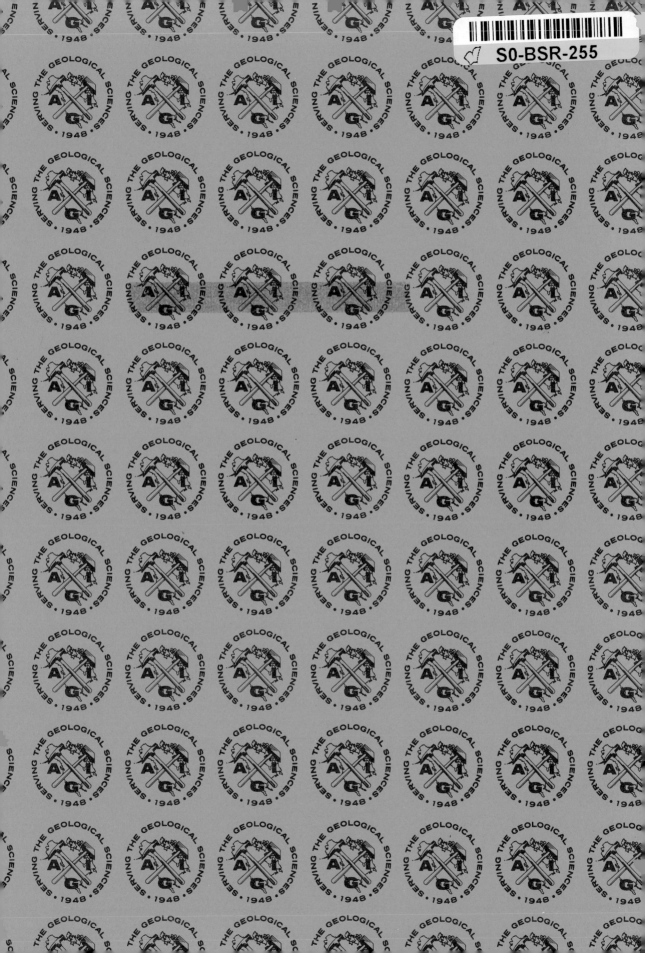

Kick at the rock, Sam Johnson, break your bones:
but cloudy, cloudy is the stuff of stones.

from "Epistemology"
by Richard Wilbur

GLOSSARY
OF
GEOLOGY

**with a foreword by
Ian Campbell**

**Margaret Gary, Robert McAfee Jr,
and Carol L. Wolf, editors**

**American Geological Institute
Washington, D.C.**

to J.V. Howell
whose monumental efforts as coordinating chairman
of a steering committee were primarily responsible
for the original edition of the American Geological
Institute's *Glossary of geology and related sciences*
published in 1957, and hence for the subsequent
work leading to this volume

and to Frank C. Calkins
who by advice and example inspired the editors with his
clarity of thought and precision of the written word.

This book was designed by Alan Nurmi.
The text was composed in Newton
by the Mack Printing Company
of Easton, Pennsylvania, and Helvetica
by Artisan Type, Inc., of Washington, D.C.;
printing (by offset lithography) and binding by
the Kingsport Press of Kingsport, Tennessee.
Production by Wendell Cochran

Foreword

It was in 1950—just a little over 20 years ago—that a group of geologists in Tulsa, Oklahoma, recognized the desirability of agreement on definitions, and for standardization of many terms commonly, but often differently, used by geologists. Not only did they recognize the need; they did something about it! They outlined a plan of action and submitted it to the Research Committee of the American Association of Petroleum Geologists. The committee, favorably impressed by the proposal but having some doubt as to its appropriateness for AAPG, recommended that the project might better be undertaken by the American Geological Institute—thereby providing for input from the many special fields represented by the AGI member societies. Thus was the first Glossary Project initiated. Under the dedicated guidance of J.V. Howell (whose name deservedly will be forever associated with the AGI *Glossary*) and with contributions from a great number of individuals, committees, and some societies, a *Glossary of geology and related sciences* embodying some 14,000 entries was published in 1957 under the imprimatur of the National Academy of Sciences—National Research Council.

Widely acclaimed as this publication was, it was recognized that—like most first attempts—it could be improved upon. Almost immediately, therefore, a Glossary Review Committee, with J. Marvin Weller as chairman, was established. This committee brought out, in 1960, a second edition, *Glossary of geology and related sciences, with supplement*—the supplement consisting of some 4,000 terms that had "come to notice since 1957".

From the first announcements of their publication, the various editions of the AGI *Glossary* have been best-sellers among publications in geology. Altogether some 4,000 copies of the first edition and more than 22,000 copies of the second edition are in print. Because of the multiplication of new terms in geology and the frustrations of having to consult two parts of the second edition to establish a definition (the supplement was separately alphabetized), it was recognized early in the 1960s that a wholly new *Glossary* would be needed before long. Moreover, although its preparation began as a revision of the second edition, it soon became apparent that the only valid basis for a comprehensive work was to gather from scratch the terminology of the various earth sciences, together with apparent meanings. To that end, several years were spent in research before the work of defining could begin.

The first editorial work was undertaken on a part-time basis by Henry R. Aldrich, who had for many years been executive secretary and editor for the Geological Society of America. Through the succeeding years many others, both staff employes and volunteers, have carried on the work. Though all these deserve credit for the final result, one in particular deserves mention, Frank Cathcart Calkins. For several years he volunteered much of his time (with the understanding coöperation of the U.S. Geological Survey) in order to edit many terms old and new. Any USGS geologist so fortunate as to have had a typescript "calkinized" knows the talent that Frank Calkins has (and still possesses although now well into his 94th year!) for improving phraseology. In some of the more pleasantly readable, and at the same time more precise, definitions in this volume (as one example, look up *roche moutonnée*) many will no doubt recognize the talent that Frank Calkins brought to this task. Unfortunately, limits on financial and clerical support precluded full use of his talents. Even so, it is eminently fitting that this edition should be dedicated not only to J.V. Howell but also to Frank C. Calkins.

Why do we need glossaries, and repeatedly revised editions of glossaries? A dramatic answer was provided just 100 years ago with the publication (in 1872) of Lewis Carroll's *Through the looking glass*. In it is this oft-quoted passage:

"There's glory for you," said Humpty Dumpty. "I don't know what you mean by glory," Alice said. Humpty Dumpty smiled contemptuously. "Of course you don't—till I tell you. I meant there's a nice knock-down argument for you!"

"But glory doesn't mean 'a nice knock-down argument,' " Alice objected.

"When *I* use a word," Humpty Dumpty said in rather a scornful tone, "it means just what I choose it to mean—neither more nor less."

"The question is," said Alice, "whether you *can* make a word mean so many different things."

"The question is," said Humpty Dumpty, "which is to be master—that's all."

Interestingly enough, some 50 years later, Arthur

Holmes used this quotation very effectively in the first edition of *The nomenclature of petrology* (1920). And Holmes was quoting in turn an earlier geological application of the quotation from the *Quantitative classification of igneous rocks!* Today it is worth quoting again, for Humpty Dumpty and Alice had dramatized inimitably one of the long-standing verities of geology: that a word frequently means just what a particular author wants it to mean, "neither more nor less." Thus, if it were not for the occasional appearance of an authoritative glossary, our language—which, as a distinguished lexicographer has said, should "mirror our culture"—would rapidly degenerate into babel. By good fortune, the efforts of lexicographers have prevented this ignominy to our science while—perhaps with equal good fortune—they have not been so completely and autocratically successful as to have frozen the meanings of words. After all, languages grow and evolve (as has been so well demonstrated by H.L. Mencken in his three volumes on *The American language),* and in their growth and evolution can provide an engrossing record of historical change and development. The language of geology is no exception.*

If I seem to have implied that on the one hand lexicographers and glossaries are needed in order to prevent a language from degenerating into babel, while on the other hand conceding that standardization of terms should not be so rigid as to stultify growth and change, that is not ambivalence. The two processes, standardization

* As a somewhat incidental yet interesting example, take the word "graywacke" which, in the second edition of the *Glossary,* required something over 500 words of explanation and definition, and which in this new edition, with some 900 words devoted to it, rates top place in size of definition among all the 33,000 entries! This is a word, first used in 1789, which by 1818 had already earned this comment from Mawe: "Geologists differ much respecting what is, and what is not, Grey Wacce." Because of the confusion that even then attached to the name, it fell somewhat into disuse for many years, particularly after Sir Roderick Murchison's pronouncement in 1839: "it has already been amply shown that this word [graywacke] should cease to be used in geological nomenclature, and I shall in the following pages give further proofs that it is mineralogically worthless." Yet in recent years the word has been retrieved from obsolescence, in a sense recycled, and again diversely redefined to the point that the current *Glossary* states that "in view of the diversity of usage, the term 'graywacke' should not be used formally without either a specific definition or a reference to a readily available published definition" and Folk (1968) has advocated discarding the term for any precise petrographic usage.

on the one hand and growth and change on the other, are not incompatible *if taken in the proper sequence.*

At various points in time, standardization becomes both desirable and necessary. Such a point may well have been achieved in the publication of this new edition of the *Glossary.* The definitions recorded here are the meanings that —in speech and in writing—we should attach to these terms for at least some little time into the future. By holding to this practice we all can the more easily, the more precisely, and the more assuredly exchange ideas and advance our science. But I should also note that the *Glossary* necessarily reports actual usages whether they are "correct" or "standard" or otherwise. For example, "tidal wave" is often used for "tsunami". Although generally condemned, that usage must be recorded in order to help the neophyte determine the author's meaning. Of course the entry "tidal wave" calls attention to the preferred term.

Inevitably, however, as the science advances, as knowledge expands, and as new and better data become available, we must modify and improve our concepts and with this must come modification and enlargement of our vocabulary.

What I am saying here is simply that the burden of proof—that is, a demonstration of need—for a new word, or for a change of meaning of an old and established word, must rest on him who proposes a new word or a modification of meaning of an old word. Can the proposer marshal sufficient evidence to show that a new word is really needed? Has he searched the *Glossary* to see whether an established word is not perhaps already available and suited to the principle or idea he wants to convey? Must an established word have its meaning modified? If so, why? Is it because of new discoveries? Because of an erroneous earlier concept? For some other truly valid reason? The evidence must be compelling before a recommended *Glossary* definition should be modified or abandoned.

Earlier I referred to the opportune timing of publication of this new *Glossary.* The preceding edition appeared in 1960, scarcely more than a decade ago. A decade, in geology, is small time indeed. Nevertheless, consider some of the changes and developments that have come about in this decade. In 1960, AGI consisted of 14 member societies. Today, with the addition of the American Institute of Professional Geologists, the Association of Earth Science Editors, the Association of Engineering Geologists, and the Geoscience Information Society, 18 societies are represented in AGI. Significantly, perhaps, two of the four new member societies (AESE and GIS) are concerned principally with the problem that looms so large over our times: communication. And a third member society (AIPG) is to a considerable degree also concerned with communication, especially as between the profession and the public. Meantime, the spectacular growth in this past decade of engineering geology as a profession has introduced new terms and new concepts into our language. In order to cover adequately such developments the new *Glossary* contains almost 33,000 entries; the second edition had only 18,-000. Certainly these comparisons are interesting and constitute an impressive record of growth as well as of the diversification so characteristic of geologists, their societies, and their vocabularies.

It is however of much greater interest to recall some of the scientific developments of the decade just past. In 1960 "sea-floor spreading" was just beginning to spread; "plate tectonics" was yet to come alive; "environmental geology" as such had scarcely been envisaged (much less enshrined as a path to a geology degree); there was scarcely a laboratory in this country where an electron microprobe was operating. "Mohole" just made it into the supplement of the 1960 *Glossary*. And though it has all but foundered since, in the meantime "JOIDES" was coined and has been at work—and with what spectacular results! The fossil record has not only been greatly enlarged, it has also been pushed back into the Precambrian by a billion years or so. A discussion of the possibility of earthquake prediction was, in 1960, something that almost no reputable seismologist would have indulged in— he would have risked excommunication from the establishment. Today this possibility is being discussed, and financed, at the highest government levels. Age-dating of rocks has been enlarged and improved by almost an order of magnitude. But of all the spectacular achievements of the decade, the success in obtaining and studying lunar samples must rank as the most spectacular. Yet, in 1960, lunar samples were no more than a dream in the minds of a few of the most optimistic NASA scientists. Has geology ever enjoyed a more exciting and a more productive decade?

Certainly no geologist can view these developments of a decade or less without enormously increased pride in his profession. Moreover, these developments, diverse as many of them seem at first sight, have nevertheless exerted a notable (and needed!) unifying influence on the sciences. For the first time, especially through the concepts of global tectonics, geologists feel that they are on the verge of knowing "just how the Earth ticks." The concept of sea-floor spreading; the contouring and sampling and drilling of the ocean basins; the ever deeper and more precise probing of the Earth's interior by seismologists; the study of lunar rocks and of Martian geomorphology—all these are contributing to a greatly improved understanding of Earth. Best of all, this new knowledge has required and has developed teamwork such as never known before. Today we see geophysicists and paleontologists not only communicating effectively but also working closely together to find solutions to problems of mutual interest; geochemists, glaciologists, paleometeorologists, and crystallographers find themselves on a single team. The study of Moon rocks brings together biologists, physicists, soils engineers, aerodynamicists, volcanologists and other specialists along with geologists.

Central to all this is geology; and central to good communication and to true understanding is this *Glossary*. Here it is—treat it critically, treat it lovingly, treat it with understanding for the terminology and with compassion for its editors. Above all, use it!

Ian Campbell
San Francisco
July 1972

Acknowledgments

The preparation of this *Glossary* was undertaken with the advice and guidance of the AGI Committee on Publications, under the chairmanship of Lewis M. Cline (1966-1968), Phillip E. La-Moreaux (1968-1970), and Charles W. Collinson (1971-1972). The work was partially funded by the National Science Foundation.

Special acknowledgment is made to Henry R. Aldrich, formerly Executive Secretary of the Geological Society of America, who began the work of revising the *Glossary of Geology and Related Sciences, with Supplement* on a part-time basis during 1964 and 1965.

Although the editors and AGI assume responsibility for content as well as for form, they respectfully thank and acknowledge the contribution of those scientists who gave their time generously and without recompense to review and write definitions relevant to their particular fields of interest:

Astrogeology: Peter Boyse, Bevan M. French, Paul D. Lowman, jr., Brian A. Mason, and Robert L. Wildey.

Cartography, surveying, map projections, photogeology: Earle J. Fennell, Arthur D. Howard, and Douglas M. Kinney.

Crystallography, mineralogy, and gems: L.G. Berry, L.L. Copeland, Michael Fleischer, Richard T. Liddicoat, jr., Kurt Servos, and George Switzer.

Economic geology: M.E. Hopkins, Thomas A. Howard, Joel J. Lloyd, Ross Shipman, Jack A. Simon, Robert B. Smith, Robert R. Wheeler, and Walter S. White.

Engineering geology and rock mechanics: Mary Hill, Eugene Kojan, and David J. Varnes.

Geochemistry: Z.S. Altschuler, William Back, P.L. Greenland, Mead LeRoy Jensen, Priestley Toulmin, III, and Irving Breger.

Geochronology and absolute age determination: Mead LeRoy Jensen, Richard F. Marvin, Zell E. Peterman, John N. Rosholt, Barney J. Szabo, Robert E. Zartman, and Thomas W. Stern.

Geomorphology: Willard Bascom, Robert F. Black, Arthur L. Bloom, Arthur D. Howard, Watson H. Monroe, J.W. Pierce, Arthur N. Strahler, and William F. Tanner.

Geophysics: Don L. Anderson, L.G. Berryman, William J. Best, John K. Costain, Jules D. Friedman, M. Charles Gilbert, Pembroke J. Hart, Douglas M. Kinney, Arthur Lachenbruch, R.J.P. Lyon, Robert Moxham, Fred Nicodemus, Otto Nuttli, Donald A. Rice, Malcolm Rigby, Frank E. Senftle, Ralph T. Shuey, John S. Summer, Ben Tsai, Stanley H. Ward, Charles A. Whitten, Frans-Erik Wickman, and George Zissis.

Glaciers, ice, snow, and glacial geology: Richard Foster Flint, C.L. McGuinness, Mark F. Meier, and Stephen C. Porter.

History and philosophy of geology: George V. Cohee and George W. White.

Hydrology: Manuel A. Benson, John Ficke, Luna Leopold, and C.L. McGuinness.

Marine geology and oceanography: Willard Bascom, A.R. Gordon, jr. and the U.S. Naval Oceanographic Office at Suitland, Maryland, and Norman J. Hyne.

Meteorology: Charles W. Collinson and Glen Stout.

Paleoclimatology and paleoecology: Robert E. Gernant and Thor N.V. Karlstrom.

Paleobotany: Anne S. Benninghoff and William S. Benninghoff.

Paleontology of the invertebrates: Richard H. Benson, Jean Berdan, William B.N. Berry, Richard S. Boardman, Raymond C. Douglass, J. Wyatt Durham, J. Thomas Dutro, jr., Robert M. Finks, Mackenzie Gordon, jr., Robert R. Hessler, Dorothy Hill, Erle G. Kauffman, Robert V. Kesling, Porter M. Kier, Bernhard Kummel, Alfred R. Loeblich, jr., Donald B. Macurda, jr., Raymond C. Moore, Norman J. Newell, William A. Newman, W.A. Oliver, John Pojeta, jr., A.J. Rowell, J.S. Ryland, K. Norman Sachs, Norman F. Sohl, Walter C. Sweet, Helen Tappan, Curt Teichert, Ruth M. Todd, J. Marvin Weller, John W. Wells, C.W. Wright, and Ellis L. Yochelson.

Palynology: Alfred Traverse.

Igneous and metamorphic petrology: Harold H. Banks, jr., Paul Bateman, Phillippa M. Black, Wilfred B. Bryan, Richard V. Dietrich, William G. Melson, Anne K. Loring, Kurt E. Lowe, Lucian B. Platt, Robert I. Tilling, and Hatten S. Yoder, jr.

Sedimentary petrology: Murray Felsher, Robert L. Folk, George deVries Klein, Roy C. Lindholm, Francis J. Pettijohn, and J.W. Pierce.

Soils: Mary Hill, Charles E. Kellog, and Arnold C. Orvedal.

Statistical methods: Peter Fenner, J.C. Gower, William C. Krumbein, Richard McCammon, Daniel F. Merriam, and G.V. Middleton.

Stratigraphy: George V. Cohee (assisted by Virginia Byers and Marjorie MacLachlan).

Structural geology and tectonics: Fred A.

Donath, William C. Gussow, Marshall Kay, Philip B. King, John M. Logan, and Robert S. Young.

Volcanology and pyroclastics: Roy A. Bailey, Wilfred B. Bryan, Richard S. Fiske, Jack Green, and Alexander McBirney.

Beverly M. Sullivan was a member of the editorial staff during 1970-1971. Her contribution to the *Glossary* is much appreciated by her fellow lexicographers.

Myrl Powell of the Library of Congress Subject Cataloging Division provided bibliographic help. Most of the difficult job of proofreading was done, and with dedication, by Lillian B. Dawson and Dorothy W. Rogers. Careful and conscientious clerical help was provided by many assistants, especially Carolyn Canion, Gail Clark, Phyllis East, Susan Gilwood, Dierdre McKenzie, Susanne Miller, and Alva Saunders.

Introduction

The terms in this *Glossary* have appeared in recognized publications and reflect, in general, North American usage, although some British and Australian terminology is included. Colloquialisms are, for the most part, restricted to those of the USA. Old and obsolete terms are defined when possible (and are distinguished as such), since the older geologic literature continues to be a valuable research tool. Foreign-language terms are included only if they appear as such in the English-language literature.

Emphasis is on the current or preferred meaning of a term rather than on its original usage or historical development, although this information is sometimes given. The *Glossary* is descriptive rather than prescriptive; the editors have hoped to present an authoritative work on the basis of completeness and a reflective, explanatory style.

The scope of the book encompasses many subdisciplines of geology:

Astrogeology, including lunar geology, impact phenomena, meteorites, and those astronomic terms which may be of interest to the astrogeologist.

Cartography, map projections, surveying, and photogeology.

Crystallography.

Economic geology, including fuels and mineral deposits, drilling and logging, materials, and the terminology of mining up to the point of removal of the ore.

Engineering geology, including rock and soil mechanics and mass movements.

Geochemistry, including phase studies, experimental geochemistry, instruments and methods, organic geochemistry, and those chemical terms and units which have a particular or special relevance to the geochemist, or a special sense or emphasis in geological usage, or which are not easily found in a general chemistry text.

Geochronology and absolute age determination.

Geomorphology, including topographic features, coasts, karst, glacial geology, patterned ground and permafrost, and speleology.

Geophysics of the solid Earth, including seismology, interior of the Earth, gravity, geodesy, paleomagnetism, radioactivity and isotope geology, remote sensing, and electrical, magnetic, and thermal prospecting.

Glaciers, ice, snow, and glacial geology.

History and philosophy of geology.

Hydrology, ground water, hydraulics, streams, limnology.

Marine geology; biological, physical, and chemical oceanography; currents, waves, and tides.

Meteorology, including the aspects of weather, and storms.

Mineralogy, including the almost 4,000 mineral names recognized by the U.S. Geological Survey (provided for this *Glossary* by Michael Fleischer), and gem terminology.

Paleoclimatology, and the relevant terminology of climatology; paleoecology.

Paleontology of the invertebrates, including morphology, taxonomy, and names of paleontologically significant animals (vertebrate paleontology is not included, in accordance with the wishes of the Society of Vertebrate Paleontology); paleobotany and palynology; evolution.

Sedimentary, igneous, and metamorphic petrology and nomenclature; soils.

Statistics and mathematical terms used in geology; also, terms used in the application of computers to geologic research.

Stratigraphy, including the geologic time scale and generally recognized European stage names and American stage and provincial series names, as well as the main cultural levels of archaeology that often appear with Quaternary terminology.

Structural geology, including experimental structural geology; tectonics.

and Volcanology.

Alphabetization is letter by letter rather than word by word.

The format of the *Glossary* makes full use of cross-references:

. . . an entry followed by an italicized term means that the entry is synonymous with that term, and that its definition may be found under that term; e.g. the entry

smaragd *emerald*

means that the term "smaragd" is a synonym of the term "emerald", and that the reader should turn to the entry "emerald" for the definition.

. . . The italicizing of a term within a definition indicates that the term is defined elsewhere in the *Glossary*.

. . . Cross-references are listed at the end of a definition, if they have not been incorporated into the text of the definition.

The full citations for references used in the

text are found in the bibliography at the end of the book. Citations are made for any of several purposes: to show original usage, to show an authoritative source other than the original, or to acknowledge a quotation or a paraphrasing of an essential concept.

When a term is entered more than once because it has a meaning in more than one subject area, an abbreviated notation of the particular subject to which each entry refers is made in brackets following the term; e.g. the annotation [sed] refers to sedimentology, sediments, or sedimentary rocks. A few of the more abbreviated forms are listed here: [cart]=cartography; [cryoped]=cryopedology; [evol]=evolution; [glac]=glaciology, glacial geology; [grd wat]= ground water; [mater]=materials; [pat grd]=pat-terned ground; [periglac]=periglacial phenomena; [shock]=shock metamorphism; [stat]=statistics and mathematical geology; [water res]= water resources; [weath]=weathering.

Many terms have multiple definitions which are prefixed by (a), (b), (c) etc. If there is a comment about the term applicable to all its meanings, it is given at the end of the definitions, following a long dash: - - - .

Occasionally a guide to pronunciation is included at the end of a definition; such a guide is presented in an informal, "sound-it-out" system.

The *Glossary* contains a very few encyclopedic rather than lexicographic entries (e.g. **International Years of the Quiet Sun**) which have been included on the advice of the reviewers.

A

aa A type of lava flow having a rough, fragmental surface; it is a blocky lava consisting of clinkers and scoria, and is characteristic of oceanic shield volcanoes and continental plateau eruptions. Cf: *pahoehoe; block lava.* Pron: ah-ah. Etymol: Hawaiian. Obs. syn: *aphrolith.*

Aalenian Stage in Great Britain: lowermost Middle Jurassic or uppermost Lower Jurassic (above Yeovilian, below Bajocian).

a axis [cryst] One of the crystallographic axes used as reference in crystal description. It is the axis that is oriented horizontally, front-to-back. In an orthorhombic or triclinic crystal, it is usually the *brachy-axis.* In monoclinic crystals, it is the *clinoaxis.* The letter *a* usually appears in italics. Cf: *b axis; c axis.*

a axis [struc petrol] In structural petrology, that *fabric axis* which is the direction of maximum displacement, e.g. of tectonic transport. It is usually assumed to be perpendicular to the fold axis, or *b* axis. Cf: *c axis.* Syn: *a direction.*

a* axis That axis of a reciprocal crystal lattice which is perpendicular to (100). Cf: *b* axis; c* axis.*

abactinal Referring to the *aboral* aspect; e.g. pertaining to the upper side of the test of an echinoid or to the side of a crinoid theca or plate opposite the oral surface. Ant: *actinal.*

abandoned channel (a) A drainage channel along which runoff no longer occurs, as on an alluvial fan. (b) *oxbow.*

abandoned cliff A sea cliff that is no longer undergoing wave attack as a result of a relative drop of sea level.

abandoned meander *cutoff meander.*

abapertural Away from the aperture of a gastropod shell. Ant: *adapertural.*

abapical Away from the apex of a gastropod shell and toward the base along the axis of spiral or slightly oblique to it.

abaxial Directed or facing away from, or situated on the outside of, the axis or center of the axis, as of an organ, plant, or invertebrate; dorsal or anterior. Also, said of the abaxial side. Ant: *adaxial.*

Abbe refractometer A *refractometer* that is used for minerals and gemstones; its function is based on the measurement of the variation of the critical angle in a hemicylinder of highly refractive glass.

abbreviation Loss of the final ontogenetic stages during the course of evolution.

ABC soil A soil having A, B, and C horizons.

ABC system A method of seismic surveying whereby the effect of irregular weathering thickness may be determined by a simple calculation from reciprocal placement of shot holes and seismometers. The method was originally used to solve refraction problems arising from irregularities in the top of the high-velocity layer.

abdomen (a) The posterior and often elongated region of the body of an arthropod, behind the thorax or cephalothorax; e.g. the tagma following the thorax of a crustacean, including the telson, and consisting of seven or fewer segments, or the complete, usually unsegmented part of the body of an arachnid or merostome, following the cephalothorax. Cf: *opisthosoma; pygidium.* (b) The third joint of the shell of a nasselline radiolarian.----Pl: *abdomens* or *abdomina.* Adj: *abdominal.*

aber The mouth of a river, or the confluence of two rivers. Etymol: Celtic.

abernathyite A mineral: $K(UO_2)(AsO_4) \cdot 4H_2O$.

AB interray Right anterior interray in echinoderms situated between A ray and B ray and clockwise of A ray when the echinoderm is viewed from the adoral side.

abiogenesis The development of living organisms from lifeless matter. Cf: *biogenesis.*

abioglyph A hieroglyph of inorganic origin (Vassoevich, 1953, p.38). Cf: *bioglyph.*

ablation [geomorph] Separation and removal of rock material,

and formation of residual deposits, esp. by wind action or the washing away of loose and soluble materials. Some writers prefer to restrict the term to wasting of glaciers by melting and evaporation.

ablation [meteorite] Removal of molten surface layers of meteorites and tektites by direct vaporization during flight.

ablation [glaciol] (a) All processes by which snow or ice is lost from a glacier, floating ice, or snow cover. These processes include melting, evaporation (sublimation), wind erosion, and calving. Sometimes calving is excluded, or the term may be restricted to surface phenomena. Cf: *accumulation* [*glaciol*]. (b) The amount of snow or ice removed by the process of ablation.----Syn: *wastage* [*glaciol*].

ablation area The part of a glacier or snowfield in which, over a year's time, ablation exceeds accumulation; the region below the equilibrium line. Cf: *accumulation area.* Syn: *zone of ablation.*

ablation breccia *solution breccia.*

ablation form A feature formed on a surface of snow, firn, or ice by melting or evaporation; e.g. *nieve penitente* or *ice pyramid.* Cf: *erosion form.*

ablation funnel A closed depression, similar to a solution channel, formed by solution processes or by removal of particulate material by circulating ground water.

ablation gradient The change in ablation with altitude on a glacier, usually expressed as millimeters of water equivalent per meter of altitude. Also used incorrectly to specify change of net balance with altitude up to the equilibrium line. Cf: *activity index.*

ablation moraine An uneven pile or continuous layer of *ablation till,* either overlying ice in the ablation area, or resting on ground moraine derived from the same glacier.

ablation rate The amount of ice or snow loss per unit time from a glacier, floating ice, or snow cover. Usually expressed in millimeters of water equivalent per hour or day.

ablation season *summer season.*

ablation till Loosely consolidated rock debris, formerly contained by a glacier, that accumulated in place as the surface ice was removed by ablation.

ablatograph An instrument that measures the distance through which the surface of snow, ice, or firn changes, because of ablation, during a given period.

ablykite A clay-mineral material consisting of an aluminosilicate of magnesium, calcium, and potassium. It resembles halloysite in its dehydration characteristics but differs from it in its thermal and X-ray-diffraction properties. Syn: *ablikite.*

Abney level A *hand level* consisting of a short telescope, a movable bubble tube, and a graduated vertical arc to which a bubble is attached, and used to measure vertical angles and esp. tree heights. Named after William de Wiveleslie Abney (1843-1920), English physicist.

abnormal [fold] Said of an anticlinorium in which the axial surfaces of the subsidiary folds converge upwards; said of a synclinorium in which the axial surfaces of the subsidiary folds converge downwards. Cf: *normal [fold]*.

aboral (a) Located opposite to or directed away from the mouth of an invertebrate; esp. applied to the *abactinal* surface (or to the structures on it) opposite that bearing the mouth and ambulacral grooves of an echinoderm, or to the part of an echinoderm theca or plate directed away from the mouth (directed downward in an edrioasteroid). Cf: *adoral.* (b) Toward the underside of a conodont element; e.g. "aboral edge" or "aboral groove" along the midline of the underside of a conodont element, or "aboral attachment scar".---Ant: *oral.*

aboral margin The trace of the aboral side of a conodont element in lateral (side) view. The term has also been used for the aboral side itself.

aboral pole The end of a flask-shaped chitinozoan that includes the chamber of the body and the base. Cf: *oral pole.*

aboral side The underside of a conodont element to which the basal plate is attached or onto which the basal cavity or at-

1

tachment scar opens. Cf: *oral side*.

aboriginal Said of the original race, fauna, or flora of a particular area, as distinguished from later immigrant or imported forms. Noun: aborigine.

abrasion The mechanical wearing, grinding, scraping, or rubbing away (or down) of rock surfaces by friction and impact, in which the solid rock particles transported by wind, ice, waves, running water, or gravity are the media doing the work of abrasion. The term *corrasion* is essentially synonymous. Also, an abraded place or the effect of abrading, such as the abrasion left by glacial action. Verb: *abrade*. Cf: *attrition*.

abrasion pH A term proposed by Stevens & Curron (1948) to designate the characteristic pH achieved by a suspension of a pulverized mineral in water, and resulting from a complex of hydrolysis and dissolution reactions.

abrasion platform An extensive, nearly horizontal, submerged surface produced by long-continued wave erosion, still in its original position at or near the wave base, with the marine or lake forces still operating on it, and representing the outward continuation of the wave-cut bench toward a flatter surface (Johnson, 1916, p.444). Cf: *erosion platform*. See also: *wave-cut platform; plain of marine erosion.*

abrasion shoreline *retrograding shoreline.*

abrasion tableland A tableland in which the outcrops of various rocks have been reduced to nearly the same level by denuding agents (Stamp, 1961, p.2).

abrasive [geomorph] A rock fragment, mineral particle, or sand grain used by natural agents in abrading rock material or land surfaces.

abrasive [mater] A natural or artificial substance suitable for grinding, polishing, cutting, or scouring. Natural abrasives include diamond, emery, silica, and diatomite.

abrolhos A term used in Brazil for a mushroom-shaped barrier reef spreading widely near the surface. Etymol: Portuguese, "caltrop; breakers; pointed rocks".

absarokite A basaltic rock, composed of olivine and augite phenocrysts in a groundmass of labradorite with orthoclase rims, ilivine, and some leucite. Absarokite grades into *shoshonite* with a decrease in the olivine content and with the presence of some dark-colored glass, and into *banakite* with a decrease in the olivine and augite. Its name is derived from the Absaroka Range, Wyoming.

abscission Separation of plant parts, e.g. of a leaf from a stem, usually by cell-wall dissolution along a certain layer (*abscission layer*).

abscission layer In a plant, the zone of cells, e.g. at the base of a petiole, along which *abscission* occurs. Syn: *separation layer.*

absite A thorian variety of brannerite.

absolute abundance The exact number of individuals of a taxon in a certain area or volume. See also: *abundance; relative abundance.*

absolute age The *geologic age* of a fossil organism, rock, or geologic feature or event given in units of time, usually years. Commonly used as a syn. of *isotopic age* or *radiometric age*, but may also refer to ages obtained from tree rings, varves, etc. Term is now in disfavor as it implies a certainty or exactness that may not be possible for an age determined by present dating methods, i.e. sometimes two absolute ages determined for the same pluton disagree by hundreds of million years. Cf: *relative age*. Syn: *actual age.*

absolute age determination Calculation of *absolute age* usually, but not always, on the basis of radioactive isotopes. The ratio of decay products to parent products in the sample is calibrated to a certain number of years as in the *age equation.*

absolute altitude *flight height.*

absolute chronology *Geochronology* in which the time-order is based on *absolute age*, usually measured in years by radiometric dating, rather than on superposition and/or fossil con-

tent as in *relative chronology*.

absolute date The date of an event usually expressed in years (*absolute age*) and related to a specific time scale.

absolute gravity instruments Devices for measuring the true value of gravity at a point. This type of measurement is much more difficult than relative determinations, because all the physical influences must be evaluated with extreme accuracy. The measurements are accomplished by various forms of reversible pendulums or by timing the motion of a body in free fall. Cf: *relative gravity instruments.*

absolute humidity The content of water vapor in air, expressed as the weight of the water per unit volume of air.

absolute pollen frequency The estimate of the actual amount of pollen deposited per unit area in a given length of time, achieved by correcting the amount of pollen per gram of sediment by factors based on rate of sedimentation. Abbrev: APF.

absolute time *Geologic time* measured in terms of years; specif. time as determined by radioactive decay of elements. Jeletzky (1956, p.681) proposed that the term be abandoned because its usage, based on criteria peculiar to the Earth and having the present part of geologic history as its starting point, is "incorrect and highly misleading". Cf: *relative time; mineral time*. Syn: *physical time.*

absolute viscosity *viscosity coefficient.*

Absonderung A term, now obsolete, applied by von Leonhard in 1823 to the parting in igneous rocks that divides them into more or less regular bodies. The parting results from fractures that developed as a cooling phenomenon (Johannsen, 1939, p.163). Etymol: German, "separation, division".

absorbed water (a) Water retained mechanically within a soil mass and having properties similar to those of ordinary water at the same temperature and pressure. (b) Water entering the lithosphere by any means. Cf: *adsorbed water.*

absorbing well *drainage well.*

absorptance [radiation] (a) The measure (ratio or percentage) of the absorbed fraction of incident flux (Nicodemus, 1971). (b) The ratio of absorbed radiant flux to incident radiant flux. Symbol: α.

absorption [chem] Taking up, assimilation, or incorporation; e.g. of liquids in solids or of gasses in liquids, sometimes incorrectly used in place of *adsorption*. Syn: *occlusion.*

absorption [phys] The process by which energy such as that of electromagnetic or seismic waves, is converted into other forms of energy, e.g. heat.

absorption [geophys] The conversion of the energy of a seismic wave into heat by the medium through which the wave passes.

absorption [optics] The reduction of light intensity in transmission through an absorbing substance or in reflection from a surface. In crystals, the absorption may vary with the wavelength of vibration direction of the transmitted light.

absorption [grd wat] The entrance of surface water into the lithosphere by any method. Verb: *to absorb*. Cf: *adsorption.*

absorption band Any of the dark bands in the *absorption spectrum* of a substance due to certain wavelengths in the spectrum being selectively absorbed on passing through a medium. Cf: *absorption line.*

absorption coefficient *linear absorption coefficient.*

absorption edge The wavelength at which there is an abrupt change in the intensity of an *absorption spectrum*. The term is usually applied to X-ray spectra.

absorption factor (a) *linear absorption coefficient*. (b) Formerly, the ratio of the intensity loss by absorption to the total original intensity of radiation.

absorption formula An expression of a crystal's relative absorption of light vibrating parallel to certain vibration directions. Cf: *pleochroic formula.*

absorption line Any of the dark lines in the *absorption spectrum* of a substance due to certain wavelengths in the spectrum being selectively absorbed on passing through a medium. Cf: *absorption band.*

absorption loss Water lost through absorption by rock and soil during the initial filling of a reservoir.

absorption spectroscopy The observation of an *absorption spectrum* and all processes of recording and measuring which go with it.

absorption spectrum The array of absorption bands or lines seen when a continuous spectrum is transmitted through a selectively absorbing medium.

absorptivity (a) A fundamental property of material, internal absorptance, through a path of unit length (Nicodemus, 1971). (b) As used in the older literature, the fraction of the radiant energy incident on the surface which is absorbed. It is a pure numeric but is often referred to as "absorptive power". From Kirchhof's Law it follows that the absorptivity is equal to the emissivity.

abstraction [water] That part of precipitation that does not become direct runoff (e.g. interception, evaporation, transpiration, depression storage, infiltration). Cf: *precipitation excess; rainfall excess.*

abstraction [streams] The merging of two or more, subparallel streams into a single stream course as a result of competition between adjacent, consequent gullies and ravines, as by the deepening and widening of one channel so that it absorbs a shallower and smaller one nearby; the simplest type of *capture.* It usually occurs at the upper end of a drainage line. Syn: *stream abstraction.*

abtragung The part of degradation not resulting directly from stream erosion, i.e. preparation and reduction of rock debris by weathering and transportation of waste (Engeln, 1942, p.265). Etymol: German *Abtragung,* "degradation; denudation".

abukumalite A mineral of the apatite group: $(Ca,Y)_5(SiO_4, PO_4)_3(OH,F)$. Cf: *britholite.*

Abukuma-type facies series A type of dynamothermal regional metamorphism named after the Central Abukuma plateau of Japan and characterized by the index minerals (in order of increasing metamorphic grade) biotite - andalusite - cordierite - sillimanite (no kyanite) representing the greenschist and cordierite-amphibolite facies. Pressures are rather low: approaching those in contact metamorphism, i.e. 2500-3500 bars (Hietanen, 1967, p.192). Cf: *Buchan-type facies series.*

abundance [geochem] A synoptic measure of large-scale average content, e.g. the abundance of Ni in meteorites, or the crustal abundance of oxygen; also, used preferentially in synoptic statements of relative average content; e.g. the order of abundance of elements in the Earth's crust is O, Si, Al, Fe, Ca, etc.; the estimated cosmic abundance of Li in atoms per 10,000 atoms of Si is 1.0 (Suess & Urey, 1966).

abundance [ecol] In ecology, the number of individuals of a particular taxon in a certain area or volume of sediment. See also: *absolute abundance; relative abundance.*

abundant In the description of coal constituents, 30-60% of a particular constituent occurring in the coal (ICCP, 1963). Cf: *rare; common; very common; dominant.*

abysmal A less-preferred syn. of *abyssal.*

abyss [geomorph] *chasm.*

abyss [oceanog] *deep* [oceanog].

abyssal [oceanog] Pertaining to the ocean environment or *depth zone* of 500 fathoms or deeper; also, pertaining to the organisms of that environment. Less-preferred syn: *abysmal.*

abyssal [intrus rocks] Pertaining to an igneous intrusion or to the rock of that intrusion which occurs at considerable depths; *plutonic.* Cf: *hypabyssal.*

abyssal [lake] Pertaining to the zones of greatest depth in a lake at which the water is "stagnant" or has a uniform temperature.

abyssal benthic Pertaining to the benthos of the abyssal zone of the ocean. Syn: *abyssobenthic.*

abyssal cone *submarine fan.*

abyssal deposits *pelagic deposits.*

abyssal fan *submarine fan.*

abyssal gap A passage that connects two abyssal plains of different levels, through which clastic sediments are transported. Syn: *gap* [*marine geol*].

abyssal hill A common low-relief feature of the ocean floor, usually found seaward of abyssal plains and in basins isolated by ridges, rises, or trenches. Its average height is 100-200 m, and average diameter is about 10 m. About 85% of the Pacific Ocean floor and 50% of the Atlantic Ocean floor are covered by abyssal hills.

abyssal injection The process of rising of plutonic magma through deep-seated contraction fissures.

abyssal pelagic Pertaining to the open ocean or pelagic environment at abyssal depths. Syn: *abyssopelagic.*

abyssal plain A flat region of the ocean floor, usually at the base of a continental rise, whose slope is less than 1:1000. It is formed by the deposition of sediments that obscure the preexisting topography.

abyssal theory A theory of mineral-deposit formation involving the separation and sinking of minerals below a silicate shell during the cooling of the Earth from a liquid stage, followed by their transport to and deposition in the crust as it was fractured (Shand, 1947, p.204). Modern thought ascribes a far more complex origin to most mineral deposits.

abyssobenthic *abyssal benthic.*

abyssolith *batholith.*

abyssopelagic *abyssal pelagic.*

acadialite A flesh-red variety of chabazite, found in Nova Scotia.

Acadian Provincial series in Canada: Middle Cambrian.

Acadian orogeny A middle Paleozoic deformation, especially in the northern Appalachians; it is named for Acadia, the old French name for the Canadian Maritime Provinces. In Gaspé and adjacent areas the climax of the orogeny can be dated by limiting strata as early as the Late Devonian, but deformational, plutonic, and metamorphic events were prolonged over a more extended period; the last two have been dated radiometrically as between 330 and 360 m.y. ago. The Acadian had best be regarded, not as a single orogenic episode, but as an orogenic era in the sense of Stille. Cf: *Antler orogeny.*

acantharian Any radiolarian belonging to the suborder Acantharina, characterized by a centrogenous skeleton composed of strontium sulfate and a central capsule enclosed by a thin simple membrane.

acanthine septum A corallite septum composed of a vertical or steeply inclined series of trabeculae and commonly marked by spinose projections along the axially directed margin of septum.

acanthite An orthorhombic mineral: Ag_2S. It is dimorphous with argentite and constitutes an ore of silver.

acanthopore A small rod-like structure of cone-in-cone layers (originally believed to be tubular) lying within the zooecial wall in Paleozoic bryozoans and forming a spine-like projection at the colony surface.

acanthostyle A monaxonic sponge spicule (style) covered with short or tiny spines over most of its surface.

acanthus A secondary deposit in the chamber floor of certain foraminifers (such as *Endothyra*), sharply pointed but not curved toward the anterior (TIP, 1964, pt.C, p.58). Pl: *acanthi.*

acarid Any arachnid belonging to the order Acarida, characterized by the absence of abdominal segmentation but with subdivision of the body into a proterosoma and hysterosoma. Their stratigraphic range is Devonian to present.

acaustobiolith A noncombustible organic rock, or a rock formed by the organic accumulation of purely mineral matter (Grabau, 1924, p.280). Cf: *caustobiolith.*

acaustophytolith An *acaustobiolith* formed by plant activity;

e.g. a pelagic ooze containing diatoms, and a nullipore reef or limestone.

accelerated development The production of a landscape where the rate of uplift is more rapid than the rate of downward erosion or where valley deepening exceeds valley widening, characterized by an increase of the relative relief and the formation of convex slopes. Cf: *declining development; uniform development.* Syn: *waxing development; ascending development.*

accelerated erosion Erosion occurring in a given region at a greater rate than *normal erosion,* usually brought about by the influence of man's activities in disturbing or destroying the natural cover, thus sharply reducing resistance of the land surface and rate of infiltration. It may result from: increase of sediment yield through deforestation, improper cultivation of soil, dry-farming, overgrazing of rangelands, burning and clearance of natural vegetation, excavation for buildings and highways, urbanization of drainage areas, strip mining, or copper smelting; and by nonhuman influences, such as lightning or rodent invasion.

acceleration During evolution, the gradual appearance of an ancestral adult characteristic in an immmature descendant.

acceleration due to gravity The acceleration of a freely falling body in a vacuum due to the Earth's gravitational attraction. Although its true value varies with altitude, latitude, and the nature of the underlying rocks, the standard value is 980.665 cm/sec^2.

accelerogram The record made by an *accelerograph.*

accelerograph An instrument that records the measurement of an *accelerometer.* See also: *accelerogram.*

accelerometer An instrument used to detect acceleration; specif. a seismometer designed to respond to the acceleration of Earth particles. This is accomplished by making its natural period much smaller than that of the ground motion. See also: *accelerograph.*

accented contour *index contour.*

accessory [paleont] adj. Said of a secondary or minor element of an ammonoid suture; e.g. "accessory lobe" or "accessory saddle". Cf: *auxiliary.*---n. An accessory lobe or an accessory saddle.

accessory [min] adj. A term applied to a mineral or minerals that occur in relatively small quantities in a rock, and whose presence or absence does not affect its analysis.----n. An accessory mineral.----Cf: *essential mineral.*

accessory [pyroclast] Said of pyroclastics that are formed from fragments of the volcanic cone or earlier lavas; it is part of a classification of volcanic ejecta based on mode of origin, and is equivalent to *resurgent* ejecta. Cf: *essential; accidental.* See also: *cognate.*

accessory aperture An opening in the test of a planktonic foraminifer that does not lead directly into a primary chamber but extends beneath or through accessory structures (such as bullae and tegilla); e.g. a *labial aperture,* an *infralaminal accessory aperture,* and an *intralaminal accessory aperture.*

accessory archeopyle suture An *archeopyle suture* that consists of a short cleft in the wall adjacent to the principal suture, or that may be more fully developed on the operculum of the dinoflagellate cyst, dividing that structure into two or more separate pieces.

accessory comb The line of large cilia within the preoral cavity in a tintinnid.

accessory element *trace element.*

accessory mineral A discrete mineral occurring in very small amounts in a sedimentary rock; it is usually a *heavy mineral.*

accessory muscle (a) A convenient noncommittal term for any muscle of a bivalve mollusk (other than an adductor muscle or a muscle withdrawing marginal parts of the mantle) of uncertain origin and having a scar of attachment to the shell. (b) One of a pair of diductor muscles branching posteriorly and ventrally from the main diductor muscles of a brachiopod

and inserted in the pedicle valve posterior to the adductor bases (TIP, 1965, pt.H, p.139).

accessory spore A spore present in a rock only in very small quantities. Accessory spores may contain types with a very restricted range and they have been used for correlation and for zoning (as of coal measures).

accident (a) A departure from the normal cycle of erosion, caused by events that occur "arbitrarily as to place and time", such as climatic changes and volcanic eruptions (Davis, 1894). Cf: *interruption.* (b) An event, such as drowning, rejuvenation, ponding, or capture, that interferes with, or entirely puts an end to, the normal development of a river system (Scott, 1922, p.188). (c) An irregular feature in, or an undulation of, a land surface.

accidental Said of pyroclastics that are formed from fragments of nonvolcanic rocks or from volcanic rocks not related to the erupting volcano; it is part of a classification of volcanic ejecta based on mode of origin, and is equivalent to *allothigenous* ejecta. Cf: *cognate; accessory; essential.* Syn: *noncognate.*

accidental error An unpredictable error that occurs without regard to any known mathematical or physical law or pattern and whose occurrence is due to chance only; e.g. an error ascribed to uncontrollable changes of external conditions. Syn: *random error.*

accidental inclusion *xenolith.*

accidented relief Rugged and irregular relief; probably a literal translation of the common French term *relief accidenté* (Stamp, 1961, p.4).

acclimation Physiologic adjustment by an organism to a change in its immediate environment. Syn: *acclimatization.*

acclimatization *acclimation.*

acclinal A syn. of *cataclinal.* Term used by Powell (1873, p.463). Not to be confused with *aclinal.*

acclivity A slope (as of a hill) that ascends from a point of reference. Ant: *declivity.*

accordance of summit levels *summit concordance.*

accordant Said of topographic features that have the same or nearly the same elevation; e.g. an *accordant* valley whose stream enters the main stream at the same elevation as that of the main stream. Ant: *discordant.*

accordant drainage Drainage that has developed in a systematic relationship with, and consequent upon, the present geologic structure. Ant: *discordant drainage.* Syn: *concordant drainage.*

accordant fold One of several folds having similar orientation.

accordant junction The joining of two streams or two valleys whose surfaces are at the same level at the place of junction. See also: *Playfair's law.* Ant: *discordant junction.* Syn: *concordant junction.*

accordant summit level A level surface indicating that the hilltops or mountain summits over a region have approximately the same elevation. Accordant summit levels in a region of high topographic relief suggest that the summits are remnants of an erosion plain formed in a previous erosion cycle. See also: *summit concordance; even-crested ridge.* Syn: *concordant summit level.*

accordion fold A fold in which the limbs are straight and maintain a constant thickness but in which there is thickening and sharpening of the hinge area. See also: *zigzag fold; chevron fold.* Syn: *angular fold; concertina fold.*

accretion [sed] (a) The gradual or imperceptible increase or extension of land by natural forces acting over a long period of time, as on a beach by the washing up of sand from the sea or on a flood plain by the accumulation of sediment deposited by a stream. Legally, the added land belongs to the owner of the land to which it is added. Cf: *avulsion; reliction.* See also: *lateral accretion; vertical accretion.* Syn: *aggradation; alluvion.* (b) The land so added or resulting from accretion. (c) *continental accretion.*

accretion [sed struc] (a) The process by which an inorganic body increases in size by the external addition of fresh particles, as by adhesion. (b) A *concretion*; specif. one that grows from the center outward in a regular manner by successive additions of material (Todd, 1903).

accretion [stream] The filling up of a stream bed, due to such factors as silting or wave action. Cf: *degradation [stream]*.

accretionary Tending to increase by external addition or accumulation; esp. said of a secondary sedimentary structure produced by overgrowth upon a preexisting nucleus, such as a rounded form that originated through rolling, or said of a limestone formed in place by slow accumulation of organic remains.

accretionary lapilli *mud ball*.

accretionary lava ball A rounded mass, varying in size from a few centimeters to several meters, formed on the surface of a lava flow such as aa by the molding of viscous lava around a core of already solidified lava.

accretion ridge A beach ridge located inland from the modern beach, representing an ancient beach deposit and showing that the coast has been built out seaward (Fisk, 1959, p.111). It is often accentuated by the development of dunes.

accretion ripple mark An asymmetric ripple mark having a gentle and curved lee slope, with a maximum angle of dip less than the angle of repose, and composed of cross-strata without conspicuous sorting of particles (Imbrie & Buchanan, 1965, p.151 & 153). Cf: *avalanche ripple mark*.

accretion topography A landscape built by accumulation of sediment.

accretion vein A type of vein in which the mineral deposits have been formed by a repeated alternation of channelway filling with reopening by fractures.

accumulated discrepancy The sum of the separate *discrepancies* that occur in the various steps of making a survey or of the computation of a survey.

accumulation (a) All processes that add snow or ice to a glacier, floating ice, or snow cover, including snowfall, condensation, avalanching, snow transport by wind, and freezing of liquid water. Syn: *nourishment [glaciol]*; *alimentation*. Cf: *ablation [glaciol]*. (b) The amount of snow and other solid precipitation added to a glacier or snowfield by the processes of accumulation.

accumulation area The part of a glacier or snowfield in which, over a year's time, accumulation exceeds ablation; the region above the equilibrium line. Cf: *ablation area*; *névé*. Syn: *firn field*; *accumulation zone*; *zone of accumulation*.

accumulation area ratio The dimensionless ratio of accumulation area divided by total area of a glacier for any given year, used as a rough guide to the balance between accumulation and ablation. Abbrev: AAR.

accumulation mountain *mountain of accumulation*.

accumulation rate The amount of ice or snow gain per unit time to a glacier, floating ice, or snow cover. Usually expressed in millimeters of water equivalent per hour or day.

accumulation season *winter season*.

accumulation zone (a) *accumulation area*. (b) The area in which the bulk of the snow contributing to an avalanche was originally deposited. Syn: *zone of accumulation*.

accumulative phase That portion of a frequency distribution curve for a magmatic series that lies toward the low-silica side of the modal peak, presumably representing accumulation of solid crystals by gravitational differentiation.

accumulative rock *cumulate*.

accumulator plant In geobotanical prospecting, a tree or plant that preferentially concentrates an element in contrast to other plants.

accuracy Closeness to true value or to a value accepted as being true; the degree of perfection attained in a measurement, computation, or estimate, or the degree of conformity to some recognized standard value (such as a conventional unit of measure, an exact mathematical value, or a survey observation determined by refined methods). Accuracy relates to the quality of a result, as distinguished from *precision*.

AC demagnetization *alternating field demagnetization*.

acequia A Spanish word, of Arabic origin, for an irrigation ditch or canal.

acervuline Heaped, or resembling little heaps; e.g. said of some foraminifers (such as *Acervulina*) having chambers in irregular clusters.

acetolysis Any chemical reaction in which acetic acid plays a role similar to that of water in hydrolysis; e.g. a reaction used in maceration in which organic material such as peat is heated in a mixture of nine parts acetic anhydride and one part concentrated sulfuric acid. It breaks down cellulose especially vigorously.

acetylenic A compound having a triple bond as in acetylene.

ACF diagram A triangular diagram showing the simplified compositional character of a metamorphic rock by plotting the character of a metamorphic rock according to the molecular quantities of the three components: $A = A_{12}O_3 + Fe_2O_3 - (Na_2O + K_2O)$; $C = CaO - 3.3P_2O_5$; and $F = FeO + MgO + MnO$. $A+C+F$ (in mols) are recalculated to 100%; the presence of excess SiO_2 is assumed. Cf: *AFM diagram*; *A'KF diagram*.

achene A dry, one-seeded, indehiscent fruit developed from a simple ovary with unfused seed coat and fruit wall. Also spelled: *akene*.

achlamydate Said of a gastropod without a mantle.

achoanitic Said of the condition in a nautiloid in which septal necks are vestigial or absent. Syn: *aneuchoanitic*.

achondrite A *stony meteorite* that lacks chondrules. Achondrites are commonly more coarsely crystallized than *chondrites*, and nickel-iron is almost completely lacking in most of them; they represent meteorites that are most like terrestrial rocks, with sizable fragments of various minerals visible to the naked eye. Adj: *achondritic*.

achroite A colorless variety of tourmaline, used as a gemstone.

acicular [cryst] Said of a crystal that is needlelike in form. Cf: *fascicular; sagenitic*.

acicular [sed] Said of a sedimentary particle whose length is more than three times its width (Krynine, 1948, p.142). Cf: *platy*.

acicular ice Freshwater ice consisting of numerous long crystals and hollow tubes having variable form, layered arrangement, and a content of air bubbles; it forms at the bottom of an ice layer near its contact with water. Syn: *fibrous ice; satin ice*.

aciculate Needle-shaped, or having a needle-like point; esp. said of a slender gastropod shell that tapers to a sharp point.

acid adj. (a) *silicic*. (b) *acidic [geol]*. (c) Said of a plagioclase that is sodic, i.e. "acid plagioclase".

acid bottle A glass, tube-like bottle partly filled with dilute hydrofluoric acid and formerly used in an *acid-dip survey* to determine the departure of a drill hole from the vertical.

acid clay A clay that yields hydrogen ions in a water suspension; e.g. "Japanese acid clay", a variety of fuller's earth occurring in Kambara, Japan.

acid-dip survey A method of determining the angular inclination of a drill hole in which an *acid bottle*, inserted in a watertight metal case, is lowered into the hole and left for 20-30 minutes, during which time the acid etches the bottle at a level plane from which the deviation from the vertical can be measured. The method was in general use before more precise techniques became available.

acidic (a) A syn. of *silicic*. "Acidic" is one of four subdivisions of a widely used system for classifying igneous rocks on the basis of silica content: (acidic), *basic, intermediate,* and *ultrabasic*. It does not imply properties as would be used by a chemist. The term is deprecated by some because of its confusing nature. (b) Applied loosely to any igneous rock composed predominantly of light-colored minerals having a low specific gravity and less than 65% silica. Partial syn: *felsic.*---Syn: *acid* [*geol*].

acidite A silicic rock.

acidity coefficient *oxygen ratio.*

acidity quotient *oxygen ratio.*

acid plagioclase A variety of plagioclase having a relatively high content of silicic acid (in the anhydrous form SiO_2); e.g. an Ab-rich member such as albite or oligoclase.

acid soil A soil whose root zone is acidic, having a pH of less than 6.6 or 7.0, i.e. less than that of a neutral soil. There is a preponderance of hydrogen over hydroxyl ions in the soil solution.

ac-joint A partial syn. of *cross joint*, used for a cross joint in folded sedimentary rock that is parallel to the fold axis.

aclinal A little-used term said of strata that have no inclination; horizontal. Not to be confused with *acclinal.*

acline [paleont] A syn. of *orthocline* as used to describe the hinge teeth or shell body of a bivalve mollusk.

acline-A twin law *Manebach-Ala twin law.*

acline-B twin law A twin law for parallel twins in feldspar with twin axis *b* and composition plane (100). Cf: *x-Carlsbad twin law.*

aclinic line *magnetic equator.*

acme That point in the phylogeny of a species, genus, family, etc. at which greatest abundance and/or variety culminates. See also: *paracme.*

acme zone *peak zone.*

acmite A brown or green mineral of the clinopyroxene group: $NaFe(SiO_3)_2$. It occurs in certain alkali-rich igneous rocks. Symbol: Ac. Syn: *aegirine.*

acmite-augite A mineral intermediate between augite and acmite; a variety of augite rich in sodium and ferric iron. Syn: *aegirine-augite.*

acolpate Said of pollen grains without furrows (colpi). In practice, such pollen grains are sometimes difficult to distinguish from *alete* spores. Cf: *inaperturate.*

acoustic impedance The product of seismic velocity and density. See also: *acoustic resistance; acoustic reactance.* Syn: *specific acoustic impedance; characteristic impedance.*

acoustic intensity *sound intensity.*

acoustic log A log that measures the physical properties of rocks around a borehole by acoustic means; a *sonic log.*

acoustic reactance The imaginary part of *acoustic impedance,* usually expressed in acoustic ohms. Cf: *acoustic resistance.*

acoustic resistance The real (as opposed to the imaginary) component of *acoustic impedance,* usually measured in acoustic ohms. Cf: *acoustic reactance.* See also: *specific acoustic resistance.*

acoustics The study of sound, including its production, transmission, reception, and utilization. With reference to Earth sciences, it is especially relevant to oceanography.

acoustic wave A wave that contains and transmits sound energy. In the solid Earth, an acoustic wave is the *P* type of seismic wave. Syn: *sound wave; sonic wave.*

acre A unit of land area used in U.S. and England, equal to 43,560 sq ft, 4840 sq yd, 160 square rods, 10 square chains, 1/640 square miles, or 0.405 hectares. It is based on an old approximate unit equal to the amount of land plowed by a yoke of oxen in a day.

acre-foot The volume of water required to cover one acre to a depth of one foot; 43,560 cu ft or 325,851 U.S. gal. Commonly used in measuring volumes of water or reservoir storage space. See also:*acre-inch.*

acre-inch The volume of water required to cover one acre to a depth of one inch. See also: *acre-foot.*

acrepid Said of a desma (of a sponge) that lacks an axial canal, implying that it was not formed about a crepis.

acritarch A unicellular, or apparently unicellular, resistant-walled microscopic organic body of unknown or uncertain biologic relationship and characterized by varied sculpture, some being spiny and others smooth. Many if not most acritarchs are of algal affinity, but the group is artificial. They range from Precambrian to Holocene, but are esp. abundant in Precambrian and early Paleozoic. The term was proposed by Evitt (1963, p.300-301) as "an informal, utilitarian, 'catch-all' category without status as a class, order, or other supra-generic unit" consisting of "small microfossils of unknown and probably varied biological affinities consisting of a central cavity enclosed by a wall of single or multiple layers and of chiefly organic composition". See also: *hystrichosphaerid; dinoflagellate.*

acrobatholithic Said of a mineral deposit occurring in or near an exposed batholith dome; also, said of the stage of batholith erosion in which that area is exposed (Emmons, 1933). The term is little used today. Cf: *cryptobatholithic; embatholithic; endobatholithic; epibatholithic; hypobatholithic.*

acron The anteriormost part of the cephalon of a crustacean, carrying the eyes and antennules.

acrotretacean Any inarticulate brachiopod belonging to the superfamily Acrotretacea, characterized by a conical to subconical, rarely convex, pedicle valve.

acrozone (a) A syn. of *range zone,* suggested by Moore (1957). (b) A term that means *peak zone* (Henningsmoen, 1961, p.68).

actinal Referring to the *oral* aspect; e.g. pertaining to the under or mouth side of the test of an echinoid or to the side of a crinoid theca or plate containing the mouth. Ant: *abactinal.*

actine (a) One of the individual branches of the triaene or triode in the ebridian skeleton. (b) A star-shaped spicule, as of a sponge.

actinodont Said of the dentition of certain bivalve mollusks of early origin having hinge teeth radiating from the beak (the outer teeth being more or less elongate).

actinolite A bright-green or grayish-green monoclinic mineral of the amphibole group: $Ca_2(Mg,Fe)_5Si_8O_{22}(OH)_2$. It may contain manganese. Actinolite is a variety of asbestos, occurring in long, slender, needle-like crystals and also in fibrous, radiated, or columnar forms in metamorphic rocks (such as schists) and in altered igneous rocks. Symbol: Ac. Cf: *tremolite.*

actinometer Any device which measures the intensity of radiation capable of effecting photochemical changes, particularly that of the Sun. Actinometers may be classified according to the quantities which they measure, as: a *pyrheliometer,* which measures the intensity of direct solar radiation; a *pyranometer,* which measures global radiation (the combined intensity of direct solar radiation and diffuse sky radiation); and a *pyrgeometer,* which measures the effective terrestrial radiation (Marks, 1969, p.34).

actinomorphic Said of an organism or organ that is radially symmetric or capable of division into essentially symmetric halves by any longitudinal plane passing through the axis. Cf: *zygomorphic.*

actinopod Any protozoan belonging to the class Actinopoda and characterized by protoplasmic extensions radiating from the spheroidal main body. Cf: *rhizopod.*

actinosiphonate Said of endosiphuncular deposits of a nautiloid, consisting of radially arranged longitudinal lamellae.

actinostele A type of *stele* consisting of alternating or radial groups of xylem and phloem within a pericycle and having a star shape in cross section.

activation [clay] The act or process of treating clay (such as bentonite) with acid so as to improve its adsorptive properties or to enhance its bleaching action, as for use in removing colors from oils.

activation [radioactivity] The process of making a substance radioactive by bombarding it with nuclear particles. The radioactivity so produced is called *induced radioactivity.*

activation analysis A method of identifying stable isotopes of elements in a sample by irradiating the sample with neutrons, charged particles, or gamma rays to render the elements radioactive, after which the elements are identified by their characteristic radiations. Cf: *neutron activation*. Syn: *radioactivation analysis*.

activation energy The extra amount of energy which any particle or group of particles must have in order to go from one energy state into another, such as changes in phase, and movement of particles in diffusion. The greater the amount of energy involved, the higher the resistance to the change, or the *potential barrier*.

active Said of a karst feature that contains running water, or that is still being developed by the action of running water.

active cave *live cave*.

active channel A channel, on an alluvial fan, in which runoff flows.

active earth pressure The minimum value of lateral *earth pressure* exerted by soil on a structure, occurring when the soil is allowed to yield sufficiently to cause its internal shearing resistance along a potential failure surface to be completely mobilized. Cf: *passive earth pressure*.

active fault A fault along which there is recurrent movement, that is usually indicated by small, periodic displacements or seismic activity. Cf: *dead fault*.

active glacier (a) A glacier that has an accumulation area, and in which the ice is flowing. Ant: *dead glacier*. (b) A glacier that moves at a comparatively rapid rate, generally occurring in a maritime environment at low latitudes where accumulation and ablation are both large. Ant: *passive glacier*.

active ice That part of a glacier showing clear evidence of movement, such as crevassing.

active layer [eng geol] The surficial deposit that undergoes seasonal changes of volume, swelling when frozen or wet and shrinking when thawing and drying.

active layer [permafrost] A surface layer of ground (or soil), above the permafrost, that is alternately frozen each winter and completely thawed each summer; it represents the *seasonally frozen ground*, only on permafrost, and its thickness varies from several centimeters to a few meters. Syn: *annually thawed layer; mollisol*.

active method A construction method in permafrost areas by which the frozen ground is thawed and removed or kept unfrozen at and near the structure. Ant: *passive method*.

active patterned ground Patterned ground that is actually growing or still developing at the present time. Ant: *fossil patterned ground*.

active permafrost Permafrost that is able to revert to a permanently frozen state under present climatic conditions after having been thawed by natural (as an unusually warm summer) or by artificial means. Ant: *passive permafrost*.

active seismometer A seismometer that is used in an active experiment, i.e. to detect artificially induced signals. Cf: *passive seismometer*.

active system An electromagnetic sensor which emits a signal and measures its returned characteristics. *Radar* is an example.

active volcano A volcano that is erupting; also, a volcano that is not presently erupting but that is expected to do so. There is no precise distinction between an active and a *dormant volcano* in this sense. Cf: *extinct volcano; inactive volcano*.

active water Water with corrosive capabilities.

activity [chem] (a) A quantity specifically defined to bear the same mathematical relationship to chemical potential in a nonideal solution that concentration bears in an ideal one. It is expressed numerically in units of concentration, and is usually symbolized by *a*. See also: *activity coefficient*. Syn: *relative fugacity*. (b) The tendency of a substance to react spontaneously and energetically with other substances.

activity [radioactivity] The intensity of radioactive emission of a substance, measured as the number of atoms decaying per unit of time. See also: *specific activity*.

activity coefficient The ratio of chemical *activity* to concentration.

activity index The rate of change with altitude of the net balance of a glacier, measured in the vicinity of the equilibrium line. High values indicate vigorous transfer of excess accumulation to lower altitudes (e.g. temperate maritime glaciers); low values indicate minimal or sluggish transfer (e.g. polar continental glaciers). Usually measured in millimeters per meter. The term was introduced by Meier in 1961. Syn: *energy of glacierization*. Cf: *ablation gradient*.

activity ratio In a sediment, the ratio of the plasticity index to the percentage of clay-sized minerals.

actual Said of faults and fault movements that are described in terms of *slip* rather than of separation. Cf: *apparent*.

actual age *absolute age*.

actual horizon A great circle on the celestial sphere whose plane is perpendicular to the direction of the plumb line (or the direction of gravity) at the point of observation. It is usually referred to as the *horizon*. Syn: *rational horizon*.

actualism *uniformitarianism*.

actual relative movement *slip* [struc geol].

actuopaleontology The paleontologic study of an area using the principle of uniformitarianism.

acute bisectrix The *bisectrix* of the acute angle between the axes of a biaxial crystal. Cf: *obtuse bisectrix*.

acyclic In plant morphology, pertaining to attachment of three or more parts, e.g. leaves, in a spiral pattern. Cf: *whorled*.

adamant A very hard mineral, stone, or metal, real or imaginary; any substance of impenetrable hardness. The term was formerly used of the diamond and of corundum.

adamantine Said of a very hard mineral, the luster of which is brilliant, e.g. diamond.

adamantine spar Silky brown *corundum*.

adamellite In English-speaking countries and in the USSR, a syn. of *quartz monzonite* of U.S. usage (i.e. quartz as 10-50% of the felsic minerals, and a plagioclase/total feldspar ratio of 35-65%). The term was first used in this sense by Brögger; it was originally used, however, by Cathrein for an orthoclase-bearing *tonalite* of Monte Adamello, Italy.

Adamic earth A term used for common clay, in reference to the material of which Adam, the first man, was made; specif. a kind of red clay (Humble, 1843, p.4).

adamite A colorless, white, or yellow orthorhombic mineral: $Zn_2(AsO_4)(OH)$. It is dimorphous with paradamite. Originally spelled: *adamine*.

adapertural (a) Toward the aperture of a gastropod shell. Ant: *abapertural*. (b) Toward the mouth of an ammonoid, or toward the aperture of an ammonoid shell; adoral or forward.

adapical (a) Toward the apex of a gastropod or cephalopod shell. (b) Toward the apical system of an echinoid.

adaptation Modification, as the result of natural selection, of an organism or of its parts so that it becomes better fitted to exist under the conditions of its environment.

adaptive grid A changing series of ecologic zones having time as one dimension.

adaptive radiation Subdivision of a group of organisms into diversified groups as a result of evolution controlled by genetic and environmental factors; occupation of equivalent niches in comparable habitats by ecologically similar but taxonomically distinct organisms, as a result of evolution. Cf: *convergence*. Syn: *divergence*.

adaptive zone A unit of environment defined in terms of its occupation by a single kind of organism.

adaxial Facing, directed toward, or situated on the same side as, the axis or center of the axis, as of an organ, plant, or invertebrate; ventral or posterior. Also, said of the adaxial side. Ant: *abaxial* [paleont].

adcumulate A cumulate formed by *adcumulus growth*, with intercumulus material comprising less than five percent of the rock. Cf: *mesocumulate*.

adcumulus growth Continued growth of cumulus crystals from material of the same composition so that the crystals are unzoned. This process reduces the intercumulus liquid by forcing it out of the intercumulus. See also: *adcumulate*.

adder stone A *serpent stone* formerly believed to be formed by an adder.

addition solid solution The addition of small atoms or ions, at random, in some interstices between atoms of a crystal structure. It may result in an *interstitial defect*. Syn: *interstitial solid solution*.

adductor *adductor muscle*.

adductor muscle (a) A muscle, or one of a pair of muscles, that contracts and thereby closes and/or tends to draw or hold together the valves of a bivalve shell (as in ostracodes, brachiopods, and pelecypods). Two adductor muscles, each dividing dorsally, are commonly present in articulate brachiopods; two pairs of adductor muscles (anterior and posterior), passing almost directly from the dorsal to the ventral side between the valves, are commonly present in inarticulate brachiopods. (b) Any transverse muscle (esp. that of the maxillary segment) for closure of aperture in a cirripede crustacean.--Cf: *diductor muscle*. Syn: *adductor*.

adductor muscle scar A *muscle scar* showing the final site where an adductor muscle was formerly attached. In articulate brachiopods, a single pair of such scars is located between diductor impressions in the pedicle valve and two pairs (anterior and posterior) in the brachial valve (TIP, 1965, pt.H, p. 139). See also:*cicatrix*. Syn: *adductor scar*.

adductor pit A depression that develops on the interior of a scutum for attachment of an adductor muscle of a cirripede crustacean.

adductor ridge A linear elevation that develops (in association with an adductor pit) on the interior of a scutum of certain cirripede crustaceans (such as Balanomorpha).

adelite A mineral: $CaMg(OH)AsO_4$. It sometimes has appreciable fluorine.

adeoniform Said of the form of a bilamellar, erect bryozoan colony such as that characterizing the family Adeonidae (order Cheilostomata).

ader wax *ozocerite*.

adfreezing The process by which two objects adhere to one another due to the binding action of ice as a result of the freezing of water; applied in permafrost studies.

adherent In plant morphology, pertaining to unlike parts that touch each other but are not fused. Cf: *adnate; connate; coherent*.

adhesion The molecular attraction between contiguous surfaces. Cf: *cohesion*.

adhesive water *pellicular water*.

adiabatic In thermodynamics, pertaining to the relationship of pressure and volume when a gas or fluid is compressed or expanded without either giving or receiving heat. In an adiabatic process, compression causes a rise in temperature, and expansion, a drop in temperature (U.S. Naval Oceanographic Office, 1966). See also: *potential density; potential temperature* [oceanog].

adiagnostic A term applied to the texture of a rock, esp. an igneous rock, in which individual components cannot be distinguished even using a microscope. Also, applied to the mineral components themselves. The term was originally used by Zirkel in German as *adiagnostisch*. Ant: *eudiagnostic*.

adinole An argillaceous sediment which has undergone albitization as a result of contact metamorphism along the margins of a sodium-rich mafic intrusion. Cf: *Schalstein; spilosite; spotted slate*. See also:*desmosite*.

adipocere A waxy or unctuous, brownish or light-colored natural substance consisting mainly of free fatty acids, hydroxy acids, and their calcium and magnesium salts, formed on decay of human, animal, or fish remains in damp areas or fresh or salt water, and whose consistency may range from soft and pliable for recent material to hard and brittle for older material. It often replaces and takes the form of the decomposing body.

adipocerite *hatchettine*.

adipocire A syn. of *hatchettine*. Not to be confused with *adipocere*.

a direction *a axis* [struc petrol].

adjusted stream A stream that carves its valley parallel to the strike of the least resistant rocks over which it flows.

adjustment The determination and application of corrections to a series of survey observations for the purpose of reducing errors, removing internal inconsistencies, and coordinating and correlating the derived results within the survey itself or with previously existing basic data; e.g. the determination and application of orthometric corrections in order to make the elevations of all bench marks consistent and independent of the circuit closures, or the positioning of public-land lines on a topographic map to indicate their true, theoretical, or approximate location relative to the adjacent terrain and culture. Adjustment may be accomplished by mathematical procedures (such as by the method of least squares) or mechanically by graphical methods (such as those employed in plane-table surveying).

adjustment of cross section The tendency in glaciers and in rivers to change the size of every cross section of the channel to accomodate the volume of ice or water that must pass through it.

adjustor muscle One of a pair of two pairs of muscles in many articulate brachiopods, branching from the pedicle, and responsible for moving the position of the shell on the pedicle. A ventral pair is attached posteriorly and laterally from the diductor muscles, and a dorsal pair is on hinge plates or floor of brachial valve behind posterior adductor muscles (TIP, 1965, pt.H, p.139).

admission *admittance*.

admittance [chem] In a crystal structure, substitution of a trace element for a common element of higher valence, e.g. Li^+ for Mg^{++}. Cf: *capture; camouflage*. Syn: *admission*.

admittance [elect] The reciprocal of *impedance*, or the ratio of complex current to voltage in a linear circuit. The practical unit is the *mho*.

admixture (a) A term applied by Udden (1914) to one of the lesser or subordinate of several particle-size grades of a sediment. See also: *coarse admixture; fine admixture, distant admixture; proximate admixture*. (b) A material that is added to another to produce a desired modification; e.g. a substance (other than aggregate, cement, or water) added during the mixing of concrete.

adnate In plant morphology, pertaining to the fusion of unlike parts. Cf: *connate; coherent; adherent*.

adobe (a) A fine-grained, usually calcareous, hard-baked clayey deposit mixed with silt, usually forming as great sheets in the central or lower parts of desert basins, as in the playas of SW U.S. and in the arid parts of Mexico and Spanish America. It is probably a windblown deposit, although it is often reworked and redeposited by running water. The term was applied originally to an unburnt, sun-dried brick made of adobe mixed with straw, and later to the clayey material constituting the brick. (b) A heavy-textured clay soil derived from adobe deposits.---Etymol: Spanish. Cf: *loess*.

adobe flat A generally narrow plain formed by sheetflood deposition of fine sandy clay or adobe brought down by an ephemeral stream, and having a smooth, hard surface (when dry) usually unmarked by stream channels.

adolescence A stage of the cycle of erosion, following youth and preceding maturity, sometimes considered "early maturity". It is characterized by: slight development of mature features; disappearance of lakes, waterfalls, and rapids; well-marked and comparatively narrow stream channels; roughly U-shaped valleys; and a well-established drainage system. Syn: *topographic adolescence*.

adolescent Pertaining to the stage of *adolescence* of the cycle of erosion; esp. said of a valley having a well-cut, smoothly graded stream channel that may reach base level at its mouth, or of a coast marked by low but nearly continuous sea cliffs.

adont Said of a class of ostracode hinges having no teeth, with a ridge or bar in one valve fitting into a groove in the opposed valve.

adoral Located or directed toward or near the mouth of an invertebrate; e.g. an "adoral surface" of an echinoderm theca bearing the mouth or ambulacral grooves, or the "adoral direction" toward the aperture of an ammonoid shell, or an "adoral fiber" representing a large thread of cytoplasm tissue leading from neuromotorium to the edge of peristome in a ciliate protozoan (such as a tintinnid). Cf: *aboral*.

adradial adj. (a) Said of the position corresponding to the boundary between ambulacral and interambulacral areas of an echinoid. (b) Directed toward the axis of an asterozoan ray. (c) Pertaining to a radius of the third order in a coelenterate.--n. (a) One of the small plates lining ambulacra in certain edrioasteroids. (b) One of a series of ossicles on the aboral surface of an asterozoan ray.

adret A mountain slope so oriented as to receive the maximum available amount of light and warmth from the Sun during the day; esp. a southward-facing slope of the Alps. Etymol: French dialect, "good side". Cf: *ubac*.

adsorbed water Water held by *adsorption*, as contrasted with absorption and chemical combination. Its physical properties are substantially different from those of *absorbed water* or chemically combined water at the same temperature and pressure.

adsorption Adherence of gas molecules or of ions or molecules in solutions to the surfaces of solids with which they are in contact. *Adsorbed water* in soil is held so strongly that it is resistant to the pull of gravity and to capillary action. Verb: *to adsorb*. Cf: *adsorption [geol]*; *absorption [grd wat]*.

adularescence A floating, billowy, white or bluish light, seen in certain directions as a gemstone (usually adularia) is turned; it is caused by diffused reflection of light from parallel intergrowths of another feldspar of slightly different refractive index from the main mass of adularia. It is often called *schiller*.

adularia A moderate-to-low temperature mineral of the alkali feldspar group: $KAlSi_3O_8$. It is a weakly triclinic (formerly regarded as an apparently monoclinic) variety of orthoclase typically occurring in well-developed, usually transparent, and colorless to milky-white (and often opalescent) pseudo-orthorhombic crystals in fissures in crystalline schists, esp. in the region of the Swiss Alps. Adularia displays pearly internal reflections and a fascinating variety of optical behavior between crossed nicols. It typically has a relatively high content of barium.

adularization Introduction of, or replacement by, adularia, as in the *poenites* of Timor.

advance [coast] A continuing seaward movement of a shoreline. Also, a net seaward movement of the shoreline during a specified period of time. Ant: *recession*. Cf: *progradation*. Syn: *progression*.

advance [glaciol] (a) The forward and downslope movement of the terminus of a glacier, resulting from a period of positive net balance, or an excess of accumulation over ablation. Its rate is usually measured in meters per year. (b) A time interval marked by an advance or general expansion of a glacier.--See also: *readvance*. Syn: *glacial advance; glacier advance*.

advance-cut meander A meander in which the outer bank of the channel is eroded so rapidly that deposition on the inner bank failed to keep pace, thereby widening the channel (Mel-

ton, 1936, p.598-599). Cf: *forced-cut meander*. Syn: *induced meander*.

advanced dune A sand dune formed on the windward side of a larger *attached dune*, remaining separate from it by the eddy motion of the wind.

advection [meteorol] Large-scale, horizontal movements within the Earth's atmosphere. Cf: *convection*.

advection [oceanog] The horizontal or vertical flow of sea water as a current.

advection [tect] Lateral mass movements of mantle material; such movement has been proposed as the cause of strike-slip movements along the midoceanic ridges. Cf: *convection*.

adventitious Said of a plant part that arises from an unusual place, e.g. a root that arises from a leaf or stem rather than from a primary root.

adventitious avicularium An avicularium (of a bryozoan) occupying some position on the external wall of an autozooid.

adventitious lobe A secondary lobe of an ammonoid suture, formed by subdivision of the first lateral saddle. Syn: *adventive lobe*.

adventitious stream A stream resulting from accidental variations of conditions, generally in an area that is approaching topographic maturity (Horton, 1945, p.341-342).

adventive *parasitic*.

adventurine A syn. of *aventurine* (esp. "aventurine feldspar").

advolute Said of a gastropod shell whose whorls barely touch one another but are not distinctly embracing or overlapping, or said of a coiled cephalopod shell in which the outer whorl touches but does not cover any part of the adjacent inner whorls. Cf: *evolute; involute; convolute*.

adyr (a) A term used in Turkmenia for a part of a desert plain devoid of sands and having soft ground. Cf: *kyr*. (b) A term used in Kazakhstan for a small flat top of relict high ground or of a mesa-like hill. (c) A term loosely applied in central Asia to a low mountain, a small hill, an eroded ridge with gentle slopes, etc.---Etymol: Turkish. The definitions above are from Murzaevs & Murzaevs (1959, p.20).

aegagropile A *lake ball* consisting of radial, outgrowing, hair-like filaments formed by algae. Etymol: Greek *aigagros*, "goat", + *pilus*, "hair". Cf: *algal biscuit*. Syn: *aegagropila; egagropile*.

aegirine A syn. of *acmite*. The term is sometimes applied to impure acmite containing calcium, magnesium, or aluminum. Syn: *aegirite*.

aegirine-augite A syn. of *acmite-augite*. Also spelled: *aegirite-augite; aegirinaugite*.

aegirite *aegirine*.

Aeneolithic Var. of *Eneolithic*.

aenigmatite A mineral: $Na_2Fe_5TiSi_6O_{20}$. Syn: *enigmatite*.

aeolian *eolian*.

aeolianite *eolianite*.

aeolotropic A syn. of *anisotropic*. Also spelled: *eolotropic*.

aeon Var. of *eon*.

aeration In a soil, the supplying of air and other gases to the pores.

aeration porosity The volume of interstices that do not hold water at a specified low moisture tension (Jacks, et al, 1960). Cf: *noncapillary porosity*.

aerial Pertaining to the air; related to, located in, or consisting of, the Earth's atmosphere. Not to be confused with *areal*.

aerial arch An anticline, the crest of which has been eroded.

aerial mapping The taking of aerial photographs for making maps and for geologic interpretation.

aerial mosaic *mosaic* [photo].

aerial photogrammetry Photogrammetry utilizing aerial photographs.

aerial photograph Any photograph taken from the air, such as a photograph of a part of the Earth's surface taken by a camera mounted in an aircraft. Syn: *aerial photo; air photograph; aerophoto; airview*.

aerial survey (a) A survey using aerial photographs as part of the surveying operation. (b) The taking of aerial photographs for surveying purposes.

aerial triangulation *aerotriangulation.*

aerobic (a) Said of an organism (esp. a bacterium) that can live only in the presence of free oxygen; also, said of its activities. Noun: *aerobe.* (b) Said of conditions that can exist only in the presence of free oxygen. Cf: *anaerobic.*

aerobic decay Decomposition of organic substances, primarily by microorganisms, in the presence of free oxygen; the ultimate decay products are carbon dioxide and water.

aerogeography The geographic study and interpretation of features by means of aerial observation and aerial photographs.

aerogeology The geologic study and interpretation of Earth features by means of aerial observation and aerial photographs. The term is loosely used as a synonym of *photogeology.*

aeroides Pale sky-blue aquamarine.

aerolite *stony meteorite.*

aerolithology An obsolete term for the science that deals with meteorites. Cf: *meteoritics.*

aeromagnetic survey A *magnetic survey* made with an *airborne magnetometer.*

aerophoto *aerial photograph.*

aerosiderite An obsolete syn. of *siderite* meteorite.

aerosiderolite An obsolete syn. of *stony-iron meteorite.*

aerosol A *sol* in which the dispersion medium is a gas (usually air) and the dispersed or colloidal phase consists of solid particles or liquid droplets; e.g. mist, haze, most smokes, and some fogs.

aerospace A mnemonic term derived from *aeronautics + space* to denote both the Earth's atmosphere and outer space as a single unit.

aerotriangulation Process of *phototriangulation* accomplished by means of aerial photographs. Syn: *aerial triangulation.*

aerugite A mineral: $Ni_9As_3O_{16}$.

aeschynite A mineral: $(Ce,Ca,Fe,Th)(Ti,Nb)_2(O,OH)_6$. It is isomorphous with priorite. Syn: *eschynite.*

aethoballism A term proposed by Grabau (1904) for local metamorphism resulting from contact with a meteorite. The term is now obsolete. Cf: *symphrattism.*

aetites A syn. of *eaglestone.* Etymol: Latin, from Greek *aetites*, "of an eagle".

affine Said of the deformation producing a tectonite, in which predeformational lineations are still linear after deformation; homogeneous deformation.

affinity In biology, the relationship that exists between two individuals or groups of individuals that closely resemble each other but that do not belong to the same taxon. Abbrev: *aff.*

affluent adj. Said of a stream flowing toward or into a larger stream or into a lake.---n. An affluent stream; esp. an *influent* flowing into a lake. The term, originally introduced by Jackson (1834, p.77-78) as distinct from a *confluent,* is obsolescent as the synonym *tributary* is more commonly used.

afflux (a) The upstream rise of water level above the normal surface of the water in a channel, caused by contraction or obstruction of the normal waterway. (b) The difference between high flood levels upstream and downstream of a weir.

afghanite A mineral: $(Na,Ca,K)_{12}(Si,Al)_{16}O_{34}(Cl,SO_4,CO_3)_4 \cdot H_2O$.

AFMAG method The use of natural electromagnetic noise to study changes in the Earth's resistivity. It is applicable in mineral prospecting. Etymol: An acronym for audio-*f*requency *mag*netic (technique).

AFM diagram A triangular diagram showing the simplified compositional character of a metamorphosed pelitic rock by plotting molecular quantities of the three components: $A = A_{12}O_3$; $F = FeO$; and $M = MgO$. Cf: *ACF diagram; A'KF diagram.*

A-form The megalospheric form of a foraminifer. Cf: *B-form.*

afterdamp The gas remaining in a coal mine after an explosion of *firedamp* or after a fire. It includes carbon monoxide and carbon dioxide. Cf: *whitedamp; blackdamp.*

aftershock An earthquake which follows a larger earthquake or *main shock* and originates at or near the focus of the larger earthquake. Generally, major earthquakes are followed by a large number of aftershocks, decreasing in frequency with increasing time. Such a series of aftershocks may last many days for small earthquakes or even many months for large earthquakes. Cf: *foreshock.*

afterworking *creep recovery.*

Aftonian Pertaining to the first interglacial stage of the Pleistocene Epoch in North America, following the Nebraskan and preceding the Kansan glacial stages. Etymol: Afton, town in Iowa. See also: *Günz-Mindel.*

afwillite A mineral: $Ca_3Si_2O_4(OH)_6$.

agalite A fine fibrous variety of talc, pseudomorphous after enstatite. Syn: *asbestine.*

agalmatolite A soft, waxy, compact mineral or stone (such as pinite, pyrophyllite, and steatite) of a grayish, greenish, yellowish, or brownish color, used by the Chinese for carving small images, miniature pagodas, and other objects. Syn: *figure stone; pagodite; lardite.*

agaric mineral *rock milk.*

agate A translucent cryptocrystalline variety of quartz, being a variegated *chalcedony* frequently mixed or alternating with opal, and characterized by colors arranged in alternating stripes or bands, in irregular clouds, or in moss-like forms. Agate is found in virtually all colors, usually of low intensity; it commonly occupies vugs in volcanic rocks and cavities in some other rocks. Cf: *onyx.* See also: *banded agate; clouded agate; moss agate.*

agate jasper An impure variety of agate consisting of jasper with veins of chalcedony. Syn: *jaspagate.*

agate opal Opalized agate. Cf: *opal-agate.*

agathocopalite *kauri.*

agatized wood A syn. of *silicified wood.* See also: *wood agate.*

age [geochron] (a) A geologic-time unit shorter than an *epoch* [geochron] and longer than a *subage,* during which the rocks of the corresponding *stage* were formed. Syn: *chron.* (b) A term used informally to designate a length of geologic time during which the rocks of any stratigraphic unit were formed. (c) A division of time of unspecified duration in the history of the Earth, characterized by a dominant or important type of life form; e.g. the "age of mammals". (d) The time during which a particular geologic event or series of events occurred or was marked by special physical conditions; e.g. the "Ice Age". (e) The position of anything in the geologic time scale; e.g. "the rocks of Miocene age". It is often expressed in years.----See also: *geologic age.*

age [ice] The stage of development of sea ice; the term usually refers to the length of time since the formation of the ice and to its thickness.

aged Said of a topographic feature, such as a shore, that is approaching reduction to base level.

age determination The evaluation of geologic age by faunal or stratigraphic means, or by physical methods involving determination of isotopic composition or radioactivity. Some methods of relative age determination are based on the extent of chemical change, like the hydration of obsidian or the fluorine uptake of bone.

age equation The relationship between radioactive decay and geologic time which, expressed mathematically, is $t=1/\lambda.1n(1+D/P)$, where t is the age of a rock or mineral specimen, D is the number of daughter isotopes, today, P is the number of parent isotopes, today, $1n$ is the natural logarithm (logarithm to base e), and λ is the decay constant. The age equation is as taken from *Geologic time* (USGS, 1970, p.13).

age of amphibians An informal designation of the late Paleozoic, i.e., the *Carboniferous* and the *Permian.*

age of coal An informal designation of the *Carboniferous*; *coal age*.

age of cycads An informal designation of the *Jurassic*.

age of ferns An informal designation of the *Pennsylvanian*.

age of fishes An informal designation of the *Silurian* and the *Devonian*.

age of gymnosperms An informal designation of the *Mesozoic*.

age of mammals An informal designation of the *Cenozoic*.

age of man An informal designation of the *Quaternary*.

age of marine invertebrates An informal designation of the *Cambrian* and the *Ordovician*.

age of reptiles An informal designation of the *Mesozoic*; *reptilian age*.

age of the Earth The age of the Earth, i.e. the length of time that the Earth has existed essentially as it is now, is 4.5 b.y. as determined from isotopic investigation of common lead relationships and uranium-lead and rubidium-strontium dating of meteorites. Oldest dated terrestrial rocks are approximately 3.5 b.y.

age of the Moon The time elapsed in days since the last new Moon.

age of the Universe Usually refers to the Hubble age of the universe which, given by the reciprocal of the Hubble constant, is 13.7 b.y. Other estimates include 10 b.y. to 15 b.y., the time when nucleosynthesis is thought to have begun, and 4.7 b.y., the age of meteorites.

age of water The length of time since a water mass was last in contact with the atmosphere at the surface of the ocean.

age ratio The ratio of daughter to parent isotope upon which the *age equation* is based. However, for a valid age determination, the isotope system must have remained closed since solidification, metamorphism, or sedimentation, the decay constant must be known, and the sample must be truly representative of the rock from which it is taken.

age-specific eruption rate In stochastic treatment of volcanologic data based on renewal theory, if the volcano has not erupted in time interval $(0,x)$, the age-specific eruption rate, or *short eruption rate*, $\phi(x)$, is the limit of the ratio to Δx of the probability of eruption in time interval $(x, x + \Delta x)$. Its physical dimension is time^{-1}. Approximately $\phi(x)$ gives the probability of almost immediate eruption, ending repose period of known age x (Wickman, 1966, p.298). Syn: *eruption rate*.

agglomerate A term originally used by Lyell in 1831 for a chaotic assemblage of coarse, angular, pyroclastic materials. The term has been variously defined since then, and it should be defined in context to avoid confusion. Cf: *volcanic breccia*.

agglomerate ice Ice that has been formed by congealing a mixture of floating ice fragments, as in a lake.

agglomerating value A measure of the binding qualities of coal determined by fusing tests in which no inert material is heated with the sample. Cf: *agglutinating value*.

agglomeroplasmic Said of an arrangement in a soil fabric whereby the plasma occurs as loose or incomplete fillings in the intergranular spaces between skeleton grains (Brewer, 1964, p.170). Cf: *porphyroskelic; intertextic*.

agglutinate n. A welded pyroclastic deposit characterized by vitric material cementing the fragments, the presence of scoria, and the absence of a tuff matrix. Also spelled: *agglutinite*.

agglutinate cone *spatter cone*.

agglutinated Said of the tests of certain protists (some foraminifers, thecamoebians, and tintinnids) composed of foreign particles (sand grains, sponge spicules, mica flakes, etc.) bound together by cement.

agglutinating value A measure of the binding qualities of a coal and an indication of its caking or coking characteristics, determined by the ability of the coal, when fused, to combine with an inert material, such as sand. Cf: *agglomerating value*. Syn: *caking index*.

agglutination A syn. of sedimentary *cementation*, esp. in regard to coarse-grained rocks, such as breccia or conglomerate.

agglutinite Var. of *agglutinate*.

aggradation [permafrost] The spread or growth of permafrost, under present climatic conditions, due to natural or artificial causes. Ant: *degradation*.

aggradation [geomorph] (a) The building up of the Earth's surface by deposition; specif. the *upbuilding* performed by a stream in order to establish or maintain uniformity of grade or slope. See also: *gradation*. Cf: *degradation*. Syn: *upgrading*. (b) A syn. of *accretion*, as in the development of a beach.

aggradational plain A broad, fan-like plain with a nearly straight longitudinal profile, formed in an arid region by stream deposition.

aggradation recrystallization Recrystallization resulting in the enlargement of crystals. Ant: *degradation recrystallization*.

aggraded valley plain An *alluvial plain*, or a stream-filled flood plain; a plain formed by a stream aggrading its valley, the in-filling with alluvium on the valley floor attaining a thickness greater than that of the stream channel (Cotton, 1958, p.193).

aggrading neomorphism A kind of *neomorphism* in which the crystal size increases (Folk, 1965, p.23); e.g. *porphyroid neomorphism* and *coalescive neomorphism*. Ant: *degrading neomorphism*.

aggrading stream (a) A stream that is actively building up its channel or flood plain by being supplied with more load than it is capable of transporting. (b) A stream that is upbuilding approximately at grade.

aggregate [geol] (a) A mass or body of rock or mineral particles, or a mixture of both, separable by mechanical means, e.g. *mineral aggregate*. (b) Sand, gravel, or any clastic material in a bedded iron ore (Arkell & Tomkeieff, 1953, p.1).

aggregate [mater] Any of several hard, inert, construction materials (such as sand, gravel, shells, slag, crushed stone, or other mineral material), or combinations thereof, used for mixing in various-sized fragments with a cementing or bituminous material to form concrete, mortar, plaster, etc., or used alone as in railroad ballast or in various manufacturing processes (such as fluxing). See also: *coarse aggregate; fine aggregate; all-in aggregate*. Syn: *mineral aggregate*.

aggregated Said of a massive sulfide or other ore deposit in which the sulfide or other valuable constituent constitutes 20% or more of the total volume. The term is little used.

aggregate polarization In crystal optics, a pattern seen between crossed nicols in a fine-grained crystal aggregate, composed of the different interference colors of variously oriented grains. The presence of aggregate polarization may define a material as crystalline that in ordinary light appears amorphous.

aggregate structure In crystallography, a mass of individual crystals or grains which extinguish under the polarizing microscope at different intervals during the rotation of the stage.

aggressive Said of a magmatic intrusion, and of the magma itself, the emplacement of which forcibly created the space in which it intruded; a forcible intrusion. Cf: *permissive*. Syn: *invasive*.

aggrotubule A *pedotubule* composed of skeleton grains and plasma which occur essentially as recognizable aggregates within which there is no directional arrangement with regard to the external form (Brewer, 1964, p.239).

aging The process by which a *young lake* becomes an *old lake* as a result of filling, *eutrophication*, vegetation encroachment, and other actions.

aglet A tiny plate pierced by a single pore in a radiolarian.

agmatite Migmatite with appearance of breccia (Dietrich & Mehnert, 1961).

agnostid Any trilobite belonging to the order Agnostida, which includes small forms with subequal cephalons and pygidia and only two segments in the thorax. Their stratigraphic range is Lower Cambrian to Upper Ordovician.

Agnotozoic *Proterozoic*.

agonic line An *isogonic line* that connects points of zero magnetic declination. Its position changes according to the secular variation of the Earth's magnetic field.

agouni A term used in Morocco for a broad, generally dry gully carved by a torrent (Termier & Termier, 1963, p.399). Etymol: Berber. Cf: *asif*.

agpaite A group of feldspathoid-bearing igneous rocks first described from Ilimausak, Greenland, that includes sodalite-bearing nepheline syenite, naujaite, lujavrite, kakortokite, etc.

agpaitic Said of crystallization in the presence of an excess of alkali (esp. sodium) so that the amount of aluminum oxide is insufficient for the formation of aluminum silicates (Thrush, 1968, p.18).

agpaitic coefficient The ratio na + k/al, where na, k and al are relative amounts of sodium, potassium, and aluminum atoms, respectively, in a rock, esp. an alkalic igneous rock.

agric Said of a soil horizon in which at least 15% of the volume is accumulated clay, silt, and humus from an overlying cultivated, fertilized layer (SSSA, 1970).

agricolite *eulytite*.

agricultural geology The application of geology to agricultural needs, e.g. formation and composition of soils, mineral deposits used as fertilizers, the location of ground water. Syn: *agrogeology*.

agrogeology *agricultural geology*.

agrology An old term for *soil science*.

aguada In the karst region of Yucatan, Mexico, a shallow depression that collects water in the rainy season. Etymol: Spanish, "watering place".

aguilarite A mineral: Ag_4SeS.

ahermatypic coral A nonreef-building coral; a coral lacking symbiontic algae; a coral capable of living in cold, deep water. Ant: *hermatypic coral*. Syn: *ahermatype*.

ahlfeldite A mineral: $(Ni,Co)SeO_3.2H_2O$.

A horizon The uppermost mineral horizon of a soil, incorporating the following subdivisions: the A horizon, characterized by an accumulation of humic material; the A2 horizon, characterized by a concentration of quartz or other minerals in the sand and silt size fractions due to leaching of clay, iron, or aluminum; and the A3 horizon, transitional to the underlying B horizon. The A horizon is also called the *zone of leaching*.

Ahren's prism A type of polarizing prism having three cut and recemented segments; the ordinary rays of the light are reflected to the sides, and the extraordinary ray forms the polarized light.

aiguille A prominent, sharp, needle-shaped rock peak or pinnacle such as is commonly found in intensely glaciated mountain regions (as near Mont Blanc in the French Alps); it is a remnant of a septum between two cirques, the rest of which has been largely or wholly removed by erosion. Cf: *gendarme*. Pron: ai-*gwee*. Etymol: French, "needle". Syn: *needle*.

aikinite (a) A mineral: $PbCuBiS_3$. Syn: *needle ore*. (b) Wolframite pseudomorphous after scheelite.

ailsyte *paisanite*.

aimless drainage Drainage without a well-developed system, such as that in an area of glacial drift or karst topography.

ainalite A mixture of cassiterite and columbite.

air [coast] A Scottish term for a beach. Etymol: Old Norse *eyrr*, "gravelly bank".

air [meteorol] A commonly used term for the Earth's *atmosphere*.

air base (a) The line joining two air stations, usually in consecutive arrangement. (b) The distance between two air stations; the length of the air base. Cf: *photobase*. (c) The distance (at the scale of the stereoscopic model) between adjacent perspective centers as reconstructed in the plotting instrument.

airborne profile recorder An electronic instrument that emits a pulsed-type radar signal from an aircraft to measure vertical distances between the aircraft and the Earth's surface. Abbrev: APR. Syn: *terrain profile recorder*.

air dome *dome* [beach].

air dry The condition of a substance whose moisture content has come into approximate equilibrium with the humidity of the surrounding air.

air entrainment The insufflation of air into moving water due to the breaking of small waves or the turbulence of the water (ASCE, 1962).

air-fall deposition Showerlike falling of pyroclastic fragments from an *eruption cloud*. See also: *ash fall; pumice fall*.

air gap A gap through which air passes; specif. a *wind gap*.

air heave The process of deformation of plastic sediments by the enlargement of a pocket of air trapped in them, such as by the accretion of small air bubbles rising through sand exposed at low tide on a beach or tidal flat. See also: *gas heave*.

air-heave structure A crumpled sedimentary structure believed to have been produced by air heave, measuring several centimeters across, and characterized by an abrupt upward doming of laminae with a core of unlaminated sandstone (Stewart, 1956, p.159). See also:*gas-heave structure*.

air mass A widespread body of air having particular characteristics, esp. of temperature and moisture content, that were acquired at its *source region* and that are modified as it moves away from its source.

air photograph An *aerial photograph*. Syn: *air photo*.

air pressure *atmospheric pressure*.

air sac A cavity or *vesicle* in a pollen grain of a pine.

air shooting In seismic prospecting, the shooting technique of applying a seismic pulse to the Earth by detonating a charge or charges in the air. See also: *Poulter seismic method*.

air shrinkage The volume decrease that a clay undergoes in drying at room temperature.

air-space ratio In a soil, the ratio of the volume of water that can be drained by gravity from a saturated soil to the total volume of the soil's voids.

air station The *camera station* from which an aerial photograph is taken.

airview *aerial photograph*.

air-void ratio In a soil, the ratio of the volume of air space to the total volume of voids in the soil.

air volcano A *mud volcano* characterized more by the gas it emanates than by the mud and rocks thrown out.

air wave An acoustic wave in the air, whose source is a seismic event. Syn: *earthquake sound*.

air well A tower of loose rock, used in some desert countries to collect water, by condensation of moisture from the warm atmosphere on the relatively cooler rock surfaces within the tower, whose temperature fluctuates about the daily and annual mean in a smaller range than does that of the air.

Airy isostasy The hypothesis of the mechanism of *isostasy*, proposed by George Bedell Airy, that postulates an equilibrium of crustal blocks of the same density but of different size; thus the topographically higher mountains would be of the same density as other crustal blocks but would have greater mass and deeper *roots*. Cf: *Pratt isostasy*.

Airy phase (a) The predominant surface-wave group corresponding to a maximum or minimum group velocity. (b) More generally, any seismic *phase* associated with a stationary group velocity.

aisle In a cave, a traversable *passage* [speleo] that is generally high and narrow.

ait A British term for a small island in a lake or river; an islet. Syn: *eyot*.

Aitoff projection (a) A map projection derived from the azimuthal equidistant projection by doubling the horizontal distances from the central meridian until the entire spherical surface is within an ellipse whose major axis (equator) is twice the length of its minor axis (central meridian). It resembles the Mollweide projection, but the parallels (except the equator) and meridians (except the central meridian) are represented by curved lines and there is less distortion at the margins. Named after David Aitoff (d.1933), Russian geographer, who introduced the projection in 1889. (b) A name commonly, but incorrectly, given to the *Hammer-Aitoff projection*.

ajkaite A pale-yellow to dark reddish-brown, sulfur-bearing fossil resin found in brown coal. Syn: *ajkite*.

ajoite A mineral: $Cu_6Al_2Si_{10}O_{29}.5.5H_2O$.

akaganéite A mineral: beta-FeO(OH).

akatoreite A mineral: $Mn_9(Si,Al)_{10}O_{23}(OH)_9$.

akdalaite A mineral: $4Al_2O_3.H_2O$.

akene Var. of *achene*.

akenobeite A granodioritic aplite composed of an aggregate of orthoclase and oligoclase, the former in excess of the latter, with an aggregate of fine quartz grains in the interstices and with little ferromagnesian material present (Johannsen, 1939, p. 238). Its name is derived from Akenobe district, Japan.

akerite An augite-bearing syenite containing orthoclase, oligoclase, biotite, a green pyroxene, and quartz. Its name is derived from Aker, Norway.

akermanite A mineral of the melilite group: $Ca_2MgSi_2O_7$. It is isomorphous with gehlenite.

A'KF diagram A triangular diagram showing the simplified compositional character of a metamorphic rock by plotting molecular quantities of the three components: $A = Al_2O_3 + Fe_2O_3 - (Na_2O + K_2O + CaO)$; $K=K_2O$; and $F=FeO + MgO + MnO$. $A'+K+F$ (In mols) are recalculated by 100%; the diagram is used in addition to the *ACF diagram* when K minerals require representation. Cf: *AFM diagram*.

akinetic surface The surface in a sedimentary rock layer that was the outer surface of the lithosphere at the place and time that oscillation of base level at that point passed through its maximum (Goldman, 1921, p.8-20).

akmolith A now obsolete term for an igneous intrusion along a zone of décollement, with tonguelike extensions into the overlying rock.

akrochordite A mineral: $Mn_4Mg(AsO_4)_2(OH)_4.4H_2O$. It occurs in reddish-brown rounded aggregates.

aksaite A mineral: $MgB_6O_{10}.5H_2O$.

aktashite A mineral: $(Cu,Hg)_2AsS_3$ (?).

aktology Study of nearshore and shallow-water areas, conditions, sediments, life, and environments.

akyrosome A term used by Niggli (1954, p.191) for a subsidiary mass (such as a vein, nodule, band, lens, or block) of a complex rock; a minor part of a migmatite. This spelling appears in the English translation of Niggli's paper and probably should have been "akyriosome". Cf: *kyriosome*.

ala A wing or wing-like process or part of an organism; e.g. a wing-like flange in the diatom *Surirella*, or the wing-like extension of the ear of a brachiopod shell, or a wing-like extension of the ventral and lateral part of the carapace of an ostracode, or the triangular lateral part of a compartmental plate of a cirripede crustacean, delimited from the paries, which is overlapped by the adjacent compartmental plate and by the radius when the latter is present. Pl: *alae*.

Ala-A twin law A twin law in plagioclase, in which the twinning axis is the *a* axis and the composition plane is (001). An Ala-A twin is usually multiple and parallel, and often occurs with the Manebach twin law. Cf: *Manebach-Ala twin law; Ala-B twin law*.

alabandite A mineral: MnS. It usually occurs in iron-black massive or granular form. Syn: *manganblende; alabandine*.

alabaster (a) A firm, very fine-grained, and massive or compact variety of *gypsum*, usually snow-white and translucent but sometimes delicately shaded or tinted with light-colored tones of yellow, brown, red, orange, or gray. It is used as an interior decorative stone (esp. for carved ornamental vases and figures) and for statuary. (b) *onyx marble*.

Ala-B twin law A twin law in plagioclase, in which the twinning

axis is the *a* axis and the composition plane is (010). It often occurs with the albite twin law. Cf: *Ala-A twin law; albite-Ala twin law*.

aladzha Impure ozokerite containing an admixture of country rocks, found in the region of the Caspian Sea (Tomkeieff, 1954, p.21). Etymol: Tatar.

alaite A dubious mineral: $V_2O_5.H_2O$.

alalite A light-green variety of diopside.

alamandine A syn. of *almandine* garnet.

alamosite A mineral: $PbSiO_3$.

alar fossula A *fossula* developed in the position of an alar septum in a rugose coral or adjoining it on the side toward the counter septum.

alar projection A wing-like extension of a foraminiferal test.

alar septum One of two protosepta of a rugose corallite located about midway between the *cardinal septum* and the *counter septum*, distinguished from other protosepta by pinnate insertion of newly formed metasepta on the side facing the counter septum (TIP, 1956, pt.F, p.245). Symbol: A.

alaskaite [**mineral**] A sulfide mineral, being a complex mixture of lead, silver, copper, and bismuth.

Alaskan band A *dirt-band ogive*. The term is not to be encouraged because such ogives are not limited to Alaska and apparently are not even the most common forms of ogives on Alaskan glaciers.

alaskite In the U.S., a commonly used term for a granitic rock containing only a few percent of dark minerals. The term was introduced by Spur (1900) for holocrystalline-granular plutonic rock characterized by essential alkali feldspar and quartz, and little or no dark component. Johannsen (1919) proposed two subdivisions of alaskites: *kalialaskites*, which lack albite, and alaskites proper, which contain albite. In the recommendations of the Terminological Commission of the Petrographic Committee of the USSR (1969), the term alaskite is used to designate granitoid rocks in which quartz constitutes 20-60% of the felsic minerals and in which the ratio of alkali feldspar to total feldspar is greater than 90%; i.e. the equivalent of alkali granite, or *kaligranite*. Cf: *aplogranite; tarantulite*.

alaskite-quartz *tarantulite*.

alate Having wings or a winged form; e.g. said of an outward lateral extension in the ventral half of an ostracode valve, usually increasing in width backward and terminating abruptly, and tending to have a triangular shape (TIP, 1961, pt.Q, p.47), or said of the form of a brachiopod shell in which the valves are drawn out at the lateral ends of the hinge line to form wing-like extensions, or said of a bivalve-mollusk shell having auricles.

A layer The seismic region of the Earth equivalent to the *crust*, extending downward to the Mohorovičić discontinuity. It is part of a classification of the interior of the Earth, having layers A to G.

alb A flat or gently inclined narrow shelf separating the nearly vertical side of an alpine glacial trough from the mountain slope above. Cf: *alp*.

albedo As used in remote sensing the term commonly refers to the reflectivity of a body compared with that of a perfectly diffusing surface at the same distance from the Sun, and normal to the incident radiation. Albedo is sometimes used to mean the flux of the reflected radiation, e.g. the Earth albedo is 0.64 calorie per square centimeter. This usage is not encouraged. Syn: *Bond albedo*.

Albers projection An equal-area conic projection having two standard parallels, each of which lies equidistant between the central and extreme parallels of the map, and along which the scale is true. Meridians are represented by equally spaced straight lines that converge to a common point beyond the map limits, and parallels are represented by arcs of concentric circles whose center is at the common point and which are at right angles to the meridians. The meridional scale is too large between the standard parallels and too small beyond them; distances between the parallels decrease north and south of the standard parallels. The projection, when used for maps of the U.S. at 1/2,500,000, has minimum scale error (1.25%). Named after Heinrich C. Albers (1773-1833), German cartographer, who devised the projection in 1805.

Albertan North American provincial series: Middle Cambrian (above Waucoban, below Croixian).

albert coal An early name for *albertite*.

albertite A dark brown to black asphaltic pyrobitumen with conchoidal fracture occurring as veins 1 to 16 ft wide in the Albert Shale of Albert County, New Brunswick. It is partly soluble in turpentine, but practically insoluble in alcohol. It was earlier called *albert coal*. See also: *libollite; stellarite; impsonite; byerite*. Syn: *asphaltic coal (obs.)*.

Albian European stage: uppermost Lower Cretaceous, or Middle Cretaceous of some authors (above Aptian, below Cenomanian). Syn: *Selbornian*.

albic Said of a soil horizon from which clay and free iron oxides have been removed (or the oxides have been segregated) so that its color is determined by the sand and silt particles, and not by their coatings (SSSA, 1970).

albite (a) A colorless or milky-white triclinic mineral of the feldspar group: $NaAlSi_3O_8$. It is a variety of plagioclase with composition ranging from $Ab_{100}An_0$ to $Ab_{90}An_{10}$; it is also an alkali feldspar, representing the triclinic modification of sodium feldspar. Albite occurs in all groups of rocks, forming a common constituent of granite and of various acid-to-intermediate igneous rocks; it is widely distributed in low-temperature metamorphic rocks (greenschist facies), and is regularly deposited from hydrothermal solutions in cavities and veins. Albite crystals frequently exhibit polysynthetic twinning, predominantly after the albite law. Cf: *analbite*. Syn: *sodium feldspar; sodaclase; white feldspar; white schorl*. (b) The pure sodium-feldspar end member in the plagioclase series.

albite-Ala twin law A complex twin law in feldspar, in which the twin axis is perpendicular to [100] and the composition plane is (010). Cf: *Ala-B twin law*.

albite-Carlsbad twin law A complex twin law in feldspar, in which the twin axis is perpendicular to [001] and the composition plane is (010).

albite-epidote-amphibolite facies Metamorphic rocks formed under intermediate pressures and temperatures by regional dynamothermal metamorphism or in outer contact-metamorphic zones. It was named by Turner (1948) to replace Eskola's *epidote-amphibolite facies*. More recently, Fyfe et al (1958) interpreted it as the highest-temperature subfacies of the greenschist facies (quartz-albite-epidote- almandine subfacies). Exact interpretation is in question.

albite-epidote-hornfels facies Metamorphic rocks formed in the outermost parts of contact aureoles under very low fluid pressures (less than 1200 bars) and temperatures of 200°-300°C (Winkler, 1967). It is part of the *hornfels facies*. Cf: *pyroxene-hornfels facies; hornblende-hornfels facies*. Syn: *knotted-hornfels facies*.

albite porphyrite *albitite*.

albite twin law A twin law in triclinic feldspars, in which the twin plane and composition plane are (010). An albite twin is usually multiple and lamellar, and shows fine striations on the (001) cleavage plane.

albitite A porphyritic igneous rock containing phenocrysts of albite in a groundmass chiefly consisting of albite. Muscovite, garnet, apatite, quartz, and opaque oxides are common accessory minerals. Syn: *albitophyre; albite porphyrite*.

albitization Introduction of, or replacement by, albite, usually replacing a more calcic plagioclase.

albitophyre *albitite*.

Alboll In U.S. Dept. of Agriculture soil taxonomy, a suborder of the soil order *Mollisol*, characterized by an albic horizon and a mollic epipedon within or below which is a seasonally perched water table. There may also be an argillic or a natric

horizon, and mottles and/or iron-manganese concretions may occur in any of the horizons (SSSA, 1970). Cf: *Aquoll; Boroll; Rendoll; Udoll; Ustoll; Xeroll*.

alboranite An olivine-free hypersthene-bearing basalt named after Alboran Island of Spain.

alcove A large, deep niche formed in a precipitous face of rock; specif. a *spring alcove*.

alcove lands An angular landscape characterized by terraced slopes consisting of resistant beds interstratified by deeply cut softer rocks (Powell, 1875, p.149-150).

alcyonarian *octocoral*.

aldanite A variety of thorianite containing lead and uranium.

Aleppo stone *eye agate*.

alete Said of a spore without a laesura. In practice, such spores are sometimes difficult to distinguish from *acolpate* pollen. Cf: *inaperturate*.

aleurite An unconsolidated sedimentary deposit intermediate in texture between sand and clay, consisting of particles having diameters in the range of 0.01 to 0.1 mm. The term is common in Russian literature, and is frequently translated as "silt". Etymol: Greek *aleuron*, "flour".

aleurolite A consolidated aleurite, intermediate in texture between sandstone and shale; esp. siltstone.

aleutite A porphyritic *belugite* having a fine-grained groundmass. Its name is derived from the Aleutian Islands.

Alexandrian North American provincial series: Lower Silurian (above Cincinnatian of Ordovician, below Niagaran). Obsolete syn: *Medinan*.

alexandrine sapphire An alexandrite-like sapphire that is blue in daylight and violet, purple, or reddish under most artificial light.

alexandrite A transparent variety of chrysoberyl that has a grass-green or emerald-green color in daylight and wine-red to brownish-red color by transmitted or incandescent artificial light.

alexoite A pyrrhotite-bearing dunite. The term is used locally, in the Alexo mine, Dundonald Township, Ontario, Canada.

Alfisol In U.S. Dept. of Agriculture soil taxonomy, a soil order characterized by an umbric or ochric epipedon, an argillic horizon, and the retention of water at less than 15 bars tension during a growing season of at least three months (SSSA, 1970). Suborders and great soil groups of this order have the suffix -alf. See also: *Aqualf; Boralf; Udalf; Ustalf; Xeralf*.

alga An individual plant of the taxon *Algae*.

algae Photosynthetic, almost exclusively aquatic plants of a large and diverse division (Algae) of the thallophytes, including seaweeds and their fresh-water allies. They range in size from simple unicellular forms to giant kelps several meters long, and display extremely varied life-cycles and physiological processes, with, for example, different complexes of photosynthetic pigments. Algae range from the Precambrian. An individual plant is called an *alga*.

algae wash Shoreline drift composed mainly of filamentous algae.

algal Of, pertaining to, or composed of algae.

algal ball *algal biscuit*.

algal biscuit Any of various hemispherical or disk-shaped, calcareous masses up to 20 cm in diameter, produced in fresh-water as a result of precipitation by various blue-green algae; e.g. a deposit of marl formed around a piece of algal material or other nucleus as a result of photosynthesis and found on the shallow bottoms of hard-water lakes of the temperate region (as in Wisconsin). Cf: *aegagropile*. Syn: *algal ball; water biscuit; lake biscuit; marl biscuit; girvanella; pycnostromid*.

algal coal *boghead coal*.

algal dust (a) Angular to subangular, medium- to dark-colored (usually brown and brownish-gray) grains or crystals of carbonate, commonly 1-5 microns in diameter, derived from breakdown of algal felts, algally precipitated aragonite needles, algal slime, and comminution of phytoplankton (Chilingar et al, 1967, p.310). Term proposed by Wood (1941). (b) Algal micrite.

algal head A large (10-12 cm in diameter), bulbous, dome-shaped or columnar mass of mechanically transported, laminated sediments collected by algae (esp. blue-green algae) on a tidal flat or in a lake, and bound together by innumerable algal filaments.

algal limestone A limestone composed largely of remains of calcium-secreting algae or one in which such algae serve to bind together the fragments of other calcium-secreting forms. They are common in early Paleozoic, esp. in the Cambro-Ordovician of the Appalachian region of U.S.

algal mound A local thickening of limestone attributed chiefly to the presence of a distinctive suite of rock types (such as massive calcilutites) containing algae.

algal paste A term used in a loose sense by Schlanger (1957) for dark-gray to black, finely divided (micrograined or micro-crystalline) flecks forming a rather dense micritic limestone or dolomite, and associated with organic frame-builders such as corals. It is common, but not restricted, to the reef core, and it may represent compact, dense, diagenetically altered algal dust.

algal pit A small depression containing, or thought to contain, algae, occurring in the ablation area of a glacier or on the surface of sea ice.

algal reef An *organic reef* largely composed of algal remains and in which algae are or were the principal organisms secreting calcium carbonate; e.g. off the coast of Yugoslavia. Dimensions may be 3 m in height and more than 4.5 m in transverse diameter.

algal ridge A low ridge or elevated margin at the seaward (outer) edge of a reef flat, composed of the calcium-carbonate secretions of actively growing calcareous algae. See also: *lithothamnion ridge.* Cf: *algal rim.*

algal rim A low, slight rim built by actively growing calcareous algae on the lagoonal side of a leeward reef or on the windward side of a reef patch in a lagoon; esp. the outer part of the main reef surface or reef top, situated behind and rising gradually from the reef front, frequently culminating in a low *algal ridge,* and varying greatly in width (sometimes more than 500 m wide).

algal structure A sedimentary structure of definite form and usually of calcareous composition, resulting from colonial secretion and precipitation by algae; it includes crusts, small pseudopisolitic and pseudoconcretionary forms, biscuit- and cabbage-like heads of considerable size, and laminated structures such as stromatolites or bedding modified by blue-green algal mats. Some so-called algal structures may be of inorganic origin.

algarite A bitumen derived from algae.

algarvite A *melteigite* having more biotite and less nepheline than the original melteigite.

Algerian onyx A distinctly banded, stalagmitic form of *onyx marble.*

algerite A pinitic pseudomorph after scapolite.

alginite A variety of *exinite* consisting of algae and characteristic of boghead coals. Cf: *cutinite; sporinite; resinite.* Syn: *algite.*

algite *alginite.*

algodonite A mineral: Cu_6As.

Algoman orogeny Orogeny and accompanying granitic emplacement which affected Precambrian rocks of northern Minnesota and adjacent Ontario about 2400 m.y. ago; it is synonymous with the *Kenoran orogeny* of the Canadian Shield and was the final event of the Archean according to the current Canadian classification.

algon The viscous, organic binding material of *vase,* consisting of finely divided remains of algae (or of land vegetation, as in the upper parts of the estuary) and iron principally in the form of FeS (Bourcart, 1939).

Algonkian *Proterozoic.*

Algophytic *Archeophytic.*

algorithm A fixed step-by-step procedure for accomplishing a given result, such as solving a mathematical problem. It frequently involves repetition of an operation.

algovite *allgovite.*

alidade (a) A rule equipped with simple or telescopic sights and used for determining the directions of objects, such as the index of an astrolabe, the movable arm of a sextant, or the straightedge ruler of a plane table; specif. a part of a surveying instrument (such as a transit) consisting of a telescope or other sighting device, with index, and reading or recording accessories to measure vertical angles and distances. (b) A surveying instrument used with a plane table for mapping; e.g. *peep-sight alidade* and *telescopic alidade.*

aligned current structure *directional structure.*

alignment (a) The placing of surveying points along a straight line. Also, the location of such points with reference to a straight line or to a system of straight lines. The term "should be limited to operations associated with straight lines" (Mitchell, 1948, p.3). (b) Representing on a map the correct direction and character of a line or a feature in relation to other lines or features.---Syn: *alinement.*

alimentation The supplying of a glacier with material, such as snow or firn, that turns to ice; the process of *accumulation.*

alimentation facies Facies characteristics that provide evidence of the source of sediments, as revealed mainly by rock composition (such as "sandstone", "clay", and "chert") (Sonder, 1956). Cf: *precipitation facies.*

alinement *alignment.*

alio A French term for an impervious, ferruginous crust formed by the precipitation of iron salts from subsurface water.

aliphatic Designating open-chain type organic structures consisting of *paraffinic,* olefinic, and *acetylenic* compounds, e.g. palmitic acid, oleic acid, and ethyl alcohol.

aliphatic hydrocarbon A straight or branched open-chain hydrocarbon having the empirical formula C_nH_{2n+2}, such as methane or phytane.

alive In ore deposits, a syn. of *quick.*

alivincular Said of a type of ligament of bivalve mollusk (e.g. *Ostrea*) whose longer axis is transverse to the hinge axis, or that is not elongated in the longitudinal direction nor necessarily situated entirely posterior to the beaks, but located between the cardinal areas (where present) of the respective valves, with the lamellar component both anterior and posterior to the fibrous component.

alkali [chem] n. Any strongly basic substance, such as a hydroxide or carbonate of an alkali metal (e.g. sodium, potassium). Plural: *alkalies; alkalis.* Adj: *alkaline; alkalic.*

alkali [mineral] Said of a feldspar or group of feldspars containing alkali metals but little calcium.

alkali [petrology] *alkalic* [petrology].

alkali basalt Basalt that contains olivine (commonly both as phenocrysts and in the groundmass), augite (commonly titaniferous), and calcic plagioclase. Feldspathoids and sodic amphiboles and/or pyroxenes are common accessory minerals. Cf: *tholeiite.* Syn: *alkalic basalt.*

alkalic [chem] An adj. of *alkali* [chem].

alkalic [petrology] (a) Said of an igneous rock that contains more alkali metals than is considered average for the group of rocks to which it belongs. (b) Said of an igneous rock that contains more sodium and/or potassium than is required to form feldspar with the available silica. (c) Said of an igneous rock having an *alkali-lime index* below 51. Cf: *calcic; alkali-calcic; calc-alkalic.* (d) Said of an igneous rock belonging to the *Atlantic suite.*----Syn: *alkali* [petrology]; *alkaline* [petrology].

alkali-calcic Said of an igneous rock, or series of igneous rocks in which the weight percentage of silica is between 51 and 56 when the weight percentages of CaO and of K_2O + Na_2O are equal. See also: *alkali-lime index.*

alkali-calc index *alkali-lime index.*

alkali charnockite According to Tobi (1971), a rock of the charnockite series in which quartz constitutes 20-60% of the felsic constituents and in which the ratio of alkali feldspar to total feldspar is greater than 90%. He uses the term as a replacement for *enderbite.*

alkali feldspar (a) A group of feldspars composed of mixtures, or mixed crystals, of potassium feldspar (Or, or $KAlSi_3O_8$) and sodium feldspar (Ab, or $NaAlSi_3O_8$) in any ratio; a group of feldspars containing alkali metals but little calcium. (b) A mineral of the alkali feldspar group, such as microcline, orthoclase, sanidine, adularia, albite, perthite, anorthoclase, and plagioclase in which the proportion of the An molecule is less than 20%.---Cf: *plagioclase.* Syn: *alkalic feldspar.*

alkali flat A level area or plain in an arid or semiarid region, encrusted with alkali salts that became concentrated by evaporation and poor drainage; a *salt flat.* Se also: *playa.*

alkali lake A *salt lake,* commonly found in an arid region, whose waters contain in solution large amounts of sodium carbonate and potassium carbonate, as well as sodium chloride and other alkaline compounds; e.g. Lake Magadi in the Eastern Rift Valley of Kenya. See also: *potash lake; soda lake.* Syn: *alkaline lake.*

alkali-lime index A means of classifying igneous rocks introduced by Peacock (1931) based on the weight percentage of silica present when the weight percentages of CaO and of K_2O + Na_2O are equal. Four chemical classes of igneous rocks based on this index are recognized: *alkalic* (when the silica percentage is less than 51), *alkali-calcic* (when the silica percentage is between 51 and 56, *calc-alkalic* (when the silica percentage is between 56 and 61), *calcic* (when the silica percentage is over 61). Syn: *alkali-calc index.*

alkaline [chem] An adj. of *alkali* [chem]; *basic* [chem].

alkaline [petrology] *alkalic* [petrology].

alkaline soil A soil whose pH value is 7.0 or higher.

alkaline-sulfide hypothesis A theory of ore-deposit formation that postulates complex sulfide ions in hydrothermal solutions as a means of ore transportation. The theory accounts for only a few of the common metals (Krauskopf, 1967, p.501-502).

alkalinity [oceanog] The number of milliequivalents of hydrogen ion that is neutralized by one liter of seawater at 20°C.

alkali soil A soil that contains sufficient sodium carbonate or other alkali salt to give it a pH of 8.5 or higher, or a high percentage (15 or more) of exchangeable sodium. See also: *black alkali; white alkali.*

alkalitrophy The quality or state of an arid-region lake exhibiting alkaline characteristics. Adj: *alkalitrophic.*

Alkemade line In a ternary phase diagram, a straight line that connects the composition points of two primary phases whose areas are adjacent and whose interface forms a boundary curve (Levin et al, 1964, p.5). Cf: *Alkemade theorem.* Incorrect syn: *conjugation line.*

Alkemade theorem The statement in chemical phase studies that, in a ternary phase diagram, the direction of falling temperature on the boundary curve between the areas of two primary phases is always away from the *Alkemade line*; also, that the temperature maximum on the boundary curve is at the point at which the Alkemade line intersects it, or, if it does not intersect, at the end of the boundary curve which, if prolonged, would intersect the Alkemade line (Levin et al, 1964, p.5).

allactite A mineral: $Mn_7(AsO_4)_2(OH)_8$.

allalinite An altered gabbro containing olivine and saussurite as the main secondary minerals, which occur as euhedral pseudomorphs after the original minerals thus maintaining the original ophitic texture. Its name is derived from Allalin, Switzerland.

allanite A monoclinic, cerium-bearing mineral of the epidote group: $(Ce,Ca,Y)(Al,Fe)_3(SiO_4)_3(OH)$. It is typically an accessory mineral in igneous rocks (granite, syenite, diorite, pegmatite) and in their metamorphic equivalents. Syn: *orthite; cerine; bucklandite; treanorite.*

allargentum A mineral: $Ag_{1-x}Sb_x$, with x = 0.09-0.16.

alleghanyite A pink mineral: $Mn_5(SiO_4)_2(OH)_2$.

Allegheny orogeny An event which deformed the rocks of the Valley and Ridge province, and those of the adjacent Allegheny Plateau in the central and southern Appalachians. The orogeny cannot be closely dated, as there are no limiting overlying strata; Pennsylvanian rocks are involved in many places, and Lower Permian (Dunkard) in a few. Most of the orogeny was probably late in the Paleozoic, but phases may have extended into the early Triassic. The term is preferable to the much more inclusive term *Appalachian Revolution.*

allemontite A mineral: AsSb. It occurs in reniform masses. Syn: *arsenical antimony.*

Allen's rule In zoology, the statement that warm-blooded animals tend to have longer protruding body parts (i.e. legs, arms, tails) in warmer parts of the environment than in cooler. Named after Joel A. Allen (d.1921), American zoologist. Cf: *Bergmann's rule.*

Allerod n. A term used primarily in Europe for an interval of late-glacial time (centered about 11,000 years ago) following the Older Dryas and preceding the Younger Dryas, during which the climate as inferred from stratigraphic and pollen data in Denmark (Iversen, 1954) ameliorated favoring accelerated retreat of the waning continental and alpine glaciers; a subunit of the late-glacial Arctic interval, characterized by birch, pine, and willow vegetation. Also spelled: *Allerôd.*---adj. Pertaining to the late-glacial Allerød interval and to its climate, deposits, biota, and events.

allevardite *rectorite.*

all face centered lattice *face-centered lattice.*

allgovite An obsolete term for a group of igneous rocks containing augite and plagioclase and ranging in composition from dolerite to gabbro, including porphyritic varieties. Its name is derived from Allgäu Alps. Also spelled: *algovite.*

alliaceous Said of a mineral that has an odor of garlic when rubbed, scratched, or heated, e.g. arsenic compounds.

all-in aggregate An *aggregate* that is a natural mixture of sand and gravel.

Alling grade scale A metric *grade scale* designed by Harold L. Alling (1888-1960), U.S. geologist, for two-dimensional measurements (such as with thin sections or polished blocks) of sedimentary rocks. It has a constant geometric ratio of 10 for the major divisions (colloid, clay, silt, sand, gravel, cobble, boulder) and one of the fourth root of 10 for the four-fold subdivisions of each major unit (Alling, 1943).

allingite A fossil resin (retinite) containing no succinic acid but considerable sulfur, found at Allinges in Haute-Savoie, France.

allite A rock name for *allitic* material; e.g. bauxite and laterite.

allitic Said of a rock or soil from which silica has been largely removed and which contains a high proportion of aluminum and iron compounds in the clay fraction.

allivalite A gabbro containing anorthite and olivine with pyroxene rare or absent and accessory augite, apatite, and opaque iron oxides. Its name is derived from Allival, Romania.

allochem A collective term introduced by Folk (1959, p.4) for one of several varieties of discrete and organized carbonate aggregates that serve as the coarser framework grains in most mechanically deposited limestones, as distinguished from sparry calcite (usually cement) and carbonate-mud matrix (micrite). Important allochems include: silt-, sand-, and gravel-size *intraclasts;* ooliths; pellets; lumps; and fossils or fossil fragments (carbonate skeletons, shells, etc.). Adj: *allochemical.* Cf: *pseudoallochem; orthochem.*

allochemical metamorphism Metamorphism that is accompanied by addition or removal of material so that bulk chemical composition of the rock is changed (Mason, 1958).

allochetite A porphyritic hypabyssal igneous rock composed of labradorite, orthoclase, titanaugite, nepheline, magnetite, and apatite phenocrysts in a fine-grained, felty groundmass of augite, biotite, magnetite, hornblende, nepheline, and orthoclase. Its name is derived from the Allochet Valley, Italy.

allochromatic Said of a mineral that, in its purest state, is colorless, but that has color due to submicroscopic inclusions, or to the presence of a closely related element that has become part of the chemical structure of the mineral. Cf: *idiochromatic*.

allochronic Said of taxa that are not contemporaneous or that occur in different segments of geologic time. Ant: *synchronic*.

allochthon [sed] A mass of redeposited sedimentary materials originating from distant sources.

allochthon [tect] A mass of rocks which has been moved from its original site of origin by tectonic forces, as in a thrust sheet or nappe. Many allochthonous rocks have been moved so far from their original sites that they differ greatly in facies and structure from those on which they now lie. Ant: *autochthone*. Adj: *allochthonous*.

allochthone Var. of *allochthon*.

allochthonous [coal] (a) Said of coal or peat that originates from accumulations of plant material transported from its place of growth and deposited elsewhere, or that is not derived from plants that grew and decayed in the place where the coal or peat is now found. (b) Pertaining to *allochthony*.

allochthonous [geol] Formed or occurring elsewhere than in place; of foreign origin, or introduced; e.g. said of a rock or sediment derived from an environment other than that in which it is found, such as a limestone composed largely or organic material transported far from the place where the base organisms lived. The term was first used by Naumann (1858, p.657) to designate rocks of distant origin; it is similar in meaning to *allogenic* (which refers to constituents, rather than whole formations). Ant: *autochthonous*. Cf: *heterochthonous*.

allochthonous [crater] Said of a fragmental unit composed of material ejected from an impact or explosion crater during formation and subsequently redeposited within and around the crater; e.g. a fallback breccia in a meteorite impact crater, containing numerous shock-metamorphosed rock fragments and perhaps meteoritic material.

allochthonous [petrology] Pertaining to a magma, or the magmatic constituent of a migma, derived from some extraneous source.

allochthonous [tect] (a) Pertaining to an *allochthon* or to the rocks of an allochthon, esp. to strata that have become detached by movement on low-angle thrust faults. (b) Said of a *displaced fold* or one that is composed of rocks that have been moved from the beds on which they originally rested.--- Cf: *parautochthonous; exotic*.

allochthonous [stream] Said of a stream flowing in an adopted channel. The term is generally applied to subterranean streams.

allochthony In coal formation, accumulation of plant remains elsewhere than in place. Ant: *autochthony*. See also: *primary allochthony; secondary allochthony; drift theory; hypautochthony*.

alloclastic breccia A breccia that is formed by volcanic activity, and consists of fragments of any pre-existing rock in a volcanic matrix; a type of *volcanic breccia*.

allocyclicity The state of cyclothemic deposition that results from changes in the supply of energy or material input into a sedimentary system (Beerbower, 1964, p.32). It involves such mechanisms as uplift, subsidence, climatic variation, eustatic change in sea level, and other changes external to the sedimentary unit. Cf: *autocyclicity*.

allogene An *allogenic* mineral or rock constituent; e.g. a xenolith in an igneous rock, or a pebble in a conglomerate, or quartz sand blown from land into an evaporite sequence, or a detrital mineral in a placer deposit. Ant: *authigene*. Syn: *allothigene*.

allogenic [geol] (a) Formed or generated elsewhere, usually from a distant place; specif. said of rock constituents and minerals that were derived from preexisting rocks and transported to their present depositional site, or that came into existence previously to the rock of which they now constitute a part and at some place other than that at which the rock is now found. Ant: *authigenic*. Cf: *chthonic; allochthonous*. Syn: *allothogenic; allothigenic; allothigenous; allothigenetic*. (b) Pertaining to a stream that derives much of its water from a distant terrain or from beyond its surface draining area, such as one originating in a humid or glacial region and flowing across an arid or desert region.

allogenic [ecol] Said of an ecologic succession that resulted from factors that arise from outside the natural community and alter its habitat, such as an allogenic drought of prolonged duration. Cf: *autogenic* [ecol].

allogenous Said of floras that have persisted from an earlier environment. Not to be confused with *allogeneous*.

allokite A clay mineral intermediate in structure between kaolinite and allophane.

allomeric *isomorphous*.

allomerism *isomorphism*.

allometry (a) The growth of a part of an organism in relation to its entirety. (b) Measurement and study of the growth of a part of an organism in relation to its entirety.

allomicrite Allochthonous *orthomicrite*.

allomorph (a) A *polymorph* or a *dimorph*. (b) A *pseudomorph*.

allomorphic *polymorphic*.

allomorphism [cryst] *paramorphism*.

allomorphism [paleont] A term used erroneously for *xenomorphism* (regarding bivalve mollusks).

allomorphite A mineral consisting of barite pseudomorphous after anhydrite.

allomorphosis Biologic evolution characterized by a rapid increase in specialization; evolutionary allometry. Cf: *aromorphosis*.

allomorphous *polymorphic*.

allopatric Said of organisms or biologic events occurring in different areas; e.g. the development of a distinct species from an isolated population. Noun: *allopatry*. Cf: *sympatric*.

allophane An amorphous clay mineral: $Al_2O_3.SiO_2.nH_2O$. It consists of a hydrous aluminum silicate gel (of highly variable composition, with minor amounts of bases and accessory acid radicals) that is or appears to be amorphous to X-ray diffraction and that changes, on standing, from a glassy or translucent constituent of clay materials to one with an earthy appearance owing to the loss of water. Allophane has various colors (snow white, blue, green, brown, yellow, or colorless) and often occurs in incrustations, thin seams, or rarely stalactitic masses. Syn: *allophanite*.

allophaneton An obsolescent term used by ceramists (esp. in Europe) for the portion of a clay that is soluble in hydrochloric acid. Cf: *kaolinton*.

allophanite An obsolete syn. of *allophane*.

allophanoid A group name for the clay minerals allophane, halloysite, and montmorillonite.

allothigene *allogene*.

allothigenetic *allogenic* [geol].

allothigenic *allogenic* [geol].

allothigenous [geol] *allogenic* [geol].

allothigenous [volc] In the classification of pyroclastics, the equivalent of *accidental* ejecta. Cf: *authigenous*.

allothimorph A constituent of a metamorphic rock which, in the new rock, has not changed its original crystal outlines (Johannsen, 1939).

allothogenic *allogenic* [geol].

allothrausmatic Said of igneous rocks with an orbicular texture in which the nuclei of the orbicules are formed of xenoliths differing in composition from the groundmass (Eskola, 1938, p.476). Cf: *isothrausmatic; crystallothrausmatic; homeothrausmatic; heterothrausmatic.*

allotrioblast *xenoblast.*

allotriomorphic A syn. of *xenomorphic.* The term *allotriomorphisch* was proposed by Rosenbusch (1887, p.11), but it lacks priority. See also: *anhedral.*

allotriomorphic-granular *xenomorphic-granular.*

allotrope A crystal form of a substance that displays *allotropism.*

allotrophic *heterotrophic.*

allotropy *Polymorphism* in an element, e.g. sulfur as both orthorhombic and monoclinic. See also *allotrope.*

alluaudite A mineral: $(Na,Ca)_{1-2}(Fe^{+3},Mn^{+2})_3(PO_4)_3$.

alluvia Seldom-used plural of *alluvium.*

alluvial [ore dep] Said of a placer formed by the action of running water; also, said of the valuable mineral, e.g. gold or diamond, associated with an alluvial placer.

alluvial [sed] adj. Pertaining to or composed of alluvium, or deposited by a stream or running water; e.g. an "alluvial clay" or an "alluvial divide". Syn: *alluvian; alluvious.*---n. *alluvium.*

Alluvial A name, now obsolete, applied by Jameson (1808) from the teachings of A.G. Werner in the 1790's to the group or series of rocks consisting of unconsolidated or poorly consolidated gravels, sands, clays, and peat that were believed to have been formed after the withdrawal of the ocean from the continents. It constituted the fourth (following Floetz) of the divisions in which Werner placed the rocks of the geologic column. Syn: *Tertiary.*

alluvial apron *bajada.*

alluvial bench A term used by Hobbs (1912, p.214) for a feature now known as a *bajada.*

alluvial channel A channel whose bed is composed of alluvium.

alluvial cone An *alluvial fan* with very steep slopes; it is generally higher and narrower than a fan, and is composed of coarser and thicker material believed to have been deposited by larger streams. The term is sometimes used synonymously with *alluvial fan.* Syn: *cone of dejection; cone of detritus; hemicone; debris cone; cone delta; dry delta; wash.*

alluvial dam A deposit of alluvium that is built by an overloaded stream and that obstructs its channel, thereby impounding water behind the deposit; esp. such a dam in a distributary on an alluvial fan.

alluvial-dam lake A lake formed behind an *alluvial dam*; esp. a lake at the apex of an alluvial fan, formed as a result of a cloudburst.

alluvial deposit [sed] *alluvium.*

alluvial fan A low, outspread, relatively flat to gently sloping mass of loose rock material, shaped like an open fan or a segment of a cone, deposited by a stream (esp. in a semiarid region) at the place where it issues from a narrow mountain valley upon a plain or broad valley, or where a tributary stream is near or at its junction with the main stream, or wherever a constriction in a valley abruptly ceases or the gradient of the stream suddenly decreases; it is steepest near the mouth of the valley where its apex points upstream, and it slopes gently and convexly outward with gradually decreasing gradient. Cf: *alluvial cone; bajada.* Syn: *detrital fan; talus fan; dry delta; talus fan.*

alluvial-fan shoreline A prograding shoreline formed where an alluvial fan is built out into a lake or sea.

alluvial fill A deposit of alluvium occupying a stream valley, conspicuously thicker than the depth of the stream. It represents a single stratigraphic unit.

alluvial flat A small *alluvial plain* bordering a river and on which alluvium is deposited during floods. Syn: *river flat.*

alluvial meander An extremely sinuous bend in an *alluvial river.*

alluvial plain (a) A level or gently sloping tract or a slightly undulating land surface produced by extensive deposition of alluvium, usually adjacent to a river that periodically overflows its banks; it may be situated on a flood plain, a delta, or an alluvial fan. Cf: *alluvial flat.* Syn: *wash plain; waste plain; river plain; aggraded valley plain.* (b) *bajada.*

alluvial-plain shoreline A prograding shoreline formed where the broad alluvial slope at the base of a mountain range is built out into a lake or sea.

alluvial river A graded river that occupies a broad flood-plain belt over which the depth of alluvium deposited by the river equals or exceeds the depth to which scour takes place in time of flood.

alluvial slope A surface, composed of alluvium, that slopes down and away from the sides of mountains and merges with a plain or a broad valley floor upon which it rests (Bryan, 1923a, p.86); an alluvial surface that lacks the distinctive form of an alluvial fan or a bajada. See also: *bajada.*

alluvial-slope spring *boundary spring.*

alluvial soil A young soil on flood plains and deltas that is actively in the process of construction and has developed no characteristics other than those of the alluvium itself.

alluvial stone A gemstone that has been transported and deposited by a stream.

alluvial terrace A *stream terrace* composed of unconsolidated alluvium (including gravel), produced by renewed downcutting of the flood plain or valley floor by a rejuvenated stream or by the later covering of a terrace with alluvium. Cf: *rock terrace.* Syn: *stream-built terrace; built terrace; fill terrace; drift terrace.*

alluvial tin *stream tin.*

alluviation (a) The subaerial deposition or formation of alluvium or alluvial features (such as cones or fans) at places where stream velocity is decreased or streamflow is checked; the process of aggradation or of building up of sediments by a stream along its course, or of covering or filling a surface with alluvium. (b) A hydraulic effect on solids suspended in a current of water, whereby the coarsest and heaviest particles are the first to settle out, and the finest muds the last, as gradient or velocity of a stream is decreased (Pryor, 1963).

alluvion [law] The formation of new land by the gradual or imperceptible action of flowing water or of waves and currents; *accretion.* Also, the land so added. Cf: *diluvion.*

alluvion [geol] (a) The wash of the sea against the shore, or the flow of a river against its bank. (b) An overflowing; an inundation or flood, esp. when the water is charged with much suspended material. (c) Material deposited by a flood; *alluvium.*

alluvion [volc] An obsolete term for a flood of volcanic-cinder mud and for a consolidated volcanic-cinder mud.

alluvium (a) A general term for clay, silt, sand, gravel, or similar unconsolidated detrital material deposited during comparatively recent geologic time by a stream or other body of running water as a sorted or semisorted sediment in the bed of the stream or on its flood plain or delta, or as a cone or fan at the base of a mountain slope; esp. such a deposit of fine-grained texture (silt or silty clay) deposited during time of flood. The term does not require permanent submergence; and it formerly included (but is not now intended to include) subaqueous deposits in seas, estuaries, lakes, and ponds. Syn: *alluvial; alluvial deposit; alluvion.* (b) A driller's term used "incorrectly" for the broken, earthy rock material directly below the soil layer and above the solid, unbroken bed or ledge rock (Long, 1960). (c) *alluvial soil.*---Etymol: Latin *alluvius*, from *alluere*, "to wash against". Pl: *alluvia; alluviums.* Cf: *eluvium; diluvium.*

almandine (a) The iron-aluminum end member of the garnet group, characterized by a deep-red to purplish color: $Fe_3Al_2(SiO_4)_3$. It occurs in mica schists and other regionally metamorphosed rocks, and is used as a gemstone. Syn: *al-*

mandite; alamandine; almond stone. (b) A violet or mauve variety of ruby spinel; a reddish-purple to purplish-red spinel. (c) A reddish-purple sapphire.

almandine-amphibolite facies Metamorphic rocks formed by regional dynamothermal metamorphism under high to very high pressures (in excess of 5000-6000 bars) and temperatures of 450°-700°C. They represent the high-pressure part of the amphibolite facies of Eskola. Characteristic index minerals include kyanite and, less commonly, andalusite but not cordierite (Winkler, 1967). Cf: cordierite-amphibolite facies.

almandite A syn. of almandine garnet.

almashite A green or black variety of amber that has a low (2.5-3%) content of oxygen.

almeriite natroalunite.

almond-shaped bomb spindle-shaped bomb.

almond stone A syn. of almandine garnet.

almost atoll An atoll with a minute noncoral island, generally of volcanic origin, in the central lagoon.

alnoite A lamprophyre chiefly composed of olivine, mica, augite, melilite, nepheline, and garnet. Perovskite may be present as an accessory. Its name is derived from Alno, Sweden. Also spelled: allnôite; alnôite.

alongshore Along the shore or coast, such as an "alongshore drift" or "alongshore current"; longshore.

alp (a) A high, rugged, steep-sided mountain, esp. one that is snow-covered, resembling topographically those in the European Alps. (b) A high pasture- or meadow-land on a mountain side, between timberline and snowline, as those in the Swiss Alps; also a high shoulder or gentle slope, esp. in the Swiss Alps, commonly above a glaciated valley at a marked change of slope. Cf: alb.

alpestrine montane.

alpha [mineral] adj. Of or relating to one of two or more closely related minerals and specifying a particular physical structure (esp. a polymorphous modification); specif. said of a mineral that is stable at a temperature lower than those of its beta and gamma polymorphs (e.g. "alpha cristobalite" or "α-cristobalite", the low-temperature tetragonal phase of cristobalite). Some mineralogists reverse this convention, using α for the high-temperature phase (e.g. "alpha carnegieite", the isometric phase of carnegieite stable above 690°C.

alpha [cryst] (a) In a biaxial crystal, the smallest index of refraction. (b) The interaxial angle between the b and c crystallographic axes.----Cf: beta [cryst]; gamma [cryst].

alpha* The interaxial angle of the reciprocal lattice between b* and c* which is equal to the interfacial angle between (010) and (001). Cf: beta* angle; gamma* angle.

alpha chalcocite digenite.

alpha decay Radioactive decay of an atomic nucleus by the emission of an alpha particle.

alpha particle (a) A particle emitted from an atomic nucleus during radioactive decay which is positively charged and has two protons and two neutrons. It is physically identical to a helium atom's nucleus. Cf: beta particle; gamma ray. (b) By extension, the nucleus of a helium atom.----Less-preferred syn: alpha ray.

alpha-particle recoil tracks The paths of radiation damage in a mica caused by the recoil nucleus accompanying the alpha-particle decay of uranium and thorium impurities. The tracks are similar to fission tracks, but are much smaller and more numerous. An age determination can be made on the basis of the track density, usually examined with an electron microscope, plus determinations of the thorium and uranium contents of the sample (Huang & Walker, 1967, p.1103-1106). Syn: alpha-recoil tracks.

alpha quartz The polymorph of quartz that is stable below 573°C, that has a vertical axis of three-fold symmetry and three horizontal axes of two-fold symmetry, and that has a higher refractive index and birefringence than those of beta quartz. It occurs in veins, geodes, and large pegmatites. Also

spelled: α-quartz. Syn: low quartz.

alpha ray A less-preferred syn. of alpha particle.

alpha-recoil tracks alpha-particle recoil tracks.

alpha-vredenburgite A homogeneous, metastable mineral: $(Mn,Fe)_3O_4$. It has the same composition as that of beta-vredenburgite, and is regarded as an iron-rich hausmannite.

alphitite A term suggested by Salomon (1915, p.398) for a clay or silt consisting largely of rock flour, such as the fine material produced by a glacier. Twenhofel (1937, p.84) recommends disuse of the term "because of inability to determine that a clay is a rock flour and not composed of particles of many origins brought together by wind or water".

Alpides A name used by Suess for the great orogenic belt, or system of young, folded mountains, including the Alps, that extends eastward from Spain into southern Asia. In a stricter sense, the term is used for structures on the north flank of the belt, those on the south flank being termed the Dinarides. Syn: Mediterranean belt; Alpine-Himalayan belt.

alpine [geomorph] Pertaining to, characteristic of, or resembling the European Alps or any lofty mountain or mountain system, esp. one so modified by intense glacial erosion as to contain aiguilles, cirques, horns, etc.; e.g. an alpine lake resulting from glacial erosion and situated in or along the border of a high mountain region. Spelled Alpine when referring specif. to the European Alps.

alpine [ecol] Characteristic or descriptive of the mountainous regions lying between timberline and snowline; said of the climate, flora, relief, ecology, etc. Less strictly, pertaining to high elevations and cold climates. Cf: montane.

alpine [struc geol] A general term for mountainous topographical, morphological, and structural features which resemble in grandeur and complexity those of the European Alps, regardless of the age and location of the mountains and features so composed.

alpine firn melt firn.

alpine glacier Any glacier in a mountain range except an ice cap or ice sheet. It usually originates in a cirque and may flow down into a valley previously carved by a stream. Syn: mountain glacier; valley glacier.

Alpine-Himalayan belt Alpides.

Alpine Meadow soil One of an intrazonal, hydromorphic group of soils that is dark and that develops under grasses above the timber line.

Alpine orogeny A name for the relatively young orogenic events of southern Europe and Asia, by which the rocks of the Alps and the remainder of the Alpide orogenic belt were strongly deformed. Stille includes in his Alpidic orogenic era all orogenic events from the Jurassic to the end of the Tertiary, but most geologists restrict the era to the Tertiary, with many episodes of varying strength from place to place, ending during the Miocene or Pliocene.

Alpine-type facies series The highest-pressure type of dynamothermal regional metamorphism at no more than moderate temperature (150° to 400°C), characterized by the presence of the pumpellyite and glaucophane schist facies. It may also involue the zeolite facies in the low-temperature/highest-pressure range and the deep-seated eclogite facies at the highest pressures with moderate temperatures (Hietanen, 1967, p.203).

alpinotype tectonics The tectonics of alpine mountain belts, regardless of their age, which are produced from orthogeosynclines. They are characterized in their internal parts by deep-seated plastic folding and plutonism, and in their external parts by lateral thrusting, which has produced nappe, thrust sheets, and closely crowded folds. Cf: germanotype tectonics. Syn: orthotectonics.

alpland An area whose topography resembles that of the Alps.

alquifou A coarse-grained galena, used by potters in preparing a green glaze.

alsbachite A porphyritic granodiorite chiefly composed of sodic plagioclase, quartz, and a small amount of orthoclase, often with accessory garnet and mica. The quartz and orthoclase commonly form the phenocrysts in a granular groundmass.

alstonite A mineral: $BaCa(CO_3)_2$. It is the orthorhombic dimorph of barytocalcite. Syn: *bromlite*.

alta A miner's term for the black, shaly, and highly sheared capping of quicksilver orebodies. Syn: *black alta.*

Altaides A name used by Suess for a late Paleozoic orogenic belt extending across the width of Eurasia, and including also the Appalachian and Ouachita belts of North America. It is named for the Altai Mountains of central Asia, where there was late Paleozoic deformation, but as these mountains are remote and still poorly known, the term is little used now by tectonic geologists; most modern references are to component parts, such as the Variscan belt in Europe.

altaite A tin-white mineral: PbTe.

alteration Any change in the mineralogic composition of a rock brought about by physical or chemical means, esp. by the action of hydrothermal solutions; also, a secondary, i.e. supergene, change in a rock or mineral. Alteration is sometimes considered as a phase of metamorphism, but is usually distinguished from it because of being milder and more localized than metamorphism is generally thought to be.

altered rock A rock that has undergone changes in its chemical and mineralogic composition since its original deposition.

alterite A general term for altered, unrecognizable grains of heavy minerals.

alternate In plant morphology, pertaining to the attachment of parts, e.g. leaves, singly at each node; also, said of plant parts in regular occurrence between other organs.

alternate folding Deflection of the surface of a brachiopod shell in which the fold of one valve is opposed by the sulcus of the other valve.

alternate pick In the interpretation of seismograph records, a secondary choice in the selection of times to be used, when two or more possibilities occur. Cf: *pick.*

alternate terrace One of several *meander-scar terraces.*

alternating field demagnetization A technique of partial *demagnetization* that preferentially removes the softer components of natural remanent magnetization; the specimen is placed in an alternating magnetic field which is gradually reduced to zero. Cf: *thermal demagnetization; chemical demagnetization.* Syn: *AC demagnetization.*

alternation A kind of vertical transition between one stratigraphic unit and another, "where the new lithology appears as separate beds, becoming more and more frequent" (Challinor, 1967, p.6).

alternation of generations The occurrence in the life cycle of a plant or animal of two or more forms having different lines of development, usually involving the regular alternation of sexual and asexual generations. Syn: *heterogony; metagenesis.*

altimeter An instrument, usually an aneroid barometer, for determining height above ground or above mean sea level, based on the fall of atmospheric pressure accompanying an increase in altitude.

altimetric frequency curve A curve showing the altitudinal distribution of the highest points in a series of small squares that divide the map of a given region.

altiplanation A group of erosion processes, involving solifluction and related mass movement, which tend to produce flat or terracelike surfaces. Such processes are especially active at high elevations and latitudes where periglacial processes predominate. Cf: *equiplanation; cryoplanation.*

altiplanation terrace A broad terrace or flattened summit or pass formed by altiplanation, and represented by an accumulation of solifluction deposits and other loose rock materials on a long smooth slope. See also: *equiplanation terrace.*

altiplano A high-lying plateau or tableland; specif. the high plateau of western Bolivia, consisting of a string of intermontane basins. Etymol: Spanish. Syn: *altiplanicie.*

altithermal adj. Pertaining or belonging to a climate characterized by rising or high temperatures; e.g. "altithermal soil" of postglacial time.

Altithermal n. A term proposed by Antevs (1948, p.176) for a dry postglacial interval (from about 7500 to 4000 years ago) following the Anathermal and preceding the Medithermal, during which temperatures were probably "distinctly warmer than at present". Karlstrom (1956, p.327) suggests that the term designates the interval of maximum postglacial climatic amelioration (centered about 5500 years ago) falling near "the culmination of an interglacial period which separates the Wisconsin glacial stage from a still unnamed glacial stage that may attain its maximum extent within the next 20,000 years". It corresponds to the *Atlantic* interval or the middle part of the *Hypsithermal.* See also: *thermal maximum; Megathermal; Xerothermic.* Syn: *Long Draught.*---adj. Pertaining to the postglacial Altithermal interval and to its climate, deposits, biota, and events.

altitude [photo] Vertical distance above the datum (usually mean sea level) of an object or point in space above the Earth's surface. See also: *flight altitude.*

altitude [surv] (a) The vertical distance of a level, a point, or an object considered as a point above or below the surface of the Earth, measured from a given datum (usually mean sea level). Altitude is positive if the point or object is above the horizon, and negative if it is below it. Cf: *elevation.* Abbrev: alt. (b) The vertical angle between the horizontal plane of the observer and any higher point (such as the summit of a peak).

alto A term used in the SW U.S. for a bluff, height, or hill. Etymol: Spanish, "high ground".

alum (a) A mineral: $KAl(SO_4)_2 \cdot 12H_2O$. It is colorless or white, and has a sweetish-sourish astringent taste. Cf: *kalinite.* Syn: *potash alum; potassium alum.* (b) A group of minerals containing hydrous aluminum sulfates, including alum, kalinite, soda alum, mendozite, and tschermigite.

alum coal A pyritic, argillaceous brown coal that forms alum as a weathering product. Cf: *alum shale.*

alum earth *alum shale.*

aluminite A mineral: $Al_2(SO_4)(OH)_4 \cdot 7H_2O$. Syn: *websterite.*

aluminocopiapite A mineral of the copiapite group: $Al_{0.67}Fe_4(SO_4)_6(OH)_2 \cdot 20H_2O$.

aluminosilicate A silicate in which aluminum substitutes for the silicon in the SiO_4 tetrahedra.

alumite *alunite.*

alumocalcite A variety of opal containing alumina and lime as impurities.

alumogel An amorphous aluminum hydroxide forming the main constituent of bauxite; *cliachite.*

alumohydrocalcite A mineral: $CaAl_2(CO_3)_2(OH)_4 \cdot 3H_2O$. It occurs as white, chalky masses consisting of radially fibrous spherulites.

alum rock *alunite.*

alum schist *alum shale.*

alum shale An argillaceous, often carbonaceous, rock impregnated with alum, originally containing iron sulfide (pyrite, marcasite) which, when decomposed, formed sulfuric acid that reacted with the aluminous and potassic materials of the rock to produce aluminum sulfates. Syn: *alum earth; alum schist; alum slate.*

alum slate *alum shale.*

alumstone *alunite.*

alunite (a) A mineral: $KAl_3(SO_4)_2(OH)_6$. It is isomorphous with natroalunite, sometimes contains appreciable sodium, generally occurs as a hydrothermal-alteration product in feldspathic igneous rocks, and is used in the manufacture of alum. Syn: *alumstone; alum rock; alumite.* (b) A group of minerals containing hydrous sulfates, including alunite, natroalunite, jarosite, natrojarosite, ammoniojarosite, argentojarosite, and plumbojarosite.

alunitization Introduction of, or replacement by, alunite.

alunogen A mineral: $Al_2(SO_4)_3.18H_2O$. It occurs as a white, fibrous incrustation or efflorescence formed by volcanic action or by decomposition of pyrite in alum shales. Syn: *feather alum; hair salt.*

alurgite A manganiferous muscovite.

alushtite A mixture of dickite with clay minerals (such as illite).

alvanite A mineral: $Al_6(VO_4)_2(OH)_{12}.5H_2O$.

alveolar weathering *honeycomb weathering.*

alveole A space or cavity, such as a *vacuole* in a foraminiferal test wall; *alveolus.*

alveolinid Any foraminifer belonging to the family Alveolinidae, characterized by an imperforate, porcelaneous, axially elongated test that may be subcylindrical, fusiform, ellipsoidal, or spherical. Their stratigraphic range is Lower Cretaceous to present. Although this group resembles the *fusulinids* in shape, the two groups are not genetically related.

alveolitoid Said of a type of reclined corallite having a vaulted upper wall and a nearly plane lower wall parallel to the surface of adherence of the colony (as in the genus *Alveolites*) (TIP, 1956, pt.F, p.245).

alveolus A small cavity or pit in an invertebrate; e.g. a minute blind cavity in the keriotheca of some fusulinids or a blind chamberlet opening only toward the back (opposite to the direction of coiling) as in the foraminiferal family Alveolinidae, or the conical cavity in the anterior end of the guard of a belemnite, or a pit located anterior to the internal face of the cardinal process in a brachiopod, or a bryozoan *cancellus.* Pl: *alveoli.* Adj: *alveolar.* Syn: *alveole.*

amagmatic Said of a structure, region, or process that doesn't involve magmatic activity.

amakinite A mineral: $(Fe,Mg)(OH)_2$.

amalgam (a) A naturally occurring alloy of silver with mercury; mercurian silver. It is found in the oxidation zone of silver deposits and as scattered grains in cinnabar ores. See also: *gold amalgam; moschellandsbergite.* Syn: *silver amalgam; argental mercury.* (b) A general term for alloys of mercury with one or more of the well-known metals (except iron and platinum); esp. an alloy of mercury with gold, containing 40-60% gold, and obtained from the plates in a mill treating gold ore.

amalgamate Said of a type of wall structure in trepostome bryozoans in which the zooecial boundaries are not visible. Cf: *integrate.*

amang A term used in Malaysia for the heavy iron and tungsten minerals (and associated minerals) found with cassiterite deposits.

amarantite A dark reddish-purple mineral: $FeSO_4(OH).3H_2O$.

amararhysis A *skeletal canal* in dictyonine hexactinellid sponges running longitudinally within the body wall, opening at intervals to the cloaca by slit-like apertures, and opening to the exterior by radial canals terminating in oscula. It is part of the exhalant system. Pl: *amararhyses.*

amargosite A syn. of *bentonite.* Etymol: *Amargosite,* a trade name for a bentonitic clay (montmorillonite) from the Amargosa River, Calif.

amarillite A pale greenish-yellow mineral: $NaFe(SO_4)_2.6H_2O$.

amatrice A green gem cut from variscite and its surrounding matrix of gray, reddish, or brownish crystalline quartz or chalcedony.

amausite A finely crystalline rock, e.g. devitrified glass (Thrush, 1968, p.33). Syn: *petrosilex.* [petrology].

amazonite An apple-green, bright-green, or blue-green laminated variety of microcline, sometimes used as a gemstone. Syn: *amazonstone.*

amazonstone The earlier name for *amazonite.* Also spelled: *Amazon stone.*

amb The contour or outline of a pollen grain (less commonly of a spore) as viewed from directly above one of the poles.

Syn: *equatorial limb.*

amber (a) A very hard, brittle, usually yellowish to brownish, translucent or transparent fossil resin that is derived from coniferous trees, that frequently encloses insects and other organisms, that takes a fine polish, and that is found in alluvial soils, clays, and recent sediments, in beds of lignite, or on some seashores (as of the Baltic Sea). It is used chiefly in making pipe mouthpieces, beads, and other small ornamental objects. Syn: *succinite; bernstein; electrum.* (b) A term applied to a group of fossil resins containing considerable succinic acid and having very variable C:H:O ratios; e.g. almashite, simetite, delatynite, and ambrosine.

amberat Dried wood-rat urine found as lustrous dark-brown or black masses on rock deposits and walls of rock shelters. Syn: *johnsonite.*

amberite *ambrite.*

amber mica *phlogopite.*

amberoid A gem material consisting of small fragments of genuine amber that have been united or reconstructed by heat and pressure. It is characterized by an obvious flow structure or by a dull spot left by a drop of ether. Syn: *ambroid; pressed amber.*

ambitus The exterior edge or periphery; e.g. the greatest horizontal circumference of an echinoid test, or the thecal outline of a dinoflagellate as viewed from the dorsal or ventral side. Pl: *ambitus.*

amblygonite A mineral: $(Li,Na)AlPO_4(F,OH)$. It commonly occurs in white or greenish cleavable masses in pegmatites, and is mined as an ore of lithium. Syn: *hebronite.*

amblyproct Said of a sponge in which the exhalant surface is in the form of an open cup.

amblystegite A dark brownish-green to black variety of hypersthene.

ambonite A group of porphyritic cordierite-bearing hornblende-biotite andesites and dacites originally described from the Indonesian island of Ambon (or Amboina); also, any rock in that group, the three main subdivisions being bronzite-bearing andesites and dacites, mica-bearing andesites and dacites; and hornblende-bearing andesites and dacites (Johannsen, 1939, p.239).

ambrite A yellowish-gray, subtransparent variety of retinite resembling amber, occurring in large masses in the coalfields of New Zealand. (Its approximate formula: $C_{40}H_{66}O_5$. Syn: *amberite.*

ambroid Var. of *amberoid.*

ambrosine A yellowish or clove-brown variety of amber containing considerable succinic acid and occurring as rounded masses in phosphate beds.

ambulacral adj. (a) Pertaining to an ambulacrum or to ambulacra collectively; e.g. a small needle-like "ambulacral spine" attached to the part of a side plate directed toward the main food groove of the ambulacrum of a blastoid. (b) Corresponding in position to an ambulacrum, or referring to the zone in which an ambulacrum is present; e.g. an "ambulacral ray" representing an area defined by direction of an ambulacrum radiating from the mouth of a crinoid. Cf: *interambulacral.* Syn: *radial.*---n. A small calcareous plate that covers part of an ambulacrum of an echinoderm; e.g. a *side plate* or the broader of the floor plates in a cystoid. Pl: *ambulacralia.*

ambulacral groove A branched passageway, furrow, or depression along the course of an ambulacrum of an echinoderm, covered by ambulacral plates, and through which food particles were conveyed to the mouth by means of ciliary currents. Syn: *food groove.*

ambulacral pore An opening, in or between ambulacral plates of an echinoderm, for the passage of tube feet or of podia or for the connection of a podium to an ampulla.

ambulacral system An organ system peculiar to echinoderms, consisting of a ring canal encircling the mouth and five radial ambulacral vessels radiating from the ring canal and lodged in the ambulacral grooves (TIP, 1966, pt.U, p.153). See also: *subvective system*; *water-vascular system*.

ambulacrum (a) One of the narrow, usually elongate areas extending radially from the mouth of an echinoderm and along which run the principal nerves, blood vessels, and canals of the water-vascular system. It encloses the ambulacral groove, and usually bears numerous tube feet. An echinoderm may exhibit as many as five such areas, as in the blastoids and edrioasteroids. See also: *ray* [paleont]. (b) A trough of the coenosteum separating collines on the surface of some meandroid coralla of a scleractinian coral.---Pl: *ambulacra*.

ameghinite A mineral: $NaB_3O_5.2H_2O$.

amelioration *climatic amelioration*.

ament A spike inflorescence of closely spaced, often intricate, apetalous flowers. It is conelike but herbaceous. Syn: *catkin*.

American cut A style of cutting for a round diamond brilliant, with proportions and facet angles that produce the maximum brilliancy consistent with a high degree of fire. It was worked out by trial and error by master American cutters and confirmed mathematically by Marcel Tolkowsky. Syn: *ideal cut*.

American jade (a) Nephrite from Wyoming. (b) A syn. of *californite* (jade-like variety of vesuvianite).

American ruby A red garnet (pyrope) found in Arizona and New Mexico.

American system *cable-tool drilling*.

amerikanite A natural glass from South America (Colombia and Peru), once classed as a tektite but now believed to be of volcanic origin.

amesite An apple-green mineral: $(Mg,Fe^{+2})_4Al_4Si_2O_{10}(OH)_8$. It may be essentially free of iron. Amesite is usually classed as a chlorite, but it is a structurally distinct, well-defined species, having some structural relations to kaolinite (Hey, 1962, 16.19.7).

amethyst (a) A transparent to translucent and purple, purple-red, reddish-purple, bluish-violet, or pale-violet variety of crystalline quartz much used as a semiprecious gemstone. The color is perhaps due to manganese or ferric iron. Syn: *bishop's stone*. (b) A term applied to a deep-purple variety of corundum and to a pale reddish-violet beryl.

amherstite A syenodiorite in which the feldspar is andesine and antiperthite. Its name is derived from Amherst County, Virginia.

Amherst stone A commercial name for a variety of *bluestone* quarried near Amherst, Ohio.

amianthus A syn. of *asbestos*, applied esp. to a fine silky variety (such as chrysotile). Syn: *amiantus*.

amictic lake A perennially frozen lake.

amino acid One of the group of organic compounds, containing both amine ($-NH_2$) and carboxyl (-COOH) groups, which are the building blocks of proteins and are, therefore, essential to life processes. All but one (glycine) are optically active and most occur in nature in the L-form.

ammersooite A clay mineral (illite?) occurring in soil.

ammite *ammonite* [sed].

ammonia alum *tschermigite*.

ammonioborite A white mineral: $(NH_4)_2B_{10}O_{16}.5H_2O$. It is found as aggregates of minute plates in fumarolic deposits.

ammoniojarosite A pale-yellow mineral of the alunite group: $(NH_4)Fe_3(SO_4)_2(OH)_6$.

ammonite [paleont] Any ammonoid belonging to the order Ammonitida, characterized by a thick, strongly ornamented shell with sutures having finely divided lobes and saddles. Their stratigraphic range is Lower Jurassic to Upper Cretaceous. Syn: *ammoniticone*.

ammonite [sed] An obsolete term, formerly applied in the 17th and 18th centuries to a sedimentary rock now known as oolite. Syn: *ammite*.

ammoniticone Any of numerous closely coiled fossil cephalopod shells of the subclass Ammonoidea; an *ammonite*.

ammonitic suture A type of suture in ammonoids characterized by complex fluting in which all of the smaller secondary and tertiary lobes and saddles (developed on a larger primary set) are denticulate or frilled; specif. a suture in ammonites. Cf: *goniatitic suture; ceratitic suture*.

ammonium alum *tschermigite*.

ammonoid Any extinct cephalopod belonging to the subclass Ammonoidea, characterized by an external shell that is symmetrical and coiled in a plane and has a bulbous protoconch, septa that form angular sutural flexures, and a small marginal siphuncle. Their stratigraphic range is Lower Devonian to Upper Cretaceous.

amnicolous Said of an organism living on sand.

amoebocyte A sponge cell of amoeboid (irregular, changing) form. It includes such cells as archaeocytes, sclerocytes, trophocytes, and collencytes. Syn: *amoeboid cell*.

amoeboid Said of a fold that has no specific shape and that has a very shallow dip, e.g. said of a placanticline.

amorphous (a) Said of a mineral or other substance that lacks crystalline structure, or whose internal arrangement is so irregular that there is no characteristic external form. The term does not preclude the existence of any degree of order. Ant: *crystalline*. (b) A term formerly used to describe a body of rock occurring in a continuous mass or without division into parts such as brought about by stratification or cleavage.

amorphous graphite Very fine-grained graphite from metamorphosed coal beds. The word "amorphous" is a misnomer because all graphite is crystalline. The term has also been applied to very fine particles of crystalline flake graphite that can be sold only for low-value uses (such as foundry facings) and to fine-grained varieties of Ceylon lump graphite.

amorphous peat Peat in which degradation of cellulose matter has destroyed the original plant structures. Cf: *pseudofibrous peat; fibrous peat*.

amosite A commercial term for an iron-rich, asbestiform variety of amphibole occurring in long fibers. It may consist of an orthorhombic amphibole (anthophyllite or gedrite) or of a monoclinic amphibole (cummingtonite or grunerite).

ampangabéite *samarskite*.

ampasimenite An igneous rock characterized by the presence of nepheline, titanaugite, hornblende, and magnetite in a fine-grained brown groundmass.

ampelite An obsolete term for a black, carbonaceous or bituminous shale.

amphiaster A siliceous sponge spicule (microsclere) composed of a straight rod with a group of radiating spines at each end.

amphiblastula A hollow, ovoid, free-swimming sponge larva composed of two morphologically distinct (small flagellated and large nonflagellated) types of cells, one grouped anteriorly and the other posteriorly. Pl: *amphiblastulae*.

amphibole (a) A group of dark, rock-forming, ferromagnesian silicate minerals closely related in crystal form and composition and having the general formula: $A_{2-3}B_5(Si,Al)_8O_{22}(OH)_2$, where A = Mg, Fe^{+2}, Ca, or Na, and B = Mg, Fe^{+2}, Fe^{+3}, or Al. It is characterized by a cross-linked double chain of tetrahedra with a silicon:oxygen ratio of 4:11, by columnar or fibrous prismatic crystals, and by good prismatic cleavage in two directions parallel to the crystal faces and intersecting at angles of about 56° and 124°; colors vary from white to black. Most amphiboles crystallize in the monoclinic system, some in the orthorhombic or triclinic systems; they constitute an abundant and widely distributed constituent in igneous and metamorphic rocks (some are wholly metamorphic or secondary), and they are analogous in chemical composition to the *pyroxenes*. (b) A mineral of the amphibole group, such as hornblende, anthophyllite, cummingtonite, tremolite, actinolite, riebeckite, glaucophane, and arfvedsonite. (c) A term sometimes used as a syn. of *hornblende*.---Etymol: Greek *amphibolos*, "ambiguous, doubtful", in reference to its many varieties.

amphibolide A general term, for use in the field, to designate any coarse-grained, holocrystalline igneous rock almost entirely composed of amphibole minerals. Syn: *amphibololite.*

amphibolite A crystalloblastic (metamorphic) rock consisting mainly of amphibole and plagioclase with little or no quartz. As the content of quartz increases, the rock grades into hornblende plagioclase gneiss. The term was originated by Brongniart. Cf: *feather amphibolite.*

amphibolite facies Metamorphic rocks formed by regional dynamothermal metamorphism under moderate to high pressures (3000-8000 bars) and temperatures (450°-700°C). The term was introduced by Eskola (1939) and was divided into the *cordierite-amphibolite facies* and the *almandine-amphibolite facies* by Winkler (1967).

amphibolization Introduction of, or replacement by, an amphibole mineral.

amphibololite *amphibolide.*

amphidetic Said of a ligament or ligamental area of certain bivalve mollusks (e.g. *Arca*) extending on both the anterior and posterior sides of the beaks. Cf: *opisthodetic.*

amphidisc A siliceous sponge spicule (microsclere) consisting of a central shaft at each end of which is a transverse, stellate disk or an umbrella-like structure (umbel) containing multiple recurved teeth. Also spelled: *amphidisk.*

amphidont Said of a class of ostracode hinges consisting of three elements of which the terminal elements are teeth (or sockets in the opposed valve) and a median element which is subdivided into an anterior socket and a bar (or a tooth and a groove in the opposed valve).

amphidromic point A geographic position in the ocean where, theoretically, there is no tide range and from which cotidal lines radiate in various directions, with the tide amplitude presumably increasing with distance from this point. See also: *amphidromic region.* Syn: *nodal point.*

amphidromic region An oceanic region whose cotidal lines radiate from one *amphidromic point.*

amphidromic system A system of tidal action in which the tide wave swings around a point or center of little or no tidal movement.

amphigene *leucite.*

amphineuran A marine mollusk belonging to the Amphineura, a group now replaced by the *polyplacophorans* and the *aplacophorans.*

amphioxea A slightly curved oxea (sponge spicule).

amphiphloic Pertaining to the siphonostele of certain vascular plants having phloem both internal and external to the xylem. Cf: *ectophloic.*

amphipod Any crustacean belonging to the order Amphipoda, the members of which resemble the *isopods* by the absence of a carapace and by the presence of unstalked sessile eyes, but differ from them in having laterally, rather than dorsoventrally, compressed bodies. Their stratigraphic range is Upper Eocene to present.

amphi-sapropel Sapropel containing coarse plant debris (Tomkeieff, 1954, p.23).

amphitheater A concave landform, generally oval or circular in outline, nearly surrounded by steep slopes, and having a relatively flat floor; e.g. a *cirque.*

amphitropous Said of an ovule whose stalk (funiculus) is curved about it so that the ovule tip and stalk base are near each other (Lawrence, 1951, p.738). Cf: *anatropous.*

amphoterite A chondritic stony meteorite composed essentially of pyroxene (bronzite) and olivine with small amounts of oligoclase and nickel-rich iron.

amplexoid septum A rugose corallite septum characterized by extreme shortness except where it is extended axially on the distal side of a tabula, as in *Amplexus.*

amplitude [ecol] The degree of adaptability exhibited by a particular kind or group of organisms to its surroundings.

amplitude [fold] The distance between the axial surfaces of adjacent folds. Syn: *axial-plane separation.*

ampulla One of the muscular vesicles of the water-vascular system of an echinoderm (such as of an asteroid), being a contractile bulb or sac-like structure of a tube foot, either seated externally in a small cup-shaped depression on the surface of an ambulacral plate or internally and connecting with the podium through the podial pore. Pl: *ampullae.*

amygdale In British usage, a partial syn. of *amygdule*; the terms are not quite synonymous, however, since the term "amygdule" is used only for a very small amygdale.

amygdaloid n. An extrusive or intrusive rock containing numerous *amygdules.* ----adj. Said of a rock having numerous amygdules. Syn: *amygdaloidal.*

amygdaloidal Said of rocks containing *amygdules* and of the structure of such rocks. Its syn. *amygdaloid* is sometimes used in the sense of an "amygdaloidal rock".

amygdule [ign] A gas cavity or vesicle in an igneous rock which is filled with such secondary minerals as zeolite, calcite, quartz, or chalcedony. The term *amygdale* is preferred in British usage, in which case "amygdule" is applied only to small amygdales. Adj: *amygdaloidal.* See also: *amygdaloid.*

amygdule [sed] An agate pebble.

Anabar block A syn. of *Angara shield*, used by Schatsky & Bogdanoff (1957).

anabatic wind A local wind that moves upward, e.g. a *valley wind.*

anabohitsite An olivine-bearing pyroxenite containing hornblende and hypersthene and a high proportion of magnetite and/or ilmenite, approximately 30 percent, according to Johannsen (1939, p.240). Its name is derived from Anabohitsy, Malagasy.

anaboly Acceleration of the ontogeny. Syn: *hypermorphosis.*

anabranch (a) A diverging branch flowing out of a main stream and later rejoining it farther downstream; esp. one of the several branches composing a braided stream. The term was coined by Jackson (1834, p.79) and is used mainly in Australia. Etymol: *anastomosing* + *branch.* Cf: *braid.* Syn: *valley braid; anastomosing branch.* (b) A branch of a stream that loses itself in sandy soil.

anaclinal Said of an *antidip stream* or of a valley that descends in a direction opposite to that of the general dip of the underlying strata it traverses; e.g. applied to an antecedent stream flowing on a surface that has been tilted in a direction opposite to that of the flow of the stream. Term introduced by Powell (1875, p. 160). See also: *obsequent.* Ant: *cataclinal.* Syn: *contraclinal.*

anacline Said of the dorsal and posterior *inclination* of the cardinal area in either valve of a brachiopod, lying in the top left or first quadrant moving clockwise from the orthocline position (TIP, 1965, pt.H, p.60, fig.61).

anadiagenesis A term used by Fairbridge (1967) for the compaction, maturation phase of diagenesis, in which the clastic sediment grains (or chemical ions) become once again lithified during deep burial (extending to 10,000 m). It is characterized by expulsion and upward migration of connate waters and other nonmagnetic fluids (such as petroleum) and often by reducing conditions; it may pass into metamorphism. See also: *epidiagenesis; syndiagenesis.* Adj: *anadiagenetic.* Syn: *middle diagenesis.*

anaerobic (a) Said of an organism (esp. a bacterium) that can live in the absence of free oxygen; also, said of its activities. Noun: *anaerobe.* (b) Said of conditions that exist only in the absence of free oxygen. -- Cf: *aerobic.*

anaerobic decay Decomposition of organic substances in the absence or near absence of oxygen; the ultimate decay products are enriched in carbon.

anaerobic sediment A highly organic sediment characteristic of some fjords and basins where restricted circulation of the water results in the absence or near absence of oxygen at the sediment surface and bottom water rich in hydrogen sulfide.

anagenesis A progressive, linear evolutionary change.

anagenite Quartz conglomerate in the northern Apennines.

anaglyph A picture produced when the overlapping parts of a *stereoscopic pair* of photographs are printed or projected one over the other in complementary colors (usually red and green), the images being oriented so that all pairs of corresponding points that lie in a given reference plane are superimposed. By viewing through filter spectacles of the corresponding complementary colors, a stereoscopic image (three-dimensional effect) is obtained.

Anahuac North American (Gulf Coast) stage: Miocene (above Chickasawhay, below Napoleonville).

anal adj. Pertaining to or situated near the anus of an animal; e.g. an "anal opening" representing a large thecal orifice marking the position of the anus in the CD interray of a blastoid.---n. *anal plate*.

analbite A triclinic *albite* with a "monoclinic" history; a triclinic modification of sodium feldspar that is not stable at any temperature and that upon heating becomes monoclinic by a displacive (reversible) transformation that takes place at once when the transformation temperature is passed. Recent information indicates that no such mineral exists. (b) A mineral: $(Na,K)AlSi_3O_8$, with the Na:K ratio equal to or greater than 9. It is a variety of anorthoclase.

analcime A white or slightly colored zeolite mineral: $NaAlSi_2O_6.H_2O$. It is closely related to albite, but contains less silica. Syn: *analcite*.

analcimite An extrusive or hypabyssal igneous rock consisting mainly of analcime and pyroxene (usually titanaugite). Feldspathoids, plagioclase, and/or olivine may be present. Apatite, sphene, and opaque oxides may be present as accessories.

analcimization Replacement of feldspars or feldspathoids by analcime, usually in igneous rocks during late- and postmagmatic stages. Syn: *analcitization*.

analcimolith An igneous rock essentially composed of primary or secondary analcime (Thrush, 1968, p.37).

analcite *analcime*.

analcitite An olivine-free analcime-bearing basalt.

analcitization *analcimization*.

anal cover plate One of many small polygonal plates that may extend over and conceal the anal opening in theca of a blastoid and that are bordered marginally by anal deltoids or deltoid plates. See also: *cover plate*.

anal deltoid An undivided interradial plate on the posterior (CD) part of blastoid theca below circlet of oral plates or mouth opening.

anal fasciole (a) A *fasciole* generated on gastropod-shell whorls by the indentation of the outer lip (such as a sinus) situated close to the adapical suture. (b) A *fasciole* adoral and lateral to the periproct of an echinoid.

analog computer A *computer* that operates with numbers represented by directly measurable quantities (such as linear lengths, voltages, or resistances) in a one-to-one correspondence; a measuring device that operates on continuous variables represented by physical or mathematical analogies between the computer variables and the variables of a given problem to be solved by the computer. Cf: *digital computer*.

analogous pole In crystallography, that pole of a crystal which becomes electrically positive when the crystal is heated or expanded by decompression. Cf: *antilogous pole*.

analogy Likeness in form or function but not in origin.

anal plate Any plate covering the anus or anal opening of an echinoderm; a plate of the CD interray of a crinoid, mostly confined to the dorsal cup but excluding fixed pinnulars incorporated in theca. Syn: *anal*.

anal pyramid A cone-shaped structure composed of several elongate, imbricate, more or less triangular plates serving to close the anus in echinoderms such as cystoids, blastoids, and edrioasteroids.

anal sac A variously shaped, generally inflated, and strongly elevated part of theca as developed among inadunate crinoids.

anal tube A conical to cylindrical structure on crinoid tegmen, usually of considerable height, bearing the anal opening at its summit. It is typically developed in many camerate, flexible, and articulate crinoids.

anal X A special anal plate in inadunate and flexible crinoids, typically located between posterior (CD) radial plates, adjacent or next adjacent to tegmen.

analysis of variance A statistical technique for simultaneously partitioning the total variance of a set of data into components which can be attributed to different sources of variation, and which can be used to test for differences among several samples. Syn: *ANOVA*.

analytical geomorphology *dynamic geomorphology*.

analytical triangulation Triangulation accomplished by computational routines using measured coordinates and appropriate formulas; e.g. aerial triangulation accomplished by computing positions and/or elevations of ground stations from measurements made on aerial photographs and known locations of control points.

analytic group A rock-stratigraphic unit formerly classed as a formation but now called a *group* because subdivisions of the unit are now considered to be formations (Weller, 1960, p.434). Cf: *synthetic group*.

analyzer In a polarizing microscope, the upper of the two *Nicol prisms* that receives and exhibits the polarized light after it has passed through the object under study; it is located above the objective. Its vibration direction is normally set at right angles to that of the *polarizer* that produced the polarized light.

anamesite A fine-grained basaltic rock intermediate in texture between fine-grained basalt and coarse-grained dolerites.

anamigmatization High-temperature, high-pressure remelting of preexisting rock to form magma. Cf: *anatexis*.

anamorphic zone A zone of intense metamorphism (*anamorphism*) in which rock flowage takes place and in which simple mineral compounds are changed into more complex ones of greater density by processes of silication, decarbonization, dehydration, and deoxidation. The term was originated in 1898 by Van Hise, and is now little used. Cf: *katamorphic zone*.

anamorphism Constructive metamorphism at considerable depth in the Earth's crust (in the *anamorphic zone*) where complex, relatively dense minerals are formed at the expense of simpler ones of lower density, and where rock flowage takes place. The term was originated by Van Hise (1904). Cf: *katamorphism*.

anamorphosis In the evolution of a group of animals or plants, the gradual change from one form to another; e.g., certain arthropods acquire an additional body segment after hatching.

anandite A mineral: $(Ba,K)(Fe,Mg)_3(Si,Al,Fe)_4O_{10}(O,OH)_2$.

anapaite A pale-green or greenish-white mineral: $Ca_2Fe(PO_4)_2.4H_2O$.

anapeirean Said of rocks of the *Pacific suite*.

anaphoresis *Electrophoresis* in which the movement of suspended negative particles in a fluid is toward the anode. Cf: *cataphoresis*.

anaplasis An evolutionary state characterized by increasing vigor and diversification of organisms; considered to be the first stage in an evolutionary cycle. Cf: *metaplasis; cataplasis*.

anaptychus One of the plates of an *aptychus*; strictly, a single plate closing the aperture of some ammonoids.

anascan Any cheilostome belonging to the suborder Anasca and characterized by the absence of a compensation sac. Cf: *ascophoran*.

anaseism Earth movement away from the focus of an earthquake; a *compression*. Cf: *kataseism*.

anastable adj. Stable, with a tendency towards upheaval. Cf: *catastable*.

anastatic water A syn. of *fringe water* proposed by Meinzer (1939) as one of the three classes of *kremastic water*.

anastomosing Said of a leaf whose veins form a netlike pattern; pertaining to an intervened leaf. Sometimes the vein branches meet only at the margin.

anastomosing branch *anabranch.*

anastomosing stream *braided stream.*

anastomosis [speleo] *spongework.*

anastomosis [streams] (a) *braiding.* (b) A product of braiding; esp. an interlacing network of branching and reuniting channels.

anastomosis tube One of many, small, irregular, repeatedly interconnected solution tubes, commonly found along bedding planes.

anatase A brown, dark-blue, or black tetragonal mineral: TiO_2. It is trimorphous with rutile (which has different facial angles) and brookite, and occurs as an alteration product of other titanium minerals. Syn: *octahedrite.*

anatectic magma The magma formed as a result of *anatexis.*

anatectite *anatexite.*

anatexis Melting of preexisting rock (this term is commonly modified by terms such as intergranular, partial, differential, selective, or complete) (Dietrich & Mehnert, 1961). Cf: *metatexis; diatexis; palingenesis; syntexis; anamigmatization.*

anatexite Rock formed by anatexis. Also spelled: *anatectite.* See also: *protectite; snytectite.*

anathermal n. A period of time during which temperatures are rising. The term was used by Emiliani (1955, p.547) for a part of a cycle as displayed in a deep-sea sediment core. Abbrev: An. Ant: *catathermal.*

Anathermal n. A term proposed by Antevs (1948, p.176) for a postglacial interval (from about 10,000 to 7500 years ago) preceding the Altithermal and representing the period of generally rising temperatures following the last major advance of the continental glaciers. It is equivalent to the Preboreal and Boreal.---adj. Pertaining to the postglacial Anathermal interval and to its climate, deposits, biota, and events.

anatomy The branch of morphology that deals with the structure of animals and plants, esp. the internal structure of organisms as revealed by the microscope.

anatriaene A sponge triaene in which the cladi are bent back toward the rhabdome.

anatropous Said of an ovule that is reversed, i.e. one whose opening (micropyle) is closed to the point of funiculous attachment (Lawrence, 1951, p.738). Cf: *amphitropous.*

anauxite A clay mineral: $Al_2(Si_3O_7)(OH)_4$. It is a high-silica variety of kaolinite (silica:alumina molecular ratio frequently approaches 3) composed of interstratified layers of kaolinite and extra silica sheets. It is probable that "much material described in the literature as anauxite is actually a discrete mixture of kaolinite and some form of silica that has not been detected" (Grim, 1968, p.37). Syn: *ionite.*

anaxial Lacking a distinct axis or axes; e.g. said of an arm or branch of a sponge spicule that has no axial filament or axial canal.

ancestral river A term applied in Australia to a major, ancient river system, older than a *prior river.*

ancestrula The primary or first-formed zooid of a bryozoan colony, derived by metamorphosis of a free-swimming larva, and from which secondary individuals are formed by budding. Pl: *ancestrulae.*

anchi- In petrologic terminology, a prefix signifying "almost".

anchieutectic Said of a rock whose minerals are almost completely in eutectic proportions.

anchimetamorphism A term introduced by Harrassowitz (1927) and still used by German authors to indicate changes in mineral content of rocks under temperature and pressure conditions prevailing in the region between the Earth's surface and the zone of true metamorphism, i.e. approximately in the zones of weathering and ground-water circulation. The term

was never accepted by Anglo-Saxon authors.

anchimonomineralic Said of an igneous rock (such as anorthosite or dunite) that consists essentially of a single mineral. Term originated by Vogt (1905). Cf: *monomineralic.* Syn: *anchimonomineral.*

anchor A holothurian sclerite in the shape of an anchor, consisting of a *shank*, two or more *flukes*, and usually a *stock*.

anchorate n. (a) A sponge spicule (hexactin) with one long ray and two (coplanar) or four recurved, short rays at one end. (b) A strongly dentate chela of a sponge.---adj. Said of a sponge spicule having one or more processes like the fluke of an anchor.

anchor branch A curved hooklet in the radiolarian suborder Phaeodarina.

anchored dune A sand dune whose movement is arrested or whose form is protected from further wind action, as by a growth of vegetation or by cementation of the sand. Cf: *attached dune; wandering dune.* Syn: *fixed dune; established dune; stabilized dune.*

anchor ice Spongy underwater ice formed on a submerged object or structure, or attached to the bottom of a body of water (as a stream, lake, or very shallow sea) which itself is not frozen; usually forms in cold, clear, and still water. Syn: *bottom ice; ground ice; depth ice; lappered ice; underwater ice.*

anchor-ice dam An accumulation of anchor ice raising the water level of a river or stream.

anchorite A nodular, veined diorite having isolated patches of mafic minerals and contemporaneous veins of felsic minerals.

ancient uvium Alluvium that has become lithified.

ancient volcano A general term for a volcano or volcanism of a past geologic age.

ancylite A mineral: $SrCe(CO_3)_2(OH) \cdot H_2O$.

andalusite A brown, yellow, green, red, or gray orthorhombic mineral: Al_2SiO_5. It is trimorphous with kyanite and sillimanite. Andalusite occurs in thick, nearly square prisms in schists, gneisses, and hornfelses; it forms at medium temperatures and pressures of a regionally metamorphosed sequence and is characteristic of contact-metamorphosed argillaceous rocks. In transparent gem quality, andalusite has a very strong pleochroism: brownish green in one direction and brownish red at 90°. See also: *chiastolite.*

Andept In U.S. Dept. of Agriculture soil taxonomy, a suborder of the soil order *Inceptisol*, characterized by formation in vitric pyroclastics or by containing much amorphous material or by having a low bulk density (SSSA, 1970). Cf: *Aquept; Ochrept; Plaggept; Tropept; Umbrept.*

andersonite A bright yellow-green secondary mineral: $Na_2Ca(UO_2)(CO_3)_3 \cdot 6H_2O$.

andesine A mineral of the plagioclase feldspar group with composition ranging from $Ab_{70}An_{30}$ to $Ab_{50}An_{50}$. It occurs as a primary constituent of intermediate igneous rocks (such as of andesites and diorites).

andesinite A coarse-grained igneous rock almost entirely composed of andesine.

andesite A dark-colored, fine-grained extrusive rock that, when porphyritic, contains phenocrysts composed primarily of zoned acid plagioclase (esp. andesine) in the range of An_{35} to An_{70} and one or more of the mafic minerals (e.g. biotite, hornblende, pyroxene) and a groundmass composed generally of the same minerals as the phenocrysts, although the plagioclase may be more acid and quartz is generally present; the extrusive equivalent of *diorite.* Andesite grades into *latite* with increasing alkali feldspar content, and into *dacite* with both more alkali feldspar and more quartz. Its name is derived from the Andes Mountains, South America.

andesite line The geographic-petrographic boundary between basalts of the *Atlantic suite* and the mainly andesitic rocks of the *Pacific suite.* The boundary on the west is generally drawn from Alaska to the east of New Zealand and Chatham Island by way of Japan, the Marianas, Palau Islands, Bismarck Archipelago, and the Fiji and Tonga groups; on the east, the boundary is less clearly defined but probably runs along the coasts of North and South America; it has not been traced in the South Pacific. Syn: *Marshall line.*

andorite A dark-gray or black mineral: $PbAgSb_3S_6$. It is closely related to ramdohrite and fizelyite. Syn: *sundtite.*

ando soil A black or dark brown soil that is formed from volcanic material, in a temperate to tropical, humid climate. It has a low bulk density. Also spelled: *andosol.*

andosol Var. of *ando soil.*

andradite The calcium-iron end member of the garnet group: $Ca_3Fe_2(SiO_4)_3$. It has a variety of colors ranging from yellow, red, and green, to brown and black; it often occurs in contact-metamorphosed limestones. Varieties include topazolite, demantoid, melanite, aplome, and bredbergite.

andrewsite A bluish-green mineral: $(Cu,Fe^{+2})Fe_3^{+3}(PO_4)_3(OH)_2.$

anegite An igneous rock characterized by the absence of feldspar and olivine, but containing pyroxene, spinel, pyrope, and hornblende.

anelasticity [seis] The effect of attenuation of a seismic wave; it is symbolized by Q. The fraction of strain energy lost per stress cycle is expressed as $\Delta E/E = 1/Q$, where $1/Q$ is the specific attenuation factor or specific dissipation function, ΔE is the strain energy lost per cycle, and $\Delta E/E$ is relative energy loss per cycle.

anelasticity [exp struc geol] The inelastic relaxation in time of very small deformations, usually studied as forced vibrations.

anemochore A plant whose seeds or spores are distributed by the wind. Cf: *anemophily.*

anemoclast A rock fragment that was broken off and more or less rounded by wind action (Grabau, 1904).

anemoclastic rock A clastic rock consisting primarily of anemoclasts.

anemolite A type of *helictite,* the erratic shape of which is thought to have been formed by air currents.

anemometer Any type of instrument designed to measure the velocity of wind.

anemophily Pollination by wind. Adj: *anemophilous.* Cf: *entomophily.*

anemousite A silica-deficient variety of albite.

aneroid barometer A nonliquid type of *barometer* that measures barometric change by its effect on the thin sides of a metal box within which there is almost no air. This type of barometer is most commonly used to measure altitude. Cf: *mercury barometer.*

aneuchoanitic achoanitic.

Angaraland Angara shield.

angaralite A mineral of the chlorite group, occurring in thin black plates. Its formula was formerly given as: $Mg_2(Al, Fe)_{10}Si_6O_{29}.$

Angara shield A name used by Suess for a small shield exposing ancient Precambrian rocks in north-central Siberia, supposed to have been the *vertex* or nucleus around which all other structures of Asia were built. Modern Soviet geologists ascribe less significance to the feature. Syn: *Anabar block; Angaraland.*

angelellite A blackish-brown mineral: $Fe_4As_2O_{11}.$

angiosperm A plant in which the seeds are enclosed in an ovary, comprising the fruit. Such plants range from the early Cretaceous or possibly earlier. Examples include grasses, orchids, elms, roses. Cf: *gymnosperm.* Syn: *flowering plant.*

angle The difference in direction between two convergent lines or surfaces; a measure of the amount of rotation required to make either of two intersecting lines coincide with or become parallel to the other, the rotation being in the plane of the lines and about the point of intersection.

angle of coverage The apex angle of the cone of rays passing through the front (incident) nodal point of a lens. A wide-angle lens has an angle of coverage between 75° and 100° and a narrow-angle lens has an angle of coverage of less than 60°.

angle of departure The acute angle between a structural plane and the vertical plane of a geologic cross section, that relates the structure to the vertical plane as the angle of dip does to the horizontal. It is measured in a plane perpendicular to the trace line of the structural plane in the cross section (Knutson, 1958). Cf: *angle of penetration.*

angle of dip dip.

angle of emergence An angle formed between a ray of energy, e.g. a seismic wave, and the horizontal. It is the complement of the *angle of incidence.* Cf: *apparent angle of emergence.* Syn: *emergence angle.*

angle of incidence The angle that a ray of energy, e.g. a seismic wave, makes with the normal to a boundary, e.g. the Earth's surface. It is the complement of the *angle of emergence.* See also: *critical angle.*

angle of penetration The minimum angle between a structural plane and the plane of a geologic cross section, comparable to the angle of plunge on a geologic map (Knutson, 1958). Cf: *angle of departure.*

angle of reflection Bragg angle.

angle of refraction In optics, the angle of a refracted ray of light, measured from a perpendicular to the surface from which the ray is refracted. Syn: *refraction angle.*

angle of repose The maximum angle of slope (measured from a horizontal plane) at which loose, cohesionless material will come to rest on a pile of similar material. This angle is somewhat less than the slope angle at which sliding will be initiated (angle of slide) and is generally 5° to 10° less than the angle of internal friction of the same material. The angle of repose commonly ranges between 33° and 37° on natural slopes, and is rarely less than 30° or more than 39°. The angle is dependent upon the frictional properties of the material and increases slightly as the size of the fragments decreases and as their angularity increases. Cf: *angle of slide.* Syn: *angle of rest.*

angle of rest angle of repose.

angle of slide The angle (usually measured from a horizontal plane) of minimum slope at which any loose material (such as earth or talus) will start to slide; it is slightly greater than the *angle of repose.*

angle of ultimate stability critical slope angle.

anglesite A white orthorhombic mineral: $PbSO_4$. It is a common secondary mineral formed by the oxidation of galena and is a valuable ore of lead. Syn: *lead vitriol; lead spar.*

angrite An achondritic stony meteorite consisting chiefly of purple titaniferous augite (more than 90%) with a little olivine and troilite.

Ångström compensation phyheliometer A *pyrheliometer* developed by K. Ångström for the measurement of direct solar radiation. The radiation receiver station consists of two identical manganin strips whose temperatures are measured by attached thermocouples. One of the strips is shaded, whereas the other is exposed to sunlight. An electrical heating current is passed through the shaded strip so as to raise its temperature to that of the exposed strip. The electric power required to accomplish this is a measure of the solar radiation (Marks, 1969, p. 54). Cf: *Ångström pyrgeometer.*

Ångström pyrgeometer A *pyrgeometer* developed by K. Ångström for measuring the effective terrestrial radiation. It consists of four manganin strips, of which two are blackened and two are polished. The blackened strips are allowed to radiate to the atmosphere while the polished strips are shielded. The electrical power required to equalize the temperature of the four strips is taken as a measure of the outgoing radiation (Marks, 1969, p. 54). Cf: *Ångström compensation pyrheliometer.*

anguclast An angular *phenoclast,* such as a large fragment of a breccia. Cf: *spheroclast.*

angular Having sharp angles or borders; specif. said of a sedimentary particle showing very little or no evidence of abrasion, with all of its edges and corners sharp, such as block with numerous (15-30) secondary corners and a roundness value between zero and 0.15 (midpoint at 0.125) (Pettijohn, 1957, p.58-59). Powers (1953) gives values between 0.17 and 0.25 (midpoint at 0.21). Also, said of the *roundness class* containing angular particles.

angular cross-bedding Cross-bedding in which the foreset beds appear in section as nearly straight lines meeting the underlying surface at high, sharp, or discordant angles; it often implies deposition by water, such as *torrential cross-bedding*. Cf: *tangential cross-bedding.*

angular discordance *angular unconformity.*

angular distance The angle, measured at the Earth's center, which subtends the great circle path between the earthquake epicenter and receiver. Cf: *epicentral distance.*

angular distortion The change in shape of a small circle on a globe when it is represented on a map projection.

angular drift "Rock debris formed by intensive frost action, derived from underlying or adjacent bedrock" (ADTIC, 1955, p.4).

angular fold *accordion fold.*

angularity [seis] *stepout time.*

angularity [sed] A term often used for the property of a sedimentary particle now commonly known as *roundness*, but used by Lamar (1928, p.148-151) for the property now referred to as *sphericity.*

angular spreading The lateral extension of ocean waves as they move out of the generating area as swell.

angular unconformity An *unconformity* between two groups of rocks whose bedding planes are not parallel or in which the older underlying rocks dip at a different angle (usually steeper) than the younger overlying strata; specif. an unconformity in which the younger overlying sediments rest upon the eroded surface of tilted or folded older rocks. It is sometimes regarded as a type of *nonconformity.* Syn: *discordance; angular discordance; clinounconformity; structural unconformity; orogenic unconformity.*

angulate drainage pattern A modified *rectangular drainage pattern* developed where streams follow joints or faults that join each other at acute or obtuse angles, rather than at right angles (Zernitz, 1932, p.517). Examples are found in the Timiskaming and Nipissing areas of Ontario, Can.

anhedral (a) Said of an individual mineral crystal, in an igneous rock, that has failed to develop its own bounding crystal faces naturally suggested by the internal molecular structure of the crystal or that has a rounded or indeterminate form determined by the crowding of adjacent mineral grains during crystallization. (b) Said of a detrital mineral grain that shows no crystal outline; also, said of a crystal, in a sedimentary rock (such as a calcite crystal in a recrystallized dolomite), characterized by the absence of crystal faces. (c) Said of the shape of an anhedral crystal.----The term was used, in reference to igneous-rock components, by Cross et al (1906, p.698) in preference to the synonymous terms *xenomorphic* and *allotriomorphic* (as these were originally defined). Cf: *euhedral; subhedral.*

anhedron An anhedral crystal. The term was introduced by Pirsson (1896) in reference to an imperfectly defined igneous-rock component (crystal). Pl: *anhedrons; anhedra.*

anhydrite A mineral consisting of an anhydrous calcium sulfate: $CaSO_4$. It represents *gypsum* without its water of crystallization, and it alters readily to gypsum, from which it differs in crystal form (anhydrite is orthorhombic) and in being harder and slightly less soluble. Anhydrite usually occurs in white or slightly colored, granular to compact masses, forming large beds or seams in sedimentary rocks or associated with gypsum and halite in evaporites. Syn: *cube spar.*

anhydrock A sedimentary rock composed chiefly of anhydrite that commonly occurs in finely granular masses and sometimes in fibrous or coarsely crystalline masses; it is typically uniformly bedded with a dense to saccharoidal texture.

anhydrous Said of a substance, e.g. magma or a mineral, that is completely or essentially without water. An anhydrous mineral contains no water in chemical combination.

anhysteretic remanent magnetization Remanent magnetization produced by simultaneous application of a constant field and an initially larger alternating field whose amplitude decreases slowly to zero.

anidiomorphic *xenomorphic.*

anilite A mineral: Cu_7S_4.

Animikean Var. of *Animikie.*

Animikie A provincial series of the Proterozoic of the Canadian Shield; it is also called the *Animikean.*

animikite A silver ore consisting of a mixture of sulfides, arsenides, and antimonides showing striking intergrowth relations and occurring in white or gray granular masses. It contains nickel and lead. Cf: *macfarlanite.*

Anisian European stage: lower Middle Triassic (above Scythian, below Ladinian). Syn: *Virglorian; Hydaspien.*

anisochela A sponge *chela* having unequal or dissimilar ends. Cf: *isochela.*

anisodesmic Said of a crystal or compound in which the ionic bonds are of unequal strength. Cf: *isodesmic.*

anisomerism (a) Repetition of parts that differ more or less importantly among themselves. (b) Reduction in number and differentiation of similar parts in organisms.

anisometric (a) An obsolete term applied to granular rocks (esp. igneous rocks) in which the grains are of different sizes. Cf: *seriate.* (b) Said of crystals having unequal axes, for which the more common and preferred term is *anisotropic.*

anisotropic Said of a medium whose physical properties vary in different directions, e.g. in crystal optics, said of a crystal whose physical properties vary according to crystallographic direction, e.g. a crystal having double refraction. All crystals except those of the isometric system are anisotropic. Ant: *isotropic.* Syn: *aeolotropic.*

anispiracle An enlarged opening in the summit part of the posterior interray of a blastoid, formed by the union of anal opening and posterior spiracle (or spiracles).

anitaxis A linear succession of crinoid anal plates. Pl: *anitaxes.*

ankaramite An olivine-bearing basalt containing numerous pyroxene and olivine phenocrysts, the former being more abundant than the latter, in a fine-grained groundmass composed of augite and titanaugite microlites, labradorite, and accessory biotite. Its name is derived from Ankaramy, Malagasy.

ankaratrite *olivine nephelinite.*

ankerite A white, red, or grayish iron-rich mineral related to dolomite: $Ca(Fe,Mg,Mn)(CO_3)_2$. It is associated with iron ores and commonly forms thin veins of secondary matter in some coal seams. Syn: *ferroan dolomite; cleat spar.*

ankylosis Fusion of columnals or other skeletal elements of a crinoid, commonly with obliteration of sutures.

annabergite An apple-green mineral: $(Ni,Co)_3(AsO_4)_2.8H_2O$. It is isomorphous with erythrite, and it usually occurs in incrustations as an alteration product of nickel arsenides. Syn: *nickel bloom; nickel ocher.*

annelid Any worm-like invertebrate belonging to the phylum Annelida, characterized by a segmented body with a distinct head and appendages. Because the annelids lack skeletal structures (except for chitinous jaws, called *scolecodonts*), they are usually known as fossils only from their burrows and trails.

annerödite A black mineral consisting of samarskite with parallel overgrowths of columbite. Also spelled: *annerodite.*

annotated photograph A photograph on which planimetric, hypsographic, geologic, cultural, hydrographic, or vegetation information has been added to identify, classify, outline, clarify, or describe features that would not otherwise be apparent in examination of an unmarked photograph (ASP, 1966, p.1127). The term generally does not apply to a photograph marked only with geodetic control or pass points.

annual balance The change in mass of a glacier from the beginning to the end of a hydrologic year (usually October 1 to September 30), or other measurement year defined by fixed calendar dates, determined at a point, as an average for an area, or as a total mass change for the glacier. Units of millimeters, meters, or cubic meters are normally used. Cf: *balance; net balance.*

annual flood (a) The highest peak discharge of a stream in a given water year. (b) That flood in a given water year which has been equalled or exceeded on the average of once a year.----(ASCE, 1962).

annual growth ring *growth ring.*

annual layer (a) A sedimentary layer deposited or presumed to have been deposited during the course of a year; e.g. a glacial varve. (b) A dark band (in a salt stock) of formerly disseminated anhydrite crystals that accumulated upon being freed by solution of the enclosing salt.

annually thawed layer *active layer.*

annual wave Annually cycling heating and cooling of the upper 3-5 meters of soils, in response to yearly weather cycles. Below this point constant annual temperatures may be observed. Cf: *diurnal wave.*

annular drainage pattern A drainage pattern in which subsequent streams follow a roughly circular or concentric path along a belt of weak rocks, resembling in plan a ring-like pattern. It is best displayed by streams draining a maturely dissected structural dome or basin where erosion has exposed rimming sedimentary strata of greatly varying degrees of hardness, as in the Red Valley which nearly encircles the domal structure of the Black Hills, S.D.

annular lobe A small secondary dorsal lobe in the center of the main *internal lobe* of a suture of some coiled nautiloid conchs.

annular space The ring-shaped space between the outer wall of a drill pipe suspended in a borehole and the casing or the side of the open hole or the inner side of a larger pipe, or a similar space between the casing and the wall of the hole.

annular tracheid The first *tracheid* to mature, characterized by the deposition of additional wall material in rings. Cf: *spiral tracheid.*

annulation A ringlike structure; e.g. a raised ring-shaped feature on the walls of a joint fissure, or a ringlike expansion of an ammonoid conch, either transverse or slightly oblique to the longitudinal axis of the conch. Cf: *ring structure.*

annulus [ceph] (a) A thin, ring-shaped endosiphuncular deposit, semicircular in cross section, on the inner side of a septal neck of a nautiloid (TIP, 1964, pt.K, p.54). (b) The *periphract* of a nautiloid.

annulus [bot] (a) In the mushrooms, a ring that is a remnant of the partial veil on the stipe. (b) In the mosses and ferns, a specialized ring of cells on the sporangium that is indirectly involved in spore release.----(Scagel, 1965, p.609).

annulus [palyn] A ring bordering a pore of a pollen grain and in which the ektexine is modified (usually thickened). Cf: *margo.* See also: *endannulus.*

anogene The original, now obsolete, form of *anogenic.*

anogenic (a) Pertaining to plutonic metamorphism or replacement. (b) Pertaining to eruptive rocks. --Obs syn: *anogene.*

anomalous [intrus rocks] Said of a magma type that is the result of *assimilation.* Cf: *hybrid.*

anomaly [geophys] A deviation from uniformity or regularity in geophysical quantities; a difference between observed and computed value.

anomaly [meteorol] A difference in the mean local value of a meteorological element from its mean value for that latitude.

anomaly [oceanog] The difference between an ocean of standard or arbitrary temperature and salinity and the conditions actually observed at a particular station.

anomite A variety of biotite differing only in optic orientation.

anomoclone A desma (of a sponge) consisting of a short arm (brachyome) and several longer arms directed at various angles away from the short arm.

anomphalous Said of a gastropod shell lacking an umbilicus. Cf: *phaneromphalous.*

anorogenic Not orogenic, lacking in or unrelated to tectonic disturbance, e.g. anorogenic area, anorogenic time, and anorogenic granite.

anorthic Said of crystals having unequal oblique axes; i.e., triclinic crystals.

anorthite (a) A white, grayish, or reddish triclinic mineral of the plagioclase feldspar group: $CaAl_2Si_2O_8$. It is the most basic member of the plagioclases, its composition ranging from $Ab_{10}An_{90}$ to Ab_0An_{100}. Anorthite occurs in basic and ultrabasic igneous rocks (gabbro, norite, anorthosite), rarely as well-developed druse mineral, sometimes in tuffs, and very rarely in metamorphic rocks (skarns). Syn: *calcium feldspar; calciclase.* (b) The pure calcium-feldspar end member in the plagioclase series.

anorthitfels *anorthitite.*

anorthitissite A hornblendite containing anorthite.

anorthitite An igneous rock almost completely composed of anorthite. Syn: *calciclasite; anorthifels.*

anorthoclase A triclinic mineral of the alkali feldspar group: $(Na,K)AlSi_3O_8$. It is a sodium-rich feldspar ($Or_{40}Ab_{60}$ to $Or_{10}Ab_{90}$) that shows deviations from monoclinic symmetry and that contains very fine-grained intergrowths; it is widespread as a groundmass constituent of slightly alkalic lavas. The term is usually applied to a mixture of several phases that may not even have a stability field of their own at any temperature. Cf: *orthoclase.* Syn: *anorthose; soda microcline.*

anorthoclasite An igneous rock almost entirely composed of anorthoclase.

anorthose *anorthoclase.*

anorthosite A group of essentially monomineralic plutonic igneous rocks composed almost entirely of plagioclase feldspar, which is usually labradorite but may also be as calcic as bytownite or as sodic as andesine or oligoclase, and little or no dark-colored minerals; also, any rock in that group. Anorthosites occur as large nonstratiform plutonic bodies, as stratiform intrusions, and have been identified in lunar rock samples. Syn: *plagioclasite; plagioclase rock.*

anorthositization Introduction of, or replacement by, anorthosite.

ANOVA *analysis of variance.*

antagonism In ecology, the relationship that exists between two organisms in which one or both are harmed, usually as a result of their trying to occupy the same ecologic niche.

antapical series The series of plates forming the terminal group behind the postcingular series in dinoflagellate theca. Cf: *apical series.*

antarctic n. The area within the Antarctic Circle; the region of the south pole.----adj. Pertaining to features, climate, vegetation, and animals characteristic of the antarctic region.

Antarctic Circle The parallel of latitude to the equator falling at approx. lat. 66°32′ S; it delineates the frigid zone of the South Pole. Cf: *Arctic Circle.*

Antarctic convergence A natural and distinct oceanographic boundary around the continent of Antarctica, more or less equivalent to the 50°F isotherm for the warmest month. The colder, denser Antarctic waters sink sharply below the warmer, lighter sub-Antarctic waters, with little mixing. The oceanographic aspect of the boundary is reflected in water and air temperatures and in the flora and fauna.

antarcticite A mineral: $CaCl_2.6H_2O$.

antecedence A postulated sequence of erosional and orogenic events that presumably led to the development of an antecedent stream or drainage system.

antecedent Said of a stream, valley, or drainage system that maintains its original course or direction despite subsequent deformation or uplift. The term was first applied by Powell (1875, p.163) to the valley thus formed.

antecedent moisture The amount of moisture present in a soil mass at the beginning of a runoff period; often expressed in terms of an *antecedent precipitation index*.

antecedent-platform theory A theory of coral-atoll and barrier-reef formation according to which reefs are built upward to the water surface from an extensive submarine platform (perhaps consisting of volcanic debris rapidly leveled by wave erosion), situated 50 m or more below sea level and predating its colonization by corals, without the intervention of relative changes in sea level (see Hoffmeister & Ladd, 1944). Cf: *glacial-control theory; subsidence theory*.

antecedent precipitation index The amount of moisture in a drainage basin before a storm. Abbrev: API. See also: *antecedent moisture*.

antecedent stream A stream that was established before local uplift or diastrophic movement was developed across it and that maintained its original course after and in spite of the deformation by incising its channel at approximately the same rate as the land was rising; a stream that existed prior to the present topography.

antecedent valley A valley eroded by or containing an *antecedent stream*.

anteclise A positive or uplifted structure of the continental platform; it is of broad, regional extent (tens to hundreds of thousands of square kilometers) and is produced by slow crustal upwarp during the course of several geologic periods. The term is used mainly in the Russian literature; e.g. the Belorussian anteclises of the Volga-Urals. Also spelled: *anticlise*. Ant: *syneclise*.

anteconsequent adj. Said of a stream, valley, or drainage system that is *consequent* in the earlier stages and *antecedent* in the later stages of an erosion or orogenic sequence. The term is rarely used "doubtless due to the practical difficulties of differentiating the effects of stages in tectonic movements in most areas" (Stamp, 1961, p.25).---n. An anteconsequent stream.

antediluvian Pertaining to or produced during the period before the flood described in the Bible; e.g. "antediluvian deposits" antedating the Noachian flood. Syn: *antediluvial; prediluvian*.

antenna (a) One of a pair of anterior sensory appendages of the cephalon of a crustacean, preceded by antennule and followed by mandible. (b) One of a pair of slender multijointed sensory appendages attached to the ventral surface of the cephalon in front of the mouth of a trilobite. (c) An obsolete term for a *chelicera* of an arachnid, used in the past to emphasize its homology with an antenna of a crustacean or of an insect.---Pl: antennae or antennas.

antennule A small antenna; specif. one of a pair of the most anterior appendages of the cephalon of a crustacean, followed by antenna. Adj: *antennular*. Syn: *first antenna; antennula*.

anter Part of the orifice in ascophoran cheilostomes (bryozoans) that is distal to the condyles. Cf: *poster*.

anterior adj. Situated toward the front of an animal, or near or toward the head or head region, as opposed to *posterior*; e.g. in a direction (in the plane of symmetry or parallel to it) away from the pedicle and toward the mantle cavity of a brachiopod, or in a direction (in the plane of bilateral symmetry) parallel to the cardinal axis of a bivalve mollusk and approximating the direction in which the mouth faces, or in a direction (typically abapical) along the midline axis of a gastropod and toward the extremity of the head, or located on the side of echinoderm theca opposite hydropore and/or gonopore (such as in a direction of the ambulacral ray of an echinoid opposite

the anal interray, where the anus has shifted from the apex of the skeleton), or in a direction toward the aperture of a foraminifer.---- n. The forward-moving or head region of an animal; e.g. the axial part of a brachiopod shell farthest from the apex or pedicle foramen (the side opposite the beak or opposite the position of the hinge line), or the end opposite the pallial sinus of a bivalve mollusk, or the aperture-bearing end of a gastropod.

anterior lateral muscle One of a pair of *retractor muscles* in some linguild brachiopods, originating on the pedicle valve posteriorly and laterally to the central muscles, and converging dorsally to their insertions anteriorly on the brachial valve (TIP, 1965, pt.H, p.139).

anterior side The front end of a conodont; e.g. the convex side of cusp (the side facing in the direction opposite that toward which the tip of cusp points) in simple conodont elements, or the convex side of cusp and denticles in compound conodont elements, or the distal end of free blade in plate-like conodont elements. Ant: *posterior side*.

anterior tubercle A polygenetic swelling or small protuberance in the anterior region of the carapace of a phyllocarid crustacean. It includes the "optic tubercle" of some authors.

antetheca The final septal face in a fusulinid; e.g. the front wall of the last-formed volution of the test of *Triticites*.

ante-turma One of two groupings in which *turmae* are classified: Sporites (for spores) and Pollenites (for pollen).

anther The pollen-bearing part of a stamen.

antheridium (a) In cryptogamous plants, the male reproductive organ, within which the male sexual cells are organized. (b) In primitive seed plants, a minute structure of only a few cells developed within the microspore.

anthill A colloquial term for *termitarium*.

anthoblast The basal portion of the zooid in certain solitary corals from which the *anthocyathus* is pinched off to form a new zooid; e.g. the stage of the ontogeny of *Acrosmilia* derived by transverse division from a solitary *Fungia* individual by extratentactular budding. Cf: *anthocaulus*.

anthocaulus The stalk-like basal portion of the zooid in certain solitary corals from which the *anthocyathus* is pinched off to form a new zooid; e.g. the stage of the ontogeny of *Acrosmilia* derived by transverse division from a solitary *Fungia* individual by sexual generation. Pl: *anthocauli*. Cf: *anthoblast*.

anthocyathus The oral disk that is pinched off from the basal portion in some solitary corals and that enlarges to become a new zooid; e.g. the neanic stage of *Fungia* after separation from an *anthocaulus* or *anthoblast*. Pl: *anthocyathi*.

anthodite In a cave, gypsum or aragonite that occurs as needlelike crystals radiating from a common base. Cf: *cave flower*.

anthoinite A white mineral: $Al_2W_2O_9 \cdot 3H_2O$.

anthonyite A lavender-colored mineral: $Cu(OH,Cl)_2 \cdot 3H_2O$.

anthophyllite A clove-brown orthorhombic mineral of the amphibole group: $(Mg,Fe)_7Si_8O_{22}(OH)_2$. It contains less iron than *cummingtonite*, and may contain manganese and calcium. Anthophyllite is a variety of asbestos, normally occurring in metamorphic rocks as lamellae, radiations, fibers, or massive. Syn: *bidalotite*.

anthozoan Any coelenterate belonging to the class Anthozoa which includes exclusively marine, polypoid, solitary or colonial, mostly sedentary forms and is characterized by the presence of a stomodaeum. Their stratigraphic range is Ordovician to present.

anthracite Coal of the highest metamorphic rank, in which fixed-carbon content is between 92% and 98%. It is hard, black, and has a semimetallic luster and semiconchoidal fracture. Anthracite ignites with difficulty and burns with a short, blue flame and without smoke. Syn: *hard coal; stone coal; kilkenny coal; black coal*.

anthracitization The metamorphic transformation of bituminous coal into anthracite.

anthracology Coal petrology; the analysis of coals by type.

anthraconite A black *bituminous limestone* (or marble) that usually emits a fetid smell on being struck or rubbed; a *stinkstone*. Syn: *swinestone; lacullan*.

anthracoxene A brownish resin which, when treated with ether, dissolves into an insoluble portion, *anthracoxenite*, and a soluble portion, *schlanite*.

anthracoxenite The insoluble resin remaining when *anthracoxene* is treated with ether. See also: *schlanite*.

anthraxolite A hard, black asphaltite with a high fixed carbon content; it occurs in veins and masses in sedimentary rocks, especially in association with oil shales.

anthraxylon A composite term for the vitreous coal components derived from woody tissues of plants and forming lustrous bands interlayered with dull *attritus* in banded coal. Etymol: Greek *anthrax*, "coal", and *xylon*, "wood".

anthraxylous-attrital coal A bright coal in which the ratio of anthraxylon to attritus varies from three to one to one to one. Cf: *attrital-anthraxylous coal; anthraxylous coal; attrital coal*.

anthraxylous coal A bright coal in which the ratio of anthraxylon to attritus is greater than three to one. Cf: *attrital coal; attrital-anthraxylous coal; anthraxylous-attrital coal*.

anthrinoid Vitrinite that occurs in noncaking anthracites and that has a reflectance higher than 2.0% (American Society for Testing and Materials, 1970, p.19). Cf: *xylinoid; vitrinoid*.

anthropic Said of an epipedon that is similar to a *mollic* epipedon but in which the content of soluble P_2O_5 is greater than 250 ppm. It develops due to long periods of cultivation and fertilization (SSSA, 1970).

anthropozoic Said of that span of geologic time since the appearance of man; also, said of the rocks formed during that time. Cf: *Diluvial*.

antibiosis Passive action by one organism which is harmful to another (Ager, 1963, p.313).

anticenter That point on the Earth's surface which is diametrically opposite the *epicenter* of an earthquake. Syn: *antiepicenter*.

anticlinal [bot] adj. At right angles to the surface or circumference of a plant organ. Cf: *periclinal*.

anticlinal [struc geol] n. An obsolete form of *anticline*. adj. Pertaining to an anticline.

anticlinal axis (a) Concerning folds, a line on a map from which the strata of a fold dip away in opposite directions; the median line of an anticline. (b) The intersection of the anticline's crest surface with a given stratum of the fold.

anticlinal nose *nose*.

anticlinal spring A contact spring occurring along the surface outcrop of an anticline, from a pervious stratum overlying one that is less pervious (ASCE, 1962).

anticline A fold, the core of which contains the stratigraphically older rocks; it is convex upward. Ant: *syncline*. See also: *antiform; anticlinal*.

anticlinorium A composite anticlinal structure of regional extent composed of lesser folds. Cf: *synclinorium*. Pl: *anticlinoria*.

anticlise Var. of *anteclise*.

anticonsequent stream *obsequent stream*.

anticusp The downward projection of the base of a conodont *cusp*. It may or may not bear denticles.

anticyclone An atmospheric high-pressure system with closed isobars, the pressure gradient being directed from the center so that the wind blows spirally outward in a clockwise direction in the Northern Hemisphere, counterclockwise in the Southern. It was named by Francis Galton in 1861. Cf: *cyclone*.

antidip stream A stream flowing in a direction opposite to that of the general dip of the strata; an *anaclinal* stream. It is frequently but not necessarily an *obsequent* stream.

antidune (a) A term used by Gilbert (1914, p.31) for an ephemeral or transient *sand wave* formed on a stream bed

(but not believed to be preserved in sediments), similar to a dune but traveling upstream as the individual sand particles move downcurrent, and characterized by erosion on the downstream slope and deposition on the upstream slope. It travels much faster and seems to be higher (heights of 1.8 m have been reported) than a subaqueous *dune* and its profile is more symmetric; it is indicated on the water surface by a regular undulating wave much like that formed behind a stern-wheel steamboat. Syn: *regressive sand wave*. (b) Any stream-bed feature (whether it moves upstream, downstream, or not at all) that is in phase with surface gravity (water) waves (Kennedy, 1963). (c) A term used by Lamont (1957) for *flame structure*.

antidune phase The part of stream traction (transitional to the *smooth phase*) whereby a mass of sediment travels in the form of a ridge-like structure having an eroded downcurrent slope and a depositional upcurrent slope (Gilbert, 1914, p.30-34); it develops when the bed load is large or the current is strong. The antidune form moves upstream as the individual particles move downstream. Cf: *dune phase*.

antiepicenter *anticenter*.

antiferromagnetism A type of *magnetic order* in which the moments of neighboring magnetic ions are aligned antiparallel, so that there is no macroscopic spontaneous magnetization. Cf: *ferromagnetism; ferrimagnetism*. See also: *weak ferromagnetism*.

antiform An anticlinal-type structure in which the stratigraphic sequence is not known. Cf: *anticline*. Ant: *synform*.

antigorite A platy or lamellar, brownish-green mineral of the serpentine group: $Mg_3Si_2O_5(OH)_4$. Cf: *chrysotile*. Syn: *picrolite; baltimorite*.

antigravitational gradation A term introduced by Keyes (1913) for wind erosion and deposition operating mainly from a lower to a higher elevation, as on broad intermont valleys of arid regions where "the wind is able to blow sands erodingly and extensively up-hill". See also: *planorasion*.

antilogous pole In crystallography, that pole of a crystal which becomes electrically negative when the crystal is heated, or is expanded by decompression. Cf: *analogous pole*.

antimagmatist *transformist*.

antimeridian The meridian that is 180 degrees of longitude from a given meridian. A meridian and its antimeridian constitute a complete great circle.

antimonate A mineral compound characterized by the presence of antimony and oxygen in the radical. An example is swedenborgite, $NaBe_4SbO_7$.

antimonite *stibnite*.

antimonpearcite A mineral: $(Ag,Cu)_{16}(Sb,As)_2S_{11}$. Cf: *arsenpolybasite*.

antimony A hexagonal mineral, the native metallic element Sb. It is brittle and commonly occurs in silvery- or tin-white and granular, lamellar, or shapeless masses.

antimony blende *kermesite*.

antimony bloom *valentinite*.

antimony glance *stibnite*.

antimony ocher Any of several native antimony oxides such as stibiconite or cervantite.

antinode That point on a standing wave at which the vertical motion is greatest and the horizontal motion is least. Ant: *node*. Syn: *loop*.

antipathetic Said of two or more minerals that are far apart from each other in a crystallization sequence and thus will not be commonly found in association. See also: *antipathies of minerals*.

antipathies of minerals An aspect of the theory of fractional crystallization, which states that minerals that are far apart in a crystallization sequence will not be found in association to any great extent. Such minerals are said to be *antipathetic*.

antiperthite A variety of alkali feldspar consisting of parallel or subparallel intergrowths in which the sodium-rich phase (albite, oligoclase, or andesine) appears to be the host from which the potassium-rich phase (usually orthoclase) exsolved. Cf: *perthite*.

antipodal bulge The tidal effect on the side of the Earth farthest from the Moon; at the point that is antipodal to the *tidal bulge*; lunar attraction is weakest and produces an apparent bulge.

antipodal point *antipode*.

antipode The opposite point with respect to any given point; specif. one of two diametrically opposite parts of the Earth. The term is usually used in the plural and often extended to include the whole region at the opposite end of a diameter of the Earth, such as Australia and New Zealand which lie roughly opposite to the British Isles. Syn: *antipodal point*.

antiripple A small, asymmetric ripple (wavelength less than 2 cm) formed by windblown silt adhering to a moist surface of unconsolidated material and characterized by a steep slope on the windward side (facing upcurrent). Cf: *wind ripple*. Syn: *antiripplet*.

antirock A meteorite composed of antimatter.

antiroot According to the Pratt isostasy hypothesis, crustal material of higher density under the oceans as isostatic compensation for its lesser mass and lower topographic elevation. Cf: *root*.

antistress mineral A term suggested by Harker (1918) for a mineral such as anorthite, potash feldspars, pyroxenes, forsterite, andalusite, etc. whose formation in metamorphosed rocks is favored by conditions that are controlled, not by shearing stress, but by thermal action and by hydrostatic pressure that is probably no more than moderate. Cf: *stress mineral*.

antithetic Pertaining to minor normal faults that are of the opposite orientation as the major fault with which they are associated. Ant: *synthetic* [*fault*].

antitrades A layer of westerly winds in the troposphere, above the *trade winds* of the tropics. Syn: *countertrades*.

antlerite An emerald-green to blackish-green mineral: $Cu_3SO_4(OH)_4$. It occurs in interlaced aggregates of needle-like crystals and constitutes an ore of copper. Syn: *vernadskite*.

Antler orogeny An orogeny which extensively deformed Paleozoic rocks of the Great Basin in Nevada during late Devonian and early Mississippian time; named by R. J. Roberts (1951) for relations in the Antler Peak quadrangle near Battle Mountain, Nevada. The main expression of the orogeny is the emplacement of eugeosynclinal western rocks over miogeosynclinal eastern rocks along the Roberts Mountains thrust. Minor orogenic pulses followed the main event, extending into the Permian. It is broadly equivalent to the *Acadian orogeny* of eastern North America.

antofagastite *eriochalcite*.

antozonite A dark-violet to black semiopaque variety of fluorite that emits a strong odor when crushed, perhaps due to free fluorine. It is produced by strong alpha bombardment, as in the inner bands of halos surrounding uraninite and thorite inclusions.

anus The posterior or terminal opening of the alimentary canal or digestive tract of an animal. In echinoderms, it includes tissues as well as the anal pyramid.

Apache tear An obsidian nodule.

apachite A phonolite characterized by the presence of abundant enigmatite and hornblende in about the same amount as the pyroxene, but of a later crystallization phase.

apalhraun An Icelandic term for both block lava and aa. Cf: *helluhraun*.

apatite (a) A group of variously colored hexagonal minerals consisting of calcium phosphate together with fluorine, chlorine, hydroxyl, or carbonate in varying amounts and having the general formula: $Ca_5(PO_4,CO_3)_3(F,OH,Cl)$. Also, any mineral of the apatite group, such as fluorapatite, chlorapatite, hydroxylapatite, carbonate-apatite, and francolite; when not specified, the term usually refers to *fluorapatite*. The apatite minerals occur as accessory minerals in almost all igneous rocks, in metamorphic rocks, and in veins and other ore deposits, and most commonly as fine-grained and often impure masses as the chief constituent of phosphate rock and of most or all bones and teeth. Syn: *calcium phosphate*. (b) A group of hexagonal minerals having the general formula: $A_5(RO_4)_3(F,OH,Cl)$, where A = Ca, Sr, or Pb, and R = P, As, V, or less commonly Si. Examples include svabite, hedyphane, mimetite, pyromorphite, and vanadinite.---Symbol: Ap.

apatotrophic Pertaining to a lake that is brackish and that contains living organisms (Termier, 1963).

aperiodic damping Damping which equals or exceeds *critical damping*.

apertural bar A fused pair of costae immediately proximal to the orifice in cribrimorph cheilostomes (bryozoans).

aperture [paleont] (a) The opening of a univalve shell; e.g. the opening at the last-formed margin of a gastropod shell and through which the head-foot mass is extended or withdrawn. (b) Any of the major openings through theca or calyx of echinoderms, such as the mouth and anus, and sometimes including hydropores and gonopores. (c) A term that is loosely used in the literature of bryozoans for an opening in the zooecium through which the living animal extends lophophore and portions of body. It is best defined as the "outermost opening of zooecium" (TIP, 1953, pt.G, p.8). Cf: *orifice*. (d) An opening in the test or shell of a foraminifer, such as a relatively large opening to the exterior in the last-formed chamber. Also, the large main opening of a radiolarian shell. (e) The opening into the mantle cavity of a cirripede crustacean.

aperture [palyn] Any of the various modifications in the exine of spores and pollen that can be used as a locus for exit of the contents; e.g. laesura, colpus, and pore. See also: *germinal aperture*.

apex [geomorph] The tip, summit, or highest or uppermost point of a landform, as of a mountain; specif. the highest point on an alluvial fan, usually the point where the stream that formed the fan emerged from the mountain or from confining canyon walls.

apex [mining] In mining, the highest point of a vein relative to the surface, whether it outcrops or not. The concept is used in mining law. See also: *apex law*.

apex [paleont] (a) The first-formed, generally pointed end of an elongate or conical form of an organism, such as the small end of the shell or spire of a gastropod. (b) The first-formed part of a brachiopod valve around which the shell has grown subsequently. The term is "usually restricted to valves having this point placed centrally or subcentrally" (TIP, 1965, pt.H, p.140). (c) The tip of the basal cavity or of a denticle of a conodont. Also, the juncture of bars, blades, or other processes of conodont elements. (d) The upper (umbonal) angle of a valve or plate of a cirripede crustacean.

apex [fold] *culmination*.

apex law In U.S. mining law, the individual whose claim contains a vein *apex* may follow and exploit that vein indefinitely along its dip, even if it enters adjoining property. It is also known as the law of *extralateral rights*.

aphanic Said of the texture of a carbonate sedimentary rock characterized by individual crystals or clastic grains whose diameters are less than 0.01 mm (Bissell & Chilingar, 1967, p.150) or 0.005 mm (Chilingar et al, 1967, p.311). The term was proposed by DeFord (1946) to replace *aphanitic*. See also: *aphanocrystalline*. Cf: *phaneric*.

aphanide An informal term used in the field to designate a wholly or partially fine-grained rock.

aphaniphyric *felsiphyric*.

aphanite Any fine-grained igneous rock whose components are not distinguishable with the unaided eye; a rock having aphanitic texture. This form of the word is now obsolete but the adjectival form, aphanitic, is still in use. Syn: *kryptomere; felsite.*

aphanitic [ign] Said of the texture of an igneous rock in which the crystalline components are not distinguishable by the unaided eye; both microcrystalline and cryptocrystalline textures are included. Also, said of a rock or a groundmass exhibiting such texture. The syn. *felsitic* is sometimes restricted to the light-colored rocks with this texture and "aphanitic" to the dark-colored (Johannsen, 1939, p.201). Ant: *phaneritic.* Syn: *fine-grained [geol].*

aphanitic [sed] A term formerly and loosely applied to a sedimentary (carbonate-rock) texture now referred to as *aphanic.*

aphanocrystalline Descriptive of an interlocking texture of a carbonate sedimentary rock having crystals whose diameters are in the range of 0.001-0.004 mm (Folk, 1959). See also: *aphanic.* Syn: *extremely finely crystalline.*

aphanophyre A syn. of *felsophyre.* Adj: *aphanophyric.*

aphanophyric *felsophyric.*

Aphebian In a three-part division of Proterozoic time, the earliest division, before the *Helikian.* Cf: *Hadrynian.*

aphodus A short canal of uniform diameter in a sponge, leading to an exhalant canal from an apopyle of approximately the same cross-sectional area. Pl: *aphodi.* Cf: *prosodus.*

aphotic zone That part of the ocean in which there isn't enough penetration of light for photosynthesis. Cf: *disphotic zone; euphotic zone.*

aphrite A foliated, lamellar, scaly, or chalky variety of calcite having a white pearly luster. Syn: *earth foam; foaming earth.*

aphrizite A black variety of tourmaline containing iron.

aphrodite *stevensite.*

aphroid Said of a massive corallum similar to *astreoid* type but with septa shortened peripherally and adjacent corallites united by a dissepimental zone.

aphrolith An obsolete syn. of *aa.* Cf: *dermolith.*

aphrosiderite *ripidolite.*

aphthitalite A white rhombohedral mineral: $(K,Na)_3Na(SO_4)_2$. Syn: *glaserite.*

aphylactic projection A map projection that does not possess any one of the three special properties of equivalence, conformality, or equidistance; e.g. a gnomonic projection.

aphyllous Said of a leafless plant.

aphyric Said of the texture of a fine-grained or aphanitic igneous rock which lacks phenocrysts. Also, said of a rock exhibiting such texture.

aphytal zone The plantless part of a lake bottom. Cf: *phytal zone.*

Aphytic A paleobotanic division of geologic time, signifying that time that preceded the first occurrence of plant life. Cf: *Archeophytic; Eophytic; Paleophytic; Mesophytic; Cenophytic.*

apical (a) Situated at or in the direction or vicinity of the apex of a shell; e.g. "apical horn", a spine at the apex of the shell of a nasselline radiolarian. (b) Located away from the mouth of an echinoderm; aboral.

apical archeopyle An *archeopyle* formed in a dinoflagellate cyst by the loss of the entire apical series of plates. See also: *haplotabular archeopyle; tetratabular archeopyle.*

apical area On an embryophytic spore, the area including the trilete suture.

apical axis The lengthwise axis of a pennate diatom. Cf: *transapical axis; pervalvar axis.*

apical papilla A dot-like thickening of the interradial area of a spore. When present, there is generally one apical papilla per interradial area, hence three per spore.

apical prominence In megaspores, mostly Paleozoic, a variously constructed projection at the intersection of the contact areas. Cf: *gula.*

apical series The series of plates grouped about an open apical pore or forming an apical cluster in the epitheca of a dinoflagellate. Cf: *antapical series.*

apical system A system of primordial plates at the aboral ter-

minus of ambulacra and interambulacra of echinoids, including when present an outer circlet of *ocular plates* surrounding an inner circlet of *genital plates* and sometimes including one or more supplementary plates. See also: *oculogenital ring.*

apiculus An open-end process extending from the valve surface in a diatom. Plural: *apiculi.*

A.P.I. gravity A standard adopted by the American Petroleum Institute for denoting the specific weight of oils, in which the lower the specific gravity, the higher the A.P.I. gravity: A.P.I. gravity=141.5 divided by the specific gravity of the liquid at 60°F-131.5. Cf: *Baumé gravity.*

apjohnite A silky-white, faintly rose-green, or yellow mineral: $MnAl_2(SO_4)_4.22H_2O$. It occurs in crusts, fibrous masses, or efflorescences. Syn: *manganese alum.*

aplacophoran A marine mollusk belonging to the class Aplacorphora and known only from living forms. See also:*amphineuran.* Cf: *polyplacophoran.*

aplite A light-colored hypabyssal igneous rock characterized by fine-grained xenomorphic-granular (i.e. aplitic) texture. Aplites may range in composition from granitic to gabbroic, but the term "aplite" with no modifier is generally understood to mean granitic aplite, consisting essentially of quartz, potassium feldspar, and acid plagioclase. Syn: *haplite.*

aplitic (a) Pertaining to the fine-grained and *saccharoidal* or xenomorphic-granular texture characteristic of aplites. See also: *autallotriomorphic.* (b) Said of an igneous rock having the characteristics and/or texture of aplite, such as a comparatively fine- and even-grained rock free from dark minerals.

aplodiorite A light-colored biotite granodiorite with little or no hornblende.

aplogranite A light-colored plutonic rock having granitic texture and essentially composed of alkali feldspar and quartz, with lesser amounts of biotite and with or without muscovite. Cf: *alaskite; two-mica granite.*

aplome A dark-brown, yellowish-green, or brownish-green variety of andradite garnet containing manganese. Syn: *haplome.*

aplowite A mineral: $(Co,Mn,Ni)SO_4.4H_2O$.

apo- In petrologic terminology, a prefix signifying metasomatic change without destruction of original texture.

apobsidian An old, devitrified obsidian.

apocarpous Said of a plant ovary whose carpels are separate rather than united; also, said of a gynoecium of separate pistils (Lawrence, 1951, p.739). Cf: *syncarpous.*

apochete An *exhalant canal* of a sponge.

apodeme One of the ingrowths from the exoskeleton of many arthropods that provide points of muscle attachment, such as an invagination of the body wall of an arachnid, an inward deflection of a sclerite of a merostome, a downward projection from the dorsal interior of a thoracic segment of a trilobite, or an infold of the exoskeleton of a crustacean. Syn: *apodema.*

apo-epigenesis Epigenesis (occurring subsequent to diagenesis) that affects sediments while they are remote from the original environment of deposition, as when they are under a relatively thick overburden (Chilingar et al, 1967, p.311). Cf: *juxta-epigenesis.*

apogean tide A tide of decreased range that occurs monthly, as the Moon is at or near the apogee of its orbit. It is a secondary modification of the tidal cycle. Ant: *perigean tide.*

apogee That point on the orbit of an Earth satellite, e.g. the Moon, which is farthest from the Earth. Cf: *perigee.*

apogrit *graywacke.*

apomagmatic Said of a hydrothermal mineral deposit of an intermediate distance from its magmatic source. Cf: *telemagmatic; perimagmatic; cryptomagmatic.*

aponeurotic band An anterior or posterior area of attachment of ligaments (mantle and visceral) on the inside of the body chamber of a nautiloid (TIP, 1964, pt.K, p.54).

apophyllite A mineral: $KCa_4Si_8O_{20}(F,OH).8H_2O$. It is a secondary mineral related to and occurring with zeolites in geodes in decomposed basalts and other igneous rocks. Syn: *fish-eye stone.*

apophysis [paleont] (a) An internal projection from interambulacral basicoronal plates of an echinoid for the attachment of muscles supporting Aristotle's lantern. (b) A lateral transverse process of a radial spine in acantharian radiolarians.---Pl: *apophyses.*

apophysis [intrus rocks] A syn. of *tongue* [intrus rocks]. Cf: *epiphysis.*

apopore The exit opening of an exhalant canal of a sponge, located either within the sponge (as on the lining of a larger exhalant canal or the cloaca) or on the surface of the sponge in which case it is equivalent to an osculum. Cf: *prosopore.*

apopyle Any opening through which water passes out of a flagellated chamber of a sponge. Cf: *prosopyle.*

aporhyolite An old rhyolite in which the once glassy groundmass has become devitrified.

aporhysis A *skeletal canal* in dictyonine hexactinellid sponges running radially through the body wall, opened at the cloacal end but closed over at the outer end. Pl: *aporhyses.*

apotaphral Descriptive of a type of tectonics involving lateral outward spreading (under gravity) of the orogenic zone away from the axis of a geosyncline (Carey, 1963, p.A6); it is characterized by nappes, thrusts, and recumbent folds. Cf: *syntaphral; diataphral.*

Appalachia One of the *borderlands* proposed by Schuchert (1923), in this case along the southeast side of North America, seaward from the Appalachian orogenic belt. Most of the evidence for Appalachia, as originally conceived, can now be otherwise interpreted. It is true that during middle and late Paleozoic time much sediment was being shed from lands in the seaward part of the Appalachian belt, but these were probably narrow, ephemeral tectonic lands. No former large extensions of this borderland into the present Atlantic Ocean basin are possible, because of the oceanic crustal structure beyond the edge of the continental shelf.

Appalachian relief A type of relief found in old mountains consisting of many anticlines and synclines, and characterized by secondary forms (such as monoclinal ridges and anticlinal valleys) that have adapted themselves to the structure and differential resistance of the rocks (Schieferdecker, 1959, term 1943). Type example: the relief of the Appalachian Mountains in North America. Cf: *Jurassian relief.*

Appalachian Revolution A widely held concept in the first part of the 20th Century that Paleozoic time was closed by a profound crustal disturbance, which especially deformed the rocks in the Central and Southern Appalachians. The term is misleading, and should be abandoned in this sense. At most, it could apply only to deformation in the Valley and Ridge province and plateau area, for which the more expressive term *Allegheny orogeny* is preferable. It is more proper to use "Appalachian orogeny" in a broad sense for the whole range of deformations, both in area and in time, in the Appalachian orogenic belt.

apparent Said of faults and fault movements that are described in terms of *separation* rather than of slip. Cf: *actual.*

apparent ablation *summer balance.*

apparent accumulation *winter balance.*

apparent angle of emergence In seismology, the angle whose tangent is equal to the ratio of the vertical and horizontal components of the ground displacement at a point on the Earth's surface, due to an arriving seismic wave (Runcorn, 1967, p.73). Cf: *angle of emergence.*

apparent crater The depression of an explosion or impact crater as it appears after modification of the original shape by postformational processes such as slumping and deposition of material ejected during crater formation; a crater that is visible on the surface and whose dimensions are measured with respect to the original ground level (Nordyke, 1962, p.3447).

The "apparent diameter" and "apparent depth" are measured using the highest points on the rim crest and the deepest part of the observable depression. Cf: *true crater.*

apparent density Obsolete syn. of *bulk density.*

apparent dip The angle that a structural surface, e.g. a bedding or fault plane, makes with the horizontal, measured in any random, vertical section rather than perpendicular to the strike. It varies from nearly zero to nearly the *true dip*, depending on whether the random section is close to the direction of the strike or or of the dip. Syn: *false dip.*

apparent horizon The somewhat irregular boundary where the visible land or water surface of the Earth appears to meet the sky as viewed from any given point; the more or less circular line along which rays from the point of observation are tangent to the Earth's surface. Strictly, it is the circle that bounds the part of the Earth's surface which would be visible from a given point if no irregularities or obstructions were present. The apparent horizon is extended slightly downward because of atmospheric refraction. In popular usage, the term *horizon* usually signifies the "apparent horizon". Cf: *true horizon.* Syn: *visible horizon; local horizon; sensible horizon; geographic horizon; topocentric horizon; natural horizon.*

apparent movement of faults The apparent movement observed in any chance section across a fault is a function of several variables: the attitude of the fault, of the disrupted strata, of the section on which the fault is observed, and of the net or actual slip of the fault.

apparent optic angle The *optic angle* as it appears under the conoscope, after being refracted upon emergence from the crystal. It is larger than the actual optic angle within the crystal.

apparent plunge The angle that a normal projection of a geologic structure will assume in the plane of a vertical cross section (Knutson, 1958).

apparent relative movement *separation.*

apparent resistivity The resistivity measured at any point on the surface of a real (inhomogeneous) Earth may be equated to the resistivity of some homogeneous Earth, for the same electrode array and the same excitation frequency. The homogeneous Earth would be characterized by the value of apparent resistivity so measured. This quantity varies from point to point over the surface of a real (inhomogeneous) Earth and its values do not relate in any simple way to the *true resistivities* of the homogeneous units making up the inhomogeneous Earth. Symbol: ρ_a.

apparent slope A vertically distorted or exaggerated slope as it appears in an aerial photograph viewed under a stereoscope.

apparent surface velocity The velocity with which a fixed phase, which is traveling through the Earth, appears to travel along the surface of the Earth.

apparent thickness The *thickness* of a stratigraphic unit or other tabular body, measured at right angles to the surface of the land. See also: *vertical.* Cf: *true thickness.*

apparent velocity The velocity with which a fixed phase of a seismic wave, usually its front or beginning, passes an observer.

apparent water table *perched water table.*

appinite A group of dark-colored hornblende-rich plutonic rocks, such as certain syenites, monzonites, and diorites, in which the hornblende occurs as large prismatic phenocrysts and also in the finer grained groundmass.

applanation All processes that tend to reduce the relief of an area, causing it to become more and more plain-like. These include lowering of the high parts by erosion and raising of the low parts by addition of material; the latter is usually more effective.

applied geology The application of various fields of geology to economic, engineering, water-supply, or military problems; geology relative to human activity.

applied geophysics The use of geophysical methods, e.g. electric, gravity, magnetic, seismic, or thermal, in the search for economically valuable deposits or water supplies. Cf: *geophysical prospecting.*

applied seismology The use of seismology in the search for economic deposits such as salt, oil and gas, or in engineering aspects, such as determining depth to bedrock or the motion of structures produced by vibratory sources; *seismic exploration; prospecting seismology.*

applied stress Downward stress exerted at an aquifer boundary.

apposed glacier A term, not in current usage, for a glacier formed by the union of two glaciers (Swayne, 1956, p.14).

apposition beach One of a series of roughly parallel beaches successively formed on the seaward side of an older beach.

apposition fabric A primary orientation of the elements of a sedimentary rock, developed or formed at the time of deposition of the material by the successive placing of particles upon those already present. See also: *depositional fabric.* Syn: *primary fabric.*

appressed [paleont] Said of very closely set conodont denticles each partly or entirely fused to adjoining denticles. Cf: *discrete.*

appressed [struc geol] Said of a fold, the limbs of which are almost closed.

approach The area of the sea extending indefinitely seaward from the shoreline at mean low-water spring tide (Wiegel, 1953, p.2).

apron [geomorph] An extensive, continuous, outspread, blanket-like deposit of alluvial, glacial, eolian, marine, or other unconsolidated material derived from an identifiable source, and deposited at the base of a mountain, in front of a glacier, etc.; e.g. a *bajada* or an *outwash plain.* Syn: *frontal apron.*

apron [glaciol] *ice apron.*

apron [ice] *ram.*

apsacline Said of the ventral and posterior *inclination* of the cardinal area in either valve of a brachiopod, lying in the bottom left or first quadrant moving counterclockwise from the orthocline position (TIP, 1965, pt.H, p.60, fig.61).

Aptian European stage: Lower Cretaceous, or Lower and Middle Cretaceous of some authors (above Barremian, below Albian). Syn: *Vectian.*

aptychus A heart-shaped structure consisting of a pair of symmetric calcareous or horny plates frequently found in the body chambers of certain ammonoids and regarded as an operculum for closure of the aperture during life. Pl: *aptychi.* Cf: *anaptychus.*

aquafact An isolated boulder or cobble, commonly on a sandy beach, that has been worn smooth on its seaward face by wave action, so that a sharp ridge parallel to the shore is developed along the exposed surface of the boulder or cobble; a *water-faceted stone* produced by *wet blasting* (Kuenen, 1947).

aquafer A var. of *aquifer*; not used by hydrogeologists.

aquagene tuff *hyaloclastite.*

Aqualf In U.S. Dept. of Agriculture soil taxonomy, a suborder of the soil order *Alfisol*, characterized by water saturation for sufficient periods of time to make cultivation without artificial drainage difficult. Aqualfs are mottled and gray (SSSA, 1970). Cf: *Boralf; Udalf; Ustalf; Xeralf.*

aquamarine (a) A transparent and pale-blue, greenish-blue, or bluish-green gem variety of beryl. The bluish color is attributed to scandium. (b) A pale-blue, greenish, light greenish-blue, or bluish-green color designation applied to mineral names; e.g. "aquamarine chrysolite" (a greenish-blue beryl), "aquamarine sapphire" (a pale-blue sapphire), "aquamarine topaz" (a greenish topaz), and "aquamarine tourmaline" (a pale-blue or pale greenish-blue tourmaline).

aquasol A soil so saturated with water that water is the medium in which the plants grow; a water soil. Cf: *hydrosol.*

aquatic (a) Living entirely or primarily in or on water. (b) Growing in or on water. (c) Living near or frequenting water.

aquatillite A term proposed by Schermerhorn (1966, p.832) for a glaciomarine or glaciolacustrine till-like deposit, such as one deposited from a melting iceberg.

aquatolysis A term proposed by Müller (1967, p.130) for the chemical and physicochemical processes that occur in a freshwater environment during transportation, weathering, and preburial diagenesis of sediments. Cf: *halmyrolysis.*

Aquent In U.S. Dept. of Agriculture soil taxonomy, a suborder of the soil order *Entisol*, characterized by water saturation for sufficient periods of time to make cultivation difficult without artificial drainage (SSSA, 1970). Cf: *Arent; Fluvent; Orthent; Psamment.*

aqueoglacial *glacioaqueous.*

aqueo-igneous Pertaining to the formation of a mineral or rock from its molten state, in which water present in the magma was influential. Syn: *hydroplutonic; hydatopyrogenic.*

aqueo-residual sand A term used by Sherzer (1910, p.627) for a sand in which the particles, produced by various residual agents, were subsequently modified by the action of water. It includes "all water-transported sand, for residual agencies have been present to some extent in the derivation of all sand from the parent material" (Allen, 1936, p.12). Cf: *residuo-aqueous sand.*

aqueous (a) Of, or pertaining to, water. (b) Made from, with, or by means of water; e.g. aqueous solutions. (c) Produced by the action of water; e.g. aqueous sediments.

aqueous fusion Melting in the presence of water, as a magma (Thrush, 1968, p.48).

aqueous ripple mark A ripple mark made by waves or currents of water, as opposed to one made by air currents (wind).

aqueous rock A sedimentary rock deposited by, in, or through the agency of water. Syn: *hydrogenic rock.*

Aquept In U.S. Dept. of Agriculture soil taxonomy, a suborder of the soil order *Inceptisol*, characterized by a water saturation for sufficient periods of time to make cultivation without artificial drainage difficult. In the upper 50cm of an Aquept there is either an umbric or histic epipedon, or an ochric epipedon with an underlying cambic horizon. Aquepts are generally gray (SSSA, 1970). Cf: *Andept; Ochrept; Plaggept; Tropept; Umbrept.*

aquic Said of a mostly reducing soil moisture regime characterized by the virtual absence of dissolved oxygen due to groundwater saturation, when soil temperature at a depth of 50cm is more than 5°C (SSSA, 1970).

aquiclude A body of relatively impermeable rock that is capable of absorbing water slowly but functions as an upper or lower boundary of an aquifer and does not transmit ground water rapidly enough to supply a well or spring. Cf: *aquifuge; aquitard; confining bed.*

aquifer A body of rock that contains sufficient saturated permeable material to conduct ground water and to yield economically significant quantities of ground water to wells and springs. The term was originally defined by Meinzer (1923, p.30) as any water-bearing formation. Syn: *water horizon; ground-water reservoir; nappe; aquafer.*

aquiferous system The entire water-conducting system between the ostia and the oscula of a sponge, including the *inhalant system* and the *exhalant system.* Syn: *canal system.*

aquifer system A heterogeneous body of intercalated permeable and less permeable material that acts as a water-yielding hydraulic unit of regional extent (Poland, et al, in press).

aquifer test A test involving the withdrawal of measured quantities of water from, or addition of water to, a well and the measurement of resulting changes in head in the aquifer both during and after the period of discharge or addition.

aquifuge An impermeable body of rock; a rock with no interconnected openings and thus lacking the ability to absorb and transmit water. Cf: *aquiclude; aquitard; confining bed.*

Aquilonian Stage in France: uppermost Jurassic. It is equivalent to Purbeckian in Great Britain.

Aquitanian European stage: lowermost Miocene (above Chattian of Oligocene, below Burdigalian). It was formerly regarded by some authors as uppermost Oligocene.

aquitard A *confining bed* that retards but does not prevent the flow of water to or from an adjacent aquifer; a *leaky confining bed.* It does not readily yield water to wells or springs, but may serve as a storage unit for ground water. Cf: *aquifuge; aquiclude.*

Aquod In U.S. Dept. of Agriculture soil taxonomy, a suborder of the soil order *Spodosol*, characterized by water saturation for sufficient periods of time to make cultivation without artificial drainage difficult. An Aquod may have a histic epipedon, an albic horizon with mottles or duripan, and mottles or gray colors in or just below the spodic horizon (SSSA, 1970). Cf: *Ferrod; Humod; Orthod.*

Aquoll In U.S. Dept. of Agriculture soil taxonomy, a suborder of the soil order *Mollisol*, characterized by water saturation for sufficient periods of time to make cultivation without artificial drainage difficult. An Aquoll may have a histic epipedon or a mollic epipedon, and there may be mottles or gray colors within or below the mollic epipedon (SSSA, 1970). Cf: *Alboll; Boroll; Rendoll; Udoll; Ustoll; Xeroll.*

Aquox In U.S. Dept. of Agriculture soil taxonomy, a suborder of the soil order *Oxisol*, characterized by the continuous presence of plinthite near the surface, or by water saturation during part of the year that requires artificial drainage for cultivation. An Aquox may contain, within the oxic horizon, indications of poor drainage such as a histic epipedon or mottles (SSSA, 1970). Cf: *Humox; Orthox; Torrox; Ustox.*

Aquult In U.S. Dept. of Agriculture soil taxonomy, a suborder of the soil order *Ultisol*, characterized by water saturation for sufficient periods of time to make cultivation without artificial drainage difficult. Aquults have characteristics associated with wetness, e.g. mottles, manganese-iron concretions, gray colors (SSSA, 1970). Cf: *Humult; Udult; Ustult; Xerult.*

arabesquitic Said of the texture of certain porphyritic rocks in which an apparently homogeneous groundmass breaks up, under crossed nicols, into irregular patches, supposedly resembling arabesques (Johannsen, 1939, p.202).

arachnid Any terrestrail *chelicerate* belonging to the class Arachnida, characterized by the presence of one pair of preoral appendages with two to three joints. Cf: *merostome.*

aragonite (a) A white, yellowish, or gray orthorhombic mineral: $CaCO_3$. It is trimorphous with calcite and vaterite. Aragonite has a greater density and hardness, and a less distinct cleavage, than calcite, and is also less stable and less common than calcite. It commonly occurs in fibrous aggregates in beds of gypsum and of iron ore, and as a deposit from hot springs, and it is a major constituent of shallow marine muds and the upper parts of coral reefs; aragonite is also an important constituent of the pearl and of some shells. Syn: *Aragon spar.* (b) A group of orthorhombic carbonate minerals, including aragonite, alstonite, witherite, strontianite, and cerussite.

Aragon spar *aragonite.*

arakawaite *veszelyite.*

aramayoite An iron-black mineral: $Ag(Sb,Bi)S_2$.

araneid Any arachnid belonging to the order Araneida, characterized by the presence of maxillary lobes and glands and by the similarity of the first pair of legs to the other legs. Their stratigraphic range is Carboniferous (possibly Devonian) to present.

arapahite A dark-colored, porous, fine-grained basalt that is microscopically holocrystalline, poikilitic, and composed of magnetite (about 50 percent), bytownite, and augite.

A ray Anterior ray in echinoderms, located opposite the CD interray.

arbitrary cutoff Vertical boundary separating two laterally in-

tergrading stratigraphic units that differ from each other in some arbitrarily defined way. See also: *cutoff* [*stratig*].

arborescent *dendritic.*

arborescent pollen Pollen of trees. Abbrev: AP. Syn: *tree pollen.*

Arbuckle orogeny A name used by Van der Gracht (1931) for the last major deformation in the Wichita orogenic belt of southern Oklahoma (Wichita and Arbuckle mountains, and subsurface). It is placed in the late Pennsylvanian (Virgil) by its relations to limiting fossiliferous strata. The nearby Ouachita Mountains were not supposed to have been materially affected by this orogeny, but to have been deformed later.

arcanite An orthorhombic mineral: K_2SO_4.

arch [**geomorph**] *natural arch.*

arch [**struc geol**] A broad, open anticlinal fold of a regional scale; it is usually a basement doming, e.g. the Cincinnati Arch. Cf: *dome.* Less-preferred syn: *swell.*

archaeocyathid Any Cambrian marine organism belonging to the phylum Archaeocyatha and characterized chiefly by a cone-, goblet-, or vase-shaped skeleton composed of calcium carbonate. The archaeocyathids have been variously classified as corals, sponges, protozoans, and calcareous algae. Syn: *pleosponge; cyathosponge.*

archaeocyte An amoebocyte of a sponge that has a large nucleus and cytoplasm rich in ribonucleic acid, that is capable of ingesting particulate material and serving as the origin of any other type of cell, and that is believed to be persistent undifferentiated embryonic cell. Syn: *archeocyte.*

archaeomagnetism The study of natural remanent magnetism of baked clays and recent lavas to determine intensity and direction of the Earth's magnetic field in the archaeologic past.

Archaeozoic Var. of *Archeozoic.*

Archaic n. In New World archaeology, a prehistoric cultural stage that follows the Lithic and is characterized in a general way by a foraging pattern of existence and numerous types of stone implements (Jennings, 1968, p.110). It is followed by the Formative. Correlation of relative cultural levels with actual age (and, therefore, with the time-stratigraphic units of geology) varies from region to region.----adj. Pertaining to the Archaic.

arch dam A dam built in the form of a horizontal arch that abuts against the side walls of a gorge and that has its convex side upstream.

Archean Said of the rocks of the *Archeozoic.*

arched iceberg An iceberg eroded in such a manner that a large opening, at the water line, extends horizontally through the ice, forming an arch.

archegonium A multicellular, female (egg-producing) sex structure of some plants, e.g. mosses and ferns.

archeocyte Var. of *archaeocyte.*

Archeophytic A paleobotanic division of geologic time, signifying the time of initial plant evolution, specif. algae. Cf: *Aphytic; Paleophytic; Mesophytic; Eophytic; Cenophytic.* Syn: *Algophytic; Proterophytic.*

archeopyle An opening in the wall of a dinoflagellate cyst by means of which the cell contents can emerge from the cyst. It is usually more or less polygonal in shape, and operculate. See also: *apical archeopyle; cingular archeopyle; precingular archeopyle; combination archeopyle.*

archeopyle suture A line of dehiscence on the dinoflagellate cyst that more or less completely separates a part of the cyst wall to form an operculum that covers the archeopyle. See also: *accessory archeopyle suture.*

Archeozoic The earlier part of Precambrian time, corresponding to *Archean* rocks. Cf: *Proterozoic.* Also spelled: *Archaeozoic.*

archetype *prototype.*

arch-gravity dam A solid-masonry arch dam that also depends upon gravity action (having sufficient mass and breadth of base) to provide stability.

archibenthic Pertaining to the benthos of the continental slope; *bathybenthic.*

archibole An obsolete syn. of *positive element.*

Archimedes' principle The statement in fluid mechanics that a fluid buoys up a completely immersed solid so that the apparent weight of the solid is reduced by an amount equal to the weight of the fluid that it displaces.

arching The transfer of stress from a yielding part of a soil mass to adjoining, less-yielding or restrained parts of the mass (ASCE, 1958, term 22).

archipelagic apron A smooth, fanlike slope or broad cone surrounding a seamount or an island. It is comparable to a continental rise and an abyssal plain in its topography and sedimentary processes.

archipelago A sea or area in a sea that contains numerous islands; also, the island group itself.

architype The type of a genus or species named in a publication prior to the time of the establishment of the current interpretation of types. Var: *arquetype.*

arc measurement The measurement, in geodesy, that follows a given meridian to determine the shape and size of the Earth along that line.

arcose Var. of *arkose.*

arc shooting (a) A method of refraction seismic prospecting in which the variation of travel time with azimuth from a shot point is used to infer geologic structure. (b) A reflection spread placed on a circle or on an arc with the center at the shot point.

arc spectrum The spectrum of light emitted by a substance at the temperature of an electric arc when it is placed in an arc or applied to one of the poles of the arc as a coating. The spectrum is representative of non-ionized atoms due to the low potential difference of the arc. Cf: *spark spectrum.*

arctic n. The area within the Arctic Circle; the area of the north pole.----adj. (a) Pertaining to cold, frigid temperatures. (b) Pertaining to features, climate, vegetation, and animals characteristic of the arctic region.

Arctic [clim] adj. Said of a climate in which the mean temperature of the coldest month is less than 0°C, and the mean temperature for the warmest month is below 10°C.

Arctic [paleoclim] n. The oldest subunit of the Blytt-Sernander climatic classification (Post, 1924), preceding the Preboreal, characterized primarily by tundra vegetation, and recording the cold climate of full and late-glacial time (prior to about 10,000 years). It is further subdivided from stratigraphic and pollen data in Denmark (Iversen, 1954) into Oldest Dryas, Bølling, Older Dryas, Allerød, and Younger Dryas, and is now dated between about 14,000 and 10,000 years ago.---adj. Pertaining to the late-glacial Arctic interval and to its climate, deposits, biota, and events.

Arctic Circle The parallel of latitude to the equator falling at approx. lat. 66°32′ N; it delineates the frigid zone of the North Pole. Cf: *Antarctic Circle.*

arctic desert *polar desert.*

arctic pack *polar ice.*

Arctic suite A group of basaltic and associated igneous rocks intermediate in composition between rocks of the *Atlantic suite* and the *Pacific suite.*

arc triangulation Triangulation designed to extend in a single general direction, following approximately the arc of a great circle. It is executed in order to connect two distinct control points or two independent and widely separated surveys. Cf: *area triangulation.*

arcuate Said of a fold, the axial trace of which is curved or bent.

arcuate delta A curved or bowed delta with its convex outer margin facing the sea or lake; the classical delta, as that formed at the mouth of the Nile River. Syn: *fan-shaped delta.*

arcuate fault A fault that has a curved trace on any given

transecting surface. Cf: *peripheral faults; plane fault.*

arculite A textural term for a bow-shaped aggregate of crystallites.

arcus A band-like thickening in the exine of a pollen grain (as in *Alnus*), running from one pore apparatus to another.

ardealite A white or light-yellow mineral: $Ca_2(HPO_4)(SO_4).4H_2O$.

Ardennian orogeny One of the 30 or more short-lived orogenies during Phanerozoic time identified by Stille, in this case late in the Silurian, within the Ludlovian Stage.

ardennite A yellow to yellowish-brown mineral: $Mn_5Al_5(VO_4)_5(SiO_4)_5(OH)_2.2H_2O$.

arduinite *mordenite.*

are A metric unit of area equal to 100 square meters, 0.01 hectares, or 119.60 square yards. Abbrev: a.

area-altitude analysis *hypsometric analysis.*

area curve In hydraulics, a curve which expresses the relation between area and some other variable, such as cross-sectional area of a stream against water-surface elevation, or surface area of a reservoir against water surface elevation (ASCE, 1962).

areal geology The geology of any relatively large area, treated broadly and primarily from the viewpoint of the spatial distribution and position of stratigraphic units, structural features, and surface forms. Cf: *regional geology.*

areal map A geologic map showing the horizontal extent and distribution of rock varieties or rock units exposed at the surface.

area of influence The area within which the potentiometric surface of an aquifer is lowered by withdrawal, or raised by injection, of water through a well or other structure designed for the purpose; the outer boundary of the *cone of depression.* Syn: *circle of influence.*

area slope The generalization of slope conditions within a given area.

area triangulation Triangulation designed to extend in every direction from a control point and to cover the region surrounding it. Cf: *arc triangulation.*

areg Arabic plural of *erg.*

areic Var. of *arheic.*

areism Var. of *arheism.*

arena A term used in Uganda for a large, undulating, relatively low-lying central area more or less completely surrounded by a hilly rim of resistant rock, and representing a dome of softer rock that has been worn away (Wayland, 1920, p.36-37).

arenaceous (a) Said of a sediment or sedimentary rock consisting wholly or in part of sand-size fragments, or having a sandy texture or the appearance of sand; pertaining to sand or arenite. Also said of the texture of such a sediment or rock. The term implies no special composition, and should not be used as a syn. of "siliceous". Syn: *psammitic; sandy; sabulous; arenarious.* (b) Said of organisms growing in sandy places.

arenarious Composed of sand; *arenaceous.*

arenated Said of a substance that is mixed with sand or that has been reduced to sand.

arendalite [mineral] A dark-green *epidote* from Arendal, southern Norway.

arendalite [rock] A French term for a garnetiferous rock.

arenicolite A sand-filled, U-shaped hole in a sedimentary rock (generally a sandstone), interpreted as a burrow of a marine worm and resembling the U-shaped burrow or trail of the modern worm *Arenicola.* It has also been regarded as the trail of a mollusk or crustacean.

Arenigian European stage: Lower Ordovician (above Tremadocian, below Llanvirnian). Syn: *Skiddavian.*

arenilitic Pertaining to, having the quality of, or resembling sandstone.

arenite (a) A general name used for consolidated sedimentary rocks composed of sand-sized fragments irrespective of composition; e.g. sandstone, graywacke, arkose, and calcarenite. The term is equivalent to the Greek-derived term, *psammite*, and was introduced as *arenyte* by Grabau (1904, p.242) who used it with appropriate prefixes in classifying medium-grained rocks (e.g. "autoarenyte", "autocalcarenyte", "hydrarenyte", and "hydrosilicarenyte"). See also: *lutite; rudite*. (b) A relatively "clean" sandstone that is relatively well-sorted, contains little or no matrix material, and has a relatively simple mineralogic composition; specif. a pure or nearly pure, chemically cemented sandstone containing less than 10% argillaceous matrix and inferred to represent a selectively and slowly deposited sediment well-washed by currents (Gilbert, 1954, p.290). The term is used for a major category of sandstone, as distinguished from *wacke*.---Etymol: Latin *arena*, "sand". Adj: *arenitic*.

arenose Full of grit or fine sand; *gritty*.

Arent In U.S. Dept. of Agriculture soil taxonomy, a suborder of the soil order *Entisol*, characterized by fragments of pedogenic horizons that have been mechanically mixed (SSSA, 1970). Cf: *Aquent; Fluvent; Orthent: Psamment*.

arenyte Var. of *arenite*.

areography [astrogeol] Description of the surface of the planet Mars.

areola (a) One of the thinner chamber- or box-like structures arranged in characteristic pattern within the shell wall of diatoms, being larger and more complex than a *puncta*, lying perpendicular to the valve surface, and permitting the diffusion of gases and nutrients. An areola may be subcircular, elliptical, or hexagonal, and wholly or partly closed on either the inner or outer surface. Syn: *areole*. (b) A space above an areolar pseudopore in some cheilostomatous bryozoans and between two projecting structures formed by secondary thickening. (c) A generally smooth, featureless area of a crinoid columnal articulum, situated between lumen and inner margin of crenularium. It may be granulose or marked by fine vermicular furrows and ridges.---Pl: *areolae*.

areolar Pertaining to an areola; e.g. an "areolar pore" representing a marginal pseudopore in the frontal wall of some cheilostomatous bryozoans.

areole (a) A *scrobicule* or depression around a boss of an echinoid for the attachment of muscles controlling the movement of spines. (b) An *areola* of a diatom.

areology The scientific study of the planet Mars.

arête A narrow, jagged, serrate mountain crest, or a narrow, rocky, sharp-edged ridge or spur, commonly present above the snowline in rugged mountains (as the Swiss Alps) sculptured by glaciers, and resulting from the continued backward growth of the walls of adjoining cirques. Etymol: French, "fish bone". See also: *horn; comb ridge; grat*. Syn: *serrate ridge; crib; arris*.

aretic *arheic*.

arfvedsonite (a) A black monoclinic mineral of the amphibole group, approximately: $Na_{2-3}(Fe,Mg,Al)_5Si_8O_{22}(OH)_2$. It may contain some calcium, and it occurs in strongly pleochroic prisms in certain (sodium-rich) igneous rocks. Syn: *soda hornblende*. (b) An end member of the amphibole group: $Na_3Fe_4^{+2}Fe^{+3}Si_8O_{22}(OH)_2$.

argental mercury Naturally occurring *amalgam*.

argentian Said of a substance that contains silver; *argentiferous*.

argentiferous Said of a substance that contains or yields silver, e.g. "argentiferous galena". Syn: *argentian*.

argentine n. A pearly-white variety of calcite with undulating lamellae.---adj. Pertaining to, containing, or resembling silver; silvery.

argentite A dark lead-gray monoclinic mineral: Ag_2S. It is dimorphous with acanthite and constitutes a valuable ore of silver. Syn: *silver glance; vitreous silver; argyrite*.

argentojarosite A yellow or brownish mineral of the alunite group: $AgFe_3(SO_4)_2(OH)_6$.

argentopyrite A mineral: $AgFe_2S_3$.

argic water A syn. of *intermediate vadose water* proposed by Meinzer (1939) as one of the three classes of *kremastic water*.

Argid In U.S. Dept. of Agriculture soil taxonomy, a suborder of the soil order *Aridisol*, characterized by the presence of an argillic or natric horizon (SSSA, 1970). Cf: *Orthid*.

argil (a) A clay; esp. a white-colored clay, such as *potter's clay*. (b) *alumina*.

argillaceous (a) Pertaining to, largely composed of, or containing clay-size particles or clay minerals, such as an "argillaceous ore" in which the gangue is mainly clay; esp. said of a sediment (such as marl) or a sedimentary rock (such as shale) containing an appreciable amount of clay. Cf: *shaly; lutaceous; pelitic; argillic*. Syn: *clayey; pelolithic; argillous*. (b) Said of the peculiar odor emitted by an argillaceous rock when breathed upon. (c) Pertaining to argillite.

argillaceous hematite A brown to deep-red variety of natural ferric oxide containing an appreciable portion of clay (or sand).

argillaceous limestone A limestone containing an appreciable (but less than 50%) amount of clay; e.g. cement rock.

argillaceous sandstone (a) A term applied loosely to an impure sandstone containing an indefinite amount of fine silt and clay. (b) A relatively weak sandstone, not suitable for building purposes, containing a considerable amount of clay that serves as the cementing material.---Cf: *clayey sandstone*.

argillan A *cutan* composed dominantly of clay minerals (Brewer, 1964, p.212); e.g. a *clay skin*. Syn: *argitan*.

argillation A term used by Keller (1958, p.233) for the development of kaolinite and other clay minerals by weathering of primary aluminum-silicate minerals.

argille scagliose A thick sheet of chaotic, allochthonous material consisting of highly plastic, churned, and slickensided clays that have been displaced many kilometers by lateral or vertical stresses aided by gravity sliding or by diapiric movement; specif. a stratigraphic unit of Jurassic to Oligocene age exposed along parts of the Apennines, on which great slabs of rock have slid. Etymol: Italian, "scaly shale".

argillic [clay] Pertaining to clay or clay minerals; e.g. "argillic alteration" in which certain minerals of a rock are converted to minerals of the clay group. Cf: *argillaceous*.

argillic [soil] Said of a soil horizon characterized by an illuvial accumulation of silicate clays. Its minimum thickness and clay content depend on the thickness and clay content of the overlying eluvial layer (SSSA, 1970).

argilliferous Abounding in clay; producing clay.

argillite (a) A compact rock, derived either from mudstone (claystone or siltstone) or shale, that has undergone a somewhat higher degree of induration than is present in mudstone or shale but that is less clearly laminated than, and without the fissility (either parallel to bedding or otherwise) of, shale, or that lacks the cleavage distinctive of slate. Flawn (1953, p.563-564) regards argillite as a weakly metamorphosed argillaceous rock, intermediate in character between a mudstone (claystone) and a *meta-argillite*, in which less than half of the constituent material (clay minerals and micaceous paste) has been reconstituted to combinations of sericite, chlorite, epidote, or green biotite, the particle size of the reconstituted material ranging from 0.01 to 0.05 mm. Cf: *clay slate*. (b) A term that has been applied to an argillaceous rock cemented by silica (Holmes, 1928, p.35) and to a claystone composed entirely of clay minerals.---Also spelled: *argillyte*.

argillith A term suggested by Grabau (1924, p.298) for a claystone. Syn: *argillyte*.

argillization Replacement or alteration of feldspars to form clay minerals, esp. that occurring in wall rocks adjacent to mineral veins.

argillutite A pure lutite. Term introduced as *argillutyte* by Grabau (1904, p.243).

argillyte (a) Var. of *argillite*. (b) *argillith*. (c) Obsolete synonym of orthoclase.

argitan *argillan*.

Argovian Substage in Great Britain: Upper Jurassic (lower Lusitanian; above Oxfordian Stage, below Rauracian Substage).

argyrite *argentite*.

argyrodite A steel-gray mineral: Ag_8GeS_6. It is isomorphous with canfieldite.

arheic Said of a drainage basin or region characterized by arheism; without flow. Syn: *areic; aretic; arhetic.*

arheism The condition of a region (such as a desert) in which runoff is nil or surface drainage is almost completely lacking, or where rainfall is so infrequent that the water sinks into the ground or evaporates. Syn: *areism.*

arich A term used in Algeria for a rocky, sand-cloaked butte on which an isolated dune is based (Capot-Rey, 1945, p.399).

arid Said of a climate characterized by dryness, variously defined as rainfall insufficient for plant life or for crops without irrigation; less than 10 inches of annual rainfall; or a higher evaporation rate than precipitation rate. Syn: *dry.*

arid cycle The cycle of erosion in an arid region; there is some doubt about the validity of the concept. Cf: *normal cycle.*

aridic Said of a soil moisture regime in which, for a soil having a temperature of 5°C at a depth of 50cm, there is no moisture for plant growth for more than half the time; or, for a soil having a temperature of 8°C at the same depth, there are no consecutive 90 days in which there is moisture for plants (SSSA, 1970).

Aridisol In U.S. Dept. of Agriculture soil taxonomy, a soil order characterized by an ochric epipedon and other pedogenic horizons, but none of them oxic. It develops in an arid climate (SSSA, 1970). Suborders and great soil groups of this soil order have the suffix -id. See also: *Argid; Orthid.*

ariegite A group of pyroxenites chiefly composed of clinopyroxene, orthopyroxene, and spinel, with pyrope and/or hornblende as possible accessories and lacking primary feldspar; a spinel pyroxenite. The term is more commonly used by European (esp. French) petrologists. Also spelled: *ariegite.*

Aristotle's lantern A complex masticatory system of forty or fewer calcareous skeletal elements that surround the mouth of an echinoid and function as jaws. Etymol: from a passage in Aristotle where the shape of a sea urchin is said to resemble the frame of a lantern (Webster, 1967, p.118). Syn: *lantern.*

arithmetic mean The sum of the values of *n* numbers divided by *n*. It is usually referred to simply as the "mean". Syn: *average.*

arithmetic mean diameter An expression of the average particle size of a sediment or rock, obtained by summing the products of the size-grade midpoints and the frequency of particles in each class, and dividing by the total frequency. Syn: *equivalent grade.*

Arizona ruby A deep-red or ruby-colored pyrope garnet of igneous origin from SW U.S.

arizonite [mineral] (a) A doubtful mineral: $Fe_2Ti_3O_9$. It is found in irregular metallic steel-gray masses in pegmatite veins near Hackberry, Ariz. Cf: *kalkowskite; pseudorutile.* (b) A mixture of hematite, rutile, ilmenite, and anatase. (c) A type of ore, discovered in Yavapai County, Ariz., whose principal vein material consists of micaceous iron, silver iodide, gold, iron sulfides, and antimony.

arizonite [rock] A light-colored hypabyssal rock composed chiefly of quartz (80 percent), and orthoclase (18 percent), with mica and apatite as possible accessories. Its name is derived from Arizona, where the rock was first found.

Arkansas stone A superior variety of *novaculite* found in the Ouachita Mountains of western Arkansas. Also, a whetstone made of Arkansas stone.

arkansite A brilliant iron-black variety of brookite from Magnet Cove, Arkansas.

arkesine A name, now obsolete, proposed by Jurine (1806, p.373-374) and for the talc- and chlorite-bearing R 2352700.003 --Adj: /----adj. hornblende granite on Mont Blanc in the French Alps. Etymol: Greek *archaios,* "ancient, primitive".

arkite A porphyritic plutonic foidite with the same general composition as *fergusite* but distinguished from it by the presence of melanite. The potassium feldspathoid usually forms the phenocrysts.

arkose A feldspar-rich, typically coarse-grained sandstone, commonly pink or reddish to pale gray or buff, composed of angular to subangular grains that may be either poorly or moderately well sorted, usually derived from the rapid disintegration of granite or granitic rocks (including high-grade feldspathic gneisses and schists), and often closely resembling or having the appearance of a granite; e.g. the Triassic arkoses of eastern U.S. Quartz is usually the dominant mineral, with feldspars (chiefly microcline) constituting at least 25%. Cement (silica or calcite) is commonly scanty, and matrix material (usually less than 15%) includes clay minerals (esp. kaolinite), mica, and iron oxide; fine-grained rock fragments are often present. Arkose is commonly a current-deposited sandstone of continental (sometimes neritic) origin, occurring as a very thick, wedge-shaped, orogenic mass of limited geographic extent (as in a fault trough or a rapidly subsiding basin); it may be strongly cross-bedded and associated with coarse granite-bearing conglomerate, and it may denote an environment of high relief and vigorous erosion of strongly uplifted granitic rocks in which the feldspars were not subjected to prolonged weathering (chemical decomposition) or to long transport before burial. Arkose may also occur at the base of a sedimentary series as a thin blanket-like residuum derived from and directly overlying granitic rock. Selected modern definitions of arkose: (1) A sandstone with more than 30% feldspar (Krynine, 1940); (2) A sandstone with more than 25% feldspars and igneous rock fragments of all kinds, and less than 10% micas and metamorphic rock fragments, and with any degree of clay content, sorting, or rounding (Folk, 1954); (3) A sandstone having a particular composition and reflecting a particular source, but exhibiting any texture and being deposited under many different conditions in a great many different environments (Gilbert, 1954, p.294-295); (4) A sandstone with less than 15% detrital clay matrix, less than 75% quartz and chert, and containing a varied assemblage (at least 25%) of unstable materials with feldspar dominant over rock fragments (Pettijohn, 1957); (5) A sandstone with more than 25% feldspar, less than 10% fine-grained rock fragments, and less than 75% quartz, quartzite, and chert (McBride, 1963, p.667); and (6) A sandstone containing less than 75% quartz and metamorphic quartzite and more than 25% feldspar and plutonic rock fragments (or whose content of feldspar and plutonic rock fragments is at least three times that of all other fine-grained rock fragments, including chert), regardless of clay content or texture (Folk, 1968, p.124). The term "arkose" was introduced by Brongniart (1823, p.497-498) in an attempt to limit use of "grés" (sandstone) and was defined by him as a rock of granular texture formed principally by mechanical aggregation and composed essentially of large grains of feldspar and glassy quartz mixed together unequally, with mica and clay as fortuitous constituents (see Oriel, 1949, p.825). Roberts (1839, p.11) attributes the term to Bonnard. Etymol: French, probably from Greek *archaios,* "ancient, primitive" (Oriel, 1949, p.826). Adj: *arkosic.* Cf: *graywacke; feldspathic sandstone; subarkose.* Syn: *arcose.*

arkose-quartzite *arkosite.*

arkosic arenite A sandstone containing abundant quartz, chert, and quartzite, less than 10% argillaceous matrix, and more than 25% feldspar (chiefly unaltered sodic and potassic varieties), and characterized by an abundance of unstable materials in which the feldspar grains exceed the fine-grained rock fragments (Gilbert, 1954, p.294). It is more feldspathic and less mature than *feldspathic arenite*. See also: *arkosic sandstone*.

arkosic bentonite A term used by Ross & Shannon (1926, p.79) for a bentonite with 25-75% sandy impurities; a bentonite derived from a volcanic ash whose abundant detrital crystalline grains remained essentially unaltered. Cf: *bentonitic arkose*.

arkosic conglomerate A poorly sorted, lithologically homogeneous *orthoconglomerate* consisting of immature gravels derived from granites in a tectonically active or sharply elevated area under conditions of fluvial transport and rapid burial in a subsiding basin; an arkose with scattered pebbles of granite or with lenses of granitic gravel. The sand and silt matrix content is high and has the composition of arkose, consisting of quartz and feldspar particles with some finer kaolinitic material. The rock forms thick, wedge-shaped deposits and is commonly interbedded with arkosic sandstones. Syn: *granite-pebble conglomerate*.

arkosic graywacke A graywacke characterized by abundant unstable materials; specif. a sandstone containing more feldspar grains than fine-grained rock fragments, the feldspar content exceeding 25% (Gilbert, 1954, p.294). It is more feldspathic than *feldspathic graywacke*.

arkosic limestone An impure limestone containing a relatively high proportion of grains and/or crystals of feldspar, either detrital or formed in place.

arkosic sandstone A sandstone with considerable feldspar, such as one containing minerals from coarse-grained quartzo-feldspathic rocks (granites, granodiorites, medium- or high-grade schists) or from older, highly feldspathic sedimentary rocks; specif. a sandstone containing more than 25% feldspar and less than 20% matrix material of clay, sericite, and chlorite (Pettijohn, 1949, p.227). It is more feldspathic than *feldspathic sandstone*. The term is used also as a general term to include *arkosic arenite*, arkosic wacke, and arkose (Gilbert, 1954, p.310), or arkose and subarkose (Pettijohn, 1954, p.364). See also: *arkosite*.

arkosic wacke A sandstone containing abundant quartz, chert, and quartzite, more than 10% argillaceous matrix, and more than 25% feldspar (chiefly sodic and potassic varieties), and characterized by an abundance of unstable materials in which the feldspar grains exceed the fine-grained rock fragments (Gilbert, 1954, p.291-292). It is more feldspathic and less mature than *feldspathic wacke*.

arkosite A quartzite with a notable amount of feldspar; e.g. a well-indurated *arkosic sandstone* (Pettijohn, 1949, p.227) or a well-cemented arkose (Tieje, 1921, p.655). Syn: *arkose-quartzite*.

arkositite A term used by Tieje (1921, p.655) for an arkose so well-cemented that the particles are interlocking.

arm [coast] A long, narrow inlet of water extending inland from another body of water, such as an "arm of the sea" or an "arm of a lake". It is usually longer and narrower than a bay.

arm [geomorph] (a) A ridge or spur extending from a mountain. (b) The trailing outer extension of a parabolic dune.

arm [paleont] (a) One of several radially disposed appendages bearing an extension of ambulacrum and mounted on the oral surface of an echinoderm, such as one of the five radial extensions of the body of an asteroid; strictly, a free, pinnule-bearing extension of a crinoid. It is a major element in the food-gathering structure of many echinoderms, and may or may not be distinct from the disc. Cf: *brachiole*. (b) A ray-like structure of a sponge spicule, whether or not it is a true ray or a pseudoactin. (c) A brachium of a brachiopod. (d) A flat extension from the central region of a radiolarian shell.

arm [stream] A tributary or branch of a stream.

armalcolite A mineral of the pseudobrookite group found in Apollo 11 lunar samples: $(Mg,Fe)Ti_2O_5$.

armangite A black rhombohedral mineral: $Mn_3(AsO_3)_2$.

armchair geology Deducing geologic conditions without intensive field work (Shepard, 1967, p.209).

Armenian bole A soft, clayey, bright-red astringent earth found chiefly in Armenia and Tuscany, used formerly (14th-18th centuries) for medical purposes and now as a coloring material. Syn: *bole Armoniac*.

armenite A mineral: $BaCa_2Al_6Si_8O_{28}.2H_2O$.

armored Said of dinoflagellates (such as those of the order Peridiniales) possessing a cellulose envelope or cell wall that is subdivided into or covered by articular plates. Ant: *unarmored*.

armored mud ball A large subspherical mass of silt or clay, which became coated or studded with coarse sand and fine gravel as it rolled along downstream; it is generally 5-10 cm in diameter, although the size may vary between 1 cm and 50 cm in diameter. See also: *clay ball; till ball*. Syn: *mud ball; pudding ball*.

armored relict An *unstable relict* that is prevented from further reaction by a rim of reaction products.

Armorican orogeny A name used by Suess for late Paleozoic deformation in western Europe; it is based on relations in Brittany (Armorica). The term is now little used; modern geologists prefer the names *Hercynian orogeny* or *Variscan orogeny*.

armoring Formation of a reaction rim resulting from a loss of equilibrium in a discontinuous reaction series. Cf: *zoning* [crystal].

arnimite A hydrous copper-sulfate mineral, perhaps identical with antlerite.

aromatic Designating cyclic organic compounds characterized by a high degree of stability in spite of their apparent unsaturated bonds and best exemplified by benzene and related structures, but also evident in other compounds.

aromatic hydrocarbon A monocyclic or polycyclic, relatively stable, hydrocarbon having the empirical formula C_nH_{2n-6} of which the simplest example is benzene. The compounds of higher molecular weight are solid and may fluoresce or be slightly colored. See also: *benzene series*.

aromatite A bituminous stone resembling myrrh in color and odor. Webster (1967, p.120) defines *aromatites* (pl: *aromatitae*) as a precious stone of ancient Arabia and Egypt.

aromorphosis Biologic evolution characterized by an increase in the degree of organization without marked specialization. Cf: *allomorphosis*.

arpent (a) An old French unit of land area, still used in certain French sections of Canada and U.S., equal to about 0.85 acre depending on local custom and usage; e.g. a unit of area in Missouri and Arkansas equal to 0.8507 acre, and in Louisiana, Mississippi, Alabama, and NW Florida equal to 0.84625 acre. (b) A unit of length equal to one side of a square having an area of one arpent (192.50 ft in Missouri and Arkansas, and 191.994 ft in Louisiana, Mississippi, Alabama, and NW Florida).---Syn: *arpen*.

arquerite A soft, malleable, silver-rich variety of native amalgam, containing about 87% silver and 13% mercury.

arquetype A var. of *architype*.

array An ordered arrangement of geophysical instruments such as magnetometers or geophones, the data from which feeds into a central point or receiver.

arris A term used in the English Lake District for *arête*; syn: *arridge* (Stamp, 1961, p.33).

arrival The appearance of seismic energy on a seismic record; the lineup of coherent energy signifying the passage of a wave front. See also: *first arrival; later arrival*. Syn: *onset; break; kick*.

arrival time In seismology, the time at which a particular wave phase arrives at a detector. It is measured from *time marks*.

arrojadite A dark-green monoclinic mineral: nearly $Na_2(Fe, Mn)_5(PO_4)_4$. It is isostructural with dickinsonite.

arrow [coast] A spit pointing seaward.

arrow [surv] A thin, sharp-pointed metal rod, peg, or pin, about a half meter in length and with a ring at one end, used in surveying to mark the ground when the measurement between two points is more than one tape or chain length apart; a chaining or taping *pin*.

arroyo (a) A term applied in the arid and semiarid regions of SW U.S. to the small, deep, flat-floored channel or gully of an ephemeral stream or of an intermittent stream, usually with vertical or steeply cut banks of unconsolidated material at least 60 cm high; it is usually dry, but may be transformed into a temporary watercourse or short-lived torrent after heavy rainfall. Cf: *dry wash.* (b) The small, discontinuous, intermittent stream, brook, or rivulet that occupies such a channel.-- Etymol: Spanish, "stream, brook; gutter, watercourse of a street". See also: *wadi; nullah.* Syn: *arroya.*

arroyo-running A term applied in the SW U.S. to the phase of local flooding characterized by a temporary mountain torrent debouching from a canyon and spreading out over a great fan (Keyes, 1910, p.572).

arsenate A mineral compound characterized by pentavalent arsenic and oxygen in the anion. An example of an arsenate is mimetite, $Pb_5Cl(AsO_4)_3$. Cf: *phosphate; vanadate.*

arsenate-belovite A mineral: $Ca_2Mg(AsO_4)_2.2H_2O$. Formerly called: *belovite.* Syn: *talmessite.*

arsenic A hexagonal mineral, the native metallic element As. It is brittle and commonly occurs in steel-gray and granular or kidney-shaped masses.

arsenical antimony *allemontite.*

arsenical nickel *nickeline.*

arsenical pyrites *arsenopyrite.*

arsenic bloom (a) *arsenolite.* (b) *pharmacolite.*

arseniopleite A brownish-red mineral consisting of a basic arsenate of manganese, calcium, iron, lead, and magnesium.

arseniosiderite A yellowish-brown mineral: $Ca_3Fe_4(AsO_4)_4(OH)_4.4H_2O$.

arsenite A mineral compound characterized by trivalent antimony and oxygen in the anion. An example is trigonite, $Pb_3MnH(AsO_3)_3$.

arsenobismite A yellowish-green mineral: $Bi_2(AsO_4)(OH)_3$.

arsenoclasite A red mineral: $Mn_5(AsO_4)_2(OH)_4$. Also spelled: *arsenoklasite.*

arsenolamprite A lead-gray polymorph of native arsenic. It was formerly regarded as a mixture of arsenic and arsenolite.

arsenolite A cubic mineral: As_2O_3. It usually occurs as a white bloom or crust, and is dimorphous with claudetite. Syn: *arsenic bloom.*

arsenopyrite A tin-white or silver-white to steel-gray orthorhombic mineral: FeAsS. It is isomorphous with loellingite. Arsenopyrite occurs chiefly in crystalline rocks and esp. in lead and silver veins, and it constitutes the principal ore of arsenic. Syn: *arsenical pyrites; mispickel; white pyrites; white mundic.*

arsenosulvanite A cubic mineral: $Cu_3(As,V)S_4$. It is isomorphous with sulvanite.

arsenpolybasite A mineral: $(Ag,Cu)_{16}(As,Sb)_2S_{11}$. Cf: *antimonpearceite.* Syn: *arsenopolybasite.*

arsenuranylite An orange-red mineral: $Ca(UO_2)_4(AsO_4)_2(OH)_4.6H_2O$.

arsoite A trachyte containing phenocrysts of sanidine, andesine, diopside, and olivine in a groundmass of sanidine, oligoclase, diopside, magnetite, and sodalite. Its name is derived from the Arso lava flow of 1302 on Monte Epomeo (Ischia, Italy).

Artenkreis A German term synonymous with *species-group.*

arterite Migmatite, the more mobile portion of which was injected magma (Dietrich & Mehnert, 1961). "This is the same as the arteritic gneiss, *injection gneiss,* and lit-par-lit gneiss of

some workers" (Dietrich, 1960, p.36). Originally proposed, along with the term *venite,* to replace *veined gneiss* with terms of genetic connotation (Mehnert, 1968, p.17). Cf: *phlebite; composite gneiss; diadysite.*

artesian An adjective referring to ground water confined under hydrostatic pressure. Etymol: French *artésien,* "of Artois", a region in northern France.

artesian aquifer *confined aquifer.*

artesian basin A terrane, often but not necessarily basin shaped, including an artesian aquifer whose potentiometric surface typically is above the land surface in the topographically lower portion of the terrane. Examples range in size from areas a few hundred feet across to several hundred miles across. Cf: *ground-water basin.*

artesian discharge Discharge of water from a well, spring, or aquifer by artesian pressure.

artesian head The *hydrostatic head* of an artesian aquifer or of the water in the aquifer.

artesian leakage Slow percolation from a confined aquifer into confining beds.

artesian pressure *Hydrostatic pressure* of artesian water, often expressed in terms of pounds per square inch at the land surface; or height, in feet above land surface, or a column of water that would be supported by the pressure.

artesian-pressure surface A potentiometric surface that is above the zone of saturation. Cf: *normal-pressure surface; subnormal-pressure surface.*

artesian province A region within which structure, stratigraphy, physiography, and climate combine to produce conditions favorable to the existance of one or, generally, several artesian aquifers; e.g. the Atlantic and Gulf Coastal Plain.

artesian spring A spring from which the water flows under artesian pressure, usually through a fissure or other opening in the confining bed above the aquifer.

artesian system (a) A structure permitting water confined in a body of rock to rise within a well or along a fissure. (b) Any system incorporating a water source, a body of permeable rock bounded by bodies of distinctly less permeable rock, and a structure enabling water to percolate into and become confined in the permeable rock under pressure distinctly greater than atmospheric.

artesian water *confined ground water.*

artesian well A well tapping confined ground water. Water in the well rises above the level of the water table under artesian pressure, but does not necessarily reach the land surface. Sometimes restricted to mean only a *flowing artesian well.* Cf: *water-table well; nonflowing artesian well.*

arthrolite A cylindrical concretion with transverse joints, sometimes found in clays or shales.

arthrophycus A sand-filled, curving and branching, rounded furrow, with faint but regularly spaced transverse ridges commonly bearing a median depression, probably representing a feeding burrow but also variously regarded as an inorganic structure or a trail produced by a worm, mollusk, or arthropod crawling over a soft mud surface. The "branches" of the trace fossil may reach 60 cm in length. It was originally described as a plant (seaweed) fossil and assigned to the genus *Arthrophycus.*

arthropod Any one of a group of solitary, marine, freshwater, and aerial invertebrates belonging to the phylum Arthropoda, characterized chiefly by jointed appendages and segmented bodies. Among the typical arthropods are trilobites, crustaceans, chelicerates, pycnogonids, and myriapods. Their stratigraphic range is Lower Cambrian to present.

arthurite An apple-green mineral: $Cu_2Fe_4[(As,P,S)O_4]_4(O,OH)_4.8H_2O$.

article An articulated *segment* of an appendage in an arthropod.

articular Pertaining to a joint or joints; e.g. "articular furrow" (groove) and "articular ridge" (linear elevation) occurring on the margins of scutum and tergum of a cirripede crustacean and together forming the articulation between these valves.

articulate [bot] Said of a plant that has nodes or joints, or places where separation may occur.

articulate [paleont] (a) n. Any brachiopod belonging to the class Articulata, characterized by calcareous valves united by hinge teeth and dental sockets.----adj. Said of a brachiopod possessing such valves, or of the valves themselves. Cf: *inarticulate*. (b) n. Any crinoid belonging to the subclass Articulata, characterized by highly differentiated brachial articulations.----adj. Said of a crinoid having highly differentiated brachial articulation.

articulation (a) The action or manner of jointing, or the state of being jointed; e.g. the interlocking of two brachiopod valves by projections along their posterior margins commonly effected in articulates by two ventral teeth fitting dental sockets of the brachial valve, or the flexible to nearly immovable union of adjoined cirrals and columnals of a crinoid stem effected by ligaments attached to articula. (b) Any movable joint between the rigid parts of an invertebrate (such as between the segments of an insect appendage).

articulite *itacolumite.*

articulum An articular facet; specif. a smooth or sculptured surface of a crinoid columnal or cirral serving for articulation with the contiguous stem. Pl: *articula.*

articulus The hinge including the hinge plate, the hinge teeth, and ligament in bivalve mollusks. Pl: *articuli.*

artificial horizon A device for indicating the horizontal and used esp. in observing altitudes; esp. a device consisting of a planar reflecting surface that can be adjusted to coincide with the astronomic horizon (i.e. can be made perpendicular to the zenith), such as a shallow dish or trough filled with mercury whose upper surface is free and assumes a horizontal position, or a level glass mirror equipped with spirit levels and leveling screws so that it can be adjusted into the plane of the horizon. It is sometimes simply called a *horizon*. Syn: *false horizon.*

artificial radioactivity The radioactivity of a synthetic nuclide; a syn. of *induced radioactivity.*

artificial recharge Recharge at a rate greater than natural, resulting from the activities of man.

artinite A snow-white mineral: $Mg_2CO_3(OH)_2.3H_2O$. It occurs in orthorhombic crystals and fibrous aggregates.

Artinskian European stage: Lower Permian (above Sakmarian, below Kungurian).

arzrunite A bluish-green mineral consisting of a sulfate and chloride of copper and lead.

ås A Swedish term for *esker.* Pron: *auss,* as in "Aussie" when pronounced by an Englishman. Pl: *åsar.* See also: *os.*

åsar Plural of *ås.*

asbecasite A mineral: $Ca_3(Ti,Sn)(As_6Si_2Be_2O_{20})$.

asbestiform Said of a mineral that is fibrous, i.e. that is like asbestos.

asbestine adj. Pertaining to or having the characteristics of asbestos.--n. A variety of talc; specif. *agalite.*

asbestos (a) A commercial term applied to a group of highly fibrous silicate minerals that readily separate into long, thin, strong fibers of sufficient flexibility to be woven, are heat resistant and chemically inert, and possess a high electric insulation, and therefore are suitable for uses (as in yarn, cloth, paper, paint, brake linings, tiles, insulation, cement, fillers, and filters) where incombustible, nonconducting, or chemically resistant material is required. (b) A mineral of the asbestos group, principally chrysotile (best adapted for spinning) and certain fibrous varieties of amphibole (esp. tremolite, actinolite, and crocidolite). (c) A term strictly applied to the fibrous variety of actinolite.---Syn: *asbestus; amianthus; earth flax; mountain flax.*

asbolane *asbolite.*

asbolite A soft, black, earthy mineral aggregate, often classed as a variety of wad, containing hydrated oxides of manganese and cobalt, the content of cobalt sometimes reaching as much as 32% (or 40% cobalt oxide). Syn: *asbolane; earthy cobalt; black cobalt; cobalt ocher.*

ascending branch Either of two ventral elements of a brachiopod loop, continuous anteriorly with ventrally recurved *descending branches* and joined posteriorly by transverse band.

ascending development *accelerated development.*

ascension theory A theory of hypogene mineral-deposit formation involving mineral-bearing solutions rising through fissures from magmatic sources in the Earth's interior. Cf: *descension theory.*

aschaffite An obsolete term originally applied to a kersantite that contains large quartz and plagioclase crystals, probably foreign inclusions (Johannsen, 1939, p.241).

ascharite *szaibelyite.*

aschistic Said of the rock of a minor intrusion which has a composition equivalent to that of the parent magma, i.e. in which there has been no significant differentiation. Such a rock may be called an *aschistite*. Cf: *diaschistic.*

aschistite An *aschistic* rock.

ascoceroid conch The mature conch of the nautiloid group Ascocerida, typically consisting of an expanded exogastric brevicone having an inflated posterior part with dorsal phragmocone, an anterior cylindric neck, and an apical end formed by a transverse partition of conch comprising specialized thick septum (septum of truncation) (TIP, 1964, pt.K, p.54 & 263). See also: *ascocone.*

ascocone A cephalopod shell (like that of *Ascoceras*) whose early portion is slender and curved and later portion is short, blunt, and wide with camerae overlying the living chamber. The initial part of the shell may be detached. See also: *ascoceroid conch.*

ascon A thin-walled sponge or sponge larva with a single flagellated chamber that is also the cloaca and lacking either inhalant or exhalant canals. Cf: *leucon; sycon*. Adj: *asconoid.*

ascophoran Any cheilostome belonging to the suborder Ascophora and characterized by the presence of a compensation sac. Cf: *anascan.*

ascopore A median *frontal pore* that serves as the inlet of the ascus in some ascophoran cheilostomes (bryozoans).

ascus A sac-like hydrostatic organ in ascophoran cheilostomes (bryozoans). Pl: *asci*. Adj: *ascan*. Syn: *compensatrix; compensation sac.*

aseismic Said of an area that is not subject to earthquakes.

aseismic ridge A submarine ridge that is a fragment of continental crust; it is so named to distinguish it from the seismically active mid-oceanic ridge. Cf: *microcontinent.*

asellate Said of certain cephalopod sutures without saddles; specif. said of the simplest known ammonoid suture. Ant: *sellate.*

asexual reproduction Reproduction that does not involve or follow the union of individuals or of germ cells of two different sexes.

ash [coal] In coal, the inorganic residue after burning. Ignition generally alters both the weight and the composition of the inorganic matter. See also: *ash content; extraneous ash; inherent ash*. Syn: *coal ash.*

ash [volc] Fine pyroclastic material (under 4.0mm diameter; under 0.25mm diameter for fine ash). The term usually refers to the unconsolidated material but is sometimes also used for its consolidated counterpart, or *tuff*. Syn: *dust; volcanic ash; volcanic dust.*

ashbed diabase An igneous rock that resembles a conglomerate but that is a scoriaceous, amygdaloidal, tabular intrusion that has incorporated sandy material.

Ashby North American stage: Middle Ordovician (upper subdivision of older Chazyan, above Marmor, below Porterfield) (Cooper, 1956).

ash cloud *eruption cloud.*

41

ash content The percentage of incombustible material in a fuel, determined by burning a sample of coal and measuring the *ash*.

ashcroftine A pink zeolite mineral: $KNaCaY_2Si_6O_{12}$-$(OH)_{10}.4H_2O$.

ash fall The descent of volcanic ash from an eruption cloud by *air-fall deposition*. Cf: *pumice fall*. Syn: *ash shower*.

ash flow A turbulent blend of unsorted, mostly fine-grained pyroclastics and high-temperature gas ejected explosively from fissures or a crater. Cf: *nùee ardente*. Syn: *pyroclastic flow; incandescent tuff flow; glowing avalanche*. Partial syn: *pumice flow*. Obsolete syn: *sand flow*.

ash-flow tuff A tuff deposited by an ash flow or gaseous cloud; a type of *ignimbrite*. It is a consolidated but not necessarily welded deposit.

Ashgillian European stage: Upper Ordovician (above upper Caradocian, below Llandoverian of Silurian).

ash shower *ash fall*.

ashstone An indurated deposit of fine volcanic ash.

ashtonite *mordenite*.

ashy grit (a) A deposit of sand-size and smaller pyroclastics. (b) A deposit of sand and volcanic ash.

asiderite An obsolete term for a meteorite that lacks metallic iron; a *stony meteorite*.

asif A term used in Morocco for a large, generally dry valley in a mountainous region. Etymol: Berber. Cf: *agouni*.

askeletal Without a skeleton; said esp. of sponges.

asparagus stone A yellow-green variety of apatite. Syn: *asparagolite*.

aspect [stratig] The general appearance of a particular geologic entity or fossil assemblage, considered more or less apart from relations in time and space; e.g. the gross or overall lithologic and/or paleontologic character of a stratigraphic unit as displayed at any single geographic point of observation (as in a borehole or outcrop section), and representing the summation or "flavor" of a facies. The aspect of a facies generally has environmental significance.

aspect [slopes] The direction toward which a slope faces with respect to the compass or to the rays of the Sun. Cf: *exposure*.

aspect angle The angle between the aspect of a slope and geographic south (if measured in the Northern Hemisphere), usually reckoned positive eastward and negative westward.

asperite Any rough vesicular lava having plagioclase as its chief feldspar (Brown & Runner, 1939, p.27).

asphalt A dark brown to black viscous liquid or low-melting solid bitumen which consists nearly entirely of carbon and hydrogen and is soluble in carbon disulfide. Natural asphalt, formed in oil-bearing rocks by the evaporation of the volatiles, occurs in Trinidad, near the Dead Sea, and in the Uinta Basin of Utah (tabbyite, liverite, and argulite); asphalt can be prepared by the pyrolysis of coals or shales. Var: *asphaltum*. Syn: *asphaltus (obs.)*; *mineral pitch (obs.)*. See also: *tabbyite*.

asphalt base *napthene base*.

asphaltene Any of the solid, amorphous, black to dark brown dissolved or dispersed constituents of crude oils and other bitumens which are soluble in carbon disulfide, but insoluble in paraffin naphthas. They consist of carbon, hydrogen, and some nitrogen and oxygen, hold most of the inorganic constituents of bitumens, and are isolated by dilution of the bitumen with 10-20 parts of petroleum ether or n-pentane, chilling overnight, and centrifugation or filtration.

asphaltic Pertaining to or containing asphalt; e.g. "asphaltic petroleum" from which asphalt may be recovered, or "asphaltic limestone" or "asphaltic sandstone" impregnated with asphalt, or "asphaltic sand" representing a natural mixture of asphalt with varying proportions of loose sand grains.

asphaltic coal An obsolete syn. of *albertite*.

asphaltic pyrobitumen A bitumen, usually black and structureless, similar in composition to *asphaltite*, but infusible and insoluble in carbon disulfide, and which contains generally less than 5 percent of oxygen. Examples are albertite, elaterite, impsonite, and wurtzilite.

asphaltite Any one of the naturally occurring, black, solid bitumens which are soluble in carbon disulfide and fuse above 230°F. Examples are uintahite, glance pitch, and grahamite.

asphalt rock A porous rock, such as a sandstone or limestone, that is impregnated naturally with asphalt. Syn: *asphalt stone; rock asphalt*.

asphalt stone *asphalt rock*.

asphaltum Var of *asphalt*.

asphaltus An obsolete syn. of *asphalt*.

aspondyl Pertaining to the irregular arrangement of the primary branches or rays of dasycladacean algae. Cf: *euspondyl*.

assay v. In economic geology, to analyse the proportions of metals in an ore; to test an ore or mineral for composition, purity, weight, or other properties of commercial interest. n. The test or analysis itself; its results.

assay foot In determining the *assay value* of an orebody, the multiplication of its *assay grade* by the number of feet along which the sample was taken. Cf: *assay inch*.

assay grade The percentage of valuable constituents in an ore, determined from assay. Cf: *assay value; value*.

assay inch In determining the *assay value* of an orebody, the multiplication of its *assay grade* by the number of inches along which the sample was taken. Cf: *assay foot*.

assay limits The geographic limits of an orebody, as determined by assays of samples taken at places whose position is recorded.

assay ton A weight of 29.166+ grams, used in assaying to represent proportionately the *assay value* of an ore. Since it bears the same ration to one milligram that a ton of 2,000 pounds does to the troy ounce, the weight in milligrams of precious metal obtained from as assay ton of ore equals the number of ounces to the ton. Abbrev: AT.

assay value (a) The quantity of an ore's valuable constituents, determined by multiplying its *assay grade* or percentage of valuable constituents by its dimensions. Cf: *assay inch: assay foot*. The figure is generally given in ounces per ton of ore, or *assay ton*. See also: *value*. (b) The monetary value of an orebody, calculated by multiplying the quantity of its valuable constituents by the market price (von Bernewitz, 1931).

assemblage [paleoecol] A group of fossils that occur at the same stratigraphic level. For its accurate interpretation, an assemblage should be relatively homogeneous or uniformly heterogeneous. Cf: *congregation*.

assemblage [petrology] (a) *mineral assemblage*. (b) *metamorphic assemblage*.

assemblage zone A formal *biostratigraphic zone* consisting of a body of strata characterized by a certain natural assemblage or association of fossil forms without regard to their ranges, and receiving its name from one or more of the fossils particularly representative of the assemblage (ACSN, 1961, art.21), although the named fossil(s) is neither necessarily restricted to the zone nor necessarily found in every part of it. It is based upon directly observable variations in the fossil taxa represented, in abundance of specimens, or in both, and it may indicate ecologic facies and/or age. The term is used without any implications as to either time or facies, and is more or less synonymous with *faunizone* or *florizone*. Cf: *range zone; peak zone*. Syn: *cenozone*.

assembled stone Any stone constructed of two or more parts of gem materials whether they be genuine, imitation, or both (Shipley, 1951, p.15); e.g. a doublet.

assimilated Said of an ore-forming fluid or mineralizer that is derived from crustal, palingenic magmas (Smirnov, 1968). Cf: *juvenile [ore dep]; filtrational*.

assimilation The process of incorporating solid or fluid foreign material, i.e. wall rock, into magma. The term implies no specific mechanisms or results. Such a magma, or the rock it produces, may be called *hybrid* or *anomalous*. See also: *hybridization; contamination; cross-assimilation; magmasklerosis.* Cf: *differentiation.* Syn: *magmatic digestion; magmatic dissolution.*

assise (a) A term approved by the 2nd International Geological Congress in Bologna in 1881 for a subordinate stratigraphic unit (rock section) next in rank below stage and equivalent to *beds.* (b) A stratum characterized by a distinctive fossil content; esp. a biostratigraphic unit consisting of a succession of two or more zones bearing typical fossils of the same species or genera. (c) A term suggested by P.F. Moore (1958) to replace *format.*---Etymol: French, "bed, course, layer, stratum, base", from the architectural usage for a continuous row or layer of stones or brick placed side by side between layers of mortar. Pron: a-*seas.*

associated natural gas Natural gas that occurs in association with oil in a reservoir. Cf: *nonassociated natural gas.*

association [ecol] A group of organisms (living or fossil) occurring together and having similar environmental requirements and patterns of development and usually having one or more dominant species.

association [petrology] *rock association.*

associes A seral community in which there are two or more dominant forms. Pron: uh-*sew*-sees.

assorted *poorly sorted.*

assortment The inverse of sorting or of the measure of dispersion of particle sizes within a frequency distribution.

assured mineral *developed reserves.*

Assyntian orogeny A name proposed by Stille and widely used in western Europe for orogenies and disturbances at the end of the Precambrian. It is named for the district of Assynt in the northwest highlands of Scotland, and based on the angular truncation of the Torridonian strata by the Lower Cambrian strata. The name is unfortunate, because the age of the Torridonian is undetermined, and may be very much older than the Cambrian.

assyntite A foyaite rich in sphene, containing augite, and with biotite and apatite among the accessories. According to Johannsen (1939, p.242), the name has been withdrawn.

Astartian *Sequanian.*

astatic Said of a geophysical instrument that has a negative restoring force which aids a deflecting force, making the instrument more sensitive and/or less stable.

astatic gravimeter An instrument of the unstable type, with a mechanical system designed to produce relatively large motions for small changes in gravity. Syn: *unstable gravimeter.*

astatic magnetometer A *gradiometer* using the principle of a torsion magnetometer.

astatic pendulum A pendulum having almost no tendency to take a definite position of equilibrium.

aster A sponge spicule (microsclere) that has a star-like appearance, with a relatively large number of rays or pseudoactins radiating from a relatively restricted central area.

asteria Any gemstone that, when cut cabochon in the correct crystallographic direction, displays a rayed figure (a star) by either reflected or transmitted light; e.g. a *star sapphire.* Syn: *starstone.*

asteriated Said of a mineral, crystal, or gemstone that exhibits asterism; e.g. "asteriated beryl". Syn: *star.*

asterism [cryst] Elongation of Laue X-ray diffraction spots produced by stationary single crystals as a result of internal crystalline deformation. The size of the Laue spot is determined by the solid angle formed by the normals to any set of diffracting lattice planes; this angle increases with increasing crystal deformation, producing progressive elongated ("asterated") spots. Measurements of asterism are used as indicators of deformation in crystals subjected to slow stress or to shock waves.

asterism [gem] The optical phenomenon of a rayed or star-shaped figure of light displayed by some crystals when viewed in reflected light (as in a star sapphire or a gemstone cut cabochon) or in transmitted light (as in some mica). It is caused by minute, multiple, oriented, acicular inclusions. See also: *star.*

asteroid [paleont] Any asterozoan belonging to the subclass Asteroidea, characterized by relatively broad arms usually not separable from the central disc; e.g. a starfish. Their stratigraphic range is Lower Ordovician to present.

asteroid [astron] One of the many small celestial bodies in orbit around the Sun. Most asteroid orbits are between those of Mars and Jupiter (NASA, 1966, p.6). Syn: *planetoid; minor planet.*

asterolith A star- or rosette-shaped coccolith with a concave face, formed of a single crystal with the vertical *c*-axis perpendicular to the plane of the disk. See also: *discoaster.*

asterozoan Any free-living echinoderm belonging to the subphylum Asterozoa, having a characteristic depressed star-shaped body composed of a central disc and symmetrical radiating arms. Their stratigraphic range is Cambrian to present.

asthenolith A body of magma that has melted at any place within the solid part of the Earth, at any time in the geologic past or present, due to heat generation by radioactive disintegration (Willis, 1938, p.603). See also: *asthenolith hypothesis.*

asthenolith hypothesis A theory of magmatic activity, both intrusive and extrusive, that postulates local *asthenoliths* or areas of melting by radioactive heat which have a repetitive cycle of melting, growth, migration, cooling, solidification, and remelting. Asthenolithic activity is postulated as the cause of uplift and subsidence, orogeny, earthquakes, and metamorphism (Willis, 1938, p.603). Cf: *blister hypothesis.*

asthenosphere The layer or shell of the Earth below the lithosphere, which is weak and in which isostatic adjustments take place, in which magmas may be generated, and in which seismic waves are strongly attenuated. It is equivalent to the *upper mantle.* Syn: *zone of mobility.*

Astian European stage: upper Pliocene (above Plaisancian, below the Pleistocene stage variously known as Villafranchian, Calabrian, or Günz).

astite A variety of hornfels in which mica and andalusite dominate (Holmes, 1928, p.37). Type locality: Cima d'Asta, Italian Alps. Cf: *aviolite; edolite.*

astogeny The life history of a colonial animal, such as a graptolite or a bryozoan.

astrakhanite A syn. of *bloedite.* Also spelled: *astrakanite.*

astreoid Said of a massive rugose corallum in which the septa of each corallite are fully developed but walls between corallites are lacking, and characterized by septa of adjacent corallites generally in alternating position. Cf: *aphroid.*

astringent (a) Said of a mineral (such as alum) having a taste that tends to pucker the tissues of the mouth. (b) Said of a clay containing an astringent salt.

astrobleme An ancient erosional scar on the Earth's surface, produced by the impact of a cosmic body, and usually characterized by a circular outline and highly disturbed rocks showing evidence of intense shock (Dietz, 1961, p.53); an eroded remnant of a meteoritic or cometary impact crater. The most spectacular example is the Vredefort Ring in the Transvaal of South Africa. The term is generally applied to *cryptoexplosion structures* of great age in which any original associated extraterrestrial fragments have been destroyed. Term introduced by Dietz (1960). Etymol: Greek *astron,* "star", + *blema,* "wound from a thrown object such as a javelin or stone". Cf: *geobleme.* Syn: *fossil meteorite crater.*

astrogeodetic The term used to describe the direct measurement of the Earth to determine the deflection of the vertical, and hence, the separation of the geoid and the ellipsoid. The method contrasts with the gravimetric, or indirect method.

astrogeology A science that applies the principles and techniques of geology, geochemistry, and geophysics to the study of the nature, origin, and history of the condensed matter and gases in the solar system (usually excluding the Earth). It includes: remote-sensing observations and in-situ manned exploration of other planetary bodies (the Moon, Mars); the study of the chemistry, mineralogy, and history of objects that occur on the Earth but are of known or possible extraterrestrial origin (such as meteorites and tektites) or that are returned to the Earth (such as lunar samples); and the study of the effects of extraterrestrial processes (such as meteorite impact, solar energy changes, and tides) on the Earth in the present and past. The term was first used by Lesevich (1877) for a branch of astronomy based primarily on the study of meteorites and secondarily on telescopic spectroscopy (see Milton, 1969). See also: *planetology; planetary geology.* Syn: *extraterrestrial geology; space geology; geoastronomy.*

astrolabe A compact optical instrument designed for measuring the altitudes or observing the positions of celestial bodies; e.g. a "prismatic astrolabe" consisting of a telescope in a horizontal position, with a prism (generally of 45 or 60 degrees) in front of the object glass, immediately underneath which is an artificial horizon (pool of mercury). It was formerly used to fix latitude precisely by observing the apparent transit of the Sun across the meridian at midday, but now has been superseded by the *sextant.*

astrolithology An obsolete term for the science that deals with meteoritic stones.

astronomic A term used in geodetic surveying to describe the method of determining latitude, longitude, or azimuth from observations on stars, referred to a plumb line, or gravity vector, at the point of observation. The sun is sometimes used in lower order surveying. Cf: *deflection of the vertical.*

astronomical position The latitude and longitude of a point on the Earth determined from measurements based on the position of the stars.

astronomical unit A unit of measuring planetary distance equivalent to the mean distance of the Earth from the Sun: 1.496×10^8 km, or approximately 93 million statute miles. This distance is also equivalent to the length of the semi-major axis of the Earth's orbit.

astronomic azimuth The horizontal angle measured from the vertical plane passing through the celestial pole to the vertical plane passing through the center of a celestial body; an arc of the horizon measured between a fixed point of observation and the point where the vertical circle passes through the center of an observed object. It results directly from observations on a celestial body, and is measured at the observer in the plane of the horizon, preferably clockwise from north through 360 degrees. It is usually simply called *azimuth.* The term *bearing,* as used in navigation, is synonymous.

astronomic equator The line on the Earth's surface whose astronomic latitude at every point is zero degrees. When corrected for station error, it becomes the *geodetic equator.*

astronomic horizon A great circle on the celestial sphere formed by the intersection of the celestial sphere and a plane passing through any point (such as the eye of an observer) and perpendicular to the zenith-nadir line; the plane that passes through the observer's eye and is perpendicular to the zenith at that point. It is the projection of a horizontal plane in every direction from the point of orientation. Cf: *celestial horizon.* Syn: *sensible horizon.*

astronomic latitude The *latitude* or angle between the plane of the celestial equator and the plumb line (direction of gravity) at a given point on the Earth's surface; the angle between the plane of the horizon and the Earth's axis of rotation. It represents the latitude resulting directly from observations on celestial bodies; when corrected for station error, it becomes the *geodetic latitude.* Symbol: Φ.

astronomic longitude The *longitude* or angle between the plane of the celestial meridian and the plane of an arbitrarily chosen prime meridian (generally the Greenwich meridian). It

represents the longitude resulting directly from observations on celestial bodies; when corrected for station error, it becomes the *geodetic longitude.* Symbol: Λ.

astronomic meridian A line on the Earth's surface having the same astronomic longitude at every point. It is an irregular line, not lying in a single plane. Syn: *terrestrial meridian.*

astronomic parallel A line or circle on the Earth's surface having the same astronomic latitude at every point. It is an irregular line, not lying in a single plane.

astronomic tide *equilibrium tide.*

astronomy The study of celestial bodies: their positions, sizes, movements, relative distances, compositions and physical conditions, interrelationships, and their history.

astrophyllite A mineral: $(K,Na)_3(Fe,Mn)_7Ti_2Si_8O_{24}(O,OH)_7$.

astrophysics That aspect of astronomy which is concerned with the physics and chemistry of celestial bodies, and with their origins.

astropyle A nipple-like projection from the central capsule of a radiolarian of the suborder Phaeodarina.

astrorhiza One of a stellate or rosette cluster of small but usually clearly defined channels, grooves, or canals diverging radially from a center on the surface of a stromatoporoid and branching toward their outer terminations. Pl: *astrorhizae.*

astrotectonic Pertaining to terrestrial deformation structures produced by cosmic bodies, such as induced by meteorite impact.

Asturian orogeny One of the 30 or more short-lived orogenies during Phanerozoic time identified by Stille, in this case late in the Carboniferous, between the Westphalian and Stephanian Stages.

asylum *refugium.*

asymmetric Said of crystals of the hemihedral class of the triclinic system, which have no symmetry elements; also, said of any irregular crystal.

asymmetric bedding Bedding characterized by lithologic types or facies that follow each other in a "circuitous" arrangement illustrated by the sequence 1-2-3-1-2-3-1-2-3. Cf: *symmetric bedding.*

asymmetric fold A fold, the limbs of which have different angles of dip relative to the axial surface, which is not vertical. Ant: *symmetrical fold.*

asymmetric ripple mark A *ripple mark* having an asymmetric profile in cross section, characterized by a short and steep slope facing downcurrent and a long and gentle slope facing upcurrent; specif. *current ripple mark.* In plan view, the crest may be relatively straight or markedly curved. Ant: *symmetric ripple mark.*

asymmetric valley A valley with one side steeper than the other.

atacamite A green orthorhombic mineral: $Cu_2Cl(OH)_3$. It is dimorphous with paratacamite, and is formed by weathering of copper lodes, esp. under desert conditions. Syn: *remolinite.*

atatschite A porphyritic igneous rock containing microscopic crystals of orthoclase, augite, and biotite in a glassy groundmass and characterized by the presence of small amounts of sillimanite and cordierite.

ataxic Said of an unstratified mineral deposit. Cf: *eutaxic.*

ataxite [meteorite] An *iron meteorite* containing more than 10% nickel and lacking the structure of either *hexahedrite* or *octahedrite.* Many ataxites show microscopic oriented plates of kamacite in a groundmass of plessite. Symbol: *D.*

ataxite A *taxite* whose components have mixed in a breccia-like manner. Cf: *eutaxite.*

atelestite A yellow mineral: $Bi_8(AsO_4)_3O_5(OH)_5$.

atexite Basic material that is unchanged during anatexis (Dietrich & Mehnert, 1961). Also spelled: *atectite.*

at grade [eng] On the same level or degree of rise; at design level or slope. The term is applied to highways, walks, culverts, etc., or combinations of these, at the point where they intersect.

at grade [geomorph] *graded* [geomorph].

athabascaite A mineral: Cu_5Se_4.

athrogenic Pertaining to pyroclastics.

Atlantic n. A term used primarily in Europe for an interval of postglacial time (from about 7500 to 4500 years ago) following the Boreal and preceding the Subboreal, during which the inferred climate was warmer than present and generally wet; a subunit of the Blytt-Sernander climatic classification, characterized by oak, elm, linden, and ivy vegetation. More recent paleoclimatic data suggest that the Atlantic in Europe, as in many other regions, includes the dryest as well as the warmest postglacial climate (Karlstrom, 1966). It corresponds to most of the *Altithermal* and the middle part of the Hypsithermal.---adj. Pertaining to the postglacial Atlantic interval and to its climate, deposits, biota, and events.

Atlantic suite One of two large groups of igneous rocks, characterized by alkalic and alkali-calcic rocks. Harker (1909) divided all Tertiary and Holocene igneous rocks of the world into two main groups (the Atlantic suite and the *Pacific suite*), the Atlantic suite being so named because of the predominance of alkalic and alkali-calcic rocks in the nonorogenic areas of crustal instability around the Atlantic Ocean. Because there is such a wide variety of tectonic environments (and in their associated rock types) in the areas of Harker's Atlantic and Pacific suites, the terms are now seldom used to indicate kindred rock types; e.g. Atlantic-type rocks are widespread in the mid-Pacific volcanic islands. Cf: *Mediterranean suite.* Syn: *intra-Pacific province.* See also: *andesite line.*

Atlantic-type coastline A *discordant coastline*, esp. one as developed in many areas around the Atlantic Ocean; e.g. the SW coastline of Ireland and the NW coastlines of France and Spain. Ant: *Pacific-type coastline.*

atlantite A dark-colored nepheline-bearing basanite having a predominance of dark-colored minerals over light.

atlas A collection of maps bound into a volume. The use of the term is derived from the figure of Atlas (a Titan of Greek mythology, often represented as supporting the heavens) used as a frontispiece to certain early collections of maps, first appearing on the general title page of Mercator's *Atlas* (1595).

atmidometer *atmometer.*

atmoclast A rock fragment broken off in place by atmospheric weathering, either chemically or mechanically.

atmoclastic rock A clastic rock consisting of atmoclasts that have been recemented without further rearrangement by wind or water (Grabau, 1924, p.292).

atmodialeima A term proposed by Sanders (1957, p.295) for an unconformity caused by subaerial processes.

atmogenic Said of a rock, mineral, or deposit derived directly from the atmosphere, as by condensation, wind action, or deposition from volcanic vapors; e.g. snow. See also: *atmolith.*

atmolith An *atmogenic* rock (Grabau, 1924, p.279).

atmometer A device that is used to measure the rate of evaporation in the atmosphere. It may be a large evaporation tank, a small evaporation pan, a porous porcelain body, or a porous paper wick device. Syn: *atmidometer; evaporimeter.*

atmophile (a) Said of those elements which are the most typical in the Earth's atmosphere: H, C, N, O, I, Hg, and inert gases (Rankama & Sahama, 1950, p.88). (b) Said of those elements which occur either in the uncombined state, or which, as volatile compounds, will concentrate in the gaseous primordial atmosphere (Goldschmidt, 1954, p.26).

atmosphere The mixture of gases that surround the Earth: chiefly oxygen and nitrogen, with some argon and carbon dioxide and minute quantities of helium, krypton, neon, and xenon. Syn: *air.*

atmospheric argon Argon in the atmosphere and argon absorbed on the surfaces or rocks and minerals that have been exposed to the atmosphere. Cf: *excess argon; inherited argon; radiogenic argon.*

atmospheric pressure The force per unit area in any part of the Earth's atmosphere. Normal pressure at sea level is variously defined as 76.0 cm or 29.92 inches of mercury, 1033.3 cm or 33.9 ft of water, 3 grams or 1,013,250.0 dynes per cm^2, 14.66 pounds per square inch, 1.01325 bars. Syn: *air pressure; barometric pressure.*

atmospheric radiation The infrared radiation emitted by the atmosphere in two directions: upward into space and downward into the Earth. The latter is known as *counterradiation.*

atmospherics *spherics.*

atmospheric tide The rhythmic and alternate vertical oscillation of the atmosphere, produced by daily temperature changes and by the gravitational action of the Moon and Sun. Syn: *tide.*

atmospheric water Water in the atmosphere, in gaseous, liquid, or solid state.

atmospheric weathering Weathering occurring at the surface of the Earth.

Atokan North American provincial series: lower Middle Pennsylvanian (above Morrowan, below Desmoinesian).

atoll (a) A ring-shaped coral reef appearing as a low, roughly circular (sometimes elliptical or horseshoe-shaped) coral island or a ring of closely spaced coral islets encircling or nearly encircling a shallow lagoon in which there is no pre-existing land or islands of noncoral origin, and surrounded by deep water of the open sea; it may vary in diameter from 1 km to more than 130 km, and is esp. common in the western and central Pacific Ocean. Etymol: native name in the Maldive Islands (Indian Ocean) which are typical examples of this structure. Syn: *lagoon island; ring reef; reef ring.* (b) An atoll-like island among the Florida Keys, composed of noncoral material.

atoll moor A peat bog that entirely surrounds a lake in the form of a ring and is in turn surrounded by a ditch or ring of open water around the original shoreline of the lake (Davis, 1907, p.128 & 154). See also: *moat lake.* Syn: *sphagnum atoll.*

atollon A term used in the Maldive Islands of the Indian Ocean for a large atoll consisting of many smaller atolls; the term "atoll" was derived from this name.

atoll structure In a metamorphic rock, porphyroblasts with hollow centers resembling atolls. The annular ring may be almost complete or a chain of granules (Joplin, 1968).

atoll texture In mineral deposits, the surrounding of one mineral by a ring of one or more other minerals. Syn: *core texture.* Cf: *tubercle texture.*

atomic absorption spectrometry Chemical analysis performed by vaporizing in a flame a sample, usually in a liquid form, and measuring the absorbance by the unexcited atoms in the vapor of various narrow resonant wavelengths of light which are characteristic of specific elements. The amount of an element present is proportional to the amount of absorption by the vapor.

atomic absorption spectrophotometer An instrument for generating and analyzing an *atomic absorption spectrum.*

atomic absorption spectroscopy The observation of an *atomic absorption spectrum* and all processes of recording and measuring which go with it.

atomic absorption spectrum The absorption spectrum seen when the unexcited atoms of a vaporized sample selectively absorb certain wavelengths of light passed through the sample.

atomic clock *radioactive clock.*

atomic plane In a crystal, any plane that contains a regular array of atomic units (atoms, ions, or molecules); it is a potential cleavage face or cleavage plane.

atomic time scale A *geologic time scale* calibrated on the basis of radioactive decay in rocks. Measurements are made in terms of hundreds to thousands of million years. Cf: *relative time scale.*

atom percent The percentage of an atomic species in a substance, calculated with reference to number of atoms rather than weight, number of molecules, or other criteria.

atopite A yellow or brown variety of romeite containing fluorine.

at rest Said of the lateral *earth pressure* when the soil is neither compressed nor allowed to yield and the structure (such as a wall) does not move. See also: *neutral pressure.*

atrium [paleont] The *cloaca* of a sponge. Pl: *atria.*

atrium [palyn] The space between the external opening (pore) and a much larger internal opening in the endexine of a pollen grain with a complex porate structure. The internal opening is so large that the endexine is missing in the pore area. Pl: *atria.* Cf: *vestibulum.*

atrium oris A preoral cavity in a crustacean, bounded ventrally by the posteriorly directed labrum, dorsally by the ventral surface of the cephalon just behind the mouth, and laterally by metastoma and mandibles.

atrypoid Any articulate brachiopod belonging to the family Atrypidae, characterized by costate or plicate shells that are unequally biconvex or convexo-plane, the brachial valve being the more convex. Their stratigraphic range is Middle Ordovician to Upper Devonian.

attached dune A dune that accumulates around a rock or other obstacle in the path of windblown sand, occurring on either the windward or the leeward sides of the obstacle, or on both sides, and varying widely in size and form. Cf: *anchored dune.*

attached ground water Ground water retained on the walls of interstices in the zone of aeration. It is considered equal in amount to the *pellicular water* and is measured by specific retention.

attached operculum The part of a dinoflagellate cyst that is completely surrounded by archeopyle sutures and hence remains joined to the main part of the theca where the suture is not developed. Cf: *free operculum.* Syn: *attached opercular piece.*

attachment scar (a) An expanded *basal cavity* of a conodont, or one that is larger than a small-sized pit. (b) The part of the aboral side of a conodont element to which the basal plate was attached.

attakolite A mineral: $(Ca,Mn,Sr)_3Al_6(PO_4,SiO_4)_7.3H_2O$. Also spelled: *attacolite.*

attapulgite *palygorskite.*

attenuation constant A constant defined by a complex relationship of angular frequency, magnetic permeability, dielectric permitivity, and electrical conductivity.

attenuation distance *depth of penetration.*

Atterberg grade scale A geometric and decimal *grade scale* devised by Albert Atterberg (1846-1916), Swedish soil scientist; it is based on the unit value 2 mm and involves a fixed ratio of 10 for each successive grade, yielding the diameter limits of 200, 20, 2.0, 0.2, 0.02, and 0.002 (Atterberg, 1905). Subdivisions are the geometric means of the grade limits. The scale has been widely used in Europe and was adopted in 1927 by the International Commission on Soil Science (but not by the U.S. Bureau of Soils).

Atterberg limits In a sediment, the water-content boundaries between the semiliquid and plastic states (known as the *liquid limit*) and between the plastic and semisolid states (known as the *plastic limit*). Syn: *consistency limits.*

attic The very small, uppermost (abaxial) chamberlet in superposed chamberlets of a foraminiferal shell volution (as in *Flosculinella* and *Alveolinella*).

Attic orogeny One of the 30 or more short-lived orogenies during Phanerozoic time, identified by Stille; in this case in the Miocene, between the Sarmatian and Pontian Stages.

attitude [photo] The angular orientation of a camera, or of the photograph taken with that camera at the instant of exposure, with respect to some external reference system.

attitude [struc geol] The position of a structural surface relative to the horizontal, expressed quantitatively by both *strike* and *dip* measurements.

attribute A qualitative *variable*, usually denoted by its presence or absence.

attrital-anthraxylous coal A bright coal in which the ratio of anthraxylon to attritus varies from one to one to one to three. Cf: *anthraxylous coal; anthraxylous-attrital coal; attrital coal.*

attrital coal A coal in which the ratio of anthraxylon to attritus varies from one to one to one to three. Cf: *anthraxylous coal; anthraxylous-attrital coal; attrital-anthraxylous coal.*

attrition The act or process of rubbing together or wearing down by friction; specif. the mutual wear and tear that loose rock fragments or particles, while being moved about by wind, waves, running water, or ice, undergo by rubbing, grinding, knocking, scraping, and bumping against one another, with resulting reduction in size and increase in roundness. Although the term has been used synonymously with *abrasion* and *corrasion,* strictly it is the wearing away suffered by the rock fragments serving as the tools of abrasion or corrasion.

attritus A composite term for dull grey to nearly black coal components of varying maceral content with granular texture that forms the bulk of some coals or is interlayered with bright bands of *anthraxylon* in others. It is formed of a tightly compacted mixture of vegetal materials, especially those that were relatively resistant to complete degradation. Syn: *durain.*

aubrite An achondritic stony meteorite whose essential mineral is enstatite and that contains diopside in minor amounts. See also: *whitleyite.* Syn: *bustite.*

auerlite A variety of thorite containing phosphorus, with a PO_4/SiO_4 ratio of about 0.8:1.

aufeis Thick (1-4 m) masses or sheets of ice (*icings*) formed on a river's flood plain in winter, when shoals in the river freeze solid or are otherwise dammed, so that water under increasing hydrostatic pressure is forced to the surface and spreads over the flood plain where it freezes in successive sheets of ice. Etymol: German. Syn: *flooding ice; flood icings; flood-plain icings; naledi.*

aufwuchs Aquatic organisms that are attached to but do not penetrate the substrate; e.g. crustaceans. Cf: *periphyton.* Etymol: German, "growth".

auganite An augite-bearing andesite; according to Johannsen (1939, p.242), an olivine-free basalt.

augelite A colorless, white, or pale-red mineral: $Al_2(OH)_3PO_4$.

augen In foliate metamorphic rocks such as schists and gneisses, large, lenticular mineral grains or mineral aggregates having the shape of an eye in cross section, in contrast to the shapes of other minerals in the rock. See also: *augen structure.* Etymol: German, "eyes".

augen-blast An augen-forming porphyroblast in dynamically metamorphosed rocks having idioblastic shapes (Bayly, 1968). Cf: *augen-clast.*

augen-clast In dynamically metamorphosed rock, augen consisting of clastic fragments which generally occur in a wholly clastic matrix (Bayly, 1968). Cf: *augen-blast.*

augen gneiss A general term for a gneissic rock containing lenticular mineral grains or aggregates representing either deformed original phenocrysts or porphyroblasts of metamorphic origin.

augen schist A mylonitic rock characterized by the presence of recrystallized minerals as augen or lenticles parallel to and alternating with schistose streaks.

augen structure In some gneissic and schistose metamorphic rocks, a structure consisting of minerals like feldspar, quartz, or garnet which have been squeezed into elliptical or lens-shaped forms resembling eyes (*augen*) which are commonly enveloped by parallel layers of contrasting constituents such as mica or chlorite. Cf: *flaser structure.* Syn: *eyed structure; phacoidal structure.*

auger (a) A screw-like boring tool resembling the conventional carpenter's auger bit but much larger, usually operated by hand and designed for use in clay, soil, and other relatively unconsolidated near-surface materials, esp. for such purposes as testing for ore minerals or drilling for oil or water, or taking any hard-rock sample below soil or alluvial cover. (b) A rotary drilling device for making seismic shotholes or geophone holes by which the cuttings are mechanically and continuously removed from the bottom of the borehole during the drilling operation without the use of fluids.

augite (a) A common mineral of the clinopyroxene group: $(Ca,Na)(Mg,Fe^{+2},Al)(Si,Al)_2O_6$. It may contain titanium and ferric iron. Augite is usually black, greenish black, or dark green, and commonly occurs as an essential constituent in many basic igneous rocks and in certain metamorphic rocks. Dana (1892) confined the name "augite" to clinopyroxenes containing appreciable $(Al,Fe)_2O_3$, but petrologists have applied it to members of the system $(Mg,Fe,Ca)SiO_3$. Cf: *pigeonite*. (b) A term often used as a syn. of *pyroxene*.---Syn: *basaltine*.

augitite A porphyritic extrusive rock containing abundant phenocrysts of augite with lesser amounts of amphibole, magnetite or ilmenite, apatite, and sometimes nepheline, haüyne, or feldspar in a dark-colored glassy groundmass probably of the composition of analcime.

augitophyre A porphyritic basaltic rock in which the phenocrysts are composed of augite.

aulacogen Schatsky's term for a fault-bounded intracratonic trough or graben (Shatsky & Bogdanoff, 1960). Cf: *taphrogeosyncline*.

aulocalycoid Said of the skeleton of a hexactinellid sponge in which presumed dictyonal strands are diagonally interwoven and connected by synapticulae.

aulos An *axial structure* in a rugose coral, consisting of a tube commonly formed by abrupt sideward deflection of the inner edges of septa and junction of them with neighbors.

aureole In geology, a zone surrounding an igneous intrusion in which the country rock shows the effects of contact metamorphism. Syn: *contact aureole; contact zone; metamorphic aureole; metamorphic zone; thermal aureole; exomorphic zone; zone [meta]*.

auric Said of a substance that contains gold, esp. gold in its trivalent state. Cf: *auriferous*.

aurichalcite A pale-green or pale-blue mineral: $(Zn,Cu)_5$ $(CO_3)_2(OH)_6$. Syn: *brass ore*.

auricle (a) An ear-like extension of the dorsal region of the shell in certain bivalve mollusks, commonly separated from the body of the shell by a notch or sinus. (b) An internal process arising from basicoronal ambulacral plates of an echinoid and serving for attachment of muscles supporting Aristotle's lantern.

auricula One of the thickened "ears" of auriculate spores. Pl: *auriculae*. Cf: *zone [palyn]*.

auricular sulcus An external furrow at the junction of an auricle with the body of the shell of a bivalve mollusk.

auriculate Said of spores having exine thickenings (auriculae) in the equatorial region that project like "ears", generally from the area of the ends of the laesura.

auricupride A mineral: Cu_3Au. Syn: *cuproauride*.

auriferous Said of a substance that contains gold, esp. said of gold-bearing mineral deposits. Cf: *auric*.

auriform Said of a mollusk shell (such as that of a gastropod) shaped like the human ear.

aurorite A mineral: $(Mn,Ag,Ca)Mn_3O_7.3H_2O$.

aurostibite A cubic mineral: $AuSb_2$.

austausch A measure of turbulent mixing, equal to the product of mass and transverse distance traveled in a unit of time by a fluid in turbulent motion as it passes through a unit area parallel to the general direction of flow (Twenhofel, 1939, p.187). Syn: *eddy conductivity, mixing coefficient, austausch coefficient, eddy coefficient; exchange coefficient*.

austausch coefficient *austausch*.

Austinian North American (Gulf Coast) stage: Upper Cretaceous (above Eaglefordian, below Tayloran).

austinite A colorless or yellowish orthorhombic mineral: $CaZnAsO_4(OH)$.

australite A jet-black, usually button-shaped or lensoid, often well-preserved tektite from southern Australia. Syn: *blackfellows' button*.

Austrian orogeny One of the 30 or more short-lived orogenies during Phanerozoic time identified by Stille, in this case at the end of the Early Cretaceous.

autallotriomorphic Pertaining to an *aplitic* texture in which all of the mineral constituents crystallized at the same time and mutually interfered.

autecology The study of the relationships between individual organisms or species and their environment. Var: *autoecology*. Cf: *synecology*.

authalic projection *equal-area projection*.

authigene An *authigenic* mineral or rock constituent; e.g. a primary or secondary mineral of an igneous rock, or the cement (of a sedimentary rock) deposited directly from solution, or glauconite in the ocean, or a mineral resulting from metamorphism. The term was introduced by Kalkowsky (1880, p.4). Ant: *allogene*.

authigenesis The process by which new minerals form in place within an enclosing sediment or sedimentary rock during or after deposition, as by replacement or recrystallization, or by secondary enlargement of quartz overgrowths. Cf: *neogenesis*.

authigenetic *authigenic*.

authigenic Formed or generated in place; specif. said of rock constituents and minerals that have not been transported or that were derived locally on the spot where they are now found, and of minerals that came into existence at the same time, or subsequently to, the formation of the rock of which they constitute a part. The term, as used, often refers to a mineral (such as quartz or feldspar) formed after deposition of the original sediment. Ant: *allogenic*. Cf: *halmeic; autochthonous*. Syn: *authigenous; authigenetic*.

authigenous [geol] *authigenic*.

authigenous [volc] In the classification of pyroclastics, the equivalent of *essential* ejecta.

authimorph A constituent of a metamorphic rock which, in the formation of the new rock, had its outlines or boundaries altered. The term is now obsolete.

autobreccia A breccia formed by some process that is penecontemporaneous with the formation or consolidation of the rock unit from which the fragments are derived; specif. a *flow breccia*. Challinor (1967, p.31) regards "the peculiar pseudobreccia" as a "sedimentary" autobreccia.

autobrecciation Formation of an autobreccia; e.g. the fragmentation process whereby portions of the first consolidated crust of a lava flow are incorporated into the still-fluid portion.

autochthon [sed] A residual deposit produced in place by decomposition.

autochthon [tect] A body of rocks that remains at its site of origin, where it is rooted to its basement. Although not moved from their original site, autochthonous rocks may be mildly to considerably deformed. Ant: *allochthon*. Cf: *parautochthonous; stationary block*. Also spelled: *autochthone*.

autochthone Var. of *autochthon*.

autochthonous [coal] (a) Said of coal or peat that originates at the place where its constituent plants grew and decayed. (b) Pertaining to *autochthony*.

autochthonous [geol] Formed or occurring in the place where found; esp. said of a rock or sediment derived in the place where it is now found, such as a salt or gypsum deposit formed by precipitation. The term was first used by Naumann (1858, p.657) to designate rock units localized at the site of their original emplacement or formation; it is similar in meaning to *authigenic* (which refers to constituents, rather than whole formations). Ant: *allochthonous.*

autochthonous [crater] Said of a fragmental unit in an impact or explosion crater, composed of shattered and brecciated material that was not ejected from its original position during crater formation, and generally occurring in the walls, and below the floor, of the original crater; e.g. an "autochthonous breccia" whose component fragments exhibit only minor rotation or translation.

autochthonous [paleont] Endemic, or indigenous or native to a region; said of a fossil representing an organism that actually lived in the place where the fossil was found.

autochthonous [petrology] (a) Pertaining to a magma, or the magmatic constituent of a migma, produced by liquefaction in place. (b) Said of a granite that remained at its place of formation. Cf: *parautochthonous.*

autochthonous [tect] (a) Pertaining to an *autochthon* or to the rocks of an autochthon, esp. to strata that have not been displaced by overthrusting. (b) Said of a fold that is still connected to its roots, composed of untraveled rocks that lie on their original basement.----Cf: *parautochthonous.*

autochthonous [stream] Said of a stream flowing in its original channel.

autochthony Accumulation of plant remains in their original environment or in the place of their growth. Ant: *allochthony.* See also: *euautochthony; hypautochthony; in-situ theory.*

autoclast A rock fragment in an autoclastic rock.

autoclastic rock A term originated by Smyth (1891) for a rock having a broken or brecciated structure, formed in the place where it is found as a result of crushing, shattering, dynamic metamorphism, orogenic forces, or other mechanical processes; e.g. an autoclastic schist produced by pressure of one rock mass upon another, or a fault breccia produced by friction of one rock mass moving over another, or a brecciated dolomite produced by diagenetic shrinkage followed by recementation, or a subglacial moraine produced by glacial abrasion. Cf: *cataclastic rock; epiclastic rock.*

autoconsequent Said of a stream whose course is guided by the slopes of material it has deposited, and of the topographic features (such as waterfalls) developed by such a stream.

autocorrelation The use of a function to measure the statistical dependence of a later value of a wave form on the present value, or the extent to which future values can be predicted from past values. It is equivalent to passing the wave form through its matched filter. Cf: *crosscorrelation.*

autocyclicity The state of cyclothemic deposition that requires no change in the total energy and material input into a sedimentary system but involves simply the redistribution of these elements within the system (Beerbower, 1964, p.32). It involves such mechanisms of deposition as channel migration; channel diversion, and bar migration. Cf: *allocyclicity.*

autodermalium A specialized sponge spicule lying within the exopinacoderm. Cf: *hypodermalium.*

autoecology *autecology.*

autogastralium A specialized sponge spicule lying within the endopinacoderm of the cloaca. Cf: *hypogastralium.*

autogenetic (a) Said of landforms that have developed or evolved under strictly local conditions, without interference by orogenic movements; esp. a topography resulting from the action of falling rains and flowing streams upon land surfaces having free drainage to the sea. (b) Said of a type of drainage (and of its constituent streams) that is determined solely by the conditions of the land surface over which the streams flow, as a drainage system developed solely by headwater erosion. See also: *self-grown stream.*---Syn: *autogenous; autogenic.*

autogenic [geomorph] *autogenetic* [geomorph].

autogenic [ecol] Said of an ecologic succession that resulted from factors originating within the natural community and altering its habitat. Cf: *allogenic* [ecol].

autogenous [geomorph] *autogenetic.*

autogeosyncline A parageosyncline without an adjoining uplifted area, containing mostly carbonate sediments (Kay, 1947, p.1289-1293); an *intracratonic basin.* Syn: *residual geosyncline.* Cf: *zeugogeosyncline.*

auto-injection *auto-intrusion.*

auto-intrusion [ign] The filling of rifts in a crystal mesh by residual magmatic liquids. The rifting, usually during a late stage of magmatic differentiation, causes a release in pressure thereby permitting gases in the liquids to expand and migrate to the areas of lower pressure. Syn: *auto-injection.*

autointrusion [sed] Sedimentary *intrusion* of rock material from one part of a bed or set of beds in process of deposition into another part.

autolith (a) An inclusion in an igneous rock to which it is genetically related. Cf: *xenolith.* Syn: *cognate inclusion; cognate xenolith; endogenous inclusion.* (b) In a granitoid rock, an accumulation of Fe-Mg minerals of uncertain origin (Balk, 1937, p. 10-12). It appears as a round, oval, or elongate segregation or clot.

autolysis (a) The process of "self-digestion", as in the albitization of plagioclase in a lava by sodium from the lava itself rather than by newly introduced sodium. (b) Return of a substance to solution, as phosphate removed from seawater by plankton and returned when these organisms die and decay.

autometamorphism (a) A process of chemical adjustment of an igneous mineral assemblage to falling temperature attributed to the action of its own volatiles, e.g. serpentinization of peridotite, spilitization of basalt. (b) The alteration of an igneous rock by its own residual liquors (Tyrrell, 1926). This process should rather be called *deuteric* because it is not considered to be metamorphic.

autometasomatism Alteration of a recently crystallized igneous rock by its own last, water-rich liquid fraction trapped within the rock, generally by an impermeable chilled border.

automicrite Autochthonous *orthomicrite.*

automolite A dark-green to nearly black variety of gahnite.

automorphic A syn. of *idiomorphic.* The term *automorphisch* was proposed by Rohrback (1885, p.17-18) and has priority, but is less used than "idiomorphic". See also: *euhedral.*

automorphic-granular *idiomorphic-granular.*

autopiracy Capture of an upper part of a stream by its lower part, as by the cutting off of a meander, generally resulting in a shortening of its own course.

autopneumatolysis Autometamorphism involving the crystallization of minerals or the alteration of a rock by gaseous emanations originating in the magma or rock itself.

autopore Tubular autozooecium in Paleozoic bryozoans.

autopotamic Said of an aquatic organism adapted to living in flowing, fresh water. Cf: *eupotamic; tychopotamic.*

autoradiograph A type of *scan* of radioactivity, e.g. a neutron, X-ray, or gamma-ray photograph. Syn: *radiograph; radioautograph.*

autoregression A method in mathematical analysis, esp. in time series.

autosite An igneous rock similar in composition to kersantite but without feldspar.

autoskeleton The endoskeleton of sponges, consisting of spicules or spongin secreted by the cells. Cf: *pseudoskeleton.*

autotheca The largest tube of three produced at each budding in the development of a graptolithine colony. It may have contained a female zooid. Cf: *bitheca; stolotheca; metatheca.*

autotrophic Said of an organism that nourishes itself by utilizing inorganic material to synthesize living matter. Green plants and certain protozoans are autotrophic. Noun: *autotroph.* Cf: *heterotrophic.*

autozooecium The skeleton of a bryozoan autozooid.

autozooid (a) A fully formed octocorallian polyp with eight well developed tentacles and septs. It is the only type of polyp in monomorphic species and it is the major type in dimorphic species. Cf: *siphonozooid*. (b) A feeding bryozoan zooid.

autumn ice Sea ice in an early stage of formation and not yet affected by lateral pressure; it is relatively salty and crystalline in appearance.

Autunian European stage: Lower Permian (above Stephanian of Carboniferous, below Saxonian).

autunite (a) A lemon-yellow or sulfur-yellow, radioactive, tetragonal mineral: $Ca(UO_2)_2(PO_4)_2.10-12H_2O$. It is isomorphous with torbernite. Autunite is commonly a secondary mineral and occurs as tabular plates or in mica-like scales. Syn: *lime uranite; calcouranite*. (b) A group of isomorphous tetragonal minerals of general formula: $R^{+2}(UO_2)_2(X-O_4)_2.nH_2O$, where R =Ca, Cu, Mg, Ba, Na$_2$, and other elements, and X = P or As. The group includes minerals such as autunite, torbernite, uranocircite, saléeite, sodium autunite, zeunerite, uranospinite, novacekite, and kahlerite.

Auversian European stage: Eocene (above Lutetian, below Bartonian). Syn: *Ledian*.

auwai An Hawaiian term for a watercourse or channel, esp. one used for irrigation.

auxiliary adj. Said of an inflection (any lateral lobe or lateral saddle) of the ammonoid suture added later than the first two or three pairs; e.g. "auxiliary lobe" springing from the umbilical lobe and occurring between the second lateral saddle and the umbilicus (TIP, 1959, pt.L, p.18). Cf: *accessory [paleont]*. -n. A auxiliary lobe or an auxiliary saddle.

auxiliary contour *supplementary contour*.

auxiliary fault A minor fault abutting against a major one. Syn: *branch fault*.

auxiliary mineral Capillary water ringing the contact points of adjacent rock or soil particles in the zone of aeration (Smith, W.O., 1961, p.2). Cf: *funicular water; pellicular water; capillary condensation*. insignificant part of the rock's volume.

auxiliary plane The plane that is perpendicular to the fault plane; the direction perpendicular to the slip. It is determined from seismic data for earthquakes.

auxotrophic Said of a microscopic organism that requires certain specific nutrients.

available moisture *available water*.

available relief The total relief available for stream dissection in a given area, equal to the vertical distance between the height of the remnants of an original upland surface and the level at which grade is first attained by adjacent streams (Glock, 1932). Cf: *local relief*.

available water Water available to plants; the difference between field capacity and wilting point. Syn: *available moisture*.

avalanche A large mass of snow, ice, soil, or rock, or mixtures of these materials, falling or sliding very rapidly under the force of gravity. Velocities may sometimes exceed 500 km/hr. Avalanches can be classified by their content, e.g. snow and ice avalanches, debris avalanches, soil or rock avalanches.

avalanche bedding Steeply inclined bedding in barchan and related dune forms, produced by an avalanche of sand down the slip face of the dune.

avalanche blast A very destructive *avalanche wind* occurring when an avalanche is stopped abruptly, as when it falls vertically onto a valley floor.

avalanche breccia A breccia formed by a rockfall.

avalanche chute A trench or trough formed by avalanches through which the snow, ice, rock and/or soil debris passes. Cf: *avalanche track*.

avalanche cone The mass of material deposited where an avalanche has fallen, consisting of snow, ice, rock, and all other material torn away and carried along by the avalanche.

avalanche ripple mark Any asymmetric ripple mark having a steep lee-side dip at or near the angle of repose and migrating by a series of small avalanches down the lee slope (Imbrie & Buchanan, 1965, p.151). Cf: *accretion ripple mark*.

avalanche track The central channel-like corridor along which

an avalanche has moved; it may take the form of an open path in a forest with bent and broken trees, or an eroded surface marked by pits, scratches, and grooves. Cf: *avalanche chute*.

avalanche wind A high wind or rush of air produced in front of a large landslide or of a fast-moving dry-snow avalanche, and sometimes causing destruction at a considerable distance from the avalanche itself. See also: *avalanche blast*.

avalanching The sudden and rapid downward movement of an avalanche.

Avalonian orogeny An orogenic event near the end of Precambrian time along the southeastern border of North America, especially well shown in the type area where a granite dated at 575 m.y. is separated from the fossiliferous Lower Cambrian by a thick, late Precambrian sequence. Indications of the event, represented by radiometric dates ranging from slightly earlier to later, occur along the southeastern edge of the Appalachian orogenic belt, and in rocks beneath the Atlantic Coastal Plain sediments as far southwest as Florida. It was named by Lilly (1966) and Rodgers (1967) for the Avalon Peninsula, southeastern Newfoundland.

avanturine Var. (error) of *aventurine*.

aven *pothole* (b). Etymol: French.

aventurescence In certain translucent minerals, a display of bright or strongly colored reflections from included crystals. Examples are aventurine quartz and aventurine feldspar (sunstone).

aventurine (a) A translucent quartz spangled throughout with tiny inclusions of another mineral; a grayish, greenish, brown, or yellowish quartzite (massive granular quartz) that exhibits aventurescence from minute crystals, platelets, flakes, or scales of minerals such as green mica, ilmenite, hematite, and limonite. Syn: *aventurine quartz*. (b) *aventurine feldspar*.--The term is also used as an adjective in referring to the brilliant, spangled appearance of a glass or mineral containing gold-colored or shiny inclusions. Syn: *avanturine; adventurine*.

aventurine feldspar A variety of feldspar (oligoclase, albite, andesine, or adularia) characterized by a reddish luster produced by fiery, golden reflections or fire-like flashes of color from numerous, thin, small but visible, disseminated mineral particles (such as flakes of hematite) oriented parallel to structurally defined planes and probably formed by exsolution; specif. *sunstone*. Syn: *aventurine*.

aventurine glass *goldstone*.

aventurine quartz *aventurine*.

average *arithmetic mean*.

average discharge As used by the U.S. Geological Survey, the arithmetic average of all complete water years of record of discharge whether consecutive or not.

average igneous rock A theoretical rock whose chemical composition is believed to be similar to the average composition of the outermost layer of the Earth.

average level anomaly Gravity anomaly related to average topographic level in an area, usually of some fixed radius. Syn: *Putnam anomaly*.

average velocity [seis] In seismology, the ratio of the distance traversed along a ray path by a seismic pulse to the time required for that traverse. The average velocity is usually measured or expressed for a ray perpendicular to the reference datum plane.

average velocity [hydraul] (a) For a stream, discharge divided by the area of a cross section normal to the flow. (b) For ground water, the volume of ground water passing through a given cross-sectional area, divided by the porosity of the material through which it moves. Syn: *mean velocity*.

avezacite A plutonic rock intermediate in composition between pyroxenite and hornblendite, with amphibole in excess of pyroxene and with ilmenite constituting approximately 20 percent of the rock.

avicennite A black cubic mineral: Tl_2O_3.

49

avicularium A specialized zooid in cheilostomatous bryozoans, resembling a bird's head, characterized by a reduced polypide but strong muscles that operate a mandible-like operculum. Pl: avicularia.

aviolite A type of hornfels whose main constituents are mica and cordierite. Type locality: Monte Aviolo, Italian Alps. Cf: *astite; edolite.*

avlakogene A term introduced by Shatsky (1955) for a broad, intraplatform depression or graben. Etymol: Russian.

avogadrite An orthorhombic mineral: $(K,Cs)BF_4$.

avon A river. Etymol: Celtic.

Avonian *Dinantian.*

avulsion (a) A sudden cutting off or separation of land by a flood or by an abrupt change in the course of a stream, as by a stream breaking through a meander or by a sudden change in current whereby the stream deserts its old channel for a new one. Legally, the part thus cut off or separated belongs to the original owner. Cf: *accretion.* (b) Rapid erosion of the shore by waves during a storm (Wiegel, 1953, p.4).

awaruite A mineral consisting of a natural alloy of nickel and iron; *nickel-iron.*

awn *seta.*

axial angle *optic angle.*

axial canal (a) A longitudinal passageway penetrating columnals of an echinoderm and connecting with the body cavity, and generally but not invariably located centrally. (b) An intrapsicular cavity left by decay of an axial filament in a sponge.

axial compression In experimental work with cylinders, a compression applied parallel with the cylinder axis.

axial cross [paleont] The orthogonal crossing of the six axial filaments at the center of a hexactinellid sponge spicule. It is used esp. where some of the axial filaments are reduced to the area about the center, as in spicules with fewer than six rays.

axial dipole field A hypothetical terrestrial magnetic field, consisting of an ideal *dipole field* centered at the Earth's center and with its axis along the Earth's rotation axis. While the actual terrestrial magnetic field does not have this ideal form, it is hypothesized that it would, after averaging over thousands of years of secular variation.

axial elements The lengths (or ratios of the lengths) and angles which define the unit cell of a crystal.

axial figure An *interference figure* in which one optic axis is centered in the figure.

axial filament An organic fiber about which the mineral substance of a sponge-spicule ray is deposited.

axial filling A deposit of dense calcite developed in the axial region of some fusulinacean foraminifers and formed probably at the same time as excavation of tunnel or foramina and formation of chomata and parachomata (TIP, 1964, pt.C, p.58).

axial furrow (a) One of the two longitudinal grooves bounding the axis of a trilobite. Syn: *dorsal furrow.* (b) A longitudinal groove separating the median lobe or axis of a merostome from the pleural area.

axial increase A type of *increase* (budding of corallites) in coralla characterized by the appearance of dividing walls between newly formed corallites approximately in position of the axis of the parent corallite.

axial jet A flow pattern characteristic of *hypopycnal inflow*, in which the inflowing water spreads as a cone with an apical angle of about $20°$ (Moore, 1966, p.88). Cf: *plane jet.*

axial lobe The *axis* of a trilobite.

axial plane [cryst] (a) The plane of the optic axes of an optically biaxial crystal. (b) A crystallographic plane that includes two of the crystallographic axes; this use is rare.

axial plane [fold] *axial surface.*

axial-plane cleavage Cleavage which is closely related to the axial planes of folds in the rock, either being rigidly parallel to the axes, or diverging slightly on each flank (*fan cleavage*).

Most axial-plane cleavage is closely related to the minor folds seen in individual outcrops, but some is merely parallel to the regional fold axes. Most axial plane cleavage is also *slaty cleavage.* Cf: *bedding-plane cleavage.*

axial-plane separation *amplitude* [fold].

axial ratio The ratio of the lengths of the crystallographic axes of a crystal, stated in terms of one axis as unity.

axial section A slice bisecting a foraminiferal test in a plane coinciding with the axis of coiling and intersecting the proloculus.

axial septulum A secondary or tertiary septum located between primary septa of a foraminifer, its plane approximately parallel to the axis of coiling and thus observable in sagittal (equatorial), parallel, and tangential sections. See also: *primary axial septulum; secondary axial septulum.*

axial stream (a) The main stream of an intermontane valley, flowing in the deepest part of the valley and parallel to its longest dimension. (b) A stream that follows the axis of a syncline or anticline.

axial structure A collective term for various longitudinal structures in the axial region of a corallite, whether a solid or spongy rod-like columella or an axial vortex. See also: *clisiophylloid; aulos.*

axial surface A surface that connects the *axes* of each stratum in a fold; in a syncline, it is the *trough surface*, and in an anticline it is the *crest surface.* Syn: *axial plane.*

axial symmetry In structural petrology, fabric symmetry having one axis. Syn: *spheroidal symmetry.*

axial trace The intersection of the axis of a fold with the surface of the Earth or other given surface; the trend of the axis. Syn: *surface axis.*

axial vortex A longitudinal structure in the axial region of a corallite, formed by twisting together of the inner edges of major septa associated commonly with the transverse skeletal elements.

axil The distal angle formed between two plant parts, specif. a stalk and the stem from which it grows.

axil angle Acute angle as shown on a map between two confluent streams, measured upstream from their junction. Symbol: ξ. Syn: *entrance angle; stream-entrance angle.*

axillary A brachial plate supporting two crinoid arm branches.

axinellid Said of a sponge skeleton built of spiculofibers in which the component spicules are all directed obliquely outward from the axes of the fibers.

axinite A brown, violet, blue, green, or gray mineral: $(Ca,Mn,Fe)_3Al_2BSi_4O_{15}(OH)$. It may contain appreciable sodium. Axinite commonly forms glassy wedge-shaped triclinic crystals. Syn: *glass schorl.*

axiolite A spherulitic aggregate elongated along a central line or axis, to which the radial structure is developed at right angles; e.g. a spherulite in a rhyolite composed of welded glass fragments to whose outlines minute acicular crystals or fibers of feldspar are approximately perpendicular and radiate inward, or a subspherical oolith or pisolith in a carbonate sediment, around which acicular needles are axially grouped. Term proposed by Zirkel (1876, p.167). Syn: *axiolith.*

axiolith *axiolite.*

axiolitic Said of the structure of a rock in which axiolites are abundant; also said of a rock containing axiolites.

axiometer A device that permits accurate location and measurement of pebble and cobble axes by means of a clamp and a track-mounted caliper (Schmoll & Bennett, 1961).

axis [geomorph] (a) The central or dominant region of a mountain chain. (b) A line that follows the trend of large land forms, such as one following the crest of a ridge or mountain range, or of the bottom or trough of a depression.

axis [cryst] *crystal axis.*

axis [paleont] (a) The median lobe of a trilobite, consisting of the central, longitudinal, raised portion of the exoskeleton lying between the pleural regions, particularly of the thorax and the pygidium. Syn: *axial lobe.* (b) The central supporting structure of certain octocorals, such as a spicular and consolidated or unconsolidated structure in a gorgonian, or a horny structure (with more or less nonspicular calcareous matter) as in the order Pennatulacea. (c) A straight line with respect to which an invertebrate is radially or bilaterally symmetric; e.g. the oral-aboral axis of a corallite, or the axis formed by ambulacral plates in the sheath of radial water vessel in an asterozoan ray (TIP, 1966, pt.U, p.28). (d) An imaginary line through the apex of a gastropod shell and about which whorls of conispiral and discoid shells are coiled. (e) An imaginary line around which a spiral or cyclical shell of a protist is coiled, transverse to the plane of coiling.---Pl: axes.

axis [fold] In a fold, a line that connects the central points of each constituent stratum, from which its limbs bend; in a syncline, it is the *trough*, and in an anticline it is the *crest.*

axis culmination *culmination.*

axis of divergence The generally vertical or oblique line in coral septum from which trabeculae incline inward and outward. See also: *fan system.*

axis of symmetry *symmetry axis.*

axis of tilt A horizontal line through the perspective center of a lens, relative to which an aerial photograph has been tilted; it is perpendicular to the principal plane.

axoblast An individual *scleroblast* that produces the axis of certain octocorals, e.g. *Holaxonia.*

axopodium A semipermanent *pseudopodium,* typically present in radiolarian and heliozoan cells, consisting of an axial rod surrounded by a protoplasmic envelope. Pl: axopodia. Syn: *axopod.*

azimuth [seis] An angle, measured from north through east between the north meridian and the arc of the great circle connecting the earthquake epicenter and the receiver. When measured at the epicenter, it is called the azimuth from epicenter to receiver. When measured at the receiver, it is called the azimuth from receiver to epicenter or back azimuth.

azimuth [surv] (a) Orientation of a horizontal line as measured on an imaginary horizontal circle; the horizontal *direction* reckoned clockwise from the meridian plane of the observer, expressed as the angular distance between the vertical plane passing through a fixed point (the point of observation) and the poles of the Earth and the vertical plane passing through the observer and the object under observation. In the basic control surveys of U.S., azimuths are measured clockwise from south, a practice not followed in all countries. Cf: *bearing.* See also: *true azimuth; magnetic azimuth.* (b) *astronomic azimuth.*

azimuthal equal-area projection *Lambert azimuthal equal-area projection.*

azimuthal equidistant projection A map projection (neither equal-area nor conformal) in which all points are placed at their true distances and true directions from the central point of the projection. Any point on the globe may be placed at the center, and a straight line radiating from this point to any other point represents the shortest distance (a great circle in its true azimuth from the center) and its length can be measured to scale. Syn: *zenithal equidistant projection.*

azimuthal projection (a) A map projection in which a portion of the sphere is projected upon a plane tangent to it at the pole or any other point (which becomes the center of the map) and on which the azimuths (directions) of all lines radiating from the central point to all other points are the same as the azimuths of the corresponding lines on the sphere. Distortion at the central point is zero and scale distortions are generated radially from the central point. All great circles through the central point are straight lines intersecting at true angles. (b) A similar projection used in structural petrology.--Syn: *zenithal projection.*

azimuthal survey A resistivity or induced-polarization survey in which an area is traversed by a voltage-measuring electrode pair along azimuths away from a fixed current electrode which may be in a drill hole or in contact with a metallic ore. The second electrode is placed at infinity. Syn: *radial array.*

azimuth angle (a) The horizontal angle, less than 180 degrees, between the plane of the celestial meridian and the vertical plane containing the observation point and the observed object (celestial body), reckoned from the direction of the elevated pole. In the astronomic triangle (composed of the pole, the zenith, and the star), it is the spherical angle at the zenith. (b) An angle in triangulation or in a traverse, through which the computation of azimuth is carried.

azimuth circle (a) An instrument for measuring azimuth, consisting of a horizontal graduated circle divided into 360 major divisions; e.g. one attached to a compass to show magnetic azimuths, or one having a telescope for accurate measurement of differences of azimuth. (b) *vertical circle.*

azimuth compass A magnetic compass, supplied with vertical sights, for measuring the angle that a line on the Earth's surface, or the vertical circle through a heavenly body, makes with the magnetic meridian; a compass used for taking the magnetic azimuth of a celestial body.

azimuth line A term used in radial triangulation for a radial line from the principal point, isocenter, or nadir point of a photograph, representing the direction to a similar point on an adjacent photograph in the same flight line.

azimuth mark A mark set at a significant distance from a triangulation or traverse station to mark the end of a line for which the azimuth has been determined and to serve as a starting or reference azimuth for later use.

Azoic (a) That earlier part of Precambrian time for which, in the corresponding rocks, there is no trace of life. Cf: *Protozoic.* (b) The entire *Precambrian.*

azoic Said of an environment that is devoid of life.

azonal peat *local peat.*

azonal soil In early U.S. classification systems, one of the three *soil orders* that embraces soils that lack well-developed horizons and resemble the parent material; also, any soil belonging to the azonal soil order. Cf: *intrazonal soil; zonal soil; Entisol.* Syn: *immature soil.*

azonate Said of spores without a zone or a similar (usually equatorial) extension.

azoproite A mineral of the ludwigite group: $(Mg,Fe^{+3},Ti)BO_5$.

azulite A translucent pale-blue variety of smithsonite often found in large masses (as in Arizona and Greece).

azurchalcedony *azurlite.*

azure quartz A blue variety of quartz; specif. *sapphire quartz.*

azure stone A term applied to lapis lazuli and to blue minerals such as lazulite and azurite.

azurite (a) A deep-blue to violetish-blue monoclinic mineral: $Cu_3(CO_3)_2(OH)_2$. It is an ore of copper and is a common secondary mineral associated with malachite in the upper (oxidized) zones of copper veins. Syn: *chessylite; blue copper ore; blue malachite.* (b) A semiprecious stone derived from compact azurite and used chiefly for ornamental objects. (c) A trade name for a sky-blue gem variety of smithsonite.

azurlite A variety of chalcedony colored blue by chrysocolla and used as a gemstone. Syn: *azurchalcedony.*

azurmalachite An intimate mixture or intergrowth of azurite and malachite, usually occurring massive and concentrically banded, and used as an ornamental stone.

azygous basal plate The smallest of the three plates of the basalia of a blastoid, normally located in the AB interray but sometimes in the DE interray. Cf: *zygous basal plate.*

azygous node A special kind of cusp located directly above the basal cavity of certain conodonts (such as *Palmatolepis* and *Panderodella*).

B

babefphite A mineral: $BaBe(PO_4)(O,F)$.

Babel quartz A variety of crystalline quartz so named for its fanciful resemblance to the successive tiers of the Tower of Babel. Syn: *Babylonian quartz.*

babingtonite A greenish-black triclinic mineral: $Ca_2(Fe^{+2},Mn).Fe^{+3}Si_5O_{14}(OH)$.

bacalite A variety of amber from Baja California, Mexico.

bache A term used in England for the valley of a small stream.

bacillite A rodlike crystallite composed of a group of parallel longulites.

back [meteorol] v. To change wind direction counterclockwise in the northern hemisphere and clockwise in the southern hemisphere.

back [eco geol] That part of a lode nearest the surface, relative to the mine workings; also, the top or roof of a mine passageway.

back bay A small, shallow bay into which coastal streams drain and which is connected to the sea through a pass between barrier islands, as along the coast of Texas. Cf: *front bay.*

backbeach *backshore.*

back bearing (a) A *bearing* along the reverse direction of a line, such as one resulting from a backsight; the reverse or reciprocal of a given bearing. If the bearing of line AB is N 58° W, the back bearing (bearing of line BA) is S 58° E. Syn: *reverse bearing; reciprocal bearing.* (b) A term used by the U.S. Public Land Survey system for the reverse direction of a line as corrected for the curvature of the line from the forward bearing at the preceding station.

backbone A ridge serving as the main axis of a mountain; the principle mountain ridge, range, or system of an area.

backdeep A syn. of *epieugeosyncline,* so named because of its relative position, away from the craton (Aubouin, 1965, p.34).

backfill (a) Earth or other material used to replace material removed during construction or mining, such as stones and gravel used to fill pipeline trenches and other excavations or placed behind structures such as bridge abutments, or waste rock used to support the roof after removal of ore from a stope. Also, material (such as sand or dirt) placed between an old structure and a new lining, as in a shaft or tunnel. (b) The process of refilling, with backfill, an excavation, a mine opening, or the space around a foundation.

background [geochem] The abundance of an element, or any chemical property of a naturally occurring material, in an area in which the concentration is not anomalous (Hawkes, 1957, p.336).

background [radioactivity] *background radiation.*

background radiation The radiation of the environment, e.g. from cosmic rays and from the Earth's naturally radioactive substances. Also, any radiation that is not part of a controlled experiment. Syn: *background.*

backhand drainage Drainage in which the general course of the tributaries on both sides is opposite to that of the main stream (Eakin, 1916, p.17).

backland [geomorph] The lowland along either side of a river, behind the natural levee; the part of a flood plain extending from the base of a valley slope and separated from the river by the natural levee. Syn: *back lands.*

backland [tect] *hinterland.*

back lead A *lead* or deposit of coastline sands above the high-water mark.

backlimb The less steep of the two limbs of an asymmetrical, anticlinal fold. Cf: *forelimb.*

backlimb thrust A thrust fault through the backlimb or more

gently dipping side of an anticline. Cf: *forelimb thrust.*

back marsh The low, wet, poorly drained areas of an alluvial flood plain.

back radiation *counterradiation.*

back reef The landward side of a reef, including the area and the contained deposits between the reef and the mainland; the terrestrial deposits connecting the reef with the land; the reef flat. The term is often used adjectivally to refer to the restricted lagoon behind a barrier reef, such as the "back-reef facies" of lagoonal deposits. Cf: *fore reef.* Also spelled: *backreef.*

back-reef moat *boat channel.*

backrush *backwash.*

back scattering *backward scatter.*

backset bed A cross-bed that dips against the direction of flow of a depositing current; e.g. an inclined layer of sand deposited on the gentle windward slope of a transverse dune, and often trapped by tufts of sparse vegetation, or a glacial deposit that formed (as the ice retreated) on the rear slope of an apron or sand plain or at the front of an esker, and consequently dipping toward the retreating ice.

backset eddy A small current revolving in the reverse direction to that of the main ocean current.

backshore (a) The upper or inner, usually dry and narrow zone of the shore or beach, lying between the high-water line of mean spring tides and the coastline, and acted upon by waves or covered by water only during exceptionally severe storms or unusually high tides. It is essentially horizontal or slopes landward, and is divided from the *foreshore* by the crest of the most seaward berm. (b) The area lying immediately at the base of a sea cliff. (c) *berm.*---Syn: *backbeach.*

backshore terrace A wave-built terrace on the backshore of a beach; a *berm.*

backsight (a) A sight or bearing on a previously established survey point (other than a closing or check point), taken in a backward direction and made in order to verify its elevation. (b) A reading taken on a level rod held in its unchanged position on a survey point of previously determined elevation when the leveling instrument has been moved to a new position. It is used to determine the height of instrument prior to making a foresight. Syn: *plus sight.*---Abbrev: B.S Ant: *foresight.*

back slope (a) A syn. of *dip slope;* the term is used where the angle of dip of the underlying rocks is somewhat divergent from the angle of the land surface. (b) The slope at the back of a scarp; e.g. the gentler slope of a cuesta or of a fault block. It may be unrelated to the dip of the underlying rocks.--Also spelled: *backslope.* Cf: *scarp slope.*

backswamp A swampy or marshy, depressed area developed on a flood plain with poor drainage due to the natural levees of the river.

backswamp deposits Thin layers of silt and clay deposited in the flood basin behind the natural levees of a river.

backswamp depression A low, usually swampy area adjacent to a leveed river. Syn: *levee-flank depression.*

back thrusting Thrust faulting in an orogenic belt, with the direction of displacement towards its interior, or contrary to the general structural trend.

backwall *headwall.*

backward scatter The scattering of radiant energy into the hemisphere of space bounded by a plane normal to the direction of the incident radiation and lying on the same side as the incident ray; the opposite of a *forward scatter.* It is sometimes referred to as *back scattering.* Atmospheric backward scatter depletes 6 to 9 percent of the incident solar beams before it reaches the Earth's surface. In radar usage, backward scatter generally refers to that radiation scattered back toward the source.

backwash The seaward return of water running down the foreshore of a beach following an *uprush* of waves; also, the seaward-flowing mass of water so moved. Cf: *uprush.* Syn: *backrush.*

backwash mark A term used by Johnson (1919, p.517) for a "criss-cross ridge" developed on a beach slope by the return flow of the uprush.

backwash ripple mark A term used by Kuenen (1950, p.292) for a broad, flat ripple mark between narrow, shallow troughs, formed on a beach by backrush above the level of maximum wave retreat; its crest is parallel to the shoreline.

backwasting (a) Wasting that causes a slope to retreat without changing its declivity. (b) The recession of the front of a glacier.---Cf: *downwasting*.

backwater (a) Water that is retarded, backed up, or turned back in its course by an obstruction (such as a dam), an opposing current, or the movement of the tide; e.g. the water in a reservoir or the water obtained at high tide to be discharged at low tide. Also, the resulting *backwater effect*. (b) A body of currentless or relatively stagnant water, parallel to a river and usually fed from it through a single channel at the lower end by the back flow of the river. Loosely, any tranquil body of water joined to a main stream but little affected by its current, such as the water collected in side channels or flood-plain depressions after it overflowed the lowland. (c) A creek, arm of the sea, or series of connected lagoons, usually parallel to the coast, separated from the sea by a narrow strip of land but communicating with it through barred outlets. (d) A backward current of water. Also, the motion of water that is turned back; a *backwash*.

backwater curve (a) The form of the water surface along a longitudinal profile, assumed by a stream above the point where depth is made to exceed the normal depth by a constriction or obstruction in the channel (ASCE, 1962). Cf: *drop-down curve*. (b) A generic term for all *surface profiles* of water; esp. *flow profile*.

backwater effect The upstream increase in height of the water surface of a stream, produced when flow is retarded above a temporary obstruction (such as a dam) or when the main stream overflows low-lying land and backs up water in its tributaries. The effect is also characterized by an expansion in width of the body of water and by a slackening in the current. Syn: *backwater*.

backwearing Erosion that causes the parallel retreat of an escarpment or of the slope of a hill or mountain, or the sideways recession of a slope without changing its declivity; a process contributing to the development of a pediment or pediplain. Cf: *downwearing*.

backweathering Weathering that contributes to slope retreat.

bacon [speleo] A type of dripstone in a cave that projects from the walls and roof in thin, translucent sheets and is characterized by parallel colored bands. Cf: *blanket* [speleo]. Syn: *bacon-rind drapery*.

bacon [sed] A quarryman's term used in Portland (southern England) for *beef*. Syn: *horseflesh*.

bacon-rind drapery *bacon*.

bacon stone (a) An old name for a variety of steatite, alluding to its greasy appearance. See also: *speckstone*. (b) A term used in Bristol, Eng., for calcite colored with iron oxide.

bacteriogenic Said of ore deposits formed by the action of anaerobic bacteria, by the reduction of sulfur or the oxidation of metals (Park & MacDiarmid, 1964, p.105-107). See also: *iron bacteria; sulfur bacteria*.

bacterium A simple, microscopic plant consisting of a single cell without an evident nucleus. It ranges from the Precambrian.

bactritoid Any straight cephalopod belonging to the subclass Bactritoidea, characterized by a shell of relatively uniform shape, having a small globular to ovate protoconch and a much larger orthoconic or cyrtoconic conch. Bactritoids have been classified both as nautiloids and as ammonoids by some workers. Their stratigraphic range is Ordovician to Permian.

baculate Said of sculpture of pollen and spores consisting of bacula.

baculite A crystallite that appears as a dark rod.

baculum One of the tiny rods, varying widely in size and either isolated or clustered, that make up the ektexine sculpture of pollen or spores. Pl: *bacula*.

baddeleyite A colorless, yellow, brown, or black monoclinic mineral: ZrO_2. It may contain some hafnium, titanium, iron, and thorium.

badenite A steel-gray mineral: $(Co,Ni,Fe)_3(As,Bi)_4$ (?).

badlands Extremely rough, high, narrowly and steeply gullied topography in arid or semiarid areas that are horizontally bedded and have a dry, loose soil. Infrequent, heavy showers cause unchecked erosion of the vegetation-free landscape. The term is esp. applied to a specific area of South Dakota which was called "mauvaises terres" by the early French fur traders.

Baer's law *von Baer's law*.

bafertisite An orthorhombic mineral: $BaFe_2TiSi_2O_9$.

Bagnold dispersive stress A shear stress between two layers in a fluid, caused by the impact between cohesionless particles that are free to collide with each other during current flow in the absence of applied body force. The stress increases as the diameter squared and the large particles, subjected to highest stress, are forced to the bed surface where the stress is zero. The term was used by Leopold et al (1966, p. 213) and named for Ralph A. Bagnold (b. 1896), British geographer, who quantified the influence of collective mutual collisions on current flow (Bagnold, 1956). The term "Bagnold effect" was proposed by Sanders (1963, p.174) for the same phenomenon. Syn: *dispersive stress*.

baguette A *step cut* used for small, narrow, rectangularly shaped gemstones, principally diamonds. Also, the gem so cut. Etymol: French, "rod".

bahada Anglicized var. of *bajada*.

bahamite A term proposed by Beales (1958, p.1851-1852) for a shallow marine deposit that consists of limestone grains closely resembling the predominant deposits (described by Illing, 1954) now accumulating in the interior of the Bahama Banks. It is very pure, generally fine-grained, massively bedded, widely extensive, and without abundant fossils; the grains are accretionary and commonly composite. The term applies to sediments that accumulate under conditions of coastal shoaling or on offshore banks, and does not imply that the limestone was formed under genetic conditions exactly comparable with those prevailing on present-day Bahama Banks. See also: *grapestone*.

bahiaite A pyroxenite containing orthopyroxene, amphibole, olivine, and a small amount of ceylonite.

bahr A body of water as found in the Saharan region; esp. a deep natural spring, often in the form of a small, crater-shaped lake of great depth, as in some oases of eastern Algeria. Etymol: Arabic, "sea". Pl: *bahar; bahrs*.

Baikalian orogeny A name widely used throughout the U.S.S.R. for orogenies late in Precambrian time. It is named after Lake Baikal in Siberia. Different phases have been recognized, some extending into the Early Cambrian, all of which are defined by relations within the many late Precambrian stratified sequences of the Soviet Union.

baikalite A dark-green variety of diopside containing iron and found near Lake Baikal, U.S.S.R.

baikerinite A thick, tarry hydrocarbon which makes up about one third of *baikerite* and from which it is separated by alcohol.

baikerite A variety of *ozocerite*. See also: *baikerinite*.

bailer A long, hollow, steel cylindrical container or pipe with a valve at the bottom (for admission of fluid), attached to a wire line and used in cable-tool drilling for recovering and removing water, cuttings, and mud from the bottom of a borehole or well. It is run inside the casing. See also: *sand pump*.

bajada A broad, continuous *alluvial slope* or gently inclined detrital surface extending along and from the base of a mountain range out into and around an inland basin, formed by the lateral coalescence of a series of separate but confluent *alluvial fans*, and having an undulating character due to the convexities of the component fans; it occurs most commonly in semiarid and desert regions, as in the SW U.S. A bajada is a surface of deposition, as contrasted with a *pediment* (a surface of erosion that resembles a bajada in surface form), and its top often merges with a pediment. Originally, the term was used in New Mexico for the gentler of the two slopes of a cuesta. Etymol: Spanish, "descent, slope". Pron: ba-*ha*-da. Syn: *bahada; apron; alluvial apron; mountain apron; fan apron; debris apron; alluvial plain; compound alluvial fan; piedmont alluvial plain; piedmont plain; waste plain; piedmont slope; gravel piedmont; alluvial bench.*

bajada breccia A term used by Norton (1917, p.167) for a wedge-shaped, imperfectly stratified accumulation of coarse, angular, poorly sorted rock fragments mixed with mud, formed in an arid region by an intermittent stream or a mudflow containing considerable water. Cf: *fanglomerate.*

bajir A term applied in the deserts of central Asia to a lake occupying a flat-bottomed basin separating sand hills or dunes.

Bajocian European stage: Middle Jurassic (above Toarcian, below Bathonian).

bakerite A mineral: $Ca_4B_4(BO_4)(SiO_4)_3(OH)_3.H_2O$. It is a variety of datolite occurring in white, compact, nodular, fine-grained masses resembling marble or unglazed porcelain.

baking In geology, the hardening of rock material by heat from magmatic intrusions or lava flows. Prolonged baking leads to contact metamorphic effects. Cf: *caustic metamorphism.*

Bala European stage: Middle and Upper Ordovician. It comprises the Caradocian and Ashgillian.

balaghat A term used in India for a tableland situated above mountain passes.

balance The change in mass (the difference between total accumulation and gross ablation) of a glacier over some defined interval of time, determined either as a value at a point, an average over an area, or the total mass change for the glacier. Units of millimeters, meters, or cubic meters of water equivalent are normally used but kilograms per square meter or kilograms are used by some. Syn: *mass balance; mass budget; regimen* [*glaciol*]; regime [*glaciol*]; economy. Cf: *net balance; annual balance.*

balanced rock (a) A large rock resting more or less precariously on its base, formed by weathering and erosion in place. See also: *pedestal rock; rocking stone.* (b) *perched block.*

balance of nature "A state of equilibrium in nature due to the constant interaction of the whole biotic and environmental complex, interference with this equilibrium (as by human intervention) being often extremely destructive" (Webster, 1967, p.165).

balance rate The rate of change of mass of a glacier at any time, equal to the difference between accumulation rate and ablation rate.

balance year The period from the time of minimum mass in one year to the time of minimum mass in the succeeding year for a glacier; the period of time between the formation of one *summer surface* and the next. See also: *net balance.* Syn: *budget year.*

balancing The process of systematically distributing corrections through any traverse to eliminate the error of closure and to obtain an adjusted position for each traverse station. Also known as "balancing a survey".

balas ruby A pale rose-red or orange gem variety of spinel, found in Badakhshan (or Balascia) province in northern Afghanistan. See also: *ruby spinel.* Syn: *balas; ballas.*

balavinskite A mineral: $Sr_2B_6O_{11}.4H_2O$.

bald n. A local term, esp. in southern U.S., for an elevated, grassy area, as a mountain top or high meadow, that is devoid of trees.

bald-headed Said of an anticline, the crust of which has been eroded.

baldite The hypabyssal equivalent of analcime-bearing basalt, composed of pyroxene phenocrysts in a groundmass of analcime, augite, and iron oxide.

balk A low ridge of earth that marks a boundary line (ASCE, 1954, p.20).

ball [coal] *coal ball.*

ball [coast] A syn. of *longshore bar.* The term is used in the expression *low and ball,* but the elongate character of the bar is "not well indicated by 'ball' which suggests a round object" (Shepard, 1952, p.1909).

ball [sed] A primary sedimentary structure consisting of a spheroidal mass of material; e.g. an armored mud ball, a slump ball, and a lake ball.

ball-and-pillow structure A primary sedimentary structure found in sandstones and some limestones, characterized by hemispherical or kidney-shaped masses resembling balls and pillows, and commonly attributed to subaqueous slump or sliding or to foundering; e.g. a *flow roll* and a *pseudonodule.* See also: *pillow structure; ball structure.*

ball-and-socket jointing In basalt columns, cross-jointing surfaces that are concave either upward or downward. Syn: *cup and ball jointing.*

ballas (a) A dense, hard, globular or nearly spherical aggregate of minute diamond crystals, having a confused radial or granular structure, and whose lack of through-going cleavage planes imparts a toughness that makes it useful as an *industrial diamond.* Cf: *bort; carbonado.* (b) A term incorrectly applied to a rounded, single-crystal form of diamond. (c) *balas ruby.*

ballast (a) Broken stone, gravel, or other heavy material used to supply weight in a ship and therefore improve its stability or control the draft. Jettisoned ballast may be found in samples of marine sediments. Also, similar material used to supply weight in equipment for use in lunar-gravity studies. (b) Rough, unscreened gravel, sand, and broken stone used as a foundation for roads, esp. that laid in the roadbed of a railroad to provide a firm surface for the track, to hold the track in line, and to facilitate drainage.

ball clay A highly plastic, sometimes refractory clay, commonly characterized by the presence of organic matter, having unfired colors ranging from light buff to various shades of gray, and used as a bonding constituent of ceramic wares; *pipe clay.* It has high wet and dry strength, long vitrification range, and high firing shrinkage. Ball clay is so named because of the early English practice of rolling the clay into balls weighing about 13-22 kg (30-50 lb) and having diameters of about 25 cm (10 in.).

ball coal Coal occurring in spheroidal masses that are probably formed by jointing. Cf: *coal ball.* Syn: *pebble coal.*

balled-up structure A term used by O.T. Jones (1937) for a knot (several centimeters to a few meters in diameter) of highly contorted silty material lying isolated in mud and produced by subaqueous slump. See also: *ball structure.*

ball ice Sea ice, either frazil ice or pancake ice, consisting of numerous soft, spongy, floating spheres (2.5-5 cm in diameter) shaped by the waves, and usually occurring in belts.

ball ironstone A sedimentary rock containing large, argillaceous nodules of ironstone.

ballistic magnetometer A type of magnetometer that uses the transient voltage induced in a coil when either the magnetized specimen or the coil is moved relative to each other.

ball jasper (a) Jasper showing a concentric banding of red and yellow. (b) Jasper occurring in spherical masses.

ballon A rounded, dome-shaped hill, formed either by erosion or by uplift. Etymol: French, "balloon".

ballstone (a) A nodule or large rounded lump of rock in a stratified unit; specif. an ironstone nodule in a coal measure.

See also: *iron ball*. Syn: *ball*. (b) A large (up to 20 m in length), more or less crystalline concretion or lenticular mass of fine, unstratified limestone surrounded by a matrix of calcareous shale or impure bedded limestone, and often containing corals in the position of growth. Examples varying greatly in size occur in the Wenlockian (Middle Silurian) limestones of Shropshire, Eng. Syn: *crog ball; woolpack*.

ball structure (a) A ball-shaped primary sedimentary structure characteristic of sandstones and some limestones; e.g. a slump ball. See also: *ball-and-pillow structure; balled-up structure*. (b) The structure of ball coal.

bally A term used in northern California for a mountain. It is thought to be a corruption of "buli", an American Indian term for mountain. Syn: *bolly*.

ballycadder A syn. of *icefoot*. Variant and shortened forms: *bellicatter; catter; cadder*.

balm A concave cliff or precipice, forming a shelter beneath the overhanging rock; a cave. Etymol: Celtic.

balneology The science of the healing qualities of baths, esp. with natural mineral waters.

balsam bog A *sphagnum bog* which also contains balsam fir.

baltimorite A grayish-green and silky, fibrous, or splintery serpentine mineral found near Baltimore, Md.; *antigorite*.

bambollaite A mineral: $CuTeSe_2$.

banakite A basaltic rock composed of olivine and augite phenocrysts in a groundmass of labradorite with orthoclase rims, olivine, augite, some leucite, and possible quartz. Banakite grades into *shoshonite* with an increase in the olivine and augite and with less sanidine, and into absarokite with more olivine and augite. Its name is derived from the Bannock (or Robber) Indians.

banalsite A mineral of the feldspar group: $BaNa_2Al_4Si_4O_{16}$.

banana hole A term used in the Bahamas for a sinkhole or doline, in which it is customary to cultivate bananas and sugar cane. This feature, when drowned as on the Bahama Banks, is known as a *blue hole*.

banatite A quartz or augite diorite containing orthoclase. The term has been variously defined since its original usage and is now obsolescent (Johannsen, 1939, p.242).

banco A term applied in Texas to the part of a stream channel or flood plain cut off and left dry by a change in the stream course; an oxbow lake. Etymol: Spanish, "sandbank, shoal, bench".

band [phys] A selection of wavelengths.

band [stratig] (a) A thin stratum with a distinctive lithology or fossil content; esp. a lamina that is conspicuous because it differs in color from adjacent layers. (b) A well-defined, widespread, thin stratum that may or may not be fossiliferous but that is useful in correlating strata; e.g. a *marine band*. (c) A deprecated term for any bed or stratum of rock (BSI, 1964, p.5).

band [glaciol] *glacier band*.

band [coal] A thin layer of shale, slate, or other rock material interstratified with coal; esp. *dirt band*.

bandaite A dacite containing labradorite or bytownite.

banded Said of a vein having alternating layers of ore that differ in color and texture but that may or may not differ in mineral composition, e.g. banded iron ore. Cf: *ribbon* [ore dep].

banded agate An *agate* whose various colors (principally different tones of gray, but also white, pale and dark brown, bluish, and other shades) are arranged in delicate parallel alternating bands or stripes of varying thicknesses. The bands are sometimes straight but more often wavy or zigzag and occasionally concentric circular; they may be sharply demarcated or grade imperceptibly into one another. Banded agate is formed by a deposit of silica (from solutions intermittently supplied) in irregular cavities in rocks, and it derives its concentric pattern from the irregularities of the walls of the cavity. Cf: *onyx*.

banded coal Heterogeneous coal, containing bands of varying luster. Banded coal is usually bituminous although banding occurs in all ranks of coal. Cf: *banded ingredients*. See also: *bright-banded coal; dull-banded coal; intermediate coal*. Syn: *common-banded coal*.

banded constituents *banded ingredients*.

banded differentiate Igneous rock that has bands of differing chemical composition, usually alternation of two types; a *layered intrusion*. The structure has been attributed to rhythmic crystal settling during convection.

banded gneiss A regularly layered metamorphic or composite rock with alternating layers of different composition and/or texture. Thickness of individual layers is usually not more than a few meters (Dietrich, 1960, p.36).

banded hematite quartzite A term used in India and Australia for *iron formation*. See also: *banded quartz-hematite*.

banded ingredients Vitrain, clarain, fusain, and durain as they appear as macroscopically visible and separable bands of varying luster in *banded coal*. Syn: *banded constituents; primary-type coal; rock type*.

banded ironstone A term used in South Africa for *iron formation* consisting essentially of iron oxides and chert occurring in prominent layers or bands of brown or red and black. This usage of the term *ironstone* is at variance with that applied in the U.S. and elsewhere.

banded peat Peat that consists of alternating bands of plant debris and sapropelic matter. Cf: *mixed peat; marsh peat*.

banded quartz-hematite A syn. of *itabirite*. See also: *banded hematite quartzite*.

banded structure [petrology] An outcrop feature developed in igneous and metamorphic rocks as a result of alternation of layers, stripes, flat lenses, or streaks differing conspicuously in mineral composition and/or texture. See also: *banding* [ign]; *banding* [meta].

banding [meta] A *banded structure* of metamorphic rocks consisting of nearly parallel bands having different textures or minerals or both. It may be produced by incomplete segregation of constituents during recrystallization or may be inherited in part from bedding in sediments or from layering in igneous rocks. Cf: *ribbon*.

banding [ign] The appearance of *banded structure* in an outcrop of igneous rock. It may be produced by flow of heterogeneous material (such as *flow banding* of rhyolites or *primary gneissic banding* of granitic rocks) or by successive deposition of layers of different materials (as in mafic plutonic rocks). Although the term strictly describes the appearance of a two-dimensional feature on a surface exposure, it "should perhaps be kept for any streaky or roughly planar heterogeneity in igneous rocks, whatever its origin" (Wager & Brown, 1967, p.5).

banding [sed] Thin bedding produced by deposition of different materials in alternating layers and conspicuous in a cross-sectional appearance of laminated sedimentary rocks; e.g. *ribbon banding*.

banding [glaciol] The occurrence of a layered structure in glacier ice, due to alternating layers of coarse-grained and fine-grained ice or bubbly and clear ice. Syn: *foliation*.

band-pass filter A *filter* in which frequency components outside of a given range (the pass band) are attenuated. It is the inverse of the *band-reject filter*.

band-reject filter A *filter* in which frequency components within a given range are attenuated, all others being passed. It is the inverse of the *band-pass filter*. Syn: *notch filter*.

band spectrum A spectrum that appears to be a number of bands because the array of intensity values in the spectrum occurs over broad ranges of wavelengths of the ordering variable. The term is also applied to cases where lines in a *line spectrum* are grouped closely together and cannot be resolved by available instruments, so that the groups of lines appear as bands. An optical band spectrum arises mainly in molecular transitions. Cf: *continuous spectrum*.

bandwidth (a) The range of frequency in an antenna within which its performance, in respect to some characteristic, conforms to a specified standard. (b) The least frequency interval, in a wave, outside of which the power spectrum of a time-varying quantity is elsewhere less than some specified fraction of its value at a reference frequency. This permits the spectrum to be less than the specified fraction within the interval. Unless otherwise stated, the reference frequency is that at which the spectrum has its maximum value.

bandylite A dark-blue tetragonal mineral: $Cu_2(B_2O_4)Cl_2.4H_2O$.

bangar *bhangar*.

bank [eng] (a) The edge of a cut or a fill. (b) An obsolete term for earthwork.

bank [coast] (a) An *embankment*; a *sandbank*. (b) A long, narrow island along the Atlantic coast of the U.S., composed of sand, and forming a barrier between the inland lagoon or sound and the ocean. (c) An obsolete term for a *seacoast*. (d) The rising ground bordering a sea.

bank [geomorph] (a) A steep slope or face (such as on a hillside), usually of sand, gravel, or other unconsolidated material, and only rarely of bedrock. (b) A term used in northern England and in Scotland for a hill or a hillside. (c) A term used in South Africa for a moderately high (up to 500 m) scarp consisting of resistant rock layers and forming a high hill or a low mountain, often occurring in groups of two or more with broad longitudinal valleys in between. The term is most often used in the plural: *banke*. Etymol: Afrikaans.

bank [mining] A coal deposit; the surface or face of a coal deposit that is being worked.

bank [oceanog] A relatively flat-topped elevation of the sea floor at shallow depth (generally less than 200 m), typically on the continental shelf or near an island.

bank [lake] (a) The sharply rising ground, or abrupt slope, bordering a lake. (b) The scarp of a littoral shelf of a lake. (c) Shoal bottom of a lake.

bank [snow] *snowbank*.

bank [streams] The sloping margin of, or the ground bordering, a stream, and serving to confine the water to the natural channel during the normal course of flow. It is best marked where a distinct channel has been eroded in the valley floor, or where there is a cessation of land vegetation. A bank is designated as right or left as it would appear to an observer facing downstream.

bank [sed] A limestone deposit consisting of nonfragmental skeletal matter, formed in place by organisms (such as crinoids and brachiopods) that lack the ecologic potential to erect a rigid, wave-resistant structure (Nelson et al, 1962, p.242). It is thinner than, and lacks the structural framework of, an organic *reef*. See also: *marine bank*. Syn: *organic bank*.

bank atoll *pseudoatoll*.

bank caving The slumping or sliding into a stream channel of masses of sand, gravel, silt, or clay, caused by a highly turbulent current undercutting or undermining the channel wall on the outside of a stream bend.

banke Plural of *bank*, an escarpment in South Africa.

banker An Australian term for a stream flowing full to the top of its banks.

banket A general term for a compact, siliceous conglomerate of vein-quartz pebbles of about the size of a pigeon's egg, embedded in a quartzitic matrix. The term was originally applied in the Witwatersrand area of South Africa to the mildly metamorphosed gold-bearing conglomerates containing muffin-shaped quartz pebbles and resembling an almond cake made by the Boers. Etymol: Afrikaans, "a kind of confectionary".

bankfull discharge The discharge at *bankfull stage*.

bankfull stage The elevation of the water surface of a stream flowing at *channel capacity*. Discharge at this stage is called *bankfull discharge*.

bank gravel Gravel found in natural deposits, usually more or less intermixed with finer material (such as sand, clay, or combinations thereof). Syn: *pit run*.

bank-inset reef A coral reef situated on a submarine flat (such as the shelf of a continent or island, or an offshore bank), well within its locally unrimmed outer margin (Kuenen, 1950, p.426). Cf: *bank reef*.

bank reef Any large reef growth, generally irregular in shape, developed over submerged highs of tectonic or other origin, and more or less completely surrounded by water too deep to support the growth of reef-forming organisms (Henson, 1950, p.227); a *bank-inset reef*. See also: *shoal reef*.

bankside The slope of a stream bank.

bank stability The quality of permanence or resistance to change in the slope and contour of the bank of a stream. It can be attained by benching, growth of vegetation, and artificial protections such as masonry walls, drainage systems, and fences. See also: *slope stability*.

bank storage Water absorbed and retained in permeable material adjacent to a stream during periods of high water and returned as effluent seepage or flow during periods of low water. Syn: *lateral storage*.

banner bank *tail* [coast].

bannisterite A mineral: $(Na,K)(Mn,Fe,Al)_5(Si,Al)_6O_{15-}(OH)_5.2H_2O$.

banquette An embankment at the toe of the land side of a levee, constructed to protect the levee from sloughing off when saturated with water.

baotite A tetragonal mineral: $Ba_4(Ti,Nb)_8Si_4O_{28}Cl$.

bar [eco geol] (a) Any band of hard rock, such as a vein or dike, crossing a lode. (b) A hard band of barren rock crossing a stream bed. (c) A mass of inferior rock in a workable deposit of granite. (d) A fault across a coal seam or orebody. (e) A banded ferruginous rock; specif. *jasper bar, jaspilite*.

bar [coast] A generic term for any of various elongate offshore ridges, banks, or mounds of sand, gravel, or other unconsolidated material, submerged at least at high tide, and built up by the action of waves or currents on the water bottom esp. at the mouth of a river or estuary, or at a slight distance from the beach. A bar commonly forms an obstruction to water navigation. Cf: *barrier*.

bar [paleont] The slender shaft of a compound conodont element, commonly bearing denticles, and having an anterior, lateral, or posterior position. Also, any conodont with discrete denticles and with a single large denticle near one end. Cf: *blade*.

bar [streams] A ridge-like accumulation of sand, gravel, or other alluvial material formed in the channel, along the banks, or at the mouth, of a stream where a decrease in velocity induces deposition; e.g. a *channel bar* or a *meander bar*. See also: *river bar*.

baraboo An ancient monadnock that had been buried and subsequently reexposed by partial erosion of overlying strata. Type locality: Baraboo Ridge, Wisc.

barachois A term used in the Gulf of St. Lawrence region for a lagoon.

bararite A hexagonal, low-temperature mineral: $(NH_4)_2SiF_6$. Cf: *cryptohalite*.

Barbados earth A fine-grained siliceous deposit of Miocene age occurring in Barbados (island in British West Indies) and containing abundant radiolarian remains. It was formed originally in deep water and later upraised above sea level.

bar beach A syn. of *barrier beach*. The term is not acceptable.

barbed drainage pattern A drainage pattern produced by tributaries that join the main stream in sharp *boathook bends* that point upstream; it is usually the result of stream piracy that has effected a reversal of flow of the main stream.

barbed tributary A stream that joins the main stream in an upstream direction, forming a sharp bend that points upstream and an acute angle that points downstream at the point of junction.

barbertonite A rose-pink to violet hexagonal mineral: $Mg_6Cr_2(CO_3)(OH)_{16}.4H_2O$. It is dimorphous with stichtite.

barbierite A name formerly applied to a hypothetical high-temperature monoclinic form of albite, and later changed to *mon-albite*. The name was originally applied to a mineral later shown to be finely twinned microcline with about 20% of un-mixed albite.

barbosalite A black mineral: $Fe^{+2}Fe_2^{+3}(PO_4)_2(OH)_2$. It is also known as "ferrous ferric lazulite".

barcan Var. of *barchan*.

barcenite A mixture of stibiconite and cinnabar.

barchan A moving, isolated, crescent-shaped sand dune lying transverse to the direction of the prevailing wind, with a gently sloping convex side facing the wind so that the wings or horns of the crescent point downwind (leeward) and an abrupt or steeply sloping concave or leeward side inside the horns; it can grow to heights of greater than 30 m and widths up to 350 m from horn to horn. A barchan forms along a flat, hard surface where the sand supply is limited and the wind is constant with only moderate velocity, and is among the commonest of the dune types, characteristic of very dry, inland desert regions the world over. Etymol: Russian *barkhan*, from Kirghiz; originally a Turki word meaning a "sand hill" in central Asia. Cf: *parabolic dune*. See also: *snow barchan; ice barchan*. Syn: *barchan dune; barcan; barchane; barkan; barkhan; horseshoe dune; crescentic dune*.

barchan dune *barchan*.

barchane A French variant of *barchan*.

bar diggings A term used in the western U.S. for gold diggings located on a bar or shallows of a stream, and worked when the water was low.

bare ice Ice without a snow cover.

bar finger A narrow, elongated, lenticular body of sand underlying, but several times wider than, a distributary channel in a bird-foot delta; it is produced and lengthened by the seaward advance of the lunate bar at the distributary mouth. Examples in the Mississippi River delta are as much as 30 km long, 8 km wide, and 80 m thick. Syn: *bar-finger sand; finger bar*.

bar-finger sand (a) A deposit of sand in the form of a bar finger. (b) *bar finger*.

bariandite A mineral: $V_2O_4.4V_2O_5.12H_2O$.

baring *overburden*.

barite A white, yellow, or colorless orthorhombic mineral: $BaSO_4$. Strontium and calcium are often present. Barite occurs in tabular crystals, in granular form, or in compact masses resembling marble, and it has a specific gravity of 4.5. It is the principal ore of barium, and is used in paints, drilling muds, and as filler for paper and textiles. Syn: *barytes; heavy spar; cawk*.

barite dollar A term used esp. in Texas and Oklahoma for a small, flat, rounded or disk-shaped mass of barite formed in a sandstone or sandy shale.

barite rosette A *rosette* consisting of a cluster or aggregate of large, tabular, sand-filled crystals of barite, usually forming in sandstone. Syn: *barite rose; petrified rose*.

bark In a woody stem, the tissue outside the cambium.

barkan Var. of *barchan*.

barkevikite A brownish or velvet-black monoclinic mineral of the amphibole group, near arfvedsonite in composition and appearance.

barkhan Var. of *barchan*. Etymol: Russian.

bar lake A lake with a sandbar across its outlet. Examples occur along the east coast of Lake Michigan.

barley coal One of the sizes of *buckwheat coal*, equivalent to number three in the series. It will pass through a 3/16 inch round mesh but not through a 3/32 inch round mesh. Cf: *rice coal*.

barnesite A mineral: $Na_2V_6O_{16}.3H_2O$. It is the sodium analogue of hewettite.

baroclinic Adj. of *baroclinity*.

baroclinity In oceanography, the condition of stratification of a fluid due to the intersection of surfaces of constant pressure with surfaces of constant density. Such a fluid is said to be *baroclinic*. See also: *barotropy*.

barodynamics Mechanics applied to the behavior of heavy structures (such as dams, mine shafts, and bridges) liable to failure because of their own weight.

barograph A barometer which makes a continuous record of changes in atmospheric pressure. It is usually an aneroid type of *barometer*.

barolite A rock composed of barite or of celestite.

barometer An instrument that is used to measure atmospheric pressure. It may be either a *mercury barometer* or an *aneroid barometer*. See also: *barograph*.

barometric altimeter *pressure altimeter*.

barometric efficiency The ratio of the fluctuation of water level in a well to the change in atmospheric pressure causing the fluctuation, expressed in the same units such as feet of water. Symbol: B. Cf: *tidal efficiency*.

barometric elevation An elevation above mean sea level estimated by the use of a barometer to measure the difference in air pressure between the point in question and some reference base of known value, whose elevation has been determined by a more precise method.

barometric leveling A type of indirect leveling in which differences of elevation are determined from differences of atmospheric pressures observed with altimeters or barometers. Cf: *thermometric leveling*.

barometric pressure *atmospheric pressure*.

barometric rate The rapidity with which the atmospheric pressure rises or falls within a given period of time.

barometric tendency On a weather map, the net change (rise or fall) of barometric pressure within a specified time (usually three hours) before a particular observation, together with a symbol showing the nature of the pressure change.

barophilic Said of marine organisms that live under conditions of high pressure.

baroque adj. Said of a pearl or tumble-polished gem material that is irregular in shape.---n. A baroque pearl.

baroseismic storm Microseisms caused by definite barometric changes.

barotropic Said of a fluid that is in a state of *barotropy*.

barotropy In oceanography, the state of zero *baroclinity*; the coincidence of surfaces of constant density with those of constant pressure. Adj: *barotropic*.

bar plain A term introduced by Melton (1936, p.594 & 596) for a relatively smooth flood plain formed by a stream with neither a low-water channel nor an alluvial cover, and characterized by a network of elongate and irregularly sized "bars" built from "the tractional and suspended load in the declining stages of the last flood". Cf: *meander plain; covered plain*.

barrage A large dam, usually of concrete but sometimes of earth, built across a major river and designed to conserve a large body of water by greatly increasing its depth or to divert water into a channel for navigation and large-scale irrigation. The term is sometimes used when no power station for electricity is included ("dam" being used if there is one), and sometimes used when the structure is concerned with annual water storage ("dam" being used when more than one year's flow is impounded).

barranca (a) Var. of *barranco*. (b) A large rift in a piedmont glacier or in an ice shelf. Cf: *donga*.

barranco [geomorph] (a) A term used in the SW U.S. for a deep, steep-sided, usually rock-walled ravine, gorge, or small canyon, and for a deep cleft, gully, arroyo, or other break or hole made by a heavy rain. (b) A term used in the SW U.S. for a steep bank or a precipice; in New Mexico it is equivalent to *cliff*.---Etymol: Spanish. Cf: *quebrada*. Syn: *barranca*.

barranco [volc] A deep, steepsided drainage valley on the slope of a volcanic cone, formed by erosion and coalescence of smaller channels. Barrancos form a radiating pattern around a volcanic cone. Etymol: Spanish *barranca*, "ravine". Syn: *sector graben*.

barrandite A pale-gray orthorhombic mineral: (Fe,Al)-$PO_4.2H_2O$. It is isomorphous with, and intermediate in composition between, strengite and variscite.

barred basin *restricted basin*.

barrel copper Pieces of native copper occurring in large enough sizes to be extracted from the gangue, and of sufficient purity to be smelted without mechanical concentration. Syn: *barrel work*.

barrel distortion A type of geometric distortion found in scanning imagery in which elements crossing the flight direction are distorted by a combination of scanner-mirror rotation, and forward motion of the aircraft. Straight lines cut obliquely appear as sigmoid curves in the resultant imagery. Cf: *scanner*. See also: *sigmoid distortion*.

barrel work A syn. of *barrel copper*, used in the Lake Superior mining region.

Barremian European stage: Lower Cretaceous (above Hauterivian, below Aptian).

barren A word, usually used in the pl., for rugged and unproductive land that is devoid of significant vegetation compared to adjacent areas because of environmental factors such as adverse climate, poor soil, or wind.

barren zone A quasi-biostratigraphic unit representing a part of the stratigraphic succession devoid of all diagnostic fossils or representatives of the taxonomic categories on which the remainder of the succession is subdivided.

barrier [coast] An elongate offshore ridge or mass usually of sand rising above the high-tide level, generally extending parallel to, and at some distance from, the shore, and built up by the action of waves and currents. Examples include *barrier beach* and *barrier island*. Cf: *bar*.

barrier [ecol] A condition, such as a topographic feature or physical quality, that tends to prevent the free movement and mixing of populations or individuals.

barrier [glaciol] *ice barrier* [glaciol].

barrier [grd wat] *ground-water barrier*.

barrier bar A syn. of *longshore bar*. The term is not acceptable.

barrier basin A basin produced by the formation of a natural dam or barrier; it may contain a *barrier lake*.

barrier beach A single, narrow, elongate sand ridge rising slightly above the high-tide level and extending generally parallel with the shore, but separated from it by a lagoon (Shepard, 1952, p.1904) or marsh; it is extended by longshore drifting and is rarely more than several kilometers long. See also: *barrier island*. This feature was termed an *offshore bar* by Johnson (1919, p.259 & 350). Syn: *offshore barrier; offshore beach; bar beach*.

barrier chain A series of barrier islands, barrier spits, and barrier beaches extending along a coast for a considerable distance (Shepard, 1952, p.1908).

barrier flat A relatively flat area, often occupied by pools of water, separating the exposed or seaward edge of a barrier and the lagoon behind the barrier.

barrier ice (a) A term used by Robert F. Scott in 1902 for the ice constituting the Antarctic ice shelf (which was then called an "ice barrier"). The syn. *shelf ice* seems to be more commonly used for the actual ice itself. (b) A term sometimes used improperly as a syn. of *ice shelf*.

barrier island (a) A long, low, narrow, wave-built sandy island representing a broadened *barrier beach* that is sufficiently above high tide and parallel to the shore, and that commonly has dunes, vegetated zones, and swampy terranes extending lagoonward from the beach. Also, a long series of barrier beaches. Examples include Miami Beach, Fla., and the Lido in Venice. This feature was termed an *offshore bar* by Johnson (1919). (b) A detached portion of a barrier beach between two inlets (Wiegel, 1953, p.5).

barrier-island marsh A salt or brackish marsh on the low inner margin of a barrier island.

barrier lagoon (a) A *lagoon* that is roughly parallel to the coast and is separated from the open ocean by a strip of land or by a barrier reef. See also: *moat*. (b) A *lagoon* encircled by coral islands or coral reefs, esp. one enclosed within an atoll.

barrier lake A lake whose waters are impounded by the formation of a natural dam or barrier, such as by a landslide, by alluvium in a delta, by glacial moraine, by an ice dam, or by lava; also, a freshwater lagoon separated from a lake by a shore dune or by a sandbank.

barrier reef (a) A long, narrow coral reef roughly parallel to the shore and separated from it at some distance by a lagoon of considerable depth and width. It typically encloses a volcanic island (either wholly or in part), or it lies at a great distance from a continental coast (such as the Great Barrier Reef off the coast of Queensland, Australia). (b) The limestone produced by the consolidation of materials in a barrier reef.

barrier spit A *barrier island* or *barrier beach* that is connected at one end to the mainland.

barrier spring A spring resulting from the diversion of a flow of ground water over or underneath an impermeable barrier in the floor of a valley (Schieferdecker, 1959, term 0308). Cf: *contact spring*.

barrier well A type of recharge well, generally one of several along a line, used to inject water of usable quality to build up a ridge of such water between wells used for water supply and a potential source of contamination, as between water-supply wells and the salt-water front in a coastal area.

barringerite A meteorite mineral: $(Fe,Ni)_2P$.

barringtonite A mineral: $MgCO_3.2H_2O$ (?).

Barrovian-type facies series The most common type of dynamothermal regional metamorphism, characterized by the appearance of metamorphic zones of the low-temperature greenschist and higher-temperature almandine-amphibolite facies. The geothermal gradient is rather low as prevailing at depth under high pressures (in excess of about 4000 bars). Typical index minerals (in order of increasing metamorphic grade) are chlorite - biotite - garnet (almandine) - staurolite - kyanite - sillimanite (but no andalusite). It is named after Barrow who first described this type of metamorphism in the Grampian Highlands of Scotland (Hietanen, 1967, p.195). Cf: *Saxonian-type facies series; Idahoan-type facies series*.

barrow pit A term used chiefly in western U.S. for a *borrow pit*; esp. a ditch dug along a roadway to furnish fill.

barshawite An andesine lugarite containing orthoclase.

bar theory A theory to account for thick evaporite deposits occurring in lagoons separated from the ocean by a bar (usually built by waves) and forming under conditions such that the water lost by evaporation from the lagoons exceeds that introduced by streams and rainfall.

barthite A variety of austinite containing copper. It was formerly thought to be a variety of veszelyite.

Bartonian European stage: Eocene (above Auversian, below Ludian). Syn: *Marinesian*.

barycenter The location of the center-of-mass of a collection of bodies.

barylite A colorless mineral: $BaBe_2Si_2O_7$.

barysilite A white mineral: $Pb_4MnSi_3O_{11}$.

barysphere The interior of the Earth, beneath the lithosphere, thus including both the mantle and the core. However, it is sometimes used to refer to only the core or only the mantle. Cf: *pyrosphere*. Syn: *centrosphere*. Nonrecommended syn: *bathysphere*.

barytes A syn. of *barite*. Also spelled: *baryte; barytine; barytite*.

barytocalcite (a) A mineral: $BaCa(CO_3)_2$. It is the monoclinic dimorph of alstonite. (b) A mixture of calcite and barite.

barytolamprophyllite A mineral: $(Na,K)_6(Ba,Ca,Sr)_3(Ti,Fe)_7$-$Si_8O_{32}(O,OH,F,Cl)_4$.

basal adj. Pertaining to, situated at, or forming the base of an animal structure; e.g. referring to the aboral part of the theca of an echinoderm, or pertaining to the under or reverse side of an incrusting or freely growing bryozoan colony.---n. A basal structure of an animal; esp. a *basal plate* of an echinoderm.

basal arkose Arkose or subarkose occurring as a thin and discontinuous blanket-like deposit at the base of a sedimentary series that overlies granitic rock. It consists of consolidated feldspathic residue resulting from the slight reworking of arkosic regolith and from the removal of the more completely decayed and finer portions.

basal cavity A pit or concavity about which a conodont element was built through accretion of lamellae, opening onto the aboral side, and present on all true conodont elements. See also: *cup; attachment scar.* Improper syn: *pulp cavity; escutcheon.*

basal cleavage A type of isometric crystal cleavage that occurs parallel to its basal pinacoid; e.g. molybdenite.

basal complex *basement.*

basal conglomerate A well-sorted, lithologically homogeneous conglomerate that forms the bottom or initial stratigraphic unit of a sedimentary series and that rests on a surface of erosion, thereby marking or immediately overlying an unconformity; esp. a coarse-grained beach deposit of an encroaching or transgressive sea. It commonly occurs as a relatively thin, widespread, or patchy sheet, interbedded with quartz-type sandstones. Cf: *marginal conglomerate.*

basal disk An expanded basal part by which certain stalked sessile organisms are attached to the substrate; specif. the aboral fleshy part of a scleractinian coral polyp, typically subcircular in outline, that closes off the lower end of the polyp. Cf: *oral disk.* Also spelled: *basal disc.*

basal funnel An excavated cone-like basal plate whose tip fits into the basal cavity of a conodont.

basal granule A dot-like body forming part of the neuromotor system in tintinnids.

basal ground water A term that originated in Hawaii and refers to a major body of ground water floating on and in hydrodynamic equilibrium with salt water. Syn: *basal water.*

basalia (a) A circlet of basal plates in theca of a blastoid, normally consisting of two large zygous plates and one small azygous plate. (b) Prostalia protruding from the base of a sponge and serving to stabilize or anchor it to the substrate.

basal lamina A usually adherent, calcified basal surface in cyclostomatous and Paleozoic bryozoans, made up from zooecial walls but extending to the growing edge of the colony. See also: *median lamina.*

basal leaf cross Broad wings on radial spines of acantharian radiolarians.

basal lobe One of two lobes set off by furrows in the posterior and lateral parts of the glabella of a trilobite, just in advance of the neck ring.

basal pinacoid In all crystals except those of the isometric system, the {001} pinacoid. Cf: *front pinacoid; side pinacoid.* Syn: *basal plane.*

basal plane *basal pinacoid.*

basal plate (a) One of a circlet of certain chiefly ventral skeletal plates of an echinoderm; e.g. a plate composing the aboral end of theca of a blastoid and articulated to the stem aborally and to radial plates on oral borders, or an interambulacral plate just below the arm-bearing radial plates of a crinoid. Syn: *basal.* (b) A thin skeletal plate formed initially as a part of a corallite immediately below the basal disk of the polyp of a scleractinian coral and from which the septa and walls begin to extend upward and outward. (c) A laminated plate-like structure of organic material attached to the aboral side of a conodont element along the attachment scar or basal cavity.

basal platform A term used by Linton (1955) as a syn. of *basal surface.*

basal pore One of the pores outlined by connector bars joining the basal ring of a radiolarian skeleton (as in the subfamily Trissocyclinae).

basal ring A ring at or below the base of the sagittal ring of a radiolarian skeleton, commonly with basal spines projecting from it (as in the subfamily Trissocyclinae).

basal sapping The undercutting, or breaking away of rock fragments, along the headwall of a cirque, due to frost action at the bottom of the bergschrund. Syn: *sapping; glacial sapping.*

basal sliding (a) The sliding of a glacier on its bed. (b) The velocity or speed of sliding of a glacier on its bed.----Syn: *basal slip.*

basal slip *basal sliding.*

basal slope *wash slope.*

basal surface The generalized boundary between weathered and unweathered rock, or the lower limit to active weathering (Ruxton & Berry, 1959). The contact, which may be regular or irregular, indicates a very rapid or sudden change upward into the base of the mass of weathering debris. Syn: *basal platform; weathering front.*

basalt [lunar] A volcanic-like rock from the Moon that is composed chiefly of nearly equal amounts of augite, plagioclase, and ilmenite. The plagioclase is characteristically highly calcic (An_{80}-An_{90}). Lunar basalt contains more titanium dioxide, rare-earth elements, zirconium, and less nickel than terrestrial basalt.

basalt (a) A dark- to medium-dark-colored, commonly extrusive (locally intrusive, as dikes), mafic igneous rock composed chiefly of calcic plagioclase (usually labradorite) and clinopyroxene in a glassy or fine-grained groundmass; the extrusive equivalent of *gabbro.* Nepheline, olivine, hypersthene, and quartz may be present, but not all simultaneously: nepheline and olivine can occur together, as can olivine and hypersthene and hypersthene and quartz, with other combinations not occurring. Apatite and magnetite are common accessories. (b) A term loosely used for dark-colored, fine-grained igneous rocks whether intrusive or extrusive. Syn: *basaltic rocks.*

basalt glass *sideromelane.*

basal thrust plane *sole fault.*

basaltic Pertaining to, composed of, containing, or resembling basalt; e.g. "basaltic lava".

basaltic dome *shield volcano.*

basaltic hornblende A black or brown variety of hornblende rich in ferric iron (ferrous iron has undergone oxidation) and occurring in basalts and other iron-rich basic igneous (volcanic) rocks; a type of *brown hornblende* characterized optically by strong pleochroism and birefringence, by high refractive indices, and by a smaller extinction angle. Syn: *lamprobolite; oxyhornblende; basaltine.*

basaltic layer A syn. of *sima,* so named for its supposed petrologic composition. It is also called the *gabbroic layer.* Cf: *granitic layer.*

basaltic rocks A general term incorporating fine-grained, compact, dark-colored, extrusive igneous rocks such as *basalt* [ign], diabase, dolerite, and dark-colored andesite.

basal till A firm clay-rich till containing many abraded stones dragged along beneath a moving glacier and deposited upon bedrock or other glacial deposits. Cf: *lodgment till.*

basaltine n. (a) *basaltic hornblende.* (b) *augite.*---adj. *basaltic.*

basaltite An older term revived by the International Geological Congress in 1900 and applied to olivine-free basalts.

basalt obsidian *sideromelane.*

basal tunnel A water-supply tunnel excavated along the basal water table in basaltic areas, esp. Hawaii. Cf: *Maui-type well*.

basaluminite A white mineral: $Al_4(SO_4)(OH)_{10}.5H_2O$. It occurs in veinlets lining crevices in ironstone. Cf: *felsôbanyite*.

basal water *basal ground water*.

basal water table The water table of a body of basal ground water.

basanite [sed] (a) A *touchstone* consisting of flinty jasper or finely crystalline quartzite. Syn: *Lydian stone*. (b) A black variety of jasper.

basanite [ign] A group of basaltic rocks characterized by calcic plagioclase, augite, a feldspathoid (nepheline, leucite), and olivine; also, any rock in that group. Without the olivine, the rock would be called a *tephrite*.

basanitoid n. A term proposed by Bücking in 1881, but never adopted, for a group of rocks intermediate in composition between basanites and basalts (Johannsen, 1939, p.243), i.e. having the chemical composition of basanite but without feldspathoids and with a glassy groundmass. --adj. Said of a basanite-like rock; having normative but no modal feldspathoids.

basculating fault *wrench fault*.

base [eng] The bottom or that part of an engineering structure resting upon the subgrade or supporting soil or solid rock; the *base course*.

base [gem] *pavilion*.

base [paleont] (a) The aboral end of cystoid theca. The term is restricted by some to the columnar facet but extended by others to include thecal plates of the basal circlet or aboral circlets. (b) The aboral end of crinoid calyx, or the part of a crinoid dorsal cup between radial plates and stem, normally composed of basal plates or of basal and infrabasal plates, but in articulate crinoids may include the centrale. (c) The area adjacent to the aboral side of a conodont element.

base [petroleum] The dominant substance held in solution in crude oil, that will be left as a residue after distillation, i.e. paraffin or napthene.

base [ign] *mesostasis*.

base [surv] *base line*.

base apparatus Any apparatus (such as wood tubes, metal wires, iron bars, steel rods, or invar tapes) used in geodetic surveying and designed to measure with accuracy and precision the length of a base line in triangulation or the length of a line in a traverse.

base bullion A crude lead that may or may not contain gold but that is recoverable for its silver content.

base-centered lattice A type of centered lattice that is centered in the {001} planes. Syn: *C-centered lattice*.

base correction A correction or adjustment of geophysical measurements to express them in terms of the values of the *base station*.

base course A bottom layer of specified material (coarse gravel, crushed stone, etc.) and thickness, constructed on the subgrade or subbase for the purpose of serving one or more functions such as distributing load, providing drainage, and minimizing frost action. Syn: *base* [eng].

base discharge As used by the U.S. Geological Survey, that discharge about which peak discharge data are published.

base exchange *ion exchange*.

base flow Sustained or fair-weather flow of a stream, whether or not affected by the works of man (Langbein & Iseri, 1960). Cf: *base runoff*.

base-height ratio The ratio between the air base and the flight height of a stereoscopic pair of aerial photographs.

base level n. (a) The theoretical limit or lowest level toward which erosion of the Earth's surface constantly progresses but seldom, if ever, reaches; esp. the level below which a stream cannot erode its bed. The general or *ultimate base level* for the land surface is sea level, but *temporary base levels* may exist locally. The base level of eolian erosion may be above or below sea level; that of marine erosion is the lowest level to which marine agents can cut a bottom. Also spelled: *baselevel*. Syn: *base level of erosion*. (b) A curved or planar surface extending inland from sea level, inclined gently upward from the sea and representing the theoretical limit of stream erosion. (c) The surface toward which external forces strive, at which neither erosion nor deposition takes place (Barrell, 1917); a surface of equilibrium.---v. To reduce by erosion to, or toward the condition of a plain at, base level.

base level of deposition The highest level to which a sedimentary deposit can be built (Twenhofel, 1939, p.8); if built of marine deposits, it coincides with the base level of erosion.

base level of erosion A syn. of *base level*. The term was introduced by Powell (1875, p.203) for an irregular "imaginary surface, inclining slightly in all its parts toward the lower end of the principal stream", below which the stream and its tributaries were supposed to be unable to erode.

base-level peneplain *peneplain*.

base-level plain A flat surface, area, or lowland of faint relief, produced by the wearing down of a region to or near its base level; a plain that cannot be materially reduced in elevation by erosion. Cf: *peneplain*.

base line (a) A surveyed line on the Earth's surface or in space, whose exact length and position have been accurately determined with more than usual care, and that serves as the origin for computing the distances and relative positions of remote points and objects or that is used as a reference to which surveys are coordinated and correlated. Syn: *base*. (b) The initial measurement in triangulation, being an accurately measured distance constituting one side of a series of connected triangles and used, together with measured angles, in computing the lengths of the other sides. (c) One of a pair of coordinate axes (along with the principal meridian) used in the U.S. Public Land Survey system, consisting of a line extending east and west along the true parallel of latitude passing through the initial point and along which standard township, section, and quarter-section corners are established. The base line is the line from which the survey of the meridional township boundaries and section lines is initiated and from which the townships, either north or south, are numbered. (d) An aeromagnetic profile flown at least twice in opposite directions and at the same level, in order to establish a line of reference of magnetic intensities on which to base an aeromagnetic survey. (e) The center line of location of a railway or highway; the reference line for the construction of a bridge or other engineering structure.---Sometimes spelled: *baseline*.

base map (a) A map of any kind showing only essential outlines necessary for adequate geographic reference and on which additional or specialized information is plotted for a particular purpose; esp. a topographic map on which geologic information is recorded. (b) *master map*. (c) Obsolete syn. of *outline map*.

basement n. (a) A *complex* of undifferentiated rocks that underlies the oldest identifiable rocks in the area. (b) The crust of the Earth below sedimentary deposits, extending downward to the Mohorovičić discontinuity. In many places the rocks of the complex are igneous and metamorphic of Precambrian age, but in some places they are Paleozoic, Mesozoic, or even Cenozoic. Syn: *basement rock; basal complex; fundamental complex; basement complex*.----adj. Said of material, processes, or structures originating or occurring in the basement.

basement complex *basement*.

basement rock *basement*.

base metal (a) Any of the more common and more chemically active metals, e.g. lead, copper. (b) The principal metal of an alloy, e.g. the copper in brass.----Cf: *noble metal*.

base net A small *net* of triangle and quadrilaterals, starting from a measured base line, and connecting with a line of the main scheme of a *triangulation net*; e.g. a triangle formed by sighting a point from both ends of a base line, or two adjacent triangles with the base line common to both. It is the initial figure in a triangulation system.

base of drift The seismic-velocity discontinuity between a glacial moraine (or drift) material and the competent formation beneath. By extension, any similar discontinuity between a layer below the weathering and the topmost competent formation layer.

base of weathering In seismic interpretation, the boundary between the low-velocity surface layer and an underlying, comparatively high-velocity layer. This may correspond to the geologic base of weathering. The boundary may vary with time, and is considered in deriving time corrections for seismic records.

base runoff Sustained or fair-weather *runoff* [water] (Langbein & Iseri, 1960). It is primarily composed of effluent ground water, but also of runoff delayed by slow passage through lakes or swamps. The term refers to the natural flow of a stream, unaffected by the works of man. Cf: *base flow; direct runoff*. Syn: *fair-weather runoff; sustained runoff*.

base station An observation point used in geophysical surveys or exploration with reference to which measurements at additional points can be compared. See also: *base correction*.

base surge A ring-shaped cloud of gas and suspended solid debris that moves radially outward at high velocity as a density flow from the base of a vertical explosion column accompanying a volcanic eruption or crater formation by an explosion or hypervelocity impact.

basic (a) Said of an igneous rock having a relatively low silica content, sometimes delimited arbitrarily as less than 54% (although the limits vary with different petrologists); e.g. gabbro, basalt. Basic rocks are relatively rich in iron, magnesium, and/or calcium and thus include most mafic rocks as well as other rocks. "Basic" is one of four subdivisions of a widely used system for classifying igneous rocks based on their silica content: *acidic, intermediate*, (basic), and *ultrabasic*. Cf: *femic*. (b) Said loosely of any igneous rock composed chiefly of dark-colored minerals.----Cf: *silicic; mafic*. (c) Said of a plagioclase that is calcic.

basic behind In granitization, a zone in which residual mafic components are concentrated.

basic border The marginal area of an igneous intrusion, characterized by a relatively more basic composition than the interior of the rock mass. Cf: *chill zone*. Syn: *mafic margin*.

basic front In granitization, an advancing zone enriched in calcium, magnesium, and iron, which is said to represent those elements in the rock being granitized that are in excess of those required to form granite. During granitization, these are displaced and move through the rock ahead of the granitization front, usually as a zone enriched in hornblende or pyroxene. Cf: *basic behind*. Syn: *mafic front; magnesium front*.

basic hydrologic data Data including inventories of land and water features that vary from place to place (as topographic and geologic maps), and records of processes that vary with time and from place to place (as precipitation, streamflow, and ground-water levels) (Langbein & Iseri, 1960). Cf: *basic hydrologic information*.

basic hydrologic information A broader term than *basic hydrologic data*, including surveys and appraisals of the water resources of an area and a study of its physical and related economic processes, interrelations, and mechanisms (Langbein & Iseri, 1960).

basicoronal Pertaining to the corona of an echinoid at the edge of peristome.

basic plagioclase A variety of plagioclase having relatively low content of silicic acid (in the anhydrous form SiO_2); e.g. an An-rich member such as bytownite or anorthite.

basic wash A driller's term for material eroded from outcrops of basic igneous rocks (gabbro, basalt) and redeposited to form a rock having approximately the same major mineral constituents as the original rock (Taylor & Reno, 1948, p.164). Cf: *granite wash*.

basidiospore A *fungal spore* produced by the basidium of a basidiomycete. Such spores with chitinous walls may occur as microfossils in palynologic preparations.

basification The development of a more basic rock, commonly richer in hornblende, biotite, and oligoclase, presumably as a result of the contamination of a granitic magma by assimilation of country rock. This phenomenon occurs chiefly at the margins of the granite mass. Syn: *thalassogenesis; oceanization*.

basimesostasis The mesostasis of a *basiophitic* rock.

basin [topog] A depressed area having no surface outlet.

basin [coast] (a) A little bay, esp. an area of water partly or wholly enclosed by land (often enlarged by excavation) and suitable for anchorage of one or more vessels, as a landlocked harbor. (b) *tidal basin*.

basin [glac geol] A local term commonly applied to a *cirque*, as in the Rocky Mountains.

basin [marine geol] A more or less equidimensional depression of the sea floor.

basin [sed] (a) An area in which sediments accumulate. (b) A sunken area caused by the accumulation of sediments on it.---See also: sedimentary basin; depocenter; cuvette.

basin [struc geol] A general term for a depressed, sediment-filled area. It may be a circular to elliptical *centrocline*, characteristic of cratonic areas, e.g. the Michigan Basin, or an elongate, fault-bordered intermontane basin within an orogenic belt, e.g. the Bighorn Basin of Wyoming. Syn: *structural basin*.

basin [water] A physiographic feature or subsurface structure that is capable of collecting, storing, and discharging water by reason of its shape and the characteristics of its confining material. Cf: *ground-water basin*.

basin [lake] (a) *lake basin*. (b) A depression in the Earth's surface, the lowest part often filled by a lake or pond.

basin [streams] (a) *drainage basin*. (b) *river basin*. (c) A part of a river or canal widened and furnished with wharves.

basin-and-range Said of a topography, landscape, or physiographic province characterized by a series of tilted fault blocks forming longitudinal, asymmetric ridges or mountains and broad, intervening basins; specif. the Basin and Range physiographic province in SW U.S., where the ranges have steep eastern faces and gentler western slopes. See also: *basin-range structure; basin range*.

basin area For a given stream order *u*, the total area, projected upon a horizontal plane, of a drainage basin bounded by the basin perimeter and contributing overland flow to the stream segment of order *u*, including all tributaries of lower order (Strahler, 1964, 4-48). Symbol: A_u. Cf: *watershed area*. See also: *law of basin areas*.

basin-area ratio Ratio of mean basin area of a given order to the mean basin area of the next lower order within a specified larger drainage basin (Schumm, 1956, p.606). Symbol: R_a.

basin-circularity ratio Ratio of the area of a drainage basin to the area of a circle with the same basin perimeter. Symbol: R_c. Syn: *circularity; circularity ratio*.

basin-elongation ratio Ratio of the diameter of a circle having the same area as a drainage basin to the maximum length of that basin. Symbol: R_e. Syn: *elongation ratio*.

basin facies A stratigraphic facies representing sediments deposited beyond the outer limits of a land-bordering submarine shelf.

basining The bending down or settling of part of the Earth's crust in the form of a basin, as by erosion or by solution and transportation of underground deposits of salt or gypsum.

basin length Horizontal distance of a straight line from the mouth of a stream to the farthest point on the drainage divide of its basin, parallel to the principal drainage line (Schumm, 1956, p.612). Symbol: L_b.

basin order The number assigned to an entire drainage basin contributing to the stream segment of a given order and bearing an identical integer designation; e.g. a first-order basin contains all of the drainage area of a first-order stream. See also: *stream order*.

basin peat *local peat*.

basin perimeter Length of the line enclosing the area of a drainage basin. Symbol: P.

basin range (a) A relatively long and narrow mountain range that owes its present elevation and structural form mainly to faulting and tilting of strata and that is isolated by alluvium-filled basins or valleys. (b) A tilted fault block.---Etymol: from the Great Basin, a region in SW U.S. characterized by fault-block mountains. See also: *basin-and-range*.

basin-range structure Regional geologic structure dominated by fault-block mountains separated by broad alluvium-filled basins; e.g. basin-range faulting characterized by antithetic movements, as in the *basin-and-range* country of SW U.S.

basin relief Difference in elevation between the mouth of a stream and the highest point within, or on the perimeter of, its drainage basin (Strahler, 1952b, p.1119); the maximum relief in the basin. Symbol: H.

basin swamp A freshwater swamp at the margin of a small lake in which there is little wave activity, or near a large lake where shallow water or a barrier serves as a protection (Twenhofel, 1939, p.78). The swamp develops from a floating mat of plants, which traps sediments and is underlain by water or black sludge.

basin valley A broad and shallow valley with gently sloping sides.

basiophitic Said of the texture of an ophitic rock whose mesostasis is composed of augite; also, said of an ophitic rock with such texture. Cf: *oxyophitic; oxybasiophitic*. See also: *basimesostasis*.

basiophthalmite The proximal segment (lowest joint) of the eyestalk of a decapod crustacean, articulating with the *podophthalmite*.

basiphytous Said of a sponge that is attached to the substrate by an encrusting base.

basipinacoderm The pinacoderm delimiting a sponge at the surface of fixation.

basipod The *basis* of a crustacean limb. Syn: *basipodite*.

basis [paleont] The limb segment of a crustacean just distal to the coxa, commonly bearing the exopod and endopod; a membranous or calcareous structure (in nonpedunculate cirripedes) that contacts the substratum. Pl: *bases*. Syn: *basipod*.

basis [ign] *mesostasis*.

basite A basic igneous rock.

basket-of-eggs topography A landscape characterized by swarms of closely spaced drumlins distributed more or less en echelon, and separated by small marshy tracts. Syn: *drumlin field*.

bass An English term for a black or bluish-black carbonaceous shale or slaty clay associated with coal and often containing pyrite; *batt*. Adj: *bassy*. Syn: *basses*.

bassanite A white mineral: $CaSO_4 \cdot 1/2H_2O$.

basset An obsolete term for the substantive "an outcrop" and the verb "to outcrop".

bassetite A yellow mineral: $Fe(UO_2)_2(PO_4)_2 \cdot 8H_2O$.

bastard adj. (a) Said of an inferior or impure rock or mineral, or of an ore deposit that contains a high proportion of uncommercial material. (b) Said of any metal or ore that gives misleading assays or values. (c) Said of a vein or other deposit close to and more or less parallel to a main vein or deposit, but less thick or extensive or of a lower grade.

bastard coal Thin partings of impure coal occurring in the lower part of shale strata immediately overlying a coal seam; any coal with a high ash content. Syn: *batt*.

bastard ganister A silica rock having the superficial appearance of a true *ganister* but characterized by an increased amount of interstitial matter, a greater variability of texture, and often an incomplete secondary silicification.

bastard quartz (a) A syn of *bull quartz*. (b) A round or spherical boulder of quartz embedded in a soft or decomposed rock.

bastard rock A term used in south Wales and North Staffordshire, Eng., for an impure sandstone containing thin lenticular layers or bands of shale or coal.

bastard shale *cannel shale*.

bastion A prominent mass of bedrock extending from the mouth of a hanging glacial trough and projecting far out into the main glacial valley; commonly found where a tributary glacier joins or has joined a trunk glacier.

bastite An olive-green, blackish-green, or brownish variety of serpentine mineral resulting from the alteration of orthorhombic pyroxene (esp. enstatite), occurring as foliated masses in igneous rocks, and characterized by a schiller (metallic or pearly luster) on the chief cleavage face of the pyroxene. Syn: *schiller spar*.

bastnaesite A greasy, wax-yellow to reddish-brown mineral: $(Ce,La)CO_3(F,OH)$. It is most commonly found in contact zones or associated with zinc lodes. Also spelled: *bastnàsite*.

bat Var. of *batt*.

Bathian A "stillborn homonym" or variant of *Bathonian* (ACSN, 1961, art.32d).

batholite Var. of *batholith*.

batholith A large, generally discordant, plutonic mass that has more than 40 sq. mi. (100 km²) in surface exposure and is composed predominantly of medium- to coarse-grained rocks of granodiorite and quartz monzonite composition. No visible floor for such a mass has yet been reported. Though a subject of controversy, its formation currently is believed by most investigators to involve magmatic processes. Also spelled: *bathylith; bathylite; batholite; batholyte; batholyth*. Syn: *abyssolith; intrusive mountain; central granite*.

batholyte Var. of *batholith*.

batholyth Var. of *batholith*.

Bathonian European stage: Middle Jurassic (above Bajocian, below Callovian). Syn: *Bathian*.

Bath's law A statement in seismology that the largest aftershock occurring within a few days of a main shock has a magnitude of 1.2 units lower than that of the main shock (Richter, 1958, p.69).

Bath stone A soft, creamy, oolitic limestone, easily quarried and used for building purposes. Type locality: near Bath, Eng.

bathvillite An amorphous, opaque, very brittle woody resin occurring as fawn-brown porous lumps in torbanite at Bathville, Scotland.

bathyal Pertaining to the ocean environment or *depth zone* between 100 and 500 fathoms; also, pertaining to the organisms of that environment.

bathybenthic Pertaining to the benthos of the bathyal zone of the ocean. Syn: *archibenthic*.

bathydermal Said of deformation or gliding of the lower part of the sialic crust. Cf: *dermal; epidermal*.

bathygenesis Negative or subsident tectonic movement; tectonic lowering of marine basins. It is analogous to *epeirogeny* or positive tectonic movements associated with the continents. Adj: *bathygenic*.

bathygenic Adj. of *bathygenesis*.

bathylimnion The deeper part of a *hypolimnion* characterized by constant rates of heat absorption at different depths. Cf: *clinolimnion*.

bathylite Var. of *batholith*.

bathylith Var. of *batholith*.

bathymetric chart A topographic map of the bottom of a body of water (such as the sea floor), with depths indicated by contours (isobaths) drawn at regular intervals.

bathymetric contour *isobath* [oceanog].

bathymetry The measurement of ocean depths and the charting of the topography of the ocean floor.

bathyorographical (a) Pertaining to ocean depths and mountain heights considered together; said of a map that shows both the relief of the land and the depths of the ocean. (b) Pertaining to the description of the relief features on the ocean floor.

bathypelagic Pertaining to the open water of bathyal depth.

bathyscaph A manned, submersible vehicle for deep-sea exploration; it is navigable, in contrast to a *bathysphere*.

bathyseism A deep-focus earthquake that is instrumentally detected worldwide. The term is little used.

bathysphere [interior Earth] A nonrecommended syn. of *barysphere*.

bathysphere [oceanog] A manned, submersible sphere that is lowered into the deep ocean by cable for observations; unlike the *bathyscaph*, it is not navigable.

bathythermogram The record (or a photographic print of it) that is made by a *bathythermograph*.

bathythermograph In oceanography, an instrument that records temperature in relation to depth. See also: *bathythermogram*.

batisite A dark-brown orthorhombic mineral: $Na_2BaTi_2(Si_2O_7)_2$.

batt [coal] (a) An English term for a compact, black, carbonaceous shale that splits into fine laminae and that is often interstratified with thin layers of coal or ironstone. Syn: *bass*. (b) *bastard coal*.----Also spelled: *bat*.

batt [clay] An English term for any hardened clay other than fireclay.

battery ore A type of manganese ore, generally a very pure crystalline manganese dioxide (pyrolusite), that is suitable for use in dry wells.

batture An elevated part of a river bed, formed by gradual accumulation of alluvium; esp. the land between low-water stage and a levee along the banks of the lower Mississippi River. Etymol: Louisiana French, from French *battre*, "to strike upon or against".

batukite A dark-colored extrusive rock composed of augite and fewer olivine phenocrysts in a groundmass of augite, magnetite, and leucite.

baulite *krablite*.

Baumé gravity The specific weight of a liquid, measured on a scale based on the weight of water; it is used in the petroleum industry for denoting the specific weight of oils. For liquids lighter than water, the relationship of specific gravity to the Baumé gravity scale is expressed as: degrees Baumé=140 divided by the specific gravity of the liquid at 60°F-130. Cf: *A.P.I. gravity*.

baumhauerite A lead- to steel-gray monoclinic mineral: $Pb_4As_6S_{13}$.

baum pot (a) A calcareous concretion in the roof of a coal seam; a *bullion*. (b) A cavity left in the roof of a coal seam due to the dropping downward of a cast of a fossil tree stump after removal of the coal.

bauxite An off-white, grayish, brown, yellow, or reddish-brown rock composed of a mixture of various amorphous or crystalline hydrous aluminum oxides and aluminum hydroxides (principally gibbsite, some boehmite), and containing impurities in the form of free silica, silt, iron hydroxides, and esp. clay minerals; a highly aluminous *laterite*. It is a common residual or transported constituent of clay deposits in tropical and subtropical regions, and occurs in concretionary, compact, earthy, pisolitic, or oolitic forms. Bauxite is the principal commercial source of aluminum; the term is also used collectively for lateritic aluminous ores. Bauxite was formerly regarded as an amorphous clay mineral consisting essentially of hydrated alumina, $Al_2O_3.2H_2O$. Named after les Baux (originally les Beaux), a locality near Arles in southern France. Syn: *beauxite*.

bauxitic Containing much bauxite; e.g. a "bauxitic clay" containing 47% to 65% alumina on a calcined basis, or a "bauxitic shale" abnormally high in alumina and notably low in silica.

bauxitization Development of bauxite from primary aluminum silicates (such as feldspars) or from secondary clay minerals under aggressive tropical or subtropical weathering conditions of good surface drainage, such as the dissolving (usually above the water table) of silica, iron compounds, and other constituents from alumina-containing material.

bavenite A white fibrous mineral: $Ca_4BeAl_2Si_9O_{24}(OH)_2$. Syn: *duplexite*.

Baveno twin law An uncommon twin law in feldspar, in which the twin plane and composition surface are (021). A Baveno twin usually consists of two individuals.

b axis [cryst] One of the crystallographic axes used as reference in crystal description. It is the axis that is oriented horizontally, right-to-left. In an orthorhombic or triclinic crystal, it is usually the *macro-axis*. In a monoclinic crystal, it is the *orthoaxis*. The letter *b* usually appears in italics. Cf: *a axis; c axis*.

b axis [struc petrol] In structural petrology, that *fabric axis* which is the direction of fold axes, bedding-plane intersections, i.e. in the plane of slip. It is assumed to be perpendicular to the direction of transport (the *a axis*). Cf: *c axis*. Syn: *b direction*.

b* axis That axis of a reciprocal crystal lattice which is perpendicular to (010). Cf: *a* axis; c* axis*.

bay [geog] (a) Any terrestrial formation resembling a bay of the sea, as a recess or extension of lowland along a river valley or within a curve in a range of hills, or an arm of a prairie extending into and partly surrounded by a forest. Also, a piece of low marshy ground producing many bay trees (such as laurel). (b) A *Carolina bay*. (c) A term used in southern Georgia and in Florida for an arm of a swamp extending into the upland as a bay-like indentation (Veatch & Humphrys, 1966, p.23).

bay [coast] (a) A wide, curving, open indentation, recess, or inlet of a sea or lake into the land or between two capes or headlands, larger than a cove, and usually smaller than, but of the same general character as, a gulf; the width of the seaward opening is normally greater than the depth of penetration into the land. (b) A large tract of water that penetrates into the land and around which the land forms a broad curve. By international agreement (for purposes of delimiting territorial waters), a bay is a water body having a baymouth less than 24 nautical miles wide and an area that is equal to or greater than the area of a semicircle whose diameter is equal to the width of the baymouth.---See also: *bight; embayment*.

bay [magnet] A simple transient magnetic disturbance, lasting typically an hour. On a magnetic record it has the appearance of a V or of a bay of the sea.

bay [ice] (a) *bight*. (b) A part of the sea partly surrounded by ice.

bay bar A syn. of *baymouth bar* and *bay barrier*. The term is "confusing" because it "fails to indicate whether or not the sand ridge is submerged or stands above the water level" (Shepard, 1952, p.1908).

bay barrier A term proposed by Shepard (1952, p.1908) to replace *bay bar*, signifying a spit that has grown "entirely across the mouth of a bay so that the bay is no longer connected to the main body of water". See also: *baymouth bar*.

bay delta A delta formed at the mouth of a stream entering, and filling or partially filling, a bay or drowned valley. See also: *bayhead delta*.

bayerite A mineral: $Al(OH)_3$. It is a polymorph of gibbsite. Not to be confused with *beyerite*.

bayhead (a) The part of a bay that lies farthest inland from the larger body of water with which the bay is in contact. (b) A local term in southern U.S. for a swamp at the bayhead.---Also spelled: *bay head*.

bayhead bar A bar formed a short distance from the shore and across a bay near its head. It commonly has a narrow inlet. Syn: *bayhead barrier*.

bayhead barrier *bayhead bar.*

bayhead beach A small, crescentic beach formed at the head of a bay by materials eroded from adjacent headlands and carried to the bayhead by longshore currents and/or storm waves. Examples occur along the coasts of Maine and Oregon. Syn: *pocket beach; cove beach.*

bayhead delta A delta at the head of a bay or estuary into which a river discharges. See also: *bay delta.*

bay ice (a) Newly formed, relatively smooth sea ice of more than one winter's growth. (b) A term sometimes used in Antarctica for thick ice floes recently broken away from an ice shelf. (c) A term used in Labrador for one-year ice that forms in bays and inlets. (d) An obsolete term for young sea ice sufficiently thick to impede navigation.

bayldonite A grass-green to blackish-green mineral: $Pb-Cu_3(AsO_4)_2(OH)_2$.

bayleyite A yellow monoclinic mineral: $Mg_2(UO_2)-CO_3)_3.18H_2O$.

baymouth The entrance to a bay; the part of a bay that is in contact, and serves as a connection, with the main body of water. Also spelled: *bay mouth.*

baymouth bar A long, narrow bank of sand or gravel deposited by waves entirely or partly across the mouth or entrance of a bay so that the bay is no longer connected or is connected only by a narrow inlet with the main body of water; it usually connects two headlands, thus straightening the coast. It can be produced by the convergent growth of two spits from opposite directions, by a single spit extending in a constant direction, or by a longshore bar being driven shoreward. See also: *bay barrier.* Syn: *bay bar.*

bayou (a) A term variously applied to many local water features in the lower Mississippi River basin and in the Gulf Coast region of the U.S., esp. Louisiana. Its general meaning is a creek, large stream, minor river, or secondary watercourse that is tributary to another river or connecting with another body of water; esp. a sluggish and stagnant stream, characterized by a slow or imperceptible current, that follows a winding course through flat alluvial lowlands, coastal swamps or marshes, or river deltas. (b) An effluent branch, esp. sluggish or stagnant, of a main river or of a lake, as a distributary flowing through a delta or enclosing a low island. Also, the distributary channel that carries floodwater or affords a passage for tidal water through swamps or marshlands. (c) An intermittent, partly closed, or disused watercourse, esp. a lake or a sluggish stream formed on a river delta or in an abandoned river channel; a *bayou lake* or an *oxbow lake.* Other examples: an outlet for a coastal lake or swamp, or a shallow inlet into a bay, lake, or river; a swampy or marshy offshoot or overflow of a lake or river; a *slough* in a salt marsh. Syn: *girt.* (d) An estuarine creek (generally tidal), or an inlet, bay, or open cove on the Gulf Coast. Also, a lagoon, lake, or bay in a salt marsh. (e) A natural canal or narrow passage connecting two bodies of open water, such as bays; a navigable channel through sandbars or mud flats. (f) A term used in northern Arkansas and southern Missouri for a clear brook or rivulet that rises in the hills (Webster, 1967, p.188).---Etymol: American French *boyau,* "gut"; from Choctaw *bayuk,* "small stream".

bayou lake A lake or pool in an abandoned and partly closed channel of a stream, as on the Mississippi River delta. Syn: *bayou.*

bay salt A kind of *solar salt* obtained by evaporating seawater in shallow bays, lagoons, or ponds.

bayside beach A beach formed along the side of a bay by materials eroded from adjacent headlands.

bazzite An azure-blue hexagonal mineral: $Be_3(Sc,Al)_2Si_6O_{18}$. It is the scandium analogue of beryl.

BC interray Right posterior interray in echinoderms situated between B ray and C ray and clockwise of B ray when the echinoderm is viewed from the adoral side.

bc-joint *longitudinal joint.*

BC soil A soil having only B and C horizons.

b direction *b axis* [struc petrol].

beach (a) A gently sloping zone, typically with a concave profile, of unconsolidated material that extends landward from the low-water line to the place where there is a definite change in material or physiographic form (such as a cliff) or to the line of permanent vegetation (usually of the effective limit of the highest storm waves); a *shore* of a body of water, formed and washed by waves or tides, usually covered by sandy or pebbly material, and lacking a bare rocky surface. See also: *strand.* (b) The relatively thick and temporary accumulation of loose water-borne material (usually well-sorted sand and pebbles, accompanied by mud, cobbles, boulders, and smoothed rock and shell fragments) that is in active transit along, or deposited on, the beach between the limits of low water and high water. The term was originally used to designate the loose wave-worn shingle or pebbles found on English shores, and is so used in this sense in some parts of England (Johnson, 1919, p.163). (c) A term used in New Jersey for a low sand island along the coast. (d) A term commonly used for a seashore or lake-shore area, esp. that part of the shore amenable to recreation.

beach berm *berm* [beach].

beach breccia A breccia formed on a beach where wave action is inefficient and angular blocks are supplied from cliffs, and resulting under conditions of rapid submergence (Norton, 1917, p.181).

beachcomber *comber.*

beach concentrate A natural accumulation in beach sand of heavy minerals selectively concentrated (by wave, current, or surf action) from the ordinary beach sands in which they were originally present as accessory minerals; esp. a *beach placer.*

beach crest A temporary ridge or berm marking the landward limit of normal wave activity (Veatch & Humphrys, 1966, p.24). Cf: *berm crest.*

beach cusp A *cusp* of sand, pebbles, gravel, or boulders, formed on the foreshore of a beach by wave action; specif. a relatively small cusp along a straight beach, measuring 10-60 m between the seaward-pointing crescentic tips with the distance generally increasing with wave height, developed by deposition on, or erosion of, the seaward face of the beach or of a beach ridge.

beach cycle The periodic retreat and outbuilding of a beach under the influence of tides and waves: cutting back occurs during periods of spring tides and of high waves produced by winter storms; building out occurs during periods of neap tides and of low waves characteristic of summer.

beach erosion The destruction and removal of beach materials by wave action, tidal currents, littoral currents, or wind.

beach face The section of the beach normally exposed to the action of wave uprush; the *foreshore* of a beach. Not to be confused with *shoreface.*

beach firmness The quality of beach sand to resist pressure; the "strength" of the sand. It is controlled by the degree of packing and sorting of sand, by moisture content, by the quality of trapped air, and by sand-particle size: the damper or finer-grained the sand, the firmer the beach.

beachline A shoreline characterized by a series of well-developed beaches.

beach mining The extraction and concentration of beach placer ore, usually by dredging.

beach ore *beach placer.*

beach placer A placer, esp. of gold, tin, or platinum, on a contemporary or ancient beach or along a coastline; a *beach concentrate* containing valuable minerals. Syn: *beach ore; seabeach placer; mineral sands.*

beach plain A continuous and level or undulating area formed by closely spaced successive embankments of wave-deposited beach material added more or less uniformly to a prograding shoreline, such as to a growing compound spit or to a cuspate foreland (Johnson, 1919, p.297 & 319); a *wave-built terrace.*

beach platform *wave-cut bench.*

beach pool (a) A small, usually temporary, body of water between two beaches or two beach ridges, or a lagoon behind a beach ridge. (b) A pool adjoining a lake and resulting from wave action.

beach profile The intersection of the beach surface by a vertical plane perpendicular to the shoreline. The beach profile of equilibrium is commonly concave upward as the slope is steeper above high water and gentler seaward.

beach ridge A low, essentially continuous mound of beach or beach-and-dune material (sand, gravel, shingle) heaped up by the action of waves and currents on the backshore of a beach beyond the present limit of storm waves or the reach of ordinary tides, and occurring singly or as one of a series of approximately parallel deposits. The ridges are roughly parallel to the shoreline and represent successive positions of an advancing shoreline. Syn: *full.*

beachrock A friable to well-cemented sedimentary rock, formed in the intertidal zone in a tropical or subtropical region, consisting of calcareous debris (detrital and/or skeletal) cemented with calcium carbonate; e.g. a thin, clearly stratified, seaward-dipping calcarenite found on a sandy coral beach. Also spelled: *beach rock.* Syn: *beach sandstone.*

beach sandstone *beachrock.*

beach scarp An almost vertical slope fronting a berm on a beach, caused by wave erosion. It may vary in height from several centimeters to a few meters, depending on the character of the wave action and the nature and composition of the beach.

beach width The horizontal dimension of the beach as measured normal to the shoreline.

beachy Covered with beach materials, esp. pebbles, shingle, or sand.

bead In blowpipe analysis of minerals, a drop of a fused material, such as a *borax bead* used as a solvent in color testing for various metals. The addition of a metallic compound to the bead will cause the bead to assume the color that is characteristic of the metal. See also: *blowpiping.*

beaded esker An esker with numerous bulges or swellings (representing fans or deltas) along its length. It may have been formed during pauses in the retreat of the glacier that nurtured the esker-forming stream.

beaded lake One of a string of lakes, as a *paternoster lake* or a long narrow lake between sand dunes.

beaded stream A stream consisting of a series of small pools or lakes connected by short stream segments; e.g. a stream commonly found in a region of paternoster lakes or an area underlain by permafrost.

beak [coast] A *promontory.*

beak [paleont] (a) The generally pointed extremity of a bivalve shell, marking the point of the initial growth of the shell; specif. the nose-like projection of the dorsal part of a pelecypod valve, located along or above the hinge line, and typically showing strong curvature and pointing anteriorly, or the tip of the umbo of a brachiopod, located adjacent or posterior to the hinge line and in the midline of the valve. Cf: *umbo.* (b) The prolongation of certain univalve shells, containing the canal (as in a gastropod). The term in this sense is not generally used by paleontologists. (c) The pair of horny jaws on either side of the mouth of a cephalopod.

beak [bot] A long, prominent point of a plant part, e.g. of a fruit or pistil.

beaked apex The upper angle of tergum of a cirripede crustacean, produced into a long narrow point (TIP, 1969, pt.R, p.91).

beak ridge A more or less angular linear elevation of a brachiopod shell, extending from each side of the umbo so as to

delimit all or most of the cardinal area (TIP, 1965, pt.H, p.140).

Beaman stadia arc A specially graduated arc attached to the vertical circle of an explorer's alidade or transit to simplify the computation of elevation differences for inclined stadia sights (without the use of vertical angles). The arc is so graduated that each division on the arc is equal to $100(0.5 \sin 2A)$, where A is the vertical angle. Named after William M. Beaman (1867-1937), U.S. topographic engineer, who designed it in 1904. Syn: *Beaman arc.*

beam balance *Westphal balance.*

bean ore A loose, coarse-grained, pisolitic iron ore; limonite occurring in lenticular aggregations. See also: *pea ore.*

bearing (a) The angular *direction* of any place or object at one fixed point in relation to another fixed point; esp. the horizontal direction of a line on the Earth's surface with reference to the cardinal points of the compass, usually expressed as an angle of less than 90 degrees east or west of a reference meridian adjacent to the quadrant in which the line lies and referred to either the north or south point (e.g. a line in the NE quadrant making an angle of 50 degrees with the meridian will have a bearing of N 50° E). Cf: *azimuth* [*surv*]. See also: *true bearing; magnetic bearing; compass bearing; back bearing.* (b) The horizontal direction of one terrestrial point from another (such as an observer on a ship), usually measured clockwise from a reference direction and expressed in degrees from zero to 360; specif. *astronomic azimuth.* (c) Relative position or direction, as in reference to the compass or to surrounding landmarks. Usually used in the plural.

bearing capacity The load per unit of area which the ground can safely support without excessive yield. See also: *ultimate bearing capacity.*

bearing tree A tree forming a *corner accessory*, its distance distance and direction from the corner being known. It is identified by prescribed marks cut into its trunk. Syn: *witness tree.*

bearsite A monoclinic mineral: $Be_2(AsO_4)(OH).4H_2O.$

beat A periodic pulsation caused by the simultaneous occurrence of waves of differing frequencies.

beat frequency That frequency which is the difference between two given frequencies.

Beaufort wind scale The most commonly used *wind scale*, in which code numbers and descriptive terms are assigned to various ranges of wind velocity, e.g. a wind velocity of 8-10 mph (or 7-10 knots) is Beaufort Code Number 3, and is called a "gentle breeze". The scale is a modernized version of that devised by Admiral Beaufort of the British Navy early in the nineteenth century.

beauxite Original spelling of *bauxite.*

beaverite A canary-yellow mineral: $Pb(Cu,Fe,Al)_3(SO_4)_2(OH)_6.$

beaver meadow An area of soft, moist ground resulting from the construction of a beaver dam; a beaver pond that has been changed into a marsh of grass or sedge upon abandonment of the dam by the beavers.

bebedourite A pyroxenite that contains biotite, with accessory perovskite, apatite, and titanomagnetite.

beck A British term for a small stream or brook, often with a stony bed, a rugged or winding course, and a rapid flow. Etymol: Old Norse.

Becke line In the *Becke test* of relative refraction indices, a bright line, visible under the microscope, that separates substances of different refractive indices.

beckelite A wax-yellow to brown isometric mineral: $Ca_3(Ce, La,Y)_4(Si,Zr)_3O_{15}.$ It may be britholite.

beckerite A brown variety of retinite having a very high (20-23%) oxygen content.

Becke test In optical mineralogy, a test under the microscope, at moderate or high magnification, for comparing the indices of refraction of two contiguous minerals, or of a mineral and a mounting medium or immersion liquid, in a thin section or other mount. If these substances differ materially in refractive index, they are separated by a bright line (the so-called *Becke line*), which moves toward the less refractive substance when the tube of the microscope is lowered, and away from that substance when the tube is raised.

beckite Original, but incorrect, spelling of *beekite*.

becquerelite An amber to yellow secondary mineral: $CaU_6O_{19}.11H_2O$. It occurs in small orthorhombic crystals and crusts on pitchblende.

Becquerel ray A term used before the terms alpha, beta, and gamma rays were introduced for the particles emitted during radioactive decay.

bed [geophys] A rock mass of relatively greater horizontal than vertical extent, characterized by a change in physical properties from those of the overlying rock.

bed [stratig] (a) A subdivision of a stratified sequence of rocks, lower in rank than a *member* or formation, internally composed of relatively homogeneous material exhibiting some degree of lithologic unity, and separated from the rocks above and below by visually or physically more or less well-defined boundary planes; "the smallest rock-stratigraphic unit recognized in classification" (ACSN, 1961, art.8). It is generally named as a formal unit only when distinctive and particularly useful to recognize; otherwise, it is an informal unit, as in "the limestone bed at the base of the formation". A bed may be of any thickness, although it commonly varies from several centimeters to a meter. The term is used primarily for a sedimentary unit, but has been applied to a metamorphic derivative of a sedimentary bed, to a layer of pyroclasts (e.g. an ash bed), to an individual lava flow in a sequence, and to a structurally defined layer in an igneous intrusion. The term is commonly used as a syn. of *stratum* and of *layer*, but has also been applied as a general term for a rock unit composed of "two or more strata or laminae" (Payne, 1942, p.1724) and as a kind of stratum whose thickness is greater than 1 cm (McKee & Weir, 1953, p.382). It originally signified a bedding plane or "surface-junction of two different strata" (Page, 1859, p.87). Cf: *lamina*. See also: *beds*. (b) A layer containing a concentration of paleontologic or anthropologic evidence; e.g. a *bone bed*.

bed [geomorph] (a) The ground upon which any body of water rests, or the land covered by the waters of a stream, lake, or ocean; the bottom of a watercourse or of a stream channel. Examples: *stream bed; seabed.* (b) The land surface marking the site of a former body of water, or representing land recently exposed by recession or by drainage.---Syn: *floor; bottom.*

bed configuration A group of *bed forms.*

bedded [ore dep] (a) Said of a vein or other mineral deposit that follows the bedding plane in a sedimentary rock. (b) Said of a layered replacement deposit. Cf: *stratiform [ore dep]; strata-bound [ore dep].*

bedded [stratig] Formed, arranged, or deposited in layers or beds, or made up of or occurring in the form of beds; esp. said of a layered sedimentary rock, deposit, or formation. The term has also been applied to nonsedimentary material that exhibits depositional layering, such as the "bedded deposits" of volcanic tuff alternating with lava in the mantle of a stratovolcano. See also: *stratified; well-bedded.*

bedded chert Brittle, close-jointed, rhythmically layered *chert* occurring in areally extensive and thick (measured in tens of meters) deposits, and containing distinct, usually even-bedded layers (3-5 cm thick) separated by partings of dark siliceous shale or by layers of siderite. Many bedded cherts are believed to be the result of primary deposition of silica under geosynclinal conditions. Examples include the Monterey (Miocene) and the Franciscan (Jurassic?) cherts of California, and the Mesozoic radiolarian cherts of the Alps and Apennines. See also: *novaculite.*

beddedness index *stratification index.*

bedded volcano A less-preferred syn. of *stratovolcano.*

bedding [mining] (a) A quarryman's term applied to a structure occurring in granite and other crystalline rocks that tend to split in well-defined planes more or less horizontally or parallel to the land surface. (b) The storing and mixing of different ores in thin layers in order to blend them more uniformly for the future reclamation.

bedding [stratig] (a) The arrangement of a sedimentary rock in beds or layers of varying thickness and character; the general physical and structural character or pattern of the beds and their contacts within a rock mass, such as "cross-bedding" and "graded bedding"; a collective term denoting the existence of beds. Also, the structure so produced. The term may be applied to the layered arrangement and structure of an igneous or metamorphic rock. See also: *stratification, layering.* (b) *bedding plane.*

bedding cave *bedding-plane cave.*

bedding cleavage *bedding-plane cleavage.*

bedding fault A fault, the fault surface of which is parallel to the bedding plane of the constituent rocks. Cf: *bedding glide.* Syn: *bedding-plane fault.*

bedding fissility The property possessed by a sedimentary rock (esp. a fine-grained rock such as shale) of tending to split more or less parallel to the bedding; *fissility* along bedding planes. It is a primary foliation that forms in a sedimentary rock while the sediment is being deposited and compacted, and is a result of the parallelism of the platy minerals to the bedding plane.

bedding glide A nearly horizontal overthrust fault produced by bedding-plane slip (Nelson, 1965). Cf: *bedding fault.* Syn: *bedding thrust.*

bedding joint In sedimentary rock, a joint that is parallel to the bedding plane; a joint that follows a bedding plane. Syn: *bed joint.*

bedding plane (a) A planar or nearly planar *bedding surface* that visibly separates each successive layer of stratified rock (of the same or different lithology) from its preceding or following layer; a plane of deposition. It often marks changes in the circumstances of deposition, and is often marked by partings, color differences, or both. (b) A term commonly applied to any bedding surface, even when the surface is conspicuously bent or deformed by folding. (c) A term commonly applied to a plane of discontinuity (usually the bedding plane) along which a rock tends to split or break readily.---Syn: *bedding; bed plane; stratification plane; plane of stratification.*

bedding-plane cave A cave passage, generally broad and flat, that has developed along a bedding plane, usually by corrosion of one of the beds. Syn: *bedding cave.*

bedding-plane cleavage Cleavage that is parallel to the bedding plane. Cf: *axial-plane cleavage.* Syn: *bedding cleavage; parallel cleavage.*

bedding-plane fault *bedding fault.*

bedding-plane parting A *parting* or surface of separation between adjacent beds or along a bedding plane.

bedding-plane slip The slipping of sedimentary strata along bedding planes during folding. It produces *disharmonic folding* and, in extreme form, *décollement.* Syn: *flexural slip.*

bedding surface A surface, usually conspicuous, within a mass of stratified rock, representing an original surface of deposition; the surface of separation or interface between two adjacent beds of sedimentary rock. If the surface is more or less regular or nearly planar, it is called a *bedding plane.*

bedding thrust *bedding glide.*

bedding void An open space between successive lava flows, formed as the overlying flow does not completely conform itself to the solidified crust of lava beneath it.

Bedford limestone A commercial name for *spergenite*, represented by a Mississippian limestone extensively quarried from Bedford (Lawrence County), Ind., and sold for building purposes. Syn: *Bedford stone.*

bed form Any deviation from a flat bed, generated by the flow on the bed of an alluvial channel (Middleton, 1965, p.247). See also: *bed configuration.*

bediasite A jet-black to brown tektite from east-central Texas. Named after the Bidai (Bedias) Indians of the Trinity River valley.

bed joint *bedding joint.*

bed load The part of the total *stream load* that is moved along (on, near, or immediately above) the stream bed, such as the larger or heavier particles (boulders, pebbles, gravel) transported by traction or saltation along the bottom; the part of the load that is not continously in suspension or solution. See also: *bed-material load; contact load; saltation load.* Also spelled: *bedload.* Syn: *bottom load; traction load.*

bed-load function The rate at which various streamflows for a given channel will transport the different particle sizes of the bed-material load. It is used to estimate the total bed-material load.

bed material The material of which the bed of a stream is composed; it may originally have been the material of suspended load or of bed load, or may in some cases be partly residual.

bed-material load The part of the total *sediment load* (of a stream) composed of all particle sizes found in appreciable quantities in the bed material; it is the coarser part of the load, or the part that is most difficult to move by flowing water. See also: *bed load.* Cf: *wash load.*

bed moisture *inherent moisture.*

Bedoulian Substage in Switzerland: Lower Cretaceous (lower Aptian; below Gargasian Substage).

bed plane *bedding plane.*

bed rock A general term for the rock, usually solid, that underlies soil or other unconsolidated, superficial material. Adj: *bedrock.* A British syn. of the adjectival form is *solid.*

bedrock valley A valley eroded in bedrock.

beds A general term used informally for stratigraphic units or complex sequences that are incompletely known as to their thickness and detailed lithologic successions or that consist of two or more beds of essentially identical lithology and following in such close succession as to constitute a unit in themselves or that are of local economic significance; e.g. "the beds of Permian age", "the sandy beds forming the hilltop", "key beds", or "coal beds". The term is not used with the proper names of stratigraphic units. See also: *bed* [*stratig*]; *assise.*

bed separation In mining, the parting of strata along bedding planes caused by differential subsidence above a mine roof. It is an important factor in the engineering structure of mines.

beechleaf marl A term used in Lancashire, Eng., for brownish, finely laminated marl of glacial origin. Cf: *toadback marl.*

beef A quarryman's term, used originally in Purbeck (southern England), for thin, flat-lying veins or layers of fibrous calcite (sometimes of anhydrite, gypsum, halite, or silica) occurring along bedding planes of shale, giving a resemblance to beef. It appears to be due to rapid crystallization in lenticular cavities. Syn: *bacon.*

beegerite A mixture of matildite and schirmerite.

beekite (a) White, opaque silica occurring in the form of subspherical, discoid, rosette-like, doughnut-shaped, or botryoidal accretions, generally intervolved as bands or layers, and commonly found on silicified fossils or along joint surfaces as a replacement of organic matter; e.g. chalcedony pseudomorphous after coral, shell, or other fossils. See also: *ooloid.* (b) Concretionary calcite commonly occurring in small rings on the surface of a fossil shell that has weathered out of its matrix.---Named after Dr. Beek, dean of Bristol. Originally spelled: *beckite.*

beerbachite A hornfels, originally described as a hypabyssal dike rock, that is often light-colored, similar to aplite in appearance but chiefly composed of fine-grained labradorite, orthopyroxene, clinopyroxene, and magnetite.

beetle stone (a) A *septarium* of coprolitic ironstone, the enclosed coprolite resembling the body and limbs of a beetle. (b) An old name for *turtle stone.*

before breast A miner's term for that part of the orebody that still lies ahead. See also: *breast.*

before present An indication of time calculation, used especially when referring to *radiometric dating.* When used with a carbon-14 date, it indicates that the quoted date is calculated from A.D. 1950, because of the increased radioactivity introduced into the atmosphere since that time by nuclear tests. Abbrev: B.P.

beforsite A dolomitic *carbonatite.* It is commonly hypabyssal.

beheaded stream The diminished lower part of a stream whose headwaters have been captured by another stream.

beheading (a) The cutting off of the upper part of a stream and the diversion of its headwaters into another drainage system by *capture.* (b) The removal of the upper part of a stream's drainage area by wave erosion. Cf: *betrunking.*

behierite A mineral: $(Ta,Nb)BO_4$.

behoite A mineral: $Be(OH)_2$.

beidellite (a) A white, reddish, or brownish-gray clay mineral of the montmorillonite group: $(Na,Ca/2)_{0.33}Al_{2.17}(Si_{3.17}Al_{0.83})$-$O_{10}(OH)_2 \cdot nH_2O$. It is an aluminian montmorillonite, characterized by replacement of Si^{+4} by Al^{+3} and by the absence or near absence of magnesium or iron replacing aluminum. Beidellite is a common constituent of soils and certain clay deposits (such as metabentonite). Cf: *montmorillonite.* (b) An old name applied to a material that is actually a mixture, esp. an interlayered mixture of illite and montmorillonite or a mixture of clay minerals and hydrated ferric oxide.---"Because of the past use of the term, it is believed desirable to drop it entirely [from clay-mineral terminology] in order to avoid confusion" (Grim, 1968, p.48).

bekinkinite An igneous rock that contains no feldspar, being composed chiefly of barkevikite, nepheline, and olivine, along with soda amphibole (e.g. hornblende), biotite, and analcime. Cf: *fasinite.*

bel [**geomorph**] A term applied in India and Pakistan to "sandy islands in the beds of rivers" (Stamp, 1961, p.60). Etymol: Panjabi. Syn: *bhel.*

Belanger's critical velocity *critical velocity* (c).

belemnite Any member of an order of coleoids characterized by a well developed internal shell consisting of a guard, phragmocone, and forward-projecting dagger- or spade-like proostracum. The body has a ten-armed crown, each arm equipped with a double row of arm hooks.

belemnoid A broad term applied to any one of the belemnitelike coleoids including, besides the belemnites, those forms having a tripartite proostracum and those with a living chamber and tentacles without arm hooks.

belite A calcium orthosilicate found as a constituent of portland-cement clinkers; specif. *larnite.* Syn: *felite.*

bell A cone-shaped nodule or concretion in the roof of a coal seam, and liable to collapse without warning. Cf: *caldron bottom; pot bottom; kettle bottom.* See also: *bell hole.*

bell hole A cavity in the roof of a coal seam, produced by the falling of a *bell.*

bellicatter Var. of ballycadder, a syn. of *icefoot.*

bellingerite A light-green or bluish-green mineral: $3Cu(IO_3)_2 \cdot 2H_2O$.

bell-jar intrusion An igneous intrusion similar to a *bysmalith* but differing in that the strata of the sunken block have become domed and severely fractured.

bell-metal ore A syn. of *stannite,* esp. the bronze-colored variety.

bell-shaped distribution A frequency distribution whose plot has the shape of a bell; usually, a *normal distribution.*

belly *pocket.*

beloeilite A granular plutonic rock composed of sodalite, less potassium feldspar, and a small amount of mafic minerals. Nepheline may or may not be present.

belonite An elongated or acicular crystallite having rounded or pointed ends.

belonosphaerite A now obsolete term for a spherulite whose minerals are in radial arrangement.

belovite (a) A mineral of the apatite group: $(Sr,Ce,Na,Ca)_5$-$(PO_4)_3(OH)_2$. (b) *arsenate-belovite*.

belt A long area or strip of pack ice, measuring from 1 km to more than 100 km in width. Cf: *strip* [*ice*].

belted coastal plain A broad, maturely dissected coastal plain on which there are a series of roughly parallel cuestas alternating with subsequent lowlands or vales; e.g. the Gulf Coastal Plain through Alabama and Mississippi.

belted metamorphics A geomorphic term proposed by Strahler (1946) to describe mountain areas of folded and faulted metamorphosed sedimentary and igneous rocks which have been differentially eroded in distinct but irregular, elongate, subparallel ridges and valleys.

belted outcrop plain A peneplain or erosion surface displaying alternating and parallel belts or bands of resistant upland rock and weaker lowland rock; e.g. a *belted coastal plain*. Syn: *belted plain*.

belteroporic fabric A rock fabric in which the preferred orientation of its mineral constituents was determined solely by the direction of easiest growth.

Beltian An approximate equivalent of *Riphean*.

Beltian orogeny A name proposed by Eardley (1962) for a supposed orogeny at the end of the Precambrian in western North America, based on unconformable relations between the Belt Series and the Cambrian in northwestern Montana. The existence of an orogeny at the time proposed is dubious; strata overlying the Belt Series are Middle Cambrian or younger, and the Belt itself has now been dated radiometrically as much older than 900 m.y.; moreover, it is overlain unconformably farther west by thick sequences of younger Precambrian strata.

belt of cementation *zone of cementation*.

belt of no erosion A zone adjacent to a drainage divide where no erosion by overland flow occurs because of insufficient depth and velocity of flow, and a slope that is too gentle to overcome the initial resistance of the soil surface to sheet erosion (Horton, 1945, p.320); its width is equal to *critical length*.

belt of soil moisture *belt of soil water*.

belt of soil water The upper subdivision of the *zone of aeration* limited above by the land surface and below by the intermediate belt. This zone contains plant roots and water available for plant growth. Syn: *belt of soil moisture; discrete-film zone; zone of soil water; soil-water zone; soil-water belt*.

belt of wandering The whole breadth of the valley floor that may be worn down by a stream.

belt of weathering *zone of weathering*.

belugite A group of intrusive rocks intermediate, in regard to the feldspar content (containing andesine and/or labradorite), between a diorite and a gabbro; also, any rock in that group. The term is rarely used. Its name is derived from the Beluga River, Alaska. Cf: *aleutite*.

belyankinite A yellowish-brown mineral: $Ca(Ti,Zr,Nb)_6O_{13}.14H_2O$.

bementite A grayish-yellow or grayish-brown mineral: $Mn_8Si_6O_{15}(OH)_{10}$ (?). It may contain small amounts of zinc, magnesium, and iron.

ben A Scottish term for a high hill or mountain, or a mountain peak; it is used only in proper names, esp. in those of the higher summits of mountains, as *Ben* Nevis or *Ben* Lomond. Etymol: Gaelic *beann* or *beinn*, "peak".

bench [*coal*] A layer of coal; either a coal seam separated from nearby seams by an intervening layer, or one of several layers within a coal seam that is mined separately from the others.

bench [*coast*] (a) *wave-cut bench*. (b) A nearly horizontal area at about the level of maximum high water on the ocean side of an artificial dike (CERC, 1966, p.A3).

bench [*geomorph*] A long, narrow, relatively level or gently inclined strip or platform of land, earth, or rock, bounded by steeper slopes above and below, and formed by differential erosion of rocks of varying resistance or by a change of base-level erosion; a small terrace or step-like ledge breaking the continuity of a slope; an eroded bedrock surface between valley walls. The term sometimes denotes a form cut in solid rock as distinguished from one (as a *terrace*) cut in unconsolidated material. See also: *berm; mesa*.

bench gravel A term applied in Alaska and the Yukon Territory to gravel beds on the side of a valley above the present stream bottom and that represents part of the stream bed when it was at a higher level. See also: *bench placer*.

benchland (a) A *bench*, esp. one along a river. Also, the land situated in, or forming, a bench. (b) A land surface composed largely of benches; e.g. a *piedmont benchland*.

bench mark (a) A relatively permanent metal tablet or other mark firmly embedded in a fixed and enduring natural or artificial object, indicating a precisely determined elevation above or below a standard datum (usually sea level) and bearing identifying information, and used as a reference in topographic surveys and tidal observations; e.g. an embossed and stamped disk of bronze or aluminum alloy, about 3.75 in. in diameter, with an attached shank about 3 in. in length, which may be cemented in natural bedrock, in a massive concrete post set flush with the ground, or in the masonry of a substantial building. Its location and position are precisely and accurately determined, and it is usually monumented to include bench-mark name and number and the responsible agency. Abbrev: B.M. See also: *permanent bench mark; temporary bench mark*. (b) A well-defined, permanently fixed point in space, used as a reference from which measurements of any sort (such as of elevations) may be made.

bench-mark soil A soil for which the full results of research are reported, and which is taken as representative of similar soils.

bench placer A *bench gravel* that is mined as a placer. Syn: *river-bar placer; terrace placer*.

bench terrace A shelf-like embankment of earth with a flat top and often a steep or vertical downhill side, constructed along the contour of sloping land to control runoff and erosion. It is used esp. in series to convert mountainous slopes to arable land.

bend [*sed*] An English miner's term for any hardened clayey substance. See also: *bind*.

bend [*geomorph*] (a) A curve or turn in the course, bed, or channel of a stream, not yet developed into a meander. Also, the land area partly enclosed by a bend or meander. (b) A curved part of a lake, inlet, or coastline.

bend folding A syn. of *false folding*. Cf: *buckle folding*.

bend gliding Bedding-plane slip along a plane that curves about an axis.

bending [*ice*] The upward or downward motion in sea ice, caused by lateral pressure of wind or tide; the first stage in the formation of deformed ice.

bendway A term applied along the Mississippi River to one of the deep-water pools occurring at the bends of meanders on alternate sides of the river. See also: *crossing*.

Benioff fault plane *Benioff seismic zone*.

Benioff seismic zone A plane beneath the trenches of the circum-Pacific belt, dipping toward the continents at an angle of about 45°, along which earthquake foci cluster. It is sometimes referred to as the *Benioff fault plane*. According to the theory of plate tectonics and sea-floor spreading, plates of the lithosphere sink into the upper mantle through this zone.

68

benitoite A blue to colorless, transparent, hexagonal mineral: $BaTiSi_3O_9$. It is strongly dichroic, resembles sapphire in appearance, and is sometimes used as a gem.

benjaminite A mineral: $Pb_2(Cu,Ag)_2Bi_4S_9$.

benstonite A rhombohedral mineral: $(Ca,Mg,Mn)_7(Ba,Sr)_6(CO_3)_{13}$.

benthic Pertaining to the *benthos*; also, said of that environment. Syn: *benthonic; demersal.*

benthogene Said of sediments derived from benthonic plants or animals, or precipitated chemically on the ocean floor (Sander, 1951, p.6).

benthograph A submersible, spherical container for photographic equipment used in photographic exploration of the deep sea.

benthonic *benthic.*

benthos Those forms of marine life that are bottom-dwelling; also, the ocean bottom itself. Certain fish that are closely associated with the benthos may be included. Adj: *benthic.*

bentonite (a) A soft, plastic, porous, light-colored rock consisting largely of colloidal silica and composed essentially of clay minerals (chiefly of the montmorillonite group) in the form of extremely minute crystals, and produced by devitrification and accompanying chemical alteration of a glassy igneous material, usually a tuff or volcanic ash. Its color ranges from white to light green and light blue when fresh, becoming light cream on exposure and gradually changing to yellow, red, or brown. The rock is greasy and soap-like to the touch (without gritty feeling), and commonly has great ability to absorb large quantities of water accompanied by an enormous increase in volume (swelling to about 8 times its original volume). The term was first used by Knight (1898) to replace *taylorite*, previously proposed for the argillaceous Cretaceous deposits occurring in the Benton Formation (formerly Fort Benton Formation) of the Rock Creek district in eastern Wyoming. Syn: *volcanic clay; soap clay; mineral soap; amargosite.* (b) A commercial term applied to any of numerous variously colored clay deposits (esp. bentonite) containing montmorillonite as the essential mineral and presenting a very large total surface area, characterized either by the ability to swell in water or to be slaked and to be activated by acid, and used chiefly (at a concentration of about 3 lb per cu ft of water) to thicken oil-well drilling muds. (c) A general term for the montmorillonitic clay minerals derived from volcanic ash.

bentonite debris flow A *debris flow* associated with the seasonal freezing and thawing and the extreme cold of the arctic region, formed where easily hydrated bentonite-rich sediments are exposed to surface water (in moderate quantities for at least several week's duration) on slopes of 5-30 degrees, and developed in a smooth-sided, fluted, leveed, and U-shaped mudflow channel 1-2 m deep and 8-10 m wide. Term proposed by Anderson et al (1969, p.173) for such features near Umiat, Alaska.

bentonitic arkose A term used by Ross & Shannon (1926, p.79) for a rock containing less than 25% bentonitic clay minerals. Cf: *arkosic bentonite.*

benzene A colorless, volatile, highly inflammable toxic liquid that is the simplest member (formula C_6H_6) of the *aromatic hydrocarbon* series. It is usually produced from coal tar or coke-oven gas or synthesized from open-chain hydrocarbons, and is used chiefly as a solvent, as a motor fuel, as a material in the manufacture of dyes, and in organic synthesis. See also: *benzol.*

benzene series The *aromatic hydrocarbon* series of liquids and solids, empirical formula C_nH_{2n-6}, containing the benzene ring; i.e. it is comprised of benzene, the simplest member, and the homologues of benzene.

benzol A commercial form of *benzene* which is at least 80 percent benzene, but also contains its homologues toluene and xylene.

beraunite A dark-red or brown mineral: $Fe^{+2}Fe_4^{+3}$-$(PO_4)_3(OH)_5.3H_2O$.

berborite A mineral: $Be_2(BO_3)(OH,F).H_2O$.

Berek compensator In a microscope used for optical analysis of minerals, a type of *compensator* used for the measurement of the path difference produced by a crystal plate; it is a calcite plate that is cut at right angles to the optic axis and mounted on a rotating axis in the tube slit above the objective. The angle through which it is rotated to reach compensation is the measurement of the path difference.

beresite An aplitic hypabyssal rock altered to a material resembling greisen, containing quartz and often pyrite. Beresite was originally described as being predominantly feldspar, later determined to be free of feldspar, and still later described as a quartz porphyry (Johannsen, 1939, p.243).

berezovite A syn. of *phoenicochroite.* Also spelled: *beresovite.*

berezovskite A variety of chromite with Fe:Mg from 3 to 1. Syn: *beresofskite.*

berg [geomorph] (a) A term used in the Hudson River valley, N.Y., for a mountain or hill. Etymol: Dutch *bergh*, akin to German *Berg*, "mountain". (b) A term used in South Africa for a mountain or mountain range. Etymol: Afrikaans.

berg [glaciol] Shortened form of *iceberg.*

bergalite A lamprophyre containing phenocrysts of melilite, haüyne, biotite, and rare augite in a fine-grained groundmass of the same minerals and also nepheline, magnetite, perovskite, apatite, and glass.

berg crystal *rock crystal.*

bergenite A yellow secondary mineral: $Ba(UO_2)_4(PO_4)_2$-$(OH)_4.8H_2O$.

Bergmann's rule In zoology, the statement that warm-blooded animals tend to be larger in colder parts of the environment than in warmer. Named after Carl Bergmann (d. 1865), German biologist. Cf: *Allen's rule.*

bergmehl (a) *diatomaceous earth.* (b) *rock milk.*---Also spelled: *bergmeal.*

bergschrund A deep and often wide gap or crevasse, or a series of closely spaced crevasses, in ice or firn at or near the head of a mountain glacier or snowfield, that separates moving ice and snow from the relatively immobile ice and snow (ice apron) adhering to the confining headwall of a cirque. It may be covered by or filled with snow during the winter, but be visible and reopened in the summer. Etymol: German *Bergschrund*, "mountain crack". Cf: *randkluft.*

bergschrund action Enlargement of a cirque occupied by a glacier through such processes as frost action and abrasion along the bergschrund.

berg till (a) A glacial till deposited intact by grounded icebergs in lakes bordering an ice sheet. (b) A lacustrine clay containing boulders and stones dropped into it by melting icebergs.-- Syn: *floe till; subaqueous till; glacionatant till.*

bergy bit A piece of floating ice, generally less than 5 m above sea level and not more than about 10 m across, larger than a *growler.* It is generally glacier ice but may be a massive piece of sea ice or disrupted hummocked ice.

bergy seltzer A sizzling sound, like that of newly uncorked seltzer water, emitted by an iceberg when melting. It is caused by the release of air bubbles that were retained in the ice under high pressure (Baker et al, 1966, p.21).

beringite A dark-colored barkevikite andesite containing albite and a smaller amount of orthoclase as the feldspars. Its name is derived from Bering Island, U.S.S.R.

berkeyite A transparent gem *lazulite* from Brazil.

berlinite A colorless to rose-red mineral: $AlPO_4$.

berm [eng] (a) A relatively narrow, horizontal, man-made strip, shelf, ledge, or bench built along an embankment, situated part way up and breaking the continuity of a slope. (b) The bank of a canal opposite the towing path. (c) The side, margin, or shoulder of a road, adjacent to and outside the paved portion.----Etymol: Dutch, "strip of ground along a dike". Syn: *berme.*

berm [beach] A low, impermanent, nearly horizontal or landward-sloping bench, shelf, ledge, or narrow terrace on the backshore of a beach, formed of material thrown up and deposited by storm waves; it is generally bounded on one side or the other by a beach ridge or beach scarp. Some beaches have no berms, others have one or several. See also: *storm berm*. Syn: *beach berm; backshore; backshore terrace*.

berm [geomorph] (a) A term introduced by Bascom (1931) for a terrace- or bench-like remnant of a surface developed to middle or late maturity in a former erosion cycle which has since been interrupted by renewed downward cutting following uplift; e.g. the undissected remnant of an earlier valley floor of a rejuvenated stream, or the remnant of an uplifted abrasion platform that underwent wave erosion along the coast. Bascom intended that the term replace *strath*, although "berm" sometimes includes the shoulder of a new valley together with the remnant of the old valley floor (Engeln, 1942, p.221). See also: *bench; strath terrace*. (b) A horizontal ledge of land bordering either bank of the Nile River and inundated when the river overflows.

bermanite A reddish-brown mineral: $Mn^{+2}Mn_2^{+3}(PO_4)_2$-$(OH)_2.4H_2O$.

berm crest The seaward or outer limit or edge, and generally the highest part, of a berm on a beach; a line representing the intersection of two berms or of a berm and the foreshore. The crest of the most seaward berm separates the foreshore from the backshore. Cf: *beach crest*. Syn: *berm edge; crest*.

berm edge *berm crest*.

bermudite A lamprophyric extrusive rock composed of small biotite phenocrysts in a groundmass presumably composed of analcime, nepheline, and sanidine.

berndtite A mineral: SnS_2.

Bernoulli effect The observation in hydrodynamics that, in a stream of fluid, pressure is reduced as velocity of flow increases. Cf: *Bernoulli's theorem*.

Bernoulli's theorem The statement in hydraulics that under conditions of uniform steady flow of water in a conduit or stream channel, the sum of the velocity head, the pressure head, and the head due to elevation at any given point along such conduit, or channel, is equal to the sum of these heads at any other point along each conduit or channel plus or minus the losses in head between the two points due to friction or other causes (being plus if the latter point is upstream and minus if downstream). It was developed by the Swiss engineer Daniel Bernoulli in 1738 (ASCE, 1962). Cf: *Bernoulli effect*.

bernstein A syn. of *amber*. Etymol: German *Bernstein*, from Old German *bôrnen*, "to burn", + *Stein*, "stone".

berondrite An igneous rock similar to *luscladite* but characterized by the presence of hornblende laths and titanaugite as mafic components. Cf: *fasinite*.

Berriasian European stage: lowermost Lower Cretaceous (above Portlandian of Jurassic, below Valanginian). It was formerly included in the Valanginian.

berryite A mineral: $Pb_2(Cu,Ag)_3Bi_5S_{11}$.

berthierine A mineral: $(Fe^{+2},Fe^{+3},Mg,Al)_{3-x}(Si,Al)_2O_5(OH)_4$, with x about 0.1 to 0.2. Much so-called *chamosite* appears to be identical with berthierine (Hey, 1962, 16.19.16a).

berthierite A dark steel-gray mineral: $FeSb_2S_4$.

berthonite *bournonite*.

bertossaite A mineral: $(Li,Na)_2(Ca,Fe,Mn)Al_4(PO_4)_4(OH,F)_4$.

bertrandite A colorless to pale-yellow mineral: $Be_4Si_2O_7(OH)_2$.

Bertrand lens A removable, plano-convex lens in the tube of a petrographic microscope that is used in conjunction with converging light to form an interference figure.

beryl (a) A mineral: $Be_3Al_2Si_6O_{18}$. It usually occurs in green or bluish-green, sometimes yellow or pink, or rarely white, hexagonal prisms in metamorphic rocks and granitic pegmatites and as an accessory mineral in acid igneous rocks.

Transparent and colored gem varieties include emerald, aquamarine, heliodor, golden beryl, and vorobyevite. Beryl is the principal ore of beryllium. (b) *green beryl*.

beryllite A mineral: $Be_3SiO_4(OH)_2.H_2O$.

beryllium detector An instrument utilizing the principles of gamma-ray *activation analysis* to detect and analyze for beryllium. An enclosed gamma-ray source, generally Sb_{124}, transforms Be_9 to Be_8 (2^α) plus a neutron (n). The measurement of neutron production rate allows for quantitative evaluation of beryllium. A popular term for the instrument is *berylometer*.

beryllonite A colorless or yellow mineral: $NaBePO_4$. It occurs in transparent, topaz-like orthorhombic crystals.

berylometer A popular term for and syn. of *beryllium detector* (*analyzer*), a commonly portable instrument used to detect and analyze for naturally occurring beryllium.

berzelianite A silver-white mineral: Cu_2Se.

berzeliite A bright-yellow mineral: $(Mg,Mn)_2(Ca,Na)_3(AsO_4)_3$. It is isomorphous with manganberzeliite. Syn: *berzelite*.

beschtauite A sodic quartz porphyry.

Bessemer ore An iron ore that contains very littel phosphorous (generally less than .045%). It is so named because it is suitable for use in the Bessemer process of refining iron for steelmaking; this process is no longer in use.

beta [mineral] adj. Of or relating to one of two or more closely related minerals and specifying a particular physical structure (esp. a polymorphous modification); specif, said of a mineral that is stable at a temperature intermediate between those of its *alpha* and *gamma* polymorphs (e.g. "beta cristobalite" or "β-cristobalite", the high-temperature isometric phase of cristobalite). Some mineralogists reverse this convention, using β for the low-temperature phase (e.g. "beta carnegieite", the triclinic phase of carnegieite produced from alpha carnegieite at temperatures below 690°C).

beta [cryst] (a) In a biaxial crystal, the intermediate *index of refraction*. (b) The interaxial angle between the a and c crystallographic axes.----Cf: *alpha [cryst]*; gamma *[cryst]*.

beta* The interaxial angle of the reciprocal lattice between a* and c* which is equal to the interfacial angle between (100) and (001). Cf: *alpha*; gamma**.

beta axis The line of intersection of two or more surfaces (Sander, 1963, p.83). It is usually written as β axis, and is equivalent to *pi axis*.

beta chalcocite *chalcocite*.

beta diagram In structural petrology, a projection of the poles of bedding planes which form a girdle, the pole of which is inferred to be a regional fold axis or b direction. Also spelled: β *diagram*. Cf: *pi diagram*.

betafite A yellow, brown, greenish, or black mineral of the pyrochlore group: $(Ca,Na,U)_2(Nb,Ta)_2O_6(O,OH)$. It is a uranium-rich variety of pyrochlore found in granitic pegmatites near Betafo, Madagascar. Betafite probably forms a continuous series with pyrochlore; the name is assigned to members of the series with uranium greater than 15%. Syn: *ellsworthite; hatchettolite; blomstrandite*.

beta particle A particle emitted from an atomic nucleus during radioactive decay which is physically identical to the electron. It may be either positively or negatively charged; if positively, it is called a positron. Cf: *alpha particle; gamma ray*. Less-preferred syn: *beta ray*.

beta quartz The polymorph of quartz that is stable from 573°C to 870°C, that has a vertical axis of six-fold symmetry and six horizontal axes of two-fold symmetry, and that has a lower refractive index and birefringence than those of *alpha quartz*. It occurs as phenocrysts in quartz porphyries, graphic granite, and granite pegmatites. Also spelled: β*-quartz*. Syn: *high quartz*.

beta ray A less-preferred syn. of *beta particle*.

beta-roselite A triclinic mineral: $Ca_2Co(AsO_4)_2.2H_2O$. It is dimorphous with roselite.

beta-uranophane A yellow, monoclinic secondary mineral:

$Ca(UO_2)_2Si_2O_7.6H_2O$. It is a dimorph of uranophane. Syn: *beta-uranotile.*

beta-vredenburgite A mineral consisting of an exsolution mixture or oriented intergrowth of jacobsite and hausmannite. Cf: *alpha-vredenburgite.*

betekhtinite An orthorhombic mineral: $Cu_{10}(Fe,Pb)S_6$.

bet lands A term used in India and Pakistan for a flood plain. Etymol: anglicization of Panjabi *bêt.*

betpakdalite A lemon-yellow mineral: $CaFe_2H_8(AsO_4)_2(MoO_4)_5.10H_2O$.

betrunked river A river that is shorn of its lower course by betrunking. Cf: *dismembered stream.*

betrunking The removal of the lower part of a stream course by submergence of a valley or by recession of the coast, leaving the several upper branches of the drainage system to enter the sea as independent streams. Cf: *dismembering; beheading.*

betwixt mountains *median mass.*

beudantite A green to black rhombohedral mineral: $PbFe_3(AsO_4)(SO_4)(OH)_6$.

beusite A mineral: $(Mn,Fe,Ca,Mg)_3(PO_4)_2$. It is related to graftonite.

bevel (a) Any surface that has or appears to have been planed off or beveled, such as the flat surface along the crest of a cuesta; esp. an inclined surface that meets another at an angle other than at right angles, such as the moderately steep and straight to convex slope produced by subaerial erosion above the vertical face of a sea cliff. (b) The angle made by a bevel. Also, the slant or inclination of a bevel.

bevel cut Any style of cutting for a gemstone having a very large table and a pavilion that may be step cut, brilliant cut, or any other style (Shipley, 1951, p.24); it is used predominantly for less valuable gems. See also: *table cut.*

beveled [geomorph] Said of a geologic structure or landform that has been cut across by *beveling.*

beveling An act or instance of cutting across a geologic structure or landform; e.g. the planation of an anticline or of the outcropping edges of strata on a mountain summit. Cf: *truncation.* Also spelled: *bevelling.*

beyerite A yellow mineral: $(Ca,Pb)Bi_2(CO_3)_2O_2$. Not to be confused with *bayerite.*

bezel (a) The portion of a brilliant-cut gem above the girdle; the *crown.* (b) More specif. the sloping surface between the girdle and the table, or only a small part (the so-called "setting edge") of that sloping surface just above the girdle.

bezel facet One of the eight large, four-sided facets on the crown of a round brilliant-cut gem, the upper points of which join the table and the lower points join the girdle. Cf: *star facet.*

B-form The microspheric form of a foraminifer. Cf: *A-form.*

bhabar A great piedmont composed of gravel and fringing the outer margin of the Siwalik Range in northern India. Etymol: Urdu-Hindi *bhâbar,* "porous". Cf: *terai.* Syn: *bhabbar.*

bhangar A term used in India for a high area (as a terrace, scarp, or hill) consisting of an old alluvial plain so situated within a river valley that it is beyond the reach of river floods. Etymol: Urdu-Hindi *bângar.* Cf: *khadar.* Syn: *bangar.*

bhel *bel.*

bhil A term applied in the Ganges delta region of India to a brackish and stagnant body of water (such as an oxbow lake or marsh) occupying an interdistributary trough that is often below sea level. Etymol: Bengali. Cf: *jheel.* Syn: *bil; bheel.*

bhit A term used in West Pakistan for a sand hill or sand ridge. Etymol: Sindhi.

B horizon A mineral horizon of a soil, below the A horizon, sometimes called the *zone of accumulation* and characterized by one or more of the following conditions: an illuvial accumulation of humus or silicate clay, iron, or aluminum; a residual accumulation of sesquioxides or silicate clays; darker, stronger, or redder coloring due to the presence of sesquioxides; a blocky or prismatic structure. B1, B2, and B3 horizons may be distinguished; B1 and B3 are transitional upward and downward, respectively. The B horizon is also called the *zone of accumulation* or the *zone of illuviation.*

bhur A term used in India and Pakistan for a hill or patch of windblown sandy soil, frequently capping the high bank of a river. Etymol: Urdu-Hindi *bhúr.*

bianchite A monoclinic mineral occurring as white crystalline crusts: $(Zn,Fe)SO_4.6H_2O$.

Biarritzian European stage: middle to upper Eocene.

bias [stat] A purposeful or accidental distortion of observations, data, or calculations in a nonrandom manner.

biaxial Said of a crystal having two optic axes and three main indices of refraction, e.g. a monoclinic, triclinic, or orthorhombic crystal. Cf: *uniaxial.*

biaxial figure An *interference figure* that may display both axes or no axes.

biaxial stress A stress system in which only one of the three principal stresses is zero.

bib A long projection of land sloping gradually into the sea (Webster, 1967, p.211).

biconvex Convex on both sides; e.g. said of a brachiopod shell having both valves convex.

bidalotite *anthophyllite.*

bideauxite A mineral: $Pb_2AgCl_3(F,OH)_2$.

bieberite A flesh-red to rose-red mineral occurring esp. in crusts and stalactites: $CoSO_4.7H_2O$. Syn: *red vitriol; cobalt vitriol.*

bielenite A peridotite that contains various pyroxenes and olivine. It differs from *lherzolite* in containing more pyroxene than olivine. Diallage, enstatite, chromite, and magnetite are commonly present.

bifoliate Pertaining to a *bilamellar* bryozoan colony.

biforaminate Said of a foraminifer (such as *Discorbis*) having both protoforamen and deuteroforamen.

biform Having the forms of two distinct kinds; e.g. said of the rhabdosome of a graptoloid (esp. of a monograptid) with thecae of two conspicuously different shapes, or said of foraminiferal shells having a growth plan that changes during ontogeny. Obsolete var: *biformed.*

biform processes In a spore, exine projections with broad bases which abruptly terminate in sharply pointed tips. The term is most commonly applied to Paleozoic spores.

bifurcation (a) The separation or branching of a stream into two parts. (b) A stream branch produced by bifurcation.

bifurcation ratio Ratio of the number of streams of a given order to the number of streams of the next higher order. According to the law of stream numbers, the ratio tends to be constant for all orders of streams in the basin. It is a measure of the degree of branching within a drainage basin. Symbol: R_b.

"big bang" hypothesis The hypothesis that the presently observed expansion of the Universe may be extrapolated back to a primeval cosmic fireball. Depending on the ratio of the initial expansion velocity to the the mass of the Universe, which is relatable to presently observable parameters (the deceleration parameter), the Universe may or may not reach a maximum distension and collapse in on itself again. Syn: *fireball hypothesis; primeval-fireball hypothesis.*

bight [coast] (a) A long, gradual bend or gentle curve, or a slight, crescent-shaped indentation, in the shoreline of an open coast or of a bay; it may be larger than a bay or it may be a segment of such a feature smaller than a bay. (b) A tract of water or a large bay formed by a bight; an *open bay.* Example: the Great Australian Bight. (c) A term sometimes, although rarely, applied to a bend or curve in a river, or in a mountain chain.

bight [ice] An extensive crescent-shaped indentation in the ice edge, formed either by wind or current. Syn: *bay [ice]; ice bay.*

71

big lime A driller's term for a thick, definitive limestone formation in an oil field.

bigwoodite A medium-grained plutonic rock consisting chiefly of microcline, microcline-microperthite, sodic plagioclase (albite), and hornblende, with acmite-augite or biotite sometimes substituting for the hornblende; an alkalic syenite.

bikitaite A white monoclinic mineral: $LiAlSi_2O_6.H_2O$.

bil *bhil*.

bilamellar (a) Said of a bryozoan colony consisting of two layers of zooids growing back to back and separated by a double *median lamina*. Syn: *bilaminar; bifoliate*. (b) Said of the walls of each chamber (in hyaline calcareous foraminifers) consisting of two primary formed layers.

bilateral symmetry The condition, property, or state of having the individual parts of an organism arranged symmetrically along the two sides of an elongate axis or having a median plane dividing the organism or part into equivalent right and left halves so that they are counterparts one of the other. Cf: *radial symmetry*. Syn: *bilateralism*.

bilinite A white to yellowish mineral: $Fe^{+2}Fe_2^{+3}(SO_4)_4.-22H_2O$. It occurs in radially fibrous masses.

bill A long, narrow promontory or headland, or a small peninsula, resembling a beak or ending in a prominent spur; e.g. Portland Bill in Dorset, England.

billabong (a) A term applied in Australia to a blind channel leading out from a river and to a usually dry stream bed that may be filled seasonally. (b) An Australian term for an elongated, stagnant backwater or pool produced by a temporary overflow from a stream, or for an *oxbow lake* which may not be permanently filled with water.----Etymol: aboriginal term meaning "dead river".

billietite An amber-yellow secondary mineral: $BaU_6O_{19}.11H_2O$. It occurs in orthorhombic plates and is closely related to becquerelite.

billitonite An Indonesian tektite from Belitung (Billiton) Island, near Sumatra; a tektite from the East Indies.

bilobite A trace fossil consisting of a two-lobed (bilobate) trail; esp. one made by a trilobite, such as a shallow pocket-like pit, passage, or burrow shoveled or scratched by a trilobite or a coffee-bean form with a median groove and transverse wrinkles representing a resting trail made by a trilobite.

biloculine Having two chambers; specif. pertaining to or shaped like *Pyrgo* ("Biloculina"), a genus of calcareous imperforate foraminifers having a two-chambered exterior part of test and found abundantly in the North Sea where their remains form much of the ooze covering the bottom.

bimaceral Said of a coal microlithotype consisting of two macerals. Cf: *monomaceral; trimaceral*.

bimagmatic Said of the texture of porphyritic rocks in which the minerals occur in two generations. A translation of the German term *bimagmatisch* (Johannsen, 1939, p.203).

bimodal distribution A frequency distribution characterized by two localized modes, each having a higher frequency of occurrence than other immediately adjacent individuals or classes. Cf: *polymodal distribution*.

bimodal sediment A sediment whose particle-size distribution shows one secondary maximum; e.g. many coarse alluvial gravels.

binary granite A syn. of *two-mica granite*, used by Keyes (1895, p.714).

binary sediment A sediment consisting of a mixture of two components or end members; e.g. a sediment with one clastic component (such as quartz) and one chemical component (such as calcite), or an aggregate containing sand and gravel.

binary system A chemical system containing two components, e.g. the $MgO-SiO_2$ system.

bind A British coal miner's term for any fine-grained, well-laminated rock (such as shale, clay, or mudstone, but not sandstone) associated with coal. See also: *bend* [*sed*]; *blaes*.

binder (a) The material that produces or promotes consolidation in loosely aggregated sediments; e.g. a mineral cement that is precipitated in the pore spaces between grains and that holds them together, or a primary clay matrix that fills the interstices between grains. (b) *soil binder*. (c) A term used in Ireland for a bed of sand in shale, slate, or clay. (d) A coal miner's term used in Pembrokeshire, Eng., for shale.

bindheimite A secondary mineral: $Pb_2Sb_2O_6(O,OH)$.

binding coal *caking coal*.

Bingham substance An idealized material showing linear-viscous behavior above a yield point. Below the yield point the material is presumed to be rigid.

binnacle A term applied in New York and Pennsylvania to a secondary channel of a stream. Etymol: Dutch *binnenkil*, "within channel". Syn: *binnekill*.

binnite A variety of tennantite containing silver.

binocular microscope A microscope adapted to the simultaneous use of both eyes.

binomen A name, as one used in *binomial nomenclature*, that consists of two words, the name of the genus followed by the name of the species itself. Syn: *binomial; specific name*.

binomial n. A syn of *binomen*.

binomial distribution A discrete frequency distribution of independent events or occurrences with only two possible outcomes, such as zero and one, or success and failure.

binomial nomenclature A system of naming plants and animals in which the name of each species consists of two words (i.e. a *binomen*), the first designating the genus followed by the name of the particular species; e.g. *Phacops rana*. Syn: *binominal nomenclature*.

bioaccumulated limestone A limestone consisting predominantly of shells and other fossil material that were derived from sedentary but noncolonial organisms and that accumulated essentially in place. It is characterized by an abundance of unsorted and unbroken fossils, diverse organic components, and scarce fine-grained matrix. Cf: *bioconstructed limestone*.

biocalcarenite A calcarenite containing abundant fossils or fossil fragments; e.g. a crinoidal limestone.

biocalcilutite A calcilutite containing abundant fossils or fossil fragments.

biocalcilyte A term used by Grabau (1924, p.297) for a calcareous biogenic clastic rock, such as coral rock, shell rock, or calcareous ooze. The modernized form "biocalcilite" might be useful (Thomas, 1960).

biocalcirudite A calcirudite containing abundant fossils or fossil fragments.

biocalcisiltite A calcisiltite containing abundant fossils or fossil fragments.

biocenology The branch of ecology concerned with all aspects of natural communities and the relationships between the members of those communities. Also spelled: *biocoenology*. Cf: *biosociology*.

biocenosis Var. of *biocoenosis*.

biochemical oxygen demand The amount of oxygen (measured in parts per million) removed from aquatic environments rich in organic material by the metabolic requirements of aerobic microorganisms. Abbrev: BOD. Cf: *chemical oxygen demand*. Syn: *biological oxygen demand*.

biochemical rock A sedimentary rock characterized by or resulting directly or indirectly from the chemical processes and activities of, or the chemical reactions in, living organisms; e.g. bacterial iron ores, and certain limestones.

biochore (a) A region with a distinctive fauna and/or flora; specif. one or more similar (ecologic) *biotopes*. (b) The part of the Earth's surface having a life-sustaining climate, characterized by a major type of vegetation. It consists largely of the *dendrochore*.

biochron (a) The total duration or time of existence of any taxonomic unit (including species). The term was coined as a time term by Williams (1901, p.579-580) for the "absolute duration of a fauna or flora or component parts of it''; it is synonymous with *biozone* of Buckman (1902), but refers to the geologic-time unit corresponding to ''biozone'' when that term is used in a time-stratigraphic sense as applied to strata. (b) 'A fossil fauna or flora of relatively short time range'' (Webster, 1967, p.218).

biochronologic unit (a) A division of time distinguished on the basis of biostratigraphic or objective paleontologic data; a *geologic-time unit* during which sedimentation of a biostratigraphic unit took place (Teichert, 1958a, p.117). Examples include a moment corresponding to a biostratigraphic zone, and a biochron equivalent to a time-stratigraphic biozone. (b) A term used by Jeletzky (1956, p.700) to replace biostratigraphic unit and time-stratigraphic unit, being a material rock unit defined in its type locality ''by agreement among specialists, elsewhere by criteria of time-correlation found in the contained rocks, which in practice means geochronologically valuable fossils''; a time-stratigraphic unit considered as a biostratigraphic unit.

biochronology *Geochronology* based on the relative dating of geologic events by biostratigraphic or paleontologic methods or evidence; i.e. the study of the relationship between geologic time and organic evolution. See also: *morphochronology; orthochronology; parachronology.*

biochronostratigraphic unit A term used by Henningsmoen (1961) for a time-stratigraphic unit based on fossil evidence.

bioclast (a) A single fossil fragment (Carozzi & Textoris, 1967, p.3). (b) Material derived from ''the supporting or protective structures of animals or plants, whether whole or fragmentary'' (Sander, 1967, p.327).

bioclastic rock (a) A rock consisting primarily of fragments that are broken from preexisting rocks, or pulverized or arranged, by the action of living organisms, such as by plant roots or earthworms (Grabau, 1904). The rock need not consist of organic material. The term includes ''rocks'' (such as concrete and cement) that owe their existence to man's activities. (b) A sedimentary rock consisting of fragmental or broken remains of organisms, such as a limestone composed of shell fragments. Cf: *biogenic rock.*

biocoenology An alternate spelling of *biocenology.*

biocoenosis A group of organisms that live closely together and form a natural ecologic unit. The term was first defined and introduced by the German zoologist Moebius in 1877. Cf: *thanatocoenosis.* Syn: *life assemblage; community.* Var: *biocenosis; biocoenose; biocenose.* Plural: *biocoenoses.* Etymol: Greek *bios,* ''mode of life'' + *koinos,* ''general, common''.

bioconstructed limestone A limestone consisting predominantly of material resulting from the vital activities of colonial and sediment-binding organisms (such as algae, corals, bryozoans, and stromatoporoids) that erect three-dimensional frameworks in place. Cf: *bioaccumulated limestone.*

biocycle A group of related biochores that comprise one of the major divisions of the biosphere; e.g. salt water, fresh water, and dry land are biocycles.

bioecology The branch of ecology concerned with the relationships between plants and animals in their common environment.

bioerosion Removal of consolidated mineral or lithic substrate by the direct action of organisms (Neumann, 1966).

biofacies [ecol] (a) A distinctive assemblage of organisms (animals or plants or both) formed at the same time but under different conditions; an association of fossils. (b) A term used by Dunbar and Rodgers (1957, p.137) for ''a local aspect of some larger life assemblage or biota, especially a modern biota, more or less limited by some environmental control'' (such as substratum of soil or of ocean sediments, or by relative abundance of other organisms). Syn: *biologic facies.*

biofacies [stratig] (a) A lateral subdivision of a stratigraphic unit, distinguished from other adjacent subdivisions on the basis of its biologic characters (fossil fauna or flora) without respect to the nonbiologic features of its lithology; esp. such a body of sediment or rock recognized by characters that do not affect lithology, such as the taxonomic identity or environmental implications of fossils (Weller, 1958, p.634). The upper and lower boundaries of biofacies, unlike those of biostratigraphic zones, correspond with the boundaries of some specific stratigraphic unit. (b) A term that has been applied to the lateral variations in the paleontologic nature of a stratigraphic unit, and to the expression or manifestation of such variations; the biologic aspect or fossil character of a facies of some definite stratigraphic unit, esp. considered as an expression of local biologic conditions; ''the total biological characteristics of a sedimentary deposit'' (Moore, 1949, p.17). (c) A term that has been applied to the ''strictly organic elements'' of a particular sedimentary deposit, to the ''total recognizable organic content'' (fossils) of a specified part of a stratigraphic unit, and to the ''fossil record'' of a biocoenosis.---Cf: *paleontologic facies.* Syn: *biologic facies.*

biofacies map A *facies map* based on paleontologic attributes, showing areal variation in the overall paleontologic character of a given stratigraphic unit. It may be based on proportions of, or ratios among, the fossil organisms present.

biogenesis (a) Formation by the action of organisms; e.g. coral reefs. (b) The doctrine that all life has been derived from previously living organisms. Cf: *abiogenesis.*

biogenetic law Ontogeny recapitulates phylogeny.

biogenetic rock *biogenic rock.*

biogenic rock An *organic rock* produced directly by the physiological activities of living organisms, either plant or animal (Grabau, 1924, p.280); e.g. coral reefs, shelly limestone, pelagic ooze, coal, and peat. Cf: *bioclastic rock; biolith.* See also: *phytogenic rock; zoogenic rock.* Syn: *biogenous rock; biogenetic rock.*

biogenous rock *biogenic rock.*

biogeochemical prospecting *Geochemical prospecting* based on the chemical analysis of systematically sampled plants in a region to detect biological concentrations of elements that might reflect hidden ore bodies. The trace-element content of one or more plant organs is most often measured. Cf: *geobotanical prospecting.*

biogeochemistry A branch of geochemistry which studies the effects of life processes on the distribution and fixation of chemical elements in the biosphere.

biogeography The science that deals with the geographic distribution of all living organisms. See also:*zoogeography; phytogeography.* Syn: *chorology.*

biogeology The application of biologic data to geology; e.g., identification of a concealed rock unit by the type of plants growing at the surface above it.

bioglyph A hieroglyph produced by an organism or of biologic origin (Vassoevich, 1953, p.38). Cf: *abioglyph.* Syn: *organic hieroglyph.*

bioherm A mound-, dome-, lens-, or reef-like or otherwise circumscribed mass of rock built up by, and composed almost exclusively of, the remains of sedentary organisms (such as corals, algae, foraminifers, mollusks, gastropods, and stromatoporoids) and enclosed or surrounded by rock of different lithology; e.g. an *organic reef* or a nonreef limestone mound. Term proposed by Cumings & Shrock (1928, p.599), and defined by Cumings (1930, p.207), as a structural term, although the term as applied often stresses the calcareous composition. Cf: *biostrome.* Syn: *organic mound.*

biohermal Pertaining to a bioherm, such as a ''biohermal limestone'' of restricted extent.

biohermite A term used by Folk (1959, p.13) for a limestone composed of debris broken from a bioherm and forming pocket-fillings or talus slopes associated with reefs; it was formerly used by Folk for a limestone now better known as *biolithite.*

biohydrology The study of the interactions of water, plants, and animals, including both the effects of water on biota and the physical and chemical changes in water or its environment caused by biota. Cf: *hydrobiology.*

biokinematic Said of sedimentary operations in which "the largest displacement vectors occur between a living organism and the unmodified deposit surrounding the structure produced" (Elliott, 1965, p.196); e.g. the activities shown by trace fossils. Also, said of the sedimentary structures produced by biokinematic operations.

biolite [mineral] A group name for minerals formed by biologic action (Hey, 1963, p.92).

biolite [sed] (a) *biolith.* (b) An old term for a concretion formed through the action of living organisms.

biolith A rock of organic origin or composed of organic remains; specif. *biogenic rock.* See also: *phytolith; zoolith.* Syn: *biolite.*

biolithite A limestone constructed by organisms (faunal or floral) that grew and remained in place, characterized by a rigid framework of carbonate material that binds allochem grains and skeletal elements. It is typical of reef cores. The major organism should be specified when using the term; e.g. "coral biolithite", "algal-mat biolithite", or "rudist biolithite". See also: *biohermite.*

biological oceanography The study of the plant and animal life of the oceans relative to the marine environment.

biological oxygen demand *biochemical oxygen demand* (Abbrev: BOD).

biologic artifact [organic] An organic compound whose chemical structure demonstrates its derivation from living matter.

biologic facies A syn. of *biofacies* as that term is used in stratigraphy and in ecology; e.g. coral reefs and shell banks are "biologic facies" characterized by the organisms themselves.

biologic time scale An uncalibrated *geologic time scale,* based on organic evolution, giving the relative order for a succession of events. Cf: *relative time scale.*

biologic weathering *organic weathering.*

biolysis Death and subsequent disintegration of the body.

biomass The amount of living organisms in a particular area, stated in terms of the weight or volume of organisms per unit area or of the volume of the environment.

biome A climax community that characterizes a particular natural region. Partial syn: *biotic formation.*

biomechanical Said of a rock or deposit formed by detrital accumulation of organic material.

biomere A term proposed by Palmer (1965, p.149-150) for "a regional biostratigraphic unit bounded by abrupt nonevolutionary changes in the dominant elements of a single phylum". The changes are not necessarily related to physical discontinuities, and they may be diachronous.

biometrics Statistics as applied to biologic observations and phenomena.

biomicrite A limestone consisting of a variable proportion of skeletal debris and carbonate mud (micrite); specif. a limestone containing less than 25% intraclasts and less than 25% ooliths, with a volume ratio of fossils and fossil fragments to pellets greater than 3 to 1, and the carbonate-mud matrix more abundant than the sparry-calcite cement (Folk, 1959, p.14). It is characteristic of environments of relatively low physical energy. The major organism should be specified when using the term; e.g. "crinoid biomicrite" or "brachiopod biomicrite". See also: *sparse biomicrite; packed biomicrite.*

biomicrosparite A biomicrite in which the carbonate-mud matrix has recrystallized to microspar (Folk, 1959, p.32); a microsparite containing fossils or fossil fragments.

biomicrudite A biomicrite containing fossils or fossil fragments that are more than one millimeter in diameter.

biomorphic Pertaining to or incorporating the forms of organisms; e.g. "biomorphic sediments" containing fossil forms.

bionomics *ecology.*

biopelite An organic pelite; specif. a *black shale.*

biopelmicrite A limestone intermediate in content between biomicrite and pelmicrite; specif. a limestone containing less than 25% intraclasts and less than 25% ooliths, with a volume ratio of fossils and fossil fragments to pellets ranging between 3 to 1 and 1 to 3, and the carbonate-mud matrix (micrite) more abundant than the sparry-calcite cement (Folk, 1959, p.14).

biopelsparite A limestone intermediate in content between biosparite and pelsparite; specif. a limestone containing less than 25% intraclasts and less than 25% ooliths, with a volume ratio of fossils and fossil fragments to pellets ranging between 3 to 1 and 1 to 3, and the sparry-calcite cement more abundant than the carbonate-mud matrix (micrite) (Folk, 1959, p.14).

biophile (a) Said of those elements which are the most typical in organisms and organic material (Rankama & Sahama, 1950, p.88). (b) Said of those elements which are concentrated in and by living plants and animals (Goldschmidt, 1954, p.26).

biopyribole A mnemonic term coined by Johannsen in 1911 in his classification of igneous rocks to indicate the presence of biotite, pyroxene, and/or amphibole. Etymol: *biotite* + *pyroxene* + *amph*ibole.

biorhexistasy The name given to a theory of sediment production related to variations in the vegetational cover of the land surface and characterized by a long-term, stable, subtropical, deep weathering conditions resulting in lateritic soils accompanied by removal of calcium, silica, alkalis, and alkaline earths (Erhart, 1956). See also: *rhexistasy; biostasy.*

biosociology The branch of ecology concerned with the social behavior of organisms in communities. Cf: *biocenology.*

biosome (a) A term proposed by Wheeler (1958a, p.648) for an ecologically controlled biostratigraphic unit that is mutually intertongued with one or more biostratigraphic units of differing character; the biostratigraphic equivalent of *lithosome.* Cf: *holosome.* (b) A term used by Sloss (in Weller, 1958, p.625) for a "body of sediment deposited under uniform biological conditions"; the record of a uniform biologic environment or of a biotope; a three-dimensional rock mass of uniform paleontologic content.---Not to be confused with *biostrome.* Cf: *biotope* [stratig].

biospace As used by Valentine (Valentine, 1969, p.686), that part of the *environmental hyperspace lattice* that actually represents conditions existing on the Earth. Syn: *realized ecological hyperspace.*

biospararenite A biosparite containing sand-sized fossils or fossil fragments.

biosparite A limestone consisting of a variable proportion of skeletal debris and clear calcite (spar); specif. a limestone containing less than 25% intraclasts and less than 25% ooliths, with a volume ratio of fossils and fossil fragments to pellets greater than 3 to 1, and the sparry-calcite cement more abundant than the carbonate-mud matrix (micrite) (Folk, 1959, p.14). It is generally characteristic of high-energy carbonate environments, with the spar being normally a pore-filling cement. According to Folk (1962), further textural subdivision may be made into "unsorted biosparite", "sorted biosparite", and "rounded biosparite". The major organism should be specified when using the term; e.g. "trilobite biosparite" or "pelecypod biosparite".

biosparrudite A biosparite containing fossils or fossil fragments that are more than one millimeter in diameter.

biospeleology The scientific study of the organisms that live in caves.

biosphere (a) All the area occupied or favorable for occupation by living organisms. It includes parts of the lithosphere, hydrosphere, and atmosphere. Cf: *ecosphere.* (b) All living organisms of the Earth and its atmosphere.

biostasy Maximum development of organisms at a time of

ectonic repose when residual soils form extensively on land and deposition of calcium carbonate is widespread in the sea. See also: *rhexistasy; biorhexistasy.*

biostratic unit *biostratigraphic unit.*

biostratigraphic unit A stratum or body of strata that is defined and identified by one or more distinctive fossil species or genera, without regard to lithologic or other physical features or relations; "a body of rock strata characterized by its content of fossils contemporaneous with the deposition of the strata" (ACSN, 1961, art.19). It is a stratigraphic unit bounded by the limits of paleontologic features such as the physical range or natural association of certain fossils or the stratigraphic position of some paleontologic evolutionary change. The fundamental unit is the biostratigraphic zone. If the fossil remains are so abundant that in themselves they become lithologically important, a biostratigraphic unit may also be a rock-stratigraphic unit. Syn: *biostratic unit.*

biostratigraphic zone A formal biostratigraphic unit, consisting of "a stratum or body of strata characterized by the occurrence of a fossil taxon or taxa from one or more of which it receives its name" and "defined solely by the fossils it contains, without reference to lithology, inferred environment, or concepts of time" (ACSN, 1961, art.20); the general basic unit in biostratigraphic classification and generally the smallest biostratigraphic unit on which intercontinental or worldwide correlations can be established. It is of limited but variable thickness, and may be based on all its fossils or solely on the fossils of one phylum, or one class, or one order, etc. The name fossil(s) may be a member of earlier or later fossil assemblages, and may in places be absent from the typical assemblage that characterizes the zone. The term is equivalent to *zone*, as defined by Oppel (1856-1858), in that the body of strata contains one or more characteristic fossils that are not characteristic of adjacent strata; but "biostratigraphic zone" also has a broader usage, comprising groups of strata now referred to as: *peak zone; assemblage zone;* and *range zone.* See also: *subzone; zonule.*

biostratigraphy Stratigraphy based on the paleontologic aspects of rocks, or stratigraphy with paleontologic methods; specif. the separation and differentiation of rock units on the basis of the description and study of the fossils they contain. The term was apparently proposed by Louis Dollo, Belgian paleontologist, in 1904 in a wider sense for the entire research field in which paleontology exercises a significant influence upon historical geology. Cf: *stratigraphic paleontology.*

biostratinomy An alternate, and currently less preferred, spelling of *biostratonomy.*

biostratonomy The branch of paleoecology concerned with the relationships between organisms and their environment after death and before and after burial. Cf: *taphonomy; Fossildiagenese.* Also spelled: *biostratinomy.*

biostromal Pertaining to a biostrome, such as a "biostromal limestone" resembling a coquina and often described as a coquinoid limestone.

biostrome A distinctly bedded and widely extensive or broadly lenticular, blanket-like mass of rock built by and composed mainly of the remains of sedentary organisms, and not swelling into a mound-like or lens-like form; an "organic layer", such as a bed of shells, crinoids, or corals, or a modern reef in the course of formation, or even a coal seam. Term proposed by Cumings (1932, p.334). Cf: *bioherm.* Not to be confused with *biosome.*

biota All living organisms of an area; the flora and fauna considered as a unit.

Biot-Fresnel law A statement in crystallography that the directions of extinction in any section of a biaxial crystal are parallel to the traces, on that section, of the planes bisecting the angles between the two planes containing the normal to the section and the optic axes.

biotic Of or pertaining to life or the mode of living of plants and animals collectively.

biotic formation (a) *biome.* (b) In botany, a broad, natural unit consisting of distinctive plants in a climax community.

biotic province A geographic area that supports one or more ecologic associations which are distinct from those of adjacent provinces.

biotite (a) A widely distributed and important rock-forming mineral of the mica group: $K(Mg,Fe^{+2})_3(Al,Fe^{+3})Si_3O_{10}(OH)_2$. It is generally black, dark brown, or dark green, and forms a constituent of crystalline rocks (either as an original crystal in igneous rocks of all kinds or a product of metamorphic origin in gneisses and schists) or a detrital constituent of sandstones and other sedimentary rocks. Biotite is useful in the potassium-argon method of age determination. (b) A general term to designate all ferromagnesian micas.----Syn: *black mica; iron mica; magnesia mica.*

biotitite An igneous rock almost entirely composed of biotite. Cf: *granitite.* Syn: *glimmerite.*

biotitize Introduction of, or replacement by, biotite.

biotope [ecol] (a) An area of uniform ecology and organic adaptation (Hesse et al, 1937, p.135); the habitat, or physical basis, of a uniform community of animals and plants adapted to its environment; a limited region characterized by certain environmental conditions under which the existence of a given biocoenosis is possible. It is more or less ephemeral and at any moment it is circumscribed by a boundary that is subject to expansion, contraction, or other shift in position. Cf: *biochore.* (b) An association of organisms characteristic of a particular geographic area. See also: *paleobiotope.*

biotope [stratig] (a) A biostratigraphic surface or area, analogous to lithotope, representing the part of a depositional interface to which an environment may be related (Wheeler, 1958a, p.653-654). (b) A faunal (or floral) unit that "may be interpreted environmentally" and that reflects "the influence of the environment on the biota" (Sloss et al, 1949, p.95-96). (c) The fossil record of an organic environment (Krumbein & Sloss, 1951, p.277); the record of an ecologic biotope that persisted for an appreciable time and that is represented by a body of a sediment or rock and its enclosed fossils. (d) A term that signifies an "organic environment" (Moore, 1949, p.17); the environment under which a biologic assemblage lived (Dunbar & Rodgers, 1957, p.317)'---The term is an ecologic term; it is used in stratigraphy in the sense of paleontologic "environment", and has been applied to an area, a zone, a fossil unit, a fossil record, a biostratigraphic unit, and the environment itself. It "is not likely to find much actual application in stratigraphic facies considerations" (Weller, 1958, p.636). Cf: *lithotope; biosome.*

bioturbation The churning and stirring of a sediment by organisms.

biotype Any one of a group of organisms having identical genetic constitutions, i.e. of the same *genotype.*

biozone (a) A term that has been used as a biostratigraphic unit and applied to rock strata; e.g. the deposits formed during the life span of a certain fossil form (the rocks deposited during a biochron), and the total mass of strata in which a certain fossil form occurs (the rocks identified by the actual occurrence of the fossil). "Whether the biozone includes all deposits equivalent in age to the life-span of the taxon or only those in which the taxon is actually found is a controversial question" (ACSN, 1961, art.22h). The term *range zone* has been proposed to replace "biozone" based on the distribution of a single taxonomic entity. (b) A better-known syn. of *biochron.* The term was coined by Buckman (1902) as a time term to indicate the total range of a particular taxon in geologic time (as reflected by its occurrence or "entombment in the strata"), and has been widely used to denote the time span of taxonomic entities of any category.---Teichert (1958a, p.114) recommends that the term be abandoned: as a chronologic term, it is preceded by "biochron"; as a stratigraphic term, it is "useless" because "no stratigraphic term is required" for rocks deposited during the survival time of a family, order, or phylum, and because for smaller units the terms "species zone", "generic zone", or "faunizone" are available.

bipartite oolith An oolith whose central part is divided into two more or less distinct fractions that differ in texture and/or grain size, so that it displays an asymmetric appearance (Choquette, 1955, p.338).

bipocillus A siliceous monaxonic sponge spicule (microsclere) consisting of a shaft at each end of which is a transverse, bowl-like, and excentrically attached expansion, the concave surfaces facing one another. Pl: *bipocilli*.

bipolarity The similarity or identity of groups of organisms occurring north and south of but not in the equatorial zone.

bipyramid *dipyramid*.

biquartz plate A type of *compensator* in a polarizing microscope, one half of which is dextrorotatory and the other half, levorotatory, with superimposed wedges. It is used to detect accurately the position of extinction.

biramous Two-branched; said of a crustacean limb in which the basis bears both exopod and endopod, or said of a trilobite appendage consisting of an outer and an inner branch.

Birch discontinuity A seismic-velocity discontinuity in the C layer or transition zone of the upper mantle at about 900 km, caused by phase or chemical change or both.

bird A geophysical measuring device such as a magnetometer that is used while airborne.

bird-foot delta A delta formed by many levee-bordered distributaries extending seaward and resembling in plan the outstretched claws of a bird; e.g. the Mississippi River delta. Syn: *digitate delta; bird's-foot delta*.

bird's-eye A spot, bleb, tube, or irregular patch of sparry calcite commonly found in limestones (such as dismicrites) and some dolomites as a precipitate that infills cavities resulting from localized disturbances, such as algal or burrowing activity, escaping gas bubbles, shrinkage cracking, soft-sediment slumping, reworking of sediments, or plant roots. Also applied to the porosity created by the presence of bird's eyes in a rock. Also spelled: birdseye. Syn: *calcite eye*.

bird's-eye coal Anthracite with numerous, small fractures that display its semiconchoidal fracture pattern.

bird's-eye limestone A syn. of *dismicrite*. The term "birdseye limestone" was applied in a titular sense in early New York State reports to the Lowville Limestone, a very fine-textured limestone containing spots or tubes of crystalline calcite or having light-colored specks due in part to a characteristic fossil supposed to be a form of coral and now known as *Tetradium cellulosum* (Wilmarth, 1938, p.192). Cf: *loferite*.

bird's-eye ore A miner's term used in Arkansas for a variety of pisolitic bauxite characteristic of residual deposits. Also spelled: birdseye ore.

bird's-foot delta *bird-foot delta*.

bird track A term applied in the mid-19th century to a dinosaur track before its true character was recognized.

bireflectance The ability of a mineral to change color in reflected polarized light with change in crystal orientation. Cf: *pleochroism*. Syn: *reflection pleochroism*.

birefracting *birefringent*.

birefraction *birefringence*.

birefractive *birefringent*.

birefringence The ability of crystals other than those of the isometric system to split a beam of ordinary light into two beams of unequal velocities; the difference between the greatest and the least indices of refraction of a crystal. Cf: *single refraction*. See also: *positive birefringence; refraction*. Adj: *birefringent*. Syn: *double refraction; birefraction*.

birefringent Said of a crystal that displays *birefringence*; such a crystal has more than one *index of refraction*. Syn: *birefractive; birefracting*.

biringuccite A monoclinic mineral: $Na_4B_{10}O_{17}.4H_2O$.

birkremite A light-colored quartz-bearing syenite containing alkali feldspar and some hypersthene; a hypersthene-bearing *kalialaskite*. Also spelled: *bjerkreimite*.

birne A syn. of *boule*. Etymol: German *Birne*, "pear".

birnessite A mineral: $(Na,Ca)Mn_7O_{14}.3H_2O$.

birotulate A sponge spicule (microsclere) having two wheel-shaped ends; e.g. an amphidisc, or one of its derivatives such as a hemidisc, staurodisc, or hexadisc. Syn: *birotule*.

birthstone A stone that has been chosen as appropriate to the time (month) of one's birth. The modern list specifies: January (garnet); February (amethyst); March (bloodstone or aquamarine); April (diamond); May (emerald); June (pearl, moonstone, or alexandrite); July (ruby); August (sardonyx or peridot); September (sapphire); October (opal or tourmaline); November (topaz); December (turquoise or zircon).

bisaccate *bivesiculate*.

bisbeeite A doubtful mineral: $CuSiO_3.H_2O$ (?). It may be the equivalent of chrysocolla.

bischofite A white to colorless mineral: $MgCl_2.6H_2O$.

biscuit *algal biscuit*.

biscuit-board topography A glacial landscape characterized by a rolling upland on the sides of which there are cirques that resemble the big bites made by a biscuit-cutter in the edge of a slab of dough; e.g. the Wind River Mountains in Wyoming. It may represent an early or partial stage in glaciation.

bisectrix In a biaxial crystal, a line that bisects either of the complementary angles between the two optic axes of a biaxial crystal. See also: *acute bisectrix; obtuse bisectrix*.

biserial Arranged in, characterized by, or consisting of two rows or series; e.g. a "biserial arm" of a crinoid composed of brachial plates arranged in a double row with interlocking sutures along the junction of the rows, or a "biserial brachiole" of a cystoid composed of plates arranged in two rows from the thecal articulation upward, or a "biserial test" of a foraminifer whose chambers are arranged in a parallel or alternating series of two rows, or a "biserial rhabdosome" of a scandent graptoloid composed of two rows of thecae in contact either back-to-back (dipleural) or side-by-side (monopleural). Cf: *uniserial*.

bishop's stone *amethyst*.

bismite A monoclinic mineral: Bi_2O_3. It is straw yellow and earthy or powdery, and is polymorphous with sillenite. Syn: *bismuth ocher*.

bismoclite A pale-grayish or creamy-white mineral: BiOCl. It is isomorphous with daubreeite.

bismuth A rhombohedral mineral, the native metallic element Bi. It is brittle and heavy and commonly occurs in silver-white or grayish-white (with a pinkish or reddish tinge) and arborescent, foliated, or granular forms.

bismuth blende *eulytite*.

bismuth glance *bismuthinite*.

bismuth gold *maldonite*.

bismuthide A mineral compound that is a combination of bismuth with a more positive element.

bismuthine *bismuthinite*.

bismuthinite A lead-gray to tin-white orthorhombic mineral: Bi_2S_3. It has a metallic luster and an iridescent tarnish, and usually occurs in foliated, fibrous, or shapeless masses associated with copper, lead, and other ore minerals. Syn: *bismuth glance; bismuthine*.

bismuth ocher A group name for earthy oxides and carbonates of bismuth; specif. *bismite*.

bismuth spar *bismutite*.

bismutite A mineral: $(BiO)_2CO_3$. It is earthy and amorphous, and usually dull white, yellowish, or gray. Syn: *bismuth spar*.

bismutoferrite A mineral: $BiFe_2(SiO_4)_2(OH)$.

bismutotantalite A black mineral: $Bi(Ta,Nb)O_4$.

bisphenoid *disphenoid*.

bistatic radar A *radar* with its transmitter and receiver spatially separated.

it *drill bit.*

itheca A small tube formed at each budding in the development of a graptolithine colony. It may have contained a male zooid. Cf: *autotheca; stolotheca.*

itter lake A *salt lake* whose waters contain in solution a high content of sodium sulfate and lesser amounts of the carbonates and chlorides ordinarily found in salt lakes; a lake whose water has a bitter taste. Examples include Carson Lake in Nevada, and the Great Bitter Lake in Egypt.

ittern (a) The bitter liquid remaining after seawater has been concentrated by evaporation until the sodium chloride has crystallized out. See also: *bittern salt.* (b) A natural solution, in an evaporite basin, that resembles a saltworks liquor, esp. in its high magnesium content.

ittern salt A dissolved salt in the *bittern* of a saltworks; e.g. magnesium chloride, magnesium sulfate, bromides, and iodides.

itter salts *epsomite.*

itter spar *dolomite* [mineral].

itumen A generic term applied to natural inflammable substances of variable color, hardness, and volatility, composed principally of a mixture of hydrocarbons substantially free from oxygenated bodies. Bitumens are sometimes associated with mineral matter, the nonmineral constituents being fusible and largely soluble in carbon disulfide, yielding water-insoluble sulfonation products. Petroleums, asphalts, natural mineral waxes, and asphaltites are all considered bitumens.

itumenite *torbanite.*

itumenization (a) *coalification.* (b) Hydrocarbon enrichment.

itumicarb Low-rank bituminous matter in coal that is derived from resins, waxes, spores, exines, etc. (Tomkeieff, 1954).

ituminous [coal] Pertaining to bituminous coal.

ituminous [mineral] Said of a mineral having an odor like that of bitumen.

ituminous [sed] (a) Said of a sedimentary rock (such as sandstone, shale, or limestone) that is naturally impregnated with, infiltrated by, containing, or constituting the source of bitumen. (b) Loosely, said of a substance containing much organic or carbonaceous matter; e.g. "bituminous ore" (iron ore whose gangue consists principally of coaly matter).

ituminous brown coal *pitch coal.*

ituminous coal Coal that ranks between subbituminous coal and semibituminous coal and that contains 15-20% volatile matter. It is dark brown to black in color and burns with a smoky flame. Bituminous coal is the most abundant rank of coal and is commonly Carboniferous in age. Syn: *soft coal.*

ituminous fermentation Fermentation of vegetable matter under conditions of no air and abundant moisture. Volatiles are retained, resulting in the formation of bitumens i.e., peat, coal.

ituminous lignite *pitch coal.*

ituminous limestone A dark, dense limestone containing abundant organic matter believed to have accumulated under stagnant conditions and emitting a fetid odor when freshly broken or vigorously rubbed; e.g. the Bone Spring Limestone of Permian age in west Texas. See also: *stinkstone; anthraconite.*

ituminous wood *woody lignite.*

iumbilicate Having a central depression (umbilicus) on each side of a foraminiferal test (as in planispiral forms).

iumbonate Said of a foraminifer having two raised umbonal bosses (as in *Lenticulina*).

valve adj. Having a shell composed of two distinct and usually movable valves that open and shut; said of a shell having two equal or subequal valves. Cf: *univalve.* Syn: *bivalved.*---n. A bivalve animal, or one that has a two-valved shell, such as a brachiopod and an ostracode; specif. a mollusk of the class Bivalvia (Pelecypoda), including the clams, oysters, scallops, and mussels, generally sessile or burrowing into soft sediment, having no distinct head, and possessing a hatchet-shaped foot and a sheet-like or lamelliform gill on each side of a bilaterally symmetric body. See also: *pelecypod.*

bivariant *divariant.*

bivariate Pertaining to or involving two mathematical variables; e.g. "bivariate distribution".

bivesiculate Said of pollen with two vescicles. Bivesiculate pollen are usually the pollen of conifers, but can also occur in other gymnosperms, e.g. Caytoniales and other seed ferns. Syn: *bisaccate; disaccate.*

bivium (a) The two posterior ambulacra of an echinoid. (b) The part of an asterozoan containing the madreporite and the rays on each side of it. The term is not recommended as applied to asterozoans (TIP, 1966, pt.U, p.29).---Pl: *bivia.* Cf: *trivium.*

bixbyite A black isometric mineral: $(Mn,Fe)_2O_3$. Syn: *partridgeite; sitaparite.*

bizardite An alnoite that contains nepheline as an essential component.

bjerezite A porphyritic igneous rock in which the phenocrysts of nepheline, pyroxene with acmite rims, elongated andesine laths, and orthoclase are contained in a fine-grained groundmass of pyroxene, brown mica, andesine, potassium feldspar, nepheline, analcime, and indeterminate zeolites.

bjerkreimite An alternate spelling of *birkremite.*

black alkali An old term for an *alkali soil* whose carbonates tend to blacken organic matter. Cf: *white alkali.*

black alta *alta.*

black amber (a) *jet* [coal]. (b) *stantienite.*

black-and-white iceberg An iceberg made up of sharply defined alternating layers of dark, opaque ("black") ice containing sand and stones, and cleaner, light-colored ("white") ice.

blackband (a) A dark, earthy variety of the mineral siderite, occurring mixed with clay, sand, and considerable carbonaceous matter, and frequently being associated with coal. Syn: *blackband ore.* (b) A thin layer (up to 10 cm in thickness) of blackband interbedded with clays or shales in blackband ironstone. (c) *blackband ironstone.*

blackband ironstone A dark variety of *clay ironstone* containing sufficient carbonaceous matter (10-20%) to make it self-calcining (without the addition of extra fuel). Syn: *blackband; blackband ore.*

blackband ore (a) *blackband.* (b) *blackband ironstone.*

blackbody (a) An ideal emitter which radiates energy at the maximum possible rate per unit area at each wavelength for any given temperature. A blackbody also absorbs all the radiant energy incident upon it. No actual substance behaves as a true blackbody, although platinum black and other soots rather closely approximate this ideal. However, one does speak of a blackbody with respect to a particular wavelength interval. This concept is fundamental to all the radiation laws, and is to be compared with the similarly idealized concepts of the *whitebody* and the *graybody.* In accordance with the Kirchhoff law, a blackbody not only absorbs all wavelengths, but emits at all wavelengths and does so with maximum possible intensity for any given temperature. (b) A laboratory device which simulates the characteristics of a blackbody, as through the use of wedge-shaped cavities. It is sometimes expressed as a subscript letter b.

black chalcedony The correct designation for most so-called *black onyx.*

black chalk A bluish-black carbonaceous clay, shale, or slate, used as a pigment or crayon.

black chert (a) Carbonaceous chert, such as that occurring in South Africa. (b) A term used in England for *flint.* Cf: *white chert.*

black cobalt *asbolite.*

black copper *tenorite.*

black cotton soil *regur.*

blackdamp A coal-mine gas that is nonexplosive and consists mainly of carbon dioxide and nitrogen, with little oxygen. Cf:

whitedamp; afterdamp; firedamp. Syn: *chokedamp.*

black diamond [coal] A syn. of *coal.*

black diamond [mineral] (a) *carbonado.* (b) A black gem diamond. (c) Dense black hematite that takes a polish like metal.

black drift *forest bed.*

black durain Durain that is high in hydrogen and volatiles; it contains many microspores and some vitrain. Cf: *gray durain.*

black earth [coal] Brown coal that is finely ground and used as a pigment. Syn: *Cologne earth; Cologne umber; Cassel brown; Cassel earth; Vandyke brown.*

black earth [soil] (a) *Chernozem.* (b) More generally, any black soil.

blackfellows' button *australite.*

black gold [mineral] (a) *maldonite.* (b) Placer gold coated with a black or dark-brown substance (such as a film of manganese oxide) so that the yellow color is not visible until the coating is removed.

black granite A *commercial granite* that when polished is dark gray to black. It may be a diabase, diorite, or gabbro.

black hematite A syn. of *romanechite.* The term is a misnomer because romanechite contains no iron. Cf: *red hematite; brown hematite.*

black ice (a) A clear and transparent thin ice layer, formed on the sea, in rivers or lakes, or on land, but appears dark because of its transparency. (b) Dark glacier ice formed by freezing of silt-laden water. Cf: *blue ice; white ice.* (c) A thin sheet of dark glazed ice formed when a light rain or drizzle falls on a surface whose temperature is below freezing; *glaze.* Cf: *verglas.*

blackjack [coal] (a) A term used in Illinois for a thin stratum of coal interbedded with layers of slate; a slaty coal with a high ash content. (b) A British term for a variety of cannel coal.----Also spelled: *black jack.*

blackjack [sed] A term used in Arkansas for a soft, black carbonaceous clay or earth associated with coal.

blackjack [mineral] A syn. of *sphalerite,* esp. a dark variety. The term was originated by miners who regarded sphalerite as an impish intrusion ("jack") of worthless material in lead ores. Also spelled: *black jack.*

blackland A term used in Texas for a Vertisol.

black lead *graphite.*

black-lead ore An old name for the black variety of cerussite.

black light (a) A prospector's and miner's term for ultraviolet light, used in exploration to detect mineral fluorescence. (b) An instrument, usually portable, that produces ultraviolet light for this purpose.

black lignite A syn of *lignite A.*

black manganese A term applied to dark-colored manganese minerals, such as pyrolusite, hausmannite, and psilomelane.

black metal A black shale associated with coal measures.

black mica *biotite.*

blackmorite A yellow variety of opal from Mount Blackmore, Mont.

black mud A type of *mud [marine geol]* whose dark color is due to hydrogen sulfide, developed under anaerobic conditions. Syn: *hydrogen sulfide mud; reduced mud; euxinic mud.*

black ocher *wad* [mineral].

black onyx The popular name for single-colored *black chalcedony,* usually artificially colored. Although the word "onyx" is not quite accurate (except for banded material), it has come to be accepted as the usual term for solid-color chalcedony.

black opal A form of *precious opal* whose internal reflections (usually red or green) are displayed against a dark (usually dark-gray, rarely black) body color; e.g. the fine Australian blue opal with flame-colored flashes. Cf: *white opal.*

black prairie A *prairie* with rich, dark soil.

Blackriverian North American substage: Middle Ordovician (lower Mohawkian Stage), below the Trentonian Substage. See also: *Wilderness.*

blacks (a) Highly carbonaceous black shale; an impure can nel coal. (b) A British term for dark coaly shale, clay, or mud stone.

black sand (a) An alluvial or beach sand consisting predom nantly of grains of heavy, dark minerals or rocks (such as c magnetite, ilmenite, chromite, cassiterite, rutile, zircon, mona zite, garnet, tourmaline, or basaltic glass) concentrated chie ly by wave, current, or surf action. It sometimes yields valu able minerals, such as gold and platinum. (b) An asphalti sand.

black-sand beach A beach containing a large quantity of blac sands concentrated by the action of waves and currents; e.g along the shores of Hawaii.

black shale (a) A dark, papery, usually thinly laminated ca bonaceous shale exceptionally rich in organic matter (5% more carbon content) and sulfide sulfur (esp. iron sulfid usually pyrite), and often containing unusual concentrations certain trace elements (U, V, Cu, Ni). It is formed by parti anaerobic decay of buried organic matter in a quiet-water, r ducing environment (such as in a stagnant marine basi characterized by restricted circulation and very slow depos tion of clastic material. Fossil organisms (principally plankto ic and nektonic forms) are preserved as a graphitic or carbo naceous film or as pyrite replacements. Syn: *biopelite.* (b) finely laminated, sometimes cenneloid, carbonaceous sha often found as a roof to a coal seam (Tomkeieff, 1954, p.29 Syn: *black metal.*

black silver *stephanite.*

black tellurium *nagyagite.*

black tin *cassiterite.*

bladder *vesicle* [palyn].

blade [speleo] A thin projection from any surface of a cave; is usually a remnant of a partition or of a bridge.

blade [mineral] A mineral that forms an elongate crystal; suc a mineral is said to be *bladed.*

blade [paleont] A laterally compressed structure of a cono dont; e.g. a denticle-bearing posterior or anterior proces (based on position with reference to the basal cavity) in compound conodont, or a generally compressed and dentic late part of the axis anterior to the basal cavity in a plate-lik conodont. Cf: *bar.*

blade [bot] The widened portion of a leaf or of a plant stru ture that resembles a leaf.

blade [sed] A bladed or triaxial shape of a sedimentary par cle, defined in *Zingg's classification* as having width/lengt and thickness/width ratios less than 2/3.

bladed Said of a mineral in the form of aggregates of flattene *blades* or elongate crystals.

blady Like a blade; e.g. "blady calcite" having elongate cry tals somewhat wider than those of fibrous calcite.

blaes (a) A Scottish term for a gray-blue carbonaceous sha that weathers to a crumbling mass and eventually to a so clay. See also: *bind.* (b) A Scottish term for a hard, joint-fre sandstone.---Syn: *blaize.*

Blagden's law The statement in chemistry that, for a give salt, the depression of the freezing point is proportional to th concentration of the solution.

blairmorite A porphyritic extrusive rock consisting predom nantly of analcime phenocrysts in a groundmass of analcim sanidine, and alkali pyroxene, with accessory sphene mela ite, and nepheline; an analcime phonolite.

blaize *blaes.*

blakeite (a) A reddish-brown mineral consisting of a ferric te lurite, found sparingly as crusts from Goldfield, Nev. (b) z conolite.

blanket [speleo] A type of dripstone in a cave that projec from the walls and roof in thick, opaque sheets. Cf: *bacon.*

blanket [sed struc] A thin, widespread sedimentary bod whose width/thickness ratio is greater than 1000 to 1, an may be as great as 50,000 to 1 (Krynine, 1948, p.146). C *ibular.* Syn: *sheet* [*sed*].

blanket bog A bog covering a large, fairly horizontal area and which depends on high rainfall or high humidity rather than local water sources for its supply of moisture. See also: *high-moor bog.*

blanket deposit [*ore dep*] A miner's term for a horizontal, tabular orebody. The term has no generic connotation.

blanket deposit [*sed*] A sedimentary deposit of great lateral or areal extent and of relatively uniform thickness; esp. a *blanket sand* and associated blanket limestones.

blanket moss An accumulation of dead algae, often forming peat. See also: *blanket peat.*

blanket peat Peat that is derived mainly from the algae, *blanket moss.*

blanket sand A *blanket deposit* of sand or sandstone of unusually wide distribution, typically an orthoquartzitic sandstone deposited by a transgressive sea advancing for a considerable distance over a broad and stable shelf area; e.g. the St. Peter Sandstone extending from Colorado to Indiana. Syn: *sheet sand; blanket sandstone.*

blast [*geophys*] The violent effect produced in the vicinity of an explosion, consisting of a wave of increased atmospheric pressure followed by a wave of decreased atmospheric pressure. Syn: *shock wave.*

blast [*meta*] In metamorphism terminology, a syllable that, when used as a prefix, signifies a relict texture, and when used as a suffix, signifies a texture formed entirely by metamorphism.

blastation A term suggested by Glock (1928) for the destructive action of windblown particles of sand and dust; *blasting.*

blastetrix In an anisotropic medium, any surface to which a direction of greatest ease of growth is perpendicular (Turner, 1948, p.223).

blastic deformation One of the processes of dynamothermal metamorphism which operates by recrystallization according to Riecke's principle in such a way that previously existing minerals are elongated perpendicular to the direction of greatest pressure and new minerals (called stress minerals) grow in the same plane. Cf: *clastic deformation; plastic deformation.*

blasting Abrasion or attrition effected by the impact of fine particles moved by wind or water against or past an exposed, stationary surface; esp. *sandblasting.* Syn: *blastation.*

blastogranitic A relict texture in a metamorphic rock in which remnants of the original granitic texture remain.

blastoid Any crinozoan belonging to the class Blastoidea, characterized chiefly by high developed quinqueradiate symmetry, a dominant meridional growth pattern, uniform arrangement of thecal plates in four cycles, specialized recumbent ambulacral areas, and by the presence of hydrospires. Their stratigraphic range is Silurian to Permian. Var: *blastid.*

blastomylonite A mylonitic rock in which some recrystallization and/or neomineralization has taken place. Cf: *mylonite.*

blastopelitic A texture of a metamorphosed argillaceous rock in which there are relicts of the parent rock.

blastophitic A relict texture in a metamorphic rock in which traces of the original ophitic texture remain.

blastoporphyritic A relict texture in a metamorphic rock in which traces of the original porphyritic texture remain.

blastopsammitic A texture of a metamorphosed sandstone which contains relicts of the parent rock.

blastopsephitic A texture of a metamorphosed conglomerate or breccia which contains relicts of the parent rock.

blast wave A sharply defined wave of increased atmospheric pressure rapidly propagated through a surrounding medium from a center of detonation or similar disturbance. See also: *shock wave.*

B layer The seismic region of the Earth from the Mohorovičić discontinuity to 410 km, equivalent to the *low-velocity zone* of the uppermost mantle. It is part of a classification of the interior of the Earth, having layers A to G.

blaze An artificial mark made on a tree trunk, usually at about breast height, in which a piece of the bark and a very small amount of the live wood tissue is removed leaving a flat scar that permanently marks the tree. It is made for the purpose of guiding the course of a survey or of a trail in wooded country.

bleached Said of a sand which has become pale due to leaching.

bleaching clay A clay or earth that, either in its natural state or after chemical activation, has the capacity for adsorbing or removing coloring matter or grease from liquids (esp. oils). Syn: *bleaching earth.*

bleaching earth *bleaching clay.*

bleach spot A greenish or yellowish area in a red rock, developed by the reduction of ferric oxide around an organic particle (Tyrrell, 1926). Syn: *deoxidation sphere.*

bleb In petrology, a small, usually rounded inclusion, e.g. olivine that is poikilitically enclosed in pyroxene.

bleeding n. (a) The process of giving off oil or gas from pore spaces or fractures; it can be observed in drill cores. (b) The exudation of small amounts of water from coal or a stratum of some other rock.

bleeding rock A sandstone that contains water.

blende (a) *sphalerite.* (b) Any of several minerals (chiefly metallic sulfides) with somewhat bright or resinous but nonmetallic luster, such as zinc blende (sphalerite), antimony blende (kermesite), bismuth blende (eulytite), cadmium blende (greenockite), pitchblende, and hornblende.----Etymol: German *Blende,* "deceiver".

blended unconformity An unconformity having no distinct surface of separation or sharp contact, such as at an erosion surface that was originally covered by a thick residual soil, which graded downward into the underlying rocks (from which it was derived) and was partly incorporated in the overlying rocks; e.g. a nonconformity between granite and overlying basal arkosic sediments derived as a product of its disintegration. Syn: *graded unconformity.*

blind Said of a mineral deposit that does not outcrop at the surface. The term is more appropriate for a deposit that terminates below the surface than for one that is simply hidden by unconsolidated surficial debris.

blind apex The near-surface end of a mineral deposit, e.g. the upper end of a seam or vein that is abutted by an unconformity. Syn: *suboutcrop.*

blind coal (a) Anthracite or other coal that burns without a flame. (b) *Natural coke* that resembles anthracite.

blind creek A creek that is dry except during a rainfall. The term is "obsolete or obsolescent and better avoided because of confusion with blind valley" (Stamp, 1961, p.66).

blind estuary A term used in Australia and South Africa for an *estuarine lagoon.*

blind island A patch of marl or organic matter covered by a shallow depth of water in a lake. Cf: *sunken island.*

blind joint In apparently massive rock that is being quarried, a plane of potential fracture along which the rock may break during excavation.

blind lake A colloquial term for a lake that has neither an influent nor an effluent.

blind lead A long narrow passage in pack ice with only one outlet. Syn: *cul-de-sac; pocket.*

blind valley (a) A type of valley that is characteristic of karst; it ends abruptly downstream at the point at which its stream disappears underground into a cave or swallow hole. See also: *half-blind valley.* (b) A steep-walled valley that terminates abruptly near the source of the stream that formed it; e.g. a valley whose precipitous walls approach each other near a karst spring. This usage is "geographically obsolete" (Stamp, 1961, p.66). See also: *pocket valley.*

blink (a) A brightening of the sky near the horizon or the underside of a cloud layer, caused by reflection of light from a snow- or ice-covered surface. See also: *iceblink; snowblink; landblink.* (b) A dark appearance of the sky near the horizon or the underside of a cloud layer, caused by the relative absence of reflected light from a water or land surface. See also: *water sky; land sky.*

blister [volc] A surficial swelling of the crust of a lava flow formed by the puffing up of gas or vapor beneath the flow. A blister is usually about one meter in diameter, and is hollow. Cf: *tumulus.*

blister [coal mining] In a coal seam, a downward protrusion of roof rock into the seam, probably formed as a filling of a streambed pothole.

blister hypothesis A theory of the cause of orogeny, that proposes that in a zone not more than 80 km deep in the crust, heat from radioactive disintegration created a large convex-upward lens of heated and expanded rock, which produced doming of the overlying crust and in turn formed orogenic structures in the near-surface rocks. The "melting spot" or *asthenolith hypothesis* of B. Willis is similar. Geophysical evidence indicates that the existence of such blisters is unlikely, and the theory is probably obsolete. Cf: *undation theory.*

blixite A mineral: $Pb_2Cl(O,OH)_2$.

blizzard A type of weather in which temperatures are low, wind velocity is high (28 knots or more), and there is much snow in the air that is mostly fine and dry and has been picked up from the ground (U.S. Naval Oceanographic Office, 1966, p.22).

block [ice] A fragment of floating sea ice varying in size from 2 m to 10 m across; the term is being replaced by *ice cake.*

block [part size] (a) A large, angular rock fragment, showing little or no modification by transporting agents, its surfaces resulting from breaking of the parent mass, and having a diameter greater than 256 mm (10 in.); it may be nearly in place or transported by gravity, ice, or other agents. Cf: *boulder.* (b) A term used by Woodford (1925) for a nearly equidimensional, angular rock fragment of any diameter greater than 4 mm. (c) A rock or mineral particle in the soil, having a diameter range of 200-2000 mm (Atterberg, 1905). (d) A layer of sedimentary rock, from 60 cm to 120 cm (2-4 ft) thick, produced by splitting (McKee & Weir, 1953, p.383).

block [tect] *fault block.*

block [volc] A pyroclast that was ejected in a solid state; it is in the size range of about 5cm in diameter to several cubic meters. It may be essential, accessory, or accidental. Cf: *lapilli; volcanic gravel; cinder.*

block clay *mélange.*

block diagram (a) A plane figure representing an imaginary rectangular block of the Earth's crust (depicting geologic and topographic features) in what appears to be a three-dimensional perspective, showing a surface area on top and including one or more (generally two) vertical cross sections. The top of the block gives a bird's-eye view of the ground surface, and its sides give the underlying geologic structure (Lobeck, 1924). (b) A sketch of a relief model; a representation of a landscape in perspective projection.

block disintegration *joint-block separation.*

blocked-out ore A syn. of *developed reserves.* Syn: *ore blocked out.*

block faulting A type of normal faulting in which the crust is divided into structural or *fault blocks* of different elevations and orientations. It is the process by which *block mountains* are formed.

block field *felsenmeer.*

block glide A translational landslide in which the slide mass remains essentially intact, moving outward and downward as a unit, most often along a preexisting plane of weakness, such as bedding, foliation, joints, faults, etc. In contrast to rotational landslides, the various points within a displaced block-glide landslide have predominantly maintained the same mutual difference in elevation in relation to points outside the slide mass. Cf: *earth slide.*

blocking out In economic geology, exposure of an orebody o three sides in order to develop it, i.e., to make estimates o its tonnage and quality. The part so prepared is an *ore block.*

blockite *penroseite.*

block lava Lava having a surface of angular blocks; it is simi lar to *aa* lava but the fragments are more regular in shape somewhat smoother and less vesicular.

blockmeer A syn. of *felsenmeer.* Etymol: German *Blockmeer* "sea of blocks".

block mountains Mountains that are formed by *block faulting* The term is not applied to mountains that are formed by thrus faulting. Syn: *fault-block mountains; block structures.*

block movement In mining, a general failure of the hangin wall.

Block-Schollen movement A type of glacier flow in which th greater portion of the glacier moves as a solid mass with nearly uniform velocity; blocks of ice are produced by move ment over irregularities in the glacier bed (Finsterwalde 1950). Etymol: German.

block sea *felsenmeer.*

block spar An industrial term for feldspar that required on hand cobbing, grinding, sizing, and often magnetic treatmer to be prepared for market (AIME, 1960).

block spread *felsenmeer.*

block stream *rock stream.*

block stripe A short and broad *sorted stripe* containing mater al that is coarser, and of less uniform size, than that in *stone stripe.*

block structures *block mountains.*

block talc A general term for any massive talc or soapston that can be worked by machines.

block waste *felsenmeer.*

blocky iceberg An iceberg with steep, precipitous sides and horizontal or nearly horizontal upper surface.

bloedite A white or colorless monoclinic mineral: Na_2Mg $(SO_4)_2.4H_2O$. Also spelled: *blödite; blodite.* Syn: *astrakhanite*

blomstrandine *priorite.*

blomstrandite *betafite.*

blood agate (a) Flesh-red, pink, or salmon-colored agate fro Utah. (b) *hemachate.*

blood rain Rain with a reddish color caused by dust-like mate rial (containing iron oxide) picked up from the air by raindrop during descent, often leaving a red stain on the ground; e. the blood rain of Italy, containing red dust carried north b great storms from the Saharan desert region. Syn: *dust fall.*

bloodstone (a) A semitranslucent and leek-green or dar green variety of chalcedony speckled with red or brownish-re spots of jasper resembling drops of blood. Cf: *plasma [mine al].* Syn: *heliotrope; oriental jasper.* (b) hematite.

bloom [ore dep] *blossom.*

bloom [mineral] *efflorescence.*

bloom [oceanog] *water bloom.*

blossom The oxidized or decomposed outcrop of a vein o coal bed, more frequently the latter. Syn: *bloom.*

blow *blowhole.*

blowhole [coast] A nearly vertical hole, fissure, or natura chimney in coastal rocks, leading from the inner end of th roof of a sea cave to the ground surface above, and throug which incoming waves and the rising tide forcibly compres the air to rush upward or spray water to spout intermittentl often with a noise resembling a geyser outburst; it is probab formed by wave erosion concentrated along planes of wea ness, as in a well-jointed rock. Also spelled: *blow-hole.* Sy *puffing hole; gloup; blow; boiler; buller.*

blowhole [volc] A minute crater on the surface of a viscou lava flow.

blowhole [glaciol] An opening that passes through a snov bridge into a crevasse or system of crevasses that are othe wise sealed by snowbridges (Armstrong et al, 1966, p.11).

blowing [sed] Transportation and deposition of sediments effected by the wind acting along the surface of the ground.

blowing cave A cave that has an alternating movement of air, or cave breathing, through its entrance. Syn: breathing cave.

blowing snow Wind-lifted snow raised from the surface of the ground to moderate or great heights (more than 2 m) and carried along by the wind in such quantity that horizontal visibility is restricted. Cf: drifting snow.

blowing well A syn. of breathing well. The term should not be confused with blow well.

blow land Land that is subject to wind erosion.

blown sand Sand that has been transported by the wind; sand consisting of wind-borne particles; eolian sand. See also: dune sand.

blowoff The removal of humus and loose topsoil by wind action. Also, the material so moved.

blowout [ore dep] (a) A prospector's term for a weathered exposure considered to be indicative of a mineral deposit. (b) A large mineral-deposit outcrop beneath which the deposit is smaller.

blowout [geomorph] (a) A general term for a small saucer-, cup-, or trough-shaped hollow, depression, basin, or valley formed by wind erosion on a preexisting dune or other sand deposit, esp. in an area of shifting sand or loose soil, or where protective vegetation is disturbed or destroyed; the adjoining accumulation of sand derived from the depression, where readily recognizable, is commonly included. Some blowouts may be many kilometers in diameter. (b) A butte, the top of which has been blown out by the wind until it resembles a volcanic crater. (c) A shallow basin formed where vegetation has been destroyed by fire or by overgrazing.---See also: deflation basin. Also spelled: blow-out. Syn: blowout basin; deflation hollow; deflation hole.

blowout [grd wat] sand boil.

blowout dune A dune consisting of a large accumulation of sand derived from the formation of a blowout. An "elongate blowout dune" is characterized by a slight migration of the blowout and its crescent-shaped rim in the direction of the prevailing wind (Stone, 1967, p.226).

blowout pond A shallow, intermittent pond occupying a blowout, as on a dune.

blowover (a) Sand blown by the wind over a barrier and deposited as a veneer in the lagoon; e.g. the deposits resulting where sand is blown by strong onshore winds over the barrier islands and into the lagoons along the Gulf coast of Texas. Cf: washover. (b) The process of forming a blowover.

blowpipe A plain brass tube that produces an intense heat by combining a flame from a bunsen burner or other heat source with a stream of air; it is used in simple qualitative analysis of minerals. See also: blowpiping.

blowpipe reaction The indicative changes of a mineral specimen as it undergoes blowpiping; e.g. color of the flame, odor, nature of the encrustation.

blowpiping In mineralogy, a qualitative test of a mineral made by heating a specimen in the flame of a blowpipe and observing its blowpipe reaction, such as color of the flame or color of the encrustation, to determine what elements may be present. See also: bead.

blow well A syn. of flowing artesian well. The term should not be used for or confused with blowing well, a syn. of breathing well.

blue amber A variety of osseous amber with a bluish tinge that is probably due to the presence of calcium carbonate.

blue asbestos crocidolite.

blue band [sed] The thin but persistent bed of bluish clay found throughout the Illinois-Indiana coal basin.

blue band [glaciol] (a) A sharply bounded lens or layer of relatively bubble-free glacier ice; a bluish band marking the appearance of such a lens or layer on the surface of a glacier. The bluish tint is due to the low content of air in the ice. Cf:

white band. (b) The dark-ribbon effect produced on the surface of the glacier by the exposure of blue bands.

blue-black ore corvusite.

blue chalcocite digenite.

blue copper ore azurite.

blue earth blue ground.

blue elvan A Cornish term for greenstone occurring in dikes.

blue-green algae A group of algae corresponding to the phylum Cyanophyta, that owes its blue color to the presence of the pigment c-phycocyanin. Cf: brown algae; green algae; red algae.

blue ground Unoxidized slate-blue or blue-green kimberlite, usually a breccia (e.g. the diamond pipes of South Africa) that is found below the surficial oxidized zone of yellow ground. Cf: hardebank. Syn: blue earth.

blue hole (a) A Jamaican term for a resurgence that is not turbulent. Cf: boiling spring. (b) A term used in the Bahamas for drowned sinkholes on the Bahama Banks, from which cooler water rises during high tide. Cf: banana hole. Syn: ocean hole.

blue ice (a) Nonbubbly, unweathered, coarse-grained glacier ice, often occurring as blue bands [glaciol]; it is distinguished by a slightly bluish or greenish color. Cf: black ice; white ice. (b) An ablation area created by wind erosion on the Antarctic Ice Sheet, characterized by bare glacier ice showing at the surface.

blue iron earth Pale-blue powdery vivianite.

blue ironstone A bluish iron-bearing mineral; specif. crocidolite and vivianite.

blue john A massive, fibrous, or columnar and blue or purple variety of fluorite found in Derbyshire, Eng. It is frequently banded, and is used esp. for the manufacture of vases. Syn: derbystone.

blue lead [mineral] A syn. of galena, esp. a compact variety with a bluish-gray color. Syn: blue lead ore.

blue lead [ore dep] A bluish, gold-bearing lead or gravel deposit found in Tertiary river channels of the Sierra Nevada, California. Pron: bloo leed.

blue malachite A misnomer used as a syn. of azurite.

blue metal [sed] A term used in England for a hard bluish-gray shale or mudstone lying at the base of a coal bed and often containing pyrite.

blue mud A hemipelagic type of mud [marine geol] whose blueish gray color is due to iron sulfides and organic matter.

blue ocher vivianite.

blue quartz (a) A faint, soft-blue or milky-blue to smoky-blue, plum-blue, or lavender-blue variety of crystalline quartz containing needle-like inclusions of rutile. It occurs as grains in metamorphic and igneous rocks. (b) sapphire quartz.

blue-rock phosphate A term used for the Ordovician, bedded, bluish-gray phosphates of Tennessee.

blueschist facies glaucophane schist facies.

blue spar lazulite.

bluestone [mineral] chalcanthite.

bluestone [rock] (a) A commercial name for a building or paving stone of bluish-gray color; specif. a dense, hard, tough, slightly metamorphosed, fine-grained, dark bluish- or slate-gray (the color is due to the presence of fine black and dark-green minerals, chiefly hornblende and chlorite) feldspathic sandstone that splits easily into thin, smooth slabs and that is extensively quarried near the Hudson River in New York State for use as flagstone. The term is applied locally to other rocks, such as dark-blue shale and blue limestone. See also: Amherst stone. (b) A highly argillaceous sandstone of even texture and bedding, formed in a lagoon or lake near the mouth of a stream (Grabau, 1920a, p.579). (c) A local term used in Great Britain for a hard shale or clay (as in south Wales), and for a basalt.

blue vitriol chalcanthite.

blue-white Said of the highest-quality diamond whose color

grade is between that grade which appears colorless in transmitted light and any grade with a yellowish tinge that is not apparent to the average inexperienced buyer (Shipley, 1951, p.28).

bluff (a) A high bank or bold headland with a broad, precipitous, almost perpendicular, sometimes rounded cliff face overlooking a plain or a body of water; esp. on the outside of a stream meander; a *river bluff*. (b) Any cliff with a steep broad face.

bluff formation Deposit of coarse *loess*, forming bluffs immediately adjacent to the edges of river flood plains, as in the Mississippi Valley region.

bluff line The side of a valley formed by a river or by ice cutting away the heads of interlocking spurs (Swayne, 1956, p.25).

blythite A hypothetical member of the garnet group: $Mn_3^{+2}Mn_2^{+3}(SiO_4)_3$.

Blytt-Sernander climatic classification A classification of late-glacial and postglacial climate inferred originally from bog stratigraphy and megascopic plant remains from Norway and Sweden, and later refined by Post (1924) from pollen evidence. This classic system remains the basic reference for worldwide research on postglacial climate. It comprises six subunits: Arctic, Preboreal, Boreal, Atlantic, Subboreal, and Subatlantic. Named after Axel Gudbrand Blytt (1843-1898), Norwegian botanist, and Johan Rutger Sernander (b. 1886-), Swedish botanist.

board coal *woody lignite*.

boar's back A *horseback* or esker in northern New England, esp. Maine.

boart Var. of *bort*.

boat channel A channel, on a reef flat, separating a fringing reef from the shore along which the channel is parallel. It is generally only a few meters in depth and width. Syn: *back-reef moat*.

boathook bend The sharp curvature of a tributary where it joins the main stream in an upstream direction in a *barbed drainage pattern*, resembling in plan a boathook.

bobierrite A mineral: $Mg_3(PO_4)_2.8H_2O$. It occurs massive or in crystals in guano.

boca The mouth of a stream, esp. the point where a stream or its channel emerges from a canyon, gorge, or other precipitous valley and flows onto or enters a plain. Etymol: Spanish, "mouth".

bocca An aperture on any part of a volcano from which magma or gas escapes. Etymol: Italian, "mouth". Pl: *bocche*.

bocche Plural of *bocca*.

bodden A broad, shallow, irregularly shaped inlet or bay along the southern Baltic coast, typically produced by partial submergence of an uneven lowland surface and characterized by seaward islands. Etymol: German *Bodden*. Cf: *fôrde*.

bodily tide *earth tide*.

bodily wave *body wave*.

body [coal] The fatty, inflammable property that makes a coal combustible; e.g. bituminous coal has more *body* than anthracite.

body [palyn] *central body*.

body [water] A separate entity or mass of water, distinguished from other water masses; e.g. an ocean, sea, stream, lake, pond, pool, and water in an aquifer are distinct "bodies of water".

body cavity A cavity or major space within an animal body, such as a *coelom*; e.g. the principal part of the coelomic space in a brachiopod, situated posteriorly, bounded by the body wall, and containing the alimentary tract, internal organs, etc.

body-centered lattice A type of *centered lattice* in which the unit cell contains two lattice points; the point at the intersection of the four body diagonals is identical with those at the corners. It is found in crystals of the cubic, tetragonal, and orthorhombic systems. Syn: *I-centered lattice*.

body chamber (a) The large, undivided, anterior space in a cephalopod shell occupied by the living body of the animal, bounded at the back by a septum and open at the front through the aperture. Syn: *living chamber; chamber* [paleont]. (b) The interior of the shell containing the soft parts of a cirripede crustacean.

body forces Any of the forces acting on a material proportional to the mass of the substance, e.g. gravity, centrifugal force, magnetic force, and measured in units of force per unit volume. Cf: *surface force*.

body wall The external surface of the body in all animals, enclosing the body cavity; e.g. the integument or external layer with any included calcareous skeleton, that encloses the disc and arms of an asteroid, or the part of a sponge between the exterior and a central cloaca, or the *perisome* of an echinoderm.

body wave A *seismic wave* that travels through the interior of the Earth and is not related to any boundary surface. A body wave may be either longitudinal (a *P wave*) or transverse (an *S wave*). A perfectly sharp distinction between body waves and *surface waves* is difficult to make unless the waves are plane or spherical. Syn: *bodily wave*.

body whorl The outer, last-formed, and typically largest *whorl* of a univalve shell; e.g. the last complete loop in the spiral of a gastropod shell, terminating in the aperture.

boehmite A grayish, brownish, or reddish orthorhombic mineral: $AlO(OH)$. It is a major constituent of some bauxites and it represents the gamma phase dimorphous with diaspore. Syn: *bôhmite*.

Boehm lamellae Planes of inclusions in quartz, oriented subparallel to the basal plane and probably developed by gliding. Cf: *Tuttle lamellae*. Also spelled: *Bôhm lamellae*.

bog (a) A waterlogged, spongy groundmass, primarily mosses, containing acidic, decaying vegetation which may develop into peat. (b) The vegetation characteristic of this environment, esp. sphagnum, sledges, and heaths.----The term is often used synonymously with *peat bog*. Cf: *fen; marsh; swamp*.

bogan *pokelogan*.

bog burst The bursting of a bog under the pressure of its swelling, due to water retention by a marginal dam of growing vegetation. The escaping water produces muddy peat flow over the surrounding area.

bog butter A substance, found preserved in Irish peat bogs, that was formerly believed to be a native hydrocarbon but is now known to be "fossil" butter that had been buried for storage and found at a much later date (Tomkeieff, 1954, p.30). Syn: *butyrellite; butyrite*.

bog coal An earthy type of brown coal.

bog flow A mudflow representing the outflow from a *bog burst*. Cf: *peat flow*.

bøggildite A mineral: $Na_2Sr_2Al_2(PO_4)F_9$.

boghead coal A *sapropelic coal* resembling *cannel coal* in its physical properties but that contains algae rather than spores. It rarely occurs in a pure state but rather in forms transitional to cannel coal. It is a source of both oil and gas. Cf: *torbanite*. Syn: *algal coal; gélosic coal*.

boghedite An old syn. of *torbanite*.

bog iron ore A general term for a soft, spongy, and porous deposit of impure hydrous iron oxides formed in bogs, marshes, swamps, peat mosses, and shallow lakes by precipitation from iron-bearing waters and by the oxidizing action of algae, iron bacteria, or the atmosphere; a *bog ore* composed principally of limonite that is often impregnated with plant debris, clay, and clastic material. It is a poor-quality iron ore found in tubular, pisolitic, nodular, concretionary, or thinly layered forms, or in irregular aggregates, in level sandy soils and is esp. abundant in the glaciated northern regions of North America and Europe (Scandinavia). See also: *murram*

Syn: *limnite; morass ore; meadow ore; marsh ore; lake ore; swamp ore.* (b) A term commonly applied to a loose, porous, earthy form of *limonite* occurring in wet ground.---Syn: *bog iron.*

bog lake A lake or small body of open water surrounded or nearly surrounded by bogs and characterized by a false bottom of organic (peaty) material, high acidity, scarcity of aquatic fauna, and vegetation growing on a firm deposit or on a semifloating mat of peat. See also: *muskeg lake.*

bog lime A soft, grayish to white, earthy or powdery, usually impure calcium carbonate precipitated on the bottoms of present-day freshwater lakes and ponds largely through the chemical action of aquatic plants, or forming deposits that underlie marshes, swamps, and bogs that occupy the sites of former (glacial) lakes. The calcium carbonate may range from 90% to less than 30%. Also spelled: *boglime.* Syn: *marl; lake marl.*

bog manganese A *bog ore* of variable composition, but consisting chiefly of hydrous manganese oxide; specif. *wad* formed in bogs or marshes by the action of minute plants.

bog-mine ore A syn. of *bog ore.* Also called: *bog mine.*

bog moat *lagg.*

bog ore A poorly stratified accumulation of earthy metallic mineral substances, consisting mainly of oxides, that are formed in bogs, marshes, swamps, and other low-lying moist places usually by direct chemical precipitation from surface or near-surface percolating waters; specif. *bog iron ore* and *bog manganese.* Cf: *lake ore.* Syn: *bog-mine ore.*

bog peat *highmoor peat.*

Bog soil One of an intrazonal, hydromorphic group of soils having a mucky or peaty surface horizon and an underlying peat horizon. Cf: *Histosol; Half-Bog soil.*

bogue A term used in Alabama and Mississippi for the mouth or outlet of a stream, or for the stream itself, or for a bayou. Etymol: American French, from Choctaw *bouk,* "stream, creek".

bogusite An intrusive rock of the same general composition as *teschenite* but of lighter color.

bohdanowiczite A mineral: $AgBiSe_2$.

Bohemian garnet A yellowish-red to dark, intense-red gem variety of *pyrope* obtained from Bohemia.

Bohemian ruby A red variety of crystalline quartz; specif. *rose quartz* cut as a gem.

Bohemian topaz *citrine.*

böhmite *boehmite.*

Böhm lamellae Var. of *Boehm lamellae.*

Bohr magneton A unit of *magnetic moment,* comparable to the actual magnetic moments of ions such as Fe^{+2}, Fe^{+3}, Mn^{+4}.

boil n. A churning agitation of water, esp. at the surface of a water body, such as a river, spring, or the sea.

boiler (a) A submerged coral reef, esp. one occurring where the sea breaks. (b) A *blowhole* along the coast.

boiling hole *boiling spring.*

boiling spring [karst] A jamaican term for a turbulent resurgence. Cf: *blue hole.* Syn: *boiling hole.*

boiling spring [grd wat] (a) A spring, the water from which is agitated by the action of heat. (b) A spring that flows so rapidly that strong vertical eddies develop.

bojite A gabbro in which primary hornblende substitutes for most of the augite, although some augite and biotite may be present; a hornblende gabbro. Cf: *evjite.*

bokite A black mineral: $KAl_3Fe_6V_6^{+4}V_{20}^{+5}O_{76}.30H_2O$.

bold coast A prominent landmass, such as a cliff or promontory, rising or sloping steeply from a body of water, esp. along the seacoast. Syn: *bold.*

bole Any of several varieties of fine, compact, friable, and earthy or unctuous clay (impure halloysite), usually colored red, yellow, or brown due to the presence of iron oxide, and consisting essentially of hydrous silicates of aluminum or less

often of magnesium; a waxy decomposition product of basaltic rocks, having the variable composition of lateritic clays. Adj: *bolar.* Syn: *bolus; terra miraculosa.*

boleite An indigo-blue mineral: $Pb_9Cu_8Ag_3Cl_{21}(OH)_{16}.H_2O$. Also spelled: *boléite.*

bolide An exploding or exploded meteor or meteorite; a detonating *fireball.*

Boliden gravimeter An electrical, stable-type gravity meter with a moving system suspended on a pair of bowed springs. The moving system carries electrical condenser plates at each end, one to measure the position of the moving system, the other to apply a balancing force to bring the system to a fixed position. Syn: *Lindblad-Malmquist gravimeter.*

Bølling n. A term used primarily in Europe for an interval of late-glacial time (centered about 12,500 years ago) following the Oldest Dryas and preceding the Older Dryas, during which the climate as inferred from stratigraphic and pollen data in Denmark (Iversen, 1954) ameliorated favoring accelerated retreat of the waning continental and alpine glaciers; a subunit of the late-glacial Arctic interval, characterized by birch and park-tundra vegetation. Also spelled: *Bôlling.*---adj. Pertaining to the late-glacial Bølling interval and to its climate, deposits, biota, and events.

bolly Var. of *bally.*

Bologna stone A nodular, concretionary, or roundish form of barite, composed of radiating fibers and being phosphorescent when calcined with charcoal. Syn: *Bolognan stone; Bologna spar.*

bolometer A thermal detector used to measure the heating effect of radiation by measuring the change in electrical resistance of a metal (e.g. platinum) or of a semiconductor (e.g. a thermistor).

bolson (a) A term applied in the desert regions of SW U.S. to an extensive, flat, saucer-shaped, alluvium-floored basin or depression, almost or completely surrounded by mountains from which drainage has no surface outlet as it runs centripetally with gentle gradients toward a playa or central depression; an interior basin, or a basin with internal drainage. See also: *semibolson.* Syn: *playa basin.* (b) A temporary lake, usually saline, formed in a bolson.---Etymol: Spanish *bolsón,* "large purse".

bolson plain A broad, intermontane plain in the central part of a bolson or semibolson, composed of deep alluvial accumulations washed into the basin from the surrounding mountains.

boltonite A greenish or yellowish granular variety of forsterite from Bolton, Mass.

boltwoodite A yellow mineral: $K_2(UO_2)_2(SiO_3)_2(OH)_2.5H_2O$.

bolus A *bole.* Etymol: Latin, "clod of earth".

bolus alba A syn. of *kaolin.* Etymol: Latin, "white clay".

bomb [geochem] A vessel in which experiments can be conducted at high temperature and pressure. It is used in geochemistry and in experimental petrology.

bomb [pyroclast] A pyroclast that was ejected while viscous and received its rounded shape while in flight. It is larger than lapilli in size, and may be vescicular to hollow inside. Actual shape or form varies greatly, and is used in descriptive classification of bombs, e.g. rotational bomb; spindle bomb.

bombiccite *hartite.*

bombite A blackish-gray aluminosilicate of ferric iron and calcium from Bombay, India. It resembles Lydian stone and is probably a glassy rock.

bomb sag Depressed and disturbed strata or laminae of tuff or other deposit in which a volcanic bomb has fallen and become buried.

bomb-type seismometer A deep-well seismometer designed as a pressure vessel. It is distinguished from other deep-well seismometers by the fact that the internal pressure of the seismometer is independent of the pressure of the surrounding medium.

bomby An Australian term for a large, submerged reef clump found in a back-reef area and constituting a hazard for navigation and fishing (Maxwell, 1968, p.133). See also: *bommy*.

bommy A coral *niggerhead*. See also: *bomby*.

bonamite A trade name for an apple-green gem variety of smithsonite, resembling the color of chrysoprase.

bonanza A miner's term for a rich body of ore or a rich part of a deposit; a mine is "in bonanza" when it is operating profitably. Etymol: Spanish, "prosperity, success". Cf: *borasca*.

bonattite A monoclinic mineral: $CuSO_4.3H_2O$.

bonchevite An orthorhombic mineral: $PbBi_4S_7$.

Bond albedo *albedo*.

bone A tough, very fine-grained, gray, white, or reddish quartz.

bone amber *osseous amber*.

bone bed Any sedimentary stratum (usually a thin bed of sandstone or gravel) in which fossil bones or bone fragments are abundant, and often containing other organic remains (such as scales, teeth, and coprolites). See also: *bone phosphate*.

bone breccia An accumulation of bones or bone fragments of vertebrates, often mixed with earth and sand, and cemented with calcium carbonate; esp. such a deposit formed in limestone caves or other animal retreats. Syn: *osseous breccia*.

bone chert A weathered, residual chert that appears chalky and somewhat porous, and that is usually white but may be stained with red or other colors. When found in insoluble residues, it is an indicator of an unconformity.

bone coal (a) Coal that has a high ash content. It is hard and compact. Syn: *bony coal*. (b) Argillaceous partings in coal, sometimes called *slate*.

bone phosphate (a) A large *bone bed* containing sufficient calcium phosphate to be designated a phosphorite. (b) Tribasic calcium phosphate obtained from bones; also, the phosphate from the phosphatic rocks of North Carolina.

bone turquoise *odontolite*.

boninite A glassy olivine-bronzite andesite that contains little or no feldspar.

Bonne projection An equal-area, modified-conic map projection having one standard parallel intersecting the central meridian (a straight line along which the scale is exact) near the center of the map. All parallels are represented by equally spaced arcs of concentric circles (divided to exact scale) and all meridians (except the central meridian) are curved lines connecting corresponding points on the parallels. The projection is commonly used for mapping compactly shaped areas in middle latitudes (such as France) and for mapping continents such as North America and Eurasia. Named after Rigobert Bonne (1727-1795), French cartographer, who is said to have introduced the projection in 1752. See also: *sinusoidal projection*.

Bononian Stage in Great Britain: Upper Jurassic (above Kimmeridgian, below Purbeckian). It is equivalent to lower Portlandian.

bony coal *bone coal*.

book *mica book*.

book clay Clay deposited in thin, leaf-like laminae. Syn: *leaf clay*.

bookhouse structure A term introduced by Sloane & Kell (1966, p.295) for a fabric found in compacted kaolin clays, consisting of parallel and random arrangements of packets of oriented clay particles (flakes). Cf: *cardhouse structure*.

book structure In ore deposits, the alternation of ore with gangue, usually quartz, in parallel sheets. Cf: *ribbon [ore dep]*.

boolgoonyakh A nonseasonal *frost mound*, "usually of a considerable size and of many years duration" (Muller, 1947, p.214). According to Stamp (1966, p.71), the Yakut term "boolyunyakh" is equivalent to "hydrolaccolith".

booming sand A *sounding sand*, found on a desert, that emits a low-pitched note of considerable magnitude and duration as it slides (either spontaneously or when induced) down the slip

face of a dune or drift (Humphries, 1966, p.135). See also: *roaring sand*.

boort Var. of *bort*.

boothite A blue monoclinic mineral: $CuSO_4.7H_2O$. Its blue color is lighter than that of chalcanthite.

boracite A white, yellow, greenish, or bluish orthorhombic mineral: $Mg_3B_7O_{13}Cl$. It is strongly pyroelectric, becomes cubic at high temperatures, and occurs in evaporites and saline deposits. See also: *stassfurtite*.

Boralf In U.S. Dept. of Agriculture soil taxonomy, a suborder of the soil order *Alfisol*, characterized by formation in frigid or cryic temperature regimes and in a udic moisture regime (SSSA, 1970). Cf: *Aqualf; Udalf; Ustalf; Xeralf*.

borasca A miner's term for an unproductive area of a mine or orebody; a mine is "in borasca" when it is exhausted. Etymol: Mexican Spanish *borrasca*, "exhaustion of a mine". Cf: *bonanza*.

borate A mineral compound characterized by a fundamental structure of BO_3^{-3}. An example of a borate is boracite, $Mg_3B_7O_{13}Cl$. Cf: *carbonate; nitrate*.

borax A white, yellowish, blue, green, or gray mineral: $Na_2B_4O_7.10H_2O$. It is an ore of boron and occurs as a surface efflorescence or in large monoclinic crystals embedded in muds of alkaline lakes. Borax is used chiefly in glass, ceramics, agricultural chemicals, and pharmaceuticals, and as a flux, cleansing agent, water softener, preservative, and fire retardant. Syn: *tincal*.

borax bead The type of *bead* commonly used in blowpipe analysis of metallic compounds.

borax lake (a) A lake whose shores are encrusted with deposits rich in borax. (b) A dry, borax-rich bed of a lake.

borcarite A mineral: $Ca_4MgH_6(BO_3)_4(CO_3)_2$.

border belt A term used by Chamberlin (1893, p.263) for superficial glacial deposits now known as a *boulder belt*.

bordered pit A *pit* [bot] in which the margin projects over the thin closing membrane, as in tracheids of coniferous wood. Cf: *simple pit*.

border facies The marginal portion of an igneous intrusion which differs in texture and composition from the main body of the intrusion, possibly due to more rapid cooling, or assimilation of material from the country rock, for example.

border fault (a) *boundary fault*. (b) *peripheral faults*.

borderland According to a concept widely held in the first part of the 20th Century, and championed by Schuchert (1923), a crystalline landmass that bordered (farther out) the Phanerozoic orogenic belts near the edges of the North American continent. The borderlands were tectonically much more active than the Canadian Shield, and were subsequently lost by foundering into the oceans. The concept is now discreditied; continental crust ends near the edges of the continental shelves, and it would be difficult to founder large areas of such crust into the ocean basins beyond. Most of the geological evicence adduced for these lands can be otherwise interpreted. Cf: *hinterland; tectonic land*. See also: *Appalachia; Cascadia; Llanoria*.

bore [drill] *borehole*.

bore [marine geol] A submarine sand ridge, in very shallow water, whose crest may rise to intertidal level.

bore [tide] (a) A large, turbulent, wall-like wave of water with a high, abrupt front, caused by the meeting of two tides or by a very rapid rise or rush of the tide up a long, shallow and narrowing estuary, bay, or tidal river where the tidal range is appreciable; it can be 3-5 m high and moves rapidly (10-15 knots) upstream with and faster than the rising tide. A bore usually occurs after low water of a spring tide. Syn: *tidal bore*.

bore [volc] The outlet of a geyser at the Earth's surface.

bore [water] A syn. of *bored well*. Sometimes applied to any deep well or shaft.

boreal (a) Pertaining to the north, or located in northern regions; northern. (b) Pertaining to the northern biotic area (or

Boreal region) characterized by tundra and taiga and by dominant coniferous forests. (c) Pertaining to the Boreal postglacial period characterized by a cool climate like that of the present Boreal region. Also, said of the climate of such a period. (d) Pertaining to a Boreal climatic zone, or to the climate of such a zone.

Boreal [clim] n. A climatic zone having a definite winter that experiences snow and a short summer that is generally hot, and characterized by a large annual range of temperature. It includes large parts of North America, central Europe, and Asia, generally between latitudes 60°N and 40°N.

Boreal [paleoclim] n. A term used primarily in Europe for an interval of postglacial time (from about 9000 to 7500 years ago) following the Preboreal and preceding the Atlantic, during which the inferred climate was relatively warm and dry; a subunit of the Blytt-Sernander climatic classification, characterized by pine and hazel vegetation. More recent paleoclimatic data (carbon-14 dating of glacial advances and sea-level shifts) indicate that the Boreal climate in many regions may have been relatively wetter as well as cooler than the following Atlantic interval.---adj. Pertaining to the postglacial Boreal interval and to its climate, deposits, biota, and events.

bored well A shallow (3-30 m, or 10-100 ft), large-diameter (20-90 cm, or 8-36 in.) water well constructed by hand-operated or power-driven augers. Syn: *bore [water]*.

borehole A circular hole made by boring; esp. a deep vertical hole of small diameter, such as a shaft, a well (an exploratory oil well or a water well), or a hole made to ascertain the nature of the underlying formations, to obtain samples of the rocks penetrated, or to gather other kinds of geologic information. Cf: *drill hole*. Syn: *bore; boring*.

borehole log A *log* obtained from a borehole; esp. a lithologic record of the rocks penetrated.

borehole survey (a) Determination of the course, and deviation from the vertical, of a borehole by precise measurements of various points along the central axis of the borehole. Also, the record of the information thus obtained. (b) Determination of the mineralogic, structural, and/or physical characteristics of the strata penetrated by a borehole. Also, the record of the information thus obtained.

bore meal Cuttings, sludge, mud, or crushed debris brought up from a borehole.

boresight camera A camera mounted in or parallel to the optical axis of an instrument to photograph the ground being tracked. It thus provides a location guide, or photographic reflectances with a radar system.

borickite A reddish-brown mineral consisting of a hydrous basic phosphate of iron and calcium.

boring [drill] (a) The *drilling* of a circular hole in rock or earth material by the rotary motion of a cutting tool, for purposes such as blasting, exploration, exploitation (as of an oil field), or drainage. (b) A *borehole*.

boring [paleont] (a) A trace fossil consisting of an etching produced by plants (fungi, algae) or animals (sponges, worms, bryozoans, barnacles) in shells, bones, or other hard parts of invertebrates and vertebrates. (b) *burrow*.

boring porosity *burrow porosity*.

bornhardt A residual peak having the characteristics of an *inselberg*; specif. a large granite-gneiss inselberg associated with the second-cycle of erosion in a rejuvenated desert region (King, 1948). Named in honor of F. Wilhelm C.E. Bornhardt (1864-1946), German explorer of Tanganyika, who first described the feature.

bornhardtite A cubic mineral: Co_3Se_4.

bornite A brittle, metallic-looking mineral: Cu_5FeS_4. It has a reddish-brown or coppery-red color on fresh fracture, but tarnishes rapidly to iridescent purple or blue. Bornite is a valuable ore of copper. Syn: *erubescite; variegated copper ore; peacock ore; horseflesh ore; purple copper ore*.

borolanite A plutonic rock composed chiefly of orthoclase and melanite with lesser amounts of nepheline, biotite, and pyroxene; a melanite nepheline syenite. The orthoclase and nepheline frequently form aggregates that resemble phenocrysts of leucite. The term was originated by Horne & Teall in 1892.

Boroll In U.S. Dept. of Agriculture soil taxonomy, a suborder of the soil order *Mollisol*, characterized by a mean annual soil temperature of less than 8°C and by never being dry for 60 consecutive days during the 90-day period following the summer solstice (SSSA, 1970). Cf: *Alboll; Aquoll; Rendoll; Udoll; Ustoll; Xeroll*.

boronatrocalcite *ulexite*.

boron metasomatism The replacement of minerals in a rock by boron-bearing minerals.

borrow Earth material (sand, gravel, etc.) taken from one location (such as a borrow pit) to be used for fill at another location; e.g. embankment material obtained from a pit when there is insufficient excavated material nearby to form the embankment.

borrow pit An excavated area where borrow has been obtained. See also: *barrow pit*.

bort (a) Granular to very finely crystalline aggregate consisting of imperfectly crystallized diamonds or of fragments produced in cutting diamonds. It often occurs as spherical forms, with no distinct cleavage, and having a radial fibrous structure. (b) A diamond of the lowest quality, so badly flawed or imperfectly crystallized or too off-color that it is suitable only for crushing into abrasive powders for industrial purposes (such as for arming saws and drill bits); an *industrial diamond*. Originally, any crystalline diamond (and later, any diamond) not usable as a gem. (c) A term formerly used as a syn. of *carbonado*.---Cf: *ballas*. Syn: *boart; boort; bortz; bowr*.

bortz *bort*.

böschung A syn. of *gravity slope*. Etymol: German *Böschung*, a term used by Penck (1924) for a rock slope maintaining constant gradient as it retreats.

boss [geomorph] A smooth and rounded mound, hillock, or other mass of resistant bedrock, usually bare of soil or vegetation.

boss [paleont] (a) A rounded and raised or knob-like ornamental structure in foraminifera. (b) The part of an echinoid tubercle, below the mamelon, shaped like a truncated cone and supporting the spheroidal summit of the tubercle. (c) A coarse, short nodule occurring on the spire of a gastropod.

boss An igneous intrusion that is less than 40 sq mi in surface exposure and that is roughly circular in plan. Cf: *stock*.

bostonite A light-colored hypabyssal rock, characterized by *bostonitic* texture and composed chiefly of albite and microcline, with accessory pyroxene; an alkalic syenite with few or no mafic components.

bostonitic Said of the texture of *bostonite*, in which microlites of rough irregular feldspar tend to form clusters of divergent laths within a trachytoid groundmass.

botallackite A bluish-green mineral: $Cu_2(OH)_3Cl.3H_2O$.

botanical anomaly A deviation which is noticeably more than the normal variation in the chemical composition, distribution, ecological assemblage, or morphology of plants indicating the possibility of the presence of an ore deposit. See also: *geobotanical prospecting*.

botn A Norwegian and Swedish term for the "bottom" of a glacial lake or of a fjord, but used as an equivalent to *cirque*.

botryogen A deep-red or deep-yellow, usually botryoidal mineral: $MgFe(SO_4)_2(OH).7H_2O$.

botryoid n. In a cave, a formation of calcium carbonate that occurs on the walls in a grapelike cluster. Syn: *clusterite; grape formation*.

botryoidal Having the form of a bunch of grapes. Said of mineral deposits, e.g. hematite, having a surface of spherical shapes; also said of crystal structure in which the spherical shapes are composed of radiating crystals. Cf: *colloform; reniform*.

botryolite A radiated, columnar variety of datolite with a botryoidal surface.

bottleneck bay A bay with a narrow entrance which is guarded from the waves by features other than barrier islands.

bottle post *drift bottle.*

bottle spring A fresh-water spring that issues through the floor of a saline lake or pool. The name is derived from the fact that fresh water can be obtained by submerging a stoppered bottle directly over the spring and then removing the stopper.

bottom [**ore dep**] (a) Syn. of *gutter.* (b) The lower limit of an orebody, either structurally or by economic grade. Syn: *root.* See also: *bottoming.*

bottom [**geog**] Low-lying, level, usually highly fertile land, esp. in the Mississippi Valley region and farther west where the term signifies a grassy lowland formed by deposition of alluvium along the margin of a watercourse; an alluvial plain or a flood plain; the floor of a valley. The term is usually used in the plural. Syn: *bottomland; flat; interval; lowland.*

bottom [**karst**] In an area of karst, a term sometimes used for a dry valley.

bottom [**geomorph**] (a) The *bed* of any body of water; the *floor* upon which any body of water rests. (b) A term used in England for the former head of a lake in a U-shaped valley, now covered with sediment deposited by inflowing streams. (c) *valley floor.*

bottom-hole pressure (a) The pressure produced at the bottom of a borehole by the weight of the column of drilling fluid or other liquid in the hole, or exerted by gas or liquids ejected from the rocks at or near the bottom of the hole. (b) The hydraulic pressure measured in a well at a test position opposite the producing formation. It may or may not be measured under flowing conditions. (c) The load applied to a drill bit or other cutting tool while drilling.---See also: *reservoir pressure.*

bottom-hole temperature The temperature of rock or rock material at or near the bottom of a borehole.

bottom ice *anchor ice.*

bottoming The downward pinching out or termination of an orebody, either structurally or by economic grade. See also: *bottom.*

bottomland A syn. of *bottom.* Also spelled: *bottom land.*

bottom load *bed load.*

bottom moraine *ground moraine* [glac geol].

bottom peat Peat that is associated with lakes or streams and is derived mainly from mosses such as *Hypnum.*

bottomset *bottomset bed.*

bottomset bed One of the horizontal or gently inclined layers of fine-grained material (silts and clays) slowly deposited on the floor of a sea or lake in front of the advancing margin of a delta amd progressively buried by *foreset beds* and *topset beds.* It appears at the bottom of a complete cross section of a delta. Also spelled: *bottom-set bed.* Syn: *bottomset.*

bottom terrace A depositional landform produced by streams having moderate or small bottom loads of coarse sand and gravel, characterized by a broad (a few meters), gently sloping surface in the direction of flow and a steep escarpment (about a meter high) facing downstream, and generally trending at right angles to the flow (Russell, 1898b, p.166-167).

bottom water [**oceanog**] The deepest and most dense *water mass,* formed by cooling at the surface in high latitudes. Cf: *deep water; intermediate water; surface water* [oceanog].

bottom water [**oil**] The water immediately underlying oil or gas in a rock stratum. Cf: *edge water.*

boudin (a) One of a series of elongate, sausage-shaped segments occurring in *boudinage* structure and lying parallel to the strike, either detached or joined by pinched connections, and having barrel-shaped cross sections. Cf: *tectonic lens.* (b) A term applied loosely, without regard to shape or origin, to any *tectonic inclusion.*----Etymol: French, "bag; blood sausage". Pron: boo-*dan.*

boudinage A structure common in strongly deformed sedimentary and metamorphic rocks, in which an original continuous competent layer or bed between less competent layers has been stretched, thinned, and broken at regular intervals into bodies resembling *boudins* or sausages, elongated parallel to the fold axes. Syn: *sausage structure.*

Bouguer anomaly A gravity anomaly calculated by allowing for the attraction effect of topography but not for that of isostatic compensation.

Bouguer plate An imaginary layer, having infinite length and a thickness equal to the height of the observation point above the reference surface, which is usually the geoid (Mueller & Rockie, 1966, p. 13).

boulangerite A bluish-gray or lead-gray metallic-looking mineral: $Pb_5Sb_4S_{11}$. It occurs in plumose masses.

boulder (a) A detached rock mass larger than a cobble, having a diameter greater than 256 mm (10 in., or -8 phi units, or about the size of a volleyball), being somewhat rounded or otherwise distinctively shaped by abrasion in the course of transport; the largest rock fragment recognized by sedimentologists. In Great Britain, the limiting size of 200 mm (8 in.) has been used. Cf: *block* [part size]. See also: *small boulder; medium boulder; large boulder; very large boulder.* (b) *glacial boulder.* (c) *boulder of weathering.* (d) *boulder stone.* (e) A general term for any rock that is too heavy to be lifted readily by hand.---Syn: *bowlder.*

boulder barricade An accumulation of many large boulders visible along a coast (such as that of Labrador) between low tide and half tide (Daly, 1902, p.260).

boulder barrier A shore ridge created by great pressure from floating ice under the influence of strong winds and gentle shore slopes, and measuring over 6 m in height and 800 m in length (Hamelin & Cook, 1967, p.97).

boulder beach A beach consisting mostly of boulders.

boulder bed (a) A boulder-bearing conglomerate. (b) A glacial deposit, such as a till or tillite, containing a wide range of particle sizes; e.g. the Talchir boulder beds of India.

boulder belt A long, narrow accumulation of glacial boulders derived from distant sources, lying transverse to the direction of movement of the glacier by which it was deposited; also, a zone of such boulders. Cf: *boulder train.* Syn: *border belt.*

boulder clay A term used in Great Britain as an equivalent to *till,* but applied esp. to glacial deposits consisting of striated subangular boulders of various sizes embedded in stiff, hard, pulverized clay or rock flour. The term "till" is preferable as a general term, applicable not only to material of the character described above but to glacial deposits that contain no boulders or that may be so sandy as to have very little clay. Syn: *drift clay.*

boulder conglomerate A consolidated rock consisting mainly of boulders.

boulder depression A type of felsenmeer situated in a shallow depression, displaying a flat surface of pure boulder material that gradually decreases in size downward, and found mainly below the timberline. Diameter: a few meters to hundreds of meters.

boulderet A term suggested by Chamberlin (1883, p.324) for a rounded, coarser fragment of glacial drift, having a diameter range of 6-15 in. (15-38 cm).

boulder facet One of the small plane surfaces on a *faceted boulder.*

boulder fan A fan-shaped assemblage of *boulder trains* diverging from the bedrock source in the direction of movement of the glacier by which they were deposited.

boulder field *felsenmeer.*

boulder flat A level tract covered with boulders.

boulder gravel An unconsolidated deposit consisting mainly of boulders.

boulder of decomposition A *boulder of weathering* produced by chemical weathering; e.g. a joint block of basalt, modified and rounded by *spheroidal weathering,* leaving a relatively fresh spherical core surrounded by shells of decayed rock.

boulder of disintegration A *boulder of weathering* produced by mechanical weathering; e.g. a boulder fashioned by exfoliation.

boulder of weathering A large, detached rock mass whose corners and edges have been rounded in place, at or somewhat below the surface of the ground, by a distinctive process of weathering; e.g. *boulder of decomposition* and *boulder of disintegration*. Cf: *boulder*. Syn: *residual boulder*.

boulder opal A miner's term applied in Queensland, Australia, to siliceous ironstone nodules of concretionary origin, containing precious opal and occurring in sandstone or clay.

boulder pavement [glac geol] (a) An accumulation of glacial boulders once contained in a moraine and remaining nearly in their original positions when the finer material has been removed by waves and currents. (b) A relatively smooth surface strewn with striated and polished boulders, abraded to flatness by the movement of a glacier. Cf: *glacial pavement*.

boulder pavement [geomorph] (a) An accumulation of boulders produced on a terrace by the eroding action of waves or river currents in removing finer material from littoral or fluvial deposits. (b) A slightly inclined surface composed of randomly spaced, flat-surfaced, usually frost-shattered blocks resulting from solifluction or other mass movement. (c) A *desert pavement* consisting of boulders.

boulder prospecting The use of boulder trains from outcrops or suboutcrops of mineral deposits as a guide to ore.

boulder quarry A quarry in which weathering has produced so much jointing in the stone that it is not possible to mine large blocks from it.

boulder rampart A *rampart* of boulders built along the seaward edge of a reef. Syn: *boulder ridge*.

boulder ridge (a) A *beach ridge* composed of boulders. (b) *boulder rampart*.

boulder size A term used in sedimentology for a volume greater than that of a sphere with a diameter of 256 mm (10 in.).

boulder stone An obsolete term for any large rock mass lying on the surface of the ground or embedded in the soil, differing from the country rock of the region, such as an erratic. Syn: *boulder*.

boulderstone A consolidated sedimentary rock consisting of boulder-size particles (Alling, 1943, p.265).

boulder stream *rock stream*.

boulder train A line or series of glacial boulders extending from the same bedrock source, often for many kilometers, in the direction of movement of the glacier by which they were deposited. Cf: *boulder belt; boulder fan*.

boulder wall A boulder-built glacial moraine.

bouldery Characterized by boulders; e.g. a "bouldery soil" containing stones having diameters greater than 60 cm (24 in.) (SSSA, 1965, p.333).

boule A pear- or carrot-shaped mass (as of sapphire, ruby, spinel, or rutile) that forms during the production of synthetic gem material by the Verneuil process. Etymol: French, "ball". Syn: *birne*.

Bouma cycle A fixed, characteristic succession, of five intervals, that makes up a complete sequence of a turbidite (Bouma, 1962). One or more of the intervals may be missing for various reasons. The five intervals, from the top: (e) pelitic; (d) upper parallel laminations; (c) current ripple laminations; (b) lower parallel laminations; and (a) graded. Named after Arnold H. Bouma, 20th-century Dutch sedimentologist.

bounce cast The cast of a bounce mark, consisting of a short ridge that fades out gradually at both ends.

bounce mark A short (up to 5 cm in length) and shallow tool mark oriented parallel to the current and produced by an object that struck or grazed against the bottom and rebounded and was carried upward; its longitudinal profile is symmetric. The mark is widest and deepest in the middle and fades out gradually in both directions. Cf: *prod mark*. See also:

brush mark.

boundary Something that serves to indicate or fix the limit or extent of anything; e.g. the surface between different types or ages of rocks, or the line separating two cartographic units on a geologic map, or a line of demarcation between adjoining parcels of land. See also: *contact* [*geol*].

boundary current A deep ocean current, esp. along the western part of the oceans, characterized by sudden changes in temperature and salinity.

boundary curve *boundary line*.

boundary fault A descriptive term used in coal-mining geology for a fault along which there has been sufficient displacement to truncate the coal-bearing strata and thus bound the coalfield. Syn: *marginal fault*. Partial syn: *border fault*.

boundary lake A lake situated on or crossed by a political boundary line, as between two states or nations.

boundary layer In a fluid, a region of concentrated velocity variation and shear stress close to a solid that is moving relatively to the fluid. In a real fluid, the boundary layer is thin and its flow may be either turbulent or laminar (Middleton, 1965, p.247).

boundary line [geochem] In a binary system, the line along which any two phase areas adjoin; in a ternary system, the line along which any two liquidus surfaces intersect. In a condensed ternary system, the boundary line represents equilibrium, typically with two solid phases and one liquid phase. See also: *reaction line*. Syn: *boundary curve; phase boundary*.

boundary line [surv] A line along which two areas meet; a line of demarcation between contiguous political or geographic entities.

boundary map A map that delineates a boundary line and the adjacent territory.

boundary monument A *monument* placed on or near a boundary line for the purpose of preserving and identifying its location on the ground.

boundary spring A type of gravity spring whose water issues from the lower slope of an alluvial cone. Syn: *alluvial-slope spring*.

boundary stratotype A type section demonstrating or designating a synchronous boundary between two time-stratigraphic units. The boundary is designated at a position within an unbroken succession in which it is unlikely that appreciable amounts of geologic time are unrepresented by rocks. Cf: *stratotype*. Syn: *type-boundary section*.

boundary survey A survey made to establish or reestablish a boundary line on the ground or to obtain data for constructing a map showing a boundary line; esp. such a survey of boundary lines between political territories. Cf: *land survey; cadastral survey*.

boundary value component *perfectly mobile component*.

boundary value problem Three boundary value problems of potential theory have great significance in geodesy. *Dirichlet's problem* is to determine a function that is harmonic outside of a given surface and assumes prescribed boundary values on the surface. *Neumann's problem* is to determine a function that is harmonic outside of a given surface and whose normal derivatives assume prescribed boundary values on the surface. The third (unnamed) problem is to determine a function that is harmonic outside of a given surface and is such that a certain linear combination of it and its normal derivative assumes prescribed boundary values on the surface.

boundary vista A lane cleared along a boundary line passing through a wooded area. It is used for readily identifying a boundary line.

boundary wave A seismic wave propagated along a free surface or an interface between definite layers.

bound gravel A hard, lenticular, cemented mass of sand and gravel occurring in the region of the water table; it is often mistaken for bedrock.

boundstone A term used by Dunham (1962) for a sedimentary carbonate rock whose original components were bound to-gether during deposition and remained substantially in the po-sition of growth (as shown by such features as intergrown skeletal matter and lamination contrary to gravity); e.g. most reef rocks and some biohermal and biostromal rocks.

bound water Water present in such materials as animal and plant cells and soils that cannot be removed without changing the structure or composition of the material and that cannot react as does *free water* in such ways as dissolving sugar and forming ice crystals.

bourne A small stream or brook; specif. an intermittent stream that flows on the chalk downs and limestone heights of southern England after a heavy rainfall. Syn: *bourn; burn; winterbourne; nailbourne; woebourne; gypsey; lavant; chalk stream.*

bournonite A steel-gray to iron-black orthorhombic mineral: $PbCuSbS_3$. It commonly occurs in wheel-shaped twin crystals associated with other copper ores. Syn: *wheel ore; cogwheel ore; endellionite; berthonite.*

bourrelet (a) An externally inflated or elevated part of an interambulacral area of an echinoid, situated adjacent to the peristome. Cf: *phyllode.* (b) Either of two parts of the ligam-ental area of a bivalve flanking the resilifer on its anterior and posterior sides. Each bourrelet comprises a growth track and a seat of the lamellar ligament.

boussingaultite A mineral: $(NH_4)_2Mg(SO_4)_2.6H_2O$.

bowenite A hard, compact, greenish-white to yellowish-green mineral of the serpentine group, representing a translucent, massive, fine-grained variety of antigorite resembling nephrite jade in appearance and composed of a dense felt-like aggre-gate of colorless fibers, with occasional patches of magnesite, flakes of talc, and grains of chromite. The term has also been applied to a serpentine rock in New Zealand. Syn: *tangiwai.*

Bowen ratio The ratio of sensible to evaporative energy (heat) loss from the surface of a body of water.

Bowen's reaction series A term used interchangeably with *reaction series* for a concept originally proposed by N. L. Bowen.

Bowie effect The indirect effect on gravity due to a warping of the geoid resulting from the application of gravity corrections. Syn: *indirect effect.*

bowlder Var. of *boulder.*

bowlingite *saponite.*

bowr *bort.*

bowralite A syenitic pegmatite composed chiefly of tabular eu-hedral sanidine crystals with lesser amounts of alkali amphi-bole and aegerine, and quartz, perovskite, zircon, and ilmenite as possible accessories.

box (a) A hollow limonitic concretion. (b) *box-stone.*

box canyon (a) A narrow gorge or canyon containing a stream following a zigzag course, characterized by high, steep rock walls and typically closed upstream with a similar wall, giving the impression as viewed from its bottom of being surrounded or "boxed in" by four almost-vertical walls. (b) A steep-walled canyon heading against a cliff; a dead-end can-yon.---Syn: *cajon.*

box core A type of *corer* that retrieves sediment samples in a block rather than in a cylinder.

box fold A fold, the crest or trough of which is broad and flat and the limbs of which are steep.

box level A spirit level in which a glass-covered box is used instead of a glass tube; specif. a *circular level.*

box-stone (a) A term used in Suffolk, Eng., for a pebble or mass of hard, brown, rounded or flattened, ferruginous or phosphatic sandstone, generally a little larger than a clenched fist, and often containing the remains of a fossil. Also, the sandstone bed containing box-stones. Syn: *box.* (b) A British term applied to a ferruginous concretion (found in Jurassic and Tertiary sands), often of rounded and rectangular or box-like form, having a hollow interior in which white, powdery

sand is sometimes present (P.G.H. Boswell in Wentworth, 1935, p.241).

box the compass To name or repeat the 32 points of the com-pass in their contact order (clockwise).

boxwork In mineral deposits, e.g. limonite, a network of inter-secting blades or plates of the mineral deposited along frac-ture planes and from which the host rock has dissolved.

braccianite A melilite-free cecilite.

brach (a) *brachial plate.* (b) *brachiopod.*

brachia (a) Plural of *brachium.* (b) A term sometimes used as a syn. of *lophophore.*

brachial adj. Pertaining to an arm or arm-like structure of an animal (such as to the rays of a starfish or the brachium of a brachiopod).---n. A brachial part; esp. *brachial plate.*

brachial plate One of the plates that form the arms of a cri-noid; any crinoid-ray plate above the radial plates (exclusive of pinnulars). Syn: *brachial; brach.*

brachial process An anteriorly directed blade- or rod-like pro-jection from the cardinalia of pentameracean brachiopods.

brachial ridge A narrow elevation of secondary shell of some articulate brachiopods, extending laterally or anteriorly as an open loop from the dorsal adductor muscle field. The brachial ridges are thought to be the region of attachment of the lo-phophore.

brachial valve The valve of a brachiopod that invariably con-tains any skeletal support (brachidium) for the lophophore and never wholly accommodates the pedicle, that is common-ly smaller than the *pedicle valve*, and that has a distinctive muscle-scar pattern (TIP, 1965, pt.H, p.141). It typically has a small or indistinguishable beak. Syn: *dorsal valve.*

brachidium The loop-like or spiraliform, internal, calcareous, skeletal support structure of the lophophore of certain brach-iopods. Pl: *brachidia.*

brachiolar facet An elliptical or subcircular *facet*, indentation, or scar-like area where a brachiole was attached, as in a cys-toid or blastoid. Also spelled: brachiole facet.

brachiolar plate One of the biserially arranged plates of a bra-chiole of a blastoid, semielliptical in cross section and subquadrangular in side view, with basal pair attached at bra-chiolar facet (TIP, 1967, pt.S, p.346).

brachiole A biserial, nonpinnulate, exothecal appendage of an echinoderm, springing independently from its surface and containing no extension of the body systems; esp. an erect, food-gathering structure arising from a cystoid thecal plate at the end or along the side of an ambulacrum and bearing an extension of the ambulacral groove and by which food was transmitted to the ambulacrum. Cf: *arm [paleont].*

brachiophore One of the short, typically stout, blade-like pro-cesses of secondary shell projecting from either side of the notothyrium and forming anterior and median boundaries of sockets in the brachial valves of certain brachiopods.

brachiophore base The basal (dorsal) part of a brachiophore which joins the floor of a brachiopod valve (TIP, 1965, pt.H, p.141).

brachiophore process A distal rod-like extension of a brachio-phore that possibly supported the lophophore in some bra-chiopods.

brachiopod Any solitary marine invertebrate belonging to the phylum Brachiopoda, characterized by a lophophore and by two bilaterally symmetrical valves that may be calcareous or composed of chitinophosphate and that are commonly at-tached to a substratum, but may also be free. Their stratigra-phic range is Lower Cambrian to present. Syn: *brach; lamp shell.*

brachistochronic path *minimum time path.*

brachitaxis A series of crinoid brachial plates. Pl: *brachitaxes.*

brachium (a) Either of the two arm-like, coiled, muscular pro-jections from either side of the mouth segment of the lopho-phore of a brachiopod, variably disposed but symmetrically placed about the mouth. (b) Any process of an invertebrate similar to an arm, such as a tentacle of a cephalopod.---Pl: *brachia*.

brachyanticline An anticline that is wider than it is long, so that it appears to be quaquaversal in outline. Cf: *brachysyncline*.

brachy-axis The shorter lateral axis of an orthorhombic or triclinic crystal; it is usually the *a axis*. Cf: *macro-axis*. Also spelled: *brachyaxis*. Syn: *brachydiagonal*.

brachydiagonal n. A syn. of *brachy-axis*.

brachydome A *first-order prism* in the orthorhombic system; it is rhombic, with four faces parallel to the brachy-axis. Its indices are {0*kl*}.

brachygeosyncline A deep, oval depression formed during the later stages of geosynclinal deformation; a type of secondary geosyncline (Peive & Sinitzyn, 1950).

brachyome The short arm of an anomoclone or ennomoclone of a sponge, or the different fourth ray of a trider of a sponge.

brachypinacoid *side pinacoid*.

brachysyncline A syncline that is wider than it is long, so that it resembles a *centrocline* in outline. Cf: *brachyanticline*.

brachyuran Any decapod belonging to the infraorder Brachyura, characterized by a carapace that becomes progressively shortened and widened, developing a lateral margin; e.g. a crab. Their stratigraphic range is Lower Jurassic to present.

brackebuschite A black to reddish mineral: $Pb_4MnFe(VO_4)_4.2H_2O$.

bracket delta A flat-topped, steep-sided delta resembling the bastion of a fort, formed rapidly by swift-flowing streams depositing coarse material (Dryer, 1910, p.258).

brackish water An indefinite term for water, the salinity of which is intermediate between that of normal sea water and of normal fresh water.

bract A modified leaf associated with a flower or inflorescence, e.g. bearing a flower on its axis or being borne on a floral axis (subtending the flower or inflorescence).

bracteate Said of a plant having *bracts*.

bradenhead A *casinghead* in an oil well, used to confine gas between the tubing and casing or between two strings of casing, and having a stuffing box usually packed with rubber to make a gastight connection. Named after Glenn T. Braden (d.1923), American oilman and inventor. Syn: *stuffing-box casinghead*.

Bradfordian (a) North American stage: uppermost Devonian (above Cassadagan, below Mississippian). Syn: *Conewangoan*. (b) Substage in Great Britain: Middle Jurassic (upper Bathonian Stage).

bradleyite A mineral: $Na_3Mg(PO_4)(CO_3)$.

bradygenesis *bradytely*.

bradytely Retardation in the development of a group of organisms which may gradually cause certain individuals to fall behind the normal rate of progress in some or all of their characteristics. Cf: *horotely; tachytely; lipogenesis*. Syn: *bradygenesis*.

Bragg angle In the *Bragg equation*, the angle of the reflected ray of light, measured from a perpendicular to the surface from which the ray is reflected. It is symbolized by θ (theta). Syn: *reflection angle; angle of reflection*.

Bragg equation A statement in crystallography that the X-ray diffractions from a three-dimensional lattice may be thought of as reflecting from the lattice planes: $n\lambda = 2d \sin\theta$, in which n is any integer, λ is the wavelength of the X-ray, d is the crystal plane separation, also known as *d-spacing*, and θ is the angle between the crystal plane and the reflected beam, also known as the *Bragg angle*. Syn: *Bragg's law*.

braggite A steel-gray mineral: $(Pt,Pd,Ni)S$.

Bragg reflection A diffracted beam of X-rays by a crystal plane according to the Bragg law.

Bragg's law *Bragg equation*.

braid v. To branch and rejoin repeatedly to form an intricate pattern or network, as of channels of an overloaded stream.-

-n. A reach of a braided stream, characterized by relatively stable branch islands and hence two or more separate channels. Cf: *anabranch*.

braided channel The channel of a braided stream, characterized by an intricate network of smaller interlacing channels that repeatedly merge and separate.

braided drainage pattern A drainage pattern consisting of braided streams. Syn: *interlacing drainage pattern*.

braided stream A stream that divides into or follows an interlacing or tangled network of several, small, branching and reuniting shallow channels separated from each other by branch islands or channel bars, resembling in plan the strands of a complex braid. Such a stream is generally believed to indicate an inability to carry all of its load, such as an overloaded and aggrading stream flowing in a wide channel on a flood plain. Syn: *anastomosing stream*.

braiding The process of successive branching and rejoining of a stream channel, resulting in the formation of elongated islands and bars that split the channel into an intricate network of smaller interlacing channels; the action of a stream that can no longer transport its load, thus producing a braided stream. Syn: *anastomosis*.

brait A rough diamond.

braitschite A mineral: $7(Ca,Na_2)O.(Ce,La)_2O_3.11B_2O_3.7H_2O$.

brammallite A micaceous clay mineral, representing the sodium analogue of illite. It is a variety of illite in which sodium replaces potassium. Syn: *sodium illite*.

branch (a) A small stream that flows into another, usually larger, stream; a tributary. (b) A term used in the southern U.S. for a creek, or a stream normally smaller than and often tributary to a river. (c) A stream flowing out of the main channel of another stream and not rejoining it, as on a delta or alluvial fan; a distributary. (d) A stream flowing out of another stream and rejoining it, such as an anabranch; a by-channel. (e) A fork of a tidal river; e.g. the fork of the Severn River, Md.

branch fault *auxiliary fault*.

branch gap An interruption in the vascular tissue of a stem at the point at which a *branch trace* occurs. It is most evident in cross section, at the point of branch-trace departure.

branchia (a) A thin-walled, finger- or leaf-like structure extending outward from a crustacean limb or secondarily from a side of the body, typically occurring in pairs, and functioning for respiration. Syn: *gill [paleont]*. (b) A slender, hollow, finger-like extension of the body wall of an asteroid.---Pl: *branchiae*.

branchial chamber The space between the body and the wall of carapace enclosing the branchiae of a crustacean. Syn: *gill chamber*.

branchial slit A *gill slit* of an echinoid.

branching bay A bay having a dendritic pattern, produced by drowning or flooding of a river valley by the sea. See also: *estuary*.

branching fault A fault that splits into two or more parts or branches.

branching ratio In the case of two competing reactions, the ratio of the probabilities of each reaction.

branchiopod Any crustacean belonging to the class Branchiopoda, characterized by the morphologic similarity of their numerous somites and limbs and by their filter-feeding mode of nourishment. Their stratigraphic range is Lower Devonian to present.

branchiostegite The extended portion of carapace covering the branchial chamber of a decapod crustacean.

branch island An island formed by the braiding of the branches of a stream; an island formed between a tributary and the main stream. Term was introduced by Jackson (1834, p.79).

branchite *hartite*.

branch trace Vascular tissue extending from a stem into a branch. Cf: *leaf trace*. See also: *branch gap*.

branch water Water from a small stream or a branch.

brandbergite A hypabyssal rock having aplitic texture and being composed of orthoclase (as whitish Carlsbad twins), quartz grains, and aggregates of biotite in a micrographic groundmass.

brandtite A mineral: $Ca_2Mn(AsO_4)_2.2H_2O$. It is isomorphous with roselite and may contain up to 3% MgO.

brannerite A mineral: $(U,Ca,Ce)(Ti,Fe)_2O_6$.

brash brash ice.

brash ice An accumulation of floating fragments (or bits) of ice, not more than 2 m across, and representing the wreckage of other forms of ice; occurs esp. near an ice pack or floe. Syn: brash [ice]; debris ice; mush [ice]; rubble [ice].

brass An English term for yellowish iron pyrites (pyrite and marcasite) found in coal or coal seams. Syn: brasses.

brassil A syn. of brazil. Also spelled: brassel.

brass ore (a) aurichalcite. (b) A mixture of sphalerite and chalcopyrite.

braunite A brittle brownish-black or steel-gray tetragonal mineral: $3Mn_2O_3.MnSiO_3$. It sometimes has appreciable ferric iron, and it represents an ore of manganese.

bravaisite A name proposed for a micaceous clay mineral having about half the potassium of muscovite, and later used as a synonym to replace illite. Material from the type locality has been shown to be a mixture of montmorillonite and illite, with illite predominating; therefore, bravaisite is not a specific mineral and has no standing as a distinct mineral species (Grim, 1953, p.36). Cf: sarospatakite.

Bravais lattice A syn. of crystal lattice; it is named after the nineteenth-century French physicist, Auguste Bravais, who demonstrated that there are only 14 possible crystal lattices.

bravoite A yellow mineral: $(Ni,Fe)S_2$. It is related to pyrite, and has a paler color.

B ray Right anterior ray in echinoderms situated clockwise of A ray when the echinoderm is viewed from the adoral side.

brazil An English dialectal term for iron pyrites, esp. associated with coal. Also, by extension, a term applied to a coal seam containing much pyrites. Adj: brazilly. Syn: brazzil; brazzle; brassil.

Brazilian emerald A transparent green variety of tourmaline occurring in Brazil and used as a gemstone.

brazilianite A yellowish-green to greenish-yellow monoclinic mineral: $NaAl_3(PO_4)_2(OH)_4$.

Brazilian ruby A reddish mineral resembling ruby in appearance and occurring in Brazil; e.g. a light rose-red spinel, or a pink to rose-red or deep-red topaz (either natural or artificially heated), or a reddish tourmaline.

Brazilian sapphire A transparent blue variety of tourmaline occurring in Brazil and used as a gemstone.

Brazilian topaz Topaz mined in Brazil and varying in color from pure white to blue; esp. true yellowish topaz.

brazilite A mixture of baddeleyite, zircon, and altered zircon. The term has also been applied to an oil shale, to a fibrous variety of baddeleyite, and as a syn. of baddeleyite.

Brazil twin law A type of twin law in quartz in which the twin plane is perpendicular to one of the a crystallographic axes; an example of optical twinning. Cf: Dauphiné twin law.

brazzil Var. of brazil.

brea A rarely used term for a viscous asphalt formed by the evaporation of volatile components from oil in seepages. In Trinidad, it is used as the name for maltha. Etymol: Spanish, "pitch".

breach v. To cut a deep opening in a landform, esp. by erosion.

breached Said of a volcanic cone or crater, the rim of which has been broken through by the outpouring lava.

breached anticline An anticline, the crest of which has been deeply eroded so that it is flanked by inward-facing erosional scarps. Syn: unroofed anticline; scalped anticline.

breachway A connecting channel.

bread-crust bomb A type of volcanic bomb that is characterized by a network of opened cracks on its surface, due to continued expansion of the interior after the solidification of the crust. See also: explosive bomb.

bread-crust surface A surface, resembling the crust of bread characterizing certain concretions formed where abundant salts are being precipitated by evaporating water in a semiarid climate (Twenhofel, 1939, p.40).

break [drill] (a) A change in penetration rate of a drill. Also, an abrupt vertical change (such as a fault or fracture) in the hardness of the rock formations that a drill is penetrating. (b) A soft layer of rock between harder strata; e.g. a "shale break".

break [geomorph] A marked variation of topography, or a tract of land distinct from adjacent land, or an irregular and rough piece of ground; e.g. a deep valley, esp. a ravine or gorge cutting through a ridge or mountain. See also: breaks.

break [seis] arrival.

break [stratig] (a) An abrupt change at a definite horizon in a chronologic sequence of sedimentary rocks, usually indicative of an unconformity (esp. a disconformity) or hiatus; esp. a marked change in lithology in a sedimentary sequence, such as one separating stray sand from a more persistent member. See also: faunal break. (b) An interruption of a normal geologic sequence, esp. of stratigraphic continuity; a discontinuity. -Syn: gap; stratigraphic break.

break [mining] A general term used in mining geology for any discontinuity in the rock, such as a fault, a fracture, or a small cavity.

break [slopes] A marked or abrupt change or inflection in a slope or profile; a knickpoint. Term is used in the expressions "break of slope" and "break of profile".

breakdown cave breakdown.

breaker A sea-surface wave that has become too steep (wave steepness of 1/7) so that the crest outraces the body of the wave and collapses into a turbulent mass on shore or over a reef or rock. See also: plunging breaker; spilling breaker; surging breaker; surf. Syn: breaking wave.

breaker depth The still-water depth at the point where a wave breaks. Syn: breaking depth.

breaker line The axis along which a wave breaks as it approaches the shore. Syn: plunge line.

breaker zone surf zone.

breaking depth breaker depth.

breaking strength fracture strength.

breaking wave breaker.

breakout The well-drilling process of pulling up drill pipes or casings from a borehole and disconnecting them for stacking in the derrick.

breakover A rounded crest that is both structurally and topographically high.

breakpoint bar A longshore bar formed at the breakpoint of waves, where there is a sudden decrease of sand moving landward outside the breakpoint but where sand is moving seaward to this point (King & Williams, 1949, p.80).

breaks (a) A term used in the western U.S. for a tract of rough or broken land dissected by ravines and gullies, as in a badlands region. (b) Any sudden change in topography, as from a plain to hilly country, or a line of irregular cliffs at the edge of a mesa or at the head of a river; e.g. Cedar Breaks, Utah.

breakthrough (a) The erosive action of water in wearing or cutting a passage. (b) The channel made by such a breakthrough.

break thrust An overthrust developed during the deformation of an anticline at that point at which folding becomes fracturing, and strata are overthrust along the fault surface.

breakup (a) The melting, loosening, fracturing, or destruction of snow or floating ice during the spring; specif. the destruction of the ice cover on a river during the spring thaw. (b) The period during the spring thaw when a breakup occurs.

breakwater An offshore structure (such as a mole, mound, wall, or jetty) that, by breaking the force of the waves, protects a harbor, anchorage, basin, beach, or shore area. Syn: *water-break*.

breast A miner's term for the *face* of a mine working. See also: *before breast*.

breast wall A wall designed to sustain the face of a natural bank of earth, such as for timber face used in the forepoling process in tunneling.

breathing *river breathing*.

breathing cave *blowing cave*.

breathing well A well, generally a water well, that, in response to changes in atmospheric pressure, alternately takes in and emits a strong current of air, often with an alternating sucking and blowing sound. It penetrates, but is uncased in at least part of, a thick zone of aeration that is porous and permeable enough to exchange air freely with the well but otherwise is poorly connected with the atmosphere because of the presence of tight soil or other low-permeability material above the unsaturated material. Syn: *blowing well*. Cf: *blow well*.

breccia (a) A coarse-grained clastic rock composed of large (greater than sand size, or 2 mm in diameter), angular, and broken rock fragments that are cemented together in a finer-grained matrix (which may or may not be similar to the larger fragments) and that can be of any composition, origin, or mode of accumulation; the consolidated equivalent of rubble. According to Woodford (1925, p.183), the rubble content must be greater than 80%. Breccia is similar to *conglomerate* except that most of the fragments have sharp edges and unworn corners; the term "breccia" formerly included conglomerate, and is still sometimes so used in Europe. The rock can be formed in many ways, chiefly by sedimentation (*sedimentary breccia*), igneous activity (*igneous breccia*), and diastrophism (*tectonic breccia*). Etymol: Italian, "broken stones or rubbish of broken walls". Pron: *bresh-ah*. Syn: *rubblerock*. (b) *volcanic breccia*.

breccia-conglomerate (a) A sedimentary rock consisting of both angular and rounded particles (Norton, 1917, p.181); a sedimentary rock that is not clearly referable to either breccia or conglomerate. (b) A term that may be applied either to a sedimentary rock in which the larger fragments are "said to be on average slightly angular" or to a breccia "conglomerated" into a rock mass (Challinor, 1967, p.53).---Syn: *breccio-conglomerate*.

breccia dike A sedimentary dike composed of breccia injected into the country rock.

breccial Pertaining to breccia.

breccia marble Any marble composed of angular fragments. The term was in use before the separate use of "breccia" in geology.

breccia pipe *pipe* [volc].

breccia porosity *Interparticle porosity* in a breccia (Choquette & Pray, 1970, p.244).

breccia-sandstone A sandstone containing "some small breccia-fragments" (King, 1899, p.105).

brecciated [geol] Converted into, characterized by, or resembling a breccia; esp. said of a rock structure marked by an accumulation of angular fragments, or of an ore texture showing mineral fragments without notable rounding.

brecciated [meteorite] A term incorrectly applied to an intermediate-type meteorite (between iron meteorite and stony-iron meteorite) in which the main mass is iron with octahedral or hexahedral structure but also containing relatively large silicate inclusions of rounded or angular form. Also, said of such a texture occurring in a meteorite, including stony meteorites.

brecciation Formation of a breccia, as by crushing a rock into angular fragments. The term should apply to the formation of a tectonic breccia only (Challinor, 1967, p.32).

breccia vein A fissure containing numerous wall-rock frag-

ments and in which mineral deposits fill the interstices.

brecciform In the form or shape of a breccia, or resembling a breccia.

breccio-conglomerate *breccia-conglomerate*.

breccioid Having the appearance of a breccia.

brecciola A well-graded, intraformational breccia consisting of small, angular limestone fragments in well-defined beds separated by dark shale, such as the breccia occurring in the northern Apennines. Etymol: Italian, diminutive of *breccia*.

bredbergite A variety of andradite garnet containing magnesium.

bredigite A mineral: Ca_2SiO_4. It is a metastable orthorhombic phase of calcium orthosilicate (but not isomorphous with olivine), stable from about 800° to 1447°C on heating and from 1447° to 670°C on cooling. Cf: *calcio-olivine; larnite*.

breithauptite A copper-red mineral: NiSb.

Breithaupt twin law A rare type of normal twin law in feldspar, in which the twin plane is $(\bar{1}11)$.

Bretonian orogeny Part of the 30 or more orogenies during Phanerozoic time identified by Stille; this orogeny consists of several phases, from the Late Devonian to the end of the Devonian; considered to be the earliest part of the Variscan orogenic era, which continued to the end of the Paleozoic.

breunnerite (a) A variety of magnesite containing 5-30% iron carbonate. (b) A mineral: $(Mg,Fe,Mn)CO_3$.

Brevaxones A group of mid-Cretaceous and younger angiosperm pollen in which the polar axis is shorter than the equatorial diameter, representing an evolutionary advance over *Longaxones*, and including such forms as Normapolles.

brevicone A typically curved shell characteristic of certain Paleozoic cephalopods, having a short, blunt form and a rapid taper from the wide living chamber to the first camera. Also, a fossil animal having such a shell. Cf: *longicone*.

Brewster angle *angle of incidence*.

brewsterite A zeolite mineral: $(Sr,Ba,Ca)Al_2Si_6O_{16} \cdot 5H_2O$. It usually contains some calcium.

brewsterlinite Liquid CO_2, found as inclusions in cavities in minerals such as quartz, topaz, and chrysoberyl. It will expand so as to fill cavities under the warmth of the hand.

Brewster's law A statement in optics that when unpolarized light is incident on a surface, it acquires maximum plane polarization at a particular angle of incidence whose tangent equals the refractive index of the substance. This angle is called the *polarizing angle*, or the Brewster angle.

brezinaite A meteorite mineral: Cr_3S_4.

Brezina's lamellae Lamellae of schreibersite oriented parallel to dodecahedral planes in parent taenite of iron meteorites. Named after M. Aristides S.F. Brezina (1848-1909), Austrian mineralogist.

brianite A mineral: $Na_2CaMg(PO_4)_2$.

briartite A mineral: $Cu_2(Fe,Zn)GeS_4$.

brick clay (a) Any clay suitable for the manufacture of bricks or coarse pottery; a *brick earth*. (b) An impure clay containing iron, calcium, magnesium, and other ingredients.

brick earth Earth, clay, or loam suitable for making bricks; specif. a fine-grained brownish deposit consisting of quartz and flint sand mixed with ferruginous clay and found on river terraces as a result of reworking by water of windblown material, such as that overlying the gravels on certain terraces of the River Thames in England. See also: *brick clay*. Also spelled: *brickearth*.

bridal-veil fall A cataract of great height and such small volume that the falling water is largely dissipated in spray before reaching the lower stream bed, and having a form that suggests a bridal veil. Type example: Bridalveil Fall in Yosemite Valley, Calif.

bridge [drill] n. A large rock fragment or fragments or other obstruction that has fallen (either accidentally or intentionally) and lodged part way down in a drill hole (as in an oil well); e.g. a plug device so arranged in a borehole as to be capable of being placed at a desired point above the bottom, or an obstruction formed in a shothole by caving or by an exploding charge.---v. To form a bridge in a drill hole.

bridge [speleo] In a cave, a solutional remnant of rock that spans a passage from wall to wall. Cf: *partition.*

bridge [geomorph] *natural bridge.*

bridge [elect] A device for comparing resistances, inductances, capacitances, or impedances. This is done by balancing two opposing voltages through a current detector and relating the unknown to the known value when a meter needle is brought to zero.

bridge [snow] A snow or ice formation joining two heavier formations. Cf: *ramp.* Not to be confused with *snowbridge.*

bridge islet An island that becomes a peninsula during low tide.

bridging [drill] The deliberate or accidental closing or plugging of a section of a drill hole at some distance above its bottom, as by sloughing of loose rock off the walls of the hole, by squeezing of plastic material such as shale, or by intentionally introducing foreign material into the hole.

bridging [photo] The extension and adjustment, by photogrammetric methods, of the positions and/or elevations of additional points between given control points on the ground. See also: *stereotriangulation.*

bridging factor A term used by Gruner (1950) for a number that expresses the manner by which SiO_4 tetrahedra are tied together in a mineral. It is equal to 0.8 plus twenty percent of the quotient of the sum of the valence bonds of a silicate divided by the number of cations. A bridging factor of 1.00 is assigned to quartz, whose SiO_4 tetrahedra are all directly tied to other tetrahedra, resulting in the highest number of bridges possible; all other structures have smaller factors. See also:*energy index [mineral].*

bridle v. To connect a group of seismic amplifiers to a common input; to constrain a group of mechanical, electrical, or electromechanical devices to operate in unison.----n. A seismogram produced with the amplifiers bridled.

brigg An English term for a headland formed by "a scarp of hard rock cropping out at or near tide marks" (Stamp, 1961, p.78). Syn: *brig.*

bright-banded coal *Banded coal* consisting mainly of vitrain and clarain, some durain, and minor fusain. Cf: *dull-banded coal; semisplint coal.*

bright coal A type of banded coal defined microscopically as consisting of more than 5% of anthraxyllon and less than 20% of opaque matter; banded coal in which translucent matter predominates. Bright coal corresponds to the microlithotypes vitrite and clarite and in part of duroclarite and vitrinerite (ICCP, 1963). Cf: *dull coal; semibright coal; semidull coal; intermediate coal.* Syn: *brights.*

brightness temperature (a) The temperature of a blackbody radiating the same amount of energy per unit area at the wavelengths under consideration as the observed body. It is sometimes called *effective temperature.* Cf: *color temperature.* (b) The apparent temperature of a non-blackbody determined by measurement with an optical pyrometer or radiometer.

brights *bright coal.*

brilliancy The total amount of light reaching the eye, after being reflected from both the exterior and interior surfaces of a gemstone. Given equal transparency and perfection of cutting, the gem species with the highest refractive index will be the most brilliant.

brilliant A *brilliant-cut* diamond. The term is less correctly applied to any brilliant-cut gemstone.

brilliant cut A style of cutting for a gemstone, and the most common style for most gemstones. The standard round brilliant consists of a total of 58 facets: 1 table, 8 bezel facets, 8 star facets, and 16 upper girdle facets on the crown; and 8 pavilion facets, 16 lower girdle facets, and usually a culet on

the pavilion. Cf: *ideal cut; step cut; single cut; mixed cut.* See also: *pear-shape cut; pendeloque.*

brim The flared or recurved portion of a cyrtochoanitic septal neck of a nautiloid, measured transverse to the longitudinal axis of the siphuncle.

brimstone A common or commercial name for *sulfur,* esp. native sulfur or fine sulfur fused into rolls, sticks, or blocks.

brine Sea water that, due to evaporation or freezing, contains more than the usual amount of dissolved salts, or about 35‰ .

brine cell A small inclusion, usually in the shape of an elongated tube about 0.05 mm in diameter, containing residual liquid more saline than seawater, formed in sea ice as it develops. Syn: *brine pocket.*

brine lake *salt lake.*

brine pit A *salt well,* or an opening at the mouth of a salt spring, from which water is taken to be evaporated for making salt.

brine pocket *brine cell.*

brine slush "A mixture of ice crystals and salt water, which retards or prevents complete freezing, often found between young sea ice and a cover of newly fallen snow" (ADTIC, 1955, p.14).

brine spring *salt spring.*

brink (a) A bank, edge, or border of a body of water, esp. of a stream. (b) The top of the slip face of a dune. It may or may not be the same as the crest.

briolette A gemstone in the shape of an oval or pear, having its entire surface cut in triangular, or sometimes rectangular, facets. Cf: *pear-shape cut; pendeloque.*

britholite A mineral of the apatite group: $(Ca,Ce)_5(SiO_4, PO_4)_3(OH,F)$. Cf: *abukumalite.*

brittle Said of a rock that fractures at less than 3-5% deformation or strain. Cf: *ductile [struc geol].*

brittle mica (a) A group of minerals resembling the true *micas* in crystallographic characters, but having the cleavage flakes less elastic and containing calcium (instead of potassium) as an essential constituent. Syn: *clintonite.* (b) A mineral of the brittle-mica group, such as clintonite, margarite, and ephesite. (c) A micaceous mineral occurring in brittle folia; e.g. chloritoid.

brittle silver ore *stephanite.*

broad A British term for a lake or wide sheet of shallow, reed-fringed freshwater, forming a broadened part of, or joined to, a sluggish river near its estuary; often used in the plural. The feature is typically found in East Anglia (Norfolk and Suffolk), and is believed to have been produced artificially by the cutting and removal of peat in the Middle Ages.

brochantite An emerald-green to dark-green mineral: $Cu_4 SO_4(OH)_6$. It is common in the oxidation zone of copper-sulfide deposits. Syn: *kamarezite.*

brock An English term for a *brook.*

brockite A red and yellow mineral: $(Ca,Th,Ce)PO_4.H_2O$.

brockram A term used in Cumberland County, NW England, for breccia whose angular blocks are believed to have accumulated as talus material; e.g. the Brockram of Appleby and Kirkby Stephen, a Permian breccia in which fragments of Carboniferous limestone are held together by gypseous cement.

brodel An *involution* characterized by a bulbous mass of silt, without horizontal continuity, intruded downward into and completely enclosed by clay (Horberg, 1954, p.1136).

bröggerite A mineral: $(U,Th)O_2$. It is a variety of uraninite bearing thorium.

broken belt The transition zone between open water and consolidated pack ice.

broken coal A size of anthracite that will pass through a 4 3/8 inch round mesh but not through a 3.25 - 3.0 inch round mesh. Cf: *egg coal; stove coal; chestnut coal; pea coal; buckwheat coal.*

broken ice An obsolete term for sea-ice concentration of 5/10 to 8/10; now replaced generally by *open pack ice* and *close* pack ice. Syn: *open pack; open ice; loose ice; slack ice.*

broken round A term used by Bretz (1929, p.507) for a round-stone (such as a pebble or cobble) that has undergone breakage and whose spalled corners are believed to indicate exceptionally high-velocity currents.

broken sand A sandstone containing a mixed sequence of deposits (such as shaly layers).

broken shoreline A shoreline characterized by many closely spaced islands, peninsulas, or jutting headlands.

broken stream A stream that repeatedly disappears and reappears, as in an arid region.

broken water Water whose surface is covered with ripples and eddies.

bromargyrite A yellow isometric mineral: AgBr. Syn: *bromyrite.*

bromellite A white hexagonal mineral: BeO.

bromlite *alstonite.*

bromoform Tribromethane: $CHBr_3$. It is used as a *heavy liquid*; its specific gravity is 2.9. Cf: *methylene iodide; Clerici solution; Sonstadt solution; Klein solution.*

bromyrite *bromargyrite.*

brontolith An obsolete syn. of *stony meteorite.* Also spelled: *brontolite.*

Bronze Age In archaeology, a cultural level that was originally the middle division of the *three-age system,* and is characterized by the technology of bronze. Correlation of relative cultural levels with actual age (and, therefore, with the time-stratigraphic units of geology) varies from region to region. The term is used mainly in Europe, since in Asian archaeology it coincides with written history and in the Americas and Africa bronze was little used (Bray and Trump, 1970, p.43).

bronzite A brown or green variety of *enstatite* containing iron and often having a bronze-like or pearly metallic luster; an orthopyroxene intermediate in composition between enstatite and hypersthene.

bronzitfels *bronzitite.*

bronzitite A pyroxenite composed almost entirely of bronzite. Syn: *bronzitfels.*

brood pouch A sac or cavity of the body of an animal where the eggs or embryos are received and undergo a part of their development; e.g. the gently to strongly swollen part of the heteromorphous (presumed female) carapace of an ostracode, thought to be used for containing the not yet independent young. Syn: *brood chamber.*

brook (a) A small *stream* or rivulet, commonly swiftly flowing in rugged terrain, of less length and volume than a *creek*; esp. a stream that issues directly from the ground, as from a spring or seep, or that is produced by heavy rainfall or melting snow. Also, one of the smallest branches or ultimate ramifications of a drainage system. (b) A term used in England and New England for any tributary to a small river or to a larger stream. Syn: *brock; bruik.* (c) A general literary term for a *creek.*

brookite A brown, reddish, or sometimes black orthorhombic mineral: TiO_2. It is trimorphous with rutile and anatase, and occurs in druses and cavities as an alteration product of other titanium minerals. Syn: *pyromelane.*

brooklet A small brook; a rill.

brookside The land adjacent to or bordering on a brook. Syn: *burnside.*

brotocrystal A crystal fragment of a previously consolidated rock that is only partially assimilated in a later magma.

brow (a) The projecting upper part or margin of a steep slope just below the crest; the edge of the top of a hill or mountain, or the place at which a gentle slope becomes abrupt. Syn: *brae.* (b) An English term for a steep slope.

brown algae A group of algae corresponding to the phylum Phaeophyta, that owes its greenish yellow to deep brown color to the presence of carotenes and xanthophylls in greater amounts than the chlorophylls. Brown algae are usually re-

stricted to salt water. Cf: *blue-green algae; green algae; red algae.*

brown body A colored spheroid formed in many bryozoan zooids by the aggregation of nonhistolyzable residue of a degenerated polypide. It is believed to have an excretory function.

brown clay *red clay.*

brown coal A brown to brownish black coal, intermediate in coalification between peat and subbituminous coal, in which original plant structures may usually be seen. The term is generally used in Europe, Australia, and Great Britain. Cf: *lignite.* Partial syn: *lignite B.*

Brown earth A syn. of *Brown Forest soil*; also, any soil of similar reaction.

Brown Forest soil An old term for one of an intrazonal, calcimorphic group of soils that develops in a temperate climate under deciduous forest. It has a calcium-rich parent material, and has a mull horizon but no clay or sesquioxides. Syn: *Brown earth.*

brown hematite A syn. of *limonite.* The term is a misnomer because true hematite (unlike limonite) is anhydrous. Cf: *red hematite; black hematite.*

brown hornblende A brown variety of hornblende rich in iron; specif. *basaltic hornblende.*

brown iron ore *limonite.*

brown lignite A syn. of *lignite B.*

brown matter *Humic degradation matter;* cell wall degradation matter that is brown and translucent in cross section.

Brown Mediterranean soil A type of soil that is similar to the *Red Mediterranean soil* but that has a brown B horizon and is usually noncalcareous. The term is little used.

brown mica *phlogopite.*

brownmillerite A mineral: Ca_2AlFeO_5. It is a constituent of portland cement. Syn: *celite.*

brown ocher A *limonite* that is used as a pigment.

brown ore A brown-colored ore; specif. the limonite group of iron ores.

Brown Podzolic soil An old term for any of a group of zonal soils that is similar to a Podzol but lacks the leached, light-colored horizon. It is considered by some to be a type of Podzol rather than a separate soil group.

brown rock A term used in Tennessee for dark (brown to black) phosphorite resulting from the weathering of phosphatic limestone. Cf: *white-bedded phosphate.*

Brown soil In early U.S. classification systems, a group of zonal soils having a brown surface and a light-colored subsurface over an accumulation of calcium carbonate. It is developed under conditions of temperate to cool aridity.

brown spar Any light-colored crystalline carbonate mineral that is colored brown by the presence of iron; e.g. ankerite, dolomite, magnesite, or siderite.

brownstone A brown or reddish-brown sandstone whose grains are generally coated with iron oxide; specif. a dark reddish-brown, ferruginous quartz sandstone of Triassic age, extensively quarried in the Connecticut River valley for use as building stone.

brucite A hexagonal mineral: $Mg(OH)_2$. It commonly occurs in thin pearly folia and in fibrous form, as in serpentines and impure limestones.

brugnatellite A flesh-red mineral: $Mg_6Fe(OH)_{13}(CO_3).4H_2O$.

bruik A Scottish var. of *brook.*

brunckite A colloidal variety of sphalerite.

Brunizem *Prairie soil.*

Brunton compass A small, compact pocket instrument that consists of an ordinary compass, folding open sights, a mirror, and a rectangular spirit-level clinometer and that can be used in the hand or upon a staff or light rod for reading horizontal and vertical angles, for leveling, and for reading the magnetic bearing of a line. It is used in sketching mine workings and in preliminary topographic and geologic surveys on the surface (such as in determining stratigraphic thickness and vertical elevations). The construction of the case enables one to measure with ease the strike and dip of rock formations. Named after its inventor, David W. Brunton (1849-1927), U.S. mining engineer. Usually called a *Brunton*. Syn: *Brunton pocket transit; pocket transit.*

brush (a) Numerous fine cytoplasmic strands radiating from the distal end of the caecum and connected to the periostracum of the punctate shells of articulate brachiopods (TIP, 1965, pt.H, p.141). (b) A bunch of fine terminal branches in phaeodarian radiolarians.

brush cast The cast of a brush mark, characterized by a crescentic depression around the downcurrent end.

brush hook A short, stout, heavy hooked blade with a sharpened iron edge, attached to an axe handle, and used by surveyors for cutting brush.

brushite A nearly colorless mineral: $CaHPO_4.2H_2O$.

brush mark A *bounce mark* whose downcurrent end has a small crescentic ridge of mud pushed up by and in front of the impinging object.

Bruxellian European stage: lower middle Eocene (above Ypresian, below Auversian). It includes Cuisian and Lutetian.

bryalgal Said of a rigid, wave-resistant limestone composed largely of materials constructed in place by frame-building bryozoans and algae that often encrust one another. The material so formed is intimately associated with reefs. Term proposed by Bissell (1964, p.586). Cf: *coralgal.*

bryochore A climatic term for the part of the Earth's surface represented by tundras.

bryophyte A nonvascular plant that may have differentiated stems and leaves, but that has no true roots. Liverworts and mosses are bryophytes. Cf: *thallophyte.*

bryozoan Any invertebrate belonging to the phylum Bryozoa and characterized chiefly by colonial growth, a calcareous skeleton, or, less commonly, a chitinous membrane, and a U-shaped alimentary canal, with mouth and anus. Their stratigraphic range is Ordovician to present, with a possible downward extension into the Upper Cambrian. Syn: *sea mat; moss animal; moss coral; moss polyp; polyzoan.*

B-tectonite A tectonite whose fabric is dominated by linear elements; a tectonite whose petrofabric diagram shows a b axis. Cf: *L-tectonite; S-tectonite; R-tectonite.*

bubble (a) A small air-filled cavity, or the globule of air or gas, in the glass tube of a spirit level. When the level is adjusted to the horizontal, the center of the bubble comes to rest under a fixed mark or etched line at the highest point possible in the tube. (b) A term sometimes applied to a bubble tube and its contents.

bubble impression A small, shallow depression (2.5 cm in diameter) formed on a beach or sedimentary surface by a bubble of gas after it has been dislodged into the air or water above. It has a smooth surface and is not margined by a raised rim, and it may pass downward into a tube. Syn: *bubble mark.*

bubble mark *bubble impression.*

bubble noise Seismic pulses generated by the collapse of gaseous cavitation in water in which an explosive charge has been detonated. One explosion can give rise to a sequence of bubbles; as each collapses, additional bubble noise is generated.

bubble pulse A pulsation attributable to the bubble produced by a seismic charge fired in deep water. The bubble pulsates several times with a period proportional to the cube root of the charge, each oscillation produces an identical unwanted seismic effect.

bubble trend A planar or linear distribution of bubbles in glacier ice.

bubble tube The circular or slightly curved glass tube containing the liquid and bubble in a spirit level and mounted with the bend convex upward.

Bubnoff unit A standard measure of geologic time-distance rates (such as for geologic movements and increments), proposed by Fischer (1969) and defined as 1 micron/year (1 mm/thousand years, or 1 m/million years). Named in honor of Serge von Bubnoff (1888-1957), Russian-born German geologist. Abbrev: *B.*

bucaramangite A pale-yellow variety of retinite that is insoluble in alcohol, found at Bucaramanga, Colombia.

buccal cavity A hollow space on the ventral side of the body of a crustacean, containing the mouthparts. In Malacostraca, it is bounded by the epistome in front and the free edges of the carapace on the sides.

buccal frame A structure enclosing the mouthparts of brachyuran decapod crustaceans, bounded laterally by free (anterior and lateral) edges of the carapace and in front by the epistome, and commonly closed maxillipeds.

buccal plate (a) One of ten small primordial plates of an echinoid, located on the tissue between peristomial margin and mouth, and containing pores for passage of tube feet. (b) *buccal shield.*

buccal shield A large, more or less triangular ossicle in the interradial position, adjoining the mouth in an ophiuroid (TIP, 1966, pt.U, p.29). Syn: *buccal plate.*

Buchan-type facies series A type of dynamothermal regional metamorphism rather similar to the *Abukuma-type facies series* but in a somewhat higher pressure environment of 3000 to 4000 bars (Hietanen, 1967, p.192).

buchite A vitrified hornfels produced by fusion of an argillaceous rock by intense, local thermal metamorphism. Cf: *hyalomylonite.*

buchonite A dark-colored extrusive rock containing hornblende and biotite in addition to plagioclase, nepheline, and augite; an orthoclase-bearing tephrite.

bucklandite (a) A black variety of epidote containing iron and having nearly symmetric crystals. (b) *allanite.*

buckle folding A syn. of *true folding.* Cf: *bend folding.*

buck quartz *bull quartz.*

buckshot A general term for a clay soil (usually alluvial) which, when dry, breaks into small particles or pellets.

buckwheat coal A series of sizes of anthracite that will pass through a 9/16 inch round mesh; there is no lower limit. The series is numbered one through five. See also: *rice coal; barley coal; mustard-seed coal.* Cf: *broken coal; egg coal; stove coal; chestnut coal; pea coal.*

bud [paleont] An asexual reproductive body (including various types of cells) that is eventually isolated from a parent sponge. See also: *common bud; initial bud.*

bud [bot] On a plant stem, a mass of meristematic tissue in which stem growth or leaf or flower production occurs. It is usually protected by bud scales, and may be apical, axillary, or adventitious in location on the stem.

budding *vegetative reproduction.*

buddingtonite A mineral: $(NH_4)AlSi_3O_8.nH_2O$, with n about 0.5. It is isostructural with orthoclase.

bud scale A protective leaf on a bud.

bud scar A scar left on a twig by the falling away of a bud or a group of bud scales (Fuller & Tippo, 1949, p.953).

buergerite A mineral of the tourmaline group: $NaFe^{+3}_3Al_6Si_6B_3O_{30}F$.

Buerger precession method The recording on film of a single level of the reciprocal lattice of an individual crystal by means of X-ray diffractions, for the purpose of determining unit cell dimensions and space groups. See also: *precession camera.* Syn: *precession method.*

buetschliite A mineral: $K_2Ca(CO_3)_2$. Its formula was previously given as: $K_6Ca_2(CO_3)_5.6H_2O$. Cf: *fairchildite.* Syn: *bütschliite.*

buffalo wallow (a) One of the small, subcircular, undrained, very shallow depressions that were once common on the Great Plains of western U.S., usually containing water after a rain (and often remaining as a stagnant water hole for most of the year), and generally believed to have been deepened or modified, and perhaps initially formed, by the trampling and wallowing of buffalo herds in mud and dust; its diameter varies from about a meter to 15-20 m, and its depth from several centimeters to a few meters. (b) A term improperly applied to one of the large natural depressions widely distributed throughout the Great Plains of western U.S. (esp. on the High Plains), often containing an intermittent pond or temporary lake (Veatch & Humphrys, 1966, p.49).

bug A colloquial syn. of *geophone*.

bug hole A miner's term for a *vug*.

bugor A hill or succession of small hills separating creeks or ravines, as on the shore of the Black Sea. Etymol: Russian *bugori*, "hillock".

buhrstone (a) A hard, compact, cellular, siliceous rock suitable for use as a material for millstones; e.g. an open-textured, porous but tough, fine-grained sandstone, or a silicified fossiliferous limestone. In some sandstones, the cement is calcareous. Syn: *millstone*. (b) A millstone cut from buhrstone.--- Syn: *burr; burrstone; burstone*.

building stone A general, nongeneric term for any massive, dense rock suitable for use in construction. Whether igneous, metamorphic, or sedimentary, a building stone is chosen for its properties of durability, attractiveness, and economy. See also: *dimension stone*.

buildup A nongenetic term used by Merriam (1962, p.73) for "any extra, stray, or super" limestone bed or beds, in addition to the "normal" sequence, as exemplified in the rhythmic (cyclic) deposits of the northern midcontinent region of U.S.; e.g. a marine bank, a bioherm, and an organic reef.

built platform *wave-built platform*.

built terrace (a) *wave-built terrace*. (b) *alluvial terrace*.

bukovskyite A mineral: $Fe(AsO_4)(SO_4)(OH).7H_2O$.

bulb glacier *expanded-foot glacier*.

bulge (a) *valley bulge*. (b) A tumescence of lava. (c) A landmass projecting beyond the general outline of the body of which it is a part; e.g. the "bulge" of Brazil. (d) A diapiric structure with a clay core beneath more competent overlying strata. Cf: *camber*. Syn: *valley bulge*.

Bulitian North American stage: Paleocene (above Ynezian, below Penutian).

bulk density The weight of an object or material divided by the volume it occupies (including the volume of its pore spaces); specif. the weight per unit volume of a soil mass that has been oven-dried to a constant weight at 105°C. Syn: *apparent density*.

bulkhead A stone, mud, wood, or concrete wall-like structure primarily designed to resist earth or water pressure, as a retaining wall holding back the ground from sliding into a channel or to protect a shore from wave erosion, or a partition preventing water from entering a working area in a mine.

bulk modulus A *modulus of elasticity* which relates a change in volume to the hydrostatic state of stress. It is the reciprocal of *compressibility*. Symbol: k. Syn: *volume elasticity; incompressibility modulus; modulus of incompressibility*.

bulla (a) One of the blister-like structures that partly or completely cover the apertures in planktonic foraminifers, that are not closely related to primary chambers, and that may be umbilical, sutural, or areal in position and may have one or more marginal accessory apertures. (b) A radially elongated tubercle of an ammonoid.---Pl: *bullae*. Adj: *bullate*.

Bullard discontinuity The seismic-velocity interface between the outer core and the inner core.

Bullard's method Computation of the effect of topography for the *Hayford zones*, by first calculating the effect of the spherical cap of height equal to the station height, then of the topographic deviations of this cap. (Schieferdecker, 1959, term 3489).

bulldust An Australian term for coarse dust or silt.

buller A *blowhole* in coastal rocks.

bullet crystal A snow crystal in the shape of a short hollow or solid hexagonal prism with one pointed end, characteristically formed at very cold temperatures.

bullette A siphonal deposit of a nautiloid, similar to annulus but flatter and more elongated in cross section in which it appears knob- or boss-like (TIP, 1964, pt.K, p.54).

bullion A smooth, ovoid, highly fossiliferous calcareous concretion or ironstone nodule (sometimes a quartzite boulder or a coal ball) in the roof of a coal seam, varying from several centimeters to a meter or more in diameter, and breaking cleanly from the enclosing rock; e.g. a *baum pot*. It may contain well preserved plant structures.

bull mica Large clusters of diversely oriented and partially intergrown crystals of muscovite with a little interstitial albite and quartz (Skow, 1962, p.169).

bull pup A miner's term for a worthless claim.

bull quartz A miner's or prospector's term for white, glassy, massive quartz, essentially free of accessory minerals and valueless as ore. Syn: *bastard quartz; buck quartz*.

bull's-eye level *circular level*.

bultfonteinite A mineral: $Ca_2SiO_2(OH,F)_4$. It is found at Bultfontein, South Africa.

bummock A downward projection from the underside of sea ice; the submariner's counterpart of a *hummock* [ice].

Bumstead head A lightweight tripod, adapted to use on foot traverses. It carries a 15-inch-square plane-table board, which can be revolved about its axis, but provides no means of leveling. It is used with a peep-sight alidade. Named after Albert H. Bumstead (1875-1940), U.S. cartographer.

bund Any artificial embankment used to control the flow of water on a river or on irrigated land. The term is applied extensively in India to large dams and dikes and also to the small ridges between rice fields. Also, an embanked causeway or thoroughfare along a river or the sea.

bundle scar In a *leaf scar*, a scar indicating the breaking of a vascular bundle that had connected the stem and the stalk.

bunker A Scottish term for a small sand hole or pit.

bunsenite A pistachio-green mineral: NiO.

Bunter European stage (esp. in Germany): Lower Triassic (above Permian, below Muschelkalk).

burbankite A pale-yellow hexagonal mineral: $(Na,Ca,Sr,Ba,Ce)_6(CO_3)_5$.

Burdigalian European stage: Miocene (above Aquitanian, below Helvetian).

Burgers circuit A loop which encloses a dislocation line in a crystal and that fails to close by an amount known as the *Burgers vector*.

Burgers dislocation In crystal structure, a screw dislocation. Cf: *Burgers circuit; Burgers vector*.

Burgers vector The vector required to complete a *Burgers circuit*. It is the same as the vector of, and independent of the position of, the crystal dislocation.

burial Covering up or concealing geologic features by the process of sedimentation.

burial metamorphism A type of regional metamorphism affecting sediments and interlayered volcanic rocks in a geosyncline without any influence of orogenesis or magmatic intrusions. Temperatures are 200°C to a maximum of 450°C. Original rock fabrics are largely preserved but mineralogical compositions are generally changed (Coombs, 1961). Cf: *dynamothermal metamorphism*.

buried channel An old channel concealed by surficial deposits; esp. a preglacial channel filled with glacial drift.

buried erosion surface An *erosion surface*, such as a peneplain, that has been covered by younger sediments; it may represent a surface of unconformity at depth. Cf: *fossil erosion surface*.

buried ice Any relatively distinct ice mass buried in the ground, esp. surface ice, as that of sea, lake, river, or glacier origin, that has been buried syngenetically by sediments, esp. in a permafrost region.

buried karst *subjacent karst.*

buried placer A placer that has been buried by a lava flow, soil, or other material.

buried river A river bed that has been concealed beneath alluvium, lava, pyroclastic rocks, or till.

buried soil A *paleosol.*

buried valley A depression in an ancient land surface, now covered by younger deposits; esp. a preglacial valley filled with glacial drift.

burkeite A white, buff, or grayish mineral: $Na_6(CO_3)(SO_4)_2$.

burl An oolith or nodule in fireclay. It may have a high content of alumina or iron oxide.

burley clay A clay containing burls; specif. a diaspore-bearing clay in Missouri, usually averaging 45-65% alumina. See also: *diaspore clay.*

burmite A dark-brown, pale-yellow, or reddish variety of retinite that resembles amber but is tougher and harder and that is found in Burma. It has also been regarded as a variety of amber low in succinic acid.

burn A term used in Scotland and northern England for a small stream or brook, such as a *bourne.*

burned Said of *slate* [coal mining] that adheres tightly to and is difficult to remove from the coal with which it is associated.

burnie A Scottish term for a little stream or brook.

burnside *brookside.*

burnt stone A gemstone whose color has been altered by heating; e.g. "burnt amethyst" whose natural brownish color has been changed to yellow.

Burozem A Russian term for a brown steppe soil.

burr (a) A term used in England for a rough or hard stone, such as a compact siliceous sandstone especially hard to drill. (b) A knob, boss, nodule, or other hard mass of siliceous rock in a softer rock; a hard lump of ore in a softer vein. (c) *buhrstone.* (d) *whetstone.*---Syn: *bur.*

burr ball *lake ball.*

burrow [paleont] A tubular or cylindrical hole or opening made by a mud-eating worm, a mollusk, or other invertebrate extending along a bedding plane or penetrating a rock, and often filled with clay or sand or preserved as a filling; it may be straight or sinuous, and vertical, horizontal, or inclined. Syn: *boring.*

burrow porosity Porosity resulting from burrowing organisms. Although this type of porosity is uncommon in ancient carbonate rocks due to collapse of burrows, other types of porosity such as interparticle porosity can develop within the burrow filling material, esp. if its permeability is greater than that of the host sediment (Choquette & Pray, 1970, p.244). Syn: *boring porosity.*

burr rock An aggregate of muscovite books and quartz (Skow, 1962).

burrstone Var. of *buhrstone.*

bursa An internal gill pouch in ophiuroids, entered by the gill slit. Pl: *bursae.*

bursaite A mineral: $Pb_5Bi_4S_{11}$.

burst [rock mech] An explosive breaking of brittle rock or coal material; e.g. a *rock burst* in a deep mine where mining has deprived the rock of support on one side. In coal mines, a burst may or may not be accompanied by a copious discharge of methane, carbon dioxide, or coal dust. See also: *outburst* [mining].

burst [rock mech] An explosive breaking of brittle rock or coal material; e.g. a *rock burst* in a deep mine where mining has deprived the rock of support on one side. In coal mines, a burst may or may not be accompanied by a copious discharge of methane, carbon dioxide, or coal dust. See also: *outburst* [mining]. Cf: *bump* [mining].

burstone Var. of *buhrstone.*

bushveld A large, flat grassy (savanna-like) area with scat-

tered trees, sometimes becoming very dense, found in tropical or subtropical areas, esp. Africa. Etymol: Afrikaans, *bosveld.*

bustamite A grayish-red mineral: $CaMnSi_2O_6$.

bustite *aubrite.*

butane A gaseous, inflammable, paraffin hydrocarbon, formula C_4H_{10}, occurring in either of two isomeric forms: n-butane, $CH_3CH_2CH_2CH_3$, or isobutane, $CH_3CH(CH_3)_2$. The butanes occur in petroleum and natural gas.

butlerite A monoclinic mineral: $FeSO_4(OH).2H_2O$. Cf: *parabutlerite.*

bütschliite *buetschliite.*

butt cleat A less-preferred syn. of *end cleat.*

butte (a) A conspicuous, usually isolated, generally flat-topped hill or small mountain with relatively steep slopes or precipitous cliffs, often capped with a resistant layer of rock and bordered by talus, and representing an erosion remnant carved from flat-lying rocks; the summit is smaller in extent than that of a *mesa*, and many buttes in the arid and semiarid regions of the western U.S. result from the wastage of mesas. Syn: *mesa-butte.* (b) An isolated hill having steep sides and a craggy, rounded, pointed, or otherwise irregular summit; e.g. a volcanic cone (as Mount Shasta, Calif., formerly known as Shasta Butte), or a *volcanic butte.*---Etymol: French, "knoll, hillock, inconspicuous rounded hill; rising ground". Pron: *bewt.*

butterball A clear-yellow, rounded segregation of very pure carnotite found in the soft sandstone of Temple Rock, San Rafael Swell, Utah.

butter rock *halotrichite.*

butte témoin [geomorph] A flat-topped hill representing the former extension of an escarpment edge or plateau, now detached by stream erosion, its surface in broadly the same plane as the main mass; an outlier. In most cases, a *butte* is a "butte témoin". Etymol: French, "witness hill". Syn: *witness butte; zeugenberg.*

buttgenbachite A sky-blue mineral: $Cu_{19}Cl_4(NO_3)_2(OH)_{32}.2H_2O$. It is isomorphous with connellite.

buttress [geomorph] A protruding rock mass on, or a projecting part of, a mountain or hill, resembling the buttress of a building; a spur running down from a steep slope. Example: a prominent salient produced in the wall of a gorge by differential weathering.

buttress [paleont] (a) An internal ridge-like projection from the shell wall of a bivalve mollusk which supports the hinge plate or chondrophore. (b) A ridge of skeletal material extending adapically from an echinoid auricle on the inner surface of the test (TIP, 1966, pt.U, p.253). (c) An aboral tongue-shaped extension of the apical denticle of a conodont, generally on the inner side and outer side.

buttress sand A sandstone body that intersects an underlying surface of unconformity, such as a sandstone flanking a buried hill or a truncated anticline. It often forms a stratigraphic trap or contains oil.

butyrellite *bog butter.*

butyrite *bog butter.*

Buys Ballot's law The statement in meteorology that describes the relationship between horizontal wind direction and barometric pressure: if an observer stands in the Northern Hemisphere with his back to the wind, the pressure to his left is lower than it is to his right; the reverse relationship holds true for an observer in the Southern Hemisphere. The law is named after the Dutch meteorologist who formulated it in 1857.

by-channel A stream or branch along one side of the main stream.

byerite A bituminous coal that resembles *albertite.*

bypassing A term applied by Eaton (1929) to sedimentary transport across areas of nondeposition, as in the case where one particle size passes another simultaneously transported, or continues in motion after the other has come to rest; e.g.

the normal decrease in average particle size of sediments away from a source area. The term is also applied to transport of coarser particles farther than finer particles ("reverse" bypassing); e.g. gravel bypassing sand along the edge of the continental shelf, probably the result of density-current deposition. Cf: *total passing*. Also spelled: *by-passing*.

bysmalith A roughly vertical, cylindrical igneous intrusion along a steep fault. It has been interpreted as a type of *laccolith*. Cf: *bell-jar intrusion*.

by-spine A small accessory spine additional to a radial spine in acantharian and spumellarian radiolarians.

byssal gape An opening between the margins of a bivalve-mollusk shell for the passage of the byssus.

byssal notch The indentation, below the anterior auricle of the right valve in many pectinacean bivalve mollusks, serving for the passage of the byssus or for the protrusion of the foot.

byssiferous Possessing a byssus.

byssolite An olive-green, fibrous variety of amphibole. The term is used in the gem trade for a variety of quartz containing greenish, fibrous inclusions of actinolite or asbestos.

byssus A tuft or bundle of long, tough, thread- or hair-like strands or filaments secreted by a gland in a groove of the foot of certain bivalve mollusks and issued from between the valves and by which a temporary attachment of the bivalve can be made to rocks or other extraneous objects. Pl: *byssi* or *byssuses*.

by-stream Said of the part of a flood plain consisting of a narrow belt of levee deposits immediately adjacent to the stream channel, and composed usually of sandy alluvium.

byströmite (a) A pale blue-gray tetragonal mineral: $MgSb_2O_6$. (b) A monoclinic polymorph of pyrrhotite.----Also spelled: *bystromite*.

by-terrace Said of the part of a flood plain consisting of a narrow belt of clayey deposits adjacent to a bounding terrace relatively distant from the stream channel.

bytownite A bluish to dark-gray mineral of the plagioclase feldspar group with composition ranging from $Ab_{30}An_{70}$ to $Ab_{10}An_{90}$. It occurs in basic and ultrabasic igneous rocks.

bytownitfels *bytownitite*.

bytownitite An igneous rock composed almost entirely of bytownite. Syn: *bytownitfels*.

by-wash A channel, spillway, or weir designed to carry surplus water from a dam, reservoir, or aqueduct and thereby to prevent overflow.

by-water A yellow-tinted diamond.

C

caballing The mixing of two water masses to produce a blend that sinks because it is denser than its original components. This occurs when the two water masses have the same density but different temperatures and salinities.

cabbage-leaf mark *frondescent cast.*

cable [drill] (a) A heavy, flexible rope composed of many steel wires or hemp fibers in groups and twisted together to form the strands, used in cable-tool drilling as the line to which cable tools are attached and as the connecting link between the drill stem and the walking beam. Syn: *drilling cable.* (b) A term used loosely to signify a *wire line.*

cable break A seismic break on a well geophone trace caused by energy being transmitted down the electric logging cable to the well geophone.

cable drilling *cable-tool drilling.*

cable tool One of a set of bottom-hole tools used in cable-tool drilling and suspended on a steel cable; e.g. a steel drill bit fastened below a drill stem, a set of jars (loosely hooked links fastened to the stem above the bit and used to loosen tools stuck in the hole by imparting a sharp jerk to the bit), or a bailer.

cable-tool drilling A method of drilling, now largely replaced by *rotary drilling*, based on a percussion principle in which the rock material at the bottom of the hole is pulverized or broken up by means of a solid-steel cylindrical bit (with a blunt, chisel-shaped cutting edge) attached to, and working vertically at, the end of a heavy string of steel tools suspended in the hole at the end of a steel cable that is activated by a walking beam, the bit chipping the rock by a series of repeated blows. It is adapted to construction of deep water wells and drilling of deep holes having diameters of 7.5-75 cm (3-30 in.). Syn: *cable drilling; cable system; rope drilling; percussion drilling; churn drilling; American system.*

cabochon (a) An unfaceted cut gemstone of domed, or convex, form. The top is unfaceted and smoothly polished; the back, or base, is usually flat or slightly convex and often is unpolished. The girdle outline may be round, oval, square, or any other shape. (b) The style of cutting such a gem. (c) A polished but uncut gem.---Etymol: French. Pron: kabb-o-*shon.* See also: *en cabochon.*

cabocle A compact rolled pebble resembling red jasper, supposed to be a hydrated phosphate of calcium and aluminum, found in the diamond-producing sands of Bahia, Brazil.

cacholong An opaque or feebly translucent and bluish-white, pale-yellowish, or reddish variety of common opal containing a little alumina. Syn: *cachalong; pearl opal.*

cacoxenite A yellow or brownish mineral: $Fe_4(PO_4)_3(OH)_3.12H_2O$. Syn: *cacoxene.*

cactolith An irregular, intrusive igneous body of obscurely cactus-like form, more or less confined to a horizontal zone and appearing to consist of irregularly related and possibly distorted branching and anastomosing dikes that fed a laccolith. Term introduced by Hunt et al. (1953, p. 151): "a quasi-horizontal chonolith composed of anastomosing ductoliths whose distal ends curl like a harpolith, thin like a sphenolith, or bulge discordantly like an akmolith or ethmolith".

cadacryst Var. of *chadacryst.*

cadastral Delineating or recording property boundaries, and sometimes subdivision lines, buildings, and other details. Etymol: French *cadastre*, an official register of the real property of a political subdivision with details of area, ownership, and value, and used in apportioning taxes.

cadastral map A large-scale map showing the boundaries of subdivisions of land, usually with the directions and lengths thereof and the areas of individual tracts, compiled for the purpose of describing and recording ownership and value. It

may also show culture, drainage, and other features relating to the value and use of the land.

cadastral survey A survey relating to land boundaries and subdivisions, made to delimit units suitable for transfer or to define the limitations of title; esp. a survey of the public lands of the U.S., such as one made to identify or restore property lines. Cf: *land survey; boundary survey.*

cadder Shortened form of *ballycadder*, a syn. of *icefoot.*

cadicone A strongly depressed, barrel-shaped cephalopod shell with more or less evolute coiling, a wide venter, and a deep, crater-like umbilicus (as in the ammonoid *Cadoceras*).

cadmia (a) *calamine.* (b) A chemical compound: CdO. (c) An impure zinc oxide that forms on the walls of furnaces in the smelting of ores containing zinc.

cadmium blende *greenockite.*

cadmium ocher *greenockite.*

cadmoselite A black hexagonal mineral: CdSe.

caducous Said of a plant whose floral parts fall away prematurely.

cadwaladerite A mineral: $Al(OH)_2Cl.4H_2O$.

caecum (a) The sac-shaped apical end of the siphuncle of a nautiloid or ammonoid. Also, a cavity associated with the digestive system in the living *Nautilus.* (b) The evagination of the outer epithelium projecting into the endopuncta of a brachiopod shell.---Pl: *caeca.* Adj: *caecal.*

Caen stone A yellowish or light cream-colored Jurassic limestone, marked with a rippled figure, and largely used for building purposes. Type locality: near Caen, city in Normandy, France.

Caerfaian European stage: Lower Cambrian (above Precambrian, below Solvan).

cafarsite A mineral: $Ca_6(Fe,Ti)_6Mn_2(AsO_4)_{12}.4H_2O$.

cafemic Said of an igneous rock or magma that contains calcium, iron, and magnesium. Etymol: a mnemonic term derived from *calcium* + *ferric* (or *ferrous*) + *magnesium* + *ic.*

cafetite An orthorhombic mineral: $Ca(Fe,Al)_2Ti_4O_{12}.4H_2O$.

cage A void in a crystal structure that is large enough to trap one or more atoms (such as argon or xenon) foreign to the structure; it is found in tektosilicates, beryl, and organic compounds.

cahemolith *humic coal.*

cahnite A tetragonal mineral occurring in white sphenoidal crystals: $Ca_2B(AsO_4)(OH)_4$.

Cainophyticum Plant life of the Cenozoic. Cf: *Palaeophyticum; Mesophyticum.*

Cainozoic Var. of *Cenozoic.*

cairn An artificial mound of rocks, stones, or masonry, usually conical or pyramidal, used in surveying to aid in the identification of a point or boundary.

cairngorm A type of *smoky quartz* from Cairngorm, a mountain southwest of Banff in Scotland. Syn: *cairngorm stone; Scotch topaz.*

cajon (a) *box canyon.* (b) A defile leading up to a mountain pass; also, the pass itself.---Etymol; Spanish *cajón*, "large box". *The term is used in the SW U.S.*

cake [drill] *mud cake.*

cake [ice] *ice cake.*

cake ice An accumulation of *ice cakes.*

caking coal Coal that softens and agglomerates when heated, and produces a hard, gray, cellular coke. Not all caking coals are necessarily good coking coals. Syn: *binding coal.*

caking index *agglutinating value.*

cal A term used in Cornwall, Eng., for iron tungstate (wolframite).

cala [coast] A short and narrow ria formed in a limestone coast; a small, semicircular, shallow bay along a *cala coast*, as along the coast of Majorca. Etymol: Spanish, "cove, small bay, inlet". See also: *caleta.*

cala [streams] A term applied in SW U.S. to a creek corresponding to a lateral stream of a main drainage. Etymol:

Spanish, "creek". See also: *caleta*.

Calabrian European stage: lower Pleistocene (above Astian of Pliocene, below middle Pleistocene). It is the marine equivalent (in France and Italy) of the terrestrial *Villafranchian*.

cala coast A coast formed by the submergence of many small valleys having steep slopes so that *calas*, separated by narrow peninsulas, are formed under the influence of breakers; examples occur along several coasts of the Mediterranean Sea. Syn: *calas coast*.

calaite *turquoise*.

calamine (a) A name used in U.S. for *hemimorphite*. This name is disapproved by the Commission on New Minerals and Mineral Names of the International Mineralogical Association (Fleischer, 1966, p.1263). (b) A name frequently used in Great Britain for *smithsonite*. (c) *hydrozincite*. (d) A commercial, mining, and metallurgical term for the oxidized ores of zinc (including silicates and carbonates), as distinguished from the sulfide ores of zinc---Syn: *cadmia*.

calamistrum A spinose comb or row of special bristles on the fourth (hind) metatarsi of spiders possessing a cribellum, used for drawing out a band of special silk from the cribellum. Pl: *calamistra*.

calanque A French term for a cove or inlet, esp. a dry valley excavated in limestone during a wet period and later submerged by the rise of sea level. Examples occur along the Mediterranean coast of France.

calaverite A pale bronze-yellow or tin-white monoclinic mineral: $AuTe_2$. It often contains silver, and is an important source of gold.

calc-alkalic (a) Said of an igneous rock, or group of igneous rocks, in which the weight percentage of silica is between 56 and 61 when the weight percentages of CaO and K_2O + Na_2O are equal. See also:*alkali-lime index*. (b) Said of an igneous rock containing plagioclase feldspar. Syn: *alkali-lime*.

calcarenaceous Said of a sandstone containing an abundant amount of calcium-carbonate detritus; e.g. "calcarenaceous orthoquartzite" in which the calcareous components constitute up to 50% of the total clastic particles (Pettijohn, 1957, p.404-405).

calcarenite A limestone consisting predominantly (more than 50%) of detrital calcite particles of sand size; a consolidated calcareous sand. The term was introduced by Grabau (1903). Cf: *calcareous sandstone*.

calcarenitic limestone A term used by Powers (1962, p.145) for a limestone composed dominantly (more than 10%) of original calcareous-mud matrix (particles with diameters less than 0.06 mm) accompanied by more than 10% coarse clastic carbonate grains (sand and gravel).

calcareous Said of a substance that contains calcium carbonate. When applied to a rock name it implies that a considerable percentage (up to 50%) of the rock is calcium carbonate (Stokes & Varnes, 1955).

calcareous alga A seaweed that builds a more or less solid structure of calcium carbonate. See also: *coralline alga*.

calcareous clay A clay containing a significant amount of calcium carbonate; specif. a *marl*.

calcareous crust An indurated soil horizon cemented with calcium carbonate; *caliche*. Syn: *calc-crust*.

calcareous dolomite A term used by Leighton & Pendexter (1962, p.54) for a carbonate rock containing 50-90% dolomite. Cf: *calcitic dolomite*.

calcareous ooze Any *ooze* whose skeletal remains are calcareous, e.g. pteropod ooze. Cf: *siliceous ooze*.

calcareous peat *eutrophic peat*.

calcareous rock A sedimentary rock containing an appreciable amount of calcium carbonate, such as limestone, chalk, tufa, or shelly sandstone. See also: *carbonate rock*.

calcareous sandstone (a) A sandstone cemented with calcite. (b) A sandstone containing appreciable calcium carbonate, but in which clastic quartz is present in excess of 50% (Petti-

john, 1957, p.381).---Cf: *calcarenite*.

calcareous shale A shale containing at least 20% calcium carbonate in the form of finely precipitated materials or small organically-fixed particles (Pettijohn, 1957, p.368-369).

calcareous sinter *travertine*.

calcareous soil A soil whose content of carbonate is sufficient to cause effervescence when tested with hydrochloric acid.

calcareous spar *calcspar*.

calcareous tufa *tufa* [sed].

calcarinate An adjective restricted by Allen (1936, p.21) to designate the calcium-carbonate cement of a sedimentary rock.

calc-crust *calcareous crust*.

calcdolomite A sedimentary rock consisting of both calcite and dolomite minerals; a *calcitic dolomite*.

calcedony *chalcedony*.

calceoloid Said of a solitary corallite shaped like the tip of a pointed slipper (as in *Calceola*), with angulated edges between flattened and rounded sides.

calc-flinta A fine-grained, calc-silicate rock of flinty appearance formed by thermal metamorphism of a calcareous mudstone with possibly some accompanying pneumatolytic action.

calcian dolomite A dolomite mineral that contains at least 8% calcium in excess of the ideal composition (Ca:Mg = 1:1 molar) of dolomite.

calciborite A white mineral: CaB_2O_4.

calcibreccia A limestone breccia, or a consolidated calcareous rubble; a *calcirudite* whose constituent particles are angular (Carozzi & Textoris, 1967, p.3).

calcic [geochem] Said of minerals and igneous rocks containing a relatively high proportion of calcium, but what proportion is required to warrent the use of the term depends on circumstances.

calcic [geol] Said of an igneous rock, or series of igneous rocks, in which the weight percentage of silica is greater than 61 when the weight percentages of CaO and of K_2O + Na_2O are equal. See also:*alkali-lime index*.

calciclase *anorthite*.

calciclasite *anorthitite*.

calciclastic Pertaining to a clastic carbonate rock (Braunstein, 1961).

calcicole A plant requiring a lime-rich, i.e. alkaline, soil. Cf: *calcifuge; calciphobe*. Syn: *calciphile* [organic].

calcicrete *calcrete*.

calciferous In a stratigraphic sense, pertaining to a series of strata containing limestone, e.g. the Calciferous Sandstone series of Scotland (Challinor, 1967).

calcification [paleont] (a) Deposition of calcium salts in living tissue. (b) Replacement of organic material by calcium salts (esp. $CaCO_3$) in fossilization.

calcification [soil] A general term used for those processes of soil formation in which the surface soil is kept sufficiently supplied with calcium to saturate the soil colloids to a high degree with exchangeable calcium and thus render them relatively immobile and nearly neutral in reaction. The process is best expressed in Chernozem and other soils having a horizon of carbonate accumulation.

calcifuge A plant surviving, but not thriving, on a lime-rich soil; it grows better on an acid soil. Cf: *calcicole; calciphobe*.

calcigranite A granite in which the plagioclase is labradorite or bytownite.

calcigravel The unconsolidated equivalent of *calcirudite*.

calcikersantite A kersantite that contains labradorite or bytownite.

calcilith (a) A term suggested by Grabau (1924, p.298) for a limestone. Syn: *calcilyte*. (b) A sedimentary rock composed principally of the calcareous remains of organisms (Pettijohn, 1957, p.429).

calcilutite A limestone consisting predominantly (more than 50%) of detrital calcite particles of silt and/or clay size; a consolidated calcareous mud. Some authors broaden the term to include calcareous rocks containing chemically precipitated crystalline components (of inorganic or organic origin). The term was introduced by Grabau (1903). See also: *micritic limestone*. Cf: *calcisiltite*. Ṣyn: *calcipelite*.

calcimicrite A term used by Schmidt (1965, p.127-128) for a limestone whose particles have diameters less than 20 microns and whose micrite component exceeds its allochem component. See also: *micritic limestone*.

calcimixtite A term proposed by Schermerhorn (1966, p.835) for a *mixtite* that is dominantly calcareous.

calcimorphic Said of an intrazonal soil whose characteristics reflect the influence of the calcification process on its development. Examples of calcimorphic soils are Brown Forest soils and Rendzina soils.

calciocarnotite tyuyamunite.

calciocopiapite A mineral of the copiapite group: Ca-Fe$_4$(SO$_4$)$_6$(OH)$_2$.19H$_2$O.

calcioferrite A yellow or green mineral: Ca$_2$Fe$_2$(PO$_4$)$_3$-(OH).7H$_2$O.

calcio-olivine An orthorhombic phase of calcium orthosilicate: γ-Ca$_2$SiO$_4$. It is stable below 780° to 830°C, and is isomorphous with olivine. The term has also been applied to a highly calciferous variety of olivine, and to any of the polymorphs of Ca$_2$SiO$_4$. Cf: *larnite; bredigite*. Syn: *lime olivine*.

calciovolborthite A green, yellow, or gray mineral: CaCu-(VO$_4$)(OH). Syn: *tangeite*.

calcipelite calcilutite.

calciphile calcicole.

calciphobe A plant requiring an acid soil; it cannot survive in a lime-rich soil. Cf: *calcicole; calcifuge*.

calciphyre A marble containing conspicuous crystals of calcium and/or magnesium silicates.

calciphyte A plant that requires large quantities of carbonate.

calcirhyolite A rhyolite that contains labradorite or bytownite as its accessory plagioclase.

calcirudite A limestone consisting predominantly (more than 50%) of detrital calcite particles larger than sand size (larger than 2 mm in diameter), and often also cemented with calcareous material; a consolidated calcareous gravel or rubble, or a limestone conglomerate or breccia. Some authors (e.g. Folk, 1968, p.162) use one millimeter as the lower limit. The term was introduced by Grabau (1903). Cf: *calcigravel; calcibreccia*.

calcisiltite A limestone consisting predominantly of detrital calcite particles of silt size; a consolidated calcareous silt. Cf: *calcilutite*.

calcisol An old term for a zonal soil developed from alluvium and characterized by a neutral or calcareous A horizon, no B horizon, and a calcareous C horizon.

calcisponge Any sponge belonging to the class Calcispongea and characterized mainly by a skeleton composed of spicules of calcium carbonate.

calcisyenite A syenite that contains labradorite or bytownite, rather than andesine or oligoclase, as its accessory plagioclase.

calcite A common rock-forming mineral: CaCO$_3$. It is trimorphous with aragonite and vaterite. Calcite is usually white, colorless, or pale shades of gray, yellow, and blue; it has perfect rhombohedral cleavage, a vitreous luster, and a hardness of 3 on Mohs' scale, and it readily effervesces in cold dilute hydrochloric acid. It is the principal constituent of limestone; calcite also occurs crystalline in marble, loose and earthy in chalk, spongy in tufa, and stalactitic in cave deposits. It is commonly found as a gangue mineral in many ore deposits and as the cementing medium in clastic sedimentary rocks; it is also a minor secondary constituent in many igneous rocks. Calcite crystallizes in a great variety of forms, such as nailhead spar, dogtooth spar, and Iceland spar. Symbol: Cc. Cf: *dolomite* [*mineral*]. Syn: *calcspar*.

calcite bubble In a cave, a formation of calcite that is formed around a gas bubble on the surface of a pond. It is a hollow sphere with a smooth interior and an exterior of jagged crystals.

calcite eye (a) One of the rounded bodies of clear calcite occurring sporadically in the radial zone and central area of foraminifers of the family Orbitolinidae. The term is usually used in the plural. (b) *bird's-eye*.

calcite flottante floe calcite.

calcite ice Deposits of calcium carbonate in a cave. Partial syn: *cave ice*.

calcite skin A partial syn. of *crust stone*.

calcitic dolomite A dolomite rock in which calcite is conspicuous, but the mineral dolomite is more abundant; specif. dolomite rock containing 10-50% calcite and 50-90% dolomite and having an approximate magnesium-carbonate equivalent of 22.7-41.0% (Pettijohn, 1957, p.418), or a dolomite rock whose Ca/Mg ratio ranges from 2.0 to 3.5 (Chilingar, 1957). Cf: *dolomitic limestone; calcareous dolomite*. Syn: *calcdolorite*.

calcitic limestone A limestone that consists essentially of calcite; specif. a limestone whose Ca/Mg ratio exceeds 10 (Chilingar, 1957).

calcitite A term used by Kay (1951) for a rock composed of calcite; e.g. a limestone.

calcitization (a) The act or process of forming calcite, such as by alteration of aragonite. (b) The alteration of existing rocks to limestones, due to the chemical replacement of mineral particles by calcite, such as of dolomite in dolomite rock or of feldspar and quartz particles in sandstones.

calcitostracum An internal layer of various mollusk shells consisting chiefly of calcite. Cf: *nacre*.

calcium carbonate compensation depth In the ocean, the level below which the rate of calcium carbonate solution exceeds the rate of its deposition. In the Pacific Ocean, the level is about 4000-5000 m; in the Atlantic and Pacific ocean it is somewhat deeper. Syn: *compensation depth; depth compensation*.

calcium catapleiite A mineral: CaZrSi$_3$O$_9$.2H$_2$O. See also: *catapleiite*.

calcium feldspar A plagioclase feldspar containing the An molecule (CaAl$_2$Si$_2$O$_8$); specif. *anorthite*. See also: *lime feldspar*.

calcium-larsenite esperite.

calcium mica margarite [mineral].

calcium phosphate A mineral name applied to naturally occurring *apatite*.

calclacite A mineral: CaCl$_2$.Ca(C$_2$H$_3$O$_2$.10H$_2$O. It is calcium chloride acetate found as a fibrous efflorescence on calcareous museum specimens.

calclithite (a) A litharenite in which carbonates constitute the most abundant rock fragment (Folk, 1968, p.124); e.g. the Oakville Sandstone (Miocene) in S Texas, derived largely from erosion of Cretaceous limestones. Not to be confused with a limestone consisting of intraclasts. Syn: *carbonate-arenite*. (b) A term originally defined by Folk (1959, p.36) as a limestone that was derived mainly from erosion of older, lithified limestones and that contains more than 50% carbonate rock fragments (extraclasts); e.g. a terrigenous rock formed as an alluvial fan in an area of intense tectonism or very dry climate.

calcolistolith A limestone olistolith.

calcouranite autunite.

calcrete (a) A term suggested by Lamplugh (1902) for a conglomerate consisting of surficial sand and gravel cemented into a hard mass by calcium carbonate precipitated from solution and redeposited through the agency of infiltrating water, or deposited by the escape of carbon dioxide from vadose water. (b) A calcareous *duricrust*; caliche.---Etymol: calcareous + concrete. Cf: *silcrete; ferricrete*. Syn: *calcicrete*.

calc-sapropel Sapropel containing calcareous algae.

calc-schist A metamorphosed argillaceous limestone with a schistose structure produced by parallelism of platy, recrystallized calcite (Holmes, 1928, p.52).

calc-silicate Said of a metamorphic rock mainly consisting of calcite and calcium-bearing silicates such as diopside and wollastonite, and formed by thermal metamorphism of impure limestone or dolomite. Syn: *lime-silicate*.

calc-silicate marble A *marble* in which calcium silicate and/or magnesium silicate minerals are conspicuous.

calc-silicate rock A metamorphic rock mainly consisting of calcium-bearing silicates such as diopside and wollastonite, and formed by thermal metamorphism of impure limestone or dolomite. Syn: *lime-silicate rock*.

calc-sinter *travertine*.

calcspar Crystalline *calcite*. Also spelled: *calc-spar*. Syn: *calcareous spar*.

calcsparite A sparry calcite crystal, as distinguished from *dolosparite*. The term is synonymous with *sparite* when the latter is understood to mean the calcareous variety.

calcspathization Widely distributed crystallization of sparry calcite (Sander, 1951, p.154).

calc-tufa *tufa* [sed].

calcurmolite A honey-yellow secondary mineral: $Ca(U-O_2)_3(MoO_4)_3(OH)_2 \cdot 11H_2O$.

caldera A large, basin-shaped volcanic depression, more or less circular or cirquelike in form, the diameter of which is many times greater than that of the included vent or vents, no matter what the steepness of the walls or form of the floor (Williams, 1941). See also: *collapse caldera; erosion caldera; explosion caldera*.

caldera lake A term that may be used for a *crater lake* when it occurs in a caldera.

calderite The manganese-iron end member of the garnet group: $Mn_3Fe_2(SiO_4)_3$.

caldron [marine geol] A small, steep-sided, round, pot-shaped depression in the ocean floor. Also spelled: *cauldron* [marine geol].

caldron [coal] *caldron bottom*.

caldron bottom The mud-filled cast of a fossil root or trunk of a tree (esp. *Sigillaria*) or fern extending vertically into the roof of a coal seam, and resembling the bottom of a caldron or pot; it represents a concretionary mass surrounded by a film of coal, and is liable to collapse without warning. Cf: *pot bottom; bell; kettle bottom*. Syn: *caldron; pot*.

Caledonian orogeny A name commonly used for the early Paleozoic deformation in Europe which created an orogenic belt, the *Caledonides*, extending from Ireland and Scotland northwestward through Scandinavia. The classical Caledonian orogeny has been dated as near the end of the Silurian, but Stille and many others use the term for an orogenic era which included pulses in Ordovician and Silurian time, or even later.

Caledonides The orogenic belt extending from Ireland and Scotland northwestward through Scandinavia, formed by the early Paleozoic *Caledonian orogeny*.

caledonite A green mineral: $Cu_2Pb_5(SO_4)_3(CO_3)(OH)_6$. Not to be confused with *celadonite*.

caleta [coast] A small *cala*; a cove or inlet. Etymol: Spanish, cove, small bay".

caleta [streams] The ultimate and smallest headwater ramification of a *cala*; a draw, drain, or coulee. Etymol: Spanish, creek".

calf [glaciol] A small mass of *calved ice*; specif. a piece of ice that has risen to the surface after breaking loose from the submerged part of an iceberg. Syn: *calf ice*.

calice [paleont] The oral (upper or distal), generally bowl-shaped surface of a corallite, on which the basal disk of a polyp rests. Pl: *calices*. See also: *calyx*.

caliche [eco geol] (a) Gravel, rock, soil, or alluvium cemented with soluble salts of sodium in the nitrate deposits of the Atacama Desert of northern Chile and Peru; it contains sodium nitrate (14-25%), potassium nitrate (2-3%), sodium iodate (up to 1%), sodium chloride, sodium sulfate, and sodium borate, mixed with brecciated clayey and sandy material in beds up to 2 m thick. It may form by leaching of bird guano, by bacterial fixation of nitrogen, by leaching from volcanic tuffs, or by drying up of former shallow lakes. (b) A term used in various geographic areas for: thin layer of clayey soil capping a gold vein (Peru); whitish clay in the selvage of veins (Chile); feldspar, or white clay, or a compact transition limestone (Mexico); a mineral vein recently discovered, or a bank composed of clay, sand, and gravel in placer mining (Colombia). The term has also been extended by some authors to quartzite and kaolinite.

caliche [soil] A term applied broadly in SW U.S. (esp. Arizona) to an opaque, reddish-brown to buff or white calcareous material of secondary accumulation (in place), commonly found in layers on, near, or within the surface of stony soils of arid and semiarid regions, but also occurring as a subsoil deposit in subhumid climates. It is composed largely of crusts or succession of crusts of soluble calcium salts in addition to impurities such as gravel, sand, silt, and clay. It may occur as a soft, thin, extremely porous and friable horizon within the soil, but more commonly it is a thick (several centimeters to a meter or more), impermeable, and strongly indurated layer near the surface or exposed by erosion; the cementing material is essentially calcium carbonate, but may contain magnesium carbonate, silica, or gypsum. The term has also been used for the calcium-carbonate cement itself. Caliche appears to form by a variety of processes, such as by capillary action in which soil solutions rise to the surface and upon evaporation, deposit their salt contents on or in the surface materials. It is called *hardpan*, or calcareous *duricrust* (or *calcrete*) in some localities, and *kankar* in parts of India. See also: *soil caliche*. Syn: *calcareous crust; croute calcaire; nari; sabach; tepetate*.---Etymol: American Spanish, from a Spanish word for almost any porous material (such as gravel) cemented by calcium carbonate. "The Spanish word originally was used for a small stone or pebble accidentally burned with the clay mass when brick or tile was made, and it also was used for a crust of lime or similar material flaking from a wall" (Cottingham, 1951, p.162).

calichification The production of caliche.

calico rock (a) A term used in South Africa for *iron formation*. (b) A local term used in eastern Pennsylvania for a quarry rock of the Helderberg Limestone.

California bearing ratio A measure of the relative resistance of a soil to penetration under controlled conditions of density and moisture content. It is the ratio of the force per unit area required to penetrate a subgrade soil to that required for corresponding penetration of a standard material (crushed-rock base material) whose resistance under standardized conditions is well established. Abbrev: *CBR*.

California onyx A dark, amber-colored or brown variety of aragonite used in ornamentation.

californite (a) A compact, massive, translucent to opaque variety of vesuvianite, characterized by a dark-, yellowish-, olive-, or grass-green color usually mottled with white or gray, closely resembling jade, and used as an ornamental stone. Principal sources are in Fresno, Siskiyou, and Tulare counties in California. Syn: *American jade*. (b) A white variety of grossular garnet from Fresno County, Calif.

caliper log A log that shows, to scale, the variations with depth in the mean diameter (or cross-sectional area) of an uncased borehole. It is produced by spring-activated caliper arms that measure the varying widths of the hole as the device is drawn upward. Syn: *section-gage log*.

calk (a) *cawk*. (b) *cauk*---Also spelled: *caulk*.

calkinsite A pale-yellow mineral: $(Ce,La)_2(CO_3)_3 \cdot 4H_2O$.

calkstone *hassock*.

callaghanite A blue mineral: $Cu_2Mg_2(CO_3)(OH)_6.2H_2O$.

callainite A massive, wax-like, translucent, apple- to emerald-green hydrated aluminum phosphate, possibly a mixture of wavellite and turquoise.

callais An ancient name used by Pliny for a precious green or greenish-blue stone, probably turquoise. The name is still sometimes used for turquoise. Pl: callaides. Syn: *callaica; callaina*.

calley stone A term used in Yorkshire, Eng., for a hard argillaceous sandstone associated with coal.

calliard Var. of *galliard*.

callose n. A carbohydrate component of cell walls in certain plants; e.g. the amorphous cell-wall substance that envelops the pollen mother cell during pollen-grain development and acts as a barrier between mother cells but that disappears as the ektexine structure is completed and impregnated with sporopollenin.---adj. Having protuberant hardened spots; e.g. "callose leaves".

Callovian European stage: lowermost Upper Jurassic (above Bathonian, below Oxfordian). The stage has been assigned by some authors to the uppermost Middle Jurassic.

calluna peat Peat that is derived mainly from the common heather *Calluna vulgaris*. Syn: *heath peat*.

callus [paleont] (a) The thickened inductura on the parietal region of a gastropod shell, or a growth of shelly material extending from the inner lip over the base of the shell and perhaps into the umbilicus in a gastropod. (b) Any excessive thickening of secondary shell (fibrous inner layer) located on the valve floor of a brachiopod and covering interior structures.

calm [meteorol] n. A region of the Earth's atmosphere in which there is no wind, or a wind velocity of less than one mile per hour.----adj. Said of such a region.

calm [sed] A Scottish term for a light-colored shale or mudstone, such as a baked shale or clay slate used for making pencils, or a shale that is easily cut with a knife.

calomel A tetragonal mineral: Hg_2Cl_2. It is a white, colorless, grayish, yellowish, or brown tasteless salt, and is used as a cathartic, fungicide, and insecticide; when fused, it has a horny appearance. Syn: *horn quicksilver; horn mercury; calomelite*.

calorie The amount of heat energy required at a pressure of one atmosphere to increase the temperature of one gram of water one Centigrade degree from 14.5 to $15.5°C$. The amount of heat energy represented by a calorie varies according to the temperature of the water, hence the "mean" calorie is one one-thousandth of the amount of heat required to raise the temperature of one gram of water from $0°C$ to $100°C$ at a pressure of one atmosphere (Zemansky, 1957, p.64). See also: *kilocalorie*. Syn: *gram-calorie; small calorie*.

calorific value For solid fuels and liquid fuels of low volatility, the amount of heat produced by combustion of a specified quantity under specified conditions. See also: *net calorific value; gross calorific value*.

calorimeter An instrument used to measure the specific heat capacity of a substance.

calotte "A French term for the ring of surrounding rock in a tunnel which has become weakened or pressure-relieved by the excavation" (Nelson & Nelson, 1967, p.57).

calthrops A tetraxonic sponge spicule in which the rays are equal or nearly equal in length. Pl: *calthrops*. See also: *candelabrum; cricocalthrops*.

caltonite A dark-colored analcime-bearing basanite that contains microphenocrysts of olivine and augite in a trachytic groundmass composed of feldspar laths, augite, iron ore, and the analcime.

calumetite An azure-blue mineral: $Cu(Cl,OH)_2.2H_2O$.

calved ice A fragment or fragments of ice, floating in a body of water after *calving* from a larger ice mass. See also: *calf*.

calving [mass move] The falling or caving in, as from under-mining, of a mass of earth or rock.

calving [glaciol] The breaking away of a mass or block of ice from a glacier, from an ice front (as from an ice shelf), or from an iceberg or floeberg; the process of iceberg formation. See also: *calved ice*.

calymma A frothy layer of cytoplasm in radiolarians.

calyptrolith A basket-shaped coccolith, opening proximally.

calyx (a) A more or less cup-shaped skeletal cover of a echinoderm including the body and internal structures, but excluding the stem and appendages such as free arms or brachioles; e.g. the plated skeletal structure surrounding the viscera of a crinoid and comprising the dorsal cup and tegmen. (b) A small cup-shaped structure or living cavity in which coral polyp sits. See also: *calice*.

calzirtite A tetragonal mineral: $CaZr_3TiO_9$.

cam A term used in the English Lake District for the crest of a mountain. Etymol: German *Kamm*.

camarophorium A spoon-shaped, adductor-bearing platform in the brachial valve of brachiopods of the superfamily Stenoscismatacea, supported by the median septum, and derived independently of the cardinalia (TIP, 1965, pt.H, p.141).

camarostome A concave space found in certain arachnids formed by a depression in the common wall of fused pedipal coxae and the convex rostrum fitting into it, and serving as a filter of liquefied food before it reaches the mouth (TIP, 1955, pt.P, p.61).

camber A superficial structure resembling an arching or ridging, caused by gravitational sagging towards topographically lower areas. Cf: *bulge*.

cambic Said of a soil horizon that is characterized by alteration or removal of mineral matter, indicated by mottling, stronger chromas, or gray or red colors. It has a fine texture and is not indurated (SSSA, 1970).

cambium In a plant, a layer, usually regarded as one cell thick, of persistent meristematic tissue (referring to vascular and cork cambia); or a persistent meristematic layer which gives rise to secondary xylem, secondary phloem, and parenchyma, usually in radial rows in vascular plants (Fernald, 1950, p.1571; Scagel, 1965, p.611). See also: *cork cambium*.

Cambrian The earliest period of the Paleozoic era, thought to have covered the span of time between 570 and 500 million years ago; also, the corresponding system of rocks. It is named after Cambria, the Roman name for Wales, where rocks of this age were first studied. See also: *age of marine invertebrates*. Obs syn: *Primordial*.

camel back A coal miner's term for a *bell, pot, kettle bottom*, or other rock mass that tends to fall easily from the roof of a coal seam. Syn: *tortoise*.

cameo A carved gem that is actually a miniature bas-relief sculpture, commonly cut from materials of differently colored layers (esp. onyx or a gastropod shell), the upper layer being used for the figure and the lower layer serving as the background. Cf: *intaglio*.

cameo mountain A mountain composed of elevated horizontal strata left behind by two or more subparallel streams carving out deep channels that eventually unite.

camera The space enclosed within a cephalopod shell between two adjacent septa but excluding the siphuncle. It represents a portion of an earlier living space now closed off by a septum. Pl: *camerae*. Syn: *chamber*.

camera axis A line perpendicular to the focal plane of the camera and passing through the interior perspective center or emergent (rear) nodal point of the lens system (ASP, 1966, p.1130).

cameral deposit A calcareous deposit secreted against the septa and/or the original walls of camerae during the life of cephalopod; e.g. *mural deposit*.

camera lucida A simple monocular instrument for hand copying or tracing a map or diagram onto a sheet of paper, consisting of a half-silvered mirror (or a prism or the optical quivalent) attached to the eyepiece of a microscope, thereby causing a virtual image of an external object to appear as if projected upon a plane. Etymol: Latin, "light chamber". Pl: *camera lucidas*.

camera station The point in space (in the air or on the ground) occupied by the camera lens at the moment of exposure. See also: *air station*. Syn: *exposure station*.

camerate [paleont] n. Any crinoid belonging to the subclass Camerata, characterized by rigidly united calyx plates.

camerate [palyn] adj. A not widely used syn. of *cavate*.

camouflage In crystal structure, substitution of a trace element for a mineral-forming element of the same valence, e.g. Ga for Al, and Hf for Zr. The trace element is then said to be camouflaged by the common element. Cf: *admittance; capture*.

campagiform Said of the loop, or of the growth stage in the development of the loop, of a dallinid brachiopod, marked by a proportionally large hood without lateral lacunae, with the position of attachment of the descending branches to the septum and hood varying in different genera (TIP, 1965, pt.H, p.141). It precedes the *frenuliniform* stage.

campagna An Italian term for a nearly level, open plain; esp. the undulating, uncultivated plain surrounding Rome, of volcanic origin, and in many places undermined by the Catacombs. Pron: cam-*pahn*-ya.

Campanian European stage: Upper Cretaceous (above Santonan, below Maestrichtian).

campanite An extrusive rock originally described as a tephrite containing large leucite crystals and later defined as a pseudoleucite-bearing nepheline syenite (Johannsen, 1939, p.245); Tröger considered it the extrusive equivalent of essexite-foidite (Streckeisen, 1967, p.208).

Campbell's law The general law of migration of drainage divides, which states that the divide tends to migrate toward an axis of uplift or away from an axis of subsidence (Campbell, 1896, p.580-581). "Where two streams that head opposite to each other are affected by an uneven lengthwise tilting movement, that one whose declivity is increased cuts down vigorously and grows in length headward at the expense of the other" (Cotton, 1948, p.342). Named in honor of its originator, Marius R. Campbell (1858-1940), American geologist.

camptonite A lamprophyre, similar in composition to diorite, being composed essentially of plagioclase (usually labradorite) and brown hornblende (usually barkevikite). Its name is derived from Campton, New Hampshire.

camptospessartite A dark-colored spessartite in which the pyroxene is titanaugite.

campylite A yellow or brown variety of mimetite sometimes crystallizing in barrel-shaped forms.

campylotropous Said of a plant ovule curved by uneven growth so that its axis is approximately at right angles to its funiculus.

camstone (a) A Scottish term for a compact whitish limestone containing much clay. (b) A Scottish term for a bluish-white pipe clay used for whitening hearths and doorsteps.

canaanite A grayish white to bluish white pyroxenite composed chiefly of large white crystals of diopside. It is associated with dolomite at Canaan, Connecticut, U.S.A. (Dana, 1892, p.356).

cañada (a) A term used in the western U.S. for a ravine, glen, or narrow valley, smaller and less steep-sided than a canyon, such as the V-shaped valley of a dry river bed; a dale or open valley between mountains. (b) A term used in the western U.S. for a small stream; a creek.---Etymol: Spanish *cañada*, "cane, reed". Pron: kan-*ya*-da.

canada balsam A natural cement used in mounting specimens for microscopic analysis; it is exuded as a viscous, yellow-green oleoresin by the balsam fir tree. Syn: *canada turpentine*.

canada turpentine *canada balsam*.

Canadian (a) North American provincial series: Lower Ordovician (above Croixian of Cambrian, below Champlainian). (b) Obsolete name applied to a system of rocks between the Ozarkian below and the Ordovician above.

Canadian pole system An obsolete system of oil-well drilling similar to cable-tool drilling but using wooden rods screwed together instead of a cable or rope.

canadite A nepheline syenite containing albite as the main feldspar, along with calcian and aluminian mafic minerals.

canal [astrogeol] Any of various dark, narrow, faint or diffuse, linear markings on the surface of Mars, generally of low contrast, that have been reported by visual observers. Very few have been confirmed by spacecraft photography.

canal [coast] (a) A long, narrow channel or arm of the sea connecting two larger stretches of water, usually extending far inland (sometimes between islands or between an island and the mainland), and approximately uniform in width; e.g. Lynn Canal in Alaska. (b) A term used along the Atlantic coast of the U.S. for a sluggish coastal stream.

canal [speleo] A passage in a cave that is partially filled with water.

canal [paleont] A hollow vessel, tube, passage, channel, or groove of an invertebrate; e.g. a *ring canal* and *stone canal* of an echinoderm, or a gutter-like extension of the lower end of a gastropod shell which carries the siphon within it, or a tube running lengthwise along the walls of a foraminifer test, or a tube leading from an external pore of a sponge to the cloaca and serving for water flow.

canal [streams] An artificial watercourse of relatively uniform dimensions, cut through an inland area, and designed for navigation, drainage, or irrigation by connecting two or more bodies of water; it is larger than a ditch.

canalarium A specialized sponge spicule lining a canal. Pl: *canalaria*.

canaliculate Grooved or channeled longitudinally; e.g. said of the sculpture of pollen and spores consisting of more or less parallel grooves, or said of a foraminifer possessing a series of fine tubular cavities.

canal system A system of passages connecting various cavities of an invertebrate body (as of a coral); e.g. the *aquiferous system* of a sponge.

canary A pale-yellow diamond.

canary stone A yellow variety of carnelian.

canasite A monoclinic mineral: $(Na,K)_5Ca_4(Si_2O_5)_5(OH,F)_3$.

Canastotan Stage in New York State: Upper Silurian (lower Cayugan; below Murderian).

cancellate Having a honeycomb-like structure, or divided into small spaces by laminae, or marked with numerous crossing plates, bars, lines, threads, etc.; e.g. said of the shell surface of a bivalve mollusk marked by subequal, intersecting concentric and radial markings.

cancellus A cylindrical, interzooidal, tube-like cavity developed in some cyclostome bryozoans (as in the family Lichenoporidae). Pl: *cancelli*. Syn: *alveolus*.

cancrinite (a) A mineral of the feldspathoid group: $(Na_2, Ca)_4(AlSiO_4)_6CO_3.nH_2O$. It occurs in igneous rocks (esp. nepheline syenites) usually as transparent to translucent masses of various colors (orange, yellow, violet, blue, green, red, white, and gray). (b) A group of feldspathoid minerals of the general formula (Hey, 1962, p.219): $(Na,K,Ca)_{6-8}(Al,Si)_{12}-O_{24}(SO_4,CO_3,Cl_2)_{1-2}.nH_2O$. The nomenclature of the group is obscure, but it includes cancrinite (the carbonate end member), vishnevite (the sulfate end member), and davyne (containing considerable chloride).

cand A term used in Cornwall, Eng., for fluorite occurring in a vein.

candela A term used in the SW U.S. for a candle-like rocky pinnacle. Etymol: Spanish, "candle".

candelabrum A *calthrops* (sponge spicule) with multiply branched rays, the branches of one ray often differing from those of the others. Pl: *candelabra*.

candite A blue spinel; *ceylonite*.

candle coal A syn. of *cannel coal*, so named because it burns with a steady flame.

candle ice Disintegrating sea or lake ice consisting of ice prisms oriented perpendicular to the ice surface; a form of *rotten ice*. Syn: *candled ice; needle ice; penknife ice*.

caneolith A heterococcolith having a central area with laths, a simple or complex wall, and petaloid upper and lower rims. See also: *complete caneolith; incomplete caneolith*.

canfieldite A black mineral: Ag_8SnS_6. It is isomorphous with argyrodite.

canga A Brazilian term for a tough, moderately hard, well-consolidated, unstratified, iron-rich rock composed of varying amounts of fragments derived from *itabirite*, high-grade hematite, or other highly ferruginous material, and cemented by limonite (which may vary from about 5 to more than 95 percent). It occurs as a near-surface or surficial deposit and blankets older rocks on or near the present or ancient erosion surface, and is very resistant to erosion and chemical weathering. Park (1959, p.580) restricts the term to a rock formed by the cementation by hematite of *rubble ore* into "a hard ironstone conglomerate". Some writers use the term to refer to a ferruginous laterite developed from any iron-bearing rock (commonly basalt or gabbro); e.g. "lateritic iron ore" is equivalent to "canga" as used in Sierra Leone.

cank ball An English term for a nodular mass of cemented sandstone and for an ironstone nodule.

cannel *cannel coal*.

cannel bass An English term for an inferior or impure carbonaceous shale approaching the character of an oil shale.

cannel coal A compact, tough *sapropelic coal* that contains spores and that is characterized by dull to waxy luster, conchoidal fracture, and massiveness. It is attrital and high in volatiles; by American standards it must contain less than 5% anthraxylon. Cf: *boghead coal; torbanite*. Syn: *candle coal; kennel coal; cannel; cannelite; parrot coal; curley cannel*. Adj: *canneloid*.

cannelite *cannel coal*.

canneloid Of or pertaining to *cannel coal*.

cannel shale A black shale or oil shale formed by the accumulation of sapropelic sediments accompanied by a considerable or an approximately equal quantity of inorganic material (chiefly silt and clay). Syn: *bastard shale*.

cannizzarite A mineral: lead bismuth sulfide.

cannonball A large, dark concretion, as much as 3 m in diameter, resembling a cannonball, as in the Cannonball Member (Paleocene) of the Fort Union Formation in the Dakotas.

canoe fold A closely folded syncline, the surface expression of which is elongate.

cañon An obsolete syn of *canyon*. Etymol: American Spanish *cañon*. Incorrectly spelled: *canon*.

canopy In a cave, a ledge of flowstone hanging from a sloping wall, fringed with stalactites on its outer edge.

cant An inclination from a horizontal, vertical, or other given line; a slope or tilt.

cantalite An old name for a glassy sodic rhyolite.

cantonite A variety of covellite occurring in cubes and probably pseudomorphous after chalcopyrite that had replaced galena.

canyon [karst] A *karst valley*.

canyon [speleo] In a cave, a stream bed that resembles a chasm.

canyon [geomorph] (a) A long, deep, relatively narrow, steep-sided valley confined between lofty and precipitous walls in a plateau or mountainous area, often with a stream at the bottom; similar to, but larger than, a *gorge*. It is characteristic of an arid or semiarid area (such as western U.S.) where stream downcutting greatly exceeds weathering; e.g. Grand Canyon. The walls of a small, young canyon are rarely so. (b)

Any valley in a region where canyons abound.---Etymol: angl[ ] cized form of American Spanish *cañon*. Cf: *cañada*. Sy[ ] cañon.

canyon bench One of a series of relatively narrow, flat lan[ ] forms occurring along a canyon wall and caused by differer[ ] tial erosion of alternating strong and weak horizontal strat[ ] See also: *step* [geomorph].

canyon dune A dune formed in a box canyon (Stone, 196[ ] p.219).

canyon fill Unconsolidated material filling a canyon, consistin[ ] of sediment in transport and sediment permanently or tempe[ ] rarily deposited.

canyonside The steeply sloping side of a canyon.

cap *ice cap*.

capacitance That property of an electrically nonconductin[ ] material which permits energy storage, due to electric di[ ] placement when opposite surfaces of the nonconductor ar[ ] maintained at a difference of potential.

capacitive coupling The capacity between two adjacent circu[ ] elements. In induced polarization surveys, capacitive couplin[ ] between a current wire and a potential wire, or between eithe[ ] wire and ground, can lead to fictitious induced polarizatic[ ] anomalies.

capacity [hydrol] (a) The ability of a soil to hold water. (b[ ] The yield of a pump, well, or reservoir.

capacity [hydraul] The ability of a current of water or wind t[ ] transport detritus, emphasizing the amount, measured at [ ] given point per unit of time. Capacity may vary according t[ ] the detrital grain size. Cf: *competence*. See also: *efficienc[ ]* [stream].

capacity curve In hydraulics, a graphic presentation of the re[ ] lationship between the water-surface elevation of a reservo[ ] and the volume of water below; also, a graphic presentatic[ ] of the rate of discharge in a pipe or conduit or through porou[ ] material (ASCE, 1962). Syn: *storage curve*.

cape [coast] An extensive, somewhat rounded irregularity [ ] land jutting out from the coast into a large body of water, ei[ ] ther as a peninsula (e.g. Cape Cod, Mass.) or as a projectin[ ] point (e.g. Cape Hatteras, N.C.); a *promontory* or *headlan[ ]* generally more prominent than a *point*. Also, the part of th[ ] projection extending farthest into the water.

cape [mineral] *Cape diamond*.

Cape diamond A diamond having a yellowish tinge. Etymo[ ] *Cape* of Good Hope. See also: *silver Cape*. Syn: *cape*.

cap effect In gravity observations of salt domes, an effec[ ] produced by the caprock material which is usually dense[ ] than the enclosing rock strate or the salt mass. Syn: *caproc[ ] effect*.

Cape ruby A misleading name applied to a ruby-colored gar[ ] net; specif. gem-quality *pyrope* that is perfectly transparer[ ] and ruby red in color, such as that obtained in the diamon[ ] mines of the Kimberley district of South Africa. Syn: *South A[ ] rican ruby*.

capilla A very fine radial ridge on the external surface of [ ] brachiopod shell (usually more than 25 capillae in a width o[ ] 10 mm). Pl: *capillae*. Cf: *costa; costella*.

capillarity (a) The state of being *capillary*. (b) The action b[ ] which a fluid, such as water, is drawn up (or depressed) i[ ] small interstices or tubes as a result of surface tension. Sy[ ] *capillary action*.

capillary [min] Said of a mineral that forms hairlike or threac[ ] like crystals, e.g. millerite. Syn: *filiform; moss; wire; wiry*.

capillary [water] adj. A term used to describe tubes or inte[ ] stices with such small openings that they can retain fluids b[ ] *capillarity*.

capillary action *capillarity*.

capillary analysis Chromatographic analysis.

capillary attraction The adhesive force between a liquid and [ ] solid in capillarity.

capillary condensation The formation of rings of *pendula[ ]* water around point contacts of grains, and, when the rings around adjacent contacts become large enough to touch, of unicular water filling clusters of interstices surrounded by a single closed meniscus.

capillary conductivity The ability of an unsaturated soil or rock to transmit water or another liquid. As the larger interstices are partly occupied by air or other gas, rather than a liquid, the liquid must move through and in bodies surrounding point contacts of rock or soil particles. For water, the conductivity increases with the moisture content, from zero in a perfectly dry material to a maximum equal to the hydraulic conductivity, or *permeability coefficient.*

capillary ejecta *Pele's hair.*

capillary flow *capillary migration.*

capillary fringe The lower subdivision of the *zone of aeration*, immediately above the water table in which the interstices are filled with water under pressure less than that of the atmosphere, being continuous with the water below the water table but held above it by surface tension. Its upper boundary with the intermediate belt is indistinct, but is sometimes defined arbitrarily as the level at which 50 percent of the interstices are filled with water. Syn: *zone of capillarity; capillary-moisture zone.*

capillary head The *capillary potential* expressed as head of water.

capillary interstice An *interstice* small enough to hold water by surface tension at an appreciable height above a free water surface, yet large enough to prevent molecular attraction from extending across the entire opening. There are no definite size limitations (Meinzer, 1923, p.18). Cf: *subcapillary interstice; supercapillary interstice.*

capillary migration The movement of water by capillarity. Syn: *capillary flow; capillary movement.*

capillary-moisture zone *capillary fringe.*

capillary movement (a) The rise of water in the subsoil above the water table by capillarity (Nelson, 1965, p.66). (b) *capillary migration.*

capillary percolation *imbibition* [water].

capillary porosity The volume of interstices in a soil mass that hold water by capillarity (Jacks, et al, 1960).

capillary potential A number representing the work required to move a unit mass of water from the soil to an arbitrary reference location and energy state (SSSA, 1965, p.348). Symbol: M. Cf: *capillary head.*

capillary pressure The difference in pressure across the interface between two immiscible fluid phases jointly occupying the interstices of a rock. It is due to the tension of the interfacial surface, and its value depends on the curvature of that surface.

capillary pyrites *millerite.*

capillary ripple *capillary wave.*

capillary rise The height above the free water level to which water will rise as a result of capillarity. Syn: *height of capillary rise.*

capillary stalagmite A hollow stalagmite that is formed by upward flow of water through cracks and capillaries in sinter covering a cave floor. The type locality is Cuba, where capillary stalagmites are composed of aragonite (Monroe, 1970).

capillary tension *moisture tension.*

capillary water (a) Water held in, or moving through, small interstices or tubes by capillarity. The term is considered obsolete by the Soil Science Society of America (1965, p.332). Syn: *water of capillarity.* (b) Water of the capillary fringe.

capillary wave A wave whose wavelength is shorter than 1.7 cm, and whose propagation velocity is controlled mainly by surface tension. Cf: *gravity wave.* Syn: *capillary ripple; ripple.*

capillary yield The amount of capillary water, in mm per day or liters per second per hectare, that rises through a plane parallel to the water table, at a given distance below the land surface (Schieferdecker, 1959, term 0301).

capitulum (a) A part of the carapace of a cirripede crustacean, enclosing trophic structures and commonly armored by calcareous plates. (b) An obsolete syn. of *gnathosoma.*---Pl: *capitula.*

capped column A type of snow crystal in the shape of a hexagonal column with thin hexagonal plates or stars at each end.

capped deflection A mountain-range *deflection* in which the arcs meet approximately at right angles and the junction consists of a *cap range* of very high mountains (Wilson, 1950, p.155).

capped quartz A variety of quartz containing thin layers of clay.

cappelenite A mineral: $(Ba,Ca,Na)(Y,La)_6B_6Si_{13}(O,OH)_{27}$.

capping (a) A syn. of *overburden*, usually used for consolidated material. (b) A syn of *gossan.*

cap range A secondary mountain arc that curves around the junction of two primary arcs, as in a *capped deflection.*

capricorn An ammonoid shell resembling a goat's horn, encircled by widely spaced blunt ribs and subequal rounded areas (interspaces) between adjacent ribs.

caprock [eco geol] (a) A syn. of overburden, usually used for consolidated material. (b) A hard rock layer, usually sandstone, overlying the shale above a coal bed.

cap rock [coast] Estuarine sandstones along the Yorkshire coast of England (Nelson & Nelson, 1967, p.58).

cap rock [tect] In a *salt dome*, an impervious body of anhydrite and gypsum, with minor calcite and sulfur, which overlies the top of the salt body, or plug. It probably results from accumulation of the less soluble minerals of the salt body during leaching of its top in the course of its ascent.

caprock effect *cap effect.*

cap-rock fall A waterfall descending over a lip of strong and resistant rock.

captor stream *capturing stream.*

capture [geochem] In a crystal structure, the substitution of a trace element for a common element of lower valence, e.g. Pb^+ for K^+. Cf: *admittance; camouflage.*

capture [streams] The natural diversion of the headwaters of one stream into the channel of another stream having greater erosional activity and flowing at a lower level; esp. diversion effected by a stream eroding headward at a rapid rate so as to tap and lead off the waters of another stream. See also: *abstraction; beheading; intercision.* Syn: *stream capture; river capture; piracy; stream piracy; river piracy; robbery; stream robbery.*

captured stream A stream whose former upper course has been diverted into the channel of another stream by capture. Syn: *pirated stream.*

capture theory The theory which holds that the Moon originated as an independent planet whose orbit around the Sun lay sufficiently close to the Earth such that it strayed into the gravitional field of, and was captured by, the Earth.

capturing stream A stream into which the headwaters of another stream have been diverted by capture. Syn: *pirate; pirate stream; captor stream.*

caracolite A colorless mineral consisting of a sulfate and chloride of sodium and lead, and occurring as crystalline incrustations.

Caradocian European stage: Middle and Upper Ordovician (above Llandeilian, below Ashgillian). It is divided into lower stage (Middle Ordovician) and an upper stage (Upper Ordovician).

carapace A bony or chitinous case or shield covering the whole or part of the back of certain animals, such as the dorsal covering of the cephalothorax in all arachnids or the resistant mineralized dorsal exoskeleton of a trilobite; esp. the commonly calcified part of the dorsal exoskeleton of a crustacean, comprising the cephalic shield and fold of integument arising from the posterior border of the maxillary somite and extending over the trunk, usually covering it laterally as well as dorsally and in many forms having a dorsal longitudinal hinge, and often fused to one or more thoracic somites. Also, the fossilized remains of an ostracode whose calcified cephalothorax covering has become divided laterally into two sub-symmetric parts (valves) joined at the dorsum.

carapace carina A narrow ridge variously located on the surface of carapace of a decapod crustacean.

carapace groove A furrow, generally dorsal, of the surface of carapace of a decapod crustacean.

carapace horn The anteriorly dorsal termination of the valves of carapace of phyllocarid crustaceans.

carapace spine A sharp projection of carapace of a decapod crustacean.

carat (a) A unit of weight for diamonds, pearls, and other gems. It formerly varied somewhat in different countries, but the metric carat or international carat equal to 0.2 gram, or 200 mg, was adopted in the U.S. in 1913 and is now standard in the principal countries of the world. Before 1913 in the U.S., the unit equaled 205.3 mg. Abbrev: c. See also: *point; grain.* (b) Var. of *karat.*

carat grain *grain* [gem].

carbankerite Any coal microlithotype containing 20-60% by volume of carbonate minerals (calcite, siderite, dolomite, and ankerite). (ICCP, 1963).

carbapatite *carbonate-apatite.*

carbargilite Any coal microlithotype containing 20-60% by volume of clay minerals, mica, and lesser proportions of quartz, all with particle size averaging 1-3 microns. (ICCP, 1963).

carbene An *asphaltene* that is insoluble in carbon tetrachloride, but soluble in carbon disulfide, benzene, or chloroform.

carbide A mineral compound that is a combination of carbon with a metal. An example is cohenite, $(Fe,Ni,Co)_3C$.

carbite A general term, now obsolete, applied to diamond and graphite.

carboborite A mineral: $Ca_2Mg(CO_3)(B_2O_5).10H_2O$.

carbocer A pitchy, ocherous, and carbonaceous mineral substance containing rare-earth elements.

carbocernaite A mineral: $(Ca,Ce,Na,Sr)(CO_3)$.

carbohumin *ulmin.*

carbohydrate A polyhydroxy aldehyde or ketone or a compound that can be hydrolyzed to such a compound. Carbohydrates, of which sugars, starches, and cellulose are examples, are produced by all green plants and form an important animal food.

carboid A bitumen formed from asphaltenes at elevated temperatures. It may occur in cracked petroleum residues, and is insoluble in benzene.

carbon (a) A nonmetallic, chiefly tetravalent chemical element (atomic number 6; atomic weight 12.01115) occurring native in the crystalline form (as the diamond and as graphite) or amorphous, and forming a constituent of coal, petroleum, and asphalt, of limestone and other carbonates, and of all organic compounds. Symbol: C. (b) *carbonado.*

carbonaceous (a) Said of a rock or sediment that is rich in carbon; coaly. (b) Said of a sediment containing organic matter.

carbonaceous chondrite A group name for friable, dull-black, chondritic stony meteorites characterized by the presence of hydrated, clay-type silicate minerals (usually fine-grained serpentine or chlorite), by considerable amounts and a great variety of organic compounds (hydrocarbons, fatty and aromatic acids, porphyrins) believed to be of extraterrestrial origin, by an absence or almost total absence of free nickel-iron, and by abnormally high contents of inert gases (esp. xenon). Much of the organic matter is a black insoluble complex of compounds of high molecular weight; the water content (usually water of hydration) is approximately 20% by weight. Carbonaceous chondrites are grouped in three types, each characterized by the amount of organic material they contain and by other

compositional features: type I contains the greatest amount o water and organic matter (3-5% combined carbon, 24-30% ignition loss); type II is intermediate in composition (12-24% ignition loss); and type III contains high-temperature mineral and some metallic constituents (2-12% ignition loss).

carbonaceous coal Coal that is intermediate in compositio between metabituminous coal and anthracite. (Tomkeieff 1954, p.35).

carbonaceous rock A sedimentary rock that consists, or con tains an appreciable amount, of original or subsequently intro duced organic material, including plant and animal residue and chemically organic derivatives greatly altered (carbonize or bituminized) from the original remains; e.g. the coal series black shale, asphaltic sediments, sapropel, certain clays, vari ous solid substances derived from altered plant remains, an esp. *carbonaceous shale.* Syn: *carbonolite.*

carbonaceous shale A dark-gray or black shale with a signifi cant content of carbon in the form of small particles or flake disseminated throughout the whole mass or mingled with inor ganic constituents; it is commonly associated with coa seams.

carbonado An impure, opaque, massive, dark (black, brown dark gray) aggregate composed of minute diamond particles forming a usually rounded mass with a granular to compac structure, and displaying a superior toughness as a result o its cryptocrystalline character and lack of cleavage planes. I is used as an *industrial diamond.* Cf: *bort; ballas.* Syn: *blac diamond; carbon diamond; carbon.*

carbonate [min] A mineral compound characterized by a fun damental anionic structure of $CO_3^{\pm}$. Calcite and aragonite $CaCO_3$, are examples of carbonates. Cf: *borate; nitrate.*

carbonate [sed] A sediment formed by the organic or inorgan ic precipitation from aqueous solution of carbonates of cal cium, magnesium, or iron; e.g. limestone and dolomite. Se also: *carbonate rock.*

carbonate-apatite (a) A mineral of the apatite group: $Ca_5(PO_4,CO_3)_3(OH,F)$. It is the principal constituent of phosphate rock formed under marine conditions. (b) An apatite minera containing a considerable amount of carbonate.---C *francolite.* Syn: *carbapatite; dahllite; podolite; collophane; tav istockite.*

carbonate-arenite *calclithite.*

carbonate-cyanotrichite A pale-blue mineral: $Cu_4Al_2(CO_3 SO_4)(OH)_{13}.2H_2O$.

carbonate cycle The biogeochemical pathways of carbonate involving transformation from or to CO_2 and HCO_3^-, and it solution, deposition in minerals, and metabolism and regen eration in biological fixation.

carbonated spring A spring whose water contains carbor dioxide gas. This type of spring is especially common in vol canic areas (Comstock, 1878, p.34).

carbonate-facies iron formation An *iron formation* character ized by alternating laminae of chert and iron-rich carbonate minerals (James, 1954, p.251-256).

carbonate-fluorapatite *francolite.*

carbonate hardness Hardness of water, expressed as $CaCO_3$ that is equivalent to the carbonate and bicarbonate alkalinity When the total alkalinity expressed as $CaCO_3$ equals or ex ceeds the total hardness, all the hardness is carbonate. It car be removed by boiling and hence is sometimes called *tempo rary hardness,* although this synonym is becoming obsolete Cf: *noncarbonate hardness; hardness.*

carbonate rock A rock consisting chiefly of carbonate miner als, such as limestone, dolomite, or carbonatite; specif. a sedimentary rock composed of more than 50% by weight o carbonate minerals. See also: *calcareous rock; carbonate* [sed].

carbonate thermometer The temperature-dependent oxygen-18/oxygen-16 isotope ratio in the carbonate shells of fossi marine animals as used to indicate the water temperature i.e. paleotemperature) existing at the time the organism was depositing its shell. Accurate determination depends on the original isotopic composition being in equilibrium with the surrounding water and upon its being preserved without further isotopic fractionation or substitution. See also: *oxygen isotope fractionation.*

carbonation (a) A process of chemical weathering involving the transformation of oxides of calcium, magnesium, potassium, sodium, and iron into carbonates or bicarbonates of these metals by carbon dioxide contained in water (e.g. a weak carbonic acid solution). (b) Introduction of carbon dioxide into a fluid. Syn: *carbonatization.*

carbonatite [ign] A carbonate rock of apparent magmatic origin, generally associated with kimberlites and alkalic rocks. The origin of these rocks is controversial, being variously explained as derived from magmatic melt, solid flow, hydrothermal solution, and gaseous transfer. A carbonatite may be calcitic (*sôvite*).

carbonatite [sed] A sedimentary carbonate rock at least 80 percent of which is calcium or magnesium; Kay (1951, p.5) used the term synonymously with "limestone". This usage of the term is rare, as it is more commonly used as an igneous rock name.

carbonatization (a) Introduction of, or replacement by, carbonates. (b) *carbonation* (b).

carbon clock A popular syn. of *carbon-14,* used in radiometric dating of rocks.

carbon cycle The continued exchange and reactions of carbon in the biosphere, atmosphere, and hydrosphere (Pettijohn, 1949, p.363).

carbon dating *carbon-14 dating.*

carbon diamond *carbonado.*

carbon-hydrogen ratio The ratio of carbon to hydrogen in coal. The ratio is used as a method of coal classification. Abbrev: C/H ratio.

Carbonic (a) *Pennsylvanian.* (b) *Carboniferous.*

Carboniferous The Mississippian and Pennsylvanian periods, combined; also, the corresponding system of rocks. In European usage, however, the Carboniferous is considered as a single period and is divided into upper and lower parts. The Permian is sometimes included. See also: *age of amphibians; age of coal.* Partial syn: *Carbonic.*

carbonification A syn. of *coalification,* suggested for standard use by the International Committee for Coal Petrology (ICCP, 1963).

carbonite *natural coke.*

carbonization (a) In the process of *coalification,* the accumulation of residual carbon by the changes in organic matter and decomposition products. (b) The accumulation of carbon by the slow, underwater decay of organic matter. (c) The conversion into carbon of a carbonaceous substance such as coal by driving off the other components, either by heat under laboratory conditions or by natural processes.

carbonolite *carbonaceous rock.*

carbonolith A term suggested by Grabau (1924, p.298) for a carbonaceous sedimentary rock. Syn: *carbonolyte.*

carbon ratio [coal] (a) The percentage of fixed carbon in a coal. (b) The ratio of fixed carbon in a coal to the fixed carbon plus the volatile hydrocarbons. Syn: *fixed carbon ratio.*

carbon ratio [isotope] The ratio of the most common carbon isotope, carbon-12, which is nonradioactive, to either of the less common isotopes, carbon-13 (nonradioactive) or carbon-14 (radioactive), or the reciprocal of one of these ratios. If unspecified, the term generally refers to the ratio (carbon-12/carbon-13).

carbon-ratio theory The hypothesis that in any area, the gravity of oil varies inversely to the carbon ratio of the coal, i.e., as the percentage of fixed carbon in coal increases due to increased temperature and pressure, the oil becomes lighter, i.e. contains more volatile hydrocarbons. The theory was first set forth by H.D. Rogers in 1860, and later by David White in 1915.

carbon spot A black fleck or flake-like carbon inclusion in the body of a diamond crystal.

carbon trash Carbon remains of plant life found in sedimentary rocks and often associated with uranium mineralization. See also: *tree ore.*

carbon-14 A heavy radioactive isotope of carbon having a mass number of 14 and a half-life of 5660±30 years, though this is constantly being refined. It is produced in nature by the reaction of atmospheric nitrogen with neutrons produced by cosmic ray collisions and artificially by atmospheric nuclear explosions. Carbon-14 is useful in dating and tracer studies of materials directly or indirectly involved with the Earth's carbon cycle during the last 50,000 years. Symbol: C^{14}. Partial syn: *radiocarbon.*

carbon-14 age A radiometric age expressed in years and calculated from the quantitative determination of the amount of carbon-14 remaining in an organic material. Syn: *radiocarbon age.* Popular syn: *carbon clock.*

carbon-14 dating A method of determining an age in years by measuring the concentration of carbon-14 remaining in an organic material, usually formerly living matter, but also water bicarbonate, etc. The method, worked out by Willard F. Libby, U.S. chemist (1908-), in the years 1946-1951, is based on the assumption that assimilation of carbon-14 ceased abruptly on removal of the material from the Earth's carbon cycle (i.e. the death of an organism) and that it thereafter remained a closed system. Most carbon-14 ages are calculated using a half-life of 5570±30 years, thus the method is useful in determining ages in the range of 500 to 30,000 or 40,000 years, although it may be extended to 70,000 years by using special techniques involving controlled enrichment of the sample in carbon-14. Syn: *radiocarbon dating; carbon dating.*

carbopyrite Any coal microlithotype containing 5-20% by volume of iron disulfide (pyrite and marcasite) (ICCP, 1963).

carborundum A synthetic substance (silicon carbide) used as an abrasive and as a refractory material. It is identical with the mineral *moissanite.*

carbuncle (a) A cabochon-cut red garnet. (b) An old name, now obsolete, for any of several precious stones of a fiery-red or scarlet-like color, such as ruby or spinel.

carbunculus A term applied to ruby, ruby spinel, almandine garnet, and pyrope.

carburan A pitchy hydrocarbon mineral containing uranium.

Cardan hinge *Galitzin hinge.*

cardella A bryozoan *condyle.*

cardhouse structure A structure found in certain marine sediments, consisting of platy aggregates of clay minerals arranged in edge-to-face fashion and resembling a "house" of playing cards (Lambe, 1953, p.38). Cf: *bookhouse structure.*

cardiac lobe The median lobe of the prosoma and the opisthosoma in merostomes.

cardinal adj. Pertaining to the hinge of a bivalve shell; e.g. "cardinal extremity" (termination of posterior hinge margin of a brachiopod shell).---n. A cardinal part; e.g. a cardinal tooth.

cardinal angle The angle formed at each of the extremities of the hinge of a bivalve shell; e.g. the angle formed between the hinge line and the anterior or posterior free margin of an ostracode valve, or the angle formed between the hinge line and the posterior and lateral margins of a brachiopod shell.

cardinal area (a) A flat or slightly concave, commonly triangular surface extending between the beak and the hinge margin in many bivalve mollusks, and partly or wholly occupied by ligament. It is set off from the remainder of the shell by a sharp angle. (b) The flattened, posterior sector of a valve of an articulate brachiopod, exclusive of the delthyrium or the notothyrium. It may be the interarea, a planarea, or the palintrope.---See also: *hinge area.*

cardinal axis The *hinge axis* in a bivalve mollusk.

cardinal fossula A *fossula* developed in the position of the cardinal septum in a rugose coral. It is most commonly due to abortion of the cardinal septum.

cardinalia A collective term for the varied internal outgrowths and structures of secondary shell located in the posterior and median region of the brachial valve near the beak of a brachiopod and associated with articulation, support of lophophore, and muscle attachment. It may include, for example, the cardinal processes, socket ridges, crural bases, and hinge plates.

cardinal margin The curved posterior margin of a brachiopod shell along which the valves are hinged, homologous with the *hinge line* of strophic shells but not parallel to the *hinge axis* (TIP, 1965, pt.H, p. 141). See also: *hinge* [*paleont*].

cardinal muscle scar A posteriorly and laterally placed muscle scar in certain brachiopods (as in the superfamilies Acrotretacea and Obolellacea).

cardinal plate A plate extending across the posterior end of the brachial valve of a brachiopod, consisting laterally of outer hinge plates and medially of either conjunct inner hinge plates or a single plate, and commonly perforated posteriorly (TIP, 1965, pt.H, p.141).

cardinal platform The *hinge plate* in a bivalve mollusk.

cardinal point One of the four principal *points* of the compass (viz: the north, south, east, or west points) that lie in the direction of the Earth's two poles and of sunrise and sunset and that indicate the four principal astronomic directions on the Earth's surface, spaced at 90-degree intervals.

cardinal process A blade or variably shaped boss, ridge, or projection of secondary shell of a brachiopod, situated medially in the posterior end of the brachial valve, and serving for separation or attachment of diductor muscles.

cardinal septum The protoseptum lying in the plane of bilateral symmetry of a rugose corallite, distinguished from other protosepta by pinnate insertion of newly formed metasepta adjacent to it on both sides. Symbol: C. Cf: *alar septum; counter septum*.

cardinal tooth A *hinge tooth*, often relatively large, situated close to and directly beneath the beak of a bivalve mollusk. Its long axis is perpendicular or oblique to the hinge line. Cf: *lateral tooth*.

cardiophthalmic region The space between the ophthalmic ridges in merostomes.

Cardium clay A Pleistocene glacial clay of northern Europe, characterized by fossil shells of the genus *Cardium*, a marine bivalve mollusk of the family Cardiidae.

carex peat Peat that is derived mainly from *Carex*, a genus of sedges of the family Cyperaceae. Cf: *eriophorum peat*. Syn: *sedge peat*.

cargneule A French term for a porous or cavernous carbonate sedimentary rock (esp. a cellular dolomite), its cavities filled with soft, friable, evaporitic material that easily dissolves or falls out, leaving a rough, corroded surface. Syn: *cornieule*.

Cariboo orogeny A name proposed by W.H. White (1959) for orogeny that is believed to have occurred during early Paleozoic time in the Cordillera of British Columbia, especially in the Selkirk and Omineca mountains, where Permian strata overlie deformed and metamorphosed Proterozoic and lower Paleozoic rocks.

caridoid adj. Said of a decapod crustacean of the infraorder Caridea containing most shrimps, prawns, and related forms; e.g. "caridoid facies", an aspect of primitive Eumalacostraca distinguished by enclosure of thorax by carapace, movably stalked eyes, biramous antennules, scaphocerite-bearing antennae, thoracopods with natatory exopods, ventrally flexed and powerfully muscled elongate abdomen, and caudal fan.-- n. A caridoid crustacean.

caries texture In mineral deposits, a replacement pattern in which the guest mineral forms in a series of scallops around and curving away from the host mineral.

carina (a) An unpaired compartmental plate adjacent to th terga of a cirripede crustacean. (b) A flange-like elevation o the side of a septum of a rugose coral, formed by thickene trabecula. See also: *yardarm carina; zigzag carina*. (c) keel-shaped structure or a flange going around the edge o some foraminiferal tests. (d) A prominent keel-like ridge o the exterior shell of a mollusk; e.g. an extended, somewh angular linear elevation on the exterior of a whorl at the edg of a gastropod shell. (e) The central and denticulated, no dose, or smooth ridge extending down the middle of the pla form or blade of a conodont. (f) A major angular elevation o the surface of a brachiopod valve, externally convex transverse profile and radial from the umbo. Cf: *fold* [*paleon* (g) A median ridge or keel-like structure on the frontal side a bryozoan branch, chiefly in Cryptostomata. Syn: *keel*. Etymol: Latin, "keel". Pl: *carinae*.

carinal adj. Pertaining to a carina; e.g. a "carinal band" repre senting an imperforate marginal area (poreless margin) be tween the carinae of a foraminiferal test.---n. One of a serie of ossicles along the midline of the aboral surface of an as erozoan ray.

carinate [**fold**] (a) Said of a fold that is so closely compresse that is it almost isoclinal. (b) Said of an anticline or synclir that is confined to incompetent strata, so that adjacent strat are undisturbed.

carinolateral A compartmental plate in certain cirripede crus taceans located on either side of the carina. Syn: *carin latus*.

carletonite A mineral: $KNa_4Ca_4Si_8O_{18}(CO_3)_4(OH,F).4H_2O$.

Carlin-type gold deposit Gold occurring as microscopic part cles (up to 30 microns) that must be identified by chemica analysis as it is not recoverable by panning. The term is take from such deposits at Carlin, Nevada. Syn: *invisible gold de posit*.

Carlsbad B twin law A twin law that is now equated with th *x-pericline twin law*.

Carlsbad twin law A twin law in feldspar, especially orthoc lase, that is a penetration twin in which the twin axis is the crystallographic axis and the composition surface is irregula Also spelled: *Karlsbad twin law*.

carmeloite A basalt or an andesite, depending on whether th plagioclase is andesine or labradorite, that contains iddingsit phenocrysts, possibly as an alteration product of olivine. Ca meloite characteristically has a high Na_2O content. Its nam is derived from Carmel Bay, California.

carminite A carmine to tile-red mineral: $PbFe_2(AsO_4)_2(OH)_2$.

carnallite A milk-white to reddish orthorhombic minera $KMgCl_3.6H_2O$. It occurs in deliquescent masses as a sali residue and represents an ore of potassium.

carnegieite An artificial mineral: $NaAlSiO_4$. It is the high-tem perature equivalent of nepheline. It is triclinic at low temper tures, isometric at high temperatures.

carnelian A translucent blood-red, flesh-red, reddish-whit orange-red, reddish-yellow, or brownish-red variety of chalc dony, pale to deep in shade, containing iron impurities. It used for seals and in the manufacture of signet rings. C *sard*. Syn: *carneol; cornelian*.

carneol *carnelian*.

Carnian European stage: Upper Triassic (above Ladinia below Norian). Syn: *Karnian*.

carnivore A heterotrophic organism that nourishes itself mai ly by feeding on other animals, living or dead. Adj: *carniv rous*. Cf: *herbivore*.

carnotite A strongly radioactive and canary-yellow to gree ish-yellow secondary mineral: $K(UO_2)_2(VO_4)_2.3H_2O$. It is a ore of uranium and vanadium, and a source of radium, and occurs as a powdery incrustation or in loosely coherent mas es chiefly in sandstones (as in western U.S.).

carobbiite A mineral: KF.

Carolina bay Any of various shallow, often oval or elliptica enerally marshy, closed depressions in the Atlantic coastal lain (from southern New Jersey to NE Florida, esp. developed in the Carolinas), ranging from about 100 m to many ilometers in length, rich in humus, and containing tree and hrub vegetation different from that of the surrounding areas. heir origin is much debated and has been attributed to fallen neteorites, upwelling springs, eddy currents, and solution. yn: *bay*.

arpathite *karpatite*.

arpedolith *stone line*.

arpel An ovule-bearing locule of the ovary of seed plants, orming a simple *pistil* or a unit of a compound pistil.

arpholite A straw-yellow mineral: $MnAl_2Si_2O_6(OH)_4$.

arphosiderite A yellow mineral consisting of a basic hydrous on sulfate. Much so-called carphosiderite is jarosite or atrojarosite. Cf: *hydronium jarosite*.

arpoid Any homalozoan having an ambulacral groove and a tereome composed of crystalline calcite with reticular microtructure.

arpolite (a) A fossil fruit, nut, or seed. Syn: *carpolith*. (b) An llipsoidal concretion or similar diagenetic structure, 1-2 cm in iameter, originally believed to be a fossil seed and assigned ne generic name *Carpolites*.

arpopod The *carpus* of a malacostracan crustacean. Syn: arpopodite.

arpus The fifth pereiopodal segment from the body of a ualacostracan crustacean, located distal to the merus and roximal to the propodus. It comprises the third segment of a ypical endopod. Pl: *carpi*. Syn: *carpopod; wrist*.

arr An isolated mass of rock found off the coast, as off the oasts of Scotland and northern England. Syn: *carrig*.

arrig A *carr* found off the coast of Ireland. Syn: *carrick*.

arrollite A light steel-gray mineral of the linnaeite group: u(Co,Ni)$_2$S$_4$. Syn: *sychnodymite*.

arrying capacity The natural *production* of a lake, as it influ-nces the population of fish and other aquatic life that the ake will support.

arrying contour A single contour line representing two or nore contours, used to show a vertical or near-vertical topo-raphic feature such as a cliff.

arse A Scottish term for low, level, and fertile land; esp. a act of alluvial land or river bottom bordering an estuary or ear a river mouth, and representing a marine terrace or a aised beach. Example: Carse of Gowrie along the coast of cotland. Syn: *carse land*.

arst Variant spelling of *karst*.

arstone A British term for a hard, firmly cemented ferrugi-ous sandstone, esp. one of Cretaceous age used as a build-g stone. Syn: *quernstone*.

artogram A small, abstracted, simplified map generally howing statistical data of various kinds in a diagrammatic ay, usually by the use of shades, curves, or dots; e.g. a *dot nap*. Syn: *diagrammatic map*.

artographic unit A rock or group of rocks that is shown on a eologic map by a single color or pattern. The standard carto-raphic unit is a *formation*.

artography (a) The art of map or chart construction, and the cience on which it is based. It includes the whole series of nap-making operations, from the actual surveying of the round to the final printing of the map. Syn: *chartology*. (b) he study of maps as scientific documents and works of art.

artology A graphic method of coal-seam correlation, involv-ıg the mapping and drawing of both vertical and horizontal ections. Cf: *composite map*.

artouche A decorative frame or scroll-shaped embellishment ı a map or chart, enclosing the title, scale, legend, and ther descriptive matter. Syn: *title box*.

aryinite A mineral: (Ca,Pb,Na)$_5$(Mn,Mg)$_4$(AsO$_4$)$_5$.

aryopsis A small, dry, one-seeded, indehiscent fruit that has completely united seed coat and pericarp and forms a single

grain, e.g. in grasses.

cascade [glac] *glacial stairway*.

cascade [streams] (a) A *waterfall*, esp. a small fall or one of a series of small falls descending over steeply slanting rocks; a shortened rapids. Also, a stepped series of small, closely spaced waterfalls or very steep rapids. Cf: *cataract*. (b) A short, rocky declivity in a stream bed over which water flows with greater rapidity and a higher fall than through a *rapids*.

cascade decay A little-used term for radioactive decay from a parent isotope through several daughter isotopes to a stable isotope.

cascade fold One of a series of folds that is formed by buckling due to gravity collapse along the limb of an anticline.

cascade stairway *glacial stairway*.

Cascadia One of the *borderlands* proposed by Schuchert (1923), in this case along the western margin of North America, partly at sea, partly inland. Most of the evidence adduced for the former existence of Cascadia can now be otherwise interpreted. Possibly there were minor offshore lands in places, and possibly some former continental material has disappeared by underthrusting at the edge of the continent, but the foundering of any extensive lands into the Pacific Ocean basin is impossible.

Cascadian Revolution A name used by Schuchert and others for a supposed profound crustal disturbance in western North America which brought an end to the Tertiary. The concept is implausible, as events at this time do not differ from those earlier or later. The type area (the Cascade Range) is unfortunate, as no notable crustal events seem to have occurred there at this time. The term should be abandoned; if names for orogenies near the Pliocene-Pleistocene boundary are needed, they should be based on areas where they can be defined by limiting fossiliferous strata, as in California.

cascading glacier A glacier passing over a steep, irregular bed, and therefore crossed by numerous crevasses and suggestive in appearance of a cascading stream. Cf: *icefall*.

cascadite A sodic minette containing biotite, olivine, and augite phenocrysts in a groundmass composed almost entirely of alkali feldspar.

cascajo Reef-derived material composed of coral detritus and other sediment, occurring in old deposits. Etymol: Spanish, "gravel". Pron: kas-*ka*-ho.

cascalho A term used in Brazil for alluvial material, including gravel and ferruginous sand, in which diamonds are found. Etymol: Portuguese, "pebbles, small stones, coarse sand, gravel, grit".

case hardening The process by which the surface of a porous rock (esp. tuff and certain sandstones) is coated with a cement or desert varnish formed by evaporation of mineral-bearing solutions. Adj: *case-hardened*. Also spelled: *casehardening*.

casing [drill] (a) A heavy metal (steel or iron) pipe or tubing of varying diameter and weight, lowered into a borehole during or after drilling in order to support the sides of the hole and thus prevent the walls from caving, to prevent loss of drilling mud to porous ground, and to prevent water, gas, or other fluid from entering the hole. It consists of sections (16-34 ft in length) that are riveted, welded, or screwed together. There may be several strings of casing, one inside the other, in a single well. See also: *surface pipe; well casing*. (b) The process of inserting casing in a borehole.

casing [mining] A term used in Ohio for a thin slab of sandstone that splits out between closely spaced joints.

casinghead A fitting attached to the top of the casing set in an oil or gas well, designed to regulate the flow of oil and gas, to separate oil from gas, and to allow pumping and the cleaning out of the well. It may have several lateral outlets through which the flow of the oil can be controlled and lead away to reservoirs by means of pipes. See also: *Christmas tree; bradenhead*. Also spelled: *casing head*.

casing point The depth in a drill hole to which a given string of casing extends.

Cassadagan North American stage: Upper Devonian (above Chemungian, below Bradfordian).

Cassel brown *black earth* [earth]. Etymol: source near Cassel, Germany.

Cassel earth *black earth* [coal]. Etymol: source near Cassel, Germany.

Casselian *Chattian*.

Cassiar orogeny A name proposed by W.H. White (1959) for orogeny that is believed to have occurred near the end of Paleozoic time in the Cordillera of British Columbia, especially in the Omineca and Cassiar districts. It was characterized by uplift, folding, and ultramafic intrusion.

cassidyite A mineral: $Ca_2(Mg,Ni)(PO_4)_2.2H_2O$.

Cassinian curve An *isochromatic curve* in a biaxial crystal.

Cassini projection A map projection constructed by computing the lengths of arcs along a central meridian and along a great circle perpendicular to that meridian, and plotting these as rectangular coordinates on a plane. The scale is preserved along the central meridian and great-circle arcs. It was formerly used as the base for much topographic and cadastral mapping. Named after C.F. Cassini de Thury (1714-1784), French astronomer, who introduced the projection in 1745.

cassinite (a) A bluish variety of orthoclase containing barium and occurring in Delaware County, Penna. (b) Perthitic intergrowths of hyalophane and plagioclase.

cassiterite A brown or black tetragonal mineral: SnO_2. It is the principal ore of tin. Cassiterite occurs in prismatic crystals of brilliant adamantine luster and also in massive forms either compact with concentric fibrous structure (wood tin) or in rolled or pebbly fragments (stream tin). Syn: *tinstone; tin ore; black tin*.

cast [paleont] Secondary rock or mineral material that fills a *natural mold*; specif. a replica or reproduction of the external details (size, shape, surface features) of a fossil shell, skeleton, or other organic structure, produced by the filling of a cavity formed by the decay or dissolution of some or all of the original hard parts of which the organism consisted. Cf: *mold* [*paleont*].

cast [sed] A sedimentary structure representing the infilling of an original mark or depression made on top of a soft bed and preserved as a solid form on the underside of the overlying and more durable stratum; e.g. a *flute cast* and a *load cast*. Cf: *mold*. Syn: *counterpart*.

castellated Said of a physiographic feature (such as a cliff, peak, or iceberg) displaying a towering or battlement-like structure.

casting [paleont] Something that is cast out or off, such as a worm casting or a fecal pellet.

casting [sed] The process of forming a sedimentary cast or casts, or the configuration of a surface characterized by such casts; e.g. "load casting". Also, a cast so formed.

castle A natural rock formation bearing a fanciful resemblance to a castle.

castle koppie A pointed *koppie*, or a small bornhardt, with a castellated profile, often occurring as a jumbled pile of joint-bounded granite blocks. Syn: *castle kopje*.

castorite A transparent variety of petalite. Syn: *castor*.

caswellite A mineral consisting of a copper-red altered biotite, resembling clintonite.

cat *cat claw*.

cataclasis Rock deformation accomplished by fracture and rotation of mineral grains or aggregates without chemical reconstitution. See also: *cataclastic metamorphism; cataclastic rock*.

cataclasite A rock formed by cataclasis; a *cataclastic rock*.

cataclasm A breaking down or rending asunder; a violent disruption.

cataclastic (a) Pertaining to the structure produced in a rock by the action of severe mechanical stress during dynam metamorphism. Characteristic features include the bending breaking, and granulation of the minerals; also said of th rocks exhibiting such structures. See also: *mortar structur* (b) Pertaining to clastic rocks, the fragments of which hav been produced by the fracture of preexisting rocks by Ear stresses, e.g. crush breccia (Teall, 1887).

cataclastic breccia A breccia produced by cataclasis; *crus breccia*.

cataclastic conglomerate *crush conglomerate*.

cataclastic flow *Flow* [exp struc geol] involving intergranul movement, i.e. mechanical displacement of particles relativ to each other.

cataclastic metamorphism A type of local metamorphism con fined to the vicinity of faults and overthrusts involving pure mechanical forces causing crushing and granulation of th rock fabric (*cataclasis*). Cf: *dislocation metamorphism; kine ic metamorphism*.

cataclastic rock A rock, such as a tectonic breccia, contain ing angular fragments that have been produced by the crust ing and fracturing of preexisting rocks as a result of mechan cal forces in the crust; a metamorphic rock produced by cata clasis. Its fabric is a structureless rock powder. Pettijoh (1957, p.281) would include glacial till as a "cataclastic de posit" as it is "an extensive gouge caused by the grindin along the base of an overthrust ice sheet". See also: *catac lasis; autoclastic rock; mylonite*. Syn: *cataclasite*.

cataclastic structure *mortar structure*.

cataclastic texture A texture in a dynamically metamorphose rock produced by severe mechanical crushing and differenti movement of the component grains and characterized b granular, fragmentary, deformed, or strained mineral crysta flattened in a direction at right angles to the mechanic stress. Syn: *pressure texture*.

cataclinal Said of a *dip stream* or of a valley that descends the same direction as that of the general dip of the underlyin strata it traverses. Term introduced by Powell (1875, p.160 Ant: *anaclinal*. Syn: *conclinal; acclinal*.

catacline Said of the *inclination* of the cardinal area in eithe valve of a brachiopod lying at right angles to the orthoclin position.

cataclysm (a) Any geologic event that produces sudden ar extensive changes in the Earth's surface; e.g. an exceptional violent earthquake. Adj: *cataclysmic; cataclysmal*. Cf: *catas trophe*. (b) Any violent, overwhelming flood of water tha spreads over the land; a deluge.

catadupe An obsolete term for a cataract or waterfall.

catagenesis [evol] Evolution leading to decadence and de creased vigor. Also spelled: *katagenesis*.

catagenesis [sed] The changes occurring in an alread formed sedimentary rock buried by a distinct (though some times thin) covering layer, characterized by pressure-tempe ature conditions that are much different from those of depos tion (Fersman, 1922); specif. the breakdown of rocks. Th term is more or less equivalent to *epigenesis* as applied b Russian geologists. Syn: *katagenesis*.

catagraphite A complex structure made up of traces of cana and cavities thought to be the result of the activity of cyand phytes and bacteria in the late Precambrian and Lower Cam brian. Cf: *oncolite*.

catalinite An agate beach pebble, used as a gem, from San Catalina Island, Calif.

catamorphism Var. of *katamorphism*.

catanorm The norm of rocks of the katazone.

cataphoresis *Electrophoresis* in which the movement of sus pended positive particles in a fluid is toward the cathode. C *anaphoresis*.

cataphorite *kataphorite*.

cataplasis An evolutionary stage in which organisms becom ecadent and display a decreasing vigor. Cf: *anaplasis; meta-plasis.*

catapleiite A yellow or yellowish-brown hexagonal mineral: $Na_2,Ca)ZrSi_3O_9.2H_2O$. See also:*calcium catapleiite.* Also spelled: *catapleite.*

catapleiite syenite A porphyritic igneous rock that has the texture of tinguaite and contains phenocrysts of catapleiite and sometimes eudialyte in a fine-grained holocrystalline groundmass of the same minerals plus alkali feldspar, nephelne, and acmite.

cataract (a) A *waterfall*, esp. one of great volume in which the vertical descent has been concentrated in one sheer drop over a precipice. Cf: *cascade.* (b) A series of steep *rapids* in a large river; e.g. in the Nile. (c) An overwhelming rush of water; a flood.

cataract lake A lake occupying the plunge basin of an extinct cataract whose stream has been diverted above the fall.

catarinite An obsolete term for an iron meteorite remarkable for high percentage of nickel.

catastable adj. Stable, with a tendency towards sinking. Cf: *nastable.*

catastrophe [geol] A sudden, violent disturbance of nature, ascribed to exceptional or supernatural causes, affecting the physical conditions and the inhabitants of the Earth's surface; e.g. the Noachian flood, or an extinction of an entire fauna. Cf: *cataclysm; paroxysm.* Syn: *convulsion.*

catastrophic advance *surge* [glaciol].

catastrophism (a) The doctrine that sudden, violent, short-lived, more or less worldwide events outside our present experience or knowledge of nature have greatly modified the Earth's crust. (b) The doctrine that changes in the Earth's fauna and flora are explained by recurring catastrophes, followed by creation of different organisms.---Cf: *uniformitarianism.* Syn: *convulsionism.*

catathermal n. A period of time during which temperatures are declining. The term was used by Emiliani (1955, p.547) for a part of a cycle as displayed in a deep-sea sediment core. Abbrev: Ct. Ant: *anathermal.*

catawberite A metamorphic rock of South Carolina that consists mainly of talc and magnetite.

catazone Var. of *katazone.*

catch basin (a) A reservoir or basin into which surface water may drain. (b) A basin to collect and retain material from a street gutter that would not readily pass through the sewer system.

catchment (a) A term used in Great Britain for an area that collects and drains rainwater; a *drainage basin.* (b) A depression that collects rainwater; a reservoir. (c) The act of catching water; also, the amount of water that is caught.

catchment area [grd wat] (a) The *recharge area* and all areas that contribute water to it. (b) An area paved or otherwise waterproofed to provide a water supply for a storage reservoir. Syn: *collecting area.*

catchment area [streams] *drainage basin.*

catchment basin [streams] *drainage basin.*

catchment glacier *drift glacier.*

catchwater drain A ditch or surface drain designed to intercept flowing water on sloping land and to divert the flow or to irigate the soil. Syn: *catchwater; catchwork.*

cat claw A miner's term used in Peoria County, Ill., for an irregular protuberance (2.5-7.5 cm in height and width) in the lower surface of a bed of marcasite overlying a coal seam (Cady, 1921, p.164). Syn: *cat.*

cat coal Coal that contains pyrites.

catena A *soil association* of a given area and developed from common parent material, whose distinguishing or differentiating features are due to local variations in the drainage pattern.

catenary The curve assumed by a perfectly flexible, inextensible cord of uniform density and cross section when suspended freely from two fixed points both at the same level; e.g. such a curve as formed by a surveyor's tape hanging between adjacent supports. See also: *sag correction.*

cateniform Said of a tabulate corallum with the corallites united laterally as palisades that appear chain-like in cross section and commonly form a network.

cat face A miner's term for glistening balls or nodules, or small discontinuous veinlets, of pyrite in the mining face of coal; also, lenticular deposits of pyrite associated with coal. Also spelled: *catface.*

cathead An English term for a nodule of ironstone; a *dogger.* Syn: *cat's-head.*

cathode-coupled amplifier *cathode follower.*

cathode follower A single-stage, vacuum-tube amplifier in which the load impedance is placed between the cathode and the ground. Syn: *cathode-coupled amplifier.*

cathole A local term used in southern Michigan for a small (less than an acre), shallow, and boggy or miry depression, esp. one formed by a glacier in a till plain (Veatch & Humphrys, 1966, p.59).

cat ice *shell ice.*

catkin *ament.*

catlinite A red, siliceous, indurated clay from the upper Missouri River valley region (SW Minnesota), formerly used by the Dakota Indians for making tobacco pipes; a *pipestone.* Named after George Catlin (1796-1872), American painter of Indians.

catoctin A residual knob, hill, or ridge of resistant material rising above a peneplain and preserving on its summit a remnant of an older peneplain. Named after Catoctin Mountain, Maryland & Virginia.

catogene Pertaining to sedimentary rocks, signifying that they were formed by deposition from above, such as of suspended material. Cf: *katogene.*

catophorite *kataphorite.*

catoptrite *katoptrite.*

Ca-Tschermak molecule A synthetic pyroxene, $CaAl(AlSi)O_6$; a hypothetical component of natural pyroxenes. Not to be confused with *tschermakite.* Syn: *Tschermak molecule.*

cat's-eye [gem] Any gemstone that, when cut cabochon, exhibits under a single strong point source of light a narrow, well-defined chatoyant band or streak that moves across the summit of the gemstone and shifts from side to side as it is turned about and that resembles the shape of a slit pupil of the eye of a cat. The phenomenon is caused by internal reflection of light from numerous parallel inclusions of minute fibrous crystals or from long, parallel cavities or tubes. Gemstones exhibiting this phenomenon include chrysoberyl, quartz, sillimanite, scapolite, cordierite, orthoclase, albite, beryl, and tourmaline.

cat's-eye [mineral] (a) A greenish gem variety of chrysoberyl that exhibits chatoyancy. Syn: *cymophane; oriental cat's-eye.* (b) A variety of minutely fibrous, grayish-green quartz (chalcedony) that exhibits an opalescent play of light. Syn: *occidental cat's-eye.* (c) A yellowish-brown silicified form of crocidolite. Cf: *tiger's-eye.*———The term, when used alone, is properly applied only to chrysoberyl.

cat's-head Var. of *cathead.* Also spelled: *catshead; cats-head.*

catstep A *terracette*, esp. one produced by slumping of deep loess deposits as in western Iowa. Also spelled: *cat step.*

catter Shortened form of *bellicatter*, a syn. of *icefoot.*

cattierite A mineral with pyrite structure: CoS_2.

cattle terrace *sheep track.*

caudal fan A powerful swimming structure in malacostracan crustaceans, consisting of a combination of laterally expanded uropods and telson, and constituting a means of steering and balancing. Syn: *tail fan.*

caudal furca A crustacean *furca* consisting of a pair of caudal rami.

caudal ramus One of a pair of appendages of the telson of a crustacean. It is usually rod- or blade-like, but may be filamentous and multiarticular. Syn: *caudal appendage; cercus.*

caudex A basal part of the axis of an erect plant where it is neither clearly stem, rhizome, or root; the persistent base of an otherwise annual herbaceous stem (Fernald, 1950, p.1571).

cauk (a) A dialectal British term for a limestone or chalk. Syn: *calk.* (b) *cawk.*

cauldron [marine geol] Var. of *caldron* [marine geol].

cauldron [volc] As used by Smith & Bailey (1968), an inclusive term for all volcanic subsidence structures regardless of shape or size, depth of erosion, or connection with the surface. The term thus includes *cauldron subsidences,* in the classical sense, and *collapse calderas.*

cauldron subsidence The sinking of an approximately cylindrical portion of the roof of a magma chamber along *ring faults,* sometimes accompanied by squeezing up of magma along the fault to form a ring dike; also, the structure so formed (Clough et al, 1909). Cf: *cauldron.* See also: *ring-fracture stoping.*

caulescent Said of a plant that is more or less stemmed or stem-bearing; having an evident stem above ground (Lawrence, 1951, p.743).

caunter lode *cross vein.*

causse A characteristic type of limestone plateau, intermediate in development between holokarst and merokarst, containing closed depressions, potholes, and caves. Type locality: the Causses of SE France, also called Les Grandes Causses. The term is sometimes used synonymously with karst. Etymol: French *cau* for *chaux,* "lime".

caustic n. (a) A curve representing a locus of points where a set of rays (as of light or sound) are brought exceptionally close together due to the refractive properties of the medium through which they travel; specif. the curve to which adjacent orthogonals of waves, refracted by a bottom whose contour lines are curved, are tangents. Syn: *caustic curve.* (b) The envelope of the system or sequence of such rays. Syn: *caustic surface.*

caustic metamorphism The indurating, baking, burning, and fritting effects of lava flows and small dikes on the rocks with which they come in contact. The term was originated by Milch in 1922. Cf: *baking.* Syn: *optalic metamorphism.*

caustobiolith A combustible organic rock (Grabau, 1924, p.280). It is usually of plant origin. Cf: *acaustobiolith.*

caustolith A rock that has the property of combustibility (Grabau, 1924, p.280). It is usually of organic origin (e.g. coal and peat) but inorganic deposits (e.g. sulfur, asphalt, graphite) also occur.

caustophytolith A *caustobiolith* formed by the direct accumulation of vegetal matter; e.g. peat, lignite, and coal.

caustozoolith A rare *caustobiolith* formed by the direct accumulation of animal matter (Grabau, 1924, p.280); e.g. some oils.

cavaedium An irregular space within a sponge, communicating directly with the exterior, but morphologically outside the sponge in that it is lined by exopinacoderm. Pl: *cavaedia.*

cavalorite A granular plutonic rock containing more orthoclase than oligoclase. See also: *oligoclasite.*

cavansite A mineral: $Ca(VO)(Si_4O_{10}).6H_2O$.

cavate (a) Descriptive of spores whose exine layers are separated by a cavity, including a rather slight separation as well as a more extensive separation producing a bladder-like protuberance (pseudosaccus). Syn: *camerate* [*palyn*]. (b) Said of a dinoflagellate cyst with space or spaces of notable size between the periphragm and endophragm (as in *Deflandrea phosphoritica*).

cave [coast] *sea cave.*

cave [speleo] (a) A naturally formed, subterranean open area or chamber, or series of chambers. The implication of size is that it is large enough for a human being to enter. A cave is commonly formed in limestone, by solution. Partial syn: *ca ern.* (b) A similar feature that was formed artificially.

cave blister An ovoid, pimplelike speleothem that has a cor of mud.

cave breakdown The process of collapse of the ceiling walls of a cave; also, the accumulation of debris thus forme Cf: *cave breccia.* Syn: *breakdown.*

cave breathing The back-and-forth movement of air in th constricted passages of caves, with a cycle of a few second to a few minutes. See also: *blowing cave.*

cave breccia The debris formed by *cave breakdown,* e.g. detached mass of angular fragments of limestone that hav fallen to the floor from the roof and sides of a cave and th are cemented with calcium carbonate or occur in a matrix cave earth. See also: *collapse breccia; solution breccia.* Sy *cavern breccia.*

cave coral *coralloid.*

cave deposits *drift* [speleo].

cave earth *drift* [speleo].

cave flower In a cave, gypsum or epsomite that occurs as curved, elongate deposit from a cave wall. Growth of the fo mation occurs at the attached end. Cf: *anthodite.* Syn: *gy sum flower; oulopholite.*

cave formation *speleothem.*

cave ice (a) Naturally formed ice in a cave. (b) Calcium ca bonate deposits in a cave; *calcite ice.*

cave-in (a) The partial or complete collapse of earth materi into a large underground opening, such as an excavation or mine. (b) The sudden slumping of wall material into a pit. (c A place where material has collapsed or fallen in or down.

cave-in lake A shallow body of water whose basin is produce by collapse of the ground following contraction due to unequ thawing of ground ice in regions underlain by permafrost; lake occupying a *thaw depression.* Syn: *kettle lake; kettl hole lake; thaw lake; thermokarst lake.*

cave inlet An opening into which a surface stream goes u derground on a gentle gradient.

cave marble *cave onyx.*

cave of debouchure *outflow cave.*

cave onyx A compact, cryptocrystalline, banded deposit calcite or aragonite found in caves, capable of taking a hig polish, and resembling true onyx in appearance. See als *onyx marble.* Syn: *cave marble.*

cave pearl A concretion of calcite formed in subterranea streams. Syn: *cave pisolite.*

cave pisolite *cave pearl.*

caver One who engages in the hobby of cave exploration, *caving.* Syn: *spelunker; potholer.*

cavern A syn. of *cave,* with the implication of large size; system or series of caves or of chambers within a cave.

cavern breccia *cave breccia.*

cavern flow Movement, usually turbulent, of ground wate through caves, coarse, sorted gravel, or large open conduit either by gravity or under pressure.

cavernous Said of an area or geologic formation, e.g. lime stone, that contains caverns, or caves.

cavernous Said of the texture of an igneous or volcanic roc that is *porous* or *cellular,* due to erosion of certain constitu ents rather than to expansion of gases.

cavernous rock Any rock that has many cavities, cells, large interstices; e.g. a cliff face pitted with shallow holes re sulting from cavernous weathering.

cavernous weathering Chemical and mechanical weatherin on a cliff face in which grains and flakes of rock are loosene so as to enlarge hollows and recesses "opened through chemically hardened shell" surfacing the cliff face (Cotto 1958, p.15). It produces the *tafoni* in seaside cliffs. Cf: *hone comb weathering.*

cavern porosity A pore system having large, cavernous oper ings. The lower size limit, for field analysis, is practically s ''about the smallest opening an adult person can enter'' Choquette & Pray, 1970, p.244).

cavern system *cave system.*

cave system (a) A complex cave, or a network of caves. (b) group of caves that are hydrologically related, although they may not be physically connected.----Syn: *cavern system.*

caving [geomorph] (a) *bank caving.* (b) A falling in; the action of caving in.

caving [speleo] The exploration of caves, rather more as a hobby or sport than a scientific study. See also: *caver; speleology.* Syn: *spelunking; potholing.*

cavings Rock fragments or splinters that fall from the walls of borehole and contaminate the drill cuttings or block the hole. They must be washed or drilled out before the borehole can be deepened. Cf: *ravelly ground.*

cavitation The usually mechanically induced growth and collapse of bubbles in a fluid, caused by the static pressure at any point in a flow being less than the fluid vapor pressure U.S. Naval Oceanographic Office, 1966).

cawk A syn. of *barite*; esp. a white, massive, opaque variety barite found in Derbyshire, Eng. Syn: *cauk; calk.*

c axis [cryst] One of the crystallographic axes used as reference in crystal description; it is oriented vertically. In tetragonal and hexagonal crystals, it is the unique symmetry axis. It usually the principal axis. The letter c usually appears in italics. Cf: *a axis; b axis.*

c axis [struc petrol] In structural petrology, that *fabric axis* which is the direction perpendicular to a cleavage or other structural surface. Cf: *b axis; c axis.* Syn: *c direction.*

c* axis That axis of a reciprocal crystal lattice which is perpendicular to (001). Cf: *a* axis; b* axis.*

cay A small, low, coastal island or emergent reef of sand or coral; a flat mound of sand and admixed coral fragments, built up on a reef flat at or just above high-tide level. Term is used esp. in the West Indies where it is pronounced ''key'' and sometimes spelled *kay.* Etymol: Spanish *cayo,* ''shoal or reef''. Cf: *key.*

cayeuxite A nodular variety of pyrite containing silicon, arsenic, antimony, and germanium.

cay sandstone A friable to firmly cemented coral sand formed near the base of a coral-reef cay and reaching above high-tide level; it is cemented by calcium carbonate deposited from freshwater.

Cayugan North American provincial series: Upper Silurian (above Niagaran, below Ulsterian of Devonian).

Cazenovian North American stage: Middle Devonian (above Chesquethawan, below Tioughniogan).

C-centered lattice *base-centered lattice.*

CD interray Posterior interray in echinoderms situated between C ray and D ray and clockwise of C ray when viewed from the oral side. It differs frequently in shape from the other interrays and contains the periproct.

c direction *c axis* [struc petrol].

cebollite A greenish to white fibrous mineral: $H_2Ca_4Al_2Si_3O_{16}$.

cecilite A basaltic rock with few phenocrysts and composed leucite, augite, melilite, nepheline, olivine, anorthite, magnetite, and apatite. Leucite comprises about 50 percent of the total rock, followed by the mafic minerals with melilite comprising about one-eighth of the total rock; nepheline constitutes about eight percent and anorthite, four (Johannsen, 1939, p.246).

cedarite *chemawinite.*

cedar-tree structure A term applied to a laccolith or volcanic rock in which sill-like intrusive layers taper away from a central intrusive mass, the whole structure resembling the outline a cedar tree in cross section.

cedricite A lamproite that contains leucite and diopside phenocrysts in a very fine-grained groundmass.

ceiling cavity A concavity in the ceiling of a cave; it is a solutional feature, the form of which is determined by jointing or bedding planes. Cf: *pocket* [speleo]. Syn: *joint pocket.*

ceiling channel *upside-down channel.*

ceiling meander *upside-down channel.*

ceja A term used in the SW U.S. for the jutting edge along the top of a mesa or upland plain, and also for the cliff at this edge; an escarpment, esp. the steeper of the two slopes of a cuesta (if the slope is a cliff) or part of this slope that is a cliff. Etymol: Spanish, ''eyebrow; mountain summit''. Pron: *say-ha.*

celadonite A soft, green or gray-green, earthy, dioctahedral mineral of the mica group, consisting of a hydrous silicate of iron, magnesium, and potassium, and generally occurring in cavities in basaltic rocks. It has a structure very similar to that of glauconite, and has been regarded as a ferruginous glauconite, a mica near glauconite, and an aluminum-poor variety of glauconite, and also as 'similar to illite'. (Hey, 1962, 14.24.6). Not to be confused with *caledonite.* Syn: *Verona earth.*

celerity The magnitude of velocity, or rate, at which a small surface wave progresses (in still water) radially outward from the point of a small disturbance. Symbol: c. See also: *wave velocity.*

celestial coordinate Any member of any system of coordinates used to locate a point on the celestial sphere; e.g. altitude, azimuth, declination, and right ascension.

celestial equator The great circle on the celestial sphere midway between the celestial poles and whose plane, passing through the center of the Earth, is perpendicular to the Earth's axis of rotation. It is formed by the intersection of the celestial sphere and the extension of the Earth's equatorial plane. In astronomic work, the plane of the celestial equator is sometimes assumed to pass through the point of observation. Usually simply called the *equator.* Syn: *equinoctial; equinoctial circle.*

celestial horizon The great circle on the celestial sphere midway between the zenith and the nadir; that circle on the celestial sphere formed by the intersection of the celestial sphere and a plane passing through the center of the Earth and perpendicular to the zenith-nadir line, or, the plane, through the center of the Earth, that is perpendicular to a radius of the Earth which passes through the point of observation on the Earth's surface. It is commonly called simply the *horizon.* Cf: *astronomic horizon; geometric horizon.* Syn: *rational horizon; true horizon.*

celestialite A variety of *ozocerite* found in some iron meteorites.

celestial latitude Angular distance north or south of the ecliptic to a point on the celestial sphere, measured along the arc of a great circle perpendicular to the ecliptic. See also: *latitude.*

celestial longitude Angular distance measured east from the vernal equinox along the ecliptic intercepted between the circle of celestial latitude of the vernal equinox and the circle of celestial latitude of a point on the celestial sphere and perpendicular to the ecliptic. See also: *longitude.*

celestial mechanics The study of the theory of the motions of celestial bodies under the influence of gravitational fields (NASA, 1966, p.11).

celestial meridian One half of a great circle of the celestial sphere passing through the zenith of a given place and terminating at the celestial poles; the hour circle that contains the zenith, or the vertical circle that contains the celestial pole. Usually called simply the *meridian.*

celestial pole Either of the two points of intersection of the celestial sphere and the extended axis of rotation of the Earth and around which the diurnal rotation of the stars appears to take place; specif. *north pole* and *south pole.* The altitude of an observer's celestial pole is equal to his geographic latitude.

celestial sphere An imaginary sphere of infinite radius, described around an assumed center (usually the center of the Earth), and upon whose ''inner surface'' all the heavenly bodies (except the Earth) appear to be projected along radii passing through the bodies and of which the apparent dome of the visible sky forms half. Since the radius of the Earth is negligible in comparison to the distances to most celestial bodies, the center of the celestial sphere is assumed for most surveying purposes to coincide with the point of observation on the Earth's surface.

celestine *celestite.*

celestite An orthorhombic mineral: $SrSO_4$. It is commonly white with an occasional pale-blue tint, and it often occurs in deposits of salt, gypsum, and residual clays. Celestite is the principal ore of strontium. Syn: *celestine; coelestine.*

celite *brownmillerite.*

cell texture In mineral deposits, a network pattern formed by exsolution or by replacement or organic structures, e.g. cell walls.

cellular Said of the texture of a rock (e.g. a cellular dolomite) characterized by moderately small to large openings or cavities which may or may not be connected. Although there are no specific size limitations, the term is usually applied to cavities larger than pore size and smaller than caverns. The syn. *vesicular* is preferred when describing igneous rocks. Cf: *porous; cavernous.*

cellular porosity A term applied originally by Howard & David (1936, p.1406) to equidimensional openings formed by solution. The term has since been applied to intraparticle porosity formed organically within fossils. Choquette & Pray (1970, p.245) suggest that the term be abandoned because of its infrequent use and diverse application.

cellular soil *polygonal ground.*

cellule A subdivision of a marginal chamberlet in the outer part of the marginal zone of a foraminifer (as in Orbitolinidae), formed by primary and secondary partitions.

cellulose A polymeric carbohydrate composed of glucose units, formula $(C_6H_{10}O_5)_x$, of which the permanent cell membranes of plants are formed, making it the most abundant carbohydrate.

cell wall A rigid wall outside the cytoplasmic membrane of most plants, most commonly composed of cellulose.

cell-wall degradation matter *humic degradation matter.*

celsian A rare, colorless, monoclinic mineral of the feldspar group: $BaAl_2Si_2O_8$. It is the barium analogue of anorthite and is dimorphous with paracelsian.

celyphytic rim Var. of *kelyphytic rim.*

cement [ore dep] Ore minerals, e.g. gold, that are part of or have replaced *cement* in the sedimentary use of the word.

cement [metal] adj. Said of a metal that is obtained by precipitation, e.g. cement copper precipitated from copper sulfate solution by iron.

cement [sed] Chemically precipitated mineral material that occurs in the spaces among the individual grains of a consolidated sedimentary rock, thereby binding the grains together as a rigid, coherent mass; it may be derived from the sediment or its entrapped waters, or it may be brought in by solution from outside sources. The most common cements are silica (quartz, opal, chalcedony), carbonates (calcite, dolomite, siderite), and various iron oxides; others include barite, gypsum, anhydrite, and pyrite. Clay minerals and other fine clastic particles should not be considered as cements.

cementation [drill] The operation whereby cement slurry is pumped into a drill hole and forced in behind the casing for such purposes as sealing the casing to the walls of the hole (thereby keeping the pipe stationary), preventing unwanted leakage of fluids into the hole or migration of oil or gas from the hole, closing the hole back to a shallower depth, sealing a dry hole, or redrilling to straighten the hole. Syn: *cementing.*

cementation [eng] The injection of cementing materials into fissured or friable rocks or rocks made permeable by the presence of solution channels for the purpose of rendering them relatively watertight or increasing their strength.

cementation [sed] The diagenetic process by which coarse clastic sediments become lithified or consolidated into hard compact rocks through the deposition or precipitation of minerals in the spaces among the individual grains of the sediment. It may occur simultaneously with sedimentation, or the cement may be introduced at a later time. Cementation may occur by secondary enlargement. Syn: *agglutination.*

cementation [soil] The binding together of the particles of soil by such cementing agents as colloidal clay, hydrates of iron or aluminum, or carbonates.

cement clay A clay with a variable amount of calcium carbonate, used in the manufacture of cement.

cement deposits In the Black Hills of the U.S., gold-bearing Cambrian conglomerates believed to be ancient beach and stream-channel deposits.

cement gravel Gravel that is consolidated by some binding material such as clay, silica, or calcite.

cement rock Any rock that is capable of furnishing cement when properly treated or without the addition of other material; specif. a massive, sparsely fossiliferous, clayey limestone that contains the ingredients (alumina, silica, lime) for cement in approximately the required proportions, such as *hydraulic limestone.* Example: the Blackjack Creek Limestone Member of the Fort Scott Limestone in Kansas. Syn: *cement stone.*

cement stone A syn. of *cement rock.* Also spelled: *cementstone.*

cemetary hummock *nonsorted circle.*

cenology An old term for the geology of surficial deposits.

Cenomanian European stage: lowermost Upper Cretaceous or Middle Cretaceous of some authors (above Albian, below Turonian).

Cenophytic A paleobotanic division of geologic time, signifying the time since the development of the angiosperms in the middle or late Cretaceous. Cf: *Aphytic; Archeophytic; Eophytic; Paleophytic; Mesophytic.* Syn: *Neophytic.*

cenosis Var. of *coenosis.*

cenosite *kainosite.*

cenospecies A group of biologic units, e.g. *ecospecies,* which there can be free exchange of genes between the units and only rarely outside the group. It is more or less equivalent to the taxonomic unit *subgenus.* Var: *coenospecies.*

cenote In Yucatán, Mexico, a *pothole* (b) that is a natural well; it is formed by collapse of an underlying cave. Etymol: Maya, *tzonot.*

cenotypal Said of a fine-grained porphyritic igneous rock having the appearance of fresh or nearly fresh extrusive rock, such as those of Tertiary and Holocene age. This term and the term *paleotypal* were introduced to distinguish Tertiary and pre-Tertiary fine-grained igneous rocks; both are now obsolete. Syn: *cenotype; kainotype.*

cenotype *cenotypal.*

Cenozoic An era of geologic time, from the beginning of the Tertiary period to the present. (Some authors do not include the Quaternary, considering it a separate era.) It is characterized paleontologically by the evolution and abundance of mammals, advanced mollusks, and birds; and paleobotanically by angiosperms. The Cenozoic is considered to have begun about 70 million years ago. Also spelled: *Cainozoic; Kainozoic.* See also: *age of mammals.*

cenozone A syn. of *assemblage zone,* suggested by Moore (1957).

cenozoology Zoology of existing animals without regard to those that are extinct.

centare A metric unit of area equal to one square meter, 0.01 are, or 10.76 square feet. Abbrev: ca.

center counter In structural petrology, an instrument used to count the number of points on a point diagram. It consists of a piece of plastic or similar material with a circular hole whose area usually equals 1.0% of that of the diagram. Cf: *peripheral counter.*

centered lattice A crystal lattice in which the axes have been hosen according to the rules for the crystal system, and in which there are lattice points at the centers of certain planes as well as at its corners; thus a centered lattice has two, three, or four lattice points per unit instead of one, as in a primitive lattice. See also: *one face centered lattice; face-centered lattice; body-centered lattice; side-centered lattice; base-centered lattice.*

center line A straight or curved line that continuously bisects feature or figure (such as a stream, a strip of land, or the bubble tube in a spirit level); specif. the line connecting opposite corresponding corners of a quarter section or quarter-quarter section, or the line extending from the true center point of overlapping aerial photographs through each of the transposed center points.

center of gravity That point in a body or system of bodies through which the resultant attraction of gravity acts when the body or system of bodies is in any position; that point from which the body can be suspended or poised in equilibrium in any position.

center-of-gravity map A *vertical-variability map,* or *moment map,* that shows the relative, weighted mean position of a lithologic type in terms of its distance from the top of a given stratigraphic unit, expressed as a percentage of total thickness of the unit. Cf: *standard-deviation map.*

center of instrument The point on the vertical axis of rotation (of a surveying instrument) that is at the same elevation as that of the collimation axis when that axis is in a horizontal position. In a transit or theodolite, it is close to or at the intersection of the horizontal and vertical axes of the instrument.

center of symmetry A point in a crystal structure through which every aspect of an array is repeated by *inversion;* it is a symmetry element. Syn: *inversion center; symmetry center.*

center point (a) The *principal point* of a photograph. (b) The central point from which a map projection is geometrically based.

central body The main part of a pollen grain or spore; e.g. the main part of a vesiculate pollen grain, as distinct from the vesicles, or the compact central part of a dinoflagellate cyst from which the projecting structures extend. Syn: *body* [palyn].

central capsule The mucoid or chitinous, perforated internal skeleton or sac of a radiolarian, enclosing the nucleus and intracapsular cytoplasm.

central cavity A narrow tube-like to broad bowl-shaped interior space enclosed by the inner wall (or rarely by the outer wall alone) in archaeocyathids (TIP, 1955, pt.E, p.6). It may be partly filled with vesicular tissue.

central complex The core or central zone in which foraminiferal chamber passages bifurcate and anastomose in a reticulate pattern (as in Orbitolinidae).

central cylinder *stele.*

centrale (a) A prominent plate at the center of the aboral surface of the disc in many asterozoans. It is the central plate of the primary circlet. Syn: *centrodorsal.* (b) A noncirriferous basal plate typically occurring inside the infrabasal circlet of some articulate crinoids (such as *Marsupites* and *Uintacrinus*).

central granite *batholith.*

centrallasite *gyrolite.*

central meridian The line of longitude at the center of a map projection; the meridian about which the geometric properties of a map projection are symmetric and which is a straight line on the map. It is used to determine the directions of axes of plane coordinates. See also: *principal meridian.*

central mound *central peak.*

central muscle One of anteriorly or medially placed pair of muscles in lingulid brachiopods, originating on the pedicle valve, and passing anteriorly and dorsally to the brachial valve (TIP, 1965, pt.H, p.142).

central peak A central area in a large crater, topographically

higher than the surrounding crater floor. One or more central peaks are commonly observed in the approximate centers of large lunar craters (such as in Copernicus and Tycho); their average height is about half the crater depth. In terrestrial craters produced by impact or by explosion, the central peaks are formed by uplift of material originally below the crater floor. Cf: *central uplift.* Syn: *central mound.*

central projection *gnomonic projection.*

central ring induction method An inductive electromagnetic method in which the transmitting and receiving coils are concentric.

central tendency Any measure or value representative or indicating the center of an entire statistical distribution; e.g. median, mode, and mean.

central uplift A central high area produced in an impact crater or in an explosion crater by inward and upward movement of material below the crater floor. It is formed at a relatively late stage during the crater-forming event and does not appear due to long-term slow adjustment. Central uplifts are characteristic of many cryptoexplosion structures believed to be produced by meteorite impact. Cf: *central peak.*

central valley *rift valley* [marine geol].

central water A surface or near-surface *water mass* in a temperate climate, characterized by high salinity and warm temperature. See also: *equatorial water.*

centric Said of the texture of a rock in which the components are arranged about a center, either radially (e.g. spherulitic) or concentrically (e.g. orbicular).

centric diatom A diatom having basically radial symmetry; a member of the diatom order Centrales. Cf: *pennate diatom.*

centrifugal drainage pattern *radial drainage pattern.*

centrifugal replacement Mineral replacement in which the host mineral is replaced from its center outward. Cf: *centripetal replacement.*

centrifuge moisture equivalent *moisture equivalent.*

centripetal drainage pattern A drainage pattern in which the streams converge more or less radially inward toward a central depression; it may be indicative of a volcanic crater or caldera, a structural basin, a breached dome, a sinkhole, or a bolson. Cf: *radial drainage pattern.* See also: *internal drainage.*

centripetal replacement Mineral replacement in which the host mineral is replaced from its periphery inward. Cf: *centrifugal replacement.*

centripetal selection Natural selection resulting in decreasing variation.

centroclinal adj. Said of strata and structures that dip towards a common center. Ant: *quaquaversal.* Cf: *periclinal.*

centroclinal fold *centrocline.*

centrocline An equidimensional basin characteristic of cratonic areas, in which the strata dip towards a central low point. The term is little used in the U.S. Cf: *pericline.* Ant: *quaquaversal* (n.) Syn: *centroclinal fold.*

centrodorsal n. (a) A commonly cirriferous crinoid columnal, or semifused to fused columnals attached to theca of some articulate crinoids. (b) The *centrale* in many asterozoans.--- adj. (a) Central and dorsal. (b) Pertaining to a centrodorsal; e.g. "centrodorsal cavity" consisting of a depression on the ventral surface of a crinoid centrodorsal and containing chambered organ and accessory structures.

centrogenous skeleton The supporting rods generated at the cell center in acantharian radiolarians.

centronelliform Said of a simple spear-shaped loop of a brachiopod (as in the subfamily Centronellinae), suspended free of valve floor, and commonly bearing median vertical plate in addition to echmidium (TIP, 1965, pt.H, p.142).

centrosphere *barysphere.*

centrosymmetric Said of a crystal having a center of symmetry.

centrotylote A monaxonic sponge spicule (tylote) with a central swelling.

centrum [**seis**] *focus* [seis].

centrum [**paleont**] (a) A differentiated central part of a sponge spicule. (b) The substance of a stem plate (columnal or cirral) of a crinoid, including luminal flanges.---Pl: *centra*.

cenuglomerate (a) A term proposed by Harrington (1946) for a rock resulting from the consolidation of mudflow material. (b) A coarse breccia formed by the accumulation of material resulting from rockfalls, landslides, or mudflows (Dunbar & Rodgers, 1957, p.171).--Etymol: Latin *coenum*, "mud", + *glomerare*, "to wind into a ball".

cephalic Pertaining to the head; esp. directed toward or situated on, in, or near the cephalon, such as the "cephalic shield" of a crustacean exoskeleton covering the head region and formed of fused tergites, or a "cephalic spine" carried by the cephalon of a trilobite.

cephalis The first or apical chamber of a nasselline radiolarian.

cephalon (a) The anterior or head region of an exoskeleton of a trilobite, consisting of several fused segments and bearing the eyes and mouth. (b) The most anterior tagma of a crustacean, bearing eyes, mouth, two pairs of antennae, mandibles, and two pairs of maxillae. See also: *head* [*paleont*].---Pl: *cephala*.

cephalopod Any marine mollusk belonging to the class Cephalopoda, characterized by a definite head, with the mouth surrounded by part of the foot that is modified into lobe-like processes with tentacles or arm-like processes with hooklets or suckers or both. The external shell, if present, as in nautiloids, is univalve and resembles a hollow cone which may be straight, curved, or coiled and divided into chambers connected by a siphuncle; the shell is internal in present-day cephalopods and their fossil ancestors, such as the belemnites. Nautiloids and ammonoids are extinct cephalopods, generally valuable as index fossils; octopuses, squids, and cuttlefishes are common living cephalopods. Cephalopods are known from Cambrian to present.

cephalothorax The fused *head* and *thorax* of certain arthropods; e.g. the anterior part of the body of a crustacean, composed of united cephalic and thoracic somites and covered by the carapace, or the fore part of the body of a merostome in front of the opisthosoma, or the anterior part of the body of an arachnid, bearing six pairs of appendages. Cf: *prosoma; gnathothorax*.

cerargyrite (a) *chlorargyrite*. (b) A group name for isomorphous isometric silver halides that includes mainly chlorargyrite, bromargyrite, and embolite.---Syn: *kerargyrite*.

ceratite Any ammonoid belonging to the order Ceratitida, characterized by a shell having sutures with serrate lobes and, in some groups, by an ornamented shell. Their stratigraphic range is Permian to Triassic.

ceratitic suture A type of suture in ammonoids characterized by small, rounded, unbroken saddles and finely denticulate lobes developed on a major set; specif. a suture in ceratites. Cf: *ammonitic suture; goniatitic suture; pseudoceratitic suture*.

ceratoid Said of a very slenderly conical, horn-shaped corallite of a solitary coral.

ceratolith A horseshoe-shaped skeletal element of the coccolithophorid *Ceratolithus*, acting optically as a single unit of calcite.

cercopod *cercus*.

cercus Either of a pair of simple or segmented appendages situated at the posterior end of certain arthropods; e.g. a *caudal ramus* of a crustacean. Pl: *cerci*. Syn: *cercopod*.

ceresine A white wax which is the product resulting from the bleaching of *ozocerite*.

cerianite A mineral: CeO_2. It usually contains some thorium.

cerine (a) *allanite*. (b) *cerite*.

cerioid Said of a massive corallum in which the walls of adjacent polygonal corallites are closely united.

cerite A mineral: $(Ca,Ce)_3Si_2(O,OH,F)_9$. Syn: *cerine*.

cerolite A yellow or greenish, wax-like mixture of serpentir and stevensite.

Cerozem Var. of *Sierozem*.

cerrito A small *cerro*. Syn: *cerrillo*.

cerro A term used in SW U.S. for a hill, esp. a craggy rocky eminence of moderate height. Etymol: Spanish.

ceruleite A turquoise-blue mineral: $CuAl_4(AsO_4)_2(OH)_8.4H_2O$ Also spelled: *cerulëite*.

cerulene (a) A trade name for a form of calcite colore blue or green by azurite or malachite and used as a gen stone. (b) A term used less correctly for a blue variety satin spar.

cerussite A colorless, white, yellowish, or grayish orthorhom bic mineral of the aragonite group: $PbCO_3$. It is a commo alteration product of galena and is a valuable ore of lead. Syn *white lead ore; lead spar*.

cervantite A white or yellow orthorhombic mineral: Sb_2O_4. was formerly regarded as identical with stibiconite.

cervical sinus An indentation at the front of the carapace of cladoceran crustacean, exposing the rear part of the head.

cesarolite A steel-gray mineral: $H_2PbMn_3O_8$. It occurs spongy masses.

cesium vapor magnetometer A type of *optically pumpe magnetometer* that uses magnetic atoms of cesium. Cf: *rub dium vapor magnetometer*.

ceylonite A dark (dark-green, brown, or black) variety of sp nel containing iron. Syn: *pleonaste; candite; ceylanite; zeyla ite*.

chabazite A zeolite mineral: $CaAl_2Si_4O_{12}.6H_2O$. It sometime contains sodium and potassium. Also spelled: *chabasite*.

chadacryst A syn. of *xenocryst*. Also spelled: *cadacryst*.

chaemolith *humic coal*.

chaetetid Any organism characterized by massive coral composed of very slender aseptate corallites with imperfora walls and complete tabulae. The chaetetids are current placed in the tabulate family Chaetetidae, but have been va iously classified as hydrozoans, anthozoans, bryozoans, an sponges.

chaff peat Peat that is derived from fragments of plants.

chagrenate Said of a smooth and translucent sculpture of po len and spores.

chain [**ore dep**] adj. In mineral deposits, e.g. chromite, said a crystal texture or structure in which a series of connecte crystals resembles a linked or chainlike pattern.

chain [**geomorph**] A general term for any series or sequenc of related, interconnected, or similar natural features arrange more or less longitudinally, such as of lakes, islands, reef dunes, barriers, seamounts, and volcanoes; esp. a *mounta chain* or other extended group of more or less parallel fea tures of high relief, including plateaus.

chain [**surv**] (a) A measuring device used in land surveyin consisting of 100 links (formerly 50 links) joined together b rings; specif. *Gunter's chain*. The term is commonly used i terchangeably with *tape* although strictly a chain is a series links and a tape is a continuous strip. (b) A unit of length pre scribed by law for the survey of U.S. public lands and equal 66 feet or 4 rods. It is a convenient length for land measure ment because 10 square chains equals one acre.

chain coral Any coral (esp. one belonging to the family Haly itidae) characterized, in plan view, by cylindrical, oval, subpolygonal corallites jointed together on two or three side to form a branching, chain-like network.

chain crater One of several small aligned depressions on th surface of the Moon, believed to be formed by either volcan activity or by secondary impacts; more commonly applied those of probable volcanic origin.

chain gage A type of *gage* used in gaging water-surface ele vation, consisting of a tagged or indexed chain, tape, or othe form of line. It is used in situations in which the water surfac difficult to reach. Cf: *staff gage.*

chaining A term that was applied originally to the operation of measuring distances on the ground by means of a surveyor's chain, but later applied to such an operation using either a chain or a surveyor's tape. The term was formerly used synonymously with *taping*, but "chaining" is now preferred (because of historical and legal reasons) for surveys of the U.S. public-lands system and "taping" for all other surveys.

chainman A surveyor's assistant who measures distances, marks measuring points, and performs related duties; specif. one who marks the tape ends in chaining or who measures distances with a tape. See also: *rodman.* Syn: *tapeman.*

chain silicate *inosilicate.*

chalazoidite *mud ball.*

chalcanthite A blue triclinic mineral: $CuSO_4.5H_2O$. It is a minor ore of copper. Syn: *blue vitriol; copper vitriol; bluestone; cyanosite.*

chalcedonic chert A transparent, translucent, vitreous, milky, smoky, waxy, or greasy variety of *smooth chert*, of any color (generally buff or blue-gray, and sometimes mottled) (Ireland et al, 1947, p.1484); it breaks into splintery fragments with smooth conchoidal surfaces. Cf: *novaculitic chert.*

chalcedonite *chalcedony.*

chalcedony (a) A cryptocrystalline variety of quartz. It is commonly microscopically fibrous, may be translucent or semitransparent, and has a nearly wax-like luster, a uniform tint, and a white, pale-blue, gray, brown, or black color; it has a lower density and lower indices of refraction than ordinary quartz. Chalcedony is the material of much chert, and often occurs as an aqueous deposit filling or lining cavities in rocks. In the gem trade, the name refers specif. to the light blue-gray or "common" variety of chalcedony. Varieties include carnelian, sard, chrysoprase, prase, plasma, bloodstone, onyx, and sardonyx. See also: *agate.* Syn: *chalcedonite; chalcedony.* (b) A general name for crystalline silica that forms concretionary masses with radial-fibrous and concentric structure and that is optically negative (unlike true quartz). (c) A trade name for a natural blue onyx.

chalcedony patch A milk-like, semitransparent blemish (sometimes occurring as a flaw) in a ruby.

chalcedonyx An onyx with alternating stripes or bands of gray and white. It is valued as a semiprecious stone.

chalchihuitl A Mexican term for any green stone that has been carved into a decorative or useful object, and sometimes any stone (regardless of color) that has been carved. It refers esp. to jadeite or chalchuite (turquoise), but sometimes to porphyry, serpentine, or smithsonite. Syn: *chalchihuite; chalchiguite.*

chalchuite A blue or green variety of turquoise.

chalcoalumite A turquoise-green to pale-blue mineral: $CuAl_4(SO_4)(OH)_{12}.3H_2O.$

chalcocite A black or dark lead-gray mineral: Cu_2S. It has a metallic luster, occurs in orthorhombic crystals or massive, and is an important ore of copper. Syn: *copper glance; chalcosine; redruthite; beta chalcocite; vitreous copper.*

chalcocyanite A white (not blue) mineral: $CuSO_4$. Syn: *hydrocyanite.*

chalcodite *stilpnomelane.*

chalcolite *torbernite.*

chalcolithic *Copper Age.*

chalcomenite A blue mineral: $CuSeO_3.2H_2O.$

chalconatronite A mineral: $Na_2Cu(CO_3)_2.3H_2O$. It occurs as greenish-blue incrustations on ancient bronze objects from Egypt.

chalcopentlandite A hypothetical high-temperature sulfide of copper, nickel, and iron, now represented by mixtures of pentlandite and chalcopyrite.

chalcophanite A black mineral: $(Zn,Mn,Fe)Mn_2O_5.nH_2O.$

chalcophile (a) Said of an element enriched in the sulfide rather than in the metallic and silicate phases of meteorites,

and that is probably enriched in the Earth's mantle relative to its crust and core (in Goldschmidt's tripartite scheme of element partition in the solid Earth). Cf: *lithophile; siderophile.* (b) Said of an element tending to concentrate in sulfide minerals and ores. Such elements have intermediate electrode potentials and are soluble in iron monosulfide.----(Goldschmidt, 1954, p.24; Krauskopf, 1967, p.580). Examples are: S, Se, As, Fe, Pb, Zn, Cd, Cu, Ag.

chalcophyllite A green mineral: $Cu_{18}Al_2(AsO_4)_3(SO_4)_3(OH)_{27}.33H_2O.$ Syn: *copper mica.*

chalcopyrite A bright brass-yellow tetragonal mineral: $CuFeS_2$. It is generally found massive and constitutes the most important ore of copper. Syn: *copper pyrites; yellow copper ore; yellow pyrites; fool's gold.*

chalcosiderite A green mineral: $Cu(Fe,Al)_6(PO_4)_4(OH)_8.4H_2O.$ It is isomorphous with turquoise.

chalcosine *chalcocite.*

chalcosphere That zone or layer of the Earth containing heavy-metal oxides and sulfides; it is the equivalent of *stereosphere.*

chalcostibite A lead-gray mineral: $CuSbS_2$. Syn: *wolfsbergite.*

chalcothallite A mineral: $Cu_3TlS_2.$

chalcotrichite A capillary variety of cuprite occurring in fine, straight, slender, interlacing fibrous crystals. Syn: *plush copper ore; hair copper.*

chalite A term, now obsolete, proposed by Pinkerton (1811, v.2, p.100) for a conglomerate containing pebbles intermediate between flint and chalcedony.

Chalk A stratigraphic term used in NW Europe for Upper Cretaceous. In Great Britain, it is divided into Lower Chalk (Cenomanian), Middle Chalk (Turonian), and Upper Chalk (Senonian).

chalk (a) A soft, pure, earthy, fine-textured, usually white to light gray or buff limestone of marine origin, consisting almost wholly (90-99%) of calcite, formed mainly by shallow-water accumulation of calcareous tests of floating microorganisms (chiefly foraminifers) and of comminuted remains of calcareous algae (such as coccoliths and rhabdoliths) set in a structureless matrix of very finely crystalline calcite (some of which may have been chemically precipitated), and marked by a porous, somewhat friable, and unindurated or slightly coherent character. It often includes remains of bottom-dwelling forms (such as of ammonites, echinoderms, and pelecypods); other chalk deposits may be almost devoid of organic remains. It is an unaltered deposit, although it may contain nodules of chert and pyrite. Best known and most widely distributed chalks are of Cretaceous age, such as the thick deposits exposed in cliffs on both sides of the English Channel. Syn: *creta.* (b) A white, pure (or nearly pure), natural calcium carbonate, breaking into crumbly or powdery pieces. (c) *chalk rock.*---Etymol: Old English *cealc,* from Latin *calx,* "lime".

chalkland A region underlain by chalk deposits, characterized by rolling hills, undulating plateaus, open expanses of pastureland, and dry valleys; e.g. SE England.

chalk rock (a) A soft, milky-colored rock resembling white chalk, such as talc, calcareous tufa, diatomaceous shale, volcanic tuff, or a bed of white limestone. Syn: *chalk.* (b) A chalky rock; specif. the Chalk Rock, a bed of hard nodular chalk, sometimes containing green-colored calcareous or phosphatic nodules, occurring at or near the base of the Upper Chalk in England (Himus, 1954, p.24).

chalk stream A stream flowing across or among the strata of a chalk deposit; a *bourne.*

chalky (a) Said of a soil or rock consisting of, rich in, or characterized by chalk. Syn: *cretaceous.* (b) Said of a limestone having the appearance of chalk. (c) Said of the porosity of such finely textured rocks as chalk and marl.

chalky chert A commonly dull or earthy, soft to hard, sometimes finely porous chert (insoluble residue) of essentially uniform composition, having an uneven or rough fracture surface, and resembling chalk (Ireland et al, 1947, p.1487). Cf: *smooth chert; granular chert*. Syn: *dead chert; cotton chert*.

chalky marl A grayish marly rock rich in chalk and containing up to 30% clayey material; specif. the Chalk Marl, the stratigraphic horizon near the base of the English Chalk.

chalmersite *cubanite*.

chalybeate An adj. applied to water strongly flavored with iron salts or to a spring yielding such water. The term is approaching obsolescence in the U.S. Etymol: Greek, an ancient tribe of ironworkers in Asia Minor.

chalybite British syn. of *siderite* (ferrous-carbonate mineral).

chamber [speleo] A *room* of a cave; in American usage it has the implication of large size. Syn: *hall*.

chamber [paleont] (a) The fundamental unit of a foraminiferal test, consisting of a cavity and its surrounding walls. It is a variously shaped enclosure that invariably is connected by pores, intercameral foramina, or other passages leading to similar enclosures or to the exterior. (b) One of the regular, juxtaposed, hollow structures formed by the skeleton of sphinctozoan sponges. Also, a term that is often used an an abbreviated form of *flagellated chamber*; this usage is not recommended. (c) A *camera* of a cephalopod. Also, the *body chamber* of a cephalopod.

chambered Said of a vein or lode of brecciated, irregular texture, e.g. *stockwork*.

chambered level A spirit-level tube with a partition near one end which cuts off a small air reservoir, so arranged that the length of the bubble can be regulated.

chamberlet A small chamber in a foraminifer, created by subdivision of chambers by axial or transverse secondary septula.

chamber passage One of the radial corridors consisting of centrally directed extensions of marginal chamberlets of foraminifers (such as Orbitolinidae).

chambersite An orthorhombic mineral: $Mn_3B_7O_{13}Cl$.

chameolith *humic coal*.

chamosite A greenish-gray or black mineral of the chlorite group: $(Mg,Fe)_3Fe_3^{+3}(AlSi_3)O_{10}(OH)_8$. It is the monoclinic dimorph of orthochamosite, and is an important constituent of many oolitic and other bedded iron ores. See also: *berthierine*.

Champlainian (a) North American provincial series: Middle Ordovician (above Canadian, below Cincinnatian). (b) Obsolete syn of *Ordovician*.

chance packing A random combination of systematically packed grains surrounded by or alternating with haphazardly packed grains (Graton & Fraser, 1935). The average porosity of a chance-packed aggregate of uniform solid spheres is slightly less than 40%.

Chandler motion *polar wandering*.

Chandler wobble An aspect of the Earth's rigid body motion departing from simple or pure spin, due to the fact that its angular momentum vector is not precisely colinear with a principle axis of inertia. The free nutation, of period 428 days, results in a variation of instantaneous astronomical latitude (as defined in accordance with an instantaneous angular velocity vector or axis of rotation) of amplitude about 0.4 sec-arc.

change of color (a) An optical phenomenon consisting of a difference in color when a mineral or gemstone is moved about; specif. *labradorescence*. (b) An optical phenomenon consisting of a difference in color from daylight to artificial light, caused by selective absorption; e.g. that shown by alexandrite.--Cf: *play of color*.

channel [drill] A cavity in a faulty cement job behind the casing in a borehole.

channel [ore dep] *channelway*.

channel [coast] (a) A relatively narrow sea or stretch of water, wider and larger than a *strait*, between two close landmasses and connecting two larger bodies of water (usually seas); e.g. the English Channel between England and France. (b) The deeper part of a moving body of water (as a bay, estuary, or strait) through which the main current flows or

which affords the best passage through an area otherwise to shallow for navigation; it is often deepened by dredging. Also a navigable waterway between islands or other obstruction as on a lake.

channel [paleont] A groove of an invertebrate, such as on that winds down the columella near its base in some gastropod shells and terminates in the siphonal notch or in th canal.

channel [sed struc] (a) A linear current mark, larger than *groove*, produced on a sedimentary surface, parallel to th current, and often preserved as a *channel cast*. It is 0.5-2 wide, 20-50 cm deep, and up to 30 m long, and is best developed in a turbidite sequence. (b) An erosional feature "tha may be meandering and branching and is part of an integrate transport system" (Pettijohn & Potter, 1964, p.288).

channel [volc] A narrow, sinuous flow channel, common formed in aa flows, that is several meters in depth. Suc channels may eventually be occupied by streams.

channel [ice] *lead*.

channel [streams] (a) The hollow bed where a natural body surface water flows or may flow; a natural passageway or de pression of perceptible extent containing continuously or pe riodically flowing water, or forming a connecting link betwee two bodies of water; a watercourse. Syn: *channelway*. (b) Th deepest or central part of the bed of a stream, containing th main current, and occupied more or less continuously b water; the *thalweg*. (c) A term used in quantitativ geomorphology for a line or pattern of lines, without regard width or depth, in the analysis of streams. Syn: *stream*. (c An abandoned or buried watercourse represented by strea deposits of gravel and sand. (e) An artificial waterway, suc as an open conduit, an irrigation ditch or canal, or a floodwa (f) An obsolete term for a stream or small river.

channel bar An elongate deposit of sand and gravel located the course of a stream, esp. of a braided stream. Cf: *poi bar*.

channel basin An obsolete term for a long, narrow, proglacia valley, trench, or channel.

channel capacity The maximum flow that a given channel ca transmit without overflowing its banks. See also: *bankfu stage*.

channel cast The cast of a *channel* that is generally cut shale and filled with sand. Cf: *washout*. Syn: *channel fil gouge channel*.

channeled scabland *Scabland* deeply eroded. On the Columbi Plateau of eastern Washington it represents intense scourin by glacial meltwater.

channeled upland *grooved upland*.

channel erosion Erosion in which material is removed b water flowing in well-defined channels; erosion caused b channel flow. Cf: *sheet erosion; rill erosion; gully erosion*.

channel fill (a) An alluvial deposit in a stream channel, esp one in an abandoned cutoff channel or where the transportin capacity of the stream is insufficient to remove material sup plied to it. (b) *channel cast*.---Syn: *channel filling*.

channel flow Movement of surface runoff in a long, narrow trough-like depression bounded by banks or valley walls tha slope toward the channel; specif. *streamflow*. Cf: *overlan flow*. Syn: *concentrated flow*.

channel frequency *stream frequency*.

channel geometry The description of the shape (form) of given cross section within a limited reach of a river channel See also: *river morphology*.

channel-gradient ratio *stream-gradient ratio*.

channel length *stream length*.

channel line The line of the fastest current or the stronges flow of a stream; it generally coincides with (and is some times known as) the *thalweg*. Cf: *thread*.

channel maintenance constant Ratio of the area of a drainag basin to the total stream lengths of all the stream orders with the basin (Schumm, 1956, p.607); the reciprocal of the *drainage density*. It expresses the minimum limiting area required for the development of a drainage channel. Symbol: C.

channel morphology *river morphology.*

channel-mouth bar A bar built where a stream enters a body of standing water; it results from a decrease in the stream's velocity.

channel net The pattern of all stream channels within a drainage basin.

channel order *stream order.*

channel pattern The configuration in plan view of a limited reach of a river channel as seen from an airplane (Leopold & Wolman, 1957, p.39-40). Recognized patterns include meandering, braided, sinuous, and relatively straight channels. See also: *river morphology.* Syn: *river pattern.*

channel porosity A system of pores in which the openings are markedly elongate and have developed independently of the textural or fabric elements of the rock (Choquette & Pray, 1970, p.245).

channel precipitation Part of direct runoff; precipitation that falls directly onto lake and stream waters. It is usually considered with surface runoff (Chow, 1964, p.14-2).

channel recording A system, chain, or cascade of interconnected devices through which geophysical data may flow from source to recorder, for example, geophone, amplifier (with filters and gain control), galvanometer, and optical system.

channel sand A sand or sandstone deposited in a stream bed or some other channel eroded into the underlying bed, and frequently containing oil, gas, gold, or other valuable minerals.

channel segment *stream segment.*

channel splay *flood-plain splay.*

channel spring A type of depression spring issuing from the bank of a stream that has cut its channel below the water table.

channel storage In a stream channel, or over its flood plains, the volume of water at a given moment.

channel wave A type of *guided wave* that is propagated in a low-velocity layer within the Earth or in the ocean or atmosphere. Cf: *Stoneley wave.*

channelway [ore dep] An opening or passage in a rock through which mineral-bearing solutions or gases may move. Syn: *channel [ore dep]; feeder; feeding channel.*

channelway [streams] *channel.*

channel width The distance across a channel or a stream, measured from bank to bank near bankfull stage. Symbol: w.

channery (a) Thin, flat *coarse fragments* of limestone, sandstone, or schist, having diameters as large as 150 mm (6 in.). (b) A term used in Scotland and Ireland for gravel.

chaoite A mineral: C.

chaos [cosmol] The disorganized state of primordial matter and infinite space before the ordered universe was created. Ant: *cosmos.*

chaos [geol] A structural term proposed by Noble (1941, p.963-977) for a gigantic breccia associated with thrusting, consisting of a mass of large and small blocks of irregular shape collected together (with very little fine-grained material) in a state of semidisorder. Type example: the Amargosa chaos, a widespread deposit in the Death Valley area of California, consisting of an extremely complex mosaic of enormous, tightly packed, often unshattered but internally coherent, random blocks and masses of formations of different ages occupying a definite zone above a major thrust fault. The blocks range in size from pods a meter in diameter to blocks more than 800 m in length. Cf: *mélange; megabreccia.*

chaotic [geomorph] Said of a surface or land area consisting of short, jumbled ridges and valleys.

chaotic [petrology] Said of a massive, unstratified tuff consisting of a mixture of equally distributed fine and coarse material such as the deposit of a glowing cloud).

chapeau de fer *gossan.*

chapeiro An isolated coral reef growing in small scattered patches, often rising like a tower to a height of 12-15 m, and sometimes spreading out in a mushroom-like top, as off the coast of Brazil. Etymol: Portuguese *chapeirão*, "broad-brimmed hat".

chapmanite A mineral: $Fe_2Sb(SiO_4)_2(OH)$.

char A term applied in India to a newly formed alluvial tract or flood-plain island formed of silt and sand deposited in the bed of a deltaic river, such as a sandbank left dry on the subsidence of a river after the flood season. Etymol: Hindu. Syn: chur; diara.

character [seis] The recognizable aspect of a seismic event, usually in the waveform, which distinguishes it from other events. It is usually a frequency or phasing effect, and is often not defined precisely and hence is dependent on subjective judgment.

characteristic fossil A fossil species or genus that is characteristic of a stratigraphic unit (formation, zone, series, etc.) or time unit. It is either confined to the unit or is particularly abundant in it. Inappropriate syn: *index fossil.* Syn: *diagnostic fossil.*

characteristic impedance *acoustic impedance.*

characterizing accessory mineral *varietal mineral.*

charco (a) A term applied in SW U.S. to a small, natural depression in which water collects, as in a desert alluvial plain; a *tank* or a water hole. Syn: *represo.* (b) A natural or artificial pool of water occupying a charco and supplied by desert floods. Also, a pool in a stream bed or a puddle in a playa.--- Etymol: Spanish, "pond, small lake".

charcoal An impure carbon residue of the burning of wood or other organic material in the absence of air. It is black, often porous, and able to absorb gases. Like *coke*, it can be used as a fuel.

charge The sediment that is carried into a channel, expressed as the ratio of the volume of sediment passing across a given cross section or portion of cross section of the channel in unit time to the portion of cross section of the channel in unit time (ASCE, 1962).

chargeability The primary unit of measurement in time-domain induced polarization surveys. It is the area under the decay curve between two delay times after cessation of the transmitted current. Usually expressed in millivolt-seconds per volt. The millivolts are a measure of the decay potential and the volts a measure of primary potential. Symbol: μ.

Charmouthian Stage in Great Britain: Lower Jurassic (above Sinemurian, below Domerian).

Charnian orogeny An orogeny that supposedly occurred late in Precambrian time in the English Midlands; the dating is questionable, however, and the term has only local significance. The Charnian folds trend NW-SE and seem to have had a posthumous influence on structures in the surrounding Paleozoic. It is named for Charnwood Forest, Leicestershire, where small inliers of Precambrian sediments emerge.

charnockite A hypersthene-bearing granite. Most classifications require that quartz constitute at least 20% of the felsic constituents and that the ratio of alkali feldspar to total feldspar fall between 40-90%. Tobi (1971) places the quartz content at 10-60% and the ratio of alkali feldspar to total feldspar at 35-90% to correspond with Streckeisen's (1967) definition of granite. Although the origin (igneous or metamorphic) of charnockite is controversial, charnockites are commonly found only in granulite facies terranes, and high temperature and pressure are generally thought to be essential to their formation. The name is derived from that of Job Charnock (d.1693), the founder of Calcutta, India, from whose tombstone the rock was first described.

charnockite series A series of plutonic rocks compositionally similar to the granitic rock series but characterized by the presence of orthopyroxene.

charophyte Said of a group of green algae corresponding to the order Charales and comprising the *stoneworts*.

chart (a) A special-purpose map; esp. one designed for purposes of navigation, such as a *hydrographic chart* or a *bathymetric chart*. (b) A base map conveying information about something other than the purely geographic. (c) *weather map*. (d) Obsolete syn. of *map*.

chart datum A standard water surface, usually low water, from which depths of soundings or tide heights are referenced. When based on the tide, it may be called a *tidal datum*.

chartology *cartography*.

chartometer An instrument for measuring distances on charts or maps, such as the length of a stream in a drainage basin on a topographic map. See also: *opisometer*. Syn: *map measurer*.

chasm (a) A deep breach, cleft, or opening in the Earth's surface, such as a yawning fissure, a narrow and steep-sided gorge or canyon, or a precipitous and impassable cavity or gap; e.g. the Ausable Chasm near Keeseville, N.Y. (b) A deep recess extending below the floor of a cave.---Syn: *abyss*.

chasmophyte A plant growing in the crevices of a rock; a *saxifragous* plant.

chassignite An achondritic stony meteorite composed almost entirely (95%) of olivine, with accessory amounts of chromite, and lacking nickel-iron. It resembles terrestrial dunite.

chathamite A variety of nickel-skutterudite containing much iron.

chatoyancy An optical phenomenon, possessed by certain minerals in reflected light, in which a movable wavy or silky sheen is concentrated in a narrow band of light that changes its position as the mineral is turned. It results from the reflection of light from minute, parallel fibers, cavities or tubes, or needle-like inclusions within the mineral. The effect may be seen on a cabochon-cut gemstone, either distinct and well-defined (as the narrow, light-colored streak in a fine chrysoberyl cat's-eye) or less distinct (as in the usual tourmaline or beryl cat's-eye). Syn: *chatoyance*.

chatoyant adj. Said of a mineral or gemstone possessing chatoyancy or having a changeable luster or color marked by a narrow band of light.--n. A chatoyant gem.

chattermark [beach] A crescent-shaped mark on a wave-worn pebble, such as flint, caused by "hammering" of a beach by wave action.

chattermark [glac geol] One of a series of small, densely packed, short curved scars or cracks (smaller than a crescentic fracture) made by vibratory chipping of a firm but brittle bedrock surface by rock fragments carried in the base of a glacier. Each mark is roughly transverse to the direction of ice movement (although a succession of such marks is parallel to that direction), and usually convex toward the direction from which the ice moved (its "horns" point in the direction of ice movement). The term has been applied loosely to any glacial crescentic mark. Also spelled: *chatter mark*.

chattermark Any mark, pit or scratch made on a rock surface by the surface of a mass that moves over it. Chattermarks can be caused by the material embedded in the bottom of a glacier, or they can occur on a fault surface. Cf: *slip-scratch; vibration mark*.

Chattian European stage: uppermost Oligocene (above Rupelian, below Aquitanian of Miocene). Syn: *Casselian*.

Chautauquan North American provincial series: upper Upper Devonian (above Senecan, below Kinderhookian of Mississippian).

Chayes point counter An instrument used for petrographic modal analysis. A pattern of regularly spaced traverses, along which are regularly spaced points, is placed over a thin section; at each point the mineral is identified and then mechanically tabulated.

Chazyan North American stage: Middle Ordovician (below

Mohawkian, above Lower Ordovician).

check dam A dam designed to retard the flow of water in channel, used esp. for controlling soil erosion.

checkerboard topography A term used by Hobbs (190 p.150) for a topography characterized by a repeating patte in the relief, such as the "diaper pattern of rectangles" due a fracture system as seen on the topographic map of the E zabethtown Quadrangle, Adirondack Mountains, N.Y. (Hobb 1911b, p.131 & plate 9).

checker coal Rectangular grains of anthracite.

cheek One of the two lateral or pleural parts of the cephalo of a trilobite, anterior to and typically much lower and flatte than the glabella. See also: *fixed cheek; free cheek*. Sy *gena*.

cheeswring (a) A mushroom-shaped rock with a narrow ste and overhanging upper block; e.g. the Cheesewring on Bo min Moor in Cornwall, Eng., a granite tor resembling an in verted muslin bag in which "curds from sour milk were one put by country people and the moisture wrung out so as leave a white cream cheese" (Stamp, 1961, p.102). (b) *gara*.

cheilostome Any ectoproct bryozoan belonging to the ord Cheilostomata and characterized by the presence of a mov ble operculum over the orifice of the zooecia. Adj: *cheiloste matous*.

cheirographic coast "A coast of folded and faulted region with complex submergences. It shows a series of deep gul and fingerlike promontories" (Swayne, 1956, p.33). Stam (1961, p.102) used "cheiragratic coast" and noted that th term is "apparently obsolete".

chela (a) The pincer-like claw or organ borne by certain the limbs of arthropods; e.g. the distal part of a crustacea limb consisting of opposed movable and immovable finger and usually involving dactylus and propodus, or the pincer an arachnid appendage formed by a rigid process of the per ultimate joint and a movable last joint. (b) A siliceou monaxonic sponge spicule (microsclere) consisting of an a cuate shaft at each end of which is a recurved, cup-like e pansion, either lobed or toothed. See also: *isochela; anis chela*.---Pl: *chelae*.

chelation Retention of a metallic ion by two atoms of a singl organic molecule. An example is magnesium being retaine by heme in hemoglobin.

chelicera One of the preoral appendages of all Chelicerat (subphylum of Arthropoda), corresponding to the second a tennae of crustaceans, but modified for piercing or biting, an composed of two or three segments (as in arachnids) or three or four(?) joints with the distal ones forming a chela (a in merostomes). Pl: *chelicerae*. Cf: *antenna*.

chelicerate Any terrestrial (*arachnid*) or aquatic (*merostome* anthropod belonging to the subphylum Chelicerata, characte ized chiefly by paired preoral appendages. Their stratigraphi range is Cambrian to present. Cf: *pycnogonid*.

cheliped Any thoracopod bearing chelae; e.g. one of the pa of legs that bears the large chelae in decapod crustaceans.

chelkarite A mineral: $CaMgB_2O_4Cl_2 \cdot 7H_2O$ (?).

chelogenic Said of a cycle of continental evolution; shield forming.

cheluviation *Eluviation* under the influence of chelating agents

chemawinite A pale-yellow to dark-brown variety of retinit found in decayed wood at Cedar Lake in Manitoba. Syn: *ceda rite*.

chemical activity *activity* [chem].

chemical composition [mineral] A chemical formula giving th relative number of atoms present in a particular mineral. Syr *composition* [mineral].

chemical composition [petrology] The weight percent of th elements (generally expressed as certain oxide molecules composing a rock. Syn: *composition* [petrology].

chemical demagnetization A technique of partial *demagnetiza tion* involving treatment by acid or other reagents to selective ly remove one magnetically ordered mineral while leaving the remanent magnetization of another unaffected. Cf: *alternating field demagnetization; thermal demagnetization.*

chemical equilibrium A state of balance between two opposing chemical reactions. The amount of any substance being built up is exactly counterbalanced by the amount being used up in the other reaction, so that concentrations of all participating substances remain constant.

chemical erosion *corrosion.*

chemical exfoliation A type of *exfoliation* caused by a volume increase induced by changes in the bulk chemical composition of the rock.

chemical gaging A type of *stream gaging* in which velocity of flow is measured by introducing a chemical of known saturation into the stream and then measuring the amount of dilution.

chemical limestone A limestone formed by direct chemical precipitation or by consolidation of calcareous ooze.

chemical magnetization *chemical remanent magnetization.*

chemical mining The extraction of valuable constituents of an orebody, either in place or within the confines of a mine, by chemical methods such as leaching (Johnson & Bhappu, 1969). See also: *solution mining.*

chemical oceanography The study of the chemistry of ocean water: its dissolved and suspended material, its acidity, and the geographic and temporal variation of its chemical features.

chemical oxygen demand The amount of oxygen required for the oxidation of all oxidizable compounds in a water body. Cf: *biochemical oxygen demand.* Var: *oxygen demand.*

chemical potential (a) In chemistry, an intensive quantity of a component that is defined as being equal to the change of the Gibbs free energy of the system with the change in the number of moles mi, the temperature, pressure, and number of moles of the other components being kept constant. It is defined at each point of the system. (b) Partial molal free energy, usually symbolized by μ; the increase in energy of a system (due to an infinitesimal addition of an element or compound, without affecting thermal and mechanical energies) divided by the amount of the element or compound added.

chemical remanence *chemical remanent magnetization.*

chemical remanent magnetization A stable remanent magnetization caused by the slow growth of magnetically ordered mineral grains in the presence of a magnetic field, e.g. during such processes as oxidation, reduction, exsolution. Syn: *chemical remanence; chemical magnetization; crystallization remanent magnetization; crystallization magnetization.* Abbrev: *CRM.*

chemical residue A *residue* formed by chemical weathering in place; e.g. a deposit of sand resulting from the removal by solution of nitrates from a Chilean niter bed.

chemical rock (a) A sedimentary rock composed primarily of material, either organic or inorganic, formed directly by precipitation from solution or colloidal suspension (as by evaporation) or by the deposition of insoluble precipitates (as by mixing solutions of two soluble salts); e.g. gypsum, rock salt, chert, tufa, and most limestones. It usually has a nonclastic (frequently crystalline) texture. (b) A sedimentary rock having less than 50% detrital material (Krynine, 1948, p.134). Cf: *detrital rock.*

chemical unconformity An unconformity or stratigraphic boundary determined by chemical analysis, such as in the case of a limestone formation whose basal part has a higher concentration of impurities (silica, magnesia, sulfur) due to the presence of inwashed fine clastic and organic material (Landes, 1957).

chemical water Water that is part of the chemical formula of a mineral. Syn: *hydration water.*

chemical weathering The process of weathering by which chemical reactions (such as hydrolysis, hydration, oxidation,

carbonation, ion exchange, and solution) transform rocks and minerals into new chemical combinations that will remain indefinitely without further change under conditions prevailing at or near the Earth's surface; e.g. the alteration of orthoclase to kaolinite, or the solution of the calcium carbonate in limestone by carbonic acid derived from rainwater containing carbon dioxide. Cf: *mechanical weathering.* Syn: *decomposition; decay.*

chemoautotrophic Said of an organism that obtains nourishment from chemical reactions of inorganic substances. Syn: *chemotrophic.*

chemocline The boundary between the circulating and the noncirculating water masses or layers of a lake; specif. the boundary separating the *mixolimnion* and the *monimolimnion* in a meromictic lake.

chemofacies A term used by Keith & Degens (1959, p.40) to designate "all the chemical elements that are collected, precipitated, or adsorbed from the aqueous environment or fixed by chemical reactions within the bottom muds" and intended "only as a convenience in discussing chemical differences among environmental groups of sediments" (as in differentiating between marine and freshwater sediments).

chemogenic Said of a rock or mineral that was deposited directly from solution without biological mediation, e.g. travertine, in contrast to clastic, bioclastic, or organogenic limestones.

chemolithotrophic Said of an organism that obtains its nourishment by oxidation of inorganic compounds. Cf: *chemoorganotrophic.*

chemoorganotrophic Said of an organism that obtains its nourishment by the oxidation of organic compounds. Cf: *chemolithotrophic.*

chemotaxis Taxis resulting from chemical stimuli. Cf: *chemotropism.*

chemotrophic *chemoautotrophic.*

chemotropism Tropism resulting from chemical stimuli. Cf: *chemotaxis.*

Chemungian North American stage: Upper Devonian (above Fingerlakesian, below Cassadagan).

chenevixite A dark-green to greenish-yellow mineral: $Cu_2Fe_2(AsO_4)_2(OH)_4 \cdot H_2O$ (?).

chenier A long, low (3-6 m high), narrow, wooded *beach ridge* or sandy hummock, forming roughly parallel to a prograding shoreline seaward of marsh and mud-flat deposits (such as along the coast of S Louisiana), enclosed on the seaward side by fine-grained sediments, and resting on peat or clay. It is well drained and fertile, often supporting large evergreen oaks or pines on higher areas; its width varies from 45 to 450 m and its length may be several tens of kilometers. Etymol: French *chêne,* "oak". Obsolete syn: *cheniere.*

chenier plain A broad plain of marshes and swamps along an open seashore, characterized by a series of cheniers, such as the 32-km wide plain extending 175 km along the coast of SW Louisiana.

cheralite A green monoclinic mineral: $(Ca,Ce,Th)(P,Si)O_4$. It is isostructural with monazite, and is essentially an intermediate member of a solid-solution series apparently extending between $CePO_4$ (monazite) and $CaTh(PO_4)_2$ (an artificial compound).

cherker A term used in Forest of Dean (SW England) for an ironstone nodule.

chernovite A mineral: $YAsO_4$.

Chernozem In early U.S. classification systems, a group of zonal soils whose surface horizon is dark and highly organic, below which is a lighter-colored horizon and an accumulation of lime. It is developed under conditions of temperate to cool, subhumid climate. Etymol: Russian *tschernosem,* "black earth". Also spelled: *Chernozyom; Tchornozem; Tschernosiom; Tschernosem.* Cf: *Chestnut soil.* Partial Syn: *black earth.*

Chernozyom Var. of *Chernozem.*

121

cherokite The dense, hard, brown or drab, silicified, residual sand constituting the cement of the chert breccias in the zinc-mining district of Joplin, Missouri.

cherry coal A soft, black, noncaking bituminous coal with a resinous luster that ignites and burns readily.

chert A hard, extremely dense or compact, dull to semi-vitreous, cryptocrystalline sedimentary rock, consisting dominantly of cryptocrystalline silica (chiefly fibrous chalcedony) with lesser amounts of micro- or crypto-crystalline quartz and amorphous silica (opal); it sometimes contains impurities such as calcite, iron oxide, and the remains of siliceous and other organisms. It has a tough, splintery to conchoidal fracture, and may be white or variously colored gray, green, blue, pink, red, yellow, brown, and black. Chert occurs principally as nodular or concretionary segregations (*chert nodules*) in limestones and dolomites, and less commonly as areally extensive layered deposits (*bedded chert*); it may be an original organic or inorganic precipitate or a replacement product. The term *flint* is essentially synonymous, although it has been used for the dark variety of chert (Tarr, 1938). Syn: *hornstone; white chert; silexite; phthanite.*

chert-arenite (a) A term used by McBride (1963, p.668) for a quartzarenite containing more than 25% chert. (b) A term used by Folk (1968, p.124) for a litharenite in which the main rock fragment is chert. It may have any clay content, sorting, rounding, or particle size.

chertification A type of silicification in which fine-grained quartz or chalcedony is introduced into limestones, as in the Tri-State mining district of the Mississippi Valley (Fowler & Lyden, 1932).

chert nodule A dense, very irregular, usually structureless, commonly fossiliferous, diagenetic segregation of *chert*, varying from regular disks up to 5 cm in diameter to large, highly irregular, tuberous bodies up to 30 cm in length, frequently occurring distributed through calcareous strata. The larger nodules, of rounded contour, are marked by warty or knobby extensions. Examples include the cherts in the Mississippian limestones of the upper Mississippi Valley region, and the flint nodules of the Cretaceous chalk of England and France. See also: *nodular chert.*

cherty Containing chert; e.g. a "cherty limestone" so siliceous as to be worthless for the limekiln, or a "cherty iron carbonate" consisting of siderite intimately interbedded with chert.

chervetite A monoclinic mineral: $Pb_2V_2O_7$.

chessman spicule *discorhabd.*

chessylite A term commonly used in France and elsewhere for *azurite.* Syn: *chessy copper.*

Chesterian North American provincial series: uppermost Mississippian (above Meramecian, below Morrowan of Pennsylvanian). Syn: *Chesteran.*

chesterlite Microcline feldspar from Chester County, Penna.

chestnut coal (a) A size of anthracite that will pass through a 1 5/8 inch round mesh but not through a 1 3/16 inch round mesh. Cf: *broken coal; egg coal; stove coal; pea coal; buckwheat coal.* (b) *nut coal;* a bituminous coal size.

Chestnut soil In early U.S. classification systems, a group of zonal soils having a dark brown surface horizon, below which is a lighter-colored horizon and an accumulation of lime. It is developed under conditions of temperate to cool, subhumid to semiarid climate, i.e. under slightly more arid conditions than that of *Chernozem.* Its characteristic vegetation is mixed tall and short grasses. Cf: *Reddish Chestnut soil.*

chevee A flat gem with a smooth, concave depression. Cf: *cuvette.*

chevron cast The cast of a chevron mark.

chevron cross-bedding Cross-bedding that dips in different or opposite directions in alternating or superimposed beds, forming a chevron or herringbone pattern. Syn: *herringbone cross-bedding; zigzag cross-bedding.*

chevron dune A V-shaped dune formed in a vegetated area where strong winds blow in a constant direction.

chevron fold An accordion fold, the limbs of which are of equal length. Cf: *zigzag fold.* Syn: *kink fold.*

chevron groove A V-shaped furrow on the cardinal area for the insertion of ligament in certain bivalve mollusks (as in some of the superfamily Arcacea and in early forms of the superfamilies Pteriacea and Pectinacea).

chevron mark One of a series of tool marks consisting of chevron-like depressions arranged in a row, the points of the chevrons presumably pointing upstream. Cf: *vibration mark.* Syn: *herringbone mark.*

Chézy equation An equation used to compute the velocity of uniform flow in an open channel: mean velocity of flow (V) equals the Chézy coefficient (C) times the square root of the product of hydraulic radius in feet (R) times the slope of the channel (S). Cf: *Manning equation.* See also: *Kutter's formula.*

chiastoclone A desma (of a sponge) in which several subequal, zygome-bearing arms radiate from a very short central shaft, giving the spicule an X-shaped profile.

chiastolite An opaque variety of *andalusite* containing black carbonaceous impurities arranged in a regular manner so that a section normal to the longer axis of the crystal shows a black Maltese cross formed as a result of the pushing aside of the impurities into definite areas as the crystal grew in metamorphosed shales. It has long been used for amulets, charms, and other inexpensive novelty jewelry. Syn: *crossstone; crucite; macle.*

chiastolite slate A rock formed by contact metamorphism of carbonaceous shale, characterized by a lack of prominent cleavage or schistosity and the presence of conspicuous chiastolite crystals in a cryptocrystalline groundmass.

chibinite A eudialyte-bearing nepheline syenite distinguished from *lujavrite* by its smaller amount of mafic components, which are in compact aggregates of thick rather than acicular crystals, and by having the eudialyte in patches in the interstices rather than as euhedral crystals. Also spelled: *khibinite.*

Chickasawhay North American (Gulf Coast) stage: Oligocene (above Vicksburgian, below Anahuac).

Chideruan *Tatarian.*

childrenite A pale-yellowish to dark-brown orthorhombic mineral: $(Fe,Mn)AlPO_4(OH)_2.H_2O$. It is isomorphous with eosphorite.

Chile saltpeter Naturally occuring sodium nitrate; *soda niter,* occurring in caliche in northern Chile. Cf: *saltpeter.* Syn: *Chile niter.*

chilidial plate One of a pair of posterior plate-like extensions of the walls of the notothyrium of certain brachiopods, commonly forming lateral boundaries of the cardinal process.

chilidium The triangular plate covering the apex of the notothyrium of certain brachiopods, commonly convex externally and extending for a variable distance ventrally over the proximal end of the cardinal process.

chillagite A variety of wulfenite containing tungsten.

chilled border *chill zone.*

chilled contact That part of a mass of igneous rock, near its contact with older rocks, that is finer grained than the rest of the mass, because of its having cooled more rapidly.

chilled margin *chill zone.*

chilled zone *chill zone.*

chill zone The border or marginal area of an igneous intrusion characterized by being of relatively finer grain than the interior of the rock mass, due to more rapid cooling. Cf: *basic border.* Syn: *chilled border; chilled margin; chilled zone.*

chimney [ore dep] *pipe.* Var. of *ore chimney.*

chimney [coast] (a) An angular and columnar mass of rock smaller than a *stack,* isolated on a wave-cut platform by differential wave erosion of a sea cliff. (b) A *blowhole;* a *spouting horn.*

chimney [speleo] In a cave, a vertical passage or opening aving rounded walls. Cf: *dome pit.*

himney [volc] A conduit through which magma reaches the arth's surface; a partial syn. of *vent* and of *pipe* [volc]. Cf: *eeder* [volc].

himney rock [geomorph] A chimney-shaped column of rock sing above its surroundings or isolated on the face of a steep ope; a small, weathered outlier shaped like a sharp pinnacle; *stack* formed by wave erosion. Syn: *pulpit rock.*

hina clay A commercial term for *kaolin* obtained from china-ay rock after washing and suitable for use in the manufac-ire of chinaware. Sometimes spelled: *China clay.*

hina-clay rock Kaolinized granite composed chiefly of quartz nd kaolin, with muscovite and tourmaline as possible ac-essories. The rock crumbles easily in the fingers. Also pelled: *China-clay rock.* Cf: *china stone* [ign].

hinaman pebble A term used in New Zealand for a pebble erived from a conglomerate consisting of quartz pebbles ce-ented with chalcedony.

hinarump A term used in the southwestern U.S. for *silicified ood.* Also spelled: *shinarump.*

hina stone [ign] Partially kaolinized granite containing quartz, aolin, and sometimes mica and fluorite. It is harder than *iina-clay rock* and is used as a glaze in the manufacture of *iina.* Syn: *petunzyte; petuntse; porcelain stone.* Cf: *Cornish one.*

hina stone [sed] A fine-grained, compact Carboniferous udstone or limestone found in England and Wales.

iine (a) A term used in England (esp. in Hampshire and in e Isle of Wight) for a narrow and deep ravine, gorge, or eft, or a small canyon, or a large fissure, cut in a soft, ear-y cliff by a stream descending steeply to the sea. (b) A ige or crest of rocks.

hinese-wall glacier A seldom used term for an ice sheet, ich as along the coast of Greenland, whose ice front is a ertical or even overhanging cliff.

iink-faceting A term applied by Wentworth (1925, p.260) to e localized grinding of smoothed, distinct, and often sharply nited facets on the surfaces of beach pebbles and cobbles at are lodged in crevices in such a way that they are bjected to a limited recurrent rubbing and to-and-fro move-ent under the continual action of waves.

iinook A term used for a *foehn* occurring on the eastern opes of the Rocky Mountains.

iiolite A snow-white tetragonal mineral: $Na_5Al_3F_{14}$.

iip A small fragment from a crystal; specif. a *diamond chip.*

iipping Abrasion of a rock fragment resulting in the flaking * of its corners.

iip yard *forest bed.*

iiral twinning *optical twinning.*

iisle A bank of shingle.

ii-square test A statistical test that employs the sum of lues given by the quotients of the squared difference be-een observed and expected (theoretical) frequencies divid- by the expected frequency. It enables assessment of odness of fit, association, or commonality in a population, d is used to determine equivalency of observed sample and pected population.

iitin A resistant organic compound with the same basic car-hydrate structure as cellulose, but nitrogenous because me hydroxyl groups are replaced by acetamide groups (i.e. is a repeating unit of N-acetylglucosamine instead of glu-se). It is a very common constituent of various invertebrate eletons such as insect exoskeletons and foraminiferal inner sts, and also occurs in hyphae and spores of fungi. Cf: *eudochitin.*

iitinous Consisting of *chitin.*

iitinozoan A pseudochitinous marine microfossil of the ex-ct group Chitinozoa, having uncertain affinity (generally as-med to represent animal remains), shaped in general like a sk, occurring individually or in chains, and ranging primarily

from uppermost Cambrian to Devonian. Chitinozoans have thin, usually black, structureless, and opaque walls, but they may be brown and translucent. Named by Eisenack (1931) who noted the resemblance of their walls to chitin.

chiton *polyplacophoran.*

chkalovite A mineral: $Na_2BeSi_2O_6$.

chladnite [meteorite] A group name for achondritic stony meteorites (aubrites and diogenites) composed essentially of orthopyroxene. The term originally applied to achondrites composed essentially of enstatite.

chladnite [mineral] Pure meteoritic *enstatite.*

chlamydospore A thick-walled, nondeciduous spore, such as a unicellular *resting spore* in certain fungi, usually borne termi-nally on a hypha and rich in stored reserves; a *fungal spore* that may have chitinous walls and therefore occur as a micro-fossil in palynologic preparations.

chloanthite *nickel-skutterudite.*

chloraluminite A mineral: $AlCl_3.6H_2O$.

chlorapatite (a) A rare mineral of the apatite group: $Ca_5(PO_4)_3Cl$. (b) An apatite mineral in which chlorine predomi-nates over fluorine and hydroxyl.

chlorargyrite A white, pale-yellowish, or grayish, isometric, wax-like mineral that darkens on exposure to light: $AgCl$. It occurs in the weathering zones of silver-sulfide deposits and it represents an important ore of silver. Syn: *cerargyrite; horn silver.*

chlorastrolite A mottled, green variety of *pumpellyite* used as a semiprecious stone, occurring as grains or small nodules of a radial, fibrous structure in geodes in basic igneous rocks. It resembles prehnite, and is found in the Lake Superior region (esp. on Isle Royale).

chlorides A miner's or prospector's term for ores containing silver chloride.

chlorine equivalent *chlorinity.*

chlorine log A neutron-gamma log designed to respond to the chlorine content of the rocks surrounding a borehole. It is used to estimate the salinity of the formation waters (since al-most all chlorine exists as soluble chlorides in pore water) and therefore to delineate a hydrocarbon-bearing zone from a salt-water zone. Syn: *salinity log.*

chlorinity The chloride content of sea water, measured by mass, or grams per kilogram of sea water, and including all the halides. Syn: *chlorine equivalent.*

chlorite (a) A group of platy, monoclinic, usually greenish minerals of general formula: $(Mg,Fe^{+2},Fe^{+3})_6AlSi_3O_{10}(OH)_8$. It is characterized by prominent ferrous iron and by the ab-sence of calcium and alkalies; chromium and manganese may also be present. Chlorites are associated with and resemble the micas (the tabular crystals of chlorite cleave into small, thin flakes or scales that are flexible, but not elastic as those of mica), and are widely distributed, esp. in low-grade meta-morphic rocks, or as alteration products of ferromagnesian minerals in igneous rocks. (b) Any mineral of the chlorite group, such as clinochlore, penninite, ripidolite, chamosite, thuringite, pennantite, and corundophilite.

chlorite schist A schist in which the main constituent, chlorite, imparts the schistosity by parallel arrangement of its flakes. Quartz, epidote, magnetite, and garnet are accessories; the latter two often as conspicuous porphyroblasts.

chloritic shale A poorly laminated shale containing a variety of angular to subrounded mineral particles (including those of unstable types) of silt size, characterized by feldspar some-times exceeding quartz in abundance and by chlorite often ab-undant in the finer matrix. It is commonly associated with graywacke (high-rank or feldspathic graywacke) and repre-sents accumulation of relatively finer detritus derived from rapidly eroded orogenic source areas and "poured" into rapid-ly subsiding depositional areas.

chloritization Introduction of, or replacement by, chlorite.

chloritoid A micaceous mineral: $Fe_2Al_4Si_2O_{10}(OH)_4$. It occurs in dull- or dark-green to gray or grayish-black masses of brittle folia in metamorphosed argillaceous sedimentary rocks, and is related to the brittle micas. Magnesium may be present.

chlormanganokalite A yellow rhombohedral mineral: K_4MnCl_6. It is isomorphous with rinneite.

chlorocalcite hydrophilite.

chloromagnesite A mineral: $MgCl_2$.

chloromelanite (a) A dark-green to nearly black variety of jadeite. (b) A solid solution of roughly equal amounts of diopside, jadeite, and acmite.

chloropal (a) A name originally applied to a deep-green, opal-like mineral that was later shown to be a crystalline clay mineral and renamed nontronite. (b) A greenish variety of common opal from Silesia.

chlorophaeite A mineraloid closely related to chlorite in composition (hydrous silicate of magnesium, iron, and calcium) and found in the groundmass of tholeiitic basalts, where it occupies spaces between feldspar laths, forms pseudomorphs after olivine, or occurs in veinlets and amygdules. It is pale green when fresh, but may be dark green, brown, or red in weathered rocks.

chlorophane A variety of fluorite that emits a bright-green light when heated.

chlorophoenicite A gray-green monoclinic mineral: $(Mn, Zn)_5(AsO_4)(OH)_7$. It is isostructural with magnesium-chlorophoenicite.

chlorophyll Generally a mixture of two waxy pigments, chlorophyll a $C_{55}H_{72}O_5N_4Mg$, blue-black, and chlorophyll b, $C_{55}H_{70}O_6N_4Mg$, yellow-green, which occurs in plasmic bodies (chloroplasts) of plants and serves as a catalyst in photosynthesis. Other forms of chlorophyll occur in diatoms, algae, etc.

chlorophyll *a* A pigment in phytoplankton that can be used to measure the abundance of phytoplankton.

chlorophyll coal A variety of dysodile which contains chlorophyll that can be extracted by alcohol.

chlorophyre A green prophyritic quartz diorite.

chlorospinel A grass-green variety of spinel containing some copper.

chlorothionite A bright-blue secondary mineral: $K_2Cu(SO_4)Cl_2$.

chlorotile A green orthorhombic mineral consisting of hydrated arsenate of copper. Cf: mixite.

chloroxiphite A dull-olive or pistachio-green monoclinic mineral: $Pb_3CuCl_2(OH)_2O_2$.

cho A rainy-season torrent carrying sand from the Himalayan foothills onto a plain below. Etymol: Panjabi, connoting "a bed of loose boulders, gravel and· sand, indicating rapid erosion" (Stamp, 1961, p.103). Also spelled: choh.

choanocyte An endoderm cell of a sponge, bearing a distinct, tabular collar-like, contractile, protoplasmic rim surrounding the base of a flagellum. Choanocytes line the inner surfaces of canals and/or cloaca. Adj: choanocytal. Syn: collar cell.

choanoderm A single layer of choanocytes in a sponge; a choanocytal membrane.

choanosome A region (inner layer) of a sponge containing choanocyte-lined cavities (flagellated chambers).

choke [drill] An aperture or constriction in the tubing of an oil or gas well used for controlling the flow or volume of oil and gas.

choke [speleo] An area in a cave that is blocked by debris.

chokedamp blackdamp.

choked stalagmite A stalagmite, the diameter of which has returned to standard after having become small.

choma A ridge-like deposit of dense shell substance delimiting a tunnel in a fusulinid. Pl: chomata. Cf: parachoma.

chondri Chondrules; plural of chondrus.

chondrite [meteorite] A stony meteorite characterized by chondrules embedded in a finely crystalline matrix consisting of orthopyroxene, olivine, and nickel-iron, with or without glass. Chondrites constitute over 80% of meteorite falls and

are usually classified according to the predominant pyroxen e.g. "enstatite chondrite", "bronzite chondrite", and "hype sthene chondrite". Adj: chondritic. Cf: achondrite.

chondrite [paleont] A common trace fossil of the "genus Chondrites, consisting of plant-like, regularly ramifying tunn structures that neither cross each other nor anastomose b radiate around a central vertical tube. It is interpreted as dwelling or feeding burrow, probably made by a marine worr It is often called a fucoid.

chondrodite A dark-red, orange-red, or yellow monoclinic mi eral of the humite group: $(Mg,Fe)_3SiO_4(OH,F)_2$. It common occurs in contact-metamorphosed dolomites. Syn: condrodite

chondrophore A relatively prominent process with a hollowe out surface for holding or attaching the internal ligament (res lium) of a bivalve mollusk. See also: resilifer.

chondrule A spheroidal granule or aggregate, often radia crystallized and usually about one millimeter in diameter, co sisting chiefly of olivine and/or orthopyroxene (enstatite bronzite), and occurring embedded more or less abundantly the fragmental bases of many stony meteorites (chondrite and sometimes free in marine sediments. Most chondrul appear to have originated as molten silicate droplets. Sy chondrus; chondre.

chondrus A syn. of chondrule. Pl: chondri.

chone An inhalant canal penetrating a cortex in a spong often leading from a vestibule to a subcortical crypt.

chonetid Any articulate brachiopod belonging to the suborc Chonetidina, characterized chiefly by a functional foramen I cated outside the delthyrium. Their stratigraphic range Lower Silurian (possibly Upper Ordovician) to Lower Jurassic

chonolite Var. of chonolith.

chonolith An igneous intrusion whose form is so obscure th it cannot be classified as a laccolith, dike, or sill. Also spelle chonolite.

chop hill A term used in Nebraska for a sand hill.

choppy cross-bedding Small-scale trough cross-bedding.

chorate cyst A spiny, encysted, unicellular alga; esp. a co densed dinoflagellate cyst bearing little morphologic reser blance to the motile theca. The ratio of the diameter of tl main body to the total diameter of the cyst is 0.6 or less. E amples: marginate chorate cyst; membranate chorate cy pterate chorate cyst; trabeculate chorate cyst. See also: pro imochorate cyst; proximate cyst.

chorismite Megascopically composite rock (mixed rock) th consists of two or more petrogenetically dissimilar materia of any origin (Dietrich & Mehnert, 1961). The term was fi introduced without genetic connotation, as a replacement f migmatite. Five types of chorismite were outlined.

chorismitization The formation of a chorismite.

choristid adj. Said of a sponge having a skeleton containi tetraxonic megascleres and lacking desmas.---n. A choris sponge; specif. a sponge of the order Choristida, class D mospongiae.

choristoporate Pertaining to a highly specialized type dasycladacean algae in which the sporangia are formed specialized gametangia, which may be specialized rays of t second or third order.

C horizon A mineral horizon of a soil, lying beneath the A a or B horizons, consisting of unconsolidated rock mater that is transitional in nature between the parent mater below and the more developed horizons above.

chorochromatic map A British term for a map in which bro distributions or variations are shown qualitatively over an ar by means of different colors, tints, or shadings. Syn: col patch map.

chorogram A generic term suggested by Wright (1944, p.65 for "any and all quantitative areal symbols" on a map.

chorographic Pertaining to chorography; specif. relating to area of regional or subcontinental extent, or said of a m representing a large region on a small scale (such as one I ween 1:500,000 and 1:5,000,000).

chorography (a) The art or practice of describing or mapping a particular region or district, esp. one larger than that considered by *topography* but smaller than that by geography. The term was widely used in the 17th and 18th centuries. (b) A broad account, description, map, or chart of a region considered by chorography. Also, the physical conformation or configuration, and the features, of such a region.---Syn: Greek *choros*, "place", + *graphein*, "to write".

chorology *biogeography.*

chott A var. of *shott*. Etymol: French, from Arabic *shatt*.

Christiansen effect In optical mineralogy, a dispersion phenomenon in which the boundary of a mineral grain that is immersed in a liquid of the same refractive index appears blue on one side and red to orange on the other.

Christmas tree The assemblage of valves, pipes, gages, and fittings at the top of the casing in a flowing oil well, used to control the flow of oil and gas from the well and to prevent blowouts. See also: *casinghead.*

christophite *marmatite.*

chromate A mineral compound characterized by the hexavalent cation CrO_4. An example is potassium chromate, K_2CrO_4. Cf: *sulfate.*

chromatic aberration In crystal optics, the production of color fringes due to the failure of rays of different wavelengths to converge at the same point.

chromatite A citron-yellow mineral: $CaCrO_4$.

chromatography A general name for several processes for separating components of a sample by moving the sample in a mixture or solution over or through a medium utilizing the property of adsorption, partition, ion exchange, or electrochromatography in such a way that the different components have different mobilities and thus become separated. One of the earliest applications was in the separation of components of dye mixtures, giving rise to bands of different colors and hence to the name chromatography. See also: *column chromatography; electrochromatography; gas chromatography; liquid chromatography; paper chromatography; thin-layer chromatography.*

chrome A term commonly used to indicate ore of chromium, consisting esp. of the mineral chromite, or chromium-bearing minerals such as chrome chlorite, chrome mica, or chrome diopside.

chrome-chert A cherty-looking rock formed by the replacement (by silica) of the silicate minerals of a chromite peridotite, the more resistant chromite grains remaining unaltered in the secondary siliceous matrix.

chrome diopside A bright-green variety of diopside containing small amount of Cr_2O_3.

chrome iron ore A syn. of *chromite*. Var: *chrome iron; chromic iron.*

chrome mica *fuchsite.*

chrome ocher A chromiferous clay; specif. a bright-green clay material containing 2-10.5% Cr_2O_3.

chrome spinel *picotite.*

chromic iron *chrome iron ore.*

chromite (a) A brownish- to iron-black mineral of the spinel group: $(Fe,Mg)(Cr,Al)_2O_4$. It occurs in octahedral crystals as a primary accessory mineral in basic and ultrabasic igneous rocks; it also occurs massive, and it forms detrital deposits. Chromite is isomorphous with magnesiochromite, and is the most important ore of chromium. Syn: *chrome iron ore.* (b) A name applied to a series of isomorphous minerals in the spinel group, consisting of magnesiochromite and chromite.

chromitite (a) An igneous rock composed chiefly of the mineral chromite. (b) A mixture of chromite with magnetite or hematite.

chromocratic *melanocratic.*

chron (a) A term preferred by Dunbar & Rodgers (1957, p.301) as a "reasonably unambiguous" and "mnemonic" syn.

---

of *moment* [stratig]. (b) A term used by Sutton (1940, p.1404) for the time interval during which a "group" (now referred to as a "stage") of rocks were formed; i.e. used as a syn of *age* [geochron] (a).----The term was introduced by Williams (1901, p.583-584) for an indefinite division of geologic time, and used by Wheeler et al (1950, p.2362) as a general geologic-time unit.

chronocline A cline involving successive changes in the members of a natural group of organisms in successive stratigraphic units.

chronofauna Geographically restricted natural assemblage of interacting animal populations that maintained its basic structure over a geologically significant period of time.

chronogenesis The time sequence of appearance of organisms in stratified rocks.

chronolith *time-stratigraphic unit.*

chronolithologic unit *time-stratigraphic unit.*

chronology Arranging events or happenings in their proper sequence in time; also, considering or measuring time in discrete units. See also: *geochronology.*

chronostratic unit *time-stratigraphic unit.*

chronostratigraphic unit A term preferred by the International Subcommission on Stratigraphic Terminology (1961, p.23) for *time-stratigraphic unit.*

chronostratigraphic zone *chronozone.*

chronostratigraphy The branch of stratigraphy that interprets geologic history by determining the age and time sequence of the Earth's rock strata (Hedberg, 1961, p.501). Syn: *time-stratigraphy.*

chronotaxial Pertaining to, characterized by, or exhibiting chronotaxy.

chronotaxis An erroneous transliteration of *chronotaxy.* The term was proposed by Henbest (1952, p.310) as a complementary term for "homotaxis".

chronotaxy Similarity of time sequence; specif. correlation of fossil or stratigraphic sequences on identity in time, or the determination of age equivalence. The term was originally proposed as *chronotaxis* by Henbest (1952, p.310). Cf: *homotaxy.*

chronozone (a) A formal time-stratigraphic unit indicating "the strata equivalent in time-span to a biostratigraphic zone or any other zone" (ACSN, 1967, p.1865-1866); e.g. "*Cardioceras cordatum* Chronozone" is the total body of rock formed anywhere during the time span of the *Cardioceras cordatum* Range Zone, regardless of whether or not *Cardioceras cordatum* is itself present. (b) An informal or general time-stratigraphic unit indicating "the body of strata representing the rocks formed during any minor interval of geologic time" (ISST, 1961, p.23). The term is used in the sense of "age zone" or "geologic-time zone", or as an informal expression of the concept of the strata equivalent in age to any indicated stratigraphic range. (c) A time-stratigraphic term proposed by Henningsmoen (1961, p.80) as equivalent to the classical "zone" and partially replacing "substage" and by Størmer (1966, p.22 & 25) as a formal substage having a local to regional, or a more limited, geographic range. Cf: *division* [stratig].---Syn: *chronostratigraphic zone.*

chrysoberyl (a) A hard mineral: $BeAl_2O_4$. It is usually yellow, pale green, or brown, contains a small amount of iron, and is used as a gem. Principal varieties are cat's-eye and alexandrite. Syn: *chrysopal; gold beryl; cymophane.* (b) An obsolete syn. of *heliodor.*

chrysocolla (a) A blue, blue-green, or emerald-green mineral: $(Cu_2H_2(Si_2O_5)(OH)_4$. It is usually cryptocrystalline or amorphous, and it occurs as incrustations and thin seams in the zone of weathering of copper ores. Its chemical composition was formerly given as: $CuSiO_3.2H_2O$. (b) An old name given to a mineral or minerals (such as chrysocolla, borax, and malachite) used for soldering gold (Hey, 1962, p.384).

chrysocolla chalcedony Translucent to semitranslucent, vivid-blue to greenish-blue chalcedony, the color of which is caused by minutely distributed chrysocolla.

chrysolite [gem] A gem (such as beryl, chrysoberyl, sapphire, spinel, or topaz) characterized by light greenish-yellow to light yellowish-green hues; specif. the pale-yellow to yellowish-green gem variety of olivine.

chrysolite [mineral] (a) A yellowish-green, reddish, or brownish variety of olivine in which the ratio of magnesium to total magnesium plus iron is between 0.90 and 0.70 or in which the Fe_2SiO content is 10-30 mole percent. The name has at times carried a wider meaning to designate a synonym of *olivine*. (b) A name that has been applied at various times to topaz, prehnite, and apatite. This usage is obsolete.———Not to be confused with *chrysotile*.

chrysomonad In some classifications of protozoans, any flagellate protozoan belonging to the order Chrysomonadina, which includes coccolithophores and silicoflagellates.

chrysopal (a) A translucent variety of common opal, colored apple green due to the presence of nickel. (b) *chrysoberyl*. (c) A gemstone-trade name for opalescent chrysolite (olivine).

chrysophyric Said of a basalt having olivine phenocrysts (Thrush, 1968, p.208).

chrysoprase (a) An apple-green or pale yellowish-green variety of chalcedony containing nickel and valued as a gem. Syn: *green chalcedony*. (b) A misleading name used in the gem trade for green-dyed chalcedony having a much darker color than natural chrysoprase.

chrysoquartz Green aventurine quartz.

chrysotile A white, gray, or greenish mineral of the serpentine group: $Mg_3Si_2O_5(OH)_4$. It is a highly fibrous, silky variety of *serpentine*, and constitutes an important type of asbestos. Not to be confused with *chrysolite*. Cf: *antigorite*. Syn: *serpentine asbestos; clinochrysotile*.

chrystocrene A term introduced as *crystocrene* by Tyrrell (1904, p.234) for a surface mass of ice formed each winter by the overflow of springs; also, the ice formed in the interstices of a mass of loose rock fragments (such as talus) by the freezing of a subjacent spring. The term is not synonymous with *rock glacier* (Tyrrell, 1910). Cf: *crystosphene*.

chthonic Said of deep-sea sediments and clastic debris derived from preexisting rocks. Ant: *halmeic*. Cf: *allogenic*.

chuco A term used in Chile for the upper part of a caliche deposit, composed mainly of sodium sulfate.

chudobaite A mineral: $(Na,K,Ca)(Mg,Zn,Mn)_2H(AsO_4)_2.4H_2O$.

chukhrovite A mineral: $Ca_3(Y,Ce)Al_2(SO_4)F_{13}.10H_2O$.

chunk mineral A term used in Wisconsin for masses of galena as they come from the mine.

chur *char*.

churchite A syn. of *weinschenkite* (phosphate mineral).

churn drilling *cable-tool drilling*.

churn hole A *pothole* in a stream bed.

chute [ore dep] Var. of *shoot*, as in *ore shoot*, q.v.

chute [speleo] An inclined channel or passage in a cave.

chute [geomorph] A term used in the Isle of Wight for a steep cutting affording a passage from the surface above a cliff to the lower undercliff ground (Stamp, 1961, p.104).

chute [hydraul] An inclined water course of either natural or artificial origin.

chute [streams] (a) A fall of water; a rapid or quick descent in a river; a steep channel by which water falls from a higher to a lower level; a rapids. See also: *shoot*. (b) A narrow channel through which water flows rapidly, esp. along an overflow river (such as the lower Mississippi River); specif. a *chute cutoff*.

chute cutoff A narrow, low-angle meander *cutoff* formed at time of flood when the main flow of a stream is diverted to the inside of a bend, along or through a trough between adjacent point bars. Cf: *neck cutoff*. Syn: *chute*.

chymogenetic Said of that portion of a composite rock formed by crystallization from an ionic- or molecular-dispersed phase or from a fluid, whether a gas, hydrothermal solution, or magma (Mehnert, 1968, p.353). Also spelled: *chymogenic*. Cf: *stereosome; metatect; mobilizate*. See also: *neosome*.

cicatricose Marked with scars; esp. said of sculpture of pollen and spores consisting of more or less parallel ridges.

cicatrisation Reconstruction of a broken or corroded crystal as a result of a secondary deposit of the same mineral in optical continuity.

cicatrix (a) A scar in an echinoderm; esp. the scar indicating the former position of the column in some echinoderms (such as cystoids) which apparently molt it. (b) A small groove or scar on the apex of some nautiloid conchs. (c) The impression on the inside of a bivalve shell caused by the insertion of the adductor muscle. See also: *adductor muscle scar*.---Pl: cicatrices. Syn: *cicatrice; scar*.

cienaga A marshy area where the ground is wet due to the presence of seepage or springs, often with standing water and abundant vegetation. The term is commonly applied in arid regions such as the southwestern U.S. Etymol: Spanish ciénaga, "marsh, bog, miry place". Pron: see-ay-nah-gah. Also spelled: cienega.

cigar-shaped mountain An anticlinal ridge pitching away at each end.

cilia Plural of *cilium*.

ciliate n. Any protozoan belonging to the class Ciliata and characterized by the presence of cilia throughout its life cycle. Their known stratigraphic range is Upper Jurassic to the present.----adj. Possessing cilia.

cilifer Said of a variant of radulifer type of brachiopod crura flattened in the plane of commissure, forming direct prolongations of horizontal hinge plates, then turning parallel to the plane of symmetry as slightly crescentic blades.

cilium One of numerous short hair-like processes found on the surface of cells, capable of rhythmic vibratory or lashing movement, and serving as organs of locomotion in free-swimming unicellular organisms and in some small multicellular forms and as producers of currents of water in higher animals. Pl: *cilia*. Cf: *flagellum*.

cima A mountain peak or dome. Etymol: Italian. Pron: che-ma.

ciminite An extrusive rock composed of basic plagioclase, olivine, augite, and labradorite with orthoclase rims in a trachytic groundmass; an olivine trachyte.

Cimmerian Var. of *Kimmerian*.

Cimmerian orogeny One of the 30 or more short-lived orogenies during Phanerozoic time and identified by Stille; in this case, two orogenies are included, the late Cimmerian at the end of the Jurassic, and the early Cimmerian late in the Triassic, between the Norian and Rhaetian stages.

cimolite A white, grayish, or reddish hydrous aluminum silicate mineral occurring in soft, clay-like masses.

Cincinnatian North American provincial series: Upper Ordovician (above Champlainian, below Alexandrian of Silurian).

cinder An uncemented, juvenile, vitric, and vesicular pyroclastic fragment that has been given various size classifications: 4-32mm diameter; 0.5-5.0cm diameter; so fine that its nature is not discernible to the naked eye; equivalent to the size of *lapilli* but vesicular; or, the same as scoria. Cf: *block [volc]; volcanic gravel*.

cinder coal *natural coke*.

cinder cone A conical hill formed by the accumulation of cinders and other scoriaceous ejecta, normally of basaltic or andesitic composition. The steepness of the slopes may differ widely depending on the coarseness of the ejecta, the height of eruption, wind velocity, and other factors, but is normally steeper than 10 degrees.

cinerite A deposit of volcanic cinders.

cingular archeopyle An *archeopyle* formed in a dinoflagellate cyst by breakage along and within the girdle.

cingular series The series of plates along the girdle in a dino-lagellate possessing theca.

cingulate Having a girdle; esp. said of a spore possessing a cingulum.

cingulum (a) Either of the two *connecting bands* forming the sides of the two valves of a diatom; a *girdle*. (b) An annular, more or less equatorial extension of a spore in which the wall is thicker than that of the main body of the spore. Cf: *zone palyn*]; crassitude.---Pl: *cingula*.

cinnabar A rhombohedral mineral: HgS. It is dimorphous with metacinnabar and represents the most important ore of mercury. Cinnabar occurs in brilliant red acicular crystals or in red, brownish, or gray masses in veins and alluvial deposits. Syn: *cinnabarite; vermilion*.

cinnamon stone Yellow-brown to reddish-brown *essonite*. Syn: *cinnamite*.

cipolin (a) *cipolino*. (b) A term used in France for any *crystalline limestone*.

cipolino A siliceous marble containing micaceous layers. Partial syn: *cipolin*.

C.I.P.W. classification A system for classifying and naming igneous rocks based on the *norm*. The initials represent the initial letters of the names of the men who devised the system, Cross, Iddings, Pirsson, and Washington (1902). Syn: *quantitative system; norm system*.

circadian Said of a time period approximately 24 hours in length or of an event occurring at roughly 24-hour intervals; e.g. "circadian rhythms". Cf: *circannian*.

circannian Said of a time period approximately one year in length or of an event that occurs annually; e.g. "circannian rhythms". Cf: *circadian*.

circinate (a) Pertaining to the unrolling of a developing fern frond. (b) Pertaining to a protist that is curled downward from the apex.

circle [pat grd] A form of horizontal patterned ground whose mesh is dominantly circular. See: *sorted circle; nonsorted circle*.

circle [surv] The graduated disk of a surveying instrument, perpendicular to and centered about an axis of rotation, and calibrated to read the amount of rotation; e.g. a horizontal circle or a vertical circle of a theodolite or transit.

circle of influence *area of influence*.

circle of latitude A meridian of the terrestrial sphere, along which latitude is measured. Cf: *parallel of latitude*.

circlet A series of plates that forms a ring entirely or partially around theca of an echinoderm.

circuit (a) A continuous series of connected survey lines that form a closed loop. (b) A line or series of lines connecting two fixed survey points.

circuit closure The *error of closure* of a level circuit, being the algebraic sum of all the junction closures in a circuit, usually reckoned counterclockwise around the circuit; hence, the accumulated error (before adjustment) of measured differences of elevation around the circuit, or the amount by which the last computed elevation fails to equal the initial elevation.

circular coal *eye coal*.

circularity ratio *basin-circularity ratio*.

circular level A *spirit level* having the inside surface of its upper part ground to a spherical shape, the outline of the bubble formed being circular, and the graduations being concentric circles. It is used where a high degree of precision is not required, as in setting an instrument in approximate position. Syn: *bull's-eye level; box level*.

circular normal distribution A frequency distribution of a polar variable analogous to a normal (Cartesian) distribution.

circular polarization In optics, circularly polarized light consisting of upward-spiraling vibration vectors that define a surface similar to the thread of a screw. It is caused by the interaction of mutually perpendicular wave motions whose path differences differ in phase by $(2n+1)/4\ \lambda$ on emergence from a crystal. Cf: *elliptical polarization*.

circular section [cryst] In a uniaxial crystal, an equatorial section perpendicular to the optic axis; in a biaxial crystal, one of two sections intersecting the beta axis of the biaxial indicatrix (Wahlstrom, 1948).

circular section [exp struc geol] *K section*.

circular slide A landslide whose slip surface follows the arc of a circle.

circulation [drill] The continuous cycling of drilling mud down the drill pipe, out through the drill bit, and up to the surface through the annular space between the drill pipe and the walls of the hole.

circulation [oceanog] In oceanography, a general term for the flow of water in a large area, usually in a closed pattern, due to wind over the surface or to varying densities of water (due to differences in salinity and water temperature).

circulation [lake] The complete mixing throughout the total depth of a lake or sea; generally it occurs when the waters are isothermal, and often at the temperature of maximum density. See also: *overturn*.

circulation fluid *drilling mud*.

circulus A cameral deposit on the concave surface of a cyrtochoanitic septal neck of a nautiloid (TIP, 1964, pt.K, p.54).

circumdenudation The denudation or erosion of a landmass such that a part of the ground is left isolated and upstanding; e.g. denudation around a resistant rock mass. Syn: *circumerosion*.

circumdenudation mountain *mountain of circumdenudation*.

circumerosion *circumdenudation*.

circumferential wave An obsolete syn. of *surface wave*.

circumferentor A type of *surveyor's compass* having vertical slit sights on projecting arms. Syn: *planchette*.

circummural budding A type of *polystomodaeal budding* in which indirectly linked stomodaea are arranged around discontinuous collines or monticules of corallum. Cf: *intramural budding*.

circumoceanic basalt Basalt that issues from volcanoes on the margins of ocean basins and that contains less than 1.75 percent TiO_2 and usually more than 15 percent Al_2O_3 (Longwell, et al, 1969, p.645).

circumoral budding A type of *polystomodaeal budding* in which directly linked stomodaea are arranged concentrically around the central parent stomodaeum.

circum-Pacific belt The *great-circle belt* that borders the Pacific Ocean along the continental margins of Asia and the Americas, and meets the *Eurasian-Melanesian belt* in the Celebes.

circum-Pacific province *Pacific suite*.

circumvallation The process whereby mountains are formed by streams incising a featureless plain (Hobbs, 1912, p.442).

cirque [lunar] *walled plain*.

cirque [glac geol] A deep, steep-walled, flat- or gently-floored, half-bowl-like recess or hollow, variously described as horseshoe- or crescent-shaped or semicircular in plan, situated high on the side of a mountain and commonly at the head of a glacial valley, and produced by the erosive activity (frost action, nivation, ice plucking) of mountain glaciers. It often contains a small round lake, and it may or may not be occupied by ice or snow. Etymol: French, from Latin *circus*. Syn: *corrie; cwm; coire; kar; basin; glacial cirque; botn; amphitheater; combe; oule; van; zanoga*.

cirque [geomorph] A term sometimes used for a semicircular, amphitheater-like, or armchair-shaped hollow of nonglacial origin but resembling a glacial cirque; e.g. a doline in a limestone region, a blowout in an arid region, or a depression formed by landslide sapping. See also: *pseudocirque*.

cirque floor The nearly flat surface at the bottom of a cirque. See also: *cirque niveau*.

cirque glacier A small glacier occupying a cirque or a rounded niche formed by the ice on a mountainside, or resting against the headwall of a cirque. It is the most common type of glacier in the mountains of the western U.S. Cf: *glacieret; niche glacier.*

cirque lake A small, deep, almost circular glacial lake occupying a cirque; it has no prominent inlet or outlet, being fed by runoff from the surrounding slopes and dammed by a lip of bedrock or by a small moraine. Syn: *tarn.*

cirque mountain *horn.*

cirque niveau The level of a *cirque floor* representing the surface of a terrace developed by preglacial erosion (Swayne, 1956, p.34); it is the approximate altitude at which most cirques in a region have excavated their floors. Etymol: *cirque* + French *niveau,* "level".

cirque platform A relatively level surface formed by the coalescence of several cirques.

cirque stairway A succession of cirques situated in a row at different levels in the same glacial valley. Cf: *glacial stairway.* Syn: *cirque steps.*

cirral adj. Pertaining to a cirrus.---n. A single segment or plate of a crinoid cirrus.

cirriped Any marine crustacean belonging to the class Cirripedia, characterized chiefly by the permanent attachment of the adult stage to some substrate; e.g. a barnacle. Their stratigraphic range is Upper Silurian to present. Also spelled: *cirripede.*

cirrus (a) Any of the flexible, root-like, jointed appendages attached to the side of the stem (and sometimes from the aboral surface) of a crinoid, exclusive of radix. It is composed of small articulated plates (cirrals). (b) A multiarticulate food-gathering thoracic appendage in a cirripede crustacean.---Pl: *cirri.*

cislunar Pertaining to phenomena, or being the space, between the Earth and the Moon or the Moon's orbit. Cf: *translunar.*

cistern (a) An artificial reservoir or tank for storing water. (b) A natural reservoir; a hollow containing water.

cistern rock *laccolith.*

citrine A transparent yellow or yellow-brown (sometimes red-orange or orange-brown) variety of crystalline quartz closely resembling topaz in color. It can be produced by heating amethyst or dark smoky quartz. Syn: *topaz quartz; false topaz; Bohemian topaz; quartz topaz; yellow quartz.*

civil engineering A branch of engineering concerned primarily with the investigation, design, construction, operation, and maintenance of civil-works projects (public and private) such as highways, bridges, tunnels, waterways, harbors, dams, water supply, irrigation, railways, airports, buildings, sewage disposal, and drainage.

clade A branch at the extremity of an actine in an ebridian skeleton that may connect adjacent actines. See also:*proclade; opisthoclade; mesoclade.*

cladi Plural of *cladus.*

cladoceran Any crustacean belonging to the order Cladocera, characterized by a univalve carapace that is bent along the back giving a bivalve appearance. Cladocerans are commonly found in fresh- to brackish-water postglacial deposits; their stratigraphic range is Oligocene to present.

cladogenesis (a) Phylogenetic splitting or branching; speciation. (b) Progressive evolutionary specialization.

cladome The group of similar rays of a diaene, triaene, or tetraene sponge spicule.

cladus One of the rays of a cladome; a branch of a ramose spicule. Term is usually used in the plural: *cladi.* Syn: *clad.*

Claibornian North American (Gulf Coast) stage: Eocene (above Sabinian, below Jacksonian).

claim In mining law, a portion of public land on which an individual may have mining rights; a *mining claim.* Size and other legal restrictions vary from country to country.

Clairaut's theorem An expression for the variation of normal gravity on the Earth that is the basis for standard gravity formulas such as the International Gravity Formula. It establishes the relationship between normal gravity and the flattering of the Earth from which it becomes possible to comput the flattening from surface gravity observations.

clam A popular term for a bivalve mollusk, commonly applied to an edible one that lives partially or completely buried in sand or mud.

Clamgulchian Floral stage in Alaska: Miocene(?) and Plice cene.

clamshell snapper A type of *grab sampler.*

clan [ecol] A small ecologic community, usually a climax community, that occupies only a few square meters of spac and has only one dominant species.

clan [petrology] A group of igneous rocks that are closely re lated in chemical composition. Clans are subdivided into fam lies. See also: *family* [petrology]. Syn: *igneous rock cla* Partial syn: *family* [petrology].

Clapeyron equation A statement in chemistry that the rate change of pressure with temperature in a phase transition of closed system is equal to the heat of the reaction divided b the product of the absolute temperature and the volum change of the reaction. It was developed by Clausius in 185 Syn: *Clausius-Clapeyron equation.* See also: *Ehrenfest rela tion; Poynting's law.*

clarain A coal *lithotype* characterized by semibright, silky lus ter and sheetlike, irregular fracture. It is distinguished from v *train* by containing fine intercalations of a duller lithotype, *d rain.* Its characteristic microlithotype is *clarite.* Cf: *fusain.*

clarinite The major maceral of clarain, according to th Stopes classification; the term is no longer in general use.

clarite A coal microlithotype that contains a combination of v trinite and exinite totalling at least 95%, and containing mo of each than of inertinite. Cf: *clarain.*

Clark degree A British unit for measuring hardness of wate equal to one grain per British gallon or 14.3 ppm as CaCO Cf: *grain* [water]; degree [water].

clarke The average percentage of an element in the crust the Earth. The measurement is sometimes compromised current differences in defining the limits and composition the crust. The term is named in honor of F.W. Clarke. Cf *clarke of concentration.* Syn: *crustal abundance.*

Clarke-Bumpus quantitative plankton sampler An instrume used for collecting plankton from the water; it also measure the quantity of water that passes through it.

clarkeite A dark-brown or reddish-brown mimeral: (Na,Ca Pb)$_2$U$_2$(O,OH)$_7$.

clarke of concentration The factor of concentration of an ele ment in a mineral or rock, compared with its crustal abu dance. The term is applied to specific occurrences as well to the average occurrence in a particular mineral or roc type.

Clarke spheroid of 1866 This is the spheroid of reference f geodetic surveys in North and Central America, the Hawaii Islands, and the Philippines. It was the *North American datu of 1927.* Cf: *ellipsoid.*

clarocollain A transitional lithotype of coal characterized b the presence of collinite with lesser amounts of other mace als. Cf: *colloclarain.* Also spelled: *clarocollite.*

clarocollite Var. of *clarocollain.*

clarodurain A transitional lithotype of coal characterized by v trinite, but more of other macerals such as micrinite and ex nite than of vitrinite; it corresponds to *semisplint coal.* Cf: *d roclarain.*

clarodurite A coal microlithotype containing at least 5% eac vitrinite, exinite, and inertinite, and in which there is mo inertinite than vitrinite. It is intermediate in composition b tween clarite and durite, but closer to durite. Cf: *duroclarite.*

clarofusain A transitional lithotype of coal characterized by th presence of fusinite and vitrinite, with other macerals; fusini

s more abundant than it is in *fusoclarain*. Also spelled: *claro-*
usite.

larofusite Var. of *clarofusain*.

larotelain A transitional lithotype of coal characterized by the
resence of telinite, and lesser amounts of other macerals.
f: *teloclarain*. Also spelled: *clarotelite*.

larotelite Var. of *clarotelain*.

larovitrain A transitional lithotype of coal characterized by
he presence of vitrinite with lesser amounts of other macer-
ls. Cf: *vitroclarain*. Also spelled: *clarovitrite*.

larovitrite Var. of *clarovitrain*.

lasmoschist A term suggested by W.D. Conybeare to replace
graywacke" (an arenaceous rock in the lower part of the
econdary strata) (Roberts, 1839, p.72).

lasolite A rock composed of fragments of other rocks; a
lastic rock.

lasper An appendage of a crustacean, modified for attach-
nent in copulation or for fixation of parasites.

lass [cryst] One of thirty-two possible combinations of the
lements of symmetry. Crystal classes are divided among the
ix crystal systems, and deal with outward symmetry. Syn:
oint group.

lass [taxon] A major unit in the taxomony of plants or ani-
nals, containing one or more orders, that is ranked above
order" but below "phylum".

lass [petrology] In the C.I.P.W. classification of igneous
ocks, a subdivision based on the relative proportions of salic
nd femic standard minerals. The classes correspond approxi-
nately to the color-based divisions leucocratic, melanocratic,
nd mesocratic. The basic unit of the class is the *order*
betrology].

lass [stat] A subdivision of the observed range of a variable,
aving stated limits.

lassic n. In New World archaeology, a cultural stage that fol-
ows the Formative and is characterized by the rise of civiliza-
ons such as the Mayan. It is followed by the Post-Classic.---
dj. Pertaining to the Classic.

lassical equilibrium constant An *equilibrium constant* that is
efined by concentrations rather than by activities.

lassification That part of *taxonomy* that deals with the actual
ssignment of organisms to the various taxonomic units.

last (a) An individual constituent, grain, or fragment of a
ediment or rock, produced by the mechanical weathering
disintegration) of a larger rock mass; e.g. a phenoclast. (b)
yroclast. (c) *bioclast.*

lastation (a) The breaking up of rock masses in situ by phys-
cal or chemical means (Grabau, 1924, p.17); *weathering.* (b)
he disrupting of rocks to form clastic sediments (Galloway,
922).

lastic adj. (a) Pertaining to or being a rock or sediment
omposed principally of broken fragments that are derived
om preexisting rocks or minerals and that have been trans-
torted individually for some distance from their places of ori-
|in; also said of the texture of such a rock. The term is often
sed to indicate a source from within the depositional basin,
s compared to *detrital.* (b) *pyroclastic.* (c) Said of a bioclas-
c rock. (d) Pertaining to the fragments (clasts) composing a
lastic rock.---n. A clastic rock. Term is usually used in the
lural; e.g. the commonest "clastics" are sandstones and
hales.

lastic breccia A breccia formed by erosion (McKinstry, 1948,
.634).

lastic deformation One of the processes of dynamothermal
netamorphism which involves the actual fracture, rupture,
nd rolling out of mineral and rock particles. Generally the ori-
ntation of the resulting fragments becomes confused al-
hough crystal structures may be preserved in some instan-
es. In the ultimate case, the rock may be thoroughly pulver-
ed (Tyrrell, 1926). Cf: *blastic deformation; plastic deforma-*
on.

clastic dike A *sedimentary dike* consisting of a variety of broken rocks derived from underlying or overlying material; esp.
a *sandstone dike* or a *pebble dike*.

clastichnic Said of a dolomite rock in which the original clastic
texture of a limestone is preserved (Phemister, 1956, p.72).

clasticity (a) The quality, state, or degree of being clastic. (b)
The maximum apparent particle size in a sediment or sedimentary rock (Carozzi, 1957). Syn: *clasticity index*.

clastic pipe A cylindrical body of clastic material, having an
irregular columnar or pillar-like shape, standing approximately
vertical through enclosing formations (usually in limestone),
and measuring a few centimeters to 50 m in diameter and a
meter to 60 m in height; esp. a *sandstone pipe*. Syn: *cylindri-*
cal structure.

clastic ratio A term introduced by Sloss et al (1949, p.100) for
the ratio of the thickness or percentage of clastic material
(conglomerate, sandstone, shale) to that of nonclastic material (limestone, dolomite, evaporites, etc.) in a stratigraphic
section; e.g. a ratio of 5 indicates that the section contains an
average of 5 m of clastics per meter of nonclastics. The ratio
is a measure of the relative contribution by materials carried
into the environment compared to sediments formed locally.
Cf: *sand-shale ratio*. Syn: *detrital ratio*.

clastic rock (a) A consolidated sedimentary rock composed
principally of broken fragments that are derived from preexisting rocks (of any origin) or from the solid products formed
during chemical weathering of such rocks, and that have been
transported mechanically to their places of deposition; e.g. a
sandstone, conglomerate, or shale, or a limestone consisting
of particles derived from a preexisting limestone. See also:
epiclastic rock. Syn: *fragmental rock.* (b) *pyroclastic rock.*
(c) *bioclastic rock.* (d) *cataclastic rock.*

clastic sediment A sediment formed by the accumulation of
fragments derived from preexisting rocks or minerals and
transported as separate particles to their places of deposition
by purely mechanical agents (such as water, wind, ice, and
gravity); e.g. gravel, sand, mud, clay. Cf: *detrital sediment*.
Syn: *mechanical sediment.*

clastic wedge The sediments of an *exogeosyncline*, derived
from the tectonic land masses of the adjoining orthogeosynclinal belt (King, 1959, p.59). Cf: *geosynclinal prism*.

clastizoic Said of a rock containing animal remains mainly in
the form of angular, little-worn debris; esp. said of a fossilifer-
ous-fragmental limestone that may often contain entire microfossils. Term introduced by Phemister (1956, p.72).

clastizoichnic Said of a dolomite or recrystallized limestone
that contains traces of original clastizoic features (Phemister,
1956, p.72).

clastomorphic Said of a *deuteromorphic* rock constituent
whose shape has been modified by erosion, such as a rounded or angular particle of a detrital sediment. The term is little
used.

clathrate A term applied by Washington in 1906 to the texture
commonly found in leucite-bearing rocks in which leucite
crystals are surrounded by tangential augite crystals giving
the appearance of a net or sponge, the augite representing
the threads or walls and the leucite the holes (Johannsen,
1939, p.205).

clatter *clitter.*

claudetite A monoclinic mineral: As_2O_3. It is dimorphous with
arsenolite.

claugh Var. of *clough*.

Clausius-Clapeyron equation *Clapeyron equation.*

clausthalite A mineral: PbSe. It resembles galena in appearance.

clavalite A belonite with a globular enlargement at each end.

clavate [paleont] (a) Club-shaped, being slender at one end
and gradually thickening near the other end, such as a long
body thickened toward the top or distal end like a baseball
bat; e.g. said of spores and pollen having sculpture consisting

of processes that widen to a knob at the end. Cf: *pilate.* (b) Pertaining to a clavus of an ammonoid.

clavicle A shelly buttress or heavy internal ridge supporting the chondrophore in some bivalve mollusks.

clavidisc A sponge spicule (microsclere) in the form of an ovate disk with a central perforation.

clavula A small ciliated spine in a fasciole of an echinoid. Pl: *clavulae.*

clavule A *sceptrule* in which the end bearing the axial cross is swollen or bears a ring of recurved teeth.

clavus An ammonoid tubercle elongated longitudinally in the direction of coiling. Pl: *clavi.*

clay [eng] Plastic material consisting mainly of particles having diameters less than 0.074 mm (passing U.S. standard sieve no.200). Cf: *silt.*

clay [geol] (a) A rock or mineral fragment or a detrital particle of any composition often a crystalline fragment of a clay mineral), smaller than a very fine silt grain, having a diameter less than 1/256 mm (4 microns, or 0.00016 in., or 8 phi units). This size is approximately the upper limit of size of particle that can show colloidal properties. See also: *coarse clay; medium clay; fine clay; very fine clay.* (b) A loose, earthy, extremely fine-grained, natural sediment or soft rock composed primarily of clay-size or colloidal particles and characterized by high plasticity and by containing a considerable amount of clay minerals (hydrous aluminum silicates) derived from feldspathic rocks by weathering (primarily decomposition) or by precipitation, and subordinate amounts of finely divided quartz, decomposed feldspar, carbonates, ferruginous matter, and other impurities; it forms a pasty, plastic, moldable, impermeable muddy mass when finely ground and mixed with water, retaining its shape upon drying, and becoming firm, rock-like, and permanently hard upon heating or firing. Some clays are nonplastic. It should have more than 50% clay-size particles (Twenhofel, 1937, p.96), and clay minerals must form at least one-fourth of the total (Pettijohn, 1957, p.341). Clays are classified by use, origin, composition, mineral constituents, and color; among their various uses are in the manufacture of tile, porcelain, and earthenware, and in filtration, oil refining, and paper manufacture. (c) A term that is commonly applied to any soft, adhesive, fine-grained deposit (such as loam or siliceous silt) or to earthy material, esp. when wet (such as mud or mire). (d) *clay mineral.*

clay [soil] (a) A term used in the U.S. and by the International Society of Soil Science for a rock or mineral particle in the soil, having a diameter less than 0.002 mm (2 microns). Prior to 1937, the term as used in the U.S. included particles less than 0.005 mm in diameter. (b) *clay soil.*

clay ball (a) A chunk of clay released by erosion of a clayey bank and rounded by wave action; esp. an *armored mud ball.* (b) A ball-like aggregate formed from clay in suspension or in a viscous state, and occurring in the bed of a lake or stream, or on the sea floor.---Also spelled: *clayball.*

clay band A light-colored, argillaceous layer in clay ironstone. Also spelled: *clayband.*

clay-band ironstone A variety of *clay ironstone* characterized by abundant clay bands.

clay boil A *mud circle* that suggests a welling-up of the central core.

clay colloid (a) A clay particle having a diameter less than 1 micron (0.001 mm) (Jacks et al, 1960, p.24). (b) A colloidal substance consisting of clay-size particles.

claycrete Weathered argillaceous material forming a layer immediately overlying bedrock.

clay doll A concretion, resembling a doll, found in clayey glacial-lake deposits.

clay dune A dune composed of clay fragments heaped up by the wind, as in the lower Rio Grande Valley, Tex. (Coffey, 1909).

C layer The seismic region of the Earth between 410km and 1000 km, equivalent to the *transition zone* of the upper mantle. It is a part of a classification of the interior of the Earth having layers A to G.

clayey Abounding in, consisting of, characterized by, or resembling clay; *argillaceous.*

clayey breccia A term used by Woodford (1925, p.183) for breccia containing at least 80% rubble and 10% clay, and no more than 10% of other material.

clayey sand (a) An unconsolidated sediment containing 50-90% sand and having a ratio of silt to clay less than 1:2 (Folk, 1954, p.349). (b) An unconsolidated sand containing 40-75% sand, 12.5-50% clay, and 0-20% silt (Shepard, 1954).

clayey sandstone (a) A consolidated *clayey sand.* (b) A sandstone containing more than 20% clay (Krynine, 1948, p.141).---Cf: *argillaceous sandstone.*

clayey silt (a) An unconsolidated sediment containing 40-75% silt, 12.5-50% clay, and 0-20% sand (Shepard, 1954). (b) An unconsolidated sediment containing more particles of silt size than of clay size, more than 10% clay, and less than 10% of all other coarser sizes (Wentworth, 1922).

clay gall (a) A small, markedly flattened, somewhat rounded pellet or curled fragment, chip, or flake of clay, generally embedded in a sandy matrix and esp. abundant at the base of sandy beds. It may arise from the drying and cracking of a thin layer of coherent mud, the fragment commonly being rolled or blown into sand and buried, and forming a lenticular bleb upon wetting. (b) An ocherous, sometimes hollow inclusion of clay or mudstone, occurring esp. in oolitic limestone (Woodward, 1894, p.340). Syn: *crick.*---Syn: *gall.*

clay gouge (a) A clayey deposit in a fault zone; *fault gouge.* (b) A thin seam of clay separating masses of ore, or separating ore from country rock. Se also: *gouge* [ore dep].

clay gravel Gravel containing fine-grained silica and clay, developing under puddling action (compaction) a dense, firm surface.

clay ironstone (a) A compact, hard, dark gray or brown, and fine-grained sedimentary rock consisting of a mixture of argillaceous material (up to 30%) and iron carbonate (siderite) occurring in layers of nodules or concretions or as relatively continous but irregular thin beds and usually associated with carbonaceous strata, esp. such a rock overlying a coal seam in the coal measures of the U.S. or Great Britain; a clayey iron carbonate, or an impure siderite ore occurring admixed with clays. The term has also been applied to an argillaceous rock containing iron oxide (such as hematite or limonite). See also: *blackband ironstone; clay-band ironstone.* (b) A siderite concretion or nodule occurring in clay ironstone and other argillaceous rocks, often displaying septarian structure. (c) A sheet-like deposit of clay ironstone.---Syn: *ironstone.*

clayite A term proposed by Mellor (1908) for a hydrous aluminum silicate thought to be the true clay substance in kaolin and considered to be an amorphous (colloidal) material of the same chemical composition as kaolinite. Cf: *pelinite.*

clay loam A soil containing 27-40% clay and 20-45% sand. See also: *silty clay loam; sandy clay loam.*

clay marl A whitish, smooth, chalky clay; a marl in which clay predominates.

clay mineral (a) One of a complex and loosely defined group of finely crystalline, metacolloidal, or amorphous hydrous silicates essentially of aluminum (and sometimes of magnesium and iron) with a monoclinic crystal lattice of the two- or three-layer type in which silicon and aluminum ions have tetrahedral coordination with respect to oxygen and in which aluminum, ferrous and ferric iron, magnesium, chromium, lithium, manganese, and other ions have octahedral coordination with respect to oxygen or hydroxyl. There may be exchangeable cations (usually calcium and sodium, sometimes potassium, magnesium, hydrogen, and aluminum) on the surfaces of the silicate layers, in amounts determined by the excess negative charge within the layer. Clay minerals are formed chiefly by alteration or weathering of primary silicate minerals such as feldspars, pyroxenes, and amphiboles, and are found in clay deposits, soils, shales, alteration zones of ore deposits, and other rocks in flake-like particles or dense, feathery aggregates of varying types; they are characterized by small particle size and ability to adsorb substantial amounts of water and ions on the surfaces of the particles. The most common clay minerals belong to the kaolin, montmorillonite, and illite groups. Syn: *clay; hydrosialite; sialite.* (b) Any crystalline substance occurring in the clay fraction of a soil or sediment.

claypan [geomorph] A term used in Australia for a shallow depression filled with clayey and silty sediment, and having a hard, sun-baked surface; a playa formed by deflation of alluvial topsoils in a desert, and in which water collects after a rain.

claypan [soil] A dense, heavy, and relatively impervious subsurface soil layer that owes its hardness to a relatively higher clay content than that of the overlying material, from which it is separated by a sharply defined boundary. It is usually hard when dry and plastic when wet, and is presumably formed by the concentration of clay by percolating waters. Cf: *hardpan; iron pan.* Also spelled: *clay pan.*

clay parting (a) Clayey material between a vein and its wall. Syn: *parting.* (b) A seam of hardened carbonaceous clay between or in beds of coal, or a thin layer of clay between relatively thick beds of some other rock (such as sandstone).

clay plug A mass of silt and clay materials containing much organic muck, deposited in and eventually filling an oxbow lake.

clay pocket A clay-filled cavity in rock; a mass of clay in rock or gravel.

clay rock An indurated clay, composed of argillaceous detrital material derived chiefly from the decomposition of feldspars, and sufficiently hardened to be incapable of being worked without grinding, but not chemically altered or metamorphosed; a *claystone.*

clay shale (a) A consolidated sediment consisting of no more than 10% sand and having a silt/clay ratio of less than 1:2 (Folk, 1954, p.350); a fissile claystone. (b) A shale that consists chiefly of clayey material and that becomes clay on weathering.

clay size A term used in sedimentology for a volume less than that of a sphere with a diameter of 1/256 mm (0.00016 in.). See also: *dust size.*

clay skin A cutan or *argillan* consisting of a coating of clay minerals on the surface of a ped or the wall of a void in a soil material.

clay slate (a) A low-grade, essentially unreconstituted slate, as distinguished from the more micaceous varieties that border on phyllite; specif. an *argillite,* less than 50% reconstituted, with a parting, slaty cleavage, or incipient foliation, or a weakly metamorphosed rock intermediate in character between a shale and a slate (Flawn, 1953, p.564). (b) A slate derived from an argillaceous rock (such as shale), in contrast to a slate derived from volcanic ash; a metamorphosed clay, with cleavage developed by shearing or pressure, as distinguished from *mica slate.* (c) An English term much used in the early 19th century for true slate.----Also spelled: *clayslate; clayslate.*

clay soil A soil containing a high percentage of fine particles and colloidal substances, becoming sticky and plastic when wet and forming very hard lumps or clods when dry; specif. a soil containing 40% or more of clay and not more than 45% of sand and 40% of silt. The term has also been used for a soil containing 30% or more of clay. Syn: *clay [soil].*

claystone [ign] (a) An obsolete term for a dull, altered, feldspathic igneous rock in which the groundmass or the whole rock has been reduced to a compact mass of earthy or clayey alteration products (Holmes, 1928, p.61-62). (b) A term used in Australia for a soft, earthy, feldspathic rock occurring in veins and having the appearance of indurated clay. --Also

spelled: *clay stone.*

claystone [sed] (a) An indurated clay having the texture and composition, but lacking the fine lamination or fissility, of shale; a massive *mudstone* in which the clay predominates over silt; a nonfissile clay shale. Flawn (1953, p.562-563) regards claystone as a weakly indurated sedimentary rock (less indurated than shale) whose constituent particles (clay minerals and/or micaceous paste) have diameters less than 0.01 mm. Shrock (1948a) regards claystone as a very fine-grained, somewhat unctuous, conchoidally fracturing sedimentary rock composed largely of clay material. Syn: *clay rock.* (b) A concretionary mass of clay found in alluvial deposits in the form of flat, rounded disks that are variously united to give rise to curious shapes (Fay, 1918, p.160). Also, a calcareous concretion frequently found in a bed of clay. (c) An old English term for an argillaceous limestone (Arkell & Tomkeieff, 1953, p.24).---Also spelled: *clay stone.*

claystone porphyry An old and indefinite name for a porphyry whose fine groundmass is more or less kaolinized "so as to be soft and earthy, suggesting hardened clay" (Kemp, 1934, p.203).

clay vein A body of clay, usually roughly tabular in form like an ore vein, which fills a crevice in a coal seam. It is believed to originate where pressure has been sufficient to force clay from the roof or floor into a small fissure, often altering or enlarging it. Syn: *dirt slip.*

clay-with-flints (a) A term used in southern England (as at North Downs and Dorset) for a residual deposit of reddish-brown, tenacious clay containing mechanically unworn flint fragments, lying unevenly and directly on the surface of chalk or occurring in funnel-shaped pipes penetrating to considerable depths. It represents in part the insoluble residue of chalk subjected to prolonged subaerial weathering and in part admixed waste material derived from formerly overlying Tertiary rocks. There is much variation in the relative proportions of flints and clay. (b) A term applied loosely to any clay-flint drift deposit that rests on chalk.

clean (a) Said of a diamond or other gemstone that is free from noticeable interior flaws. (b) Said of a mineral that is virtually free of undesirable nonore or waste-rock material.

clean sandstone A relatively pure or well-washed sandstone containing little matrix; specif. an arenite with less than 10% argillaceous matrix (Gilbert, 1954) or an orthoquartzite with less than 15% detrital clay matrix (Pettijohn, 1954). The particles are held together by a mineral cement or an introduced precipitated mineral that fills the pore spaces between grains and holds them together. It is usually deposited by fluids of low density. Cf: *dirty sandstone.*

clearing *polynya.*

cleat In a coal seam, a joint or system of joints along which the coal fractures. There are usually two cleat systems developed perpendicular to each other. See also: *face cleat; end cleat.* Also spelled: *cleet.*

cleating A syn. of *jointing,* used with reference to coal.

cleat spar Crystalline mineral matter occurring in the cleat planes of a coal seam; specif. *ankerite.*

cleavage [mineral] The breaking of a mineral along its crystallographic planes, thus reflecting crystal structure. The types of cleavage are named according to the structure, e.g. prismatic cleavage. Cf: *fracture [mineral]; parting [mineral].*

cleavage [struc geol] The property or tendency of a rock to split along secondary, aligned fractures or other closely spaced, planar structures or textures, produced by deformation or metamorphism. See also: *schistosity.* Obs. syn: *secondary cleavage.*

cleavage banding A compositional banding that is parallel to the cleavage rather than to the bedding. It results from the mechanical movement of incompetent material, such as argillaceous rocks, into the cleavage planes in a more competent rock, such as sandstone. The argillaceous bands are commonly only a few millimeters thick (Billings, 1954). Cf: *segregation banding.*

cleavage face In a crystal, a smooth surface produced by cleavage. It may be almost planar, e.g. in mica.

cleavage fan *fan cleavage.*

cleavage fold A *shear fold* in which the shear occurs along cleavage planes of secondary foliation. Syn: *shear-cleavage fold.*

cleavage fragment A fragment of a crystal that is bounded by cleavage faces.

cleavage mullion A type of *mullion* formed by the intersection of cleavage planes with bedding (Wilson, 1953). Cf: *pencil cleavage; fold mullion.*

cleavage plane That surface along which a rock tends to split, due to *cleavage* texture or structure. Cleavage planes in a rock are parallel to subparallel.

cleavelandite A white, lamellar or leaf-like variety of albite having an almost pure Ab content and often forming fanshaped aggregates of tabular crystals that frequently show mosaic development and present a bent appearance. It is formed as a late hydrothermal product deposited in veins or as masses during the late pegmatitic stages of granite formation. Also spelled: *clevelandite.*

cleavings The partings in a coal seam that separate it into beds.

cleet A less-preferred spelling of *cleat.*

cleft An abrupt chasm, cut, breach, or other sharp opening, such as a craggy fissure in a rock, a wave-cut gully in a cliff, a trench on the ocean bottom, a notch in the rim of a volcanic crater, or a narrow recess in a cave floor. Syn: *clift.*

cleft deposit A *pocket,* specifically a fissure filling, in alpine regions.

cleft girdle On a fabric diagram, an annular maximum occupying a small circle of the net (Turner and Weiss, 1963, p.58). Cf: *girdle; maximum.* Syn: *small-circle girdle.*

cleftstone *flagstone.*

cleithral Said of a bryozoan ovicell closed by the operculum of the parent zooid.

cleme A long hexactinellid-sponge spicule (monactin) with alternating thorn-like lateral spines arranged in two opposite rows. Cf: *uncinate.*

Clerici solution A solution of thallium malonate and thallium formate in water that is used as a *heavy liquid;* its specific gravity is 4.15. Cf: *Sonstadt solution; Klein solution; bromoform; methylene iodide.*

cleuch A Scottish var. of *clough.*

cleugh A Scottish var. of *clough.*

cleve (a) An English syn. of *cliff* in the sense of a hill or steep slope. (b) An English term for *brae* or steeply sloping ground.

cleveite A variety of uraninite containing rare earths (cerium).

clevelandite *cleavelandite.*

cliachite (a) A ferruginous bauxite. (b) A group name for colloidal aluminum hydroxides occurring as generally brownish constituents of bauxite. Syn: *kliachite; alumogel.*

cliff [geomorph] (a) *sea cliff.* (b) Any high, very steep to perpendicular or overhanging face of rock (sometimes earth or ice) occurring in the mountains or rising above the shore of a lake or river; a precipice. A cliff is usually produced by erosion, less commonly by faulting. (c) A British term for a steep slope or declivity, or a hill. Syn: *cleve.*

cliff [sed] (a) *clift.* (b) An old term used in SW England for rock lying directly above or between coal seams. Pl: *clives.*

cliffed headland A headland characterized by a cliff, such as one formed by erosion during the early development of an embayed coast.

cliff erosion *sapping.*

cliff glacier A short glacier that occupies a niche or hollow on a steep slope and does not reach a valley, such as a glacier perched on a ledge or bench on the face of a cliff. Cf: *hanging glacier.* Nonpreferred syn: *cornice glacier.*

cliffline The *coastline* on a steep coast, represented by an imaginary line along the base of the cliffs.

cliff of displacement *fault scarp.*

cliffordite A mineral: UTe_3O_8.

cliff overhang A rock mass jutting out from a slope; esp. the upper part or edge of an eroded cliff projecting out over the lower, undercut part, as that above a wave-cut notch.

cliffside The steep side of a cliff.

cliffstone A hard chalk found in England and used in paint, as a filter for wood, and in the manufacture of rubber (Thrush, 1968, p.218).

clift [geomorph] (a) Obsolete var. of *cleft.* (b) Dialectal var. of *cliff.*

clift [sed] A term used in southern Wales for various kinds of shale, esp. a strong, usually silty, mudstone. Syn: *cliff.*

cliftonite A black polycrystalline aggregate of graphite with cubic morphology, representing a minutely crystalline form of carbon occurring in meteorites. It is considered by some to be a pseudomorph after diamond.

climate [clim] A characteristic condition of the various elements of weather of a given region, such as temperature, humidity, rainfall, or other atmospheric elements that prevail in a given area. See also: *climatic province.*

climate classification An arrangement or description of the various climate types by a particular descriptive factor, such as temperature, rainfall, temperature-rainfall relationship, humidity, wind, or position relative to land and sea; e.g. *Thornthwaite's classification of climate; Köppen's classification of climate.* See also: *climatic zone; temperature zone.*

climate-stratigraphic unit A term used by the American Commission on Stratigraphic Nomenclature (1959, p.669) for a time interval now known as a *geologic-climate unit.* It is not strictly a stratigraphic unit.

climatic [ecol] Said of ecologic formations resulting from or influenced by differences in climate. Cf: *edaphic.*

climatic [clim] (a) Of or pertaining to a climate. (b) According to Langbein & Iseri (1960, p.13), any event, process, or change that is completed over several centuries or a few millenia. Cf: *secular.*

climatic accident A departure from the normal cycle of erosion, caused by marked changes in the climate, such as those effected by glaciation or by a change to aridity, independent of the normal climatic change due to loss of relief from youth to old age. "The term has become virtually obsolete since the idea of humid temperate climate being 'normal' no longer holds credence" (Monkhouse, 1965, p.3).

climatic amelioration A term designating warmer and/or dryer climate, applied specif. to the primary and secondary climatic trends of late glacial and postglacial time; an "improvement" of climate. Cf: *climatic deterioration.* Syn: *amelioration.*

climatic deterioration A term designating colder and/or wetter climate, applied specif. to the primary and secondary climatic trends of late glacial and postglacial time such as occurred during the Little Ice Age; a "degeneration" or "worsening" of climate. Cf: *climatic amelioration.* Syn: *deterioration.*

climatic optimum An informal term, frequently used in the past, designating the postglacial interval of most equitable climate and warmest temperatures. The concept is derived from the postglacial warm interval of Post (1924) who stratigraphically defined it as comprising the Boreal, Atlantic, and Subboreal climatic intervals, or from about 9000 to 2500 years ago. See also: *thermal maximum; Hypsithermal.* Also spelled: *Climatic Optimum.*

climatic peat Peat that characteristically occurs in a certain climatic zone.

climatic province A region characterized by a particular *climate.*

climatic snowline (a) The average line or altitude above which horizontal surfaces have more than 50 percent snow cover, averaged over a long time period of climatic significance (e.g.

10 or 30 years). Cf: *snowline.* (b) The same line as observed in late summer so that it approximately coincides with the firn line or *equilibrium line* on glaciers.----See also: *regional snow-line.*

climatic terrace A stream terrace whose formation is controlled by climatic changes that may induce aggradation or degradation of a valley.

climatic zone A general term for a latitudinal region characterized by a relatively homogeneous climate, e.g. any of the zones delimited by the Tropics of Cancer and Capricorn and the Arctic and Antarctic circles, or a zone or province in some *climatic classification.* Cf: *temperature zone.*

climax In ecology, the final stable stage of development that a community, species, flora, or fauna attains in a given environment. The major world climaxes correspond to formations and biomes. Cf: *pioneer.*

climax avalanche A snow avalanche of maximum size, containing a large portion of old snow, and arising from conditions that developed over a period of time longer than one year.

climbing bog An elevated boggy area on the margin of a swamp, usually in a region characterized by a short summer and considerable rainfall, caused by the upward growth of sphagnum from the original level of the swamp to higher ground.

climbing dune A dune formed by the piling up by wind of sand against a cliff or mountain slope. Syn: *rising dune.*

climbing ripple One of a series of cross-laminae produced by superimposed migrating ripples, in which the crests of vertically succeeding laminae appear to be advancing upslope. See also: *ripple drift.*

cline A gradational series of variant forms (e.g. morphologic or physiologic variations) in a group of related organisms, usually developing as a result of environmental or geographic transition.

clinker [coal] (a) Coal that has been altered by igneous intrusion. Cf: *natural coke.* Syn: *scoria coke; clinker bed.* (b) Masses of coal ash that are a byproduct of combustion. Cf: *coke.*

clinker [volc] A rough, jagged pyroclastic fragment such as aa that resembles the clinker or slag of a furnace. It is usually smaller than 15 cm in diameter. Adj: *clinkery.*

clinker bed *clinker* [coal]. Also spelled: *klinker bed.*

clinkertill Glacial till baked by the burning of lignite beds.

clinkery Adj. of *clinker* [volc]; it is used to describe the surface of a lava flow.

clinkstone An obsolete syn. of *phonolite* (in its broadest sense).

clinoamphibole (a) A group name for amphiboles crystallizing in the monoclinic system. (b) Any monoclinic mineral of the amphibole group, such as hornblende, cummingtonite, grunerite, tremolite, actinolite, riebeckite, glaucophane, and arfved-sonite.———Cf: *orthoamphibole.*

clinoaugite *clinopyroxene.*

clinoaxis In a monoclinic crystal, the lateral axis that is oblique to the vertical; it is the *a* axis. Cf: *orthoaxis.* Syn: *clinodiagonal.*

clinobronzite A variety of clinoenstatite containing iron; a clinopyroxene intermediate in composition between clinoenstatite and clinohypersthene, having less than 20 mole percent of $FeSiO_3$.

clinochlore A greenish mineral of the chlorite group: (Mg, Fe^{+2},Al)$_3$(Si,Al)$_2O_5$(OH)$_4$. It is sometimes substantially free from iron. Cf: *ripidolite.*

clinochrysotile A syn. of *chrysotile.* The term is used to denote its monoclinic form. Cf: *orthochrysotile.*

clinoclase A dark-green mineral: $Cu_3(AsO_4)(OH)_3$. Syn: *clinoclasite.*

clinodiagonal n. A syn. of *clinoaxis.*

clinodome A *first-order prism* in the monoclinic system. Its indices are {0kl} and its symmetry is 2/m. Cf: *brachydome.*

clinoenstatite A mineral of the clinopyroxene group: (Mg, Fe)SiO_3; specif. the monoclinic magnesium silicate $MgSiO_3$.

clinoenstenite A group name for the clinopyroxenes of the $MgSiO_3$-$FeSiO_3$ isomorphous series. It includes clinoenstatite, clinohypersthene, and clinoferrosilite.

clinoferrosilite A mineral of the clinopyroxene group: (Fe, Mg)SiO_3; specif. a mineral consisting of the monoclinic iron silicate $FeSiO_3$. See also: *ferrosilite.* Cf: *orthoferrosilite.*

clinograde Pertaining to the decreasing concentration of oxygen or other chemicals in the hypolimnion of a lake. Cf: *ortho-grade.*

clinograph An instrument for determining the departure from the vertical of a borehole, well, or shaft; e.g. the "Surwell clinograph" using a gyroscope and a spherical level with a photographic record. Cf: *driftmeter.*

clinographic Pertaining to a representation of a crystal in which no crystal face is projected as a line.

clinographic curve A curve representing the slope or slopes of an area of the Earth's surface as it varies with altitude; in practice it is designed to show the actual variation of the average slope within each contour interval.

clinographic projection An oblique projection used for representing crystals in such a manner that no crystal face will be projected as a line.

clinohedral class *domatic class.*

clinohedrite A colorless, white, or purplish monoclinic mineral: $CaZnSiO_3(OH)_2$.

clinoholmquistite A monoclinic mineral of the amphibole group: (Na,Ca)(Al,Li,Mg,Fe)$_7Si_8O_{22}$(OH,F)$_2$.

clinohumite A monoclinic mineral of the humite group: $Mg_9Si_4O_{16}$(F,OH)$_2$.

clinohypersthene A mineral of the clinopyroxene group: (Mg, Fe)SiO_3. It has a higher iron content (20-50 mole percent of $FeSiO_3$) than that of clinoenstatite.

clinolimnion The upper part of a *hypolimnion* where the rate of heat absorption falls off almost exponentially with depth. Cf: *bathylimnion.*

clinometer Any of various instruments used for measuring angles of slope, elevation, or inclination (esp. the dip of a geologic stratum or the slope of an embankment); e.g. a simple hand-held device consisting of a tube with cross hair, with a graduated vertical arc and an attached spirit level so mounted that the inclination of the line of sight can be read on the circular scale by centering the level bubble at the instant of observation. A clinometer is usually combined with a compass (e.g. the Brunton compass). See also: *inclinometer* [drill].

clinopinacoid In a monoclinic crystal, a pinacoid that is parallel to the mirror plane of symmetry and perpendicular to the axis of symmetry.

clinoplain An inclined plain projecting from the mountains and forming a low bluff on the side of a flood plain, as in the Rio Grande valley (Herrick, 1904, p.379).

clinoptilolite A zeolite mineral: (Na,K,Ca$_{2-3}$Al$_3$(Al,Si)$_2$Si$_{13}$-O$_{36}$.12H$_2$O. It is a potassium-rich variety of *heulandite,* but was formerly believed to be a silica-rich variety of heulandite.

clinopyroxene (a) A group name for pyroxenes crystallizing in the monoclinic system and sometimes containing considerable calcium with or without aluminum and the alkalies. (b) Any monoclinic mineral of the pyroxene group, such as diopside, hedenbergite, clinoenstatite, clinohypersthene, clinoferrosilite, augite, acmite, pigeonite, spodumene, jadeite, and omphacite.———Cf: *orthopyroxene.* Syn: *monopyroxene; clinoaugite.*

clinounconformity An obsolete syn. of *angular unconformity.* Term proposed by Crosby (1912, p.296). Syn: *clinunconform-ity.*

clinoungemachite A monoclinic mineral consisting of a sulfate of ferric iron, sodium, and potassium. It is probably dimorphous with ungemachite.

clinozoisite A grayish-white, pink, or green mineral of the epidote group: $Ca_2Al_3Si_3O_{12}$(OH). It is the monoclinic dimorph of *zoisite* and grades into, but is lighter in color than, epidote.

clint [karst] A slab of *limestone pavement* that is separated from adjacent clints by solutional activity along joints. See also: *grike*.

clint [geomorph] A Scottish term used in a general sense for any hard or flinty rock, such as a rock or ledge projecting from a hillside or in a stream bed or between crevices; also, a rocky cliff.

clintonite (a) A mineral of the brittle-mica group: Ca(Mg, Al)$_3$(Al$_3$Si)O$_{10}$(OH)$_2$. It has a reddish-brown, copper-red, or yellowish color, and occurs in monoclinic crystals and foliated masses. Syn: *seybertite; xanthophyllite.* (b) A group name for the *brittle micas.*

Clinton ore A red, fossiliferous sedimentary iron ore, e.g. the Clinton formation (Middle Silurian) or correlative rocks of the east-central U.S., containing lenticular or oolitic grains of hematite. Cf: *fossil ore; flaxseed ore.*

clinunconformity Var. of *clinounconformity.*

clisere A sere that develops as a result of great physiographic changes.

clisiophylloid Said of an *axial structure* in a rugose coral (such as in *Clisiophyllum*) characterized by a short medial plate joining the cardinal and counter septa and resembling a spider web in transverse section.

clitter A term used in Devonshire, England, for a *scree* or mass of loose boulders, esp. the subangular and rounded granite boulders derived from tors. Syn: *clatter; clitter-clatter.*

clives Plural of *cliff* [sed], so called because of the "easy cleavage" of the rocks overlying the coal.

cloaca An exhalant chamber in an invertebrate; esp. the large central cavity of a sponge into which pores and/or canals empty and which communicates through the osculum directly and externally with the surface of the sponge. Pl: *cloacae.* Syn: *atrium; spongocoel; paragaster.*

clockwise inclination The inclination to the right of a heterococcolith suture as it proceeds to the periphery. Ant: *counterclockwise inclination.*

clod [mining] A miner's term for soft shale associated with a coal bed.

clod [sed] A miner's term applied to a soft, weak, or loosely consolidated shale (or to a hard, earthy clay), esp. one found in close association with coal or immediately overlying a coal seam. It is so called because it falls away in lumps when worked.

clod [soil] An artificially formed structural unit of soil. Cf: *ped.*

clone A group of individuals descended asexually from a single parent.

Cloosian dome An elliptical uplift or upwarping, described by Hans Cloos and exemplified in the East African and Rhine rift valleys (1939).

closed basin An enclosed area having no drainage outlet and from which water escapes only by evaporation, as in an arid region. Cf: *interior basin.*

closed bay A bay indirectly connected with the sea through a narrow pass.

closed contour A continuous contour represented on a map by a contour line that forms a closed loop and that does not intersect the edge of the arbitrary map area on which it is drawn; e.g. a *depression contour* indicating a closed depression, or a contour indicating a hilltop.

closed depression An area of lower ground indicated on a topographic map by a hachured *depression contour* line forming a closed loop; e.g. a fault sag, or a hollow below the general land surface, with no surface outlet. Syn: *topographic depression.*

closed drainage *internal drainage.*

closed fault A fault, or a section of a fault, the two walls of which remain in contact. Cf: *open fault.*

closed fold A fold, the limbs of which have been so compressed that they are parallel. The structure contour lines of

such a fold form a closed loop. A fold can be described on a relative scale from closed to open. Ant: *open fold.* Syn: *tight fold.*

closed form A crystal form whose faces enclose space, e.g. a dipyramid. Cf: *open form.*

closed lake A lake that does not have a surface effluent and that loses water by evaporation (as in an arid or semiarid region where the lake is usually saline or brackish) or by seepage (e.g. a *seepage lake*). Cf: *enclosed lake.* Ant: *open lake.*

closed ridge A circular, elliptical, oval, or irregularly shaped ridge of glacial material surrounding a central depression (or sometimes a mound of glacial material or a moraine plateau), and resulting from the melting of a block of stagnant ice (Gravenor & Kupsch, 1959, p.52-53).

closed structure A structure which, when represented on a map by contour lines, is enclosed by one or more closed contours; e.g. a closed anticline or a closed syncline. Ant: *open structure.*

closed system [chem] A chemical system in which, during the process under consideration, no transfer of matter (either into or out of the system) takes place. Cf: *open system.*

closed system [permafrost] A condition of freezing of the ground in which no additional ground water is available (Muller, 1947, p.214), exemplified by the pingos of the Mackenzie Delta in Canada. Ant: *open system.*

closed traverse A surveying traverse that starts and terminates upon the same station or upon a station of known position. Cf: *open traverse.*

close-grained Said of a rock, and of its texture, characterized by fine and tightly packed particles.

close ice An obsolete term for sea-ice concentration of 8/10 to 10/10; now replaced generally by *close pack ice* and *very close pack ice.* Syn: *close pack; packed ice.*

close-joints cleavage An old term for both *slip cleavage* and *fracture cleavage.*

close-packed structure A type of crystal structure that provides the tightest possible packing: a first layer of atoms in which each atom has six similar atoms touching it, a second layer of atoms fitting into the indentations of the first, and a third layer either as a repetition of the first or in a third position. Cf: *open-packed structure.*

close pack ice Pack ice in which the ice cover or concentration is 7/10 to 9/10 and composed of floes mostly in contact. See also: *close ice; broken ice.*

close packing The manner of arrangement of uniform solid spheres packed as closely as possible so that the porosity is at a minimum; e.g. the packing of a face-centered cubic lattice or of a close-packed hexagonal lattice. See also: *rhombohedral packing.* Ant: *open packing.*

close sand *tight sand.*

close suture A suture between immovably united but not fused crinoid ossicles. Cf: *loose suture.*

closing error *error of closure.*

closing the horizon Measuring the last angle of a series of adjacent horizontal angles at a station, required to make the series complete around the horizon. See also: *horizon closure.*

closterite A dense, laminated, brownish-red canneloid material from the Irkutsk River basin of Siberia. The organic matter is largely *Pila bibractensis* colonies (Twenhofel, 1950, p.475).

closure [drill] The difference in the relative positions of the bottom and the collar of a borehole, expressed in horizontal distance in a specific compass direction (Long, 1960).

closure [struc geol] In a fold, dome, or other structural trap, the vertical distance between the structure's highest point and its lowest contour that encloses itself. It is used in the estimation of petroleum reserves. Syn: *structural closure.* Cf: *fault closure.*

closure [surv] A cumulative measure of the various individual errors in survey measurements; the amount by which a series of survey measurements fails to yield a theoretical or previously determined value for a survey quantity. See also: *error of closure; discrepancy.*

clot A group of ferromagnesian minerals in igneous rock, from a few inches to a foot or more in size, commonly drawn out longitudinally, that may be an altered foreign inclusion or a segregation (Balk, 1937).

clotted *grumous.*

cloud An atmospheric mass of droplets of water (or ice crystals) condensed from water vapor around such nuclei as dust, pollen, negative ions. Clouds are classified according to height and shape.

cloudburst flood An ephemeral flood commonly occurring during an abrupt summer rain of high intensity, usually in an arid or semiarid region.

clouded agate A transparent or semitransparent, light-gray *agate* with irregular, indistinct, or more or less rounded spots or patches of darker gray resembling dark clouds. Syn: *cloud agate; cloudy agate.*

clouding The effect produced in crystals (such as of plagioclase) by the presence of numerous minute dark particles (microlites, dust-like specks, short rods, thin hair-like growths, needles) distributed through the crystals. The particles consist of one or more minerals recognized with difficulty and seldom with certainty.

clough (a) A British term for a cleft in a hill; esp. a ravine, gorge, or other narrow and rugged valley or glen with precipitous and rocky sides. (b) The cliff or precipitous face of a clough.---Pron: *kluf; klow* (as in "now"). Syn: *cleugh; cleuch; claugh.*

clove A term used in the Catskill Mountains, N.Y., for a narrow, deep valley; esp. a ravine or gorge. Etymol: Dutch *kloof,* "cleft".

Clovelly North American (Gulf Coast) stage: Miocene (above Duck Lake, below Foley).

clunch (a) A term used in England for various stiff, tough, or indurated clays, esp. one forming the floor of a coal seam; a fireclay or underclay, or a bluish hard clay. Also, a soft, fine-grained, often clayey rock (shale) that breaks readily into irregular layers and that does not make a good roof during coal mining. See also: *stone clunch.* (b) A term used in England for a soft limestone; specif. a marly chalk.

cluse A narrow, steep-sided, deep gorge or trench, or water gap, cutting transversely through an otherwise continuous ridge; esp. an antecedent valley crossing an anticlinal limestone ridge in the Jura Mountains of the European Alps. Etymol: French. Pron: *klooz.* Cf: *combe; val.*

cluster A group of criss-crossing dikes demonstrably related to an exposed pluton. Cf: *dike swarm; dike set.*

cluster analysis A statistical procedure for arranging a number of objects in homogeneous subgroups based on their mutual similarities and hierarchical relationships.

clusterite *botryoid.*

clymenid Any ammonoid belonging to the order Clymeniida, characterized by the dorsal, rather than ventral, position of the marginal siphuncle. Clymenids are found only in the Upper Devonian.

clypeus (a) The labrum-carrying part of the cephalon of a crustacean. (b) The part of carapace of an arachnid between its anterior edge and the eyes.---Pl: *clypei.*

CM pattern A sample-point pattern designed to distinguish different depositional environments of sediments and to define, compare, and correlate clastic sediments (Passega, 1957, p.1952). It represents the variations in a sedimentary deposit of the maximum or one-percentile particle size (C) and the median particle size (M), as plotted on a diagram (C on the ordinate, M on the abscissa). Syn: *CM diagram.*

cnidarian Any coelenterate belonging to the subphylum Cnidaria, characterized mainly by the presence of nematocysts and simple muscles. All known fossil coelenterates belong to this subphylum.

cnidoblast A cell that produces a coral nematocyst or that develops into a nematocyst.

Coahuilan North American provincial series: Lower Cretaceous (above Upper Jurassic, below Comanchean).

coak An obsolete var. of *coke.*

coal A readily combustible rock containing more than 50% by weight and more than 70% by volume of carbonaceous material including inherent moisture, formed from compaction and induration of variously altered plant remains similar to those in peat. Differences in the kinds of plant materials (type), in degree of metamorphism (rank), and in the range of impurity (grade) are characteristic of coal and are used in classification (American Society for Testing and Materials, 1970, p.70). Syn: *black diamond.*

coal age An informal designation of the Carboniferous; *age of coal.*

coal apple *coal ball.*

coal ash *ash.*

coal ball Concretions of mineralized plant debris, occurring in coal seams or in adjacent rocks. Cf: *ball coal; sulfur balls.* Syn: *ball; coal apple; negrohead.*

coal basin A coalfield with a synclinal basin structure, e.g. the Carboniferous coal measures of England.

coal bed A *coal seam.* Also spelled: *coalbed.*

coal blende *coal brass.*

coal brass Iron pyrites (pyrite) found in coal or coal seams. Syn: *coal blende.*

coal breccia Naturally fragmented coal in a seam. It often shows polished or slickensided surfaces (Stutzer & Noe, 1940, p.248).

coal classification (a) The analysis or grouping of coals according to a particular property, such as degree of metamorphism (*rank*), constituent plant materials (*type*), or degree of impurity (*grade*). (b) The analysis or grouping of coals according to the percentage of volatile matter, caking properties, and coking properties.

coal clay *underclay.*

coal equivalent The heat energy of fuels other than coal, expressed in terms of comparable heat energy of coal.

coalescing fan One of a series of confluent alluvial fans that form a bajada.

coalescing pediment One of a series of expanding pediments that join to produce a continuous pediment surrounding a mountain range or whose merging levels over a broad region may ultimately reduce a desert mountain mass to an approximately continuous level; one of a number of pediments that make up a *pediplain.*

coalescive neomorphism A term introduced by Folk (1965, p.22) for *aggrading neomorphism* in which small crystals are converted to large ones by gradual enlargement maintaining a uniform crystal size at all given times (all crystals are consuming or being consumed); e.g. the process that forms most microspar calcite. Cf: *porphyroid neomorphism.*

coalfield A region in which coal deposits of possible economic value occur.

coal gas The fuel gas produced from gas coal; its average composition, by volume, is: 50% hydrogen, 30% methane, 8% carbon monoxide, 4% other hydrocarbons, and 8% carbon dioxide, nitrogen, and oxygen (Nelson, 1965, p.89).

coal gravel A secondary coal deposit consisting of transported and redeposited fragments. Cf: *float coal.*

coalification The alteration or metamorphism of plant material into coal; the biochemical processes of diagenesis and the geochemical process of metamorphism in the formation of coal. See also: *carbonization; incorporation; vitrinization; fusinization; peat-to-anthracite theory; coalification break.* Syn: *carbonification; incarbonization; incoalation; bitumenization.*

coalification break That point in the process of *coalification* at which the quantity of gas produced begins to decrease instead of increase, relative to the increase in content of volatile matter. It is defined as the point at which the constituent vitrinite contains 29.5% volatile matter; this is the boundary between gas coal and fat coal. Microspores change from very dark to pale, due to their loss of methane (Stach, 1965, p.11-12).

coalingite A mineral: $Mg_{10}Fe_2(CO_3)(OH)_{24}.2H_2O$.

coal land An area containing coal beds which falls within the public domain.

Coal Measures A stratigraphic term used in Europe (esp. in Great Britain) for Upper Carboniferous, or for the sequence of rocks (typically, but not necessarily, coal-bearing) occurring in the upper part of the Carboniferous System. It is broadly synchronous with the Pennsylvanian of North America. Abbrev: C.M.

coal measures (a) A succession of sedimentary rocks (or *measures*) varying in thickness from a meter or so to a few thousand meters, and consisting of clays or shales, sandstones, limestones, and conglomerates with interstratified beds of coal and occasionally iron ore. (b) A group of coal seams.

coal-measures unit A sequence (from oldest to youngest) of coal, shale, and sandstone, occurring in coal measures.

coal pipe A cylindrical extension of a coal seam into the overlying rock, representing a tree stump that was rapidly buried.

coal plant A fossil plant found in association with, or contributing by its substance to the formation of, beds of coal, esp. in the coal measures.

coal seam A stratum or bed of coal, usually of such quality and thickness (at least 30 cm) as to be mined with profit. Syn: *coal bed*.

coal seat *seat earth*.

coal split *split*.

coaly Covered with coal, or containing or resembling coal; e.g. "coaly rashings", small pieces of soft, dark shale containing much carbonaceous material.

coarse [ore dep] A little-used syn of *low-grade* [ore dep].

coarse [sed] Composed of or constituting relatively large particles; e.g. "coarse sandy loam". Ant: *fine*.

coarse admixture A term applied by Udden (1914) to an *admixture* (in a sediment of several size grades) whose particles are coarser than those of the dominant or maximum grade; material coarser than that found in the maximum histogram class.

coarse aggregate The portion of an *aggregate* in which the larger particles are retained on a certain numbered sieve, such as those particles with diameters greater than 1/4 or 3/16 inch or 4.76 mm. Cf: *fine aggregate*.

coarse clay A geologic term for a *clay* particle having a diameter in the range of 1/512 to 1/256 mm (2-4 microns, or 9 to 8 phi units). Also, a loose aggregate of clay consisting of coarse clay particles.

coarse fragment A rock or mineral particle in the soil with an equivalent diameter greater than 2 mm; it may be gravelly, cobbly, stony, flaggy, cherty, slaty, or shaly. See also: *channery*. Cf: *soil separate*.

coarse-grained [geol] (a) Said of a crystalline rock, and of its texture, in which the individual minerals are relatively large; specif. said of an igneous rock whose particles have an average diameter greater than 5 mm (0.2 in.). Johannsen (1931, p.31) earlier used a minimum diameter of 1 cm, and referred to igneous rocks having walnut- to coconut-size grains as "very coarse-grained". Syn: *phaneritic*. (b) Said of a sediment or sedimentary rock, and of its texture, in which the individual constituents are large enough to see with the naked eye; specif. said of a sediment or rock whose particles have an average diameter greater than 2 mm (0.08 in., or granule size and larger). Cf: *coarsely crystalline*.----The term is used in a relative sense, and various size limits have been suggested and used. Cf: *fine-grained; medium-grained*.

coarse-grained [soil] Said of a soil in which gravel and/or sand predominate. In the U.S., the minimum average diameter of the constituent particles is 0.05 mm (0.002 in.), or as used by engineers, 0.074 mm (retained on U.S. standard sieve no.200); the International Society of Soil Science recognizes a diameter limit of 0.02 mm. Cf: *fine-grained*.

coarse gravel An engineering term for *gravel* whose particles have a diameter in the range of 19-76 mm (3/4 to 3 in.).

coarsely crystalline Descriptive of an interlocking texture of a carbonate sedimentary rock having crystals whose diameters are in the range of 0.25-1.0 mm (Folk, 1959), or exceed 0.2 mm (Carozzi & Textoris, 1967, p.5) or 4 mm (Krynine, 1948, p.143). Cf: *coarse-grained* [geol].

coarse pebble A geologic term for a *pebble* having a diameter in the range of 16-32 mm (0.6-1.3 in., or -4 to -5 phi units) (AGI, 1958).

coarse sand (a) A geologic term for a *sand* particle having a diameter in the range of 0.5-1 mm (500-1000 microns, or 1 to zero phi units). Also, a loose aggregate of sand consisting of coarse sand particles. (b) An engineering term for a *sand* particle having a diameter in the range of 2 mm (retained on U.S. standard sieve no.10) to 4.76 mm (passing U.S. standard sieve no.4) (c) A soil term used in the U.S. for a *sand* particle having a diameter in the range of 0.5-1 mm. The diameter range recognized by the International Society of Soil Science is 0.2-2 mm. (d) Soil material containing 85% or more of sand-size particles (percentage of silt plus 1.5 times the percentage of clay not exceeding 15), 25% or more of very coarse sand and coarse sand, and less than 50% of any other one grade of sand (SSSA, 1965, p. 347).

coarse silt A geologic term for a *silt* particle having a diameter in the range of 1/32 to 1/16 mm (31-62 microns, or 5 to 4 phi units). In Great Britain, the range 1/20 to 1/10 mm has been used. Also, a loose aggregate of silt consisting of coarse silt particles.

coarse topography A topography with coarse *topographic texture*, characterized by low drainage density and widely spaced streams, and common in regions of resistant rocks where the surface is incompletely dissected or the erosional features are on a large scale.

coast (a) A strip of land of indefinite width (may be many kilometers) that extends from the *coastline* or the base of the cliffs inland to the first major change in landform features; it excludes the *shore* (or the beach), which is a narrower zone seaward or lakeward from the coast. (b) The part of a country regarded as near the coast, often including the whole of the coastal plain; a littoral district having some specific feature, as the Gold *Coast*. (c) The border or bank of a river or other large body of water. (d) An obsolete term for a point of the compass, and for a region or tract of the Earth or a part of the world.

coastal Pertaining to a coast; bordering a coast, or located on or near a coast, as *coastal* waters or *coastal* lowlands.

coastal area The areas of land and sea bordering the shoreline, and extending seaward through the breaker zone (CERC, 1966, p.A6).

coastal beach A beach along a coast, containing well-sorted sediments chiefly of sand size, and characteristic of areas where the continental shelf is wide or where a pre-existing coastal plain is well established.

coastal desert Generally, any desert area bordering an ocean. See also: *west coast desert*.

coastal dune A sand dune found chiefly on low-lying land recently abandoned or built up by the sea; the dune may ascend a cliff and travel inland.

coastal energy The total energy, including that of waves, tides, and currents, available for work along the coast. Most well-known coasts are characterized as dominated by *wave energy*.

coastal lake A lake produced by shoreline processes, as by the formation of a bar across a bay or by the joining of an offshore island to the mainland by a double tombolo.

coastal marsh A marsh bordering a seacoast, generally formed under the protection of a barrier beach, or enclosed in the sheltered part of an estuary. Cf: *open-coast marsh.*

coastal plain (a) A low, generally broad but sometimes narrow plain that has its margin on the shore of a large body of water (esp. the ocean) and its strata either horizontal or very gently sloping toward the water, and that generally represents a strip of recently emerged sea floor or continental shelf; e.g. the coastal plain of SE U.S. extending from New Jersey to Texas for 3000 km along the Atlantic Ocean and the Gulf of Mexico. (b) Less restrictedly, any lowland area bordering a sea or ocean, extending inland to the nearest elevated land, and sloping very gently seaward; it may result from the accumulation of material, as along the Adriatic coast of northern Italy.--Not to be confused with *coast plain.*

coastal-plain shoreline The typical shoreline of emergence, resulting from the emergence of a submarine or sublacustrine coastal plain (Johnson, 1919, p.186).

coastland Land along a coast; esp. a section of seacoast.

coastline (a) The seaward limit of the permanently exposed land, or the extreme landward limit of the highest storm waves on the beach; the line that forms the boundary between the coast and the shore or the beach. Syn: *cliffline.* (b) Commonly, the line that forms the boundary between the land and the water, esp. the water of a sea or ocean. (c) A general term to describe the appearance or configuration of the land along a coast, esp. as viewed from the sea; it includes bays, but crosses narrow inlets and river mouths. (d) A broad zone of land and water extending indefinitely both landward and seaward from a shoreline.---Cf: *shoreline.*

coast plain (a) A wave-cut plain of denudation. (b) A base level marking the sea level to which the land has been reduced by subaerial forces (Reusch, 1894, p.349).---Not to be confused with *coastal plain.*

Coast Range orogeny (a) A name proposed by W.H. White (1959) for major deformation, metamorphism, and plutonism during Jurassic and Early Cretaceous time in the Coast Mountains of the Cordillera of British Columbia. It is broadly equivalent to the *Nevadan orogeny* of the U.S (b) A term sometimes used for the late Cenozoic orogenic events in southern California; *Pasadenean orogeny.*

coast shelf *submerged coastal plain.*

coated grain A term used by Wolf (1960, p.1414) for a sedimentary particle possessing concentric or enclosing layers of calcium carbonate disposed about a central nucleus; e.g. an oolith or pisolith, or a dolomitized remnant, or a skeletal grain encrusted by algae or foraminifers.

coated stone A gemstone partially or entirely covered by some transparent substance to heighten color, improve phenomenal effects, or conceal defects.

coba Uncemented rock or gravel underlying the nitrate (caliche) deposits of Chile. Cf: *congela.*

cobalt bloom *erythrite.*

cobalt glance *cobaltite.*

cobaltite A grayish to silver-white isometric mineral with a reddish tinge: CoAsS. It usually occurs massive and in association with smaltite, and represents an important ore of cobalt. Syn: *cobalt glance; white cobalt; gray cobalt.*

cobaltocalcite (a) A red variety of calcite containing cobalt. (b) *spherocobaltite.*

cobalt ocher (a) *erythrite.* (b) *asbolite.*

cobaltomenite A mineral: $CoSeO_3.2H_2O$.

cobalt pentlandite A mineral: $(Co,Ni)_9S_8$. It is the cobalt analogue of pentlandite.

cobalt pyrite A variety of pyrite containing cobalt.

cobalt pyrites *linnaeite.*

cobalt vitriol *bieberite.*

cobb *promontory.*

cobble [geomorph] A term used in NE U.S. for a rounded hill of moderate elevation.

cobble [part size] (a) A rock fragment larger than a pebble and smaller than a boulder, having a diameter in the range of 64-256 mm (2.5-10 in., or -6 to -8 phi units, or a size between that of a tennis ball and that of a volleyball), being somewhat rounded or otherwise modified by abrasion in the course of transport; in Great Britain, the range of 60-200 mm has been used. Also, a similar rock fragment rounded in place by weathering at or somewhat below the surface of the ground; e.g. a "cobble of exfoliation" or a "cobble of spheroidal weathering". See also: *large cobble; small cobble.* (b) A rock or mineral fragment in the soil, having a diameter in the range of 20-200 mm (Atterberg, 1905). In the U.S., the term is used for a soil particle having a diameter in the range of 75-250 mm (3-10 in.) (SSSA, 1965, p.333). Syn: *cobblestone.* (c) An engineering term for a particle having a diameter greater than 76 mm. (d) *cobblestone.*

cobble beach *shingle beach.*

cobble conglomerate A consolidated rock consisting mainly of cobbles.

cobble size A term used in sedimentology for a volume greater than that of a sphere with a diameter of 64 mm (2.5 in.) and less than that of a sphere with a diameter of 256 mm (10 in.).

cobblestone (a) A naturally rounded, usually waterworn stone suitable for use in paving a street or in other construction. Syn: *cobble; roundstone.* (b) A *cobble* in the soil. (c) A consolidated sedimentary rock consisting of cobble-size particles (Alling, 1943, p.265).

cobbly Characterized by cobbles; e.g. a "cobbly soil" or a "cobbly land" containing an appreciable quantity of cobbles (SSSA, 1965, p.333).

Coblenzian European stage: upper Lower Devonian (above Gedinnian, below Eifelian). It includes Siegenian and Emsian.

cocarde ore *cockade ore.*

coccinite A mercury mineral, supposedly HgI_2.

coccolite A granular variety of diopside of various colors.

coccolith (a) A general term applied to various microscopic calcareous structural elements or button-like plates having many different shapes and averaging about 3 microns in diameter (some have diameters as large as 35 microns), constructed of minute calcite or aragonite crystals, and constituting the outer skeletal remains of a coccolithophore. Coccoliths are found in chalk and in deep-sea oozes of the temperate and tropical oceans, and were probably not common before the Jurassic. See also: *rhabdolith.* (b) Two shields connected by a central tube in a coccolithophore. (c) A term loosely applied to a coccolithophore.

coccolithophore Any of numerous, minute, mostly marine, planktonic biflagellate protists having brown pigment-bearing cells that at some phase of their life cycles are encased in a sheath of coccoliths to form a complex calcareous shell; a flagellate organism that produces coccoliths. Coccolithophores have been variously classified as algae and as protozoans. Var: *coccolithophorid.*

coccosphere (a) The entire spherical or spheroidal test or skeleton of a coccolithophore composed of an aggregation of interlocking coccoliths that are external to or embedded within an outer gelatinous layer of the cell. (b) A coccolithophore.

cocinerite A doubtful silver-gray mineral: Cu_4AgS.

cocite A fine-grained hypabyssal rock containing olivine and diopside phenocrysts in a groundmass of leucite, orthoclase, biotite, and magnetite.

cockade ore A metallic ore in which the minerals are deposited in successive crusts around rock fragments, e.g. around vein breccia fragments. Syn: *cocarde ore; ring ore; sphere ore.*

cockpit (a) An originally Jamaican term for steep-sided, closed depression associated with karst. (b) The star-shaped depression associated with *cone karst.*

cockpit karst A syn. of *cone karst*, referring to the depressions.

cockscomb barite A comb-like variety of barite displaying tabular crystals disposed roughly parallel to one another.

cockscomb pyrites A crest-like form of *marcasite* in twin crystals. Cf: *spear pyrites.*

cockscomb ridge A term used in South Africa for a wind-chiseled ridge, similar to a yardang.

cockscomb structure *hacksaw structure.*

coconinoite A light creamy-yellow secondary mineral: $Fe_2Al_2(UO_2)_2(PO_4)_4(SO_4)(OH)_2.20H_2O$.

cod A bag-shaped area of water or land; e.g. the inmost recess of a bay or meadow.

coda The latter part of a seismogram following the early, identifiable surface waves and in which arrive long trains of waves. It may last for hours, especially if long oceanic paths are involved.

coefficient of acidity *oxygen ratio.*

coefficient of anisotropy The square root of the ratio of the true transverse resistivity to true longitudinal resistivity in an anisotropic material. Symbol: λ.

coefficient of earth pressure The principal-stress ratio at a given point in a soil.

coefficient of fineness The ratio of suspended solids to turbidity; a measure of the size of particles causing turbidity, the particle size increasing with coefficient of fineness (ASCE, 1962).

coefficient of kinematic viscosity *kinematic viscosity.*

coefficient of permeability *permeability coefficient.*

coefficient of storage *storage coefficient.*

coefficient of thermal expansion The relative increase of the volume of a system with increasing temperature in an isobaric process.

coefficient of transmissibility *transmissibility coefficient.*

coefficient of variation The standard deviation of a set of data divided by its arithmetic mean. Syn: *coefficient of variability.*

coefficient of viscosity The ratio of the rate of shear stress in a substance to the rate of shear strain; *viscosity.*

coefficient of volume compressibility The compression of a lithologic unit per unit of original thickness per unit increase of pressure.

coelenterate Any multicelled invertebrate belonging to the phylum Coelenterata, characterized by a body wall composed of two layers of cells connected by a structureless mesogloea, by a single body cavity with a single opening for ingestion and egestion, and by radial or biradial symmetry. Their stratigraphic range is Precambrian to present.

coelenteron The spacious, internal cavity enclosed by the body wall of a coelenterate and opening externally through the mouth. Pl: *coelentera.*

coelestine *celestite.*

coeloconoid Said of a gastropod shell that approaches a conical shape but has concave sides. Cf: *cyrtoconoid.*

coelom The general *body cavity* occurring in multicelled animals other than the sponges and coelenterates, as distinguished from all special cavities such as the *enteron.* Where well developed, the coelom forms a large space between the alimentary viscera and the body walls. Adj: *coelomic.* Syn: *coelome.*

coelome Var. of *coelom.*

coenenchyme The complex mesogloea uniting the polyps of a compound coral; a collective term for both the coenosteum and the coenosarc. Adj: *coenenchymal.* Syn: *coenenchym; coenenchyma.*

coenobium A colony of algal cells having a definite cell arrangement and not increasing in number at maturity.

coenocyte An organism such as certain green algae that consists of continuous, multinucleate protoplasm, lacking walls to separate protoplasts.

coenosarc Common soft tissue connecting coral polyps in a colony (TIP, 1956, pt.F, p.246).

coenosis A population that is held together by ecologic factors in a state of unstable equilibrium (Stamp, 1966, p. 115). Var: *cenosis.* Plural: *coenoses.*

coenospecies A var. of *cenospecies.*

coenosteum (a) Calcareous skeletal deposits formed between the individual corallites of a colonial coral. (b) Vesicular or dense calcareous skeletal material between zooecia of some cyclostomatous and cryptostomatous bryozoans, esp. in the exozone.---Pl: *coenostea.*

coercive force The opposing applied magnetic field H required to reduce the remanent magnetization of a substance to zero. See also: *hysteresis.* Syn: *coercivity.*

coercivity *coercive force.*

coeruleolactite A mineral: $(Ca,Cu)Al_6(PO_4)_4(OH)_8.4-5H_2O$. It has a milk-white to sky-blue color, and is related to turquoise.

coesite A monoclinic mineral: SiO_2. It is a very dense (2.93 g/cm^3) polymorph of *quartz* and is stable at room temperature only at pressures above 20,000 bars; the silicon has a coordination number of 4. Coesite is found naturally only in structures that are presently best explained as impact craters or in rocks (such as suevite) associated with such structures. Cf: *stishovite.*

coetaneous A suggested replacement for the term *isochronous*, in the sense of being equal in duration or uniform in time.

coffee rock An indurated horizon containing iron and organic material in a coastal dune deposit, produced from the leaching action of percolating rainwater.

coffee shale A driller's term, used in the Appalachian basin of the U.S., for well cuttings consisting of dark-colored shale chips mixed with lighter-colored mud.

coffinite A black mineral: $U(SiO_4)_{1-x}(OH)_{4x}$. It is an important ore of uranium and occurs in many sandstone deposits and in hydrothermal veins.

cogeoid *compensated geoid.*

cognate [struc geol] Said of fractures or joints in a system that have the same time and deformational type of origin. Cf: *conjugate; complementary.*

cognate [pyroclast] Said of pyroclastics that are *essential* or *accessory.* Cf: *accidental.*

cognate inclusion *autolith.*

cognate xenolith *autolith.*

cogwheel ore The mineral *bournonite* esp. when occurring in wheel-shaped twin crystals.

cohenite A tin-white, isometric, meteorite mineral: $(Fe,Ni,Co)_3C$. It occurs as an accessory mineral in iron meteorites. The artificial material is *cementite.*

coherent [geochem] Said of a group of elements which, owing to similarity in radius and valency, occur intimately associated in nature, such as entering into the same minerals at about the same stage of fractional crystallization; e.g. zirconium and hafnium form a "coherent pair" (Goldschmidt, 1937, p.662).

coherent [geol] Said of a rock or deposit that is consolidated, or that is not easily shattered.

coherent [bot] In plant morphology, pertaining to like parts that are touching but not fused. Cf: *connate; adnate; adherent.*

cohesion *Shear strength* in a sediment not related to interparticle friction. Cf: *adhesion.*

cohesionless Said of a soil that has relatively low shear strength when air-dried, or cohesion when wet, e.g. a sandy soil. Cf: *cohesive.* Syn: *noncohesive; frictional.*

cohesive Said of a soil that has relatively high shear strength when air-dried, and cohesion when wet, e.g. a clay-bearing soil. Cf: *cohesionless.*

cohesiveness A mass property of unconsolidated, fine-grained sediments by which like or unlike particles (having diameters less than 0.01 mm) cohere or stick together by surface forces.

cohesive strength Inherent strength of a material when normal stress across the prospective surface of failure is zero.

coiling direction The direction (dextral or sinistral) in which a gastropod shell or a foraminiferal test is coiled. Changes in the coiling directions of planktonic foraminiferal tests are applied in stratigraphy to interpret paleoclimates or to determine correlations.

coire A var. of *corrie*. Etymol: Gaelic, "large kettle".

coke (a) A combustible material consisting of the fused ash and fixed carbon of bituminous coal, produced by driving off by heat the coal's volatile matter, i.e., by carbonization. It is grey, hard, and porous, and as a fuel it is practically smokeless.----Also spelled: *coak* (obsolete). Cf: *clinker* [*coal*]; natural coke; charcoal.

coke coal *natural coke.*

cokeite *natural coke.*

coking coal A bituminous coal containing about 90% carbon and suitable for the production of coke.

col (a) A high, narrow, sharp-edged *pass* or depression in a mountain range, generally across a ridge or through a divide, or between two adjacent peaks; esp. a deep pass formed by the headward erosion and intersection of two cirques, as in the French Alps. Also, the highest point on a divide between two valleys. (b) A marked, saddle-like depression in the crest of a mountain ridge; the lowest point on a ridge. Syn: *saddle*. (c) A short ridge or elevated neck of land connecting two larger and higher masses.---Etymol: French, from Latin *collum*, "neck". Cf: *gap; notch.* Syn: *joch.*

cold avalanche A snow avalanche involving the movement of dry snow and occurring during the time of greatest cold, usually coinciding with a drop in temperature; e.g. a *dry-snow avalanche.* Ant: *warm avalanche.*

cold desert A desert in a high latitude or high altitude area whose low temperature restricts or prohibits plant and animal life. The term is often used for tundra areas.

cold flow Permanent deformation of a solid under sustained pressure.

cold front The sloping *front* between an advancing cold air mass and a warmer air mass beneath which the colder air pushes like a wedge. Its passage is usually accompanied by a rise of pressure, a fall in temperature, a veer of wind, and a heavy shower or line squall. Cf: *warm front.*

cold fumarole A fumarole whose steam is less than 100°C in temperature.

cold lahar A flow of cooled volcanic materials down the slope of a volcano, produced by heavy rains. Cf: *hot lahar.* Syn: *cold mudflow.*

cold loess Periglacial loess derived from glacial outwash and formed in garlands about the Pleistocene ice sheets, as in northern Europe and in north-central U.S. Cf: *warm loess.*

cold mudflow *cold lahar.*

cold region An area where the temperature is sufficiently low to effect engineering design, construction, and operation.

cold spring A spring whose water has a temperature appreciably below the mean annual atmospheric temperature in the area; also but not preferred usage, any *nonthermal spring* in an area having thermal springs (Meinzer, 1923, p.55).

colemanite A colorless or white monoclinic mineral: $Ca_2B_6O_{11} \cdot 5H_2O$. It is an important source of boron, occurring in massive crystals or as nodules in clay.

coleoid Any member of a subclass of cephalopods having a muscular mantle, internal shell, fins, ink bag, chromatophores, suction cups, closed funnel, and camera-like eyes (Jeletsky, 1966). See also: *decapod.* Syn: *dibranchiate; endocochlian.*

colina A term used in the SW U.S. for a hillock or other small eminence. Etymol: Spanish. Syn: *collado.*

colk A *pothole* in a stream bed. Etymol: Dutch, "hollow", such as a hole eroded by outrushing water at the base of a broken dike. Cf: *kolk.*

collabral Conforming to the shape of the outer lip of a gastro-

pod shell, as indicated by growth lines.

collado (a) *colina.* (b) A term used in the SW U.S. for a saddle, gap, or pass.---Etymol: Spanish.

collain A kind uf *euvitrain* that consists of ulmin compounds that are redeposited by precipitation from solution. Cf: *ulmain.* Also spelled: *collite.*

collapse breccia A breccia formed by the collapse of rock overlying an opening, as by foundering of the roof of a cave or of the roof of country rock above an intrusion; e.g. a *solution breccia.* Syn: *founder breccia.*

collapse caldera A type of *caldera* produced by collapse of the roof of a magma chamber due to removal of magma by voluminous pyroclastic eruptions or by subterranean withdrawal of magma. Most calderas are of this type. Cf: *erosion caldera; explosion caldera.*

collapse crater A large lunar crater believed to have formed by roof subsidence of lava-filled cavities. The type is not well-established.

collapse sink A type of *doline* that is formed by collapse of an underlying cave.

collapse structure Any rock structure resulting from removal of support and consequent collapse by the force of gravity, e.g. gravitational sliding on fold limbs, salt solution causing collapse of overlying rocks in salt basins, sink-hole collapse, or collapse into mine workings.

collar [drill] (a) *drill collar.* (b) The mouth or open end of a drill hole or mine shaft; the rock surrounding the mouth of a borehole.

collar [geol] A band of rocks encircling rocks of another kind or of different structure; e.g. a discontinuous "collar" of downfaulted arcuate slices in a volcanic vent.

collar [paleont] The smooth tapering part of an echinoid spine located above the milled ring.

collar cell *choanocyte.*

collar pore A tiny aperture that occurs in a horizontal plate at the base of the cephalis in some nasselline radiolarians.

collecting area *catchment area* [grd wat].

collective group An assemblage of identifiable species of which the generic assignment is uncertain. For taxonomic convenience the assemblage is treated as a genus but requires no type species (ICZN, 1964, p.43).

collective species *species-group.*

collector well A well constructed by sinking a concrete cylinder, sealed at the bottom and with perforated pipes extending radially, into an aquifer. "The radial pipes are jacked hydraulically into the formation. Fine material around the pipes is removed by washing during construction. Collector wells are most often constructed in alluvial formations adjoining rivers. The radial pipes extend toward and under the river, thereby inducing movement of water downward through the stream bed to the pipes" (Chow, 1964, p.13-31). This well was developed by the engineer Leo Ranney and is also known as the *Ranney collector.*

collenchyma A strengthening tissue in a plant, composed of cells with walls usually thickened at the angles of the walls (Fuller & Tippo, 1949, p.955).

collencyte An amoebocyte of a sponge, often stellate or fusiform, that forms the cellular web of the mesohyle.

collenia A markedly convex or slightly arched or turbinate laminated algal structure (stromatolite), about 10 cm in diameter and less than 3 cm in height, produced by the late Precambrian blue-green (calcareous) algae of the genus *Collenia* by the addition of external layers of varying thicknesses. It is associated with beds of flat-pebble conglomerate.

collimate (a) To make refracted or reflected rays of light parallel to a certain line or direction, such as by means of a lens or concave mirror. (b) To adjust the line of sight of a surveying instrument or the lens axis of an optical instrument so that it is in its proper position relative to other parts of the instrument, such as by means of a collimator. (c) To adjust the fiducial marks of a surveying camera so that they define the principal point.

collimating mark *fiducial mark*.

collimation axis The straight line passing through the rear nodal point of the objective lens, perpendicular to the axis of rotation of the telescope of a surveying instrument. It is perpendicular to the horizontal axis of the telescope in a transit or theodolite, and perpendicular to the vertical axis in a leveling instrument.

collimation error The angle by which the line of sights of an optical instrument differs from what it should be; the angle between the line of collimation of a telescope and its collimation axis. Syn: *error of collimation*.

collimation line *line of collimation*.

collimation plane The plane described by the collimation axis of the telescope of a transit when the telescope is rotated about its horizontal axis.

collimator An optical device for producing a beam of parallel rays of light or for artificially creating an infinitely distant target that can be viewed without parallax, usually consisting of a tube having an objective converging lens with an arrangement of cross hairs placed in the plane of its principal focus. It is used in testing and adjusting certain optical surveying instruments. See also: *vertical collimator*.

colline A protuberant ridge of corallum surface between corallites of a scleractinian coral. Cf: *monticule* [*paleont*].

collinite (a) A variety of euvitrinite characteristic of collain and consisting of cell-wall material that has been jellified and precipitated plant material. Cf: *ulminite*. (b) A suggested, preferable syn. of *euvitrinite*.

collinsite A light-brown triclinic mineral: $Ca_2(Mg,Fe)(PO_4)_2 \cdot 2H_2O$. It is isomorphous with fairfieldite.

collite Var. of *collain*.

collobrierite A metamorphic rock composed of fayalite, garnet (almandine-spessartine), grunerite, magnetite, and some feldspar. The term was originated by LaCroix in 1917.

colloclarain A transitional lithotype of coal characterized by the presence of collinite, but more of other macerals than of collinite. Cf: *clarocollain*. Also spelled: *colloclarite*.

colloclarite Var. of *colloclarain*.

colloclast A term used by Sander (1967, p.327-328) for a weakly cemented, accretionary aggregate of mud or fine silt, typically possessing surficial irregularities with a lobate outline, and having a texture similar to the stratum in which it occurs; e.g. such an aggregate attaining silt or sand size, formed in place on the sea floor, or an aggregate attaining sand or gravel size, formed by the transport and redeposition of fragments made up of silt-, sand-, or gravel-size carbonates weakly cemented prior to movement.

collocryst A crystal formed by recrystallization of aggregated colloidal parent material, as in a mobilized sediment.

colloform Said of the rounded, globular texture of a colloidal mineral deposit. Syn: *gel texture*. Cf: *botryoidal; reniform*.

colloid (a) A particle-size range of less than 0.00024 mm, i.e. smaller than clay size (U.S. Naval Oceanographic Office, 1966). (b) Originally, any finely divided substance that doesn't occur in crystalline form; in a more modern sense, any fine-grained material in suspension, or any such material that can be easily suspended (Krauskopf, 1967).

colloidal complex In a soil, a mixture of humus and clay.

colloid plucking A mechanical-weathering process in which small fragments are pulled off or loosened from rock surfaces by soil colloids in contact with them (Reiche, 1945, p.14).

colloidstone A consolidated sedimentary rock consisting of colloid-size particles (Alling, 1943, p.265).

collophane Any of the massive cryptocrystalline varieties of apatite, often opaline, horny, dull, colorless, or snow-white in appearance, that constitute the bulk of phosphate rock and fossil bone and that are used as a source of phosphate for fertilizers; esp. *carbonate-apatite* or a hydroxylapatite containing carbonate, and sometimes francolite. The chemical formula, $Ca_3P_2O_8 \cdot 2H_2O$, is sometimes given for collophane, but is probably not a true mineral. Syn: *collophanite*.

collophanite *collophane*.

colluvial Pertaining to colluvium; e.g. "colluvial deposits" or "colluvial soil".

colluviation The formation of colluvium.

colluvium (a) A general term applied to any loose, heterogeneous, and incoherent mass of soil material or rock fragments deposited chiefly by mass-wasting, usually at the base of a steep slope or cliff; e.g. talus, cliff debris, and avalanche material. (b) Alluvium deposited by unconcentrated surface runoff or sheet erosion, usually at the base of a slope.---Cf: *slope wash*. Etymol: Latin *colluvies*, "collection of washings, dregs".

colmatage A term used in New Zealand for the artificial impounding of silt-laden water in order to build up the banks in the lower part of a river by subsequent deposition of alluvium; originally the term referred to the natural process of bank growth. Cf: *warping*.

Cologne earth *black earth* [coal]. Etymol: source near Cologne, Germany.

Cologne umber *black earth* [coal]. Etymol: source near Cologne, Germany.

cololite A trace fossil now assigned to the "genus" *Lumbricaria*, consisting of a cylindrical, string-like, and tortuous or convoluted body approximately 3 mm wide, probably representing a fossil cast of a worm, but formerly regarded as the petrified intestines of a fish or the contents of such intestines. It occurs esp. in lithographic limestones such as Solenhofen stone.

colombianite A ball-like glass object found near Cali in Colombia, once identified as a tektite but now believed to be of volcanic origin.

colonial Said of an animal that lives in close association with others of the same species and that usually cannot exist as a separate individual; esp. a "colonial coral" in which the individuals are attached together as a unit. Cf: *solitary coral*.

colonnade In columnar jointing, the lower zone that has thicker and better formed columns than does the upper zone, or *entablature*.

colony (a) A group of living or fossil organisms found in an area or rock unit other than that of which they are characteristic or that migrate into and become established in a barren area. (b) A physiologically connected aggregation of organisms; e.g. certain anthozoan corals.

colophonite (a) A coarse, cloudy, yellow-brown variety of andradite garnet. (b) A nongem variety of vesuvianite.

color [min] A phenomenon of light or visual perception by which otherwise identical objects may be differentiated (Webster, 1967). In mineral analysis, color is an important diagnostic feature; it is affected or formed by the response of light to composition and structure. Color of a metallic mineral is usually constant, whereas that of a nonmetallic mineral is more apt to vary. The color of a mineral's streak may differ from that of a massive sample.

color [sed] A mass property of a sediment, represented by the overall hue caused by combinations of the colors of the particles, surface coating, matrix, and cement, and controlled in part by the degree of fineness of the particles.

coloradoite A grayish-black isometric mineral: HgTe.

color center In crystal optics, a defect in the atomic structure that selectively absorbs a component of visible light. See also: *F center*.

colored stone A gemstone of any species other than diamond. This usage illogically classifies all varieties of such species as colored stones, including colorless varieties, but it does not include colored diamonds; however, it has proved a practicable and satisfactory classification.

color grade The grade classification into which a gem is placed by examination of its color in comparison to the color f other gems of the same variety (Shipley, 1951, p.52).

colorimeter An instrument for measuring and comparing the intensity of color of a compound for quantitative chemical analysis, usually based on the relationship between concentration of a chemical solution and the amount of absorption of certain characteristic colors of light. See also: *Dubosq colorimeter; spectrocolorimeter.*

colorimetric Pertaining to *colorimetry*, e.g. *colorimetric analysis.*

colorimetric analysis Quantitative chemical analysis performed by adding a certain amount of a substance to both an unknown and a standard solution and then comparing color intensities.

colorimetry The art or process of measuring and/or comparing colors, usually with a *colorimeter*, for quantitative chemical analysis.

color index In petrology, esp. in the classification of igneous rocks, a number that represents the percent, by volume, of dark or colored (i.e. mafic) minerals in a rock. According to this index, rocks may be divided into "leucocratic" (color index, 0-30), "mesocratic" (color index, 30-60), melanocratic (color index, 60-100). Syn: *color ratio.*

color-patch map *chorochromatic map.*

color ratio *color index.*

color temperature (a) An estimate of the temperature of an incandescent body, determined by observing the wavelength at which it is emitting with peak intensity (its color) and applying the Wien law. For an ideal blackbody, the temperature so estimated would be its true temperature and would agree with its effective temperature; but for actual bodies, the color temperature is generally only an approximate value. (b) The temperature to which a blackbody radiator must be raised in order that the light it emits may match a given light source in color. It is usually expressed in degrees Kelvin. Cf: *brightness temperature.*

colpa A nonrecommended syn. of *colpus.* Pl: *colpae.*

colpate Said of pollen grains having more or less elongated, longitudinal furrows (colpi) in the exine.

colpi Plural of *colpus.*

colporate Said of pollen grains having colpi in which there is a pore or some other modification of the exine (such as a transverse furrow), usually at the equator.

colpus A longitudinal furrow- or groove-like modification in the exine of pollen grains, associated with germination (either enclosing a germ pore or serving directly as the place of emergence of the pollen tube) and often also with harmomegathic swelling. It may be distal (as in monocolpate pollen), meridional (as in tricolpate pollen), or otherwise disposed. The membrane of the colpus consists of exine in which ektexine and/or endexine are thinned or absent. Pl: *colpi.* Cf: *pore* [*palyn*]; pseudocolpus. Syn: *furrow; germinal furrow; colpa sulcus* [*palyn*].

colpus transversalis *transverse furrow.*

columbite A black mineral: $(Fe,Mn)(Nb,Ta)_2O_6$. It is isomorphous with *tantalite*, occurs in granites and pegmatites, and is the principal ore of niobium as well as a source of tantalum. Syn: *niobite; dianite; greenlandite.*

columbotantalite A noncommittal term for minerals of the columbite-tantalite series.

columbretite An extrusive rock composed of sanidine and altered hornblende laths in a dense groundmass of corroded oligoclase microlites with interstitial sanidine enclosing rounded leucite grains. The embayments of the microlites are filled with analcime, augite, and magnetite (Johannsen, 1939, p.247).

columella [*paleont*] (a) A solid or nonsolid, pillar-like, calcareous axial structure of a corallite, formed by various modifications of the inner edges of septa. It commonly projects into the central part of the calice of many corals in the form of a sharp-pointed protuberance. See also: *trabecular columella;* *lamellar columella; fascicular columella; styliform columella.* (b) The solid or hollow, medial pillar surrounding the axis of a spiral gastropod shell, formed by the coalescence of the inner (adaxial) walls of the whorls. (c) A vertical rod between two horizontal rings, or within the shell cavity, in certain radiolarians.---Pl: columellae.

columella [*palyn*] One of the rodlets of ektexine that may branch and fuse distally to produce a tectum on pollen grains with complex exine structure. Pl: *columellae.*

columellar fold A *fold* or spirally wound ridge on the columella of a gastropod, projecting into the shell interior.

columellar lip The adaxial part of the *inner lip* of a gastropod shell.

columellate Possessing or forming a columella or columellae; e.g. said of pollen grains with a complex ektexine structure consisting of columellae.

column [*speleo*] *stalacto-stalagmite.*

column [*paleont*] (a) A cylindrical structure consisting of a series of disk-like plates mounted one on top of the other and attached to the aboral end of theca of crinoids, blastoids, and most cystoids, and presumably used for anchoring or as a means of support. The distal end is known to be variously modified in some species. Syn: *stem.* (b) The smooth cylindrical body wall of a scleractinian coral polyp between the basal disks and the oral disks.

column [*stratig*] The vertical or chronologic sequence of geologic formations in a region; specif. *geologic column.*

columnal One of the numerous individual vertical segments (ossicles or plates) that make up the column or stem of an echinoderm. Columnals are circular or polygonal, discoid, or button-shaped.

columnar A crystal habit that is a subparallel arrangement of columnar individuals.

columnar coal Coal that has developed a columnar fracture structure, usually due to metamorphism by an igneous intrusion.

columnar facet A normally circular indentation in the basal plates of cystoid theca to accommodate the proximal end of the column.

columnar ice Ice that has been built by *columnar ice crystals*, mostly broader in the lower part than in the upper part of the ice cover. Black ice is columnar ice.

columnar ice crystal A vertical ice column. Massed together columnar ice crystals form *columnar ice.*

columnar jointing Parallel, prismatic columns, either hexagonal or pentagonal in cross section, in basaltic flows and sometimes in other extrusive and intrusive rocks. It is formed as the result of contraction during cooling. Syn: *columnar structure; prismatic jointing; prismatic structure.*

columnar section A *vertical section*, or graphic representation in a vertical strip, of the sequence and original stratigraphic relations of rock units that occur throughout a given area or at a specific locality. Thicknesses of rock units are drawn to scale, and their lithology indicated by standard or conventional symbols, which are usually supplemented by brief descriptive notes (indicating also age, rock classification, fossil contents, etc.) in a parallel column. See also: *geologic column.*

columnar structure [*mineral*] An aggregate of slender, elongate, individual crystals in a columnar, nearly parallel arrangement.

columnar structure [*sed*] A primary sedimentary structure found in some calcareous shales or argillaceous limestones, consisting of columns (9-14 cm in diameter, and 1-1.4 m in length) perpendicular to bedding and oval to polygonal in section (Hardy & Williams, 1959).

columnar structure [*struc geol*] *columnar jointing.*

column chromatography A chromatographic technique for separating components of a sample by moving it in a mixture or solution through tubular structures, packed with appropriate substrates, in such a way that the different components have different mobilities and thus become separated. See also: *chromatography.*

column crystal A snow crystal in the shape of a hexagonal prism, either solid or hollow.

colusite A bronze-colored tetrahedral mineral: $Cu_3(As,Sn,V,Fe,Te)S_4$.

comagmatic Said of igneous rocks (or of the region in which they occur) having a common set of chemical, mineralogic, and textural features, and hence regarded as having been derived from a common parent magma. See also: *consanguinity.* Less preferred syn: *consanguineous.*

comagmatic region *petrographic province.*

Comanchean (a) North American provincial series: Lower and Upper Cretaceous (above Coahuilan, below Gulfian). (b) Obsolete name applied to a period (or system of rocks) between the Jurassic below and the Cretaceous above. Syn: *Comanchian.*

comb [ore dep] n. A mineral-deposit vein filling which individual crystals have grown perpendicular to the vein walls. ---- adj. Said of that crystal texture or structure in a vein.

comb [geomorph] (a) The crest of a mountain or hill; a mountain ridge. Syn: *combe.* (b) A var. of *combe,* a valley. (c) A var. of *cwm.*

comb [paleont] (a) An arachnid structure resembling a comb; e.g. a row of serrated bristles on the fourth tarsi of an araneid, found only in the family Theridiidae, or a pair of abdominal appendages situated on the sternite following upon the genital opercula, present in all scorpions but not in any other arachnids. (b) A radial series of knobs or projections in acantharian radiolarians.

combe [geomorph] (a) A British term, also used in France, for a small, deep, and narrow valley running down to the sea. Also, a short, bowl-shaped, generally unwatered valley, basin, hollow, or cove on the flank of a hill, esp. a dry, closed-in valley on the sides of and under the chalk downs of southern England. Etymol: Celtic. Syn: *comb; coom; coomb; coombe.* (b) A large, high-lying longitudinal depression or valley along the crest or side of an anticline in the folded Jura Mountains of the European Alps, formed either by downfaulting or more generally by differential erosion, often occurring along the line of junction of a hard crystalline rock with one that is soft. Cf: *cluse.* See also: *val; crêt.* (c) The amphitheater-like steep bank of an incised meandering stream. (d) Var. of *comb,* a mountain crest. (e) A term used in England and southern Scotland for a glacial valley and for a *cirque.* Also, a var. of *cwm.*

combeite A rhombohedral mineral: $Na_4Ca_3Si_6O_{16}(OH,F)_2$.

comber (a) A long, curling, deep-water ocean wave whose high, breaking crest (much larger than a whitecap) is pushed forward by a strong wind. (b) A long-period *spilling breaker.*---Cf: *roller.* Syn: *beachcomber.*

combe rock *coombe rock.*

combination Any set that can be made by using all or part of a given collection of objects without regard to sequence. See also: *permutation.*

combination archeopyle An *archeopyle* formed by the release of a part of the dinoflagellate-cyst wall that corresponds to plates of more than one thecal plate series (such as combining the plates of the apical series and the precingular series).

combination well An *open well* connected to one or more other wells.

combined twinning A rare type of twinning in quartz in which there appears to be a 180° rotation around c with reflection over 1120 or over |0001|. The crystal axes are parallel but the electrical polarity of the a-axes is not reversed in the twinned parts.

combined water "Water of solid solution and water of hydration which does not freeze even at the temperature of -78°C" (Muller, 1947, p.214).

comb ridge A jagged, sharp-edged, steep-sided mountain ridge whose crest resembles a cockscomb because it bea pinnacles alternating with notches; an *arête* marked by a se ries of *aiguilles.* It commonly separates adjacent cirques glaciated mountain regions.

combustible shale *tasmanite* [coal].

comendite A white *pantellerite.* Its name is derived from Com ende, Sardinia. Cf: *taurite.*

comitalia Small megascleres (spicules) adherent to the ra of larger megascleres in lyssacine hexactinellid sponges.

commensal Said of organisms living in a state of *commensa ism.*

commensalism The relationship that exists between two org nisms in which one benefits from the other, the other bein neither benefitted nor harmed. Adj: *commensal.* Cf: *paras ism; inquilinism; mutualism; symbiosis.*

commercial dust Impure *gold dust.*

commercial granite A general term for a decorative buildir stone that is hard and crystalline. It may be a granite, gneis syenite, monzonite, granodiorite, or anorthosite or laurvikit See also: *black granite.*

comminution (a) The gradual diminution of a substance to fine powder or dust by crushing, grinding, or rubbing; spec the reduction of a rock to progressively smaller particles weathering, erosion, or tectonic movements. (b) The brea ing, crushing, or grinding by mechanical means of stone, coa or ore, as for road metal, aggregate, or ballast.----Syn: *pulve ization; trituration.*

commissural plane The plane containing the cardinal marg of a brachiopod and either the commissure of a rectimarg nate shell or points on the anterior commissure midway be tween crests of folds in both valves (TIP, 1965, pt.H, p.142).

commissure [paleont] (a) The line of junction between th edges or margins of valves in a brachiopod or bivalve mollus (b) In a plant, the surface of joining of two mericarps, or appressed stigmas or style branches. (c) The groove of th laesura along which an embryophytic spore germinates. It essentially equivalent to *suture.*

common In the description of coal constituents, 5-10% of particular constituent occurring in the coal. (ICCP, 1963). C *rare; very common; abundant; dominant.*

common-banded coal *banded coal.*

common bud The growing edge or surface of a colony cyclostomatous and Paleozoic bryozoans.

common canal A tubular cavity lying along the dorsal side of stipe of a graptolithine, formed by the sum of the commo portions of all thecae. It is approximately the sum of all pre thecae along a stipe.

common depth point A *depth point* that produces seismic re flections at different offset distances on several profiles. Se also: *common-depth-point shooting.* Abbrev: CDP.

common-depth-point shooting A type of seismic shooting d signed for multiple subsurface coverage to establish commo *depth points.*

common feldspar *orthoclase.*

common lead Lead isotopes which, at the time considere because of radioactive decay, have only insignificant amoun of uranium and thorium with them. They are composed of p meval lead plus *radiogenic lead.* Syn: *ordinary lead.*

common-lead method The determination of an age in yea for a lead by isotopically determining the lead ratios Pb^{20} Pb^{204}, Pb^{207}/Pb^{204}, and Pb^{208}/Pb^{204} which can be plotte and compared to proposed growth curves (time curves).

common mica *muscovite.*

common opal A variety of opal that never exhibits play color, that is found in a wide variety of colors and pattern that sometimes occurs as an earthy form, and that general is not suitable for gem use. Cf: *precious opal.*

common pyrites *pyrite.*

common salt A colorless or white crystalline compound co sisting of sodium chloride (NaCl), occurring abundantly in n ture as a solid mineral (*halite*) or in solution (constituting about 2.6% of seawater), or as a sedimentary deposit (such as in salt domes or as a crust around the margin of a salt lake).

communication pore An opening in the interzooidal wall of a bryozoan. Syn: *mural pore*.

communtiy *biocoenosis*.

commutate To reverse, mechanically, the direction of a unidirectional electric current by a change of connections, or to reverse every other cycle of an alternating current so as to form a unidirectional current.

compact (a) Said of any rock or soil that has a firm, solid, or dense texture, and whose particles are firmly united or closely packed with very slight intervening space. (b) Said of a close-grained rock in which no component particles or crystals can be recognized by the unaided eye. (c) Said of a finely textured rock with low matrix porosity. (d) *massive* [ign].

compactability A property of a sedimentary material that permits it to decrease in volume or thickness under pressure; it is a function of the size, shape, hardness, and brittleness of the constituent particles.

compaction [soil] Any process (such as by weight of overburden or desiccation) by which a soil mass loses pore space and achieves a higher density, thereby increasing the bearing capacity, reducing the tendency to settle or deform under load, and increasing the general stability of the soil; the densification of a soil by mechanical means, accomplished by rolling, tamping, or vibrating. See also: *consolidation*.

compaction [sed] (a) Reduction in bulk volume or thickness of, or the pore space within, a body of fine-grained sediments in response to the increasing weight of overlying material that is continually being deposited or to the pressures resulting from earth movements within the crust. It is expressed as a decrease in porosity brought about by a tighter *packing* of the sediment particles. See also: *differential compaction*. (b) The process whereby fine-grained sediment is converted to consolidated rock, such as a clay lithified to a shale.

compaction curve The curve showing the relationship between the density (dry unit weight) and the water content of a soil for a given compactive effort. Syn: *moisture-density curve*.

compaction fold A type of *supratenuous fold* developed by differential compaction of sedimentary material over more resistant rock, over a subsurface structure such as a buried hill, or over an active fault or fold.

compaction test A laboratory compacting procedure to determine the optimum water content at which a soil can be compacted so as to be most stable and best suited for use in fills and embankments and for other engineering purposes. The method involves placing (in a specified manner) a soil sample at a known water content into a mold of given dimensions, subjecting it to a compactive effort of controlled magnitude, and determining the resulting unit weight (ASCE, 1958, term 74). The procedure is repeated for various water contents sufficient to establish a relation between water content and unit weight. Syn: *moisture-density test*.

compartmental plate A rigid articulated skeletal element forming part of the shell wall in certain cirriped crustaceans; e.g. a lateral and a carinolateral. Syn: *mural plate*.

compass [paleont] A slender, arched, radial rod in ambulacral position at the top of Aristotles's lantern in an echinoid.

compass [surv] (a) An instrument or device for indicating horizontal reference directions relative to the Earth by means of a magnetic needle or group of needles; specif. *magnetic compass*. Also, a nonmagnetic device that serves the same purpose; e.g. a *gyrocompass*. (b) A simple instrument for describing circles, transferring measurements, or subdividing distances, usually consisting of two pointed, hinged legs (one of which generally having a pen or pencil point) joined at the top by a pivot.

compass bearing A *bearing* expressed as a horizontal angle

measured clockwise from north as indicated by a magnetic compass.

compass error The amount by which a compass direction differs from the true direction, usually expressed as the number of degrees east or west of true azimuth north and marked plus or minus according to whether the compass direction is less or greater than true azimuth. It combines the effects, or is the algebraic sum, of the deviation and variation of the compass.

compass rose A graduated circle, usually marked in degrees, printed or inscribed on a chart for reference, and indicating directions. It may be oriented with respect to true north or to magnetic north, or to both. Compass roses are common to nautical charts, but are also used to give directions of prevailing winds and of geologic structures (such as faults and joints). Syn: *rose*.

compass traverse A surveying traverse in which a number of straight lines are accurately measured by tape or pace and their bearings are taken by a prismatic compass. It is executed where lines are not shown or not clearly defined on a map, or when a base map of an area is not available.

compensated geoid One of various surfaces approximating the *geoid* and obtained from *Stoke's formula* or similar equations using gravity anomalies in the calculations. (The compensated geoid is sometimes referred to in an abbreviated form, *cogeoid*). A different cogeoid is obtained for each system of reduction; free-air cogeoid, isostatic cogeoid, etc.

compensation depth [oceanog] *calcium carbonate compensation depth*.

compensation depth [tect] Var. of *depth of compensation*.

compensation level [oceanog] The depth of the ocean at which the consumption and production of oxygen are equal. It is the deepest level at which phytoplankton, who produce oxygen, can exist; i.e., it is the bottom of the *euphotic zone*.

compensation level [tect] Var. of *depth of compensation*.

compensation method A procedure for determining the voltage difference between two points in the ground by compensating this voltage difference with a voltage difference derived from a source and adjusted in phase and amplitude to effect the compensation. See also: *compensator*.

compensation point In crystal optics, the point at which an interference color is compensated by introduction of a quartz wedge or by rotation of a Berek compensator.

compensation sac The *ascus* of an ascophoran cheilostome.

compensator [elect] An instrument to determine the voltage difference between two points in the ground by the *compensation method*.

compensator [optics] An apparatus in a polarizing microscope that measures the phase difference between two components of polarized light, e.g. a *Berek compensator* or a *biquartz plate*.

compensatrix The *ascus* of an ascophoran cheilostome.

competence The ability of a current of water or wind to transport detritus, emphasizing the particle size rather than the amount, measured as the diameter of the largest particle transported. It depends, therefore, on velocity: a small but swift stream, for example, may have greater competence than a larger but slower-moving stream. Cf: *capacity*. Adj: *competent*.

competent [struc geol] (a) Said of a bed or stratum that is able to withstand the pressures of folding without flowage or change in original thickness. (b) Said of a fold in which the strata have not flowed or changed their original thickness; i.e. *concentric fold*.----It is a relative term. Ant: *incompetent*.

competent [hydraul] Pertaining to the *competence* of a stream or current of air.

competent rock A volume of rock which under a specific set of conditions is able to support a tectonic force. Such a volume may be competent or incompetent a number of times in its deformational history depending upon the environmental conditions, degree and time of fracturing, etc. Cf: *incompetent rock.*

compilation The selection, extraction, and assembly of map detail from various source materials (such as existing maps, aerial photographs, surveys, and new data), and the preparation and production of a new or improved map (or a part of a map) based on this detail. See also: *delineation.*

compiled map A map (esp. a small-scale map of a large area) incorporating information collected from various source materials and not developed from original survey data for the map in question; a map prepared by compilation.

complementary [petrology] Said of different rocks or groups of rocks differentiated from the same magma and whose average composition is that of the parent magma.

complementary [struc geol] Said of sets of fractures that are considered to be *conjugate* although their origin is unknown. Cf: *cognate.*

complete caneolith A *caneolith* having upper and lower rim elements and a wall. Cf: *incomplete caneolith.*

complete flower A flower having all four types of floral appendages: sepals, petals, stamens, and carpels. Cf: *incomplete flower.*

complete overstep A term proposed by Swain (1949, p.634) for an *overstep* in which an unconformity (partly angular, partly parallel) is universal throughout a basin of deposition or in which the older rocks of the basin are entirely blanketed with an unconformable relationship by younger rocks.

complete tabula A coral *tabula* consisting of a single platform. Cf: *incomplete tabula.*

complex n. (a) A large-scale association or assemblage of different rocks of any age or origin, having structural relations that are so intricately involved or otherwise complicated that the rocks cannot be differentiated in mapping; e.g. a "volcanic complex". See also: *igneous complex; injection complex; metamorphic complex; basement complex.* (b) A rock-stratigraphic unit that includes a mass of rock "composed of diverse types of any class or classes or...characterized by highly complicated structure" (ACSN, 1961, art.6j). The term may be used as part of the formal name instead of a lithologic or rank term; e.g. Crooks Complex (Precambrian) in central Arizona. (c) A time-stratigraphic unit corresponding to the rocks formed during an eon (Kobayashi, 1944a, p.478).

complex crater A meteorite impact crater of large diameter and relatively shallow depth, characterized by formation of a central uplift and a peripheral ring depression which apparently develop during the late stages of the crater-forming event as a result of yielding of the rock beneath the crater (Dence, 1968, p.182). Cf: *simple crater.*

complex cuspate foreland A cuspate forelnad in which erosion on one side of the cusp has truncated beach ridges and swales but which is later prograded so that beach ridges, swales, and other symmetrical lines of growth are parallel to the new shoreline (Johnson, 1919, p.325); e.g. Cape Kennedy, Fla. Cf: *simple cuspate foreland.*

complex drainage pattern A drainage pattern that shows variations among component parts, as one in an area of complicated geologic structure and geomorphic history (Thornbury, 1954, p.123).

complex dune A dune formed by multidirectional winds, resulting in the intersection of two or more dunes.

complex fold A fold that contains a *cross fold.*

complex mountain A mountain that includes a combination of intimately admixed structures and that shows a great variety of landforms; a mountain whose structures defy simple classification.

complex ore (a) An ore that yields several metals. Cf: *simple ore.* (b) An ore that is difficult to utilize because it contains more than one metal or because of the presence of unusual metals. In this sense, the term is not precise.

complex ripple mark A syn. of *cross ripple mark.* Term used

by Kelling (1958, p.121) for an "interference ripple-pattern" of any kind.

complex spit A large *recurved spit* with minor or secondary spits developed at the end of the main spit (such as by minor currents). The lines of growth of the minor and the main spits do not merge or curve into each other, but intersect at distinct angles. Example: Sandy Hook, N.J. Cf: *compound spit.*

complex stream A stream that has entered a second or later cycle of erosion (Davis, 1889b, p.218); e.g. the headwaters of a compound stream.

complex tombolo The system of islands and beaches that result when several islands are united with each other and with the mainland by a complicated series of tombolos. Syn: *tombolo series; tombolo cluster.*

complex twin A twin in feldspar that is the result of both normal twinning and parallel twinning.

complex valley A valley in which part of its course may be parallel and part transverse to the general structure of the underlying strata. Term introduced by Powell (1874, p.50). Cf: *simple valley; compound valley.*

component One of a set of chemical compositions of a thermodynamic system, the relative masses of which may be varied to describe all compositions within it. A component need not be physically realizable, and may be so defined as to have negative concentration for some components of the system. In terms of the phase rule, the number of components is the minimum number of components required to describe the composition of every phase in the system.

componental movement In the deformation of a rock, the relative movements of component particles; *tectonic transport.*

composite An association of algae and fungi in an aqueous environment.

composite coast A term used by Cotton (1958, p.441) for an initial coast resulting from deformation (either upwarping or subsidence) occurring along lines transverse to the coast, characterized by salients and embayments developed on a very large scale; a "coast of transverse deformation" consisting of alternate zones of submergence and emergence, as along the coast of New Zealand near Wellington.

composite cone A less-preferred syn. of *stratovolcano.*

composite fault scarp A fault scarp whose height results from the combined effects of differential erosion and fault movement.

composite fold *compound fold.*

composite gneiss Gneiss which once was constituted by at least two phase-wise different materials (Dietrich & Mehnert, 1961). Syn: *mixed gneiss.* Cf: *arterite; phlebite; venite.*

composite grain A sedimentary particle formed by aggregation of two or more discrete particles; esp. a carbonate particle resulting from clustering of lumps, pellets, coated grains, or detrital, skeletal, or algal particles (Bissell & Chilingar, 1967, p.153).

composite intrusion Any type of igneous intrusion which is composed of more than one injection of varying chemical composition; a differentiated intrusion. Cf: *multiple intrusion.* See also: *partial pluton.*

composite map In mining, a map that shows several levels of a mine on one sheet; a map that horizontally projects data from different elevations in the mine. Cf: *cartology.*

composite photograph A photograph made by assembling the separate photographs (made by the several lenses of a multiple-lens camera in simultaneous exposure) into the equivalent of a photograph taken with a single wide-angle lens.

composite profile A profile consisting of the highest points of a series of profiles drawn along several regularly spaced and parallel lines on a map (Monkhouse & Wilkinson, 1952); it represents the surface of any relief area as viewed in the horizontal plane of the summit levels from an infinite distance. Cf: *superimposed profile; projected profile.* Syn: *zonal profile.*

composite ripple mark A term used by Tanner (1960, p.484

or an *oscillation cross ripple mark* formed by two intersecting sets of waves neither of which is parallel with the crests of the ripple marks. It is best developed on shoals where wave refraction produces two sets of waves not quite at right angles from each other.

composite seam A thick coal seam (greater than a meter in thickness) composed of two or more layers that are in contact where dirt bands or other intervening strata have wedged out.

composite section A single inclined or vertical section prepared by projecting data from various, more or less parallel sections.

composite sequence A term used by Duff & Walton (1962) for a sequence of beds that comprises all lithologic types (in a succession displaying cyclic sedimentation) in the order in which they tend to occur. It is constructed from statistical data based on actual rock successions.

composite set A term proposed by McKee & Weir (1953, p.384) for a large sedimentary unit of similar or gradational lithology, "compounded from both stratified and cross-stratified units", and consisting of horizontal strata together with cosets of cross-strata.

composite species (a) A species represented by a group of specimens that were obtained from more than one locality and that are not all of the same age. (b) A species consisting of two or more subspecies; a polytypic species.

composite stream A stream whose drainage basin receives waters from areas of different geomorphologic structures (Davis, 1889b).

composite topography A landscape whose topographic features have developed in two or more cycles of erosion.

composite unconformity An unconformity representing more than one episode of nondeposition and possible erosion.

composite volcano A syn. of *stratovolcano*. Cf: *compound volcano*.

compositing *mixing*.

composition [mineral] *chemical composition* [mineral].

composition [petrology] (a) *chemical composition* [petrology]. (b) The makeup of a rock in terms of the species and number of minerals present; mineralogic composition.

compositional maturity A type of sedimentary *maturity* in which a clastic sediment approaches the compositional end product to which it is driven by the formative processes that operate upon it. It may be expressed as a ratio between chemical compounds (e.g. alumina/soda) or between mineral components (e.g. quartz/feldspar) (Pettijohn, 1957, p.286 & 508). Cf: *mineralogic maturity; textural maturity*.

composition face *composition surface*.

composition plane In a crystal twin, a *composition surface* that is planar.

composition point In a plot of phase equilibria, that point whose coordinates represent the chemical composition of a phase or mixture.

composition surface The surface along which the individuals of a crystal twin are joined. It may or may not be a plane, and it is usually identical with the twin plane. See also: *composition plane*. Syn: *composition face*.

compost A mass of rotted organic matter made from waste plant residues. Inorganic fertilizers, especially nitrogen, and a little soil usually are added to it. The whole pile is kept moist and allowed to decompose; it is usually turned once or twice. The material is used to improve garden soils.

compound alluvial fan *bajada*.

compound conodont element A blade- or bar-like *conodont element*, commonly bearing denticles.

compound coral The skeleton of a colonial coral.

compound cross-bedding A term used by Gilbert (1899, p.139) for a laminated structure "produced by deposition and partial erosion associated with shifting sand ripples". Cf: *compound foreset bedding*.

compound cuspate bar A bar formed by a *compound spit* that unites with the shore at its distal end (Johnson, 1919, p.319).

compound eye An eye of an arthropod consisting essentially of a great number of minute eyes closely crowded together; e.g. an array of contiguous visual units (ommatidia) of a crustacean having a common optic-nerve trunk, or the lateral eye of a merostome composed of many facets, or a *holochroal eye* of a trilobite. See also: *ocellus* [paleont].

compound fault A zone or series of closely spaced, approximately parallel faults.

compound fold A fold upon which minor folds with similar axes have developed. Cf: *simple fold.* Syn: *composite fold.*

compound foreset bedding Cross-bedding characterized by a concave base and by several foreset beds dipping in more than one direction, as where tangential foresets recently deposited may have their upper ends truncated and may then be overlain by new beds of similar nature dipping in the same or in a different direction (Lahee, 1923, p.80). It may develop by changes in the stream or by fluctuations of water level of a lake, such as interference among adjacent lobes on a delta front. Cf: *compound cross-bedding.*

compound laccolith A laccolith having several parts that are separated from each other by thick layers of country rock but that were formed by a single intrusion.

compound leaf A leaf blade composed of several distinct parts or *leaflets*.

compound operculum A dinoflagellate operculum that is divided into two or more pieces that are completely separable from one another. Cf: *simple operculum.*

compound pellet A pellet of silt, sand, or granule size or larger, originating from pelleted limestone with micritic or sparry cement, and sometimes containing matrix or interstitial material (Bissell & Chilingar, 1967, p.153).

compound plate An ambulacral-plate unit of an echinoid, composed of two or more individual ambulacral plates, each with a pore for tube foot, bound together by a single large tubercle articulating with the *primary spine* (TIP, 1966, pt.U, p.253).

compound ripple mark (a) A *cross ripple mark* resulting from the simultaneous interference of wave oscillation with current action and characterized by a systematic breaking or offsetting of the crests of the current ripples (Bucher, 1919, p.195). (b) A ripple mark produced by modification of a preexisting set of ripples by a later set of either wave or current origin (Kelling, 1958, p.122); e.g. *wave-current ripple mark.*

compound shoreline A shoreline showing a very marked development of the features characteristic of at least two of the following: a shoreline of emergence, a shoreline of submergence, a neutral shoreline (Johnson, 1919, p.172); e.g. where a formerly submerged shoreline is elevated slightly but not enough to destroy the effect of submergence, or where a dissected shoreline of emergence undergoes a slight submergence so that the coastal-plain valleys are drowned.

compound spit A *recurved spit* whose inner side shows a series of intermittent landward-deflected points representing successive recurved termini. Cf: *complex spit.*

compound stream A stream that is of different ages in its different parts (Davis, 1889b); e.g. a stream with old headwaters rising in the mountains and a young lower course traversing a coastal plain.

compound structure A term used by Cotton (1948) for an arrangement of rocks characterized by a simple *cover mass* resting unconformably on more complex *undermass*.

compound trabecula A *trabecula* of a scleractinian coral, composed of bundles of sclerodermites. Cf: *simple trabecula.*

compound valley A valley whose main course may be a *simple valley* or a *complex valley*, but whose tributary valleys are of a different kind. Term introduced by Powell (1874, p.50).

compound valley glacier A glacier composed of two or more individual valley glaciers emanating from tributary valleys.

compound volcano A volcanic structure having more than one cone. Cf: *composite volcano.*

compreignacite A yellow secondary mineral: $K_2U_6O_{19}.11H_2O$.

compressed (a) Said of a nautiloid conch whose dorsal/ventral diameter is greater than its lateral diameter. (b) Said of an ammonoid whorl section that is higher than it is wide.--- Ant: *depressed.*

compressibility The reciprocal of *bulk modulus*; it equals $1/k$ where k is the bulk modulus. Its symbol is β. Syn: *modulus of compression.*

compressing flow A flow pattern on glaciers in which the velocity decreases with distance downstream; thus the longitudinal strain rate (velocity gradient) is compressive. This condition requires a transverse or vertical expansion or a negative net balance on the surface because ice is almost incompressible. Ant: *extending flow.*

compression [seis] *anaseism.*

compression [paleont] The remains of a fossil plant that have been flattened by the vertical pressure of overlying rocks.

compression [exp struc geol] A system of forces or stresses that tends to decrease the volume or shorten a substance; also, the change of volume produced by such a system of forces.

compression [tect] An adjustment of the Earth's crust to stress of contraction, as in some rift valleys, or to pressure of overlying sediment, as in geosynclines. Syn: *compressional movement; compressive settling; compression-subsidence.*

compressional movement *compression.*

compressional wave *P wave.*

compression fault A general term for a fault that has been produced by lateral crustal compression, e.g., a reverse fault. Cf: *tension fault.*

compression-subsidence *compression.*

compression tests *triaxial compression test.*

compressive settling *compression.*

compressive strength The maximum *compressive stress* that can be applied to a material, under given conditions, before failure occurs.

compressive stress A *normal stress* that tends to push together material on opposite sides of a real or imaginary plane. See also: *compressive strength.* Cf: *tensile stress.*

compromise boundary A surface of contact, not corresponding to a crystal face, between two mutually growing but differently oriented crystals.

comptonite An opaque variety of thomsonite from the Lake Superior region.

computer [cart] A person who calculates (from notes made by survey parties) latitudes, longitudes, and areas for map making.

computer [science] A calculator designed for performing simple and complex calculations; specif. an automatic electronic device capable of accepting information, applying prescribed processes to the information, and supplying the results of these processes. See also: *analog computer; digital computer.*

conate Said of sculpture of pollen and spores consisting of coni.

conca A term introduced by Tanakadate (1929) for a conchoidal collapse structure in the area of repeated volcanic activity. Pl: *conche.*

concave bank The outer bank of a curved stream, with the center of the curve toward the channel; e.g. an *undercut slope.* Ant: *convex bank.*

concave cross-bedding (a) Cross-bedding with concave (downward-arching), generally tangential, foreset beds. This type of cross-bedding is very common and is used as a criterion for distinguishing top from bottom in sedimentary rocks. Cf: *convex cross-bedding.* (b) Cross-bedding deposited on a lower concave surface, as in festoon cross-bedding.

concave slope *waning slope.*

concavo-convex Concave on one side and convex on the other; e.g. said of a brachiopod shell having a concave brachial valve and a convex pedicle valve. Cf: *convexo-concave.*

concealed coalfield Deposits of coal that do not outcrop. Cf: *exposed coalfield.*

concealed pediment A pediment that is buried by a thin layer of alluvium from an encroaching bajada; it is generally caused by a local rise of base level. Cf: *suballuvial bench; fan-topped pediment.*

concentrated flow *channel flow.*

concentrated wash *channel erosion.*

concentration [sed] *sediment concentration.*

concentration [ice] The ratio in eighths or tenths of the area extent actually covered by sea ice to the total area of the sea surface, both covered by ice and ice free, at a specific location or over a defined area. See also: *ice cover.*

concentration boundary "A line approximating the transition between two areas of pack ice with distinctly different concentrations" (U.S. Naval Oceanographic Office, 1968, p.B32).

concentration time (a) The time required for water to flow from the farthest point in a watershed to a gaging station. (b) That time at which the rate of runoff equals the rate of precipitation of a storm. Syn: *time of concentration.*

concentric Said of a system of fractures having a common axis. See also: *funnel joints.*

concentric fold A fold in which the strata have not changed their original thickness during deformation; a *competent* fold. Cf: *similar fold.* Syn: *parallel fold.*

concentric shearing surface Var. of *surface of concentric shearing.*

concentric weathering Weathering characterized by the loosening and separation of successive rounded or spherical shells; specif. *spheroidal weathering.*

concertina fold *accordion fold.*

conch (a) The portion of a cephalopod shell developed after the embryonic shell; e.g. the complete shell of an ammonoid exclusive of the protoconch, or all the hard calcareous parts of a nautiloid (including the external shell, septa, and siphuncle) exclusive of cameral deposits and any structures within the siphuncle. (b) Any of various large spiral-shelled marine gastropods, often of the genera *Strombus* or *Cassis.* Also, the shell of such a conch, often used for making cameos and formerly made into horns. (c) Any of various marine shells of invertebrates, including bivalve mollusks and brachiopods.---Pl: *conchs.*

conchal furrow A shallow, midventral groove on the inside wall of a nautiloid conch.

conche Plural of *conca.*

conchiform Shell-shaped; esp. shaped like one half of a bivalve shell.

conchilite A small, bowl-shaped body of limonite or goethite growing in an inverted position on mineralized bedrock and resembling the shell of an oyster or giant clam coated with a rusty deposit (Tanton, 1944). It is roughly oval or circular in plan, with a smooth or irregular and scalloped outline; it varies from 2.5 cm to 1 m in diameter and from 2 cm to 7.5 cm in height.

conchiolin A fibrous protein, $C_3H_{48}N_9O_{11}$, that constitutes the organic basis of most mollusk shells; e.g. the material of which the periostracum and organic matrix of the calcareous parts of a bivalve-mollusk shell are composed. Syn: *conchyolin.*

conchitic Abounding in fossil shells; esp. said of a rock composed of or containing many shells.

conchoidal Said of a type of mineral or rock fracture that gives a smoothly curved surface. It is a characteristic habit of quartz and of obsidian.

conchology The study of shells of both fossil and existing animals. Cf: *malacology.*

conchostracan Any branchiopod belonging to the order Conchostraca, characterized by a translucent bivalve shell with

146

aw-like furca developed at the posterior end. Their stratigraphic range is Lower Devonian to present.

onclinal A syn. of *cataclinal*. Term introduced by Powell 874, p.50).

oncordance of summit levels *summit concordance*.

oncordant [intrus rocks] Said of a contact between an igneus intrusion with the country rock which parallels the foliation bedding of the latter. Cf: *discordant; conformable*. See so: *interjection*.

oncordant [stratig] Structurally *conformable*; said of strata splaying parallelism of bedding or structure. The term may e used where a hiatus cannot be recognized but cannot be smissed. Ant: *discordant*.

oncordant [hydraul] Said of flows, e.g. floods, at different oints along a channel which have the same frequency of ocrrence (Langbein & Iseri, 1960, p.6).

oncordant [geochron] (a) Agreement of radiometric ages etermined by more than one method, agreement being within nalytical precision for the determining methods. (b) Agreement of radiometric ages given by coexisting minerals determined by the same method. (c) In a more restricted sense, e term has been used to indicate agreement of U^{238}-Pb^{206}, ^{235}-Pb^{207}, Pb^{207}-Pb^{206}, and Th^{232}-Pb^{208} ages determined, thin experimental error, for the same mineral sample. Ant: scordant.

oncordant bedding A sedimentary structure marked by beds at are parallel and without angular junctions. Ant: *discordant dding*. Syn: *parallel bedding*.

oncordant coastline A coastline that is broadly parallel to the ain trend of the land structure (such as mountain ranges or d belts) forming the margin of the ocean basin; it is generly linear and regular. Cf: *Dalmatian coastline*. Ant: *discornt coastline*. Syn: *Pacific-type coastline; longitudinal coaste*.

oncordant drainage *accordant drainage*.

oncordant junction *accordant junction*.

oncordant summit level *accordant summit level*.

oncordia The graphed curve formed when the Pb^{206}/U^{238} tio is plotted against the Pb^{207}/U^{235} ratio as both increase value due to nuclear decay of uranium to lead with passage time, assuming a closed U-Pb system. The curve (concora) is the locus of all concordant U-Pb ages, thus it is a time rve. Syn: *concordia plot; concordia diagram; concordia rve*.

oncordia curve *concordia*.

oncordia diagram *concordia*.

oncordia intercept The intersections of the *concordia* curve me curve) and a straight line (chord) which depicts the plot cus) of discordant U-Pb ages. It may indicate the age of o important events in the life of the U-Pb system under udy, the older age representing the start of the U-Pb system d the younger age the time when this system was disrbed.

oncordia plot *concordia*.

oncrescence A growing together, or a coalescence of origilly separate parts; e.g. union of radial spines in a radiolarian keleton.

oncretion (a) A hard, compact, rounded, normally subspheril (but commonly very oblate or disk-shaped, and sometimes irregular form with odd or fantastic outlines) mass or agegate of mineral matter generally formed by orderly and lolized precipitation from aqueous solution (often about a nueus or center, such as a leaf, shell, bone, or fossil) in the ores of a sedimentary or fragmental volcanic rock and usual of a composition widely different from that of the rock in hich it is found and from which it is rather sharply sepated. It represents a concentration of some minor constituent f the enclosing rock) or of cementing material such as sili (chert), calcite, dolomite, iron oxide, pyrite, or gypsum, d it varies in size from a small pellet-like object to a great

spheroidal body as much as 3 m in diameter. Concretions formed in limestone and shale appear to be contemporaneous with deposition and those found in sandstone appear to have been formed postdepositionally by cementation of sand grains. Cf: *nodule; secretion*. See also: *accretion; incretion; intercretion; excretion*. (b) A collective term applied loosely to various primary and secondary mineral segregations of diverse origin, including irregular nodules, spherulites, crystalline aggregates, geodes, septaria, and related bodies.

concretionary Characterized by, consisting of, or producing concretions; e.g. a ''concretionary ironstone'' composed of iron carbonate with clay and calcite, or a zonal ''concretionary texture'' (of an ore) characterized by concentric shells of slightly varying properties due to variation during growth.

concurrent-range zone A formal *range zone* or body of strata defined by the overlapping ranges of two or more specified taxa from one or more of which it takes its name (ACSN, 1961, art.23); the part of the section where two range zones overlap. It is recognized by the lowest occurrence of certain fossils and the highest occurrence of others, and it is the biostratigraphic zone generally used in attempting time-correlation of strata. The term is equivalent to Oppel's (1856-1858) usage of *zone*. Syn: *Oppelzone; multifossil range zone*.

concussion crack A *crack* in sea ice, produced by the impact of one ice cake upon another.

concussion fracture One of a system of fractures in individual grains of a shock-metamorphosed rock which are generally radial to the grain surface and related to the contacts with adjacent grains. They are apparently formed by violent grain-to-grain contacts in the initial stages of passage of a shock wave and are formed by the resulting tensile stresses parallel to the surfaces of the impacting grains (Kieffer, 1971, p.5456, 5468).

concyclothem A cyclic sequence of strata resulting from the local coalescence of two or more cyclothems. Term introduced by Gray (1955).

condensate The liquid hydrocarbon that emanates from a gas well or from the gas-cap of an oil well.

condensation [stratig] A stratigraphic process in which the thinning of a sedimentary deposit or succession takes place contemporaneously, such as subjecting a crystalline rock to strong hydrostatic pressures resulting in solution along grain boundaries.

condensation [phys] The process by which a vapor becomes a liquid or solid; the opposite of *evaporation*.

condensation room In a cave, an area in which water vapor condenses on the ceiling and walls due to a difference in temperature between the room or passage and the air that enters it.

condensed deposit A sedimentary deposit in a *condensed succession*.

condensed succession A relatively thin, but uninterrupted, stratigraphic succession in which the deposits accumulated very slowly. Ant: *extended succession*.

condensed system (a) A chemical system in which the vapor pressure is negligible, and can thus be ignored. (b) A chemical system in which the pressure maintained on the system is greater than the vapor pressure of any portion.

condrodite *chondrodite*.

conductance The electrical conductance, measured in mhos, is the inverse of electrical resistance.

conduction (a) The transport of electrical current through a material by electrons, holes, or ions. (b) *heat conduction*.

conductivity (a) *electrical conductivity*. (b) *thermal conductivity*.

conductivity log *induction log*.

conductor A material which offers a low resistance to the passage of an electric current.

conduit A subterranean passage that is completely filled with water and that is always under hydrostatic pressure. Cf: *siphon.*

condyle (a) One of a pair of oppositely placed protuberances on which the operculum pivots in the orifice of ascophoran cheilostomes (bryozoans). Syn: *cardella.* (b) A swollen knob on the shell surface of an acantharian radiolarian. (c) An articular prominence on a bone. See also: *occipital condyle.*

cone [glac geol] A steep-sided pile of sand, gravel, and sometimes boulders, with a fan-like outwash base, deposited against the front of a glacier by meltwater streams.

cone [geomorph] A mountain, hill, or other landform shaped like a cone, having relatively steep slopes and a pointed top; specif. an *alluvial cone.*

cone [marine geol] *submarine fan.*

cone [bot] A branch that bears a group of spore-bearing leaves or sporophylls.

cone [volc] *volcanic cone.*

cone cup The inside of an enclosing cone of a sedimentary cone-in-cone structure.

cone delta *alluvial cone.*

cone dike *cone sheet.*

cone-in-cone coal Coal exhibiting cone-in-cone structure. Syn: *crystallized coal.*

cone-in-cone structure [sed] (a) A minor sedimentary structure in thin, generally calcareous layers of some shales and in the outer parts of some large concretions (esp. septaria), resembling a set of concentric, right circular cones fitting one into another in inverted positions (base upward, apex downward) and commonly separated by clay films, and consisting usually of fibrous calcite and rarely of siderite or gypsum. The apical angles are commonly between 30 and 60 degrees and the cone axes are normal to the bedding; the height of the cones usually vary between 10 mm and 10 cm, and their sides are often ribbed, fluted, or grooved and marked by annular depressions and ridges that are more pronounced near the bases and finer and more obscure near the apices. The structure appears to be due to pressure aided by crystallization and weathering (solution) along conical shear zones that intersect one another. (b) A similar structure developed in coal and consisting of a set of interpenetrating cones packed closely together.---Syn: *cone-in-cone.*

cone-in-cone structure [volc] A volcanic structure in which a younger cone or cones have developed within the original one; also, a similar pattern of crater development. Adj: *nested.*

cone karst A type of karst typical of tropical regions, characterized by a pattern of steep, conical hills and adjacent star-shaped depressions, sometimes known as *cockpits*, that have steep, convex sides and slightly concave floors. Syn: *cockpit karst; Kegelkarst.* Etymol: German, *Kegelkarst.*

cone of dejection An *alluvial cone* consisting of coarse material, formed where a mountain torrent emerges from a narrow valley upon a plain or plunges over a valley-side bench. Syn: *dejection cone.*

cone of depression A depression in the potentiometric surface of a body of ground water that has the shape of an inverted cone and develops around a well from which water is being withdrawn. It defines the *area of influence* of a well. The shape of the depression is due to the fact that the water must flow through progressively smaller cross sections as it nears the well, and hence the hydraulic gradient must be steeper. Cf: *drawdown [grd wat]* (*b*); *cone of exhaustion.*

cone of detritus *alluvial cone.*

cone of exhaustion A cone-shaped depression in the potentiometric surface of a body of ground water that develops around a well when water is being withdrawn more rapidly than it can percolate laterally. Cf: *cone of depression.*

cone of pressure relief Cone of depression in the potentiometric surface of a body of confined ground water. As defined by Tolman (1937, p.562), "An imaginary surface indicating pressure-relief conditions in a confined aquifer due to pumping".

cone penetration test A soil *penetration test* in which a standard steel cone is pushed into the soil by hand (mechanicall) or hydraulically and the force required to advance the cone a slow constant rate or a specified distance, or in some d signs the penetration at various loads, is recorded.

cone sheet A dike that is arcuate in plan and dips gen (30°-45°) toward the center of the arc. It occurs in concentr sets which presumably converge (by projection) at the ma matic center. It is associated with *ring dikes* to form a ri complex. Syn: *cone dike.*

Conewangoan *Bradfordian.*

conferva peat Peat that is derived mainly from filamento algae of the genus *Tribonema.*

confidence interval The interval between confidence limits.

confidence limit Either the upper or the lower value betwee which an actual measurement or parameter will fall with stated probability.

configuration (a) The form or shape of a part of the Earth surface with regard to its horizontal outline, its elevation, a its relative disposition or arrangement with other parts of t surface. (b) The topography of a region as shown by the typ cal contour map, where the contour interval is well suited express the slope of the terrain of the area (ASCE, 195 p.39-40). Syn: *configuration of terrain.*

confined aquifer An aquifer bounded above and below by in permeable beds or beds of distinctly lower permeability tha that of the aquifer itself; an aquifer containing confined grou water. Syn: *artesian aquifer.*

confined ground water Ground water under pressure signi cantly greater than that of the atmosphere and whose upp surface is the bottom of an impermeable bed or a bed of di tinctly lower permeability than the material in which the wat occurs. Ant: *unconfined ground water.* Syn: *artesian wate confined water; piestic water.*

confined water *confined ground water.*

confining bed A body of impermeable or distinctly less perm able material stratigraphically adjacent to one or more aqu ers. Cf: *aquitard; aquifuge; aquiclude.*

confining pressure An equal, all-sided pressure; e.g. geostat pressure resulting from the load of overlying rocks, or hydr static pressure resulting from the weight of water in a zone saturation.

confluence [glaciol] (a) A flowing together of two or more gl ciers. (b) A junction or place where confluence occurs.---An *diffluence.*

confluence [streams] (a) A place of meeting of two or mo streams; the point where a tributary joins the main stream; fork. (b) A flowing together of two or more streams. (c) T stream or other body of water produced by confluence; combined flood.---See also: *junction.*

confluence plain A plain formed by the merging of the vall floors of two or more streams.

confluence step A rock step that rises upstream toward t heads of two glacial valleys at their place of confluence. It probably caused by the strengthening of glacial action at th point. Ant: *diffluence step.*

confluent adj. Said of a stream, glacier, vein, or other geolo ic feature that combines or meets with another like feature form one stream, glacier, vein, etc. Ant: *diffluent.*---n. A co fluent stream, usually a stream uniting with another of nea equal size; a fork or branch of a river. The term is sometim loosely applied to an *affluent.*

conformability The quality, state, or condition of being co formable, such as the relationship of conformable strata; co formity.

conformable [intrus rocks] Said of the contact of an intrusi body when it is aligned with the intrusion's internal structure Cf: *disconformable; concordant.*

conformable [stratig] Said of strata or stratification characte ized by an unbroken sequence in which the layers are forme he above the other in parallel order by regular, uninterrupted eposition under the same general conditions; also said of the ontacts (abrupt, gradational, or intercalated) between such trata. The term is often applied to a later formation having edding planes that are parallel with those of an earlier forma-on and showing an arrangement in which disturbance or ero-ion did not take place at the locality during deposition. Den-'s (1967, p.26) regards the term as descriptive, primarily re-erring to succession without disturbance and not necessarily nplying parallelism of succeeding beds or continuity of depo-tion; e.g. cross-bedding without intervening erosion is con-rmable, but not parallel. Cf: *unconformable; concordant.*

onformably superimposed stream A term suggested by Küm-el (1893, p.380) for a stream superimposed from a con-rmable cover.

onformality The unique property of a conformal map projec-n in which all small or elementary figures on the surface of e Earth retain their original shapes on the map. Cf: *equival-nce.* Syn: *orthomorphism.*

onformal projection A map projection on which the shape of ny very small area of the surface mapped is preserved un-hanged on the map and the scale at any point is the same in very direction (as along the meridian and the parallel at that oint), although the scale may vary from point to point. It al-ays shows a right-angle intersection of any parallel with any eridian. Conformal projections suffer from severe scale hanges so that areas are not correctly represented. Exam-es include: stereographic projection, Mercator projection, d Lambert conformal conic projection. Cf: *equal-area pro-ction.* Syn: *orthomorphic projection.*

onformity (a) The mutual and undisturbed relationship be-ween adjacent sedimentary strata that have been deposited orderly sequence with little or no evidence of time lapses; ue stratigraphic continuity in the sequence of beds without idence that the lower beds were folded, tilted, or eroded be-re the higher beds were deposited. Syn: *conformability.* (b) n uninterrupted sequence of strata displaying conformity.---: *unconformity.*

onfused sea A rough sea with indeterminate direction and riod of wave travel. It is caused by the superimposition of rious wave trains.

ongela A term used in Chile for *coba* with a high salt con-nt.

ongelation The change from a fluid to a solid state; freezing. so, the product of such a change. Syn: *gelation.*

ongelifluction The progressive and lateral flow of earth mate-al under periglacial conditions; solifluction in a region under-in by frozen ground. Syn: *gelifluction; gelisolifluction.*

ongelifract An angular rock fragment split off by frost action *congelifraction*), ranging from very large blocks to finely mminuted material. Syn: *gelifract.*

ongelifractate A mass of rock fragments (*congelifracts*) of fferent sizes produced by frost action.

ongelifraction The mechanical disintegration, splitting, or eakup of a rock or soil due to the great pressure exerted by e freezing of water contained in cracks or pores, or along edding planes; term introduced by Bryan (1946, p.640). Syn: *ost shattering; frost splitting; frost riving; frost bursting; frost eathering; frost wedging; gelivation; gelifraction.*

ongeliturbate A mass of soil or other unconsolidated earth aterial moved or disturbed by frost action, and usually coar-r than the underlying material; esp. a rubbly deposit formed solifluction. See also: *head; rubble drift; coombe rock.* Syn: *ost soil; cryoturbate; warp; trail.*

ongeliturbation A collective term suggested by Bryan (1946, 640) to describe the stirring, churning, modification, and all her disturbances of soil, resulting from frost action; it in-lves frost heaving, solifluction, and differential and mass ovements, and it produces patterned ground. Syn: *cryotur-tion; frost stirring; frost churning; geliturbation.*

conglomerate A coarse-grained, clastic sedimentary rock composed of rounded (to subangular) fragments larger than 2 mm in diameter (granules, pebbles, cobbles, boulders) set in a fine-grained matrix of sand, silt, or any of the common natu-ral cementing materials (such as calcium carbonate, iron oxide, silica, or hardened clay); the consolidated equivalent of gravel both in size range and in the essential roundness and sorting of its constituent particles. The rock or mineral frag-ments may be of varied composition and range widely in size, and are usually rounded and smoothed from transportation by water or from wave action. Conglomerates may be classified according to nature or composition of fragments, proportion of matrix, degree of size sorting, type of cement, and agent or environment responsible for their formation. Etymol: Latin *conglomeratus*, "heaped, rolled, or pressed together". Abbrev: cgl. Cf: *breccia.* Syn: *puddingstone.*

conglomerated ice All forms of floating ice compacted into one mass. The term refers to the contents of an ice mass, not to the concentration.

conglomeratic Pertaining to a conglomerate; composed or having the properties of conglomerate. Syn: *conglomeritic.*

conglomeratic mudstone A mudstone with a sparse to liberal sprinkling of pebbles or cobbles; e.g. a consolidated gravelly mud containing 5-30% gravel and having a ratio of sand to mud less than 1:1 (Folk, 1954, p.346), such as a *pebbly mud-stone* or a *tilloid.* Pettijohn (1957, p.261) considered the term synonymous with *paraconglomerate.* Cf: *mudstone conglom-erate.*

conglomeratic sandstone (a) A sandstone containing 5-30% gravel and having a ratio of sand to mud (silt + clay) greater than 9:1 (Folk, 1954, p.347); a consolidated gravelly sand. (b) A sandstone containing more than 20% pebbles (Krynine, 1948, p.141). Cf: *pebbly sandstone.*

conglomerite A term suggested by Willard (1930) for a con-glomerate that has reached the same stage of induration as that of a quartzite, characterized by the welding together of matrix and clasts as evidenced by fractures passing through both. Examples include the partially metamorphosed or "stretched" Carboniferous conglomerates of Rhode Island, esp. those east of Newport and at Natick. Cf: *metaconglom-erate.*

Congo copal A hard, yellowish to colorless *copal* derived from certain trees of the genus *Copaifera*, found as a fossil resin in the Congo, and used in making varnish. Syn: *Congo gum.*

congregation In paleoecology, those fossil species occurring together in and characterizing the rocks of a particular zone. A congregation may include all of, part of, or more than one *assemblage.*

congressite A light-colored coarsely granular igneous rock consisting chiefly of nepheline, with minor amounts of soda-lite, plagioclase, mica, and calcite. Cf: *craigmontite.*

congrous Var. of *congruent* [fold].

congruent Said of a drag fold, the axis and axial surface of which are parallel to the axis and axial surface of the main fold to which it is related. Also spelled: *congrous* [fold]. Ant: *incongruous* [fold].

congruent melting Melting of a substance directly to a liquid that is of the same composition as the solid.

coni Plural of *conus* [palyn].

Coniacian European stage: Upper Cretaceous (above Turo-nian, below Santonian). Syn: *Emscherian.*

conical fold A fold model that can be described geometrically by the rotation of a line about one of its ends, which is fixed. Cf: *cylindrical fold.*

conical wave *head wave.*

conichalcite A pistachio- to emerald-green mineral: $CaCu(AsO_4)(OH)$. It often contains phosphorus.

conic projection One of a group of map projections produced by projecting the geographic meridians and parallels onto the surface of a cone that is tangent to, or intersects, the surface

149

of the sphere at a selected latitude or latitudes (which become one or more standard parallels of the projection), and then developing (unrolling and laying flat) the cone as a plane. All parallels are arcs of concentric circles, and the meridians are straight (sometimes curved) lines radiating from the apex of the cone to the divisions of the standard parallel. True distances are measured along the standard parallels but the scale is distorted to the north and south of these parallels. Examples include: Lambert conformal conic projection, Albers projection, and Bonne projection. See also: *polyconic projection.* Syn: *conical projection.*

conic scale A lamina that extends over only a part of a cone of a sedimentary cone-in-cone structure.

conidiophore A structure that bears conidia; specif. a specialized, typically erect *hypha* that produces successive conidia in certain fungi.

conidiospore A *conidium*; a *fungal spore* that may have chitinous walls and therefore occur as a microfossil in palynologic preparations.

conidium An asexual spore produced from the tip of a conidiophore; broadly, any asexual spore not borne within an enclosing structure, such as one not produced in a sporangium. Pl: *conidia.* See also: *conidiospore.*

conifer A gymnosperm, usually a large tree, having needlelike or scalelike leaves and seeds borne in cones. Conifers include pines, firs, and spruces.

coniferous Said of a plant whose reproductive structures are cones.

coning The process by which water underlying an oil reservoir invades upward into the oil column and enters the well, as when the well is overproduced or when insufficient back pressure is maintained on the well.

conispiral adj. Said of a gastropod shell with a cone-shaped spire; said of a cephalopod shell characterized by a spiral coiled on the surface of a cone, the whorls not in a single plane.---n. A conispiral shell.

conjugate [fault] Said of faults that are of the same age and deformational episode.

conjugate [joint] Said of a joint system, the sets of which are related in deformational origin, usually compression. Conjugate fracture sets usually intersect at right angles, sharing either the strike or the dip. Also, said of the mineral deposits which may form in such joints. Cf: *cognate; complementary.*

conjugate fold system Two sets of minor folds, the axial surfaces of which are inclined towards each other. Cf: *kink band.*

conjugate image point A term formerly used to denote a *corresponding image point.*

conjugate solutions Two solutions coexisting in equilibrium whose compositions are separated by a miscibility gap in a potentially continuous compositional field. The possibility of a critical point at which the two phases would become identical is implicit.

conjugation line A *tie line* in a two-liquid field. The term is also sometimes used incorrectly as a synonym of *Alkemade line* and of *join.*

conjunct Said of a pore rhomb in which externally visible slits are continuous across suture between plates bearing the rhomb. Cf: *disjunct.*

conker Var. of *kankar.*

connate [bot] In plant morphology, pertaining to the fusion of like parts. Cf: *adnate; adherent; coherent.*

connate [petrology] Said of fluids derived from the same magma.

connate [sed] Originating at the same time as adjacent material; esp. pertaining to waters and volatile materials (such as carbon dioxide) entrapped in sediments at the time the deposits were laid down.

connate water Water entrapped in the interstices of a sedimentary or extrusive igneous rock at the time of its deposition; White (1957, p.1661) has recommended that "connate

water" be defined as "water that has been out of contact with the atmosphere for at least an appreciable part of a geologic period," including both interstitial water of metamorphosed rocks and water that has been driven from them; only "fossil water" would be used to refer to water trapped in rocks from the time of their deposition. Commonly misused by reservoir engineers and well-log analysts to mean *interstitial water* or *formation water.* Syn: *fossil water; fossilized brine; native water.*

connected map *migrated map.*

connecting band (a) One of the two hoop-like bands at the edge of the flange or valve mantle in a diatom frustule. See also: *girdle* [paleont]; *cingulum.* (b) A part of the loop of terebratellacean brachiopods that joins descending branches to the median septum or that joins ascending branches and descending branches posterior to their anterior curvature (TIP, 1965, pt.H, p.142).

connecting bar A bar that is connected at both ends to a land mass; esp. a *tombolo.*

connecting lobe A rounded linear elevation of the valve surface of an ostracode, confluent with two or more subvertical-trending *lobes* (TIP, 1961, pt.Q, p.49).

connecting ring The partly calcareous and partly conchiolinous delicate tubular membrane forming between septa the wall of a cephalopod siphuncle, such as one that connects the septal neck of an ellipochoanitic nautiloid conch with the septum immediately behind it. Vestiges of the connecting ring are also found in holochoanitic forms.

connecting tubule A subhorizontal tubular connection between neighboring corallites in a fasciculate corallum.

connector bar A bar that joins the sagittal ring of a radiolarian skeleton to the lattice shell (as in the subfamily Trissocyclinae.

connellite A deep-blue mineral: $Cu_{19}Cl_4(SO_4)(OH)_{32}.3H_2O$ (?). It is isomorphous with buttgenbachite. Syn: *footeite.*

conode *tie line.*

conodont A small, generally between 0.3 and 1.5 mm in length, tooth- or plate-like structure composed of laminae of translucent, amber-colored, crystalline apatite. Their biologic function and systematic position are unknown, but they are useful biostratigraphically. The known geologic range is Middle Cambrian to Upper Triassic.

conodont element A unit or complete specimen of a conodont. Examples: *compound conodont element; plate-like conodont element; lamellar conodont element; fibrous conodont element.*

conoplain A rarely used syn. of *pediment.* The term was used by Ogilvie (1905, p.28) for an erosion surface peripheral to and sloping radially away in all directions from a laccolithic mountain mass.

conoscope A polariscope that tests the interference figures of a crystal in polarized light; it is a polarizing microscope that contains a Bertrand lens. Cf: *orthoscope.*

Conrad discontinuity The seismic-velocity discontinuity that is equivalent to the boundary between the sial and the sima, at which velocities increase from 6.1 km/sec to 6.4-6.7 km/sec. It is not always discernible; i.e. there is a general increase in velocity with depth in the crust, with layering of the increase occurring in some areas. See also: *Conrad layer.* A possible equivalent is the *Riel discontinuity.*

Conrad layer The seismic region of the Earth between the *Conrad discontinuity* and the Mohorovicic discontinuity.

consanguineous [petrology] The adj. of consanguinity and a less preferred syn. of *comagmatic.*

consanguineous [sed] Said of a natural group of sediments or sedimentary rocks related to one another by origin; e.g. "consanguineous association" (such as flysch, molasse, or paralic sediments) interrelated by common ancestry, environment, and evolution, representing a "facies" in a broad sense but "not equivalent to a single lithologic type" (Fairbridge

958, p.319).

consanguinity In petrology, the genetic relationship that exists between igneous rocks that are presumably derived from the same parent magma. Such rocks are closely associated in space and time and commonly have similar geologic occurrence and chemical, mineralogic, and textural characteristics. Adj: *consanguineous*. See also: *comagmatic*.

consequent [geomorph] Said of a geologic or topographic feature that originated as a result of and in harmony with preexisting conditions or features; e.g. a *consequent* ridge (such as an anticlinal arch that retained its axial eminence), or a *consequent* island in a lake basin (such as an elevation that remained above the water level at the time of the formation of the lake), or a *consequent* waterfall (such as one resulting from the irregularities of the surface over which the stream flows). Cf: *subsequent; resequent; obsequent.*

consequent [streams] adj. Said of a stream, valley, or drainage system whose course or direction is dependent on or controlled by the general form and slope of an existing land surface. The term was first applied by Powell (1875, p.163-66) to the valley thus formed.---n. *consequent stream.*

consequent divide A divide between two consequent streams.

consequent fault scarp A fault scarp whose face has been rapidly changed by mass-wasting shortly after the initial formation of the scarp.

consequent lake A lake occupying a depression that represents an original inequality in any new land surface; e.g. a lake existing in a depression in glacial deposits or among sand dunes, or in an irregularity of a recently uplifted sea floor. Syn: *newland lake.*

consequent stream A stream that originates on a newly exposed or recently formed surface and that flows along a course determined entirely by the initial slope and configuration of that surface; a stream whose direction of flow is directly related to and a necessary consequence of the original up-slope surface of the land and the geologic structure of the area. Syn: *consequent; original stream.*

consequent valley A valley eroded by or containing a *consequent stream.*

consertal An obsolete syn. of *sutured.*

conservative elements In sea water, elements that are either so abundant or so inert that only minute quantities enter the particulate state; they form most of the salts in sea water. Their abundance relative to each other remains constant; this principle is the basis for determining salinity. Cf: *nonconservative elements.* Syn: *conservative solutes; conservative ions.*

conservative ions *conservative elements.*

conservative solutes *conservative elements.*

consistency The relative ease with which a soil can be deformed. The term expresses the degree of firmness or cohesion of soil particles and their resistance to rupture or deformation.

consistency index The ratio of the difference between the liquid limit and the natural water content (of a soil) to the plasticity index (the difference between the liquid limit and the plastic limit). Syn: *relative consistency.*

consistency limits *Atterberg limits.*

consociation An ecologic community, within an association, that has one dominant species, usually one of several dominants.

consolidated pack ice Pack ice in which the concentration is 8/8 (or 10/10) and the floes are frozen together. Obsolete syn: *consolidated pack; consolidated ice; field ice.*

consolidation [soil] Gradual or slow reduction in volume and increase in density of a soil mass in response to increased load or compressive stress; e.g. the adjustment of a saturated soil involving the squeezing of water from pore spaces. See also: *compaction [soil].*

consolidation [geol] Any process whereby loosely aggregated, soft, or liquid earth materials become firm and coherent rock; specif. the *solidification* of a magma to form an igneous rock, or the *lithification* of loose sediments to form a sedimentary rock.

consolidation [tect] A term that is sometimes used for diastropic processes by which orthogeosynclines are converted from mobile belts into rigid cratons. The word has many connotations besides this special one, and the more specific terms "orogeny" or "orogenesis" are preferable.

consolute Said of liquids that are miscible in all proportions.

consolute point A point that represents the composition and maximum or minimum temperature or pressure of a miscibility gap. Syn: *critical solution point.*

consortium An intimately associated group of different kinds of organisms.

constancy of composition *constancy of relative proportions.*

constancy of interfacial angles *law of constancy of interfacial angles.*

constancy of relative proportions The law that the ratios between the more abundant dissolved solids in sea water are virtually constant, regardless of the absolute concentration of the total dissolved solids. Syn: *law of constancy of relative proportions; constancy of composition; rule of constant proportion.*

constant error A *systematic error* that is the same in both magnitude and sign throughout a given series of observations (the observational conditions remaining unchanged) and that tends to have the same effect upon all the observations of the series or part thereof under consideration; e.g. the index error of a precision instrument.

constant slope The straight part of a hillside surface, lying below the *free face* and determined by the angle of repose of the material eroded from it (Wood, 1942); it merges downslope into the *waning slope.* Cf: *gravity slope.* Syn: *debris slope.*

constitutional ash *inherent ash.*

constructional (a) Said of a landform that owes its origin or general character (form, position, direction, etc.) to the processes of upbuilding, such as accumulation by deposition (e.g. a plain or an alluvial terrace), volcanic eruption (e.g. a volcanic cone), and diastrophism or orogenic activity (e.g. a mountain, a fault block, or a tectonic valley); said of a surface whose form was not acted upon by erosion. Ant: *destructional.* Cf: *initial landform.* (b) Said of a stream or drainage pattern that is formed by runoff from a constructional landform or surface (Davis, 1894, p.74).

constructional void porosity Primary porosity in a carbonate framework (Murray, 1960, p.61). Cf: *growth-framework porosity.*

constructive metamorphism *temperature-gradient metamorphism.*

constructive waterfall In a cave, a waterfall over a rimstone dam.

constructive wave A wave that builds up a beach by moving material (esp. shingle) landward, as a gentle wave with a more powerful uprush than backwash. Ant: *destructive wave.*

consumer An organism that is unable to manufacture its food from nonliving matter but is dependent on the energy stored in other living things. Cf: *producer.*

consumptive use The difference between the total quantity of water withdrawn from a source for use and the quantity of water, in liquid (and, rarely, solid) form, returned to the source. It includes mainly water transpired by plants and evaporated from the soil, but may also include water diverted from one watershed to another, to the ocean, etc.

consumptive waste Water returned to the atmosphere without having benefited man (Langbein & Iseri, 1960).

contact [eco geol] Adj. Said of a mineral deposit, sometimes veinlike, occurring at the contact of two unlike rock types, usually an igneous-sedimentary contact.

contact [geol] n. A plane or irregular surface between two different types or ages of rocks. See also: *boundary*.

contact [oil] *interface*.

contact area One of the areas of the proximal side of a spore or pollen grain that contacted the other members of the tetrad. Contact areas are seldom visible in mature pollen grains but are frequently apparent in spores. Trilete spores have three contact areas; monolete spores have two contact areas.

contact aureole *aureole*.

contact breccia A breccia around an igneous intrusion, caused by wall-rock fragmentation and consisting of both intrusive material and wall rock; *intrusion breccia*.

contact erosion valley A valley eroded along a line of weakness at the contact between two different types of rock.

contact goniometer A *goniometer* that measures the solid angle between two crystal planes by contact with the surface. It is a 180° protractor, divided in degrees and accurate to half a degree, with a straight edge pivoted at the center. Cf: *two-circle goniometer; reflection goniometer.*

contact load The part of the *bed load* that is in substantially continuous contact with the stream bed.

contact log A log that places electrodes in direct contact with the borehole wall (in order to eliminate the effects of the drilling mud) and that uses short electrode spacings (in order to measure the resistivity of a small volume of the formation adjacent to the borehole); a *microlog*.

contact-metamorphic Said of a rock or mineral that has originated through the process of *contact metamorphism*.

contact metamorphism One of the principal local processes of thermal metamorphism genetically related to the intrusion (and extrusion) of magmas and taking place in rocks or at near their contact with a body of igneous rock. Metamorphic changes are effected by the heat and materials emanating from the magma and by some deformation connected with the emplacement of the igneous mass (Holmes, 1920). Cf: *thermal metamorphism*. Adj: *contact-metamorphic*. See also: *exomorphism; endomorphism.*

contact metasomatism A mass change in the composition of rocks in contact with an invading magma from which "fluid" constituents are carried out to combine with some of the country-rock constituents to form a new suite of minerals. The term was originated by Barrell (1907). Cf: *regional metasomatism.*

contact mineral A mineral that was formed by contact metamorphism.

contact resistance The resistance observed between a grounded electrode and the ground, or between an electrode and a rock specimen.

contact spring A type of gravity spring whose water flows to the land surface from permeable strata over the outcrop of less permeable or impermeable strata that prevent or retard the downward percolation of ground water, deflecting it to the surface (Meinzer, 1923, p.51). Syn: *hillside spring; stratum spring; outcrop spring*. Cf: *barrier spring.*

contact twin A twinned crystal, the two individuals of which are symmetrically arranged about a twin plane. Syn: *juxtaposition twin.*

contact zone *aureole*.

contaminated In igneous petrology, a syn. of *hybrid*.

contamination [ign] The process whereby the chemical composition of a magma is altered as a result of the *assimilation* of country rock. Cf: *hybridization*.

contamination [water] The addition to water of any substance or property preventing the use or reducing the usability of the water for ordinary purposes such as drinking, preparing food, bathing, washing, recreation, and cooling. Sometimes arbitrarily defined differently from *pollution*, but generally considered synonymous.

contemporaneous Formed, existing, or originating at the same time. Also, said of lava flows interbedded in a single time-stratigraphic unit and, generally, of any rock or facies that de-

velops during the formation of the enclosing rocks. Cf: *penecontemporaneous*.

contemporaneous deformation Deformation that takes place in sediments during or immediately following their deposition. Includes many varieties of soft-sediment deformation, such as small-scale slumps, crumpling and brecciation, but in some areas (as the northern Apennines, Italy) features of large dimensions. Syn: *penecontemporaneous deformation*.

contemporaneous erosion Local, slight, or insignificant erosion that goes on while elsewhere deposition is occurring generally or continuously.

contemporaneous fault *growth fault*.

contemporary carbon The isotopic carbon content of living matter, implying maximum natural proportion of carbon-14 and a radioactivity of about 16 disintegrations per gram-minute. Usage should be limited as isotopic composition of living matter can change with time due to changes in proportions in the atmosphere.

continens An orange, yellow, or reddish area on the surface of Mars. Cf: *mare*. Pl: *continentes*.

continent [lunar] A rugged, heavily cratered highlands region of the Moon; *terra*.

continent [Earth] One of the Earth's major land masses of sialic rock, including both dry land and its continental shelves. About one third of the Earth's surface is covered by continents.

continental accretion A theory, originally proposed by J.D. Dana in the 19th Century, that continents have grown at the expense of the ocean basins by the gradual addition of new continental material around an original nucleus. Most of the new material was believed to have accumulated in concentric geosynclinal belts, each in turn consolidated by orogeny and succeeded by a new belt farther out. There is good evidence along some continental borders, as in western North America, that much new continental crust (now land area) has been added to the continent during Phanerozoic time, but most geologists now reject the larger implications of the concept. It is directly opposed to several other larger concepts, such as *oceanization* and *continental displacement*.

continental alluvium Alluvium produced by the erosion of a highland area and deposited by a network of rivers to form an extensive plain.

continental apron *continental rise*.

continental basin A region, in the interior of a continent, that comprises one or more closed basins.

continental borderland That area of the *continental margin* between the shoreline and the continental slope which is topographically more complex than the *continental shelf*. It is characterized by ridges and basins, some of which are below the depth of the continental shelf. An example is the southern California continental borderland.

continental climate The climate of the interior of a continent characterized by seasonal temperature extremes and by the occurrence of maximum and minimum temperature soon after summer and winter solstice, respectively. Cf: *oceanic climate*.

continental crust That type of the Earth's *crust* which underlies the continents and the continental shelves; it is equivalent to the *sial* and the *sima* below, and is about 35 km deep with depths under mountain ranges as much as 60 km. The density of the upper layer of the continental crust is 2.7 g/cm^3, and the velocities of compressional seismic waves through it are less than 6.2 km/sec. Cf: *oceanic crust*.

continental deposit A sedimentary deposit laid down on land (whether a true continent or only an island) or in bodies of water (whether fresh or saline) not directly connected with the ocean, as opposed to a marine deposit; a glacial, fluvial, lacustrine, or eolian deposit formed in a nonmarine environment. See also: *terrestrial deposit*.

continental displacement A general term, which can be used for many aspects of the theory originally propounded at length y Wegener (1912); also, less appropriately, called *continen-al drift*. Wegener postulated the displacement of large plates f continental (sialic) crust, moving freely across a subs-atum of oceanic (simatic) crust, but the mechanisms in-olved were so implausible to most geologists that the con-ept was generally discredited for many decades. Now, new vidence has been found and more acceptable mechanisms ave been proposed, so that the original theory has gained a vider measure of acceptance: (1) the continents have re-nained relatively fixed but the Earth has expanded (see *ex-anding Earth*), leaving progressively wider gaps of oceanic reas between; (2) the continents have moved away from ach other by *sea-floor spreading* along a median ridge or rift see *world rift system*), producing new oceanic areas be-ween the continents; (3) or, the masses propelled away from he ridges consist of thick plates (see *plate tectonics*), com-osed of both continental and oceanic crust, which have noved in various directions independently of each other. *robably a true explanation of world tectonics will combine ome or all of the newer explanations that have been pro-osed. Syn: *displacement theory; Wegener hypothesis; peirophoresis theory; continental migration*.

:ontinental divide A drainage *divide* that separates streams lowing toward opposite sides of a continent, often into differ-nt oceans; e.g. in North America, the divide separating the vatersheds of the Pacific Ocean from those of the Atlantic)cean, and extending from the Yukon Territory, along the 3ritish Columbia-Alberta boundary, through western Montana, Wyoming, Colorado, western New Mexico, and into Mexico.

:ontinental drift *continental displacement*. Var: *drift* [tect].

:ontinental flexure A hinge-line structure along the contact of :ontinent and sea floor, in which warping steepens the angle)f the slope of the continental shelf, causing relative uplift of he continent and eventual formation of coastal ranges.

:ontinental glacier (a) A glacier of considerable thickness :ompletely covering a large part of a continent or an area of at least 50,000 sq km, obscuring the relief of the underlying surface, such as the ice sheets covering Antarctica and 3reenland. See also: *inland ice*. Syn: *continental ice sheet; :ontinental ice; ice sheet*. (b) Any glacier in a continental, as)pposed to a maritime, climatic environment; this usage is not ecommended.

:ontinental ice sheet *continental glacier*.

:ontinental margin The ocean floor that is between the shore-ine and the abyssal ocean floor, including various provinces; he *continental shelf; continental borderland; continental slope*, and the *continental rise*.

:ontinental migration *continental displacement*.

:ontinental nucleus *shield*.

:ontinental platform *continental shelf*.

:ontinental rise That part of the *continental margin* that is be-ween the continental slope and the abyssal plain, except in areas of an oceanic trench. It is a gentle incline with slopes of l:40 to 1:2000, and generally smooth topography, although it nay bear submarine canyons. Syn: *continental rise*.

:ontinental river A river with no outlet to the sea, its water disappearing by percolation or evaporation.

:ontinental sea *epicontinental sea*.

:ontinental shelf That part of the *continental margin* that is)etween the shoreline and the continental slope (or, when here is no noticeable continental slope, a depth of 200 m). It s characterized by its very gentle slope of 0.1°. Cf: *insular shelf; marginal plateau; continental borderland*. Syn: *continen-tal platform; shelf* [marine geol].

:ontinental shield *shield*.

:ontinental slope That part of the *continental margin* that is)etween the continental shelf and the continental rise (or oce-anic trench). It is characterized by its relatively steep slope of 3-6°. Cf: *insular slope*.

:ontinental time A term used by Kobayashi (1944a, p.477) for

fossil time as indicated by nonmarine organisms. Cf: *marine time*.

continental transgression Any enlargement of the area of con-tinental deposition (such as of piedmont deposition on a ba-jada) in which previously erosional areas (or neutral areas of equilibrium of erosion and deposition) are covered with sedi-ments. Cf: *transgression* [stratig].

continentes Plural of *continens*.

continental period Time, of variable duration, when a specific area was above sea level, forming part of a continent; it was a continental period for that particular area.

continuation Determining from a set of measurements of a gravity or magnetic field at one elevation the values the field will have at another elevation.

continuity The quality or state of being continuous, such as the uninterrupted persistence of a sedimentary stratum as ob-served when traveling across country from one locality to an-other. See also: *law of original continuity*.

continuity equation An axiom stating that the rate of flow past one section of a conduit is equal to the rate of flow past an-other section of the same conduit plus or minus any additions or subtractions between the two sections (ASCE, 1962).

continuous cleavage Pervasive cleavage, as in shale; the planes of cleavage are continuous rather than spaced. It is a *penetrative texture*. Cf: *spaced cleavage*.

continuous-creation hypothesis *steady-state theory*.

continuous deformation Deformation by flow rather than by fracture. Cf: *discontinuous deformation*.

continuous permafrost A zone of permafrost that, for the most part, is uninterrupted by pockets or patches of unfrozen ground. Cf: *discontinuous permafrost; sporadic permafrost*.

continuous porosity A term originally proposed by Murray (1930, p.452) for systems of interconnected pores, as oppos-ed to *discontinuous porosity*. The term is little used now and not recommended (Choquette & Pray, 1970, p.245). Cf: *effec-tive porosity*.

continuous profiling A seismic method in which geophone groups are placed uniformly along the length of the line and so spread that continuous (or 100%) coverage is obtained (Sheriff, 1968). Cf: *correlation shooting*.

continuous reaction series A *reaction series* in which reaction of early-formed crystals with later liquids takes place without abrupt phase changes; e.g., the plagioclase feldspars for a continuous reaction series. Cf: *discontinuous reaction series*.

continuous spectrum The array of intensity values in the spec-trum are distributed evenly over the entire range of wave-lengths of the ordering variable. An optical continuous spec-trum is most often derived from incandescent solids, liquids, or gases. Cf: *line spectrum; band spectrum*.

continuous stream A stream that does not have interruptions in space; it may be perennial, intermittent, or ephemeral, but it does not have wet and dry reaches. Ant: *interrupted stream*.

continuous velocity log A log that records continuously the ve-locity of acoustic or seismic waves over small intervals as the logging device traverses a borehole; a *sonic log*. Abbrev: CVL.

contorted bedding *convolute bedding*.

contorted drift A glacial deposit that exhibits folding, thrusting, and other irregularities resulting from pressure during ice movement. The contortions are usually aligned in the direction of ice movement.

contortion (a) The intricate folding, bending, or twisting to-gether of laminated sediments on a considerable scale, the laminae being drawn out or compressed in such a manner as to suggest kneading more than simple folding; esp. *intrafor-mational contortion*. Also, the state of being contorted. It oc-curs on a larger scale than *corrugation*. (b) A structure pro-duced by contortion.

contour [cart] n. (a) An imaginary line on a land surface, all points of which are at the same elevation (vertical distance) above or below a specified datum surface (usually mean sea level). It is the line of intersection of a level surface with the surface of the ground, as illustrated by the shoreline of an imaginary body of water whose surface is at the elevation represented by the contour. The term has been extended to various underground surfaces. Syn: *topographic contour.* (b) *contour line.*---v. To provide a map with contour lines; to draw a contour line.

contour [geog] The outline, periphery, or horizontal configuration of a figure or body, seen two-dimensionally; esp. the outline of a coast, mountain pass, or other topographic feature, such as the "contour" of any portion of the land as defined by the waters that surround it.

contour [geol] An imaginary line or surface along which a certain quantity, otherwise variable, has the same value, e.g. a *structure contour.*

contour current An ocean current flowing along isopycnic lines approximately parallel to the bathymetric contours; e.g. a density current flowing parallel to the submarine slopes at the margin of an ocean basin.

contour diagram A *fabric diagram* in which there is a contouring of points for easier visual comprehension. Cf: *point diagram.*

contour horizon *datum horizon.*

contour interval The difference in value between two adjacent contours; specif. the vertical distance between the elevations represented by two successive contour lines on a topographic map. It is generally a regular unit chosen according to the amount of vertical distance involved and the scale of the map, but it need not be constant over the entire map (a variable contour interval is often used for optimum portrayal of relief features). Syn: *vertical interval.*

contour line [cart] A line, drawn on a map or chart, representing a *contour.* In ordinary usage, "contour" and "contour line" are synonymous terms, although it is sometimes desirable to distinguish between "contour" as applied to the surface of the Earth and "contour line" as applied to a topographic map (USGS, 1958, p.155). In a map legend, where there is no possibility of misunderstanding, the term "contour" may stand for "contour line". Cf: *form line.* Syn: *isohypse.*

contour line [geol] A term used loosely in the general sense of an *isopleth;* e.g. a line (on a map) connecting points of equal magnitude of a mass property of a sediment (as of porosity, permeability, color, or thickness, or of size, shape, or roundness of sedimentary particles).

contour map A map that portrays surface configuration by means of contour lines; esp. a topographic map that shows surface relief by means of contour lines drawn at regular intervals above mean sea level, or a *structure-contour map* that shows the configuration of rock surfaces underground.

contour sketching Freehand delineation, on a map, of the surface relief as seen in perspective view and controlled by map locations corresponding to salient points on the ground. See also: *field sketching.*

contour value A numerical value placed upon a contour line, such as a figure denoting elevation relative to mean sea level.

contraclinal A syn. of *anaclinal.* Term used by Powell (1873, p.463).

contracting Earth A theory widely believed in the 19th and first part of the 20th centuries that the orogenic and other structures of the Earth were produced by compression of the crust during its gradual contraction on the surface of a cooling but originally molten globe (a familiar textbook illustration of the time was a dried apple). The theory is now discredited, as the Earth is not cooling and contracting in the manner believed. Cf: *expanding Earth; tetrahedral hypothesis; wedge theory.*

contraction fault A term used by Norris (1954) for a fault in which there has been a compression-type displacement in the plane of bedding. It is partially synonymous with reverse fault or *thrust fault.*

contraction fissure A fissure that is formed as a result of cool-

ing or drying and consequent contraction of the rock. Cf *cooling crack.*

contraction stripe One of the parallel lines on a playa or along its margin, consisting of vegetation growing along large-scale cracks caused by contraction of muds upon drying (Stone, 1967, p.228). Syn: *giant contraction stripe.*

contragradation Stream aggradation caused by an obstruction (Shaw, 1911). Syn: *dam gradation.*

contraposed shoreline A term introduced by Clapp (1913) for a shoreline "placed against" a previously buried (usually resistant) land surface that has been exhumed by wave erosion removing a marine cover of softer sediments; e.g. along the coast of Maine.

contrast In photogrammetry, the actual difference in density between the shadows and the highlights (darkest parts) of a photographic negative, or the ratio of reflecting power between the shadows and the highlights (lightest parts) of a photographic print. Also, the quality or rating of a photograph as determined by contrast.

contrasted differentiation Differentiation of magma into basic and acidic magmas. Reactions between these contrasted magmas may produce intermediate rock types which resemble intermediate types usually considered as the product of progressive fractionation (Nockolds, 1934).

contratingent Said of a minor septum (of a rugose coral) that leans against the adjoining major septum on the side toward the counter septum.

contributory An obsolete syn. of *tributary.* Also spelled: *contributary.*

control [surv] (a) The coordinated and correlated dimensional data that are used to establish the position, scale, and orientation of the detail of a map and that are responsible for the interpretations placed on the map. (b) The assemblage of accurately located points that determines the degree of accuracy of a map and with which local secondary surveys may be tied to insure their essential accuracy; a system of relatively precise field measurements of points, marks, or objects on the ground, whose horizontal and/or vertical positions have been (or will be) more or less accurately determined by surveying instruments. A map that includes many such points is said to have "good control".

control [hydraul] (a) A section or reach of an open channel in which natural or artificial conditions make the water level above it a stable index of discharge. It may be either complete (i.e., water-surface elevation above the control is completely independent of downstream water-level fluctuations) or partial; it may also shift. (b) That waterway cross-section which is the bottleneck for a given flow and determines the energy head required to produce the flow. In an open channel, it is the point at which flow is at critical depth; in a closed conduit, it is the point at which hydrostatic pressure and cross-sectional area of flow are definitely fixed, except where the flow is limited at some other point by a hydrostatic pressure equal to the greatest vacuum that can be maintained unbroken at that point.----(ASCE, 1962).

control base A surface upon which a map projection and ground control are plotted and upon which templets have been assembled or aerotriangulation has been accomplished and the control points thus determined have been marked.

controlled mosaic An aerial *mosaic* in which the photographs have been adjusted, oriented, and scaled to horizontal ground control in order to improve the accuracy of representation with respect to distances and distortions. It is usually assembled from rectified photographs that have been corrected for tilt and for variations in flight altitude.

control point Any station identified on a photograph and used to aid in fixing the attitude and/or position of a photograph or group of photographs. See also: *pass point.*

control station An accurately located point, mark, or object on the ground, whose horizontal and/or vertical position is used s a base for a dependent survey; any surveyed point used for horizontal and/or vertical control.

control survey A survey that provides horizontal- and/or vertical-position data for the support or control of subordinate surveys or for mapping; e.g. a survey that provides the geographic positions (and plane coordinates) of triangulation and traverse stations and the elevations of bench marks. Control surveys are classified according to their precision and accuracy: the highest prescribed order is designated first order, the next lower prescribed order is second order, and so on.

conulariid Any scyphozoan belonging to the order Conulariida, characterized by a tetramerous cone-shaped to elongate pyramidal or subcylindrical, chitinophosphatic periderm which may be smooth or have longitudinal markings. Their stratigraphic range is Middle Cambrian to Lower Triassic.

conule A cone-shaped projection from the body surface of certain sponges, generally over the end of a fiber.

conulite A type of conical speleothem that is formed downward in mud of a cave floor by the action of dripping water.

conus One of the pointed projections making up the sculpture of certain pollen and spores, being more or less rounded at the base and less than twice as high as the basal diameter. Pl: coni.

convection [meteorol] Large-scale, vertical movements within the Earth's atmosphere. Cf: advection.

convection [oceanog] A general term for the movement and mixing of water masses in the ocean.

convection [tect] A supposed mass movement of subcrustal or mantle material, either laterally, or in upward- or downward-directed convection cells, mainly as a result of heat variations. Theories have been proposed utilizing convection currents to explain deep-sea trenches, island arcs, geosynclines, orogeny, and the like. Cf: advection.

convection cell In tectonics, a pattern of mass movement of mantle material in which the central area is uprising and the outer area is downflowing, due to heat variations. See also: convection.

convergence [meteorol] An increase of an air mass within a given volume; greater inflow than outflow of air. Cf: divergence [meteorol].

convergence [currents] The meeting of ocean currents or water masses, having differing densities, temperatures, or salinities, resulting in the sinking of the denser, colder, or more saline water; also, the line or area in which convergence occurs. See also: polar convergence. Cf: divergence [currents].

convergence [evol] The acquisition or possession of similar characteristics by animals or plants of different groups as a result of similarity in habits or environment. Cf: adaptive radiation; parallelism; radiation [evol]. See also: convergent evolution.

convergence [petrology] The production, esp. during metamorphism, of two or more petrographically similar rocks along different lines of petrogenesis and showing evidence only of the last process to have acted upon them (Challinor, 1967, p.57).

convergence [photo] The principle that "the many separate items of photogeological data, all indicating the presence of the same structure, give the interpretation of this structure a high order of reliability" (Allum, 1966, p.36-37). Syn: convergence of evidence.

convergence [stratig] The gradual decrease in the vertical distance or interval between two specified rock units or geologic horizons as a result of the thickening and thinning of intervening strata; e.g. the reduction in thickness of sedimentary beds (as measured in a given direction and at right angles to the bedding planes), caused by variable rates of deposition or by unconformable relationship.

convergence [surv] convergence of meridians.

convergence map isochore map.

convergence of meridians (a) The angular drawing together of the geographic meridians in passing from the equator to the poles. (b) The difference between the two angles formed by the intersection of a great circle with two meridians. Also, the relative difference of direction of meridians at specific points on the meridians.---Syn: convergence.

convergent evolution The development of similar-appearing forms in genetically unrelated lineages: convergence. Cf: parallel evolution.

converter plant A plant that incorporates into its structure an insoluble element from the soil and later, when the plant decays, returns that element to the soil in a soluble form.

convex bank The inner bank of a curved stream, with the center of the curve away from the channel; e.g. a slip-off slope. Ant: concave bank.

convex cross-bedding Cross-bedding with convex (upward-arching) foreset beds. Cf: concave cross-bedding.

convexo-concave Convex on one side and concave on the other; e.g. said of a resupinate brachiopod shell having a convex brachial valve and a concave pedicle valve. Cf: concavo-convex.

convexo-plane Convex on one side and flat on the other; e.g. said of a brachiopod shell having a convex brachial valve and a flat pedicle valve. Cf: plano-convex.

convex slope waxing slope.

convolute Coiled or wound together one part upon another; e.g. said of a coiled foraminiferal test in which the inner part of the last whorl extends to the center of the spiral and covers the inner whorls, or said of a coiled gastropod shell whose inner or earlier whorls are entirely concealed or embraced by the outer or later whorls. Cf: involute; evolute; advolute.

convolute bedding A descriptive term used by Kuenen (1953) for the wavy, extremely disorganized, and markedly and intricately crumpled, twisted, or folded laminae that are confined within a single, relatively thin, well-defined, undeformed layer, that die out both upward and downward, and that are overlain and underlain by parallel undisturbed layers. It is characteristic of some coarse-silt or fine-sand beds and involves only the internal laminae of the bed (and not the bed itself, which remains undeformed). The structure appears to result from pre-consolidation deformation or from deformation during deposition of highly mobile or plastic sediments; e.g. formation by subaqueous slumping, gliding, or sliding, by load deformation, or by the joint action of eddies, surface drag, and upward migration of pore waters within the laminae. Hills (1963, p.32) restricts the term to a structure formed only during deposition. See also: slip bedding; slump bedding; glide bedding; intraformational contortion. Syn: convolute lamination; convolution; contorted bedding; crinkled bedding; curly bedding; hassock structure; gnarly bedding; intrastratal flow structure.

convoluted organ A loose, calcareous, spicular structure around the axial sinus in many camerate crinoids.

convolute lamination convolute bedding.

convolution (a) The process of producing convolute bedding; the state of being convoluted. (b) A structure produced by convolution, such as a small-scale but intricate fold. (c) convolute bedding.

convolutional ball A comparatively small, concentric sedimentary structure formed in association with convolute bedding. Syn: roll-up structure.

convulsion catastrophe.

convulsionism catastrophism.

cookeite A mineral of the chlorite group: $LiAl_4(AlSi_3)O_{10}(OH)_8$.

cooking snow water snow.

coolgardite A mixture of coloradoite, sylvanite, and calaverite, found at Kalgoorlie in Western Australia.

cooling crack A joint that is formed as a result of cooling of an igneous rock. Cf: contraction fissure.

cooling unit A flow, or rapid multiple flows, of volcanic material having characteristic patterns of welding and crystallization.

coom (a) Var. of *cwm*. (b) An English term for a hollow in the side of a hill or mountain; a cove; a *combe*.

coomb (a) A var. of *combe*, a valley. (b) A var. of *cwm*.

coombe A var. of *combe*, a valley.

coombe rock An irregular mass of unstratified rock debris of any type accumulating as a result of solifluction; esp. the angular mass of unrolled and unweathered flints mixed with chalk rubble and other earthy matrix material, partly filling a dry valley (coombe), and spreading out onto the coastal plain, as in SW England. See also: *head; rubble drift.* Syn: *combe rock; elephant rock.*

coon-tail ore A term used in the Cave-in-Rock district of southern Illinois for a light- and dark-banded ore of sphalerite and fluorite.

cooperite A steel-gray tetragonal mineral: (Pt,Pd)S. It occurs in minute irregular grains in igneous rocks.

coordinate n. Any one of a set of numbers designating linear and/or angular quantities that specify the position of a point on a line, in space, or on a given plane or other surface in relation to a given reference system; e.g. latitude and longitude are coordinates of a point on the Earth's surface. The term is usually used in the plural, esp. to designate the particular kind of reference system (such as "spherical coordinates", "plane coordinates" and "polar coordinates").

coordinate system A reference system for defining points in space or on a particular surface by means of distances and/or angles with relation to designated axes, planes, or surfaces.

coordination number [cryst] In crystallography, the number of ions that can be directly attached to an ion in the crystal lattice; e.g., cubic coordination, tetrahedral coordination, octahedral coordination.

coorongite [min] A soft, brown, rubbery variety of *elaterite*, originating from deposits of *Elaeophyton* algae in salt water bodies near the Coorong in Australia; it may represent a stage in the formation of boghead coal. See also: *n'hangellite.*

cop A term used in England for the steep-sided top or summit of a hill; also, a small hill with a rounded top.

copal An inclusive term for a wide variety of hard, brittle, semitransparent, yellowish to red fossil resins from various tropical trees (such as *Copaifera* and *Agathis*), being nearly insoluble in the usual solvents, and resembling amber in appearance; e.g. *Congo copal* and *kauri*. Copal also occurs as modern resinous exudations. Syn: *gum copal.*

copalite A clear, pale-yellow, dirty-gray, or dirty-brown fossil resin resembling copal in hardness, color, transparency, and difficult solubility in alcohol, containing succinic acid, and being much poorer in oxygen than most amber; e.g. the "Highgate resin" found in the blue Tertiary clay of Highgate Hill in London. Syn: *copaline; fossil copal.*

Copenhagen water *normal water.*

copepod Any crustacean belonging to the class Copepoda, characterized by the absence of both a carapace and compound eyes. The only known fossil copepods have been found in Miocene lake deposits.

Copernican (a) Pertaining to the youngest lunar topographic features and lithologic map units constituting a system of rocks formed during the period of formation of ray craters (such as Copernicus). (b) Said of the stratigraphic period during which the Copernican System was developed.

Copernican system A concept of planetary motion according to which the Earth rotates on an axis once a day and revolves about the Sun once a year, and according to which all other planets have Sun-centered orbits. It is named after the Polish astronomer Copernicus (d.1543) who maintained it.

copiapite (a) A yellow mineral: $(Fe,Mg)Fe_4^{+3}(SO_4)_6(OH)_2 \cdot 20H_2O$. Syn: *yellow copperas; ihleite; knoxvillite.* (b) A group of minerals containing hydrous iron sulfates, including copiapite, aluminocopiapite, calciocopiapite, cuprocopiapite,

ferricopiapite, ferrocopiapite, and magnesiocopiapite.

coppaelite An olivine-free melilite-bearing extrusive rock containing small diopside phenocrysts in a holocrystalline groundmass of melilite, diopside, and phlogopite.

copper A reddish or salmon-pink isometric mineral, the native metallic element Cu. It is ductile and malleable, a good conductor of heat and electricity, usually dull and tarnished, and formerly an important ore. Copper is the only metal that occurs native abundantly in large masses; it frequently occurs in dendritic clusters or mossy aggregates, in sheets, or in plates filling narrow cracks or fissures. It has many uses, notably as an electric conductor and as the base metal in brass, bronze, and other alloys.

Copper Age In archaeology, a cultural level that is sometimes discernible between the *Bronze Age* and the *Iron Age* of the *three-age system*. It is characterized by the experimentation and use of copper for such technological purposes as weapons and tools. Correlation of relative cultural levels with actual age (and, therefore, with the time-stratigraphic units of geology) varies from region to region. Syn: *Chalcolithic; Eneolithic.*

copperas (a) *melanterite.* (b) A name sometimes applied to other sulfate minerals, such as copiapite ("yellow copperas") and goslarite ("white copperas").

copper glance *chalcocite.*

copper mica *chalcophyllite.*

copper nickel *nickeline.*

copper pyrites *chalcopyrite.*

copper uranite *torbernite.*

copper vitriol *chalcanthite.*

coppice mound A small mound of fine-grained desert material stabilized around shrubs or coppices.

coprogenic Said of a deposit formed of animal excrement. Etymol: Greek *kopros*, "dung".

coprolite (a) The fossilized excrement of vertebrates such as fishes, reptiles, and mammals, larger than a *fecal pellet*, measuring up to 20 cm in length, characterized by an ovoid to elongate form, a surface marked by annular convolutions, and a brown or black color, and composed largely of calcium phosphate. The term is incorrectly used to refer to desiccated or fresh fecal remains. Term introduced by Buckland (1829a). (b) An English term applied commercially and popularly to any *phosphatic nodule* mined for fertilizer.

copropel A term introduced by Swain & Prokopovich (1954, p.1184) to replace the vaguely applied term *gyttja*, and signifying a "dark-brown or gray coprogenic ooze, containing chitinous exoskeletons of benthonic arthropods in addition to reworked organic matter".

copula A short, hollow, funnel-like tube in the position of the mucron at the base of the chamber of chain-forming chitinozoans. Copulas connect to prosomes of preceding chambers to form chains.

coquimbite A white or slightly colored hexagonal mineral $Fe_2(SO_4)_3 \cdot 9H_2O$. It sometimes contains appreciable aluminum, and it is dimorphous with paracoquimbite. Syn: *white copperas.*

coquina A detrital limestone composed wholly or chiefly of mechanically sorted fossil debris that experienced abrasion and transport before reaching the depositional site and that is weakly to moderately cemented but not completely compacted and indurated; esp. a soft, porous, friable, coarse-grained whitish limestone made up of loosely aggregated shells and shell fragments (with or without broken corals and calcareous parts of other marine organisms), such as the relatively recent deposits of shell heaps occurring in Florida and used for roadbeds and building purposes. Wentworth (1935, p.244) recommended a particle size greater than 2 mm in diameter (gravel size). Etymol: Spanish, "cockle, shellfish". Cf: *coquinoid limestone; microcoquina.*

coquinite Compact, well-indurated, and firmly cemented equivalent of coquina.

coquinoid limestone A limestone consisting of coarse, unsorted, and often unbroken shelly materials that have accumulated in place without subsequent transportation or agitation, and generally having a fine-grained matrix. It is autochthonous, unlike the allochthonous *coquina*; under certain conditions it can develop into a biostrome. See also: *microcoquinoid limestone*. Syn: *coquinoid*.

coracite *uraninite*.

coral (a) A general name for any of a large group of bottom-dwelling, sessile, marine invertebrate organisms (polyps) that belong to the class Anthozoa (phylum Coelenterata), that are common in warm intertropical modern seas and abundant in the fossil record in all periods later than the Cambrian, that produce external skeletons of calcium carbonate, and that exist as solitary individuals or grow in colonies. (b) A hard calcareous substance consisting of the continuous skeleton secreted by coral polyps for their support and habitation, and found in single specimens growing plant-like on the bottom or in extensive solidified accumulations (coral reefs). Also, any marine deposit like coral resulting from vital activities of various organisms (such as certain algae, or bryozoans and worms). (c) A piece of coral; e.g. "precious coral", a semi-translucent to opaque mass usually found in light to dark tones of red to orangy red, but may be more nearly orange, flesh-colored, white, cream, brown, blue, or black.

coral cap A thick deposit of coral-reef material overlying material of noncoral origin. Cf: *reef cap; coral crust*.

coral crust A thin layer of coral-reef material overlying material of noncoral origin. Cf: *coral cap*.

coral formation *coralloid*.

coralgal Said of a firm carbonate rock formed by an intergrowth of frame-building corals and algae (esp. coralline algae). The material so formed is an excellent sediment binder in a coral reef. Cf: *bryalgal*.

coral head A rounded, massive, often knobby or mushroom-shaped protuberance or growth of coral, usually forming on the submerged part of a coral reef, and frequently large enough to be dangerous to navigation; a small *reef patch* of coralline material. Syn: *coral knoll*.

coral horse A long remnant of a former reef tract, characterized by a flat top (or by a flat side if it has been undercut and tilted); a *niggerhead* formed by solution and dismemberment of a former reef platform.

coral island (a) A coral reef that appears above sea level, situated far from any other kind of land. (b) An oceanic island formed from coral accumulations lying atop volcanic peaks. (c) A mound of sand resting on a flat coral reef.

coral knoll A *coral head* in the form of a small knoll. The term was used by Emery (1948) for a *pinnacle*.

coralla Plural of *corallum*.

coral limestone A limestone consisting of the calcareous skeletons of corals, often containing fragments of other organisms and often cemented by calcium carbonate. Se also: *coral-reef limestone*. Syn: *coralline limestone*.

coralline n. Any organism that resembles a coral in forming a massive calcareous skeleton or base, such as certain algae or stromatoporoids.

coralline alga A type of *calcareous alga* that forms encrustations.

corallite The calcareous exoskeleton formed by an individual coral polyp and consisting of walls, septa, and accessory structures such as tabulae and dissepiments. It is embedded in the general structure of the *corallum*.

coralloid In a cave, a nodular formation of calcite. Syn: *cave coral; coral formation*.

corallum The calcareous exoskeleton of a coral colony, or the *corallite* of a solitary coral; the entire skeleton of a coral. Pl: *coralla*.

coral pinnacle *pinnacle* [reef].

coral rag A well-cemented, rubbly limestone composed largely of broken and rolled fragments of coral-reef deposits; e.g. the Coral Rag of the Jurassic, used locally in Great Britain as a building stone. Syn: *reef-rock breccia; corraline rag*.

coral reef (a) A calcareous *organic reef* or ridge of limestone, often of great extent, composed chiefly of fragments of coral (esp. the accumulated skeletons of coral-polyp colonies), coral sands, and the solid limestone resulting from organic secretion of calcium carbonate, and often built up around a framework of other organic remains, esp. those of calcareous algae (coral may constitute less than half of the reef material). Coral reefs occur along continents and islands in the Pacific Ocean where the temperature is suitable, generally a minimum of about 18°C (b) A popular term for a reef composed of organisms (such as algae, crinoids, bryozoans) other than coral.

coral-reef limestone A *reef limestone* made up in large part of the skeletons of corals, but which may contain the remains of other organisms; a fossil coral reef. See also: *coral limestone*. Syn: *coral rock*.

coral-reef shoreline A shoreline formed by deposits of coral and algae, partly exposed at low tide, and characterized by reefs built upward from a submarine floor or outward from the margin of a land area.

coral rock *coral-reef limestone*.

corange line A line on a chart joining points of equal tide range.

corbiculoid Said of heterodont dentition of a bivalve mollusk with three cardinal teeth in each valve, the middle tooth of the right valve occupying a median position below the beaks. Cf: *lucinoid*. Obsolete syn: *cyrenoid*.

corcass A low marsh or mud flat along the bank of a tidal river in Ireland. Syn: *corcagh*.

cordate Said of a leaf, esp. the basal portion, that is heart-shaped or gently lobed. Cf: *obcordate*.

corded pahoehoe The typical kind of pahoehoe, having a surface resembling coils of rope. Cf: *elephant-hide pahoehoe; entrail pahoehoe; festooned pahoehoe; filamented pahoehoe; sharkskin pahoehoe; shelly pahoehoe; slab pahoehoe*.

cordierite A light- to dark-blue or violet-blue orthorhombic mineral: $(Mg,Fe)_2Al_4Si_5O_{18}$. It exhibits strong pleochroism, is easily altered by exposure, and is an accessory mineral in granites and a common constituent in metamorphic rocks formed under low pressure. Syn: *iolite; dichroite*.

cordierite-amphibolite facies Metamorphic rocks formed by regional dynamothermal metamorphism under low to intermediate pressures (from 2000 to 5000 or 6000 bars) and temperatures (450°-700°C). It represents the low-pressure part of the *amphibolite facies* of Eskola. Characteristic index minerals are andalusite - cordierite - muscovite - sillimanite and almandine, but not kyanite (Winkler, 1967). Cf: *almandine-amphibolite facies*.

cordillera (a) A comprehensive term for an extensive series or broad assemblage of more or less parallel ranges, systems, and chains of mountains (together with their associated valleys, basins, plains, plateaus, rivers, and lakes), the component parts having various trends but the mass itself having one general direction; esp. the main mountain axis of a continent, as the great mountain region of western North America from the eastern face of the Rocky Mountains to the Pacific Ocean, or the parallel chains of the Andes in South America; a mountain province. (b) An individual *mountain chain* with closely connected, distinct summits resembling the strands of a rope or the links of a chain; e.g. one of the parallel chains of the Rocky Mountains. (c) A term also used in South America for an individual *mountain range*.---Etymol: Spanish, "chain or range of mountains", from Latin *chorda*, "cord".

cordylite A colorless to wax-yellow mineral: $(Ce,La)_2Ba(CO_3)_3F_2$.

core [drill] n. A cylindrical or columnar piece of solid rock or section of soil, usually 5-10 cm in diameter and from a centimeter up to 15 m or so in length, taken as a sample of an underground formation by a special hollow-type drill bit, and brought to the surface for geologic examination and/or chemical analysis. It records a section of the rock or soil penetrated. Cf: *core [oceanog]*. Syn: *drill core*.---v. To take a core.

core [eng] A wall or structure of impervious material forming the central part of an embankment, dike, or dam the outer parts of which are pervious. Syn: *core wall*.

core [interior Earth] The central zone or nucleus of the Earth's interior, below the Gutenberg discontinuity at a depth of 2900 km. It is divided into an *inner core* and an *outer core*, with a transition zone in between, and is equivalent to the E, F, and G layers. Apparently, only compression waves traverse the core. The magnetic field originates within the core.

core [oceanog] An undisturbed, vertical, cylindrical section of ocean-bottom sediment collected by an oceanographic corer.

core [petrology] The innermost completely granitized part of a regional body of rock undergoing granitization processes (Challinor, 1967).

core [fold] The inner or central part of a fold, especially of a folded structure that includes some sort of structural break. Cf: *envelope*.

core-and-shell structure A term used by McKee (1954, p.65) for a sedimentary structure resembling a concretion, developed in massive silty mudstone, and characterized by a rounded lump (core) of variable shape and size (ovoid to elongate and 3-30 cm in diameter) surrounded by a series of concentric layers (shells) each with a thickness ranging from 3 mm to 12 mm and appearing to be due to shrinkage in uniform, structureless masses of mud. It is characteristic of subaerial conditions following flooding in a region of considerable aridity.

core barrel (a) A hollow tube or cylinder above the bit of a core drill, used to recieve and preserve a continuous section or core of the material penetrated during drilling. The core is recovered from the core barrel. See also: *double core barrel*. (b) The tubular section of a corer, in which ocean-bottom sediments are collected either directly in the tube or in a plastic liner placed inside the tube.

core bit A hollow, cylindrical *drill bit* for carving, removing, and holding a core or sample of rock or soil material from the drill hole; the cutting end of a core drill. Syn: *coring bit*.

core box A large and lidded wooden, metal, or cardboard box or other container, divided into narrow parallel partitions, and used to store the cores at the surface as they are extracted from a borehole or corer.

cored bomb A type of volcanic bomb that has a core of non-volcanic rock or already solidified lava, around which the lava has molded itself. Syn: *perelith*.

core drill (a) A drill (usually a rotary drill, rarely a cable-tool drill) that cuts, removes, and brings to the surface a cylindrical rock sample (core) from the drill hole. It is equipped with a core bit and a core barrel. (b) A lightweight, usually mobile drill that uses drill tubing instead of drill pipe and that can (but need not) core down from grass roots.

core drilling Drilling with a core drill; the act or process of obtaining a core by drilling. Syn: *coring*.

core hole Any hole that is made for the purpose of obtaining core samples; e.g. a well drilled for information only, or a borehole made by a core drill.

core method A method for tracing a water mass or type from its origin across an area through it is spreading by variations in a characteristic parameter such as temperature, salinity or oxygen content.

corer An ocean-bottom sampler that is a metal or plastic cylinder (or, sometimes, a box) lowered by cable and driven into the ocean floor by impact. It usually consists of the cylinder or barrel, a removable plastic liner, a cutter on the lower end, and a valve or core catcher to retain the sample. See also:

piston corer; gravity corer; box core. Cf: *dredge; grab sampler*.

core record A record showing the depth, character, lithology, porosity, permeability, and fluid content of cores.

core recovery The amount of the drilled rock or soil material withdrawn as core in core drilling, generally expressed as a percentage of the total length of core recovered compared with the total length of the interval cored.

core sample One or several pieces of whole or split parts of an undisturbed core selected as a sample for subsequent analysis; a sample obtained by a corer.

core-stone An ellipsoidal or broadly rectangular joint block of granite formed by subsurface weathering in the same manner as a *tor*, but entirely separated from bedrock (Linton, 1955).

core test A hole drilled with a core drill for the purpose of obtaining information about the stratigraphy, nature, and in some cases the structure, either of deeply buried rocks or of formations at shallow depths just below the topsoil.

core texture *atoll texture*.

core wall *core [eng]*.

corindon *corundum*.

coring [drill] n. (a) The process of obtaining an undisturbed core of ocean-bottom sediments by means of a corer. (b) *core drilling*.

coring [paleont] adj. Said of spicules, or foreign bodies such as sand grains, that occupy the axial region of a skeletal fiber of a sponge.

coring bit *core bit*.

Coriolis effect The effect produced by the *Coriolis force*, viz. the tendency of all particles of matter in motion on the Earth's surface to be deflected to the right (with respect to their compass direction of motion) in the Northern Hemisphere and to the left in the Southern Hemisphere; the magnitude of the effect is proportional to the velocity and latitude of the moving particles.

Coriolis force The apparent deflective component in the centrifugal force produced by the rotation of the Earth. Named for Gustave Gaspard Coriolis (1792-1843), French mathematician who studied the effects of the force. Cf: *pole-fleeing force*. See also: *Ferrel's law*. Also spelled: *coriolis force*. Syn: *geostrophic force*.

cork A secondary suberized tissue formed from the *cork cambium* of stems and roots, chiefly in woody plants. See also: *periderm*. Cf: *phelloderm*.

cork cambium A *cambium* which functions in secondary growth by forming *cork* cells toward the outside of the stem and cork parenchyma or *phelloderm* toward the inside (Bold, 1967, p.521). See also: *periderm*. Syn: *phellogen*.

corkite A rhombohedral mineral: $PbFe_3(PO_4)(SO_4)(OH)_6$. It is isomorphous with svanbergite, woodhouseite, and hinsdalite.

corkscrew flute cast A corkscrew-shaped flute cast, with a "twisted" beak at the upcurrent end.

corneite A biotite hornfels formed during the deformation of a shale.

cornelian *carnelian*.

corner (a) A point of intersection of two boundary lines of a tract of land; esp. a point on the Earth's surface, determined by surveying, that marks an extremity of a boundary of a subdivision of the public lands, usually at the intersection of two or more surveyed lines. Corners are described in terms of the points they represent; e.g. "township corner" located at the extremity of a township boundary. See also: *witness corner*. (b) A term that is often incorrectly used to denote the physical station, or *monument*, erected to mark the corner.

corner accessory A physical object adjacent to a corner, to which such a corner is referred for its future identification of restoration; e.g. a *bearing tree*. Corner accessories include mounds, pits, ledges, rocks and other natural features to which distances or directions, or both, from the corner of monument are known.

corner frequency For a seismic wave generated from an earthquake, the frequency at which the theoretical spectral field begins to decrease. It is related theoretically to the dimensions of the source.

cornetite A peacock-blue mineral: $Cu_3(PO_4)(OH)_3$.

cornice An overhanging ledge or mass of snow or ice on the edge of a steep ridge or cliff face. Syn: *snow cornice.*

cornice glacier *cliff glacier.*

corniculate Said of a plant part that terminates in a small, hornlike process. Cf: *cornutate.*

cornieule *cargneule.*

Cornish diamond A rock crystal (clear quartz) from Cornwall, Eng.

Cornish stone A variety of *china stone* composed of feldspar, mica, and quartz and used as a bond in the manufacture of pottery. Syn: *Cornwall stone.*

corn snow *spring snow.*

cornstone (a) A calcareous concretion embedded in marl and grading into concretionary limestone. Its presence indicates fertile soil suitable for corn growing. (b) A calcareous conglomerate, consisting of fragments of marl and limestone embedded in a sandy or calcareous matrix.---The term is used in England and refers to two distinct rock types, both typically associated with the Old Red Sandstone and the New Red Sandstone. Allen (1960) suggests that the use of "concretionary cornstone" and "conglomeratic cornstone" would avoid confusion.

cornubianite A hornfels formed by contact metamorphism, and consisting of micas, quartz, and feldspar (Holmes, 1928, p.69). Cf: *leptynolite; proteolite.*

cornubite A mineral: $Cu_5(AsO_4)_2(OH)_4$. Cf: *cornwallite.*

cornulitid An invertebrate animal known only by a tapering, flexuous tube composed of calcium carbonate, with circular cross section, and with transverse rings developing in later growth stages. Cornulitids belong to the family Cornulitidae and are taxonomically unassigned in the Treatise on Invertebrate Paleontology (1962, pt. W, p.137), but were originally described as being closely related to certain annelids. Their stratigraphic range is Middle Ordovician to Mississippian.

cornuspirine Having a tube-like, planispirally coiled test; specif. pertaining to the foraminifer *Cyclogyra* ("*Cornuspira*").

cornutate Said of a diatom valve that has hornlike extensions. Cf: *corniculate.*

cornwallite An emerald-green mineral: $Cu_5(AsO_4)_2(OH)_4 \cdot H_2O$. Cf: *cornubite.*

Cornwall stone *Cornish stone.*

corona [paleont] The principal skeletal structure or main part of the calcareous test of an echinoid, including all ambulacra and interambulacra, but excluding the apical system of plates, the periproctal and peristomial systems, Aristotle's lantern, and appendages.

corona [bot] In charophyte algae, the outer layer of the nucule, formed by cells cut off from the sheath cells of the female sex organ.

corona [palyn] A more or less equatorial extension of a spore, similar in disposition to a *zone* but divided in fringe-like fashion.

corona [petrology] A zone of minerals, usually radial around another mineral, or at the contact between two minerals. It is a general term and has been applied to reaction rims, corrosion rims, and originally crystallized minerals. Cf: *kelyphytic rim.*

coronadite A black mineral: $Pb(Mn^{+2},Mn^{+4})_8O_{16}$. It is isostructural with hollandite and cryptomelane.

coronal (a) Pertaining to a corona. (b) Pertaining to certain openings of organisms, such as fringing an osculum of a sponge; e.g. "coronal pores" or tiny openings at the periphery of shields in acantharian radiolarians.

coronate Having or resembling a crown; esp. said of a cephalopod whorl section resembling a crown as viewed from the side, or said of a gastropod shell when the spire is surrounded by a row of spines or tubercles.

coronite A rock containing mineral grains surrounded by coronas.

coronula In charophyte algae, one or two tiers of small cells resting on the apical ends of the enveloping cells to form a more or less erect, elevated ring around the summit.

corrasion (a) A process of erosion whereby rocks and soil are mechanically removed or worn away by the abrasive action of solid materials moved along by wind, waves, running water, glaciers, or gravity; e.g. the wearing away of the bed and banks of a stream by the cutting, scraping, scratching, and scouring effects of a sediment load carried by the stream, or the sawing and grinding action of sand, gravel, and boulders hurled by waves and currents against a shore. The term has also been used for the loosening of rock material by the impact of rushing water itself, and was used by Penck (1953, p.112) for the "freeing of loosened rock fragments from their place of origin". The term *abrasion* is essentially synonymous. Syn: *mechanical erosion.* (b) A term sometimes used as a syn. of *attrition.* (c) A term formerly used as a syn. of *corrosion,* or as including the work of corrosion.---The term was first used by Powell (1875, p.205) for channel cutting or the deepening of any valley floor, and extended by Gilbert (1877, p.101) to the work of all running water (including lateral corrasion). Verb: *corrade.*

corrasion valley An elongated hollow or furrow excavated by the corrading action of a moving mass of material (Penck, 1953, p.112).

correction [phys] In the measurement of physical properties, a quantity that is applied to a measured quantity in order to negate the effects of interference and obtain the true quantity or value, or to reduce the measurement to some arbitrary standard.

correction line A *standard parallel* in the U.S. Public Land Survey system.

correlatable Capable of being correlated. Incorrectly spelled: *correlable.*

correlate v. To establish or explain the position in time or space of one geologic phenomenon, feature, or event in terms of other phenomena, features, or events; e.g. to show definite correspondence in character and in stratigraphic position between geologic formations or fossil faunas of two or more separated areas.---adj. Belonging to the same stratigraphic position or level.

correlation [geomorph] The concept applied by Penck (1953, p.419) whereby strata formed from the products of denudation are related to the period during which the denudation occurred.

correlation [seis] (a) The identification of a phase of a seismic record as representing the same phase on another record, to relate reflections from the same stratigraphic sequence or refractions from the same marker. (b) The measurement of the linear relationship between a pair of traces, or of the extent to which one can be considered as a linear function of the other.----(Sheriff, 1968).

correlation [stat] The intensity of association or interdependence between two or more mathematical variables; e.g. *linear correlation.*

correlation [geol] (a) The determination, investigation, or demonstration of the dependency or association of one geologic phenomenon upon or with other geologic phenomena, and the establishment of these phenomena in a logical and complete chronologic system (such as the geologic time scale); esp. the determination of the spatial position or the time of occurrence of one geologic feature or event in relation to others, or of several such features or events in the geologic histories of two or more separated areas. See also: *stratigraphic correlation; tectonic correlation; metamorphic correlation.* (b) The condition or fact of being correlated, such as the correspondence between or relation of two or more geologic phenomena, features, or events.

correlation coefficient A number that expresses the degree of correlation between two mathematical variables. Many different correlation coefficients are in use. Symbol: r.

correlation shooting A seismic shooting method in which isolated profiles are shot and correlated to obtain relative structural positions of the horizons mapped. Cf: *continuous profiling*.

correlative estimate In hydraulics, a likely discharge value estimated by correlation for a particular span of time.

corrensite A clay-mineral mixture with a regular interstratification of chlorite and vermiculite. It shows swelling behavior in glycerol. See also: *swelling chlorite*.

corresponding image point Any of the *image points* of the same object point on two or more photographs (such as overlapping photographs). Formerly called *conjugate image point*.

corridor [karst] A solution trench developed along joint planes in limestone. Cf: *grike*. Syn: *zanjón*.

corridor [speleo] In a cave, a traversable *passage* [speleo] that is generally narrow and straight.

corrie A term used in Scotland as a syn. of *cirque*, esp. a *hanging cirque*. Etymol: Gaelic *coire*, "kettle". Syn: *coire; corry*.

corrom A Scottish term for a drainage divide developed on a delta by a stream whose drainage basin was altered by glacial action (Kendall & Bailey, 1908, p.25).

corrosion [geomorph] (a) A process of erosion whereby rocks and soil are removed or worn away by natural chemical processes, esp. by the solvent action of running water, but also by other reactions such as hydrolysis, hydration, carbonation, and oxidation. Syn: *chemical erosion*. (b) A term formerly used interchangeably with *corrasion* for the erosion ("gnawing away") of land or rock, including both mechanical and chemical processes. The mechanical part is now properly restricted to "corrasion" and the chemical to "corrosion".---Verb: *corrode*.

corrosion [petrology] The partial resorption, dissolution, fusion, modification, or eating away of the outer parts of early-formed crystals (such as quartz phenocrysts) or xenoliths by the solvent action of the residual magma in which they are contained, resulting from a progressive change in conditions affecting the solubilities of the preexisting phases or the composition of the residual solution. It sometimes results in the formation of corrosion borders. See also: *embayment*.

corrosion border One of a series of borders of one or more secondary minerals around a central, larger, and original crystal, and representing the irregular modification of the outline of a phenocryst due to the corrosive action of a magma upon the mineral which, under different conditions, was previously stable. Cf: *reaction rim*. Syn: *corrosion rim; corrosion zone; resorption border*.

corrosion rim A *corrosion border* as seen in section.

corrosion spring swamp An unusual circumstance occurring when springs, carrying a considerable amount of dissolved material to the surface, cause a local, rather than a general lowering of the water resulting in a small swamp.

corrosion surface A pitted, irregular bedding surface found only in certain carbonate sediments, characterized by a black manganiferous stain, and presumed to result from cessation of lime deposition and from submarine solution or resorption of some of the previously deposited materials. Syn: *corrosion zone*.

corrosion zone (a) *corrosion surface*. (b) *corrosion border*.

corrugated ripple mark A *longitudinal ripple mark* with a sigmoidal profile, an equally rounded and usually symmetric crest and trough, and a ridge that rarely branches (Kelling, 1958, p.124). Cf: *mud-ridge ripple mark*.

corrugation (a) The process of deforming or crumpling sedimentary beds into small-scale folds, wrinkles, or furrows; esp.

intraformational corrugation. Also, the state of being corrugated. It occurs on a smaller scale than *contortion*. (b) A structure produced by corrugation.

corry Var. of *corrie*.

corsilite A rock composed of smaragdite replacing diallage, some diallage, saussurite, and feldspars altered to a mixture of albite, zoisite, epidote, quartz, muscovite, calcite, etc. (Thrush, 1968, p.269). Also spelled: *corsilyte*.

corsite An orbicular gabbro. Syn: *napoleonite; miagite*.

cortex [paleont] (a) A layer of the ectosome of a sponge, consolidated by a distinctive skeleton, either organic or mineral, or both. Also, a layer of specialized sponge spicules or modified structure at the outer surface of the skeleton. (b) An outer coenenchymal layer of certain octocorals (esp. the Gorgonacea); the outer horny layer of the axis in the Holaxonia, as opposed to its medulla. (c) The dense, differentiated outer layer of an echinoid spine, usually bearing ornamentation. It is nonliving material on a mature spine. (d) The tissue, composed mostly of parenchyma cells, between the central vascular cylinder and the epidermis of a stem or root (Cronquist, 1961, p.873).

cortical Pertaining to, located in or on, or consisting of a cortex or outer part of an invertebrate; e.g. "cortical shell" (outermost of the concentric shells of a spumellinid radiolarian), or "cortical tissue" (outer layer of laminated material of the periderm of a graptolite).

cortlandtite A peridotite that contains hornblende and olivine. Syn: *hudsonite*.

corundolite *emery rock*.

corundophilite A mineral of the chlorite group: $(Mg,Fe)_3(Al,Fe)_3(Si,Al)_4O_{10}(OH)_8$.

corundum A mineral: Al_2O_3. It occurs as shapeless grains and masses, or as variously colored rhombohedral crystals (such as prisms or tapering hexagonal pyramids), including the gem varieties such as ruby and sapphire. Corundum is extremely tough, has a hardness of 9 on Mohs' scale, and is used industrially as an abrasive. See also: *emery*. Syn: *adamantine spar; diamond spar; corindon*.

corvusite A blue-black, brown, or purplish mineral: $V_2O_4 \cdot 6V_2O_5 \cdot nH_2O$ (?). It is an ore of vanadium. Syn: *blue-black ore*.

corynite A variety of gersdorffite containing antimony.

cosalite A lead-gray or steel-gray mineral: $Pb_2Bi_2S_5$. It often contains copper.

cosedimentation Contemporaneous sedimentation, such as the precipitation of iron-bearing minerals at the same time as deposition of fine argillaceous sediment to produce a shale or mudstone with a high iron content (Pettijohn, 1957, p.368).

coseism *coseismal line*.

coseismal Said of points which have the same seismic intensity.

coseismal line A line connecting points on the Earth's surface at which an earthquake wave has arrived simultaneously. Cf: *isoseismal line*. Syn: *coseism; coseismic line; homoseismal line; homoseism*.

coseismic line *coseismal line*.

coset A term proposed by McKee & Weir (1953, p.384) for a sedimentary unit composed of two or more *sets*, either of strata or of cross-strata, "separated from other strata or cross-strata by original flat surfaces of erosion, nondeposition, or abrupt change in character". See also: *composite set*.

cosmecology A science that treats the Earth in its relation to cosmic phenomena (Webster, 1967, p.514).

cosmic Of, from, or relating to the cosmos; esp. pertaining to phenomena or features that occur, exist, or originate beyond the Earth's atmosphere or in the universe in contrast to the Earth alone, such as "cosmic sediment" found in the oceans and containing particles of extraterrestrial origin.

cosmic dust (a) Very finely divided particles of solid matter moving about in interplanetary space or in any other part of the universe. (b) The smallest particles that invade the Earth's atmosphere from interplanetary space, reaching the Earth's surface (as on the sea floor or in polar ice) in an essentially unaltered state at an estimated accumulation rate of 1000 to one million tons a year. Their composition and structure are similar to those of meteorites, and they are believed to represent primordial condensation or sublimation products, cometary debris, or debris resulting from collisions between meteorites and asteroids or from collisions of meteorites, asteroids, and comets with the Moon and Earth. Cf: *meteoric dust; meteoritic dust.* Syn: *zodiacal dust.*

cosmic erosion The gradual degradation or catastrophic destruction of rocks in space on a planetary surface as a result of shock-wave interactions produced by hypervelocity impacts of particles against the exposed rock surfaces. The term includes both the gradual wearing away of rock surfaces due to spallation processes caused by interactions of the shock waves with the free surfaces of the rock as well as the catastrophic rupture and breakup of entire rocks (Hörz, et al, 1971, p.5789).

cosmic radiation The bombardment of the Earth's atmosphere by very high-energy, subatomic particles from outer space. Primary cosmic rays (atomic nuclei) are almost all absorbed in the upper atmosphere; secondary cosmic rays are those which reach the Earth's surface. They have less energy than those of the primary type. Cosmic radiation is part of the natural background radiation.

cosmic spherule A small particle formed when a molten droplet ablates from a meteorite entering the Earth's atmosphere and first recognized in deep-sea sediments. See also: *magnetic spherule.*

cosmic water Juvenile water that is brought to the Earth from space in meteorites.

cosmochemistry The study of the origin, distribution, and abundance of elements in the universe.

cosmoclastic rock One of the original rocks of the Earth (Fairchild, 1904).

cosmogenic radioisotope A radioisotope that was produced by the action of cosmic radiation.

cosmogony Either a scientific theory or a cultural mythology regarding the creation or origin of the Universe, either as a whole or in more limited scope, such as the solar system of the Earth. There is partial overlap in definition on both sides with *cosmology,* which is the study of the entire space-time structure of the Universe, but as a whole.

cosmolite *meteorite.*

cosmological principle The principle that the universe presents essentially the same picture throughout all space and has done so throughout all time.

cosmology The study, both theoretical and observational, of the space-time structure of the Universe in a global rather than a local sense.

cosmopolitan Said of a kind of organism or a species that is widely distributed throughout the world in various geographic and ecologic provinces. Noun: *cosmopolite.*

cosmopolite A *cosmopolitan* organism.

cosmos The universe considered as an orderly and harmonious system. Ant: *chaos.*

cossyrite Sodium-rich variety of enigmatite, occurring in minute black crystals in lava.

costa [paleont] (a) One of the modified spines overarching the frontal membrane in cribrimorph bryozoans, usually united with neighboring costae to form a frontal shield. Syn: *costula.* (b) A round-topped elevation of moderate width and prominence disposed collabrally on the surface of a gastropod shell. (c) One of several moderately broad and prominent elevations of the surface of a bivalve-mollusk shell, directed radially or otherwise from the beak. Syn: *rib.* (d) A radial ridge on the external surface of a brachiopod shell, originating at the margin of the protegulal node. Also, any coarse rib of a

brachiopod, without reference to origin (usually fewer than 15 costae in a width of 10 mm). Cf: *costella; capilla.* (e) A ridge on the external surface of a foraminiferal test. It may run along a suture or be transverse to it. (f) The prolongation of septum on the outer side of a corallite wall. (g) A long narrow raised area or ridge of a conodont.---Pl: *costae.*

costa [bot] (a) In a simple leaf, the midrib. (b) In a pinnately compound leaf, the rachis. (c) In a diatom, a wall mark formed by two well defined ridges and containing fine pores.

costa [palyn] One of the rib-like thickenings in the endexine of pollen, associated with colpi or pores. Costae are most often meridional and border colpi, but they may be transverse in association with transverse furrows.

costella A radial ridge on the external surface of a brachiopod shell, not extending to the margin of the protegulal node but arising by bifurcation of the existing costae or costellae or by intercalation between them. Also, a fine rib of a brachiopod, without reference to origin (usually 15-25 costellae in a width of 10 mm). Pl: *costellae.* Cf: *costa; capilla.*

costibite A mineral: CoSbS.

costula A small ridge of an invertebrate; e.g. a marking that makes up the sculpture of a mollusk shell, or a *costa* of a cribrimorph bryozoan. Pl: *costulae.* Syn: *costule.*

coteau A French word used in parts of the U.S. for a variety of features: a range or sharp ridge of hills; a high plateau; a hilly upland, including the divide between two valleys; a morainal hill; an elevated pitted plain of rough surface (as in Missouri); a low, dry ridge within a swampy area (as in Louisiana); a side of a valley, esp. a prominent and dissected escarpment forming the edge of a plateau (as in Missouri). Etymol: Canadian French, "slope of a hill; hillock". Pron: co-toe.

cotectic Said of conditions of temperature, pressure, and composition under which two·or more solid phases crystallize simultaneously and without resorption from a single liquid over a finite range of falling temperature; also, said of the geometric form (e.g. line or surface) representing the corresponding phase boundary on the liquidus of a phase diagram.

cotidal line A line on a chart connecting points at which high water occurs simultaneously. The lines show the time lapse, in lunar-hour intervals, between the Moon's passage over a reference meridian, usually Greenwich, and high water.

cotterite A variety of quartz having a peculiar metallic pearly luster.

cotton ball *ulexite.*

cotton chert *chalky chert.*

cotton rock (a) A term used in Missouri for a soft, fine-grained, siliceous, white to slightly gray or buff magnesian limestone having a chalky or porous appearance that is suggestive of cotton. (b) The white or light-colored, decomposed exterior surrounding the dense black interior of a chert nodule.

cotton stone *mesolite.*

cotunnite A soft, white to yellowish, orthorhombic mineral: $PbCl_2$.

cotyledon One of the first leaves of a young seed plant, developed by the embryo within the seed. It may be a storage organ. Syn: *seed leaf.*

cotype A term originally used for either a *syntype* or a *paratype* but now recommended for rejection by the International Commission on Zoological Nomenclature because of its dual meaning.

coulee [mass move] A tongue-like mass of debris moved by solifluction (Monkhouse, 1965, p.81).

coulee [volc] A flow of viscous lava that has a blocky, steep-fronted form. Also spelled: *coulée.* Etymol: French, "outflowing".

coulee [geomorph] (a) A term applied in western U.S. to a small stream, often intermittent. Also, the bed of such a stream when dry. (b) A term applied in NW U.S. to a dry or intermittent stream valley, gulch, or wash of considerable extent; esp. a long, steep-walled, trench-like gorge or valley representing an abandoned overflow channel that temporarily carried meltwater from an ice sheet, as the Grand Coulee (formerly occupied by the Columbia River) in Washington State. (c) A small valley or a low-lying area.---Etymol: French coulée, "flow or rush of a torrent". Pron: koo-lee. Syn: coulie.

coulee lake A lake produced by the damming of a water course by lava.

coulisse A term introduced by Scrivenor (1921, p.354) for one of the prominent features, formed by the erosion of folded stratified rocks and of igneous intrusions, that are arranged en echelon on the Earth's surface "like the wings of the stage in a theatre". Etymol: French, "a side scene or wing on a theater stage". Pron: koo-*lease*.

couloir (a) A deep, narrow, steep-sided valley; esp. a gorge or gully on a mountain side in the Alps. (b) A French term for a passage in a cave, or a vertical cleft in a cliff.---Etymol: French, "passage". Pron: kool-*whar*.

coulomb The unit of electric charge equal to the quantity of electricity transported in one second by a current of one ampere.

Coulomb criterion A criterion of brittle shear failure based upon the concept that shear failure will occur along a surface when the shear stress acting in that plane is sufficiently large to overcome the cohesive strength of the material plus the frictional resistance to movement. Cohesive strength is equal to inherent shear strength when the stress normal to the shear surface is zero; frictional resistance to movement is equal to stress normal to the shear surface multiplied by the coefficient of internal friction of the material. See also: *Coulomb equation*.

Coulomb's equation An equation describing the *Coulomb criterion*, or the failure of a material in shear fracture: critical shear stress for failure equals cohesion plus the coefficient of internal friction times normal stress across potential failure surface.

Coulombs' modulus *modulus of rigidity.*

coulometry A method of quantitative chemical analysis utilizing the number of coulombs necessary to release a substance during electrolysis to determine the amount of substance released. One faraday (96,500 coulombs) will release or deposit 1 gram-equivalent weight of an ion.

coulsonite A mineral of the spinel group: FeV_2O_4. Syn: *vanado-magnetite.*

coulter The cusp (of a river bar) lying between two converging currents (Haupt, 1883, p.147).

counterclockwise inclination The inclination to the left of a heterococcolith suture as it proceeds to the periphery. Ant: *clockwise inclination.*

countercurrent A secondary current flowing in a direction opposite to that of the main or an adjacent current.

counter fossula A *fossula* developed in the position of the counter septum in a rugose coral.

counter-lateral septum One of two protosepta of a rugose corallite that adjoin the *counter septum* on either side. Symbol: KL.

counterlode *cross vein.*

counterpart A *cast* of a bedding-plane mark preserved on the underside of the overlying bed (Middleton, 1965, p.248).

counterradiation The downward component of infrared *atmospheric radiation* that passes through a given surface, specif. sea level. Cf: *terrestrial radiation; effective terrestrial radiation.* Syn: *back radiation.*

counterscarp In landslides underlain by noncircular shear surfaces, a scarp that is parallel to the crown scarp and that occurs on the downslope side. Together, the scarps bound the structural trough or graben that is formed by such a landslide.

counter septum The protoseptum lying directly opposite the cardinal septum in the plane of bilateral symmetry of a rugose

corallite. Symbol: K. Cf: *alar septum; counter-lateral septum.*

countertrades *antitrades.*

countervein *cross vein.*

country rock (a) The rock enclosing or traversed by a mineral deposit. Syn: *mother rock.* Cf: *wall rock.* (b) The rock intruded by and surrounding an igneous intrusion.

coupled wave A type of *surface wave* that is continuously generated by another wave which has the same phase velocity. Syn: *C wave.*

course [stratig] (a) An old term used in Great Britain and applied to a stratum or outcrop, and to stratification. (b) A British term for a coal seam.

course [struc geol] The horizontal direction of a geologic structure; e.g. the strike of a lode or vein.

course [surv] (a) A term used in surveying with several meanings: the bearing of a line; the length of a line, and the bearing (or azimuth) and length of a line, considered together. Also, the line connecting two successive stations in a traverse. (b) A term used in navigation for the azimuth or bearing of a line along which a ship or aircraft is to or does travel, without change of direction; the line drawn on a chart or map as the intended track.

course [streams] The path followed by, or the channel through which, water flows; a *watercourse.*

Couvinian *Eifelian.*

covariance The arithmetical mean or the expected mean value of the product of the deviations of two variables from their respective mean values.

cove [coast] (a) A small, narrow, sheltered bay, inlet, creek, or recess in a coast, often inside a larger embayment; it usually affords anchorage to small craft. (b) A small, often circular, wave-cut indentation in a cliff; it usually has a restricted or narrow entrance. (c) A fairly broad, looped embayment in a lake shoreline (Veatch & Humphrys, 1966, p.73). (d) A shallow tidal river, or the backwater near the mouth of a tidal river.

cove [geomorph] (a) Any precipitously walled and rounded or cirque-like opening, as at the head of a small steep valley; specif. a deep recess, hollow, or nook in a cliff or steep mountainside, or a small, straight valley extending into a mountain or down a mountainside. Examples: among the foothills of the Blue Ridge in Virginia, and in the English Lake District. Cf: *rincon.* (b) A term used in the southern Appalachian Mountains for a relatively level area sheltered by hills or mountains; e.g. Cades Cove in eastern Tennessee. (c) A gap or pass between hills. (d) A basin or hollow where the land surface has undergone differential weathering or subsidence, as from solution of underlying rock. (e) A Scottish term for a hollow or recess in rock, as a cave or cavern.

cove beach A *bayhead beach* formed in a cove.

covelline *covellite.*

covellite An indigo-blue hexagonal mineral: CuS. It is a common secondary mineral and represents an ore of copper. Syn: *covelline; indigo copper.*

cover [stratig] (a) The sedimentary accumulation over the crystalline basement. Syn: *sedimentary cover; sedimentary mantle.* See also: *cover mass.* (b) The vertical distance between any position in strata and the surface or any other position used as a reference.

cover [snow] *snow cover.*

covered flagellar field The part of a flagellate coccolithophore having a complete cover of coccoliths in the flagellar region. Cf: *naked flagellar field.*

covered karst Karst that is developed in a region with a soil cover that subdues its topographic features. Cf: *naked karst.*

covered plain A term introduced by Melton (1936, p.594) for an alluvial flood plain through which a low-water channel does not actively meander; it usually has a natural levee and a thick alluvial cover derived from the suspended load. Cf: *meander plain; bar plain.*

cover head A thick accumulation of debris, consisting of talus cones and alluvial fans, resting on an elevated marine terrace; the material is deposited during and after the emergence of the terrace.

cover mass A *cover* of softer rock material with relatively simple structure, overlying the *undermass*; the material above the surface or an angular unconformity. See also: *compound structure.*

cover plate Any of small, polygonal, biserially arranged plates forming the walls and roof of ambulacral groove of an echinoderm (such as of a blastoid). See also: *anal cover plate.* Syn: *covering plate.*

cover rock The thickness of rock between mine workings and the Earth's surface.

cover sand An eolian deposit of fine to very fine sand, usually containing more than 90% quartz, and believed to have been deposited by heavy snowstorms during the glacial epoch.

covite A nepheline syenite in which the mafic minerals are acmite-augite and hornblende. The term is more frequently used by Russian and European petrographers.

cow-dung bomb A type of volcanic bomb whose flattened shape is due to its impact while still viscous. Its surface is somewhat scoriaceous. It is an intermediate type between the Hawaiian and Strombolian types.

coxa The first segment of the leg of an arthropod by which the leg articulates with the body; e.g. the proximal (basal) segment of the limb of a crustacean (except rarely where precoxa is distinguishable), or the basal segment of all cephalothoracic appendages of an arachnid (the name is rarely used in the case of the chelicerae), or the proximal (basal) joint of the thoracic appendage of a merostome, directly attached to the body. Pl: *coxae.* Adj: *coxal.* See also: *coxopod; maxilla.*

coxopod The *coxa* of a crustacean limb. Syn: *coxopodite.*

crab The condition indicated on a vertical photograph by the lateral edges not being parallel to the air-base lines, caused by failure to orient the camera with respect to the track of the aircraft. Also, the angle between the edge of the photograph and the flight line.

crab hole A low spot or depression in *gilgai* (a microrelief of heavy clay soil). Cf: *puff.*

crack [struc geol] A partial or incomplete fracture.

crack [ice] Any *fracture [ice]*, in sea ice, not sufficiently wide to be described as a lead, and usually narrow enough to jump across. See also: *tide crack.*

cracking The breaking up of more complex chemical compounds into simpler ones, usually by heating; i.e. the subjecting of compounds to pyrolysis.

crackle breccia (a) An incipient breccia having fragments parted by planes of rupture but suffering little or no displacement (Norton, 1917, p.161). It is commonly a chemical deposit. (b) *shatter breccia.*

crag [geomorph] (a) A steep, rugged, often precipitous point or eminence of rock, as one projecting from the side of a mountain. Syn: *craig.* (b) An obsolete term for a sharp, rough, detached or projecting fragment of rock.

crag [sed] A shelly sandstone or a compacted fossiliferous sandy marl of marine origin, found in eastern England (Norfolk, Suffolk, and Essex) and used for manure, and of Pliocene and Pleistocene age; e.g. the Coralline Crag.

crag and tail An elongate hill or ridge resulting from glaciation, having at the stoss end a steep, often precipitous, face or knob of ice-smoothed, resistant bedrock (the "crag") obstructing the movement of the glacier, and at the lee end a tapering, streamlined, gentle slope (the "tail") of intact weaker rock and/or till protected by the "crag". Cf: *knob and trail.*

craig A Scottish term for *crag;* e.g. Ailsa Craig. Etymol: Celtic, "rock".

craigmontite A light-colored nepheline syenite composed of nepheline, oligoclase, and muscovite with minor amounts of calcite, magnetite, corundum, and biotite. Craigmontite has more nepheline and less plagioclase and corundum than does *congressite* or *raglanite*. It is named for its occurrence in Craigmont Mountain, Ontario.

craignurite A glass-rich variety of rhyodacite.

cranch In mining, a part of a vein that is left unworked (Nelson, 1965, p.110).

crandallite A white to light-grayish mineral: $CaAl_3(PO_4)_2$-$(OH)_5.H_2O$. It sometimes contains appreciable strontium, barium, iron, or rare earths. Syn: *pseudowavellite.*

craniacean Any inarticulate brachiopod belonging to the superfamily Craniacea, characterized generally by a strongly punctate, calcareous shell.

cranidium The central part of the cephalon of a trilobite, consisting of the glabella and its two fixed cheeks, and bounded by the facial sutures. Pl: *cranidia.* Adj: *cranidial.*

crassitude A more or less local exine thickening of a spore. Cf: *cingulum; zone [palyn].*

crater [lunar] An approximately circular or polygonal depression in the surface of the Moon, having a diameter that may range from a few centimeters to hundreds of kilometers and a depth that is small relative to its diameter. Large lunar craters often have lofty, rugged rim crests, terraced walls, and prominent central peaks. Lunar craters are believed to have formed in several ways, including meteoritic or cometary impact (e.g. Kepler and Tycho), secondary debris impact, volcanic activity, or subsidence. Syn: *lunar crater.*

crater [geophys] A typically bowl-shaped pit, depression, cavity, or hole, generally of considerable size and with steep slopes, formed on a surface or in the ground by the explosive release of chemical or kinetic energy; e.g. an *impact crater* and an *explosion crater.*

crater [volc] A basinlike, rimmed structure that is usually at the summit of a volcanic cone; its floor is approximately the diameter of the vent. It may be formed by an explosive eruption or by the gradual accumulation of pyroclastic material into a surrounding rim. Cf: *caldera.*

crater [grd wat] geyser crater.

crater chain A distinctive, chain-like, linear group of small lunar craters, observed in most regions of the Moon's surface (such as east of the crater Copernicus). A chain may be as much as 325 km long. See also: *chain crater.*

crater depth (a) In an artificial crater, the maximum depth measured from the deepest point of the depression to the original ground surface (Flanders and Sauer, 1960, p.5). (b) In a natural crater, in which the original ground level may be uncertain, the depth measured from the highest point on the rim crest to the deepest part of the observable depression.

crater fill Solidified lava at the bottom of a volcanic crater, with associated cinders and weathering debris.

cratering (a) The dynamic process or mechanism of formation of an individual crater. (b) The process of modification of a planetary or lunar surface by repeated crater formation.

crater lake A lake, usually of fresh water, that has formed in a volcanic crater or caldera by the accumulation of rain and ground water. See also: *caldera lake.*

crater lip The elevated rim at the edge of an explosion or impact crater, formed by the uplifting of strata and by the deposition of ejecta from the crater.

crater radius (a) In an artificial crater, the average radius, measured at the level corresponding to the original ground surface (Flanders and Sauer, 1960, p.6). (b) In a natural crater, in which the original ground level may be uncertian, the crater radius measured to the rim crest.

crater ring A low-relief rim of fragmental material surrounding a *maar.*

cratogene shield.

craton A part of the Earth's crust which has attained stability, and which has been little deformed for a prolonged period. As originally defined, cratons included parts of both continents and ocean basins, but modern knowledge of the ocean basins indicates that existence of cratons there is unlikely, so the term is now restricted to continental areas. The extensive central cratons of the continents (*central stable regions*) include both *shields* and *platforms*; these have been called *hedreocratons* but the prefix seems superfluous. Parts of the more mature Phanerozoic foldbelts have now achieved, or are approaching, a cratonic condition. Also spelled: *kratogen*; *kraton*. See also: *thalassocraton*.

cratonic basin *intracratonic basin.*

crawl In a cave, a *passage* [speleo] that is traversable only by crawling. Syn: *crawlway.*

crawlway *crawl.*

C ray Right posterior ray in echinoderms situated clockwise of B ray when the echinoderm is viewed from the adoral side.

craze plane A planar void in a soil material, having a highly complex conformation of the walls due to the interconnection of numerous, short, flat and/or curved planes (Brewer, 1964, p.198).

cream ice *sludge.*

crease [glac geol] A term used by Woodworth (1901, p.665) for an *overflow channel* that formerly contained meltwater.

crease [mining] A limestone quarry in a mountainside.

crednerite A steel-gray to iron-black mineral: $CuMnO_2$.

creedite A white or colorless monoclinic mineral: $Ca_3Al_2(SO_4)(F,OH)_{10} \cdot 2H_2O$.

creek [coast] (a) A British term for a small inlet, narrow bay, or arm of the sea, longer than it is wide, and narrower and extending farther into the land than a cove. The term is used in the U.S. (as in Maryland and Virginia) in names given during the earliest period of English colonization for a narrow recess in the shore of the sea, a river, or a lake, and often offering port or anchorage facilities for vessels. (b) A small and narrow tidal inlet or estuary of a small river, esp. occurring on a low-lying coast or emptying into the lower reaches of a wide river; e.g. Napa Creek, Calif. Syn: *tidal creek.*

creek [stream] (a) A term generally applied over most of the U.S. (except New England) and in Canada and Australia to any natural stream of water, normally larger than a *brook* but smaller than a *river*; a branch or tributary of a main river; a lowland watercourse of medium size; a flowing rivulet. Also, a wide or short arm of a river, such as one filling a short ravine that joins the river. (b) A term used in the SW U.S. and in Australia for a long, shallow stream of intermittent flow; an *arroyo.*

creep [mass move] The slow, gradual, more or less continuous, nonrecoverable (permanent) deformation sustained by ice, soil, and rock materials under gravitational body stresses. Many types of creep have been distinguished and described in the literature on the basis of material properties, stress level, stage and rate of deformation, fundamental mechanics of failure, geometric patterns, and rates of deformation and cause. However, the general term should not be limited by a presumption of mechanism, depth, velocity profile, thickness of creep zone, or lateral extent. Syn: *creeping.*

creep [mining] *heave.*

creep [struc geol] Continuously increasing, usually slow, deformation (strain) of solid rock resulting from a small, constant stress acting over a long period of time.

creeping [streams] The slow *shifting* of a divide from one position to another, as where a stream, because of its greater steepness and volume or the weaker nature of the rocks over which it flows, cuts down more rapidly than another stream on the opposite side of the divide (Cotton, 1958, p.69-70). Cf: *leaping.*

creeping [mass move] A syn. of *creep.*

creep limit The maximum stress that a material can withstand without observable creep.

creep recovery The gradual recovery of elastic strain when stress is released. Syn: *elastic aftereffect; elastic afterworking; afterworking; transient strain.*

creep strength The load per unit area leading to a specified steady creep strain rate at a given temperature.

creep wrinkle One of a series of small-scale corrugations of a bedding-plane surface, oriented at right angles to the direction of movement (slumping or creep) (McIver, 1961, p.227). See also: *crinkle mark; pseudo ripple mark.*

crenate Having the edge, margin, or crest cut into rounded scallops or shallow rounded notches; e.g. "crenate costae" of bivalve mollusks.

crenella (a) A narrow furrow between culmina of a crinoid columnal articulum. (b) A small radially disposed groove on the stem impression at the base of blastoid theca and on the distal and proximal surfaces of columnals (TIP, 1967, pt.S p.346).---Pl: *crenellae.*

crenula A ridge (culmen) combined with the adjacent furrow (crenella) of a crinoid columnal articulum. Pl: *crenulae.*

crenularium The entire area of a crinoid columnal articulum bearing crenulae. Pl: *crenularia.*

crenulate shoreline A minutely irregular shoreline, characterized by crenulate lines and sharp headlands, developed during a youthful stage of submergence by differential wave erosion acting upon less resistant rocks; e.g. the shoreline of SW Ireland.

crenulation Small-scale folding that is superimposed on larger-scale folding. Crenulations may occur along the cleavage planes of a deformed rock. Cf: *plication.*

crenulation cleavage *slip cleavage.*

creoline A purple epidotized basalt.

creolite (a) Red-and-white banded jasper from Shasta and San Bernardino counties, Calif. (b) A silicified rhyolite from Baja California.

crepis The initial mineral body about which a desma of a sponge is secreted. Pl: *crepides.*

crescent beach A crescent-shaped beach, concave toward the sea, formed along a hilly or mountainous coast at a bayhead or at the mouth of a stream entering a bay.

crescent cast *current crescent.*

crescentic crack *crescentic fracture.*

crescentic dune *barchan.*

crescentic fracture A *crescentic mark* in the form of a hyperbolic crack, of larger size (up to 10-12 cm long) than a chattermark; it is convex toward the direction from which the ice moved (its "horns" point in the direction of ice movement) and consists of a single fracture without removal of any rock. Cf: *lunate fracture; crescentic gouge.* Syn: *crescentic crack.*

crescentic gouge A *crescentic mark* in the form of a groove or channel with a somewhat rounded bottom, formed by glacial plucking on a bedrock surface; it is concave toward the direction from which the ice moved (its "horns" point away from the direction of ice movement) and consists of two fractures from between which rock has been removed. Cf: *crescentic fracture.* Syn: *gouge; gouge mark; lunoid furrow.*

crescentic lake A lake occupying a crescent-shaped depression; e.g. an *oxbow lake.*

crescentic levee lake A lake confined between old natural levees inside a meander and resulting from the enlargement of the meander.

crescentic mark [glac] Any curved or lunate mark produced by a glacier moving over a bedrock surface; e.g. a *crescentic fracture* and a *crescentic gouge.* Syn: *lunate mark.*

crescentic mark [sed] A crescent-shaped scour mark; specif *current crescent.*

crescent-type cross-bedding *trough cross-bedding.*

crescumulate n. A plutonic rock formed by crystal accumulation and exhibiting crescumulate texture.

crescumulate texture The texture of certain igneous rocks in which large, elongated crystals are oriented roughly at right angles to cumulate layering in the rock. The term *harrisitic* was originally applied to this texture which was thought to be limited to olivine crystals, as observed in the igneous rock arrisite; "crescumulate" has been suggested as a more general term without mineralogic restrictions (Wager, 1968, .579).

crest [beach] (a) *berm crest.* (b) *beach crest.*

crest [geomorph] The highest point or line of a landform, from which the surface slopes downward in opposite directions; esp. the highest point of a mountain or hill, or the highest line or culminating ridge of a range of mountains or hills. See also: *summit.*

crest [struc geol] A line that connects the highest points or culminations of a fold; the *axis* of an anticline. Cf: *crest surface.* Syn: *crest line.*

crest line *crest.*

crest plane *crest surface.*

crest surface A surface that connects the *crests* of the beds of an anticline. Syn: *crest plane.*

crêt In the French Jura, an in-facing cliff or escarpment; the wall of a *combe.*

creta (a) *chalk.* (b) *fuller's earth.*---Etymol: Latin, "chalk".

Cretaceous The final period of the Mesozoic era (after the Jurassic and before the Tertiary period of the Cenozoic era), thought to have covered the span of time between 136 and 65 million years ago; also, the corresponding system of rocks. It is named after the latin word for chalk ('creta') because of the English chalk beds of this age.

cretaceous A seldom-used syn. of *chalky.*

cretification The process or an instance of converting a rock into chalk, as by infiltration with calcium salts.

crevasse [geomorph] (a) A wide breach or crack in the bank of a river or canal; esp. one in a natural levee or an artificial bank of the lower Mississippi River. Etymol: American French. (b) A wide and deep break, opening, or fissure in the Earth after an earthquake.

crevasse [glaciol] A deep and nearly vertical split, fissure, crack, or rift in a glacier or other mass of land ice, or in a snowfield, caused by stresses resulting from differential movement over an uneven surface. Crevasses may be concealed beneath snowbridges, and some are as much as 100 m in depth. Etymol: French. See also: *transverse crevasse; longitudinal crevasse; marginal crevasse; splaying crevasse.* Syn: *crevass; fissure [glaciol]; crevice [glaciol].*

crevasse filling A short, straight ridge of stratified sand and gravel believed to have been deposited in a crevasse of a wasting glacier and left standing after the ice melted; a variety of kame (Flint, 1928, p.415). Cf: *till crevasse filling.*

crevasse hoar A type of hoarfrost consisting of large leaf-, plate-, or cup-shaped ice crystals that form and grow below the surface of a snowfield or glacier in a crevasse or other large open space where water vapor can condense under calm, still conditions. Cf: *depth hoar.*

crevasse ridge A mass of fluvial material originally deposited in a crevasse and now forming a more or less straight ridge that stands above the general land surface and extends parallel to the direction of ice movement (Leighton, 1959, p.340).

crevice [geol] A narrow opening or slit-like recess or enlarged joint, having some depth, and caused by a variety of processes; e.g. an excavation formed in a sea cliff by wave erosion. A crevice may be filled by mineral deposits.

crevice [glaciol] A colloquial syn. of *crevasse.*

crevice karst A karst pattern of deep solution along closely spaced joints. Cf: *limestone pavement.*

crib [eng] (a) A bin-type retaining wall consisting of interlocking members of steel, concrete, or wood (logs), and used to stabilize slopes and protect road cuts. (b) An engineering structure enclosing a water intake and filter offshore in a lake.

crib [glac geol] A Welsh term for *arête.*

cribellum A single or paired perforated plate in a small group of spiders, corresponding to the anterior and medial spinnerets of Liphistiina and serving as an outlet for special silk glands (TIP, 1955, pt.P, p.61). Pl: *cribella.*

cribrate Like a sieve; e.g. said of a foraminifer perforated with round holes.

cribrilith A discolith with numerous central perforations and a lamellar rim.

cribrimorph adj. Pertaining to the cheilostomatous bryozoan Cribrimorpha, characterized by a frontal shield of costae.---n. A cribrimorph bryozoan.

crichtonite A mineral: $Fe_{16}^{+2}Fe_{14}^{+3}Ti_{66}O_{169}$. It was long supposed to be identical with ilmenite.

crick [sed] An English term for a *clay gall* in oolite.

cricocalthrops A *calthrops* (sponge spicule) bearing a series of annular ridges on each ray.

cricolith A heterococcolith having units arranged in a simple ring; any elliptical ring coccolith.

crimp A marginal band on the aboral side of a plate in a platelike conodont element, representing the area covered by the last lamella accreted to the element.

crinanite An olivine-analcime diabase in which the ophitic texture is well developed.

crinkled bedding (a) *convolute bedding.* (b) Bedding that displays minute wrinkles; in carbonate rocks, it is believed to be related to algal mats.

crinkle mark One of a series of subparallel corrugations of a bedding-plane surface, related to very small-scale and crumpled internal laminae, and produced by subaqueous solifluction (Williams & Prentice, 1957, p.289). See also: *creep wrinkle; pseudo ripple mark.*

crinoid Any pelmatozoan echinoderm belonging to the class Crinoidea, characterized by quinqueradiate symmetry, by a disk-shaped or globular body enclosed by calcareous plates and from which appendages, commonly branched, extend radially, and by the presence of a stem, or column, more common in fossil than in living forms. Syn: *crinite; encrinite.*

crinoidal limestone A limestone consisting almost entirely of the fossil skeletal parts of crinoids in which the plates, ossicles, or joints (representing single crystals of calcite) are often cemented with clear calcite in crystallographic continuity with the crinoid fragment. The hard parts are allochthonous and commonly show evidence of sorting. Examples are common in the Osagian (Lower Mississippian) of the Illinois basin. See also: *criquinite.* Syn: *encrinite; encrinal limestone.*

crinozoan Any attached echinoderm belonging to the subphylum Crinozoa, characterized by a partial meridional growth pattern tending to produce an aboral cup-shaped or globoid plated theca and a partial radially divergent growth pattern forming appendages. Among the major groups included in the subphylum are blastoids, crinoids, and cystoids.

cripple A swampy or low, wet area in the Pine Barrens, N.J., supporting a growth of Atlantic white cedar; sometimes defined as always having flowing water. Cf: *spong.*

criquina A coquina composed of crinoid fragments.

criquinite Compact, well-indurated, and firmly cemented equivalent of criquina. See also: *crinoidal limestone.*

crisscross-bedding A kind of *cross-bedding* characteristic of eolian deposits, in which the layers dip in opposite directions.

crista One of the elevations making up the sculpture of certain pollen and spores, characterized by long, curved bases (sometimes irregularly fused) and variously bumpy apices. Pl: *cristae.*

cristate Crested, or having a crest; esp. said of sculpture of pollen and spores consisting of cristae.

cristobalite A mineral: SiO_2. It is a high-temperature polymorph of *quartz*, and occurs as white octahedrons in the cavities and fine-grained groundmass of acidic volcanic rocks. Cristobalite is stable only above 1470°C; it has a tetragonal structure (alpha-cristobalite) at low temperatures and an isometric structure (beta-cristobalite) at higher temperatures. Cf: *tridymite.*

critical angle The least *angle of incidence* at which there is total reflection when electromagnetic radiation passes from one medium to another and less refracting medium.

critical damping *Damping* to the point at which the displaced mass just returns to its original position without oscillation. Cf: *aperiodic damping*.

critical density [phase] The density of a substance at its critical temperature and under its critical pressure.

critical density [exp struc geol] That density of a saturated, granular material below which, under rapid deformation, it will lose strength and above which it will gain strength.

critical depth In a channel of water, the depth at which flow is at a *critical velocity*, e.g. the depth at which a flow is at its minimum energy with respect to the channel bottom. Cf: *critical flow*.

critical distance In refraction seismic work, that distance at which the direct wave in an upper medium is matched in arrival time by that of the refracted wave from the medium below having greater velocity.

critical end point A point at which two of three or more phases participating in a univariant equilibrium become identical, thus terminating the line representing the equilibrium.

critical flow Fluid flow at a *critical velocity*, e.g. flow at the point at which it changes from laminar to turbulent. Cf: *critical depth*.

critical gradient *critical slope* [hydraul].

critical height The maximum height at which a vertical or sloped bank of soil will stand unsupported under a given set of conditions (ASCE, 1958, term 98).

critical hydraulic gradient In a cohesionless soil, that *hydraulic gradient* at which intergranular pressure is reduced to zero by the upward flow of water (ASCE, 1962).

critical length Maximum horizontal distance over which sheet erosion does not occur, measured in the direction of overland flow from the drainage divide to a point at which the eroding stress becomes equal to the resistance of the soil to erosion (Horton, 1945, p.320); it determines the width of the *belt of no erosion*. Symbol: x_c. Cf: *length of overland flow*.

critically undersaturated Said of a rock having feldspathoids and olivine (but no hypersthene) in its norm. Cf: *undersaturated*.

critical materials Materials vital to the security of a nation, their main source being within that nation, but which may not be produced in sufficient quality and quantity to meet the nation's wartime needs. Cf: *strategic materials*.

critical mineral A mineral or one of a mineral association that is stable only under the conditions of one metamorphic facies and that will change upon change of facies. Cf: *typomorphic mineral; index mineral*.

critical moisture In a soil, that degree of moisture below which, under constant load increase, deformation will increase, and above which it will decrease.

critical point [phase] A point representing a set of conditions (pressure, temperature, composition) at which two phases become physically indistinguishable; in a system of one component, the temperature and pressure at which a liquid and its vapor become identical in all properties.

critical point [photo] A peak, or high ground with abrupt local relief, requiring investigation (in planning aerial photography) for possible lack of stereoscopic coverage.

critical pressure The pressure required to condense a gas at the critical temperature, above which, regardless of pressure, the gas cannot be liquified.

critical resolved shear stress Shear stress in a crystal at which slip begins in a particular plane.

critical slope [hydraul] The slope or grade of a channel that is exactly equal to the loss of head per foot resulting from flow at a depth that will give uniform flow at critical depth; the slope of a conduit which will produce critical flow (ASCE, 1962). Syn: *critical gradient*.

critical slope angle The local maximum slope inclination which the soil and rock materials underlying the slope can support

without failure under existing climate, vegetation, and land use. Syn: *angle of ultimate stability*.

critical solution point *consolute point*.

critical temperature That temperature above which a substance can exist only in the gaseous state, no matter what pressure is exerted.

critical tractive force The minimum *tractive force* required to set sediment particles of a stream bed moving.

critical velocity (a) That velocity of fluid flow at which the flow changes from laminar to turbulent. (b) That velocity of fluid flow at which the flow changes from laminar to turbulent, and at which friction becomes proportional to a power of the velocity higher than the first power. Syn: *Reynolds critical velocity*. (c) That velocity of fluid flow at which the fluid's minimum energy value is attained. Syn: *Belanger's critical velocity*. (d) In an open channel, that velocity of fluid flow at which silt is neither picked up nor deposited. Syn: *Kennedy's critical velocity*. (e) In an open channel, that velocity of fluid flow at which the velocity head equals one half the mean depth, and at which the energy head is at a minimum. Syn: *Unwin's critical velocity*.----See also: *critical flow; critical depth*.

crocidolite A lavender-blue, indigo-blue, or leak-green asbestiform variety of *riebeckite*, occurring in silky fibers and in massive and earthy forms. Syn: *blue asbestos; krokidolite*.

crocoite A bright-red, yellowish-red, or orange monoclinic mineral: $PbCrO_4$. Syn: *red lead ore; crocoisite*.

crocus A term used in New Hampshire quarries for gneiss or other rock type in contact with granite.

crocydite Migmatite with flake- or fluff-like light-colored parts (Dietrich & Mehnert, 1961). Var: *krokydit*.

crog ball An English term for a large to immense, concretionary *ballstone* occurring in limestone. Syn: *crog*.

Croixian North American provincial series: Upper Cambrian (above Albertan, below Canadian of Ordovician). Syn: *Croixan; Saint Croixian*.

cromaltite A pyroxenite in which acmite-augite is the predominant mineral, with melanite as a characteristic component and smaller amounts of biotite, perovskite, and ore minerals.

cromfordite *phosgenite*.

Cromwell current *equatorial undercurrent*.

cronstedtite A black to brownish-black mineral: $Fe_4^{+2}Fe_2^{+3}(Fe_2^{+3}Si_2)O_{10}(OH)_8$. It is structurally related to kaolinite; it is not a chlorite.

crooked hole A borehole that is deflected beyond the allowable limit from the vertical or from its intended course.

crookesite A lead-gray mineral: $(Cu,Tl,Ag)_2Se$.

crook-veined Said of a rock that is traversed by short crooked channels of recrystallization with the multiplicity of channels produced by a breccioid structure.

crop n. Deprecated syn. of *outcrop*.----v. To appear at the surface of the ground; to *outcrop*.

crop coal (a) That part of a coal seam that is nearest the surface. (b) A coal deposit that outcrops. (c) Low grade coal occurring near the surface.----Cf: *deep coal*.

crop out v. *outcrop*.

cropping n. Deprecated syn. of *outcrop*.

cross-assimilation Simultaneous *assimilation* of country rock into magma and of magma into country rock, so that the same phases develop in both.

cross bar (a) A short, bifurcating "ridge" of a transverse ripple mark. (b) A low ridge trending across a blind valley.

cross-bed A single, thin-bedded, often lenticular (sometimes tabular or wedge-shaped) layer of homogeneous or gradational lithology, deposited at an angle to the original surface of deposition, confined to a single bed or sedimentation unit, and separated from adjacent layers by surfaces of erosion, nondeposition, or abrupt change in character. It is generally truncated at the top, but approaches the bottom contact in a broad tangential curve. The term is restricted by McKee & Weir (1953, p.382) to a foreset bed that is more than 1 cm in thickness. See also: *cross-stratum; cross-lamina.* Also spelled: *crossbed.*

cross-bedding An internal arrangement of the layers in a stratified rock, characterized by minor beds or laminae inclined more or less regularly in straight sloping lines or concave forms at various angles (but less than the angle of repose) to the original depositional surface or principal bedding plane, or to the dip or contact of the formation. It is produced by swift, local, changing currents of air or water, and is characteristic of granular sedimentary rocks (esp. sandstone) and sand deposits (as in a dune, stream channel, or delta). The term is restricted by McKee & Weir (1953, p.382) to bedding characterized by foreset beds that are more than 1 cm in thickness. Also, a cross-bedded structure; a cross-bed. See also: *current bedding; inclined bedding; discordant bedding; crisscross-bedding.* Also spelled: *crossbedding.* Syn: *cross-stratification; cross-lamination; false bedding; diagonal bedding; oblique bedding; foreset bedding.*

cross channel A transverse drainageway cutting across an interstream area or connecting two successive low areas.

crosscorrelation A type of seismic correlation which measures the similarity of two wave forms, or the degree of linear relationship between them, or the extent to which one is a linear function of the other. Cf: *autocorrelation.*

cross coupling A term applied to the use of gravity meters aboard ship; the cross-coupling effect is produced by vertical movements of the gravity meter beam which are correlated with the components of horizontal movement in the direction of the beam. It is an important source of error if not eliminated or if corrections are not made for it.

cross course *cross vein.*

crossed-lamellar Said of the type of bivalve-mollusk shell structure composed of primary and secondary lamellae, the latter inclined in alternate directions in successive primary lamellae.

crossed nicols In a polarizing microscope, two nicol prisms or polaroid plates that are oriented so that the transmission planes of polarized light are at right angles; light that is transmitted from one will be intersected by the other, unless there is an intervening substance. Syn: *crossed polars.*

crossed polars *crossed nicols.*

crossed twinning *cross-hatched twinning.*

cross fault (a) A fault, the strike of which is perpendicular to that of the constituent strata or general trend of the regional structure. (b) A minor fault that intersects a major fault.

cross-fiber adj. Said of veins of *fibrous* minerals, esp. asbestos, in which the fibers are at right angles to the vein.

cross fold A fold that intersects a pre-existing fold of different orientation; the resulting structure is a *complex fold.* Syn: *superimposed fold; transverse fold; subsequent fold.*

cross fracture A small-scale joint structure developed between *fringe joints.*

cross-grading The process of slope dissection by rills and gullies in which an original slope (alongside, and parallel with, a stream) is replaced by a new slope deflected toward the stream (Horton, 1945, p.335). See also: *micropiracy.*

cross-groove One of two or more intersecting sets of groove casts.

cross-hatched twinning Repeated twinning after two laws, e.g. microcline twin law, with intersecting composition planes. Syn: *grating structure; gridiron twinning; crossed twinning.*

cross-hatching (a) The process of drawing or shading (on a map) with a pattern consisting of two sets of parallel lines crossing each other at a predetermined angle (obliquely or at right angles) so that the space between the lines in one set is identical to the space between the lines in the other set. (b) The effect produced by cross-hatching, such as a pattern indicating abrupt gradients.---See also: *hatching.*

crossing A term applied along the Mississippi River to a shallow part of the channel, separating deeper pools of water (or bendways) at the bends of meanders.

crossing canal A proximal (prothecal) part of graptoloid theca, growing across the sicula to develop on that side of the sicula opposite from the side from which it originated.

crossite A blue monoclinic mineral of the amphibole group, intermediate in composition between glaucophane and riebeckite; a variety of glaucophane rich in iron.

cross joint A joint that is perpendicular to the major lineation of the rock. Syn: *transverse joint; Q-joint; ac-joint.*

cross-joint fan In igneous rock, a fanlike pattern of cross joints that follow the arching of the flow lineation.

cross-lamina A *cross-bed.* The term is restricted by McKee & Weir (1953, p.382) to a *cross-stratum* that is less than 1 cm in thickness. Syn: *cross-lamination.*

cross-lamination (a) A term that is generally used as a syn. of *cross-bedding,* but restricted by McKee & Weir (1953, p.382) to *cross-stratification* characterized by foreset beds that are less than 1 cm in thickness. (b) A cross-laminated structure; a *cross-lamina.*---See also:*flow-and-plunge structure.* Syn: *oblique lamination; diagonal lamination.*

cross lode *cross vein.*

crossopodium A trace fossil of the "genus" *Crossopodia,* consisting of a sinuous or meandering marking about one centimeter wide with a median furrow, and believed to be a trail left by a creeping marine animal. Pl: *crossopodia.*

cross profile A profile of a stream or valley, drawn along a line on the stream floor at right angles to the direction of the valley. Cf: *longitudinal profile.* Syn: *transverse profile.*

cross ripple mark A ripple mark resulting from the interference of at least two sets of ripples, one set forming after the completion of, or simultaneously with, the other; e.g. *current cross ripple mark* and *oscillation cross ripple mark.* Term introduced by Bucher (1919, p.190) as "cross-ripple". See also: *compound ripple mark; tadpole nest.* Syn: *interference ripple mark; dimpled current mark; complex ripple mark.*

cross sea A confused, irregular, choppy state of the ocean, occurring where waves from two or more different directions meet. Wave direction may appear to be the same as one of the original directions, or it may be a new direction.

cross section (a) A diagram or drawing that shows features transected by a given plane; specif. a vertical section drawn at right angles to the longer axis of a geologic feature, such as a structure section drawn perpendicularly to the strike of the strata or to the trend of an orebody, or a section at right angles to the mean direction of flow of a stream and bounded by the stream's wetted perimeter and free surface. Cf: *longitudinal section.* Syn: *transverse section.* (b) An actual exposure or cut that shows transected geologic features.----Adj: *cross-sectional.* Also spelled: *cross-section.*

cross spread (a) A type of seismic spread which makes a large angle to the line of traverse. (b) A type of seismic spread which is laid out in the pattern of a cross.

cross-stone (a) *chiastolite.* (b) *staurolite.* (c) *harmotome.*

cross-stratification A term that is generally used as a syn. of *cross-bedding,* but considered by McKee & Weir (1953, p.382) as a general term that includes *cross-bedding* and *cross-lamination.* Also, a cross-stratified structure; a cross-stratum. Syn: *false stratification; diagonal stratification.*

cross-stratum A *cross-bed.* McKee & Weir (1953, p.382) consider it a general term that includes *cross-bed* and *cross-lamina.*

cross valley *transverse valley.*

cross vein (a) A vein or lode that intersects another, larger or more important one. Syn: *cross course; cross lode; countervein; counterlode; caunter lode.* (b) A vein that crosses the bedding planes in a sedimentary sequence.

croute calcaire Hardened *caliche,* often found in thick masses or beds overlain by several centimeters of earth (SSSA, 1965, p.334). Etymol: French *croûte calcaire,* "calcareous crust".

crowfoot Obsolete syn. of *stylolite.*

crown [**drill**] *diamond bit.*

crown [**mass move**] The practically undisturbed material still in place and adjacent to the highest parts of the scarp along which a landslide moved.

crown [**geomorph**] The top or highest part of a mountain or an igneous intrusion; the summit.

crown [**gem**] The portion of any faceted gemstone above the girdle. Cf: *pavilion.* Syn: *bezel; top.*

crown [**paleont**] The whole of a crinoid exclusive of pelma; the part of a crinoid skeleton above the column, including the dorsal cup, tegmen, and arms.

crown-in In mining, a falling of the mine roof or a heave of the mine floor due to the pressure of overlying strata. Cf: *flash; inbreak; heave.*

crown scarp The outward-facing scarp, bordering the upper portion of a landslide. It is almost always concave in a downslope direction. The scarp surface may have a slickensided surface, indicating the emergence of the sublandslide shear surface; or, it may have a rough, suggary surface, indicating a tension or cleavage fracture formed by downslope displacement of the slide mass (Varnes, 1958).

crowstone An English term for a very hard, compact, siliceous sandstone representing the floor of a coal seam; a *ganister.*

crucite (a) *chiastolite.* (b) Pseudomorph of hematite or limonite after arsenopyrite.

crude Said of a mineral deposit in its natural, unrefined state, e.g. crude oil, crude lead.

crude oil Petroleum in its natural state as it emerges from a well, or after passing through a gas-oil separator but prior to any refining or distilling process.

crude petroleum *petroleum.*

crumble coal An incoherent brown coal that lacks cementing material. Syn: *formkohle.*

crumble peat Peat that is friable and earthy.

crumbling The breaking or falling into small pieces, such as the disintegration of a rock.

crumb structure A type of soil structure in which the peds are spheroids of polyhedrons that have little or no accommodation to surrounding peds, are porous, and range in size from less than 1.0 mm to 5 mm. Cf: *granular structure.*

crumina A sac-like, semienclosed space developed in the ventral part of the carapace of female ostracodes belonging to some Paleozoic species. Pl: *cruminae.*

crump Ground movement, perhaps violent, due to failure under stress of the ground surrounding an underground working (usually in coal), and so named because of the sound produced.

crumpled *plicated.*

crumpled ball A highly irregular, crumpled-up mass of laminated sandstone, measuring 5-25 cm across, and flattened parallel to the bedding (Kuenen, 1948, p.371). Cf: *slump ball.*

crumpled mud-crack cast A mud-crack cast that displays tortuous and contorted crumpling produced by the adjustment of the sand filling to the compaction of the enclosing mud matrix (Bradley, 1930).

crura Plural of *crus.*

crural base A part of the crus of a brachiopod, united to a hinge plate, and separating the inner hinge plates and the outer hinge plates when present.

cruralium A spoon-shaped structure of the brachial valve of a pentameracean brachiopod, formed by the dorsal union of outer plates (or homologues) and bearing adductor muscles.

crural plate A plate extending from the inner edge of an outer hinge plate or crural base to the floor of the brachial valve of a brachiopod (TIP, 1965, pt.H, p.143). See also: *septalial plate.*

crural process The pointed part of the crus of a brachiopod directed obliquely inward and ventrally.

crus Either of a pair of short, curved, calcareous basal processes that extend from the cardinalia or septum of a bra-

chiopod to give support to the posterior end of the lophophore. The distal end may also be prolonged into primary lamella of spiralium or descending branch of loop (TIP, 1965, pt.H, p.142). Pl: *crura.*

crush belt A belt of intensely crushed rock characterized by intense cataclasis and localized mylonitization.

crush border A microscopic, granular metamorphic structure sometimes characterizing adjacent feldspar particles in granite due to their having been crushed together during or subsequent to their crystallization (Dale, 1933).

crush breccia (a) A breccia formed in place, or nearly in place, by mechanical fragmentation of rocks during crustal movements; a *tectonic breccia* associated with planes of movement and formed as a result of folding or faulting. (b) A term used by Norton (1917, p.186-188) for a tectonic breccia in which the brecciation was accomplished without either faulting or folding "except so far as the rupture planes of the breccia may be considered as minute faults"; e.g. a breccia produced by lateral pressure without any further mass deformation than that exhibited by gentle warpings.----Syn: *cataclastic breccia.*

crush burst A major *rock burst* in which masses of rock actually fail at the face of a mine working and are suddenly projected from the roof, sides, and/or pillars of a mine. Cf: *strain burst.*

crush conglomerate A rock formed essentially in place by deformation (folding or faulting) of brittle, closely jointed rocks, containing lozenge-shaped fragments produced by granulation of rotated joint blocks and rounded by attrition, and closely simulating a normal (sedimentary) conglomerate; a rock similar to a crush breccia but having fragments that are more rounded. It is characterized by similarity in composition (generally one rock type) of fragments and matrix. Term was introduced by Lamplugh (1895, p.563). Syn: *tectonic conglomerate; cataclastic conglomerate.*

crushing strength Unconfined compressive strength of a material.

crushing test *unconfined compression test.*

crush zone An area of fault breccia or fault gouge.

crust [**interior Earth**] The outermost layer or shell of the Earth, defined according to various criteria: that part of the Earth above the Mohorovičić discontinuity; the *sial,* or the sial and the *sima.* It represents less than 0.1% of the Earth's total volume. See also: *continental crust; oceanic crust.* Cf: *tectonosphere.*

crust [**sed**] A laminated, commonly crinkled deposit of algal dust, filamentous or bladed algae, or clots (from slightly arched forms to bulbous cabbage-like heads) of algae, formed on rocks, fossils, or other particulate matter by accretion, aggregation, or flocculation.

crust [**ice**] *ice rind.*

crust [**snow**] *snow crust.*

crustacean Any arthropod belonging to the superclass Crustacea, characterized chiefly by the presence of two pairs of antennae on the head. The majority occurs in marine environments, and they are second only to the insects in numbers of individuals. Their stratigraphic range is Cambrian to present.

crustaceous (a) Having, suggesting, or of the nature of a crust or shell. (b) Belonging to the Crustacea; crustacean.

crustal abundance *clarke.*

crustal plate One of the six blocks into which the lithosphere is divided, according to the scheme of global tectonics. It is about 100 km thick.

crustified Said of a vein in which the mineral filling is deposited in layers on the wall rock. Syn: *healed.*

crust reef A coral reef formed on a submerged bank.

crust stone In a cave, flaky incrustations of calcite or other minerals that form on the walls. Partial syn: *calcite skin.*

cryergic A term recommended by Baulig (1956, paragraph 77) as a syn. of the broad meaning of *periglacial* as when the atter word is commonly but "rather incorrectly" applied to processes and deposits "in some regions not actually peripheral to glaciated regions". The term is currently used "to denote phenomena due to cold conditions" (Stamp, 1966, .141).

ryergy (a) A term used in Europe as a loose syn. of *cryology*. (b) A process under which cryergic (periglacial) deposits re formed.

ryic Said of a soil temperature regime in which the mean annual temperature (measured at 50cm depth) is more than 0°C ut less than 8°C, with cold summer temperatures, and a ummer-winter variation of more than 5°C (SSSA, 1970). Cf: *rigid* [*soil*].

ryochore A climatic term for the part of the Earth's surface covered with perpetual snow and ice.

ryoclinometer A device that measures the horizontal dimensions of a field of sea ice from an aircraft (Baker et al, 1966, .42).

ryoconite [**mineral**] A mixture of garnet, sillimanite, zircon, yroxene, quartz, and other minerals.

ryoconite [**glaciol**] A dark, powdery dust, once thought to be f cosmic origin, transported by the wind and deposited on a now or ice surface (e.g. on the Greenland Ice Cap). It is ound mainly in cryoconite holes, but may form long stripes or n almost continuous cover. Syn: *kryokonite*.

ryoconite hole A cylindrical *dust well* of varying size containing particles of cryoconite that absorb solar radiation and cause excessive melting of glacier ice around and below hem.

ryogenic lake A lake produced by local thawing in a permafrost region.

Cryogenic period An informal designation for a period in geologic history "when large bodies of ice formed at or near he poles and the climate was generally suitable for the growth of continental glaciers" (ADTIC, 1955, p.22).

ryogenics The branch of physics pertaining to the production and effects of very low temperatures; the science of extreme cold. Adj: *cryogenic*. Obsolete syn: *cryogeny*.

ryokarst *thermokarst*.

ryolaccolith *hydrolaccolith*.

cryolite A white or colorless monoclinic mineral: Na_3AlF_6. It may contain iron and is found chiefly in Greenland usually in cleavable masses of waxy luster. Cryolite is used in the manufacture of aluminum (extraction from bauxite) and for making sodium and aluminum salts and porcellaneous glass. Syn: *Greenland spar; ice stone*.

ryolithionite A colorless isometric mineral: $Na_3Li_3Al_2F_{12}$.

ryolithology The study of the development, nature, and structure of underground ice, esp. ice in permafrost regions; a branch of *geocryology*.

ryology (a) In the U.S., "cryology" is the science of refrigeration, and although the term frequently has been proposed for he study of glaciers, it has not attained wide usage outside he field of low-temperature physics. Cf: *cryergy*. (b) *glaciology*.

ryometer An instrument for the measurement of low temperatures.

ryometry The measurement of low temperatures.

ryomorphology The part of geomorphology "pertaining to the various processes and products of cold climates" (Black, 1966, p.332). See also: *periglacial geomorphology*.

ryonival (a) Pertaining to the combined action of frost and snow. (b) *periglacial* (Hamelin, 1961, p.200).

ryopedology The study of the processes of intensive frost action and the occurrence of frozen ground, esp. permafrost, including the civil engineering methods used to overcome or minimize the difficulties involved; term introduced by Bryan 1946).

cryophilic Said of an organism that prefers low temperatures, esp. below 10°C. Syn: *psychrophilic*.

cryophyllite A variety of zinnwaldite with some deficiency in the (Li,Fe,Al) group and containing some ferric iron.

cryoplanation The reduction and modification of a land surface by processes associated with intensive frost action, such as solifluction, supplemented by the erosive and transport actions of running water, moving ice, and other agents (Bryan, 1946, p.640). Cf: *altiplanation*.

cryosphere The part of the Earth's surface that is perennially frozen; the zone of the Earth where ice and frozen ground are formed.

cryostatic Descriptive of frost-induced hydrostatic phenomena (Washburn, 1956, p.842), such as *cryostatic* movement, the supposed upward injection of uniform material confined between the ground surface and an internal impermeable surface (such as a frost table or bedrock) by progressive downward freezing that generates large hydrostatic pressures in pockets of such material.

cryotectonic Said of complicated and deranged features and deposits found at glacier borders, and consisting of material that has been overturned, inverted, folded, and transported by the shoving action of glaciers. Syn: *glaciotectonic*.

cryoturbate A *congeliturbate*, resulting from or the product of congeliturbation.

cryoturbation A syn. of *congeliturbation*. Variant: *kryoturbation*.

cryptacanthiiform Said of a brachiopod loop composed of descending branches fused distally to form echmidium, which bears a hood on the ventral anterior end (TIP, 1965, pt.H, p.143). With continued growth, the echmidium becomes deeply cleft anteriorly but still connected with descending branches.

crypthydrous Said of vegetal matter deposited on a wet substratum. Cf: *phenhydrous*.

cryptic layering That type of *layering* in an igneous intrusion in which there is a regular vertical change in chemical composition, so named because it isn't as obvious as *rhythmic layering*.

cryptobatholithic Said of a mineral deposit occurring in the roof rocks of an unexposed batholith (Emmons, 1933). The term is little used today. Cf: *acrobatholithic; embatholithic; endobatholithic; epibatholithic; hypobatholithic*.

cryptoclastic rock (a) A clastic rock whose extremely fine constituents can be seen only under the microscope. Ant: *macroclastic rock*. Cf: *microclastic rock*. (b) A carbonate sedimentary rock having an aphanic clastic texture and discrete particles whose diameters are less than 0.001 mm, and displaying little or no crystallinity under high-power magnification (Bissell & Chilingar, 1967, p.154); e.g. extremely finely comminuted carbonate dust. Cf: *cryptograined*.

cryptocrystalline (a) Said of the texture of a rock consisting of or having crystals that are too small to be recognized and separately distinguished even under the ordinary microscope (although crystalline features may be shown by use of the electron microscope); indistinctly crystalline, as evidenced by the confused aggregate effect under polarized light. Also, said of a rock with such a texture. Cf: *microcrystalline*. Syn: *microcryptocrystalline; phanerocrystalline-adiagnostic*. (b) Said of the texture of a crystalline rock in which the crystals are too small to be recognized megascopically. This usage is not recommended "since it cannot be known that an aphanitic rock is cryptocrystalline until the microscope has shown that it is actually microscopically crystalline" (Johannsen, 1939, p.206). (c) Descriptive of a crystalline texture of a carbonate sedimentary rock having discrete crystals whose diameters are less than 0.001 mm (Bissell & Chilingar, 1967, p.103) or less than 0.01 mm (Pettijohn, 1957, p.93). Some petrographers use an upper limit of 0.004 mm.

cryptocyst A more or less horizontal calcareous lamina on the basal side of the frontal membrane in some cheilostomatous bryozoans (such as Anasca), developed from the vertical walls of the zooid but not completely dividing its body cavity.

cryptodepression A lake basin whose bottom lies below sea level (although the water surface may be above sea level).

cryptodont Said of the dentition of certain bivalve mollusks of early origin lacking hinge teeth.

cryptoexplosion structure A nongenetic, descriptive term suggested by Dietz (1959, p.496-497) to designate a roughly circular structure formed by the sudden, explosive release of energy and exhibiting intense, often localized rock deformation with no obvious relation to volcanic or tectonic activity. Such structures typically show some or all of the following: wide variation in diameter (less than 1.5 km to more than 50 km); a central dome-shaped uplift with intense structural deformation, often surrounded by a concentric ring depression; complex faulting and subordinate folding; widespread brecciation and shearing; and occurrence of shatter cones. Many cryptoexplosion structures are believed to be the result of hypervelocity impact of crater-forming meteorites of asteroidal dimensions; others may have been produced by obscure volcanic activity. The term, as presently used, largely replaces the earlier term *cryptovolcanic structure*. See also: *astrobleme*. Syn: *cryptoexplosive structure*.

cryptogam A plant that reproduces by spores rather than by seeds. Examples of cryptogams include thallophytes, bryophytes, and pteridophytes.

cryptogenic Of obscure or unknown origin; e.g. said of a species that appears and disappears suddenly (in the geologic column) and that has neither a known ancestry nor descendants, or said of a rock (such as limestone) that was subjected to such considerable changes that its original composition and conditions of formation cannot be established with certainty or remain completely unknown. Ant: *phanerogenic*.

cryptograined Said of the texture of a carbonate sedimentary rock having discrete clastic (or precipitated or flocculated) particles whose diameters are less than 0.001 mm (Bissell & Chilingar, 1967, p.103; and DeFord, 1946) or less than 0.01 mm (Thomas, 1962). Some petrographers use an upper limit of 0.004 mm. Cf: *cryptoclastic rock*.

cryptographic Said of the texture of a rock whose individual components are so fine that they cannot be clearly distinguished under the microscope; a cryptocrystalline granophyric texture.

cryptohalite A gray, cubic, high-temperature mineral: $(NH_4)_2SiF_6$. Cf: *bararite*.

cryptolite *monazite*.

cryptomagmatic Said of a hydrothermal mineral deposit without demonstrable relationship to igneous processes. Cf: *apomagmatic; telemagmatic; perimagmatic*.

cryptomelane A mineral: $K(Mn^{+2},Mn^{+4})_8O_{16}$. It is isostructural with hollandite and coronadite.

cryptomere An alternate spelling of *kryptomere*.

cryptonelliform Said of a long brachiopod loop (as in the superfamily Cryptonellacea) unsupported in adults by median septum and having a narrow transverse band.

cryptoolitic Pertaining to an oolitic texture of such fine grain that it can be recognized only under the microscope. Also, said of a rock with such a texture.

cryptoperthite An extremely fine-grained variety of perthite in which the lamellae are of submicroscopic dimensions (1-5 microns wide) and are detectable only by X-rays or with the aid of the electron microscope. The potassium-rich host can be sanidine, orthoclase, or microcline; the sodium-rich lamellae can be analbite or albite. Cryptoperthite frequently displays a bluish to whitish milky luster. Cf: *microperthite*.

cryptorheic Said of drainage by subterranean streams. Syn: *cryptoreic*.

cryptosiderite A stony meteorite poor in nickel-iron.

cryptostome Any ectoproct bryozoan belonging to the order Cryptostomata and resembling a trepostome but having a short endozone and an aperture at the bottom of the vesti-

bule. Adj: *cryptostomatous*.

cryptovolcanic structure (a) A term introduced by Branco & Fraas (1905) and originally applied to a highly deformed, strongly brecciated, generally circular structure believed to have been produced by volcanic explosions, but lacking any direct evidence (volcanic rocks, hydrothermal alteration, contact metamorphism, mineralization) of volcanic activity (type example: Steinheim Basin, Germany). Many of these structures are now believed to have been formed by meteorite impact, and the nongenetic term *cryptoexplosion structure* is preferred. Also, the term "has tended to become a 'waste basket' term and now includes many structures which are unquestionably of volcanic origin" (Dietz, 1959, p.496). (b) A circular structure lacking evidence of shock metamorphism or of meteorite impact and therefore presumed to be of igneous origin, but lacking exposed igneous rocks or obvious volcanic features; a rock structure produced by completely concealed volcanic activity.

Cryptozoic That part of geologic time for which, in the corresponding rocks, evidence of life is only slight and of primitive forms. Cf: *Phanerozoic*.

cryptozoon (a) A structure in Precambrian rocks, believed to be the remains of primitive life. (b) A hemispherical or cabbage-like algal structure of variable size, spreading somewhat above its base, composed of irregular and concentric laminae of calcite of very unequal thicknesses traversed by minute canals that branch irregularly, produced by the problematical Cambrian and Ordovician reef-forming calcareous algae of the genus *Cryptozoon*.---Pl: *cryptozoa*.

crystal (a) A homogeneous, solid body of a chemical element, compound, or isomorphous mixture having a regularly repeating atomic arrangement that may be outwardly expressed by plane faces. (b) *rock crystal*. (c) A colorless, transparent diamond.

crystal accumulation In a magma, the development of *layering* by the process of *crystal settling*.

crystal axial indices *indices of lattice row*.

crystal axis (a) *crystallographic axis*. (b) One of the three edges of the chosen unit cell in a crystal lattice. (c) Any lattice row; it can be considered a zone axis.----Syn: *axis* [*cryst*].

crystal cast The filling of a crystal mold; e.g. *ice-crystal cast* and *salt-crystal cast*.

crystal chemistry The study of the relations among chemical composition, internal structure, and the physical properties of crystalline matter. Syn: *stereochemistry*.

crystal class *class* [*cryst*].

crystal defect An imperfection in the ideal crystal structure. See also: *line defect; point defect; plane defect*. Syn: *lattice defect*.

crystal face The outward, planar surfaces that bound a crystal, and reflect its internal orderly structure. The symmetries of crystal faces are used to classify crystals into systems. Syn: *face* [*cryst*].

crystal flotation In petrology, the floating of lighter weight crystals in a body of magma. Cf: *crystal settling*. Syn: *flotation*.

crystal form (a) The geometric shape of a crystal. (b) An assemblage of symmetrically equivalent crystal planes making up a form which displays the symmetry of a crystal class. A crystal may be bounded by one or more forms, each consistent with the internal symmetry of the crystal. Crystal form may be symbolized by Miller indices enclosed in braces, e.g. {hkl}.

crystal fractionation Magmatic differentiation resulting from the settling, by gravity, of the heavier crystals as they form. Cf: *fractional crystallization*.

crystal gliding Slip along a plane in a crystal in response to plastic deformation. It often produces crystal twins. See also: *twin gliding*. Syn: *gliding; translation gliding; slip* [*cryst*].

crystal indices *Miller indices.*

crystallaria A general term proposed by Brewer (1964, p.284) or a group of soil features consisting of single crystals or arrangements of crystals of relatively pure fractions of the soil plasma that do not enclose the matrix of the soil material but form cohesive masses whose morphology (shape and internal fabric) is "consistent with their formation and present occurence in original voids in the enclosing soil material"; e.g. spherulites, rosettes, crystal tubes, and intercalary crystals embedded in a dense soil matrix. See also: *crystal tube.*

crystal lattice The three-dimensional, regularly repeating atomic arrangement of a crystal, each point of which has identical surroundings. The lattice is built by the regular, parallel translation in space of the *unit cell.* There are fourteen possible lattice patterns. Syn: *Bravais lattice; space lattice; translation lattice.*

crystalline [cryst] (a) Pertaining to or having the nature of a crystal, or formed by crystallization; specif. having a crystal structure or a regular arrangement of atoms in a space lattice. Ant: *amorphous.* (b) Said of a mineral particle of any size, having the internal structure of a crystal but lacking well-developed crystal faces or an external form that reflects the internal structure. (c) Resembling a crystal; clear, transparent, pure.

crystalline [petrology] adj. (a) Said of a rock consisting wholly of crystals or fragments of crystals; esp. said of an igneous rock developed through cooling from a molten state and containing no glass, or of a metamorphic rock that has undergone recrystallization as a result of temperature and pressure changes. The term may also be applied to certain sedimentary rocks (such as quartzite, some limestones, evaporites) composed entirely of contiguous crystals. (b) Said of the texture of a crystalline rock characterized by closely fitting or interlocking particles (many having crystal faces and boundaries) that have developed in the rock by simultaneous growth.----n. A crystalline rock. Term is usually used in the plural; e.g. Precambrian "crystallines". This usage is not recommended.

crystalline carbonate A term used by Dunham (1962) for a carbonate sedimentary rock in which the depositional texture is not recognizable, owing to recrystallization and replacement; e.g. dolomite rock and dolomitic limestone.

crystalline chert Obsolescent syn. of *granular chert* (Ireland et al, 1947, p.1486).

crystalline chondrite A hard, crystalline stony meteorite containing firm, round, radial chondrules that break with the matrix.

crystalline flake *flake graphite.*

crystalline-granular texture [ign] The granular texture of an igneous rock, characterized by crystallized particles or angular grains held together by crystal intergrowth and without cement. Cf: *fine-granular.*

crystalline-granular texture [sed] (a) A primary granular texture of a sedimentary rock, produced by crystallization from an aqueous medium; it may be exhibited by rock salt, gypsum, or anhydrite. (b) A texture of a carbonate sedimentary rock produced by dolomitization of a limestone containing packed granules (Thomas, 1962, p.197).

crystalline limestone [meta] (a) A calcarenite with crystalline calcite cement formed in optical continuity with crystalline fossil fragments by diagenesis (Pettijohn, 1957, p.407-408). Syn: *sedimentary marble.* (b) A metamorphosed limestone; a *marble* formed by recrystallization of sedimentary limestone as a result of metamorphism. Syn: *cipolin.*

crystalline limestone [sed] A limestone formed by recrystallization as a result of diagenesis; specif. a limestone in which calcite crystals larger than 20 microns in diameter are the predominant components (Schmidt, 1965, p.128). Examples include the crinoidal limestones whose fragments (ossicles, plates, etc.) have been enlarged by growth of calcite. Syn: *marble* [sed].

crystalline rock (a) An inexact but convenient term designating an igneous rock or a metamorphic rock, as opposed to a sedimentary rock. (b) A rock consisting wholly of crystals or fragments of crystals; e.g. a plutonic rock or an igneous rock lacking glassy material, or a metamorphic rock. The term may be applied to a sedimentary rock (such as a sandstone) composed of crystalline materials or exhibiting a crystalline texture.

crystalline structure *crystal structure.*

crystallinity (a) The degree to which a rock (esp. an igneous rock) is crystalline, whether holocrystalline, glassy, or cryptocrystalline. (b) The degree to which the crystalline character of an igneous rock is developed (e.g. macrocrystalline, microcrystalline, or cryptocrystalline) or is apparent (e.g. phaneritic or aphanitic) (Challinor, 1967, p.62).

crystallinoclastic rock A clastic rock containing abundant crystalline material, such as one having a crystalline cement.

crystallinohyaline *hyalinocrystalline.*

crystallite (a) A broad term applied to a minute body of unknown mineralogic composition or crystal form and which does not polarize light. Crystallites represent the initial stage of crystallization of a magma. Adj: *crystallitic.* Cf: *microlite; crystalloid.* (b) An obsolete syn. of *stylolite.*

crystal-lithic tuff A tuff that is intermediate between *crystal tuff* and *lithic tuff* but is predominantly the former. Cf: *lithic-crystal tuff.*

crystallitic Of, pertaining to, or composed of *crystallites.*

crystallization The process of matter becoming crystalline, from a gaseous, fluid, or dispersed state.

crystallization banding Banding developed in an igneous rock (esp. a plutonic rock) by successive deposition of crystals, such as those settling at the bottom of a pool of magma.

crystallization differentiation *fractional crystallization* [petrology].

crystallization fabric A term used by Friedman (1965, p.643) for the size and mutual relations of mineral crystals in sedimentary rocks such as evaporites, chemically deposited cements, and recrystallized limestones and dolomites. Cf: *crystallization texture.*

crystallization index In igneous petrology, specif. igneous differentiation, the number that is calculated from the system anorthite-diopside-forsterite and represents the sum (in weight percent) of normative anorthite, magnesian diopside, normative forsterite, normative enstatite converted to forsterite, and magnesian spinel calculated from normative corundum in ultramafic rocks (Poldervaart & Parker, 1964, p.281). Abbrev: CI.

crystallization interval (a) The interval of temperature (or, less frequently, of pressure) between the formation of the first crystal and the disappearance of the last drop of a magma upon cooling, usually excluding late-stage aqueous fluids. (b) More specifically, with reference to a given mineral, the temperature range or ranges over which that particular phase is in equilibrium with liquid. In the case of equilibria along reaction lines or reaction surfaces, crystallization intervals defined in this way include temperature ranges in which certain solid phases are actually decreasing in amount with temperature decrease.---- Syn: *freezing interval.*

crystallization magnetization *chemical remanent magnetization.*

crystallization remanent magnetization *chemical remanent magnetization.*

crystallization texture A term used by Friedman (1965, p.643) for the shape of mineral crystals in sedimentary rocks such as evaporites, chemically deposited cements, and recrystallized limestones and dolomites. Cf: *crystallization fabric.*

crystallized coal *cone-in-cone coal.*

crystallizing force The expansive force of a crystal that is forming within a solid medium. The force varies according to crystallographic direction. Syn: *force of crystallization.*

crystalloblast A crystal of a mineral produced entirely by metamorphic processes. See also: *idioblast; holoblast; hypidioblast; xenoblast.* Adj: *crystalloblastic.*

crystalloblastesis Deformation accomplished by metamorphic recrystallization (Knopf and Ingerson, 1938).

crystalloblastic (a) Pertaining to a *crystalloblast.* (b) Said of a crystalline texture produced by metamorphic recrystallization under conditions of high viscosity and directed pressure, in contrast to igneous rock textures that are the result of successive crystallization of minerals under conditions of relatively low viscosity and nearly uniform pressure (Becke, 1903). See also: *homeoblastic; heteroblastic.*

crystalloblastic order *crystalloblastic series.*

crystalloblastic series An arrangement of metamorphic minerals in order of decreasing form energy so that crystals of any of the listed minerals tend to assume idioblastic outlines at surfaces of contact with simultaneously developed crystals of all minerals occupying lower positions in the series (Becke, 1913). Syn: *idioblastic series; crystalloblastic order.*

crystalloblastic strength *form energy.*

crystallogeny That branch of crystallography which deals with crystal growth.

crystallographic Pertaining to crystallography or to the properties of a crystal.

crystallographic axis One of three (four in a hexagonal crystal) imaginary lines in a crystal that pass through its center; it is used as a reference in describing crystal structure and symmetry. One or all of the crystallographic axes may coincide with axes of symmetry. Syn: *crystal axis.*

crystallographic orientation The relation of the axes or planes of a given crystal to some other established directions in space, e.g. geographic or geologic lines or planes.

crystallographic plane Any plane, crystal face, cleavage or lattice plane which can be described mathematically in terms of the lengths and directions of the crystallographic axes.

crystallographic texture A texture of replacement or exsolution mineral deposits in which the distribution and form of the inclusions are controlled by the crystallography of the host mineral.

crystallography The study of crystals, including their growth, structure, physical properties, and classification by form.

crystalloid n. A microscopic crystal which, when examined under a microscope, polarizes light but has no crystal outline or determinable optical properties. Cf: *crystallite; microlite.*

crystallolith A crystalline coccolith; e.g. a disciform holococcolith.

crystallothrausmatic Said of igneous rocks with an orbicular texture in which early phenocryst minerals form the nuclei of the orbicules (Eskola, 1938, p.476). Cf: *allothrausmatic; isothrausmatic; heterothrausmatic; homeothrausmatic.*

crystal mold A cavity left by solution or sublimation of a crystal (as of salt, ice, or pyrite) embedded in soft, fine-grained sediment.

crystal mush Partially crystallized magma; "an aggregate of solid crystals lubricated by compressed water vapor" (Krauskopf, 1967, p.419).

crystal optics The study of the transmission of light in crystals; the concern of *optical crystallography.*

crystal pool In a cave, standing water that contains crystals of calcite.

crystal recovery Recovery of original properties in a crystal that has been distorted by stress, resulting from continued relief from stress, heating, or decrease in speed of deformation (Knopf, 1938).

crystal sandstone (a) A siliceous sandstone in which silica is precipitated upon quartz grains in crystalline position, thereby outwardly converting the grains into crystals. (b) A sandstone with coarsely crystalline carbonate cement that envelops one or more of the detrital grains as inclusions within the crystalline unit.

crystal sedimentation *crystal settling.*

crystal seeding The use of a *seed crystal* or foreign particle in a solution to initiate crystallization of the solute.

crystal settling In a magma, the sinking of crystals due to their greater density, and sometimes aided by magmatic convection. It results in *crystal accumulation*, which develops *layering.* Cf: *crystal flotation.* Syn: *crystal sedimentation.*

crystal sorting The separation, by any process, of crystals from a magma, or of one crystal phase from another during crystallization of the magma.

crystal structure The regular, orderly, and repeated arrangement of atoms in a crystal, described by the crystal lattice or space lattice. Syn: *crystalline structure.*

crystal system One of six groups or classifications of a crystal according to the symmetry of its crystal faces, and having characteristic dimensional equivalences in the lattices or axes of reference. The systems are: *isometric system, hexagonal system, tetragonal system, orthorhombic system, monoclinic system,* and *triclinic system.* Within the six systems there is a total of 32 crystal classes. Syn: *system [cryst].*

crystal tube A type of *crystallaria* consisting of masses of crystals filling or partly filling relatively large tube-shaped or acicular voids in soil material. It is usually formed by crystallization from the walls inward.

crystal tuff A tuff that consists predominantly of crystals and fragments of crystals. Cf: *crystal-vitric tuff; crystal-lithic tuff.*

crystal-vitric tuff A tuff that consists of 50-75% crystal fragments and the remainder, of volcanic glass fragments. Cf: *crystal tuff; vitric tuff.*

crystal zone (a) Three or more nonparallel crystal faces, the edges of intersection of which are parallel to a common line or lattice row called the *zone axis.* (b) *zoning.*

crystocrene *chrystocrene.*

crystosphene A buried mass or sheet of clear ice developed by a wedging growth between beds of other material (Tyrrell, 1904, p.234), as the freezing of springwater rising and spreading laterally beneath alluvial deposits or under swamps in a tundra region. Cf: *chrystocrene.*

ctenodont Said of the dentition of certain bivalve mollusks of early origin having numerous short hinge teeth transverse to the margin.

ctenoid cast A very rare, tooth-like sole mark having the form of an obliquely cut, longitudinally ribbed cylinder and probably representing a bounce cast made in mud by plant stems of *Equisetites* that intermittently touched bottom as they were carried along by a current of water. The term was introduced as "ctenoid marking" by Beasley (1914) who likened the structure to the large tortoise-shell comb worn in women's hair during the early Victorian period.

ctenolium A comb-like row of small teeth on the lower side of the byssal notch in some pectinacean mollusks.

ctenostome Any ectoproct bryozoan belonging to the order Ctenostomata and characterized by the presence of comb-like protrusions at the mouth.

cubanite a bronze-yellow orthorhombic mineral: $CuFe_2S_3$. Syn: *chalmersite.*

cube A crystal form of six equivalent (not necessarily square) and mutually perpendicular faces, with indices of {100}.

cube ore *pharmacosiderite.*

cube spar *anhydrite.*

cubic cleavage A type of isometric crystal cleavage that occurs parallel to the faces of a cube; e.g. galena.

cubic close packing The packing of a face-centered cubic crystal lattice. Cf: *hexagonal close packing.*

cubic coordination An atomic structure or arrangement in which an ion is surrounded by eight ions of opposite sign whose centers form the points of a cube. An example is the structure of cesium chloride. Partial syn: *hexahedral coordination.*

cubic packing The "loosest" manner of systematic arrangement of uniform solid spheres in a clastic sediment or crystal lattice, characterized by a unit cell that is a cube whose eight corners are the centers of the spheres involved (Graton & Fraser, 1935). An aggregate with cubic packing has the maximum porosity (47.64%). Cf: *rhombohedral packing*. See also: *open packing*.

cubic plane In a crystal of the cubic system, any plane at right angles to a crystallographic axis.

cubic system *isometric system*.

cuboctahedron A cubic crystal form bounded by both the six equal squares of the cube and the eight equal triangles of the octagon, the latter cutting off the corners of the former.

cubo-dodecahedron A crystal in the cubic system that is bounded by cube and dodecahedron forms.

cucalite A chlorite-rich diabase. In the Rhaetian Alps it grades locally into chlorite schist.

cuchilla A term used in the SW U.S. for a sharply edged crest of a sierra. Etymol: Spanish, "large knife".

cuesta (a) A hill or ridge with a gentle slope on one side and a steep slope on the other; specif. an asymmetric ridge (as in the SW U.S.) with one face (dip slope) long and gentle and conforming with the dip of the resistant bed or beds that form it, and the opposite face (scarp slope) steep or even cliff-like and formed by the outcrop of the resistant rocks, the formation of the ridge being controlled by the differential erosion of the gently inclined strata. Originally, the term applied to the steep slope or scarp that terminates a gently sloping plain at its upper end; the term has also been used to denote the sloping plain itself, such as the top of a mesa. (b) A ridge or belt of low hills formed between lowlands in a region of gently dipping sedimentary rocks (as on a coastal plain), having a gentle slope conforming with the dip of the rocks and a relatively steep slope descending abruptly from its crest.---Etymol: Spanish, "flank or slope of a hill; hill, mount, sloping ground". Cf: *hogback*. Syn: *wold; scarped ridge; escarpment*.

Cuisian European stage: lower Eocene (above Ypresian, below Lutetian).

cul-de-sac [coast] An inlet with a single small opening.

cul-de-sac [karst] A closed, abondoned swallow hole that has been partially filled.

cul-de-sac [speleo] A passage in a cave that has only one entrance; a dead end.

cul-de-sac [ice] *blind lead*.

culet The small facet that is polished parallel to the girdle plane across what would otherwise be the sharp point or ridge that terminates the pavilion of a diamond or other gemstone. Its principal function is to reduce the possibility of damage to the gem.

culm (a) *kolm*. (b) The anthracite contained in the series of shale and sandstones of North Devon, England known as the Culm measures. (c) Fine particles (smaller than rice coal) of anthracite; coal dust; waste.

culmen A narrow ridge between adjoining crenellae on the articular surface of a columnal of a crinoid or blastoid. Pl: *culmina*.

culmination The highest point of a structural feature, e.g. of a dome, anticlinal axis, or a nappe. The axis of an anticline may have several culminations that are separated by saddles. See also: *crest*. Syn: *axis culmination; apex*.

cultural feature An artificial or man-made feature as shown on a map.

culture The details of a map, representing the works of man such as roads, railroads, buildings, canals, trails, towns, and bridges), as distinguished from natural features; they are usually printed in black on a topographic map. The term also includes political boundary lines, meridians, parallels, all place names, and the legends on a map.

culvert Any covered structure, not classified as a bridge, that constitutes a transverse drain, waterway, or other opening under a road, railroad, canal, or similar structure.

cumacean Any malacostracan belonging to the order Cumacea, characterized chiefly by the long, slim, subcylindrical pleon which is usually strongly differentiated from the broad, commonly inflated pereion and cephalon (TIP, 1961, pt.R, p.368).

cumberlandite A coarse-grained ultramafic rock with olivine crystals (approximately 50 percent of the total rock) in a groundmass of ilmenite and magnetite (together, 40 percent), labradorite, and accessory spinel; the state rock of Rhode Island, U.S.A

cumbraite A porphyritic extrusive rock composed of basic plagioclase (bytownite, anorthite) phenocrysts in a groundmass of labradorite, pyroxene (enstatite, augite), and abundant glass; its chemical composition is andesitic rather than basaltic. See also: *inninmorite*.

cumengite A deep-blue or light indigo-blue tetragonal mineral: $Pb_4Cu_4Cl_8(OH)_8 \cdot H_2O$. Syn: *cumengeite; cumengèite*.

cummingtonite A brownish monoclinic mineral of the amphibole group: $(Fe,Mg)_7Si_8O_{22}(OH)_2$. It contains more iron than *anthophyllite*, and may contain zinc and manganese, and it usually occurs in metamorphic rocks as lamellae or fibers. Cf: *grunerite*.

cumulate n. An igneous rock formed by the accumulation of crystals that settle out from a magma by the action of gravity. Syn: *accumulative rock*.

cumulative curve *cumulative frequency distribution*.

cumulative frequency distribution A curve drawn to represent the percentage of occurrences of a number of observations of a variable less and greater than any given value for an entire sample. Syn: *cumulative curve*.

cumulite A cloudy aggregate of globulites commonly found in glassy igneous rocks.

cumulo-dome *volcanic dome*.

cumulophyre An igneous rock characterized by *cumulophyric* texture.

cumulophyric Said of the texture of a porphyritic igneous rock in which the phenocrysts, not necessarily of the same mineral, are clustered in irregular groups; also, said of a rock exhibiting such texture, e.g. a *cumulophyre* (Cross, et al 1906, p.703). Cf: *glomerophyric; synneusis; gregaritic*.

cumulose Pertaining to a soil material consisting chiefly of partly decomposed vegetable matter, accumulated in situ. An example is peat. Cf: *residual soil material*.

cumulo-volcano *volcanic dome*.

cumulus The accumulation of crystals that precipitated from a magma without having been modified by later crystallization. See also: *cumulus crystal*.

cumulus crystal A unit of the *cumulus*. Syn: *primary precipitate crystal*. See also: *primocryst*.

cuneate Said of a leaf or petal that is wedge-shaped or triangular, with the narrower end at the base.

cunette A small channel dug in the bottom of a much larger channel or conduit for the purpose of concentrating the flow at low-water stages.

cuniculus A continuous tunnel-like cavity formed in foraminifers (such as Verbeekinidae) by strong septal fluting, the opposed folds of adjacent septa meeting to form continuous spiral sutures with vaulted arches between, and serving to connect adjoining chambers from one foramen to the next. Pl: *cuniculi*.

Cunnersdorf twin law A rare type of normal twin law in feldspar, in which the twin plane is $(\bar{2}01)$.

cup A greatly expanded *basal cavity* beneath the anterior or posterior half of some conodont elements.

cup and ball jointing *ball and socket jointing*.

cup crystal An exquisitely fragile, transparent ice crystal with stepped surfaces in the form of a hollow hexagonal cup, one side of which may be undeveloped and appear to be rolled up. It is a common form of *depth hoar*.

cuphole A term proposed by Hudson (1909, p.161) for a

shoreline excavation developed by sand- and silt-laden water agitated by waves, its depth being greater than its width (5-15 cm in diameter), and having a hyperbolic or parabolic outline in cross section; it is smaller and shallower than a stream pothole. Cf: *dentpit*.

cupid's dart *flèche d'amour*.

cupola [paleont] A large vaulted dome in nassellarian and spumellarian radiolarians.

cupola [intrus rocks] An upward projection of an igneous intrusion into its roof. Cf: *pendant*.

cupolate Said of a button-shaped scleractinian corallite with a flat base and a highly convex oral surface.

cupped pebble A pebble whose upper side has been subject to solution, often being so corroded that it becomes a mere shell (Scott, 1947). Cf: *pitted pebble*.

cuprite A red (crimson, scarlet, vermilion, brownish-red) isometric mineral: Cu_2O. It is an important ore of copper, and occurs as a secondary mineral in the zone of weathering of copper lodes. Syn: *red copper ore; red oxide of copper; ruby copper; octahedral copper ore*.

cuproauride *auricupride*.

cuprobismutite A monoclinic mineral: $CuBiS_2$. It is dimorphous with emplectite, and was formerly regarded as a mixture of bismuthinite and emplectite.

cuprodescloizite *mottramite*.

cupropiapite A mineral of the copiapite group: $CuFe_4(SO_4)_6(OH)_2.20H_2O$.

cuprorivaite A mineral: $CaCuSi_4O_{10}$.

cuproscheelite A mixture of scheelite and cuprotungstite.

cuprosklodowskite A strongly radioactive, greenish-yellow or grass-green, orthorhombic secondary mineral: $Cu(UO_2)_2Si_2O_7.6H_2O$. It is isostructural with sklodowskite and uranophane.

cuprostibite A mineral: $Cu_2Sb(Te)$.

cuprotungstite A mineral: $Cu_2(WO_4)(OH)_2$.

cuprouranite *torbernite*.

cupule (a) A cuplike *involucre*, esp. of the acorn, whose bracts are adherent at the base and may be free or not, upwards. (b) A free sheathing structure from the peduncle, investing one or more seeds.

curely cannel A syn. of *cannel coal*, so named for its conchoidal fracture.

curie A unit of measurement of radioactivity, defined as the equivalent of 37 billion disintegrations per second, which is approximately equal to the radioactivity of one gram of radium.

Curie balance A *magnetic balance* used to determine saturation magnetization as a function of temperature, hence to determine Curie point.

Curie point The temperature above which thermal agitation prevents spontaneous magnetic ordering. Specifically, the temperature at which there is a transition from ferromagnetism to paramagnetism. Cf: *Néel point*. Syn: *Curie temperature*.

Curie's law The statement that magnetic susceptibility is inversely proportional to absolute temperature. It is applicable to substances which do not show spontaneous magnetic order at low temperatures. Cf: *Curie-Weiss Law*.

Curie temperature *Curie point*.

Curie-Weiss law The statement that the magnetic susceptibility of a ferromagnet above its Curie point is inversely proportional to the difference between actual temperature and the Curie point. Cf: *Curie's law*.

curio stone A stone of little intrinsic value, but which combines uniqueness or souvenir value with a reasonable amount of beauty and durability; e.g. fairy stone and Niagara spar. Cf: *ornamental stone*.

curite An orange-red, radioactive mineral: $Pb_2U_5O_{17}.4H_2O$. It is an alteration product of uraninite.

curl (a) A term used in southern U.S. for a bend in a stream. (b) An eddy in a stream.

curly bedding *convolute bedding*.

current (a) The part of a fluid body (as air or water) that moving continuously in a definite direction, often with a veloci ty much swifter than the average, or in which the progress the fluid is principally concentrated. (b) A horizontal move ment or continuous flow of water in a given direction with more or less uniform velocity, producing a perceptible mas transport, set in motion by winds, waves, gravity, or diffe ences in temperature and density, and of a permanent seasonal nature; esp. an *ocean current*. (c) The velocity flow of a fluid in a stream.

current bedding Any bedding or bedding structure produce by current action; specif. cross-bedding resulting from wate or air currents of variable direction. The term is used (esp. Great Britain) as a syn. of *cross-bedding*, but such usage not recommended (Middleton, 1965, p.248). See also: *ripp bedding*. Syn: *false bedding*.

current crescent (a) A small, semicircular or U-shape rounded ridge, convex upcurrent, commonly with a pit in th center, and developed on a muddy surface by current actic (McKee, 1954, p.63). (b) A flute cast of a horseshoe-shape moat eroded on the upcurrent side of a pebble, shell, or oth obstacle. Syn: *horseshoe flute cast; crescent cast; crescent mark*.

current cross ripple mark A *cross ripple mark* resulting fro the intersection at any angle of a preexisting current rippl mark by a later current moving in a different direction an being "sufficiently weak and only of very short duration" so a not to destroy the first set of ripples (Bucher, 1919, p.19 195).

current density The current per unit area perpendicular to th direction of current flow.

current direction *set* [current].

current electrode A metal contact with the ground used facilitate current flow through the ground.

current-focused log *focused-current log*.

current lineation A term used by Stokes (1947) for *parting neation*.

current mark (a) Any structure formed by the action of a cu rent of water, either directly or indirectly, on a sedimentar surface; e.g. a *scour mark* made by the current itself, and *tool mark* formed by solid objects swept along by the curren See also: *flow mark*. Syn: *current marking*. (b) An irregula structure made by a tidal current in the beach zone, consis ing of a small depression extending toward the shore from th lee side of an obstruction. (c) A *current ripple mark*. Kindl (1917, p.36) used the term to designate the linguoid variety current ripple mark.

current meter Any one of numerous instruments for measu ing the speed alone, or both the speed and direction, of flow ing water, as in a stream or the ocean; it is usually operate by a wheel equipped with a set of revolving vanes or cup whose rate of turning is proportional to the velocity of the cu rent.

current rip A *rip* consisting of small waves formed on a wate surface by the meeting of currents flowing from different di rections.

current ripple *current ripple mark*.

current-ripple cast A term used by Kuenen (1957, fig.6) for sedimentary structure now known as a *transverse scou mark*.

current ripple mark An *asymmetric ripple mark* with a shar or rounded crest between rounded troughs, formed by cu rents of air or water moving more or less constantly in a uni form direction over a sandy surface (such as a stream ba tidal flat, beach, or sand dune), the ripple slowly migratin downcurrent much like a miniature sand dune. The sand i coarsest in the trough and finest on the crest. It is a commo shallow-water feature but may form at any depth. See als *linguoid ripple mark; rhomboid ripple mark; normal ripp*

174

mark. Cf: *oscillation ripple mark*. Syn: *current ripple; current mark; parallel ripple mark*.

current rose A graphic diagram that indicates (for a given ocean area over a period of time) the percentage frequency of current setting toward each of the principal compass points by means of proportional radiating arrows that may be further subdivided (by thickness or pattern) to designate current speeds.

curtain [speleo] A thin sheet of dripstone that hangs from the wall of a cave or that projects from the wall. Syn: *drapery; drip curtain*.

curtain [geomorph] (a) A rock formation that connects two neighboring bastions. (b) One of a series of steps cut in a valley side and exaggerated by cultivation (Swayne, 1956, p.45).

curtisite idrialite.

curvatura A visible line of some (mostly mid-Paleozoic) trilete spores that connects the extremities of the ends of the laesura and outlines the contact areas; e.g. a "curvatura perfecta" having three lines complete all around the spore's proximal face, or a "curvatura imperfecta" having fork-like projections from the radial ends of the laesura but not joining with their neighbors. Pl: *curvaturae*.

curvature (a) *terminal curvature*. (b) *outcrop curvature*. (c) *earth curvature*.

curvature correction An adjustment applied to an observation or computation (such as of difference in elevation) to allow for *earth curvature*. In geodetic spirit leveling, the effects of curvature and of atmospheric refraction are considered together, and tables have been prepared from which combined corrections can be taken.

curvature of gravity A vector of quantity calculated from torsion-balance data indicating the shape of the equipotential surface. It points in the direction of the longer radius of curvature.

curve *trajectory* [seism].

curved fracture cleavage A pattern of cleavage surfaces in graded beds which cut more directly across the lower, coarser parts and curve to a more diagonal direction in the upper, finer parts, thus being convex outward from an anticlinal axis (Muller, 1965).

curved-path theory The statement in exploration seismology that ray paths curve as velocity increases with depth. It is applied to the analysis and plotting of seismic data.

curved ripple mark A ripple mark whose crest appears curved or crescentic in plan view; it may occur where two or more sets of simple ripple marks cross or intersect.

curve of erosion A theoretical curve produced as the result of erosion and portraying the profile of a stream, valley, hill, coast, or skyline; esp. a single, continuously descending curve representing a stream course at grade, generally concave upward and flattening out as it descends from the source to the sea.

curvette A common misspelling of *cuvette*.

curviplanar Pertaining to a surface or form derived from the curving of a plane about one or more axes.

cuselite A biotite-augite-bearing lamprophyre intermediate in composition between a minette and a vogesite and between a versantite and a spessartite.

cushion An artificial pool designed to absorb the kinetic energy of falling water and so prevent erosion.

cusp [coast] (a) Any of a series of low, crescent-shaped mounds or ridges of beach material built by wave action and separated by smoothly curved shallow depressions, spaced at more or less regular intervals along and generally at right angles to the shoreline, and varying in length across their seaward-pointing apices from less than a meter to many kilometers; e.g. a *beach cusp*, a *storm cusp*, a *giant cusp*, a *cuspate spit*, and a *cuspate foreland*. The term "beach cusp" is frequently and loosely used as a synonym. (b) The triangular deposit representing the ends of two cusps, with the point of the triangle extending toward the sea or lake and the base of the triangle on the shore.

cusp [geomorph] A landform characterized by a projection with indentations of crescent shape on either side; e.g. a *meander cusp*.

cusp [paleont] A spine- or fang-like or cone-shaped structure (large denticle) located above the basal cavity of conodont elements. It comprises the entire element in simple conodont elements. See also: *anticusp*.

cuspate bar A crescent-shaped bar uniting with the shore at each end. It may be formed by a single spit growing from the shore and then turning back to meet it, or by two spits growing obliquely from the shore and converging to form a bar of sharply cuspate form. Cf: *V-bar; looped bar*.

cuspate delta A tooth-shaped delta in which a single dominant river builds the delta forward into a lake or sea while vigorous wave action spreads the deltaic deposits uniformly on either side of the river mouth to form two curving beaches, each concave toward the water; e.g. the delta of the Tiber River on the Mediterranean Sea.

cuspate foreland The largest *cusp*, occurring as a cape or as a broadly triangular point of sand or shingle with the apex pointing seaward along an open coast, measuring in some places many kilometers from apex to apex and extending seaward for several kilometers; it is formed by long-continued shore drifting of sediment, as by the convergence to a meeting point of separate spits or beach ridges from opposed directions, or by the progradation of cuspate bars. Term originated by Gulliver (1896, p.401). Examples: Cape Kennedy, Fla., and Cape Hatteras, N.C cuspate-foreland bar (20) (a) A transition form between a *compound cuspate bar* and a *cuspate foreland*. (b) A bar produced where a cuspate bar enclosing a triangular lagoon or marsh is prograded by the addition of successive beach ridges on its seaward side (Johnson, 1919, p.324-325).

cuspate reef A wall reef whose ends curve leeward and border the passages between adjacent reefs (Maxwell, 1968, p.99 & 101).

cuspate ripple mark *linguoid ripple mark*.

cuspate sandkey A *cuspate bar* that has been built up above the water surface to form an island, as along the west coast of Florida.

cuspate spit A prominent point commonly extending from a barrier island into a bay or lagoon (Shepard, 1952, p.1911); the distance between the crescentic tips is about 2 km or more.

cusp cast A term introduced by Spotts & Weser (1964, p.199) for a crescentic, asymmetric sole mark that lacks a deeper upstream end and that is not elongated in the downstream direction. The original depression responsible for the cast may represent current scour.

cuspidine A mineral: $Ca_4Si_2O_7(F,OH)_2$. Syn: *custerite*.

cusplet A minor *beach cusp*, measuring about 1.5 m or less between the tips of the crescent, occurring in the swash zone; it has a short life-span, appearing and disappearing with the turn of the tide. Syn: *beach cusplet*.

cusp-ripple A term used by McKee (1954, p.60) for a crescent-shaped current ripple resulting from an "irregular and fluctuating" stream. See also: *linguoid ripple mark*.

custerite *cuspidine*.

cut [geol] v. To excavate or hollow out a depression, channel, furrow, etc., by erosion.---n. (a) A notch, depression, channel, inlet, or other incision produced by erosion or natural excavation, as by water or waves. (b) A passage or space from which material has been excavated, such as a road *cut*. Also, the material excavated.

cut [gem] The style or form in which a gem has been fashioned; e.g. *brilliant cut* and *step cut*.

cutan A modification of the texture, structure, or fabric of a soil material (such as a soil aggregate, ped, skeleton grain) along a natural surface within it and caused by a concentration of a particular soil constituent. It can be composed of any of the component substances of the soil material. Examples: *argillan*; *sesquan*; *mangan*; *soluan*; *silan*; *skeletan*. Etymol: Latin *cutis*, "a coating, rind, or surface of a thing; skin".

cut and fill [eng] The excavating of material in one place and the depositing of it as fill in an adjacent place, as in the building of a road, canal, or embankment, or in stope mining.

cut and fill [geomorph] A process of leveling whereby material eroded from one place by waves, currents, streams, or winds is deposited nearby until the surfaces of erosion and deposition are continuous and uniformly graded; esp. lateral erosion on the concave banks of a meandering stream accompanied by deposition within its loops. Cf: *scour and fill*.

cut and fill [sed struc] A sedimentary structure consisting of a small erosional channel that is subsequently filled; a small-scale *washout*.

cutbank A local term in the western U.S. for a steep bare slope formed by lateral erosion of a stream; esp. *undercut slope*.

cuticle A layer of *cutin* covering the outer walls of a plant's epidermal cells.

cutin The waxy material of the *cuticle* covering external cell surfaces of vascular plants and some mosses (Scagel et al, 1965, p.614).

cutinite A variety of *exinite* consisting of plant cuticles. Cf: *sporinite; resinite; alginite*.

cutoff [eng] An impermeable wall, collar, or other structure placed beneath the base or within the abutments of a dam to prevent or reduce losses by seepage along otherwise smooth surfaces or through porous strata.

cutoff [stratig] A boundary, oriented normal to bedding planes, that marks the areal limit of a specific stratigraphic unit where the unit is not defined by erosion, pinch-out, faulting, or other obvious means. Cutoffs are applicable to map, cross-sectional, and three-dimensional views, and are in effect specialized facies boundaries. See also: *arbitrary cutoff*. Syn: *stratigraphic cutoff*.

cutoff [streams] (a) The new and relatively short channel formed when a stream cuts through a narrow strip of land and thereby shortens the length of its channel. See also: *neck cutoff; chute cutoff*. Syn: *meander cutoff; cutoff channel*. (b) A channel constructed to straighten a stream or to bypass large bends, thereby relieving an area normally subjected to flooding or channel erosion. See also: *pilot channel*. (c) The crescent-shaped body of water separated from the main stream by a cutoff. (d) The formation of a cutoff.

cutoff channel A meander *cutoff*.

cut-off filter A filter shaped to remove unwanted radiation, either above or low-pass, or below or high-pass a desired band of radiation. Cf: *filter; infrared filter*.

cutoff grade In economic geology, the lowest grade of mineralized material that qualifies as ore in a given deposit; ore of the lowest assay value that is included in an ore estimate. Cf: *economic limit*.

cutoff lake *oxbow lake*.

cutoff meander A meander that has been abandoned by its stream after the formation of a neck cutoff. See also: *oxbow*. Syn: *abandoned meander*.

cutoff spur The remnant of a meander spur, formed when a vigorously downcutting stream breaks through a narrow strip of land between adjacent curves in the stream course; it usually stands as an isolated hill. Syn: *meander core*.

cutout *horseback*.

cut plain A *stratum plain* of any hard rock that has been much dissected by erosion, but whose original surface is approximately represented by the summits of the least-eroded parts (Hill, 1900, p.7).

cut platform *wave-cut platform*.

cut point Point of intersection of a drill hole and a fault plane.

cutter A term used in Tennessee for solution crevices in limestone underlying residual phosphate deposits.

cut terrace (a) *wave-cut terrace*. (b) *rock terrace*.

cuttings Rock chips or fragments produced by drilling and brought to the surface. The term does not include the core recovered from core drilling. See also: *well cuttings; sludge; drillings*. Syn: *drill cuttings*.

cutty clay Plastic clay formerly used in England for making tobacco pipes; *pipe clay*.

cuvette [gem] An *intaglio* with a raised cameo-like figure in a polished depression. Cf: *chevee*.

cuvette [sed] A large-scale *basin* in which sedimentation has occurred or is taking place, as distinguished from a tectonic basin due to folding of preexisting rocks; e.g. the Anglo-Parisian cuvette of SE England and NE France in which Cenozoic rocks accumulated and were later folded into several distinct but smaller basins, such as the Paris Basin and the London Basin. Etymol: French, "small tub or vat". Sometimes misspelled *curvette*.

Cuvier's principle The theory that certain very different characteristics are commonly associated, e.g. kinds of feet and teeth among the vertebrates. This principle is named after the French naturalist Georges Cuvier (1769-1832).

cuyamite A teschenite composed of labradorite, analcime, haüyne, hornblende, augite, and magnetite. Its name is derived from Cuyamas Valley, California, U.S.A.

C wave *coupled wave*.

cwm Welsh term for *cirque*. The term is occasionally used in Wales for a narrow, deep valley of nonglacial origin in a mountain region (Stamp, 1961, p.142). Pron: *koom*. Pl: *cwms*. Syn: *coom; coomb; combe; comb*.

cyanite *kyanite*.

cyanochroite A blue mineral: $K_2Cu(SO_4)_2.6H_2O$.

cyanosite A syn. of *chalcanthite*. Also spelled: *cyanose*.

cyanotrichite A bright-blue or sky-blue mineral: $Cu_4Al_2(SO_4)(OH)_{12}.2H_2O$. Syn: *lettsomite; velvet copper ore*.

cyatholith *placolith*.

cyathosponge *archaeocyathid*.

cycad *cycadophyte*.

cycadophyte A gymnosperm having compound leaves consisting of simple leaflets arranged on both sides of a common stem, and naked seeds borne in simple cones. Such plants range from the Permian. Syn: *cycad*.

cycle [geol] (a) A series of events or changes that are normally, but not inevitably, considered to be recurrent and to return to a starting point, that are repeated in the same order several or many times at more or less irregular intervals, and that operate under conditions which at the end of the series are the same as they were at the beginning; e.g. the cycle of the seasons, a geochemical cycle, or a cycle of sedimentation. (b) An interval of time during which one sequence of a regularly recurring succession of events or phenomena runs to completion, each stage or event occurring in the same order in succeeding periods of the same or varying length, with the last stage or event being quite different from the first; e.g. a cycle of erosion, or an igneous cycle. (c) A group of rock units, the units of which occur in a certain order, with one unit being repeated frequently throughout the succession (Duff & Walton, 1962, p.239); esp. a *cyclothem*. ----Cf: *rhythm*. Adj: *cyclic*.

cycle [paleont] (a) A ring of segments in heterococcoliths. (b) A set of septa or tentacles of like age in a coral.

cycle of denudation *cycle of erosion*.

cycle of erosion (a) The complete, progressive, and systematic sequence of natural changes or stages in a landscape from the start of its erosion on a newly uplifted or exposed surface through its dissection into mountains and valleys until it has been reduced in the final stage to a low, featureless plain or to a base level (such as sea level) that limits the activity of the agents involved; according to some authors, a complete ycle is from base level back to base level. The cycle, usually ubdivided into youthful, mature, and old-age stages, is large-y hypothetical because it is normally interrupted before it runs o completion; the landforms produced and destroyed during he sequence are a function of climate, geography, and geoloic structure. The concept was first developed and formalized y Davis (1889b). See also: *normal cycle*. (b) The interval of ime during which the cycle of erosion is completed; the time nvolved in the reduction of a newly uplifted land area to a ase level.---Syn: *geomorphic cycle; geographic cycle; eroion cycle; cycle of denudation; physiographic cycle*.

ycle of fluctuation *phreatic cycle*.

ycle of sedimentation (a) A sequence of related processes nd conditions, repeated in the same order, that is recorded n a sedimentary deposit; e.g. the processes and conditions hat determine the ordered sequence of orthoquartzite, grayvacke, and arkose. (b) The deposition of sediments in a asin between the beginnings of two successive marine transressions, comprising the deposits formed initially on dry land, ollowed by shallow-water and then deep-water deposits that n turn gradually become of shallow-water and then dry-land ype during a marine regression. (c) A cyclothem.---Syn: *sedmentary cycle*.

ycleology A term used by Maxim K. Elias (1965, p.339) to esignate the detection and study of cycles in paleontologic nd geologic phenomena. The basic unit in cycleology is the hase.

yclic Adj. of *cycle*; recurrent rather than *secular*.

yclic evolution Evolution, supposed by some to have ocurred in many lineages, involving successively initial rapid nd vigorous expansion, a long stable or slowly changing hase, and a final short episode in which overspecialized, legenerate, or inadaptive forms led to extinction.

yclic salt Salt that is lifted from the sea as spray, is blown inand, and returns to the sea via drainage.

yclic sedimentation A syn. of *rhythmic sedimentation*. The erm is used esp. for sedimentation involving a circuitous seuence of conditions, such as found in a megacyclothem exibiting asymmetric bedding.

yclic terrace One of several stream terraces representing ormer valley floors formed during periods when downcutting ad essentially stopped for a time and lateral erosion had beome dominant; e.g. a *valley-plain terrace*. Terraces on opposite sides of the valley are paired or correspond in altitude long any given section of the valley. Cf: *noncyclic terrace*.

yclic twinning *Repeated twinning* of three or more individual rystals according to the same twin law but with the twin axes r twin planes not parallel. Cyclic twinning often results in hreefold, fourfold, fivefold, sixfold, or eightfold twins which, if qually developed, would display geometrical symmetry not ormed in single crystals. Cf: *polysynthetic twinning*.

yclocystoid Any small, discoid echinozoan belonging to the lass Cyclocystoidea, characterized by a theca composed of alcareous plates separable into central oral and aboral discs, ubmarginal ring, and marginal ring, by a flat aboral surface, nd by a multiple branching ambulacral system. Their stratigaphic range is Middle Ordovician to Middle Devonian.

yclodont Said of the dentition of bivalve mollusks characterzed by arched hinge teeth curving out from below the hinge nargin and by a small or absent hinge plate.

yclographic projection A term used in structural geology for he representation of planes on a stereogram by means of reat circles (Dennis, 1967, p.140).

yclohexane A colorless, liquid, saturated homocyclic hydrocarbon of the cycloparaffin series, formula C_6H_{12}, which has pungent odor and is found occurring naturally in petroleum. yn: *hexamethylene*.

yclolith Any elliptical or circular ring coccolith. The term hould be restricted to circular forms.

yclone An atmospheric low-pressure system that is circular or nearly so, around which the wind blows in a counterclockwise direction in the Northern Hemisphere and clockwise in the Southern. The term is usually restricted to the small, violent *tropical cyclone*. See also: *extratropical cyclone*. Cf: *anticyclone*.

cycloparaffin A saturated homocyclic hydrocarbon having the empirical formula CnH_{2N}. Examples are cyclopentane and cyclohexane, both of which are found in petroleum. Syn: *naphthene*.

cyclopean texture *mosaic texture*.

cyclosilicate A class or structural type of *silicate* characterized by the linkage of the SiO_4 tetrahedra in rings, with a ratio of $Si:O = 1:3$. An example of a cyclosilicate is beryl, $Be_3Al_2(Si_6O_{18})$. Cf: *nesosilicate; sorosilicate; inosilicate; phyllosilicate; tectosilicate*. Syn: *ring silicate*.

cyclostome Any ectoproct bryozoan belonging to the order Cyclostomata and characterized by calcareous tubular zooecia with circular apertures and no operculum. Adj: *cyclostomatous*.

cyclosystem A gastrozooid of a hydrozoan, surrounded by a circular row of five to seven individual dactylozooids.

cyclothem (a) A term proposed by Weller (in Wanless & Weller, 1932, p.1003) for a series of beds deposited during a single sedimentary cycle of the type that prevailed during the Pennsylvanian Period. It is an informal rock-stratigraphic unit equivalent to "formation". Because of extremely variable development, a cyclothem cannot be defined rigidly in terms of the members actually present at any locality (Weller, 1956, p.27-28). Cyclothems are typically associated with unstable shelf or interior basin conditions in which alternate marine transgressions and regressions occur; nonmarine sediments usually occur in the lower half of a cyclothem, marine sediments in the upper half. The term has also been applied to rocks of different ages and of different lithologies from the Pennsylvanian cyclothems. The cyclothem concept was developed by Weller (1930). See also: *ideal cyclothem; megacyclothem; rhythmite*. (b) A *cycle* applied to sedimentary rocks (Duff & Walton, 1962).

cylindrical fault A fault, the plane of which is curved and the displacement of which is rotational about an axis, parallel with the fault plane (Dennis, 1967).

cylindrical fold A fold model that can be described geometrically by the rotation of a line parallel to itself at a fixed distance from a central point. Cf: *conical fold*. Syn: *cylindroidal fold*.

cylindrical projection A projection on the surface of a cylinder; esp. any of numerous map projections of the Earth, produced by projecting the geographic meridians and parallels onto the surface of a cylinder that is tangent to, or intersects, the surface of the sphere, and then developing (unrolling and laying flat) the cylinder as a plane. The principal scale is preserved along the arc of a great circle, the meridians and parallels appear as straight lines perpendicular to each other (if the cylinder is tangent to the sphere at the equator), and all parallels are the same length as the equator. Example: Mercator projection.

cylindrical structure A vertical sedimentary structure with an irregular columnar or pillar-like shape; e.g. a *clastic pipe*.

cylindrite A blackish lead-gray mineral: $Pb_3Sn_4Sb_2S_{14}$. It occurs in cylindrical forms that separate under pressure into distinct shells or folia.

cylindroidal fold *cylindrical fold*.

cymatogeny Undulating movement or warping of the Earth's crust to produce regional, linear arching or doming, but with minimal deformation. The concept was introduced by Lester King (1959, p.117) and according to it, most mountain ranges are cymatogenic rather than orogenic. Cf: *orogeny; epeirogeny*.

cymoid adj. In economic geology, said of a vein that in cross section forms a reverse curve; a vein that swerves from its course, then returns. A pair of veins in such a pattern forms a cymoid loop.

cymophane A syn. of *chrysoberyl*. The name is applied esp. to chrysoberyl exhibiting a girasol or chatoyant effect, and more specif. to *cat's-eye* only.

cymrite A zeolite mineral: $Ba_2Al_5Si_5O_{19}(OH).3H_2O$.

cyphonautes Pelagic bivalved larva of some bryozoans (such as in Membranipora).

cyprine A light-blue variety of vesuvianite containing a trace of copper.

cyrenoid Obsolete syn. of *corbiculoid*.

cyrilovite A brown mineral: $NaFe_3(PO_4)_2(OH)_4.2H_2O$.

cyrstal-body playa A playa with one or more thick salt bodies at or near the surface, formed by the evaporation of a Pleistocene lake that once occupied the area (Stone, 1967, p.220); e.g. Searles Lake in California.

cyrtochoanitic Said of a comparatively short, retrochoanitic septal neck of a nautiloid, curved so as to be concave outward (TIP, 1964, pt.K, p.55).

cyrtocone A curved, slender cephalopod conch (like that of *Cyrtoceras*) that completes less than one whorl. Syn: *cyrtoceracone*.

cyrtoconoid (a) Said of a gastropod shell in which the spire approaches a conical shape but has convex sides. Cf: *coeloconoid*. (b) Resembling a cyrtocone.

cyrtolite An altered variety of zircon containing uranium, beryllium, and rare earths.

cyrtolith A basket- or calotte-shaped heterococcolith with laths and a projecting central structure.

cyst [paleont] A sac or capsule that is secreted by many protozoans and other minute animals as a prelude to a resting or a specialized reproductive phase and that envelops protoplasm and protects it from adverse environmental conditions.

cyst [palyn] A microscopic *resting spore* with a resistant wall formed in dinoflagellates and many algae (such as blue-green algae and desmids) by the breaking up of parts of the filaments or by the enclosing of a cell group and their investment by a sheath or envelope. Dinoflagellate cysts exist abundantly as fossils. See also: *statospore*.

cystid (a) The combined cellular and skeletal layers of the wall of a bryozoan zooid. (b) A *cystoid*.

cystiphragm A blister-like calcareous partition wholly or partly encircling the zooecial wall in a bryozoan (chiefly in trepostomes), from which it extends part way across the zooecial tube in a downcurving manner until rejoining the zooecial wall. Cystiphragms are commonly arranged in vertical, overlapping series.

cystocarp In the red algae, the structure comprising the sex structure and the surrounding pericarp.

cystoid Any crinozoan belonging to the class Cystoidea, characterized by diplopores, pore rhombs, and brachioles. Their stratigraphic range is Lower Ordovician to Upper Devonian. Var: *cystid*.

D

dachiardite A white to colorless zeolite mineral: $(Ca,Na_2, K_2)_5Al_{10}Si_{38}O_{96}.25H_2O$.

Dacian European stage: Pliocene.

dacite A fine-grained extrusive rock with the same general composition as *andesite* but having a less calcic feldspar. Syn: *quartz andesite*.

dacitoid An extrusive rock with the same chemical composition as a dacite but lacking modal quartz.

dactylite An igneous rock characterized by *dactylitic* texture.

dactylitic A term applied to a rock texture produced by a *symplectic* intergrowth in which one mineral is penetrated by finger-like projections from another mineral; also, said of a rock exhibiting such texture, e.g. a *dactylite*; also, said of the intergrowth itself, i.e. a *dactylotype intergrowth*.

dactylopod The *dactylus* of a malacostracan crustacean. Syn: *dactylopodite*.

dactylopore A relatively small or narrow tubular cavity of certain hydrozoans occupied by a dactylozooid. Cf: *gastropore*.

dactylotype intergrowth A symplectic intergrowth in which finger-like projections of one mineral penetrate the second mineral. See also: *dactylitic*.

dactylous (a) Pertaining to a dactylus of a crustacean. (b) Said of a pedicellaria of an echinoid having spoon-shaped jaws mounted on individual stalks.

dactylozooid An elongate, slender, mouthless polyp housed in a dactylopore of a hydrozoan, equipped with numerous stinging cells, and performing protective, food-capturing, and tactile functions for the colony. Cf: *gastrozooid*.

dactylus The seventh and most distal segment of the pereiopod of a malacostracan crustacean. Pl: *dactyli*. Syn: *dactyl; dactylopod*.

dagala A syn. of *steptoe*, used in the region of Mt. Etna, Italy.

dahamite An albite-rich paisanite.

dahllite A resinous, yellowish-white *carbonate-apatite* mineral in association sometimes occurring as concretionary spherulites. Incorrectly spelled "dahlite" by Pettijohn (1957, p.472).

daily variation Oscillation of the Earth's magnetic field with a period of one day. It has an amplitude of about 20 gamma and is most rapid near local noon. A small lunar daily variation is superposed on the solar daily variation. Syn: *diurnal variation*.

daily wave *diurnal wave*.

dakeite *schroeckingerite*.

dal A Swedish term for a *valley*.

dale (a) A term used in northern England and southern Scotland for a broad, open, river valley situated between enclosing hills or running through high land. (b) A poetic var. of *valley*; *vale* or a small valley.---Cf: *dell*.

dalles (a) The rapids in a deep, narrow stream confined between the rock walls of a canyon or gorge; e.g. The Dalles of the Columbia River where it flows over columnar basalt. (b) A steep-sided part of a stream channel, near the dalles proper, marked by clefts, ravines, or gorges; e.g. along the Wisconsin River, Wisc.---Etymol: French plural of *dalle*, "gutter". Pron: *dalz*. Syn: *dells*.

dallinid Any articulate brachiopod belonging to the family Dallinidae, characterized by loop development passing through precampagiform, campagiform, frenuliform, terebrataliiform, and dalliniform, or variations of these.

dalliniform Said of the loop, or of the growth stage in the development of the loop, of a dallinid brachiopod (as in *Dallina*), consisting of long descending branches recurved into ascending branches that meet in transverse band, all free of valve floor (TIP, 1965, pt.H, p.143). The dalliniform loop is morphologically similar to the *magellaniform* and cryptonelliform loops.

dallol A term applied in Nigeria to a very broad, flat-bottomed, steep-sided, dry valley, usually many kilometers wide; specif. a part of an old drainage system on the left bank of the Niger River.

Dalmatian coastline A *concordant coastline* developed where drowning or a rise of sea level has produced lines of narrow islands (representing the outer mountain ranges) separated by long inlets or straits (representing parallel valleys) lying peripheral, and trending roughly parallel, to the coast. Type region: the eastern coastline of the Adriatic Sea in the region of Dalmatia, Yugoslavia.

Dalradian A division of the Precambrian. Syn: *Grampian*.

dalyite A mineral: $K_2ZrSi_6O_{15}$.

dam [eng] n. (a) An artificial barrier or wall constructed across a watercourse or valley for one or more of the following purposes: creating a reservoir (pond or lake) for storage of water; diverting water from a watercourse into a conduit or channel; creating a hydraulic head that can be used to generate power; improving river navigability; controlling floods; or retention of debris. It may be constructed of wood, earth materials, rocks, or solid masonry. (b) A body of standing water confined or held by a dam.----v. To provide a dam; to obstruct or confine a flow of water.

dam [geomorph] An obstruction formed by a natural agent (such as a glacier or a lava flow, or the activity of animals such as beavers) across a stream so as to produce a lake or pond; e.g. an *ice dam*.

damascened Said of the interwoven texture, observed in some volcanic glasses, that resembles the markings on Damascus sword blades.

dambo A term used in central Africa (esp. Zambia) for a small, ill-defined flood plain or drainage channel that is extremely flat with broad, grassy clearings, swampy during the wet season but dry for the greater part of the year. Etymol: Bantu.

dam gradation *contragradation*.

damkjernite A hypabyssal rock composed of biotite and pyroxene phenocrysts in a fine-grained groundmass of pyroxene, biotite, and magnetite, with interstitial nepheline, microcline, and calcite, possibly primary.

damourite A variety of muscovite that gives off water more readily and that has less elastic folia and a more pearly or silky luster. It is unctuous or talc-like to the touch. Syn: *talcite*.

damp A general term for coal-mine gases.

damped-wave structure A series of ring-shaped uplifts and depressions of rapidly diminishing amplitude, surrounding the central uplift of a cryptoexplosion structure (Dietz, 1959, p.496-497).

damping A term in seismology for the resistance to vibration that causes a decay of motion with time or distance, e.g. the diminishing amplitude of an oscillation. Damping may be produced by air, as in inductive geophones; by electromagnetism, as in induction seismometers; and by oil. See also: *critical damping; dynamic damping; viscous damping*.

damping capacity The ability of a material to dissipate strain within itself.

damping constant In a damped seismograph, one half the ratio of the damping resistance (force per unit velocity) to the moving mass. It has the dimension of a frequency.

damping factor The ratio of the damped and the undamped frequency of a seismograph or seismometer.

damping ratio The ratio of two equiphase peak amplitudes within one period of a damped seismograph or seismometer. The ratio is always greater than unity since the greater amplitude is divided by the succeeding amplitude.

damsite A site for a dam.

danaite A variety of arsenopyrite containing cobalt.

danalite A mineral: $(Fe,Mn,Zn)_4Be_3(SiO_4)_3S$. It is the iron end member isomorphous with helvite and genthelvite.

danburite An orange-yellow, yellowish-brown, grayish, or colorless transparent to translucent orthorhombic mineral: $CaB_2(SiO_4)_2$. It resembles topaz in crystal habit, physical properties, and appearance, and is used as an ornamental stone.

dancalite An extrusive rock containing phenocrysts of oligoclase, augite, and rare amphibole in a trachytic groundmass composed of plagioclase laths with interstitial analcime; an analcime trachyandesite.

Danian European stage: lowermost Paleocene (above Maestrichtian of Cretaceous, below Montian). It is regarded by some authors as uppermost Cretaceous.

dannemorite A yellowish-brown to greenish-gray monoclinic mineral of the amphibole group: $(Fe,Mn,Mg)_7Si_8O_{22}(OH)_2$. It is a manganiferous cummingtonite.

dans A broad, shallow valley in South Africa. Etymol: Afrikaans.

d'ansite A mineral: $Na_{21}Mg(SO_4)_{10}Cl_3$.

daphnite A mineral of the chlorite group: $(Mg,Fe)_3(Fe,Al)_3(Si,Al)_4O_{10}(OH)_8$.

darapskite A mineral: $Na_3(NO_3)(SO_4) \cdot H_2O$.

darcy A standard unit of permeability, equivalent to the passage of one cubic centimeter of fluid of one centipoise viscosity flowing in one second under a pressure differential of one atmosphere through a porous medium having an area of cross-section of one square centimeter and a length of one centimeter. A millidarcy is one one-thousandth of a darcy.

Darcy's law A derived formula for the flow of fluids on the assumption that the flow is laminar and that inertia can be neglected. The numerical formulation of this law is used generally in studies of gas, oil, and water production from underground formations. For example, in gas flow, the velocity of the flow is proportional to the pressure gradient multiplied by the ratio of permeability times density, divided by the viscosity of the gas.

Darcy-Weisbach formula A formula for determining the loss of head in flowing water: loss of head is equal to a coefficient that varies with the surface roughness of the conduit and the Reynolds number, times the length of the conduit divided by its diameter, times the velocity head of the flowing fluid. In the case of a noncircular conduit or a circular conduit not flowing full, four times the hydraulic radius is substituted for the diameter (ASCE, 1962).

dark-colored Said of a rock-forming mineral having a dark color as viewed megascopically, but being transparent in thin section; also, said of the rock such minerals form; *melanocratic*. Cf: *light-colored; mafic*.

dark halo crater A small lunar crater surrounded by material with a lower albedo than that of the adjacent terrain.

dark red silver ore *pyrargyrite*.

dark ruby silver *pyrargyrite*.

Darwin glass A highly siliceous, vesicular, frothy glass found in the Mt. Darwin range in western Tasmania (near Queenstown) in the form of blobs, drops, and twisted shreds. It is probably an impactite, although believed by some to be a tektite or a glassy meteorite. Syn: *queenstownite*.

Darwinism The theory that evolution resulted from variation and the selection of favored individuals through natural selection. This doctrine is named after the English naturalist Charles Darwin (1809-1882).

dashkesanite A monoclinic mineral of the amphibole group: $(Na,K)Ca_2(Fe,Mg)_5(Si,Al)_8O_{22}Cl_2$. It contains a high content (7.2%) of chlorine.

dasycladacean Pertaining to a type of green algae (type genus *Dasycladus*) whose filaments are whorled about a central axis which is often encrusted with lime.

date To assign a specific or approximate position on the geologic time scale for past geologic events.

date line A longitudinal line on the Earth's surface which closely follows the 180° meridian from Greenwich, England, and is taken as the point at which the calendar day begins. Syn: *international date line*.

dating Age determination of naturally occurring substances o relics by any of a variety of methods based on the amount o change, happening at a constant rate and which can be mea sured, in a component. The changes can be chemical or ei ther induced or spontaneous nuclear processes and take place over a period of time.

datolite A greenish monoclinic mineral: $CaBSiO_4(OH)$. It com monly occurs in cracks and cavities in diabase or basalt; it i sometimes used as a minor gem. Syn: *humboldtite; dystom spar*.

datum [geol] (a) The top or bottom of a bed of rock on whic structure contours are drawn. (b) *datum horizon*. ----Pl: da tums.

datum [geodesy] The latitude and longitude of an initial poin the azimuth of a line from this point, the *deflection of the ve tical*, the *geoidal separation* (frequently these quantities rela ing to the geoid are assumed to be zero for lack of mor complete information), and the two constants necessary t define the reference ellipsoid on which horizontal control su veys are to be computed. See also: *geodetic coordinate* Syn: *geodetic datum*.

datum [photo] A direction, level, or position from which ar gles, heights, depths, speeds, or distances are conventionall measured; e.g. in photographic mapping, the assumed hor zontal plane on which the map is constructed.

datum [surv] Any numerical or geometric quantity or valu that serves as a base or reference for other quantities o values; any fixed or assumed position or element (such as point, line, or surface) in relation to which others are dete mined, such as a *level surface* to which depths or heights a referred in leveling. Pl: *datums*; the plural "data" is used for group of statistical or inclusive references, such as "geo graphic data" for a list of latitudes and longitudes. See als *datum plane; mean sea level*.

datum horizon A geologic horizon used as a reference plan for the position of rock strata or for the comparative measure ment of the thickness of strata; the *key horizon* or bed o which elevations are taken or to which all elevations are fina ly referred in making a structure-contour map. Syn: *datum structural datum; contour horizon*.

datum level Any level surface (such as mean sea level) take as a surface of reference from which elevations are reckone a *datum plane*.

datum line A *reference line* representing a horizontal line fro which heights and depths are reckoned.

datum plane [seism] An arbitrary reference surface, used seismic mapping to minimize or eliminate local topographi effects, to which seismic times and velocity determination are referred.

datum plane [surv] A permanently established horizonta plane, surface, or level to which soundings, ground elevations water-surface elevations, and tidal data are referred; e.g mean sea level is a common datum plane used in topographi mapping. See also: *tidal datum; chart datum; sounding datum* Syn: *datum level; reference level; reference plane*.

datum point An assumed or fixed point used as a referenc from which calculations or measurements may be taken.

daubrée A unit of intensity of wear of a sedimentary particl equivalent to the removal of 0.1 gram from a 100-gra sphere of quartz (Wentworth, 1931, p.25). Named in honor Gabriel Auguste Daubrée (1814-1896), French mining eng neer and mineralogist, who showed experimentally (Daubré 1879) that quartz grains lost only one part in 10,000 per kilo meter of travel.

daubreeite A yellowish mineral: $BiO(OH,Cl)$. It is isomorphou with bismoclite. Syn: *daubréeite; daubreite*.

daubreelite A black meteorite mineral: $FeCr_2S_4$. Also spelle *daubréelite*.

daughter Any one of the intermediate members of nuclides of a *radioactive series*, between the *parent* and the *end product*.

daunialite A sedimentary rock consisting of a siliceous montmorillonitic clay, as distinct from bentonite of volcanic origin. It contains 25% organic silica (opal, chalcedony, quartz) and small amounts of sericite, chlorite, and kaolinite.

Dauphiné twin law A twin law in quartz in which two right-handed or two left-handed individuals form an interpenetration twin by a 180° rotation about the *c* crystal axis. Syn: *electrical twinning*.

davainite An ultramafic rock containing brown hornblende, secondary after pyroxene, some orthopyroxene, and accessory plagioclase. According to Johannsen (1939, p.248), it may be metamorphic.

davidite A dark-brown to brownish-black, uraniferous, iron-titanate mineral: $A_6B_{15}(O,OH)_{36}$, where $A = Fe^{+2}$, rare earths, U, Ca, Zr, & Th, and $B = Ti$, Fe^{+3}, V, & Cr. Ideal end member is $FeTi_3O_7$. Davidite is a primary mineral in high-temperature hydrothermal lodes; it occurs in pegmatites and basic igneous rocks, and in all stages of intergrowth and exsolution with ilmenite and hematite.

davidsonite A greenish or greenish-yellow variety of beryl.

Davisian Pertaining to the "American" school of geomorphology based on the teachings and writings of William Morris Davis (1850-1934), Harvard geologist and geographer; esp. said of the concepts of peneplanation and of the cycle of erosion, and of the genetic method of landform description whereby differences are largely explainable in terms of geologic structure, geomorphic process, and stages of development.

davisonite A white mineral: $Ca_3Al(PO_4)_2(OH)_3.H_2O$.

davyne A chlorine-bearing feldspathoid mineral of the cancrinite group: $(Na,Ca,K)_8(Al_6Si_6O_{24})(Cl,SO_4,CO_3)_{2-3}$.

dawsonite A white bladed mineral: $NaAl(CO_3)(OH)_2$.

daylighting In engineering geology, the exposure of strata by a cut whose angle is steeper than that of the dip of the beds. Such exposure increases the likelihood of landsliding, if such a tendency already exists.

D-coal Microscopic coal particles that are predominantly durain, as found in miners' lungs. Cf: *F-coal; V-coal*.

dead (a) In economic geology, said of an economically valueless area, in contrast to a *quick* area or ore; barren ground. (b) In mining, said of an area of subsidence that is thought to be completely settled and will not move again; dead ground.

dead cave A cave in which there is no longer any moisture or growth of mineral deposits that is associated with the presence of moisture. Cf: *live cave*. Syn: *dry cave*.

dead chert chalky chert.

dead cliff A former sea cliff, ordinarily elevated and no longer subject to alteration by marine processes.

dead coral reef A coral reef or a part of a coral reef with no living corals.

dead fault A fault along which movement has ceased. Cf: *active fault*.

dead glacier A glacier that has ceased to flow and is without an accumulation area or is no longer receiving material from the accumulation area. It may continue to spread or creep downhill due to its bulk and topographic setting. Ant: *active glacier*. Cf: *dead ice*. Syn: *stagnant glacier*.

dead ground Land unseen by an observer because of intervening hills or ridges, etc.; it is a term used mainly in surveying.

dead ice (a) Ice that is not flowing forward and is not receiving material from an accumulation area; the ice of a *dead glacier*. (b) Detached blocks of ice left behind by a retreating glacier, usually buried in moraine and melting very slowly without the production of large quantities of water.----See also: *fossil ice*. Syn: *stagnant ice*.

dead lake An *extinct lake*; esp. one that has become filled with vegetation.

dead line The level above which a batholith is metalliferous and below which it is economically barren. It is exposed during the epibatholithic stage of batholith erosion (Emmons, 1933).

dead sea (a) A body of water from which evaporites (such as rock salt or gypsum) have been precipitated. Type locality: Dead Sea in the Near East. (b) A lake whose surface is calm.

dead trace A trace on a seismic record that is devoid of information or noise due to instrument failure.

dead valley *dry valley*.

dead water In a stream, water that is or appears to be standing or still.

death assemblage *thanatocoenosis*.

debacle (a) A *breakup* on a river, esp. on the great rivers of the U.S.S.R. and of North America. (b) The rush of water, broken ice, and debris in a stream immediately following a breakup. Syn: *ice run*. (c) Any sudden, violent, destructive flood, deluge, or rush of water that breaks down opposing barriers and sweeps before it debris of all kinds.---Etymol: French débâcle.

debitumenization *devolatilization*.

debouchment (a) The issuing forth of a stream, as the *debouchment* of a tributary into the main stream. (b) The mouth of a river or channel. Syn: *debouchure*.---Etymol: French *débouchement*.

debouchure (a) *debouchment*. (b) The place where an underground stream reaches the surface; the opening from which a spring issues. See also: *resurgence*. (c) The point in a cave where a tubular passage connects with a larger passage or chamber.

debris [astron] Interplanetary material, ranging in size from particles less than one micron in diameter to chunks many kilometers across, including asteroids, comets, meteors, meteorites, and cosmic dust.

debris [glaciol] The rocks, earth, and other material lying on the surface, or incorporated in the body, of a glacier, or pushed ahead of the glacier front. Syn: *glacial debris*.

debris [geol] Any surficial accumulation of loose material detached from rock masses by chemical and mechanical means, as by decay and disintegration, and occurring in the place where it was formed, or transported by water or ice and redeposited. It consists of rock fragments, earth, soil material, and sometimes organic matter. The term is often used synonymously with *detritus*, although "debris" has a broader connotation. Etymol: French *débris*. Pl: *debris*. Syn: *rock waste*.

debris apron *bajada*.

debris avalanche The very rapid and usually sudden sliding and flowage of masses of initially incoherent, unsorted mixtures of soil and rock material.

debris cone [geomorph] (a) *alluvial cone*. (b) A conical mound of tightly packed, fine-grained debris piled at the angle of repose atop certain boulders moved by a landslide.

debris cone [glaciol] A cone or mound of ice or snow on a glacier, covered with a veneer of debris thick enough to protect the underlying material from the ablation that has lowered the surrounding surface. Cf: *dirt cone; sand cone*. Syn: *glacier cone*.

debris dam A mass of coarse alluvium deposited at the mouth of a tributary stream, commonly during a flash flood, and forming an obstruction in the main valley.

debris fall The relatively free falling or rapid movement of predominantly unconsolidated or weathered material from a vertical or overhanging cliff; a landslide that is esp. common along the undercut banks of streams. Syn: *soilfall*.

debris flood A disastrous flood intermediate between the turbid flood of a mountain stream and a true mudflow, as that which occurs in Los Angeles, Calif. (Strahler, 1963, p.465-466).

debris flow A mass movement involving rapid flowage of debris of various kinds under various conditions; specif. a high-density *mudflow* containing abundant coarse-grained materials and resulting almost invariably from an unusually heavy rain.

debris ice (a) Sea ice containing soil, mud, stones, shells, and other material. (b) *brash ice.*

debris island A *sorted circle* having a diameter of about 1 m and consisting of an isolated patch of fine-textured, compact material surrounded by blocks of frost-shattered boulders; term introduced by Washburn (1956, p.827). Syn: *earth island; rubble island.*

debris line A *swash mark* composed of debris washed up on a beach by storm waves. Cf: *trash line.*

debris plain A plain covered with rock waste.

debris slide A landslide involving a slow-to-rapid downslope movement of comparatively dry and predominantly unconsolidated and incoherent earth, soil, and debris ·in which the mass does not show backward rotation (as in a slump) but slides or rolls forward, forming an irregular hummocky deposit resembling a moraine (Sharpe, 1938, p.74). It is often called an *earth slide* but this is incorrect because the moving mass of a debris slide is greatly deformed or consists of many small units.

debris slope A *constant slope* with a surface of talus; a *talus slope.*

decalcified Said of a soil which has been *leached* of its calcium carbonate.

decapod (a) Any eumalacostracan belonging to the order Decapoda, characterized by the presence of five pairs of uniramous limbs behind the maxillipeds. Their stratigraphic range is Permian to present. (b) Any early name for a *coleoid*, but now discontinued owing to its prior use for an order of crustaceans.

decay [organic] Decomposition by oxidation; chemical weathering in the presence of moisture and oxygen.

decay [waves] The attenuation or loss of energy from wind-generated ocean waves after they leave the generating area and pass into a region of lighter winds; it is accompanied by a gradual increase in wavelength and a gradual decrease in wave height.

decay [radioactivity] *radioactive decay.*

decay [weath] The general weathering or wasting away of rock; specif. *chemical weathering.*

decay constant [elect] The time τ for an exponentially decaying quantity to decrease to $1/e$ of its initial value (e is the base of the Naperian logarithm).

decay constant [radioactivity] In radioactive decay, the constant ratio between the number of decaying atoms per unit of time and the existing number of atoms. Symbol: l. Syn: *disintegration constant; radioactive constant.*

Decca A brand name for an electronic positioning system for ships at sea, using a technique of comparison of low-frequency, continuous wave phases from two synchronized stations.

Deccan basalt A fine-grained, nonporphyritic, tholeiitic lava covering an area of about 200,000 square miles in the Deccan region of southeast India and consisting essentially of labradorite, clinopyroxene, and iron ore. Olivine is generally absent, or is present in minor amount, usually near the bottom of flows. It corresponds to the plateau basalt of the Pacific Northwest of the U.S.A and the Thulean province of western Scotland, northeast Ireland, and Iceland.

decementation The dissolving out or leaching of the cement of a sedimentary rock, as in a sandstone whose void-filling fluids and solid grains do not form a closed system, thereby permitting fluids to move (or ions to diffuse) in and out (Pettijohn, 1957, p.659); e.g. the removal of carbonates from a calcareous sandstone.

deciduous (a) Said of a plant that loses its leaves annually. (b) Said of plant parts which are lost within a year of their production.

decke The German equivalent of *nappe*, sometimes used in the English-language literature.

deckenkarren A general term for the various solutional features of a cave ceiling (Monroe, 1970). Etymol: German.

declination The horizontal angle in any given location between true north and magnetic north; it is one of the *magnetic elements*. Syn: *magnetic variation.*

declinator *declinometer.*

declined Said of a graptoloid rhabdosome with stipes hanging below the sicula and enclosing an angle less than 180 degrees between their ventral sides. Cf: *deflexed; reclined; reflexed.*

declining development The production of a landscape where the rate of downward erosion is more rapid than the rate of uplift or where valley widening exceeds valley deepening, characterized by a decrease of the relative relief and the formation of concave slopes. Cf: *accelerated development; uniform development.* Syn: *waning development; descending development.*

declinometer An instrument that measures magnetic declination. Syn: *declinator.*

declivity (a) A slope (as of a hill) that descends from a point of reference; esp. a steep or overhanging slope (as of a cliff). Ant: *acclivity.* (b) A gradient of a surface; a deviation downward from the horizontal; an inclination.

décollement Detachment structure of strata due to deformation, resulting in independent styles of deformation in the rocks above and below. It is associated with folding and with overthrusting, but it is merely a descriptive term. Etymol: French, "unsticking, detachment". Cf: *tectonic denudation; disharmonic folding.* See also: *bedding-plane slip.* Syn: *detachment.* Obsolete syn: *strip thrust.*

décollement fault *sole fault.*

décollement fold A fold, the strata of which are independent of the basement due to décollement. Syn: *superficial fold; Jura-type fold.*

decomposition A syn. of *chemical weathering*, or the breaking down of rocks and minerals by chemical processes; e.g. *humic decomposition.* Usually complex compounds are broken down into simpler ones that are more stable under the prevailing conditions at or near the Earth's surface. Cf: *disintegration.*

decorative stone A stone used for architectural decoration, as in mantles, columns, and store fronts. It is sometimes set in silver or in gold-filled jewelry, but usually as curio stones.

decrement *ground-water discharge.*

decrepitation The breaking up of a mineral, usually violently and noisily, when heated.

decurrent Said of a leaf that extends downward and adnate to the stem.

decussate structure A microstructure in thermally metamorphosed rocks in which axes of contiguous crystals lie in diverse, criss-cross directions which are not random but rather are part of a definite mechanical expedient for minimizing internal stress. It is most noticeable in rocks composed largely of minerals with a flaky or columnar habit (Harker, 1939).

dedolomitization A process whereby, presumably during contact metamorphism at low pressure, part or all of the magnesium in a dolomite or dolomitic limestone is used for the formation of magnesium oxides, hydroxides, and silicates (e.g. brucite, forsterite) resulting in the enrichment in calcite (Teall, 1903). The term was originally used by Morlot (1847) for the replacement of dolomite by calcite during diagenesis or chemical weathering.

deduction Reasoning from the general to the particular; inferring consequences from evidence; deriving applications from general principles. "It is a mental process not always given its proper priority in considering geological questions. An explanation or hypothesis in accord with deduction must be preferred to any other" (Challinor, 1967, p.68). Ant: *induction.*

deep [**oceanog**] n. A clearly discernible depression of the ocean floor. Syn: *abyss*.

deep coal Coal that is far enough below the surface to require underground mining methods. Cf: *crop coal*.

deep earthquake *deep-focus earthquake*.

deep-focus earthquake An earthquake whose focus is at a depth of 300-700 km, or deeper than 300 km. Cf: *shallow-focus earthquake; intermediate-focus earthquake*. Syn: *deep earthquake*.

deep lead A *lead* or alluvial placer that is buried under soil or rock.

deep percolation Precipitation moving downward, below the root zone, toward storage in subsurface strata. Cf: *shallow percolation*.

deep scattering layer A stratified area of marine organisms in the open ocean that scatter sound waves from an echo sounder. Cf: *shallow scattering layer; surface scattering layer*. Syn: *false bottom; phantom bottom*.

deep-sea channel A trough-shaped, low-relief valley on the deep-sea floor beyond the continental rise. It has few tributaries, and may be either parallel or at an angle to the continental margins. Syn: *mid-ocean canyon*.

deep-sea fan *submarine fan*.

deep-sea sediments *pelagic deposits*.

deep-seated Said of geologic features and processes that originate or are situated at depths of about one kilometer or more below the Earth's surface; *plutonic*.

deep soil (a) Generally, a soil deeper than 40 inches. (b) Also, a soil with a deep black surface layer.

deep water A very dense *water mass* that is formed along the Antarctic by cooling, then sinking and spreading at great depth. Cf: *intermediate water; bottom water; surface water* [*oceanog*].

deep-water wave A wave on the surface of a body of water, the wave length of which is less than twice the depth of the water, and for which the water depth is not an influence on the velocity or on the shape of the orbital. Cf: *shallow-water wave; transitional-water wave*. Syn: *short wave; surface wave*.

deep well (a) A water well, generally drilled, extending to a depth greater than that characteristic of shallow wells, which are generally dug. The term may be applied to a well 50 ft deep in an area where shallow wells average 20 or 25 ft deep, or to a much deeper well in an area where the shallowest aquifer supplies wells as much as a few hundred feet deep. (b) A well whose water level is deep enough to prevent use of a shallow-well (suction) pump.----Cf: *shallow well*.

deerite A mineral: $(Fe,Mn)_{13}(Fe,Al)_7Si_{13}O_{44}(OH)_{11}$.

Deerparkian North American stage: Lower Devonian (above Helderbergian, below Onesquethawan).

defeated stream A stream that, owing to uplift or other cause, is unable to degrade as fast as the land rises and thereby fails to maintain its original course, becomes ponded and diverted into a new course, and resumes as a consequent stream.

defect lattice *Schottky defect*.

defect-lattice solid solution *omission solid solution*.

deferred junction The place of meeting on a flood plain of a main stream and a tributary whose course (parallel to the main stream) is prolonged downstream for a considerable distance by a barrier along the main stream; esp. the junction of a *yazoo stream* with the main stream, as along the convex side of a major meander. Syn: *yazoo; delayed junction; deferred tributary junction*.

deferred tributary *yazoo stream*.

defile A very long, narrow, steep-sided pass or passage through hills or mountains, often forming the approach to a larger pass; esp. a passage enclosed between high, precipitous walls, as a gorge.

definition In photogrammetry, the degree of clarity and sharpness of an image in a photograph; also, the ability of a lens to record fine detail.

deflation The sorting out, lifting, and removal of loose, dry, fine-grained particles (clay and silt sizes) by the turbulent eddy action of the wind, as along a sand-dune coast or in a desert; a form of *wind erosion*.

deflation armor A *desert armor* whose surface layer of coarse particles is concentrated chiefly by deflation.

deflation basin A topographic basin excavated and maintained by wind erosion which removes unconsolidated material and commonly leaves a rim of resistant rock surrounding the depression. See also: *blowout*. Syn: *wind-scoured basin*.

deflation hollow *blowout*.

deflation lake A lake occupying a basin formed mainly by wind erosion, esp. in an arid or semiarid region; it is usually very shallow and may contain water only during certain seasons. See also: *dune lake*.

deflation residue *desert pavement*.

deflation ripple *granule ripple*.

deflection [**drill**] An intentional or accidental change in the intended course of a borehole, produced by various rock conditions (dipping strata, fault planes, etc.) encountered in the hole or by the operational characteristics (imperfections) of the drilling equipment that prevent the drilling of a completely straight hole; e.g. deflection by the use of a whipstock. Syn: *deviation*.

deflection [**mtns**] A sudden change in the trend of one or more branches of a mountain range, as shown in plan view. See also: *capped deflection; fractured deflection*. Cf: *linkage*.

deflection [**geodesy**] *deflection of the vertical*.

deflection [**streams**] A relatively spontaneous diversion of a stream, as by warping, alluviation, glaciation, lateral corrasion, volcanic action, or along a shore by the shifting downcurrent of an inlet at the stream mouth.

deflection angle [**drill**] The angular change in the course of a deflected borehole.

deflection angle [**photo**] A vertical angle, measured in the vertical plane containing the flight line, by which the datum of any model in a stereotriangulated strip departs from the datum of the preceding model.

deflection angle [**surv**] A horizontal angle measured from the forward prolongation of the preceding line to the following line; the angle between one survey line and the extension of another survey line that meets it. A deflection angle to the right is positive; one to the left is negative.

deflection basin A basin hollowed out by the erosive action of ice in front of a barrier obstructing the path of a glacier (Geikie, 1898, p.297).

deflection of the vertical A term used in geodesy to describe the angle at a given point on the Earth between the vertical, defined by gravity, and the direction of the normal to the reference spheroid through that point. It is sometimes referred to as *deviation of the vertical* or deflection of the plumb line, but is most commonly called the *deflection*. Cf: *astronomic; deviation; datum*. Syn: *station error*.

deflection pool A pool occupying a depression scooped out by a stream in its obstructing bed at a curve in its course (Miller, 1883, p.275).

deflexed Said of a graptoloid rhabdosome with initial part of stipes hanging below the sicula and enclosing an angle less than 180 degrees between their ventral sides and distal parts of stipes tending to the horizontal. Cf: *declined; reflexed; reclined*.

defluent A stream that flows down, as from a lake or glacier.

deformation A general term for the process of folding, faulting, shearing, compression, or extension of the rocks as a result of various Earth forces; a diastrophic phase.

deformation band *slipband*.

deformation ellipsoid *strain ellipsoid*.

deformation fabric Rock fabric or that fabric element or elements resulting from deformation, e.g. metamorphism; the fabric of a *secondary tectonite*. Cf: *depositional fabric*. Syn:

tectonic fabric.

deformation lamella A type of *slipband* that is produced esp. in quartz by active slip within a mineral grain during tectonic deformation; also, a similar feature produced by shock.

deformation path *deformation plan.*

deformation pattern *deformation plan.*

deformation plan A synthesis of the sequence of deformation of a rock. It is the result of analysis of the rock's fabric. Syn: *deformation path; deformation pattern; movement picture; movement plan.*

deformation plane In structural petrology, the plane that is perpendicular to the *flow surface* and parallel to the direction of movement. Syn: *displacement plane.*

deformation twin A crystal twin that is produced by gliding, i.e. deformation within a preexisting crystal. Syn: *glide twin; mechanical twin.*

deformation twinning In a crystal, twinning that is produced by gliding. Syn: *secondary twinning.* Cf: *growth twinning.*

deformed cross-bedding Cross-bedding in which the foreset beds are "overturned or buckled in the downcurrent direction usually prior to deposition of the overlying bed" (Pettijohn & Potter, 1964, p.299). The foreset dip angle may also be altered by subsequent tectonic folding.

deformed ice A term now used by the U.S. Naval Oceanographic Office (1968, p.B32) for *pressure ice.*

degenerative recrystallization *degradation recrystallization.*

Deglacial n. A term used by Antevs (1953) for a time unit that covers in North America the time from the greatest extension of Wisconsin glaciation to the beginning of recession from the Cochrane moraines.

deglaciation The uncovering of a land area from beneath a glacier or ice sheet by the withdrawal of ice due to shrinkage by melting. As used in Great Britain, the term is restricted to a process that occurred in the past, in contrast to *deglacierization.* Also, the result of deglaciation.

deglacierization A term used in Great Britain for gradual withdrawal of a glacier or an ice sheet from a land area going on at the present time. Cf: *deglaciation.*

degradation [geomorph] (a) The wearing down or away, and the general lowering or reduction, of the Earth's surface by the natural processes of weathering and erosion; e.g. the deepening by a stream of its channel. The term sometimes includes the process of transportation; and sometimes it is used synonymously with *denudation,* or used to signify the results of denudation. See also: *gradation.* (b) Less broadly, the vertical erosion or *downcutting* performed by a stream in order to establish or maintain uniformity of grade or slope. Cf: *aggradation.*

degradation [permafrost] The shrinkage or disappearance of permafrost due to natural or artificial causes. Ant: *aggradation.* Cf: *depergelation.*

degradation [soil] Change in a soil (decrease in exchangeable bases and destruction of layer-silicate clay) due to leaching (Jacks, 1960, p.162).

degradation [stream] The lowering of a stream bed, due to such factors as increased scouring. Cf: *accretion [stream].*

degradation recrystallization Recrystallization resulting in a relative decrease in the size of crystals. Ant: *aggradation recrystallization.* Syn: *grain diminution; degenerative recrystallization.*

degradation vacuity The space-time value of the degradationally removed part of a transgressive-regressive depositional succession; e.g. the part of a *lacuna* resulting from degradation of formerly existing rocks at an unconformity. The term was used by Wheeler (1964, p.602) to replace *erosional vacuity.* Cf: *hiatus.*

Degraded Chernozem In early U.S. classification systems, a group of zonal soils having a dark brown to black surface horizon and an underlying, gray leached horizon which in turn rests on a brown horizon. It is developed under an area in which forest vegetation has encroached upon grassland, and is intermediate between chernozemic and podzolic types.

degraded illite Illite that has had much of its potassium removed from the interlayer position as a result of prolonged leaching. Syn: *stripped illite.*

degrading neomorphism A kind of *neomorphism* in which the crystal size decreases (Folk, 1965, p.23). Ant: *aggrading neomorphism.*

degrading stream (a) A stream that is actively cutting down its channel or valley and that is capable of transporting more load than it is supplied with. (b) A stream that is downcutting approximately at grade.

degree In hydrologic terminology, a unit for expressing the hardness of water. Cf: *Clark degree.*

degree-day One of the degrees between mean daily temperature and an arbitrary temperature such as $65°F$. Degree-days are used to indicate patterns of deviation from a given temperature standard.

degree of curve A measure of the sharpness of curvature; e.g. the angle at the center of a circle subtended by a chord 100 ft long (as for U.S. railroads) or by an arc 100 ft long (as in highway surveying).

degree of freedom The capability of variation in a chemical system. The number of degrees of freedom in a system may be defined as the number of independent, intensive variables (temperature, pressure, and concentration in the various phases) necessary to completely define the system, or as the number of variables which may be changed independently without causing a change in phase.

degree of slope The angular measurement (expressed in degrees) of slope from a horizontal plane (Van Riper, 1962, p.603).

degree of sorting A measure of the spread or range of variation of the particle-size distribution in a sediment. It is defined statistically as the extent to which the particles are dispersed on either side of the average: the wider the spread, the poorer is the sorting. It may be expressed by *sigma phi.*

dehiscent Said of certain fruits and sporangia which split open along definite seams at maturity.

dehrnite A grayish to greenish-white mineral of the apatite group: $(Ca,Na,K)_5(PO_4)_3(OH)$. Cf: *lewistonite.*

DE interray Left posterior interray in echinoderms situated between D ray and E ray and clockwise of D ray when the echinoderm is viewed from the adoral side.

dejection cone *cone of dejection.*

delafossite A mineral: $CuFeO_2.$

delatynite A variety of amber rich in carbon, low in succinic acid, and lacking sulfur, and found at Delatyn in the Carpathian Mountains of Galicia.

delawarite Pearly orthoclase from Delaware County, Penna.

delayed runoff Water from precipitation that sinks into the ground and discharges later into streams through seeps and springs (Tarr & Engeln, 1926, p.70). As defined above, delayed runoff is a syn. of *ground-water runoff,* but it could also be defined as runoff delayed by any means, such as temporary storage of precipitation in the form of snow and ice.

delay time The additional time, for any segment of a ray path, over the time which would be required to traverse the horizontal component of that segment at highest velocity encountered on the trajectory. Syn: *intercept time.*

deldoradite A light-colored cancrinite syenite.

delessite A mineral of the chlorite group: $(Mg,Fe^{+2},Fe^{+3},Al)_6(Si,Al)_4O_{10}(OH)_8$. It occurs in cavities and seams in basic igneous rocks.

deleveling An alteration in the level of a part of the Earth's surface: it is positive when the land is depressed and negative when it is elevated, in relation to sea level.

delhayelite A mineral: $(Na,K)_{10}Ca_5Al_6Si_{32}O_{80}(Cl_2,F_2,SO_4)_3.18H_2O.$

delineation A step in map *compilation* in which mapworthy features are distinguished and outlined on various possible source materials or are visually selected (as when operating a stereoscopic plotting instrument).

dell (a) A small, deep, secluded, wooded valley or natural hollow. The term is often used in a literary sense with no definite meaning. Cf: *dale*. (b) A depression upvalley from the source of a stream (Penck, 1953, p.421). Etymol: German *Delle*, "dent".

dellaite A mineral: $Ca_6Si_3O_{11}(OH)_2$.

dellenite *rhyodacite*.

dells A corruption of *dalles*, as applied along the Wisconsin River, Wisc.

Delmontian North American stage: uppermost Miocene (above Mohnian, below Repettian).

delorenzite A syn. of *tanteuxenite*. It was originally described as: $(Y,U,Fe)(Ti,Sn)_3O_8$.

delphinite A yellowish-green *epidote* from France.

delrioite A mineral: $CaSrV_2O_6(OH)_2 \cdot 3H_2O$.

delta The low, nearly flat, alluvial tract of land deposited at or near the mouth of a river, commonly forming a triangular or fan-shaped plain of considerable area enclosed and crossed by many distributaries of the main river, perhaps extending beyond the general trend of the coast, and resulting from the accumulation in a wider body of water (usually a sea or lake) of sediment supplied by a river in such quantities that it is not removed by tides, waves, and currents. Most deltas are partly subaerial and partly below water. The term was introduced by Herodotus in the 5th century B.C. for the tract of land, at the mouth of the Nile River, whose outline broadly resembled the Greek capital letter "delta", Δ, with the apex pointing upstream.

delta bar A "bar" formed by a tributary stream that is building a delta into the channel of the main stream.

delta bedding The bedding characteristic of a delta, consisting of comparatively flat topset beds and bottomset beds, between which are steeper foreset beds leading from close to the delta surface to the bottom of surrounding water but which are commonly deformed later; the inclined bedding "presumed to originate as foresets of small deltas" (Pettijohn & Potter, 1964, p.299).

delta cap An alluvial cone built on the delta plain, and having an apex that migrates upstream (Russell, 1898b, p.127).

delta cycle A term used by Barrell (1912, p.397) for a two-phase tectonic cycle dependent on stream erosion and changing sea level, and involving deposition with a stationary crust followed by vertical movement (normally subsidence) of the bottom. For a geosynclinal fill that is essentially a deltaic accumulation, the cycle is one of increasing coarseness from the base upward, of an increasing volume of clastic material, and of eventual overtake of subsidence by sedimentation.

delta fan A deposit formed by the merging of an alluvial fan with a delta. Syn: *fan delta*.

deltafication The formation of a delta. Syn: *deltation*.

delta front A narrow zone where deposition in deltas is most active, consisting of a continuous sheet of sand, and occurring at the effective depth of wave erosion (about 10 m or less). It is the break in slope separating the *prodelta* from the *intradelta*, and it may or may not be steep.

delta-front platform A zone of shallows, up to about 5 km wide, in front of the advancing distributaries of a delta.

delta-front trough A trough-shaped submarine valley formed off a large river delta on the continental shelf and slope. It has straight walls of soft rock, few if any tributaries, and a flat, seaward-sloping floor.

deltageosyncline *exogeosyncline*.

deltaic Pertaining to or characterized by a delta; e.g. "deltaic sedimentation". Also, constituting a delta; e.g. a "deltaic coast".

deltaic coastal plain A coastal plain composed of a series of more or less coalescing deltas; it consists initially of natural levee ridges separated by basins.

deltaic deposit A sedimentary deposit laid down in a delta, characterized by well-developed local cross-bedding and by a mixture of sand, clay, and the remains of brackish-water organisms and of organic matter. Cf: *estuarine deposit*.

deltaic tract An extension of the *plain tract* of a stream, characterized by the formation of a delta and the deposition of fine sediments.

deltaite A mixture of crandallite and hydroxylapatite.

delta kame A flat-topped, steep-sided hill of well-sorted sand and gravel deposited by a meltwater stream flowing into a proglacial lake; the upper margin of the delta was built in contact with glacier ice. Syn: *kame delta*.

delta lake A lake formed along the margin of or within a delta, as by the building of bars across a shallow embayment or by the enclosure of part of the sea by the growth of deltaic deposits.

delta levee lake A lake on an advancing delta, formed between sandbars or natural levees deposited at the mouths of distributaries. Examples: Lake Pontchartrain on the Mississippi River delta.

deltalogy The study of deltas and delta formation (Stamp, 1961, p.150). Syn: *deltology*.

delta-mooreite *torreyite*.

delta moraine *ice-contact delta*.

delta plain The level or nearly level surface composing the landward part of a large delta; strictly, an alluvial plain characterized by repeated channel bifurcation and divergence, multiple distributary channels, and interdistributary flood basins.

delta plateau A raised or abandoned delta plain.

delta ray An electron that was ejected by ionizing radiation.

delta shoreline A prograding shoreline produced by the advancing of a delta into a lake or sea.

delta structure The sedimentary structure produced by the three sets of beds in a delta: bottomset beds, foreset beds, and topset beds.

delta T In the description of seismic reflection and refraction stepout times, the time difference between two time values either observed or interpolated from observed values, or interpolated from computed values.

delta terrace A fan-shaped terrace composed of a delta that remained after the disappearance of the stream that produced it (Chamberlin, 1883, p.304).

deltation *deltafication*.

delta top *intradelta*.

delta unit A syn. of *delta value*. Abbrev: d-unit.

delta value The *isotope ratio* given as the per mil deviation of the difference between the ratio of the heavy isotope divided by the light isotope of the sample and the similar ratio of the standard sample, divided by the latter, and multiplied by 1000. Syn: *delta unit*. Abbrev: δ-unit.

delthyrial chamber The cavity beneath the umbo of the pedicle valve of a brachiopod, bounded by dental plates (if present) or by posterior and lateral shell walls (if dental plates are absent). It corresponds to the *notothyrial chamber* of the brachial valve.

delthyrial plate A plate within the delthyrial chamber of some spiriferoid brachiopods, extending a variable distance from the apex between dental plates. It is probably homologous with the pedicle collar (TIP, 1965, pt.H, p.143).

delthyrium The median triangular or subtriangular opening beneath the beak of the pedicle valve of a brachiopod, bisecting the ventral cardinal area or pseudointerarea, and commonly serving as a pedicle opening. Pl: *delthyria*. Cf: *notothyrium*.

deltidial plate One of two plates growing medially (inward) from the margins of the delthyrium of a brachiopod and partly or completely closing it.

deltidium The cover of the delthyrium of a brachiopod, formed by conjunct deltidial plates (in contact anteriorly and dorsally of the pedicle) whose line of junction is visible. Pl: *deltidia*. Cf: *pseudodeltidium*.

deltohedron A *tristetrahedron* whose faces are quadrilateral, rather than triangular, as in the *trigonal tristetrahedron*. Syn: *deltoidal dodecahedron; tetragonal tristetrahedron*.

deltoid [lunar] A delta-shaped raised feature on the surface of the Moon, found in association with certain ring structures (Fielder, 1965, p.170).

deltoid [paleont] *deltoid plate*.

deltoidal cast A term used by Birkenmajer (1958, p.143) for *frondescent cast*.

deltoidal dodecahedron *deltohedron*.

deltoid branch A branch of a stream, enclosing a whole delta (Jackson, 1834, p.79).

deltoid island A *branch island* formed on a delta (Jackson, 1834, p.79).

deltoid plate One of a circlet of interradial, subtriangular plates of a blastoid, situated near the summit (oral end) of theca but aboral to oral plates, between adjacent ambulacra, and above radial plates. Syn: *deltoid*.

deluge A great inundation or overflowing of the land by water; specif. the Deluge (the Noachian flood).

deluvium *diluvium*.

delvauxite A mineral, perhaps: $Fe_4^{+3}(PO_4)_2(OH)_6.nH_2O$.

delve A surface depression or hollow.

demagnetization The reduction of remanent magnetization, often selectively only of the softer or less stable components of natural remanent magnetization. Techniques employed include *alternating field demagnetization, thermal demagnetization*, and *chemical demagnetization*.

demantoid A bright-green to yellowish-green, transparent gem variety of andradite garnet, characterized by a brilliant luster, a dispersion stronger than that of diamond, and a hardness that is less than that for other garnets. Syn: *Uralian emerald*.

demarcation line An imaginary line on the surface of a valve of a bivalve mollusk, originating at the beak and marking the locus of points on successive positions of the margin where the transverse-growth component has had maximum effect. It forms a dorsal/ventral profile when the valve is viewed from one end.

deme A randomly interbreeding population smaller than the species population. The term is usually preceded by a prefix indicating the nature or cause of the separation of such a group; e.g., "topodemic," applied to a population restricted to a particular geographic area.

demersal Said of bottom-dwelling organisms; *benthic*.

demesmaekerite A bottle-green to clear olive-green mineral: $Pb_2Cu_5(UO_2)_2(SeO_3)_6(OH)_6.2H_2O$.

demiplate An ambulacral plate of an echinoid touching an adradial suture but not touching a perradial suture.

demipyramid One of the ten elements that support the teeth in Aristotle's lantern of an echinoid. See also:*pyramid [paleont]*.

demkha A term used in Algeria for an almond-shaped *dune massif* (Capot-Rey, 1945, p.395).

demoiselle A term used in the French Alps for an *earth pillar* capped by a large boulder, esp. one formed by weathering of volcanic breccia or of glacial till. Etymol: French, "young lady". Pron: duh-mwa-zel.

demoiselle hill A rounded, symmetric, beehive-shaped elevation with a grassy surface, bordered by a shallow or deep cauldron-like depression, occurring on the Magdalen Islands in the Gulf of St. Lawrence (Clarke, 1911, p.145); it may vary from a knob or knoll to a hill 175 m high.

demorphism An obsolete syn. of *weathering*.

demosponge Any sponge belonging to the class Demospongea and characterized chiefly by the presence of spongin as all or part of its skeleton. "Most demosponges contain siliceous spicules, with or without spongin. A few contain only spongin, and still fewer produce neither spicules nor spongin" (TIP, 1955, pt.E, p.34).

dendrite A mineral, e.g. a surficial deposit of an oxide of manganese, or an inclusion, that has crystallized in a branching pattern. Adj: *dendritic*. Syn: *dendrolite*.

dendritic Said of a mineral that has crystallized in a branching pattern; pertaining to a *dendrite*. Syn: *arborescent*.

dendritic drainage pattern A drainage pattern in which the streams branch irregularly in all directions and at almost any angle, resembling in plan the branching habit of certain trees (such as oaks or maples), and produced where a consequent stream receives several tributaries which in turn are fed by smaller tributaries. It is indicative of insequent streams flowing across horizontal and homogeneous strata or complex crystalline rocks offering uniform resistance to erosion. Cf: *pinnate drainage pattern*.

dendritic glacier A *trunk glacier* joined by many tributary glaciers to form a pattern that resembles (in plan) the profile of a tree with branches.

dendritic tufa *dendroid tufa*.

dendrochore The part of the Earth's surface having a tree-sustaining climate. It constitutes the bulk of the *biochore*.

dendrochronology The study of annual *growth rings* of trees for dating of the recent past. Cf: *dendroclimatology*. Syn: *tree-ring chronologyy*.

dendroclimatology The study of the patterns and relative sizes of annual *growth rings* of trees for paleoclimatologic data of the recent past. Usually dry years yield thin rings; wet years wide rings. Cf: *dendrochronology*.

dendroclone A desma (of a sponge) having a smooth, straight shaft with a tree-like, branching process at each end.

dendrodate A date calculated by *dendrochronology*.

dendrogram A tree-like diagram depicting the mutual relationships of a group of items sharing a common set of variables, the variables representing either samples on which multiple measurements have been made or measured attributes for a group of samples; esp. a diagram illustrating the interrelations (based on degrees of similarity) throughout geologic time of fossil organisms connected by a common ancestral form. Cf: *dendrograph*.

dendrograph A tree-like, two-dimensional correlation diagram depicting the mutual relationships between and within groups of items sharing a common set of variables (McCammon, 1968). It permits a more accurate interpretation of geologic data involving large numbers of variables. Cf: *dendrogram*.

dendroid adj. (a) Said of certain invertebrates that form many-branched colonies or that have a tree-like habit of growth, such as the irregular, bushy colonies formed by graptolites. (b) Said of an irregularly branching type of fasciculate corallum.---n. A sessile graptolite of the order Dendroidea ranging from Middle Cambrian to Carboniferous, characterized by a typically erect, dendroid rhabdosome having many stipes which are composed of autothecae, bithecae, and stolothecae arranged in regularly alternating triads along each branch.

dendroid tufa Gray tufa occurring as spheroidal, mushroom, or dome-shaped masses with concentric macrostructure and less pronounced internal dendritic structure; e.g. the tufa along the shore of the extinct Lake Lahontan in Nevada, where it constitutes the major part of the dome-like mass. Cf: *thinolitic tufa; lithoid tufa*. Syn: *dendritic tufa*.

dendrolite *dendrite*.

denivellation A variation in the level of a body of water, esp. of a lake; e.g. "wind denivellation" is a rise of water level due to wind drift, as along the windward shore of a lake. Etymol: French *dénivellation*, "difference of level".

denningite A colorless to pale-green tetragonal mineral: $(Mn, Zn)Te_2O_5$.

dense [geol] (a) Said of a fine-grained, aphanitic igneous rock whose particles generally average less than 0.05 to 0.1 mm in diameter, or whose texture is so fine that the individual particles cannot be recognized by the unaided eye. (b) Said of a compact rock whose particles or various parts are massed or

186

rowded close together. The rock may be fine- or coarse-grained. (c) Said of a rock or mineral possessing a relatively high density or specific gravity.

dense [optics] Said of a substance that is highly refractive.

dense [photo] Said of a photographic negative or a positive transparency having a high or relatively high density and in which the silver deposit per unit area is greater than normal due to overdevelopement.

densilog density log.

densitometer [photo] A photoelectric or visual instrument for measuring the opacity (or photographic density) of a developed photographic plate or film by measuring the density of the silver deposit in a photographic image. The device constantly reduces the intensity of a standard light until it matches an identical beam of light that has been transmitted through or reflected from the material to be measured.

density [oceanog] The mass of water per unit volume, usually stated in grams per cubic centimeter. Density of water ranges from about 1.0 for fresh water to 1.07 for water in the deep ocean. Density may be written as *sigma-t* or σ_T. See also: *in situ density; potential density.*

density [optics] A measure of the degree of opacity of any translucent medium, defined strictly as the common logarithm of the opacity; e.g. the degree of blackening of an exposed photographic film, plate, or paper after development, or of the direct image. Cf: *contrast.* Syn: *optical density.*

density current A gravity-induced flow of relatively more dense water under surrounding water. Factors affecting density differences include temperature, salinity, and concentration of suspended sediments. See also: *salinity current, turbidity current.*

density log A log showing the varying densities of rocks penetrated in drilling; specif. a *gamma-gamma log* used to indicate density by recording the amount of backscattering of gamma rays (this amount being a simple function of bulk density of the rocks). It is applied only to uncased holes. Syn: *densilog.*

density profile A series of closely spaced gravity stations over a local topographic irregularity, preferably a hill or valley or both, reduced for different densities of the hill material and plotted as a family of curves which are compared to the elevation profile for the same stations. The criterion for actual density is that which gives the smoothest reduced gravity profile across the topographic irregularity.

density stratification The *stratification* of a lake produced as a result of density differences, the lightest layer occurring near the top and the heaviest layer at the bottom. It is usually brought about by temperature changes, but may also be caused by differences at different depths in the amount of suspended or dissolved material (e.g. where a surface layer of freshwater overlies salt water). See also: *thermal stratification.*

densofacies A term used by Vassoevich (1948) for *metamorphic facies.*

densospore A trilete spore, chiefly Paleozoic, with a pronounced cingulum which has a tendency to be "doubled", i.e. thicker part toward the center of the spore and an inner, more external part; e.g. the genus *Densosporites* and other similar genera such as *Cristatisporites.*

dental plate One of variably disposed plates of secondary shell underlying and supporting the hinge teeth of a brachiopod and extending to the floor of the pedicle valve.

dental socket A shallow excavation in the posterior margin of the brachial valve of a brachiopod for the reception of a hinge tooth of the pedicle valve. Syn: *socket.*

dentate Toothed, or having teeth or small, pointed, conical or tooth-like projections; e.g. "dentate chela" of a sponge with toothed terminal expansions.

denticle (a) A small spine-, needle-, or sawtooth-like structure of compound and plate-like conodont elements, similar to a cusp but commonly smaller. (b) One of the small, sharp, pro-

truding ridges that alternate with complementary dental sockets located along the cardinal margin or the hinge line of both valves of a brachiopod. (c) A primary or secondary toothlet on the sutural edge of the radius of a compartmental plate of a cirripede crustacean, serving to strengthen the articulation of the plates; a small, delicate, spine- or tooth-like projection on the carapace of an ostracode.

denticulate (a) Having small teeth, or bearing a series of small spine- or tooth-like projections; e.g. said of a shell covered with small pointed projections. (b) Minutely or finely dentate, as of a leaf.

denticulation The state of being denticulate; also, a series of small tooth-like structures, such as the denticles on the anterior and posterior margins of the shell of a cytherid ostracode.

dentition The number, kind, and arrangement of teeth or tooth-like structures; e.g. the hinge teeth and sockets of a bivalve mollusk, considered collectively.

dentpit A term proposed by Hudson (1910, p.167) for a shallow cavity developed on calcareous rocks where relatively sand-free water is strongly agitated by waves, its width (1-5 cm in diameter) greatly exceeding its depth, and having a circular outline in cross section. Cf: *cuphole.*

denudation (a) The sum of the processes that result in the wearing away or the progressive lowering of the Earth's surface by various natural agencies that include weathering, erosion, mass-wasting, and transportation; also the combined destructive effects of such processes. The term is wider in its scope than *erosion,* although it is commonly used as a syn. of that term. It is also used as a syn. of *degradation,* although some authorities regard "denudation" as the actual processes and "degradation" as the results produced; Davis (1909, p.408) distinguished between "denudation" as the active processes operating early in the cycle of landform development, and "degradation" as the more leisurely processes operating later. (b) Strictly, the laying bare, uncovering, or exposure of bedrock or a designated rock formation through the removal of overlying material by erosion. This is the original, and etymologically correct, meaning of the term, which was often applied in a catastrophic sense, e.g. the "Great Denudation" resulting from the universal flood.

denudation chronology The study of the sequence of events leading to the formation and evolution of an existing landscape.

deoxidation sphere *bleach spot.*

departure (a) The projection of a line onto an east-west axis of reference, such as the projection of a traverse course in a plane survey onto a line perpendicualr to the meridian, being equal to the length of the course multiplied by the sine of its bearing. (b) The length of such a projection; e.g. *easting* (positive departure) and *westing* (negative departure). Cf: *latitude difference; meridional difference.* Syn: *longitude difference.*

dependable yield The minimum water supply to a given area that is available on demand and which may decrease on the average of once every *n* number of years (Langbein & Iseri, 1960).

dependent variable The *variable* whose magnitude is plotted as a function of fixed consecutive values of a second or *independent variable.*

depergelation The act or process of thawing permafrost (Bryan, 1946, p.640). Cf: *degradation.*

depletion [eco geol] The exhaustion of a natural resource, i.e. ore, by commerical exploitation, measured quantitatively in monetary terms (Brown & Runner, 1939).

depletion [water] Loss of water from surface- or ground-water reservoirs at a rate greater than that of recharge.

depletion curve A hydrograph showing the loss of water from ground-water storage, by seepage or flowage into streams, or from any storage or channel. See also: *recession curve.*

depoaxis The axis of maximum sediment deposition during a geologic epoch.

depocenter An area or site of maximum deposition; the thickest part of any specified stratigraphic unit in a depositional *basin*.

deposit n. In sedimentology, anything laid, placed, or thrown down; specif. material of any type or from any source that has accumulated by some natural process or agent, either in the form of consolidated or unconsolidated material. The term originally applied to material left by water, but has been broadened to include matter accumulated by any agent, such as wind, ice, volcanoes, chemical action, etc. Cf: *sediment.*-- v. To lay down or let drop by a natural process; to become precipitated.

deposited snow All snow lying on the ground or on other snow, firn, or ice, subject to the several types of metamorphism and alteration by wind action. Syn: *fallen snow.*

deposit feeder An animal that obtains its food from detrital material and associated microorganisms on the sea floor; e.g., sea cucumber, scaphopod. Syn: *detritovore.*

deposition (a) The laying, placing, or throwing down of any material; specif. the constructive process of accumulation into beds, veins, or irregular masses of any kind of loose, solid, rock material by any natural agent, such as the mechanical settling of sediment from suspension in water, the chemical precipitation of mineral matter by evaporation from solution, or the accumulation of organic material through the processes of death of plants and animals. See also: *sedimentation.* (b) Material that is deposited; a deposit or sediment.

depositional (a) Pertaining to the process of deposition; e.g. a "depositional basin" or a "depositional surface". (b) Formed by the process of deposition; e.g. a "depositional topography"

depositional dip *primary dip.*

depositional fabric Rock fabric or that fabric element or elements that result from deposition during the rock's formation, e.g. sedimentary grains in an unmetamorphosed, current-laid sediment or crystals deposited by crystal settling in a magma chamber; the fabric of a *primary tectonite.* Cf: *deformation fabric.*

depositional fault *growth fault.*

depositional interface The interface between the water and the bottom where sediments are deposited in relation to the energy level at the interface (Plumley et al, 1962, p.86).

depositional mark An irregularity formed on the bedding plane of a sediment during deposition; e.g. a scour mark and a tool mark. Syn: *depositional marking.*

depositional remanent magnetization Remanent magnetization resulting from mechanical orientation of ferrimagnetic mineral grains along the ambient field during sedimentation. Its inclination is generally less than that of the ambient field. Abbrev: DRM.

depositional topography Topography formed in a body of standing water by the processes of deposition, as by variations in the kind and amount of energy from place to place (esp. variations in turbulence generated principally by waves and wind-driven currents).

depressed In invertebrate morphology, said of a nautiloid conch whose lateral diameter is greater than its dorsal/ventral diameter; also, said of an ammonoid whorl section that is wider than it is high.---Ant: *compressed.*

depressed flute cast A flat or weakly developed flute cast.

depressed moraine An irregular moraine "developed along the ice front in line with the normal morainal ridge, but failing to rise above the adjacent outwash", and due to "the nonaccumulation of marginal material than to an excess of outwash" (Fuller, 1914, p.33-34), as those on Long Island, New York State.

depression [geomorph] Any hollow in or relatively sunken part of the Earth's surface; esp. a low-lying area completely surrounded by higher ground and having no natural outlet for surface drainage, as an interior basin or a karstic sink.

depression [meteorol] An atmospheric region of relatively low pressure. The term may refer to the extratropical-type cyclone which is larger and less violent than a tropical-type cyclone since low pressure is usually associated with cyclonic circulation. Syn: *low [meteorol].*

depression [tect] A structurally low area in the crust, produced by negative movements that depress, downsink or downthrust the rocks. The term includes *basins* and *furrows.* Cf: *uplift.*

depression [surv] (a) The angular distance of an object beneath the horizontal plane that passes through the observer. (b) The angular distance of a celestial object below the horizon.

depression contour A *closed contour*, inside of which the ground is at a lower elevation than that outside, and distinguished on a map from other contour lines by hachures marked on the downslope side. See also: *closed depression.*

depression spring A type of gravity spring, with its water flowing onto the land surface from permeable material as a result of the land surface sloping down to the water table (Meinzer 1923, p.51).

depression storage Accumulation of water from precipitation in depressions in the land surface (Langbein & Iseri, 1960); accumulation of rainwater or snowmelt in depressions when the soil has reached its infiltration capacity (Chow, 1964, p.20-5). Cf: *detention.*

depressor muscle crest The elevated denticles, on the interior of a tergum of a cirripede crustacean, for attachment of muscles that depress or draw down.

depth In oceanography, the vertical distance from a specified datum, to the bottom of the body of water.

depth contour see *isobath* [oceanog].

depth hoar Ice crystals formed below the surface and usually near the bottom of a snowpack by *temperature-gradient metamorphism.* The crystals are stepped or layered, usually delicate, and may assume various complex shapes such as cup crystals, scrolls, or columns. Syn: *sugar snow.* Cf: *surface hoar; crevasse hoar.*

depth ice (a) *anchor ice.* (b) Small particles of ice formed below the surface of a sea churned by wave action (ADTIC 1955, p.24).

depth of compensation [oceanog] *calcium carbonate compensation depth.*

depth of compensation [tect] According to the concept of *isostasy*, the depth above which the rock material is brittle and below which there is a slow movement of plastic rock material to adjust to changes in load of the Earth's surface. Estimates of the depth of compensation range from 35 miles to 75 miles (Strahler, 1963, p.400). Var: *compensation level; compensation depth; level of compensation; isostatic depth of compensation.* See also: *isostatic compensation.*

depth of exploration The depth to which a source of a geophysical anomaly can be placed and still have the anomaly recognizable above noise.

depth of focus The distance from the focus of an earthquake to the epicenter. Syn: *focal depth.*

depth of frictional influence *friction depth.*

depth of frictional resistance *friction depth.*

depth of penetration [elect] The depth at which a plane electromagnetic wave penetrates a dissipative medium before the electric field intensity associated with the wave decreases to $1/e$ of its value at the surface of the medium.

depth of penetration [remote sensing] (a) The depth below the surface of a material by which the incident radiation has been attenuated to $1/e$, or to 37 percent. This is sometimes called *skin depth*, or *attenuation distance.* (b) The depth at which the integrated temperature differentials of adjacent spatial resolution cells are equivalent to the temperature resolution of the sensor. This is also called *effective depth of penetration.* (c) The depth from which radiation may still reach e receiver. Such radiation is integrated over the distance ownwards from the surface.----See also: *interfacial geology.*

epth point In seismic work, a position at which a depth determination of a mapped horizon has been calculated. For eismic reflections it is generally midway between shotpoint nd geophone. See also: *common depth point.*

epth sounder *echo sounder.*

epth-velocity curve *vertical-velocity curve.*

epth zones [oceanog] Four oceanic environments, or ranges oceanic depths: the *littoral* zone, between high and low des; the *neritic* zone (between low-tide level and 100 fathms); the *bathyal* zone (between 100 and 500 fathoms); and e *abyssal* zone (500 fathoms and deeper).

epth zones [met] Characteristic physico-chemical environents at various depths in the Earth which give rise to differnt metamorphic phenomena (Grubenmann, 1904): *epizone; esozone; katazone.*

eranged drainage pattern A distinctively disordered drainage attern in a recently glaciated area whose former surface and reglacial drainage has been remodeled and effaced, and in hich the new drainage system shows a complete lack of nderlying structural and bedrock control. It is characterized irregular streams that flow into and out of lakes, by only a w short tributaries, and by swampy interstream areas. Syn: *acially disturbed drainage pattern.*

erangement The process by which changes in the drainage course of a stream are effected by agents other than reams, such as by glaciation, wind deposition, or diastrohism. Cf: *diversion.*

erbylite A black or brown orthorhombic mineral: Fe_6Ti_6-O_2O_{23}.

erbyshire spar A popular name for *fluorite* from Derbyshire, ng. Syn: *Derby spar.*

erbystone *blue john.*

erelict (a) A tract of dry land formed by *dereliction.* (b) Any operty abandoned at sea, often of sufficient size as to be a enace to navigation.

ereliction A recession of water from the sea or other body of ater so that land above high-water mark is left dry. Cf: *relic-n.*

erivate A general term, now obsolete, for a rock derived om the products of destruction of older rocks; a sedimentary ck. Cf: *ingenite.*

erivative map A map of one of the derivatives of a potential eld, such as the Earth's gravity or magnetic field. It is usual- of the second vertical derivative, or a *second-derivative ap.*

erivative rock A rock composed of materials derived from e weathering of older rocks; a *sedimentary rock,* or a rock rmed of material that has not been in a state of fusion imediately before its accumulation.

erivative structure A crystal structure having a multiple unit ell and/or a suppression of some symmetry elements, rmed by the substitution of a simple fraction of one metal by nother; e.g. the structure of chalcopyrite as derivative after halerite.

erived Said of geologic materials that are not native or that ave been displaced or brought from elsewhere; esp. said of a fossil washed out from its original site and redeposited in a ter formation at a different locality. Also said of geologic aterials that are not primary or original. Cf: *reworked; re-anié.*

erived till A till-like deposit "formed from the erosion of sub-antially older tillites, not necessarily with any glacial trans-rt involved in the second formation"; e.g. the deposit of the -called "Cretaceous glaciation" of South Australia (Harland al, 1966, p.232).

ermal [paleont] Pertaining to the exopinacoderm, a cortex, the ectosome of a sponge; e.g. "dermal skeleton" (differ-tiated skeleton at the outer surface of a sponge, normally

connected with the exopinacoderm, cortex, or ectosome), or "dermal membrane" (exopinacoderm, cortex, or ectosome roofing over a vestibule), or "dermal pores" (minute openings, or prosopores, in the surface of a sponge, giving access to the inhalant canals).

dermal [tect] Said of deformation or gliding in the upper part of the sialic crust. Cf: *epidermal; bathydermal.*

dermalium A usually specialized sponge spicule supporting all or part of the ectosome. Pl: *dermalia.*

dermolith An obsolete syn. of *pahoehoe.* Cf: *aphrolith.*

dermoskeleton *exoskeleton.*

derrick A steel-framework tower erected over a deep drill hole (such as an oil well), used to support the various tools and tackle employed in the boring of the hole and for hoisting and lowering equipment. See also: *rig.*

desalination The removal of dissolved salts from sea water in order to make it potable. The most common method is distillation.

descendant A topographic feature carved from the mass beneath an older topographic form that has been removed (Willis, 1903, p.74).

descending branch Either of two dorsal elements of a brachiopod loop, extending distally from crura and recurved ventrally at anterior ends. Cf: *ascending branch.*

descending development *declining development.*

descensional deposit A sedimentary deposit produced by the disintegration of rock and the aggregation of the resulting particles in beds.

descension theory A theory of supergene mineral-deposit formation involving the descent from above of mineral-bearing solutions. The theory originated with the Neptunian school of thought of the 18th Century that postulated an aqueous origin for all rocks. Cf: *ascension theory.*

descloizite A brown to black mineral: $Pb(Zn,Cu)(VO_4)(OH)$. It is isomorphous with mottramite. Syn: *eusynchite.*

desert An area of low moisture due to low rainfall, i.e. less than ten inches annually, high evaporation, or extreme cold and which supports only specialized vegetation, not that typical of the latitudes, and is generally unsuitable for human habitation under natural conditions. Deserts are not characterized by uniformity of elevation but wind often produces distinctive erosional features, e.g. dunes. Adj: *desertic; eremic.*

desert arch An elongate *desert dome.*

desert armor A *desert pavement* whose surface stony fragments protect the underlying finer-grained material from further wind erosion; a common feature of *stony deserts.* See also: *pebble armor.* Syn: *deflation armor.*

desert crust (a) A hard layer, containing calcium carbonate, gypsum, or other binding matter, exposed at the surface in a desert region. (b) *desert varnish.* (c) *desert pavement.*

desert dome A convex rock surface with very uniform and smooth slopes, representing the result of prolonged exposure to desert erosion of a mountain mass; e.g. Cima Dome in the Mojave Desert, Calif. See also: *desert arch; granite dome.* Syn: *pediment dome.*

desert lacquer *desert varnish.*

desert mosaic A *desert pavement* characterized by tightly interlocking and evenly set fragments, covering the surface in the manner of a mosaic; e.g. a *pebble mosaic.*

desert patina *desert varnish.*

desert pavement A thin, natural, smooth or sheet-like, residual concentration of wind-polished, closely packed pebbles, boulders, gravel, and other rock fragments, mantling a desert surface (such as an area of reg) where wind action and sheetwash have continually removed all smaller particles (sand and dust), and usually protecting the underlying finer-grained material from further deflation. The fragments commonly are cemented by mineralized solutions. See also: *desert armor; desert mosaic; lag gravel; boulder pavement; reg.* Syn: *desert crust; deflation residue.*

desert peneplain A syn. of *pediplain*, esp. in reference to a wind-erosion surface in southern Africa. The term is inappropriate because such a surface is produced under different conditions and by different processes than those of a humid-land peneplain.

desert plain (a) A general term used by Blackwelder (1931) for any plain commonly found in a desert; e.g. a river flood plain, a structural plain, a playa, a bajada, and a pediment. (b) *pediplain*.

desert polish (a) A smooth, shiny or glossy surface imparted to rocks of desert regions by windblown sand and dust. Syn: *wind polish*. (b) A term sometimes used as a syn. of *desert varnish*.

desert rind *desert varnish*.

desert ripple One of a system of slightly arcuate ridges produced by the wind, arranged roughly en echelon about 15 m apart with crests supporting vegetation and troughs plated with caliche; it may be as high as 1 m and as long as 150 m.

desert rose A radially symmetric group of crystals with a fanciful resemblance to a rose, formed in sand, soft sandstone, or clay. The crystals are usually of calcite, less commonly of barite, gypsum, or celestite.

Desert soil In early U.S. classification systems, a group of zonal soils having a light-colored surface horizon overlying calcareous material and, commonly, a hardpan. It is developed under conditions of aridity, warm to cool climate, and scant scrub vegetation.

desert varnish A thin, dark, hard, shiny or glazed iridescent (red, brown, black) film, coating, stain, or polish composed of iron oxide accompanied by traces of manganese oxide and silica, formed in desert regions after long exposure upon the surfaces of pebbles, boulders, and other rock fragments, as well as upon the cracked walls of ledges and other rock outcrops; it is believed to be caused by exudation of mineralized solutions from within and deposition by evaporation on the surface. A similar appearance produced by wind abrasion is properly known as *desert polish*. Syn: *desert patina; desert lacquer; desert crust; desert rind*.

desiccation A complete or nearly complete drying out or drying up, or a deprivation of moisture or of water not chemically combined; e.g. the loss of water from pore spaces of soils or sediments as a result of compaction, or the formation of evaporites as a result of direct evaporation from bodies of water in an arid region, or the progressive increase in aridity of an area as a result of a climatic change (such as decreasing rainfall) or of accelerated erosion (such as deforestation). Cf: *dehydration; exsiccation*.

desiccation breccia A breccia formed where irregular dried-out and mud-cracked polygons have broken into angular fragments which have then been deposited with other sediments. Syn: *mud breccia*.

desiccation conglomerate A term used by Shrock (1948, p.208) for a conglomerate consisting of fragments eroded from a mud-cracked layer of sediment and rounded by transportation. Syn: *mudstone conglomerate*.

desiccation crack A crack in sediment, produced by drying; esp. a *mud crack*. Syn: *drying crack; desiccation fissure; desiccation mark; klizoglyph*.

desiccation mark A *desiccation crack*; esp. a *mud crack*.

desiccation polygon A small *nonsorted polygon* formed in a nonfrigid environment (as on a mud flat) by drying of moist, fine-grained, clayey soil or sediment, thus producing contortion resulting in cracking. The polygon normally has three to five sides which may measure 2-30 cm in length. See also: *giant desiccation polygon*. Syn: *mud-crack polygon; mud-flat polygon; drought polygon; shrinkage polygon*.

designation Statement of the type of a nominal genus or species.

design flood A flood against which protective measures are taken.

desilication [petrology] The removal of silica from a rock c magma, by the breakdown of silicates and the resultar freeing of silica or by reaction with the wall rock of a body c magma.

desilication [soil] The removal of silica from soils of a warm humid climate by the percolation of large amounts of rair water, resulting in a pedalfer type of soil relatively rich in hy droxides or iron, aluminum, and manganese.

desilting basin A *settling basin* consisting of an enlargement i a stream where silt carried in suspension may be deposited.

desma An irregularly branched, siliceous sponge spicule tha bears knotty outgrowths (zygomes) which interlock with adja cent spicules. Pl: *desmas* or *desmata*.

desmid A unicellular, microscopic green alga that sometime produces an encysted, resistant, walled zygospore.

desmine *stilbite*.

desmite *Residuum* that is transparent; it is characteristic c higher coal grades.

desmodont adj. Said of the dentition of a bivalve mollus characterized by the prominence of a large chondrophore in side the hinge line.----n. Any bivalve mollusk belonging to th order Desmodonta, characterized by the presence of tw equal muscle scars and a pallial sinus and by the absence c hinge teeth or with irregular hinge teeth intimately connecte with the chondrophore.

desmoid A siliceous sponge spicule bearing outgrowths lik those of a desma but which do not interlock with adjacer spicules.

Desmoinesian North American provincial series: upper Middl Pennsylvanian (above Atokan, below Missourian). Syn: *De Moines*.

desmosite A banded *adinole*.

desquamation An obsolescent syn. of *exfoliation* characterize by the peeling off or detachment of scaly rock fragments.

dess A term used in Morocco for silt deposited by a nev stream flowing in an arid region (Termier & Termier, 1963 p.403).

destinezite *diadochite*.

destructional Said of a landform that owes its origin or genera character (form, position, direction, etc.) to the removal c material by erosion and weathering; e.g. a mesa, canyor cliff, or plain resulting from the wearing down or away of th land surface. Ant: *constructional*. Cf: *sequential landform*.

destructive metamorphism *equitemperature metamorphism*.

destructive wave A wave that erodes a beach by moving ma terial seaward, as a storm wave with a more powerful back rush than uprush. Ant: *constructive wave*.

detached core The inner bed or beds of a fold that becom separated or pinched off from their source due to extrem folding and compression.

detachment [soils] Separation of transportable particles fror a soil layer, usually by running water, raindrop impact, c wind. Cf: *dispersion*.

detachment [fault] *décollement*.

detachment fault *sole fault*.

detachment thrust *sole fault*.

detail log An electric log plotted on a scale larger than th conventional scale (one inch per 100 ft of depth) in order t portray minor variations in the formations penetrated by th borehole; specif. an electric log plotted on a scale of 5 inche per 100 ft of depth.

detectivity In infrared detector terminology, the reciprocal c *noise equivalent power* (Bernard, 1970, p.58). Symbol: *D*.

detector *radiation detector*.

detention The amount of water from precipitation existing a overland flow. *Depression storage* is not considered part c the detention (Rechard & McQuisten, 1968). Syn: *detentio storage; surface detention*.

detention storage *detention*.

deterioration *climatic deterioration*.

determinism "The theory that all occurrences in nature are determined by antecedent causes or take place in accordance with natural laws" (Webster, 1967, p.616).

deterministic process A process in which there is an exact mathematical relationship between the independent and dependent variables in the system. Ant: *stochastic process.*

detrital adj. Pertaining to or formed from *detritus*; said esp. of rocks, minerals, and sediments. The term is often used to indicate a source from outside the depositional basin, as compared to *clastic.*---n. A detrital rock. Term is usually used in the plural.

detrital deposit A deposit containing *detrital minerals*, such as a placer deposit. See also:*detrital sediment.*

detrital fan *alluvial fan.*

detrital mineral Any mineral grain resulting from mechanical disintegration of parent rock; esp. a *heavy mineral* found in a sediment or weathered and transported from a vein or lode and found in a placer or alluvial deposit.

detrital ratio *clastic ratio.*

detrital rock A rock composed primarily of particles or fragments detached from preexisting rocks either by erosion or by weathering; specif. a sedimentary rock having more than 50% detrital material (Krynine, 1948, p.134). Cf: *chemical rock.*

detrital sediment A sediment formed by the accumulation of detritus, esp. that derived from preexisting rocks and transported to the place of deposition. Cf: *clastic sediment.* See also:*detrital deposit.*

detrition A general term for the processes involved in producing detritus; a wearing off or away by breaking or rubbing of rock masses.

detritovore *deposit feeder.*

detritus A collective term for loose rock and mineral material that is worn off or removed directly by mechanical means, as by disintegration or abrasion; esp. fragmental material, such as sand, silt, and clay, derived from older rocks and moved from its place of origin. Cf: *debris.* Pl: *detritus.* See also: *reef detritus.*

deuteric Said of a process or of an effect in an igneous rock that takes place in the later stages and as a direct result of consolidation of a magma; e.g. corrosion of quartz grains, formation of reaction rims. Syn: *paulopost; epimagmatic.* Cf: *multopost.* See also: *synantectic; synantexis; autometamorphism.*

deuteroconch The chamber in larger foraminifers immediately adjoining the proloculus and formed next after it.

deuteroforamen A secondary aperture independent of the tooth plate in some enrolled foraminifers. Cf: *protoforamen.*

deuterogene An old term for a secondary rock; a rock formed from a pre-existing rock. Adj: *deuterogenic.* Cf: *protogene.*

deuterogenic Adj. of *deuterogene.* Syn: *deuterogenous.*

deuterogenous *deuterogenic.*

deuteroglacial Pertaining to the last great glaciation following the *proteroglacial* period (Hansen, 1894, p.128).

deuterolophe A spirally coiled part of a brachiopod lophophore, bearing a double brachial fold and a double row of paired filamentar appendages. It is homologous with the side arms of *plectolophe.* Cf: *spirolophe.*

deuteromorphic A general term applied to crystals whose shapes have been acquired or modified by mechanical or chemical processes acting on the original forms (Loewinson-Lessing, 1899). Depending on the nature of the secondary agent, they may be described as *clastomorphic, lytomorphic, schizomorphic, tectomorphic,* or *neomorphic.* These terms are now obsolete.

deuteropore One of a group of *protopores* fusing into a single larger pore cavity in the outer wall of foraminifers.

deuterosomatic A now obsolete term originated by Loewinson-Lessing in 1893 that pertains to a regenerated rock, either clastic or crystalline, e.g. phyllite, schist, contact-metamorphic rock.

deutonymph The second developmental stage in the arachnid order Acarida.

developed ore *developed reserves.*

developed reserves Ore that has been exposed on three sides and for which tonnage and quality estimates have been made. Cf: *positive ore; proved reserves.* Syn: *developed ore; measured ore; ore in sight; blocked-out ore; assured mineral.*

development [eco geol] (a) The preparation of a mining property or area so that an orebody can be opened up for analysis. (b) The preparation or opening up of an orebody to estimate its tonnage and quality. Development is an intermediate stage between *exploration* and mining.

development [grd wat] (a) In the construction of a water well, to remove fine-grained material adjacent to a drill hole, enabling water to enter the hole more freely. (b) Exploitation of ground water.

development well A well drilled within the known or proved productive area of an oil field, with the expectation of obtaining oil or gas from the producing formation or formations in that field. Cf: *exploratory well.*

deviation [drill] *deflection.*

deviation [geodesy] *deflection of the vertical.*

deviation [stat] (a) *mean deviation.* (b) *standard deviation.*

deviation of the vertical *deflection of the vertical.*

devilline A dark-green mineral: $Cu_4Ca(SO_4)_2(OH)_6.3H_2O$. Syn: *devillite.*

devil's slide (a) An *avalanche track* down a steep slope. (b) A long narrow mass of talus material descending a steep mountain.

devil's toenail [sed] Obsolete syn. of *stylolite.*

devitrification Conversion of the glassy texture of a rock to a crystalline texture after its solidification.

devolatilization In coal, the loss of volatile constituents and resulting proportional increase in carbon content during coalification. It is a process of metamorphism; the higher the rank of coal, the higher the level of devolatilization. Syn: *debitumenization.*

Devonian A period of the Paleozoic era (after the Silurian and before the Mississippian), thought to have covered the span of time between 395 and 345 million years ago; also, the corresponding system of rocks. It is named after Devonshire county, England, where rocks of this age were first studied. See also: *age of fishes.*

devonite A porphyritic diabase containing large potassium-rich labradorite phenocrysts in a groundmass of altered plagioclase and augite.

dew Condensation of atmospheric water vapor on a surface that is below the dew point but above the freezing point. If the surface were below the freezing point, the condensed water vapor would take the form of *frost.*

deweylite A mixture of clinochrysotile (or sometimes lizardite) and stevensite. It was formerly regarded as a mineral of the serpentine group. Syn: *gymnite.*

dewindtite A canary-yellow secondary mineral: $Pb(UO_2)_2^{\cdot}(PO_4)_2.3H_2O$.

dew point [phase] The temperature to which air must be cooled, at constant pressure and constant water-vapor content, in order for *saturation* to occur; the temperature at which the saturation pressure is the same as the existing vapor pressure.

dextral Pertaining, inclined, or spiraled to the right; specif. pertaining to the normal or clockwise direction of coiling of gastropod shells. A dextral gastropod shell in apical view (apex toward the observer) has the whorls apparently turning from the left toward the right; when the shell is held so that the axis of coiling is vertical and the apex or spire is up (as in orthostrophic shells) or down (as in hyperstrophic shells), the aperture is open toward the observer to the right of the axis. Actually, the definition depends on features of soft anatomy: with genitalia on the right side of the head-foot mass, the soft parts and shell are arranged such that the aperture is on the right when viewed with the apex (of orthostrophic shells) uppermost (TIP, 1960, pt.I, p.130). Ant: *sinistral*. Syn: *right-handed* [*paleont*].

dextral fault *right-lateral fault*.

dextral fold An asymmetric fold, the long limb of which appears to have a righthand offset, viewed along its dip. Cf: *sinistral fold*.

dextral imbrication The condition in a heterococcolith in which each segment overlaps the one to the right when viewed from the center of the cycle. Ant: *sinistral imbrication*.

dextrorotatory *right-handed*.

dhand A term used in Sind (region of West Pakistan) for a salt lake, esp. an alkali lake. Etymol: Sindhi.

D horizon A soil horizon that may be present below a B or a C horizon, consisting of unweathered rock.

diabantite A mineral of the chlorite group: $(Mg,Fe^{+2},Al)_6(Si,Al)_4O_{10}(OH)_8$. It occurs in cavities in basic igneous rocks.

diabase (a) In the U.S., an intrusive rock whose main components are labradorite and pyroxene and which is characterized by ophitic texture. As originally applied by Brongniart in 1807, the term corresponded to what is now recognized as *diorite*. Syn: *dolerite*. (b) In British usage, an intrusive igneous rock of the composition of diabase as defined in the U.S. but which has been highly altered by the decomposition of feldspars and mafic minerals.

diabasic (a) A syn. of *ophitic*. Kemp (1900, p.158-159) considered that "diabasic" applied to textures in which there was a predominance of plagioclase, with augite filling the interstices, while "ophitic" indicated a predominance of augite over plagioclase. (b) Composed of or resembling diabase.

diablastic Pertaining to a texture in metamorphic rock that consists of intricately intergrown and interpenetrating constituents with usually rodlike shapes (Becke, 1903).

diaboleite A sky-blue mineral: $Pb_2CuCl_2(OH)_4$.

diabrochite Metamorphic rock which owes its new mineralogical composition to intensive penetration by ascending solutions or vapors (ichor) or by partial fusion but without injection of visible granitic material as in migmatite (Dunn, 1942, p.37,234).

diachronic Pertaining to, or during, the time of the Earth's existence; considering events or changes as they happen or develop over time. Ant: *prochronic*. Cf: *synchronous*.

diachronism A term introduced by Wright (1926) for the transgression, across time planes or biostratigraphic zones, by a rock unit whose age varies from place to place; the state or condition of being diachronous.

diachronous Said of a rock unit that is of varying age in different areas or that cuts across time planes or biostratigraphic zones; e.g. said of a sedimentary formation related to a narrow depositional environment that shifted geographically (laterally) with advancing time and whose age consequently varies from place to place within the area of deposition (such as a marine sand that was formed during an advance or recession of a shoreline and that becomes younger in the direction in which the sea was moving). Syn: *time-transgressive*.

diachyte Rock product of marked mechanical and/or chemical contamination of anatectic magma by cognate basic material (Dietrich & Mehnert, 1961).

diaclinal Said of a stream or valley that passes through or across a fold, with a direction at right angles to the strike of the underlying strata it traverses. Also said of a region having diaclinal streams. Term introduced by Powell (1874, p.50). Ant: *paraclinal*.

diacrystallic Pertaining to the texture of a diagenetically recrystallized and essentially monomineralic rock in which contiguous crystals interpenetrate in a complicated manner (Phemister, 1956, p.72).

diactin A sponge spicule having two rays, usually monaxonic. See also: *rhabdodiactin*. Syn: *diact; diactine*.

diad Said of a symmetry axis that requires a rotation of 180 to repeat the crystal's appearance. It refers to *twofold symmetry*. Also spelled: *dyad*. Cf: *triad*. Syn: *digonal*.

diadactic Var. of *diatactic*.

diadochite A brown or yellowish mineral: $Fe_2(PO_4)(SO_4)(OH).5H_2O$. It is isomorphous with sarmientite. Syn: *destinezite*.

diadochy The replacement of one atom or ion in a crystal structure by another; *ionic substitution*; *proxying*.

diadysite A migmatite consisting of granitic-composition veins and metamorphic parent rock (Mehnert, 1968, p.354). Cf: *arterite; phlebite; venite*.

diaene A sponge spicule with two rays of equal length and one of a different length, usually longer; a triaene with one ray reduced or absent.

diagenesis [**mineral**] Recombination or rearrangement of a mineral resulting in a new mineral; specif. the geochemical, mineralogic, or crystallochemical processes or transformations affecting clay minerals before burial in the marine environment of sedimentation, such as illitization, glauconitization, or any transformation affecting the lattice of a clay mineral before burial. For a discussion of diagenesis in clay minerals see Keller (1963). Cf: *halmyrolysis*.

diagenesis [**sed**] All the chemical, physical, and biologic changes, modifications, or transformations undergone by a sediment after its initial deposition (i.e. after it has reached its final resting place in the current cycle of erosion, transportation, and deposition), and during and after its lithification, exclusive of surficial alteration (weathering) and metamorphism. This is the popular definition of the term as applied by most geologists in the U.S. (Twenhofel, 1939, p.254-255) and in Germany (Correns, 1950). It embraces those nondestructive or reconstructive processes (such as consolidation, compaction, cementation, reworking, authigenesis, replacement, solution, precipitation, crystallization, recrystallization, oxidation, reduction, leaching, hydration, dehydration, polymerization, adsorption, bacterial action, and formation of concretions) that occur under conditions of pressure (up to 1 kb) and temperature (maximum range of 100° C to 300° C) that are normal to the surficial or outer part of the Earth's crust, and may include changes occurring after lithification under the same conditions of temperature and pressure. The father of this concept, as defined above, was Walther (1893-1894, p.693-711), although the term "Diagenese" was first used by Gümbel (1868, p.838) for a postsedimentary transformation of sediments into individual crystalline minerals, leading to the creation of metamorphic rocks such as gneiss and schist. Russian (and some U.S.) geologists restrict the term to the initial phase of postsedimentary changes occurring in the zone where the sediment is still unconsolidated, the process being complete when the sediment has been converted to a more or less compact sedimentary rock (Fersman, 1922); in this usage, the term is equivalent to *early diagenesis* as used in the U.S. There is no universally accepted definition of the term, and no delimitation (such as the boundary with metamorphism), as it has been used in more and in less restricted senses by various authors; for a historical discussion and review of the usage of the term, see Larsen & Chilingar (1967) and Dunoyer de Segonzac (1968). Cf: *epigenesis* Syn: *diagenism*.

diagenetic Pertaining to or caused by diagenesis; e.g. a "diagenetic change" resulting from compaction, a "diagenetic structure" (such as a stylolite) formed after deposition, a "diagenetic deposit" (such as dolomitized limestone or one consisting of manganese nodules), or a "diagenetic environment" of rock consolidation. Syn: *postdepositional*.

diagenetic differentiation The redistribution of material within a sediment by solution and diffusion toward centers or nuclei where reprecipitation occurs, leading to segregation of minor constituents into diverse forms and structures, such as chert nodules in limestone or concretions in shale (Pettijohn, 1957, p.672).

diagenetic facies A facies that includes all rocks or sedimentary material that have, by a process of diagenesis, developed "mineral assemblages that are the result of adjustment to a particular diagenetic environment" (Packham & Crook, 1960, p.400). A "low-rank" facies corresponds to an early stage of alteration, a "high-rank" facies to a late stage. Cf: *parfacies*.

diagenic metamorphism *diagenism*.

diagenism A term used by Grabau (1904, p.235) as a syn. of *diagenesis* as defined by Walther (1893-1894, p.693-711). Syn: *diagenic metamorphism; static metamorphism*.

diagenite A diagenetic rock.

diagenodont Said of the dentition of a bivalve mollusk (e.g. *Astarte*) having differentiated cardinal teeth and lateral teeth located on the hinge plate, with the lateral teeth not exceeding two or the cardinal teeth not exceeding three in either valve. Cf: *teleodont*.

diaglomerate A conglomerate in which individual fragments are recognized as related fragments.

diaglyph A hieroglyph formed during diagenesis (Vassoevich, 1953, p.33).

diagnosis A description of a taxon that gives the characteristics which distinguish it from other taxa.

diagnostic fossil *characteristic fossil*.

diagnostic mineral A mineral, such as olivine or quartz, whose presence, in an igneous rock, indicates whether the rock is undersaturated or oversaturated. Syn: *symptomatic mineral*.

diagonal bedding An archaic syn. of *inclined bedding*, or bedding diagonal to the principal surface of deposition; specif. *cross-bedding*.

diagonal fault *oblique fault*.

diagonal joint A joint, the strike of which is oblique to that of the strike of the strata of the sedimentary rock, or to the cleavage plane of the metamorphic rock, in which it occurs. Syn: *oblique joint; (hkO) joint*.

diagonal lamination *cross-lamination*.

diagonal scour mark One of a series of scour marks arranged diagonally to the main direction of flow and formed by concentration of smaller scour marks (usually longitudinal flutes) into distinct rows that alternate with areas where scour marks are absent or less abundant.

diagonal-slip fault *oblique-slip fault*.

diagonal stratification *cross-stratification*.

diagrammatic map *cartogram*.

dial n. A compass used for surface and underground surveying, fitted with sights, spirit levels, and vernier, and mounted on a tripod.---v. To survey or measure with a dial and chain.

diallage (a) A dark-green or grass-green, brown, gray, or bronze-colored clinopyroxene (usually a variety of augite or of aluminum-bearing diopside) occurring in lamellae or in foliated masses and often having a metallic or brassy luster (schiller). It is characterized by a conspicuous parting parallel to the front pinacoid, and is typically found in basic igneous rocks such as gabbros. (b) A term applied to various poorly defined alteration products of pyroxene.

diallagite A pyroxenite composed almost entirely of diallage. Other pyroxenes, hornblende, ceylonite, and garnet may be present in very small amounts as accessories.

dialogite A syn. of *rhodochrosite*. Also spelled: *diallogite*.

dialysis A method of separating compounds in solution by their differing rates of diffusion through a semipermeable membrane, some colloidal particles not moving through at all, and others diffusing quite readily, some more slowly. Cf: *osmosis*. See also: *electrodialysis*.

diamagnetic Having a small, negative magnetic susceptibility. All materials which do not show paramagnetism or magnetic order are diamagnetic. Typical diamagnetic minerals are quartz and feldspar. Cf: *paramagnetic*.

diamantiferous *diamondiferous*.

diametral spine A basally fused spine opposite radial spines and passing through the diameter of the central capsule of an acantharian radiolarian.

diamict A general term proposed by Harland et al (1966, p.229) to include *diamictite* and *diamicton*.

diamictite A comprehensive, nongenetic term proposed by Flint et al (1960b) for a nonsorted or poorly sorted, noncalcareous, terrigenous sedimentary rock that contains a wide range of particle sizes, such as a rock with sand and/or larger particles in a muddy matrix; e.g. a tillite or a pebbly mudstone. Cf: *diamicton*. Syn: *mixtite*. Originally termed *symmictite* by Flint et al (1960a).

diamicton A general term proposed by Flint et al (1960b) for the nonlithified equivalent of a *diamictite*; e.g. a till. Originally termed *symmicton* by Flint et al (1960a).

diamond [mineral] (a) An isometric mineral, representing a naturally occurring crystalline form of carbon dimorphous with graphite and being the hardest substance known (hardness of 10 on Mohs' scale). It often occurs in octahedrons with rounded edges or curved faces. Diamonds form under extreme temperatures and pressures and are found in ultrabasic breccias, pipes in igneous rocks, and alluvial deposits. Pure diamond is colorless or nearly so, becoming tinted yellow, brown, reddish, orange, green, blue, violet, or black by impurities. When transparent and more or less free from flaws, it is the most cherished and among the most highly valued gemstones; its high refractive index and dispersive powers result in remarkable brilliance and play of prismatic color when faceted. Off-color or flawed diamonds are used for industrial purposes (such as in rock drills, abrasive powder, and cutting tools). (b) Artificially produced crystallized carbon similar to the native form. (c) A crystalline mineral that resembles diamond in brilliance, such as "Alencon diamond" (a smoky quartz sometimes valued as a jewel); esp. any of various kinds of rock crystal such as "Bristol diamond", "Herkimer diamond", "Lake George diamond", and "Arkansas diamond".

diamond [snow] A colloquial term for a glint of sunlight reflected from snow; usually used in the plural.

diamond bit A rotary-drilling bit studded with diamonds (usually borts). It is used for drilling extremely hard rock or in a program of continuous coring. Syn: *crown*.

diamond chip A thin, tabular *chip* of an uncut diamond crystal, weighing less than 0.75 carat.

diamond drilling A variety of rotary drilling in which diamond bits are used as the rock-cutting tool. It is a common method of prospecting for mineral deposits, esp. in development work where core samples are desired.

diamond dust [ice] Minute ice crystals, usually solid columns or bullet-shaped, precipitated out of clear air at very low temperatures as on the polar plates of Antarctica.

diamond dust [mater] Powdered, crushed, or finely fragmented diamond material used as a cutting, grinding, and polishing abrasive or medium. Syn: *diamond powder*.

diamondiferous Said of any substance (esp. rock or alluvial material) containing or yielding diamond. Syn: *diamantiferous*.

diamond powder *diamond dust*.

diamond spar *corundum*.

diamond structure A type of crystal structure in which there is fourfold coordination; minerals of this structure are characteristically rigid and have low electrical conductivity.

diancistra A C-shaped siliceous sponge spicule (microsclere) having sharply recurved ends and bearing blade-like lamellae on the inner side so that it resembles a partly opened penknife. Pl: *diancistrae*.

dianite *columbite*.

diaphaneity The light-transmitting quality of a mineral.

diaphanotheca The relatively thick, light-colored to transparent, intermediate layer of the spirotheca next below the *tectum* in fusulinids.

diaphorite A gray-black orthorhombic mineral: $Pb_2Ag_3Sb_3S_8$.

Syn: *ultrabasite*.

diaphragm Any of various more or less rigid partitions in the bodies or shells of invertebrates; e.g. a transverse, flat or gently curved, calcareous plate extending across the zooecial tube in cyclostomatous and Paleozoic bryozoans, dividing the zooecia into two chambers, or a thin crescentic plate of secondary brachiopod shell developed around the visceral disc (part of shell posterior to geniculate bend) of the brachial valve and restricting the gap between the brachial valve and the trail of the pedicle valve, or an imperforate partition crossing the siphuncle of a nautiloid, or a partial septum just below the aperture in a thecamoebian and perforated for protrusion of pseudopodia.

diaphthoresis *retrograde metamorphism*.

diaphthorite A crystalline schist in which minerals characteristic of a lower metamorphic grade have developed by retrograde metamorphism at the expense of minerals peculiar to a higher metamorphic grade. The term was originated by Becke in 1909.

diapir A dome or anticlinal fold, the overlying rocks of which have been ruptured by the squeezing out of the plastic core material. Diapirs in sedimentary strata usually contain cores of salt or shale; igneous intrusions may also show diapiric structure. See also: *diapirism*. Syn: *piercement; diapiric fold; piercing fold*.

diapiric fold *diapir*.

diapirism The process of piercing or rupturing of domed or uplifted overlying rocks by core material heated to the plastic state, either by tectonic stresses as in anticlinal folds, or by the effect of geostatic load in sedimentary strata as in salt domes or shale diapirs, or as in the case of igneous intrusions, forming diapiric structures such as plugs or batholiths. The concept was first applied to salt structures, which are the most common type of *diapir*. Obs. syn *tiphon*.

diaplectic Said of glass-like mineralogic features produced by shock waves in such a way that the characteristics of the liquid state are lacking; e.g. a "diaplectic mineral" whose disordered and deformed crystals have been modified by shock waves without melting and have characteristics such as planar features, lowered refractive indices, and lowered birefringence, or a "diaplectic glass" (of quartz, feldspar, or other minerals) representing an amorphous phase produced by shock waves without melting. Diaplectic materials represent intermediate stages of structural order between the crystalline and the normal glassy phases. Term proposed by Wolf von Engelhardt in 1966 (Engelhardt & Stöffler, 1968, p.163). Etymol: Greek *diaplesso*, "to destroy by beating or striking". Cf: *thetomorphic*.

diapositive A photographic *positive* on a transparent medium (usually glass or film), such as a small *transparency* of an aerial photograph on a glass plate and used in a plotting instrument or projector.

diara *char*.

diaresis A transverse groove on the posterior part of the exopod (rarely also the endopod) of a uropod of a malacostracan crustacean.

diarhysis A radial *skeletal canal* in dictyonine hexactinellid sponges penetrating the body wall completely, opened at each end, and containing a flagellated chamber. Pl: *diarhyses*.

diaschistic Said of a rock of a minor intrusion in which differentiation has occurred, so that its composition isn't the same as that of the parent magma. Such a rock may be called a *diaschistite*. Cf: *aschistic*.

diaschistite A *diaschistic* rock.

diaspore A white, gray, yellowish, or greenish orthorhombic mineral: AlO(OH). It represents the alpha base dimorphous with boehmite. Diaspore is found in bauxite and is associated with corundum and dolomite; it occurs in lamellar masses with pearly luster or in prismatic crystals. Syn: *diasporite*.

diaspore clay A high-alumina refractory clay consisting essentially of the mineral diaspore and believed to have developed at the expense of clay minerals by the leaching action of carbonate solutions passing through fractures in flint clays. Commercial diaspore clay commonly contains more than 70% alumina after calcination. See also: *burley clay*.

diastem A minor or relatively short interruption in sedimentation, involving only a brief or temporary interval of time, with little or no erosion before deposition is resumed; a depositional break of less magnitude than a *paraconformity*, or a paraconformity of very small time value. Diastems are not ordinarily susceptible of individual measurement, even qualitatively, because the lost intervals are too short; they are often deduced solely on paleontologic evidence. The term was introduced by Barrell (1917, p.794) for a slight discontinuity or small break in the marine sedimentary record, involving a "bed or series of beds" (represented elsewhere by deposits of less than formation rank), as contrasted with an *unconformity* representing a major break of longer duration. The synonymous term *non-sequence* is preferred in Great Britain. Adj: *diastemic*. Etymol: Greek *diastema*, "interval".

diastrophic Adj. of *diastrophism*. Cf: *orographic*.

diastrophic eustatism The worldwide change in sea level produced by a change in the capacity of the ocean basins because of diastrophic changes (Thornbury, 1954, p.142). Cf: *glacio-eustatism; sedimento-eustatism*. Se also: eustacy. Syn *tectono-eustatism*.

diastrophic plateau A plateau formed by the upheaval of a plain and cut, broken, or divided into parts by rivers, faults, or flexures (Powell, 1895, p.39-40).

diastrophism A general term for all movement of the crust produced by Earth forces, including the formation of ocean basins, continents, plateaus, and mountain ranges, etc. *Orogeny* and *epeirogeny* are major subdivisions. The use of this general term for small-scale features (e.g. diastrophic event, diastrophic ridge, diastrophic structure) is vague and undesirable; more specific terms should be substituted. Adj: *diastrophic*. Syn: *tectonism*.

diatactic Said of a sedimentary structure, as that shown by varves, characterized by the repetition of a pair of unlike laminae showing a gradation in grain size from coarse below to fine above. Syn: *diadactic*.

diataphral Descriptive of a type of tectonics in which *syntaphral* folds and faults are refolded by upward diapiric regurgitation of the axial zone of a geosyncline (Carey, 1963, p.A6). Cf: *apotaphral*.

diatexis High-grade (i.e. nearly, but not complete) *anatexis* involving rock components with high melting points (Dietrich & Mehnert, 1961). Cf: *metatexis; anamigmatization*.

diatexite Rock formed by diatexis. Also spelled: *diatectite*.

diathermancy The ability of a "wall" between two thermodynamic systems to transmit heat. See also: *diathermic*.

diathermic Said of a substance or "wall" between two thermodynamic systems that is able to transmit heat. Such a property is *diathermancy*. Two systems separated by a diathermic barrier will eventually reach thermal equilibrium with each other (Zemansky, 1957, p.5). Cf: *adiabatic*.

diatom A microscopic, single-celled plant growing in marine or fresh water. Diatoms secrete siliceous frustules of a great variety of forms that may accumulate in sediments in enormous numbers.

diatomaceous Composed of or containing numerous diatoms or their siliceous remains; e.g. "diatomaceous earth".

diatomaceous chert A diatomite that has a well-developed siliceous cement or groundmass.

diatomaceous earth A white, yellow, or light-gray *siliceous earth* composed predominantly of the opaline frustules of diatoms, accumulated esp. in lakes or swamps, and containing a great variation in the amount and nature of impurities such as sponge spicules, radiolarian remains, clay minerals, silica sand, and alkaline earths; the unconsolidated equivalent of *di-*

atomite. It is used as an absorbent (as in the manufacture of high explosives), filtering agent, light aggregate, insulator, and abrasive. See also: *tripoli; infusorial earth.* Syn: *tripolite; tripoli-powder; kieselguhr; fossil flour; fossil farina; rock meal; bergmehl; mountain meal.*

diatomaceous shale An impure diatomite with much clayey matter and with shaly partings.

diatomite (a) The dense, chert-like, consolidated equivalent of diatomaceous earth. (b) Indurated diatom ooze. (c) A term that is often applied as a syn. of *diatomaceous earth* or to the pulverulent siliceous deposit formed by the accumulation of diatom frustules; dried diatomaceous earth.

diatom ooze An *ooze* whose skeletal remains are the frustules of diatoms; it is a siliceous ooze.

diatom-saprocol *dysodile.*

diatreme A breccia-filled volcanic *pipe* that was formed by a gaseous explosion.

dibranchiate *coleoid.*

dicalycal Said of graptolithine theca from which two others originate.

dice mineral A term used in Wisconsin for galena occurring in small cubes.

dicentric Said of a corallite formed by a polyp retaining distomodaeal condition permanently.

dichotomy A repeated, twofold branching of the main axis of a plant, e.g. in liverworts and in some seaweeds.

dichotriaene A sponge *triaene* with dichotomously branched cladi.

dichroic Said of a mineral that displays *dichroism.*

dichroism *Pleochroism* of a crystal that is indicated by two different colors. A mineral showing dichroism is said to be *dichroic.* Cf: *trichroism.*

dichroite *cordierite.*

dichroscope An optical instrument that is used to analyze the colors of a pleochroic crystal; it consists of a calcite rhomb and a lens.

dickinsonite A green mineral: $H_2Na_6(Mn,Fe,Ca,Mg)_{14}(PO_4)_{12}.H_2O$. It is isostructural with arrojadite.

dickite A well-crystallized clay mineral of the kaolin group: $Al_2Si_2O_5(OH)_4$. It is polymorphous with kaolinite and nacrite. Dickite is structurally distinct from other members of the kaolin group, having a more complex order of stacking in the *c*-axis direction than kaolinite. It usually occurs in hydrothermal veins.

dicolpate Said of pollen grains having two colpi.

dicolporate Said of pollen grains having two colpi, with as least one colpus provided with a pore or transverse furrow. Dicolporate pollen are rare.

dicot *dicotyledon.*

dicotyledon An angiosperm whose seeds contain two embryonic, net-veined leaves. Such a plant has flower parts in fours or fives, reticulate leaf venation, and tricolpate pollen. Examples include roses, thistles, and oaks. Dicotyledons range from the Jurassic. Cf: *monocotyledon.* Syn: *dicot.*

dicranoclone (a) A tuberculate monaxonic desma (of a sponge) of dipodal to polypodal form with root-like terminal zygomes. (b) A desma (of a sponge) with arms that diverge from one side of a central point.

dictyonal framework The rigid interior skeleton of dictyonine sponges between the dermalia and gastralia, built of fused dictyonal strands, commonly (but not always) forming a cubic lattice.

dictyonalia The sponge spicules of a dictyonal framework.

dictyonal strand A linear series of hexactins (sponge spicules) in parallel orientation and fused to form a continuous strand.

dictyonine adj. Said of a hexactinellid sponge whose parenchymalia (spicules) form a rigid framework composed of dictyonal strands; more loosely, said of a sponge whose skeleton is composed of subparallel hexactins rigidly fused so that the limits of individual spicules are not apparent. Ant: *lyssacine.*---n. A dictyonine sponge.

dictyonite Migmatite with reticulated character, i.e. with a veinlet network (Dietrich & Mehnert, 1961). Var: *diktyonite.*

dictyospore A multicellular *fungal spore* that has both cross septa and longitudinal chitinous walls. Such spores may occur as microfossils in palynologic preparations.

dictyostele A *stele* consisting of separate vascular bundles, or of a network of bundles (Fuller & Tippo, 1949, p.956). See also: *meristele.*

dicyclic (a) Said of a crinoid having two circlets of plates proximal to radial plates or (in some inadunate crinoids that lack radial plates) proximal to oral plates. (b) Said of the apical system of an echinoid in which ocular and genital plates are arranged in two concentric circles, the genital plates alone in contact with periproctal margin.---Cf: *monocyclic.*

didodecahedron *diploid* [cryst].

diductor muscle A muscle that opens the valves in articulate brachiopods; commonly, one of a pair of two pairs of muscles attached to the brachial valve immediately anterior to the beak, usually to the cardinal process. The principal pair is commonly inserted in the pedicle valve on either side of the *adductor muscles* and the accessory pair is inserted posterior to them. Syn: *diductor; divaricator.*

didymoclone A desma (of a sponge) consisting of a short, straight shaft, from the enlarged ends of which several zygome-bearing arms project, predominantly on one side of the spicule.

didymolite A dark-gray monoclinic mineral: $Ca_2Al_6Si_9O_{29}$.

die-back A large area of exposed and unprotected swamp or marsh deposits which are the result of the salinity of a coastal lagoon being increased by either natural or artificial means and, subsequently, the natural freshwater reed-swamp vegetation diminishing or dying off completely. These unprotected deposits are then eroded by wave action.

diel Of or pertaining to a 24-hour day.

dielectric Said of a material in which displacement currents predominate over conduction currents, i.e. an insulator.

dielectric constant A measure of the displacement currents resulting from application of an electric field. Symbol: K_e.

dielectric displacement The displacement of one charge center relative to another, upon application of an electric field.

dielectric loss The time rate of energy loss in a dielectric material due to conduction and the hysteresis in polarization.

dielectric strength The maximum electric field that a dielectric can sustain without destructive breakdown.

dienerite A gray-white isometric mineral: Ni_3As.

diesel squeeze A *squeeze job* in which dry cement mixed with diesel oil is pumped through casing perforations to recement water-bearing areas and leave oil-bearing areas unaffected.

Diestian European stage: lower Pliocene (above Pontian, below Plaisancian).

dietella A small enclosed space near the base of the vertical walls in the distal part of the zooid in certain cheilostomatous bryozoans. Its walls contain communication pores traversed by mesodermal fibers. Syn: *pore chamber.*

dietrichite A mineral: $(Zn,Fe,Mn)Al_2(SO_4)_4.22H_2O$.

dietzeite A dark golden-yellow mineral: $Ca_2(IO_3)_2(CrO_4)$.

differential compaction A kind of *compaction* produced by uneven settling of homogeneous earth material under the influence of gravity (as where thick sediments in depressions settle more rapidly than thinner sediments on hilltops) or by differing degrees of compactability of sediments (as where clay loses more interstitial water and comes to occupy less volume than sand).

differential curvature A quantity represented by the acceleration due to a gravity times the difference in the curvatures in the two principal planes, that is, g(1/p1 - 1/p2), where p1 and p2 are the radii of curvature of the two principal planes.

differential entrapment The control of oil and gas migration and accumulation by selective trapping or gas flushing in interconnecting reservoirs. A trap filled with oil is an effective gas trap but a trap filled with gas is not an effective oil trap. As a result, gas is trapped downdip and oil is flushed updip (Gussow, 1954).

differential erosion Erosion that occurs at irregular or varying rates, caused by the differences in the resistance and hardness of surface materials: softer and weaker rocks are rapidly worn away while harder and more resistant rocks remain to form ridges, hills, or mountains.

differential fault *scissor fault*.

differential force Stress that causes distortion of a body because it isn't equal at all points. Cf: *hydrostatic pressure*. Syn: *directed pressure*.

differential infrared line-scan system An image-forming system which produces quantitative imagery, i.e., the gray scale of the images produced is not directly related to the incident flux striking the detector by a known energy-transfer function. Temperature calibration points on the terrain are required to establish a crude relationship between gray-scale increments and radiation temperature of the terrain sensed (Friedman, 1970, p.35). Cf: *radiometric line-scan system*.

differential leveling The process of measuring the difference of elevation between any two points by spirit leveling.

differential melting Partial melting of a rock, in which part of the rock remains solid, as a function of differences in melting points of minerals.

differential nutation The small difference between the *nutation* of a moving object and the nutation of the fixed background stars.

differential pressure The difference in pressure between the two sides of an orifice; the difference between reservoir and sand-face pressure; between pressure at the bottom of a well and at the well head; between flowing pressure at the well head and that in the gathering line; any difference in pressure between that upstream and downstream where a restriction to flow exists.

differential settlement Nonuniform *settlement*; the uneven lowering of different parts of an engineering structure, often resulting in damage to the structure.

differential solution *intrastratal solution*.

differential stress In experimental rock deformation, the maximum principal stress minus the least principal stress.

differential thermal analysis *Thermal analysis* carried out by uniformly heating or cooling a sample material which undergoes chemical and physical changes while simultaneously heating or cooling in identical fashion a reference material which undergoes no changes. The temperature difference between the sample and the reference material is measured as a function of the temperature of the reference material. Symbol: DTA. Syn: *thermography*.

differential weathering Weathering that occurs at irregular or different rates, caused by variations in composition and resistance of a rock or by differences in intensity of weathering, and usually resulting in an uneven surface where harder material stands higher or protrudes above softer parts. Syn: *selective weathering*.

differentiate n. A rock formed by magmatic differentiation.

differentiated Said of an igneous intrusion in which there is more than one rock type, due to *differentiation*.

differentiation [intrus rocks] (a) The process of developing more than one rock type, in situ, from a common magma. Cf: *assimilation*. See also: *differentiated*. Syn: *magmatic differentiation*. (b) In a broader sense, crystallization or recrystallization phenomena that occur in a magma as it cools, e.g. *magmatic segregation*.

differentiation [sed] (a) *sedimentary differentiation*. (b) *diagenetic differentiation*.

differentiation index In igneous petrology, the number that represents the sum of the weight percentages of normative quartz, orthoclase, albite, nepheline, leucite, and kalsilite

(Thornton & Tuttle, 1960); a numerical expression of the extent of differentiation of a magma.

diffission Hobb's term for the natural process whereby rocks are broken into fragments and blocks (as much as 8 m in diameter in western Texas) by a cloudburst or downpour of rain (1912, p.204).

diffluence (a) A lateral branching or flowing apart of a glacier in its ablation area. This separation may result from the glacier's spilling over a preglacial divide or through a gap made by basal sapping of a cirque wall, or from downvalley blocking at the junction of a tributary glacier. (b) A place at which diffluence occurs.---Ant: *confluence*.

diffluence pass The lower part of a trough end, where a distributary glacier has left the main valley.

diffluence step A rock step that rises downstream away from the main glacial valley at the place of diffluence. It is probably caused by the weakening of glacial action at that point. Ant: *confluence step*.

diffluent Said of a stream or glacier that flows away or splits into two or more branches. Ant: *confluent*.

diffraction [phys] The process by which the direction of light and other forms of radiant energy (as X-rays, electrons, and neutrons) is modified by the bending of radiation around the edges of obstacles, and the resultant formation of an interference pattern within the *geometric shadow* of the obstacle. Also, the scattering of light rays when reflected from a ruled surface having scattering centers separated by distances that are comparable to the wavelength of the radiation.

diffraction [seis] Any deviation of a ray, specif. a seismic ray, that cannot be interpreted as *reflection* or *refraction*.

diffraction [water waves] The bending of a wave in a body of water around an obstacle, e.g. the interruption of a wave train by a breakwater or other barrier.

diffraction grating *grating*.

diffraction pattern The interference pattern of lines obtained when waves of rays, such as X-rays, light rays, or particle rays, are passed through a small opening or around the edge of a particle. Each substance has a characteristic diffraction pattern, which, when found, is taken to be evidence that that substance is present.

diffraction spacing In a crystal lattice, interplanar spacings given by a diffraction pattern.

diffractogram A record of diffraction of a crystalline sample, obtained by electronic detectors and recorded on a paper chart.

diffractometer In mineral analysis, an instrument that records X-ray powder diffraction patterns as an inked trace on a strip chart.

diffuse layer The outer, mobile layer of ions, in an electrolyte, required to satisfy a charge unbalance within a solid with which the electrolyte is in contact. It constitutes part of the *double layer* of charge adjacent to the electrolyte-solid interface. Cf: *fixed layer*.

diffuse-porous wood A type of wood in which the vessels are more or less uniform in size and distribution throughout each annual ring (Fuller & Tippo, 1949, p.957).

diffusivity *thermal diffusivity*.

digenite A blue to black mineral: Cu_9S_5. It is isometric and occurs with chalcocite. Syn: *blue chalcocite; alpha chalcocite*.

digestion Partial or complete chemical incorporation of wall rock into a magma.

digital computer A *computer* that operates with numbers expressed directly as digits in a decimal, binary, or other system; a counting device that operates on discrete or discontinuous variables represented by digits of numbers and performs arithmetic by manipulating the digits and executing the basic arithmetic operations in a manner similar to a human mathematician. Cf: *analog computer*.

digital log A well log that has been discretely sampled and recorded on a magnetic tape for use in computer-processed interpretation and plotting.

digital seismograph A seismograph whose data are recorded in digital form, so that they can be processed directly by a digital computer.

digitate delta A *bird-foot delta* whose seaward-extending margin has a finger-like outline in plan.

digitation The emanation of subsidiary recumbent anticlines from a larger recumbent anticline.

digitize To sample a continuous function at discrete time intervals and to record the values as a sequence of numbers.

digonal *diad*.

digue *dike* [eng].

dihexagonal Said of a symmetrical, twelve-sided figure, the alternate angles of which are equal. Such a figure is common in crystals of the hexagonal system.

dihexagonal dipyramid A crystal form that is a dipyramid of 24 faces, in which any section perpendicular to the sixfold axis is dihexagonal. Its indices are $\{hkl\}$ and its symmetry is $6/m\ 2/m\ 2/m$.

dihexagonal dipyramidal class That crystal class of the hexagonal system having symmetry $6/m\ 2/m\ 2/m$.

dihexagonal prism A crystal form of twelve faces parallel to the symmetry axis, in which any cross section perpendicular to the prism axis is dihexagonal. Its indices are $\{hk0\}$ with symmetry $6/m\ 2/m\ 2/m$.

dihexagonal pyramid A crystal form consisting of a pyramid of 12 faces, in which any cross section perpendicular to the sixfold axis is dihexagonal. Its indices are $\{hkl\}$ or $\{hkl\}$ in symmetry $6mm$.

dihexagonal-pyramidal class That crystal class in the hexagonal system having symmetry $6mm$.

dihydrite *pseudomalachite*.

dike [eng] An artificial wall, embankment, ridge, or mound, usually of earth, stones, or riprap, built around a relatively flat, low-lying area to protect it from flooding by checking, stopping, or deflecting the water currents of a sea, lake, or stream; a levee. A dike may also be constructed on the shore or border of a lake to prevent inflow of undesirable water into the lake. Syn: *digue; dyke*.

dike [intrus rocks] A tabular igneous intrusion that cuts across the planar structures of the surrounding rock. Also spelled: *dyke*. Cf: *sill* [intrus rocks]; *sheet* [intrus rocks]. See also: *dikelet*.

dike [sed] *sedimentary dike*.

dike [streams] (a) An artificial watercourse; esp. a deep drainage ditch. The term has also been applied to any channel, including those formed naturally. (b) A pool or small pond.

dike chamber A *magma chamber* whose width is relatively small in relation to its length and height. Syn: *fissure chamber*.

dike compartment A crudely rectangular body of permeable basalt bounded by relatively impermeable dikes and sills. It contains high-level ground water which eventually works its way downward and outward to become basal water. The term originated in the Hawaiian Islands.

dikelet A small *dike*. There is no agreement on specific size distinctions.

dike ridge (a) *dike wall*. (b) A small wall-like ridge (as one along a shore) produced by differential erosion.

dike rock The intrusive rock comprising a dike, generally considered to be of hypabyssal texture or character. Syn: *dykite*.

dike set A group of linear or parallel dikes. Cf: *dike swarm; cluster*.

dike spring A spring issuing from the contact between a dike composed of an impermeable rock, such as basalt or dolerite, and a permeable rock into which the dike was intruded.

dike swarm A group of dikes, either radial from a single source or in parallel, linear arrangement. Their relationship with the plutonic body may not be directly observable. Cf:

cluster; *dike set; swarm*.

dike wall A ridge, such as a *hogback*, consisting of a dike that formed in a more or less vertical crevice and that was left standing after the rocks on either side were removed by erosion. Syn: *dike ridge*.

diktyonite *dictyonite*.

diktytaxitic Said of the rock texture occurring in some olivine basalts in the Pacific northwest, that is characterized by numerous jagged, irregular vesicles bounded by crystals, some of which protrude into the cavities (Dickinson, 1965, p.101).

dilatancy An increase in the bulk volume during deformation, caused by a change from close-packed structure to open-packed structure, accompanied by an increase in the pore volume. The latter is accompanied by rotation of grains, microfracturing, grain boundary slippage, etc.

dilatated septum A partly or wholly thickened septum of a rugose coral.

dilatation [seis] *kataseism*.

dilatation [exp struc geol] *dilation* [exp struc geol].

dilatational transformation In a crystal, usually rapid thermal dilation and a rearrangement of the anion from cubic coordination to octahedral coordination, due to heating, e.g. the transformation of CsCl to the NaCl structure at $460°$ C.

dilatational wave *P wave*.

dilatation theory The theory that attributed glacier movement to infiltration and freezing of water in cracks and other openings.

dilation Deformation by a change in volume but not shape. Also spelled: *dilatation*.

dilation dike An igneous dike intruded into a fracture, causing the walls of the fracture to move away from each other.

dilation vein A mineral deposit in a vein space formed by bulging of the walls, contrasted with veins formed by wall-rock replacement.

dillnite A mineral: $Al_{15}Si_6O_{25}(OH,F)_{18}Cl$. It is a variety of zunyite rich in fluorine.

Diluvial That period of geologic time since the appearance of man. Cf: *anthropozoic*.

diluvial (a) Pertaining to, produced by, or resembling a flood, esp. the Noachian flood; e.g. *diluvial* deposits. (b) Pertaining to *diluvium*.

diluvialist A believer in *diluvianism*. Cf: *fluvialist*.

diluvianism The doctrine that the widespread surficial deposits (now known to be glacial drift) and other geologic phenomena can be explained by a former worldwide flood or deluge.

diluvion (a) *diluvium*. (b) A term used in India as an ant. of *alluvion*; "it appears to mean loss of land by river erosion after flooding" (G.T. Warwick in Stamp, 1961, p.157).

Diluvium A term used in continental Europe equivalent to *Pleistocene*.

diluvium (a) An archaic term applied during the early 1800s to certain widespread surficial deposits that could not be explained by the normal action of rivers and seas but were believed to be produced by extraordinary floods of vast extent, esp. the Noachian flood; these deposits are now known to be mostly glacial drift. (b) A general term used in continental Europe for the older Quaternary, or Pleistocene, glacial deposits, as distinguished from the younger *alluvium*. Syn: *drift*.---Syn: *deluvium; diluvion*.

dimble An English term for a *dingle* with a watercourse.

dimensional analysis A calculation in which physical quantities are expressed qualitatively in terms of their fundamental dimensional units. It is used when accurate data are lacking and to spot-check calculations involving several types of units.

dimensional orientation In structural petrology, a tendency of elongate (planar or linear) fabric elements to be so oriented so that their longer axes are approximately parallel.

dimension stone Building stone that is quarried and prepared in blocks according to specifications.

197

dimictic Said of a lake with two yearly overturns or periods of circulation, such as a deep, freshwater lake in a temperate climate, with overturns in the spring and fall. Cf: *monomictic.*

dimorph Either of two crystal forms displaying *dimorphism.* Partial syn: *allomorph.*

dimorphic [**cryst**] *dimorphous.*

dimorphic [**paleobot**] Said of a plant or plant part that occurs in two forms, e.g. both juvenile and adult foliage types. Syn: *dimorphous.*

dimorphism [**cryst**] That type of *polymorphism* in which there occur two crystal forms, known as *dimorphs.* Adj: *dimorphous.* Cf: *trimorphism; tetramorphism.*

dimorphism [**biol**] The characteristic of having two distinct forms in the same species, such as male and female, megaspheric and microspheric stages.

dimorphite An orange-yellow mineral: As_4S_3. It was originally described as one of two dimorphous substances (the other, however, being orpiment).

dimorphous [**cryst**] Adj. of *dimorphism.* Syn: *dimorphic.*

dimorphous [**paleobot**] *dimorphic.*

dimple crater A small, almost circular crater-like feature restricted to a mare region of the Moon's surface and attributed to volcanic activity (possibly to withdrawal of molten subsurface lava). It lacks the raised rim of most lunar impact craters.

dimpled current mark An obsolete syn. of *cross ripple mark.*

dimyarian adj. Said of a bivalve mollusk or its shell with two adductor muscles, whether equal (isomyarian) or unequal (anisomyarian) in size. Cf: *monomyarian.*---n. A dimyarian mollusk.

Dinantian European stage: Lower Carboniferous. It includes Tournaisian and Viséan. Syn: *Avonian.*

Dinarides The southern flank of the great orogenic belt called the *Alpides.*

Dinas rock A disintegrated sandstone of high silica content, formerly used for making refractory brick. Type locality: Craig-y-Dinas, a crag in south Wales. Syn: *Dinas clay.*

dinite A yellowish-crystalline hydrocarbon mineral which has a low melting point (30°C) and is found in lignite.

dinoflagellate A one-celled, microscopic, chiefly marine, usually solitary flagellated organism with resemblances to both animal (motility, ingestion of food) and plant (photosynthesis) kingdoms, characterized by one transverse flagellum encircling the body and usually lodged in the girdle and one posterior flagellum extending out from a similar median groove. Certain dinoflagellates have a theca (test) that is resistant to decay and that may be simple and smooth or variously sculptured and divided into characteristic plates and grooves; others produce a resting stage or cyst with a resistant organic wall that is often very spiny and may differ markedly from the theca of the same species. Both thecae and cysts exist abundantly as fossils, and have a range primarily Triassic to present (dinoflagellates are known from the Paleozoic, but are mainly important for correlating and dating Jurassic, Cretaceous, and Tertiary deposits). Dinoflagellates inhabit all water types and are capable of extensive diurnal vertical migrations in response to light; they constitute a significant element in marine plankton, including certain brilliantly luminescent forms and those that cause red tide. See also: *hystrichosphaerid.*

dinosaur leather A local term applied by Chadwick (1948) to complex sole marks, probably including both flute casts and load casts.

dioctahedral Said of a layered-mineral structure in which only two of the three available octahedrally coordinated positions are occupied. Cf: *trioctahedral.*

dioecious Said of a taxonomic unit of plants whose members have staminate and pistillate flowers on different plants (Lawrence, 1951, p.749).

diogenite An achondritic stony meteorite composed essentially of bronzite or hypersthene. Syn: *rodite.*

diopside A mineral of the clinopyroxene group: $CaMg(SiO_3)_2$. It contains little or no aluminum and may contain some iron, and varies in color from white to green; transparent varieties are used in jewelry. Diopside occurs in some metamorphic rocks, and is found esp. as a contact-metamorphic mineral in crystalline limestones. Symbol: Di. Syn: *malacolite.*

diopside-jadeite *tuxtlite.*

diopsidite A hypabyssal rock composed almost entirely of diopside, with iron ore, ceylonite, and garnet as common accessories.

dioptase A rare emerald-green hexagonal mineral: $CuSiO_2(OH)_2$. It occurs in the zone of weathering of copper lodes in Chile and Siberia. Syn: *emerald copper.*

diorite A group of plutonic rocks intermediate in composition between acidic and basic rocks, characteristically composed of dark-colored amphibole (esp. hornblende), acid plagioclase (oligoclase, andesine), pyroxene, and sometimes a small amount of quartz; also, any rock in that group; the approximate intrusive equivalent of *andesite.* Diorite grades into *monzonite* with an increase in the alkali feldspar content. In typical diorite, plagioclase contains less than 50% anorthite, hornblende predominates over pyroxene, and mafic minerals total less than 50% of the rock. Etymol: Greek *diorizein,* "to distinguish", in reference to the fact that the characteristic mineral, hornblende, is usually identifiable megascopically. Cf: *dolerite; gabbro.* See also: *diabase.*

dip [**geomorph**] (a) A low place or marked depression in the land surface; e.g. a steep-sided hollow among hills or a gap in a ridge. (b) A pronounced depression in a highway at the point of intersection with a dry stream bed; it is esp. common in the western U.S.

dip [**magnet**] *magnetic inclination.*

dip [**seis**] (a) The angle between a reflecting or refracting seismic wave and the horizontal. (b) The angle between a seismic discontinuity surface and the horizontal.

dip [**struc geol**] n. The angle that a structural surface, e.g. a bedding or fault plane, makes with the horizontal, measured perpendicular to the *strike* of the structure. See also: *attitude; hade; inclination.* Syn: *true dip; angle of dip.*

dip [**surv**] (a) The vertical angle, at the eye of an observer, between the plane of the horizon and the line of sight tangent to the apparent (visible or sensible) horizon; the angular distance of the apparent horizon below the horizontal plane through the observer's eye. See also: *dip angle.* Also called: *dip of horizon.* (b) The apparent depression of the visible horizon due to the observer's elevation and to the convexity of the Earth's surface. (c) The first detectable decrease in the altitude of a celestial body after reaching its maximum altitude on or near the meridian transit.

dip angle The vertical angle, measured at an observation point in surveying or at an exposure station in photogrammetry, between the plane of the true horizon and a line of sight to the apparent horizon. See also: *dip.*

dip calculation *migration of dips.*

dip circle An obsolete type of *inclinometer.*

dip-corrected map A map that shows strata in their original position before movement.

dip equator *magnetic equator.*

dip fault A fault that strikes parallel with the dip of the strata involved. Cf: *strike fault; oblique fault.*

diphyletic Said of an organism that developed along two lines of descent.

dip joint A joint, the strike of which is approximately perpendicular to the bedding or cleavage of the constituent rock. Cf: *strike joint.*

dipleural Said of the arrangement of the two rows of thecae in the biserial rhabdosome of a scandent graptoloid in which the rows are in contact back-to-back. Cf: *monopleural.*

diploblastic Said of the structure of lower invertebrates (sponges, coelenterates) having ectodermal and endodermal layers but lacking a true mesoderm.

diploclone A sublithistid desma (of a sponge) consisting of a shaft bearing expansions at each end but which does not articulate with neighboring spicules.

diploconical Said of a radiolarian shell formed by fusion of bases of two cones opposite in one axis (TIP, 1954, pt.D, p.14).

diplodal Said of a flagellated chamber of a sponge that has both an aphodus and a prosodus.

dip log A log that shows the dips of the formations traversed by a borehole; e.g. a *dipmeter log*.

diplogenetic Said of a mineral deposit that is in part *syngenetic* and in part *epigenetic* in origin. (Lovering, 1963, p.315-316).

diplohedron *diploid* [cryst].

diploid n. A crystal form of the isometric system having 24 similar quadrilateral faces in a paired arrangement. Each face intersects the crystallographic axes at unequal lengths. Its indices are $\{hkl\}$ in symmetry $2/m\overline{3}$. Syn: *didodecahedron; diplohedron; dyakisdodecahedron.*

diploidal class That crystal class in the isometric system having symmetry $2/m\overline{3}$.

diplopore Any of double pores piercing a thecal plate in certain cystoids and mostly confined to that plate. It usually consists of a Y-shaped branching (but sometimes unbranched) canal or tube that is oblique or perpendicular to the surface of the plate and that has two external openings at the outer end. Cf: *pore rhomb; haplopore.*

diplorhysis The condition of a dictyonine hexactinellid sponge in which both epirhyses and aporhyses are present.

diplotype The holotype of a type species. Syn: *genoholotype; genotype.*

diploxylonoid Said of bisaccate pollen, in which the outline of the sacci in distal-proximal view is discontinuous with the body outline so that the grain appears to consist of three distinct, more or less oval figures. Cf: *haploxylonoid.*

dipmeter An instrument that measures the amount (angle) and direction of dip of geologic formations, esp. those exposed in the sides of a borehole. Also spelled: *dip meter.*

dipmeter log A *dip log* produced by dipmeter measurements, in which three units (such as three electrodes), oriented with respect to magnetic north and spaced 120 degrees apart in a plane perpendicular to the borehole, define the slope of specific resistive layers within the hole.

dip needle An obsolete type of magnetometer used for iron-ore exploration. It consists of a magnetized needle pivoted to rotate freely in a vertical plane, with an adjustable weight on one side of the pivot.

dipole Two poles of opposite charge an infinitesimal distance apart.

dipole-dipole array An electrode array in which one dipole (a connected pair of electrodes) provides current to the ground and an adjacent dipole facilitates measurement of potential in the ground. The separation between dipoles is usually comparable to or greater than the spacing within each electrode pair constituting a dipole. The potential dipole lies entirely outside the current dipole.

dipole field A mathematically simple magnetic field, having an axis of symmetry, with magnetic field lines pointing outward along one half of the axis (*positive pole*) and inward along the negative half of the axis (*negative pole*). Most magnetic fields that are sufficiently remote from their source resemble a dipole field. See also: *axial dipole field.*

diporate Said of pollen grains having two pores.

dipping rod *divining rod.*

dip plain A *stratum plain* coincident in slope with the dip of the underlying resistant rock (Hill, 1891, p.522).

dip poles The locations on the Earth where the horizontal magnetic field is zero and the magnetic inclination is ±90°. Partial syn: *magnetic poles.*

dip reversal A growth-fault structure in which the dip of the beds of the downthrown wall curve towards the fault surface in a way that is exactly opposite to that produced by drag. Syn: *reverse drag; rollover; turnover.*

dip separation The distance of *separation* of formerly adjacent beds on either side of a fault surface, measured along the dip of the fault. Cf: *dip slip; strike separation.* See also: *dip-separation fault.*

dip-separation fault A fault. the displacement of which has been *dip separation.* Cf: *lateral fault.*

dip shift In a fault, the *shift* or relative displacement of the rock units parallel to the dip of the fault, but outside the fault zone itself; a partial syn. of *dip slip.* Cf: *strike shift.*

dip shooting Any system of seismic surveying in which the primary concern is the registration and computation of reflections for dip values.

dip slip In a fault, the component of the movement or slip that is parallel to the dip of the fault. Cf: *dip separation; strike slip; oblique slip.* Partial syn: *dip shift.* Syn: *normal displacement.*

dip-slip fault A fault, the actual movement of which is parallel to the dip of the fault. Cf: *strike-slip fault.*

dip slope A slope of the land surface, roughly determined by and approximately conforming with the direction and the angle of dip of the underlying rocks; specif. the long, gently inclined face of a cuesta. Cf: *scarp slope.* Syn: *back slope; outface.*

dip stream A consequent stream flowing in the direction of the general dip of the strata it traverses; a *cataclinal* stream.

dip throw The component of the slip of a fault measured parallel with the dip of the strata.

dip valley A valley trending in the direction of the general dip of the strata of a region; a valley at right angles to a subsequent stream.

dipyramid A closed crystal form consisting of two *pyramids* that are arranged base to base so that they appear as mirror images across the plane of symmetry. Adj. *dipyramidal.* Syn: *bipyramid.*

dipyramidal Having the symmetry of a *dipyramid.*

dipyre A syn. of *mizzonite;* specif. a term applied to a variety of scapolite with the components marialite and meionite in a ratio of about 3:1 to 3:2. Syn: *dipyrite.*

dipyrite (a) *dipyre.* (b) *pyrrhotite.*

dipyrization *scapolitization.*

direct angle An angle measured directly between two lines; e.g. a horizontal angle measured clockwise from a preceding surveying line to a following one.

direct intake Recharge to the aquifer directly through the zone of saturation.

direction (a) The position of one point relative to another without reference to the distance between them. It may be three-dimensional or two-dimensional. (b) The angle between a line or plane and an arbitrarily chosen reference line or plane; specif. the angle between a great circle passing through both the position of the observer and a given point on the Earth's surface and a true north-south line passing through the observer. When the reference line is north and the angle is designated east or west, the direction is called the *bearing;* when the reference line is south and the angle is reckoned clockwise, the direction is called the *azimuth.* (c) A syn. of *trend.*

direction-action avalanche A snow avalanche that occurs during or immediately after a storm and is the direct result of that storm. Ant: *delayed-action avalanche.*

directional drilling The intentional drilling of a borehole or well at an angle with the vertical along a predetermined course, and often controlled by a whipstock. It is used to drill a well under a park or swamp, to tap an oil field that is located beneath a town or that extends offshore, to shut off a blowout, to deflect a crooked hole back on course, or to bypass an obstruction in the hole. Syn: *slant drilling.*

directional load cast A term originally applied to a structure interpreted as a flowage cast, but now regarded as a flute cast (Pettijohn & Potter, 1964, p.301).

directional log A log that shows the drift (deviation from the vertical) of a borehole or well, and the direction of the drift; e.g. a *photoclinometer log*.

directional structure Any sedimentary structure that indicates the direction of the current that produced it; e.g. cross-bedding, current marks, and ripple marks. Syn: *paleocurrent structure; aligned current structure; vector structure.*

directional well A well produced by directional drilling. Syn: *slant well.*

direction of dip *line of dip.*

direction of the wind That point of the compass from which the wind blows, e.g. a "westerly" wind is blowing from (rather than toward) the west. It may also be stated in degrees, measured clockwise from the north, e.g. an east wind has a direction of 90°.

direction of tilt (a) The azimuth of the principal plane of a photograph. (b) The direction of the principal line on a photograph.

direction theodolite A theodolite in which the graduated horizontal circle remains fixed during a series of observations, the telescope being pointed on a number of signals or objects in succession, and the direction of each read on the circle, usually by means of micrometer microscopes. Syn: *direction instrument.*

directive couple A pair of mesenteries in the so-called dorsoventral (extending from the dorsal toward the ventral side) plane of a coral polyp, characterized by pleats on the opposite rather than the facing sides of the mesenteries.

direct lattice A syn. of *crystal lattice,* used when comparison is made with the *reciprocal lattice.*

direct leveling A type of *leveling* in which differences of elevation are determined by means of a continuous series of short horizontal lines, the vertical distances from these lines to adjacent ground marks being determined by direct observations on graduated rods with a leveling instrument equipped with a spirit level. Cf: *indirect leveling.*

direct linkage A type of *linkage* in scleractinian corals with mesenterial strands connecting the adjacent stomodaea. See also: *lamellar linkage.* Cf: *indirect linkage.*

direct runoff The *runoff* reaching stream channels immediately after rainfall or snow melting (Langbein & Iseri, 1960). Cf: *base runoff.* Syn: *direct surface runoff; immediate runoff; stormflow; storm runoff; storm water.*

direct stratification *primary stratification.*

direct surface runoff *direct runoff.*

direct tide An oceanic tide that is in phase with the apparent motions of the tide-producing body, so that high tide is directly under the tide-producing body and is accompanied by a high tide on the opposite side of the Earth. Cf: *reversed tide; opposite tide.*

Dirichlet's problem One of three well known *boundary value problems.* Cf: *Neumann's problem.*

dirt band [glac] (a) Any dark layer in a glacier, usually the trace of silt or debris along a *summer surface.* (b) A dark band below an icefall that may be related to dirt collected in the broken ice of the icefall or between the ridge of one *wave ogive* and another.----Syn: *dust band.* (c) A term that was originally applied to a *Forbes band.*

dirt band [coal] A thin stratum of shale, mudstone, or soft, earthy material in a coal seam or interbedded with coal seams. Syn: *dirt bed; dirt parting; stone band.*

dirt-band ogive A curved band or *ogive* composed of debris- or dirt-laden ice that may be related to dirt collected in the broken ice of an icefall or between the ridge of one *wave ogive* and another. Cf: *dirt band [glaciol];* Forbes band. See also: *Alaskan band.*

dirt bed [soil] A *paleosol* whose organic material is only partially decayed. It sometimes occurs in glacial drift.

dirt bed [coal] *dirt band.*

dirt cone A cone or mound of ice or snow on a glacier, covered with a veneer of silt thick enough to protect the underlying material from the ablation that has lowered the surrounding surface. Cf: *debris cone.*

dirt parting *dirt band* [coal].

dirt slip *clay vein.*

dirty arkose *impure arkose.*

dirty sand A term used in electrical prospecting for a sandstone, with abundant clay, which exhibits appreciable membrane polarization and abnormally high electrical conductivity because of surface conduction along the clay minerals.

dirty sandstone An impure sandstone containing much matrix; specif. a wacke with more than 10% argillaceous matrix (Gilbert, 1954) or a graywacke with more than 15% detrital clay matrix (Pettijohn, 1954). The particles are held together by primary, fine-grained interstitial detritus or matrix of clay-like nature, or by authigenic derivatives of such material. It is usually deposited by fluids of high density (or of high viscosity). Cf: *clean sandstone.*

disaccate *bivesiculate.*

disaggregation Separation or reduction of an aggregate into its component parts; specif. *mechanical weathering.*

disappearing stream *lost stream.*

disc (a) The central part of the body of an echinoderm, more or less distinctly separable from its arms. Sometimes spelled *disk.* (b) A discoidal, typically imperforate sclerite of a holothurian.

discharge [sed] *sediment discharge.*

discharge [hydraul] In hydraulics, the rate of flow at a given moment in time, expressed as volume per unit of time. See also: *specific discharge.*

discharge area An area in which subsurface water, including both ground water and vadose water, is discharged to the land surface, to bodies of surface water, or to the atmosphere. Cf: *recharge area.*

discharge coefficient That coefficient by which a theoretical discharge must be multiplied to obtain the actual discharge. It is the product of the contraction coefficient and the velocity coefficient (ASCE, 1962).

discharge efficiency *drainage ratio.*

discharge-rating curve *stage-discharge curve.*

discharge velocity The rate of discharge of water through a porous medium, measured per unit of total area perpendicular to the direction of flow (ASCE, 1962).

disciform Of round or oval shape; e.g. "disciform holococcolith" having a discolith-like shape and a raised margin two or more cycles of microcrystals high.

discinacean Any inarticulate brachiopod belonging to the superfamily Discinacea, characterized by holoperipheral growth of the brachial valve, with the beak marginal to central.

discoaster One of the tiny star- or rosette-shaped calcareous plates, 10-35 microns in diameter, that are generally agreed to be the remains of a planktonic organism and that may be isolated coccolith-like bodies either of a motile cell or cyst. Discoasters are common in Tertiary deposits but are apparently absent in the Pleistocene; the level where they disappear has been suggested as a Pliocene-Pleistocene boundary. See also: *asterolith.*

discohexaster A hexactinal sponge spicule (microsclere) in which the ray tips bear branches terminated by umbels.

discoid Adj. Having the shape of a disk; e.g. a solitary corallite. n. An object having such a shape. ----Syn: *discoidal.*

discoidal *discoid.*

discolith A discoidal coccolith with a single, apparently imperforate elliptical or circular shield and with a thickened margin. Cf: *tremalith.*

disconformable [intrus rocks] Said of the contact of an intrusive body when it is not aligned with the intrusion's internal structures. Cf: *conformable; discordant.*

disconformable [stratig] Pertaining to a stratigraphic disconormity. Term proposed by Grabau (1905, p.534) to refer to ormations that exhibit parallel bedding but "compromise beween them a time break of greater or less magnitude".

disconformity An *unconformity* in which the bedding planes above and below the break are essentially parallel, indicating a significant interruption in the orderly sequence of sedimentary rocks, generally by a considerable interval of erosion (or sometimes of nondeposition), and usually marked by a visible and irregular or uneven erosion surface of appreciable relief; e.g. an unconformity in which the older rocks remained essentially horizontal during erosion or during simple vertical rising and sinking of the crust (without tilting or faulting). The tendency is to apply the term to breaks represented elsewhere by rock units of formation rank (Stokes & Varnes, 1955, p.157). The term formerly included what is now known as *paraconformity*. Syn: *parallel unconformity; erosional unconformity; nonangular unconformity; stratigraphic unconformity; paraunconformity*.

discontinuity [seism] A surface at which seismic-wave velocities abruptly change; a boundary between seismic layers of the Earth. Syn: *interface; seismic discontinuity; velocity discontinuity*.

discontinuity [stratig] Any interruption in sedimentation, whatever its cause or length, usually a manifestation of nondeposition and accompanying erosion; an unconformity. Syn: *break*.

discontinuity [struc geol] A surface separating two unrelated groups of rocks; e.g. a fault or an unconformity. See also:*discrete*.

discontinuity layer A *thermocline* in a lake or ocean.

discontinuous deformation Deformation by fracture rather than flow. Cf: *continuous deformation*.

discontinuous gully A gully with a vertical headcut at the upstream end and a fan at the point where the plane of the gully floor intersects the more steeply sloping plane of the original valley floor (Leopold et al, 1964, p.448-449). The depth of its channel decreases downstream.

discontinuous permafrost A zone of permafrost containing patches (*taliks*) of unfrozen ground, as beneath large rivers or lakes; it occurs in an intermediate zone between the northerly *continuous permafrost* and the southerly *sporadic permafrost*.

discontinuous porosity A term originally proposed by Murray (1930, p.452) for poorly connected or isolated pores, as opposed to *continuous porosity*. The term is little used now and not recommended (Choquette & Pray, 1970, p.245).

discontinuous reaction series A *reaction series* in which reaction of early-formed crystals with later liquid represents an abrupt phase change; e.g., the minerals olivine, pyroxene, amphibole, and biotite form a discontinuous reaction series. Cf: *continuous reaction series*.

discordance (a) Lack of parallelism between adjacent strata. The term was used by Willis (1893, p.222) in cases where the process resulting in absence of parallelism is in doubt. Although the term has not been widely adopted, "it appears eminently suitable for descriptive use where there is insufficient evidence to decide between stratigraphic discordance (unconformity) and tectonic discordance (e.g. overthrust, slide, detachment fault)" (Dennis, 1967, p.36). (b) *angular unconformity*.

discordance index A numeric statistic used by Pearn (1964, p.401) to represent the amount of deviation of any actual rock sequence from the *ideal cyclothem*. It is defined as the minimum value of the number of missing lithologic units. Symbol: G.

discordant [geomorph] Said of topographic features that do not have the same or nearly the same elevation; e.g. a *discordant* valley whose stream enters the main stream via a waterfall, or a *discordant* lip over which the floor of a hanging valley passes into the floor of the main valley. Ant: *accordant*.

discordant [intrus rocks] Said of a contact between an igneous intrusion with the country rock which does not parallel the foliation or bedding planes of the latter. Cf: *concordant; disconformable*.

discordant [stratig] Structurally *unconformable*; said of strata lacking conformity or parallelism of bedding or structure. Ant: *concordant*.

discordant [geochron] (a) Disagreement, beyond experimental error, of radiometric ages determined by more than one method for the same sample or coexisting minerals. (b) Disagreement of radiometric ages given by coexisting minerals determined by the same method. (c) In a more restricted sense, the term has been used to indicate disagreement of U^{238}-Pb^{206}, U^{235}-Pb^{207}, Pb^{207}-Pb^{206}, and Th^{232}-Pb^{208} ages determined for the same mineral sample. Discordant ages usually imply that one or more of the isotopic systems used for dating purposes has been disturbed by some geologic event (metamorphism, weathering) following the initial formation of the geologic material or by inadvertent laboratory procedures. Ant: *concordant*.

discordant bedding A sedimentary structure in which parallelism of beds is lacking or in which sedimentary layers are inclined to the major lines of deposition, such as bedding developed by rapid deposition of material from heavily laden currents of air or water; specif. *cross-bedding*. Ant: *concordant bedding*. See also: *inclined bedding*.

discordant coastline A coastline that develops where the general structural grain of the land (such as mountain chains or folded belts) is transverse or at right angles to the margin of the ocean basin, and that represents faulting, subsidence, or a general interruption of a formerly continuous and harmonious structure; it is generally irregular, with many inlets. Ant: *concordant coastline*. Syn: *Atlantic-type coastline; transverse coastline*.

discordant drainage Drainage that has not developed in a systematic relationship with, and is not consequent upon, the present geologic structure. Ant: *accordant drainage*.

discordant fold A fold, the axis of which is inclined to that of the *longitudinal fold* axis of the area.

discordant junction The joining of two streams or two valleys whose surfaces are at markedly different levels at the place of junction, as the abrupt entry of a tributary flowing at a high level into a main stream at a lower level. Ant: *accordant junction*.

discordant margin A margin of closed valves (of a bivalve mollusk) not in exact juxtaposition, but with one overlapping the other.

discordogenic fault A fault in a tectonic belt that separates zones of uplift and subsidence, and that remains active during several geologic periods (Nikolaev, 1959).

discorhabd A sponge spicule (streptaster) consisting of a straight shaft bearing whorls of spines or transverse discoidal flanges. Syn: *chessman spicule*.

discotriaene A sponge *triaene* in which the cladome is represented by a transverse disk containing the axial canals of the three cladi.

discovery The actual finding of a valuable mineral, indicative of a deposit (lode or placer or coal seam). Legally, a discovery is a prerequisite to making a mining claim on an area.

discovery claim A claim containing the original discovery of mineral deposits, and which may lead to claims being made on adjoining lands.

discovery vein The original mineral deposit on which a mining claim is based. Cf: *secondary vein; discovery claim*.

discovery well The first well to encounter gas or oil in a hitherto unproven area.

discrepancy A term used in surveying for the difference in computed values of a quantity obtained by different processes using data from the same survey; e.g. the difference in the length of two measures of the same line, or the amount by which the values of a position of the third point of a triangle as computed from the two other points may fail to agree when the triangle has not been corrected for *closure*. See also: *accumulated discrepancy.*

discrete [paleont] Said of conodont denticles that are not closely set, each denticle being separated from adjacent denticles by open space. Cf: *appressed.*

discrete [struc geol] Said of any body of rock which has a definite boundary with adjacent rocks in space. See also:*discontinuity.*

discrete [weath] A term proposed by Gilbert (1898) as an adjective that is descriptive of the surficial, weathered, and unconsolidated material composing the regolith.

discrete-film zone *belt of soil water.*

discriminant analysis A statistical procedure for classifying subsequent samples into categories previously defined and differentiated on the basis of samples from known populations. Syn: *discriminant function analysis.*

disembogue To discharge water through an outlet or into another body of water, such as a stream *disemboguing* into the ocean.

disequilibrium assemblage An association of minerals not in thermodynamic equilibrium (Fyfe et al, 1958).

disharmonic fold A fold, the form of which is modified by the varying strata through which it passes. Ant: *harmonic fold.*

disharmonic folding Folding of sedimentary strata in which *bedding-plane slip* along the less competent layers cause structural discordance between the layers. It is a characteristic type of fold deformation and in fact is virtually demanded by the geometry of concentric folds. An associated structure is *décollement*. Ant: *harmonic folding.*

dish structure A primary sedimentary structure developed in the interior of coarse turbidites, consisting of small meniscus-shaped lenses (4-50 cm long and one to a few centimeters thick) that are oval in plan, oriented parallel to the bedding, and defined by slightly finer-grained, concave-up bottoms each of which truncates the underlying lenses. It is thought to form during rapid aggradation in antidune flow, or possibly to result from internal shear within aggrading sand beds. Examples occur in Cretaceous and Paleocene sandstones exposed in sea cliffs in the northern Coast Ranges of California. Term introduced by Wentworth (1967).

disintegration [coal] The decomposition of vegetable matter by slow combustion, in which there is no formation of carbon compounds and in which only volatile substances (carbon dioxide and water) are produced. Cf: *moldering; peat formation; putrefaction.*

disintegration [glaciol] *ice disintegration.*

disintegration [radioactivity] *radioactive decay.*

disintegration [weath] A syn. of *mechanical weathering*, or the breaking down or reduction of a rock to smaller fragments mainly by physical forces; e.g. *granular disintegration*. Sometimes the term includes chemical action, in which case it is practically synonymous with "weathering". Cf: *decomposition.*

disintegration constant *decay constant.*

disjunct (a) Said of a pore rhomb in which externally visible slits forming parts of the rhomb are separated by solid (slit-free) areas of plates. Cf: *conjunct.* (b) Said of the apical system of an echinoid whose anterior part is separated from the posterior part.

disjunctive fold A fold in which the more brittle strata have fractured and separated and the more plastic beds have flowed under the forces of deformation.

disk [paleont] (a) The flattened circumoral part of a coelenterate (such as a sea anemone). See also: *oral disk; basal disk.* (b) *disc.*

disk [sed] A notably discoidal (flat and circular), or oblate or tabular, shape of a sedimentary particle, defined in *Zingg's classification* as having a width/length ratio greater than 2/3 and a thickness/width ratio less than 2/3.

disk hardness gage A device for measuring snow hardness in terms of the resistance of snow to the horizontal pressure exerted by a disk attached to a spring-loaded rod.

dislocation [cryst] *line defect.*

dislocation [struc geol] A syn. of *displacement*, with reference to fault movement.

dislocation breccia *fault breccia.*

dislocation metamorphism A form of dynamic regional metamorphism concentrated along narrow belts of shearing or crushing without an appreciable rise in temperature. The term was originated by Lossen in 1883 and is considered to be equivalent to *dynamometamorphism*. Cf: *dynamic metamorphism; cataclastic metamorphism.*

dismal *pocosin.*

dismembered drainage A complex drainage system that has been altered by dismembering, thus creating a series of independent streams that enter the sea by separate mouths.

dismembered stream A tributary that is left as an independent stream after the lower part of the drainage system to which it formerly belonged was submerged by an invasion of the sea. Cf: *betrunked river.*

dismembering The making of a tributary into an independent stream by a change of geologic conditions, esp. by the submergence of the lower part of a valley by the invasion of the sea. Cf: *betrunking.*

dismicrite A fine-textured limestone with less than 1% allochems, consisting mainly of lithified carbonate mud (micrite), and containing irregular patches or bird's-eyes of sparry calcite filling cavities caused by local disturbances (Folk, 1959 p.28). Syn: *bird's-eye limestone.*

disomatic A term applied to a crystal now called a *xenocryst.*

disorder in minerals In a substitutional solid solution, the random occupation of one atom site in a crystal by two or more different atoms of similar size and charge, or of similar size and different charge if there is a concomitant substitution to balance charges, as in plagioclase, in which (Na and Si) in albite is substituted by (Ca and Al) as the composition approaches anorthite. Cf: *order in minerals.*

dispellet limestone A term used by Wolf (1960, p.1416) for a pelleted limestone with tubules or irregular patches of sparite.

dispersal [glac geol] *glacial dispersal.*

dispersal [ecol] The spreading of a species by migration into new areas having conditions favorable for its existence.

dispersal center The place on a delta at which the first stream distributary branches off from the main channel (Moore, 1966 p.92).

dispersal map A stratigraphic map that shows the inferred source area and the direction or distance of transportation of clastic materials (Krumbein & Sloss, 1963, p.484).

dispersal shadow An accumulation of sediments formed downcurrent from a generating source (Pettijohn, 1957, p.574); e.g. a boulder train on the lee side of a resistant knob overridden by ice. Cf: *sedimentary petrologic province.*

dispersed element An element that is generally too rare and unconcentrated to become an essential constituent of a mineral and which therefore occurs principally as a substituent of the more abundant elements.

dispersed phase Colloidal material suspended in another phase, which in turn is known as the *dispersion medium.*

dispersion [soils] Breaking down or separation of soil aggregates into single grains. Cf: *detachment.*

dispersion [phys] The dependence of an observed quantity upon frequency, e.g. the variation of the velocity of seismic surface waves with frequency, due to the varying elastic properties of the layers of the Earth. See also: *normal dispersion, inverse dispersion.*

dispersion [seis] The variation of the velocity of seismic surface waves with frequency, due to the varying elastic properties of the layers of the Earth. See also: *normal dispersion; inverse dispersion.*

dispersion [optics] In optics, the differences in the optical constants, e.g. wavelengths and indices of refraction, of a ven mineral for different wavelengths of the spectrum. See so: *dispersion curve*.

spersion [stat] The range or scatter of values about a central tendency; a statistical spread or variability. Common measures of dispersion are standard deviation and sorting.

spersion curve [seis] A plot of seismic-wave velocity versus riod or frequency.

spersion curve [optics] The plotting on a logarithmic scale of crystal's *dispersion*.

spersion ellipse The ground area, usually elliptical in shape, vered by a meteorite shower. Syn: *strewn field*.

spersion flow Flow of granular sediment in which collisions tween particles maintain the fluidity of the material.

spersion medium That material (solid, liquid, or gas) in hich colloidal material, known as the *dispersed phase*, is spended.

spersion ratio In a soil, the ratio of the percentage of silt d clay that remains suspended (after a standard agitation ocedure) to the percentage of the soil's clay and silt as alyzed mechanically.

spersive power The refractive ability of a transparent substance, usually isotropic; it is symbolized by δ and equals n_F- $/n_{D-1}$, in which n is the refractive index for Fraunhofer lines C, and D: C = hydrogen discharge at wavelength 656.3, F hydrogen discharge at wavelength 486.1, and D = sodium ame at wavelength 589.3, measured in nannometers. Syn: *lative dispersion*.

spersive stress *Bagnold dispersive stress*.

sphenoid A closed crystal form consisting of two *sphenoids*, which the two faces of the upper sphenoid alternate with ose of the lower sphenoid. Adj: *disphenoidal*. Syn: *bisphenoid*.

sphenoidal Having the symmetry of a disphenoid, e.g. rhom-c 222.

sphotic zone That part of the ocean in which there is only m light and little photosynthesis; it lies between the *aphotic* ne and the *euphotic zone*, or may be considered as the wer part of the euphotic zone. Syn: *twilight zone*.

splacement [photo] Any shift in the position of an image on photograph which does not alter the perspective character-tics of the photograph. It may be caused by the relief of the jects photographed, the tilt of the photograph, changes of ale in the photograph, or atmospheric refraction. Cf: *distor-n*.

splacement [struc geol] The relative movement of the two des of a fault, measured in any chosen direction; also, the ecific amount of such movment. Displacement in an apparntly lateral direction includes strike slip and strike separa-n; displacement in an apparently vertical direction includes p slip and dip separation. Syn: *dislocation*.

splacement meter A seismometer designed to respond to e displacement of Earth particles. This is accomplished by aking its natural period much larger than that of the ground otion.

splacement plane *deformation plane*.

splacement shear A fracture surface which develops on earing, parallel or subparallel to the direction of relative ovement. It often occurs in the lateral, marginal zones of rthflows.

splacement theory *continental displacement*.

splacive transformation A high-low type of crystal *transformation*, e.g. in high-low quartz at 573°C, involving rotation of O$_4$ tetrahedra with no breaking of bonds. It is usually a rapid ansformation. Cf: *dilatational transformation; reconstructive ansformation; rotational transformation; substitutional trans-rmation*.

srupture Separation along a fracture in a rock. The term is tle used.

ssected pediment A pediment that exhibits a considerable gree of dissection; it is generally regarded as a product of

second-cycle erosion of an originally nearly flat pediment, although it may also be "born dissected"

dissected peneplain An ancient and uplifted peneplain that has become the initial surface upon which erosion begins to cut the forms of a new cycle; a partially destroyed peneplain represented in a maturely dissected region only by a few remnants, such as plateaus or occasional flat-topped mountains and ridges.

dissected plateau A plateau in which a large part of the original level surface has been deeply cut into by streams.

dissection The process of erosion whereby the continuity of a relatively even topographic surface is gradually sculptured or destroyed by the formation of gullies, ravines, canyons, or other kinds of valleys; esp. the work of streams in cutting or dividing the land into hills and ridges, or into flat upland areas, separated by fairly close networks of valleys. The process is applicable esp. to surfaces, such as plains and peneplains, that have been uplifted. Adj: *dissected*.

disseminated Said of a mineral deposit (esp. metals) in which the minerals occur as scattered particles in the rock, but in sufficient quantity to make the deposit a worthwhile ore. There is no genetic connotation. Cf: *impregnated*.

dissepiment (a) A small domed calcareous plate forming a *vesicle* or cyst-like enclosure typically occurring between radiating septa in the peripheral region of a corallite. Its convex surface faces inward and upward. (b) A crossbar between connecting branches of a fenestrated bryozoan colony. (c) A curved or flat, subhorizontal, imperforate plate forming a cyst-like connection between septa in the intervallum of an archaeocyathid. (d) A crossbar or strand of periderm uniting adjacent branches (stipes) in a dendroid graptolite colony or rhabdosome, such as in *Dictyonema*.

dissepimentarium The peripheral zone of the interior of a corallite, occupied by dissepiments. See also: *regular dissepimentarium*.

dissociation constant The equilibrium constant for a dissociation reaction, defined as the product of activities of the products of dissociation divided by the activity of the original substance. When used for ionization reactions, it is called an *ionization constant*; when it refers to a very slightly soluble compound, it is called a *solubility product*.

dissociation point That temperature at which a compound breaks up reversibly to form two or more other substances, e.g. CaCO$_3$ becoming CaO plus CO$_2$. All variables should be stated in order to define the point precisely. The term dissociation refers to the breakup itself, and covers a wide variety of types, such as the breakup of molecular groupings in gases or liquids.

dissociation temperature A presumed fixed temperature point at which a given dissociation occurs; in fact, however, it is usually a range of temperature due to the compositional or pressure variance, and may refer merely to the temperature at which the rate of a given dissociation becomes appreciable, under stated conditions.

dissoconch The postlarval shell of a bivalve mollusk.

dissolution A space or cavity in or between rocks, formed by the solution of part of the rock material. Cf: *discission*.

dissolved load The part of the total *stream load* that is carried in solution, such as chemical ions yielded by erosion of the landmass during the return of rainwater to the ocean. Syn: *dissolved solids; solution load*.

dissolved oxygen The amount of dissolved oxygen, in parts per million by weight, present in water, now generally expressed in mg/l. A critical factor for fish and other aquatic life, and for self-purification of a surface-water body after inflow of oxygen-consuming pollutants. Abbrev: D.O.

dissolved solids (a) *dissolved load*. (b) A term that expresses the quantity of dissolved material in a sample of water, "either the residue on evaporation, dried at 180°C, or, for many waters that contain more than about 1000 parts per million, the sum of determined constituents" (USGS, 1958, p.50).

distal [**paleont**] Remote or away from the point of attachment or place of reference, a point conceived of as central, or the point of view. Examples in invertebrate morphology: "distal direction" away from crinoid theca toward holdfast or free lower extremity of column or away from the mouth or center of disc of an asterozoan, or opposite to the proloculus in the growth of a foraminiferal shell; "distal (last-formed) portion" of the rhabdosome of a graptolite colony, farthest away from the point of origin; "distal ray" of a sponge spicule directed toward a bounding surface of the sponge; "distal shield" on the convex side of a placolith; and "distal side" away from the ancestrula or origin of growth of a bryozoan colony. Ant: *proximal*.

distal [**palyn**] Said of the parts of pollen grains or spores away from the center of the original tetrad; e.g. said of the side of a monocolpate pollen grain upon which the colpus is borne, or said of the side of a spore opposite the laesura. Ant: *proximal*.

distal [**sed**] Said of a sedimentary deposit consisting of fine clastics and formed farthest from the source area; e.g. a "distal turbidite" consisting of thin silty varves, or the most remote foreland deposit derived from the borderland in a geosynclinal region. Cf: *proximal*.

distance-function map A term used by Krumbein (1955) for a map now known as *facies-departure map*.

distant admixture A term applied by Udden (1914) to an *admixture* (in a sediment of several size grades) whose particles are most different in size from those of the dominant or maximum grade; material in one of the two classes at the extreme ends of a histogram.

disthene *kyanite*.

disthenite A metamorphic rock composed almost entirely of kyanite (disthene) and some quartz, often associated with magnetiferous quartzite and amphibolite (LaCroix, 1922, p.497).

distichous Said of a plant that has leaves, leaflets, or flowers on opposite sides of the same point on a stem; two-ranked.

distillation [**paleont**] A process of fossilization whereby the liquid and/or gaseous components of an organic substance are removed leaving a carbonaceous residue.

distillation [**water**] Conversion of a liquid to a vapor by the addition of heat and returning the vapor to a liquid again by cooling, as in the purification of water.

distinctive mineral *varietal mineral*.

dististele The distal region of a crinoid column. Cf: *proxistele*.

distomodaeal budding A type of budding in scleractinian corals in which two stomodaea are developed within a common tentacular ring and two interstomodaeal couples of mesenteries are located between the original and each new stomodaeum.

distortion [**cart**] The change in shape on a map projection due to lack of conformality; specif. *angular distortion*. See also: *zero distortion*.

distortion [**photo**] Any shift in the position of an image on a photograph which alters the perspective characteristics of the photograph. It may be caused by lens aberration, differential shrinkage or expansion of film or paper, motion of the film or camera, and irregular motions (roll, pitch, yaw, speed changes) of the aircraft. Cf: *displacement*.

distortional wave *S wave*.

distributary [**stream**] (a) An irregular, divergent stream flowing away from the main stream and not returning to it, as in a delta or on an alluvial plain. It may be produced by stream deposition choking the original channel. Ant: *tributary*. (b) One of the channels of a braided stream; a channel carrying the water of a stream distributary. Syn: *distributary channel*.

distributary glacier Any ice stream or lobe that flows away or forks off from the lower part of a glacier; a subsidiary terminus or outlet of a trunk glacier. Cf: *outlet glacier*. See also: *glacial lobe*.

distributed fault *fault zone*.

distribution [**ecol**] *range* [ecol].

distribution [**stat**] An array of the instances of a variable arranged by classes according to their value; e.g. *frequency distribution*.

distribution coefficient *distribution ratio*.

distribution ratio The ratio of concentrations of a solute in two immiscible solvents. Syn: *distribution coefficient*.

distributive fault *step fault*.

distributive province The environment embracing all rocks that contribute to the formation of a contemporaneous sedimentary deposit and the agents responsible for their distribution (Milner, 1922, p.366). Cf: *provenance*.

distromatic Said of a thallus that is composed of two layers of cells; said of a blade of a plant that is two cells in thickness. Cf: *monostromatic; oligostromatic; polystromatic*.

disturbance A term used by some geologists for a minor orogeny; Schuchert (1924), for example, used *revolution* for major orogeny at the end of an era, and the term disturbance for an orogeny within an era; this usage is dubious and should be discontinued. Cf: *event; phase* [tect]; pulsation.

disturbing potential The difference between the *geopotential* and the spheropotential at a given point. It is also referred to as *potential disturbance, potential of random masses*, and *potential of disturbing masses*.

ditch [**drill**] A trough used in rotary drilling for conducting drilling mud from the collar of the borehole to the slush pit.

ditch [**water**] A long, narrow excavation artificially dug in the ground; esp. an open and usually unpaved waterway, channel, or trench for conveying water for a drainage or irrigation, and usually smaller than a canal. Some ditches may be natural watercourse.

ditetragonal Said of a crystal form having eight similar faces the alternate interfacial angles of which are equal.

ditetragonal dipyramid A crystal form that is a dipyramid of 16 faces in which any section perpendicular to the fourfold axis is ditetragonal. Its indices are $\{hkl\}$ in symmetry $4/m\,2/m\,2/m$.

ditetragonal-dipyramidal class That crystal class in the tetragonal system having symmetry $4/m\,2/m\,2/m$.

ditetragonal prism A crystal form of eight faces parallel to the symmetry axis in which any section perpendicular to the prism axis is ditetragonal. Its indices are $\{hk0\}$ with symmetry $4/m\,2/m\,2/m$.

ditetragonal pyramid A crystal form consisting of eight faces in a pyramid, in which any section perpendicular to the fourfold symmetry axis is ditetragonal. Its indices are $\{hkl\}$ $\{hk\bar{l}\}$ in symmetry $4mm$.

ditetragonal-pyramidal class That crystal class in the tetragonal system having symmetry $4mm$.

ditrigonal Said of a symmetrical, eight-sided figure, the alternate angles of which are equal. Such a figure is characteristic of certain crystal forms in the hexagonal system.

ditrigonal dipyramid A crystal form that is a dipyramid of twelve faces in which any section perpendicular to the threefold or sixfold symmetry axis is ditrigonal. Its indices are $\{hkl\}$ or $\{khl\}$ in symmetry $\bar{6}\,m2$.

ditrigonal-dipyramidal class That class in the hexagonal system having symmetry $\bar{6}\,m2$.

ditrigonal prism A crystal form of six faces parallel to the symmetry axis, in which any section perpendicular to the axis is ditrigonal. Its indices are $\{hk0\}$ or $\{kh0\}$ in symmetry $\bar{6}\,m$ or $3m$.

ditrigonal pyramid A crystal form consisting of six faces in a pyramid, in which any section perpendicular to the symmetry 3 axis is ditrigonal. Its indices are $\{hkl\}$, $\{hk\bar{l}\}$, $\{khl\}$, $\{kh\bar{l}\}$ in symmetry 3.

ditrigonal-pyramidal class That class in the rhombohedral division of the hexagonal system having symmetry $3m$.

ditrigonal-scalenohedral class *hexagonal-scalenohedral class*.

ditroite A *nepheline syenite* containing alkali feldspar and less amounts of sodalite, biotite, and cancrinite. The term was introduced by Zirkel in 1866, and later Brögger proposed applying it to nepheline syenites having granular texture (Johannsen, 1939, p.249). Its name is derived from Ditrau (Ditró Hungarian), Romania. Cf: *foyaite.*

Dittonian Series in the Old Red Sandstone of England: Lower Devonian (upper Gedinnian; above Downtonian).

diurnal current A tidal current that has only one flood period and one ebb period during a tidal day.

diurnal inequality The difference between the heights and durations of the two successive high waters or of the two successive low waters of a tidal day.

diurnal tide A tide with only one high water and one low water occurring during a tidal day, as in the Gulf of Mexico.

diurnal variation *daily variation.*

diurnal vertical migration A pattern of daily movement of certain marine organisms that is upward at sunset and downward at sunrise, in response to changes in light and other factors.

diurnal wave The daily temperature rise of soil, which under the heating of the sun progresses downward from the surface as a heavily dampened wave and dies out about 30 cm below the surface. Below this point, relatively constant daily temperature may be experienced. Cf: *annual wave.* Syn: *daily wave.*

divagation The lateral shifting of a stream course as a result of extensive deposition of alluvium in its bed, esp. accompanied by the development of meanders.

divariant Said of a chemical system having two degrees of freedom. Syn: *bivariant.*

divaricator A muscle that causes divergence or separation of parts; specif. a *diductor muscle* of a brachiopod.

divergence [glac geol] Interruption of a drainage pattern by the advance of a glacier or ice sheet.

divergence [meteorol] A decrease of an air mass within a given volume; greater outflow than inflow of air. Cf: *convergence [meteorol].*

divergence [currents] The separation of ocean currents by horizontal flow of water in different directions from a common source, usually upwelling; also; the area in which divergence occurs. Cf: *convergence.*

divergence [evol] *adaptive radiation.*

diversion (a) The process by which a stream actively effects changes in the drainage or course of another stream, as by degradation or capture. Cf: *derangement.* (b) The artificial draining, pumping, siphoning, or other removal of water from a stream, lake, or other body of water, into a canal, pipe, or other conduit. (c) A channel designed to divert water from a body of water for purposes such as prevention of flooding, reduction of erosion, or promotion of infiltration.

diverted stream A stream whose course or drainage has been affected by another stream; e.g. a captured stream.

diverter *diverting stream.*

diverting stream A stream that effects diversion; e.g. a capturing stream. Syn: *diverter.*

Divesian *Oxfordian.*

divide [grd wat] A ridge in the water table or other potentiometric surface from which the ground water represented by that surface moves away in both directions. Water in other aquifers above or below, and even in the lower part of the same aquifer, may have a potentiometric surface lacking the ridge, and so may flow past the divide. Syn: *water-table divide; ground-water divide; ground-water ridge.*

divide [streams] (a) The line of separation, or the ridge, summit, or narrow tract of high ground, marking the boundary between two adjacent drainage basins or dividing the surface waters that flow naturally in one direction from those that flow in the opposite direction; the line forming the rim of or enclosing a drainage basin; a line across which no water flows. An "anomalous" divide is one that does not follow the crest of the highest mountain range of a mountain chain. See also: *continental divide.* British syn: *watershed.* Syn: *drainage divide;*

water parting; height of land; topographic divide; watershed line. (b) A tract of relatively high ground between two streams; a line that follows the summit of an interfluve.(c) The highest summit of a pass or gap. (d) *ground-water divide.*

diviner *dowser.*

divining *dowsing.*

divining rod Traditionally, a forked wooden stick used in *dowsing* which supposedly dips downward sharply when held over a body of ground water or a mineral deposit, thus revealing its presence. Syn: *witching stick; wiggle stick; dowsing rod; mineral rod; dipping rod; twig; dowser.* Cf: *water witch; waterfinder.*

division [bot] A major unit in the taxonomy of plants, ranking below "kingdom" and above "class". Equivalent to *phylum* in the classification of animals.

division [stratig] A term proposed by Størmer (1966, p.25) as a time-stratigraphic unit equivalent to stage (and possibly series), but having a local to regional, or a more limited, geographic range. Cf: *chronozone.*

divisional plane A general term that includes joints, cleavage, faults, bedding planes, etc.

dixenite A black hexagonal mineral: $Mn_5(SiO_3)(AsO_3)_2(OH)_2$.

djalmaite *microlite* [mineral].

djebel Var. of *jebel.*

djerfisherite A meteorite mineral: $K_3(Cu,Na)(Fe,Ni)_{12}S_{14}$. It occurs in enstatite chondrites.

djurleite A mineral: $Cu_{1.96}S$. Its X-ray pattern is near to, but distinct from, that of chalcocite.

D layer The seismic region of the Earth between 1000 km and 2900 km, equivalent to the *lower mantle.* At a depth of 2700 km, there is a change from chemical homogeneity to inhomogeneity; the upper division is the D' layer, and the lower is the D'' layer. It is a part of a classification of the interior of the Earth, having layers A to G.

dneprovskite *wood tin.*

doab [sed] An Irish term for a dark sandy clay or shale found in the vicinity of bogs (Power, 1895).

doab [streams] (a) A term applied in the Indo-Gangetic Plain of northern India to the tongue of low-lying, alluvial land between two confluent rivers; specif. the Doab, the tract between the Ganges and Jumna rivers. The term is commonly restricted to the alluvial-plains portion characterized by very little relief. Cf: *interfluve.* (b) The confluence of two rivers.--- Etymol: Persian, "two waters".

dock (a) The berthing space or waterway between two wharves or two piers, or cut into the land. Syn: *slip.* (b) A basin or enclosure (usually artificial) in connection with a harbor or river, designed to receive vessels and provided with means for controlling the water level. (c) An erroneous syn. of *pier.*

docrystalline A term, now obsolete, suggested by Cross et al (1906, p.694) for porphyritic rocks which are dominantly crystalline, the ratio of crystals to glass being less than 7 to 1 but greater than 5 to 3.

dodd A term used in the English Lake District for a rounded summit, esp. a lower summit or blunt shoulder or boss attached to another hill. Syn: *dod.*

dodecahedral cleavage A type of isometric crystal cleavage that occurs parallel to the faces of the dodecahedron; e.g. sphalerite.

dodecahedron A crystal form with 12 faces that are either pentagonal or rhombic; if rhombic, the faces are equal, but if pentagonal, they are not regular. Each face is parallel to one crystallographic axis and intersects the other two. See also: *pyritohedron; rhombic dodecahedron.*

dodecant In the hexagonal crystal system, one of the 12 spatial divisions made by the four reference axes.

dodging The process of holding back light (passing through certain parts of a photographic negative) from certain areas of sensitized material in order to avoid overexposure. It is used during the enlargement of photographs by projection.

dofemic One of five classes in the C.I.P.W. (Cross, *et al*, 1902, p.585) chemical-mineralogic classification of igneous rocks and in which the ratio of salic minerals to femic is less than three to five but greater than one to seven. Cf: *salfemic; perfemic.*

Dogger Middle European series: Middle Jurassic (above Lias, below Malm).

dogger (a) A large, irregular nodule of ironstone (usually of clay ironstone), sometimes containing fossil remains, found in a sedimentary rock, such as in the Jurassic rocks in Yorkshire (Eng.). Syn: *cathead.* (b) An English term for any large, lumpy mass of sandstone longer than it is broad, with steep rounded sides.

dogger stone A miner's term for a brown, compact, relatively pure, nonoolitic clay ironstone interbedded with oolitic ironstones in the British Middle Jurassic.

doghole A term used in western U.S. for a small inlet on the coast where ships load lumber (Webster, 1967, p.668).

dogleg An abrupt angular change in course or direction, as of a borehole or in a survey traverse. Also, a deflected borehole, survey course, or anything with an abrupt angle felt to resemble the hind leg of a dog.

dogtooth spar A variety of calcite in sharply pointed crystals of acute scalenohedral form resembling the tooth of a dog. Syn: *hogtooth spar.*

dohyaline In the C.I.P.W. (Cross, *et al*, 1906, p.694) classification of igneous rocks, those rocks in which the ratio of crystals to glassy material is greater than three to five but less than one to seven. This term is rarely used. Cf: *perhyaline; hyalocrystalline.*

dolarenaceous Said of the texture of a dolarenite.

dolarenite A dolomite rock consisting predominantly of detrital dolomite particles of sand size (Folk, 1959, p.16); a consolidated dolomitic sand.

doldrums The equatorial region of calms or light winds over the oceans between the northeast and southeast tradewinds. It tends to move north and south with the Sun (with about a six-week lag), but is more often north of than south of the equator.

dolerine A type of talc schist containing feldspar and chlorite, found in the Pennine Alps. The term is now obsolete.

dolerite (a) In the U.S., a syn. of *diabase.* (b) In British usage, the preferred term for what is called *diabase* in the U.S.---Etymol: Greek *doleros* "deceitful", in reference to the fine-grained character of the rock which makes it difficult to identify megascopically. Cf: *diorite.*

doleritic (a) Of or pertaining to dolerite. (b) *ophitic.*

dolerophanite A brown monoclinic mineral: $Cu_2(SO_4)O$. Syn: *dolerophane.*

Dolgellian European stage: Upper Cambrian (above Festiniogian, below Tremadocian of Ordovician).

dolimorphic Said of an igneous rock in which released minerals are prominent; e.g. a lamprophyre composed chiefly of biotite and quartz, with a little hornblende.

dolina Var. of *doline.*

doline A general term for a closed depression in an area of karst topography that is formed either by solution of the surficial limestone or by collapse of underlying caves. It is measured in meters, and its form is generally basinlike or funnel-shaped. See also: *swallow hole; collapse sink.* Cf: *karst valley.* Partial syn: *sinkhole; pothole; solution basin.* Pl: *dolinen.* Also spelled *dolina.*

dolinen Plural of *doline.*

doll A concretion (commonly calcareous) found in argillaceous sediments and resembling a doll; e.g. *loess doll* and *clay doll.*

Dollo's law A syn. of *irreversibility.* This law is named after the Belgian paleontologist Louis Dollo (d.1931).

dolocast A cast or impression of a dolomite crystal, preserved

in an insoluble residue. Adj: *dolocastic.* Cf: *dolomold.*

doloclast A lithoclast derived by erosion from an older dolomite rock; also, an intraclast disrupted from partly consolidate dolomitic mud on the bottom of a sea or lake.

dololithite A dolomite rock containing 50% or more of fragments of older dolomitic rocks that have been eroded an redeposited (Hatch & Rastall, 1965, p.223).

dololutite A dolomitic rock consisting predominantly of detrit dolomite particles of silt and/or clay size (Folk, 1959, p.16); consolidated dolomitic mud. It is commonly interlayered wi dense primary dolomites in evaporitic sequences. Cf: *dolos tite.*

dolomicrite A sedimentary rock consisting of clay-sized dolomite crystals, interpreted as a lithified dolomite mud (analgous to calcite mud or micrite), and containing less than 1 allochems (Folk, 1959, p.14). See also: *primary dolomit* Syn: *dolomite mudstone.*

dolomilith *dolomith.*

dolomite [mineral] A common rock-forming rhombohedr mineral: $CaMg(CO_3)_2$. Part of the magnesium may be replaced by ferrous iron and less frequently by manganes Dolomite is white, colorless, or tinged yellow, brown, pink, gray; it has perfect rhombohedral cleavage and a pearly vitreous luster, feebly effervesces in cold dilute hydrochlor acid, and forms curved, saddle-like crystals. Dolomite found in extensive beds as a compact limestone or dolomi rock; it is also precipitated directly from seawater, possib under warm shallow conditions. Cf: *calcite.* Syn: *bitter spa pearl spar; magnesian spar; rhomb spar.*

dolomite [sed] A carbonate sedimentary rock consisting chiely (more than 50% by weight or by areal percentages und the microscope) of the mineral dolomite or approximating th mineral dolomite in composition, or a variety of limestone marble rich in magnesium carbonate; specif. a carbonate seimentary rock containing more than 90% dolomite and le than 10% calcite, or one having a Ca/Mg ratio in the range 1.5-1.7 (Chilingar, 1957), or one having an approximate Mg equivalent of 19.5-21.6% or magnesium-carbonate equivale of 41.0-45.4% (Pettijohn, 1957, p.418). Dolomite occurs crystalline and noncrystalline forms, is clearly associated an often interbedded with limestone, and usually represents postdepositional replacement of limestone. Pure dolomite (u less finely pulverized) will effervesce very slowly in cold h drochloric acid. Named after Déodat Guy de Dolomie (1750-1801), French geologist, and first applied to certa carbonate rocks of the Tyrolean Alps. See also: *primary dol mite; magnesian limestone.* Syn: *dolostone; dolomite rock.*

dolomite limestone (a) *dolomitic limestone.* (b) A term suggested by Grout (1932, p.288) for a carbonate rock composed predominantly of the mineral dolomite. The term in th usage is not recommended (Pettijohn, 1957, p.416).

dolomite mudstone *dolomicrite.*

dolomite rock *dolomite* [sed].

dolomith A term suggested by Grabau (1924, p.298) for dolomite rock. Syn: *dolomilith; dolomyte.*

dolomitic (a) Dolomite-bearing, or containing dolomite; es said of a rock that contains 5-50% of the mineral dolomite the form of cement and/or grains or crystals. (b) Containir magnesium; e.g. "dolomitic lime" containing 30-50% magnsium.

dolomitic conglomerate (a) A conglomerate consisting limestone pebbles and dolomite cement (Nelson & Nelso 1967, p.112). (b) A breccia-conglomerate of Keuper age Somerset, Eng. (Arkell & Tomkeieff, 1953, p.38).

dolomitic limestone (a) A limestone in which the mineral dolomite is conspicuous, but calcite is more abundant; specif. limestone containing 10-50% dolomite and 50-90% calci and having an approximate magnesium-carbonate equivale of 4.4-22.7% (Pettijohn, 1957, p.418), or a limestone whos Ca/Mg ratio ranges from 4.74 to 60 (Chilingar, 1957). Cf: *ca

tic dolomite; magnesian limestone. Syn: dolomite limestone.
b) A limestone that has been incompletely dolomitized (Chilgar et al, 1967, p.314).

olomitic marble A variety of marble composed largely of olomite and formed by the metamorphism of dolomitic or agnesian limestone. Cf: magnesian marble.

olomitic mottling A textural feature resulting from incipient or rrested dolomitization of limestones, characterized by preferntial alteration that leaves patches, blotches, bird's-eyes, minae, allochems and/or other structures unaffected. Also, similar phenomenon resulting from arrested or incomplete edolomitization.

olomitite A term used by Kay (1951) for a rock composed of e mineral dolomite; a dolostone.

olomitization The process whereby limestone is wholly or artly converted to dolomite rock or dolomitic limestone by e replacement of the original calcium carbonate (calcite) by agnesium carbonate (mineral dolomite), usually through the ction of magnesium-bearing water (seawater or percolating eteoric water). It can occur penecontemporaneously or hortly after deposition of the limestone, or during lithification t a later period. It is commonly accompanied by recrystalliation and by shrinkage of volume (as much as 11% of the riginal limestone) leading to the formation of pores, cavities, nd fissures. Syn: dolomization.

olomization dolomitization.

olomold A rhombohedral opening (of any size) in an insolue residue, formed by the solution of a dolomite (or calcite) rystal. Adj: dolomoldic. Cf: dolocast.

olomorphic A term used by Ireland et al (1947, p.1483) to escribe an insoluble residue characterized by replacement or teration of dolomite or calcite by an insoluble mineral which ls a dolomoldic cavity and assumes the crystal form of the ssolved mineral; it replaces "dolocastic" as used by Cloud et l (1943, p.135).

oloresite A dark-brown monoclinic mineral: $H_8V_6O_{16}$.

olorudite A dolomite rock consisting predominantly of detrital olomite particles larger than sand size (Folk, 1959, p.15); a onsolidated dolomitic gravel.

olosiltite A dolomite rock consisting predominantly of detrital olomite particles of silt size; a consolidated dolomitic silt. Cf: ololutite.

olosparite A sparry dolomite crystal. Cf: calcsparite.

olostone A term proposed by Shrock (1948a, p.126) for the edimentary rock dolomite, in order to avoid confusion with e mineral of the same name. Syn: dolomitite.

omain [mag] A region within a grain of magnetically ordered ineral, within which the spontaneous magnetization has a onstant value characteristic of the mineral composition and emperature.

omain [struc petrol] fabric domain.

omain [sed] The areal extent of a given lithology or environent; specif. the area in which a given set of physical conols combined to produce a distinctive sedimentary facies.

omatic class That crystal class in the monoclinic system aving symmetry m. Syn: clinohedral.

ome [lunar] A small, almost circular surface bulge, generally everal kilometers wide and a few hundred meters high, found a mare region of the Moon's surface. Domes often have pparently smooth summits capped by craters; they are genrally believed to be formed by local extrusive or intrusive igeous activity.

ome [beach] A miniature elevation (2.5 cm or more high, nd 5 to 30 cm in diameter), composed of sand, with a hollow enter, formed on beaches by the rush of waves entrapping nd confining air (Shepard, 1967, p.58). Syn: air dome; sand ome.

ome [geomorph] (a) A general term for any dome-shaped ndform or rock mass, such as a smoothly rounded rockapped mountain summit, roughly resembling the dome of a

building; e.g. the rounded granite peaks of Yosemite, Calif. The term is also applied to a dome-shaped region, as the English Lake District or the Black Hills of South Dakota. (b) A rounded snow peak; e.g. in the French Alps.

dome [cryst] An open crystal form composed of two nonparallel faces that intersect along and astride a symmetry plane, regardless of the orientation of the line of their intersection. Cf: sphenoid.

dome [marine geol] A general, nonrecommended term for such ocean-floor features as a seamount or a knoll.

dome [petrology] A large igneous intrusion whose surface is convex upward with the sides sloping away at low but gradually increasing angles. Intrusive domes include laccoliths and batholiths; the term is used when the evidence as to the character of the lower parts of the intrusion is insufficient to distinguish between a laccolith or a batholith.

dome [fold] n. An uplift or anticlinal-type structure, either circular or elliptical in outline, in which the rock dips gently away in all directions. A dome may be small, e.g. the Gulf Coast salt domes, or many kilometers in diameter. The origin of a dome may vary; it may be, for instance, a salt dome or diapir, a volcanic dome, or a cratonic dome. Type structure: Nashville Dome, Tennessee. See also:pericline; arch. Syn: dome structure; structural dome; quaquaversal fold. Less-preferred syn: swell. Ant: basin.----v. To bend, push, or thrust up into a dome, e.g. underlying magma doming the surface by upward pressure.

dome [volc] (a) volcanic dome. (b) lava dome.

dome mountain A mountain produced where a region of flatlying sedimentary rocks is warped or bowed by upward pressure from beneath to form a structural dome; a mountain or mountain range resulting from dissection of a structural dome. Examples: the Black Hills in South Dakota, and the Weald uplift in S England. Syn: domal mountain; domed mountain.

dome pit In a cave, a vertical passage or high chamber formed by solution. It has a domed ceiling and vertical walls. Cf: chimney.

Domerian Stage in Great Britain: middle Lower Jurassic (above Charmouthian, below Whitbian).

dome structure dome.

dome volcano volcanic dome.

domeykite A tin-white or steel-gray mineral: Cu_3As.

domicilium The main part of the carapace of an ostracode exclusive of alae or other accessory projecting structures.

dominant In the description of coal constituents, more than 60% of a particular constituent occurring in the coal (ICCP, 1963). Cf: rare; common; very common; abundant.

dominant discharge That discharge of a natural channel which determines the characteristics and principal dimensions of the channel. It depends on the sediment characteristics, the relationship between maximum and mean discharge, duration of flow, and flood frequency (ASCE, 1962).

domite An altered, decomposed oligoclase-biotite trachyte that contains tridymite.

domoikic In the C.I.P.W. (Cross, et al, 1906, p.704) textural classification of igneous rocks, those rocks in which the ratio of oikocrysts to xenocrysts is less than seven to one but greater than five to three. This term is rarely used. Cf: peroikic; xenoikic.

donathite A tetragonal mineral: $(Fe,Mg)(Cr,Fe)_2O_4$. It is a dimorph of chromite.

donga [glaciol] A small, steep-walled rift in a piedmont glacier or in an ice shelf. Cf: barranca.

donga [streams] (a) A term used in South Africa for a small, narrow, steep-sided ravine or gorge formed by turbulent water flow; it is usually dry except in the rainy season. A donga is similar to a wadi or a mullah. (b) A term used in South Africa for a gully formed by soil erosion.---Etymol: Afrikaans, from Bantu (Zulu).

doodlebug A popular term for any of various kinds of geophysical prospecting equipment.

dopatic In the C.I.P.W. (Cross, *et al*, 1906, p.701) textural classification of igneous rocks, those rocks in which the ratio of groundmass to phenocrysts is less than seven to one but greater than five to three. This term is rarely used. Cf: *perpatic; sempatic.*

doppler effect A change in the observed frequency of electromagnetic or other waves, caused by relative motion between the source and the observer. See also: *doppler signal.*

dopplerite An amorphous, brownish-black, gelatinous calcium salt of a humic acid that is found at depth in marsh and bog deposits. See also: *phytocollite.*

doppler signal The difference in frequency of electromagnetic or other waves produced by the *doppler effect.*

dorbank A term used in southern Africa for a calcareous and siliceous concretion occurring beneath the surface layer of sandy loam. Etymol: Afrikaans, "dry layer".

dore A term used in the English Lake District for a narrow, door-like opening or fissure between walls of rock, as a pass through a narrow gorge; it is often a widespread joint.

doreite An andesitic lava containing approximately equal amounts of potassium and sodium; the extrusive equivalent of *mangerite* (Streckeisen, 1967, p.209).

dorgalite A basalt in which the phenocrysts are exclusively olivine; an olivine basalt.

dormant volcano A volcano that is not presently erupting but that has, perhaps, done so within historic time and is considered likely to do so in the future. There is no precise distinction between a dormant and an *active volcano.* Cf: *extinct volcano; inactive volcano.* Syn: *subactive volcano.*

dornick A colloquial term for a boulder of iron ore found in a limonite mine.

dorr A glacial trough, open at both ends, across a ridge or mountain range, lying in a pass; it is formed through overdeepening by the crowding of ice through the pass. The term was introduced by Chadwick (1939, p.362) to describe a fjord-like trough that may or may not have been submerged, and is named in honor of George B. Dorr (b.l853), executive of the Hancock County (Me.) Trustees of Public Reservations. Example: Somes's Sound on Mount Desert Island, Me.

dorsal (a) Pertaining or belonging to or situated near or on the back or upper surface of an animal or of one of its parts; e.g. in the direction toward the brachial valve from the pedicle valve of a brachiopod, or pertaining to the hinge region of the shell of a bivalve mollusk where the valves are connected by ligament, or pertaining to the side of the uniserial stipe opposite the thecal apertures of a graptoloid or to the spiral side of a trochoid foraminifer. (b) Referring to the direction or side of an echinoderm away from the mouth, normally downward and outward, such as referring to the direction toward the point of attachment of the column with blastoid theca or to the part of the crinoid or cystoid calyx located toward the column; aboral.---Ant: *ventral.*

dorsal area The part of the whorl of a coiled nautiloid conch in contact with the preceding whorl (TIP, 1964, pt.K, p.55).

dorsal cup The cup-shaped part of crinoid theca forming the aboral and lateral walls about the viscera. It does not include the free arms, tegmen, and column.

dorsal furrow (a) An *axial furrow* of a trilobite. (b) A *septal furrow* of a nautiloid.

dorsal lobe The median primary lobe of a suture on the dorsum of a cephalopod shell. See also: *internal lobe.* Cf: *ventral lobe.*

dorsal shield (a) One of a series of ossicles along the midline of the aboral surface of an arm in an ophiuroid. Cf: *ventral shield.* (b) The entire dorsal test of a trilobite, including the cephalon, thorax, and pygidium.

dorsal valve The *brachial valve* of a brachiopod.

dorsomyarian Said of a nautiloid in which the retractor muscles of the head-foot mass are attached to the shell along the interior areas of the body chamber adjacent to, or coincide with, its dorsal midline (TIP, 1964, pt.K, p.55). Cf: *pleurom arian; ventromyarian.*

dorsum The back or entire dorsal surface of an animal; e. the dorsal side of a cephalopod conch, opposite the ventr side and equivalent to the impressed area in slightly involut shells but referring only to part of the conch adjacent to th venter of the preceding whorl in deeply involute shells, or th more or less flattened area of the carapace surface of an os tracode, adjacent to the hinge line and set off from the latera surface (flattened side) of the valves. The term is somewha loosely applied among fossil nautiloids to the concave side a whorl (in coiled forms) and to the side farthest remove from the siphuncle (in straight or curved conchs with eccen tric siphuncle) (TIP, 1964, pt.K, p.55). Pl: *dorsa.* Cf: *venter.*

dosalic One of five classes in the C.I.P.W. (Cross, *et al*, 190 p.585) chemical-mineralogic classification of igneous rock and in which the ratio of salic minerals to femic is less tha seven to one but greater than five to three. Cf: *persalic; sa femic.*

dosemic In the C.I.P.W. (Cross, *et al*, 1906, p.701) textur classification of igneous rocks, those rocks in which the rati of groundmass to phenocrysts is less than three to five b greater than one to seven. The term is rarely used. Cf: *sen patic; persemic.*

dot chart (a) A graphic aid used in the correction of static gravity for terrain effect, or for computing gravity effects o irregular masses. It can also be used in magnetic interpreta tion. (b) A transparent graph-type chart used in the calcula tion of the gravity effects of various structures. The dots o the chart represent unit areas.

dot map A *cartogram* utilizing dots (usually of uniform size each dot representing a specific number of the objects whos distribution is being mapped.

double core barrel A *core barrel* consisting of two nestin tubes, the inner tube holding the core and remaining station ary during drilling while the outer tube rotates. It is designe for taking a core from delicate material with a minimum o damage. Syn: *double-tube core barrel.*

double layer A layer of ions, in an electrolyte, required to sa isfy a charge unbalance within a solid with which the electro lyte is in contact. See also: *diffuse layer; fixed layer.*

double-line stream A watercourse drawn to scale (on a map by two lines representing the banks. Cf: *single-line stream.*

double-refracting spar *Iceland spar.*

double refraction *birefringence.*

double-serrate Said of a serrate leaf having coarse serration with minute teeth on their edges.

doublet A gem substitute composed of two pieces of gen material, or one of gem material and a second of glass o synthetic, fused or cemented together with a colorless or co ored cement; e.g. a glass imitation with a thin layer of genu ine garnet fused on the top. Cf: *triplet.*

double tide (a) A high tide consisting of two high-water maxi ma of nearly the same height separated by a slight lowering o water. (b) A low tide consisting of two low-water minima sep arated by a slight rise of water.

double tombolo Two separate bars connecting an island (usu ally of large extent and close to the shore) with the mainlanc Cf: *single tombolo; triple tombolo.*

double valley A valley with a low divide on its floor, fror which one stream flows in one direction and a second strear flows in another.

doublure An infolded margin of the exoskeleton of an arthro pod, such as the reflexed ventral margin of the carapace inte gument of a crustacean or the inwardly deflected margina part of the dorsal exoskeleton of a merostome; esp. a gene ally narrow band extending around the border of the dorsa exoskeleton of a trilobite, turned or bent under to the ventra side.

doubly plunging fold A fold, either an anticline or a syncline, that reverses its direction of plunge within the observed area (Billings, 1954, p.46). Use of the term is not recommended, because of its ambiguity. Cf: *plunging fold*.

doughnut (a) A small, well developed, circular *closed ridge* of glacial origin (Gravenor & Kupsch, 1959, p.52). (b) *rock doughnut*.

douglasite A mineral: $K_2FeCl_4.2H_2O$.

Douglas scale A series of numbers formerly used to indicate swell and the state of the sea, ranging from zero (calm) to 9 (confused). Named for H.P. Douglas, British naval officer, who devised the scale in 1921.

doup A term used in northern England for a rounded depression or cavity in a rock or hillside.

dousing *dowsing*.

doverite A mineral: $CaY(CO_3)_2F$. It is the yttrium analogue of synchysite.

down An obsolete term applied to a hill; esp. a *dune* on or near a shore, consisting of sand thrown up by the sea or by the wind.

downbuckle A compressional downfolding of sialic crust, associated with oceanic trenches. Syn: *tectogene in its restricted sense*.

downbuilding A theory of salt-dome formation, based on the fact that the top of the salt body in a dome is near the level at which the salt was originally deposited, and that the thick sequence of strata around it was formed by subsidence of the surroundings of the dome. While these facts are evident, the theory itself is mechanically implausible.

downcoast Said of the coastal direction generally trending toward the south (CERC, 1966, p.A10). Ant: *upcoast*.

downcutting Stream erosion in which the cutting is directed in a downward direction (as opposed to lateral erosion). Cf: *degradation*. Syn: *vertical erosion*.

downdip A direction that is downwards and parallel to the dip of a structure or surface. Cf: *updip*.

downdip block The rocks on the *downfaulted* side of a fault. Cf: *updip block*.

downfaulted Said of the rocks on the downdip side of a fault, the *downdip block*. Cf: *upfaulted*.

downhole n. A borehole drilled at any angle inclined downward in a direction below the horizon.---adj. In a borehole; e.g. "downhole equipment".

downslope n. A slope that lies downward; downhill.---adj. In a downward direction, or descending; e.g. a *downslope* ripple that migrated down a sloping surface.

downstream Toward, at, or from a point near the mouth of a stream; in a direction toward which a stream or glacier is flowing. Similarly, *downriver*.

down-structure method The method of examining structures on a geologic map by orienting the map so as to look down "into" it along the direction of pitch. The structures then appear in much the same attitude as they would in vertical cross section, since any plane, e.g. one parallel to the ground surface, that intersects plunging structures produces such a pattern, though with different proportions. The method is very useful for quick interpretation of complex folding and faulting (Mackin, 1950).

down-the-hole Pertaining to measurements or methods performed in a drill hole; e.g. "down-the-hole logging" (a subsurface procedure that measures geophysical characteristics of materials penetrated by the drill), or "down-the-hole extensometer" (a device that measures differential strains in a drill hole), or "down-the-hole drill" (a hard-rock drill whose bit-driven mechanism follows the drill bit down into the hole being drilled).

downthrow adj. A syn. of *downthrown*, e.g. a downthrow fault. (a) The downthrown side of a fault. (b) The amount of downward vertical displacement of a fault.----Cf: *upthrow; heave*.

downthrow fault *downthrow side*.

downthrown Said of that side of a fault that appears to have moved downward, compared with the other side. Cf: *upthrown*. Syn: *downthrow (adj)*.

downthrown block *downthrow side*.

downthrown side Var. of *downthrow side*.

downthrow side The downthrown side of a fault; a *downthrow*. Syn: *downthrow fault; downthrown side; downthrown block*.

down-to-basin fault A term used in petroleum geology for a fault whose downthrown side is toward the basin. A "down-to-coast fault" is one whose downthrown side is toward the coast. Syn: *down-to-the-basin fault*.

Downtonian Series in the Old Red Sandstone of England: Lower Devonian (lowermost Gedinnian; below Dittonian). It was originally assigned to uppermost Silurian (upper Ludlovian).

downward bulge *root* [tect].

downward-continuation method A method of interpreting gravity values at lower levels from those at the surface.

downwarping The downwards *warping* [tect] or subsidence of a regional area of the Earth's crust, usually as the result of isostatic pressure, e.g. geosynclinal sedimentation. Cf: *upwarping*.

downwash Fine-grained surface material (such as soil) moved down a mountain slope or hillside by rain, esp. where there is little vegetation. Syn: *wash*.

downwasting (a) *mass-wasting*. (b) The thinning of a glacier during ablation.---Cf: *backwasting*.

downwearing Erosion that causes the flattening out of a hill or mountain and the decline of its slope; a process contributing to the development of a peneplain. Cf: *backwearing*.

downwelling [currents] *sinking* [currents].

dowser (a) One who practices *dowsing*. Syn: *diviner*. (b) *divining rod*.----Syn: *water witch; waterfinder*.

dowsing The purported art or practice of locating ground water, mineral deposits, or other objects by means of a *divining rod* or a pendulum. A *dowser* may claim also to be able to diagnose diseases, determine the sex of unborn babies, etc. Syn: *dousing; divining; water witching*. Cf: *rhabdomancy; pallomancy*.

dowsing rod *divining rod*.

doxenic In the C.I.P.W. (Cross, *et al*, 1906, p.704) textural classification of igneous rocks, those rocks in which the ratio of oikocrysts to phenocrysts is less than three to five but greater than one to seven. This term is rarely used. Cf: *xenoikic; perxenic*.

Drachenfels trachyte A syn. of *drakonite*. Its name is derived from Drachenfels, Germany.

draft A term used in eastern U.S. for a gully or gorge, and for a small stream or creek.

drag [eco geol] *drag ore*.

drag [struc geol] The bending of strata on either side of a fault, caused by the friction of the moving blocks along the fault surface; also, the bends or distortions so formed. Drag flexures are better developed in relatively weaker or softer beds and also in the relatively more active fault block. Drag adjacent to a normal fault is upward on the downthrown side and downward on the upthrown side; thus it can indicate the direction of displacement. Less-preferred syn: *terminal flexure*.

drag [hydraul] (a) The friction of moving air against a water surface which tends to pull the water-surface layer in the direction of the wind. (b) The force exerted by a flowing fluid on an object in or adjacent to the flow. Cf: *push* [hydraul].

drag-and-slippage zone The zone bordering the crust and the Earth's interior, along which the entire crust may have shifted relative to the interior (Weeks, 1959, p.378). See also: *phorogenesis*.

drag cast A more appropriate term for *drag mark* as used by Kuenen (1957, p.243-245).

drag coefficient The ratio of the force per unit area exerted on a body by a flowing liquid to the pressure at the stagnation point (ASCE, 1962).

drag fold (a) A minor fold, usually one of a series, formed in a thinly laminated or soft (incompetent) bed lying between harder or more rigid (competent) beds, and produced by relative movement of the competent beds in opposite directions relative to one another. A drag fold may also develop beneath a sheet. The axis of the drag fold is perpendicular to the direction in which the beds slip; the acute angle between the main bedding and the axial plane of the drag fold indicates the direction of movement. The folds are usually small; a centimeter to a few meters in size. Cf: *subsidiary fold; intrafolial fold.*

drag groove A *drag mark* consisting of a long, narrow, and even groove.

drag line A short, feeble glacial striation formed on the lee side of an older glacial groove.

dragma A siliceous monaxonic sponge spicule (microsclere) occurring in bundles and produced within a single sclerocyte. Pl: *dragmata.*

drag mark (a) A long, even *groove* or *striation* made by a solid body dragged over a soft sedimentary surface, as by a stone or shell pulled along the mud bottom by attached algae; it tends to be narrower and deeper than a typical *slide mark.* See also: *drag groove; drag striation.* (b) A term used by Kuenen (1957, p.243-245) for the structure called *groove cast* by Shrock (1948, p.162-163), being a broad and rounded or flat-topped or sharp-crested ridge, commonly with longitudinal striations, formed on the underside of an overlying bed by the filling of a *drag groove* probably under turbidity-current conditions. A more appropriate term for this feature would be *drag cast.*

dragonite A rounded quartz pebble representing a quartz crystal that has lost its brilliancy and angular form that was formerly believed to be a fabulous stone obtained from the head of the flying dragon.

drag ore Crushed and broken fragments of rock or ore torn from the faulted ends of an orebody and contained in and along the fault zone. See also: *trail of a fault.* Syn: *drag.*

drag striation A *drag mark* consisting of a short, narrow, and curved or straight striation. Dzulynski & Slaczka (1958, p.234) used "drag stria" for a feature that is essentially a striation cast.

drain n. (a) A small, narrow, natural watercourse. (b) A channel, conduit, or waterway, either natural or artificial, for draining or carrying off excess water from an area, such as a surface ditch designed to lower the water table so that land may be farmed; a sewer or trench. v. To carry away the surface water or discharge of streams in a given direction or to an outlet.

drainage (a) The manner in which the waters of an area pass or flow off by surface streams or subsurface conduits. (b) The processes of surface discharge of water from an area by streamflow and sheet flow, and the removal of excess water from soil by downward flow. Also, the natural and artificial means for effecting this discharge or removal, such as a system of surface and subsurface conduits. (c) A collective term for the streams, lakes, and other bodies of surface water by which a region is drained; a *drainage system.* (d) The water features of a map, such as seas, lakes, ponds, streams, and canals. (e) An area or district drained of water, as by a stream; a *drainage area.* (f) The act or an instance of removing water from a previously marshy land area.

drainage area [oil] That area from which one well can produce all of the contained hydrocarbons without the assistance of secondary recovery methods.

drainage area [streams] The horizontal projection of the area whose surface directs water toward a stream above a specified point on that stream; a *drainage basin.*

drainage basin A region or area bounded peripherally by a

drainage divide and occupied by a drainage system; spec[i] the whole area or entire tract of country that gathers wat[e] originating as precipitation and contributes it ultimately to [a] particular stream channel or system of channels, or to a lak[e] reservoir, or other body of water. Cf: *river basin.* Syn: *basi[n] watershed; drainage area; catchment; catchment area; catc[h] ment basin; gathering ground; feeding ground; hydrograph[ic] basin.*

drainage coefficient The amount of runoff (expressed in wat[er] depth or other units) removed or drained from an area in 2[4] hours.

drainage composition Quantitative description of a drainag[e] basin in terms of stream order, drainage density, bifurcati[on] ratio, and stream-length ratio. Term introduced by Hort[on] (1945, p.286) to imply "the numbers and lengths of strea[ms] and tributaries of different sizes or orders, regardless of the[ir] pattern".

drainage density Ratio of the total stream lengths of all th[e] stream orders within a drainage basin to the area of that bas[in] projected to the horizontal; the reciprocal of the *chann[el] maintenance constant.* It is an expression of *topograph[ic] texture*: high density values are favored in regions of weak [or] impermeable surface materials, sparse vegetation, mountai[n] ous relief, and high rainfall intensity. Term introduced by Ho[r] ton (1932, p.357) to represent the average stream leng[th] within the basin per unit area. Symbol: D.

drainage district A governmental corporation or other publ[ic] body created by a state to provide drainage of specified ter[ri] tory; it functions under legal regulations for financing, co[n] structing, and operating a drainage system.

drainage divide The boundary between adjacent drainage b[a] sins; a *divide.*

drainage lake An *open lake* that loses water through a surfac[e] outlet or whose level is largely controlled by the discharge [of] its effluent. Cf: *seepage lake.*

drainage line The course or channel of a major stream in [a] drainage system.

drainage network The actual arrangement of a drainage sy[s] tem whose streams interlace or cross like the fabric of a ne[t]; a *drainage pattern.* Syn: *drainage net.*

drainage pattern The configuration or arrangement in pla[n] view of the natural stream courses in an area. It is related [to] local geologic and geomorphologic features and history. Sy[n:] *drainage network.*

drainage ratio The ratio between runoff and precipitation in [a] given area for a given period of time. Syn: *discharge efficie[n] cy.*

drainage system A surface stream or a body of impounde[d] surface water, together with all other surface streams a[nd] bodies of impounded surface water that are tributary to it a[nd] by which a region is drained. An artificial drainage system i[n] cludes also surface and subsurface conduits.

drainage varve An abnormally thick and sandy varve forme[d] by the drainage of lakes ponded between the ice edge a[nd] higher land, or by dams of glacial deposits.

drainageway A channel or course along which water moves [in] draining an area.

drainage well A type of *inverted well* used to drain excess so[il] or surface water, where the aquifer penetrated is permeabl[e] enough, and has a head far enough below the land surface, [to] remove the water at a satisfactory rate. Drainage wells hav[e] been used to dispose of some untreated domestic and othe[r] wastes, but such uses are now largely prohibited. Syn: *a[b] sorbing well.* Cf: *relief well.*

drakonite An extrusive rock composed of phenocrysts of san[i] dine, plagioclase, and biotite and/or hornblende in a trachyt[ic] groundmass of alkali feldspar microlites and interstitial alka[li] amphibole or pyroxene. The plagioclase ranges from olig[o] clase to labradorite. Sodalite, acmite-augite, apatite, titanit[e,] magnetite, and zircon may be present as accessories. Sy[n]

drachenfels trachyte.

drape fold An open, *plains-type fold.*

draper point The temperature, 977°F (525°C), at which red light first becomes visible from a heated object in darkened surroundings; hence, the minimum temperature of incandescent lava (Draper, 1847; Siegel & Howell, 1967).

drapery *curtain.*

draping The general structural concordance of warped strata lying above a limestone reef or other hard core to the surface of that reef or core, due either to initial dip or to differential compaction, or to both.

Dravidian The Cambrian to Carboniferous of India.

dravite A brown, magnesium-rich mineral of the tourmaline group: $NaMg_3Al_6(BO_3)_3(Si_6O_{18})(OH)_4$.

draw [mining] The horizontal distance, measured on the surface ahead of an underground coalface, over which the rocks are influenced by subsidence.

draw [geomorph] (a) A small, natural watercourse or gully, generally shallower or more open than a ravine or gorge; a shallow gulch; a valley or basin. (b) A usually dry stream bed; a coulee whose water results from periodic rainfall. (c) A sag or trough-like depression leading up from a valley to a gap between two hills.

drawdown [water] The distance by which the level of a reservoir is lowered by the withdrawal of water.

drawdown [grd wat] (a) The difference between the height of the water table and that of the water in a well. (b) Reduction of the pressure head as a result of the withdrawal of free water. Cf: *cone of depression.*

drawdown [hydraul] In a stream or conduit, the difference in elevation between the water-surface elevation at a constriction and what the elevation would be if there were no constriction (ASCE, 1962).

draw slate In coal mining, soft shale that occurs above a coal seam and that collapses with or after removal of the coal. See also: *slate [coal mining].*

draw works The hoisting mechanism (used in drilling) for handling (lifting and lowering) the drilling tools and for supplying driving power (such as transmitting power to the rotary table of a rotary-drilling rig).

D ray Left posterior ray in echinoderms situated clockwise of C ray when the echinoderm is viewed from the adoral side.

dredge [eng] (a) A large floating machine for scooping up or excavating earth material at the bottom of a body of water, raising it to the top, and discharging it on a bank through a floating pipeline or by conveyors or into a scow for removal to a distant point; e.g. a "hydraulic dredge" using a centrifugal pump to draw mud or saturated sand from a river channel. Other dredges use dippers, clamshells, bucket chains, and scrapers; they may or may not be self-propelled. Dredges are used to excavate or deepen harbor channels, to raise the level of lowland areas, and to dig ditches and improve drainage. (b) A ship designed to remove sediment from a channel.

dredge [oceanog] An ocean-bottom sampler that scoops sediment and benthonic organisms as it is dragged behind a moving ship. It is usually a heavy, metal container; one variety is made of chain mail affixed to a metal collar. Cf: *corer; grab sampler.*

dredge peat *sedimentary peat.*

dreikanter (a) A doubly pointed *windkanter* or stone having three curved faces intersecting in three sharp edges, resembling a Brazil nut. Syn: *pyramid pebble.* (b) A term loosely applied as a syn. pf *ventifact*, as for a wind-worn pebble having more than three edges.---Etymol: German *Dreikanter*, "one having three edges". Pl: *dreikanters; dreikanter.*

Dresbachian North American stage: Upper Cambrian (below Franconian).

dresserite A mineral: $Ba_2Al_4(CO_3)_4(OH)_8 \cdot 3H_2O$.

drewite A white, neritic, fine-grained, impalpable, calcareous mud or ooze, consisting chiefly of minute aragonite needles a few microns in length, and believed to have been precipitated directly from seawater through the action of nitrate- and sulfate-reducing bacteria. Named after George Harold Drew (1881-1913), British(?) scientist, who studied the marine bacteria associated with this sediment in the shallow lagoons in the Bahamas (Drew, 1911).

driblet adj. A syn. of *spatter.*

driblet cone *hornito.*

dried ice Sea ice whose whitened surface contains cracks and thaw holes following the disappearance of meltwater.

dries "An area of a reef or other projection from the bottom of a body of water which periodically is covered and uncovered by the water" (Baker et al, 1966, p.51). Syn: *uncovers.*

drift [drill] (a) The deviation of a borehole from the vertical or from its intended course. (b) The angle that a borehole has been deflected from the vertical.

drift [coasts] Detrital material moved and deposited by waves and currents; e.g. *littoral drift.* Also, floating material (as driftwood or seaweed) that has been washed ashore by waves and left stranded on a beach.

drift [glac geol] A general term applied to all rock material (clay, sand, gravel, boulders) transported by a glacier and deposited directly by or from the ice, or by running water emanating from a glacier. Drift includes unstratified material (till) that forms moraines, and stratified deposits that form outwash plains, eskers, kames, varves, glaciofluvial sediments, etc. The term is generally applied to Pleistocene glacial deposits in areas (as large parts of North America and Europe) that no longer contain glaciers. The term "drift" was introduced by Murchison (1839, v.1, p.509) for material, then called *diluvium*, that he regarded as having drifted in marine currents and accumulated under the sea in comparatively recent times; this material is now known to be a product of glacial activity. Partial syn: *glacial drift; fluvioglacial drift.*

drift [speleo] Detrital material accumulated in a cave. Syn: *wash [speleo]; fill [speleo]; cave earth; cave deposits.*

drift [geophys] A gradual change in the reference reading, or in any quantitative characteristic that is supposed to remain constant, of an instrument (such as a gravimeter) due to slow changes in the properties of the equipment (such as the gravimeter spring) due to elastic aging, long-term creep, hysteresis, temperature, or other factors. See also: *drift correction.*

drift [oceanog] (a) One of the wide, slower movements of surface oceanic circulation under the influence of, and subject to diversion or reversal by, prevailing winds; e.g. the easterly drift of the North Pacific. Syn: *drift current; wind drift; wind-driven current.* (b) The slight motion of ice or vessels resulting from ocean currents and wind stress. (c) The speed of an ocean current or ice floe, usually given in nautical miles per day or in knots.

drift [photo] Apparent rotation of aerial photographs with respect to the true flight line, caused by the displacement of the aircraft due to the action of cross winds or other causes, and by the failure to orient the camera to compensate for the angle between the flight line and the direction of the aircraft's heading. The photograph edges remain parallel to the intended flight line, but the aircraft itself drifts further and further from that line.

drift [sed] n. (a) A general term, used esp. in Great Britain, for all surficial, unconsolidated rock debris transported from one place and deposited in another, and distinguished from solid bedrock; e.g. *river drift.* It includes loess, till, river deposits, etc., although the term is often used specif. for glacial deposits. (b) Any surface movement of loose incoherent material by the wind; also, an accumulation of such material, such as a *snowdrift* or a *sand drift.*---v. To accumulate in a mass or be piled up in heaps by the action of wind or water.

drift [tect] *continental drift.*

drift [hydraul] "The effect of the velocity of fluid flow upon the velocity (relative to a fixed external point) of an object moving within the fluid" (Huschke, 1959, p.178).

drift [river] In South Africa, a ford in a river. The term is used in many parts of Africa to indicate a ford or a sudden dip in a road over which water may flow at times (Stamp, 1961, p.162). Syn: *drif (Afrikaans)*.

drift avalanche *dry-snow avalanche*.

drift-barrier lake A glacial lake formed upstream from a moraine that has blockaded a valley or a drainage course (Fairchild, 1913, p.153). Cf: *valley-moraine lake*.

drift bed A layer of drift "of sufficient uniformity to be distinguished from associated ones of similar origin" (Fay, 1918, p.231).

drift bedding An old term used by Sorby (1857, p.279) to replace *false bedding*. See also: *ripple drift*.

drift bottle A bottle containing a record that identifies the date and place at which it was released into the sea and that is to be returned by the finder with the date and place of recovery; it is used in studying surface currents. Syn: *bottle post*.

drift clay *boulder clay*.

drift coal Coal formed according to the *drift theory*; allochthonous coal.

drift copper A native, *float* copper transported from its source by the action of a glacier.

drift correction Adjustment to remove the effects of geophysical *drift*, usually by repeated observations at a base station.

drift current (a) *drift* [oceanog]. (b) "A current defined by assuming that the wind stress is balanced by the sum of the Coriolis and frictional forces" (Baker et al, 1966, p.52).----Cf: *stream current*.

drift curve A graph of a series of gravity values read at the same station at different times and plotted in terms of instrument reading versus time.

drift dam A dam formed by the accumulation of glacial drift in a pre-existing stream valley.

drift epoch A syn. of *glacial epoch*; specif. the *Drift epoch*, also known as the Pleistocene Epoch. See also: *Drift period*.

drifter An air-driven, percussive rock drill used for excavating horizontal underground passages and tunnels.

drift glacier A small mass of flowing ice in a mountain area nourished primarily with windblown snow from adjacent snowfields, slopes, or ridges. Syn: *Ural-type glacier; snowdrift glacier; glacieret*. Nonpreferred syn: *catchment glacier*.

drift gravel A gravel placer of gold or tin, lying on slate or granite and covered with basalt (von Bernewitz, 1931).

drift ice (a) Any ice that has broken apart and drifted from its place of origin by winds and currents, such as a fragment of a floe or a detached iceberg; loose, unattached pieces of *floating ice* with open water preponderating over ice and navigable with ease. (b) A syn. of *pack ice* as that term is used in a broad sense.

drifting ice station An oceanographic research base established on the ice in the Arctic Ocean.

drifting snow Snow particles raised to heights of less than 2 m above the ground by the wind. Visibility at eye level is not appreciably affected. Cf: *blowing snow*. Syn: *drift snow; snowdrift; driving snow*.

drift lake A glacial lake occupying a depression left in the surface of glacial drift after the disappearance of the ice (White, 1870, p.70).

driftless area A region that was surrounded, but apparently not covered, by continental ice sheets of the Pleistocene Epoch, and in which there are no glacial deposits; specif. the "Driftless Area" occuping SW Wisconsin and parts of Illinois, Iowa, and Minnesota.

drift line A line of drifted material washed ashore and left stranded. It marks the highest stage of water, such as of a flood.

drift map A British term for a geological map representing a true picture of the visible ground, including all surficial deposits and only those rock outcrops exposed at the surface (Nel-

son & Nelson, 1967, p.114). Cf: *solid map*.

driftmeter An instrument for determining the deviation of drill pipe from the vertical and the depth at which it occur Cf: *clinograph*.

drift mining (a) The extraction of placer ore by undergroun horizontal or inclined tunneling methods rather than by th use of running water. Cf: *placer mining*. (b) The extraction near-surface coal seams by underground, inclined tunnelir methods rather than by opencut mining or by vertical-sha methods.

drift peat Peat that occurs in association with glacial drift.

Drift period A term formerly used to designate the Pleistocer Epoch. See also: *drift epoch*.

drift plain A plain, as a *till plain*, underlain with glacial drift.

drift scratch *glacial striation*.

drift sheet A widespread sheet-like body of glacial drift, de posited, continuously or discontinuously, during a single glac ation (e.g. the Cary drift sheet) or during a series of close related glaciations (e.g. the Wisconsin drift sheet).

drift terrace A term used in New England for an *alluvial te race*.

drift theory [coal] The theory that coal originates from acc mulations of plant material which has been transported fro its place of growth and deposited in another locality wher coalification occurs. Ant: *in-situ theory*. See also: *allochthony*.

drift theory [glac geol] A theory of the early 19th centu which attributed the origin of widespread surficial deposits, i cluding the erratic boulders, to the action of marine curren and floating ice. Cf: *glacier theory*.

driftwood Woody material, such as parts of trees, drifted floated by water and cast ashore or lodged on beaches b storm waves.

drill [drill] n. A tool, machine, instrument, or other form of a paratus with an edged or pointed end, used for making circ lar holes in rock or earth material by a succession of blows by rotation of the drill; specif. a *drill bit*.---v. To make a circ lar hole with a drill or other cutting tool.

drill [stream] An archaic term for a small trickling stream c rill.

drill bit Any device attached to, or being an integral part o the lower end of a drill stem, used as a cutting or boring to in drilling a hole; the cutting edge of a drill. See also: *core b* Syn: *bit; drilling bit; drill*.

drill collar A length of extra-heavy, thick-walled drill pipe co nected to a rotary drill string directly above either the cor barrel or the drill bit and used to concentrate weight and giv rigidity so that the bit will cut properly (vertically). It is usual of nearly the same outside diameter as that of the bit or cor barrel on which it is used. Syn: *collar [drill]; drilling collar*.

drill column *drill pipe*.

drill core *core [drill]*.

drill cuttings *cuttings*.

drilled well A large, deep well of high capacity, constructed b drilling methods (e.g. cable-toll or rotary) for industrial, irriga tion, or municipal use (such as for obtaining water or petrole um). It may extend to a depth of more than 6000 m (20,00 ft).

driller's log A record, filled out on a tabulated form by th chief of the crew that drills an oil or gas well, showing dr progress (such as number of feet drilled each day), bit de scriptions, drilling tools used, size of the hole, rock encour tered (including its character, color, and description), person nel present during the tour of duty, and other pertinent fact having to do with the drilling (including any unusual event).

drill hole A circular hole made by drilling; esp. one made b cable tools, or one made to explore for valuable minerals or obtain geologic information. Cf: *borehole*.

drilling (a) The act or process of making a circular hole with drill or other cutting tool, for purposes such as blasting, explo ration, prospecting, valuation, or obtaining water. Cf: *boring*

) A hole drilled in the ground.

rilling bit *drill bit.*

rilling cable *cable* [drill].

rilling fluid *drilling mud.*

rilling in The act or process of completing a well by drilling to the producing (oil-, water-, or gas-bearing) formation.

rilling mud A heavy suspension, usually in water but sometimes in oil, used in rotary drilling, consisting of various substances in a finely divided state (commonly bentonitic clays nd chemical additives such as barite), introduced continuously down the drill pipe under hydrostatic pressure, out rough openings in the drill bit, and back up in the annular pace between the pipe and the walls of the hole and to a urface pit where it is purified and reintroduced into the pipe. is used to lubricate and cool the bit, to carry the cuttings up om the bottom, and to prevent blowouts and cave-ins by astering and consolidating the walls with a clay lining, thereby making casing unnecessary during drilling and also offsetng pressures of fluid and gas that may exist in the formation. ee also: *oil-base mud.* Syn: *mud; drilling fluid; drill mud; mud ush; circulation fluid.*

rilling rig The derrick, power supply (engine and engineouse), draw works, drill machine, and other surface equipent and auxiliary tools necessary for the operation of a drill-g unit (including boring and pumping operations). Syn: *rig.*

rillings The material removed by a drill in making a hole. See lso: *cuttings.*

rilling time (a) The time required for a rotary drill bit to penerate a specified thickness (usually one foot) of rock. (b) The apsed time required to drill a well, excluding periods when ot actually drilling.

rilling-time log A log of the varying rate of drilling penetraon, plotted on a log strip divided into units of depth to indiate how fast a formation is drilled (such as the faster drilling ate that may be produced by a porous reservoir rock).

rill pipe A long, heavy, steel pipe or hollow rod that drives e drill bit in rotary drilling by transmitting the motion from e rotary table at the top to the attached bit at the bottom of e hole and that conducts the drilling mud from the surface the bottom. It is normally formed of 30-ft sections conected end to end. Syn: *drill rod; drill column; drill stem.*

rill rod A *drill pipe,* esp. a thin, lightweight drill pipe, such as sed in shallow-depth core drilling.

rill stem (a) A term used in rotary drilling for the *drill pipe* nd also for the *drill string.* (b) A term used in cable-tool drillg for a solid shaft or cylindrical bar of steel or iron attached the drill bit to give it weight.---Also spelled: *drillstem.* Syn: *tem.*

rill-stem test A procedure for determining productivity of an il or gas well by measuring reservoir pressures and flow caacities while the drill pipe is still in the hole and the well is till full of drilling mud and usually before the well is cased; .g. releasing reservoir gas and fluids into a perforated pipe ttached to an empty drill pipe which measures pressures and ow. Abbrev: DST. Cf: *wire-line test.*

rill string (a) A term used in rotary drilling for the assemlage in a borehole of drill pipes, drill bit, and either core barel or drill collars, connected to and rotated by the drill mahine at the surface. Syn: *drill stem.* (b) A term used in able-tool drilling for the assemblage in a borehole of drill bit, rill stem, cable, and other tools, connected to the walking eam at the surface.---Syn: *string; drilling string.*

rip curtain *curtain.*

riphole (a) A small hole or niche in clay or rock beneath a oint where water drips. (b) The center hole in a feature built p beneath dripping water.

ripstone A general term for any cave of calcium carbonate r other mineral formed by dripping water, including stalactes and stalagmites. Cf: *flowstone.* Partial syn: *stagmalite.*

riven well A shallow, usually small-diameter (3-10 cm) well

constructed by driving a series of connected lengths of pipe into unconsolidated material to a water-bearing stratum, without the aid of any drilling, boring, or jetting device. Syn: *drivewell; tube well.*

drivewell *driven well.*

drop In hydraulics, the difference in water-surface elevations measured upstream and downstream from a constriction in the stream (ASCE, 1962).

drop-down curve The form of the water surface along a longitudinal profile, assumed by a stream or open conduit upstream from a sudden fall. In a uniform channel, the curve is convex upward (ASCE, 1962). Cf: *backwater curve.*

drops In the roof of a coal seam, funnel-shaped downward intrusions of sedimentary rock, usually sandstone.

dropstone An obsolete syn. of *stagmalite.*

drought polygon *desiccation polygon.*

drowned Said of a land surface or land feature that has undergone *drowning;* e.g. a *drowned* coast or a *drowned* shore.

drowned atoll An atoll occurring in an area of relatively rapid subsidence so that continued upgrowth of the reef does not reach the water surface. See also: *drowned reef.*

drowned reef A reef situated at such great depth that reef growth is excluded or greatly hampered. See also: *drowned atoll.*

drowned river mouth The lower end of a river that is widened or submerged by seawater invading the coast; an *estuary.* Example: Chesapeake Bay.

drowned stream A stream that has been inundated from the sea. Syn: *flooded stream.*

drowned valley A valley that is partly submerged (at its lower end) by the advance of a sea or a lake. Syn: *submerged valley.*

drowned-valley lake A lake formed at the mouth of a *drowned valley.*

drowning The submergence of a land surface or topography beneath water, either by a rise in the water level or by a sinking or subsidence of the land margin.

druid stone A *sarsen* used in the building of ancient stone (druidical) circles in Stonehenge and in other regions of Great Britain. Syn: *druidical stone.*

drum (a) A Scottish term for a long narrow ridge. (b) *drumlin.*

drumlin (a) A low, smoothly rounded, elongated and oval hill, mound, or ridge of compact glacial till, built under the margin of the ice and shaped by its flow, or carved out of an older moraine by readvancing ice; its longer axis is parallel to the direction of movement of the ice. It usually has a blunt nose pointing in the direction from which the ice approached, and a gentler slope tapering in the other direction. Height is 8-60 m, average 30 m; length is 400-2000 m, average 1500 m. Syn: *drum.* (b) *rock drumlin.---*Etymol: Irish & Gaelic, diminutive of *druim,* "back, ridge".

drumlin field *basket-of-eggs topography.*

drumlinoid A *rock drumlin* whose form approaches that of a true drumlin but does not fully attain it because of superficial modification by moving ice.

drumloid An oval hill or ridge of glacial till whose shape resembles that of a *drumlin* but is less regular and symmetrical.

drupe A stone fruit, e.g. a plum.

druse (a) An irregular cavity or opening in a vein or rock, having its interior surface or walls lined (encrusted) with small projecting crystals usually of the same minerals as those of the enclosing rock, and sometimes filled with water; e.g. a small solution cavity, a steam hole in lava, or a lithophysa in volcanic glass. Cf: *geode; vug* [eco geol]; *miarolitic cavity.* (b) A mineral surface covered with small projecting crystals; specif. the crust or coating of crystals lining a druse in a rock, such as sparry calcite filling pore spaces in a limestone. ---- Etymol: German *druos,* "gland, bump". Adj: *drusy.*

drusy (a) Pertaining to a *druse,* or containing many druses. Cf: *miarolitic.* (b) Pertaining to an insoluble residue or encrustation, esp. of quartz crystals; e.g. a "drusy oolith" covered with subhedral quartz.

dry In climatology, *arid*.

dry assay Any type of assay procedure that does not involve liquid as a means of separation. Cf: *wet assay*.

dry avalanche *dry-snow avalanche*.

dry basin An *interior basin* (as in an arid region) containing no perennial lake due to the drainage being "occasional only and not continuous" (Gilbert, 1890, p.2).

dry beach The upper part of a beach that is covered only by storm waves. Ant: *wet beach*.

dry bed *dry wash*.

dry-bone ore An earthy, friable, honeycombed variety of *smithsonite*, usually found in veins or beds in stratified calcareous rocks, accompanying sulfides of zinc, iron, and lead. The term is sometimes applied to hemimorphite. Syn: *dry bone*.

dry bulk density The specific gravity of a substance, e.g. a sediment, without interstitial water. Cf: *natural density*.

dry calving The breaking away of a mass of ice from a glacier on dry land.

dry cave *dead cave*.

dry delta (a) *alluvial fan*. (b) *alluvial cone*.

dry digging *dry placer*.

drydock iceberg *valley iceberg*.

dry firn *polar firn*.

dry frozen ground Relatively loose and crumbly ground (or soil) that has a temperature below freezing but contains no ice.

dry gap A gap that is not occupied by a stream; specif. a *wind gap*.

dry gas A natural gas containing very little liquid hydrocarbons. Cf: *wet gas*.

dry hole [drill] A drill hole in which no water is used for drilling, such as a hole driven upward through rock in a quarry.

dry ice (a) Ice at a temperature below the freezing point; specif. bare glacier ice on which there is no slush or standing water (ADTIC, 1955, p.26). (b) A substance consisting of solidified carbon dioxide.

drying crack *desiccation crack*.

dry lake (a) A lake basin that formerly contained a lake. (b) A *playa*; a tract of salt-encrusted land in an arid or semiarid region, occasionally covered by an intermittent lake.

dry peat Peat derived from humic matter and formed under drier conditions than those of a moor.

dry permafrost Loose and crumbly permafrost containing little or no ice or moisture.

dry placer A placer that cannot be mined due to lack of the necessary water supply. Syn: *dry digging*.

dry playa A playa that is normally hard, buff in color, and smooth as a floor (Thompson, 1929); the water table is at a considerable distance beneath the dense, sun-baked surface. Cf: *wet playa*.

dry quicksand A sand accumulation that offers no support to heavy loads because of alternating layers of firmly compacted sand and loose, soft sand.

dry snow Deposited snow retained below the freezing point. Cf: *wet snow*.

dry-snow avalanche An avalanche composed of dry, loose or powdery snow that is set in motion by the wind and is sometimes drifted but not wind-packed; the driving ahead of a column of compressed air creates a vacuum in its wake. It is the fastest-moving of the snow avalanches, capable of reaching a speed of 450 km/hr. Syn: *dry avalanche; dust avalanche; drift avalanche; wind avalanche; powdery avalanche*.

dry snow line The boundary on a glacier or ice sheet between the *dry snow zone* and an area where surface melting occurs.

dry snow zone The area on a glacier or ice sheet where no surface melting occurs even in summer, delimited by the *dry snow line*.

dry unit weight The *unit weight* of soil solids per unit of total volume of soil mass. See also: *maximum unit weight*. Syn: *unit dry weight*.

dry valley A valley that is devoid or almost devoid of running water; a streamless valley. It may be the result of stream capture, a climatic change, or a fall in the water table. Dry valleys are common in areas underlain by chalk and limestone; other examples include wind gaps and glacial overflow channels. Syn: *dead valley*.

dry wash A *wash* that carries water only at infrequent intervals and for short periods, as after a heavy rainfall. Cf: *arroyo*. Syn: *dry bed*.

dry weathering Mechanical weathering of a rock without the action of running water, as in an arid region.

d-spacing in refraction of X-rays by a crystal, the distance of separation between the successive and identical parallel planes in the crystal lattice. It is expressed as d in the *Bragg equation*.

dubiocrystalline Said of the texture of a rock whose crystallinity can be determined with difficulty or which cannot be determined with certainty; e.g. said of the texture of a porphyry whose groundmass is too fine to be resolved into its constituents under a microscope, but shows faint anisotropism or polarizes light as an aggregate. Also, said of a rock with such texture. Term introduced by Zirkel (1893, p.455) as *dubio krystallinisch*. Cf: *microcryptocrystalline*.

Dubosq colorimeter An instrument which compares visually the color intensity of solution of unknown strength with that of a variable depth of standard solution. From the depth of standard solution required to obtain a visual match, the strength of the unknown can be determined. See also: *colorimeter*.

Duck Lake North American (Gulf Coast) stage: Miocene (above Napoleonville, below Clovelly).

ducktownite A term used in Tennessee for an intimate mixture of pyrite and chalcocite, or for the matrix of a blackish copper ore containing grains of pyrite.

ductile [exp struc geol] Said of a rock that is able to sustain, under a given set of conditions, 5-10% deformation before fracturing or faulting. Cf: *brittle [struc geol]*.

ductolith A horizontal igneous intrusion that resembles a tear drop in cross section; a headed dike.

duff A type of organic surface horizon of forested soils, consisting of matted, peaty organic matter that is only slightly decomposed. It is a constituent of the *forest floor*. Cf: *litter; leaf mold*.

dufrenite A blackish-green mineral: $Fe^{+2}Fe_4^{+3}(PO_4)_3(OH)_5 \cdot 2H_2O$. Syn: *kraurite; green iron ore*.

dufrenoysite A lead-gray orthorhombic mineral: $Pb_2As_2S_5$.

duftite An orthorhombic mineral: $PbCu(AsO_4)(OH)$.

dug well A shallow, large-diameter well constructed by excavating with hand tools or power machinery instead of by drilling or driving, such as a well for individual domestic water supplies (and yielding considerably less than 100 gpm).

Duhem's theorem The statement in chemistry that the state of any closed system is completely defined by the values of any two independent variables, extensive or intensive, provided the initial masses of each component are given. The choice of variables, however, must not conflict with the phase rule.

dull A quality of luster of a mineral or rock surface that diffuses rather than reflects light, even though the surface may feel smooth; *earthy*.

dull-banded coal Banded coal consisting mainly of vitrain and durain, some clarain, and minor fusain. Cf: *bright-banded coal*.

dull coal A type of banded coal defined microscopically as consisting mainly of clarodurain and durain and of 20% or less of bright materials such as vitrain, clarain, and fusain. Cf: *bright coal; semidull coal; semibright coal; intermediate coal*. Syn: *dulls*.

dulls *dull coal*.

dumalite A trachyandesite characterized by intersertal texture and a glassy mesostasis which possibly has the composition nepheline.

dumbbell Two land areas connected by a relatively narrow isthmus of sand which is never below the high-water mark in any part of its length, and whose highest points are higher above sea level than any part of the isthmus (Schofield, 1920); e.g. off the coast of southern China. Syn: dumbbell land.

dumontite A yellow orthorhombic mineral: $Pb_2(U-O_2)_3(PO_4)_2(OH)_4.3H_2O$.

dumortierite A bright-blue or greenish-blue mineral of the sillimanite group: $Al_7(BO_3)(SiO_4)_3O_3$. It may contain iron, and it occurs principally in schists and gneisses.

dumortierite-quartz A massive, opaque, intense-blue to greenish-blue or violet-blue variety of crystalline quartz colored by intergrown crystals of dumortierite.

dump [mass move] A heap or pile of avalanche material.

dumped deposit An unsorted sediment deposited directly below wave base or current base, or brought in at a rate too rapid for waves and currents to distribute it (Weeks, 1952, 2107); e.g. a shaly sand or a sandy shale.

dump moraine A terminal moraine consisting of englacial and superglacial material dropped by a glacier at its front.

dumpy level A leveling instrument in which the telescope is permanently attached (either rigidly or by a hinge) to the vertical spindle or leveling base and that is capable only of rotary movement in a horizontal plane. The dumpy level takes its name from the dumpy appearance of the early type of this instrument, the telescope of which was short and had a large object glass. Cf: wye level.

dun An inconspicuous hill in the English Lake District. Etymol: Gaelic.

dundasite A white mineral: $PbAl_2(CO_3)_2(OH)_4.2H_2O$.

dune [geomorph] A low mound, ridge, bank, or hill of loose, windblown granular material (generally sand, sometimes volcanic ash), either bare or covered with vegetation, capable of movement from place to place but always retaining its own characteristic shape. Etymol: French. See also: sand dune; own.

dune [stream] A term used by Gilbert (1914, p.31) for a sand wave formed on a stream bed and usually transverse to the direction of flow, traveling downstream by the erosion of sand from the gentle upstream slope and its deposition on the steep downstream slope, and having an approximately triangular cross section (in a vertical plane in the direction of flow); a large-scale mound or ridge of sand, similar to an eolian sand dune, but formed in a subaqueous environment. Cf: antidune. Syn: subaqueous sand dune.

dune complex An aggregate of moving and fixed sand dunes in a given area, together with sand plains and the ponds, lakes, and swamps produced by the blocking of streams by the sand.

dune lake (a) A lake occupying a basin formed as a result of the blocking of the mouth of a stream by sand dunes migrating along the shore; e.g. Moses Lake, Wash. (b) A deflation lake occupying a blowout on a dune.

dune massif A large, irregular, cone- or pyramid-shaped dune with curved slopes and steep sides consisting of small hollows and terraces (Stone, 1967, p.225).

dune movement In hydraulics, the movement of sediment along the bed of a stream in the form of a wave or dune which travels downstream. The upstream face of the wave is eroded and the eroded material is deposited on the downstream face of the wave. The water surface has only a slight undulation (ASCE, 1962).

dune phase The part of stream traction whereby a mass of sediment travels in the form of a small, dune-like body having a gentle upcurrent slope and a much steeper downcurrent slope (Gilbert, 1914, p.30-34); it develops when the bed load

is small or the current is weak. The dune form moves downstream. Cf: smooth phase; antidune phase.

dune ridge A series of parallel dunes (whose movements are arrested by the growth of vegetation) built along the shore of a retreating sea. See also: foredune.

dune rock An eolianite consisting of dune sand.

dune sand A type of blown sand that has been piled up by the wind into a sand dune, usually consisting of rounded quartz grains having diameters varying from 0.1 to 1 mm.

dune slack A damp slack between dunes or dune ridges on a shore; a dune valley.

dune valley A hollow, furrow, or depression between dunes or dune ridges. Syn: dune slack.

dungannonite An alkalic corundum-bearing diorite. Its name is derived from Dungannon, Ontario, Canada.

dunite Peridotite in which the mafic mineral is almost entirely olivine, with accessory chromite almost always present. Syn: olivine rock.

duns A term used in SW England for a shale or massive clay associated with coal.

dunstone [ign] An amygdaloidal spilite; a local name in the Plymouth area of England.

dunstone [sed] (a) A term used near Matlock, Eng., for a hard, granular, yellowish or cream-colored magnesian limestone. (b) A term used in Wales for a hard fireclay or underclay, and in England for a shale.

duplexite bavenite.

duplicature A doubling or a fold; e.g. the distal portion of an ostracode carapace that is folded toward the interior of the valve around the free margin to form a doubling of the lamella (if fused the radial pore canals can be seen extending from the inner to the outer margin, otherwise a cavity or vestibule is present esp. in the anterior and posterior regions). See also: skeletal duplicature.

duplivincular Said of a type of ligament of a bivalve mollusk (e.g. Arca) in which the lamellar component is repeated as a series of bands, each with its two edges inserted in narrow grooves in the cardinal areas of the respective valves.

durability index The relative resistance to abrasion exhibited by a sedimentary particle in the course of transportation, represented by the ratio of the reduction index of a standard (such as quartz) to that of a given rock or mineral under the same conditions (Wentworth, 1931, p.26). Abbrev: D.I.

durain A coal lithotype characterized by dull, matte luster, grey to brownish black color, and granular fracture. It occurs in bands up to many centimeters in thickness, and its characteristic microlithotype if durite. Cf: vitrain; clarain; fusain.

durangite An orange-red monoclinic mineral: $NaAl(AsO_4)F$.

Durangoan North American (Gulf Coast) stage: Lower Cretaceous (above LaCasitan of Jurassic, below Nuevoleonian).

duration (a) The interval of time in which a tidal current is either ebbing or flooding, reckoned from the middle of slack water. (b) The interval of time from high water to low water (falling tide), or from low water to high water (rising tide).

duration-area curve A curve which shows the area beneath a duration curve, and any value of the flow, and is therefore the integral of duration with respect to stream flow. When the duration curve is plotted as a percentage of time, the resulting duration area shows the average flow available below a given discharge (ASCE, 1962).

duration curve A graphic illustration of how often a given quantity is equaled or exceeded during a given span of time, e.g. a flow-duration curve. It is used in hydraulics.

durbachite A plutonic rock composed chiefly of orthoclase, biotite, and hornblende, with a smaller amount of plagioclase and accessory quartz, apatite, sphene, zircon, and opaque oxides; a dark-colored biotite-hornblende syenite. The orthoclase phenocrysts form Carlsbad twins in a groundmass that is essentially an aggregate of biotite flakes and orthoclase (Johannsen, 1939, p.249). Its name is derived from Durbach, in the Black Forest, Germany.

duricrust A general term for a hard crust on the surface of, or layer in the upper horizons of, a soil of a semiarid climate. It is formed by the accumulation of soluble minerals deposited by mineral-bearing waters that move upward by capillary action and evaporate during the dry season. It may consist of concentrations of aluminous and ferruginous material *(ferricrete)* upon feldspathic rocks, or of siliceous material *(silcrete)* upon argillaceous and arenaceous rocks, or of calcareous material (*caliche* or *calcrete*) upon carbonate rocks. The term sometimes excludes (implicitly) calcareous crusts. Etymol: Latin *dursu*, "hard", + *crust*. Cf: *hardpan.*

durinite The major maceral of durain, according to the Stopes classification; the term is no longer in general use.

durinode In a soil, a nodule that has been cemented or indurated with silica.

duripan A horizon in a mineral soil that is characterized by cementation by silica (esp. opal or microcrystalline forms) and, possibly, by accessory cements (SSSA, 1970).

durite A coal microlithotype that contains a combination of inertinite and exinite totalling at least 95%, and containing more of each than of vitrinite. Cf: *durain.*

duroclarain A transitional lithotype of coal characterized by the presence of vitrinite with lesser amounts of other macerals such as micrinite and exinite; it corresponds to *semisplint coal.* Cf: *clarodurain.*

duroclarite A coal microlithotype containing at least 5% each vitrinite, exinite, and inertinite, and in which there is more vitrinite than inertinite. It is intermediate in composition between clarite and durite, but closer to clarite. Cf: *clarodurite.*

durofusain A coal lithotype transitional between fusain and durain, but predominantly fusain. Cf: *fusodurain.*

durotelain A coal lithotype transitional between telain and durain, but predominantly telain. Cf: *telodurain.*

durovitrain A coal lithotype transitional between vitrain and durain, but predominantly vitrain. Cf: *vitrodurain.*

durovitreous Said of a solid that is amorphous, uniform, and very strong. The term is not recommended. Cf: *liquevitreous.*

dussertite A mineral: $BaFe_3(AsO_4)_2(OH)_5$.

dust [sed] (a) Dry, solid, organic or inorganic matter consisting of clay- and silt-size earthy particles (diameters less than 1/16 mm, or 62 microns), larger than molecules but so finely divided or comminuted that they can be readily lifted and carried considerable distances in suspension by turbulent eddies in the wind, freely mixing with atmospheric gases, and staying aloft almost indefinitely but eventually falling back to the Earth's surface when winds become weak. Terrestrial sources of atmospheric dust include: volcanic eruptions; salt spray from the seas; mineral particles blown by the wind (esp. in desert regions); pollen and bacteria; and smoke and ashes produced industrially or in forest fires. See also: *volcanic dust.* (b) Small extraterrestrial particles that invade the Earth's atmosphere, such as *cosmic dust* and *meteoric dust.* (c) *gold dust.* (d) *diamond dust.* (e) A syn. of volcanic *ash,* esp. the finer fractions of ash.

dust [volc] A syn. of volcanic *ash,* esp. the finer fractions of ash.

dust avalanche *dry-snow avalanche.*

dust band *dirt band* [glaciol].

dust basin A large, shallow *dust well.*

dust cloud *eruption cloud.*

dust-cloud hypothesis A theory of the formation of the planets by the accretion of a cloud of small, cold bodies which are sometimes called "planetesimals". Syn: *planetesimal hypothesis.*

dust fall (a) *dusting.* (b) *blood rain.*

dust hole A small *dust well.*

dusting The process by which dust and dust-like particles are deposited from the atmosphere; e.g. the deposition of opal phytoliths in the ocean. Syn: *dust fall.*

dust ring A ring of tiny inclusions seen in thin section and marking the original surface of a detrital sand grain that has grown by secondary enlargement.

dust sand A term used by Searle (1923, p.1) for a material whose particles have diameters in the range of 0.025-0.0 mm and are washed out by a stream having a velocity of 1 mm/sec. The particle sizes correspond to medium silt and coarse silt.

dust size A term used in sedimentology for a volume less than that of a sphere with a diameter of 1/16 mm (0.0025 in.); includes *silt size* and *clay size.*

dust snow Deposited snow of low density, less than 30 kg m^3. Syn: *wild snow.*

dust tuff A tuff of very fine fraction; an indurated deposit of volcanic dust. Syn: *mud tuff.*

dust veil The stratospheric pall which results from ejection of volcanic dust into the stratosphere as an implicit consequence of paroxysmal eruptions (Lamb, 1970, p.425-533).

dust well A small hollow or pit on the surface of glacier ice or sea ice, produced by the gradual sinking into the ice of a patch of small dark windblown particles, which absorbs solar radiation and causes the surrounding ice to melt more rapidly. Cf: *cryoconite hole; dust basin; meridian hole; dust hole.*

dusty Characterized by, consisting of, or covered with dust.

duttonite A pale-brown monoclinic mineral: $VO(OH)_2$.

duty of water The quantity (or depth) of irrigation water required for a given area for the purpose of producing a particular crop; it is commonly expressed in acre-inches or acre-feet per acre, or simply as depth in inches or feet. Syn: *duty.*

duxite An opaque, dark-brown variety of retinite containing about 0.5% sulfur, found in lignite at Dux in Bohemia, Czechoslovakia.

dwip The basic unit of deposition in a tidal channel, consisting of a circular bank with a horseshoe outline and a hollow center, the toe pointing upstream (Strickland, 1940). It is caused primarily by channel bifurcation and reversing tidal currents.

dy A dark, jelly-like, freshwater mud consisting largely of unhumified or peaty organic matter (such as that derived from an acidic peat bog) that was brought to a nutrient-deficient lake in colloidal form and precipitated there. Etymol: Swedish "silt". Cf: *gyttja; sapropel.*

dyad An uncommon grouping in which mature pollen grains are shed in fused pairs. Cf: *tetrad; polyad.*

dyakisdodecahedron *diploid* [cryst].

Dyassic An old equivalent of Permian.

dying lake A lake nearing extinction from any cause.

dyke Var. of *dike.*

dykite *dike rock.*

dynamic breccia *tectonic breccia.*

dynamic damping *Damping* which is proportional to the velocity of the moving mass.

dynamic ellipticity A ratio expressed as the difference between the moments of inertia about the polar and equatorial axes, and the moment of inertia about the polar axis.

dynamic equilibrium A condition of a system in which there is a balanced inflow and outflow of materials. Cf: *stable equilibrium.*

dynamic geology A general term for the branch of geology that deals with the causes and processes of geologic phenomena; physical geology.

dynamic geomorphology The quantitative analysis of geomorphic processes treated as "gravitational or molecular shear stresses acting upon elastic, plastic, or fluid earth materials to produce the characteristic varieties of strain, or failure, that constitute weathering, erosion, transportation, and deposition" (Strahler, 1952a, p.923); the processes are considered in terms of steady-state operations that are self-regulatory to a large degree. Syn: *analytical geomorphology.*

dynamic head That head of fluid which would produce statically the pressure of a moving fluid (ASCE, 1962).

ynamic height The distance above the geoid, measured from ea level, of points on the same equipotential surface expressed in linear units measured along a plumb line at a specied latitude, usually 45°. Since gravity is assumed to be constant along the specified plumb line, there is generally no trict geometric relationship between the dynamic heights of ench marks in a leveling network. Cf: *height; orthometric eight, geopotential number.* Syn: *geopotential height.*

ynamic metamorphism The total of the processes and effects of orogenic movements and differential stresses in producing new rocks from old with marked structural and minerlogical changes due to crushing and shearing at low temperatures and extensive recrystallization at higher temperatures. usually involves large areas of the Earth's crust; thus it is egional in character. Cf: *dynamothermal metamorphism; reional metamorphism; dislocation metamorphism.* Syn: *dynaometamorphism.*

ynamic pressure The pressure of a flowing fluid against a urface. Reaction to dynamic pressure affects direction and elocity of flow (ASCE, 1962).

ynamic range In seismic frequency filtering, the ratio of the aximum to minimum signal (for a given distortion level) that an be handled. It is usually specified over a certain frequeny bandwidth and is measured in decibels.

ynamic rejuvenation A kind of a *rejuvenation* caused by peirogenic uplift of a landmass, with accompanying tilting nd warping.

ynamic theory A theory of tides that considers the horizontal de-producing forces to be the most important factor in causg movement of water and that regards the vertical tide-producing forces as small periodical variations in the acceleration f gravity (Baker et al, 1966, p.53).

ynamic viscosity *viscosity coefficient.*

ynamofluidal Pertaining to a texture in dynamometamorhosed rocks showing parallel arrangement in one direction nly. Obs. syn: *metafluidal.*

ynamogranite An augen gneiss containing much microcline nd orthoclase (Krivenko & Lapchik, 1934). Use of the term s not recommended.

ynamometamorphism The equivalent of *dislocation metamorhism* and a syn. of *dynamic metamorphism.*

ynamometer In oceanography, an instrument that is used in conjunction with a bottom sampling to indicate that the bottom has been reached. Its operation is based on wire tension.

ynamo theory The statement that the Earth's main magnetic ield is thought to be sustained by self-exciting dynamo action n the fluid core. The conducting liquid is supposed to flow in uch a pattern that the electric current induced by its motion hrough the magnetic field sustains that field.

ynamothermal metamorphism A common type of metamorphism involving the effects of directed pressures and shearing stress as well as a wide range of confining pressures (3000-10,000 bars) and temperatures of 400°C to a maximum of about 800°C. It is related both geographically and genetically to large orogenic belts, hence is regional in character. Cf: *burial metamorphism; regional metamorphism; dynamic metamorphism.*

dypingite A mineral: $Mg_5(CO_3)_4(OH)_2.5H_2O$.

dysànalyte A variety of perovskite containing niobium and tantalum.

dyscrasite A silver-white mineral: Ag_3Sb.

dyscrystalline Said of the texture of an igneous rock whose mineral grains are too small to be seen without a microscope, as in the groundmasses of certain porphyritic rocks. Also, said of a rock having such texture. Cf: *eucrystalline.*

dysgeogenous Not easily weathered; said of a rock that produces by weathering only a small amount of detritus. Ant: *eugeogenous.*

dysluite A brown variety of gahnite containing manganese and iron.

dysodile (a) An inflammable (burning with a highly fetid odor), flexible, slightly elastic, yellow or greenish-gray hydrocarbon from Melili, Sicily, and from certain German lignite deposits (Rice, 1945). See also: *chlorophyll coal.* (b) A sapropelic coal of lignitic rank derived from diatomaceous sediments formed under anaerobic conditions. It burns readily and with a bad odor. Dysodile occurs in Tertiary limestones and lignites. Syn: *diatom-saprocol.*

dysodont Said of the dentition of a bivalve mollusk (e.g. some Mytilacea) characterized by small, weak hinge teeth close to the beaks.

dystome spar *datolite.*

dystrophic lake A lake that is characterized by a deficiency in nutrient matter and by a notably high oxygen consumption in the hypolimnion; its water is brownish or yellowish with much unhumified or dissolved humic matter and a small bottom fauna. It is often associated with acidic peat bogs. Juday and Birge (1966, p.33) rejected this term, and Hutchinson (1967, p.380) called it "unfortunate". Cf: *oligotrophic lake.*

dysyntribite A hydrated aluminosilicate of sodium and potassium, probably a variety of pinite or an impure muscovite.

dzhalindite A yellow-brown mineral: $In(OH)_3$. It is an alteration product of indite.

dzhetymite A term proposed by Dzholdoshev (1964; see Schermerhorn, 1966, p.833) as a quantitative designation for a nonsorted rock composed of approximately equal proportions (25-35% each) of angular "gravel" (1-10 mm diameter in Russian literature), sand (0.1-1 mm), and mud (under 0.1 mm).

dzhezkazganite A mineral: a lead rhenium sulfide (?).

E

ea An English term for a stream.

Eaglefordian North American (Gulf Coast) stage: Upper Cretaceous (above Woodbinian, below Austinian).

eaglestone A walnut-sized concretionary nodule (usually of clay ironstone or flint), often containing a loose stone in its hollow interior, and believed by the ancients to be taken by an eagle to her nest to facilitate egg-laying. Syn: *aetites*.

EA interray Left anterior interray in echinoderms situated between E ray and A ray and clockwise of E ray when the echinoderm is viewed from the adoral side.

eakerite A mineral: $Ca_2SnAl_2Si_6O_{16}(OH)_6$.

ear The flattened or pointed extremity of a brachiopod shell, subtended between the hinge line and the lateral part of the commissure.

earlandite A pale-yellow to white mineral consisting of a hydrous calcium citrate: $Ca_3(C_6H_5O_7)_2.4H_2O$. It is found in ocean-bottom sediments from the Weddell Sea in the south Atlantic Ocean.

early Pertaining to or occurring near the beginning of a segment of time. The adjective is applied to the name of a geologic-time unit (era, period, epoch) to indicate relative time designation and corresponds to *lower* as applied to the name of the equivalent time-stratigraphic unit; e.g. rocks of a Lower Jurassic batholith were intruded in Early Jurassic time. The initial letter of the term is capitalized to indicate a formal subdivision (e.g. "Early Devonian") and is lowercased to indicate an informal subdivision (e.g. "early Miocene"). The informal term may be used for eras and epochs, and for periods where there is no formal subdivision. Cf: *middle* [*geochron*]; *late*.

early diagenesis Diagenesis occurring immediately after deposition or immediately after burial. It is equivalent to *diagenesis* as interpreted by Russian geologists. Syn: *syngenesis; syndiagenesis*.

early wood *springwood*.

earth [eng geol] Material that can be removed and handled economically with pick and shovel or by hand, or which can be loosened and removed with a power shovel.

earth [chem] (a) A difficultly reducible metallic oxide (such as alumina) formerly regarded as an element. (b) One of the four elements of the alchemists (the others: air, water, fire). Aristotle emphasized the state rather than the material, but the latter concept is so deeply embedded in literature that it is ineradicable.

earth [geog] A general term for the solid materials that make up the physical globe, in distinction from water and air. Also, the firm land or ground of the Earth's surface, uncovered by water.

earth [mater] A syn. of *earth color*.

earth [sed] (a) An organic deposit that has remained unconsolidated although it is no longer in the process of accumulation; e.g. *radiolarian earth* and *diatomaceous earth*. (b) An amorphous fine-grained material, such as a clay or a substance resembling clay; e.g. *fuller's earth*.

earth [soil] A term for the loose, softer, or fragmental material of the Earth's surface, formed by rock disintegration, as distinguished from firm or solid rock (bedrock); not synonymous with the term *soil*. Much of it has been moved by wind, ice, or water.

Earth That planet of the solar system which is fifth in size of the 9 major planets, and the third (between Venus and Mars) in order of distance from the Sun (about 150 x 10^6 km, or 93 million mi). Large-scale data for the Earth: equatorial radius: 6,378 km (3,963.5 mi); polar radius: 6,357 km (3,941 mi); mean radius: 6,371 km (3,950 mi); equatorial circumference: 40,075 km (24,902 mi); surface area: 5.101 x 10^8 km^2 (197 x 10^6 sq mi).

earth circle *soil circle*.

earth color A mineral used as a pigment. Syn: *earth*. See also: *mineral paint; mineral pigment*.

Earth current Static or alternating electric current flowing through the ground and arising either in natural or artificial electric or magnetic fields. Syn: *ground current; telluric current*.

earth curvature The divergence of the surface of the Earth (spheroid) from a plane. See also: *curvature correction*.

earth dam A dam constructed of earth material (such as gravel, broken rock, sand silt, and soil). It has an impervious clay core and facing.

earth dike An artificial *levee* built of random earth fill.

earth finger A miniature "earth pillar" in a nearly horizontal attitude, produced by wind-driven rain falling upon clayey material (Cotton, 1958, p.31 & 33).

earth flax A fine silky *asbestos*.

earthflow A mass-movement landform and process characterized by downslope translation of soil and weathered rock over a discrete basal shear surface (landslide) within well defined lateral boundaries. The basal shear surface is more or less parallel with the ground surface in the downslope portion of the flow, which terminates in lobelike forms. Overall, little or no rotation of the slide mass occurs during displacement, although, in the vicinity of the crown scarp, minor initial rotation is usually observed in a series of slump blocks. Earthflows grade into mudflows through a continuous range in morphology associated with increasing fluidity. Also spelled: *earth flow*.

earth foam Soft or earthy *aphrite*.

Earth history *geologic history*.

earth hummock A low, knob-like, dome-shaped *frost mound* consisting of a fine-textured earthen core covered by a tight mass of vegetation, esp. mosses, but also humus, plant roots, grasses, sedges, and scrubby plants, and produced by hydrostatic pressure of ground water in arctic and alpine regions; the average height is 120 mm and the diameter varies from 1/2 to 1 m. Earth hummocks form in groups to produce nonsorted patterned ground. Syn: *earth mound; thufa*.

Earth inductor A type of *inclinometer*, based on the principle that a voltage is induced in a coil rotating in the Earth's field whenever the rotation axis does not coincide with the field direction.

earth island *debris island*.

earthlight The faint illumination of the dark part of the Moon produced by sunlight reflected onto the Moon from the Earth's surface and atmosphere. It is best seen during the Moon's crescent phases. Syn: *earthshine*.

earth lurch A distinctive and uncommon earthquake motion that is perpendicular to a stream bank or cliff and that results in yielding of the material in its unsupported direction.

earth mound *earth hummock*.

earth pillar A tall, conical column of unconsolidated to semiconsolidated earth materials (as clay, till, or landslide debris) produced by differential erosion in a region of sporadic heavy rainfall (as in a badland or a high alpine valley), and usually capped by a flat, hard boulder that shielded the underlying, softer material from erosion; it often measures 6-9 m in height, and its diameter is a function of the width of the protective boulder. Cf: *hoodoo*. Syn: *earth pyramid; demoiselle; fairy chimney; hoodoo column; penitent*.

earth pitch *maltha*.

earth pressure The pressure acting between earth material (such as soil or sediments) and a structure (such as a wall); the pressure exerted by soil on any boundary. See also: *active earth pressure; passive earth pressure; at rest*.

earth pyramid A less steeply sided and more conical *earth pillar*, occurring esp. where rainwash is the chief agent of erosion.

earthquake A sudden motion or trembling in the Earth caused by the abrupt release of slowly accumulated strain (by faulting or by volcanic activity). Partial syn: *seismic event; quake.* Syn: *shock; seism; macroseism; temblor.*

earthquake engineering The study of the behavior of foundations and structures relative to seismic ground motion, and the attempt to mitigate the effect of earthquakes on such structures. Syn: *engineering seismology; earthquake seismology.*

earthquake intensity A measure of the effects of an earthquake at a particular place on humans and/or structures. The intensity at a point depends not only upon the strength of the earthquake, or the *earthquake magnitude*, but also upon the distance from the earthquake to the epicenter and the local geology at the point. See also: *intensity scale.*

earthquake magnitude A measure of the strength of an earthquake or the strain energy released by it, as determined by seismographic observations. The concept was introduced by the seismologist C.F. Richter, who first applied it to southern California earthquakes. For that region he defined local magnitude, to the logarithm, to the base 10, of the amplitude in microns of the largest trace deflection that would be observed on a standard torsion seismograph (static magnification = 2800, period = 0.8 sec, damping constant = 0.8) at a distance of 100 km from the epicenter. Magnitudes determined at teleseismic distances using the logarithm of the amplitude to period ratio of body waves are called body-wave magnitudes, and using the logarithm of the amplitude of 20-sec period surface waves are called surface-wave magnitudes. The local, body-wave, and surface-wave magnitudes of an earthquake will have approximately the same numerical value. See also: *Richter scale.* Cf: *earthquake intensity.*

earthquake period That time during which a region affected by an earthquake continues to receive shocks without any long respite between them.

earthquake prediction That aspect of seismology which deals with the physical conditions or indications that precede an earthquake, in order to extrapolate the size, time, and location of an impending shock.

earthquake record *seismogram.*

earthquake rent Popular syn. of *reverse scarplet.*

earthquake scarplet A low, nearly straight fault scarp or step, often continuous for many kilometers, formed simultaneously with an earthquake either as cause or effect. Cf: *reverse scarplet.*

earthquake sea wave *tsunami.*

earthquake seismology *earthquake engineering.*

earthquake sound *air wave.*

earthquake swarm A series of minor earthquakes, none of which may be identified as the main shock, occurring in a limited area and time. Syn: *swarm [seism]; swarm earthquakes.*

earthquake tremor *tremor.*

earthquake volume The volume of an earthquake's major potential energy content and chosen equal to the total volume of aftershocks following each shock (Baåth, 1966, p.79). It is expressed in cm^3.

earthquake wave *seismic wave.*

earthquake zone An area of the Earth's crust in which crustal movements and sometimes associated volcanism occur; a *seismic area.*

earth radiation *terrestrial radiation.*

earth rotation The turning of the Earth about its axis, described either as counterclockwise about the north pole or as eastward according to the movement of the equator. A proposed rate of rotation is 0.000072921 radians per second.

earth run A lobe of earth material that has flowed downhill beneath the vegetation cover, forming a sloping step whose front is often 1/3-2 m high (Sharpe, 1938, p.42).

earth science A term frequently used as a synonym for the more exact term, *geology.* The word may be misleading as, in its wider scope, it can be considered to include the sciences of agronomy, meteorology, soil chemistry, etc., which are not

geological sciences.

earth sculpture *sculpture.*

Earth shell Any one of the concentric *shells* that constitute the structure of the Earth.

earthshine *earthlight.*

Earth's orbit The path through space of the Earth in its annual journey around the Sun. This path is an ellipse, with semimajor axis of about 92,700,000 miles, an eccentricity of 0.03, and the Sun at one focus.

Earth stretching A method for calculating dispersion of surface waves in a spheroidal Earth using a computer program developed for flat anisotropic layers.

earth stripe *soil stripe.*

Earth tide The response of the solid Earth to the forces that produce the tides of the sea; semidaily earth tides have a fluctuation of between seven and fifteen centimeters (Strahler, 1963, p.110-111). Syn: *bodily tide.*

earth tremor *tremor.*

earth wave An obsolete syn. of *seismic wave.*

earth wax *ozocerite.*

earthwork (a) The processes or operations involved with excavations and embankments of earth, such as in preparing foundations for buildings or in constructing railroads. (b) An embankment or other construction made of earth.

earthy [mineral] (a) Said of *dull* luster; the mineral surface may feel rough to the touch. (b) Said of a type of fracture similar to that of a hard clay.

earthy [geol] Composed of or resembling earth, or having the properties or nature of earth or soil; e.g. an "earthy limestone" containing argillaceous material and characterized by high porosity, loosely aggregated particles, and close association with chalk.

earthy breccia A term used by Woodford (1925, p.183) for a breccia in which rubble, sand, and silt + clay each constitute more than 10% of the rock.

earthy calamine *hydrozincite.*

earthy cobalt *asbolite.*

earthy lignite Lignite that is structurally amorphous and is soft and friable. Cf: *woody lignite.*

earthy manganese *wad* [mineral].

east (a) The general direction of sunrise; the direction toward the right of one facing north. (b) The place on the horizon where the Sun rises when it is near one of the equinoxes. (c) The cardinal point directly opposite to west. Abbrev: E. (d) The direction of the Earth's daily rotation; the direction of revolution of the Earth around the Sun. (e) The point of the horizon having an azimuth of 90 degrees measured clockwise from north.

easting A *departure* (difference in longitude) measured to the east from the last preceding point of reckoning; e.g. a linear distance eastward from the north-south (vertical) grid line that passes through the origin of the grid.

eastonite A variety of biotite: $K_2Mg_5Al_4Si_5O_{20}(OH)_4$.

eating The gradual wasting or wearing away of a part of the Earth's surface; e.g. the *eating* of a cliff by waves.

ebb current The tidal current associated with the decrease in the height of a tide, generally moving seaward or down a tidal river or estuary. Cf: *flood current.* Erroneous syn: *ebb tide.*

ebbing-and-flowing spring *periodic spring.*

ebbing well A well in permeable rocks near the coast in which the water level fluctuated with the tide.

ebb tide (a) *falling tide.* (b) An erroneous syn. of *ebb current.*

èboulement A *landslide.* Etymol: French, "falling in, fall"; shortened from *èboulement de terre,* "landslide".

ebridian A marine protist characterized by the presence of flagella, a skeleton of solid silica rods, and by the absence of chromatophores.

ecardinal Said of an inarticulate brachiopod (or of its shell) without a hinge.

eccentricity (a) The condition, degree, amount, or instance of deviation from a center or of not having the same center; e.g. the horizontal displacement of a surveying instrument or signal from a triangulation station mark at the time an observation is made, or an effect seen in a surveyor's compass when the line of sight fails to pass through the vertical axis of the compass or when a straight line through the ends of the magnetic needle fails to pass through the center rotation of the needle. (b) The distance of the center of figure of a body from an axis about which it turns; e.g. the "eccentricity of alidade" represented by the distance between the center of figure of the index points on the alidade and the center of figure of the graduated circle. (c) The ratio of the distances from any point of a conic section to a focus and the corresponding directrix; e.g. the "eccentricity of ellipse" represented by the ratio of the distance between the center and a focus of an ellipse to the length of its semimajor axis, or $e^2 = (a^2 - b^2)/a^2$, where e = eccentricity, a = semimajor axis, and b = semiminor axis. Eccentricity is less than one in the ellipse, greater than one in the hyperbola, equal to one in the parabola, and equal to zero in the circle.

eccentricity correction The correction that must be applied to an observed direction made with an eccentric instrument or signal, or both, in order to reduce the observed value to what it would have been if there had been no eccentricity. Also known as "eccentric reduction".

eccentric signal A survey signal (target) which is not in the same vertical line with the station which it represents, such as a signal placed in triangulation at some point other than directly over the triangulation station and not in line with the station and the instrument.

eccentric station A survey point over which an instrument is centered and observations are made, and which is not in the same vertical line with the station which it represents and to which the observations will be reduced before being combined with observations at other stations (Mitchell, 1948, p.26). It is established and occuppied when it is not practicable to set up directly over the actual station center or when it becomes necessary in order to see points that are not visible from the station center.

ecdemite A bright-yellow to green mineral, perhaps: Pb_6As_2-O_7Cl_4. Syn: *ekdemite.*

ecesis The establishment of a plant in a new location as a result of successful germination, growth, and reproduction. Syn: *establishment.*

echinate Spiny-surfaced, or densely covered with stiff, stout, or bluntish bristles, prickles, or spines; e.g. "echinate pollen" having a sculpture consisting of spines.

echinating Said of sponge spicules that protrude at an angle from the surface of a skeletal fiber.

echinoderm Any solitary marine benthic (rarely pelagic) invertebrate, belonging to the phylum Echinodermata, characterized, generally, by radial symmetry, an endoskeleton formed of plates or ossicles composed of crystalline calcite, and a water-vascular system. Echinozoans, asterozoans, crinozoans, and homalozoans are echinoderm subphyla.

echinoid Any echinozoan belonging to the class Echinoidea, characterized by a subspherical to modified spherical shape, interlocking calcareous plates, and by movable appendages; e.g. a *sea urchin.*

echinozoan Any free-living, globoid, discoid, or cylindroid echinoderm belonging to the subphylum Echinozoa, characterized by the absence of arms, brachioles, and outspread rays and by the dominance of a meridional growth pattern over bilateral symmetry. Their stratigraphic range is Lower Cambrian to present.

echmidium A spear-shaped plate formed during ontogeny of a brachiopod loop by fusion of anterior ends of descending branches.

echodolite *phonolite.*

echogram The graphic record made by an *echo sounder,* in the form of a continuous profile. See also: *fathogram.*

echo sounder In oceanography, a *sounding* instrument that measures water depth by measurement of the time that it takes a sonic or supersonic sound signal to travel to and return from the sea floor. See also: *echogram; fathometer, pinger; precision depth recorder.* Syn: *sonic depth finder, depth sounder.*

eckermannite A monoclinic mineral of the amphibole group: $Na_3(Mg,Li)_4(Al,Fe)Si_8O_{22}(OH,F)_2$.

Eckert projection One of a series of six map projections of the entire Earth in each of which the geographic poles are represented by parallel straight lines that are one half the length of the equator. The parallels are rectilinear and the meridians may be rectilinear or curved. They are broadly similar in appearance to the Mollweide projection. Named after Max Eckert (1868-1938), German cartographer, who developed the projections in 1906.

eckrite A monoclinic mineral of the amphibole group: $NaCa(Mg,Fe^{+2})_4Fe^{+3}Si_8O_{22}(OH)_2$.

eclipse An obscuration of the light of one celestial body with another, e.g. of the Sun by the Moon.

eclogite facies Eclogites formed by regional dynamothermal metamorphism (according to Eskola, 1939) at extremely high pressures (in excess of 10,000 bars) and temperatures of 600°-700°C, considered to be required for the origin of the high-density mineral association of omphacite and garnet plus rutile, kyanite, enstatite, and diamond. Recent experimental work indicates that, under certain conditions, eclogites may form over a large range of pressures and temperatures (750C to over 10,000 bars and 350°-750°C), transgressing upon conditions in other facies (e.g. amphibolite). Hence the eclogite facies concept may have to be abandoned (Winkler, 1967).

eclogitic (a) Pertaining to an *eclogite.* (b) Said of a rock having an association of clinopyroxene and garnet with a proportion of jadeite molecule in the clinopyroxene. In this case no genetic connotation is implied nor bulk composition considered (Church, 1968, p.757).

ecochronology A term introduced as *Oekochronologie* by Schindewolf (1950, p.35) for *geochronology* based on the ecology of life forms.

ecocline A cline related to the gradation between two different niches.

ecography The descriptive part of *ecology.*

ecologic facies (a) *facies* [ecol]. (b) *environmental facies.*

ecologic niche *niche.*

ecologic potential A term used by Lowenstam (1950) for the capability of an organism to directly control or modify its environment, such as that possessed by a reef-building organism that is able to erect a rigid and resistant framework in the zone of wave action.

ecology The study of the relationships between organisms and their environments, including the study of communities, patterns of life, natural cycles, relationships of organisms to each other, biogeography, and population changes. See also *paleoecology; ecography.* Adj: *ecologic; ecological.* Syn *bionomics.*

economic geology The analysis and exploitation of geologic bodies and materials that can be utilized profitably by man, including fuels, metals and nonmetallic minerals, and water.

economic limit The boundary of an orebody by grade; the extent of a mineralized area that is economically valuable. Cf *cutoff grade.*

economic mineral The valuable metallic, nonmetallic, or industrial constituent in a deposit (Park & MacDiarmid, 1970 p.1).

economic yield The maximum estimated rate at which water may be withdrawn from an aquifer without creating a deficiency or affecting the quality of the supply. Cf: *safe yield.*

economy The input and consumption of energy within a system (such as a stream), and the changes that result; e.g. the *balance* of a glacier.

ecospace As used by Valentine (1969, p.687), that volume within the *environmental hyperspace lattice* corresponding to the environmental conditions under which a particular organism may live. See also: *realized ecospace; prospective ecospace.*

ecospecies A unit, within a *cenospecies*, that contains individuals capable of successful gene exchange with other members of the unit but only occasionally with members of other ecospecies in the cenospecies. It is more or less equivalent to the taxonomic unit *species*. See also: *ecotype.*

ecosphere Portions of the universe favorable for the existence of living organisms; esp. the *biosphere.*

ecostratigraphic unit A stratigraphic unit based on the mode of origin or the environment of deposition of rocks (Hedberg, 1958, p.1893); e.g. a marine zone, a brackish-water zone, or a glacially deposited zone. The terminology of most ecostratigraphic units has not reached formal status. Syn: *ecozone.*

ecostratigraphy (a) A term introduced as *Oekostratigraphie* by Schindewolf (1950, p.35) for stratigraphy based on the ecology of life forms; the stratigraphic occurrence of local or regional faunal or floral assemblages that are valuable for ecologic considerations. (b) A term used by Hedberg (1958, p.1893) for the study and classification of stratified rocks according to their mode of origin or their environment of deposition.---Syn: *oecostratigraphy.*

ecosystem A unit in ecology consisting of the environment with its living elements and the factors which exist in and effect it.

ecotone A transition zone that exists between two ecologic communities. Members of both communities may compete within this zone. Syn: *tension zone.*

ecotope The habitat of a particular organism. See also: *ecotopic.*

ecotopic Having the tendency to adjust to the specific conditions of the *ecotope.*

ecotype [ecol] An ecologic variant of a species that has adapted to local environmental conditions.

ecotype [taxon] A unit, within an *ecospecies*, that contains individuals capable of interbreeding both with other members of that ecotype and with members of other ecotypes in the same ecospecies but that remains distinctive through selection and isolation. If it is morphologically distinct, it is more or less equivalent to the taxonomic unit *subspecies.*

ecoulement A syn. of *gravitational sliding.* Etymol: French, "flowing".

ecozone (a) *ecostratigraphic unit.* (b) An assemblage zone that demonstrates a particular ecology (facies).

ectexine Var. of *ektexine.*

ectexis Migmatization with in situ formation of the mobile part (Dietrich & Mehnert, 1961). Var: *ektexis.* Cf: *entexis.*

ectexite Rock formed by ectexis. Also spelled: *ectectite.*

ectinite Rock formed as a result of essentially isochemical regional metamorphism, i.e. with no notable associated metasomatism (Dietrich & Mehnert, 1961).

ectocyst An external, membranous layer covering a zooecium or an entire zoarium of a bryozoan, with or without incorporated calcification. Cf: *endocyst.*

ectoderm The outer body layer of an organism; e.g. the outer layer of the oral and basal disks, tentacles, and column wall of a coral polyp. Cf: *endoderm; mesoderm.*

ectoderre The principal layer in the external covering of a chitinozoan, thicker than the *endoderre; periderre.*

ectodynamic *ectodynamomorphic.*

ectodynamomorphic An old term applied to a soil whose characteristics were produced in part by external forces, e.g. vegetation, climate. Also spelled: *ektodynamomorphic.* Cf: *endodynamomorphic.* Syn: *ectodynamic.*

ectoexine *ektexine.*

ectogene Said of external factors that influence the texture of a rock (Sander, 1951, p.11). Cf: *entogene.*

ectooecium The outer, generally calcified layer of the wall of a bryozoan ovicell. Cf: *entooecium.*

ectophloic Pertaining to the siphonostele of certain vascular plants having phloem external to the xylem. Cf: *amphiphloic.*

ectophragm A thin membrane lying between distal ends of processes on a dinoflagellate cyst. Cf: *endophragm; periphragm.*

ectoproct Any bryozoan belonging to the subphylum Ectoprocta and characterized by a circular or horseshoe-shaped lophophore around the mouth but not the anus. Their stratigraphic range is Ordovician to present, with a possible downward extension into the Upper Cambrian. Cf: *entoproct.*

ectosiphuncle The wall of the siphuncle of certain cephalopods, consisting generally of septal necks and connecting rings. Cf: *endosiphuncle.* Syn: *ectosiphon.*

ectosolenian Said of a foraminifer (e.g. *Lagena*) having an external tube-like neck. Cf: *entosolenian.*

ectosome The peripheral region of a sponge beneath the inhalant surface and devoid of flagellated chambers; the cortical part of a sponge. Cf: *endosome.*

ectyonine Said of a sponge skeleton built of spiculofibers made up of both coring and echinating monaxons.

ecumeme That part of the Earth which is permanently inhabited.

edaphic Said of ecologic formations resulting from or influenced by conditions of the soil or other substratum; also, an old term applied to any soil characteristic that affects plant growth, e.g. acidity, alkalinity.

edaphon All of the animals and plants living in the soil.

eddy A water current that is generally a circular motion with a different direction from that of the main current. It is a temporary current, usually formed at a point at which a current passes some obstruction, or between two adjacent currents flowing in opposite directions, or at the edge of a permanent current. Cf: *whirlpool; maelstrom.*

eddy-built bar A bar presumably built by currents rotating as an eddy in a tidal lagoon; e.g. one of the ridges surrounding some of the Carolina bays. Syn: *Neptune's racetrack.*

eddy coefficient *austausch.*

eddy conductivity *austausch.*

eddy diffusion Mixing by turbulent flow. Syn: *turbulent diffusion.*

eddy diffusivity The coefficient of the proportionality of the rate of transfer of mass to the gradient of the average concentration. It depends on the nature of the turbulent motion (Fairbridge, 1966, p.230).

eddy flux The rate of transport or flux of fluid properties, e.g. momentum or suspended matter, by turbulent flow. Syn: *turbulent flux.*

eddy mark One of numerous superimposed or overlapping loops (0.3-1 m in diameter) forming a spiral impression on a sedimentary (sandstone) surface, believed to result from "dragging of a small limb of a larger floating log caught in a vortex or eddy current" of a stream or from movement of a pebble or stick caught in circular winds of a dust devil after the sands had been exposed along the stream bank (Rigby, 1959).

eddy mill A *pothole* in a stream bed.

eddy spectrum Turbulent flow described in terms of the frequency distribution of eddy size, or of the partition of kinetic energy between eddies of various sizes. Syn: *turbulence spectrum.*

eddy viscosity The transfer coefficient for momentum, corresponding to *kinematic viscosity.*

Edenian North American stage: Upper Ordovician (above Mohawkian, below Maysvillian).

edenite (a) A light-colored and iron-free variety of hornblende. (b) An end member in the amphibole mineral group: $NaCa_2$-

$Mg_5AlSi_7O_{22}(OH)_2$. Cf: *pargasite*.

edentulous Said of a bivalve mollusk lacking hinge teeth.

edge (a) A sharply pointed ridge; also the crest of such a ridge. (b) The escarpment that terminates a plateau; the extreme margin of a cliff. (c) The highest part of an elevated tract of land of great extent; esp. a ridge or divide between two streams.

edge dislocation In a crystal, a row of atoms marking the edge of a crystallographic plane and extending only part way; it is a type of *line defect*.

edge effect In ecology, the influence of the presence of adjacent plant communities on the number and kind of animals present in the immediate vicinity.

edge line A heavy line on a relief map, depicting a sudden and sharp change or break of slope.

edge water The water surrounding, bordering, or underlying oil or gas in a pool or oil-bearing formation. Cf: *bottom water*. Also spelled: *edgewater*.

edge-water line *water level* [oil].

edgewise conglomerate A conglomerate exhibiting *edgewise structure*; e.g. an intraformational conglomerate containing elongated calcareous pebbles that are transverse to the bedding.

edgewise structure A primary sedimentary structure characterized by an arrangement of flat, tabular, or disc-shaped fragments whose long axes are set at varying and steep angles to the bedding. It may be due to running water or to sliding or slumping soon after deposition. See also: *edgewise conglomerate*.

edge zone A fold of the body wall of a coral polyp, extending laterally and/or downward over the edge of the wall.

edingtonite A white or grayish-white zeolite mineral: $BaAl_2Si_3O_{10}.4H_2O$. It sometimes contains appreciable calcium.

edolite A type of hornfels consisting mainly of feldspar and mica. There are varieties that contain also cordierite (aviolite) or andalusite (astite) (Holmes, 1928, p.87). Type locality: Edolo, Italian Alps.

edrioasteroid Any many-plated, attached echinozoan belonging to the class Edrioasteroidea, having a well-developed quinqueradial endothecal ambulacral system. Their stratigraphic range is Lower Cambrian to Lower Carboniferous.

eel In seismic reflection profiling at sea, a device for towing a hydrophone array in a manner that tends to cancel the towing noises and to reinforce the reflection signals. It consists of a buoyant, liquid-filled tube, and it does not have to be slacked after each shot. (Fairbridge, 1966, p.818).

effective depth of penetration *depth of penetration*.

effective diameter (a) The diameter of the particles in an assumed rock or soil that would transmit water at the same rate as the rock or soil under consideration, and that is composed of spherical particles of equal size and arranged in a specified manner. (b) The approximate diameter of a rock or soil particle equal to the sieve size that allows 10% (by weight) of the material to pass through; the particle diameter of the 90 percent line of a cumulative curve, or the maximum diameter of the smallest 10% of the particles of sediment.---Syn: *effective size*.

effective drainage porosity *effective porosity*.

effective force The force transmitted through a soil mass by effective stresses.

effective permeability The observed permeability of a porous medium to one fluid phase under conditions of physical interaction between this phase and other fluid phases present.

effective pore volume The pore space available for the free circulation of water. This excludes pore space taken up by air and pellicular water.

effective porosity The percent of the total volume of a given mass of soil or rock that consists of interconnecting interstices. The use of this term as a syn. of *specific yield* is to be discouraged. Syn: *effective drainage porosity*. Cf: *porosity*

[grd wat]; *continuous porosity*.

effective precipitation (a) That part of precipitation producing runoff. (b) That part of precipitation falling on an irrigated area that meets the demands of consumptive use. Cf: *precipitation excess*.

effective pressure *effective stress*.

effective radiation *effective terrestrial radiation*.

effective shot depth *shot depth*.

effective size *effective diameter*.

effective stress The average normal force per unit area transmitted directly from particle to particle of a soil or rock mass. It is the stress that is effective in mobilizing internal friction. In a saturated soil in equilibrium, the effective stress is the difference between the total stress and the neutral stress of the water in the voids; it attains a maximum value at complete consolidation of the soil. Syn: *effective pressure; intergranular pressure*.

effective temperature *brightness temperature*.

effective terrestrial radiation The difference between the Earth's outgoing infrared *terrestrial radiation* and the downward infrared *counterradiation* from the Earth's atmosphere (U.S. Naval Oceanographic Office, 1966, p.55). Syn: *effective radiation*.

effective unit weight The *unit weight* of a soil that, when multiplied by the height of the overlying column of soil, yields the effective stress due to the weight of the overburden.

effective velocity The actual velocity of ground water percolating through water-bearing material. "It is measured by the volume of ground-water passing through a unit cross-sectional area divided by effective porosity" (Tolman, 1937, p.593). The velocity is the average velocity for water moving through the interstices.

efficiency The *capacity* (maximum load a stream can carry) per unit discharge and unit stream gradient, or the quotient of capacity by the product of discharge and gradient (Gilbert, 1914, p.36). It is a measure of the stream's potential work or transportation in relation to its potential energy. Symbol: E.

efflata A little-used term for pyroclastics.

efflorescence (a) A whitish, mealy (or fluffy) or crystalline powder produced as a surface incrustation on a rock in an arid region by evaporation of water brought to the surface by capillary action or by loss of water of crystallization on exposure to the air. It may consist of one or several minerals commonly soluble salts such as gypsum, calcite, natron, and halite. Syn: *bloom*. (b) The process by which an efflorescence is formed.

effluent adj. Flowing forth or out; emanating.---n. (a) A surface stream that flows out of a lake (e.g. an outlet), or a stream or branch that flows out of a larger stream (e.g. a distributary). Ant: *influent*. Cf: *effluent stream*. (b) A liquid discharged as waste, such as contaminated water from a factory or the outflow from a sewage works; water discharged from a storm sewer or from land after irrigation.

effluent cave *outflow cave*.

effluent flow Flow of water from the ground into a body of surface water; e.g. the flow of water to an effluent stream.

effluent lava flow A lava flow that is discharged from a volcano by way of a lateral fissure (Dana, 1890); an obsolete term. Cf: *interfluent lava flow; superfluent lava flow*.

effluent seepage Diffuse discharge of ground water to the land surface; *seepage* of water from out of the ground.

effluent stream (a) A stream or reach of a stream that receives water from the zone of saturation and provides base flow; its channel lies below the water table. Syn: *gaining stream*. (b) *effluent*.

efflux *outflow*.

effusive *extrusive*.

egeran A brown or yellowish-green variety of vesuvianite.

egg In an embryophytic plant, the female gamete; a nonmotile gamete which can fuse with a sperm to form a zygote (Cron quist, 1961, p.874).

egg coal A size of anthracite that will pass through a 3 1/4 inch round mesh but not through a 2 7/16 inch round mesh. Cf: *broken coal; stove coal; chestnut coal; pea coal; buckwheat coal.*

eggstone *oolite.*

eglestonite A brownish-yellow isometric mineral: Hg_4Cl_2O.

Egyptian jasper A brown or banded jasper occurring as pebbles or small boulders scattered over the desert surface between Cairo and the Red Sea. Syn: *Egyptian pebble.*

Ehrenfest relation A modified *Clapeyron equation* that is used for second-order transitions.

ehrwaldite An augitite that contains both orthopyroxene and clinopyroxene.

Eifelian European stage: Middle Devonian (above Emsian, below Givetian). Syn: *Couvinian.*

eightling A crystal twin, either cyclic or interpenetrating, that consists of eight individuals. Cf: *twoling; trilling; fourling; fiveling.*

einkanter A *windkanter* or stone having only one face or a single sharp edge; it implies a steady, unchanging wind direction. Etymol: German *Einkanter*, "one having one edge". Pl: *einkanters; einkanter.*

eiscir An Irish term for ridge; esp. *esker.*

eitelite A hexagonal mineral: $Na_2Mg(CO_3)_2$.

ejecta [crater] Glass, shock-metamorphosed rock fragments, and other material thrown out of an explosion or impact crater during formation. Such material may be distributed around a crater in distinctive patterns forming "ejecta rays" or "ejecta loops".

ejecta [pyroclast] Material thrown out by a volcano; *pyroclastics.* Syn: *ejectamenta.*

ejecta blanket A deposit surrounding an impact crater or explosion crater, consisting of material (such as base-surge deposits, throwout, and fallout breccias) ejected from the crater during formation; e.g. lunar-crater material, probably chiefly crushed rock with large blocks, occurring on a mare region and often forming hummocky to smooth layers ranging from about a meter to several hundred meters in thickness.

ejectamenta *ejecta.*

ekanite A green mineral: $(Th,U)(Ca,Fe,Pb)_2Si_8O_{20}$.

ekdemite *ecdemite.*

ekerite A granite that contains arfvedsonite as an essential component along with acmite, microperthite, and soda microcline, and with little quartz.

Ekman layer A layer in the ocean, situated above a certain depth in the *Ekman spiral*, at which both the current and the frictional forces associated with it become negligibly small. The average flow of water is at right angles to the wind driving it (on the Northern Hemisphere). It may be produced near the surface by wind stresses (upper Ekman layer) or near the bottom by a pressure gradient (lower Ekman layer).

Ekman spiral A theoretical, graphic description of the effect that a wind blowing uniformly and steadily over a homogeneous ocean of unlimited depth and extent and of constant viscosity would cause currents in the surface layers to vary with depth, the water at the very surface drifting at an angle of $45°$ to the right of the wind direction in the Northern Hemisphere (and to the left in the Southern Hemisphere) and water at successive depths drifting in directions farther to the right (as a spiral) with a rapidly decreasing speed until at the *friction depth* it would move in the direction opposite to the wind; the net water transport (Ekman transport) is $90°$ to the right of the wind direction in the Northern Hemisphere. It is named for Vagn Walfrid Ekman, Swedish oceanographer, who in 1902 developed the theory of the spiral, which has also been applied to atmospheric motion.

eksedofacies Facies of the weathering environment (Vassoevich, 1948).

ektexine The outer layer of the two layers of the *exine* of spores and pollen, normally more densely or deeply staining than the *endexine*, and characterized by richly detailed external sculpture and often by complex internal structure of granules, columellae, and other elements. Syn: *ectexine; ectoexine; sexine.*

ektexis *ectexis.*

ektodynamomorphic Var. of *ectodynamomorphic.*

elaeolite *eleolite.*

elastic Said of a body in which strains are instantly and totally recoverable and in which deformation is independent of time. Cf: *plastic [struc].*

elastic aftereffect *creep recovery.*

elastic afterworking *creep recovery.*

elastic bitumen *elaterite.*

elastic compaction Compaction that is proportional to the change in effective stress and that is fully recoverable if the stress returns to its original condition (Poland, et al, in press). Cf: *inelastic compaction.*

elastic compliance The reciprocal of *Young's modulus.*

elastic constant One of various coefficients that define the elastic properties of a body, including the *Lamé constants, Poisson's ratio,* or one of the *moduli of elasticity.*

elastic deformation Deformation of a substance which disappears when the deforming forces are removed. Specifically, that type of deformation in which stress and strain are linearly related, according to *Hooke's law.* Cf: *plastic deformation.*

elastic discontinuity A boundary between strata of different elastic moduli and/or density at which seismic waves are reflected and refracted.

elastic energy The energy stored within a solid during elastic deformation.

elastic limit The greatest stress that can be developed in a material without permanent deformation remaining when the stress is released.

elastic modulus *modulus of elasticity.*

elasticoplastic Said of deformation that has a perfectly elastic phase and a perfectly plastic phase. It is demonstrated by the model of a *Saint Venant substance.*

elasticoviscous Said of a material in which instantaneous elastic strain, under stress below the elastic limit, is followed by continuously developed permanent strain under long sustained stress of constant magnitude. If the strain is kept constant at some point beyond the elastic limit, the stress is reduced exponentially. See also: *Maxwell liquid.* Syn: *viscoelastic.*

elastic rebound Elastic recovery from strain.

elastic rebound theory The statement that movement along a fault is the result of an abrupt release of a progressively increasing elastic strain between the two rock masses on either side of the fault. Such a movement returns the rocks to a condition of little or no strain. The theory was proposed by Harry Fielding Reid in 1911. Syn: *Reid mechanism.*

elastic strain The strain developed during the elastic behavior of a material.

elastic wave A wave which is propagated through a medium by some kind of elastic deformation. A seismic wave is an elastic wave.

elater The ribbon-like, filamentous appendage of certain spores (as of *Equisetum*), consisting of more or less coiled strips of exine. It aids in spore dispersal.

elaterite A brown, soft and elastic, infusible, asphaltic pyrobitumen that becomes hard and brittle upon exposure to air and which is derived from the metamorphosis of petroleum. See also: *coorongite.* Syn: *elastic bitumen; liverite (as used locally in Utah); mineral caoutchouc.*

elatolite A supposedly high-temperature modification of calcite, but probably crystal casts of calcium carbonate after villiaumite (Hey, 1962, p.412).

E layer [seism] The seismic region of the Earth from 2900 km to 4710 km, equivalent to the *outer core.* It is divided into an upper (E') and a lower (E'') part at 4560 km, at which level the velocity gradient of the P wave is reduced to zero. It is a part of a classification of the interior of the Earth, having layers A to G.

elb A *transverse dune* in the desert of Algeria.

elbaite A mineral of the tourmaline group: $Na(Li,Al)_3Al_6-(BO_3)_3(SiO_3)_6(OH)_4$.

elbasin A term used by Taylor (1951, p.613) for an "elevated basin, often wrongly called a plateau"; e.g. the British Columbia plateau.

Elbe The term applied in northern Europe to the first glacial stage of the Pleistocene Epoch, followed by the Elster; probably equivalent to the *Günz* and *Nebraskan*.

elbow A sharp bend or change in direction of a stream channel, coastline, shoal, etc.

elbow of capture The point at which capture was effected along a stream course, characterized by an abrupt or sharp bend where the course turns from the captured part of its valley into the valley of the capturing stream.

elbow twin *geniculate twin*.

electrical conductivity A measure of the ease with which a conduction current can be caused to flow through a material under the influence of an applied electric field. It is measured in mhos per meter and is the reciprocal of resistivity.

electrical log *electric log*.

electrical method A geophysical prospecting method which depends upon the electrical or electrochemical properties of rocks. The resistivity, spontaneous polarization, induced polarization, and inductive electromagnetic methods are the principal electrical methods. See also: *electrical prospecting; electrical survey*.

electrical overlap *mixing*.

electrical prospecting Prospecting for minerals by use of one or more of the *electrical methods*.

electrical resistivity The longitudinal electrical resistance per unit length and per unit reciprocal cross-sectional area of a material. Symbol: ρ.

electrical resistivity sounding A procedure for determining depths to geological interfaces wherein separations of electrodes in an array are increased by increments. A plot of observed apparent resistivity versus electrode separation, when compared with similar plots for theoretically computed cases, yields estimates of the depths to the interfaces and the resistivities of the strata.

electrical survey A survey or mapping of a portion of the Earth's interior by use of one of the *electrical methods*.

electrical twinning *Dauphiné twinning*.

electric calamine *hemimorphite*.

electric field The domain of the electric field intensity.

electric field intensity The strength of an electric field at any point. It is measured by the force exerted on a unit positive charge placed at that point. Syn: *voltage gradient*.

electric log A *geophysical log* of an uncased part of a well or borehole, obtained by lowering and raising an electrode device on a wire line and making in-situ measurements (continuously recorded at the surface) of the electrical properties of the geologic formations encountered at various depths. The log usually consists of a series of traces representing one or more *resistivity curves* (on the right side of the log) and a *spontaneous-potential curve* (on the left side of the log). It is useful for locating changes in composition, making local correlations, and indicating the nature and amount of fluids in the pores of the rock. Syn: *electrical log; electrolog; Schlumberger; E-log*.

electric logging The process of recording the resistivity, spontaneous polarization, and induced polarization versus depth within a well or borehole.

electric potential A mathematically convenient quantity from which the electric field intensity may be obtained by differentiation.

electrochemical induration A method of strengthening saturated and poorly consolidated rock and earth materials by passing a direct current of electricity through them (see Titkov et al, 1965).

electrochemical potential The sum of the chemical potential and the electric potential. Symbol: U.

electrochromatography *Chromatography* wherein an applied electric potential is used to produce differential electrical migration.

electrode array A distribution or configuration of electrodes on or in the ground for the purpose of effecting an electrical survey. Syn: *electrode configuration*.

electrode configuration *electrode array*.

electrodeless discharge Emission of light from matter energized by induced electrical currents.

electrodiagenesis Diagenesis affected or stimulated by electric currents and potentials.

electrodialysis *Dialysis* assisted by the application of an electromotive force to electrodes adjacent to the semipermeable membranes. Two important uses of electrodialysis are as a technique in water desalination and in removing electrolytes from naturally occurring colloids such as proteins. Cf: *electroosmosis*.

electrofiltration *electrostatic precipitation*.

electrofiltration potential An electrical potential that is caused by movement of fluids through porous formations. Syn: *streaming potential; electrokinetic potential*.

electrographic A method for analyzing minerals and metals by transferring a small amount of the sample by electrical means to a prepared surface where the ions are identified.

electrokinetic potential *electrofiltration potential*.

electrolog *electric log*.

electrolysis A method of breaking down a compound in its natural form or in solution by passing an electric current through it, the ions present moving to one electrode or the other where they may be released as new substances.

electromagnetic detector An instrument used in electrical surveys by the electromagnetic method and usually consisting of an induction coil followed by a suitable means of voltage amplification; a device for converting an alternating magnetic field to a measurable alternating voltage.

electromagnetic energy *electromagnetic radiation*.

electromagnetic field The domain of the five vectors: magnetic induction, magnetic field intensity, electric field intensity, dielectric displacement, and conduction current density.

electromagnetic induction The generation of electric field or current in an electric conductor when that body is in a changing magnetic field or is moving through a magnetic field. It is a phenomenon of *electromagnetism*. Nonrecommended syn: *magnetic induction*.

electromagnetic methods A subgroup of methods of electrical exploration based on the measurement of alternating magnetic fields associated with currents artificially or naturally maintained in the subsurface. If the subsurface currents are induced by a primary alternating magnetic field, the name inductive electromagnetic method applies, whereas if the subsurface currents are conducted into the ground via electrodes, then the name conductive electromagnetic method applies.

electromagnetic radiation Emission or transfer of energy in the form of electromagnetic waves or particles; also, these electromagnetic waves or particles.

electromagnetism The totality of electric and magnetic phenomena, or their study; particularly those phenomena with both electric and magnetic aspects, such as *electromagnetic induction*.

electromagnetometer A device for detecting conductive bodies such as massive sulfide mineral deposits at the surface or at depths down to several hundred feet beneath the surface. They depend upon generating audio frequency electromagnetic inductive fields in coils carried by low-flying aircraft, inducing eddy currents in subsurface conductive bodies and detecting the presence of these bodies by sensing the secondary field which the eddy currents re-radiate.

electromigration A method of separating isotopes or ions by their differing rates of movement during electrolysis.

electromotive force The voltage or potential which drives a current through a conductor. Abbrev: *emf.*

electron capture A mode of radioactive decay in which an atomic electron is captured by the nucleus.

electron diffraction The bending at the edge of the plate of a beam of electrons of very short wavelength sent through a thin layer of a substance, so that the waves show up on an interference pattern as light and dark bands. Each *electron diffraction pattern* is characteristic for that particular substance, thus the basis for *electron diffraction analysis.*

electron diffraction analysis A method of analysis using the diffraction pattern of a beam of electrons sent through a substance to determine its structure and identity, each substance having a characteristic *electron diffraction pattern.*

electron diffraction pattern The interference pattern seen when a beam of electrons is sent through a substance, each pattern being characteristic for that particular substance.

electronic log A *gamma-ray log* obtained by using an electronic device (such as a Geiger counter) to detect and record the intensity of gamma rays.

electron magnetic resonance *electron spin resonance.*

electron microprobe An analytical instrument which utilizes a finely focused beam of electrons to selectively excite X-ray emission from selected portions of a sample. From the emitted X-ray spectrum the composition of the sample at the point of excitation can be determined. Spots as small as 1 micron in diameter can be analyzed, with sensitivities around 50ppm or less for most metals. Syn: *electron probe; microanalyzer.*

electron microscope An electron-optical instrument in which a beam of electrons, focused by systems of electrical or magnetic lenses, is used to produce enlarged images of minute objects on a fluorescent screen or photographic plate in a manner similar to that in which a beam of light is used in a compound microscope. The electron microscope, because of the very short wavelength of the electrons, is capable of resolving much finer structures than the optical instrument, and is capable of magnifications on the order of 100,000X. See also: *scanning electron microscope.*

electron microscopy Determining and identifying the structure of substances by using the *electron microscope.*

electron paramagnetic resonance A syn. of *electron spin resonance.* Symbol: EPR.

electron probe *electron microprobe.*

electron spin resonance *Resonance* occurring when electrons which are undergoing transitions between energy levels in a substance are irradiated with electromagnetic energy of a proper frequency to produce maximum absorption. Symbol: ESR. Syn: *electron magnetic resonance; electron paramagnetic resonance (Symbol: EPR); paramagnetic resonance.*

electro-osmosis The motion of liquid through a membrane under the influence of an applied electric field.

electroosmosis The movement of a conducting liquid through a solid medium such as a porous diaphragm or a capillary tube assisted by the application of an electromotive force to electrodes on opposite sides of the medium. Cf: *electrodialysis.* See also: *osmosis.*

electrophoresis The movement toward electrodes of suspended charged particles in a fluid by applying an electromotive force to the electrodes which are in contact with the suspension. Se also: cataphoresis; anaphoresis.

electroprecipitation *electrostatic precipitation.*

electrostatic precipitation A method for removing suspended solid or liquid particles from a gas by applying a strong electric field to the mixture which charges the particles and precipitates them. Syn: *electrofiltration; electroprecipitation.*

electrostatics That branch of physics which deals with attractions and repulsions between static electrical charges.

electrostriction Deformation induced in materials upon their polarization by an applied electric field.

electroviscosity The viscosity of a fluid as influenced by electric properties, e.g. greater viscosity of a low-conductivity fluid than of a high-conductivity fluid flowing through narrow capillaries.

electrum (a) A naturally occurring, deep- to pale-yellow alloy of gold with silver; argentiferous gold, containing more than 20% silver. Also spelled: *elektrum.* (b) An ancient Greek name, now obsolete, for *amber.* Also spelled: *elektron.*

elements of symmetry *symmetry elements.*

eleolite A syn. of *nepheline,* esp. of a translucent, massive or coarsely crystalline, and dark (grayish, bright-green, or brown to brownish-red) variety having a greasy luster and sometimes used as an ornamental stone. Syn: *elaeolite; elaolite.*

eleolite syenite An obsolescent syn. of *nepheline syenite,* esp. of a coarse-grained variety containing eleolite.

elephant-head dune A sand shallow or small sand dune resembling the head of an elephant, having a rounded windward face covered with vegetation and a long, tapering snout of bare sand on the leeward side; examples occur in the Coachella Valley of the Colorado Desert in California.

elephant-hide pahoehoe A type of pahoehoe having a wrinkled and draped surface. Cf: *corded pahoehoe; entrail pahoehoe; festooned pahoehoe; filamented pahoehoe; sharkskin pahoehoe; shelly pahoehoe; slab pahoehoe.*

elephant rock (a) A term used in SE Missouri for a *rocking stone,* not necessarily delicately balanced, formed in place by the weathering and removal of surrounding material. (b) *coombe rock.*

eleutheromorph A new mineral in a metamorphic rock that has been freely developed and that is independent of pre-existing minerals for its form, in contrast with a pseudomorph.

eleutherozoan n. Any echinoderm that does not live attached to a substrate.----adj. Said of an echinoderm having a free mode of life. Var: *eleutherozoic.*----Cf: *pelmatozoan.*

elevated shoreline A shoreline whose development has been interrupted by a relatively sudden rise of the coast or by a rapid lowering of the water level; it is not a true shoreline because it is no longer being shaped by waves and currents. Examples: a broad marine terrace (common along the continental and insular coasts of the Pacific Ocean), or a narrow strandline. Not to be confused with *shoreline of elevation.*

elevation [geomorph] A general term for a topographic feature of any size that rises above the adjacent land or the surrounding ocean bottom; a place or station that is elevated.

elevation [surv] The vertical distance from a datum (usually mean sea level) to a point or object on the Earth's surface; esp. the *height* of a ground point above the level of the sea. The term is used synonymously with *altitude* in referring to distance above sea level, but in modern surveying practice the term "elevation" is preferred to indicate heights on the Earth's surface whereas "altitude" is used to indicate the heights of points in space above the Earth's surface. Abbrev: elev.

elevation correction In seismic measurements, the correction applied to observed reflection time values due to differences of station elevation, in order to reduce the observations to an arbitrary reference datum. A similar correction is made for refracted wave arrivals.

elevation head Hydrostatic pressure expressed as the potential energy of elevation of a flowing liquid; the head due to the elevation of the point in question above a datum. See also: *static head; total head.* Syn: *potential head.*

elevation meter A mechanical or electromechanical device on wheels that measures slope and distance and that automatically and continuously integrates their product into difference of elevation.

elevator tectonics A term used by Dietz & Holden (1966, p.353) for the rise and fall of blocks of sialic crust from atmospheric to abyssal levels.

elision An act or instance in which the continuity of the sedimentary record has been disturbed by the omission of sediments, such as produced by their removal and redeposition in adjacent depressions.

elkhornite A hypabyssal labradorite-bearing augite syenite. Its name is derived from the Elkhorn district of Montana.

ellipochoanitic Said of a relatively short, retrochoanitic septal neck of a nautiloid that does not reach as far as the preceding septum.

ellipsoid A mathematical figure, in geodesy, closely approaching the geoid in form and size. It is generally defined by its equatorial radius and by the reciprocal of the flattening, a/(a-b), where a and b are the equatorial and polar radii. A task of the geodesists is the determination of more and more exact parameters of the ellipsoid. Cf: *spheroid; Clarke spheroid of 1866; reference spheroid.*

ellipsoidal lava (a) *pillow lava.* (b) An inclusive term for any lava flow with an ellipsoidal pattern, i.e. pillow lava and the toes of pahoehoe. (MacDonald, 1953).

ellipsoid of revolution The simple mathematical figure that would be produced by an ellipse revolving around its minor axis. It is often used as reference surface for the Earth and is sometimes referred to as a *meridional ellipse.* See also: *spheroid.*

elliptical polarization The terminus of an instantaneous electric or magnetic field vector, of an electromagnetic field, varying with time so as to trace out an ellipse.

elliptical polarization In optics, elliptically polarized light consisting of upward-spiraling vibration vectors, the surface of which is elliptical rather than circular, as in *circular polarization.* It is caused by the inconstant lengths of vibration vectors of mutually perpendicular plane-polarized waves whose path differences differ in phase by amounts other than $(n+1)/4\lambda$ on emergence from a crystal.

elliptical projection One of several map projections showing the Earth's surface upon the interior of an ellipse; e.g. Mollweide projection and Aitoff projection.

ellipticity [elect] The ratio of minor to major axes of an ellipse of polarization.

ellipticity [geodesy] The degree of flattening of the reference spheroid as expressed by the equation $e=(a-b)/a$, where a and b are the equatorial and polar radii. Cf: *equatorial bulge.*

ellipticone A coiled cephalopod shell having elliptic coiling of the last whorl or half whorl which breaks away from the spiral or slightly breaks the regularity of the spiral form.

ellsworthite *betafite.*

E-log *electric log.*

elongation ratio *basin-elongation ratio.*

elongation sign *sign of elongation.*

elpasolite A colorless isometric mineral: K_2NaAlF_6.

elphidiid Any foraminifer belonging to the family Elphidiidae, characterized by having a sutural canal system opening into rows of sutural pores. Their stratigraphic range is Paleocene to present.

elpidite A white to brick-red mineral: $Na_2ZrSi_6O_{15}.3H_2O$.

Elster The term applied in northern Europe to the second glacial stage of the Pleistocene Epoch, following the Elbe and preceding the Saale glacial stages; equivalent to the *Mindel* and *Kansan.*

elutriation (a) A method of mechanical analysis of a sediment, in which the finer, lightweight particles are separated from the coarser, heavy particles by means of a slowly rising current of air or water of known and controlled velocity, carrying the lighter particles upward and allowing the heavier ones to sink. (b) Purification, or removal of material from a mixture or in suspension in water, by washing and decanting, leaving

the heavier particles behind. (c) The washing away of the lighter-weight or finer particles in a soil by the splashing of raindrops.

eluvial [eco geol] Said of an incoherent ore deposit, such as a placer resulting from the decomposition or disintegration of rock in place. The material may have slumped or washed downslope for a short distance but has not been transported by a stream.

eluvial [sed] Pertaining to or composed of wind-deposited eluvium; e.g. the "eluvial (or passive) phase" of a dune cycle, marked by sufficient vegetation to check deflation. Cf: *eolian.*

eluvial [soil] Said of a soil horizon from which material has been removed by the process of *eluviation*; pertaining to the process of illuviation. Cf: *illuvial.*

eluvial [weath] Pertaining to eluvium formed by the weathering of rock in place; *residual.*

eluviated Said of a soil horizon or of materials that have been subjected to the process of *eluviation.*

eluviation The downward movement of soluble or suspended material in a soil, from the A horizon to the B horizon, by ground-water percolation. The term refers especially but not exclusively to the movement of colloids, whereas the term *leaching* refers to the complete removal of soluble materials. Adj: *eluvial; eluviated.* Cf: *illuviation.* See also: *cheluviation.*

eluvium [sed] Fine soil or sand moved and deposited by the wind, as in a sand dune. Cf: *alluvium.*

eluvium [weath] An accumulation of rock debris produced in place by the decomposition or disintegration of rock to a greater or less depth; a weathering product; a *residue.*

elvan A Cornish term for hypabyssal rocks having the composition of granite, esp. a quartz porphyry. Tourmaline, fluorite, and topaz may be accessories. Syn: *elvanite.*

emarginate [paleont] Having a notched margin; e.g. said of a gastropod having a notched or variously excavated margin of the outer lip, or said of a bivalve mollusk whose margin is interrupted by a notch or sinus, or said of the posteriorly deflected median segment of the anterior commissure of a brachiopod, or said of a leaf with a shallow notch at the apex.

embankment [eng] An artificial structure, usually of earth or gravel, constructed above the natural ground surface and designed to hold back water from overflowing a level tract of land, to retain water in a reservoir or a stream in its bed, or to carry a roadway or railroad; e.g. a dike, a seawall, and a fill.

embankment [coast] A narrow depositional feature, as a spit, barrier, or bar, built out from the shore of a sea or lake by the action of waves and currents that deposit excess material at its deep end; it may be emerged or submerged. Syn: *bank.*

embatholithic Said of a mineral deposit occurring in a batholith in which exposure of the batholith and of the country rock is about equal; also, said of that stage of batholith erosion (Emmons, 1933). The term is little used today. Cf: *acrobatholithic; cryptobatholithic; endobatholithic; epibatholithic; hypobatholithic.*

embayed coast An ungraded coast with many projecting headlands, bays, and outlying islands, usually resulting from submergence.

embayed mountain A mountain, near the sea, that has been depressed so that seawater enters the valleys; e.g. on the coast of SW Ireland.

embayment [coast] (a) The formation of a bay, as by the sea overflowing a depression of the land near the mouth of a river. (b) A *bay*, either the deep indentation or recess of a shoreline, or the large body of water (as an open bay) thus formed.

embayment [petrology] (a) An irregular *corrosion* or modification of the outline of a crystal by the magma from which it previously crystallized or in which it occurs as a foreign inclusion; esp. the deep corrosion into the sides of a phenocryst. (b) The penetration of a crystal by another, generally euhedral crystal.-----Such a crystal is called an "embayed crystal".

embayment [struc geol] (a) A geosynclinal or downwarped

rea containing stratified rocks, either sedimentary or volcan-
: or both, that extends into a terrain of other rocks, e.g. the
Mississippi Embayment of the U.S. Gulf Coast. (b) *recess*
old].

mbedded Covered or enclosed by sediment in a matrix, such
s gravel *embedded* in silt.

mbolite A yellow-green isometric mineral: Ag(Cl,Br). It is in-
ermediate in composition between chlorargyrite and bromar-
yrite.

mbossed rock A term introduced by Hitchcock (1843, p.180)
s a syn. of *roche moutonnée*.

mbouchure (a) The mouth of a river, or that part where it
nters the sea. (b) An expansion of a river valley into a plain.-
Etymol: French. Syn: *embouchement*.

mbrechite Migmatite in which some textural components of
ne pre-existing rocks are preserved (Dietrich & Mehnert,
961); a migmatite with pre-existing parallel layering, often in-
luding feldspar phenoblasts or granitic layers and lenses
Mehnert, 1968, p.354).

mbryo A young sporophytic plant; the *germ* of a *seed*.

mbryonic Said of the earliest growth stage in the life history
f an animal; the stage preceding the *nepionic* stage.

mbryonic apparatus A group of chambers at the center of
ome megalospheric tests of foraminifers, larger in size and
lifferent in shape and arrangement from other chambers. See
ilso: *juvenarium*. Syn: *nucleoconch*.

mbryonic volcano A breccia-filled volcanic pipe without sur-
ace expression and considered to be produced by phreatic
xplosions. Examples of Permian age occur in Scotland.

mbryophytic Said of a plant in which a diploid embryo is pro-
luced as a part of the life cycle, due to fusion of haploid egg
ind sperm nuclei. Liverworts, mosses, and vascular plants
ire of this type.

merald [gem] n. Any of various gemstones having a green
:olor, such as "oriental emerald" (sapphire), "copper emer-
ıld" (dioptase), "Brazilian emerald" (tourmaline), and "Ural-
an emerald" (demantoid).---adj. Said of a gemmy and richly
green-colored mineral, such as "emerald jade" (jadeite),
emerald spodumene" (hiddenite), and "emerald malachite"
dioptase).

merald [mineral] A brilliant- or grass-green variety of beryl,
ighly prized as a gemstone. The rich, full green color is
:aused by the presence of chromium, and it varies from me-
lium-light or medium-dark tones of slightly bluish-green to
hose of slightly yellowish-green. Syn: *smaragd*.

merald copper *dioptase*.

merald cut A *step cut* in which the finished gem is square or
ectangular and the rows (steps) of elongated facets on the
:rown and pavilion are parallel to the girdle with sets on each
of the four sides and at the corners. It is commonly used on
liamonds and on emeralds and other colored stones. See
ılso: *square emerald cut*.

merald nickel *zaratite*.

merged bog A bog which tends to grow vertically, i.e. in-
:rease in thickness, by drawing water up through the mass of
llants to above the water table where the growth takes place.
:f: *immersed bog*.

merged shoreline *shoreline of emergence*.

mergence [coast] A change in the relative levels of water
ind land such that the land is higher (in relation to the water
evel) than it was before the change and areas formerly under
water become dry land; it results either from an uplift of the
and or from a fall of the water level. Ant: *submergence*.

mergence [streams] *resurgence*.

mergence [bot] Any outgrowth of cortical and epidermal
llant tissues, lacking vascular tissues. See also: *enation*.

mergence angle *angle of emergence*.

mergence velocity A vertical component of glacier motion
neasured at the surface, representing the difference in verti-
:al displacement of a stake or marker fixed in the ice and the

product of horizontal displacement times the tangent of ice
slope; the rate the surface would rise if there were no ablation.

emergent [seis] Said of an *emersio*, or gradual arrival of a
seismic phase. Cf: *impulsive*.

emergent [ecol] A plant that rises above its substrate; e.g., an
emergent aquatic plant. Syn: *emersed*.

emergent aquatic plant A rooted plant growing in shallow
water, with part of its stem and leaves above the water surface; e.g., bulrush; cattail.

emergent evolution Evolution characterized by the appearance
of completely new and unpredictable characteristics or qualities at different levels due to a rearrangement of preexisting
entities.

emersed *emergent*.

emersio A gradual onset or arrival of a seismic phase on a
seismogram. Adj: *emergent*. Cf: *impetus*.

emery (a) A dark (gray-black), granular, impure variety of
corundum that contains varying amounts of iron oxides (usually magnetite or hematite) and that is used in the form of
coarse or fine powder or grains for polishing and grinding and
as an abrasive. It occurs naturally as masses in limestone
and as segregations in igneous rocks. (b) A natural abrasive
composed essentially of pulverized impure corundum. Also,
the commercial product obtained by crushing emery rock. (c)
emery rock.

emery rock A granular rock that is composed essentially of an
impure mixture of corundum, magnetite, and spinel, and that
may be formed by magmatic segregation or by metamorphism
of highly aluminous sediments. Syn: *emery; corundolite*.

emigrant In ecology, a migrant plant or animal.

emildine A variety of spessartine garnet containing yttrium.
Syn: *emilite*.

eminence (a) An elevated area of any size, shape, or height;
a mass of high land; a mountain or a hill. (b) The high point of
an elevated feature.

eminent cleavage Perfectly displayed crystal cleavage, with
smooth surfaces, as often seen in mica or calcite.

emission spectroscopy The observation of an *emission spectrum* and all processes of recording and measuring which go
on with it.

emission spectrum A general term for any spectrum issuing
from a source.

emissivity (a) The ratio, as a function of direction, of the thermal radiance of the radiator in each direction to that of a
blackbody at the same temperature (Nicodemus, 1971). (b)
In the older literature, the ratio of the radiant flux density of a
surface of a temperature radiator to that of a blackbody at the
same temperature.----Cf: *emittance*. Syn: *directional emissivity*.

emittance (a) Formerly, the quotient of the flux leaving an element of the surface containing the point by the area of that
element (Nicodemus, 1971). (b) The ratio of the rate of radiant energy emission from a body, as a consequence of its
temperature only, to the corresponding rate of emission from
a blackbody at the same temperature.----In U.S. usage, there
has been a tendency to pair "emittance" with *emissivity* in the
same way as are the terms reflectance and reflectivity and
transmittance and transmissivity, with the "-ance" term denoting values measured on actual samples of a material or surface, and the "-ivity" term denoting values of the same quantity, as a material property, measured for samples of pure
material, and clean and smooth surfaces. Because of the possible ambiguity, it has been suggested that the term, if used
at all, be defined at least once where it is first introduced in
the text (Nicodemus, 1971). ----The term *exitance* has been
suggested as a replacement for the term "emittance".

emmonsite A yellow-green mineral: $Fe_2Te_3O_9 \cdot 2H_2O$.

emplacement [ore dep] The localization of ore minerals, by
whatever process; ore deposition.

emplacement [intrus rocks] A term used for the process of *intrusion*.

emplectite A grayish or white orthorhombic mineral: $CuBiS_2$. It is dimorphous with cuprobismutite.

empolder v. To reclaim land by the creation of polders; to make low-lying or periodically flooded land cultivable by adequate drainage and the erection of dikes to prevent or control inundation. See also: *polderization*. Syn: *impolder*.----n. A tract of empoldered land; a polder.

empoldering *polderization*.

empressite A pale-bronze mineral: AgTe.

Emscherian *Coniacian*.

Emsian European stage: Lower Devonian (above Siegenian, below Eifelian).

emulsion stage That stage in the crystallization of some magmas in which the concentration of water exceeds the solubility and a new, water-rich phase is formed, either as a gas or as liquid droplets (Shand, 1947).

emulsion texture An ore texture showing minute blebs or rounded inclusions of one mineral irregularly distributed in another mineral.

enalite A variety of thorite containing uranium.

enantiomorph Either of two crystals that display *enantiomorphism*.

enantiomorphism The characteristic of two crystals to be mirror images of each other, e.g. right-handed and left-handed quartz. Such crystals are called *enantiomorphs*. Adj: *enantiomorphous*.

enantiomorphous Adj. of *enantiomorphism*.

enantiotropy The relationship between polymorphs that possess a stable transition point, and that therefore can be stably interconverted by changes of temperature and/or pressure. Although the term was originally applied only in systems with a vapor present, modern usage seems to give the term the more general meaning above. Cf: *monotropy*.

enargite A grayish-black or iron-black orthorhombic mineral: Cu_3AsS_4. It is isomorphous with famatinite and dimorphous with luzonite. Enargite is an important ore of copper, occurring in veins in small crystals or granular masses, and often containing antimony (up to 6%) and sometimes iron and zinc.

enation An epidermal outgrowth on a plant structure; an *emergence*.

en cabochon adv. Cut in a style characterized by a smooth-domed, but unfaceted, surface; e.g. a ruby cut "en cabochon" in order to bring out the star. Etymol: French. See also: *cabochon*.

enclave An *inclusion*. This usage is not common in the U.S. (Holmes, 1928, p.28).

enclosed lake A lake that has neither surface influent nor effluent and that never overflows the rim of its basin; e.g. a kettle lake or a crater lake.

enclosed meander Var. of *inclosed meander*.

enclosure An *inclusion* in an igneous rock.

encrinal Pertaining to or made up of encrinites; specif. said of a carbonate rock or sediment containing the stem and/or plate fragments of a crinoid or crinoids. Syn: *encrinic; encrinital; encrinoid; encrinitic*.

encrinal limestone A *crinoidal limestone*; specif. a limestone in which the crinoidal fragments constitute more than 10%, but less than 50%, of the bulk (Bissell & Chilingar, 1967, p.156). Cf: *encrinite*.

encrinite [paleont] A syn. of *crinoid*, esp. a fossil crinoid belonging to the genus *Encrinus*.

encrinite [sed] A *crinoidal limestone*; specif. a limestone in which crinoidal fragments constitute more than 50% of the bulk (Bissell & Chilingar, 1967, p.156). Cf: *encrinal limestone*.

encroachment [petroleum] (a) The movement of edge water into a petroleum reservoir after much of the oil and gas has been recovered and the pressure has been greatly reduced. (b) *salt-water encroachment*.

encrustation Var. of *incrustation*.

endannulus An *annulus* in the endexine of a pollen grain.

end cleat The minor *cleat* system or jointing in a coal seam. See also: *end of coal*. Cf: *face cleat*. Syn: *butt cleat*.

endellionite *bournonite*.

endellite A name used in U.S. for a clay mineral: $Al_2Si_2O_5(OH)_4.4H_2O$. It is the more hydrous form of halloysite and is synonymous with *halloysite* of European authors. Syn: *hydrated halloysite; hydrohalloysite; hydrokaolin*.

endemic Said of an organism or group of organisms that is restricted to a particular region or environment. Syn: *indigenous; native*.

enderbite A plagioclase-rich member of the charnockite series containing quartz, plagioclase (commonly antiperthitic) hypersthene, and a small amount of magnetite. Most classification systems require that quartz constitute 10-65% of the felsic constituents and that the ratio of alkali feldspar to total feldspar be greater than 87.5%. Tobi (1971) has abandoned the term in favor of *alkali charnockite*.

endexine The inner, usually homogeneous and smooth layer of the two layers of the *exine* of spores and pollen, normally less deeply staining than the *ektexine*. Syn: *intexine; nexine*.

endite One of the appendages of the inner side of the limb of an arthropod, such as the medially directed lobe of precoxa, coxa, basis, or ischium of a crustacean, or the median or inner lobe of segment of biramous appendage of a trilobite. Cf: *exite*.

end lap The *overlap* between successive aerial photographs of adjoining areas in the same flight line. Syn: *forward lap*.

endlichite A variety of vanadinite in which the vanadium is partly replaced by arsenic; a mineral intermediate in composition between mimetite and vanadinite.

end member (a) One of the two or more simple compounds of which an isomorphous (solid-solution) series is composed. For example, the end members of the plagioclase feldspar series are albite, $NaAlSi_3O_8$, and anorthite, $CaAl_2Si_2O_8$. Syn: *minal*. (b) One of the two extremes of a series, e.g. types of sedimentary rock or of fossils.

end moraine A moraine that is being produced at the front of an actively flowing glacier at any given time; a moraine that has been deposited at the lower or outer end of a glacier. Cf: *terminal moraine*. Syn: *frontal moraine*.

endoadaptation Adjustment of one part of an organism to its other parts. Cf: *exoadaptation*.

endobatholithic Said of a mineral deposit occurring in or near an island or roof pendant of batholithic country rock; also said of the stage of batholith erosion in which that area is exposed. (Emmons, 1933). The term is little used today. Cf: *acrobatholithic; cryptobatholithic; embatholithic; epibatholithic; hypobatholithic*.

endobiontic Said of an organism living in bottom sediments. Cf: *epibiontic*.

endoblastesis Deuteric crystallization of minerals in an igneous rock. This term is not currently in common use. Adj: *endoblastic*.

endoblastic (a) The adj. of *endoblastesis*. Cf: *metablastic*. Sometimes applied to a minute, oriented inclusion in a crystal. Cf: *poikilitic*.

endocarp (a) The inner layer of a *pericarp*, when the pericarp has two or more different layers. Cf: *mesocarp; epicarp*. (b) The fossilized stone of a drupe.

endocast *steinkern*.

endocochlian *coleoid*.

endocoel A cavity in the capsule formed by the endophragm in a dinoflagellate cyst. Cf: *pericoel*.

endocone One of a series of concentric conical calcareous deposits or structures formed within the posterior or adapical part of the siphuncle of certain cephalopod conchs (e.g. of *Endoceras*). The apices of the cones point toward the apex of the conch.

endocyclic Said of a *regular* echinoid whose periproct is located within the oculogenital ring. Ant: *exocyclic*.

endocyst The soft layer of the body wall of a bryozoan, lining the interior of a zooecium and enclosing the polypide, and giving rise to the *ectocyst* (TIP, 1953, pt.G, p.10).

endoderm The inner body layer of an organism; e.g. the inner layer of the outer body walls of a coral polyp, occurring as a double lamina in mesenteries. Cf: *ectoderm; mesoderm*.

endodermis A layer of specialized cells in many roots and some stems, delimiting the inner margin of the cortex (Cronquist, 1961, p.874). Cf: *epidermis*.

endoderre The wall of the prosome of a chitinozoan. It is sometimes regarded as a third layer of the wall. Cf: *ectoderre; periderre*.

endodynamomorphic An old term applied to a soil whose characteristics reflect those of the parent material more than those of external agents. Cf: *ectodynamomorphic*.

end of coal The plane or surface of the *end cleat* in a coal seam. Cf: *face of coal*.

endogastric (a) Said of a cephalopod shell that is curved or coiled so that the venter is on or near the inner or concave side or area of whorls. (b) Said of a gastropod shell that is coiled so as to extend backward from the aperture over the extruded head-foot mass, as in most adult forms (TIP, 1960, pt.I, p.130).---Cf: *exogastric*.

endogene effects The effects of igneous intrusion upon the margin of the intrusive body itself (Bateman, 1950). Cf: *exogene effects*.

endogenetic *endogenic*.

endogenic Said of a geologic process; or of its resultant feature or rock, that originates within the Earth, e.g. volcanism, volcanoes, extrusive rocks; derived from within. Cf: *exogenic*. Syn: *endogenetic; endogenous*. Noun: *endogeny; endogenesis*.

endogenous *endogenic*.

endogenous dome A volcanic dome that has grown primarily by expansion from within and is characterized by a concentric arrangement of flow layers (Williams, 1932). Cf: *exogenous dome*.

endogenous inclusion *autolith*.

endoglyph A hieroglyph occurring within a single sedimentary bed (Vassoevich, 1953, p.37). Cf: *exoglyph*.

endokinematic Said of sedimentary operations in which "the largest displacement vectors occur between some matter within that part of the deposit destined to form the structure and the unmodified deposit" (Elliott, 1965, p.196); e.g. translational slumping, and horizontal or vertical transposition. Also, said of the sedimentary structures produced by endokinematic operations. Cf: *exokinematic*.

endokinetic Said of a fissure in a rock that is the result of strain within the rock unit itself. Cf: *exokinetic*.

endolithic Said of an organism that lives within rock or other stony matter. Syn: *pertricolous*. Cf: *epilithic*.

endolithic breccia A breccia formed by forces acting within the Earth's crust, as by tectonic movements, by swelling or hydration, or by foundering.

endometamorphism *endomorphism*.

endomorph A crystal that is surrounded by a crystal of a different mineral species. Adj: *endomorphic*.

endomorphic metamorphism *endomorphism*.

endomorphism Changes within an igneous rock produced by the complete or partial assimilation of country-rock fragments or by reaction upon it by the country rock along the contact surfaces. It is a form of *contact metamorphism* with emphasis upon changes produced within the igneous body rather than in the country rock. The term was originated by Fournet (1847). Cf: *exomorphism*. Syn: *endometamorphism; endomorphic metamorphism*.

endopelos Animals that lie on or burrow in soft mud.

endophragm (a) The complex internal skeletal structure of a crustacean, formed by the fusion of apodemes, and providing a framework for muscle attachment. Syn: *endophragmal skeleton*. (b) The inner-wall layer of a dinoflagellate cyst. Cf: *ectophragm; periphragm*.

endopinacoderm The *pinacoderm* lining the inhalant and exhalant systems of a sponge. Cf: *exopinacoderm*.

endopleura *tegmen*.

endopod The medial or internal ramus of a limb of a crustacean, arising from the basis. Cf: *exopod*. Syn: *endopodite*.

endopore The internal opening in the endexine of a pollen grain with a complex porate structure. See also: *vestibulum*. Cf: *exopore*. Syn: *os [palyn]*.

endopsammon Animals that lie on or burrow in sand.

endopuncta A *puncta* of a brachiopod shell not extending to its external surface, occupied by a caecum, and found only in the secondary layer. These pores are common over the whole inner surface of the shell but are not visible on the outer shell surface if the primary layer is present. Cf: *exopuncta*. Pl: *endopunctae*. Syn: *endopunctum*.

endorheic Said of a basin or region characterized by internal drainage; relating to endorheism. Syn: *endoreic*.

endorheism (a) *internal drainage*. (b) The condition of a region in which little or none of the surface drainage reaches the ocean.---Ant: *exorheism*. Syn: *endoreism*.

endosiphuncle The space within the *ectosiphuncle* of certain cephalopods, including all organic tissues and calcareous structures. Syn: *endosiphon*.

endoskeleton An internal *skeleton* in an animal, serving as a supporting framework; e.g. any internal hard parts serving for the attachment of muscles in a crustacean, or the internal system of articulated bones in a vertebrate. Cf: *exoskeleton*.

endosome The inner part of the body of various sponges; e.g. the choanosome with few if any supporting spicules, or the part of a sponge internal to a cortex, or a part of a sponge surrounding the cloaca and devoid of flagellated chambers. Because of conflicting usage, the term is not recommended. Cf: *ectosome*.

endosperm A food-storage tissue in a seed. In the gymnosperms, it a part of the female gametophyte and is haploid; in the angiosperms, it results from the fusion of a sperm with two polar nuclei and is thus triploid (Fuller & Tippo, 1949, p.958).

endosphere All that part of the Earth below the lithosphere.

endospore (a) A syn. of *intine*. The term is mostly applied to the sporoderm of spores, rather than to pollen. Syn: *endosporium*. (b) An asexual spore developed within the cell, esp. in bacteria. (c) A thin-walled spore of blue-green algae.---Cf: *exospore*.

endosternite A part of the endoskeleton of an arthropod; e.g. a tendinous endoskeletal plate in the cephalon of a crustacean.

endostratic Bedded within; e.g. said of the formation of bedding in clays as a result of "alternating desiccation and saturation by ground water" (Becker, 1932, p.85), or said of a breccia bedded within a distinct stratum (Norton, 1917).

endotheca A collective term for the dissepiments inside the wall of a scleractinian corallite. Cf: *exotheca*.

endothermic Pertaining to a chemical reaction that occurs with an absorption of heat. Cf: *exothermic*.

endotomous Characterized by bifurcation in two main crinoid arms that give off branches only on their inner sides. Ant: *exotomous*.

endozone Basal or axial part of a bryozoan colony, usually characterized by thinness of zooecial walls and by relative sparseness of diaphragms. It is typically present in cryptostomes and trepostomes. Cf: *exozone*. Syn: *immature region*.

end peneplain A syn. of *endrumpf*, originally a German word with no adequate English equivalent. The term is unsatisfactory because the Davisian peneplain is developed by a different process and has different characteristics.

end product As applied to radioactivity, the stable nuclide at the end of a *radioactive series*. Cf: *parent; daughter.*

endrumpf A term proposed by W. Penck (1924) for the final landscape or plain that results from the erosion of a landmass that had high relief; it represents the end product of a period of degradation marked by waning uplift. Although Penck considered the Davisian *peneplain* as an equivalent term, it differs from "peneplain" in that it does not imply a particular sequence of development leading up to leveling of the original relief; "endrumpf" may be more likened to an extended use of "pediplain". Etymol: German *Endrumpf*, "terminal torso". Cf: *primàrrumpf.* Syn: *end peneplain.*

endurance limit That stress below which a material can withstand hundreds of millions of repetitions of stress without fracturing. It is considerably lower than rupture strength. Syn: *fatigue limit.*

endurance ratio The ratio of the endurance limit of a material to its static, tensile strength. Syn: *fatigue ratio.*

en echelon adj. Said of geologic features that are in an overlapping or staggered arrangement, e.g. faults. Each is relatively short but collectively they form a linear zone, in which the strike of the individual features is oblique to that of the zone as a whole. Etymol: French *en échelon*, "in steplike arrangement".

Eneolithic A syn. of *Copper Age.* Also spelled: *Aeneolithic.*

energy flux Seismic energy transmitted along the wave, per unit area of wavefront units; it equals energy per time, divided by area.

energy grade line *energy line.*

energy gradient The slope of the *energy line* of a body of flowing water, with reference to any plane. Syn: *energy slope.*

energy index [mineral] A term used by Gruner (1950) for a number that expresses the stability of a silicate. It is equal to the *bridging factor* multiplied by the electronegativity. Quartz has the highest energy index (1.80).

energy index [sed] The inferred degree of water agitation in the sedimentary environment of deposition. Abbrev: EI.

energy level The kinetic energy (due to wave or current action) that existed or exists in the water of a sedimentary environment, either at or one or two meters above the interface of deposition. See also: *high-energy environment; low-energy environment.*

energy line In hydraulics, a line joining the elevations of the energy heads of a stream when referred to the stream bed. It lies above the water surface at any cross section; the vertical distance is equal to the velocity head at that cross section (ASCE, 1962). See also: *energy gradient.* Syn: *energy grade line.*

energy loss The difference between energy input and output as a result of transfer of energy between two points. In the flow of water, the rate of energy loss is represented by the slope of the *hydraulic grade line.*

energy of glacierization *activity index.*

energy slope *energy gradient.*

engineering geology Geology as applied to engineering practice, esp. mining and civil engineering. As defined by the Association of Engineering Geologists (1969), it is the application of geologic data, techniques, and principles to the study of naturally occurring rock and soil materials or ground water for the purpose of assuring that geologic factors affecting the location, planning, design, construction, operation, and maintenance of engineering structures, and the development of ground-water resources, are properly recognized and adequately interpreted, utilized, and presented for use in engineering practice. Syn: *geologic engineering.*

engineering seismology *earthquake engineering.*

englacial Contained, embedded, or carried within the body of a glacier or ice sheet; said of meltwater streams, till, drift, moraine, etc. Syn: *intraglacial.*

englishite A white mineral: $K_2Ca_4Al_8(PO_4)_8(OH)_{10}.9H_2O$. Also spelled: *Englishite.*

engrafted stream A stream composed of the waters of severa previously independent streams that unite with each other be fore reaching the sea; esp. a main stream consisting of sever al separate extended streams flowing from an oldland an merging with each other on an uplifted coastal plain. Syn: *in grafted stream.*

engysseismology That aspect of seismology which deals wit records made near to the point of origin of the impulses. Cf *teleseismology.*

enhydrite (a) A mineral or rock having cavities containin water. (b) *enhydros.*

enhydros A hollow nodule or geode of chalcedony containin water, sometimes in large amount. Syn: *enhydrite; wate agate.*

enhydrous Said of certain crystalline minerals containin water or having drops of included fluid; e.g. "enhydrou chalcedony". Cf: *anhydrous.*

enigmatite *aenigmatite.*

enneri A term used in northern Africa (esp. Libya) for a wad or dry river valley.

ennomoclone A desma (of a sponge) consisting of one shor distal arm (brachyome) and three or six longer proximal arm directed symmetrically away from it; e.g. a *tricranoclone* or *sphaeroclone.*

enrichment The supergene processes of mineral depositio including near-surface oxidation, downward migration, an precipitation, e.g. sulfide enrichment. See also: *oxidized zone.*

enrockment A mass of large stones thrown into water to form a base, such as for a pier.

ensialic geosyncline A geosyncline, the geosynclinal prism o which contains clastics accumulating on sialic crust (Wells 1949). Cf: *ensimatic geosyncline.* See also: *miogeosyncline.*

ensimatic geosyncline A geosyncline, the geosynclinal prism of which contains effusive rocks accumulating on simati crust (Wells, 1949, p.1927). Cf: *ensialic geosyncline.* Se also: *eugeosyncline.*

enstatite A common rock-forming mineral of the orthopyrox ene group: $MgSiO_3$. It is isomorphous with hypersthene, an may contain a little iron replacing the magnesium. Enstatit varies from grayish white to yellowish, olive green, and brown It is an important primary constituent of intermediate an basic igneous rocks. Symbol: En. Cf: *bronzite.* Syn: *chladnite.*

enstatolite A pyroxenite that is composed almost entirely o enstatite.

enstenite A group name for the orthopyroxenes of the $MgSiO_3$-$FeSiO_3$ isomorphous series. It includes enstatite, hyper sthene, and orthoferrosilite.

entablature In columnar jointing, the upper zone that has thin ner and less regular columns than does the lower zone, o *colonnade.*

enterolithic (a) Said of a sedimentary structure consisting o ribbons of intestine-like folds that resemble those produced b tectonic deformation but that originate through chemica changes involving an increase or decrease in the volume o the rock; e.g. said of a small fold or local crumpling formed i an evaporite by flowage or by the swelling of anhydrite durin hydration. See also: *tepee structure.* (b) Said of the deforma tion or folding that produces enterolithic structures.

enteron The digestive cavity, or alimentary system, of an ani mal, generally consisting of esophagus, stomach, and intes tine. Cf: *coelom.*

entexis Migmatization with introduction from without of th more mobile part (Dietrich & Mehnert, 1961). Cf: *ectexis.*

entexite Rock formed by entexis. Also spelled: *entectite.*

enthalpy A thermodynamic quantity that is defined as the sum of a body's internal energy plus the product of its volume multiplied by the pressure. Syn: *heat content.*

entire Said of a leaf or other plant organ with a continuou smooth margin, not indented in any way; whole.

Entisol In U.S. Dept. of Agriculture soil taxonomy, a soil order characterized by a lack of distinct horizons within a depth of one meter (SSSA, 1970). Suborders and great soil groups of this soil order have the suffix -ent. See also: *Aquent; Arent; Fluvent; Orthent; Psamment.* Cf: *azonal soil.*

entocoele The space within a pair of mesenteries of a coral. Cf: *exocoele.*

entogene Said of conditions within a depositional basin that influence the texture of a sedimentary rock formed in that basin (Sander, 1951, p.11). Cf: *ectogene.*

entomodont Said of a class of ostracode hinges intermediate in form between merodont hinge and amphidont hinge, having denticulate terminal and median elements with a partial subdivision of the median element.

entomophily Pollination by insects. Adj: *entomophilous.* Cf: *anemophily.*

entomostracan An obsolete term originally applied to insect-shelled crustaceans. Cf: *malacostracan.*

entooecium The inner, generally membranous layer of the wall of a bryozoan ovicell. Cf: *ectooecium.*

entoolitic Pertaining to oolitic structures or grains that have formed or grown inward by the filling of small cavities, as by the deposition of successive coats on the cavity walls. Ant: *extoolitic.*

entoproct Any bryozoan belonging to the subphylum Entoprocta and lacking hard parts and a body cavity. These bryozoans are not found as fossils. Cf: *ectoproct.*

entoseptum A scleractinian-coral septum developed within an entocoele. Cf: *exoseptum.*

entosolenian Said of a foraminifer (e.g. *Oolina*) having an internal tube-like apertural extension (siphon). Cf: *ectosolenian.*

entotoichal Said of a bryozoan ovicell that appears to be immersed in the distal zooid while opening independently to the exterior.

entozooidal Said of a bryozoan ovicell that appears to be immersed in the distal zooid while opening below the operculum of the parent zooid.

entrail pahoehoe A type of pahoehoe that has a surface resembling an intertwined mass of entrails, formed on steep slopes as dribbles around and through cracks in the flow crust. Cf: *corded pahoehoe; elephant-hide pahoehoe; festooned pahoehoe; filamented pahoehoe; sharkskin pahoehoe; shelly pahoehoe; slab pahoehoe.*

entrainment The process of picking up and carrying along, as the collecting and movement of sediment (as bed load or in suspension) by currents, or the transfer and mixing of water between opposing currents by friction.

entrance angle *axil angle.*

entrapment burrow A term used by Kuenen (1957, p.253) for a burrow occupied by an animal buried below the sandy deposit of a passing turbidity current.

entrenched meander (a) An *incised meander* carved vertically downward below the surface of the valley in which the meander originally formed; it exhibits a symmetric cross profile (little or no contrast between the steep slopes of the valley sides of the meander). Such a deepened meander, which preserves its original pattern with little modification or enlargement, suggests rejuvenation of a meandering stream, as when there has been a rapid vertical uplift, or a tilt of the land, or a lowering of base level. Cf: *ingrown meander.* Syn: *inherited meander.* (b) A generic term used as a syn. of *incised meander.*--Syn: *intrenched meander.*

entrenched stream A stream, often meandering, flowing in a narrow trench or valley cut into a plain or relatively level upland; e.g. a stream that has inherited its course from a previous cycle of erosion and that cuts into bedrock with little modification of the original course. Syn: *intrenched stream.*

entrenchment The process whereby a stream erodes downward so as to form a trench or to develop an entrenched meander. Also, the results of such a process. Cf: *incision.*

entropy A macroscopic, thermodynamic quantity, ultimately reflecting the degree of microscopic randomness or disorder of a system. It increases in all natural processes by an amount at least as great as the quotient of the heat absorbed by the system, divided by the absolute temperature. The entropy of a system is equal to the negative of the partial derivative of its Gibbs free energy with respect to temperature at constant pressure.

entropy [stratig] A measure of the degree of "mixing" of the different kinds of rock components in a stratigraphic unit (Pelto, 1954). The entropy value of a given component is the product of its proportion in the unit and the natural logarithm of that proportion. A stratigraphic unit with equal parts of each component has an entropy value of 100; as the composition approaches that of a single component, the entropy value approaches zero.

entropy [streams] A quantity that is expressed by the probability of a given distribution of energy utilization within or along a stream from headwaters to a downstream point, the most probable condition existing when the stream is graded or the energy is as uniformly distributed as may be permitted by physical constraints (Leopold & Langbein, 1962).

entropy map A *facies map* that is based on the degree of "mixing" of three end members (rock components) of a given stratigraphic unit, but that does not distinguish the natures of these end members. Cf: *entropy-ratio map.* Syn: *isentropic map; entropy-function lithofacies map.*

entropy-ratio map A *facies map* that is based on the degree of "mixing" of three end members (rock components) of a given stratigraphic unit and that indicates by map pattern the nature of the lithologic "mixture" through which a given end member is approached (Forgotson, 1960, p.93). Cf: *entropy map.*

entropy unit A unit of measurement defined as one calorie per mole-degree. It is essentially equivalent to the *gibbs.*

entry [streams] The mouth of a river.

envelope [petrology] The migmatized and/or metamorphosed part of a regional body or rock undergoing granitization (Challinor, 1967).

envelope [fold] The outer or covering part of a fold, especially of a folded structure that includes some sort of structural break. Cf: *core.*

enveloping cells In charophyte green algae, the usually spirally twisted cells arising from the node cell and elongating to form a cover for the oospore. See also: *equatorial angle.*

environgeology *environmental geology.*

environment [biol] All those external factors and conditions which may influence an organism or a community. Syn: *habitat.*

environment [sed] A geographically restricted complex where a sediment accumulates, described in geomorphic terms and characterized by physical, chemical, and biological conditions, influences, or forces; e.g. a lake, swamp, or flood plain.

environmental facies Facies that are concerned solely with environment or determined by the nature of environment; e.g. lithotopes, biotopes, and tectotopes (Weller, 1958, p.628). They are not material units or bodies of sediments or rocks, but are areas inferred from the results of a combination of mutually interacting influences and conditions as these are exhibited in the form of sedimentary types and organic communities. See also: *facies [stratig].* Syn: *ecologic facies.*

environmental geochemistry The distribution and interrelations of the chemical elements and radioactivity among surficial rocks, water, air, and biota.

environmental geology The collection, analysis, and application of geologic data and principles to problems created by human occupancy and use of the physical environment, including the maximization of a rapidly shrinking living space and resource base to the needs of man, the minimization of the deleterious effects of man's interaction with the Earth, and the accomodation of the exponentially increasing human population to the finite resources and terrain of the Earth. It involves studies of hydrogeology, topography, engineering geology, and economic geology, and is concerned with Earth processes, Earth resources, and engineering properties of Earth materials. It involves problems concerned with construction of buildings and transportation facilities, installation of utility facilities, safe disposal of solid and liquid waste products, development and management of water resources, evaluation and mapping of rock and mineral resources, and overall long-range physical planning and development of the most efficient and beneficial use of the land. See also:*urban geology*. Syn: *geoecology; environgeology.*

environmental hyperspace lattice In a geometric model of the environment and its ecologic units, a multidimensional space containing as many dimensions as there are possible environmental factors (Valentine, 1969, p.685). See also: *biospace; ecospace.*

environmental resistance Factors in the environment that tend to restrict the development of an organism or group of organisms and to limit its numerical increase.

environmental science (a) An earth science applied to the human habitat, consisting essentially of geomorphology, meteorology, climatology, soil science, and physical and applied oceanography. (b) A science that describes and understands "all of nature we perceive or can observe, that is our physical environment-a composite of Earth, Sun, sea, and atmosphere, their interactions, and the hazards they present" (ESSA, 1968).

Eocambrian An approximate equivalent of *Riphean.*

Eocene An epoch of the lower Tertiary period, after the Paleocene and before the Oligocene; also, the corresponding worldwide series of rocks. It is sometimes considered to be a period, when the Tertiary is designated as an era.

Eogene *Paleogene.*

eogenetic A term proposed by Choquette & Pray (1970, p.219-220) for the period of time between final deposition of a sediment and burial of that sediment below the depth to which surface or near-surface processes are effective. The upper limit of the eogenetic zone is the land surface; the lower boundary, less clearly defined due to the gradual diminishing of surface-related processes, is the *mesogenetic* zone. Also applied to the porosity that develops during the eogenetic stage. Cf: *telogenetic.*

eohypse A contour line of former land surface, reconstructed or "restored" on a map by plotting the surviving portions of the land and by extrapolation of the original contours. Syn: *eohyps; eoisohypse.*

eoisohypse *eohypse.*

eolation The gradational work performed by the wind in modifying the land surface, such as the transportation of sand and dust, the formation of dunes, the effects of sandblasting, and the indirect action of water waves (building of beaches, carving of sea cliffs) generated by wind currents.

eolian (a) Pertaining to the wind; esp. said of rocks, soils, and deposits (such as loess, dune sand, and some volcanic tuffs) whose constituents were transported (blown) and laid down by atmospheric currents, or of landforms produced or eroded by the wind, or of sedimentary structures (such as ripple marks) made by the wind, or of geologic processes (such as erosion and deposition) accomplished by the wind. (b) Said of the active phase of a dune cycle, marked by diminished vegetal control and increased dune growth. Cf: *eluvial [sed].*---Etymol: Aeolus, god of the winds. Syn: *aeolian; eolic.*

eolianite A consolidated sedimentary rock consisting of clastic material deposited by the wind; e.g. dune sand cemented below ground-water level by calcite. Syn: *aeolianite; dune rock.*

eolic *eolian.*

Eolithic n. In archaeology, a cultural level that is pre-Paleolithic or the very beginning of the *Paleolithic*. It is characterized

by eoliths, or stones that appear to be man-made but were naturally formed.----adj. Pertaining to the Eolithic.

eolomotion A relatively slow downwind or downhill movement of sand due to direct or indirect wind action on surface rock particles (Kerr & Nigra, 1952).

eometamorphism Early or initial metamorphism (Murray, 1965).

eon (a) Any grand division or large part of geologic time; specif. the longest geologic-time unit, next in order of magnitude above *era*, such as the Phanerozoic Eon which is the Paleozoic Era, the Mesozoic Era, and the Cenozoic Era, or the Cryptozoic Eon which is the time prior to the beginning of the Cambrian Period. (b) One billion (10^9) years.----Also spelled *aeon.*

eophyte (a) A fossil plant of the earliest fossiliferous rocks. (b) One of a series of straight, parallel, or curved drag marks on a bedding plane, produced by organisms or comprising inorganic objects, but long supposed to represent a fossil plant and assigned the generic name *Eophyton.*

Eophytic A paleobotanic division of geologic time, signifying that time during which algae were abundant. Cf: *Aphytic; Archeophytic; Cenophytic; Mesophytic; Paleophytic.*

eoposition A term suggested by Glock (1928) for "wind placed" deposition, as where transported materials encounter a "stagnant zone of air".

eosphorite A pink to rose-red mineral: $(Mn,Fe)AlPO_4(OH)_2 \cdot H_2O$. It is isomorphous with childrenite.

Eötvös correction In gravity measurement, a correction for centripetal acceleration caused by east-west velocity over the surface of the rotating Earth. Syn: *Eötvös effect.*

Eötvös effect *Eötvös correction.*

Eötvös torsion balance A geophysical prospecting instrument that is used to determine distortions in the gravitational field. It consists of a pair of masses suspended by a sensitive torsion fiber and so supported that they are displaced both horizontally and vertically from each other. A measurement is made of the rotation of the suspended system about the fiber; the rotation is caused by slight differences in the gravitational attraction on the two masses. Syn: *torsion balance.*

Eötvös unit The unit of measurement in work with Eötvös torsion balance having the dimensions of acceleration divided by length, for the gradient and differential curvatures values. For the gradient, 1×10^{-9} gal per horizontal centimeter (Barton, 1929, p. 421).

eozoan An obsolete term for a *protozoan*, or a one-celled animal of the subkingdom Eozoa. Syn: *eozoon.*

Eozoic The earliest part of the Precambrian.

eozoon (a) An inorganic banded structure of various ophicalcites (coarsely crystalline calcite and serpentine) occurring in the Grenville Series of Canada, but originally interpreted as the remains of a gigantic foraminifer, *Eozoon canadense*. Pl: *eozoons; eozoa.* Adj: *eozoonal.* (b) *eozoan.*

epanticlinal fault A longitudinal or transverse fault that is associated with a doubly plunging minor anticline and formed concurrently with the folding (Irwin, 1926). Also spelled: *epi-anticlinal fault.*

epaulet A five-sided step cut of a gem, resembling a shoulder ornament in outline.

epeiric sea *epicontinental sea.*

epeirocratic (a) Adj. of *epeirocraton.* (b) Said of a period of low sea level in the geologic past. Cf: *thalassocratic.*

epeirocraton A craton of the continental block. Cf: *hedreocraton; thalassocraton.* Adj: *epeirocratic.*

epeirogenesis *epeirogeny.*

epeirogenetic Var. of *epeirogenic.*

epeirogenic Adj. of *epeirogeny.* Cf: *orographic.* Also spelled *epeirogenetic.*

epeirogeny As defined by Gilbert (1890), a form of *diastrophism* which has produced the larger features of the continents and oceans, for example, plateaus and basins; in con-

...ast to the more localized process of *orogeny*, which has pro-
...uced the mountain chains. Epeirogenic movements are pri-
...arily vertical, either upward or downward, which have af-
...ected large parts of the continents, not only in the cratons,
...ut also in stabilized former orogenic belts, where they have
...roduced most of the present mountainous topography. Some
...oeirogenic and orogenic structures grade into each other in
...etail, but most of them contrast strongly. Adj: *epeirogenic.*
...yn: *epeirogenesis.* Cf: *bathygenesis; cymatogeny.*

...peirophoresis theory *continental displacement.*

...pharmone An organism that has undergone morphologic
...nange as a result of changes in the environment and therefor
...ffers from the normal or usual form. See also: *ecad.*

...phebic Said of the adult stage in the life history of an animal;
...e., the stage when the animal is normal in size and able to
...eproduce; the stage following the *neanic* stage and preceding
...e *gerontic* stage.

...phemeral lake A short-lived lake. Cf: *intermittent lake; evan-
...scent lake.* Syn: *temporary lake.*

...phemeral stream A stream or reach of a stream that flows
...riefly only in direct response to precipitation in the immediate
...ocality and whose channel is at all times above the water
...ible. The term "may be arbitrarily restricted" to a stream that
...oes "not flow continuously during periods of as much as one
...nonth" (Meinzer, 1923, p.58). Cf: *intermittent stream.*

...phemeris time A uniform measure of time determined by rel-
...tive changes in the positions of Earth, Moon, and stars.

...phesite A mineral of the brittle-mica group: $NaLi-
I_2(Al_2Si_2)O_{10}(OH)_2$. It is related to margarite.

...phippium The dorsal brood pouch of various cladoceran
...rustaceans that is shed with the eggs and serves for protec-
...ion until hatching. Pl: *ephippia.*

...pi- In petrologic terminology, a prefix signifying "alteration"

...pi-anticlinal fault Var. of *epanticlinal fault.*

...pibatholithic Said of a mineral deposit occurring in the pe-
...pheral area of a batholith; also, said of the stage of batholith
...rosion in which that area is exposed (Emmons, 1933). The
...erm is little used today. Cf: *acrobatholithic; cryptobatholithic;
...mbatholithic; endobatholithic; hypobatholithic.* See also:*dead
...ne.*

...pibiontic Said of an organism living on the surface of bottom
...ediments. Cf: *endobiontic.*

...pibole A syn. of *peak zone.* The term was proposed by True-
...nan (1923, p.200) as a stratigraphic term (time-stratigraphic
...nit) for the rocks accumulated during a hemera.

...pibolite A term introduced by Jung & Roques for a migmatite
...vith granitic layers concordant with the gneissosity of its non-
...ranitic parent rock (Roques, 1961).

...picarp The external layer of a *pericarp.* Cf: *endocarp; meso-
...arp.* Syn: *exocarp;*

...picenter That point on the Earth's surface which is directly
...bove the *focus* of an earthquake. Cf: *anticenter.* Syn: *epi-
...entrum.*

...picentral distance The distance from the epicenter to the re-
...eiver. It may be measured in angular units (*angular dis-
...ance*) or in units of linear distance, measured along the
...reat-circle path between the epicenter and receiver.

...picentrum *epicenter.*

...piclastic rock (a) A rock formed at the Earth's surface by
...onsolidation of fragments of preexisting rocks; a sedimentary
...ock whose fragments are derived by weathering or erosion.
...b) Any *clastic rock* other than the "pyroclastic" variety,
...orming a new rock in a new place, as contrasted with *auto-
...lastic rock* (Challinor, 1967, p.85).

...picontinental Pertaining to the continental shelf.

...picontinental sea A sea on the continental shelf or within a
...ontinent. See also: *mediterranean sea.* Syn: *inland sea; con-
...nental sea; epeiric sea.*

...picycle A minor or secondary cycle within a major or primary
...ycle; specif. a subdivision of a cycle of erosion, initiated by a

small change of base level, such as an episode of standstill of
sufficient duration to rank as a part of a cycle and recorded
as a terrace.

epidermal Said of shallow or surficial deformation of the sialic
crust. Cf: *dermal; bathydermal.*

epidermis [paleont] Any of various animal integuments; e.g.
the periostracum of a mollusk, or imperforate outer layer in
foraminifera, or the external cellular layer in the body wall of a
coelenterate. Cf: *hypodermis.*

epidermis [bot] The characteristic outermost tissue of leaves
and of young roots and stems (Cronquist, 1961, p.874). Cf:
endodermis.

epidermis [geol] The sedimentary part of the Earth's crust
(Bemmelen, 1949, p.285).

epidiabase A name proposed as a replacement for *epidiorite.*

epidiagenesis A term used by Fairbridge (1967) for the final
emergent phase of diagenesis, in which sediments are lithified
during and after uplift or emergence but before erosion. It is
characterized by modification of connate solutions (by deeply
penetrating and downward-migrating ground waters) and by
reintroduction of oxidizing conditions; near the Earth's surface, it passes into a zone where weathering processes become dominant. It is equivalent to *late diagenesis.* See also:
syndiagenesis; anadiagenesis. Cf: *epigenesis.* Adj: *epidi-
agenetic.*

epididymite A colorless, orthorhombic mineral: $NaBe-
Si_3O_7(OH)$. It is dimorphous with eudidymite.

epidiorite A metamorphosed gabbro or diabase in which generally fibrous amphibole (uralite) has replaced the original
clinopyroxene (commonly augite). It is usually massive but
may have some schistosity. See also: *epidiabase.*

epidosite A metamorphic rock consisting of epidote and
quartz, and generally containing other secondary minerals
such as uralite and chlorite.

epidote (a) A yellowish-, pistachio-, or blackish-green mineral: $Ca_2(Al,Fe)_3Si_3O_{12}(OH)$. It commonly occurs associated
with albite and chlorite as formless grains or masses or as
monoclinic crystals in low-grade metamorphic rocks (derived
from limestones) or as a rare accessory constituent in igneous rocks where it represents alteration products of ferromagnesian minerals. Syn: *pistacite; arendalite; delphinite; thallite.*
(b) A mineral group, including minerals such as epidote, zoisite, clinozoisite, piemontite, and hancockite.

epidote-amphibolite facies Metamorphic rocks formed by regional dynamothermal metamorphism under intermediate
pressures (3000-7000 bars) and temperatures (250°-450°C)
with conditions that are intermediate between those of the
greenschist and amphibolite facies (Eskola, 1939). The name
was changed to the *albite-epidote-amphibolite facies* by Turner (1948) which in turn was altered by Fyfe et al (1958) to
the quartz-albite-epidote- almandine subfacies. Exact interpretation is in question.

epidotite A rock that consists mainly of epidote. The term has
no textural or genetic connotation (Bayly, 1968, p.230).

epidotization The hydrothermal introduction of epidote into
rocks or the alternation of rocks in which plagioclase feldspar
isalbitized, freeing the anorthite molecule for the formation of
epidote (and zoisite) often accompanied by chloritization.
These processes are characteristically associated with metamorphism.

epieugeosyncline A postorogenic geosyncline without associated volcanism whose detrital source is an uplifted eugeosynclinal area (Kay, 1947, p.1289-1293). Syn: *backdeep.* Cf: *secondary geosyncline.* See also: *nuclear basin.*

epifauna (a) Fauna living upon rather than below the surface
of the sea floor. Cf: *infauna.* (b) Fauna living attached to
rocks, seaweed, pilings, or to other organisms in shallow
water and along the shore.

epigene [geol] adj. Said of a geologic process, and of its resultant features, occurring at or near the Earth's surface. Cf:

hypogene [*geol*]. Syn: *epigenic*.

epigene [**cryst**] Pertaining to a crystal that is not natural to its enclosing substance, e.g. a pseudomorph.

epigenesis [**meta**] The change in the mineral character of a rock as a result of external influences operating near the Earth's surface, e.g. mineral replacement during metamorphism.

epigenesis [**sed**] The changes, transformations, or processes, occurring at low temperatures and pressures, that affect sedimentary rocks subsequent to their compaction, exclusive of surficial alteration (weathering) and metamorphism; e.g. postdepositional dolomitization. The term is equivalent to *late diagenesis* (as used in the U.S.) and to *metharmosis*, but is considered by Russian geologists to include those changes occurring subsequent to *diagenesis* (as defined in a restricted sense). Cf: *epidiagenesis*. Syn: *metagenesis; metadiagenesis; catagenesis*.

epigenesis [**streams**] *superimposition*.

epigenetic [**eco geol**] Said of a mineral deposit of later origin than the enclosing rocks. Cf: *syngenetic* [*ore dep*]; *diplogenetic*. Syn: *xenogenous*.

epigenetic [**sed**] (a) Said of a secondary sedimentary structure (such as a fold, fault, or other tectonic feature, or certain concretions) formed after the deposition of the sediment. (b) Pertaining to sedimentary epigenesis.---Cf: *syngenetic*. Syn: *epigenic*.

epigenetic [**streams**] *superimposed*.

epigenic *epigene*.

epigenite A steel-gray mineral: $(Cu,Fe)_5AsS_6$ (?).

epiglacial bench A terrace cut by the lateral erosion of a supraglacial meltwater stream originating on a large glacier; also, the valley-side channel developed by such a stream (Stamp, 1961, p.179).

epiglacial epoch A period of time that closes a "great glacial series", representing a "constant phase" of glacial activity (Hansen, 1894, p.131).

epiglyph A hieroglyph on the top of a sedimentary bed (Vassoevich, 1953, p.37). Cf: *hypoglyph*.

epigone A volcanic cone that is parasitic.

epigynous Said of a plant in which the sepals, petals, and stamens appear to arise from the top of the ovary. Such an ovary is called an *inferior ovary*. Cf: *hypogynous*.

epiianthinite A yellow pseudomorphous alteration product of ianthinite, identical with *schoepite*.

epiliminal *epilimnetic*.

epilimnetic Pertaining to an *epilimnion*. Syn: *epilimnial*.

epilimnion The uppermost layer of water in a lake, characterized by an essentially uniform temperature that is generally warmer than elsewhere in the lake and by a relatively uniform mixing caused by wind and wave action; specif. the light (less dense), oxygen-rich layer of water that overlies the *metalimnion* in a thermally stratified lake. The oceanographic equivalent is *mixed layer*. Cf: *hypolimnion*.

epilithic Said of an organism that lives on or attached to rock or other stony matter. Syn: *petrophilous*. Cf: *endolithic*.

epimagma A vesicular magmatic residue that is relatively gas free and of semisolid, pasty consistency, commonly formed by the cooling of lava in a lava lake. Cf: *hypomagma; pyromagma*.

epimagmatic *deuteric*.

epimere A lateral downfold of a crustacean tergite. Syn: *epimeron; pleurite; pleuron; tergal fold*.

epinorm Theoretical calculation of minerals in metamorphic rock of the epizone as indicated by chemical analyses (Barth, 1959). Cf: *mesonorm*.

epipedon In U.S. soil classification, a diagnostic surface layer of soil, about one foot in thickness. An epipedon may be *mollic, umbric, ochric, anthropic,* or *plaggen* according to its properties.

epipelagic Pertaining to the *pelagic* environment of the ocean

to a depth of 100 fathoms. Cf: *mesopelagic*.

epipelic Said of an organism growing on sediment, esp. so mud.

epiphysis [**paleont**] An interambulacral element at the top Aristotle's lantern in echinoids.

epiphysis [**intrus rocks**] An *apophysis* or *tongue* of an intrusio which is detached from its source.

epiphyte A plant not growing from the soil but living attache to another plant or some inanimate object such as a pole c wire. Adj: *epiphytic*.

epipsammon Animals that live on a sandy surface.

epirhysis A *skeletal canal* in dictyonine hexactinellid sponge corresponding to an inhalant canal. Pl: *epirhyses*.

episeptal deposit A proximal cameral deposit on the conca (adoral) side of septum of a nautiloid. Ant: *hyposeptal depos it*.

episkeletal Above or outside the endoskeleton of an animal.

episode (a) A term used informally and without time implica tion for a distinctive and significant event or series of event in the geologic history of a region or feature; e.g. "glacial ep sode" and "volcanic episode". (b) A term preferred by Stø mer (1966, p.25) for *subage*, or the geologic-time unit durin which the rocks of a *substage* were formed. Syn: *phase* [*gec chron*] (*a*); *time (d)*.

episome The anterior part of the cell body above the girdle c an unarmored (naked) dinoflagellate. Cf: *hyposome*.

epistilbite A white or colorless zeolite mineral: $CaAl_2Si_6 O_{16}.5H_2O$.

epistome (a) The region between the antennae and the mout of a crustacean. Also, a plate covering this region, such a that between the labrum and the bases of the antennae i brachyuran decapods. (b) The *rostrum* of a trilobite. (c) small lobe or labiate organ overlapping the mouth in variou bryozoans, such as in Phylactolaemata.---Pl: *epistomes* c *epistomae*. Syn: *epistoma*.

epitactic Recommended adj. of *epitaxy*.

epitaxial Adj. of *epitaxy*.

epitaxic Adj. of *epitaxy*.

epitaxy Orientation of one crystal with that of the crystallin substrate on which it grew. It is a type of overgrowth in whic the two nets in contact share a common mesh. Adj: *epitactic epitaxic; epitaxial*. Cf: *topotaxy; syntaxy*.

epitheca (a) An external calcareous layer or sheath of skele tal tissue laterally surrounding a corallite and comprising a extension of the basal plate. (b) A dark, secondary deposit i the inner wall of a fusulinid foraminifer; a tectorium. (c) Th thin outermost calcareous layer of a thecal plate of a cystoid It is thinner than *stereotheca*. (d) The anterior part of a dino flagellate theca, above the girdle. Cf: *hypotheca*. (e) *epivalve*.

epithelium (a) In an animal, a cellular tissue that forms a cov ering or lining and that may serve various functions such a protection, secretion. (b) In a plant, one or more layers c parenchyma cells, e.g. lining a resin duct and excreting resin.

epithermal Said of a hydrothermal mineral deposit forme within 3,000ft of the Earth's surface and in the temperature range of 50°-200°C, occurring mainly as veins (Park & Mac Diarmid, 1970, p. 344). Also, said of that environment. Cf: *hy pothermal; mesothermal; leptothermal; teletothermal; xenother mal*. See also: *Tertiary-type ore deposit*.

epithermal neutron log *neutron-neutron log*.

epithyridid Said of a brachiopod pedicle opening lying wholl within the ventral umbo and ventral from the beak ridges (TIP 1965, pt.H, p.144).

epitract The part of a dinoflagellate cyst anterior to the girdl region. Ant: *hypotract*.

epityche An excystment (emergence from a cyst) aperture i the acritarch genus Veryhachium. It originated as an arche slit between two processes, and in which rupture allowed th folding back of a relatively large flap.

epivalve The outer half of a diatom frustule. Cf: *hypovalv*

Syn: *epitheca*.

epixenolith A xenolith that is derived from the adjacent wall rock (Goodspeed, 1947, p.1251). Cf: *hypoxenolith*.

epizoic Said of an organism growing on the body surface of an animal.

epizone According to Grubemann's classification of metamorphic rocks (1904), the uppermost *depth zone* of metamorphism, characterized by low to moderate temperatures (less than 300°C) and hydrostatic pressures with low to high shearing stress. Mechanical and chemical metamorphism produce hydrous silicates (e.g. sericite, chlorite, talc) and carbonates (e.g. calcite, dolomite). Typical rocks are slates, phyllites, and sericite and chlorite schists. The concept was modified by Grubenmann and Niggli (1924) to include effects of low-temperature contact metamorphism and metasomatism. Modern usage stresses pressure-temperature conditions (low metamorphic grade) rather than depth of zone. Cf: *mesozone; katazone*.

epizygal The distal brachial plate of a pair joined by syzygy in a crinoid. Cf: *hypozygal*.

epoch [geochron] (a) A geologic-time unit longer than an *age* [geochron] and a subdivision of a *period* [geochron], during which the rocks of the corresponding *series* were formed. (b) A term used informally to designate a length (usually short) of geologic time; e.g. *glacial epoch*.

epoch [paleomag] (a) A date to which measurements of a time-varying quantity are referred, e.g. "a chart of magnetic declination for epoch 1965.0". (b) *polarity epoch*.

epoch [tide] *tidal epoch*.

epontic Said of an organism that grows attached to some substratum.

epsomite [mineral] A mineral: $MgSO_4.7H_2O$. It consists of native Epsom salts, and occurs in colorless prismatic crystals, botryoidal masses, incrustations in gypsum mines or limestone caves, or mineral waters (in solution). Syn: *Epsom salt; bitter salts; hair salt*.

epsomite [sed] Obsolete syn. of *stylolite*.

Epsom salt (a) A chemical: $MgSO_4.7H_2O$. It is a bitter, colorless or white, crystalline salt with cathartic qualities. (b) *epsomite*.---Syn: *Epsom salt*.

equal-area projection (a) A map projection on which a constant ratio of areas is preserved, so that any given part of the map has the same relation to the area on the sphere it represents as the whole map has to the entire area represented. Equal-area projections suffer from severe distortions of shapes and directions; they are useful for showing the areal extent of a given surface property (such as relative areas of oceans and continents). Examples include: Bonne projection, Albers projection, and Mollweide projection. Cf: *conformal projection*. Syn: *authalic projection; equivalent projection; homolographic projection*. (b) *equiareal projection*.

equant (a) Said of a crystal, in an igneous or sedimentary rock, having the same (or nearly the same) diameters in all directions. Syn: *equidimensional*. (b) Said of a sedimentary particle whose length is less than 1.5 times its width (Krynine, 1948, p.142).

equant element A fabric element all of whose dimensions are approximately equal. Cf: *linear element; planar element*.

equation of state An equation interrelating the thermodynamic variables that define the state of a system. It is classically applied to simple gases and liquids in terms of pressure, volume, and temperature, but in modern geochemistry and petrology it is commonly extended to solids and to solutions, in which case the equations must contain terms describing the composition of the phase.

equator [palyn] An imaginary line connecting points midway between the poles of a spore or pollen grain.

equator [surv] (a) The great circle formed on the surface of a sphere or spheroid by a plane drawn through its center and perpendicular to its polar axes, such as the great circle mid-way between the poles of rotation of a celestial body; specif. the Equator, or the great circle on the Earth's surface that is everywhere equally distant from the two poles and whose plane is perpendicular to the Earth's axis of rotation, that divides the Earth's surface into the northern and southern hemispheres, and that is the line from which latitudes are reckoned, its own latitude being everywhere zero degrees. It is the largest of the parallels of latitude, having a length on the Earth's surface of 40,075.76 km (24,901.92 miles). Syn: *terrestrial equator*. (b) *celestial equator*.

equator The great circle of a celestial sphere, having a plane that is perpendicular to the axis of the Earth.

equatorial [paleont] Pertaining to or located in the median plane normal to the axis of coiling or symmetry in a foraminifer; e.g. "equatorial section" representing a slice through a foraminiferal test passing through the proloculus. See also: *sagittal*.

equatorial [clim] Said of a climate characterized by uniformly high temperature and humidity and heavy rainfall, and occurring in lowland areas within five to ten degrees of the equator. Cf: *tropical*.

equatorial angle In charophyte green algae, the angle formed with the horizontal by the *enveloping cells* as they cross the equator of the oogonium.

equatorial aperture A symmetric aperture in a planispiral foraminiferal test. It is commonly an *interiomarginal aperture*, but it may be areal or peripheral.

equatorial bulge An expression used to describe the *ellipticity* or flattening of the earth. It does not accurately describe the ellipticity and may lead to incorrect concepts. Its use should be avoided.

equatorial countercurrent An ocean-surface current near the equator, flowing eastward between the westward-flowing equatorial currents to the north and south. Cf: *equatorial undercurrent*.

equatorial current (a) Any of the ocean-surface currents in the tropical areas just north or just south of the equator, driven southwest or west in the Northern Hemisphere by northeast trade winds (North Equatorial Current), or northwest or west in the Southern Hemisphere by southeast trade winds (South Equatorial Current). (b) A tidal current occurring twice monthly when the Moon in near or over the Earth's equator.

equatorial limb A term sometimes applied to the *amb* of a pollen grain or spore. It is less desirable because of possible confusion. Syn: *limb [palyn]*.

equatorial projection One of a group of projections that have their center points on the equator and their polar axes vertical; e.g. an "equatorial cylindrical conformal map projection" (also known as the Mercator projection).

equatorial space A four-sided region resulting from formation of basal leaf cross in an acantharian radiolarian.

equatorial spine A radial spine arising on the shell equator in an acantharian radiolarian.

equatorial tide A tide occurring twice monthly when the Moon is near or over the Earth's equator, and displaying the least diurnal inequality. Cf: *tropic tide*.

equatorial undercurrent A narrow *undercurrent* in the ocean, flowing from west to east beneath or sometimes embedded in the westward-flowing equatorial currents. Cf: *equatorial countercurrent*. Syn: *Cromwell current*.

equatorial view The view of a spore or pollen grain from an aspect more or less midway between the poles.

equatorial water A surface *water mass* separating *central waters* in the Pacific and Indian oceans.

equator system of coordinates A system of curvilinear celestial coordinates (usually declination and right ascension) based on the celestial equator as the primary great circle. Cf: *horizon system of coordinates*.

equiareal projection A term used in structural petrology for an *equal-area projection* developed from the center of a sphere through points on its surface to a plane that is tangent at the south pole of the sphere and so constructed that areas between meridians and parallels on the plane are equal to corresponding areas on the surface of the sphere.

equidimensional *equant.*

equidistance The property of a map projection in which the principal scale is preserved radially from the point of zero distortion (as for an azimuthal equidistant projection) or perpendicularly to the line of zero distortion (as for an equidistant cylindrical projection or an equidistant conic projection).

equidistant projection A map projection in which distances are represented true to scale and without length distortion in all directions from a given point or along or at right angles to a given meridian or parallel.

equiform Said of the crystals, in an igneous rock, having the same (or nearly the same) shape.

equiglacial line A line drawn on a map or chart to show coincidence of ice conditions, as in a lake or river, at a given time. See also: *isopag; isopectic; isotac.*

equigranular (a) Said of a rock texture having or characterized by crystals of the same (or nearly the same) size. (b) Said of a rock with equigranular texture. The term was originally applied by Cross et al (1906, p.698) to igneous rocks, but it is also used for sedimentary rocks (such as recrystallized carbonate rocks) whose constituent crystals are of approximately the same size.----Ant: *inequigranular.* Syn: *even-grained; homeocrystalline; isomeric; isometric.* Cf: *isogranular.*

equilateral Bilaterally symmetric; specif. said of a bivalve-mollusk shell whose parts anterior and posterior to the beaks are subequal or equal in length and nearly symmetric. Cf: *equivalve.* Ant: *inequilateral.*

equilibrium [phase] That state of a chemical system in which the phases do not undergo any change of properties with the passage of time, provided they have the same properties when the same conditions, with respect to the variants, are again reached by a different procedure.

equilibrium [geol] In geology, a balance between form and process, e.g. between the resistance of rocks along a coast and the erosional force of waves.

equilibrium constant A number representing equilibrium of a chemical reaction, defined as the result of multiplying the activities of the equation and dividing by a similar product of the activities of the reactants. It may be referred to as the *thermodynamic equilibrium constant*; when concentrations instead of activities are used, it may be called the *classical equilibrium constant.* See also: *van't Hoff equation.*

equilibrium diagram *phase diagram.*

equilibrium limit *equilibrium line.*

equilibrium line The level on a glacier where the net balance equals zero, and accumulation equals ablation; the line separating the superimposed ice zone of the accumulation area (above) from the ablation area (below). For some temperate mountain glaciers, it is very nearly coincident with the *firn line*, in which case it is common practice to use the latter term; but on subpolar glaciers, the equilibrium line is lower than the firn line because freezing of meltwater occurs below the firn line forming superimposed ice. Cf: *climatic snowline; snowline.* Syn: *equilibrium limit.*

equilibrium moisture content The moisture content of a soil mass at a time when there is no moisture movement (Nelson & Nelson, 1967, p.127).

equilibrium path On a phase diagram, the crystallization sequence in which all crystals react continuously and completely with the liquid, so that adjustment of crystal composition throughout the crystallization interval is perfect.

equilibrium profile *profile of equilibrium.*

equilibrium shoreline A shoreline that has a local vertical profile of equilibrium and also an equilibrium shape in plan view; a *graded shoreline.*

equilibrium stage In hypsometric analysis of drainage basins the stage in which a steady state is developed and maintained as relief slowly diminishes, and corresponding to maturity and old age in the geomorphic cycle (Strahler, 1952b, p.1130) the hypsometric integral is stable between 35% and 60%. Cf *inequilibrium stage; monadnock phase.*

equilibrium theory A tidal hypothesis that assumes an Earth with no continental barriers and with a uniform and deep ocean cover; the *equilibrium tide* would respond instantly to the gravitational forces of the Sun and the Moon.

equilibrium tide The hypothetical tide as described by the *equilibrium theory.* Syn: *gravitational tide; astronomic tide.*

equinoctial circle The *celestial equator.* Syn: *equinoctial line.*

equinoctial tide A tide occurring when the Sun is near equinox, characterized by greater-than-average ranges of the spring tide.

equiplanation Those processes that operate at high latitudes and tend toward reduction of the land without reference to base-level control and without involving any loss or gain of material (Cairnes, 1912, p.76). Cf: *altiplanation.*

equiplanation terrace A syn. of *altiplanation terrace* (G. Warwick in Stamp, 1961, p.21).

equipotential line A contour line of a potentiometric surface; line along which the pressure head of a body of ground water is the same. Syn: *isopiestic line; isopotential line; piezometric contour.*

equipotential-line method An early electrical-survey method wherein lines of equal potential, near the current electrode were searched for and mapped using a pair of potential electrodes, one of which was held stationary for each line mapping.

equipotential surface A surface on which the potential is everywhere constant for the attractive forces concerned. The gravity vector is everywhere normal to a gravity equipotential surface; the geoid is an 'equipotential'. Syn: *gravity equipotential surface; niveau surface; level surface.*

equirectangular projection *plate carrée projection.*

equitemperature metamorphism A process of modification of ice crystals in deposited snow, characterized by vapor transfer from regions of high surface energy to regions of low surface energy in a relatively constant-temperature, below-freezing environment, and leading to the formation of uniform, well rounded grains. Syn: *destructive metamorphism.* Cf: *temperature-gradient metamorphism.*

equivalence [cart] The unique property of an equal-area map projection in which the ratio between areas on the map is the same as that between the corresponding areas on the surface of the Earth. Cf: *conformality.* Syn: *orthembadism.*

equivalence [stratig] Geologic contemporaneity, esp. as indicated by identical fossil content. Syn: *equivalency.*

equivalent [stratig] adj. Corresponding in geologic age or stratigraphic position; esp. said of strata or formations (in regions far from each other) that are contemporaneous in time of formation or deposition or that contain the same fossil forms.----n. A stratum that is contemporaneous or equivalent in time or character.

equivalent diameter Twice the *equivalent radius.*

equivalent grade A term used by Baker (1920, p.367), and synonymous with *arithmetic mean diameter.*

equivalent projection *equal-area projection.*

equivalent radius A measure of particle size, equal to the computed radius of a hypothetical sphere of specific gravity 2.65 (quartz) having the same settling velocity and same density as those calculated for a given sedimentary particle in the same fluid; one half of the *equivalent diameter.* Cf: *nominal diameter; sedimentation diameter.*

equivalve Having valves equal in size and form; specif. said of a bivalve mollusk or its shell in which the right valve and the left valve are subequal or equal and symmetric about the plane of commissure. Cf: *equilateral.* Ant: *inequivalve.*

equivoluminal wave *S wave.*

era A geologic-time unit next in order of magnitude below an *on* and including two or more *periods*, during which the rocks of the corresponding *erathem* were formed; e.g. the Paleozoic era, the Mesozoic Era, and the Cenozoic Era, during each of which a specified combination of systems was developed. Because of difficulties in establishing a chronologic succession of period for Precambrian eras, they are formed independently (Archeozic Era and Proterozoic Era).

radiation *terrestrial radiation.*

erathem The largest recognized time-stratigraphic unit, next in rank above *system*; the rocks formed during an *era* of geologic time, such as the Mesozoic Erathem composed of the Triassic System, the Jurassic System, and the Cretaceous System. The term was favored by the International Subcommission on Stratigraphic Terminology (1961, p.13 & 27) and recognized by the American Commission on Stratigraphic Nomenclature (1967, p.1865). See also: *supersystem.* Obsolete syn: *group; sequence.*

eratosthenian (a) Pertaining to lunar topographic features and lithologic map units constituting a system of rocks formed during the period of formation of large craters (such as Eratosthenes) whose rays are no longer visible. Eratosthenian rocks are older than those of the Copernican System but younger than those of the Imbrian System. (b) Said of the stratigraphic period during which the Eratosthenian System was developed.

e ray [cryst] In a uniaxial crystal, the ray of light which vibrates in a plane containing the optic axis and at an angle with the basal pinacoid and whose velocity or refraction approaches that of the O ray as the angle approaches zero; the *extraordinary ray.* Cf: *O ray.* Syn: *extraordinary wave.*

e ray [paleont] Left anterior ray in echinoderms situated clockwise of D ray when the echinoderm is viewed from the dorsal side.

eremacausis The gradual transformation by oxidation of plant material into humus from exposure to air and moisture.

eremeyevite A syn. of *jeremejevite.* Also spelled: *eremeevite.*

eremic Pertaining to a *desert* or deserts, or to sandy regions.

eremology The scientific study of the desert and its phenomena.

eremophyte *xerophyte.*

erg A vast region in the Sahara Desert deeply covered with shifting sand and occupied by complex eolian sand dunes; an extensive tract of *sandy desert*; a *sand sea.* Etymol: Hamitic. Pl: *areg; ergs.* See also: *koum; nefud.* Syn: *ergh.*

ergeron A French term for a very fine argillaceous sand, or variety of loess, containing a substantial amount of calcium carbonate, commonly occurring in northern France and the Belgian province of Hainaut.

ergh *erg.*

erian North American provincial series: Middle Devonian (above Ulsterian, below Senecan).

erian orogeny One of the 30 or more short-lived orogenies during Phanerozoic time identified by Stille, in this case at the end of the Silurian; the last part of the Caledonian orogenic era. Syn: *Hibernian orogeny.*

ericaite A mineral: $(Fe,Mg,Mn)_3B_7O_{13}Cl$.

ericophyte A plant growing on a heath or moor.

ericssonite A mineral: $BaMn_3Fe(Si_2O_7)(OH)$.

erikite A yellow-green, greenish-yellow, or brown mineral: $(La,Ce)_x(P,Si)O_4 \cdot H_2O$. It is perhaps a silicate-rich rhabdophane. The type erikite from Greenland has been shown to be monazite.

eriochalcite A bluish-green to greenish-blue mineral: $Cu_2 \cdot 2H_2O$. Syn: *antofagastite.*

erionite A zeolite mineral: $(Ca,Na,K,Mg)_5Al_9Si_{27}O_{72} \cdot 27H_2O$.

eriophorum peat Peat formed mainly from Eriophorum, also known as cotton grass, a genus of sedges of the family Cyperaceae. Cf: *carex peat.*

ernstite A mineral: $(Mn^{+2}_{1-x}Fe^{+3}_x)Al(PO_4)(OH)_{2-x}O_x$. It is an oxidation product of eosphorite.

erode (a) To wear away the land, as by the action of streams, waves, wind, or glaciers. (b) To produce or modify a landform by the wearing away of the land.

erodibility (a) The quality, degree, or capability of being eroded or yielding more or less readily to erosion. (b) The tendency of soil to be detached and carried away; the rate of soil erosion.---Cf: *erosiveness.* Adj: *erodible.* Syn: *erodability; erosibility.*

eroding channel A channel in which sufficient amounts of sediment are not available for stream transport according to a bed-load function.

eroding stress Shear stress of overland flow available to dislodge or tear loose soil material per unit area. Originally defined by Horton (1945, p.319) as the "eroding force" exerted parallel with the soil surface per unit of slope length and width. Symbol: F_1.

erosibility *erodibility.*

erosion (a) The general process or the group of processes whereby the earthy and rocky materials of the Earth's crust are loosened, dissolved, or worn away, and simultaneously removed from one place to another, by natural agencies that include weathering, solution, corrasion, and transportation, but usually exclude mass-wasting; specif. the mechanical destruction of the land and the removal of material (such as soil) by running water (including rainfall), waves and currents, moving ice, or wind. The term is sometimes restricted by excluding transportation (in which case "denudation" is the more general term) or weathering (thus making erosion a dynamic or active process only). Cf: *denudation.* (b) An instance or product, or the combined effects, of erosion.

erosional Pertaining to or produced by the work of erosion.

erosional flood plain A *flood plain* produced by the lateral erosion and gradual retreat of the valley walls.

erosional unconformity An unconformity made manifest by erosion, or a surface that separates older rocks that have been subjected to erosion from younger sediments that cover them; specif. *disconformity.*

erosional vacuity An inaccurate term formerly used by Wheeler (1958, p.1057) and now replaced by *degradation vacuity.*

erosion caldera A type of *caldera* that is developed by the erosion and resultant widening of a caldera or by erosion of a volcanic cone, resulting in a large, central cirquelike depression. It is considered by some not to be a true caldera type, since it is not formed by volcanic processes. Cf: *explosion caldera; collapse caldera.*

erosion crater *makhtésh.*

erosion cycle *cycle of erosion.*

erosion fault scarp *fault-line scarp.*

erosion groove A sedimentary structure formed by "closely spaced lines of straight-sided scour marks" (Dzulynski & Sanders, 1962, p.66). The scouring may be initially concentrated by a preexisting groove.

erosion integral An expression of the relative volume of a landmass removed by erosion at a given contour; the inverse of the *hypsometric integral.*

erosion intensity Quantity (or depth) of solid material actually removed from the soil surface by sheet erosion per unit of time and area; originally termed *erosion rate* by Horton (1945, p.324). Symbol: E_a.

erosionist A believer in the obsolete theory that the irregularities of the Earth's surface are mainly due to erosion.

erosion lake A lake occupying a basin excavated by erosion.

erosion pavement A surficial concentration of pebbles, gravel, and other rock fragments that develops after sheet erosion or rill erosion has removed the finer soil particles and that tends to protect the underlying soil from further erosion.

erosion plain A general term for any plain produced by erosion, such as a peneplain, a pediplain, a panplain, or a plain of marine erosion.

erosion platform (a) A relatively level surface of limited extent formed by erosion. (b) A *wave-cut platform* along a coast. Cf: *abrasion platform.*

erosion proportionality factor Ratio of *erosion intensity* to *eroding stress.* It expresses the resistance of the ground surface to erosion by surface runoff by representing the quantity of solid material removed per unit of time and surface area. Symbol: k_e.

erosion ramp A sloping belt of reef rock immediately above the reef flat on an atoll islet, where marine erosion is active.

erosion remnant A topographic feature that is remaining or left standing above the general land surface after erosion has reduced the surrounding area; e.g. a monadnock, a butte, or a stack. Syn: *residual; relic; remnant.*

erosion ridge One of a series of small ridges on a snow surface, formed by the corrasive action of windblown snow, either parallel or perpendicular to the direction of the wind (ADTIC, 1955, p.27). Cf: *sastruga.*

erosion ripple A minor wave-like feature produced by the cutting action of the wind on a lower and somewhat more coherent layer of a sand dune.

erosion scarp A *scarp* produced by erosion; e.g. a *fault-line scarp* or a *beach scarp.*

erosion surface A land surface shaped and subdued by the action of erosion, esp. by running water. The term is generally applied to a level or nearly level surface; e.g. a *stripped structural surface.* Syn: *planation surface.*

erosion terrace A terrace produced by erosion; specif. a *rock terrace.*

erosion thrust A thrust fault, the displacement of which is along an erosion surface at the surface of the Earth.

erosiveness (a) The quality, degree, or capability of accomplishing erosion; the power or tendency to effect erosion. (b) A term sometimes used as a syn. of *erodibility.* (c) The exposure of soil to erosion.---Adj: *erosive.*

erosive velocity That velocity of water in a channel above which erosion of the bed or banks will occur.

erpoglyph *worm casting.*

erratic [geol] n. A relatively large rock fragment lithologically different from the bedrock on which it lies, either free or as part of a sediment, and that has been transported, sometimes over a considerable distance, from its place of origin; e.g. *sarsen; exotic block; glacial boulder.* The term excludes a pebble in a conglomerate (Challinor, 1967, p.88).

erratic [glac geol] n. A rock fragment carried by glacier ice or by floating ice, and deposited when the ice melted at some distance from the outcrop from which the fragment was derived. Generally of boulder size, although the fragments range from pebbles to house-sized blocks. Cf: *perched block; exotic.* Syn: *erratic block; erratic boulder; glacial erratic; glacial boulder.*---adj. Transported by a glacier from its place of origin, as *erratic* boulder. Syn: *traveled.*

erratic [coal] A pebble, boulder, or fossil tree trunk found in or associated with a coal seam.

erratic block *erratic.*

erratic boulder *erratic.*

error (a) The difference between an observed, calculated, or measured value of a quantity and the ideal or true value of that quantity or some conventional or standard value determined by established procedure or authority and used in lieu of the true value. (b) An inaccuracy or variation in the measurements, calculations, or observations of a quantity due to mistakes, imperfections in equipment or techniques, human limitations, changes of surrounding conditions, or other uncontrollable factors. (c) The amount of deviation of a measurement from some standard, arbitrary, estimated, or other reference value.---See also: *random error; systematic error; personal error.*

error of closure The amount by which a value of a quantity

obtained by a series of related surveying measurements differs from another value (of the same quantity) such as the true or theoretical value of the quantity or a fixed value obtained from previous determinations; esp. the amount by which the final value of a series of survey observations made around a closed loop differs from the initial value. The surveyed quantities may be angles, azimuths, elevations, or traverse-station positions. See also: *closure; circuit closure; triangle closure; horizon closure; mis-tie.* Syn: *misclosure; closing error.*

error of collimation *collimation error.*

erubescite *bornite.*

eruption The ejection of volcanic materials (lava, pyroclastics, and volcanic gases) onto the Earth's surface. It is usually a violent phenomenon, although an eruption along a fissure may be relatively calm.

eruption breccia *explosion breccia.*

eruption cloud A gaseous cloud of volcanic ash and other pyroclastic fragments that forms by volcanic explosion; the ash may fall from it by *air-fall deposition.* Syn: *explosion cloud; ash cloud; dust cloud; volcanic cloud.* See also: *eruption column.*

eruption column The initial form that an eruption cloud takes at the time of explosion; the lower portion of an *eruption cloud.*

eruption rain A rain following a volcanic eruption that results from condensation of the eruption's associated steam. Syn: *volcanic rain.*

eruption rate *age-specific eruption rate.*

eruption-time In stochastic treatment of volcanologic data based on renewal theory, the age at which a repose period is ended by an outbreak (Wickman, 1966, p.293).

eruptive Said of a rock formed by the solidification of magma, i.e. either an *extrusive* or an *igneous* rock. Some writers restrict the term to its extrusive or volcanic sense.

eruptive evolution Evolution characterized by the sudden appearance of varied new stocks from a common ancestral line.

erythraean Pertaining to the ancient sea that occupied the Arabian Sea, the Red Sea, and the Persian Gulf areas. Also spelled: *erythrean.*

erythrean Var. of *erythraean.*

erythrine *erythrite.*

erythrite A rose-, crimson-, peach-, or pink-red mineral, $Co_3(AsO_4)_2.8H_2O$. It is isomorphous with annabergite, and may contain some nickel. Erythrite occurs in monoclinic crystals, in globular and reniform masses, or in earthy forms as a weathering product of cobalt ores in the upper (oxidized) parts of veins. Syn: *erythrine; cobalt bloom; red cobalt; cobalt ocher; peachblossom ore.*

erythrosiderite A mineral: $K_2FeCl_5.H_2O$. It may contain some aluminum.

erythrozincite A variety of wurtzite containing manganese.

erzbergite Calcite and aragonite in alternate layers.

Erzgebirgian orogeny One of the 30 or more short-lived orogenies during Phanerozoic time identified by Stille, in this case in the early Upper Carboniferous (Namurian or Westphalian).

esboite An orbicular diorite in which andesine or oligoclase is the dominant plagioclase and forms the orbicules. Its name is derived from Esbo, Finland.

esboitic crystallization The process by which the orbicules of an esboite attain oligoclasitic composition (Eskola, 1938, p.449).

escar Var. of *esker.*

escarpment [geomorph] (a) A long, more or less continuous cliff or relatively steep slope facing in one general direction, breaking the general continuity of the land by separating two level or gently sloping surfaces, and produced by erosion or by faulting. The term is often used synonymously with *scarp,* although "escarpment" is more often applied to a cliff formed by differential erosion. (b) A high, steep, abrupt face of rock often presented by the upward termination of strata in a line of cliffs, and generally marking the outcrop of a resistant layer occurring in a series of gently dipping softer strata; specif the steep face of a cuesta. See also: *scarp slope.* (c) A term used loosely in Great Britain as a syn. of *cuesta.*---Etymol: French *escarpment,* "steep face or slope".

eschar Var. of *esker.*

escharan Said of bryozoans with free bilamellar colonies, as in *Eschara.*

eschwegeite A syn. of *tanteuxenite.* Also spelled: *eschwegeite.*

eschynite *aeschynite.*

escutcheon (a) The flat or simply curved, typically lozenge-shaped, dorsal differentiated area extending posteriorly from the beaks of certain bivalve mollusks and sometimes bordered by a ridge in each valve. It corresponds to the posterior part of the cardinal area and is separated from the remainder of the shell surface by a sharp change in angle. (b) An improper term for the *basal cavity* of a conodont.

eskar Var. of *esker.*

eskebornite A mineral: $CuFeSe_2$.

esker A long, low, narrow, sinuous, steep-sided ridge or mound composed of irregularly stratified sand and gravel that was deposited by a subglacial or englacial stream flowing between ice walls or in an ice tunnel of a continuously retreating glacier, and was left behind when the ice melted. It may be branching and is often discontinuous, and its course is usually at a high angle to the edge of the glacier. Eskers range in length from less than a kilometer to more than 160 km, and in height from 3 to 30 m. Etymol: Irish *eiscir,* "ridge". Cf: *kame.* Syn: *äs; os; eskar; eschar; escar; serpent kame; Indian ridge.*

esker delta A stratified deposit of sand formed by and at the mouth of a glacial stream as it issued from an ice tunnel and flowed into a lake or sea. Syn: *sand plateau; sand plain.*

esker fan A small plain of gravel and sand built at the mouth of a subglacial stream, and associated with an esker forming at the same time.

eskerine Characteristic of an esker; e.g. *eskerine* topography.

esker lake A lake enclosed or dammed by an esker (Dryer, 1901, p.129).

esker trough A term applied in Michigan to a shallow valley, cut in till, that contains an esker (Leverett, 1903, p.118).

eskolaite A rhombohedral mineral: Cr_2O_3. It is isomorphous with hematite.

esmeraldite A coarse- to medium fine-grained granitic rock having hypidiomorphic-granular texture and in which quartz and muscovite are the essential components. Its name is derived from Esmeralda County, Nevada. Syn: *northfieldite; nordfieldite.*

espalier drainage pattern *trellis drainage pattern.*

esperite A mineral: $(Ca,Pb)ZnSiO_4$. Syn: *calcium-larsenite.*

espichellite A lamprophyric rock, similar to camptonite, in which hornblende, augite, olivine, magnetite, and pyrite phenocrysts are contained in a compact groundmass of magnetite, hornblende, augite, mica, and labradorite with orthoclase rims. Analcime may also occur in the groundmass. Espichellite also resembles teschenite but has less analcime and is porphyritic rather than granular. Its name is derived from Cape Espichel, Portugal.

esplanade (a) A term used in the SW U.S. for a rather broad bench or terrace bordering a canyon, esp. in a plateau region. (b) A level stretch of open and grassy or paved ground, often designed for providing a vista.

essential Said of pyroclastics that are formed from magma; it is part of a classification of ejecta based on mode of origin, and is equivalent to the terms *juvenile* and *authigenous.* Cf: *accessory; accidental; cognate.*

essential element An element whose presence is necessary in order for an organism to carry out its life processes.

essential mineral A mineral component of a rock that is necessary to its classification and nomenclature, but that is not necessarily present in large amounts. Cf: *accessory.*

essexite An alkali gabbro primarily composed of plagioclase, hornblende, biotite, and titanaugite, with lesser amounts of alkali feldspar and nepheline. Essexite grades into theralite with a decrease in potassium feldspar and an increase in the feldspathoid minerals. Its name is derived from Essex County, Massachusetts, from where it was originally defined by Sears in 1891 (Johannsen, 1939, p.250).

essonite A yellow-brown, orangy-brown, or reddish-brown transparent gem variety of grossular garnet containing iron. Syn: *hessonite; cinnamon stone; hyacinth; jacinth.*

established dune *anchored dune.*

establishment *ecesis.*

esterellite A porphyritic quartz diorite also containing zoned andesine and hornblende. Its name is derived from Esterel, France.

Estérel twin law A twin law for parallel twins in feldspar with twin axis *a* and composition plane $(0kl)$ parallel to *a.*

estero (a) An estuary or inlet, esp. when marshy. (b) Land adjoining an estuary inundated by the tide.---Etymol: Spanish.

estuarine Pertaining to or formed or living in an estuary; esp. said of deposits and of the sedimentary or biological environment of an estuary. Syn: *estuarial.*

estuarine delta A long, narrow delta that has filled, or is in the process of filling, an estuary; e.g. the delta of the Susquehanna River.

estuarine deposit A sedimentary deposit laid down in the brackish water of an estuary, characterized by fine-grained sediments (chiefly clay and silt) of marine and fluvial origin mixed with a high proportion of decomposed terrestrial organic matter; it is finer-grained and of more uniform composition than a *deltaic deposit.*

estuarine lagoon A lagoon produced by the temporary sealing of a river estuary by a storm barrier. Such lagoons are usually seasonal and exist until the river breaches the barrier; they occur in regions of low or spasmodic rainfall. Syn: *blind estuary.*

estuarine salinity Salinity that varies according to tidal or seasonal conditions, as in an estuary.

estuary (a) The seaward end or the widened funnel-shaped tidal mouth of a river valley where freshwater mixes with and measurably dilutes seawater and where tidal effects are evident; e.g. a *tidal river,* or a partially enclosed coastal body of water where the tide meets the current of a stream. (b) A portion of an ocean, as a *firth* or an arm of the sea, affected by freshwater; e.g. the Baltic Sea. (c) A *drowned river mouth* formed by the subsidence of land near the coast or by the drowning of the lower portion of a nonglaciated valley due to the rise of sea level.---See also: *ria; branching bay; liman; fjord.* Etymol: Latin *aestus,* "tide".

étang A French term for a shallow pool, pond, or lake, esp. one lying among sand dunes, formed by the ponding of inland drainage by beach material thrown up by the sea, and gradually becoming filled with silt, as those along the Mediterranean coast of France (e.g. in Languedoc).

etched pothole *solution pan.*

etch figure The pattern of pitting produced on the face of a crystal by a solvent. It reflects crystal structure.

etching [geomorph] (a) The gradation of the Earth's surface by the slow processes of differential weathering, mass-wasting (esp. creep), sheetwash, and deflation, so that areas underlain by more resistant rocks are brought into relief as the less resistant rocks are lowered (Rich, 1951). (b) A general term for the formation of a landform by erosion or chiseling, as the *etching* of a canyon by a stream.

etchplain A relatively inextensive erosion surface, believed to develop by the comparatively rapid but local differential lowering, during uplift, of a peneplain surface kept at or near base level by the removal of a deep overlying cover of weathered rock. The feature was originally described as an "etched plain" by Wayland (1934).

ethane A colorless, odorless, water-insoluble, gaseous paraffin hydrocarbon, formula C_2H_6, which occurs in natural gas or can be produced as a by-product in the cracking of petroleum.

ethmolith A discordant pluton that is funnel-like in cross section.

etindite A dark-colored extrusive rock intermediate in composition between leucitite and nephelinite, with phenocrysts of augite in a dense groundmass of leucite, nepheline, and augite.

etnaite An alkali olivine basalt (Streckeisen, 1967, p.185).

Etroeungtian *Strunian.*

ettringite A mineral: $Ca_6Al_2(SO_4)_3(OH)_{12}.26H_2O$.

euaster A sponge spicule (microsclere) having the form of a modified aster in which the rays arise from a common center. Cf: *streptaster.*

euautochthony Accumulation of plant remains (such as roots, stumps, tree trunks) which are now found in the exact place, and more or less in the correct relative positions, in which they grew. Cf: *hypautochthony.*

eu-bitumen A collective name for those fluid, viscid, or solid bitumens which are easily soluble in organic solvents. Examples are petroleum, ozokerite, elaterite, and asphalt (Tomkeieff, 1954).

eucairite A silver-white to lead-gray isometric mineral: CuAgSe. Syn: *eukairite.*

euchlorin An emerald-green mineral: $(K,Na)_8Cu_9$-$(SO_4)_{10}(OH)_6$. Also spelled: *euchlorine; euchlorite.*

euchroite An emerald- or leek-green mineral: $Cu_2(AsO_4)$-$(OH).3H_2O$.

euclase A brittle monoclinic mineral: $BeAlSiO_4(OH)$. It occurs in pale tones of blue, green, yellow, or violet, and sometimes colorless; the blue variety is greatly esteemed by gem collectors.

eucolite A variety of *eudialyte* that is optically negative. Syn: *eukolite.*

eucrite [meteorite] An achondritic stony meteorite composed essentially of calcic plagioclase and pigeonite. It has a higher content of iron and calcium than that of howardite. Eucrites were originally regarded to be anorthite-augite meteorites. Syn: *eukrite.*

eucrite [ign] A very basic gabbro composed chiefly of calcic plagioclase (bytownite, anorthite) and augite, with accessory olivine.

eucryptite A colorless or white hexagonal mineral: $LiAlSiO_4$.

eucrystalline Said of the texture of igneous rocks whose mineral grains are large enough to be visible to the unaided eye. Also, said of a rock having such texture. Cf: *dyscrystalline; eudiagnostic.*

eudiagnostic Said of the texture of a rock (esp. an igneous rock) in which all mineral components are of such size and shape as to be identifiable. The term was originally used by Zirkel in German as *eudiagnostich.* Ant: *adiagnostic.* Cf: *eucrystalline.*

eudialyte A pale-pink to brownish-red mineral: $Na_4(Ca, Fe^{+2})_2ZrSi_6O_{17}(OH,Cl)_2$. It is optically positive. Cf: *eucolite.*

eudidymite A white, glassy, monoclinic mineral: $NaBeSi_3O_7(OH)$. It is dimorphous with epididymite.

eudiometer An instrument such as a graduated glass tube for measuring the amounts of different gases in a gas mixture by exploding the gases one at a time by passing an electric spark through the mixture.

eugeiite A yellowish-brown mineral consisting of a hydrous phosphate of ferric iron with a little calcium and aluminum and occurring in small nodules in clay.

eugeogenous Easily weathered; said of a rock that produces by weathering a large amount of detritus. Ant: *dysgeogenous.*

eugeosyncline A geosyncline in which volcanism is associated with clastic sedimentation; the volcanic part of an *orthogeosyncline,* located away from the craton (Stille, 1940). Cf: *miogeosyncline.* Syn: *pliomagmatic zone.* See also: *ensimatic geosyncline.*

eugranitic *granitic.*

euhedral (a) Said of an individual mineral crystal, in an igneous rock, that is completely bounded by its own regularly developed crystal faces and whose growth was not restrained or interfered with by adjacent crystals. (b) Said of a crystal, in a sedimentary rock (such as a calcite crystal in a recrystallized dolomite), characterized by the presence of crystal faces. (c) Said of the shape of a euhedral crystal.----The term was proposed, in reference to igneous-rock components, by Cross et al (1906, p.698) in preference to the synonymous terms *idiomorphic* and *automorphic* (as these were originally defined). Cf: *anhedral; subhedral.*

euhedron A euhedral crystal. Pl: *euhedrons; euhedra.* Cf: *idiomorph.*

eukairite *eucairite.*

eukolite A syn. of *eucolite.* Also spelled: *eukolyte.*

eukrite *eucrite* [meteorite].

euktolite *venanzite.*

eulerhabd A sinuous, stout, U-shaped oxea (sponge spicule); a more sharply curved variety of ophirhabd.

Eulerian (a) Pertaining to a system of coordinates or equations of motion in which the properties of a fluid are assigned to various points in space at each given time, without attempt to identify individual fluid parcels from one time to the next; e.g. a sequence of synoptic charts is an Eulerian representation. (b) Said of a direct method of measuring the speed and/or direction of an ocean current that flows past a geographically fixed point (as an anchored ship) where a current meter is stationed.----Cf: *Lagrangian.* Named in honor of Leonhard Euler (1707-1783), Swiss mathematician.

Euler number Inertial force divided by pressure-gradient force.

Euler's theorem A theorem concerning continental displacement that postulates that if continents have drifted as rigid bodies, their positions prior to drift can be calculated by rotation about a radius through some point on the surface of the Earth.

eulysite A pyroxene peridotite containing manganese-rich fayalite, garnet, and magnetite and having a granular texture.

eulytite A mineral: $Bi_4Si_3O_{12}$. Syn: *eulytine; bismuth blende; agricolite.*

eumalacostracan Any malacostracan belonging to the subclass Eumalacostraca and differing mainly from other malacostracans by the nonbivalve nature of the carapace and the presence of biramous thoracic appendages with a single joint in the protopod; a shrimp-like crustacean. Their stratigraphic range is Middle Devonian to present.

eumorphism A property of an equal-area map projection in which undue distortion of shapes is not shown, as in the arithmetic mean of the sinusoidal and Mollweide projection (BNCG, 1966, p.17 & 47).

eumycete A plant of the subdivision Eumycetes, which comprises the true fungi. Cf: *myxomycete; schizomycete.*

euosmite A brownish-yellow resin, with a low oxygen content and a characteristic pleasant odor, found in brown coal..

eupholite A euphotide that contains talc.

euphotic zone That part of the ocean in which there is sufficient penetration of light to support photosynthesis. The actual depth varies, but averages about 80 m. Its lower boundary is the *compensation depth.* Cf: *disphotic zone; aphotic zone.* Syn: *photic zone.*

euphotide A gabbro in which the feldspar has been saussuritized. The term was originally applied by Haüy as a synonym of gabbro (Johannsen, 1939, p.251). It is obsolete in the U.S.A. but is still used by French petrologists.

eupotamic Said of an aquatic organism adapted to living in

both flowing and still, fresh water. Cf: *autopotamic; tychopo-*
amic.

Eurasian-Melanesian belt The *great-circle belt* that extends
from the Mediterranean, across southern Asis to the Celebes,
where it meets the *circum-Pacific belt.*

euretoid Said of the skeleton of a dictyonine hexactinellid
sponge in which the dictyonal strands occur in more than one
layer and are not parallel to the sponge surface. Cf: *farreoid.*

euripus A strait or narrow channel of water where the tide or
a current flows and reflows with turbulent force. Pl: *euripi.*

eurite (a) A compact, fine-grained, porphyritic igneous rock
that contains quartz phenocrysts. (b) Any fine-grained granitic
rock. ---Adj: *euritic.*

euritic *microgranitic.*

eurybathic Said of a marine organism that tolerates a wide
range of depth. Cf: *stenobathic.*

eurybiontic Said of an organism that can tolerate a wide range
of a particular environmental factor.

euryhaline Said of a marine organism that tolerates a wide
range of salinities. Cf: *stenohaline.*

euryplastic Having great capacity for modification and adaptation to a wide range of environmental conditions; capable of
major evolutionary differentiation. Cf: *stenoplastic.*

euryproct Said of a sponge in which the cloaca is conical with
the widest part forming the osculum.

eurypterid Any merostome, typically brackish or freshwater,
belonging to the subclass Eurypterida, characterized by an e-
longate lanceolate body that is only rarely trilobed. Their stratigraphic range is Ordovician to Permian. Cf: *xiphosuran.*

eurypylous Said of a flagellated chamber of a sponge that has
very large apopyle.

eurysiphonate Said of nautiloids with relatively large siphu-
cles. Cf: *stenosiphonate.*

eurythermal Said of a marine organism that tolerates a wide
range of temperatures. Cf: *stenothermal.*

eurytopic Said of an organism occurring in many different ha-
itats.

eu-sapropel Mature sapropel.

euspondyl Pertaining to the regular arrangement of the primary branches or rays of dasycladacean algae around a cen-
tal axis. Cf: *aspondyl.*

eustacy n. The worldwide sea-level regime and its fluctuations, caused by absolute changes in the quantity of seawater,
e.g. by continental icecap fluctuations. See also: *glacio-eusta-
ism; sedimento-eustatism; diastrophic eustatism.* Adj: *eusta-
ic.* Syn: *eustatism.*

eustatic Pertaining to worldwide changes of sea level that af-
ect all the oceans. Eustatic changes may have various
changes, but those in the last few million years were caused
by additions of water to, or removal of water from, the conti-
ental icecaps. n. eustacy.

eustatic rejuvenation A kind of *rejuvenation* resulting from
causes that produce worldwide lowering of sea level.

eustatism *eustacy.*

eustratite A dense lamprophyric rock having rare phenocrysts
of olivine, corroded hornblende, augite, and possibly oligo-
lase, in a groundmass composed of idiomorphic augite and
magnetite with interstitial feldspar, mica, and colorless glass.

eusynchite *descloizite.*

eutaxic Said of a stratified mineral deposit. Cf: *ataxic.*

eutaxite A *taxite* whose components have aggregated into
separate bands. Cf: *ataxite.*

eutaxitic Said of the banded structure in certain extrusive
rocks, resulting from the parallel arrangement and alternation
of layers of different textures, mineral composition, or color.
Also, said of a rock exhibiting such structure, e.g. *eutaxite.*

eutectic Said of a system consisting of two or more solid
phases and a liquid whose composition can be expressed in
terms of positive quantities of the solid phases, all coexisting
at an (isobarically) invariant point, which is the minimum

melting temperature for that assemblage of solids. Addition or
removal of heat causes an increase or decrease, respectively,
of the proportion of liquid to solid phases, but does not
change the temperature of the system or the composition of
any phases. See also: *eutectoid.*

eutectic point The lowest temperature at which a eutectic
mixture will melt. Syn: *eutectic temperature.*

eutectic ratio The ratio of solid phases forming from the eutectic liquid at the eutectic point; it is such as to yield a gross
composition for the crystal mixture that is identical with that
of the liquid. It is most frequently stated in terms of weight
percent.

eutectic temperature *eutectic point.*

eutectic texture In mineral deposits, the pattern of crystal intergrowth of the different minerals, as of a *eutectic mixture.*
See also: *exsolution texture.*

eutectofelsite *eutectophyre.*

eutectoid The equivalent of *eutectic,* when applied to a system
all of whose participating phases are crystalline.

eutectoid texture *eutectic texture.*

eutectoperthite *mesoperthite.*

eutectophyre A light-colored tuffaceous igneous rock composed of interlocking quartz and orthoclase crystals. Syn: *eutectofelsite.*

eutrophication The process by which waters become more eutrophic; esp. the artificial or natural enrichment of a lake by
an influx of nutrients required for the growth of aquatic plants
such as algae that are vital for fish and animal life.

eutrophic lake A lake that is characterized by an abundance
of dissolved plant nutrients and by a seasonal deficiency of
oxygen in the hypolimnion; its deposits usually have considerable amounts of rapidly decaying organic mud and its water is
frequently shallow. Cf: *oligotrophic lake; mesotrophic lake.*

eutrophic peat Peat containing abundant plant nutrients, such
as nitrogen, potassium, phosphorus, and calcium. Cf: *mesotrophic peat; oligotrophic peat.* Syn: *calcareous peat.*

eutrophy The quality or state of a *eutrophic lake.*

euvitrain Structureless, amorphous *vitrain.* It is the more common type. Cf: *provitrain.* See also: *collain; ulmain.* Syn: *xylovitrain.*

euvitrinite A variety of the maceral *vitrinite* characteristic of
euvitrain and including the varieties ulminite and collinite.
Plant material has been completely jellified and shows no cell
structure. The term *collinite* has been proposed as a preferable synonym. Cf: *provitrinite.*

euxenite A brownish-black mineral: $(Y,Ca,Ce,U,Th)(Nb,Ta,O)_2O_6$. It is isomorphous with polycrase and occurs in granite
pegmatites.

euxinic (a) Pertaining to an environment of restricted circulation and stagnant or anaerobic conditions, such as a fjord or a
nearly isolated or silled basin with toxic bottom waters. Also,
pertaining to the material (such as black organic sediments
and hydrogen-sulfide muds) deposited in such an environment
or basin, and to the process of deposition of such material (as
in the Black Sea). (b) Pertaining to a rock facies that includes black shales and graphitic sediments of various kinds.-
-Etymol: Greek *euxenos,* "hospitable". Cf: *pontic.*

evaluation map A stratigraphic map that summarizes the results of stratigraphic analyses made for economic purposes
(Krumbein & Sloss, 1963, p.484).

evanescent lake A short-lived lake formed after a heavy rain.
Cf: *ephemeral lake.*

evansite A colorless or milky-white to brown or reddish-brown
mineral: $Al_3(PO_4)(OH)_6.6H_2O$ (?). It has a bluish, greenish,
or yellowish tinge. Evansite may contain small amounts of
uranium and thorium.

evapocryst An individual crystal of a primary mineral in an
evaporite (Greensmith, 1957). Cf: *neocryst.*

evapocrystic texture A primary texture of an evaporite in
which no lamination or linearity of evapocrysts is evident.

evapolensic texture A primary, nonporphyritic, roughly laminated texture of an evaporite.

evapoporphyrocrystic texture A texture of an evaporite in which large evapocrysts are embedded in a finer-grained matrix.

evaporates Goldschmidt's name for the evaporite group of sediments; sedimentary salts precipitated from acqueous solutions and concentrated by evaporation. The synonymous term *evaporite* is more commonly used. Cf: *reduzates; oxidates; resistates; hydrolyzates.*

evaporation The process, also called *vaporization*, by which a substance passes from the liquid or solid state to the vapor state. Limited by some to vaporization of a liquid, in contrast to *sublimation*, the direct vaporization of a solid. Also limited by some (e.g. hydrologists) to vaporization that takes place below the boiling point of the liquid. The opposite of *condensation* (Langbein & Iseri, 1960).

evaporation discharge The release of water from the zone of saturation by evaporation from the soil ("soil discharge") or by the transpiration of plants ("vegetal discharge").

evaporimeter *atmometer.*

evaporite A nonclastic sedimentary rock composed primarily of minerals produced from a saline solution that became concentrated by evaporation of the solvent; esp. a deposit of salt precipitated from a restricted or enclosed body of seawater or from the water of a salt lake. Examples include: gypsum, anhydrite, rock salt, chemically precipitated limestone, primary dolomite, and various rare nitrates and borates. The term sometimes includes rocks developed by metamorphism of other evaporites. Syn: *evaporate; saline deposit; saline residue.*

evaporite mineral A mineral precipitated as a result of evaporation; e.g. halite, anhydrite, and aragonite.

evaporite ratio A term used by Krumbein & Sloss (1963, p.463) for the ratio of the thickness or percentage of evaporites (anhydrite, gypsum, salt) to that of carbonates (limestone, dolomite) in a stratigraphic section.

evaporite-solution breccia A term used by Sloss & Laird (1947, p.1422-1423) for a *solution breccia* formed where soluble evaporites (rock salt, anhydrite, gypsum, etc.) have been removed.

evapotranspiration Loss of water from a land area through transpiration of plants and evaporation from the soil. Also, the volume of water lost through evapotranspiration.

eveite A mineral: $Mn_2(AsO_4)(OH)$.

even-crested ridge One of the high, folded ridges, as in the Appalachian Mountains of Pennsylvania, whose tops all rise to an approximately uniform elevation, indicating that a plain reconstructed by filling the surface depressions to the level of the ridgetops is an old peneplain. See also: *summit concordance; accordant summit level.*

even fracture A type of mineral fracture that gives a more or less flat surface.

even-grained *equigranular.*

evening emerald Olivine (peridot or chrysolite) that loses some of its yellow tint by artificial light, appearing more greenish (like an emerald) and used as a gem. Syn: *night emerald.*

evenkite A hydrocarbon mineral: $C_{21}H_{44}$ (= n-tetracosane). It was formerly regarded to be a paraffin wax: $C_{21}H_{42}$.

even-pinnate Said of a pinnately compound leaf having an even number of leaflets. Cf: *odd-pinnate.*

event [paleomag] *polarity event.*

event [seis] *seismic event.*

event [tect] A noncommittal term used for any incident of probable tectonic significane that is suggested by geologic, radiometric, or other evidence, but whose full implications are unknown. It is used especially for minor clusters of radiometric dates whose relations to geologic structures or processes

have not been precisely evaluated. Cf: *phase [tect]; pulsation; disturbance.*

everglade A term used esp. in the southern U.S. for a large expanse of marshy land, e.g. the Florida Everglades.

evergreenite A nordmarkite that contains sulfide ores, such as chalcopyrite and bornite.

Evian water Noneffervescent alkaline mineral water. The term is derived from Evian-les-Bains, a town in southeastern France where it is found.

evjite A hornblende gabbro in which the only light-colored mineral is labradorite or bytownite and in which the hornblende must be of primary origin and not the result of uralitization. Cf: *bojite.*

evolute Loosely coiled or tending to uncoil; e.g. said of a foraminiferal test with nonembracing chambers so that all the whorls are visible, or said of a gastropod shell whose whorls are not in contact, or said of a cephalopod conch with little or no overlapping of the whorls and hence having a wide or broad umbilicus. Cf: *involute; advolute; convolute.*

evolution (a) The development of an organism toward perfect or complete adaptation to environmental conditions to which has been exposed with the passage of time. (b) The theory that life on Earth has developed gradually, from one or a few simple organisms to more complex organisms. Syn: *organic evolution.*

evolutional paleontology The study of fossils from an evolutionary point of view (Challinor, 1967, p.89).

evolutionary plexus A complex lineage of organisms that consists of minor lines that repeatedly divide and reunite. Syn: *plexus.*

evolutionary series In paleontology, a morphologic series which corresponds with time to a significant degree. Cf: *lineage.*

evorsion The formation of potholes in a stream bed by the erosional action of vortices and eddies. Etymol: German(?). Syn: *pothole erosion.*

evorsion hollow A *pothole* in a stream bed.

ewaldite A mineral: $Ba(Ca,Y,Na,K)(CO_3)_2$.

Ewing corer The most commonly used variety of *piston corer.*

exaration The general process of glacial erosion. Grabau (1924, p.263-264) suggested that the term be restricted to "glacial denudation, i.e. the removal and transport of weathered material" by glaciers.

excavation (a) The act or process of removing soil and/or rock materials from one location and transporting them to another. It includes digging, blasting, breaking, loading, and hauling. (b) A pit, cavity, hole, or other uncovered cutting produced by excavation. (c) The material dug out in making a channel or cavity.

excentric Not centrally located; e.g. said of an ammonoid umbilicus characterized by abrupt opening up of the spiral described by the umbilical seam or by tendency to closing of this spiral while the peripheral spiral is relatively unchanged (TIP, 1959, pt.L, p.4).

excess argon Argon-40 trapped in rocks or minerals at the time of their crystallization or formation; i.e. not formed by radioactive decay. Cf: *atmospheric argon; radiogenic argon; inherited argon.*

excess pore pressure Transient pore pressure at any point in an aquitard or aquiclude in excess of the pressure that would exist at that point if steady-flow conditions had been attained throughout the bed (Poland, et al, in press).

excess water *rainfall excess.*

exchange In glaciology, the arithmetic sum of accumulation plus ablation, averaged over a glacier for a balance year or hydrologic year; a measure of the intensity of mass exchange with the atmosphere or hydrosphere. Usually given in millimeters or meters of water equivalent.

exchange capacity The ability of a substance, e.g. a soil or clay, to have *ion exchange.* It is measured by the quantity of exchangeable ions in a given unit of the material.

exchange coefficient *austausch.*

exchange force A quantum-mechanical apparent interaction between electrons that is the cause of *magnetic order.*

excitation potential The characteristic minimum energy required to remove an electron from an atom.

excretion A term proposed by Todd (1903) for a *concretion* that grows progressively inward from the exterior; e.g. a shell of sand cemented by iron oxide and generally filled by unconsolidated sand or containing other shells of cemented sand.

excurrent Said of a plant part that extends beyond the margin or tip, e.g. a midrib developing into a mucro or awn; also, said of a plant having a continuous unbranched axis, e.g. the excurrent habit of spruces and firs (Lawrence, 1951, p.752).

excurrent canal *exhalant canal.*

exfoliation The process by which thin (from less than a centimeter to several meters), concentric shells, slabs, sheets, scales, flakes, or plates of rock are successively broken loose, spalled, peeled, or stripped from the bare outer surface of a larger rock mass; it is caused by the action of physical, thermal, or chemical forces producing differential stresses within an expanding rock, such as by rapid temperature changes in a desert region, or by the release of confining pressure of a once deeply buried rock as it is brought nearer to the surface by erosion (*pressure-release jointing*). It often results in a rounded rock mass or dome-shaped hill. Cf: *spheroidal weathering; spheroidal parting.* Syn: *spalling; scaling; desquamation; sheeting; sheet jointing.*

exfoliation cave A cave formed by the partial destruction of a plate, sheet, or slab of rock produced by exfoliation, having a planar back wall and a continuation of the joint (of exfoliation) up into the roof of the cave (Bradley, 1963, p.525). Examples are found on the Colorado Plateau in SW U.S.

exfoliation dome A large, rounded, dome-shaped form developed in massive, homogeneous, coarse-grained rocks (esp. granite) by exfoliation: well-known examples occur in Yosemite Valley, Calif.

exfoliation joint *sheeting structure.*

exhalant canal Any canal forming part of the exhalant system of a sponge. Syn: *excurrent canal; apochete.*

exhalant system The part of the *aquiferous system* of a sponge between the apopyles and the oscula, characterized by water flowing outward toward the oscula. Cf: *inhalant system.*

exhumation The uncovering or exposure by erosion of a preexisting surface, landscape, or feature that had been buried by later deposition. See also: *resurrected.*

exhumed *resurrected.*

exine The outer, very resistant layer of the two major layers forming the wall (sporoderm) of spores and pollen, consisting of sporopollenin, and situated immediately outside the *intine.* It is divided into two layers (*ektexine* and *endexine*) on the basis of staining characteristics. See also: *perisporium.* Syn: *extine; exospore.*

exinite A hydrogen-rich maceral group including *sporinite, cutinite, alginite,* and *resinite,* and consisting of spores, cuticular matter, resins, and waxes. They are a common component of attrital coal. Cf: *inertinite; vitrinite.* Syn: *liptinite.*

exinoid A maceral group that includes the macerals in the exinite series.

exinonigritite A type of *nigritite* that is derived from spore exines. Cf: *humonigritite; exinonigritite; keronigritite.*

exinous Consisting of exine.

exite A movable lobe on the exterior of the limb of an arthropod, such as a lateral ramus (e.g. exopod) of protopodal limb segments of a crustacean, or the lateral or outer lobe of joint of biramous appendage of a trilobite. Cf: *endite.*

exoadaptation Adaptation of organisms to their external environments. Cf: *endoadaptation.*

exocarp *epicarp.*

exocast *external cast.*

exocoele The space between adjacent pairs of mesenteries of a coral. Cf: *entocoele.*

exocyclic Said of an *irregular* echinoid whose periproct is located outside of the oculogenital ring. Ant: *endocyclic.*

exodiagenesis A term used by Shvetsov (1960) for diagenesis in subaerial environments and shallow stable seas. It is characterized by dehydration, coagulation of colloids, rapid growth of crystals (recrystallization), formation of concretions, and preservation of textural properties of sediments.

exogastric (a) Said of a cephalopod shell that is curved or coiled so that the venter is on or near the outer or convex side or area of whorls. (b) Said of a gastropod shell that is coiled so as to extend forward from the aperture over the front of the extruded head-foot mass, as in the earliest forms (TIP, 1960, pt.I, p.130).---Cf: *endogastric.*

exogene effects The effects of an igneous mass upon the rock it invades (Bateman, 1950). Cf: *endogene effects.*

exogenetic *exogenic.*

exogenic Said of a geologic process, or of its resultant feature or rock, that originates externally to the Earth, e.g. weathering, erosional features, clastic rocks; derived from without. Cf: *endogenic.* Syn: *exogenetic; exogenous; exogene.*

exogenite A little-used term for an epigenetic mineral deposit differing in composition from the enclosing rock.

exogenous *exogenic.*

exogenous dome A volcanic dome that is built by surface effusion of viscous lava, usually from a central vent or crater (Williams, 1932). Cf: *endogenous dome.*

exogenous inclusion *xenolith.*

exogeology *extraterrestrial geology.*

exogeosyncline A parageosyncline accumulating clastic sediments from the uplifted orthogeosynclinal belt adjacent to it but outside the craton (Kay, 1947, p.1289-1293). Syn: *deltageosyncline; foredeep; transverse basin.* Cf: *secondary geosyncline.*

exoglyph A hieroglyph occurring at the bounding surface of a sedimentary bed (Vassoevich, 1953, p.37). Cf: *endoglyph.*

exogyrate Shaped like the shell of *Exogyra* (a genus of Upper Jurassic and Cretaceous bivalve mollusks having a thick shell and spirally twisted beak); i.e. with the left valve strongly convex and its dorsal part coiled in posterior direction, and with the right valve flat and spirally coiled.

exokinematic Said of sedimentary operations in which "the largest displacement vectors occur between matter outside the deposit and the unmodified deposit surrounding the structure produced" (Elliott, 1965, p.196); e.g. types of streamflow. Also, said of the sedimentary structures produced by exokinematic operations. Cf: *endokinematic.*

exokinetic Said of a fissure in rock that is the result of strain in an adjacent rock unit. Cf: *endokinetic.*

exometamorphism *exomorphism.*

exomorphic metamorphism *exomorphism.*

exomorphic zone *aureole.*

exomorphism Changes in country rock produced by the intense heat and other properties of magma or lava in contact with them; *contact metamorphism* in the usual sense. The term was originated by Fournet (1847). Cf: *endomorphism.* Syn: *exometamorphism; exomorphic metamorphism.*

exoolitic *extoolitic.*

exopinacoderm The *pinacoderm* covering the free surface of a sponge. Cf: *endopinacoderm.*

exopod The lateral or external ramus of a limb of a crustacean, arising from the basis. Cf: *endopod.* Syn: *exopodite.*

exopore The external opening in the ektexine of a pollen grain with a complex porate structure. See also: *vestibulum.* Cf: *endospore.*

exopuncta A *puncta* of the external shell surface of a brachiopod, commonly restricted to the primary layer, and never penetrating to the internal surface. Cf: *endopuncta.* Pl: *exopunc-*

tae. Syn: *exopunctum.*

exorheic Said of a basin or region characterized by external drainage; relating to exorheism. Syn: *exoreic.*

exorheism (a) *external drainage.* (b) The condition of a region in which its water reaches the ocean directly or indirectly.--- Ant: *endorheism.* Syn: *exoreism.*

exoseptum A scleractinian-coral septum developed within an exocoele. Cf: *entoseptum.*

exoskeleton An external *skeleton* of an animal, serving as a protective and supportive covering for its softer parts; e.g. the outer shell of a brachiopod or pelecypod, the system of sclerites covering the body of an arthropod, or the bony plates covering an armadillo. Cf: *endoskeleton .* Syn: *dermoskeleton.*

exosphere The upper region of an atmosphere in which the distance between collisions of particles approaches the dimension of the planet's gravitational potential well for a significant fraction of particles, i.e. the region in which some of the particles are escaping the planet's gravity.

exospore (a) A syn. of *exine.* The term is mostly applied to the sporoderm of spores, rather than to pollen. Syn: *exosporium.* (b) One of the asexual spores formed by abstriction from a parent cell (as in certain fungi). (c) One of the spores formed from above downward and one at a time in certain blue-green algae.---Cf: *endospore.*

exostome The outer peristome of a protist.

exotheca A collective term for the dissepiments outside the wall of a scleractinian corallite. Cf: *endotheca.*

exothermic Pertaining to a chemical reaction that occurs with a liberation of heat. Cf: *endothermic.*

exotic [ecol] Said of an organism that has been introduced into a new area from an area where it grew naturally. Ant: *indigenous.*

exotic [struc geol] Applied to a boulder, block, or larger rock body unrelated to the rocks with which it is now associated, and which has been moved far from its place of origin by one of several possible processes. Exotic masses of tectonic origin are also *allochthonous;* those of glacial or ice-rafted origin are generally called *erratics.*

exotic limonite Limonite precipitated in rock that did not formerly contain any iron-bearing sulfide. Cf: *indigenous limonite.*

exotic stream A stream that derives much of its waters from a drainage system in another region; e.g. a stream that has its source in a humid or well-weathered area but that flows across a desert before reaching the sea. Example: the Nile. Cf: *indigenous stream.*

exotomous Characterized by bifurcation in two main crinoid arms that give off branches only on their outer sides. Ant: *endotomous.*

exozone Peripheral region of a bryozoan colony, characterized by thick zooecial walls and often by abundant diaphragms. It is typically present in cryptostomes and trepostomes. Cf: *endozone.* Syn: *mature region.*

expanded foot A broad, bulb-shaped lobe or fan-like mass of ice formed where the lower part of a valley glacier leaves its confining walls and extends onto an adjacent lowland at the foot of a mountain slope. Syn: *piedmont bulb.* Nonpreferred syn: *ice fan; glacier bulb.*

expanded-foot glacier A small *piedmont glacier* consisting of an expanded foot. Syn: *foot glacier; bulb glacier.*

expanding Earth A theory, favored by many geologists, that the diameter of the Earth has grown progressively larger through time, and perhaps by a third or more during recorded geologic time, as a result of changes in atomic and molecular structure in the core and lower mantle, without change in actual mass. The theory has been linked with *continental displacement* and *sea-floor spreading* although these have also been otherwise explained. Arguments and evidence for an expanding Earth were eloquently presented by Holmes (1965, chap. 27). Cf: *contracting Earth.*

expanding-lattice clay A clay mineral whose crystal lattice is

expandable according to the amount of water it takes on; e.g a three-layer clay (such as montmorillonite) in which diffus negative charges originating in the central octahedral sheet result in less tendency for successive layers to be tight bound by cations, thereby causing the layers to be readi pushed apart by adsorbed water.

expanding Universe The interpretation of the universally extra galactic red shift as a Doppler shift in which the observation may be interpreted as indicating that the Universe is expanc ing at a uniform rate.

expansion breccia A breccia formed by increase of volum due to chemical change, as by recrystallization or by hydra tion (Norton, 1917, p.191).

expansion fissure In petrology, one of a system of fissure that radiate irregularly through feldspar and other mineral adjacent to olivine crystals that have been replaced by se pentine. The alteration of olivine to serpentine involves consic erable increase in volume, and the stresses so produced ar relieved by the fissuring of the surrounding minerals. This phe nomenon is common in norite and gabbro (Tyrrell, 1950).

expansion joint *sheeting structure.*

expectation The expected number of statistical occurrences a given observation for a specified number of trials. It is e: pressed by the number of trials times the probability of occu rence of the given observation.

experimental petrology A branch of petrology concerned wit the laboratory study of reactions designed to elucida petrological processes. The term includes experiments dealir with the physical properties or physical chemistry of mineral rocks, rock melts, or of vapors, gases, or solutions coexistir with solid or molten materials (Wyllie, 1966).

experimental structural geology The study of high-pressu deformation of samples of rock; also, the construction of d namic models that illustrate structural processes.

explanation A term used by the U.S. Geological Survey preference to *legend* (except on international maps).

exploding-bomb texture In mineral deposits, a pattern of pyri replacement by copper sulfides in which scattered, residu pyrite fragments are surrounded by the copper minerals. Als spelled: *exploded-bomb texture.*

exploration (a) The search for deposits of useful minerals fossil fuels; *prospecting.* It may include geologic reconnai sance, e.g. photogeology, geophysical and geochemical met ods, and both surface and underground workings. (b) Esta lishing the nature of a known mineral deposit, preparatory *development.* In the sense that exploration goes beyond di covery, it is a broader term than prospecting.

exploratory well A well drilled in unproven territory either search of a new and as yet undiscovered field of oil or gas with the expectation of extending the known limits of a fie already partly developed; e.g. a *test well* and a *wildcat we* Cf: *development well.*

explorer's alidade A lightweight, compact alidade with a lo pillar and a reflecting prism through which the ocular may k viewed from above. Syn: *Gale alidade.*

explosion breccia A type of volcanic breccia that is formed l a volcanic explosion. Syn: *eruption breccia; pyroclastic bre cia.*

explosion caldera A type of *caldera* that is formed by expl sion removal of the upper part of a volcanic cone. It is e tremely rare, and is small in size. Cf: *collapse caldera; er sion caldera.*

explosion cloud *eruption cloud.*

explosion crater (a) A saucer-shaped to conical *crater* pr duced experimentally by detonation of a nuclear device or chemical explosive. (b) A meteorite crater formed by hype velocity impact.

explosion seismology That branch of seismology which utiliz manmade explosions in the study of the interior of the Earth.

explosion tuff A tuff whose pyroclastic fragments are in tl place in which they fell, rather than having been washed into place after they landed.

explosive bomb A *bread-crust bomb* that throws off fragments of its crust due to the tension of continued expansion of its interior after solidification of its crust.

explosive evolution The splitting of a group of organisms into numerous different lines of descent within a relatively short period of geologic time. Syn: *explosive radiation*.

explosive index The percentage of pyroclastics among the total products of a volcanic eruption. Syn: *explosivity index*.

explosive radiation *explosive evolution*.

explosivity index *explosive index*.

exponential distribution A frequency distribution whose ordinate is proportional to the value of the dependent variable and plots as the variable exponent of a constant.

exposed coalfield Deposits of coal that outcrop at the surface, e.g. a coal basin that outcrops at its rim. Cf: *concealed coalfield*.

exposure [geol] (a) A continuous area in which a rock formation or geologic structure is visible ("hammerable"), either naturally or artifically, and is unobscured by soil, vegetation, water, or the works of man. Cf: *outcrop*. (b) The condition or act of being exposed to view at the Earth's surface.

exposure [photo] (a) The total quantity of light received per unit area on a sensitized plate or film, usually expressed as the product of the light intensity and the time during which the light-sensitive material is subjected to the action of light. (b) A loosely used term "generally understood to mean the length of time during which light is allowed to act on a sensitive surface" (Smith, 1968, p.496). (c) The act of exposing a light-sensitive material to a light source. (d) An individual picture of a strip of photographs.

exposure [slopes] The nature and degree of openness of a slope or place to wind, sunlight, weather, oceanic influences, etc. The term is also sometimes regarded as synonymous with *aspect*.

exposure interval The time interval between the taking of successive photographs.

exposure station *camera station*.

exsert adj. Projecting beyond an enclosing part or organ, such as having the ocular plates of an echinoid not in contact with the periproctal margin. Ant: *insert*. Syn: *exserted*.---v. To protrude, stick out, or thrust forth, such as the lophophore of a bryozoan projecting through an orifice in the zooid wall.

exsiccation The drying up of an area due to a change that drives out, or decreases the amount of, available moisture without reducing appreciably the average rainfall; e.g. the draining of a swamp or marsh, or the migration of sand dunes across cultivated ground, or increased evaporation. Cf: *desiccation*.

exsolution The process whereby an initially homogeneous solid solution separates into two (or possibly more) distinct crystalline phases without addition or removal of material to or from the system, i.e., without change in the bulk composition. It generally, though not necessarily, occurs on cooling. Syn: *unmixing*.

exsolutional Pertaining to sediments or sedimentary rocks that solidified from solution either by precipitation or by secretion.

exsolution lamella A *lamella* produced by exsolution, such as one consisting of diopside associated with enstatite augen, or one contained within crystals of tremolite, hornblende, or cummingtonite.

exsolution texture In mineral deposits, a general term for the texture of any mineral aggregate or intergrowth formed by exsolution. See also: *eutectic texture*.

exsudation (a) A kind of *salt weathering* by which rock surfaces are scaled off due to growth of salines by capillary action (Thornbury, 1954, p.39). (b) Var. of *exudation*.

exsurgence *resurgence* [*streams*].

extended consequent stream A consequent stream that flows seaward across a newly emerged coastal plain and that forms an extension of an earlier, larger stream with headwaters in the older land behind the coastal plain.

extended stream A stream lengthened by the extension of its course downstream across newly emerged land (such as a coastal plain, a delta, or a plain of glacial deposition).

extended succession A relatively thick and uninterrupted stratigraphic succession in which the deposits accumulated rapidly. Ant: *condensed succession*.

extended valley A valley that is lengthened downstream by a regression of the sea or by uplift of the coastal area; a valley eroded by or containing an *extended stream*.

extending flow A flow pattern on glaciers in which the velocity increases with distance downstream; thus the longitudinal strain rate (velocity gradient) is extending. This condition requires a transverse or vertical compression or a positive net balance on the surface to maintain the continuity of the ice. Ant: *compressing flow*.

extensiform Said of a graptoloid (such as Didymograpti) with two stipes that are horizontal.

extension [photo] Execution of additional control points (horizontal or vertical) from existing control points, as by *phototriangulation* from a controlled area to an area without control.

extension [streams] The lengthening of a stream by headward erosion and the multiplication of tributaries, or by regression of the sea or uplift of the coastal area.

extensional fault *tension fault*.

extension fault A term used by Norris (1954) for a fault in which there has been a tension-type displacement in the plane of bedding. It is partially synonymous with *tension fault*.

extension fracture A fracture that develops perpendicular to the direction of greatest stress; and parallel to the direction of compression; a *tension fracture*. See also: *extension joint*; *tension crack*.

extension joint A joint that forms parallel to the direction of compression; a joint that is an *extension fracture*.

extension ore The *possible ore* ahead of or beyond an exposure in a mine. Cf: *probable ore*.

extension test *triaxial extension test*.

extensive quantity A thermodynamic quantity such as volume or mass that depends on the total quantity of matter in the system.

extensometer An instrument used for measuring small deformations, deflection or displacement, as in tests of stress.

exterior orientation The *orientation* of a photograph (or of the camera that took the photograph) at the instant of exposure with respect to the position of the camera station.

extermination The local or even regional disappearance of a species, that still exists elsewhere, as a result of changing environmental conditions, disease, competitors, or other adverse conditions. Cf: *extinction*.

external cast An improper term sometimes used as a syn. of *external mold*. Syn: *exocast*.

external drainage Drainage whereby the water reaches the ocean directly or indirectly. Ant: *internal drainage*. Syn: *exorheism*.

external furrow One of the shallow, linear, and axial depressions or grooves on the outer surface of a fusulinid test, formed at the point of union between successive chambers and corresponding in position to a septum, and dividing the outer surface into melon-like lobes. Syn: *septal furrow*.

external lobe The *ventral lobe* in normally coiled cephalopod conchs.

external magnetic field The relatively small and varying portion of the natural *magnetic field* near the Earth's surface which is due to electric currents in the upper atmosphere.

external mold A *mold* or impression in the surrounding earth or rock, showing the surface form and markings of the outer hard parts of a fossil shell or other organic structure; also, the surrounding rock material whose surface receives the external mold. Cf: *internal mold; external cast*. Syn: *Incrustation* [*paleont*].

external rotation In structural petrology, rotation of the bulk of the material that changes its external form, regardless of what *internal rotation* there may or may not be; rotational strain.

external structure The morphology or form of a rock mass, including its size and shape, the nature of its boundaries (sharp or gradational, conformable or unconformable), and the types of folds and faults it exhibits. Ant: *internal structure*.

external suture The part of a suture of a coiled cephalopod conch exposed on the outside of whorls between the umbilical seams. Cf: *internal suture*.

externides The external, or outer part of an orogenic belt, nearest to the craton or foreland, commonly the site of a miogeosyncline during its early phases and subjected to marginal deformation (folding and lateral thrusting) during the orogenic phase. Cf: *internides*. See also: *secondary orogeny*. Syn: *secondary arc*.

extinction [*optics*] The more or less complete darkness obtained in a birefringent mineral at two positions during a complete rotation of a section between crossed nicols. Also, a mineral section will remain in darkness for a complete rotation if the line of sight is parallel to the optic axis. See also: *extinction angle; extinction direction; inclined extinction; parallel extinction; undulatory extinction*.

extinction [*evol*] The total disappearance of an organism so that it no longer exists anywhere. Cf: *extermination*.

extinction [*lake*] The drying up of a lake by the loss of water (*temporary extinction*) or by destruction of the lake basin (*permanent extinction*).

extinction angle The angle through which a section of a birefringent mineral must be rotated from a known crystallographic plane or direction to the position at which it gives maximum *extinction* or darkness under a polarizing microscope. The extinction angle can be diagnostic in the identification of a particular crystal.

extinction coefficient In oceanography, a measure of the attenuation of downward radiation in the sea (U.S. Naval Oceanographic Office, 1966).

extinction direction One of the two positions at which a section of a birefringent crystal shows *extinction* between crossed nicols.

extinct lake (a) A lake that has lost all of its water, either temporarily or permanently. (b) A lake whose surface expanse of water in its basin no longer exists; e.g. a lake whose open water has been replaced by vegetation and whose status has reached that of a bog, marsh, or swamp.----Syn: *dead lake*.

extinct volcano A volcano that is not presently erupting and that is not considered likely to do so in the future. Cf: *active volcano; dormant volcano; inactive volcano*.

extine Var. of *exine*. The term is not in good or current usage in palynology.

extoolitic Pertaining to oolitic structures or grains that have formed or grown outward by deposition of material around a core or center, as in the formation of a small concretion. Ant: *entoolitic*. Syn: *exoolitic*.

extraclast A fragment of calcareous sedimentary material, produced by erosion of an older rock outside the area in which it accumulated; a component of calclithite. Cf: *intraclast*.

extractable metal A metal that is susceptible to beneficiation from a particular ore.

extraction In mining, (a) the removal of ore from the Earth, and (b) the removal or separation of the ore from the total material that is mined.

extragalactic nebula *galaxy*.

extraglacial Said of glacial deposits formed by meltwater beyond the farthest limit of the ice, or of glacial phenomena displayed in an area never covered by ice. Ant: *intraglacial*.

extralateral rights The *apex law*.

extramorainal Said of deposits and phenomena occurring outside the area occupied by a glacier and its lateral and end moraines. Ant: *intramorainal*. Syn: *extramorainic*.

extraneous ash *Ash* in coal that is derived from inorganic material introduced during formation of the seam, such as sedimentary particles, or filling cracks in the coal. Cf: *inherent ash*. Syn: *secondary ash; sedimentary ash*.

extraordinary ray *E ray*.

extrapolation Estimation of the value of a variate based on at least two known values on only one side of the unknown value. It is used to extend a curve by determining points on it beyond those for which data are available.

extra river A diamond (*river*) of the very highest grade.

extratentacular budding Formation of new scleractinian coral polyps by invagination of the edge zone or coenosarc outside of the ring of tentacles surrounding the mouth of the parent. Cf: *intratentacular budding*.

extraterrestrial Existing, occurring, or originating beyond the Earth or its atmosphere; e.g. "extraterrestrial radiation" of solar radiation received "on top of" the Earth's atmosphere.

extraterrestrial geology A science that applied geologic principles and techniques to the study of celestial bodies other than the Earth; *astrogeology*. Syn: *exogeology*.

extraumbilical aperture An aperture in the final chamber of foraminiferal test not connecting with the umbilicus. It is commonly sutural midway between the umbilicus and the periphery. See also: *interiomarginal aperture*.

extraumbilical-umbilical aperture An aperture in the final chamber of a foraminiferal test that extends along its forward margin from the umbilicus toward the periphery, thus reaching a point outside the umbilicus (as in *Globorotalia*).

extravasation The eruption of molten or liquid material onto the surface of the Earth, e.g. lava, or water from a geyser.

extreme desert An area without a regular season of rainfall and where 12 consecutive months without rainfall have been recorded.

extremely coarsely crystalline Descriptive of an interlocking texture of a carbonate sedimentary rock having crystals whose diameters exceed 4 mm (Folk, 1959).

extremely finely crystalline *aphanocrystalline*.

extrinsic ionic conduction Electrical conduction arising in transport of charge through a solid as a result of movement of ions or vacancies through a crystal lattice. The ions involved are impurity ions. It can take place at room temperature. Cf: *intrinsic ionic conduction*.

extrinsic semiconduction Electronic conduction in a solid facilitated by energy bands due to impurities. Takes place at room temperatures. Cf: *intrinsic semiconduction*.

extrusion In volcanology, the igneous process of emitting lava and other ejectamenta onto the Earth's surface; also, the rock so formed.

extrusion flow A discredited hypothesis for a type of glacier flow in which the pressure of overlying ice is supposed to force the basal part of the glacier to flow faster than the upper part. *Gravity flow*, originally contrasted with "extrusion flow", is now considered to account for all glacier flow.

extrusive adj. Said of igneous rock that has been ejected onto the surface of the Earth. Extrusive rocks include lava flows and detrital material such as volcanic ash.----n. An extrusive rock.----Cf: *intrusive*. Syn: *effusive; volcanic; eruptive*.

exudation basin A spoon-shaped depression on the ice surface at the head of an *outlet glacier*. Examples are found on the Greenland and Antarctic ice sheets.

exudation vein *segregated vein*.

exuviation The removal of the theca of a dinoflagellate, either plate by plate or as small groups of plates.

eye [*meteorol*] In meteorology, the approximately circular area of relatively light wind and good weather in the center of tropical-type cyclone. Such an area may be 4-40 miles in diameter.

eye [paleont] The ring-like part of a *hook* of a holothurian, sometimes partly closed by a bar. Also, a ring-like end of a *rod* of a holothurian.

eye [grd wat] The opening from which the water of a spring flows out onto the land surface.

eye agate Agate displaying concentric bands (usually of various alternating colors) about a dark center (suggesting an eye). Syn: *Aleppo stone.*

eye-and-eyebrow structure Said of the texture of certain rhyolites having crescent-shaped bodies of quartz with pea-like bodies of quartz on the concave side; the convex side is toward the top of the flow.

eye base A term used in photogrammetry for the distance between the pupils of the eyes of an observer. Syn: *interocular distance.*

eyebrow scarp A fault scarp that crosses a piedmont alluvial fan near its apex and that seldom maintains the dip of the fault surface in the unconsolidated gravels of the fan (Davis, 1927, p.62).

eye coal Coal that contains structural disks in circular or elliptical shapes, either parallel or normal to the bedding, with concentric, bending rims and radiating striae. They reflect light in a mirrorlike way. Syn: *circular coal.* Etymol: German *Augenkohle*, "eye coal".

eyed structure *augen structure.*

eye line A raised line or narrow band extending from the forward and inner part of a trilobite eye to the anterior part of the glabella.

eyepiece The lens (or lenses) in a microscope or telescope through which the image formed by the *objective* is viewed. Syn: *ocular.*

eyestalk One of the movable peduncles in a decapod crustacean, carrying the eye at its distal extremity. Syn: *ophthalmite.*

eye tubercle A polished, transparent, and rounded protuberance in the anterior and dorsal region of an ostracode valve, forming the lens of the eye. See also: *tubercle.*

eyot Var. of *ait.*

ezcurrite A triclinic mineral: $Na_4B_{10}O_{17}.7H_2O$. It is dimorphous with nasinite.

F

fabianite A monoclinic mineral: $CaB_3O_5(OH)$.

fabric [struc petrol] The sum of all the structural and textural features of a rock. The term incorporates the notion of function or behavior (the correlative physical properties) as well as of form (arrangement of structural and textural components). The term was originally defined by Sander (1930) who used the German word Gefuge. The term "orientation" is sometimes used synonymously although in a strict sense it refers only to the aspect of form. See also: *subfabric*. Syn: *rock fabric; petrofabric; structural fabric*.

fabric [ign] A term suggested by Cross et al (1902, p.614) as one of the component factors of the texture of a crystalline (esp. igneous) rock, representing the external appearance of pattern produced by the shapes and orientation of the crystalline and noncrystalline parts of the rock. It is dependent on the relative sizes and shapes of the parts and their positions with respect to one another and to the glassy groundmass (when present).

fabric [sed] The orientation (or lack of it) in space of the elements (discrete particles, crystals, cement) of which a sedimentary rock is composed. Cf: *packing*. See also: *crystallization fabric*.

fabric [soil] The physical nature of a soil according to the spatial arrangement of its particles and voids.

fabric analysis *structural petrology*.

fabric axis One of three orthoponal axes used in structural petrology as reference in the orientation of fabric elements, and in the description folding and of the movement symmetry of deformed rocks. See also:*tectonic axis; a axis [struc petrol]; b axis [struc petrol]; c axis [struc petrol]; pi axis; f axis*. Syn: *reference axis*.

fabric diagram In structural petrology, a graphic representation of the data of fabric elements, e.g. a stereogram, equal-area net; an *orientation diagram*. See also: *point diagram; contour diagram*. Syn: *petrofabric diagram*.

fabric domain A three-dimensional area or volume, defined by boundaries such as structural or compositional discontinuities, within which the rock fabric is uniform. Syn: *domain*.

fabric element In a rock fabric, a surface or line of structural discontinuity. It may be a crystallographic feature such as a lattice plane, or noncrystallographic, such as foliation. A spatial array of a particular fabric element is called a *subfabric*.

face [geomorph] The principal side or surface of a landform; esp. *rock face*.

face [cryst] *crystal face*.

face [mining] Any surface on which mining operations are in progress. Syn: *breast; highwall*.

face [struc petrol] A term used in fabric analysis for a natural or artificial exposure plane.

face [struc geol] n. A term used by Shrock (1948, p.17-18) for the original top or upper surface of a layer of rock, esp. if it has been raised to a vertical or steeply inclined position. Syn: *facing*.----v. To be directed toward or to present an aspect of. Sedimentary beds are said "to face" in the direction of the stratigraphic top of the succession (or to be directed toward the younger rocks or to the side that was originally upward), so that an overturned bed facing to the east may have a dip of 45 degrees to the west. Folds are said "to face" in the direction of the stratigraphically younger rocks along their axial surfaces and normal to their axes (Shackleton, 1958, p.363); this coincides with the direction toward which the beds face at the hinge (a normal upright fold faces upward, an overturned anticline faces downward, and an asymmetric fold faces its steeper flank). Faults are said "to face" in the direction of the structurally lower unit. Syn: *young*.

face-centered lattice A type of *centered lattice* in which each unit cell has a lattice point at the center of each face as we as those at each corner, i.e. it has four lattice points per ur cell. It is found in crystals of the cubic and orthorhombic sy tems. Syn: *F-centered lattice; all face centered lattice*.

face cleat The major *cleat* system or jointing in a coal sear Cf: *end cleat*. See also: *face of coal*.

facellite *kaliophilite*.

face of coal The plane or surface of the *face cleat* in a co seam. Cf: *end of coal*.

face pole In crystallography, a point on a projection surfac that represents the intersection of a crystal *pole* with the cry tal face.

facet [geog] Any part of an ancient or present-day intersec ing surface forming a unit of geographic study; e.g. a flat or slope.

facet [geomorph] (a) A small, nearly plane surface produce on a rock fragment by abrasion, as by wind or by the grindir action of a glacier; e.g. a rough, flat face produced by stream that differentially removed material from the upstrea side of a boulder or pebble, and inclined at an angle of 50 de grees or less to the direction of the impinging current (Ma son, 1940, p.721). (b) Any plane surface produced by erosic or faulting, and intersecting a general slope of the land; e.g. *triangular facet*.

facet [gem] One of the small, plane, polished surfaces on gemstone, such as on a brilliant-cut diamond.

facet [paleont] (a) A nearly flat surface on an echinoder plate, serving for articulation with contiguous skeletal el ment; e.g. *brachiolar facet*. (b) A small, circular or oval pr tuberance set within a channel of a cyclocystoid and co nected with the ambulacral system.

faceted boulder A boulder that has been ground flat on one more sides by the action of natural agents, such as by glaci ice (examples of such boulders are commonly found in Gre Britain and Scandinavia) or by streams. Cf: *faceted pebble*.

faceted pebble A pebble on which facets have been deve oped by natural agents, such as by wave erosion on a beac or by the grinding action of a glacier; specif. a *windkanter*. C *faceted boulder*.

faceted spur A spur or ridge with an inverted-V face that wa produced by faulting or by the trimming, beveling, or trunca ing action of streams, waves, or glaciers. See also: *truncate spur*.

facial (a) Pertaining to facies. The term is used by some no English European geologists. (b) Pertaining to an outer su face.

facial suture One of the two symmetric sutures that opens the time of molting of a trilobite, extending from the anteri margin of the cephalon around the eye and outward or bacl ward to the lateral or posterior margin. It separates the fre cheek from the fixed cheek.

faciation Part of an ecologic association, usually a large con munity, which is characterized by the dominance of two more but not all of the more abundant organisms.

facieology The study of facies.

facies [geol] (a) A term whose basic and original meanir signifies an aspect, appearance, or expression of somethir having two or more groups of attributes in different portion e.g. any observable attribute or attributes (such as overall a pearance, composition, or conditions of formation) of one pa of a rock as contrasted with another or several other parts the same rock, and the changes that may occur in these a tributes over a geographic area. Facies has no significanc "except as it is in contrast to one or more other related fa cies", and therefore implies variation, comparison, and diffe entiation, and also a certain degree of constancy and contin ity within whatever rock mass is being compared with or di tinguished from others (Weller, 1958, p.609-610). The ter was introduced into geologic literature by Steno (1669, p.6 75) for the entire aspect of a part of the Earth's surface du

...g a certain interval of geologic time, and was first used systematically in stratigraphy. (b) A relative term that implies relations in time and space" and that is not restricted as to "kind, size, propinquity, succession, or time" although it can be qualified "to bear on any of these in a given instance" (Eaton, 1951, p.79). (c) A term used to refer to a distinguished part or parts of a single geologic entity, differing from other parts in some general aspect; e.g. any two or more significantly different parts of a recognized body of rock or stratigraphic unit, distinguished from other parts of the same rock or unit by appearance or composition. The term implies physical closeness and genetic relation or connection between the parts.----Etymol: Latin (and French), "face, form, aspect, appearance, look, condition, figure, shape". Pron. *fay*-sheez or *ay*-seez in North America. Pl: *facies*. The term is "inherently clear and general but, in geologic use, nearly always somewhat restricted and made to carry certain (but varying) implications", and attempts to define it "consistently and permanently...are doomed to failure as the term itself refuses to bring to mind any particular specialized meaning" (Challinor, 1967, p.91-92). Usages and definitions of "facies" have been reviewed by Moore (1949), Weller (1958), and Teichert (1958).

facies [ecol] (a) A local assemblage of animals or plants or both, constituting part of the larger fauna, flora, or biota of a region; an association of living or fossil forms. The term is used esp. in marine ecology for an association of animals or plants that live under certain types of bottom conditions; e.g., the term "sandy facies" refers to an association of organisms that live on a sandy sea floor, not to the sandy character of the sediment on the floor. In plant ecology, the term signifies "the developmental unit of associes characterized ...by the grouping of dominants" (Weaver & Clements, 1938, p.101). The term "facies" as used in ecology is conceptionally different from its usage in stratigraphy: an ecologic facies may be devoid of sediments, and a stratigraphic facies may be without organic remains. (b) A term used by Hesse (1924) for variants of biotopes (or of the habitats or physical bases of a biologic association). (c) The general appearance of a fauna or flora; a particular local aspect or change of an ecologic community. Syn: *ecologic facies*.

facies [petrology] (a) *mineral facies*. (b) *metamorphic facies*. (c) *igneous facies*. (d) *tectonic facies*.

facies [stratig] (a) The sum of all primary lithologic and paleontologic characteristics exhibited by a sedimentary rock and from which its origin and environment of formation may be inferred; the general aspect, nature, or appearance of a sedimentary rock produced under or affected by similar conditions; a distinctive group of characteristics that differs from other groups within a stratigraphic unit. The term is used here in an abstract, descriptive sense for something (such as the "composite character") that a rock has, rather than to designate a particular kind of rock (Teichert, 1958, p.2739). This usage corresponds closely to the original concept of stratigraphic facies first recognized and defined by Gressly (1838, p.10-12, 20-25) who, in his studies of Jurassic sedimentary rocks in the eastern Jura Mountains near Solothurn (Switzerland), used the term "facies" to designate the sum total of the very distinct lateral changes in the interdependent lithologic and paleontologic characteristics of a definite stratigraphic unit (restricted to sedimentary rocks), these characteristics being quite different from those of other facies of the same unit. Later, the concept was broadened (esp. in Europe) in a stratigraphically unconfined sense to include vertical changes, so that the term referred to the sum of the lithologic and paleontologic characters of a sedimentary deposit at a given place (Haug, 1907, p.145). (b) An exclusive, mappable, and areally restricted part of a defined stratigraphic rock body, such as a stratum or group of strata differing in lithologic character or fossil contents from other beds deposited at the same time and in physical continuity; a lateral subdivision of a specified stratigraphic unit; a *lithofacies*. It may be a formally recognized part of a formation, such as the facies provided with geographic names by Stockdale (1939). Facies of this sort are applied in rock-stratigraphic nomenclature to distinguish lateral variants within the regional distribution of a major mappable stratigraphic unit. See also: *sedimentary facies*. (c) A term applied to intertonguing sedimentary rock masses of differing lithologic and paleontologic characteristics, occurring within a stratigraphic unit, having irregular lateral boundaries, and exhibiting upper and lower limits that do not necessarily correspond with the boundaries of the stratigraphic unit that contains them or with the upper and lower limits of other facies. It may occur singly or repeatedly in a vertical section. (d) A rock or group of rocks distinguished from other more or less related or comparable rocks of different appearance or composition, and identified by any observable feature (except those produced during weathering, metamorphism, or structural disturbance, but sometimes including secondary features resulting from diagenesis); specif. any actual body of sediment or rock of some particular kind or combination of kinds, irrespective of age, form, local geologic relations, and geographic or stratigraphic occurrence (although broadly corresponding to a certain type of environment or common mode of origin), such as "red-bed facies", "black-shale facies", "limestone facies", and any other informally designated facies that includes all similar rocks, wherever found, that may appear at any time. (e) A term applied to the composition or actual nature of a rock body; e.g. "the particular composition or condition of a formation in a given region" (Williams, 1895, p.69), such as "argillaceous facies" or "calcareous facies" as distinguished from "arenaceous facies". (f) A term applied (esp. in Europe) to the environment in which a rock was found or to the environment recorded by a rock body, and defined according to assumed or generalized environmental conditions or situations; e.g. "eolian facies", "volcanic facies", "marine facies", and "glacial facies", and facies of geographic-climatic provinces ("tropical facies", "boreal facies") and of depth zonation ("littoral facies", "bathyal facies"). The term has also been used to designate the totality of local geographic and biologic conditions that determine the lithologic and paleontologic nature of the sedimentary rock. See also: *environmental facies*. (g) A term used in a broad, even vague, sense for paleogeographically or geotectonically defined belts of rock units or sequences; e.g. "geosynclinal facies", "foredeep facies", "orogenic facies", "transgressive facies", and "Appalachian facies". The term has also been used to refer to the characteristics of large belts or basins of sedimentation, such as the "nappe facies" of Alpine geology. See also: *tectofacies*. (h) A generic term that includes all restricted or specialized types or kinds of facies used in stratigraphy.--- See also: *operational facies; stratigraphic facies; petrographic facies*.

facies change A lateral or vertical variation in the lithologic or paleontologic characteristics of contemporaneous sedimentary deposits. It is caused by, or reflects, a change in the depositional environment. Cf: *facies evolution*.

facies contour The trace (on a map) of a vertical surface that cuts a three-dimensional rock body into facies segments; a line indicating equivalence in lithofacies development.

facies-departure map A *facies map* based on the degree of similarity to some particular sedimentary rock composition (optimum facies or single component end member). Term suggested by Forgotson (1960, p.94) to replace *distance-function map*.

facies evolution A gradual change of facies over a period of time, indicating gradually changing depositional conditions (Teichert, 1958, p.2723). Cf: *facies change*.

facies family A term used by Teichert (1958, p.2737) for several genetically interconnected *facies tracts*; e.g. coral-atoll deposits and desert deposits. See also: *facies suite*.

facies fauna A group of animals characteristic of a given facies or adapted to life in a restricted environment; e.g. the black-shale fauna of the Middle and Upper Devonian of the Appalachian region of U.S.

facies fossil A fossil, usually a single species or a genus, that is restricted to a defined facies or is adapted to life in a restricted environment. It prefers certain ecologic surroundings and will exist in them from place to place with little change for long periods of time.

facies map A broad term for a stratigraphic map showing the gross areal variation or distribution (in total or relative content) of observable attributes or aspects of different rock types occurring within a designated stratigraphic unit, without regard to the position or thickness of individual beds in the vertical succession; specif. a *lithofacies map*. Conventional facies maps are prepared by drawing lines of equal magnitude through a field of numbers representing the observed values of the measured rock attributes. Cf: *vertical-variability map*. See also: *biofacies map; isofacies map; isolith map; percentage map; ratio map; entropy map; entropy-ratio map; facies-departure map*.

facies plane A term used by Caster (1934, p.19 & 24) for the boundary between magnafacies (or between parvafacies), although it is usually never sharp enough to be recognizable as a plane in any section. Cf: *plane of contemporaneity*.

facies sequence A term used by Teichert (1958, p.2723) for a vertical succession of different (heteropic) facies formed at different times.

facies strike The compass direction of a facies contour at a given point on a map.

facies suite (a) A term suggested by S.S. Oriel (in Teichert, 1958, p.2737) for several genetically interconnected *facies families*; e.g. all marine deposits or all continental deposits. (b) A collection or group of rocks that shows variations within a single rock mass.

facies tract A system of different, but genetically interconnected sedimentary facies of the same age (Teichert, 1958, p.2723); e.g. the outer-slope deposits of a coral atoll, or dry channel deposits. It includes the areas of erosion from which the sediments of these facies are derived, so that an erosional interval represents part of a facies tract. The concept was developed by Walther (1893-1894). See also: *facies family*. Syn: *macrofacies*.

facing [struc geol] (a) *face*. (b) The direction toward which the face is presently facing. It is usually specified geographically (e.g. east or west) or with reference to a major structure (e.g. updip or downdip). Cf: *regard; vergence*. Syn: *younging*.

faciostratotype A supplemental stratotype designated for local reference or reference to different facies, and which distinguishes different ecologic conditions that existed during the time of the time-stratigraphic unit represented by the stratotype (Sigal, 1964).

facsimile crystallization *mimetic crystallization*.

factor analysis A statistical method for identifying the minimum number of influences necessary to account for the maximum observed variation in a set of data and for indicating the extent to which each influence accounts for the variance observed in the data.

facultative Said of an organism capable of growth under a number of specific conditions; adaptable to alternate environments. Cf: *obligate*.

fadama A term used in western Africa for a flood plain in a wide, flattish river valley, subject to annual inundation and characterized by savanna vegetation (J.C. Pugh in Stamp, 1961, p.187). Etymol: Hausa.

faecal pellet Var. of *fecal pellet*.

faheyite A white hexagonal mineral: $(Mn,Mg)Fe_2^+{}^3Be_2(PO_4)_4.6H_2O$.

fahlband Bands of sulfide impregnation in metamorphic rocks Etymol: German *Fahlband*, "light-colored band", so named because weathering darkens the country rock relative to the mineralized area.

fahlerz A syn. of *fahlore*. Etymol: German *Fahlerz*, "pale ore".

fahlore Any gray-colored ore mineral consisting essentially of sulfantimonides or sulfarsenides of copper; specif. *tetrahedrite* and *tennantite*. Syn: *fahlerz*.

fahlunite An altered form of cordierite.

faikes Var. of *fakes*.

failure Fracture or rupture of a rock or other material that has been stressed beyond its *ultimate strength*. Syn: *rock failure*.

fairchildite A mineral: $K_2Ca(CO_3)_2$. It is found in fused wood ash in partly burned trees. Cf: *buetschliite*.

fairfieldite A white, pale-yellow, or greenish-white triclinic mineral: $Ca_2(Mn,Fe)(PO_4)_2.2H_2O$. It is isomorphous with collinsite.

fairway (a) The main navigable channel (usually buoyed) of a river or bay through which vessels enter or leave a port or harbor; the part of a waterway that is kept open and unobstructed for navigation. (b) *midway*.

fair-weather runoff *base runoff*.

fairy-castle structure Intricate microtopography of the lunar surface, believed to be responsible for the Moon's optical properties. Its existence is doubtful.

fairy chimney A translation of the French term *cheminée de fée*, used in the French Alps for *earth pillar*.

fairy stone (a) A cruciform-twinned crystal of staurolite, used as a *curio stone* without fashioning for adornment. The term is also applied as a syn. of *staurolite*, and esp. to the variety occurring in the form of a twinned crystal. (b) Any of various odd or fantastically shaped calcareous or ferruginous concretions formed in alluvial clays. (c) A fossil sea urchin. (d) A stone arrowhead.

fakes A British vernacular term for a platy rock, such as a fissile sandy shale, a shaly sandstone, a cross-bedded or well-laminated siltstone, or a flaggy sandstone or micaceous flagstone. Adj: *fakey*. See also: *flaikes*. Syn: *faikes*.

falaise An old and low sea cliff, on an emergent coast, that re-establishes contact with the open sea; the type example is the coast of Normandy. Etymol: French, "cliff".

falcate Hooked or curved like a sickle; e.g. said of a sickle-shaped cephalopod rib. Syn: *falciform*.

falcifer Said of brachiopod crura that arise on the dorsal side of hinge plates and project into the brachial valve as broad blade-like processes.

falcon's-eye *hawk's-eye*.

fall [mass move] (a) A very rapid downward movement of a mass of rock or earth that travels mostly through the air by free fall, leaping, bounding, or rolling, with little or no interaction between one moving unit and another; e.g.: *rockfall; debris fall*. (b) The mass of material moved by a fall.

fall [meteorite] One or more meteorites that are picked up immediately after they have reached the Earth's surface and for which information (place and time of fall) are known. Cf: *find*.

fall [streams] A *waterfall* or other precipitous descent of water. The plural "falls" is commonly used in place names, esp. where applied to a series of waterfalls.

fall [slopes] (a) The descent of land or of a hill; a slope or declivity. (b) The distance to which a stream or physiographic feature slopes.

fallback Fragmental material ejected from an impact or explosion crater during formation and redeposited within, and partly filling, the true crater almost immediately after formation. It includes slide-block deposits, talus material, and aerially transported dust. Cf: *fallout [crater]; throwout*.

fallback breccia An allochthonous breccia composed of fallback.

fall diameter The diameter of a sphere that has a specific gravity of 2.65 and the same uniform settling velocity as that of a given particle having any specific gravity "when each is allowed to settle alone in quiescent distilled water of infinite extent and at a temperature of 24°C" (Simons et al, 1961, p.vii).

fallen snow *deposited snow.*

falling dune An accumulation of sand that is formed as sand is blown off a mesa top or over a cliff face or steep slope, forming a solid wall sloping at the angle of repose of dry sand or a fan extending downward from a reentrant in the mesa wall.

falling star *shooting star.*

falling tide That part of a tide cycle between high water and the following low water, characterized by seaward or receding movement of water. Also, an outgoing tidal river. Ant: *rising tide.* Syn: *ebb tide.*

fall line An imaginary line or narrow zone connecting the waterfalls on several successive and near-parallel rivers and marking the points where these rivers make a sudden descent from the upland to the lowland, as at the edge of a plateau; specif. the Fall Line marking the boundary between the ancient and resistant crystalline rocks of the Piedmont Plateau and the younger and softer sediments of the Atlantic Coastal Plain in the eastern U.S. It also marks the limit of navigability of the rivers. Syn: *fall zone.*

fallout [crater] Fragmental material ejected from an impact or explosion crater during formation and eventually redeposited in and around the crater. It may have undergone considerable atmospheric sorting before deposition. Cf: *throwout; fallback.*

fallout [radioactivity] The descent of usually radioactive particles through the Earth's atmosphere, following a nuclear explosion; also, the particles themselves.

fallout breccia An allochthonous breccia composed of fallout from a crater. It is generally one of the last ejecta units to be deposited, and it characteristically contains small amounts of glass fragments and a limited range of fragment sizes.

fall overturn A seasonal *overturning* in lakes during the fall and winter seasons, when the cool weather creates uniform temperature and density throughout the lake.

fall velocity *settling velocity.*

fall zone *fall line.*

false beach A bar above water level, located a short distance offshore (Veatch & Humphrys, 1966, p.111).

false bedding An old term for *cross-bedding* and *current bedding*, or bedding affected by currents that were often erratic and frequently changed directions. The term is not recommended in this sense because it might refer to *pseudobedding* Hills, 1963, p.10). See also: *drift bedding.*

false body *thixotropic clay.*

false bottom [eco geol] An apparent bedrock (underlying an alluvial deposit) that conceals another alluvial deposit lying beneath it; e.g. beds of clay or sand cemented by iron, on which gold placer deposits accumulate, and under which there is an alluvial deposit resting directly upon bedrock.

false bottom [oceanog] *deep scattering layer.*

false bottom [lake] The poorly defined bottom of a lake occurring where a firm bottom grades upward to a suspended or soupy mass of muck, colloidal sludge, soft marl, or organic matter, through which a weight easily sinks; e.g. in a bog lake.

false cleavage A quarryman's term for minor cleavage in a rock, e.g. slip cleavage, to distinguish it from the dominant or *true cleavage*. Geologically, the term is misleading and should be avoided.

false diamond A colorless mineral (such as zircon, white sapphire, white topaz, and quartz) that makes a brilliant gemstone when cut and polished.

false dip *apparent dip.*

false drumlin *rock drumlin.*

false esker A term introduced by Woodworth (1894b, p.216) for a feature resembling an esker but "composed of till instead of water-laid drift".

false floor In a cave, a more or less horizontal layer of flowstone that has open space beneath it.

false folding Folding that is not generically related to lateral compression, e.g. shear folding, supratenuous folding. Ant: *true folding.* Syn: *bend folding.*

false form *pseudomorph* [cryst].

false galena *sphalerite.*

false gossan A displaced rather than indigenous iron oxide zone. It is so named because of its confusion with indigenous iron oxide of a *gossan*, which is weathered from underlying sulfide deposits.

false horizon (a) *artificial horizon.* (b) A line resembling the apparent horizon, but situated above or below it.

false lapis (a) *lazulite.* (b) Agate or jasper artificially dyed blue.

false mud crack A sedimentary structure resembling a mud crack, such as a polygonal pattern formed in soil or a fucoidal network. See also: *pseudo mud crack.*

false oolith *pseudo-oolith.*

false origin An arbitrary reference point chosen to the south and west of the true origin of a grid system and from which grid distances are measured eastward and northward to insure that all points have positive coordinate values.

false shoreline The line of contact between the open water of a lake and the front or edge of a floating mat of vegetation built out from the true shore (Veatch & Humphrys, 1966, p.111).

false stratification An old term for *cross-stratification.* It was used by Lyell (1838, p.38) for the diagonal arrangement of "minor layers placed obliquely to the general planes of stratification". The term is rarely used today. Cf: *pseudostratification.*

false stream An accumulation of water in a hollow along the side of a flood plain that slopes away from the main stream toward the side of the valley (Swayne, 1956, p.58).

false topaz (a) A yellow transparent variety of quartz resembling the color of *topaz*; specif. *citrine.* (b) A yellow variety of fluorite.

falun A French term for *shell marl* composed of an unconsolidated accumulation of sand-sized shell fragments.

famatinite A gray to copper-red orthorhombic mineral: Cu_3SbS_4. It is isomorphous with enargite.

Famennian European stage: uppermost Devonian (above Frasnian, below Tournaisian of Carboniferous).

family [ecol] An ecologic community composed of only one kind of organism, usually occupying a small area and representing an early stage in a succession.

family [taxon] A group of organisms, either plant or animal, that contains several related genera or sometimes a single distinctive genus. As a unit in taxonomy, it ranks above "genus" and below "order". In zoology, the name of a family characteristically has the ending -idae; e.g. Cytheridae.

family [petrology] (a) The basic unit of the *clan.* (b) *clan.*

fan [geomorph] (a) A gently sloping, fan-shaped mass of detritus forming a section of a very low cone commonly at a place where there is a notable decrease in gradient; specif. an *alluvial fan.* (b) A fan-shaped mass of congealed lava that formed on a steep slope by the continually changing direction of effusions.

fan [marine geol] *submarine fan.*

fan apron *bajada.*

fan bay The head of an alluvial fan that extends a considerable distance into a mountain valley. Cf: *fanhead.*

fan cleavage A type of *axial-plane cleavage* in which the cleavage planes fan out at small to large angles on each flank of the axial planes of folds. Syn: *cleavage fan.*

fan coral Any coral that forms flat, fan-like colonies.

fancy cut Any style of diamond cutting other than the round brilliant cut or single cut. It includes, among others, the marquise, emerald cut, pear-shape cut, baguette, and half moon.

Syn: *modern cut.*

fancy diamond Any diamond with a natural body color strong enough to be attractive, rather than off-color. Red, blue, and green are very rare; orange and violet, rare; strong yellow, yellowish-green, brown, and black stones are more common.

fancy sapphire A *sapphire* of any hue other than blue or colorless, although colorless is sometimes included.

fan delta (a) A gently sloping alluvial deposit produced where a mountain stream flows out onto a lowland. (b) *delta fan.---* Cf: *fan-shaped delta.*

fan fold A fold, both limbs of which are overturned.

fanglomerate A sedimentary rock consisting of slightly water-worn, heterogeneous fragments of all sizes, originally deposited in an alluvial fan and subsequently cemented into a firm rock, and characterized by a considerable persistence parallel to the depositional strike but by a rapid downdip thinning. The term was proposed by Lawson (1913, p.329) for the coarser, consolidated rock material occurring in the upper part of an alluvial fan. Cf: *bajada breccia.*

fanhead The area on an alluvial fan close to its apex. Cf: *fan bay.*

fanhead trench A linear depression formed by a drainage line that is incised considerably below the surface of an alluvial fan.

fan mesa An alluvial-fan remnant left standing after dissection of the fan.

fan scarp A *piedmont scarp* formed by faulting, occurring entirely within alluvium and not observed to cross bedrock in any part of its course (Billings, 1954, p.156).

fan-shaped delta An *arcuate delta.* Cf: *fan delta.*

fan shooting A refraction type of seismic shooting in which a fan of detectors is laid out from a single shot point. It is used to detect the presence of shallow salt domes.

fan structure The fold structure of a *normal* anticlinorium.

fan system A fan-shaped pattern formed by diverging trabeculae in the plane of septum of a scleractinian coral. See also: *axis of divergence.*

fan-topped pediment A pediment with a thin (15-60 m) covering of alluvial fans built upon it in response to some minor change of climate or of other controlling conditions (Blackwelder, 1931, p.139). Cf: *concealed pediment.*

fan valley A valley in a submarine fan; it is a continuation of the submarine canyon. It is either V-shaped or trough-shaped, and has natural levees and distributaries.

faradaic path One of the two available paths for transfer of energy across an electrolyte-metal interface. Energy is carried in it by conversion of atom to ion or vice versa, due to electrochemical reaction and ion diffusion. Cf: *nonfaradaic path.*

faratsihite A pale-yellow clay mineral: $(Al,Fe)_2Si_2O_5(OH)_4$. It has been regarded as an iron-bearing variety of kaolinite, a mixture of kaolinite and nontronite, and identical with nontronite.

farinaceous Pertaining to a texture or structure of a rock or sediment that is mealy, soft, and friable, e.g. a limestone or a pelagic ooze; also, said of a lava flow in which its particles seem to be in a state of mutual repulsion. Syn: *mealy.*

far infrared Pertaining to the longer wavelengths of the infrared region, beyond 25 micrometers to 1 mm. The longer end adjoins the microwave region around 1 millimeter. This is severely limited in terrestrial use, as the atmosphere transmits very little radiation beyond 25 micrometers until the millimeter regions. Cf: *near infrared.*

faro A small, atoll-shaped or oblong reef with a lagoon up to 30 m deep, forming part of the rim of a barrier reef or of an atoll. Etymol: political subdivision of an atollon in the Maldive Islands of the Indian Ocean.

farreoid Said of the skeleton of a dictyonine hexactinellid sponge in which the dictyonal strands occur in a single layer parallel to the sponge surface. Cf: *euretoid.*

farringtonite A colorless, wax-white, or yellow meteorite mineral: $Mg_3(PO_4)_2$.

farrisite A fine-grained hypabyssal rock composed primarily [of] a melilite-like mineral (e.g. diopside, composing approximate[ly] one third of the rock) and barkevikite, with smaller amoun[ts] of lepidomelane, olivine, and magnetite.

farsundite A hypersthene-hornblende- bearing granite. C[f: *o]paldite.*

fascicular Said of an aggregate of *acicular* crystals.

fascicular columella A *columella* in scleractinian coral[s] formed by twisted vertical ribbons or rods resembling pali [or] paliform lobes.

fasciculate Arranged in fascicles; e.g. said of a corallum wi[th] cylindrical corallites that are somewhat separated from on[e] another but may be joined by connecting tubules, or descri[p]tive of ribbing in coiled nautiloid conchs having bunched [or] bundled ribs near the umbilical margin.

fasciole (a) A heavily ciliated tract in an echinoderm; esp. [a] narrow band of small tubercles bearing densely ciliated spine[s] (clavulae) on the denuded test of certain echinoids. The ter[m] is also applied to a narrow band of such spines in which cil[ia] beat to create currents. (b) A band generated on a gastropo[d] shell by a narrow sinus or notch in, or a lamellose projecti[on] of, successive growth lines (TIP, 1960, pt.I, p.131).---Se[e] also: *anal fasciole.*

fashioning The sawing, cleaving, rounding up, facet-grindin[g] and polishing, and other operations employed in preparin[g] rough gem material for use in jewelry.

fasibitikite A medium-colored riebeckite-acmite granite th[at] also contains eucolite and zircon.

fasinite A coarse-grained plutonic rock that contains augi[te] and nepheline as its main components along with soda m[i]crocline, olivine, and biotite. It has the same chemical com[position as *berondrite* and differs from *bekinkinite* by the a[b]sence of hornblende and analcime.

fassaite A pale- to dark-green variety of augite containin[g] considerable calcium: $(Ca,Mg,Fe^{+2},Al)(Si,Al,Fe^{+3})O_3$. Als[o] spelled: *fassaite.*

fast ice Any sea ice that forms along and remains attached [to] the coast (e.g. icefoot), an ice front, an ice wall, or a groun[d]ed iceberg, or fastened to the bottom in shallow water (e.[g. anchor ice). Fast ice may form in situ from seawater or b[y] freezing of pack ice to the shore. It may extend a few mete[rs] to several hundred kilometers from the coast.

fast-ice belt *icefoot.*

fastigate Said of an ammonoid with a roof-shaped venter, th[e] periphery of the shell being sharpened but not keeled.

fastland A *mainland*; esp. one that is high and dry near wate[r] such as an *upland.*

fast ray In crystal optics, that component of light in any bir[e]fringent crystal section which travels with the greater veloci[ty] and has the lower index of refraction. Cf: *slow ray.*

fat clay A cohesive and compressible clay of high plastici[ty] and containing a high proportion of minerals that make it gre[a]sy to the feel. It is difficult to work when damp, but stron[g] when dry. Ant: *lean clay.* Syn: *long clay.*

fat coal *long-flame coal.*

fathogram The graphic record produced by a *fathometer*; [a] type of *echogram.*

fathometer A copyrighted name for a type of *echo sounde[r].* See also: *fathogram.*

fatigue Failure of a material after many repetitions of a stre[ss] that of itself is not strong enough to cause failure.

fatigue limit *endurance limit.*

fatigue ratio *endurance ratio.*

fatty A syn. of *greasy* mineral luster.

fatty acid Any one of a group of organic acids that occur [in] animal and vegetable oils and fats. Common examples a[re] butyric acid ($C_4H_8O_2$); palmitic acid ($C_{16}H_{32}O_2$); stearic ac[id] ($C_{18}H_{36}O_2$); and oleic acid ($C_{18}H_{34}O_2$).

fauces terrae A term used in international law for headlan[ds]

nd promontories that enclose territorial water that is not part
the high seas. Etymol: Latin, "gulf".

ujasite A cubic zeolite mineral: $(Na_2,Ca)Al_2Si_4O_{12}.6H_2O$. Cf:
melinite.

ult [cryst] (a) A general term for a dislocation in a crystal.
) A *stacking fault*.

ult [struc geol] A surface or zone of rock fracture along
hich there has been displacement, from a few centimeters in
a few kilometers in scale. See also: *faulting*. Obsolete syn:
araclase.

ult-angle valley The original or subsequently modified valley
ong the junction made by a fault angle. Syn: *tilt-block basin*.

ult apron A mass of rock waste deposited along the base of
fault scarp, formed by numerous coalescing alluvial cones.

ult basin A depression separated from the surrounding area
faults.

ult bench A small *fault terrace*.

ult block A crustal unit formed by *block faulting*; it is bound-
d by faults, either completely or in part. It behaves as a unit
uring block faulting and tectonic activity. An example is the
erra Nevada of California. See also: *tilt block*. Syn: *block
ect]*.

ult-block mountains *block mountains*.

ult breccia (a) A *tectonic breccia* composed of angular
agments resulting from the crushing, shattering, or shearing
rocks during movement on a fault, from friction between
e walls of the fault, or from distributive ruptures associated
ith a major fault; a *friction breccia*. It is distinguished by its
oss-cutting relations, by the presence of *fault gouge*, and by
ackened and slickensided blocks. Syn: *dislocation breccia*.
) A term sometimes used as a syn. of *fault rubble*.

ult cliff A cliff formed by faulting; esp. a *fault scarp*.

ult closure In petroleum geology, the bounding of a reservoir
ck by a fault surface. Cf: *closure*.

ult coast A coast formed directly by faulting (see Cotton,
16), as one along a fault line, a fault scarp, a fault-line
arp, or a narrow arm of the sea that floods a fault trough
.g. the coast of the Red Sea). See also: *fault shoreline*.

ult complex A group of faults that interconnect and inter-
ct, having the same or different ages.

ult-dam spring *fault spring*.

ult embayment A fault trough, or other depressed region in
fault zone or between two faults, invaded by the sea; e.g.
e Red Sea.

ult escarpment *fault scarp*.

ult fissure A fissure that is the result of faulting. It may or
ay not be filled with vein material.

ult-fold A structure that is associated with a combination of
lding and nearly vertical faulting, in which crustal material
at has been fractured into elongate strips tends to drape
ver the uplifted areas to resemble anticlines, and to crumple
to the downthrown areas to resemble synclines. The struc-
re has been described in parts of Germany (Hills, 1963).

ult gap A depression between the offset ends of a ridge,
rmed by a transverse fault that laterally displaces the ridge
that the two parts are no longer continuous (Lahee, 1961,
356). Cf: *fault-line gap*.

ult gouge Soft, uncemented, pulverized, clayey or clay-like
aterial, commonly a mixture of minerals in finely divided
rm, found along some faults or between the walls of a fault
d filling or partly filling a fault zone; a slippery mud that
ats the fault surface or cements the *fault breccia*. It is
rmed by the crushing and grinding of rock material as the
ult developed as well as by subsequent decomposition and
teration caused by underground circulating solutions. Syn:
uge; *clay gouge*; *selvage*.

ult growth Intermittent, small-scale movement along a fault
rface that, accumulated, results in considerable displace-
ent.

ulting The process of fracturing and displacement that pro-

duces a *fault*.

fault ledge *fault scarp*.

fault-line adj. Said of a secondary or subsequent landform or
feature created solely by processes (such as erosion) acting
upon faulted materials; e.g. a "fault-line outlier", an isolated
hillock or ridge capped by resistant rock and created by differential erosion of a low-angle thrust mass (Sharp, 1954,
p.27).

fault line The trace of a fault plane with the surface or with a
horizontal plane. Syn: *fault trace*.

fault-line gap A gap produced solely by erosion of a resistant
ridge laterally offset by earlier faulting; e.g. such a gap located along the line of outcrop of a dip fault or of a diagonal fault
that intersects the rock layer of the ridge (Lahee, 1961,
p.367). Cf: *fault gap*.

fault-line saddle A saddle created by rapid erosion of a ridge
crest where it is crossed by a fault (Sharp, 1954, p.27). Examples occur along the San Gabriel Fault within the San Gabriel Range in southern California. Cf: *fault saddle*.

fault-line scarp (a) A steep slope or cliff formed by differential
erosion along a fault line, as by the more rapid erosion of soft
rock on one side of a fault as compared to that of more resistant rock on the other side; e.g. the east face of the Sierra
Nevada in California. See also: *obsequent fault-line scarp*; *resequent fault-line scarp*. Syn: *erosion fault scarp*. (b) A *fault
scarp* that has been modified by erosion. This usage is not
recommended because the scarp is usually not located on the
fault line (Washburne, 1943, p.496).

fault-line-scarp shoreline A shoreline formed by the partial
submergence of a fault-line scarp so that the waters of a sea
or lake rest against the scarp.

fault-line valley A valley that is formed along or follows a fault
line; e.g. a subsequent valley developed by headward erosion
in the soft, crushed, relatively weak material along a fault
zone. Cf: *fault valley*.

fault-line-valley shoreline A shoreline formed by the partial
submergence of a valley that has been eroded along the
crushed zone of a fault or along a narrow strip of faulted
weak rock, such as along the coast of northern Nova Scotia.

fault plane A *fault surface* that is more or less planar.

fault rubble An assemblage of detached, jumbled, and
crushed or shattered angular fragments torn from the walls of
a fault; an unconsolidated *fault breccia*.

fault saddle A particular type of kerncol, being a notch, col, or
saddle in a ridge, created by actual displacement of the ridge
crest by faulting (Sharp, 1954, p.25). Cf: *fault-line saddle*.

fault sag A small, enclosed depression along an active or recent fault. It is caused by differential movement between
slices and blocks within the fault zone or by warping and tilting associated with differential displacement along the fault,
and it forms the site of a *sag pond*. Term introduced by Lawson et al (1908, p.33). Syn: *sag* [struc geol].

fault scarp (a) A steep slope or cliff formed directly by movement along one side of a fault and representing the exposed
surface of the fault before modification by erosion and weathering. It is an initial landform. Cf: *fault-line scarp*. Syn: *fault
escarpment*; *fault cliff*; *fault ledge*; *cliff of displacement*. (b) A
term used in England for any scarp that is due to the presence of a fault, even though the relief may be erosional.

fault-scarp shoreline A *fault shoreline* produced by recent
faulting, characterized by deep water near the shore and no
continental shelf.

fault-scarp shoreline A *fault shoreline* produced by recent
faulting and characterized by deep water near the shore and
often by no continental shelf.

fault set A group of faults that are parallel or nearly so, and
that are related to a particular deformational episode. Cf: *fault
system*.

fault-slice ridge *slice ridge*.

fault splinter A narrow, ramplike connection between the opposite ends of two parallel, normal faults. The feature occurs in major fault zones (Strahler, 1963, p.596). Cf: *fault step.*

fault spring A spring flowing onto the land surface from a fault that brings a permeable bed into contact with an impermeable bed. Cf: *fracture spring; fissure spring.* Syn: *fault-dam spring.*

fault step Along a normal fault expressed at the surface, one of a series of thin rock slices along which the fault's total displacement is dispersed (Strahler, 1963, p.596).

fault striae *striations* [fault].

fault surface In a fault, the surface along which displacement has occurred. Partial syn: fault plane.

fault system (a) Two or more interconnecting *fault sets.* (b) A syn. of *fault set.*

fault terrace An irregular, terrace-like tract between two fault scarps, produced on a hillside by step faulting in which the downthrow is systematically on the same side of two approximately parallel faults. Cf: *fault bench.*

fault trace *fault line.*

fault trap The sealing of a porous and permeable reservoir formation by the abutment of an impermeable bed juxtaposed by fault action.

fault-trellis drainage pattern A *trellis drainage pattern* developed where a series of parallel faults have brought together alternating bands of hard and soft rocks (Thornbury, 1954, p.123).

fault trench A cleft or crack formed on the Earth's surface. It is a smaller-scale feature than a fault trough (rift valley).

fault trough *rift valley.*

fault-trough lake *sag pond.*

fault valley A linear depression produced by faulting; e.g. a small, narrow valley created within a major fault zone by relative depression of narrow slices, or a large graben situated between tilted block mountains, or a valley created by relative uplift on opposite sides of two parallel thrust faults. Cf: *fault trough; fault-line valley.*

fault wall *wall* [fault].

fault wedge A wedge-shaped rock mass bounded by two faults.

fault zone A fault that is expressed as a zone of numerous small fractures or of breccia or fault gouge. A fault zone may be as wide as hundreds of meters. Cf: *step fault.* Syn: *distributed fault; distributive fault.* Less-preferred syn: *shatter belt.*

fauna (a) The entire animal population, living or fossil, of a given area, environment, formation, or time span. Cf: *flora.* (b) Sometimes broadly used to include both the animal and plant fossils of a particular rock unit. Adj: *faunal.*

faunal break An abrupt change or *break* from one fossil assemblage to another at a definite horizon in a stratigraphic sequence, usually produced by an unconformity or hiatus or sometimes by a change in bottom ecology without interruption of deposition; e.g. a gap in the orderly evolution of a single organism through a vertical series of beds.

faunal diversity The number of species, either estimated or counted, whose combined totals comprise 95% of the total population.

faunal dominance That percentage of a population constituted by the most common species.

faunal evolution Change in the composition of a fauna with time.

faunal stage A time-stratigraphic unit (stage) based on a faunizone.

faunal succession The observed chronologic sequence of life forms (esp. animals) through geologic time. See also: *law of faunal succession.*

faunal zone *faunizone.*

faunichron A term used by Dunbar & Rodgers (1957, p.300) for the geologic-time unit corresponding to faunizone of Buckman (1902).

faunizone (a) A biostratigraphic unit or body of strata characterized by a particular assemblage of fossils (specif. fossil

faunas), regardless of whether it is inferred to have time or only environmental significance. (b) A term sometimes used for the strata equivalent in age to a certain overlap of "bio zones" and having dominantly time-stratigraphic significance. (c) A term that has been used in the sense of "zone" regarded as a time unit of biochronologic significance.---The term was introduced by Buckman (1902) for "belts of strata, each of which is characterized by an assemblage of organic remains", and has been generally regarded as the animal-based variety of (biostratigraphic) *zone* of Oppel (1856-1858). The American Commission on Stratigraphic Nomenclature (1961, art.21d) states that the term is "not generally accepted" and that its correct definition is "in dispute". See also:*assemblage zone.* Cf: *florizone.* Syn: *faunal zone.*

faunula (a) A set of animal species found in a relatively small and poorly isolated region, and not peculiar to it. (b) *Faunule.*

faunule A term used by Fenton & Fenton (1928) for a diminutive fauna or an assemblage of fossil animals associated in single stratum or a few contiguous strata of limited thickness and dominated by the representatives of one community; the faunal assemblage of a *zonule.* Cf: *florule.* Syn: *faunula.*

faustite An apple-green mineral: $(ZnCu)Al_6(PO_4)_4(OH)_8 \cdot 5H_2O$. It is the zinc analogue of turquoise.

favositid Any tabulate belonging to the family Favositidae, characterized by massive colonies (usually without coenenchyme) of slender corallites with mural pores, short, equal spinose septa, and complete tabulae. Their stratigraphic range is Upper Ordovician to Permian, possibly Triassic also.

***f* axis** In structural petrology, the *fabric axis* of rotation around which a gliding plane may be bent, normal to the *t direction.*

fayalite A brown to black mineral of the olivine group: Fe_2SiO_4. It is isomorphous with forsterite, and occurs chiefly in igneous rocks. Symbol: Fa. Syn: *iron olivine.*

F center A type of *color center* in a crystal that is formed by negative ion vacancy with two bound electrons. The F stands for Farbe, the German word for "color".

F-centered lattice *face-centered lattice.*

F-coal Microscopic coal particles that are predominantly fusain, as found in miners' lungs. Cf: *V-coal; D-coal.*

F-distribution test *F test.*

feather [gem] (a) A series of elongated and irregular liquid inclusions in a gemstone, grouped together in orderly proximity to each other so as to resemble the overall pattern of a bird feather. (b) Any flaw inside a gemstone, such as a jagged fracture that is white in appearance.

feather alum (a) *halotrichite.* (b) *alunogen.*

feather amphibolite A metamorphic rock in which porphyroblastic crystals of amphibole (usually hornblende) tend to form stallate or sheaflike groups on the planes of foliation or schistosity. Cf: *amphibolite.* Syn: *Garbenschiefer.*

feather edge The very thin, fine, sharp edge of a bed or rock (usually sedimentary) where it thins, pinches, or wedges out to extinction. Also spelled: *featheredge.* Syn: *knife edge.*

featheredging *feathering.*

feather fracture A less-preferred syn. of *plume structure.* Although it was the original term, its use would lead to confusion with the term feather jointing.

feather ice *pipkrake.*

feathering [cart] The technique of progressively dropping contour lines (to avoid congestion on steep slopes) and tapering the line weight near the end of the contour line to be dropped. Syn: *featheredging.*

feathering [photo] The thinning or tearing of overlapping edges of aerial photographs (prior to assembly into a mosaic) in order to make a better matching and to reduce or eliminate shadows and sharp changes in contrast. Syn: *featheredging.*

feather jointing A joint pattern formed in a fault zone by shear and tension. The joints appear to the fault as the barbs of feather to its shaft. Syn: *pinnate jointing.*

feather ore A capillary, fibrous, or feathery form of an anti mony-sulfide mineral such as stibnite and boulangerite; specif. *jamesonite.*

feather out To end irregularly. The term is applied to lenticular bodies of rock.

feather quartz Quartz in imperfect crystals, the bases of which meet at an angle along a crystal plane so that a cross section looks somewhat like a feather.

feather zeolite *hair zeolite.*

fecal pellet An organic excrement, mainly of invertebrates, occurring esp. in modern marine deposits but also fossilized in some sedimentary rocks, usually of a simple ovoid form less than a millimeter in length or more rarely rod-shaped with longitudinal or transverse sculpturing, devoid of internal structure, and smaller than a *coprolite.* Syn: *faecal pellet.*

fedorite A mineral: $(Na,K)CaSi_4(O,OH)_{10} \cdot 1.5H_2O$.

Fedorov stage *universal stage.*

feedback Partial reversal of a certain process to its beginning or to a preceding stage as a means of reinforcement or modification, esp. in biologic, psychologic, and social systems.

feeder [eco geol] *channelway.*

feeder [intrus rocks] The conduit through which magma passes from the magma chamber to the localized intrusion. It may be a dike.

feeder [volc] A conduit through which magma reaches the Earth's surface; a partial syn. of *vent.* Cf: *chimney [volc].*

feeder [streams] *tributary.*

feeder beach An artificially widened beach serving to nourish downdrift beaches by natural littoral currents.

feeder current The part of a rip current that flows parallel to the shore (inside the breakers) before converging with other feeder currents to form the *neck* of the rip current.

feeding channel *channelway.*

feeding esker A small esker joining a larger esker.

feeding ground *drainage basin.*

feidj A term used in the Saharan region for a sand-covered interdune passage. Cf: *gassi.* Other spellings: *feidsh; feij; fejj.*

feitknechtite A mineral: β-MnO(OH). Cf: *manganite [mineral].*

fei ts'ui An emerald- or bluish-green variety of jadeite from Burma, esp. one resembling the color of the brilliant blue-green back of the kingfisher. Etymol: Chinese, "kingfisher jade".

felder Crustal blocks of a polygonal, mosaic pattern that are produced by taphrogeny. Syn: *tesserae.*

feldmark *fell-field.*

feldspar (a) A group of abundant rock-forming minerals of general formula: $MAl(Al,Si)_3O_8$, where M = K, Na, Ca, Ba, Rb, Sr, and Fe. Feldspars are the most widespread of any mineral group and constitute 60% of the Earth's crust; they occur as components of all kinds of rocks (crystalline schists, migmatites, gneisses, granites, most magmatic rocks) and as fissure minerals in clefts and druse minerals in cavities. Feldspars are usually white or nearly white and clear and translucent (they have no color of their own but are frequently colored by impurities), have a hardness of 6 on Mohs' scale, frequently display twinning, exhibit monoclinic or triclinic symmetry, and possess good cleavage in two directions (intersecting at 90° as in orthoclase and at about 86° as in plagioclase). On decomposition, feldspars yield a large part of the clay of soil and also the mineral kaolinite. (b) A mineral of the feldspar group, such as alkali feldspar (orthoclase, microcline), plagioclase (albite, anorthite), and celsian.---Syn: *felspar; feldspath.*

feldsparphyric Said of a rock that contains feldspar phenocrysts.

feldspath *feldspar.*

feldspathic Said of a mineral aggregate containing feldspar.

feldspathic arenite A sandstone containing abundant quartz, chert, and quartzite, less than 10% argillaceous matrix, and 10-25% feldspar (generally fresh and limpid), and characterized by an abundance of unstable materials in which the feld-

spar grains exceed the fine-grained rock fragments (Gilbert, 1954, p.294 & 316). It is less feldspathic and more mature than *arkosic arenite.* The rock is roughly equivalent to *subarkose.* See also: *feldspathic sandstone.*

feldspathic graywacke (a) A graywacke characterized by abundant unstable materials; specif. a sandstone containing a variable content (generally less than 75%) of quartz and chert and 15-75% detrital clay matrix, and having feldspar grains (chiefly sodic plagioclase, and indicating a plutonic provenance) in greater abundance than rock fragments (indicating a supracrustal provenance) (Pettijohn, 1954; 1957, p.303). Gilbert (1954, p.294) gives a feldspar content of 10-25% and an argillaceous matrix greater than 10%; it is less feldspathic than *arkosic graywacke.* The rock is equivalent to *high-rank graywacke* of Krynine (1945). (b) A term used by Folk (1954, p.354) for a sandstone containing 25-90% micas and metamorphic rock fragments, 10-50% feldspars and igneous rock fragments, and 0-65% quartz and chert. Cf: *impure arkose.* (c) A term used by Hubert (1960, p.176-177) for a sandstone containing 25-90% micas and micaceous metamorphic rock fragments, 10-50% feldspars and feldspathic crystalline rock fragments, and 0-65% quartz, chert, and metamorphic quartzite. Cf: *micaceous arkose.*---Cf: *lithic graywacke.*

feldspathic litharenite (a) A term used by McBride (1963, p.667) for a litharenite containing appreciable feldspar; specif. a sandstone containing 10-50% feldspar, 25-90% fine-grained rock fragments, and 0-65% quartz, quartzite, and chert. (b) A term used by Folk (1968, p.124) for a sandstone containing less than 75% quartz and metamorphic quartzite and having a "F/R ratio" between 1:1 and 1:3, where "F" signifies feldspars and fragments of gneiss and granite, and "R" signifies all other fine-grained rock fragments.---Cf: *lithic arkose.*

feldspathic quartzite (a) A term used by Hubert (1960, p.176-177) for a sandstone containing 70-95% quartz, chert, and metamorphic quartzite, 5-15% feldspars and feldspathic crystalline rock fragments, and 0-15% micas and micaceous metamorphic rock fragments. Cf: *micaceous quartzite.* (b) A term used by Pettijohn (1949, p.227) for a well-indurated *feldspathic sandstone,* and later by Pettijohn (1954, p.364) as a syn. of *subarkose.*

feldspathic sandstone A feldspar-rich sandstone; specif. a sandstone intermediate in composition between an *arkosic sandstone* and a quartz sandstone, containing 10-25% feldspar and less than 20% matrix material of clay, sericite, and chlorite (Pettijohn, 1949, p.227). Petttijohn (1957, p.322) redefined the term and used it as a less-preferred syn. of *subarkose.* Krumbein & Sloss (1963, p.170) used the term for a quartzose sandstone with 10-25% feldspar (mainly potassic feldspar), and Gilbert (1954, p.316) used it as a general term to include *feldspathic arenite* and feldspathic wacke. See also: *feldspathic quartzite; arkose.*

feldspathic shale A usually well-laminated shale characterized by a feldspar content greater than 10% in the silt size and by a finer matrix of kaolinitic clay minerals, commonly associated with arkose, and representing the removal of finer material from coarser arkosic debris (Krumbein & Sloss, 1963, p.175). Syn: *kaolinitic shale.*

feldspathic subgraywacke A term used by Folk (1954, p.354) for a sandstone composed of subequal amounts of rock fragments of igneous and metamorphic derivation; specif. a sandstone containing 10-25% feldspars and igneous rock fragments, 10-25% micas and metamorphic rock fragments, and 50-80% quartz and chert.

feldspathic sublitharenite *lithic subarkose.*

feldspathic wacke A sandstone containing abundant quartz, chert, and quartzite, more than 10% argillaceous matrix, and 10-25% feldspar (esp. sodic plagioclase), and characterized by an abundance of unstable materials in which the feldspar grains exceed the fine-grained rock fragments (Gilbert, 1954, p.292 & 316). It is less feldspathic and more mature than *ar-*

kosic wacke.

feldspathide *feldspathoid.*

feldspathization The formation of feldspar in a rock usually as a result of metamorphism leading toward granitization. Material for the feldspar may come from the country rock or may be introduced by magmatic or other solutions (in the liquid or gaseous state).

feldspathoid (a) A group of comparatively rare rock-forming minerals consisting of aluminosilicates of sodium, potassium, or calcium and having too little silica to form feldspar. Feldspathoids are chemically related to the feldspars, but differ from them in crystal form and physical properties; they take the places of feldspars in igneous rocks that are undersaturated with respect to silica or that contain more alkalies and aluminum than can be accommodated in the feldspars. Feldspathoids may be found in the same rock with feldspars but never with quartz or in the presence of free magmatic silica. See also: *foid; lenad.* (b) A mineral of the feldspathoid group, including leucite, nepheline, sodalite, nosean, hauyne, lazurite, cancrinite, and melilite.—Syn: *felspathoid; feldspathide.*

feldspathoidite A rarely used name applied to a group of igneous rocks that contains the most feldspathoid-rich of the *foidites.*

felite *belite.*

fell (a) A term used in Scotland and northern England for a bare, uncultivated, open hillside or mountain. (b) A term used in Great Britain for an elevated tract of wasteland or a mountain moorland; a *fell-field.*---Etymol: Scandinavian. See also: *fjeld; fjåll.*

fell-field An open, treeless, rock-strewn area that is above the timberline or in a high latitude and that has a sparse ground cover of low plants or grasses and sedges. Syn: *fell; feldmark; fjeldmark.*

fellside A hillside or mountainside.

feloid A group name for the feldspar and feldspathoid minerals.

fels Massive metamorphic rock lacking schistosity or foliation, e.g. calcsilicate fels (Winkler, 1967). Cf: *granofels.*

felsenmeer A flat or gently sloping area covered with a continuous veneer of large angular and subangular blocks of rock derived from well-jointed underlying bedrock by intensive frost action and usually occurring in situ on high, flat-topped mountains or plateaus above timberline in middle and high latitudes. Etymol: German *Felsenmeer*, "sea of rock". Syn: *block field; block spread; block waste; boulder field; stone field; mountain-top detritus; sorted field; block sea; rock sea; blockmeer.*

felsic A mnemonic adjective derived from *feldspar + lenad* (feldspathoid) + *silica + c*, and applied to an igneous rock having light-colored minerals in its mode; also, applied to those minerals. It is the opposite of *mafic.*

felside An informal term proposed for field use and applied to any fine-grained light-colored nonporphyritic igneous rock, e.g. nonporphyritic rhyolites, trachytes, phonolites, latites, and light-colored andesites.

felsiphyric A term applied by Cross et al (1906, p.703) to the texture of a rock that is microscopically aphanitic; also, said of a rock exhibiting such texture. Cf: *felsophyric; microcryptocrystalline.* Syn: *aphaniphyric.*

felsite A light-colored, fine-grained extrusive or hypabyssal rock with or without phenocrysts and composed chiefly of quartz and feldspar; a rock characterized by *felsitic* texture. Syn: *aphanite; felstone.* Cf: *felsitoid.*

felsitic A syn. of *aphanitic.* The term is sometimes applied only to the light-colored dense rocks, with "aphanitic" being reserved for the dark-colored; of, or pertaining to, a *felsite.* The term is not recommended because in its original use it was applied to a mineral substance now known to be a mixture of quartz and feldspar.

felsitoid An informal term applied to any igneous rock in which the mineral grains are too small to be distinguished by the unaided eye. Cf: *felsite.*

felsoandesite An andesite having a felsitic groundmass; a felsitic andesite.

felsobanyite A snow-white mineral: $Al_4(SO_4)(OH)_{10}.5H_2O$. It has the same formula as, but a different X-ray pattern than, that of *basaluminite.* Incorrectly spelled: *felsobanyite.*

felsophyre A porphyritic igneous rock characterized by *felsophyric* texture. Syn: *aphanophyre.* Cf: *vitrophyre; granophyre.*

felsophyric A term applied by Cross, et al (1906, p.703) to the texture of a rock that is megascopically aphanitic; also, said of a rock exhibiting such texture, e.g. a *felsophyre.* Cf: *felsiphyric; microcrystalline.* Syn: *aphanophyric.*

felsosphaerite A spherulite composed of a felsitic substance. The term is now obsolete.

felspar Chiefly British spelling of *feldspar.* The spelling was introduced by Kirwan (1794) who mistakenly believed the term to be derived from German *Fels*, "rock", rather than from Swedish *feldt*, "field".

felspathoid *feldspathoid.*

felstone An obsolete syn of *felsite.*

felty Said of *pilotaxitic* texture in which the microlites are randomly oriented.

femag *mafic.*

femic Said of an igneous rock having one or more normative dark-colored iron-, magnesium-, or calcium-rich minerals as the major components of the norm; also, said of such minerals. Etymol: a mnemonic term derived from *ferric + magnesium + ic.* Cf: *basic; salic; mafic; felsic.*

femur (a) The third segment of a leg of an arachnid, forming the "hip" articulation with the preceding segment (trochanter) and the "knee" articulation with the following segment (patella) (TIP, 1955, pt.P, p.61). (b) A joint belonging to the proximal part of a prosomal appendage of a merostome.

fen A waterlogged, spongy groundmass containing alkaline decaying vegetation characterized by reeds and which may develop into peat. It sometimes occurs in the sinkholes of karst regions. Cf: *bog.*

fenaksite A pale-rose monoclinic mineral: $(K,Na)_4(Fe Mn)_2(Si_4O_{10})_2(OH,F)$. Not to be confused with *phenakite.*

fence diagram [geochem] A diagram of chemical factors influencing mineral stability, having discrete fields defined by boundary representing the presence or absence of particular phases in an assemblage of minerals, rocks, or compounds. Such boundaries are called *geochemical fences.*

fence diagram [illustration] A drawing in perspective of three or more geologic sections with their relationships to one another.

fenestra [paleont] A small opening in an invertebrate; e.g. an open space in a reticulate bryozoan colony, or an open or closed window in the wall of lorica of a tintinnid. Pl: *fenestrae.*

fenestra [sed] A term used by Tebbutt et al (1965, p.4) for *shrinkage pore*, or "primary or penecontemporaneous gap in rock framework, larger than grain-supported interstices". may be an open space in the rock, or be completely or partly filled with secondarily introduced sediment or cement. Also used to describe the porosity and fabric of rocks with fenestral features (Choquette & Pray, 1970, p.246). Pl: *fenestrae.*

fenestrate A syn. of *fenestrated.* The term has been applied to bryozoans possessing small window-like openings, to coral having regularly perforated septa, and to pollen exhibiting large isodiametric, noncolpate, and geometrically arranged holes in the exine.

fenestrated Having one or more openings or transparent areas, such as having an open meshwork in certain radiolarian shells; perforated or reticulated. Syn: *fenestrate.*

fenestrula An uncalcified area in the ectooecium of a bryozoan through which the entoooecium is discharged.

fenêtre A syn. of *window.* Etymol: French, "window".

fenite A quartzo-feldspathic rock which has been altered b

kali metasomatism at the contact of a carbonatite intrusive
omplex. The process is called *fenitization*. Fenite is mostly
kalic feldspar, with some aegerine, subordinate alkali-horn-
ende, and accessory sphene and apatite.

nitization As generally used today, widespread alkali meta-
matism of quartzo-feldspathic country rocks in the immedi-
e environs of carbonatite complexes. The name fenite for
e altered rock was originated by Brogger (Turner & Verhoo-
en, 1960).

n peat *lowmoor peat.*

nster A syn. of *window*. Etymol: German, "window".

ral Said of a landform or landscape in early maturity when
e crests of ridges and spurs are shaped by the intersection
f the slopes of valley sides that are for the most part still
eep, so that the ridges are at first sharp and serrate (Cot-
n, 1948, p.214-215). Cf: *subdued.*

rberite A grayish to black mineral of the wolframite series:
eWO$_4$. It is isomorphous with huebnerite, and may contain
o to 20% manganese tungstate.

rdisilicite A mineral: FeSi$_2$.

rghanite A sulfur-yellow secondary mineral, possibly:
JO$_2$)$_3$(VO$_4$)$_2$.6H$_2$O. It is perhaps a leached or weathered
roduct of tyuyamunite. Also spelled: *ferganite.*

rgusite A plutonic foidite containing a potassium feldspa-
oid (leucite) and 30 to 60 percent mafic minerals, such as
livine, apatite, and biotite, with accessory opaque oxides. Its
ame is derived from Fergus County, Montana. Cf: *arkite;
issourite; italite.*

rgusonite A brownish-black mineral: Y(Nb,Ta)O$_4$. It is isom-
rphous with formanite, and may also contain erbium, cerium,
on, titanium, and uranium.

ermat's principle The statement that the path of a ray or
eismic wave between two points is that which takes the least
mount of time. Such a path is called a *least-time path.*

rmorite A white mineral of the apatite group: (Ca,Sr)$_5$[(As,
)O$_4$]$_3$(OH).

rn A vascular, nonflowering plant of the class Filicineae. It
as complex fronds growing from an underground rhizome,
nd its sporangia are grouped on the surface of the leaf.

rnandinite A dull-green mineral: CaV$_2$$^{+4}V_{10}$$^{+5}O_{30}$.14H$_2$O
?).

rralite (a) A term used in the formerly French parts of North
frica for a soil that originated from basic crystalline rocks
hich have undergone chemical change and that consist of a
ixture of hydrates of iron, aluminum, and sometimes manga-
ese and titanium. (b) A humid, tropical soil formed by the
eaching of silica and bases by mildly acidic or neutral solu-
ons and characterized by a large content of iron oxide. A
erralitic soil" has a silica/sesquioxide ratio of less than 2
Van Riper, 1962, p.83). Also spelled: *ferrallite.*

rrazite A mineral: (Pb,Ba)$_3$(PO$_4$)$_2$.8H$_2$O (?).

errel's law The statement that the centrifugal force (*Coriolis
orce*) produced by the rotation of the Earth causes a *rota-
onal deflection* of currents of water and air to the right in the
orthern Hemisphere and to the left in the Southern Hemi-
ohere.

rretto zone A term used by early geologists for a reddish or
eddish-brown B-horizon produced under conditions of free
ubsurface drainage in permeable near-surface material such
s loess or sand and gravel extending to the surface or over-
ain by thin till (Flint, 1957, p.213). It is the well-drained
quivalent of *gumbotil.*

rrian Containing ferric iron. Cf: *ferroan.*

rricopiapite A mineral of the copiapite group: Fe$_5$$^{+3}$-
O_4)$_6$O(OH).20H$_2$O.

rricrete (a) A term suggested by Lamplugh (1902) for a
onglomerate consisting of surficial sand and gravel cement-
d into a hard mass by iron oxide derived by the oxidation of
ercolating solutions of iron salts. (b) A ferruginous *duricrust.*
Etymol: *ferruginous* + con*crete*. Cf: *calcrete; silcrete.*

ferricrust (a) A general term for an indurated soil horizon cemented with iron oxide (mainly hematite). (b) The hard crust
of an iron concretion.

ferride One of a group of elements that is related to iron: cobalt, nickel, manganese, titanium, chromium, and vanadium.

ferrierite A zeolite mineral: (Na,K)$_2$MgAl$_3$Si$_{15}$O$_{36}$(OH).9H$_2$O.

ferriferous Iron-bearing; said esp. of a mineral containing iron,
or of a sedimentary rock that is much richer in iron than is
usually the case, such as a shale whose iron-oxide content is
greater than 15%. Cf: *ferruginous.*

ferrilith A term suggested by Grabau (1924, p.298) for an
iron-rich sedimentary rock (ironstone). Syn: *ferrilyte.*

ferrimagnetism A type of *magnetic order*, macroscopically resembling *ferromagnetism*. Magnetic ions at different crystal
sites are opposed, i.e., antiferromagnetically coupled. There is
nevertheless a net magnetization because of inequality in the
number or magnitude of atomic magnetic moments at the two
sites. This type of magnetic order occurs in magnetite. Cf:
antiferromagnetism.

ferrimolybdite A yellowish mineral: Fe$_2$(MoO$_4$)$_3$.8H$_2$O (?). It
occurs as an earthy powder or incrustation, or as silky, fibrous, and radiating crystals, and is formed by the oxidation
of molybdenite. Cf: *molybdite.* Syn: *molybdic ocher.*

ferrinatrite A grayish-white or whitish-green mineral: Na$_3$-
Fe^{+3}(SO$_4$)$_3$.3H$_2$O.

ferrisicklerite A dark-brown mineral: Li(Fe^{+3},Mn^{+2})PO$_4$. It is
isomorphous with sicklerite.

ferrisymplesite An amber-brown mineral: Fe$_3$(AsO$_4$)$_2$-
(OH)$_3$.5H$_2$O. Cf: *symplesite.*

ferrite [ign] A general term applied to grains, scales, and
threads of unidentifiable, more or less transparent or amorphous, red brown, or yellow iron oxide in the groundmass of a
porphyritic rock (Johannsen, 1939, p.177). Cf: *opacite; viridite.*

ferrite [sed] A term used by Tieje (1921, p.655) for a cemented iron-rich sediment whose particles do not interlock.

ferritungstite A mineral: Ca$_2$Fe$_2$$^{+2}Fe_2$$^{+3}$(WO$_4$)$_7$.9H$_2$O. It occurs as a pale-yellowish to brownish-yellow earthy powder.
Syn: *tungstic ocher.*

ferriturquoise A variety of turquoise containing 5% Fe$_2$O$_3$.

ferroactinolite A monoclinic mineral representing an end
member of the amphibole group: Ca$_2$Fe$_5$$^{+2}Si_8O_{22}(OH)_2$. It is
a variety of actinolite containing no magnesium, and is isomorphous with tremolite. Syn: *ferrotremolite.*

ferroan Containing ferrous iron. Cf: *ferrian.*

ferroan dolomite A mineral that is intermediate in composition
between dolomite and ferrodolomite; specif. *ankerite.*

ferroaxinite A mineral: Ca$_2$(Fe,Mn)Al$_2$BSi$_4$O$_{15}$(OH).

ferrocarpholite A mineral: (Fe,Mg)Al$_2$Si$_2$O$_6$(OH)$_4$.

ferrocopiapite A mineral of the copiapite group: Fe^{+2}-
Fe$_4$$^{+3}$(SO$_4$)$_6(OH)_2$.20H$_2$O.

Ferrod In U.S. Dept. of Agriculture soil taxonomy, a suborder
of the soil order *Spodosol*, characterized by having at least six
times as much elemental iron as organic carbon in its spodic
horizon. A Ferrod has none of the characteristics associated
with wetness (SSSA, 1970). Cf: *Aquod; Humod; Orthod.*

ferrodolomite A mineral component: CaFe(CO$_3$)$_2$. It is isomorphous with dolomite, but probably does not occur naturally
except in ankerite.

ferroelectricity The spontaneous alignment of electric dipoles
by mutual interaction. It occurs in barium titanate.

ferrohexahydrite A monoclinic mineral: FeSO$_4$.6H$_2$O.

ferromagnesian Containing iron and magnesium; applied to
mafic minerals.

ferromagnetism A type of *magnetic order* in which all magnetic atoms in a domain have their moments aligned in the same
direction; loosely, any type of magnetic order. Cf: *ferrimagnetism; antiferromagnetism.*

ferroselite An orthorhombic mineral: FeSe$_2$. It resembles
marcasite.

ferrosilite (a) A mineral component in the orthopyroxene group: $FeSiO_3$. It is the iron analogue of enstatite and occurs in hypersthene, but it does not exist separately in nature. Symbol: Fs. Syn: *iron hypersthene*. (b) A mineral group consisting of *clinoferrosilite* and *orthoferrosilite*.

ferrospinel (a) *hercynite*. (b) A synthetic magnetic substance of spinel structure, containing iron, and being a poor conductor of electricity.

ferrotremolite *ferroactinolite*.

ferruccite An orthorhombic mineral: $NaBF_4$.

ferruginate adj. A term restricted by Allen (1936, p.22) to designate the iron-bearing cement of a sedimentary rock.---v. To stain a rock with an iron compound.

ferruginous (a) Pertaining to or containing iron, e.g. a sandstone that is cemented with iron oxide or a schist that contains enough iron to be economically valuable. Cf: *ferriferous; siderose*. (b) Said of a rock having a rusty color due to the presence of ferric oxide. The quantity of iron does not have to be large; neither do all red rocks derive their color from iron.

fersilicite A mineral: FeSi.

fersmanite A brown mineral: $Na_4Ca_2Ti_4Si_3O_{18}(O,F)_2$. Sometimes misspelled: *fersmannite*.

fersmite A black mineral: $(Ca,Ce,Na)(Nb,Ti,Fe,Al)_2(O,OH,F)_6$.

fervanite A golden-brown mineral: $Fe_4(VO_4)_4.5H_2O$. It occurs with radioactive minerals but is not itself radioactive.

Festiniogian European stage: Upper Cambrian (above Maentwrogian, below Dolgellian).

festoon The upfolded or pointed part of a layer in a congeliturbate. Ant: *pocket*.

festoon cross-bedding A variety of *trough cross-bedding* consisting of elongate, semiellipsoidal, eroded, plunging troughs or scoop-like structures that are filled by sets of thin laminae conforming in general to the shapes of the troughs and that crosscut each other so that only parts of each unit are preserved, resulting in a festoon-like (a hanging open loop or curve) appearance in section. The cross-beds are deposited on concave surfaces so that both the lower bounding surfaces and the cross-beds are trough-shaped.

festooned pahoehoe A type of pahoehoe, the ropy surface of which has been dragged by flow of underlying molten lava into festoon patterns. Cf: *corded pahoehoe; elephant-skin pahoehoe; entrail pahoehoe; filamented pahoehoe; sharkskin pahoehoe; shelly pahoehoe; slab pahoehoe*.

fetch (a) A term used in wave-forecasting for the area of the open ocean over the surface of which the wind blows with constant speed and direction, thereby creating a wave system. Syn: *generating area*. (b) The extent of the fetch, measured horizontally in the direction of the wind. Syn: *fetch length*.

fetch length *fetch*.

feuerstein A syn. of *firestone*. Etymol: German *Feuerstein*, "flint".

fiamme Dark, vitric lenses in *piperno*, averaging a few centimeters in length, perhaps formed by the collapse of fragments of pumice. The presence of fiamme may be called *flame structure*. Etymol: Italian *fiamma*, "flame".

fiard Var. of *fjard*.

fiasconite An anorthite-bearing leucitite-basanite that also contains augite, olivine, nepheline, and iron ores. Its name is derived from Montefiascone, Italy.

fiber An elongated, tapering, thick-walled strengthening cell occurring in various parts of vascular plants (Esau, 1953).

fiber tracheid A type of *tracheid* that is a wood cell with a thick secondary wall, pointed ends, and small bordered pits having lenticular to slitlike, usually extended, inner apertures (Record, 1934, p.30).

Fibrist In U.S. Dept. of Agriculture soil taxonomy, a suborder of the soil order *Histosol*, characterized by containing much undecomposed plant fiber and by being saturated with water for sufficient periods of time to make cultivation without artificial drainage difficult. Fibrists have a bulk density of less than 0.1 (SSSA, 1970). Cf: *Folist; Hemist; Saprist*.

fibroblastic Pertaining to a *homeoblastic* type of texture of a metamorphic rock due to the development during recrystallization of minerals with a fibrous habit. Cf: *nematoblastic*.

fibrocrystalline Characterized by the presence of fibrous crystals.

fibroferrite A yellowish mineral: $FeSO_4(OH).5H_2O$.

fibrolite *sillimanite*.

fibrous Said of the habit of a mineral, and of the mineral itself (e.g. asbestos), that crystallizes in elongated thin, needle-like grains, or fibers.

fibrous conodont element A *lamellar conodont element* in which the lamellae are thick and "white matter" is absent or reduced to a thin column along cusp and denticle axes, and whose broken edges are typically frayed.

fibrous ice *acicular ice*.

fibrous layer The *secondary layer* in articulate brachiopods secreted intracellularly as fibers bounded by cytoplasmic sheaths.

fibrous ligament The part of a ligament of a bivalve mollusk characterized by fibrous structure and in which conchiolin is commonly impregnated with calcium carbonate. It is secreted by epithelium of the mantle and is elastic chiefly to compressional stresses. Cf: *lamellar ligament*.

fibrous peat Peat in which original plant structures are only slightly altered by degradation of cellulose matter. It is tough and nonplastic. Cf: *pseudofibrous peat; amorphous peat*. Syn: *woody peat*.

fibrous texture In mineral deposits, a pattern of bladelike, serrated crystals. See also: *cross-fiber*.

fichtelite A white, translucent, crystalline, non-aromatic hydrocarbon, with an approximate hydrogen to carbon ratio of 1.6, which is a resin acid found in fossil woods.

fiducial mark An index or point used as a basis of reference, e.g. one of usually four index marks connected with a camera lens (as on the metal frame that encloses the negative) and that forms an image on the negative or print such that lines drawn between opposing points intersect at and thereby define the principal point of the photograph. Syn: *collimating mark*.

fiducial time A time on a seismograph record which may be marked to correspond, by employing necessary corrections, to a datum plane in space.

fiedlerite A colorless monoclinic mineral: $Pb_3(OH)_2Cl_4$.

field [eco geol] A region or area that possesses or is characterized by a particular mineral resource, e.g. gold field, coal field.

field [geol] A broad term for the area, away from the laboratory and esp. outdoors, in which a geologist makes firsthand observations, as of actual phenomena, and collects data such as measurements of strikes and dips, rock and mineral samples, and fossils.

field [geophys] That area or space in which a given geophysical effect, e.g. gravity or magnetism, occurs and is measureable.

field [ice] (a) *ice field*. (b) A very large floe or other unbroken area of sea ice.

field capacity The quantity of water held by soil or rock against the pull of gravity. It is sometimes limited to a certain drainage period, thereby distinguishing it from *specific retention* which is not limited by time. Syn: *field-moisture capacity; normal moisture capacity*.

field classification A preliminary analysis of fossils or hand specimens of rocks or minerals in the field, usually with the aid of a hand lens.

field coefficient of permeability *field permeability coefficient*.

field completion Obtaining additional information in the field in order to edit and publish a topographic map from a compiled manuscript or to fill in and confirm that part of a map manuscript prepared by stereocompilation. It includes: a comprehensive examination of the compilation for completeness, quality, and topographic expression; the addition, deletion, or correction of map features; the classification of buildings, roads, and drainage; the mapping of public-land subdivision lines and civil boundaries; obtaining names; and checking the map for compliance with vertical-accuracy standards.

field contouring Contouring of a topographic map by field methods accomplished by plane-table surveys on a prepared base. It is usually done for terrain unsuitable for contouring by photogrammetric methods. Cf: *field sketching*.

field focus The total area or volume which the source of an earthquake occupies. If a fault is the source, the focus is the local fault surface, and is called the "field" because it is inferred from the area of shaking as observed in the field. The concept is inexact and the term is not commonly used.

field geology The direct observation of the geology of an area or of a particular geologic aspect of an area; original, primary reconnaissance; *field work*.

field ice (a) An obsolete term for *consolidated pack ice* consisting of very large, relatively flat floes many kilometers across. (b) A general term used for all types of *sea ice* except that newly formed.

field intensity [grav] The force of attraction exerted on a unit mass particle at a point by the matter causing the force field.

field map A preliminary or original map of the geology of an area, made in the field and upon which a final map may be based.

field moisture Water present in the ground above the water table.

field-moisture capacity *field capacity*.

field-moisture deficiency The amount of water required to raise the moisture content of the soil to field capacity.

field moisture equivalent The minimum water content of a soil mass, expressed as a percentage of its dry weight, at which a drop of water placed on a smoothed surface of the soil will not be absorbed but will spread out, giving a shiny appearance to the soil.

field of force A region of space within which there exists a value of the force at each point in the space.

field of view The solid angle through which an instrument is sensitive to radiation. Due to various effects, diffraction, etc., the edges are not sharp. In practice they are defined as the *half-power points*, i.e. the angle outwards from the optical axis, at which the energy sensed by the radiometer will see out to 4° or 5° off the axis because of the diffraction effect, and should be corrected accordingly. See also: *instantaneous field of view*.

field permeability coefficient The *permeability coefficient* defined for prevailing conditions rather than for a temperature of 60°F. Syn: *coefficient of field permeability*.

field reversal *geomagnetic reversal*.

field sketching The art of drawing contours based on the elevations of selected features located on a plane-table sheet usually without a prepared base. Cf: *field contouring*. See also: *contour sketching*.

field work *field geology*.

figure of the Earth The geoid, or the surface of the Earth, as it would be determined from mean sea level over the oceans and the sea level surface extended continuously through the continents. When referring to the irregularities or undulations of the *geoid*, some authors refer to humps and hollows but in no case can the surface be considered to be concave.

figure stone *agalmatolite*.

filament (a) In a flower, the stalk of a stamen which bears an anther at its apex. (b) In a protist, a branched or unbranched row of cells joined end to end.

filamented pahoehoe A type of pahoehoe, the surface of which displays threadlike strands formed by escaping gas

bubbles, and that are recumbent and aligned with the direction of flow. It is a common type and is often found superimposed on other forms. Cf: *corded pahoehoe; elephant-hide pahoehoe; entrail pahoehoe; festooned pahoehoe; sharkskin pahoehoe; shelly pahoehoe; slab pahoehoe*.

filiform *capillary*.

filiform lapilli *Pele's hair*.

fill [eng geol] (a) Man-made deposits of natural earth materials (such as soil, sand, gravel) and waste materials (such as tailings or spoil from dredging), used to fill a receptacle, cavity, or passage (such as an old stope or chamber in a mine) or to improve shore land (such as extending beaches, or building dams to raise lake levels). See also: *backfill*. (b) Soil or loose rock used to raise the surface of low-lying land, such as an embankment, to fill a hollow or ravine in railroad construction. Also, the place filled by such an embankment. (c) The depth to which material is to be placed (filled) to bring the surface to a predetermined grade.

fill [speleo] *drift* [speleo].

fill [sed] Any sediment deposited by any agent so as to fill or partly fill a valley, sink, or other depression.

filled-lake plain A swampy plain formed by the filling of a lake by sediments, aided by the growth of plants (Tarr, 1902, p.82).

filled valley A wide-basin valley, in an arid or semiarid region, that contains abundant alluvium in the form of fans, flood plains, and lake deposits.

filler [streams] A stream that empties into and fills a lake.

fill-in fill terrace A terrace left by a stream that, having incised its valley fill, partly fills up the new valley and incises anew (Schieferdecker, 1959, term 1514).

fillowite A brown, yellow, or colorless mineral: $H_2Na_6(Mn,Fe,Ca)_{14}(PO_4)_{12}.H_2O$ (?).

fillstrath terrace A *fill terrace* whose surface has been eroded to a level below the original depositional surface (Howard, 1959, p.242); it consists of alluvial material as contrasted with a *strath terrace* formed in bedrock.

fill terrace (a) A term used by Bucher (1932, p.131) for a remnant, resulting from stream rejuvenation, of a flat valley bottom (or of an alluvial plain) that had been produced by stream aggradation; e.g. an *alluvial terrace* or a glacial terrace. (b) The part of a former alluvial valley floor built upward by deposition of valley-filling sediments (Howard, 1959, p.242); it includes *filltop terrace* and *fillstrath terrace*.---Leopold et al (1964, p.460) find that the term is "confusing" and "should probably be abandoned".

filltop terrace A *fill terrace* whose flat surface is the original depositional surface (Howard, 1959, p.242).

film crust A type of *snow crust* consisting of a continuous film or very thin layer of ice formed on the surface of snow by the freezing of meltwater. It is not as thick as an *ice crust* [glaciol]. Cf: *firnspiegel*.

film water *pellicular water*.

filter [seis] In seismic profiling, an electrical or mechanical device used for *filtering*, i.e. the attenuation of some components of a signal and the enhancement of others. See also: *band-pass filter; band-reject filter*.

filter [photo] Any transparent material that selectively absorbs and transmits radiant energy through an optical system; specif. a glass or gelatin plate placed in front of, in, or behind a camera lens for the purpose of reducing or eliminating the effect of light of a certain color or colors on the film or plate, such as an "infrared filter" that holds back almost all visible light but transmits the infrared rays or an "ultraviolet filter" that cuts down ultraviolet rays but allows all visible light to pass.

filter bridge A narrow *land bridge* which permits the selective migration of some organisms.

filter cake *mud cake* [drill].

filter feeder An animal that obtains its food by straining out organic matter from water as it passes through some part of its body; e.g. a coral.

filtering In seismic prospecting, the attenuation of certain frequency components of a signal and the enhancement of others. It may be done mechanically or electrically by a *filter* or by a numerical procedure after the signal has been recorded.

filter pressing Magmatic differentiation involving the separation of crystals from a residual liquid as a result of compression, usually during a late stage of consolidation. The liquid flows to areas of lower pressure and thereby is separated from the crystals. Syn: *filtration differentiation.*

filtration Removal of suspended and/or colloidal material from a liquid by passing it through a relatively fine porous medium.

filtrational Said of an ore-forming fluid or mineralizer that is a nonmagmatic, underground water (Smirnov, 1968). Cf: *juvenile* [*ore dep*]; *assimilated.*

filtration differentiation *filter pressing.*

filtration spring A spring whose water percolates from numerous small openings in permeable material. It may have either a small or a large discharge (Meinzer, 1923, p.50). Cf: *fracture spring; seepage spring.*

filum aquae The thread of a stream. Etymol: Latin, "thread of water", Pl: *fila aquarum.*

fimmenite A peat that is derived mainly from spores.

finandranite A coarse-grained potassium-rich kalisyenite composed of microcline, amphibole, and some biotite, ilmenite, and apatite.

find [**meteorite**] A meteorite not seen to fall, but recognized as such by its composition and structure. Cf: *fall.*

fine [**sed**] Composed of or constituting relatively small particles; e.g. "fine sandy loam". Ant: *coarse.*

fine admixture A term applied by Udden (1914) to an *admixture* (in a sediment of several size grades) whose particles are finer than those of the dominant or maximum grade; material finer than that found in the maximum histogram class.

fine aggregate The portion of an *aggregate* in which the smaller particles are able to pass a certain numbered sieve, such as those particles with diameters smaller than 1/4 or 3/16 inch or 4.76 mm. Cf: *coarse aggregate.*

fine clay A geologic term for a *clay* particle having a diameter in the range of 1/2048 to 1/2024 mm (0.5-1 microns, or 11 to 10 phi units). Also, a loose aggregate of clay consisting of fine clay particles.

fine earth (a) A soil classification based on particle size: a soil which can be passed through a No. 10 (2.0mm) sieve without grinding its primary particles. Cf: *fine-grained* [*soil*]. (b) A general term for loose earth.

fine-grained [**geol**] (a) Said of a crystalline or glassy rock, and of its texture, in which the individual minerals are relatively small; specif. said of an igneous rock whose particles have an average diameter less than 1 mm (0.04 in.). Syn: *aphanitic.* (b) Said of a sediment or sedimentary rock, and of its texture, in which the individual constituents are too small to distinguish with the unaided eye; specif. said of a sediment or rock whose particles have an average diameter less than 1/16 mm (62 microns, or silt size and smaller). Cf: *finely crystalline.*----The term is used in a relative sense, and various size limits have been suggested and used. Cf: *coarse-grained; medium-grained.*

fine-grained [**soil**] Said of a soil in which silt and/or clay predominate. In the U.S., the maximum average diameter of the constituent particles is 0.05 mm (0.002 in.), or as used by engineers, 0.074 mm (passing U.S. standard sieve No. 200); the International Society of Soil Science recognizes a diameter limit of 0.02 mm. Cf: *coarse-grained; fine earth.*

fine-granular Said of the *crystalline-granular texture*, of an igneous rock, that can be observed with the unaided eye.

fine gravel (a) A soil term used in the U.S. for *gravel* whose particles have a diameter in the range of 2-12.5 mm (1/12 to 1/2 in.); it was formerly applied to soil particles (now called

very coarse sand) having diameters of 1-2 mm. (b) An engineering term for *gravel* whose particles have a diameter in the range of 4.76 mm (retained on U.S. standard sieve no.4) to 19 mm (3/4 in.).

finely crystalline Descriptive of an interlocking texture of a carbonate sedimentary rock having crystals whose diameters are in the range of 0.016-0.062 mm (Folk, 1959) or 0.01-0.1 mm (Carozzi & Textoris, 1967, p.5) or 0.01-0.05 mm (Bissell & Chilingar, 1967, p.103), or are less than 1.0 mm (Krynine, 1948, p.143). Cf: *fine-grained* [*geol*].

fineness The state of subdivision of a substance; the size of the constituent particles of a substance. The term is applied in describing sedimentary texture.

fineness factor A measure of the average particle size of clay and ceramic material, computed by summing the products of the reciprocal of the size-grade midpoints and the weight percentage of material in each class (expressed as a decimal part of the total frequency) (Purdy, 1908). The measure is based on the assumption that the surface areas of two powders are inversely proportional to their average particle sizes. Syn: *surface factor..*

fine pebble A geologic term for a *pebble* having a diameter in the range of 4-8 mm (1/6 to 0.3 in., or -2 to -3 phi units) (AGI, 1958).

fines [**mining**] Very small, finely crushed or powdered material, such as the smallest particles of coal, crushed rock, or ore, as contrasted with the coarser particles; esp. material smaller than the minimum for any specified size or grade (such as coal with a maximum particle size less than 1.6 mm or less than 3.2 mm, or ores in too pulverulent a condition to be smelted in the ordinary way), or material passing through a given screen or sieve.

fines [**sed**] (a) Very small particles, esp. those smaller than the average in a mixture of particles of various sizes; e.g. the silt and clay fraction in glacial drift, or the fine-grained sediment that settles slowly to the bottom of a body of water. (b) An engineering term for the clay- and silt-sized soil particles (diameters less than 0.074 mm) passing U.S. standard sieve no.200.

fine sand (a) A geologic term for a *sand particle having a diameter in the range of 0.125-0.25 mm (125-250 microns, or 3 to 2 phi units). Also, a loose aggregate of sand consisting of fine sand particles.* (b) An engineering term for a *sand* particle having a diameter in the range of 0.074 mm (retained on U.S. standard sieve no.200) to 0.42 mm (passing U.S. standard sieve no.40). (c) A soil term used in the U.S. for a *sand* particle having a diameter in the range of 0.10-0.25 mm. The diameter range recognized by the International Society of Soil Science is 0.02-0.2 mm. (d) Soil material containing 85% or more of sand-size particles (percentage of silt plus 1.5 times the percentage of clay not exceeding 15) and 50% or more of fine sand or less than 25% of very coarse sand, coarse sand, and medium sand together with less than 50% of very fine sand (SSSA, 1965, p.347).

fine silt A geologic term for a *silt* particle having a diameter in the range of 1/128 to 1/64 mm (8-16 microns, or 7 to 6 phi units). In Great Britain, the range 1/100 to 1/20 mm has been used. Also, a loose aggregate of silt consisting of fine silt particles.

fine topography A topography with fine *topographic texture*, characterized by high drainage density and closely spaced streams, and common in regions of weak rocks. An "ultrafine" topography is characterized by the extremely fine dissection of badlands topography.

finger One of two pincer-like blades of the distal end of a cheliped of a crustacean or of a chela of an arachnid. One finger is movable and the other is rigid or fixed.

finger bar *bar finger.*

finger coal *Natural coke* occurring as small, hexagonal columns associated with igneous intrusion.

finger gully One of a group of very small gullies that forms a fan-shaped extension at the head of a system of gullies.

finger lake A long, relatively narrow lake, usually of glacial origin, which may occupy a rock basin in the floor of a glacial trough or be held in by a morainal dam across the lower end of the valley; esp. one of a group of such lakes disposed somewhat like the fingers of a hand, such as the Finger Lakes in central New York State.

Fingerlakesian North American stage: lower Upper Devonian (above Taghanican, below Chemungian). Syn: *Finger Lakes.*

fingertip channel One of the smaller, unbranched stream channels of a drainage network.

finite strain theory A theory of material deformation which considers displacements and strains too large to be evaluated through *infitesimal strain theory*, where the displacements are still continuous and the strains vary gradually.

finnemanite A gray, olive-green, or black hexagonal mineral: $Pb_5(AsO_3)_3Cl$.

fiord Anglicized variant of *fjord.*

fiorite A syn. of *siliceous sinter;* a variety of opal occurring as a gray or white, fibrous, pearly incrustation near a hot spring.

fire Flashes of the different spectral colors seen in diamonds and other gemstones as the result of dispersion. Cf: *play of color.*

fire assay Any type of assay procedure that involves that heat of a furnace.

fireball A bright or brilliant meteor with luminosity that equals or exceeds that of the brightest planets. Cf: *bolide.*

fireball hypothesis *"big bang" hypothesis.*

fireblende A syn. of *pyrostilpnite.* Also spelled: *fire blende.*

fireclay (a) A siliceous clay rich in hydrous aluminum silicates, capable of withstanding high temperatures without deforming (either disintegrating or becoming soft and pasty), and useful for the manufacture of refractory ceramic products (such as crucibles, or firebrick for lining furnaces). It is deficient in iron, calcium, and alkalies, and approaches kaolin in composition, the better grades containing at least 35% alumina when fired. (b) A term formerly, but inaccurately, used for *underclay.* Although many fireclays commonly occur as underclays, not all fireclays carry a roof of coal and not all underclays are refractory.---Also spelled: *fire clay.* Syn: *firestone; refractory clay; sagger.*

fireclay mineral A disordered variety of kaolinite. See also: *mellorite.*

firedamp A coal-mine gas that is explosive and consists mainly of *methane.* Cf: *blackdamp; whitedamp; afterdamp.*

fire fountain Rhythmic, rise-and-fall eruption of incandescent lava either from a central volcanic vent or along a fissure.

fire marble *lumachelle.*

fire opal A transparent to translucent and orangy-yellow, brownish-orange, hyacinth-red, or brownish-red variety of opal that gives out fiery reflections in a bright light and that may or may not have play of color. See also: *gold opal.* Syn: *sun opal; pyrophane.*

firestone (a) Any fine-grained siliceous stone formerly used for striking fire; specif. *flint.* Syn: *feuerstein.* (b) A nodule of pyrite formerly used for striking fire. (c) A fine-grained siliceous rock that can resist or endure high heat and that is used for lining furnaces and kilns, such as certain Cretaceous and Jurassic sandstones in southern England. (d) *fireclay.*

firm ground Materials in which a tunnel can be advanced without any roof support and the permanent lining can be constructed before the walls begin to move.

firmoviscosity The *elasticoviscous* state, as modeled by a Kelvin body. In response to a given stress, elastic strain is produced only over a finite period of time; unloading is time dependent, although all of the strain is recoverable. At a constant strain, stress may be supported indefinitely (Turner and Weiss, 1963, p.279).

firn (a) A material that is transitional between snow and glacier ice, being older and denser than snow, but not yet transformed into glacier ice. Snow becomes firn after existing through one summer melt season; firn becomes glacier ice when its permeability to liquid water drops to zero. The term has also been defined, although rarely, on the basis of certain physical properties, such as density (snow becomes firn at a density greater than 0.4 g/cc, as in the older literature, or 0.55 g/cc, the greatest density of ice if its grains were shifted around so that they fit most snugly), but this criterion is difficult to measure and variable in specific areas. Syn: *névé; firn snow.* (b) A geographic term applied to the accumulation area or upper region of a glacier. This usage is being supplanted by *firn field*, or by *névé* (in Great Britain).----Etymol: German, adjective meaning "old, of last year".

firn basin *firn field.*

firn edge The boundary on a glacier between glacier ice and firn during the ablation season.

firn field The *accumulation area* of a glacier; a broad expanse of glacier surface over which snow accumulates and firn is created; an area of *firn.* Syn: *firn basin; névé.*

firn ice *iced firn.*

firnification The process whereby snow is transformed into firn and then into ice in a glacier.

firn limit *firn line.*

firn line (a) The highest level to which the winter snow cover retreats on a glacier; the *snowline.* (b) The edge of the snow cover at the end of the summer season, thus the boundary between the *superimposed ice zone* below and the soaked zone above. See also: *equilibrium line.* Syn: *firn limit.*

firn mirror *firnspiegel.*

firn snow (a) *firn.* (b) *old snow.*

firnspiegel A thin sheet or film of clear ice on a snow surface, bridging across hollows in the snow, formed under certain meteorologic conditions when surface meltwater is immediately refrozen as a thin ice film, and the snow below continues to melt by radiation passing through the transparent ice sheet. Etymol: German *firn*, "old, last year's", + *Spiegel*, "mirror". Syn: *firn mirror.* Cf: *film crust.*

first antenna *antennule.*

first arrival The first recorded signal or *arrival* attributable to seismic-wave travel from a known source. First arrivals on reflection records are used for information about the low-velocity or weathering layer; refraction studies are based on first arrivals (Sheriff, 1968). Cf: *later arrival.* Syn: *first break.*

first bottom The normal flood plain of a river. Cf: *second bottom.*

first break *first arrival.*

first-class ore An ore of sufficiently high grade to be acceptable for market without preliminary treatment. Syn: *shipping ore.* Cf: *second-class ore.*

first law of thermodynamics The statement describing the *internal energy* of a system, which says that the change of energy of a system equals the amount of energy received from the external world, which in turn equals the heat taken in by the system and the work done on the system.

first maxilla *maxillule.*

first meridian *prime meridian.*

first-order leveling Spirit leveling of high precision and accuracy in which, for a section of 1-2 km in length, the maximum allowable discrepancy in results obtained by running the line first forward to the objective point and then backward to the starting point is 4.0 mm times the square root of the distance in kilometers separating the ends of the line (or 0.017 ft times the square root of the distance in miles). Cf: *second-order leveling; third-order leveling.*

first-order pinacoid In a triclinic crystal, the $\{0kl\}$ pinacoid and the $\{0\bar{k}l\}$ pinacoid. Cf: *second-order pinacoid; third-order pinacoid; fourth-order pinacoid.*

first-order prism A crystal form: in a tetragonal crystal, the $\{110\}$ prism; in a hexagonal crystal, the $\{10\bar{1}0\}$ prism; in an orthorhombic crystal, any |0*kl*| prism; and in a monoclinic crystal, any |0*kl*| prism. Cf: *second-order prism; third-order prism; fourth-order prism.* See also: *clinodome; brachydome.*

first water The highest quality (value) or the purest luster of a gemstone, such as that of a diamond that is flawless, perfectly clear and transparent, and colorless or almost blue-white. Cf: *second water.*

first-year ice Sea ice not more than one winter's growth, developing from young ice; it is subdivided on the basis of thickness: "thin" (30-70 cm; also known as *white ice*); "medium" (70-120 cm); and "thick" (120 cm to 2 m). See also: *winter ice: one-year ice.*

firth A long, narrow arm of the sea; also, the opening of a river into the sea. Along the Scottish coast, it is usually the lower part of an *estuary* (e.g. Firth of Forth), but sometimes it is a fjord (e.g. Firth of Lorne) or a strait (e.g. Pentland Firth). Etymol: Scottish. Syn: *frith.*

fischerite A mineral, consisting of a green hydrous aluminum phosphate, that is probably identical with *wavellite.*

fish [drill] (a) Broken or lost equipment in a drill hole, recoverable only by fishing. (b) Any foreign material, in a well, that cannot be removed at will.

fish [oceanog] Any oceanographic sensing device that is towed behind a ship.

fish-eye stone *apophyllite.*

fishhook dune A dune consisting of a long, sinuous, sigmoidal ridge forming the "shaft" and a well-defined crescent forming the "hook" (Stone, 1967, p.228). Syn: *hooked dune.*

fishing Searching for and attempting to recover (by the use of specially prepared tools) a piece or pieces of drilling equipment (such as sections of pipe, cables, or casing) that has become detached, broken, or lost from a drilling tool and left in the drill hole or that has been accidentally dropped into the hole.

fish kill Destruction of fish in lakes or ponds due to a decrease in oxygen resulting from snow or from excessive amounts of suspended organic matter; to toxic pollutants; or to the total freezing of shallow lakes or ponds.

fishtail structure In a coal seam, the ragged termination of the seam, probably caused by the washing away of peat and the wedging in of sediment parallel to the bedding.

fissiculate Said of a blastoid having exposed or partly exposed hydrospire slits or spiracular slits.

fissile (a) Capable of being easily split along closely spaced planes; exhibiting *fissility.* (b) Said of bedding that consists of laminae less than 2 mm in thickness (Payne, 1942).

fissility A general term for the property possessed by some rocks of splitting easily into thin sheets or layers along closely spaced, roughly planar, and approximately parallel surfaces, such as along bedding planes (as in shale) or along cleavage planes (as in schist) induced by fracture or flowage; its presence distinguishes shale from mudstone. The term is not applied to minerals, but is analogous to cleavage in minerals, and includes phenomena such as *bedding fissility* and *fracture cleavage.* Etymol: Latin *fissilis,* "that which can be cleft or split". Adj: *fissile.*

fission [isotope] The spontaneous or induced splitting, by particle collision, of heavy nucleus into a pair (only rarely more) of nearly equal fission fragments plus some neutrons. The splitting is accompanied by the release of a large quantity of energy. Cf: *fusion.* See also: *spallation.* Syn: *nuclear fission.*

fission [evol] Asexual reproduction occurring when a single cell divides into two theoretically equal parts.

fissionable Said of nuclei, such as uranium and plutonium, which are capable of *fission.*

fission-track dating A method of calculating an age in years by determining the ratio of the spontaneous fission-track density to induced fission tracks. The method, which has been used for ages from 20 to 1.4×10 years, works best for micas, tektites, and meteorites, and is also useful for deter-

mining the amount and the distribution of the uranium concentration in the sample. Syn: *fission-track method; spontaneous fission-track dating.*

fission-track method *fission-track dating.*

fission tracks The paths of *radiation damage* made by nuclear particles in a mineral or glass by the spontaneous fission of uranium-238 impurities. They are similar in occurrence and formation to *alpha-particle recoil tracks,* but are larger and less numerous. Fission-track density is established by etching techniques and subsequent microscopic examination.

fissure [joints] A surface of fracture or a crack in rock along which there is a distinct separation. It is often filled with mineral-bearing material. Syn: *joint fissure; open joint.*

fissure [glaciol] *crevasse.*

fissure chamber *dike chamber.*

fissure eruption An eruption that takes place from an elongate fissure, rather than from a central vent. It is generally basaltic. Cf: *plateau eruptions.* See also: *fissure flow.*

fissure flow A flow formed as a result of a *fissure eruption.*

fissure-flow volcano *fissure volcano.*

fissure polygon A *nonsorted polygon* marked by intersecting grooves or fissures producing a gently convex polygonal surface pattern and by the absence of a well-defined stone border. The term is inadequate because some polygons with fissures have stone borders coincident with the fissures (Washburn, 1956, p.825). The feature is typical of broad areas of NW Canadian lowlands. See also: *mud polygon; ice-wedge polygon.*

fissure spring A spring issuing from a crack or joint. Several springs of this type may flow out along the same fissure line. Cf: *fault spring; fracture spring.*

fissure system A group of fissures of the same age and of more or less parallel strike and dip.

fissure theory An early theory, now partially discredited, that oil and gas migrate through fissures resulting from the arching of beds to form anticlines. Oil and gas have been known to migrate through cracks and fissures in complexly faulted areas and to have accumulated into economically viable reservoirs of igneous rock under such conditions.

fissure vein A type of mineral deposit of veinlike shape with the implication of clearly defined walls rather than extensive host-rock replacement.

fissure volcano One of a series in a pattern of eruption that is along a fissure plane as a series of vents; such a flow characteristically forms as extensive sheets. Cf: *areal eruption.* Syn: *fissure-flow volcano.*

fistulose Said of a foraminifer having tubular irregular growth in the apertural region.

fitzroyite A lamproite composed of leucite and phlogopite phenocrysts in a very fine-grained groundmass.

fiveling A crystal twin consisting of five individuals. Cf: *twoling; trilling; fourling; eightling.*

fix (a) A relatively accurate geographic position determined without reference to any former position by visual observations of terrestrial objects or heavenly bodies or by radio or electronic means; e.g. the position on a map of a point of observation obtained by surveying processes. (b) The act of determining a fix.

fix-bitumens All authigenic, nonfluid bitumens. They are divided into stabile protobitumens and stabile metabitumens (Tomkeieff, 1954).

fixed ash *inherent ash.*

fixed carbon In coal, coke, and bituminous materials, the remaining solid, combustible matter after removal of moisture, ash, and volatile matter, expressed as a percentage.

fixed carbon ratio *carbon ratio.*

fixed cheek The part of the *cheek* of a trilobite inside of the facial suture, remaining attached to the glabella at the time of molting. Cf: *free cheek.* Syn: *fixigene.*

fixed dune *anchored dune.*

fixed elevation An elevation that has been adopted either as a result of tide observations or previous adjustment of spirit leveling and that is held at its accepted value in any subsequent adjustment (Mitchell, 1948, p.27).

fixed form A crystal form whose indices are fixed relative to length, e.g. cube |100|, or octahedron |111| of the cubic system. Syn: *singular crystal form.*

fixed ground water Ground water in material having interstices so small that the water is held permanently to the walls of the interstices, or moves so slowly that it is not available for withdrawal at useful rates. Outside the zone of saturation material with infinitely small openings can hold water indefinitely against the pull of gravity, while there is apparently always movement, even though at very low rates, within the zone of saturation.

fixed layer The inner, relatively immobile layer of ions in an electrolyte, required to satisfy a charge unbalance within a solid with which the electrolyte is in contact. It constitutes part of the *double layer* of charge adjacent to the electrolyte-solid interface. Cf: *diffuse layer.*

fixed moisture Moisture retained in the soil in a quantity that is less than the hygroscopic coefficient.

fixigene The *fixed cheek* of a trilobite. Syn: *fixigena.*

fizelyite A lead-gray mineral: $Pb_5Ag_2Sb_8S_{18}$. It is closely related to andorite and ramdohrite. Also spelled: *fizélyite.*

fjall A Swedish word for a mountain rising above the timberline and having flat undissected areas (Stamp, 1961, p.193). See also: *fjeld; fell.*

fjard A small, narrow, irregular inlet or bay, typically formed by submergence of a glacial valley excavated in a lowland along the margin of a flat rocky coast, as that of S Sweden; often accompanied by numerous fringing islands. A fjard is shorter and shallower, and broader in profile, than a *fjord,* but deeper than a ria. Pron: *fyard.* Etymol: Swedish *fjård,* a large continuous area of water surrounded by skerry-guard islands (Stamp, 1961, p.193); usage in English therefore has taken a more specialized meaning not apparent in Sweden. Syn: *fiard.*

fjeld A Norwegian word for "field" having a wide meaning but when used in English refers to an elevated, rocky, bleak, almost barren plateau above the timberline, covered with snow during the winter, as in the Scandinavian upland. See also: *fell; fjäll.* Syn: *fjell.*

fjeldbotn A Norwegian term for a cirque carved by an ice field (Termier & Termier, 1963, p.405).

fjeldmark *fell-field.*

fjord (a) A long, narrow, winding, U-shaped and steep-walled, generally deep (often several hundred fathoms) inlet or arm of the sea between high rocky cliffs or slopes along a mountainous coast, typically with a shallow sill or threshold of solid rock or earth material submerged near its mouth, and becoming deeper far inland; it usually represents the seaward end of a deeply excavated glacial-trough valley that is partially submerged by drowning after the melting of the ice. Examples: along the glaciated coasts of Alaska, Greenland, and Norway. (b) Any embayment of the seacoast in a Scandinavian country regardless of the adjacent topography, as a *fjard* in the low flat Swedish coast or a *fôrde* in eastern Denmark.---Pron: *fyord.* Etymol: Norwegian. Cf: *estuary; ria.* Syn: *fiord; fyord; sea loch.*

fjord coast A deeply indented, glaciated coast characterized by a partial submergence of glacial troughs and by the presence of steep parallel walls, truncated spurs, and hanging valleys. Its development is favored by marine west-coast climates combined with strong relief, as in Alaska or southern Chile.

fjorded Dissected or partially divided by fjords.

fjord ice Ice formed during the winter in a fjord and melting in place during the summer. Cf: *sikussak.*

fjord lake A lake in a glacially excavated rock basin of a U-shaped valley near sea level.

fjord shoreline A shoreline of submergence characterized by the development of numerous *fjords;* e.g. along the west coast of Norway.

fjord strait A strait formed by the adjoining of two fjords opening in opposite directions; e.g. the Straits of Magellan.

flabellate Fan-like or resembling a fan in shape; e.g. said of a fan-shaped corallite, or of a meandroid corallum with a single continuous, laterally free, linear series of corallites.

flabellum A body part resembling a fan; esp. the thin, most distal exite of the limb of a branchiopod crustacean. Pl: *flabella.*

Fladen (a) A distinctive, generally flat, pancake-shaped body resembling a volcanic bomb and composed of glass and fragments of rocks and minerals, found in the suevite breccias at the Ries basin in Germany. It exhibits flow structure and surface sculpturing apparently produced by aerodynamic forces. See also: *impact bomb.* (b) Any similar glass-rich, aerodynamically shaped body, formed by meteorite impact, and found associated with other meteorite impact structures.----Etymol: German, "flat cake". Pl: *Fladen.* Syn: *Flådle.*

flag A syn. of *flagstone.* The term is often used in the plural, such as the "Lingula flags" of the European Upper Cambrian.

flagellar field The area around the flagella of a coccolithophore; e.g. *covered flagellar field* and *naked flagellar field.*

flagellar pore One of the pores in a dinoflagellate for extrusion of flagella, usually located at the anterior or the posterior junction of girdle and sulcus.

flagellate n. An organism, esp. a protozoan, that bears flagella. Syn: *mastigophoran.*----adj. Possessing flagella.

flagellated chamber Any cavity, in a sponge, lined by choanocytes. See also: *chamber* [*paleont*].

flagellum (a) Any of various elongated and thread-like appendages of animals, such as the slender, multiarticulate, distal part of the limb of a crustacean, or a long, slender, whiplike extension in a sponge. (b) A long, whiplike, tapering, protoplasmic process that projects singly or in groups from a cell or microorganism, that is possibly equivalent to a much enlarged *cilium,* and that is the primary organ controlling the movement (through water) of a flagellated protozoan and of many algae, bacteria, and zoospores.---Pl: *flagella.*

flaggy (a) Splitting or tending to split into layers of suitable thickness for use as flagstones; specif. descriptive of a sedimentary rock that splits into layers from 1 cm to 5 cm in thickness (McKee & Weir, 1953, p.383). (b) Said of bedding that consists of layers from 1 cm to 10 cm in thickness (Payne, 1942). (c) Pertaining to a flag or flagstone. (d) Said of a soil full of flagstone fragments.

flagstaffite A mineral: $C_{10}H_{22}O_3$. It occurs in colorless, transparent, orthorhombic crystals identical with *cis*-terpin hydrate, and is found with resin in the radial cracks of fossil pine trees.

flagstone (a) A hard, evenly- and thin-bedded, usually micaceous and fine-grained sandstone (or firm shale or sandy limestone) that splits readily and uniformly along bedding planes or joints into large, thin, flat slabs suitable for making pavements or covering the side of a house. Cf: *freestone.* (b) A flat slab of flagstone used for paving; esp. a thin piece split from flagstone. Also, a surface of such stone. (c) A relatively thin and flat fragment (of limestone, sandstone, shale, slate, or schist) occurring in the soil, having a length in the range of 15-38 cm (6-15 in.) (SSSA, 1965, p.336).---Syn: *flag; slabstone; cleftstone.*

flaikes A Scottish term for a shaly or fissile sandstone that splits along the grain. Se also: *fakes.* Syn: *flakes.*

flajolotite *tripuhyite.*

flake [**mass move**] A large slab-like unit of a landslide.

flake [**snow**] *snowflake.*

flake graphite In economic geology, graphite disseminated in metamorphic rock as thin, megascopically visible flakes that are separated from the rock by mechanical means. Syn: *crystalline flake.*

flake mica Finely divided mica recovered from mica schist or sericite schist or obtained as a byproduct of beneficiation of feldspar or kaolin.

flame coal Bituminous coal that contains more than 40% volatile matter. Etymol: German *Flammkohle*, "flame coal".

flame emission spectrometry *flame photometry.*

flame photometer A *spectrophotometer* utilizing flame excitation of samples, usually compounds in solution, to provide spectra for analysis. See also: *photometer.*

flame photometry Measurement of the intensity of the lines in a *flame spectrum* by a *flame photometer.* Syn: *flame emission spectrometry.*

flame spectroscopy The observation of a *flame spectrum* and all processes of recording and measuring that go with it.

flame spectrum The spectrum of light emitted by a substance by heating it in a nonluminous flame.

flame structure [sed] A term introduced by Walton (1956, p.267) for a sedimentary structure consisting of wave- or flame-shaped plumes of mud that have been squeezed irregularly upward into an overlying layer. It is probably formed by load casting accompanied by horizontal slip or drag. The term *antidune* as used by Lamont (1957) is synonymous. See also: *load wave; streaked-out ripples.*

flame structure [pyroclast] The presence of *fiamme* in a welded tuff, e.g. piperno.

flame test A qualitative analysis of a mineral made by intensely heating a sample in a flame and observing the flame's color, which will be indicative of the element involved, e.g. green from copper.

flange (a) A projecting rim or edge of an organism; e.g. a plate-like marginal extension along foraminiferal chambers (as in *Sphaeroidinella*) or bordering the aperture as a highly developed foraminiferal lip (as in *Hantkenina*), or the part of a coccolith that spreads out like a rim, or a shelf-like structure along the inner side or outer side of a blade or bar of a conodont, or a lateral projection from a brachiopod crus, formed by anterior extension of part of the outer hinge plate adjacent to the crural base. (b) In palynology, an equatorial extension of the exine of a spore. It is a less precisely defined term than *cingulum* or *zona.* (c) A spreading out, such as of a vein of ore.

flank [paleont] (a) The lateral side of a cephalopod conch between the venter and the dorsum; the *whorl side.* (b) The sloping surface between the venter and the lateral margin of a brachiopod valve. (c) In a bivalve mollusk, the median part of the surface of the valve.

flank [struc geol] *limb.*

flanking moraine A moraine left by a glacial lobe or by a tongue-like projection of an ice sheet (Fairchild, 1932, p.629). Cf: *lateral moraine.*

flaser The streaky layers of parallel, scaly aggregates surrounding the lenticular bodies of granular material in *flaser structure.* Etymol: German, "streak".

flaser gabbro Coarse-grained blastomylonite formed by dislocation metamorphism of a gabbro. Flakes of mica or chlorite sweep around augen of feldspar and/or quartz with much recrystallization and neomineralization (Joplin, 1968, p.21). Cf: *gabbro schist; zobtenite.*

flaser structure [met] A structure in dynamically metamorphosed rock in which lenses and layers (*flaser*) of original or relatively unaltered granular minerals are surrounded by a matrix of highly sheared and crushed material, giving the appearance of a crude flow structure, e.g. flaser gabbro. Cf: *mylonitic structure; augen structure.* Syn: *phacoidal structure.*

flaser structure [sed] A primary sedimentary structure consisting of fine-sand or silt lenticles that are commonly aligned and usually cross-bedded and that superficially resemble metamorphic flaser structure (Pettijohn & Potter, 1964, p.305).

flash [mining] (a) A subsidence of the surface resulting from underground mining, esp. from the working of rock salt and pumping of brine. Cf: *inbreak; heave; crown-in.* (b) A small lake or shallow reach of water occupying a hollow produced by such a subsidence.

flash [water] (a) A sudden rise of water in a stream, as where water is held back by a dam. (b) A pool of water; a marsh or marshy place.---Cf: *flush.*

flash figure An optic analysis of uniaxial crystals under the conoscope, a vague black cross that appears when the optic axis is parallel to the plane of polarization of either prism. Only slight rotation from this position causes the cross to separate into two hyperbolic segments and leave the field (Wahlstrom, 1948).

flash flood A local and sudden flood or torrent or relatively great volume and short duration, overflowing a stream channel in a usually dry valley (as in a semiarid area), carrying an immense load of mud and rock fragments, and generally resulting from a rare and brief but heavy rainfall over a relatively small area having steep slopes. It may also be caused by ice jams and by dam failure. See also: *freshet.* Syn: *flashy stream.*

flash opal Opal in which the play of color is pronounced only in one direction (Shipley, 1951, p.82).

flashy stream *flash flood.*

flat [eco geol] n. A horizontal orebody, regardless of genetic type.

flat [geog] (a) A tract of low-lying, level wetland; e.g. a marsh or swamp in a river valley. (b) A term used in northern and central U.S. as a syn. of *bottom,* or low-lying land along a watercourse.

flat [geomorph] adj. Having or marked by a continuous surface or stretch of land that is smooth, even, or horizontal, or nearly so, and that lacks any significant curvature, slope, elevations, or depressions.---n. A general term for a level or nearly level surface or small area of land marked by little or no relief, as a plain; specif: *mud flat; valley flat.* Also, a nearly level region that visibly displays lower relief than its surroundings.

flat [lake] (a) The low-lying, exposed, flat land of a lake delta or of a lake bottom. (b) The flat bottom of a desiccated lake in the arid parts of western U.S.----Commonly used in the plural.

flatiron One of a series of short, triangular-shaped hogbacks terminating a spur or ridge on the flank of a mountain, having a narrow apex at the top and a broad base below, resembling (when viewed from the side) a huge flatiron standing on its heel; it usually consists of a plate of steeply inclined resistant rock adhering to the dip slope.

flat joint In igneous rock, a randomly oriented joint dipping at an angle of 45° or less.

flatland A region, or tract of land, characterized by predominant levelness or by no significant variation in elevation, as along a river or a coast.

flatness A term used by Wentworth (1922a) to express the shape of a pebble, defined as the ratio of the radius of curvature of the most convex portion of the flattest developed face to the mean radius of the pebble. Wentworth (1922b) also expressed flatness as the arithmetic mean of the long and intermediate diameters (length and width) of the pebble divided by the short diameter (thickness). Cf: *roundness.*

flats and pitches (a) A phrase descriptive of the structure of the lead and zinc deposits in dolomite of the Upper Mississippi region of the U.S. The "flats" are the nearly horizontal solution openings; the "pitches" are the inclined, interconnecting joints. (b) A slump structure of both horizontal and steeply inclined cracks in sedimentary strata. Syn: *pitches and flats.*

flattening The flattening of the Earth is the ellipticity of the spheroid and equals the ellipticity of an ellipse forming a meridional section of the spheroid. If a and b represent the major and minor semi-axes of the spheroid, and f is the flattening of the Earth, then $f = a-b/a$.

flat-topped ripple mark A ripple mark with a flat, wide crest between narrow troughs; e.g. a shallow-water ripple mark whose crest was planed off during ebb tide or fall in water level.

flaw [gem] A faulty part of a diamond or other gemstone, such as a crack, an inclusion, visible imperfect crystallization, or internal twinning or cleavage. Syn: *imperfection*.

flaw [ice] (a) A narrow separation zone (*fracture* [ice]) between pack ice and fast ice, formed when pack ice at this boundary undergoes shearing due to a strong wind or current, and characterized by pieces of ice in a chaotic state; it is not wide enough to permit passage of a navigable vessel. (b) Obsolete syn. of *flaw lead*.

flaw [struc geol] An old term for a strike-slip fault.

flaw lead A navigable passage between pack ice and fast ice. See also: *shore lead*.

flawless Said of a diamond or other gemstone that is free from all internal and external blemishes or *flaws* of every description as observed under 10-power magnification.

flaxseed ore A term for an iron-bearing sedimentary deposit, e.g. *Clinton ore*, composed of disk-shaped hematitic oolites that have been somewhat flattened parallel to the bedding plane. Cf: *fossil ore*.

F layer [seism] The seismic region of the Earth from 4710 km to 5160 km, equivalent to the *transition zone* between the *outer core* and the inner core. It is a part of a classification of the interior of the Earth, having layers A to G. Together with the G layer, it is the equivalent of the *lower core*.

F layer [soil] A surficial layer of humus or partially decomposed organic matter over a forest soil. It lies above the *H layer*.

flèche d'amour An acicular, hair-like crystal of rutile embedded in sagenitic quartz. The term is used loosely as a syn. of "sagenitic quartz", and was used formerly for amethyst containing brown needles of goethite. Etymol: French, "arrow of love". Syn: *cupid's dart; love arrow*.

Fleckschiefer A type of *spotted slate* characterized by minute flecks or spots of indeterminate material. Etymol: German. Cf: *Knotenschiefer; Garbenschiefer; Fruchtschiefer*.

fleet (a) A term used in England for a small, shallow inlet, estuary, creek, or arm of the sea. Also, a place where water flows; a small rapid stream. (b) A small, usually salty or brackish lagoon behind the coastline, separated from the open sea by a long bank of sand or shingle parallel to the coast (Monkhouse, 1965, p.125).

fleischerite A mineral: $Pb_3Ge(SO_4)_2(OH)_6 \cdot 3H_2O$.

fleshy sponge Any sponge that lacks a skeleton, i.e. one of the demosponges.

fleur-de-lis A sedimentary structure consisting of frondescent spatulate elevations and resembling the appearance of an iris. Etymol: French *fleur de lis*, "lily flower". Syn: *fleur-de-lys*.

flexible Said of a mineral, the tenacity of which allows it to be bent without breaking but without returning to its original form; e.g. talc.

flexible crinoid Any crinoid belonging to the subclass Flexibilia, characterized by the incorporation, but not firm attachment, of the lower brachials in the dorsal cup and by a flexible tegmen.

flexible sandstone A fine-grained, thin-layered variety of *itacolumite*.

flexible silver ore *sternbergite*.

flexostyle The tubular enrolled chamber of a foraminiferal test immediately following the proloculus (as in *Amphisorus*).

flexural fold A fold in which the mechanism of folding, either flow within layers (*flexure-flow fold*) or slip between them (*flexure-slip fold*) is controlled by the layering of the strata. Cf: *passive fold*.

flexural slip *bedding-plane slip*.

flexure (a) A *fold*; a *monocline* formed by warping or other gentle deformation rather than by compression. (b) *hinge*.

flexure correction In pendulum observations of gravity, a necessary correction for the influence of the rather complex coupled vibration phenomena caused by the oscillation of the receiver case, the pillar, and the surface soil. Such vibrations cause the period of the pendulum itself to change.

flexure fault (a) *growth fault*. (b) A flexure that changes into a fault (Dana, 1895, p.109).

flexure-flow fold A flexural fold in which the mechanism of folding is flow within layers, resulting in thickening of hinge areas and thinning of limbs. Cf: *flexure-slip fold*.

flexure line A line, extending from the beak to the anterior border of both ventral propareas in some lingulacean brachiopods, marked by deflection of growth lines (TIP, 1965, pt.H, p.145).

flexure-slip fold A *flexure fold* in which the mechanism of folding is slip along bedding planes or along surfaces of foliation. There is no change in thickness of individual strata. Cf: *flexure-flow fold*.

flight *terrace flight*.

flight altitude The vertical distance above a given datum (usually mean sea level) of an aircraft in flight or during a specified portion of a flight. See also: *flight height*.

flight height A term used in aerial photography for the *flight altitude* when the datum is mean ground level of the area being photographed. Syn: *absolute altitude*.

flight line A line drawn on a map or chart to represent the planned or actual track of an aircraft during the period of taking aerial photographs.

flight map A map on which are indicated the desired flight lines and/or the positions of air stations previous to the taking of aerial photographs, or a map on which are plotted, after photography, selected air stations and the tracks between them.

flight strip A succession of overlapping aerial photographs taken along a single course.

flimmer *mastigoneme*.

flinkite A greenish-brown mineral: $Mn_3(AsO_4)(OH)_4$.

flint [mineral] (a) A term that has been considered as a mineral name for a massive, very hard, somewhat impure variety of chalcedony, usually black or of various shades of gray, breaking with a conchoidal fracture, and striking fire with steel. Syn: *firestone*. (b) Pulverized quartz of any kind; e.g. "potter's flint" in the form of powdered quartz, made by pulverizing flint pebbles.

flint [sed] A term that is widely used as a syn. of *chert* or for the homogeneous, dark-gray or black variety of chert. According to Tarr (1938), the term "flint" should either be discarded or be reserved for siliceous artifacts (such as the "flint arrowheads" used by primitive man) because rocks described as flint are identical with chert in texture and composition, despite the fact that the term "flint" has been in use since about A.D. 700 for "anything hard" and since A.D. 1000 for "a variety of stone" and that it antedates "chert" by almost 1000 years. Flint has been described as having a denser texture, a more perfect (smooth) conchoidal or less splintery fracture, a smaller quartz content, and greater infusibility than chert, and as having thin translucent splinters or sharp cutting edges. The term is commonly used in southern England for one of the siliceous nodules occurring in the Cretaceous chalk beds, and elsewhere in England for any hard rock (such as in Shropshire County for a fine-grained sandstone suitable for building purposes). Syn: *black chert; silex; hornstone; petrosilex*.

flint clay A very hard, smooth, flint-like fireclay that breaks with a conchoidal fracture and that develops no plasticity when ground up. Its principal clay mineral is halloysite.

flint curtain A concentration of silica derived from flints, occurring along a vertical joint plane in the chalk beds of eastern Denmark; it results from dehydration of silica gel and flowage of silica, immediately subsequent to jointing.

flint meal Fine, flour-like material consisting primarily of minute fossils (such as sponge spicules) and occurring in an enclosed cavity of a flint nodule from the chalk beds of southern England.

flinty (a) Composed of flint, or containing more than the normal percentage of silica; e.g. a "flinty slope". (b) Resembling flint in hardness or fracture; flint-like.

flinty crush rock *ultramylonite.*

flinty slate A *touchstone* consisting of siliceous slate.

float A general term for an isolated, displaced fragment of rock within another rock or on the surface, e.g. *floating ore.* Cf: *floating reef.* Syn: *floater.*

float coal Isolated bodies of coal in sandstone or shale, probably deposited as pieces of peat that were eroded and transported from the original deposit. Cf: *coal gravel.* Syn: *raft.*

float coccolith A modified coccolith serving as a suspension organ in nonmotile coccolithophores exhibiting dimorphism (such as *Thorosphaera*).

floater *float.*

floating ice Any form of ice floating in water, including ice that is stranded or grounded and ice formed on land but drifting in the sea. The term formerly excluded icebergs and other forms of land ice. See also: *drift ice.*

floating island A mass or mat of vegetation with little or no soil, floating freely in a lake or tropical sea, usually due to detachment from a marshy or boggy shore during a storm or a rise in the water level.

floating marsh *flotant.*

floating peat Peat that is derived from floating plants.

floating reef An isolated, displaced rock mass in alluvium. Cf: *float.*

floating sand [oil] *running sand.*

floating sand [sed] An isolated grain of quartz sand that is not, or does not appear to be, in contact with neighboring sand grains scattered throughout the finer-grained matrix of a sedimentary rock, esp. of a limestone; e.g. a grain entirely surrounded in three dimensions by coarse mosaic of calcite cement. Syn: *floating sand grain.*

float ore Scattered fragments of vein material broken off from outcrops; a type of float.

floatstone [mineral] A lightweight, porous, friable variety of opal that floats on water and occurs in white or grayish, spongy, and concretionary or tuberous masses. Also spelled: *float stone.* Syn: *swimming stone.*

floatstone [mining] A miner's term for cellular or honeycomb quartz detached from a lode.

floc (a) A loose, open-structured mass formed in a suspension by flocculation; e.g. a small aggregate of tiny sedimentary grains or of colloidal clay particles. (b) A term used by Brewer (1964, p.367) for soil *plasma* that has a relatively low luster and a rough surface that gives it the appearance of clusters of silt-size grains in reflected light under a hand lens up to magnifications of approximately 20 times. Cf: *lac* [*soil*].

flocculation The act or process by which a number of individual, minute, suspended particles are tightly held together in clot-like masses or are loosely aggregated or precipitated into small lumps, clusters, or granules; e.g. the clumping together of soil colloids into a small group of soil particles, or the coalescence of a finely divided precipitate (in a liquid) into small, soft, woolly, cloud-like flakes, or the deposition or settling out of suspension of clay particles in salt water, or the agglomeration of colloidal particles in a suspension by the addition of an electrolyte.

floe A piece of floating ice other than fast ice or glacier ice, larger than an *ice cake* and smaller than an *ice field.* Floes are subdivided according to horizontal extent and many varying size limits have been used in the past; the U.S. Naval Oceanographic Office (1968, p.A27) gives the following dimensions: "giant" (over 10 km); "vast" (2-10 km); "big" (500 m to 2 km); "medium" (100-500 m); "small" (20-200 m). Syn: *ice floe.*

floeberg (a) A massive piece of sea ice composed of a *hummock* or a group of hummocks, frozen together and separated from any ice surroundings, and floating with its highest point up to 5 m above sea level (U.S. Naval Oceanographic Office 1968, p.B33). It resembles a small iceberg. (b) In the older literature, a thick mass of well-hummocked sea ice originating from an ice floe, and sometimes projecting more than 15 m above sea level.

floe calcite In a cave, a very thin film of calcite on the surface of a body of standing water. Syn: *calcite flottante.*

floe till *berg till.*

Floetz A name applied by A.G. Werner in the 1790's to the group or series of rocks that included most of the obviously stratified, comparatively flat, fossiliferous rocks (and certain associated trap rocks) and that were believed to represent the emergence of mountains from beneath the receding ocean, with products of the resulting erosion deposited on the mountain flanks. The rocks succeeded the *Transition* series and included the whole range of strata from the Devonian through the Tertiary. Etymol: German *Flötz* (now *Flöz*), "flat layer, stratum, seam, bed". Syn: *Secondary.*

floitite A rock consisting of biotite and those minerals that are typical of the greenschist facies. The term was originated by Becke in 1922.

flokite *mordenite.*

flood [sed] A term used by Milner (1940, p.457) to describe the occurrence, in a sedimentary rock, of a mineral species "so far in excess of all others as to constitute almost a pure concentrate".

flood [water] (a) A rising body of water (as in a stream, lake, or sea, or behind a dam) that overtops its natural or artificial confines and that covers land not normally under water; esp. any relatively high streamflow that overflows its banks in any reach of the stream, or that is measured by gage height or discharge quantity. (b) A flood of special severity or local interest; specif. the Flood, or the *Noachian flood.* (c) An archaic term for a large body of moving water, such as a river.

flood absorption A reduction in discharge resulting from the storage of flowing water in a reservoir, channel, or lake (ASCE, 1962).

flood basalt *plateau basalt.*

flood basin (a) The tract of land actually covered by water during the highest known flood. (b) The broad, flat area between a sloping, low plain and a natural levee of a river, "occupied by heavy soils and commonly having either no vegetation or strictly swamp vegetation" (Bryan, 1923b, p.39). Syn: *tule land.*

flood control The prevention or reduction as far as possible of damage caused by flooding, such as containing water upstream from areas where it would do damage, improving channel capacity to convey water past or through critical areas with the least amount of damage, and diverting excess water into bypasses or floodways.

flood crest The highest stage of a flood. The term is nearly synonymous with *flood peak*, but does not refer to discharge since it connotes the top of the flood wave (Langbein & Iseri, 1960, p.10).

flood current The tidal current associated with the increase in the height of a tide, generally moving toward the shore or up a tidal river or estuary. Cf: *ebb current.* Erroneous syn: *flood tide.*

flood dam A dam to store floodwaters temporarily or to supply a surge of water (such as for clearing a channel). Syn: *flooding dam.*

flooded stream *drowned stream.*

flood frequency The average occurrence of flooding of a given magnitude, over a period of years.

flood-frequency curve A graphic illustration of the number of times per year that a flood of a given magnitude is equaled or exceeded.

flood fringe *pondage land.*

floodgate (a) A gate for shutting out, admitting, releasing, or otherwise regulating a body of water, such as excess water in times of flood; specif. the lower gate of a lock. See also: *sluice.* (b) The stream stopped by or allowed to pass by a floodgate.

flood icings *aufeis.*

flooding The covering or causing to be covered with a fluid, such as the covering of flat lands with a thin sheet of water; the filling or becoming full with water, esp. to excess.

flooding ice *aufeis.*

floodland The land, bordering a river, that may be submerged by floodwaters; a *flood plain.*

flood peak The highest discharge or stage value of a flood. Cf: *flood crest.* Syn: *peak discharge.*

flood plain (a) The surface or strip of relatively smooth land adjacent to a river channel, constructed (or in the process of being constructed) by the present river in its existing regimen and covered with water when the river overflows its banks at times of high water. It is built of alluvium carried by the river during floods and deposited in the sluggish water beyond the influence of the swiftest current. A river has one flood plain and may have one or more terraces representing abandoned flood plains. Cf: *valley flat; erosional flood plain.* (b) Any flat or nearly flat, usually dry lowland that borders a stream and that may be covered by its waters at flood stages; the land, beyond a stream channel, described by the perimeter of the maximum probable flood. Syn: *floodland.* (c) The part of a lake-basin plain between the shoreline and the shore cliff, subject to submergence during a high stage of the lake.--- Also spelled: floodplain; flood-plain.

flood-plain deposit Sandy and clayey sediment deposited by river water that was spread out over a flood plain; a deposit beneath and forming a flood plain, being thickest near the river and thinning out toward the valley slopes. See also: *overbank deposit.* Syn: *vertical-accretion deposit.*

flood-plain icings *aufeis.*

flood-plain lobe The part of a flood plain enclosed by a stream meander.

flood-plain meander scar A crescentic mark indicating the former position of a river meander on a flood plain.

flood-plain scroll One of a series of short, crescentic, slightly sinuous strips or patches of coarse alluvium formed along the inner bank of a stream meander and representing the beginnings of a flood plain. Syn: *flood scroll.*

flood-plain splay A small alluvial fan or other outspread deposit formed where an overloaded stream breaks through a levee (artificial or natural) and deposits its material (often coarse-grained) on the flood plain. Syn: *sand splay; channel splay.*

flood plane (a) The position occupied by the water surface of a stream during a specific flood. (b) Loosely, the elevation of the water surface at any of various points along the stream during a specific flood.

flood probability The probability, determined statistically, that a flood of a given size will be equaled or exceeded in a given period, e.g. a 10% probability would be called a ten-year flood.

flood routing Progressive determination of the timing and shape of a flood wave at successive points along a river (Langbein & Iseri, 1960, p.10).

flood scroll *flood-plain scroll.*

flood series A listing of flood events for a given period of time, arranged in order of magnitude.

flood stage (a) The height of the gage at the lowest bank of the reach (other than an unusually low place or break). (b) The stage at which stream overflow begins to cause damage.

flood tide (a) *rising tide.* (b) A tide at its greatest height. (c) An erroneous syn. of *flood current.*

flood tuff A syn. of *ignimbrite.*

floodwall A wall, often of reinforced concrete, built to prevent

flooding. Cf: *levee.*

floodwater (a) Water that has overflowed its confines; the water of a flood. (b) The flooded area in back of a dam; an impoundment.

flood wave A rise in the stage of a stream that culminates in a crest before receding.

floodway (a) A large-capacity channel constructed to divert floodwaters or excess streamflow from populous or damageable areas, such as a bypass route marked out by levees. (b) The part of a flood plain kept clear of encumbrances and reserved for emergency diversion of floodwaters. (c) *flowage land.*

flood zone [stratig] *peak zone.*

flood zone [stream] (a) The land bordering a stream, subject to floods of about equal frequency; e.g. a strip of the flood plain, subject to flooding more often than once but not as frequently as twice in a century (Langbein & Iseri, 1960, p.11). (b) The land bordering a reservoir or stream impoundment, subject to inundation above the normal operating level.

floor [eco geol] The *footwall* of a horizontal orebody.

floor [intrus rocks] The country rock bordering the lower surface of an igneous intrusion. Cf: *roof* [intrus rocks].

floor [strat] A rock surface, usually an eroded surface, upon which sedimentary strata have been deposited.

floor [geomorph] (a) The *bed* of any body of water; esp. the continuous and gently curved or essentially horizontal surface of the ground beneath the water of a stream, lake, or ocean. Syn: *bottom.* (b) *valley floor.*

floor [coal mining] The bed immediately underlying a coal seam. Cf: *footwall.*

floor plate Any of double- or single-row plates forming an ambulacral groove in an echinoderm.

flora The entire plant population of a given area, environment, formation, or time span. Cf: *fauna.*

floral stage A time-stratigraphic unit (stage) based on a florizone or commonly on a floral assemblage; e.g. the Ravenian (upper Eocene) of Washington State.

floral zone *florizone.*

florencite A pale-yellow mineral: $CeAl_3(PO_4)_2(OH)_6$.

floricome A sponge spicule (hexaster) with petal-like terminations on the rays.

Florida earth A variety of fuller's earth from Florida (esp. from Quincy and Jamieson) or resembling that from Florida.

Florida phosphate *floridite.*

floridite An obsolescent term for phosphorite from Florida, usually fluorapatite. Syn: *Florida phosphate.*

florizone A biostratigraphic unit or body of strata characterized by a particular assemblage of fossil floras, regardless of whether it is inferred to have time or only environmental significance. Although the term, like *faunizone*, has been given different meanings, it is close in concept to *assemblage zone* and has been generally regarded as the plant-based variety of (biostratigraphic) zone. The American Commission on Stratigraphic Nomenclature (1961, art.21d) states that the term is "not generally accepted" and that its correct definition is "in dispute". Syn: *floral zone.*

florule A term used by Fenton & Fenton (1928,p.15) for a diminutive flora or an assemblage of fossil plants associated in a single stratum or a few contiguous strata of limited thickness and dominated by the representatives of one community; the floral assemblage of a *zonule.* Cf: *faunule.* Syn: *florula.*

floscelle A star-shaped area around the peristome of an echinoid, formed by phyllodes and bourrelets.

flos ferri An arborescent variety of aragonite occurring in delicate white coralloid masses that commonly encrust hematite, forming picturesque snow-white pendants and branches.

floss A British term for a stream.

flotant A coastal marsh formed along an abandoned channel or in a low basin between natural levees of active and inactive stream channels, as in Louisiana south of New Orleans; it is not as firm as marshland. Syn: *floating marsh.*

flotation *crystal flotation.*

flour A finely powdered rock or mineral mass, resulting from pulverization and grinding; e.g. chalky-appearing, finely comminuted carbonate mud in limestone, formed by disintegration and abrasion of fossil debris and algal growths under intense wave or current action in shoal areas. See also: *rock flour.*

flour sand *very fine sand.*

flow [coast] A Scottish term for an arm of the sea; used chiefly in place names.

flow [mass move] (a) A mass movement of unconsolidated material that exhibits a continuity of motion and a plastic or semifluid behavior resembling that of a viscous fluid; e.g. *creep; solifluction; earthflow; mudflow; debris avalanche.* Water is usually required for most types of flow movement. (b) The mass of material moved by a flow.

flow [exp struc geol] Any rock deformation that is not instantly recoverable without permanent loss of cohesion. Various types of flow in which the mechanism is known include *cataclastic flow, gliding flow,* and *recrystallization flow.* Syn: *flowage; rock flowage.*

flow [volc] *lava flow.*

flow [glaciol] *glacier flow.*

flow [hydraulics] The movement of water, and the moving water itself; also, the rate of movement.

flowage [exp struc geol] *flow* [exp struc geol].

flowage [streams] (a) An act of flowing or flooding, such as the overflowing of a stream onto adjacent land; the state of being flooded. (b) A body of water resulting from flowage; the floodwater of a stream. The term is used locally in Wisconsin for the backwater of an artificial lake. (c) The area affected by a previous flooding.

flowage cast A term used by Birkenmajer (1958, p.141) for a sole mark believed to result from the flowage of mobile, hydroplastic sand over the uneven bottom in the direction of slope; it may be transverse, longitudinal, or multidirectional. See also: *flow cast.*

flowage differentiation Magmatic differentiation resulting from the retarding effect of relatively stable wall rock on the movement, under pressure, of crystal mush in a magma (Schieferdecker, 1959, term 4807).

flowage fold *flow fold.*

flowage land The part of a flood plain that will be covered by the water impounded by a proposed dam, exclusive of the river bed; the principal flow-carrying part of the natural cross section of a stream. Syn: *floodway.*

flowage line *flow line.*

flow-and-plunge structure A variety of *cross-lamination,* consisting of short, obliquely laminated beds deposited irregularly at various angles of slope, resulting from tidal action accompanied by plunging waves.

flow banding The structure of an igneous rock, characterized by alternating layers of different mineralogic composition and texture, formed as a result of the flowing of magma or lava. Syn: *fluxion banding.* See also: *banding* [ign].

flow bog A peat bog whose surface level fluctuates with rain and the tides.

flow breccia A breccia that is formed contemporaneously with the movement of a lava flow; the cooling crust becomes fragmented while the flow is still in motion. It is a type of *autobreccia.*

flow cast (a) A term introduced by Shrock (1948, p.156) for a sole mark consisting of a lobate ridge or other raised feature and representing the filling of a depression produced by the flowage or warping of the soft and hydroplastic underlying sediment. Kuenen (1953, p.1058) applied the term *load cast* to such a structure produced by vertical adjustments. Prentice (1956) revived "flow cast" for a sole mark resulting from a combination of load casting and current-oriented flow, such as a load cast modified by horizontal flowage of sediment during or after settling. See also: *flowage cast.* (b) *flute cast.*

flow chart A graphic representation or schematic diagram of steps in a sequence of operations that are represented by symbols, as for a computer program.

flow cleavage A syn. of *slaty cleavage,* so called because of the assumption that recrystallization of the slaty minerals is accompanied by rock flowage.

flow-duration curve A type of *duration curve* showing how often a particular stream discharge is equaled or exceeded.

flow earth *solifluction mantle.*

flower The reproductive structure of an angiosperm. In a morphologic sense, it is considered to be a specialized branch system.

flowering plant An informal designation of an *angiosperm.*

flow fold A fold composed of relatively plastic rocks that have flowed towards the synclinal trough. In this type of deformation, there are no apparent surfaces of slip. Syn: *flowage fold.* Cf: *reverse- flowage fold; rheid fold.*

flowing artesian well An *artesian well* whose head is sufficient to raise the water above the land surface. Cf: *nonflowing artesian well; flowing well.* Syn: *blow well.*

flowing well (a) A well that yields water at the land surface without pumping or other means of raising it to the surface. It is distinguished from a *flowing artesian well* by the possibility that the flow may be due to gas rather than artesian pressure. (b) *flowing oil well.*

flow joint A joint parallel to the flow layers of a plutonic rock (Tomkeieff, 1943).

flow layer A layer in an igneous rock, distinguished from adjacent layers by compositional or textural differences. The layering is a result of the flow of magma or lava prior to solidification. Cf: *flow line* [petrology].

flow line [petrology] An alignment of crystals, mineral streaks or inclusions in an igneous rock, the long axes of the crystals or grains being parallel. Cf: *flow layer.*

flow line [hydraul] (a) The position of the surface of a flowing fluid. (b) A water-level contour around a body of water, e.g. maximum or mean flow line of a lake.----*flowage line.* (c) In an open channel, the *hydraulic grade line* (ASCE, 1962).

flow mark (a) A small channel or gouge cut in a sedimentary surface by a current of water; a flute. (b) A cast of a flow mark, preserved in overlying sediment (Rich, 1950); specif. a *flute cast.* (c) A small ridge formed on the upper surface of a muddy sediment by a current of water (McKee, 1954, p.63).-See also: *current mark.* Syn: *flow marking.*

flow net In the study of seepage phenomena, a graph of flow lines and equipotential lines.

flow profile The form of the water surface of a *gradually varied flow;* it is commonly known as the *backwater curve.*

flow regime A range of streamflows with similar bed forms, resistance to flow, and mode of sediment transport (Middleton, 1965, p.249).

flow roll A rounded, pillow-like body or mass of sandstone occurring within or just above finer-grained sediment or commonly within the basal part of a sandstone overlying shale or mudstone, having a shape approaching that of an elongate flattened ellipsoid (short axis more or less vertical), and presumed to form by deformation, such as by large-scale load casting or mud flowage accompanied by subaqueous slump or foundering of sand channels. The term was used by Pepper et al (1954, p.88) in reference to the characteristic rolled appearance of the structure and because deformation of strata occurred prior to complete lithification of the rocks. See also: *ball-and-pillow structure; pseudonodule; storm roller.*

flow slide A landslide of waterlogged material in which there is no well-defined slip surface.

flowstone A general term for any cave formation of calcium carbonate or other mineral formed by flowing water on the walls or floor of a cave. Cf: *dripstone.*

flow stretching Orientation of crystals during plastic flow, with their long axes parallel to the direction of flow.

flow string The *string* of casing, tubing, or pipe through which oil or gas reaches the surface.

flow structure [ign] *flow texture.*

flow structure [sed] A primary sedimentary structure resulting from subaqueous slump or flow (Cooper, 1943, p.190).

flow surface In structural petrology, the plane separating adjacent flow layers. Cf: *deformation plane.* Syn: *slip surface [struc petrol].*

flow system In hydrodynamics, a set of flow lines in which any two flow lines are always adjacent, and can be intersected in one direction only by an uninterrupted surface across which flow takes place (Meyboom, 1962, p.355).

flow texture The texture, of an igneous rock, characterized by a wavy or swirling pattern in which platy or prismatic minerals are oriented along planes of lamellar flowage in fine-grained and glassy igneous rocks. Syn: *flow structure [ign]; fluidal texture; fluidal structure; flowage texture; fluxion structure; fluxion texture; rhyotaxitic texture.*

flowtill A supraglacial till that is modified and transported by plastic mass flow (Hartshorn, 1958, p.481).

flow unit A group of sheets or beds of lava that were formed by a single eruption or outpouring.

flow velocity In soil, a vector point function used to indicate rate and direction of movement of water through soil, per unit of time and perpendicular to the direction of flow.

fluctuation [water] The alternate rising and lowering of the water table or of the level of a body of surface water, either regularly or periodically.

flue [intrus rocks] A pipelike igneous intrusion.

flue [sed] A hard, sandy shale in the Lancashire (Eng.) coalfield, probably so named in reference to its splitting or fissile quality.

fluellite A colorless or white mineral: $Al_2(PO_4)F_2(OH).7H_2O$. It was previously formulated: $AlF_3.H_2O$.

fluent An obsolete term for a stream or other current of water.

fluidal structure *flow texture.*

fluidal texture *flow texture.*

fluid dynamics That aspect of *fluid mechanics* which deals with motion of fluids.

fluid inclusion In an igneous rock, a tiny (1.0-100.0 microns in diameter) cavity containing liquid and/or gas, formed by entrapment in crystal irregularities of the liquid from which the rock crystallized. Partial syn: *liquid inclusion.*

fluidity index The ratio of sand detritus to the interstitial detrital matrix of a sandstone (Pettijohn, 1954, p.362-363). It is a measure of the fluidity (density and viscosity) of the depositing medium and it partly determines the sorting of transported sediment; e.g. a high ratio indicates a poorly sorted sediment deposited from a medium of high density and high viscosity. Syn: *fluidity factor.*

fluidization The mixing process of gas and loose, fine-grained material so that the whole flows like a liquid, e.g. the formation of an ash flow or núee ardente during a volcanic eruption.

fluid mechanics The study of the mechanics or behavior of liquids and gases. It is broad in scope, and includes such disciplines as hydraulics and aerodynamics. See also: *fluid dynamics.*

fluid potential With reference to ground water, the mechanical energy per unit mass of a fluid (here, water) of any given point in space and time, with respect to an arbitrary state and datum. At a given point in a body of liquid, the fluid potential is proportional to the *head*; it is the head multiplied by the acceleration due to gravity.

fluke One of two or more recurved components of an *anchor* of a holothurian.

flume [eng] An artificial, inclined channel used for conveying water for industrial purposes, such as irrigation, transportation, mining, logging, and power production, or for diverting the water of a stream from its channel for the purpose of

washing the sand and gravel in the bed left dry.

flume [geomorph] A ravine, gorge, or other deep, narrow, steep-sided valley, with a stream flowing through in a series of cascades; e.g. in the White Mountains, N.H.

fluoborite A colorless hexagonal mineral: $Mg_3(BO_3)(F,OH)_3$. Syn: *nocerite.*

fluocerite A pale- or reddish-yellow hexagonal mineral: $(Ce, La,Nd)F_3$. Syn: *tysonite.*

fluolite *pitchstone.*

fluor (a) The original form of *fluorite,* still used chiefly in Great Britain. (b) An obsolete name for a mineral belonging to a group including fluorite and characterized by the alchemists as resembling gems and usable as metallurgic fluxes (Webster, 1967, p.877). Etymol: German *Fluss,* "flux".

fluorapatite (a) A very common mineral of the apatite group: $Ca_5(PO_4)_3F$. It is a common accessory mineral in igneous rocks. Syn: *apatite.* (b) An apatite mineral in which fluorine predominates over chlorine and hydroxyl.

fluorescence A type of luminescence in which the emission of light ceases when the external stimulus ceases; also, the light so produced. Cf: *phosphorescence.*

fluorescence spectrum The spectrum produced when a fluorescent material is induced to emit radiation of one kind when irradiated with radiation of another kind as in *X-ray fluorescence spectroscopy.*

fluorine dating Determination of relative age on the basis of fluorine content of Pleistocene or Recent bones. The method depends on the gradual combination with time of fluorine in ground water with the calcium phosphate of bone. In areas where this method has been used, the fluorine content averages 2% in lower Pleistocene bone, 1% in middle Pleistocene bone, 0.5% in upper Pleistocene bone, and 0.3% in Recent bone.

fluorite A transparent to translucent mineral: CaF_2. It is found in many different colors (often blue or purple) and has a hardness of 4 on Mohs' scale. Fluorite occurs in veins usually as a gangue mineral associated with lead, tin, and zinc ores, and is commonly found in crystalline cubes with perfect octahedral cleavage. It is the principal ore of fluorine, and is used as a flux, in the preparation of glasses and enamels and the manufacture of hydrofluoric acid, and for carved ornamental objects. Syn: *fluorspar; fluor; Derbyshire spar.*

fluorspar *fluorite.*

flurosion A term proposed by Glock (1928, p.477-478) for the work of transportation and erosion carried on by streams.

flush [water] (a) A sudden increase in the volume of a stream; a sudden flow or rush of water, as down a stream and filling the channel or overflowing the banks. Syn: *fresh.* (b) A British term for a pool or for a low swampy place.---Cf: *flash.*

flushing period The interval of time necessary for an amount of water equal to the volume of a lake to pass through its outlet, computed by dividing lake volume by flow rate (usually mean) flow of the outlet.

flute [geomorph] A small, longitudinal, shallow channel formed by *fluting* and running nearly vertically down the face of a rock surface.

flute [speleo] (a) A term used by Bretz (1942, p.731) for one of the small, shallow, oval-shaped, regularly and closely spaced hollows formed by a turbulent flow of water and found on the walls, floor, or (rarely) ceiling of a cave, having an asymmetric cross section along its main axis (being invariably steeper on the upstream side); also, one of a series of unsymmetric, irregularly placed, short ridges, in the ground plan of which continuity of crest lines is more nearly attained across the current than with it. Flutes strikingly resemble aqueous current ripple marks in sand, and are believed to be solutional in origin. They are used to determine direction of flow. See also: *solution ripple.* British syn: *scallop.* (b) An incised, vertical channel or groove developed in a cave shaft by solution.

flute [**sed**] (a) A primary sedimentary structure, commonly seen as a flute cast, consisting of a discontinuous scoop-shaped or lobate depression or groove 2-10 cm in length, usually formed by the scouring action of a turbulent, sediment-laden current of water flowing over a muddy bottom, and having a steep or abrupt upcurrent end where the depth of the mark tends to be the greatest. Its long axis is generally parallel to the current. (b) A term that is sometimes used loosely as a syn. of *flute cast*. (c) A scalloped or rippled rock surface. The term is usually used in the plural.---See also: *fluting*.

flute cast A term suggested by Crowell (1955, p.1359) for a spatulate or lingulate sole mark consisting of a raised, oblong, and subconical bulge on the underside of a siltstone or sandstone bed, characterized by a steep or blunt and rounded, bulbous, or beaked upcurrent (deeper) end from which the structure flattens or flares out in the downcurrent direction and merges with the bedding plane. It is formed by the filling of a flute. See also: *lobate rill mark*. Syn: *fluting; flute; flow cast; flow mark; scour cast; scour finger; vortex cast; linguoid sole mark; turboglyph; lobate plunge structure*.

fluting [**geomorph**] A process of differential weathering and erosion by which an exposed, well-jointed, coarse-grained rock (such as granite or gneiss) develops a corrugated surface of *flutes*; esp. the formation of small-scale ridges and depressions by wave action.

fluting [**glac geol**] The formation by glacial action of large, smooth, deep, gutter-like channels or furrows on the stoss side of a rocky hill obstructing the advance of a glacier; the furrows are larger than glacial grooves, and they do not extend around the hill to the lee side (Chamberlin, 1888, p.246). Also, a furrow so formed.

fluting [**sed**] (a) The process of forming a *flute* by the cutting or scouring action of a current of water flowing over a muddy surface. (b) Scalloped or rippled rock surfaces. (c) *flute cast*.

Fluvent In U.S. Dept. of Agriculture soil taxonomy, a suborder of the soil order *Entisol*, characterized by formation in alluvial deposits, recent with a carbon content that decreases with depth (SSSA, 1970). Cf: *Aquent; Arent; Orthent; Psamment*.

fluvial (a) Of or pertaining to a river or rivers. (b) Existing, growing, or living in or about a stream or river. (c) Produced by the action of a stream or river.---The term is used by geologists esp. in regard to river flow and river action. See also:*fluviatile*. Etymol: Latin *fluvius*, "river".

fluvial cycle of erosion *normal cycle*.

fluvial denudation Denudation of a land surface by the scouring action of streams, assisted by weathering, mass-wasting, overland flow, and the action of rainwater flowing over the ground.

fluvial deposit A sedimentary deposit consisting of material transported by, suspended in, or laid down by a river stream. Syn: *fluviatile deposit*.

fluvialist A believer in the doctrine that the widespread surficial deposits (now known to be glacial drift) and other geologic phenomena can be explained by ordinary stream action. Cf: *diluvialist*.

fluvial lake (a) A lake with a perceptible flow of water, e.g. a body of water connecting two larger bodies whose differences in elevation may be sufficient to create a flow from one to another. (b) A slowly moving part of the river, such as occurring where the width of the river has been expanded. See also: *fluviatile lake*.

fluviatile A syn. of *fluvial* (Fowler, 1937, p. 184). Geologists tend to use the term for the results of river action (e.g. *fluviatile* dam, or *fluviatile* sands) and for river life (e.g. *fluviatile* fauna). Syn: *fluviatic*.

fluviatile lake A lake formed as a result of the action of a river or stream, or a lake occupying a basin produced by running water capable of erosion or deposition; e.g. an oxbow lake on the flood plain of a meandering river, or a lake occupying a plunge basin at the foot of a waterfall, or a lake formed by the damming action of excess sediment at the confluence of tributary and the main river. See also: *fluvial lake*.

fluviation The activities engaged in, and the processes employed, by streams (Glock, 1928, p.477). Syn: *stream action*.

fluvicoline Said of an animal that lives in or frequents streams.

fluvioclastic rock A hydroclastic rock containing current- or river-worn fragments (Grabau, 1924, p.295). Syn: *potamoclastic rock*.

fluvioeolian Pertaining to the combined action of streams and wind; e.g. a *fluvioeolian* deposit.

fluvioglacial A syn. of *glaciofluvial*. The term "glaciofluvial" is preferred in U.S "since logically the *glac-* precedes the *fluv-*" (Monkhouse, 1965, p.127), but see *fluvioglacial drift*.

fluvioglacial drift A general term for *drift* transported and deposited by running water emanating from a glacier. Cf: *glacial drift*.

fluviograph A device for measuring and recording automatically the rise and fall of a river. Syn: *fluviometer*.

fluviolacustrine Pertaining to sedimentation partly in lake water and partly in streams, or to sediments deposited under alternating or overlapping lacustrine and fluvial conditions.

fluviology The science of rivers.

fluviomarine Said of marine sediments that contain resorted and redistributed fluvial material along with the remains of marine organisms.

fluviometer *fluviograph*.

fluviomorphology *river morphology*.

fluvioterrestrial Consisting of or pertaining to the land and its streams.

fluviraption A term introduced by Malott (1928a) for *hydraulic action* performed by rivers.

flux [**streams**] A stream of flowing water; a flood or an outflow.

flux [**radioactivity**] The number of radioactive particles in a given volume times their mean velocity.

fluxgate magnetometer A magnetometer that measures the component of the magnetic field along the axis of its sensor.

fluxie A field term generally used by sedimentologists for *fluxoturbidite*.

fluxion banding British usage for *flow banding*.

fluxion structure [**ign**] British usage for "flow structure", a syn. of *flow texture*.

fluxion texture British usage for *flow texture*.

fluxoturbidite A term proposed by Dzulynski et al (1959, p.1114) for a sediment produced by a mechanism related both to deposition from turbidity currents and to submarine slumping or sliding. It is characterized by coarse grain, thick bedding, and poor development of grading and of sole marks. Cf: *undaturbidite*. Syn: *fluxie*.

fly A Norwegian term for the moderately steep slope lying between a scarp face and the high, treeless area below it, as in eastern Norway.

flying bar A looped bar or spit formed on the landward side of an island that is subsequently reduced below sea level by wave erosion before the bar or spit is destroyed. Term originated by Gulliver (1899, p.190).

flying veins A pattern of mineral-deposit veins overlapping and intersecting in a branchlike pattern.

fly leveling Spirit leveling in which some of the restrictions of precise leveling (such as limiting the length of sight, or balancing backsights and foresights) are relaxed in order to obtain elevations of moderate accuracy more rapidly; e.g. running a level line (a line over which leveling operations are accomplished) at the close of a working day in order to check the results of an extended line run in one direction only.

flysch (a) A descriptive term for a marine sedimentary facies characterized by a thick sequence of poorly fossiliferous, thinly bedded, graded deposits composed chiefly of marls and andy and calcareous shales and muds, rhythmically interbed-ed with conglomerates, coarse sandstones, and graywackes. ee also: *wildflysch; macigno*. (b) An extensive, preorogenic sedimentary formation representing the totality of the flysch facies deposited in different troughs, during the later stages of filling of a geosynclinal system, by rapid erosion of an adja-cent and rising mountain belt at a time directly previous to the main paroxysmal (diastrophic) phase of the orogeny or when initial diastrophism had already developed interior ridges ex-posed to erosion; specif. the Flysch strata of late Cretaceous to Oligocene age along the borders of the Alps, deposited in the foredeeps in front of northward-advancing nappes rising from beneath the sea, before the main phase (Miocene) of the Alpine orogeny. (c) A term that has been loosely applied to any sediment with nearly all of the lithologic and stratigra-phic characteristics of a flysch, such as almost any turbidite.- Etymol: dialectal term of German origin used in Switzerland for a crumbly or fissile material that slides or flows. Pron: *flish*. Pl: *flysches*. Adj: *flyschoid*. Cf: *molasse*.

f-number The ratio of the equivalent focal length of a photo-graphic lens to the *relative aperture*, or a number expressing the relative aperture of the lens; e.g. the f-number of a lens with a relative aperture of $f/4.5$ is 4.5. The smaller the num-ber, the brighter the image and therefore the shorter the ex-posure required. Etymol: *f*, symbol for focal length. See also: *speed*.

foam *pumice*.

foam crust A snow-surface feature produced during ablation and resembling small overlapping waves, like sea foam on a beach. Cf: *plowshare*.

foaming earth Soft or earthy *aphrite*.

foam mark A surface sedimentary structure consisting of a pattern of almost imperceptible ridges and hollows, formed where foam produced by wind action on seawater is driven over a surface of moist or wet sand.

focal depth *depth of focus*.

focal distance *focal length*.

focal length A general term for the distance from the principal point or center of a lens to the principal focus. In photogram-metry, the term "equivalent focal length" is the distance mea-sured along the lens axis from the rear nodal point of the lens to the position of the focal plane that provides the best aver-age definition in the aerial negative, and the term "calibrated focal length" is an adjusted value of the equivalent focal length computed to distribute the effect of lens distortion over the entire field of the negative. Symbol: f. Syn: *focal distance*.

focal plane The plane, perpendicular to the axis of a lens, in which images of points in the object field are brought to a focus; a plane that passes through a principal focus.

focal point *principal focus*.

focal sphere An arbitrary reference sphere drawn about the hypocenter or focus of an earthquake, to which body waves recorded at the Earth's surface are projected for studies of earthquake mechanisms.

focus [seism] That point within the Earth which is the center of an earthquake and the origin of its elastic waves. Cf: *epi-center*. Syn: *hypocenter; seismic focus; centrum*.

focus [photo] (a) The point at which rays of light converge to form an image after passing through a lens or optical system or after reflection by a mirror. See also: *principal focus*. (b) The condition of sharpest imagery.---Pl: *foci*.

focused-current log A resistivity log obtained in borehole surveying by means of a multiple-electrode arrangement (such as a ring-shaped central measuring electrode shielded by long upper and lower metal bars or "guard" electrodes maintained at equal potentials) and an automatic control sys-tem whereby an electric current is focused or confined to flow radially through the formations in a horizontal, disk-shaped sheet of predetermined thickness and perpendicular to the borehole (thus insuring deeper current penetration in the bed

at the level of the central electrode and minimizing the influ-ence of adjacent beds and of the borehole itself on the mea-surement). The measurement involves a part of ground of lim-ited vertical extent, and the measured values are practically unaffected by the mud column. The log is useful for detailed definition of thin beds, correlation work, reservoir evaluation, and resistivity measurements where the drilling mud is highly saline or the formation is of high resistivity. Syn: *laterolog; guard-electrode log; current-focused log*.

foehn A warm and dry *katabatic wind* occurring on the lee-ward side of a mountain ridge, specif. in the alps. See also: *chinook*. Also spelled: *föhn*.

fog A cloud at the Earth's surface, formed by condensation of atmospheric water vapor into a dense mass of tiny (about 40 microns in diameter) droplets of water or perhaps, crystals of ice.

fog desert A west coast desert having a cold-water marine current just offshore of the warm continental land mass; the coolness and moist air combine to produce fog.

foggara A term used in the Saharan desert region (esp. in Morocco and Mauritania) for a gently inclined, underground conduit or tunnel designed to intercept ground water near the foot of mountains and to conduct it by gravity to a neighboring lowland for irrigation purposes; a horizontal well. Etymol: Ara-bic. Cf: *ganat; karez*.

fohrde A local name in Germany for a *förde*. Pl: *föhrden*. Etymol: German *Föhrde*.

foid A collective term coined by Johannsen (1917, p.69-70) to denote the *feldspathoid* group of minerals. Etymol: *feldspa-thoid*.

foidite A plutonic or volcanic rock in which feldspathoids con-stitute 60-100 percent of the light-colored components; e.g. urtite, ijolite, melteigite, italite. Sometimes restricted to those igneous rocks in which feldspathoids represent 90-100 per-cent of the light-colored constituents. Cf: *feldspathoidite*.

fold [geomorph] A British term for an undulation in the land surface, either a low rounded hill or a shallow depression.

fold [paleont] (a) A major rounded elevation on the surface of a brachiopod valve (generally the brachial valve), externally convex in transverse profile and radial from the umbo, and usually median in position. It is typically associated with the *sulcus*. Cf: *carina*. (b) A spirally wound ridge on the interior of the wall of a gastropod shell; e.g. *columellar fold* and *parietal fold*.

fold [struc geol] n. A curve or bend of a planar structure such as rock strata, bedding planes, foliation, or cleavage. A fold is usually a product of deformation, although its definition is de-scriptive and not genetic and may include primary structures. See also: *folding*.

foldbelt A widely used syn. of *orogenic belt*, corresponding to the French term "region de plissement". Also spelled: *fold belt; fold-belt*.

fold breccia A local *tectonic breccia* composed of angular fragments resulting from the sharp folding of thin-bedded, brit-tle rock layers between which are incompetent plastic beds; e.g. a breccia formed where interbedded chert and shale are sharply folded. Syn: *reibungsbreccia*.

fold coast A coast whose configuration is controlled by folded rocks.

folded fault A thrust fault, the hanging wall of which has be-come slightly folded due to the development of step thrusting or *step faults* beneath it (Jones, P.B., 1971); a *warped fault*.

fold fault An overfold, the middle limb of which is replaced by a fault surface. Cf: *lag fault; slide [fault]*.

folding The curving or bending of a planar structure such as rock strata, foliation, or cleavage by deformation. The term is generally used for the compression of strata in the formation of fold structures on a broad scale, and sometimes has the connotation of general deformation of which the actual folding is only a part. See also: *fold*.

fold mountains Mountains that have been formed by the compression and uplift of large-scale folding.

fold mullion A type of *mullion* formed by the cylindrical undulations of bedding; internal structure displays conformable bedding laminations (Wilson, 1953). Cf: *cleavage mullion.*

fold system A group or area of folds that are produced by the same tectonic episode.

Foley North American (Gulf Coast) stage: Pliocene (above Clovelly, below Pleistocene).

folgerite *pentlandite.*

folia Thin, leaflike layers or laminae, specif. cleavable folia of gneissic or schistose rocks. Singular: *folium.*

foliate Adj. of *foliation.*

foliated Adj. of *foliation.* Obsolete syn: *parafoliate.*

foliated ground ice A large mass of ice commonly occupying thermal contraction cracks in permafrost, and characterized by parallel or subparallel structures marked by air bubbles, films of organic or inorganic matter, or boundary surfaces between ice layers of different composition. It is usually, but not always, wedge-shaped. Syn: *wedge ice.*

foliation [struc geol] A general term for a planar arrangement of textural or structural features in any type of rock, e.g. *cleavage* in slate or *schistosity* in a metamorphic rock. It is most commonly applied to metamorphic rock. Adj: *foliate; foliated.*

foliation [glaciol] The planar or layered structure produced in the ice of a glacier by plastic shear strain, manifest as alternating layers of coarse-grained and fine-grained ice, or bubbly and clear ice. Nonpreferred syn: *banding.*

foliose Leafy, or resembling a leaf; esp. said of a corallum with laminar branches.

Folist In U.S. Dept. of Agriculture soil taxonomy, a suborder of the soil order *Histosol*, characterized by accumulations of organic material, e.g. forest litter, that is less than one meter deep (SSSA, 1970). Cf: *Fibrist; Hemist; Saprist.*

folium The singular form of *folia.*

follicle A simple, dry, dehiscent fruit that produces either several or numerous seeds and that has one carpel and splits along one seam only.

following wind A wind whose direction is the same as that of ocean waves. Cf: *opposing wind.*

Foner magnetometer *vibration magnetometer.*

font An archaic term for a stream or a spring, fountain, or source of a stream; it forms part of place names, such as Chalfont.

Fontainebleau sandstone (a) A desilicified quartz sandstone (or an uncemented quartz sand) whose calcareous cement forms a crystalline aggregate of calcite in which sand grains are embedded; the cement is continuous and the easily fractured surfaces of the rock indicate that the calcite is in crystallographic continuity. Type locality: Fontainebleau, in the Paris Basin of France. Cf: *meulerization.* (b) A name given to a variety of calcite; specif. *sand-calcite.* Syn: *Fontainebleau limestone.*

food chain The passage of energy and materials from producers through a progressive sequence of plant-eating and meat-eating consumers. Cf: *food cycle.*

food cycle All of the *food chains* in an association of organisms; the food relations between the members of a population that make it possible for the population to survive.

food groove An *ambulacral groove* in an echinoderm; e.g. a furrow running along the adoral surface of crinoid ray ossicles and traversing tegmen to mouth.

fool's gold A popular term for pyrites resembling gold in color; specif. *pyrite* and *chalcopyrite.*

foot [paleont] (a) The ventral part of the body of a mollusk, consisting chiefly of a muscular surface or process, and used for locomotion; e.g. a broad, flattened muscular sole used for creeping in most gastropods, or a broad or narrow, protrusible, tapering, blade-like muscular structure extending from the

midline of the body of a bivalve mollusk (anteriorly and ventrally in more typical bivalves), and used for burrowing or locomotion. (b) A limb of an arthropod. (c) One of the radial appendages extending from the ultimate joint of the shell of certain radiolarians (such as of the suborders Nassellina and Phaeodarina).---Cf: *head.*

foot [struc geol] The lower bend of a fold or structural terrace. Cf: *head.* Syn: *lower break.*

foot [geomorph] (a) The bottom of a slope, grade, or declivity. (b) A term loosely applied to the lower part of any elevated landform.

footeite *connellite.*

foot glacier *expanded-foot glacier.*

foothill A lower, subsidiary hill at the base of a mountain or higher hills. See also: *foothills.*

foothills A region of relatively low, rounded hills at the base of or fringing a mountain range; e.g. the low, undulating region along the western base of the Sierra Nevada in California.

footing A relatively shallow foundation by which concentrated loads of a structure are distributed directly to the supporting soil or rock through an enlargement of the base of a column or wall. Its ratio of base width to depth of foundation commonly exceeds unity. Cf: *pier [eng].*

foot layer A downward extension of ektexine of a pollen grain partly surrounded by endexine.

foot slope A general term for a hillside surface whose top part is the *wash slope* and that includes "all the slopes of diminishing gradient" (Penck, 1953, p.419). Syn: *fusshang.*

footwall The underlying side of a fault, an orebody, or of mine workings. Syn: *heading wall; heading side; lower plate.* Cf. *floor [coal mining]; hanging wall; wall.*

foralite An inorganic structure resembling a worm tube, found in stratified rock.

foramen A small opening, perforation, pore, or orifice, such as the *pedicle foramen* of a brachiopod or the *septal foramen* of a nautiloid; esp. an opening that connects adjacent chambers in the test of a foraminifer, located at the base of septa or areal in position, and often representing a previous aperture or a secondarily formed aperture (but not equivalent to a pore of a perforate test). Pl: *foramina.*

foramina Plural of *foramen.*

foraminifer Any protozoan belonging to the order Foraminiferida, characterized by the presence of a test composed of agglutinated particles or of secreted calcite (and rarely of silica or aragonite) and commonly found in marine to brackish environments from the Cambrian to the present. Precambrian forms and some freshwater forms are also known. Syn: *foram; foraminiferan; foraminifera.*

foraminiferal Pertaining to or derived from the Foraminifera or their shells; e.g. "foraminiferal test".

foraminiferal limestone A limestone composed chiefly of the remains of bottom-dwelling and floating foraminifers, and commonly lacking a fine-grained matrix; e.g. *fusulinid limestone* and *nummulitic limestone.*

foraminiferal ooze An *ooze* whose skeletal remains are the tests of foraminifera; it is a calcareous ooze. See also: *globigerina ooze.*

foraminite A sedimentary rock composed predominantly of the remains of foraminifers.

foram number In quantitative foraminiferal studies, the total number of all foraminiferal species in a one-gram sample (dry weight) of material greater than one tenth of a millimeter in diameter.

forb A noncultivated, dicotyledonous, herbaceous plant; a herb other than grass; a weed. The term appears in some palynologic literature dealing with Quaternary sediments.

Forbes band An obsolescent term for one of a group of bands forming a type of *ogive* pattern that occurs on valley glaciers below icefalls and is characterized by alternating dark and light curved bands which cross the glacier. These bands nor-nally occur in a regular succession at roughly equal intervals. This type of band was described by James D. Forbes (1809-1868), English physicist, who originally called it a *dirt band* [glaciol]. Syn: *Forbes ogive*. Cf: *dirt-band ogive; wave ogive*.

orbesite A dull grayish-white mineral: $H_2(Ni,Co)_2(As-O_4)_2 \cdot 7H_2O$.

orce [phys] That which tends to put a stationary body in motion or to change the direction or speed of motion of a moving body.

orce [streams] A name given in northern England to a waterfall or cascade. Etymol: Old Norse *fors*. Cf: *fors*. Syn: *foss; fosse*.

orce-apart A sedimentary structure similar to ball-and-pillow structure (Pettijohn & Potter, 1964, p.308).

orce couple Two equally intense forces acting in opposite directions but not in the same line on a body, creating a tendency for the body to rotate. Syn: *couple*.

orced-cut meander A meander in which deposition on the inner bank proceeds at the same rate as lateral erosion on the outer bank, thereby maintaining a channel of constant width (Melton, 1936, p.596-597). Cf: *advance-cut meander*. Syn: *scroll meander; forced meander*.

orced oscillation An oscillation that is imposed on a body, e.g. the Earth, by an external force. Cf: *free oscillation*. Syn: *forced vibration*.

orced vibration *forced oscillation*.

orced wave A wave that is generated and maintained by a continuous force, e.g. wind. Cf: *free wave*.

orce of crystallization *crystallizing force*.

orceps A C-shaped siliceous sponge spicule (sigma) having the form of a tongs, with subparallel extremities.

orcherite An orange-yellow variety of opal colored with orpiment.

ord (a) A shallow and usually narrow part of a stream, estuary, or other body of water that may be crossed, as by wading or by a wheeled land vehicle. It usually has a firm, level, and relatively boulder-free bottom. Syn: *fording; wath*. (b) An archaic term for a stream or other body of water.

orde A Danish term for a long, narrow, straight-sided inlet of the sea in a coastline consisting of till or surrounded by terminal moraines, and typically produced by drowning of a subglacial valley along a low-lying coast, as that of the SE Jutland Peninsula. Pl: *fôrden*. Cf: *fjord; bodden*. Syn: *fôhrde*.

oredeep An elongate depression bordering an island arc or other orogenic belt. Cf: *trench*. (b) A syn. of *exogeosyncline*, so named because of its relative position, near the craton.

oredune A coastal dune or *dune ridge* oriented parallel to the shoreline of an ocean or large lake, occurring at the landward margin of the beach (or along the shoreward face of a beach ridge) or at the landward limit of the highest tide, and more or less completely stabilized by vegetation.

oreign inclusion A fragment of country rock enclosed in an igneous intrusion.

oreland [geog] The land lying in front of or adjoining other land and physiographically related to it.

oreland [coast] (a) An extensive area of land, either high ground or low land, jutting out from the coast into a large body of water (usually the sea); a *headland*; a *promontory*. See also: *cuspate foreland*. (b) A prograded strip of low, flat land built by waves and currents at the base of a cliff; an initial stage in the development of a *strand plain*. (c) A part of the natural shore, located seaward of an embankment, that receives the shock of sea waves and deadens their force.

oreland [glac geol] A lowland area covered by piedmont glaciers that moved down from adjacent mountain ranges.

oreland A stable area marginal to an orogenic belt, toward which the rocks of the belt were thrust or overfolded. Generally the foreland is a continental part of the crust, and is the edge of the craton or platform area.

foreland facies *shelf facies*.

forelimb The steeper of the two limbs of an asymmetrical, anticlinal fold. Cf: *backlimb*.

forelimb thrust A thrust fault through the forelimb or more steeply dipping side of an anticline. Cf: *backlimb thrust*.

forellenstein *troctolite*.

Forel scale A color scale of yellows, greens, and blues that is used against a white Secchi disk to measure the color of sea water.

forepoling A method of advancing an excavation in loose, caving, weak, or watery ground (such as quicksand) by driving sharp-pointed poles, planks, timbers, boards, or slabs into the ground ahead of, or simultaneously with, the excavating; a method of supporting a very weak roof. It is useful in tunneling and in extracting coal from under shale or clay.

fore reef The seaward side of a reef, commonly a steeply dipping slope with deposits of reef talus. Cf: *back reef; off-reef*. Also spelled: *forereef*.

forerunner A low, long-period ocean swell representing the gradual change of water level that commonly begins several hours before the arrival of a distant storm, esp. a tropical hurricane.

foreset n. A *foreset bed*.---adj. Pertaining to or forming a steep and advancing frontal slope, or the sediments deposited on such a slope.

foreset bed One of the inclined, internal, and systematically arranged layers of a cross-bedded unit; specif. one of the gently inclined layers of sandy material deposited upon or along an advancing and relatively steep frontal slope, such as the outer margin of a delta or the lee side of a dune, and progressively covering the *bottomset bed* and in turn being covered or truncated by the *topset bed*. Foreset beds represent the greater part of the bulk of a delta. Also spelled: *fore-set bed*. Syn: *foreset*.

foreset bedding A syn. of *cross-bedding*; an internal bedding inclined to the principal surface of accumulation. See also: *compound foreset bedding*.

foreshock A small tremor that commonly precedes a larger earthquake or *main shock* by seconds to weeks and that originates at or near the focus of the larger earthquake. Cf: *aftershock*.

foreshore (a) The lower or outer, gradually seaward-sloping zone of the shore or beach, lying between the crest of the most seaward berm on the *backshore* (or the upper limit of wave wash at high tide) and the ordinary low-water mark; the zone regularly covered and uncovered by the rise and fall of the tide, or the zone lying between the ordinary tide levels. Sometimes referred to as the *shore*. Syn: *beach face*. (b) A term loosely applied to a strip of ground lying between a body of water, as a lake or stream, and land that is cultivated or otherwise occupied.

foreside A stretch of country fronting the sea, as Falmouth Foreside, Maine.

foresight (a) A sight or bearing on a new survey point, taken in a forward direction and made in order to determine its elevation. In a transit traverse, it is a point set ahead on line to be used for reference when resetting the transit. Also, a sight on a previously established survey point, taken to close a circuit. (b) A reading taken on a level rod to determine the elevation of the point on which the rod rests when read. Syn: *minus sight*.---Abbrev: F.S. Ant: *backsight*.

foreslope The steeply sloping part of an organic reef, extending from its outer margin to an arbitrary depth of 10 fathoms (formerly 6).

forest bed An interglacial deposit consisting of soil and woody remains of trees and other vegetation. Syn: *black drift; chip yard; woodyard*.

forest floor The highly organic surface layer of a forest soil, including *litter, duff*, and *leaf mold*.

forest marble *landscape marble*.

forest-moss peat Peat formed in forested swamps.

forest peat A highmoor peat formed of the remains of trees.

fork (a) A place where two or more streams join to form a larger waterway; a confluence. (b) The smaller of two streams that unite at a fork; any uniting stream at a fork where the resulting stream is much larger; a branch. (c) The land bounded by, adjoining, or lying in the angle made by, a fork. The term is often used in the plural in place names.

form In geomorphology, a syn. of *landform*.

forma The smallest category used in ordinary taxonomic works. It is generally applied to trivial variations occurring among individuals of any population (Lawrence, 1951, p.56).

formal unit A stratigraphic unit that is "defined and named in accordance with the rules of an established or conventional system of classification and nomenclature" (ISST, 1961, p.18); e.g. a unit that is established in conformance with Article 13 of the Code of Stratigraphic Nomenclature prepared by the American Commission on Stratigraphic Nomenclature (1961). The initial letter of each word used in forming the name of a formal unit should be capitalized (although a name is not necessarily formal because it is capitalized). Cf: *informal unit*.

formanite A black mineral: $Y(Ta,Nb)O_4$. It is isomorphous with fergusonite, and may also contain uranium, thorium, calcium, titanium, and zirconium.

format A term introduced by Forgotson (1957, p.2110) for an informal, laterally continuous rock-stratigraphic unit that includes two or more lithologically dissimilar units but is suitable for regional mapping, "defined in general terms as a rock unit which is related at one point (or in one area) to a unit of formal stratigraphy, but which crosses facies boundaries and cutoffs to reach other areas where other formal units are employed". The term is applied to an operational unit representing strata sandwiched between observable *markers* that are believed to be isochronous surfaces. Formats are useful for correlations (particularly in the subsurface) between areas where the stratigraphic section is divided into different formations that do not correspond in time value. See also: *assise*.

formation [cart] A persistent body of igneous, sedimentary, or metamorphic rocks, having easily recognizable upper and lower boundaries that can be traced in the field without recourse to detailed paleontologic or petrologic analysis, and large enough to be represented on a geologic map as a practical or convenient unit for mapping and description; the basic *cartographic unit* in geologic mapping.

formation [drill] A general term applied by drillers without stratigraphic connotation to a sedimentary rock that can be described by certain drilling or reservoir characteristics; e.g. "hard formation", "cherty formation", "porous formation".

formation [speleo] *speleothem*.

formation [geomorph] A naturally formed topographic feature, commonly differing conspicuously from adjacent objects or material, or being noteworthy for some other reason; esp. a striking erosional form on the land surface.

formation [ecol] A group of associations that exist together as a result of their closely similar life pattern, habits, and climatic requirements.

formation [stratig] (a) The basic or fundamental rock-stratigraphic unit in the local classification of rocks, consisting of a body of rock (usually a sedimentary stratum or strata, but also igneous and metamorphic rocks) generally characterized by some degree of internal lithologic homogeneity or distinctive lithologic features (such as chemical composition, structures, textures, or gross aspect of fossils), by a prevailingly but not necessarily tabular shape, and by mappability at the Earth's surface (considered as practicability of delineation at scales of the order of 1:25,000) or traceability in the subsurface (ACSN, 1961, art.6); a convenient unit, of considerable thickness (from a few meters to several thousand meters) and extent, used in mapping, describing, or interpreting the geology of a region, and the only formal unit that is used for completely dividing the whole geologic column all over the world into named units on the basis of lithology. A formation is a genetic unit, or a product of uniform or uniformly alternating conditions, and may contain between its upper and lower limits rock of one lithologic type, or repetitions of two or more lithologic types, or extreme heterogeneity of constitution which in itself may constitute a form of unity compared to the adjacent strata. Also, it may represent a long or short time interval, be composed of materials from one or several sources, and include breaks in the time-stratigraphic sequence; its age or time value may not necessarily be the same wherever it is recognized. Formations may be combined in *groups* or subdivided into *members*. A formation name should preferably consist of a geographic name followed by a descriptive lithologic term (usually the dominant rock type) or by the word "formation" if the lithology is so variable that no single lithologic distinction is appropriate. Abbrev: fm. (b) A general term used in England for any rock unit having some character (age, origin, composition) in common, but never applied to a specific unit. (c) A term that has been used for a time-stratigraphic unit representing a rock of constant time span; specif. a stage, or the rocks corresponding to the time interval of an age. This usage is not recommended. In Germany, the term "Formation" is equivalent to the time-stratigraphic term "system".--- Syn: *geologic formation*.

formation evalution The process of evaluating gas- or oil-bearing formations penetrated by a well, and of appraising their commercial significance.

formation factor The ratio of conductivity of an electrolyte to the conductivity of a rock saturated with that electrolyte. Symbol: F. Syn: *resistivity factor; formation resistivity factor*.

formation resistivity factor *formation factor*.

formation water Water present in a water-bearing formation under natural conditions as opposed to introduced fluids, such as drilling mud. Syn: *native water*. Cf: *connate water*.

Formative n. In New World archaeology, the cultural stage that follows the Archaic and is characterized by the development of agriculture and a settled population. It is followed by the Classic. Correlation of relative cultural levels with actual age (and, therefore, with the time-stratigraphic units of geology) varies from region to region. Syn: *Pre-Classic*.----adj. Pertaining to the Formative.

form contour A topographic contour determined by stereoscopic examination of aerial photographs without ground control or by some other means not involving conventional surveying.

form energy The potentiality of a mineral to develop its own crystal form against the resistance of the surrounding solid medium (Eskola, 1939). Syn: *power of crystallization; crystalloblastic strength*.

form genus (a) Most commonly applied to a unit in taxonomy for those fossilized remains, esp. of plants, that are only part of the whole individual and that usually occur widely separated from other parts; e.g. stems, leaves, or seeds described and classified according to the same rules as complete individuals even though the identity of the whole plant is unknown or cannot be determined positively. Cf: *parataxon*. (b) Less commonly applied to a genus in a series of related genera which have resulted from the splitting up of an old familiar genus. (c) Also, a genus containing several species with the same general form but probably with different ancestors; *genomorph*.----Also spelled: *form-genus*.

formkohle *crumble coal*.

formkreis (a) One of a series of related landforms that owe their existence to the same natural agent (such as running water or ice action). (b) *morphogenetic region*.---Etymol: German *Formkreis*, "form cycle". Pl: *formkreise*. Cf: *formenkreis*.

form line A line (usually broken) on a map, sketched or drawn by visual observation, depicting the general surface configuration or shape of the terrain without regard to a true vertical datum and regular spacing and usually without indicating elevations; an uncontrolled or interpolated *contour line*, or one that is not instrumentally or accurately surveyed. Also spelled: *formline*.

form ratio The ratio of mean stream depth to width of stream measured from bank to bank (Gilbert, 1914, p.35-36); it expresses the deepness or shallowness of a stream channel. The hydraulic radius is sometimes substituted for the mean depth if the stream is broad and shallow. Symbol: R.

fornacite An olive-green mineral: $(Pb,Cu)_3[(Cr,As)O_4]_2(OH)$. It is isomorphous with vauquelinite.

fors A Swedish term for a rapids or cataract, or a waterfall of low inclination. Cf: *force*.

forsterite A whitish or yellowish mineral of the olivine group: Mg_2SiO_4. It is isomorphous with fayalite, and occurs chiefly in metamorphosed dolomites and crystalline limestones. Symbol: Fo. Syn: *white olivine*.

forsterite marble A variety of magnesian marble in which the principal magnesian silicate impurity is olivine (forsterite) whose hydration may result in multicolored serpentine marble or ophicalcite.

fortification agate An agate having angular markings or parallel zigzag lines resembling the plan of a fortification.

fortunite A verite that contains olivine and phlogopite phenocrysts in a very fine-grained groundmass that is megascopically undeterminable but under the microscope can be identified as a combination of ortho- and clinopyroxene, mica, feldspar, and some glass. Its name is derived from Fortuna, Spain.

forward lap *end lap*.

forward scatter The scattering of radiant energy into the hemisphere of space bounded by a plane normal to the direction of the incident radiation and lying on the side toward which the incident radiation was advancing; the opposite of *backward scatter*. In *Rayleigh scattering*, forward scatter accounts for half of the total. As the particle size increases above the Rayleigh limit, an increasing fraction of the total scattering is forward scattering.

foshagite A white mineral: $Ca_4Si_3O_9(OH)_2$.

foshallassite A snow-white mineral: $Ca_3Si_2O_7 \cdot 3H_2O$ (?). Syn: *foschallasite*.

foso A term used in SW U.S. for a stream channel without conspicuous banks or bluffs. Etymol: Spanish, "moat, ditch".

foss (a) Var. of *fosse*. (b) *force*.

fossa [paleont] A depression on an articular face of a crinoid ossicle for the attachment of muscles or ligaments. Pl: *fossae*.

fossa [struc geol] A trench-type geosyncline bordering and surrounding a continent. The term is used for the structural trench crossing the ranges of Honshu, Japan: the Fossa Magna.

fosse [glac geol] A long, narrow depression or trough-like hollow between the edge of a retreating glacier and the containing wall of its valley, or between the front of a moraine and its outwash plain. It may result from local acceleration of melting due to absorbed or reflected heat from the valley sides.

fosse [streams] (a) A long, narrow waterway; a canal, ditch, or trench. Etymol: Latin *fossa*, "ditch". Syn: *foss*. (b) *force*.

fosse lake A glacial lake occupying a fosse.

fossette (a) One of the slit-like pits, depressions, or grooves paralleling the periphery on the surface of the tests of some foraminifers (such as *Elphidium*). (b) A depression for the resilium in the shell of a bivalve mollusk.

fossil n. Any remains, trace, or imprint of a plant or animal that has been preserved, by natural processes, in the Earth's crust since some past geologic time; any evidence of past life.----adj. Said of any object that existed in the geologic past and of which there is still evidence.

fossil copal *copalite*.

fossildiagenese A German term applied to the branch of

paleoecology concerned with the history of organic remains after burial. Cf: *biostratonomy; taphonomy*.

fossil erosion surface An *erosion surface* that had been buried by younger sediments and later exposed by the removal of the covering sediments. The term is sometimes used synonymously with *buried erosion surface*. Syn: *exhumed erosion surface; resurrected erosion surface*.

fossil farina (a) *diatomaceous earth*. (b) *rock milk*.

fossil flood plain A flood plain that is beyond the reach of the highest flood (Bryan, 1923a, p.88). Cf: *living flood plain*.

fossil flour *diatomaceous earth*.

fossil fuel A general term for any hydrocarbon deposit that may be used for fuel: petroleum, natural gas, and coal.

fossil geochronometry Measurement of *growth lines* on fossil shells as a means of estimating the length of days and lunar months in *geologic time*. Cf: *lichenometry*. See also: *geochronometry*.

fossil ice (a) Ice formed in, and remaining from, the geologically recent past. It is preserved in cold regions, such as the coastal plains of northern Siberia where Pleistocene ice remains have been found. See also: *dead ice*. (b) Relatively old *ground ice* in a permafrost region. Also, underground ice in a region where present-day temperatures are not low enough to create it (Huschke, 1959, p.230).

fossil ice wedge A sedimentary structure formed by the filling of the space formerly occupied by an *ice wedge* that had melted; the sediment fill may be wedge-shaped or very irregular. Syn: *ice-wedge pseudomorph; ice-wedge fill; ice-wedge cast*.

fossilization All processes involving the burial of a plant or animal in accumulating sediment and in the preservation of all, part, or a trace of it.

fossilized brine *connate water*.

fossil karst *paleokarst*.

fossil meteorite crater An ancient (pre-Pliocene), deeply eroded, meteorite impact structure; an *astrobleme*.

fossil ore A term for an iron-bearing sedimentary deposit, e.g. *Clinton ore*, in which shell fragments have been replaced and cemented together by hematite and carbonate. Cf: *flaxseed ore*.

fossil patterned ground Patterned ground that is inactive or no longer developing at the present time; e.g. ice-wedge casts and involution structures found in a region of mild climate but formed during colder periods of the past (as the Pleistocene epoch) when the region was under periglacial conditions. Ant: *active patterned ground*.

fossil peneplain A peneplain that had been buried by younger sediments and later exposed by the removal of the covering sediments. Syn: *exhumed peneplain; resurrected peneplain; revealed peneplain; stripped peneplain*.

fossil permafrost *passive permafrost*.

fossil pingo The remains of a pingo. See also: *pingo-remnant*.

fossil plain A plain that had been buried by younger sediments and later exposed by the removal of the covering sediments.

fossil resin Any of various natural *resins* found in geologic deposits as exudates of long-buried plant life; e.g. amber, retinite, and copal.

fossil ripple A *ripple mark* preserved on a sedimentary rock surface.

fossil soil *paleosol*.

fossil time *Geologic time* estimated on the basis of organic evolution (Kobayashi, 1944a, p.476). See also: *marine time; continental time*. Cf: *relative time*.

fossil turquoise *odontolite*.

fossil water *connate water*.

fossil wax *ozocerite*.

fossula An unusually wide or relatively prominent space between septa of a rugose coral, distinguished by its shape and size, and caused by failure of one or more septa to develop as rapidly as others. See also: *cardinal fossula; counter fossu-*

la; alar fossula.

fossulate Said of sculpture of pollen and spores consisting of grooves that anastomose.

fouling The attachment and growth of aquatic plants and/or animals on submerged surfaces.

foundation (a) The lower, man-made, supporting part of an engineering structure, in contact with the underlying soil or rock and transmitting the weight of the structure and its included loads to the underlying earth material (Ireland, 1969). It is usually below ground level. (b) A term that is sometimes applied to the upper part of the soil or rock mass in contact with, and supporting the loads of, and engineering structure; the subsoil.

foundation bed The rock or soil layer immediately beneath the foundation that receives the load of an engineering structure.

foundation coefficient A coefficient expressing how many times stronger is the effect of an earthquake in a given rock than would have been the case in an undisturbed crystalline rock under the same conditions (Schieferdecker, 1957, p.197).

foundation soil The upper part of the surficial earth materials directly carrying the load of an engineering structure.

founder breccia *collapse breccia.*

fount A *fountain* or spring of water.

fountain (a) A spring of water issuing from the Earth. Syn: *fount.* (b) The *source* or head of a stream.

fountain head According to Tolman (1937, p.559), "The elevation of water surface in a conduit if the overlying confining stratum extends above the water table, or elevation of water table above the upper termination of the confining stratum where the latter is below the water table".

fountainhead [stream] The fountain or spring that is the source of a stream. Syn: *springhead; wellspring.*

fourchite An olivine-free *monchiquite.* Its name is derived from the Fourch Mountains, Montana.

fourfold coordination *tetrahedral coordination.*

Fourier analysis A method for representing a periodic mathematical function as an infinite series of summed sine and cosine terms. It involves comparison of observed periodic data with this theoretical form, and also all harmonic (period, amplitude, phase) relationships of the series. Named after Jean Baptiste Joseph Fourier (1768-1830), French physicist. Syn: *harmonic analysis.*

fourling A crystal twin consisting of four individuals. Cf: *twoling; trilling; fiveling; eightling.*

fourmarierite An orange-red to brown secondary mineral: $PbU_4O_{13}.4H_2O$.

fourth-order pinacoid In a triclinic crystal, the $\{\bar{h}\,kl\}$, $\{h\bar{k}\,l\}$, or $\{hk\bar{l}\}$ pinacoid. Cf: *first-order pinacoid; second-order pinacoid; third-order pinacoid.*

fourth-order prism A crystal form: in a monoclinic crystal, the $\{hkl\}$ prism or the $\{\bar{h}\,kl\}$ prisms of four faces. Cf: *first-order prism; second-order prism; fourth-order prism.* Obsolescent syn: *hemipyramid.*

four-way dip In seismic prospecting, dip determined by spreads placed at points in four directions from a shot point. Three of the locations are essential; the fourth serves as a check.

foveolate Pitted; e.g. said of sculpture of pollen and spores consisting of pits in the ektexine.

fowlerite A variety of rhodonite containing zinc.

foyaite A *nepheline syenite* containing a predominance of potassium feldspar. Originally described by Blum as synonymous with nepheline syenite, and later applied by Brögger to nepheline syenite with trachytic texture (Johannsen, 1939, p.252). Its name is derived from Foya, Portugal. Cf: *ditroite.*

fractional crystallization [petrology] Separation of a cooling magma into parts by the successive crystallization of different minerals at progressively lower temperatures. Syn: *fractionation; fractionalization; crystallization differentiation.* Cf: *crys-*

tal fractionation.

fractional crystallization [salt] Controlled precipitation from a saline solution of salts of different solubilities, as affected by varying temperatures or by the presence of other salts in solution (Bateman, 1950, p.183).

fractional section A *section* containing appreciably less than 640 acres, usually due to invasion by a segregated body of water or by other land which cannot properly be surveyed as part of that section, or due to closing of the public-land survey on the north and west boundaries of the township.

fractional township A *township* containing appreciably less than 36 normal sections, usually due to invasion by a segregated body of water or by other land which cannot properly be surveyed as part of that township, or due to closing of the public-land survey on a State boundary or other limiting line.

fractionation [geochem] The differentiation of a mixture by separation of one or more of its component minerals, elements, or isotopes, or by separation of water masses differing in their properties.

fractionation [petrology] *fractional crystallization.*

fractoconformity The relation between conformable strata where faulting of the older beds proceeds contemporaneously with deposition of the newer.

fractography The study of the surfaces of fractures, esp. microscopic study.

fracture [mineral] The breaking of a mineral other than along planes of *cleavage.* A mineral can be described in part by its characteristic fracture, e.g. uneven, fibrous, conchoidal.

fracture [struc geol] A general term for any break in a rock, whether or not it causes displacement, due to mechanical failure by stress. Fracture includes cracks, joints, and faults.

fracture [exp struc geol] Deformation due to a momentary loss of cohesion or loss of resistance to differential stress and a release of stored elastic energy. Cf: *flow* [exp struc geol]. Syn: *rupture.*

fracture [ice] Any break or rupture through fast ice, a single floe, or highly concentrated pack ice, and resulting from deformation caused by tides, temperature changes, currents, or wind. Its length may vary from a few meters to many kilometers, and its width from zero to more than 500 m. Includes *crack* [ice]; *flaw* [ice]; *lead* [ice].

fracture cleavage A type of cleavage that occurs in deformed but only slightly metamorphosed rocks and that is based on closely spaced, parallel joints and fractures. Syn: *close-joint cleavage.*

fractured deflection A mountain-range *deflection* in which the arcs meet at large obtuse angles, and from the junction two lineaments appear to cross the ranges and mark major structural changes (Wilson, 1950, p.151).

fracture porosity Porosity resulting from the presence of openings produced by the breaking or shattering of an otherwise less pervious rock.

fracture spring A spring whose water flows from joints or other fractures, in contrast to the numerous small openings from which a *filtration spring* flows (Meinzer, 1923, p.50). Cf: *fissure spring; fault spring.*

fracture strength In experimental structural geology, the differential stress at the moment of fracture. Syn: *fracture stress; breaking strength.*

fracture stress *fracture strength.*

fracture system A set or group of contemporaneous fractures related by stress.

fracture zone On the deep-sea floor, an elongate zone of unusually irregular topography that often separates regions of different depths. Such a zone often crosses and apparently displaces the mid-oceanic ridge by faulting.

fragipan A dense, natural subsurface layer of soil whose hardness and relatively slow permeability to water are chiefly due to extreme density or compactness rather than to high clay content (as in *claypan*) or to cementation (as in *hardpan*).

ppears indurated when dry, but shows a moderate to weak
ittleness when moist; it contains much silt and sand, but lit-
e clay and organic matter.

agment (a) A rock or mineral *particle* larger than a *grain*.
) A piece of rock that has been detached or broken from a
eexisting mass; e.g. a clast produced by volcanic, dynamic,
weathering processes.

agmental rock (a) *clastic rock.* (b) *pyroclastic rock.* (c) *bio-
astic rock.*

agmental texture (a) A texture of sedimentary rocks, char-
cterized by broken, abraded, or irregular particles in surface
ntact, and resulting from the physical transport and deposi-
n of such particles; the texture of a clastic rock. The term
used in distinction to a "crystalline" texture. (b) The texture
a pyroclastic rock, such as that of a tuff or a volcanic brec-
a.

agmentary Consisting of or characterized by clastic or detri-
l material; fragmental. The term was formerly applied to
cks composed of fragments of older rocks, or to rocks hav-
g an inhomogeneous texture; as used in geology, the term is
solete.

agmentation The act or process of breaking into pieces or
actionating, or the state of being fragmentated or fractionat-
; esp. the breaking up of a sponge into several others with-
t concomitant formation of specialized reproductive bodies.

amboid A microscopic aggregate of pyrite grains, often in
heroidal clusters. It was considered to be the result of col-
dal processes but is now linked with the presence of organ-
materials; sulfide crystals fill chambers or cells in bacteria
ark & MacDiarmid, 1970, p.133). Adj: framboidal.

amesite A South African variety of bort showing minute bril-
nt points possibly due to included diamonds.

amework [sed] (a) The rigid arrangement created in a sedi-
ent or sedimentary rock by particles that support one anoth-
at their points of contact; e.g. the clasts of a fragmental
ck (esp. a sandstone), constituting a mechanically firm
ructure capable of supporting open pore spaces, although
terstices may be occupied by cement or matrix. (b) The
jid, wave-resistant, calcareous structure built by sedentary
ganisms (such as sponges, corals, and bryozoans) in a
gh-energy environment.

amework [tect] *tectonic framework.*

amework silicate *tectosilicate.*

ancevillite A yellow secondary mineral $(Ba,Pb)(UO_2)_2$-
$O_4)_2.5H_2O$.

anckeite A dark-gray or black mineral: $Pb_5Sn_3Sb_2S_{14}$.

ancolite A colorless mineral of the apatite group: $Ca_5(PO_4,$
$O_3)_3(F,OH)$. Cf: *carbonate-apatite.* Syn: *carbonate-fluorapa-
e; staffelite; kurskite.*

anconian North American stage: Upper Cambrian (above
resbachian, below Trempealeauan).

angite A comprehensive term proposed by Bastin (1909,
450) for all sedimentary rocks (unconsolidated or ce-
ented), and their dynamically metamorphosed representa-
ves, formed from the disintegration of igneous rocks without
tensive decomposition or mechanical sorting; e.g. arkose,
aywacke, gneiss. Etymol: Latin *frangere,* "to break". Adj:
angitic.

anklinian Floral stage in Washington State: lower Eocene.

anklinite An iron-black mineral of the magnetite series in the
inel group: $(Zn,Mn^{+2},Fe^{+2})(Fe^{+3},Mn^{+3})_2O_4$. It resembles
agnetite but is less strongly magnetic, and it represents an
e of zinc.

asch process A hydraulic method of sulfur mining in which
perheated water that has been forced into the in-place de-
sits to dissolve the sulfur is pumped to the surface and
eated to recover the sulfur.

asnian European stage: Upper Devonian (above Givetian,
low Famennian).

aunhofer line Any of the absorption lines in the spectrum of

the sun caused by certain bright lines in the identical positions
in the spectrums of chemical elements present being absorbed by the gases around the sun.

frazil (a) A group of individual ice crystals having the form of
small discoids or spicules which are formed in supercooled
turbulent water. Syn: *frazil ice.* (b) *frazil crystal.*----Etymol:
Canadian French *frasil,* from French *fraisil,* "forge cinders".

frazil crystal A small discoid or needle-like spicule of ice
formed by freezing of supercooled turbulent water. Syn: *frazil; ice crystal.*

frazil ice A spongy, slushy, cinder-like mass or aggregate of
frazil crystals collected by adhesion or regelation and suspended in supercooled turbulent water, esp. common in a rapidly flowing stream, but also found in turbulent seawater
(where it is called *lolly ice*). Syn: *frazil; needle ice.*

frazil slush An agglomerate of loosely packed frazil floating on
the water surface which can accumulate under the ice cover.
Cf: *slush* [*snow*].

freboldite A hexagonal mineral: CoSe.

Fredericksburgian North American (Gulf Coast) stage: Lower
Cretaceous (above Trinitian, below Washitan).

free Said of a native element.

free-air anomaly A gravity anomaly which is calculated from a
theoretical model and elevation above sea level, but without
allowance for the attraction effect of topography and isostatic
compensation.

free arm The freely mobile part of a crinoid ray not incorporated in theca.

free blade The portion of a blade of a plate-like conodont element not flanked by platforms.

freeboard (a) The additional height that is above the recorded
or design high-water mark of an engineering structure (such
as a dam, seawall, flume, or culvert) associated with a body
of water and that represents an allowance against overtopping
by transient disturbances. (b) The vertical distance between
the water level at a given time and the top of an engineering
structure, such as the distance between the normal operating
level of a reservoir and the crest of the associated dam.

free-burning coal *noncaking coal.*

free cheek A lateral part of the cephalon of a trilobite outside
of the facial suture, separated from the cranidium at the time
of molting and including the visual surface of the eye. See
also: *cheek* [*paleont*]. Cf: *fixed cheek.* Syn: *librigene.*

free corer A type of *gravity corer* that operates from a float
rather than from a ship.

free degradation Degradation of a slope from the foot of
which no debris is removed, e.g. an abandoned cliff (Hutchinson, 1967).

free energy A thermodynamic function of the state of a system, providing a measure of the maximum work obtainable
from the system under specified conditions. The functions
most commonly used are the *Helmholtz* and *Gibbs free energies.* Because these functions (among others) also measure
the driving forces for processes occurring under specified
conditions, they are often referred to as *thermodynamic potentials.*

free enthalpy *Gibbs free energy.*

free face The part of a hillside surface consisting of an outcrop of bare rock (such as a scarp or cliff) that stands more
steeply than the angle of repose of the *constant slope* immediately below (Wood, 1942); a rock wall from which weathered material falls to the bottom to form a talus.

free field The field of a sound wave that would propagate in an
infinite medium.

free flow In hydraulics, flow that is not disturbed by submergence or backwater (Brown & Runner, 1939, p.155).

free ground water *unconfined ground water.*

free margin The peripheral distal border of an ostracode
valve, exclusive of the hinge.

free meander A stream meander that displaces itself very

easily by lateral corrasion, esp. where vertical corrasion is of no importance. Syn: *free-swinging meander.*

free moisture *free water.*

free operculum The part of a dinoflagellate cyst that is completely surrounded by archeopyle sutures, with no unsutured connection to the rest of the cyst. Cf: *attached operculum.* Syn: *free opercular piece.*

free oscillation An oscillation of a body, e.g. the Earth, that occurs without external influence other than the initiating force, and that has its own natural frequency. Such oscillations follow major earthquakes. Cf: *forced oscillation.* Syn: *free vibration.*

free period The time for one complete swing of the seismograph mass when all damping is removed and the Earth is quiet.

freestone [rock] Any stone (esp. a thick-bedded, even-textured, fine-grained sandstone) that breaks freely and can be cut and dressed with equal ease in any direction without splitting or tending to split. The ease with which it can be shaped into blocks makes it a good building stone. The term was originally applied to limestone, and is still used for such rock. See also: *konite.* Cf: *flagstone.*

freestone [water] Water containing little or no dissolved material. Syn: *freestone water.*

free surface The upper surface of a layer of fluid where the pressure on it is equal to the external atmospheric pressure.

free vibration *free oscillation.*

free water (a) Water in the soil in excess of field capacity that is free to move in response to the pull of gravity. Syn: *gravity water; gravitational water; infiltration water.* Cf: *gravity ground water.* (b) Water that can be removed from another substance, as in ore analysis, without changing the structure or composition of the substance. Ant: *bound water.* Syn: *free moisture.*

free-water content The fraction by weight of liquid water in wet snow; some workers prefer fraction by volume. Not to be confused with *water equivalent.* Syn: *snow moisture; liquid water content.* See also: *water content* [snow].

free-water elevation *water table.*

free-water level *free-water surface.*

free-water surface The surface of a body of water at which the pressure is atmospheric and below which the pressure is greater than atmospheric; the surface of any pond, reservoir, etc., that is open to the atmosphere, or a *water table.* Syn: *free-water level.*

free wave A wave that is created by a sudden rather than continuous impulse and that continues to exist after the generating force is gone. Cf: *forced wave.*

freeze-out lake A very shallow lake subject to being deeply frozen over for long periods.

freeze-thaw action *frost action.*

freezeup (a) The formation of a continuous ice cover, generally restricted to the hardening of locally formed young ice, but sometimes including the freezing together of pieces of drift ice. (b) The period during which a body of water in an area is frozen over, esp. when marking the beginning of winter.

freezing The process by which a liquid becomes a solid, involving the removal of heat. Commonly applied to water but also used by solidification upon cooling of molten metals, magma, etc.

freezing interval *crystallization interval.*

freibergite A steel-gray variety of tetrahedrite containing silver.

freieslebenite A steel-gray to dark-gray mineral: $Pb_3Ag_5Sb_5S_{12}$.

freirinite A lavender to turquoise-blue mineral: Na_3Cu_3-$(AsO_4)_2(OH)_3 \cdot H_2O$. Cf: *lavendulan.*

fremontite *natromontebrasite.*

French chalk A soft, white, compact variety of talc, steatite,

or soapstone, finely ground into powder and used for crayons, as a grease remover in dry cleaning, or other special purposes.

Frenkel defect In a crystal lattice, the displacement of an atom from its original position to an interstice; it is a type of *point defect.* Cf: *interstitial defect; Schottky defect.*

frenuliniform Said of the loop, or of the growth stage in the development of the loop, of a dallinid brachiopod (as in the subfamily Frenulininae), marked by lateral resorption gaps (lacunae) occurring in the hood but before resorption of shell occurs posterior to the gaps (TIP, 1965, pt.H, p.145). It is subsequent to the *campagiform* stage.

frenulum A small cylinder connected to the internal part of the nasal tube of a style near the base of a galea in a phaeodarian radiolarian. Pl: *frenula.*

frequency analysis The interpretation of a past record of events in terms of the future probabilities of occurrence; e.g. an estimate of the frequencies of hydrologic events such as floods and droughts.

frequency curve A curve that graphically represents a frequency distribution; e.g. a smooth line drawn on a histogram if the class interval is made smaller and the steps between several bars grows smaller.

frequency distribution A *distribution* or systematic arrangement of statistical data (such as a graphic or tabular display of the number of observations of a variable) that exhibits the division of the values of the variable into mutually exclusive (but closely related), usually ranked, and exhaustive discrete categories or classes and that indicates the frequencies or relative frequencies that correspond to each of the categories or classes. It is generally selected on the basis of some progressively variable physical character, such as diameters of sedimentary particles.

frequency domain Transmission of a continuous wave (usually sinusoidal) and simultaneous reception of electromagnetic energy as a function of frequency. It is used with induced electrical polarization and electromagnetic methods.

frequency response The percentage response of a seismic amplifier for various frequencies at a given filter setting.

fresh [water] adj. Said of water that does not contain or is not composed of salt water.---n. (a) An increased, sudden flow or rush of water; a *freshet* or a *flush.* (b) A stream, spring, or pool of freshwater. (c) A freshwater stream flowing into the sea, or the part of a stream or its shores above the flow of tidal seawater. (d) The mingling of freshwater and salt water.

fresh [weath] Said of a rock or rock surface that has not been subjected to or altered by surface weathering, such as a rock newly exposed by fracturing. Syn: *unweathered.*

freshening Making water less salty; separating water from saline constituents.

freshet (a) A great rise in, or a sudden overflowing of, a small stream, usually caused by heavy rains or rapidly melting snow in the highlands at the head of the stream; a rapidly rising flood, usually of minor severity and short duration. See also *flash flood.* Syn: *fresh; spate; high water.* (b) A small, clear freshwater stream or current flowing swiftly into the sea; an area of comparatively fresh water at or near the mouth of a stream flowing into the sea. (c) A small stream flowing swiftly into a lake (as in the spring) and often carrying a heavy silt load during its peak flow.

fresh ice (a) *young ice.* (b) *freshwater ice.* (c) Ice that was formed on salt water but is now salt-free.

fresh snow *new snow.*

fresh water n. Water containing only small quantities of dissolved minerals, such as the water of streams and lakes unaffected by salt water or salt-bearing rocks; water that lacks saline or mineral taste, though it may contain suspended sediment, pathogenic organisms, and/or small quantities of dissolved constituents not detectable by taste but, nevertheless, toxic, and may have harmless but objectionable taste, odor or color. Cf: *potable water*. Syn: *sweet water*. Also spelled: *freshwater*; *fresh-water*.

freshwater estuary An estuary into which river water pours with sufficient volume to exclude salt water. See also: *positive estuary*.

freshwater ice Ice formed by the freezing of freshwater in lakes or streams, or in the ground. Syn: *fresh ice*.

freshwater limestone A limestone formed by accumulation or precipitation in a freshwater lake, a stream, or a cave. It is often algal and sometimes nodular. See also: *underclay limestone*.

freshwater sediment A sediment that accumulates, or has accumulated, in a freshwater environment; e.g. a sediment resulting from lacustrine, fluvial, or glaciofluvial activity.

freshwater swamp A swamp that depends on nontidal fresh water rather than a saltwater source.

Fresnian North American stage: upper Eocene (above Narizian, below Refugian).

fresnoite A mineral: $Ba_2TiSi_2O_8$.

fret A spot worn or eroded by *fretting*, as on a limestone surface.

fretted upland A preglacial upland surface completely consumed by the intersection of cirques from opposite sides; the "ultimate product of cirque sculpture by glaciers" (Hobbs, 1912, p.373). Cf: *grooved upland; scalloped upland*.

fretting (a) The wearing away of a surface, such as a stream making a channel for itself; the eating into of a substance, such as rainwater corroding a rock surface. See also: *fret*. (b) The agitation or disturbance of running water, such as the rippling of a brook over rocks.

fretum An arm of the sea; a strait. Pl: *freta*.

fretwork weathering *honeycomb weathering*.

freudenbergite A black hexagonal mineral: $Na_2Fe_2Ti_7O_{18}$.

freyalite A variety of thorite high in rare earths (esp. cerium).

friability The condition of being friable.

friable (a) Said of a rock or mineral that crumbles naturally or is easily broken, pulverized, or reduced to powder, such as a soft or poorly cemented sandstone. (b) Said of a soil consistency in which moist soil material crushes easily under gentle to moderate pressure (between thumb and forefinger) and coheres when pressed together.

friction Mechanical resistance to the relative motion of contiguous bodies or of a body and a medium. Cf: *internal friction*.

frictional As applied to a soil, a syn. of *cohesionless*.

frictional layer The layer of the ocean that is affected by the action of the wind.

friction breccia A breccia composed of broken or crushed rock fragments resulting from friction, such as a "volcanic friction breccia" formed where a rising column of nearly congealed lava was shattered against the walls of the volcanic vent and later cemented by newly rising magma; specif. a *fault breccia* produced by friction of the two walls of the fault rubbing against each other.

friction crack A short *crescentic mark* that is transverse to the direction of ice movement and that includes a distinct fracture that dips forward into the bedrock, indicating the direction of ice movement (Harris, 1943). It presumably results from local increase in frictional pressure between ice and bedrock.

friction depth The depth in the ocean where the velocity vector in an *Ekman spiral* is exactly opposite to the wind direction; commonly about 100 m, rarely deeper than 200 m. Syn: *depth of frictional influence; depth of frictional resistance*.

friction head That head of fluid which is lost because of friction. See also: *friction slope*. Syn: *head loss; friction loss*.

friction loss *friction head*.

friction mark An obsolete term for a ripple mark or an undulating sedimentary structure that originates as a surface of least friction.

friction slope The *friction head* or loss per unit length of con-duit. For most conditions of flow the friction slope coincides with the energy gradient, but where a distinction is made between energy losses due to bends, expansions, impacts, etc., a distinction must also be made between the friction slope and the energy gradient. In uniform channels, the friction slope is equal to the bed or surface slope only for uniform flow (ASCE, 1962).

friedelite A rose-red mineral: $Mn_8Si_6O_{18}(OH,Cl)_4.3H_2O$.

friendly ice A submariner's term for an ice canopy containing more than 10 large skylights (or other features that permit a submarine to surface) per 30 nautical miles (56 km) along the submarine's track (U.S. Naval Oceanographic Office, 1968, p.B33). Ant: *hostile ice*.

frigid Said of a soil temperature regime in which the mean annual temperature (measured at 50cm depth) is more than 0°C but less than 8°C with a summer-winter variation of more than 5°C and with warm summer temperatures (SSSA, 1970). Cf: *isofrigid; cryic*.

frigid climate A type of climate characteristic of a region in which there is a more or less permanent snow and ice cover over the permanently frozen surface (Twenhofel, 1939). Cf: *polar climate*.

frill A relatively large lamella projecting well beyond the general contour of a brachiopod valve, deposited by margin of highly retractile mantle (TIP, 1965, pt.H, p.145). Cf: *growth lamella*.

fringe [coast] The line beyond which detritus from the delta-forming river is no longer the major fraction in the sea-floor sediment (Bayly, 1968, p.151-152).

fringe [glac geol] A thin sprinkling of erratic boulders in front of the terminal moraine of a glacier.

fringe joint A small-scale joint peripheral to a master joint, usually at a 5-25° angle from the face of the main joint. It is formed by tension or shear. See also: *cross fracture*.

fringe ore Ore of the outer limits of the mineralization pattern or halo. Syn: *halo ore*.

fringe water Water of the capillary fringe. Syn: *anastatic water*.

fringing reef A coral reef that is directly attached to or borders the shore of an island or continent, having a rough, table-like surface that is exposed at low tide; it may be more than 1 km wide, and its seaward edge slopes sharply down to the sea floor. There may be a shallow channel or lagoon between the reef and the mainland, although strictly there is no body of water between the reef and the land upon which it is attached. Cf: *barrier reef*. Syn: *shore reef*.

frith A syn. of *firth*. Etymol: alteration of *firth*.

fritting The partial melting of quartz, etc. grains so that they are now each surrounded by a zone of glass. Fritting is due to contact action of basalt and other lavas on other rocks (Johannsen, 1931).

frohbergite A mineral: $FeTe_2$. It is isomorphous with marcasite.

frolovite A white mineral: $CaB_2O_4.4H_2O$.

frond The expanded, leaflike portion of ferns and some other cryptogams (Fernald, 1950, p.1575).

frondelite A mineral: $MnFe_4^{+3}(PO_4)_3(OH)_5$.

frondescent cast A term used by Haaf (1959, p.30) for a feather-like sole mark resembling certain shrubs or large cabbage leaves, the spreading "foliage" always directed downcurrent; it is usually several decimeters in length. Syn: *cabbage-leaf mark; deltoidal cast*.

front [meteorol] The contact at the Earth's surface between two different air masses, commonly cold and warm, that generally moves in an easterly direction. See also: *cold front; warm front*.

front [paleont] The part of the carapace of a brachyuran decapod crustacean (crab) between the orbits.

front [petrology] A metamorphic zone of changing mineralization developed outward from a large expanding igneous intrusion.

front [glaciol] (a) *ice front.* (b) *snout* [glaciol].

front [geomorph] (a) The more or less linear outer slope of a mountain range that rises above a plain or plateau. (b) Land that faces or abuts, esp. on a body of water. Syn: *frontage.*

frontal [paleont] adj. Pertaining or belonging to the front part; esp. pertaining to the exposed or orifice-bearing side of a bryozoan zooid or colony (particularly of cheilostomes). Syn: *obverse.*

frontal apron *apron* [geomorph].

frontal area Space occupied in life by the frontal membrane in anascan cheilostomes (bryozoans).

frontal kame A kame that consists of a steep alluvial fan against the edge of an ice sheet.

frontal membrane Uncalcified part of frontal wall in cheilostomatous bryozoans lacking asci. It may be exposed or covered by a frontal shield.

frontal moraine (a) *end moraine.* (b) "A moraine rampart at the front of a former glacier" (Schieferdecker, 1959, term 0918).

frontal plain *outwash plain.*

frontal plate The modified rostrum of a brachyuran decapod crustacean, bearing a process that projects ventrally between antennules to unite with the episteme.

frontal pore (a) A pore on the front of the lattice shell of a radiolarian and adjacent to the basal ring. It is similar in appearance to the sternal pore. (b) A pore on the exposed or orifice-bearing side of a bryozoan zooid or colony; e.g. *ascopore.*

frontal scar The scar on the interior of the carapace of an ostracode, just anterior and dorsal to the adductor muscle scars.

frontal shield Calcareous cover above the frontal membrane in some cheilostomatous bryozoans, commonly formed by the fusion of costae. Syn: *pericyst.*

frontal terrace *outwash terrace.*

frontal wall Calcareous, frontal body wall covering the ascus in ascophoran cheilostomes (bryozoans).

front bay A large, irregular, shallow bay connected with the sea through a pass between barrier islands, as along the coast of Texas. Cf: *back bay.*

front pinacoid In an orthorhombic, monoclinic, or triclinic crystal, the {100} pinacoid. Cf: *basal pinacoid; side pinacoid.* Syn: *macropinacoid; orthopinacoid.*

front range The outermost range of a mountain system; e.g. the Front Range of the Rocky Mountains, extending southward from Casper, Wyo., through Colorado and into New Mexico, including the Sangre de Cristo Mountains.

front slope The *scarp slope* of a cuesta.

froodite A monoclinic mineral: $PdBi_2$. Cf: *michenerite.*

frost (a) A light, feathery deposit of ice caused by the condensation of water vapor on a surface that is below the freezing point. Cf: *dew.* (b) In weather description, a below-freezing temperature.

frost action (a) The mechanical weathering process caused by alternate or repeated cycles of freezing and thawing of water in pores, cracks, and other openings, usually at the surface. It includes *congelifraction* and *congeliturbation.* (b) The resulting effects of frost action on materials and structures.-- Syn: *freeze-and-thaw action; freeze-thaw action.*

frost-active soil A fine-grained soil that undergoes changes in volume and bearing capacity due to frost action (Nelson & Nelson, 1967, p.151).

frost belt A ditch designed to assist the early and rapid freezing of the soil in order to obstruct seepage of shallow ground water. It is commonly placed upslope from foundations in permafrost areas. Syn: *frost dam.*

frost blister A low *frost mound,* usually less than 8 m high, caused by seasonal growth of segregated layers of ice or by

hydrostatic pressure of ground water. Cf: *gravel mound.*

frost boil (a) An accumulation of excess water and mud liberated from ground ice by accelerated spring thawing, commonly softening the soil and causing a quagmire. (b) A low mound developed by local differential frost heaving at a place most favorable for the formation of segregated ice and accompanied by an absence of an insulating cover of vegetation (Taber, 1943, p.1458-1459). (c) A break in a surface pavement due to swelling frost action; as the ice melts, soupy subgrade materials issue from the break.

frost bursting *congelifraction.*

frost churning *congeliturbation.*

frost circle A term used by Williams (1936) in referring to a *sorted circle* developed in horizontal thin-bedded limestones in Ontario.

frost crack A nearly vertical fracture developed in rock or in frozen ground with an appreciable ice content by contraction of ice as the temperature is lowered below the freezing point. Syn: *ice crack.*

frost cracking The contraction of frozen ground at very low temperatures; the formation of *frost cracks.*

frost-crack polygon A *nonsorted polygon* formed by intersecting frost cracks resulting from low-temperature contraction of frozen ground (Hopkins et al, 1955, p.139). It is similar to an *ice-wedge polygon,* but lacks a border of ice wedges and therefore is not necessarily associated with permafrost.

frost creep Soil creep resulting from frost action (Kerr, 1881).

frost drift The movement, by frost action, of debris on a slope (Kerr, 1881).

frost flower A type of *surface hoar,* usually growing on an ice surface, and characterized by leafy or dendritic blades oriented at various random angles to the surface. Syn: *ice flower.*

Frost gravimeter An astatic, balance-type gravity meter consisting of a mass at the end of a nearly vertical arm, supported by a main spring inclined to the vertical at about a 45° angle. The beam rises and falls with gravity variation, but is restored to its normal position by a sensitive weighing spring tensioned by a micrometer screw.

frost heave *frost heaving.*

frost heaving The uneven lifting or upward movement, and general distortion, of surface soils, rocks, vegetation, and other structures, such as pavements, due to internal frost action resulting from subsurface freezing of water and growth of ice masses (esp. ice lenses), and usually producing a frost mound; any upheaval of ground caused by freezing. See also *frost thrusting.* Syn: *frost heave.*

frost hillock The marked upward bulging sometimes present in the center of a mud polygon (Sharpe, 1938, p.36).

frosting (a) A lusterless, finely pitted, ground-glass, or mat surface imposed by wind action on the surface of a rounded grain (usually a quartz grain) by innumerable impacts of other similar grains. (b) The etching or pitting process that produces a frosting.

frost line (a) The maximum depth of frozen ground (or soil) in areas where there is no permafrost; it may be expressed for a given winter or the average of several winters or the greatest depth on record. Cf: *frost table.* (b) The bottom limit of permafrost. Cf: *permafrost table.* (c) The altitudinal limit below which frost never occurs; applied esp. in tropical regions.

frost mound A general term for a knoll, hummock, hill, or conical mound in a permafrost region, containing a core of ice, and representing a generally seasonal and localized upwarp of the land surface, caused by frost heaving and/or hydrostatic pressure of ground water. See also: *pingo; palsa; hydrolaccolith; earth hummock; frost blister; ground-ice mound; ice mound; gravel mound.* Syn: *soil blister; soffosion knob.*

frost-pattern soil A term used by Troll (1944) for what is now known as *patterned ground;* it is a misleading term because patterned ground need not consist of soil, nor need it involve a periglacial origin.

frost-point temperature That temperature to which a sample of moist air must be cooled (at constant pressure and water-vapor content) in order to reach the point of saturation with respect to ice.

frost polygon One of the network polygons forming *polygonal ground*.

frost riving *congelifraction*.

frost scar A *nonsorted circle* or irregular form representing a small patch of bare soil produced by local frost heaving intense enough to disturb the vegetation cover. See also: *mud circle*. Syn: *mud spot; spot medallion*.

frost shattering *congelifraction*.

frost shifting *frost thrusting*.

frost soil *congeliturbate*.

frost splitting *congelifraction*.

frost stirring A syn. of *congeliturbation* involving no mass movements.

frost table An irregular surface that represents, at any given time, the penetration of thawing in frozen ground; the upper limit of frozen ground. It coincides with the *permafrost table* when the active layer is completely thawed. Cf: *frost line*.

frost weathering *congelifraction*.

frost wedge A term used loosely for any *ice wedge*, either active (or seasonal) or fossil; any wedge-shaped mass whose origin involves cold or freezing conditions.

frost wedging A type of *congelifraction* by which jointed rock is pried and dislodged by ice acting as a wedge.

frost zone *seasonally frozen ground*.

Froude number A dimensionless numerical quantity used as an index to characterize the type of flow in a hydraulic structure that has the force of gravity (as the only force producing motion) in conjunction with the resisting force of inertia. It is the ratio of inertia forces to gravity forces, and is equal to the square of a characteristic velocity (mean, surface, or maximum velocity) of the system divided by the product of a characteristic linear dimension (e.g. diameter or depth) and the gravity constant, acceleration due to gravity, all of which are expressed in consistent units in order that the combinations will be dimensionless. The number is used in open-channel flow studies or where the free surface plays an essential role in influencing motion (ASCE, 1962).

frozen [ore dep] Said of the contact between a vein wall and the mineral deposit filling the vein in which the vein material adheres closely to the wall; also, said of the vein material and of the wall.

frozen ground Ground (or soil) that has a temperature below freezing and generally contains a variable amount of water in the form of ice. Terms inadvisedly used as syns: *frost; ground frost; permafrost*. Syn: *tjaele; gelisol; merzlota; tele; taele*.

fruchtschiefer A type of *spotted slate* characterized by concretionary spots having shapes suggestive of grains of wheat. Etymol: *German*. Cf: *Fleckschiefer; Garbenschiefer; Knotenschiefer*.

fructification A reproductive organ or fruiting structure of a plant.

fruit (a) In a strict sense, the pericarp and its seeds, the fertilized and developed ovary. (b) More broadly, the matured pericarp and its contents, with any integral, external part (Jackson, 1953, p.153).

frustule The siliceous skeleton of a diatom, consisting of two halves (the epivalve and the hypovalve). It is ornate, microscopic, and boxlike.

F test A statistical test for equality or comparison of sample variances, expressed as the ratio between sample variances. Syn: *F-distribution test*.

fuchsite A bright-green, chromium-rich variety of muscovite. Syn: *chrome mica*.

fucoid n. (a) An informal name now applied loosely to any indefinite tunnel-like sedimentary structure identified as a trace fossil but not referred to a described genus. It was once considered to be the remains of the marine alga *Fucus*, and later was regarded as a cylindrical, U-shaped, regularly branching feeding burrow of a marine animal and assigned to the plant-like "genus" *Fucoides*. The term has been applied to crustacean tracks, worm burrows, molluscan trails, marks made by the tide or waves, and rill marks. Fucoids are distinguished from hieroglyphs by being within sedimentary layers and formed of material more or less unlike the matrix (Vassoevich, 1953, p.21). See also: *chondrite*. (b) A fossil of an alga, or a fossil resembling an alga or the remains or impression of a seaweed. (c) A seaweed of the order Fucales (brown algae).----adj. Pertaining to or resembling a fucoid. Syn: *fucoidal*.

fuel ratio In coal, the ratio of fixed carbon to volatile matter. It is sometimes a factor in the analysis or classification of coals. Cf: *carbon ratio*.

fugacity A thermodynamic function directly related to the chemical potential and defined in such a way that thermodynamic equations describing the behavior of ideal gases apply equally well to nonideal gases when the fugacities of the nonideal gases are substituted in the equations for the pressures of the ideal gases. Numerical values of fugacity are expressed in units of pressure.

fugacity coefficient The ratio of the fugacity of a gas to its pressure; hence, a measure of the nonideality of the gas.

fugitive In ecology, said of an organism, esp. a plant species, that is not permanently established and is likely to disappear.

fugitive constituent A substance originally present in magma but which was lost during crystallization of the magma, so that it does not commonly appear as a rock constituent (Shand, 1947, p.34).

fukuchilite A pyrite-type mineral: Cu_3FeS_8.

fulcral plate A small plate raised above the floor of the brachial valve of a brachiopod, extending between the posterior margin and the brachiophore base, and bounding the dental socket anteriorly and laterally (TIP, 1965, pt.H, p.145).

fulcral ridge A linear elevation on an articular face of a crinoid ossicle serving as an axis of differential movement.

fulcrum The intersection of the end of a recurved spit with the next succeeding stage in development of a compound spit.

fulgurite An irregular, glassy, often tubular or rod-like structure or crust produced by the fusion of loose sand (or rarely, compact rock) by lightning, and found esp. on exposed mountain tops or in dune areas of deserts or lake shores. It may measure 40 cm in length and 5-6 cm in diameter. Etymol: Latin *fulgur*, "lightning". Syn: *lightning stone; lightning tube; sand tube*.

fulji A term used in northern Arabia for a depression between barchans, occurring esp. where the dunes are pressing closely on one another; it has a steep slope on the windward side and a gentle slope on the lee side. Etymol: Arabic. Pl: *fuljis*. Syn: *fulje*.

full A British term for *beach ridge*.

full-cut brilliant A brilliant-cut diamond or colored stone with the usual total of 58 facets, consisting of 32 facets and a table above the girdle, and 24 facets and a culet below. On colored stones, the girdle is usually polished, but seldom on diamonds.

full dip *true dip*.

fuller's earth A very fine-grained, naturally occurring earthy substance (such as a clay or clay-like material) possessing a high adsorptive capacity, consisting largely of hydrated aluminum silicates (chiefly the clay minerals montmorillonite and palygorskite), and used originally in England for whitening, degreasing, or fulling (shrinking and thickening by application of moisture) woolen fabrics but now extensively as an adsorbent in refining and decolorizing oils (mineral, vegetable, animal) and fats; a natural bleaching agent. Its color ranges from light brown through yellow and white to light and dark green, and it differs from ordinary clay by having a higher percentage of water and by lacking (or showing very little) plasticity, tending to break down into a muddy sediment in water. Fuller's earth probably forms as a residual deposit by decomposition of rock in place, such as by devitrification of volcanic glass. The term is applied without reference to any particular chemical or mineral composition, texture, or origin. Syn: *creta; walker's earth.*

full meander A stream meander consisting of two loops, one in a clockwise direction and the other in a counterclockwise direction.

fully arisen sea *fully developed sea.*

fully developed sea A sea in which all possible wave frequencies in the wave spectrum for prevailing wind speed have developed the maximum energies. The ocean waves are at the maximum height for the given wind blowing over sufficient fetch and with sufficient duration. Syn: *fully arisen sea.*

fuloppite A lead-gray mineral with a bluish or bronzy tarnish: $Pb_3Sb_8S1_5$. Also spelled: *fuloppite.*

Fultonian Floral stage in Washington State: middle Eocene.

fulvic acid That organic matter of indefinite composition which remains in solution when an aqueous alkaline extract of soil is acidified.

fulvurite An old synonym for *brown coal.*

fumarole A vent, usually volcanic, from which gases and vapors are emitted; it is characteristic of a late stage of volcanic activity. It is sometimes described by the composition of its gases, e.g. chlorine fumarole. Fumaroles may occur along a fissure or in apparently chaotic clusters or fields. Se also: solfatara; fumarolic stage. Also spelled: fumerole.

fumarolic stage A later or decadent type of volcanic activity characterized by the emission of gases and vapors from a volcanic vent or *fumarole.* Cf: *solfataric stage.*

fume cloud A vaporous cloud of volcanic gases risen from a body of molten lava.

fumerole Var. of *fumarole.*

functional morphology The study of the form and structure of an organism in relation to its adaptation to a specific environment and/or survival under specific conditions; the *morphology* of an animal or plant as it responds or responded to environmental changes and conditions.

fundamental complex *basement.*

fundamental jelly ulmin.

fundamental strength The maximum stress that a body can withstand, under given conditions but regardless of time, without creep.

fundamental substance *ulmin.*

fungal spore A spore of the usually multicellular, nonvascular, heterotrophic plants (fungi). Such spores include a wide variety of types, from simple unicellular to multicellular sclerotia; they have a range of Precambrian to Holocene, and those that are preserved in sediments and studied in coal petrology and palynology are chitinous. Examples: *basidiospore; chlamydospore; conidiospore; dictyospore; phragmospore; teleutospore; urediospore.*

fungus A multicelled thallophyte plant, probably polyphyletic, that feeds on organic matter rather than performing photosynthesis; the informal term for the class Fungi. It ranges from the Devonian.

funicle A spirally wound narrow ridge extending upward from the inner lip of a gastropod shell into the umbilicus.

funicular water Capillary water contained in a cluster of rock or soil particles in the zone of aeration, the interstices of the cluster being completely filled with water bounded by a single closed capillary meniscus (Smith, W.O., 1961, p.2). Cf: *pendular water; pellicular water; sejunction water; capillary condensation.*

funiculus [paleont] A band of mesodermal tissue connecting the polypide with the zooidal wall in a bryozoan (or with the communication pores in cheilostomes and ctenostomes). Pl: *funiculi.*

funiculus [bot] In a fruit, the stalk by which an ovule is at-

tached to the ovary wall or placenta.

funnel filling The red-brown to dark-brown, opaque to translu cent, coarsely laminated portion of the basal plate of som conodont elements, occupying the cavity in the basal funnel.

funnel intrusion An igneous intrusion of an inverted conica shape; generally layered and mafic or ultramafic in compos tion.

funnel joints A joint set that is *concentric,* with the joints dip ping towards a common center.

funnel sea A gulf or bay that is narrow at its head and wide a its mouth, and that deepens rapidly from head to mouth, thu resembling one half of a funnel split lengthwise; e.g. the Gu of California.

furca A two-forked last abdominal segment of certain crusta ceans; specif. *caudal furca.* Pl: *furcae.*

fur-cap rock *mushroom rock.*

furcula A wishbone-shaped sponge spicule. Pl: *furculae.*

furious cross-bedding Bedding that is doubly cross-beddec characterized by foreset beds that are themselves cross-bec ded (Reiche, 1938, p.926).

furrow [geol] A linear marking or depression, wider than line, produced by the digging out and removal of rock materia from a surface, as by glacial action or by faulting; e.g. *groove* or a *striation.*

furrow [beach] *swale.*

furrow [palyn] *colpus.*

furrow [tect] A nongeneric term used by Bucher (1933) for depressed part of the crust of any size with a distinct linea development. Cf: *welt.*

furrow [streams] A term applied in Africa to a natural or artif cial watercourse used for drainage or irrigation.

furrow cast A term introduced by McBride (1962, p.58) for sole mark consisting of a cast of a group of closely spacec parallel, and linear indentations separated by long, narrow round or flat-topped, slightly sinuous septa which appear a depressions in the cast; it lacks the steep or blunt upcurrer end of a flute cast. The term was suggested hesitantly b Kuenen (1957, p.244) to replace the ambiguous "groov cast". Cf: *furrow flute cast.*

furrow flute cast A *furrow cast* with an upcurrent terminatio similar to that of a flute cast. Syn: *sludge cast; rill cast.*

fusain A coal lithotype characterized by its silky luster, fibrou structure, friability, and black color. It occurs in strands c patches and is soft and dirty when not mineralized. Its cha acteristic microlithotype if *fusite.* Cf: *vitrain; clarain; durain.*

fusainisation Var. of *fusinization.*

fusellar Said of the inner layer or tissue of the periderm of graptolite, exhibiting growth increments, and situated insid the outer layer of cortical tissue.

fusibility A characteristic of metals by which they can be mea sured on a temperature scale. See also:*fusibility scale.* Cf: in *fusible.*

fusibility scale A temperature scale based on the *fusibility* of standard group of metals, with which other metals may b compared. An analysis that can be made with a burner and blowpipe is based on the following series: stibnite, $550°C$ chalcopyrite, $800°C$; almandine garnet, $1050°C$; actinolite $1200°C$; orthoclase, $1300°C$; enstatite (bronzite), $1400°C$; an quartz, infusible.

fusiform Shaped like a spindle or cigar, tapering toward eac end from a swollen middle.

fusiform bomb A volcanic bomb that tapers at both ends fror a swollen middle, it includes both *rotational bombs* and *spir dle-shaped bombs.*

fusil A spindle-shaped siliceous concretion.

fusinite A variety of inertinite characteristic of fusain and con sisting of carbonized woody material.

fusinization A process of *coalification* in which fusain i formed. Cf: *incorporation; vitrinization.* Also spelled: *fusainisa tion.*

fusinoid Fusinite that has a reflectance that is distinctly higher than that of associated xylinoids, vitrinoids, or anthrinoids, and that has well-developed cellular structure (American Society for Testing and Materials, 1970, p.175).

fusion [isotope] The combination, or fusion, of two light nuclei to form a heavier nucleus. The reaction is accompanied by the release of a large amount of energy as in the hydrogen bomb. Cf: *fission*. Syn: *nuclear fusion*.

fusion [petrology] The process whereby a solid becomes liquid by the application of heat; melting; also, the unification or mixing of two or more substances as by melting together.

fusion [photo] *stereoscopic fusion*.

fusion crust A thin glassy coating, usually black and rarely exceeding one millimeter in thickness, formed by ablation on the surface of a meteorite. Due to differing effects of the atmosphere upon different meteorite surfaces, fusion crusts may be knobby, striated, ribbed, net, porous, warty, or scoriaceous.

fusion tectonite An igneous rock whose alignment of early-formed crystals was by continuous movement in an enclosing melt; a type of *primary tectonite* (Turner and Weiss, 1963, p.39).

fusite A coal microlithotype that contains a combination of fusinite, semifusinite, and sclerotinite totalling at least 95%. Cf: *fusain*.

fusoclarain A transitional lithotype of coal characterized by the presence of fusinite and vitrinite, with other macerals; fusinite is less abundant than it is in *clarofusain*. Also spelled: *fusoclarite*.

fusoclarite Var. of *fusoclarain*.

fusodurain A coal lithotype transitional between durain and fusain, but predominantly durain. Cf: *durofusain*.

fusotelain A coal lithotype transitional between telain and fusain, but predominantly telain. Cf: *telofusain*.

fusovitrain A coal lithotype transitional between vitrain and fusain, but predominantly vitrain. Cf: *semifusain*. Also spelled: *fusovitrite*.

fusovitrite Var. of *fusovitrain*.

fusshang A syn. of *foot slope*. Etymol: German *Fusshang*, "foot slope".

fusulinacean Any fusulinid belonging to the superfamily Fusulinacea, characterized by a spindle-shaped, spheroidal, or discoid test with a complex internal structure.

fusulinid Any foraminifer belonging to the suborder Fusulinina, characterized by a calcareous microgranular test. Their known stratigraphic range is Ordovician to Triassic. Syn: *fusuline*. See also: *alveolinid*.

fusulinid limestone A *foraminiferal limestone* composed chiefly of fusulinid tests; e.g. the numerous Missourian and Virgilian (Upper Pennsylvanian) limestones of midwestern U.S.

future ore *possible ore*.

fyord Var. of *fjord*.

G

gabbride A term used in the field for any igneous rock having pyroxene as the only dark mineral, which forms over 50 percent of the rock, with a smaller amount of feldspar; e.g. augite diorite, gabbro, norite.

gabbro A group of dark-colored, basic intrusive igneous rocks composed principally of basic plagioclase (commonly labradorite or bytownite) and clinopyroxene (augite, generally diallage), with or without olivine and orthopyroxene; also, any member of that group. It is the approximate intrusive equivalent of *basalt*. Apatite and magnetite or ilmenite are common accessory minerals. Gabbro grades into *monzonite* with increasing alkali-feldspar content. According to Streckeisen (1967, p.171, 198), plagioclase with more than 50% anorthite distinguishes gabbro from *diorite*; quartz is 0-20% of the light-colored constituents, and the plagioclase/total feldspar ratio is 90/100.

gabbroic layer *basaltic layer.*

gabbroid Said of the texture of a rock similar to that of gabbro; a nonpreferred syn. of *ophitic*. Also, said of a rock resembling gabbro.

gabbrophyre A porphyritic hypabyssal rock composed of labradorite and augite phenocrysts in a groundmass of calcic plagioclase and hornblende.

gabbro schist A gabbro that has become schistose due to cataclasis. Cf: *flaser gabbro*.

gabion A specially designed earth-filled basket, cylinder, or box of corrosion-resistant wire used to hold rock and other coarse aggregate, as in forming a groin or seawall, or to assist in developing a bar or dike in a harbor.

gabrielsonite A mineral: $PbFe(AsO_4)(OH)$.

gadolinite A black, greenish-black, or brown mineral: $Be_2FeY_2Si_2O_{10}$. It is a source of rare earths.

gagarinite A creamy, yellowish, or rosy hexagonal mineral: $NaCaY(F,Cl)_6$.

gagatite Coalified woody material, resembling jet. See also: *gagatization*.

gagatization In coal formation, the impregnation of wood fragments with dissolved organic substances. Se also: gagatite.

gage n. In hydraulics, a device for measuring such factors as water-surface elevation, velocity of flow, water pressure, or precipitation. See also: *staff gage; chain gage*.

gage height *stage* [hydraulics].

gageite A mineral: $(Mn,Mg,Zn)_8Si_3O_{14}.2H_2O$.

gaging *stream gaging.*

gaging station A particular site on a stream, canal, lake, or reservoir where systematic observations of gage height, discharge, or water quality parameters (or any combination of these) are obtained.

gahnite A dark-green (sometimes yellowish, gray, or black) mineral of the spinel series: $ZnAl_2O_4$. It often contains some magnesium. Syn: *zinc spinel*.

gahnospinel A blue or greenish variety of spinel containing zinc.

gain control In a seismic amplifier, control of the amplification or attenuation which varies with time. It may be automatic by individual channel or for all channels together, programmed prior to the shot as a function of time, or binary (allowed to vary by factors of two, with programmed changes) (Sheriff, 1968). Syn: *volume control*.

gaining stream *effluent stream.*

gaize A porous and fine-grained, micaceous and glauconitic sandstone containing much soluble silica, occurring among the Cretaceous rocks of France and Belgium; a calcareous clastic sediment (of Cretaceous age in France), cemented by chert or flint. Se also: opoka.

gal An acceleration of one centimeter per second. A milligal is

0.001 gal. The term gal is not an abbreviation: it was invented to honor the memory of Galileo.

galactic cluster *star cluster.*

galactite (a) A variety of white natrolite, occurring in colorless acicular crystals. (b) An obsolete syn. of *novaculite*. (c) An unidentified stone (possibly of calcium nitrate) whose milky solution gave rise to several medieval legends and superstitions.

galaxite A black mineral of the spinel series: $MnAl_2O_4$. The manganese is often replaced in part by ferrous iron or magnesium, and the aluminum by ferric iron.

galaxy One of billions of large systems of stars, nebulae, star clusters, globular clusters, and interstellar matter that make up the Universe. When the term is capitalized, it refers to the Milky Way stellar system. Syn: *extragalactic nebula*.

galea (a) Conical process in the skeleton of a phaeodarian radiolarian. (b) The spinning tube on the movable finger of the chelicera of certain arachnids (pseudoscorpions). (c) The outer distal hood-like lobe of the second segment of maxillule of a crustacean, adjacent to the *lacinia* (TIP, 1969, pt.R, p.96).

Gale alidade *explorer's alidade.*

galeite A trigonal mineral: $Na_{15}(SO_4)_5F_4Cl$. Cf: *schairerite*.

galena A bluish-gray to lead-gray mineral: PbS. It almost always contains silver. Galena occurs in cubic or octahedral crystals, in masses, or in coarse or fine grains; it is often associated with sphalerite as disseminations in veins in limestones, dolomites, and sandstones. It has a shiny metallic luster, exhibits highly perfect cubic cleavage, and is relatively soft and very heavy. Galena is the most important ore of lead and one of the most important sources of silver. Syn: *galenite; lead glance; blue lead*.

galenite *galena.*

galenobismutite A lead-gray or tin-white mineral: $PbBi_2S_4$.

Galitzin hinge An almost frictionless hinge consisting of a flexible metal strip; it is used in gravity meters, seismographs, and in other precision instruments. Syn: *Cardan hinge*.

Galitzin-type seismograph A type of *vertical seismograph* consisting of a horizontal beam hinged at one end and weighted at the other, supported by a spring attached below the line connecting the hinge and center of gravity of the weight in order to lengthen the period (increase the sensitivity) of the system.

gall [sed] (a) *clay gall.* (b) A sand pipe.

gall [soil] A small, barren or infertile surface spot or area from which the original surface soil has been removed by erosion or excavation.

galleried cave A cave that contains *galleries* at two or more levels.

gallery [speleo] A large, more or less horizontal passage in a cave. It may have been formed by a subterranean stream. See also: *galleried cave*.

gallery [grd wat] *infiltration gallery.*

galliard A hard, smooth, close-grained, siliceous sandstone; a *ganister*. Syn: *calliard*.

gallite A tetragonal mineral: $CuGaS_2$.

Gall projection A stereographic, modified-cylindrical map projection in which the cylinder intersects the globe along the parallels 45°N and 45°S. The scale is preserved along these parallels, but is too small between them and too large poleward of them; there is less distortion of areas and shapes in high latitudes than in the Mercator projection. It was introduced in 1885. Named(?) after Johann G. Galle (1812-1910), German astronomer.

galmei *hemimorphite.*

galt Var. of *gault*.

galvanometer An instrument for detecting or measuring small electric currents.

gamagarite A dark-brown monoclinic mineral: $Ba_4(Fe,Mn)_2V_4O_{15}(OH)_2$.

gametophyte The individual or sexual generation of a plant that produces gametes; e.g. the haploid generation of an embryophytic plant, produced by germination of the spores. In lower vascular plants and bryophytes, the gametophyte is a separate plant, but in seed plants, it is confined to the several cells of the *microgametophyte* in the pollen grain and the multicellular *megagametophyte* in the ovule (with the seed developing from the fertilized ovule). Cf: *sporophyte*. See also: *prothallus*.

gamma [**mag**] A unit of magnetic field strength, symbolized by small Greek letter ν. It is equal to 10^{-5} oersted.

gamma [**mineral**] adj. Of or relating to one of three or more closely related minerals and specifying a particular physical structure (esp. a polymorphous modification); specif. said of a mineral that is stable at a temperature higher than those of its *alpha* and *beta* polymorphs (e.g. "gamma quartz" or γ-quartz").

gamma [**cryst**] (a) In a biaxial crystal, the largest *index of refraction*. (b) The interaxial angle between the *a* and *b* crystallographic axes.----Cf: *alpha* [*cryst*]; *beta* [*cryst*].

gamma* The interaxial angle of the reciprocal lattice between a* and b* which is equal to the interfacial angle between (100) and (010). Cf: *alpha*; beta*.

gamma decay The transformation of an atomic nucleus by the emission of a gamma ray during radioactive decay, without a change in atomic number.

gamma-gamma log A *gamma-ray log* that records the intensity of induced gamma rays emitted by a subsurface formation as a result of gamma-ray bombardment. It is obtained by lowering into a borehole a source that emits gamma rays and a detector (shielded from the source) that receives signals indicative of the gamma rays scattered back to it from the surrounding rocks. Syn: *scattered gamma-ray log; density log*.

gamma radiation *gamma ray*.

gamma ray Electromagnetic radiation from an atomic nucleus, usually accompanying emission of *alpha particles* and *beta particles*. Syn: *gamma radiation*.

gamma-ray log A *radioactivity log* obtained by recording the natural radioactivity of the rocks traversed by a cased or uncased borehole or well, and expressed by measuring the intensity of naturally emitted gamma rays and plotting the data as a function of depth. It is used for definition of strata and determination of interfaces, and extensively with the *neutron log* and laterolog for correlation work and for delineating degrees of shale content (shales and marine clays generally contain the greatest concentration of radioactive salts of all formations encountered in drilling). See also: *gamma-gamma log; spectral log*. Syn: *electronic log*.

gamma-ray spectrometer An instrument for measuring the energy distribution, or spectrum, of gamma rays, whether from natural or artificial sources. It is used as an airborne remote-sensing technique for potassium, thorium, and uranium. Cf: *scintillation spectrometer*.

gamma-ray spectrometry Determinations of gamma-ray energies and states of polarization, and also studies of the correlations between gamma rays emitted in sequence from a nucleus.

gamma-ray spectroscopy The observation of a gamma-ray spectrum and all processes of recording and measuring which go with it.

gamma-sulfur *rosickyite*.

gangmylonite An ultramylonite or mylonite which shows intrusive relations to the adjacent rock without any evidence of fusion (Hammer, 1914).

gangue The valueless rock or mineral aggregates in an ore; that part of an ore that is not economically desirable. Syn: *matrix*. Cf: *ore mineral*.

ganister (a) A hard, compact, even-grained, refractory, and highly siliceous sedimentary rock (fine-grained quartzose sandstone) used in the manufacture of silica brick, composed of subangular quartz particles (0.15-0.5 mm in diameter, although some authors use a lower limit of 0.05 mm, thereby extending into the silt-size range), cemented with secondary silica, and possessing a characteristic splintery fracture that gives rise to smooth, subconchoidal surfaces and sharp edges. It is normally pale to dark gray (sometimes black when showing streaks of carbonaceous matter), and is often (but not necessarily) found as part of the *seat earth* of a coal seam. Ganister is distinguished from chert by its more granular texture and by the relatively small quantity of chalcedonic or amorphous silica. The Sheffield ganister of the Lower Coal Measures of Yorkshire, Eng., is considered the type. See also: *pencil ganister; bastard ganister; silica rock; crowstone; galliard*. (b) A mixture of ground quartz and fireclay used as a furnace lining.---Also spelled: gannister.

ganomalite A colorless to gray tetragonal mineral: $Ca_2Pb_3Si_3O_{11}$.

ganophyllite A brown mineral: $(Na,K)(Mn,Fe,Al)_5(Si,Al)_6O_{15}(OH)_5 2H_2O$.

gap [**coast**] (a) A narrow passage or channel between an island and the shore. (b) A break in a levee through which a stream distributary may flow; a tidal inlet.

gap [**geomorph**] (a) A term used in Pennsylvania and farther south for a sharp break or opening in a mountain ridge, or for a short *pass* through a mountain range; e.g. a *wind gap*. Cf: *notch; col*. (b) A ravine or gorge cut deeply through a mountain ridge, or between hills or mountains; e.g. a *water gap*.---Cf: *gate*.

gap [**marine geol**] *abyssal gap*.

gap [**photo**] Any space where aerial photographs fail to meet minimum coverage requirements, such as a space not covered by any photograph or a space where the minimum specified overlap was not obtained.

gap [**stratig**] A *break* in continuity, such as a lacuna or an unconformity.

gap [**fault**] In a fault, the horizontal component of separation measured parallel to the strike, with the faulted bed absent from the measured interval. Cf: *overlap (fault)*. Obsolete syn: *stratigraphic heave*.

gape [**paleont**] A localized opening remaining between the margins of a bivalve shell (of a mollusk, brachiopod, or ostracode) when the valves are shut or drawn together by adductor muscles; e.g. the anterior gap between the valves when a productoid is feeding.

gara A mushroom-shaped rock occurring in arid or desert regions, resulting from undercutting by wind-driven sand of soft rock, esp. if overlain by more resistant strata. Pl: *gour*. Syn: *cheesewring*.

G/A ratio The principle of the geometric increase in ore tonnage relative to the arithmetic decrease of ore grade, as applied mainly to sedimentary, some residual, and disseminated orebodies (Lasky, 1950).

Garbenschiefer (a) A type of *spotted slate* characterized by concretionary spots whose shape resembles that of a caraway seed. Etymol: German. Cf: *Knotenschiefer; Fleckschiefer; Fruchtschiefer*. (b) *feather amphibolite*.

gardening A phenomenon in which the lunar regolith is constantly and very slowly churning due to successive impacts whereby bottom material works its way to the top and surface material gets buried.

garéwaite A nearly feldspar-free lamprophyre, in the same series with vogesite, spessartite, and odinite, composed of corroded diopside phenocrysts in a fine-grained holocrystalline groundmass of olivine, pyroxene, chromite, and magnetite, with green spinel and labradorite as accessories.

garganite A vogesite that contains both augite and hornblende.

Gargasian Substage in Switzerland: Lower Cretaceous (upper Aptian; above Bedoulian Substage).

gargulho A Brazilian term used in the plateau region of Bahia for a comparatively coarse, clay-cemented ferruginous conglomerate in which diamonds are found.

garland *stone garland.*

garnet (a) A group of minerals of formula: $A_3B_2(SiO_4)_3$, where A = Ca, Mg, Fe^{+2}, and Mn^{+2}, and B = Al, Fe^{+3}, Mn^{+3}, and Cr. (b) Any of the minerals of the garnet group, such as the end members almandine (Fe-Al), andradite (Ca-Fe), grossular (Ca-Al), pyrope (Mg-Al), spessartine (Mn-Al), and uvarovite (Ca-Cr).———Garnet is a brittle and transparent to subtransparent mineral, having a vitreous luster, absence of cleavage, and usually red color (but also green, brown, black, and most any other color except possibly blue). It occurs as an accessory mineral in a wide range of igneous rocks, but is most commonly found as distinctive euhedral isometric crystals in metamorphic rocks (gneiss, mica schist, eclogite); it may also be massive or granular. Garnet is used as a semiprecious stone and as an abrasive.

garnetiferous Containing garnets; e.g. "garnetiferous peridotite".

garnetite A metamorphic rock consisting chiefly of an aggregate of interlocking garnet grains. Cf: *tactite.*

garnetization Introduction of, or replacement by, garnet. This process is commonly associated with contact metamorphism.

garnet jade A light-green variety of grossular garnet, closely approaching fine jadeite in appearance, esp. that found in Transvaal, South Africa.

garnetoid A group name for minerals with structures similar to that of garnet; e.g. griphite and berzeliite.

garnierite (a) An apple-green or pale-green mineral, probably: $(Ni,Mg)_3Si_2O_5(OH)_4$. The ratio Ni:Mg is very variable. Garnierite has been regarded as having no crystal structure, but it may be monoclinic; it is sometimes considered to be a nickel-rich antigorite. It is an important ore of nickel and is used as a gemstone. (b) A group name for poorly defined hydrous nickel-magnesium silicates.---See also: *genthite.* Syn: *noumeite; nepouite.*

garrelsite A monoclinic mineral: $(Ba,Ca)_4H_6Si_2B_6O_{20}$.

garronite A mineral: $Na_2Ca_5Al_{12}Si_{20}O_{64}\cdot27H_2O$.

gas Naturally occurring gaseous hydrocarbon produced in association with petroleum or as marsh gas.

gas barren An area, as large as several acres, that is characterized by a lack of vegetation due to fumarolic activity and acid leaching of the surface rocks.

gas cap Gas overlying liquid hydrocarbons in a reservoir under trap conditions.

gas chromatography A process for separating gases or vapors from one another by passing them over a solid (*gas-solid chromatography*) or liquid (*gas-liquid chromatography*) phase. The gases are repeatedly adsorbed and released at differential rates resulting in separation of their components. Symbol: GC. Cf: *liquid chromatography.* See also: *chromatography.*

gas coal Bituminous coal that is suitable for the manufacture of flammable gas because it contains about 33-38% volatile matter. Cf: *high-volatile bituminous coal.* See also: *coal gas.* Syn: *gas flame coal.*

gas-cut mud Drilling mud returned from the bottom of a drill hole, characterized by a fluffy texture, gas bubbles, and reduced density due to the retention of entrained natural gas rising from the strata traversed by the drill.

gaseous transfer Separation, in a magma, of a gaseous phase which moves relative to the magma and releases dissolved substances, usually in the upper levels of the magma, when it enters an area of reduced pressure. See also: *pneumatolytic differentiation.* Syn: *volatile transfer.*

gaseous transfer differentiation *pneumatolytic differentiation.*

gas field *gas pool.*

gas flame coal *gas coal.*

gas fluxing In volcanology, a rapid upstreaming of free, juvenile gas through a column of molten magma in the conduit of a volcano; the gas acts as a flux to promote melting of the wall rocks. Syn: *volcanic blowpiping.*

gash breccia A term used in Pembrokeshire, Eng., for a rock believed to have originated from the collapse of walls and roofs of huge caves that had been eroded by the solvent action of underground water.

gas heave The distortion of prodelta sediments, produced by the weight of a distributary-mouth bar compacting the underlying sediment and causing carbon dioxide to escape (Moore, 1966, p. 101). See also: *air heave.*

gas-heave structure A sedimentary structure, restricted to the smaller distributaries of the Mississippi River delta, produced by gas heave. See also: *air-heave structure.* Cf: *mudlump.*

gash fracture Small-scale tension fractures occurring at an angle to a fault, that tend to remain open. Syn: *open gash fracture.*

gash vein A vein that is wide above and narrow below, and that terminates within the formation it traverses; a nonpersistent vein that begins as fairly wide but soon narrows. The term was originally applied to vein fillings of solution joints in limestone.

gas inclusion A gas bubble within a gemstone, often visible to the unaided eye, such as a clouded structure in a diamond; esp. one within a synthetic stone, often enabling it to be distinguished from a genuine gem.

gas-liquid chromatography A gas, such as helium, argon, hydrogen, or nitrogen, carrying a gaseous mixture to be resolved is passed over a nonvolatile liquid coated on a porous inert solid support (May & Cuttitta, 1967, p.115) where the components are separated by differential mobility rates. Symbol: GLC Cf: *gas-solid chromatography.* See also: *gas chromatography.*

gasoclastic sediment A sediment resulting from sedimentary volcanism, such as mud ejected by enormous volumes of gas (Bucher, 1952, p.87).

gas-oil interface A surface that forms the boundary between an accumulation of oil and an overlying accumulation of natural gas.

gas-oil ratio (a) The quantity of gas produced with oil from an oil well, usually expressed as the number of cubic feet of gas produced per barrel of oil. Abbrev: GOR. (b) *reservoir gas-oil ratio.*

gaspeite A mineral: $(Ni,Mg,Fe)CO_3$.

gas phase In volcanology, that stage in a volcanic eruption that is characterized by the release of large amounts of *volcanic gases.*

gas pit A small circular pit (2.5-30 cm in diameter, and from less than 3 cm to more than 30 cm deep) surrounded by a mound of mud and produced by the escape of gas bubbles (as of methane generated during the decomposition of organic matter) rising from the surface of a mud bar (Maxson, 1940a).

gas pool An area of subsurface formations, under adequate closure, which will yield gas in economic quantities. Syn: *gas field.*

gas sand An oil sand or sandstone formation containing a large quantity of natural gas.

gassi A term used in the Saharan region for a sand-free interdune passage, in some cases traversing an entire erg from end to end. Etymol: Arabic, "neck, closed ground"'. Pl: *gassis.* Cf: *feidj.*

gas skin The envelope of hot, volcanic gas surrounding each of the particles in an ash flow or núee ardente.

gas-solid chromatography A gas, such as helium, argon, hydrogen, or nitrogen, carrying a gaseous mixture to be resolved is passed over a porous adsorbing solid (May & Cuttitta, 1967, p.114) where the components are separated by differential mobility rates. Symbol: GSC. Cf: *gas-liquid chromatography.* See also: *gas chromatography.*

gas streaming A process of magmatic differentiation involving the formation of a gaseous phase, usually at a late stage in consolidation of the magma, which results in partial expulsion, by the escaping gas bubbles, of residual liquid from the crystal network.

gasteropod *gastropod.*

gastral Pertaining to the surface of the cloaca of a sponge; e.g. "gastral membrane" (endopinacoderm lining the cloaca) and "gastral cavity" (the cloaca itself).

gastralium A specialized sponge spicule lining the cloaca. Pl: *gastralia.*

gastrolith A polished stone or pebble from the stomach of some vertebrates and thought to have been used in grinding up their food. Syn: *stomach stone; gizzard stone.*

gastropod Any mollusk belonging to the class Gastropoda, characterized by a distinct head with eyes and tentacles and, in most, by a single calcareous shell that is closed at the apex, sometimes spiralled, not chambered, and generally asymmetrical; e.g. a snail. Their stratigraphic range is Upper Cambrian to present. Also spelled: *gasteropod.*

gastropore A relatively large tubular cavity of certain hydrozoans providing lodgment for a gastrozooid. Cf: *dactylopore.*

gastrovascular cavity The interior space (coelenteron) of a coral polyp, radially partitioned by septa, and functioning in both digestion and circulation.

gastrozooid A short, cylindrical feeding and digestive polyp housed in a gastropore of a hydrozoan. Cf: *dactylozooid.*

gas-water contact *gas-water interface.*

gas-water interface A *water level* or surface that forms the boundary between ground water and an overlying body of natural gas. Syn: *gas-water contact.*

gas well A *well* that is capable of producing natural gas or that produces chiefly natural gas. Some statutes define the term on the basis of the gas-oil ratio.

gat (a) A natural or artificial opening, as a strait, channel, or other passage, extending inland through shoals, or between sandbanks, or in the cliffs along a coast. (b) A strait or channel from one body of water to another, as between offshore islands or shoals, or connecting a lagoon with the sea.---Syn: *gate.*

gate [coast] (a) An entrance to a bay or harbor, located between promontories; e.g. the Golden Gate in San Francisco, Calif. (b) *gat.*

gate [geomorph] (a) A mountain pass affording an entrance into a country. (b) A broad, low valley or opening between highlands, generally wider than a *gap.* Syn: *geocol.* (c) A restricted passage along a river valley; e.g. the Iron Gate on the River Danube.

gate [paleont] A large opening or fissure in the skeleton of a spumellarian radiolarian.

gathering ground (a) An area over which water is collected from precipitation, percolation from springs, and drainage; esp. an area that supplies water to a reservoir. (b) *drainage basin.*

gathering zone A term suggested for the area between the land surface and the water table.

gating In marine seismic profiling and echo sounding, a method of eliminating near-surface background noise from bottom and sub-bottom echoes.

gaudefroyite A mineral: $Ca_4Mn_{3-x}^{+3}(BO_3)_3(CO_3)(O,OH)_3.$

gaufrage A syn. of *plaiting.* Etymol: French, "stamping, goffering".

gault A stiff, firm, compact clay, or a heavy, thick clay soil; specif. the Gault, a Lower Cretaceous clay formation in Great Britain. Syn: *galt.*

Gause's principle In ecology, the statement that two identical species cannot coexist in the same area of space at the same time. Named after G.F. Gause (d:1855) German geneticist. Var: *Gause's rule; Gause's axiom.*

gauss The cgs (centimeter gram second) unit for magnetic induction or magnetic flux density, the magnetic field conventionally symbolized by B. The field one centimeter from a

straight wire carrying 5 amps is one gauss.

gaussbergite An extrusive rock similar to orendite but with a glassy groundmass and containing augite and olivine in place of phlogopite. Its name is derived from Gaussberg volcano, Kaiser Wilhelm II Land, Antarctica.

Gaussian distribution A syn. of *normal distribution.* Named after Karl Friedrich Gauss (1777-1855), German mathematician.

Gauss projection Any of several conformal map projections used or developed by Karl Friedrich Gauss (1777-1855), German mathematician; esp. the "Gauss-Kruger projection", a special case of the *transverse Mercator projection* derived by the direct conformal representation of the spheroid on a plane.

Gauss' theorem The statement that the volume integral of the divergence of a vector function taken over any volume is equal to the surface integral taken over the closed surface surrounding the volume.

gauteite A porphyritic hypabyssal rock characterized by a bostonitic groundmass predominantly of plagioclase, more than in a typical bostonite, along with magnetite, augite, hornblende, and biotite and with phenocrysts of hornblende, augite, some biotite, and abundant plagioclase feldspar. There is a small amount of colorless interstitial material in the groundmass that is frequently decomposed and replaced by a yellow fibrous material. Analcime is also a common constituent.

gaylussite A yellowish-white to gray mineral: $Na_2Ca(CO_3)_2.5H_2O.$

geanticlinal n. The original, now obsolete form of *geanticline.* adj. Pertaining to a geanticline.

geanticline (a) A mobile upwarping of the crust of the Earth, of regional extent. Ant: *geosyncline.* (b) More specifically, an anticlinal structure that develops in geosynclinal sediments, due to lateral compression.----Var: *geoanticline.*

gearksutite An earthy mineral: $CaAl(OH)F_4.H_2O.$ It occurs with cryolite. Syn: *gearksite.*

gedanite A brittle, wine-yellow variety of amber with very little succinic acid.

Gedinnian European stage: Lower Devonian (above Ludlovian of Silurian, below Siegenian).

gedrite An aluminum-bearing variety of anthophyllite.

gedroitzite A clay mineral of the vermiculite group: $6(K,Na)_2O.5Al_2O_3.14SiO_2.12H_2O.$ It is characteristic of many alkali soils from the Ukraine. Also spelled: *gedroizite.*

geest In general, alluvial material that is not of recent origin lying on the surface. A example is the sandy region of the North Sea coast in Germany.

gehlenite A mineral of the melilite group: $Ca_2Al_2SiO_7.$ It is isomorphous with akermanite.

geic acid An obsolete syn. of *humic acid.*

Geiger counter *Geiger-Müller counter.*

Geiger-Müller counter An instrument consisting of a *Geiger-Müller tube* plus the electronic equipment necessary to record the tubes' electric pulses. Syn: *G-M counter; Geiger counter.*

Geiger-Müller tube A radiation detector consisting of a gas-filled tube with a cathode envelope and an axial wire electrode. It functions by detecting momentary current pulses caused by ionizing radiation. It is a part of the *Geiger-Müller counter.* Syn: *G-M tube.*

geikielite A bluish- or brownish-black mineral: $MgTiO_3.$ It is isomorphous with ilmenite, and often contains much iron. It is usually found in rolled pebbles.

gel (a) A translucent to transparent, semisolid, apparently homogeneous substance in a colloidal state, being elastic and jelly-like (or sometimes more or less rigid), offering little resistance to liquid diffusion, and containing a dispersion or network of fine particles that have coalesced to some degree. (b) A nonhomogeneous gelatinous precipitate; e.g. a *coagel.* (c) A liquified mud that has become firm and readsorbed most of the water released earlier.----A gel is in a more solid form than a *sol*, and can sustain limited shear stress.

gelation [chem] The formation of a gel from a sol, as by coagulation (as by cooling, by evaporation, or by precipitation with an electrolyte).

gelation [ice] *congelation.*

geli- Of or pertaining to cold, frost action, or permafrost; e.g. *geli*vation.

Ge:Li detector *lithium-drifted germanium detector.*

gelifluction A syn. of *congelifluction*. Also spelled: *gelifluxion*.

gelifraction *congelifraction.*

gelisol *frozen ground.*

gelisolifluction *congelifluction.*

gelite A name for opal (or chalcedony?) as a secondary deposit in rocks (Hey, 1962, p.435).

geliturbation *congeliturbation.*

gelivation *congelifraction.*

gelivity The sensitivity or susceptibility of rock to congelifraction; the property of being readily split by frost (Hamelin & Clibbon, 1962, p.219).

gel mineral *mineraloid.*

gélose *ulmin.*

gélosic coal Coal that is rich in gélose or ulmin; a syn. of algal coal or *boghead coal*.

gelosite A microscopic constituent of torbanite, consisting of squashed, translucent, pale-yellow spheres of birefringent material (Dulhunty, 1939).

gel texture *colloform.*

gem (a) A cut-and-polished stone that has intrinsic value and possesses the necessary beauty, durability, rarity, and size for use in jewelry as an ornament or for personal adornment; a jewel whose value is not derived from its setting. (b) An especially fine or superlative specimen (as compared to others cut from the same species), generally of superb color, unusual internal quality, and finely cut; e.g. a gem turquoise having a pure uniform blue of the highest possible intensity. In this usage, the meaning depends on the ethics and the range of qualities handled by the seller. (c) A semiprecious stone that possesses value because it is carved or engraved. (d) An uncut diamond of which all or a part can be used for cutting and polishing into a diamond usable in jewelry.

gem color The finest or most desirable color for a particular gem variety.

gem crystal A crystal from which a gem can be cut.

gem gravel A sediment of gravel grade contining appreciable amounts of gem minerals; a placer deposit resulting from washing away of finer material, leaving a residual concentration of valuable minerals.

gemma An asexual propagule sometimes appearing as, but not homologous with, a vegetative bud (Lawrence, 1951, p.753).

gemmary (a) The science of gems. (b) A collection of gems; gems considered collectively.

gemmate adj. (a) Said of sculpture of pollen and spores consisting of more or less spherical projections. (b) Having gemmae.---v. To produce or propagate by a bud.

gem material Any rough material, either natural or artificial, that can be fashioned into a jewel.

gemmation Asexual reproduction involving the origination of a new organism as a localized area of growth on or within the body of the parent and subsequently differentiating into a new individual, as in some corals.

gemmiferous Producing or containing gems.

gem mineral Any mineral species that yields varieties that have sufficient beauty and durability to be classed as gemstones.

gemmology British var. of *gemology*.

gemmule An internal resistant asexual reproductive body of a sponge, consisting of a mass of archaeocytes charged with reserves and enclosed in a noncellular protective envelope.

gemmy Having the characteristics (such as hardness, brilliance, and color) desired in a gemstone.

gemology The study of gems and gemstones, including their identification, grading, and appraisal. Syn: *gemmology*.

gemstone Any mineral, rock, or other natural material (including organic materials such as pearl, amber, jet, shell, ivory and coral) that, when cut and polished, has the necessary beauty (brilliance, color) and durability or hardness for use as a personal adornment or other ornament.

gem variety The variety of a mineral species that yields gemstones.

gena A *cheek* of a trilobite. Pl: *genae*.

genal angle The posterior lateral corner of the cephalon of a trilobite (typically terminating in genal spine, but may be rounded), or of the prosoma of a merostome.

genal spine A spine extending backward from the genal angle of the cephalon of a trilobite, produced on the outer posterior margin of the free cheek.

gendarme [glac geol] A sharp rock pinnacle on an arete, such as will retard or prevent progress along the crest of the aréte; it is less pointed and more tower-like than an *aiguille*, and is commonly found in the Alps. Pron: zhan-*darm*. Etymol: French, "policeman".

gendarme [gem] "A flaw in a diamond or other precious stone" (Webster, 1967, p.944).

gene The fundamental unit governing the transmission and development or determination of hereditary characteristics. Genes occur in a linear sequence on the chromosomes of a cell nucleus.

gene complex The system comprising all the interacting genetic factors of an organism.

genera The plural of *genus*.

general base level *ultimate base level.*

general form The crystal form in each crystal class which has different intercepts on each crystal axis, and has the largest number of equivalent crystal faces for the symmetry present. The form displays and is characteristic of the point group symmetry; other forms may display more symmetry.

general-purpose map A map designed to provide a large amount of general information for many purposes (land-use planning, petroleum geology, highway locations, etc.). Its scale is usually 1/62,500 or larger.

generating area *fetch.*

generation All the crystals of the same mineral species that crystallized at the same time. If there are olivine phenocrysts in a groundmass containing olivine, there are said to be two "generations" of olivine.

generic The adj. form of *genus*.

genetic drift Gradual change with time in the genetic composition of a continuing population resulting from the elimination of some genetic features and the appearance of others.

genetic pan A natural soil *pan* of low or very low permeability, with a high concentration of small particles, and differing in certain physical and chemical properties from the soil immediately above or below it (SSSA, 1965, p.341); e.g. claypan, fragipan, and hardpan. Cf: *pressure pan*.

genetic physiography *geomorphogeny.*

genetic type A sedimentary body representing a complex of genetically related facies formed in the same environment (on land or in the sea) and mostly as the effect of a single leading process (Krasheninnikov, 1964, p.1245); e.g. alluvial deposits, deltaic deposits, lagoonal deposits, or marine deposits. The concept of genetic types, long established in Russian geology, was introduced by Pavlov (1889) to demonstrate the diversity of continental deposits according to their origin.

genicular spine A spine originating from the geniculum of a graptolite.

geniculate Bent abruptly at an angle; specif. said of a brachiopod shell characterized by an abrupt and more or less persistent change in direction of valve growth, producing an angular bend in lateral profile.

geniculate twin A type of crystal twin that bends abruptly, e.g. crystals of rutile that are twinned on the second-order pyramid 101). Syn: *elbow twin.*

geniculum A small knee-shaped structure or abrupt bend in an organism; e.g. a distinct bend in the free ventral wall of graptolithine theca, or the portion between two successive segments of erect, jointed, coralline algae. Pl: *genicula.*

genital plate (a) One of the inner circlet of primordial plates of the *apical system* of an echinoid, arranged in an interradial position and usually perforated by one or more pores for discharge of reproductive products. Cf: *ocular plate.* (b) One of the circlet of five plates on the aboral surface immediately around the centrale of an asterozoan.

genobenthos All terrestrial organisms.

genocline A cline caused by hybridization between adjacent populations that are genetically distinct.

genoholotype *diplotype.*

genomorph (a) A member of a genus which differs from the genotype in exhibiting some basic evolutionary trend over a considerable length of time. (b) A group of superficially similar but not closely related species; a *form genus.*

genotype (a) *type species.* (b) *diplotype.* (c) In genetics, the genetic constitution of an organism or a species in contrast to its observable physical characteristics. See also: *biotype.*

gens A syn. of *species-group.* Plural: *gentes.*

gentes The plural of *gens.*

genthelvite A mineral: $(Zn,Fe,Mn)_4Be_3(SiO_4)_3S$. It is the zinc end member isomorphous with helvite and danalite.

genthite A soft, amorphous, pale-green or yellowish mixture of hydrous nickel-magnesium silicates, representing a mineral near: $(Ni,Mg)_4Si_3O_{10}.6H_2O$. See also: *garnierite.*

gentnerite A mineral: $Cu_8Fe_3Cr_{11}S_{18}$.

genus A unit in taxonomy composed of species that are similar in structure and phylogeny or composed of a single, highly distinctive species. It ranks below "family" and above "species". Adj: *generic.* Plural: *genera.*

geo (a) A term used in northern Scotland for a long, deep, narrow, parallel-sided coastal inlet or cove, walled in by steep, rocky cliffs, and formed by marine erosion of the cliffs along a line of weakness, esp. in a well-jointed rock, as the Old Red Sandstone of Caithness and Orkney. Syn: *gio; goe; gja.* (b) Broadly, an opening to the sea.---Etymol: Scandinavian, akin to Old Norse *gja,* "chasm".

geoacoustics The study of the sounds of the Earth, e.g. those of a volcanic eruption.

geoanticline Var. of *geanticline.*

geoastronomy (a) *astrogeology.* (b) *geocosmology.*

geobarometry Any method, such as the interpretation of pressure-sensitive chemical reactions, for the direct or indirect determination of the pressure conditions under which a rock or mineral formed. The term is analogous to *geothermometry.*

geobasin Rich's term for a geosynclinal-type structure but without folding of its sedimentary filling (1938).

geobiology The study of the biosphere.

geobios That area of the Earth occupied by terrestrial plants and animals. Cf: *hydrobios.*

geobleme A term used by Bucher (1963) for a cryptoexplosion structure caused from within the Earth. Cf: *astrobleme.*

geobotanical prospecting The visual study of plants, their morphology, and their distribution as indicators of such things as soil composition and depth, bedrock lithology, the possibility of ore bodies, and climatic and groundwater conditions. Cf: *biogeochemical prospecting.* See also: *botanical anomaly.*

geocentric A reference to the center of mass of the Earth used in defining coordinate systems. When combined with latitude, such as geocentric latitude, it is the angle at the center of the Earth between the plane of the *celestial equator* and a line to a point on the surface of the Earth.

geocentric horizon The plane passing through the center of the Earth, parallel to the apparent horizon.

geocentric latitude (a) The *latitude* or angle at the center of the Earth between the plane of the celestial equator and a line (radius) to a given point on the Earth's surface. It would be identical with *geodetic latitude* only for a truly spherical Earth. Symbol: ψ. (b) The celestial latitude of a body based on or as seen from the Earth's center.

geocentric longitude (a) *geodetic longitude.* (b) The celestial longitude of a body based on or as seen from the Earth's center.

geocerain *geocerite.*

geocerite A mineral consisting of a white, flaky, wax-like resin of approximate composition: $C_{27}H_{53}O_2$. It is found in brown coal. Syn: *geocerain; geocerin.*

geochemical analysis The application of chemical analysis to geological problems.

geochemical anomaly A concentration of one or more elements in rock, soil, sediment, vegetation, or water that is markedly higher than background. The term may also be applied to hydrocarbon concentrations in soils.

geochemical balance The study of the proportional distribution, and the rates of migration, in the global fractionation of a particular element, mineral, or compound; e.g. the distribution of quartz in igneous rocks, its liberation by weathering, and its redistribution to sediments and, in solution, to terrestrial waters and the oceans.

geochemical cycle The sequence of stages in the migration of elements during geologic changes. Rankama and Sahama distinguish a major cycle, proceeding from magma to igneous rocks to sediment to sedimentary rocks to metamorphic rocks, and possibly through migmatites and back to magma, and a minor or exogenic cycle proceeding from sediment to sedimentary rocks to weathered material and back to sediments again.

geochemical exploration The search for economic mineral deposits or petroleum by detection of abnormal concentrations of elements or hydrocarbons in surficial materials or organisms, usually accomplished by instrumental, spot-test, or "quickie" techniques which may be applied in the field. Syn: *geochemical prospecting.*

geochemical facies Any areal geological entity that is distinguishable on the basis of a compositional variable or a related physical property. Any lithofacies would be definable chemically, but a geochemical facies could be delimited by features which remain imperceptible lithologically, as, for example, trace-element composition or radioactivity.

geochemical fence In a *fence diagram,* a boundary between discrete fields that represents the presence of a particular phase on one side of it and its absence on the other.

geochemical prospecting *geochemical exploration.*

geochemical survey The mapping of geochemical facies.

geochemistry As defined by Goldschmidt (1954, p.1), the study of the distribution and amounts of the chemical elements in minerals, ores, rocks, soils, water, and the atmosphere, and the study of the circulation of the elements in nature, on the basis of the properties of their atoms and ions; also, the study of the distribution and abundance of isotopes, including problems of nuclear frequency and stability in the universe. More generally, all geological study involving chemical change (Clarke, 1924, p.10). A major concern of geochemistry is the synoptic evaluation of the abundances of the elements in the Earth's crust and in major classes of rocks and minerals.

geochron An interval of geologic time corresponding to a *rock-stratigraphic unit.* It may, like the age of that unit, vary from place to place.

geochrone A standard unit of geologic time, now obsolete. It was proposed by Williams in 1893 and was set equal to the duration of the Eocene.

geochronic An obsolete syn. of *geochronologic.*

geochronologic Pertaining or relating to *geochronology.* Syn:

geochronic (obs.)

geochronologic unit geologic-time unit.

geochronology Study of time in relationship to the history of the Earth, esp. by the absolute age determination and relative dating systems developed for this purpose. Cf: geochronometry. See also:chronology. Syn: geologic chronology; geochrony (obs.)

geochronometer A physical feature, material, or element whose formation, alteration, or destruction can be calibrated or related to a known interval of time. See also: radioactive clock.

geochronometry Measurement of geologic time by geochronologic methods, esp. radiometric dating. Cf: geochronology. See also: fossil geochronometry.

geochrony An obsolete syn. of geochronology.

geocline A cline related to a geographic transition.

geocol A term introduced by Taylor (1951, p.614) for a broad, low gap between highlands; a gate.

geocosmogony The science of the origin of the Earth. See also: geogony.

geocosmology The science that deals with the origin and geologic history of the Earth, including its planetary attributes (shape, mass, density, physical fields, rotation, location of poles), the influence of the solar system, the galaxy, and the universe on the geologic development of the Earth, and the material interaction (exchange) between the Earth and the universe (Galkiewicz, 1968). Syn: geoastronomy.

geocronite A lead-gray mineral: Pb_5SbAsS_8.

geocryology The study of ice and snow on the Earth, esp. the study of permafrost.

geocyclic (a) Pertaining to or illustrating the rotation of the Earth. (b) Circling the Earth periodically.

geocyclicity The quality or state of being geocyclic, as in defining all events of cyclic nature recorded in sedimentary rocks for which a specific causal mechanism could not be inferred.

geode (a) A hollow or partly hollow and globular or subspherical body, from 2.5 cm to 30 cm or more in diameter, found in certain limestone beds and also, but rarely, in shales, and characterized by a thin and sometimes incomplete outermost primary layer of dense chalcedony, by a cavity that is partly filled by an inner drusy lining of inward-projecting crystals (often perfectly formed and usually of quartz or calcite and sometimes of barite, celestite, and various sulfides) deposited from solution on the cavity walls, and by evidences of growth by expansion in the cavities of fossils or along fracture surfaces of shells. Unlike a druse, a geode is separable (by weathering) as a discrete nodule or concretion from the rock in which it occurs and its inner crystals are not of the same minerals as those of the enclosing rock. A geode tends to be slightly flattened with its equatorial plane parallel to the bedding. (b) The crystal-lined cavity in a geode. (c) A term applied to a rock cavity and its lining of crystals that is not separable as a discrete nodule from the enclosing rock.---Cf: vug.

geodepression Haarmann's term for a long, narrow depression on the scale of a geosyncline, but not necessarily filled with sediments (Glaessner & Teichert, 1947, p.476).

geodesic A line of shortest distance between any two points on a mathematically defined surface.

geodesy (a) A science concerned with the determination of the size and shape of the Earth and the precise location of points on its surface. (b) The determination of the gravitational field of the Earth and the study of temporal variations such as Earth tides, polar motion, rotation of the Earth, etc.

geodetic azimuth The angle between the tangent to the meridian at a given point and the tangent to the geodesic line from the given point to the azimuth point. See also: Laplace azimuth.

geodetic coordinates Quantities defining the horizontal position of a point on a spheroid of reference with respect to a specific geodetic datum, usually expressed as latitude and longitude. These may be referred to as geodetic positions or geographic coordinates. The elevation of a point is also a geodetic coordinate and may be referred to as a height above sea level.

geodetic datum datum.

geodetic engineering geodetic surveying.

geodetic equator The circle on the spheroid midway between its poles of revolution and connecting points of zero degrees geodetic latitude. Its plane, which contains the center of the spheroid and is perpendicular to its axis, cuts the celestial sphere in a line coinciding with the celestial equator because the axis of the spheroid is by definition parallel to the axis of rotation of the Earth. Cf: astronomic equator.

geodetic latitude The latitude or angle which the normal at a point on the spheroid makes with the plane of the geodetic equator (the equatorial plane of the spheroid). It is equivalent to the astronomic latitude corrected for station error and to true geographic latitude. Latitude as shown on topographic maps is geodetic latitude. Symbol: ϕ. Cf: geocentric latitude.

geodetic leveling Spirit leveling of a high order of accuracy, usually extended over large areas (and with proper applications of orthometric corrections), to furnish accurate vertical control for surveying and mapping operations.

geodetic longitude The longitude or angle between the plane of the geodetic meridian and the plane of an arbitrarily chosen prime meridian (generally the Greenwich meridian). It is equivalent to the astronomic longitude corrected for station error and to true geographic longitude. Symbol: λ. Syn: geocentric longitude.

geodetic meridian A line (ellipse) on the spheroid having the same geodetic longitude at every point.

geodetic parallel A line or circle on the spheroid having the same geodetic latitude at every point; a small circle whose plane is parallel with the plane of the geodetic equator.

geodetic position The latitude and longitude of a point on the Earth's surface determined by triangulation, trilateration, or traverse and referred to an adopted spheroid of reference. geodetic coordinates.

geodetic sea level mean sea level.

geodetic survey (a) A survey of a large land area accomplished by the processes of geodetic surveying and used for the precise location of basic points suitable for controlling other surveys. (b) An organization engaged in making geodetic surveys.

geodetic surveying Surveying in which account is taken of the figure and size of the Earth and corrections are made for earth curvature; the applied science of geodesy. It is used where the areas or distances involved are so great that results of desired accuracy and precision cannot be obtained by plane surveying. Syn: geodetic engineering.

geodic Pertaining to a geode; e.g. a "geodic cavity or vein" in which iron sulfide may precipitate.

geodiferous Containing geodes.

geodimeter An electronic-optical device that measures ground distances precisely by electronic timing and phase comparison of modulated light waves that travel from a master unit to a reflector and return to a light-sensitive tube where an electric current is set up. It is normally used at night and is effective with first-order accuracy up to distances of 5-40 km (3-25 miles). The ultimate precision of the geodimeter over that of the tellurometer is roughly by a factor of 3. Etymol: Geodimeter (acronym for "geodetic-distance meter"), a trade name for a product manufactured by Geodimeter Co., a division of Berg, Hedstrom & Co. Inc., New York City.

geodynamic Pertaining to physical processes within the Earth as they affect the features of the crust.

geodynamics That branch of science which deals with the forces and processes of the interior of the Earth.

Geodynamics Project An international program of research

(1971-1977) on the dynamics and dynamic history of the Earth with emphasis on deep-seated foundations of geological phenomena, esp. movements and deformations of the lithosphere. The program is coordinated by the Inter-Union Commission on Geodynamics which was established by the International Council of Scientific Unions (ICSU) at the request of the International Union of Geodesy and Geophysics (IUGG) and the International Union of Geological Sciences (IUGS).

geoecology *environmental geology.*

geoelectricity The Earth's natural electric fields and phenomena. It is closely related to geomagnetism.

geoelectric section A vertical distribution of layer thicknesses, electrical conductivities, dielectric permittivities, and magnetic permeabilities which are descriptive of the subsurfaces. It is usually obtained by sounding or logging.

geoevolutionism A term introduced by Goode (1969) for the "idea of evolution of geologic processes" in which some geologic changes foster new geologic processes that in turn bring about new changes.

geoexploration The application of geophysics to the search for deposits of economic value; *geophysical prospecting.*

geoflex An earlier syn. of *orocline.*

geofracture *geosuture.*

geogeny *geogony.*

geognosy An antiquated term used by Werner for "knowledge of the Earth": the study of the Earth's rock masses, structure, and constitution. As originally used, when very little was known about the Earth, the term was generally synonymous with *geology*; it later became restricted to absolute knowledge of the Earth as distinct from the theoretical and speculative reasoning of "geology".

geogony The science or theory of the formation of the Earth, esp. a speculative study of its origin. Syn: *geogeny.* See also: *geocosmogony.*

geogram A term used by Marr (1905, p.lxii) for a hypothetical geologic column, connoting principally lateral variations in lithology and in organic assemblages that could be traced in a deposit over an area of any width up to that of the circumference of the Earth.

geographic Pertaining or relating to geography.

geographical position An inclusive term used to designate both geodetic and astronomic coordinates.

geographic center The point on which an area on the Earth's surface would balance if it were a plate of uniform thickness (i.e. the center of gravity of such a plate). The geographic center of the conterminous U.S. is in the eastern part of Smith County, Kansas (lat. 39°50′ N, long. 98°35′W); the geographic center of North America is in Pierce County, N.D., a few miles west of Devils Lake.

geographic coordinates *geodetic coordinates.*

geographic cycle *cycle of erosion.*

geographic grid A system of parallels and meridians used to locate points on the Earth's surface.

geographic horizon The boundary line of that part of the Earth's surface visible from a given point of view; the *apparent horizon.*

geographic latitude The *latitude* or angle between the plane of the Earth's equator and the line perpendicular to the standard spheroid at a given point on the Earth's surface; *geodetic latitude.* It is also equal to the angular elevation of the celestial pole above the surface tangential to the spheroid representing the Earth.

geographic longitude The *longitude* or angle between the line perpendicular to the standard spheroid at the observer's position on the Earth's surface and the plane of an arbitrarily chosen prime meridian (generally the Greenwich meridian); *geodetic longitude.*

geographic meridian A general term for a line on the Earth's surface having the same longitude at every point. It is applied to astronomic meridian and geodetic meridian.

geographic name The proper name or expression by which a particular geographic entity is known; esp. the name of a river, town, or other natural or artificial feature at or near which a rock-stratigraphic unit is typically developed.

geographic north *true north.*

geographic parallel A general term for a line or circle on the Earth's surface having the same latitude at every point. It is applied to astronomic parallel and geodetic parallel.

geographic pole Either of the two *poles* or points of intersection of the Earth's surface and its axis of rotation; specif. *north pole* and *south pole.* Syn: *terrestrial pole.*

geographic province An extensive region all parts of which are characterized by similar geographic features. Cf: *physiographic province.*

geographic race Part of a biologic species coinciding with a particular geographic area and probably resulting from specific characteristics of the environment; a geographic subspecies or variety.

geography The study of all aspects of the Earth's surface including its natural and political divisions, the distribution and differentiation of areas and, often, man in relationship to his environment. See also: *physical geography.*

geohistory *geologic history.*

geohydrologic unit An aquifer, a confining unit (aquiclude or aquitard), or a combination of aquifers and confining units, comprising "a framework for a reasonably distinct hydraulic system" (Maxey, 1964, p.126). Cf: *hydrostratigraphic unit.*

geohydrology A term, often used interchangeably with *hydrogeology*, referring to the hydrologic, or flow characteristics of subsurface waters. Also used in reference to all hydrology on the Earth without restriction to geologic aspects (Stringfield, 1966, p.3). The term was first suggested in 1942 (p.4) by Meinzer for the branch of hydrology dealing with subterranean waters. Syn: *ground-water hydrology.*

geoid The *figure of the Earth* considered as a sea level surface extended continuously through the continents. It is a theoretically continuous surface that is perpendicular at every point to the direction of gravity (the plumb line). It is the surface of reference for astronomical observations and for geodetic leveling. See also: *compensated geoid; datum.*

geoidal horizon A circle on the celestial sphere formed by the intersection of the celestial sphere and a plane tangent to the sea-level surface of the Earth at the zenith-nadir line (DOD, 1969, p.97).

geoidal map The contoured development of the separations of the geoid and a specified reference ellipsoid.

geoidal section A section profile of the geoid with relation to the reference ellipsoid.

geoidal separation The distance between the mathematical reference spheroid and the geoid. Cf: *datum.*

geoisotherm The locus of points of equal temperature within the Earth, referring to a line in two dimensions or to a surface in three dimensions. Syn: *geotherm; isogeotherm.*

geolith *rock-stratigraphic unit.*

geologese (a) Literary style or speech peculiar to geologists. (b) Geological language that is "progressing rapidly" toward the construction of "sentences in such a way that their meaning is not apparent on first reading" (Vanserg, 1952, p.221).

geologic Pertaining to or related to geology.

geologic age (a) The age of a fossil organism or of a particular geologic event or feature referred to the geologic time scale and expressed in terms either of years or centuries (*absolute age*) or of comparison with the immediate surroundings (*relative age*); an age datable by geologic methods. (b) The term is also used to emphasize the long-past periods of time in geologic history, as distinct from present-day or historic times. See also: *age* [*geochron*].

geologic age determination Determination of the *relative* or *absolute age* of a geologic event or feature.

geological Pertaining to or related to geology.

geological oceanography That aspect of the study of the ocean which deals specifically with the ocean floor and the ocean-continent border, including submarine relief features, the geochemistry and petrology of the sediments and rocks of the ocean floor, and the influence of sea water and waves on the ocean bottom and its materials. Syn: *marine geology; submarine geology.*

geological ore *possible ore.*

geological science Any of the subdisciplinary specialties that are part of the science of *geology;* e.g. geophysics, geochemistry, paleontology, petrology, etc. See also: *geoscience.* Cf: *Earth science.*

geologic body A loose term denoting "any volume of rock selected for study or comment, without restriction as to size" (Turner & Weiss, 1963, p.15); e.g. the great granite pluton of the Sierra Nevada in California, a nodule of actinolite schist enclosed in serpentinite, or the aggregate of quartz particles comprising a thin section of any sandstone.

geologic chronology *geochronology.*

geologic climate *paleoclimate.*

geologic-climate unit A term used by the American Commission on Stratigraphic Nomenclature (1961, art.39) for "an inferred widespread climatic episode defined from a subdivision of Quaternary rocks"; e.g. glaciation, interglaciation, stade, and interstade. It is strictly not a stratigraphic unit. The different stratigraphic boundaries that define the limits of the geologic-climate unit in different latitudes are not likely to be isochronous. Syn: *climate-stratigraphic unit.*

geologic column (a) A composite diagram that shows in a single column the subdivisions of part or all of geologic time or the sequence of stratigraphic units of a given locality or region (the oldest at the bottom and the youngest at the top, with dips adjusted to the horizontal) so arranged as to indicate their relations to the subdivisions of geologic time and their relative positions to each other. See also:*columnar section.* (b) The vertical or chronologic arrangement or sequence of rock units portrayed in a geologic column. See also: *geologic section.*---Syn: *stratigraphic column.*

geologic cycle The sum total or general cyclic sequence of the major internal and external geologic processes and conditions acting on the materials of the Earth's crust or throughout a region of the crust, such as: deposition in a geosyncline; diastrophism, with deformation, uplift, metamorphism, and igneous intrusion; erosion; submergence; and deposition of a further set of deposits. Cf: *geostrophic cycle.*

geologic engineering *engineering geology.*

geologic erosion A syn. of *normal erosion,* or erosion caused naturally by geologic processes.

geologic formation *formation* [stratig].

geologic hazard A naturally occurring or man-made geologic condition or phenomenon that presents a risk or is a potential danger to life and property. Examples of geologic hazards: landsliding, flooding, earthquakes, ground subsidence, coastal and beach erosion, faulting, dam leakage and failure, mining disasters, pollution and waste disposal, and seawater intrusion.

geologic high An oil-field term for a relatively young (later) formation, regardless of its elevation. Cf: *topographic high.*

geologic history The history of the Earth and its inhabitants throughout geologic time, considered for a certain area or duration of time, or in certain aspects. It comprises all chemical, physical, and biologic conditions that have existed on and in the Earth, all processes that have operated to make and modify these conditions, and all events that have affected any part of the Earth (including its inhabitants) during time from the beginning of the planet onward to the present, and is circumscribed "in no way by what we think we know about it" (Moore, 1949, p.3). Syn: *geohistory; Earth history.*

geologic horizon *horizon* [geol].

geologic informatics The science and technology of geological information, including automatic data processing.

geologic low An oil-field term for a relatively old (earlier) formation, regardless of its elevation. Cf: *topographic low.*

geologic map A map on which is recorded geologic information such as the distribution and nature of rock units (the surficial deposits may or may not be mapped separately) and the occurrence of structural features (folds, faults, joints, etc.) mineral deposits, and fossil localities. It may indicate geologic structure by means of formational outcrop patterns and by conventional symbols giving the directions and amounts of dip, cleavage, etc., at certain points.

geologic name The name of a geologic formation or other rock-stratigraphic unit.

geologic norm The condition resulting from *normal erosion* of the land, undisturbed by the activity of man and his agents.

geologic province An extensive region all parts of which are characterized by similar geologic history or by particular structural, petrographic, or physiographic features. Cf: *physiographic province.*

geologic range *stratigraphic range.*

geologic record The "documents" or "archives" of the history of the Earth, represented by bedrock, regolith, and the Earth's morphology; the rocks and the accessible solid part of the Earth. Also, the geologic history based on inferences from a study of the geologic record. See also: *stratigraphic record.*

geologic section (a) Any sequence of rock units found in a given region either at the surface or below it (as in a drilled well or mine shaft); a local *geologic column.* Syn: *stratigraphic section.* (b) *section* [geol].

geologic thermometer A mineral or mineral assemblage whose composition, structure, or inclusions are fixed within known thermal limits under particular conditions of pressure and composition and whose presence thus denotes a limit or a range for the temperature of formation of the enclosing rock. The α to β inversion of quartz, the occurrence of wollastonite, the selenium content of galena, and the homogenization of liquid inclusions are examples of possible geologic thermometers.

geologic thermometry Measurement or estimation, by direct or indirect methods, of the maximum, minimum, or actual temperatures at which geologic processes occur or have occurred in the past; e.g. the determination of the temperatures at which rocks and minerals crystallized within the Earth's crust. See also:*paleothermometry.* Syn: *geothermometry.*

geologic time The period of time dealt with by historical geology, or the time extending from the end of the formative period of the Earth as a separate planetary body to the beginning of written or human history; the part of the Earth's history that is represented by and recorded in the succession of rocks. The term implies extremely long duration or remoteness in the past, although no precise limits can be set. See also: *time (a).*

geologic time scale An arbitrary, chronologic arrangement or sequence of geologic events, used as a measure of the relative or absolute duration or age of any part of geologic time, and usually presented in the form of a chart showing the names of the various rock-stratigraphic, time-stratigraphic, or geologic-time units, as currently understood; e.g. the geologic time scales published by Holmes (1959) and Kulp (1961). See also: *atomic time scale; relative time scale; biologic time scale.* Syn: *time scale.*

geologic-time unit A span of continuous time in geologic history, during which a corresponding *time-stratigraphic unit* was formed; a division of time "distinguished on the basis of the rock record, particularly as expressed by time-stratigraphic units" (ACSN, 1961, art.36). It is not a material unit or body of strata, and therefore it is strictly not a stratigraphic unit. Geologic-time units in order of decreasing magnitude are eon, era, period, epoch, and age. The names of periods, epochs, and ages are identical with those of the corresponding time-stratigraphic units; the names of eras and eons are independently formed. Cf: *biochronologic unit.* Syn: *geochronologic unit; time unit.*

geologist One who does or participates in any of the geological sciences.

geologize v. To participate in or talk about geology; to practice geology.

geograph A device that automatically records the rate of penetration of a drill bit during rotary drilling.

geology The study of the planet Earth. It is concerned with the origin of the planet, the material and morphology of the Earth, and its history and the processes that acted (and act) upon it to affect its historic and present forms. In the pursuit of that knowledge, the science considers the physical forces that influenced, and continue to influence, change; the chemistry of its constituent materials; the record and age of its past as revealed by the organic remains that are preserved in the layers of its crust or by interpretation of relic morphology and environment. Clues to the origin of the Earth are sought through the study of extraterrestrial bodies and their atmospheres that may reflect an earlier stage of this planet, or whose history may share the events and forces that created the Earth. All of the knowledge obtained through the study of the planet is placed at the service of man, to discover useful materials within the Earth; to identify stable environments for the support of his constructed arts and utilities; and to provide him with a foreknowledge of dangers associated with the mobile forces of a dynamic Earth, that may threaten his welfare or being. See also: *geological science; Earth science; geoscience; geonomy; historical geology; physical geology.*

geomagnetic axis The axis of the dipole magnetic field most closely approximating the actual magnetic field of the Earth.

geomagnetic electrokinetograph A shipboard instrument containing electrodes and designed to measure ocean-surface currents in depths greater than 100 fathoms by recording the electric current generated by the movement of a high-conductivity electrolyte (ocean water) through the Earth's magnetic field. Abbrev: GEK.

geomagnetic equator The great circle of the Earth whose plane is perpendicular to the geomagnetic axis; the line connecting points of zero geomagnetic latitude.

geomagnetic latitude The magnetic latitude a location would have if the Earth's field were replaced by the dipole field most closely approximating it. It is latitude reckoned relative to the geomagnetic axis instead of to the Earth's rotation axis. Cf: *geomagnetic longitude.*

geomagnetic longitude Longitude reckoned around the geomagnetic axis instead of around the Earth's rotation axis. The prime geomagnetic meridian is the one extending to the geographic south from the north geomagnetic pole. Cf: *geomagnetic latitude.*

geomagnetic meridian A great circle of the Earth through the geomagnetic poles.

geomagnetic polarity epoch *polarity epoch.*

geomagnetic polarity event *polarity event.*

geomagnetic polarity reversal *geomagnetic reversal.*

geomagnetic polarity time scale A chronology based on counting reversals of the Earth's magnetic field. From youngest to oldest, named epochs are Brunhes normal (now to 0.7 m.y.), Matuyama reversed (0.7 my to 2.5 m.y.), Gauss normal (2.5 m.y. to 3.3 m.y.) and Gilbert reversed (3.3 my to 5 m.y.). See also: *polarity epoch.*

geomagnetic poles The expression on the Earth's surface of that axis of the geocentric magnetic dipole which most closely approximates the Earth's magnetic field. Partial syn: *magnetic poles.*

geomagnetic reversal A change of the Earth's magnetic field between *normal polarity* and *reversed polarity.* Syn: *field reversal; geomagnetic polarity reversal; reversal; polarity reversal.*

geomagnetic secular variation *secular variation.*

geomagnetism The magnetic phenomena exhibited by the Earth and its atmosphere; also, the study of such phenomena. Syn: *terrestrial magnetism.*

geomalism Equal lateral growth of an organism in response to gravitational force. Cf: *geotaxis; geotropism.*

geomathematics Mathematics as applied to geology; "the closely interrelated system involving mathematical and applied statistics, experimental design, and associated computer techniques as applied to geological sciences" (Osborne, 1969, p.121). See also: *mathematical geology.*

geomatics The mathematics of the Earth. Adj: *geomatic.*

geomechanics A branch of geology that embraces the basic fundamentals of structural geology and a knowledge of the response of natural materials to deformation or changes due to the application of stress and/or strain energy.

geometric factor The factor by which the ratio of voltage to current is multiplied, for resistivity data, in order to obtain the resistivity. It is dependent upon the electrode array and spacing.

geometric grade scale A *grade scale* having a constant ratio between size classes; e.g. the Wentworth grade scale, each size class of which differs from its predecessor by the constant ratio 1/2.

geometric horizon A term originally applied to the *celestial horizon,* but now more commonly to the intersection of the celestial sphere and an infinite number of straight lines tangent to the Earth's surface and radiating from the eye of the observer. It would coincide with the apparent horizon if there were no terrestrial refraction.

geometric latitude *reduced latitude.*

geometric mean The *n*th root of the product of the values of *n* positive numbers; the antilogarithm of the mean of the logarithms of individual values.

geometric mean diameter An expression of the average particle size of a sediment or rock, obtained by taking the antilogarithm of the *phi mean diameter,* the diameter equivalent of the arithmetic mean of the logarithmic frequency distribution.

geometric projection *perspective projection.*

geometrics Measurements (through space and time) in regard to the Earth.

geometric sounding That form of electromagnetic *sounding* in which separation between transmitting and receiving coils is the variable. Cf: *parametric sounding.*

geometry number A dimensionless constant representing the ratio of the product of maximum basin relief and drainage density within a given drainage basin to the tangent of the stream gradient; it summarizes the essentials of landform geometry, particularly the relations between planimetric and relief aspects of a drainage basin (Strahler, 1958, p.287 & 295). Symbol: N_G.

geomonocline A broad flank of a geosyncline, the beds of which have a uniform dip; a unilateral, marginal geosyncline.

geomorphic (a) Pertaining to the form of the Earth or of its surface features; e.g. a *geomorphic* province. (b) Pertaining to geomorphology; geomorphologic.

geomorphic cycle *cycle of erosion.*

geomorphic geology *geomorphology.*

geomorphogeny The part of geomorphology that deals with the origin, development, and changes of the Earth's surface features or landforms. Cf: *geomorphography.* Syn: *genetic physiography.*

geomorphography The part of geomorphology that deals with the description of the Earth's surface features or landforms. Cf: *geomorphogeny.*

geomorphology (a) The science that treats the general configuration of the Earth's surface; specif. the study of the classification, description, nature, origin, and development of present landforms and their relationships to underlying structures, and of the history of geologic changes as recorded by these surface features. The term is esp. applied to the genetic interpretation of landforms, but has also been restricted to features produced only by erosion or deposition. The term was applied widely in Europe before it was used in the U.S. where it has come to replace the term *physiography* and is usually considered a branch of geology (in Great Britain, it is usually regarded as a branch of geography). See also: *physical geography.* Syn: *physiographic geology; geomorphic geology; geomorphy.* (b) Strictly, any study that deals with the form of the Earth (including geodesy, and structural and dynamic geology). This usage is more common in Europe, where the term has even been applied broadly to the science of the Earth. (c) The features dealt with in, or a treatise on, geomorphology; e.g. the *geomorphology* of Texas.

geomorphy (a) *geomorphology.* (b) A syn. of *topography* "in the broader application of that term" (Lawson, 1894, p.241).

geomyricite A mineral consisting of a white, waxy resin of approximate composition: $C_{32}H_{62}O_2$. It is found in brown coal. Syn: *geomyricin.*

geon A hypothetical entity postulated on the basis of electromagnetic and gravitational theories and consisting of an electromagnetic field maintained by the gravitational attraction of the field's mass.

geonomy A term variously recommended as a synonym for *geology,* as the science of the dynamic Earth, as the science concerned exclusively with the physical forces relating to the Earth, and to denote the study of the upper mantle of the Earth. The word has never been given broad recognition or acceptance, and the variety of proposed definitions stated by its proponents suggest that it should be used with great care, if at all.

geop A surface within the gravity field of the Earth in which all points have equal and constant *geopotential.* Its numerical value is the *geopotential number.* It is sometimes called a *geopotential surface.*

geopetal Pertaining to any rock feature that indicates the relation of top to bottom at the time of formation of the rock; e.g. a "geopetal fabric" or the internal structure or organization that indicates original orientation of a stratified rock, such as cross-bedding or grains on a boundary surface. Term introduced by Sander (1936, p.31; 1951. p.2).

geophagous Said of an organism that feeds on soil.

geophone A trade name for a *seismic detector* that is located on or in the ground. Various synonyms are: *bug; jug; geotector; tortuga; receptor.* Cf: *seismometer; seismograph.*

geophoto A photograph for use in geologic investigations; esp. an aerial photograph so used.

geophysical exploration The use of geophysical instruments and methods to determine subsurface conditions by analysis of such properties as specific gravity, electrical conductivity, or magnetic susceptibility. See also: *geophysical prospecting; geophysical survey.*

geophysical log A *log* obtained by lowering an instrument into a borehole or well and recording continuously on a meter at the surface some physical property of the rock material being logged. Examples include: *electric log; radioactivity log; sonic log; temperature log.*

geophysical prospecting *Geophysical exploration* that usually has an economic objective, e.g. discovery of fuel or mineral deposits. Cf: *applied geophysics.* Syn: *geoexploration.*

geophysical survey The use of one or more geophysical techniques in *geophysical exploration,* such as earth currents, electrical, infra-red, heat flow, magnetic, radioactivity, and seismic. Geochemical techniques may sometimes be included.

geophysicist One who studies the Earth as a planet, according to its various physical phenomena; a specialist in geophysics. A geophysicist may also study the Moon and the other planets.

geophysics The study of the Earth as a planet. The three basic divisions of geophysics are solid-earth, atmosphere and hydrosphere, and magnetosphere (solar-terrestrial physics). The term is also used, as a convenience, for the study of the Moon and planets, although there are more specific terms such as selenophysics.

geophysiography The synthesis of all knowledge available about the Earth; "the combination...of relevant parts of geophysics, geochemistry, geomorphology, and geoecology" (Strøm, 1966, p.8).

geopiezometry Measurement of pressure and compressibility of rocks in metamorphism (Chinner, 1966).

geoplanetology A term proposed by Nayak (1970, p.1279) for the study of "geological and related aspects of the Moon and other planets which are likely to be investigated in the future".

geoplosics The systematic study of explosion effects in the Earth (Flanders & Sauer, 1960, p.iii).

geopolar Pertaining or relating to a pole of the Earth.

geopotential The potential energy of a unit mass relative to sea level, numerically equal to the work that would be done in lifting the unit mass from sea level to the height at which the mass is located. It is commonly expressed in terms of *dynamic height.* Cf: *disturbing potential, spheropotential.*

geopotential height *dynamic height.*

geopotential number The numerical value, C, assigned to a given *geop* or geopotential surface when expressed in geopotential units (g.p.u.) where 1 g.p.u. = 1 m x 1 kgal. Since g = 1.98 kgal, C = gH = 0.98 H. Thus the geopotential number in g.p.u. is almost equal to the height above sea level in meters. Geopotential numbers are generally preferred to *dynamic heights* because the physical meaning is more correctly expressed.

geopotential surface *geop.*

geopressured aquifer A term used by Charles A. Stuart for an aquifer, esp. in the Gulf Coast, in which fluid pressure exceeds normal hydrostatic pressure of 0.465 pound per square inch per foot of depth (Jones, P.H., 1969, p.34).

georgiadesite A white or brownish-yellow orthorhombic mineral: $Pb_3(AsO_4)Cl_3$.

georgiaite A greenish tektite from Georgia, U.S.

Georgian Obsolete syn. of *Waucoban.*

geoscience A short form denoting the collective disciplines of the *geological sciences.* The term, as such, is synonymous with *geology.*

geoselenic Relating to the Earth and the Moon.

geosere A series of climax communities that succeed each other as a result of changes in the physical and climatic characteristics of the environment.

geosophy A term coined by Wright (1947, p.12) for "the study of geographical knowledge from any or all points of view" dealing with "the nature and expression of geographical knowledge both past and present", and covering "the geographical ideas, both true and false, of all manner of people".

geosphere (a) The *lithosphere.* (b) The lithosphere, hydrosphere, and atmosphere combined. (c) Any of the so-called spheres or layers of the Earth.

geostatic Pertaining to, or capable of sustaining, the pressure exerted by earth; e.g. a "geostatic load" representing the weight of the overburden at a particular depth, or a "geostatic structure" resulting from gravity phenomena. Syn: *stereostatic.*

geostatic pressure The vertical pressure at a point in the Earth's crust, equal to the pressure that would be exerted by a column of the overlying rock or soil. Syn: *lithostatic pressure; overburden pressure; ground pressure; rock pressure.*

geostatistics Statistics as applied to geology; the application of statistical methods or the collection of statistical data for use in geology.

geostratigraphic Pertaining to worldwide stratigraphy; e.g. "geostratigraphic standards" or "geostratigraphic stage".

geostrome A term, now obsolete, proposed by Patrin "to deote the strata of the Earth" (Pinkerton, 1811, v.l, p.542).

geostrophic current A wind or ocean current in which the horizontal *pressure force* is exactly balanced by the equal but opposite *Coriolis force*. The geostrophic current is neither accelerated nor affected by friction. It flows to the right of the pressure gradient force along the isobars in the northern hemisphere. Cf: *gradient current.*

geostrophic cycle A term suggested by Tomkeieff (1946, p.326) for the Huttonian concept of one great cycle of dynamic changes occurring in the Earth; the cycle embraces both the organic and the inorganic spheres, and consists alternately of the complementary processes of destruction and construction. Cf: *geologic cycle.*

geosuture A boundary zone between contrasting tectonic units of the crust; in many places a fault which probably extends through the entire thickness of the crust. Syn: *geofracture.*

geosynclinal n. The original, now obsolete, term for *geosyncline*, first used by Dana in 1873. adj. Pertaining to a geosyncline.

geosynclinal couple Aubouin's concept of the true geosyncline as miogeosynclinal and eugeosynclinal furrows linked by geanticlinal ridges (1965, p.34); an *orthogeosyncline.*

geosynclinal cycle *tectonic cycle.*

geosynclinal facies A sedimentary facies characterized by great thickness, predominantly argillaceous character, and paucity of carbonate rocks; it consists of uniform, rhythmic, and graded beds of shale or silty shale regularly interbedded with graywackes, deposited rapidly in a strongly subsiding geosyncline of a deep-water marine environment. Cf: *shelf facies.* See also: *orogenic facies; graptolitic facies.*

geosynclinal prism The load of sediments which accumulates, often to great thicknesses, in the downwarped part of a geosyncline, having a shape similar to that of a long, plano-convex prism whose convexity is at the floor. Cf: *nepton; clastic wedge.*

geosynclinal trough A linear depression or basin that subsides as it receives clastic material, located not far from the source supplying the sediment.

geosyncline A mobile downwarping of the crust of the Earth, either elongate or basin-like, measured in scores of kilometers, which is subsiding as sedimentary and volcanic rocks accumulate to thicknesses of thousands of meters. A geosyncline may form in part of a tectonic cycle in which orogeny follows. The concept was presented by Hall in 1859, and the term, *geosynclinal*, was proposed by Dana in 1873. The differing opinions of the origin, mechanics, and essential features of geosynclines are reflected in the various schemes that have been used to define aspects of the term. Some are based on the tectonic relationship of crustal units, some emphasize mountain-building processes, and others are concerned with the relationship of geosynclinal sedimentation to subsidence. Cf: *mobile belt.* Se also: *synclinorium.* Ant: *geanticline.*

geotaxis Taxis resulting from gravitational attraction. Cf: *geotropism; geomalism.*

geotechnical Pertaining to geotechnics; e.g. a "geotechnical process" that changes the properties of soil such as compaction, stabilization, and electrochemical induration.

geotechnics The application of scientific methods and engineering principles to the acquisition, interpretation, and use of knowledge of materials of the Earth's crust to the solution of civil-engineering problems; the applied science of making the Earth more habitable. It embraces the fields of soil mechanics and rock mechanics, and many of the engineering aspects of geology, geophysics, hydrology, and related sciences. Syn: *geotechnique.*

geotechnique A syn. of *geotechnics.* Etymol: French *géotechnique.*

geotechnology The application of scientific methods and engineering techniques to the exploitation and use of natural resources (such as mineral resources).

geotectocline The geosynclinal accumulation of sediments formed above a downbuckle; the basin between the limbs of a downbuckle (Hess, 1938, p.79). Syn: *tectocline.*

geotectogene *tectogene.*

geotectonic *tectonic.*

geotectonic cycle A sequence of geosynclinal, orogenic, and cratonic stages. Such a cycle may be repeated (Stille, 1940).

geotectonics *tectonics.* Cf: *megatectonics.*

geotector *geophone.*

geotexture The texture of the Earth's surface as manifested by the largest features of relief (such as continental massifs and ocean basins), the formation of which is connected with worldwide processes (I.P. Gerasimov & J.A. Mescherikov in Fairbridge, 1968, p.731).

geotherm *geoisotherm.*

geothermal Pertaining to the heat of the interior of the Earth. Syn: *geothermic.*

geothermal brine A brine that is overheated with respect to its observed depth, as a result of its prior association with an anomalous heat source, such as heat transferred from a fault zone.

geothermal energy Useful energy that can be extracted from naturally occurring steam and hot water found in the Earth's volcanic and young orogenic zones, whose surface manifestations include hot springs, fumaroles, and geysers. All such areas under development are in areas of late Cenozoic volcanic activity. (McNitt, 1965, p.240).

geothermal flux *Geothermal heat flow* per unit time.

geothermal gradient The rate of increase of temperature in the Earth with depth; the *thermal gradient* of the Earth. The gradient near the surface of the Earth varies from place to place depending upon the heat flow in the region and on the thermal conductivity of the rocks. The approximate average geothermal gradient in the Earth's crust is about $25°C/km$.

geothermal heat flow The amount of heat energy leaving the Earth per cm_2/sec, measured in calories/cm_2/sec. The mean heat flow for the Earth is about 1.5 ± 0.15 microcalories/cm_2/sec, or about 1.5 heat flow units (HFU). Heat-flow measurements in igneous rocks have recently shown a linear correlation between heat production in rocks and surface heat flow. The heat production is due to the presence of uranium, potassium, and thorium. See also: *geothermal flux.*

geothermal metamorphism A type of deep-seated static metamorphism in which a regular downward increase in temperature attributed to deep burial by overlying rocks (geothermal gradient) is the controlling factor. Cf: *thermal metamorphism; load metamorphism; static metamorphism.*

geothermal prospecting Exploration for sources of *geothermal energy.* Cf: *thermal prospecting.*

geothermic *geothermal.*

geothermometer [geol] *geologic thermometer.*

geothermometry (a) The science of the Earth's heat. It includes the temperature of the Earth, the effect of temperature on physical and chemical processes, the source of the Earth's heat, and volcanology. (b) *geologic thermometry.*

geotomical axis A minor axis with small spines in the shell of an acantharian radiolarian. Cf: *hydrotomical axis.*

geotraverse The multidisciplinary exploration or survey of a selected area.

geotropism Tropism resulting from gravitational attraction. Cf: *geotaxis; geomalism.*

geoundation An upward or downward warping of the Earth's crust, on a continental or oceanic scale; it is an epeirogenic movement.

gerasimovskite A mineral: $(Mn,Ca)_2(Nb,Ti)_5O_{12}.9H_2O$.

gerhardtite An emerald-green mineral: $Cu_2(NO_3)(OH)_3$.

germ In botany, a syn. of *embryo.*

germanite A reddish-gray mineral: $Cu_3(Ge,Ga,Fe)(S,As)_4$.

germanotype tectonics The tectonics of the cratons and stabilized foldbelts, typified by structures in Germany north of the Alps. The milder phases of germanotype tectonics are epeirogenic, but they also include broad folds dominated by vertical uplift and high-angle faults, block-faulted terranes, and sedimentary basins deformed within a frame of surrounding massifs. Cf: *alpinotype tectonics*. Syn: *paratectonics*.

germinal aperture An *aperture* (such as a colpus or germ pore) of a pollen grain through which the pollen nuclei emerge on germination of the grain. The term is sometimes used to include the laesura of spores.

germinal furrow *colpus*.

germ pore A *pore* or thin area in the exine of a pollen grain through which the pollen tube emerges on germination.

gerollton A term used by Pettijohn (1957, p.265) for a nonglacial conglomeratic mudstone (or as a syn. of *tilloid*). It was introduced as *Geröllton* (German for "gravel clay") by Ackerman (1951) who applied it to a pebble-bearing clay in which the pebbles and clay were deposited simultaneously.

gerontic The senile or old-age growth stage in ontogeny; the stage following the *ephebic* stage.

gerontomorphosis Evolutionary changes involving modifications of the adult characteristics of organisms.

gersdorffite A silver-white to steel-gray isometric mineral: NiAsS. It closely resembles cobaltite, and may contain some iron and cobalt. Syn: *nickel glance*.

gerstleyite A red mineral: $(Na,Li)_4As_2Sb_8S_{17}.6H_2O$.

getchellite A mineral: $AsSbS_3$.

geversite A mineral: $PtSb_2$. It has pyrite-type structure.

geyser A type of hot spring that intermittently erupts jets of hot water and steam, the result of ground water coming into contact with rock or steam hot enough to create steam under conditions preventing free circulation; a type of intermittent spring. Syn: *pulsating spring; gusher* [grd wat].

geyser basin A valley that contains numerous springs, geysers, and steaming fissures fed by the same ground-water flow (Schieferdecker, 1959, term 4524).

geyser cone A large or small, low hill or mound built up of siliceous sinter around the orifice of a geyser. The term is sometimes mistakenly applied to an algal growth on objects (such as wooden snags) occurring along the shores of some Tertiary lakes. Syn: *geyser mound*.

geyser crater The bowl- or funnel-shaped opening of the geyser pipe that often contains a *geyser pool*. Syn: *crater* [grd wat].

geyserite A syn. of *siliceous sinter*, used esp. for the compact, loose, concretionary, scaly, or filamentous incrustation of opaline silica deposited by precipitation from the waters of a geyser.

geyser jet The plume of water and steam emitted during the eruption of a geyser.

geyser mound *geyser cone*.

geyser pipe The narrow tube or well of a geyser extending downward from the *geyser pool*. Syn: *pipe* [grd wat]; *geyser shaft*.

geyser pool The comparatively shallow pool of heated water ordinarily contained in a *geyser crater* at the top of a *geyser pipe*.

geyser shaft *geyser pipe*.

ghat A term used in India originally for a mountain pass, or a path leading down from a mountain, but now commonly and erroneously applied (by Europeans) to a mountain range, specif. the mountain ranges parallel to the east and west coasts of India. Etymol: Hindi *ghāt*. Syn: *ghaut*.

ghizite An analcime- and olivine-bearing basalt characterized by the presence of biotite.

G horizon A soil horizon of intense reduction characterized by ferrous iron and gray or olive colors.

ghost [petrology] A visible outline of a former crystal shape, fossil, or other rock structure that has been partly obliterated (as by diagenesis or replacement) and that is bounded by in-

clusions and outlined by bubbles or foreign material. Syn: *phantom*.

ghost [seis] Seismic energy that travels upward from a seismic profiling shot and then is reflected downward at the base of the weathering or at the surface. Such energy usually unites with a downward-traveling wave train, although it may sometimes be distinguished as a separate wave (Sheriff, 1968). Syn: *secondary reflection*.

ghost coal A coal that burns with a bright white flame.

ghost crystal *phantom crystal*.

ghost member Any part, of the ideal or typical cyclothem, that is absent in a particular cyclothem.

ghost stratigraphy A term used by Challinor (1967, p.113) for "the lingering relict traces, in highly metamorphosed strata, of the original lithology and stratification".

Ghyben-Herzberg ratio A ratio describing the static relation of fresh ground water and seawater in coastal areas. The term was formally defined by Badon-Ghyben in 1889 and later (1901) independently by Herzberg and credited to these two (Chow, 1964, p.1-8). It expresses the principle of the U-tube in which a column of seawater is balanced by a column of fresh water. For each foot of fresh water head above sea level the salt-water surface will be displaced to 40 feet below sea level. The static relationship is modified by dynamic factors that cause mixing of fresh and salt water, esp. the seaward flow of fresh water and tide-induced fluctuations of the interface. Syn: *Ghyben-Herzberg relation; Ghyben-Herzberg principle; Ghyben-Herzberg formula*. See also: *salt-water encroachment*.

ghyll Obsolescent syn. of *gill*.

giant cusp A slightly protruding *cusp*, commonly 300-500 m between the crescentic tips (some range up to 1500 m), with a submarine ridge continuing seaward as a transverse bar along one or both sides of which is a deep channel. A giant cusp appears to be characteristic of areas or times of relatively strong littoral currents.

giant desiccation polygon A *desiccation polygon* formed on a playa by contraction of muds upon drying, consisting of a fissure or crack measuring several meters in depth, up to 1 m in width, and extending over an area of several hundred meters (Stone, 1967, p.228).

giant granite *pegmatite*.

giant ripple *megaripple*.

giant's cauldron *giant's kettle*.

giant's kettle A cylindrical hole bored in bedrock beneath a glacier by water falling through a deep moulin or by boulders rotating in the bed of a meltwater stream; it may contain a pond or marsh. Syn: *moulin pothole; glacial pothole; pothole; potash kettle; giant's cauldron*.

giant stairway *glacial stairway*.

gibber An Australian term for a pebble or boulder; esp. one of the wind-polished or wind-sculptured stones that compose a desert pavement or the lag gravels of an arid region. stones that compose a desert pavement or the lag gravels of an arid region. It is pronounced with a hard "g".

gibber plain A desert plain strewn with *gibbers*; a gravelly desert in Australia.

gibbs A unit of measurement of entropy, heat capacity, and various commonly used thermodynamic functions that is essentially equivalent to *entropy unit*.

Gibbs free energy A *thermodynamic potential* associated with the variables pressure and temperature. In an irreversible thermodynamic process at constant temperature and pressure, the Gibbs free energy of a system decreases; in a reversible process it remains constant. It is one of the commonly used functions describing *free energy*. Cf: *Helmholtz free energy*. Syn: *Gibbs function; free enthalpy*.

Gibbs function *Gibbs free energy*.

gibbsite A white or tinted monoclinic mineral: $Al(OH)_3$. It is polymorphous with bayerite. Gibbsite is formed by weathering

f igneous rocks and is the principal constituent of bauxite; it occurs in mica-like crystals or in stalactitic and spheroidal forms. Syn: *hydrargillite*.

Gibbs phase rule *phase rule*.

gibelite A trachyte characterized by the presence of abundant large soda microcline phenocrysts, sometimes microperthite, and a small amount of augite in a groundmass composed of soda microcline and exhibiting flow texture. Augite and amphibole are present as accessories.

Gibraltar stone A light-colored *onyx marble* found at Gibraltar and elsewhere.

giessenite An orthorhombic mineral: $Pb_9CuBi_6Sb_{1.5}S_{30}$.

gigantism (a) In animals, development to abnormally large size as a result of the excessive growth of certain hard parts accompanied by weakness and sexual impotence. Cf: *nanism*. (b) In plants, excessive vegetative growth.

gigayear A term proposed by Rankama (1967) for one billion (10^9) years. Abbrev: *Gyr*.

gilgai n. The microrelief of heavy clay soils with high coefficients of expansion and contraction according to changes in moisture. Gilgai is typical of vertisols.----adj. Said of a soil that displays gilgai.----Pl: *gilgaies*. See also: *crab hole; puff*. Syn: *melon hole*.

gill [paleont] An organ for obtaining oxygen from water; e.g. a *branchia* of a crustacean.

gill [streams] (a) A term used in the English Lake District for a deep, narrow, steep-sided, rocky valley, esp. a wooded ravine with a rapid stream running through it. (b) A narrow mountain stream or brook flowing swiftly through a gill. Also, a term used in Yorkshire, Eng., for a stream flowing in a shallow valley, sometimes ending in a pothole.---Syn: *ghyll; ghyl; gil*.

gill chamber *branchial chamber*.

gillespite A red mineral: $BaFeSi_4O_{10}$.

gill slit An opening in an echinoderm, such as a fissure in the disc of an ophiuroid along the side of the base of an arm and leading into the bursa, or an indentation of the peristomial margin of echinoid interambulacra for the passage of the stem of an external branchia. Syn: *branchial slit*.

gilpinite *johannite*.

gilsonite *uintahite*.

ginko A member of the gymnosperm subclass Gingkoales, characterized by parallel venation, fan-shaped or regularly bifurcating leaves, and terminally borne seeds. Ginkoes range from the Permian.

ginorite A white monoclinic mineral: $Ca_2B_{14}O_{23}.8H_2O$.

ginzburgite A group name for iron-rich clay minerals of the kaolin group.

gio Var. of *geo*.

giobertite *magnesite*.

gipfelflur (a) An imaginary, relatively smooth surface touching the tops of the accordant summits of a region. See also: *summit plane; peak plain*. Syn: *hilltop surface*. (b) The concept of the uniformity or accordance of summit levels; such uniformity is independent of geologic structure and of rock type, and is believed to result from erosion of a former uplifted surface.--Etymol: German *Gipfelflur*, "summit plain". The term was originated by A. Penck (1919).

gipsy *gypsey*.

girasol adj. Said of any gem variety (such as of sapphire and chrysoberyl) that exhibits a billowy, gleaming, round or elongated area of light that "floats" or moves about as the stone is turned or as the light source is moved.---n. A name that has been applied to many gemstones with a girasol effect, such as moonstone; specif. a translucent variety of fire opal with reddish reflections in a bright light and a faint bluish-white floating light emanating from the center of the stone.

girdle [gem] The outer edge or periphery of a fashioned gemstone; the portion that is usually grasped by the setting or mounting; the dividing line between the *crown* and the *pavilion*.

girdle [paleont] (a) The region of overlap of the two valves of a diatom frustule, consisting of the connecting band from each valve. Also, either of the two *connecting bands* forming the girdle. Syn: *cingulum*. (b) A transverse furrow around the theca or body of a dinoflagellate; the part of the shell lying between the epivalve and hypovalve in certain dinoflagellates. (c) The muscular, flexible marginal band of uniform width of the mantle of a chiton, encircling the shell plates and differentiated from the central portion of the back. It belongs with the soft parts of the chiton, although its covering may be studded with needle-like or scale-like calcareous spicules. (d) A spiral or annular shelf in the skeleton of a spumellarian radiolarian.--See also: *perignathic girdle*.

girdle [struc petrol] On a fabric diagram or equal-area projection net, a belt of concentration of points approximately coincident with a great circle of the net and which represents orientation of fabric elements (Turner and Weiss, 1963, p.58). Cf: *maximum; cleft girdle*.

girdle axis On a fabric diagram, the pole of the areas of least point population.

girdle facet One of the 32 triangular facets that adjoin the girdle of a round brilliant-cut gem; there are 16 facets above the girdle and 16 below.

girdle list A high membranous ridge perpendicular to the wall and bordering the girdle of a dinoflagellate.

girdle view The side view of a diatom frustule, showing only the edges of the valves.

girdle zone The circular central region with shelves in a radiolarian shell.

girt *bayou*.

girvanella An *algal biscuit* characterized by a complex of microscopic filaments.

gisement The angle between the *grid meridian* and the *geographic meridian*. It is a term used primarily in connection with military grids and is referred to as the *mapping angle*. Cf: *grid azimuth*.

Gish-Rooney method In electrical prospecting, the use of a double commutator to reverse periodically the direction of flow of current in both power and potential leads to eliminate Earth-current potentials.

gismondite A zeolite mineral: $CaAl_2Si_2O_8.4H_2O$. It sometimes contains potassium. Syn: *gismondine*.

giumarrite An amphibole-bearing monchiquite.

Givetian European stage: Middle Devonian (above Eifelian, below Frasnian).

gizzard The last part of the foregut (anterior part of the alimentary canal) of an arachnid, developed as a pumping organ. Its dorsal dilative muscle is attached to an apodeme visible on the external surface of the carapace (TIP, 1955, pt.P, p.62).

gizzard stone *gastrolith*.

gja [coast] *geo*.

gja [volc] A gaping, dilation fissure of the Icelandic rift system, from which volcanic eruptions take place. Etymol: Icelandic, "chasm". Pl: *gjàr*.

glabella The raised (convex) axial part of the cephalon of a trilobite. It represents the anterior part of the axis or axial lobe. Pl: *glabellae*.

glabellar furrow A narrow groove extending transversely across the glabella of a trilobite. It is generally incomplete or interrupted. Also called "lateral glabellar furrow".

glabellar lobe A transverse lobe on the glabella of a trilobite, more or less bounded by complete or partial glabellar furrows. It represents a remnant of the original segments fused in the cephalon. Also called "lateral glabellar lobe".

glacial adj. (a) Of or relating to the presence and activities of ice or glaciers, as *glacial* erosion. (b) Pertaining to distinctive features and materials produced by or derived from glaciers and ice sheets, as *glacial* lakes. (c) Pertaining to an ice age or region of glaciation. (d) Suggestive of the extremely slow movement of glaciers. (e) Used loosely as descriptive or suggestive of ice, or of below-freezing temperature.---n. A *glacial age*, or *glacial stage*, of a glacial epoch, esp. of the Pleistocene Epoch; e.g. the Wisconsin *glacial*.

glacial action All processes due to the agency of glacier ice, such as erosion, transportation, and deposition. The term sometimes includes the action of meltwater streams derived from the ice. See also: *glacial erosion.*

glacial advance *advance* [glaciol].

glacial age A subdivision of a glacial epoch, esp. of the Pleistocene Epoch.

glacial basin A *rock basin* caused by erosion of the floor of a glacial valley.

glacial block A large, markedly angular rock fragment that has not been greatly modified during glacial transport.

glacial boulder A *boulder* or large rock fragment that has been moved for a considerable distance by a glacier, being somewhat modified by abrasion but not always "rounded"; an *erratic.* Syn: *ice boulder.*

glacial boundary The position occupied, in a given region or during a given glacial stage, by the outer or lower margin of an ice sheet; this may extend beyond the terminal moraine.

glacial canyon A canyon eroded by a glacier, usually occupying the site of an older stream valley and having a U-shaped cross profile.

glacial chute A term suggested by Harvey (1931, p.231) for one of a group of narrow, closely spaced, steeply plunging glacial troughs, with vertical or nearly vertical walls and gently curving U-shaped bottoms, that acted as valves in passing or stopping snow, ice, and rock material. Examples occur on the flanks of Mount Puy Puy in Peru.

glacial-control theory A theory of coral-atoll and barrier-reef formation according to which marine erosion and lowering of the sea level during the Ice Age destroyed existing coral reefs and left an extensive, level rock surface from which coral reefs were built up continuously during the gradual postglacial rise of sea level when the oceans became rapidly warmed. Theory was proposed by Reginald A. Daly in 1910. Cf: *antecedent-platform theory; subsidence theory.*

glacial cycle A term used by Davis (1911, p.56) for the ideal case of glaciation operating for a long period of time under fixed climatic conditions such that glacial erosion would be complete and replaced by normal erosion. The term is obsolescent and should be abandoned as it embodies a concept that lacks reality.

glacial debris (a) *glacial drift.* (b) *debris* [glaciol].

glacial deposit *glacial drift.*

glacial-deposition coast A coast with partly submerged moraines, drumlins, and other glacial deposits.

glacial dispersal A center of "radial outflow" from which glacial erratics originated.

glacial drainage (a) The flow system of glacier ice. (b) The system of meltwater streams flowing from a glacier or ice sheet.

glacial drainage channel A "safer expression" for *overflow channel,* because "the exact mode of origin of many channels supposedly due to overflow is uncertain" (Challinor, 1967, p.176); a channel formed by an englacial or subglacial stream.

glacial drift A general term for *drift* transported by glaciers or icebergs, and deposited directly on land or in the sea. Cf: *fluvioglacial drift.* Syn: *glacial deposit; glacial debris.*

glacial epoch Any part of geologic time, from Precambrian onward, in which the climate was notably cold in both the northern and southern hemispheres, and characterized by widespread glaciers that moved toward the equator and covered a much larger total area than those of the present day; specif. the *Glacial Epoch,* or the latest of the glacial epochs, also

known as the Pleistocene Epoch. Syn: *glacial period; ice age; drift epoch.*

glacial erosion The grinding, scouring, plucking, gouging grooving, scratching, and polishing effected by the movemen of glacier ice armed with rock fragments frozen into it, togeth er with the erosive action of meltwater streams. Syn: *ice ero sion.*

glacial erratic *erratic* [glac geol].

glacial eustasy *glacio-eustatism.*

glacial flour *rock flour.*

glacial geology (a) The study of the geologic features and ef fects resulting from the erosion and deposition caused by gla ciers and ice sheets. Cf: *glaciology.* (b) The features of a re gion that has undergone glaciation.---Syn: *glaciogeology.*

glacial groove A deep, wide, usually straight furrow cut i bedrock by the abrasive action of a rock fragment embedde in the bottom of a moving glacier; it is larger and deeper tha a *glacial striation,* ranging in size from a deep scratch to glacial valley.

glacialism (a) The *glacier theory.* (b) *glaciation.*

glacialist (a) A believer in the *glacier theory.* (b) A student o glaciology; a glaciologist.---Syn: *glacierist.*

glacial lake (a) A lake that derives much or all of its wate from the melting of glacier ice, as one fed by meltwater, o one lying entirely on glacier ice and due to differential melting (b) A lake occupying a basin produced by glacial deposition as one held in by a morainal dam. (c) A lake occupying basin produced in bedrock by glacial erosion (scouring quarrying), as a *cirque lake* or *fjord lake.* (d) A lake occupy ing a basin produced by collapse of outwash material sur rounding masses of stagnant ice. (e) *glacier lake.*

glacial lobe A large, rounded, tongue-like projection from th margin of the main mass of an ice cap or ice sheet; a short broad *distributary glacier.* Cf: *tongue* [glaciol]; *outlet glacie* Syn: *lobe* [glac].

glacial-lobe lake A lake occupying a depression that was ex cavated by a glacial lobe as it advanced over the drainage basin of a former river.

glacially disturbed drainage pattern *deranged drainage pat tern.*

glacial marine Said of marine sediments that contain glacia material. Syn: *glaciomarine.*

glacial maximum The time or position of the greatest advance of a glacier, or of glaciers (such as the greatest extent o Pleistocene glaciation). Ant: *glacial minimum.* Syn: *glaciatio limit.*

glacial meal *rock flour.*

glacial milk *glacier milk.*

glacial mill *moulin.*

glacial minimum The time or position of the greatest retreat o a glacier. Ant: *glacial maximum.*

glacial pavement A polished, striated, relatively smooth planed-down rock surface produced by glacial abrasion. Cf *boulder pavement.* Syn: *glacier pavement; ice pavement.*

glacial period (a) A syn. of *glacial epoch;* specif. the *Glacia Period,* also known as the Pleistocene Epoch. Usage of this term is not strictly correct "as glacial intervals during earth history were not of the period rank, but of shorter duration" (ADTIC, 1955, p.34). (b) A geologic *period,* such as the Quaternary Period, that embraced an interval of time marked by a major advance of ice.

glacial plain A plain formed by the direct action of the glacie ice itself. Cf: *outwash plain.*

glacial polish A smoothed surface produced on bedrock by glacial abrasion.

glacial pothole A syn. of *giant's kettle.* The term is misleading because the feature is not produced by the direct scouring action of glacier ice, but by a stream of water falling through a moulin.

glacial pressure ridge *ice-pushed ridge.*

glacial recession *recession* [glaciol].

glacial refuge *refugium*.

glacial retreat *recession* [glaciol].

glacial scour The eroding action of a glacier, including the removal of surficial material and the abrasion, scratching, and polishing of the bedrock surface by rock fragments dragged along by the glacier. Cf: *grinding* [glac geol]. Syn: *scouring*.

glacial scratch *glacial striation*.

glacial spillway *overflow channel*.

glacial stage A major subdivision of a glacial epoch, esp. one of the cycles of growth and disappearance of the Pleistocene ice sheets; e.g. the "Wisconsin Glacial Stage". Syn: *glacial*.

glacial stairway A glacial valley whose floor is shaped like a broad staircase composed of a series of irregular step-like benches (*treads*) separated by steep *risers*. Cf: *cirque stairway*. Syn: *giant stairway; cascade stairway; cascade* [glac].

glacial stream A flow of water that is supplied by melting glacier ice; a *meltwater* stream.

glacial stria *glacial striation*.

glacial striation One of a series of long, delicate, finely cut, usually straight and parallel furrows or lines inscribed on a bedrock surface by the rasping and rubbing of rock fragments embedded at the base of a moving glacier, and usually oriented in the direction of ice movement; also formed on the rock fragments transported by the ice. Cf: *glacial groove*. Syn: *glacial scratch; glacial stria; drift scratch*.

glacial terrace A terrace formed by glacial action, either by rearranging glacial materials in terrace form (such as a remnant of a valley train), or by cutting into bedrock. See also: *kame terrace*.

glacial theory *glacier theory*.

glacial till *till* [glac geol].

glacial trough A deep, flat-floored, steep-sided *U-shaped valley* leading down from a cirque, and excavated by an alpine glacier that has widened, deepened, and straightened a preglacial river valley; e.g. Yosemite Valley, Calif.

glacial valley A U-shaped, steep-sided valley showing signs of glacial erosion; a glaciated valley, or one that has been modified by a glacier.

glaciated Said of a formerly glacier-covered land surface, esp. one that has been modified by the action of a glacier or an ice sheet, as a *glaciated* rock knob. Cf: *glacier-covered*.

glaciated coast A coast whose features were modelled by continental glaciers of the Pleistocene epoch, or a coast covered by glaciers at the present time.

glaciation (a) The formation, movement, and recession of glaciers or ice sheets. (b) The covering of large land areas by glaciers or ice sheets. Syn: *glacierization*. (c) The geographic distribution of glaciers and ice sheets. (d) A collective term for the geologic processes of glacial activity, including erosion and deposition, and the resulting effects of such action on the earth's surface. (e) Any of several minor parts of geologic time during which glaciers were more extensive than at present; a *glacial epoch*, or a *glacial stage*. "A climatic episode during which extensive glaciers developed, attained a maximum extent, and receded" (ACSN, 1961, art.40).---The verb *glaciate* is apparently derived from *glaciation*. Syn: *glacialism*.

glaciation limit (a) The lowest altitude in a given locality at which glaciers can develop or form, usually determined as below the minimum summit altitude of mountains on which glaciers occur but above the maximum summit altitude of mountains having topography favorable for glaciers but on which none occur. (b) *glacial maximum*.

glacieluvial Said of sand and gravel, sometimes with a loamy base, deposited in wide spreads by the irregular wash of water from melting ice. The term was introduced by Gregory (1912, p.175) to describe a glacial deposit that corresponds to colluvium, as distinct from alluvium; e.g. "glacieluvial kame" formed of gravel and sand deposited as a "bank of wash along the edge of a melting glacier".

glacier (a) A large mass of ice formed, at least in part, on land by the compaction and recrystallization of snow, moving slowly by creep downslope or outward in all directions due to the stress of its own weight, and surviving from year to year. Included are small mountain glaciers as well as ice sheets continental in size, and ice shelves which float on the ocean but are fed in part by ice formed on land. See also: *ice stream* [glaciol]; *ice sheet; ice cap*. (b) Nonpreferred syn. of *alpine glacier*. (c) A term used in Alaska for flood-plain *icing* or a mass of *ground ice* [ice]. (d) A stream-like landform having the appearance of, or moving like, a glacier; e.g. a "rock glacier".----Etymol: French *glace*, "ice", alteration of Latin *glacies*.

glacier advance *advance* [glaciol].

glacier band The appearance of one of a series of more or less extensive zones, layers, or lenses, on or within a glacier, that differ visibly (such as in color or texture) from the adjacent material. It may consist of ice, firn, snow, rock, debris, dirt, organic matter, or any mixture of these materials, and it may originate by infilling, plastic flow, or shear. Syn: *band* [glaciol].

glacier bed The channel occupied by a glacier.

glacier berg *glacier iceberg*.

glacier breeze *glacier wind*.

glacier bulb *expanded foot*.

glacier burst *glacier outburst flood*.

glacier cap *ice cap*

glacier cave A cave that is formed within a glacier. Partial syn: *ice cave* [speleo].

glacier cone A rarely used syn. of *debris cone*.

glacier corn Glacier ice broken into irregular crystals of various sizes (Brigham, 1901, p.92).

glacier cornice A mass of glacier ice projecting into an open crevasse. It was formerly underlain by ice containing numerous rock fragments which, when warmed by solar radiation, melted the ice around them and caused the ice above them to project like a *cornice* (Hobbs, 1912, p.397).

glacier-covered Said of a land surface overlaid by glacier ice at the present time. Cf: *glaciated*. Syn: *glacierized; ice-covered; glaciered*.

glacière A French term applied to a natural undergound ice formation, but now used as a syn. of *ice cave*.

glaciered *glacier-covered*.

glacieret (a) A very small glacier on a mountain slope or in a cirque, as in the Sierra Nevada, Calif.; a miniature alpine glacier. Cf: *cirque glacier*. (b) A tiny mass of ice or firn in high mountains, resembling a glacier but defying a precise definition; a *drift glacier*.---Also spelled: *glacierette*.

glacier flood *glacier outburst flood*.

glacier flow The slow downward or outward movement of the ice in a glacier, due to the force of gravity (*gravity flow*). Deformation within the ice, by intragranular gliding, grain-boundary migration, and recrystallization, is involved together with sliding of the glacier on its bed in some situations. Usually expressed in meters per day or year or rarely in furlongs per fortnight. Syn: *ice flow; flow* [glaciol].

glacier grain (a) An individual ice crystal in a glacier. (b) A mechanically separate particle of ice in a glacier.----Cf: *grain* [glaciol].

glacier ice Any ice that forms in or was once a part of a glacier, including land ice that is flowing or that shows evidence of having flowed, and glacier-derived ice floating in the sea.

glacier iceberg A greenish, irregularly shaped iceberg consisting of ice detached from a coastal glacier, typically found in the Arctic. Syn: *glacier berg*.

glacierist *glacialist*.

glacierization A term used in Great Britain for *glaciation* in the sense of the gradual covering or "inundation" of a land surface by glaciers or ice sheets.

glacierized A term used in Great Britain for *glacier-covered*.

glacier lake A lake held in place by the damming of natural drainage by the edge or front of a glacier or ice sheet, as a lake ponded by glacier ice advancing across a valley, or a lake occurring along the margin of a continental ice sheet. Cf: *proglacial lake*. Syn: *glacial lake; marginal lake; ice-dammed lake*.

glacier meal *rock flour.*

glacier milk A stream of turbid, whitish meltwater containing *rock flour* in suspension. Syn: *glacial milk.*

glacier mill *moulin.*

glacier mouse A small, rounded, moss-covered stone, 7-10 cm in diameter, found on certain glaciers, that either lies on morainal material or has rolled off it onto the adjacent ice. Glacier mice were described and named by Eythórsson (1951, p.503). Cf: *polster.*

glacier outburst flood A sudden, often annual, release of meltwater from a glacier or glacier-dammed lake, sometimes resulting in a catastrophic flood, formed by melting of a drainage channel or buoyant lifting of ice by water or by subglacial volcanic activity. Syn: *jõkulhlaup; outburst* [*glaciol*]; *glacier burst; glacier flood.*

glacier pothole *moulin.*

glacier recession *recession* [glaciol].

glacier remanié A glacier formed by regelation of accumulated ice blocks brought down by avalanches and icefalls from the ends of glaciers at higher levels. Etymol: French, "reworked glacier". Syn: *recemented glacier; reconstructed glacier; regenerated glacier; reconstituted glacier*. See also: *remanié* [*glaciol*].

glacier snout *terminus* [glaciol].

glacier surge *surge* [glaciol].

glacier table A boulder or large block of rock supported by an *ice pedestal* that rises from the surface of a glacier. It occurs where the melting of the glacier is retarded by the insulation effect afforded by the rock which thus becomes perched on the pedestal.

glacier theory The theory, first propounded about 1840 and now universally accepted, that the drift was deposited through the agency of glaciers and ice sheets moving slowly from higher to lower latitudes during the Pleistocene Epoch. Cf: *drift theory*. See also: *glacialist*. Syn: *glacial theory; glacialism.*

glacier tongue *tongue* [glaciol].

glacier tongue afloat The *tongue* [glaciol] of a glacier that extends so far into the sea that its lower end floats on the ocean. It is esp. common in the Antarctic. Nonpreferred syn: *ice tongue afloat.*

glacier wave A syn. of *wave ogive*; also used loosely and nonspecifically for *kinematic wave* and the active rapid-flow phase of a surging glacier.

glacier well *moulin.*

glacier wind (a) A cold *katabatic wind* blowing off a glacier. (b) A cold wind blowing out of ice caves in a glacier front, due to the difference in density between the colder air inside and the warmer air outside. Syn: *glacier breeze.*

glacification (a) The change of state from liquid water to ice. (b) A rarely used, almost obsolete, syn. of *glaciation* in the senses of the formation of, and the covering of land areas by, glaciers and ice sheets.

glacigene Of glacial origin, as *glacigene* deposits. Syn: *glacigenous.*

glacioaqueous Pertaining to or resulting from the combined action of ice and water; the term is often used as a syn. of *glaciofluvial*. Syn: *aqueoglacial.*

glacio-eustatism The worldwide change in sea level produced by the successive withdrawal and return of water to the oceans accompanying the formation and melting of ice sheets. Syn: *glacial eustacy; glacio-eustacy; glacio-eustatic change.*

glaciofluvial Pertaining to the meltwater streams flowing from wasting glacier ice and esp. to the deposits and landforms produced by such streams, as kame terraces and outwash plains; relating to the combined action of glaciers and streams. Syn: *fluvioglacial; glacioaqueous.*

glaciogeology *glacial geology.*

glacio-isostasy The state of hydrostatic equilibrium in the Earth's crust as influenced by the weight of glacier ice.

glaciokarst A glaciated limestone region that has both karstic and glacial characteristics (Monroe, 1970). Cf: *thermokarst.*

glaciolacustrine Pertaining to, derived from, or deposited in glacial lakes; esp. said of the deposits and landforms composed of suspended material brought by meltwater streams flowing into lakes bordering the glacier, as delta kames and varved sediments.

glaciology (a) The study of all aspects of snow and ice or the science that treats quantitatively the whole range of processes associated with all forms of solid existing water. Syn *cryology*. (b) The study of existing glaciers and ice sheets and of their physical properties. This definition is not internationally accepted.

glaciomarine *glacial marine.*

glacionatant Relating to or derived from floating ice of glacial origin, as *glacionatant* till.

glaciosolifluction Gravitative sliding, on the surface of a melting glacier, of heterogeneous material mixed with water, generally into crevasses or potholes.

glaciotectonic *cryotectonic.*

glacis A gently inclined slope or bank, less steep than a talus slope; e.g. a piedmont slope, or an easy slope on a mountain side. Etymol: from its resemblance to a glacis used in fortifications as a defense against attack; originally from Middle French *glacer*, "to freeze".

glacon A fragment of sea ice ranging in size from brash ice (2 m across) to a floe of medium- to big-sized dimensions (about 1 km across). Etymol: French *glaçon*, "piece of ice".

glade [geog] A term that usually indicates a clearing between slopes; it can be a high meadow, sometimes marshy and forming the headwaters of a stream, or it can be a low, grassy marsh, which is periodically inundated.

glade [karst] (a) A type of depression associated with karst topography that is elongate with steep sides; this usage is Jamaican. Cf: *karst valley*. (b) Limestone pavement in a karst area that has little soil cover.

glade [ice] A *polynya*, esp. an opening or a stretch of open water in the ice of a lake or of a river.

gladite A lead-gray mineral: $PbCuBi_5S_9$.

gladkaite A fine-grained gray hypabyssal rock composed of a granular aggregate of sodic plagioclase, abundant quartz, hornblende, and some biotite; a quartz lamprophyre.

glady [clay] A term used in Devon, Eng., for a variegated black and white clay having a slippery or smooth texture and often associated with stoneware clays. Also spelled: *gladii.*

glady [soil] Said of a limestone-outcrop area having shallow soil.

glaebule A term proposed by Brewer (1964, p.259-260) for a three-dimensional unit, usually prolate to equant in shape within the matrix of a soil material, recognizable by its greater concentration of some constituent, by its difference in fabric as compared with the enclosing soil material, or by its distinct boundary with the enclosing soil material; e.g. a nodule, concretion, septarium, pedode, or papule. Etymol: Latin *glaebula* "a small clod or lump of earth". Pl: *glaebules*. See also: *papule.*

glamaigite An intrusive rock composed of dark patches of marscoite in a lighter groundmass.

glance A mineral that has a resplendent luster; e.g. chalcocite, or copper glance.

glance coal *pitch coal.*

glance pitch A variety of *asphaltite* with a brilliant conchoidal fracture and which is sometimes called *manjak*. It is similar o gilsonite, but has a higher specific gravity and percentage of fixed carbon, it fuses between 230°F and 250°F, and it is ound in Barbados, Cuba, Mexico, Argentina, Utah, Russia, yria, and elsewhere.

glaposition A term suggested by Glock (1928, p.477) to include "deposition by glaciers".

lare ice A smooth, glassy or bright, highly reflective sheet of ice on a surface of water, land, or glacier.

lareous Said of an organism that lives in gravelly soil. Syn: *lareal*.

glarosion A term suggested by Glock (1928, p.476) to include all destructive activities of glaciers plus the transportation of material produced or supplied"; *glacial erosion*.

laserite *aphthitalite*.

lass [phase] A state of matter intermediate between the close-packed, highly ordered array of a crystal and the poorly packed, highly disordered array of a gas. Most glasses are supercooled liquids, i.e. are metastable, but there is no break in the change in properties between the metastable and stable states. The distinction between glass and liquid is made solely on the basis of viscosity, and is not necessarily related, except indirectly, to the difference between metastable and stable states.

lass [ign] In igneous petrology, an amorphous noncrystalline product of the rapid cooling of a magma. It may constitute the whole rock (e.g. obsidian) or only part of the groundmass.

lass porphyry *vitrophyre*.

lass rock A term used in northern Illinois and southern Wisconsin for a pure cryptocrystalline limestone of Trentonian age.

lass sand A sand that is suitable for glassmaking because its silica content is 98-100% and its content of iron oxide impurities is less than 1%.

lass schorl *axinite*.

lass sponge *hyalosponge*.

lassy [min] *vitreous*.

lassy [ign] Said of the texture of certain extrusive igneous rocks that is similar to that of broken glass or quartz and developed as a result of rapid cooling of the lava, without distinct crystallization. Also, said of any of the other properties of a volcanic rock that resemble those of glass, such as hardness, luster, or composition. Syn: *hyaline [ign]*; *vitreous [ign]*.

lassy feldspar *sanidine*.

glauberite A brittle, light-colored, monoclinic mineral: $Na_2Ca(SO_4)_2$. It has a vitreous luster and saline taste, and occurs in saline residues.

Glauber's salt A syn. of *mirabilite*. Named after Johann R. Glauber (1604-1668), German chemist. Also spelled: *Glauber salt*.

glaucocerinite *glaucokerinite*.

glaucochroite A bluish-green, violet, or pale-pink orthorhombic mineral: $CaMnSiO_4$.

glaucodot A mineral: $(Co,Fe)AsS$. Syn: *glaucodote*.

glaucokerinite A sky-blue mineral: $(Cu,Zn)_{10}Al_4(SO_4)(OH)_{30} \cdot 2H_2O$ (?). Syn: *glaucocerinite*.

glauconite (a) A dull-green, amorphous, and earthy or granular mineral of the mica group: $(K,Na)(Al,Fe^{+3},Mg)_2(Al,Si)_4O_{10}(OH)_2$. It has often been regarded as the iron-rich analogue of illite. Glauconite occurs abundantly in greensand, and seems to be forming in the marine environment at the present time; it is the most common sedimentary (diagenetic) iron silicate and is found in marine sedimentary rocks from the Cambrian to the present. Glauconite is an indicator of very slow sedimentation. (b) A name applied to a group of green minerals consisting of hydrous silicates of potassium and iron.

glauconitic Said of a mineral aggregate that contains glauconite, resulting in the characteristic green color, e.g. glauconitic shale or clay.

glauconitic sand *greensand*.

glauconitic sandstone A sandstone containing sufficient grains of glauconite to impart a marked greenish color to the rock; *greensand*.

glauconitization A submarine metamorphic process whereby a mineral is converted to glauconite under anaerobic conditions.

glaucophane A blue, bluish-black, or grayish-blue monoclinic mineral of the amphibole group: $Na_2(Mg,Fe^{+2})_3Al_2Si_8O_{22}(OH)_2$. It is a fibrous or prismatic mineral that occurs only in certain crystalline schists resulting from regional metamorphism of sodium-rich igneous rocks (such as spilites).

glaucophane schist A type of amphibole schist in which glaucopnane rather than hornblende is an abundant mineral. Epidote frequently occurs, and there are quartz and mica varieties (Holmes, 1928, p.106).

glaucophane schist facies A term originated by Eskola (1939) for rocks formed by dynamothermal metamorphism and having a broad mineral association of amphiboles (e.g. glaucophane, actinolite), pyroxenes (e.g. jadeite, aegirine), lawsonite, pumpellyite, stilpnomelane, etc., representing hydrostatic pressures in excess of 5000 bars and temperatures of 300° to 400°C (Turner and Verhoogen, 1960). It is generally interpreted as the high-pressure, glaucophanitic phase of the *greenschist facies* (in excess of 5000-6000 bars) at the higher temperature range (over 400°C). At lower temperatures (250°-400°C) it is now called the *lawsonite-glaucophane-jadeite facies* (Winkler, 1967). Syn: *blueschist facies*.

glaucopyrite A variety of loellingite containing cobalt.

G layer The seismic region of the Earth below 5160 km, equivalent to the *inner core*. It is a part of a classification of the interior of the Earth, having layers A to G. Together with the F layer, it is the equivalent of the *lower core*.

glei Var. of *gley*.

gleization The formation of *gley* in a soil. Syn: *gleying*.

glen A long, narrow, steep-sided, flat-bottomed, and secluded valley, usually wooded, often containing a stream or lake at its bottom; esp. a narrow-floored, glaciated mountain valley in Scotland and Ireland. It is narrower and more steep-sided than a *strath*. Syn: *glyn*.

Glenarm A provincial series of the late Precambrian in New Jersey, Pennsylvania, Delaware, Maryland, and Virginia.

glendonite A pseudomorph of a carbonate (calcite or esp. siderite) after glauberite.

Glen flow law An empirical relation relating the shear strain rate (ℓ) of ice to the shear stress (σ), as $\epsilon = k\sigma^n$, where the parameter k varies with temperature, type of ice, and geometry of stress, and n is a number between 1.5 and 4.5 (Glen, 1955, p.528). This relation is basic to most analyses of glacier flow.

glenmuirite An orthoclase-bearing teschenite.

glessite A brown variety of retinite found on the shores of the Baltic Sea.

gletscherschlucht A gorge sculptured by meltwater streams, often initiated where a moulin empties onto jointed rock. Etymol: German *Gletscherschlucht*, "glacier gorge".

gley Soil mottling, caused by partial oxidation and reduction of its constituent ferric iron compounds, due to conditions of intermittent water saturation. The process is called *gleization*. Also spelled: *glei*.

gleying *gleization*.

gley soil A soil in which gley has formed; it is usually gray in color, with yellow and brown mottling.

glidder A local British term used synonymous with *scree* in the sense of a loose stone, as on a hillside.

glide A gently flowing and calm reach of shallow water.

glide bedding A variety of *convolute bedding* produced by subaqueous gliding. Cf: *slip bedding*.

glide breccia A breccia formed by rapid and slow subaqueous gravitational movements that deform, shatter, or crush newly formed or partly consolidated bottom sediments deposited under somewhat unstable conditions at higher levels. It may be caused by overloading, earthquakes, or deformation.

glide direction The crystallographic direction of glide translation.

glide fold *shear fold.*

glide line In structural petrology, the direction of movement on an s surface.

glide plane A plane in a crystal along which translation of one part of the crystal may take place due to plastic deformation. Syn: *translation plane; gliding plane; slip plane; T plane.*

glide twin *deformation twin.*

gliding [cryst] *crystal gliding.*

gliding [tect] *gravitational sliding.*

gliding flow *Flow* [exp struc geol] involving gliding parallel to the preferred crystallographic orientation, e.g. intragranular deformation in a crystal by twin gliding or translation gliding. Syn: *secondary flowage.*

gliding plane *glide plane.*

gliding surface *slip surface* [mass move].

glimmer A syn. of *mica.* Etymol: German *Glimmer.*

glimmergabbro A biotite-bearing gabbro.

glimmerite *biotitite.*

glimmerton An early name for *illite.*

glinite A group name for clay minerals from clay deposits.

glint An escarpment or steep cliff, esp. one produced by erosion of a dipping resistant formation. See also: *klint.* Etymol: Norwegian, "boundary".

glint lake A lake formed along a *glint line,* esp. a long, narrow glacial lake occupying a basin excavated in bedrock where a glacier is dammed by an escarpment ("glint"), as certain lakes in Norway and Scotland. Syn: *glint-line lake.*

glint line An extensive erosional escarpment produced by the denudation of a very gently dipping resistant formation, as the Silurian limestone of the Great Lakes region. The term is used specif. for the boundary between an ancient shield and younger rocks, e.g. in Russia where Paleozoic rocks rocks rise above the Baltic Shield.

glitter A British syn. of *scree* or a mass of loose stones. Also spelled: *glitters.*

gloap *gloup.*

global scale A map scale (smaller than 1/5,000,000) involving all or a major part of the Earth's surface.

global tectonics Tectonics on a global scale, such as tectonic processes related to very large-scale movement of material within the Earth; specif. *new global tectonics.* Cf: *megatectonics.*

globe (a) A body having the form of a sphere; specif. a spherical, typically hollow ball that has a map of the Earth drawn on it and that is usually rotatable at an angle corresponding to the inclination of the Earth's axis. Also, a chart of the celestial sphere, depicted on a sphere. (b) A planet; esp. the Earth.

globigerina ooze An *ooze* whose skeletal remains are foraminiferal tests, predominantly of the genus *Globigerina.* It is calcareous, and a particular type of *foraminiferal ooze.*

globigerinid Any planktonic foraminifer belonging to the superfamily Globigerinacea, characterized by a perforate test with bilamellid septa and walls of radial calcite crystals. Their stratigraphic range is Middle Jurassic to present. Var: *globigerine.*

globosphaerite A more or less spherical cumulite in which the globulites have a somewhat radial arrangement.

globular *spherulitic.*

globularite A speleothem or cave formation consisting of small calcite crystals tipped with spheres of radiating fibers (Monroe, 1970).

globular projection A map projection (neither conformal nor equal-area) representing a hemisphere upon a plane parallel to its base, the point of projection being removed to a point outside of the opposite surface of the sphere. The equator and central meridian are straight lines intersecting at right angles; all other meridians and parallels are circular arcs. The projec-

tion is an arbitrary distribution of curves conveniently constructed; distance and directions can neither be measured nor plotted. It is commonly used in pairs in atlases.

globule The male reproductive structure, having both sterile and fertile cells, of a Charophyceae.

globulite A spherical, or globular, crystallite commonly found in volcanic glass.

globulith An intrusive body, or group of closely associated bodies, having a globular or botryoidal shape and almost concordant contacts resulting from the effects of the intrusion(s) on the immediate surroundings (Berthelsen, 1970, p.73).

glomeroclastic Pertaining to lumpal particles grouped together in clusters in a carbonate sedimentary rock. Also said of the texture characterized by lumps.

glomerocryst An aggregate of crystals of the same mineral. Cf: *polycrystal.*

glomerophyric An obsolete term applied to the texture of porphyritic igneous rocks containing closed clusters of equant crystals, usually of the same mineral. Cf: *cumulophyric.* Syn: *glomeroporphyritic.*

glomeroplasmatic An obsolete term applied to the texture of granites and gneisses containing open clusters of individual crystals or grains of the same mineral. Cf: *glomerophyric.*

glomeroporphyritic *glomerophyric.*

glomospirine Having an irregularly wound coiled tubular chamber; specif. pertaining to the foraminifer *Glomospira.*

glory-hole mining A type of opencut mining in which the ore body is worked from the top down in a conical excavation and is removed by an underground system beneath the orebody. Syn: *mill-hole mining.*

gloss *polish.*

gloss coal The highest ranking lignite. It is deep black, compact, with definite conchoidal fracture and glossy luster. Cf: *subbituminous coal.*

glossopterid n. The informal name for the fossil genus *Glossopteris,* a fern or fernlike plant common in the Permian and Triassic.----adj. Pertaining to such a plant or plant assemblage.

glossothyropsiform Said of a brachiopod loop developed from the cryptacanthiiform stage by final resorption of the posterior part of the echmidium and consisting of two descending branches unconnected posteriorly, bearing two broad ascending elements joined by a wide transverse band (TIP, 1965, pt.H, p.145).

gloup A Scottish term for a *blowhole* along the coast. Syn: *gloap.*

glowing avalanche *ash flow.*

glowing cloud *nuée ardente.*

glucine A mineral: $CaBe_4(PO_4)_2(OH)_4.1/2H_2O.$

glume A chaffy bract, specif. at the base of a grass spikelet. See also: *tegmen.*

glutenite A collective term, now obsolete, for breccias, conglomerates, and sandstones (Holmes, 1928, p.7). Syn: *glutinite.*

glyders A term used in North Wales for a *scree* or mass of loose stones. Syn: *glydrs.*

glyn A Welsh syn. of *glen.*

glyptogenesis The *sculpture* of the Earth's surface by erosion.

glyptolith A term proposed by Woodworth (1894a, p.70) for a wind-cut stone or *ventifact.* Etymol: Greek *glyptos,* "carved" + *lithos,* "stone".

G-M counter *Geiger-Müller counter.*

gmelinite A hexagonal zeolite mineral: $(Na_2,Ca)Al_2Si_4O_{12}.6H_2O.$ Cf: *faujasite.*

G-M tube *Geiger-Müller tube.*

gnamma hole A term used in the deserts of Western Australia for a rounded hollow eroded or indented in solid rock (usually at the intersections of joints in granite) and frequently containing water; it has a narrow orifice, but widens out below.

gnarly bedding *convolute bedding.*

gnathal lobe The masticatory endite of the mandible of a crustacean.

gnathosoma The anterior part of the body in the arachnid order Acarida, bearing the mouthparts. Obsolete syn: *capitulum*. Syn: *gnathosome*.

gnathothorax The thorax and the part of the head bearing the feeding organs of an arthropod; e.g. the tagma of a crustacean resulting from fusion of mandibular and two maxillary somites with one or more thoracic somites, limbs of which are modified to act as mouthparts (TIP, 1969, pt.R, p.96). Cf: *cephalothorax*.

gneiss A foliated rock formed by regional metamorphism in which bands or lenticles of granular minerals alternate with bands and lenticles in which minerals having flaky or elongate prismatic habits predominate. Generally less than 50% of the minerals show preferred parallel orientation. Although a gneiss is commonly feldspar- and quartz-rich, the mineral composition is not an essential factor in its definition (American usage). Varieties are distinguished by texture (e.g. augen gneiss), characteristic minerals (e.g. hornblende gneiss), or general composition and/or origins (e.g. granite gneiss). See also: *gneissic; gneissoid; gneissose*.

gneissic Pertaining to the texture or structure typical of gneisses, having foliation that is wider spaced, less marked, and often more discontinuous than that of a *schistose* texture or structure (Johannsen, 1931). Cf: *gneissoid; gneissose*.

gneissic structure In a metamorphic rock, commonly *gneiss*, the coarse, textural lineation or banding of the constituent minerals into alternating silicic and mafic layers. Syn: *gneissosity; gneissose structure*. Cf: *primary gneissic banding*.

gneissoid Pertaining to a gneisslike rock structure or texture which is the result of nonmetamorphic processes, e.g. viscous magmatic flow forming a gneissoid granite. Cf: *gneissic; gneissose*.

gneissose (a) Said of a rock, or of its structure, that resembles gneiss but that is not the result of metamorphic processes. Cf: *gneissoid*. (b) Said of a rock whose structure is composite, having alternating schistose and granulose bands and lenses which differ in mineral composition and texture. Cf: *gneissic*.----Use of this ambiguous term is discouraged.

gneissose structure *gneissic structure*.

gneissosity *gneissic structure*.

gnomonic projection (a) A perspective azimuthal map projection (of a part of a hemisphere) on a plane tangent to the surface of the sphere, having the point of projection at the center of the sphere. All straight lines on the tangent plane represent arcs of great circles on the Earth's surface; all great circles appear as straight lines. The point of tangency may be at a pole, on the equator, or at any point in between (oblique gnomonic projection). It is used, in conjunction with the Mercator projection, to plot great-circle courses in navigation. Syn: *central projection; great-circle projection; great-circle chart*. (b) A similar projection used in optical mineralogy to plot data obtained by measurements of crystals with a two-circle goniometer, characterized by a plane of projection that is tangent to the north pole of the sphere, with the poles of the faces parallel to the vertical axis of the sphere lying at infinity.

gobi (a) A Mongolian term introduced by Berkey & Morris (1924, p.105) for a small, open, level-surfaced basin within a *tala*. (b) A lenticular mass of sedimentary deposits occupying a gobi.

goblet valley *wineglass valley*.

godlevskite A mineral: $(Ni,Fe)_7S_6$.

goe Var. of *geo*.

goethite A yellowish, reddish, or brownish-black mineral: α-$FeO(OH)$. It is dimorphous with lepidocrocite. Goethite is the commonest constituent of many forms of natural rust or of limonite, and it occurs esp. as a weathering product in the gossans of sulfide-bearing ore deposits. Syn: *göthite; xanthosiderite*.

gold A soft, heavy, yellow, isometric mineral, the native metallic element Au. It is commonly naturally alloyed with silver or copper and occasionally with bismuth, mercury, or other metals, and is widely found in alluvial deposits (as nuggets and grains) or in veins associated with quartz and various sulfides. Gold is very malleable and ductile, and is used chiefly for jewelry and as the international standard for world finance.

gold amalgam A variety of native gold containing mercury; a naturally occurring *amalgam* composed of gold, silver, and mercury, the gold averaging about 40%. It is usually associated with platinum, and occurs in yellowish-white grains that crumble readily.

gold beryl A syn. of *chrysoberyl*. Not to be confused with *golden beryl*.

gold dust Fine particles, flakes, or pellets of gold, such as those obtained in placer mining. Cf: *commercial dust*.

golden beryl A clear, golden-yellow or yellowish-green gem variety of beryl. See also: *heliodor*. Not to be confused with *gold beryl*.

goldfieldite A dark lead-gray mineral: $Cu_3(Sb,As)(Te,S)_4$. It is a variety of tetrahedrite containing tellurium.

goldichite A pale-green monoclinic mineral: $KFe(SO_4)_2.4H_2O$.

goldmanite A mineral of the garnet group: $Ca_3(V,Al,Fe)_2(SiO_4)_3$.

gold opal A *fire opal* that exhibits only an overall color of golden yellow..

gold quartz Milky quartz containing small inclusions of gold.

goldschmidtine *stephanite*.

goldschmidtite *sylvanite*.

Goldschmidt's mineralogical phase rule *mineralogical phase rule*.

goldstone A translucent, reddish-brown glass containing a multitude of tiny, thin, metallic-copper tetrahedra or hexagonal platelets that exhibit bright reflections, producing a popular but poor imitation of aventurine (quartz or feldspar). Syn: *aventurine glass*.

goletz terrace A terrace "in which bedrock is exposed in the scarp and is close to the surface of the bench" (Bird, 1967, p.248).

gompholite *nagelfluh*.

gonal spine A spine situated only at plate corners on a dinoflagellate cyst.

gonatoparian adj. Of or concerning a trilobite having facial sutures, the posterior sections of which reach the cephalic margin at the genal angles; also, said of the sutures themselves.

gondite A metamorphic rock consisting of spessartite and quartz, probably derived from manganese-bearing sediment. It is named after Gonds (the Gondite Series), central India. Cf: *collobrierite; eulysite*.

Gondwana The hypothetical protocontinent of the Southern Hemisphere. It is named for the Gondwana system of India with an age range from Carboniferous to Jurassic and containing glacial tillite in its lower part and coal measures higher up. Similar sequences of the same age are found in all the continents of this hemisphere; this similarity, along with much compelling evidence in the older rocks, indicates that all these continents were once joined into a single larger mass. Earlier geologists attempted to connect the present continental fragments by assuming that large parts of the ocean areas of the Southern Hemisphere were originally continental, and that the inferred continental connections afterwards foundered into the ocean basins. The preponderance of modern evidence indicates that the present continents are fragments which have been separated from each other by some form of continental displacement with the aid of sea-floor spreading. The hypothetical counterpart of Gondwana in the Northern Hemisphere is *Laurasia; the hypothetical supercontinent from which both were derived is Pangea*. Var: *Gondwanaland*.

Gondwanaland Var. of *Gondwana*.

gongylodont Said of a class of ostracode hinges consisting of three elements wherein the terminal elements are opposites in the same valve, as in the genus *Loxoconcha*.

goniatite Any ammonoid belonging to the order Goniatitida, characterized generally by a shell having sutures with eight undivided lobes. Their stratigraphic range is Middle Devonian to Upper Permian.

goniatitic suture A type of suture in ammonoids characterized by simple fluting in which most or all of the lobes and saddles are entire or plain (not denticulate or frilled), the only common exception being the ventral lobe, which is subdivided and may be denticulate; specif. a suture in goniatites. Cf: *ammonitic suture; ceratitic suture*.

goniometer (a) An instrument used in optical crystallography for measuring the angles between crystal faces. Types of goniometers are the *contact goniometer*, the *reflection goniometer*, and the *two-circle goniometer*. (b) An instrument that measures x-ray diffractions; a *diffractometer*.

gonnardite A zeolite mineral: $Na_2CaAl_4Si_6O_{20}.7H_2O$. It occurs in finely fibrous, radiating spherules.

gonoecium An ambiguous term applied to certain brood chambers in cheilostomatous bryozoans. Pl: *gonoecia*.

gonopore (a) A simple opening that serves as an exit from the genital system of an echinoderm (such as a cystoid and edrioasteroid); a genital pore. Cf: *hydropore*. (b) The outlet of the genital ducts in crustaceans; a sexual pore.

gonozooid A bryozoan zooid modified as a brood chamber, esp. in the order Cyclostomata.

gonyerite A mineral of the chlorite group: $(Mn, Mg)_6Si_4O_{10}(OH)_8$.

gooderite A plutonic rock similar to nepheline syenite but with a predominance of albite rather than potassium feldspar.

goodness of fit A statistical test or method used to ascertain equivalency of observed data with theoretical distributions, with other observed data, or with some mathematical functions.

Goodsprings twin law A rare type of normal twin in feldspar, in which the twin plane is $(\bar{1}\,12)$.

goongarrite A lead-bismuth sulfide consisting of a mixture of cosalite and galena, found at Lake Goongarrie in Western Australia.

gooseberry stone A syn. of *grossular*, esp. used for the yellow-green to yellowish-green varieties.

gooseneck The part of a winding valley resembling in plan the curved neck of a goose; esp. a part formed by an entrenched meander.

gorceixite A brown mineral: $BaAl_3(PO_4)_2(OH)_5.H_2O$.

gordonite A colorless mineral: $MgAl_2(PO_4)_2(OH)_2.8H_2O$.

gordunite A peridotite composed chiefly of olivine and less pyroxene, with small amounts of pyrope, picotite, and opaque oxides; a garnet-bearing wehrlite.

gore [cart] One of the series of related and triangular or lune-shaped sections (pieces of paper or thin cards) of a map or chart, usually bounded by meridians and tapering to the poles, applied to the surface of a sphere (with a negligible amount of distortion) to form a globe.

gore [surv] A small irregularly shaped tract of land; esp. a triangular piece of land left between two adjoining surveyed tracts, often because of inaccuracies in the boundary surveys or as a remnant of a systematic survey. It is an officially recognized tract in some States (such as Maine and Vermont).

gorge (a) A small, narrow, deep valley with nearly vertical rocky walls, enclosed among mountains, smaller than a *canyon*, and more steep-sided than a *ravine*; esp. a narrow, restricted, steep-walled part of a canyon. (b) A narrow defile or passage between hills or mountains.---Etymol: French, "throat".

gorgeyite A mineral: $K_2Ca_5(SO_4)_6.1H_2O$.

goshenite A colorless, white, or bluish beryl from Goshen, Mass.

goslarite A white mineral: $ZnSO_4.7H_2O$. It forms by oxidation of sphalerite and it usually occurs massive. Syn: *white vitriol; zinc vitriol; white copperas*.

gossan An iron-bearing, weathered product overlying a sulfide deposit. It is formed by the oxidation of sulfides and the leaching out of the sulfur and most metals, leaving hydrated iron oxide. Syn: *capping; iron hat; chapeau de fer*. Also spelled: *gozzan*. Cf: *oxidized zone; false gossan*.

gossany Pertaining to or comprising gossan.

gote A British term for a watercourse.

gothite *goethite*.

Gothlandian Var. of *Gotlandian*.

Gotlandian An alternative name of the Silurian, specif. the late Silurian, used in Europe. Also spelled: *Gothlandian*.

gotzenite A mineral: $(Ca,Na)_7(Ti,Al)_2Si_4O_{15}(F,OH)_3$.

gouffre A French term for a gulf, chasm, or pit, sometimes applied in English-language publications to a natural gorge, a karstic depression, or a pit cave.

gouge [ore dep] (a) A thin layer of soft, earthy, putty-like rock material along the containing wall of a mineral vein or between the country rock and the vein, so named because a miner is able to "gouge" it out and thereby facilitate the mining of the vein itself. Syn: *selvage; pug*. See also: *clay gouge*. (b) A term used in Nova Scotia for a narrow band of gold-bearing slate next to a vein, extractable by a thin, long pointed stick.

gouge [glac geol] *crescentic gouge*.

gouge channel A term used by Kuenen (1957, p.242) for a large sole mark (larger than a flute cast) now known as a *channel cast*. Syn: *megaflow mark*.

gouge mark *crescentic gouge*.

gouging [glac geol] (a) The formation of *crescentic gouges*. (b) The local basining of a bedrock surface by the action of glacier ice (Thornbury, 1954, p.48).

gouging [mining] The working of a mine without plan or system in which only the high-grade ore is mined.

gour [speleo] A syn. of *rimstone dam* and of *rimstone pool*; it is taken from the French. See also: *microgour; grand gour*.

gour [geomorph] Plural of *gara*.

gowerite A monoclinic mineral: $CaB_6O_{10}.5H_2O$.

goyazite A yellowish-white mineral: $SrAl_3(PO_4)_2(OH)_5.H_2O$. Syn: *hamlinite*.

goyle An English term for a ravine or other steep, narrow valley.

goz A term used in Sudan for a long, gentle, dune-like accumulation of sand, varying in thickness from a few decimeters to tens of meters; also. a large-scale, undulating tract containing such accumulations. Etymol: Arabic. Pl: *gozes*.

gozzan Var. of *gossan*.

grab An implement for gripping and extricating broken tools (such as a drill or cable) from a borehole or well.

graben An elongate, relatively depressed crustal unit or block that is bounded by faults on its long sides. It is a structural form that may or may not be geomorphologically expressed as a *rift valley*. Etymol: German, "ditch". Cf: *horst*. Syn: *trough* [fault].

grab sampler An ocean-bottom sampler that operates by enclosing the material between its two jaws upon contact with the bottom. See also: *Shipek bottom sampler; Peterson grab; clamshell snapper*. Cf: *dredge; corer*. Syn: *snapper*.

gradation [geomorph] (a) The leveling of the land, or the bringing of a land surface or area to a uniform or nearly uniform grade or slope through erosion, transportation, and deposition; specif. the bringing of a stream bed to a slope at which the water is just able to transport the material delivered to it. See also: *gradation; degradation; aggradation*. (b) A term often used as a syn. of *degradation*.

gradation [part size] The proportion of material of each particle size, or the frequency distribution of various sizes, constituting a particulate material such as a soil, sediment, or sedinentary rock. The limits of each size are chosen arbitrarily. Cf: *sorting; grading.*

gradation period "The entire time during which the base-level remains in one position; that is, the interval between two elevations of the Earth's surface of sufficient magnitude to produce a marked change in the position of sea level" (Hayes, 1899, p.22).

grade [coal] A *coal classification* based on degree of purity, i.e., quantity of inorganic material or ash left after burning. Cf: *type [coal]; rank [coal].*

grade [ore dep] The relative quantity or percentage of ore-mineral content; e.g. *high-grade, low grade.* Syn: *tenor.*

grade [meta] *metamorphic grade.*

grade [part size] A particular size (diameter), size range, or size class of particles of a soil, sediment, or rock; a unit of a grade scale, such as "clay grade", "silt grade", "sand grade", or "pebble grade".

grade [surv] (a) A datum level; a level of reference. (b) Height above sea level; actual elevation. Also, the elevation of the finished surface of an engineering project (such as of a canal bed, embankment top, or excavation bottom).

grade [streams] (a) The condition of balance, achieved by a stream, between erosion and deposition, brought about by the adjustments between the capacity of the stream to do work and the quantity of work that the stream has to do (Davis, 1902, p.86). It is represented by the continuously descending curve (the longitudinal profile of the stream) which everywhere is just steep enough to allow the stream to transport the load of sediment made available to it. Grade involves an equilibrium among slope, load, volume, and velocity of the stream. The term was used by Gilbert (1876) but the concept was first formally introduced by Davis (1894) who admits that it "cannot be understood without rather careful thinking"; although a precise definition is difficult, the concept is useful as it implies both an adjustability of the channel to changes in independent variables and a stability in form and profile. (b) A term sometimes used as a syn. of *gradient* of a given length of a stream. This usage is confusing and not recommended.

grade [eng] (a) A degree of inclination, or a rate of ascent or descent, with respect to the horizontal, of a road, railroad, embankment, conduit, or other engineering structure; it is expressed as a ratio (vertical to horizontal), a fraction (such as m/km or ft/mi), or a percentage (of horizontal distance). (b) A graded part of a road, embankment, or other engineering structure that is ascending, descending, or level.---The synonymous term *gradient* is used in geomorphology.

grade correction *slope correction.*

graded [geomorph] Said of a surface or feature when neither degradation nor aggradation is occurring, or when both erosion and deposition are so well balanced that the general slope of equilibrium is maintained. Cf: *in regime.* Syn: *at grade.*

graded [part size] (a) A geologic term pertaining to an unconsolidated sediment or to a cemented detrital rock consisting of particles of essentially uniform size or of particles lying within the limits of a single grade. Syn: *sorted.* (b) An engineering term pertaining to a soil or an unconsolidated sediment consisting of particles of several or many sizes or having a uniform or equable distribution of particles from coarse to fine; e.g. a "graded sand" containing coarse, medium, and fine particle sizes. See also: *well-graded.*---The term is "rarely used in geology to refer to the sorting of the sediment, although this is common among engineers" (Middleton, 1965, p.249). Ant: *nongraded.*

graded bed A sedimentary bed, usually thin, exhibiting *graded bedding,* generally having an abrupt contact with the fine material of the underlying bed but a gradational or indefinite contact near the top; e.g. sand or coarse silt grading upward into shaly material.

graded bedding A type of bedding in which each layer displays

a gradual and progressive change in particle size, usually from coarse at the base of the bed to fine at the top. It may form under conditions in which the velocity of the prevailing current declined in a gradual manner, such as by deposition from a single and short-lived turbidity current. See also: *sorted bedding.*

graded profile *profile of equilibrium.*

graded reach (a) A part (of a stream) characterized by a condition of balance between erosion and deposition, as where a stream crossing the outcrops of weak rocks is in equilibrium while the profile of the stream across resistant rocks remains for a long time irregular and steep. (b) A reach of a graded stream.

graded shoreline A smooth shoreline that has been straightened by the formation of bars across embayments and by the cutting back of headlands, and that possesses a broadened sloping surface so adjusted in slope that the energy of incoming waves is completely absorbed and the shifting of the shoreline is reduced to a very slow rate; a shoreline with a vertical *profile of equilibrium,* typical of an advanced stage of development. Syn: *equilibrium shoreline.*

graded slope The downstream gradient of a graded stream; it permits the most effective transport of load and is represented by the profile of equilibrium.

graded stream (a) A stream in equilibrium, showing a balance between its transporting capacity and the amount of material supplied to it, and thus between degradation and aggradation in the stream channel. "A graded stream is one in which, over a period of years, slope is delicately adjusted to provide, with available discharge and with prevailing channel characteristics, just the velocity required for the transportation of the load supplied from the drainage basin. ... its diagnostic characteristic is that any change in any of the controlling factors will cause a displacement of the equilibrium in a direction that will tend to absorb the effect of the change" (Mackin, 1948, p.471). A graded stream is not a stream that is loaded to capacity (streams probably never attain this condition), and neither is it a stream that is neither eroding nor depositing (erosion may occur in one part of the channel and deposition in another part). The term is not to be confused with *gradient,* which is possessed by all streams. Syn: *steady-state.* (b) A stream characterized by the absence of waterfalls and rapids (Kesseli, 1941).---Cf: *poised stream; regime stream.*

graded unconformity *blended unconformity.*

grade level The level attained by a stream when its "whole course" has been reduced to a uniform gradient, or when its longitudinal profile is a straight line (Park, 1914, p.42).

grade scale A systematic, arbitrary division of an essentially continuous range of particle sizes (of a soil, sediment, or rock) into a series of classes or scale units (or grades) for the purposes of standardization of terms and of statistical analysis; it is usually logarithmic. Examples include: *Udden grade scale; Wentworth grade scale; phi grade scale; Atterberg grade scale; Tyler Standard grade scale; Alling grade scale.* See also: *geometric grade scale.*

gradient [stream] A *stream gradient.* Cf: *grade.*

gradient [geophys] The change in value of one variable with respect to another variable, especially vertical distance with respect to horizontal distance, or geophysical properties such as gravity, temperature, magnetic susceptibility, or electrical potential with respect to horizontal distance. See also: *gradiometer.*

gradient [hydraul] *hydraulic gradient.*

gradient [geomorph] (a) A degree of inclination, or a rate of ascent or descent, of an inclined part of the Earth's surface with respect to the horizontal; the steepness of a slope. It is expressed as a ratio (vertical to horizontal), a fraction (such as m/km or ft/mi), a percentage (of horizontal distance), or an angle (in degrees). Syn: *slope.* (b) A part of a surface feature that slopes upward or downward; a slope, as of a stream channel or of a land surface.---The synonymous term *grade* is used in engineering.

gradient array An electrode array used in lateral search in which both current electrodes are fixed while the two potential electrodes are sufficiently close together so as to measure the gradient of the potential. The potential probes are moved along traverse lines normal to geologic lineation and parallel to a line joining the current electrodes. A square area of dimensions one third of the separation of the current electrodes, and situated midway between the current electrodes, is surveyed. It is used in resistivity and induced polarization surveys.

gradient current A wind or ocean current in which the horizontal *pressure force* is exactly balanced by the sum of the *Coriolis force* and the surface or bottom frictional forces. It flows to the right of the pressure gradient force in the Northern Hemisphere, but not along the isobars. Cf: *geostrophic current.*

gradienter An attachment to a surveyor's transit with which an angle of inclination is measured in terms of the tangent of the angle instead of in degrees and minutes. It may be used as a telemeter in observing horizontal distances.

gradient of the head *Hydraulic gradient* for which the specified direction is that of maximum rate of increase in head.

grading [geomorph] The reduction of the land to a level surface, such as erosion to base level by streams or the shaping of an excavation slope with graders.

grading [part size] The gradual reduction, in a progressively upward direction within an individual stratification unit, of the upper particle-size limit. It implies pulsatory turbulent-fluid deposition. Cf: *gradation.*

grading factor A *sorting index* developed by Baker (1920, p.368) and defined as the difference between unity and the quotient of mean deviation divided by arithmetic mean diameter (equivalent grade); a measure of how nearly the degree of sorting approaches perfection. Abbrev: G.F.

gradiometer Any instrument that is used to measure the *gradient* of a physical quantity, e.g. a device consisting of two magnetometers, one above the other, that measures the vertical gradient of the magnetic field by the difference in their readings. See also: *astatic magnetometer.*

gradually varied flow The flow in an open channel where the velocity changes slowly along the channel and the flow is assumed uniform for an increment of length. See also: *flow profile.*

graduation (a) The method or system of dividing into degrees or quantity, such as the placing of equally spaced intermediate marks on an instrument or device (such as a tape or thermometer) to represent standard or conventional values. (b) A mark or the marks so placed; one of the equal divisions or dividing lines on a graduated scale.

Graf sea gravimeter A balance-type gravity meter that is heavily damped in order to attenuate shipboard vertical accelerations. It consists of a mass at the end of a horizontal arm, supported by a torsion spring rotational axis. The mass rises and falls with gravity variation, but is restored to near its null position by a horizontal reading spring, tensioned with a micrometer screw. The difference between actual beam position and null position gives indication of gravity value after the micrometer screw position has been taken into account. (U.S. Naval Oceanographic Office, 1966, p.72).

graftonite A salmon-pink monoclinic mineral: (Fe,Mn, Ca)$_3$(PO$_4$)$_2$. It occurs in laminated intergrowths with triphylite.

grahamite [meteorite] *mesosiderite.*

grahamite [mineral] A black asphaltite with a variable luster, black streak, high specific gravity, and high fixed carbon content.

grail Coarse- or medium-grained sediment particles; specif. gravel or sand.

grain [petrology] (a) A mineral or rock *particle*, smaller than a

fragment, having a diameter of less than a few millimeter and generally lacking well-developed crystal faces; esp. small, hard, more or less rounded mineral particle, such as sand grain. Also, a general term for sedimentary particles c all sizes (from clay to boulders), as used in the expression "grain size", or "fine-grained" and "coarse-grained". (b) The factor of rock texture that depends upon the absolute size (fineness or coarseness) of the distinct particles composing the rock. Also, the factor of rock texture that is due to the arrangement or trend of constituent particles, such as a linea tion or stratification; e.g. the "magnetic grain" in the crusta structure of a region.

grain [geomorph] (a) The broad, linear arrangement of the to pographic features (such as mountain ranges and valleys) o underlying geologic structures (such as folds and bedding) o a country or region; e.g. the arrangement of roughly paralle ridges and valleys often displayed in regions of tilted strata (b) The general direction or trend of such physical or structur al features; e.g. the *grain* of northern Scotland runs NW an SE.---Syn: *grain of the country.*

grain [gem] A unit of weight commonly used for pearls an sometimes for other gems, equal to $1/4$ *carat*, or 0.050 gram. Syn: *carat grain.*

grain [palyn] *pollen grain.*

grain [water] A unit of hardness of water, expressed in term of equivalent CaCO$_3$. A hardness of one grain per U.S. ga equals 17.1 ppm by weight as CaCO$_3$. Cf: *Clark degree.*

grain [glaciol] An individual particle in snow, ice, or glacie material, consisting of a single ice crystal or a mechanicall separate particle of ice. Cf: *snow grain; glacier grain.*

grain boundary In a polycrystalline solid, the boundary be tween two crystals. See also: *plane defect.*

grain density Specific gravity of the grains composing a sed ment or sedimentary rock.

grain diminution (a) *degradation recrystallization.* (b) *micrit zation.*

grain-foliated Said of a structure of a monomineralic meta morphic rock in which foliation is shown by parallel arrange ment of lenticular bands of different grain size, each ban being of uniform grain.

grain growth (a) A metallurgical term for the solid-stat growth, coalescence or enlargement of a crystal at the ex pense of another, occurring between unstrained or unde formed grains. (b) Applied by Bathurst (1958, p.24) to car bonate sediments, e.g. calcite mud or fibers changing to ca cite mosaic with a coarser texture; in this sense it is equiva lent to recrystallization; Folk (1965, p.16-20) objects to th usage of this term in carbonate petrology. (c) The growth of crystal, as from solution on the walls of a container, in ope pore space or in a magma chamber; *crystal growth.*

grain-micrite ratio A ratio that expresses the relative propor tion of larger particles to smaller particles in a carbonate sed mentary rock, defined as the sum of the percentages c grains (detrital grains, skeletal grains, pellets, lumps, coate grains, and mineral grains) divided by the percentage of m crite (calcareous mud or its consolidated equivalent. It ex cludes diagenetic or postdepositional features such as ce ment, vugs, fractures, vein fillings, and recrystallized areas Abbrev: GMR.

grain plane A quarryman's term for a plane of parting in a me tamorphic rock, e.g. slate, that is perpendicular to the flov cleavage (Nevin, 1949, p.162).

grain shape *particle shape.*

grain size (a) *particle size.* (b) *granularity.*

grainstone A term used by Dunham (1962) for a mud-fre (less than 1% of material with diameters less than 20 m crons), grain-supported, carbonate sedimentary rock. It ma be current-laid or formed by mud being washed out from pre viously deposited sediment, or it may result from mud bein bypassed while locally produced particles accumulated. C backstone; mudstone.

grain-supported A term used by Dunham (1962) to describe a sedimentary carbonate rock with little or no muddy matrix and whose sand-size particles are so abundant that they are in three-dimensional contact and able to support one another. Cf: mud-supported.

gralmandite A garnet intermediate in chemical composition between grossular and almandine (almandite).

gram-calorie calorie.

gramenite nontronite.

Grampian Dalradian.

granat (a) A term used in Ireland for quartzose grit. It is presumably an obsolete variant of "granite" (Arkell & Tomkeieff, 1953, p.50). (b) An obsolete form of "garnet". Etymol: German Granat, "garnet".

Grand Canyon A provincial series of the Proterozoic in Arizona.

grand gour A gour or rimstone dam or pool on the scale of a meter. Cf: microgour.

grandidierite A mineral: $(Mg,Fe)Al_3(BO_4)(SiO_4)O$.

grandite A garnet intermediate in chemical composition between grossular and andradite.

granide A syn. of granitic rock, proposed by Johannsen 1939, p.253) for light-colored medium- to coarse-grained quartz- and feldspar-bearing plutonic rocks also containing biotite or hornblende.

granilite An obsolete term formerly applied to a crystalline igneous rock having more than three components.

graniphyric Said of the texture of a porphyritic igneous rock having a microcrystalline groundmass (Cross, et al, 1906, p.704). Cf: granophyric.

granite [drill] Any igneous or metamorphic rock occurring below the sedimentary sequence in a particular area.

granite [seism] In early seismologic work, any rock in which velocity of the compressional wave is about 5.5-6.2 km/sec.

granite [petrology] (a) A plutonic rock in which quartz constitutes 10 to 50 percent of the felsic components and in which the alkali feldspar/total feldspar ratio is generally restricted to the range of 65 to 90 percent. Rocks in this range of composition are scarce in nature, and sentiment has been growing to expand the definition to include rocks designated as adamellite or quartz monzonite, which are abundant in the U.S. Streckeisen (1967) recommends that granite include plutonic rocks in which quartz constitutes 20 to 60 percent of the felsic components and in which the alkali feldspar/total feldspar ratio is between 35 and 90 percent. He also suggests two subdivisions of the granite group, one in which alkali feldspar predominates over plagioclase and one in which quartz, alkali feldspar, and plagioclase are present in nearly equal amounts. The origin of granite is in dispute, with some petrologists regarding it as igneous, having crystallized from a magma, and others considering it as the product of intense metamorphism of pre-existing rocks. (b) Broadly applied, any holocrystalline, quartz-bearing plutonic rock. Syn: granitic rock.----Etymol: Latin granum, "grain".

granite dome A term introduced by Davis (1933) for what was later (Davis, 1938) termed a desert dome because the feature was not always developed across granites.

granite gneiss (a) A gneiss derived from a sedimentary or igneous rock and having a granite mineralogy. (b) A metamorphosed granite.

granitelle An obsolete term formerly applied to a compound of quartz and feldspar; originally used as a syn. of two-mica granite.

granitello An obsolete term formerly applied to fine-grained granite.

granite-pebble conglomerate A term used by Krumbein & Sloss (1963, p.164) for arkosic conglomerate.

granite porphyry A hypabyssal rock differing from a quartz porphyry by the occasional presence of mica, amphibole, or pyroxene phenocrysts in a medium- to fine-grained groundmass.

granite series A sequence of products that evolved continuously during crustal fusion, earlier products tending to be deep-seated, syntectonic, and granodioritic and later products tending to be shallower, late syntectonic, or post-tectonic, and more potassic (Turner & Verhoogen, 1960, p.388).

granite tectonics The study of the structural features, such as foliation, lineation, and faults, of plutonic rocks and the reconstruction of the movements that created them.

granite wash A driller's term for material eroded from outcrops of granitic rocks and redeposited to form a rock having approximately the same major mineral constituents as the original rock (Taylor & Reno, 1948, p.164); e.g. an arkose consisting of granitic detritus in a kaolinitic matrix. Cf: basic wash.

granitic (a) Said of the holocrystalline texture of nonporphyritic igneous rocks in which all the constituents, apparently the product of continuous crystallization, are anhedral and of approximately the same size. See also: granular texture. (b) Pertaining to or composed of granite.----Syn: granitoid; eugranitic.

granitic layer A syn. of sial, so named for its supposed petrologic composition. Cf: basaltic layer.

granitic rock A term loosely applied to any light-colored coarse-grained plutonic rock containing quartz as an essential component, along with feldspar and mafic minerals. Syn: granite [petrology]; granitoid; granide.

granitification granitization.

granitine A crystalline rock containing any three minerals other than those of a granite.

granitite A granite that contains biotite but no muscovite or other ferromagnesian mineral. The term, now obsolete, was mainly used by French petrologists; it has also been used for a granite rich in oligoclase or containing hornblende or accessory minerals (Johannsen, 1939, p.254). Cf: biotitite.

granitization An essentially metamorphic process or group of processes by which a solid rock is converted or transformed to a granitic rock by the entry and exit of material and without passing through a magmatic stage. Some authors include in this term all granitic rocks formed from sediments by any process, regardless of the amount of melting or any evidence of movement. The precise mechanism, frequency, and magnitude of the processes are still in dispute. See also: transformism. Cf: magmatism; transfusion. Syn: transformation; granitisation; granitification.

granitizer transformist.

granitogene Said of a sediment composed of granitic fragments.

granitoid n. A granitic rock. --adj. A syn. of granitic.

granitotrachytic ophitic.

granoblastic (a) Pertaining to a homeoblastic type of texture in a nonschistose metamorphic rock upon which recrystallization formed essentially equidimensional crystals with normally well sutured boundaries (Harker, 1939). See also: sutured. (b) Pertaining to a secondary texture due to diagenetic change either by crystallization or recrystallization in the solid state, in which the grains are of equal size (Pettijohn, 1949). This usage is discouraged on the basis that diagenesis is not a metamorphic process.

granodiorite A group of coarse-grained plutonic rocks intermediate in composition between quartz diorite and quartz monzonite (U.S. usage), containing quartz, plagioclase (oligoclase or andesine), and potassium feldspar, with biotite, hornblende, or, more rarely, pyroxene, as the mafic components; also, any member of that group; the approximate intrusive equivalent of rhyodacite. The ratio of plagioclase to total feldspar is at least two to one but less than nine to ten. With less alkali feldspar it grades into quartz diorite, and with more alkali feldspar, into granite or quartz monzonite. The term first appeared in print in 1893 in a paper by Lindgren and was applied to all rocks intermediate in composition between granite and diorite. The term has the connotation that the rock is a diorite with granitic characteristics, i.e. with quartz and a certain amount of orthoclase (Johannsen, 1939, p.254).

granodolerite A dolerite that contains quartz and orthoclase, usually as micropegmatitic interstitial filling.

granofels A field name for a medium- to coarse-grained granoblastic metamorphic rock with little or no foliation or lineation (Goldsmith, 1959). Cf: *fels*.

granogabbro A rare granodiorite that contains more than 50 percent basic plagioclase.

granolite A plutonic rock characterized by granitic texture rather than porphyritic.

granomerite A holocrystalline rock that contains no kryptomerous groundmass.

granophyre (a) A porphyritic extrusive rock characterized by a micrographic holocrystalline groundmass. Syn: *pegmatophyre*. (b) A porphyritic rock characterized by a crystalline-granular groundmass.---Adj: *granophyric*. Cf: *felsophyre; vitrophyre.*

granophyric (a) As defined by Rosenbusch, a term applied to the texture of a porphyritic igneous rock in which the phenocrysts and groundmass mutually penetrate each other, having crystallized simultaneously; of or pertaining to a *granophyre* (Cross et al, 1906, p.703). Syn: *micropegmatitic; pegmatophyric*. Cf: *graphophyric; graphiphyric*. (b) As defined by Vogelsang, a term applied to a porphyritic igneous rock having a crystalline- granular groundmass (Johannsen, 1939, p.214). Cf: *graniphyric*.

granoschistose Pertaining to a structure of a monomineralic metamorphic rock produced by the parallel elongation of grains of a mineral which is normally equidimensional or nearly so.

granosphaerite A spherulite composed of radially or concentrically arranged grains.

granotubule A *pedotubule* composed essentially of skeleton grains without plasma (Brewer, 1964, p.239).

grantsite A dark olive-green to greenish-black mineral: Na_4-$Ca_xV_{2x}^{+4}V_{12-2x}^{+5}O_{32}.8H_2O$.

granular [geol] (a) Said of a rock (sedimentary, igneous, or metamorphic) consisting of grains or granules representing mineral particles of any kind; esp. said of a rock (such as sandstone) made up of small, hard, somewhat rounded particles. (b) Said of a structure or texture (of a rock, ore, or soil) consisting of or appearing to consist of grains, esp. grains of approximately equal size. (c) Pertaining to a grain or granule.

granular [paleont] Covered with very small grains or having numerous small protuberances; e.g. "granular pattern" of ornamentation on the walls of spores and pollen grains, or "granular hyaline wall" representing a perforate and lamellar part of a foraminiferal test composed of minute, equidistant, and variously oriented grains of calcite and seen between crossed nicols as a multitude of tiny flecks of color (TIP, 1964, pt.C, p.60).

granular cementation Chemical deposition of material from solution onto a free surface between detrital grains of a sediment, resulting in outward growth of crystalline material adhering to that surface (Bathurst, 1958, p.14); e.g. growth of calcite in the pores of an unconsolidated sand. Cf: *rim cementation.*

granular chert A compact, homogeneous, hard to soft chert (insoluble residue) composed of distinguishable and relatively uniform-sized grains, granules, or druses, characterized by an uneven or rough fracture surface and by a dull to glimmering luster (Ireland et al, 1947, p.1486); it may appear saccharoidal. See also: *granulated chert*. Cf: *smooth chert; chalky chert*. Syn: *crystalline chert.*

granular disintegration A type of mechanical weathering consisting of grain-by-grain breakdown of rock masses composed of discrete, and usually coarse, mineral crystals that separate from one another along their natural contacts to produce a coarse sand or gravel, each grain having much the same shape and size as in the original rock. It develops esp. in coarse-grained rocks (such as granite, gneiss, sandstone, and conglomerate) occurring in regions of great temperature extremes (such as deserts). Syn: *mineral disintegration; granular exfoliation.*

granular exfoliation *granular disintegration.*

granular ice Ice made of small crystals of irregular form but having somewhat rounded forms like sand particles.

granularity The quality, state, or property of being granular; specif. one of the component factors of the texture of a crystalline (esp. igneous) rock, representing the appearance or effect due to the actual size or magnitude of the constituent particles. Friedman (1965, p.647) used the term to refer to the "size and mutual relations of crystals" in a sedimentary rock such as an evaporite, a chemically deposited cement, or a recrystallized limestone or dolomite. Syn: *grain size.*

granular snow Relatively coarse, uniform-grained snow, resulting from prolonged equitemperature or melt-freeze metamorphism, with most grains in the shape of spheres. Syn: *snow grains*. Cf: *spring snow.*

granular structure A type of soil structure in which the peds are spheroids of polyhedrons that have little or no accommodation to surrounding peds, are relatively nonporous, and range in size from less than 1.0mm to more than 10.0mm. Cf: *crumb structure.*

granular texture A rock texture due to the aggregation of mineral grains of approximately equal size. The term may be applied to a sedimentary or metamorphic rock, but is esp. used to describe an equigranular, holocrystalline igneous rock whose particles range in diameter from 0.05 to 10 mm. See also: *granitic.*

granulated chert A type of *granular chert* composed of rough, irregular grains or granules of chert tightly or loosely held together in small irregular masses or fragments (Hendricks, 1952, p.12 & 18).

granulation The act or process of being formed into grains, granules, or other small particles; specif. the crushing of rock under such conditions that no visible openings result. Also, the state or condition of being granulated.

granule [paleont] A minute, more or less spherical skeletal element situated on the surface of asterozoan ossicles, generally in a pit or distributed in covering skin.

granule [sed] (a) A term proposed by Wentworth (1922, p.380-381) for a rock fragment larger than a very coarse sand grain and smaller than a pebble, having a diameter in the range of 2-4 mm (1/12 to 1/6 in., or -1 to -2 phi units, or size between that of the head of a small wooden match and that of a small pea) being somewhat rounded or otherwise modified by abrasion in the course of transport. The term *very fine pebble* has been used as a synonym. (b) A little grain or small particle, such as one of a number of the generally round or oval, nonclastic (precipitated), internally structureless grains of glauconite of other iron silicate in iron formation; pseudo-oolith.

granule gravel An unconsolidated deposit consisting mainly of granules.

granule ripple A large *wind ripple* consisting in part of granule-sized particles. Syn: *deflation ripple.*

granule texture A texture of iron formation in which precipitated or nonclastic granules are separated by a fine-grained matrix.

granulite [ign] A muscovite-bearing granite, esp. in the French literature.

granulite [meta] (a) A metamorphic rock consisting of even-sized, interlocking mineral grains (granoblastic texture) less than 10% of which have any preferred orientation. (b) A relatively coarse, granular rock formed at the high pressures and temperatures of the granulite facies which may exhibit a rude gneissic structure due to the parallelism of flat lenses of quartz and/or feldspars. The texture is typically granuloblastic.

granulite [sed] A sedimentary rock composed of sand-sized aggregates of constructional (nonclastic) origin, simulating in texture an arenite of clastic origin; e.g. a rock formed of lapilli or of oolitic grains. The term was introduced by Grabau (1911, p.1007). Syn: *granulyte*.

granulite facies A group of gneissic (and probably polymetamorphic) rocks formed by deep-seated regional, dynamothermal metamorphism at high temperatures (above 650°C) and high pressures (3000-12,000 bars) marked by a distinctive fabric (granoblastic) and mineral composition, and found in Precambrian shield areas (Winkler, 1967).

granulitic [meta] (a) Pertaining to a *granoblastic* texture having xenoblastic crystal development. (b) Pertaining to a structure resulting from the production of granular fragments in a rock by crushing.----Use of this term requires precise definition of its meaning.

granulitic [ign] (a) Said of the texture of a granular, holocrystalline igneous rock in which most or all of the components are xenomorphic. When Michel-Lévy (1874, 1889) first used the term, he considered *panidiomorphic-granular* as a synonym since Rosenbusch has applied that term to rocks with xenomorphic components. (b) A term applied by Judd (1886) to the texture of basaltic or doleritic rocks in which discrete crystals of augite and/or olivine occupy the interstices of a network of plagioclase laths (Johannsen, 1939, p.215). The term *intergranular* is a synonym and now preferentially used. (c) Of, pertaining to, or composed of *granulite* [ign].

granulitic [sed] Said of the rock structure resulting from the production of flattened or granular fragments in a rock by crushing (Stokes & Varnes, 1955, p.66).

granulitization In regional metamorphism, reduction of the components of a solid rock such as a gneiss to grains. The extreme result of the process is the development of mylonite.

granuloblastic Pertaining to a texture of a metamorphic rock, consisting of a mosaic of xenoblasts with smooth intergranular boundaries or straight edges approximating a polygonal arrangement. It is common in rocks of the granulite facies and in monomineralic rocks (Joplin, 1968). See also: *granulitic* [meta].

granulometric facies An interpretative term introduced by Rivière (1952) for semilogarithmic cumulative curves representing grain-size analyses of sediments. The facies is subdivided as "linear", "parabolic", "logarithmic", and "hyperbolic" according to the shape of the curve. "The use of the term facies to describe such statistical representations of one single property of a sedimentary rock seems inadmissible" (Teichert, 1958, p.2726).

granulometry The measurement of grains, esp. of grain sizes.

granulose [paleont] Having a surface roughened with granules; e.g. having very small grains on the tests of certain foraminifers or on the epitheca or tabulae in some corals.

granulose [met] Pertaining to the structure that is typical of granulite and that is due to the presence of granular minerals, e.g. quartz, feldspars, garnet, pyroxene, in alternating streaks and bands developed on a megascopic or microscopic scale. No typical foliation is developed due to the absence of lamellar or prismatic minerals.

granulyte *granulite* [sed].

grape formation *botryoid*.

grapestone A term used by Illing (1954) for a cluster of small calcareous pellets or other grains commonly of sand size and stuck together by incipient cementation shortly after deposition. The cluster has a lumpy outer surface that resembles a bunch of grapes. Grapestones occur in modern carbonate environments, such as on the Bahama Banks. See also: *bahamite*.

grapevine drainage pattern *trellis drainage pattern*.

graphic Said of the texture, of an igneous rock, that results from the regular intergrowth of quartz and feldspar crystals. The quartz commonly occupies triangular areas, producing the effect of cuneiform writing on a background of feldspar. Intergrowths of other minerals are less common. See also: *graphic intergrowth*. Syn: *runic*.

graphic granite A pegmatite characterized by graphic intergrowths of quartz and alkali feldspar. Syn: *Hebraic granite; runite*. See also: *pegmatite*.

graphic intergrowth An intergrowth of crystals, commonly feldspar and quartz, that produces a type of poikilitic texture in which the larger crystals have a fairly regular geometric outline and orientation, resembling cuneiform writing. See also: *graphic*.

graphic log A graphic record on which the formations penetrated in drilling a well or borehole are drawn to a uniform vertical scale; e.g. a *strip log*. It usually indicates the points at which oil, gas, or water was found, and the lengths of casing used; also, conventional colors and symbols are usually added in order to abbreviate the record.

graphic tellurium An old name for *sylvanite* occurring in monoclinic crystals that are arranged in more or less regular lines, having a fanciful resemblance to writing (such as to runic characters).

graphiocome A sponge spicule (hexaster) with fine brush-like terminations on the rays. Syn: *graphiohexaster*.

graphiphyre A rock having a granophyric groundmass in which the constituents are of microscopic size (Cross et al, 1906, p.704). Cf: *graphophyre*. Adj: *graphiphyric*.

graphite A hexagonal mineral, representing a naturally occurring crystalline form of carbon dimorphous with diamond. It is opaque, lustrous, very soft, greasy to the touch, and iron black to steel gray in color; it occurs as crystals or as flakes, scales, laminae, or grains in veins or bedded masses or as disseminations in metamorphic rocks. Graphite conducts electricity and heat, and is used in lead pencils, paints, and crucibles, as a lubricant and an electrode, and as a moderator in nuclear reactors. Syn: *plumbago; black lead*.

graphitic Pertaining to, containing, derived from, or resembling graphite; e.g. "graphitic rock".

graphitite A variety of shungite or type of graphitic rock that does not give the so-called nitric-acid reaction (Tomkeieff, 1954, p.52).

graphitization The formation of graphite-like material from organic compounds.

graphitoid n. (a) A variety of shungite that will burn in the Bunsen flame. It may represent merely impure graphite. (b) A term applied to meteoritic graphite.———adj. Resembling graphite.

graphocite The end product of coal metamorphism, comparable to *meta-anthracite* and composed mainly of graphitic carbon (ASTM, 1970, p.185).

graphoglypt A trace fossil consisting of a presumable worm trail appearing as a relief on the undersurface of flysch beds (mostly sandstones) and having an ornamental (meandering, spiral, net-like) pattern related to a highly organized foraging behavior; e.g. *Paleodictyon*. It was interpreted by Fuchs (1895) as a string of spawn of gastropods. Cf: *rhabdoglyph; vermiglyph*.

graphophyre A rock having a granophyric groundmass in which the constituents are of megascopic size (Cross et al, 1906, p.704). Cf: *graphiphyre*. Adj: *graphophyric*.

graptolite Any colonial marine organism belonging to the class Graptolithina, characterized by a chitinous cup- or tube-shaped exoskeleton, arranged with other individuals along one or more branches (stipes) to form a colony (rhabdosome). Graptolites commonly occur in black shales (graptolitic facies). They have been variously allied with cephalopods, coelenterates, and bryozoans, but current evidence seems to indicate pteropod affinities (TIP, 1955, pt.V, p.20). Their known stratigraphic range is Middle Cambrian to Carboniferous. Adj: *graptolithine*.

graptolitic facies A term applied to a *geosynclinal facies* containing an abundance of graptolites.

graptoloid Any graptolite belonging to the order Graptoloidea, characterized by a planktonic or epiplanktonic mode of life and by a colony consisting of a few branches with only one kind of theca, the autotheca. Their stratigraphic range is Lower Ordovician to Lower Devonian.

grass opal An *opal phytolith* derived from a grass.

grat A term used in the Alps for a small, lateral *arête*. Etymol: German *Grat*, "ridge".

graticule [cart] The network of lines representing meridians of longitude and parallels of latitude on a map or chart and upon which the map or chart was drawn. Not to be confused with *grid*.

graticule [geophys] A template divided into blocks or cells that is used to graphically integrate a geophysical quantity such as gravity. It is used in computing terrain corrections and gravitational or magnetic attractions of irregular bodies. Syn: *grating*.

graticule [optics] An accessory to an optical instrument such as a microscope to aid in measurement of the object under study; it is a thin glass disk bearing a scale which is superimposed upon the object.

grating [geophys] *graticule* [geophys].

grating [cryst] (a) In optical spectroscopy, equidistant and parallel lines that are used in producing spectra by diffraction. Syn: *diffraction grating*. (b) The grate-like pattern of lines observed in some serpentinized hornblende crystals, resulting from the occurrence of the initial alteration along cleavage cracks.

gratonite A rhombohedral mineral: $Pb_9As_4S_{15}$.

graupel A soft, usually spherical snow crystal which has been completely enveloped by frozen water droplets. Etymol: German *Graupel*, "sleet, soft hail". Syn: *pellet snow; soft hail*.

grauwacke Var. of *graywacke*. Etymol: German *Grauwacke*.

gravel (a) An unconsolidated, natural accumulation of rounded rock fragments resulting from erosion, consisting predominantly of particles larger than sand (diameter greater than 2 mm, or 1/12 in.), such as boulders, cobbles, pebbles, granules, or any combination of these fragments; the unconsolidated equivalent of conglomerate. In Great Britain, the range of 2-10 mm has been used. Cf: *rubble; pebble*. (b) A popularly used term for a loose accumulation of rock fragments, such as a detrital sediment associated esp. with streams or beaches, composed predominantly of more or less rounded pebbles and small stones, and mixed with sand that may compose 50-70% of the total mass. (c) A soil term for rock or mineral particles having a diameter in the range of 2-20 mm (Jacks et al, 1960, p.14); in this usage, the term is equivalent to *pebbles*. The term has also been used in Great Britain for such particles having a diameter in the range of 2-50 mm. In the U.S., the term is used for rounded rock or mineral soil particles having a diameter in the range of 2-75 mm (1/6 to 3 in.); formerly the term applied to fragments having diameters ranging from 1 to 2 mm. See also: *fine gravel*. (d) An engineering term for rounded fragments having a diameter in the range of 4.76 mm (retained on U.S. standard sieve no.4) to 76 mm (3 in.). See also: *fine gravel; coarse gravel*. (e) A stratum of gravel. (f) An obsolete term for sand. (g) *volcanic gravel*.

gravel bank A natural mound or exposed face of gravel, esp. such a place from which gravel is dug. Cf: *gravel pit*.

gravel deposit In economic geology, an alluvial deposit consisting mainly of gravel but commonly including sand and clay. It is used as a construction material.

gravel desert *reg*.

gravelly layer A thin layer of gravel that separates surficial

material from the underlying beds in a periglacial region characterized by solifluction.

gravelly mud An unconsolidated sediment containing 5-30% gravel and having a ratio of sand to mud (silt + clay) less than 1:1 (Folk, 1954, p.346).

gravelly sand (a) An unconsolidated sediment containing 5-30% gravel and having a ratio of sand to mud (silt + clay) greater than 9:1 (Folk, 1954, p.346). (b) An unconsolidated sediment containing more particles of sand size than of gravel size, more than 10% gravel, and less than 10% of all other finer sizes (Wentworth, 1922, p.390).

gravelly soil A soil whose fragments are rounded to angular, not much flattened, and have an upper size limit of three inches diameter.

gravel mound Any mound of gravel; restricted by Muller (1947, p.217) to a low *frost mound* of sand and gravel formed by hydrostatic pressure of ground water. Cf: *frost blister*.

gravel packing The placing of gravel or coarse sand opposite an oil-producing sand to prevent or retard the movement of loose sand grains (along with the oil) into the oil well and to reduce resistance offered to the flow of fluid into the well. It is usually forced through perforations under pressure.

gravel piedmont A term used by Hobbs (1912, p.214) for a feature now known as a *bajada*.

gravel pipe A *sand pipe* filled predominantly with gravel.

gravel pit A pit or other surface working from which gravel is obtained. Cf: *gravel bank*.

gravel rampart A *rampart* of loosely compacted reef rubble built along the seaward edge of a reef. Syn: *gravel ridge*.

gravel ridge *gravel rampart*.

gravelstone (a) A rounded rock fragment or constituent of gravel. (b) Consolidated gravel; a conglomerate.

gravel train A *valley train* composed chiefly of gravel.

gravimeter (a) A weighing device of sufficient sensitivity to register variations in the weight of a constant mass when the mass is moved from place to place and thereby subjected to the influence of gravity at those places. Gravimeters measure differences in the intensity of gravity between an initial station at which the value of gravity is known or assumed, and other points for which values of gravity are required. Syn: *gravity meter*. (b) *gravitometer*.

gravimetric (a) Pertaining or relating to measurement by weight, e.g. *gravimetric analysis*. (b) Pertaining to measurements of variations of the gravitational field.

gravimetric analysis Quantitative chemical analysis in which the different substances of a compound are measured by weight.

gravimetry The measurement of gravity or gravitational acceleration, especially as used in geophysics, applied geophysics and geodesy.

graviplanation A term used by Coleman (1952, p.455) for the process whereby lunar material is transported and deposited under the "morphological influence" of the Moon's gravitational force.

gravitation The mutual attraction between two masses. See also: *law of universal gravitation*.

gravitational acceleration The mutual acceleration between any two masses resulting from gravitational attraction.

gravitational constant The constant G in the law of universal gravitation: its value is $6.670 \pm 0.005 \times 10^{-8}$ cm^3/gm sec (NASA SP-7012, 1964, p.5).

gravitational differentiation Magmatic differentiation involving the settling out, by gravity, of the earlier formed, denser crystals, or the settling of a heavier, still liquid portion of the magma.

gravitational field A region associated with any mass distribution which gives rise to gravitational attraction forces. See also: *gravity field*.

gravitational gliding *gravitational sliding*.

gravitational intensity The measure of gravitational force, expressed as a vector quantity, exerted on a unit mass at a particular point.

gravitational method gravity prospecting.

gravitational separation The stratification of oil, gas, and water in a reservoir rock according to their relative gravities; the segregation of water from oil and of heavier hydrocarbons from lighter hydrocarbons by the force of gravity, either in the producing horizon or by gravity separators after production. See also: gravity separation.

gravitational sliding Downward movement of rock masses on slopes by the force of gravity, e.g. along a thrust-fault plane. See also: gravity tectonics. Syn: gravity sliding; gravity gliding; gravitational gliding; gliding; écoulement; sliding.

gravitational tide equilibrium tide.

gravitational water free water.

gravitational wave A hypothetical wave which travels at the speed of light and by means of which the gravitational attraction effect is propagated.

gravitometer An instrument that measures the specific gravity of a solid, liquid, or gas. Partial syn: gravimeter.

gravity (a) The resultant effect upon any body of matter in the universe of the inverse square law attraction between it and all other matter lying within the frame of reference and of any centrifugal force which may act on the body because of its motion in any orbit. (b) The resultant force on any body of matter at or near the Earth's surface due to the attraction by the Earth and to its rotation about its axis. (c) The force exerted by the Earth and by its rotation on unit mass or the acceleration imparted to a freely falling body in the absence of frictional forces.

gravity anomaly The difference between the observed value of gravity at a point and the theoretically calculated value is based on a simple gravity model, usually modified in accordance with some generalized hypothesis of subsurface density variation as related to surface topography.

gravity compaction Compaction of sediment resulting from pressure of overburden.

gravity corer An oceanographic corer that penetrates the ocean floor solely by its own weight. It isn't as efficient as a piston corer. There are several varieties, including the Phleger corer and the free corer.

gravity correction gravity reduction.

gravity dam A dam so proportioned that it will resist overturning and sliding forces by its own weight; e.g. Grand Coulee Dam on the Columbia River and Aswan Dam on the Nile.

gravity equipotential surface equipotential surface.

gravity erosion mass erosion.

gravity fault normal fault.

gravity field A term used instead of gravitational field when other influences are also involved, such as centrifugal force.

gravity flow Movement of glacier ice as a result of the inclination of the slope on which the glacier rests; glacier flow. See also: extrusion flow.

gravity fold A fold that is generically related to isostatic movements.

gravity formula A formula expressing normal gravity on the surface of a specified reference ellipsoid in terms of latitude.

gravity gliding gravitational sliding.

gravity gradient The partial derivative of the acceleration of gravity with respect to distance in a particular direction, for which purpose the acceleration of gravity in considered as a scalar.

gravity ground water The water that would be withdrawn from a body of rock or soil by the influence of gravity should the zone of saturation and capillary fringe be moved downward entirely below that body, remaining there for a specific length of time, no water being lost or received by the body except through the force of gravity (Meinzer, 1923, p.27). Cf: free water.

gravity meter gravimeter.

gravity orogenesis A concept proposed by Bucher (1956) for mountain building that results entirely from gravitational stresses. Others believe that such forces may account for folding and buckling but not for entire mountains. Cf: sedimentary tectonics.

gravity prospecting The determination of specific-gravity differences of rock masses by mapping the force of gravity of an area, using a gravimeter. Syn: gravitational method.

gravity reduction A modification of the theoretical value of gravity to determine an anomaly in accordance with some hypothesis such as the free-air, Bouguer, or isostatic anomalies. Syn: gravity correction.

gravity separation Separation of mineral particles, with the aid of water or air, according to the differences in their specific gravities. See also: gravitational separation.

gravity sliding gravitational sliding.

gravity slope The upper, relatively steep slope of a hillside, commonly lying at the angle of repose of the material eroded from it; it is steeper than the wash slope below. Term introduced by Meyerhoff (1940). Cf: constant slope. Syn: steilwand; böschung.

gravity solution A solution used to separate the different mineral particles of rock by exploiting their specific-gravity differences; e.g. a solution of mercuric iodide in potassium iodide, having a maximum specific gravity of 3.19.

gravity spring A spring issuing from the point where the water table and the land surface intersect; an outcrop of the water table.

gravity tectonics Tectonics in which the dominant propelling mechanism is down-slope gliding under the influence of gravity. In the northern Apennines of Italy gravity tectonics seems to dominate; in the Helvetic Alps downslope gliding by gravity may have aided in the forward travel of the nappes; in North America, emplacement of near-surface rocks along the Heart Mountain thrust was probably by downslope gliding by gravity. In general, however, the extent of structures produced mainly by gravity remains controversial; probably all gravity movements were triggered by deeper-seated crustal forces, and probably many structures produced by dominantly deep-seated forces were modified to some extent by gravity. See also: gravitational sliding.

gravity unit In prospecting, one tenth of a milligal. Abbrev: G unit.

gravity water (a) free water. (b) Water delivered in canals or pipelines by gravity instead of by pumping, as for irrigation or a public water supply.

gravity wave A wave whose propagation velocity is controlled mainly by gravity, and whose wavelength is 1.7 cm or more. Cf: capillary wave.

gravity wind katabatic wind.

gray antimony (a) stibnite. (b) jamesonite.

grayband Sandstone used for sidewalks; flagstone.

graybody (a) A nonselective radiator whose constant spectral emissivity is less than unity (Nicodemus, 1971). (b) A radiating surface such that while its radiation has the same spectral-energy distribution, its emissive power is less than that of a blackbody at the same temperature; and such that, while not black its absorptivity is nonselective. Cf: whitebody.

Gray-Brown Podzolic soil An old term for any of a group of zonal soils that is transitional between Podzol and Brown Forest soil.

gray cobalt (a) smaltite. (b) cobaltite.

gray copper ore (a) tetrahedrite. (b) tennantite. Syn: gray copper.

Gray Desert soil Sierozem.

gray durain Durain that is low in hydrogen and volatiles; it contains a small amount of microspores and some fusain. Cf: black durain.

gray earth Sierozem.

gray hematite *specularite.*

gray ice A type of *young ice* (10-15 cm thick) that is less elastic than nilas; it breaks on swell and usually rafts under pressure. Cf: *gray-white ice.*

grayite A yellow powdery mineral: (Th,Pb,Ca)PO$_4$.H$_2$O.

gray manganese ore (a) *manganite* [mineral]. (b) *pyrolusite.*

gray mud A type of *mud* [marine geol] that is intermediate in composition between globigerina ooze and red clay.

gray scale A monochrome strip of continuous tones ranging from white to black with intermediate tones of gray, used to determine the density of a color photograph. Cf: *step wedge.*

graystone A dense gray-green rock, resembling basalt and composed of feldspar and augite (Thrush, 1968, p.508).

graywacke An old rock name that has been variously defined but is now generally applied to a dark (usually gray or greenish gray, sometimes black) and very hard, tough, and firmly indurated, coarse-grained sandstone that has a subconchoidal fracture and consists of poorly sorted and extremely angular to subangular grains of quartz and feldspar with an abundant variety of small, dark rock and mineral fragments embedded in a preponderant and compact, partly metamorphosed clayey matrix having the general composition of slate and containing an abundance of very fine-grained micaceous (illite and sericite) and chloritic minerals; e.g. the Jackfork Sandstone (Mississippian) in Oklahoma, parts of the Franciscan Formation (Mesozoic) in western California, and certain Ordovician rocks in the Taconic region of New York and Vermont. This description is similar to Naumann's (1858, p.663) definition of the type graywacke, the Tanner Graywacke (Upper Devonian and Lower Carboniferous) of the Harz Mountains, Germany: a predominantly gray rock containing angular or subrounded quartz grains and small fragments of siliceous slates, phyllites, and other rocks, and in some cases feldspar grains, all bound together by a clay matrix or a chert cement that imparts a great toughness and hardness to the rock. Graywacke is very abundant within the sedimentary section (esp. in the older strata), usually occurring as a thick, extensive body with sole marks of various kinds and exhibiting massive or obscure stratification in the thicker units but marked graded bedding in the thinner layers. It generally requires an environment in which erosion, transportation, deposition, and burial are so rapid that complete chemical weathering does not occur, as in an orogenic belt where sediments derived from recently elevated source areas were "poured" into a geosyncline. Graywackes are typically interbedded with marine shales or slates and associated with submarine lava flows and bedded cherts; they are generally of marine origin and believed to have been deposited by submarine turbidity currents (Pettijohn, 1957, p.313). Selected modern definitions of graywacke: (1) A sandstone that may or may not be intensely indurated or metamorphosed and that is composed of more than 33% of easily destroyed minerals and rock fragments derived by rapid disintegration of basic igneous rocks, slates, and dark-colored rocks (Allen, 1936, p.22); (2) A sandstone constituting the basic equivalent of arkose, composed of slightly decomposed particles derived from the disintegration of basic igneous rocks of granular nature and their metamorphic equivalents, thus having a large content of ferromagnesian minerals, plagioclase, and magnetite (Twenhofel, 1939, p.289); (3) A sandstone composed of angular quartz (and chert) grains and abundant metamorphic rock fragments, with little or no cement and feldspar, and containing more than 12-17% micas and chlorite (either in the clay matrix or as metamorphic rock fragments) (Krynine, 1948); (4) A sandstone with more than 25% metamorphic rock fragments and coarse micas, and less than 10% feldspars and igneous rock fragments, and with any degree of clay content, sorting, or rounding (Folk, 1954); (5) A deeply buried wacke (more than 10% clayey matrix) derived from any source and varying widely in composition, having been accumulated in a rapidly subsiding marine basin or geosyncline (Gilbert, 1954, p.293-297); (6) A sandstone characterized by a dominant detrital clay matrix (15-75%) and no chemical or mineral cement, having a variable quartz content (generally less than 75%), and containing a varied assemblage (at least 25%) of unstable materials consisting of feldspar grains (at least 5%) and sand-sized rock fragments (at least 10%) (Pettijohn, 1957); (7) An indurated sandstone with more than 15% matrix of chlorite and sericite, more than 10% unstable fine-grained rock fragments, and more than 5% feldspar (McBride, 1962a); and (8) A sandstone containing 30-40% quartz, 10-50% feldspar, and 5-10% rock fragments and detrital chert, and having more than 15-20% clay matrix (Krumbein & Sloss, 1963, p.171-172). The first recorded use of the term was by Lasius (1789, p.132-152) who referred to "Grauewacke" as a German miner's term for barren country rock of certain ore veins in the Harz Mountains, and who described the rock as a gray or dark quartz "breccia" with mica flakes and fragments of chert or sandstone in a clay cement (see Dott, 1964). The term "greywacke" was probably first used in English by Jameson (1808) who adopted it from the teachings of Werner and applied it to coarse- and fine-grained deposits of fragments of Primitive rocks, including all the upper Precambrian and lower Paleozoic sedimentary rocks (other than limestones) constituting a part of the Transition series (or what were then regarded as the lowest members of the Secondary strata), i.e. the hard conglomerates, sandstones, siltstones, and shaly and slaty sandstones. Early usage was wide and vague: "geologists differ much respecting what is, and what is not, Grey Wacce" (Mawe, 1818, p.92), and "it has already been amply shown that this word should cease to be used in geological nomenclature, and I shall in the following pages give further proofs that it is mineralogically worthless" (Murchison, 1839). In view of the diversity of usage, the term "graywacke" should not be used formally without either a specific definition or a reference to a readily available published definition. Folk (1968, p.125) advocates discarding the term for any precise petrographic usage, and relegating it to nonquantitative field usage (like that of "trap rock") for a very hard, dark, clayey, impure sandstone "that you can't tell much about in the field". Etymol: German *Grauwacke,* "gray stone", probably so named because the original graywackes resembled partly weathered basaltic residues (wackes). See also: *wacke.* Cf: *arkose; subgraywacke.* Syn: *greywacke; grauwacke; apogrit.*

graywacke slate An old rock name, now obsolete, for a variety of graywacke in which "the grains are so minute as to be scarcely perceptible by the naked eye" (Humble, 1843, p.112); e.g. the hard, micaceous shaly and slaty mudstones of the Transition series.

graywether Var. of *greywether.* Also spelled: *gray weather.*

gray-white ice A type of *young ice* (15-30 cm thick) that is more likely to be ridged than rafted under pressure. Cf: *gray ice.*

Gray Wooded soil A Canadian term for a *Gray Podzolic soil.*

grazing The feeding of zooplankton upon phytoplankton.

grease ice A soupy layer of *new ice* formed on a water surface (esp. in the sea) by the coagulation of frazil crystals; it reflects little light, giving the sea a matte, greasy appearance. Syn: *ice slush.*

greasy A type of mineral luster that seems oily to the touch or by sight. Syn: *fatty.*

greasy quartz A type of *milky quartz* with a greasy luster.

great circle A curve formed on the surface of a sphere by the intersection of any plane that passes through the center of the sphere; specif. a circle on the Earth's surface, the plane of which passes through the center of the Earth and an arc of which constitutes the shortest distance between any two terrestrial points. It is the largest possible circle that can be inscribed on a given sphere. On the Earth, the equator is a great circle, and each meridian is half of a great circle. Cf:

mall circle. Syn: *orthodrome.*

great-circle belt The distribution pattern of *primary arcs* on the Earth's surface, into belts of the Earth's major tectonic activity, i.e. the *circum-Pacific belt* and the *Eurasian-Melanesian belt* (Strahler, 1963, p.403).

great-circle chart (a) A chart on a gnomonic projection, on which a great circle appears as a straight line. (b) *gnomonic projection.*

great-circle projection *gnomonic projection.*

great divide A drainage divide between major drainage systems; specif. the Great Divide (Continental Divide) of the North American continent.

great elliptic The point of intersection of the ellipsoid and a plane containing two points on the ellipsoid and its center.

Great Ice Age The Pleistocene Epoch.

great soil group A group of soils having common internal soil characteristics. Great soil groups are subdivisions of soil orders, and are themselves divided into subgroups and families. Names of great soil groups, as well as of soil orders, are usually but not invariably capitalized. See also: *soil type.*

Greco-Latin square An array of size *n* x *n*, sometimes used in design of experiments, that is composed of two *Latin squares* superimposed such that each resulting array location has a unique two-letter identification. Syn: *Graeco-Latin square.*

green algae A group of algae corresponding to the phylum *Chlorophyta*, that owes its grassy green color to the dominance of chlorophyll pigmentation. Such algae occur in a great variety of forms, from a unicellular type to a complex halloid type. Cf: *brown algae; blue-green algae; red algae; yellow-green algae.*

greenalite An earthy- or pale-green mineral: $(Fe^{+2}, Fe^{+3})_{5-6}$ $Si_4O_{10}(OH)_8$. It occurs in small ellipsoidal granules in cherty rock associated with the iron ores of the Mesabi district, Minn. Greenalite resembles glauconite in appearance, but contains no potassium.

greenalite rock A dull, dark-green rock, uniformly fine-grained with conchoidal fracture, containing grains of greenalite in a matrix of chert, carbonate minerals, and ferruginous amphiboles (Van Hise & Leith, 1911, p.165 & 474).

green beryl A term applied to the light-green or pale-green gem variety of *beryl*, as distinguished from the full-green or richly green-colored emerald and the light blue-green aquamarine.

green chalcedony (a) Chalcedony that has been artificially colored green. (b) *chrysoprase.*

green earth Any of various naturally occurring silicates (esp. of iron) used chiefly as bases for green basic dyes and for green pigments; specif. glauconite and celadonite. Syn: *terre verte; terra verde.*

greenhouse effect The almost perfect transparency of the Earth's lower atmosphere to incoming solar radiation, and its partial opacity to outgoing terrestrial radiation. It is due to the fact that the absorption bands of water vapor, ozone, and carbon dioxide are more prominent in the relatively long wavelengths of terrestrial radiation than they are in the relatively short wavelengths of solar radiation.

green iron ore *dufrenite.*

green john A green variety of fluorite.

greenlandite *columbite.*

Greenland spar *cryolite.*

green lead ore *pyromorphite.*

Green Mountains disturbance A name used by Schuchert (1924) for a supposed time of deformation at the end of the Cambrian, on the dubious assumption that all geological periods were closed by at least minor orogenic events. Evidence for this disturbance is unconvincing, even in the type area of the Green Mountains of Vermont, and the term should be discarded.

green mud A type of *mud* [marine geol] whose greenish color is due to the presence of chlorite or glauconite minerals.

greenockite A yellow or orange hexagonal mineral: CdS. It is dimorphous with hawleyite and usually occurs as an earthy incrustation or coating on sphalerite and other zinc ores. Syn: *cadmium blende; cadmium ocher; xanthochroite.*

greenovite A red, pinkish, or rose-colored variety of sphene containing manganese.

greensand (a) A sand having a greenish color, such as due to attached algae along a lake beach; specif. an unconsolidated marine sediment consisting largely of dark greenish grains of glauconite often mingled with clay or sand (quartz may form the dominant constituent), found between the low-water mark and the inner mud line. The term is loosely applied to any glauconitic sediment. Syn: *glauconitic sand.* (b) A sandstone consisting of greensand that is often little or not at all cemented, having a greenish color when unweathered (but an orange or yellow color when weathered), and forming prominent deposits in Cretaceous and Eocene beds (as in the Coastal Plain areas of New Jersey and Delaware); specif. either or both of the Greensands (Lower and Upper) of the Cretaceous System in England, whether containing glauconite or not. Syn: *glauconitic sandstone.*---Also spelled: *green sand.*

greensand marl A marl containing sand-size grains of glauconite.

greenschist A schistose metamorphic rock whose green color is due to the abundance of chlorite, epidote, or actinolite present in it. Cf: *greenstone.*

greenschist facies A term originated by Eskola (1939) for schistose rocks containing an abundance of green minerals, e.g. chlorite, epidote, or actinolite, which are produced by regional and dislocation metamorphism at low and intermediate temperatures (300°-500°C) and at low to moderate, hydrostatic pressures (3000-8000 bars) (Turner and Verhoogen, 1960, p.534). See also: *glaucophane schist facies.*

green snow A general name for green-tinted snow colored by a growth of green microscopic algae, such as the species of *Stichococcus.* Cf: *red snow.*

greenstone [mineral] (a) *nephrite.* (b) An informal name for a greenish gemstone, such as fuchsite or chiastolite.

greenstone [mining] Freshly quarried stone containing natural moisture called *quarry water.*

greenstone [ign] In Scotland, any intrusion of igneous rock in the coal measures.

greenstone [met] (a) An old field term applied to any compact dark-green altered basic to ultrabasic igneous rock (e.g. spilite, basalt, gabbro, diabase, serpentinite) owing its color to the presence of chlorite, hornblende, and epidote and commonly found in folded mountain ranges.

greenstone [sed] A compact, nonoolitic, relatively pure chamosite mudstone interbedded with oolitic ironstone in the Lower Jurassic of Great Britain.

green vitriol *melanterite.*

Greenwich meridian The astronomic meridian that passes through the original site of the Royal Astronomical Observatory at Greenwich, near London, Eng. Its adoption as the worldwide reference standard, or *prime meridian*, for the Earth was approved almost unanimously at an International Meridian Conference in Washington, D.C., in 1884. Cf: *national meridian.*

greet stone A term used in Yorkshire, Eng., for a coarse-grained or gritty sandstone.

gregaritic Said of the texture of a porphyritic igneous rock in which independently oriented grains (esp. of augite) in the groundmass occur in clusters. Cf: *synneusis; cumulophyric; glomeroporphyritic.*

greigite A dark mineral with spinel-like structure: Fe_3S_4. Syn: *melnikovite.*

greisen A granitic rock composed chiefly of quartz, mica, and topaz, with accessory tourmaline, fluorite, rutile, cassiterite, and wolframite. See also: *greisenization.*

greisenization A process of hydrothermal alteration in which feldspar and muscovite are converted to an aggregate of quartz, topaz, tourmaline, and lepidolite (i.e., *greisen*) by the action of water vapor containing fluorite. Syn: *greisenisation; greisening.*

grenatite (a) *staurolite.* (b) *leucite.*

Grenville A provincial series of the Precambrian of lanada and New York.

Grenville orogeny A name that is widely used for a major plutonic, metamorphic, and probably deformational event during the Precambrian, dated radiometrically as between 880 and 1000 m.y. ago which affected a broad province along the southeastern border of the Canadian Shield. Originally, the name Grenville was used for a metasedimentary series in the southern part of the province, and the name *Laurentian* was used for the associated plutonic rocks. Pertinent objections have been raised (Osborne, 1956; Gilluly, 1966) to use of Grenville for the orogeny, the province, and for its northwestern structural "front", but these uses will be continued until generally acceptable alternatives are proposed.

grenz A syn. of *recurrence horizon.* Etymol: German "Grenzhorizont", recurrence horizon.

greywacke Var. of *graywacke.*

greywether A popular term for a *sarsen* on the English chalk downs, so named from its fancied resemblance to a sheep lying down on a distant hillside. Syn: *graywether; gray weather.*

grid (a) An orderly network composed of two sets of uniformly spaced parallel lines usually intersecting at right angles and forming squares, superimposed on a map, chart, or aerial photograph to permit identification of ground locations by means of a system of coordinates and to facilitate computation of direction and distance to other points. The term is frequently used to designate a plane-rectangular coordinate system superimposed on a map projection, and usually carries the name of the projection; e.g. "Lambert grid". Not to be confused with *graticule.* (b) A systematic array of points or lines; e.g. a rectangular pattern of pits or boreholes used in alluvial sampling.

grid azimuth The angle at a given point in the plane of a rectangular coordinate system between the central meridian or a line parallel to it, and a straight line to the azimuth point. Cf: *gisement.*

gridiron twinning *cross-hatched twinning.*

grid line One of the lines used to establish a grid.

grid meridian A line through a point parallel to the central meridian or Y axis of a system of plane-rectangular coordinates. Cf: *gisement.*

grid method A method of plotting detail from oblique photographs by superimposing a perspective of a map grid on a photograph and transferring the detail by eye (using the corresponding lines of the map grid and its perspective as placement guides).

grid north The northerly or zero direction indicated by a meridional line of a rectangular map grid. It is coincident with true north only along the meridian of origin.

grief stem A *kelly.* Also spelled: *griefstem.* Syn: *grief joint.*

griffel schiefer A phyllite which cleaves into tiny stick- or pencil-like fragments.

griffithite An iron-rich clay mineral of the montmorillonite group; specif. a variety of saponite containing ferrous iron. It was formerly regarded as identical with nontronite.

grike A solutional, vertical fissure developed along a joint, separating *limestone pavement* into *clints.* Also spelled *gryke.* Cf: *corridor.*

grinder A spherical or discus-shaped stone rotated by the force of helical water currents in a stream pothole, the rotation producing a deepening of the pothole.

grinding [geomorph] The process of erosion by which rock fragments are worn down, crushed, sharpened, or polished through the frictional effect of continued contact and pressure by larger fragments.

grinding [glac geol] Abrasion by rock fragments embedded i a glacier and dragged along the bedrock floor; it produce gouges and grooves, and chips out fragments of bedrock. C *glacial scour.*

griotte A French quarryman's term for a marble or fine grained limestone of red color, often variegated with sma dashes of purple and spots or streaks of white or brown. It in cludes goniatite shells and is often used as an ornaments stone. Etymol: French, "morello cherry".

griphite A mineral: $(Na,Al,Ca,Fe)_6Mn_4(PO_4)_5(OH)_4$. Its crys tal structure is related to that of garnet.

griquaite A coarse-grained garnet- and diopside-bearing hypa byssal rock that may or may not contain olivine or phlogopite It occurs as nodular xenoliths in kimberlite pipes and dikes; garnet-bearing ariegite.

grit (a) A coarse-grained sandstone, esp. one composed c angular particles; e.g. a breccia composed of particles varyin in diameter from 2 mm to 4 mm (Woodford, 1925, p.183). (b A sand or sandstone made up of angular grains that may b coarse or fine. The term has been applied to any sedimentar rock that looks or feels gritty on account of the angularity c the grains. (c) *gritstone.* (d) A sandstone composed of parti cles of conspicuously unequal sizes (including small pebble or gravel). (e) A sandstone with a calcareous cement. Th term has been applied incorrectly to any nonquartzose roc resembling a grit; e.g. *pea grit* or a calcareous grit. (f) small particle of a stone or rock; esp. a hard, angular granul of sand. Also, an abrasive composed of such granules. (g The structure or "grain" of a stone that adapts it for grindin or sharpening; the hold of a grinding substance. Also, the siz of abrasive particles, usually expressed as their mesh. (h) A obsolete term for sand or gravel, and for earth or soil.---Th term is vague and has been applied widely with many differer connotations. It was first used as a provincial name in En gland for coarse-grained sandstone, esp. a hard, siliceous coarse sandstone used for millstones and grindstones. Later the term was applied to a sharp-grained sandstone regardles of the particle size. Allen (1936, p.22) proposed to restrict th term to a coarse-grained sandstone composed of angular par ticles varying in diameter from 0.5 mm to 1 mm. Etymol: Ol English *greot,* "gravel, sand".

gritrock *gritstone.*

gritstone A hard, coarse-grained, siliceous sandstone; esp one used for millstone and grindstones. Syn: *grit; gritrock.*

gritty (a) Said of the feel of a soil or of a loose or cemente sediment containing enough angular particles of sand to im part a roughness to the touch. The actual quantity of sand i such a soil is usually small. (b) Containing or resembling san or grit. Syn: *arenose.*

grivation The angular difference in direction between gri north and magnetic north at any point, measured east or wes from grid north. It is used esp. in aerial navigation.

groin A low, narrow, rigid jetty constructed of timber, stone concrete, or steel, usually extending roughly perpendicular t the shoreline, designed to protect the shore from erosion b currents, tides, or waves, or to trap sand and littoral drift fo the purpose of building up or making a beach. It may be per meable or impermeable. Syn: *groyne.*

gronlandite A hypersthene hornblendite containing more horn blende than hypersthene.

groove [glac geol] *glacial groove.*

groove [sed] A long, straight, and narrow depression pro duced on a sedimentary surface (such as mud or shale) by simultaneous and rectilinear advance of objects propelled b a continuous current, and often preserved as a *groove cast;* has a uniform depth and cross section, and is larger an wider than a *striation* but smaller than a *channel.* See also *drag mark; slide mark.*

groove [fault] One of a series of parallel scratches develope along a fault surface. A groove is a larger structure than a *striation*. Cf: *slickensides; mullion structure; slip-scratch.*

groove-and-spur structure *spur-and-groove structure.*

groove cast A term used by Shrock (1948, p.162-163) for a rounded or sharp-crested rectilinear ridge, a few millimeters high and many centimeters in length and width, produced on the underside of a sandstone bed by the filling of a *groove* on the surface of an underlying mudstone. This structure was called a *drag mark* by Kuenen (1957, p.244) who considered "groove cast" as a general term including drag marks and slide marks. Cf: *striation cast; mud furrow.* See also: *ruffled groove cast.* Syn: *proglyph.*

grooved upland An upland surface largely unaffected by feeble cirque-cutting that has left extensive undissected remnants of the preglacial surface (Hobbs, 1911a, p.30). Cf: *fretted upland.* Syn: *channeled upland.*

groove lake A glacial lake occupying a *glacial groove.*

groove spine One of a cluster or row of short, blunt, generally recumbent spines bordering ambulacral grooves in many asteroids.

grooving (a) The formation of grooves on a rock surface. (b) A groove, or a set of grooves.

grorudite A hypabyssal rock composed of phenocrysts of microcline or microcline-perthite, acmite, and less kataphorite in a tinguaitic groundmass of microcline or microperthite, acmite, and abundant quartz; an acmite-rich sodic granite.

grospydite An igneous rock containing garnets (esp. grossular), plagioclase, pyroxene, and, at high pressures, kyanite, rather than spinel or olivine. Its name is derived from the initial letters of *grossular, pyroxene,* and *disthene* (kyanite).

gross calorific value A *calorific value* calculated on the assumption that the water in the products is completely condensed. Cf: *net calorific value.*

gross heat of combustion *gross calorific value.*

grossular The calcium-aluminum end member of the garnet group, usually characterized by a green color: $Ca_3Al_2(SiO_4)_3$. It may be colorless, yellow, orange, brown, rose, or red, and it often occurs in contact-metamorphosed impure limestones. The principal variety is essonite. Syn: *grossularite; grossularia; gooseberry stone.*

grossularite *grossular.*

grothite *sphene.*

grotto A small cave, or one of the rooms or chambers of a cave system, developed in limestone by solution.

ground [geog] (a) The surface or upper part of the Earth. (b) Land, particularly a region or area.

ground [elect] In electrical prospecting, the voltage level of the ground; also, the reference voltage level of an electrical system or instrument.

ground acceleration The acceleration of an Earth particle, usually as measured by an *accelerometer.*

ground air Air in the ground, principally in the zone of aeration, but including any bubbles trapped in the zone of saturation. Cf: *subsurface air; soil atmosphere; included gas.*

ground avalanche A snow avalanche that glides over a rock or earth surface; a *wet-snow avalanche.*

ground control The marking of survey, triangulation, or other key points or system of points on the Earth's surface so that they may be recognized in aerial photographs.

ground current *Earth current.*

ground data A general term for geophysical, geochemical, geologic, geographic, pedologic, hydrologic, oceanographic or other data collected on or near the surface of the Earth, in contrast to the information collected by airborne or satellite remote sensing systems. Nonrecommended syn: *ground truth.*

grounded hummock A hummocked formation of *grounded ice,* appearing singly or in a line or chain of grounded hummocks.

grounded ice Floating ice that is aground in shallow water. Cf: *stranded ice; grounded hummock.*

ground frost [permafrost] (a) Any frozen soil including permafrost; a deprecated syn. of *frozen ground.* (b) *ground ice.*

ground frost [meteorol] An occurrence of below-freezing temperature on the surface of the ground, while the air temperature remains above the freezing point.

ground ice [permafrost] All ice, of whatever origin or age, found below the surface of the ground, esp. a lens, sheet, wedge, seam, or irregular mass of clear nonglacial ice enclosed in permanently or seasonally frozen ground, often at considerable depth. Syn: *fossil ice; subsurface ice; subsoil ice; subterranean ice; underground ice; stone ice; ground frost; glacier (in Alaska).*

ground ice [ice] (a) A deprecated syn. of *anchor ice.* (b) Glacier ice, sea ice, or lake ice that has been covered with soil (ADTIC, 1955, p.37). (c) Ice formed on the ground by freezing of rain or snow, or by compaction of a snow layer.

ground-ice layer A layer of ground ice; a syn. of *ice layer.*

ground-ice mound A *frost mound* containing bodies of ice, esp. a *pingo* whose bodies of ice are segregated.

ground-ice wedge *ice wedge.*

grounding The temporary dropping and lodgement of sedimentary particles carried in saltation, typically in sandbars, natural levees, or gravel beds (McGee, 1908, p.199).

groundmass [ign] The interstitial material of a porphyritic igneous rock; it is relatively more fine-grained than the phenocrysts and may be glassy or both. Cf: *mesostasis.* Syn: *matrix [ign].*

groundmass [sed] A term sometimes used for the *matrix* of a sedimentary rock.

ground moraine [glac geol] The rock debris dragged along in and beneath a glacier or ice sheet; also, this material after it has been deposited or released from the ice during ablation, to form an extensive, fairly even thin layer of till, having a gently rolling surface. Syn: *bottom moraine; subglacial moraine.*

ground-moraine shoreline An irregular shoreline formed where masses of glacial drift abut against the sea as a result of submergence.

ground motion A general term for all seismic motion, including ground acceleration, displacement, stress and strain. See also: *strong motion.*

ground nadir The point on the ground or at sea-level datum that is vertically beneath the perspective center of the camera lens. See also: *map nadir.* Syn: *plumb point.*

ground noise In exploration seismology, a seismic disturbance that is not caused by the shot. Cf: *ground roll.*

ground plane The assumed horizontal plane passing through the ground nadir of a camera station.

ground pressure The vertical pressure exerted on underlying rocks by the weight of superincumbent rocks and rock materials or by the diastrophic forces created by movements in the rocks forming the Earth's crust; *geostatic pressure.*

ground roll In exploration seismology, the seismic surface wave caused by the shot. It is of low frequency and of low velocity. Cf: *ground noise.*

ground slope *valley-side slope.*

ground survey A survey made by ground methods, as distinguished from an aerial survey.

ground swell A long and high ocean *swell.*

ground truth A nonrecommended syn. of *ground data.* It is a misleading term, since it implies that the truth may be found only on the ground; the whole truth is preferred.

ground water (a) That part of the subsurface water that is the zone of saturation, including underground streams. See also: *phreatic water.* Syn: *plerotic water.* (b) Loosely, all *subsurface water* (excluding internal water) as distinct from surface water.---- Also spelled: groundwater; ground-water. Syn: *subterranean water; underground water.*

ground-water artery A roughly tubular body of permeable material surrounded by impermeable or less permeable material and saturated with water confined under artesian pressure.

"The term is especially applicable to deposits of gravel along ancient stream channels that have become buried in less permeable alluvial material under alluvial fans" (Meinzer, 1923, p.42).

ground-water barrier A natural or artificial obstacle, such as a dike or fault gouge, to the lateral movement of ground water, not in the sense of a confining bed. It is characterized by a marked difference in the level of the ground water on opposite sides. Syn: *barrier* [grd wat]; *hydrologic barrier; ground-water dam.* Cf: *ground-water cascade; interrupted water table.*

ground-water basin (a) A subsurface structure having the character of a basin with respect to the collection, retention, and outflow of water. (b) An aquifer or system of aquifers, whether or not basin shaped, that has reasonably well defined boundaries and more or less definite areas of recharge and discharge. Cf: *basin* [water]; *artesian basin.*

ground-water budget A numerical account, the *ground-water equation,* of the recharge, discharge, and changes in storage of an aquifer, part of an aquifer, or system of aquifers. Syn: *ground-water inventory.*

ground-water cascade The near-vertical or vertical flow of ground water over a *ground-water barrier.* Cf: *interrupted water table.*

ground-water cement A secondary concentration of calcium carbonate, usually in the desert, resulting from evaporation of ground water at the surface or in shallow soil; a type of *water-table rock.* Syn: *water-table cement.*

ground-water dam *ground-water barrier.*

ground-water decrement *ground-water discharge.*

ground-water discharge (a) Release of water from the zone of saturation by any means. (b) The water or the quantity of water released. Syn: *ground-water decrement; decrement; phreatic-water discharge.*

ground-water divide *divide* [grd wat].

ground-water equation (a) The equation that balances the *ground-water budget;* $R = E + S - I$, where R is rainfall, E is evaporation and transpiration loss, S *is water discharged from the area as streamflow, and I* is the recharge. (b) A mathematical statement of ground-water losses and gains in a specified area.----(Tolman, 1937, p.560).

ground-water flow (a) *ground-water movement.* (b) *ground-water runoff.*

ground-water geology The science of subsurface water, with emphasis on the geologic aspects; *hydrogeology.*

ground-water hydrology *geohydrology.*

ground-water increment *recharge.*

ground-water inventory *ground-water budget.*

ground-water lake A body of surface water that represents an exposure of the upper surface of the zone of saturation, or of the water table.

Ground-Water Laterite soil One of an intrazonal, hydromorphic group of soils having characteristic lateritic concretions or a ferruginous layer above the water table.

ground-water level (a) A syn. of *water table.* (b) The elevation of the water table or another potentiometric surface at a particular place or in a particular area, as represented by the level of water in wells or other natural or artificial openings or depressions communicating with the zone of saturation.

ground-water mining The process, deliberate or inadvertent, of extracting ground water from a source at a rate so in excess of the replenishment that the ground-water level declines persistently, threatening actual exhaustion of the supply or at least a decline of pumping levels to uneconomic depths.

ground-water mound A rounded, mound-shaped elevation in a water table or other potentiometric surface that builds up as a result of the downward percolation of water, through the zone of aeration or an overlying confining bed, into the aquifer represented by the potentiometric surface. Syn: *water-table mound.*

ground-water movement The movement, or flow, of water in the zone of saturation, whether naturally or artificially induced. Syn: *ground-water flow.*

ground-water outflow The discharge from a drainage basin, or from any area, occurring as ground water.

Ground-Water Podzol soil One of an intrazonal, hydromorphic group of soils having a prominent, light-colored leached A_2 horizon overlain by thin organic material and underlain by dark hardpan. It develops under various types of forest vegetation, in humid climates of various temperature.

ground-water province An area or region in which geology and climate combine to produce ground-water conditions consistent enough to permit useful generalizations.

ground-water recession curve The part of a stream hydrograph supposedly representing the inflow of ground-water at a decreasing rate after surface runoff to the channel has ceased. Because the base runoff to the stream may include some water that had been stored in lakes and swamps rather than in the ground, the lower recession curve cannot be assumed to represent ground water only.

ground-water recharge *recharge.*

ground-water replenishment *recharge.*

ground-water reservoir (a) *aquifer.* (b) A term used to refer to all the rocks in the zone of saturation, including those containing permanent or temporary bodies of perched ground water.----Syn: *ground-water zone; reservoir* [grd wat].

ground-water ridge (a) A linear elevation in the water table that develops beneath an influent stream. Cf: *interstream ground-water ridge.* (b) *divide* [grd wat].

ground-water runoff The *runoff* that has entered the ground, become ground water, and been discharged into a stream channel (Langbein & Iseri, 1960). Cf: *surface runoff; storm seepage; delayed runoff.* Syn: *ground-water flow.*

ground-water storage (a) The quantity of water in the zone of saturation. (b) Water available only from storage as opposed to capture.

ground-water surface *water table.*

ground-water table *water table.*

ground-water trench A trough-like depression in the water table or other potentiometric surface caused by flow of ground water into a stream, drainage ditch, or a thalweg beneath a stream.

ground-water wave A high in the water table or other potentiometric surface that moves laterally, with a wave-like motion away from a place where a substantial quantity of water has been added to the zone of saturation within a brief period. Syn: *phreatic wave.*

ground-water withdrawal The process of withdrawing ground water from a source; also, the quantity of water withdrawn. Syn: *offtake; recovery.*

ground-water zone *ground-water reservoir.*

ground wave A seismic wave whose path of propagation is through both the material beneath the ocean floor and through the water.

group (a) A major rock-stratigraphic unit next higher in rank than *formation,* consisting wholly or partly of two or more (commonly two to five) contiguous or associated formations having significant lithologic features in common (ACSN, 1961, art.9). A group name customarily combines a geographic name with the word "group", and no lithologic designation is included (ACSN, 1961, art.10d). Abbrev: gr. See also: *subgroup; supergroup.* (b) A term applied in reconnaissance work to an informally recognized succession of strata too thick or inclusive to be considered a formation or to a stratigraphic unit that appears to be divisible into formations but has not yet been so divided. The term "formation" is recommended for this usage and "when future work demonstrates that the sequence can be divided into formally named formations, the unit can be raised to group status, the same name being retained for the group" (Cohee, 1962, p.4). See also: *analytic group.* (c) A general term for an assemblage or consecutive sequence of related layers of rock, such as of igneous rocks or of sedimentary beds. (d) A term proposed at the 2nd International Geological Congress in Bologna in 1881 as the time-stratigraphic equivalent of an era, and subsequently used quite widely for the rocks formed during an era. This usage is not recommended, the synonymous term *erathem* having been formally adopted (ISST, 1961, p.28). (e) An obsolete term for a time-stratigraphic unit representing a local or provincial subdivision of a system (usually less than a standard series, or the equivalent of "stage" as that term is presently used) and containing two or more formations.

group velocity The velocity of the transport of energy with which an observable wave group (consisting of individual waves having their own *phase velocities*) is propagated through a medium. For deep-water waves, it is equal to one-half the phase velocity; for shallow-water waves, it is equal to the phase velocity. Symbol: C. Cf: *particle velocity*.

grout (a) A cementitious component of high water content, fluid enough to be poured or injected into spaces and thereby fill or seal them (such as the fissures in the foundation rock of a dam, or the interstices between fragments in a brecciated rock, or the space between the lining of a tunnel and the surrounding earth); specif. a pumpable slurry or mixture of portland cement, sand, and water forced under pressure into a borehole during oil-well drilling to seal crevices and prevent contamination of the oil by seepage or flow of ground water, to provide a protective wall around the metal casing, or to improve the strength and elastic properties of the rock. Syn: *grouting*. (b) The stony waste material, of all sizes, obtained in quarrying.

groutite A jet-black mineral: $HMnO_2$. It is polymorphous with manganite.

grovesite A mineral: $(Mn,Fe,Al)_{13}(Al,Si)_8O_{22}(OH)_{14}$.

growan (a) An English term for a granite or any coarse grit or sandstone. (b) A *grus* developed by the disintegration of a granite.----Syn: *grouan*.

growler A small fragment of massive floating ice of glacier or sea-ice origin, extending less than 1 m above sea level and smaller than a *bergy bit*.

growth axis The line formed by tips of lamellae in cusp and denticles of conodont elements and commonly emphasized by concentration of "white matter".

growth band A *growth line* on the surface of a bivalve-mollusk shell.

growth fabric Orientation of fabric elements independent of the influences of stress and resultant movement, i.e. characteristic of the manner in which the rock or crystal was formed. Syn: *growth-zone fabric*.

growth fault A fault in sedimentary rock that forms contemporaneously and continuously with deposition, so that the throw increases with depth and the strata of the downthrown side are thicker than the correlative strata of the upthrown side. Such a structure occurs in the Gulf Coast region. See also: *hinge-line fault*. Syn: *contemporaneous fault*. Less-preferred syn: *depositional fault; flexure fault; Gulf Coast-type fault; progressive fault; sedimentary fault; slump fault; synsedimentary fault*.

growth-framework porosity Primary porosity developed from organic and/or inorganic processes during the in-place growth of a carbonate rock framework (Choquette & Pray, 1970, p.246-247). Intraparticle porosity of individual organisms or particles that were clastic components of the rock are excluded, a more restrictive meaning than that of *constructional void porosity*, which includes these openings.

growth island An irregular layer or patch on a crystal face due to *spiral growth* along an internal screw dislocation.

growth lamella A concentric outgrowth of a brachiopod shell, smaller than a *frill*, deposited by margin of retractile mantle TIP, 1965, pt.H, p.145).

growth lattice The rigid, reef-building, in-situ framework of an organic reef, consisting of the skeletons of sessile organisms and excluding reef-flank and other associated fragmental deposits (MacNeil, 1954, p.390). Syn: *organic lattice*.

growth layers *growth rings*.

growth line (a) One of a series of fine to coarse ridges on the outer surface of a brachiopod shell, concentric about the beak and parallel or subparallel to the margins of the valves, and indicating the former positions of the margins when the anterior and lateral growth of the shell temporarily was in abeyance. (b) One of a usually irregularly arranged and more or less obscure series of concentric lines on the surface of a bivalve-mollusk shell, approximately parallel to the borders of the valve, and representing successive advances of the shell margin at earlier growth stages. Cf: *growth ruga*. Syn: *growth band*. (c) One of a series of lines on the surface of a cephalopod conch, denoting periodic increases in size and hence former positions of the aperture. (d) One of a series of collabrally disposed surface markings (low ridges) on the outer surface of a gastropod shell, parallel to and indicating the former positions of the outer lip. (e) An irregular marking on epitheca of rugose corallites, such as a slight ridge or depression parallel to the upper edge of the corallite, defining a former position of this margin during growth. Syn: *growth ring*.

growth ring [geochron] Layer of wood produced in a tree or woody plant during its annual growth period. It is seen in cross section as a ring or concentric rings which can be analyzed for chronologic and climatic data based on number and relative sizes. See also: *dendrochronology; dendroclimatology*. Syn: *annual growth ring; tree ring*.

growth ring [paleont] A *growth line* on a rugose coral.

growth ruga An irregular *ruga* or wrinkle on the surface of a bivalve-mollusk shell, having a similar origin to that of a *growth line* but corresponding to a more pronounced halt in growth.

growth twin A twinned crystal that developed accidentally during its formation by change in lattice orientation during growth.

growth-zone fabric *growth fabric*.

groyne Var. of *groin*.

grumous Formed of clustered, aggregated, or flocculated grains; esp. said of a secondary texture in a microcrystalline, carbonate sedimentary rock that has experienced pervasive recrystallization (such as a diagenetic dolomite), characterized by patches of coarse crystals or limy particles irregularly invading shell fragments, ooliths, and matrix, and by dark, dense, and fine-grained unrecrystallized areas that are ultimately surrounded by sparry (clear, coarse, crystalline) calcite. Syn: *clotted*.

Grumusol Formerly, a general U.S. term for a *vertisol*.

grunerite A monoclinic mineral of the amphibole group: $Fe_7Si_8O_{22}(OH)_2$. Cf: *cummingtonite*. Also spelled: *grünerite*.

grus An accumulation of waste consisting of angular, coarse-grained fragments resulting from the granular disintegration of crystalline rocks (esp. granite), generally in an arid or semiarid region. Etymol: German *Grus*, "grit, fine gravel, debris". Also spelled: *gruss; grush*. Syn: *slack; growan*.

grush Var. of *grus*.

gruss Var. of *grus*.

gryke Var. of *grike*.

gryphaeate Shaped like the shell of *Gryphaea* (a genus of fossil bivalve mollusks); i.e. with the left valve strongly convex and its dorsal part incurved, and with the right valve flat.

Gshelian Var. of *Gzhelian*.

guadalcazarite A variety of metacinnabar containing zinc.

Guadalupian North American provincial series: Lower and Upper Permian (above Leonardian, below Ochoan).

guanajuatite A bluish-gray mineral: Bi_2Se_3.

guano (a) A phosphate deposit formed by the leaching of bird excrement accumulated in arid regions, (e.g., islands of the eastern Pacific Ocean and the West Indies). It is processed for use as a fertilizer. Syn: *ornithocopros.* (b) Similar deposits of bat excrement worked for phosphate and found in caves, as in Malaya.

guard The thick, hard, cigar-shaped (fusiform or subcylindrical) calcareous structure that ensheathes the phragmocone of a belemnite and forms the rear end of the shell. Syn: *rostrum.*

guard cells Specialized epidermal cells, two of which bind the stoma, that in surface view are generally crescent-shaped with blunt ends (kidney-shaped) and function to change the size of the stomatal aperture (Esau, 1965, p.160-161). They are most common on epidermal surfaces of leaves of plants.

guard-electrode log A *focused-current log* in which "guard" electrodes focus the current in order to facilitate deeper penetration of current into the strata on either side of the hole. Syn: *guard log.*

guayaquilite A soft, pale-yellow, amorphous fossil resin with a high (15%) oxygen content, soluble in alcohol and alkalies, and found near Guayaquil, Ecuador. Its approximate formula: $C_{40}H_{26}O_6$. Syn: *guayaquillite; guyaquillite.*

gubbin A variety of ironstone; a clunch or clod with ironstone nodules.

gudmundite A silver-white to steel-gray orthorhombic mineral: $FeSbS$.

guerinite A mineral: $Ca_5H_2(AsO_4)_4.9H_2O$.

guern *khurd.*

guest A mineral introduced into and usually replacing a preexistent mineral or rock; a *metasome.* Ant: *host.*

guest element *trace element.*

guettardite A mineral: $Pb_9(Sb,As)_{16}S_{33}$.

guhr (a) A white (sometimes red or yellow), loose, earthy, water-laid deposit of a mixture of clay or ocher, occurring in the cavities of rocks. (b) *kieselguhr.*

guidebook A road log of a field trip, summarizing the geology of the particular stops; also, a guide to the minerals or fossils available in an area.

guided wave Any seismic wave that is propagated along some surface or discontinuity, e.g. a *surface wave, Stoneley wave,* or a *channel wave.*

guide fossil (a) Any fossil that has actual, potential, or supposed value in identifying the age of the strata in which it is found or in indicating the conditions under which it lived; a fossil used esp. as an index or guide in the local correlation of strata. (b) A fossil that is most characteristic of an assemblage zone, but that is not necessarily restricted to the zone or found throughout every part of it (ACSN, 1961, art. 21e).-- See also: *zonal guide fossil.* Cf: *index fossil.*

guide meridian A north-south line used for reference in surveying; specif. one of a set of auxiliary governing lines of the U.S. Public Land Survey system, projected north or south from points established on the base line or a standard parallel, usually at intervals of 24 miles east or west of the principal meridian, and on which township, section, and quarter-section corners are established.

guildite A dark chestnut-brown mineral: $CuFe(SO_4)_2$-$(OH).4H_2O$.

guilielmite A subaqueous sedimentary structure formed in mud by collapse around a fossil and characterized by small, polished slip surfaces arranged with radial or orthorhombic symmetry around the fossil (Wood, 1935).

guilleminite A canary-yellow secondary mineral: $Ba(U-O_2)_3(SeO_3)_2(OH)_4.3H_2O$.

gula A projecting, rather ornate extension of the trilete laesura of fossil megaspores. Pl: *gulae.* Cf: *apical prominence.*

gulch A term used esp. in the western U.S. for a small, narrow, deep ravine with steep sides, larger than a *gully;* esp. a short, precipitous cleft in a hillside, formed and occupied by a torrent, and containing gold (as in California).

gulf [coast] A relatively large part of an ocean or sea extending far into the land, partly enclosed by an extensive sweep of the coast, and opened to the sea through a strait; the largest of various forms of inlets of the sea. It is usually larger, more enclosed, and more deeply indented than a bay.

gulf [karst] In an area of karst, a closed depression having steep walls and a flat alluvial floor, often with a disappearing or a resurgent stream.

gulf [geomorph] A deep and narrow hollow, depression, gorge, or chasm; e.g. one of the long, narrow, precipitous, stream-worn excavations west of the Adirondack Mountains in northern New York State.

Gulf Coast-type fault *growth fault.*

gulf-cut island An island formed by the cutting backward of two parallel inlets into opposite sides of a piece of subsiding land (Powell, 1895, p.64).

Gulfian North American provincial series: Upper Cretaceous (above Comanchean, below Paleocene of Tertiary).

Gulf-type gravimeter A meter consisting of a mass suspended at the end of a spring, the latter so designed that its extension will cause the mass to rotate. By this means the linear displacement of the spring is converted into an angular deflection which is more easily measured (Wyckoff, 1941, p.13). The design also minimizes the sensitivity to seismic disturbances and the basic instrument is therefore well suited for underwater observations (Pepper, 1941, p. 34) Syn: *Hoyt gravimeter.*

gull A structure formed by mass-movement processes, consisting of widened, steeply inclined tension fissures or joints, resulting from lateral displacement of a slide mass and filled with debris derived from above. Gulls generally trend parallel to surface contours, and are usually associated with camber structure (Hollingworth, *et al,* 1944).

gullet [paleont] (a) A variably tubular invagination of the cytoplasm of various protists (such as tintinnids) that sometimes functions in the intake of food. (b) A longitudinal groove present in certain algae (such as some Cryptophyceae and Euglenophyceae).

gullet [streams] A narrow opening or depression, such as a defile or ravine; a *gully* or other channel for water.

gull hummock A conical, arched- or domed-shaped, peaty mound, formed by accretion of well-manured grasses near the nest of the great black-backed gull, on islands in the Arctic (ADTIC, 1955, p.61). Syn: *pingo.*

gully [coast] A wave-cut chasm in a cliff, or a minor channel incised in a mud flat below the high-water level (Schieferdecker, 1959, terms 1149 & 1233).

gully [geomorph] (a) A very small valley, such as a small *ravine* in a cliff face, or a long, narrow hollow or channel worn in earth or unconsolidated material (as on a hillside) by running water and through which water runs only after a rain or the melting of ice or snow; it is smaller than a *gulch.* Syn: *gulley; gullet.* (b) Any erosion channel so deep that it cannot be crossed by a wheeled vehicle or eliminated by plowing, esp. one excavated in soil on a bare slope. (c) A small, steep-sided wooded hollow.

gully erosion Erosion of soil or soft rock material by running water that forms distinct, narrow channels that are larger and deeper than rills and that usually carry water only during and immediately after heavy rains or following the melting of ice or snow. Cf: *sheet erosion; rill erosion; channel erosion.* Syn: *gullying; ravinement.*

gully gravure A term used by Bryan (1940) for the process or processes whereby the steep slopes of hills and mountains retreat by "repeated scoring or graving", each groove (gully) "so disposed as to reduce rather than emphasize inequalities" (p.92); the development of rills into gullies.

gullying *gully erosion.*

gum An organic, viscid juice extracted from, or exuded by certain trees and plants. It hardens in the air and is soluble in water.

gumbo A term used locally in the U.S. for a fine-grained, clay soil that becomes sticky, impervious, and plastic when wet.

gumbotil (a) A gray to dark-colored, leached, deoxidized clay representing the B-horizon of fully mature soils, developed from profoundly weathered clay-rich till under conditions of low relief and poor subsurface drainage (as beneath broad, flat uplands). It consists chiefly of beidellite, and may contain altered rock fragments originally mixed with the clay; it is very sticky and plastic when wet, extremely firm when dry. Term introduced by Kay (1916). Cf: *silttil; mesotil.* See also: *ferretto zone.* (b) A term used for a fossilized soil beneath a deposit of later till.

gum copal An inferior resin or amber; *copal.*

gummite A general term for yellowish, orange, reddish, or brownish secondary minerals consisting of a mixture of hydrous oxides of uranium, thorium, and lead, and occurring as alteration products of uraninite and not otherwise identified. It includes silicates, phosphates, and oxides; much of the material is probably mixtures or amorphous gels, but some consists perhaps largely of curite. Syn: *uranium ocher.*

gunite n. A mixture of portland cement, sand, and water applied by pneumatic pressure through a specially adapted hose and used as a fireproofing agent and as a sealing agent to prevent weathering of mine timbers and roadways. Etymol: *Gunite,* a trademark. Syn: *shotcrete.*----v. To apply gunite; to cement by spraying gunite.

gunkhole A nearly unnavigable shallow cove or channel containing mud, rocks, or vegetation.

gunningite A monoclinic mineral: $(Zn,Mn)(SO_4).H_2O$.

Gunnison River A provincial series of the Precambrian in Colorado.

Gunter's chain A surveyor's *chain* that is 66 feet long, consisting of a series of 100 metal links each 7.92 inches long and fastened together with rings. It served as the legal unit of length for surveys of U.S public lands, but has been superseded by steel or metal tapes graduated in chains and links. Named after Edmund Gunter (1581-1626), English mathematician and astronomer, who invented the device about 1620. Syn: *pole chain.*

Günz (a) European stage: Pleistocene (above Astian of Pliocene, below Mindel). (b) The first glacial stage of the Pleistocene Epoch in the Alps. Se also: Nebraskan; Elbe.---Etymol: Günz River, Bavaria. Adj: *Günzian.*

Günz-Mindel The term applied in the Alps to the first interglacial stage of the Pleistocene Epoch, following the Günz and preceding the Mindel glacial stages. See also: *Aftonian.*

gurhofite A snow-white variety of dolomite mineral, containing a large proportion of calcium. Syn: *gurhofian.*

gusher [grd wat] *geyser.*

gushing spring *Vauclusian spring.*

gustavite A mineral: $Pb_5Ag_3Bi_{11}S_{24}$.

gut (a) A very narrow passage or channel connecting two bodies of water; e.g. a contracted strait, or a small creek in a marsh or tidal flat, or an inlet. Also, "a channel in otherwise shallow water, generally formed by water in motion" (CERC, 1966, p.A14). (b) A tidal stream connecting two larger waterways. (c) A term used in the Virgin Islands and elsewhere for a gully, ravine, small valley, or narrow passage on land.

Gutenberg discontinuity The seismic-velocity discontinuity marking the mantle-core boundary, at which the velocities of *P* waves are reduced and *S* waves disappear. It probably reflects the change from a solid to a liquid phase. It is named after Beno Gutenberg, seismologist. Syn: *Oldham-Gutenberg discontinuity; Weichert-Gutenberg discontinuity.*

Gutenberg low-velocity zone The *low-velocity zone* of the upper mantle.

gutsevichite A mineral: $(Al,Fe)_3(PO_4,VO_4)_2(OH)_3.8H_2O$ (?).

guttation The process by which water in liquid form is exuded from an uninjured surface of a plant. Cf: *transpiration.*

gutter [ore dep] The lowest and usually richest portion of an alluvial placer. The term is used in Australia for the dry bed of a buried Tertiary river. Syn: *bottom.*

gutter [**streams**] (a) A shallow, natural channel, furrow, or gully worn by running water. (b) A shallow, steep-sided valley that drains a marshy upland; it usually marks an area where the drainage is about to be rejuvenated. (c) An artificially paved watercourse, such as a roadside ditch for carrying off excess surface water to a sewer. (d) An archaic term for a brook.

guyaquillite *guayaquillite.*

guyot A type of *seamount* that has a platform top. Etymol: Arnold Guyot, nineteenth century Swiss-American geologist. Syn: *tablemount; tableknoll.*

G wave A long-period (1-4 min) *Love wave* in the upper mantle, usually restricted to an oceanic path. Corresponding waves over continental paths are depressed and not pulselike. It is named after Gutenberg.

gymnite *deweylite.*

gymnocyst Part of the frontal wall of a cheilostomatous bryozoan, lying between the space occupied in life by the frontal membrane and the free edges of the vertical walls. It is usually most developed on the proximal side of the space occupied by the frontal membrane.

gymnosolen A finger-like or digitate form of stromatolite, splitting off into two or more upward directions from algal structures resembling a series of stacked inverted thimbles or (if large) of soup bowls similarly arranged (Pettijohn, 1957, p.222). It is produced by blue-green algae of the genus *Gymnosolen.*

gymnosperm A plant whose seeds are not enclosed in an ovary. Examples include cycads, ginkgo, pines, firs, and spruces. Such plants range from the upper Devonian. Cf: *angiosperm.*

gymnospore A naked spore, or one not developing in a sporangium. The term is not in good or current usage in palynology.

gyp A syn. of *gypsum.* Also spelled: *gyps.*

gyparenite A sandstone composed of discrete, wind-drifted particles of gypsum.

gypcrete A gypsum cement, found in some playa-lake beachrock environments in an arid climate (Fairbridge, 1968, p.555).

gyprock [drill] A driller's term for any rock in which he has trouble in drilling a well.

gyprock [sed] A sedimentary rock composed chiefly of gypsum varying from coarsely crystalline to finely granular masses; it usually displays disturbed bedding due to expansion during hydration of anhydrite. See also: *rock gypsum.*

gypsey A syn. of *bourne.* Also spelled: *gipsy; gypsy.*

gypsic Said of a soil horizon that is characterized by enrichment in calcium sulfate.

gypsification Development of, or conversion into gypsum; e.g. the hydration of anhydrite.

gypsinate Cemented with gypsum.

gypsite (a) An earthy variety of gypsum containing dirt and sand, found only in arid regions as an efflorescent deposit occurring over the ledge outcrop of gypsum or of a gypsum-bearing stratum. Syn: *gypsum earth.* (b) *gypsum.*

gypsolith A term suggested by Grabau (1924, p.298) for a gypsum rock. Syn: *gypsolyte.*

gypsum A widely distributed mineral consisting of hydrous calcium sulfate: $CaSO_4.2H_2O$. It is the commonest sulfate mineral, and is frequently associated with halite and *anhydrite* in evaporites or forming thick, extensive beds interstratified with limestones, shales, and clays (esp. in rocks of Permian and Triassic age). Gypsum is very soft (hardness of 2 on Mohs' scale), and is white or colorless when pure, but can be tinted grayish, reddish, yellowish, bluish, or brownish. It occurs massive (*alabaster*), fibrous (*satin spar*), or in monoclinic crystals (*selenite*). Gypsum is used chiefly as a soil amendment, as a retarder in portland cement, and in making plaster of Paris. Etymol: Greek *gypsos*, "chalk". Syn: *gypsite; gyp; plaster stone; plaster of Paris.*

gypsum cave (a) A cave that is formed in gypsum by solution. (b) A cave containing abundant gypsum incrustations.

gypsum earth *gypsite.*

gypsum flower *cave flower.*

gypsum plate In a polarizing microscope, a plate of clear gypsum (selenite) that gives a first-order red interference color; it is used to determine optical sign with crystals or interference figures and to determine the position of vibration-plane traces in crystal plates.

gyral *gyre.*

gyrate Winding or coiled round; convolute, like the surface of the brain.

gyre A great, closed, circular motion of water in each of the major ocean basins, centered on a subtropical high-pressure region; its movement is generated by convective flow of warm surface water poleward, by the deflective effect of the Earth's rotation, and by the effects of prevailing winds. The water within each gyre turns clockwise in the Northern Hemisphere and counterclockwise in the Southern Hemisphere. Syn: *gyral.*

gyrocompass A nonmagnetic *compass* that functions by virtue of the couples generated in a rotor when the latter's axis of rotation is displaced from parallelism with that of the Earth and that consists of a continuously driven gyroscope whose supporting ring confines the spring axis to a horizontal plane. It automatically aligns itself in the celestial meridian (thus pointing to the true north) by the Earth's rotation which causes it to assume a position parallel to the Earth's axis. The gyrocompass is used in underground and borehole surveying. Syn: *gyroscopic compass; gyrostatic compass.*

gyrocone A loosely coiled cephalopod shell (like that of *Gyroceras*) in which the successive whorls are not in contact with each other or in which only a single whorl is approximately completed. Syn: *gyroceracone.*

gyroid An isometric crystal form consisting of 24 crystal faces with indices |*hkl*| and symmetry 432. A gyroidal crystal may be right- or left-handed.

gyroidal class That crystal class in the isometric system having symmetry 432.

gyrolite A white mineral with a micaceous cleavage: $Ca_2Si_3O_7(OH)_2 \cdot H_2O$. Syn: *centrallasite.*

gyroscopic compass (a) *gyrocompass.* (b) A magnetic compass whose equilibrium is maintained by the use of gyroscopes.

gyrostatic compass *gyrocompass.*

gyttja A dark, pulpy, freshwater mud characterized by abundant organic matter that is more or less determinable and deposited or precipitated in a lake whose waters are rich in nutrients and oxygen, or in a marsh. It is an anaerobic sediment laid down under conditions varying from aerobic to anaerobic and is capable of supporting aerobic life.

Gzhelian Stage in Russia: upper Upper Carboniferous (above Moscovian, below Uralian). Syn: *Gshelian.*

H

Haalck gravimeter A gravimeter in which the change in weight of a mercury column is balanced by a gas spring (Heiland, 1940, p.124).

habit [geol] A general term for the outward appearance of a mineral or rock.

habit [cryst] The characteristic crystal form or combination of forms, including characteristic irregularities, of a mineral.

habit [ecol] The characteristic appearance of an organism.

habitat *environment* [biol].

habitat form *ecad.*

hachure n. One of a series of short lines used on a topographic map for shading and for indicating surfaces in relief (such as steepness of slopes), drawn perpendicular to the contour lines; e.g. an inward-pointing "tick" trending downslope from a depression contour. Hachures are short, broad (heavy), and close together for a steep slope, and long, narrow (light), and widely spaced for a gentle slope, and they enable minor details to be shown but do not indicate elevations above sea level. Etymol: French. Syn: *hatching; hatchure.*---v. To shade with or show by hachures.

hackly Said of a mineral or rock fracture that gives a jagged surface.

hackmanite A variety of sodalite containing a little sulfur and usually fluorescing orange or red under ultraviolet light.

hacksaw structure The irregular, *saw-toothed* or saw-shaped termination of a crystal (such as of augite) or mineral particle due to intrastratal solution. Syn: *hacksaw termination; cockscomb structure.*

hadal Pertaining to the deepest oceanic environment, specifically that of oceanic trenches, i.e., over 6.5 km in depth.

hade n. In structural geology, the angle that is the complement of the *dip*; the angle that a structural surface makes with the vertical, measured perpendicular to the strike of the structure. It is little used. Syn: *rise.* v. To incline from the vertical.----Syn: *underlay.*

Hadley cell A thermally driven unit of atmospheric circulation that extends in both directions from the equator to about 30°. Air rises at the equator, flows poleward, descends, and then flows toward the equator. It is named after G. Hadley who described it in 1735.

Hadrynian In a three-part division of Proterozoic time, the latest division, above the *Helikian*. Cf: *Aphebian.*

Haeckel's law *recapitulation theory.*

haematite Original spelling of *hematite.*

haff A long, shallow, freshwater, coastal lagoon separated from the open sea by a sandspit (*nehrung*) across a river mouth; esp. such a lagoon on the East German coast of the Baltic Sea. Pl: *haffs; haffe.* Etymol: German *Haff*, "lagoon".

hagendorfite A greenish-black mineral: (Na,Ca)(Fe, Mn)$_2$(PO$_4$)$_2$.

häggite A black monoclinic mineral: V$_2$O$_3$(OH)$_3$.

haidingerite A white or colorless mineral: CaHAsO$_4$.H$_2$O.

hail Precipitation from cumulonimbus clouds in the form of pellets of ice. The pellets vary in size and shape and may be opaque or clear or have an alternation of clear and opaque layers. Hail may occur in association with rain, thunderstorms, or snow.

hail imprint A small, shallow, circular to elliptical depression or crater-like pit formed by a hailstone falling on a soft sedimentary surface. It is generally larger, deeper, and more irregular than a *rain print*. Syn: *hail pit; hailstone imprint.*

hail pit *hail imprint.*

hailstone imprint *hail imprint.*

hair ball *lake ball.*

hair copper *chalcotrichite.*

hairpin dune A greatly elongated *parabolic dune* that has mi-

grated downwind, its horns drawn out parallel to each other, formed where a constant wind is in conflict with vegetation.

hair pyrites (a) *millerite.* (b) Capillary pyrite.

hair salt (a) *alunogen.* (b) Silky or fibrous *epsomite.*

hairstone A variety of clear crystalline quartz thickly penetrated with fibrous, thread-like, or acicular inclusions of other minerals, usually crystals of rutile or actinolite; esp. sagenitic quartz. See also: *Venus hairstone; Thetis hairstone.* Also spelled: *hair stone.* Syn: *needle stone.*

hair zeolite A group of fibrous zeolite minerals, including natrolite, mesolite, scolecite, thomsonite, and mordenite. See also: *needle zeolite.* Syn: *feather zeolite.*

haiweeite A pale- to greenish-yellow secondary mineral: Ca(UO$_2$)$_2$Si$_6$O$_{15}$.5H$_2$O.

hakite A mineral of the tetrahedrite group: (Cu,Hg)$_{12}$Sb$_3$(S, Se)$_{13}$.

haldenhang A syn. of *wash slope.* Etymol: German *Haldenhang*, "under-talus rock slope of degradation" (Penck, 1924).

half-blind valley A *blind valley* whose stream may flood and overflow when the swallow hole cannot accept all the water (Monroe, 1970).

Half-Bog soil One of an intrazonal, hydromorphic group of soils that is similar to a *Bog soil* but has an underlying gray mineral soil.

half-height width A measure of the shape of a narrow band filter. Also referred to as *half width.* Cf: *field of view.*

half island A peninsula.

half-life *radioactive half-life.*

half-life period *radioactive half-life.*

half moon A style of gem cutting that produces a stone shaped as a half circle (Shipley, 1951, p.99).

half-power points The outward angle from the optical axis at which the energy sensed by the *radiometer* will see out to 4° or 5° off the axis. Cf: *field of view.*

half section A half of a normal *section* of the U.S. Public Land Survey system, representing a piece of land containing 320 acres as nearly as possible; any two quarter sections within a section which have a common boundary. It is usually identified as the north half, south half, east half, or west half of a particular section.

half-space A region bounded by a plane uppermost surface which extends to infinity. The bounding plane usually coincides with a model of the Earth's surface.

half-tide level *mean tide level.*

half tube A remnant or trace of a *tube* visible on the roof or walls of a cave.

half-value thickness The thickness of an absorbing medium which will reduce any incident radiation to half its initial density.

half width According to Dobrin (1952) and Nettleton (1940), half the width of a simple anomaly (esp. a gravity or magnetic anomaly) at the point of half the maximum value of the anomaly. It is usually confined to symmetrical or almost symmetrical anomalies. There is some confusion about the term; half width has also been defined as the full width at half maximum value of the anomaly.

halide A mineral compound characterized by a halogen such as fluorine, chlorine, iodine, or bromine as the anion. Halite, NaCl, is an example of a halide. Syn: *halogenide.*

halilith A term suggested by Grabau (1924, p.298) for rock salt (a sedimentary rock). Syn: *halilyte.*

halite A mineral: NaCl. It is native salt, occurring in massive, granular, compact, or cubic-crystalline forms, and having a distinctive salty taste. Symbol: Hl. Syn: *common salt; rock salt.*

halitic Pertaining to halite; esp. said of a sedimentary rock containing halite as cementing material, such as "halitic sandstone".

hall *chamber.*

hälleflinta Fine-grained, compact and horny granular hornfels formed by contact metamorphism of an acid igneous rock such as rhyolite, quartz porphyry, or acid tuff resulting in a banded and/or blastoporphyritic rock. Cf: *leptite*.

hälleflintgneiss An obsolete syn. of *leptite*.

Hallian North American stage: Pleistocene (above Wheelerian, below Holocene).

hallimondite A yellow secondary mineral: $Pb_2(UO_2)(AsO_4)_2$.

halloysite (a) A name used in U.S. for a porcelain-like clay mineral: $Al_2Si_2O_5(OH)_4.2H_2O$. It is synonymous with *metahalloysite* of European authors. Halloysite is made up of minute slender and hollow tubes, observed with the electron microscope. The term has also been used to designate a nonhydrated mineral having the chemical composition of, but structurally distinct from, kaolinite. (b) A name used in Europe for a clay mineral: $Al_2Si_2O_5(OH)_4.4H_2O$. It is synonymous with *endellite* of U.S. authors, and designates a more highly hydrated mineral than metahalloysite. (c) A general term proposed by MacEwan (1947) and used by Grim (1968, p.39) for all the naturally occurring halloysite minerals (hydrated, nonhydrated, and intermediate) and for artificially prepared complexes. It is a group name used with significant self-explanatory qualifications to denote the type of halloysite (e.g. "glycerol-halloysite" and "fully hydrated halloysite") that is being described.

halmeic Said of a mineral or sediment that is derived directly from sea water. Syn: *halmyrogenic; halogenic*.

halmeic Said of a deep-sea sediment formed directly from solution or around an organic nucleus, e.g. barite, phosphorite, manganese nodules. Cf: *authigenic*. Ant: *chthonic*.

halmyrogenic *halmeic*.

halmyrolysis The geochemical reaction of sea water and sediments in an area of little or no sedimentation. Examples include modification of clay minerals, formation of glauconite from feldspars and micas, and the formation of phillipsite and palagonite from volcanic ash. Alternate spelling: halmyrosis. Cf: *diagenesis*. Syn: *submarine weathering*.

halo A circular or crescentic distribution pattern about the source or origin of a mineral, ore, mineral association, or petrographic feature. It is encountered principally in magnetic and geochemical surveys.

halocline A locally steepened, vertical gradient of salinity of sea water. Cf: *thermocline*.

halogenic *halmeic*.

halogenide *halide*.

halokinesis *salt tectonics*.

halomorphic Pertaining to an intrazonal soil whose characteristics have been strongly influenced by the presence of neutral or alkali salts or both.

halo ore *fringe ore*.

halophilic Said of an organism that prefers a saline environment. Noun: *halophile*. Cf: *haloxene*.

halophreatophyte A plant receiving its water supply from saline ground water.

halophyte A plant growing in soil or water with a high content of salts.

halosere A sere that develops in a saline environment.

halotrichite (a) A mineral: $FeAl_2(SO_4)_4.22H_2O$. It occurs in yellowish fibrous crystals. Syn: *feather alum; iron alum; mountain butter; butter rock*. (b) Any of several sulfates similar to halotrichite in construction and habit; e.g. alunogen.

haloxene Said of an organism that can tolerate saline conditions but does not prefer them. Cf: *halophilic*.

hals A British term for a pass or col. Cf: *hause*. Syn: *halse*.

halurgite A mineral: $Mg_2B_8O_{14}.5H_2O$.

hamada Var. of *hammada*.

hambergite A grayish-white or colorless mineral: Be_2BO_3OH.

hamlinite *goyazite*.

hammada An extensive, nearly level, upland desert surface that is either bare bedrock or bedrock thinly veneered by pebbles, smoothly scoured and polished and generally swept clear of sand and dust by wind action; a *rock desert* of the plateaus, esp. in the Sahara. The term is also used in other regions, as in Western Australia and the Gobi Desert. Etymol: Arabic, *hammadah*. See also: *reg; serir*. Syn: *hamada; hammadah; hammadat; hamadet; nejd*.

hammarite A reddish steel-gray mineral: $Pb_2Cu_2Bi_4S_9$ (?).

hammer A hand tool used for breaking rock, consisting of a firmly fixed head of hardened steel with a blunt surface at one end and either a sharpened point (or pick, for breaking hard rock) or a sharpened edge (or chisel, for breaking soft rock) at the other end, and set crosswise on a handle. It is the recognized symbol of the geologist, and his chief field instrument.

Hammer-Aitoff projection An equal-area map projection derived from the equatorial aspect of the Lambert azimuthal equal-area projection by doubling the horizontal distances along each parallel from the central meridian until the entire spherical surface can be represented within an ellipse whose major axis (equator) is twice the length of its minor axis (central meridian). It resembles the Mollweide projection, but all parallels (except the equator) are represented by curved lines and there is less angular distortion near the margins. The projection was introduced in 1892 by H.H. Ernst von Hammer (1858-1925), German geodesist, but is often attributed to David Aitoff (d.1933), Russian geographer, who previously introduced a similar-appearing projection based on the azimuthal equidistant projection. Incorrect syn: *Aitoff projection*.

hammock (a) *hummock* [geog]. (b) A term applied in the SE U.S. to a fertile area of deep, humus-rich soil, generally covered by hardwood vegetation and often rising slightly above a plain or swamp; esp. an island of dense, tropical undergrowth in the Florida Everglades. Syn: *hummock*.

hammock structure The intersection of two vein or fracture systems at an acute angle.

hampshirite Steatite pseudomorphous after olivine.

hamrongite A dark violet-gray fine-grained lamprophyre containing phenocrysts of black mica in a groundmass characterized by intersertal texture and composed of mica, andesine, and some quartz; a quartz kersantite.

hamulus A hook-shaped secondary deposit on the chamber floor in foraminifers of the family Endothyridae. The point of the hook is directed toward the aperture of the test. Pl: *hamuli*.

hancockite A mineral of the epidote group: $(Pb,Ca,Sr)_2(Al,Fe)_3(SiO_4)_3(OH)$.

hand lens A small magnifying glass for use in the field or in other preliminary investigations of a mineral, fossil, or rock. Syn: *pocket lens*.

hand level A small, hand-held leveling instrument in which the spirit level is so mounted that the observer can view the bubble at the same time that he sights an object through the telescope. The viewing of the bubble is accomplished by means of a prism or mirror in the telescope tube: when the cross hair bisects the bubble and the object in view, that object is on a level with the eye. The hand level is used where a high degree of precision and accuracy is not required, such as in reconnaissance surveys. See also: *Abney level; Locke level*.

hand specimen A piece of rock of a size that is convenient for megascopic study and for preserving in a study collection.

hanger *hanging wall*.

hanging adj. Situated or lying on steeply sloping ground (such as a *hanging* meadow) or on top of other ground (such as a *hanging* wall), or jutting out and downward (such as a *hanging* rock), or situated at or having a discordant junction (such as a *hanging* valley).---n. (a) A downward slope; a declivity. (b) *hanging wall*.

hanging cirque A cirque on a mountainside, excavated by a former hanging glacier and not continued in a valley. Cf: *valley-head cirque*. See also: *corrie*.

hanging drumlin A drumlin on a valley slope, consisting of subglacial debris pushed laterally into its present position by a thin overriding glacier.

hanging glacier A glacier, generally small, protruding from a basin or niche on a mountainside above a cliff or very steep slope, from which ice may break off occasionally and abruptly to form an ice avalanche. Cf: *cliff glacier.*

hanging side *hanging wall.*

hanging tributary A tributary stream or tributary glacier occupying a hanging valley.

hanging trough A glacial *hanging valley.*

hanging valley [glac geol] A glacial valley whose mouth is at a relatively high level on the steep side of a larger glacial valley. The larger valley was eroded by a trunk glacier and the smaller one by a tributary glacier, and the discordance of level of their floors, as well as their difference in size, is due to the greater erosive power of the trunk glacier. Syn: *hanging trough; hanging glacial valley; perched glacial valley.*

hanging valley [streams] (a) A tributary valley whose floor at the lower end is notably higher than the floor of the main valley in the area of junction, produced where the more rapid deepening of the main valley results in the creation of a cliff or steep slope over which a waterfall may develop. (b) A coastal valley whose lower end is notably higher than the shore to which it leads, produced where betrunking or rapid cliff recession causes the mouths of streams to "hang" along the cliff front. Syn: *valleuse.*

hanging wall The overlying side of an orebody, fault, or other structure. also, of mine workings. Syn: *hanging side; hanger.* Cf: *footwall; wall; upper plate.*

hanksite A white or yellow hexagonal mineral: $Na_{22}K(SO_4)_9(CO_3)_2Cl.$

hannayite A mineral: $Mg_3(NH_4)_2H_4(PO_4)_4.8H_2O.$ It occurs as slender yellowish crystals in guano.

hanusite A mixture of stevensite and pectolite. It was formerly regarded as a mineral: $Mg_2Si_3O_7(OH)_2.H_2O.$

haplite *aplite.*

haplome A "more correct", but "not generally accepted", spelling of *aplome* (Hey, 1962, p.446).

haplophyre A granite found in the Alps and characterized by large quartz and feldspar grains in a mortar structure (Thrush, 1968, p.526).

haplopore An unbranched pore lying normally within one thecal plate of a cystoid. If connected in pairs, haplopores are designated *diplopores.*

haplotabular archeopyle An *apical archeopyle* in a dinoflagellate cyst consisting of a single plate.

haploxylonoid Said of bisaccate pollen, in which the outline of the sacci in distal-proximal view is more or less continuous with the outline of the body, the sacci appearing more or less crescent-shaped, and the outline of the whole grain presenting a more or less smooth ellipsoidal form. Cf: *diploxylonoid.*

haplozoan Any one of a small group of supposedly free-living echinoderms belonging to the subphylum Haplozoa, comprising only two genera, and characterized by a thick calcareous skeleton composed of a few plates arranged around a median, crater-like depression. They are known only from the Middle Cambrian.

hapteron An attaching structure in some brown algae; it is usually multicellular, branched, and rootlike.

haptonema A thread- to club-like part of a cell in a coccolithophorid, located between, but more rigid than, the flagella. It can be contracted into a spiral or extended and used as an organ of attachment to substrate material.

haptotypic character A feature of spores and pollen grains that is a product of its contact with other members of the tetrad in which it was formed; e.g. the laesura and contact areas of spores.

haradaite A mineral: $SrVSi_2O_7.$

harbor (a) A small bay or a sheltered part of a sea, lake, or other large body of water, usually well protected either naturally or artificially against high waves and strong currents, and deep enough to provide safe anchorage for ships; esp. such a place in which port facilities are furnished. (b) An inlet; e.g. Pearl Harbor, Hawaii.---British spelling: *harbour.*

harbor bar A bar built across the exit to a harbor.

hardcap A term used in bauxite mining for the uppermost foot or two of a bauxite deposit. Since it is harder and tougher than the material below it, it is usually used as a roof during mining.

hard coal *anthracite.*

hardbank Unaltered kimberlite below the zone of *blue ground.*

hard ground A nonrecommended syn. of *surface of unconformity.*

hardhead A syn. of *negrohead.* Also spelled: *hard head.*

hard magnetization Magnetization that is not easily destroyed; specifically, remanent magnetization with a large coercivity. Cf: *soft magnetization; stable magnetization.*

hard mineral A mineral that is as hard as or harder than quartz, i.e. seven or higher according to Mohs' scale. CF: *soft mineral.*

hardness [mineral] The resistance of a mineral to scratching; it is a property by which minerals may be described, relative to a standard scale of ten minerals known as *Mohs' scale,* to the *technical scale* of fifteen minerals, or to any other standard.

hardness [water] A property of water causing formation of an insoluble residue when used with soap and causing formation of a scale in vessels in which water has been allowed to evaporate. It is primarily due to the presence of calcium and magnesium ions but also to ions of other alkali metals, ions of other metals (e.g. iron), and even hydrogen ion. Hardness of water is generally expressed as parts per million as $CaCO_3$ (40 ppm Ca produced a hardness of 100 ppm as $CaCO_3$); also, milligrams per liter; the combination of *carbonate hardness* and *noncarbonate hardness.* Syn: *total hardness.* Cf: *soft water.*

hardness points Small, pointed pieces of minerals of different hardness, affixed to small metal handles and used for testing the hardness of another mineral by ascertaining which point will scratch it. Minerals of hardness 6 to 10 on Mohs' scale are usually used as the points for testing gemstones.

hard ore A term used in the Lake Superior region for a compact, massive iron ore mainly composed of specular hematite and/or magnetite and containing more than 58% iron. Cf: *soft ore.*

hardpan (a) A general term for a relatively hard, impervious, and often clayey layer of soil lying at or just below the surface, produced as a result of cementation of soil particles by precipitation of relatively insoluble materials such as silica, iron oxide, calcium carbonate, and organic matter, offering exceptionally great resistance to digging or drilling, and permanently hampering root penetration and downward movement of water; e.g. a subsoil horizon of secondary accumulation of mineral matter (such as sulfates and carbonates in an arid or semiarid region, or clays and iron oxides in a humid region) leached from the topsoil. Its hardness does not change appreciably with changes in moisture content, and it does not slake or become plastic when mixed with water; it can be shattered by explosives. The term is not properly applied to hard clay layers that are not cemented, or to layers that may seem indurated but which soften when soaked in water. See also: *lime pan; iron pan; ortstein.* Cf: *duricrust; claypan; fragipan.* (b) A layer of gravel encountered in the digging of a gold placer, occurring one or two meters below the ground surface and partly cemented with limonite. (c) A term commonly applied in NW U.S. to a compact, subglacial till which must be drilled or blasted before removal. Also, a cemented layer of sand or gravel enclosed within till, or a capping of partly cemented material at the top of a waterbearing layer of sand or gravel in till (Wentworth, 1935, p.243). (d) A popular term used loosely to designate any relatively hard layer that is difficult to excavate or drill; e.g. a thin resistant layer of limestone interbedded with easily drilled soft shales. (e) A crust of "alkali" deposited in irrigated soils. (f) A compact subsurface material cemented with iron hydroxides and iron silicates. (g) *caliche*. (h) *plow sole*.----Legget (1962, p.798) suggests that the term be avoided "in view of its wide and essentially popular local use for a wide range of materials".

hard rock (a) A term used loosely for an igneous rock or a metamorphic rock, as distinguished from a sedimentary rock. (b) A rock that is relatively resistant to erosion; a firm, solid, or compact rock that does not yield easily to pressure. (c) Rock that requires drilling and blasting for its economical removal. (d) A term used loosely by drillers for a pre-Cretaceous sedimentary rock (such as a limestone, dolomite, or anhydrite, in addition to consolidated sandstone or shale) that is drilled relatively slowly and that produces samples usually readily identified as to depth interval.----Ant: *soft rock*.

hard-rock geology A colloquial term for geology of igneous and metamorphic rocks, as opposed to *soft-rock geology*.

hard-rock phosphate A term used in Florida for a hard, massive, close-textured, homogeneous, light-gray phosphate, showing irregular cavities that are usually lined with secondary mammillary incrustations of calcium phosphate. It is essentially equivalent to the term *white-bedded phosphate* that is used in Tennessee.

hard shore A shore composed of sand, gravel, cobbles, boulders, or bedrock. Ant: *soft shore*.

hard spar A name applied to corundum and andalusite.

hard water Water that does not lather readily when used with soap and that forms a scale in containers in which it has been allowed to evaporate; water with more than 60 mg/l of hardness-forming constituents, expressed as $CaCO_3$ equivalent. See also: *hardness*. Cf: *soft water*.

hardwood The wood of an angiospermous tree. Actually, such wood may be either hard or soft. Cf: *softwood*.

hardystonite A white mineral: $Ca_2ZnSi_2O_7$.

Harker diagram *variation diagram*.

harkerite A colorless mineral: $Ca_{48}Mg_{16}Al_3(BO_3)_{15}(CO_3)_{18}$-$(SiO_4)_{12}(OH,Cl)_8.3H_2O$.

Harlechian European stage: Lower Cambrian.

harlequin opal Opal with small, close-set, angular (mosaic-like) patches of play of color of similar size. Cf: *pinfire opal*.

harmomegathus The *membrane* of a pore or colpus (of a pollen grain) when it serves to accommodate, by expansion, an increase in volume of the grain, which usually results from the taking up of water. Pl: *harmomegathi*. Adj: *harmomegathic*.

harmonic analysis *Fourier analysis*.

harmonic fold A fold, the form of which is constant throughout its constituent strata. Ant: *disharmonic fold*.

harmonic folding Folding in which the strata remain parallel or concentric, without structural discordances between them, and in which there are no sudden changes in the form of the folds at depth. Ant: *disharmonic folding*.

harmonic mean The reciprocal of the arithmetic mean of the reciprocals of *n* numbers.

harmotome A zeolite mineral: $(Ba,K)_2(Al,Si)_2Si_6O_{16}.6H_2O$. It forms cruciform twin crystals. Syn: *cross-stone*.

harpolith A large, sickle-shaped igneous intrusion that was injected into previously deformed strata and subsequently deformed with the host rock by horizontal stretching or orogenic forces.

harrisite [mineral] Chalcocite pseudomorphous after galena.

harrisite [petrology] A granular igneous rock composed chiefly of olivine and a smaller amount of anorthite and characterized by *harrisitic* texture.

harrisitic Said of the texture observed in certain olivine-rich rocks (e.g. *harrisite* [petrology]) in which the olivine crystals are oriented approximately at right angles to the cumulate layering of the rock. This phenomenon is now known to occur

with other minerals and is called *crescumulate* (Wager, 196?, p.579).

harrow mark One of a group of parallel fine-grained ridges of sand, silt, and clay about 1-10 cm high and 5-50 cm apart with intervening trough-like strips of coarser sediments, occurring in stream channels and extending for distances as great as 100 m. It has been ascribed to the action of regular systems of longitudinal helical flow patterns with alternating senses of rotation (Karcz, 1967).

harstigite A mineral: $Ca_6(Mn,Mg)Be_4Si_6(O,OH)_{24}$.

hartite A white, crystalline, fossil resin found in lignites. Syn: *bombiccite; branchite; hofmannite; josen*.

Hartley gravimeter An early form of stable-type gravimeter consisting of a weight suspended from a spiral spring, a hinged lever, and a compensating spring for restoring the system to a null position.

Hartmann's law The statement that the acute angle between two sets of intersecting shear planes is bisected by the axis of greatest principal stress, and the obtuse angle by the axis of least principal stress.

Hartmann's lines A group of *slipbands*.

hartschiefer A metamorphic rock of compact, dense, cherty or felsitic texture having a banded structure in which the bands are of approximately even thickness, have rigid parallelism, and differ considerably in mineral and chemical composition. It is formed by intense dynamic metamorphism from ultramylonites and is associated with other rocks of mylonitic habit (Holmes, 1920, p.116). The term was originated by Quensel in 1916. Etymol: German.

harzburgite A peridotite composed chiefly of olivine and orthopyroxene. See also: *saxonite*.

hassock A term used in England for a soft, somewhat calcareous sandstone containing glauconite. Syn: *calkstone*.

hassock structure A variety of *convolute bedding* in which the laminae resemble tufts of grass or sedge. Syn: *hassock bedding*.

hastate Said of a leaf that is arrow-shaped, and has pointed basal lobes that point away from the petiole.

hastingsite A monoclinic mineral of the amphibole group: Na$Ca_2(Fe,Mg)_5Al_2Si_6O_{22}(OH)_2$. It generally contains a little potassium.

hastite An orthorhombic mineral: $CoSe_2$. It is dimorphous with trogtalite.

hatchettine A soft, yellow-white, mineral paraffin wax, perhaps $C_{38}H_{78}$, having a melting point of $55°-65°C$ in the natural state and $79°C$ after purification. It occurs as vein-like masses in ironstone nodules associated with coal-bearing strata (as in south Wales) or in cavities in limestone (as in France). Syn: *hatchettite; adipocire; adipocerite; mineral tallow; mountain tallow; naphthine*.

hatchettolite *betafite*.

hatching (a) The drawing of hachures on a map to give an effect of shading. See also: *cross-hatching*. (b) *hachure*.

hatchite A lead-gray triclinic mineral: $(Pb,Tl)_2AgAs_2S_5$.

hatchure Var. of *hachure*.

hatherlite An anorthoclase-bearing biotite-hornblende syenite. Syn: *leeuwfonteinite*.

hauerite A reddish-brown or brownish-black mineral: MnS_2. It occurs in octahedral or pyritohedral crystals.

haughtonite A black, iron-rich variety of biotite.

hause An English term for a pass, or a ridge connecting two higher elevations, or a narrow gorge (Whitney, 1888, p.137). Cf: *hals*. Syn: *haws*.

hausmannite A brownish-black, opaque mineral: Mn_3O_4.

haustorium A food-absorbing organ of a parasitic plant. It is often rootlike.

Hauterivian European stage: Lower Cretaceous (above Valanginian, below Barremian).

haüyne A blue feldspathoid mineral of the sodalite group: (Na,Ca)$_{4-8}(Al_6Si_6O_{24})(SO_4,S)_{1-2}$. It is related to nosean and oc urs in rounded and subangular grains embedded in various volcanic rocks. Pron: ah-ween. Also spelled: hauyne. Syn: haüynite.

haüynite A syn. of hauyne. Also spelled: hauynite.

haüynitite A plutonic or hypabyssal rock composed chiefly of haüyne and pyroxene, usually titanaugite. Small amounts of feldspathoids and sometimes plagioclase and/or olivine are present. Apatite, sphene, and opaque oxides occur as accessories. See also: haüynophyre.

haüynolith A monomineralic extrusive rock composed entirely of haüyne.

haüynophyre An extrusive rock similar in composition to a leucitophyre but containing haüyne in place of some of the leucite. Other possible components include nepheline, augite, magnetite, apatite, melilite, and mica. A partial syn. of haüynitite; some rocks are called "haüynophyre" when haüyne is a conspicuous mineral but not necessarily a major component. Syn: haüynporphyr.

haüynporphyr haüynophyre.

haüy's law law of rational indices.

haven A small bay, recess, or inlet of the sea affording anchorage and protection for ships; a harbor.

havsband A Swedish term for the outermost or seaward part of a skerry-guard, constituting bare skerries and the smallest rock islets (Stamp, 1961, p.230 & 419).

Hawaiian-type bomb A general type of volcanic bomb produced from very fluid lava, characterized by its relatively small size (from a millimeter to about ten centimeters) and elongated form. Cf: Strombolian-type bomb.

Hawaiian-type eruption A type of volcanic eruption characterized by fountaining of basaltic lava and the formation of active lava lakes. Cf: Pelée-type eruption; Strombolian-type eruption; Vulcanian-type eruption.

hawaiite [mineral] A pale-green, iron-poor gem variety of olivine from the lavas of Hawaii.

hawaiite [petrology] An andesine-bearing alkali olivine basalt typically found in the Hawaiian Islands.

hawk's-eye A transparent to translucent, colorless variety of quartz that contains minute, parallel, closely packed, bluish fibrous crystals of partly replaced crocidolite (from which a pale-blue to greenish-blue sheen is produced by reflection of light) and that resembles the eye of a hawk when cut cabochon; a blue variety of tiger's-eye. Cf: sapphire quartz. Syn: hawk-eye; falcon's-eye.

hawleyite A yellow isometric mineral: CdS. It is dimorphous with greenockite.

haws hause.

haxonite A meteorite mineral: $(Fe,Ni)_{23}C_6$.

Hayford zones A subdivision of the globe into areas, used in the calculation of topographic and isostatic reductions around a gravity station. They are named after the U.S geodesian J.F. Hayford. See also: Bullard's method.

haystack hill A less-preferred syn. of mogote, used in Puerto Rico.

haze (a) Fine dust, salt particles, smoke, or water particles (finer and more scattered than those of fog) dispersed through a part of the atmosphere, causing a lack of transparency of the air (which assumes a characteristic opalescent appearance that subdues all colors) and reducing the horizontal visibility of distant objects to more than one, but less than two, kilometers. (b) The obscuration, or a lack of transparency, of the atmosphere near the Earth's surface, caused by haze or by heat refraction (shimmering).

head [eng] (a) A body of water kept in reserve at a height, such as that used for supplying a mill or held in a reservoir or high-altitude lake. (b) The bank, dam, or wall by which such a head is contained or kept up.

head [phys sci] The upper or higher end of a geomorphic feature, as the head of a valley or the head of a slope.

head [coast] (a) A headland, usually coupled with a specific place name. (b) The inner part of a bay, creek, or other coastal feature extending farthest inland. (c) The apex of a triangular-shaped delta.

head [mass move] A term used in southern England for a thick, poorly stratified, compact mass of locally derived angular rubble mixed with sand and clay, formed by solifluction under periglacial conditions, and mantling the high ground or occurring on slopes and in valley bottoms; a congeliturbate. See also: coombe rock. Syn: rubble drift.

head [currents & tide] The part of a rip current that has slackened and widened out seaward of the breaker line. Cf: neck [currents]. Syn: rip head.

head [paleont] (a) The anterior tagma of a crustacean, consisting of the cephalon alone or comprising the cephalon and one or more anterior thoracomeres (having limbs that are modified as mouthparts) fused to it. See also: cephalothorax. (b) The anterior dorsal part of the body of a mollusk, bearing the mouth, sensory organs, and major nerve ganglia.---Cf: foot.

head [sed] algal head.

head [struc geol] The upper bend of a fold or structural terrace. Cf: foot. Syn: upper break.

head [hydraul] (a) The pressure of a fluid on a given area, at a given point caused by the height of the fluid surface above the point. Cf: fluid potential. (b) Water-level elevation in a well, or elevation to which the water of a flowing artesian well will rise in a pipe extended high enough to stop the flow. (c) When not otherwise specified, it usually refers to static head.

head [lake] The influent end of a lake, or the end of a lake opposite the outlet.

head [streams] The source, beginning, or upper part of a stream. (b) The farthest upstream point reached by a vessel; the limit of river navigation.

head [slopes] The upper part or end of a slope; a valley head.

headcut A vertical face or drop on the bed of a stream channel, occurring at a knickpoint.

head dune A dune that accumulates on the windward side of an obstacle. Cf: tail dune.

headed dike A dike which has a terminal expansion of teardrop shape.

head erosion headward erosion.

heading [navigation] The compass direction (azimuth) of the longitudinal axis of a ship or aircraft.

heading [grd wat] A horizontal tunnel, into an aquifer, that taps ground water penetrating fissures, for the purpose of supplying wells and reservoirs.

heading side footwall.

heading wall footwall.

headland [coast] (a) An irregularity of land, esp. of considerable height with a steep cliff face, jutting out from the coast into a large body of water (usually the sea or a lake); a bold promontory or a high cape. Syn: head; mull. (b) The high ground flanking a body of water, such as a cove. (c) The steep crag or cliff face of a promontory.

headland [soil] A term used in soil conservation for the source of a stream.

headland beach A narrow beach formed at the base of a cliffed headland.

headland mesa A part of a general plateau that projects into a meander loop of a large river (Lee, 1903, p.73). See also: island mesa.

head loss friction head.

headpool A pool near the head of a stream.

heads Low-grade material overlying a wash or alluvial placer.

headstream A stream that is the source or one of the sources of a larger stream or river.

headwall A steep slope at the head of a valley; esp. the rock cliff at the back of a cirque. Syn: backwall.

headwall recession The steepening and backward movement

of the headwall of a cirque, caused by alternate thawing and refreezing.

headward erosion The lengthening and cutting back upstream of a young valley or gully above the original source of its stream, effected by erosion of the upland at the valley head; it is accomplished by rainwash, gullying, spring sapping, and the slumping of material into the head of the growing valley. Syn: *head erosion; headwater erosion; retrogressive erosion.*

headwater (a) The *source* (or sources) and upper part of a stream (esp. of a large stream or river), including the upper drainage basin; a stream from this source. The term is usually used in the plural. Syn: *waterhead.* (b) The water upstream from a structure, as behind a dam.

headwater erosion *headward erosion.*

headwater opposition The position of, or relationship shown by, two valleys facing in opposite directions, each valley growing upstream by headward erosion and separated from the other by a ridge-like divide (Fenneman, 1909, p.35-36).

head wave A wave resulting from a mechanical disturbance impressed on a medium and traveling through it at a speed greater than the characteristic speed of the medium. In seismology, the wave travels along an interface which separates two media of different elastic moduli and density. Syn: *conical wave; von Schmidt wave.*

healed *crustified.*

heaped dune *star dune.*

heat Energy in transit from a higher-temperature system to a lower-temperature system. The process ends in thermal equilibrium between the two systems, at which point there is no longer any occasion to use the word "heat" (Zemansky, 1957, p.62).

heat balance (a) Equilibrium which exists on the average between the radiation received by a planet and its atmosphere from the Sun and that emitted by the planet and atmosphere. That the equilibrium does exist in the mean is demonstrated by the observed long-term constancy of the Earth's surface temperature. On the average, regions of the Earth nearer the equator than about 35° latitude receive more energy from the Sun than they are able to radiate, whereas latitudes higher than 35° receive less. The excess of heat is carried from low latitudes to higher latitudes by atmospheric and oceanic circulations and is reradiated there (Marks, 1969, p.245).

heat balance (a) Equilibrium between the radiation emitted by the Earth and its atmosphere and that which they receive from the Sun. (b) The state of equilibrium in a *heat budget* of a system.

heat budget (a) The amount of heat required to raise the water of a lake from its minimum winter temperature to its maximum summer temperature; it is usually expressed as gram calories of heat per square centimeter of lake surface. (b) The accounting for the total amount of heat received and lost by a particular system, such as a lake, a glacier, or the entire Earth during a specific period. See also: *heat balance.*

heat capacity That quantity of heat required to increase the temperature of a system or substance one degree of temperature. It is usually expressed in calories per degree centigrade (Weast, 1970, p.F-80). The average heat capacity of a system is the ratio of the heat absorbed to the temperature difference before and after the change; the instantaneous heat capacity is this ratio in the limit as the final temperature approaches the initial temperature (Zemansky, 1957, p.70). See also: *specific heat capacity.*

heat conduction The process of heat transfer through solids, from a higher-temperature to a lower-temperature region, by molecular impact without transfer of the matter itself, i.e. without convection (McIntosh, 1963, p.60). See also: *thermal*

conductivity. Syn: *thermal conduction.*

heat conductivity *thermal conductivity.*

heat content *enthalpy.*

heat crack A crack in a rock, formed by vigorous temperatur changes, and causing the rock to fall apart into two or mor pieces.

heated stone A gemstone that has been heated to change it color, such as biue zircon, or to improve its color, such a many aquamarines. Cf: *stained stone.*

heat equivalent of fusion The quantity of heat necessary t change one gram of solid to a liquid with no temperatur change (Weast, 1970). Syn: *latent heat of fusion.*

heat flow The product of the thermal conductivity of a sub stance and the thermal gradient in the direction of the flow c heat. See also: *heat flux.*

heat-flow measurement In geophysics, the measurement c the amount of heat leaving the Earth. It involves the measure ment of the *geothermal gradient* of rocks by accurate resis tance thermometers in drill holes (preferably more than 30 meters deep), and the measurement of the *thermal conduc tivity* of rocks, usually in the laboratory, on rock samples fror the drill holes. Heat flow is the product of the gradient and th conductivity. Heat-flow measurements on the ocean floor use slightly different techniques.

heat-flow unit A measurement of terrestrial heat flow equiva lent to 10^{-6} cal/cm^2/sec.

heat flux *Heat flow* per unit time.

heath peat *calluna peat.*

heat sink An area in which energy loss due to terrestrial ra diation exceeds energy gain due to insolation.

heave [soil] A predominantly upward movement of a surface caused by expansion or displacement, such as due to swellin clay, removal of overburden, seepage pressure, or frost ac tion; esp. *frost heaving.* See also: *air heave; gas heave.* Syn *heaving.*

heave [mining] A slow rising of a mine floor, esp. in a coa mine, due to its softness and the weight of the supporting pil lars. Syn: *creep [mining].* Cf: *flash; crown-in; inbreak.*

heave [struc geol] In a fault, the horizontal component of sep aration or displacement. It is an old, imprecise term. Cf *throw.* Syn: *horizontal throw.*

heaving shale An incompetent shale that runs, falls, swells, o squeezes into a drill hole, such as the shale (under consider able pressure) adjacent to a faulted trap.

heavy-bedded Said of a shale whose splitting property is inter mediate between that of a thin-bedded shale (easy to split and that of a platy or flaggy shale (hard to split) (Alling, 1945 p.753).

heavy crop A collective term used in Great Britain for th *heavy minerals* of a sedimentary rock.

heavy gold Gold occurring as large particles. Cf: *nugget.*

heavy isotope Isotopes of an element having a greater tha normal mass; e.g. deuterium.

heavy liquid In analysis of minerals, a liquid of high densit such as *bromoform* in which specific gravity tests can be made, or in which mechanically mixed minerals can be sepa rated. When a mineral grain is placed in the liquid, the liquids specific gravity is adjusted by the addition of a lighter o heavier liquid until the mineral neither rises nor sinks; the spe cific gravity of the liquid and of the mineral are then equal See also: *Klein solution; Sonstadt solution; Clerici solution Westphal balance; methylene iodide.* Syn: *specific-gravity liq uid.*

heavy metal Any of the metals that react readily with dithi zone, e.g. zinc, copper, cobalt, lead, bismuth, gold, cadmium iron, manganese, nickel, tantalum, tellurium, platinum, silver and others.

heavy mineral [petrology] A rock-forming mineral generall having a specific gravity greater than 2.8; e.g. a mafic miner al. Cf: *light mineral.*

heavy mineral [sed] A *detrital mineral* from a sedimentary rock, having a specific gravity higher than a standard (usually 2.85), and commonly forming as a minor constituent or *accessory mineral* of the rock (less than 1% in most sands); e.g. magnetite; ilmenite, zircon, rutile, kyanite, garnet, tourmaline, sphene, apatite, biotite. Cf: *light mineral*. See also: *heavy crop*.

heavy oil Crude oil that has a low Baumé gravity or A.P.I. gravity. Cf: *light oil*.

heavy spar *barite*.

heazlewoodite A mineral: Ni_3S2.

hebraic granite *graphic granite*.

Hebridean *Lewisian*.

hebronite *amblygonite*.

hecatolite Orthoclase *moonstone*.

hectare A metric unit of land area equal to 10,000 square meters, 100 ares, or 2.471 acres. Abbrev: ha.

hectorite A trioctahedral, lithium-rich clay mineral of the montmorillonite group: $(Ca/2,Na)_{0.33}(Mg_{2.67}Li_{0.33})Si_4O_{10}(OH,F)_2 \cdot nH_2O$. It represents an end member in which the replacement of aluminum by magnesium and lithium in the octahedral sheets is essentially complete.

hedenbergite A black mineral of the clinopyroxene group: $CaFeSi_2O_6$. It occurs as a skarn mineral at the contacts of limestones with granitic masses.

hedgehog stone Quartz with needle-shaped inclusions of goethite.

hedleyite A mineral: Bi_7Te_3. It is an alloy consisting of a solid solution of Bi_5 in Bi_2Te_3.

hedreocraton A stable, continental craton, including both continental shield and platform. Cf: *thalassocraton; epeirocraton*.

hedrumite A coarse-grained, light-colored porphyritic hypabyssal rock characterized by trachytoid texture and containing accessory nepheline; a pulaskite porphyry. Its name is derived from Hedrum, Norway.

hedyphane A yellowish-white mineral of the apatite group: $(Ca,Pb)_5(AsO_4)_3Cl$. It may contain barium.

heidornite A monoclinic mineral: $Na_2Ca_3B_5O_8(SO_4)_2Cl(OH)_2$.

height [geomorph] (a) A landform or area that rises to a considerable degree above the surrounding country, such as a hill or plateau. The term is often used in the plural. (b) The highest part of a ridge, plateau, or other high land.

height [geodesy] The distance between an equipotential surface through a point and a reference surface, measured along a line of force or along its tangent. Cf: *dynamic height; orthometric height*.

height [paleont] (a) The maximum distance, measured normal to the length in the plane of symmetry, between a concavoconvex or convexo-concave shell of a brachiopod and the line joining the beak and the anterior margin. Also, the *thickness* of a biconvex, plano-convex, or convexo-plane shell of a brachiopod. (b) The distance between two planes parallel to the hinge axis of a bivalve-mollusk shell and perpendicular to the plane of symmetry, and which just touch the most dorsal and ventral parts of the shell. Cf: *length*.

height [surv] The vertical distance above a datum (usually the surface of the Earth); *altitude* or *elevation* extending upward above a given level or surface.

height of capillary rise *capillary rise*.

height of instrument A surveying term used in spirit leveling for the height of the line of sight of a leveling instrument above the adopted datum, in trigonometric leveling for the height of the center of the theodolite above the ground or station mark, in stadia surveying for the height of the center of the telescope of the transit or telescopic alidade above the ground or station mark, and in differential leveling for the height of the line of sight of the telescope at the leveling instrument when the instrument is leveled. Abbrev: H.I.

height of land The highest part of a plain or plateau; specif. a drainage *divide*, or a part thereof.

heinrichite A yellow to green secondary mineral: $Ba(UO_2)_2(AsO_4)_2 \cdot 10\text{-}12H_2O$. Cf: *metaheinrichite*.

heintzite *kaliborite*.

hekistotherm A plant that can grow at low temperatures, esp. in areas where the warmest month has a mean temperature below 50°F.

helatoform Shaped like a nail; e.g. "helatoform cyrtolith" having a nail-shaped central structure.

Helderbergian North American stage: lowermost Devonian (above Upper Silurian, below Deerparkian).

held water Water retained within the soil as a liquid or a vapor.

helenite A variety of *ozocerite*.

helical flow *helicoidal flow*.

helicitic Pertaining to a metamorphic rock texture consisting of bands of inclusions which indicate original bedding or schistosity of the parent rock, and which cut through later-formed crystals of the metamorphic rock. The relict inclusions commonly occur in porphyroblasts as curved and contorted strings. The term was originally, but is no longer, confined to microscopic examination. Also spelled: *helizitic*. Cf: *poikiloblastic*.

helicoid Forming or arranged in a spiral; specif. said of a gastropod shell having the form of a flat coil or flattened spiral, or said of an ammonoid coiled in regular three-dimensional spiral form with a constant angle.

helicoidal flow At the bend of a river, a coiling type of flow motion that results in erosion of the concave, outer bank and deposition along the convex, inner bank. Syn: *helical flow*.

helicoplacoid Any echinozoan belonging to the class Helicoplacoidea, characterized by a fusiform to pyriform placoid body with a spirally pleated expansible and flexible test (TIP, 1968, pt.U, p.131). They are known only from the Lower Cambrian.

helictite A speleothem that resembles a stalagmite in origin but that angles or twists erratically. Cf: *heligmite*. See also: *anemolite*.

heligmite A speleothem that resembles a stalagmite in origin but that angles or twists erratically. Cf: *helictite*.

Helikian In a three-part division of Proterozoic time, the middle division, between the *Aphebian* and the *Hadrynian*.

heliodor A golden-, greenish-, or brownish-yellow transparent gem variety of beryl found in southern Africa. See also: *golden beryl*. Obsolete syn: *chrysoberyl*.

heliolite *sunstone*.

heliolith (a) A coccolith constructed of many tiny calcite crystals, commonly in radial arrangement. (b) An individual of the Heliolithae, a subdivision of the family Coccolithophoridae.-- Cf: *ortholith*.

heliolitid Any coral belonging to the family Heliolitidae, characterized by massive coralla with slender tabularia separated by coenenchyme and commonly having 12 equal spinose septa and complete tabulae. Heliolitids are considered to be tabulates by some workers. Their stratigraphic range is Middle Ordovician to Middle Devonian.

heliophyllite A mineral representing an orthorhombic polymorph of ecdemite.

heliostat A geodetic heliotrope.

heliotrope [mineral] *bloodstone*.

heliotrope [surv] An instrument used in geodetic surveying to aid in making long-distance (up to 320 km) observations and composed of one or more plane mirrors so mounted and arranged that a beam of sunlight may be reflected by it in any desired direction (such as toward a distant survey station where it can be observed with a theodolite). It is sometimes used as an observation signal in triangulation.

heliozoan Any actinopod belonging to the subclass Heliozoa, characterized by pseudopodia that are not stiff or rigid, being strengthened only by an axial rod of fibrils.

helium age method Determination of the age of a mineral in years based on the known radioactive decay rates of uranium and thorium isotopes to helium. This oldest method of radioactive age measurement is not in common use as both uranium and helium are known for their ease of movement through geologic materials thus preventing the determination of reliable ages. Magnetite and pyrite seem most amenable to the method and fossil shells of original aragonite with incorporated uranium isotopes can also be dated. Syn: *helium dating.*

helium dating *helium age method.*

helium index An obsolete term for the experimental age obtained by substituting helium and radioactivity values into the age equation.

helizitic Var. of *helicitic.*

hellandite A mineral: $Ca_3Y_4B_4Si_6O_{27}.3H_2O$.

helluhraun An Icelandic term for pahoehoe. Cf: *apalhraun.*

hellyerite A mineral: $NiCO_3.6H_2O$.

Helmert's formula Friedrich Robert Helmert (1843-1917), Director of the Geodetic Institute at Potsdam, developed many formulae relating to all aspects of geodosy. Among the better known and more frequently referenced are those relating to gravity, astrogeodetic leveling, precise leveling, and isostatic reductions.

Helmholtz coil In a magnetometer, a pair of coaxial coils with their distance apart from each other equal to their radius. Electric current in the coils produces an unusually uniform magnetic field between the coils.

Helmholtz free energy A *thermodynamic potential* that is a function of temperature and volume. It is one of the commonly used functions describing *free energy*, and is useful in determining the course of constant-volume isothermal processes. Cf: *Gibbs free energy.*

helminthite An obsolete term for a doubtfully distinguished trace fossil consisting of a long, sinuous surface trail or filled-up burrow of a supposed marine worm, without impressions of lateral appendages. Syn: *helmintholite.*

heloclone A sinuous, monaxonic sponge spicule of irregular outline, often bearing articulatory notches along its length or at its end.

helophyte A perennial marsh plant that has its overwintering buds beneath the water. See also: *hydrophyte.*

helsinkite A hypidiomorphic-granular hypabyssal rock composed primarily of albite and epidote. Its name is derived from Helsinki, Finland.

Helvetian European stage: Miocene (above Burdigalian, below Tortonian).

helvite A mineral: $(Mn,Fe,Zn)_4Be_3(SiO_4)_3S$. It is the manganese end member isomorphous with danalite and genthelvite. Syn: *helvine.*

hemachate A light-colored agate spotted with red jasper. Syn: *blood agate.*

hemafibrite A brownish- to garnet-red mineral: $Mn_3(AsO_4)-(OH)_3.H_2O$.

hematite A common iron mineral: $\alpha\text{-}Fe_2O_3$. It is dimorphous with maghemite. Hematite occurs in splendent, metallic-looking, steel-gray or iron-black rhombohedral crystals, in reniform masses or fibrous aggregates, or in deep-red or red-brown earthy forms, and it has a distinctive cherry-red to reddish-brown streak and a characteristic brick-red color when powdered. It is found in igneous, sedimentary, and metamorphic rocks both as a primary constituent and as an alteration product. Hematite is the principal ore of iron. Symbol: Hm. See also: *specularite.* Syn: *haematite; red hematite; red iron ore; red ocher; rhombohedral iron ore; oligist iron; bloodstone.*

hematophanite A mineral: $Pb_4Fe_4O_9(Cl,OH)_2$.

hemera (a) The geologic-time unit corresponding to *peak zone* (or epibole); the time span of the acme or greatest abundance, in a local section, of a taxonomic entity. Also, the period of time during which a race of organisms is at the apex of its evolution. The term was proposed by Buckman (1893, p.481-482) for the time of acme of development of one or more species, but later used by him (Buckman, 1902) (intrazonal time) in the sense of *moment* or the time during whic a biostratigraphic zone was deposited and by Jukes-Brown (1903, p.37) for the duration of a subzone. (b) A term sometimes incorrectly applied to a biostratigraphic zone (body c strata) comprising the time range of a particular foss species.---Etymol: Greek, "day". Pl: *hemerae; hemera.* Ad *hemeral.*

hemichoanitic Said of a retrochoanitic septal neck of a naut loid that extends one-half to three-fourths of the distance t the preceding septum.

hemicone *alluvial cone.*

hemicrystalline *hypocrystalline.*

hemicyclothem Half of a cyclothem. The term is generally ap plied either to the lower nonmarine part, or to the upper ma rine part, of a Pennsylvanian cyclothem.

hemidisc A sponge spicule consisting of an unequal-ende amphidisc.

hemihedral Said of the *merohedral* crystal class (or classes in a system, the general form of which has half the number c equivalent faces of the corresponding holohedral form. Syn *hemisymmetric.*

hemihedrite A mineral: $Pb_{10}Zn(CrO_4)_6(SiO_4)_2F_2$.

hemimorph A crystal having polar symmetry, i.e. displayin *hemimorphism.*

hemimorphism The characteristic of a crystal to have *pola* symmetry, so that the two ends of a biterminated crystal hav different forms. Such a crystal is a *hemimorph.* Adj: *hemimor phic.*

hemimorphite (a) A white or colorless to pale-green, blue, o yellow orthorhombic mineral: $Zn_4Si_2O_7(OH)_2.2H_2O$. It is sim lar to smithsonite, but is distinguished from it by strong pyroe lectric properties. Hemimorphite is a common secondary min eral, and is an ore of zinc. Syn: *calamine; electric calamine galmei.* (b) A term sometimes used (esp. in the gem trade as a syn. of *smithsonite.*

hemiopal *semiopal.*

hemipelagic deposits Deep-sea sediments containing a sma amount of terrigenous material as well as remains of pelagi organisms. Cf: *terrigenous deposits; pelagic deposits.*

hemiperipheral growth Growth of brachiopod shells in whic new material is added anteriorly and laterally but not pos teriorly.

hemiphragm A transverse, calcareous, shelf-like platform ex tending from the proximal wall part way across the zooecia tube in Paleozoic bryozoans.

hemipyramid A crystal form that is a prism with two pairs o parallel faces, rhombic in cross section and parallel to an in clined direction in monoclinic crystals. Its indices are $\{hkl\}$ o $\{\bar{h}kl\}$. It is an old name for the *fourth-order prism.*

hemiseptum (a) A shelf-like platform extending part wa across the zooecial vestibule in a cryptostome bryozoan. A "superior hemiseptum" may be present on the proximal wal and an "inferior hemiseptum" may be present on the dista wall. (b) A partial septum, between normal ones, subdividing a foraminiferal chamber, as in some Lituolacea.---Pl: hemi septa.

hemisphere Half a sphere; usually refers to half of the Earth as divided by the equator into a northern and a southern hemisphere, by the 20°W and 160°E meridians into an eastern (the Old World) and a western (the New World) hemisphere or by land surface into a land and a water hemisphere.

Hemist In U.S. Dept. of Agriculture soil taxonomy, the subor der of the soil order *Histosol,* characterized by having an in termediate degree of decomposition of plant fiber, a bulk den sity of 0.1 to 0.2, and by being saturated with water for suffi cient periods of time to make cultivation without artificia drainage difficult (SSSA, 1970). Cf: *Fibrist; Folist; Saprist.*

hemisymmetrical *hemihedral.*

hendersonite A black mineral: $Ca_2V^{+4}V_8^{+5}O_{24}.8H_2O$.

hendricksite A mineral of the mica group: $K(Zn,Mn)_3(Si_3-Al)O_{10}(OH)_2$.

henritermierite A mineral of the hydrogarnet group: $Ca_3(Mn,Al)_2(SiO_4)_2(OH)_4$.

hepatic cinnabar A liver-brown or black variety of cinnabar. Syn: *liver ore*.

hepatite A variety of barite that emits a fetid odor when rubbed or heated.

heptane Any of nine colorless, liquid, isomeric paraffin hydrocarbons of formula C_7H_{16}. n-heptane, $CH_3(CH_2)_5CH_3$, occurs in crude oils and in some pine oils.

heptaphyllite (a) A group of mica minerals that contain seven cations per ten oxygen and two hydroxyl ions. (b) Any mineral of the heptaphyllite group, such as muscovite and other light-colored micas; a dioctahedral clay mineral.---Cf: *octaphyllite*.

heptorite A dark-colored basanite composed of barkevikite, tianaugite, and haüyne phenocrysts in a glassy groundmass containing labradorite microlites.

herb A flowering plant whose stem does not become woody.

herbaceous Said of a green, vascular plant whose stem does not become woody.

herbivore A heterotrophic organism that feeds on plants. Cf: *carnivore*.

Hercules stone A syn. of *lodestone* [magnet]. Also called: *Heraclean stone*.

Hercynian orogeny By present usage, the late Paleozoic orogenic era of Europe, extending through the Carboniferous and Permian, hence synonymous with the *Variscan orogeny*; European usage today is about equally divided between one term or the other. Many German geologists regard "Hercynian" as a NW folding direction without time significance, hence they prefer the name Variscan; many French and Swiss geologists, following M. Bertrand, prefer "Hercynian" in the time sense; thus, the crystalline massifs of the northern Alps are said to be Hercynian, rather than Variscan. Cf: *Armorican orogeny*.

Hercynides A name used for the fold-belt created by the *Hercynian orogeny*, extending from southern Ireland and Wales to northern France, Belgium, and northern Germany.

hercynite A black mineral of the spinel series: $FeAl_2O_4$. It often contains some magnesium. Syn: *iron spinel; ferrospinel*.

herderite A colorless to pale-yellow or greenish-white monoclinic mineral: $CaBe(PO_4)(F,OH)$. It is isomorphous with hydroxyl-herderite.

heredity All of the qualities and potentialities that an individual has acquired genetically from its ancestors.

Herkimer diamond A quartz crystal from Herkimer County, N.Y. See also: *Lake George diamond*.

Hermann-Mauguin symbols An internationally accepted shorthand-type notation system of the elements of symmetry of crystal classes, which expresses both outward and inward symmetry. An example is $4/m\,\overline{3}\,2/m$ for the hexoctahedral class of the isometric system, in which the numbers with a bar are axes of rotoreflection, *m* is a symmetry plane, and a number over *m* indicates an axis of symmetry with a plane of symmetry perpendicular to it. Cf: *Schoenflies notation*.

hermatobiolith An organic *reef rock*.

hermatolith *reef rock*.

hermatopelago A *reef cluster* of submerged reefs. Etymol: Greek *hermato*, "submerged reefs", + *pelagos*, "sea".

hermatypic coral A reef-building coral; a coral characterized by the presence within their endodermal tissue of many symbiontic algae; a coral incapable of adjusting to aphotic conditions. Ant: *ahermatypic coral*. Syn: *hermatype*.

heronite A hypabyssal rock composed of spheroidal phenocrysts of orthoclase in a groundmass composed of radiating bundles of labradorite and acmite with interstitial analcime. The rock appears to be an altered tinguaite (Johannsen, 1939, p.256). Its name is derived from Heron Bay, Ontario, Canada.

herrerite A blue and green variety of smithsonite containing copper.

herringbone cross-bedding *chevron cross-bedding*.

herringbone mark *chevron mark*.

herringbone texture In mineral deposits, a pattern of adjacent rows of parallel crystals, each row in a somewhat reverse direction from the adjacent one.

herschelite A zeolite mineral: $(Na,Ca,K)AlSi_2O_6 \cdot 3H_2O$.

hervidero A syn. of *mud volcano*. Etymol: Spanish *hervir*, "to boil".

herzenbergite A mineral: SnS. Syn: *kolbeckine*.

hessite A lead-gray cubic mineral: Ag_2Te. It is sectile, usually massive, and often auriferous.

hessonite *essonite*.

hetaerolite A black mineral: $ZnMn_2O_4$. It is found with chalcophanite.

heterad crystallization Adcumulus growth in which cumulus crystals and poikilitic crystals of the same composition continue to develop until little or no interstitial liquid remains.

heteradcumulate A cumulate in which cumulus crystals and unzoned poikilitic crystals have the same composition.

heteroaxial Said of the deformation producing a tectonite, in which the fabric symmetry differs from that of the external form of the body of the rock.

heteroblastic Pertaining to a type of *crystalloblastic* texture in a metamorphic rock in which the essential mineral constituents are of two or more distinct orders of magnitude of size. The term was originated by Becke (1903). Cf: *homeoblastic*.

heterochronism The phenomenon by which two analogous geologic deposits may not be of the same age although their processes of formation were similar.

heterochronous [evol] Said of a fauna or flora appearing in a new region at a time that is quite different from the time it appeared in the region which it previously inhabited.

heterochronous [stratig] Said of a sequence of sediments representing lateral development of a similar lithofacies in successively younger stages. Term introduced by Nabholz (1951).

heterochronous homeomorphs Homeomorphs, of which one from a later geologic time resembles one from an earlier geologic time.

heterochthonous (a) Said of a transported rock or sediment, or one that was not formed in the place where it now occurs. Also, said of certain fossils removed by erosion from their original deposition site and re-embedded. Cf: *allochthonous*. (b) Said of a fauna or flora that is not indigenous.

heterococcolith A coccolith constructed of differing elements. Cf: *holococcolith*.

heterocoelous Said of sponges whose cloacae are not lined with choanocytes; specif. pertaining to syconoid or leuconoid sponges having calcium-carbonate spicules. Cf: *homocoelous*.

heterocolpate Said of pollen grains having large elongate holes (pseudocolpi) geometrically arranged in the exine.

heterocyst A sporelike structure produced by some blue-green algae; it may be a spore or a degenerate cell (Scagel, 1965, p.619). It differs from the other cells in the filament in that it is enlarged, clear, colorless, and thick-walled (Haupt, 1953, p.436).

heterodesmic Said of a crystal or other material that is bonded in more than one way. Cf: *homodesmic*.

heterodont adj. (a) Said of the dentition of a bivalve mollusk having a small number of distinctly differentiated cardinal teeth and lateral teeth that both fit into depressions on the opposed valve. (b) Said of hingement of ostracode valves effected by combination of tooth-and-socket and ridge-and-groove types, characterized by pointed or slightly crenulate teeth in one or both valves associated with a ridge in one valve and a groove in the other valve (TIP, 1961, pt.Q, p.50). ---n. A heterodont mollusk; specif. a bivalve mollusk of the order Heterodonta, having few hinge teeth but usually with both lateral teeth and cardinal teeth and with unequal adductor muscles. Cf: *taxodont.*

heterogeneous equilibrium Equilibrium in a system consisting of more than one phase. Cf: *homogeneous equilibrium.*

heterogenite A black mineral occurring in mammillary masses: $CoO(OH)$. It may contain some copper and iron. Syn: *stainierite.*

heterogony *alternation of generations.*

heterolithic unconformity A term proposed by Tomkeieff (1962, p.412) to replace *nonconformity* in the sense of an unconformity developed between "unlike rocks".

heteromesic Said of sediments or rocks formed in different media or under different conditions. Ant: *isomesic.*

heteromorph An aberrant form, such as an organism or part that differs from the normal form of a group; specif. an ammonoid or ammonoid shell of any form that deviates from the normal (planispiral) mode of coiling and/or whose walls of the coil are not in contact.

heteromorphic [evol] Deviating from the usual form or having diversity of form. Syn: *heteromorphous.* Cf: *isomorphic.*

heteromorphic [petrology] Said of igneous rocks having similar chemical composition but different mineralogic composition.

heteromorphism The crystallization of two magmas of identical chemical composition into two different mineral aggregates as a result of different cooling histories.

heteromorphite A mineral: $Pb_7Sb_8S_{19}.$

heteromorphosis The production by an organism of an abnormal or misplaced part esp. as the result of regeneration.

heteromyarian adj. Said of a bivalve mollusk, or of its shell, having the anterior adductor muscle conspicuously smaller than the posterior adductor muscle.----n. A heteromyarian mollusk. Cf: *anisomyarian; monomyarian.*

heteropic Said of sedimentary rocks of different facies, or said of facies characterized by different rock types. The rocks may be formed contemporaneously or in juxtaposition in the same sedimentation area or both, but the lithologies are different; e.g. facies that replace one another laterally in deposits of the same age. Also, said of a map depicting heteropic facies or rocks. Cf: *isopic.*

heteropod Any prosobranch belonging to the suborder Heteropoda, a group of pelagic forms with shells of aragonite.

heterosis The high capacity for growth and activity frequently displayed by crossbred organisms as compared with those that are inbred.

heterosite A mineral: $(Fe^{+3},Mn^{+3})PO_4$. It is isomorphous with purpurite.

heterosporous Characterized by heterospory; specif. said of plants that produce both microspores and megaspores.

heterospory The condition in embryophytic plants in which spores are of two types: microspores and megaspores. Cf: *homospory.*

heterostrophy The quality or state of being coiled in a direction opposite to the usual one; specif. the condition of a gastropod protoconch in which the whorls appear to be coiled in a direction opposite to those of the teleoconch. Adj: *heterostrophic.*

heterotactic [struc petrol] Said of a tectonite fabric, the subfabrics of which do not conform to a common symmetry. Cf: *homotactic [struc petrol].*

heterotactic [stratig] *heterotaxial.*

heterotaxial Pertaining to, characterized by, or exhibiting heterotaxy. Syn: *heterotactic; heterotactous; heterotaxic.*

heterotaxis An erroneous transliteration of *heterotaxy.*

heterotaxy Abnormal or irregular arrangement; specif. the condition of strata that are widely separated and not equivalent as to their relative positions in the geologic sequence, or that are lacking uniformity in stratification or arrangement. Ant: *homotaxy.* Syn: *heterotaxis; heterotaxia.*

heterotherm *poikilotherm.*

heterothermic *poikilothermic.*

heterothrausmatic Said of igneous rocks with an orbicular texture in which the nuclei of the orbicules are formed of various kinds of rock or mineral fragments. Cf: *allothrausmatic; crystallothrausmatic; isothrausmatic; homeothrausmatic.*

heterotomous Characterized by division of a crinoid arm in unequal branches. Ant: *isotomous.*

heterotopic Said of rocks formed in different environments, such as in different sedimentary basins or geologic provinces. Ant: *isotopic [sed].*

heterotrichy The occurrence of erect filaments arising from the prostrate portion in some algae and bryophytes.

heterotrophic Said of an organism that nourishes itself by utilizing organic material to synthesize living matter. Most animals are heterotrophic. Noun: *heterotroph.* Cf: *autotrophic.* Syn: *metatrophic; allotrophic; zootrophic.*

heterozooid A specialized bryozoan zooid, such as a *keno-zooid;* a zooid that is not an autozooid.

Hettangian European stage: lowermost Jurassic (above Rhaetian of Triassic, below Sinemurian).

heulandite A zeolite mineral: $(Na,Ca)_{4-6}Al_6(Al,Si)_4Si_{26}O_{72}.24H_2O$. It often occurs as foliated masses or as coffin-shaped monoclinic crystals in cavities in decomposed basic igneous rocks. See also: *clinoptilolite; stilbite.*

heumite A dark-colored, fine-grained hypabyssal rock characterized by granular texture and composed of alkali feldspar, barkevikite, biotite, and smaller amounts of nepheline, sodalite, diopside, and minor accessories. Its name is derived from Heum, Norway.

hewettite A deep-red mineral: $CaV_6O_{16}.9H_2O$. It occurs in silky, slender orthorhombic crystal aggregates.

hexacoral *scleractinian.*

hexactin A siliceous sponge spicule having six rays arising from a common center at right angles to one another.

hexactinellid A nonpreferred syn. of *hyalosponge.*

hexadisc A hexactinellid-sponge spicule (microsclere) composed of three interpenetrating amphidiscs at right angles to one another about a common center. Cf: *staurodisc.*

hexagonal close packing In a crystal, close packing of spheres by stacking close-packed layers in the sequence ABAB etc. Cf: *cubic close packing.*

hexagonal cross ripple mark An *oscillation cross ripple mark* formed by parallel ripples arranged in zigzag fashion and characterized by obtuse angles in adjoining ripples facing in opposite directions, by crossbars connecting apexes on opposite sides of the ripple, and by an enclosed pit that tends to be bounded by six sides. It appears to be formed by waves that oscillate at some angle between $45°$ and $90°$ to the direction of the original ripple mark.

hexagonal dipyramid A crystal form of 12 faces consisting of a hexagonal pyramid repeated across a mirror plane of symmetry. A cross section perpendicular to the sixfold axis is hexagonal. Its indices are $\{h0l\}$ or $\{hhl\}$ with symmetry $6/m$ $2/m$ $2/m$, and 622, $\{hhl\}$ only with symmetry $\bar{6}$ m2, also $\{hkl$ with symmetry $6/m$.

hexagonal-dipyramidal class That class of the hexagonal system having symmetry $6/m$.

hexagonal indices *Miller-Bravais indices.*

hexagonal prism A crystal form of six faces parallel to the symmetry axis, whose interfacial angles are $60°$. Its indices are $\{110\}$ or $\{100\}$ in several hexagonal classes, or $\{hk0\}$ in symmetry $6/m$, 6, and $\bar{3}$.

hexagonal pyramid A crystal form consisting of six faces in a pyramid, in which any cross section perpendicular to the sixfold axis is hexagonal. Its indices are $\{h01\}$ and $\{hhl\}$ in classes 6mm and 6, only $\{hhl\}$ in 3m, and $\{hkl\}$ in 6.

hexagonal-pyramidal class That crystal class in the hexagonal system having symmetry 6.

hexagonal-scalenohedral class That class crystal in the rhombohedral division of the hexagonal system having symmetry 3

330

2/m. Syn: *trigonal-scalenohedral class; ditrigonal-scalenohedral class.*

hexagonal scalenohedron A *scalenohedron* of twelve faces and having symmetry $\bar{3}2/m$. It resembles a ditrigonal pyramid. Cf: *tetragonal scalenohedron.*

hexagonal system One of the six *crystal systems,* characterized by one unique axis of threefold or sixfold symmetry that is perpendicular and unequal in length to three identically long axes which intersect at angles of 120°. This definition includes the *trigonal system* of threefold symmetry; however, the two systems of threefold and sixfold symmetries may be defined separately. Cf: *isometric system; tetragonal system; orthorhombic system; monoclinic system; triclinic system.*

hexagonal-trapezohedral class That crystal class in the hexagonal system having symmetry 622.

hexagonal trapezohedron A crystal form with 12 faces, a sixfold axis, and three twofold axes, but neither mirror planes nor a center of symmetry. It is composed of top and bottom hexagonal pyramids, one of which is twinned less than 30° about c with respect to the other. It may be right-handed or left-handed, and its indices are $\{hkl\}$ or $\{hk\bar{l}\}$ in symmetry 622.

hexahedral Adj. of *hexahedron.*

hexahedral coordination An atomic structure or arrangement in which an ion is surrounded by eight ions of opposite sign, whose centers form the points of a hexahedron (which may or may not be a cube). It may be synonymous with *cubic coordination.*

hexahedrite An *iron meteorite* made up of large single crystals or coarse aggregates of kamacite (which has cubic structure), usually containing 4-6% nickel in the metal phase, and characterized upon etching by the presence of Neumann bands caused by twinning parallel to the octahedral planes. Symbol: *H.* Cf: *octahedrite; ataxite.*

hexahedron A polyhedron of six equivalent faces, e.g. a cube or a rhombohedron. Adj: *hexahedral.*

hexahydrite A white or greenish-white monoclinic mineral: $MgSO_4 \cdot 6H_2O.$

hexamethylene *cyclohexane.*

hexane Any of five colorless, liquid, volatile, isomeric paraffin hydrocarbons of formula C_6H_{14}. The hexanes, especially n-hexane, $CH_3(CH_2)_4CH_3$, occur in crude oil.

hexarch Said of a stele having six strands or origins (Jackson, 1953, p.180).

hexastannite A mineral: Cu_6FeSnS_{10} (?).

hexaster A sponge spicule (microsclere) having the form of a hexactin with anaxial branches or extensions at the ray tips.

hexatetrahedron Var. of *hextetrahedron.*

hexoctahedral class That crystal class in the isometric system having symmetry $4/m\bar{3}\,2/m.$

hexoctahedron An isometric crystal form of 48 equal, triangular faces, each of which cuts the three crystallographic axes at different distances. Its indices are $\{hkl\}$ and its symmetry is $4/m\bar{3}\,2/m.$

hextetrahedral class That crystal class in the isometric system having symmetry $\bar{4}\,3m.$

hextetrahedron An isometric crystal form of 24 faces, with indices $\{hkl\}$ and symmetry $\bar{4}\,3m.$

heyrovskite A mineral: $(Pb,Ag,Bi)_6Bi_2S_9.$

hiatal [ign] Said of the inequigranular texture of an igneous rock in which the sizes of the crystals are not in a continuous series but are broken by hiatuses, or where there are grains of two or more markedly different sizes, as in *porphyritic* rocks (Johannsen, 1939, p.216). Cf: *seriate.*

hiatal [stratig] Pertaining to or involving a stratigraphic hiatus.

hiatus (a) A break or interruption in the continuity of the geologic record, such as the absence in a stratigraphic sequence of rocks that would normally be present but are missing, either because they were never deposited or because they were eroded before the deposition of the beds immediately overlying the break. (b) A lapse in time, such as the time interval not represented by rocks at an unconformity. The term has been used for the *time value* of an episode of nondeposition and for the *time value* of nondeposition and erosion together. Wheeler (1958, p.1057) regards the term as "the space-time value of nondeposition during a regressive-transgressive episode" (the part of a *lacuna* resulting from nondeposition), or as the "nondepositional and erosional cyclic phase" during which the *degradation vacuity* was "developed" (Wheeler, 1964, p.607).

hibernaculum A dormant zooid with protective walls, produced mainly at the onset of winter by freshwater and brackish-water bryozoans (such as those in the class Gymnolaemata) and developed as part of a colony in the spring. Pl: *hibernacula.* See also: *statoblast.*

Hibernian orogeny *Erian orogeny.*

hibonite A dark-brown mineral: $(Ca,Ce)(Al,Ti,Mg)_{12}O_{18}.$

hibschite A syn. of *hydrogrossular.* It was formerly believed to be a hydrogarnet with A = Ca, B = Al, and x = 1.

hidalgoite A white mineral: $PbAl_3(SO_4)(AsO_4)(OH)_6.$

hiddenite An intense, light-to-medium green, yellowish-green, or emerald-green transparent gem variety of *spodumene* containing chromium.

hielmite *hjelmite.*

hieratite A grayish, cubic, high-temperature mineral of fumaroles: $K_2SiF_6.$

hieroglyph Any sedimentary mark or structure found on a bedding plane; esp. a sole mark. A classification of hieroglyphs has been proposed by Vassoevich (1953). The term was first used by Fuchs (1895) for a problematic fossil whose appearance is suggestive of a drawing or ornament.

high [geophys] *maximum* [geophys].

high [oil] (a) The uppermost part of a geologic structure where petroleum is likely to be discovered. (b) *geologic high.* (c) *topographic high.*

high [struc geol] n. A general term for such features as a crest, culmination, anticline, or dome. Cf: *low [struc geol].* Syn: *structural high.*

high albite High-temperature albite, stable above 450°C. Natural high albite is almost always contaminated with appreciable amounts of potassium and calcium in solid solution. Cf: *low albite.*

high-angle cross-bedding Cross-bedding in which the cross-beds have an average maximum inclination of 20° or more (McKee & Weir, 1953, p.38). NXF: *low-angle cross-bedding.*

high-angle fault A fault, the dip of which is greater than 45°. Cf: *low-angle fault.*

high-calcium limestone A limestone that contains very little magnesium; specif. a limestone in which the approximate MgO equivalent is less than 1.1%, or in which the approximate magnesium-carbonate equivalent is less than 2.3% (Pettijohn, 1957, p.418). Dolomite is not present (the magnesium carbonate is in solid solution in calcite). According to Cooper (1945, p.9), the calcium-carbonate content is greater than 95%. Cf: *magnesian limestone.*

high chalcocite Hexagonal chalcocite, stable above 105°C.

high-energy coast A coast exposed to ocean swell and stormy seas and characterized by average breaker heights of greater than 50 cm. Cf: *moderate-energy coast; low-energy coast.*

high-energy environment An aqueous sedimentary environment characterized by a high *energy level* and by turbulent action (such as that created by waves, currents, or surf) that prevents the settling and accumulation of fine-grained sediment; e.g. a beach or a river channel, containing sand deposits. Cf: *low-energy environment.*

high-grade adj. Said of an ore with a relatively high ore-mineral content. Cf: *low-grade.* See also: *grade.*

high island In the Pacific Ocean, a volcanic rather than a coralline island. Cf: *low island.*

highland [geomorph] (a) A general term for a relatively large area of elevated or mountainous land standing prominently above adjacent low areas; a mountainous region. The term is often used in the plural in a proper name; e.g. the Highlands of Scotland. (b) A relative term denoting the higher land of a region; it may include mountains, valleys, and plains. Cf: *upland*. (c) A lofty headland, cliff, or other high landform.

highland [tect] A dissected mountain region composed of old folded rocks; e.g. one of two anticlines produced in Colorado during the late Paleozoic.

highland glacier A semicontinuous ice cap or glacier system covering the highest or central position of a mountainous area, partly reflecting irregularities of the land surface beneath; e.g. a *plateau glacier*. Syn: *highland ice*. Cf: *ice field*.

high-level ground water Ground water occurring above the basal water table and separated from it by impermeable or less permeable material such as ash beds, intrusive igneous rocks, or ice. It makes its way through, over, and around the low-permeability materials to join the basal water (Stearns & Macdonald, 1942, p.132).

high marsh A syn. of *salting*. The term is a "less correct and rather obsolete" syn. of *salt marsh* (Schieferdecker, 1959, term 1243). Cf: *low marsh*.

highmoor bog A bog, often on the uplands, whose surface is largely covered by sphagnum mosses which, because of their high degree of water retention, make the bog more dependent on relatively high rainfall than on the water table. The bog often occurs as a *raised peat bog* or *blanket bog*. Cf: *lowmoor bog*.

highmoor peat Peat occurring on high moors and formed predominantly of moss, such as sphagnum. Its moisture content is derived from rain water rather than from ground water, and is acidic. Ash and nitrogen contents are low, and cellulose content is high. Cf: *lowmoor peat*. Syn: *moorland peat; moor peat; sphagnum peat; bog peat; moss peat*.

high-oblique photograph An *oblique photograph* in which the camera is sufficiently inclined to the vertical so that the apparent horizon is included within the field of view. Syn: *high oblique*.

high plain An extensive area of comparatively level land not situated near sea level; e.g. the High Plains, a relatively undissected section of the Great Plains of the U.S., extending along the eastern side of the Rocky Mountains at elevations above 600 m.

high-polar glacier A *polar glacier* in whose accumulation area the firn is at least 100 m thick, and which does not melt appreciably even during the summer (Ahlmann, 1933); e.g. most of the glaciers in Antarctica. Cf: *subpolar glacier*.

high quartz High-temperature quartz; specif. *beta quartz*.

high-rank graywacke A term introduced by Krynine (1945) for a graywacke containing abundant (20%) feldspar, usually a sodic plagioclase. It is formed in eugeosynclines. The rock is equivalent to *feldspathic graywacke* of Pettijohn (1954) and is regarded as "graywacke proper" by Pettijohn (1957, p.320). Cf: *low-rank graywacke*.

high-rank metamorphism Metamorphism that is accomplished under conditions of high temperature and pressure. Cf: *low-rank metamorphism*.

high-speed layer A layer of a medium in which the speed of elastic-wave propagation is relatively greater than that of an adjacent layer. See also: *seismic stringer*.

high tide The tide at its highest; the accepted popular syn. of *high water* in the sea.

high-volatile A bituminous coal A nonagglomerating, bituminous, *high-volatile coal* that has 14,000 or more BTU/pound. Cf: *high-volatile B bituminous coal; high-volatile C bituminous coal*.

high-volatile B bituminous coal A nonagglomerating, *bituminous*, *high-volatile coal* that has 13,000 or more but less than 14,000 BTU/pound. Cf: *high-volatile A bituminous coal; high-volatile C bituminous coal*.

high-volatile bituminous coal Bituminous coal that contains

more than 31% volatile matter, analyzed on a dry, mineral-matter-free basis. See also: *high-volatile A bituminous coal; high-volatile B bituminous coal; high-volatile C bituminous coal*. Cf: *low-volatile bituminous coal; medium-volatile bituminous coal; gas coal*.

high-volatile C bituminous coal A nonagglomerating, *high-volatile bituminous coal* that has 11,500 or more but less than 13,000 BTU/pound; or, an agglomerating, high volatile bituminous coal that has 10,500 or more but less than 11,500 BTU/pound. Cf: *high-volatile A bituminous coal; high-volatile B bituminous coal*.

highwall In coal mining, the *face* of an opencast mine.

high water Water at the maximum level reached during a tidal cycle. Abbrev: HW. Cf: *low water*. Syn: *high tide*.

high-water platform A *wave-cut bench* developed a little below the high-water level, commonly on a rock surface.

highway geology Engineering geology as applied to the planning, design, construction, and maintenance of public roads.

highwoodite A dark-colored intrusive rock composed of loxoclase, labradorite, pyroxene, biotite, iron ore, apatite, and possibly a small amount of nepheline. Its name is derived from the Highwood Mountains, Montana, U.S.A.

hilairite A porphyritic crystalline igneous rock composed of large albite, nepheline, sodalite, acmite, and eudialyte phenocrysts in a trachytic groundmass of acmite, nepheline, albite and orthoclase.

hilate Said of a spore or pollen grain possessing a hilum.

hilgardite A mineral: $Ca_2B_5ClO_8(OH)_2$.

hill [geog] Dry or solid ground surrounded by wet land or by water.

hill [geomorph] (a) A natural elevation of the land surface rising rather prominently above the surrounding land, usually of limited extent and having a well-defined outline (rounded rather than peaked or rugged), and generally considered to be less than 300 m (1000 ft) from base to summit; the distinction between a hill and a *mountain* is arbitrary and dependent on local usage. See also: *mount*. (b) Any slightly elevated ground or other conspicuous elevation in a relatively flat area. (c) An eminence of inferior elevation in an area of rugged relief. (d) A range or group of hills, or a region characterized by hills or by a highland. Term usually used in the plural; e.g. the Black Hills of South Dakota.

hill creep Slow downhill movement, on a steep hillside and under the influence of gravity, of soil and rock waste flowing toward the valleys; it is an important factor in the wasting of hillsides during dissection, as in the Alps. See also: *terminal creep*. Syn: *hillside creep*.

hillebrandite A white mineral: $Ca_2SiO_3(OH)_2$.

hill-island A glacial moraine, mainly of sand and of variable size, rising as a mature hill from an outwash plain of a late glacial epoch (P.R. Barham in Stamp, 1966, p.235).

hillock A small, low hill; a mound. Adj: *hillocky*.

hillock moraine A moraine consisting of a series of hillocks.

hill of planation A term applied by Gilbert (1877, p.130-131) to a bedrock erosion surface now described as a *pediment* (although such a surface is not a hill in any sense).

hill peat Peat occurring in cold, temperate areas and derived from mosses, heaths, pine trees, and related plant forms. Syn: *subalpine peat*.

hill shading (a) A method of showing relief on a map by simulating the appearance of sunlight and shadows, assuming an oblique light from the NW so that slopes facing south and east are shaded (the steeper slopes being darker), thereby giving a three-dimensional impression similar to that of a relief model. The method is widely used on topographic maps in association with contour lines. (b) The pictorial effect (of contoured topographic features) emphasized by hill shading, in which the features are shown by the shadows they cast.-- Syn: *hillwork; relief shading; plastic shading; shading*.

hillside A part of a hill between its crest and the drainage line at the foot of the hill. Syn: *hillslope.*

hillside creep *hill creep.*

hillside spring *contact spring.*

hillslope (a) The sloping surface that forms a hillside. (b) *hillside.*

hilltop surface *gipfelflur.*

hillwash The process of *rainwash* operating on a hillslope. Also spelled: *hill wash.*

hillwork *hill shading.*

hilly (a) Descriptive of a region characterized by an abundance of hills. (b) Resembling the inclination or character of a hill.

Hilt's law The generalization that in a vertical succession at any point in a coal field, coal rank increases with depth.

hilum [bot] A scar on a seed coat that marks the place of attachment of the seed stalk to the seed.

hilum [palyn] A germinal aperture of a spore or pollen grain, formed by the breakdown of the exine in the vicinity of one of the poles. The hilum in the spore *Vestispora* is associated with an operculum that may become separated from the spore.

hinge [paleont] A collective term for the structures of the dorsal region which function during the opening and closing of the valves of a bivalve shell; esp. a flexible ligamentous joint. The term is often used loosely for *hinge line*, and for the *cardinal margin* of a brachiopod (TIP, 1965, pt.H, p.145).

hinge [fold] The point of maximum curvature or bending of a fold. See also: *hinge line.* Syn: *flexure.*

hinge area (a) The flattened area margining the hinge of a brachiopod or pelecypod shell. See also: *cardinal area.* (b) The surface involved in the hingement of ostracode valves, commonly differentiated into anterior and posterior areas containing more complex elements and between these an inter-terminal area with simpler structures (TIP, 1961, pt.Q. p.50-51).

hinge axis (a) An imaginary straight line about which the two valves of a bivalve shell are hinged. Syn: *cardinal axis.* (b) The line joining the points of articulation about which the valves of a brachiopod rotate when opening and closing.---Cf: *hinge line* [paleont].

hinge crack A *crack* in sea ice, parallel and adjacent to a pressure ridge, and believed to be caused by the weight of the ridge.

hinge fault [fault] A fault, the movement of one side of which hinges about an axis perpendicular to the fault plane; displacement increases with distance from the hinge. It is a questionable term. Partial syn: *scissor fault; rotational fault; rotary fault; pivotal fault.*

hinge line [paleont] (a) A line along which articulation takes place; e.g. the middorsal line of junction of two valves of a crustacean carapace, permitting movement between them, or the line along which the two valves of an ostracode articulate, seen when the carapace is complete. (b) The straight posterior margin, edge, or border of a brachiopod shell, parallel to the *hinge axis.* The term is also used as a syn. of *cardinal margin.* (c) A term applied loosely to the part of a bivalve-mollusk shell bordering the dorsal margin and occupied by or situated close to the hinge teeth and ligament. The term is sometimes used as a syn. of *hinge axis.*---Syn: *hinge.*

hinge line [fault] That line along a fault surface at which the direction of apparent displacement changes. See also: *scissor fault.*

hinge line [fold] A line connecting the points of flexure or maximum curvature of the bedding planes in a fold. See also: *hinge.*

hinge line [struc geol] A line or boundary between a stable region and a region undergoing upward or downward movement. In Pleistocene geology, it is the boundary between regions undergoing postglacial uplift, and those of no uplift (for example, in Great Lakes area).

hinge-line fault A fault that is caused by sedimentary overload,

as in an area of embayment, with downthrow and sedimentary thickening in a seaward direction. It is an alternative interpretation to *growth fault* for the Gulf Coast region of the U.S.

hingement The area of the juncture and articulation of the two halves or valves of the carapace of an ostracode.

hinge node A localized thickened part of the hinge of the right valve of phyllocarid crustaceans, serving to strengthen the hinge.

hinge plate (a) The shelly internal plate bearing the hinge teeth in a bivalve mollusk, situated below the beak and the adjacent parts of the dorsal margins of each valve, and lying in a plane parallel to that of commissure. Syn: *cardinal platform.* (b) A plate, simple or divided, typically nearly parallel to the plane between the valves of a brachiopod, lying along the hinge line in the interior of the brachial valve and bearing its dental sockets, and joined to crural bases. See also: *inner hinge plate; outer hinge plate.*

hinge tooth An articulating projection of one valve of a bivalve shell, located near the hinge line or adjacent to the dorsal margin, and fitting into an accompanying indentation in the opposite valve for the purpose of holding the valves in position when closed; e.g. a *cardinal tooth* or a *lateral tooth* of a bivalve mollusk, or one of a pair of small or stout, wedge-shaped processes situated at the base of the delthyrium of the pedicle valve of a brachiopod and articulating with the dental sockets in the brachial valve. Syn: *tooth.*

hinge trough A V- or U-shaped depression formed by fusion of bifurcated median septum with combined socket ridges and crural bases of some terebratellacean brachiopods. Syn: *trough* [paleont].

hinsdalite A dark-gray or greenish rhombohedral mineral: $(Pb,Sr)Al_3(PO_4)(SO_4)(OH)_6$. It is isomorphous with svanbergite, corkite, and woodhouseite.

hinterland An area bordering, or within, an orogenic belt on the internal side, away from the direction of overfolding and thrusting; it is related to the *internides* and to the discredited *borderlands* of Schuchert. Syn: *backland.*

hintzeite *kaliborite.*

hiortdahlite A pale-yellow triclinic mineral: $(Ca,Na)_{13}Zr_3Si_9(O, OH,F)_{33}$.

hipotype *hypotype.*

hirnantite An intrusive rock composed of lath-shaped sodic plagioclase (albitized andesine) with interstitial chlorite and small amounts of quartz, leucoxene, hematite, and calcite.

hirst Var. of *hurst.*

hisingerite A black or brownish-black amorphous mineral: $Fe_2^{+3}Si_2O_5(OH)_4.2H_2O.$

hislopite A bright grass-green variety of calcite in which the color is due to admixed glauconite.

hispid Rough or covered with minute, fine, short, hair-like spines.

histium The near-ventral ridge confluent with the connecting lobe of the carapace of some heteromorphic ostracodes.

histogram A vertical bar-graph representation of a frequency distribution in which the height of bars is proportional to frequency of occurrence within each class interval and, due to the subdivision of the x-axis into adjacent class intervals, there are no empty spaces between bars when all classes are represented in a sample so graphed. Histograms are commonly used to depict particle-size distribution in sediments.

histometabasis Preservation of structure by the minerals that have replaced organic tissues, as in silicified wood and in many other fossils.

historical geology A major branch of *geology* that is concerned with the evolution of the Earth, its atmosphere, and its environment from its origins to its present-day forms. The study of historical geology therefore involves investigations into stratigraphy, paleontology, and geochronology, as well as the consideration of paleoenvironments, glacial periods, and polar wandering as evidenced by remanent magnetism. It is complementary to *physical geology*.

historical geomorphology A branch of geomorphology concerned with the actual series of events within a particular geologic period and in a particular geographic region.

history of geology That branch of the history of science which treats of the development of geologic knowledge, including the history of observations of geologic features, the history of theories to explain their origin, and the history of the organization and development of geologic institutions and societies. The biographical study of geologists is included. Not to be confused with *historical geology*.

Histosol In U.S. Dept. of Agriculture soil taxonomy, an organic soil order characterized by being more than half organic in its upper 80cm, or by having its interstices filled with organic material (SSSA, 1970). Suborders and great soil groups of this order have the suffix -ist. Cf: *Bog soil*. See also: *Fibrist; Folist; Hemist; Saprist*.

hjelmite A black, often metamict mineral of formula: AB_2O_6 or $A_2B_3O_{10}$, where $A = Y$, Fe^{+2}, U^{+4}, Mn, or Ca, and $B =$ Nb, Ta, Sn, or W. It may be equivalent to pyrochlore + tapiolite. Syn: *hielmite*.

hkl indices The *Miller indices*, in general terms that represent integral numbers.

(hk0) joint A partial syn. of *diagonal joint*.

H layer In a forest soil, a layer of amorphous organic material below the litter and the partially decomposed *F layer*. Syn: *humus layer*.

hoar *hoarfrost*.

hoarfrost A deposit of thin ice crystals formed as a result of radiational cooling of a surface (McIntosh, 1963, p.129). It may be partly frost and partly dew that was frozen after deposition. Cf: *rime*. Syn: *hoar*.

Hochmoor *raised bog*.

hodgkinsonite A bright-pink to reddish-brown mineral: $MnZn_2SiO_5 \cdot H_2O$.

hodochrone *traveltime curve*.

hodograph [seis] *traveltime curve*.

hodograph [oceanog] The locus of one end of a variable vector as the other end remains fixed, and representing the linear velocity of a moving point; used specif. in oceanography to describe the Ekman spiral and the motion of the mean tidal-current cycle.

hodrushite A mineral: $Cu_4Bi_6S_{11}$.

hoe A promontory or a point of land stretching into the sea; a spur of a hill, or a projecting ridge of land; a cliff. Term is obsolete except in English place names; e.g. Plymouth Hoe. Syn: *howe*.

hoegbomite *hôgbomite*.

hoelite A yellow mineral: $C_{14}H_8O_2$ (anthraquinone).

hoernesite A white monoclinic mineral: $Mg_3(AsO_4)_2 \cdot 8H_2O$. Its crystals resemble those of gypsum. Syn: *hôrnesite*.

hofmannite *hartite*.

hogback [glac geol] A term applied in New England to a *drumlin* (western Massachusetts) and to a *horseback* or esker (Maine).

hogback [geomorph] Any ridge with a sharp summit and steep slopes of nearly equal inclination on both flanks, and resembling in outline the back of a hog; specif. a long, narrow, sharp-crested ridge formed by the outcropping edges of very steeply inclined or highly tilted resistant rocks (such as igneous dikes), and produced by differential erosion. The term is usually restricted to ridges carved from beds dipping at angles greater than $20°$ (Stokes & Varnes, 1955, p.71). Cf: *cuesta*. See also: *dike wall; razorback*. Also spelled: *hog-back*. Syn: *hog's back; stone wall; swine back*.

hôgbomite A black mineral: $Mg(Al,Fe,Ti)_4O_7$. Syn: *hoegbomite*.

hogbomitite A hogbomite-rich magnetite.

hog's back Var. of *hogback*. Also spelled: *hog's-back; hogsback*.

hogtooth spar *dogtooth spar*.

hog wallow (a) A *wallow* made by swine. Also, a similar depression believed to be formed by heavy rains. (b) A faintly billowing land surface characterized by many low, coalescent or rounded mounds (such as *Mima mounds*) that are slightly higher than the basin-shaped depressions between them. Also spelled: *hogwallow*.

hohmannite A mineral: $Fe_2(SO_4)_2(OH)_2 \cdot 7H_2O$.

holacanth A trabecula of a rugose coral, seemingly consisting of a clear rod of calcite, as in septa of *Tryplasma* (TIP, 1956, pt.F, p.248).

holarctic The arctic regions as a whole.

holaspis A juvenile trilobite at any stage having the number of thoracic segments typical of the species. Pl: *holaspides*.

holdenite A red orthorhombic mineral: $(Mn,Zn)_6(AsO_4)-(OH)_5O_2$.

holdfast Something that supports or holds in place; e.g. a basal discoid or root-like structure by which the thallus of many algae is attached to a solid object in water, or any structure, at the distal extremity of a crinoid column, serving for fixation.

hole [drill] (a) *drill hole*. (b) *borehole*. (c) A mine, well, or other shaft dug in earth material.

hole [coast] A term used in New England for a small bay, cove, or narrow waterway; e.g. Woods Hole, Mass.

hole [geomorph] (a) A term used in the western U.S. for a comparatively level, grassy valley shut in by the mountains; e.g. Jackson Hole, Wyo. Cf: *park*. (b) An abrupt hollow in the ground, such as a pothole, a kettle, or a cave.

hole [cryst] *vacancy*.

hole [ice] (a) An opening through a piece, or between pieces, of sea ice. (b) *thaw hole*.

hole [stream] (a) A deep place in a stream. (b) A *water hole* in the bed of an intermittent stream.

hole fatigue A phenomenon caused by the delay between the detonation of a shot and the initiation of the seismic impulse from it because of changes in the shot environment (usually cavity formation) produced by an earlier shot in the same hole (Sheriff, 1968). Syn: *shothole fatigue*.

holism The theory that in nature organisms develop from individual structures acting as "whole" units. Cf: *vitalism*.

holisopic Said of an *isopic* condition that includes both lithologic and paleontologic similarities.

hollandite A silvery-gray to black mineral: $Ba(Mn^{+2}, Mn^{+4})_8O_{16}$. It is isostructural with coronadite and cryptomelane.

hollingworthite A mineral: $(Rh,Pt,Pd)AsS$

hollow (a) A low tract of land surrounded by hills or mountains; a small, sheltered valley or basin, esp. in a rugged area. (b) A term used in the Catskill Mountains of New York for a notch or pass. (c) A landform represented by a depression, such as a cirque, a cave, a large sink, or a blowout, or a volcanic structure formed by a lack of volcanic material.

Holmes' classification A classification of igneous rocks based chiefly on the degree of saturation of a rock, with other aspects of the mineralogy as secondary considerations. The system was proposed in 1928 by Arthur Holmes (1890-1965).

Holmes effect The effect leading to overestimation of the relative area of an opaque grain in a thin section of a rock viewed in transmitted light, due to the fact that the apparent area of an opaque grain will always be that of its maximum cross section in the slide, whereas the desired reference is the surface of the thin section. This term was coined by F. Chayes (1956, p.95) for the effect described by the petrographer Arthur Holmes (1890-1965) in 1927.

holmite An igneous rock similar to monchiquite but with a groundmass of melilite rather than analcime.

holmquistite A bluish-black orthorhombic mineral of the amphibole group: $(Na,K,Ca)Li(Mg,Fe)_3Al_2Si_8O_{22}(OH)_2$. It is related to anthophyllite.

holoaxial *holohedral.*

holoblast A *crystalloblast* that is newly and completely formed during metamorphism. The term was first used by Sander (1951).

Holocene An epoch of the Quaternary period, from the end of the Pleistocene to the present time; also, the corresponding series of rocks and deposits. When the Quaternary is designated as an era, the Holocene is considered to be a period. Syn: *Recent; Postglacial.*

holochoanitic Said of a retrochoanitic septal neck of a nautiloid that extends backward through the length of one camera.

holochroal eye A trilobite eye whose external visual surface is smooth (covered by a continuous cornea) and does not reveal the pattern of numerous underlying lenses. Cf: *schizochroal eye.* Syn: *compound eye.*

holoclastic rock An ordinary (sedimentary) clastic rock, as distinguished from a *pyroclastic rock.*

holococcolith A coccolith consisting entirely of microcrystals of usual crystallographic shape, whether or not they are identical. Cf: *heterococcolith.*

holocrystalline Said of an igneous rock composed entirely of crystals, having no part glassy. Syn: *pleocrystalline.*

holocrystalline-porphyritic Said of a porphyritic igneous rock having an entirely crystalline groundmass.

holocyst A smooth frontal wall in some cheilostomatous bryozoans, often overlain by secondary calcification. Cf: *pleurocyst; tremocyst.* Syn: *olocyst.*

holohedral Said of that crystal class having the maximum symmetry possible in each crystal system. Such a class is also called the *normal class.* Cf: *merohedral; tetartohedral.* Syn: *holosystematic; holosymmetric.*

holohedron Any crystal form in the holohedral class of a crystal system.

holohyaline Said of an igneous rock that is composed entirely of glass or whose texture is completely glassy.

holokarst Karst that is completely developed, characterized by thick limestone bedrock, little or no surface drainage, and a bare surface with well-formed depressions and caves. Cf: *merokarst.*

hololeims Coalified remains of entire plants (Kristofovich, 1945, p.138). Cf: *meroleims.* See also: *phytoleims.*

holomictic lake A lake that undergoes a complete circulation of its waters during periods of circulation or overturn, the mixing extending throughout the entire depth of the lake. Cf: *meromictic lake.*

holomixis The process leading to, or the condition of, a *holomictic lake.*

holoperipheral growth Increase in size of a brachiopod valve all around the margins (in posterior, anterior, and lateral directions). Cf: *mixoperipheral growth.*

holophyte A plant which derives its nourishment entirely from its own organs. Adj: *holophytic.*

holophytic The obsolescent adj. of *holophyte;* plant-type nutrition. Cf: *holozoic.* Syn: *phototrophic; photoautotrophic.*

holoplankton Plankton that are planktonic during their complete life cycles, as opposed to the temporarily planktonic *meroplankton.*

holosiderite A meteorite consisting of metallic iron without stony matter. Cf: *oligosiderite.*

holosome A term introduced by Wheeler (1958, p.1061) for an intertongued time-stratigraphic unit that may be either depositional (comprising one or more contiguous holostromes) or hiatal (consisting of combined contiguous hiatuses). Cf: *lithosome; biosome.*

holostomatous Said of a gastropod shell with a more or less circular apertural margin uninterrupted by a siphonal canal or siphonal notch. Cf: *siphonostomatous.*

holostratotype The originally defined stratotype (Sigal, 1964).

holostrome A term introduced by Wheeler (1958, p.1055-1056) for a time-stratigraphic unit "embodying the space-time value of a complete (restored) transgressive-regressive depositional sequence", including strata that may later have been removed by erosion. Cf: *lithostrome.*

holosymmetric *holohedral.*

holosystematic *holohedral.*

holothuroid Any cylindroid echinozoan, usually free-living, belonging to the class Holothuroidea, and characterized by the absence of an articulated test and by the reduction of skeletal elements to microscopic sclerites; e.g. a *sea cucumber.* Var: *holothurian.*

holotype The single specimen designated as the *type specimen* of a species or subspecies at the time its original description is published. Cf: *lectotype; neotype.*

holozoic Said of an organism that is nourished by the ingestion of organic matter; animal-type nourishment. Cf: *holophytic.*

holtite A mineral that is related to dumortierite: $(Al,Sb,Ta)_7(B,Si)_4O_{18}$ (?).

Holweck-Lejay inverted pendulum An instrument for measuring differences in gravity in which a mass is suspended from below by a weak flat-leaf spring. The instrument is used near its instability configuration and, as used, its period of oscillation varies with changes in gravity by a much greater percentage than is the case in the gravity pendulum (Nettleton, 1940, p. 30).

holyokeite An albitite characterized by ophitic texture.

homalographic projection *homolographic projection.*

homalozoan Any echinoderm belonging to the subphylum Homalozoa, characterized by the absence of radial symmetry, having a basically asymmetrical body. The subphylum comprises the carpoids and possibly the Machaeridia.

homeoblastic Pertaining to a type of *crystalloblastic* texture in a metamorphic rock in which the essential mineral constituents are approximately of equal size. Depending on the habit of the minerals involved, this texture may also be called more specifically *granoblastic, lepidoblastic, nematoblastic,* or *fibroblastic.* The term was originated by Becke (1903). Cf: *heteroblastic.*

homeochilidium An externally convex triangular plate closing almost all or only the apical part of the notothyrium in the brachiopod order Paterinida. Cf: *homeodeltidium.* Also spelled: *homoeochilidium.*

homeocrystalline *equigranular.*

homeodeltidium An externally convex triangular plate closing almost all or only the apical part of the delthyrium in the brachiopod order Paterinida. Cf: *homeochilidium.* Also spelled: *homoeodeltidium.*

homeomorph [**cryst**] A crystal that displays *homeomorphism* with another.

homeomorph [**evol**] An individual that bears a superficial resemblance to another organism although the two have different ancestors.

homeomorphic Adj. of *homeomorphism.*

homeomorphism The characteristic of crystalline substances of dissimilar chemical composition to have similar crystal form and habit; such crystals are known as *homeomorphs.* Adj: *homeomorphic; homeomorphous.*

homeomorphous Adj. of *homeomorphism.*

homeomorphy The phenomenon in which species having superficial resemblance are unlike in structural details; general similarity but dissimilarity in detail.

homeostasis The trend toward a relatively stable internal condition in the bodies of the higher animals as a result of a sequence of interacting physiologic processes; e.g., the ability to maintain relatively constant body heat during widely varying external temperatures.

homeothrausmatic Said of igneous rocks with an orbicular texture in which the nuclei of the orbicules are formed of inclusions of the same generation as the groundmass (Eskola, 1938, p.4776). Cf: *isothrausmatic; allothrausmatic; heter-*

othrausmatic; crystallothrausmatic.

Homerian Floral stage in Alaska: Miocene and Pliocene(?).

homilite A black or blackish-brown mineral: $Ca_2(Fe, Mg)B_2Si_2O_{10}$.

Hommel's classification A classification of igneous rocks in which a rock is represented by a two-part formula, one giving the molecular proportions of the oxides and the other, the percentages of the normative minerals. The system was proposed in 1919 by W. Hommel.

homoaxial (a) Said of a fold system in which the axes are parallel. (b) Said of a petrographic fabric in which the lineation of the mineral constituents is parallel.

homoclinal Adj. of *homocline*.

homoclinal shifting A term used by Cotton (1922, p.392) as a syn. of *monoclinal shifting*.

homocline A general term for a rock unit in which the strata have the same dip, e.g. one limb of a fold, a tilted fault block, a monocline, or an isocline. Cf: *monocline*. Adj: *homoclinal*.

homocoelous Said of sponges whose flagellated chambers are also cloacae; specif. pertaining to asconoid sponges having calcium-carbonate spicules. Cf: *heterocoelous*.

homodesmic Said of a crystal or other material that is bonded in only one way. Cf: *heterodesmic*.

homogeneous equilibrium Equilibrium in a system consisting of only one phase, typically liquid or gaseous. Cf: *heterogeneous equilibrium*.

homogeneous strain A state of *strain* in which the resultant form and orientation of the body are similar to the original form and orientation, because the strain is the same at all points. Syn: *uniform strain*.

homogeny *homology*.

homoiolithic Said of a sedimentary rock containing fragments of similar rock material or composed of two similar rock materials, and having a structure that indicates "contemporary erosion and redistribution" (Phemister, 1956, p.73).

homoiothermic Said of an organism whose body temperature remains relatively uniform and independent of the temperature of the environment; warm-blooded. Syn: *homothermous; homothermal; homeothermic*. Cf: *poikilothermic*.

homologous [paleont] The adj. of *homology*.

homologous [geol] (a) Said of strata, in separated areas, that are correlatable (contemporaneous) and that are of the same general lithologic character or facies and/or occupy analogous structural positions along the strike. (b) Said of faults, in separated areas, that have the same relative position or structure.

homolographic projection An *equal-area projection*. The term is sometimes given to a particular map projection, such as the "Mollweide homolographic projection". Syn: *homalographic projection*.

homologue An organism or part of an organism exhibiting homology. Also spelled: *homolog*.

homology (a) In biology, similarity, but not identity, between parts of different organisms as a result of evolutionary differentiation from the same or corresponding part of an ancestor. Syn: *true homology; homogeny*. Cf: *homoplasy*. (b) Similarity of position, proportion, structure, etc. without restriction to common ancestry. ----Adj: *homologous*.

homolosine projection An equal-area map projection consisting of a sinusoidal projection (between parallels 40°N and 40°S) combined with a Mollweide homolographic projection (between these parallels and the poles); specif. an *interrupted projection* that allows continental masses to be recentered on several meridians in order that they be shown with a minimum of shape distortion, leaving gaps in the interrupted ocean areas between each section to accommodate errors. Also, a similar projection showing the oceans to best advantage.

homomorphosis The regeneration by an organism of a part having a form similar to that of a part that has been lost.

homomyarian *isomyarian*.

homonym In taxonomy, any one of two or more identical names applied to different taxa of the same rank. See also *homonymy*.

homonymy In taxonomy, identity in spelling of the names applied to different taxa of the same rank. See also: *law of homonymy*.

homoplasy Similarity or correspondence of parts or organs that developed as a result of convergence or parallelism. Cf: *homology*. Adj: *homoplastic*.

homopycnal inflow Flowing water of the same density as the body of water it enters, resulting in easy mixing (Moore, 1966, p.89). Cf: *hyperpycnal inflow; hypopycnal inflow*.

homoseism *coseismal line*.

homoseismal line *coseismal line*.

homospore One of the spores of an embryophytic plant which reproduced by homospory. Its range is Silurian to Holocene. Syn: *isospore*.

homosporous Characterized by homospory.

homospory The condition in embryophytic plants in which all spores produced are of the same kind; the production by various plants of homospores. Cf: *heterospory*. Syn: *isospory*.

homotactic [struc petrol] Said of a tectonite fabric, all the subfabrics of which are conformable in symmetry. Cf: *heterotactic [struc petrol]*.

homotactic [stratig] *homotaxial*.

homotaxial Pertaining to, characterized by, or exhibiting homotaxy; e.g. said of rock-stratigraphic units or biostratigraphic units that have a similar order of arrangement in different locations but are not necessarily contemporaneous (ACSN, 1961, art.2a). Syn: *homotaxeous; homotactic*.

homotaxis An erroneous transliteration of *homotaxy*.

homotaxy Similarity of serial arrangement; specif. taxonomic similarity between stratigraphic or fossil sequences in separate regions, or the condition of strata characterized by similar fossils occupying corresponding positions in different vertical sequences, without connotation of similarity of age. The term was originally proposed as *homotaxis* by Huxley (1862, p.xlvi) to avoid the common fallacy of confusing taxonomic similarity with synchroneity. Etymol: Greek. Cf: *chronotaxy*. Ant: *heterotaxy*. Syn: *homotaxia*.

homothetic [geomorph] Said of geomorphologic features that show geometric similarity (similar in shape though perhaps differing in size) and that have corresponding points that are collinear (Strahler, 1958, p.291).

hondo A term used in the SW U.S. for a broad, low-lying arroyo. Etymol: Spanish, "bottom".

hondurasite *selen-tellurium*.

honessite A mineral consisting of a basic sulfate of iron and nickel.

honeycomb structure [ice] A sea ice structure consisting of soft, spongy ice filled with pockets of meltwater or seawater and characteristic of *rotten ice*.

honeycomb structure [weath] A rock or soil structure having cell-like forms suggesting a honeycomb; e.g. a *stone lattice*. See also: *tafone*.

honeycomb weathering A type of chemical weathering whereby innumerable small pits are produced on a rock exposure by the decomposition of individual mineral grains, the pits becoming deeper, larger, and more numerous until they unite at the surface while still separate below. The deeply pitted surface resembles an enlarged honeycomb and is characteristic of finely granular rocks (such as tuffs and sandstones) in an arid region. Cf: *cavernous weathering*. Syn: *fretwork weathering; alveolar weathering*.

honey stone A syn. of *mellite*. Also spelled: *honeystone*.

Honkasalo correction A term added to the conventional Earth tide correction formula to reduce observed gravity values at a point to a common average value, instead of eliminating the luni-solar effect altogether.

hood [paleont] (a) An arched plate of secondary shell of a brachiopod, arising from the echmidium of *Cryptacanthia* or from the median septum of dallinids. (b) The tough fleshy structure located above the head of *Nautilus* and covering the aperture when the head is withdrawn into the living chamber.

hood [intrus rocks] The *metallized hood* of a batholith.

hoodoo (a) A fantastic column, pinnacle, or pillar of rock produced in a region of sporadic heavy rainfall by differential weathering or erosion of horizontal strata, facilitated by joints and by layers of varying hardness, and occurring in varied and often eccentric or grotesque forms. Cf: *earth pillar*. Syn: *rock pillar*. (b) *hoodoo column*.---Etymol: African; from its fancied resemblance to animals and embodied evil spirits.

hoodoo column A term sometimes applied to an *earth pillar*. Syn: *hoodoo*.

hoodoo rock One of several topographic forms of bizarre shape, developed or modified by differential weathering; e.g. *pedestal rock; earth pillar; hoodoo.*

hook [geomorph] (a) A sandy or gravelly spit or narrow cape turned sharply landward at the outer end, and resembling a hook in form; e.g. a low peninsula or bar ending in a recurved spit and formed at the end of a bay. Also, a *recurved spit*. (b) A sharp bend, curve, or angle in a stream.

hook [paleont] A holothurian sclerite in the form of a common hook used in angling, consisting of an *eye*, a *shank*, and a *spear*.

Hookean substance A material that deforms in accordance with *Hooke's law*.

hooked bay A bay similar to a bight but having a headland at only one end.

hooked dune *fishhook dune*.

hooked spit *recurved spit*.

Hooke's law A statement of *elastic deformation*, that the strain is linearly proportional to the applied stress. See also: *Hookean substance*.

hook valley The valley containing a barbed tributary.

hope [geog] A piece of dry, arable land surrounded by swamp or marsh.

hope [coast] A term used in Scotland for a small bay or inlet.

hope [geomorph] A British term (used esp. in southern Scotland) for a small enclosed valley; esp. the broad upper end of a narrow mountain valley, or a blind valley branching from a larger or wider valley. It is usually rounded and often has a stream flowing through it.

hopeite A gray orthorhombic mineral: $Zn_3(PO_4)_2.4H_2O$. It is dimorphous with parahopeite.

hopper crystal A cubic crystal of salt in which the faces of the cube have grown more at the edges than in the center.

horizon [soil] *soil horizon*.

horizon [geol] (a) A plane of stratification assumed to have been once horizontal and continuous; a particular stratigraphic level in the geologic column or the systematic position of a stratum in the geologic time scale, such as an imaginary isochronous surface that denotes a certain position with the strata over a wide area and that is without actual thickness. (b) A thin stratum belonging to a particular time and characterized by distinctive features; esp. a biostratigraphic zone, or bed or group of beds identified by a certain fossil or fossils peculiar to it. It easily recognized and is persistent over a wide area. The term may also indicate igneous rocks of a particular time, or an identifiable stratum regionally known to contain or be associated with rock containing valuable minerals.---The term, in the strict sense, designates the surface separating two beds and hence having no thickness. The use of the term to denote a bed or stratum is improper (USGS, 1958, p.162). Syn: *geologic horizon*.

horizon [surv] One of several lines or planes used as reference for observation and measurement relative to a given location on the Earth's surface and referred generally to a horizontal direction (Huschke, 1959, p.283); esp. *apparent horizon*. The term is also frequently applied to: *celestial horizon; actual horizon;* and *artificial horizon*.

horizon A The uppermost reflecting horizon of the ocean floor. Cf: *horizon beta; horizon B*.

horizon B The lowermost *reflecting horizon* of the ocean floor. Cf: *horizon A; horizon beta*.

horizon beta A *reflecting horizon* of the ocean floor, occurring between *horizon A and horizon B*.

horizon circle A circle, in an azimuthal projection, defined by points equidistant from the center of the projection. The maximum horizon circle on a polar projection is the equator.

horizon closure The amount by which the sum of a series of adjacent measured horizontal angles around a point fails to equal exactly the theoretical sum of 360 degrees; the *error of closure* of horizon. Also known as "closure of horizon". See also: *closing the horizon*.

horizon system of coordinates A system of curvilinear celestial coordinates (usually altitude and azimuth or azimuth angle) based on the celestial horizon as the primary great circle. Cf: *equator system of coordinates*.

horizontal adj. In geodesy, said of a direction that is tangent to the geoid at a given point. Cf: *vertical [geophys]*.

horizontal angle An angle in a horizontal plane. Cf: *vertical angle*.

horizontal axis The axis about which the telescope of a theodolite or transit rotates when moved vertically. It is the axis of rotation that is perpendicular to the *vertical axis* of the instrument.

horizontal circle A graduated disk affixed to the lower plate of a transit or theodolite by means of which horizontal angles can be measured.

horizontal control A system of points whose horizontal positions and interrelationships have been accurately determined for use as fixed references in positioning and correlating map features.

horizontal dip slip *horizontal slip*.

horizontal direction A direction in a horizontal plane; an observed horizontal angle at a triangulation station, reduced to a common initial direction.

horizontal displacement *strike slip*.

horizontal equivalent The distance between two points on a land surface, projected onto a horizontal plane; e.g. the shortest distance between two contour lines on a map. Abbrev: H.E. Cf: *vertical interval*.

horizontal fault A fault, the dip of which is *zero*. Cf: *vertical fault*.

horizontal field balance An instrument that measures the horizontal component of the magnetic field by means of the torque that the field component exerts on a vertical permanent magnet. The two most common types are the *Schmidt field balance* and the *torsion magnetometer*. Cf: *vertical field balance*.

horizontal fold *nonplunging fold*.

horizontal form index A term used by Bucher (1919, p.154) to express the degree of asymmetry of a current ripple mark, defined as the ratio of the horizontal length of the steep (downcurrent) side to that of the gentle (upcurrent) side. Twenhofel (1950, p.568) used the ratio of the length of the upcurrent side to that of the downcurrent side. Cf: *vertical form index*. See also: *ripple symmetry index*.

horizontal intensity The horizontal component of the vector magnetic field intensity; it is one of the *magnetic elements*, and is symbolized by H. Cf: *vertical intensity*.

horizontal-loop method An electromagnetic method in which the plane of the transmitting coil is horizontal (and in which incidentally the plane of the receiving coil is also horizontal).

horizontal section (a) A *section* representing a horizontal segment of the Earth's crust, such as one drawn parallel to the general direction of the structure. (b) A reconstruction of the vertical section of geologic structure along a given line, drawn from the information on a geologic map or from a special traverse across the land (Challinor, 1967, p.222).

horizontal separation *strike slip*. Cf: *vertical separation*.

horizontal slip In a fault, the horizontal component of the new slip. Cf: *vertical slip*. Syn: *horizontal dip slip*.

horizontal throw The *heave* of a fault.

hormites A group name suggested (but not approved) for the sepiolite and palygorskite clay minerals.

hormogonium A multicellular segment of filamentous blue-green algae, fragmented at random or at separation disks.

horn [geog] (a) A body of land (such as a spit) or of water shaped like a horn. (b) The pointed end of a dune or beach cusp; e.g. the forward extending, outer end of a barchan crescent. Syn: *wing*.

horn [glac geol] A high, rocky, sharp-pointed, steep-sided, pyramidal mountain peak with prominent faces and ridges, bounded by the intersecting walls of three or more cirques that have been cut back into a mountain by headward erosion of glaciers; e.g. the Matterhorn of the Pennine Alps. See also: *tind*. Syn: *glacial horn; matterhorn; cirque mountain; pyramidal peak; monumental peak; horn peak*.

hornblende (a) The commonest mineral of the amphibole group: $Ca_2Na(Mg,Fe^{+2})_4(Al,Fe^{+3},Ti)(Al,Si_8O_{22}(O,OH)_2$. It has a variable composition, and may contain potassium and appreciable fluorine. Hornblende is commonly black, dark green, or brown, and occurs in distinct monoclinic crystals or in columnar, fibrous, or granular forms. It is a primary constituent in many acid and intermediate igneous rocks (granites, syenites, diorites, andesites) and less commonly in basic igneous rocks, and it is a common metamorphic mineral in gneisses and schists. Symbol: Ho. (b) A term sometimes used (esp. by the Germans) to designate the *amphibole* group of minerals. The term "Hornblende" is an old German name for any dark, prismatic crystal found with metallic ores but containing no valuable metal (the word "Blende" indicates "a deceiver").---Obsolete syn: *hornstone*.

hornblende andesite *hungarite*.

hornblende-hornfels facies Rocks formed in the middle grades of thermal (contact) metamorphism at temperatures between 350°C and 550°C and at low pressures not exceeding about 2500 bars (Turner and Verhoogen, 1960, p.511). It is part of the *hornfels facies*. Cf: *pyroxene-hornfels facies; albite-epidote-hornfels facies*.

hornblendite An igneous rock composed almost entirely of hornblende. The term has been equated incorrectly by some authors with the metamorphic rock amphibolite.

hornesite *hoernesite*.

hornfels A fine-grained rock composed of a mosaic of equidimensional grains without preferred orientation and typically formed by contact metamorphism. Porphyroblasts or relict phenocrysts may be present in the characteristically granoblastic (or decussate) matrix (Winkler, 1967). Cf: *skleropelite*.

hornfels facies Rocks formed by thermal (contact) metamorphism at relatively shallow depths in the Earth's crust (not exceeding about 10 km or pressure of 3000 bars) and at temperatures ranging from 250°C to 800°C, depending on the distance from the intrusive contact and the source of heat (Winkler, 1967). It is now considered to be inclusive of three facies: the *albite-epidote-hornfels facies*; the *pyroxene-hornfels facies*; and the *hornblende-hornfels facies*.

hornito A *spatter cone*, esp. around a rootless vent on a lava flow. Syn: *driblet cone*.

horn lead *phosgenite*.

horn mercury *calomel*.

horn peak *horn*.

horn quicksilver *calomel*.

horn silver *chlorargyrite*.

hornstein A syn. of *hornstone*. Etymol: German *Hornstein*, "chert".

hornstone [mineral] (a) A compact, flinty, brittle variety of chalcedony; "a siliceous mineral substance, sometimes ap-

proaching nearly to flint, or common quartz" (Lyell, 1854, p.807). (b) An obsolete name formerly applied to *hornblende*.

hornstone A general term for a compact, tough, and siliceous rock having a splintery or subconchoidal fracture (Holmes, 1920). The term has been used to describe flint or chert as well as hornfels, and has also been confused with hornblende. According to Tarr (1938, p.26), the term should be abandoned because it is doubtful that it can designate readily identifiable material. Syn: *hornstein; petrosilex*.

horny sponge Any demosponge that possesses spicules but has a spongin skeleton.

horotely A phylogenetic phenomenon characterized by a normal or average rate of evolution. Cf: *bradytely; tachytely*.

horse [ore dep] A miners' term for a barren rock mass equivalent to the country rock but occurring within the vein.

horse [fault] A displaced rock mass that has been caught along the fault between the walls.

horse *horseback*.

horseback [glac geol] A low, sharp ridge of sand, gravel, or rock; specif. an esker or esker-like deposit, or a kame, in northern New England, esp. Maine. Syn: *hogback; boar's back*.

horseback [mining] (a) In a coal seam, a mass of shale or sandstone filling a natural channel cut by flowing water in a coal seam. (b) A bank or ridge of foreign matter in a coal seam.----Syn: *kettleback; horse; symon fault; cutout; roll; swell; washout*.

horseflesh *bacon* [sed].

horseflesh ore Cornish syn. of *bornite*.

horse latitudes Oceanic areas between 30°-35°N and S, characterized by calms or light winds, heat, and dryness. These belts move north and south about 5° following the Sun.

horseshoe A topographic feature, such as a valley or a mountain range, shaped like a horseshoe.

horseshoe bend An *oxbow* in the course of a stream.

horseshoe dissepiment One of a single vertical series of dissepiments of a rugose coral, characterized by a horizontal base and a strongly arched top part. Cf: *lateral dissepiment*.

horseshoe dune *barchan*.

horseshoe flute cast *current crescent*.

horseshoe lake A lake occupying a horseshoe-shaped basin; specif. an *oxbow lake*.

horseshoe moraine A terminal moraine, markedly convex on the down-valley side, usually formed at the end of a valley glacier that never advanced beyond the mountain front.

horseshoe reef A horseshoe-shaped reef that develops from a reef pinnacle or table reef parallel to the dominant wave action on a platform reef. Its cusp extends downwind and its interior parts often become densely vegetated to produce a small wooded island of low relief.

horsetail [ore dep] adj. Said of veins that are smaller fissures emanating from a major vein; also, said of an ore comprising a series of such veins.

horsetail [bot] *sphenopsid*.

horsetailing A feathery or frond-like fluting, grooving, or other structure developed on the surfaces of shatter-coned rocks by the distinctive striations that radiate from the apex of each shatter cone and extend along its length. The presence of multiple nested and parasitic shatter cones produces a distinctive horsetail-like effect.

horsfordite A silver-white mineral: Cu_5Sb.

horst [speleo] A type of *pendant*, the connecting part of which is smaller than the projection.

horst [struc geol] An elongate, relatively uplifted crustal unit or block that is bounded by faults on its long sides. It is a structural form and may or may not be expressed geomorphologically. Etymol: German, no direct English equivalent. Cf: *graben*.

hortite An obsolete name applied to a dark-colored syenite that may have formed from gabbro as a result of the assimilan of limestone.

orton number A dimensionless number formed by the product of *runoff intensity* and *erosion proportionality factor*. It expresses the relative intensity of the erosion process on the slopes of a drainage basin. Named in honor of Robert E. Horton (1875-1945), U.S. hydraulic engineer. Symbol: N_H.

rtonolite A mineral of the olivine group: $(Fe,Mg,Mn)_2SiO_4$. is a variety of fayalite containing magnesium and manganese.

st A rock or mineral that is older than other rocks or minerals introduced into it or formed within or adjacent to it, such a *host rock* and a large crystal that contains the inclusions smaller crystals of a different mineral species; a *paleosome* a *palasome*. Ant: *guest*.

st element An essential element replaced by a trace, or est, element in a mineral.

stile ice A submariner's term for an ice canopy containing large skylights or other features that permit a submarine to rface (U.S. Naval Oceanographic Office, 1968, p.B33). Ant: *endly ice*.

st rock A body of rock serving as a *host* for other rocks or mineral deposits; e.g. a pluton containing xenoliths, or any ck susceptible to attack by mineralizing solutions and in ich ore deposits occur (such as the wall rock of an igenetic ore deposit).

t As related to radioactivity, said of a highly radioactive bstance.

t brine Warm and very saline water with an abnormally high ntent of metal ions, such as are found on the bottom of the ed Sea. Temperature may be as high as 56°C, and salinity high as 256‰.

t desert An arid area where the mean annual temperature higher than 18°C (64.4°F) (Stone, 1967, p.230).

t lahar A flow of hot volcanic materials down the slope of a lcano, produced by heavy rains after an eruption. Cf: *cold har*. Syn: *hot mudflow*.

t mudflow *hot lahar*.

t spring A *thermal spring* whose temperature is above that the human body (Meinzer, 1923, p.54). Cf: *warm spring*.

urglass structure A type of zoning, especially common in nopyroxenes, in which the core, distinguished from the ter part by a difference of color or optical properties, has a oss section resembling that of an hourglass.

urglass valley (a) A valley whose pattern in plan view resembles an hourglass; e.g. a valley extending without interruption across a former divide, toward which it narrows from both ections (Engeln, 1942, p.377). (b) *wineglass valley*.

ver A floating island of vegetation.

w An English term for a low, small hill in a valley or dale; a und or hillock.

wardite An achondritic stony meteorite consisting largely of lcic plagioclase and orthopyroxene (commonly hyperhene). It has a lower content of iron and calcium than that eucrite.

we (a) A Scottish term for a hollow or depression, esp. one the Earth's surface, as a basin or a valley. (b) *hoe*.

wieite A mineral: $Na(Fe,Mn)_{10}(Fe,Al)_2Si_{12}O_{31}(OH)_{13}$.

wlite A white nodular or earthy mineral: $Ca_2Bi_5SiO_9(OH)_5$.

ya A stream bed, valley, or basin in a rugged mountainous gion, as the Peruvian Andes. Etymol: Spanish, "large hole, vity, pit".

yt gravimeter *Gulf-type gravimeter*.

ngitudinal joint.

ianghualite An isometric mineral: $Ca_3Li_2Be_3(SiO_4)_3F_2$.

angho deposit A general term applied by Grabau (1936, 253) to a coastal-plain deposit that consists of alluvium that spread out over a level surface (as a flood plain or a delta) ove the normal reach of the sea but that passes laterally o marine beds of equivalent age. Type locality: the loessrived alluvial deposits at the mouth of the Huang Ho (Yel-

low River) in northern China. See also:*shantung*.

huanghoite A hexagonal mineral: $BaCe(CO_3)_2F$.

hub The cylindrical or hemispherical projection on the central part (and usually on the lower surface) of a *wheel* of a holothurian.

Hubble constant The amount by which the distance to a galaxy must be multiplied in order to get its velocity or red shift: 75 km sec^{-1} 10^{-6} PSC^{-1}. It represents the present expansion rate of the Universe. It has the unit of reciprocal time and would represent the age of the Universe were there no gravitational deceleration.

hubnerite *huebnerite*.

Hudsonian orogeny A name proposed by Stockwell (1964) for a time of plutonism, metamorphism, and deformation during the Precambrian in the Canadian Shield (especially in the Churchill, Bear, and Southern provinces), dated radiometrically as between 1640 and 1820 m.y. ago. It is synonymous with the *Penokean orogeny* of Minnesota and Michigan.

hudsonite A nonpreferred syn. of *cortlandtite*, since "hudsonite" had been used earlier for a variety of pyroxene.

huebnerite A brownish-red to black mineral of the wolframite series: $MnWO_4$. It is isomorphous with ferberite, and may contain up to 20% iron tungstate. Syn: *hübnerite*.

huemulite A triclinic mineral: $Na_4MgV_{10}O_{28}.24H_2O$.

huerfano A term used in the SW U.S. for a hill or mountain of older rock entirely surrounded, but not covered, by any kind of later sedimentary material; esp. a solitary eminence separated by erosion from the mass of which it once formed a part. Etymol: Spanish *huérfano*, "orphan". Pron: ware-fa-no. Cf: *lost mountain; tejon*.

hügelite A brown to orange-yellow secondary mineral: $Pb_2(UO_2)_3(AsO_4)_2(OH)_4.3H_2O$.

Hugoniot The locus of points describing the pressure-volume-energy relations or states that may be achieved within a material by shocking it from a given initial state. Named after Pierre Henri Hugoniot (1851-1887), French physicist. Pron: ewe-go-neeoh. Syn: *Hugoniot curve*.

hühnerkobelite A mineral: $(Na_2,Ca)(Fe^{+2},Mn^{+2})_2(PO_4)_2$. It is isomorphous with varulite.

hullite A soft, black, waxy-appearing aluminosilicate of ferric iron, magnesium, calcium, and alkalies, occurring as interstitial matter and amygdaloidal infillings in certain basalts. It is perhaps identical with chlorophaeite.

hulsite A black mineral: $(Fe^{+2},Mg)_2(Fe^{+3},Sn)(BO_3)O_2$.

hum A residual limestone hill in a region of karst; a karst inselberg. It has steep sides and is surrounded by flat alluvial plains. Syn: *mogote*.

humanthracite Humic coal of anthracitic rank; it is the highest stage in the *humolith series*.

humanthracon Humic coal of bituminous rank; it is the fifth stage in the *humolith series*.

humate A salt or ester of *humic acid*.

humberstonite A mineral: $Na_7K_3Mg_2(SO_4)_6(NO_3)_2.6H_2O$.

Humble gravimeter A gravimeter consisting of a mass, hinged lever, and several springs. The gravity force is therefore balanced by an elastic force. The instrument depends for its sensitivity on proximity to an instability configuration (Bryan, 1937, p. 301-308).

humboldtine A mineral: $FeC_2O_4.2H_2O$. It occurs in capillary or botryoidal forms in brown coal and black shale. Syn: *humboldtite; oxalite*.

humboldtite (a) *datolite*. (b) *humboldtine*.

humic Pertaining to or derived from *humus*.

humic acid Black, acidic, organic matter extracted from soils, low rank coals, and other decayed plant substances by alkalis. It is insoluble in acids and organic solvents. Syn: *geic acid (obs.)*.

humic-cannel coal *pseudocannel coal*.

humic coal Coal that is derived from peat by the process of humification. Most coal is of this type, including brown coal and lignite, and bituminous coal and anthracite. Cf: *sapropelic coal*. See also: *humolith series*. Syn: *cahemolith; chameolith; chaemolith; humulith; humus coal; humulite; humolite; humite [coal]; humolith*.

humic decomposition Chemical breakdown of rocks and minerals by the action of vegetable acids.

humic degradation matter Organic degradation matter that is cellulosic and in which the individual particles are still recognizable; it is similar to anthraxylon. It is classified according to the type of constituent plant material. See also: *translucent humic degradation matter*. Syn: *cell-wall degradation matter; brown matter*.

Humic Gley soil One of an intrazonal, hydromorphic group of soils having a dark surface horizon underlain by gley. It occurs in wet meadows and in swamps with forest vegetation.

humidity The relationship between the atmosphere and the water vapor it contains. The unmodified term often signifies *relative humidity*. See also: *absolute humidity; specific humidity*.

humification The process of development of humus or humic acids, essentially by slow oxidation. Adj: *humified*. See also: *mor*.

humin *ulmin*.

huminite A variety of oxidized bitumen, resembling brown coal, found in a granite-pegmatite vein in Sweden.

humite [coal] *humic coal*.

humite [mineral] (a) A white, yellow, brown, or red orthorhombic mineral: $Mg_7Si_3O_{12}(F,OH)_2$. It sometimes contains appreciable iron, and it is found in the masses ejected from volcanoes. (b) A group of isomorphous magnesium-silicate minerals frequently containing fluorine and closely resembling one another in chemical composition, physical properties, and crystallization. It consists of olivine, humite, clinohumite, chondrodite, and norbergite.

hummerite A mineral: $KMgV_5O_{14}.8H_2O$.

hummock [permafrost] A *frost mound*, esp. an *earth hummock*.

hummock [geog] A rounded or conical knoll, mound, hillock, or other small elevation, generally of equidimensional shape and not ridge-like. Also, a slight rise of ground above a level surface. Syn: *hammock*.

hummock [ice] A mound, hillock, or pile of broken floating ice, either fresh or weathered, that has been forced upward by pressure, as in an ice field or ice floe. Cf: *bummock*.

hummocked ice Sea ice having a rugged, uneven surface due to the formation of hummocks; it has the appearance of smooth hillocks when weathered. A form of *pressure ice*.

hummocky Abounding in hummocks, or uneven; said of topographic landforms, as a hummocky dune, and of hummocked ice.

hummocky moraine An area of knob-and-kettle topography that may have been formed either along a live-ice front or around masses of stagnant ice (Gravenor & Kupsch, 1959, p.52).

humocoll Humic material of the rank of peat; it is the second stage in the *humolith series*. Cf: *saprocol*.

Humod In U.S. Dept. of Agriculture soil taxonomy, a suborder of the soil order *Spodosol*, characterized by accumulations of organic carbon and aluminum (but not of iron) in its spodic horizon. A Humod has none of the characteristics associated with wetness (SSSA, 1970). Cf: *Aquod; Ferrod; Orthod*.

humodil Humic coal of lignitic rank; it is the third stage in the *humolith series*. Cf: *saprodil*.

humodite Humic coal of subbituminous rank; it is the fourth stage in the *humolith series*. Cf: *saprodite*.

humodurite *translucent attritus*.

humogelite *ulmin*.

humolite *humic coal*.

humolith *humic coal*.

humolith series Humic material and coals in order of meta-

morphic rank: *humopel, humocoll, humodil, humodite, huma thracon* and *humanthracite* (Heim & Potonié, 193 p.146). Cf: *sapropelite series; humosapropelic serie saprohumolith series*. See also: *humic coal*.

humonigritite A type of *nigritite* that occurs in sediments. C *polynigritite; exinonigritite; keronigritite*.

humopel Organic matter, or *ulmin*, of humic coals; it is th first stage in the *humolith series*. Cf: *sapropel*.

humosapropelic series Organic materials and coals interme ate between the *humolith series* and the *sapropelite serie* with humolithic materials predominating. Cf: *saprohumol series*.

humosite A dark brownish-red, translucent, isotropic, micr scopic constituent of torbanite (Dulhunty, 1939).

humovitrinite Vitrinite in vitrain of humic coal. Cf: *saprovitr ite*.

Humox In U.S. Dept. of Agriculture soil taxonomy, a subord of the soil order *Oxisol*, characterized by being moist almc all or all of the time, by having a high content of organic co tent within the first meter of depth, and by having a mean a nual soil temperature of less than 22°C (SSSA, 1970). C *Aquox; Orthox; Torrox; Ustox*.

humpy A small morainal mound with a central depressi (Gravenor & Kupsch, 1959, p.53).

humulite *humic coal*.

humulith *humic coal*.

Humult In U.S. Dept. of Argiculture soil taxonomy, a subord of the soil order *Ultisol*, characterized by having a high c ganic-carbon content (SSSA, 1970). Cf: *Aquult; Udult; Ustu Xerult*.

humus The generally dark, more or less stable part of the c ganic matter of the soil so well decomposed that the origir sources cannot be identified (sometimes used incorrectly the total organic matter of the soil, including relatively und composed material). Syn: *soil ulmin*.

humus coal *humic coal*.

humus layer *H layer*.

hungarite An andesite containing abundant hornblende. Sy *hornblende andesite*.

hungchaoite A mineral: $MgB_4O_7.9H_2O$.

hungry (a) Said of a rock, lode, or belt of country that is ba ren of ore minerals (such as white quartz) or of geologic in cations of ore, or that contains very low-grade ore. Ant: *like* (b) Said of a soil that is poor or not fertile.

huntite A white mineral: $CaMg_3(CO_3)_4$.

hureaulite A monoclinic mineral: $H_2Mn_5(PO_4)_4.4H_2O$. It c curs in yellowish, orangy, reddish, rose, or grayish prisma crystals or massive.

hurlbutite A mineral: $CaBe_2(PO_4)_2$.

Huronian A division of the Proterozoic of the Canadian Shield.

hurricane (a) A *tropical cyclone*, esp. in the West Indies. C *typhoon*. (b) On the Beaufort wind scale, a wind having a v locity of 73-80 mph.

hurricane delta A delta formed by storm waves carrying sa across a reef or barrier island and depositing it in a lagoon.

hurricane surge *storm surge*.

hurricane tide (a) *storm surge*. (b) The height of a stor surge above the astronomically predicted level of the sea.

hurricane wave *storm surge*.

hurst (a) A wooded knoll, hill, or other small eminence; grove or a thick wood; a copse. (b) A bank or piece of risi ground; esp. a sandbank in or along a river.----The term very common in place names. Syn: *hirst; hyrst*.

husebyite A plagioclase-bearing nepheline syenite.

hushing *hydraulic prospecting*.

hussle A term used in England for soft clay associated w coal and for contorted carbonaceous shale immediately bel a coal seam. Etymol: Yorkshire dialect for "rubbish".

hutchinsonite A scarlet to deep cherry-red orthorhombic m eral: $(Pb,Tl)_2(Cu,Ag)As_5S_{10}$.

Huttonian Of or relating to James Hutton (1726-1797), Scottish geologist, who advocated the theory of *plutonism*, introduced the concepts of uniformitarianism and the geologic cycle, and emphasized the length of geologic time.

Huttonite A colorless to pale-cream monoclinic mineral: SiO_4. It is dimorphous with thorite and isostructural with monazite.

Huygens' principle The statement in physics that any point or particle excited by the impact of wave energy becomes a new point source of energy (Wahlstrom, 1948, p.30). It is named after the Dutch astronomer and mathematician Christian Huygens (d. 1695).

H wave hydrodynamic wave.

hyacinth (a) A transparent orange, red, reddish-brown, or brownish *zircon*, sometimes used as a gem. The term has been used interchangeably with *jacinth* to designate yellow-orange, yellow, or brown zircon, and loosely to signify any zircon. (b) Yellow, orange, red-orange, or brownish *essonite* used as a gem. Syn: *hyacinth garnet; hyacinthoid.* (c) A term applied as a syn. of various orange-red to orange minerals, and also of minerals such as harmotome, vesuvianite, and meionite. (d) A precious stone believed by the ancients to be a sapphire.

hyaline [mineral] Said of a mineral that is amorphous.

hyaline [paleont] Said of the glassy clear or transparent fine-textured outer wall of a foraminifer.

hyaline [ign] A syn. of *glassy* [ign]; sometimes used as a prefix ("hyalo-") to names of volcanic rocks with a glassy texture, e.g. "hyalobasalt".

hyalinocrystalline Said of the texture of a porphyritic igneous rock in which the phenocrysts are in a glassy groundmass. Also, said of a rock having such texture. Syn: *crystallohyaline.*

hyalite A colorless variety of common opal that is sometimes clear as glass and sometimes translucent or whitish and that occurs as globular concretions (resembling drops of melted glass) or botryoidal crusts lining cavities or cracks in rocks. Syn: *water opal; Müller's glass.*

hyalithe An opaque glass resembling porcelain and frequently black, green, brown, or red.

hyalo- A prefix that is used with rock names to indicate a glassy nature. Cf: *vitr-.*

hyalobasalt tachylite.

hyaloclastite A deposit resembling tuff that is formed by the flowing of basalt under water or ice and its consequent granulation or shattering into small, angular fragments. Syn: *aquagene tuff.*

hyalocrystalline Said of the texture of a porphyritic igneous rock in which the phenocrysts and glassy groundmass are equal or nearly equal in amount, the ratio of crystals to groundmass being between five to three and three to five (Cross, et al, 1906, p.694).

hyalomelane A volcanic glass, commonly porphyritic, differing from *tachylite* in its insolubility in acids; a balsatic vitrophyre.

hyalomylonite A glassy rock formed by fusion of granite, arkose, etc. by frictional heat in zones of intense differential movement. Cf: *buchite.*

hyalo-ophitic Said of the texture of an igneous rock in which the groundmass is glassy and composes a higher proportion of the rock than with *intersertal* texture. Cf: *hyalopilitic.*

hyalophane A colorless monoclinic mineral of the feldspar group: $(K,Ba)Al(Al,Si)_3O_8$. It is intermediate in composition between celsian and orthoclase.

hyalopilitic Said of the *intersertal* texture of an igneous rock in which needle-like microlites of the groundmass are set in a glassy mesostasis, the groundmass forming a greater proportion of the rock than in *hyalo-ophitic* textures.

hyalopsite obsidian.

hyalosiderite A rich olive-green variety of olivine containing considerable iron (30-50 mole percent of Fe_2SiO_4).

hyalospongea Any sponge belonging to the class Hyalospongea and characterized chiefly by a skeleton composed of siliceous spicules and without calcium carbonate or spongin. Syn: *hexactinellid; glass sponge.*

hyalotekite A white or gray mineral: $(Pb,Ca,Ba)_4BSi_6O_{17}(OH,F)$.

hybrid [evol] An individual having parents belonging to different species.

hybrid [ign petrol] adj. Pertaining to a rock, the chemical composition of which is the result of *assimilation.* Cf: *anomalous.* Syn: *contaminated.*----n. A rock whose composition is the result of assimilation.----See also: *hybridism.*

hybrid age The radiometric age given by an isotopic system that has lost radiogenic isotopes due to thermal, igneous, or tectonic activity some time after the start of the isotopic system. See also: *overprint* [geochron]; *updating; mixed ages.*

hybridism (a) *hybridization.* (b) The condition of being *hybrid.*

hybridization The process whereby rocks of different composition from that of the parent magma are formed, by *assimilation.* Cf: *contamination.* Syn: *hybridism.*

Hydaspien Anisian.

hydatogenesis [sed] The crystallization or precipitation of salts from normal aqueous solutions; the formation of an evaporite.

hydatogenic Said of a rock or mineral deposit formed by an aqueous agent, e.g. a mineral deposit in a vein from a magmatic solution or an evaporite from a body of salt water. Cf: *pneumatogenic; hydatopneumatogenic.*

hydatomorphic hydatogenic.

hydatopneumatogenic Said of a rock or mineral deposit formed by both aqueous and gaseous agents. Cf: *hydatogenic; pneumatogenic.*

hydatopyrogenic aqueo-igneous.

hydnophorid Said of a scleractinian corallum with corallite centers arranged around protuberant collines or monticules (TIP, 1956, pt.F, p.248).

hydrarch adj. Said of an ecologic succession (i.e. a sere) that develops under *hydric* conditions. Cf: *mesarch; xerarch.* See also: *hydrosere.*

hydrargillite (a) *gibbsite.* (b) A name that has been applied to various aluminum-bearing minerals, including aluminite, wavellite, and turquoise.

hydrate n. A mineral compound that is produced by hydration; a mineral compound in which water is part of the chemical composition. v. To cause to incorporate water into the chemical composition of a mineral.

hydrated halloysite endellite.

hydration rind dating obsidian hydration dating.

hydration water chemical water.

hydraulic [eng] Conveying, acting, operated, effected, or moved by means of water or other fluids, such as a "hydraulic dredge" using a centrifugal pump to draw mud or saturated sand from a river channel.

hydraulic [hydraul] Pertaining to a fluid in motion, or to movement or action caused by water.

hydraulic [mater] Hardening or setting under water; e.g. "hydraulic lime" or "hydraulic cement".

hydraulic action The mechanical loosening and removal of loose or weakly resistant material solely by the pressure and *hydraulic force* of flowing water, as by a stream surging into rock cracks or impinging against the bank on the outside of a bend, or by ocean waves and currents pounding the base of a cliff. See also: *fluviraption.*

hydraulic conductivity permeability coefficient.

hydraulic current A local current produced by differences in water level (at the two ends of a channel) set up by the rising and falling tide at constrictions in a baymouth or in the narrow strait connecting two bodies of water having tides that differ in time or range; e.g. in The Hell Gate, where Long Island Sound joins the East River, N.Y.

hydraulic element A quantity pertaining to a particular stage of flowing water in a particular cross section of a conduit or stream channel, e.g. depth of water, cross-sectional area, hydraulic radius, wetted perimeter, mean depth of water, velocity, energy head, friction factor (ASCE, 1962).

hydraulic equivalent The number of Udden size grades between the size of a given mineral grain of a sediment and the size of the quartz grain with which it was deposited or to which it is hydraulically equivalent (the size of a larger or smaller grain that settles with the given mineral grain under the same conditions) (Rittenhouse, 1943). See also: *hydraulic ratio*.

hydraulic fill Earth or waste material that has been excavated, transported, and flushed into place by moving water.

hydraulic-fill dam A dam composed of hydraulic-fill material in which the sorting of particle sizes into an impervious central core supported by outer zones of coarser material is accomplished by arrangement of peripheral discharge outlets and flow in the central pool.

hydraulic force The eroding and shearing force of flowing water, involving no sediment load and resulting in *hydraulic action*.

hydraulic friction The resistance to flow exerted on the perimeter or contact surface between a stream and its containing conduit, due to the roughness characteristic of the confining surface, which induces a loss of energy. Energy losses arising from excessive turbulence, impact at obstructions, curves, eddies, and pronounced channel changes are not ordinarily ascribed to hydraulic friction (ASCE, 1962).

hydraulic geometry The description, at a given cross section of a river channel, of the graphical relationships among plots of hydraulic characteristics (such as width, depth, velocity, channel slope, roughness, and bed particle size, all of which help to determine the shape of a natural channel) as simple power functions of river discharge (Leopold & Maddock, 1953).

hydraulic grade line In a closed channel, a line joining the elevations that water would attain in atmospheric pressure; in an open channel, the free water surface. Its slope represents *energy loss*. See also: *hydraulic gradient; hydraulic head*.

hydraulic gradient (a) In an aquifer, the rate of change of *pressure head* per unit of distance of flow at a given point and in a given direction. Cf: *pressure gradient*. See also: *gradient of the head*. Syn: *potential gradient*. (b) In a stream, the slope of the *hydraulic grade line*. See also: *critical hydraulic gradient*.

hydraulic head (a) The height of the free surface of a body of water above a given subsurface point. (b) The water level at a point upstream from a given point downstream. (c) The elevation of the *hydraulic grade line* at a given point above a given point of a pressure pipe.----(ASCE, 1962).

hydraulic jump In fluid flow, a change in flow conditions accompanied by a stationary, abrupt turbulent rise in water level in the direction of flow. It is a type of stationary wave.

hydraulic limestone An impure limestone that contains silica and alumina (usually as clay) in varying proportions and that yields, upon calcining, a quicklime that will harden under water to form a firm, strong, and solidified mass. See also:*cement rock*. Syn: *water lime*.

hydraulic mean depth *hydraulic radius*.

hydraulic mining The extraction of ore by means of strong jets of water. The ore may be surficial, such as a placer, or subterranean, such as a soft coal. Cf: *placer mining*.

hydraulic permeability The ability of a rock or soil to transmit water under pressure. It may vary according to direction.

hydraulic plucking A process of stream erosion by which rock fragments are forcibly removed by the impact of water entering cracks in a rock. Syn: *quarrying*.

hydraulic profile A vertical section of the potentiometric surface of an aquifer.

hydraulic prospecting The use of water to clear away surficial deposits and debris to expose outcrops, for the purpose of exploring for mineral deposits. Syn: *hushing*.

hydraulic radius In a stream, the ratio of the area of its cross section to its *wetted perimeter*. Symbol: R. Syn: *hydraulic mean depth*.

hydraulic ratio A value expressing the quantity of any given heavy mineral in a sediment, equal to the weight of a heavy mineral in a given size class divided by the weight of light minerals in the *hydraulic-equivalent* class (Rittenhouse, 1943). The value is commonly multiplied by 100 to reduce the number of decimal places.

hydraulics That aspect of engineering which deals with the flow of water or other liquids; the practical application of *hydromechanics*.

hydraulic wedging Pressure produced in a cavity within a reef or other body of rock by pounding surf (Cloud, 1957, p.1016).

hydric Said of a habitat that has or requires abundant moisture; also, said of an organism or group of organisms occupying such a habitat. Cf: *xeric; mesic*. See also: *hydrarch*.

hydroamphibole A mixture of hornblende and chlorite.

hydrobasaluminite A mineral: $Al_4(SO_4)(OH)_{10}.36H_2O$.

hydrobiolite An organic rock formed by the simple accumulation and drying out of organisms.

hydrobiology The biology of bodies of water, esp. of lakes and other bodies of fresh water. Cf: *biohydrology*.

hydrobios That area of the Earth occupied by aquatic plants and animals. Cf: *geobios*.

hydrobiotite (a) A light-green, trioctahedral, mixed-layer clay mineral composed of interstratification of biotite and vermiculite. (b) A term applied originally to a biotite-like material high in water.

hydroboracite A white mineral: $CaMgB_6O_{11}.6H_2O$.

hydrocalcite (a) A mineral name applied to material that is perhaps $CaCO_3.2H_2O$ or $CaCO_3.3H_2O$. (b) A mineral name used by Marschner (1969) for a compound now known as *monohydrocalcite*.

hydrocalumite A colorless to light-green mineral: $Ca_2Al(OH)_7.3H_2O$ or $Ca_4Al_2O_7.12H_2O$.

hydrocarbon Any organic compound, gaseous, liquid, or solid, consisting solely of carbon and hydrogen. They are divided into groups of which those of especial interest to geologists are the *paraffin, cycloparaffin, olefin*, and *aromatic* groups. Crude oil is essentially a complex mixture of hydrocarbons.

hydro cast *hydrographic cast*.

hydrocerussite A colorless hexagonal mineral: $Pb_3(CO_3)_2(OH)_2$. It occurs as a secondary product as an encrustation on native lead or on galena.

hydrochemical facies (a) The diagnostic chemical character of ground-water solutions occurring in hydrologic systems (Back, 1966, p.11). It is determined by the flow pattern of the water and by the effects of chemical processes operating between the ground water and the minerals within the lithologic framework. (b) A term used by Chebotarev (1955, p.199) to indicate concentration of dissolved solids (facies may be low-, transitional-, or high-saline).

hydrochore A plant whose seeds or spores are distributed by water.

hydroclast A rock fragment that is transported and deposited in an aqueous environment.

hydroclastic rock (a) A clastic rock deposited by the agency of water. Syn: *hydrolith*. (b) A rock broken by wave or current action. (c) A volcanic rock broken or fragmented during chilling under water or ice. Syn: *hydroclastic volcanic rock*.

hydroclimate The physical and often the chemical factors that characterize a particular aquatic environment.

hydrocyanite *chalcocyanite*.

hydrodialeima A term proposed by Sanders (1957, p.295) for an unconformity caused by subaqueous processes.

hydrodolomite A mixture of hydromagnesite and calcite.

hydrodynamic jetting Directional ejection of molten or vaporized material at very high velocities as a result of shock-wave

teractions at the interface between projectile and target in
e early stages of hypervelocity impact (Gault et al, 1968,
90). Such a process is believed to operate to form tektites
' meteorite impacts.

ydrodynamics That aspect of *hydromechanics* which deals
ith forces that produce motion. Cf: *hydrostatics; hydrokinet-
s.*

ydrodynamic wave An obsolete term for a type of *surface
ave* that is similar to a Rayleigh wave but having a particle
otion in the direction of the advance of the wave front at the
aximum "up" position of the trajectory.

ydroelectric power Electrical energy generated by means of
power generator coupled to a turbine through which water
asses. Cf: *waterpower; hydropower; white coal* [*water*].

ydroexplosion An explosion caused by the contact of a mol-
n lava flow with water. It may be a submarine explosion or a
toral explosion.

ydrogarnet (a) A group of garnet minerals of the general for-
ula: $A_3B_2(SiO_4)_{3-x}(OH)_{4x}$. (b) A mineral of the hydrogarnet
oup, such as hydrogrossular.

ydrogenesis The natural condensation of moisture in the air
aces of surficial soil or rock material.

ydrogenic Said of a soil whose dominant formative influence
water, as in a cold, humid area. Such a soil may be called
hydrosol.

ydrogenic rock A sedimentary rock formed by the agency of
ater; an *aqueous rock*. The term was restricted by Grabau
924, p.280) to an aqueous rock or *hydrolith* "wholly of
emical origin", such as a precipitate from solution in water.

ydrogenous (a) Said of coals high in moisture content, such
s brown coals. (b) Said of coals high in volatiles, such as
propelic coals.

ydrogen sulfide mud *black mud.*

ydrogeochemistry The chemistry of ground and surface wa-
rs, particularly the relationships between the chemical char-
cteristics and quality of waters and the areal and regional
eology.

ydrogeology The science that deals with subsurface waters
nd related geologic aspects of surface waters. Also used in
e more restricted sense of *ground-water geology* only. In
919, the term was defined by Mead (p.2) as the study of the
ws of the occurrence and movement of subterranean wa-
rs. In later usage it has been used interchangeably with
eohydrology.

ydroglauberite A mineral: $Na_4Ca(SO_4)_3.2H_2O$.

ydrograph A graph showing stage, flow, velocity, or other
aracteristics of water with respect to time (Langbein &
eri, 1960). A stream hydrograph commonly shows rate of
ow, a ground-water hydrograph, water level, or head.

ydrographic basin (a) The *drainage basin* of a stream. (b)
n area occupied by a lake and its drainage basin.

ydrographic cast A hydrographic survey station at which
mperature and salinity measurements are made at standard
cean depths, in order to compute densities. Syn: *hydro cast,
ansen cast; oceanographic cast.*

ydrographic chart A map used in navigation, showing water
epth, bottom relief, tides and currents, adjacent land, and
stinguishing surface features. Syn: *nautical chart.*

ydrography (a) The science that deals with the physical as-
ects of all waters on the Earth's surface, esp. the compila-
on of navigational charts of bodies of water. (b) The body of
cts encompassed by hydrography.

ydrogrossular A mineral of the hydrogarnet group: Ca_3-
$l_2(SiO_4)_{3-x}(OH)_{4x}$, with x near $1/2$. Syn: *hibschite; plazo-
te; hydrogrossularite.*

ydrohalite A mineral: $NaCl.2H_2O$. It is formed only from salty
ater at or below the freezing temperature of pure water.

ydrohalloysite *endellite.*

ydrohematite *turgite.*

ydroherderite *hydroxyl-herderite.*

hydrohetaerolite A black mineral of uncertain composition:
$Zn_2Mn_4O_8.H_2O$ (?).

hydroid Any one of a group of hydrozoans belonging to the
order Hydroida, among which the polypoid (usually colonial)
generation is dominant, and the skeleton is commonly composed of a horn-like material. Cf: *millepore; stylaster.*

hydrokaolin (a) *endellite.* (b) A fibrous variety of kaolinite
from Saglik in Transcaucasia, U.S.S.R.

hydrokinetics That aspect of *hydromechanics* which deals
with forces that cause change in motion. Cf: *hydrodynamics; hydrostatics.*

hydrolaccolith An ice-cored *frost mound*, roughly 0.1 to 6 m
high and resembling a laccolith in section, produced in a tundra or periglacial area by the freezing of water trapped between the frozen surface and the underlying permafrost, and
upwarped into a lenticular body of ice. The term is used often
as a syn. of *pingo*, but a hydrolaccolith can be a seasonal
mound, whereas a pingo is perennial. Syn: *cryolaccolith.*

hydrolite A term variously applied to enhydros, to siliceous
sinter, and to the zeolite mineral gmelinite.

hydrolith (a) A term proposed by Grabau (1904) for an aqueous rock that is chemically precipitated, such as rock salt or
gypsum; a *hydrogenic rock*. (b) A rock that is "relatively free
from organic material" (Nelson & Nelson, 1967, p.185). (c) A
hydroclastic rock consisting of carbonate fragments (Bissell &
Chilingar, 1967, p.158).

hydrologic balance *hydrologic budget.*

hydrologic barrier *ground-water barrier.*

hydrologic budget An accounting of the inflow to, outflow
from, and storage in a hydrologic unit such as a drainage
basin, aquifer, soil zone, lake, or reservoir (Langbein & Iseri,
1960); the relationship between evaporation, precipitation,
runoff, and the change in water storage, expressed by the
hydrologic equation. Syn: *water balance; water budget;
hydrologic balance.*

hydrologic cycle The constant circulation of water from the
sea, through the atmosphere, to the land, and its eventual return to the atmosphere by way of evaporation from the sea
and the land surfaces. Syn: *water cycle.*

hydrologic equation The equation that balances the *hydrologic
budget*; $P = E + R + \Delta S$, with P as precipitation, E as evaporation, R as runoff, and ΔS as the change in water storage,
whether negative or positive. Syn: *water-balance equation.*

hydrologic properties Those properties of a rock that govern
the entrance of water and the capacity to hold, transmit, and
deliver water; porosity, effective porosity, specific retention,
permeability, and direction of maximum and minimum permeability.

hydrologic regimen (a) *regimen* [*water*]. (b) *regimen* [*lake*].

hydrologic system A complex of related parts—physical,
conceptual, or both—forming an orderly working body of
hydrologic units and their man-related aspects such as the
use, treatment and reuse, and disposal of water and the costs
and benefits thereof, and the interaction of hydrologic factors
with those of sociology, economics, and ecology.

hydrology (a) The science that deals with continental water
(both liquid and solid), its properties, circulation, and distribution, on and under the Earth's surface and in the atmosphere,
from the moment of its precipitation until it is returned to the
atmosphere through evapotranspiration or is discharged into
the ocean. In recent years the scope of hydrology has been
expanded to include environmental and economic aspects. At
one time there was a tendency in the U.S (as well as in Germany) to restrict the term "hydrology" to the study of subsurface waters (DeWiest, 1965, p.1). Syn: *hydroscience.* (b)
The sum of the factors studied in hydrology; the hydrology of
an area or district.

hydrolysate *hydrolyzate.*

hydrolyzates Sediments characterized by elements which are
readily hydrolyzed and concentrate in the fine-grained alter-

ation products of primary rocks and are thus enriched in clays, shales, and bauxites. Hydrolyzate elements are aluminum and associated silicon, potassium, and sodium. It is one of Goldschmidts' groupings of sediments as analogues of differentiation stages in rock analysis. Also spelled: *hydrolysates.* Cf: *resistates; oxydates; reduzates; evaporates.*

hydromagnesite A white, earthy mineral: $Mg_4(OH)_2$ $(CO_3)_3 \cdot 3H_2O$. It occurs in small monoclinic crystals (as in altered ultrabasic rocks) or in amorphous masses or chalky crusts (as in the temperate caves of eastern U.S.).

hydromagniolite A general term for hydrous magnesium silicates.

hydromechanics The theoretical, experimental, or practical study of the action of forces on water. See also: *hydrodynamics; hydrokinetics; hydrostatics; hydraulics.*

hydrometamorphism Alteration of rock by material that is added, removed, or exchanged by water solutions, without the influence of high temperature and pressure. Syn: *hydrometasomatism.*

hydrometasomatism *hydrometamorphism.*

hydrometer An instrument that is used to measure the specific gravity of a liquid such as sea water.

hydrometry (a) The use of the hydrometer to measure specific gravity of a fluid. (b) The study of the flow of water, esp. measurement.

hydromica Any of several varieties of muscovite that are less elastic and more unctuous than mica, that have a pearly luster, and that sometimes contain less potash and more water than ordinary muscovite; e.g. a common micaceous clay mineral resembling sericite but having weaker double refraction. The term is practically synonymous with *illite.* Syn: *hydrous mica.*

hydromolysite A mineral: $FeCl_3 \cdot 6H_2O$.

hydromorphic Said of an intrazonal soil having characteristics that were developed in the presence of excess water all or part of the time, i.e. under poor drainage conditions; e.g. a bog soil.

hydromuscovite A term applied loosely to any fine-grained, muscovite-like clay mineral commonly but not always high in water content and deficient in potassium. It is probably an illite.

hydronium jarosite A mineral: $(H_3O)Fe_3(SO_4)_2(OH)_6$. Cf: *carphosiderite.*

hydrophane A semitranslucent or almost opaque and white, yellowish, brownish, or greenish variety of common opal that becomes more translucent or transparent when immersed in water.

hydrophilite A white mineral: $CaCl_2$. Syn: *chlorocalcite.*

hydrophone An electroacoustic transducer that responds to sound transmitted through water.

hydrophyte (a) A plant growing in water, either submerged, emergent, or floating; esp. a *helophyte.* (b) A plant that requires large quantities of water for its growth. Syn: *hygrophyte.* --Cf: *mesophyte; xerophyte.*

hydroplasticity Plasticity that results from the presence of pore water and absorbed water films in a sediment, so that it yields easily to changes of pressure.

hydroplutonic *aqueo-igneous.*

hydropore A pore, slit, or small external opening that serves as an adit to the water-vascular system of an echinoderm. It may be covered by the madreporite. Cf: *gonopore.*

hydropower Literally, *waterpower,* but now generally considered a syn. of *hydroelectric power.* Syn: *white coal.*

hydropsis *synoptic oceanography.*

hydroscarbroite A mineral: $Al_{14}(CO_3)_3(OH)_{36} \cdot nH_2O$.

hydroscience *hydrology.*

hydroscopic water *hygroscopic water.*

hydrosere A sere that develops in an aquatic (i.e. hydric) environment; a *hydrarch* sere. Cf: *mesosere; xerosere.*

hydrosialite A syn. of *clay mineral.* Also spelled: *hydrosyalite.*

hydrosilicate inclusion A fluid inclusion in a crystal repre senting the late-silicate fraction of magmatic crystallization.

hydrosol [soil] A water soil; a major group of soils, as distin guished from mineral soils and organic soils; a *hydrogen* soil. Cf: *aquasol.*

hydrospace *inner space.*

hydrosphere The waters of the Earth, as distinguished fro the rocks (lithosphere), living things (biosphere), and the a (atmosphere). Includes the waters of the ocean; rivers, lake and other bodies of surface water in liquid form on the con nents; snow, ice, and glaciers; and liquid water, ice, ar water vapor in both the unsaturated and saturated zone below the land surface. Included by some, but excluded b others, is water in the atmosphere, which includes wate vapor, clouds, and all forms of precipitation while still in th atmosphere.

hydrospire An infolded, thin-walled, calcareous linear stru ture in the interior of blastoid theca beneath and parallel ambulacral border. Its function is apparently respiratory.

hydrospire pore One of numerous, minute, rounded *pores* b tween side plates or outer side plates near the margin ambulacrum, leading into a hydrospire of a blastoid or co necting the space enclosed by a hydrospire with the exterior.

hydrospire slit A longitudinal opening of the thin calcareou wall surrounding a hydrospire of a blastoid, excavated in th substance of deltoid and radial plates.

hydrostatic equilibrium [oceanog] In a fluid, the horizontal c incidence of the surfaces of constant pressure and consta mass; gravity and pressure are in balance.

hydrostatic head The height of a vertical column of water, th weight of which, if of unit cross section, is equal to the hydr static pressure at a point; static head, as applied to wate See also: *artesian head.*

hydrostatic level The level to which the water will rise in a we under its full pressure head. It defines the potentiometric su face. Syn: *static level.*

hydrostatic pressure [exp struc geol] In experimental structu al geology, stress that is uniform in all directions, e.g. benea a homogeneous fluid, and causes dilation rather than disto tion. Cf: *differential force.*

hydrostatic pressure [hydraul] The pressure exerted by th water at any given point in a body of water at rest. The hydr static pressure of ground water is generally due to the weig of water at higher levels in the zone of saturation (Meinze 1923, p.37). See also: *artesian pressure.*

hydrostatics That aspect of *hydromechanics* which deals wi forces that produce equilibrium. Cf: *hydronamics; hydr kinetics.*

hydrostatic stress A state of stress in which the norm stresses acting on any plane are equal and where shearin stresses do not exist in the material.

hydrostatic weighing The determination of the specific gravi of a substance (such as a gemstone) by weighing it im mersed in water and also in air. The specific gravity is state as the ratio between the weight in air and the loss of weight water.

hydrostratigraphic unit A term proposed by Maxey (196 p.126) for a body of rock having considerable lateral exte and composing "a geologic framework for a reasonably di tinct hydrologic system". Cf: *geohydrologic unit.*

hydrotachylite A volcanic glass similar to *tachylite* but co taining as much as 13 percent water.

hydrotalcite A pearly-white rhombohedral. mineral: Mg_6Al $(CO_3)(OH)_{16} \cdot 4H_2O$. It is dimorpous with manasseite.

hydrothermal Of or pertaining to heated water, to the action heated water, or to the products of the action of heate water, such as a mineral deposit precipitated from a h aqueous solution, with or without demonstrable associatio with igneous processes (also, said of the solution itself). "H drothermal" is generally used for any heated water but ha o been restricted to heated water of magmatic origin.

drothermal alteration Alteration of rocks or minerals by the action of hydrothermal water with preexisting solid phases.

drothermal metamorphism A local type of metamorphism used by the percolation of hot solutions or gases through ctures, causing mineralogic changes in the neighboring ck (Coombs, 1961). Syn: *hydrothermal metasomatism.*

drothermal metasomatism *hydrothermal metamorphism.*

drothermal stage That stage in the cooling of a magma conning volatiles during which the residual fluid is strongly enhed in water and other volatiles. The exact limits of the ige are variously defined by different authors, in terms of her phase assemblage, temperature, composition, or vapor essure; most definitions consider it as the last stage of igous activity, presumably coming at a later time (and hence ver temperature) than the *pegmatitic stage.*

drothermal synthesis Mineral synthesis in the presence of iter at elevated temperatures.

drothermal water Subsurface water whose temperature is ph enough to make it geologically or hydrologically signifint, whether or not it is hotter than the rock containing it. It iy include magmatic and metamorphic water, water heated radioactive decay or by energy release associated with lting, meteoric water that descends slowly enough to acire the temperature of the rocks in accordance with the rmal geothermal gradient but then rises more quickly so as retain a distinctly above-normal temperature as it approaches the surface, meteoric water that descends to and is ated by cooling intrusive rocks, water of geopressured aquis, and brine that accumulates in an area of restricted cirlation at the bottom of a sea.

drotomical axis A major axis with large spines in the skelen of an acantharian radiolarian. Cf: *geotomical axis.*

drotroilite A black, finely divided colloidal material: ·S.nH_2O. It is perhaps formed by bacteria on bottoms of irine basins characterized by reducing conditions and reicted circulation; it quickly changes to more stable pyrite.

drotungstite A mineral: $H_2WO_4.H_2O$.

drougrandite A mineral of the garnet group: $(Ca,Mg,$ $)_3(Fe,Al)_2(SiO_4)_{3-x}(OH)_{4x}$.

drous Said of a mineral compound containing water.

drous mica *hydromica.*

droxide A type of *oxide* characterized by the linkage of a ietallic element or radical with the ion OH, such as brucite, g$(OH)_2$.

droxyapatite *hydroxylapatite.*

droxylapatite (a) A rare mineral of the apatite group: Ca_5- $^3O_4)_3(OH)$. (b) An apatite mineral in which hydroxyl edominates over fluorine and chlorine.---Syn: *hydroxyapa-* e.

droxyl-bastnaesite A wax-yellow to dark-brown mineral: Ce,La)$CO_3(OH,F)$.

droxyl-herderite A monoclinic mineral: $CaBe(PO_4)(OH)$. It isomorphous with herderite. Syn: *hydroherderite.*

drozincite A white, grayish, or yellowish mineral: $n_5(CO_3)_2(OH)_6$. It is a minor ore of zinc and is found in the pper (oxidized) zones of zinc deposits as an alteration prodct of sphalerite. Syn: *zinc bloom; calamine; earthy calamine.*

drozoan Any coelenterate belonging to the class Hydrozoa, iaracterized by forms that are both polypoid and medusoid by exclusively medusoid forms and by the absence of ematocysts and a stomodaeum. Their stratigraphic range is recambrian or Lower Cambrian to present.

vetal Pertaining to rain, rainfall, or rainy regions; e.g. *hyetal* terval, or the difference in rainfall between two isohyets. Cf: *uvial* [*meteorol*].

vetometer *rain gage.*

vgrograph A self-recording *hygrometer.*

vgrometer An instrument that is used to measure the humidiof the air. See also: *hygrograph; psychrometer.*

hygrophilous Said of an organism that lives in moist areas. Syn: *hygrophile; hygrophilic.*

hygrophyte *hydrophyte.*

hygroscope A type of *hygrometer* that measures changes in humidity.

hygroscopic capacity *hygroscopic coefficient.*

hygroscopic coefficient The ratio of the weight of water a completely dry mass of soil will absorb if in contact with a saturated atmsophere until equilibrium is reached, to the weight of the dry soil mass, expressed as a percentage. See also: *hygroscopic water* [*soil*]. Syn: *hygroscopic capacity.*

hygroscopicity "The quantity of water absorbed by dry soil in a secluded space above 10 percent sulphuric acid at room temperature (about $18°C$), expressed as a percentage of the weight of dry soil" (Schieferdecker, 1959, term 0356); the ability of a soil for absorbing and retaining water.

hygroscopic moisture *hygroscopic water.*

hygroscopic water [**soil**] Moisture held in the soil that does not evaporate at ordinary temperatures (i.e. temperatures below the boiling point); moisture in equilibrium with that in the atmosphere to which the soil is exposed. Syn: *hygroscopic moisture; hydroscopic water.* See also: *hygroscopic coefficient.*

hyolithid An invertebrate animal characterized by bilateral symmetry, a closed tapering shell, triangular in cross section, with a conical embryonic chamber not separated from the rest of the shell, and with an operculum possessing two pairs of bilaterally symmetrical muscle scars. Hyolithids belong to the order Hyolithida and are questionably assigned to the mollusks, but have been variously identified as worms, pteropods, cephalopods, and incertae sedis. Their stratigraphic range is Lower Cambrian to Middle Permian.

hypabyssal [**intrus rocks**] Pertaining to an igneous intrusion, or to the rock of that intrusion, whose depth is intermediate between that of *abyssal* or *plutonic* and the surface. This distinction is not considered relevant by some petrologists (Stokes and Varnes, 1955). Syn: *subvolcanic.*

hypautochthony (a) Accumulation of plant remains which no longer occur in the exact place, but still within the same general region, of their growth, as in a peat bog. Cf: *euautochthony.* (b) A term sometimes used as a syn. of *allochthony.*

hypautomorphic A syn. of *hypidiomorphic.* The term *hypautomorphisch* was proposed by Rohrbach (1885, p.17-18) and has priority, but is less used than "hypidiomorphic". See also: *subhedral.*

hypautomorphic-granular *hypidiomorphic-granular.*

hyperborean Pertaining or relating to the far north; of a frigid northern region.

hypercline Said of the dorsal and anterior *inclination* of the cardinal area in the brachial valve of a brachiopod, lying in the top right or second quadrant moving clockwise from the orthocline position (TIP, 1965, pt.H, p.60, fig.61).

hypercyclothem A term proposed by Weller (1958a, p.203-204) for a great cyclic sequence consisting of four *megacyclothems* and an alternating detrital sequence "of more than ordinary thickness and complexity".

hyperfusible n. Any substance capable of lowering the melting ranges in end-stage magmatic fluids. Syn: *hyperfusible component.*

hypergene *supergene.*

hypergenesis A term introduced by Fersman (1922), and persisting to the present day in Russian geology, for surficial alteration (weathering) of sedimentary rocks. Syn: *retrograde diagenesis; regressive diagenesis; retrodiagenesis.*

hypergeometric distribution A frequency distribution resulting from sampling a population without subsequent replacement of that sample.

hyperglyph A hieroglyph formed during weathering (Vassoevich, 1953, p.33).

hyperite A plutonic rock composed of hypersthene, plagioclase, olivine, and augite or diallage, being intermediate in composition between a gabbro and a norite. See also: *hyperite texture.*

hyperite texture The texture characteristic of *hyperite* in which a fibrous amphibole reaction rim is formed at the contact of the olivine and plagioclase crystals (Johannsen, 1939, p.257).

hypermorphosis *anaboly.*

hyperpiestic water A class of *piestic water* including waters that rise above the land surface. Cf: *hypopiestic water; mesopiestic water.*

hyperpycnal inflow Flowing water that is denser than the body of water it enters, resulting in formation of a turbidity current. Its flow pattern is that of a *plane jet* (Moore, p.89). Cf: *hypopycnal inflow, homopycnal inflow.*

hypersolvus Said of those granites, syenites, and nepheline syenites that are characterized by the absence of plagioclase except as a component of perthite (Tuttle and Bowen, 1958). Cf: *solvus; subsolvus.*

hypersthene A common rock-forming mineral of the orthopyroxene group: $(Mg,Fe)SiO_3$. It is isomorphous with enstatite. Hypersthene is grayish, greenish, black, or dark brown, and often has a bronze-like or greenish-brown luster (schiller) on the cleavage surface. It is an essential constituent of many igneous rocks (gabbros, andesites). Symbol: Hy.

hypersthenfels *norite.*

hypersthenite Originally defined as a syn. of *norite*, but now commonly used to mean a rock composed entirely of hypersthene.

hyperstomial Said of a bryozoan ovicell that rests on or is partly embedded in the distal zooid and that opens above the operculum of the parent zooid.

hyperstrophic Said of a rare gastropod shell in which the whorls are coiled on an inverted cone so that the apex points forward rather than backward, and the spire is depressed instead of elevated. A hyperstrophic shell is not easily distinguished from an *orthostrophic* shell unless the aperture shows the siphon pointed in the same direction as the apex (Beerbower, 1968, p.341).

hypertely Evolution to the extreme of being a disadvantage.

hyperthermic Said of a soil temperature regime in which the mean annual temperature (measured at 50cm) is at least 22°C, with a summer-winter variation of at least 5°C (SSSA, 1970). Cf: *isohyperthermic.*

hypertrophy Excessive growth of a species.

hypervelocity impact The impact of a projectile onto a surface at a velocity such that the stress waves produced upon contact are orders of magnitude greater than the static bulk compressive strength of the target material. The minimum required velocities vary for different materials, but are generally 1-10 km/sec, and about 4-5 km/sec for most crystalline rocks. In such an impact, the kinetic energy of the projectile is transferred to the target material in the form of intense shock waves whose interactions with the surface produce a crater much larger in diameter than the projectile. Meteorites striking the Earth at speeds in excess of about 5 km/sec are examples of large hypervelocity impacts and produce correspondingly large craters (Dietz, 1959, p.499).

hypha One of the individual tubular filaments or threads that make up the *mycelium* of a fungus; e.g. a *conidiophore*. Pl: *hyphae.*

hypidioblast A mineral grain that is newly formed by metamorphism and which is bounded only in part by its characteristic crystal faces. It is a type of *crystalloblast.* Cf: *idioblastic; xenoblastic.* Syn: *subidioblast.*

hypidioblastic Pertaining to a *hypidioblast* of a metamorphic rock; also, said of such a texture. It is analogous to the term hypidiomorphic (subhedral) in igneous rocks. Cf: *idioblastic.*

hypidiomorphic (a) Intermediate between *idiomorphic* and *xenomorphic*; esp. said of the texture or fabric of an igneous rock characterized by crystals bounded in part by the crystal

faces peculiar to the mineral species to which the crys belongs and in part by surfaces formed against preexisti crystals or characterized by crystals only some of which ha distinct crystalline forms. Also, said of an igneous rock (su as a granitic rock) with hypidiomorphic texture. (b) An obs lescent syn. of *subhedral.*----The term *hypidiomorphisch* w proposed by Rosenbusch (1887, p.11) originally to describe an igneous rock the individual mineral crystals (now known subhedral crystals) that are bounded only in part by th characteristic crystal faces. Current usage tends to apply "h pidiomorphic" to an igneous-rock texture or fabric characte ized by subhedral crystals. Syn: *hypautomorphic; subidiome phic.*

hypidiomorphic-granular Said of the granular texture, of a pl tonic rock, characterized by a hypidiomorphic fabric in whi most of the essential minerals have subhedral crystals or which some of the mineral crystals are euhedral, some su hedral, and the rest anhedral; e.g. a granitic texture. Als said of the rock with such a texture. Syn: *hypautomorph granular.*

hypidiotopic Intermediate between *idiotopic* and *xenotop* esp. said of the fabric of a crystalline sedimentary rock which the majority of the constituent crystals are subhedr Also, said of the rock (such as an evaporite, a chemically d posited cement, or a recrystallized limestone or dolomit with such a fabric. The term was proposed by Friedm (1965, p.648).

hypobatholithic Said of a mineral deposit occurring in th deeply eroded region of a batholith; also, said of that stage batholith erosion (Emmons, 1933). Cf: *acrobatholithic; crypt batholithic*; embatholithic; *endobatholithic; epibatholithic.*

hypocenter *focus* [seism].

hypocotyl The portion of a seed-plant embryo axis below th attachment of the cotyledons (seed leaves); the growing tip a hypocotyl forms the primary root of a seedling.

hypocrystalline Said of the texture of an igneous rock whi has crystalline components in an amorphous groundmas Syn: *semicrystalline; hemicrystalline; merocrystalline; hyp hyaline; miocrystalline.*

hypocrystalline-porphyritic Said of the texture of a porphyri igneous rock having a hypocrystalline groundmass.

hypodermalium A specialized sponge spicule of the corte lying largely beneath the exopinacoderm. Cf: *autodermalium.*

hypodermis (a) A reticulate layer beneath the *epidermis* in th walls of certain foraminifera. (b) The cellular layer that unde lies and secretes the external chitinous membrane of arthr pods.

hypodigm A group consisting of all the specimens used as th basis for the description of a species, i.e. all specimens r ferred to the species at the time of its original description; th *type material.*

hypogastralium A specialized sponge spicule lying largely b neath the endopinacoderm of the cloaca. Cf: *autogastralium.*

hypogeal *hypogene.*

hypogeic *hypogene.*

hypogene [ore dep] Said of a mineral deposit or enrichme formed by ascending solutions; also, said of those solutio and of that environment. Cf: *supergene; mesogene.*

hypogene [geol] adj. Said of a geologic process, and of its re sultant features, occurring within and below the crust of th Earth. Cf: *epigene.* Syn: *hypogenic; hypogeal; hypogeic.*

hypogene [intrus rocks] A rarely used syn. of *plutonic.*

hypogenesis The direct development of an organism witho alternation of generations.

hypogenic *hypogene.*

hypoglyph A hieroglyph on the bottom of a sedimentary be (Vassoevich, 1953, p.37). Cf: *epiglyph.*

hypogynous Said of a plant in which the ovary surmounts receptacle, so that the sepals, petals, and stamens radia from below the locules. Cf: *epigynous; perigynous.*

hypohyaline *hypocrystalline*.

hypolimnetic Pertaining to a *hypolimnion*. Syn: *hypolimnial*.

hypolimnion The lowermost layer of water in a lake, characterized by an essentially uniform temperature (except during a turnover) that is generally colder than elsewhere in the lake and often by relatively stagnant or oxygen-poor water; specif. the dense layer of water below the metalimnion in a thermally stratified lake. See also: *clinolimnion; bathylimnion*. Cf: *epilimnion*.

hypolithic Said of a plant that grows beneath rocks.

hypomagma Relatively immobile, viscous, lava that forms at depth beneath a shield volcano, is undersaturated with gases, and activates volcanic activity. Cf: *epimagma; pyromagma*.

hyponome The muscular tube, nozzle, or swimming funnel just below the head of a cephalopod, extending externally from the mantle cavity, and through which water is confined and expelled from the mantle cavity.

hyponomic sinus The large, concave, invariably ventral notch or reentrant in the middle of the aperture of a cephalopod, marking the location through which the hyponome protrudes.

hypoparian adj. Of or concerning a trilobite that lacks facial sutures and that is generally blind.---n. A hypoparian trilobite; specif. a trilobite of the order Hypoparia known from the Cambrian through the Lower Ordovician.

hypopiestic water A class of *piestic water* including waters that rise above the bottom of the upper confining bed but not as high as the water table. Cf: *hyperpiestic water; mesopiestic water*.

hypopycnal inflow Flowing water that is less dense than the body of water it enters, e.g. a river entering the ocean. Its flow pattern is that of an *axial jet*. (Moore, 1966, p.89). Cf: *hyperpycnal inflow; homopycnal inflow*.

hyposeptal deposit A distal cameral deposit on the convex (adapical) side of septum of a nautiloid. Ant: *episeptal deposit*.

hyposome The posterior part of the cell body below the girdle of an unarmored (naked) dinoflagellate. Cf: *episome*.

hypostega (a) The coelomic cavity of a bryozoan between a cryptocyst and the overlying membrane. (b) The space between the frontal membrane and the overlying frontal shield in some bryozoans. This usage is not recommended. Syn: *hypostege*.

hypostoma A syn. of hypostome. Pl: *hypostomata*.

hypostomal suture The line of junction in a trilobite between the anterior margin of the hypostome and the posterior margin of the frontal doublure or rostrum.

hypostome (a) A ventral plate of the head region behind and above which the mouth of a trilobite is located. Cf: *metastoma*. Syn: *labrum*. (b) A vase-shaped or conical process bearing the mouth of a hydrozoan.---Syn: *hypostoma*.

hypostracum A term used originally for the inner layer of the shell wall of a bivalve mollusk, secreted by the entire epithelium of the mantle, but also applied in a later sense to the myostracum. Cf: *ostracum*.

hypotheca (a) The posterior part of a dinoflagellate theca, below the girdle. Cf: *epitheca*. (b) *hypovalve*.

hypothermal Said of a hydrothermal mineral deposit formed at great depth and in the temperature range of 300°-500°C (Park MacDiarmid, 1970, p.293). Also, said of that environment. Syn: *katathermal*. Cf: *mesothermal, epithermal, leptothermal; telethermal; xenothermal*.

hypothermal n. A term proposed by Cooper (1958, p.944) for postglacial interval (the last 2600 years) characterized by a moderate decrease in temperature, by limited glacial expansion, and by supposed eustatic lowerings of sea level. It is roughly synonymous with *Little Ice Age*.

hypothesis A conception or proposition that is tentatively assumed, and then tested for validity by comparison with observed facts and by experimentation; e.g. the "planetesimal hypothesis" and the "nebular hypothesis" to explain the evolu-

tion of the planets. It is less firmly founded than a *theory*.

hypothyridid Said of a brachiopod pedicle opening located below or on the dorsal side of the beak ridges with the umbo intact (TIP, 1965, pt.H, p.146).

hypotract The part of a dinoflagellate cyst posterior to the girdle region. Ant: *epitract*.

hypotype A specimen not in the group of specimens upon which the original description of a particular species is based but known by some other published description or illustration that extends or corrects the knowledge of the species. Also spelled: *hipotype*.

hypovalve The inner valve of a diatom frustule. Cf: *epivalve*. Syn: *hypotheca*.

hypoxenolith A xenolith that is derived from a source further than the adjacent wall rock (Goodspeed, 1947, p.1251). Cf: *epixenolith*.

hypozygal The proximal brachial plate of a pair joined by syzygy in a crinoid. Cf: *epizygal*.

Hypsithermal n. A term proposed by Deevey & Flint (1957) as a substitute for *climatic optimum* and *thermal maximum*, and representing the postglacial interval when "most of the world entered a period when mean annual temperatures exceeded those of the present". As defined stratigraphically, it includes the Boreal, Atlantic, and Subboreal climatic intervals, or from about 9000 to 2500 years ago. It is distinguished from *Altithermal* by including the Subboreal and by suggesting that the major arid period in the west occurred in post-Atlantic time; this suggestion has not gained general acceptance (see Baumhoff & Heizer, 1965). See also: *Megathermal; Xerothermic*.---adj. Pertaining to the postglacial Hypsithermal interval and to its climate, deposits, biota, and events. Also spelled: *hypsithermal*.

hypsographic curve A cumulative-frequency profile representing the statistical distribution of the absolute or relative areas of the Earth's solid surface (land and sea floor) at various elevations above, or depths below, a given datum, usually sea level. Syn: *hypsometric curve*.

hypsography (a) A branch of geography dealing with the observation and description of the varying elevations of the Earth's surface with reference to a given datum, usually sea level. Cf: *hypsometry*. (b) Topographic relief. Also, the parts of a map, collectively, that represent topographic relief. (c) The portrayal of topographic relief on maps.

hypsoisotherm A line connecting points of equal temperature on a vertical section of the atmosphere. Cf: *isallotherm; isotherm*.

hypsometer (a) An instrument used in measuring or estimating the elevation of a point on the Earth's surface in relation to sea level by determining atmospheric pressure through observation of the boiling point of water at that point. It is useful in mountainous or high-altitude regions. Syn: *thermobarometer*. (b) An instrument used in forestry for determining heights of trees.

hypsometric Pertaining to hypsometry or to elevation above a datum; esp. relating to elevations above sea level determined with a hypsometer.

hypsometric analysis The measurement of the distribution of ground surface area (or horizontal cross-sectional area) of a landmass with respect to elevation (Strahler, 1952b, p.1118). Syn: *area-altitude analysis*. Cf: *hypsometry*.

hypsometric curve A *hypsographic curve*; e.g. *percentage hypsometric curve*.

hypsometric integral Proportionate area below the *percentage hypsometric curve*; it expresses the relative volume of a landmass at a given contour. In the study of drainage basins, Strahler (1952b, p.1121) used this term to express the ratio of the volume of earth material to the volume of the solid reference figure having a base equal to basin area and a height equal to maximum basin relief. Symbol: I. Cf: *erosion integral*.

hypsometric map Any map showing relief by means of contours, hachures, shading, tinting, or any other convention.

hypsometric tint A color applied to the area between two selected contour lines on a map of an area whose relief is depicted by layer tinting. Syn: *layer tint.*

hypsometry The science of determining, by any method, height measurements on the Earth's surface with reference to sea level; e.g. "barometric hypsometry" in which elevations are determined by means of mercurial or aneroid barometers. Cf: *hypsography; hypsometric analysis.*

hyrst Var. of *hurst.*

hysteresis [elect] A phase lag of dielectric displacement behind electric field intensity, due to energy dissipation in polarization processes.

hysteresis [magnet] A lagging of magnetization M or induction B behind an alternating applied magnetic intensity H. The value of M when H=0 is called *remanent magnetization.* The opposing H required to reduce M to zero is called *coercive force.*

hysterobase A diabase composed of plagioclase, quartz, biotite, and brown hornblende that is paramorphic after augite.

hysterocrystalline An obsolete term applied to a mineral produced in an igneous rock by secondary crystallization.

hysterogenetic Said of the last crystallization products of magma; e.g. dikes, zoning.

hysterogenous A little-used term pertaining to a mineral deposit on the Earth's surface formed from the debris of other rocks. Syn: *hysteromorphous.* Cf: *idiogenous; xenogenous.*

hysteromorphous *hysterogenous.*

hysterosoma That section of the body of an acarid that is behind the second pair of legs.

hystrichosphaerid A general term formerly used for a great variety of resistant-walled organic microfossils ranging from Precambrian to Holocene and characterized by spherical to ellipsoidal, usually spinose remains found among fossil microplankton. These are now divided among the *acritarchs* and *dinoflagellate* cysts. The term has no formal taxonomic status. Syn: *hystrichosphere.*

hystrichosphere *hystrichosphaerid.*

I

ianthinite (a) A violet-black, orthorhombic, secondary mineral: UO_2-$5UO_3$.$10H_2O$. (b) A mineral name that was erroneously given to *wyartite*.

ice (a) Water in the solid state; specif. the dense substance formed in nature by the freezing of liquid water, by the condensation of water vapor directly into ice crystals, or by the recrystallization or compaction of fallen snow. It is colorless to pale blue or greenish blue, usually white from included gas bubbles. At standard atmospheric pressure, it is formed at and has a melting point of 0°C, in freezing it expands about one eleventh in volume. Ice commonly occurs as hexagonal crystals, and in large masses is classed as a rock. (b) A term often substituted for *glacier*, as in "continental ice".

ice age A loosely used syn. of *glacial epoch*, or time of extensive glacial activity; specif. the *Ice Age*, or the latest of the glacial epochs, also known as the Pleistocene Epoch.

ice apron [eng] A wedge-shaped structure placed on the upstream end of a bridge pier to protect it from floating ice.

ice apron [glaciol] The thin mass of snow and ice attached to the headwall of a cirque above the bergschrund (but not present at a randkluft). Syn: *apron* [glaciol].

ice avalanche A sudden fall, down a steep slope, of ice broken from an ice sheet or glacier (most commonly a hanging glacier). Syn: *icefall*.

ice bar An ice edge consisting of floes compacted by wind, sea, and swell, and difficult to penetrate.

ice barchan A small crescentic dune composed of ice crystals.

ice barrier [glaciol] A syn. of *ice front* and *ice shelf*. The term was introduced by Sir James C. Ross in 1841 to designate the high, steep, seaward cliff face or edge of a great ice mass in Antarctica because it obstructed navigation; the syn. "ice front" is now preferred. The term was later applied to the entire mass of ice and was widely adopted for similar polar morphologic features, esp. for an immense ice formation of very great extent; the term "ice shelf" is now used for this type of ice mass. Syn: *barrier* [glaciol].

ice barrier [ice] *ice dam*.

ice-barrier lake A lake formed in a mountain valley whose lower end is dammed by a glacier descending another valley.

ice-basin lake A lake, pond, or pool on sea ice or on glacier ice (ADTIC, 1955, p.41).

ice bay *bight* [ice].

iceberg A large, massive piece of floating or stranded glacier ice of any shape detached (calved) from the front of a glacier into a body of water. An iceberg extends more than 5 m above sea level and has the greater part of its mass ($4/5$ to $8/9$) below sea level. It may reach a length of more than 80 km. Syn: *berg* [glaciol]. Cf: *floeberg*.

iceberg tongue "A major accumulation of icebergs projecting from the coast, held in place by grounding, and joined together by fast ice" (U.S. Naval Oceanographic Office, 1968, p.B34).

ice blade A "crest" or "spire" of ice, 0.5-1.5 m high, rising from a surface of firn, and formed by unequal melting (Russell, 1885, p.318).

iceblink [meteorol] A relatively bright, usually yellowish or whitish glare in the sky near the horizon or on the underside of a cloud layer, produced in a polar region by light reflected from a large ice-covered surface (as an ice sheet) that may be too far away to be visible; not as bright as *snowblink*. Also spelled: *ice blink*. Syn: *ice sky*.

iceblink [glaciol] A cliff extending along the seaward margin of a mass of inland ice. Examples are found on the coast of Greenland. Syn: *isblink*.

ice blister *ice mound*.

ice-block ridge A ridge, either closed or linear, surrounding or separating depressions in a moraine (Deane, 1950, p.14).

ice boulder [glac geol] An obsolete syn. of *glacial boulder*.

ice boulder [ice] A large fragment of sea ice shaped by wave action into a nearly spherical form and then stranded on the shore.

ice boundary "The demarcation at any given time between fast ice and pack ice or between areas of pack ice of different concentrations" (U.S Naval Oceanographic Office, 1968, p.B34). Cf: *ice edge*.

ice breccia Pieces of ice of different ages frozen together. Syn: *ice mosaic*.

ice cake A *floe* or piece of floating sea ice less than 10 m across. Syn: *block; cake* [ice].

ice canopy A submariner's term for *pack ice*.

ice cap A dome-shaped or plate-like cover of perennial ice and snow covering all of the summit area of a mountain mass so that no peaks emerge through it, or covering a flat landmass such as an Arctic island, spreading due to its own weight outwards in all directions, and having an area of less than 50,000 sq km. Cf: *ice field; ice sheet; glacier*. Nonpreferred syn: *ice carapace; cap* [glaciol]; *glacier cap*. Also spelled: *icecap*.

ice carapace *ice cap*.

ice cascade *icefall*.

ice cast A shell of ice formed around a beach pebble as a result of the wetting action of spray, tides, and waves, and subsequent freezing; the ice is sometimes separated from the pebble.

ice cauldron A wide area in a valley, upon which glacier ice once piled up so high as to flow radially outward through pre-existing passes that were deepened by glacial scour.

ice cave (a) An artificial or natural cave, in a temperate climate, in which ice forms and persists throughout all or most of the year. Syn: *ice grotto; ice glen; glacière*. (b) *glacier cave*.

ice clearing (a) The end phase of breakup. (b) *polynya*.

ice cliff Any vertical wall of ice; e.g. a very steep surface bounding a glacier or a mass of shelf ice. Disapproved syn: *ice front; ice wall*. Syn: *ice face*.

ice cluster A concentration of sea ice covering hundreds of square kilometers and found in the same region every summer. See also: *ice pack*. Syn: *ice massif*.

ice column *ice pillar*.

ice concrete A dense frozen mixture of sand, rock fragments, and ice. Syn: *icecrete*.

ice cone *ice pyramid*.

ice-contact delta A delta built by a stream flowing (into a lake) between a valley slope and the margin of glacier ice. Syn: *morainal delta; delta moraine*.

ice-contact deposit Stratified drift deposited in contact with melting glacier ice, such as an esker, kame, kame terrace, or a feature marked by numerous kettles.

ice-contact plain *kame plain*.

ice-contact slope The steep slope of material deposited against glacier ice, marking the position of the ice front; an irregular scarp against which glacier ice once rested.

ice-contact terrace *kame terrace*.

ice-contorted ridge *ice-pushed ridge*.

ice cover (a) The extent of glacier ice on a land surface at the present time, with special reference to its thickness. (b) The ratio of an area of sea ice of any *concentration* to the total area of sea surface within some large geographic locale that may be global, hemispheric, or specific to a given study (U.S. Naval Oceanographic Office, 1968, p.B34). (c) The extent of ice on a lake surface.

ice-covered *glacier-covered*.

ice crack *frost crack*.

ice-crack moraine A linear ridge consisting of very sandy, unstratified drift believed to have been deposited between blocks of dead ice in a disintegrating glacier (Sproule, 1939, p.104).

icecrete *ice concrete.*

ice crust [ice] (a) *ice rind.* (b) A thin layer of ice on a rock surface, formed by freezing of water condensed from the air.

ice crust [glaciol] A type of *snow crust,* formed when surface water (such as meltwater or rainwater) freezes to form a continuous layer of ice upon the surface of deposited snow. It is thicker than a *film crust.* See also: *ice layer* [glaciol].

ice crystal (a) A macroscopic particle of ice with a regular (usually hexagonal) structure; it is anisotropic. (b) *frazil crystal.*

ice-crystal cast A *crystal cast* formed by the filling of an ice-crystal mark with mud or sand; it commonly appears as a straight, slightly raised ridge on the underside of a sandstone bed.

ice-crystal mark A crack formed on a sedimentary surface by the sublimation of a crystal of ice.

ice dam A river obstruction formed of floating blocks of ice that may cause ponding and widespread flooding during spring and early summer. Syn: *ice barrier.*

ice-dammed lake *glacier lake.*

ice dendrite A thin branching ice crystal. Dendritic ice forms the first skim of ice over still water, and also may grow under water from an existing ice surface.

iced firn A mixture of ice and firn; firn permeated with meltwater and then refrozen. Syn: *firn ice.*

ice dike A secondary formation of ice, usually made up of columnar crystals, filling a crevasse or other gash in glacier ice.

ice disintegration The process of breaking up a stagnant and wasting glacier into numerous small blocks: it is said to be "controlled" where the blocks are separated along fractures or other lines of weakness to form linear or lobate features, and "uncontrolled" where equal forces broke up the glacier along cracks extending in all directions to produce round, oval, or rudely polygonal features (Gravenor & Kupsch, 1959, p.48-49). Syn: *disintegration* [glaciol].

ice disintegration The process of breaking up a stagnant and wasting glacier into numerous small blocks: it is said to be "controlled" where the blocks are separated along fractures or other lines of weakness to form linear or lobate features, and "uncontrolled" where equal forces broke up the glacier along cracks extending in all directions to produce round, oval, or rudely polygonal features (Gravenor & Kupsch, 1959, p.48-49). Syn: *disintegration* [glaciol].

ice dome An accumulation of glacier ice in a caldera.

ice edge The demarcation at any given time between open water and sea, lake, or river ice of any kind, whether fast or drifting. It may be "compacted" by wind or current, or it may be "diffuse" or "open" when dispersed or poorly defined. Cf: *ice boundary; ice limit.*

ice erosion (a) Erosion resulting from the freezing of water in cracks or fractures in rock. (b) *glacial erosion.*

ice face *ice cliff.*

icefall (a) The part of a glacier that is highly crevassed because of a very steep slope of the glacier bed. Syn: *ice cascade.* Cf: *cascading glacier.* (b) *ice avalanche.*----Also spelled: *ice fall.*

ice fan *expanded foot.*

ice field [permafrost] *icing.*

ice field [glaciol] (a) An extensive mass of land ice covering a mountain region consisting of many interconnected valley and other types of glaciers, covering all but the highest peaks and ridges. Cf: *ice cap; highland glacier.* (b) A general, but not recommended, designation for a large and irregular body of glacier ice.----Also spelled: *icefield.*

ice field [ice] An extensive, flat sheet of unbroken pack ice, consisting of any size floes, and greater than 10 km (5 mi) across; the largest areal subdivision of sea ice. Ice fields are subdivided according to horizontal extent as follows: "large"

(over 20 km); "medium" (15-20 km); "small" (10-15 km). Cf: *ice patch.* Syn: *field.*

ice floe (a) A large fragment or extensive sheet of ice, detached and floating freely in open water. Syn: *ice raft.* (b) *floe.*

ice flow *glacier flow.*

ice flower *frost flower.*

ice foot (a) The ice at the lower end or front of a glacier. (b) A mass or wall of ice formed by the freezing of snow that accumulated along the foot of a mountain slope. It is not formed from converging glaciers. (c) *icefoot.*

icefoot A narrow strip, belt, or fringe of ice formed along and firmly attached to a polar coast, unmoved by tides, and remaining after the fast ice has broken away; it is usually formed by the freezing of wind-driven spray, or of seawater during ebb tide. A true icefoot has its base at or below the low-water mark. Also spelled *ice foot.* Syn: *ballycadder; bellicatter; cadder; catter; collar ice; fast-ice belt; shore-ice belt; ice ledge.*

ice-foot niche A hollow created at the base of a soft cliff (as of limestone) by floating sea ice; during spring it permits the collapse of the overhanging wall (Hamelin & Cook, 1967, p.101).

ice free "No sea ice present. There may be some ice of land origin" (U.S Naval Oceanographic Office, 1968, p.B34). Cf: *open water.*

ice fringe (a) A very narrow *ice piedmont,* extending less than about one kilometer inland from the sea. (b) A belt of sea ice that extends a short distance offshore. (c) *ice ribbon.*

ice front (a) The floating vertical cliff forming the seaward face or edge of an ice shelf or other glacier that enters water varying in height from 2 to 50 m or more above sea level. Cf: *ice wall.* Syn: *ice barrier* [glaciol]; *front* [glaciol]. (b) Deprecated syn. of *ice cliff* in the sense of any vertical wall of ice (c) The *snout* of a glacier.

ice gang A rush of water following a breakup; a *debacle.*

ice gland A rudely cylindrical vertical column of ice or of iced firn in a firn field.

ice glen *ice cave* [speleo].

ice gneiss Frozen ground with ice segregated in laminae; term used by Taber (1943).

ice gorge (a) The vertical-walled opening left after an ice jam has broken through. (b) An *ice jam* in a river channel.

ice grass *ice stalk.*

ice grotto *ice cave* [speleo].

ice gruel A type of *sludge* floating on the sea surface and formed by the irregular freezing together of frazil crystals.

ice hummock *hummock* [ice].

ice island A form of large *tabular iceberg* broken away from an ice shelf and found in the Arctic Ocean, having a thickness of 15-50 m and an area between a few thousand square meters and 500 sq km or even more. The surfaces of ice islands are usually marked by broad, shallow, and regular undulations that give them a ribbed appearance from the air.

ice jam (a) An accumulation of broken river ice lodged in a narrow or obstructed part of the channel; it frequently produces local floods during a spring breakup. Syn: *ice gorge.* (b) An accumulation of large fragments of lake ice or sea ice thawed loose from the shore during early spring and subsequently piled up on or blown against the shore by the wind, often exerting great pressures.

ice keel The submerged mass of broken ice under a *pressure ridge,* forced downward by pressure, and extending as much as 50 m below sea level.

ice-laid drift *till* [glac geol].

Iceland agate A syn. of *obsidian,* applied to gem-quality varieties.

Iceland spar A very pure and transparent variety of calcite the best of which is obtained in Iceland, that cleaves easily and perfectly into rhombohedrons and that exhibits strong double refraction; an *optical calcite*. It occurs in vugs and cavities in volcanic rocks and as nodules in residual clays in limestone regions. Syn: *Iceland crystal; double-refracting spar*.

ice layer [permafrost] An approximately horizontal layer of ground ice, sometimes lenticular (see *ice lens*). Syn: *ground-ice layer*.

ice layer [glaciol] A layer of solid ice or iced firn in a mass of snow or firn, either a remnant *ice crust* [glaciol] that has been covered by snow or a layer formed by the freezing of water trapped on a relatively impermeable horizon such as a snow crust.

ice lens [permafrost] A discontinuous layer of ground ice tapering at the periphery; ice lenses in soil commonly occur parallel to each other in repeated layers.

ice lens [glaciol] A discontinuous, horizontal ice layer that tapers out in all directions.

ice limit (a) The extreme minimum or the extreme maximum extent of the *ice edge* in any given time period, based on observations over several years. The term should be prefaced by the word "minimum" or "maximum". (b) Obsolete syn. of *mean ice edge*.

ice mantle *ice sheet*.

ice-marginal drainage Stream drainage along the side or the front of a glacier.

ice-marginal lake *proglacial lake*.

ice-marginal terrace *kame terrace*.

ice-marginal valley A valley parallel to the front of a glacier, serving for the draining away of meltwaters.

ice massif *ice cluster*.

ice mosaic *ice breccia*.

ice mound (a) A *frost mound* containing bodies of ice. (b) *icing mound*. Syn: *ice blister*.

ice mountain A popular term for a large iceberg.

ice-centered lattice *body-centered lattice*.

ice pan A large flat piece of first-year ice protruding from several centimeters to a meter above the sea surface. Syn: *pan [ice]*.

ice patch An area of pack ice smaller than an *ice field*. Syn: *patch [ice]*.

ice pavement *glacial pavement*.

ice pedestal A pinnacle, column, or cone of ice projecting from the surface of a glacier and supporting, or formerly supporting, a large rock (*glacier table*) or mass of debris which protects the ice underneath it from solar radiation, so that it melts less rapidly than the ice around it. See also: *mushroom ice*. Syn: *ice pillar*.

ice penitente A *nieve penitente* consisting of glacier ice.

ice period The period of time from freezeup to breakup of ice.

ice piedmont A mass of ice, sloping gently seaward that covers a coastal strip of low-lying land backed by mountains (Armstrong & Roberts, 1956, p.7). It may be anywhere from one kilometer to 50 km wide, and its outer edge may be marked by a line of ice cliffs. Ice piedmonts frequently merge into ice shelves. See also: *ice fringe*.

ice pillar Any tall, narrow mass of ice, such as an *ice pedestal* or *mushroom ice*. Syn: *ice column*.

ice pipe An ice mass of cylindrical shape.

ice plateau (a) An ice-covered highland area whose upper surface is nearly level and whose sides descend steeply to lowlands or to the ocean. See also: *plateau glacier*. (b) Any ice sheet with a level or gently rounded surface, as the Polar Plateau in the Antarctic.

ice pole The approximate center of the most consolidated part of the arctic pack ice, and therefore a difficult point to reach by surface travel; it is near lat. 83°-84°N and long. 160°W. Syn: *pole of inaccessibility*.

ice potential "The potential amount of ice that would be formed in a given water mass if surface heat loss provided the thermohaline circulation" (Baker, et al, 1966, p. 84).

ice push (a) The lateral pressure exerted by the expansion of shoreward-moving ice, esp. of lake ice. Syn: *ice shove; ice thrust*. (b) The ridge of material formed by an ice push. Syn: *lake rampart; ice-push ridge*.

ice-pushed ridge An asymmetric ridge of local, essentially nonglacial material (such as deformed bedrock, with some drift incorporated in it) that has been pressed up by the shearing action of an advancing glacier. It is typically 10-60 m high, about 150-300 m wide, and as long as 5 km. Examples are common on the Great Plains of North America where they occur on the sides of escarpments formed of relatively incompetent rocks that face the direction from which the ice moved. The term is sometimes used "not quite correctly" as a synonym of *push moraine*, which is an accumulation of glacial drift (Schieferdecker, 1959, term 0924). Syn: *ice-contorted ridge; ice-thrust ridge; glacial pressure ridge; pressure ridge [glaciol]*.

ice-push ridge *lake rampart*.

ice-push terrace A terrace-like accumulation of coarse material pushed up along a shore; esp. a terrace consisting of successive *lake ramparts*. Syn: *ice-pushed terrace*.

ice pyramid A roughly conical mound of ice on the surface of a glacier, formed by differential ablation; e.g. an ice pedestal whose sides have been melted back, making a cone shape. Syn: *ice cone*.

ice raft *ice floe*.

ice-rafted Said of material, such as boulders or till, deposited by the melting of floating ice containing this material; esp. said of till distributed widely in marine sediments.

ice-rafting The transporting of rock and other minerals, of all sizes, on or within icebergs, ice floes, river drift, or other forms of floating ice. See also: *rafting*.

ice rampart A syn. of *lake rampart*. The term is misleading because the rampart is a wall of boulders and other coarse material, not of ice.

ice receiving area The portion, generally near the terminus, of a surging glacier that is periodically refilled by glacier surges.

ice reservoir area That portion of a glacier that is periodically drained by glacier surges. This reservoir, which is refilled by direct snow accumulation or by normal ice flow between surges, may be located at nearly any point in the glacier system.

ice ribbon A thin, white, curly deposit of ice growing to a length of 10 cm or more, formed by the freezing of moisture exuded from the dead stem of a plant just above the ground in the early frosty period of winter before the ground is thoroughly frozen. Syn: *ice fringe*.

ice rind A brittle, thin but hard crust of sea ice formed on a quiet surface by direct freezing or from grease ice, usually in water of low salinity; its thickness is generally less than 5 cm, and it is easily broken into rectangular pieces by wind or swell (U.S. Naval Oceanographic Office, 1968, p.B34). Syn: *ice crust; crust [ice]*.

ice rise An ice mass, usually dome-shaped, resting on unexposed rock and surrounded either by an ice shelf, or partly by an ice shelf and partly by sea or ice-free land or both.

ice run (a) Movement of ice floes with the current in a river at breakup. The ice run may be characterized as thin, close, or compact. (b) A rush of water following a breakup; a *debacle*.

ice-scour lake A glacial lake occupying a rock basin eroded by a glacier; e.g. a *finger lake*.

ice sheet A *glacier* of considerable thickness and more than 50,000 sq km in area forming a continuous cover of ice and snow over a land surface, spreading outward in all directions and not confined by the underlying topography; a *continental glacier*. Ice sheets are now confined to polar regions (as on Greenland and Antarctica), but during the Pleistocene Epoch they covered large parts of North America and northern Europe. Armstrong & Roberts (1956, p.7) also apply the term to any extensive body of floating sea ice, but the term should be restricted to land ice. Not to be confused with *sheet ice*. See also: *inland ice.* Cf: *ice cap.* Syn: *ice mantle.*

ice shelf A sheet of very thick ice, with a level or gently undulating surface, which is attached to the land along one side but most of which is afloat and is bounded on the seaward side by a steep cliff (ice front) rising 2-50 m or more above sea level. Ice shelves have been formed along polar coasts (e.g. those of Antarctica, the Canadian Arctic islands, and Greenland), and they are generally of great breadth, some of them extending several hundreds of kilometers seaward from the coastline. They are nourished by annual snow accumulation and by seaward extension of land glaciers; limited areas may be aground. Term used by Sir Douglas Mawson in 1912. Disapproved syn: *shelf ice; barrier; ice barrier; barrier ice.*

ice shove *ice push.*

ice sky *iceblink.*

ice slush *grease ice.*

ice spar A white or colorless, glassy, transparent variety of orthoclase; specif. *sanidine.*

ice spicule A small needle-like ice crystal that grows in water.

ice stalk A fibrous or spiky efflorescence of *Taber ice,* developed on the surface of freezing sediments. Syn: *ice grass.*

ice stone *cryolite.*

ice stream [glaciol] (a) A current of ice in an ice sheet or ice cap that flows more rapidly than the surrounding ice, usually flowing to the ocean or to an ice shelf and not constrained by exposed rock. Cf: *outlet glacier.* See also: *glacier.* Syn: *stream* [glaciol]. (b) One component of a *valley glacier;* e.g. an *inset ice stream.* (c) An obsolete syn. of *outlet glacier* and *valley glacier.* (d) A term sometimes popularly applied to a glacier of any kind, esp. a *valley glacier.* Syn: *stream* [glaciol].

ice structure An imperfection in a diamond or other gem, usually consisting of a group of cracks about an included foreign particle.

ice table A mass of *level ice.*

ice thrust *ice push.*

ice-thrust ridge (a) *ice-pushed ridge.* (b) *lake rampart.*

ice tongue *tongue* [glaciol].

ice tongue afloat *glacier tongue afloat.*

ice vein *ice wedge.*

ice wall (a) A cliff of ice forming the seaward margin of a glacier that is not afloat, such as of an ice sheet, an ice piedmont, or an ice rise. It is aground, the rock basement being at or below sea level. Cf: *ice front.* (b) Deprecated syn. of *ice cliff* in the sense of any vertical wall of ice.

ice-walled channel A term used by Gravenor & Kupsch (1959, p.56) for a meltwater channel, either an open trench or a *tunnel valley,* containing a stream that may have flowed beneath a glacier.

ice wedge A large, usually wedge-shaped mass of foliated ground ice produced in permafrost, occurring as a vertical or inclined sheet, dike, or vein tapering downward, and generally measuring from a few millimeters to 3 m wide (some massive wedges are 6 m wide) and 1-10 m high (sometimes reaching 30 m) when seen in transverse cross section. It originates by the growth of hoar frost or by the freezing of water in a narrow crack or fissure produced by thermal contraction of the permafrost. Syn: *ground-ice wedge; ice vein.*

ice-wedge cast A syn. of *fossil ice-wedge.* Other syn: *ice-wedge fill; ice-wedge pseudomorph.*

ice-wedge polygon A large *nonsorted polygon* characterized by borders of intersecting ice wedges occupying fissures in permafrost regions and formed by contraction of frozen ground. The fissured borders delineating the polygon may be ridges (low-centered polygon in which sediments are being upturned) or shallow troughs (high-centered polygon in which erosion and thawing are prevalent), and are generally underlain by irregular masses of wedge-shaped ice. Diameter: up to 150 m, averaging 10-40 m. In plan, the pattern tends to be tetragonal, but three-, five-, and six-sided forms also occur.

See also: *fissure polygon.* Cf: *frost-crack polygon.* Syn: *tundr polygon; Taimyr polygon.*

ice-worn Abraded by ice; rubbed, striated, grooved, polished scoured, or plucked by glacial action.

ice yowling A long, high-pitched sound accompanying the for mation of contraction cracks in ice. Syn: *yowling.*

ichn A combining form signifying preservation of an origina feature after alteration, e.g. a "clastichnic rock".

ichnite A fossil footprint or track. Syn: *ichnolite.*

ichnocoenosis An association of trace fossils.

ichnofacies A sedimentary facies characterized by evidence of the life activities of fossil animals.

ichnofossil *trace fossil.*

ichnolite (a) *ichnite.* (b) The rock containing an ichnite.

ichnology The study of trace fossils; esp. the study of foss tracks.

ichor A fluid thought to be responsible for such processes a granitization. Originally the term carried the connotation o derivation from a magma (Dietrich & Mehnert, 1961). Syn *residual magma.*

ichthammol A brownish-black viscous liquid which is a distilla tion product of bituminous schists and is used as an emollien and as an antiseptic. See also: *Ichthyol.*

Ichthyol The trademark for *ichthammol* which is a distillate o bituminous schists and is used as an emollient and as an anti septic.

icicle [pat grd] A small *ice wedge* hanging downward from horizontal sheet of ice joining larger ice wedges, thus produc ing a polygonal or honeycombed pattern of ground ice.

icicle [ice] A narrow cone-shaped spike or shaft of clear ice hanging with its point downward, formed by the freezing c dripping water; its length varies from finger length to 8 m.

icing (a) A surface ice mass formed during the winter in permafrost area by successive freezing of sheets of wate that may seep from the ground, or from a spring or river. Se also: *aufeis.* Syn: *glacier (colloquial in Alaska); ice field.* (b The accumulation of an ice deposit on exposed objects.

icing mound A thick, localized surface mound on an *icing;* may form by the upwarp of a layer of ice (as in a river) by th hydrostatic pressure of water. Syn: *ice blister.*

icosacanthic law *Müllerian law.*

icositetrahedron A term for the isometric *trapezohedron.*

Idahoan-type facies series A type of dynamothermal regiona metamorphism characteristic of Boehls Butte, Idaho in whic the polymorphs andalusite, kyanite, and sillimanite occur to gether with some staurolite and cordierite. Thus the Al_2SiO triple point is characteristically involved in this type of regiona dynamothermal metamorphic series. At pressures betwee 3000 and 6000 bars, it lies approximately between the *Pyre neean-type* and the *Barrovian-type facies series* (Hietanen 1967, p.195).

idaite A mineral: Cu_3FeS_4, but perhaps Cu_5FeS_6.

idd A term applied in northern Sudan to a place in the bed o an intermittent stream where water may be obtained in shal low wells for most if not all of the dry season (J.H.G. Lebon i Stamp, 1961, p.254).

Idding's classification A classification of igneous rocks ir which the mineralogic classifications of Rosenbush and Zirke are correlated with the C.I.P.W. (i.e. *normative* classifica tion). The system was proposed in 1913 by J.P. Iddings.

iddingsite A reddish-brown mixture of silicates (of ferric iron calcium, and magnesium) formed by the alteration of olivine It forms rust-colored patches in basic igneous rocks.

ideal cut *American cut.*

ideal cyclothem A theoretical *cyclothem* that represents, in a given region and within a given stratigraphic interval, the optimum succession of deposits during a complete sedimentary cycle. It is constructed from theoretical considerations and from accumulated data from modern environments and experimental evidence. An ideal cyclothem of ten members for western Illinois consists of the following sequence (Weller & others, 1942, p.10): (10) marine shale with ironstone concretions; (9) clean marine limestone; (8) black laminated shale with limestone concretions or layers; (7) impure, lenticular, fine-grained marine limestone; (6) gray marine shale with pyritic nodules; (5) coal; (4) underclay; (3) freshwater, usually nonfossiliferous limestone; (2) sandy shale; and (1) fine-grained micaceous sandstone, locally unconformable on underlying beds. See also: *discordance index.*

ideal section A hypothetical geologic cross section that gives both the evidence or actual stratigraphy and/or structure as well as the interpretation of what is not present.

ideal solution A solution in which the molecular interaction between components is the same as that within each component; a solution which conforms to *Raoult's law.* Cf: *nonideal solution.*

idioblast A mineral constituent of a metamorphic rock formed by recrystallization which is bounded by its own crystal faces. It is a type of *crystalloblast.* The term was originated by Becke (1903). Cf: *hypidioblast; xenoblast.*

idioblastic Pertaining to an *idioblast of a metamorphic rock. It is analogous to the term idiomorphic (euhedral) in igneous rocks.* Cf: *hypidioblastic.*

idioblastic series *crystalloblastic series.*

idiochromatic Said of a mineral, the color of which is due to its chemical composition, i.e. it is inherent rather than accidental. Cf: *allochromatic.*

idiogenous A little-used syn. of *syngenetic.* Cf: *hysterogenous; xenogenous.*

idiogeosyncline A type of late-cycle geosyncline between stable and mobile areas of the crust, the sediments of which are only weakly folded, such as the marginal basins of the East Indian island arc (Umbgrove, 1933, p.33-43). Cf: *parageosyncline (b).*

idiomorph A crystal in an igneous rock, sharply bounded by some or all of the crystal faces characteristic of the mineral species to which the crystal belongs. Cf: *euhedron.*

idiomorphic (a) Said of the texture or fabric of an igneous rock having or characterized by crystals completely bounded by the crystal faces peculiar to the mineral species to which the crystal belongs. Also, said of an igneous rock with idiomorphic texture. Cf: *xenomorphic; hypidiomorphic.* See also: *panidiomorphic.* (b) An obsolescent syn. of *euhedral.*----The term *idiomorphisch* was proposed by Rosenbusch (1887, p.11) originally to describe in an igneous rock the individual mineral crystals (now known as euhedral crystals) that had sufficient room to form their own proper crystal faces. Current usage tends to apply "idiomorphic" to an igneous-rock texture or fabric characterized by euhedral crystals. Syn: *automorphic; idiomorphous.*

idiomorphic-granular Said of the granular texture, of an igneous rock, characterized by an idiomorphic fabric; also said of the rock with such a texture. Syn: *automorphic-granular.*

idiophanous Said of a crystal that exhibits an interference figure to the naked eye, without the help of optical instruments.

idiotopic Said of the fabric of a crystalline sedimentary rock in which the majority of the constituent crystals are euhedral. Also, said of the rock (such as an evaporite, a chemically deposited cement, or a recrystallized limestone or dolomite) with such a fabric. The term was proposed by Friedman (1965, p.648). Cf: *xenotopic; hypidiotopic.*

idocrase *vesuvianite.*

idrialite A hydrocarbon mineral: $C_{22}H_{14}$. It was previously formulated $C_{24}H_{18}$. Idrialite is often found mixed with cinnabar and clay. Syn: *curtisite.*

igneous Said of a rock or mineral that solidified from molten or partly molten material, i.e., from a magma; also, applied to processes leading to, related to, or resulting from the formation of such rocks. "Igneous" rocks constitute one of the three main classes into which all rocks are divided (i.e. igneous, metamorphic, sedimentary). Etymol: Latin *ignis,* "fire". See also: *plutonic; hypabyssal; extrusive.* Obs. syn: *ignigenous; pyrogenous.* Deprecated syn: *eruptive.*

igneous breccia (a) A breccia that is composed of fragments of igneous rock. (b) Any breccia produced by igneous processes, e.g. volcanic breccia, intrusion breccia.

igneous complex An assemblage of intimately associated and roughly contemporaneous igneous rocks differing in form or in petrographic type; esp. a plutonic body consisting of a diverse association of layered igneous rocks.

igneous-contact shoreline A shoreline formed by the partial submergence of the relatively steep slope left by the removal of weak beds from one side of a straight igneous contact (Johnson, 1925, p.24-27).

igneous cycle The usual sequence of events in which volcanic activity is followed by major plutonic intrusions, and then minor intrusions (e.g. dikes).

igneous facies A part or variety of a single igneous rock body, differing in some attribute (structure, texture, or mineralogic or chemical composition) from the normal or typical rock of the main mass; e.g. a granite mass may grade into a porphyritic "igneous facies" near its borders. It is due to differentiation in place. Syn: *facies [petrology].*

igneous lamination In plutonic rocks, the parallel arrangement of tabular crystals to each other and to any layering the rocks may have.

igneous metamorphism A high-temperature metamorphic process which includes the effects of magma upon adjacent rocks as well as those due to igneous injection pegmatitization (Lindgren, 1933). The term is no longer in common use. Cf: *pyrometamorphism.*

igneous province *petrographic province.*

igneous quartz *silexite.*

igneous rock clan *clan.*

igneous rock series An assemblage of related igneous rocks of the same general form of occurrence (plutonic, hypabyssal, or volcanic), found in a single district and formed in a single period of igneous activity, and characterized by possessing in common certain chemical, mineralogic, and textural features or properties so that the rocks together exhibit a continuous variation from one extremity of the series to the other. Syn: *series; rock series.*

ignigenous An obs. syn. of *igneous.*

ignimbrite The rock formed by the deposition and consolidation of ash flows and núees ardentes. The term originally implied dense welding but there is no longer such a restriction, so that the term is only partially synonymous with *welded tuff,* and may include the nonwelded *sillar* as well. See also: *tufflava.* Partial syn: *ash-flow tuff.* Syn: *flood tuff.*

ignispumite A type of rhyolite characterized by lenticles and banding and which is believed to have been deposited as an acid, foamy lava and to be transitional with true ignimbrite. Cf: *tufflava.*

ihleite *copiapite.*

ijolite A series of plutonic rocks containing nepheline and 30-60% mafic minerals, generally pyroxene, and including sphene, apatite, and melanite; also, any rock of that series. *Melteigite* and *jacupirangite* are more mafic members of the series; *urtite* is a type rich in nepheline. Cf: *tawite.*

ijussite An igneous rock intermediate in composition between teschenite and pyroxenite, being composed of abundant titanaugite and barkevikite with smaller amounts of bytownite, anorthoclase, and analcime.

ikaite A chalky mineral: $CaCO_3.6H_2O.$

ikunolite A mineral: $Bi_4(S,Se)_3.$ Cf: *laitakarite.*

ilesite A green mineral: $(Mn,Zn,Fe)SO_4.4H_2O.$

ilimaussite A mineral: $Ba_2Na_4CeFeNb_2Si_8O_{28}.5H_2O.$

illidromica A hydromica, low in potassium and high in water; *illite.* Also, a clay mineral intermediate in composition between illite and montmorillonite.

Illinoian Pertaining to the third glacial stage of the Pleistocene Epoch in North America, beginning 115,000(?) years ago, after the Yarmouth interglacial stage and before the Sangamon. See also: *Riss.* Syn: *Illinoisan.*

illite (a) A general name for a group of three-layer, mica-like, and gray, light-green, or yellowish-brown clay minerals that are widely distributed in argillaceous sediments (esp. marine shales and soils derived from them), that are intermediate in composition and structure between muscovite and montmorillonite, that have 10-angstrom c-axis spacings that show substantially no expanding-lattice characteristics, and that have the general formula: $(H_3O,K)_y(Al_4.Fe_4.Mg_4.Mg_6)(Si_{8-y}.Al_y)O_{20}(OH)_4$, with y less than 2 and frequently equal to 1 to 1.5. The term is generally used for "clay-mineral micas of both dioctahedral and trioctahedral types and of muscovite and biotite crystallizations" (Grim, 1968, p.42). Illite contains less potassium and more water than true micas, and more potassium than kaolinite and montmorillonite; it appears intermediate between kaolin and montmorillonite clays in cation-exchange capacity, in ability to absorb and retain water, and in physical characteristics (such as plasticity index). Much so-called illite may be a mechanical mixture of fine-grained montmorillonite and muscovite, or a clay containing alternate layers having a montmorillonite and a muscovite structure. (b) A mineral of the illite group; esp. the mineral having the complex chemical composition of *muscovite* or of a hydrated muscovite but giving a line-poor X-ray powder pattern (Hey, 1962, 16.3.20).---The term was proposed by Grim, et al (1937) as a general term (not a specific mineral name) in recognition of the state (Illinois) in which clay study has received much encouragement. See also: *bravaisite.* Syn: *hydromica; illidromica; glimmerton.*

illuvial Said of a soil horizon to which material has been added by the process of *illuviation*; pertaining to the process of illuviation. Cf: *eluvial [soil].*

illuviated Said of a soil horizon or of materials that have been subjected to the process of *illuviation.*

illuviation The accumulation of soluble or suspended material in a lower soil horizon that was transported from an upper horizon by the process of *eluviation.* Adj: *illuvial; illuviated.*

illuvium Material from a soil that is leached and redeposited in another horizon.

ilmenite An iron-black, opaque, rhombohedral mineral: $FeTiO_3$. It is the principal ore of titanium. Ilmenite occurs as a common accessory mineral in basic igneous rocks (esp. gabbros and norites), and is also concentrated in mineral sands. See also: *menaccanite.* Syn: *titanic iron ore; mohsite.*

ilmenitite A hypabyssal rock composed almost entirely of ilmenite, with accessory pyrite, chalcopyrite, pyrrhotite, hypersthene, and labradorite.

ilmenomagnetite (a) Magnetite with microintergrowths of ilmenite. (b) Titanian maghemite with exsolution ilmenite.---Cf: *magnetoilmenite.*

ilmenorutile A black mineral: $(Ti,Nb,Fe)_3O_6$.

ilsemannite A black, blue-black, or blue mineral: $Mo_3O_8.nH_2O$ (?).

ilvaite A brownish to black orthorhombic mineral: $CaFe_2^{+2}-Fe^{+3}(SiO_4)_2(OH)$. It is related to epidote, and usually contains manganese in small amounts. Syn: *lievrite; yenite.*

image [photo] (a) A pictorial representation of a subject on photographic film whether produced by direct photography or by imagery. (b) The optical counterpart of an object, produced by the reflection or refraction of light when focused by a lens, mirror, or other optical system.

image [sed] (a) A term introduced by Wadell (1932, p.449) for a binomial expression of the shape of a sedimentary particle, expressed as a fraction giving the roundness of the particle in the numerator and the sphericity in the denominator. (b) A two-dimensional (plane) projection or cross section of a sedimentary particle, obtained by photography or by tracing; it is useful in determining *roundness.*

image-forming system An airborne, satellite and ground (tripod-mounted) image-forming system which preserves and records all radiation-intensity information in the field of view, providing a comprehensive picture of the scene observed. Three types are commonly used to reproduce an image faithfully: a film or extensive mosaic-type system; an electronic scanning system similar to the vidicon or image orthicon tube used in television reproduction; and a mechanical raster-scan system. In the third type a detector or detectors systematically examine a field of view by mechanical action of a mirror or prism. The ultimate resolution of all three systems depends on an equivalent detection element. All three systems have a frame time or exposure time associated with the action of surveying the scene of interest (Jamieson, 1963, p.620).

image/frame That data from one spectral band of one sensor for a nominal framing area of the Earth's surface (NASA, 1971, p.P-1).

image motion The smearing or blurring of imagery on an aerial photograph due to the relative movement of the camera with respect to the ground. An "image-motion compensator" installed with the camera intentionally imparts movement to the film at such a rate as to compensate for the forward motion of the aircraft during exposure time.

image point An image on a photograph, corresponding to a definite object or specific point on the ground. See also: *corresponding image point.*

imager A colloquial and nonrecommended term for any of a variety of optical-mechanical imaging systems.

imagery The pictorial and indirect representation of a subject produced by electromagnetic radiation emitted or reflected from, or transmitted through, the subject, and detected electronically by "a reversible-state physical or chemical transducer whose output is capable of providing an image" (Robinove, 1963, p.880), such as one displayed on a television type tube, or a photograph of same.

imandrite A rock composed chiefly of quartz and albite formed by the combination of a nepheline syenite magma with a graywacke.

imatra stone *marlekor.*

imbibition [rock] Formation of feldspathic minerals by the penetration of alkaline solutions of magmatic origin into aluminum-rich metamorphic rocks.

imbibition [water] (a) The absorption of a fluid, usually water, by a granular rock or any other porous material, under the force of capillary attraction, and in the absence of any pressure. Syn: *capillary percolation.* (b) Absorption of water by plants.

imbibometry A method of analysis involving measuring water uptake, or the uptake of other fluids, in a solid substance.

Imbrian (a) Pertaining to lunar topographic features and lithologic map units constituting a system of rocks formed during the period of formation of the Mare Imbrium basin and of deposition of mare material of the Procellarum Group, or during any time between these two events. Imbrian rocks are older than the post-mare craters and associated ejecta of the Eratosthenian and Copernican systems. (b) Said of the stratigraphic period during which the Imbrium System was developed.

imbricated Said of the overlapping pattern of leaves on a bud.

imbricated fault zone A zone of closely spaced faults exhibiting *imbricate structure*; it is a décollement, underlain by a sole fault.

imbricated texture A texture resembling overlapping plates, seen in certain minerals (such as tridymite) under the microscope.

imbricate structure [sed] A sedimentary structure characterized by *imbrication* of pebbles all sloping in the same direction, and whose flat sides commonly display an upstream dip (pebbles "lean" downstream) or a seaward overlapping (as in beach deposits). Syn: *shingle structure.*

imbricate structure [tect] A tectonic structure displayed by a series of nearly parallel and overlapping minor thrust faults, high-angle reverse faults, or slides and characterized by rock slices, sheets, plates, blocks, or wedges that are approximately equidistant and have the same displacement and that are all steeply inclined in the same direction (toward the source of stress); a superimposition of nappe structures. It is generally attributable to the passage overhead of a major fault movement, causing a succession of many small, compressed overfolds that find relief through dislocation and resulting in lower (older) rocks overlapping higher (younger) rocks in the manner of tiles on a roof. See also: *imbricated fault zone.* Syn: *schuppen structure; shingle-block structure.*

imbrication [sed] (a) The slanting, overlapping arrangement of tabular or platy fragments or flat pebbles in a stream bed or on a beach, in the manner of tiles or shingles on a roof. It is caused by flowing water. (b) Formation of *imbricate structure.*---Syn: *shingling.*

imbrication [tect] (a) The steeply inclined, overlapping arrangement of thrust sheets in imbricate structure. (b) Formation of such structure.

imerinite A colorless to blue monoclinic mineral of the amphibole group: $Na_2(Mg,Fe)_6Si_8O_{22}(O,OH)_2$. It is related to richterite.

imhofite A mineral: $Tl_6CuAs_{16}S_{40}$ (?).

imitation Any material that simulates a genuine, natural gem; specif. glass, plastic, or other amorphous material, as well as crystalline material, that simulates the appearance of a natural gem, as distinguished from a synthetic stone and an assembled stone. Syn: *simulated stone; imitation stone.*

immature [geomorph] Said of a topography or region, and of its features, not having attained maturity; esp. said of a valley or drainage system that is well above base level and whose side slopes descend steeply to the riverbanks, or of a drainage system that is actually expanding into an imcompletely drained upland.

immature [sed] (a) Pertaining to the first stage of textural maturity (Folk, 1951); said of a clastic sediment that has been differentiated or evolved from its parent rock by processes acting over a short time and/or with a low intensity and that is characterized by relatively unstable minerals (such as feldspar), abundance of mobile oxides (such as alumina), presence of weatherable material (such as clay), and poorly sorted and angular grains. Example: an "immature sandstone" containing over 5% clay and commonly occurring in deeper marine, flood-plain, swamp, and mudflow deposits. Cf: *submature; mature; supermature.* (b) Said of an argillaceous sedimentary material intermediate in character between a clay and a shale; e.g. an "immature shale".

immature region *endozone.*

immature soil *azonal soil.*

immaturity [geomorph] A stage in the cycle of erosion characterized by *immature* features.

immediate runoff *direct runoff.*

immersed bog A bog which tends to grow horizontally by growth of plants under water. Cf: *emerged bog.*

immersion cup A cup-like accessory to a microscope, used for examining inclusions in a gem that is immersed in a liquid of high refractive index, and designed to eliminate reflections from highly polished facets.

immersion liquid A liquid of known refractive index that is used in the *immersion method* of determining a mineral's refractive index. An example is acetone. Syn: *index liquid.*

immersion method A method of determining the relative refractive index of a mineral in order to identify it, by immersing the sample in a liquid of known refractive index (an *immersion liquid*).

immigrant In ecology, an organism that becomes established in a region where it was previously unknown.

immiscibility gap A term that is incorrectly used for *miscibility gap.*

immiscible Said of two or more phases that, at mutual equilibrium, cannot dissolve completely in one another, e.g. oil and water. Cf: *miscible.*

impact A forceful contact or collision between bodies, such as that involved in the production of a meteorite crater or cryptoexplosion structure. Also, the degree or concentration of force in a collision.

impact bomb A porous mass of impactite formed by splattering and exhibiting aerodynamic sculpturing. See also: *Fladen.*

impact cast *prod cast.*

impact crater A *crater* formed on a surface by the impact of an unspecified projectile; esp. a terrestrial or lunar crater in which the nature of the impacting body (meteorite, asteroid, comet, etc.) is not known. See also: *meteorite crater; penetration funnel; primary crater; secondary crater.*

impact erosion The wearing away of rocks or fragments through the effect of definite blows of relatively large fragments.

impact glass Glassy *impactite.*

impactite (a) A vesicular, glassy to finely crystalline material produced by fusion or partial fusion of target rock by the heat generated from the impact of a large meteorite and occurring in and around the resulting crater typically as individual bodies composed of mixtures of melt and rock fragments, often with traces of meteoritic material; a rock (such as suevite) from a presumed impact site. Syn: *impact slag; impact glass.* (b) A term used incorrectly for any shock-metamorphosed rock.

impact lava *impact melt.*

impact law A physical law governing the settling of coarse particles, in which (for a given particle density, fluid density, and fluid viscosity) the settling velocity is directly proportional to the square root of the particle diameter. Cf: *Stokes' law.*

impact mark *prod mark.*

impact melt Molten material produced by fusion of target rock during a meteorite impact and emplaced in and around the resulting crater as discrete, partly to completely crystalline dike- or sill-like bodies, as the matrix of fragmental breccias, or as discrete fragments ejected from the crater. Syn: *impact lava.*

impact metamorphism A type of *shock metamorphism* in which the shock waves and the observed changes in rocks and minerals result from the hypervelocity impact of a body such as a meteorite (Chao, 1967, p.192). The term was used by Dietz (1961) to describe a field of geology involving the study of minerals (such as coesite and minute diamonds) created by meteorite impact.

impact slag *impactite.*

impact structure A generally circular or crateriform structure produced by impact (usually extraterrestrial) onto a planetary surface. The stage of erosion of the structure and the nature of the impacting body need not be specified.

imparipinnate *odd-pinnate.*

impedance [elect] The complex ratio of voltage to current in an electrical circuit, or more generally, the complex ratio of electric field intensity to magnetic field intensity in an electromagnetic field. It is the reciprocal of *admittance.*

impedance [seis] *acoustic impedance.*

impedance matching The process of matching the output impedance of a transmitter or the input impedance of a receiver to the characteristic impedance of the wave-transmitting medium so as to insure a maximum transfer of energy to or from the medium. It is applicable to both electric and acoustic *impedance.*

imperfect flower A flower having either stamens or carpels but not both. Cf: *perfect flower.*

imperfection *flaw* [gem].

imperforate (a) Not perforated, or lacking a normal opening; esp. descriptive of foraminiferal-test walls without pores or tiny perforations. (b) Said of a spiral mollusk shell having the umbilicus obliterated by the later whorls.---Cf: *perforate*.

Imperial jade A translucent to semitransparent variety of jadeite characterized by the finest, highly intense emerald-green color; "true jade". Also spelled: *imperial jade*.

impermeability The condition of a rock, sediment, or soil that renders it incapable of transmitting fluids under pressure. Syn: *imperviousness*. Ant: *permeability [geol]*. Adj: *impermeable*.

imperviousness *impermeability*.

impetus A sharp onset or arrival of a seismic phase on a seismogram. Adj: *impulsive*. Cf: *emersio*.

impingement The mechanism or process in dolomitization whereby dolomite crystals replace calcareous particles (commonly skeletal particles such as crinoid ossicles and plates) but not in optical continuity with the calcite of the original particle (Lucia, 1962).

implication A rarely used syn. of *symplectite*.

impolder Var. of *empolder*.

impregnated Said of a mineral deposit (esp. metals) in which the minerals are epigenetic and are diffused in the host rock. Cf: *disseminated; interstitial*.

impressed area The concave dorsum of a coiled cephalopod conch in contact with the venter of the preceding (next-older) whorl and tending to overlap it. Syn: *impressed zone*.

impression (a) The form, shape, or indentation made on a soft sedimentary surface (as of mud or sand) by an organic or inorganic body (usually a harder structure, such as a fossil shell or the strengthened surface of a leaf) that has come in contact with it; a *mold*. It usually occurs as a negative or concavity found on the top of a bed, and a cast of it may then be found on the base of the overlying bed. (b) A small, shallow, circular pit or depression formed by rain, hail, drip, or spray. (c) A fossil footprint, trail, track, or burrow.----Syn: *imprint*.

imprint [sed] An *impression*, esp. one made by a thin object (such as a leaf) or one made by a falling hailstone or raindrop.

imprint [struc petrol] *overprint*.

imprisoned Said of a boulder or block of rock resting intimately against others with common or closely fitting interfaces, such as those found along a rocky coast (as that of Victoria, Australia). Term used by Baker (1959, p.206).

imprisoned lake A term used by Dana (1895, p.199) for a lake occupying a basin that had been cut off from a river system, or a crater of an extinct volcano, or a basin or depression formed by earth movements or glacial action.

impsonite A dull, black, nearly infusible, asphaltic pyrobitumen with a hackly fracture and high fixed carbon content that closely resembles *albertite*, but which is almost insoluble in turpentine; it is derived from the metamorphism of petroleum.

impulse In seismology, a sudden and short time-duration force caused by an explosion or other mechanical means.

impulsive Said of an *impetus*, or sharp arrival of a seismic phase. Cf: *emergent*.

impunctate Lacking pores, perforations, or punctae; specif. said of a brachiopod shell without endopunctae or pseudopunctae and in which the shell substance is dense. Impunctate brachiopods are by far the most numerous of all brachiopods. Cf: *punctate*.

impure arkose A term commonly applied to a sandstone (esp. a graywacke) that is highly feldspathic, but not an arkose; specif. a sandstone containing 25-90% feldspars and igneous rock fragments, 10-50% micas and metamorphic rock fragments, and 0-65% quartz and chert (Folk, 1954, p.354). The term is roughly equivalent to *micaceous arkose* of Hubert (1960), and was used by Krynine (1948, p.137) for a transitional rock between arkose and high-rank graywacke. Cf: *lithic arkose; feldspathic graywacke*. Syn: *dirty arkose*.

inactive volcano A volcano that has not been known to erupt. Cf: *active volcano; dormant volcano; extinct volcano*.

inadunate Any crinoid belonging to the subclass Inadunata,

characterized by firmly jointed calyx plates, a mouth concealed by the tegmen, and arms that are free above the radials.

in-and-out channel A crescentic valley excavated on a hillside by meltwater flowing around a projecting glacial lobe (Kendall, 1902, p.483).

inaperturate Said of pollen and spores having no germinal, harmomegathic, or other openings. Cf: *acolpate; alete*.

inarticulate n. Any brachiopod belonging to the class Inarticulata, characterized by valves that are calcareous or composed of chitinophosphate and are commonly held together by muscles rather than hinge teeth and dental sockets.----adj. Said of a brachiopod having such valves, or of the valves themselves. Cf: *articulate [brach]*.

inbreak A subsidence of the surface over a mine due to subterranean shattering of rock material. Cf: *crown-in; inbreak heave*.

incandescent Said of an ash flow or núee ardente or of any pyroclastic matter that is glowing or fiery.

incandescent tuff flow A term that is essentially synonymous with *ash flow* and was originally used to describe the fragmental outbursts of fine-grained rhyolitic material in the Arequipa region of Peru (Fenner, 1948, p.879).

incarbonization *coalification*.

Inceptisol In U.S. Dept. of Agriculture soil taxonomy, a soil order characterized by the alteration or removal of material other than carbonates or amorphous silica from the pedogenic horizon(s) (SSSA, 1970). Suborders and great soil groups of this order have the suffix -ept. See also: *Andept; Aquept; Ochrept; Plaggept; Tropept; Umbrept*.

incertae sedis In taxonomy, a term applied to a fossil or group of fossils whose exact position in the classification system cannot be determined with certainty. Etymol: Latin, "of uncertain place".

incidental vein A vein discovered after the original vein or which a claim is based.

incipient peneplain A syn. of *strath* in the restricted sense proposed by Bucher (1932, p.131) of a fluvial-degradation surface consisting of a broad valley floor and of extensive valley floor side strips. Syn: *partial peneplain; local peneplain*.

incipient species A natural group of individuals (e.g. a subspecies) that is capable of interbreeding with another related group but that is prevented from doing so because of some specific barrier.

incised In geomorphology, said of a stream meander or notch that has downcut or entrenched into the surface during, and because of, relative uplift of the surface. See also: *incision*.

incised meander (a) A generic term for an old stream meander that has become deepened by rejuvenation and that is more or less closely bordered or enclosed by valley walls. Two types are usually recognized: *entrenched meander* and *ingrown meander*. Syn: *inclosed meander*. (b) A term used more restrictedly as a syn. of *entrenched meander*.

incision (a) The process whereby a downward-eroding stream deepens its channel or produces a narrow, steep-walled valley; esp. the downcutting of a stream during, and as a result of, relative movement (uplift) of the crust. Also, the product of such a process, e.g. an incised notch or meander. Cf: *entrenchment*. (b) The process whereby a deep, narrow, steep-sided trench or notch intersects a plane surface or slope; e.g. current erosion of the continental slope to produce a submarine canyon.

incisor A process with a biting surface on the gnathal lobe of the mandible of a crustacean. Cf: *molar [paleont]*.

inclination [magnetism] In magnetic inclination, the angle at which magnetic field lines dip; it is one of the *magnetic elements*. Syn: *magnetic dip*.

inclination [paleont] The attitude of the cardinal area (or pseudointerarea) in either valve of a brachiopod, based on the convention of viewing the specimen in lateral profile with

eaks to the left and brachial valve uppermost, referring the
ardinal area to its position within one of four quadrants de-
ned by the commissural plane and the plane normal to it and
he plane of symmetry, touching the base of the cardinal
reas (TIP, 1965, pt.H, p. 146). See also: *orthocline; cata-*
line; anacline; hypercline; apsacline; procline.

inclination [struc geol] A general term for the slope of any geological body or surface, measured in the upward or downward direction and from the horizontal or the vertical. It is
ften used synonymously with *dip.*

inclination [slopes] (a) A deviation from the true vertical or
orizontal. Also, the amount of such deviation; the rate of
ope, or grade. (b) An inclined surface; a slope.

inclinator *inclinometer.*

inclined bedding (a) An inclusive term for bedding inclined to
he principal surface of deposition. The term is not recommended for use as a syn. of *cross-bedding* because it may
equally refer ... to any initial dip" (Hills, 1963, p.10). See
lso: *discordant bedding.* Archaic syn. *diagonal bedding;*
blique bedding. (b) Bedding laid down with primary or initial
ip (Dennis, 1967, p.12).

inclined extinction A type of *extinction* seen in birefringent
rystal sections in which the vibration directions are inclined
o a crystal axis or direction of cleavage. Cf: *parallel extinc-*
ion; undulatory extinction. Syn: *oblique extinction.*

inclined fold A fold whose axial surface is inclined from the
ertical, and in which one limb is steeper than the other. The
erm sometimes includes the restriction that the steeper of
he two limbs not be overturned.

inclined polarization Polarization that is inclined with reference
o either the linear dimensions of a magnetized body or the
lumb line or horizon.

inclinometer [drill] Any of various instruments for indicating
nclination; esp. an instrument that indicates the amount
and direction of departure from the vertical of a borehole or
well. See also: *clinometer.*

inclinometer [mag] An instrument that measures magnetic inclination. See also: *Earth inductor, dip circle.* Syn: *inclinator.*

inclosed meander A syn. of *incised meander*; it was proposed
as a generic term by Moore (1926). Syn: *enclosed meander.*

included gas Gas in isolated interstices in either the zone of
aeration or the zone of saturation (Meinzer, 1923, p.21). It
may also be applied to bubbles of air or other gas, not in iso-
ated interstices, that are surrounded by water in either zone
and that act as obstacles to water flow until the gas disappears by dissolving in the water. Cf: *ground air; subsurface*
air; natural gas.

inclusion A fragment of older, previously crystallized rock
within an igneous rock to which it may or may not be genetically related. See also: *xenolith; autolith.* Syn: *enclave; enclo-*
sure.

incoalation *coalification.*

incoherent Said of a rock or deposit that is loose or unconsol-
dated, or that is unable to hold together firmly or solidly.

incompetent (a) Said of a bed or stratum that is deformed by
flowage under the stresses of folding. (b) Said of a fold in
which the strata have flowed and changed their original thickness, e.g. a *similar fold*; an *injection fold.*----It is a relative
term. Ant: *competent.*

incompetent rock A volume of rock which at a specific time
and under specific conditions is not able to support a tectonic
orce. Cf: *competent rock.*

incomplete caneolith A *caneolith* having upper and lower rim
elements but lacking a wall. Cf: *complete caneolith.*

incomplete flower A flower which lacks one or more of the
four floral appendages (stamens, carpels, corolla or petals,
and calyx or sepals). Cf: *complete flower.*

incomplete ripple mark A ripple-marked surface characterized
by isolated crests of ripple marks, such as one receiving an
nsufficient supply of sand. Syn: *starved ripple mark.*

incomplete tabula A coral *tabula* consisting of several *tabellae*
joined together. Cf: *complete tabula.*

incompressibility modulus *bulk modulus.*

incongruous [fold] Said of a drag fold, the axis and axial surface of which are not parallel to the axis and axial surface of
the main fold to which it is related. Ant: *congruent [fold].*

incongruent melting Melting accompanied by decomposition
or by reaction with the liquid, so that one solid phase is converted into another; melting to give a liquid different in composition from the original solid. An example is orthoclase
melting incongruently to give leucite and a liquid richer in silica than the original orthoclase.

incongruent solution Dissolution accompanied by decomposition or by reaction with the liquid so that one solid phase is
converted into another; dissolution to give dissolved material
in different proportions from those in the original solid.

inconsequent A syn. of *insequent.* The term was used by Gilbert (1877, p.143-144), but is not now in general use.

incorporation A process of *coalification* in which there is no
modification of material. Cf: *vitrinization; fusinization.*

increase The addition of corallites to colonies by budding. Examples: *axial increase; lateral increase; intermural increase;*
peripheral increase.

increment *recharge.*

incretion (a) A term proposed by Todd (1903) for a cylindrical
concretion with a hollow core; e.g. a rhizocretion. (b) A concretion whose growth has been directed inward from without.

incrop A former outcrop concealed by or buried beneath
younger unconformable strata.

incrustation [geol] (a) A crust or hard coating of minerals
formed on a rock surface; e.g. carnotite on sandstone, calcium carbonate on cave objects, and soluble salts on a playa.
(b) The process by which an incrustation is formed.----Syn:
encrustation.

incrustation [paleont] An *external mold* of a plant, usually in
some incompressible rock such as sandstone or tufa (Walton,
1940).

incumbent Lying above; said of an overlying or superimposed
stratum.

incurrent canal *inhalant canal.*

incurve A gentle landward sweep or curving in of the coast,
as at the head of a bay.

indehiscent Said of a fruit or other plant structure that is not
regularly opening, e.g. a seed pod or anther.

indelta A term used in Australia for an inland area where a
river subdivides (Taylor, 1951, p.615).

independent ovicell A bryozoan ovicell that develops independently of the distal zooid. Syn: *recumbent ovicell.*

independent variable The *variable* whose magnitude changes
systematically. Cf: *dependent variable.*

inderborite A monoclinic mineral: $CaMgB_6O_{11} \cdot 11H_2O$.

inderite A mineral: $Mg_2B_6O_{11} \cdot 15H_2O$.

index bed *key bed.*

index contour A contour line shown on a map in a distinctive
manner for ease of identification, being printed more heavily
than other contour lines and generally labeled with values
(such as figures of elevation) along its course. It appears at
regular intervals, such as every fourth or fifth contour line
(depending on the contour interval). Syn: *accented contour.*

index ellipsoid The *indicatrix* of a crystal other than an isotropic crystal.

index error (a) An instrument error, constant in behavior,
caused by the displacement of the zero or index mark of a
vernier; e.g. an error resulting from inclination of the upper
plate in a transit having a fixed vertical vernier. (b) An instrument error in the magnetic bearing given by readings of the
needle of a compass, such as an error arising from oblique
magnetization of the needle or from the disturbance of the
line of sight.

index fossil (a) A fossil that identifies and dates the strata or succession of strata in which it is found; esp. any fossil taxon (generally a genus, rarely a species) that combines morphologic distinctiveness with relatively common occurrence or great abundance and that is characterized by a broad, even worldwide geographic range and by a narrow or restricted stratigraphic range (range zone) that may be demonstrated to approach isochroneity. The best index fossils include swimming or floating organisms that were evolved rapidly (short-lived) and distributed widely, such as graptolites and ammonites. The fossil need not necessarily be either confined to, or found throughout every part of, the strata for which it serves as an index. Syn: *key fossil; type fossil*. (b) A term used, esp. in U.S., for *characteristic fossil*; e.g. a fossil that is characteristic of an assemblage zone and so far as known is restricted to the zone (ACSN, 1957, p.1881). This usage is not recommended because it implies that the fossil found anywhere in the unit will be present throughout and that, because the fossil has time value, the unit also has time value.---Cf: *guide fossil*.

index horizon A structural surface used as a reference point in analyzing the geologic structure of an area. Syn: *index plane*.

index liquid *immersion liquid*.

index map (a) A map, usually of small scale, that depicts the location of, or specific data regarding, one or more small areas in relation to (or within) a larger area and that typically points up special features in the larger area; e.g. a map showing the main surface features (towns, roads, streams, etc.) with reference to a mine property. It often encloses a small area in a rectangle on a large map. (b) A map showing the location and numbers of flight strips and aerial photographs; a map showing the outline of the area covered by each aerial photograph. Cf: *photoindex*.

index mineral A mineral developed under a particular set of temperature and pressure conditions, thus characterizing a particular degree of metamorphism. When dealing with progressive metamorphism, it is a mineral whose first appearance (in passing from low to higher grades of metamorphism) marks the outer limit of the zone in question (Turner and Verhoogen, 1960, p.491). Cf: *critical mineral; typomorphic mineral*.

index of refraction In crystal optics, a number that expresses the ratio of the velocity of light in vacuo to the velocity of light within the crystal. Its conventional symbol is *n*. Modifying factors include wavelength, temperature, and pressure. *Birefringent* crystals have more than one index of refraction. See also: *relative index of refraction; alpha* [cryst]; *beta* [cryst]; *gamma* [cryst]. Syn: *refractive index*.

index plane *index horizon*.

index species (a) A species of plant or animal that is characteristic of a particular set of environmental conditions and therefore whose presence in a particular area indicates the existence of those conditions in that area. (b) An index fossil of species rank.

index surface A two-shelled geometric surface that represents the indices of refraction in a biaxial crystal in the direction of propagation. Cf: *indicatrix*.

index zone A stratum or body of strata, recognizable by paleontologic or lithologic characters, that can be traced laterally and identifies a reference position in a stratigraphic section.

indialite A hexagonal mineral: $Mg_2Al_4Si_5O_{18}$.

indianaite A white, porcelain-like clay mineral representing an impure variety of halloysite from Lawrence County, Indiana.

Indiana limestone A commercial name for *spergenite*, represented by a uniform and gray or buff Mississippian limestone easily quarried in southern Indiana and widely used for building purposes. Syn: *Indiana oolitic limestone*.

indianite A variety of anorthite occurring as gangue of the corundum of the Carnatic of India.

Indian ridge A term used in New England for a sinuous *esker*.

Syn: *serpent kame; Indian road*.

indicated ore A known mineral deposit for which quantitative estimates are made partly from inference and partly from specific sampling. Cf: *inferred ore; possible ore; potential ore* Syn: *probable ore*.

indicator [eco geol] A geologic or other feature that suggest the presence of a mineral deposit, e.g. a carboniferous shale indicative of coal, or a pyrite-bearing seam which may lead to gold ore at its intersection with a quartz vein.

indicator [glac geol] *indicator stone*.

indicator [ecol] A plant or animal peculiar to a specific environment and which can therefore be used to identify that environment.

indicator fan A pattern formed by the distribution of *indicator stones* derived from a restricted source.

indicator horizon *marker band*.

indicator plant (a) A plant whose occurrence is broadly indicative of the soil of an area, e.g. its salinity or alkalinity, level of zone of saturation, and other soil conditions. (b) A plant that grows exclusively or preferentially on soil rich in a given metal or other element.----Var: *plant indicator*.

indicator stone A glacial erratic whose source and direction of transportation are known because of its identity with bedrock in a certain small or restricted area. Syn: *indicator*.

indicatrix In optics, a geometric figure that represents the refractive indices of a crystal: it is formed by drawing, from a central point representing the center of the crystal, lines in all directions, whose lengths represent the refractive indices for those vibration directions. The figure for an isotropic crystal is a sphere; for a uniaxial crystal, an ellipsoid of revolution; and for a biaxial crystal, it is a triaxial ellipsoid (Berry and Mason, 1959, p.192). Cf: *index surface*. Partial syn: *index ellipsoid* Syn: *optic indicatrix*.

indices of lattice row Integral numbers, enclosed in square brackets, symbolized by [*uvw*], and determined, with any lattice point on the *lattice row* as origin, by the coordinates ua vb, and wc, of the next lattice point on the row in terms of the unit cell edges a, b, and c. Indices enclosed in the following manner indicate a symmetrical set of axes: <*uvw*>. Syn *crystal axial indices*.

indicolite An indigo-blue (light-blue to bluish-black) variety of tourmaline, used as a gemstone. Syn: *indigolite*.

indifferent point In a system having two or more components that point at which two phases become identical in composition, with the result that the system loses one degree of freedom, e.g. the maximum and minimum in a solid-solution series, and the melting point of a congruently melting compound (Levin et al, 1964, p.6).

indigene An indigenous organism. Var: *indigen*.

indigenous Said of an organism originating in a specific place; native. Syn: *endemic*. Ant: *exotic*.

indigenous coal Coal formed according to the *in-situ theory;* autochthonous coal.

indigenous limonite Sulfide-derived limonite that remains fixed at the site of the parent sulfide. Cf: *exotic limonite; relief limonite*.

indigenous stream A stream that lies wholly within its drainage basin. Cf: *exotic stream*.

indigirite A mineral: $Mg_2Al_2(CO_3)_4(OH)_2.15H_2O$.

indigo copper *covellite*.

indigolite *indicolite*.

indirect effect *Bowie effect*.

indirect intake Recharge to the aquifer by way of another body or rock.

indirect leveling A type of *leveling* in which differences of elevation are determined indirectly, as from vertical angles and horizontal distances (trigonometric leveling), from atmospheric pressures (barometric leveling), or from the boiling point of water (thermometric leveling). Cf: *direct leveling*.

indirect linkage A type of *linkage* in scleractinian corals with ne or more couples of mesenteries between each pair of ∎eighboring stomodaea. See also: *trabecular linkage*. Cf: *direct linkage*.

∎ndirect stratification [sed] *secondary stratification*.

∎ndite An iron-black mineral: FeIn$_2$S$_4$.

∎ndium A silvery-white tetragonal mineral, the native metallic ∎lement In. It is soft and malleable, and occurs in very small ∎uantities in zinc and other ores.

∎ndochinite A tektite from southeast Asia (Cambodia, Laos, ∎iet-Nam, China, Thailand).

∎nduced infiltration Recharge to ground water by infiltration, ∎ither deliberate or inadvertent, from a body of surface water ∎s a result of the ground-water withdrawal and subsequent ∎wering of the ground-water head below the surface-water ∎vel. Syn: *induced recharge*.

∎nduced magnetization That component of a rock's *magnetization* which is proportional to the ambient magnetic field, and ∎as the same direction. Cf: *remanent magnetization*. See ∎lso: *susceptibility* [mag].

∎nduced meander *advance-cut meander*.

∎nduced polarization The production of a double layer of ∎harge at a mineral interface, or production of changes in ∎ouble layer density of charge, brought about by application of ∎n electric field (induced electrical polarization) or of a mag-∎etic field (induced magnetic polarization); used almost exclu-∎ively for induced electrical polarization wherein it is manifest-∎d either by a decay of voltage in the Earth following the ces-∎ation of a excitation current pulse, or by a frequency depen-∎ence of the apparent resistivity of the Earth. Abbrev: *IP*.

∎nduced radioactivity Radioactivity that is produced by *activation*. Syn: *artificial radioactivity*.

∎nduced recharge *induced infiltration*.

∎nductance That property of an electric circuit by which an ∎lectromotive force is induced in it by a current variation (ei-∎her in it or in a neighboring circuit).

∎nduction [magnet] (a) *magnetic induction*. (b) *electromag-∎etic induction*.

∎nduction [philos] Reasoning from the particular to the gen-∎ral, or from the individual to the universal; deriving general ∎rinciples from the examination of separate facts. Ant: *deduc-∎on*.

∎nduction log An electric log obtained, without the use of elec-∎rodes, by lowering into the uncased borehole a generating ∎oil (fed with a constant alternating current) that induces, in ∎he rocks surrounding the borehole, eddy currents that are ∎oncentric with the hole and that are detected by a receiver ∎oil. The magnitude of the currents is proportional to the con-∎uctivity of the surrounding rocks, and the log gives a contin-∎ous record of the conductivity with depth. It is appropriate ∎or measurements in empty holes or in holes drilled with oil or ∎il-base mud. Syn: *conductivity log*.

∎nduction coupling The mutual impedance between a transmit-∎ing wire and a potential wire, arising in induction. This effect ∎an lead to fictitious anomalies in induced electrical polariza-∎ion surveys.

∎nductive methods Electrical exploration methods in which ∎lectric current is introduced in the ground by means of elec-∎romagnetic induction and in which one determines the mag-∎etic field that is associated with the current (Schieferdecker, ∎959, term 3674).

∎nductura The smooth shelly layer of a gastropod shell se-∎reted by the general surface of the mantle, commonly ex-∎ending from the inner side of the aperture over the parietal ∎egion, columellar lip, and part or all of the shell exterior (TIP, ∎960, pt.I, p.131).

∎ndurated (a) Said of a compact rock or soil hardened by the ∎ction of pressure, cementation, and esp. heat. (b) Said of an ∎mpure, hard, slaty variety of talc.

∎nduration (a) The hardening of a rock or rock material by the ∎ction of heat, pressure, or the introduction of some cement-

ing material not commonly contained in the original mass; esp. the process by which relatively consolidated rock is made harder or more compact. See also: *lithification*. (b) The hardening of a soil horizon by chemical action to form a hard-pan. (c) A hardened, compact mass of rock or soil produced by induration.

indusium The covering of a sorus on a fern.

industrial diamond A general term for a nongem-quality dia-mond (e.g. one having flaws or poor color) that is suitable only for tools, abrasives, drills, and other industrial applica-tions where its superior hardness is desirable. It also includes gem-quality crystals used for tools and esp. for dies where lack of both internal strain and flaws is required. See also: *ballas; bort; carbonado*. Syn: *industrial stone*.

industrial mineral Any rock, mineral, or other naturally occur-ring substance of economic value, exclusive of metallic ores, mineral fuels, and gemstones; one of the *nonmetallics*. See also: *economic mineral*.

industrial stone *industrial diamond*.

inelastic compaction Compaction that is roughly proportional to the logarithm of stress increase but is not recoverable when the stress is decreased. Cf: *elastic compaction*.

inequigranular (a) Said of a rock texture having or character-ized by crystals of different sizes. (b) Said of a rock with ine-quigranular texture. The term was originally applied by Cross et al (1906, p.698) to igneous rocks, but it is also used for sedimentary rocks (such as recrystallized carbonate rocks) whose constituent crystals vary in size.----Ant: *equigranular*.

inequilateral Having the two ends unequal; specif. said of a bi-valve-mollusk shell whose parts anterior and posterior to the beaks differ appreciably in length (such as the anterior part of the valves being much shorter than the posterior part). Cf: *in-equivalve*. Ant: *equilateral*.

inequilibrium stage In hypsometric analysis of drainage ba-sins, the stage of early development corresponding to youth in the geomorphic cycle (Strahler, 1952b, p.1130); the hypso-metric integral is greater than 60%. Cf: *equilibrium stage*.

inequivalve Having valves unequal in size and form; specif. said of a bivalve mollusk or its shell in which one valve is flat-ter (and often smaller) than the other. Cf: *inequilateral*. Ant: *equivalve*.

inert component A component whose amount in a rock after a metasomatic process depends on its initial concentration rath-er than on its chemical potential as externally fixed by the en-vironment. Cf: *mobile component; perfectly mobile compo-nent*. Syn: *initial value component*.

inertinite A carbon-rich maceral group including *micrinite*, *sclerotinite*, *fusinite*, and *semifusinite*. They are relatively inert during the carbonization process. Cf: *exinite; vitrinite*. Syn: *in-erts*.

inerts An informal term, synonymous with *inertinite*.

inesite A rose- to flesh-red mineral: Ca$_2$Mn$_7$Si$_{10}$O$_{28}$-(OH)$_2$.5H$_2$O.

inface The steeper of the two slopes of a cuesta; the *scarp slope*. The term is an abbreviation of "inward-facing escarp-ment", referring to the cliff portion (of a cuesta) facing the oldland, as on a coastal plain.

infancy The initial or very early stage of the cycle of erosion, presumably following an uplift or equivalent change with re-spect to base level, when a region is freshly exposed to the action of surface waters. It is characterized by: smooth, near-ly level erosion surfaces imperfectly dissected by narrow stream gorges; numerous original and slight depressions occupied by marshy lakes and ponds; shallow streams; and imperfect drainage systems. Cf: *youth*. Syn: *topographic in-fancy*.

infantile Pertaining to the stage of *infancy* of the cycle of ero-sion; esp. said of a stream that has just begun its work of erosion, or of a landscape with a smooth surface and numer-ous shallow lakes. Cf: *youthful*.

infauna Those aquatic animals that live within rather than on the bottom sediment. Cf: *epifauna*.

inferior ovary An *epigynous* plant ovary.

inferred ore A known mineral deposit for which quantitative estimates are made in only a general way, based on geologic relationships and on past mining experience, rather than on specific sampling. Cf: *indicated ore; possible ore; potential ore*.

infilling A process of deposition by which sediment falls or is washed into depressions, cracks, or holes, as the filling in of crevasses upon the melting of glacier ice.

infiltration The movement of water or solutions, esp. ore-bearing solutions, into a rock through its interstices or fractures or into the soil, from another area. Cf: *percolation*.

infiltration capacity The maximum or limiting *infiltration rate*. The term is considered an obsolete syn. of *infiltration rate* by the Soil Science Society of America (1965, p.338). Symbol: f.

infiltration coefficient The ratio of infiltration to precipitation for a specific soil under specified conditions (Nelson & Nelson, 1967, p.192).

infiltration front *pellicular front*.

infiltration gallery A horizontal conduit constructed for the purpose of intercepting ground water. The galleries often parallel rivers, which provide a perennial water supply to the conduit. Syn: *gallery* [grd wat].

infiltration index The average rate of infiltration, expressed in inches per hour. It equals the average rate of rainfall "such that the volume of rainfall at greater rates equals the total direct runoff" (Langbein & Iseri, 1960, p.12).

infiltration rate The rate at which a soil under specified conditions can absorb falling rain or melting snow; expressed in depth of water per unit time (cm/sec; in/hr). Syn: *infiltration velocity*. Cf: *infiltration capacity*.

infiltration vein An interstitial mineral deposit formed by the action of percolating waters. Cf: *segregated vein*.

infiltration velocity *infiltration rate*.

infiltration water Nonpreferred syn. of *free water*.

infiltrometer An instrument used to measure the infiltration of water into soil.

infinitesimal strain theory A theory of material deformation in which during elastic behavior, small displacements and small strains of closely spaced elements in very small volumes are analyzed. Cf: *finite strain theory*.

inflammable cinnabar A mixture of cinnabar, idrialite, and clay.

inflation [mollusk] The distance, measured normal to the plane of symmetry, between the right and left sides of a bivalve mollusk; the distance of the middorsal-midventral line, or the "width" of a bivalve mollusk. Syn: *thickness* [paleont].

inflation [volc] *tumescence*.

inflection angle The angle at which a contour line diverges downstream from a stream channel. Symbol: ψ.

inflorescence A cluster of flowers. See also: *panicle*.

inflow (a) The act or process of flowing in; e.g. the flow of water into a lake. Syn: *influx*. (b) Water that flows in; e.g. ground water and rainfall flowing into the streams of a drainage basin. Also, the amount of water that has flowed in.

inflow cave A cave into which a stream flows, or is known to have flown. Cf: *outflow cave; through cave*. Syn: *influent cave*.

influent adj. Flowing in.---n. (a) A surface stream that flows into a lake (e.g. an inlet), or a stream or branch that flows into a larger stream (e.g. a tributary). Ant: *effluent*. Cf: *influent stream*. Syn: *affluent*. (b) A stream that flows into a cave.

influent cave *inflow cave*.

influent flow Flow of water into the ground from a body of surface water; e.g. the flow of water from an influent stream.

influent seepage Movement of gravity water in the zone of aeration, from the ground surface toward the water table;

seepage of water into the ground.

influent stream (a) A stream or reach of a stream that con‐ tributes water to the zone of saturation and develops ban‐ storage; its channel lies above the water table. Syn: *losin‐ stream*. (b) *influent*.

influx (a) *inflow*. (b) *mouth* [geol] (a).

informal unit A stratigraphic unit that is used in "a broad c‐ free sense without precise connotation or without being a par‐ of an organized system of terminology" (ISST, 1961, p.18)‐ e.g. a unit that is not established as a *formal unit*. Units re‐ garded as informal by the American Commission on Stratigra‐ phic Nomenclature (1961) include: aquifers, oil sands, coa‐ beds, and quarry layers (marble, slate, stone, etc.), even ‐ named (although certain such units are recognized formall‐ as beds, members, or formations by ACSN, 1965, p.297); un‐ named units such as "formation A" or "map unit 2"; zones a‐ applied to rock-stratigraphic units (e.g. "producing zone" c‐ "mineralized zone"); marker horizons established on electri‐ and other mechanically recorded logs; soils in a rock-stratig‐ raphic classification; several minor units in one vertical se‐ quence having identical geographic names; the several unit‐ constituting a cycle of sedimentation; general rock units (e.g‐ "Lower Cambrian strata"); and a unit within another (e.g‐ "Chinle sandstone" referring to "sandstone of the Chinle For‐ mation"). Names of informal units follow the same rules o‐ capitalization as ordinary common nouns (although failure t‐ capitalize a unit name does not necessarily render the nam‐ informal).

infrabasal plate Any plate of proximal circlet in crinoid dorsa‐ cup having two circlets of plates (dicyclic) below the radia‐ plates. Syn: *infrabasal*.

Infracambrian Eocambrian.

infraglacial *subglacial*.

infralaminal accessory aperture An *accessory aperture* in th‐ test of a planktonic foraminifer that leads to a cavity beneat‐ accessory structures and that is at the margin of these struc‐ tures (as in *Catapsydrax*). Cf: *intralaminal accessory aper‐ ture*.

inframarginal sulcus *scrobis septalis*.

infraorder A less commonly used taxonomic unit rankin‐ below "suborder" but above "superfamily" and containing on‐ or more superfamilies.

infrared (a) Pertaining to or designating the portion of th‐ electromagnetic spectrum with wavelengths just beyond th‐ red end of the visible spectrum. (b) Pertaining to the range o‐ wavelengths of from 0.7 to about 1.0 micrometer, i.e. the use‐ ful limits of film sensitivity.

infrared absorption spectroscopy The observation of an ab‐ *sorption spectrum* in the infrared frequency region and all pro‐ cesses of recording and measuring which go with it. Absorp‐ tion of radiant energy for transitions within molecules at th‐ lowest energy levels produce an infrared absorption spectrum.

infrared atmospheric transmission window In the atmosphere‐ a spectral band in which there is minimal absorption, henc‐ maximum transmission of radiation. Water vapor is the singl‐ most important attenuator of infrared radiation. The larges‐ transmission window in the middle infrared region occurs be‐ tween 8 and 14 m. Syn: *transmission window*.

infrared film Film that has a response of about 0.9 μm. Unles‐ elements of the scene are very hot, these films respond t‐ variations in reflected illumination (as do ordinary films) rathe‐ than to thermally emitted radiation. Such films have prove‐ useful in aerial photography. Infrared film is not to be con‐ fused with infrared radiometer and line-scan systems utilizin‐ solid-state detectors and raster-scan devices to record in‐ frared radiation in the middle or thermal infrared region be‐ yond 0.9 μm.

infrared filter (a) Any material of high infrared transmittance‐ which by selective absorption or reflection, or by the additio‐ of surficial interference coatings, transmits only a predeterʰined band of wavelengths. These may be narrow band, wide ᵇand, low pass, or short pass by design. Cf: *filter; cut-off filᵗⁱ ᵉr; narrow band filter.* (b) A photographic filter which passes ᵉear infrared radiation but is almost or completely opaque to ᵛisible light.

ⁿfrared image *thermographic image.*

ⁿfrared photography A type of aerial photography utilizing a ʰlm that is more sensitive to infrared than to visible light rays.

ⁿfrared radiation *Electromagnetic radiation* in the wavelength ⁱⁿterval from about 0.75 micrometer to an indefinite upper ᵇoundary sometimes arbitrarily set at 1 μm. Syn: *long-wave ᵃadiation.*

ⁿfrared thermography A technique or process by which a ⁱⁿne-scan image-forming system detects middle or thermal in-ʳared radiation and produces a two-dimensional image of the ˢcene recorded, called a thermographic image.

ⁿfrasculpture A kind of structure of spores and pollen con-ˢisting of organized internal modifications of exine.

ⁿfrastructure Structure produced at a deep crustal level, in a ᵖlutonic environment, under conditions of elevated tempera-ᵘure and pressure, which is characterized by plastic folding, ᵃnd the emplacement of granite and other migmatitic and ᵐagmatic rocks. This environment occurs in the internal parts ᵒf most orogenic belts, but the term is used especially where ᵗhe infrastructure contrasts with an overlying, less disturbed ᵃayer, or *superstructure.*

ⁿfundibulum (a) A deep indentation of *scrobis septalis* or a ᵇasal indentation of the apertural face of a foraminiferal test ᵃas in *Alabamina*) (TIP, 1964, pt.C, p.61). (b) The apertural ʳegion of a tintinnid.---Pl: *infundibula.*

ⁿfusible Said of a mineral that ranks with quartz on the fusibi-ⁱty scale; i.e., that will not fuse in temperatures up to about ˡ500°C. Cf: *fusibility.*

ⁿfusorial earth A term that was formerly and commonly, but ⁿcorrectly, applied to the white and earthy or powdery sub-ˢtance or soft rock composed essentially of the siliceous re-ᵐains of diatoms, such as *diatomaceous earth.* Syn: *infusorial ˢilica.*

ⁿfusorial silica *infusorial earth.*

ⁿgenite A general term, now obsolete, for a rock originating ᵇelow the Earth's surface; an igneous or metamorphic rock. Ϲf: *derivate.*

ⁿglenook A term proposed by Hobbs (1901, p.152) for a ᵖrism-like indentation of a basin wall, caused by downfaulting ᵃnd characterized by steep walls, a wide and nearly level ˡloor, and "the inadequacy of ... present drainage" to explain ⁱts formation. It is not much used.

ⁿgrafted stream Var. of *engrafted stream.*

ⁿgression The entering of the sea at a given place, as the ᵈrowning of a river valley (Schieferdecker, 1959, terms 1260 & 1840).

ⁿgrown meander A term proposed by Rich (1914, p.470) for ᵃ continually growing or expanding *incised meander* formed ᵈuring a single cycle of erosion by the enlargement or accent-ᵘation of an initial minor sinuosity while the stream was ac-ᵗively downcutting; a meander that "grows in place". It exhib-ⁱts a pronounced asymmetric cross profile (a well-developed, ˢteep undercut slope on the outside of the meander, a gentle ˢlip-off slope on the inside) and is produced when the rate of ᵈowncutting is slow enough to afford time for lateral erosion. Ϲf: *entrenched meander.*

ⁿgrown stream A stream that has enlarged its original course ᵇy undercutting the outer (concave) banks of its curves.

ⁿhalant canal Any canal forming part of the inhalant system ᵒf a sponge. Syn: *incurrent canal; prosochete.*

ⁿhalant system The part of the *aquiferous system* of a sponge ᵇetween the ostia and the prosopyles, characterized by water ʰlowing inward from the ostia. Cf: *exhalant system.*

ⁿherent ash *Ash* in coal derived from inorganic material that ᵂas structurally part of the original plant material. It cannot

be separated mechanically from the coal, constitutes not more than 1%. Cf: *extraneous ash.* Syn: *constitutional ash; fixed ash; intrinsic ash; plant ash.*

inherent mineral matter *Mineral matter* in coal that was structurally part of the original organic material.

inherent moisture In coal, that fraction of the *moisture content* that is structurally contained in the material. It cannot be removed by natural drying but requires heat to 220°F. Syn: *bed moisture.*

inherited (a) Said of a geologic structure, feature, or land-scape that owes its character to conditions or events of a for-mer period; esp. said of a *superimposed* stream, valley, or drainage system. (b) Also, said of a soil characteristic that is directly related to the nature of the parent material rather than to formative processes. See also: *lithomorphic.*

inherited argon Argon-40 present in a metamorphosed mineral as a result of premetamorphic radioactive decay. Also, ra-dioactive decay argon-40 in xenoliths which was produced prior to solidification of the rock as a whole (Damon, 1968, p.13). Cf: *atmospheric argon; excess argon; radiogenic argon.*

inherited flow control The control of glacial drainage by the disposition of blocks of dead ice that had broken along thrust planes or open crevasses when the ice was in motion (Grave-nor & Kupsch, 1959, p.49).

inherited meander *entrenched meander.*

inhomogeneity In electrical prospecting, a local region differ-ing in electrical properties to its surroundings.

inhomogeneity breccia A term used by Sander (1951, p.2) for a breccia that forms paradiagenetically by the rupture of rela-tively friable layers occurring within a more plastic sediment. It contains sharp fragments with broken borders that some-times can be matched to each other.

initial bud An outgrowth through a hole in the wall of a grapto-lithine sicula, producing the first theca of the rhabdosome.

initial dip (a) A syn. of *primary dip.* (b) The dip that a bedded deposit attains due to compaction after sedimentation, but be-fore tectonic deformation. Cf: *primary dip.*

initial landform A landform that is produced directly by epeiro-genic, orogenic, or volcanic activity, and whose original fea-tures are only slightly modified by erosion; it is dominant in the initial and youthful stages of the erosion cycle. Cf: *se-quential landform; ultimate landform; constructional.* Syn: *ini-tial form.*

initial meridian *prime meridian.*

initial point The point from which any survey is initiated; esp. the point from which a survey within a given area of the U.S. public-land system begins and from which a base line is run east and west and a principal meridian is run north and south.

initial shoreline A shoreline brought about by regional tectonic activity (subsidence, uplift, faulting, folding), by volcanic ac-cumulation, or by glacial action; it may have any slope (from almost vertical to nearly horizontal) and may be either smooth or irregular.

initial value component *inert component.*

injection [ign] *intrusion* [ign].

injection [sed] (a) The forcing, under abnormal pressure, of sedimentary material (downward from above, upward from below, or laterally) into a preexisting deposit or rock, either along some plane of weakness or into a preexisting crack or fissure; e.g. the transformation of wet sands and silts to a fluid-like state and their emplacement in adjacent sediments, producing structures such as sandstone dikes or sand volca-noes. See also: *intrusion [sed].* (b) A sedimentary structure or rock formed by injection.---Syn: *sedimentary injection.*

injection breccia A fragmental rock formed by the introduction of largely foreign rock fragments into veins and fractures in the host rock (Speers, 1957). Some examples (notably the Sudbury breccias at Sudbury, Canada) are associated with structures of probable meteorite impact origin and may have formed as a result of the impact process itself.

injection complex An assemblage or association of rocks consisting of igneous intrusions in intricate relationship to sedimentary and metamorphic rocks, such as the ancient rocks underlying the oldest sedimentary formations in eastern U.S.

injection dike A sedimentary dike formed by abnormal pressure of injection from below or above or from the side. Cf: *neptunian dike*.

injection fold An *incompetent* fold.

injection gneiss A composite rock whose banding is wholly or partly caused by *lit-par-lit* injection of granitic magma into layered rock (Holmes, 1928, p.124). Cf: *arterite; phlebite; venite.*

injection ice Shallow ground ice, either seasonal or perennial, formed when water intrudes beneath rock layers; it can occur in beds several meters thick and hundreds of meters long.

injection metamorphism Metamorphism accompanied by intimate injection of sheets and streaks of liquid magma (usually granitic) in zones near deep-seated instrusive contacts (Turner, 1948). Cf: *plutonic metamorphism.*

injection well (a) *recharge well.* (b) A well into which water or a gas is pumped for the purpose of increasing the yield of other wells in the area. Syn: *input well.*

inland Pertaining to or lying in the interior part of a country or continent, or not bordering on the sea; e.g. an *inland* lake.

inland basin *interior basin.*

inland drainage *internal drainage.*

inland ice (a) The ice forming the inner part of a *continental glacier* or large *ice sheet*. The term is applied esp. to the ice on Greenland. (b) A continental glacier or ice sheet its entirety.

inland sea *epicontinental sea.*

inland waters The *territorial waters* (such as lakes, canals, rivers, inlets, bays) within the territory of a state, but excluding high seas and marginal waters that are subject to sovereign rights of bordering states; waters that are above the rise and fall of the tides.

inland waterway One of a system of navigable inland bodies of water (such as a river, canal, or sound).

inlet (a) A small, narrow opening, recess, indentation, or other entrance into a coastline or a shore of a lake or river, and through which water penetrates into the land. Cf: *pass.* (b) A surface waterway into a sea, lake, or river; a creek; an inflowing stream. (c) A narrow strip of water running into the land; e.g. an arm of the sea extending a considerable distance inland. (d) A short, narrow waterway running between islands or connecting a bay, lagoon, or similar body of water with a larger body of water, such as a sea or lake; e.g. a waterway through a coastal obstruction (as a reef or a barrier island) leading to a bay or lagoon. Syn: *tongue.* (e) *tidal inlet.*

inlier An area or group of rocks surrounded by outcrops of younger age, e.g. a window, or an eroded anticlinal crest. Cf: *outlier.*

in-line offset That component of *offset shotpoint* which is the distance from the projection of the shot point onto the line of the spread. Cf: *perpendicular offset.*

innate Said of certain igneous rocks that have undergone transformation without intrusion or other change of position, such as rocks formed by simple fusion in place (Medlicott & Blanford, 1879, p.752).

innelite A mineral: $Na_2(Ba,K)_4(Ca,Mg,Fe)Ti_3Si_4O_{18}(OH,F)_{1.5}(SO_4)$.

inner bar A bar formed at the upper bend of a flood channel, or where the waters of a river are checked by a flood tide. Ant: *outer bar.*

inner beach The part of a sandy beach that is covered by the wash of gentle waves and is ordinarily saturated. Cf: *foreshore.*

inner core The central part of the Earth's *core*, extending from a depth of about 5100 km to the center (6371 km) of the Earth; its radius is about one third of the whole core. The inner core is probably solid because compressional wave travel noticeably faster through it than through the outer core. The chemical composition is probably pure iron. Density ranges from 10.5 to 15.5 g/cm^3. It is equivalent to the *G layer.* Cf *outer core.* Syn: *lower core.*

inner hinge plate Either of a pair of subhorizontal *hinge plate* in the cardinalia of some brachiopods (such as rhynchonelloids, spiriferoids, and terebratuloids), located median of the crural bases and fused laterally with them. Cf: *outer hinge plate.*

inner lamella The thin layer covering an ostracode body in the anterior, ventral, and posterior parts of the carapace, chitinous except for calcified marginal parts forming the duplicature (TIP, 1961, pt.Q, p.51). Cf: *outer lamella.*

inner lamina The inner shell layer of a compartmental plate of certain cirripede crustaceans, separated from an outer lamina by parietal tubes.

inner lead An area of calm water between a line of parallel offshore islands (such as a string of skerries) and the mainland.

inner lip The adaxial (inner) margin of the aperture of a gastropod shell, extending from the foot of the columella to the suture. It consists of the *columellar lip* and the *parietal lip.* Cf *outer lip.*

inner lowland The innermost of the lowland belts of a belted coastal plain, formed in less-resistant rocks that separate the oldland from the cuesta landscape (the first cuesta scarp descending to the bottom of the lowland). Syn: *inner vale.*

inner mantle *lower mantle.*

inner plate One of a pair of subvertical plates in the cardinalia of some pentameracean brachiopods, lying on the ventral side of the base of the brachial process and fused dorsally with it. Cf: *outer plate.*

inner side The portion of a conodont element on the concave side of the anterior-posterior midline. Ant: *outer side.*

inner space The region involved in marine research; the ocean or the ocean environment. Cf: *outer space.* Syn: *hydrospace.*

inner vale *inner lowland.*

inninmorite An igneous rock composed of augite and plagioclase (anorthite to labradorite) in a groundmass of sodic plagioclase, augite, and abundant glass. It is similar in composition to *cumbraite.*

inoperculate adj. Having no operculum; e.g. said of an irregular tear that serves as the opening of a sporangium through which spores are discharged.---n. An inoperculate animal or shell; e.g. an inoperculate gastropod shell.

inorganic Pertaining or relating to a compound that contains no carbon. Cf: *organic.*

inosculation The union of tributaries to form a main stream.

inosilicate A class or structural type of *silicate* characterized by the linkage of the SiO_4 tetrahedra into linear chains by the sharing of oxygens. In a simple chain, e.g. pyroxenes, two oxygens are shared; in a double chain or band, e.g. amphiboles, half the SiO_4 tetrahedra share three oxygens and the other half share two. The Si:O ratio of the former type is 1:3 and for the latter it is 4:11. Cf: *nesosilicate; sorosilicate; cyclosilicate; phyllosilicate; tectosilicate.* Syn: *chain silicate.*

input well *injection well.*

inquilinism A form of *commensalism* in which one organism lives inside another, usually in the digestive tract or respiratory chamber. Adj: *inquiline.*

in regime Said of a stream or channel that has attained an average equilibrium or that is capable of adjusting its cross-sectional form or longitudinal profile by means of alterations imposed by the flow, and in which the average values of the quantities that constitute regime show no definite trend over a period of years (such as 10-20 years). Cf: *graded* [*geomorph*].

inselberg A prominent, isolated, steep-sided, usually smoothed and rounded, residual knob, hill, or small *mountain* of circumdenudation rising abruptly from and surrounded by an extensive and nearly level, lowland erosion surface in a hot, dry region (as in the deserts of southern Africa or Arabia), generally bare and rocky although partly buried by the debris derived from and overlapping its slopes; it is characteristic of an arid or semiarid landscape in a late stage of the erosion cycle. The term was originated by W. Bornhardt. Etymol: German Inselberg, "island mountain". Pl: inselberg; inselberge. Cf: monadnock; bornhardt. Syn: island mountain.

insequent adj. Said of a stream, valley, drainage system, or type of dissection that is seemingly uncontrolled by the associated rock structure or surface features, being determined by minor inequalities not falling into any larger-scale pattern. Etymol: "in" + "consequent". Syn: inconsequent.---n. insequent stream.

insequent stream A stream developed on the present surface but not consequent upon it and apparently not controlled or adjusted by the rock structure and surface features; a self-guided stream that develops under accidental or chance controls and whose resulting drainage pattern is dendritic, as a young stream wandering irregularly on a nearly level plain underlain by homogeneous or horizontally stratified rocks. The term was proposed by Davis (1897, p.24). Syn: insequent.

insequent valley A valley eroded by or containing an insequent stream; a valley whose direction is not explainable by determinable factors.

insert adj. Having the ocular plates of an echinoid in contact with the periproctal margin. Ant: exsert. Syn: inserted.

inset [cart] inset map.

inset [petrology] A term proposed by Shand (1947) to replace the term phenocryst.

inset [streams] A channel where water flows in.

inset ice stream An ice stream [glaciol] from a tributary glacier that is set into the surface of a larger glacier and does not extend to the bed; e.g. a superimposed ice stream set into the surface of a trunk glacier a short distance from their confluence. Cf: juxtaposed ice stream.

inset map A small, separate map that is positioned within the neat line, and in an unimportant part, of a larger map for economy of space or for legibility; e.g. a map of an area geographically outside a map sheet but included therein for convenience of publication, or a part of a larger map drawn at an enlarged or reduced scale. It may or may not be at the same scale as the larger map. Syn: inset.

inset terrace A stream terrace formed during successive periods of vertical and lateral erosion such that remnants of the former valley floor are left on both sides of the valley (Schieferdecker, 1959, term 1512).

inshore (a) Situated close to the shore or indicating a shoreward position; specif. said of a zone of variable width extending from the low-water shoreline through the breaker zone. See also: offshore; nearshore. (b) In a narrow sense, said of a zone that is equivalent to the shoreface.

inshore water (a) Water that is adjacent to land and whose physical properties are influenced considerably by continental conditions. Ant: offshore water. (b) A strip of open water located seaward of an icefoot or land, produced by the melting of fast ice along the shore.

inside pond A body of freshwater enclosed or partly enclosed by sediment deposited from bifurcating distributaries, and lying inland from the delta front of the Mississippi River delta. Cf: outside pond.

in situ density The density of a unit of water, measured at its original depth. Cf: potential density.

in situ temperature The temperature of a unit of water, measured at its original depth. Cf: potential temperature [oceanog].

in situ theory The theory that coal originates at the place where its constituent plants grew and decayed. Ant: drift theory. See also: autochthony. Syn: swamp theory.

insoak The absorption of free surface water by unsaturated soil.

insolation [meteorol] Solar radiation received by a given body, e.g. the Earth; also, the rate at which it is received, per unit of horizontal surface.

insolation [weath] (a) Exposure to the Sun's rays. (b) The geologic effect of the Sun's rays on the Earth's surficial materials; specif. the effect of changes of temperature on the mechanical weathering of rocks. See also: shadow weathering.

insolilith A relatively rounded pebble with a rough or cracked surface produced by exfoliation or granular disintegration resulting from insolation.

insoluble residue The material remaining after a more soluble part of a specimen has been dissolved in hydrochloric acid or acetic acid. It is chiefly composed of siliceous material (such as chert or quartz) and various insoluble detrital minerals (such as anhydrite, glauconite, pyrite, and sphalerite). Abbrev: IR. See also: siliceous residue.

inspissation [oil] The thickening (in consistency) of an oil deposit, as by evaporation or oxidation, in which, after long exposure, gases and lighter fractions escape and pitch, gum, asphalt, and heavier oils remain.

inspissation [stratig] A term used incorrectly for the increase in thickness of sedimentary units, such as commonly occurs where they are traced inward from the edge of a depositional basin.

instant (a) A term defined cryptically by Kobayashi (1944, p.742) as a "stationary point in time-current" and specifically by him (p.750) as a zone time involving a million years or, more precisely, ranging from 300,000 to 3 million years; the time value corresponding to the deposition of a biostratigraphic subzone. (b) A theoretical time interval "of no appreciable length", representing the beginning or the end of a moment or other biochronologic unit (Teichert, 1958a, p.113); a true time plane. It may not necessarily be worldwide.

instantaneous field of view (a) A term specifically denoting the narrow field of view designed into line-scan systems, so that, while about 120° may be under scan, at any one instant only a fraction of that is being recorded. (b) The field of view of a line-scan with the scan motion stopped.

instant rock A colloquial term for a fragile rock produced from originally fragmental materials by the shock waves associated with explosions or meteorite impacts, i.e. by shock-lithification.

instar (a) The ontogenetic stage in the life of an arthropod occurring between two successive molts (periods of ecdysis); an immature molted or shed carapace of an ostracode. Also, an individual in a specified instar. (b) A single episode of shell formation in foraminifers, commonly of a single chamber.

instrument error A systematic error resulting from imperfections in, or faulty adjustment of, the instrument or device used. Such an error "may be accidental or random in nature and result from the failure of the instrument to give the same indication when subjected to the same input signal" (ASP, 1966, p.1138). Syn: instrumental error.

instrument station A station at which a surveying instrument is set up for the purpose of making measurements; e.g. the point over which a leveling instrument is placed for the purpose of taking a backsight or foresight. Syn: setup.

insular [ecol] Said of an organism that has a limited or isolated range or habitat.

insular [clim] Said of a climate in which there is little seasonal temperature variation, e.g. an oceanic climate.

insular shelf An area of the ocean floor analogous to the continental shelf, but surrounding an island. Syn: island shelf.

insular slope An area of the ocean floor analogous to a continental slope, but surrounding an island. Syn: island slope.

insulated stream A stream or reach of a stream that neither contributes water to the zone of saturation nor receives water from it (Meinzer, 1923, p.56); it is separated from the zone of saturation by an impermeable bed.

insulosity The percentage, of the area within the shoreline of a lake, occupied by islands.

intaglio A carved gem that may be used as a seal, in which the design has been engraved into the stone. Cf: *cameo.* See also: *cuvette.*

intake [eng] An opening of an apparatus (such as of a pipe) through which a fluid (air, water, steam, etc.) enters an enclosure from the source of supply; e.g. the place at which water is received into a channel, pipe, or pump from a stream or other body of water for the purpose of driving a mill, supplying a canal, etc.

intake [grd wat] (a) *recharge.* (b) The openings in water-bearing materials through which water passes into a well.

intake area *recharge area.*

intectate Said of a pollen grain lacking a tectum.

integraph Any device used in carrying out a mathematical integration by graphical means, e.g., dot charts for computing terrain effects used in the reduction of gravity data.

integrate Said of a type of wall structure in trepostome bryozoans in which the zooecial boundaries appear as narrow, well-defined lines. Cf: *amalgamate.*

integrated drainage Drainage developed during maturity in an arid region, characterized by coalescence (across intervening ridges and mountains) of drainage basins as a result of headward erosion in the lower basins or of spilling over from the upper basins due to aggradation (Lobeck, 1939, p.12-13); drainage developed where various higher local base levels are replaced by a single lower base level.

integration In petrology, the formation of larger crystals from smaller crystals by recrystallization. See also: *regenerated crystal.*

integripalliate Said of a bivalve mollusk devoid of a pallial sinus. Cf: *sinupalliate.*

integrometer Any mechanical device designed to perform a mathematical integration; e.g. a planimeter.

integument In a flowering plant, the covering layer of the ovule. It is the forerunner of the seed coat or *testa.*

intensity [seism] *earthquake intensity.*

intensity scale A standard of relative measurement of *earthquake intensity.* Three such systems are: the *Mercalli scale,* the *modified Mercalli scale,* and the *Rossi-Forel scale.* The *Richter scale* is a measurement of magnitude rather than of intensity.

intensive variable A thermodynamic variable that is independent of the total amount of matter in the system. Examples include temperature, pressure, and mole fraction.

interambulacral adj. Situated between ambulacra of echinoderms, such as an "interambulacral ray" between two ambulacral rays of a crinoid; esp. referring to the thecal plates making up the area between ambulacra. Cf: *ambulacral.*---n. Any plate situated between the ambulacral plates of an echinoderm.

interambulacrum One of the areas between two ambulacra in an echinoderm; any of the five interradial sections of most echinoderms.

interamnian Situated between or enclosed by rivers. See also: *interfluvial.*

interantennular septum A plate that separates two antennular cavities in some malacostracan crustaceans. Syn: *proepistome.*

interarea The posterior sector of a brachiopod shell with the growing edge at the hinge line. The term is more commonly used for any plane or curved surface lying between the beak and the posterior margin of a brachiopod valve and bisected by the delthyrium or the notothyrium (TIP, 1965, pt.H, p.146). The interarea is generally distinguished by a sharp break in angle from the remainder of the valve and by the absence of costae, plicae, or coarse growth lines. Cf: *pseudointerarea; planarea.*

interbasin area A roughly triangular area, within a drainage basin, that has not developed a drainage channel but contributes drainage directly into a higher-order channel (Schumm, 1956, p.608). Symbol: A_0.

interbasin length The maximum horizontal length of the *interbasin area,* measured from the apex of the triangular ground surface to the adjacent channel (Strahler, 1964, 4-47). Symbol: L_0.

interbed A bed, typically thin, of one kind of rock material occurring between or alternating with beds of another kind.

interbedded Said of beds laid between or alternating with others of different character; esp. said of rock material laid down in sequence between other beds, such as a contemporaneous lava flow "interbedded" with sediments. Cf: *interstratified.*

interbrachial adj. Situated between arms; e.g. an "interbrachial margin" of the disc of an asterozoan.---n. A crinoid plate occurring in the dorsal cup between brachial plates of adjacent ambulacral rays or between braches of any single ray.

intercalary adj. Interposed or inserted between, or introduced or existing interstitially; e.g. "intercalary apical system" of an echinoid in which ocular plates II and IV meet at midline so as to separate anterior and posterior portions, or hoop-like "intercalary bands" located between the valves and connecting bands in a diatom frustule and often projected inward as incomplete partitions parallel to the valves, or "intercalary rock" lying between other strata.---n. One of many thecal plates occurring between radial and basal circlets of some crinoids (e.g. *Acrocrinus*).

intercalated Said of layered material that exists or is introduced between layers of a different character; esp. said of relatively thin strata of one kind of material that alternate with thicker strata of some other kind of material, such as lava flows, beds of shale, or intrusive sills that are "intercalated" in a large body of sandstone.

intercalation (a) The existence of a layer or layers between other layers; e.g. the presence of sheets of lava between sedimentary strata, or the occurrence of a particular fossil horizon between fossil zones of a different character, or the inclusion of lamellar particles of one mineral in another in such a way that the inclusions are oriented more or less exactly in planes related to the crystal structure of the host mineral. The term "sometimes refers ... to the fact that ... beds of sandstone occur at several levels within a predominantly shaly series, with no implication that they are not in the normal order of deposition" (Challinor, 1967, p.135). (b) The introduction of a layer (bed or stratum) between other layers already formed, such as the insertion of rock by intrusion or thrusting in a preexisting series of stratified rock. (c) An intercalated body of material, such as a bed in an intertongued zone or a lens of volcanic ash in a sedimentary deposit.

intercamarophorial plate A short, low median septum on the posterior midline of the camarophorium in brachiopods of the superfamily Stenoscismatacea, extending to the underside of the hinge plate but independent of the median septum duplex (TIP, 1965, pt.H, p.146).

intercameral Located between the chambers of a foraminiferal test; e.g. "intercameral foramen" representing a primary or secondary opening between successive chambers.

intercardiophthalmic region A small rectangular area of the prosoma of a merostome, embracing the cardiac lobe and a minor part of the interophthalmic region.

intercellular space The space resulting from a separation of adjacent plant cell walls from each other along more or less extended areas of contact. In some cases it may result from the splitting of the *middle lamella* in plant cell walls (Esau, 1965, p.62).

intercept [sed] One of the three linear dimensions or diameters of a sedimentary particle: the longest dimension is the "maximum" intercept, and the shorter dimensions are the "intermediate" and "short" intercepts.

intercept [surv] The part of the rod seen between the upper and lower stadia hairs of a transit or telescopic alidade; e.g. a stadia interval.

interception The process by which water from precipitation is caught and stored on plant surfaces and eventually returned to the atmosphere without having reached the ground. Also, the amount of water intercepted. Cf: *throughfall*.

intercept time The sum of the *delay times* at the shot and receiver ends of the path.

intercision (a) A type of *capture* characterized by sidewise swinging of mature streams (Lobeck, 1939, p.201). (b) A type of diversion accomplished by the cutting back of bluffs along a lake shore such that the lake advances inland and cuts into a bend of a river valley some distance above its mouth (Goldthwait, 1908).

intercretion A term proposed by Todd (1903) for a *concretion* that grows by accretion (on the exterior) and by irregular and interstitial addition, causing a circumferential expansion and resultant cracking and wedging apart of the interior of the concretion; e.g. a septarium.

intercrystal porosity Porosity between individual equant, equal-sized crystals (Choquette & Pray, 1970, p.247).

intercumulus The space between crystals of a cumulus.

intercumulus liquid Magmatic liquid that surrounds the crystals of the cumulus, i.e. that occupies the intercumulus. See also: *intercumulus material*. Syn: *interprecipitate liquid; interprecipitate material*.

intercumulus material Material that crystallized from *intercumulus liquid*. Syn: *interprecipitate material*.

interdigitation *intertonguing*.

interdistributary bay A pronounced indentation of the delta front between advancing stream distributaries, occupied by shallow water, and either open to the sea or partly enclosed by minor distributaries.

interdune Pertaining to the relatively flat surface (sand-free or sand-covered) between dunes; e.g. said of the long, trough-like, wind-swept passage between parallel longitudinal dunes, such as a gassi or a feidj.

interestuarine Situated between two estuaries.

interface [seism] *discontinuity*.

interface [petroleum] The two-dimensional contact plane between water and oil or between oil and gas. Syn: *contact*.

interface [sed] A depositional boundary separating two different physicochemical regions; specif. the surface separating the top of the uppermost layer of sediment and the medium (usually water) in which the sedimentation is occurring.

interfacial angle In crystallography, the angle between two faces of a crystal.

interfacial geology The detailed study of the uppermost soil layers, which determine the *depth of penetration* of electromagnetic radiations.

interfelted Said of strata that have been so intimately forced together by pressure and heat as to produce interlocking of structure along contiguous surfaces.

interference The condition occurring when the area of influence of a water well comes into contact with or overlaps that of a neighboring well, as when two wells are pumping from the same aquifer or where two wells are located near each other.

interference colors In crystal optics, the colors displayed by a birefringent crystal in crossed polarized light. Thickness and orientation of the sample and the nature of the light are factors that affect the colors and their intensity.

interference figure The pattern or figure that a crystal displays in polarized light under the conoscope. It is a combination of the *isogyre* and the *isochromatic curve*, and is used to distinguish axial from biaxial crystals and to determine optical sign. See also: *axial figure; biaxial figure*.

interference ripple mark *cross ripple mark*.

interfingering *intertonguing*.

interflow *Storm seepage*.

interfluent lava flow A lava flow that is discharged into and through subterranean fissures and cavities in a volcano and may never reach the surface (Dana, 1890); an obsolete term. Cf: *effluent lava flow; superfluent lava flow*.

interfluminal *interfluvial*.

interfluve The area between rivers; esp. the relatively undissected upland or ridge between two adjacent valleys containing streams flowing in the same general direction. Cf: *doab*. Syn: *interstream area*.

interfluve hill A relatively flat-topped remnant of an antecedent slope on which gradation was arrested when the adjacent streams and valleys were developed (Horton, 1945, p.360); it occurs along the divides in drainage basins approaching maturity. Syn: *interfluve plateau*.

interfluvial Lying between streams; pertaining to an interfluve. See also: *interstream; interamnian*. Syn: *interfluminal*.

interfolding The simultaneous development of discrete fold systems with different orientations.

interformational Formed or existing between one geologic formation and another; e.g. "interformational unconformity".

interformational conglomerate A conglomerate that is present within a formation, and of which the constituents have a source external to the formation. Cf: *intraformational conglomerate*.

intergelisol *pereletok*.

interglacial adj. Pertaining to or formed during the time interval between two successive *glacial epochs* or between two *glacial stages*. The term implies both the melting of ice sheets to about their present level, and the maintenance of a warm climate for a sufficient length of time to permit certain vegetational changes to occur (Suggate, 1965, p.691).---n. *interglacial stage*.

interglacial stage A subdivision of a glacial epoch separating two glaciations, characterized by a relatively long period of warm or mild climate during which the temperature rose to that of the present day; esp. a subdivision of the Pleistocene Epoch, as the "Sangamon Interglacial Stage". Syn: *interglacial; thermal*. Less-preferred syn: *interglacial period*.

interglaciation A climatic episode "during which the climate was incompatible with the wide extent of glaciers that characterized a glaciation" (ACSN, 1961, art.40).

intergrade adj. Said of a soil that is transitional between two types.

intergranular [struc petrol] Said of tectonic transport or movement that takes place by rotation and displacement of grains relative to each other. Cf: *intragranular*.

intergranular [ign] Said of the ophitic texture of an igneous rock in which the augite occurs as an aggregation of grains (rather than in large crystals), not in optical continuity, in the interstices of a network of feldspar laths which may be diverse, subradial, or subparallel (Johannsen, 1939, p.219). It is distinguished from *intersertal* by the absence of interstitial glass. Syn: *granulitic [ign]* (in the sense of Judd).

intergranular movement [glaciol] A process that goes on within a glacier when grains of ice rotate and slide over each other like grains of corn do in a chute. It is a significant factor in glacier flow only near the surface of a glacier. Cf: *intragranular movement*.

intergranular porosity The porosity between the grains or particles of a rock, such as that between the lithoclasts or the bioclasts of a carbonate sedimentary rock. Cf: *interparticle porosity*.

intergranular pressure *effective stress*.

intergrowth The state of interlocking of crystals of two different minerals as a result of their simultaneous crystallization.

interio-areal aperture An aperture in the face of the final chamber of a foraminiferal test, not at its base.

interiomarginal aperture A basal aperture in a foraminiferal test at the margin of the final chamber and along the final suture. In coiled forms it may be an *equatorial aperture* or an *extraumbilical aperture*. See also: *spiroumbilical aperture*.

interior The central or inner area of a mass, region, or structure.

interior basin (a) A depression entirely surrounded by higher land and from which no stream flows outward to the ocean. Cf: *closed basin*. Syn: *inland basin*. (b) *intracratonic basin*.

interior drainage *internal drainage*.

interior orientation The *orientation* of the principal distance and the position of the principal point of a photograph with respect to the fiducial marks of the camera; the determination of the interior perspective of the photograph as it was at the instant of exposure.

interior plain A plain that is situated far from the borders of a continent, as contrasted with a *coastal plain*.

interior valley A closed depression characteristic of karstic topography; it is as large as several kilometers in diameter and has a flat floor, sometimes with a detrital cover, and steep walls. Syn: *polje*.

interjection An intrusion that is *concordant*.

interlacing drainage pattern *braided drainage pattern*.

interlacustrine Situated between lakes; e.g. an "interlacustrine overflow stream" spilling over from one lake to another.

interlaminated Said of laminae occuring between or alternating with others of different character; intercalated in very thin layers. Syn: *interleaved*.

interlayer A layer placed between others of a different nature; e.g. an interbed.

interlayering The regular or random arrangement of structural units of clay minerals in a clay, each unit differing from the adjacent unit either in composition or in crystallographic orientation. Syn: *interstratification*.

interleaved *interlaminated*.

interlobate deposit Drift lying between two adjacent glacial lobes.

interlobate moraine A lateral moraine formed along the line of junction and roughly parallel to the axes of two adjacent glacial lobes which have pushed their margins together. Syn: *intermediate moraine*.

interlocking seismic recording *mixing*.

interlocking spur One of several projecting ridges extending alternately from the opposite sides of the wall of a young, V-shaped valley down which a river with a winding course is flowing, each lateral spur extending into and separating a concave bend of the river so that viewed upstream the spurs seem to "interlock" or "overlap" with each other. Syn: *overlapping spur*.

interlocking texture A rock texture in which particles with irregular boundaries interlock by mutual penetration, as in a crystalline limestone.

intermediate Said of an igneous rock that is transitional between *basic* and *silicic*, generally having a silica content of 54 to 65 percent; e.g. syenite and diorite. "Intermediate" is one subdivision of a widely used system for classifying igneous rocks on the basis of their silica content; the other subdivisions are *acidic*, *basic*, and *ultrabasic*. Syn: *mediosilicic*.

intermediate belt That part of the *zone of aeration* that lies between the capillary fringe and the belt of soil water. Syn: *intermediate zone*.

intermediate coal A type of banded coal defined microscopically as consisting of between 60% and 40% of bright ingredients such as vitrain, clarain, and fusain, with clarodurain and durain composing the remainder. Cf: *semibright coal; semidull coal; bright coal; dull coal*.

intermediate contour A contour line drawn between index contours.

intermediate earthquake *intermediate-focus earthquake*.

intermediate-focus earthquake An earthquake whose focus occurs between a depth of 60 or 70 km and 300 km. Cf: *shallow-focus earthquake; deep-focus earthquake*. Syn: *intermediate earthquake*.

intermediate layer A syn. of *sima*.

intermediate moraine *interlobate moraine*.

intermediate plain A plain intermediate in altitude between the highest summits of an erosion surface and the bottoms of the deepest valleys (Trowbridge, 1921, p.31-33). The term is misleading because it implies a plain intermediate in position between two other plains.

intermediate vadose water Water of the intermediate belt. Syn: *argic water*.

intermediate water A cold, relatively fresh *water mass* originating at arctic and antarctic convergences. It lies above *deep water* and *bottom water*. Cf: *surface water* [*oceanog*].

intermediate wave *transitional-water wave*.

intermediate zone *intermediate belt*.

intermittent island A patch of the shallow bottom of a lake exposed during a period of low lake level and covered during periods of higher level.

intermittent lake A lake that normally contains water for only part of the year or that is only seasonally dry; e.g. a deflation lake. Cf: *ephemeral lake*. Syn: *temporary lake*.

intermittent spring A spring that discharges only during certain periods and at other times is dry or does not exist as a spring. A geyser is a special type of intermittent spring (Meinzer, 1923, p.54). Syn: *intermitting spring*. Cf: *perennial spring; periodic spring*.

intermittent stream (a) A stream or reach of a stream that flows only at certain times of the year, as when it receives water from springs or from some surface source. The term "may be arbitrarily restricted" to a stream that flows "continuously during periods of at least one month" (Meinzer, 1923, p.58). (b) A stream that does not flow continuously, as when water losses from evaporation or seepage exceed the available streamflow.---Cf: *ephemeral stream*. Syn: *temporary stream; seasonal stream*.

intermitting spring *intermittent spring*.

intermont adj. *intermontane*.---n. A hollow between mountains.

intermontane Situated between or surrounded by mountains, mountain ranges, or mountainous regions; e.g. the Great Basin of western U.S., between the Sierra Nevada and the Wasatch Mountains. Syn: *intermont; intermountain*.

intermontane glacier A glacier formed by the confluence of several valley glaciers and occupying a depression between separate mountain ranges or ridges.

intermontane plateau A plateau that is partly or completely enclosed by mountains, and that is formed in association with them; e.g. the Tibetan plateau.

intermontane trough (a) A subsiding area in an island-arc region of the ocean, lying between stable or uprising regions (b) A basin-like area between mountain ranges, sometimes occupied by an intermontane glacier.

intermorainal Situated between moraines, as an *intermorainal* lake occupying a narrow depression between parallel moraines of a retreating glacier. Syn: *intermorainic*.

intermountain adj. *intermontane*.---n. *intermontane basin*.

intermural increase A type of *increase* (budding of corallites) in cerioid coralla characterized by the sideward outgrowth of offsets, the initial parts of which become surrounded by the growing wall of the parent corallite.

internal cast A syn. of *steinkern*. The term should not be used for an *internal mold*.

internal drainage Surface drainage whereby the water does not reach the ocean, such as drainage toward the lowermost or central part of an interior basin. It is common in arid and semiarid regions, as in western Utah. Ant: *external drainage*. See also: *centripetal drainage pattern*. Syn: *interior drainage; inland drainage; closed drainage; endorheism*.

internal energy That energy of a system which is described by the *first law of thermodynamics*.

internal erosion Erosion effected within a compacting sediment by movement of water through the larger pores (Bathurst, 1958, p.33). Cf: *internal sedimentation.*

internal friction The *viscosity* of a substance, e.g. a sediment. Cf: *friction.*

internal lobe The *dorsal lobe* in normally coiled cephalopod conchs. See also: *annular lobe.*

internal mold A *mold* or impression showing the form and markings of the inner surfaces of a fossil shell or other organic structure; it is made on the surface of the rock material filling the hollow interior of the shell or organism. It is sometimes called incorrectly a "cast of the interior", but can be so called only if the shell or structure itself be regarded as a mold. Cf: *external mold; internal cast.*

internal oblique muscle One of a pair of muscles in some inarticulate brachiopods, originating on the pedicle valve between the anterior adductor muscles, and passing posteriorly and laterally to insertions on the brachial valve located anteriorly and laterally from the posterior adductor muscles (TIP, 1965, pt.H, p.146). Cf: *lateral oblique muscle.*

internal rotation In structural petrology, fine-scale rotation producing distortion of the crystal lattice, regardless of what *external rotation* there may or may not be; e.g. calcite twinning. See also: *tectonic rotation.*

internal sedimentation Accumulation of clastic or chemical sediments derived from the surface of, or within, a more or less consolidated carbonate sediment (mud or silt), and deposited in secondary cavities formed in the host rock (after its deposition) by bending of laminae or by *internal erosion* or solution (Bathurst, 1958, p.31).

internal seiche A free oscillation of a submerged layer in a stratified body of water occupying an enclosed or semienclosed basin; esp. an oscillation of the thermocline in a lake. It is believed to be initiated by the same factors that produce a surface *seiche.*

internal structure A structure occurring within the body of a rock mass, such as bedding, ripple marks, molds, and casts. Cf: *external structure.*

internal suture The part of a suture of a coiled cephalopod conch situated on the dorsum (or within the impressed area), extending between umbilical seams, and hidden from view unless the conch is broken. Cf: *external suture.*

internal tide Submerged vertical oscillations on density surfaces in the sea with tidal periods.

internal water Water in the interior of the Earth, below the zone of saturation, where interstices cannot exist due to the pressure of overlying rocks (Meinzer, 1923, p.22).

internal wave A submerged wave occurring on a density surface, i.e. the thermocline, in density-stratified water. Because of the small density gradients involved in internal waves compared with external or surface waves, the internal wave heights, periods, and wavelengths are usually large. See also: *wave.*

International Active Sun Years An internationally cooperative program of studying solar-terrestrial phenomena during an active-sun, i.e. sunspot-maximum, period. It is related to the *International Geophyscial Year* and to the *International Years of the Quiet Sun.* Abbrev: IASY.

international date line *date line.*

International Geophysical Cooperation A continuation of the major part of the programs of the *International Geophysical Year,* from January 1, 1958 to January 1, 1959.

International Geophysical Year An internationally cooperative program of observation of geophysical phenomena from July 1, 1957 to December 31, 1958, near sunspot maximum. See also: *International Active Sun Years; International Years of the Quiet Sun; International Geophysical Cooperation.* Abbrev: IGY.

International Hydrological Decade A ten-year program, 1965-74, patterned after the International Geophysical Year, aimed at training hydrologists and technicians and at establishment of networks for measuring hydrologic data. The idea originated in the United States, but the program is sponsored by UNESCO, and a large proportion of the membership of the United Nations is participating.

international low water A datum plane so low that the tide will seldom fall below it: a plane below mean sea level by an amount equal to half the range between mean lower low water and mean higher high water multiplied by 1.5 (Baker et al, 1966, p.87). Abbrev: ILW.

international map of the world A map series at a scale of $1/1,000,000$ (one inch to 15.78 miles), having a uniform set of symbols and conventional signs, using the metric system for measuring distances and elevations, and printed in modified polyconic projection on 840 sheets each covering an area of $4°$ lat. and $6°$ long. except above the 60th parallel where the longitude covered is $12°$ on each sheet; specif. the International Map of the World on the Millionth Scale, first suggested at the 5th International Geographical Congress in 1891 and accepted in principle in 1909, consisting of an incomplete series of map sheets (many needing revision) generally published by national mapping agencies of concerned countries under the auspices of the United Nations. Abbrev: IMW. Syn: *millionth-scale map of the world.*

International Years of the Quiet Sun An internationally cooperative program during 1964-1965 of studying solar-terrestrial phenomena during a quiet-sun, i.e. sunspot-minimum, period. It is related to the *International Geophysical Year* and to the *International Active Sun Years.* Abbrev: IQSY.

internides The internal part of an orogenic belt, farthest away from the craton, commonly the site of a eugeosyncline during its early phases, and subjected later to plastic folding and plutonism. Cf: *externides.* See also: *primary orogeny; hinterland.* Syn: *primary arc.*

internode An interval or part between two successive nodes; e.g. a segment of a jointed bryozoan colony between the surfaces of articulation, or a segment of a jointed algal thallus.

interocular distance *eye base.*

interophthalmic region The space between the cardiac lobe and ophthalmic ridge of a merostome.

interparticle porosity The porosity between particles in a rock; e.g. *breccia porosity.* Choquette & Pray (1970, p.247) recommend use of this term rather than the term *intergranular porosity* which suggests limitation to grain-size particles.

interpenetration twin A twinned crystal, the individuals of which appear to have grown through one another. Syn: *penetration twin.*

interpluvial adj. Said of an episode of time that was dryer than the pluvial periods between which it occurred. It may correspond to an interglacial period.---- n. Such an episode or period of time.

interpolation Estimation of the value of a variate based on two or more known surrounding values; a method used to determine intermediate values between known points on a curve.

interprecipitate liquid *intercumulus liquid.*

interprecipitate material *intercumulus material.*

interpretative log A *sample log,* based on rotary drill cuttings, in which only the rock cut by the drill bit at the level indicated is portrayed, ignoring the admixed material from a higher level. It is used in the Mid-Continent region of the U.S. Cf: *percentage log.* Also spelled: interpretive log.

interradial [paleont] adj. Situated midway between the axes of adjacent rays of an echinoderm; e.g. an "interradial suture" representing a common line or division between adjacent radial plates of a blastoid.---n. (a) A structure in the interradial area; e.g. a crinoid plate above a basal. (b) *interray.*

interradial [palyn] Pertaining to areas of the proximal face or equator of trilete spores, lying between the arms of the laesura. Cf: *radial.*

interray The area between two adjacent rays of an echinoderm; e.g. the part of a theca between any two adjacent crinoid rays. Syn: *interradial*.

interreef Situated between reefs; e.g. the "interreef region" characterized by relatively unfossiliferous rock, or "interreef sediments" deposited between reefs.

interrupted profile A normal profile that has been altered by an *interruption*; e.g. a longitudinal profile of a stream, where, after rejuvenation, the head of the second-cycle valley touches the first-cycle valley.

interrupted projection A map projection lacking continuous outlines, having several central meridians instead of one, or whose origin is repeated, in order to reduce the peripheral shape distortion and the linear scale discrepancy; e.g. a *homolosine projection* split along several meridians. Syn: *recentered projection*.

interrupted stream A stream that contains perennial reaches with intervening intermittent or ephemeral reaches, or a stream that contains intermittent reaches with intervening ephemeral reaches (Meinzer, 1923, p.58). Ant: *continuous stream*.

interrupted water table A water table that slopes steeply over a *ground-water barrier*, with pronounced difference in elevation above and below the barrier, but not as steep as a *ground-water cascade*.

interruption A break in, or the cutting short of, the cycle of erosion, characterized by a change in the position of base level relative to a landmass or terrain, and resulting in the initiation of a new erosion cycle. It is commonly caused by earth movements (involving deformation, dislocation, or tilting) or by fluctuations of sea level. Cf: *accident*.

intersecting peneplain One of two peneplains forming a *morvan* landscape.

intersection (a) A method in surveying by which the horizontal position of an unoccupied point is determined by drawing lines of known direction from two or more points of known position. Cf: *resection*. (b) Determination of positions by triangulation.

intersection shoot An ore shoot located at the intersection of one vein or vein system with another. It is a common type of ore deposit.

interseptal ridge A longitudinal elevation on the outer surface of the wall of a corallite, corresponding in position to the space between a pair of adjacent septa on the inner surface of the wall. Cf: *septal groove*.

intersequent stream A stream following a consequent course in a depression between the margins of opposing alluvial fans, as on a bajada.

intersertal Said of the texture of a porphyritic igneous rock in which the groundmass, composed of a glassy or partly crystalline material other than augite, occupies the interstices between unoriented feldspar laths, the groundmass forming a relatively small proportion of the rock. Cf: *hyalo-ophitic; hyalo-pilitic; intergranular*.

interstade A warmer substage of a glacial stage, marked by a temporary retreat of the ice; "a climatic episode within a glaciation during which a secondary recession or a stillstand of glaciers took place" (ACSN, 1961, art.40). Example: the Alleröd interstade of Denmark. Syn: *interstadial; oscillation*.

interstadial adj. Pertaining to or formed during an *interstade*.-n. *interstade*.

interstice An opening or space between one thing and another, as an opening in a rock or soil that is not occupied by solid matter. On the basis of origin, it may be classified as an *original interstice* or a *secondary interstice*; on the basis of size, as a *capillary interstice*, a *subcapillary interstice*, or a *supercapillary interstice*. Syn: *void; pore*. Adj: *interstitial*.

interstitial Said of a mineral deposit in which the minerals fill the pores of the host rock. Cf: *impregnated*.

interstitial defect In a crystal lattice, the filling of a normally void interstice with an extra atom; it is a type of *point defect*.

See also: *addition solid solution*. Cf: *Frenkel defect; Schottky defect*.

interstitial solid solution *addition solid solution*.

interstitial water Subsurface water in an interstice. Syn: *pore water*. Cf: *connate water*.

interstratification (a) The state or condition of being interstratified or occurring between strata of a different character. (b) *interlayering*.

interstratified Said of strata laid between or alternating with others of different character; esp. said of sedimentary rocks laid down in sequence in an alternating arrangement. Cf: *interbedded*.

interstream Said of an area, divide, or topographic feature situated or lying between streams, such as an *interstream* area (or *interfluve*). See also: *interfluvial*.

interstream ground-water ridge A residual ridge in the water table that develops between two effluent streams as a result of the percolation of ground water toward the streams. Cf: *ground-water ridge*.

intertentacular Situated between tentacles; e.g. "intertentacular organ" representing a flask-shaped, tubular bryozoan structure providing passageway for extrusion of ova between two tentacles located toward the back and near the midline.

intertextic Said of an arrangement in a soil fabric whereby the skeleton grains are linked by intergranular braces or are embedded in a porous matrix (Brewer, 1964, p.170). Cf: *porphyroskelic; agglomeroplasmic*.

interthecal Between thecae; e.g. "interthecal septum" separating the adjacent thecal cavities in graptoloids.

intertidal A syn. of *littoral*, in one of its senses.

intertidalite A *tidalite* that is demonstrably known to be deposited by tidal processes in the intertidal zone.

intertongued lithofacies A *lithofacies* whose irregular boundaries separate intertonguing stratigraphic bodies of contrasting characteristics (such as shale and sandstone) (Weller, 1958, p.633). It is not a unit that can be mapped in the normal manner. The term *lithosome*, as originally defined, is synonymous. Cf: *statistical lithofacies*.

intertonguing The disappearance of sedimentary bodies in laterally adjacent masses owing to splitting into many thin units (tongues), each of which reaches an independent pinch-out termination; the intergradation of markedly different rocks through a vertical succession of thin interlocking or overlapping wedge-shaped layers. Syn: *interfingering; interdigitation*.

intertrappean Pertaining to a deposit that occurs between two lava flows.

intertrough The median, narrowly triangular furrow dividing the pseudointerarea of the pedicle valve of some acrotretacean brachiopods.

interval [cart] *contour interval*.

interval [geog] A term used in New England as a syn. of *bottom*, or a tract of low-lying, alluvial land along a watercourse between the river and the hills or higher ground by which the level part of the valley is bounded. Also spelled: *intervale*.

interval [glac geol] An informal term for a subdivision of an interstade.

interval [stratig] The distance between any two given geologic horizons, measured normal to them; specif. *stratigraphic interval*.

interval change The increase or decrease of the time interval between two seismic reflection events, suggesting basinward thickening of the corresponding geological section or thinning often encountered on the flanks of a local structure.

interval correlation Stratigraphic correlation based on well-defined stratigraphic intervals identified by their positions between marker horizons (Krumbein & Sloss, 1963, p.343).

intervale Var. of *interval*.

interval-entropy map A multicomponent *vertical-variability map* that expresses the degree of vertical alternations (homogeneity or heterogeneity) of rock types in a given succession of beds or within a given stratigraphic unit. The map was introduced by Forgotson (1960).

intervallum The space between the outer and inner walls of an archaeocyathid (TIP, 1955, pt.E, p.7). It may contain various structures, esp. the septa.

intervalometer An electrical timing device, used on an aerial camera, that automatically operates the shutter at predetermined intervals for the purpose of obtaining a desired end lap between successive photographs.

interval velocity Seismic-wave velocity measured over a depth interval, e.g. in a sonic log or borehole survey. It usually refers to compressional velocity and usually implies measurement across bedding.

interzonal time *Geologic time* represented by a diastem or stratigraphic hiatus (Kobayashi, 1944, p.745).

interzone A term used by Henningsmoen (1961, p.83) for the barren or nonfossiliferous rocks between two local range zones.

interzooecial Existing between or among zooecia; e.g. "interzooecial space" being that part of a bryozoan zoarium between zooecia.

interzooidal Existing between or among zooids; e.g. "interzooidal avicularium" extending to the basal surface of a bryozoan colony but wedged in between autozooids rather than replacing one of them in a series.

intexine A syn. of *endexine*. Also spelled: *intextine*.

intine The thin, inner layer of the two major layers forming the wall (sporoderm) of spores and pollen, composed of cellulose and pectates, and situated inside the *exine*, surrounding the living cytoplasm. Syn: *endospore*.

intracapsular Said of cell materials within the central capsule of a radiolarian; e.g. "intracapsular layer" consisting of protoplasm exclusive of nucleus.

intraclast A broad and general term introduced by Folk (1959, p.4) for a component of a limestone, representing a torn-up and reworked fragment of a penecontemporaneous sediment having any degree of lithification, but usually weakly consolidated) that has been eroded within the basin of deposition (such as an adjoining part of the sea floor or an exposed carbonate mud flat) and redeposited there to form a new sediment; an *allochem* derived from the same formation (thereby excluding fragments derived from older carbonate outcrops). The fragment may range in size from fine sand to gravel (smaller grains are *pellets*), and it is generally rounded but may be equant to very discoidal. Cf: *protointraclast; extraclast*.

intracoastal Being within or near the coast; esp. said of inland waters near the coast.

intracontinental geosyncline *intrageosyncline*.

intracratonic basin *autogeosyncline*.

intracrystal porosity The porosity within individual crystals, pores in large crystals of echinoderms, and fluid inclusions (Choquette & Pray, 1970, p.247).

intracyclothem A cyclic sequence of strata resulting from the splitting of a cyclothem. Term introduced by Gray (1955).

intradeep A geosynclinal trough appearing within a geosynclinal belt at the end of or following uplift of the belt; a type of *secondary geosyncline*. See also: *foredeep; backdeep*.

intradelta The landward part of a delta, largely subaerial but extending for a short distance below the water level, marked by a great diversity of environments and commonly covered by marshes and swamps; it contains the distributary channels, flanked by levees. Cf: *delta front; prodelta*. Syn: *delta top*.

intrafacies A term used by Cloud & Barnes (1957, p.169) to denote a minor or subordinate facies occurring within a differing major facies.

intrafolial fold A minor fold pattern along the foliation of an otherwise unfolded rock. It resembles a *drag fold* but there is

no generic connotation.

intraformational Formed or existing within a geologic formation, or originating more or less contemporaneously with the enclosing geologic formation. The term is esp. used in regard to convolute bedding; e.g. "intraformational deformation" or "intraformational fragments". See also: *intrastratal*.

intraformational breccia A rock formed by brecciation of partly consolidated material, followed by practically contemporaneous sedimentation. It is similar in nature and origin to an intraformational conglomerate but contains fragments showing greater angularity.

intraformational conglomerate (a) A conglomerate in which the clasts are essentially contemporaneous in origin with the matrix, developed by the breaking up and rounding of fragments of a newly formed or partly consolidated sediment (usually shale or limestone) and their nearly immediate incorporation in new sedimentary deposits; e.g. an edgewise conglomerate. Fragmentation is commonly caused by shoaling and temporary withdrawal of water, followed by desiccation and mud cracking. Examples abound in the lowest Paleozoic limestones and dolomites of the Appalachian region of eastern U.S. (b) A conglomerate occurring in the midst of a geologic formation, such as one formed during a brief interruption in the orderly deposition of strata. It may contain "foreign" clasts external to the formation, and is therefore not necessarily formed of broken pieces of partly consolidated local sediment. The term is used in this sense esp. in England.--- Cf: *interformational conglomerate*.

intraformational contortion Intricate and complicated folding exhibited by *convolute bedding*; esp. such deformation resulting from the subaqueous slumping or sliding of unconsolidated sediments under the influence of gravity. See also: *intraformational corrugation*. Syn: *intrastratal contortion*.

intraformational corrugation A term applied to *intraformational contortion* on a small scale.

intraformational fold A minor fold confined to a sedimentary layer lying between undeformed beds; it results from processes (such as sliding or slumping) that are responsible for the layer in which it is found or that occur prior to complete lithification.

intrageosyncline DuToit's term for a *parageosyncline* (1937).

intraglacial (a) Said of glacial deposits formed on ground actually covered by the ice, or of glacial phenomena pertaining to a region covered by the ice at any given time. Ant: *extraglacial*. (b) *englacial*.

intragranular Said of tectonic transport or movement that takes place along a glide plane. Cf: *intergranular*.

intragranular movement A gliding movement by which favorably oriented ice crystals are deformed by slip along layers, without breaking the continuity of the crystal lattice. It is an important mechanism in glacier flow. Cf: *intergranular movement*.

intragranular porosity The porosity existing within individual grains or particles of a rock, esp. within skeletal material of a carbonate sedimentary rock. Cf: *intraparticle porosity*.

intralaminal accessory aperture An *accessory aperture* in the test of a planktonic foraminifer that leads through accessory structures into a cavity beneath them and not directly into the chamber cavity (as in *Rugoglobigerina*). Cf: *infralaminal accessory aperture*.

intramicarenite An intramicrite containing sand-sized intraclasts.

intramicrite A limestone containing at least 25% intraclasts and in which the carbonate-mud matrix (micrite) is more abundant than the sparry-calcite cement (Folk, 1959, p.14).

intramicrudite An intramicrite containing gravel-sized intraclasts.

intramorainal Said of deposits and phenomena occurring within a lobate curve of a moraine. Ant: *extramorainal*. Syn: *intramorainic*.

intramural budding A type of *polystomodaeal budding* in which the stomodaea are directly or indirectly linked in a single linear series. Cf: *circummural budding*.

intra-Pacific province *Atlantic suite*.

intraparticle porosity The porosity within individual particles of a rock. Choquette & Pray (1970, p.247) recommend use of this term rather than the term *intragranular porosity* which suggests limitation to grain-size particles.

intrapermafrost water Unfrozen ground water in layers or lenses within the permafrost.

intrapositional deposit Sediments deposited by the process of *stratigraphic leak*, e.g. as crevice or erosional channel fillings (Foster, 1966).

intraspararenite An intrasparite containing sand-sized intraclasts.

intrasparite A limestone containing at least 25% intraclasts and in which the sparry-calcite cement is more abundant than the carbonate-mud matrix (micrite) (Folk, 1959, p.14). It is common in environments of high physical energy, where the spar usually represents pore-filling cement.

intrasparrudite An intrasparite containing gravel-sized intraclasts.

intrastratal Formed or occurring within a stratum or strata; e.g. formation of iron-rich authigenic clay by "intrastratal alteration" of hornblende. The term is esp. used in regard to convolute bedding; e.g. "intrastratal crumpling" or "intrastratal flowage". See also: *intraformational*.

intrastratal contortion *intraformational contortion*.

intrastratal flow structure A variety of *convolute bedding* formed by flowage.

intrastratal solution Removal by chemical solution of certain mineral species from within a sedimentary bed following deposition. Syn: *differential solution*.

intratabular Said of features of a dinoflagellate cyst that correspond to more or less the central parts of thecal plates rather than to the lines of separation between them. Cf: *nontabular; peritabular*.

intratelluric (a) Said of a phenocryst, of an earlier generation than its groundmass, that formed at depth, prior to extrusion of a magma as lava. (b) Said of that period of crystallization occurring deep within the Earth just prior to the extrusion of a magma as lava. (c) Located, formed, or originating deep within the Earth.

intratentacular budding Formation of new scleractinian coral polyps by invagination of the oral disk of the parent inside the ring of tentacles surrounding its mouth. Cf: *extratentacular budding*.

intraumbilical aperture An aperture in a foraminiferal test located in the umbilicus but not extending outside of it.

intrazonal soil In early U.S. classification systems, one of the *soil orders* including soils with more or less well-developed soil characteristics that reflect the dominating influence of some local factor of relief, parent material, or age over the normal effects of the climate and vegetation; also, any soil belonging to the intrazonal soil order. Cf: *zonal soil; azonal soil*.

intrazonal time *Geologic time* represented by a biostratigraphic zone (Kobayashi, 1944, p.745); the *hemera* of Buckman (1902).

intrenched meander Var. of *entrenched meander*.

intrenched stream Var. of *entrenched stream*.

intrinsic ash *inherent ash*.

intrinsic geodesy The aim of intrinsic geodesy is the study of the gravity field of the Earth by considering only natural quantities, endowed with physical reality, and thus capable of actual observation and measurement.

intrinsic ionic conduction Electrical conduction arising in transport of charge through a solid as a result of a movement of ions through a crystal lattice. It usually takes place at elevated temperature and does not depend upon the presence of impurities or vacancies. Cf: *extrinsic ionic conduction*.

intrinsic permeability *specific permeability*.

intrinsic semiconduction Electronic conduction in solids due to thermal (or other) agitation of electrons from a valence band to a conduction band and drift of the conduction band electrons due to an applied electric field. Cf: *extrinsic semiconduction*.

intrusion [ign] The process of emplacement of magma in pre-existing rock; magmatic activity; also, the igneous rock mass so formed within the surrounding rock. See also: *pluton*. Syn: *injection; emplacement; invasion; irruption*.

intrusion [sed] (a) A sedimentary *injection* on a relatively large scale; e.g. the forcing upward of clay, chalk, salt, gypsum, or other plastic sediment, and its emplacement under abnormal pressure in the form of a diapiric plug. See also:*autointrusion*. (b) A sedimentary structure or rock formed by intrusion. (c) *stone intrusion*.---Syn: *sedimentary intrusion*.

intrusion [grd wat] *salt-water encroachment*.

intrusion breccia *contact breccia*.

intrusion displacement Faulting due to an igneous intrusion. See also: *marginal thrust; trap-door fault*.

intrusive adj. Of or pertaining to *intrusion*, both the processes and the rock so formed. n. A *intrusive* rock. -- Cf: *extrusive*. Syn: *irruptive*.

intrusive mountain *batholith*.

intrusive tuff *tuffisite*.

intrusive vein An igneous intrusion resembling a *sheet* and rich in volatiles.

intumescence [mineral] A property or characteristic that some minerals have when heated, of swelling or frothing due to the release of gases.

inundation A rising of water and its spreading over land not normally submerged.

invar An alloy of nickel and iron, containing about 36% nickel, and having an extremely low coefficient of thermal expansion (approximately 1×10^{-6} inch per inch per degree centigrade at ordinary temperatures). It is used in the construction of surveying instruments such as pendulums, level rods, first-order leveling instruments, and tapes. Etymol: *Invar*, a trademark.

invariant equilibrium Equilibrium of a phase assemblage which has zero degrees of freedom. See also: *invariant point*.

invariant point A point representing the conditions of *invariant equilibrium*.

invasion [ign] *intrusion* [ign].

invasion [stratig] *transgression*.

invasive *aggressive* [intrus rocks].

inver (a) A place where a river flows into the sea or into an arm of the sea; e.g. Inverness, Scotland. (b) The confluence of two streams.---Etymol: Gaelic.

invernite A granite-like holocrystalline intrusive rock characterized by orthoclase and plagioclase phenocrysts in a groundmass of euhedral feldspar and rare hornblende or mica with interstitial quartz.

inverse dispersion The *dispersion* of seismic surface waves in which the recorded wave period decreases with time. Cf: *normal dispersion*.

inverse estuary An "estuary" in which evaporation exceeds the influx of freshwater (land drainage and precipitation) so that the salinity rises above that of seawater. Ant: *positive estuary*. Syn: *negative estuary*.

inverse projection *transverse projection*.

inverse thermoremanent magnetization An artificial remanent magnetization acquired during a temperature increase from subzero temperature. Cf: *partial thermoremanent magnetization*. Abbrev: *ITRM*.

inverse zoning reversed zoning.

inversion [geomorph] (a) The development of inverted relief whereby anticlines are transformed into valleys and synclines into mountains; e.g. the formation of a deep basin in an area formerly occupied by land delivering vast quantities of sediment. (b) The occupancy by a lava flow of a former ravine or valley in the side of a volcano, thereby producing a divide over the former valley and forcing the stream to develop a new valley on its former divide (Cotton, 1958, p.366-367).---Syn: inversion of relief.

inversion [meteorol] In meteorology, a reversal of the gradient of a meteorologic element, e.g. an increase rather than a decrease of temperature with height.

inversion [cryst] (a) transformation [cryst]. (b) Rotation about a center of symmetry.

inversion center center of symmetry.

inversion layer In a body of water, a water layer whose temperature increases rather than decreases with depth.

inversion point [phase] (a) A point representing the temperature at which one polymorphic form of a substance, in equilibrium with vapor, reversibly changes into another under invariant conditions. (b) The temperature at which one polymorphic form of a substance inverts reversibly to another under univariant conditions and a specific pressure. (c) More closely, the lowest temperature at which a monotropic phase inverts at an appreciable rate into a stable phase, or at which a given phase dissociates at an appreciable rate, under given conditions. (d) A single point at which different phases are capable of existing together at equilibrium.----transition point; transition temperature.

invert The floor, bottom, or lowest part of the internal cross section of a closed conduit (aqueduct, sewer, tunnel, drain). The term originally referred to the inverted arch used to form the bottom of a masonry-lined sewer or tunnel.

invertebrate n. An animal belonging to the Invertebrata which includes all animals without backbones such as the mollusks, arthropods, and coelenterates.----adj. Of, or pertaining to, an animal that lacks a backbone.

invertebrate paleontology The branch of paleontology dealing with fossil invertebrates.

inverted overturned.

inverted pendulum pendulum (b).

inverted plunge The plunge of folds, or sets of folds, whose inclination has been carried past the vertical, so that the plunge is now less than 90° in a direction opposite from the original attitude. It is a rather common feature in excessively folded or refolded terranes.

inverted relief A topographic configuration that is the inverse of the geologic structure, as where mountains occupy the sites of synclines and valleys occupy the sites of anticlines. See also: inversion [geomorph]. Ant: uninverted relief.

inverted siphon A portion of a water conduit that is depressed in a U shape. Cf: siphon [hydraul].

inverted stream (a) A beheaded stream whose drainage turns back into the capturing stream (Lobeck, 1939, p.199). (b) A term proposed by Davis (1889b, p.210) but later abandoned (Davis, 1895, p.134) in favor of obsequent stream.

inverted tide reversed tide.

inverted unconformity (a) An unconformity in which the younger strata end abruptly against the older rocks, such as one produced by intense folding of a complex region (Grabau, 1924, p.826). (b) Truncation of the upper parts of laminae in sediment, shown in some load casts where underlying plastic material has been squeezed upward and intruded laterally into overlying sediment (Kuenen, 1957, p.250).

inverted well A well that takes in water near its top and discharges it at lower levels, into permeable material; e.g. a drainage well.

visible gold deposit A syn. used in the popular press for Car-

lin-type gold deposit.

involucre One or more whorls of small leaves or bracts subtending a flower or inflorescence. See also: cupule.

involute Coiled or rolled inward; e.g. said of a foraminiferal test having closely coiled and strongly overlapping whorls in which the inner part of the last whorl extends in toward the center of the coil to cover part of the adjacent inner whorl, or said of a coiled gastropod shell having the last whorl enveloping the earlier whorls which are more or less visible in the umbilici, or said of a coiled cephalopod conch with considerably overlapping whorls and hence having a narrow umbilicus. Cf: convolute; evolute; advolute.

involution [sed] (a) A highly irregular, sack-like, aimlessly contorted sedimentary structure consisting of local folds and interpenetrations of fine-grained material in clayey strata, and developed by the formation, growth, and melting of ground ice (congeliturbation) in the active layer overlying permafrost. See also: brodel. (b) An irregularly contorted and penetrating structure, as a wave or fold, in a soil deposit.

involution [struc geol] The refolding of nappes, resulting in complex patterns of association.

inwash Alluvium deposited against the margin of a glacier by a stream of nonglacial origin.

inyoite A colorless monoclinic mineral: $Ca_2B_6O_{11} \cdot 13H_2O$.

iodargyrite A yellowish or greenish hexagonal mineral: AgI. Syn: iodyrite.

iodate A mineral compound that is characterized by the radical IO_3^-. An example is salesite, $Cu(IO_3)(OH)$.

iodobromite An isometric mineral: $Ag(Br,Cl,I)$.

iodyrite iodargyrite.

iolanthite A banded reddish jasper-like mineral from Oregon.

iolite A syn. of cordierite. The name is used esp. for the gem variety.

ion exchange The reversible replacement of certain ions by others, without loss of crystal structure. It occurs in interlayer exchangeable ions, e.g. of clay minerals, in channelways of weak bonding, e.g. in zeolites, and also in the hydrocarbon networks of resins. See also: exchange capacity. Syn: base exchange.

ionic substitution The spatial replacement of one or more ions in a crystal structure by others of generally of similar size and charge. Syn: diadochy.

ionite (a) anauxite. (b) An earthy, resinous, brownish-yellow fossil hydrocarbon found in the lignite of Ione Valley, Amador County, Calif. It is not a recognized mineral species.

ionium A naturally occurring radioactive isotope of thorium in the uranium series and daughter of uranium-234.

ionium-deficiency method Calculation of an age in years for fossil coral or shell from 10,000 to 250,000 years old, based on the growth of ionium (thorium-230) toward equilibrium with uranium-238 and uranium-234 which entered the carbonate shortly after its formation or burial. The age depends on the change in equilibrium ratio which is directly related to the passage of time. See also: uranium-series age methods; thorium-230 to protactinium-231 deficiency method.

ionium-excess method The calculation of an age in years for deep-sea sediments formed during the least 300,000 years, based on the assumptions that the initial ionium (thorium-230) content of accumulating sediments has remained constant for the total section of sediments under study and that this initial ionium content is "ionium-excess", i.e. uranium-unsupported thorium-230. The age depends on this ionium excess content which decreases with the passage of time. See also: ionium-thorium age method; uranium-series age methods.

ionium-thorium age method The calculation of an age in years for deep-sea sediments formed during the last 300,000 years, based on the assumption that the initial thorium-230 ratio for accumulating sediments has remained constant for the total section of sediments under study. The age depends on the thorium-230 to thorium-232 ratio which gradually decreases with the passage of time. See also: *ionium-excess method; uranium-series age methods.* Syn: *thorium-230 to thorium-232 age method.*

ionization constant As applied to ionization reactions, a syn. of *dissociation constant.*

ionization potential The voltage required to drive an electron from an atom or molecule, leaving a positive ion.

ionizing radiation Radiation, e.g. alpha, beta, or gamma radiation, in which electrons are displaced.

ionography Ion-exchange *electrochromatography* wherein the ions migrate by electrostatic attraction, usually over or through an ion-exchange resin.

iowaite A mineral: $Mg_4Fe^{+3}(OH)_8OCl.2-4H_2O.$

Iowan The earliest substage of the Wisconsin glacial stage, formerly regarded as a separate stage; it occurred more than 30,000 years ago.

iozite *wüstite.*

iranite A saffron-yellow triclinic mineral: $PbCrO_4.H_2O.$

irarsite A mineral: $(Ir,Ru,Rh,Pt)AsS.$

irhzer A term used in northern Africa for a straight groove carved in a mountainside by a stream (Termier & Termier, 1963, p.408). Etymol: Berber.

iridescence The exhibition of prismatic colors (producing rainbow effects) in the interior or on the surface of a mineral, caused by interference of light from thin films or layers of different refractive index.

iridosmine A tin-white or steel-gray rhombohedral mineral: $(Os,Ir).$ It is a native alloy containing 20-68% iridium and 32-80% osmium; it usually contains some rhodium, platinum, ruthenium, iron, and copper. Cf: *osmiridium.* Syn: *iridosmium.*

iridosmium *iridosmine.*

iriginite A canary-yellow secondary mineral: $U(MoO_4)_2\cdot(OH)_2.2H_2O.$

iris (a) A transparent quartz crystal containing minute air- or liquid-filled internal cracks that produce iridescence by interference of light. The cracks may occur naturally or be caused artificially by heating and sudden cooling of the specimen. Syn: *iris quartz; rainbow quartz.* (b) An iridescent mineral; e.g. "California iris" (kunzite).

iron [meteorite] An *iron meteorite.*

iron [mineral] A heavy, magnetic, malleable and ductile, and chemically active mineral, the native metallic element Fe. It has a silvery or silver-white color when pure, but readily oxidizes in moist air. Native iron occurs rarely in terrestrial rocks (such as disseminated grains in basalts), but is common in meteorites; it occurs combined in a wide range of ores and in most igneous rocks. Iron is the most widely used of the metals.

Iron Age In archaeology, a cultural level that is the final age in the *three-age system,* and is characterized by the technology of iron. Correlation of relative cultural levels with actual age (and, therefore, with the time-stratigraphic units of geology) varies from region to region; e.g. the Iron Age began in Europe about 1100 B.C. at the earliest, but in the Americas, there was no iron technology until contact with European culture was made. In general, the Iron Age falls within historic time (Bray and Trump, 1970, p.115).

iron alum *halotrichite.*

iron bacteria Anaerobic bacteria that precipitate iron oxide from solution, either by oxidizing ferrous salts or by releasing oxidized metals from organic compounds. Accumulations of iron developed in this way are *bacteriogenic* ore deposits. Cf: *sulfur bacteria.*

iron ball A term used in Lancashire, Eng., for an ironstone nodule. See also: *ballstone.*

iron-bearing formation *iron formation.*

iron cordierite *sekaninaite.*

iron cross twin law A twin law in crystals of the diploidal class

of the isometric system that is formed by interpenetratio twinning of two pyritohedrons. The twin axis is perpendicula to a face of the rhombic dodecahedron.

iron formation A chemical sedimentary rock, typically thin bedded and/or finely laminated, containing at least 15% iro of sedimentary origin, and commonly but not necessarily con taining layers of chert (James, 1954, p.239). Various primar facies (usually not weathered) of iron formation are distin guished on the basis of whether the iron occurs predominant as oxides, silicates, carbonate, or sulfide. It is usually of Pre cambrian age. In mining usage, the term refers to a low grade sedimentary iron ore in which the iron mineral(s) is se gregated in bands or sheets irregularly mingles with chert o fine-grained quartz (Thrush, 1968, p.590), Cf: *ironstone jaspilite; See also: oxide-facies iron formation; carbonate facies iron formation; iron ore; silicate-facies iron formatior sulfide-facies iron formation.* Essentially synonymous terms *itabirite; banded hematite quartzite; taconite; quartz-bande ore; banded ironstone; calico rock; jasper bar; iron-bearin formation.* Also spelled: *iron-formation.*

iron froth A fine, spongy or micaceous variety of hematite.

iron glance A variety of hematite; specif. *specularite.*

iron hat *gossan.*

iron hypersthene (a) An iron-rich hypersthene. (b) *ferrosilite.*

iron meteorite A general name for meteorites consisting es sentially of nickeliferous iron (solid solution of iron with 4% t 30% or more of nickel); e.g. a *hexahedrite,* an *octahedrite* and an *ataxite.* Syn: *iron; siderite; meteoric iron.*

iron mica (a) *lepidomelane.* (b) *biotite.* (c) Micaceous hema tite.

iron-monticellite *kirschsteinite.*

iron olivine *fayalite.*

iron ore Ferruginous rock containing one or more distinct na ural chemical compounds from which metallic iron may b profitably extracted when properly treated. The chief ores o iron consist mainly of the oxides: red hematite (Fe_2O_3 brown hematite, limonite, or goethite ($Fe_2O_3\cdot 3H_2O$); magne tite (Fe_3O_4); and the carbonate: siderite or chalybit ($FeCO_3$).

iron pan A general term for a *hardpan* in a soil in which iro oxides are the principal cementing agents; several types o iron pans are found in dry and wet areas and in soils of widel varying textures. Also spelled: *ironpan.* See also: *moorpar* Cf: *claypan.*

iron pitch A variety of Lake Trinidad *land asphalt* that has ov erflowed the lake onto the land and hardened on weatherin to such an extent that it resembles refined *lake asphalt* (Abra ham, 1960, p.177).

iron pyrites (a) *pyrite.* (b) *marcasite* [mineral].---Sometime incorrectly spelled: *iron pyrite.*

iron range A term used in the Great Lakes region of the U.S and Canada for a productive belt of iron formations. The ter implies a linear region rather than a topographical elevation.

iron sand A sand containing particles of iron ore (usuall magnetite), as along a coastal area.

iron shale A material, usually with a laminated structure, con sisting of iron oxides and produced by the weathering of a iron meteorite.

ironshot adj. (a) Said of a mineral that is streaked, speckle or marked with iron or an iron ore. (b) Containing small nod ules or oolitic bodies of limonite or hematite; e.g. an "ironsho sand" or an "ironshot rock" in which the ooliths are essentia ly composed of limonite. n. A limonitic oolith in an ironsho rock.

iron spar A syn. of *siderite* (ferrous-carbonate mineral).

iron spinel *hercynite.*

ironstone (a) Any rock containing a substantial proportion o an iron compound, or any iron ore from which the metal ma be smelted commercially; specif. an iron-rich sedimentar rock, either deposited directly as a ferruginous sediment or resulting from chemical replacement. The term is customarily applied to a hard, coarsely banded or nonbanded, and noncherty sedimentary rock of post-Precambrian age, in contrast with *iron formation*. The iron minerals may be oxides (limonite, hematite, magnetite), carbonate (siderite), or silicate (chamosite); most ironstones containing iron oxides or chamosite are oolitic. (b) *clay ironstone*. (c) *banded ironstone*.

ironstone cap A surficial or near-surface sheet or cap of concretionary *clay ironstone*.

iron-stony meteorite *stony-iron meteorite*.

iron talc *minnesotaite*.

iron vitriol *melanterite*.

irradiance The radiant energy per unit time per unit area incident upon a surface. The preferred symbol for this quantity is H, and it is expressed in watts per cm^2 (Billings, 1963).

irregular Pertaining to an echinoid of the Exocycloida order displaying an *exocyclic* test in which the periproct is located outside of the oculogenital ring or in a posterior or oral position. Cf: *regular*.

irregular crystal A snow crystal not otherwise classifiable as a plate, stellar crystal, column, etc.

irreversibility In evolution, the theory that an evolving group of organisms, or part of an organism, does not return to the ancestral condition. Syn: *Dollo's law*.

irreversible process Any process which proceeds in one direction spontaneously, without external interference.

irrigation The artificial distribution of water over the ground surface, as by canals, pipes, or flooding, in order to promote plant growth.

irrotational strain Strain at a point, in which the orientation of the axes of strain remains unchanged. Cf: *rotational strain*. Syn: *nonrotational strain*.

irrotational wave *P wave*.

irruption [ecol] An abrupt sharp increase in a natural population, usually connected with favorable environmental changes. Adj: *irruptive*.

irruption [intrus rocks] A syn. of *intrusion*.

irruptive *intrusive*.

Irvingtonian Stage in southern California: lower Pleistocene (below Rancholabrean).

isabnormal *isanormal*.

isallobar A line on a weather map connecting points of equal barometric tendencies.

isallotherm A line connecting points of equal temperature variation in a given time interval. Cf: *isotherm; hypsoisotherm*.

isanomal A line connecting points of equal anomalies of a given normal value. Syn: *isabnormal*.

isanomalic line Var. of *isoanomalous line*; an *isoanomaly*.

isanomaly *isoanomaly*.

isarithm An *isopleth*, esp. one drawn through points on a graph at which a given quantity has the same numerical value.

isblink A term used in Greenland for *iceblink* or seaward cliff of ice. Etymol: Danish.

ischium The third pereiopodal segment from the body of a malacostracan crustacean, distal to the basis and proximal to the merus. It comprises the first segment of the endopod. Pl: *ischia*. Syn: *ischiopod; ischiopodite*.

isedemes On a map or diagram, lines connecting points of equal swelling characteristics. Cf: *isocals; isocarbs; isohumes; isovols*.

isenite A feldspathoid-bearing hornblende trachyandesite containing andesine, soda microcline, hornblende, and biotite phenocrysts in a groundmass of oligoclase, orthoclase, and nosean, with smaller amounts of augite, apatite, and iron ore.

isentrope A line or surface that represents the locus of points of a given entropy.

isentropic Said of a process that is at constant entropy.

isentropic map *entropy map*.

iserine A variety of ilmenite found as loose rounded crystals or grains in the sands at Iserwiese in Bohemia. It was formerly regarded as probably a ferruginous rutile. Syn: *iserite*.

iserite (a) A doubtful variety of rutile with considerable amounts of FeO. (b) *iserine*.

I-shaped valley An extremely young valley (such as a canyon) in which downcutting greatly exceeds lateral erosion (Lane, 1923).

ishikawaite A black mineral: $(U,Fe,Y,Ce)(Nb,Ta)O_4$.

ishkulite A chromium-bearing variety of magnetite.

ishkyldite A mineral: $Mg_{15}Si_{11}O_{27}(OH)_{20}$. It may be a variety of chrysotile high in silica. Syn: *ishkildite*.

Ising gravimeter An astatic, balance-type gravity meter consisting of a quartz frame and thread, with an attached mass at the end of a short vertical beam attached to the center of the thread, the size of the mass and the stiffness of the thread being chosen so that the system is in nearly neutral equilibrium. When the frame is given a small measured tilt, the change in the equilibrium position of the beam is dependent on the tilt and on the value of gravity.

isinglass A syn. of *mica*, esp. muscovite in thin transparent sheets and superficially resembling certain kinds of gelatin.

island (a) A tract of land, smaller than a continent, completely surrounded by water, under normal conditions, in an ocean, sea, lake, or stream. The term has been loosely applied to land-tied and submerged areas, such as in a lake, and to land cut off on two or more sides by water, such as a peninsula. (b) An elevated piece of land surrounded by a swamp, marsh, or alluvial land, or isolated at high water or during floods. (c) Any isolated and distinctive tract of land surrounded by terrain with other characteristics; e.g. a woodland surrounded by prairie or flat open country.

island arc A chain of islands, e.g. the Aleutians, rising from the deep sea floor and near to the continents; a *primary arc* expressed as a curved belt of islands. Its curve is generally convex toward the open ocean. According to Bucher (1965), an island-arc pattern results from shrinkage. This is a typical tension pattern which would result in the brittle crust by rotating a polar circumference into an equitorial circumference such as by sliding the Earth's crust on the interior. Syn: *volcanic arc*.

island hill An isolated, partly buried, bedrock hill standing island-like in the midst of alluvium (as of a *sandy*) or of the silt of a lake plain (Shaw, 1911, p.489). Type example: the hill bearing the town of Island in Kentucky.

island mesa A *headland mesa* that has been cut off from the main plateau by a river, so that it stands as an isolated mass (Lee, 1903, p.73).

island mountain (a) A mountain more or less completely encircled by valleys that separate it from other mountains or drainage-divide ridges. (b) *inselberg*.

island shelf *insular shelf*.

island slope *insular slope*.

island-tying The formation of a *tombolo*.

island volcano *volcanic island*.

isle An *island*, generally but not necessarily of small size; e.g. the British Isles. The term is a diminutive of "island".

islet A small or minor island.

isoanomalous line A syn. of *isoanomaly*. Also spelled: *isanomalous line*.

isoanomaly A line connecting points of equal geophysical anomalies. Syn: *isanomaly; isoanomalous line*.

isoanomaly curve A curve representing equal gravity anomaly values.

isoanthracite lines (a) On a map or diagram, lines connecting points of equal carbon-hydrogen ratio in anthracite. (b) A nonpreferred synonym of *isovols* as applied to anthracite.

isobar A line on a chart (meteorologic, thermodynamic) connecting points of equal pressures.

isobaric surface A surface, all points of which have equal pressure; it is not necessarily horizontal.

isobase A term used for a line which connects all areas of equal uplift or depression, as in a structure contour; it is used especially in Quaternary geology as a means for expressing movements related to postglacial uplift.

isobath (a) In oceanography, a line on a map or chart that connects points of equal water depth. Syn: *bathymetric contour; depth contour.* (b) An imaginary line on a land surface all points along which are the same vertical distance above the upper or lower surface of an aquifer or above the water table.

isobath [grd wat] An imaginary line on a land surface all points along which are the same vertical distance above the upper or lower surface of an aquifer or above the water table.

isobed map *isostratification map.*

isobenth In biological oceanography, a line that connects points of equal fertility (*primary production*) of the ocean bottom.

isobiolith A para-time-rock unit defined by fossils (Wheeler et al, 1950, p.2362).

isocals On a map or diagram, lines connecting points of equal calorific value in coal. Cf: *isocarbs; isodemes; isohumes; isovols.*

isocarbon map A coal-deposit map showing points of equal fixed-carbon content by contour lines, or *isocarbs.*

isocarbs On a map or diagram, lines connecting points of equal fixed-carbon content in coal. Cf: *isocals; isodemes; isohumes; isovols.* See also: *isocarbon map.*

isocenter (a) The unique point common to the principal plane of a tilted photograph and the plane of an assumed truly vertical photograph taken from the same camera station and having an equal principal distance. It is the center of radial displacement of images due to tilt. (b) The point on an aerial photograph intersected by the bisector of the angle between the plumb line and the perpendicular to the photograph. (c) The point of intersection (on a photograph) of the principal line and the isometric parallel.

isochela A sponge *chela* having equal or similar ends. Cf: *anisochela.*

isochemical metamorphism Metamorphism that involves no bulk change in chemical composition (Eskola, 1939). It is a theoretical concept which is only approached in nature. Syn: *treptomorphism.*

isochemical series Rocks displaying the same bulk chemical composition throughout a sequence of mineralogic or textural changes, as in a sequence of weathered or metamorphic phases.

isochore [phase] In a phase diagram, a line connecting points of constant volume.

isochore [stratig] A line drawn on a map through points of equal interval between two stratigraphic units or horizons. Cf: *isopach.*

isochore map A map that shows the varying interval between two designated stratigraphic units or horizons by means of isochores drawn at right angles to the bedding planes; e.g. a map prepared directly from two structure-contour maps by subtracting the elevations of the lower surface from those of the upper surface at each control point, and drawing contours of equal interval between the two surfaces. Isochore maps are useful in projecting surface structure into lower horizons through beds of variable thickness, and in serving as a basis for estimating oil-reservoir content. The term is often used synonymously with *isopach map*, although strictly an "isochore map" expresses the variations in many units and the effects of one or more unconformities, whereas an "isopach map" expresses variation within a single unit. Syn: *convergence map.*

isochromatic curve In optics of biaxial and uniaxial crystals, a band of color indicating the emergence of those components of light having equal path difference. It is a part of the *interference figure.* Cf: *isogyre.* See also: *Cassinian curve.*

isochrome map A contour map that depicts the continuity and extent of color stains on geologic formations.

isochron [seism] A line, on a map, connecting points at which a characteristic time or interval has the same value; e.g. in seismology, a line passing through points at which difference between arrival times of seismic waves from two reflecting surfaces is equal. See also: *isotime curve.*

isochron [geochron] The term denotes a straight line containing all the points representing the same time or age. In geochronology it usually refers to a line constructed either by plotting the ratio Sr^{87}/Sr^{86} as a function of ratio Rb^{87}/Sr^{86} or by plotting the ratio Pb^{207}/Pb^{204} or Pb^{208}/Pb^{204} as a function of ratio Pb^{206}/Pb^{204} for minerals or rocks of the same age. See also: *isochrone.*

isochronal (a) *isochronous.* (b) *isochronic.*

isochrone A line, on a map or chart, connecting all points at which an event or phenomenon occurs simultaneously or which represent the same time value or time difference; e.g. a line along which duration of travel is constant, or a line indicating the places at which rain begins at a specified time. See also: *isochron.*

isochroneity The state or quality of being *isochronous*; equivalence in duration. Syn: *isochronism.*

isochronic Having isochrones; e.g. an isochronic map. Syn: *isochronal.*

isochronism *isochroneity.*

isochronous [geol] (a) Equal in duration or uniform in time; e.g. an "isochronous interval" between two synchronous surfaces, or an "isochronous unit" of rock representing the complete rock record of an isochronous interval. Mann (1970, p.750) has proposed that the word *coetaneous* would be more appropriate for isochronous used in this sense. (b) A term that is frequently applied in the sense of *synchronous*, such as an "isochronous surface" having everywhere the same age or time value within a body of strata (Hedberg, 1958, p.1890; and ACSN, 1961, art.28c).----Syn: *isochronal.*

isoclasite A white mineral: $Ca_2(PO_4)(OH).2H_2O.$

isoclinal Adj. of *isocline.*

isocline A fold, the limbs of which have been so compressed that they have the same dip. Isoclines occur in homogeneous rocks such as slates, and are characteristic of strong deformation. Adj: *isoclinal.*

isoclinic line An *isomagnetic line* connecting points of equal magnetic inclination.

isocommunity A natural community that closely resembles another community in morphology and ecology.

isocon A line connecting points of equal geochemical concentration, e.g. salinity.

isodesmic Said of a crystal or other material having ionic bonding of equal strength, e.g. NaCl. Cf: *anisodesmic.*

isodietic Pertaining to a line or surface along which similar conditions of deposition have occurred. Term proposed by Hind & Howe (1901). Etymol: Greek *diaita*, "way of living".

isodiff A line on a map or chart connecting points of equal correction or difference in datum; e.g. *isolat* and *isolong.*

isodimorphism The characteristic of two crystalline substances to be both dimorphous and isomorphous, e.g. calcite and aragonite. Adj: *isodimorphous.*

isodimorphous Said of two crystalline substances displaying *isodimorphism.*

isodont Said of the dentition of a bivalve mollusk (e.g. *Spondylus* and *Plicatula*) characterized by a small number of symmetrically arranged hinge teeth (such as two large subequal teeth in one valve and corresponding sockets in the other).

isodynamic line An *isomagnetic line* connecting points of equal magnetic intensity. It is used for maps of total, horizontal, or vertical magnetic intensity. Syn: *isogam*.

isofacial [petrology] Pertaining to rocks belonging to the same facies and having reached equilibrium under the same set of physical conditions. As applied to metamorphic facies, the term is synonymous with *isogradal*.

isofacial [stratig] Pertaining to rocks belonging to the same facies; e.g. an "isofacial line" on a map, along which the thickness of a stratum of the same lithologic composition is constant.

isofacies map A map showing the distribution of one or more facies within a designated stratigraphic unit. See also: *facies map*.

isofract A graphic representation of the locus of all compositions in a system having a given value of the index of refraction.

isofrigid Said of a soil temperature regime having the same temperature range as the *frigid* regime, but with a summer-winter variation of less than 5°C (SSSA, 1970).

isogal In gravity prospecting, a contour line of equal gravity values (after gal, the common unit of gravity measurement; one gal = lcm/sec/sec).

isogam *isodynamic line.*

isogeolith A para-time-rock unit defined by lithology (Wheeler et al, 1950, p.2362).

isogeotherm *geoisotherm.*

isogon [mag] *isogonic line.*

isogon [meteorol] A line connecting points of equal wind direction. Cf: *isotach.*

isogonic line An *isomagnetic line* connecting points of equal magnetic declination. See also: *agonic line.* Syn: *isogon.*

isograd A line on a map joining points at which metamorphism proceeded at similar values of pressure and temperature as indicated by rocks belonging to the same metamorphic facies. Such a line represents the intersection of an inclined isograd surface with the Earth's surface and marks the boundary between two contiguous facies or zones of metamorphic grade as defined by the appearance of specific index minerals, e.g. garnet isograd, staurolite isograd.

isogradal Pertaining to rocks which have reached the same grade of metamorphism irrespective of their initial compositions. Cf: *isophysical series.* Syn: *isogradal; isofacial.*

isograde adj. A syn. of *isogradal.*

isogram A general term proposed by Galton (1889, p.651) for any line on a map or chart connecting points having an equal numerical value of some physical quantity (such as of temperature, pressure, or rainfall); an *isopleth.*

isogranular Said of the hypidiomorphic to idiomorphic texture of an igneous rock having pyroxene grains in the interstices between, and of the same size as, plagioclase crystals. Cf: *equigranular.*

isogriv A line on a map or chart connecting points of equal grivation.

isogyre In crystal optics, a black or shadowy part of an *interference figure* that is produced by extinction and indicates the emergence of those components of light having equal vibration direction. It may look like one arm of a black cross. Cf: *isochromatic curve.* Syn: *polarization brush.*

isohaline adj. Of equal or constant salinity.----n. A line on a chart that connects points of equal salinity in the ocean.----Cf: *isopycnic.*

isoheight *isohypse.*

isohumes On a map or diagram. lines connecting points of equal moisture content in coal. Cf: *isocals; isocarbs; isodemes; isovols.*

isohyet A line connecting points of equal precipitation.

isohyperthermic Said of a soil temperature regime having the same temperature range as the *hyperthermic* regime, but with a summer-winter variation of less than 5°C (SSSA, 1970).

isohypse An isopleth for height or elevation; a *contour line* on a topographic map. Syn: *isohyps; isoheight.*

isokite A white monoclinic mineral: $CaMg(PO_4)F$. It is isomorphous with tilasite.

isolat An *isodiff* connecting points of equal latitude corrections.

isolated porosity The property of rock or soil of containing noncommunicating interstices, expressed as the percent of bulk volume occupied by such interstices; the numerical difference between total porosity and effective porosity. Interstices such as vesicles in lava and fluid inclusions in mineral grains.

isolation In biology, any process or condition by which a group of individuals is cut off and separated for a considerable length of time from other areas or groups, as a result of geographic, feeding, or other factors.

isolation of outcrops A method of geologic mapping that outlines all areas of exposed rock to distinguish them from areas where the rock is buried or otherwise concealed. Syn: *multiple-exposure method.*

isoline [phys sci] *isopleth.*

isoline [photo] A line of common scale representing the intersection of the planes of two overlapping aerial photographs having a common perspective center and equal principal distances; e.g. a line representing the intersection of the plane of a vertical photograph with the plane of an overlapping oblique photograph taken from the same camera station. Cf: *isometric parallel.*

isolith (a) An imaginary line connecting points of similar lithology and separating rocks of differing nature, such as of color, texture, or composition (Kay, 1945a, p.427). The term "isolithic boundary" was used by Grossman (1944, p.48) for a zone of lithofacies change separating rocks of different grain sizes. (b) An imaginary line of equal aggregate thickness of a given lithologic facies or particular class of material within a formation, measured perpendicular to the bedding at selected points (which may be on outcrops or in the subsurface).

isolith map A map that depicts isoliths; esp. a *facies map* showing the absolute (net) thickness of a single rock type or selected rock component in a given stratigraphic unit.

isolong An *isodiff* connecting points of equal longitude corrections.

isomagnetic line A line connecting points of equal value of some magnetic element. See also: *isoclinic line; isogonic line; isodynamic line; isopors.*

isomegathy A term introduced by Shepard & Cohee (1936) for a line, on a map, connecting points of equal median size of sedimentary particles.

isomerite An obsolete term originally applied to an igneous rock composed entirely of crystals without finer interstitial material (i.e. groundmass).

isomesic [sed] Said of sediments or rocks formed in the same medium or under identical conditions, although not necessarily at the same time; e.g. all accumulations formed in a particular sea, lake, or stream. Ant: *heteromesic.*

isomesic [soil] Said of a soil temperature regime having the same temperature range of a *mesic* regime, but with a summer-winter variation of less than 5°C (SSSA, 1970).

isometric line [cart] A term introduced by Wright (1944) for a line, drawn on a map, representing a constant value obtained from measurement at a series of points along its course; an *isopleth.*

isometric parallel An *isoline* in which the vertical photograph is truly vertical (tilt-free); a line passing through the isocenter, parallel to the horizon and perpendicular to the principal plane.

isometric projection A projection in which the plane of projection is equally inclined to the three spatial axes of a three-dimensional object, so that equal distances along the axes are drawn equal. It gives a bird's-eye view, combining the advantages of a ground plan and elevation; e.g. as in a block diagram showing three faces.

isometric system One of the six crystal systems, characterized by four threefold axes of symmetry as body diagonals in a cubic unit cell of the lattice. It comprises five crystal classes or point groups. Cf: *hexagonal system; tetragonal system; orthorhombic system; monoclinic system; triclinic system*. Obsolete syn: *tesseral*. Syn: *cubic system*.

isomicrocline An optically positive variety of microcline.

isomodal layering Layering in a cumulate in which the layers are characterized by a uniform proportion of one or more cumulus minerals.

isomorph An organism, or a part of an organism, that is similar to another but unrelated to it.

isomorphic *isomorphous*.

isomorphic Having identical or similar form. Syn: *heteromorphic*.

isomorphism [cryst] The characteristic of two or more crystalline substances to have similar chemical composition, axial ratios, and crystal forms, and to crystallize in the same crystal class. Such substances form an *isomorphous series*. Adj: *isomorphous*. Cf: *isostructural*. Syn: *allomerism*.

isomorphism [evol] The similarity that develops in organisms of different ancestry as a result of convergence.

isomorphous Adj. of *isomorphism*. Syn: *isomorphic; allomeric*.

isomorphous mixture *isomorphous series*.

isomorphous series Two or more crystalline substances that display *isomorphism*; their physical properties vary along a smooth curve. An example is olivine, usually found in nature as a solid solution of Mg_2SiO_4 and Fe_2SiO_4, i.e. an isomorphous series between forsterite and fayalite. The exact lattice dimensions and other physical properties vary along a smooth curve with change of the Mg/Fe ratio.

isomudstone map A contour map that depicts areas having equal quantities of mudstone within assumed stratigraphic intervals.

isomyarian adj. Said of a bivalve mollusk or its shell having two adductor muscles of equal or nearly equal size. Syn: *homomyarian*.---n. An isomyarian mollusk.

isontic line Obsolete syn. of *isopleth*. Term proposed by Lane (1928, p.37).

iso-orthoclase An optically positive variety of orthoclase. It has been found in granitic gneiss. Syn: *isorthoclase; isorthose*.

isopach A line drawn on a map through points of equal thickness of a designated stratigraphic unit or group of stratigraphic units. Cf: *isochore*. Syn: *isopachous line; isopachyte; thickness line; thickness contour*.

isopach map A map that shows the thickness of a bed, formation, sill, or other tabular body throughout a geographic area, based on a variety of types of data; a map that shows the varying true thickness of a designated stratigraphic unit or group of stratigraphic units by means of isopachs plotted normal to the bedding or other bounding surface at regular intervals. Cf: *isochore map*. Syn: *isopachous map; thickness map*.

isopachous Of, relating to, or having an isopach; e.g. an "isopachous contour".

isopachous line *isopach*.

isopachous map *isopach map*.

isopach strike The compass direction of an isopach line at a given point on a map.

isopachyte British term for *isopach*.

isopag An *equiglacial line* connecting points where ice is present for approximately the same number of days per year.

isopectic An *equiglacial line* connecting points where ice begins to form at the same time in winter. Cf: *isotac*.

isoperimetric curve A line on a map or map projection (as on an equal-area projection) along which there is no variation from exact scale.

isoperthite A variety of alkali feldspar consisting of perthitic intergrowths of the same kind of feldspar or of two kinds of feldspar belonging to the same isomorphous series.

isophysical series A series of rocks of different chemical composition which were metamorphosed under identical physical conditions. Cf: *isogradal*.

isopic Said of sedimentary rocks of the same facies, or said of facies characterized by identical or closely similar rock types. The rocks may be formed in different sedimentation areas or at different times or both, but the lithologies are the same; e.g. a facies repeated in vertical succession. Also, said of a map depicting isopic facies or rocks. Cf: *heteropic, holisopic*.

isopiestic line *equipotential line*.

isopleth [phys sci] (a) A general term for a line, on a map or chart, along which all points have a numerically specified constant or equal value of any given variable, element, or quantity (such as a line of equal size, abundance, or magnitude), with respect to space or time; esp. a *contour line*. Etymol: Greek *isos*, "equal", + *plethos*, "fullness, quantity, multitude". Syn: *isogram; isoline; isontic line; isometric line*. (b) A line drawn through points on a graph at which a given quantity has the same numerical value (or occurs with the same frequency) as a function of two coordinate variables. It is often used of a meteorologic element that varies with the time of the year (month) and the time of day (hour).---Syn: *isarithm*.

isopleth [geochem] In a strict sense, a line, surface, etc. on which some mathematical function has a constant value. It is sometimes distinguished from a *contour* by the fact that an isopleth need not refer to a directly measurable quantity characteristic of each point in the map area, e.g. maximum temperature of a particular point. More generally, the term is sometimes used as a synonym of *isocompositional section*.

isopleth map A general term for any map showing the areal distribution of some variable quantity in terms of lines of equal or constant value; e.g. an isopach map.

isopod Any amalcostracan belonging to the order Isopoda, characterized generally by the absence of a carapace and the presence of sessile eyes and a compressed body. Their stratigraphic range is Triassic to present. Cf: *amphipod*.

isopollen A line on a map connecting locations with samples having the same percentage or amount of pollen of a given kind. Syn: *isopoll*.

isopors *Isomagnetic lines* of equal secular change, e.g. equal annual change of isogonic or isoclinic lines.

isopotal Having equal infiltration capacities; e.g. an isopotal area in a watershed.

isopotential level The level to which artesian water can rise. Syn: *potentiometric surface*.

isopotential line *equipotential line*.

isoprenoid A substance, such as isoprene, containing or relating to a characteristic branched-chain grouping of five carbon atoms. This particular molecular grouping occurs in many natural compounds such as rubber, terpenes, and vitamins A and E.

isopycnic adj. Of constant or equal density, measured in space or in time.----n. A line on a chart that connects points of equal density.----Cf: *isohaline*. See also: *isostere*.

isorad A line connecting points of equal radioactivity.

isorat Line connecting points of equal isotope ratios.

isoseism *isoseismal line*.

isoseismal n. A syn. of *isoseismal line*.

isoseismal line A line connecting points on the Earth's surface at which earthquake intensity is the same. It is usually a closed curve around the epicenter. Cf: *coseismal line*. Syn: *isoseism; isoseismic line; isoseist; isoseismal*.

isoseismic line *isoseismal line*.

isoseist *isoseismal line.*

isosinal map A *slope map* whose contour lines are lines of equal slope represented by sines of slope angles read from a topographic map.

isospore A plant spore functioning as either male or female in reproduction; a *homospore.*

isospory The quality or state of having isopores; *homospory.*

isostasy The condition of equilibrium, comparable to floating, of the units of the brittle crust above the plastic mantle. The two differing hypotheses of the mechanism of isostasy are called *Airy isostasy* and *Pratt isostasy.* See also: *isostatic compensation; depth of compensation.*

isostatic adjustment Var. of *isostatic compensation.*

isostatic anomaly A gravity anomaly calculated on a hypothesis that the gravitational effect of masses extending above sea level is approximately compensated by a deficiency of density of the material beneath those masses; the effect of deficiency of density in ocean waters is compensated by an excess of density in the material under the oceans.

isostatic compensation The adjustment of the crust of the Earth to maintain equilibrium among units of varying mass and density; excess mass above is balanced by a deficit of density below, and vice versa. See also: *depth of compensation; isostasy.* Var: *isostatic adjustment.*

isostatic depth of compensation Var. of *depth of compensation.*

isostatic isocorrection line map A contour map showing lines connecting places for which the isostatic correction has the same value.

isostere A line connecting points of equal density of the Earth's atmosphere; an *isopycnic* of the atmosphere.

isostratification map A map that shows the number or thickness of beds in a stratigraphic unit by means of contour lines representing equal stratification indices (Kelley, 1956, p.299). Syn: *isobed map.*

isostructural Said of crystalline substances that have corresponding atomic positions, whether or not they form a chemical series, i.e. are *isomorphous.* Within an isostructural group of minerals there may be much ionic substitution.

isotac An *equiglacial line* connecting points where ice melts at the same time in spring. Cf: *isopectic.*

isotach A line connecting points of equal wind velocity. Cf: *isogon.*

isotangent map A *slope map* whose contour lines are lines of equal slope represented by tangents of slope angles read from a topographic map.

isotaque In crystal optics, one of several curves representing equal wave-normal velocities; in a uniaxial crystal it is a circle concentric with the optic axis, and in a biaxial crystal, it is a spherical ellipse.

isotherm A line connecting points of equal temperature. Cf: *isallotherm; hypsoisotherm.*

isothermal Pertaining to the process of changing the thermodynamic state of a substance, i.e., its pressure and volume, e.g., while maintaining the temperature constant.

isothermal remanent magnetization Remanent magnetization due solely to application of a magnetic field, and in particular without change of temperature. Abbrev: IRM.

isothermic Said of a soil temperature regime that has the characteristics of a *thermic* regime except for a summer-winter variation of less than 5°C (SSSA, 1970).

isothism A line joining points of equal, lateral orogenic displacement. See also: *isothismic map.*

isothismic map A map that shows lateral orogenic displacement by means of *isothisms*; it is a type of paleogeologic map.

isothrausmatic Said of igneous rocks with an orbicular texture in which the nuclei of the orbicules are composed of the same rock as the groundmass (Eskola, 1938, p.476). Cf: *allothraus-*

matic; crystallothrausmatic; homeothrausmatic; heterothrausmatic.

isotime curve A line of equal time difference; specif. an *isochron* used in seismic surveying.

isotomous Characterized by division of a crinoid arm in equal branches. Ant: *heterotomous.*

isotope One of two or more species of the same chemical element, i.e. the same number of protons in the nucleus, but differing from one another by having different atomic weights, i.e. a different number of neutrons. Though the isotopes of an element have basically the same chemical properties, due to the similarity in electron configuration, they have slightly different physical properties by which they can be separated. See also: *radioisotope.*

isotope dilution An analytical technique wherein a known amount of radioactive isotope (spike) is added to a sample, the total amount of that element in the sample is then separated out, and its radioactivity measured and subsequently used to determine how much of the element was present originally.

isotope effect The slight changes in chemical and physical properties, such as rates of reaction and diffusion, density, and equilibrium distribution, observed in a molecule when one isotope of an element is substituted for another, and caused by the differences in mass.

isotope geochemistry *isotope geology.*

isotope geology The application of the study of radioactive and stable isotopes, especially their abundances, to geology. It includes the calculation of geologic time, and the determination of the origin, mechanisms, and conditions of geologic processes by isotopic means. Syn: *isotope geochemistry; nuclear geology; nuclear geochemistry.*

isotope ratio Isotope abundances (e.g. for oxygen-18 and oxygen-16) given as a ratio relative to a standard rather than as an absolute quantity and expressed mathematically by a *delta value.*

isotopic [**isotope**] Pertaining or relating to an *isotope.*

isotopic [**sed**] Said of rocks formed in the same environment, such as in the same sedimentary basin or geologic province. Ant: *heterotopic.*

isotopic age *radiometric age.* A common syn. is *absolute age.*

isotopic age determination *radiometric dating.*

isotopic fractionation The relative enrichment of one isotope over another in a system caused by the differential effects of temperature, mainly, but also kinetic effects, activity coefficients, etc. on the slight mass differences of the isotopes.

isotopic number The number of excess neutrons, i.e. the number of neutrons minus the number of protons, in an atomic nucleus which is usually an indication of the radioactivity of the nucleus.

isotropic Said of a medium whose properties are the same in all directions, e.g. in crystal optics, said of a crystal whose physical properties do not vary according to crystallographic direction, e.g. light travels with the same speed in any direction. Cubic crystals and amorphous substances are usually isotropic. Ant: *anisotropic.*

isotropization The solid-state conversion of an originally birefringent mineral such as quartz or feldspar into a more or less isotropic phase at temperatures below the melting point, as a result of destruction of the crystallinity by such processes as shock-wave action or neutron bombardment.

isotubule A *pedotubule* composed of skeleton grains and plasma that are not organized into recognizable aggregates and within which the basic fabric shows no directional arrangement with regard to the external form (Brewer, 1964, p.241).

isotypic Said of crystalline substances that have analogous crystal structures and chemical compositions, e.g. zircon and xenotime.

isovols On a map or diagram, lines connecting points of equal volatile content in coal. Cf: *isocarbs; isocals; isodemes; isohumes.* See also: *isoanthracite lines.*

ispatinow One of the steep, narrow, sharp-crested ridges or hills aligned parallel to the direction of ice movement and occurring in basins of large postglacial lakes, as those in Cree Lake of northern Saskatchewan. It is composed of loose, unsorted rock flour mixed with boulders, and has a summit that is rounded from the crest downward; average height is about 40 m, and length varies from 0.4 to 1.6 km. It was probably formed in a narrow gorge in the ice sheet when the glacier front was bounded by a deep lake. Etymol: Cree, "a conspicuous hill". Term introduced by Tyrrell & Dowling (1896, p.23).

issite A dark-colored hypabyssal rock characterized by xenomorphic-granular texture and composed chiefly of hornblende with smaller amounts of green pyroxene and even less labradorite and accessory magnetite and apatite. Its name is derived from the Issa River, Penza oblast, U.S.S.R.

issue The place where a stream flows out into a larger body of water.

isthmus A narrow strip or neck of land, bordered on both sides by water, connecting two larger land areas, as a peninsula and the mainland (e.g. Isthmus of Suez) or two continents (e.g. Isthmus of Panama). Etymol: Greek *isthmos.* See also: *submarine isthmus.*

itabirite A laminated, metamorphosed, *oxide-facies iron formation* in which the original chert or jasper bands have been recrystallized into megascopically distinguished grains of quartz and in which the iron is present as thin layers of hematite, magnetite, or martite (Dorr & Barbosa, 1963, p. 18). The term was originally applied in Itabira, Brazil, to a pure or high-grade, massive specular-hematite ore (66% iron) associated with a schistose rock composed of granular quartz and scaly hematite. The term is now widely used outside Brazil. Cf: *jacutinga; canga.* Syn: *banded-quartz hematite.*

itacolumite A micaceous sandstone (sedimentary) or a schistose quartzite (metamorphic), containing interstitial, loosely interlocking grains of mica, chlorite, and talc, that exhibits flexibility (bends noticeably without breaking) when split into thin slabs. Type locality: Itacolumi Mountain in the state of Minas Gerais, Brazil. Syn: *flexible sandstone; articulite.*

italite A volcanic foidite rich in potassium feldspathoid (leucite) and 0 to 30 percent mafic minerals, such as melilite, biotite, and apatite. Its name is derived from Italia (Italy). Cf: *fergusite; missourite.*

iterative evolution Repeated development of new forms from a limited stock; repeated, independent evolution.

itoite An orthorhombic mineral: $Pb_3GeO_2(SO_4)_2(OH)_2$.

itsindrite A potassium-rich nepheline syenite hypabyssal rock containing microcline, nepheline, biotite, acmite, and zoned melanite. Its name is derived from the Itsindra Valley, Malagasy.

ivorite A black tektite from the Ivory Coast, western Africa.

ivory The translucent to opaque, fine-grained, creamy-white substance that composes the tusks of the elephant and certain other large animals; it has long been esteemed for a wide variety of ornamental articles.

I wave A longitudinal or *P* wave in the Earth's inner core. Cf: *K wave.*

ixiolite A mineral: $(Ta,Nb,Sn,Fe,Mn)_4O_8$. It was previously considered to be a mixture of cassiterite with columbite or tapiolite, and to be a manganese-tantalate isomorph of tapiolite and of mossite.

J

jacinth (a) A *zircon*; specif. a yellow or brown zircon. The term was originally an alternate spelling of *hyacinth*, and has been used to designate a red or orange zircon and sometimes a gem zircon more nearly pure orange in color than a hyacinth. (b) An orange-red to orange *essonite*.---The term, having become meaningless", is obsolete in the American gem trade (Shipley, 1951, p.116).

jack [coal] (a) Cannel coal interstratified with shale. (b) Coaly, often canneloid, shale. (c) A large ironstone nodule in the coal measures of Wales.

jack [mineral] A zinc ore; specif. *sphalerite.*

jack iron A term used in the zinc-mining area of Missouri for solid flint rock containing disseminated sphalertie, or blackjack.

Jacksonian North American (Gulf Coast) stage: Eocene (above Claibornian, below Vicksburgian).

jacobsite A black magnetic mineral of the magnetite series in the spinel group: $(Mn^{+2}, Fe^{+2}, Mg)(Fe^{+3}, Mn^{+3})_2O_4$.

Jacob's staff A single, straight rod, staff, or pole, pointed and shod with iron at its lower end for insertion in the ground and fitted with a ball-and-socket joint at its upper end for adjustment to a level position, and used instead of a tripod for mounting and supporting a surveyor's compass or other instrument. Named after Jacob St. James, symbolized in religious art by a pilgrim's staff. Syn: *Jacob staff.*

jacupirangite An ultramafic plutonic rock that is part of the ijolite series, composed chiefly of titanaugite and magnetite, with a smaller amount of nepheline; a nepheline-bearing pyroxenite. Its name is derived from Jacupiranga, Brazil.

jacutinga A term used in Brazil for disaggregated (powdery) itabirite and for variegated, thin-bedded, high-grade hematite iron ores associated with and often forming the matrix of gold ore. Etymol: from its resemblance to the colors of the plumage of *Pipile jacutinga*, a Brazilian bird.

jade (a) A hard, extremely tough, compact gemstone consisting of either the pyroxene mineral jadeite or the amphibole mineral nephrite, and having an unevenly distributed color ranging from dark or deep green to dull or greenish white. It takes a high polish, and has long been used for jewelry, carved articles, and various ornamental objects. Syn: *jadestone*. (b) A term that is often applied to various hard green minerals; e.g. "California jade" (or californite, a green compact variety of vesuvianite), "Mexican jade" (or tuxtlite, and also green-dyed calcite), saussurite, and green varieties of amazonite, pectolite, garnet, and serpentine.

jadeite A high-pressure mineral of the clinopyroxene group, essentially: $Na(Al,Fe)Si_2O_6$. It occurs in various colors (esp. green) and is found chiefly in Burma; when cut, it furnishes the most valuable and desirable variety of jade and is used for ornamental purposes.

jadeitite A metamorphic rock consisting of jadeite associated with small amounts of feldspar or feldspathoids. It is probably derived from an alkali-rich igneous rock by high-pressure metamorphism.

jadeolite A deep-green chromiferous syenite cut as a gemstone and resembling jade in appearance.

jadestone *jade.*

jager A high-quality bluish-white diamond.

jagoite A yellow-green mineral: $Pb_3FeSi_3O_{10}(OH,Cl)$.

jalpaite A lead-gray mineral: Ag_3CrS_2. It is a cupriferous argentite.

jamesonite A lead-gray to gray-black orthorhombic mineral: $Pb_4FeSb_6S_{14}$. It is a minor ore of lead, and sometimes contains copper and zinc. Jamesonite has a metallic luster and commonly occurs in acicular crystals with fibrous or feather-like forms. Syn: *feather ore; gray antimony.*

Jamin effect The restrictive force exerted upon the flow of fluids through narrow tubes or passages by successive bubbles of air or other gas. If a narrow tube is expanded at several places, and each bulb or enlargement contains a gas, the liquid can support a pressure of several atmospheres before it begins to flow.

Jänecke diagram A square phase diagram whose corners represent two reciprocal salt pairs (e.g., NaCl - KCl - NaBr - KBr) and on which are plotted the configuration of the surface representing the aqueous solution saturated with the salts. It is particularly useful in the study of phase equilibria relevant to evaporites. Syn: *reciprocal salt-pair diagram.*

Japanese twin law A twin law in quartz crystals having two individuals with a composition plane of (1122); four varieties are possible.

jardang Var. of *yardang.*

jargoon A colorless, pale-yellow, or smoky-tinted gem variety of zircon from Ceylon. Syn: *jargon.*

jarlite A colorless to brownish mineral: $NaSr_3Al_3F_{16}$.

jarosite (a) An ocher-yellow or brown mineral of the alunite group: $KFe_3(SO_4)_2(OH)_6$. Syn: *utahite.* (b) A group of minerals consisting of hydrous iron sulfates, including jarosite, natrojarosite, ammoniojarosite, argentojarosite, plumbojarosite, and hydronium jarosite.

jaspachate *jaspagate.*

jaspagate A syn. of *agate jasper*. The term is sometimes applied to agate jasper in which the jasper predominates. Syn: *jaspachate.*

jasper A dense, cryptocrystalline, opaque (to slightly translucent) variety of chert (always quartz) associated with iron ores and containing iron-oxide impurities that give the rock various colors, characteristically red, although yellow, green, grayish-blue, brown, and black cherts have also been called jaspers. The term has also been applied to any red chert or chalcedony irrespective of associated iron ore. Syn: *jasperite; jaspis; jasperoid.*

jasper bar A term used in Australia for *iron formation.* Syn: *bar; jaspilite.*

jasperine Banded jasper of varying colors.

jasperite *jasper.*

jasperization The conversion or alteration of igneous or sedimentary rocks into banded, jaspilite-like rocks by metasomatic introduction of iron oxides and cryptocrystalline silica.

jasperoid n. (a) A dense, usually gray, chert-like, siliceous rock in which chalcedony or cryptocrystalline quartz has replaced the carbonate minerals of limestone or dolomite; a silicified limestone. It typically develops as the gangue of metasomatic sulfide deposits of the lead-zinc type (as those of Missouri, Oklahoma, and Kansas). (b) *jasper.* adj. Resembling jasper.

jasper opal A yellow or yellow-brown, almost opaque common opal containing iron oxide and other impurities, having the color of yellow jasper but the luster of common opal. Some varieties are almost reddish brown to red. Syn: *jaspopal; opal jasper.*

jaspery Resembling or containing jasper; e.g. "jaspery iron ore" (impure hematite interbedded with jasper), or "jaspery chert" (a silicified radiolarian ooze associated with volcanic rocks in Ordovician strata of southern England). Syn: *jaspidean.*

jaspidean Resembling or containing jasper; jaspery. Syn: *jaspideous.*

jaspilite (a) A banded, ferruginous (at least 25% iron), compact, siliceous rock occurring with iron ores and resembling jasper, e.g. the rock of the Precambrian iron-bearing district of the Lake Superior region. (b) A general term (used esp. in Australia) for banded *iron formation.* Syn: *jasper bar.* Also spelled: *jaspilyte.* See also: *jasperization.*

jaspis A syn. of *jasper*. Etymol: German *Jaspis*.

jaspoid (a) Resembling jasper (Thrush, 1968, p.598). (b) A syn. of *tachylite* (Hey, 1962, p.470).

jasponyx An opaque *onyx*, part or all of whose bands consist of jasper.

jaspopal A syn. of *jasper opal*. Also spelled: *jasp-opal*.

javaite An Indonesian tektite from Java. Syn: *javanite*.

jebel A hill, mountain, or mountain range in northern Africa. Etymol: Arabic. Syn: *jabal; djebel*.

jefferisite A variety of vermiculite.

jeffersonite A dark-green or greenish-black mineral of the clinopyroxene group: $Ca(Mn,Zn,Fe)Si_2O_6$.

jelly *ulmin*.

jenkinsite A variety of antigorite containing iron.

jennite A mineral: $Na_2Ca_8(SiO_3)_3(Si_2O_7)$.

jeremejevite A colorless to pale yellowish-brown hexagonal mineral: $Al_6B_5O_{15}(OH)_3$. Syn: *eremeyevite*.

jeromite A mineral: $As(S,Se)_2$ (?).

jet [coal] A hard, lustrous, pure black variety of lignite with a conchoidal fracture and taking a high polish. It occurs as isolated masses in bituminous shale and is probably derived from waterlogged pieces of driftwood. Jet is used for jewelry and other ornamentation. Syn: *black amber*. See also: *jet shale; pitch coal*.

jet [hydraul] A sudden and forceful rush or gush of fluid (water, gas, vapor, etc.) through a narrow or restricted opening either in spurts or in a continuous flow; e.g. a stream of water or air used to flush cuttings from a borehole.

jet flow A type of streamflow characterized by water moving in plunging, jet-like surges, produced where a stream reaches high velocity along a sharply inclined stretch or moves swiftly over a waterfall, or where a turbulent stream enters a body of standing water. Syn: *shooting flow*.

jetonized wood Lamellae of vitrain in coal.

jet rock *jet shale*.

jet shale Bituminous shale containing *jet* [coal]. Syn: *jet rock*.

jetted well A shallow water well constructed by a high-velocity stream of water directed downward into the ground.

jetting (a) The process of sinking a borehole, or of flushing cuttings or loosely consolidated materials from a borehole, by using a directed, forceful stream (jet) of air or water. (b) Injection of gas into a stratum for purposes such as maintaining pressure in an oil reservoir or of secondary recovery of oil.

jetty (a) An engineering structure (such as a breakwater, groin, seawall, or small pier) extending out from the shore into a body of water, designed to direct and confine the current or tide, or to protect a harbor, or to induce scouring, or to prevent shoaling of a navigable passage by littoral materials; it is often built in pairs on either side of a harbor entrance, or at the mouth of a river. (b) A British term for a landing wharf or pier used as a berthing place for vessels.

jew's-stone [mineral] A piece of marcasite used in making ornaments (esp. costume jewelry).

jew's-stone [paleont] A large fossil clavate spine of a sea urchin.

jezekite *morinite*.

jheel A term applied in the Ganges flood plain of India to a backwater, such as a pool, marsh, or lake, remaining from inundation, existing during the cold weather at about the same level as that of the river and rising with the river during the rainy season. Etymol: Hindi. Cf: *bhil*. Pron: *jeel*. Syn: *jhil*.

jhil *jheel*.

jimboite An orthorhombic mineral: $Mn_3(BO_3)_2$. It is isostructural with kotoite.

joaquinite A honey-yellow mineral: $NaBa_2Ce_2Fe(Ti, Nb)_2Si_8O_{26}(OH,F)_2$.

Job's tears Rounded grains of olivine (peridot) found associated with garnet in Arizona and New Mexico.

joch A mountain pass with a long, approximately level summit between two parallel slopes (Stamp, 1961, p.491); a *col*. Etymol: German *Joch*, "yoke". Pron: *yuhkh*. Syn: *yoke-pass*.

joesmithite A mineral: $PbCa_2(Mg,Fe)_4Fe^3Si_6O_{12}(OH)_4(O, OH)_8$.

johachidolite A colorless and transparent mineral: $Na_2Ca_3-Al_4B_6O_{14}(F,OH)_{10}$.

johannite A green secondary mineral: $Cu(UO_2)_2(SO_4)_2-(OH)_2.6H_2O$. Syn: *gilpinite*.

johannsenite A clove-brown, grayish, or greenish mineral of the clinopyroxene group: $CaMnSi_2O_6$.

Johannsen number A number composed of three or four digits and which defines the position of any igneous rock in *Johannsen's classification*. The first digit represents the class, the second the order, and the third and fourth, the family.

Johannsen's classification A quantitative mineralogic classification of igneous rocks developed by the petrographer Albert Johannsen (1939).

johnsonite *amberat*.

johnstrupite A brownish-green mineral, approximately: $(Ca, Na)_3(Ce,Ti,Zr)(SiO_4)_2F$. Cf: *mosandrite*.

join The line or plane drawn between any two or three composition points. There is no special phase significance to join; it need not be a limiting binary or ternary subsystem. Incorrect syn: *conjugate line*.

joint [paleont] (a) An articulation in a crustacean; commonly, the movable connection of an individual *segment* of an appendage with its neighbors or the body, or the movable connection of body parts. (b) A connection between any pair of contiguous crinoid ossicles. (c) A segment of the shell of a nasselline radiolarian.

joint [struc geol] A surface of actual or potential fracture or parting in a rock, without displacement; the surface is usually plane and often occurs with parallel joints to form part of a *joint set*. See also: *jointing*.

joint block A body of rock that is bounded by joints; the rock that occurs between adjacent joints.

joint-block separation A type of mechanical weathering in which the rock breaks down or comes apart along well-defined joint planes. Syn: *block disintegration*.

joint drag A less-preferred syn. of *kink band*.

joint fissure *fissure*.

joint frequency *joint spacing*.

jointing n. The condition or presence of *joints* in a body of rock. Partial syn: *cleating*.

joint plane The surface of fracturing or potential fracture of a joint.

joint-plane fall A waterfall whose crest is irregularly broken by the falling away of joint blocks (Tarr & Engeln, 1926, p.83).

joint pocket *ceiling cavity*.

joint set A regional pattern of groups of parallel *joints*; in sedimentary rocks, one group is usually parallel to the dip and another to the strike. In massive igneous or metamorphic rock a third set of joints may occur. See also: *joint system*.

joint spacing The interval between joints of a particular joint set, measured on a line perpendicular to the joint planes. Syn: *joint frequency*.

joint system Two or more *joint sets* that intersect. They may be of the same age or of different ages.

joint valley A valley whose drainage pattern is controlled by master joint systems, e.g. a *rectangular drainage pattern*.

jokul An Icelandic mountain permanently covered with ice and snow; an Icelandic snow-capped peak. The term has been applied as well to an Icelandic glacier or small ice cap. Etymol: Icelandic *jökull*, "icicle, glacier". Pron: *yo-kool*. Pl: *jokul*. Syn: *jökull; jökul*.

jökulhlaup An Icelandic term for *glacier outburst flood*. Pron: *yo-kool-loup* (the last syllable as in "out").

Jolly balance In mineral analysis, a delicate spring balance used to measure specific gravity.

Jordanite A lead-gray mineral: $(Pb,Tl)_{13}As_7S_{23}$.

Jordan's law A theory in evolutionary biology stating that closely related organisms have the tendency to occupy adjacent rather than identical or distant ranges. This theory is named after the American biologist David Jordan (1851-1931).

Jordisite An amorphous mineral: MoS_2. Cf: *molybdenite*.

Josefite An altered hypabyssal rock having microgranular texture and composed of augite, olivine, serpentine, and calcite.

Joseite A mineral: $Bi_3Te(Se,S)$.

Josen *hartite*.

Josephinite A mineral consisting of a natural alloy of iron and nickel occurring in stream gravel from Josephine County, Oregon; *nickel-iron*.

Jotnian A division of the Precambrian.

Jotunite A plutonic rock, intermediate between mangerite and jorite, containing plagioclase and microperthite, and attributed in part to monzonite, in part to monzonite-diorite, and in part to monzonite-gabbro (Streckeisen, 1967, p.169). Syn: *jotun-norite*.

Jotun-norite *jotunite*.

Jouravskite A mineral: $Ca_6Mn_2(SO_4,CO_3)_4(OH)_{12}.24H_2O$.

Jug A colloquial syn. of *geophone*.

Jugum (a) A medially placed connection of secondary shell between two primary lamellae of brachiopod spiralia; a more or less complex skeletal crossbar linking the right and left halves of the brachidium of certain brachiopods. (b) A transverse structure crossing the center of a heterococcolith and connecting one side of the cycle with the other.---Pl: *juga* or *jugums*. Adj. *jugal*.

Julgoldite A mineral: $Ca_2Fe^{+2}(Fe,Al)_2(SiO_4)(Si_2O_7)(OH)_2.H_2O$. It is related to pumpellyite.

Julienite A blue mineral occurring in needle-like crystals: $Na_2Co(SCN)_4.8H_2O$ (?).

Jumillite An extrusive rock, commonly fine grained, composed of barium-bearing sanidine, olivine, and phlogopite phenocrysts in a fine-grained groundmass of altered leucite, sanidine, and iron-rich diopside mantled with acmite-augite, with interstitial kataphorite.

Junction [stratig] The contact between two rock masses, esp. where there is an upward sequence of stratified rocks. It may be conformable, disconformable, or unconformable, or it may be of tectonic nature (such as a "faulted junction" in which the contact is between two layers through a fault).

Junction [surv] A point common to two or more survey lines.

Junction [streams] The meeting of two or more streams; also, the place of such a meeting; a *confluence*. Examples: accordant junction; deferred junction.

junction closure The amount by which a new survey line into a junction fails to give the previously determined position or elevation for the junction point.

jungle A wild, tangled, densely vegetated region in an equatorial area.

junior In taxonomy, said of the later published of two synonyms or homonyms. Cf: *senior*.

Jura *Jurassic*.

Jurassian relief A type of relief found in young mountains consisting of many parallel anticlines and synclines, and characterized by primary structural forms or by features upon which erosion has had relatively little influence (Schieferdecker, 1959, term 1944). Type example: the relief of the Jura Mountains in Switzerland. Cf: *Appalachian relief*.

Jurassic The second period of the Mesozoic era (after the Triassic and before the Cretaceous), thought to have covered the span of time between 195-190 and 136 million years ago; also, the corresponding system of rocks. It is named after the Jura Mountains between France and Switzerland, in which rocks of this age were first studied. See also: *age of cycads*. Syn: *Jura*.

Jura-Trias The Jurassic and Triassic periods, combined.

Jura-type fold *décollement fold*.

jurupaite A variety of xonotlite containing magnesium.

juvenarium The proloculus and first few chambers of a foraminifer. See also: *embryonic apparatus*.

juvenile [ore dep] Said of an ore-forming fluid or mineralizer that is derived from subcrustal, basaltic magmas (Smirnov, 1968). Cf: *assimilated; filtrational*.

juvenile [geomorph] *youthful*.

juvenile [water] A term applied to water and gases that are known to be derived directly from magma and that come to the Earth's surface for the first time. Cf: *resurgent*.

juvenile [volc] In the classification of pyroclastics, the equivalent of *essential* ejecta.

juvite A light-colored nepheline syenite in which the feldspar is exclusively or predominantly orthoclase and in which the potassium-oxide content is higher than the sodium oxide.

juxta-epigenesis Epigenesis (occurring subsequent to diagenesis) that affects sediments while they are near the original environment of deposition, either under a relatively thin overburden or while exposed above sea level (Chilingar et al, 1967, p.316). Cf: *apo-epigenesis*.

juxtaposed ice stream Ice from a tributary glacier that is set into the surface of a glacier and extends to the bed. Cf: *inset ice stream*.

juxtaposition twin *contact twin*.

K

kaersutite A black variety of hornblende containing titanium.

kahlerite A yellow to yellow-green secondary mineral of the autunite group: $Fe(UO_2)_2(AsO_4)_2.nH_2O$.

kaimoo A stratified ice and sediment rampart built during the autumn on an Arctic beach by wave action. Etymol: Eskimo.

kainite A usually whitish monoclinic mineral: $MgSO_4.KCl.3H_2O$. It is a natural salt occurring in irregular granular masses, and is used as a source of potassium and magnesium compounds.

kainosite A yellowish-brown mineral: $Ca_2(Ce,Y)_2(SiO_4)_3(CO_3).H_2O$. Syn: *cenosite*.

kainotype *cenotypal*.

Kainozoic Var. of *Cenozoic*.

kaiwekite A trachytic extrusive rock composed of phenocrysts of barkevikite, small acmite-augite crystals, and anorthoclase with inclusions of acmite-augite and other minerals, and with a few pseudomorphs of serpentine after olivine, in a groundmass composed chiefly of oligoclase with some augite and magnetite; the approximate extrusive equivalent of laurvikite.

kajanite A feldspar-free extrusive rock composed of bronze-colored mica and olivine phenocrysts in a groundmass of leucite, diopside, and titaniferous iron ore.

kakirite A megascopically sheared and brecciated rock in which fragments of original material are surrounded by gliding surfaces along which has occurred intense granulation and some recrystallization (Holmes, 1928, p.129). It was named by Svenonius after Lake Kakir, Swedish Lapland.

kakortokite A banded nepheline syenite of varied composition, having light-colored layers rich in feldspar and nepheline (white) or in eudialyte and nepheline (red) and dark-colored layers rich in acmite and arfvedsonite (black). The rock was originally described from Julianehaab (Kakortok), Greenland.

kalahari A term used in SW Africa for a salt pan; specif. the Kalahari Desert of Botswana.

kali- A prefix which, when in an igneous rock name, signifies an absence of plagioclase or a plagioclase content of less than 5.0%.

kalialaskite An *alaskite* with no modal albite; the intrusive equivalent of kalitordrillite. Cf: *birkremite*.

kaliborite A colorless to white mineral: $HKMg_2B_{12}O_{21}.9H_2O$. Syn: *heintzite; hintzeite; paternoite*.

kalicinite A colorless to white or yellowish mineral: $KHCO_3$. Syn: *kalicine; kalicite*.

kaligranite A plagioclase-free granite that frequently contains soda pyriboles; the intrusive equivalent of kalirhyolite. Cf: *alaskite*. Syn: *orthogranite*.

kalikeratophyre An albite-free keratophyre containing potassium feldspar instead.

kaliliparite An igneous rock having a chemical composition of approximately 68 percent SiO_2, 16 percent Al_2O_3, 1 percent CaO, 1 percent MgO, 1 percent Fe_2O_3, 11 percent K_2O, and 2 percent Na_2O. It is used for low-alkali glass (Thrush, 1968, p.605).

kalinite A mineral of the alum group: $KAl(SO_4)_2.11H_2O$. Cf: *alum*. Syn: *potash alum*.

kaliophilite A hexagonal mineral of volcanic origin: $KAlSiO_4$. It is dimorphous with *kalsilite*. Syn: *facellite; phacellite*.

kalirhyolite An extrusive rock composed of quartz, alkali feldspar, and a ferromagnesian mineral; the extrusive equivalent of kaligranite.

kalistronite A hexagonal mineral: $K_2Sr(SO_4)_2$. It is isostructural with palmierite.

kalisyenite A syenite containing less than 5 percent of the total feldspar as plagioclase.

kalitordrillite A plagioclase-free hypabyssal or extrusive rock composed chiefly of quartz and alkali feldspar; the extrusive equivalent of kalialaskite.

kalitrachyte An alkalic trachyte in which potassium feldspar is the only feldspar and in which the dark minerals are usually alkalic.

kalkowskite A very rare, brownish or black mineral: $Fe_2Ti_3O_9$ (?). It may be ilmenite. It usually contains small amounts of rare-earth elements, niobium, and tantalum. Cf: *arizonite* [*mineral*]. Syn: *kalkowskyn*.

kallar A term used in India (west of the Indo-Gangetic region) as a syn. of reh. Etymol: Panjabi. Also spelled: *kalar*.

kalsilite A mineral: $KAlSiO_4$. It is dimorphous with kaliophilite and sometimes contains sodium.

kam A mound or short ridge developed at right angles to the direction of ice movement (Giles, 1918, p.163). Term is obsolete.

kamacite A meteorite mineral consisting of the body-centered cubic alpha-phase of a *nickel-iron* alloy, with a fairly constant composition of 5-7% nickel. It occurs in iron meteorites as bars or "girders" flanked by lamellae of *taenite*.

kamarezite *brochantite*.

kame A long, low, steep-sided hill, mound, knob, hummock or short irregular ridge, composed chiefly of poorly sorted and stratified sand and gravel deposited by a subglacial stream as an alluvial fan or delta against or upon the terminal margin of a melting glacier, and generally aligned parallel to the ice front. The term has undergone several changes in meaning but can still be usefully applied to a deposit of poorly sorted glaciofluvial sand and gravel whose precise mode of formation is uncertain (Thornbury, 1954, p.378-379). Etymol: a Scottish variant of "comb", a long, steep-sided ridge. Cf: *esker*.

kame-and-kettle topography *knob-and-kettle topography*.

kame complex An assemblage of kames, constituting a hilly landscape.

kame delta *delta kame*.

kame field A group of closely spaced kames, interspersed in places with kettles and eskers, and having a characteristic hummocky topography.

kame moraine (a) A terminal moraine that contains numerous kames. (b) A group of kames along the front of a stagnant glacier.---See also: *moraine kame*.

kamenitza *solution pan*.

kame plain A flat-topped outwash plain originally entirely bounded by ice-contact slopes. Syn: *ice-contact plain*.

kame terrace A terrace-like ridge consisting of stratified sand and gravel deposited by a meltwater stream between a melting glacier or a stagnant ice lobe and a higher valley wall or lateral moraine, and left standing after the disappearance of the ice; a filling of a *fosse*. A kame terrace terminates a short distance downstream from the terminal moraine; it is commonly pitted with kettles and has an irregular ice-contact slope. Syn: *ice-contact terrace; ice-marginal terrace*.

kämmererite (a) A reddish variety of penninite containing chromium. Its formula is near: $Mg_5(Al,Cr)_2Si_3O_{10}(OH)_8$. (b) A hypothetical end member of the chlorite group: $Mg_2Cr_2SiO_5(OH)_4$.

kammgranite A porphyritic hornblende granite (Thrush, 1968, p.605).

kamperite A fine-grained black hypabyssal rock composed of small euhedral orthoclase crystals and a small amount of oligoclase in a groundmass of dark-colored mica.

kamptomorph A mineral in a metamorphic rock which is not fractured by the metamorphism and which has unbroken outlines, but which has been bent and shows undulatory extinction. The term is now obsolete.

kanat *ganat*.

kandite A name suggested (but not approved) for the *kaolin* group of clay minerals, including kaolinite, nacrite, dickite, and halloysite.

kankan-ishi A black, resinous, flinty andesite composed of hypersthene, oligoclase, and hornblende microphenocrysts in a groundmass of colorless glass and a network of acicular crystals of colorless bronzite (Johannsen, 1939, p.259).

kankar (a) A term used in India for masses or layers of concretionary calcium carbonate, usually occurring in nodules, found embedded in the older alluvium or stiff clay of the Indo-Gangetic plain, or for precipitated calcium carbonate in the form of cement in porous sediments or as a coating on pebbles. (b) A limestone containing kankar and used for making lime and building roads.---Etymol: Hindi. The term was used originally for gravel, stone, or any small rock fragments, whether rounded or not; it is occasionally applied in the U.S. to a residual calcareous deposit, such as *caliche*. Inappropriate syn: *travertine*. Syn: *kunkur; kunkar; conker.*

Kansan Pertaining to the second glacial stage of the Pleistocene Epoch in North America, beginning 400,000(?) years ago, after the Aftonian interglacial stage and before the Yarmouth. See also: *Mindel.*

kansite *mackinawite.*

kantography The depiction of *edge lines* on relief maps. Etymol: German *Kantographie.*

kaoleen A colloquial term used in south-central Missouri for a chalky, porous, weathered chert with a white to tan or buff color. Etymol: corruption of *kaolin*, to which the material bears a slight resemblance.

kaolin [petrog] A soft, fine, earthy, nonplastic, usually white or nearly white clay (rock) composed essentially of clay minerals of the kaolin group, principally kaolinite, derived from in-situ decomposition (extreme weathering or pneumatolysis) of aluminous minerals (such as feldspars in a granitic rock), containing a variable proportion of other constituents (quartz, mica flakes) derived from the parent rock, and remaining white or nearly white on firing; a *porcelain* clay, or natural (unwashed) *china clay*. It is used in the manufacture of ceramics, refractories, and paper. Type locality: Kao-ling (meaning "high hill"), hill in Kiangsi province, SE China. Syn: *kaoline; white clay; bolus alba.*

kaolin [mineral] (a) A group of clay minerals characterized by a two-layer crystal lattice in which each silicon-oxygen sheet is alternately linked with one aluminum-hydroxyl sheet and having approximate composition: $Al_2O_3.2SiO_2.2H_2O$. The kaolin minerals include kaolinite, nacrite, dickite, and anauxite; although structurally and chemically different, the minerals halloysite, endellite, and allophane are sometimes included. The kaolin minerals are generally derived from alteration of alkali feldspars and micas. Compared to montmorillonite and illite, they have lower base-exchange capacities, and they absorb less water and thus have lower plasticity indexes, lower liquid limits, and less shrinkage when drying from a wet state. See also: *kandite.* Syn: *kaolinite.* (b) A mineral of the kaolin group; specif. *kaolinite.* The term was once applied to a single clay mineral which later was known to comprise at least four minerals of the kaolin group.

kaolinic Pertaining to or resembling kaolin.

kaolinite (a) A common, white to grayish or yellowish clay mineral of the kaolin group: $Al_2Si_2O_5(OH)_4$. It is the characteristic mineral of most kaolins, and is polymorphous with dickite and nacrite. Kaolinite consists of sheets of tetrahedrally coordinated silicon joined by an oxygen shared with octahedrally coordinated aluminum; it is a high-alumina clay mineral that does not appreciably expand under varying water content and does not exchange iron or magnesium. The mineral was formerly known as *kaolin.* (b) A name sometimes applied to the *kaolin* group of clay minerals, and formerly applied to individual minerals of that group (such as to dickite and nacrite).

kaolinitic shale *feldspathic shale.*

kaolinization (a) Formation of kaolin by the weathering of aluminum silicate minerals or other clay minerals. (b) Some-

times loosely used for the formation of kaolin by hydrothermal processes.----Var: *kaolinisation; kaolisation.*

kaolinton An obsolescent term used by ceramists (esp. in Europe) for the portion of a clay that is soluble in sulfuric acid but not soluble in hydrochloric acid. Cf: *allophaneton.*

kar A syn. of *cirque.* Etymol: Swiss-German *Kar*, "cirque".

K-Ar Potassium-argon, e.g. *potassium-argon age method.*

karang A term used in Indonesia for an emerged terrace composed of ancient fringing-reef material, and also for the coral limestone itself (Termier & Termier, 1963, p.408). Etymol: Malay, "reef, coral reef".

karelianite A black mineral: V_2O_3. It may contain some iron, chromium, and manganese.

karewa A term applied in Kashmir to the level surface between the incised streams dissecting a terrace. Etymol: Kashmiri.

karez A term used in Pakistan for a gently inclined, underground channel dug so as to conduct ground water by gravity from alluvial gravels and the foot of hills to an arid lowland or basin; a horizontal well. Etymol: Baluchi. Pl: *karezes.* Cf: *qanat; foggara.*

karite A grorudite containing approximately 50 percent quartz.

karling (a) A high, dissected region, characterized by cirques, as Mount Anne in Tasmania. (b) A cluster or group of cirques.

Karlsbad twin law Var. of *Carlsbad twin law.*

karnaite A rock found on the island Kärnä in Lake Lappajärvi, central Finland, originally described as a volcanic rock with glassy groundmass and considerable inclusions, consisting of agglomerate-like tuff, and having a composition similar to dacite, the principal feldspar phenocrysts being monoclinic (probably sanidine). The rock has also been interpreted as an impactite with bedrock fragments (Svensson, 1968).

karnasurtite A honey to pale-yellow metamict mineral: $(Ce,La,Th)(Ti,Nb)(Al,Fe)(Si,P)_2O_7(OH)_4.3H_2O$. It gives a monazite-like X-ray pattern when heated, but it may be equivalent to rhabdophane.

Karnian Var. of *Carnian.*

karoo Var. of *karroo.*

karpatite A hydrocarbon mineral: $C_{24}H_{12}$. Syn: *carpathite; pendletonite.*

karpinskite A greenish-blue mineral: $(Mg,Ni)_2Si_2O_5(OH)_2$ (?). Not to be confused with *karpinskyite.*

karpinskyite A white hexagonal mineral: $Na_2(Be,Zn,Mg)Al_2Si_6O_{16}(OH)_2$. Not to be confused with *karpinskite.*

karren In karst topography, a general term for solutional furrows or channels formed on the surface of limestone, ranging in depth from a few millimeters to more than a meter, and usually separated by knifelike ridges. The term is always used in this plural form. Etymol: German, "furrow". Se also: karrenfeld. Syn: *lapei s.*

karrenfeld A karstic surface limestone characterized by *karren.* Cf: *limestone pavement.*

karroo A tableland found esp. in South Africa and which often rises to a considerable height in terraces, that does not support vegetation in the dry season, but which becomes a grassy plain or pastureland during the wet season. Also spelled: *karoo.*

karst A type of topography that is formed over limestone, dolomite, or gypsum by dissolving or solution, and that is characterized by closed depressions or sinkholes, caves, and underground drainage. Etymol: German, from the Slavic *kras*, "a bleak, waterless place". Type locality: Karst, a limestone plateau in the Dinaric Alps of Yugoslavia and the Free Territory of Trieste. Also spelled *carst.* Adj: *karstic.*

karst base level In an area of karst, that level below which karstification ceases (Monroe, 1970).

karst bridge A *natural bridge* in a limestone terrane.

karst fenster *karst window.*

karst hydrology The drainage pattern and features that are characteristic of karstic topography.

karstic Adj. of *karst*.

karstification The formation of the features of a karstic topography by the solutional, and sometimes mechanical, action of water in a region of limestone, dolomite, or gypsum bedrock.

karst lake *karst pond*.

karstland An area of karst topography.

karst margin plane *marginal karst plane*.

karst plain A plain upon which karst features are developed; a region of karst topography, usually of limestone. See also: *marginal karst plain*. Syn: *karst plateau*.

karst plateau *karst plain*.

karst pond A body of standing water occurring in a closed depression of a karst region. Syn: *karst lake; sink lake; solution lake*.

karst spring A spring in which original surface waters, supplied from precipitation that fell only on limestone and that remained undiluted by runoff from noncalcareous areas, reappear after having been diverted through underground routes (such as fractures or caves). See also: *Vauclusian spring*. Syn: *resurgence; rise; rising*.

karst valley An elongate, closed depression in a karst area that is formed by the coalescence of several *dolines* and measures about one kilometer in diameter. Cf: *valley sink; glade*. Syn: *uvala; solution valley; canyon* [*speleo*]; *nested sinkholes*.

karst window An unroofed area of a subterranean stream; a depression at the bottom of which can be seen a subterranean stream. Syn: *karst fenster*.

kasoite A variety of celsian containing potassium.

kasolite A yellow to brown monoclinic mineral: $Pb(UO_2)$-$SiO_4.H_2O$.

kassaite A fine-grained hypabyssal rock composed of phenocrysts of haüyne, labradorite with oligoclase rims, barkevikite, and augite in a holocrystalline tinguaitic groundmass of acicular hastingsite crystals and andesine with oligoclase and orthoclase rims.

kassite A mineral: $CaTi_2O_4(OH)_2$.

katabatic wind A local wind that moves downward, e.g. as a result of surface cooling during the night. Cf: *anabatic wind*. See also: *mountain wind; foehn; glacier wind*. Syn: *gravity wind*.

kataclastic Var. of "cataclastic". Also spelled: *kataklastic*.

katagenesis *catagenesis*.

kataglyph A hieroglyph formed during catagenesis, or under a covering set of beds (Vassoevich, 1953, p.33).

katamorphic zone The zone of shallow depth in the Earth's crust in which complex mineral compounds are broken down into simple ones by the processes of weathering (in the upper belt of the zone) and ground-water effects, including cementation (in the lower belt of the zone). The term was originated by Van Hise. It has been suggested that use of the term be discontinued in order to avoid confusion with the term *katazone*. Cf: *anamorphic zone*.

katamorphism The destructive process of *metamorphism* whereby rocks at or near the Earth's surface (in the *katamorphic zone*, i.e. in the zones of weathering and cementation) are broken down and altered mainly through oxidation, hydration, solution etc. with characteristic production of simpler, less dense minerals from more complex ones (Van Hise, 1904). It has been suggested that use of the term be discontinued since this type of rock alteration is no longer considered to be a metamorphic process, and because the prefix kata- is now commonly used to denote the deepest zone of the Earth's crust, i.e. the *katazone*. Also spelled: *catamorphism*. Cf: *anamorphism*.

kataphorite A brownish monoclinic mineral of the amphibole group: $Na_2Ca(Fe^{+3},Al)_5AlSi_7O_{22}(OH)_2$. Syn: *catophorite; cataphorite*.

kataseism Earth movement toward the focus of an earthquake; a *dilatation* or *rarefaction*. Cf: *anaseism*.

katatectic layer A sedimentary layer that is "built downward" specif. a distinct, generally horizontal or slightly dipping layer of solution residue (gypsum and/or anhydrite) formed by intermittent compaction of sulfate accumulating on top (in the cap rock) of a salt stock by solution of salt. The term was introduced by Goldman (1933, p.84). Etymol: Greek *kata* "down", + *tekton*, "builder".

katatectic surface A surface separating two katatectic layers (Goldman, 1952, p.v).

katathermal *hypothermal*.

Katathermal n. *Little Ice Age*.

katazone According to Grubenmann's classification of metamorphic rocks (1904), the lowermost *depth zone* of metamorphism, characterized by high temperatures ($500°-700°C$) mostly strong hydrostatic pressure, and low or no shearing stress. Long-continued reconstitution and recrystallization often without deformation, and deep-seated metamorphism associated with igneous action, produce such rocks as high-grade schists and gneisses, granulites, eclogites, and amphibolites. The concept was modified by Grubenmann and Niggli (1924) to include effects of high-temperature contact metamorphism and metasomatism. Modern usage stresses temperature-pressure conditions (highest metamorphic grade) rather than depth of zone. Cf: *mesozone; epizone*. Also spelled: *catazone*. See also: *katamorphism*.

katogene A now obsolete term pertaining to a breaking down of a rock by atmospheric or other agents, or to shallow-depth replacements.

katoptrite A black monoclinic mineral: $(Mn,Mg,Fe)_{14}(Al,Fe)_4Sb_2Si_2O_{29}$. Syn: *catoptrite*.

katothermal Said of a lake (such as a polar lake) whose temperature increases with depth. The term is "apparently defunct" (Stamp, 1961, p.278). Cf: *katathermal*.

katungite An extrusive rock composed chiefly of melilite, with a smaller amount of olivine and magnetite and minor leucite and perovskite; a pyroxene-free melilitite.

katzenbuckelite A hypabyssal rock with tinguaitic texture and composed of phenocrysts of nepheline, biotite, olivine, nosean, leucite, and apatite in a fine-grained groundmass of nepheline, leucite, and acmite.

kauaiite An orthoclase-bearing olivine-augite diorite in which the feldspar is zoned, with calcic labradorite in the inner rims grading outward into alkali feldspar. Its name is derived from the Hawaiian island of Kauai.

kauri A light-colored, whitish-yellow, or brown *copal* usually found as a fossil resin from the kauri pine (a tree of the genus *Agathis*), esp. from *Agathis australis*, a tall timber tree of New Zealand. Etymol: Maori *kawri*. Syn: *kauri resin; kauri gum; kauri copal; agathocopalite*.

kavir (a) A term used in Iran for a *salt desert*; specif. the Great Kavir of inner Iran, a series of closed basins noted for marshy conditions and high salinities. (b) A *playa* on a kavir. Syn: *kewire; kevir*. (c) A term used in Iran for a salt marsh.

kay A variant of *key* and *cay*.

Kazanian European stage: Upper Permian (above Kungurian, below Tatarian).

kazanskite A hypabyssal plagioclase-bearing dunite, with accessory hornblende, plagioclase, magnetite (approximately one-fourth of the rock), and green spinel.

K-bentonite *potassium bentonite*.

keatite A tetragonal polymorph of SiO_2, synthesized hydrothermally at high pressures.

keazoglyph A hieroglyph consisting of small transverse displacements along cracks (Vassoevich, 1953, p.64).

kedabekite An igneous rock similar to eucrite and composed of bytownite (originally identified as anorthite), calcium-iron garnet, and hedbergite (originally identified as violaite).

keel (a) A costa or rib-like structure on the aboral (under) side of plate-like conodont elements. (b) A continuous sharp ridge along the venter of a nautiloid conch, esp. in coiled forms; a distinct, longitudinal, continuous ridge, either solid or hollow, on the venter of an ammonoid shell. (c) A *carina* of a bryozoan. (d) A vertical sail-like plate in a radiolarian; a keel-like ridge along the outer margin of the test of a foraminifer. (e) A canal or cleft in the valve of some pennate diatoms.

Keewatin A division of the Archeozoic rocks of the Canadian Shield. Also spelled: *Kewatinian*.

Kegelkarst A synonym common in the older literature for *cone karst*.

Kegelsonde A spring-driven penetrometer for measuring the horizontal resistance of snow to penetration. Etymol: German.

kehoeite An amorphous mineral: $(Zn,Ca)_4Al_8(PO_4)_8$-$(OH)_8 \cdot 20H_2O$ (?).

keilhauite A radioactive variety of sphene containing aluminum, iron, and yttrium and other rare earths.

keldyshite A mineral: $(Na,H)_2ZrSi_2O_7$.

kelly A hollow, 40-ft (12-m) tube or rod of square or hexagonal cross section, forming the top section of the rotary-drill shaft in an oil well, its lower end screwed into and supporting the drill pipe. It is fitted into and passes through the rotary table and is turned by it during drilling, thereby transmitting the rotary motion of the table to the drill pipe. Syn: *kelly joint; kelly stem; kelly bar; grief stem.*

Kelvin wave A tide progression developing in a relatively confined body of water (such as the English Channel or the North Sea) in which, because of the Coriolis force, the tide range in the Northern Hemisphere increases to the right (and decreases to the left) of the direction of travel.

kelyphytic rim (a) In some igneous rocks, a peripheral zone of pyroxene or amphibole developed around olivine where it would otherwise be in contact with plagioclase or around garnet where it would otherwise be in contact with olivine or other magnesium-rich minerals. Cf: *reaction rim.* (b) A secondary *reaction rim.* (c) *reaction rim.----Also spelled: celyphytic rim.* Syn: *kelyphytic border.*

kemmlitzite A mineral: $SrAl_3(AsO_4)(SO_4)(OH)_6$.

kempite An emerald-green orthorhombic mineral: $Mn_2(OH)_3$-$Cl.$

kennedyite A mineral: $MgFe_2^{+3}Ti_3O_{10}$.

Kennedy's critical velocity *critical velocity* (d).

kennel (a) A Scottish term for a hard sandstone, often with calcareous cement; *kingle.* (b) *kennel coal.*

kennel coal *cannel coal.*

Kenoran orogeny A name proposed by Stockwell (1964) for a time of plutonism, metamorphism, and deformation during the Precambrian of the Canadian Shield (especially in the Superior and Slave provinces), dated radiometrically at 2390-2600 m.y. ago, or at the end of the Archean of the present Canadian classification. It is synonymous with *Algoman orogeny* of Minnesota.

kenozooid A bryozoan *heterozooid* without a polypide and usually without either orifice or muscles.

kentallenite A dark-colored monzonite composed of approximately equal amounts of augite, olivine, orthoclase, and plagioclase, with smaller amounts of biotite, apatite, and opaque oxides; an olivine-bearing monzonite. Its name is derived from Kentallen, Argyllshire, Scotland.

kentrolite A dark reddish-brown mineral: $Pb_2Mn_2Si_2O_9$.

kentsmithite A local name used in the Paradox Valley, Colo., for a black vanadium-bearing sandstone.

kenyaite A mineral: $Na_2Si_{22}O_{41}(OH)_8 \cdot 6H_2O$.

kenyte An olivine-bearing phonolite composed of anorthoclase, nepheline, acmite-augite, sodic amphibole, olivine, apatite, and opaque oxides. The groundmass may have a trachytic or hyalopilitic texture. Its name is derived from Mount Kenya, Kenya.

Keppler's laws of planetary motion The statements that each planet moves in an elliptical orbit with the Sun at one focus of the ellipse; that the line from the Sun to any planet sweeps out equal areas of space in equal intervals of time; and that the squares of the sidereal periods of the several planets are proportional to the cubes of their mean distances from the Sun. Although Keppler's laws are a mathematical consequence of Newton's laws, which are more fundamental, they preceded Newton's laws, were empirically based on the observations of Tycho Brahe, and were a significant extension of Copernican philosophy.

kerabitumen *kerogen.*

keralite A quartz-biotite hornfels. The term was originated by Cordier in 1868.

kerargyrite *cerargyrite.*

keratophyre A name originally applied by Gümbel (1874, p.43) to trachytic rocks containing highly sodic feldspars, but now more generally applied to all salic extrusive and hypabyssal rocks characterized by the presence of albite or albite-oligoclase and chlorite, epidote, and calcite, generally of secondary origin. Originally the term was restricted to lavas of pre-Tertiary age but this distinction is not recognized in current usage. Some varieties of keratophyre contain sodic orthoclase and sodic amphiboles and pyroxenes. Keratophyres commonly are associated with spilitic rocks and interbedded with marine sediments.

keratose Said of a horny sponge whose skeleton consists entirely of organic fibers without spicules (although it sometimes may contain foreign particles including spicules of other sponges).

keriotheca The relatively thick shell layer with honeycomb-like structure in the wall of some fusulinids (such as schwagerinids), occurring next below the tectum, and forming part of the spirotheca. It may be divisible into *lower keriotheca* and *upper keriotheca.*

kermesite A cherry-red mineral: Sb_2S_2O. It usually occurs as tufts of capillary crystals resulting from the alteration of stibnite. Syn: *antimony blende; red antimony; purple blende; pyrostibite.*

kernbut A projecting ridge or buttress created by displacement on a fault traversing a hillslope and separated from the hill by a *kerncol*; the outer ridge-like edge of a fault terrace or fault bench. The term was introduced by Lawson (1904, p.332) for a primary feature occurring in Kern Canyon, Calif., but the type locality has since been shown to be one of fault-line forms (Webb, 1936). Etymol: *Kern* Canyon + *butt*ress.

kerncol A low sag or trough separating a *kernbut* from the hillside, occurring where a faulted block joins the hill. The term was introduced by Lawson (1904, p.332) for a primary feature occurring in Kern Canyon, Calif., but the type locality has since been shown to be one of fault-line forms (Webb, 1936). Etymol: *Kern* Canyon + *col.*

kernite A colorless to white monoclinic mineral: Na_2B_4-$O_7 \cdot 4H_2O$. It is an important source of boron. Syn: *rasorite.*

kerogen Fossilized, insoluble, organic material found in sedimentary rocks, usually shales, which can be converted by distillation to petroleum products. Syn: *kerabitumen; petrologen.*

kerogenite Fissile and laminated rock containing organic material that will yield hydrocarbons on distillation.

kerogen shale *oil shale.*

keronigritite A type of *nigritite* that is derived from kerogen. Cf: *polynigritite; humonigritite; exinonigritite.*

kerosene shale Var. of *kerosine shale.*

kerosine shale (a) A syn. of *torbanite.* (b) Any bituminous oil shale.----Also spelled: *kerosene shale.*

kersantite A lamprophyre containing biotite, plagioclase (usually oligoclase or andesine), and augite, with or without diopside and olivine. Obs syn: *kersanton.*

kersanton An obsolete syn. of *kersantite.*

kerzinite In the Urals, lignite that is impregnated with hydrated nickel silicate and thus is mined for nickel.

kess-kess An Arabic term used in Morocco for a reef knoll isolated by erosion.

kesterite A mineral: $Cu_2(Zn,Fe)SnS_4$. It is the zinc analogue of stannite. Also spelled: *kesterite*.

kettle [glac geol] A steep-sided, usually basin- or bowl-shaped hole or depression without surface drainage in glacial-drift deposits (esp. outwash and kame), often containing a lake or swamp, and believed to have formed by the melting of a large, detached block of stagnant ice (left behind by a retreating glacier) that had been wholly or partly buried in the glacial drift. A kettle is usually 10-15 m deep, and 30-150 m in diameter. Cf: *pothole* [glac geol]. Syn: *kettle hole; kettle basin*.

kettle [streams] A *pothole* in a stream bed.

kettleback *horseback*.

kettle basin *kettle*.

kettle bottom A part of the roof of a coal seam, such as a piece of slate, resembling the bottom of a kettle and easily loosened without warning, leaving a smooth cavity in the roof. Cf: *caldron bottom; pot bottom; bell*. Syn: *saddle* [coal mining].

kettle drift A mound or ridge of gravelly drift; an esker or a kame.

kettle hole *kettle*.

kettle lake (a) A body of water occupying a kettle in a pitted outwash plain or in a kettle moraine. Syn: *kettle-hole lake; pit lake*. (b) *cave-in lake*.

kettle moraine A terminal moraine whose surface is marked by many kettles.

kettle plain A *pitted outwash plain* marked by many kettles.

kettnerite A brown to yellow tetragonal mineral: $CaBi(CO_3)OF$.

Keuper European stage (esp. in Germany): Upper Triassic (above Muschelkalk, below Jurassic).

kevir Var. of *kavir* (a playa).

Kewatinian Var. of *Keewatin*.

Keweenawan A provincial series of the Precambrian in Michigan and Wisconsin.

kewire Var. of *kavir* (a playa).

key [cart] A *legend* on a map.

key [coast] A *cay*, esp. one of the coral islets or barrier islands off the southern coast of Florida. See also: *sandkey*. Syn: *kay*.

key [taxon] An orderly arrangement of the distinguishing features of a group of plants or animals or of species or genera, constructed to facilitate classification and determination of taxonomic relationships of unidentified members of the group.

key bed (a) A well-defined, easily identifiable stratum or body of strata that has sufficiently distinctive characteristics (such as lithology or fossil content) to facilitate correlation in field mapping or subsurface work. (b) A bed the top or bottom of which is used as a datum in making structure-contour maps.--Syn: *key horizon; index bed; marker bed*.

keyed *sutured*.

key fossil *index fossil*.

key horizon (a) The top or bottom of an easily recognized, extensive bed or formation (such as a layer containing certain fossils) that is so distinctive as to be of great help in stratigraphy and structural geology; e.g. a *datum horizon*. (b) A term that is used interchangeably with *key bed*.

keystone fault A grabenlike structure developed on the crest of an anticline.

K-feldspar *potassium feldspar*.

K-feldspar-cordierite-hornfels facies A new name given to the *pyroxene-hornfels facies* (Winkler, 1967) in order to indicate a more diagnostic mineral assemblage that would distinguish it from the hornblende-hornfels facies. Characteristic orthopyroxenes form in the higher-temperature phases (in excess of

about $700°C$ of this facies.

khadar A term used in India for a low-lying area (as a strath or flood plain) consisting of a new alluvial plain that is liable to be flooded by the waters of a river. Etymol: Urdu-Hindi *khādar*. Cf: *bhangar*. Syn: *khaddar; khuddar*.

khagiarite A black pantellerite characterized by a glassy microlitic groundmass exhibiting flow texture.

khal (a) A term used in East Pakistan for a narrow stream channel. (b) A sluggish creek on the lower delta of the Ganges.---Etymol: Bengali.

kharafish A limestone plateau in the Libyan desert, formed by wind erosion.

khari A term used in East Pakistan for a small, deep stream of local origin. Etymol: Bengali.

kheneg A term used in the Atlas Mountains of northern Africa for a canyon. Etymol: Arabic.

khibinite [mineral] *mosandrite*.

khibinite [petrology] An alternate spelling of *chibinite*.

khlopinite A tantalian variety of samarskite.

khoharite A hypothetical end member of the garnet group: $Mg_3Fe_2(SiO_4)_3$.

khondalite A group of metamorphosed aluminous sediments consisting of garnet-quartz-sillimanite rocks with garnetiferous quartzites, graphite schists, and marbles (Walker, 1902, p.11). It is named after Khonds (the Khondalite series), India.

khor (a) A term used in Sudan for an intermittent stream. (b) A term used in northern Africa for a watercourse or ravine, esp. one that is dry.---Etymol: Arabic *khawr*, "wadi, dry wash".

khud A term used in India for a ravine or precipice. Etymol: Hindi *khad*.

khurd A term used in Algeria for a high (80-100 m), pyramid-shaped sand dune with curved slopes, formed by the intersection of seif dunes (Capot-Rey, 1945, p.393); some reach 150 m in height. Cf: *rhourd*. Syn: *guern*.

kick [drill] (a) Loss of normal fluid circulation in an oil well, caused by the fluid pressure in the well exceeding the pressure exerted by the drilling mud being pumped into the well. (b) A quick snap of the drill stem caused by the core breaking in a blocked core barrel. (c) A small sidewise displacement in a borehole, caused by the deflection of the drill bit when entering a hard, dipping stratum underlying softer rock.

kick [seis] *arrival*.

kickout The lateral distance from a drilling site reached by a directional well.

kidney ore A variety of hematite occurring in compact kidney-shaped masses, concretions, or nodules, together with clay, sand, calcite, or other impurities; concretionary ironstone. Syn: *kidney iron ore*.

kidneys A miner's term for a mineral zone that contracts, expands, and again contracts downwards.

kidney stone [mineral] *nephrite*.

kidney stone [sed] A pebble or nodule roughly resembling the shape of a kidney; e.g. a small, hard, red-coated, ironstone nodule common in the Oxford Clay of England.

kies A general term for the sulfide ores. Etymol: German *Kies*, "finer gravel".

kieselguhr A syn. of *diatomaceous earth*. Etymol: German *Kieselguhr*. Syn: *guhr; kieselgur*.

kieserite A white monoclinic mineral: $MgSO_4 \cdot H_2O$. It occurs in saline residues.

kiirunavaarite *magnetitite*.

kilchoanite A mineral: $Ca_3Si_2O_7$. It is dimorphous with rankinite.

kilkenny coal *anthracite*.

kill A creek, channel, stream, or river. The term is used chiefly in place names in Delaware and New York State; e.g. Peekskill, N.Y. Etymol: Dutch *kil*.

Killarney Revolution A name proposed by Schuchert (1924

for a supposed major orogeny at the end of Precambrian time in North America; based on the Killarney Granite north of Lake Huron in Ontario, supposed to be of post-Keweenawan age. Radiometric data now indicate that the Keweenawan is 1000 m.y. old, and that the Killarney Granite is older, and probably equivalent to the Penokean Granite (see *Penokean orogeny*) of Michigan and Minnesota. Actually, no notable tectonic events are now known to have occurred in this part of North America at the end of the Precambrian. The term Killarney Revolution is obsolete, and should be abandoned.

killas A name used in Devon and Cornwall for any rock that has been metamorphosed by contact with granite.

kilocalorie One thousand calories. Cf: *calorie*. Syn: *great calorie; large calorie; kilogram calorie*.

kilogram calorie *kilocalorie*.

kimberlite A porphyritic alkalic peridotite containing abundant phenocrysts of olivine (commonly serpentinized or carbonatized) and phlogopite (commonly chloritized) and possibly geikielite and chromian pyrope in a fine-grained groundmass of calcite and second-generation olivine and phlogopite and with accessory ilmenite, serpentine, chlorite, magnetite, and perovskite.

Kimeridgian *Kimmeridgian*.

Kimmerian [stratig] European stage (Black Sea area): Pliocene (above Meotian). Syn: *Cimmerian*.

Kimmeridgian European stage: Upper Jurassic (above Oxfordian, below Portlandian). The spelling *Kimeridgian* was used by Arkell (1956, p.20) on the basis that the type locality, the village of Kimmeridge in the Isle of Purbeck, southern England, was spelled with one "m" until 1892.

kimzeyite A mineral of the garnet group: $Ca_3(Zr,Ti)_2(Al, Si)_3O_{12}$.

kin A headland. Also, a term used in Ireland for the highest point of anything. Etymol: Gaelic.

kindchen A nodule or concretion that resembles the head of a child; specif. *loess kindchen*. Etymol: German *Kindchen*, "little child, infant, baby".

Kinderhookian North American provincial series: lowermost Mississippian (above Chautauquan of Devonian, below Osagian).

K index A measure of intensity of magnetic disturbance. It is a figure ranging from zero to nine, indicative of range of magnetic intensity in a three-hour interval, after subtraction of normal daily variation.

kindly *likely*.

kindred *rock association*.

kinematic viscosity The ratio of the viscosity coefficient (in poises) to density at room temperature (in g/sq cm). See also: *eddy viscosity*. Syn: *coefficient of kinematic viscosity*.

kinematic wave A disturbance in the steady-state flow of a glacier that is propagated downstream at a velocity greater than the ice velocity. The disturbance can be in the form of a change of thickness or net balance due to a change in climate. Cf: *glacier wave*.

kinetic metamorphism A type of metamorphism that produces deformation of rocks without chemical reconstitution or recrystallization to form new minerals (Turner and Verhoogen, 1951, p.370). Cf: *cataclastic metamorphism*. Syn: *mechanical metamorphism*.

kingdom (a) The highest ranking unit in the classification of living and fossil organisms. "Phylum" or "division" is the next lower rank. (b) Any one of the three major divisions into which all natural objects are commonly classified, viz. animal kingdom, plant kingdom, mineral kingdom.

kingeniform Said of the loop of an adult dallinid brachiopod (as in the subfamily Kingeninae) in which the tendency to retain the campagiform hood during loop development leads to a broad sheet-like transverse band with connecting bands leading to the septum "in addition to normal ones joining descending branches with septum" (TIP, 1965, pt.H, p.147).

kingite A white mineral: $Al_3(PO_4)_2(OH,F)_3.9H_2O$.

kingle A Scottish term for a very hard rock, esp. a siliceous or calcareous sandstone occurring in oil shales but destitute of bituminous matter; e.g. *kennel*.

kink band A type of *deformation band* occurring microscopically in crystals and megascopically in foliated rocks, in which the orientation of the lattice or of the foliation is changed or deflected by gliding or slippage and shortening along slippage planes. Kink bands are associated with shock-wave action as well as with normal metamorphism. See also: *kink plane*. Cf: *conjugate fold system*. Syn: *knick band; knick zone; joint drag*.

kink fold A *chevron fold* formed by kink bands.

kinoite A mineral: $Ca_2Cu_2Si_3O_{10}.2H_2O$.

kinradite A name used in California and Oregon for jasper containing spherical inclusions of colorless or nearly colorless quartz.

kinzigite A coarse-grained metamorphic rock of pelitic composition occurring in the granulite facies. Essential minerals are garnet and biotite with varying amounts of quartz, K-feldspar, oligoclase, muscovite, cordierite, and sillimanite. The term was originated by Fischer in 1860 who named it after Kinzig, Schwarwald, Germany.

kipuka A syn. of *steptoe*, used in Hawaii.

kirovite A mineral: $(Fe,Mg)SO_4.7H_2O$. It is a variety of melanterite containing magnesium.

kirschsteinite A mineral: $Ca(Fe,Mg)SiO_4$. It is isomorphous with monticellite. Syn: *iron-monticellite*.

kitkaite A mineral: $NiTeSe$.

kivite A dark-colored leucite-bearing basanite. Its name is derived from Lake Kivu in east-central Africa.

kjelsasite A syenodiorite similar to larvikite but with more calcium oxide and fewer alkalies.

kladnoite A mineral: $C_6H_4(CO)_2NH$ (phthalimide). It occurs as monoclinic crystals formed in burning waste heaps in the Kladno coal basin of Bohemia.

klapperstein A *rattle stone* that results from the weathering of a box-stone. Etymol: German *Klapperstein*.

klebelsbergite A mineral consisting of a basic antimony sulfate (?) and occurring in the interstices between crystals in columnar aggregates of stibnite.

kleinite A yellow to orange mineral: $Hg_2N(Cl,SO_4).nH_2O$. Cf: *mosesite*.

Klein solution A solution of cadmium borotungstate that is used as a *heavy liquid*; its specific gravity is 3.6. Cf: *bromoform; Clerici solution; Sonstadt solution; methylene iodide*.

kliachite Var. of *cliachite*, the amorphous material constituting most bauxites.

klimakotopedion A term introduced by Schwarz (1912, p.95) as a syn. of *stepped plain*. Etymol: translation into Greek of "stepped plain".

klingstein An obsolete syn. of *phonolite*.

klinker bed Var. of *clinker bed*.

klint [coast] A term used in Denmark and Sweden for a vertical mountain wall or abrasion precipice, several meters high and 100 m or more long; esp. a steep cliff along the shore of the Baltic Sea. Pl: *klintar*. See also: *glint*.

klint [reef] An exhumed bioherm or coral reef, its softer surrounding rocks having been eroded, leaving the more resistant reef core standing in relief as a prominent knob, ridge, or hill. Pl: *klintar*. Not to be confused with *clint*.

klintite The rock composing a klint (Pettijohn, 1957, p.397); e.g. a loosely knit and cavernous or reticulating network of dense, hard, tough dolomite, which because of its rigid framework gives, to the massive reef core of an exhumed bioherm, its strength and resistance to denudation.

klippe An isolated rock unit that is an erosional remnant or outlier of a *nappe*. The original sense of the term, however, was merely descriptive, i.e. included any isolated rock mass such as an erosional remnant. Plural: *klippen*. Etymol: German, "rock protruding from a sea or lake floor".

klizoglyph *desiccation crack*.

klockmannite A reddish-violet to slate-gray mineral: CuSe. It tarnishes blue-black and is found in granular aggregates.

kloof A term used in South Africa for a deep, narrow, rugged gorge, ravine, glen, or other short and steep-sided valley, and also for a mountain pass. In some place names, the term may refer to a wide, open valley. Etymol: Afrikaans.

knap (a) A crest or summit of a hill. (b) A small hill or slight rise of ground.

kneaded (a) Said of a vague sedimentary structure resembling kneaded dough, such as a variety of flow roll or ball-and-pillow structure, or a structure formed by intrastratal slippage. (b) Said of a sediment or sedimentary particles transported by mudflows; e.g. "kneaded gravel".

knebelite A mineral: $(Fe,Mn)_2SiO_4$. It is a manganiferous fayalite.

knee fold A *zigzag fold* occurring in gravity-collapse structures.

knick (a) A *knickpoint*; esp. the place of junction where a gently inclined pediment and the adjacent mountain slope meet at a sharp angle. Syn: *knickpunkte*. (b) *nick*.

knick band *kink band*.

knickline A line formed by the angle of a *knick* in a slope, esp. in a desert region where there is an abrupt transition from a pediment surface to the mountain slope.

knickpoint Any interruption or break of slope; esp. a point of abrupt change or inflection in the longitudinal profile of a stream or of its valley, occurring where a new curve of erosion (graded to a new base level after a relative lowering of the former level) intersects an earlier curve, and resulting from rejuvenation, glacial erosion, or the outcropping of a resistant bed. Etymol: German *Knickpunkt*. Also spelled: *knick point*. Syn: *knick; nick; nickpoint; knickpunkt; break; rejuvenation head; rock step*.

knickpunkt (a) A *knickpoint* in a stream profile, esp. one resulting from rejuvenation or from an uplift. (b) The sharp angle made by the *haldenhang* and the *steilwand*; a *knick*.---Pl. *knickpunkte*. Etymol: German Knickpunkt, "bend point".

knick zone *kink band*.

knife edge (a) A narrow ridge of rock or sand. (b) *feather edge*.

knipovichite A mineral consisting of a hydrous carbonate of calcium, aluminum, and chromium.

knitted A texture characteristic of serpentinite, involving the interlacing of scaly crystals parallel to the original mineral's cleavage.

knitted texture A texture that is typical for the mineral serpentine in a rock when it replaces a clinopyroxene. Cf: *lattice texture*.

knob (a) A rounded eminence, as a knoll, hillock, or small hill or mountain; esp. a prominent or isolated hill with steep sides, commonly found in the southern U.S. See also: *knobs*. (b) A peak or other projection from the top of a hill or mountain. Also, a boulder or group of boulders or an area of resistant rocks protruding from the side of a hill or mountain.

knob-and-basin topography *knob-and-kettle topography*.

knob-and-kettle topography An undulating morainal landscape in which a disordered assemblage of knolls, mounds, or ridges of glacial drift is interspersed with irregular depressions, pits, or kettles that are commonly undrained and may contain swamps or ponds. Syn: *knob-and-basin topography; kame-and-kettle topography*.

knob and trail A structure, found in glaciated areas, that is made up of a protruding mass of resistant rock (the "knob") and a ridge of softer rock (the "trail") extending from the lee

side of the knob (Chamberlin, 1888, p.244-245). Cf: *crag and tail*.

knobs An area marked by a group of rounded, isolated hills ("knobs").

knock A hill in the English Lake District or in Scotland. Etymol: Gaelic.

knoll [geomorph] (a) A small, low, rounded hill; a hillock or mound. (b) The rounded top of a hill or mountain.---Syn: *knowe; knowle*.

knoll [marine geol] A mound-like relief form of the sea floor, less than 1000 m in height. Syn: *seaknoll*.

knoll [reef] A *reef knoll*. See also: *knoll reef*.

knoll reef (a) A reef from which a *reef knoll* has been exhumed or exposed by later denudation. (b) A term used by Tiddeman (1890) interchangeably with *reef knoll* in the sense of an original knoll-like feature of a reef. Also, a reef in the form of a knoll or a chain of knolls.

knopite A variety of perovskite containing cerium.

knot [geomorph] (a) A term used in the English Lake District for a hill of moderate height; esp. one having a bare-rock surface. (b) An elevated land area formed by the meeting of two or more mountainous regions; e.g. the structural-junction area of ridges of folded mountains.

knot [mining] A miner's term for small concretions, e.g. galena in sandstone, or for segregations of darker minerals in granite and gneiss.

Knotenschiefer A type of *spotted slate* characterized by conspicuous subspherical or polyhedral clots which are often individual minerals (Holmes, 1928). Etymol: German. Cf: *Garbenschiefer; Fruchtschiefer; Fleckschiefer*.

knotted With reference to metamorphic rocks, a syn. of *maculose*.

knotted-hornfels facies Metamorphic rocks formed in the lowest grades of thermal (contact) metamorphism at temperatures between 200° and 350°C and at pressures not exceeding 2500 bars (Hietanen, 1967). Syn: *albite-epidote-hornfels facies*.

knotted schist *spotted slate*.

knotted slate *spotted slate*.

knowe A Scottish syn. of *knoll*. Syn: *know*.

knoxvillite *copiapite*.

Knudsen formula A formula that expresses the relationship between salinity and chlorinity of sea water: $S‰ = 0.030 + 1.8050\ Cl‰$; in which S = salinity and Cl = chlorinity. See also: *Knudsen's tables*.

Knudsen's tables Hydrographic tables to aid in the computation of salinity, density, and sigma-t from chlorinity trirations and hydrometer readings, based on the *Knudsen formula*.

kobeite A black mineral: $(Y,U)(Ti,Nb)_2(O,OH)_6$. Also spelled: *kobeite*.

kobellite A blackish-gray mineral: $Pb_2(Bi,Sb)_2S_5$.

kodurite A coarsely crystalline rock of doubtful origin consisting of potassium feldspar, garnet (spessartite, andradite) and apatite.

koechlinite A greenish-yellow orthorhombic mineral: Bi_2MoO_6.

koenenite A very soft mineral: $Na_4Mg_9Al_4Cl_{12}(OH)_{22}$.

Koenigsbergen ratio The ratio of the remanent magnetization to magnetization induced by the Earth's field. Its symbol is Q.

koenlinite *könlite*.

koettigite A carmine mineral: $Zn_3(AsO_4)_2.8H_2O$. Syn: *köttigite*.

köfelsite A frothy, pumiceous high-silica glass occurring as small veins in fractured gneisses in the Köfels structure, Austria and apparently formed by vesiculation of an impact melt. The material is extremely heterogeneous, ranging in color from white to dark brown, and contains shock-metamorphosed mineral fragments and shock-melted glasses which provide definite evidence of origin by meteorite impact.

köflachite A dark-brown variety of retinite found in brown coal

Köflach in Styria, Austria.

kohalaite An andesite that contains normative oligoclase and may or may not contain modal olivine. Its name is derived from the Kohala Mountains, Hawaii.

koktaite A mineral: $(NH_4)_2Ca(SO_4)_2.H_2O$.

kolbeckine *herzenbergite*.

kolbeckite A blue to gray mineral: $ScPO_4.2H_2O$. It was formerly described as a hydrous phosphate and silicate of aluminum, beryllium, and calcium. Syn: *sterrettite*.

kolk A deep, isolated hole or depression, scoured out by eddying water in soft rock. Etymol: German *Kolk*, "deep pool, eddy, scour". Cf: *colk*.

kollanite A term, now obsolete, proposed by Pinkerton (1811, v.2, p.98) to distinguish the English puddingstones from other conglomerates.

kolm Nodular or concretionary bodies of coal found in the Paleozoic alum shales of Sweden and that contain rare metals, especially uranium. Syn: *culm*.

kolskite A mineral: $Mg_5Si_4O_{13}.4H_2O$ (?). It may be antigorite.

kona A term used in Hawaii for the leeward side, i.e. one away from the trade winds (Stamp, 1966, p.282).

kongsbergite A silver-rich variety of native amalgam, containing about 95% silver and 5% mercury.

koninckite A yellow mineral: $FePO_4.3H_2O$ (?).

konite A term, now obsolete, introduced by Pinkerton (1811, v.1, p.429) for a *freestone* consisting of limestone. Etymol: Greek *konia*, a term used by Theophrastus for "lime".

konlite A brown to yellow mineral consisting of a hydrocarbon found in brown coal and having an approximate composition of 91.75% carbon, 7.50% hydrogen, and 0.75% oxygen. Syn: *könleinite; koenlinite*.

kop A mountain or large hill that stands out prominently. Etymol: Dutch, "head".

kopje A var. of *koppie*. Etymol: Dutch, "small head".

Köppen's classification of climate A *climate classification*, formulated by W. Köppen in 1918, that is based on the climatic requirements of certain types of vegetation. Cf: *Thornthwaite's classification of climate*.

koppie A small but prominent hill occurring on the veld of South Africa, sometimes reaching 30 m above the surrounding land; esp. an isolated, elongate, scrub-covered hillock or knob composed of igneous rock and representing an erosion remnant, such as a small inselberg. See also: *castle koppie*. Etymol: Afrikaans, from Dutch *kopje*, "small head". Syn: *kopje*.

koppite A variety of pyrochlore containing iron, potassium, and cerium and lacking titanium.

koris A term used in northern Africa for a dry valley.

korite A variety of palagonite (Hey, 1962, p.485).

kornelite A colorless to brown mineral: $Fe_2(SO_4)_3.7H_2O$.

kornerupine A colorless, yellow, brown, or sea-green mineral: $Mg_3Al_6(Si,B,Al)_5O_{21}(OH)$. It resembles sillimanite in appearance.

korzhinskite A mineral: $CaB_2O_4.H_2O$.

kosmochlor *ureyite*.

kostovite A mineral: $CuAuTe_4$.

koswite A peridotite composed of olivine, diallage, and hornblende in a groundmass of magnetite; a magnetite peridotite.

kotoite An orthorhombic mineral: $Mg_3(BO_3)_2$. It is isostructural with jimboite.

kotschubeite A rose-red variety of clinochlore containing chromium.

köttigite *koettigite*.

kotulskite A mineral: $Pd(Te,Bi)_{1-2}$.

koum A *sandy desert* or continuous tract of sand dunes in central Asia, equivalent to an *erg*. Etymol: French. See also: *kum*.

koutekite A hexagonal mineral: Cu_2As.

kozulite A manganese-rich mineral of the amphibole group:

$(Na,K)_3(Mn,Mg,Fe)_5Si_8O_{22}(OH,F)_2$.

krablite A rhyolite containing plagioclase grains enclosed in orthoclase phenocrysts, along with smaller amounts of augite and quartz. The rock was originally identified as the mineral feldspar. It occurs as ejecta from Mount Krafla, Iceland. Syn: *baulite; kraflite*.

kraflite *krablite*.

kragerite *kragerôite*.

kragerôite A rutile-bearing albite aplite, with minor amounts of quartz, potassium feldspar, and ilmenite. Its name is derived from Kragerö, Norway. Syn: *kragerite*.

Krakatoan caldera A type of caldera formed in the summit region of a volcano following evacuation of the underlying magma chamber by a voluminous outpouring of pyroclastic ejecta, normally of silicic composition.

krans *krantz*.

krantz A term used in southern Africa for a precipitous rock face or sheer cliff. Etymol: Afrikaans *krans*, "wreath". Pl: *krantzes*. Syn: *krans*.

krantzite A variety of retinite found in small yellowish grains disseminated in brown coal.

Krasnozem A Russian term for a zonal red soil developed in a Mediterranean climate.

kratochvilite A hydrocarbon mineral: $C_{13}H_{10}$ (fluorene).

kratogen An early variation of *craton*.

kraton An early variation of *craton*.

kraurite *dufrenite*.

krausite A yellowish-green mineral: $KFe(SO_4)_2.H_2O$.

krauskopfite A mineral: $BaSi_2O_5.3H_2O$.

KREEP An acronym for a basaltic lunar rock type first found in Apollo 12 fines and breccias and characterized by unusually high contents of potassium (K), rare-earth elements (REE), phosphorous (P), and other trace elements in comparison to other lunar rock types. The material, which is found in a variety of crystalline and glassy (shock-melted?) rock types, is distinctly different from the iron-rich mare basalts. The term *nonmare basalt* is equivalent.

kreittonite A black variety of gahnite containing ferrous iron or ferric iron, or both.

kremastic water A syn. of *vadose water* proposed by Meinzer (1939, p.676), including *rhizic water*, *argic water*, and *anastatic water*.

kremersite A ruby-red mineral: $[(NH_4),K]_2FeCl_5.H_2O$. It is a variety of erythrosiderite containing ammonium.

krennerite A silver-white to pale-yellow mineral: $AuTe_2$. It often contains silver. Syn: *white tellurium*.

kribergite A white, chalk-like mineral: $Al_5(PO_4)_3(SO_4)-(OH)_4.2H_2O$ (?).

krinovite A meteorite mineral: $NaMg_2CrSi_3O_{10}$.

krohnkite An azure-blue monoclinic mineral: $Na_2Cu-(SO_4)_2.2H_2O$. Also spelled: *kroehnkite*.

krokidolite *crocidolite*.

krotovina An irregular tubular or tunnel-like structure in soil, made by a burrowing animal and subsequently filled in; a worm cast in soil. Etymol: Russian.

kryokonite Var. of *cryoconite* [glaciol].

kryomer A relatively cold period within the Pleistocene Epoch, such as a *glacial stage* (Lüttig, 1965, p.582). Ant: *thermomer*.

kryoturbation Var. of *cryoturbation*.

kryptogene adj. Said of a rock whose origin cannot be determined.

kryptomere A rock of such fine grain that its individual components cannot be distinguished megascopically. Adj: *kryptomerous*. Syn: *aphanitic; cryptomere*.

krystic Pertaining to ice, in all of its forms, as a surface feature of the Earth. Term is rarely used.

kryzhanovskite A mineral: $MnFe_2(PO_4)_2(OH)_2.H_2O$. Syn: *kruzhanovskite*.

K section One of the two circular cross sections through a strain ellipsoid. Syn: *circular section.*

ksimoglyph A hieroglyph produced by a solid object that is dragged or propelled by a continuous current over a sedimentary surface (Vassoevich, 1953, p.61 & 72); e.g. a drag mark or a groove cast.

ktenasite A blue-green mineral: $(Cu,Zn)_3(SO_4)(OH)_4.2H_2O$.

ktypéite A mineral substance intermediate between calcite and aragonite.

kukersite An organic sediment rich in alga *Gloexapsamorpha prisca* remains found in the Ordovician of Estonia.

kulaite An extrusive rock containing both orthoclase and calcic plagioclase, a small amount of nepheline and olivine, and hornblende as the dominant mafic mineral.

kullaite A porphyritic diorite containing altered plagioclase and sodic orthoclase phenocrysts in an ophitic groundmass of plagioclase, sodic orthoclase, and pseudomorphs of chlorite after augite, with minor amounts of quartz, apatite, and opaque oxides.

Kullenberg corer A type of *piston corer* that has many modifications or varieties.

kullerudite A mineral: $NiSe_2$.

kum A Turkish term for "sand", applied to the sandy deserts of central Asia; e.g. Kizil Kum. See also: *koum.*

Kummerian Floral stage in Washington State: lower Oligocene.

Kungurian European stage: Lower Permian or Middle Permian of some authors (above Artinskian, below Kazanian).

kunkar (a) Var. of *kankar.* (b) A term used in Australia for a calcareous duricrust, or caliche.

kunkur Var. of *kankar.*

kunzite A pinkish, light-violet, or lilac-colored transparent gem variety of *spodumene.*

kupfernickel *nickeline.*

kupletskite A mineral: $(K,Na)_3(Mn,Fe)_7(Ti,Nb)_2Si_8O_{24}(O,OH)_7$.

kurchatovite A mineral: $Ca(Mg,Mn)B_2O_4$.

kurgantaite A mineral: $(Sr,Ca)_2B_4O_8.H_2O$ (?).

kurnakovite A white mineral: $Mg_2B_6O_{11}.15H_2O$.

kurskite An alkali-bearing *francolite.*

kurtosis (a) The quality, state, or condition of peakedness or flatness of the graphic representation of a statistical distribution. (b) A measure of the peakedness of a frequency distribution; e.g. a measure of concentration of sediment particles about the median diameter: $(Q_3-Q_1)/2(P_{90}-P_{10})$, where Q_3 and Q_1 are the particle diameters, respectively, at the 75% and 25% intersections on the cumulative frequency distribution, P_{90} is the particle diameter such that 90% of the particles are larger and 10% smaller, and P_{10} is the particle diameter such that 10% of the particles are larger and 90% smaller. Various approximations or coefficients of kurtosis have been devised in an attempt to assign genetic significance to sediment distributions. Abbrev: K.---Cf: *skewness.*

kurumsakite A mineral: $(Zn,Ni,Cu)_8Al_8V_2Si_5O_{35}.27H_2O$ (?).

kuskite A light-colored hypabyssal rock originally thought t contain scapolite, quartz, and some decomposed plagioclas phenocrysts in a fine-grained groundmass of quartz, orth clase, and muscovite, but later the scapolite was identified quartz and the name was withdrawn (Johannsen, 193 p.261). Its name is derived from the Kuskokwim River, Ala ka.

kutinaite A mineral: Cu_2AgAs.

kutnahorite A mineral: $Ca(Mn,Mg,Fe)(CO_3)_2$. It is isomo phous with dolomite. Syn: *kutnohorite.*

Kutter's formula A formula that expresses the value of th Chézy coefficient in the *Chézy equation* in terms of the frictio slope, hydraulic radius, and a roughness coefficient (Brown Runner, 1939, p.199).

kvellite A very dark-colored, ultramafic hypabyssal rock cor taining lepidomelane, olivine, barkevikite, apatite, ilmenit and magnetite phenocrysts in a groundmass of lath-shape anorthoclase.

K wave A longitudinal or *P* wave in the Earth's outer core. C *l wave.*

kyanite A blue or light-green triclinic mineral: Al_2SiO_5. It is tr morphous with andalusite and sillimanite. Kyanite occurs long, thin, bladed crystals and crystalline aggregates schists, gneisses, and granite pegmatites, and has a hardnes of 4-5 along the length of the crystal and 6-7 across it. forms at medium temperatures and high pressures in a reg ionally metamorphosed sequence. Syn: *cyanite; sappare disthene.*

kyanophilite A mineral: $(K,Na)Al_2Si_2O_7(OH)$ (?).

kyle A Scottish term for a narrow channel, sound, or stra between two islands or an island and the mainland, or for narrow inlet into the coast. Etymol: Gaelic.

kylite An olivine-rich theralite.

kymoclastic rock A hydroclastic rock containing marine c other wave-formed fragments (Grabau, 1924, p.295).

kyphorhabd A curved monaxonic sponge spicule wit transverse swellings on the convex side.

kyr A term used in central Asia for flat land, a plateau, or th top of a small hill or mountain, and indiscriminately for a lo hill or small mountain, but applied specif. in Turkmenia t stony, hard ground as contrasted to *adyr* (Murzaevs & Mur zaevs, 1959, p.131).

kyriosome A term used by Niggli (1954, p.191) for the funda mental mass or framework fraction of a complex rock; th major part of a migmatite. Cf: *akyrosome.*

kyrtome A triradiate, more or less thickened area associate with and bordering the laesura of a trilete spore. Cf: *torus.*

kyschtmite A medium- to fine-grained hypabyssal rock com posed of euhedral corundum crystals and some biotite in groundmass of calcic plagioclase. Cf: *luscladite.*

L

laagte A broad, almost level drainage course in the veld of southwestern Africa, less well-defined than a valley, and dry for most of the year except after a rain. Etymol: Afrikaans. Pron: loch-ta. Syn: leegte.

lanilite A coarse-grained pegmatoid composed chiefly of garnet, biotite, quartz, and iron ores.

lavenite lavenite.

labial aperture An accessory aperture formed in a foraminiferal test by the free parts of the apertural lip, not leading directly to a chamber.

labial pore A median or submedian pore in the proximal wall of the peristome in some cheilostomatous bryozoans and resulting from closure of a sinus.

labiate Having lips; e.g. said of an exaggerated marginate foramen of a brachiopod in which the dorsal edge is prolonged lip-like (TIP, 1965, pt.H, p.147).

labiatiform cyrtolith A cyrtolith coccolith with a central structure shaped like a double lip (as in Anthosphaera robusta).

labile [geol] (a) Said of rocks and minerals that are mechanically or chemically unstable; e.g. a "labile sandstone" or "labile graywacke" containing abundant unstable fragments of rocks and minerals and less than 75% matrix of fine silt and clay (Packham, 1954), or "labile constituents" (such as feldspar and rock fragments in a sandstone) that are easily decomposed. Cf: unstable [sed]. (b) Said of protobitumen that represents easily decomposable plant and animal products (such as fat, oil, or protein) in peat and sapropel. Ant: stabile.

labilizing force In an unstable gravimeter, a force acting in the same direction as the force being measured and, therefore, opposite to the direction of the ordinary restoring force.

labite A mineral: $MgSi_3O_6(OH)_2.H_2O$. It may be chrysotile.

labium (a) The lower lip of an arthropod (such as of an insect); the metastoma of a crustacean. Cf: labrum. (b) The columellar part of the aperture of a gastropod shell.---Pl: labia.

labor A Spanish term used in early land surveys in Texas for unit of area equal to about 177.14 acres (representing a tract 1000 varas square). Pron: la-bore.

labradite labradoritite.

labradophyre An anorthosite composed of labradorite phenocrysts in a groundmass of the same mineral.

labradorescence An optical phenomenon consisting of flashes of a laminated iridescence of a single bright hue that changes gradually as a mineral or gemstone is moved about in reflected light, caused by internal structures that selectively reflect only certain colors; specif. the light-interference effect exhibited by labradorite and set up in thin plates of feldspar (produced by repeated twinning), resulting in a series of vivid colors (usually brilliant blue or green) spread over large areas. Syn: change of color.

labradorfels labradoritite.

labradorite [mineral] A dark (gray, blue, green, or brown) mineral of the plagioclase feldspar group with composition ranging from $Ab_{50}An_{50}$ to $Ab_{30}An_{70}$. It commonly shows a rich, beautiful play of vivid colors (commonly brilliant blue or green), and is therefore much used for ornamental purposes. Labradorite is common in igneous rocks of intermediate to low silica content. Pron: lab-ra-daw-rite. Syn: Labrador spar.

labradorite [ign] A name applied by French petrologists to light-colored labradorite-rich basalt and by Soviet petrologists to a light-colored gabbro or norite (i.e. anorthosite).

labradoritite A rock composed almost entirely of labradorite. Syn: labradorfels; labradite.

Labrador spar A syn. of labradorite [mineral]. Also called: Labrador stone; Labrador rock.

labrum (a) An unpaired outgrowth of an arthropod, consisting of a single median piece or flap immediately in front of or above the mandibles and more or less covering the mouth; e.g. the upper lip in front of the mouth of a crustacean or of a merostome, or the hypostome of a trilobite. Cf: labium. (b) The external margin of a gastropod shell. (c) A more or less enlarged and modified lip-like primordial plate of an echinoid, bordering the peristome in interambulacrum 5.---Adj: labral.

labuntsovite A mineral: $(K,Ba,Na)(Ti,Nb)(Si,Al)_2(O,OH)_7.H_2O$. It was originally described as titaniferous elpidite but is now shown to contain only a trace of ZrO_2.

labyrinthic Said of some agglutinated foraminifers having a complex spongy wall with interlaced dendritic channels perpendicular to the surface.

lac [soil] A term used by Brewer (1964, p.366) for soil plasma that has a high luster and smooth surface in reflected light under a hand lens up to magnifications of approximately 20 times. Cf: floc.

lac [lake] The French term for lake; it appears in proper names in parts of the U.S. where the influence of early French settlement remains, and in French-speaking regions of Canada.

LaCasitan North American (Gulf Coast) stage: Upper Jurassic (above Zuloagan, below Durangoan of Cretaceous; it is equivalent to European Portlandian and Kimmeridgian) (Murray, 1961).

laccolite The original, now superceded, term for a laccolith.

laccolith A concordant igneous intrusion with a known or assumed flat floor and a postulated dikelike feeder somewhere beneath its thickest point. It is generally lenslike in form and roughly circular in plan, less than five miles in diameter, and from a few feet to several hundred feet in thickness. See also: bysmalith. Syn: laccolite; cistern rock.

lacine One of a series of detached, tongue-shaped, ridge-like meander scrolls, frequently found spread apart like the rays of a fan. Lacines are not as long, as smoothly curved, or as closely spaced as the more symmetric meander scrolls. Term introduced by Melton (1936, p.599). Etymol: Latin lacinia, "flap, tongue".

lacine meander A detached scroll meander in which lateral erosion of the outer bank is somewhat retarded, thereby producing a low-water channel of unequal width (Melton, 1936, p.599-600). Examples occur in the Mississippi River between Cairo, Ill., and Baton Rouge, La.

lacinia The inner distal spiny lobe of the second segment of maxillule of a crustacean, adjacent to the galea (TIP, 1969, pt.R, p.97).

lacinia mobilis A small, generally toothed process articulated with the incisor process of the mandible of a malacostracan crustacean.

laciniate Said of a leaf having narrow, pointed lobes, as if slashed.

LaCoste-Romberg gravimeter A long-period vertical seismograph suspended system adapted to the measurement of gravity differences. Sensitivity is achieved by adjusting the system to proximity to an instability configuration. Syn: zero-length spring gravimeter.

lacroixite A pale yellowish-green mineral, approximately: $Na(Ca,Mn)AlPO_4(OH)_3$. It often contains considerable fluorine.

lacullan anthraconite.

lacuna [paleont] (a) A perforation or true pore between bryozoan costae making up the frontal shield in cribrimorph cheilostomes. (b) A space, in the lorica of a tintinnid, lacking reticulation or other surface marking. (c) A lateral hole or gap in the hood of dallinid brachiopods, produced by resorption during the frenuliniform growth stage in loop development.---Pl: lacunae or lacunas.

lacuna [palyn] A rarely used term for a depressed space, pit, or hole on the outer surface of a pollen grain.

lacuna [stratig] A time-stratigraphic unit representing a gap in the stratigraphic record; specif. the missing interval at an unconformity, representing the interpreted space-time value of both *hiatus* (period of nondeposition) and *degradation vacuity* (period of erosion) (Wheeler, 1964, p.599).

lacuster The central part of a lake (Veatch & Humphrys, 1966, p.162).

lacustrine (a) Pertaining to, produced by, or formed in a lake or lakes; e.g. "lacustrine sands" deposited on the bottom of a lake, or a "lacustrine terrace" formed along the margin of a lake. (b) Growing in or inhabiting lakes; e.g. "lacustrine fauna". (c) Said of a region characterized by lakes; e.g. a "lacustrine desert" containing the remnants of numerous Pleistocene lakes that are now dry.---Cf: *limnic.* Syn: *lacustral; lacustrian.*

lacy residue An insoluble residue containing irregular openings and having constituent material comprising less than 25% of the volume (Ireland et al, 1947, p.1482). Cf: *skeletal residue.*

ladder lodes *ladder veins.*

ladder reefs *ladder veins.*

ladder veins Mineral deposits in transverse, roughly parallel fractures that have formed along foliation planes perpendicular to the walls of a dike during its cooling. They may also be formed along shrinkage joints in basaltic rocks or dikes. Syn: *ladder lodes; ladder reefs.*

lade (a) The mouth of a river. (b) A watercourse.

Ladinian European stage: upper Middle Triassic (above Anisian, below Carnian).

laesura The line or scar on the proximal face of an embryophytic spore, marking the contact with other members of the tetrad. It may be trilete or monolete. Pl: *laesurae.* See also: *Y-mark; suture [palyn].* Syn: *tetrad scar.*

laevigate A syn. of *psilate.* The term is more often applied to spores than to pollen.

lag (a) *lag gravel.* (b) *sedimentary lag.*

lag deposit *lag gravel.*

lag fault An overthrust, the thrusted rocks of which move differentially so that the upper part of the geologic section is left behind; the replacement of the upper limb of an overturned anticline by a *fold fault.* Syn: *tectonic gap.*

lagg The depressed marginal areas, often moatlike, i.e. covered with water, surrounding some convex-shaped bogs. Etymol: Swedish. Syn: *bog moat.*

lag gravel (a) A residual accumulation of coarse, usually very hard rock fragments remaining on a surface after the finer material has been blown away by winds. See also:*desert pavement.* (b) Coarse-grained material that is rolled or dragged along the bottom of a stream at a slower rate than the finer material, or that is left behind after currents have winnowed or washed away the finer material.---Syn: *lag; lag deposit.*

lag mound A term used by Packer (1965) for a residual remnant of thin, unconsolidated surface materials on a limestone pavement during the early stages of its development. It is only an apparent mound, as depressions develop around it.

lagoon [geog] A closed depression in a high, grass-covered tableland of the cordilleras of the western U.S.

lagoon [coast] (a) A shallow stretch of seawater, such as a sound, channel, bay, or salt-water lake, near or communicating with the sea and partly or completely separated from it by a low, narrow, elongate strip of land, such as a reef, barrier island, sandbank, or spit; esp. the sheet of water between an offshore coral reef and the mainland. It often extends roughly parallel to the coast, and it may be stagnant. (b) A shallow freshwater pond or lake near or communicating with a larger lake or a river; a stretch of freshwater cut off from a lake by a barrier, as in a depression behind a shore dune; a barrier lake. (c) A shallow body of water enclosed or nearly enclosed

within an atoll. (d) The term has been widely applied to other coastal features, such as an estuary, a slough, a bayou, marsh, and a shallow pond or lake into which the sea flows.- Etymol: Latin *lacuna,* "pit, pool, pond". Syn: *lagune; laguna.*

lagoon [water res] Any shallow artificial pond or other water filled excavation for the natural oxidation of sewage or disposal of farm manure, or for some decorative or aesthetic purpose.

lagoon [water] (a) The basin of a hot spring; also, the pool formed by a hot spring in such a basin. Etymol: Italian *lagone* "large lake". (b) A perennial brine pool near the margin of a alkaline lake; e.g. near Lake Magadi in southern Kenya.

lagoonal Pertaining to a lagoon, esp. *lagoonal* deposition.

lagoon atoll *pseudoatoll.*

lagoon beach The sandy fringe on the inner, protected side o a reef island, facing toward the lagoon.

lagoon channel (a) The stretch of deep water separating reef from the neighboring land (mainland or island). (b) *pass* through a reef, and into and through a lagoon.

lagoon cliff *lagoon slope.*

lagoon cycle The sequence of events, and the time required involved in the filling of a lagoon with sediments followed b eventual erosion by wave action and refilling by deposition.

lagoon flat The nearly horizontal reef flat located lagoonwar of the lagoon beach.

lagoon floor The undulating to nearly level bottom of a lagoon often encircled by the *lagoon slope.*

lagoon island (a) One of many scattered islets rising from within the lagoon of a composite atoll or a large barrier reef generally marking former fringing reefs that grew up with th postglacial eustatic rise. (b) *atoll.*

lagoonlet A small lagoon; esp. a shallow pool of water on platform reef. Syn: *moat; miniature lagoon.*

lagoon margin The lagoonward margin of the *lagoon shelf* o of the reef flat along those parts of the reef that lack islands.

lagoon phase The strata or stratigraphic facies formed by ac cumulation of sediment in a shallow coastal area that is separated and isolated from the open sea by a barrier.

lagoon plain A flat landform produced by the filling of lagoon with sediments.

lagoon scarp *lagoon slope.*

lagoon shelf The part of a reef that borders the lagoonside o a reef island; the sand-covered, lagoonward-sloping shel commonly found where sedimentation conspicuously exceed organic growth.

lagoonside The land bordering on a lagoon.

lagoon slope The seaward border zone or side of a lagoor sloping downward from the *lagoon margin* to the *lagoon floor* It is usually an abrupt boundary to the interior of an atoll. Syr *lagoon cliff; lagoon scarp.*

Lagorio's rule An approximate rule according to which quart usually begins to crystallize early from highly siliceous rhy olites and porphyries and late from the less siliceous ones The rule was proposed in 1887 by A. Lagorio.

Lagrangian (a) Pertaining to a system of coordinates or equa tions of motion in which the properties of a fluid are identifie for all time by assigning them coordinates that do not var with time. (b) Said of a direct method of measuring the spee and/or direction of an ocean current by tracking the move ment of the same water mass through the ocean by means o tracers, drift bottles, buoys, deep drogues, current poles, an other devices.----Cf: *Eulerian.* Named in honor of Josep Louis Lagrange (1736-1813), French mathematician.

laguna [coast] A var. of *lagoon.* Etymol: Spanish and Italian.

laguna [lake] A term used in areas of Spanish influence, in cluding the southwestern U.S., for a lake or lagoon; esp. shallow ephemeral lake in the lower part of a bolson, fed b streams rising in the neighboring mountains and flowing onl as a result of rainstorms. Etymol: Spanish, "pond, small lake".

lagune (a) Var. of *lagoon*. Etymol: French. (b) A term used in the SW U.S. for a small lake.

lahar A *mudflow* [mass move] composed chiefly of volcaniclastic materials on the flank of a volcano. The debris carried in the flow includes pyroclastic material, blocks from primary lava flows, and epiclastic material. Etymol: Indonesian. Syn: *mudflow* [volc].

laitakarite A mineral: $Bi_4(Se,S)_3$. Cf: *ikunolite*.

lakarpite A syenite composed of orthoclase or microcline, calcic plagioclase, and an arfvedsonite-like amphibole, with accessory acmite, altered nepheline, rosenbuschite, and secondary natrolite.

lake [coast] A term loosely applied to a sheet of water lying along a coast and connected with the sea; e.g. one of the shallow (1-2 m deep), interconnected bodies of water in the Florida Bay area, Fla. See also: *seashore lake*.

lake [speleo] In a cave, any body of standing water that is too deep to traverse by walking.

lake [ice] A submariner's term for a *polynya* during the summer. Cf: *skylight*.

lake [lake] (a) Any inland body of standing water occupying a depression in the Earth's surface, generally of appreciable size (larger than a *pond*) and too deep to permit vegetation (excluding subaqueous vegetation) to take root completely across the expanse of water; the water may be fresh or saline. The term includes an expanded part of a river, a reservoir behind a dam, or a lake basin intermittently or formerly covered by water. (b) An inland area of open, relatively deep water whose surface dimensions are sufficiently large to sustain waves capable of producing somewhere on its periphery a barren wave-swept shore (Welch, 1952).---Syn: *lac; lago; loch; lough*.

lake [streams] An English term for a brook or small stream; also, a channel.

lake asphalt Soft asphalt, rich in bitumen, from the pitch lake of Trinidad. See also: *land asphalt*. Syn: *lake pitch*.

lake ball A spherical mass of tangled, waterlogged fibers and other filamentous material of living or dead vegetation (such as blue-green algae, moss, spruce needles, and fragments of peat, grass, or twigs), produced mechanically along a lake bottom by wave action, and usually impregnated with sand and fine-grained mineral fragments. It may vary in size up to that of a man's head. See also: *aegagropile; peat ball*. Cf: *sea ball*. Syn: *hair ball; burr ball*.

lake basin (a) The depression in the Earth's surface occupied or formerly occupied by a lake and containing its shore features. (b) The area from which a lake receives drainage.---See also: *basin* [lake].

lakebed (a) The flat to gently undulating ground underlain by fine-grained sediments deposited in a formerly existing lake. (b) The ground upon which a lake presently rests; the bottom of a lake; a lake basin.

lake biscuit *algal biscuit*.

lake delta A delta, usually arcuate with a steep front, built out by a river into a freshwater lake; e.g. the delta of the Rhône River in Lake Geneva.

lake deposit A sedimentary deposit laid down conformably on the floor of a lake, usually consisting of coarse material near the shore and sometimes passing rapidly into clay and limestone in deeper water; most of it is of fluvial or glacial origin mixed with freshwater or terrestrial organic matter. It may show a clearly marked seasonal layering, as in varved clays.

lake district A region marked by the grouping together of lakes.

lake gage A gage for measuring the elevation of the water surface of a lake.

lake George diamond Colorless, doubly terminated quartz crystal from Herkimer County, N.Y. See also: *Herkimer diamond*.

lake gun A term used on Seneca Lake, New York, for a phenomenon that is the apparent production of sounds like distant thunder or gunfire. The phenomenon is known to occur on several European lakes and has a variety of names (Hutchinson, 1957, p.361-362).

lake-head delta A delta built at the mouth of a river at the head of a lake. Syn: *lake delta*.

lake ice Ice formed on a lake, regardless of observed location; it is usually *freshwater ice*.

lake inlet A stream that flows into a lake.

lakelet A small lake.

lake loam A term applied to loess that may have been formed by deposition in lakes (Veatch & Humphrys, 1966, p.171).

lake marl *bog lime*.

lake marsh (a) A part of the bottom of a lake, supporting a dense growth of emergent aquatic plants. (b) A marsh that occupies the site of a former lake.----See also: *marsh lake*.

lake ocher Ocherous deposits formed on the bottom of a lake by bacteria capable of precipitating ferric hydroxide, or found in a marsh or swamp that was formerly the site of a lake.

lake ore (a) A flat, disk-like or irregular, concretionary mass of ferric hydroxide less than a meter thick, or a layer of soft-to-hard, porous, yellow bedded limonite, formed along the borders of certain lakes. Cf: *bog ore*. (b) *bog iron ore*.

lake outlet A stream that flows out of a lake.

lake peat *sedimentary peat*.

lake pitch *lake asphalt*.

lake plain (a) The nearly level surface marking the floor of an extinct lake, filled in by well-sorted deposits from inflowing streams. (b) A flat lowland or a former lake bed bordering an existing lake. See also: *lake terrace*.

lake rampart An irregular, conspicuous, wall-like ridge composed of unconsolidated coarse material along a lake shore, produced by shoreward movement of lake ice, as by winds, waves, or currents, and esp. by expansion of ice that caused a ridging of yielding lake-shore deposits, or that shoved, pushed, and then stranded bottom lake deposits as it overrode the shore. Its height may reach 2 m. Examples occur along the shores of the Great Lakes. See also: *walled lake*. Syn: *ice rampart; rampart; ice-push ridge; ice thrust ridge*.

lakescape The entire area of a lake or a part of a lake, including its water surface, islands, and shoreline features, that can be viewed from an observation point.

lakeshore The narrow strip of land in contact with or bordering a lake; esp. the beach of a lake. Also spelled: *lake shore*. Syn: *lakeside*.

lakeside *lakeshore*.

lake terrace A narrow shelf, partly cut and partly built, produced along a lake shore in front of a nip or line of low cliffs, and later exposed when the water level falls (Cotton, 1958, p.489). See also: *lake plain*.

lakmaite A dense, dark-colored (green to black) igneous rock containing small feldspar phenocrysts in a glassy groundmass (Thrush, 1968, p.622).

lallan Scottish var. of *lowland*. Syn: *lalland*.

Lamarckism A theory of evolution stating that changes in the environment cause structural changes in an organism especially by inducing new or increased use of organs or parts as a result of adaptive modification or greater development, and also cause disuse and eventual atrophy of other parts, and that these changes are passed on to offspring. This theory is named after the French naturalist J.B. de Monet Lamarck (1744-1829).

Lambert azimuthal equal-area projection An azimuthal map projection having the pole of the projection at the center of the area mapped, the azimuths of great circles radiating from this pole (center) and being truly represented on the map but the scale along such great-circle lines so varying with distance from the center that an equal-area projection is produced. The pole (center) of the projection may be at the pole of the sphere, on its equator, or at any point in between. The projection is useful for representing a single hemisphere or continental masses, but extreme distortion of areas is encountered near the map periphery. See also: *Schmidt projection.*

Lambert conformal conic projection A conformal conic map projection on which all meridians are represented by equally spaced straight lines that radiate from a common point outside the map limits and the parallels (of which one or two are standard parallels along which the scale is exact) are represented by circular arcs having this common point for a center and intersecting the meridians at right angles. The scale is the same in every direction at any point on the map, but increases north and south from the standard parallel(s); where there are two standard parallels, the scale is too small between them and too large beyond them. The projection is used for maps of middle latitudes (for maps of the contermious U.S., smallest distortion occurs when the standard parallels represent latitudes 33°N and 45°N) and as a base for sets of large-scale aeronautical charts produced by the U.S. Coast & Geodetic Survey. Named after Johann H. Lambert (1728-1777), German physicist, who introduced the projection in 1772. Syn: *Lambert conformal projection.*

Lambert's law The statement that the intensity of blackbody radiation emerging from an aperture is greatest in the direction perpendicular to the plane of the aperture, and decreases with the cosine of the angle between the perpendicular and the direction of observation. Such a reflective body is called a *perfectly diffuse reflector*; real bodies seldom approach this condition.

Lamb's problem An investigation in seismology that is concerned with disturbances in a semi-infinite, perfectly elastic medium that are initiated at a point or along a line on the surface of the medium.

Lamé constants Two *elastic constants* or parameters, λ and μ, which express the relationships between the components of stress and strain for linear elastic behavior of an isotropic solid; λ is identical with rigidity, and μ is equivalent to the bulk modulus minus $2\mu/3$.

lamella [geol] A thin plate, scale, flake, leaf, lamina, or layer; e.g. one of the units of a polysynthetically twinned mineral (such as plagioclase), or a thin plate compressed between the bands of narrow plates seen in cross section on the polished surface of an iron meteorite exhibiting Widmanstatten figures. Pl: *lamellae* or *lamellas.* See also: *deformation lamella; exsolution lamella.*

lamella [biol] An organ, process, or part of an organism resembling a plate; e.g. a primary lamella of a brachiopod, or a thin plate of the gill of a bivalve mollusk, or the inner lamella or the outer lamella of an ostracode, or a thin sheet or flap-like plate of tissue on the dorsal surface of the thallus of a bryophyte, or a continuous thin layer in the membranelle of a tintinnid. Pl: *lamellae.*

lamellar Composed of or arranged in lamellae; disposed in layers like the leaves of a book. Syn: *lamellate.*

lamellar columella A plate-like coral *columella.* In rugose corals, it is generally in the plane of the cardinal septum and the counter septum; in scleractinian corals, it is oriented parallel with the longer axis of the calice.

lamellar conodont element A *conodont element* consisting of numerous thin layers or sheaths, being most obvious in specimens that also contain opaque "white matter". See also: *fibrous conodont element.*

lamellar flow Flow of a liquid in which layers glide over one another. Cf: *laminar flow.*

lamellar layer The *primary layer* of a brachiopod.

lamellar ligament The part of a ligament of a bivalve mollusk characterized by lamellar structure and containing no calcium

carbonate. It is secreted at the edge of the mantle and is elastic to both compressional and tensional stresses. Cf: *fibrous ligament.*

lamellar linkage The joining, by lamellar septal plates, of corallite centers in scleractinian corals, corresponding to *direct linkage* of stomodaea.

lamellar pyrites *marcasite* [mineral].

lamellar wall A foraminiferal test constructed of thin plate-like layers of aragonite or calcite, one layer being formed with addition of each new chamber, and covering whole previously formed test.

lamellibranch *pelecypod.*

lamina [paleont] A thin plate-, scale-, flake-, layer-, or sheet-like structure in an organism; e.g. a uniform thin sheet of wall substance in the lorica of a tintinnid, or a coccolith with two large dimensions and a very small third dimension, or a sheet-like structure in some corals, formed by the juxtaposition of two layers of skeletal material in septa and the column, or a basal lamina and median lamina in a bryozoan.

lamina [bot] The blade or expanded portion of a leaf.

lamina [sed] The thinnest or smallest, recognizable unit layer of original deposition in a sediment or sedimentary rock, differing from other layers in color, composition, or particle size, and resulting from variations in the rate of supply or deposition of different material during a momentary or local fluctuation in the velocity of the depositing current; specif. such a sedimentary layer less than 1 cm in thickness (commonly 0.05-1.00 mm thick). As defined by Otto (1938, p.575), a lamina "may vary from microscopic size for clays to many inches for coarse gravel deposits". It may be parallel or oblique to the general stratification. Several laminae may constitute a *bed* (Payne, 1942, p.1724) or a *stratum* (McKee & Weir, 1953, p.382). Pl: *laminae.* See also: *phase.* Syn: *lamination; striacule.*

laminar Consisting of, arranged in, or resembling laminae; e.g. "laminar corrugation" representing small-scale intraformational folding, or "laminar structure" produced by alternation of very thin sedimentary layers of differing composition.

laminar flow [glaciol] A type of glacier flow in which the surface, bed, and flow vectors are all parallel; there is neither extending flow nor compressing flow.

laminar flow [hydraul] Water flow in which the stream lines remain distinct and in which the flow direction at every point remains unchanged with time. It is characteristic of the movement of ground water. Cf: *turbulent flow; mixed flow; lamellar flow.* Syn: *streamline flow; sheet flow.*

laminarian Pertaining to a large family of kelps, Laminariaceae, of the order Laminariales.

laminarite A straight and parallel structure on bedding planes, supposedly a fossil seaweed assigned to the "genus" *Laminarites* comprising very heterogeneous "species" related or similar to the kelps of the genus *Laminaria*, but seemingly in part a ripple, a flow cast, or a trail.

laminar velocity That velocity of water in a stream below which the flow is laminar and above which it may be either laminar or turbulent. Cf: *turbulent velocity.*

laminar wall In a diatom, a single layer of silica in the wall of a frustule. It may be either of uniform thickness or have local thickenings that form ribs or costae.

laminate adj. Consisting of or containing laminae. Syn: *laminated.*

laminated [ign] Said of the texture imparted to a lava by lamination; also, said of a rock with such texture.

laminated [sed struc] (a) Said of a rock (such as shale) that consists of laminae or that can be split into thin layers. Syn: *laminate.* (b) Said of a substance that exhibits lamination; e.g. "laminated clay" formed in a lake. (c) Said of the sedimentary structure possessed by a laminated rock.

laminated quartz Vein quartz characterized by slabs or films of other material.

lamination [sed] (a) *lamina.* (b) The formation of a lamina or laminae. (c) The state of being laminated; specif. the finest stratification or bedding, typically exhibited by shales and fine-grained sandstones. (d) A laminated structure.

lamination [ign] The spreading out of the constituents of a lava parallel to the underlying rocks.

laminite (a) A term used by Lombard (1963) for a finely laminated, detrital rock of the flysch lithofacies, frequently occurring in geosynclinal successions in natural sequences complementary to typical turbidites. It is finer-grained and thinner-bedded than a turbidite, ranging in thickness from a few millimeters to 30 cm, and is believed to form seaward from turbidites as a bottomset bed of a large delta. (b) A term suggested by Adolph Knopf (in Sander, 1951, p.135) to replace *rhythmite* in order to avoid the positive implication of perfect periodicity in the recurrence of laminae.

laminoid Laterally elongate parallel to stratification; e.g. "laminoid-fenestral fabric" of a limestone, characterized by particulate carbonate interrupted by horizontally elongate gaps (fenestrae) that tend to outline lamination (Tebbutt et al, 1965, p.4).

lampadite A variety of wad containing as much as 18% copper oxide and often containing cobalt. The term is often used for all hydrous manganese oxides containing copper.

lamprobolite *basaltic hornblende.*

lamproite A group of dark-colored hypabyssal or extrusive rocks that represent the end members of the syenites; also, any rock in that group, such as madupite, verite, cedricite, or wyomingite (Thrush, 1968, p.623).

lamprophyllite A mineral: $Na_2(Sr,Ba)_2Ti_3(SiO_4)_4(OH,F)_2$.

lamprophyre A group of dark-colored, porphyritic, hypabyssal igneous rocks characterized by panidiomorphic texture (i.e. *lamprophyric*), a high percentage of mafic minerals (esp. biotite, hornblende, and pyroxene), which form the phenocrysts, and a fine-grained groundmass with the same mafic minerals in addition to light-colored minerals (feldspars or feldspathoids); also, any rock in that group, e.g. minette, vosgesite, kersantite, spessartite, camptonite, monchiquite, fourchite, alnoite. Lamprophyres are frequently highly altered and commonly associated with carbonatites. Cf: *leucophyre.*

lamprophyric Said of the *panidiomorphic* texture exhibited by *lamprophyre*, in which phenocrysts of mafic minerals are contained in a fine-grained crystalline groundmass.

lamproschist Metamorphosed lamprophyre with a schistose structure containing brown biotite and green hornblende.

lamp shell A syn. of *brachiopod*, esp. a terebratuloid.

Lanarkian European stage: lower Upper Carboniferous (above Lancastrian, below Yorkian). It is equivalent to lowermost Westphalian.

lanarkite A white, greenish, or gray monoclinic mineral: Pb_2SO_5 or $PbO.PbSO_4$.

Lancastrian European stage: lower Upper Carboniferous (above Viséan, below Lanarkian). Cf: *Namurian.*

lanceolate Spear-shaped, or shaped like a lance head, such as a leaf or prism that is much longer than broad, widening above the base and tapering to a point at the apex; e.g. said of the form of a lobe of an ammonoid suture, or said of a nautiloid whorl section with an acute periphery.

lancet plate An elongate and spear-shaped or triangular plate located along the midline of ambulacrum of blastoids.

land In a general sense, that part of the Earth's surface that stands above sea level. The inclusion of Antarctica's permanent ice in calculating the land surface of the Earth is controversial. The term should not be confused with the term "soil".

land accretion Reclamation of land from the sea or other low-lying or flooded areas by draining and pumping, dumping of fill, or planting of marine vegetation.

land asphalt Hard asphalt, containing less bitumen and more

impurities than *lake asphalt*, from areas outside the pitch lake of Trinidad. It is divided into cheese, slate, stone, and iron varieties depending on the depth at which it is found. Syn: *land pitch.*

landauite A mineral: $(Zn,Mn,Fe)Ti_3O_7$.

landblink A yellowish reflection on the underside of a cloud layer over snow-covered land in a polar region; yellower than iceblink. Also spelled: *land blink.*

land bridge A land connection, often subject to temporary or permanent submergence, between continents or landmasses and which permits the migration of organisms, e.g. the Bering Land Bridge. See also: *neck; filter bridge.*

land compass *surveyor's compass.*

Landenian European stage: upper Paleocene (above Montian, below Ypresian of Eocene). It includes Thanetian and Sparnacian.

landerite A pink to rose-pink variety of grossular garnet. Syn: *rosolite; xalostocite.*

landesite A brown mineral: $Mn_{10}Fe_3{}^{+3}(PO_4)_8(OH)_5.11H_2O$ (?). It occurs as an alteration product of reddingite. Cf: *salmonsite.*

landfill Disposal of waste by burying it under layers of earth materials in low-lying ground.

land floe An unusually thick fragment of fast ice that has become detached and is now adrift.

landflood An overflowing of inland water onto the land.

landform Any physical, recognizable form or feature of the Earth's surface, having a characteristic shape, and produced by natural causes; it includes major forms such as a plain, plateau, or mountain, and minor forms such as a hill, valley, slope, esker, or dune. Taken together, the landforms make up the surface configuration of the Earth. Also spelled: *land form.* See also: *physiographic form; topographic form.* Syn: *relief feature.*

landform map *physiographic diagram.*

land hemisphere That half of the Earth containing the bulk (about six-sevenths) of dry land surface; it is mostly north of the equator with Paris as its approx. center. Cf: *water hemisphere.*

land ice Any ice mass formed from snow, rain, or other freshwater on land, as an ice shelf or a glacier, even though it may be floating in the sea, as an iceberg. Ant: *sea ice.*

landlocked Said of a body of water that is enclosed or nearly enclosed by land; e.g. a landlocked bay separated from the main body of water by a bar, or a landlocked lake having no surface outlet.

landmark (a) Any conspicuous object, natural or artificial, located near or on land and of sufficient interest or prominence in relation to its surroundings to make it outstanding or useful in determining a location or a direction. (b) Any monument, material mark, or fixed object (such as a river, tree, or ditch) used to designate the location of a land boundary on the ground.

landmass A land area studied as a unit, without regard necessarily to size or relief, on the basis of the sediments derived from it or the paleogeographic evidence indicated by the change in shorelines (Eardley, 1962, p.6).

landmass volume The volume, beneath a land surface, of a body with vertical sides and a base equal to *basin area* at the elevation of the stream's mouth (Strahler, 1952b, p.1120). Symbol: V.

land pebble *land-pebble phosphate.*

land-pebble phosphate A term used in Florida for a residual *pebble phosphate* occurring in clayey, gravelly, or compacted beds below the surface of the ground. Cf: *river-pebble phosphate.* Syn: *land pebble; land rock; matrix rock.*

land pitch *land asphalt.*

land rock A syn. used in South Carolina for *land-pebble phosphate.*

landscape The distinct association of landforms, esp. as modified by geologic forces, that can be seen in a single view, e.g. glacial landscape.

landscape marble A close-grained limestone characterized by dark, conspicuous, dendritic markings that suggest natural scenery (woodlands, forests); e.g. the argillaceous limestone in the Cotham Marble near Bristol, Eng. Syn: *forest marble.*

land sculpture *sculpture.*

landside (a) That part of a near-water feature that is facing toward the land. (b) An obsolete term for *shore.*

land sky Dark or gray streaks or patches in the sky near the horizon or on the underside of low clouds, caused by the absence of reflected light from the bare ground (as a land surface that is not snow-covered); not as dark as *water sky.*

landslide A general term covering a wide variety of mass-movement landforms and processes involving the moderately rapid to rapid (on the order of one foot per year or greater) downslope transport, by means of gravitational body stresses, of soil and rock material en masse. Usually, but not always, the displaced material moves over a relatively confined zone or surface of shear. The wide range in site variables, structure, and in the material properties affecting resistance to shear, as well as in the factors affecting the driving forces, results in a great range in landslide morphology, rates, and patterns of movement, and in scale. Landsliding is usually preceded, accompanied by, and followed by perceptible creep deformation, along the surface of sliding and/or within the slide mass. Terminology designating particular landslide types generally refers to the landform as well as the process responsible for the landform, e.g. rockfall, talus, translational slide, rockslide, block glide, debris slide, avalanche, earth-flow, mudflow, quick clay slide, liquefaction slide, slump, rotational slide, etc. Syn: *landsliding; slide.*

landslide breccia A breccia that is largely fragmented and wholly assembled by the force of gravity, as by a rockfall or a rockslide.

landslide lake (a) A lake resulting from the damming of a stream valley by the material in a landslide. (b) A long, narrow lake between the back slope of a landslide terrace and a valley wall.

landslide sapping The process of causing landslides by a stream undermining a canyon wall (Freeman, 1925, p.78).

landslide scar A bare or relatively bare surface or niche on the side of a mountain or other steep slope, left by the removal of earth material from the place where a landslide started.

landslide shear surface *slip surface* [mass move].

landslide terrace A short, rough-surfaced terrace resulting from a landslide.

landslide track The exposed path in rock or earth formed by a landslide. Syn: *slide.*

landsliding The downward movement of a landslide. Syn: *landsliding.*

land survey A survey made to determine boundaries and areas of tracts of land, esp. of privately owned parcels of land. Cf: *cadastral survey; boundary survey.*

land-tied island A *tied island* connected with the mainland by a tombolo.

lane (a) A narrow, not necessarily navigable, fracture or channel of water through sea ice; it may widen into a *lead.* (b) A term often used as a syn. of *lead* [ice].

langbanite An iron-black hexagonal mineral: (Mn,Sb,Ca,Fe, Mg)O$_8$(SiO$_4$) (?).

langbeinite A colorless, yellowish, reddish, or greenish isometric mineral: K$_2$Mg$_2$(SO$_4$)$_3$. It is much used in the fertilizer industry as a source of potassium sulfate.

langite A blue to green mineral: Cu$_4$(SO$_4$)(OH)$_6$.2H$_2$O.

lansfordite A colorless mineral: MgCO$_3$.5H$_2$O. It alters to nesquehonite on exposure to air.

lantern *Aristotle's lantern.*

lantern-node The central octahedron of a lychnisc in a sponge.

lanthanite A colorless, white, pink, or yellow earthy or crystalline mineral: (La,Ce)$_2$(CO$_3$)$_3$.8H$_2$O.

lapiaz Var. of *lapieś,* a syn. of *karren.*

lapidary (a) A cutter, grinder, and polisher of colored stones, or of precious stones other than diamonds. Syn: *lapidist.* (b) The art of cutting gems. (c) An obsolete term for a short treatise on metals, stones, and gems, describing their supposed medicinal, magical, or mythical characteristics.

lapidification An obsolete term signifying the conversion into stone or stony material, such as the process of petrifaction or of lithification.

lapidofacies Facies related to diagenesis (Vassoevich, 1948).

lapiés A syn. of *karren.* Also spelled: lapiaz. Etymol: French.

lapilli Pyroclastics that may be either essential, accessory, or accidental in origin, of a size range that has been variously defined within the limits of 1-64 mm diameter. The fragments may be either solidified or still viscous when they land (though some classifications restrict the term to the former); thus there is no characteristic shape. An individual fragment is called a *lapillus.* Cf: *volcanic gravel; block* [volc]; cinder.

lapilli tuff An indurated deposit that is predominantly lapilli, with a matrix of tuff.

lapillus The singular form of *lapilli.*

lapis lazuli (a) A blue, semitranslucent to opaque, granular crystalline rock used as a semiprecious stone for ornamental purposes and composed essentially of lazurite and calcite but also containing haüyne, sodalite, spangles of pyrite inclusions, and other minerals. It usually has a rich azure-blue color, but may be dark blue, violetish blue, greenish blue, and light blue, depending upon the amount of inclusions. It is probably the original sapphire of the ancients. Syn: *lazuli.* (b) An old name for *lazurite,* still used esp. for the gem variety. (c) An ultramarine-colored serpentine from India.

Laplace azimuth A *geodetic azimuth* derived from an astronomic azimuth by means of the *Laplace equation,* expressing the relationship between astronomic and geodetic azimuths in terms of astronomic and geodetic longitudes and geodetic latitude. Cf: *azimuth.*

Laplace equation An equation, used to derive the *Laplace azimuth,* which expresses the relationship between astronomic and geodetic azimuths in terms of astronomic and geodetic longitudes and geodetic latitude. See also: *Laplace station.*

Laplace station A triangulation or traverse station at which the *Laplace equation* can be formulated through observation of longitude and astronomic azimuths.

lap-out map A map showing the areal distribution of formations immediately overlying an unconformity. Syn: *worm's-eye map.*

lapparentite (a) A mineral: Al$_2$(SO$_4$)$_2$(OH)$_2$.9H$_2$O. (b) *tamarugite.*

lappered ice *anchor ice.*

lapse rate The rate at which temperature decreases with height; its vertical gradient. Its average is about 0.6°C/10 meters; however, it varies according to the moisture content of the air.

laqueiform Said of the loop pattern in a dallinid brachiopod (as in the family Laqueidae) in which posterior connecting bands from the ascending branches to the descending branches "are retained during enlargement and proportional thinning during change from frenuliform to terebrataliiform loop" (TIP, 1965, pt.H, p.147).

Laramian orogeny *Laramide orogeny.*

Laramic orogeny *Laramide orogeny.*

Laramide orogeny A time of deformation typically developed in the eastern Rocky Mountains of the United States, whose several phases extended from late Cretaceous until the end of the Paleocene. Intrusives and accompanying ore deposits enaced about this time in the mountain states are commonly alled Laramide (example, Boulder batholith, Montana). Geolgists differ as to whether to restrict the Laramide closely in me and space, as to a single event near the end of the Creceous, and to deformations near the type area -- or whether apply it broadly to all orogenies from early in the Cretaous through the Eocene or later, and to deformations in the hole Cordilleran belt of western North America. Best usage probably somewhere between the extremes, and the Larmide can properly be considered as an orogenic era in the ense of Stille. It is named for the Laramie Formation of yoming and Colorado, probably a synorogenic deposit. Also pelled: *Laramic orogeny*; *Laramian orogeny*. Syn: *Laramide evolution*.

aramide Revolution *Laramide orogeny*.

rdalite *laurdalite*.

rderellite A white mineral: $(NH_4)B_5O_8.2H_2O$.

rdite (a) White hydrated silica, probably a variety of opal, ccurring in clay in central Russia. (b) Massive talc; *steatite*.) *agalmatolite*.

rd stone Massive talc; *steatite*.

rge boulder A *boulder* having a diameter in the range of 024-2048 mm (40-80 in., or -10 to -11 phi units).

rge calorie *kilocalorie*.

rge cobble A geologic term for a *cobble* having a diameter the range of 128-256 mm (5-10 in., or -7 to -8 phi units).

rger foraminifera An informal term generally used to desigate those foraminifers that are studied without the aid of thin ectioning. Cf: *smaller foraminifera*.

rge-scale map A map drawn at a scale (in the U.S., larger an 1/62,500) such that a small area can be shown in fine etail and with great accuracy; a map whose representative action has a small denominator (such as 1/24,000).

rge wave An obsolete syn. of *surface wave*.

rnite A gray mineral: β-Ca_2SiO_4. It is a metastable monoclic phase of calcium orthosilicate, stable from 520° to 670°C, nd tending to break down to the stable *calcio-olivine*. Cf: *redigite*. Syn: *belite*.

rsenite A colorless or white orthorhombic mineral: oZnSiO4.

arsen method *lead-alpha age method*. The lead-alpha age ethod was suggested and developed under the guidance of sper Signius Larsen, Jr. (1879-1961), U.S. mineralogist and etrologist, and is sometimes referred to by his name.

arsen variation diagram A diagram in which the weight perent of each oxide constituent of a rock is plotted as the ordiate against the abscissa which is one-third of the SiO_2 + 2O - FeO - MgO - CaO. The diagram was devised by Larsen 938).

rvikite An alkalic syenite composed of phenocrysts of two, ten intimately intergrown, feldspars (esp. oligoclase and alali feldspar) which comprise up to eight- or nine-tenths of e rock, with diopsidic augite and titanaugite as the chief aafic minerals and accessory apatite (generally abundant), ilienite, and titaniferous magnetite, and even less commonly, ivine, bronzite, lepidomelane, and feldspathoids (less than 0 percent by volume). Its name is derived from Larvik, Noray. Also spelled: *laurvikite*. Syn: *blue granite*.

ssenite A name, now considered obsolete, that was formerly pplied to a volcanic glass originally thought to have had the omposition of trachyte but now known to be dacitic. Cf: *meibolite*.

te Pertaining to or occurring near the end of a segment of me. The adjective is applied to the name of a geologic-time nit (era, period, epoch) to indicate relative time designation nd corresponds to *upper* as applied to the name of the quivalent time-stratigraphic unit; e.g. rocks of an Upper Jurssic batholith were intruded in Late Jurassic time. The initial tter of the term is capitalized to indicate a formal subdivision

(e.g. "Late Devonian") and is lowercased to indicate an informal subdivision (e.g. "late Miocene"). The informal term may be used for eras and epochs, and for periods where there is no formal subdivision. Cf: *middle* [*geochron*]; early.

late diagenesis Deep-seated diagenesis, occurring a long time after deposition, when the sediment is more or less compacted into a rock, but still in the realm of pressure-temperature conditions similar to those of deposition; it represents a transition from diagenesis to metamorphism. Syn: *epigenesis; epidiagenesis; metharmosis*.

late-glacial Pertaining to the time of the waning of the last glaciation; specif. the "Late Glacial period" of the Pleistocene Epoch, immediately preceding the Preboreal phase.

latent heat of fusion *heat equivalent of fusion*.

latent magma A highly viscous magma that exists under high pressure beneath the Earth's crust and reacts as a solid body, e.g. with respect to the propagation of earthquake waves. With a decrease in pressure, the magma becomes sufficiently fluid to flow (Schieferdecker, 1959, term 3827).

latera Plural of *latus*.

lateral [*paleont*] n. (a) A compartmental plate in certain cirripede crustaceans bounded by a carinolateral and a rostrolateral. In other cirripedes, the term is synonymous with *latus*. (b) One of a series of ossicles along the side of an arm in an ophiuroid. Also, one of a circlet of five plates in certain cystoids. (c) A lateral part; e.g. a lateral tooth.

lateral [*stratig*] Said of the direction of extension of strata, measured at right angles to the *vertical* direction.

lateral [*volc*] *parasitic*.

lateral [*streams*] n. A *lateral stream*.

lateral accretion Outward or horizontal sedimentation; e.g. digging away of the outer bank of a stream meander and building up of the inner bank to water level by deposition of material brought there by rolling or pushing along the bottom. Cf: *vertical accretion*.

lateral bud In seed plants, an axillary bud.

lateral channel A channel formed by a meltwater stream flowing laterally away from a glacier through a notch in bordering hills (Rich, 1908, p.528).

lateral consequent stream A *secondary consequent stream* flowing down the flank of an anticline or syncline.

lateral corrasion Corrasion of the banks of a stream.

lateral crevasse *marginal crevasse*.

lateral depressor pit A small hollow near one or both basal angles of the scutum of a cirripede crustacean, serving for attachment of the lateral muscle that depresses or draws down.

lateral dissepiment A dissepiment of a rugose coral, characterized by blister-like form and developed in isolated manner on the sides of septa. Cf: *horseshoe dissepiment*.

lateral dune A sand dune flanking a larger dune, formed around an obstacle.

lateral erosion The erosion performed on its banks by a meandering stream as it swings from side to side, impinging against and undercutting the banks as it flows downstream; it results in *lateral planation*.

lateral fault A fault along which there has been strike separation. Cf: *strike-slip fault; dip-separation fault*. See also: *wrench fault; right-lateral separation; left-lateral separation*. Syn: *strike-separation fault*.

lateral increase A type of *increase* (budding of corallites) in fasciculate coralla characterized by sideward outgrowth of offsets.

lateral lake A fluviatile lake formed in the valley of a tributary stream by the silting up of the channel of the main stream, thereby producing embankments or levees that impound the water of the tributary.

lateral levee lake A lake occupying a depression behind a natural levee.

lateral lobe (a) Any primary lobe (other than the ventral lobe) of the external suture of an ammonoid shell. The "first lateral lobe" is next to the ventral lobe, usually on the whorl side but in depressed whorls commonly on the venter, and the "second lateral lobe" is next to the first lateral lobe, commonly on the whorl side and morphogenetically part of the umbilical lobe (TIP, 1959, pt.L, p.4). (b) Any adapical inflection of a suture of a nautiloid shell between the ventral lobe and the dorsal lobe (in coiled conchs, the lateral lobes may be external or internal according to whether they are on the flanks or dorsal area) (TIP, 1964, pt.K, p.57).

lateral log A *resistivity log* consisting of three electrodes in the borehole (two potential electrodes and one current electrode). The spacings between the current electrode and the point midway between the potential electrodes is generally 18 ft 8 in.; this large spacing allows the determination of resistivities up to 30 ft from the borehole. The lateral log gives good depth determinations of bed bottoms. Cf: *normal log*.

lateral moraine [glac geol] (a) A low ridge-like moraine carried on, or deposited at or near, the side margin of a mountain glacier. It is composed chiefly of rock fragments loosened from the valley walls by glacial abrasion and plucking, or fallen onto the ice from the bordering slopes. (b) An end moraine built along the side margin of a glacial lobe occupying a valley. Cf: *flanking moraine*.---Syn: *side moraine; valley-side moraine*.

lateral oblique muscle One of a pair of muscles in some inarticulate brachiopods, originating on the pedicle valve anteriorly and laterally from the posterior adductor muscles, and passing anteriorly and dorsally to insertions either on the brachial valve and anterior body wall against the anterior adductor muscles (as in the family Discinidae) or entirely on the anterior body wall (as in the family Craniidae) (TIP, 1965, pt.H, p.147). Cf: *internal oblique muscle*.

lateral planation The reduction of the land in an interstream area to a plane or a nearly flat surface by the *lateral erosion* of a meandering stream; the creation and development by a stream of its flood plain.

lateral saddle (a) Any primary saddle of the external suture of an ammonoid shell (other than the median saddle located on the venter). The "first lateral saddle" is a forward (adoral) deflection that separates the ventral lobe from the first lateral lobe, and the "second lateral saddle" is a forward deflection that separates the first lateral lobe from the second lateral lobe (TIP, 1959, pt.L, p.4). (b) Any adoral inflection of a suture of a nautiloid shell, separating lateral lobes from each other or from external lobes or internal lobes (in coiled conchs, the lateral saddles may be external or internal according to whether they are on the flanks or dorsal area) (TIP, 1964, pt.K, p.57).

lateral search *profiling*.

lateral secretion A theory of ore genesis formulated in the 18th Century that postulates the formation of ore by the leaching of the adjacent wall rock. A contemporary term for such a mineral deposit is the adjective *lithogene*. See also: *segregated vein*.

lateral shift The offset of the position of the peak of an anomaly, in either a gravitational or a magnetic field, with respect to the crest of the mass which produces it, arising from asymmetry of the mass or of the magnetization.

lateral sinus A notch or reentrant in the lateral part of the apertural margin (peristome) of a cephalopod.

lateral storage *bank storage*.

lateral stream A stream situated on, directed toward, or coming from the side; e.g. a stream flowing along the edge of a lava flow that recently filled part of a valley. Syn: *lateral*.

lateral tooth A *hinge tooth* partly or wholly located some distance from the beak of a bivalve mollusk, situated anterior or posterior to the middle of the hinge, and lying ahead of or behind the *cardinal teeth*. Its long axis is parallel to the hin line.

lateral valley A *longitudinal valley* developed parallel to the r gional structure.

lateral variation A facies change of the sedimentary chara teristics within a formation in a horizontal direction. It is pa ticularly significant in petroleum geology when porosity a permeability are affected, as in an alternation from sandsto to siltstone to shale (a *shale-out*).

later arrival An *arrival* on a seismogram following the *first a rivals*. It may be reflected, refracted, diffracted, or som combination.

laterite A highly weathered, red subsoil or material rich in se ondary oxides of iron, aluminum, or both, nearly void of bas and primary silicates, and maybe containing large amounts quartz and kaolinite. It develops in a tropical or foreste warm to temperate climate, and is a residual or end produ of weathering. Laterite is capable of hardening after a trea ment of wetting and drying, and can be cut and used f bricks; hence its etymology: Latin, *latericius*, "brick". S also: *lateritic soil*.

lateritic soil A soil containing *laterite*; also, any reddish, trop cal soil developed from much weathering. Syn: *latosol*.

lateritization Var. of *laterization*.

laterization A general term for the process that converts rock or soil to *laterite*. Also spelled: *lateritization*.

laterolog A trade name for a conductively *focused-current lo* See also: *microlaterolog*.

late wood *summerwood*.

lath [paleont] The part of a heterococcolith with one large d mension, one intermediate, and one very small.

lath-shaped Said of a habit of a crystal as long and thin, an of moderate to narrow width. In thin section, lath-shape crystals often cross sections of platy or tabular crystals.

Latin square An array of size $n \times n$, containing n letters di played n times in such a manner that each letter appea once only in each row and in each column of the array. Se also: *Greco-Latin square*.

latite A porphyritic extrusive rock having plagioclase and p tassium feldspar (probably mostly sanidine) present in near equal amounts as phenocrysts, little or no quartz, and a fine crystalline to glassy groundmass, which may contain obscu potassium feldspar; the extrusive equivalent of *monzonite*. La tite grades into trachyte with an increase in the alkali feldspa content, and into andesite or basalt, depending on the pres ence of acid or calcic plagioclase, as the alkali feldspar co tent decreases. It is usually considered synonymous with *tra chyandesite* and *trachybasalt*, depending on the color.

latitude (a) Angular distance from some specified circle plane of reference, such as the angle measured at the cente of a sphere or spheroid between the plane of the equator an the radius to any point on the surface of the sphere or sphe oid; specif. angular distance of a point on the Earth's surfac north or south of the equator, measured along a meridia through 90 degrees (the equator being latitude zero degree the North Pole lat. 90°N, and the South Pole lat. 90°S). A de gree of latitude on the Earth's surface varies in length fro 68.704 statute miles at the equator to 69.407 statute miles the poles. Abbrev: lat. Symbol: ϕ. See also: *astronomic la itude; geodetic latitude; geocentric latitude; geographic la itude; celestial latitude*. Cf: *parallel*. (b) The projection on th meridian of a given course in a plane survey equal to th length of the course multiplied by the cosine of its bearing (c) A linear coordinate distance measured north or sout from a specified east-west line of reference; e.g. northing an southing.---Cf: *longitude*.

latitude correction [grav] The amount of the adjustment observed gravity values to an arbitrarily chosen base latitude $K = 0.8122 \sin 2\phi$ (mg/km) and $K = 1.307 \sin 2\phi$ (mg/m

ere ϕ is the latitude angle.
itude correction [mag] The north-south corrections made to served magnetic intensities in order to remove the Earth's mal field (leaving, as the remainder, the anomalous field). *longitude correction.*
itude difference The length of the projection of a line onto a ridian of reference in a plane survey, being equal to the gth of the line multiplied by the cosine of its bearing; e.g. *rthing* (positive difference) and *southing* (negative differce). Cf: *meridional difference; departure.*
iumite A mineral: $(Ca,K)_8(Al,Mg,Fe)(Si,Al)_{10}O_{25}(SO_4)$.
osol A soil that contains abundant hydroxide minerals; a *eritic soil.*
rappite A mineral: $(Ca,Na)(Nb,Ti,Fe)O_3$. Cf: *perovskite.*
tice [cryst] *crystal lattice.*
tice [reef] *growth lattice.*
tice bar A bar composing the lattice shell of a radiolarian.
tice constant In a crystal lattice, the length of one edge of unit cell; also, the angle between two edges of the cell. n: *lattice parameter; parameter (a).*
tice defect *crystal defect.*
tice drainage pattern *rectangular drainage pattern.*
tice parameter *lattice constant.*
tice pore One of the open spaces surrounded by lattice rs in a radiolarian skeleton of the subfamily Trissocyclinae.
tice row A series of lattice points coincident with the interction of two nonparallel crystal faces; and lattice axis or ne axis. See also: *indices of lattice row.*
tice shell A porous sheath that surrounds all or a part of the gittal ring of some radiolarians or that is divided into symtric halves by the sagittal ring in other radiolarians; a shwork radiolarian skeleton.
tice spine One of the spines that project from the lattice r in a radiolarian skeleton, either distributed randomly or nfined to the junctions of two or more lattice bars.
tice texture [eco geol] In mineral deposits, a texture proced by exsolution in which elongate crystals are arranged ng structural planes.
tice texture [met] A texture that is typical for the mineral rpentine in a rock when it replaces an amphibole. Cf: *knit- texture.*
torfian A syn. of *Tongrian.* Also spelled: *Latdorfian.*
us (a) Any of paired plates forming part of the shell in cern cirripede crustaceans, but not including tergum and scun; e.g. "carinal latus" (carinolateral) or "rostral latus" strolateral). Syn: *lateral [paleont].* (b) The surface of a crid columnal or cirral, exclusive of articular facets.---Pl: *lat-a.*
bmannite A mineral: $Fe_3{}^{+2}Fe_6{}^{+3}(PO_4)_4(OH)_{12}$. It may ntain a little manganese or calcium.
ue camera The instrument used in the *Laue method* of ay diffraction analysis. A pinhole defines an X-ray beam rpendicular to flat film. When the back-reflection method is ed, a pinhole is made through the film which is then mount- between tube and crystal.
ue equations Simultaneous equations that represent the cessary conditions for radiation diffraction by a three-dinsional lattice.
uegram The diagram of X-ray diffraction made according to *Laue method.* See also: *Laue spot.* Syn: *Laue pattern.*
eite A honey-brown triclinic mineral: $MnFe_2(PO_4)_2(OH)_2$. $_2O$. It is polymorphous with strunzite. Cf: *pseudolaueite.*
ue method A technique of X-ray diffraction analysis using a gle, fixed crystal irradiated by a beam of a continuous ectrum of X-rays. The patterns, or *Lauegram,* are observed er X-ray transmission or by reflection back to their source, lled back reflection. See also: *Laue camera.*
ue pattern *Lauegram.*
ue spot A single spot on a *Lauegram.*

laugenite An oligoclase diorite.
laumontite A white zeolite mineral: $CaAl_2Si_4O_{12}.4H_2O$. It sometimes contains appreciable sodium, and upon exposure to air it loses water and becomes opaque and crumbles. It occurs as prismatic crystals in veins in schists and slates and in cavities in igneous rocks. Syn: *laumonite; lomonite; lomontite.*
laumontite-prehnite-quartz facies A term introduced by Winkler (1967) to replace the term *zeolite facies.*
launayite A mineral: $Pb_{22}Sb_{26}S_{61}$.
launder A trough, channel, gutter, flume, or chute by which water or powdered ore is conveyed in a mining operation.
lauoho o pele *Pele's hair.*
Laurasia A combination of Laurentia, a paleographic term for the Canadian Shield and its surroundings, and Eurasia. It is the protocontinent of the Northern Hemisphere, corresponding to Gondwana in the Southern Hemisphere, from which the present continents of the Northern Hemisphere have been derived by separation and continental displacement. The hypothetical supercontinent from which both were derived is Pangea. The protocontinent included most of North America, Greenland, and most of Eurasia, excluding India. The main zone of separation was in the North Atlantic, with a branch in Hudson Bay, and geologic features on opposite sides of these zones are very similar.
laurdalite A larvikite containing more than 10 percent modal feldspathoids and characterized by pseudoporphyritic texture. Also spelled: *lardalite.*
Laurentian A name that is widely and confusingly used for granites and orogenies of Precambrian age in the Canadian Shield. It is named for the Laurentian Highlands northwest of the St. Lawrence River in eastern Canada (a part of the Grenville province of current usage) where Logan (1863) recognized the Laurentian granites, now dated radiometrically at about 1000 m.y. The term was misapplied by Lawson (1885) to the oldest granites near the U.S.-Canadian border northwest of Lake Superior, from which Schuchert subsequently derived his Laurentian Revolution, or orogeny, that was supposed to have terminated the Archeozoic. Modern work shows that Lawson's Laurentian is older than the 2400 m.y.-old Algoman orogeny and granites, at the end of the Archean of the present Canadian classification, but no radiometric dates for it survive, and its significance and extent are uncertain. It has been suggested that the term Laurentian be restored to Logan's original meaning. See also: *Grenville orogeny.*
laurionite A colorless mineral: $Pb(OH)Cl$. It is dimorphous with paralaurionite.
laurite An iron-black mineral: RuS_2. It often contains osmium. Laurite is found in association with platinum in placer deposits and usually occurs in minute octahedrons resembling those of magnetite.
laurvikite *larvikite.*
lausenite A white silky or fibrous mineral: $Fe_2(SO_4)_3.6H_2O$.
lautarite A monoclinic mineral: $Ca(IO_3)_2$.
lautite A mineral: $CuAsS$.
lava A general term for a molten extrusive; also, for the rock that is solidified from it.
lava ball A globular mass of lava that is scoriaceous inside and compact on the outside; it is formed by the coating of a fragment of scoria by fluid lava. Syn: *pseudobomb; volcanic ball; volcanic dumpling.*
lava breccia *volcanic breccia.*
lava-dam lake A lake formed where a lava flow obstructs or has obstructed a watercourse. Syn: *lava-dammed lake.*
lava dome (a) A dome-shaped mountain of solidified lava in the form of many individual flows, formed by the extrusion of highly viscous lava, as Mauna Loa, Hawaii. Cf: *volcanic dome.* Syn: *dome [volc].* (b) *shield volcano.* (c) A tumulus devel-

oped on a lava flow.

lava flow A lateral, surficial outpouring of molten lava from a vent or a fissure; also, the solidified body of rock that is so formed. Syn: *flow* [*volc*]; *nappe* [*volc*].

lava lake A lake of usually basaltic molten lava in a volcanic crater or depression. The term refers to solidified and partly solidified stages as well as to the molten, active lava lake.

lava levee The scoriaceous sheets of lava that overflowed their natural channels of flow and solidified to form a levee, similar to a levee formed by an overflowing stream of water.

lavant *bourne.*

lava rag A bit of scoriaceous material ejected from a volcano.

lava toe One of a series of small, bulbous projections that develop at the front of a moving pahoehoe flow, formed by the breaking open of the crust and the emergence of fluid lava. Syn: *toe.*

lava tree A *lava tree mold* that projects above the surface. Syn: *lava tree cast.*

lava tree cast A *lava tree mold* that projects above the surface; a syn. of *lava tree.*

lava tree mold A cylindrical hollow in a lava flow formed by the envelopment of a tree by the flow, solidification of the lava in contact with the tree, and disappearance of the tree by burning and subsequent removal of the charcoal and ash. The inside of the mold preserves the surficial features of the tree. See also: *lava tree; lava tree cast.*

lava trench A collapsed *lava tube.*

lava tube A hollow space beneath the surface of a solidified lava flow, formed by the withdrawal of molten lava after the formation of the surficial crust. See also: *lava trench; volcanic flow drain.* Syn: *lava tunnel.*

lava tunnel *lava tube.*

lavendulan A lavender mineral: $NaCaCu_5(AsO_4)_4Cl.5H_2O$. Cf: *freirinite.*

lavenite A mineral: $(Na,Ca)_3Zr(Si_2O_7)(O,OH,F)_2$. Cf: *wöhlerite.* Syn: *laavenite.*

lavialite A metamorphosed basaltic rock with relict phenocrysts of labradorite in an amphibolitic groundmass. The term was originated by Sederholm in 1899, who named it after Lavia, Finland.

lavrovite A green variety of diopside containing small amounts of vanadium and chromium. Syn: *lavroffite.*

law [**geomorph**] A Scottish term for a more or less rounded or conical hill or mound. Syn: *low.*

law [**science**] A formal statement of the invariable and regular manner in which natural phenomena occur under given conditions; e.g. the "law of superposition" or a "law of thermodynamics".

law of acceleration A theory in biology stating that the order of development of a structure or organ is directly related to its importance to the organism.

law of accordant junctions *Playfair's law.*

law of basin areas A general law expressing the direct geometric relation between stream order and the mean basin area of each order in a given drainage basin, originally stated by Schumm (1956, p.606). The law is expressed as a linear regression of logarithm of mean basin area on stream order, the positive regression coefficient being the logarithm of the basin-area ratio.

law of constancy of interfacial angles The statement in crystallography that the angles between corresponding faces on different crystals of one substance are constant. It was first noted by the Danish scientist Nicolaus Steno in 1669. Syn: *constancy of interfacial angles.*

law of constancy of relative proportions *constancy of relative proportions.*

law of correlation of facies A ruling principle in European stratigraphy, enunciated by Walther (1893-1894): within a given sedimentary cycle, the same succession of facies that occurs laterally is also present in vertical succession.

law of crosscutting relationships A stratigraphic princip whereby relative ages of rocks can be established: a roc (esp. an igneous rock) is younger than any other rock acros which it cuts.

law of equal declivities Where homogeneous rocks are m turely dissected by consequent streams, all hillside slopes the valleys cut by the streams tend to develop at the sam slope angle, thereby producing symmetric profiles of ridge spurs, and valleys. The principle was formulated by Gilbe (1877, p.141).

law of equal volumes *Lindgren's volume law.*

law of faunal assemblages A general law of geology: Simil assemblages of fossil organisms (faunas and floras) indica similar geologic ages for the rocks that contain them.

law of faunal succession A general law of geology: Fossil c ganisms (faunas and floras) succeed one another in a defini and recognizable order, each geologic formation having a d ferent total aspect of life from that in the formations above and below it; or, the age of rocks can be determined fro their fossil content.

law of homonymy A principle in taxonomy stating that a name that is a junior homonym of another name must be r jected and replaced. See also: *homonymy.*

law of minimum lateral thrust The statement that the relati displacement of overlying strata to underlying strata surroun ing an inclined, concordant intrusion can be expressed as $Acot\Theta$, in which B=the horizontally measured width of the i clined part of the intrusive body, A=the vertically measure width of the horizontal part of the intrusive body, and $\Theta = t$ angle of inclination (DuToit, 1920, p.28).

law of nature A generalization of science, representing an i trinsic orderliness of natural phenomena or their necessa conformity to reason. Syn: *natural law.*

law of original continuity A general law of geology: A wate laid stratum, at the time it was formed, must continue latera in all directions until it thins out as a result of nondepositio until it abuts against the edge of the original basin of depo tion. The law was first clearly stated by Steno (1669).

law of original horizontality A general law of geology: Wate laid sediments are deposited in strata that are horizontal nearly horizontal, and parallel or nearly parallel to the Earth surface. The law was first clearly stated by Steno (1669).

law of priority A principle in taxonomy stating that the olde available name applied to a taxon is the valid name, unle that name has been invalidated because it does not meet o or more of the rules of nomenclature. See also: *priority.*

law of rational indices The statement in crystallography th crystal faces make simple rational intercepts on suitable cry tal axes. The axes are the axes of reference or the three ax forming the edges of the unit cell of each crystal lattice. Sy *Haüy's law.*

law of reflection The statement in physics that the angle b tween the reflected ray (normal to the wave front) and t normal to the reflecting surface is the same as the angle b tween this normal and the incident ray, provided the wa travels with the same velocity as the incident wave. See als *reflection.*

law of refraction The statement in physics that when a wa crosses a boundary between two isotropic substances, t wave normal changes direction in such a manner that the si of the angle of incidence between wave normal and bounda normal divided by the velocity in the first medium equals t angle of refraction divided by the velocity in the second me um. Syn: *Snell's law.*

law of stream gradients A general law expressing the inver geometric relation between stream order and the me stream gradient of a given order in a given drainage bas originally stated by Horton (1945, p.295).

law of stream lengths A general law expressing the direct geometric relation between stream order and the main stream lengths of each order in a given drainage basin, originally stated by Horton (1945, p.291). The law is expressed as a linear regression of logarithm of mean stream length on stream order, the positive regression coefficient being the logarithm the stream-length ratio.

law of stream numbers A general law expressing the inverse geometric relation between stream order and the number of streams of each order in a given drainage basin, originally stated by Horton (1945, p.291). The law is expressed as a linear regression of logarithm of number of streams on stream order, the negative regression coefficient being the logarithm the bifurcation ratio. See also: *number of streams*.

law of superposition A general law upon which all geologic chronology is based: In any sequence of sedimentary strata (or of extrusive igneous rocks) that has not been subsequently disturbed by overthrusting or overturning, the youngest stratum is at the top and the oldest at the base, the older strata being successively covered or overlain by younger and younger layers; or, each bed is younger that the bed beneath, but older than the bed above it. The law was first clearly stated by Steno (1669).

law of surface relationships A stratigraphic principle developed by Wheeler (1964, p.602-603): "time as a stratigraphic dimension has meaning only to the extent that any given moment in the Earth's history may be conceived as precisely coinciding with a corresponding worldwide lithosphere surface and all simultaneous events either occurring thereon or directly related thereto". The "lithosphere surfaces" (surfaces of deposition or surfaces of erosion) are envisaged as "the only universal physical geologic 'datum' surfaces with direct stratigraphic implication".

law of unequal slopes A stream flowing down the steeper slope of an asymmetric ridge or divide erodes its valley more rapidly than one flowing down the gentler slope, thereby causing the crest of the divide to migrate away from the more actively eroding stream toward the less actively eroding one. The principle was first recognized by Gilbert (1877, p.140).

law of universal gravitation The statement, explaining *gravitation*, that every mass particle in the universe attracts every other mass particle with a force which is directly proportional to the product of the two masses and inversely proportional to the square of the distance between them, the direction of the force being in the line joining the two particles. The law applies only to particles, and not to bodies of finite size.

lawrencite A green or brown meteorite mineral: $(Fe,Ni)Cl_2$. It occurs as an abundant accessory mineral in iron meteorites.

lawsonite A colorless to grayish-blue orthorhombic mineral: $CaAl_2(Si_2O_7)(OH)_2 \cdot H_2O$.

lawsonite-albite facies A term introduced by Winkler (1967) for rocks formed by burial metamorphism by the same temperatures (250°-400°C) as, but lower pressures (6000-7500 bars) than, those of the *lawsonite-glaucophane-jadeite facies*.

lawsonite-glaucophane-jadeite facies A term introduced by Winkler (1967) for that part of the *glaucophane schist facies* formed at lower temperatures (250°-400°C). The rocks are formed by burial metamorphism at very high pressures (above 6500 bars), and the coexistence of lawsonite and glaucophane as index minerals is required. Cf: *lawsonite-albite facies*.

laxite An old name for unconsolidated fragmental rocks.

layer [seism] One of a series of concentric zones or belts of the Earth, delineated by seismic discontinuities. A classification of the interior of the Earth designates layers A to G, e.g. layer, B layer, etc.

layer [stratig] A general term for any tabular body of rock (igneous, metamorphic, or sedimentary), ice, or unconsolidated material (sediment or soil), lying in a position essentially parallel to the surface or surfaces on or against which it was formed or laid, and more or less distinctly limited above and below; specif. a single *bed* or *stratum* of rock, with no limitation as to thickness.

layer-cake Said of the geologic concept of successive layers of strata, each separated by an unconformity and completely independent of or structurally different from other layers above and below (Levorsen, 1943, p.907-912).

layer depth In the ocean, the depth to the top of the thermocline; i.e., the depth of the mixed layer.

layered intrusion An intrusive body in which there are layers of varying chemical composition, formed by some type of *layering*. An example is the Bushveld Igneous Complex in southern Africa. See also: *banded differentiate*.

layered permafrost Ground consisting of permanently frozen layers alternating with unfrozen layers or taliks (Muller, 1947, p.218).

layered series A body of igneous rocks showing banding that simulates the stratification of a sedimentary sequence and having a parallelism of platy minerals that resembles the bedding of sediments (Wager & Deer, 1939, p.36).

layering [cart] *layer tinting*.

layering [petrology] A tabular structure, or the tabular succession of the different components (mineralogic, textural, structural), of an igneous, sedimentary, or metamorphic rock, or the formation of layers of material, one upon the other, in a particular rock; esp. the phenomenon occurring in plutonic rocks as a result of *crystal settling* in magma during slow cooling or of congelation of successive sheets around the top and sides of a pool of magma. Wager & Brown (1967, p.v) suggest that the term "be kept for the high-temperature sedimentation features of igneous rocks", leaving *bedding* and *stratification* for use in regard to the "usual" sedimentary rocks and *banding* for use in regard to metamorphic rocks. See also: *rhythmic layering; cryptic layering; crystal accumulation*.

layer silicate *phyllosilicate*.

layer stripping A procedure in seismology which removes the effects of the outer layers on the travel times and distances of seismic rays. It effectively enables one to place the seismic source and receivers at the base of the stripped layers. Syn: *stripping the Earth*.

layer structure A type of crystal structure built up by distinct layer units, e.g. micas, clay, graphite. See also: *two-layer structure; three-layer structure*.

layer tint *hypsometric tint*.

layer tinting A method of depicting relief on a map by the distinctive shading or coloring of the areas between contour lines in a manner suggestive of progressive change so that the pattern of distribution of high and low land areas is revealed or emphasized at a glance. Syn: *layering*.

lay of the land *topography*.

layover Displacement of the top of an elevated terrain feature with respect to its base on the radar linage. The peaks look like dip slopes.

lazuli A syn. of *lapis lazuli*. Also spelled: *lazule*.

lazulite An azure-blue to violetish-blue mineral: $(Mg,Fe^{+3})Al_2(PO_4)_2(OH)_2$. It is isomorphous with scorzalite, and occurs in small masses or in monoclinic crystals. Syn: *blue spar; false lapis; berkeyite*.

lazurite An intense blue or violetish-blue feldspathoid mineral of the sodalite group: $(Na,Ca)_8(Al,Si)_{12}O_{24}(S,SO_4)$. It is the principal constituent of *lapis lazuli*. See also: *ultramarine*.

leachate A solution obtained by leaching; e.g. water that has percolated through soil containing soluble substances and that contains certain amounts of these substances in solution.

leached Said of a soil in which *leaching* has taken place. Partial syn: *decalcified*.

leach hole *sinkhole*.

leaching (a) The separation, selective removal, or dissolving out of soluble constituents from a rock or orebody by the natural action of percolating water. (b) The removal in solution of nutritive or harmful constituents (such as mineral salts and organic matter) from an upper to a lower soil horizon by the action of percolating water, either naturally (by rainwater) or artificially (by irrigation). Cf: *eluviation*. (c) The extraction of soluble metals or salts from an ore by means of slowly percolating solutions; e.g. the separation of gold by treatment with a cyanide solution.----Syn: *lixiviation*.

lead [eco geol] (a) A syn of *lode*. (b) A placer deposit. See also: *back lead; blue lead; deep lead*.----Pron: leed.

lead [mineral] (a) A soft, heavy, malleable, and ductile isometric mineral, the native metallic element Pb. It is silvery bluish white when freshly cut, but tarnishes readily in moist air to dull gray. Lead rarely occurs in native form, being found mostly in combination (as in galena, cerussite, and anglesite). It is used often in the form of alloys. (b) A term sometimes applied to graphite.---Pron: *led*.

lead [ice] Any *fracture* [ice], *water opening*, or long narrow strip of ocean water through sea ice (esp. pack ice), navigable by surface vessels, and sometimes covered by young ice; wider than a *lane*. Cf: *polynya*. Syn: *channel* [ice]; *lane*. Pron: leed.

lead [streams] An open watercourse, usually artificial, leading to or from a mill, mine, reservoir, etc. Syn: *leat*.

lead-alpha age method A method of calculating an age in years by spectrographically determining the total lead content and the alpha-particle activity of a zircon, monazite, or xenotine concentrate, the alpha-particle activity representing the uranium-thorium content. Precision of this age method is less than of the potassium-argon or rubidium-strontium age methods, and is best used for rocks younger than Precambrian. Syn: *Larsen method; lead-alpha dating*.

lead-alpha dating *lead-alpha age method*.

lead glance *galena*.

leadhillite A yellowish or greenish- or grayish-white monoclinic mineral. $Pb_4(SO_4)(CO_3)_2(OH)_2$. It is dimorphous with susannite.

leading stone *lodestone* [magnet].

lead-isotope age *lead-lead age*.

lead-lead age An age in years calculated from the ratio of lead-207 to lead-206, a by-product of the *uranium-thorium-lead age method*. Syn: *lead-isotope age*.

lead line A weighted line of wire or cord that is used in *sounding*. The line is lowered from a ship until it reaches bottom; then its length is measured. Syn: *sounding line*.

lead ocher A yellowish or reddish, and scaly or earthy lead monoxide; specif. *massicot* and *litharge*. Syn: *plumbic ocher*.

lead ratio The ratio of lead-206 to uranium-238, of lead-207 to uranium-235, or of lead-208 to thorium-232, formed by the radioactive breakdown of the uranium and/or thorium within a mineral. The ratios are the basis of the *uranium-thorium-lead age method*.

lead spar (a) *cerussite*. (b) *anglesite*.

lead-uranium age method *uranium-lead age method*.

lead-uranium ratio The ratio of lead-206 to uranium-238 and/cr lead-207 to uranium-235, formed by the radioactive decay of uranium within a mineral. The ratios are frequently used as part of the *uranium-thorium-lead age method*.

lead vitriol *anglesite*.

leaf A very thin sheet or folium of a metal such as gold.

leaf clay *book clay*.

leaf gap A parenchymatous opening into a stele, left by the departure of a leaf trace (Cronquist, 1961, p.877).

leaf gold Gold occurring naturally as thin flakes or sheets; not to be confused with man-made gold leaf.

leaflet One of the parts of a *compound leaf*.

leaf mold A general term for an accumulation on the soil surface that is composed chiefly of partially decomposed vegeta-

ble matter (usually fallen leaves and the remains of herba ceous plants). It is a constituent of the *forest floor*. Cf: *litte duff*.

leaf peat *paper peat*.

leaf primordium The outgrowth of a bud which develops into leaf.

leaf scar A scar on a twig following the abscission of a lea See also: *bundle scar*.

leaf trace Vascular tissue extending from a stem into a lea Cf: *branch trace*.

leafy Pertaining to a sedimentary structure resembling a lea or said of a rock containing such a structure; e.g. "leafy post a thinly laminated sandstone containing micaceous laye (as in Durham, Eng.).

league (a) Any of various linear units of distance, varyin from about 2.42 to 4.6 statute miles; esp. "land league" (a English land unit equal to 3 statute miles) and "marin league" (a marine unit equal to 3 nautical miles). (b) Any various units of land area equal to a square league; esp. a old Spanish unit for the area of a tract 5000 varas squar equal to 4428.4 acres (1792.1 hectares) in early Texas lan descriptions or equal to 4439 acres (1796 hectares) in o California surveys.

leak *stratigraphic leak*.

leaking mode A surface seismic wave which is imperfect trapped, so that its energy leaks or escapes across a laye boundary, causing some attenuation. Syn: *leaky wave*.

leaky aquifer A confined aquifer whose confining beds w conduct significant quantities of water into or out of the aqu fer. Cf: *leaky confining bed*.

leaky confining bed A confining bed through which water ca move into or out of the adjacent aquifer. Cf: *aquitard; leak aquifer*.

leaky wave *leaking mode*.

lean *low-grade* [ore dep].

lean cannel coal Cannel coal that is low in hydrogen an transitional to bituminous coal in rank. Cf: *subcannel coa* Syn: *semicannel coal*.

lean clay A clay of low to medium plasticity owing to a rela tively high content of silt or sand. Ant: *fat clay*.

lean coal *short-flame coal*.

leaping The sudden and radical *shifting* of a divide from on position to another, as where the valley system drained by a abstracted stream is "transferred and added in a moment t that of the master stream" (Cotton, 1958, p.69). Cf: *creepin*

least squares Any of several statistical methods for fitting line, curve, or higher-degree surface to a set of data suc that the sum of the squares of the distances of points to th fitted surface is minimized. Syn: *method of least squares*.

least-time path *minimum time path*.

leat An English dialectal syn. of *lead* [streams].

leatherstone *mountain leather*.

lebensspur A sedimentary structure left by a living organism a *trace fossil*. The term is also applied to a Holocene track c burrow. Etymol: German *Lebensspur*, "life mark". Pl: *le bensspuren*.

lechatelierite Naturally fused amorphous silica, occurring fulgurites and impact craters as a vitreous or glassy produc formed by the melting of quartz sand as a result of lightnin or of the heat generated by the impact of a meteorite; a natu ral *silica glass* formed at high temperatures. Also spellec *lechatelièrite*.

Le Chatelier's rule The statement in chemistry that, if condi tions of a system that is initially at equilibrium are changec the equilibrium will shift in such a direction as to tend to re store the original conditions.

lechosos opal A variety of precious opal exhibiting a deep green play of color; esp. a Mexican opal exhibiting emeral green play of color and flashes of carmine, dark violet, dar e, and purple.

ontite A colorless mineral found in bat guano: $(NH_4,K)Na-$
$O_4).2H_2O$.

toparatype *paralectotype*.

tostratotype A stratotype chosen among less clearly de-
ed sections or occurrences (Sigal, 1964).

totype One of the syntypes of a species or subspecies that
s been designated as *type specimen* after the original publi-
ion of the description of the species or subspecies. Cf: *hol-
pe; neotype*.

ge [mining] In mining, a quarry exposure or natural outcrop
a mineral deposit.

ge [geol] (a) A narrow, usually horizontal, shelf-like ridge
projection of rock, much longer than it is high, formed in a
k wall or on a cliff face. Also, the surface of such a rocky
ss. (b) A rocky outcrop; rock solid enough to form a ledge.
n: *ledge rock*.

ge [coast] (a) A platform composed of resistant rock,
med along a coast by differential wave erosion of softer
ks. (b) An underwater ridge of rocks, esp. one near the
re or connecting with and fringing the shore; also, a near-
re reef.

dian *Auversian*.

ikite A clay mineral: $K(Fe,Mg)_3(Si,Al)_8O_{20}(OH)_4$. It is the
octahedral analogue of illite.

morite A melanite-bearing nepheline syenite with more py-
ene and less melanite than borolanite with which it occurs
hannsen, 1939, p.262). Its name is derived from the Led-
re River, Scotland.

[glac geol] Said of the side or slope of a hill, knob, or
minent rock located away from which an advancing glacier
ice sheet moved; facing the downstream side of a glacier,
d relatively protected or sheltered from its abrasive action.
t: *stoss*.

[wind] n. (a) The part or side (as of a hill or prominent
k) sheltered or turned away from the wind. (b) The direc-
n or region toward which the wind is blowing, or the direc-
n or region opposite to that from which the wind is blowing.
adj. Pertaining to or located on the side sheltered from the
nd; leeward.

e configuration A configuration employing electrodes, the
ter two of which are the current and the inner three of
ich are the potential electrodes. Syn: *partitioning method*.

dune A general term for a dune formed to the leeward of a
rce of loose sand or of an obstacle of any kind, and gener-
y under a wind of constant direction. See also: *umbracer
ne; umbrafon dune*.

gte *laagte*.

shore A shore crossed by wind from the land and thereby
tected from strong wave action. Ant: *weather shore*.

-source dune *umbrafon dune*.

uwfonteinite A syn. of *hatherlite*. This name was suggested
cause the rock does not occur at Hatherley factory but at
euwfontein in the Bushveld (Transvaal, South Africa).

ward adj. (a) Said of the side or slope (as of a hill or
minent rock) sheltered or located away from the wind;
nwind. (b) Said of a tidal current running onshore and set-
g in the same direction as that in which the wind is blowing.
n. The lee side, or the lee direction. Ant: *windward*.

wave An *internal wave* occurring on the downstream side
a submarine ridge.

bank The bank of a stream situated to the left of an ob-
ver who is facing downstream.

-handed [cryst] Said of an optically active crystal which ro-
es the plane of polarization of light to the left. Cf: *right-
nded*. Syn: *levorotatory*.

-handed [paleont] *sinistral*.

-handed separation *left-lateral separation*.

-lateral fault A fault, the displacement of which is *left-later-*

al separation. Syn: *sinistral fault; left-lateral slip fault; left-slip
fault*.

left-lateral separation Movement of a *lateral fault* along which,
in plan view, the side opposite the observer appears to have
moved to the left. See also: *left-lateral fault*. Cf: *right-lateral
separation*. Syn: *left-handed separation*.

left-lateral slip fault *left-lateral fault*.

left-slip fault *left-lateral fault*.

left valve The valve lying on the left-hand side of a bivalve
mollusk when the shell is placed with the anterior end pointing
away from the observer, the commissure being vertical and
the hinge being uppermost. Abbrev: LV. Ant: *right valve*.

leg On a seismogram, a single cycle of more or less periodic
motion in a wave train. Cf: *leggy*.

legal geology The application of expert knowledge of geology,
esp. structural and economic geology, in testifying in litigation
concerning ownership of or other rights to economic deposits
or geologically significant areas or structures.

legato injection An igneous intrusion that was formed in a single stage or event.

legend A brief explanatory list of the symbols, cartographic
units, patterns (shading and color hues), and other cartographic conventions appearing on a map, chart, or diagram. On a
geologic map, it shows the sequence of rock units, the oldest
at the bottom and the youngest at the top. The legend formerly included a textural inscription of, and the title on, the map
or chart. Syn: *explanation; key*.

leggy Said of a wave train that contains a number of similar
cycles or *legs*. Such a wave train is caused by a filter band-
pass that is too narrow. Syn: *tailing*.

legrandite A yellow to nearly colorless mineral: $Zn_2(AsO_4)-$
$(OH).H_2O$.

legume A dry, dehiscent fruit that is developed from a single
carpel and that splits along two seams.

lehiite A white mineral: $(Na,K)_2Ca_5Al_8(PO_4)_8(OH)_{12}.6H_2O$ (?).

lehm A term used in Alsace, France, for *loess*. Etymol: German *Lehm*, "loam".

lehmanite An obsolete term, proposed by Pinkerton but never
widely used, for an igneous rock containing feldspar and
quartz (Johannsen, 1939, p.262).

leidleite A glassy variety of dacite or rhyodacite containing
microlites, not phenocrysts, of calcic plagioclase and pyroxene, with accessory apatite and opaque oxides.

leifite A colorless mineral: $Na_2(Si,Al,Be)_7(O,OH,F)_{14}$.

leightonite A pale-blue mineral: $K_2Ca_2Cu(SO_4)_4.2H_2O$.

leiosphaerid A thin-walled, more or less spherical body of
probable algal relationship, characterized by the genus *Leios-
phaeridia*, and usually referred to the acritarchs. It is mostly
Ordovician to Silurian in age.

lekolith A term proposed by Coats (1968, p.71) for "a mass of
extrusive igneous rock more or less equant in plan, with a
nearly level upper surface, commonly a lower surface determined by the shape of the basin that it filled, and a diameter
greater than its depth"; e.g. a mass formed by a congealed
lava lake. Etymol: Greek *lekos*, "dish", + *lithos*, "stone".

Lemberg's stain A test used to distinguish calcite from dolomite. A solution of logwood in an aqueous solution of aluminum
chloride is used to stain the minerals by boiling; calcite and
aragonite become violet, whereas dolomite does not change
color. Cf: *Meigen's solution*.

Lemnian bole A gray to yellow or red clay obtained from Lem-
nos (island in the Aegean Sea) and formerly used for medicinal purposes. Syn: *Lemnian earth; terra Lemnia*.

lemoynite A mineral: $(Na,Ca)_3Zr_2Si_8O_{22}.8H_2O$.

Lemuria An imaginary continent, beloved by science-fiction
writers, alleged to have occupied most of the central Pacific
Ocean until historic time, when it sank, leaving only the Pacific islands as tiny remnants. The dispersal of Polynesian peo-
ples and cultures is supposed to have been facilitated by the

existence of Lemuria, but this dispersal is easily explained otherwise. Geologically, the existence of such a continent, either modern or ancient, is impossible.

lenad (a) A group name for the *feldspathoid* standard minerals. (b) A mnemonic term for leucite and nepheline.----Etymol: *leucite* + *nepheline* + *ad.*

lengenbachite A steel-gray mineral: $Pb_6(Ag,Cu)_2As_4S_{13}$.

length [paleont] (a) The distance from the most posterior part (normally the umbo) of a brachiopod valve to the farthest point on the anterior margin, measured on or parallel with the commissural plane (containing the cardinal margin and either commissure of rectimarginate shell or points on anterior commissure midway between crests of folds in both valves) in the plane of symmetry (at right angles to the width and thickness). (b) The distance between two planes perpendicular to the hinge axis of a bivalve-mollusk shell and just touching the anterior and posterior extremities of the shell. Cf: *height.*

length [lake] The shortest distance, through the water or on the water surface, between the most distant points on a lake shore.

length of overland flow Distance along the ground surface, projected to the horizontal, of nonchannel flow from a point on the drainage divide to a point of contact with a definite stream channel; the length is always measured at right angles to the contour lines in the drainage basin. Symbol: L_g. Cf: *critical length.* Syn: *slope length.*

lennilite (a) A green variety of feldspar (orthoclase) from Lenni Mills, Delaware County, Penna. (b) A vermiculite mineral.

lenoblite A mineral: $V_2O_4.2H_2O$.

lens n. A geologic deposit bounded by converging surfaces (at least one of which is curved), thick in the middle and thinning out toward the edges, resembling a convex lens; e.g. an orebody having a length many times greater than its width and pinching out laterally at its extremities. Geologic lenses may be double-convex or plano-convex. See also: *sand lens; lentil.* Syn: *lense.*---v. (a) To deposit or form a geologic lens. (b) To disappear laterally in all directions; e.g. a lentil is said to "lens out" in both directions within a mapped area.

lensing The thinning out of a stratum in one or more directions; e.g. the disappearing laterally of a lentil.

lentelliptical Lenticular and elliptical; e.g. said of a lens-shaped radiolarian shell with elliptical outline.

lenticel In a plant, a pore through which the exchange of gases occurs. In a woody stem, lenticels occur in the bark (Fuller & Tippo, 1949, p.963).

lenticle (a) A large or small lens-shaped stratum or body of rock; a *lentil.* (b) A lens-shaped rock fragment of any size.

lenticular (a) Resembling a lens in shape, esp. a double-convex lens. The term may be applied to a body of rock, to a sedimentary structure, to a geomorphologic feature, or to a mineral habit. (b) Pertaining to a geologic lens or lentil. --Syn: *lentiform.*

lenticule A small *lentil.*

lentiform *lenticular.*

lentil (a) A minor rock-stratigraphic unit of limited geographic extent, being a subdivision of a formation and similar in rank to a *member*, and thinning out in all directions; "a geographically restricted member that terminates on all sides within a formation" (ACSN, 1961, art.7). Term originated by Keith (1895). Cf: *tongue* [stratig]. (b) A thin-edged, lens-shaped body of rock, enclosed by strata of different material; a geologic *lens.* See also: *lenticule; lenticle.*

lentil ore *liroconite.*

lentocapillary point An obsolete term for the moisture content at which water movement through the soil becomes slow (Jacks, et al, 1960, p.54).

Leonardian North American provincial series: Lower Permian (above Wolfcampian, below Guadalupian).

leonhardite A zeolite mineral: $Ca_2Al_4Si_8O_{24}.7H_2O$. It is a variety of laumontite altered by partial loss of water.

leonhardtite *starkeyite.*

leonite A colorless, white, or yellowish monoclinic mineral: $K_2Mg(SO_4)_2.4H_2O$.

leopardite An igneous rock composed of small quartz phenocrysts in a microgranitic groundmass of quartz, orthoclase, albite, and mica. Iron and manganese hydroxide stains give the rock a characteristic streaked or spotted appearance.

leopoldite *sylvite.*

leperditiid Any ostracode belonging to the order Leperditicopida, characterized by a large, strongly calcified, thick-walled shell that is usually smooth, but sometimes finely ornamented to nodose and that has a long, straight hinge, large muscle scar pattern, and secondary shell layers. Leperditiids are commonly four or five times larger than other ostracodes. Their stratigraphic range is Lower Ordovician (possibly Upper Cambrian) to Upper Devonian.

lepidoblastic Pertaining to a *homeoblastic* type of texture of a foliated or schistose rock that is due to the development during recrystallization of minerals with a flaky or scaly habit, e.g. mica, chlorite.

lepidocrocite A ruby-red or blood-red to reddish-brown mineral: γ-FeO(OH). It is dimorphous with, but less common than goethite, and is associated with limonite in iron ores.

lepidodendrid n. An arborescent lycopod of the genus *Lepidodendron* that is well known from Carboniferous deposits.---adj. Pertaining to the genus *Lepidodendron* or to related genera.----Cf: *sigillarid.*

lepidolite A mineral of the mica group: $K(Li,Al)_3(Si,Al)_4O_{10}(OH)_2$. It commonly occurs in rose- or lilac-colored masses made up of small scales, as in pegmatites. Syn: *lithium mica; lithia mica; lithionite.*

lepidolith A thin, apparently homogeneous, elliptical coccolith, e.g. a surface plate of the coccolithophorid *Thorosphaera flagellata.*

lepidomelane A black variety of biotite with a high content of ferric iron. Syn: *iron mica.*

lepidote Said of a plant part that is covered with fine scales. Cf: *squamose.*

leptite A Fennoscandian term for a granular, quartzo-feldspathic, metamorphic rock formed by regional metamorphism of the highest grade (granulite facies). The terms *granulite* (meaning b), *leptynite*, and *hälleflinta* are approximately synonymous.

leptochlorite (a) A group name for chlorites of indistinct crystallization. (b) A group name for chlorites with a composition corresponding to: $(Mg,Fe^{+2},Al)_n(Si,Al)_4O_{10}(OH)_8$, where n is less than 6 (Hey, 1962, p.495).---Cf: *orthochlorite.*

leptogeosyncline An oceanic trough containing only minor sedimentation and associated with volcanism (Trumpy, 1955).

leptokurtic (a) Said of a frequency distribution that has a concentration of values about its mean greater than for the corresponding normal distribution. (b) Said of a narrow frequency distribution curve that is more peaked than the corresponding normal distribution curve.---Cf: *platykurtic; mesokurtic.*

leptoma A thin region of exine situated at the distal pole of a pollen grain and usually functioning as the point of emergence of the pollen tube. See also: *pseudopore* [palyn].

leptomorphic An obsolete syn. of *xenomorphic.*

leptopel Finely particulate, mainly colloidal organic and inorganic matter (such as silicates, hydrous oxides, or insoluble carbonates) occurring suspended in natural waters (Fox, 1957, p.383). Cf: *pelogloea.*

leptosporangiate A type of sporangial development occurring in the higher ferns, in which the sporogenous tissue arises from the outer segment of the initial cell (Haupt, 1953, p.437).

leptothermal Said of a hydrothermal mineral deposit formed mperature and depth conditions intermediate between *esothermal* and *epithermal*; also, said of that environment. *: hypothermal; xenothermal; telethermal.*

ptynite A French term for *leptite*.

ptynolite A fissile or schistose variety of hornfels containing ica, quartz, and feldspar with or without accessories such s andalusite and cordierite. The term was originated by Corer in 1868 (Holmes, 1928, p.139). Cf: *cornubianite; proteoe.*

rmontovite A mineral: $(U,Ca,Ce)_3(PO_4)_4.6H_2O(?)$.

stiwarite A syenite-aplite composed chiefly of microperthite, cmite, arfvedsonite, and accessory sphene. Its name is deved from Lestiware, Finland.

tdown The natural lowering (in the stratigraphic section) of abs and fragments of a resistant formation by weathering nd erosion of more vulnerable underlying rock; e.g. the stage " brecciation in which bedding is but little disturbed (Landes, 958, p.125).

tovicite A mineral: $(NH_4)_3H(SO_4)_2$.

ttsomite *cyanotrichite.*

ucaugite A white or grayish variety of augite resembling diopside: $CaMgSi_2O_6$.

uchtenbergite A mineral (clinochlore) of the chlorite group, ften resembling talc and containing little or no iron.

ucite A white or gray mineral of the feldspathoid group: $AlSi_2O_6$. It is an important rock-forming mineral in alkalic cks (esp. lavas), and usually occurs in trapezohedral crysls with a glassy fracture. Syn: *amphigene; grenatite; white arnet; Vesuvian garnet; vesuvian.*

ucitite A fine-grained or porphyritic extrusive or hypabyssal neous rock chiefly composed of pyroxene (esp. titanaugite) nd leucite, with little or no feldspar and without olivine.

ucitohedron *trapezohedron.*

ucitolith An extrusive rock composed almost entirely of leuite.

ucitophyre A porphyritic extrusive rock composed chiefly of ucite, nepheline, and augite. Cf: *haüynophyre.*

ucochalcite *olivenite.*

ucocratic Light colored; applied to a light-colored igneous ock relatively poor in mafic minerals. The percentage of nafic minerals permissible for a rock to be classified as "leuocratic" varies among petrologists, but is usually given as ess than 30 to 35 percent. Cf: *melanocratic; mesocratic.* Joun: *leucocrate.* Syn: *light-colored.*

ucon A sponge or sponge larva in which the flagellated hambers are connected to both exhalant and inhalant canals nd do not open directly either to the cloaca or to the exterior xcept through a canal. Cf: *ascon; sycon.* See also: *rhagon.* dj: *leuconoid.*

ucophanite A glassy greenish to pale-yellow mineral: $(Na, a)_2BeSi_2(O,F,OH)_7$. Syn: *leucophane.*

ucophoenicite A light purplish-red mineral: $Mn_7Si_3O_{12}(OH)_2$.

ucophosphite A white mineral: $K_2Fe_4(PO_4)_4(OH)_2.9H_2O$.

ucophyre A term originally applied to altered diabase in vhich the feldspar has been altered to saussurite, kaolin, and hlorite. This usage is obsolete, but the term is occasionally sed for a light-colored hypabyssal rock, being the antithesis f *lamprophyre.*

ucophyride A term used in the field for any light-colored porhyritic igneous rock with a fine-grained groundmass.

ucopyrite *loellingite.*

ucosome The light-colored part of a migmatite, usually rich n quartz and feldspar (Mehnert, 1968, p.355). Cf: *melanoome.*

ucosphenite A white to grayish-blue monoclinic mineral: aNa_4Ti_2B_2Si_{10}O_{30}.

ucotephrite A tephrite containing leucite as the only sodic eldspathoid. The term does not mean "light-colored tephrite".

ucoxene A general term for fine-grained, opaque, whitish alteration products of ilmenite, commonly consisting mostly of rutile and partly of anatase or sphene, and occurring in some igneous rocks. The term has also been applied to designate a variety of sphene.

leurodiscontinuity A term proposed by Sanders (1957, p.295) for an unconformity characterized by a regular surface. Cf: *trachydiscontinuity.*

levee [marine geol] An embankment of sediment, bordering one or both sides of a fan valley or a deep-sea channel. It is similar to a river-channel levee in the subaerial environment.

levee [stream] (a) *natural levee.* (b) An artificial embankment, usually of random earth fill, built along the bank of a watercourse or an arm of the sea and designed to protect land from inundation or to confine streamflow to its channel. Cf: *floodwall.* Syn: *earth dike.* (c) A landing place along a river; a pier or quay. (d) *mudflow levee.*---Etymol: French *levée.*

levee delta A delta having the form of a long narrow ridge, resembling a natural levee (Dryer, 1910).

levee-flank depression *backswamp depression.*

levee lake A lake formed by a natural levee that acts as a barrier or enclosure for holding water. Types of levee lakes include: *lateral levee lake; delta levee lake.*

levee ridge The elevated strip of land upon which a river flows, produced by the building up of the stream bed and the natural levees on each side.

level [geog] n. Any large expanse of relatively flat, usually (but not necessarily) low-lying country, unbroken by noticeable elevations or depressions; specif. any flat, alluvial tract of recent formation, such as the Bedford Level in Lincolnshire, Eng.

level [surv] n. (a) A *leveling instrument.* (b) A device or attachment for finding a horizontal line or plane or for adjusting an instrument to the horizontal; specif. a *spirit level.* (c) A measurement of the difference of altitude of two points on the Earth's surface by means of a level.---v. To find the heights of different points by means of a level.

level [water] (a) An open reach of water in a stream or canal, such as between two canal locks. (b) The elevation of the surface of a body of water; a water table.

level fold *nonplunging fold.*

level ice Sea ice that is, or has been, unaffected by deformation, displaying a flat surface, and typically occurring in undisturbed waters. Ant: *pressure ice.*

leveling The operation of determining the comparative altitude of different points on the Earth's surface, usually by sighting through a leveling instrument at one point to a level rod at another point. Also, the finding of a horizontal line or the establishing of grades (such as for a railway roadbed) by means of a level. See also: *spirit leveling; direct leveling; indirect leveling.* Also spelled: *levelling.*

leveling instrument An instrument for establishing a horizontal line of sight, usually by means of a spirit level or a pendulum device; e.g. a *surveyor's level* and a *pendulum level.* It is used, with a level rod, to determine differences in elevation between two widely separated points on the Earth's surface. Syn: *level [surv].*

leveling rod A syn. of *level rod.* Also known as a "leveling pole" or "leveling staff".

leveling screw One of three or more adjusting screws for bringing an instrument (such as a surveyor's level) to the horizontal.

levelman A surveyor who operates a leveling instrument.

level of compensation Var. of *depth of compensation.*

level of saturation *water table.*

level of zero amplitude The depth (below the Earth's surface) below which the temperature gradient of permafrost is relatively stable throughout the year.

level rod A straight rod or bar, with a flat face graduated in plainly visible linear units with zero at the bottom, used in measuring the vertical distance between a point on the Earth's surface and the line of sight of a leveling instrument that has been adjusted to a horizontal position. It is usually made of metal or well-seasoned wood. See also: *target rod; speaking rod.* Syn: *rod; leveling rod; surveyor's rod.*

level surface *equipotential surface.*

level trier An apparatus for use in measuring the angular value of the divisions of a spirit level.

leverrierite A discredited name for a clay mineral known to be kaolinite, or a mixture of alternating plates of kaolinite and muscovite, or a mixture of kaolinite and illite.

levorotatory *left-handed.*

levynite A white or light-colored zeolite mineral: $(Na,Ca)_2(Al, Si)_9O_{18}.8H_2O$. Syn: *levyne; levyine; levyite.*

Lewisian A division of the Precambrian in Scotland. Syn: *Hebridean.*

lewisite A mineral: $(Ca,Fe,Na)_2(Sb,Ti)_2O_7$. It is a titanian romeite.

Lewistonian Stage in New York State: Middle Silurian (Niagaran). It was formerly regarded as upper Lower Silurian.

lewistonite A white mineral of the apatite group: $(Ca,K, Na)_5(PO_4)_3(OH)$. It is a potassium-rich variety of hydroxylapatite. Cf: *dehrnite.*

lexicon An alphabetic compilation of geologic names, accompanied by formal definitions that state the lithology, thickness, age, underlying and overlying formations, type locality, and original reference; e.g. the "Wilmarth lexicon" containing 13,090 names (Wilmarth, 1938) and the "Keroher lexicon" containing 14,634 names (Keroher et al, 1966).

***Lg* wave** Short-period, higher-mode Love and Rayleigh waves with a group velocity of about 3.5 km/sec that travel over long paths in the continental crust only. Cf: *Rg wave.*

lherzite A hornblendite composed chiefly of brown hornblende, with minor amounts of biotite, ilmenite, and, occasionally, garnet.

lherzolite Peridotite composed chiefly of olivine, orthopyroxene, and clinopyroxene, and in which olivine is generally most abundant; a two-pyroxene peridotite. Cf: *bielenite.*

Lias Middle European series: Lower Jurassic (above Triassic, below Dogger). Syn: *Liassic.*

lias A bluish or whitish, compact, argillaceous limestone or cement rock, typically interbedded with shale or clay; esp. such a limestone quarried in Somerset and other parts of SW England. Syn: *lyas.*

Liassic *Lias.*

liberite A mineral: Li_2BeSiO_4.

libethenite An olive-green to dark-green orthorhombic mineral: $Cu_2(PO_4)(OH)$.

libollite A variety of *albertite* from Angola.

libration The small, angular change in the face that a body, e.g. the Moon, presents toward the Earth. Only a tiny part is due to dynamic rotational motion (physical libration). In the case of the Moon, it is due primarily to the fact that although the Moon is in synchronous rotation, its rotation is uniform while its rate of revolution varies due to orbital eccentricity, producing longitudinal geometric libration; and also the fact that its rotational axis is not exactly perpendicular to the plane of its orbit (producing latitudinal geometric libration).

libriform Said of wood fibers that are thick-walled, elongate, and have simple pits.

librigene The *free cheek* of a trilobite. Syn: *librigena.*

Libyan Desert glass Silica glass from the Libyan Desert, possibly a tektite or an impactite.

lichen A thallophytic plant of the order Lichenes that is composed of a fungus and an alga living in a symbiotic relationship. The alga is protected by the fungus, which in turn relies upon the alga for the production of food.

lichenometry Measurement of the diameter of lichens growing on exposed rock surfaces as a method of dating geomorph features. Cf: *fossil geochronometry.*

lichen polygon A highly specialized *vegetation polygon* which the pattern appears to be confined to the thick vegeta tion (reindeer moss, as in northern Quebec) border itse (Rousseau, 1949, p.50). The polygon sides are 30-50 cm length.

lick *salt lick.*

lido (a) An Italian term for a barrier beach; e.g. the one pre tecting the lagoon of Venice. (b) A bathing beach at a sea side resort, but now extended to include those at freshwat and artificial-lake resorts. Type example: the Lido near Ve ice. Syn: *plage.*

lie A British term for the disposition of topographic features for the slope of the land surface.

liebenerite A variety of pinite containing alkalies, iron, and ca cium.

liebigite An apple-green or yellow-green mineral: $Ca_2U (CO_3)_4.10H_2O$. It occurs as secondary concretions or coa ings. Syn: *uranothallite.*

Liesegang banding *Liesegang rings.*

Liesegang rings Secondary, nested rings or bands caused b rhythmic precipitation within a fluid-saturated rock. Syn *Liesegang banding.*

lievrite *ilvaite.*

life assemblage *biocoenosis.*

lift [geog] A slight rise or elevation of the ground.

lifting Process by which a stream deepens its bed, involvin the upward movement of bottom rock particles into the turbu lent part of the stream where they are carried away or floate off in suspension (Cleland, 1916, p.84).

lift joint A horizontal tension joint in massive rock such a granite, probably formed as a result of removal of load pres sure during quarrying; a type of *strain break.*

ligament A tough structure of connecting tissue in an anima esp. a chitinous or horny elastic band in bivalve mollusks co necting the valves of the shell dorsally along a line adjacent the umbones and acting as a spring to open the valves whe the adductor muscles relax.

ligamentary articulation A type of articulation of crinoid o sicles effected solely by ligaments but sometimes supplemen ed by calcareous deposition. Cf: *muscular articulation.*

ligament field A concave or flat part of a crinoid articular fac for the attachment of ligaments.

ligament groove A narrow depression in the cardinal area of bivalve mollusk for the attachment of the fibers of a ligamen Cf: *ligament pit.*

ligament pit (a) A relatively broad depression in the cardina area of a bivalve mollusk for the attachment of the ligamen Cf: *ligament groove.* (b) A generally steep-sided small de pression in a crinoid dorsal-ligament fossa adjoining the cen ter of the transverse ridge.

light-colored Said of a rock-forming mineral that is light color and generally also light in weight; also, said of the roc such minerals form; *leucocratic.* Cf: *dark-colored.* Syn: *ligh mineral.*

lighthouse A term used in Kentucky for a *natural bridge.*

light mineral [sed] (a) A rock-forming mineral of a detrita sedimentary rock, having a specific gravity lower than a stan dard (usually 2.85); e.g. quartz, feldspar, calcite, dolomite muscovite, feldspathoids. Cf: *heavy mineral.* (b) A light-co ored mineral.

lightning stone *fulgurite.*

lightning tube A tubular *fulgurite.*

light oil Crude oil that has a high Baumé gravity or A.P. gravity. Cf: *heavy oil.*

light red silver ore *proustite.*

light ruby silver *proustite.*

lignilite Obsolete syn. of *stylolite.*

lignin An organic substance somewhat similar to carbohydrates in composition that occurs with cellulose in woody plants.

lignite (a) A brownish-black coal that is intermediate in coalification between peat and subbituminous coal; consolidated coal with a calorific value less than 8300 BTU/pound, on a moist, mineral-matter-free basis. Cf: *brown coal; lignite A; lignite B.*

lignite A *Lignite* that contains 6,300 or more BTU/pound but less than 8,300 BTU/pound. Cf: *lignite B.* Syn: *black lignite.*

lignite B *Lignite* that contains less than 6,300 BTU/pound; essentially synonymous with *brown lignite* or *brown coal.*

ligule A term used for various straplike plant structures, e.g. a membranous structure internal to the leaf base in the lycopods *Isoetes* and *Selaginella*, or the limb of the ray flowerets of a composite plant.

ligurite An apple-green variety of sphene.

likasite A sky-blue orthorhombic mineral: $Cu_{12}(NO_3)_4(PO_4)_2(OH)_{14}$.

likely Said of a rock, lode, or belt of country that gives indications of containing valuable minerals. Syn: *kindly.* Ant: *hungry.*

lillianite A steel-gray mineral: $Pb_3Bi_2S_6$.

lily pad *stool stalagmite.*

lily-pad ice A term used for *pancake ice* consisting of circular pieces of ice that are not more than about 50 cm in diameter.

liman (a) A shallow, muddy, branching, and isolated lagoon, bay, or marshy lake, formed at the mouth of a river behind the seaward deposits of a delta and protected by a barrier or a spit; an *estuary* or broad freshwater bay of the sea. Etymol: Russian, from Greek *limen*, "harbor". (b) An area of mud or slime deposited near the mouth of a river.

liman coast A coast with many lagoons (*limans*) and drowned valleys, protected from the open sea by a barrier or a spit; e.g. the northern coast of the Black Sea.

limb [palyn] *equatorial limb.*

limb [fold] The side of a fold; each limb is common to its adjacent fold. The term may be modified according to its structural relations. Syn: *flank.* Obs. syn: *shank.*

limb [surv] (a) The graduated margin of an arc or circle in an instrument for measuring angles, such as the part of a marine sextant carrying the altitude scale. (b) The graduated staff of a leveling rod.

limb [astron] The outer edge of a lunar or planetary disk (Baldwin, 1965, p.141).

limbate Having a thickened border or edge of a foraminiferal chamber, commonly at the suture but sometimes elevated.

limb darkening A photometric function of the Moon in which, at full moon, the center of the disk is much brighter than the limbs, evidently because of the Moon's roughness at optical dimensions.

limburgite A dark-colored, porphyritic extrusive igneous rock having olivine and clinopyroxene as phenocryst minerals in an alkali-rich glassy groundmass which may have microlites of clinopyroxene, olivine, and opaque oxides; some nepheline and/or analcime may be present, and feldspars are typically absent. Its name is derived from Limburg, Germany. Partial syn: *magma basalt.*

limbus A crease at the edge of the vesicle of a vesiculate pollen grain (or at the edge of the pseudosaccus of a pseudosaccate grain) in which the proximal and distal exine layers are more or less fused.

lime (a) Calcium oxide, CaO; specif. quicklime and hydraulic lime. The term is used loosely for calcium hydroxide (as in *hydrated lime*) and incorrectly for calcium carbonate (as in *agricultural lime*). (b) A term commonly misused for calcium in such deplorable expressions as "carbonate of lime" or "lime feldspar". (c) A limestone. The term is sometimes used by drillers for any rock consisting predominantly of calcium carbonate.

limeclast A lithoclast derived by erosion from an older limestone; also, an intraclast disrupted from partly consolidated calcareous mud on the bottom of a sea or lake.

lime concretion A concretion in soil, having a variable shape and size, consisting of an aggregate of precipitated calcium carbonate or of other material cemented by precipitated calcium carbonate.

lime feldspar A misnomer for *calcium feldspar.*

lime mica *margarite* [mineral].

lime mud The unconsolidated micritic component of a limestone.

lime mudstone A term proposed by Dunham (1962) for a fairly pure (93-99% calcium carbonate), mainly nonporous and impermeable, texturally uniform limestone whose main constituent (75-85%) is calcite mud (micrite). See also: *micritic limestone.*

lime olivine *calcio-olivine.*

lime pan [geomorph] A playa with a smooth, hard surface of calcium carbonate, commonly tufa.

lime pan [soil] A type of *hardpan* consisting of a thick, hard soil layer cemented chiefly with calcium carbonate. Also spelled: *limepan.*

lime rock A term used in SE U.S. (esp. Florida and Georgia) for an unconsolidated or partly consolidated form of limestone, usually containing shells or shell fragments, with a varying percentage of silica. It hardens on exposure and is sometimes used as road metal. Also spelled: *limerock.*

lime-silicate *calc-silicate.*

lime sink Var. of *sinkhole.*

lime-soda feldspar A misnomer for *sodium-calcium feldspar.*

limestone (a) A sedimentary rock consisting chiefly (more than 50% by weight or by areal percentages under the microscope) of calcium carbonate, primarily in the form of the mineral calcite, and with or without magnesium carbonate; specif. a carbonate sedimentary rock containing more than 95% calcite and less than 5% dolomite. Common minor constituents include silica (chalcedony), feldspar, clays, pyrite, and siderite. Limestones are formed by either organic or inorganic processes, and may be detrital, chemical, oolitic, earthy, crystalline, or recrystallized; many are highly fossiliferous and clearly represent ancient shell banks or coral reefs. Limestones include chalk, calcarenite, coquina, and travertine, and they effervesce freely with any common acid. Abbrev: ls. (b) A general term used commercially (in the manufacture of lime) for a class of rocks containing at least 80% of the carbonates of calcium or magnesium and which, when calcined, gives a product that slakes upon the addition of water.

limestone log A log that uses a resistivity measuring device consisting of four current electrodes (symmetrically arranged so that two sets contain two electrodes that are 4-5 in. apart) and of a measuring electrode placed in the middle of the device at a distance of 30-35 in. from the two sets of current electrodes. It is used to give an accurate resistivity log in borehole surveying of hard formations (such as limestones). Syn: *limestone lateral.*

limestone pavement (a) A karst surface on limestone. Cf: *karrenfeld.* (b) A limestone bedding-plane surface in a karst area that is divided into *clints* by *grikes.* Cf: *crevice karst.*

limestone sink Var. of *sinkhole.*

lime uranite *autunite.*

limewater Natural water with large amounts of dissolved calcium bicarbonate or calcium sulfate.

limiting beds The oldest strata immediately above and the youngest strata immediately below an angular unconformity; they are used to date the folding and erosion (Spieker, 1956).

limn A combining form signifying fresh water.

limnal Pertaining to a body or bodies of freshwater, esp. to a lake or lakes.

limnetic (a) Relating to the pelagic or open part of a body of fresh water. (b) Said of lake-dwelling organisms and communities that are free from direct dependence on the bottom or shore.----Syn: *limnic*.

limnic [coal] (a) Said of coal deposits formed inland in freshwater basins, peat bogs, or swamps, as opposed to *paralic* coal deposits. (b) Said of peat formed beneath a body of standing water. Its organic material is mainly planktonic.

limnic [lake] (a) Pertaining to a body of freshwater. (b) *limnetic*.

limnite *bog iron ore.*

limnogenic rock A sedimentary rock formed by precipitation from freshwater, esp. that of a lake (Grabau, 1924, p.329).

limnogeology The geology of lakes.

limnogram A record of lake-level variations as recorded by a water-level gage, such as a record made on a limnimeter. Syn: *limnograph*.

limnograph *limnogram.*

limnokrene *spring lake.*

limnology The scientific study of the physical, chemical, meteorological, and esp. the biological and ecological, conditions and characteristics in pools, ponds, and lakes and by extension in all inland waters. Etymol: Greek *limne*, "marsh, lake, pool".

limon (a) Viscous mud deposited during floods by rivers of the Mediterranean basin, the Atlantic coast of Morocco, and western Africa, and characterized by a binder of fine iron-hydroxide grains. (b) A widespread, fine-grained, surficial deposit of periglacial loam in northern France, from which brown loamy soils have developed. It is probably of windblown and wind-deposited origin, but different from loess in that it is formed under a more humid climate. (c) A term sometimes used as a French syn. of *loess*.---Etymol: French, "loam, silt, ooze, mud". Pron: lee-*mone*.

limonite A general field term for a group of brown, amorphous, naturally occurring hydrous ferric oxides whose real identities are unknown. Limonite was formerly thought to be a distinct mineral ($2Fe_2O_3.3H_2O$), but is now considered to have a variable composition (and variable chemical and physical properties) and to consist of any of several iron hydroxides (commonly goethite) or of a mixture of several minerals (such as hematite, goethite, and lepidocrocite) with or without presumably adsorbed additional water. It is a common secondary material formed by oxidation (weathering) of iron or iron-bearing minerals, and it may also be formed as an inorganic or biogenic precipitate in bogs, lakes, springs, or marine deposits; it occurs as coatings (such as ordinary rust), as loose or dense earthy masses, as pseudomorphs after other iron minerals, and in a variety of stalactitic, fibrous, reniform, botryoidal, or mammillary forms, and it represents the coloring material of yellow clays and soils. Limonite is commonly dark brown or yellowish brown, but may be yellow, red, or nearly black; it is a minor ore of iron. See also: *bog iron ore.* Syn: *brown iron ore; brown hematite; brown ocher.*

limurite A metasomatic rock found at the contact of calcareous rocks with intruded granite and consisting of over 50% axinite. Other minerals include diopside, actinolite, zoisite, albite, and quartz. The term was originated by Zirkel in 1879.

limy (a) Containing a significant amount of lime or limestone; e.g. "limy soil". (b) Containing calcite; e.g. "limy dolomite" (a calcitic dolomite rock).

lin A var. of *linn.*

linarite A deep-blue monoclinic mineral: $PbCu(SO_4)(OH)_2$.

lindackerite A light-green or apple-green mineral: H_2Cu-$(AsO_4)_4.8-9H_2O$. It may contain a little nickel or cobalt.

Lindblad-Malmquist gravimeter *Boliden gravimeter.*

lindgrenite A green mineral: $Cu_3(MoO_4)_2(OH)_2$.

Lindgren's volume law The statement that during metasomatic formation of ore, there is not change in rock volume or form (Lindgren, 1933).

lindinosite An igneous rock composed of more than 50 percent riebeckite, with quartz and microcline (Thrush, 196?, p.644).

lindoite A light-colored hypabyssal rock characterized b? bostonitic texture and similar in composition to *solvsbergit?* but being quartz-rich and poor in dark-colored minerals; th? extrusive equivalent of an alkalic granite.

lindstromite A lead-gray to tin-white mineral: $PbCuBi_3S_6$. Also spelled: *lindstrômite*.

line [cart] A mark on a map, indicating a boundary, division or contour.

line [seism] A linear array of seismologic observation points.

lineage In evolution, a line of descent. Although it is some times used synonymously with *evolutionary series*, it usuall? refers to a particular line of descent within the evolutiona? plexus.

lineage boundary The surface along which *plane defects* in crystal occur.

lineament [lunar] A conspicuous linear feature on the surfac? of the Moon; e.g. a rille, wrinkle ridge, crater chain, ray, an? fault. Also, a less distinctive lunar feature, such as an elon? gated valley, a mountain ridge, and a straight section of crater wall. Lineaments may represent surface manifestation? of regional stresses within the Moon.

lineament [photo] Any line, on an aerial photograph, that ? structurally controlled, including any alignment of separat? photographic images such as stream beds, trees, or bushe? that are so controlled. The term is widely applied to lines rep? resenting beds, lithologic horizons, mineral bandings, veins faults, joints, unconformities, and rock boundaries (Allum 1966, p.31).

lineament [tect] Straight or gently curved, lengthy features ? the Earth's surface, frequently expressed topographically a? depressions or lines of depressions; these are prominent o? relief models, high-altitude air photographs, and radar imag? ery. Their meaning has been much debated; some certain? express valid structural features, such as faults, aligned vo? canoes, and zones of intense jointing with little displacemen? but the meaning of others is obscure, and their origins may b? diverse, or purely accidental. Syn: *linear.*

linear [geol] adj. Arranged in a line or lines, as a linear dik? swarm. It is a one-dimensional arrangement, in contrast t? the two-dimensional *planar* arrangement.

linear [tect] n. A syn. of *lineament.*

linear absorption coefficient A quotient of the internal absorp? tance of a path element traversed by the radiation, by th? length dx of that element (Nicodemus, 1971, p.129). Syn: *ab? sorption coefficient.* Partial syn: *absorption factor.*

linear correlation The *correlation* of two or more variable? measured by means of their straight-line relationship. If th? correlation is perfect (unity or negative unity), the value ? one variable is proportional to that of the others; if the corre? lation is absent (zero), there is no predictability of one valu? given that for any other.

linear element A fabric element having one dimension that ? much greater than the other two. Cf: *planar element; equan? element.*

linear flow structure *platy flow structure.*

linear parallel texture The *parallel texture* of a rock in whic? the constituents are parallel to a line and not just to a plane as in *plane parallel texture.*

linear scale ratio In model analysis, a ratio of the length in th? prototype to the length in the model (Strahler, 1958, p.291) Symbol: λ.

linear selection Natural selection favoring variation in a partic? ular direction.

lineation [sed] Any linear structure, of megascopic or micro? scopic nature, on or within a sedimentary rock, and esp? characterizing a bedding plane; e.g. a ripple mark, a sol?

mark, or a linear parallelism in fabric caused by preferred alignment of long axes of clasts or fossils at the time of deposition. It is largely the product of current action. See also: parting lineation.

lineation [struc geol] A general, nongeneric term for any linear structure in a rock, of whatever scale; e.g. flow lines, slickensides, linear arrangements of components in sediments, or axes of folds. Lineation in metamorphic rocks includes mineral streaking and stretching in the direction of transport, crinkles and minute folds parallel to fold axes, and lines of intersection between bedding and cleavage, or of variously oriented cleavages.

line defect A type of *crystal defect* occurring along certain lines in the crystal structure. Cf: *plane defect; point defect.* See also: *screw dislocation.* Syn: *dislocation [cryst].*

line map *planimetric map.*

line of collimation The *line of sight* of the telescope of a surveying instrument, defined as the line through the rear nodal point of the objective lens of the telescope and the center of the reticle when they are in perfect alignment. Syn: *collimation line.*

line of concrescence The proximal line of the junction or fusion in an ostracode of the duplicature with the outer lamella.

line of dip The direction of the angle of dip, measured in degrees by compass direction. It generally refers to true dip, but can be said of apparent dip as well. Syn: *direction of dip.*

line of force [phys] In a *field of force,* a line that is perpendicular to every equipotential surface it intersects.

line of force [mag] *magnetic field line.*

line of induction *magnetic field line.*

line of section A line on a map, indicating the position of a *profile section.* It is the *profile line* of the section as seen in plan.

line of seepage *seepage line.*

line of sight (a) A line extending from an observer's eye or an observing instrument to a distant point (such as on the celestial sphere) toward which the observer is looking or directing the instrument; e.g. *line of collimation.* (b) The straight line between two points. It is in the direction of a great circle but does not follow the curvature of the Earth. (c) A line joining the Earth or the Sun and a distant astronomic body.

line of strike *strike.*

line rod *range rod.*

line scanner An airborne, orbital, or ground (tripod-mounted) image-forming system, generally incorporating either an electronic scanning system similar to the vidicon or image orthicon tube used in television reproduction, or a mechanical raster-scan system. Syn: *line-scan system; scanner.*

line-scan system *line scanner.*

line source A straight current element of infinite extent but infinitesimal cross section.

line spectrum The array of intensity values in the spectrum occurs in very short, distinct ranges (i.e. only certain wavelengths) of the ordering variable so that the spectrum appears to be a number of discrete lines with spaces between. An optical line spectrum results from electron transitions within atoms. Cf: *band spectrum; continuous spectrum.*

line squall A cold front as long as 300 miles that is characterized by *squalls* and their associated precipitation, thunder, and lightning. Syn: *squall line.*

line-up n. On a seismogram trace, alignment in phase.

lingulacean n. Any inarticulate brachiopod belonging to the superfamily Lingulacea, characterized by subequal, generally phosphatic valves, with the pedicle valve being slightly larger. Their stratigraphic range is Lower Cambrian to present.----adj. Said of a brachiopod having subequal phosphatic valves, or of the valves themselves. Var: *linguloid.*

lingulid Any lingulacean brachiopod belonging to the family Lingulidae, characterized mainly by an elongate oval to spatulate outline and a biconvex shell. Their stratigraphic range is Silurian (possibly Ordovician) to present. The genus *Lingula* belongs to this family and has frequently been used loosely for any Ordovician species in the family.

linguloid *lingulacean.*

linguloid ripple mark *linguoid ripple mark.*

linguoid current ripple *linguoid ripple mark.*

linguoid ripple mark An aqueous *current ripple mark* characterized by a tongue-shaped outline or having a barchan-like shape whose horns point into the current; it is best developed on the bottoms of shallow streams where it shows a highly irregular pattern with a wide range in the variety of forms. The term "linguoid" applied to a ripple or ripple mark was introduced by Bucher (1919, p.164). See also: *cusp-ripple.* Syn: *linguoid current ripple; linguloid ripple mark; cuspate ripple mark.*

linguoid sole mark *flute cast.*

lining (a) A brick, concrete, cast-iron, or steel casing placed around a tunnel or shaft to provide support. (b) A cover of clay, concrete, polythene, or other material, placed over the whole or part of the perimeter of a conduit or a reservoir to resist erosion, minimize seepage losses, withstand pressure, and improve flow.

lining pole *range rod.*

link (a) One of the 100 standardized divisions of a surveyor's chain, each consisting of iron rods or heavy steel wire looped at both ends and joined together by three oval rings, and measuring 7.92 inches in length. (b) A unit of linear measure equal to 7.92 inches or one one-hundredth of a chain.

linkage [mtns] The joining at a sharp angle of two branches of a mountain range, as shown in plan view. Cf: *deflection.*

linkage [paleont] A type of intratentacular budding in scleractinian corals, characterized by development of two or more mouths with stomodaea inside the same tentacular ring. See also: *direct linkage; indirect linkage.*

linked veins An ore-deposit pattern in which adjacent, more or less parallel veins are connected by diagonal veins or veinlets.

links [coast] A Scottish term for a narrow area of flat or undulating land built up along a coast by drifting sand, and covered with turf or coarse grass; in Scotland, such land is often used as a golf course.

links [stream] A winding of a river. Also, the ground along such a winding.

linn (a) A pool of water, esp. a deep one below a fall of water. (b) A torrent running over rocks; a waterfall, cataract, or cascade. (c) A precipice or a steep ravine.---The term is used chiefly in Scotland and northern England. Etymol: Gaelic *linne,* "pool". Syn: *lin; lyn; lynn.*

Linnaean Of or pertaining to the method of Carl von Linné (in Latin, Carolus Linnaeus) (d. 1778), the Swedish botanist who established the system of binomial nomenclature.

Linnaean species A species defined entirely on the basis of its morphology. Also spelled: *Linnean species.*

linnaeite (a) A pale steel-gray isometric mineral: $(Co,Ni)_3S_4$. It has a coppery-red tarnish and constitutes an ore of cobalt. Syn: *linneite; cobalt pyrites.* (b) A group of isomorphous nickel-bearing sulfides, including linnaeite, carrollite, siegenite, violarite, and polydymite.

linophyre An igneous rock characterized by *linophyric* texture.

linophyric A term, now obsolete, applied to porphyritic igneous rocks with the phenocrysts arranged in lines or streaks (Cross et al, 1906, p.703); of or pertaining to a *linophyre.*

linosaite A basaltic rock having alkalic affinities, by the presence of sodic pyribole or minor feldspathoid or by being associated with feldspathoid-bearing rocks. Its name is derived from Linosa, one of the Pelagie islands.

linsey A term used in Lancashire (county in NW England) for a strong, striped shale and a streaky, banded sandstone or siltstone, interbedded in such a manner as to resemble a mixed linen and woollen fabric ("linsey-woolsey"). Syn: *lin and wool.*

lintonite A greenish, agate-like variety of thomsonite from the Lake Superior region.

lip [eng] A low parapet erected on the downstream edge of a millrace or dam apron for the purpose of minimizing scouring of the river bottom.

lip [geomorph] (a) A projecting or overhanging edge, rim, or margin, such as of a rock on a mountainside. (b) A steep slope or abyss. (c) *crater lip.*

lip [paleont] (a) A margin of the aperture of a gastropod shell; e.g. *inner lip* and *outer lip.* (b) An elevated border of the aperture of a foraminiferal test. It may be small and at one side of the aperture, or completely surround it. (c) Either the labrum (upper lip) or labium (lower lip) of an arthropod.

Lipalian A name that was formerly used for the interval of time represented by a widespread unconformity separating Precambrian and Cambrian strata.

liparite A syn. of *rhyolite* used by German and Soviet authors. Its name is derived from Lipari Islands, in the Tyrrhenian Sea.

lip height The height above the ground surface to which earth materials have been piled around a crater formed by an explosion. See also: *rim height.*

lipid Any of several saponifiable oxygenated fats or fatty-acid containing substances such as waxes, exclusive of hydrocarbon and certain other nonsaponifiable ether-soluble compounds, which in general are soluble in organic solvents, but barely soluble in water. They, along with proteins and carbohydrates, are the principal structural components of living cells. Also spelled: *lipide.*

lipide Var of *lipid.*

lipogenesis In evolution, accelerated development as a result of the omission of certain ancestral stages. Cf: *bradytely; tachytely.*

lipotexite Nonliqufied basic material within anatectic magma (Dietrich & Mehnert, 1961). Also spelled: *lipotectite.*

lipscombite A mineral: $(Fe,Mn)Fe_2(PO_4)_2(OH)_2$.

liptinite *exinite.*

liptite *sporite.*

liptobiolite (a) A resistant plant material that is left behind after the less resistant parts of the plant have wholly decomposed and that is characterized by relative stability of composition; e.g. resin, gum, wax, amber, copal, and pollen. (b) *liptobiolith.*

liptobiolith A combustible organic rock formed by an accumulation of liptobiolites; e.g. spore coal and pollen peat. Syn: *liptobiolite.*

liptocoenosis In paleontology, an assemblage of dead organisms together with the traces and products of their life prior to burial. Its syn., *necrocoenosis*, is used more commonly in biology.

liquation In a magma, the separation of the residual liquid from earlier formed crystals.

liquefaction [soil] The sudden large decrease of the shearing resistance of a cohesionless soil, caused by a collapse of the structure by shock or strain, and associated with a sudden but temporary increase of the pore fluid pressure (ASCE, 1958, term 205). It involves a temporary transformation of the material into a fluid mass.

liquefaction [sed] The transformation of loosely packed sediment into a fluid mass preliminary to movement of a turbidity current by subaqueous slumping or sliding.

liquefaction slide The rapid and often catastrophic failure of a loose mass of predominantly cohesionless material which is generally at or near full saturation. The essential mechanism of such a slide is the sudden transfer of load from the particle contacts to the pore fluid, with resultant high transient pore fluid pressures and consequent loss of strength. Flow slides usually follow upon a disturbance (e.g. by earthquake or con-

ventional slide) and can occur both subaqueously and subae ially (Koppejan, *et al*, 1948). Syn: *flow slide.*

liquefied petroleum gas A compressed hydrocarbon gas o tained through distillation and usable as a motor fuel or in ce tain industrial processes. It is commonly abbreviated and r ferred to as LPG.

liquevitreous Said of a solid that is amorphous, uniform, a weak. The term is not recommended. Cf: *durovitreous.*

liquid chromatography A process for separating componen in a liquid phase from one another by passing them over solid or liquid stationary phase where the components a separated by their differential mobility rates. The techniqu used, based on the nature of the stationary phase, is ofte *column chromatography, paper chromatography,* or *thin-lay chromatography.* Cf: *gas chromatography.* See also: *chrom tography.*

liquid flow Movement of a liquid that is usually of low visco ity, involving either or both laminar and turbulent flow. Cf: *vi cous flow [exp struc geol]; solid flow.*

liquid immiscibility A process of magmatic differentiation i volving division of the magma into two or more separate liqui phases which are then separated from each other by gravi or other processes.

liquid inclusion A partial syn. of *fluid inclusion.*

liquidity index An expression for the consistence of a soil at i natural moisture content: its water content minus the wate content at the plastic limit, all divided by the plasticity index the liquid limit (Nelson and Nelson, 1967). Syn: *water-plasti city ratio; relative water content.*

liquid limit The water-content boundary between the semiliqu and the plastic states of a sediment, e.g. a soil. It is one o the *Atterberg limits.* Cf: *plastic limit.*

liquidus The locus of points in a temperature-composition d agram representing the maximum solubility (saturation) of solid component or phase in the liquid phase. In a binary sys tem it is a line; in a ternary system, a curved surface; in quaternary system, a volume. In an isoplethal study, at tem peratures above the liquidus, the system is completely liquid and at the intersection of the liquidus and the isopleth, the liq uid is in equilibrium with one crystalline phase.

liquid water content *free water content.*

lira A fine raised line or linear elevation on the surface o some shells, resembling a thread or a hair; e.g. one of th parallel fine ridges on the surface of a nautiloid conch sepa rated by *striae* and not easily discernible with the naked eye or a fine linear elevation within the outer lip or on the she surface of a gastropod. Pl: *lirae* or *liras.*

liroconite A sky-blue to verdigris-green monoclinic minera $Cu_2Al(AsO_4)(OH)_4.4H_2O.$ It usually contains some phospho rus. Syn: *lentil ore.*

liskeardite A soft, white mineral: $(Al,Fe)_3(AsO_4)(OH)_6.5H_2O.$

list A rod that strengthens the periderm in graptolites.

listric surface A curvilinear, usually concave-upward surfac of fracture that curves, at first gently and then more steeply from a horizontal position. Listric surfaces form wedge-shape masses, appearing to be thrust against or along each other Etymol: Greek, *listron*, "shovel".

listrium A plate closing the anterior end of the pedicle openin (which has migrated posteriorly) in some discinacean bra chiopods.

listwanite A schistose rock of yellowish green color compose of various combinations of the minerals quartz, dolomite magnesite, talc, and limonite (Holmes, 1928, p.143). It i found at Beresowsk, Ural Mountains.

litchfieldite A plutonic rock composed chiefly of albite, wit smaller amounts of nepheline, lepidomelane, and sometimes cancrinite and sodalite. Its name is derived from Litchfield Maine.

-lyte A rock-name suffix derived from the Greek word fo

410

"stone". Cf: -lyte.

lith A combining form which, as a prefix, means stone or stonelike, and which, as a suffix, means rock or rocklike.

litharenite (a) A term introduced by McBride (1963, p.667) as a shortened form of *lithic arenite* and used by him for a sandstone (regardless of texture) containing more than 25% fine-grained rock fragments, less than 10% feldspar, and less than 75% quartz, quartzite, and chert. See also: *sublitharenite*. (b) A general term used by Folk (1968, p.124) for a sandstone containing less than 75% quartz and metamorphic quartzite and more than 25% fine-grained volcanic, metamorphic, and sedimentary rock fragments, including chert (or whose content of such rock fragments is at least three times that of feldspar and plutonic rock fragments), regardless of clay content or texture.

litharge A reddish or yellow, tetragonal mineral: PbO. Cf: *massicot*. Syn: *lead ocher*.

lithia mica *lepidolite*.

lithia water Mineral water containing lithium salts (e.g. lithium bicarbonate, lithium chloride).

lithic (a) Said of a medium-grained sedimentary rock, and of a pyroclastic deposit, containing abundant fragments of previously formed rocks; also, said of such fragments. (b) Pertaining to or made of stone; e.g. "lithic artifacts" or "lithic architecture". (c) A term sometimes used as a syn. of *litholoic*, as in "lithic unit" (a rock-stratigraphic unit).

Lithic n. In New World archaeology, the basal prehistoric cultural stage, characterized by the appearance, by migration, of man in the New World and the hunting of big game. It is followed by the Archaic. Correlation of relative cultural levels with actual age (and, therefore, with the time-stratigraphic units of geology) varies from region to region.----adj. Pertaining to the Lithic.

lithic arenite (a) A term used by Gilbert (1954, p.294 & 304) for a sandstone containing abundant quartz, chert, and quartzite, less than 10% argillaceous matrix, and more than 10% feldspar, and characterized by an abundance of unstable materials in which the fine-grained rock fragments exceed feldspar grains. It is better sorted and more porous and permeable, and contains better-rounded grains, than lithic wacke. The rock is roughly equivalent to Pettijohn's (1957) redefinition of "subgraywacke". See also:*litharenite*. (b) A term used by Pettijohn (1954, p.364) as a syn. of *lithic sandstone*.

lithic arkose (a) A term used by McBride (1963, p.667) for an arkose containing appreciable rock fragments; specif. a sandstone containing 10-50% fine-grained rock fragments, 25-90% feldspar, and 0-65% quartz, quartzite, and chert. (b) A term used by Folk (1968, p.124) for a sandstone containing less than 75% quartz and metamorphic quartzite and having a "F/R ratio" between 1:1 and 3:1, where "F" signifies feldspars and fragments of gneiss and granite, and "R" signifies all other fine-grained rock fragments.---Cf: *feldspathic litharenite; impure arkose*.

lithic-crystal tuff A tuff that is intermediate between *crystal tuff* and *lithic tuff* but is predominantly lithic. Cf: *crystal-lithic tuff*.

lithic graywacke A graywacke characterized by abundant unstable materials; specif. a sandstone containing a variable content (generally less than 75%) of quartz and chert and 5-75% detrital clay matrix, and having rock fragments (primarily of sedimentary or low-rank metamorphic origin) in greater abundance than feldspar grains (chiefly sodic plagioclase, and indicating a plutonic provenance) (Pettijohn, 1957, p.304). Example: some of the gray sandstones of the Siwalik series (India) with little or no feldspar and 40-45% metamorphic rock fragments (mainly phyllite or schist). The rock is equivalent to *low-rank graywacke* of Krynine (1945) and to *subgraywacke* as originally defined by Pettijohn (1949). The

term was introduced by Pettijohn (1954, p.364) and by Gilbert (1954, p.294). Cf: *feldspathic graywacke*.

lithiclast *lithoclast*.

lithic sandstone A sandstone containing rock fragments in greater abundance than feldspar grains. The term was used by Pettijohn (1954, p.364) for such a sandstone with less than 15% detrital clay matrix (e.g. subgraywacke and protoquartzite), by Gilbert (1954, p.310) to include lithic arenite and lithic wacke, and by Hatch & Rastall (1965, p.111-112) to include the sublitharenite of McBride (1963). See also: *lithic arenite*.

lithic subarkose A term used by McBride (1963, p.667) for a sandstone composed of subequal amounts of feldspar and rock fragments; specif. a sandstone containing 10-25% feldspar, 10-25% rock fragments, and 50-80% quartz, quartzite, and chert. Syn: *feldspathic sublitharenite*.

lithic tuff A tuff that consists predominantly of sediments other than pyroclastics. Cf: *crystal-lithic tuff; crystal tuff; lithic-crystal tuff*.

lithic wacke (a) A sandstone containing abundant quartz, chert, and quartzite, more than 10% argillaceous matrix, and more than 10% feldspar (esp. sodic plagioclase), and characterized by an abundance of unstable materials in which the fine-grained rock fragments exceed feldspar grains (Gilbert, 1954, p.291-292 & 301). (b) A quartz wacke containing abundant (up to 40-50%) fine-grained rock fragments (bits of shale, coal, etc.) (Krumbein & Sloss, 1963, p.172-173).

lithidionite A mineral: $(Cu,Na_2,K_2)Si_3O_7$.

lithifaction Var. of *lithification*.

lithification [coal] A compositional change in a coal seam from coal to bituminous shale or other rock; the lateral termination of a coal seam due to a gradual increase in impurities.

lithification [sed] (a) The conversion of a newly deposited, unconsolidated sediment into a coherent and solid rock, involving processes such as cementation, compaction, desiccation, crystallization, recrystallization, and compression. It may occur concurrent with, or shortly or long after, deposition. (b) A term that is sometimes applied to the *solidification* of a molten lava to form an igneous rock.---See also: *consolidation; induration*. Syn: *lithifaction*.

lithify To change to stone, or to petrify; esp. to consolidate from a loose sediment to a solid rock.

lithionite *lepidolite*.

lithiophilite A salmon-pink or clove-brown orthorhombic mineral: $Li(Mn^{+2},Fe^{+2})PO_4$. It is isomorphous with triphylite.

lithiophorite A mineral: $(Al,Li)MnO_2(OH)_2$.

lithiophosphate A white or colorless mineral: Li_3PO_4. It is a hydrothermal alteration product of montebrasite. Syn: *lithiophosphatite*.

lithistid adj. Said of a stone-like or stony sponge whose rigid skeletal framework consists of interlocking or fused siliceous spicules (desmas).----n. Any demosponge belonging to the order Lithistida and characterized by the presence of desmas, interlocked and cemented to form a rigid framework.

lithium-drifted germanium detector A *semiconductor radiation detector* containing germanium rather than silicon and in which lithium is diffused into the semiconductor to compensate for impurities. Syn: *Ge:Li detector*.

lithium mica *lepidolite*.

lithizone A para-time-rock unit representing a zone or succession of strata possessing common lithologic characteristics (Wheeler et al, 1950, p.2364). Cf: *monothem*. Syn: *lithozone*.

lithocalcarenite A calcarenite containing abundant limeclasts.

lithocalcilutite A calcilutite containing abundant limeclasts.

lithocalcirudite A calcirudite containing abundant limeclasts.

lithocalcisiltite A calcisiltite containing abundant limeclasts.

lithoclast A mechanically formed and deposited fragment of a carbonate rock, normally larger than 2 mm in diameter, derived from an older, lithified limestone or dolomite within, ad-

411

jacent to, or outside the depositional site. Syn: *lithiclast*.

lithodesma A small calcareous plate reinforcing the internal ligament (resilium) in many shells of bivalve mollusks. Pl: *lithodesmata*. Syn: *ossiculum*.

lithodolarenite A dolarenite containing abundant doloclasts.

lithodololutite A dololutite containing abundant doloclasts.

lithodolorudite A dolorudite containing abundant doloclasts.

lithodolosiltite A dolosiltite containing abundant doloclasts.

lithodomous *lithotomous*.

lithofacies (a) A lateral, mappable subdivision of a designated stratigraphic unit of any kind (or of any arbitrarily limited body of sedimentary deposits), distinguished from other adjacent subdivisions on the basis of noteworthy lithologic characters, including all physical and chemical (mineralogic and petrographic) characters and those biologic (paleontologic) characters that influence the appearance, composition, or texture of the rock (some definitions consider only the inorganic characters); a stratigraphic *facies* characterized by particular lithologic features. Laterally equivalent lithofacies may be separated by vertical arbitrary-cutoff planes, by intertonguing surfaces, or by gradational changes. See also: *statistical lithofacies; intertongued lithofacies*. (b) A term used by Moore (1949, p.17 & 32) to signify any particular kind of sedimentary rock or distinguishable rock record formed under common environmental conditions of deposition, considered without regard to age or geologic setting or without reference to designated stratigraphic units, and represented by the sum total of the lithologic characteristics (including both physical and biologic characters) of the rock. This usage closely parallels Wells' (1947) definition of *lithotope*. (c) The general aspect or appearance of the lithology of a sedimentary bed or formation, esp. considered as the expression of the local depositional environment; the collective lithologic characters of any sedimentary rock; the lithologic aspect of a facies of some definite stratigraphic unit. Cf: *physiofacies*. (d) A term that has been applied to "lithology", "lithologic type", "lithologic unit", "formation", and the "manifestation" of lithologic characters.

lithofacies map A *facies map* based on lithologic attributes, showing areal variation in the overall lithologic character of a given stratigraphic unit. The map may emphasize the dominant, average, or specific lithologic aspect of the unit, and it gives information on the changing composition of the unit throughout its geographic extent.

lithofraction The fragmentation of rocks during transportation in streams or by wave action on beaches.

lithogene adj. Said of a mineral deposit formed by the process of mobilization of elements from a solid rock and their transportation and redeposition elsewhere. On a local scale the process may be called a product of *lateral secretion*; on a larger scale, the deposit may be called a product of regional metamorphism (Lovering, 1963, p.315-316).

lithogenesis (a) The origin and formation of rocks, esp. of sedimentary rocks. Also, the science of the formation of rocks. Cf: *petrogenesis*. Syn: *lithogeny*. (b) The first stage of mountain building, during which sediment is accumulated in the sea (esp. in a sinking geosyncline) and later compacted to form sedimentary rock.---Adj: *lithogenetic*.

lithogenetic unit A term used by Schenck & Muller (1941) for a local mappable assemblage of rock strata (such as a formation, member, or bed), considered without regard to time; a cartographic unit. See also: *rock-stratigraphic unit*.

lithogenous Said of stone-secreting organisms, such as a coral polyp.

lithogeny *lithogenesis*.

lithographic limestone A compact, dense, homogeneous, exceedingly fine-grained limestone having a pale creamy yellow or grayish color and a conchoidal or subconchoidal fracture; a *micritic limestone*. It was formerly much used in lithography for engraving and the reproduction of colored plates. See also: *Solenhofen stone*. Syn: *lithographic stone*.

lithographic stone *lithographic limestone*.

lithographic texture A sedimentary texture of certain calcareous rocks, characterized by uniform particles of less than clay size and by an extremely smooth appearance resembling that of the stone used in lithography.

lithoid Pertaining to or resembling a rock or stone, e.g. lithoid tufa.

lithoidal Said of the texture of dense, microcrystalline igneous rocks, of devitrified glass, or of a microcrystalline groundmass, in which individual constituents are too small to be distinguished with the naked eye.

lithoidite A nonporphyritic, cryptocrystalline rhyolite composed of felsitic minerals.

lithoid tufa Gray, compact, bedded tufa, occasionally containing gastropod shells, occurring in the core of dome-like masses in the desert basins of NW Nevada, as along the shore of the extinct Lake Lahontan. It is older and more stone-like than the overlying *thinolitic tufa* and *dendroid tufa*.

lithologic Adj. of *lithology*. Syn: *lithic*.

lithologic correlation A kind of *stratigraphic correlation* based on the correspondence in lithologic characters such as particle size, color, mineral content, primary structures, thickness, weathering characteristics, and certain physical properties.

lithologic guide In mineral exploration, a kind of rock known to be associated with an ore. Cf: *stratigraphic guide*. See also: *ore guide*.

lithologic log A log that shows the distribution of lithology with depth in a borehole or well.

lithologic map A type of geologic map showing the rock types of a particular area.

lithologic unit *rock-stratigraphic unit*.

lithology (a) The description of rocks, esp. sedimentary clastics and esp. in hand specimen and in outcrop, on the basis of such characteristics as color, structures, mineralogic composition, and grain size. As originally used, "lithology" was essentially synonymous with *petrography* as currently defined. (b) The physical character of a rock. ----Adj: lithologic. Cf: *petrology*.

lithomarge A smooth, indurated or firm, and compact variety of common kaolin, consisting of a mixture at least in part of kaolinite and halloysite.

lithomorphic Said of a soil whose characteristics are mainly *inherited*.

lithophagous Said of an organism that feeds on rock material.

lithophile (a) Said of an element that is enriched in the silicate rather than the metal or sulfide phases of meteorites. Such elements concentrate in the Earth's silicate crust in Goldschmidt's tripartite division of elements in the solid Earth. Cf: *chalcophile; siderophile*. (b) Said of an element with a greater free energy of oxidation per gram of oxygen than iron. It occurs as an oxide and more often as an oxysalt, esp. in silicate minerals.----(Goldschmidt, 1954, p.24). Examples are: Se, Al, B, La, Ce, Na, K, Rb, Ca, Mn, U. Syn: *oxyphile*.

lithophilous *rupestral*.

lithophyl A petrified leaf or its impression; also, the rock containing it.

lithophysa A large, spherulitic cavity in glassy basalts, e.g. rhyolites, usually having a radial or concentric structure. Plural: *lithophysae*. Adj: *lithophysal*. Syn: *stone bubble*.

lithophyte A plant living on the surface of a rock. Adj: *lithophytic*.

lithorelic Said of a soil feature that is derived from the parent material. Cf: *pedorelic*.

lithosere A sere that develops on a rock surface.

lithosiderite *stony-iron meteorite*.

Lithosol One of an azonal group of soils characterized by recent and imperfect weathering. It usually develops on steep slopes. See also: *mountain soil*. Syn: *skeletal soil*.

lithosome (a) A three-dimensional rock mass of essentially uniform (or uniformly heterogeneous) lithologic character, having intertonguing relationships in all directions with adjacent and similar masses of different lithologic character. The term was introduced by Wheeler & Mallory (in Fischer et al, 1954, p.929) and defined by them (1956, p.2722) as a rock-stratigraphic body or vertico-laterally segregated unit that is "mutually intertongued with one or more bodies of differing lithic constitution". It is essentially identical with Caster's (1934) *magnafacies*. Cf: *biosome; holosome; intertongued lithofacies*. (b) The sedimentary record of a physicochemical environment or of a more or less uniform lithotope; a body of sediment deposited under uniform physicochemical conditions (Sloss, in Weller, 1958, p.624). (c) A term defined by Moore (1957a, p.1787-1788) as "an independent body of genetically related sedimentary deposits of any sort", or, alternatively, "a spatially segregated part of any genetically related body of sedimentary deposits".---Cf: *lithostrome*.

lithospar A naturally occurring mixture of spodumene and feldspar.

lithosphere The solid portion of the Earth, as compared with the *atmosphere* and the *hydrosphere*; the *crust* of the Earth, as compared with the *barysphere*. Partial syn: *geosphere*. Syn: *oxysphere*.

lithostatic pressure *geostatic pressure*.

lithostratic unit *rock-stratigraphic unit*.

lithostratigraphic unit A term preferred by the International Subcommission on Stratigraphic Terminology (1961, p.19) for *rock-stratigraphic unit*.

lithostratigraphic zone A term used by the International Subcommission on Stratigraphic Terminology (1961, p.19-20) for an informal rock-stratigraphic unit indicating "a body of strata which is unified in a general way by lithologic features but for which there is insufficient need or insufficient information to justify designation as a formal named unit"; e.g. the "shaly żone" in the lower part of the Parker Formation.

lithostratigraphy Preliminary stratigraphy based only on the physical and petrographic features of rocks; delineation and classification of strata as three-dimensional, lithologically unified bodies. Syn: *petrostratigraphy; rock-stratigraphy*.

lithostrome A term introduced by Wheeler & Mallory (1956, p.2721-2722) for a rock-stratigraphic layer "consisting of one or more beds of essentially uniform or uniformly heterogeneous lithologic character" and representing the "three-dimensional counterpart of a lithotope" (an area); esp. an individual tongue projecting from a lithosome. The term is regarded as essentially synonymous with *lithosome* as defined by some, and with *lithotope* as defined by Wells (1947). Weller (1958, p.636) would reject the term because it represents "nothing more than a rock-stratigraphic unit" such as "bed", "member", "tongue", "stratum", or "layer". Cf: *holostrome*.

lithothamnion A plant of the genus *Lithothamnion*, an encrusting or nodular, red calcareous (calcitic) alga of the family Corallinaceae, abundant in post-Jurassic rocks, and reported as a living form from considerable depths and very cold waters. It is most abundant on the seaward edge of the reef flat where it acts as a cementing medium of some coral reefs. Syn: *lithothamnioid; lithothamnium*.

lithothamnion ridge An *algal ridge* built by *Lithothamnion* and other red calcareous algae, rising about 1 m above the surrounding reef and extending to depths of 6-7 m below sea level.

lithotomous Said of an organism that burrows in rock. Syn: *lithodomous*.

lithotope (a) An area or surface of uniform sediment or sedimentation; an area of uniform sedimentary environment, or a place distinguished by relative uniformity of the principal environmental conditions of rock deposition (including occurrence of kinds of organisms associated with these conditions). (b) A paleoecologic term originally proposed by Wells (1944, p.284) for "the sedimentary rock record of a biotope" whose life community or biocoenosis was preserved, but was later defined by Wells (1947, p.119) as "the rock record of the environment" (including both its physical and biologic expressions). The term has subsequently been used for a stratigraphic unit, a part of a stratigraphic section, a particular kind of sediment or rock, and a body of uniform sediments formed by persistence of the depositional environment, and also in an intangible sense for a sedimentary rock environment and a physical environment. Cf: *lithofacies; lithostrome; biotope* [*stratig*].

lithotype In coal, a macroscopically visible band in humic coals, analyzed by physical characteristics rather than by botanical origin. The four lithotypes of banded, bituminous coal are: *vitrain; clarain, durain*, and *fusain*. These were originally described by Stopes (1919). Lithotypes are still used in descriptions of coal seams, although coal analysis by *microlithotype* is now more common.

lithoxyl A term applied to *wood opal* in which the original woody structure is observable. The term is also used to designate petrified (opalized) wood. Syn: *lithoxyle; lithoxylite; lithoxylon*.

lithozone A "more euphonious" syn. of *lithizone* (P.F. Moore, 1958, p.449).

lit-par-lit adj. Having the characteristic of a layered rock, the laminae of which have been penetrated by numerous thin, roughly parallel sheets of igneous material, usually granitic. Etymol: French, bed-by-bed. Cf: *injection gneiss*.

litter In forestry, a general term for the layer of loose organic debris, composed of freshly fallen or only slightly decayed material, that accumulates in wooded areas. It is a constituent of the *forest floor*. Cf: *duff; leaf mold*.

Little Ice Age A period of limited but pulsatory expansion of mountain glaciers in many parts of the world (including the Alps, Scandinavia, and Alaska) that marked a deterioration of climate and the termination of postglacial trends toward climatic amelioration. The shift toward climatic deterioration in most regions is dated by carbon-14 about 5500 years ago and was followed by glacial expansions that attained maximum extensions between 4000 and 2000 years ago in many areas and as late as A.D. 1550 to A.D. 1850 in others. Most glaciers seen today are products of this readvance. The term "little ice-age" was introduced informally by Matthes (1939, p.520) for "an epoch of renewed but moderate glaciation that began about 4000 years ago". It comprises the Subboreal and the Subatlantic, and in some areas is roughly synonymous with the *Medithermal*, the *Hypothermal*, and the *Katathermal*.

littoral (a) Pertaining to the benthic ocean environment or *depth zone* between high water and low water; also: pertaining to the organisms of that environment. Syn: *intertidal*. (b) Pertaining to the depth zone between the shore and a depth of about 200 m. In this meaning, the term includes the *neritic* zone. See also: *sublittoral*. Cf: *supralittoral*.

littoral cone A small volcanic cone of a rootless vent, formed where a lava flow enters a body of water.

littoral current An ocean current caused by the approach of waves to a coast at an angle. It flows parallel to and near to the shore. See also: *littoral drift*. Syn: *longshore current*.

littoral drift Material (such as shingle, gravel, sand, and shell fragments) that is moved along the shore by a *littoral current*. Syn: *longshore drift; shore drift*.

littoral explosion An explosion that is the result of the contact of a flow of molten lava at the edge of a body of water; a *hydroexplosion*.

littoral shelf A shallow, nearshore, terrace-like part of a submerged lake bed, produced by the combined effects of wave erosion and current deposition, and often extending a considerable distance lakeward from the beach.

lituicone A nautiloid conch (like that of *Lituites*) that completes a few whorls (becomes coiled) in the early stages of development and then becomes straight in the mature stages. Syn: *lituiticone*.

live cave A cave in which there is moisture and growth of speleothems or mineral deposits that is associated with the presence of moisture. Cf: *dead cave*. Syn: *active cave*.

liveingite A mineral: $Pb_9As_{13}S_{28}$.

liverite Name used locally in Utah for *elaterite*.

liver opal *menilite*.

liver ore *hepatic cinnabar*.

liver rock A sandstone that breaks or cuts as readily in one direction as in another and that can be worked without being affected by stratification; a dense freestone that lacks natural division planes.

liverwort A member of the bryophyte class Hepaticae, characterized by a round, creeping, or branched thallus, no leaves, and unicellular rhizoids. Liverworts range from the Carboniferous. Cf: *moss*.

livesite A clay mineral intermediate between kaolinite and halloysite; a disordered kaolinite.

live stream *perennial stream*.

living chamber The *body chamber* housing the soft parts of a cephalopod.

living flood plain A flood plain that is overflowed in times of high water (Bryan, 1923a, p.88). Cf: *fossil flood plain*.

living fossil An animal or plant that lives at the present time, is also known as a fossil from an earlier geologic time, and that has generally undergone little modification since that earlier time.

livingstonite A lead-gray mineral: $HgSb_4S_9$.

lixiviation *leaching*.

lixivium *leachate*.

lizardite A platy mineral of the serpentine group: $Mg_3Si_2O_5(OH)_4$. It is a polymorph of chrysotile, distinct from clinochrysotile, orthochrysotile, and parachrysotile.

L-joint *primary flat joint*.

Llandeilian European stage: Middle Ordovician (above Llanvirnian, below lower Caradocian).

Llandoverian European stage: Lower Silurian (above Ashgillian of Ordovician, below Tarannon). Syn: *Valentian*.

llanite An igneous rock composed of red feldspar and blue quartz phenocrysts in a fine-grained groundmass of quartz, microcline, and albite.

Llano A provincial series of the Precambrian in Texas.

llano A term for an extensive tropical plain, with or without vegetation, applied esp. to the generally treeless plains of northern South America and the southwestern U.S Etymol: Spanish.

Llanoria One of the *borderlands* proposed by Schuchert (1923), in this case south of North America, between the Ouachita geosyncline and the Gulf of Mexico. Evidence proposed for Llanoria was much more tenuous than for Appalachia and Cascadia, and modern knowledge of the substructure of the Gulf Coastal Plain and Gulf of Mexico virtually precludes its former existence.

Llanvirnian European stage: Middle Ordovician (above Arenigian, below Llandeilian).

L layer A surficial layer of leaf litter over a soil.

llyn A Welsh term for a pool or lake.

load [sed] (a) The material that is actually moved or carried by a natural transporting agent, such as by a stream, a glacier, the wind, or by waves, tides, and currents; specif. *stream load*. (b) The actual quantity or amount of such material at any given time.---Syn: *sediment load*.

load [struc geol] Vertical pressure caused by the weight of the overlying rocks, as a result of gravity alone. Cf: *standard state*.

load cast A sole mark, usually measuring less than a meter in any direction, consisting of a swelling in the shape of a slight bulge, a deep or shallow rounded sack, a knobby excrescence, a highly irregular protuberance, or a bulbous, mammillary, or papilliform protrusion of sand or other coarse clastics extending downward into finer-grained, softer, and originally hydroplastic underlying material (such as wet clay, mud, or peat) containing an initial depression. It is produced by the exaggeration of the depression as a result of unequal settling and compaction of the overlying material and by the partial sinking of such material into the depression, such as during the onset of deposition of a turbidite on unconsolidated mud. A load cast is more irregular than a flute cast (it is usually not systematically elongated in the current direction), and is characterized by an absence of a distinction between the upcurrent and downcurrent ends. The term was proposed by Kuenen (1953, p.1058) to replace *flow cast* used by Shrock (1948, p.156), although Kuenen excluded the phenomenon of warping of underlying laminae and applied the term to a feature resulting from vertical adjustment only. See also: *load-flow structure*. Syn: *load casting*; *teggoglyph*.

load-casted Said of a current mark (such as a groove or flute) that is exaggerated, misshapen, or entirely obscured by the development of a load cast. Also said of a sole mark (such as a flute cast or groove cast) that is similarly modified by load casting.

load casting (a) The formation or development of a load cast or load casts; also, the configuration of the underside of a stratum characterized by load casts. (b) *load cast*.

load-cast lineation A small-scale, poorly defined, irregular linear structure that appears as a cast on the underside of a sandstone bed, and that is attributed to a dense, sluggish turbidity current moving over soft mud (Crowell, 1955, p.1358).

load-cast striation A rill-like sedimentary structure of uncertain origin (Pettijohn & Potter, 1964, p.319).

loaded stream A stream that has all the sediment it can carry. A partly loaded stream is one carrying less than full capacity. See also: *overloaded stream; underloaded stream*.

load-flow structure A term sometimes used for *load cast* because the structure forms by downsinking of overlying material and not by infilling of a depression (as implied by the term "cast").

load fold A plication of an underlying stratum, believed to result from unequal pressure and settling of overlying material (Sullwold, 1959).

load metamorphism A type of *static metamorphism* in which high temperature has been a controlling influence, as well as overhead pressure due to deep burial (Daly, 1917). Cf: *geothermal metamorphism; thermal metamorphism*.

load mold The mold of a load cast; the depression in an underlying stratum, occupied by a load pocket, such as the sea-floor surface beneath a depositing turbidite (Sullwold, 1960).

load pocket The material within a load cast, consisting of a "bulge of sand" pressing into an underlying stratum (Sullwold, 1959, p.1247).

loadstone *lodestone* [magnet].

load wave The "salient curved unevenness" of underlying material that appears to have been "squirted up" into a superjacent turbidity-current deposit as a result of unequal settling of the overlying material (Sullwold, 1959); it resembles a ripple mark or other wave-like structure. The term "refers to smooth upward bulges as well as to tenuous breaking wave or flame shapes" (Sullwold, 1960, p.635). See also: *flame structure*.

loam (a) A rich, permeable soil composed of a friable mixture of relatively equal and moderate proportions of clay, silt, and sand particles, and usually containing organic matter (humus) with a minor amount of gravelly material; specif. a soil consisting of 7-27% clay, 28-50% silt, and 23-52% sand. It has a somewhat gritty feel yet is fairly smooth and slightly plastic.

Loam may be of residual, fluvial, or eolian origin, and includes many loesses and many of the alluvial deposits of flood plains, alluvial fans, and deltas. It usually implies a fertile soil, and is sometimes called *topsoil* in contrast to the subsoils that contain little or no organic matter. (b) A term used in the old English literature for a mellow soil rich in organic matter, regardless of texture. (c) A obsolete term formerly used in a broad sense for clay, impure clay, clayey earth, and mud.

loaming A method of geochemical prospecting in which samples of surficial material are tested for traces of the metal desired, its presence presumably indicating a near-surface orebody.

loamy (a) Said of a soil (such as a clay loam and a loamy sand) whose texture and properties are intermediate between a coarse-textured or sandy soil and a fine-textured or clayey soil. (b) Pertaining to, consisting of, or characterized by loam.

loamy sand A soil containing 70-90% sand, 0-30% silt, and 0-15% clay, or a soil containing 85-90% sand at the upper limit and having the percentage of silt plus 1.5 times the percentage of clay not less than 15, or a soil containing 70-85% sand at the lower limit and having the percentage of silt plus twice the percentage of clay not exceeding 30 (SSSA, 1965, p.347); specif. such a soil containing at least 25% very coarse sand, coarse sand, and medium sand, and less than 50% fine sand or very fine sand. It is subdivided into loamy coarse sand, loamy fine sand, and loamy very fine sand. Cf: *sandy loam.*

lobate delta A delta formed where the current of a river distributary predominates over shore currents and wave attack.

lobate plunge structure *flute cast.*

lobate rill mark A term used by Clarke (1918) and Shrock (1948, p.131) for a spatulate or lingulate sedimentary structure (cast) resembling the bowl of an inverted spoon and believed to develop on a beach by ebbing tidal currents or retreating storm waves of the intertidal zone. The structure is now considered to be a *flute cast* formed by current action. See also: *rill mark.*

lobate soil *step* [pat grd].

lobe [paleont] (a) An element or undulation of a suture line in a cephalopod shell that forms an angle or curve whose convexity is directed backward or away from the aperture (or toward the apex). Ant: *saddle.* (b) One of the longitudinal divisions of the body, or one of the lateral divisions of the glabella, in a trilobite. (c) A rounded major protuberance of the valve surface of an ostracode, generally best developed in the dorsal part of the carapace (TIP, 1961, pt.Q, p.52); e.g. *connecting lobe.*

lobe [bot] Any part or segment of a plant organ; specifically a part of a petal or calyx or leaf that represents a division to about the middle (Lawrence, 1951, p.759).

lobe [glac] (a) A rounded, tongue-like projection of glacial drift beyond the main mass of drift. (b) *glacial lobe.*

lobe [lake] A long, rounded indentation of a lake.

lobe [streams] (a) *meander lobe.* (b) *flood-plain lobe.*

local A general term applied in geology to a geologic feature or process that occurs in a relatively small, restricted area. It is contrasted with *regional* [geol].

local base level *temporary base level.*

local correlation Correlation of geologic features over areas of comparatively small extent, such as the correlation of strata or orebodies across a fault.

local gravity map A gravity map from which regional changes of gravity have been eliminated.

local horizon (a) *apparent horizon.* (b) The actual lower boundary of the observed sky or the upper outline of terrestrial objects including nearby obstructions or irregularities.

local metamorphism Metamorphism caused by a local process which may be contact metamorphism or metasomatism near an igneous body or dislocation metamorphism in a fault

zone. Cf: *regional metamorphism.*

local peat Peat developed by ground water. Syn: *basin peat; azonal peat.*

local peneplain (a) *incipient peneplain.* (b) *partial peneplain.*

local range zone The *range zone* or body of strata of a specified taxon or group of taxa in any single geographically located section or local area (ACSN, 1961, art.22g). The summation of all the local range zones is the true range zone of the taxa. Syn: *teilzone; topozone; partial range zone.*

local relief The vertical difference in elevation between the highest and the lowest points of a land surface within a specified horizontal distance or in a limited area. Syn: *relative relief.* Cf: *available relief.*

local sorting The action responsible for the size-frequency distribution of sedimentary particles, and of its homogeneity, at a given place (Pettijohn, 1957, p.540-541).

local unconformity An unconformity that is strictly limited in geographic extent and that usually represents a relatively short period, such as one developed around the margins of a sedimentary basin or along the axis of a structural trend that rose intermittently while continuous deposition occurred in an adjacent area. It may be similar in appearance to, but lacks the regional importance of, a disconformity. Cf: *regional unconformity.*

location [drill] (a) The spot or place where a borehole is to be drilled; e.g. a *well site.* (b) The spacing unit between two boreholes or wells.

lochan A Scottish term for a small lake (loch) or pond, usually lying in a cirque.

lock A stretch of water in a canal, stream, or dock, enclosed by gates at each end, and used in raising or lowering boats as they pass from one water level to another.

Locke level A *hand level* with a fixed bubble tube that can be used only for horizontal sighting.

Lockportian Stage in New York State: upper Middle Silurian.

locomorphic stage A term introduced by Dapples (1962) for the middle geochemical stage of diagenesis characterized by prominent mineral replacement (without reactions). It is typical of lithification of a clastic sediment, and is more advanced than the *redoxomorphic stage* and precedes the *phyllomorphic stage.*

locular wall A wall of a diatom frustule having separate inner and outer laminae connected by vertical partitions that form areolae.

locule A compartment, cavity or chamber in a plant; e.g. in Ascomycetes, a stromatic chamber containing asci; in flowering plants, a cavity in the ovary containing ovules, or in anthers containing pollen grains (Scagel, et al, 1965, p.622). Syn: *loculus.*

loculus [paleont] One of the chambers in a foraminiferal test. Pl: *loculi.* Adj: *locular.*

loculus [bot] *locule.*

lode [eco geol] A mineral deposit consisting of a zone of veins; a mineral deposit in consolidated rock as opposed to *placer* deposits. Syn: *lead.* Cf: *vein; vein system.*

lode [streams] A local English term for a channel or watercourse, usually partly artificial and embanked above the surrounding country.

lode claim A mining claim on an area containing a vein or lode. Cf: *placer claim.*

lode country *ore channel.*

lodestone [magnet] (a) A magnetic variety of natural iron oxide (Fe_3O_4) or of the mineral magnetite; specif. a piece of magnetite possessing polarity like a magnet or magnetic needle and hence one that, when freely suspended, will attract other iron objects to itself. Syn: *loadstone; leading stone; Hercules stone.* (b) An intensely magnetized rock or ore deposit.

lodestuff Both the gangue and the economically valuable minerals of a lode; the contents of an *ore channel*.

lode tin Cassiterite occurring in veins, as distinguished from *stream tin*.

lodge moraine A terminal moraine of billowy relief, consisting of subglacial debris lodged under a thin margin of a glacier; widespread in North America. Syn: *submarginal moraine*.

lodgment The plastering beneath a glacier of successive layers of basal till upon bedrock or other glacial deposits. Cf: *plastering-on*.

lodgment till A *basal till* commonly characterized by compact fissile structure and containing stones oriented with their long axes generally parallel to the direction of ice movement.

lodranite A stony-iron meteorite composed of a mixture of bronzite and olivine, enclosed within a fine network of nickel-iron.

loess A widespread, homogeneous, commonly nonstratified, porous, friable, unconsolidated but slightly coherent, usually highly calcareous, fine-grained, blanket deposit (generally less than 30 m thick) of marl or loam, consisting predominantly of silt with subordinate grain sizes ranging from clay to fine sand, and covering areas extending from north-central Europe to eastern China as well as in the Mississippi Valley and Pacific Northwest of the U.S. It is buff to light yellowish or yellowish brown in color (locally gray, brown, or red), often contains shells, bones, and teeth of mammals, and is traversed by networks of many small, narrow, vertical tubes (frequently lined with calcium-carbonate concretions) left by successive generations of grass roots that allow the loess (when undisturbed) to stand in steep or nearly vertical faces. Although source and origin is still a controversial question, loess is now generally believed to be windblown dust of Pleistocene age, carried from desert surfaces, alluvial valleys, and outwash plains lying south of the limits of the ice sheets, or from unconsolidated glacial or glaciofluvial deposits uncovered by successive glacial recessions but prior to invasion by a vegetation mat. The mineral grains, composed mostly of silica and associated heavy minerals, are fresh and angular, and are generally held together by calcareous cement. Etymol: German *Löss*, from dialectal (Switzerland) *lösch*, "loose", so named by peasants and brickworkers along the Rhine valley where the deposit was first recognized. Pron: *luehss*. Cf: *limon: adobe*. Syn: *löss; lehm; bluff formation*.

loessal Pertaining to or consisting of loess. Syn: *loessial*.

loess doll A spheroidal or irregular, compound nodule or concretion of calcium carbonate found in loess and resembling a doll or potato, or a childs's head. It is often hollow but may contain a loose stone. Syn: *loess nodule; loess kindchen; puppet*.

loess flow A fluid suspension of dry porous silt in air, as that which occurred following the 1920 earthquake in Kansu Province, China (Close & McCormick, 1922).

loessification Formation and development of loess.

loess kindchen A *loess doll* resembling the head of a child. Etymol: German *Lösskindchen*.

loessland Land whose surface is composed of loess.

loess nodule *loess doll*.

loessoïde A Dutch term for deposits in southern Limburg (a province in southern Netherlands), believed to be of loessal origin, but reworked and redeposited by streams, possibly with an admixture of residual material from in-situ decomposition.

loeweite A white to pale-yellow mineral: $Na_{12}Mg_7(SO_4)_{13} \cdot 15H_2O$. Syn: *löweite*.

Loewinson-Lessing classification A chemical classification of igneous rocks (into the four main types acid, neutral, basic, and ultrabasic) based on acidity quotients and calculated from molecular values.

loferite A term suggested by Fischer (1964, p.124) for a limestone or dolomite riddled by shrinkage pores, such as the carbonate sediments in the Triassic Dachstein Formation (Lofer facies) in Salzburg, Austria. The term is partly synonymous with *bird's-eye limestone*.

log (a) A detailed, systematic, and sequential record (book, sheet, chart) of the progress made in drilling a well or borehole. It may include notes on the depths, lithologies, and thicknesses of the rocks and earth materials penetrated, geologic structure (dips), fossil content, water conditions, depths at which mineral substances or fluids (oil, gas, water) are found, diameter and length and the kinds of casing or pipe used, rate of drilling, and other pertinent facts having to do with the drilling; a chronologic record of what was found in sinking a well or borehole. It is obtained by examining cuttings and cores and by using various geophysical devices, and it may be arranged in a narrative, tabular, graphic, or symbolic form. See also: *geophysical log*. Syn: *well log; borehole log*. (b) A graph showing the variation with depth in a well or borehole of some physical property such as electrical resistivity, spontaneous potential, gamma-ray intensity, density, or acoustic velocity. (c) The device used in making a log.

logan [*geog*] Shortened form of *pokelogan*.

logan [*geomorph*] *logan stone*.

Logan's Line A structural discontinuity along the northwestern edge of the Northern Appalachians, between complexly deformed geosynclinal rocks on the southeast and undisturbed cratonic and shield rocks on the northwest. The name commemorates its discovery by Logan (1863) near Quebec City. For part of its distance the line is a major low-angle thrust fault, but northeast of Quebec City the line is beneath the St. Lawrence Estuary and its nature is undetermined; southward in Vermont the frontal fault changes into a succession of discontinuous breaks. It is interpreted by many geologists as having been formed during the Taconic orogeny of early Paleozoic time.

logan stone An English name for a *rocking stone* consisting of a large mass of granite or gneiss chemically weathered along horizontal joints and so balanced on its base as to "log" or rock from side to side; e.g. the stone weighing about 80 tons near Land's End in Cornwall. Syn: *logan; loggan stone; logging stone*.

logarithmic mean diameter An expression of the average particle size of a sediment or rock, obtained by taking the arithmetic mean of the particle-size distribution in terms of logarithms of the class midpoints. Cf: *phi mean diameter*.

loggan stone Var. of *logan stone*.

logging (a) The act or process of making or recording a log. (b) The method or technique in which a subsurface formation is analyzed by the making of a log.

logging stone *logan stone*.

lognormal distribution A frequency distribution whose logarithm follows a normal distribution.

log strip A long, narrow piece of paper on which a *strip log* is plotted.

loipon A term proposed by Shrock (1947) for a residual surficial layer produced by intense and prolonged chemical weathering and composed largely of certain original constituents of the source rock. Typical accumulations of loipon are the gossans over orebodies, bauxite deposits in Arkansas, terra rossa deposits of Europe, and duricrust of Australia. Etymol: Greek, "residue". Pron: *loy-pon*. Adj: *loiponic*.

lokkaite A mineral: $(Y,Ca)_2(CO_3)_3 \cdot 2H_2O$.

lollingite A syn. of *loellingite*. Also spelled: *lollingite*.

lolly ice Soft *frazil ice* formed in turbulent seawater. Syn: *lolly*.

loma A term used in the SW U.S. for an elongated, gentle swell or rise of the ground (as on a plain), or a rounded, broad-topped, inconspicuous hill. Etymol: Spanish, "hillock, rising ground, slope".

lomita A small, low *loma*.

lomonite Original spelling of *laumontite*.

lomonosovite A dark cinnamon-brown to black or rose-violet mineral: $Na_2Ti_2Si_2O_9.Na_3PO_4$. Cf: *murmanite*.

lomontite *laumontite*.

lonchiole A *sceptrule* with a single spine opposite the single ray.

Londinian *Ypresian*.

Longaxones A group of primitive, usually lightly sculptured, tricolpate, Cretaceous and younger angiosperm pollen in which the polar axis is as long as, or longer than, the equatorial diameter. Cf: *Brevaxones*.

long clay A highly plastic clay; a *fat clay*.

Long Draught *Altithermal*.

long-flame coal Coal that is high in volatiles. Syn: *fat coal*. Cf: *short-flame coal*.

longicone A long, slender, conical or gradually tapering shell characteristic of certain cephalopods. Also, a fossil animal having such a shell. Cf: *brevicone*.

longitude (a) An angular distance between the plane of a given meridian through any point on a sphere or spheroid and the plane of an arbitrary meridian selected as a line of reference, measured in the plane of a great circle of reference or in a plane parallel to that of the equator; specif. the length of the arc or portion of the Earth's equator or of a parallel of latitude intersected between the meridian of a given place and the prime meridian (or sometimes from a national meridian), expressed either in time or in degrees east or west of the prime or national meridian (which has longitude zero degrees) to a maximum value of 180 degrees. A degree of longitude on the Earth's surface varies in length as the cosine of the latitude, being 69.95 statute miles at the equator, 53.43 miles at lat. 40°, and zero at the poles; in time, it represents 4 minutes so that 15 degrees of longitude is equivalent to a difference of one hour of local time. Longitude may also be measured as the angle at the poles lying between the two planes that intersect along the Earth's axis to produce the two meridians. Abbrev: long. Symbol λ. See also: *astronomic longitude; geodetic longitude; geographic longitude; celestial longitude;* Cf: *meridian.* (b) A linear coordinate distance measured east or west from a specified north-south line of reference; e.g. easting and westing.---Cf: *latitude*.

longitude correction The north-south corrections made to observed magnetic intensities by subtracting the Earth's normal field. Cf: *latitude correction [mag]*.

longitude difference *departure*.

longitudinal Said of an entity that is extended in a lengthwise direction; esp. said of a topographic feature that is oriented parallel to the general strike of a region. Ant: *transverse [geomorph]*.

longitudinal band Foliation in a glacier that is parallel to the direction of ice movement.

longitudinal coastline *concordant coastline*.

longitudinal consequent stream A consequent stream whose direction is determined by the pitch of a fold; esp. a stream flowing in a synclinal trough.

longitudinal crevasse A crevasse roughly parallel to the direction of ice movement. This type of *crevasse* in a valley glacier is longitudinal only in the center of the glacier; away from the center it becomes a *splaying crevasse* (the preferred term).

longitudinal drift A long, tapered, sharp-crested *sand drift* formed on the lee side of a narrow gap in a ridge or scarp oriented transversely to the prevailing wind, esp. in a desert or steppe region where ridges interrupt flat plains or plateaus; it may extend one-half kilometer in length.

longitudinal dune A long, narrow, usually symmetrical (in profile) sand dune oriented parallel with the direction of the prevailing wind responsible for its construction, being wider and steeper on the windward side but tapering to a point on the leeward side, and commonly forming behind obstacles in an area where sand is abundant and the wind is strong and con-

stant; it may be a few meters high and up to 100 km long in some inland regions. See also: *seif dune*.

longitudinal fault A fault, the strike of which is parallel with that of the general structural trend of the region.

longitudinal flagellum A thread-shaped flagellum in a dinoflagellate, trailing after the body and arising from the posterior pore in the sulcus if two are present, its proximal part lying in the ventral sulcus near the major axis.

longitudinal fold A fold, the axis of which trends in accordance with the general strike of the area's structures. Cf: *discordant fold*. Syn: *strike fold*.

longitudinal joint A steeply dipping joint plane in a pluton that is oriented parallel to the lines of flow. Syn: *S-joint; (hO1) joint; bc-joint*.

longitudinal moraine A moraine rampart consisting of a medial moraine and an englacial moraine of a former glacier (Schieferdecker, 1959, term 0920).

longitudinal profile (a) The profile of a stream or valley, drawn along its length from the source to the mouth of the stream; it is the straightened-out, upper edge of a vertical section that follows the winding of the valley. See also: *thalweg*. Cf: *cross profile*. Syn: *long profile; valley profile; stream profile; river profile*. (b) A similar profile of a landform, such as of a pediment.

longitudinal resistivity Resistivity of rock measured along the direction of bedding. Cf: *transverse resistivity*.

longitudinal ripple mark A ripple mark with a relatively straight crest, formed parallel to the direction of the current, such as one related to oscillatory wave action (Straaten, 1951); its profile may be asymmetric or symmetric. See also: *corrugated ripple mark; mud-ridge ripple mark*.

longitudinal section A diagram drawn on a vertical or inclined plane and parallel to the longer axis of a given feature; e.g. a section drawn parallel to the strike of a vein. Cf: *cross section*.

longitudinal septum A septum in certain cirripede crustaceans disposed normal to the inner and outer laminae of a compartmental plate and separating the parietal tubes. Syn: *parietal septum*.

longitudinal stream A *subsequent stream* that follows the strike of the underlying strata.

longitudinal valley (a) A *subsequent valley* developed along or in the same direction as the general strike of the underlying strata; a valley at right angles to a consequent stream. This is the current usage of the term, as used by Powell (1873, p.463). (b) A term originally applied by Conybeare & Phillips (1822, p.xxiv) to a long valley developed parallel to the general trend of a ridge, range, or chain of mountains or hills. According to current usage, the term is correctly used only "where the mountain or hill ranges are parallel to the strike" (Stamp, 1961, p.300). Syn: *lateral valley*.---Cf: *transverse valley*.

longitudinal wave *P wave*.

Longmyndian A division of the Precambrian in Shropshire, England.

long period A period of seismic activity that is more than six seconds in duration. Cf: *short period*.

long profile *longitudinal profile*.

long-range fossil A fossil taxon that possesses an extensive vertical range and that may be expected to occur through a great thickness of strata.

long-range order A strong tendency for the random atoms in a random solid solution to order as the solution cools from the elevated temperature at which it was formed. Cf: *long-range order*.

longshore Pertaining or belonging to the shore or coast, or a seaport; littoral. Syn: *alongshore*.

longshore bar A low, elongate sand ridge, built chiefly by wave action, occurring at some distance from, and extending generally parallel with, the shoreline, being submerged at least by high tides, and typically separated from the beach by an intervening *trough.* Syn: *ball; offshore bar; submarine bar; barrier bar.*

longshore current *littoral current.*

longshore drift *littoral drift.*

longulite A cylindrical or conical belonite thought to have formed by the coalescence of globulites.

longwall mining A method of coal mining in which the coal is mined in a single, continuous operation. The emptied space is either filled with some packing material or is allowed to collapse. Cf: *pillar mining.*

long wave [seis] An obsolete syn. of *surface wave.*

long wave [atmos] *planetary wave.*

long wave [water] *shallow-water wave.*

long-wave radiation *infrared radiation.*

lonsdaleite A meteorite mineral consisting of a form of carbon.

lonsdaleoid septum A rugose corallite septum characterized by discontinuity toward the peripheral edge of the septum, as in *Lonsdaleia.*

loop [coast] *looped bar.*

loop [glac geol] *loop moraine.*

loop [geophys] A pattern of field observations which begin and end at the same point with a number of intervening observations. Such a pattern is useful in correcting for drift in gravity-meter observations, diurnal variation in magnetometer surveys, and to detect faults or other causes of misclosure in seismic dip shooting. See also: *looping.*

loop [waves] *antinode.*

loop [paleont] A support (brachidium) for a brachiopod lophophore, composed of secondary shell and extending anteriorly from crura as a closed apparatus, variably disposed and generally ribbon-like with or without supporting septum from the floor of the brachial valve (TIP, 1965, pt.H, p.147).

loop bar *looped bar.*

loop bedding Bedding characterized by small groups of laminae that are sharply constricted or that end abruptly at intervals, giving the effect of long, thin loops or links of a chain; it is found in fine calcareous sediments and in oil shale.

looped bar A curved bar on the leeward or landward side of an offshore island undergoing wave erosion, formed by the union of two separate spits that have trailed off behind and joined together to form a loop that encloses or nearly encloses a body of water. Cf: *cuspate bar.* Syn: *loop; loop bar.*

looping Making geophysical observations around a closed *loop* or a traverse.

loop lake *oxbow lake.*

loop moraine An end moraine of a valley glacier, shaped like an arc or half-loop, concave toward the direction from which the ice approached; it is usually steep on both sides and extends across the valley. Syn: *valley-loop moraine; moraine loop; loop.*

loose ice *broken ice.*

loose-snow avalanche A snow avalanche that starts at a point and widens downhill, in snow lacking cohesion. Cf: *wind-slab avalanche; slab avalanche.*

loose suture An externally visible suture between movably united crinoid ossicles. Cf: *close suture.*

loparite A brown to black mineral: $(Ce,Na,Ca)_2(Ti,Nb)_2O_6$. It was formerly regarded as a variety of perovskite containing alkalies and cerium.

lopezite An orange-red mineral: $K_2Cr_2O_7$.

lophophore A feeding organ usually consisting of a circular or horseshoe-shaped fleshy ridge surrounding the mouth and bearing the tentacles and serving to engulf food particles and provide a respiratory current in bryozoans and brachiopods. In brachiopods, it is a feeding organ with filamentous appendages, symmetrically disposed about the mouth, typically sus-

pended from the anterior body wall but may be attached to the dorsal mantle, and occupying the mantle cavity (TIP, 1965, pt.H, p.147). See also: *brachia.*

lophophytous Said of a sponge that is fastened to the substrate by a root-tuft (tuft of spicules).

lophotrichous Said of a bacterial cell having a tuft of flagella at one or both ends. Cf: *monotrichous; peritrichous.*

lopodolith A basket-shaped coccolith opening distally.

lopolith A large, concordant igneous intrusion of planoconvex or lenticular shape, that is sunken in its central part due to sagging of the underlying country rock.

lorac A precision radio surveying technique or hyperbolic radio navigational system similar to loran in which two or more fixed transmitters emit continuous waves and the position of a mobile receiver in the resulting standing-wave pattern is determined by measuring the phase difference of the waves emanating from two of the transmitters. The useful range is about 200 nautical miles. Etymol: *long-range accuracy.*

loran Any of various long-range electronic navigational systems by which hyperbolic lines of position are determined by measuring the difference in arrival times of synchronized pulse signals from two or more fixed transmitting radio stations of known geographic position. Loran fixes may be obtained at a range of 1400 nautical miles at night. Cf: *shoran.* Etymol: *long-range navigation.*

lorandite A cochineal- to carmine-red or dark lead-gray monoclinic mineral: $TlAsS_2$.

loranskite A black mineral: $(Y,Ce,Ca,Zr)TaO_4$ (?).

lorenzenite A dark-brown to black mineral: $Na_2Ti_2Si_2O_9$. Syn: *ramsayite.*

lorettoite A honey-yellow mineral: $Pb_7O_6Cl_2$.

lorica (a) A hard, protective, commonly tubular or vase-like, external organic covering, case, or shell secreted or built with agglutinated foreign matter by tintinnids, thecamoebians, certain algae, and other protists, and having a calcareous or siliceous composition. (b) The cell wall or two valves of a diatom.---Pl: *loricae.*

loseyite A bluish-white mineral: $(Mn,Zn)_7(CO_3)_2(OH)_{10}$.

losing stream *influent stream.*

löss Var. of *loess.* Etymol: German *Löss.*

lost circulation The condition during a drilling operation when the drilling mud escapes into the porous or cavernous sidewalls of the borehole and does not return to the surface.

lost mountain An isolated mountain standing in a desert and so far removed from the main mass of mountains as to have no apparent connection with them; e.g. an outlier or a monadnock that has resisted erosion more effectively than the surrounding land. A smaller feature is called a *lost hill.* Cf: *huerfano.*

lost stream (a) A surface stream that disappears into an underground channel and that does not reappear in the same or even adjacent drainage basin; e.g. a stream in a karst region that disappears into a sinkhole and follows a definite channel through limestone caves. (b) A dried-up stream in an arid region.----Syn: *sunken stream; disappearing stream; sinking creek.*

Lotharingian European stage: uppermost lower Lower Jurassic (above Sinemurian, below Pliensbachian).

lotrite *pumpellyite.*

lottal A field term used by King (1962, p.179) for the aqueous clayey mixtures formed by mass movement down hillslopes. Etymol: the jingle "Careful with that catsup bottle; none'll come and then a lot'll".

louderback A remnant of a lava flow appearing in a tilted fault block and bounded by a dip slope. It is named after George D. Louderback, a North American geologist, who used it as evidence of block faulting in basin and range topography.

loughlinite A pearly-white, asbestiform mineral: $Na_2Mg_3Si_6O_{16}\cdot8H_2O$.

loupe Any small magnifying glass mounted for use in the hand so that it can be held in the eye socket or attached to spectacles, and used to study gemstones.

louchorrite mosandrite.

love arrow flèche d'amour.

lovénian system A numbering system in which the individual ambulacral and interambulacral areas of the test of echinoids are designated by Roman (I-V) and Arabic (1-5) numerals, respectively. It is based on bilateral symmetry with respect to a plane passing through the apical system, peristome, and periproct in irregular echinoids, and chiefly by the position of the madreporite in regular echinoids. Named after Sven L. Lovén (1809-1895), Swedish zoologist.

love wave A major type of surface wave having a horizontal motion that is shear or transverse to the direction of propagation. Its velocity depends only on density and rigidity modulus, and not on bulk modulus. It is named after A.E.H. Love, the English mathematician who discovered it. See also: G wave. Syn: Q wave. Obsolete syn: Querwellen wave.

lovozerite A mineral: $(Na,K)_2(Mn,Ca)ZrSi_6O_{16}.3H_2O$ (?).

low [beach] (a) swale. (b) trough.

low [geomorph] law.

low [geophys] minimum [geophys].

low [meteorol] depression [meteorol].

low [oil] (a) geologic low. (b) topographic low.

low [struc geol] n. A general term for such features as a structural basin, a syncline, a saddle, or a sag. Cf: high [struc geol]. Syn: structural low.

low albite Low-temperature albite common in nature, stable below 450°C. It takes almost no calcium or potassium into solid solution, and has a completely ordered lattice. Cf: high albite.

low and ball A descriptive name for a longshore bar ("ball") separated by a distinct longitudinal trough ("low") lying parallel to the shoreline in the shoreface or offshore region along a seashore or a lake shore; esp. such a feature as described from Lake Michigan.

low-angle cross-bedding Cross-bedding in which the cross-beds have an average maximum inclination of less than 20° (McKee & Weir, 1953, p.388). Cf: high-angle cross-bedding.

low-angle fault A fault, the dip of which is 45° or less. Cf: high-angle fault.

low-angle thrust overthrust.

low chalcocite Orthorhombic chalcocite, stable below 105°C.

loweite loeweite.

low-energy coast A sheltered coast protected from strong wave action by headlands, islands, or reefs and characterized by average breaker heights of less than 10 cm. Cf: high-energy coast; moderate-energy coast; zero-energy coast.

low-energy environment An aqueous sedimentary environment characterized by a low energy level and by standing water or general lack of wave or current action, thereby permitting very fine-grained sediment to settle and accumulate; e.g. a coastal lagoon or an alluvial swamp, containing shale. Cf: high-energy environment.

lower [stratig] Pertaining to rocks or strata that are normally below those of later formations of the same subdivision of rocks. The adjective is applied to the name of a time-stratigraphic unit (system, series, stage) to indicate position in the geologic column and corresponds to early as applied to the name of the equivalent geologic-time unit; e.g. rocks of the Lower Jurassic System were formed during the Early Jurassic Period. The initial letter of the term is capitalized to indicate a formal subdivision (e.g. "Lower Devonian") and is lowercased to indicate an informal subdivision (e.g. "lower Miocene"). The informal term may be used where there is no formal subdivision of a system or of a series. Cf: upper; middle.

lower break foot.

lower Carboniferous In European usage, the approximate equivalent of the Mississippian. Cf: Upper Carboniferous.

lower core A term that is inclusive of the inner core and the transitional zone of the outer core, i.e. the equivalent of the F and G layers.

lower keriotheca The adaxial (lower) part of keriotheca in the wall of a fusulinid, characterized by coarse alveolar structure (as in Schwagerina). Cf: upper keriotheca.

lower low-water datum An approximation to the plane of mean lower low water, adopted as a standard reference plane for a specific area (as the Pacific coast of the U.S.) and retained for an indefinite period although it may differ slightly from a later (and better) determination. Cf: low-water datum.

lower mantle That part of the mantle which lies below a depth of about 1000 km and has a density of 4.7 g/cm^3, in which the rates of increase of seismic velocities are attenuated. It is equivalent to the D layer. Syn: inner mantle; mesosphere; pallasite shell.

lower Paleolithic n. The first and oldest division of the Paleolithic, characterized by Australopithecus and Homo erectus. Correlation of cultural levels with actual age (and, therefore, with time- stratigraphic units of geology) varies from region to region. Cf: middle Paleolithic; upper Paleolithic.----adj. Pertaining to the lower Paleolithic.

lower plate The footwall of a fault. Cf: upper plate.

Lower Silurian An old syn. of Ordovician.

lower tectorium The adaxial secondary dark layer of spirotheca in the wall of a fusulinid, next below the diaphanotheca or tectum (as in Profusulinella). Cf: tectorium; upper tectorium.

low-flow frequency curve A graphic illustration of both the magnitude and frequency of minimum flows in a given time span.

low-grade [ore dep] Said of an ore with a relatively low ore-mineral content. Syn: lean; coarse. Cf: high-grade. See also: grade.

low island In the Pacific Ocean, a coralline rather than a volcanic island. Cf: high island.

lowland (a) A general term for low-lying land or an extensive region of low land, esp. near the coast and including the extended plains or country lying not far above tide level. (b) The low and relatively level ground of a region, in contrast with the adjacent, higher country; e.g. a vale between two cuestas. (c) A low or level tract of land along a watercourse; a bottom.----The term is usually used in the plural. Ant: upland.

low-latitude desert tropical desert.

low marsh The flat, usually bare ground situated seaward of a salt marsh and regularly covered and uncovered by the tide (Carey & Oliver, 1918, p.166); e.g. a mud flat. Syn: slob land.

lowmoor bog A bog that is at or only slightly above the water table and which is dependent on it for accumulation and preservation of peat which is usually composed of sedge, reed, and shrub remains and various mosses. Cf: highmoor bog.

lowmoor peat Peat occurring on low-lying moors or swamps and containing little or no sphagnum. Its moisture is standing surface water and is low in acidity. Ash and nitrogen content are high; cellulose content is low. Cf: highmoor peat. Syn: fen peat.

low-oblique photograph An oblique photograph in which the camera is insufficiently inclined to the vertical so that the apparent horizon is excluded from the field of view and the entire picture is below the horizon. Syn: low oblique.

low quartz Low-temperature quartz; specif. alpha quartz.

low-rank graywacke A term introduced by Krynine (1945) for a graywacke in which feldspar is almost absent. It is related to miogeosynclines. The rock is equivalent to subgraywacke as originally defined by Pettijohn (1949) and to lithic graywacke of Pettijohn (1954). Cf: high-rank graywacke.

low-rank metamorphism Metamorphism that is accomplished under conditions of low to moderate temperature and pressure. Cf: high-rank metamorphism.

low tide The tide at its lowest; the accepted popular syn. of *low water* in the sea.

low-tide delta A delta formed at the step by drainage of water from the beach onto the tidal flat. It is associated with enlarged and accentuated rills.

low-tide terrace A relatively horizontal zone of the foreshore near the low-water line.

low-velocity correction *weathering correction.*

low-velocity layer *low-velocity zone.*

low-velocity zone (a) Any layer or shell of the Earth in which seismic-wave velocities are lower than in adjacent layers or zones, because of its reduced strength. (b) Specifically, the zone in the upper mantle, variously defined as from 60 to 250 km in depth, in which velocities are about 6% lower than in the outermost mantle. It is probably caused by the near-melting-point temperature of the material. Syn: *Gutenberg low-velocity zone; B layer.* (c) Specifically, a region inside the core boundary below a depth of 2900 km which produces a *shadow zone* at the Earth's surface.

low-volatile bituminous coal Bituminous coal that contains 15-22% volatile matter, analyzed on a dry, mineral-matter-free basis. Cf: *high-volatile bituminous coal; medium-volatile bituminous coal.*

low water Water at the minimum level reached during a tide cycle. Abbrev: LW. Cf: *high water.* Syn: *low tide.*

low-water datum An approximation to the plane of *mean low water,* adopted as a standard reference plane for a specific area (as the Atlantic coast of the U.S.) and retained for an indefinite period although it may differ slightly from a later (and better) determination. Cf: *lower low-water datum.*

loxochoanitic Said of a short, straight, retrochoanitic septal neck of a nautiloid that points obliquely toward the interior of the siphuncle.

loxoclase A variety of orthoclase containing considerable sodium: $(K,Na)AlSi_3O_8$. It has a green tinge due to small inclusions of diopside. The loxoclase series ranges from Or_1Ab_1 to Or_1Ab_4, with K_2O in the range of 4-7%. Syn: *soda orthoclase.*

loxodrome *rhumb line.*

loxodromic curve *rhumb line.*

L-tectonite A tectonite whose fabric is dominated by constricted linear elements, e.g. a stretched-pebble conglomerate. Cf: *S-tectonite; R-tectonite; B-tectonite.*

lublinite A very soft, cheesy or spongy mixture of calcite and water. See also: *moonmilk.* Syn: *rock milk; mountain milk.*

lubricating layer In a décollement, that stratum which acted as a lubricant for the gliding of the overthrust. See also: *sole* [*fault*].

lucinoid Said of heterodont dentition of a bivalve mollusk with two cardinal teeth in each valve, the anterior tooth in the left valve occupying a median position below the beaks. Cf: *corbiculoid.*

luclite A fine-grained diorite composed chiefly of plagioclase, hornblende, and occasionally a small amount of quartz. It is somewhat coarser grained than *malchite,* which it otherwise resembles.

Luders lines A group of *slipbands.*

Ludian European stage: uppermost Eocene (above Bartonian, below Tongrian of Oligocene).

ludlamite A green monoclinic mineral: $(Fe,Mg,Mn)_3$-$(PO_4)_2.4H_2O$.

Ludlovian European stage: Upper Silurian (above Wenlockian, below Gedinnian of Devonian).

ludwigite A blackish-green orthorhombic mineral: $(Mg,Fe^{+2})_2Fe^{+3}BO_5$. It is isomorphous with vonsenite.

lueneburgite A colorless mineral: $Mg_3B_2(PO_4)_2(OH)_6.5H_2O$.

lueshite An orthorhombic mineral: $NaNbO_3$. It has perovskite-type structure and is dimorphous with natroniobite.

lugarite A coarse-grained porphyritic ijolite that contains analcime in place of nepheline. Phenocrysts of barkevikite prism,

titanaugite, and zone labradorite occur in an analcime groun[ ] mass.

luhite An igneous rock transitional between polzenite ar[ ] melilite-nepheline basalt; a haüyne-melilite damkjernite (J[ ] hannsen, 1939, p.264).

Luisian North American stage: Miocene (above Relizia[ ] below Mohnian).

lujavrite A coarse-grained, trachytic, eudialyte-bearing neph[ ] line syenite containing thin parallel feldspar crystals with inte[ ] stitial nepheline grains and acicular acmite crystals. The roc[ ] was originally described from Luijaur (Lujavr), Lapland. Als[ ] spelled: *lujaurite; luijaurite; lujauvrite.* Cf: *chibinite.*

lumachelle (a) A compact, dark-gray or dark-brown limestor[ ] or marble, composed chiefly of fossil mollusk shells, ar[ ] characterized by a brilliant iridescence or chatoyant reflectic[ ] from within. Syn: *fire marble.* (b) Any accumulation of shel[ ] (esp. oysters) in stratified rocks.---Etymol: French, "coquina[ ] oyster bed", from Italian *lumachella,* "little snail".

lumb A term used in Sheffield, Eng., for a steep-sided valley.

lumen (a) A small, round central open space through a ce[ ] lumnal of a crinoid, blastoid, or cystoid; e.g. the wide cavity i[ ] a short, ring-like, proximal columnal of many cystoids. (I[ ] One of the spaces between muri of pollen and spores exhi[ ] iting reticulate sculpture.---Pl: *lumina.* Adj: *luminal.*

lumina Plural of *lumen.*

luminescence The emission of light by a substance that ha[ ] received energy or electromagnetic radiation of a differe[ ] wavelength from an external stimulus; also, the light so pr[ ] duced. It occurs at temperatures lower than those require[ ] for incandescence. See also: *phosphorescence; fluorescence[ ]*

lump (a) A descriptive term applied to a composite, loba[ ] grain in recent carbonate sediments, believed to have forme[ ] by aggregation, flocculation, or clotting of two or more pellet[ ] ooliths, skeletons, etc., or fragments thereof, or by disruptic[ ] of newly deposited or partly indurated carbonate mud. It typ[ ] cally possesses surficial reentrants. See also: *megalump.* (l[ ] *mudlump.*

lump coal A size of bituminous coal that passes through a [ ] inch round mesh in an initial screening, preparatory to pr[ ] cessing.

lump graphite Cryptocrystalline or very finely crystalline natu[ ] ral graphite from vein deposits, occurring in particle size[ ] ranging from that of walnuts to finer than 60-mesh.

lumping In taxonomy, the practice of ignoring minor diffe[ ] ences in the recognition or definition of species and genera. [ ] taxonomist known for his frequent lumping of taxa is called [ ] "lumper". Cf: *splitting.*

lump limestone A limestone containing numerous lumps (suc[ ] as aggregates of pellets or ooliths) in a matrix of micrite [ ] such as some of the Cenozoic limestones of the western inte[ ] rior of the U.S. Syn: *lumpal limestone.*

lumpy [gem] Said of a gemstone characterized by a thick cu[ ] (too great a depth in proportion to its width).

lumpy [sed] (a) Said of bedding consisting of scattered t[ ] loosely packed nodular bodies. (b) Pertaining to a lump o[ ] lumps.

lunabase A general term for dark, lunar surface rocks o[ ] basic (basaltic?) composition; e.g. *marebase.* Term intro[ ] duced by Spurr (1944). Cf: *lunarite.*

lunar (a) Pertaining to or occurring on the Moon, such as [ ] "lunar probe" designed to pass close to the Moon, or "luna[ ] dust" consisting of fine-grained material produced by meteori[ ] tic bombardment. (b) Resembling the surface of the Moon[ ] such as the "lunar landscape" of certain glaciers.

lunar crater *crater* [lunar].

lunar day The time required for the Earth to rotate once wit[ ] respect to the Moon, or the interval between two successive[ ] upper transits of the Moon across the meridian of a place; [ ] is about 24.84 hours (24 hours and 50 minutes). Cf: *tidal day.*

lunar geology A science that applies geologic principles and techniques to the study of the Moon, esp. its composition and the origin of its surface features. See also: *selenology*.

lunarite A general term for light-toned, brightly reflecting surface rocks of the lunar highlands (terrae). Term introduced by Spurr (1944). Cf: *lunabase*.

lunarium A hood-like projection from the wall on the proximal side of orifice in certain bryozoans (as in cryptostomes). Pl: *lunaria*.

lunar microcrater *micrometeorite crater*.

lunar playa A relatively small (as much as a few kilometers long), level area on the Moon's surface, occupying a low place in the ejecta blankets surrounding lunar craters such as Tycho and Copernicus. It is believed to be either a fallback deposit or a small lava flow.

lunar regolith A thin, gray layer on the surface of the Moon, perhaps several meters deep, consisting of partly cemented or loosely compacted fragmental material ranging in size from microscopic particles to blocks more than a meter in diameter. It is believed to be formed by repeated meteoritic and secondary fragment impact over a long period of time. Syn: *lunar soil; soil*.

lunar soil *lunar regolith*.

lunar tide The part of the tide caused solely by the tide-producing force of the Moon. Cf: *solar tide*. Syn: *moon tide*.

lunar transient phenomenon A temporarily abnormal appearance of a small area of the Moon, generally involving brightening, darkening, obscuration of surface features, or significant color changes, especially in the red and blue. Durations range from a few seconds to about a day, with a typical duration of about 30 minutes. The areas involved are generally part of a single crater or an isolated mountain but may occasionally include a whole crater or larger areas. These phenomena are strongly distributed around mare margins, as in the occurrence of dark-floored and dark-haloed craters, domes, and sinuous rilles. They are generally thought to arise from some type of internal volcanic action involving outgassing or extrusion of lava. Abbrev: *LTP*.

lunar varnish A hypothetical substance forming a dark coating on lunar particles and suggested as responsible for the low albedo of shallow subsurface lunar regolith.

lunate bar A crescent-shaped bar commonly found off a pass between barrier islands or the entrance to a harbor, or at a stream mouth.

lunate fracture A *crescentic mark* that is similar to a *crescentic fracture* but consists of two fractures from between which rock has been removed.

lunate mark *crescentic mark*.

lunate sandkey A *lunate bar* that has been built up above the water surface to form a crescent-shaped island, as along the west coast of Florida.

lundyite An intrusive rock characterized by orthophyric texture, a high content of alkali minerals, and a kataphorite-like amphibole.

lunette A term proposed by Hills (1940) for one of the broad, low (rarely more than 6-9 m high), even-crested, smooth, crescentic mounds or ridges of clay loam or silty clay bordering the leeward (eastern) shore of almost every lake and swamp in the plains of northern Victoria, Australia; it is produced by dust-laden winds.

lunitidal interval The interval between the Moon's transit over the local or Greenwich meridian and the time of the following high water or low water. Syn: *retardation*.

lunker A Scottish term for a lenticular mass of sandstone or clay ironstone; a big nodule.

lunoid furrow *crescentic gouge*.

lunule (a) A flat or curved, commonly cordate, area in front of the beak on the outside of many bivalve (pelecypod) shells, corresponding to the anterior part of the cardinal area, and

distinguished from the remainder of the shell surface by a sharp change in angle. (b) One of the openings in an echinoid test from the aboral surface through the oral surface at a perradial or interradial suture.

lunulitiform Said of bryozoans with unattached conical colonies, as in *Lunulites*.

lurain Lunar terrain.

lusakite A variety of staurolite containing cobalt.

luscladite An olivine theralite or essexite having hyperite texture and characterized by the absence of hornblende and by the presence of biotite as well as olivine. Orthoclase forms the reaction rim around the plagioclase. Nepheline is not abundant and fills the interstices. Cf: *berondrite; kyschtmite*.

Lusitanian European stage: Upper Jurassic (above Oxfordian, below Kimmeridgian). It includes the Argovian, Rauracian, and Sequanian substages.

lusitanite A dark-colored albite syenite composed of riebeckite, acmite, orthoclase, microcline-microperthite, albite, and minor amounts of quartz and osannite. Its name is derived from Lusitania (i.e. Portugal).

luster The reflection of light from the surface of a mineral, described by its quality and intensity; the appearance of a mineral in reflected light. Terms such as metallic or resinous are used for general appearance; terms such as bright or dull are used for intensity.

luster mottling [ign] A syn. of *poikilitic* originated by Raphael Pumpelly (Johannsen, 1939, p.183).

luster mottling [sed] The shimmering appearance of a broken surface of a sandstone cemented with calcite, produced by the brilliant reflection of light from the cleavage faces of conspicuously large (a centimeter or more in diameter) and independently oriented calcite crystals incorporating colonies of detrital sand grains; e.g. the luster displayed by Fontainebleau sandstone. It may also develop locally in barite, gypsum, or dolomite cements.

lusungite A rhombohedral mineral: $(Sr,Pb)Fe_3(PO_4)_2(OH)_5 \cdot H_2O$.

lutaceous Said of a sedimentary rock formed from mud (clay-and/or silt-size particles) or having the fine texture of impalpable powder or rock flour; pertaining to a lutite. Also said of the texture of such a rock. Term introduced by Grabau (1904, p.242). Cf: *argillaceous; pelitic*.

lutalite A tephritic leucite nephelinite with more than 50 percent mafic minerals.

lutecite Fibrous chalcedony characterized by inclined extinction and by fibers that are seemingly elongated about 30° to the c-axis. Syn: *lutecin*.

luteous Having an essential proportion of muddy sediment in a limestone, recognized by the presence of many particles of clastic quartz of silt size. Term introduced by Phemister (1956, p.73).

Lutetian European stage: Eocene (above Cuisian, below Auversian).

lutetium-hafnium age method The determination of an age in years based on the known radioactive decay rate of lutetium-176 (half-life approximately 2.2×10^{10} years) to hafnium-176. The method can be used, under favorable conditions, for dating minerals containing rare earths.

lutite A general name used for consolidated rocks composed of muds (silts and/or clays) and of the various associated materials which, when mixed with water, form mud; e.g. shale, mudstone, and calcilutite. The term is equivalent to the Greek-derived term, *pelite*, and was introduced as *lutyte* by Grabau (1904, p.242) who used it with appropriate prefixes in classifying fine-grained rocks (e.g. "anemolutyte", "anemosilicilutyte", "hydrolutyte", and "hydrargillutyte"). Etymol: Latin *lutum*, "mud". See also: *rudite; arenite*.

lutyte Var. of *lutite*.

luxullianite A granite characterized by phenocrysts of orthoclase and quartz which enclose clusters of radially arranged acicular tourmaline crystals in a groundmass of quartz, tourmaline grains, orthoclase crystals, brown mica, and cassiterite. Its name is derived from Luxulyan, Cornwall. Also spelled: *luxulianite; luxulyanite.* Var: *luxuliane.*

luzonite An isometric mineral: Cu_3AsS_4. It is dimorphous with enargite, and was formerly regarded as a variety of famatinite containing arsenic. Cf: *sinnerite.*

L wave *surface wave.*

lyas Var. of *lias.*

lychnisc A dictyonal hexactin (sponge spicule) in which the central crossing of the rays is replaced by an open octahedron with the rays arising from the octahedral angles.

lycopod A pteridophyte characterized by dense, simple, spirally arranged leaves, spore-bearing organs situated in the leaf axils which, in some forms, create specialized stromboli or cones; a member of the Lycopodineae, coextensive with the subdivision Lycopsida. Lycopods range from the Silurian, and include the club mosses. See also: *lycopsid.*

lycopsid A member of the subdivision Lycopsida; a *lycopod.*

Lydian stone A *touchstone* consisting of a compact, extremely fine-grained, velvet- or gray-black variety of jasper. It occurs in chert-like bands in older formations where it may represent a shale, limestone, or tuff that has been silicified. Etymol: Greek *Lydia*, ancient country in Asia Minor. Syn: *lydite; basanite.*

lydite A syn. of *Lydian stone.* Also spelled: *lyddite.*

lyell A block of rock transported and released by an iceber an ice-rafted block. Term proposed by Hamelin (1961, p.202

lyn A var. of *linn.*

lyndochite A variety of aeschynite, relatively high in calciu and thorium.

lyrula A subopercular, often anvil-shaped, median tooth on t proximal side of the orifice in some cheilostomatous bry zoans.

lysimeter A structure used to measure quantities of wa used by plants, evaporated from soil, and lost by deep perc lation. It consists of a basin, having closed sides and a bo tom fitted with a drain, in which soil is placed and plants a grown. Quantities of natural and/or artificial precipitation a measured, the deep percolate is measured and analyze water taken up by plants is weighed, etc.

lysocline The level or ocean depth at which the rate of so tion of calcium carbonate increases significantly.

lyssacine adj. Said of a hexactinellid sponge whose mega cleres (spicules) are unfused or separate, or are incomplete fused or so fused that their individual boundaries are app ent. Ant: *dictyonine.*---n. A lyssacine sponge.

lytomorphic A *deuteromorphic* crystal modified by aqueo solutions. The term is obsolete.

M

maacle *macle* [cryst].

maar A low-relief, coneless volcanic crater formed by a single explosive eruption. It is surrounded by a *crater ring*, and is commonly filled by water. Its type occurrence is in the Eifel area of Germany. Pl: *maars*.

maars Pl. of *maar*. Syn: *megata*.

maastrichtian Var. of *Maestrichtian*.

macadam effect Cementation of calcareous fragments by wetting and partial solution followed by deposition of cement on evaporation.

macallisterite A mineral: $Mg_2B_{12}O_{20}\cdot15H_2O$. Syn: *mcallisterite*.

macaluba A syn. of *mud volcano*. The name is taken from that of a low mud volcano, Macaluba, in Sicily.

macdonaldite A mineral: $BaCa_4Si_{15}O_{35}\cdot11H_2O$.

macedonite [min] A mineral: $PbTiO_3$.

macedonite [pet] A fine-grained basaltic rock composed of anorthoclase, sodic plagioclase, biotite, olivine, and rare pyroxenes; an olivine trachyte.

maceral Organic units that comprise the coal mass; all petrologic units seen in microscope thin sections of coal. Macerals are to coal as minerals are to rock. Maceral names bear the suffix "-inite". Cf: *phyteral*. Syn: *micropetrological unit*.

maceration The act or process of disintegrating sedimentary rocks (such as coal and shale) by various chemical and physical techniques in order to extract and concentrate acid-insoluble microfossils (including palynomorphs) from them. It includes mainly chemical treatment by oxidants and alkalies and use of other separating techniques that will remove extraneous mineral and organic constituents. Maceration is widely used in palynology.

macfarlanite A silver ore consisting of a mixture of sulfides, arsenides, etc., and containing cobalt, nickel, and lead. Cf: *animikite*.

macgovernite A mineral: $(Mn,Mg,Zn)_{15}As_2Si_2O_{17}(OH)_{14}$. Syn: *mcGovernite*.

machaeridia A class questionably assigned to the homalozoans and characterized by an elongate, bilaterally symmetrical test composed of an even number of longitudinal columns of plates. They have been variously identified as mollusks, annelids, and arthropods.

macigno The classical *flysch* facies in the northern Apennines, consisting of alternating strata of sandstone and mudstone, and showing graded bedding. Etymol: Italian, "millstone". Pron: mah-*cheen*-yo.

mackayite A green mineral: $FeTe_2O_5(OH)$ (?).

mackelveyite A dark-green or black mineral, approximately: $Na_2Ba_2Ca(Y,U)_2(CO_3)_9\cdot5H_2O$. Syn: *mckelveyite*.

mackereth sampler A variety of *piston corer* that operates by air hoses rather than by cable. It is used in shallow water.

mackinawite A tetragonal mineral: $(Fe,Ni)_{1.1}S$. It occurs as a corrosion product on iron pipes. Syn: *kansite*.

mackinstryite A mineral: $(Ag,Cu)_2S$. Syn: *mckinstryite*.

mackintoshite *thorogummite*.

mackle *macle* [cryst].

macle [mineral] (a) A dark or discolored spot in a mineral. (b) *chiastolite*.

macle [cryst] A twinned crystal; esp. a flat, often triangular, diamond composed of two flat crystals. Etymol: French, "wide-meshed net". Syn: *maacle; mackle*.

macled [mineral] (a) Said of a mineral that is marked like chiastolite. (b) Said of a mineral that is spotted.

macled [cryst] Said of a crystal having a twin structure.

maconite A vermiculite from North Carolina.

macro- A less preferred syn. of the prefix *mega-*.

macro-axis The longer lateral axis of an orthorhombic or triclinic crystal; it is usually the *b* axis. Cf: *brachy-axis*. Syn: *macrodiagonal*.

macrochoanitic Said of a retrochoanitic septal neck of a nautiloid that reaches backward beyond the preceding septum and is invaginated into the preceding septal neck.

macroclastic [coal] Said of coal that contains many recognizable fragments. Cf: *microclastic*.

macroclastic rock A clastic rock whose constituents are visible to the naked eye without magnification. Ant: *cryptoclastic rock*.

macrococcolith One of the larger coccoliths in coccolithophores exhibiting dimorphism but with the dimorphic coccoliths irregularly placed. Cf: *micrococcolith*.

macrocrystalline Said of the texture of a rock consisting of or having crystals that are large enough to be distinctly visible to the unaided eye or with the use of a simple lens; also, said of a rock with such a texture. Howell (1922) applied the term to the texture of a recrystallized sedimentary rock having crystals whose diameters exceed 0.75 mm, and Bissell & Chilingar (1967, p.103) to the texture of a carbonate sedimentary rock having crystals whose diameters exceed 1.0 mm. Syn: *megacrystalline; macromeritic*.

macrodiagonal n. A syn. of *macro-axis*.

macrodome A crystal form of either two or four faces which are parallel to the macroaxis in an orthorhombic crystal. A macrodome with four faces is a rhombic prism, or a *second-order prism* of the orthorhombic system.

macroevolution Evolution occurring in large, complex stages, such as the development of one species from another. Cf: *microevolution*.

macrofabric *megafabric*.

macrofacies *facies tract*.

macrofauna *megafauna*.

macroflora *megaflora*.

macrofossil A fossil large enough to be studied without the aid of a microscope. Cf: *microfossil*. Syn: *megafossil*.

macrofragmental Said of a coal composed of recognizable fragments or lenses of vegetal matter. Cf: *microfragmental*.

macrograined Said of the texture of a carbonate sedimentary rock having clastic particles whose diameters are greater than one millimeter (Bissell & Chilingar, 1967, p.103). See also: *megagrained*.

macrolithology The study, or characteristics, of rocks considered collectively as a part or the whole of the stratigraphic column in a given area. Cf: *microlithology*.

macromeritic An obsolete syn. of *macrocrystalline*.

macromutation Mutation that is large and easily observed.

macronucleus A relatively large vegetative nucleus that is believed to exert a controlling influence over the trophic activities in the body of a tintinnid. Cf: *micronucleus*.

macrophagous Said of an organism that feeds on relatively large particles. Cf: *microphagous*.

macrophyric *megaphyric*.

macrophyte A megascopic plant, esp. in an equatic environment.

macropinacoid *front pinacoid*.

macroplankton Plankton of the size range one millimeter to one centimeter. They are larger than *ultraplankton*, *nannoplankton*, and *microplankton*, but smaller than *megaloplankton*.

macropolyschematic Said of mineral deposits having megascopically distinguishable textural elements.

macropore A pore too large to hold water by capillarity. Syn: *megapore*.

macroporphyritic *megaphyric*.

macrorelief A term applied to surface irregularities only when it is necessary to distinguish them from *microrelief*.

macrosclere *megasclere.*

macroscopic (a) *megascopic.* (b) According to Dennis (1967, p.152), a term introduced to describe tectonic features that are too large to be observed directly in their entirety. Cf: *mesoscopic.*

macroseism A syn. of *earthquake*, as opposed to *microseism.*

macroseismic observations Measurements of an earthquake that are other than instrumental. Cf: *microseismic data.*

macrospecies A large group of organisms, usually a polymorphic species, distinctly different from related forms. Cf: *microspecies.*

macrospore A seldom-used and unsatisfactory syn. of *megaspore.*

macrotectonics A term used by Tomkeieff (1943, p.348) as a syn. of *megatectonics.*

macrotherm *megatherm.*

macula [paleont] One of the clusters of small, modified zooecia which may be regularly spaced throughout a Paleozoic bryozoan colony, appearing at the surface as a flattened or slightly depressed area typically surrounded by unusually large zooecia. Pl: *maculae.* Cf: *monticulus.*

macula [intrus rocks] A local pocket of magma that is formed by the fusion of shale and that acts as a type of *magma chamber* (Hobbs, 1953). Pl: *maculae.*

maculae The plural of *macula.*

maculose Said of a group of contact-metamorphic rocks, e.g. spotted slates, that have a spotted or knotted character; also, said of the structure itself. Syn: *spotted; knotted.*

macusanite *amerikanite.*

madeirite An intrusive rock composed of large augite and olivine phenocrysts in a fine-grained groundmass, less abundant than the phenocrysts and composed of plagioclase microlites in secondary calcite and magnetite. Its name is derived from Madeira, Spain.

made land Man-made land; an area of artificial fill consisting of earth materials more or less mixed with waste, refuse, and debris, such as on the marshy borders of a lake shore or on a shallow lake bottom bordering the shoreline. Also spelled: *madeland.*

madocite A mineral: $Pb(Sb,As)_{16}S_{25}$.

madreporite A perforated or porous, sieve-like structure that is situated at the distal end of the stone canal in an echinoderm and that provides access to the water-vascular system from the exterior; e.g. a conspicuous plate on the aboral surface of an asteroid or in the right anterior genital plate of an echinoid. Syn: *nucleus* [paleont].

madupite An extrusive rock composed of phlogopite, diopside, and perovskite phenocrysts in a brown glassy groundmass with the composition of leucite; a lamproite.

maelstrom A rapid, confused current formed by the combination of strong, wind-generated waves and a strong, opposing tidal current, and that may display *eddy*-type or *whirlpool*-type characteristics. It characteristically occurs along the south shore of the Lofoten Islands of Norway.

maenaite A hypabyssal calcic bostonite differing from bostonite in being rich in calcium oxide and poor in potassium oxide.

Maentwrogian European stage: Upper Cambrian (above Menevian, below Festiniogian).

Maestrichtian European stage: Upper Cretaceous (above Campanian, below Danian of Tertiary). Syn: *Maastrichtian.*

mafelsic Said of an igneous rock in which the felsic and mafic minerals are present in approximately equal amounts.

mafic Said of an igneous rock composed chiefly of one or more ferromagnesian, *dark-colored* minerals in its mode; also, said of those minerals. The term was proposed by Cross, et al (1912, p.561) to replace the term *femag*, which they did not consider to be euphonious. Etymol: a mnemonic term derived from *ma*gnesium + *f*erric + *ic*. It is the opposite of *felsic*. Cf: *femic; salic; basic.* Partial syn: *ferromagnesian.*

mafic front A term preferred by some petrologists over synonym *basic front.*

mafic margin *basic border.*

mafite A mafic mineral.

mafraite A hypabyssal rock containing labradorite with so[d] sanidine rims, pyroxene, magnetite, and euhedral hornblend[e] and without nepheline although its chemical components a[re] present in the groundmass.

mafurite A variety of olivine leucitite in which kalsilite is pre[s]ent instead of leucite.

magadiite A mineral: $NaSi_7O_{13}(OH)_3 \cdot 3H_2O$. It is found in la[ke] beds at Lake Magadi, Kenya.

magadiniform Said of the loop, or of the growth stage in th[e] development of the loop, of a terebratellid brachiopod (as [in] the subfamily Magadinae), marked by completed descendi[ng] branches from the cardinalia to the median septum, with [a] ring-like structure (on the septum) representing an early a[s]cending portion of the loop (TIP, 1965, pt.H, p.147). C[f:] *premagadiniform.*

magbasite A mineral: $KBa(Al,Sc)(Mg,Fe^{+2})_6Si_6O_{20}F_2$.

magellaniform Said of the free loop, or of the growth stage [in] the development of the loop, of a terebratellid brachiopod ([as] in *Magellania*), consisting of long descending branches r[e]curved into ascending branches that meet in transverse ba[nd] (TIP, 1965, pt.H, p.147). The magellaniform loop is morphol[o]gically similar to the *dalliniform* loop.

magelliform Said of the loop, or of the growth stage in the d[e]velopment of the loop, of a terebratellid brachiopod (as in *M*[a]*gella*), in which the bases of the septal ring on the medi[an] septum meet and fuse with the attachments of complete[d] descending branches (TIP, 1965, pt.H, p.147).

maghemite A strongly magnetic mineral of the magnetite s[e]ries in the spinel group: γ-Fe_2O_3. It is dimorphous with hem[a]tite. Syn: *oxymagnite.*

magma Naturally occurring mobile rock material, generate[d] within the Earth and capable of intrusion and extrusion, fro[m] which igneous rocks are thought to have been derived throu[gh] solidification and related processes. It may or may not conta[in] suspended solids (such as crystals and rock fragments) an[d] or gas phases. Adj: *magmatic.*

magma basalt A partial syn. of *limburgite*, also applied to [a] porphyritic, glassy basaltic rock with more resemblance [to] ordinary basalt.

magma blister A pocket of magma whose formation ha[s] raised the overlying land surface.

magma chamber A reservoir of magma in the shallow part [of] the lithosphere (to a few thousand meters), from which vo[l]canic materials are derived; the magma has ascended in the crust from an unknown source below. See also: *macul*[a]; *dike chamber.* Syn: *magma reservoir.*

magmagranite A granite produced by crystallization of [a] magma.

magma province *petrographic province.*

magma reservoir *magma chamber.*

magmasklerosis The magmatic *assimilation* of limestone. Th[e] term is not recommended.

magmatic Of, pertaining to, or derived from *magma*. Syn: o[r]*thotectic.*

magmatic assimilation *assimilation.*

magmatic corrosion *corrosion* [petrology].

magmatic differentiation *differentiation.*

magmatic digestion *assimilation.*

magmatic dissolution The solution of country rock by magma[;] *assimilation.* Syn: *magmatic solution.*

magmatic evolution The continuing change in composition of [a] magma as a result of magmatic differentiation, assimilatio[n,] or mixing of magmas.

magmatic ore deposit An ore deposit formed by *magmatic se*[g]regation, as crystals of metallic oxides or from an immiscib[le] lfide liquid. Syn: *magmatic segregation deposit.*

magmatic segregation Concentration of crystals of a particular mineral (or minerals) in certain parts of a magma during its cooling and crystallization. Some economically valuable deposits (i.e. *magmatic ore deposits*) are formed in this way. See also: *differentiation.* Syn: *segregation [petrology].*

magmatic segregation deposit *magmatic ore deposit.*

magmatic solution *magmatic dissolution.*

magmatic stoping A term originated by Daly for a process of magmatic emplacement or intrusion by detaching and engulfing pieces of the country rock. The engulfed material presumably sinks downward and/or is assimilated. See also: *piecemeal stoping; ring-fracture stoping; overhead stoping; overhand stoping; underhand stoping.*

magmatic water Water contained in or expelled from magma. Cf: *juvenile water; plutonic water.*

magmatism (a) The development, movement, and solidification to igneous rock, of magma. (b) The theory that much granite has crystallized from magma of any origin; opposed to *transformism.* A proponent of this theory is called a *magmatist.*

magmatist A proponent of the theory of *magmatism.*

magmatite A rock formed from magma.

magma type Categorization of magma having a distinctive chemical composition.

magmosphere *pyrosphere.*

magnacycle A term proposed by Merriam (1963, p.106) for a large, complex rock unit that follows a repetitious pattern and that can be considered cyclic in nature.

magnacyclothem A magnacycle that is larger than a *megacyclothem* (Merriam, 1963, p.106).

magnafacies A term proposed by Caster (1934, p.19) for a major, continuous, and homogeneous belt of deposits that is distinguished by similar lithologic and paleontologic characters and that extends obliquely across time planes or through several defined time-stratigraphic units; a complete "lithic member" or perfect rock-stratigraphic unit of the same facies but formed at different times; an isopic facies of European usage. It represents a distinct depositional environment that persisted with more or less shifting of geographic placement during time, and it may be divisible into, or assignable to, several noncontemporaneous *parvafacies.* The term is very nearly synonymous with *lithosome* as defined by Wheeler & Mallory (1956). Etymol: Latin *magna*, "great", + *facies.* See also: *megafacies.*

magnesia alum *pickeringite.*

magnesia mica (a) *phlogopite.* (b) *biotite.*

magnesian calcite A variety of calcite: $(Ca,Mg)CO_3$. It consists of randomly substituted magnesium carbonate in a disordered lattice of calcite. Low-magnesian calcite has less than 4% $MgCO_3$ in solid substitution, and is essentially the common form of calcite. High-magnesian calcite has 4-19% $MgCO_3$ in solid substitution; it is metastable and during limestone formation converts to low-magnesian calcite or to dolomite. Syn: *magnesium calcite.*

magnesian dolomite A dolomite rock with an excess of magnesium; specif. a dolomite rock whose Ca/Mg ratio ranges from 1.0 to 1.5 (Chilingar, 1957), or a dolomite rock containing 50-75% dolomite and 25-50% magnesite (Bissell & Chilingar, 1967, p.108).

magnesian limestone (a) A limestone that contains appreciable magnesium; specif. a limestone having at least 90% calcite, no more than 10% dolomite, an approximate MgO equivalent of 1.1-2.1%, and an approximate magnesium-carbonate equivalent of 2.3-4.4% (Pettijohn, 1957, p.418), or a limestone whose Ca/Mg ratio ranges from 60 to 105 (Chilingar, 1957), or a limestone containing 5-15% magnesium carbonate but in which dolomite cannot be detected (Holmes, 1928, p.149). Some petrographers use the term for a limestone with

some MgO but containing no dolomite; others for a rock with all possible mixtures of dolomite and calcite. Cf: *high-calcium limestone; dolomitic limestone.* (b) A dolomitic limestone; specif. the Magnesian Limestone, a facies of the Permian of NE England. (c) A term commonly, but loosely, used to indicate *dolomite* rock.

magnesian marble A type of metamorphosed magnesian limestone containing some dolomite (generally less than 10-15%). Cf: *dolomitic marble.*

magnesian spar *dolomite [mineral].*

magnesiochromite (a) A mineral of the spinel group: $(Mg,Fe)(Cr,Al)_2O_4$. It is isomorphous with chromite. Syn: *magnochromite.* (b) *picrochromite.*

magnesiocopiapite A mineral of the copiapite group: $MgFe_4(SO_4)_6(OH)_2.20H_2O$. It is a magnesium-rich variety of copiapite.

magnesioferrite A mineral of the magnetite series in the spinel group: $(Mg,Fe)Fe_2O_4$. It is strongly magnetic and usually black. Syn: *magnoferrite.*

magnesioriebeckite A mineral of the amphibole group: $Na_2(Mg,Fe^{+2},Fe^{+3})_5Si_8O_{22}(OH)_2$.

magnesite A white to grayish, yellow, or brown mineral: $MgCO_3$. It is isomorphous with siderite. Magnesite is generally found as earthy masses or irregular veins resulting from the alteration of limestones and dolomite rocks by magmatic solutions or of rocks rich in magnesium silicates (such as olivines). It is an ore of magnesium, and is used chiefly in making refractories and magnesia. Syn: *giobertite.*

magnesium calcite *magnesian calcite.*

magnesium-chlorophoenicite A monoclinic mineral: $(Mg,Mn)_5(AsO_4)OH)_7$. It is isostructural with chlorophoenicite.

magnesium front *basic front.*

magnet A magnetized body, especially a *permanent magnet.*

magnetic aftereffect *magnetic viscosity.*

magnetic anisotropy (a) *susceptibility anisotropy.* (b) *magnetocrystalline anisotropy.*

magnetic azimuth The *azimuth* measured clockwise from magnetic north through 360 degrees; the angle at the point of observation between the vertical plane through the observed object and the vertical plane in which a freely suspended magnetized needle, influenced by no transient artificial magnetic disturbance, will come to rest.

magnetic balance (a) An instrument in which the translational force on a magnetic moment in a nonuniform magnetic field is balanced against a spring, torsional or gravitational force. See also: *Curie balance.* (b) *Schmidt field balance.*

magnetic bearing The *bearing* expressed as a horizontal angle between the local magnetic meridian and a line on the Earth; a bearing measured clockwise from magnetic north. It differs from a *true bearing* by the amount of magnetic declination at the point of observation.

magnetic cleaning Partial demagnetization of natural remanent magnetization, by the removal of less stable, secondary components of magnetization or viscous remanent magnetization. Syn: *magnetic washing.*

magnetic compass A *compass* whose operation depends upon an element that senses the Earth's magnetic field; e.g. an instrument having a magnetic needle that turns freely on a pivot in a horizontal plane and that always swings to such a position that one end points to the magnetic north. See also: *prismatic compass.*

magnetic diffusivity The constant relating the rate of change of the intensity of a magnetic field to its gradient.

magnetic dip *inclination.*

magnetic elements The characteristics of a magnetic field which can be expressed numerically. The seven magnetic elements are *declination* D, *inclination* I, *total intensity* F, *horizontal intensity* H, *vertical intensity* Z, *north component* X, *east component* Y. Only three elements are needed to give a complete vector specification of the magnetic field.

magnetic equator The line on the Earth's surface at which magnetic inclination is zero; the locus of points with zero magnetic latitude. Syn: *aclinic line; dip equator.*

magnetic field (a) A region in which *magnetic forces* would be exerted on any magnetized bodies or electric currents present; the region of influence of a magnetized body or an electric current. Se also: *external magnetic field.* (b) *magnetic field intensity.* (c) *magnetic induction.*

magnetic field intensity A vector quantity, symbolized by H. Its direction is that sought by a freely suspended magnetized needle, and its magnitude is proportional to the torque on a magnetized body or the deflecting force of an electric current. Syn: *magnetic field; magnetic field strength.* Nonrecommended syn: *magnetic force.*

magnetic field line A curve whose tangent at any point is in the magnetic field direction at that point. Syn: *line of force; line of induction; magnetic flux line.*

magnetic field strength *magnetic field intensity.*

magnetic flux The surface area times the normal component of magnetic induction B; the number of magnetic field lines crossing the surface of a given area.

magnetic flux line *magnetic field line.*

magnetic force (a) The mechanical forces exerted by a *magnetic field*, or between magnetized bodies and electric currents. (b) A nonrecommended syn. of *magnetic field intensity.*

magnetic induction (a) Magnetic flux density, symbolized by B. In a magnetic medium the vector sum of the inducing field H and the magnetization M, according to the equation $B = H + \pi M$ in cgs units or $B = \mu_0 (H + M)$ in mks units. Syn: *magnetic field.* (b) A nonrecommended syn. of *electromagnetic induction.* (c) The process of magnetizing a body by applying a magnetic field. This usage is not recommended.

magnetic iron ore A syn. of *magnetite.* Var: *magnetic iron.*

magnetic latitude The angle whose tangent is one half the tangent of the magnetic inclination. It would equal geographic latitude if Earth's actual magnetic field were an axial dipole field.

magnetic meridian *magnetic north.*

magnetic moment A vector quantity characteristic of a magnetized body or an electric current system; it is proportional to the magnetic field intensity produced by this body and also to the force experienced in the magnetic field of another magnetized body or electric current. See also: *Bohr magneton.* Syn: *magnetic dipole moment.*

magnetic needle A short, slender, wire-like length of magnetic material (such as a bar magnet) that is used as a compass and that is so suspended at its midpoint as to indicate the direction of the magnetic field in which it is placed by orienting itself toward the Earth's magnetic north. Usually referred to as *needle.*

magnetic north The uncorrected direction indicated by the north-seeking end of the needle of a magnetic compass; the direction from any point on the Earth's surface of the horizontal component of the Earth's magnetic lines of force connecting the observer with the north magnetic pole; the northerly direction of the magnetic meridian at any given point. It is the common zero-degree (or 360-degree) reference in much of navigational practice. Cf: *true north.* Syn: *magnetic meridian.*

magnetic order A repetitive arrangement of the magnetic moments of ions in mineral crystals, analogous to the repetitive arrangement of the positions of the ions. It is applicable only for ions with an intrinsic magnetic moment, such as Fe^{+3}, Fe^{+2}, or Mn^{+2}. See also: *ferromagnetism; ferrimagnetism; antiferromagnetism; magnetic order.*

magnetic permeability *permeability* [magnet].

magnetic poles (a) Two areas near opposite ends of a magnet where the magnetic intensity is greatest. The magnetic lines of force leave the magnet at the positive or north-seeking pole

and enter at the negative or south-seeking pole. See als *negative pole; positive pole.* (b) *dip poles.* (c) *geomagnet poles.*

magnetic potential A scalar quantity whose negative gradie is the vector magnetic field intensity H. It is often symbolize by W.

magnetic prospecting A technique of applied geophysics: survey is made with a magnetometer on the ground or fro the air of local spatial variations in total intensity or vertic intensity. These measurements are interpreted as to th depth, size, shape, and magnetization of geologic feature causing any anomalies disclosed.

magnetic pyrites *pyrrhotite.*

magnetic resonance Interaction between the magnetic m tion, electron spin, and nuclear spin of certain atoms with a external magnetic field.

magnetic signature A shape of a magnetic anomaly, useful fc comparison with known or model anomalies.

magnetic spherule A black *cosmic spherule* consisting c magnetite and sometimes including a metal core.

magnetic storm A world-wide disturbance of the Earth's mag netic field. It generally lasts several days, and is thought to b caused by charged particles ejected by solar flares.

magnetic stratigraphy *paleomagnetic stratigraphy.*

magnetic survey Measurement of a component or element c the geomagnetic field at different locations. It is usually mad to map either the broad patterns of the Earth's main field c local anomalies due to variation in rock magnetization. Se also: *aeromagnetic survey.*

magnetic susceptibility *susceptibility* [mag].

magnetic variation (a) Changes of the magnetic field in tim or in space. (b) Magnetic *declination.*

magnetic viscosity A slow change of magnetization toward the direction of the ambient magnetic field. See also:*viscou magnetization.* Syn: *magnetic after effect.*

magnetic washing *magnetic cleaning.*

magnetism A class of physical phenomena associated wit moving electricity, including the mutual mechanical force among magnets and electric currents.

magnetite (a) A black, isometric, strongly magnetic, opaqu mineral of the spinel group: $(Fe,Mg)Fe_2O_4$. It often contain variable amounts of titanium oxide, and it constitutes an im portant ore of iron. Magnetite commonly occurs in octahe drons and also granular or massive; it is a very common an widely distributed accessory mineral in rocks of all kinds (i orebodies as a magmatic segregation, in lenses enclosed i schists and gneisses, in igneous rocks as a primary minera or as a secondary alteration product, in placer deposits, an as a constituent or heavy mineral in sands). Syn: *magneti iron ore; octahedral iron ore.* (b) A name applied to a serie of isomorphous minerals in the spinel group, consisting o magnetite, magnesioferrite, franklinite, jacobsite, trevorite and maghemite.--Symbol: Mt.

magnetitite An igneous rock composed chiefly of the minera magnetite. It has an iron content of at least 65-70%, and apa tite may be present. Syn: *kiirunavaarite.*

magnetization The magnetic moment per unit volume; a vec tor quantity symbolized by M, I or J. The magnetization of a rock is the sum of its two types: *induced magnetization* an *remanent magnetization.* Syn: *volume magnetization.* Non recommended syn: *polarization.*

magnetocrystalline anisotropy Dependence of the electronic energy of a magnetically ordered crystal upon the direction in which the atomic magnetic moments are aligned. Those crys tallographic directions for which the energy is lowest are called "easy" directions. In magnetite these are the (111) di rections. Syn: *magnetic anisotropy.*

magnetogram A continuous record produced by a *magneto graph* of temporal variations in magnetic elements.

magnetograph Instrument to automatically and continuously record temporal variations in the magnetic elements; the record it produces is a *magnetogram*.

magnetohydrodynamics The study of the relationship between a magnetic field and an electrically conducting fluid. It is relevant to studies of the Earth's core.

magnetoilmenite A high-temperature solid solution of magnetite in ilmenite. Cf: *ilmenomagnetite*.

magnetometer An instrument that measures the Earth's magnetic field and its changes, or the magnetic field of a particular rock (from which its magnetization is deduced). There are many types, of varying methods.

magnetoplumbite A black hexagonal mineral: $(Pb, Mn)_2Fe_6O_{11}$. Cf: *plumboferrite*.

magnetosphere The region around the Earth to which the Earth's magnetic field is confined, due to interaction between the solar wind and the geomagnetic field. On the sunlit side, the magnetosphere is approximately hemispherical, with a radius of about ten Earth radii under quiet conditions; it may be compressed to about six Earth radii by magnetic storms. On the side opposite the sunlit side, the magnetosphere extends in a "tail" of several hundred Earth radii.

magnetostriction Elastic strain or deformation accompanying magnetization. Cf: *piezomagnetism*.

magnetotelluric method An electromagnetic method in which natural electric and magnetic fields are measured; usually the two horizontal electric field components plus the three magnetic field components are recorded; orthogonal pairs yield elements of the tensor impedance of the Earth. This impedance is thereby measured at frequencies within the range 0^{-5}hz to 10 hz.

magniotriplite A mineral: $(Mg,Fe,Mn)_2(PO_4)F$. It is a magnesium-rich variety of triplite.

magniphyric Said of the texture of a microphyric igneous rock in which the greatest dimension of the phenocrysts is between 0.2 mm and 0.4 mm (Cross, et al, 1906, p.702); also, said of a rock having such texture. Cf: *mediophyric; magnophyric*.

magnitude *earthquake magnitude*.

magnocalcite Dolomitic calcite; a mixture of dolomite and calcite.

magnochromite *magnesiochromite*.

magnocolumbite A black orthorhombic mineral: $(Mg,Fe,Mn)(Nb,Ta)_2O_6$. It is the magnesium analogue of columbite.

magnoferrite *magnesioferrite*.

magnophorite A monoclinic mineral of the amphibole group: $NaKCaMg_5Si_8O_{23}OH$.

magnophyric Said of the texture of a porphyritic igneous rock in which the greatest dimension of the phenocrysts is more than 5 mm (Cross et al, 1906, p. 702); also, said of a rock having such texture. Cf: *magniphyric; mediophyric*.

magnussonite A green isometric mineral: $Mn_5(AsO_3)_3(OH,Cl)$. It may contain some magnesium and copper.

main joint *master joint*.

mainland A continuous body of land constituting the chief part of a country; e.g. a continent, or a main island relative to an adjacent smaller island. Syn: *fastland*.

main partition A radial wall of a foraminiferal test, extending from the marginal zone toward the center of the chamber (as in Orbitolinidae). It may be a simple transverse septum.

main shock The largest earthquake in a sequence. See also: *aftershock; foreshock*. Syn: *principal shock; principal earthquake*.

main stem The principal course of a stream.

main stream The principal, largest, or dominating stream of any given area or drainage system. Syn: *master stream; trunk stream*.

major earthquake An earthquake having a magnitude of seven or greater on the Richter scale. Such a limit is arbitrary, and

may vary according to the user. Cf: *microearthquake; ultramicroearthquake*.

major fold A large-scale or dominant fold or an area, with which *minor folds* are usually associated.

majorite A meteorite mineral of the garnet group: $Mg_3(Fe,Al, Si)_2Si_3O_{12}$.

major joint *master joint*.

major septum One of the initial or secondary septa of a corallite; specif. a protoseptum or a metaseptum. Major septa are of subequal length and extend most of the distance from the wall to the axis. Cf: *minor septum*.

makatea A Polynesian term used in the south Pacific Ocean for a raised rim of a coral reef, of for a broad, uplifted coral reef surrounding an island. Etymol: Tuamotu.

makatite A mineral: $Na_2Si_4O_9.5H_2O$.

make n. (a) A formation or accumulation of ore in a vein; esp. the wide or thick part of a lode or orebody. Cf: *pinch*. (b) The output, actual yield, or amount produced by an oil or gas well or a mine over a specified period. The term is colloquial.

makhésh A term used in Israel for a huge, cirque-like hollow somewhat resembling an elongated meteorite crater, produced by erosion of a structural dome (Amiran, 1950-1951). Etymol: Hebrew, "mortar". Pl: *makhteshim*. Syn: *erosion crater*.

mäkinenite A mineral: gamma-NiSe.

making hole The act of, or the portion of work time spent in, actual drilling and advancement of a drill hole or well.

malachite A bright-green monoclinic mineral: $Cu_2CO_3(OH)_2$. It is an ore of copper and is a common secondary mineral associated with azurite in the upper (oxidized) zones of copper veins. Malachite occurs in masses having smooth mammillated or botryoidal surfaces, and it is often concentrically banded in different shades of colors. It is used to make ornamental objects.

malacolite A syn. of *diopside*. The term originally designated a light-colored (pale-green or yellow) translucent variety of diopside from Sweden.

malacology The study of mollusks. Cf: *conchology*.

malacoma Collective name for the soft parts of radiolarians.

malacon A brown altered or hydrated variety of zircon. Syn: *malakon; malacone*.

malacostracan (a) Any crustacean belonging to the class Malacostraca, characterized by the presence of compound eyes, a thorax composed of eight somites, typically with a carapace, and by an abdomen composed of six or seven somites. Their stratigraphic range is Lower Cambrian to present. (b) In very early usage, a soft-shelled crustacean. Cf: *entomostracan*.

malayaite A mineral: $CaSnSiO_5$.

malaysianite A tektite from the Malay Peninsula.

malchite A fine-grained lamprophyre, generally porphyritic with small, somewhat rare phenocrysts of hornblende, and labradorite, and sometimes biotite, in a groundmass of hornblende, andesine, and a small amount of quartz. Its name is derived from Malchen, Germany. Cf: *luciite*.

maldonite A mineral: approximately Au_2Bi. It consists of a pinkish to silvery-white alloy of gold and bismuth. Syn: *black gold; bismuth gold*.

malenclave A body of contaminated or unusable ground water surrounded by uncontaminated water. Classification of malenclaves depends on whether their volume expands, diminishes, or is constant with time (Legrand, 1965, p.88).

malezal swamp A swamp due to drainage of water over an extensive plain which has only a slight, almost imperceptible slope.

malgachite An igneous-rock facies that includes granite, granodiorite, diorite, and gabbro (Thrush, 1968, p.675); also, any rock in that facies.

malignite A mafic nepheline syenite which has more than 5%

nepheline and roughly equal amounts of pyroxene and potassium feldspar. The name is derived from the Maligne River, Ontario, Canada.

malinowskite A variety of tetrahedrite containing lead.

malladrite A hexagonal, low-temperature mineral of fumaroles: Na_2SiF_6. Not to be confused with *mallardite*.

mallardite A pale-rose monoclinic mineral: $MnSO_4.7H_2O$. Not to be confused with *malladrite*.

Mallard's constant In *Mallard's law*, the constant for any combination of lenses on a given microscope; it is written as K.

Mallard's law A statement in crystallography that planes or axes of pseudosymmetry in a space lattice may become twinning planes or axes. The formula is $D=K \sin E$, in which D equals half the distance between the points of emergence of the optic axes, E equals one half the optic axial angle in air, and K equals the constant for any combination of lenses on a given microscope; this constant is *Mallard's constant*. It is named after the French crystallographer and mineralogist of the nineteenth century, Ernest Mallard.

malleable Said of a mineral, e.g. gold, silver, copper, platinum, which can be plastically deformed under compressive stress, e.g. hammering.

malloseismic Said of an area that is likely to be visited several times in a century by destructive earthquakes.

Malm Middle European series: Upper Jurassic (above Dogger, below Cretaceous).

malmstone (a) A hard, cherty, grayish-white sandstone whose matrix contains minute opaline globules derived from sponge spicules that once filled now-empty casts; specif. the Malmstone from the upper part of the Upper Greensand (Cretaceous) of Surrey and Sussex in England, used as a building and paving material. (b) A marly or chalky rock; *malm*.---Syn: *malm rock*.

malpais A term used in the southwestern U.S. for a region of rough and barren lava flows. The connotation of the term varies according to the locality. Etymol: Spanish, *mal païs*, "bad land".

maltha A term used to designate the softer, more viscid varieties of native asphalt. In Trinidad, it is called *brea*. Syn: *earth pitch; mineral tar; malthite; pissasphalt.*

malthacite A scaly, sometimes massive, white or yellowish clay related to fuller's earth, having a Si/Al ratio of about 4.

malthite *maltha*.

mamelon A raised, rounded top of an echinoid tubercle on which the spine articulates.

mamelon A small, rounded volcano formed over a vent by slow extrusion of viscous, siliceous lava.

mamlahah A term used on the Arabian peninsula for an interior salt-incrusted playa. Cf: *sebkha*.

mammillary Said of an aggregate of crystals taking the form of rounded masses.

mammillary hill A smooth, rounded, more or less elongate drumlin having an elliptical base.

mammillary structure *pillow structure* [sed].

mammillated surface A hummocky rock surface characterized by smoothed and rounded mounds alternating with hollows, esp. a streamlined surface formed by glacial erosion in mountainous areas, as in the Adirondack Mountains, N.Y.

manaccanite *menaccanite* [mineral].

manandonite A white mineral: $Li_4Al_{14}B_4Si_6O_{29}(OH)_{24}$ (?).

manasseite A hexagonal mineral: $Mg_6Al_2(CO_3)(OH)_{16}.4H_2O$. It is dimorphous with hydrotalcite.

mandible (a) An articulated part of a bryozoan avicularium, moved by muscles, and homologous with the operculum of an autozooid. (b) Any of various invertebrate mouthparts serving to hold or bite into food materials and/or to move food into the mouth; e.g. one of the third pair of cephalic appendages of a crustacean. (c) An obsolete term used by some arachnologists for chelicera and by other for pedipalpal coxa.

mandibular muscle scar The place of attachment of the muscle leading to the mandible of an ostracode from the inner surface of the carapace just anterior and ventral to the adductor muscle scars.

mandibular palp The distal articulated part of the mandible of a crustacean, aiding in feeding and cleaning.

mandshurite A nepheline basanite.

Manebach-Ala twin law A complex twin law in triclinic feldspar in which the twin axis is perpendicular to [001] and the composition plane is (001). Cf: *Ala-A twin law*. Syn: *acline-A twin law*.

Manebach pericline twin law A complex twin law in feldspars in which the twin axis is at right angles to [010], and the composition plane is (001).

Manebach twin law A twin law in feldspars, both monoclinic and triclinic, usually simple, with the twin plane and composition plane of (001).

mangan A *cutan* consisting of manganese oxides or hydroxides (Brewer, 1964, p.215).

manganandalusite *viridine*.

manganaxinite A mineral: $Ca_2(Mn,Fe)Al_2BSi_4O_{15}(OH)$.

manganbabingtonite A mineral: $Ca_2(Mn,Fe^{+2})Fe^{+3}Si_5O_1$ (OH).

manganbelyankinite A mineral: $(Mn,Ca)(Ti,Nb)_5O_{12}.9H_2O$.

manganberzeliite A mineral: $(Mn,Mg)_2(Ca,Na)_3(AsO_4)_3$. It is isomorphous with berzeliite.

manganblende *alabandite*.

manganese alum *apjohnite*.

manganese epidote *piemontite*.

manganese-hoernesite A mineral: $(Mn,Mg)_3(AsO_4)_2.8H_2O$. Syn: *manganese-hörnesite*.

manganese nodule A small, irregular, black to brown, friable laminated concretionary mass consisting primarily of manganese salts and manganese-oxide minerals (Mn content is 15 30%) alternating with iron oxides, abundant on the floors of the world's oceans (and also of the Great Lakes) as a result of pelagic sedimentation or precipitation esp. in an area of slow deposition, and occurring suspended in sediments (esp. in red clay and sometimes in organic ooze) or as a surface coating on rocks and other minerals, as a rounded ball with a small nucleus (such as a shark's tooth), or as an intergrowth forming a large slab. Manganese nodules vary in size from a few microns to 25 cm in diameter (generally 3-5 cm) and have an average weight of 115 grams, although larger ones exist (a nodule weighing 770 kg has been found). Syn: *pelagite*.

manganese spar (a) *rhodonite*. (b) *rhodochrosite*.

manganite A brilliant steel-gray or iron-black orthorhombic mineral: $\gamma-MnO(OH)$. It is polymorphous with groutite, and represents an ore of manganese. Cf: *feitknechtite*. Syn: *gray manganese ore*.

mangan-neptunite A dark-red mineral: $(Na,K)_2(Mn,Fe^{+2})TiSi_4O_{12}$. Cf: *neptunite*.

manganocalcite (a) A variety of rhodochrosite containing calcium. (b) A variety of calcite containing manganese.

manganolangbeinite A rose-red isometric mineral: $K_2Mn_2(SO_4)_3$.

manganolite [rock] A general term for rocks composed of manganese minerals (esp. manganese oxides such as wad and psilomelane).

manganolite [mineral] *rhodonite*.

manganomelane A field term used synonymously for *psilomelane* to designate hard, massive, botryoidal, colloform manganese oxides not specifically identified. The term was rejected by the International Mineralogical Association.

manganophyllite (a) A manganiferous variety of biotite: $K(Mn,Mg,Al)_{2-3}(Al,Si)_4O_{10}(OH)_2$. (b) A hypothetical biotite end member: $K_2Mn_5Al_4Si_5O_{10}(OH)_4$.

manganosiderite A variety of siderite (ferrous-carbonate mineral)

428

ral) containing manganese. It is an intermediate member of
the isomorphous series siderite-rhodochrosite.

manganosite An isometric mineral: MnO. It occurs in small
emerald-green octahedrons that turn black on exposure.

manganotantalite A variety of tantalite, with manganese substituting for most of the ferrous iron: $(Mn,Fe)(Ta,Nb)_2O_6$. Syn:
mangantantalite.

manganpyrosmalite A mineral: $(Mn,Fe)_8Si_6O_{15}(OH,Cl)_{10}$. Cf:
pyrosmalite.

mangerite A plutonic rock of the charnockite series, corresponding to monzonite. Typically it contains microperthite as
the dominant feldspar with varying amounts of mafic minerals,
esp. hypersthene; a hypersthene-bearing alkalic syenite containing a predominance of perthitic feldspars; the intrusive
equivalent of *doreite* (Streckeisen, 1967, p. 209). See also:
pyroxene monzonite.

mangrove coast A tropical or subtropical low-energy coast the
shoreline of which is overgrown by mangrove vegetation, such
as in southern Florida.

mangrove swamp A tropical or subtropical *marine swamp*
characterized by abundant mangrove trees.

maniculifer Said of brachiopod crura derived from the radulifer
type, with hand-like processes at the end of straight, ventrally
directed crura.

manjak A variety of *glance pitch* (i.e. it is an asphaltite) found
in Barbados which contains 0.7 to 0.9 percent sulfur and 1 to
percent mineral matter.

manjakite An igneous rock exhibiting equigranular texture and
containing garnet, biotite, pyroxene, and variable amounts of
feldspar, magnetite, hypersthene, and labradorite. It resembles kentallenite but contains less calcium oxide (Thrush,
1968, p.678).

man-made shoreline A shoreline consisting of the works of
man, such as harbor areas, breakwaters, causeways, piers,
seawalls, and docks.

Manning equation An equation used to compute the velocity of
uniform flow in an open channel: $V = 1.486/n \ R^{2/3} \ S^{1/2}$,
where V is the mean velocity of flow (in cfs units), R is the
hydraulic radius in feet, S is the slope of the channel or sine
of the slope angle, and n is the Manning roughness coefficient. Cf: *Chézy equation.*

mansfieldite A white to pale-gray orthorhombic mineral: Al-
$sO_4.2H_2O$. It is isomorphous with scorodite.

mantle [interior Earth] The zone of the Earth below the crust
and above the core (to a depth of 3480 km), which is divided
into the *upper mantle* and the *lower mantle*, with a transition
zone between.

mantle [cryst] The outer zone in a zoned crystal; *an overgrowth.*

mantle [paleont] (a) The fold, lobe, or pair of lobes of the
body wall in a mollusk or brachiopod, lining the shell and
bearing the shell-secreting glands, and usually forming a mantle cavity; e.g. a prolongation of the body wall of a brachiopod, such as the two folds of ectodermal epithelium lying
above or below the viscera and lining the inner surface of
each valve, or the integument surrounding the vital organs of
a bivalve mollusk, consisting of a pair of folds from the dorsal
body wall extending laterally and ventrally over the sides of
the animal. Syn: *pallium.* (b) The fleshy structure of cirripede
crustaceans, strengthened by five calcified plates (carina,
terga, and scuta) (TIP, 1969, pt.R, p.98). (c) Variously
formed covering or coat in a radiolarian.

mantle [geol] A general term for an outer covering of material
of one kind or another, such as a *regolith*; specif. *waste mantle.*

mantle canal Any of the flattened, tube-like, branching extensions of the body cavity into the mantle of a brachiopod,
through which fluids circulate in the mantle. Syn: *pallial sinus.*

mantle cavity The cavity, between the mantle and the body

proper, holding the respiratory organs of a mollusk or brachiopod; e.g. the anterior space between brachiopod valves,
bounded by the mantle and the anterior body wall, and containing the lophophore. Syn: *pallial chamber.*

mantle-crust mix Rock whose properties are between those of
the crust and those of the mantle, e.g. *P*-wave velocities between 7.4 and 7.7 km/sec.

mantled Said of bedding, e.g. of pyroclastic deposits, that
conforms to the underlying, irregular topography.

mantled gneiss dome A term used by Eskola (1948) for a
dome in metamorphic terranes that has a core of gneiss that
was remobilized from an original basement and that has risen
through its cover of younger, also metamorphosed rocks. The
gneiss is surrounded by a concordant sheath of the basal part
of the overlying metamorphic sequence. Examples are the
Baltimore gneiss domes in the eastern Maryland Piedmont,
which yield 2 radiometric dates, that of the original basement,
and that of the later remobilization.

mantle rock A syn. of *regolith*. Also spelled: *mantlerock.*

manto A flat-lying, bedded deposit; either a sedimentary or an
igneous, strata-bound orebody, Etymol:· Spanish, "vein, stratum".

map n. A diagram, drawing, or other graphic representation
usually on a plane (horizontal, flat) surface of selected physical features (natural, artificial, or both) of a part or the whole
of the surface of the Earth, some other planet, or the Moon,
or of any desired surface or subsurface, by means of signs
and symbols and with the means of orientation indicated, so
that the relative position and size of each feature on the map
corresponds to the correct geographic situation according to a
definite and established scale or projection. The type of information that a map is primarily designed to convey is frequently designated by an adjective to distinguish it from maps of
other types; e.g. "geologic map", "topographic map", or
"structure map". Etymol: Latin *mappa*, "napkin, cloth". Cf:
chart; plan.---v. To produce or prepare a map; to represent or
delineate on a map; to engage in a mapping operation.

map face The area on a map, enclosed by the neat line.

map measurer *chartometer.*

map nadir The map position of the *ground nadir.*

mapping The process of making a map of an area; esp. the
field work necessary for the production of a map.

mapping angle *gisement.*

map projection (a) Any orderly system or arrangement of
lines drawn on a plane surface and representing a corresponding system of imaginary lines on an adopted terrestrial
or celestial datum surface; esp. a graticule formed by two intersecting systems of lines (representing parallels of latitude
and meridians of longitude) that portray upon a flat surface
the whole or any part of the curved surface of the Earth, or a
grid based on such parallels and meridians. It is frequently referred to as a *projection.* (b) Any systematic method by which
a map projection is made; the process of transferring the outline of surface features of the Earth onto a plane. (c) The
mathematical concept of a map projection.

map reading The interpretation of the information shown on a
map.

map scale *scale* [cart].

map series A group of maps generally conforming to the
same cartographic specifications or having some common unifying characteristic, such as the same scale or the same size
of area covered. It usually has a uniform format and is identified by a name, number, or a combination of both. Examples
are the National Topographic Map Series and the Geologic
Quadrangle Map Series published by the U.S. Geological Survey. Syn: *series.*

map sheet An individual map, either complete in itself or part
of a map series.

mar A Swedish term for a bay or creek whose entrance is

filled with silt so that the water is almost fresh (Stamp, 1961, p.308). Pl: *marer*.

marais A French term for swamp used in place names in certain localities of the U.S.

marble [met] (a) A metamorphic rock consisting predominantly of fine- to coarse-grained recrystallized calcite and/or dolomite usually with a granoblastic (saccharoidal) texture. (b) In commerce and industry it is a term applied to any more or less crystallized limestone capable of taking a polish or of being used for fine architectural work or ornamental purposes. See also: *crystalline limestone*.

marble [sed] *crystalline limestone*.

marble [eco geol] Any calcareous rock or other rock of comparable hardness that is used for decorative purposes because it takes a good polish. Pleasing color and texture are also considerations.

marcasite [gem] A popular term used in the gemstone trade to designate any of several minerals with a metallic luster (esp. crystallized pyrite, as used in jewelry) and also polished steel and white metal.

marcasite [mineral] A common, very light brownish-yellow or grayish, orthorhombic mineral: FeS_2. It is dimorphous with pyrite and resembles it in appearance, but marcasite has a lower specific gravity, less chemical stability, and usually a paler color. Marcasite often occurs in sedimentary rocks (such as chalk) in the form of nodules or concretions with a radiating fibrous structure. Syn: *white iron pyrites; iron pyrites; white pyrite; white pyrites; cockscomb pyrites; spear pyrites; lamellar pyrites*.

marchite A pyroxenite composed of enstatite and diopside.

mare (a) One of the several dark, low-lying, level, relatively smooth, plains-like areas of considerable extent on the surface of the Moon, having fewer large craters than the highlands, and composed of mafic or ultramafic volcanic rock; e.g. Mare Imbrium (a circular mare) and Mare Tranquillitatis (a mare with an irregular outline). It is completely waterless. Cf: *terra*. (b) A dark area, on the surface of Mars, whose origin is not definitely known. Cf: *continens.*--Etymol: Latin, "sea", from Galileo's belief that lunar maria represented great seas of water. Pron: *mah*-rey. Pl: *maria*. Syn: *sea*.

marebase Lunar rock of basic composition specific to the maria. See also: *lunabase*.

mare basin A large, approximately circular or elliptical topographic depression in the lunar surface, filled or partly filled with mare material; e.g. the Imbrium basin. See also: *thalassoid [lunar]*.

marekanite Obsidian that occurs as rounded to subangular bodies, usually less than two inches in diameter and having indented surfaces. These bodies occur in masses of perlite and are of special interest because of their low water content as compared with the surrounding perlite, the ratio often being as small as one to ten. Partial syn: *obsidianite*.

mare material Dark, relatively smooth, heavily cratered igneous rock (chiefly of mafic or ultramafic composition) underlying the lunar maria.

maremma A low, marshy or swampy tract of coastland. Etymol: Italian.

mareogram Var. of *marigram*.

mareograph Var. of *marigraph*.

mare ridge *wrinkle ridge*.

mareugite A bytownite- and hauyne-bearing plutonic rock; a hauyne gabbro. Its name is derived from Mareuges, France.

margarite [mineral] A mineral of the brittle-mica group: $CaAl_2(Al_2Si_2)O_{10}(OH)_2$. It has a pale-pink, reddish-white, or yellowish color, and is marked by a pearly luster. Syn: *lime mica; calcium mica; pearl mica*.

margarite [ign] A bead-like string of globulites, commonly found in glassy igneous rocks.

margarodite A pearly variety of muscovite resembling talc and

affording a small percentage of water upon ignition.

margarosanite A colorless or snow-white triclinic mineral: $Pb(Ca,Mn)_2(SiO_3)_3$.

marginal chamberlet A simple subdivision of a primary chamber of a foraminiferal test, located in the marginal zone of the chamber, and formed by main partitions only (as in Orbitolinidae).

marginal channel A channel formed by a meltwater stream flowing along the margin of a glacier or an ice sheet (Rich, 1908, p.528).

marginal conglomerate A conglomerate that forms along a shore along the landward margins of sediments of other types into which it grades (Twenhofel, 1939, p.30). It lies at different stratigraphic levels in the section (as seen over a large area) and thereby diagonally transects time intervals. If sea level is rising, the conglomerate is a *basal conglomerate*.

marginal cord A thick spiral structure beneath the surface at the periphery of a foraminiferal test (as in Nummulitidae) (TIP, 1964, pt.C, p.61).

marginal crevasse A *crevasse* near the margin of a glacier. It normally extends obliquely upstream from either side toward its middle at an angle of about 45° (as seen in plan). Cf: *transverse crevasse; splaying crevasse*. Syn: *lateral crevasse*.

marginal fault *boundary fault*.

marginal fissure A fracture bordering an igneous intrusion which has become filled with magma.

marginal granule A dot-like body in a lamella of a tintinnid.

marginalia Sponge spicules (prostalia) around or of an oscular margin.

marginal karst plain That part of a *karst plain* that borders a higher region so that it receives detrital material that allows surface drainage (Monroe, 1970). Var: *karst margin plain*.

marginal lagoon A lagoon that is adjacent to a shore or coast line.

marginal lake *glacier lake*.

marginal moraine A term formerly used as a syn. of *terminal moraine* (Hobbs, 1912, p.279).

marginal nunatak A nunatak that is partly bounded by the sea or by land; e.g. Jensen Nunatak of western Greenland.

marginal plain An obsolete term for an *outwash plain* flanking the margin of a terminal moraine. Also, a vague term loosely applied to various topographic features around the margins of glaciers.

marginal plateau A relatively flat shelf adjacent to a continent and similar topographically to, but deeper than, a *continental shelf*.

marginal ring The distal part of a cyclocystoid, bordering *sub marginal ring* and composed of small imbricating plates that distally decrease in size.

marginal salt pan A natural salt pan along a coast, such as the Great Rann of Kutch in the Gujarat region of western India; a salt marsh along a coast.

marginal sea A semienclosed sea adjacent to a continent, floored by submerged continental mass. See also: *shelf sea*.

marginal spine One of a series of spines, often jointed at the base, surrounding the frontal area of a cheilostomatous bryozoan; e.g. a *scutum*.

marginal suture (a) The ecdysial (molting) junction between exoskeleton elements at the prosomal margin in a merostome. (b) A suture running along the edge of the cephalon of certain trilobites (TIP, 1959, pt.O, p.122).

marginal thrust One of a series of faults bordering an igneous intrusion and crossing both the intrusion and the wall rock; *intrusion displacement*. The total displacement is dispersed along the thrusts in small amounts. Syn: *marginal upthrust*.

marginal trench *trench [marine geol]*.

marginal upthrust Var. of *marginal thrust*.

marginal zone The peripheral portion of foraminiferal chambers where chamberlets are subdivided by primary and secon ary partitions (as in Orbitolinidae).

marginarium The peripheral part of the interior of a corallite, characterized by generally abundant dissepiments or by a dense deposit of skeletal tissue producing a stereozone. Adj: *marginarial*. Cf: *tabularium*.

marginate chorate cyst A dinoflagellate *chorate cyst* whose outgrowths are characteristically localized on the lateral margins, leaving the dorsal and more often the ventral surfaces free of outgrowths.

margination texture The texture of a granite characterized by sinuous contacts between quartz and feldspar grains, resulting from the corrosion of earlier formed crystals by a later crystallization.

margo (a) A modified margin of the colpus of a pollen grain, consisting of a thickening or thinning in the ektexine. Cf: *annulus [palyn]*. (b) A term sometimes used for similar marginal features associated with the laesura of spores.

maria Plural of *mare*.

marialite A mineral of the scapolite group: $3NaAlSi_3O_8 \cdot NaCl$ or three albite plus sodium chloride). It is isomorphous with meionite. Symbol: Ma.

marienbergite A plagioclase-bearing phonolite containing natrolite instead of nepheline.

marignacite A variety of pyrochlore containing appreciable amounts of rare earths (esp. cerium).

marigram A *tide curve*; esp. the autographic record traced by a *marigraph*. Syn: *mareogram*.

marigraph A self-registering *tide gage*, usually actuated by a float in a tube or pipe communicating with the sea through a small hole that filters out short-period waves. See also: *marigram*. Syn: *mareograph*.

marine abrasion The erosion of the ocean floor by sediment that is moved by wave energy. Syn: *wave erosion*.

marine arch *sea arch*.

marine bank A general term for limestone deposits that are locally abnormally thick and that appear to have formed over submerged shallow areas that rose above the general level of the surrounding sea floor (Harbaugh, 1962, p.13). Marine banks lack the hard, wave-resistant character of organic reefs. See also: *bank [sed]*.

marine bench *marine-cut bench*.

marine biology The study of marine fauna and flora.

marine bridge *sea arch*.

marine-built Constructed or built up by the action of waves and currents of the sea. See also: *wave-built*.

marine-built platform A syn. of *marine-built terrace*. The term is inconsistent because a platform is usually regarded as an erosional feature.

marine-built terrace A *wave-built terrace* produced by marine processes.

marine cave (a) *sea cave*. (b) A cave formed on the bottom of the sea.

marine cliff *sea cliff*.

marine climate *oceanic climate*.

marine-cut Carved or cut away by the action of waves and currents of the sea. See also: *wave-cut*. Syn: *sea-cut*.

marine-cut bench A *wave-cut bench* of marine origin. Syn: *marine bench*.

marine-cut platform A *wave-cut platform* produced by marine processes.

marine-cut terrace A syn. of *marine-cut platform*. The term is inconsistent because a terrace is usually regarded as a constructional feature.

marine delta plain A nearly flat plain built in a bay by stream deposits at the place where the current is checked upon entering quiet water (Tarr, 1902, p.73-74); it is built a slight distance above sea level.

marine-deposition coast A coast whose configuration results chiefly from marine deposition, as a coast straightened by the formation of spits or bars, or a coast prograded by wave and current deposits.

marine ecology The study of the relationships between marine organisms and their environments, and among the organisms themselves.

marine-erosion coast A coast whose configuration results chiefly from marine erosion, as the straightening of sea cliffs by waves.

marine erratics Sedimentary particles of anomalous size or lithology, transported and deposited in marine sediments by ice raffing, plants, or animals.

marine geodesy The precise determination of positions at sea and the establishing of boundaries and boundary markers at sea. It also includes the measurement of gravity at sea and the study of all the associated physical characteristics of the sea environment that effect such measurements.

marine geology *geological oceanography*.

marine invasion The spreading of the sea over a land area.

marine limit The limit of the sea; a coastline. See also: *marin gräns*.

marine marsh A flat, vegetated, savanna-like land surface at the edge of the sea, and usually covered by the sea during high tide. Cf: *salt marsh*.

marine onlap A term proposed by Melton (1947, p.1869) for *onlap* in connection with marine strata that are progressively pinched out landward above an unconformity. Example: the relations of the Cambrian rocks of the Grand Canyon.

marine peneplain An abrasion platform of large areal extent, uplifted above the reach of the waves before wave erosion had succeeded in perfecting a smooth plane; an almost-plane surface of uncompleted marine denudation. Cf: *plain of marine erosion*.

marine plain (a) *plain of marine erosion*. (b) A coastal plain of marine sediments.

marine plane A level wave-cut surface produced during the ultimate stage of marine erosion; a *plain of marine erosion*.

marine platform *marine-cut platform*.

marine salina A body of salt water found along an arid coast, separated from the sea by a sand or gravel barrier through which seawater enters, and having little or no inflow of freshwater; e.g. at Larnaca on Cyprus. Some salt may be deposited in it.

Marinesian *Bartonian*.

marine snow *sea snow*.

marine stack *stack [coast]*.

marine swamp A low, salty or brackish water area along the shore, characterized by an abundant growth of grass, reeds, mangrove trees, and similar types of vegetation. See also: *mangrove swamp*. Syn: *paralic swamp*.

marine terrace (a) A narrow, constructional, coastal strip, sloping gently seaward, veneered by a marine deposit (typically silt, sand, fine gravel). See also: *wave-built terrace*. (b) A narrow coastal plain whose margin has been strongly cliffed by marine erosion. (c) Loosely, a wave-cut platform that has been exposed by uplift along a seacoast or by the lowering of the sea level, and measuring as little as 3 m to more than 40 m above mean sea level; an elevated marine-cut bench. Cf: *raised beach*. (d) A terrace formed along a seacoast by the merging of a wave-built terrace and a wave-cut platform.--- Syn: *sea terrace; shore terrace*.

marine time A term used by Kobayashi (1944a, p.477) for *fossil time* as indicated by marine organisms. Cf: *continental time*.

marine transgression *transgression [stratig]*.

marin gräns Any maximum stand of the sea against the coast; esp. the highest *marine limit* or coastline of the postglacial sea. Etymol: Swedish, "marine border (or limit)". Abbrev: M G.

marining A term proposed by Grabau (1936, p.254) for a temporary or short-lived flooding of a level coastal plain or deltaic deposits by an epicontinental sea; e.g. a momentary submergence accompanying a tsunami.

mariposite A bright-green, chromium-rich variety of muscovite (or phengite), having a high silica content.

maritime Bordering on the sea, as a *maritime* province.

maritime climate *oceanic climate.*

maritime plant A plant growing in salty conditions of the foreshore.

mariupolite An albite-nepheline diorite containing acmite and lepidomelane, with zircon and beckelite as the main accessories.

mark A sedimentary structure along a bedding plane. The term usually signifies a *mold* or primary sedimentary structure (depression), such as a slide mark or a tool mark, but is also frequently applied to a cast (filling), such as a sole mark or a drag mark. Syn: *marking.*

marker (a) An easily recognized geologic feature having characteristics distinctive enough for it to serve as a reference or datum or to be traceable over long distances, esp. in the subsurface (such as a guide in well drilling or in a mine working); e.g. a stratigraphic unit readily identified by characteristics recognized on an electric log, or any recognizable rock surface such as an unconformity or a corrosion surface. See also: *format.* Syn: *marker bed; marker horizon.* (b) A term used in South Africa for an outcrop.

marker band An identifiable thin bed that has the same stratigraphic position throughout a considerable area (Wills, 1956, p.14). Syn: *indicator horizon.*

marker bed [seis] (a) A layer which accounts for a characteristic segment of a seismic refraction time-distance curve and which can be followed over reasonably extensive areas. (b) A layer which yields characteristic reflections over a more or less extensive area. Syn: *marker horizon.*

marker bed [stratig] (a) A geologic formation serving as a *marker.* (b) *key bed.*

marker horizon [seis] *marker bed.*

marker horizon [stratig] A *marker* represented by a rock surface or stratigraphic level, such as a vertical or lateral boundary based on electric or other mechanically recorded logs and that may serve to delineate rock-stratigraphic units.

markfieldite A granite containing plagioclase phenocrysts in a micrographic groundmass. Its name is derived from Markfield, England.

marking *mark.*

Markov process A stochastic process in which the state of a system at time *t(n)* depends on the state of the system at time *t(n-1).* It assumes that in a sequence of random events, the outcome or probability of each event is influenced by or depends upon the outcome of the immediately preceding event. Process introduced by Andrei A. Markov (1856-1922), Russian mathematician. Syn: *Markov chain; Markoff process.*

marl (a) An old term loosely applied to a variety of materials most of which occur as soft, loose, earthy, and semifriable or crumbling unconsolidated deposits consisting chiefly of an intimate mixture of clay and calcium carbonate in varying proportions, formed under either marine or esp. freshwater conditions; specif. an earthy substance containing 35-65% clay and 35-65% carbonate (Pettijohn, 1957, p.410). It is usually gray, although yellow, green, blue, and black varieties are not uncommon. Marl is used esp. as a fertilizer for acid soils deficient in lime. In the Coastal Plain area of SE U.S., the term has been used for little-indurated sedimentary deposits such as: fine-grained calcareous sands; calcareous clays and silts; clays, silts, and sands containing calcareous matter and glauconite (greensand marls); and newly formed deposits of shells mixed with clay. The term has also been used to designate a soft, friable clay with very little content of calcium carbonate and also to designate a very fine, loose, almost pure calcium carbonate with little clay or silt. Syn: *calcareous clay.* (b) A term used erroneously in the interior of U.S. for a calcareous lake deposit more correctly known as *bog lime.* (c) A term occasionally used (as in Scotland) for a compact, impure, argillaceous limestone. (d) A term loosely applied to any soil that falls readily to pieces on exposure to the air. (e) A literary term for clay or earthy material.---Etymol: French *marle.*

marlaceous Resembling or abounding with marl.

marl ball *marl biscuit.*

marl biscuit An *algal biscuit* found on the shore or shallow bottom of a lake (esp. in northern U.S. and southern Canada), consisting of a hard, flattish, rounded (biscuit- or pebble-shaped) concretion of marl formed around a shell fragment or other nucleus. Syn: *marl ball; marl pebble.*

marlekor A calcareous concretion of certain glacial clays, as of the varved lake clays of Scandinavia and in the Connecticut River valley of New England. Syn: *imatra stone.*

marlite (a) A hardened marl resistant to the action of air; *marlstone.* (b) A semi-indurated sheet or crust formed on the bottoms and shores of lakes by the intergrowth or cementation together of a considerable number of marl biscuits.---Syn: *marlyte.*

marl lake (a) A lake whose bottom deposits contain large quantities of marl. (b) A lake that has been mined or dredged as a commercial source of marl, esp. for the manufacture of portland cement.---Syn: *merl.*

marloesite A pale-gray, fine-grained extrusive rock composed of phenocrysts of feldspar and lath-shaped pseudomorphs of mica after olivine in a groundmass characterized by glomerophyric texture and composed of augite, sodic plagioclase, and iron ore. Its name is derived from Marloes on Skomer Island, Wales.

marl pebble *marl biscuit.*

marl slate An English term for fissile calcareous rock (shale); it is not a true slate.

marlstone (a) A consolidated or better-indurated rock of about the same composition as marl, more correctly called an earthy or impure argillaceous limestone rather than a calcareous shale. It has a blocky subconchoidal fracture, and is less fissile than shale. Syn: *marlite.* (b) A hard ferruginous rock (ironstone) of the Middle Lias in England, worked as an iron ore; specif. the Marlstone, a calcareous and sideritic oolite made up of ooliths, shell chips, and crinoid ossicles, set in a carbonate cement. (c) A term originally applied by Bradley (1931) to slightly magnesian calcareous mudstones or muddy limestones in the Green River Formation of the Uinta Basin, Utah, but subsequently applied to associated rocks (including conventional shales, dolomites, and oil shales) whose lithologic characters are not readily determined. Picard (1953) recommends abandonment of the term as used in the Uinta Basin.

marly Pertaining to, containing, or resembling marl; e.g. "marl limestone" containing 5-15% clay and 85-95% carbonate, or "marly soil" containing at least 15% calcium carbonate and no more than 75% clay (in addition to other constituents).

marlyte (a) *marlite.* (b) An obsolete term for a shale having imperfect lamination and so slightly indurated as to be fragile (Dana, 1874).

marmarization Var. of *marmorization.*

marmarosis Var. of *marmorosis.*

marmatite A dark-brown to black, iron-rich variety of sphalerite. Syn: *christophite.*

marmolite A thinly laminated, usually pale-green serpentine mineral; a variety of chrysotile.

Marmor North American stage: Middle Ordovician (lower subdivision of older Chazyan, above Whiterock, below Ashby) (Cooper, 1956).

marmoraceous Pertaining to or resembling marble.

marmorization The conversion of limestone into marble by an process of metamorphism. Also spelled: *marmarization*. Syn: *marmorosis*.

marmorosis A syn. of *marmorization*. Also spelled: *marmarosis*.

marne A French term for a marl or calcareous clay containing more than 50% clay and not less than 15% calcium carbonate.

marokite A black orthorhombic mineral: $CaMn_2O_4$.

marosite An intrusive rock intermediate in composition between shonkinite and nepheline diorite, being composed of biotite, augite with hornblende rims, sanidine, calcic plagioclase, nepheline, sodalite, apatite, and iron ore.

marquise A style of diamond cutting in which the girdle outline is boat-shaped. The shape and placement of the facets is of the brilliant style.

marrite A monoclinic mineral: $PbAgAsS_3$.

marscoite An intrusive rock containing quartz and feldspar phenocrysts in a gabbroid groundmass. The term was intended for use only in the Marsco area of Skye, Scotland.

Marsden chart A chart that is used to show meteorologic data over the oceans. It is based on a Mercator map projection with systematically numbered squares.

Marsden square One of a system of numbered areas each 10 degrees latitude by 10 degrees longitude, based on the Mercator projection, and used chiefly for identifying geographic positions and showing distribution of worldwide oceanographic and meteorologic data on a chart. Each square is subdivided into 100 one-degree subsquares which are numbered from 00 to 99 starting with 00 nearest the intersection of the equator and the Greenwich meridian. The system was introduced in 1831 by William Marsden (1754-1836), Irish orientalist.

marsh A watersaturated, poorly drained area, intermittently or permanently water-covered, having aquatic and grasslike vegetation, essentially without peatlike accumulation. Cf: *swamp; bog*.

Marshall line A syn. of *andesite line*, named after the New Zealand geologist P. Marshall.

marsh bar A narrow ridge of sand piled up at the seaward edge of a marsh undergoing wave erosion, as along the Delaware Bay shores of New Jersey.

marsh basin The depression occurring between stream banks in a salting.

marsh creek A drainage channel developed on a salt marsh.

marsh gas *Methane* produced during the decay of vegetable substances in stagnant water.

marshite A reddish, oil-brown isometric mineral: CuI.

marsh lake (a) An area of open water in a marsh, surrounded by wide expanses of marshland. (b) A lake covered completely or nearly so by emergent aquatic plants, esp. sedge and grasses.----See also: *lake marsh.*

marsh ore *bog iron ore.*

marsh pan A *salt pan* in a marsh.

marsh peat Peat that is derived from both plant debris and sapropelic matter. Cf: *banded peat.*

marsh shore A lake shore consisting of marsh vegetation which often merges with the emergent aquatic vegetation of the lake.

marthozite A mineral: $Cu(UO_2)_3(SeO_3)_3(OH)_2.7H_2O$.

martinite [mineral] A variety of whitlockite containing carbonate.

martinite [petrology] A leucite-bearing orthoclase-labradorite extrusive rock characterized by fine-grained, vesicular texture and composed of leucite, feldspar, and augite phenocrysts in a felty groundmass of soda labradorite, orthoclase, augite, leucite, olivine, magnetite, and apatite.

martite Hematite occurring in iron-black octahedral crystals pseudomorphous after magnetite.

masafuerite A dark-colored hypabyssal rock containing olivine as the only phenocryst mineral, and comprising over 50 percent of the rock, in a groundmass of pleochroic augite and calcic plagioclase, ilmenite, and magnetite.

masanite A quartz monzonite containing phenocrysts of zoned plagioclase and corroded quartz in a micrographic groundmass.

masanophyre A masanite containing phenocrysts of oligoclase with orthoclase rims in a groundmass of blue-green hornblende and sphene.

mascagnite A yellowish-gray mineral: $(NH_4)_2SO_4$. It occurs as powdery crusts in volcanic districts and with other ammonium sulfates in guano deposits.

mascon A large-scale, high-density, lunar mass concentration below a ringed mare (Muller & Sjogren, 1968, p.680). Etymol: *mass + concentration.*

maskeeg *muskeg.*

maskelynite Thetomorphic plagioclase glass; a colorless meteorite mineral consisting of a shock-formed noncrystalline phase (glass) that results from vitrification of plagioclase in rocks transfigured by shock waves and that retains the external features of crystalline plagioclase.

masonite A variety of chloritoid occurring in broad dark-green plates.

mass balance *balance.*

mass budget *balance.*

mass defect The divergence of the atomic mass of an isotope from its mass number, an integer; i.e. the difference between the sum of the atomic weights of the particles of the isotope and its atomic weight as a whole.

mass erosion A term which includes all processes by which soil and rock materials fail and are transported downslope predominantly en masse by the direct application of gravitational body stresses. Syn: *gravity erosion.*

mass heaving The all-sided, general expansion of the ground during freezing, involving significant horizontal forces over a considerable area (Washburn, 1956, p.840). Syn: *mass heave.*

massicot A yellow, orthorhombic mineral: PbO. Cf: *litharge.* Syn: *lead ocher.*

massif A massive topographic and structural feature in an orogenic belt, commonly formed of rocks more rigid than those of its surroundings. These rocks may be protruding bodies of basement rocks, consolidated during earlier orogenies, or younger plutonic bodies. (The term is also used in general for a plutonic rock body and its underlying area.) Examples are the crystalline massifs of the Helvetic Alps whose rocks were consolidated during the Hercynian orogeny, and before the Alpine orogeny.

massive [eco geol] Said of a mineral deposit (esp. sulfide) characterized by a great concentration of ore in one place, as opposed to a disseminated or veinlike deposit.

massive [rock mech] Said of a competent rock (such as granite, marble, and some sedimentary rocks) that is considered to be elastically perfect, isotropic, and homogeneous, and to possess a strength that does not vary appreciably from point to point.

massive [mineral] (a) Said of a mineral without internal structure, such as one lacking a platy or fibrous structure. (b) Said of an amorphous mineral, or one without apparent crystalline structure. This usage is not recommended.

massive [paleont] Said of corallum composed of corallites closely in contact with one another, or of a bryozoan colony consisting of thick heavy zoarium, generally hemispherical or subglobular in shape.

massive [meta] Said of a metamorphic rock whose constituents are not oriented in parallel position or not arranged in layers; said of a metamorphic rock that does not have schistosity, foliation, or any similar structure.

massive [ign] (a) Said of granite, diorite, and other igneous rocks that have a homogeneous structure over wide areas and that display a lack of layering, foliation, cleavage, or similar features. Also. said of the structure of such rocks. The term is often, but incorrectly, used as a syn. of "igneous". Syn: *compact*. (b) Said of a pluton that is not tabular in shape.

massive [sed] (a) Said of a stratified rock that occurs in very thick, homogeneous beds, or of a stratum that is imposing by its thickness; specif. said of a bed that is more than 10 cm (4 in.) in thickness (Payne, 1942) or more than 1.8 m (6 ft) in thickness (Kelley, 1956, p. 294). (b) Said of a stratum or stratified rock that is obscurely bedded, or that is or appears to be without internal structure (such as a rock free from minor joints, fissility, or lamination), regardless of thickness. The massive appearance may be deceiving as many "massive" beds display laminae and other structures when X-rayed. See also: *unstratified*. (c) Descriptive of a sedimentary rock (such as shale) that is difficult to split, or that splits into layers greater than 120 cm (4 ft) in thickness (McKee & Weir, 1953, p.383).

massive unit weight *wet unit weight*.

mass movement A unit movement of a portion of the land surface; specif. *mass-wasting* or the gravitative transfer of material down a slope. Cf: *mass transport*.

mass property A property of a sediment considered as an aggregate, such as porosity, permeability, color, density, elasticity, plasticity, compactibility, packing, hygroscopicity, adsorption, moisture equivalent, shrinkage, tensile and crushing strength, slaking, fusibility, cohesiveness, electrical resistivity, magnetic susceptibility, thermal conductivity, and radioactivity.

mass spectograph Strictly, a recording *mass spectrometer*, but commonly used only for those instruments which record on a photographic plate as contrasted to those which record numerically or graphically. The latter are usually simply called mass spectrometers.

mass spectrometer An instrument for producing and measuring, usually by electrical means, a *mass spectrum*. It is especially useful for determining molecular weights and relative abundances of isotopes within a compound. See also: *mass spectrograph.*

mass spectrometry The art or process of using a *mass spectrometer* to study mass spectra.

mass spectroscopy The observation of a *mass spectrum* and all processes of recording and measuring that go with it.

mass spectrum The spectrum of intensity values of a substance ordered according to mass or mass-to-charge ratio.

mass susceptibility *specific susceptibility*.

mass transport [oceanog] The movement of water by orbital wave motion.

mass transport [sed] The carrying of material in a moving medium such as water, air, or ice. Cf: *mass movement.*

massula (a) A more or less irregular, coherent mass of many pollen grains shed from the anther and fused together. Cf: *pollinium*. (b) A term sometimes applied to a structure associated with the laesura and the attached nonfunctional spores of certain megaspores.---Pl: *massulae*.

mass wasting A general term for the dislodgement and downslope transport of soil and rock material under the direct application of gravitational body stresses. In contrast to other erosion processes, the debris removed by mass wasting processes is not carried within, on, or under another medium possessing contrasting properties. The mass strength properties of the material being transported depend on the interaction of the soil and rock particles with each other. It includes slow displacements such as creep and solifluction and rapid movements such as earthflows, rockslides, avalanches, and falls. Cf: *mass erosion*. Syn: *mass movement.*

master cave An area in or a portion of a cave that seems to be the largest, most level part and to which the auxiliary passages seem to lead.

master joint A persistent joint plane of greater than average extent; the dominant jointing of an area. Syn: *main joint; major joint.*

master map An original map, usually of large scale, containing all the information from which other maps showing specialized information can be compiled; a primary source map. Syn: *base map.*

master stream *main stream.*

mastigoneme One of the delicate, hair-like, lateral threads, filaments, or processes along the length of some flagella. Syn: *flimmer.*

mastigophoran *flagellate.*

matched terrace *paired terrace.*

mathematical geography That branch of geography that is concerned with the figures and motions of the Earth and their representation on maps and charts using various projection methods.

mathematical geology Mathematics as applied to geology "the name given to the discipline devoted to the investigation of probability distributions of values of random variables with the object of obtaining information concerning geological processes" (Vistelius, 1967, p.9). These investigations are based on the methods of mathematical probablity, statistical theory and special computational mathematics developed in work on the solutions of various particular problems. See also *geomathematics.*

matildite A gray mineral: $AgBiS_2$. Syn: *schapbachite; plenargyrite.*

matlockite A mineral: $PbFCl$.

matraite A mineral: ZnS.

matrix [ore dep] *gangue.*

matrix [gem] A gemstone cut from material consisting of a mineral (such as opal or turquoise) and the surrounding rock material in which the mineral is contained.

matrix [paleont] The natural rock or earthy material in which a fossil is embedded or surrounded, as opposed to the actual fossil itself.

matrix [ign] The fine-grained interstitial material of an igneous rock; e.g. the material surrounding the phenocrysts of a porphyritic rock. The syn. *groundmass* is more commonly used.

matrix [sed] The smaller or finer-grained, continuous material enclosing, or filling the interstices between, the larger grains or particles of a sediment or sedimentary rock; the natural material in which a sedimentary particle is embedded. The term refers to the relative size and disposition of the particles and no particular particle size is implied. In carbonate sedimentary rocks, the matrix usually consists of clay minerals or micritic components surrounding coarser material; although the term should be used in a descriptive, nongenetic, and noncompositional manner, it has been applied (inappropriately) as a syn. of *micrite*. Syn: *groundmass.*

matrix limestone *micritic limestone.*

matrix porosity The porosity of the matrix or finer part of a carbonate rock, as opposed to the porosity of the coarse constituents (Choquette & Pray, 1970, p.247).

matrix rock *land-pebble phosphate.*

matrosite Black, opaque, microscopic material forming the matrix of torbanite (Dulhunty, 1939).

matterhorn A glacial *horn* resembling the Matterhorn, a peak in the Pennine Alps. Syn: *Matterhorn peak.*

matteuccite A mineral: $NaHSO_4.H_2O$.

Matura diamond Colorless to faintly smoky gem-quality zircon from the Matara (Matura) district of southern Ceylon. It either occurs colorless or is decolorized by heating.

mature [geomorph] Pertaining to the stage of *maturity* of the cycle of erosion; esp. said of a topography or region, and of its landforms (such as a plain or plateau), having undergone maximum development and accentuation of form, or of a stream (and of its valley) with a fully developed profile of equilibrium, or of a coast that is relatively stable.

mature [sed] Pertaining to the third stage of textural maturity (Folk, 1951); said of a clastic sediment that has been differentiated or evolved from its parent rock by processes acting over a long time and with a high intensity and that is characterized by stable minerals (such as quartz), deficiency of the more mobile oxides (such as soda), absence of weatherable material (such as clay), and well-sorted but subangular to angular grains. Example: a clay-free "mature sandstone" on a beach, with a standard deviation of less than 0.5 phi units (a range of less than 1 phi unit between the 16th and 84th percentiles of the particle-size distribution). Cf: *immature; subature; supermature*.

matureland The land surface of the mature stage of the cycle of erosion, varying from surfaces having attained maximum relief to those of reduced "but not low" relief (Maxson & Anderson, 1935, p.90). The term was introduced by Willis (1928, p.493) in a broader sense to include eroded surfaces "qualified as vigorous, advanced, or subdued, according to the stage of development", a subdued matureland approaching the flatness of a peneplain. Davis (1932, p.429), noting that a "subdued" surface is neither mature nor old, but senescent, proposed the term *senesland* for this kind of matureland.

mature region *exozone*.

mature soil *zonal soil*.

mature stream A stream developed during the stage of *maturity*; a graded stream.

maturity [topog] The second of the three principal stages of the cycle of erosion in the topographic development of a landscape or region, intermediate between youth and old age (or following adolescence), lasting through the period of greatest diversity of form or maximum topographic differentiation, during which nearly all the gradation resulting from operation of existing agents has been accomplished. It is characterized by: numerous, closely spaced mature streams; disappearance of initial level surfaces, as the land is completely dissected and reduced to slopes; large, well-defined drainage systems with numerous and extensive tributaries and sharp, narrow divides, and an absence of swamps or lakes on the uplands; greatest degree of ruggedness possible, with a new plain of erosion just beginning to appear; and pedimentation (in an arid cycle). Syn: *topographic maturity*.

maturity [coast] A stage in the development of a shore, shoreline, or coast that begins when a profile of equilibrium is attained, and that is characterized by: decrease of wave energy; creation of beaches; disappearance of lagoons and marshes; straightening of the shoreline by bridging of bays and cutting back of headlands so as to produce a smooth, regular shoreline consisting of sweeping curves; and retrogradation of the shore beyond the bayheads so that it lies against the mainland as a line of eroded cliffs throughout its course (this process does not necessarily occur at the same rate everywhere due to varying rock resistance). See also: *secondary [coast]*.

maturity [sed] The extent to which a clastic sediment texturally and compositionally approaches the ultimate end product to which it is driven by the formative processes that operate upon it (Pettijohn, 1957, p.508 & 522). See also: *textural maturity; mineralogic maturity; compositional maturity*.

maturity [streams] The stage in the development of a stream in which it has reached its maximum vigor and efficiency, having attained a profile of equilibrium and a velocity that is just sufficient to carry the sediment delivered to it by tributaries. It is characterized by: a load that is just about equal to the ability of the stream to carry it; lateral erosion predominating over downcutting, with the formation of a broad, open, flat-floored valley having a regular and moderate or gentle gradient and gently sloping, soil-covered walls with few out-

crops; absence of waterfalls, rapids, and lakes; a steady but deliberate current, and muddy water; numerous and extensive tributaries, some of whose headwaters may still be in the youthful stage; development of flood plains, alluvial fans, deltas, and meanders, as the stream begins to deposit material; and a graded bed.

maturity index A measure of the progress of a clastic sediment in the direction of chemical or mineralogic stability; e.g. a high ratio of alumina/soda, of quartz/feldspar, or of quartz + chert/feldspar + rock fragments, indicates a highly mature sediment (Pettijohn, 1957, p.509).

maucherite A reddish silver-white mineral: $Ni_{11}As_8$. It tarnishes to gray copper-red.

Maui-type well A type of *basal tunnel* characterized by a vertical or inclined shaft dug from the land surface to the basal water table, and by one or more tunnels dug along the water table to skim off the uppermost basal ground water to avert possible salt-water encroachment (Stearns & Macdonald, 1942, p.126). This procedure was first used on the island of Maui, Hawaii.

mawsonite A mineral: $Cu_7Fe_2SnS_{10}$.

maxilla One of the first or second pairs of mouthparts posterior to the mandibles in various arthropods; e.g. the last cephalic appendage of a crustacean, following the maxillule and serving for feeding and respiration, or the *coxa* of a pedipalpus of an arachnid. Pl: *maxillae*. Adj: *maxillary*.

maxilliped One of the three pairs of appendages of a crustacean, situated next behind the maxillae; an anterior thoracopod modified to act as a mouthpart, its somite usually fused to the cephalon.

maxillule The fourth cephalic appendage of a crustacean, between the mandible and the maxilla, serving as a mouthpart. Pl: *maxillulae*. Syn: *first maxilla; maxillula*.

maximum [glac geol] *glacial maximum*.

maximum [geophys] n. An anomaly in a given area that is relatively greater than those in neighboring areas; e.g. a gravity maximum or a geothermal maximum. Cf: *minimum [geophys]*. Syn: *high [geophys]*.

maximum [struc petrol] On a fabric diagram, a single area of concentration of points representing orientation of fabric elements (Turner and Weiss, 1963, p.58). Cf: *girdle; cleft girdle*. Syn: *point maximum*.

maximum slope A slope that is steeper than the slope units above or below it.

maximum unit weight The *dry unit weight* defined by the peak of a compaction curve.

maximum water-holding capacity The average moisture content of a disturbed soil sample, one cm high, after equilibration with a water table at its lower surface (Jacks, 1960, p.45). The retained water represents the lower part of the capillary fringe.

maxwell The cgs (centimeter gram second) unit of magnetic flux. Maxwell = 10^{-8} *weber*; the flux through one square centimeter normal to a field of magnetic induction one gauss.

Maxwell liquid A model of *elasticoviscous* behavior. During the application of stress the body deforms both elastically and viscously. When the stress is released the elastic strain is recovered, releasing the stored energy. If the body is retained in a strained condition, the stress is relaxed as the elastic strain is slowly recovered.

mayaite A white to gray-green or yellow-green mineral grading from tuxtlite to a nearly pure albite, found in the ancient tombs of the Mayans and elsewhere in Central America.

mayenite An isometric mineral: $Ca_{12}Al_{14}O_{33}$.

Maysvillian North American stage: Upper Ordovician (above Edenian, below Richmondian).

mboziite A mineral of the amphibole group: $Na_2CaFe_3^{+2}-Fe_2^{+3}Al_2Si_6O_{22}(OH)_2$.

mbuga A term used in SW Africa for a temporary swamp or

black claypan (playa) marking the last stand of a now desiccated lake.

mcallisterite *macallisterite.*

McGovernite A syn. of *macgovernite.* Also spelled: *mcgovernite.*

m-charnockite A name proposed by Tobi (1971, p.202) in his classification of the charnockite suite for that member which contains mesoperthite as the only feldspar.

mckelveyite *mackelveyite.*

mckinstryite *mackinstryite.*

M-discontinuity Syn: of *Mohorovičić discontinuity,* suggested by Vening Meinesz (1955, p.321). Also spelled: *M discontinuity.*

meadow ore *bog iron ore.*

meadow peat Peat that is derived from grasses.

meadow soil A general term for an intrazonal soil developed in a humid climate on flood plains and low terraces that border streams. It is a wet soil under grass and has a dark, organic upper layer and an underlying gley horizon. Syn: *Weisenboden.*

mealy *farinaceous.*

mean An arithmetic average of a series of values; esp. *arithmetic mean.* See also: *geometric mean; harmonic mean.* Cf: *mode; median.*

mean depth In hydraulics, the cross-sectional area of a stream divided by its width at the surface. Cf: *mean hydraulic depth.*

meander [surv] v. To survey a meander line on or along.---n. *meander line.*

meander [streams] n. (a) One of a series of somewhat regular, sharp, freely developing, and sinuous curves, bends, loops, turns, or windings in the course of a stream. It is produced by a mature stream swinging from side to side as it flows across its flood plain or shifts its course laterally toward the convex side of an original curve. Etymol: Greek *maiandros,* from Maiandros River in western Asia Minor (now known as Menderes River in SW Turkey), proverbial for its windings. (b) *valley meander.*---v. To wind or turn in a sinuous or intricate course; to form a meander.

meander amplitude The distance between points of maximum curvature of successive meanders of opposite phase, measured in a direction normal to the general course of the meander belt (Langbein & Iseri, 1960, p.14).

meander bar A deposit of sand and gravel located on the inside, and extending into the curve, of a meander; specif. a *point bar.*

meander belt The zone along a valley floor across which a meandering stream shifts its channel from time to time; specif. the area of the flood plain included between two lines drawn tangentially to the extreme limits of all fully developed meanders. It may be from 15 to 18 times the width of the stream.

meander breadth The distance between the lines used to define the *meander belt* (Langbein & Iseri, 1960, p.14).

meander core (a) A central hill encircled or nearly encircled by a stream meander. Syn: *rock island.* (b) *cutoff spur.*

meander cusp A projection on the eroded edge of a meander-scar terrace, formed by the intersection of two or more meander scars. See also: *two-swing cusp; three-swing cusp; two-sweep cusp.* Syn: *terrace cusp.*

meander cutoff A *cutoff* formed when a stream cuts through a meander neck.

meandering stream A stream having a pattern of successive meanders; it is capable of carrying on lateral erosion. Syn: *snaking stream.*

meandering valley A valley having a pattern of successive windings broadly resembling the trace of a meandering stream. The windings, or *valley meanders,* are of the same general order of size.

meander length (a) The distance between corresponding part of successive meanders of the same phase, measured along the general course of the meanders (Langbein & Iseri, 1960 p.14). (b) Twice the distance between successive points inflection of the meander (Leopold & Wolman, 1957, p.55).

meander line A surveyed line, usually of irregular course, tha is not a boundary line; esp. a metes-and-bounds traverse the margin or bank of a permanent natural body of water, ru approximately along the mean-high-water line for the purpos of defining the sinuosities of the bank or shoreline and as means of providing data for computing the area of land remaining after the water area has been segregated. Syn: *meander.*

meander lobe The more or less elevated, tongue-shaped are of land enclosed within an acute stream meander. Syn: *tongue.*

meander neck The narrow strip of land, between the tw limbs of a meander, that connects a meander lobe with th mainland.

meander niche On the wall of a cave, a conical or crescent opening formed by downward and lateral stream erosion. Sy *wall niche.*

meander plain A term introduced by Melton (1936, p.594) f a plain built by the meandering process, or a plain of later accretion; it is seldom or never subject to overbank floo and thus lacks any alluvial cover. Cf: *covered plain; bar plain.*

meander scar (a) A crescentic, concave mark on the face a bluff or valley wall, produced by the lateral planation of meandering stream which undercut the bluff, and indicati the abandoned route of the stream. See also: *flood-plain m ander scar.* Syn: *meander scarp.* (b) An abandoned meande often filled in by deposition and vegetation, but still discernib (esp. from the air).

meander scarp *meander scar.*

meander-scar terrace A local terrace formed by the shifting meanders during the slow and continuous excavation of a va ley (Schieferdecker, 1959, term 1519). Syn: *alternate terrace*

meander scroll (a) One of a series of long, parallel, close fitting, arcuate ridges and troughs formed along the inn bank of a stream meander as the channel migrated lateral down-valley and toward the outer bank. Cf: *point bar; lacir* (b) A small, elongate lake occurring on a flood plain in a we defined part of an abandoned stream channel, commonly an oxbow.

meander spur An undercut projection of high land extendi into the concave part of, and enclosed by, a meander.

meander-spur terrace A terrace on a meander spur.

meander terrace A small, relatively short-lived *stream terra* formed by a freely swinging meander cutting into a form and higher flood plain; an *unpaired terrace.*

mean deviation The arithmetic mean of the absolute devi tions of observations from their mean.

mean diameter (a) *arithmetic mean diameter.* (b) *geometr mean diameter.* (c) *logarithmic mean diameter.* (d) *phi me diameter.*

meandroid Said of a corallum characterized by meanderi rows of confluent corallites with walls only between the rows.

mean ground elevation Average elevation above mean se level of the terrain to be photographed from the air.

mean higher high water The average height of all the high high waters recorded at a given place over a 19-year peri or a computed equivalent period. Abbrev: MHHW.

mean high water The average height of all the high waters corded at a given place over a 19-year period or a comput equivalent period. Abbrev: MHW.

mean high-water neap The average high-water height duri quadrature, recorded over a 19-year (or computed equivaler period. Abbrev: MHWN. Cf: *mean low-water neap.*

mean high-water spring The average high-water height

zygy, recorded over a 19-year (or computer equivalent) period. Abbrev: MHWS. Cf: *mean low-water spring*.

mean hydraulic depth The cross-sectional area of a stream divided by the length of its wetted perimeter. Cf: *mean depth*.

mean ice edge The average position of the *ice edge* in any given time period (usually a month), based on observations over several years. Formerly known as *ice limit*.

mean lower low water The average height of all the lower low waters recorded at a given place over a 19-year period or a computed equivalent period. Abbrev: MLLW. See also: *lower low-water datum*.

mean low water The average height of all the low waters recorded at a given place over a 19-year period or a computed equivalent period. Abbrev: MLW. See also: *low-water datum*.

mean low-water neap The average low-water height during quadrature, recorded over a 19-year (or computed equivalent) period. Abbrev: MLWN. Cf: *mean high-water neap*.

mean low-water spring The average low-water height at syzygy, recorded during a 19-year (or computed equivalent) period. Abbrev: MLWS. Cf: *mean high-water spring*.

mean range The difference in height between mean high water and mean low water. Abbrev: Mn. Cf: *tide range*.

mean refractive index (a) The median *index of refraction* for any crystalline substance, with variation due to zoning. (b) The median index of refraction in any microcrystalline substance for which specific index values related to crystal directions are not determinable. (c) In a biaxial crystal, the beta, y or N_m index of refraction (in which y or m = mean). This not the average index.

mean sea level The average height of the surface of the sea for all stages of the tide over a 19-year period, usually determined from hourly height observations on an open coast or in adjacent waters having free access to the sea; the assumed actual sea level at its mean position midway between high water and mean low water. It is adopted as a *datum plane* or tidal datum for the measurement of heights, such as *sea-level datum*. Cf: *mean tide level*. Abbrev: MSL. Popular n: *sea level*. Syn: *geodetic sea level*.

mean spheroid The hypothesized spheroid assumed to be an ellipsoid of revolution which coincides most closely with the actual figure of the earth at sea level.

mean spring range The average semidiurnal range of the tide at syzygy. Abbrev: Sg. Syn: *spring range*.

mean stress The algebraic average of the three *principle stresses*.

mean tide level The plane or surface that lies exactly midway between mean high water and mean low water; the average of observed heights of high water and low water. Abbrev: TL. Cf: *mean sea level*. Syn: *ordinary tide level; half-tide level*.

mean velocity *average velocity*.

mean velocity curve *vertical-velocity curve*.

mean water level The average height of the surface of water, determined at equal (usually hourly) intervals of time. Abbrev: WL.

measured ore *developed reserves*.

measures A group or series of sedimentary rocks having some characteristic in common, such as the strata of a mineral; specif. *coal measures*. The term apparently refers to the old practice of designating the different seams of a coalfield by their "measure" or thickness.

meat earth A term sometimes used in mining for the topsoil of an opencut mine, which may be saved and used for restoration of the area.

mechanical analysis Determination of the particle-size distribution of a soil, sediment, or rock by screening, sieving, or other means of mechanical separation; "the quantitative expression of the size-frequency distribution of particles in granular, fragmental, or powdered material" (Krumbein & Petti-

john, 1938, p.91). It is usually expressed in percentage by weight (and sometimes by number or count) of particles within specific size limits. See also: *particle-size analysis*.

mechanical clay A clay formed from the products of abrasion of rocks.

mechanical equivalent of heat A dimensionless conversion factor relating joules and calories. Its numerical value is equal to the number of joules of work, i.e. energy, necessary to produce the same change of state, e.g. pressure, temperature, volume, in a system as that produced by the absorption of one calorie of heat (Zemansky, 1957, p.65). Symbol: J.

mechanical erosion *corrasion*.

mechanical metamorphism *kinetic metamorphism*.

mechanical sediment *clastic sediment*.

mechanical seismograph A seismic detector in which all amplification of ground motion is done by mechanical means.

mechanical stage A microscope *stage* that allows exact recording of the position of the object, e.g. a thin section, and that has a device for moving the object sideways and forward and backward.

mechanical twin *deformation twin*.

mechanical weathering The process of weathering by which physical forces (such as frost action, salt-crystal growth, absorption of water, and temperature changes) break down or reduce a rock to smaller and smaller fragments, and involving no chemical change. Cf: *chemical weathering*. Syn: *physical weathering; disintegration; disaggregation*.

mechanoglyph A hieroglyph of mechanical origin (Vassoevich, 1953, p.38).

medano A Spanish term for a *sand dune*, esp. one occurring along a seashore, as in Chile or Peru.

medial *middle* [geochron].

medial moraine [glac geol] (a) An elongate moraine carried in or upon the middle of a glacier and parallel to its sides, usually formed by the merging of adjacent and inner lateral moraines below the junction of two coalescing valley glaciers. (b) A moraine formed by glacial abrasion of a rocky protuberance near the middle of a glacier and whose debris appears at the glacier surface in the ablation area. (c) The irregular ridge left behind in the middle of a glacial valley, when the glacier on which it was formed has disappeared.---Syn: *median moraine*.

median The value of the middle item in a set of data arranged in rank order. If the set of data has an even number of items, the median is the arithmetic mean of the middle two ranked items. Cf: *mean; mode*.

median diameter An expression of the average particle size of a sediment or rock, obtained graphically by locating the diameter associated with the midpoint of the particle-size distribution; the middlemost diameter that is larger than 50% of the diameters in the distribution and smaller than the other 50%.

median dorsal plate An elongate plate that posteriorly and dorsally separates the carapace valves of a phyllocarid crustacean.

median lamina The central, fused *basal laminae* of two layers of zooids growing back to back in a *bilamellar* bryozoan colony. Syn: *mesotheca*.

median mass A less disturbed block, or structural unit, in the midst of an orogenic belt, bordered on each side by orogenic structures that are thrust away from it toward forelands on each flank. Median masses occur at intervals along the Alpine-Himalayan belt in the Eastern Hemisphere, between the Alpides in the strict sense on the north and the Dinarides on the south. A good example is the Hungarian Plain; no certain examples can be identified in North America. Syn: *Zwischengebirge; betwixt mountains*.

median moraine *medial moraine* [glac geol].

median section A slice in the central sagittal part and perpendicular to the axis of coiling of a foraminiferal test.

median septum A calcareous ridge built along the midline of the interior of a brachiopod valve (Beerbower, 1968, p.284).

median sulcus A prominent vertical depression in the anterior and median surface of an ostracode valve.

median valley *rift valley.*

medical geology The application of geology to medical and health problems, involving such studies as the occurrence of toxic elements in unusual quantities in parts of the Earth's crust, or the distribution of trace elements as related to nutrition, or the geographic patterns of disease. The medical syn. is "regional pathology". See also: *environmental geochemistry.*

medicinal spring A spring of reputed therapeutic value due to the substances contained in its waters. Cf: *spa.*

mediglacial Relating to or formed between glaciers, or situated in the midst of glaciers.

mediiphyric A term applied by Cross et al (1906, p.702) to porphyritic rocks in which the longest dimension of the phenocrysts is between 0.04 mm and 0.008 mm.

Medinan Obsolete syn. of *Alexandrian.*

mediophyric Said of the texture of a porphyritic rock in which the longest dimension of the phenocrysts is between 1 mm and 5 mm (Cross et al, 1906, p.702); also, said of a rock having such texture. Cf: *magnophyric; magniphyric.*

mediosilicic A term proposed by Clarke (1908, p.357) to replace *intermediate*. Cf: *subsilicic; persilicic.*

mediterranean n. *mesogeosyncline.*

Mediterranean belt *Alpides.*

Mediterranean climate A climate characterized by hot and dry summers and by mild, rainy winters.

mediterranean delta A delta built out into a landlocked sea that is tideless or has a low tidal range (Lyell, 1840, v.1. p.422).

mediterranean sea A type of *epicontinental sea* that is deep and that connects with the ocean by a narrow opening.

Mediterranean suite A major group of igneous rocks, characterized by potassium-rich rocks. This suite was so named because of the predominance of potassium-rich lavas around the Mediterranean Sea; specif. of Vesuvius and Stromboli. Cf: *Atlantic suite; Pacific suite.*

Medithermal A term used by Antevs (1948, p.176) for a period of time now more generally known as *Little Ice Age.*

medium bands In banded coal, vitrain bands from 2.0 to 5.0 mm thick (Schopf, 1960, p.39). Cf: *thin bands; thick bands; very thick bands.*

medium-bedded A relative term applied to a sedimentary bed whose thickness is intermediate between *thin-bedded* and *thick-bedded.*

medium boulder A *boulder* having a diameter in the range of 512-1024 mm (20-40 in., or -9 to -10 phi units).

medium clay A geologic term for a *clay* particle having a diameter in the range of 1/1024 to 1/512 mm (1-2 microns, or 10 to 9 phi units). Also, a loose aggregate of clay consisting of medium clay particles.

medium-crystalline Descriptive of an interlocking texture of a carbonate sedimentary rock having crystals whose diameters are in the range of 0.062-0.25 mm (Folk, 1959) or 0.1-0.2 mm (Carozzi & Textoris, 1967, p.5) or 1-4 mm (Krynine, 1948, p.143). Cf: *medium-grained.*

medium-grained (a) Said of an igneous rock, and of its texture, in which the individual crystals have an average diameter in the range of 1-5 mm (0.04-0.2 in.). Johannsen (1931, p.31) earlier used the range of 1-10 mm. (b) Said of a sediment or sedimentary rock, and of its texture, in which the individual particles have an average diameter in the range of 1/16 to 2 mm (62-2000 microns, or sand size). Cf: *medium-crystalline.-- --*The term is used in a relative sense to describe rocks that are neither *coarse-grained* nor *fine-grained.*

medium pebble A geologic term for a *pebble* having a diameter in the range of 8-16 mm (0.3-0.6 in., or -3 to -4 phi units) (AGI, 1958).

medium sand (a) A geologic term for a *sand* particle having a diameter in the range of 0.25-0.5 mm (250-500 microns, or 2 to 1 phi units). Also, a loose aggregate of sand consisting of medium sand particles. (b) An engineering term for a *sand* particle having a diameter in the range of 0.42 mm (retained on U.S. standard sieve no.40) to 2 mm (passing U.S. standard sieve no.10). (c) A soil term used in the U.S. for a *sand* particle having a diameter in the range of 0.25-0.5 mm.

medium silt A geologic term for a *silt* particle having a diameter in the range of 1/64 to 1/32 mm (16-31 microns, or 6 to 5 phi units). Also, a loose aggregate of silt consisting of medium silt particles.

medium-volatile bituminous coal Bituminous coal that contains 23-31% volatile matter, analyzed on a dry, mineral-matter-free basis. Cf: *high-volatile bituminous coal; low-volatile bituminous coal.*

medmontite A clayey mineral: $CuAl_2(Al,Si)_4O_{10}(OH)_2$. It is related to montmorillonite, and is believed to be a mixture of chrysocolla and mica.

medulla [paleont] (a) The central zone of certain octocorals e.g. the central chord of the axis of the Holaxonia (TIP, 1956 pt.F, p.174). (b) The internal part of some protozoans.

medullary shell The internal concentric shell of spumellarian radiolarians.

meerschaum Massive *sepiolite.* Etymol: German *Meerschaum*, "sea froth".

mega- A prefix signifying a characteristic visible to the unaided eye. It is preferred to its synonymous prefix, *macro-.* Cf *micro-.*

megabarchan A giant *barchan*, up to 100 m or more in height (Stone, 1967, p.232).

megabreccia (a) A term used by Landes (1945) for a rock produced by brecciation on a very large scale, containing blocks that are randomly oriented and invariably inclined at angles from 6° to 25° and that range from a meter to more than 100 m in horizontal dimension. (b) A term used by Longwell (1951) for a coarse breccia containing individual blocks as much as 400 m long, developed downslope from large thrusts by gravitational sliding. It is partly tectonic and partly sedimentary in origin, containing blocks that are shattered but little rotated.---Cf: *chaos.*

megacell A cell (or group of cells) in some algae that grow much larger than the surrounding ones.

megaclast (a) One of the larger fragments in a variable matrix of a sedimentary rock (Crowell, 1964). Cf: *phenoclast.* (b) A constituent of a mixtite (Schermerhorn, 1966).

megaclone A large, smooth, monaxonic desma (of a sponge) having branches that bear cup-like articular facets, mostly terminal.

megacryst A nongenetic term introduced by Clarke (1958 p.12) for "any crystal or grain", in an igneous or metamorphic rock, that is "significantly larger" than the surrounding groundmass or matrix; e.g. a large microcline crystal in porphyritic granite. It may be a phenocryst, a porphyroblast, or a porphyroclast.

megacrystalline *macrocrystalline.*

megacyclothem A term introduced by Moore (1936, p.29) to designate a combination of related *cyclothems*, or a cycle of cyclothems, such as in the Pennsylvanian of Kansas. Also, a cyclothem on a large scale, comprising minor cyclothems. Cf *hypercyclothem; magnacyclothem.*

megafabric The fabric of a rock, as seen on a large scale (in hand specimen or outcrop), or as capable of being distinguished without the aid of a microscope. Cf: *microfabric.* Syn *macrofabric.*

megafacies (a) A term used by Cooper & Cooper (1946 p.68) apparently for a large intertonguing lithologic body. (b

A term used mistakenly for *magnafacies*.

megafauna (a) Living or fossil animals large enough to be seen with the naked eye. (b) The animals occupying a broad area of uniform characteristics; a large or widespread group of animals. Cf: *microfauna; megaflora.* Syn: *macrofauna.*

megaflora (a) Plants large enough to be seen with the naked eye. (b) The plants of a large habitat; a large, widespread group of plants. Cf: *microflora; megafauna.* Syn: *macroflora.*

megaflow mark A term used by Kuenen (1957, p.243) as a syn. of *gouge channel.*

megafossil *macrofossil.*

megagametophyte The female *gametophyte* or haploid generation that develops from the megaspore of a heterosporous embryophytic plant. In lower vascular plants, it may be a small free-living plant bearing archegonia, but in seed plants, it is contained within the ovule, and the egg is produced in it (the embryo which develops from fertilization of the egg, plus the developing maternal tissues of the ovule, then comprise the seed). Cf: *microgametophyte.*

megagrained Said of the texture of a carbonate sedimentary rock having clastic particles whose diameters are greater than one millimeter (DeFord, 1946). See also: *macrograined.*

megagroup A term used by Swann & William (1961) for a rock-stratigraphic unit that is next higher in rank than group and that represents a major event in the course of geologic history. It is not recognized as a formal unit by the American Commission on Stratigraphic Nomenclature (1961). Cf: *supergroup.*

megaloplankton The largest plankton; they are more than one centimeter in size. Cf: *ultraplankton; nannoplankton; microplankton; macroplankton.*

megalospheric Said of a foraminiferal test or shell produced asexually and characterized by a large initial chamber (proloculus), relatively few chambers, small size of the adult test, and incomplete ontogeny. Cf: *microspheric.*

megalump A gravel-sized *lump* in a limestone. It usually originates by disruption (by high-energy waves or currents or possibly by turbidity currents) of newly deposited or partly indurated carbonate mud, which is then incorporated within the sedimentary unit from which it was derived.

megaphyric Said of the texture of a porphyritic igneous rock in which the greatest dimension of the phenocrysts is more than 2 mm (Cross et al, 1906, p.702); also, said of a rock having such texture. Cf: *microphyric.* Syn: *macroporphyritic; megaporphyritic; macrophyric.*

megapore (a) *macropore.* (b) In the pore-size classification of Choquette & Pray (1970, p.233), an equant to equant-elongate pore or a tubular or platy pore with an average diameter or thickness greater than 4 mm.----Cf: *mesopore; micropore.*

megaporphyritic *megaphyric.*

megarhizoclone A large *rhizoclone* approaching the form of a megaclone.

megaripple A large *sand wave* or ripple-like feature having a wavelength greater than 1 m (Straaten, 1953) or a ripple height greater than 10 cm (Imbrie & Buchanan, 1965, p.155), composed of sand, and formed in very shallow water in a fluvial, tidal, or marine environment. Wavelengths may reach 100 m and ripple heights about 1 m. Not to be confused with *metaripple.* Syn: *giant ripple.*

megasclere A large *sclere;* specif. one of the primary spicules forming the principal skeletal support in a sponge. It usually differs in form from a *microsclere.* Syn: *macrosclere.*

megascopic Said of an object or phenomenon or of its characteristics that can be observed with the naked eye or with a hand lens. Syn: *macroscopic.*

megashear A term used by Carey (1958) for a strike-slip fault of such extent that it underlies the entire orogen.

megasporangium A *sporangium* that develops or bears megaspores; e.g. the nucellus in a gymnospermous seed plant.

Cf: *microsporangium.*

megaspore (a) One of the spores of a heterosporous embryophytic plant that germinates to produce a megagametophyte (multicellular female gametophyte) and that is ordinarily larger than the *microspore.* Its range is mid-Devonian to Holocene. (b) A term arbitrarily defined in paleopalynology as a spore or pollen grain greater than 200 microns in diameter (although it may not have been biologically a megaspore in function). Cf: *miospore.*---Syn: *macrospore.*

megata A Japanese term for *maars* that appears in English-language geologic literature.

megatectonics The tectonics of the very large structural features of the Earth, or of the whole Earth. Similar terms are *geotectonics* and *global tectonics,* but all these large, vague words seem superfluous since the subject of *tectonics* itself differs from the subject of *structural geology* in dealing only with the very large structural features. Cf: *microtectonics; global tectonics.* Syn: *macrotectonics.*

megatherm A plant requiring high temperatures and large quantities of water for its existence. Syn: *macrotherm.* Cf: *microtherm; mesotherm.*

megathermal Pertaining to a climate characterized by high temperature. Cf: *mesothermal; microthermal.*

Megathermal n. An infrequently used term proposed by Judson (1953, p.59) for *Altithermal,* but now regarded as a syn. of *Hypsithermal.*

megayear A term proposed by Rankama (1967) for one million (10^6) years. Abbrev: Myr.

megazone *superzone.*

megerliiform Said of the loop of a terebratellacean brachiopod with descending branches joining anterior projections from the large ring on a low median septum, differing from similar dallinid and terebratellid loops by the appearance of a well-developed ring before growth of descending branches (TIP, 1965, pt.H, p.148).

Meigen's reaction A test used to distinguish calcite from aragonite. A cobalt nitrate solution is used to stain the minerals by boiling; aragonite becomes lilac and retains this color in thin section, whereas calcite and dolomite becomes pale blue but without showing the color in thin section. Cf: *Lemberg's stain.*

meimechite An alternate spelling of *meymechite.*

meinzer A syn. of *permeability coefficient,* named for O.E. Meinzer (1876-1948), a hydrogeologist with the USGS. Syn: *Meinzer unit.*

meionite A mineral of the scapolite group: $3CaAl_2Si_2O_8 \cdot CaCO_3$ (or three anorthite plus calcium carbonate). It is isomorphous with marialite, and may contain other anions (sulfate, chloride). Symbol: Me.

meizoseismal Said of or pertaining to the maximum destructive force of an earthquake.

meizoseismal curve A curved line connecting the points of the maximum destructive energy of an earthquake shock around its epicentrum.

melaconite *tenorite.*

melanchym A complex humic substance separated into two fractions by alcohol: insoluble *melanellite* and soluble *rochlederite.* It is found in the brown coal of Bohemia.

melane Any mafic mineral.

melanellite The insoluble portion remaining when *melanchym* is treated with alcohol. See also: *rochlederite.*

mélange A heterogeneous medley or mixture of rock materials; specif. a mappable body of deformed rocks consisting of a pervasively sheared, fine-grained, commonly pelitic matrix, thoroughly mixed with angular and poorly sorted inclusions of native and exotic tectonic fragments, blocks, or slabs (of diverse origins and geologic ages) that may be as much as several kilometers in length. Examples include argille scagliose and wildflysch. Etymol: French, "mixture". Cf: *chaos.* Syn:

block clay.

melanic *melanocratic.*

melanide A term proposed for any igneous rock in which the dark-colored mineral, such as an amphibole or a pyroxene, cannot be identified megascopically.

melanite A black variety of andradite garnet containing titanium. Cf: *schorlomite.* Syn: *pyreneite.*

melanized Said of a soil whose dark color is due to its content of humus.

melanocerite A brown or black rhombohedral mineral: $(Ca, Ce,Y)_8(BO_3)(SiO_4)_4(F,OH)_4$ (?).

melanocratic Dark colored; applied to a dark-colored igneous rock rich in mafic minerals. The percentage of mafic minerals required for a rock to be classified as "melanocratic" varies among petrologists, but is usually given as at least 50 to 60 percent. Cf: *leucocratic; mesocratic.* Syn: *chromocratic; melanic; dark-colored.* Noun: *melanocrate.*

melanophlogite A mineral consisting of silicon dioxide (SiO_2) and containing carbon and sulfur. It was formerly believed to be a partly oriented pseudomorph of alpha quartz after cristobalite containing H_2SO_4.

melanophyride A broad term used in the field for any dark-colored porphyritic igneous rock having a fine-grained groundmass.

melanosome The dark-colored part of a migmatite, rich in mafic minerals (Mehnert, 1968, p.355). Cf: *leucosome.* Also spelled: *melasome.*

melanostibite A mineral: $Mn(Sb,Fe)O_3$.

melanotekite A black or dark-gray mineral: $Pb_2Fe_2^{+3}Si_2O_9$.

melanovanadite A black mineral: $2CaO.2V_2O_4.3V_2O_5.nH_2O$.

melanterite A green or greenish-blue monoclinic mineral: $FeSO_4.7H_2O$. It usually results from the decomposition of iron sulfides. Syn: *copperas; green vitriol; iron vitriol.*

melaphyre A term originally applied to any dark-colored porphyritic igneous rock but later restricted to altered basalt, esp. of Carboniferous and Permian age.

melasome *melanosome.*

melatope In an *interference figure*, a point indicating the crystal's optic axis.

melee (a) A collective term for small, round, faceted diamonds, such as those mounted in jewelry. The term is sometimes applied to colored stones of the same size and shape of the diamonds. (b) A small diamond cut from a fragment of a larger size.---Etymol: French. Anglicized pron; *mell-ee.*

melikaria (a) Skeletal structures of quartz formed in place by deposition of silica from rising waters in the bottoms of deep shrinkage cracks in septaria or other concretions, the enclosing rock having been removed by solution (Burt, 1928). They resemble septarian veins in form, and may be as large as 45 x 20 x 10 cm (as in the Quaternary alluvial deposits of Brazos County, Tex.). Cf: *septarium.* (b) A term applied to the vein skeletons of septaria (Twenhofel, 1939, p.552).---Etymol: Greek, "honeycombs".

melilite (a) A group of minerals of general formula: $(Na, Ca)_2(Mg,Al)(Si,Al)_2O_7$. It consists of an isomorphous solid-solution series, and may contain some iron. (b) A tetragonal, often honey-yellow mineral of the melilite group, such as the end members gehlenite and akermanite. It occurs as a component of certain recent basic volcanic rocks.---The melilites of volcanic rocks are usually classed as feldspathoids, but have also been considered as "undersaturated pyroxenes". Syn: *mellilite.*

melilitholith An extrusive rock composed entirely of melilite.

melilitite A generally olivine-free extrusive rock composed of melilite and augite (or other mafic mineral, usually comprising more than 90 percent of the rock), with minor amounts of feldspathoids and sometimes plagioclase.

melilitolite A group of rare plutonic mafic rocks with a predominance of melilite (Streckeisen, 1967, p.174); also, any

rock in that group, e.g. *uncompahgrite.*

meliphanite A yellow, red, or black mineral: $(Ca,Na)_2Be(S, Al)_2(O,OH,F)_7$. Syn: *meliphane.*

melkovite A mineral: $CaFeH_6(MoO_4)_4(PO_4).6H_2O$.

mellilite (a) *melilite.* (b) *mellite.*

mellite A honey-colored mineral: $Al_2[C_6(COO)_6].18H_2O$. has a resinous luster, usually occurs as nodules in brow coal, and is in part a product of vegetable decompositior Syn: *honey stone; mellilite.*

mellorite A name suggested for a poorly crystallized materia of the kaolin group of clay minerals in which randomness e stacking of the layer packets in the *c*-axis direction is presen Because there is considerable range of disorder in the les well-crystallized kaolinites, there is no need for a specifi mineral name such as "mellorite"; the term in general use i *fireclay mineral.*

mellow Said of a soil structure that is porous, without a ter dency toward compaction; also, said of the soil itself. Mellow ness is an optimum condition for plant growth, and is usuall associated with soil fertility.

melnikovite *greigite.*

melon hole An Australian syn. of *gilgai.*

melonite A reddish-white mineral: $NiTe_2$.

melt n. In petrology, a liquid, fused rock.

melteigite A dark-colored plutonic rock that is part of the *ijo ite* series and contains nepheline and 60-90% mafic mineral: esp. green pyroxene. Cf: *turjaite; urtite; algarvite; microme teigite.*

melt firn Firn formed under conditions of melting and freezin characterized by a comparatively rapid transformation fror snow to firn to ice. Syn: *alpine firn; alpine-type firn.* Cf: *pole firn.*

melt-freeze metamorphism A process of modification of ic crystals in deposited snow, involving wetting of grains by me twater followed by freezing, together with accelerated equi emperature metamorphism, leading to a rise in density an strength of the snowpack; an important early part of the prc cess of firnification.

meltwater Water derived from the melting of snow or ice, esp the stream water flowing in, under, or from melting glacie ice. Also spelled: *melt water.*

member A rock-stratigraphic unit of subordinate rank, con prising some specially developed part of a varied *formatic* (such as a subdivision of only local extent, or a lithologicall unified subdivision distinguished from adjacent parts of th formation by color, hardness, composition, or similar features and not defined by specified shape or extent (ACSN, 196 art.7). It may be formally defined and named, or informall named, or unnamed. It is not necessarily mappable, and named member may extend from one formation into anothe Laterally equivalent parts of a formation that differ recogniz ably may be considered members; e.g. the gravel membe and the silt member of the Bonneville Formation (Pleistocene of west-central Utah. A member name combines a geographi name followed by the word "member"; where a lithologic des ignation is useful, it should be included (e.g. the Wedingto Sandstone Member of the Fayetteville Shale). It is higher i rank than a *bed.* Abbrev: mbr. Cf: *lentil; tongue.*

membranate chorate cyst A dinoflagellate *chorate cyst* with prominent membrane (e.g. *Membranilarnacia*).

membrane The thinned, generally delicate and elastic exinou floor of a pore or colpus of a pollen grain; e.g. *harmomega thus.*

membranelle A flattened, blade-like vibrating organ in a tinti nid, comprised of a row of fused cilia and fringed with lame lae, and used for locomotion. Syn: *membranella.*

menaccanite [**mineral**] A variety of *ilmenite* found as a san near Manaccan (Menachan) in Cornwall, Eng. Syn: *menacha nite; manaccanite.*

menaccanite [sed] A black, volcanic sand from near Manac-an (Menachan) in Cornwall, Eng., from which the element titanium was first isolated. Syn: *menachanite; menachite.*

mendeleyevite A titanium- and rare-earth-bearing betafite. Also spelled: *mendeleyeevite; mendeleevite.*

m-enderbite A name proposed by Tobi (1971, p.202) in his classification of the charnockite suite for that member in which mesoperthite and free plagioclase are both present.

mendip (a) A buried hill that is exposed (by the cutting of a valley across a cuesta) as an inlier. (b) A coastal-plain hill that at one time was an offshore island.---Type locality: Mendip Hills in Somerset, England.

mendipite A white orthorhombic mineral: $Pb_3Cl_2O_2$.

mendozite A monoclinic mineral of the alum group: $NaAl(SO_4)_2 \cdot 11H_2O$ (?). Cf: *soda alum.*

meneghinite A blackish lead-gray mineral: $CuPb_{13}Sb_7S_{24}$. Its formula was previously given as: $Pb_{13}Sb_7S_{23}$.

Menevian European stage: Middle Cambrian (above Solvan, below Maentwrogian).

mengwacke A wacke with 33-90% unstable materials (Fischer, 1934). Etymol: German *Mengwacke,* "mixed wacke".

menilite An opaque, impure, dull-grayish or brown (liver-colored) variety of opal found in rounded or flattened concretions at Menilmontant near Paris, France. Syn: *liver opal.*

Meotian European stage (Black Sea area): lowermost Pliocene (above Sarmatian of Miocene, below Kimmerian). It has also been regarded as uppermost Miocene.

Meramecian North American provincial series: Upper Mississippian (above Osagian, below Chesterian).

meraspis A juvenile trilobite that does not yet have the number of thoracic segments typical of the species; a late trilobite larva in which the pygidium is beginning to form. Pl: *meras-ides.*

Mercalli scale An arbitrary scale of earthquake intensity, ranging from I (detectable only instrumentally) to XII (causing almost total destruction). It is named after Giuseppi Mercalli (d.1914), the Italian geologist who devised it in 1902. Its adaptation to North American conditions is known as the *modified Mercalli scale.*

mercallite A sky-blue mineral: $KHSO_4$.

Mercator chart A chart or map drawn on the Mercator projection. It is commonly used for marine navigation.

Mercator equal-area projection *sinusoidal projection.*

Mercator projection An equatorial, cylindrical, conformal map projection derived by mathematical analysis (not geometrically) in which the equator is represented by a straight line true to scale, the meridians by parallel straight lines perpendicular to the equator and equally spaced according to their distance apart at the equator, and the parallels by straight lines perpendicular to the meridians and parallel with (and the same length as) the equator. The parallels are spaced so as to achieve conformality, their spacing increasing rapidly with their distance from the equator so that at all places the degrees of latitude and longitude have the same ratio to each other as to the sphere itself, resulting in great distortion of distances, areas, and shapes in the polar regions (above 80° lat.), the scale increasing poleward as the secant of the latitude. Because any line of constant direction (azimuth) on the sphere is truly represented on the projection by a straight line, the Mercator projection is of great value in navigation and is used for hydrographic charts, and also to show geographic variations of some physical property (such as magnetic declination) or to plot trajectories of Earth satellites in oblique orbits. Named after Gerhardus Mercator (1512-1594), Flemish mathematician and geographer, whose world map of 1569 used this projection. See also: *transverse Mercator projection.*

Mercator track A *rhumb line* constructed on a Mercator projection.

mercury A heavy, silver- to tin-white hexagonal mineral, the native metallic element Hg. It is the only metal that is liquid at ordinary temperatures. Native mercury is found as minute fluid globules disseminated through cinnabar or deposited from the waters of certain hot springs, but it is unimportant as a source of the metal. It usually contains small amounts of silver. Mercury combines with most metals to form alloys or amalgams. Syn: *quicksilver.*

mercury barometer A type of *barometer* that measures barometric change by its effect on the mercury or other liquid in a U-shaped glass tube closed at one end. Cf: *aneroid barometer.*

mer de glace A general term applied to any of the large glaciers or ice sheets of the Pleistocene Epoch. Type example: Mer de Glace, the largest glacier on the Mont Blanc massif in the Alps. Etymol: French, "sea of ice".

mere [coast] An obsolete term for an estuary, creek, inlet, or other arm of the sea, and for the sea itself.

mere [lake] (a) A sheet of standing water; esp. a large pond or a small, shallow lake, occupying a hollow among drumlins, or often occupying a basin resulting from subsidence caused by the removal of subsurface salt or by the solvent action of ground water on the salt. (b) A levee lake behind a barrier consisting of sediment carried upstream by the tide.

merenskyite A mineral: $(Pd,Pt)(Te,Bi)_2$.

mergifer Said of a variant of radulifer type of long brachiopod crura, very close together and parallel, arising directly from the swollen edge of a high dorsal median septum.

meridian (a) One of the imaginary great circles on the surface of the Earth passing through the poles and any given place and perpendicular to the equator, connecting all points of equal longitude; a north-south line of constant longitude, or a plane, normal to the geoid or spheroid and passing through the Earth's axis, defining such a line. Also, a half of such a great circle included between the Earth's poles with a plane coinciding with that of the astronomic meridian of the place. Syn: *terrestrial meridian.* (b) Any one of a series of lines, corresponding to meridians, drawn on a globe, map, or chart at intervals due north and south and numbered according to the degrees of longitude east or west from the prime meridian. (c) *celestial meridian.*---Cf: *parallel.*

meridian hole A term introduced by Agassiz (1866, p.293-294) for a shallow, crescent-shaped *dust well* that accurately registers on the surface of a glacier the position of the Sun during the day. It has a steeper wall on its southern side than on its northern side.

meridian line A line running accurately north and south through any given point on or near the Earth's surface; specif. a line used in plane surveying and defined by the intersection of the plane of the celestial meridian and the plane of the horizon.

meridional (a) Pertaining to a movement or direction between the poles of a thing, e.g. the Earth's north-south water or air circulation patterns, or the alignment of colpi on a pollen grain. (b) Southern.

meridional difference The difference (distance) between the meridional parts of any two given parallels of latitude. It is found by subtraction if the two parallels are on the same side of the equator, and by addition if they are on opposite sides. Cf: *latitude difference; departure.* Also called: *meridional difference of latitude.*

meridional ellipse *ellipsoid of revolution.*

meridional part The linear length of the arc of a meridian between the equator and a given parallel of latitude on a Mercator chart, expressed in units of one minute of longitude at the equator.

meridional projection A projection of a sphere onto a plane

that is parallel to the plane of a meridian passing through the point of projection; e.g. "meridional orthographic map projection" or "Lambert meridional equal-area projection".

merismite Chorismite in which there is irregular (shape-wise) penetration of the diverse units (Dietrich & Mehnert, 1961).

meristele The intervening portion of vascular tissue between two *dictyosteles*.

meristem A relatively undifferentiated plant tissue with thin-walled cells characterized by persistent, active cell division, producing daughter cells capable of differentiating into other plant tissues and organs, thus contributing to the growth and development of a plant.

merocrystalline *hypocrystalline*.

merodont Said of a class of ostracode hinges having three elements and characterized by crenulate terminal elements with either a positive or negative, and a crenulate or smooth, median element.

merohedral Said of crystal classes in a system, the general form of which has only one half, one fourth, or one eighth the number of equivalent faces of the corresponding form in the *holohedral* class of the same system. This condition is known as *merohedrism*. Cf: *tetartohedral*. Syn: *merosymmetric*. See also: *hemihedral; ogdosymmetric*.

merohedrism The condition of being *merohedral*. Syn: *merohedry*.

merohedry *merohedrism*.

merokarst Karst that is imperfect or incomplete, characterized by thin or impure limestone bedrock and the presence of surface drainage. Cf: *holokarst*.

meroleims Coalified remains of plant debris (Kristofovich, 1945, p.138). Cf: *hololeims*. See also: *phytoleims*.

meromictic lake A lake that undergoes incomplete mixing of its waters during periods of circulation; specif. a lake in which the bottom, noncirculating water mass (*monimolimnion*) is adiabatically isolated from the upper, circulating layer (*mixolimnion*). Cf: *holomictic lake*.

meromixis The process leading to, or the condition of, a *meromictic lake*.

meroplankton An organism that is temporarily planktonic, e.g. eggs and larvae of benthic and nektonic organisms. Cf: *holoplankton*. Syn: *temporary plankton*.

meropod The *merus* of a malacostracan crustacean. Syn: *meropodite*.

merostome Any aquatic *chelicerate* belonging to the class Merostomata, characterized by the presence of one pair of preoral appendages with three, possibly four, joints. Cf: *arachnid*.

merosymmetric *merohedral*.

merosyncline Bubnoff's term for that part of a geosynclinal belt having independent mobility (Glaessner & Teichert, 1947, p.588).

meroxene A variety of biotite with its axial plane parallel to the crystallographic *b*-axis.

Merriam effect The relationship between mountain mass and the vertical distribution of animals and plants. The term was designated by Lowe (1961, p.45-46) for the indirect factor of environmental conditions such as the total elevation of a mountain, the size or mass of the mountain, and the elevation of the basin or plain from which the mountain rises on the vertical placement and displacement of species and communities of plants and animals. Named after Clinton Hart Merriam (1855 - 1942), U.S. biologist, who first recognized the relationship (Merriam, 1890).

merrihueite A mineral: $(K,Na)_2(Fe,Mg)_5Si_{12}O_{30}$.

merrillite A colorless meteorite mineral: $Na_2Ca_3(PO_4)_2O$ (?). It is related to whitlockite.

Mersey yellow coal *tasmanite* [coal].

merumite A mineral: $Cr_2O_3.H_2O$ (?). Its exact nature is doubtful: it may be a mixture containing eskolaite and perhaps

a hydrated oxide (Hey, 1963, 7.14.1a).

merus The fourth pereiopodal segment from the body of a malacostracan crustacean, bounded proximally by the ischium and distally by the carpus. Pl: *meruses*. Syn: *meropod*.

merwinite A colorless to pale-green monoclinic mineral: $Ca_3Mg(SiO_4)_2$.

merzlota A Russian term for *frozen ground*.

mesa (a) An isolated, nearly level landmass standing distinctly above the surrounding country, bounded by abrupt or steeply sloping erosion scarps on all sides, and capped by layers of resistant, nearly horizontal rocks (usually lavas). Less strictly, a very broad, flat-topped, usually isolated hill or mountain of moderate height bounded on at least one side by a steep cliff or slope and representing an erosion remnant. A mesa is similar to, but has a more extensive summit area than, a *butte*, and is a common topographic feature in the arid and semiarid regions of the U.S. See also: *table mountain*. (b) A broad terrace or comparatively flat plateau along a river valley, marked by an abrupt slope or escarpment on one side. See also: *bench*.---Etymol: Spanish, "table". Pron: *may-sa*.

mesabite An ocherous variety of goethite from the Mesabi Range in Minnesota.

mesa-butte A *butte* formed by the erosion and reduction of a mesa. Cf: *volcanic butte*.

mesa plain The flat summit of a hilly mountain or of a plateau (Hill, 1900, p.6). Cf: *plateau plain*.

mesarch adj. Said of an ecologic succession (i.e. a sere) that develops under *mesic* conditions. Cf: *hydrarch; xerarch*. See also: *mesosere*.

mesa-terrace An obsolete term used by Lee (1900, p.504-505) for an alluviated, planate rock surface contained within a valley, lying between the flood plain of a nearby stream and the steeper slope leading up to a mesa.

mesenchyme (a) The *mesohyle* of a sponge. (b) A term used by zoologists for the fleshy connective tissue in coelenterates, but applied by paleontologists to the stony skeletal structures between corallites secreted by the common fleshy connective tissue (Shrock & Twenhofel, 1953, p.133). Cf: *sclerenchyme*.---Also spelled: *mesenchyma*.

mesentery One of several radially disposed fleshy laminae or sheets of soft tissue that are attached to the inner surface of the oral disk and column wall of a coral polyp and that partition the internal body cavity by extending inward from the body wall. Adj: *mesenterial*.

meseta (a) A small *mesa*. (b) An extensive plateau or flat upland, often with an uneven or eroded surface, forming the central physical feature of a region; e.g. the high, dissected tableland of the interior of Spain.---Etymol: Spanish, "tableland".

mesh [pat grd] The unit component of patterned ground (excepting steps and stripes), as a circle, a polygon, or an intermediate form (Washburn, 1956, p.825).

mesh [part size] One of the openings or spaces between the wires of a sieve or screen. See also: *mesh number*.

mesh number The size of a sieve or screen, or of the material passed by a sieve or screen, in terms of the number of meshes per linear inch; e.g. mesh number 20 indicates that the sieve or screen has 20 holes per linear inch (this takes no account of the diameter of the wire, so that the mesh number does not always have a definite relation to the size of the hole).

mesh texture *reticulate* [petrology].

mesic [ecol] Said of a habitat receiving a moderate amount of moisture; also, said of an organism or group of organisms occupying such a habitat. Cf: *hydric; xeric*. See also: *mesarch*.

mesic [soil] Said of a soil temperature regime in which the mean annual temperature (measured at 50cm) is at least 8°C but less than 16°C, with a summer-winter variation of more than 5°C (SSSA, 1970). Cf: *isomesic*.

mesilla A term used in SW U.S. for a small *mesa*. Etymol: Spanish, "small table". Syn: *mesita; meseta.*

mesistele The intermediate part of a crinoid column, between proxistele and dististele. It is doubtfully distinguishable in pluricolumnals.

mesita *mesilla.*

mesitis Transformation (tending toward homogenization) between chemically different rocks under the same temperature and pressure conditions (Sørensen, 1961, p.61).

mesitite A white variety of magnesite containing 30-50% iron carbonate. Syn: *mesitine; mesitine spar.*

mesoautochthon An autochthon or basement formed temporarily where a nappe has ceased movement, and on which the sediments of a *parallochthon* are deposited. Cf: *paleoautochthon; neoautochthon.*

mesocarp The middle layer of *pericarp*, when the pericarp consists of more than two different layers. Cf: *endocarp; epicarp.*

mesoclade One of the median *clades* or skeletal branches that connect the bifurcating parts of the actines in an ebridian skeleton.

mesoconch The part of a dissoconch of a bivalve mollusk formed at an intermediate stage of growth and separated from earlier- and later-formed parts by pronounced discontinuities.

mesocoquina A term used by Bissell & Chilingar (1967, p.153) for a detrital limestone composed of weakly cemented shell detritus of sand size (2 mm in diameter) or less. Cf: *microcoquina.*

mesocratic Composed of almost equal amounts of light and dark constituents; applied to igneous rocks intermediate in color between *leucocratic* and *melanocratic*. The percentage of mafic minerals permissible for a rock to be classified as "mesocratic" varies among petrologists but usually ranges between 30 and 60 percent. Syn: *mesotype.*

mesocrystalline Said of the texture of a rock consisting of or having crystals whose diameters are intermediate between those of a microcrystalline and a macrocrystalline rock; also, said of a rock with such a texture. Howell (1922) applied the term to the texture of a recrystallized sedimentary rock having crystals whose diameters are in the range of 0.20-0.75 mm, and Bissell & Chilingar (1967, p.103) to the texture of a carbonate sedimentary rock having crystals whose diameters are in the range of 0.05-1.0 mm.

mesocumulate A cumulate containing a small amount of intercumulus material; a cumulate intermediate between an *orthocumulate* and an *adcumulate.*

mesoderm An intermediate body layer present in animals more highly organized than the coelenterates. Cf: *endoderm; ectoderm.*

mesogene Said of a mineral deposit or enrichment of mingled *hypogene* and *supergene* solutions; also, said of those solutions and of that environment.

mesogenetic A term proposed by Choquette & Pray (1970, p.220) for the period between the time when newly buried deposits are affected mainly by processes related to the depositional interface (*eogenetic* stage) and the time when long-buried deposits are affected by processes related to the erosional interface (*telogenetic* stage). Also applied to the porosity that develops during the mesogenetic stage.

mesogeosyncline A geosyncline between two continents and receiving clastics from both of them (Schuchert, 1923). Syn: *mediterranean.*

mesogloea A gelatinous substance between the endoderm and ectoderm of certain invertebrates; e.g. an extracellular material, containing proteins and carbohydrates, found in the mesohyle of many sponges, or a noncellular jelly-like middle layer of the outer walls and mesenteries of coral polyps. Syn: *mesoglea.*

mesograined Said of the texture of a carbonate sedimentary rock having clastic particles whose diameters are in the range of 0.05-1.0 mm (Bissell & Chilingar, 1967, p.103) or 0.1-1.0 mm (DeFord, 1946).

mesogyrate Said of the umbones (of a bivalve mollusk) curving toward the center. Cf: *orthogyrate.*

mesohyle Loosely organized material constituting a sponge between the pinacoderm and the choanoderm, commonly consisting of spongin, spicules, and various types of cells (mainly amoebocytes), embedded in mesogloea, although one or more of these elements may be missing. Syn: *mesenchyme; parenchyma.*

mesokurtic Closely resembling a normal frequency distribution; e.g. said of a distribution curve that is neither *leptokurtic* (very peaked) nor *platykurtic* (flat across the top).

mesolimnion A *metalimnion* in a lake. Adj: *mesolimnetic.*

mesolite A zeolite mineral: $Na_2Ca_2Al_6Si_9O_{30}.8H_2O$. It is intermediate in chemical composition between natrolite and scolecite, and is usually found in white or colorless tufts of very delicate acicular crystals in amygdaloidal basalts. Syn: *cotton stone.*

Mesolithic n. In archaeology, the middle division of the *Stone Age*, characterized by the change from glacial to postglacial climate and the absence of agriculture. Correlation of relative cultural levels with actual age (and, therefore, with the time-stratigraphic units of geology) varies from region to region. Cf: *Paleolithic; Neolithic.* Syn: *Transitional; Middle Stone Age.*---adj. Pertaining to the Mesolithic.

mesolithion Animals that live in cavities in rock.

mesomicrocline A pseudomonoclinic mineral of the alkali feldspar group: $KAlSi_3O_8$. It is intermediate in degree of ordering between microcline and orthoclase.

mesonorm A theoretical calculation of normative minerals in metamorphic rocks of the mesozone from chemical analyses (Barth, 1959). Cf: *epinorm.*

mesopelagic Pertaining to the *pelagic* environment of the ocean between 100-500 fathoms. Cf: *epipelagic.*

mesopeltidium A sclerite (commonly one of a pair) of segmented carapace of an arachnid, situated immediately behind the *propeltidium* and in front of the *metapeltidium.*

mesoperthite A variety of perthitic feldspar consisting of an intimate mixture of about equal amounts of potassium feldspar and plagioclase (usually albite, sometimes oligoclase). It is intermediate in composition between perthite and antiperthite. Syn: *eutectoperthite.*

mesophilic Said of an organism that prefers a moderate environment; e.g. mesothermal conditions. Noun: *mesophile.*

mesophyll The chlorophyllous tissues in the interior of a leaf.

mesophyte A plant that cannot survive extreme conditions of temperature or water supply. Cf: *hydrophyte; xerophyte.* Adj: *mesophytic.*

Mesophytic A paleobotanic division of geologic time, signifying the time between the first occurrence of the angiosperms. Cf: *Aphytic; Archeophytic; Eophytic; Paleophytic; Cenophytic.*

Mesophyticum Plant life of the Mesozoic. Cf: *Palaeophyticum; Cainophyticum.*

mesopiestic water A class of *piestic water* including water that rises above the water table but not to the land surface. Cf: *hypopiestic water; hyperpiestic water.*

mesoplankton (a) Plankton of the size range 0.5-1.0 mm; a type of *microplankton.* (b) Plankton that live at middle depths. The term is rarely used because it is confusing.

mesopore [petrology] In the pore-size classification of Choquette & Pray (1970, p.233), an equant to equant-elongate pore or a tubular or platy pore with an average diameter or thickness between 4 and 1/16 mm. Cf: *megapore; micropore.*

mesopore [paleont] An unusually small and irregular or angular skeletal (zooecial) tube found in Paleozoic bryozoans, parallel to but smaller in diameter than the autozooecia, and generally restricted to the exozone. It occupies the space between the normal larger zooecia and is characterized by numerous diaphragms.

mesopsammon Animals that live in cavities in sand.

mesoscopic According to Dennis (1967, p.152), a term introduced to describe a tectonic feature large enough to be observed without the aid of a microscope yet small enough that it can still be observed directly in its entirety. Cf: *macroscopic.*

mesosere A sere that develops in an environment having a moderate amount of moisture, i.e. in a mesic environment; a *mesarch* sere. Cf: *hydrosere; xerosere.*

mesosiderite A *stony-iron meteorite* in which the silicates are mainly pyroxene (usually orthopyroxene) and calcic plagioclase. Mesosiderites often appear to be breccias made up of fragments of widely different chemical and mineralogical composition, cemented together by a nickel-iron matrix. Olivine is sometimes present, generally as separately enclosed crystals of fairly large size. Syn: *grahamite.*

mesosilexite A silexite in which the dark-colored components constitute more than five percent of the rock.

mesosoma The middle region of the body of some invertebrates, esp. when this cannot be readily analyzed into its primitive segmentation (as in arachnids and most mollusks); specif. the anterior part of a merostome opisthosoma carrying appendages. Cf: *metasoma.* Syn: *mesosome.*

mesosphere The *lower mantle*; it is not involved in the Earth's tectonic processes.

mesostasis The last-formed interstitial material, either glassy or aphanitic, of an igneous rock. Cf: *groundmass* [ign]. Syn: *basis* [ign]; *base* [ign].

mesotectonics Tectonics of features about 100-1000 km in size.

mesotheca *median lamina.*

mesotherm A plant that requires moderate temperatures for successful growth. Cf: *microtherm; megatherm.*

mesothermal [eco geol] Said of a hydrothermal mineral deposit formed at considerable depth and in the temperature range of 200°-300°C (Park & MacDiarmid, 1970, p. 317). Also, said of that environment. Cf: *hypothermal; epithermal; leptothermal; teletothermal; xenothermal.*

mesothermal [ecol] Said of an organism that prefers moderate temperatures, i.e. in the 25-37°C range.

mesothermal [clim] Pertaining to a climate characterized by moderate temperature. Cf: *megathermal; microthermal.*

mesothyridid Said of a brachiopod pedicle opening when the foramen is located partly in the ventral umbo and partly in the delthyrium, with the beak ridges appearing to bisect the foramen (TIP, 1965, pt.H, p.148).

mesotil A semiplastic or semifriable derivative of chemically weathered till, developed beneath a partially drained area, and intermediate in texture between *gumbotil* and *silttil* (Leighton & MacClintock, 1930, p.42-43).

mesotourmalite A *tourmalite* in which tourmaline comprises 5-50 percent of the rock.

mesotrophic lake A lake that is characterized by a moderate supply of nutrient matter, neither notably high nor low in its total production; it is intermediate between a *eutrophic lake* and an *oligotrophic lake.*

mesotrophic peat Peat containing a moderate amount of plant nutrients. Cf: *oligotrophic peat; eutrophic peat.*

mesotrophy The quality or state of a *mesotrophic lake.*

mesotype [mineral] (a) A group of zeolite minerals, including natrolite, mesolite, and scolecite. Syn: *needle zeolite.* (b) A term used, mainly in France, in the restricted meaning of *natrolite*, because its form is intermediate between those of stilbite and analcime.

mesotype [petrology] *mesocratic.*

Mesozoic An era of geologic time, from the end of the Paleo-

zoic to the beginning of the Cenozoic, See also: age of *gy nosperms; age of reptiles.* Obs syn: *Secondary.*

mesozone According to Grubenmann's classification of met morphic rocks (1904), the intermediate *depth zone* of met morphism, characterized by temperatures of 300°500° moderate hydrostatic pressure and shearing stress (som times lacking). Chemical and regional metamorphism pr dominate; association of some epizone and katazone minera is characteristic. The concept was modified by Grubenma and Higgli (1924) to include effects of intermediate-temper ture contact metamorphism. Modern usage stresses temper ture-pressure conditions (medium to high metamorphic grad rather than depth of zone. Cf: *katazone; epizone.*

mesquitelite A mineral, approximately: $(Mg,Ca)Al_4S$ $O_{25}.5H_2O$. It is a clay-like alteration product of feldspar.

messelite A mineral: $Ca_4Fe_2(PO_4)_4.5H_2O$.

messenger A metal weight on the cable of an oceanograp device that acts as an activator.

meta- A prefix that, when used with a name of a sediment or igneous rock, indicates that the rock type has been me morphosed, e.g. metabasalt, metasediment.

meta-aluminite A mineral: $Al_2(SO_4)(OH)_4.5H_2O$.

meta-alunogen A mineral: $Al_4(SO_4)_6.27H_2O$.

meta-ankoleite A yellow secondary mineral: $K_2(UO_2 (PO_4)_2.6H_2O$.

meta-anthracite Coal having a fixed-carbon content of 98% more; the highest rank of anthracite. Cf: *graphocite.* Syn: s *peranthracite; subgraphite.*

meta-argillite An *argillite* that has been weakly metamo phosed. The term was used by Flawn (1953, p.564) for a lo grade metamorphic rock, without cleavage or parting, which more than half of the constituent material (clay min als and micaceous paste) has been reconstituted to combir tions of sericite, chlorite, epidote, or green biotite, the parti size of the reconstituted material ranging from 0.01 to 0. mm.

meta-arkose Arkose (often in Precambrian terrane) which h been "welded" by metamorphism so that it resembles a tr granite or a granitized sediment (Pettijohn, 1957, p.325). (*recomposed granite.*

meta-autunite A yellow secondary mineral: $Ca(UO_2)_2(PO_4 26H_2O$. It is apparently not formed directly in nature, but m field and museum specimens of autunite have been par dehydrated to this phase. Cf: *para-autunite.*

metabasite A collective term, first used by Finnish geologis for metamorphosed mafic rock which had lost all traces original texture and mineralogy due to complete recrystalliz tion (Myashiro, 1968, p.800).

metabentonite (a) Metamorphosed, altered, or somewhat durated bentonite, characterized by clay minerals (es beidellite) that no longer have the property of absorbing or a sorbing large quantities of water; nonswelling bentonite, bentonite that swells no more than do ordinary clays. T term has been applied to certain Ordovician clays of the A palachian region and upper Mississippi River valley. See al potassium bentonite. Syn: *subbentonite.* (b) A mineral of t montmorillonite group with SiO_2 layers in the montmorillor structure.

metabituminous coal Coal that contains 89-91.2% carbo analyzed on a dry, ash-free basis. Cf: *semibituminous coal.*

metablastesis (a) Recrystallization and growth of a mineral group of minerals. (b) Essentially isochemical recrystallizat without evidence of a separate mobile phase.---- (Dietrich Mehnert, 1961, p.61). Adj: *metablastic.*

metablastic The adj. of *metablastesis.* This term has son times been applied to crystals formed during metamorphis to distinguish them from those formed from magma, i.e. *doblastic.*

metaboghead coal High-ranking torbanite.

metabolism of rocks A term proposed by Barth (1962) for the distribution of granitizing materials within sediments by mobilization, transfer, and reprecipitation, as opposed to metasomatism involving addition of new materials.

metabolite [meteorite] An iron meteorite showing metamorphic effects due to reheating.

metabolite [ecol] An excretion or external secretion (e.g., an enzyme, hormone, or vitamin) of an organism which affects the associated organisms by inhibiting their activities or even killing them.

metabolite [ign] A term proposed, but not used, for altered essenite.

metaboly The capability of an organism to change its shape.

metaborite A white mineral: HBO_2. It is the cubic modification of metaboric acid.

metacannel coal Cannel coal of high metamorphic rank. Cf: subcannel coal.

metacinnabar A black isometric mineral: HgS. It is dimorphous with cinnabar and represents an ore of mercury. Syn: metacinnabarite.

metaclase Leith's term for a rock possessing secondary cleavage, or cleavage in its current meaning (1905, p.12). Cf: protoclase.

metacolloid An originally colloidal substance that has become crystalline, e.g. serpophite.

metacryst Any large crystal developed in a metamorphic rock by recrystallization, such as garnet or staurolite in mica schist; a syn. of porphyroblast. Syn: metacrystal.

metacrystal metacryst.

metadiagenesis epigenesis [sed].

metafluidal An obsolete syn. of dynamofluidal.

metagenesis [evol] alternation of generations.

metagenesis [sed] A term applied by Russian geologists to epigenesis (changes occurring in a more or less compact sedimentary rock) or to late epigenesis.

metagenic Said of a sediment or sedimentary rock formed through diagenetic alteration of other sediments (Grabau, 1920, p.2).

metaglyph A hieroglyph formed during metamorphism (Vassoevich, 1953, p.33).

metahaiweeite A secondary mineral: $Ca(UO_2)_2Si_6O_{15} \cdot nH_2O$, where n is less than 5. It is apparently a dehydration product of haiweeite.

metahalloysite A name used in Europe for the less hydrous form of halloysite. It is synonymous with halloysite of U.S. authors. The term has also been used to designate the nonhydrated form of halloysite.

metaharmosis Var. of metharmosis.

metaheinrichite A yellow to green secondary mineral: $Ba(UO_2)_2(AsO_4)_2 \cdot 8H_2O$. Cf: heinrichite.

metahewettite A red mineral: $CaV_6O_{16} \cdot 9H_2O$. It resembles hewettite but differs slightly from it in its behavior during hydration; it is found in highly oxidized ore as coatings and fracture fillings.

metahohmannite An orange mineral: $Fe_2(SO_4)_2(OH)_2 \cdot 3H_2O$. It constitutes a partly dehydrated hohmannite.

metajennite A mineral: $Na_2Ca_8Si_5O_{19} \cdot 7H_2O$.

metakahlerite A yellow to yellowish-green secondary mineral: $(UO_2)_2(AsO_4)_2 \cdot 8H_2O$.

metakaolinite An intermediate product obtained when kaolinite is heated between about 500°C and 850°C artificially dehydrated kaolinite. Syn: metakaolin.

metakirchheimerite A pale-rose mineral: $Co(UO_2)_2(AsO_4)_2 \cdot 8H_2O$.

metal In the older geologic literature, a now obsolete term for any hard rock.

metal factor A derived parameter used to represent induced polarization anomalies. Abbrev: MF.

metalignitous coal Coal that contains 80-84% carbon, analyzed on a dry, ash-free basis. Cf: subbituminous coal.

metalimnion The horizontal layer of a thermally stratified lake in which the temperature decreases rapidly with depth. The metalimnion lies between the epilimnion and the hypolimnion, and includes the thermocline. Less preferred syn: thermocline (b) discontinuity layer; mesolimnion.

metallic (a) Pertaining to a metal. (b) A brilliant type of mineral luster that is characteristic of metals. Cf: nonmetallic; submetallic.

metalliferous Pertaining to a mineral deposit from which a metal or metals can be extracted by metallurgical processes.

metallization The process or processes by which metal or metals are introduced into a rock, resulting in an economically valuable deposit; the mineralization of metal(s).

metallized hood The upper shell or roof of a batholith that is the first area to solidify after intrusion and that contains virtually all the metalliferous lodes of a batholith. (Emmons, 1933). Syn: hood.

metallogenetic metallogenic.

metallogenic Adj. of metallogeny. Syn: metallogenetic; minerogenic; minerogenetic.

metallogenic element An element that occurs as a native element or that occurs in sulfides, selenides, tellurides, arsenides, antimonides, or sulfosalts. It is one of H.S. Washington's bipartate groupings of elements of the lithosphere, now obsolete. Cf: petrogenic element.

metallogenic epoch A unit of geologic time favorable for the deposition or ores, or characterized by a particular assemblage of mineral deposits. Several metallogenic epochs may be represented within a single area, or metallogenic province.

metallogenic map A map, usually on a regional scale, on which is shown the distribution of particular assemblages or provinces of mineral deposits and their relationship to such geologic features as tectonics and petrography.

metallogenic province An area characterized by a particular assemblage of mineral deposits, or by one or more characteristic types of mineralization. A metallogenic province may contain more than one episode of mineralization, or metallogenic epoch. Syn: metallographic province.

metallogeny The study of the genesis of mineral deposits, with emphasis on their relationship in space and time to regional petrographic and tectonic features of the Earth's crust. Although the term is used for both metallic and nonmetallic mineral deposits, it is more common than the alternative term, mineralogenesis. Adj: metallogenic.

metallographic province A little-used syn. of metallogenic province.

metallo-organic Describes a compound in which an atom of a metal is bound to an organic compound through an atom other than carbon, such as oxygen, nitrogen, or sulfur, to form a coordination compound. Cf: organometallic.

metallotect A term used in metallogenic studies for any geologic features (tectonic, lithologic, geochemical, etc) considered to have influenced the concentration of elements to form mineral deposits; an ore control, but without the implication of economic value.

metallurgy The science and art of separating metals from their ores by mechanical and chemical processes; the preparation of metallic raw materials from ore.

metaluminous Said of an igneous rock in which the molecular proportion of aluminum oxide is greater than that of sodium and potassium oxides combined but generally less than of sodium, potassium, and calcium oxides combined; one of Shand's (1947) groups of igneous rocks, classified on the basis of the degree of aluminum-oxide saturation. Cf: peralkaline; peraluminous; subaluminous.

metamict Said of a radioactive mineral (or one containing substituted radioactive elements) in which various degrees of lattice disruption and changes have taken place as a result of radiation damage while at the same time retaining its original external morphology. Examples occur in zircon, thorite and several other minerals. All minerals containing radioactive elements are not necessarily metamict (the process is also structure-dependent); e.g. the minerals xenotime and apatite.

metamorphic adj. Pertaining to the process of metamorphism or to its results. n. A *metamorphic rock*, usually used in the plural, e.g. "the metamorphics" of an area.

metamorphic assemblage (a) *metamorphic complex*. (b) A metamorphic *mineral assemblage*.

metamorphic aureole *aureole*.

metamorphic complex The metamorphic rocks constituting a whole group closely related on a regional and/or stratigraphic basis, e.g. the Dalradian metamorphic complex of Scotland. Syn: *metamorphic assemblage*.

metamorphic convergence A term introduced by Bayly (1968) to indicate two metamorphic processes converging from opposite directions but resulting in the same metamorphic product, e.g. at the same temperature a diorite may be converted retrogressively and a dolomitic marl progressively into the identical epidote-chlorite-actinolite rock.

metamorphic correlation The correlation or determination of equivalence of metamorphic features, either between the metamorphic grades of rocks of different original composition, or between a metamorphic unit and its unmetamorphosed representative elsewhere (Challinor, 1967).

metamorphic differentiation A collective term for the various processes by which contrasted mineral assemblages develop in some sequence from an initially uniform parent rock during metamorphism, e.g. garnet porphyroblasts in fine-grained mica schist (Stillwell, 1918).

metamorphic diffusion Migration, by diffusion, of materials from one part of a rock mass to another during metamorphism. Diffusion may involve chemically active fluids from magmatic sources, hot pore fluids or fluids released from hydrous minerals or carbonates. Ionic diffusion in the solid state may also occur (Joplin, 1968). Cf: *solid diffusion*.

metamorphic facies All the rocks of any chemical composition and varying mineralogical composition that have reached chemical equilibrium during metamorphism within the limits of a certain pressure-temperature range characterized by the stability of specific index minerals. The concept was introduced by Eskola (1915) and modified after Fyfe et al (1958). Cf: *mineral facies*. See also: *metamorphic facies series; metamorphic subfacies*. Syn: *densofacies*.

metamorphic facies series A group of *metamorphic facies* characteristic of an individual area or terrane, and represented by a curve or a group of curves in a pressure-temperature diagram illustrating the range of the different types of metamorphism and metamorphic facies (Hietanen, 1967). The term was introduced by Miyashiro (1961).

metamorphic grade The extent or rank of metamorphism, measured by the amount or degree of difference between the original parent rock and the metamorphic rock. It indicates in a general way the P, T environment or facies in which the metamorphism took place. For example, conversion of shale to slate or phyllite would be low-grade dynamothermal metamorphism (greenschist facies) while its continued metamorphism to a garnet-sillimanite schist would be called high-grade (almandine-amphibolite facies). Syn: *metamorphic rank*.

metamorphic overprint *overprint*.

metamorphic rank *metamorphic grade*.

metamorphic rock (a) In its original usage (Lyell, 1833), the group of gneisses and crystalline schists. (b) In current usage, any rock derived from pre-existing rocks by mineralogical, chemical, and structural changes, essentially in the solid state, in response to marked changes in temperature, pressure, shearing stress, and chemical environment at depth

in the Earth's crust, i.e. below the zones of weathering an cementation. Syn: *metamorphic (n.)*; metamorphite.

metamorphic subfacies A subdivision of a *metamorphic facie* based on compositional and mineralogical differences rathe than on pressure-temperature differences. Such subdivisior must be used with care, since knowledge of the precise sta bility fields of the metamorphic index minerals is incomplete abolition of all subfacies has been suggested (Winkler, 1967).

metamorphic water Water that is driven out of rocks by th process of metamorphism and that is commonly involved the process (Tolman, 1937).

metamorphic zone *aureole*.

metamorphic zoning *zoning* [meta].

metamorphism The mineralogical and structural adjustment solid rocks to physical and chemical conditions which hav been imposed at depth below the surface zones of weatherir and cementation, and which differ from the conditions unde which the rocks in question originated (Turner and Verhoo gen, 1960, p.450). In an older and generally obsolete sens the scope of the term included *katamorphism*, i.e. the proces of cementation and weathering (Van Hise, 1904).

metamorphite A *metamorphic rock*.

metamorphosis (a) In biology, a process involving marked abrupt reorganization of an animal during post-embryonic de velopment, such as the transformation of a larva into a suc ceeding stage of development and growth. (b) Any change form, structure, substance, etc.

metanauplius A postnaupliar crustacean larva with the sam general body and limb morphology as *nauplius*, but having ad ditional limbs (about seven pairs).

metanovacekite A yellow mineral: $Mg(UO_2)_2(AsO_4)_2 \cdot 4-8H_2O$ It is a partly dehydrated form of novacekite.

metaparian Of or concerning a trilobite that appears to hav nonfunctional facial sutures both beginning and ending at th posterior margin of the cephalon.

metapeltidium The last sclerite (usually single, rarely one of pair) of a segmented carapace of an arachnid, following upo the *mesopeltidium*.

metapepsis A now obsolete term originated by Kinaha (1878) for regional metamorphism as seemingly due to in tensely heated water or steam that "stews" the rocks. Ad *metapeptic*. Cf: *paroptesis*.

metapeptic Adj. of *metapepsis*.

metaphyte A multicellular plant. Cf: *protophyte*.

metaplasis That stage in evolution in which organisms obtai maximum vigor and diversification. Cf: *anaplasis; cataplasis*.

metapodosoma A section of the body of an acarid arachni bearing the third and fourth pairs of legs. Cf: *propodosoma*.

metaquartzite A quartzite formed by metamorphic recrystal zation, as distinguished from an *orthoquartzite*, whose crysta line nature is of sedimentary origin.

metaripple A term introduced by Bucher (1919, p.190) for large, asymmetric, ripple-like feature whose surface config ration and internal structure show that the final form was pro duced under conditions quite distinctive from those unde which it was initiated; e.g. a ripple that is transformed into an other pattern (by waves acting in the same direction as th preceding current) when the velocity of the current changed. Metaripples are common in the Waddenzee, a sha low sea in NW Netherlands. Not to be confused with *megarip ple*. Cf: *para-ripple*.

metarossite A light-yellow or pale greenish-yellow minera $CaV_2O_6 \cdot 2H_2O$. It is a dehydration product of rossite.

metaschoderite A monoclinic mineral: $Al_2(PO_4)(VO_4) \cdot 6H_2O$ is a dehydration product of schoderite.

metaschoepite A mineral: $UO_3 \cdot nH_2O$, where n is less than It is a dehydration product of *schoepite*.

metasediment (a) A sediment or sedimentary rock whic shows evidence of having been subjected to metamorphisr

446

) A metamorphic rock of sedimentary origin.

metaseptum One of the main septa of a corallite other than a *protoseptum*, generally distinguished by its extension axially much beyond that of minor septa (TIP, 1965, pt.F, p.249).

metasicula The distal part of the *sicula* of a graptolithine, formed of normal growth increments of fuselar tissue overlain by cortical tissue. Cf: *prosicula*.

metasideronatrite A yellow mineral: $Na_4Fe_2{}^{+3}(SO_4)_4$- $(OH)_2.3H_2O$. It is a partly dehydrated form of sideronatrite.

metasilicate According to the now obsolete classification of silicates as oxyacids of silicon, a salt of the hypothetical metasilicic acid, H_2SiO_3. Cf: *orthosilicate*.

metasom Var. of *metasome*.

metasoma The hind region of the body of some invertebrates, esp. when this cannot be readily analyzed into its primitive segmentation (as in some mollusks and arachnids); specif. the posterior part of a merostome opisthosoma lacking appendages, or the *metasome* of a copepod crustacean. Cf: *mesosoma*.

metasomasis *metasomatism*.

metasomatic Pertaining to the process of metasomatism and to its results. The term is especially used in connection with the origin of ore deposits.

metasomatic rock (a) A rock whose chemical composition has been substantially changed by the metasomatic alteration of its original constituents (Lindgren, 1912). (b) More generally, any rock formed essentially by the process of metasomatism; a *metasomatite*.

metasomatic texture A secondary texture produced by dynamic metamorphism or chemical replacement. The term is vague and now obsolete.

metasomatism The process of practically simultaneous capillary solution and deposition by which a new mineral of partly or wholly different chemical composition may grow in the body of an old mineral or mineral aggregate (Lindgren, 1928). In current usage, the presence of interstitial, chemically active pore liquids or gases contained within the rock body or introduced from external sources are essential for the replacement process which commonly occurs at constant volume with little disturbance of textural or structural features. Obs. syn: *metasomatosis*.

metasomatite A *metasomatic rock*.

metasomatosis An obsolete syn. of *metasomatism*.

metasome [geol] (a) The replacing mineral where one mineral grows in size at the expense of another mineral (or *palasome*); a mineral grain formed by metasomatism. Syn: *guest*. (b) The newly formed part of a migmatite or composite rock, introduced during metasomatism. Cf: *neosome*. ----Also spelled: *metasom*.

metasome [paleont] The posterior part of the prosome of a copepod crustacean, consisting of free thoracic somites in front of the major articulation. Syn: *metasoma*.

metaspondyl In dasycladacean algae, one of the primary branches or rays that occur in a cluster of three or six and that are regularly arranged in whorls.

metastable (a) Said of a phase that is stable with respect to small disturbances but that is capable of reaction with evolution of energy if sufficiently disturbed. (b) Said of a phase that exists in the temperature range in which another phase of lower vapor pressure is stable. A vapor phase need not be present. It is not to be confused with instability. In general, metastability is due to the reluctance of a system to initiate the formation of a new, stable phase.

metastable relict *unstable relict*.

metastasis [met] Changes of a paramorphic character, such as the recrystallization of a limestone or the devitrification of a glassy rock (Bonney, 1886).

metastasis [tect] *metastasy*.

metastasy A term used by Gussow (1958) for regular, lateral

adjustments of the Earth's crust, as opposed to such vertical movements (isostasy). Syn: *metastasis*.

metaster The portion of a migmatite that remained solid (immobile or less mobile) during migmatization. Cf: *restite; stereosome*. See also: *paleosome*.

metastoma A median plate-like process behind the mouth in certain arthropods; e.g. a small ventral plate of the head region behind the position of the mouth of a trilobite, or a plate at the posterior edge of the mouth of a merostome. Also, the lower lip behind the mandibles of a crustacean, usually cleft into paragnaths. Pl: *metastomata*. Cf: *hypostome*. Syn: *metastome; labium*.

metastome *metastoma*.

metastrengite *phosphosiderite*.

metatarsus The proximal (typically the sixth) segment of a leg of an arachnid, following upon the tibia and preceding the tarsus. Pl: *metatarsi*.

metatect The fluid or more mobile part of a migmatite, formed as a result of the remelting of pre-existing rock (anatexis). Cf: *chymogenetic; mobilizate*. See also: *neosome*.

metatectite Lipotexite whose mineralogy and texture have been changed mainly through metasomatism accompanying anatexis. Synonymous with the *metatexite* of some workers (Dietrich & Mehnert, 1961).

metatexis Low-grade *anatexis*, i.e. partial or differential melting of rock components with low melting points (Mehnert, 1968, p.355). Cf: *diatexis; anamigmatization*.

metatexite The rock resulting from metatexis. Synonymous with the *metatectite* of some workers (Dietrich & Mehnert, 1961).

metatheca The distal part of graptoloid theca. It is equivalent to *autotheca* in those graptolites with more than one type of theca.

metathenardite A mineral representing a high-temperature polymorph (perhaps hexagonal) of thenardite and occurring in fumaroles on Martinique Island.

metatorbernite A green secondary mineral: $Cu(UO_2)_2(PO_4)_2.8H_2O$. It contains less water than torbernite.

metatrophic *heterotrophic*.

metatyuyamunite A yellow secondary mineral: $Ca(UO_2)_2(VO_4)_2.35H_2O$.

meta-uranopilite A yellow, grayish, brown, or green mineral: $(UO_2)_6(SO_4)(OH)_{10}.5H_2O$. It is partly dehydrated uranopilite.

meta-uranospinite A yellow secondary mineral: $Ca(UO_2)_2(AsO_4)_2.8H_2O$. It is partly dehydrated uranospinite.

metavandendriesscheite A mineral: $PbU_7O_{22}.nH_2O$, with n less than 12.

metavanuralite A mineral: $Al(UO_2)_2(VO_4)_2(OH).8H_2O$.

metavariscite A green monoclinic mineral: $AlPO_4.2H_2O$. It is dimorphous with variscite and isomorphous with phosphosiderite.

metavauxite A colorless mineral: $Fe^{+2}Al_2(PO_4)_2(OH)_2.8H_2O$. It has more water than vauxite but less water than paravauxite.

metavoltine A mineral: $(K,Na,Fe)_5Fe_3(SO_4)_6(OH)_2.9H_2O$ (?).

metaxite [mineral] A fibrous serpentine mineral; a variety of chrysotile.

metaxite [sed] *micaceous sandstone*.

metaxylem Primary xylem which matures after the *protoxylem*, concomitantly with or after the surrounding tissues (Cronquist, 1961, p.877).

metazellerite A yellow secondary mineral: $Ca(UO_2)(CO_3)_2.3H_2O$.

metazeunerite A green secondary mineral: $Cu(UO_2)_2(AsO_4)_2.8H_2O$. It has less water than zeunerite.

meteor (a) A phenomenon or appearance in the Earth's atmosphere, such as lightning, rainbow, and snowfall; specif. the visible streak of light resulting from the entry into the atmosphere of a solid particle from space. (b) A term commonly applied to any physical object or relatively small fragment of solid material associated with a meteor and made luminous as a result of friction during its passage through the Earth's atmosphere; a *meteoroid*. Syn: *shooting star*.

meteor crater *meteorite crater*.

meteoric [meteorite] (a) Pertaining to, dependent on, derived from, or belonging to the Earth's atmosphere; e.g. "meteoric erosion" caused by rain, wind, or other atmospheric forces. (b) Relating to or composed of meteors or meteoroids.

meteoric [water] Pertaining to water of recent atmospheric origin.

meteoric dust Small particles (diameters ranging from a few microns to 100 microns) representing the product of melting and oxidation of meteors in the Earth's atmosphere. Cf: *meteoritic dust; cosmic dust*.

meteoric iron (a) Iron of meteoric origin. (b) An *iron meteorite*.

meteoric stone (a) A stone of meteoric origin; a *stony meteorite*. (b) A meteorite having the appearance of a stone.

meteorite Any *meteoroid* that has fallen to the Earth's surface in one piece or in fragments without being completely vaporized by intense frictional heating during its passage through the atmosphere; a stony or metallic meteoroid large enough to survive passage through the Earth's atmosphere and reach the ground. Most meteorites are believed to be fragments of asteroids and to consist of primitive solid matter similar to that from which the Earth was originally formed. Adj: *meteoritic*. Syn: *cosmolite; skystone*.

meteorite crater An *impact crater* formed by the falling of a large meteorite onto a surface; e.g. Barringer Crater (Meteor Crater) in Coconino County, Ariz., and Chubb Crater in Quebec, Canada. Cf: *penetration funnel*. Syn: *meteor crater; meteorite impact crater*.

meteorite impact crater *meteorite crater*.

meteoritic dust Small, angular or flat particles representing the product of fragmentation or crushing of meteorites. The particles maintain the composition and structure peculiar to meteorites. Cf: *meteoric dust; cosmic dust*.

meteoritics A science that deals with meteors and meteorites. Cf: *aerolithology*.

meteoroid One of the countless solid objects moving in interplanetary space, distinguished from asteroids and planets by their smaller size but considerably larger than an atom or molecule. Cf: *meteor; meteorite*.

meteorolite An obsolete term for a meteorite, esp. a *stony meteorite*. Syn: *meteorlithe*.

meteorologic tide A change in water level due to such factors as strong winds or barometric pressure. See also: *wind setup*.

meteorology The study of the Earth's atmosphere, including its movements and other phenomena, especially as they relate to weather forecasting.

meteor shower A large concentration of falling meteors; also, the phenomenon observed when members of a meteor swarm encounter the Earth's atmosphere and their luminous paths appear to diverge from a single point in the sky.

meteor swarm A group of meteoroids that have closely similar orbits around the sun.

meter rod A precise leveling rod graduated in whole and fractional meters.

metes and bounds The boundaries or limits of a tract of land; esp. the boundaries of irregular pieces of land (such as claims, grants, and reservations) in which the bearing and length of each successive line is given and in which the lines may be described by reference to local natural or artificial monuments along it (such as a stream, ditch, road, or fence). Such boundaries have been established for much of the land in non-public-land surveys, and are distinguished from those established by beginning at a fixed starting point and runni[ng] therefrom by stated compass courses and distances.

methane A colorless, odorless, inflammable gas which is t[he] simplest paraffin hydrocarbon, formula CH_4. It is the princi[pal] constituent of natural gas and is also found associated w[ith] crude oil. See also: *marsh gas; firedamp*.

methane series The homologous series of saturated alipha[tic] hydrocarbons, empirical formula CnH_2n+2, of which metha[ne] is the lowest and representative member followed by etha[ne,] propane, the butanes, etc. Syn: *paraffin series*.

metharmosis The changes occurring in a sediment after burial (after uplift or consolidation) but before weathering b[e]gins; in this usage, the term is equivalent to *late diagenesis [or] epigenesis*. The term was proposed by Kessler (1922) in [a] less restricted sense to designate all changes that a sedime[nt] may undergo, including diagenesis proper and metamorphis[m.] Syn: *metaharmosis*.

method of least squares *least squares*.

methylene iodide A liquid compound that is used as a *hea[vy] liquid*; its specific gravity is 3.33. Cf: *Clerici solution; Sonsta[d] solution; Klein solution; bromoform*.

methylosis A now obsolete term for change by chemical a[c]tion (Kinahan, 1878).

metroscope An instrument for measuring distances and ina[c]cessible heights and for leveling.

meulerization Local cementation, and replacement (in pa[rt]) by opaline or chalcedonic silica carried by ground water, o[f a] carbonate sandstone or a limestone, such as the reaction o[c]curring in certain sedimentary rocks of the Paris Basin. Ety[m]ol: French *meule*, "millstone". Cf: *Fontainebleau sandstone*.

Mexican onyx Yellowish- or greenish-brown *onyx marb[le]* found chiefly in Tecali, Mex.

meyerhofferite A colorless triclinic mineral: $Ca_2B_6O_{11}.7H_2O$. [It] is an alteration product of inyoite.

meymacite A resinous, light-brown mineral: $WO_3.2H_2O$.

meymechite An ultramafic igneous rock composed of abun[d]ant olivine phenocrysts (usually altered) in a serpentine-ri[ch] groundmass; according to Russian petrologists, the extrusi[ve] equivalent of kimberlite. Also spelled: *meimechite*.

miagite *corsite*.

mianthite Dark-colored enclosures, patches, streaks, etc. [in] an anatexite (Dietrich & Mehnert, 1961).

miargyrite An iron-black to steel-gray monoclinic miner[al:] $AgSbS_2$. It has a cherry-red powder.

miarolithite A chorismite having miarolitic cavities or remnan[ts] thereof; a variety of *ophthalmite*.

miarolitic A term applied to small irregular cavities in igneo[us] rocks, esp. granite, into which small crystals of the roc[k-] forming mineral protrude; characteristic of, pertaining to, [or] occurring in such cavities. Also, said of a rock containi[ng] such cavities, e.g. a miarolite.

miaskite A biotite-bearing mepheline syenite also containi[ng] oligoclase, microperthite, and mica. Its name is derived fro[m] Miask, in the Urals, U.S.S.R. Also spelled: *miascite*.

mica (a) A group of minerals of general formula: $(K,N[a,] Ca)(Mg,Fe,Li,Al)_{2-3}(Al,Si)_4O_{10}(OH,F)_2$. It consists of co[m]plex phyllosilicates (with sheet-like structures) that crystalli[ze] in forms apparently orthorhombic or hexagonal (such as tab[u]lar six-sided prisms) but really monoclinic, that are characte[r]ized by low hardness and by perfect basal cleavag[e] readily splitting into very thin, tough, and somewhat elast[ic] laminae or plates which have a splendent pearly luster o[n] their surfaces (none of the elastic micas contain calcium) and that vary in color from colorless, silvery white, pa[le] brown, or yellow to green or black. Micas are prominent roc[k-] forming constituents of many igneous and metamorph[ic] rocks, and commonly occur as flakes, scales, or shreds. The[y] are used for optical purposes and as electric insulators. (

brittle mica. Syn: *isinglass; glimmer.* (b) Any mineral of the mica group, including muscovite, biotite, lepidolite, phlogopite, zinnwaldite, roscoelite, paragonite, and sericite.

mica book A crystal of mica, usually large and irregular. It is so named because of the resemblance of its cleavage plates to the leaves of a book. Syn: *book.*

micaceous (a) Consisting of, containing, or pertaining to mica; e.g. a "micaceous sediment". (b) Resembling mica; e.g. a "micaceous rock" composed of thin plates or scales, or a "micaceous mineral" capable of being easily split into thin sheets, or a "micaceous luster".

micaceous arkose A term used by Hubert (1960, p.176-177) for a sandstone containing 25-90% feldspars and feldspathic crystalline rock fragments, 10-50% micas and micaceous metamorphic rock fragments, and 0-65% quartz, chert, and metamorphic quartzite. The term is roughly equivalent to *impure arkose* of Folk (1954). Cf: *feldspathic graywacke.*

micaceous iron ore A soft, unctuous variety of hematite having a foliated structure resembling that of mica.

micaceous quartzite A term used by Hubert (1960, p.176-177) for a sandstone containing 70-95% quartz, chert, and metamorphic quartzite, 5-15% micas and micaceous metamorphic rock fragments, and 0-15% feldspars and feldspathic crystalline rock fragments. Cf: *feldspathic quartzite.*

micaceous sandstone A quartz or feldspar sandstone containing conspicuous layers or flakes of mica, usually muscovite. Syn: *metaxite.*

micaceous shale A gray or brownish-gray, usually well-laminated shale containing abundant muscovite flakes along its lamination planes and finer-grained sericite in its clay matrix, commonly associated with subgraywacke (quartz wacke or low-rank graywacke), and representing detrital deposition under moderately unstable conditions in the sedimentary basin.

mica plate In a polarizing microscope, a phase plate consisting of a sheet of muscovite that is used to determine optical sign from interference figures. Its interference color in white light is a light, neutral gray. Syn: *quarter-wave plate.*

micarelle Mica pseudomorphous after scapolite.

mica schist A schist whose essential constituents are mica and quartz, and whose schistose foliation is mainly due to the parallel arrangement of the mica flakes.

micatization Introduction of, or replacement by, a mica.

michenerite An isometric mineral: $(Pd,Pt)BiTe$. Cf: *froodite.*

micrinite A variety of *inertinite* that is granular but shows no plant-cell structure and that is characteristic of durain. It is opaque and of medium hardness, and is divided into massive (10-100 microns) and fine (1-10 microns) according to grain size. See also: *residuum [coal].*

micrinoid A maceral group that includes the macerals in the micrinite series.

micrite (a) A descriptive term used by Folk (1959) for the semiopaque, crystalline, interstitial component (matrix) of limestones, consisting of chemically precipitated carbonate (calcite) mud whose crystals have diameters of less than 4 microns (generally 1-3 microns), and interpreted as a lithified ooze. Leighton & Pendexter (1962) used a diameter limit of 31 microns. Chilingar et al (1967, p.317) and Bissell & Chilingar (1967, p.161) extended the usage of the term to include unconsolidated material that may be of either chemical or mechanical origin (and possibly of biologic, biochemical, or physicochemical origin). It is finer-textured than *sparite.* See also: *matrix [sed].* (b) A limestone with less than 1% allochems and consisting dominantly of micrite matrix (Folk, 1959, p.14); e.g. lithographic limestone. See also: *micritic limestone.*

micritic limestone A limestone consisting of more than 90% micrite (Leighton & Pendexter, 1962, p.60) or less than 10%

allochems (Wolf, 1960, p.1415); a *micrite.* See also:*calculutite; lithographic limestone; lime mudstone; calcimicrite.* Syn: *micrite limestone; matrix limestone.*

micritization Decrease in the size of sedimentary particles, possibly due to boring algae. Syn: *grain diminution.*

micro- A prefix signifying a microscopic characteristic. Cf: *mega-.*

microaerophilic Said of an organism that can exist with very little free oxygen present. Noun: *microaerophile.*

microanalyzer *electron microprobe.*

microaphanitic *microcryptocrystalline.*

microatoll A ring-shaped organic reef consisting usually of a colony or circular growth of corals or serpulids and characterized by a central depression and a breadth of 1-6 m. They are commonly found within the intertidal belt in relatively warm seas or scattered across a reef flat. Syn: *miniature atoll.*

microbiofacies The biologic aspect of a *microfacies* (Fairbridge, 1954).

microbiostratigraphy Biostratigraphy based on microfossils.

microbreccia (a) A poorly sorted sandstone containing relatively large and sharply angular particles of sand set in a very fine silty or clayey matrix; e.g. a graywacke. It is somewhat less micaceous than a siltstone. (b) A breccia within fragments of a coarser breccia (Sander, 1951, p.28).

microchemical tests Chemical tests made on minute grains or polished surfaces under a microscope. They are often combined, in identifying a substance, with observations on form, color, and optical properties.

microclastic Said of coal that is composed mainly of fine particles, e.g. cannel coals. Cf: *macroclastic.*

microclastic rock A clastic rock whose constituents are very minute. Cf: *cryptoclastic rock.*

microcline A clear, white to light-gray, pale-yellow, brick-red, or green mineral of the alkali feldspar group: $KAlSi_3O_8$. It is the fully ordered, triclinic modification of potassium feldspar and is dimorphous with *orthoclase*, being stable at lower temperatures; it usually contains some sodium in minor amounts. Microcline is a common rock-forming mineral of granitic rocks and pegmatites, and is often secondary after orthoclase. It is generally characterized by cross-hatch twinning.

microcline-perthite A perthite consisting of an intergrowth of microcline and plagioclase.

microcline twin law A twin law in feldspar that is a combination of the albite and pericline twin laws and that produces a *cross-hatched twinning.* Syn: *M twin law.*

microclinite An igneous rock composed entirely of microcline.

micrococcolith One of the smaller coccoliths in coccolithophores exhibiting dimorphism but with the dimorphic coccoliths irregularly placed. Cf: *macrococcolith.*

microconglomerate A poorly sorted sandstone containing relatively large and rounded particles of sand set in a very fine silty or clayey matrix.

microcontinent A submarine plateau that is an isolated fragment of continental crust. Cf: *aseismic ridge.*

microcoquina (a) A detrital limestone composed wholly or chiefly of weakly cemented shell detritus of sand size (2 mm in diameter) or less. (b) A variety of chalk (Bissell & Chilingar, 1967, p.153).---Cf: *coquina; mesocoquina.*

microcoquinoid limestone A *coquinoid limestone* composed of small shells. Syn: *microcoquinoid.*

microcosmic salt *stercorite.*

microcrater *micrometeorite crater.*

micro cross-bedding A small but distinctive cross-lamination, similar to a small-scale trough cross-bedding. See also: *rib and furrow.* Also spelled: *micro-cross-bedding.*

microcryptocrystalline Microscopically *cryptocrystalline*; said of the texture of a rock that megascopically does not show crystallinity but microscopically reveals a crystalline nature (although individual crystals are too small to be separately distinguished). Also, said of a rock with such a texture. Cf: *dubiocrystalline; felsiphyric.* Syn: *microaphanitic; microfelsitic.*

microcrystal A crystal, the crystalline nature of which is discernible only under the microscope; such crystals form a *microcrystalline* substance.

microcrystalline Said of the texture of a rock consisting of or having crystals that are small enough to be visible only under the microscope; also, said of a rock with such a texture. In regard to carbonate sedimentary rocks, various diameter ranges are in use: 0.01-0.20 mm (Pettijohn, 1957, p.93); less than 0.01 (Carozzi & Textoris, 1967, p.6); and 0.001-0.01 mm (Bissell & Chilingar, 1967, p.161, who note that some petrographers use 0.004-0.062 mm). Cf: *cryptocrystalline; felsophyric.* See also: *microcrystal.* Syn: *micromeritic.*

microdistributive fracture One of a pattern of numerous, tiny fractures along which slight movement has taken place. Such movement in the aggregate can have a strong effect on the form and structure of a large rock body (Rubey & Hubbert, 1959, p.197).

microearthquake An earthquake having a magnitude of two or less on the Richter scale. Such a limit is arbitrary, and may vary according to the user. Cf: *major earthquake; ultramicroearthquake.*

microelement *trace element.*

microeutaxitic Said of the texture of certain extrusive igneous rocks that are microscopically eutaxitic.

microevolution Evolution that occurs within a continuous population but does not result in the development of genetic discontinuities. The evolutionary changes, brought about by selective accumulation of minute variations, are thought to be chiefly responsible for evolutionary differentiation. Cf: *macroevolution.*

microfabric The fabric of a rock, as seen under the microscope. Cf: *megafabric.*

microfacies A petrologic term suggested by Brown (1943) for the features, composition, or appearance of a rock or mineral as seen in thin section or under the microscope. It was used by Cuvillier (1951) and Fairbridge (1954) as a textural term to refer to lithologic and paleontologic properties, or a "certain look", recognizable in thin sections of sedimentary rocks. Although the term might indicate a "fine-grained facies" or a facies on a small scale, it has not been applied in this sense to a rock body. See also:*microbiofacies; microlithofacies.*

microfauna (a) Living or fossil animals too small to be seen with the naked eye. (b) A very localized or small group of animals; animals occupying a very small habitat. Cf: *microflora; megafauna.*

microfelsitic *microcryptocrystalline.*

microflora (a) Living or fossil animals too small to be seen with the naked eye. The term is sometimes used synonymously with *palynoflora* but is more frequently applied to living microscopic algae and fungi. (b) A very localized or small group of plants; plants occupying a very small habitat. Cf: *microfauna; megaflora.*

microfluidal *microfluxion.*

microfluxion Said of the flow texture of an igneous rock that is visible only with the aid of a microscope. Syn: *microfluidal.*

microforaminifera (a) The chitinous inner tests of certain, almost always spiral foraminifers, frequently found in palynologic preparations of marine sediments, and generally much smaller than "normal" whole foraminifers but displaying recognizable characteristics of "normal" species. (b) Tiny foraminifers much smaller than those generally observed and studied.

microfossil A fossil too small to be studied without the aid of a microscope. It may be the remains of a microscopic organism or a part of a larger organism. Cf: *macrofossil.*

microfragmental Said of a coal composed of macerated vegetal matter. Cf: *macrofragmental.*

microgametophyte The male *gametophyte* or haploid generation that develops from the microspore of a heterosporous embryophytic plant. In lower vascular plants, a few-celled microgametophyte as well as the sperm cells are produced entirely within the microspore; in seed plants, the microgametophyte plus the surrounding microspore wall is the pollen grain, in which the microgametophyte is further reduced, consisting of only three cells in the angiosperms. Cf: *megagametophyte.*

micro-gas survey Soil analysis to determine the presence of hydrocarbon gases that have presumably seeped upwards into the overburden from buried sources.

microgeography The detailed analysis of the natural features of a very limited area.

microgeology (a) Study of the geologic (and geochemical) role of microorganisms (Ehrenberg, 1854). (b) Study of microscopic features of rocks.

microgour A *gour* or rimstone dam or pool on the scale of a few centimeters. Cf: *grand gour.*

micrograined (a) Said of the texture of a carbonate sedimentary rock having clastic particles whose diameters are in the range of 0.001-0.01 mm (Bissell & Chilingar, 1967, p.103) or 0.001-0.004 mm (DeFord, 1946). Some petrographers use the limits of 0.004-0.062 mm. (b) Said of the texture of a carbonate sedimentary rock wherein the particles are mostly 0.01-0.06 mm in diameter, are poorly sorted, and are admixed with clay-sized calcareous mud (Thomas, 1962). Also said of a sedimentary rock with such a texture. Cf: *microgranular.*

microgranitic Said of the texture of a megascopically aphanitic igneous rock that is crystalline-granular (granitic) under the microscope. Also said of an igneous rock with such a texture. Syn: *euritic; microgranular; microgranitoid.*

microgranitoid *microgranitic.*

microgranular [paleont] Said of a foraminiferal wall (as in Endothyracea) composed of minute calcite crystals, probably originally granular but possibly recrystallized. The granules may be aligned in rows perpendicular to the outer wall, resulting in fibrous structure.

microgranular [ign] *microgranitic.*

microgranular [sed] Minutely granular; specif. said of the texture of a carbonate sedimentary rock wherein the particles are mostly 10-60 microns in diameter and are well-sorted, and the finer clay-sized matrix is absent (Thomas, 1962). Also said of a sedimentary rock with such a texture. Cf: *micrograined.*

microgranulitic Said of the granulitic texture of an igneous rock that is distinguishable only with the aid of a microscope; also, said of a rock with such texture.

micrograph A graphic recording, e.g. a *photomicrograph*, of that which is seen through the microscope, e.g. a petrologic thin section.

micrographic Said of the graphic texture of an igneous rock that is distinguishable only with the aid of a microscope; also said of a rock having such texture. Syn: *micropegmatitic.*

microgroove cast A term used by McBride (1962, p.56) for a *striation cast* of a striation less than 2.5 cm in length.

microhill A very rough, miniature sand column raised by the formation of pipkrakes, varying from a few millimeters to several centimeters high and having a height-diameter ratio of 2/5 (Otterman & Bronner, 1966, p.56).

microite A coal microlithotype that contains at least 95% inertinite, with micrinite as the dominant maceral.

microlaterolog A resistivity log obtained with a miniaturized electrode arrangement similar in geometry to that of a *laterolog* but with the electrodes embedded in concentric fashion in an insulating pad that is pressed and held by means of springs against the wall of the borehole. The current from a central electrode is focused and flows out in a pattern reminiscent of the shape of a trumpet. The log investigates only a small volume of rock immediately adjacent to the hole (depth of investigation is about 7.5 cm from the borehole). Cf: *microlog.* Syn: *trumpet log.*

microlite [**mineral**] A pale-yellow, reddish, brown, or black isometric mineral of the pyrochlore group: $(Na,Ca)_2(Ta, Nb)_2O_6(O,OH,F)$. It is isomorphous with pyrochlore, with Ta greater than Nb, and it often contains small amounts of other elements (including uranium and titanium). Microlite occurs in granitic pegmatites and in pegmatites related to alkalic igneous rocks, and it constitutes an ore of tantalum. Syn: *djalmaite.*

microlite [**cryst**] A microscopic crystal which polarizes light and which has some determinable optical properties. Cf: *crystallite; crystalloid.* Syn: *microlith.*

microlith *microlite.*

microlithofacies The lithologic aspect of a *microfacies* (Fairbridge, 1954).

microlithology The study, or characteristics, of rocks as they appear under the microscope. Cf: *macrolithology.*

microlithon The rock material between cleavage planes that is folded or kinked (DeSitter, 1954).

microlithotype A typical association of macerals in humic coals, occurring in bands at least 50 microns wide. Microlithotype names bear the suffix "-ite". Cf: *lithotype.*

microlitic Said of the texture of a porphyritic igneous rock in which the groundmass is composed of an aggregate of differently oriented or parallel microlites in a generally glassy base. Hyalopilitic, pilotaxitic, and trachytic are microlitic textures.

microlog A *resistivity log* obtained in borehole surveying with three closely spaced electrodes (one inch apart) mounted on an insulating rubber pad that is pressed and held by means of springs against the wall of the hole in order to reduce the short-circuiting action of the drilling mud. The electrodes measure the average resistivity of a small volume of highly resistant rock in front of the pad. The log distinguishes porous and permeable strata from impervious layers by detecting the mud cake deposited on the borehole wall by invading drilling mud. Cf: *microlaterolog.* Syn: *contact log.*

micromelteigite A fine-grained hypabyssal *melteigite.*

micromeritic An obsolete syn. of *microcrystalline.*

micromeritics The study of the characteristics and behavior of small particles. It is applicable to soil physics.

micrometeorite A very small meteorite or meteoritic particle with a diameter generally less than a millimeter; a meteorite so small that it undergoes atmospheric entry without vaporizing or becoming intensely heated and hence without disintegration.

micrometeorite crater A small crater produced by hypervelocity impact of primary micrometeorite particles on exposed surfaces of lunar rocks on the lunar surface. The craters are typically small (less than a few millimeters in diameter) and are characterized by a central glass-lined pit, a concentric lightened area of shock-fractured minerals, and a roughly circular spall area approximately 4.5 times larger in diameter than the central pit. The informal term *zap crater* is equivalent (Hörz et al, 1971, p.5785). Syn: *microcrater; lunar microcrater.*

micromineral Any crystalline matter of the clay fraction in sediments and soils, including the iron and aluminum oxides, allophane, fine-grained carbonates, etc, in addition to the phyllosilicates, to which the term *clay mineral* is usually restricted (Yaalon, 1965).

micronucleus A small nucleus that is concerned with reproductive functions in the body of a tintinnid. Cf: *macronucleus.*

micro-oil A term used by Vernadskiy for hydrocarbons occurring in a diffused state in sedimentary rock; the nascent oil, still within and sorbed to its source rock (Vassoevich, N.B., 1965, p.510).

micro-ophitic Said of the ophitic texture, of an igneous rock, that is distinguishable only with the aid of a microscope. Also, said of a rock with such texture.

micropaleontology A branch of paleontology that deals with the study of fossils too small to be observed without the aid of a microscope; the study of microfossils.

micropedology The study of the microscopic phenomena of soils.

micropegmatitic A syn. of *micrographic*; not recommended because it implies a restriction to the combination of quartz and feldspar crystals.

micropellet A pellet or pellet-like sedimentary particle of a fine to very fine grade size, "possibly smaller than 0.01 mm in diameter" (Bissell & Chilingar, 1967, p.161). Adj: *micropelletoid.*

microperthite A variety of perthite in which the lamellae (5-100 microns wide) are visible only with the aid of the microscope. Cf: *cryptoperthite.*

microperthitite An igneous rock composed entirely of microperthite.

micropetrological unit *maceral.*

microphagous Said of an organism that feeds on relatively minute particles. Cf: *macrophagous.*

microphotograph A less-preferred syn. of *photomicrograph.*

microphyllous Said of a plant having small leaves with one unbranched midvein (Scagel, et al, 1965, p.623).

microphyric Said of the texture of a porphyritic igneous rock in which the phenocrysts are of microscopic size, i.e. their longest dimension not exceeding 0.2 mm (Cross et al, 1906, p.702); also, said of a rock having such texture. Cf: *macrophyric.* Syn: *microporphyritic.*

micropiracy The overtopping and breaking down of the narrow ridges between adjacent rill channels, and diversion of flow from the higher, shorter, and shallower rill channels to the lower, longer, and deeper ones more closely adjacent to the initial rill (Horton, 1945, p.335). See also: *cross-grading.*

microplankton Plankton of the size range 60 microns--one millimeter, e.g. most phytoplankton. They are larger than *ultraplankton* and *nannoplankton*, but smaller than *macroplankton* and *megaloplankton.* Syn: *net plankton.*

micropoikilitic Said of the poikilitic texture of an igneous rock that can be distinguished only with the aid of a microscope. Also, said of a rock having such texture.

micropore (a) A pore small enough to hold water against the pull of gravity and to inhibit the flow of water. (b) In the pore-size classification of Choquette & Pray (1970, p.233), an equant to equant-elongate pore or a tubular or platy pore with an average diameter or thickness of less than 1/16 mm.---- Cf: *mesopore; megapore.*

microporphyritic *microphyric.*

micropulsation A short-period geomagnetic variation in the range of about 0.2-600 seconds, typically exhibiting an oscillatory waveform.

micropyle The minute opening in the integument of an ovule or seed, through which a pollen tube grows to reach the female gametophyte (Fuller & Tippo, 1949, p.964).

microquartz Nonclastic anhydrous crystalline silica occurring in sediments and having particle diameters usually less than 20 microns.

microradiograph A picture produced by X-rays or rays from a radioactive source showing the minute internal structure of a substance.

microrelief (a) Local and slight irregularities of a land surface, including such features as low mounds, swales, and shallow pits, generally about a meter in diameter, and causing variations amounting to no more than 3 m. (b) Relief features that are too small to show on a topographic map; e.g. small gullies, mounds, boulders, pinnacles, or other features less than 60 m in diameter and less than 6 m in elevation, in an area whose topographic map has a scale of 1:50,000 or smaller and a contour interval of 3 m (10 ft) or larger.---Cf: *macrorelief.*

microrhabd A rod-shaped monaxonic sponge spicule (microsclere).

microsclere A small *sclere;* specif. one of the minute secondary spicules scattered throughout a sponge or concentrated in the cortex or elsewhere. It usually differs in form from a *megasclere.*

microscope An optical instrument that is used to produce an enlarged image of a small object; it consists of the lens (or lenses) of the objective and of the eyepiece set into a tube, and held by an adjustable arm over a stage on which the object is placed. Types of microscopes vary according to intended use and to types of illumination used, e.g. natural light, polarized light, transmitted light, reflected light, electrons, or x-rays.

microscopic (a) Said of an object or phenomenon or of its characteristics that cannot be observed without the aid of a microscope. (b) Of or pertaining to a microscope.

microsection (a) Any thin section used in microscopic analysis. (b) A *polished section.* (c) A *polished thin section.*

microseism A more or less continuous motion in the Earth that is unrelated to an earthquake and that has a period of 1.0-9.0 sec. It is caused by a variety of natural and artificial agents, esp. atmospheric agents, and is a weak, oscillatory disturbance for which sensitive detectors are required. Cf: *macroseism.* See also: *microseismology.* Syn: *seismic noise.*

microseismic data Earthquake measurement or observation by instrumental means, as opposed to *macroseismic observations.* The term is not to be confused with the connotation of the term microseism.

microseismograph A seismic detection system that is specifically designed for the detection and recording of microseismic motion. Cf: *microseismometer.*

microseismology The study of *microseisms.*

microseismometer A seismometer designed for the detection of microseisms. Cf: *microseismograph.*

microsere A sere of a very small habitat, usually failing to attain climax and ending with the loss of identity of the habitat. Syn: *serule.*

microsolifluction The frost movements that produce patterned ground (Troll, 1944).

microspar Calcite matrix in limestones, occurring as uniformly sized and generally loafish-shaped crystals ranging from 5 to more than 20 microns in diameter. It develops by recrystallization or neomorphism of carbonate mud (micrite). Cf: *microsparite.*

microsparite (a) A term used by Folk (1959, p.32) for a limestone whose carbonate-mud matrix has recrystallized to microspar. (b) A term used by Chilingar et al (1967, p.320) for a sparry crystal of calcite whose diameter ranges from 5 to 20 microns. Cf: *microspar.*

microspecies A small usually localized group of organisms that are only slightly but positively different from other related organisms. Cf: *macrospecies.*

microspheric Said of a foraminiferal test or shell produced sexually and characterized by a very small initial chamber (proloculus), many chambers, often large size of the adult test, and more complete ontogeny. Cf: *megalospheric.*

microspherulitic Said of the spherulitic texture of an igneous rock that is distinguishable only with the aid of a microscope. Also, said of a rock having such texture.

microsporangium A *sporangium* that develops or bears microspores; e.g. the anther in an angiosperm or the pollen sac in all other seed plants. Cf: *megasporangium.*

microspore One of the spores of a heterosporous embryophytic plant that germinates to produce a microgametophyte (male gametophyte) and that is ordinarily smaller than the *megaspore* of the same species. In seed plants, pollen grains consist of a microspore wall or exine with a microgametophyte contained inside. See also: *small spore.*

microstriation A microscopic scratch developed on the polished surface of a rock or mineral as a result of abrasion.

microstylolite A *stylolite* in which the relief along the surface is less than a millimeter, such as one indicating differential solution between two mineral grains.

microtectonics A syn. of *structural petrology.*

microtektite A small (less than one millimeter in diameter), usually spherical, glassy object found in some deep-sea sediments and possibly related to tektites in outward form and composition.

microtherm A plant that requires low temperatures for successful growth. Cf: *mesotherm; megatherm.*

microthermal Pertaining to a climate characterized by low temperature. Cf: *mesothermal; megathermal.*

microtinite A light-colored, coarse-grained igneous rock characterized by monzonitic texture and by the presence of vitreous plagioclase.

microtopography Topography on a small scale. The term has been applied to features having relief as small as 1-10 cm as well as to those involving amplitudes of 50-100 meters and wavelengths of a few kilometers.

microvermicular Said of the texture of a rock having worm-like intergrowths of crystals that are visible in thin section under the polarizing microscope.

microvitrain Vitrain bands occurring in clarain, .05-2.0 mm thick.

microwave That region of the electromagnetic spectrum in the millimeter and centimeter wavelengths which is bounded on the short wavelength sides by the far infrared and by the atmosphere absorption bands near 0.5 mm. It is equivalent to the 1,000-300,000 megacycle frequency region. Passive sensing systems operating at these wavelengths are called microwave systems; active systems are called radar.

mictite Coarsely composite rock formed as the result of contamination, i.e. incorporation and partial or complete assimilation of country rock fragments, of a magma under conditions of relatively low temperature and probably at relatively high levels in the crust (Dietrich & Mehnert, 1961).

midalkalite A rarely used syn. of *nepheline syenite.*

mid-bay bar A bar built across a bay at some point between its mouth and its head.

midden [sed] A mound-like accumulation of calcareous sediment trapped or bound together by algal growth.

midden [soil] A mass of highly organic soil formed by an earthworm around its burrow; also, any organic debris on soil, deposited by an animal.

middle [geochron] Pertaining to a segment of time intermediate between *late* and *early.* The adjective is applied to the name of a geologic-time unit (era, period, epoch) to indicate relative time designation and corresponds to *middle* as applied to the name of the equivalent time-stratigraphic unit; e.g. rocks of a Middle Jurassic batholith were intruded in Middle Jurassic time. The initial letter of the term is capitalized to indicate a formal subdivision (e.g. "Middle Devonian") and is lowercased to indicate an informal subdivision (e.g. "middle Miocene"). The informal term may be used for eras and epochs, and for periods where there is no formal subdivision. Syn: *medial.*

middle [stratig] Pertaining to rocks or strata that are intermediate between *upper* and *lower.* The adjective is applied to the name of a time-stratigraphic unit (system, series, stage) to indicate position in the geologic column and corresponds to *middle* as applied to the name of the equivalent geologic-time unit; e.g. rocks of the Middle Jurassic System were formed during the Middle Jurassic Period. The initial letter of the term is capitalized to indicate a formal subdivision (e.g. "Middle

Devonian") and is lowercased to indicate an informal subdivision (e.g. "middle Miocene"). The informal term may be used where there is no formal subdivision of a system or of a series.

middle diagenesis *anadiagenesis.*

middle ground A bar deposit or shoal formed in the middle of a channel or fairway at the entrance and exit of a constricted passage (as a strait) by the rise and fall of the tide, and characterized by a flow of water on either side of the deposit.

middle lamella In plants, the intercellular substance between the primary walls of two contiguous cells, composed chiefly of calcium pectate (Esau, 1965, p.34). See also: *intercellular space.*

middle lateral muscle One of a pair of *protractor muscles* in some lingulid brachiopods, originating on the pedicle valve between the central muscles, and diverging slightly posteriorly before insertion on the brachial valve (TIP, 1965, pt.H, p.148). Cf: *outside lateral muscle.*

middle latitude n. The latitude of the point situated midway on a north-south line between two parallels; half the arithmetic sum of the latitudes of two places on the same side of the equator.

middle-latitude desert A vast desert area occurring within lat. 30°-50° north or south of the equator in the deep interior of a large continental mass, usually situated with high mountains across the path of prevailing winds (thus, a *rain-shadow desert*), and commonly characterized by a cold, dry climate.

middle Paleolithic n. The second division of the *Paleolithic*, characterized by Neanderthal man in Eurasia. Correlation of cultural levels with actual age (and, therefore, with the time-stratigraphic units of geology) varies from region to region. Cf: *lower Paleolithic; upper Paleolithic.*----adj. Pertaining to the middle Paleolithic.

middle Stone Age *Mesolithic.*

midfan The area between the fanhead and the outer, lower margins of an alluvial fan.

midfan mesa A much eroded, island-like remnant of an old upfaulted alluvial fan, commonly a primary product of piedmont faulting (Eckis, 1928, p.243-246).

mid-ocean canyon *deep-sea channel.*

mid-oceanic ridge A continuous, seismic, median mountain range extending through the North and South Atlantic Oceans, the Indian Ocean, and the South Pacific Ocean. It is a broad, fractured swell with a central rift valley and usually extremely rugged topography, 1-3 km in elevation, about 1500 km in width, and over 84,000km in length. According to the hypothesis of sea-floor spreading, the mid-oceanic ridge is the source of new crustal material. See also: *rift valley [marine geol]; sea-floor spreading.* Syn: *mid-ocean rise; oceanic ridge.*

mid-ocean rift *rift valley.*

mid-ocean rise *mid-oceanic ridge.*

midrange The arithmetic mean of the smallest and largest values in a sample. Syn: *range midpoint.*

midrib The central rib of leaf venation. It is a continuation of the petiole.

midstream (a) The part of a stream well removed from both sides or from the source and the mouth. (b) A line along a stream course, midway between the sides of the stream.

midwater trawl A towed, netlike device that is used to gather marine organisms anywhere between the bottom and the water surface.

midway The middle, deepest, or best navigable channel used in defining water boundaries between states, as in a sound, bay, strait, estuary, or other arm of the sea, and in a lake or landlocked sea. Syn: *fairway; thalweg.*

Midwayan North American (Gulf Coast) stage: Paleocene (above Navarroan of Cretaceous, below Sabinian)

miemite A yellowish-brown, fibrous dolomite mineral occurring at Miemo in Tuscany, Italy.

miersite A canary-yellow isometric mineral: $(Ag,Cu)I$.

Mie scattering Scattering produced by the interaction of electromagnetic radiation and spherical particles usually expressed in terms of a dimensionless parameter involving the ratio of particle diameter to *wavelength.*

Mie theory A complete mathematical-physical theory of the scattering of electromagnetic radiation by spherical particles, developed by G. Mie in 1908. In contrast to *Rayleigh scattering*, the Mie theory embraces all possible ratios of diameter to wavelength. The Mie theory is very important in meteorological optics, where diameter-to-wavelength ratios of the order of unity and larger are characteristic of many problems regarding haze and cloud scattering.

migma Mobile, or potentially mobile, mixture of solid rock material(s) and rock melt (the rock melt could have been injected into or melted out of the rock material) (Dietrich & Mehnert, 1961). Etymol: Greek, "mixture".

migmatite A composite rock composed of igneous or igneous-looking and/or metamorphic materials which are generally distinguishable megascopically (Dietrich, 1960, p.50). Its formation (*migmatization*) may involve solid-state reconstruction in the presence of fluids (Dietrich, 1969, p.557). Injection of magma, or in situ melting, or both, may take place. The term was introduced by Sederholm in 1907 (p.88-89). Cf: *chorismite.*

migmatitization *migmatization.*

migmatization Formation of a *migmatite*. It may involve either injection or in situ melting. Also spelled: *migmatitization.*

migrated map A *seismic map* whose z coordinate is depth. Syn: *connected map.*

migrating dip A dipping event in a reflection seismogram which is mapped to its true position in space. Such a process is called *migration of dips.* Syn: *offsetting dip; swinging dip.*

migrating dune *wandering dune.*

migrating inlet A tidal inlet, such as that connecting a coastal lagoon with the open sea, that shifts its position laterally in the direction in which the dominant longshore current flows. It results from deposition on one side of the inlet, accompanied by erosion on the other.

migration [geomorph] The movement of a topographic feature from one locality to another by the operation of natural forces; specif. the movement of a dune by the continual transfer of sand from its windward to its leeward side.

migration [ecol] A broad term applied to the movements of plants and animals from one place to another over long periods of time.

migration [oil] The movement of liquid and gaseous hydrocarbons from their source or generating beds, through permeable formations into reservoir rocks.

migration [streams] (a) *shifting.* (b) The slow, downstream movement of a system of meanders, accompanied by enlargement of the curves and widening of the meander belt.

migration of dips The process by which dipping events in a reflection seismogram are mapped to their true spatial positions. It requires a priori knowledge of the true velocity. Such an event is called a *migrating dip.* Syn: *dip calculation.*

miharaite An alboranite in which the groundmass is olivine-free but contains free silica; a quartz-bearing hypersthene basalt.

mijakite A manganese-rich basalt composed of augite and bytownite, and sometimes biotite, hypersthene, and apatite phenocrysts in a groundmass having intersertal texture and composed of lath-shaped feldspar, magnetite grains, and a nearly opaque red-brown mineral identified as pyroxene.

Milankovitch theory An astronomical theory of glaciation, proposed by Milutin Milankovitch (b. 1879), Yugoslav mathematician, in which climatic changes result from solar-radiation fluctuations determined by variations of the Earth's orbital elements (eccentricity and longitude of perihelion) and axial inclination (Milankovitch, 1920). Recent radiometrically dated reconstructions of ocean temperature and glacial sequences suggest parallelisms with the theoretical radiation curves and have stimulated serious reconsideration of the theory.

milarite A colorless to greenish, glassy, hexagonal mineral: $K_2Ca_4Be_4Al_2Si_{24}O_{60}.H_2O$.

mile Any of various units of distance that were derived from the ancient Roman marching unit of 1000 double paces (a double pace = 5 ft) and that suffered many changes as the term came into use among the western nations (e.g. a mile = 1620 English yards or 1482 meters); specif. *statute mile* and *nautical mile*. Etymol: Latin *mille*, "thousand".

milieu A French term used in paleontology, sedimentation, and stratigraphy for environment, surroundings, or setting; e.g. the environment characteristic of a stratigraphic facies.

miliolid A foraminifer belonging to the family Miliolidae, characterized by a test that usually has a porcelaneous and imperforate wall and has two chambers to a whorl variably arranged about a longitudinal axis.

milioline Pertaining or belonging to or resembling the foraminiferal genus *Miliola* or suborder Miliolina; e.g. formed as in the foraminiferal test of the superfamily Miliolacea, commonly with narrow elongate chambers (two to a whorl) added in differing planes of coiling (TIP, 1964, pt.C, p.61).

miliolite A fine-grained limestone of eolian origin, consisting chiefly of the tests of *Miliola* and other foraminifers.

military geology The application of the earth sciences, especially soil science and climatology, to such military concerns as terrain analysis, water supply, cross-country movements, foundations, location of construction materials, and construction of roads and airfields.

milk opal A translucent and milk-white to greenish, yellowish, or bluish variety of common opal.

milky quartz A milk-white, nearly opaque variety of crystalline quartz often having a greasy luster. The milkiness is usually due to the presence of innumerable very small cavities containing air. Syn: *greasy quartz*.

milled ring A flange near the base of an echinoid spine for the attachment of muscles controlling the movement of the spine.

millepore Any one of a group of hydrozoans belonging to the order Milleporina, characterized by a calcareous skeleton and free-swimming sexual individuals. Cf: *hydroid; stylaster*.

Miller-Bravais indices A four-index type of *Miller indices*, useful but not necessary in order to define structures of crystals in the hexagonal system; the symbols are hkil, in which i = h + k. Syn: *hexagonal indices*.

Miller indices A set of three or four symbols (letters or integers) used to define the position and orientation of a crystal face or internal crystal plane. The indices are determined by expressing, in terms of lattice constants, the reciprocals of the intercepts of the face or plane on the 3 crystallographic axes, and reducing (clearing fractions) if necessary to the lowest integers retaining the same ratio. When the exact intercepts are unknown, the general symbol (*hkl*) is used for the indices, where *h*, *k*, and *l* are respectively the reciprocals of rational but undefined intercepts along the *a*, *b*, and *c* crystallographic axes. In the hexagonal system, the Miller indices are (*hkil*); these are known as the *Miller-Bravais indices*. Indices designating individual crystal faces are enclosed in parentheses; complete crystal forms, in braces; crystal zones, in square brackets; and crystallographic lines, in greater than/less than symbols. To denote the interception at the negative end of an axis, a line is placed over the appropriate index, as (*111*). The indices were proposed by William H. Miller (1801-1880), English mineralogist. See also: *indices of lattice row*. Syn: *crystal indices; hkl indices*.

millerite A brass-yellow to bronze-yellow rhombohedral mineral: NiS. It usually has traces of cobalt, copper, and iron, and is often tarnished. Millerite generally occurs in fine hair-like or capillary crystals of extreme delicacy, chiefly as nodules in clay ironstone. Syn: *capillary pyrites; nickel pyrites; hair pyrites*.

millet-seed sand Sand that consists essentially of smoothly and conspicuously rounded, spheroidal grains about the size of a millet (cereal or forage grass) seed; specif. a desert sand whose grains have a surface like that of ground glass and are very perfectly rounded as a result of wind action that caused them to be constantly impacting against each other.

mill-hole mining *glory-hole mining*.

millidarcy The customary unit of fluid permeability, equivalent to 0.001 darcy. Abbrev: md.

milling ore *second-class ore*.

millionth-scale map of the world *international map of the world*.

millisite A white mineral: $(Na,K)CaAl_6(PO_4)_4(OH)_9.3H_2O$.

mill ore Var. of *milling ore*.

millstone A *buhrstone*; e.g. a coarse-grained sandstone or a fine-grained quartz conglomerate. Also, one of two thick disks of such material used for grinding grain, cement rocks, and other materials fed through a center hole in the upper stone.

millstone grit Any hard, siliceous rock suitable for use as a material for millstones; specif. the Millstone Grit of the British Carboniferous, a coarse conglomeratic sandstone.

Mima mound A term used in the NW U.S. for one of numerous low, circular or oval domes composed of loose, unstratified, gravelly silt and soil material, built upon glacial outwash on a *hog-wallow* landscape; the basal diameter varies from 3 m to more than 30 m, and the height from 30 cm to about 2 m. The mounds are probably built by pocket gophers (Arkley & Brown, 1954). Named after the Mima Prairie in western Washington State. Pron: *my*-ma. Cf: *pimple mound*. Also spelled: *mima mound*.

mimetene *mimetite*.

mimetesite *mimetite*.

mimetic [**cryst**] Pertaining to a twinned or malformed crystal that appears to have a higher grade of symmetry than it actually does.

mimetic [**evol**] Said of an organism that exhibits or is characterized by *mimicry*.

mimetic [**struc petrol**] Said of a tectonite whose deformation fabric, formed by recrystallization or neomineralization, reflects and is influenced by pre-existing anisotropic structure; also, said of the fabric itself.

mimetic crystallization Recrystallization and/or neomineralization in metamorphism which reproduces any pre-existent anisotropy, bedding, schistosity, or other structures (Knopf and Ingerson, 1938). Syn: *facsimile crystallization*.

mimetite A yellow to yellowish-brown mineral of the apatite group: $Pb_5(AsO_4)_3Cl$. It is isomorphous with pyromorphite, and commonly contains some calcium or phosphate. Mimetite usually occurs in the oxidized zone of lead veins, and is a minor ore of lead. Syn: *mimetene; mimetesite*.

mimicry The superficial similarity that exists between organisms or between an organism and its surroundings as a means of concealment, protection, or other advantage. See also: *mimetic*.

mimosite A dark-colored dolerite containing abundant augite and ilmenite. Cf: *soggendalite*.

minable Said of a mineral deposit for which extraction is economically worthwhile and structurally to technically feasible.

minal *end member*.

minasragrite A blue efflorescent mineral: $(VO)_2H_2(SO_4)_3.15H_2O$.

Mindel (a) European stage: Pleistocene (above Günz, below Riss). (b) The second glacial stage of the Pleistocene Epoch in the Alps, after the Günz-Mindel interglacial stage. See also: *Kansan; Elster*.---Etymol: Mindel River, Bavaria. Adj: *Mindelian*.

Mindel-Riss The term applied in the Alps to the second inter-glacial stage of the Pleistocene Epoch, after the Mindel glacial stage and before the Riss. See also: *Yarmouth*.

mine n. (a) Any opening or excavation for the extraction of mineral deposits or building stone. The term implies men working in a mine, therefore a well is excluded from the sense of the word. In general, a mine is assumed to be subterranean; the term is usually modified to show that it is otherwise, e.g. opencut mine. The term is also modified according to its valuable commodity, e.g. coal mine. (b) A subterranean excavation for the extraction of mineral deposits, in contrast to surficial excavations such as quarries, placer-deposit workings, and various types of opencut mines. (b) Any opening or excavation for the development of a mineral deposit prior to extraction, as well as for extraction. (d) The area or property of a mineral deposit that is being excavated; a mining claim. v. To excavate for and extract mineral deposits or building stone.

mineragraphy An obsolescent syn. of *ore microscopy*.

mineral (a) A naturally formed chemical element or compound having a definite chemical composition and, usually, a characteristic crystal form. A mineral is generally considered to be inorganic, though organic compounds are classified by some as minerals. Those who include the requirement of crystalline form in the definition of a mineral would consider an amorphous compound such as opal to be a *mineraloid*. (b) Any naturally formed, inorganic material, i.e. a member of the mineral kingdom as opposed to the plant and animal kingdoms. A naturally occurring, usually inorganic, crystalline substance with characteristic physical and chemical properties that are due to its atomic arrangement. See also: *mineraloid*.

mineral aggregate (a) An *aggregate* or assemblage of more than one crystal grain (which may be of several mineral species or all of one species) and containing more than one crystal lattice. It can occur as sediment if loosely bound, or as rock if tightly bound. (b) *aggregate* [mater].

mineral assemblage (a) The minerals that compose a rock, esp. an igneous or metamorphic rock. The term includes the different kinds and relative abundances of minerals, but excludes the texture and fabric of the rock. See also: *metamorphic assemblage*. (b) *mineral association*.

mineral association A group of minerals found together in a rock (esp. in a sedimentary rock). Syn: *mineral assemblage*.

mineral belt An elongated region of mineralization; an area containing several mineral deposits.

mineral blossom Drusy quartz.

mineral caoutchouc *elaterite*.

mineral charcoal *fusain*.

mineral deposit A mass of naturally occurring mineral material, e.g. metal ores or nonmetallic minerals, usually of economic value, without regard to mode of origin. The organic fuels (coal and petroleum) may or may not be considered as mineral deposits; usage varies and should be defined in context.

mineral disintegration Separation of a rock into its component minerals by natural forces; *granular disintegration*.

mineral facies Rocks of any origin whose constituents have been formed within the limits of a certain pressure-temperature range characterized by the stability of certain index minerals. It is a broader concept than that of *metamorphic facies*.

mineralization [ore dep] The process or processes by which a mineral or minerals are introduced into a rock, resulting in an economically valuable or potentially valuable deposit. This is a general term, incorporating various types (e.g. metallization) and modes (e.g. fissure filling, impregnation, replacement) of mineralization.

mineralization [paleont] A process of fossilization whereby the organic components of an organism are replaced by inorganic material.

mineralize To convert to a mineral substance; to impregnate with mineral material. The term is applied to the processes of ore vein deposition as well as to the process of fossilization (Challinor, 1967, p.161).

mineralizer (a) Magmatic gases and fluids, as in hydrothermal solutions, that aid the crystallization of ore minerals; a mineralizing agent that combines with metal to form an ore. Syn: *ore-forming fluid*. (b) A gas that is dissolved in a magma and that aids in the concentration and crystallization of certain minerals and in the development of certain textures as it is released from the magma by decreasing temperature and/or pressure.

mineral lands Legally, areas considered economically valuable, more for their ore deposits than for agriculture or other purpose. Cf: *stone land*.

mineral matter The inorganic material in coal. See also: *inherent mineral matter*.

mineral occurrence Any mineral in any concentration found in bedrock or as float; esp. a mineral in sufficient concentration as to suggest further exploration. Cf: *mineral deposit*.

mineralogic Adj. of *mineralogy*.

mineralogical Adj. of *mineralogy*.

mineralogical phase rule Any of several modifications of the fundamental Gibbs phase rule, taking into account the number of degrees of freedom consumed by the fixing of physical-chemical variables in the natural environment. The most famous such rule, due to Goldschmidt, assumes that two variables (taken as pressure and temperature) are fixed externally and that consequently the number of phases (minerals) in a system (rock) will not generally exceed the number of components. The Korzhinskiy-Thompson version of the mineralogical phase rule takes into account the external imposition of chemical potentials of perfectly mobile components, and thereby reduces the maximum expectable number of minerals in a generally located rock to the number of inert components. Syn: *Goldschmidt's mineralogical phase rule*.

mineralogic maturity A type of sedimentary *maturity* in which a clastic sediment approaches the mineralogic end product to which it is driven by the formative processes that operate upon it. The ultimate sand is a concentration of pure quartz, and the mineralogic maturity of sandstones is commonly expressed by the quartz/feldspar ratio; this ratio is not so appropriate for sand derived from feldspar-poor rocks and the ratio of quartz + chert/feldspar + rock fragments may be substituted as more generally applicable. Cf: *compositional maturity; textural maturity*.

mineralogist One who studies the formation, occurrence, properties, composition, and classification of minerals; a geologist whose field of study is *mineralogy*.

mineralography A syn. of *mineragraphy*; both are obsolescent terms for *ore microscopy*.

mineralogy (a) The study of minerals: their formation and occurrence, their properties and composition, and their classification. See also: *mineralogist*. Adj: *mineralogic; mineralogical*. Obs. syn: *oryctology; oryctognosy*. (b) An obsolete use of the term is for the general geology of a region.

mineraloid A naturally occurring, usually inorganic substance that is not considered to be a *mineral* because it is amorphous and thus lacks characteristic physical and chemical properties; e.g. opal. Syn: *gel mineral*.

mineral paint An inorganic material used as a paint pigment, e.g. ocher, iron oxide, barite. See also: *earth color; mineral pigment*.

mineral pathology Study of the changes undergone by unstable minerals in an environment whose conditions of temperature, pressure, and composition are different from those under which the minerals originally formed (Pettijohn, 1957, p.502).

mineral pigment An inorganic pigment, either natural or synthetic, used to give color, opacity, or body to a paint, stucco, plaster, or similar material. See also: *earth color; mineral paint.*

mineral pitch An obsolete syn. of *asphalt.*

mineral reserves Known mineral deposits that are recoverable under present conditions but that are as yet undeveloped. The term excludes *potential ore.* Cf: *mineral resources.*

mineral resin Any of a group of resinous (usually fossilized) mineral hydrocarbon deposits; e.g. bitumen and asphalt. See also: *resin.*

mineral resources The valuable minerals of an area that are presently recoverable or may be so in the future; both the known orebodies (*mineral reserves*) and *potential ore* of a region.

mineral rod *divining rod.*

mineral sands A *beach placer,* e.g. of zircon, ilmenite, rutile, etc.

mineral sequence *paragenesis.*

mineral soap *bentonite.*

mineral soil A soil that is composed mainly of mineral matter but having some organic material also.

mineral spring A spring whose water contains enough mineral matter to give it a definite taste, in comparison to ordinary drinking water, esp. if the taste is unpleasant or if the water is regarded as having therapeutic value. This type of spring is often described in terms of its principal characteristic constituent; e.g. *salt spring.*

mineral streaking In metamorphic rocks, lineation of grains of a mineral. Cf: *stretching.* Syn: *streaking.*

mineral survey The marking of legal boundaries of ore deposits or mineral-bearing formations on public land, when such boundaries are not the normal subdivisions of public land.

mineral tallow *hatchettine.*

mineral tar *maltha.*

mineral time *Geologic time* estimated on the basis of radioactive minerals (Kobayashi, 1944a, p.476). Cf: *absolute time.*

mineral water Water which contains naturally or artificially supplied mineral salts or gases (e.g. carbon dioxide).

mineral wax *ozocerite.*

mineral zone An informal term for a stratigraphic unit classified on the basis of mineral content (usually detrital minerals) and usually named from characteristic minerals (ISST, 1961, p.29).

mineral zoning *zoning of ore deposits.*

minerocoenology The study of mineral associations in the broadest sense, such as the correlation of igneous rocks or magmatic provinces with their ore deposits (Thrush, 1968, p.712).

minerogenetic *metallogenic.*

minerogenic *metallogenic.*

miner's inch A measure of water flow equal to 1.5 cu ft/min.

minette A lamprophyre primarily composed of biotite phenocrysts in a groundmass of orthoclase and biotite.

minguzzite A green monoclinic mineral (oxalate): $K_3Fe(C_2O_4)_3.3H_2O$.

miniature atoll *microatoll.*

miniature lagoon (a) *lagoonlet.* (b) *pseudolagoon.*

minimum [glac geol] *glacial minimum.*

minimum [geophys] n. An anomaly in a given area that is relatively lesser than those in neighboring areas; e.g. a gravity minimum or a geothermal minimum. Cf: *maximum [geophys].* Syn: *low [geophys].*

minimum detectable power In infrared detector technology, as applied to the Earth sciences, the incident power which will give a signal-to-noise ratio equal to unity at the output of the detector (Smith et al, 1968, p.250). Syn: *noise equivalent power.*

minimum pendulum A pendulum used in gravity measurements so designed that changes in period resulting from small length changes are at a minimum. Among factors which may tend to change the length are temperature, creep, or knife-edge wear.

minimum slope A slope that is flatter in gradient than the slope units above or below it.

minimum time path A *Fermat path;* the path between two points along which the time of travel of a ray is a true minimum. See also: *Fermat's principle.* Syn: *least-time path; brachistochronic path.*

mining The process of extracting mineral deposits or building stone from the Earth. The term may also include preliminary treatment of the ore or building stone, e.g. cleaning, sizing, dressing. Cf: *mining geology; mining engineering.*

mining claim A *claim* on mineral lands.

mining engineering The structural evaluation of mineral deposits; the planning and design of mines; the supervision of extraction and preliminary refinement of the raw material. Cf: *mining; mining geology.*

mining geology The study of mineral-deposit occurrence and structure, and the geologic aspects of mine planning and development. Cf: *mining; mining engineering.*

miniphyric Said of the texture of a porphyritic igneous rock in which the greatest dimension of the phenocrysts does not exceed 0.008 mm (Cross et al, 1906, p.702); also, said of a rock having such texture.

minium A bright-red, scarlet, or orange-red mineral: Pb_3O_4. Syn: *red lead.*

minnesotaite A mineral: $(Fe,Mg)_3Si_4O_{10}(OH)_2$. It is probably isomorphous with talc. It occurs abundantly in the iron ores of Minnesota. Syn: *iron talc.*

minophyric Said of the texture of a porphyritic igneous rock in which the greatest dimension of the phenocrysts is between 0.2 mm and 1 mm (Cross et al, 1906, p.702); also, said of a rock with such texture.

minor element (a) A syn. of *trace element.* (b) A term that is occasionally used for an element that normally comprises between one and five percent of a rock; it is not quantitatively defined.

minor fold A small-scale fold that is associated with or related to the *major fold* of an area. Cf: *subsidiary fold.*

minor planet *asteroid.*

minor septum One of the relatively short, third-cycle septa of a corallite, commonly inserted between, and much shorter than, adjacent *major septa.*

minus-cement porosity The porosity that a sedimentary material would have if it contained no chemical cement.

minus sight *foresight.*

minute (a) A unit of time equal to 1/60 of an hour and containing 60 seconds. Abbrev: min; m (in physical tables). (b) A unit of angular measure equal to 1/60 of a degree and containing 60 seconds of arc. Symbol: ′. Syn: *minute of arc.*

minverite A mafic intrusive rock, diabasic in nature, containing hornblende and albite. According to Johannsen (1939, p.267), the albite is in part primary and in part secondary, and the rock may be metamorphic.

minyulite A white mineral: $KAl_2(PO_4)_2(OH,F).4H_2O$.

Miocene An epoch of the upper Tertiary period, after the Oligocene and before the Pliocene; also, the corresponding worldwide series of rocks. It is sometimes considered to be a period, when the Tertiary is designated as an era.

miocrystalline *hypocrystalline.*

miogeosyncline A geosyncline in which volcanism is not associated with sedimentation; the nonvolcanic aspect of an orthogeosyncline, located near the craton. (Stille, 1941). Syn: *miomagmatic zone.* Cf: *eugeosyncline.* See also: *ensialic geosyncline.*

miomagmatic zone *miogeosyncline.*

miospore A term arbitrarily defined in paleopalynology as a spore or pollen grain less than 200 microns in diameter. Cf: *megaspore.* See also: *small spore.*

miothermic Pertaining to or characterized by prevailing temperature conditions on the Earth as opposed to warmer or colder periods. Cf: *pliothermic*.

mirabilite A white or yellow monoclinic mineral: $Na_2SO_4.10H_2O$. It occurs as a residue from saline lakes, playas, and springs, and as an efflorescence. Syn: *Glauber's salt*.

mire (a) A small and muddy piece of marshy, swampy, or boggy ground; wet spongy earth. (b) Soft, heavy, often deep mud or slush.----Obsolete syn: *slough* [geol].

mirror glance *wehrlite* mineral.

mirror plane of symmetry *plane of mirror symmetry*.

mirror stone *muscovite*.

miry ground Ground that is deeply wet, generally sticky, and not having sufficient bearing strength to support loads.

mischungskorrosion Limestone solution at depth by calcite-undersaturated water. Such water is formed by the mixing of two saturated waters of differing equilibrium carbon dioxide pressures; it is undersaturated because of the nonlinear relationship between calcite solubility and carbon dioxide partial pressure. It is used as an English word, although its translated form, *mixture dissolving*, is also used. Etymol: German.

miscibility gap A compositional range intermediate between phases of variable composition, in which the assemblage of those phases is stable relative to a single phase. It is sometimes, and incorrectly, called an *immiscibility gap*.

miscible Said of two or more phases that, when brought together, have the ability to mix and form one phase. Cf: *immiscible*.

misclosure *error of closure*.

mise a la masse A drill-hole resistivity or induced-polarization survey technique in which a buried conductor is directly energized and thereby serves as a large buried electrode. Potentials are measured on surface, in bore holes, or underground. A great depth of exploration is expected with the technique.

misenite A white mineral: $KHSO_4$.

miserite A pink mineral: $K(Ca,Ce)_4Si_5O_{13}(OH)_3$.

misfit stream (a) A stream whose meanders are obviously not proportionate in size to the meanders of the valley or to the meander scars preserved in the valley wall; a stream that is either too large (an *overfit stream*) or too small (an *underfit stream*) to have eroded the valley in which it flows. (b) A term that is often incorrectly used as a syn. of *underfit stream*.

mispickel *arsenopyrite*.

Mississippian A period of the Paleozoic era (after the Devonian and before the Pennsylvanian), thought to have covered the span of time between 345 to 320 million years ago; also, the corresponding system of rocks. It is named after the Mississippi River valley in which there are good exposures of rocks of this age. It is the approximate equivalent of the *Lower Carboniferous* of European usage.

Mississippi Valley type deposit A strata-bound deposit of lead and/or zinc in carbonate rocks, as occurring in the Mississippi valley, and possibly including associated deposits of fluorite and barite.

Missourian North American provincial series: lower Upper Pennsylvanian (above Desmoinesian, below Virgilian).

missourite A plutonic rock containing a potassium feldspathoid (leucite) and 60 to 90 percent mafic minerals, such as pyroxene and olivine. Its name is derived from Missouri River in Montana. Cf: *fergusite; italite*.

mis-tie A term used in surveying for the failure of the first and final observations around a closed loop to be identical, or for the failure of the values at identical points on intersecting loops to be the same. See also: *error of closure*.

misy A term for various poorly defined iron sulfates.

mitridatite A mineral: $Ca_2Fe_3(PO_4)_{3-x}(OH)_{4+3x}.nH_2O$.

mitscherlichite A greenish-blue tetragonal mineral: $K_2CuCl_4.2H_2O$.

Mitscherlich's law A statement in crystallography that isomorphous substances have similar chemical compositions and analogous formulas. It is named after a German chemist of the nineteenth century, Eilhardt Mitscherlich.

mixability The degree with which different soils will mix together; e.g. clay and sand have low mixability, but sand will mix readily with silt or gravel.

mix-crystal *solid solution*.

mixed ages Discordant ages given by various dating methods (e.g. potassium-argon or rubidium-strontium) for the same igneous or metamorphic body which are the result of thermal and/or dynamic changes that affected the body at some time after its formation. See also: *hybrid age; overprint* [geochron]; *updating*.

mixed base A crude oil in which both paraffin and naphthene are present in approximately equal proportion. Cf: *paraffin base; naphthene base*.

mixed crystal *solid solution*.

mixed current A tidal current with two flood periods and two ebb periods during a tidal day, and characterized by a conspicuous diurnal inequality.

mixed cut A combination of *brilliant cut* above the girdle, with usually 32 facets, and *step cut* below, with the same number of facets. It is often used for colored stones, esp. fancy sapphires, to improve color and retain brilliancy.

mixed flow Water flow that is partly *turbulent flow* and partly *laminar flow*.

mixed gneiss *composite gneiss*.

mixed layer The layer of ocean water above the thermocline; it is mixed by wave action. It is equivalent to the *epilimnion* in a lake.

mixed-layer mineral A mineral, the structure of which is interstratified, consisting of alternating layers of two clays or of a clay with some other mineral; e.g. mica and brucite in chlorites.

mixed ore An ore of both oxidized and unoxidized minerals.

mixed peat Peat that is stratified according to varying plant associations. Cf: *banded peat*.

mixed rock *chorismite*.

mixed tide A tide with two high waters and two low waters occurring during a tidal day, and having a marked diurnal inequality (as in parts of the Pacific and Indian oceans). The term is usually applied to a tide that is intermediate between a predominantly diurnal tide and a predominantly semidiurnal tide, or to a tide with alternating periods of diurnal and semidiurnal components.

mixed water A term used by White (1957a, p.1639) for any mixture of volcanic and meteoric water in any proportion. White recommends discontinuing use of the term for chloride- and sulfide-rich acid waters.

mixing The combination of energy of different channels, generally to cancel noise. The technique is used in seismograph recording, in which the result is similar to that of a system of multiple geophones. Syn: *compositing; electrical overlap; interlocking seismic recording*.

mixing coefficient *austausch*.

mixing length In turbulent fluid flow, the length, normal to the flow direction, over which a small volume of fluid is assumed to retain its identity in the mass exchange process in turbulent flow. It is related to the coefficient of eddy viscosity and the rate of change of velocity normal to the line of flow (Middleton, 1965, p.250).

mixing ratio The ratio of water-vapor mass to the mass of air with which it is associated.

mixite An emerald- or blue-green to whitish mineral: $Bi_2Cu_{12}(AsO_4)_6(OH)_{12}.6H_2O$.

mixolimnion The upper, low-density, freely circulating layer of a *meromictic lake*. Cf: *monimolimnion; chemocline*.

mixoperipheral growth Growth of a brachiopod valve in which the posterior part increases in size anteriorly and toward the other valve. Cf: *holoperipheral growth.*

mixotrophic Said of an organism that is nourished by both autotrophic and heterotrophic mechanisms.

mixtite A descriptive group term proposed by Schermerhorn (1966, p.834) for a coarsely mixed, nonsorted or poorly sorted (fragments range in size from clay to blocks larger than boulders), clastic sedimentary rock, without regard to composition or origin; e.g. a tillite. Syn: *diamictite.*

mixtum A term proposed by Schermerhorn (1966, p.834) for an unconsolidated *mixtite.*

mixture dissolving *mischungskorrosion.*

miyakite A porphyritic basalt containing bytownite and augite phenocrysts in a groundmass of plagioclase, magnetite, manganian pyroxene, and glass.

mizzonite A mineral of the scapolite group intermediate between meionite and marialite, and containing 54-57% silica; esp. such a variety of scapolite occurring in clear crystals in ejected masses on volcanoes. Syn: *dipyre.*

MM scale *modified Mercalli scale.*

mo [glac geol] A Swedish term for "glacial silts or rock flour having little plasticity" (Stokes & Varnes, 1955, p.93). Pron: *moo.*

moat [glac] A glacial channel resembling a moat; e.g. a deep trench in glacier ice, surrounding a nunatak, and produced by ablation; or a channel at the margin of a dwindling glacier.

moat [marine geol] A ring-like depression around the base of many seamounts. It may be discontinuous. Syn: *sea moat.*

moat [volc] A valleylike depression around the inner side of a volcanic cone, between the rim and the lava dome.

moat [streams] A syn. of *oxbow lake.* The term is used in New England and was also applied by Shaler (1890, p.277) to the waters in abandoned channels in the Mississippi River flood plain.

moat lake A senescent lake characterized by a peripheral or outer ring of water enclosing a filled interior (Veatch & Humphrys, 1966, p.202). Se also: atoll moor.

mobile belt A long, relatively narrow crustal region of tectonic activity, measured in scores of miles. The term *geosyncline* is applied to its phase of sedimentation and subsidence. See also: *orogenic belt; orogenic cycle.*

mobile component (a) A component whose amount in a system changes during a given process. Cf: *perfectly mobile component; inert component.* (b) An element (or group of elements) that can migrate beyond the limits of a single mineral (Mehnert, 1968, p. 356).

Mobilisat *mobilizate.*

mobility A term used by W. Penck (1924) for the concept that the relative rate of uplift of the Earth's crust primarily determined the nature of the landforms produced by erosional processes.

mobilizate English translation of the German word *Mobilisat,* introduced to refer to the mobile phase, of any consistency, that existed during migmatization. Cf: *chymogenetic; metatect.* See also: *neosome.*

mobilization Any process that renders a solid rock sufficiently soft and plastic to permit it to flow or to permit geochemical migration of the mobile components. Cf: *rheomorphism.*

Mocha stone A white, gray, or yellowish form of moss agate containing brown to red iron-bearing or black manganese-bearing dendritic inclusions. The term is also used as a syn. of *moss agate.* Named for the city of Mocha (Al Mukhā) in Yemen. Also spelled: *mocha stone; mochastone.* Syn: *Mocha pebble.*

mock lead *sphalerite.*

mock ore *sphalerite.*

moctezumite A bright-orange mineral: $Pb(UO_2)(TeO_3)_2$.

modal The adj. of *mode.*

modal cycle A term proposed by Duff & Walton (1962) for a particular group of beds that occurs most frequently through a succession displaying cyclic sedimentation.

modal diameter An expression of the average particle size of a sediment or rock, obtained graphically by locating the highest point of the frequency curve or by finding the point of inflection of the cumulative curve; the diameter that is most frequent in the particle-size distribution.

mode [seis] A stationary vibration pattern of an oscillatory system.

mode [petrology] The actual mineral composition of a rock, usually expressed in weight or volume percentages. Adj: *modal.* Cf: *norm.*

mode [stat] The value or group of values that occurs with the greatest frequency in a set of data; the most typical observation. Cf: *mean; median.*

model An accurate simulation, by means of description, statistical data, or analogy, of a thing or process that cannot be observed directly or that is difficult to observe directly. Models may be derived by various methods, e.g. by computer, from stereoscopic photographs.

model scale The relationship existing between a distance measured in a model (such as in a stereoscopic image) and the corresponding distance on the Earth.

moder Plant material in a state intermediate between living and decayed.

moderate-energy coast A coast protected from strong wave action by headlands, islands, or offshore reefs and characterized by average breaker heights of 10-50 cm. Cf: *high-energy coast; low-energy coast.*

moderately sorted Said of a *sorted* sediment that is intermediate between a well-sorted sediment and a poorly sorted sediment and that has a sorting coefficient in the range of 2.5 to 4.0. Based on the phi values associated with the 84 and 16 percent lines, Folk (1954, p.349) suggests sigma phi limits of 0.50-1.00 for moderately sorted material.

modern cut A syn. of *fancy cut.* Also spelled: *moderne cut.*

modified Mercalli scale One of the earthquake *intensity scales,* having twelve divisions ranging from I (not felt by people) to XII (damage nearly total). It is a revision of the *Mercalli scale* made by Wood and Neumann in 1931. Cf: *Rossi-Forel scale; Richter scale.* Abbrev: *MM scale.*

modified polyconic projection A projection used for maps of large areas, derived from the regular *polyconic projection* by so altering the scale along the central meridian that the scale is exact along two standard meridians, one on either side of the central meridian and equidistant therefrom. It is used for the International Map of the World, the scale being preserved along two extreme parallels (4° apart) and along two meridians each of which lies 2° from the central meridian.

modlibovite A monticellite-free *polzenite* with anomite phenocrysts together with olivine, melilite, lazurite, phlogopite, biotite, and interstitial nepheline. Cf: *vesecite.*

modulus of compression *compressibility.*

modulus of deformation A term used instead of *modulus of elasticity* for materials that deform other than according to Hooke's law.

modulus of elasticity The ratio of stress to its corresponding strain under given conditions of load, for materials that deform elastically, according to Hooke's law. It is one of the *elastic constants.* See also: *Young's modulus; modulus of rigidity; modulus of deformation; static modulus; bulk modulus.* Syn: *elastic modulus; modulus of volume elasticity.*

modulus of incompressibility *bulk modulus.*

modulus of rigidity A *modulus of elasticity* in shear. Symbol: μ or G. Syn: *torsion modulus; shear modulus; rigidity modulus; Coulomb's modulus.*

modulus of volume elasticity *modulus of elasticity.*

modumite A calcium- and aluminum-rich essexite.

moel A term used in Wales for a rounded hill with a vegeta-n-clad summit (Marr, 1901). Etymol: Welsh, "bare field".

ofette The exhalation of carbon dioxide in an area of late-age volcanic activity; also, the small opening from which the s is emitted. An example is Yellowstone National Park in e U.S. Etymol: French, "noxious gas".

ofettite A natural carbon-dioxide gas.

ogote A syn. of *hum* that is used for tropical karst regions. n: *haystack hill; pepino hill*.

ohavite *tincalconite*.

ohawkian North American stage: Middle Ordovician (above azyan, below Edenian). See also: *Trentonian*.

ohnian North American stage: Miocene (above Luisian, low Delmontian).

oho Abbreviated form of *Mohorovičić discontinuity*, sug-sted by Birch (1952, p.229).

ohole project A now discontinued project to penetrate the arth's crust and sample the mantle, i.e. to drill through the ohorovičić discontinuity. The drill hole(s) itself may be alled the mohole.

ohorovičić discontinuity The boundary surface or sharp eismic-velocity discontinuity that separates the Earth's crust om the subjacent mantle. It marks the level in the Earth at hich *P*-wave velocities change abruptly from 6.7-7.2 km/sec n the lower crust) to 7.6-8.6 km/sec or average 8.1 km/sec t the top of the upper mantle); its depth varies from about -10 km beneath the ocean floor to about 35 km below the ontinents, although its depth below the geoid may reach 70 n under some mountain ranges. The discontinuity probably presents a chemical change between the basaltic materials oove to periodotitic or dunitic materials below, rather than a nase change (basalt to eclogite); however, the discontinuity nould be defined by seismic velocities alone. It is variously stimated to be between 0.2 and 3 km thick. It is named in onor of its discoverer. Andrija Mohorovičić (1857-1936), roatian seismologist. Abbrev: M. Syn: *Moho; M-discontinuity*.

ohr envelope An envelope of a series of *Mohr circles*; the cus of points whose coordinates represent the stresses ausing failure. Syn: *rupture envelope*.

ohrite A mineral: $(NH_4)_2(Fe,Mg)(SO_4)_2.6H_2O$.

ohr-Knudsen method In oceanography, a chemical method r estimating the chlorinity of sea water.

ohr's circle A graphic representation of the state of stress at particular point at a particular time. See also: *Mohr enve-pe*.

ohsite *ilmenite*.

ohs' scale. A standard of ten minerals by which the *hardness* a mineral may be rated. The scale includes, from softest to ardest and numbered one to ten: talc; gypsum; calcite; fluo-te; apatite; orthoclase; quartz; topaz; corundum; and dia-ond. Cf: *technical scale*.

oinian A division of the Precambrian in Scotland.

oire Said of feldspars having the appearance of watered lk.

oissanite A meteoritic mineral: SiC. It is identical with the rtificial *carborundum*.

oist playa *wet playa*.

oisture Water diffused in the atmosphere or the ground, in-luding soil water.

oisture content [coal] In coal, both the surface or free mois-ure that can be removed by natural drying, and the *inherent* noisture that is structurally contained in the substance.

oisture content [soil] The amount of moisture in a given soil ass, expressed as weight of water divided by weight of ven-dried soil, multiplied by 100 to give a percentage. See so: *water content [sed]*.

oisture-density curve *compaction curve*.

oisture-density test *compaction test*.

oisture equivalent The ratio of weight of water that a satu-ated soil will retain against a centrifugal force 1000 times the

force of gravity to the weight of dry soil (Meinzer, 1923, p.25). Syn: *centrifuge moisture equivalent*.

moisture index A means for classifying climates devised by Thornthwaite; wet-season surplus minus 0.6 times dry-season deficiency, divided by total need--all expressed in the same unit, such as inches--multiplied by 100 to get a percentage.

moisture meter An instrument for determining the percentage of moisture in a substance such as timber or soil, usually by measuring its electrical resistivity.

moisture tension In a soil, negative gage pressure of the water, equal to the equivalent pressure necessary to bring the soil water to hydraulic equilibrium through a porous wall, with a pool of water of equivalent composition (SSSA, 1970). Syn: *soil-moisture tension; capillary tension*.

molar n. A process with a grinding surface on the gnathal lobe of the mandible of a crustacean. Cf: *incisor*.

molar specific heat capacity The heat capacity of a system per gram-mole, numerically equal to about 6 cal/mole-°C for the common pure metals including lead (atomic weight=207) and aluminum (atomic weight=27). Since the number of mol-ecules in one gram-mole is the same all substances, it fol-lows that virtually the same quantity of heat is needed per molecule to raise the temperature of each of the common pure metals by a given amount, even though their masses are quite different (Sears, 1958, p.516).

molasse (a) A descriptive term for a paralic (partly marine, partly continental or deltaic) sedimentary facies consisting of a very thick sequence of soft, ungraded, cross-bedded, fossil-iferous conglomerates, sandstones, shales, and marls, char-acterized by primary sedimentary structures and sometimes by coal and carbonate deposits. It is more clastic and less rhythmic than the preceding flysch facies. (b) An extensive, postorogenic sedimentary formation representing the totality of the molasse facies resulting from the wearing down of ele-vated mountain ranges during and immediately succeeding the main paroxysmal (diastrophic) phase of an orogeny, and deposited considerably in front of the preceding flysch; specif. the Molasse strata, mainly of Miocene and partly of Oligocene age, deposited on the Swiss Plain and Alpine foreland of southern Germany subsequent to the rising of the Alps.--- Etymol: French *mollasse*, "soft". Pron: mo-*laas*. Adj: *mo-lassic*. Cf: *flysch*.

mold [paleont] (a) An *impression* made in the surrounding earth or rock material by the exterior or interior of a fossil shell or other organic structure. A complete mold would be the hollow space with its boundary surface. Cf: *cast*. See also: *external mold; internal mold*. (b) *natural mold*. (c) A cast of the inner surface of a fossil shell or other organic structure.----Syn: *mould*.

mold [sed] An original mark or primary depression made on a sedimentary surface; e.g. a flute, striation, or groove. Strictly, a mold is the filling of such a depression, and although usage has applied the term *cast* to such a filling, "some authors re-verse the usage and regard the structures on the bottoms of beds as molds" (Middleton, 1965, p.247); others regard "cast" and "mold" as synonymous. Syn: *mark*.

mold [soil] (a) An old term for a soft, friable soil rich in humus and suited to plant growth, e.g. leaf mold. (b) An old term for surface soil; the surface of the Earth; the ground.---- Etymol: Old English, *molde*, "earth; dust; soil".

moldavite [tektite] A translucent, olive- to brownish-green or pale-green tektite from western Czechoslovakia (southern Bo-hemia and southern Moravia), characterized by marked sculpturing on its surface due to solution etching. Named after the Bohemian river Moldau (German name for Vltava), in whose valley moldavites are found. Syn: *moldauite; vltavite; pseudochrysolite*.

moldavite [mineral] A variety of *ozocerite* from Moldavia.

moldering The decomposition of organic matter under conditions of insufficient oxygen, so that a carbon-rich residue is formed. Cf: *disintegration* [*coal*]; *peat formation; putrefaction.*

moldic porosity Porosity resulting from the removal, usually by solution, of a former individual constituent of a rock, such as a shell (Choquette & Pray, 1970, p.248-249).

mole A massive, solid-fill, protective structure extending from the shore into deep water, formed of masonry and earth or large stones, and serving as a breakwater or a pier.

molecular heat *molar specific heat capacity.*

molecular percent *molecular proportion.*

molecular proportion The ratio of the weight percentage of a particular rock component, esp. an oxide, to its molecular weight. Syn: *molecular percent.*

mole track A small, geologically short-lived ridge, 30-60 cm high, formed by the humping up and cracking of the ground where movement along a large strike-slip fault occurred in heavily alluviated terrain. It resembles the track of a gigantic mole, or a line of disturbed earth turned by a great plowshare.

mollic Pertaining to a dark, thick *epipedon* having at least 0.58% organic carbon, a base saturation of at least 50% when measured at a pH of 7, and less than 250 ppm P_2O soluble in citric acid (SSSA, 1970). Cf: *umbric; ochric.* See also: *anthropic.*

mollisol *active layer.*

Mollisol In U.S. Dept. of Agriculture soil taxonomy, a soil order characterized by a mollic epipedon with an underlying horizon having a base saturation of 50% or more, measured at pH 7. It has no oxic or spodic horizon but may contain a histic epipedon or a natric, albic, argillic, cambic, gypsic, calcic, or petrocalcic horizon (SSSA, 1970). Suborders and great soil groups of this soil order have the suffix -oll. See also: *Alboll; Aquoll; Boroll; Rendoll; Udoll; Ustoll.*

mollition The thawing of the active layer (Bryan, 1946, p.640).

molluscoid In some classifications, any invertebrate animal possessing a lophophore; i.e. a brachiopod or bryozoan.

mollusk A solitary invertebrate belonging to the phylum Mollusca, characterized by a nonsegmented body which is bilaterally symmetrical and by a radially or biradially symmetrical mantle and shell. Among the classes included in the mollusks are the gastropods, pelecypods, and cephalopods. Also spelled: *mollusc.* Adj: *molluscan.*

Mollweide projection An equal-area map projection on which the entire surface of the Earth is enclosed within an ellipse whose major axis (the equator, representing 360° of longitude) is twice the length of the minor axis (the central meridian, representing 180° of latitude). All parallels are represented by straight lines at right angles to the central meridian and more widely spaced at the equator than at the poles, and all meridians are represented by equally spaced elliptical arcs with the exception of the central meridian (a straight line) and the meridian 90° from the center (a full circle, representing the hemisphere centered at the origin of the projection). The meridional curvature increases away from the central meridian. There is excessive angular distortion (shearing) at the margins of the map. Named after Karl B. Mollweide (1774-1825), German mathematician and astronomer, who introduced the projection in 1805. Also known as "Mollweide homolographic projection".

molten Reduced to the fluid state as a result of heating; fused; melted.

moluranite A black amorphous mineral: $UO_2.3UO_3.7MoO_3.20H_2O$.

molybdate A mineral compound characterized by the radical MoO_4, in which the six-valent molybdenum ion and the four oxygens form a flattened square rather than a tetrahedron. Tungsten and molybdenum may substitute for each other. An example of a molybdate is wulfenite, $PbMoO_4$. Cf: *tungstate.*

molybdenite A greenish lead-gray hexagonal mineral: MoS_2. It

is the principal ore of molybdenum. Molybdenite generally occurs in foliated masses or scales and is found in pegmatite dikes and quartz veins; it resembles graphite in appearance and to the touch, but has a bluer color. Cf: *jordisite.*

molybdic ocher (a) *ferrimolybdite.* (b) *molybdite.*

molybdite A mineral: MoO_3. Much so-called molybdite is *ferr molybdite.* Syn: *molybdine; molybdic ocher.*

molybdomenite A colorless to yellowish-white mineral $PbSeO_3$.

molybdophyllite A colorless, white, or pale-green mineral $(Pb,Mg)_2SiO_4.H_2O$ (?).

molysite A brownish-red or yellow mineral: $FeCl_3$.

moment [stat] The average or sum of the deviations or some power of the deviations of the elements of a frequency distribution from a specified norm (Webster, 1967, p.1456, def.8) See also: *moment measure.*

moment [stratig] (a) A term recommended by Teicher (1958a, p.113-115, 117) for the time interval during which biostratigraphic zone was deposited; the geologic-time unit corresponding to Oppel's (1856-1858) "zone". It is, for a practical purposes, the shortest perceptible time interval int which geologic time can be subdivided, and is of the general order of between half a million and 5 million years. Term wa formerly proposed by Renevier et al (1882) in 1881. Cf: *ir stant; hemera.* Syn: *phase* [geochron]; *secule; chron; zon time.* (b) Any short time interval during which some geologi event or process occurs, and for which further temporal anal ysis is not possible; e.g. the period during which shells in thin bed of rock were fossilized.

moment map A stratigraphic map that expresses the position al relations of beds as a continuous variable (Krumbein & Libby, 1957, p.200); e.g. a *center-of-gravity map* and a *stan dard-deviation map.*

moment measure The expected value of each of the power of a random variable that has a given distribution; a weighte measure of central tendency. In sedimentology, moment mea sures are related to the center of gravity of the particle-siz distribution curve and are defined about the mean value of the variable. See also:*moment* [stat].

monacanth A trabecula in a rugose coral in which the fiber are related to a single center of calcification and radiate up ward and outward from the axis formed by upward shifting c the center (TIP, 1956, pt.F, p.235). Cf: *rhabdacanth.*

monactin A sponge spicule having a single ray. Syn: *monact monactine.*

monadnock An upstanding rock, hill, or *mountain of circum denudation* of resistant rock rising conspicuously above the general level of a peneplain in a temperate climate, repre senting an isolated remnant of a former erosion cycle in a mountain region that has been largely beveled to its base level. Type locality: Mount Monadnock in SW New Hampshire Cf: *unaka; inselberg.* Syn: *torso mountain.*

monadnock phase In hypsometric analysis of drainage basins the transitory stage characterized by abnormally low hypso metric integrals (less than 35%): removal of the monadnock by fluvial erosion will restore the distorted hypsographic curve to equilibrium form (Strahler, 1952b, p.1130). Cf: *equilibrium stage.*

monalbite Monoclinic albite; a monoclinic, high-temperature modification of sodium feldspar. It forms a complete solid-so lution series with sanidine. Formerly called: *barbierite.*

monaxon A simple uniaxial sponge spicule with a single axia filament or axial canal, or one developed by growth along a single axis. It may be curved or straight and may bear expan sions at one or both ends. Obsolete syn: *rhabd.*

monazite A yellow, brown, or reddish-brown monoclinic mineral: $(Ce,La,Nd,Th)(PO_4,SiO_4)$. It is a rare-earth phosphate with appreciable substitution of thorium for rare earths and licon for phosphorus; thorium-free monazite is rare. It is widely disseminated as an accessory mineral on granites, gneisses, and pegmatites, and it is often naturally concentrated in detrital sand, gravel, and alluvial tin deposits. Monazite is the principal ore of the rare earths and of thorium. Syn: cryptolite.

moncheite A steel-gray hexagonal mineral: (Pt,Pd)(Te,Bi)$_2$.

monchiquite A lamprophyre containing olivine, pyroxene, and usually mica or amphibole (barkevikite) phenocrysts in a groundmass of glass or analcime, often highly altered. Its name is derived from Serra de Monchique, Portugal. Cf: *fourchite*.

mondhaldeite A hypabyssal rock similar in composition to camptonite and characterized by the presence of long acicular hornblende phenocrysts along with augite, bytownite, and leucite, in a glassy groundmass with felty texture.

moneron In some classifications, any one of a group of organisms (i.e. Monerozoa) made up of a protoplasmic mass without a nucleus and including bacteria, blue-green algae, and certain fungi. Their known stratigraphic range is Early to Middle Precambrian. Var: *monera*.

monetite A yellowish-white mineral: CaHPO$_4$.

moniliform Bead-like, or jointed at regular intervals so as to resemble a string of beads; e.g. "moniliform antennae".

monimolimnion The deep, usually salty, stable, high-density, and perennially stagnant, unmixed, or noncirculating layer of a meromictic lake. Cf: *mixolimnion; chemocline*.

monimolite A yellowish, brownish, or greenish mineral: (Pb, Ca)$_3$Sb$_2$O$_8$ (?). It may contain ferrous iron.

monk rock *penitent rock*.

monmouthite An urtite containing hastingsite in place of pyroxene. It is named after Monmouth Township, Ontario, Canada.

monocentric Said of a corallite formed by a monostomodaeal polyp.

monochromatic illuminator *monochromator*.

monochromatic light Electromagnetic radiation of a simple wavelength or frequency. It is used in crystal optics to determine indices of refraction.

monochromator An instrument for selecting a narrow portion of a spectrum. In optics, a variable filter, grating, or prism which can isolate light of only one wavelength (color) or of a very narrow range of wavelengths. Syn: *monochromatic illuminator*.

monoclinal Adj. of *monocline*.

monoclinal coast A coast resulting from monoclinal flexure at the shoreline (Cotton, 1958, p.475); e.g. the west coast of South Island, New Zealand.

monoclinal scarp A scarp resulting from a steep downward flexure between an upland block and a tectonic basin (Cotton, 1958, p.174).

monoclinal shifting The downdip migration of a divide (and of a stream channel) resulting from the tendency of streams in a region of inclined strata to flow along the strike of less resistant strata, as where differential erosion proceeds more rapidly along the steeper slope of a cuesta or monoclinal ridge. The process was first noted by Gilbert (1877, p.135-140). See also: *shifting*. Syn: *homoclinal shifting; uniclinal shifting*.

monocline A unit of strata that dips or flexes from the horizontal in one direction only, and is not part of an anticline or syncline. It is generally a large feature of gentle dip. Cf: *homocline; flexure*. Adj: *monoclinal*. Obs. syn: *unicline*.

monoclinic system One of the six *crystal systems*, characterized by either a single, twofold axis of symmetry, a single plane of symmetry, or a combination of the two. Of the three unequal axes, two are obliquely inclined to each other and the third is perpendicular to the plane formed by them. Cf: *isometric system; hexagonal system; tetragonal system; orthorhombic system; triclinic system*.

monocolpate Said of pollen grains having a single, normally distal colpus. Syn: *monosulcate*.

monocot *monocotyledon*.

monocotyledon An angiosperm whose seeds contain a single, parallel-veined embryonic leaf. Such a plant usually has flowering parts in threes, parallel leaf venation, and monocolpate pollen. Examples include grasses, palms, and lilies. Monocotyledons range from the Cretaceous. Cf: *dicotyledon*. Syn: *monocot*.

monocrepid Said of a desma (of a sponge) with a monaxial crepis.

monocyclic (a) Said of a crinoid having only a single circlet of plates proximal to radial plates. (b) Said of the apical system of an echinoid in which genital plates and ocular plates are arranged in a single ring around the periproct.---Cf: *dicyclic*.

monoecious Said of a plant that has both staminate and pistillate flowers, or both male and female gametangia.

monogene adj. (a) *monogenetic*. (b) A term applied specif. by Naumann (1850, p.433) to an igneous rock (such as dunite) composed essentially of a single mineral. Cf: *polygene; monomineralic; monomictic [sed]*. Syn: *monogenic*.

monogenetic (a) Resulting from one process of formation or derived from one source, or originating or developing at one place and time; e.g. said of a volcano built up by a single eruption. (b) Consisting of one element or type of material, or having a homogeneous composition; e.g. said of a gravel composed of one type of rock. --Cf: *polygenetic*. Syn: *monogene; monogenic*.

monogenic (a) *monogenetic*. (b) *monogene*.

monogeosyncline A single geosynclinal trough along the continental margin and receiving sediments from a borderland on its oceanic side (Schuchert, 1923). Cf: *polygeosyncline*.

monoglacial theory The belief that the Pleistocene ice sheet made only one general advance and one general recession, without any substantial "interglacial" activity of recession followed by advance (Wright, 1914, p.124-125).

monograptid n. Any graptoloid belonging to the family Monograptidae, characterized by scandent uniserial rhabdosomes with thecae of variable form and "monograptid" growth. They are known from the Silurian and Lower Devonian.----adj. Said of the upward direction of growth in graptoloid thecae.

monohydrocalcite A rare mineral: CaCO$_3$.H$_2$O. It was first observed in lake-bottom sediments, and it may be formed by precipitation from cold water in contact with air. Cf: *hydrocalcite*.

monolete adj. Said of an embryophytic spore having a laesura consisting of a single line or mark. Cf: *trilete*.---n. A monolete spore. The usage of this term as a noun is improper.

monolith [eng] One of many large stones or blocks of stone forming the component parts of an engineering structure (such as of concrete gravity dam).

monolith [geol] (a) A flawless piece of unfractured bedrock, usually more than a few meters across; e.g. an unweathered joint block lifted by a glacier. (b) A large upstanding mass of rock, such as a volcanic spine or a mass produced by differential erosion. (c) A mountain or large hill apparently composed of one kind of rock, usually of a coarse-grained igneous rock.

monolith [soil] A vertical soil section, taken to illustrate the soil profile.

monomaceral Said of a coal microlithotype consisting of a single maceral. Cf: *bimaceral; trimaceral*.

monomict breccia A brecciated meteorite in which all the fragments have essentially the same composition. Cf: *polymict breccia*.

monomictic [sed] Said of a clastic sedimentary rock composed of but a single mineral species. Cf: *oligomictic; polymictic; monogene [geol]*. Syn: *monomict*.

monomictic [lake] Said of a lake with only one yearly overturn

461

or period of circulation, such as a subtropical lake with a winter overturn. Cf: *dimictic*.

monomineralic Said of a rock composed wholly or almost wholly of a single mineral; esp. said of an igneous rock (such as anorthosite or dunite) consisting or one essential mineral. The amounts of other minerals tolerated under the definition vary with different authors. Cf: *polymineralic; anchimonomineralic; monogene*. Syn: *monomineral*.

monomorphic Said of a taxon that has but one form; a taxon comprising individuals with essentially the same characteristics. Syn: *monomorphous*.

monomyarian adj. Said of a bivalve mollusk or its shell with only the posterior adductor muscle. Cf: *dimyarian; anisomyarian*.---n. A monomyarian mollusk, such as an oyster or scallop.

monophyletic Evolving from a single ancestral stock. Cf: *polyphyletic*.

monoplacophoran Any mollusk belonging to the class Monoplacophora, characterized by nearly bilateral symmetry and internal serial repetition. Originally thought to be represented only by Paleozoic forms, it is now known to exist in present-day marine environments.

monopleural Said of the arrangement of the two rows of thecae in the biserial rhabdosome of a scandent graptoloid in which the rows are in contact side-by-side. Cf: *dipleural*.

monopodial Having one main axis of growth.

monoporate Said of pollen grains provided with a single pore, as in grasses.

monopyroxene *clinopyroxene*.

monosaccate Said of pollen with a single vesicle, usually extending all around the pollen grain more or less at the equator.

monoschematic Said of mineral deposits having a uniform texture. Cf: *polyschematic*.

monosomatic chondrule A chondrule consisting of a single crystal. Cf: *polysomatic chondrule*.

monostatic radar A *radar* with the transmitter and receiver located at the same place.

monostomodaeal Said of stomodaea of a scleractinian coral polyp, each having its own tentacular ring after originating by distomodaeal budding or by tristomodaeal budding.

monostratum A simple layer, as in a first-order laminite (Lombard, 1963, p.14).

monostromatic Said of a thallus that is composed of only one layer of cells; said of a blade of a plant that is only one cell in thickness. Cf: *distromatic; oligostromatic; polystromatic*.

monosulcate A term essentially equivalent to *monocolpate* in ordinary usage.

monothalamous *unilocular*.

monothem A term proposed by Caster (1934, p.18) for a noncyclic, or not obviously cyclic, time-stratigraphic unit of genetically related strata, representing a "more ordinary, and perhaps more normal major subdivision" of a stage, but interpreted by Moore (1949, p.19) as a "local deposit having essentially uniform lithologic character" and corresponding to formation or member of rock-stratigraphic classification. Weller (1958, p.636) regards the term as "superfluous because its meaning is the same as 'substage'". Cf: *lithizone*.

monothermite A clay-mineral material that shows a single high-temperature endothermal reaction at about 550°C. It appears to be a mixture in which illite and kaolinite are important components (Grim, 1968, p.48).

monotopism The origin of a species or other systematic group in only one geographic area. Cf: *polytopism*.

monotrichous Said of a bacterial cell with a single flagellum occurring at one pole. Cf: *lophotrichous; peritrichous*.

monotrophic Said of an organism that feeds on one kind of food only.

monotropy The relationship between two different forms of the

same substance, e.g. pyrite and marcasite, that have no de[fi]nite transition point, since only one of the forms, e.g. pyrite, [is] stable, and the change from the unstable to the stable form [is] irreversible. Cf: *enantiotropy*.

monotypic In taxonomy, said of a taxon that includes only o[ne] taxon of the next lower rank, e.g. a genus or subgenus w[ith] only one originally included species. Cf: *polytypic*.

monsmedite A mineral: $K_2O.Tl_2O_3.8SO_3.15H_2O$.

monsoon A seasonal type of wind system in which its dire[c]tion changes with the seasons, e.g. over the Arabian S[ea] where the winds are from the northeast for six months a[nd] then from the southeast for the next six months.

monster An organism with extreme departure in form or stru[c]ture from the usual type of its species.

monstrosity A part of an organism exhibiting considerable d[e]viation in the structure or form which may be injurious or u[se]useful to the species and is usually not propagated.

montane Of, pertaining to, or inhabiting cool upland slope[s] below the timber line, characterized by the dominance of e[v]ergreen trees. Cf: *alpine [ecol]*. Syn: *subalpine; alpestrine*.

montanite A mineral: $Bi_2O_3.TeO_3.2H_2O$.

Mont Blanc ruby Reddish quartz or *rubasse* from Mont Blan[c,] a mountain in SE France.

montbrayite A tin-white triclinic mineral: Au_2Te_3.

montebrasite A mineral: $LiAlPO_4(OH)$. It is isomorphous wi[th] amblygonite and natromontebrasite.

Monte Carlo method A procedure in statistics by which ra[n]dom numbers are used to approximate the solution to intra[c]table mathematical or physical problems.

monteponite A black mineral: CdO.

montesite A mineral: $PbSn_4S_5$.

montgomeryite A green to colorless mineral: Ca_2Al $(PO_4)_3(OH).7H_2O$. It was formerly regarded as: Ca_4Al $(PO_4)_6(OH)_5.11H_2O$.

Montian European stage: Paleocene (above Danian, belo[w] Thanetian).

monticellite A colorless or gray mineral related to olivin[e:] $CaMgSiO_4$. It is isomorphous with kirschsteinite, and usual[ly] occurs in contact-metamorphosed limestones.

monticle *monticule*.

monticule [geomorph] (a) A little mound; a hillock, moun[d,] knob, or other small elevation. (b) A small, subordinate vo[l]canic cone developed on the flank or about the base of [a] larger volcano. ----Etymol: French. Syn: *monticle*.

monticule [paleont] (a) A protuberant part of corallum surfac[e] of a scleractinian coral, produced in circummural budding. C[f:] *colline*. (b) One of the small rounded nodes or swellings of [a] brachiopod shell, commonly bearing spines. (c) *monticulus*.

monticulus One of the clusters of small, modified zooeci[a] which may be regularly spaced throughout a Paleozoic bry[o]zoan colony, appearing at the surface as a small protube[r]ance or elevation. Cf: *macula*. Syn: *monticule [paleont]*.

montiform Having the shape of a mountain; mountain-like.

montmartrite A variety of gypsum from Montmartre (a sectio[n] of Paris, France).

montmorillonite (a) A group of expanding-lattice clay minera[ls] of general formula: $R_{0.33}Al_2Si_4O_{10}(OH)_2.nH_2O$, where R in[-]cludes one or more of the cations Na^+, K^+, Mg^{+2}, Ca^{+2}, an[d] possibly others. The minerals are characterized by a three[-] layer crystal lattice (one sheet of aluminum and hydroxyl be[-]tween two sheets of silicon and oxygen), by deficiencies i[n] charge in the tetrahedral and octahedral positions balanced b[y] the presence of cations (most commonly calcium and sod[i]um) subject to base exchange, and by swelling on wettin[g] (and shrinking on drying) due to introduction of considerab[le] interlayer water in the c-axis direction. Magnesium or iro[n] may proxy for aluminum, and aluminum for silica. The mon[t]morillonite minerals are generally derived from alteration [of] ferromagnesian minerals, calcic feldspars, and volcanic glass[,] es; they are the chief constituents of bentonite and fuller's earth, and are very common in soils, sedimentary rocks, and associated with some mineral deposits. Syn: *smectite*. (b) A dioctahedral clay mineral of the montmorillonite group: $Na_{0.33}Al_{1.67}Mg_{0.33}Si_4O_{10}(OH)_2.nH_2O$. It is usually white, grayish, pale red, or blue, and represents a high-alumina end member that has some slight replacement of Al^{+3} by Mg^{+2} and substantially no replacement of Si^{+4} by Al^{+3}. Cf: *beidellite*. (c) Any mineral of the montmorillonite group, such as montmorillonite, nontronite, saponite, hectorite, sauconite, beidellite, volkonskoite, and griffithite.

montmorillonite-saponite An alternative name preferred by the International Mineralogical Association for the *smectite* group of clay minerals.

montrealite A peridotite containing, in order of decreasing abundance, pyroxene, hornblende, and olivine, with little or no feldspar or nepheline.

montroseite [mineral] A black mineral: $(V,Fe)O(OH)$.

montroseite [sed] A uranium-bearing sandstone.

montroydite A mineral: HgO.

monument [geomorph] (a) An isolated pinnacle, column, or pillar of rock resulting from erosion and resembling a man-made monument or obelisk, usually extremely regular in form and of grand dimensions. (b) *tind*.

monument [surv] A natural or artificial (but permanent) physical structure that marks the location on the ground of a *corner* or other survey point; e.g. a pile of stones indicating the boundary of a mining claim, or a road or fence marking the boundary of real property. See also: *boundary monument.*

monumental peak *horn* [glac geol].

monumented upland A term proposed by Hobbs (1921, p.373) for "the extreme type of mountain sculpture ... believed to be due to continued glacial action upon a fretted upland like that of the Alps", characterized by enlargement of cirques and reduction of horns. Example: Glacier National Park, Mont.

monzonite A group of plutonic rocks intermediate in composition between *syenite* and *diorite*, containing approximately equal amounts of orthoclase and plagioclase, little or no quartz, and commonly augite as the main mafic mineral; also, any rock in that group; the intrusive equivalent of *latite*. With a decrease in the alkali feldspar content, monzonite grades into *diorite* or *gabbro*, depending on the composition of the plagioclase; with an increase in alkali feldspar, it grades into syenite. Syn: *syenodiorite.*

monzonitic (a) Said of the texture of an igneous rock containing euhedral plagioclase crystals and some interstitial orthoclase. (b) Of, pertaining to, or composed or monzonite.

monzonorite In Tobi's (1971, p.202) classification of the charnockite suite, a quartz-poor member containing more plagioclase than microperthite.

Moody diagram A diagram showing the variation of the Darcy-Weisbach coefficient against the Reynolds number.

moon Any natural satellite of a planet; specif. the Moon, the Earth's only known natural satellite and next to the Sun the most conspicuous object in the sky, deriving its light from the Sun and reflecting it to the Earth. The Moon revolves about the Earth from west to east in about 29.53 days with reference to the Sun (interval from new moon to new moon) or about 27.32 days with reference to the stars; it has a mean diameter of 3475.9 km (2160 miles, or about 27% that of the Earth), a mean distance from the Earth of about 384,400 km (238,857 miles), a mass of 7.354 x 10^{25} g (about 1/81 that of the Earth), a volume about 1/49 that of the Earth, and a mean density of 3.34 g/cm^3. The Moon rotates once on its axis during each revolution in its orbit and therefore it always presents nearly the same face (41%) to the Earth; it has essentially no detectable atmosphere or water, and no life forms are believed to exist there.

moonmilk (a) A soft, white, plastic or putty-like, calcareous deposit that occurs on the walls of limestone caves and that may consist of aragonite, huntite, or esp. calcite. See also: *lublinite*. (b) A group of carbonate minerals occurring in soft, cottage-cheese-like masses and found frequently in caves. It includes hydromagnesite, nesquehonite, huntite, aragonite, calcite, magnesite, and dolomite.----Etymol: Swiss dialect *moonmilch*, "elf's milk".

moonquake An agitation or disturbance of the Moon's surface, analogous to a terrestrial earthquake.

moonscape The surface of the Moon, as observed in photographs or through a telescope or as delineated on the basis of photographic or telescopic evidence.

moonstone (a) A semitransparent to translucent alkali feldspar (adularia) or cryptoperthite that exhibits a bluish to milky-white, pearly, or opaline luster; an opalescent variety of orthoclase. Flawless moonstones are used as gemstones. Cf: *sunstone*. Syn: *hecatolite*. (b) A name incorrectly applied to *peristerite* or to opalescent varieties of plagioclase (esp. albite). (c) A name incorrectly applied (without proper prefix) to milky or girasol varieties of chalcedony, scapolite, corundum, and other minerals.

moon tide *lunar tide.*

moor coal A lignite or brown coal that is friable.

mooreite A glassy white mineral: $(Mn,Zn,Mg)_8(SO_4)$-$(OH)_{14}.4H_2O$. Cf: *torreyite.*

moorhouseite A mineral: $(Co,Ni,Mn)(SO_4).6H_2O$.

moorland pan *moorpan.*

moorland peat *highmoor peat.*

moorpan An *iron pan* occurring in a peaty soil or forming at the bottom of a bog, containing compact redeposited iron and humus compounds. Syn: *moorland pan.*

moor peat *highmoor peat.*

mor A raw type of humus, usually on the surface with a sharp lower boundary, and developed under cool, moist conditions. Cf: *mull.* Syn: *raw humus.*

moraesite A white mineral: $Be_2(PO_4)(OH).4H_2O$.

morainal Of, relating to, forming, or formed by a moraine. Syn: *morainic.*

morainal apron *outwash plain.*

morainal channel A meltwater-stream channel formed during the construction of a moraine (Rich, 1908, p.528).

morainal-dam lake A glacial lake impounded by a *drift dam* left in a preexisting valley by a retreating glacier.

morainal delta An obsolete syn. of *ice-contact delta.*

morainal lake A glacial lake occupying a depression resulting from irregular deposition of drift in a terminal or ground moraine of a continental glacier.

morainal plain *outwash plain.*

morainal stuff An obsolete term for the material carried upon the surface of a glacier.

morainal topography An irregular landscape produced by deposition of drift and characterized by irregularly scattered hills and undrained depressions.

moraine [glac geol] A mound, ridge, or other distinct accumulation of unsorted, unstratified glacial drift, predominantly till, deposited chiefly by direct action of glacier ice in a variety of topographic landforms that are independent of control by the surface on which the drift lies. The history of the term is confused: it was probably used originally, and is still often used, in European literature as a petrologic name for *till* that is being carried and deposited by a glacier, but is now more commonly used as a geomorphologic name for a landform composed mainly of till that has been deposited by either a living or an extinct glacier. Etymol: French, a term used by Alpine peasants in the 18th century for any heap of earth and stony debris; neither the exact origin of the term nor its first use in glacial geology can be traced.

moraine [volc] The solidified volcanic debris carried on the surface of a lava flow.

moraine bar A terminal moraine serving as a bar, rising out of deep water some distance from the shore (Tarr & Martin, 1914, p.294).

moraine kame A term applied by Salisbury et al (1902, p.118) to a kame that forms one of a group having the characteristics, topography, constitution, position, and "the same general significance" of a terminal moraine. See also: *kame moraine*.

moraine loop *loop moraine*.

moraine plateau A relatively flat area within a hummocky moraine, generally at the same elevation as, or a little higher than, the summits of surrounding knobs (Gravenor & Kupsch, 1959, p.50).

moraine rampart An elongated ridge or row of lateral and terminal moraines, sometimes forming an amphitheatrical arrangement (Schieferdecker, 1959, terms 0916 & 0922).

morainic *morainal*.

moralla (a) Poorly crystallized or massive opaque-appearing greenish material from Colombian emerald mines. (b) Any of the poorer grades of emerald.---Syn: *morallion; morallon*.

morass A general, mainly literary term for low-lying, swampy, boggy, or marshy land with decaying vegetation and, often, a muddy, offensive appearance.

morass ore *bog iron ore*.

mordenite A zeolite mineral: $(Ca,Na_2,K_2)_4Al_8Si_{40}O_{96}.28H_2O$. Syn: *ashtonite; flokite; arduinite; ptilolite*.

morel basin A cavity or depression developed by solution on the surface of a limestone pebble, smaller than a tinajita but larger than a pit, and having a diameter/height ratio of about 1:0.75 (Scott, 1947, p.147). See also: *solution-morel*.

morencite *nontronite*.

morenosite An apple-green or light-green mineral: $NiSO_4.7H_2O$. It may contain appreciable magnesium, and it occurs in secondary incrustations. Syn: *nickel vitriol*.

Morey bomb A simple reaction vessel for hydrothermal experimentation at modest pressures, in which pressure is produced by sealing a known quantity of water or other volatile material in a fixed-volume bomb, which is then heated.

morganite *vorobyevite*.

morinite A mineral: $Na_2Ca_4Al_4(PO_4)_4O_2F_6.5H_2O$. Syn: *jezekite*.

Morin transition A magnetic transition in hematite, occurring at about $250°K$, below which "the feeble ferromagnetism with a molecular saturation moment of about one-hundredth of a Bohr magnetron disappears" (Runcorn, 1967, p.867-868).

morion A nearly black, opaque variety of smoky quartz or cairngorm.

morlop A mottled variety of jasper found in New South Wales, Australia. It often occurs as pebbles associated with diamonds.

morphogenesis [geomorph] The origin and subsequent early growth or development of landforms or of a landscape.

morphogenetic region A region in which, under a given set of climatic conditions, the predominant geomorphic processes will give to the landscape certain regional characteristics that contrast with those of other areas developed under different climatic conditions. Peltier (1950) has postulated nine morphological regions based on temperature and moisture conditions. See also: *formkreis*.

morphogeny [geomorph] The interpretative morphology of a region; specif. *geomorphogeny*.

morphographic map *physiographic diagram*.

morphography The descriptive morphology of a region, or the phenomena so described; specif. *geomorphography*.

morphologic region A distinctive region delimited according to form, rock structure, and evolutionary history. Cf: *physiographic province*.

morphologic series In paleontology, a graded series of fossils showing variation either in individuals as a whole or in some particular variable feature.

morphologic species A species based solely on morphologi characteristics and consisting of individuals included betwee more or less arbitrarily selected limits.

morphologic unit [geomorph] A surface, either depositional e erosional, that is recognized by its topographic character.

morphologic unit [stratig] *morphostratigraphic unit*.

morphology [phys sci] The scientific study of form, and of th structures and development that influence form. The term widely used in most sciences.

morphology [geomorph] (a) The shape of the Earth's surface geomorphology, or "the morphology of the Earth" (Kin 1962). (b) The external structure, form, and arrangement e rocks in relation to the development of landforms.

morphology [meteorite] The study of the dimensions, form surface relief, fusion crust, and inner macrostructure e meteorites.

morphology [paleont] (a) A branch of biology or of paleontolo gy that deals with the form and structure of animals an plants or of their fossil remains; esp. a study of the form relations, and phylogenetic development of organs apart fro their functions. See also: *paleomorphology; function morphology*. (b) The features comprised in the form an structure of an organism or any of its parts.

morphology [soil] The study of the distribution patterns of so horizons in the soil profile, and of the soil's properties.

morphology [streams] *river morphology*.

morphometry [geomorph] "The measurement and mathemat cal analysis of the configuration of the Earth's surface and e the shape and dimensions of its landforms. The main aspect examined are the area, altitude, volume, slope, profile, an texture of the land as well as the varied characteristics of riv ers and drainage basins" (Clarke, 1966, p.235).

morphometry [lake] (a) The measurement of the form char acteristics (area, depth, length, width, volume, bottom grad ents) of lakes and their basins. (b) The branch of limnolog dealing with such measurements.

morphosculpture A topographic feature of less magnitud than, and often developed within or on, a *morphostructure* e.g. a ripple mark, ledge, or knoll on the ocean floor.

morphosequent Said of a surface feature that does not reflec the underlying geologic structure. Ant: *tectosequent*.

morphostratigraphic unit A distinct stratigraphic unit, define by Frye & William (1960, p.7) as comprising "a body of roc that is identified primarily from the surface form it displays that may or may not be distinctive lithologically from contigu ous units, and that may or may not transgress time through out its extent. The term is used in stratigraphic classificatio of surficial deposits such as glacial moraines, alluvial-terrac deposits, alluvial fans, lake plains, beach ridges, and othe such deposits where landforms serve to give identity to a bod of clastic sediments. Syn: *morphologic unit*.

morphostructure A major topographic feature that coincide with or is an expression of a geologic structure (e.g. a trenc or ridge on the ocean floor) or that is formed directly by tec tonic movements (e.g. a basin or dome). It is produced b the interaction between endogenetic and exogenetic forces the endogenetic factor being predominant in the tectoni movements of the Earth's crust. Cf: *morphosculpture*.

morphotectonics The tectonic interpretation of the morpholog ical, or present topographic features, of the Earth's surface; deals thus with their tectonic or structural relations and ori gins, rather than with their more obvious origins by surficia processes of erosion and sedimentation. Cf: *orogeny*.

morriner *esker*.

morro A term used in Latin America for an isolated hill e ridge which may or may not be on a coastal plain near th present shoreline; esp. a headland or bluff. Etymol: Spanis and Portuguese.

Morrowan North American provincial series: lowermost Penn ylvanian (above Chesterian of Mississippian, below Atokan).

mortar bed A valley-flat deposit (occurring in Nebraska and Kansas) consisting of sand, or of a mixture of clay, silt, sand, and gravel, firmly cemented by calcium carbonate and resembling hardened mortar; a type of caliche.

mortar structure A *cataclastic* structure produced by dynamic metamorphism of crystalline rocks (specif. granites and gneisses) and characterized by a mica-free aggregate of small, finely crushed grains of quartz and feldspar occupying the interstices between, or forming borders on the edges of, much larger and rounded relicts (of the same minerals) having unbroken cores or being more resistant to granulation. See also: *pseudogritty structure.* Syn: *cataclastic structure; murbruk structure; porphyroclastic structure.*

mortar texture A texture found in crystalline sedimentary rocks in which relatively large crystalline grains are separated by a microcrystalline mosaic.

mortlake A British syn. of *oxbow lake.* The term is "practically obsolete" in Britain (Stamp, 1961, p.327). Etymol: probably from Mortlake, a parish in a SW suburb of London, situated near a drained oxbow lake of the River Thames.

morvan (a) The intersection of two peneplains, as where an exhumed, tilted peneplain is cut across obliquely by a younger surface that has more nearly retained its original horizontal attitude; e.g. the intersection between the stripped and distinctly sloped Fall Zone peneplain (along the eastern Piedmont Plateau of the U.S.) with the late Tertiary Harrisburg peneplain. Also, the "problem" of the intersection of two peneplains. (b) A region that exhibits a *morvan* relationship, marked by a hard-rock upland bordered by a sloping land of older rock. The term was introduced by Davis (1912, p.115): "a region of composite structure, consisting of an older undermass, usually made up of deformed crystalline rocks, that had been long ago worn down to small relief and that was then depressed, submerged, and buried beneath a heavy overmass of stratified deposits, the composite mass then being uplifted and tilted, the tilted mass being truncated across its double structure by renewed erosion, and in this worn-down condition rather evenly uplifted into a new cycle of destructive evolution". Type locality: Morvan region of central France. Syn: *kiou.*

mosaic [geomorph] *desert mosaic.*

mosaic [paleont] (a) A pattern formed on the interior of a brachiopod valve by outlines of the adjacent fibers of the secondary layer of shell. (b) Arrangement of plates in edrioasteroids and cyclocystoids, more or less in plane and not imbricating, and presumably rather rigid.

mosaic [photo] An assembly of aerial photographs whose edges have been feathered (torn) or cut and matched to form a continuous photographic representation of a part of the Earth's surface; e.g. a composite photograph formed by joining together parts of several overlapping vertical photographs of adjoining areas of the Earth's surface. See also: *controlled mosaic; uncontrolled mosaic.* Syn: *aerial mosaic; photomosaic.*

mosaic breccia A breccia having fragments that are largely but not wholly disjointed and displaced.

mosaic structure Slight irregularity of orientation of small, angular, and granular fragments of varying sizes in a crystal, the fragments appearing in polarized light like pieces of a mosaic.

mosaic texture [met] A granoblastic texture in a dynamically metamorphosed rock in which the individual grains meet with straight or only slightly curved, but not interlocking or sutured, boundaries. Syn: *cyclopean texture.*

mosaic texture [sed] A texture in a crystalline sedimentary rock, characterized by more or less regular grain-boundary contacts; e.g. a texture in a dolomite or recrystallized limestone in which the mineral dolomite forms rhombs of uniform size in such a manner that in section contiguous crystals appear to dovetail, or a texture in an orthoquartzite in which secondary quartz is deposited in optical continuity on detrital nuclei.

mosandrite A reddish-brown or yellowish-brown mineral: (Na, Ca,Ce)$_3$Ti(SiO$_4$)$_2$F. Cf: *johnstrupite.* Syn: *rinkite; rinkolite; lovchorrite; khibinite.*

moschellandsbergite A mineral: Ag$_2$Hg$_3$. It consists of a naturally occurring alloy of silver with mercury, and was formerly included with *amalgam.*

Moscovian Stage in Russia: middle Upper Carboniferous (above Namurian, below Gzhelian).

moscovite *muscovite.*

mosesite A yellow mineral: Hg$_2$N(SO$_4$,MoO$_4$).H$_2$O. Cf: *kleinite.*

mosor A monadnock that has survived because of remoteness from the main drainage lines; esp. a *hum* in a karstic region. Etymol: originally a German term named by Penck (1900) after the Mosor Mountains in Dalmatia, Yugoslavia. Pl: *mosore.*

moss [eco geol] adj. A syn. of *capillary*, e.g. moss gold.

moss [gem] A fracture, fissure, or other flaw in a gemstone, having the appearance of moss; specif. such a fracture in an emerald.

moss [bot] A bryophyte of the class Musci, characterized by a protonema that produces a leafy, upright, gametophyte and by multicellular rhizoids. Cf: *liverwort.*

moss agate (a) A general term for any translucent chalcedony containing inclusions of any color arranged in dendritic patterns resembling trees, ferns, leaves, moss, and similar vegetation; specif. an *agate* containing brown, black, or green moss-like markings due to visible inclusions of oxides of manganese and iron. See also: *Mocha stone; tree agate.* (b) A moss agate containing green inclusions of actinolite or of other green minerals.

moss animal *bryozoan.*

Mössbauer effect The almost recoil-free emission and absorption of nuclear gamma rays by atoms tightly bound in a solid. It is a special case of *nuclear resonance* characterized by an extremely sharply defined resonant frequency. The emission of gamma rays by unbound atoms involves recoil of the emitting atom, to conserve momentum, and consequent varying of the energy and frequency of the emitted gamma ray. If, on the other hand, the atom is tightly bound in a crystal lattice, the atomic recoil is minimized and the emitted gamma ray is limited to a narrow frequency range - the Mössbauer effect, named in honor of Rudolf L. Mössbauer (1929-), German physicist. So sharp is the Mössbauer resonance that very slight Doppler shifts and even gravitational shifts of the emitted frequency can be detected.

Mössbauer spectrometry The art or process of using a spectrometer to analyze a *Mössbauer spectrum*, mainly for determining chemical structure. Measurement is made of the nuclear resonant absorption of gamma rays passing from a radioactive source to an absorber, usually the material being studied (DeVoe & Spijkerman, 1966, p.382R).

Mössbauer spectrum The spectrum seen when Mössbauer gamma-ray intensity is plotted as a function of relative velocity between the radioactive source and the absorber, usually the material being studied (DeVoe & Spijkerman, 1966, p.382R).

moss coral *bryozoan.*

mossite A mineral: Fe(Nb,Ta)$_2$O$_6$. It is isomorphous with tapiolite.

moss land An area with abundant moss, yet not wet enough to be a bog.

moss peat *highmoor peat.*

moss polyp *bryozoan.*

mother cell A cell from which new cells are formed; e.g. *spore mother cell* and *pollen mother cell.*

mother crystal A mass of raw quartz (faced or rough) as found in nature.

mother geosyncline Stille's term for a geosyncline that matured by evolving into a folded mountain system (Glaessner & Teichert, 1947, p.588). See also: *orogeosyncline*.

motherham An old syn. of *fusain*.

mother liquor In crystallization, the liquid which remains after the substances readily and regularly crystallizing have been removed. Syn: *mother liquid; mother water*.

mother lode (a) A main mineralized unit that may not be economically valuable in itself but to which workable veins are related, e.g. the Mother Lode of California. (b) An ore deposit from which a placer is derived; the *mother rock* of a placer.

mother of coal A syn. of *fusian*. Also spelled: *mother-of-coal*.

mother-of-emerald (a) *prase*. (b) Green fluorite.

mother-of-pearl The *nacre* of a pearl-bearing mollusk, extensively used for making small ornamental objects.

mother rock [eco geol] (a) A general term for the rock in which an ore deposit originated, e.g. *source rock, mother lode*. (b) A syn. of *country rock*.

mother rock [sed] A rock from which other rocks or sediments are derived; *source rock*.

motile Exhibiting or capable of movement, such as by cilia; e.g. "motile phase" of the life cycle of a flagellate coccolithophorid, producing holococcoliths, caneoliths, cyrtoliths, cricoliths, and cribriliths. Ant: *nonmotile*.

motion (a) A quarry from which small paving blocks are mined. (b) That part of a quarry in which work is actually being done.

mottle (a) A spot, blotch, or patch of color or shade of color, occurring on the surface of a sediment or soil. (b) A small, irregular body of material in a sedimentary matrix of different texture (difference in color not being essential) (Moore & Scruton, 1957, p.2727).

mottled [sed] Said of a sediment or sedimentary rock (such as a clay or sandstone) marked with spots of different colors, usually as a result of oxidation of iron compounds. Cf: *variegated*.

mottled [soil] Said of a soil that is irregularly marked with spots or patches of different colors, usually indicating poor aeration or seasonal wetness.

mottled limestone Limestone with narrow, branching, fucoidlike cylindrical masses of dolomite, often with a central tube or hole (Van Tuyl, 1916, p.345); it may be organic or inorganic in origin.

mottled structure Discontinuous lumps, tubes, pods, and pockets of a sediment, randomly enclosed in a matrix of contrasting textures, and usually formed by the filling of animal borings and burrows (Moore & Scruton, 1957, p.2727). Syn: *mottling*.

mottling (a) Variation of coloration in sediments and soils as represented by localized spots, patches, or blotches of color or shades of color. Also, the formation of mottles or of a mottled appearance. (b) *mottled structure*. (c) *luster mottling* (d) *dolomitic mottling*.

mottramite A mineral: $Pb(Cu,Zn)(VO_4)(OH)$. It is isomorphous with descloizite. Syn: *cuprodescloizite; psittacinite*.

Mott Smith gravimeter An instrument in which the moving system consists entirely of fused quartz, the restoring and labilizing forces being provided by quartz fibers.

motu A Polynesian term for a small coral island with vegetation. Pl: *motu; motus*.

mould Var. of *mold*.

moulin A roughly cylindrical, nearly vertical, well-like opening, hole, or shaft in the ice of a glacier, scoured out by swirling meltwater as it pours down from the surface. Etymol: French, "mill", so called because of the loud roaring noise made by the vortical motion of the water falling down the hole. Pron: moo-lanh. Syn: *glacier mill; glacial mill; glacier pothole; pot-*
hole [*glaciol*]*; glacier well*.

moulin kame A conical hill of glaciofluvial material formed in large circular hole ("moulin") in glacier ice.

moulin pothole *giant's kettle*.

mounanaite A mineral: $PbFe_2(VO_2)_2(OH)_2$.

mound (a) A general term for a low, isolated, rounded natural hill, generally of earth; a knoll. Syn: *tuft*. (b) An organic structure built by fossil colonial organisms, such as crinoids.

mound spring A spring characterized by a mound at the place where it flows onto the land surface. According to Meinzer (1923, p.55) "mound springs may be produced, wholly or in part, by the precipitation of mineral matter from the spring water; or by vegetation and sediments blown in by the wind-- method of growth common in arid regions". Cf: *pool spring*. See also: *spring mound*.

mount (a) An abbreviated form of the term *mountain*, esp. used preceding a proper name and usually referring to a particular summit within a group of elevations; e.g. Mount Marcy in the Adirondack Mountains. Abbrev: mt. (b) A high *hill*; esp. an eminence rising abruptly above the surrounding land surface, such as Mount Vesuvius. (c) *seamount*.

mountain (a) Any part of the Earth's crust higher than a *hill*, sufficiently elevated above the surrounding land surface of which it forms a part to be considered worthy of a distinctive name, characterized by a restricted summit area (as distinguished from a plateau), and generally having comparatively steep sides and considerable bare-rock surface; it can occur as a single, isolated eminence, or in a group forming a long chain or range, and it may form by earth movements, erosion, or volcanic action. Generally, a mountain is considered to project at least 300 m (1000 ft) above the surrounding land, although older usage refers to an altitude of 600 m (2000 ft) or more above sea level. When the term is used following a proper name, it usually signifies a group of elevations, such as a range (e.g. the Adirondack Mountains) or a system (e.g. the Rocky Mountains). Abbrev: mt.; mtn. Syn: *mount*. (b) An conspicuous or prominent elevation in an area of low relief, esp. one rising abruptly from the surrounding land and having a rounded base. (c) A term used in structural geology for large-scale, disordered or disturbed landmass, having thick crumpled strata, regionally metamorphosed rocks, and granitic batholiths; e.g. a *block mountain* or a *folded mountain*. (d) A region characterized by mountains; term usually used in the plural.

mountain and bolson desert A desert area made up of elongated mountain ranges and intervening alluvium-filled fault basins (Stone, 1967, p.233).

mountain apron *bajada*.

mountain blue A blue copper mineral; specif. azurite and chrysocolla.

mountain building *orogeny*.

mountain butter A term used for various salts; esp. *halotrichite*.

mountain chain A complex, connected series of several more or less parallel *mountain ranges* and *mountain systems* grouped together without regard to similarity of form structure, and origin, but having a general longitudinal arrangement or well-defined trend; e.g. the Mediterranean mountain chain of southern Europe. See also: *cordillera*.

mountain climate A climate of high altitudes, characterized by extremes of surface temperature, low atmospheric temperature, strong winds, and rarefied air (Swayne, 1956, p.98).

mountain cork (a) A white or gray variety of asbestos consisting of thick interlaced fibers and resembling cork in texture and lightness (floats on water). Syn: *rock cork*. (b) A fibrous clay mineral such as sepiolite or palygorskite.

mountain crystal *rock crystal*.

mountain flax A fine silky *asbestos*.

mountain glacier An *alpine glacier*; a glacier formed on a mountain slope.

mountain green A green mineral; specif. malachite, green earth, and chrysocolla.

mountain group An assemblage of several mountain peaks or of short mountain ridges; e.g. the Catskill Mountains, N.Y.

mountainite A monoclinic mineral: $(Ca,Na_2,K_2)_2Si_4O_{10}.3H_2O$. Cf: *rhodesite.*

mountain leather (a) A tough variety of asbestos occurring in thin flexible sheets made of interlaced fibers. Syn: *rock leather; mountain paper.* (b) A fibrous clay mineral such as sepiolite or palygorskite.---Syn: *leatherstone.*

mountain limestone A term used in England for a carboniferous limestone occurring in the hills and mountains; specif. the Early Carboniferous limestone forming the Pennine Chain of northern England.

mountain mahogany *obsidian.*

mountain meal *diatomaceous earth.*

mountain milk *lublinite.*

mountain of accumulation A symmetric mountain, frequently of great height, formed by the accretion of material on the Earth's surface, esp. by the ejection of material from a volcano; it tends to occur as an isolated peak. Syn: *accumulation mountain.*

mountain of circumdenudation A mountain consisting of resistant rock that remains after the surrounding, less resistant rock has been worn away, or a mountain representing the remains of a preexisting plateau; e.g. a *monadnock or an inselberg.* Syn: *relict mountain; subsequent mountain; remainder mountain; circumdenudation mountain; mountain of circumerosion.*

mountain of denudation A remnant of "undisturbed and otherwise continuous strata, that have been in part removed by erosion" (Gilbert, 1875, p.21).

mountain of dislocation A mountain "due to the rearrangement of strata, either by bending or fracture" (Gilbert, 1875, p.21); a *fold mountain* or a *fault mountain.*

mountainous (a) Descriptive of a region characterized by mountains or mountain ranges. (b) Resembling a mountain, such as a *mountainous* dome that is strongly elevated and around whose flanks the strata are steeply dipping.

mountain paper A paper-like variety of asbestos occurring in thin sheets; specif. *mountain leather.*

mountain pediment (a) A term introduced by Bryan (1923a, p.30,52-58,88) for a plain of combined erosion and transportation at the foot of a desert mountain range, similar in form to an alluvial plain, and surrounding a mountain in such a manner that at a distance the plain appears to be broad triangular mass (resembling a pediment or gable of a low-pitched roof) above which the mountain projects. This usage is similar to *piedmont pediment.* (b) A pediment occurring within a mountain mass as a relatively high-altitude surface truncating a mountain structure (Tator, 1953, p.51).

mountain range A single, large mass consisting of a succession of mountains or narrowly spaced mountain ridges, with or without peaks, closely related in position, direction, formation, and age; a component part of a *mountain system* or of a *mountain chain.*

mountainside A part of a mountain between the summit and the foot. Syn: *mountain slope.*

mountain slope (a) The sloping surface that forms a mountainside. (b) *mountainside.*

mountain soap A dark clay mineral having a greasy feel and streak; specif. *saponite.* Syn: *rock soap.*

mountain soil An old term for a skeletal soil or *Lithosol,* formed by physical weathering processes in mountainous areas.

mountain system A group of *mountain ranges* exhibiting certain unifying features, such as similarity in form, structure, and alignment, and presumably originating from the same general causes; esp. a series of ranges belonging to an orogenic belt, as many of the major structural provinces in North América. Cf: *mountain chain.*

mountain tallow *hatchettine.*

mountain tract The narrow, upper part of a stream near its source in the mountains, characterized by a steep gradient and a narrow, V-shaped valley in which the water is flowing swiftly. Cf: *valley tract; plain tract.* Syn: *torrent tract.*

mountain wall A very steep mountainside.

mountain wind A nighttime *katabatic wind,* flowing down a mountain slope. It often alternates with a *valley wind.*

mountain wood (a) A compact, fibrous, gray to brown variety of asbestos resembling dry wood in appearance. Syn: *rock wood.* (b) A fibrous clay mineral such as sepiolite or palygorskite.

mourite A violet mineral consisting of a hydrous uranium molybdate: $U^{+4}Mo^{+6}{}_5O_{12}(OH)_{10}$.

mouth [geol] (a) The place of discharge of a body of water into a larger body of water, as where a tributary enters the main stream or where a river enters a sea or lake. Syn: *influx.* (b) *baymouth.* (c) The opening of a geomorphic feature, resembling a mouth and affording entrance or exit, such as of a cave, valley, or canyon. Also, the surface outlet of a subsurface passageway, such as of a volcano.

mouth [paleont] The entrance to the digestive tract through which food passes into the body of an animal; e.g. the central opening at the summit of theca leading to the alimentary system of a blastoid, or the external opening of the body cavity of a coelenterate through which indigestible material is also discharged, or the open end or aperture of the body chamber of an ammonoid shell.

mouth frame The angulated girdle of ossicles surrounding the mouth of an asterozoan.

moutonnée A French adjective meaning "fleecy" or "curled", but often used (incorrectly) as a shortened form of *roche moutonnée.* The term was introduced into geologic literature by Saussure (1786, par.1061, p.512-513) in describing an assemblage of rounded, Alpine hills whose contiguous and repeated curves, taken as a whole and as seen from a distance, resemble a thick fleece and also a curly or wavy wig ("perruque moutonnée") that was fashionable in the late 18th century. The term later implied a fancied resemblance between the general form of a roche moutonnée and that of a grazing sheep whose head is represented by the stoss side.

movable bed A stream bed consisting of readily transportable materials.

movement picture *deformation plan.*

movement plan *deformation plan.*

moveout *stepout time.*

moyite A granite in which quartz comprises more than 50 percent of the light-colored components, and potassium feldspar is the only feldspar present. Its name is derived from the Moyie Sill, British Columbia, Canada.

mozarkite The state rock of Missouri: a varicolored, easily polished Ordovician chert.

M twin law *microcline twin law.*

muck [sed] Dark, finely divided, well decomposed, organic material intermixed with a high percentage of mineral matter, usually silt, which forms surface deposits in some poorly drained areas, e.g. areas of permafrost and lake bottoms.

muck [mining] *waste rock.*

muckite A yellow variety of retinite found in minute particles in coal in a region of central Europe about the upper valley of the Oder River. Named after its discoverer, H. Muck, 19th century German mineralogist.

muck soil A soil that contains 20-50% organic matter; in the U.S., a well-decomposed organic soil (SSSA, 1970, p.11).

mucro An abrupt, sharp terminal point or process of an animal part or a plant part; e.g. a blunt or spinous elevation of the proximal lip of the orifice in some cheilostomatous bryozoans, or the terminal segment of the springing appendage of certain arthropods, or a short abrupt spur or spiny tip of some leaves. Pl: *mucrones* or *mucros*. Syn: *mucron.*

mucron (a) A perforate central scar or button-like projection on the aboral end of a chitinozoan test, serving for attachment. (b) A bryozoan *mucro.*

mucronate Ending in an abrupt, sharp terminal point or process, such as by a distinct and obvious mucro; e.g. said of the cardinal margin of a brachiopod in which the extremities extend into sharp points.

mud [drill] (a) *drilling mud.* (b) The aqueous suspension consisting of cuttings produced by the bit when drilling a borehole.

mud [marine geol] A sticky, fine-grained, marine detrital sediment, either pelagic or terrigenous. Muds are usually described by color: *blue mud; black mud; gray mud; green mud; red mud.*

mud [sed] (a) A slimy and sticky or slippery mixture of water and finely divided particles (silt size or smaller) of solid or earthy material, with a consistency varying from that of a semifluid to that of a soft and plastic sediment, such as that produced by sediment from turbid waters, by injection from a spring or volcano, or when rain falls on an earthy surface; a very wet and soft soil or earthy mass; mire, sludge. (b) An unconsolidated sediment consisting of clay and/or silt, together with material of other dimensions (such as sand), mixed with water, without connotation as to composition; e.g. a recently exposed lake-bottom clay in a soft, ooze-like condition. (c) A mixture of silt and clay; the silt-plus-clay portion of a sedimentary rock, such as the finely divided calcareous matrix of a limestone.

mud aggregate An aggregate of mud grains, commonly having the size of a sand or silt particle, and usually mechanically deposited.

mud ball (a) A spherical mass of mud or mudstone in a sedimentary rock, developed by weathering and breakup of clay deposits. It may measure as much as 20 cm in diameter. (b) *armored mud ball.*

mudbank A submerged or partly submerged ridge of mud along a shore or in a river, usually exposed during low tide.

mud breccia A term used by Ransome & Calkins (1908, p.31) for a *desiccation breccia* containing angular or slightly rounded fragments of "fine-grained argillite embedded in somewhat coarser-grained and more arenaceous material".

mud-buried ripple mark Ripple mark covered by mud settling out of water, characterized by the filling up of troughs and little or no accumulation on the crests (Shrock, 1948, p.109-110).

mud cake [drill] A clay lining or layer of concentrated solids adhering to the walls of a well or borehole, formed where the drilling mud lost water by filtration into a porous formation during rotary drilling. Syn: *filter cake.*

mud cake [sed] A clast formed by desiccation and occurring in an intraformational breccia.

mud circle A *nonsorted circle* characterized by a central core of upwardly injected clay, silt, or sometimes fine sand, surrounded by vegetation; the center is round and generally 10 cm to 2 m in diameter. See also: *frost scar.* Syn: *clay boil; tundra ostiole.*

mud column The length, measured from the bottom of a borehole, of drilling mud standing in a borehole either while being circulated during drilling or when the drill string is not in the hole.

mud cone A small cone of sulfurous mud built around the opening of a *mud volcano* or mud geyser. Syn: *puff cone.*

mud crack (a) An irregular fracture in a crudely polygonal pattern, formed by the shrinkage of clay, silt, or mud in the course of drying under the influence of atmospheric surface

conditions. Also referred to as a *sun crack*, a *shrinkage crack*, and a *desiccation crack.* (b) *mud-crack cast.*---Also spelled: *mudcrack.*

mud-crack cast A mud crack after it has been filled and the filling material (generally sand) has been hardened into rock; often occurs on the underside of a bed immediately overlying a mudstone. Syn: *mud crack.*

mud-crack polygon A *desiccation polygon* bounded by mud cracks. It commonly has three, four, or five sides, although in some instances it may have as many as eight. Syn: *mud polygon.*

mudding off The introduction of drilling mud into a borehole or oil well for the purpose of sealing the walls against natural gas or water during drilling.

muddy [geomorph] n. A topographic feature consisting of a very fine-grained stream deposit produced when water has been ponded, and marked by a top that is "nearly horizontal though more or less concave" (Shaw, 1911, p.489). Type example: the bottom of Muddy River valley in southern Illinois. Cf: *sandy.*

muddy [sed] adj. Pertaining to or characterized by mud; esp. said of water made turbid by sediment, or of sediment consisting of mud.

muddy gravel An unconsolidated sediment containing 30-80% gravel and having a ratio of sand to mud (silt + clay) less than 1:1 (Folk, 1954, p.346).

muddy sand An unconsolidated sediment containing 50-90% sand and having a ratio of silt to clay between 1:2 and 2:1 (Folk, 1954, p.349).

mud field An area of ground saturated with ground water due to the presence of fumaroles (Schieferdecker, 1959, term 4511).

mud flat A relatively level area of fine silt along a shore (as in a sheltered estuary) or around an island, alternately covered and uncovered by the tide, or covered by shallow water; a muddy *tidal flat* barren of vegetation. Cf: *sand flat.* Also spelled: *mudflat.* Syn: *flat.*

mud-flat polygon A *desiccation polygon* developed on a mud flat.

mudflow [mass move] A general term for a mass-movement landform and a process characterized by a flowing mass of predominantly fine-grained earth material possessing a high degree of fluidity during movement. The degree of fluidity is revealed by the observed rate of movement or by the distribution and morphology of the resulting deposit. If more than half of the solid fraction of such a mass consists of material larger than sand size, the term *debris flow* is preferable (Sharp & Nobles, 1953; Varnes, 1958). Mudflows are intermediate members of a gradational series of processes characterized by varying proportions of water, clay, and rock debris. The water content of mudflows may range up to 60%. The degree of water bonding, determined by the clay content and mineralogy, critically affects the viscosity of the matrix and the velocity and morphology of the flow. With increasing fluidity, mudflows grade into loaded and clear streams; with a decrease in fluidity, they grade into earthflows. Also spelled: *mud flow.*

mudflow [sed] A minor sedimentary structure found in fine grained rocks and indicative of local flowage while the material was still soft. Syn: *mud mark.*

mud flow [volc] *lahar.*

mudflow levee A sharp linear ridge marking the edge of mudflow type of mass movement, and consisting of boulders shoved aside by the force of the flow.

mud flush *drilling mud.*

mud furrow An obsolete term (used by Hall, 1843, p.234) for the structure called *groove cast* by Shrock (1948, p.162-163).

mud geyser A geyser that erupts sulfurous mud; a type of mud volcano.

mud glacier A viscous mass of surficial material moving slow downslope by solifluction, as a glacier.

mud lava [volc] The water-saturated volcanic debris of mud-flows or lahars.

mud lava [grd wat] The sulfurous and sometimes carbona-ous material contained in mud pots or erupted from mud volcanoes or mud geysers.

mud log A continuous analysis of the drilling mud and cuttings to determine the presence or absence of oil, gas, or water in the formations penetrated by the drill bit, and to ascertain the depths of any oil- or gas-bearing formations.

mudlump A diapiric sedimentary structure that forms a small (1000 square meters in area), short-lived island near the mouth of a major distributary of the Mississippi River delta and that consists of a broad, low (2-4 m above sea level) mound or swelling of silt or thick plastic clay. It is created by the loading action of rapidly deposited delta-front sands upon inter-weight prodelta clays, causing the clays to be intruded or thrusted upward into and through the overlying sandbar deposits. Cf: *gas-heave structure*. Also spelled: *mud lump*.

mud mark *mudflow* [sed].

mud pellet A small (3-13 mm in diameter), flattened to rounded, and irregularly shaped mass of mud or mudstone in a sedimentary rock (such as in a siltstone, limestone, or con-glomerate). It is developed by weathering and breakup of clay deposits (small particles of which are washed about locally or transported short distances) and by penecontemporaneous compaction and induration of the mud.

mud pit *slush pit*.

mud polygon (a) A *nonsorted polygon* whose center is bare of vegetation but whose outlining reticulate fissures contain peat and plants. The term was suggested by Elton (1927, p.165) to replace *fissure polygon*, but it is ambiguous because forms without a stone border do not invariably consist of "mud" but may consist of sand, gravel, or a nonsorted mixture of sand, clay, and silt with stones (Washburn, 1956, p.825-826). (b) *mud-crack polygon*.

mud pot A type of hot spring containing boiling mud, usually sulfurous and often multicolored, as in a *paint pot*. Mud pots are commonly associated with geysers and other hot springs in volcanic areas, esp. Yellowstone National Park, Wyo., U.S.A. Syn: *sulfur-mud pool*.

mud-ridge ripple mark A *longitudinal ripple mark* with a regu-lar profile, a usually symmetric crest, and a narrow and angu-lar ridge that is situated between much wider and relatively flat troughs and that frequently branches (always converging downcurrent). Cf: *corrugated ripple mark*.

mud ring A ring of solid material on the walls of a well or borehole, formed where the drilling mud lost water by filtration to a porous formation during rotary drilling.

mud rock A syn. of *mudstone*. Also spelled: *mudrock*.

mud-rock flood A violent and destructive rush of water gener-ated by a cloudburst and laden with rocks, mud, and debris engulfed along its path.

mudrush The sudden inflow of waterlogged surface material into shallow mine workings. See also: *running ground*.

mud shale A consolidated sediment consisting of no more than 10% sand and having a silt/clay ratio between 1:2 and 2:1 (Folk, 1954, p.350); a fissile mudstone.

mudslide A relatively slow-moving type of mudflow in which movement occurs predominantly by sliding upon a discrete boundary shear surface (Hutchinson & Bhandari, 1971). Cf: *earthflow*.

mud spot *frost scar*.

mud stalagmite A stalagmite that is composed of clay or sandy clay, with less than 30% calcite.

mudstone (a) An indurated mud having the texture and com-position, but lacking the fine lamination or fissility, of shale; a blocky or massive, fine-grained sedimentary rock in which the proportions of clay and silt are approximately the same; a

nonfissile mud shale. Shrock (1948a) regards mudstone as a partly indurated mud that slakes upon wetting. See also: *clay-stone; siltstone*. (b) A general term that includes clay, silt, claystone, siltstone, shale, and argillite, and that should be used only when the amounts of clay and silt are not known or specified or cannot be precisely identified, or "when a deposit consists of an indefinite mixture of clay, silt, and sand parti-cles, the proportions varying from place to place, so that a more precise term is not possible" (Twenhofel, 1937, p.98), or when it is desirable to characterize the whole family of finer-grained sedimentary rocks (as distinguished from sands-tones, conglomerates, and limestones). Syn: *mud rock*. (c) A term used by Dunham (1962) for a mud-supported carbonate sedimentary rock containing less than 10% grains (particles with diameters greater than 20 microns); e.g. a calcilutite. The term specifies neither mineralogic composition nor mud of clastic origin. Cf: *wackestone; packstone; grainstone*.---The term was apparently first used by Murchison (1839) for cer-tain massive, dark-gray, fine-grained Silurian shales of Wales, which on exposure and wetting rapidly decompose or disinte-grate into muds.

mudstone conglomerate (a) A conglomerate containing mud-stone clasts, such as one produced by penecontemporaneous compaction and induration of muds. (b) *desiccation conglom-erate*.---Cf: *conglomeratic mudstone*.

mudstone ratio A uranium-prospector's term for the ratio of the total thickness of red mudstone to that of green mudstone within an assumed stratigraphic horizon. "Its value is based upon the premise that uranium-bearing solutions will bleach red mudstone containing ferric iron to green mudstone con-taining ferrous iron" (Ballard & Conklin, 1955, p.194).

mud stream A mass of moving sediment mixed with water, such as a mudflow, turbidity current, or mud glacier. Also spelled: *mudstream*.

mud-supported A term used by Dunham (1962) to describe a sedimentary carbonate rock whose sand-size particles (at least 10% of the total bulk, but not in sufficient amount to be able to support one another) are embedded or "floating" in, and supported by, the muddy matrix. Cf: *grain-supported*.

mud tuff *dust tuff*.

mud volcano An accumulation, usually conical, of mud and rock ejected by volcanic gases; also, a similar accumulation formed by escaping petroliferous gases. The term has also been used for a *mud cone* not of eruptive origin. Syn: *hervid-ero; macaluba*. Cf: *air volcano; salinelle*.

mugearite A dark-colored, fine-grained rock similar to tra-chyte in texture but in which the chief feldspar is oligoclase, and containing orthoclase and olivine, with accessory apatite and opaque oxides.

muirite A mineral: $Ba_{10}Ca_2MnTiSi_{10}O_{30}(OH,Cl,F)_{10}$.

mukhinite A mineral of the epidote group: $Ca_2(Al_2V)(Si_2O_8)-(OH)$.

mule's-ear cave A driller's term, used in the Appalachian re-gion of the U.S., for a hard shale that caves into wells in the form of splinters that bear a fancied resemblance to the ears of mules.

mull [coast] A Scottish term for a *headland*; e.g. *Mull* of Gal-loway.

mull [soil] A type of humus usually developed in the forest that is incorporated with underlying mineral matter. Cf: *mor*.

Müllerian law The law that expresses the regularity in distribu-tion of 20 radial spines on the shells of radiolarians of the su-border Acantharina (four spines on each of the five circles that are comparable to the equatorial, two tropical, and two polar circles of the terrestrial globe): "between two poles of a spineless axis are regularly disposed five parallel zones, each with four radial spines; the four spines of each zone are equi-distant one from another, and also equidistant from each pole; and the four spines of each zone are so alternating with those of each neighboring zone, that all twenty spines together lie in four meridian planes, which transect one another at an angle of 45°'' (translated from Haeckel, 1862, p.40). Named by Haeckel in honor of Johannes Müller (1801-1858), German physiologist and zoologist, who first recognized the regularity in the disposition of the 20 radial spines (Müller, 1858, p.12 & 37). Syn: *icosacanthic law*.

Müller's glass *hyalite*.

mullicite A variety of vivianite occurring in cylindrical masses.

mullion A columnar structure in folded sedimentary and metamorphic rocks in which columns of rock appear to intersect. Mullions may be formed parallel to the direction of movement, as along fault planes, or perpendicular to it, as in *fold mullions, cleavage mullions*, or other *irregular mullions* (Spencer, 1969, p.245). Cf: *rodding*.

mullion structure A wavelike pattern of parallel grooves and ridges, measuring as much as several feet from crest to crest, and formed along a fault surface due to the fault movement. Etymol: Old French, *moienel*, "medial". Cf: *striation* [*fault*]; *groove* [*fault*]; *slickenside; slip-scratch*.

mullite A rare orthorhombic mineral: $Al_6Si_2O_{13}$. It is used as a refractory. Mullite is often found as a synthetic material in ceramic products. Syn: *porcelainite*.

mullitization The formation of mullite from minerals of the sillimanite group by heating.

mullock *waste rock*.

mulola An *oshana* in Angola (Stamp, 1961, p.347).

multi-band system A multiple image-forming system for simultaneously observing the same target utilizing several filtered spectral bands, through which data can be recorded. The term is usually applied to line-scan image-forming systems which utilize dispersant optics to split wavelength bands apart for viewing by several filtered detectors. Cf: *spectra-zonal system*. Less preferred syn: *multi-channel system*; *multi-spectral system*.

multi-channel system A nonrecommended syn. of *multi-band system*.

multicycle adj. Said of a landscape or landform produced during or passing through more than one cycle of erosion, and bearing the traces of the former condition(s); e.g. a *multicycle* coast with a series of elevated sea cliffs separated from each other in stair-like fashion by narrow wave-cut benches, each sea cliff representing a separate shoreline cycle (Cotton, 1922, p.426); or a *multicycle* valley showing on its sides a series of straths resulting from successive uplifts. Syn: *multicyclic; multiple-cycle; polycyclic*.

multifossil range zone *concurrent-range zone*.

multigelation Often-repeated freezing and thawing by any process (Washburn, 1956, p. 838). See also: *regelation*.

multilocular Having, divided into, or composed of many small chambers or vesicles; specif. said of a many-chambered test of a unicellular organism (such as a foraminifer). See also: *polythalamous*.

multipartite map A *vertical-variability map* that shows the degree of distribution of one lithologic type within certain parts (such as top, middle, and bottom thirds) of a given stratigraphic unit. The map was introduced by Forgotson (1954).

multiple-cycle *multicycle*.

multiple detectors Two or more seismic detectors, e.g. seismometers, geophones, whose combined energy is fed into a single amplifier-recorder circuit in order to reduce undesirable near-surface waves. Syn: *multiple recording group*.

multiple-exposure method *isolation of outcrops*.

multiple fault *step fault*.

multiple glaciation The alternating advance and recession of glacier ice during the Pleistocene Epoch.

multiple intrusion Any type of igneous intrusion which has been produced by several injections separated by periods of crystallization. Chemical composition of the various injections

is approximately the same. Cf: *composite intrusion*.

multiple recording group *multiple detectors*.

multiple reflection A seismic wave which has been reflected more than once. Syn: *repeated reflection; secondary reflection*.

multiple shotholes In the seismic shooting technique of seismic prospecting, two or more shotholes which are used simultaneously. They are so spaced as to minimize near-surface interferences.

multiple tunnel One of a series of openings in the chamber a fusulinid test, produced by resorption of lower (adaxial parts of septa.

multiple twin A twinned crystal that is formed by repeated twinning.

multiple working hypotheses The name given by Chamberl (1897) to a method of "mental procedure" applicable geologic studies in which several rational and tenable explanations of a phenomenon are developed, coordinated, ar evaluated simultaneously in an impartial manner.

multiplex A stereoscopic (anaglyphic) plotting instrument use in preparing topographic maps from aerial photographs.

multisaccate Said of pollen with more than two vesicles.

multiserial Arranged in, characterized by, or consisting of several or many rows or series; e.g. "multiserial ambulacrum" an echinoid with pore pairs arranged in more than two longitudinal series, or said of a protist composed of numerous rows of cells or other structural features.

multi-spectral system A nonrecommended synonym of *multiband system*. All such systems sense radiation in the electromagnetic spectrum.

multisystem A set of phases more numerous than can coexist stably at any set of conditions, thus formally possessing, the phase-rule sense, a negative number of degrees of freedom; the equilibrium relationships among the phases of a multisystem can be represented by an array of invariant points connected by univariant, bivariant, etc., equilibria, some which in general will be stable, some metastable.

multivariate Pertaining to, having, or involving two or more independent mathematical variables; e.g. "multivariate analysis that separates and defines the effects of a number of statistically independent variables.

multivincular Said of a type of ligament of a bivalve mollusc (e.g. *Isognomon*) consisting of serially repeated elements alivincular type. Also, said of the hinge of various bivalves having several small separate ligaments.

multi-zonal system *spectra-zonal system*.

multipost Said of a process involving an igneous rock that occurs some time after consolidation of the magma, i.e. later than a *deuteric* process.

mundic A syn. of *pyrite*. Drillers often use "mundick" to designate pyrite.

muniongite A hypabyssal rock resembling *tinguaite*, being composed of loxoclase, nepheline, acmite, and sometimes cancrinite. It contains more nepheline than orthoclase, the distinguishing it from tinguaite.

munro A Scottish term for a hill more than 900 m in height separated from another by a "dip" of more than 150 m (Dalling & Boyd, 1964, p.23). Named after H.T. Munro, Scottish mountaineer.

Munsell color system A system of color classification that applied in geology to the colors of rocks and soils. Color defined by its hue, value or brilliance, and chroma (purity).

muntenite A variety of amber from Rumania.

mural deposit A *cameral deposit* along the wedge-like extension of each septum attached to the wall of a nautiloid conch.

mural escarpment A rocky cliff with a face nearly vertical, like a wall (Lee, 1840, p.350).

muralite A *phyteral* of coal that represents the structure plant cell walls and that occurs in some types of vitrain.

mural joint structure A pattern of cubical or rectangular blocks of rock formed by numerous right-angle joints. Syn: *rectangular joint structure*.

mural plate *compartmental plate*.

mural pore (a) A circular or oval small hole in the wall between adjoining corallites, as in some tabulates. (b) An opening in the shell wall of a foraminifer, as distinguished from a *septal pore*. (c) A *communication pore* in a bryozoan.

mural rim The raised edge of the gymnocyst, where it meets the frontal area, in many anascan cheilostomes (bryozoans). It often bears marginal spines.

murambite A leucite basanite containing abundant mafic minerals.

murasakite A schistose rock composed essentially of piedmontite and quartz. The term was originated by Kato in 1887 who named it after Murasako, Japan.

murbruk structure *mortar structure*.

murchisonite (a) A flesh-red perthitic variety of orthoclase with good cleavage and often gold-yellow reflections in a direction perpendicular to (010). (b) A name applied to moonstone and the iridescent feldspar from Frederiksvaern, Norway.

Murderian Stage in New York State: Upper Silurian (upper Cayugan; above Canastotan).

murdochite A black isometric mineral: $PbCu_6O_8$.

muri The walls of the positive reticulate sculpture in pollen and spores.

murite A dark-colored feldspathoid-rich phonolite in which the mafic minerals comprise about 50 percent of the rock.

murmanite A violet mineral: $Na_2(Ti,Nb)_2Si_2O_9.nH_2O$. Cf: *monosovite*.

muromontite A mineral: $Be_2FeY_2(SiO_4)_3$ (?). It is perhaps identical with gadolinite or is a member of the clinozoisite group.

murram Deposits of *bog iron ore* in the tropics of Africa.

murus reflectus A sutural indentation of the apertural face of foraminiferal test, longitudinally and obliquely folded below the aperture (as in *Osangularia*). Etymol: Latin.

Muschelkalk European stage (esp. in Germany): Middle Triassic (above Bunter, below Keuper).

muscle field (a) An area of a brachiopod valve where muscle scars are concentrated. (b) A concave or flat area on the ventral (inner) side of an articular face of a muscularly articulated plate of a crinoid ray, serving for the attachment of muscle fibers.

muscle platform A relatively broad and solid or undercut elevation of the inner surface on either valve of some brachiopods, commonly bearing muscles or to which the muscles are attached. Syn: *platform [paleont]*.

muscle scar (a) One of the differentiated, more or less well-defined impressions or elevations on the inner surface of a bivalve shell (as in an ostracode, brachiopod, or pelecypod), marking the former place of attachment of a muscle; e.g. an *adductor muscle scar*. Syn: *muscle mark; scar*. (b) Smooth or slightly depressed paired areas in the external surface of the axial region of a trilobite exoskeleton, interpreted as areas of muscle attachment (TIP, 1959, pt.O, p.123).

muscle track The path of successive muscle impressions formed in a brachiopod by migration of the muscle base during growth. Syn: *track [paleont]*.

muscovadite A cordierite-biotite norite, formed as a result of the partial absorption of country rock fragments by the magma.

muscovado A term applied in Minnesota to rusty-colored outcropping rocks (such as gabbros and quartzites) that resemble brown sugar. Etymol: Spanish, "brown sugar".

muscovite (a) A mineral of the mica group: KAl_2-$(AlSi_3)O_{10}(OH)_2$. It is usually colorless, whitish, or pale brown, and is a common mineral in metamorphic rocks (gneisses and schists), in most acid igneous rocks (such as granites and pegmatites), and in many sedimentary rocks (esp. sandstones). See also: *sericite*. Syn: *white mica; potash mica; common mica; Muscovy glass; mirror stone; moscovite*. (b) A term applied in clay mineralogy to *illite*.

Muscovy glass *muscovite*.

muscular articulation A type of articulation of crinoid ossicles effected by muscle fibers in addition to ligaments. Cf: *ligamentary articulation*.

mush *brash ice*.

mush frost *pipkrake*.

mushroom ice An *ice pedestal* with a round and expanded top.

mushroom rock A table-like rock mass formed by wind abrasion in an arid region, consisting of a upper layer of resistant rock underlain by a softer, partially eroded layer, thereby forming a thin "stem" supporting a wide mass of rock, the whole feature itself resembling a mushroom in shape. See also: *pedestal rock; cheesewring; zeuge; gara*. Syn: *toadstool rock; fur-cap rock*.

mushroom stalagmite *stool stalagmite*.

musical sand A *sounding sand* that emits a definite musical note or tone when stirred, trodden on, or otherwise disturbed; esp. *whistling sand*.

muskeg (a) A bog, usually a sphagnum bog, frequently with tussocks of deep accumulations of organic material, growing in wet, poorly drained, boreal regions, often areas of permafrost. Tamarack and black spruce are commonly associated with muskeg areas. Syn: *maskeeg*. (b) A term sometimes used in Michigan for a *bog lake*.

mussel (a) *mytilid*. (b) Any of the common freshwater pelecypods belonging to the superfamily Unionacea.

mustard gold A spongy, free type of gold deposit associated with the gossan above telluride deposits.

mustard-seed coal The smallest size of *buckwheat coal*, equivalent to number five in the series. It will pass through a 3/64 inch round mesh.

mutant The offspring bearing the *mutation*.

mutation A spontaneously occurring fundamental change in heredity thought to result in the development of new individuals that are basically unlike their parents and that can be acted upon by natural selection to effect desirable changes and to establish new species. See also: *mutant*.

muthmannite A gray-white mineral: $(Ag,Au)Te$.

mutual inductance The complex of voltage in one circuit to current in another to which it is inductively coupled.

mutualism The relationship that exists between two organisms in which both are benefitted. Cf: *parasitism; commensalism; symbiosis*.

mycelium The whole vegetative mass of a fungus, composed of *hyphae*.

mycobiont The fungal partner, or component, of a lichen. Cf: *phycobiont*.

mycorrhiza A symbiotic association of root and/or rhizome with a fungus.

mylonite As introduced by Lapworth in 1885, a compact, chertlike rock without cleavage, but with a streaky or banded structure, produced by the extreme granulation and shearing of rocks which have been pulverized and rolled during overthrusting or by action of intense dynamic metamorphism in general. Mylonite may also be described as a microbreccia with fluxion structure (Holmes, 1920). See also:*protomylonite; ultramylonite; blastomylonite*.

mylonite gneiss A metamorphic rock that is intermediate in character between mylonite and schist. Felsic minerals show cataclastic phenomena without much recrystallization, and often occur as augen surrounded by and alternating with schistose streaks and lenticels of recrystallized mafic minerals (Holmes, 1928, p.164).

mylonitic structure A structure characteristic of mylonites, produced by intense microbrecciation and shearing which gives the appearance of a flow structure. Cf: *flaser structure.*

mylonitization Deformation of a rock by extreme microbrecciation, due to mechanical forces applied in a definite direction, without noteworthy chemical reconstitution of granulated minerals. Characteristic features of the mylonites thus produced have a flinty, banded, or streaked appearance, and undestroyed augen and lenses of the parent rock in a granulated matrix (Schieferdecker, 1959). Also spelled: *mylonization.*

mylonization *mylonitization.*

myocyte A fusiform contractile cell in sponges.

myodocope Any marine ostracode belonging to the order Myodocopida, characterized by a shell with subequal valves that may be ornamented or smooth and by a well developed rostrum. Most planktonic ostracodes are myodocopes. Their stratigraphic range is Ordovician to present.

myophore A part of a shell adapted for the attachment of a muscle; e.g. a process for attachment of an adductor muscle of a pelecypod, or the distal expanded part of the differentiated cardinal process of a brachiopod to which the diductor muscles were attached.

myophragm A median ridge of secondary shell of a brachiopod, secreted between muscles and not extending beyond the muscle field.

myostracum The part of the shell wall of a bivalve mollusk secreted at the attachments of the adductor muscles. Syn: *hypostracum.*

myriapod Any terrestrial arthropod belonging to the superclass Myriapoda, which includes insects. They are characterized by a body that is divided into a head and trunk with one pair of antennae on the head and uniramous appendages. They are rarely preserved as fossils but are known from the Upper Silurian to the present.

myrickite (a) A whitish or grayish chalcedony, opal, or massive quartz unevenly colored by or intergrown with pink reddish inclusions of cinnabar, the color of which tends to become brownish. The opal variety is known as *opalite.* (b) Cinnabar intergrown with common white opal or transluce chalcedony.

myrmekite A wart-like intergrowth of plagioclase feldsp (generally oligoclase) and vermicular quartz, generally repla ing potassium feldspar, formed during the later stages of co solidation in an igneous rock or during a subsequent period plutonic activity. The quartz occurs as blobs, drops, or verm cular shapes within the feldspar.

myrmekite-antiperthite A myrmekite-like intergrowth of pr dominant plagioclase and vermicular orthoclase (Schiefe decker, 1959, term 5177).

myrmekite-perthite A myrmekite-like intergrowth of microclir and vermicular plagioclase (Schieferdecker, 1959, ter 5176).

myrmekitic (a) Said of a texture characterized by inte growths of feldspar and vermicular quartz. (b) Pertaining to characteristic of myrmekite.

mytilid Any bivalve mollusk belonging to the family Mytilida characterized by an equivalve, inequilateral shell with pros gyrate umbones. Syn: *mussel.*

mytiliform Said of a slipper-shaped shell of a bivalve mollus specif. shaped like a mussel shell, such as the elongated ar equivalve shell of *Mytilus* (a genus of marine bivalve mc lusks).

myxomycete An organism of the class Myxomycetes, con prising the slime molds, of uncertain systematic position b usually associated with the fungi. It exists as complex, mobi plasmodia and reproduces by spores. Cf: *eumycete; schizor ycete.*

myxosponge Any demosponge whose only skeleton is meso loea, being without spicules or spongin.

N

nab A British term for a projecting part of an eminence; e.g. a headland or promontory (a *ness*), or a spur of an escarpment.

nacre The hard, iridescent internal layer of various mollusk shells, having unusual luster and consisting chiefly of calcium carbonate in the form of aragonite deposited organically in a succession of thin overlapping laminae parallel to the growth lines of the shell and interleaved with thin sheets of organic matrix (such as conchiolin). Cf: *calcitostracum.* Syn: *mother-pearl.*

nacreous A type of mineral luster resembling that of mother-pearl. Syn: *pearly.*

nacrite A well-crystallized clay mineral of the kaolin group: $Al_2Si_2O_5(OH)_4$. It is polymorphous with kaolinite and dickite. Nacrite is structurally distinct from other members of the kaolin group, being the most closely stacked in the c-axis direction.

nadir [photo] (a) *photograph nadir.* (b) *ground nadir.*

nadir [geodesy] The point on the celestial sphere that is directly beneath the observer and directly opposite to the *zenith.*

nadir point A nadir; specif. *photograph nadir.*

nadorite A brownish-yellow mineral: $PbSbO_2Cl$.

naegite A variety of zircon containing thorium and uranium. Also spelled: *naegite.*

nafud *nefud.*

nagatelite A phosphatian variety of allanite.

nagelfluh A massive and variegated Miocene conglomerate accompanying the molasse of the Alpine region in Switzerland. It contains pebbles that appear like flights or swarms of nailheads. Etymol: German *Nagel*, "nail", + *Fluh*, "mass of rock, stratum, layer". Syn: *gompholite.*

nagyagite A dark lead-gray mineral: $Pb_5Au(Te,Sb)_4S_{5-8}$. Syn: *black tellurium; tellurium glance.*

nahcolite A white monoclinic mineral: $NaHCO_3$.

naif Said of a gemstone having a true or natural luster when cut. Syn: *naife.*

nailbourne *bourne.*

nailhead spar A variety of calcite in crystals showing a combination of hexagonal prisms with flat rhombohedrons.

nailhead striation A glacial striation with a definite or blunt head or point of origin, generally narrowing or tapering in the direction of ice movement and coming to an indefinite end. Syn: *nailhead scratch.*

naked flagellar field The area around the flagella in which no coccoliths are present in coccolithophores lacking a complete cover of coccoliths. Cf: *covered flagellar field.*

naked karst Karst that is developed in a region without soil cover, so that its topographic features are well exposed. Cf: *covered karst.*

naked pole The end in a nonflagellate coccolithophore that is free of coccoliths.

nakhlite An achondritic stony meteorite consisting of a holo-crystalline aggregate of diopside (75%) and olivine.

nala Var. of *nullah.*

naledi A Russian term for *aufeis.* Sing: *naled.*

nallah Var. of *nullah.*

Namurian European stage: Lower and Upper Carboniferous (above Viséan, below Westphalian). It is divided into a lower stage (Lower Carboniferous or Upper Mississippian) and an upper stage (Upper Carboniferous or Lower Pennsylvanian). Cf: *Lancastrian.*

nanism The development of abnormally or exceptionally small size; dwarfishness. Cf: *gigantism.*

nannofossil (a) A collective term for fossil discoasters and coccoliths, both primarily calcareous microfossils, mostly rather near the limit of resolution of the light microscope and hence best studied with electron microscopy. (b) A term sometimes used in a more general sense for other small marine (usually algal) fossils.

nannoplankton Plankton of the size range 5-60 microns. They are larger than *ultraplankton* but smaller than *microplankton*, *macroplankton*, and *megaloplankton*.

nanozooid A dwarf bryozoan zooid containing reduced polypide (as in some bryozoans of the order Cyclostomata).

Nansen bottle A device used in oceanography to obtain sub-surface sea-water samples and in-situ temperature measurements. It is a bottle, open at both ends, to which are attached a pair of *reversing thermometers.* When the device is in place, it is reversed, which encloses the water sample and records its temperature. Nansen bottles are usually used in a series on a line.

Nansen cast A syn. of *hydrographic cast*, so called because of the use of a series of Nansen bottles to obtain the samples and measurements.

nant A little valley with a stream. Etymol: Celtic, "brook".

nantokite A colorless, white, or grayish isometric mineral: $CuCl$.

naotic septum A rugose corallite septum characterized by development peripherally in a series of closely spaced dissepiment-like plates, as in *Naos.*

naphtha An archaic term for liquid petroleum. It is now used to designate those hydrocarbons of the lowest boiling point (under $250°C$) that are liquid at standard conditions, but easily vaporize and become inflammable, and which are used as cleaners and solvents.

naphthalene A white, crystalline, bicyclic aromatic hydrocarbon, formula $C_{10}H_8$, which has a characteristic odor and occurs in coal tar and some crude oils.

naphthene *cycloparaffin.*

naphthene base Crude oil with a high carbon and low oxygen content that leaves an asphaltic residue after refining. Cf: *paraffin base; mixed base.* Syn: *asphalt base.*

naphthine A syn. of *hatchettine.* Also spelled: *naphtine; naphtein.*

napoleonite *corsite.*

Napoleonville North American (Gulf Coast) stage: Miocene (above Anahuac, below Duck Lake).

nappe [struc geol] A sheetlike, allochthonous rock unit of whatever internal structure or origin. The mechanism may be thrust faulting or recumbent folding or both. The term was first used as "nappe de recouvrement" (1893) for the large, allochthonous sheets of the western Alps, and it has been adopted into English. The German equivalent, *Decke*, is also sometimes used in English. Etymol: French, "cover sheet, tablecloth". See also: *klippe.*

nappe [volc] *lava flow.*

nappe [hydraul] A sheet of water overflowing a dam.

nappe outlier *klippe.*

nari A variety of *caliche* that forms by surface or near-surface alteration of permeable calcareous rocks (dissolution and redeposition of calcium carbonate) and that occurs in the drier parts of the Mediterranean climatic region. It is characterized by a fine network of veins surrounding unreplaced remnants of the original rock, and it often contains clastic particles (rocks and shells). Etymol: Arabic *nar*, "fire", in allusion to its use in limekilns.

Narizian North American stage: upper Eocene (above Ulatisian, below Fresnian).

narrow A constricted section of a mountain pass, of a valley, or of a cave; a gap or narrow passage between mountains. See also: *narrows.*

narrow band filter A material of high infrared transmittance which transmits only a narrow band of predetermined wavelength. Cf: *infrared filter.*

narsarsukite A yellow mineral: $Na_2(Ti,Fe)Si_4(O,F)_{11}$.

nasal tube A curved cylinder or prismatic tube in a phaeodarian radiolarian, embracing the central capsule on one side and a galea on the other side. Syn: *rhinocanna*.

nase *naze*.

nasinite A monoclinic mineral: $Na_4B_{10}O_{17}.7H_2O$. It is dimorphous with ezcurrite.

nasledovite A mineral: $PbMn_3Al_4(CO_3)_4(SO_4)O_5.5H_2O$.

nasonite A white mineral: $Ca_4Pb_6Si_6O_{21}Cl_2$.

nassellarian *nasselline*.

nasselline Any radiolarian belonging to the suborder Nassellina, characterized by a central capsule perforated only at one pole and enclosed by a single membrane. Syn: *nassellarian*.

nasturan *pitchblende*.

national meridian A meridian chosen in a particular nation as the reference datum for determining longitude for that nation. It is commonly defined in several European countries with respect to a key point in the capital city. Cf: *Greenwich meridian*.

native adj. A syn. of *endemic*.

native asphalt Liquid or semiliquid asphalt exudations or seepages including surface flows and lakes. Native asphalt deposits containing more than 10 percent by dry weight of mineral matter are called native asphalts associated with mineral matter. Syn: *natural asphalt*.

native coke *natural coke*.

native element Any element found uncombined in a nongaseous state in nature. Nonmetallic examples are carbon, sulfur, and selenium; semimetal examples are antimony, arsenic, bismuth, and tellurium; *native metals* include silver, gold, copper, iron, mercury, iridium, lead, palladium, and platinum.

native metal A metallic *native element*.

native paraffin *ozocerite*.

native water (a) *connate water*. (b) *formation water*.

natric Said of a soil horizon that has the same properties as an argillic horizon, but that also displays a blocky, columnar, or prismatic structure and has a subhorizon with an exchangeable-sodium saturation of over 15% (SSSA, 1970).

natroalunite A mineral of the alunite group: $NaAl_3(SO_4)_2$-$(OH)_6$. It is isomorphous with alunite. Syn: *almerite*.

natroborocalcite A syn. of *ulexite*. Also spelled: *natronborocalcite*.

natrochalcite An emerald-green mineral: $NaCu_2(SO_4)_2$-$(OH).H_2O$.

natrojarosite A yellowish-brown to golden-yellow mineral of the alunite group: $NaFe_3(SO_4)_2(OH)_6$. Syn: *utahite*.

natrolite A zeolite mineral: $Na_2Al_2Si_3O_{10}.2H_2O$. It sometimes contains appreciable calcium, and usually occurs in slender, acicular or prismatic crystals. Partial syn: *mesotype* [mineral]; *needle zeolite*.

natromontebrasite A mineral: $(Na,Li)AlPO_4(OH,F)$. It is isomorphous with amblygonite and montebrasite. Syn: *fremontite*.

natron A white, yellow, or gray monoclinic mineral: $Na_2CO_3.10H_2O$. It is very soluble in water, and occurs mainly in solution (as in the soda lakes of Egypt and western U.S.) or in saline residues.

natroniobite A monoclinic mineral: $NaNbO_3$. It is dimorphous with lueshite.

natron lake *soda lake*.

natrophilite A mineral: $NaMn(PO_4)$.

natural arch (a) A *natural bridge* resulting from erosion. (b) A landform similar to a natural bridge but not formed by erosive agencies (Cleland, 1910, p.314). (c) *sea arch*.---Syn: *arch*.

natural area (a) An area of land or water that has retained its wilderness character, although not necessarily completely natural and undisturbed, or that has rare or vanishing flora, fauna, archaeological, scenic, historical, or similar features of scientific or educational value (Ohio Legislative Service Commission, 1969, p.3); e.g. a "research natural area" where "natural processes are allowed to predominate and which is preserved for the primary purposes of research and education" (U.S. Federal Committee on Research Natural Area 1968, p.2). (b) Any outdoor site that contains an unusu biologic, geologic, or scenic feature or that illustrates "common principles of ecology uncommonly well" (Lindsey et a 1969, p.4). ----See also: *wilderness area*.

natural asphalt *native asphalt*.

natural bridge (a) Any arch-like rock formation created erosive agencies and spanning a ravine or valley; an openin found where a stream abandoned a meander and broke through the narrow meander neck, as at Rainbow Bridg Utah. (b) In a limestone terrane, the remnant of the roof an underground cave or tunnel that has collapsed. Syn: *ka bridge*. (c) *sea arch*.---Syn: *natural arch*.

natural coke Coal that has been naturally carbonized by co tact with an igneous intrusion or by natural combustion. Sy *carbonite*; *coke coal*; *cokeite*; *native coal*; *finger coal*; *bli coal*; *black coal*; *cinder coal*. Cf: *clinker* [coal]; *coke*.

natural gas [eco geol] Any of the gaseous hydrocarbons ge erated below the Earth's surface. Methane is usually found a product of decaying vegetable matter; ethane, propane, a butane in association with liquid hydrocarbons and distill within the lithified crust.

natural gas [grd wat] Gas trapped in the zone of saturatio under pressure from, and partially dissolved in, underlyi water or petroleum (Meinzer, 1923, p.21). Cf: *subsurface a included gas*.

natural gas liquids Hydrocarbons that occur naturally in ga eous form or in solution with oil in the reservoir, and that a recoverable as liquids by condensation or absorption pr cesses; e.g. natural gasoline, condensate, and liquefied petr leum gases.

natural gasoline The liquid fraction recovered from natural g emerging from a well.

natural glass A vitreous, amorphous, inorganic substance o curring in nature that has solidified from magma too quickly crystallize. Granitic or acid natural glass includes pumice a obsidian; an example of a basaltic natural glass is tachyli Quartz may also occur as a natural glass on sandy deserts.

natural history The study of the nature and history of all a mal, vegetable, and rock and mineral forms. The term sometimes considered old-fashioned except when applied the animal world, but is still used by some in its geolog sense.

natural horizon *apparent horizon*.

natural landscape A landscape that is unaffected by the a tivities of man (in contrast to the "cultural landscape" resu ing from man's settlement); it includes landforms and the natural plant cover, and the contrast between land and wate Syn: *physical landscape*.

natural law *law of nature*.

natural levee (a) A long, broad, low ridge or embankment sand and coarse silt, built by a stream on its flood plain a along both banks of its channel, esp. in time of flood whe water overflowing the normal banks is forced to deposit th coarsest part of its load. It has a gentle slope (about 60 cm km) away from the river and toward the surrounding floo plain, and its highest elevation (about 4 m above the floo plain) is closest to the river bank, at or near normal floo level. Syn: *levee*; *raised bank*; *spill bank*. (b) Any natura produced low ridge resembling a natural levee; e.g. a la levee, or a sediment ridge bordering a fan-valley.

natural load The quantity of sediment that a stable strea carries.

natural mold The empty space or cavity left after solution an original fossil shell or other organic structure, bounded the external impression (external mold) and the surface of th internal filling (or of the steinkern) (Shrock & Twenhofe 1953, p.19). See also: *mold* [paleont]; *cast* [paleont].

natural region (a) A part of the Earth's surface characterize by relatively uniform and distinctive physical features (relief, structure, climate, vegetation) within its borders, and therefore possessing (to a certain extent) a uniformity in human activities. (b) A region that possesses a unity based on significant geographic characteristics (physical, biological, cultural), in contrast to an area marked out by boundaries imposed for political or administrative purposes.----The term provides a convenient regional basis for nomenclature and integration of the whole landscape.

natural remanence *natural remanent magnetization.*

natural remanent magnetism *natural remanent magnetization.*

natural remanent magnetization The entire *remanent magnetization* of a rock in situ. Abbrev: NRM. Syn: *natural remanence; natural remanent magnetism.*

natural resin An unmodified *resin* from a natural source (such as a tree) and distinguished from synthetic resin; e.g. a copal.

natural scale (a) The *scale* of a map, expressed in the form of a fraction or ratio, independent of the linear units of measure; specif. *representative fraction.* The term is not recommended. (b) True scale, as it exists in nature, without magnification or reduction.

natural selection The process by which organisms are eliminated or preserved according to their fitness or adaptation to their surroundings and the changes therein. Syn: *selection.* See also: *struggle for existence.*

natural slope (a) The slope assumed by a mass of loose heaped-up material, such as earth. (b) *angle of repose.*

natural stone A gemstone that occurs in nature, as distinguished from a man-made substitute (such as an imitation or a synthetic stone).

natural tunnel A cave that is nearly horizontal and that is open at both ends. It may contain a stream. Syn: *tunnel cave; tunnel [speleo].*

natural well A sinkhole or other natural opening resembling a well that extends below the water table and from which ground water can be withdrawn.

naujaite A coarse, hypidiomorphic-granular sodalite-rich nepheline syenite that contains microcline and small amounts of albite, analcime, acmite, and sodium amphiboles and is characterized by poikilitic texture. The rock was first described from Naujakasik on the southwest coast of Greenland.

naujakasite A silvery-white or grayish mineral: $(Na,K)_6(Fe,Mn,Ca)(Al,Fe)_4Si_8O_{26}.H_2O$.

naumannite An iron-black isometric mineral: Ag_2Se.

naupliar eye An unpaired median eye appearing in nauplius and retained in some mature stages.

nauplius A crustacean larva in the early stage after leaving the egg, having only three pairs of limbs (corresponding to antennules, antennae, and mandibles), a median (naupliar) eye, and little or no segmentation of the body. Pl: *nauplii.* Cf: *metanauplius.*

nautical chart *hydrographic chart.*

nautical distance The length in nautical miles of the rhumb line joining any two places on the Earth's surface.

nautilicone A strongly involute nautiloid conch (like that of *Nautilus*) coiled in a plane spiral with the outer whorls embracing the inner whorls.

nautiloid Any cephalopod belonging to one of the following subclasses: Nautiloidea, Endoceratoidea, Actinoceratoidea, characterized by a straight, curved, or commonly, coiled chambered external shell, with the siphuncle located centrally, rather than marginally as in ammonoids and with less elaborate sutural flexures than in ammonoids. Nautiloids are known today only from the genus *Nautilus*, having reached their peak in the Ordovician and Silurian; their stratigraphic range is Upper Cambrian to present.

navajoite A dark-brown mineral: $V_2O_5.3H_2O$.

Navarroan North American (Gulf Coast) stage: Upper Cretaceous (above Tayloran, below Midwayan of Tertiary).

Navier-Stokes equations Equations of motion for a viscous fluid.

navite A dark-colored porphyritic igneous rock containing phenocrysts of basic plagioclase (labradorite) and olivine frequently altered to iddingsite, and some augite and rare enstatite in a holocrystalline groundmass composed chiefly of feldspar and minor second-generation augite.

naze A promontory or headland; a *ness.* Etymol: perhaps from the Naze, a promontory in Essex, Eng. Syn: *nase.*

neanic Said of a youthful or immature growth stage of an organism; the stage following the *nepionic* stage and preceding the *ephebic stage.*

neap tide A tide occurring at the first and third quarters of the Moon when the gravitational pull of the Sun opposes (or is at right angles to) that of the Moon, and having an unusually small or reduced tide range (usually 10-30% less than the mean range). Cf: *spring tide.*

near earthquake An earthquake whose epicenter is within about 1000-1200 km of the detector.

near infrared Pertaining to the shorter wavelengths in the infrared region extending from about 0.7 micrometer (visible red), to around 2 or 3 micrometers. The longer wavelength end grades into the middle infrared. It is also called *solar infrared*, as it is only available for use during the daylight hours. Cf: *far infrared.*

nearshore Extending seaward or lakeward an indefinite but generally short distance from the shoreline; specif. said of the indefinite zone extending from the low-water shoreline well beyond the breaker zone, defining the area of *nearshore currents*, and including the *inshore* zone and part of the *offshore* zone.

neat line The innermost of a series of lines that frame or bound the topographic or planimetric detail of a map. Also spelled: *neatline.* Cf: *sheet line.*

neat model The portion of the gross overlap (of a pair of photographs) that is actually used in photogrammetric procedures, generally approximating a rectangle whose width equals the air base and whose length equals the width between flight lines.

Nebraskan Pertaining to the first glacial stage of the Pleistocene Epoch in North America, beginning about 1,000,000 years ago; followed by the Aftonian interglacial stage. See also: *Günz.*

nebula Historically, any faintly luminous, diffuse object seen in the heavens. In modern usage, an interstellar cloud of gas or dust. The other diffuse objects are clusters of stars, or galaxies. Galaxies are still often referred to as extragalactic nebulae (Stokes & Judson, 1968, p.512).

nebular hypothesis A model or the origin of the universe (by Laplace in 1796) which supposes a rotating, primeval nebula of gas and dust which, as it contracted, increased its rotation. This led to a flattening of the mass and, as centrifugal forces exceeded gravity, ejections of matter from its equator. These castoffs formed into planets around the original mass, the Sun. The model has been abandoned since it was discovered that the angular momentum of the Sun is too low.

nebulite A chorismite in which one textural element occurs in indistinct or *nebulitic* lenticular masses (*schlieren*).

nebulitic (a) Lacking distinct boundaries between textural elements. (b) Of, or pertaining to, a *nebulite.*

neck [ore dep] pipe [ore dep].

neck [geog] (a) A narrow stretch or strip of land connecting two larger areas; e.g. the lowest part of a level mountain pass between two ridges, or a narrow isthmus joining a peninsula with the mainland. See also: *land bridge.* (b) Any narrow strip of land such as a cape, promontory, peninsula, bar, or hook. (c) *meander neck.*

neck [currents] The narrow band or "rip" of water forming the part of a rip current where *feeder currents* converge and flow swiftly through the incoming breakers or surf and out to the *head.*

neck [**paleont**] (a) The constricted anterior part of the body chamber in specialized brevicones between the aperture and the inflated portion (TIP, 1964, pt.K, p.57). (b) *septal neck.*

neck [**bot**] The tapering portion of an archegonium.

neck [**volc**] A vertical, pipelike intrusion that represents a formed volcanic *vent.* The term is usually applied to the form as an erosional remnant. Cf: *plug* [*volc*].

neck cutoff A high-angle meander *cutoff* formed where a stream breaks through or across a narrow meander neck, as where downstream migration of one meander has been slowed and the next meander upstream has overtaken it. Cf: *chute cutoff.*

neck ring The most posterior segment of the cephalon of a trilobite, generally set off at the front by a prominent occipital furrow. Syn: *occipital ring.*

necrocoenosis *liptocoenosis.*

necrology As used by Hecker (1865, p.21), the study of the processes which act on plant and animal remains under various conditions.

necronite A blue pearly variety of orthoclase that emits a fetid smell upon hammering. It occurs in limestone near Baltimore, Md.

necrophagous Said of an organism that feeds on dead matter.

needle [**geol**] A pointed, elevated, and detached mass of rock formed by erosion, such as an *aiguille* and a *stack.*

needle [**cryst**] A needle-shaped or acicular mineral crystal.

needle [**surv**] *magnetic needle.*

needle [**snow**] A long, slender snow crystal that is at least five times as long as it is broad.

needle ice (a) *pipkrake.* (b) *frazil ice.* (c) *candle ice.*

needle ironstone A variety of goethite occurring in fibrous aggregates of acicular crystals. Syn: *needle iron ore.*

needle ore (a) Iron ore of very high metallic luster, found in small quantities, which may be separated into long slender filaments resembling needles. (b) *aikinite.*

needle stone (a) *needle zeolite.* (b) *hairstone.*---Also spelled: *needlestone.*

needle tin ore A variety of cassiterite with acute pyramidal forms.

needle zeolite A syn. of *mesotype* [mineral]; specif. a syn. of *natrolite.* See also: *hair zeolite.* Syn: *needle stone.*

Nèel point (a) The temperature at which the susceptibility of an antiferromagnetic mineral has a maximum. Above this point, thermal agitation prevents antiferromagnetic ordering. (b) The temperature at which thermal agitation overcomes magnetic order in a ferrimagnetic mineral. *Curie point* is also used with this meaning.----Syn: *Nèel temperature.*

Nèel temperature *Nèel point.*

neftdegil *neft-gil.*

neft-gil A mixture of paraffins and a resin found in the Caspian area on Cheleken Island and which is related to *pietricikite.* Syn: *neftdegil.*

nefud (a) A deep or large sandy desert in Arabia, equivalent to an *erg.* Syn: *nafud.* (b) A high sand dune in the Syrian desert (Stone, 1967, p.267).

negative [**optics**] (a) Said of anisotropic crystals: of a uniaxial crystal, in which the extraordinary index of refraction is greater than the ordinary index; and of a biaxial crystal in which the intermediate index of refraction β is closer to η than to α. Cf: *positive* [*optics*]. (b) Said of a crystal containing a cavity, the form of which is one of the possible crystal forms of the mineral.

negative [**photo**] A photographic image (on exposed film, plate, or paper) that reproduces the bright parts of the subject as dark areas and the dark parts as light areas; it is used to print a *positive.*

negative area [**geog**] An area that is almost uncultivable or uninhabitable.

negative area [**tect**] *negative element.*

negative center The region of negative potential in the center of an observed anomaly in spontaneous polarization.

negative confining bed The upper confining bed of an aquifer whose head is below the upper surface of the zone of saturation, i.e. below the water table. Currently little used by hydro geologists.

negative delta A term used by Playfair (1802, p.430) for an *estuary.*

negative element A large structural feature or portion of the Earth's crust, characterized through a long period of geologic time by frequent and conspicuous downward movement (subsidence, submergence) or by extensive erosion, or by an uplift that is considerably less rapid or less frequent than those of adjacent *positive elements.* Syn: *negative area.*

negative elongation In a section of an anisotropic crystal, *sign of elongation* that is parallel to the faster of the two plane-polarized rays. Cf: *positive elongation.*

negative estuary *inverse estuary.*

negative landform A relatively depressed or low-lying topographic form, such as a valley, basin, or plain, or a volcanic feature formed by a lack of material (as a caldera). Ant: *positive landform.*

negative movement (a) A downward movement of the Earth's crust relative to an adjacent part of the crust, such as produced by subsidence; a negative movement of the land may result in a *positive movement* of sea level. (b) A relative lowering of the sea level with respect to the land, such as produced by a *positive movement* of the Earth's crust or by a retreat of the sea.

negative pole The south-seeking member of the *magnetic poles.* Cf: *positive pole.* See also: *dipole field.*

negative shoreline A shoreline resulting from a positive movement of the land or by a negative movement of the sea level; a *shoreline of emergence.* Ant: *positive shoreline.*

negative strip *Vening Meinesz zone.*

negrohead A syn. of *niggerhead.* Also spelled: *negro head.*

nehrung A long, narrow sandpit, sandbar, or barrier beach enclosing or partially enclosing a lagoon (*haff*), formed across a river mouth by longshore drifting of sand; esp. such a feature along the East German coast of the Baltic Sea. Pl: *nehrungs* *nehrungen.* Etymol: German *Nehrung,* "sandbar, baymouth bar, spit".

neighborite An orthorhombic mineral: $NaMgF_3$.

nejd A syn. of *hammada.* Var: *nijd.*

nek A term used in South Africa for a low place in a mountain range; a saddle or col. Etymol: Afrikaans, "neck".

nekoite A triclinic mineral: $Ca_3Si_6O_{15}.8H_2O$. Cf: *okenite.*

nektobenthos Those forms of marine life that live just above the ocean bottom and occasionally rest on it.

nekton Aquatic animals that are actively free-swimming, e.g. cephalopods, fish. Adj: *nektonic.*

nektonic Said of that type of *pelagic* organism which actively swims; adj. of *nekton.* Cf: *planktonic.*

nelsonite A group of hypabyssal rocks composed chiefly of ilmenite and apatite, with or without rutile. It is named after Nelson County, Virginia, U.S.A.

nema A hollow thread-like prolongation of the apex of the prosicula of a graptolite; the delicate tube to which the base of a graptolite colony is attached. The term is used where the prolongation is "exposed" as in all except scandent rhabdosomes. Cf: *virgula.*

nemalite A fibrous variety of brucite containing ferrous oxide.

nematath A term used by Carey (1958) for a submarine ridge across an Atlantic-type ocean basin which is not an orogenic structure, but which is composed of otherwise undeformed continental crust that has been stretched across a sphenochasm or rhombochasm. Carey cites as an example the Lomonosov ridge that extends across the Arctic Ocean basin from North America to Asia.

nematoblastic Pertaining to a *homeoblastic* type of texture of metamorphic rock due to the development during recrystallization of slender prismatic crystals. Cf: *fibroblastic*.

nematocyst One of the minute stinging cells or organs of hyrozoans, scyphozoans, and anthozoans; e.g. a "thread cell" formed within a cnidoblast of a coral.

nenadkevichite An orthorhombic mineral: (Na,Ca,K) (Nb,Ti)₂O₇.2H₂O.

neoautochthon A stable basement or autochthon formed where a nappe has ceased movement and has become denct. Cf: *paleoautochthon; mesoautochthon*.

Neocene An obsolete syn. of *Neogene*.

neocomian European stage: Lower Cretaceous (above Portlandian of Jurassic, below Aptian). It includes: Berriasian (lowermost Cretaceous), Valanginian, Hauterivian, and Barremian (although some authors omit Barremian).

neocryst An individual crystal of a secondary mineral in an evaporite (Greensmith, 1957). Cf: *evapocryst*.

neocrystic texture A secondary, nonlaminated texture of an evaporite.

neoformation *neogenesis*.

Neogene An interval of time incorporating the Miocene and Pliocene of the Tertiary period; the upper Tertiary. When the Tertiary is designated as an era, then the Neogene, together with the *Paleogene*, may be considered to be its two periods. Obs Syn: *Neocene*.

neogenesis The formation of new minerals, as by diagenesis or metamorphism. Cf: *authigenesis*. Syn: *neoformation*.

neogenic Said of newly formed minerals; pertaining to neogenesis.

neoglaciation The readvance of glacier ice during the Little Ice Age, subsequent to its shrinkage or disappearance during the warmer Altithermal interval. The term refers only to mountain areas that experienced renewal of glacier growth (Moss, 1951, p.62).

neoichnology The study of Holocene tracks, burrows, and other structures left by living organisms, as opposed to *paleoichnology*.

neokaolin Kaolinite artificially produced from nepheline.

neolensic texture A secondary, nonporphyritic, roughly laminated texture of an evaporite.

neolithic n. In archaeology, the last division of the *Stone Age*, characterized by the development of agriculture and the domestication of farm animals. Correlation of relative cultural levels with actual age (and, therefore, with the time-stratigraphic units of geology) varies from region to region. Syn: *New Stone Age*.----adj. Pertaining to the Neolithic.

neomagma A magma formed by partial or complete refusion of pre-existing rocks under the conditions of plutonic metamorphism.

neomineralization Chemical interchange within a rock whereby its mineral constituents are converted into entirely new mineral species (Knopf and Ingerson, 1938, p.14); a type of *recrystallization*.

neomorphic A *deuteromorphic* crystal modified by secondary growth. The term is obsolete.

neomorphism An inclusive term suggested by Folk (1965, p.20-21) for all transformations between one mineral and itself or a polymorph, whether the new crystals are larger or smaller or simply differ in shape from the previous ones, or present a new mineral species. It includes the processes of inversion, recrystallization, and strain recrystallization, in which the gross composition remains essentially constant. The term is appropriate where it is not possible to distinguish between recrystallization and inversion, or where the mechanism of change is not known. See also: *aggrading neomorphism; degrading neomorphism*.

neontology The study of existing organisms, as opposed to *paleontology*.

neophytic *Cenophytic*.

neoporphyrocrystic texture A texture of an evaporite in which large neocrysts are embedded in a finer-grained matrix.

neosome A geometric element of a composite rock or mineral deposit, appearing to be younger than the main rock mass (or *paleosome*); e.g. an injection in country rock, or the metasomatically introduced or newly formed material of a migmatite. Sometimes used in place of the term *metatect, mobilizate*, or *chymogenetic*. Cf: *metasome* [*geol*].

neostratotype A stratotype established after the holostratotype has been destroyed or is otherwise not usable (Sigal, 1964).

neotectonic map A map portraying *neotectonics* or the last structures and structural history of the Earth's crust. It is a kind of *paleotectonic map*. Because most neotectonic movements were epeirogenic, such a map emphasizes the broad upwarps and downwarps of the crust.

neotectonics The study of the last structures and structural history of the Earth's crust, after the Miocene and during the later Tertiary and the Quaternary. Although some deformational and even orogenic structures were formed during this time, most neotectonic features are epeirogenic and were produced by vertical upward or downward movements. Soviet geologists and geomorphologists pay particular attention to this field. Neotectonic features are presented on a *neotectonic map*.

neoteny (a) Arrested development such that youthful characteristics are retained by the adult organism. Syn: *paedomorphism; paedomorphosis; proterogenesis*. (b) Acceleration in the attainment of sexual maturity relative to general body development. Syn: *paedogenesis*.

Neothermal n. A term introduced by Antevs (1948, p.176) designating the climatic interval since the culmination of the latest major advance of Wisconsin glaciation to the present (approximately the past 10,000 years) and comprising the subunits Anathermal, Altithermal, and Medithermal.---adj. Pertaining to the postglacial Neothermal interval and to its climate, deposits, biota, and events.

neotocite A mineral consisting of a hydrous silicate of manganese and iron of uncertain formula. It may be an alteration product of rhodonite, possibly an opal with disseminated manganese and iron oxides.

neotype A single specimen designated as the *type specimen* of a species or subspecies when the *holotype* (or *lectotype*) and all paratypes or all syntypes have been lost or destroyed.

neovolcanic Said of extrusive rocks that are of Tertiary or younger age. Cf: *paleovolcanic*.

nepheline A hexagonal mineral of the feldspathoid group: (Na, K)AlSiO₄. It occurs as glassy crystals or colorless grains, or as coarse crystals or green to brown masses of greasy luster without cleavage, in alkalic igneous rocks, and it is an essential constituent of some sodium-rich rocks. Syn: *nephelite; eleolite*.

nepheline basalt An older syn. of *olivine nephelinite*.

nepheline syenite A plutonic rock composed essentially of alkali feldspar and nepheline. It may contain an alkali ferromagnesian mineral, e.g. an amphibole (riebeckite, arfvedsonite, barkevikite) or a pyroxene (acmite or acmite-augite); the intrusive equivalent of *phonolite*. Sodalite, cancrinite, hauyne, and nosean, in addition to apatite, sphene, and opaque oxides, are common accessories. Rare minerals are also frequent accessories. Cf: *foyaite; ditroite*. Obs syn: *eleolite syenite; midalkalite*.

nephelinite A fine-grained or porphyritic extrusive or hypabyssal rock, of basaltic character, but primarily composed of nepheline and pyroxene, esp. titanaugite, and lacking feldspar.

nephelinitoid A nepheline groundmass in an igneous rock; the glassy groundmass in nepheline rocks.

nephelite *nepheline*.

nepheloid zone A layer of water near the bottom of the continental rise and slope of the western North Atlantic Ocean that contains suspended sediment of the clay fraction and organic matter. It is from 200 m to 1,000 m thick.

nephelometer An instrument used in nephelometry and designed to measure the amount of cloudiness of a medium.

nephelometry The measurement of the cloudiness of a medium; esp. the determination of the concentration or particle sizes of a suspension by measuring, at more than one angle, the scattering of light transmitted or reflected by the medium. Cf: *turbidimetry.*

nephlinolith An extrusive igneous rock composed entirely of nepheline.

nephrite An exceptionally tough, compact, fine-grained, greenish or bluish amphibole (specif. tremolite or actinolite) constituting the less rare or valuable kind of jade and formerly worn as a remedy for kidney diseases. Syn: *kidney stone; greenstone.*

nepioconch The earliest-formed part of a dissoconch of a bivalve mollusk, separated from the later part by a pronounced discontinuity. Cf: *mesoconch.*

nepionic Said of the stage or period in which the young shell of an invertebrate does not yet show distinctive specific characteristics, i.e. of the stage following the *embryonic* stage and preceding the *neanic* stage.

nepouite *garnierite.*

nepton A term used by Makiyama (1954) for a body of sedimentary rock filling a basin; e.g. a *geosynclinal prism.*

Neptune's racetrack *eddy-built bar.*

neptunian adj. (a) Pertaining to neptunism and the rocks whose origin was explained by neptunism. (b) Formed by the agency of water.---n. *neptunist.*

neptunian dike A sedimentary dike formed by infilling of sediment, generally sand, in an undersea fissure or hollow. Cf: *injection dike.*

neptunic rock (a) A rock formed in the sea. (b) A general term proposed by Read (1944) for all sedimentary rocks. Cf: *plutonic rock; volcanic rock.*

neptunism The theory, advocated by Werner and long since obsolete, that the rocks of the Earth's crust (including basalt and granite) all consist of material deposited from, or crystallized out of, water. Etymol: Neptune, Roman god of waters. See also: *Wernerian.* Ant: *plutonism.* Syn: *neptunianism; neptunian theory.*

neptunist A believer in the theory of neptunism. Ant: *plutonist.* Syn: *neptunian.*

neptunite A black mineral: $(Na,K)_2(Fe^{+2},Mn)TiSi_4O_{12}$. Cf: *mangan-neptunite.*

nereite A trace fossil of the "genus" *Nereites*, consisting of a meandering feeding trail (1-2 cm wide) with a narrow central axis and regularly spaced lateral, leaf-shaped, or lobe-like projections, and formed perhaps by a worm or a gastropod.

neritic Pertaining to the ocean environment or *depth zone* between low-tide level and 100 fathoms, or between low-tide level and approximately the edge of the continental shelf; also, pertaining to the organisms living in that environment. It is called by some the *sublittoral zone,* i.e. is considered by some to be part of the *littoral* zone.

Nernst distribution law The statement that the ratio of the molar concentration of a substance dissolved in two immiscible liquids is constant and depends only on temperature. The ratio is called the *partition coefficient.*

nesophitic Said of the ophitic texture, of an igneous rock, in which pyroxene is interstitial to plagioclase and occurs in isolated areas (Walker, 1957, p.2). Cf: *sporophitic.*

nesosilicate A class or structural type of *silicate* characterized by the linkage of the SiO_4 tetrahedra by ionic bonding only, rather than by sharing of oxygens. An example of a nesosilicate is olivine, $(Mg,Fe)_2SiO_4$. Cf: *sorosilicate; cyclosilicate; inosilicate; phyllosilicate; tectosilicate.*

nesquehonite A colorless or white mineral: $MgCO_3.3H_2O$. It occurs in radiating groups of prismatic crystals.

ness A British term used esp. in Scotland for a *promontory,*

headland, or cape, or any point or projection of the land int the sea; commonly used as a suffix to a place name, e.g Fife*ness.* Syn: *naze; nose; nore; nab.*

nest A concentration of some relatively conspicuous elemen of a geologic feature, such as a "nest" of pebbles or inclu sions within a sand layer or igneous rock; esp. a small, iso lated, pocket-like mass of ore or mineral within another for mation.

nested (a) Said of volcanic cones, craters, or calderas tha occur one within another; cones, craters, or calderas showin *cone-in-cone structure.* (b) Said of two or more calderas tha intersect, having been formed at different times or by differen explosions.

nested sinkholes An American syn. of *karst valley.*

net [pat grd] A form of horizontal patterned ground whos mesh is intermediate between a *circle* and a *polygon.* See *sorted net; nonsorted net.*

net [struc petrol] A coordinate system or network of mer dians and parallels, projected from a sphere at suitable inter vals (usually 2°), used to plot points whose spherical coord nates (latitude and longitude) are known and to study orienta tion and distribution of planes and points; e.g. stereographi net, equal-area net. It is used in structural petrology. Syr *projection net; stereographic net.*

net [surv] A series of surveying (leveling) stations that hav been interconnected in such a manner that closed loops c circuits have been formed or that are so arranged as to pro vide a check on the consistency of the measured values; e.g a *base net* and a *triangulation net.* Syn: *network.*

net ablation A nonrecommended term with various meanings such as *summer balance* and net balance of the ablatio area.

net accumulation A nonrecommended term with variou meanings, such as *winter balance* and net balance of the ac cumulation area.

net balance The change in mass of a glacier from the time c minimum mass in one year to the time of minimum mass i the succeeding year (*balance year*); the mass change be tween one *summer surface* and the next. It can be deter mined at a point, as an average for an area, or as a tota mass change for the glacier. Units of millimeters, meters, c cubic meters of water equivalent are generally used. Syn: *ne budget.* Cf: *annual balance; balance.*

net budget *net balance.*

net calorific value A *calorific value* calculated from *gross ca orific value* under conditions such that all the water in th products remains in the form of vapor.

net heat of combustion *net calorific value.*

net plankton *microplankton.*

net primary production The amount of organic matter pro duced by living organisms within a given volume or area in given time, minus that which is consumed by the respirator processes of the organisms. Cf: *primary production.*

net slip On a fault, the distance between two formerly adja cent points on either side of the fault, measured on the fau surface or parallel to it. It defines both the direction and rela tive amount of displacement. Syn: *total slip.*

net venation In a leaf, a type of *venation* in which the vein branch repeatedly to form a network through the leaf. Cf: *pa allel venation.* See also: *pinnate venation; palmate venation.*

network Especially in surveying and gravity prospecting, pattern or configuration of stations, often so arranged as t provide a check on the consistency of the measured values e.g., a level network, a gravity network based on the integra tion of torsion-balance gradients. Syn: *net.*

network [surv] *net.*

network deposit *stockwork.*

neudorfite A waxy, pale-yellow variety of retinite containing little nitrogen, found in coal at Neudorf in Moravia, Czechoslo vakia.

Neumann bands Fine, straight lines observed on etched surfaces of iron meteorites (hexahedrites) and caused by mechanical twinning on (211) planes in kamacite. Named after Franz E. Neumann (1798-1895), German mineralogist. Syn: *Neumann lines; Neumann lamellae*.

Neumann's problem One of three well known *boundary value problems*. Cf: *Dirichlet's problem*.

neuromotorium A ganglion-like granular body forming the dynamic center of ciliates (as in tintinnids).

neuston A community of aquatic organisms whose environment is the surface film of the water. Such forms occur mostly in fresh water. See also: *pleuston*.

neutral axis In a two-dimensional structural model, the equivalent of a *surface of no strain*.

neutral depth *normal depth*.

neutral dune A small, irregular sand dune (Wolfe et al, 1966, p.614).

neutral estuary An estuary in which neither freshwater inflow nor evaporation dominates.

neutral pressure (a) The hydrostatic pressure of the water in the pore space of a soil; *neutral stress*. (b) The lateral earth pressure when the soil is *at rest*.

neutral shoreline A shoreline whose essential features are independent of either the submergence of a former land surface or the emergence of a former underwater surface (Johnson, 1919, p.172 & 187); a shoreline resulting without a change in the relative level of land and water. It includes shorelines of deltas, alluvial plains, outwash plains, volcanoes, coral reefs, and those produced by faulting.

neutral soil A soil whose pH value is 7.0. In practice, however, the pH value of a neutral soil varies from 6.6 to 7.3.

neutral stress The stress transmitted through the fluid that fills the voids between particles of a soil or rock mass; e.g. that part of the total normal stress in a saturated soil due to the presence of interstitial water. Syn: *pore pressure; pore-water pressure; neutral pressure*.

neutral surface *surface of no strain*.

neutron activation *Activation analysis* using neutrons to irradiate the sample.

neutron-gamma log A *neutron log* that records the varying intensity of gamma rays resulting from synthetic neutron bombardment. The induced gamma radiation is related to the hydrogen content (and hence the fluid content) of the rocks penetrated.

neutron log A *radioactivity log* that measures the intensity of radiation (neutrons or gamma rays) artificially produced when the rocks around a borehole or well are bombarded by neutrons from a synthethic source. It indicates the presence of fluids (but does not distinguish between oil and water) in the rocks, and is used with the *gamma-ray log* to differentiate porous and nonporous formations; it is also used to determine the presence of oxygen and chlorine. See also: *neutrongamma log; neutron-neutron log*.

neutron-neutron log A *neutron log* that detects neutrons produced artificially by neutron bombardment. It is sensitive to hydrogen content and is used for porosity determination. Syn: *epithermal neutron log; n-n log*.

neutron soil-moisture meter An instrument for measuring water content of soil and rocks as indicated by the scattering and absorption of neutrons emitted from a source, and resulting gamma radiation received by a detector, in a probe lowered into an access hole.

Nevadan orogeny A time of deformation, metamorphism, and plutonism during Jurassic and Early Cretaceous time in the western part of the North American Cordillera, typified by relations in the Sierra Nevada, California. In that area, deformation of the supracrustal rocks can be closely dated by limiting fossiliferous strata as late in the Jurassic (between the Kimmeridgian and Portlandian Stages), but earlier and later Nevadan deformation occurs elsewhere. In the Sierra Nevada itself, emplacement of granite and other plutonism were more prolonged than the deformation, and have been dated radiometrically between 180 m.y. and 80 m.y., or from Early Jurassic to Early Cretaceous. Geologists differ as to whether to restrict the Nevadan closely in time and space, or to use it broadly; it can most properly be considered as an orogenic era, in the sense of Stille. Also spelled: *Nevadian orogeny; Nevadic orogeny*. Cf: *Coast Range orogeny*.

Nevada twin law A rare, parallel twin law in feldspar, with a twin axis of [112].

Nevadian orogeny Var. of *Nevadan orogeny*.

Nevadic orogeny Var. of *Nevadan orogeny*.

nevadite A term, now obsolete, that was applied to rhyolite containing abundant large phenocrysts of quartz, feldspar, biotite, and hornblende in a small amount of groundmass.

névé A French term meaning a mass of hardened snow at the source or head of a glacier; it refers to the overall snow cover which exists during the melting period and sometimes from one year to another. The term was originally used in English as an exact equivalent of *firn* (the material), and is still frequently so used, but it is perhaps best to restrict the term, as proposed by Bristish glaciologists, to a geographic meaning, such as an area covered with perennial snow or an area of firn (a *firn field*), or more generally the *accumulation area* above or at the head of a glacier.

nevyanskite A tin-white variety of iridosmine containing 35-50% osmium or more than 40% iridium and occurring in flat scales.

newberyite A white orthorhombic mineral: $HMgPO_4.3H_2O$.

new global tectonics A general térm introduced by Isacks et al (1968) for *global tectonics* based on the related concepts of continental drift, sea-floor spreading, transform faults, and underthrusting of the lithosphere (crust and uppermost mantle) at island arcs, as they are jointly applied to an integrated global analysis of the relative motions of crustal segments delineated by the major seismic belts.

new ice A general term for recently formed ice (esp. floating sea ice) less than 5 cm thick, composed of ice crystals that are only weakly, if at all, frozen together and that have a definite form only while they are afloat; e.g. frazil ice, grease ice, sludge, shuga, ice rind, nilas, and pancake ice.

newlandite A griquaite containing garnet, enstatite, and chrome diopside.

newland lake A term used by Hobbs (1912, p.401) for a *consequent lake*, esp. one occupying a depression on a newly emerged ocean bottom.

New Red Sandstone The red sandstone facies of the Permian and Triassic systems, well-developed in NW England. See William Buckland in Phillips (1818, p.71-79).

new snow (a) Fallen snow in which the original crystalline structure is retained and therefore recognizable; it may or may not be recently fallen snow. Ant: *old snow*. (b) Snow that has fallen in a single day; its depth is measured from one morning to the next.---Cf: *fresh snow*.

new Stone Age *Neolithic*.

Newtonian flow In experimental structural geology, flow in which the rate of shear strain is directly proportional to the shear stress; flow of a *Newtonian liquid*. Cf: *non-Newtonian flow*. Syn: *viscous flow*.

Newtonian liquid A substance in which the rate of shear strain is proportional to the shear stress. This constant ratio is the *viscosity* of the liquid. See also: *Newtonian flow*.

Newton's law of gravitation The statement in physics that every particle of matter in the universe attracts every other particle with a force whose magnitude is proportional to the product of their masses and inversely proportional to the square of the distance between them, and whose direction is that of the line between them.

nexine The inner division of the exine of pollen, more or less equivalent to *endexine*. Cf: *sexine*.

neyite A mineral: $Pb(Cu,Ag)_2Bi_6S_{11}$.

ngavite A chondritic stony meteorite composed of bronzite and olivine in a friable, breccia-like mass of chondrules.

n'hangellite A green, elastic bitumen, similar to *coorongite*, that represents deposits of the alga cf. Coelosphaerium.

Niagaran North American provincial series: Middle Silurian (above Alexandrian, below Cayugan).

Niagara spar A name applied in the vicinity of Niagara Falls, N.Y., to fibrous gypsum imported through Canada from England, and to fibrous calcite found in veins in limestone near Niagara Falls, Ont. (Shipley, 1951, p.152).

niccolite A syn. of *nickeline*. Also spelled: *nicolite*.

niche [geomorph] A shallow cave or reentrant produced by weathering and erosion near the base of a rock face or cliff, or beneath a waterfall.

niche [ecol] The position of an organism or a population in the environment as determined by its needs, contributions, potential, and interaction with other organisms or populations. Syn: *ecologic niche*.

niche glacier A common type of small mountain glacier, occupying a funnel-shaped hollow or irregular recess in a mountain slope. Cf: *cirque glacier*.

nick (a) A place of abrupt inflection in a stream profile; a *knickpoint*. (b) A sharp angle cut by waves, currents, or ice at the base of a cliff.---Syn: *knick*.

nickel A nearly silver-white hard mineral, the native metallic element Ni. It occurs native esp. in meteorites and also alloyed with iron in meteorites. Nickel is used chiefly in alloys and as a catalyst.

nickel-antimony glance *ullmannite*.

nickel bloom A green hydrated and oxidized patina, coating, film, or incrustation on outcropping rocks, indicating the existence of primary nickel minerals; specif. *annabergite* (a nickel arsenate). The term is also applied to zaratite (a nickel carbonate) and to morenosite (a nickel sulfate).

nickel glance *gersdorffite*.

nickelhexahydrite A mineral: $(Ni,Mg,Fe)SO_4.6H_2O$.

nickeline A pale copper-red hexagonal mineral: NiAs. It is one of the chief ores of nickel, and may contain antimony, cobalt, iron, and sulfur. Syn: *niccolite; arsenical nickel; copper nickel; kupfernickel*.

nickel-iron An alloy of nickel and iron (Ni,Fe) occurring native terrestrially in pebbles and grains (as in stream gravel) and in meteorites. See also: *kamacite; taenite*. Syn: *awaruite; josephinite*.

nickel ocher *annabergite*.

nickel pyrites *millerite*.

nickel-skutterudite A tin-white to steel-gray isometric mineral: $(Ni,Co)As_3$. It may contain iron, and it represents a valuable ore of nickel, often associated with smaltite and skutterudite. Syn: *chloanthite; white nickel*.

nickel vitriol *morenosite*.

nickpoint A syn. of *knickpoint*. Also spelled: *nick point*.

nicol (a) *Nicol prism*. (b) Any apparatus that produces polarized light, e.g. Nicol prisms or polaroid; a *polarizer*.

nicolite *niccolite*.

nicolo A variety of onyx with a black or brown base and a bluish-white or faint-bluish top layer.

Nicol prism In a polarizing microscope, a pair of prisms that polarize and analyze the light used for illumination of the thin section under study. The lower nicol, or *polarizer*, is located below the stage; it consists of a rhombohedron of optically clear calcite so cut and recemented that the ordinary ray produced by double refraction in the calcite is totally reflected and the extraordinary ray transmitted. The upper nicol, or *analyzer*, is located above the objective and receives the polarized light after it has passed through the object under study. Its vibration direction is normally set at right angles to that of the polarizer. Partial syn: nicol. Syn: *polarizing prism*.

nicopyrite *pentlandite*.

nieve penitente (a) A jagged pinnacle or spike of snow or fir up to several meters in height, resulting from differential abla tion under conditions of strong insolation, especially in hig altitude-low latitude environments; an advanced stage of s cup development. (b) An assemblage of nieve penitentes.-- Etymol: Spanish, "penitent snow", shortened from "nieve c los penitentes", from the illusion of human figures hangin their heads in penitence; first used in South America. Syn *penitent* [glaciol]; penitente; ice penitente; snow peniten sun spike.

nife A petrologically descriptive name of the material of th core of the Earth, as a combination of nickel and iron. Etym Ni + Fe. Also spelled: *nifel*.

nifel Var. of *nife*.

nifontovite A mineral: $CaB_2O_4.3H_2O$.

nigerite A dark-brown mineral: $(Zn,Mg,Fe^{+2})(Sn,Zn)_2(A Fe^{+3})_{12}O_{22}(OH)_2$.

niggerhead (a) A large block or boulder of coral torn from th outer face of a reef and thrown onto the reef flat by stor waves and wind action, and quickly overgrown by a crust black lichens, as on the Great Barrier Reef off the NE coa of Australia; a blackened lump of dead coral on a reef. Th term is sometimes applied to a growing coral head. Syn *coral horse; bommy*. (b) Any dark, round, shaggy tussoc hummock, peat mass, or clump of organic or soil materi found in far northern regions, resulting from the differenti frost heaving of wet soil, and often covered with mosses, chens, or other low plant growth; specif. one of black, matt sedge tussocks with a peaty base, formed by the growth cotton grass, and separated from other tussocks by wet ho lows filled with snow, esp. noticeable in many permafrost r gions. Also, any more or less uniform hummock of tundra. (A syn. of *coal ball*, associated with metamorphism from an i neous intrusion. Syn: *hardhead*.----Syn: *negrohead*.

niggliite A silver-white mineral: PtSn or PtTe (?).

Niggli's classification A chemical classification of igneou rocks that is essentially a modification and simplification Ossan's classification. This system was proposed in 1920 b the Swiss mineralogist Paul Niggli (1888-1953).

night emerald *evening emerald*.

nigrine A black variety of rutile containing iron.

nigritite Coalified, carbon-rich fix bitumens. See also: *po nigritite; humonigritite; exinonigritite; keronigritite*.

nijd A var. of *nejd* which is a syn. of *hammada*.

niklesite A pyroxenite containing the three pyroxenes diopsid enstatite, and diallage.

nilas A thin elastic crust of gray-colored ice formed on a cal sea, having a matte surface, and easily bent by waves an thrusted into a pattern of interlocking "fingers"; it is subdivi ed by color into "dark nilas" (less than 5 cm thick) and "lig nilas" (5-10 cm thick). Etymol: Russian.

niligongite A plutonic foidite intermediate in composition be tween fergusite and ijolite, containing approximately equ amounts of nepheline and leucite and 30 to 60 percent maf minerals. Its name is derived from Niligongo, East Africa.

nimite A mineral of the chlorite group: $(Ni,Mg,Fe,Al)_6AlSi O_{10}(OH)_8$.

ningyoite A brownish-green to brown mineral: $(U,CA,Ce)_2 (PO_4)_2.1-2H_2O$. It occurs as coatings or filling cavities in urar ium ore.

niningerite A meteorite mineral: $(Mg,Fe,Mn)S$.

niobite *columbite*.

niobophyllite A mineral: $(K,Na)_3(Fe,Mn)_6(Nb,Ti)_2Si_8(O,OI 3)_{31}$.

niocalite A pale-yellow orthorhombic mineral: $Ca_4NbSi_2O_{1} (O,F)$.

nip [coast] A small, very low cliff or break in slope produce at the high-water mark by wavelets, and often cited as an in tial feature in the development of a shoreline of emergenc The term has also been applied in a broader sense to th small *notch* resulting from the formation of such a cliff.

nip [streams] The place on the bank of a meander lobe where erosion occurs as a result of the crowding of the stream current toward the lobe (Tower, 1904, p.593).

nip [coal] A *pinch* or thinning of a coal seam, esp. as a result of tectonic movements. Syn: *want.*

nisbite A mineral: $NiSb_2$.

nissonite A mineral: $Cu_2Mg_2(PO_4)_2(OH)_2.5H_2O$.

niter (a) A white orthorhombic mineral: KNO_3. It is a soluble crystalline salt that occurs as a product of nitrification in most arable soils in hot, dry regions, and in the loose earth forming the floors of some natural caves. Cf: *soda niter.* Syn: *saltpeter.* (b) A term that was formerly used for a variety of saline efflorescences, including natron and soda niter.---Syn: *nitre.*

nitrate A mineral compound characterized by a fundamental anionic structure of NO_3^-. Soda niter, $NaNO_3$, and niter, KNO_3, are nitrates. Cf: *carbonate; borate.*

nitratine *soda niter.*

nitre *niter.*

nitride A mineral compound that is a combination of nitrogen with a more positive element. An example is osbornite, TiN.

nitrification The formation of nitrates by the oxidation of ammonium salts to nitrites (usually by bacteria) followed by oxidation of nitrites to nitrates. It is one of the processes of soil formation.

nitrobarite A colorless mineral: $Ba(NO_3)_2$.

nitrocalcite A mineral: $Ca(NO_3)_2.4H_2O$. It occurs as an efflorescence, as on walls and in limestone caves. Syn: *wall saltpeter.*

nitrogen balance In a soil, the net loss or gain of nitrogen.

nitrogen fixation In a soil, the conversion of atmospheric nitrogen to a combined form by the metabolic processes of algae and, possibly, other organisms.

nitromagnesite A mineral: $Mg(NO_3)_2.6H_2O$. It occurs as an efflorescence in limestone caverns.

nitrophyte A plant that requires nitrogen-rich soil for growth.

nival Characterized by or living in or under snow, or pertaining to a snowy environment; e.g. *nival* fauna or *nival* climate.

nival gradient The angle between a nival plane and the horizon (Young, 1910, p.252).

nival plane The imaginary planar surface containing all of the different snowlines of the same time period (Young, 1910, p.252).

nivation (a) Erosion of rock or soil beneath a snowbank or snow patch and around its fluctuating margin, caused mainly by frost action but also involving chemical weathering, solifluction, and meltwater transport of weathering products; the "digging-in" and "hollowing-out" effected by a snowbank. Nivation is most active behind snowbanks in the summer when nightly freezing alternates with daytime thawing. (b) More generally, the work of snow and ice beyond or outside the limits of glacial action.---Syn: *snow-patch erosion.*

nivation cirque *nivation hollow.*

nivation glacier A small, "new-born" glacier, representing the initial stage of glaciation. Syn: *snowbank glacier.*

nivation hollow A small, shallow recess, depression, or cirquelike basin formed, and occupied during part of the year, by a small snow patch or snowbank that, through nivation, supposedly initiates the process of glaciation in mountainous regions. Syn: *nivation cirque; snow niche.*

nivation ridge *winter-talus ridge.*

niveal Said of features and effects "due to the action of snow and ice" (Scheidegger, 1961, p.24). See also: *niveoglacial.*

niveau surface *equipotential surface.*

nivenite A velvet-black variety of uraninite containing rare earths (cerium) and yttrium.

niveo-eolian *niveolian.*

niveoglacial Pertaining to the combined action of snow and ice. See also: *niveal.*

niveolian Pertaining to simultaneous accumulation and inter-

mixing of snow and airborne sand at the side of a gentle slope; e.g. said of material deposited by snowstorms under periglacial conditions. Syn: *niveo-eolian.*

nivo-karst "A characteristic of periglacial areas" (Hamelin & Cook, 1967, p.73) whereby differential chemical weathering beneath snowbanks produces a karst-like topography, as the solution of limestone fragments by snowmelt containing carbonic acid.

n-n log *neutron-neutron log.*

Noachian flood The flood, described in Genesis 5:28-10:32, during which the patriarch Noah was said to have saved his family and representative creatures. Early writers believed that the waters of this flood deposited material now known as *drift.* Also known as "The Deluge".

no-basement interpretation *thin-skinned structure.*

nobleite A monoclinic mineral: $CaB_6O_{10}.4H_2O$.

noble metal Any metal or alloy of comparably high economic value, or one that is relatively superior in certain desired properties. Cf: *base metal.*

nocerite *fluoborite.*

nodal line A line on any oscillating surface (such as on a standing wave) along which the oscillation has zero amplitude. Cf: *node.*

nodal point *amphidromic point.*

node [waves] That point on a standing wave at which the vertical motion is least and the horizontal velocity is greatest. It is also associated with seiches. Cf: *nodal line; partial node.* Ant: *antinode.*

node [paleont] (a) A place of articulation in a jointed bryozoan colony. (b) A carinal spine in cryptostomatous bryozoans. (c) A knob, protuberance, or thickened or swollen body part of an animal, such as a small boss at the end of a foraminiferal pillar.

node [bot] The place on a plant stem from which a leaf and bud emerge; the region between two successive joints of a jointed algal thallus.

node [fault] That point along a fault at which the direction of apparent displacement changes. It can occur, for instance, at the intersection of a lateral fault with a fold. See also: *scissor fault.*

nodular (a) Composed of nodules; e.g. "nodular bedding" consisting of scattered to loosely packed nodules in matrix of like or unlike character, or "nodular limestone" characterized by lumps, flocculated material, roundish aggregations, or large coated grains, often composed of the same material that encloses them. (b) Having the shape of a nodule, or occurring in the form of nodules; e.g. "nodular ore" such as a colloform mineral aggregate with a bulbed surface. Syn: *nodulated.* (c) *orbicular.*

nodular chert (a) Chert in the form of *chert nodules.* (b) A term used in Missouri for chalky chert containing small irregular grains (Grohskopf & McCracken, 1949, pl.3).

nodulated Occurring in the form of nodules; *nodular.*

nodule [sed] (a) A small, hard, and irregular, rounded, or tuberous body (knot, mass, lump) of a mineral or mineral aggregate, normally having a warty or knobby surface and no internal structure, and usually exhibiting a contrasting composition from and a greater hardness than the enclosing sediment or rock matrix in which it is embedded; e.g. a nodule of ironstone, a rounded mass of pyrite in a coal bed, a *chert nodule* in limestone, or a *phosphatic nodule* in marine strata. Most nodules appear to be secondary structures: in sedimentary rocks they are primarily the result of postdepositional replacement of the host rock and are commonly elongated parallel to the bedding. Nodules can be separated as discrete masses from the host material. (b) One of the widely scattered concretionary lumps of manganese, cobalt, iron, and nickel found on the floors of the world's oceans; esp. a *manganese nodule.*---Etymol: Latin *nodulus*, "small knot". Cf: *concretion.*

nodule [ign] A fragment of a coarse-grained igneous rock, apparently crystallized at depth, occurring as an *inclusion* in an extrusive rock; e.g. a "peridotite nodule" in a flow of olivine basalt. Syn: *plutonic nodule*.

noise (a) Any undesired sound, and by extension, any unwanted disturbance within a useful frequency band. Cf: *signal*. (b) An erratic, intermittent, or statistically random oscillation. (c) In electrical circuit analysis, that portion of the unwanted signal which is statistically random, as distinguished from hum, which is an unwanted signal occurring at multiples of the power-supply frequency.

noise equivalent input In infrared detector terminology, the incident radiation on an infrared detector that will produce a signal/noise ratio of one (Bernard, 1970, p.58). Abbrev: *NEI*. Cf: *noise equivalent power*.

noise equivalent power A syn. of *minimum detectable power*, measured in watts rather than watts per cm^2 as is *noise equivalent input*. Abbrev: *NEP*. Cf: *detectivity*.

nolanite A black hexagonal mineral: $Fe_3V_7O_{16}$.

nomenclature In biology, a system of Latin names for plants and animals (living or fossil) that has been standardized by an international commission and is internationally accepted and employed.

nomen nudum A scientific name that is not valid because it does not meet one or more of the requirements expressed by the rules of nomenclature. Etymol: Latin, "nude name, mere name". Plural: *nomina nuda*.

nominal In taxonomy, a term applied to a particular taxon that is objectively defined by its type; e.g. "the nominal genus *Musca* is always that to which its type species, *Musca domestica*, belongs" (ICZN, 1964, p.153).

nominal diameter The computed diameter of a hypothetical sphere having the same volume as that calculated for a given sedimentary particle; it is a true measure of particle size independent of either the shape or the density of the particle. Cf: *equivalent radius; sedimentation diameter*.

nomogenesis A theory of evolution stating that evolutionary change is governed by predetermined natural processes and is independent of environmental influences.

nomogram A type of line chart that graphically represents an equation of three variables, each of which is represented by a graduated straight line. It is used to avoid lengthy calculations; a straight line connecting values on two of the lines automatically intersects the third line at the required value. Syn: *nomograph*.

nomograph *nomogram*.

nonangular unconformity *disconformity*.

nonarborescent pollen Pollen of herbs and shrubs. Abbrev: NAP. Syn: *nontree pollen*.

nonartesian ground water *unconfined ground water*.

nonasphaltic pyrobitumen Any of a group of pyrobitumens including peat, coal, and nonasphaltic pyrobitumenous shales which are dark-colored, relatively hard and nonvolatile solids, composed of hydrocarbons containing oxygenated bodies. They are sometimes associated with mineral matter, the nonmineral constituents being infusible and largely insoluble in carbon disulfide (Abraham, 1960, p.57).

nonassociated natural gas (a) Natural gas that occurs in a reservoir without oil. Cf: *associated natural gas*. (b) Natural gas that occurs with oil but is not dissolved in it.

nonbanded coal Coal without bands of lustrous material, consisting mainly of clarain or durain or intermediate material, without vitrain.

noncaking coal Coal that does not cake or agglomerate when heated; it is usually a hard or dull coal. Syn: *free-burning coal*.

Noncalcic Brown soil An old term for a group of zonal soils having a slightly acidic, light pink or reddish brown A horizon and a light brown or dull red B horizon. It is developed under a mixture of grass and forest vegetation, in a subhumid climate. Syn: *Shantung soil*.

noncapillary porosity The volume of large interstices in a rock or soil that do not hold water by capillarity (Jacks, et al, 1960). Cf: *aeration porosity*.

noncarbonate hardness Hardness of water, expressed as $CaCO_3$, that is in excess of the $CaCO_3$ equivalent of the carbonate and bicarbonate alkalinity. It cannot be removed by boiling and hence is sometimes called *permanent hardness*, although this synonym is becoming obsolete. Cf: *carbonate hardness; hardness*.

nonclastic (a) Said of a sedimentary texture showing no evidence that the sediment or rock was derived from a preexisting rock or was deposited mechanically. (b) Pertaining to a chemically or organically formed sediment or sedimentary rock.---Syn: *nonmechanical*.

noncognate *accidental* [pyroclast].

noncohesive *cohesionless*.

nonconformable Pertaining to a nonconformity.

nonconformity (a) An *unconformity* developed between sedimentary rocks and older rocks (plutonic igneous or massive metamorphic rocks) that had been exposed to erosion before the overlying sediments covered them. The restriction of the term to this usage was proposed by Dunbar & Rodgers (1957, p.119). Although the term is "well known in the classroom", it is "not commonly used in practice" (Dennis, 1967, p.160). Syn: *heterolithic unconformity*. (b) A term that formerly was widely, but now less commonly, used as a syn. of *angular unconformity*, or as a generic term that includes angular unconformity.---Term proposed by Pirsson (1915, p.291-293).

nonconservative elements In sea water, elements that are uncommon. Large proportions of their total quantities enter and leave the particulate phase. Cf: *conservative elements*.

noncyclic terrace One of several stream terraces representing former valley floors formed during periods when continued valley deepening accompanied lateral erosion. Terraces on opposite sides of the valley are unpaired. Cf: *cyclic terrace*.

nondepositional unconformity A term used by Tomkeieff (1962, p.412) for a surface of nondeposition in the case of marine sediments. It is equivalent to *paraconformity*.

nondetrital Pertaining to sedimentary material derived from solution by chemical, physical, physicochemical, biochemical, or biologic means, including authigenic minerals formed in the sediment after deposition. In the next erosion cycle, nondetrital material may become detrital.

noneroding velocity That velocity of water in a channel which will maintain silt in movement but which will not scour the bed. Cf: *transporting erosive velocity*.

nonesite A porphyritic basalt composed of enstatite, labradorite, and augite phenocrysts in a groundmass of plagioclase and augite.

nonfaradaic path One of the two available paths for transfer of energy across an electrolyte-metal interface. Energy is carried in it by capacitive transfer, i.e. charging and discharging of the double-layer capacitance. Cf: *faradaic path*.

nonferrous Said of metals other than iron, usually the base metals.

nonflowing artesian well An *artesian well* whose head is not sufficient to raise the water above the land surface. Cf: *flowing artesian well; nonflowing well*.

nonflowing well A well that yields water at the land surface only by means of a pump or other lifting device. It may be either a *water-table well* or a *nonflowing artesian well*.

nonfoliate Pertaining to a metamorphic rock lacking foliation on the scale of hand specimens.

nongraded (a) A rarely used geologic term pertaining to an unconsolidated sediment or to a cemented detrital rock consisting notably of particles of more than one size or of particles lying within the limits of more than one grade; e.g. a loam or a till. Syn: *poorly sorted*. (b) An engineering term pertaining to a soil or an unconsolidated sediment consisting of particles of essentially the same size. See also: *poorly graded*. -Ant: *graded* [part size].

nonideal solution A solution in which the molecular interaction between components is not the same as that between each component. Cf: *ideal solution.*

nonmare basalt An equivalent term to *KREEP.*

nonmechanical *nonclastic.*

nonmetal (a) A naturally occurring substance that does not have metallic properties, such as high luster, conductivity, and, for the most part, opaqueness and ductility. (b) In economic geology, any rock, mineral, or other naturally occurring substance mined for its nonmetallic element, such as sulfur, fuels, salt, gemstones. Syn: *nonmetallic (n.)* See also:*industrial mineral.*

nonmetallic adj. Of or pertaining to a nonmetal. ----n. A *nonmetal*; usually plural.

nonmetallic Said, in general, of mineral lusters other than *metallic* luster; Cf: *submetallic.*

nonmotile Not *motile*; e.g. "nonmotile phase" of the life cycle of a nonflagellated coccolithophorid, producing a coccolith cover of scapholiths, rhabdoliths, placoliths, osteoliths, and pentaliths.

non-Newtonian flow Flow in which the relationship of the shear stress to the rate of shear is nonlinear, i.e. flow of a substance in which viscosity is not constant. Cf: *Newtonian flow.*

nonparametric statistics Statistics which do not assume specific distributions.

nonpareil A large, specially cut gemstone; esp. a *solitaire.*

nonpenetrative Said of a texture of deformation that affects only part of a rock, e.g. kink bands. Ant: *penetrative.* Cf: *spaced cleavage.*

nonplunging fold A fold, the axial surface of which is horizontal. Cf: *plunging fold.* Syn: *horizontal fold; level fold.*

nonrotational strain *irrotational strain.*

nonsaline alkali soil A soil whose percentage of exhangeable sodium is greater than 15, and whose pH value is usually between 8.5 and 10.0. Cf: *saline alkali soil.*

non-sequence A term used in Great Britain for a *diastem*, or for a break or gap in the continuity of the geologic record, representing a time during which no permanent deposition took place. A non-sequence usually can be detected only by a study of successive fossil contents. Cf: *paraconformity.*

nonsilting velocity That velocity of water in a channel which maintains silt in movement. Syn: *transportation velocity.*

nonsorted Said of a nongenetic group of patterned-ground features displaying an absence of a border of stones surrounding or alternating with finer material such as that characterizing *sorted* patterned ground; often there is a border of vegetation-covered ground between relatively bare ground or finer material.

nonsorted circle A form of patterned ground "whose mesh is dominantly circular and has a nonsorted appearance due to the absence of a border of stones" (Washburn, 1956, p.829); developed singly or in groups. Vegetation characteristically outlines the pattern by forming a bordering ridge. When well-developed, it has a distinctly domed central area. Diameter: 0.5 m. Syn: *cemetary hummock; mud circle; frost scar; peat ring; tussock ring.*

nonsorted crack A very rare form of patterned ground representing the boulder-free variant of a *sorted crack.*

nonsorted field A structureless, but clearly frost-affected, ground that can be identified by determining the long axes of stones oriented in the direction of slope.

nonsorted net A form of patterned ground "whose mesh is intermediate between that of a nonsorted circle and a nonsorted polygon and has a nonsorted appearance due to the absence of a border of stones" (Washburn, 1956, p.830); e.g. an *earth hummock.*

nonsorted polygon A form of patterned ground "whose mesh is dominantly polygonal and has a nonsorted appearance due to the absence of a border of stones" (Washburn, 1956,

p.831-832); never developed singly. Its borders commonly, but not invariably, are marked by wedge-shaped fissures narrowing downward; it typically results from infilling of these fissures. Diameter: a few centimeters to tens of meters. See also: *fissure polygon; mud polygon; ice-wedge polygon; vegetation polygon; sand-wedge polygon; frost-crack polygon; desiccation polygon.*

nonsorted step A form of patterned ground "with a steplike form and a nonsorted appearance due to a downslope border of vegetation embanking an area of relatively bare ground upslope" (Washburn, 1956, p.834); formed in groups. See also: *turf-banked terrace.*

nonsorted stripe One of the alternating bands comprising a form of patterned ground characterized by "a striped pattern and a nonsorted appearance due to parallel lines of vegetation-covered ground and intervening strips of relatively bare ground oriented down the steepest available slope" (Washburn, 1956, p.837). Vegetation characteristically outlines the pattern as the absence of lines of stones is an essential feature; the bare ground consists of finer-grained material or a nonsorted mixture of fines and stones. See also: *solifluction stripe; vegetation stripe; stripe hummock.*

nonsteady flow *unsteady flow.*

nonstrophic Said of a brachiopod shell whose posterior margin is not parallel with the hinge axis. Cf: *strophic.*

nonsystematic joints Joints that are not part of sets. They do not cross other joints and they often terminate at bedding surfaces, their surfaces are strongly curved, and the structures on their faces are not oriented. Cf: *systematic joints.*

nontabular Said of projecting surface features of a dinoflagellate cyst that are neither sutural nor *intratabular* and that have a random arrangement or show no apparent relation to a tabulate scheme. Cf: *peritabular.*

nontectonite Any rock whose fabric shows no influence of movement of adjacent grains, e.g. a rock formed by mechanical settling. Some rocks are transitional between a tectonite and a nontectonite (Turner and Weiss, 1963, p.39).

nonthermal spring A spring the temperature of whose water is not appreciably above the mean atmospheric temperature in the vicinity. A spring whose temperature approximates the mean annual temperature or a *cold spring* is considered a nonthermal spring (Meinzer, 1923, p.55).

nontree pollen A syn. of *nonarborescent pollen.* Abbrev: NTP.

nontronite A pale-yellow to apple- or pistachio-green or yellowish-green, dioctahedral, iron-rich clay mineral of the montmorillonite group: $Na_{0.33}Fe_2^{+3}(Al_{0.33}Si_{3.67})O_{10}(OH)_2.nH_2O$. It represents an end member in which the replacement of aluminum by ferric iron in the octahedral sheets is essentially complete. Nontronite commonly occurs in weathered basaltic rocks, where it may occupy vesicles or veins or may occur between lava flows. Syn: *chloropal; gramenite; morencite; pinguite.*

nonuniform flow In hydraulics, a type of steady flow in an open channel in which velocity varies at different points along the channel.

nonuniformist One who believes that past changes in the Earth's structure have proceeded from cataclysms or processes more violent than are now operating; a believer in the doctrine of catastrophism. Syn: *nonuniformitarian.*

nonvascular plant A plant without a vascular system or well differentiated roots, stems, and leaves, e.g. a thallophyte or bryophyte.

nonwetting sand Sand that resists infiltration of water, consisting of angular particles of varying sizes, and occurring as a tightly packed lens.

nook An obsolete syn. of *promontory.*

norbergite A yellow or pink orthorhombic mineral of the humite group: $Mg_3SiO_4(F,OH)_2$.

nordenskioldine A mineral: $CaSn(BO_3)_2$.

nordfieldite *esmeraldite.*

nordite A pale brown mineral: $(La,Ce)(Sr,Ca)Na_2(Na,Mn)(Zn,Mg)Si_6O_{17}$.

nordmarkite [**mineral**] A variety of staurolite containing manganese.

nordmarkite A quartz-bearing alkalic syenite that has microperthite as its main component with smaller amounts of oligoclase, quartz, and biotite, and is characterized by granitic or trachytoid texture.

nordsjoite A juvite that contains abundant nepheline and orthoclase but no microperthite.

nordstrandite A mineral: $Al(OH)_3$ or $Al_2O_3.3H_2O$. It is distinct from gibbsite and bayerite.

nore *ness*.

Norian European stage: Upper Triassic (above Carnian, below Rhaetian).

norilskite A mineral consisting of platinum with high contents of iron and nickel.

norite A coarse-grained plutonic rock containing basic plagioclase (labradorite) as the chief component and differing from gabbro by the presence of orthopyroxene (hypersthene) as the dominant mafic mineral. Cf: *hypersthenite*. Syn: *hypersthenfels*.

norm The theoretical mineral composition of a rock expressed in terms of *standard mineral* molecules that have been determined by specific chemical analyses for the purpose of classification and comparison; "the theoretical mineral composition that might be expected had all chemical components crystallized under equilibrium conditions according to certain rules" (Stokes & Varnes, 1955, p.94). Adj: *normative*. Cf: *mode*. See also: *C.I.P.W. classification*.

normal [**geodesy**] adj. Forming a right angle, or situated at right angles to; perpendicular.----n. (a) A straight line perpendicular to a given surface or to another line; e.g. a straight line perpendicular to the surface of the spheroid. (b) The condition of being perpendicular to a surface or line.

normal [**meteorol**] n. The average value of a meteorological element (such as pressure, temperature, rainfall, or duration of sunshine) over any fixed period of years that is recognized as standard for a given country or element. The period 1901-1930 was selected by the International Meteorological Organization at a Warsaw conference in 1935 as the international standard period for climatological normals; the U.S. Weather Bureau's temperature and precipitation normals, however, are computed from the period 1921-1950.----adj. Approximating the statistical norm or average, such as the "normal rainfall" of a region and for a definite time.

normal [**fold**] Said of an anticlinorium in which the axial surfaces of the subsidiary folds converge downwards; said of a synclinorium in which the axial surfaces of the subsidiary folds converge upwards. Cf: *abnormal* [*fold*]. See also: *fan structure*.

normal class The *holohedral class* of a crystal system.

normal consolidation Consolidation of sedimentary material in equilibrium with overburden pressure. Cf: *overconsolidation*.

normal cross section *profile* [struc petrol].

normal curve A bell-shaped curve that graphically represents a normal distribution.

normal cycle A *cycle of erosion* whereby the complete reduction or lowering of a region to base level is effected largely by running water, specif. the action of rivers as the dominant erosion agent. Cf: *arid cycle*. Syn: *fluvial cycle of erosion*.

normal depth (a) Water depth in an open channel that corresponds to uniform velocity for a given flow. It is the hypothetical depth in a steady, nonuniform flow; the depth for which the surface and bed are parallel. Syn: *neutral depth*. (b) Water depth measured perpendicular to the bed.----(ASCE, 1962).

normal dip *regional dip*.

normal dispersion The *dispersion* of seismic surface waves in which the recorded wave period increases with time. Cf: *inverse dispersion*.

normal displacement A syn. of *dip slip*. Cf: *total displacement*.

normal distribution A frequency distribution whose plot is continuous, infinite, bell-shaped curve that is symmetrical about its arithmetic mean, mode, and median (which in this distribution are numerically equivalent). Syn: *Gaussian distribution; bell-shaped distribution*.

normal earth A mass whose bounding equipotential surface is the earth-spherop, and its gravity given by prescribed gravity formulae. See also: *normal gravity*.

normal earthquake *shallow-focus earthquake*.

normal erosion (a) Erosion that is effected by prevailing agencies and that is mainly responsible for the present modification of the habitable land surface; specif. subaerial erosion by running water, rain, and certain physical and organic weathering processes. The term, used originally for stream erosion in a temperate climate, is open to criticism because erosion as found in temperate areas may in fact be "abnormal" (esp. in regard to past geologic conditions) or because one mode of erosion is just as "normal" as another. Cf: *special erosion*. (b) Erosion of rocks and soil under natural environmental conditions, undisturbed by human activity. It includes erosion by running water, rain, wind, ice, waves, gravity, and other geologic agents. Cf: *accelerated erosion*. See also: *geologic norm*. Syn: *geologic erosion*.

normal fault A fault in which the hanging wall appears to have moved downward relative to the footwall. The angle of the fault is usually 45-90°. There is dip separation but there may or may not be dip slip. Cf: *thrust fault*. Syn: *gravity fault; normal slip fault; slump fault*.

normal fold *symmetrical fold*.

normal geopotential number *spheropotential number*.

normal gradient *normal gravity*.

normal gravity The gravity caused by the attraction of the normal earth combined with the centrifugal force due to its rotation. Syn: *normal gradient*.

normal horizontal separation *offset*.

normal hydrostatic pressure In porous strata or in a well, pressure at a given point that is approximately equal to the weight of a column of water extending from that point to the surface.

normal log A *resistivity log* consisting of two electrodes in the borehole (a potential electrode and a current electrode). It has been a standard resistivity measuring device and when used in conjunction with the spontaneous-potential log and the microlog gives porosity and saturation values of a formation. The "short normal log" has an electrode separation of 16 in.; the "long normal log" has an electrode separation of 64 in. Cf: *lateral log*.

normal moisture capacity *field capacity*.

normal moveout The increase in stepout time due to an increase in the distance from source to detector, when there is no dip. Abbrev: NMO. Syn: *spread correction; stepout correction*.

normal polarity (a) A natural remanent magnetization closely parallel to the present ambient geomagnetic field direction. See also: *geomagnetic reversal*. (b) A configuration of the Earth's magnetic field with the magnetic negative pole, where field lines enter the Earth, located near the geographic north pole.----Cf: *reversed polarity*.

normal pore canal A tubule or *pore canal* piercing an ostracode carapace at right angles, and believed to serve as receptors of sensory setae. Cf: *radial pore canal*.

normal-pressure surface A potentiometric surface that coincides with the upper surface of the zone of saturation (Meinzer, 1923, p.39). It is usually the same as the water table. Cf: *subnormal-pressure surface; artesian-pressure surface*.

normal projection (a) A projection in which a three-dimensional object is projected onto two mutually perpendicular planes. (b) A projection whose surface axes coincide with nose of the sphere.

normal ripple mark An aqueous *current ripple mark* consisting of a "simple asymmetrical ridge" that may have "various round plans" (Shrock, 1948, p.101).

normal sandstone A term used by Shrock (1948a) for a sandstone composed almost exclusively of quartz and subordinate amounts of other minerals.

normal section A line between two points on the surface of an ellipsoid formed by the intersection of the ellipsoid and plane containing the normal at one point and the other point.

normal shift In a fault, the horizontal component of the shift, measured perpendicular to the strike of the fault. Cf: *offset.*

normal slip fault *normal fault.*

normal soil A soil whose profile is more or less in equilibrium with the environment, and which shows the effects of the environment on its development from the parent material.

normal stress That component of *stress* which is perpendicular to a given plane. It may be either *tensile stress* or *compressive stress.* Symbol: σ. Cf: *shear stress.*

normal twin A twinned crystal, the twin axis of which is perpendicular to the composition surface. Cf: *parallel twin.*

normal water A standardized sea water, the chlorinity of which is between 19.30‰ and 19.50‰ and that has been analyzed to within 0.001‰. Syn: *Copenhagen water; standard sea water.*

normal zoning *Zoning* in a crystal, in which the zones become progressively more sodic outward. Cf: *reversed zoning.*

Normapolles A group of Cretaceous and lower Paleogene porate (usually triporate) pollen with a complex pore apparatus (e.g. an oculus) and sometimes other peculiarities such as double Y-marks. Cf: *Postnormapolles.*

normative The adj. of *norm.*

normative mineral *standard mineral.*

norm system *C.I.P.W. classification.*

norsethite A rhombohedral mineral: $BaMg(CO_3)_2$.

north (a) The primary reference direction relative to the Earth; the direction of the north terrestrial pole, or the direction to the left of one facing east or of one facing the sunrise when the Sun is near one of the equinoxes. (b) The cardinal point directly opposite to south. Abbrev: N. (c) The direction along any meridian toward that pole of the Earth viewed from which the Earth's rotation is counterclockwise. (d) The direction to the left when one faces the direction of revolution of the Earth around the Sun. (e) The point of the horizon having an azimuth of zero degrees.

North American datum Prior to 1913 the N.A.D. was known as the United States standard datum. At that time it was adopted by Canada and Mexico and the name was changed accordingly. It was defined by the data of the station at Meade's Ranch; latitude 39°13′ 28.686″N., longitude 98°32′30.506″W., azimuth to Waldo 75°28′14.52″. See also: *North American datum of 1927.*

North American datum of 1927 The entire triangulation network of the United States, on which the *North American datum* is based, was readjusted between 1925 and 1930. The latitude and longitude of Meade's Ranch was not changed, but its azimuth to Waldo was adjusted to 75°28′09.64″. The triangulation networks of Canada and Mexico were then adjusted to the North American datum of 1927. Cf: *Clarke spheroid of 1866.*

northfieldite *esmeraldite.*

north geographic pole *north pole.*

northing A *latitude difference* measured toward the north from the last preceding point of reckoning; e.g. a linear distance northward from the east-west line that passes through the origin of a grid.

north pole [geog] The *geographic pole* in the northern hemisphere of the Earth at lat. 90°N, representing the northernmost point of the Earth or the northern extremity of its axis of rotation. Also spelled: North Pole. Syn: *north geographic pole.*

north pole [astron] The north *celestial pole* representing the zenith of the heavens as viewed from the north geographic pole.

northupite A colorless, white, yellow, or gray isometric mineral: $Na_3Mg(CO_3)_2Cl$.

nose [geomorph] (a) A projecting and generally overhanging buttress of rock. (b) The projecting end of a hill, spur, ridge, or mountain. (c) The central forward part of a parabolic dune. (d) *ness.*

nose [sed] The forward part of a turbidity current, which is more dense than the *tail* and carries coarser material.

nose [fold] A short, plunging anticline without closure. Syn: *structural nose; anticlinal nose.*

nosean A feldspathoid mineral of the sodalite group: Na_8Al_6-$Si_6O_{24}(SO_4)$. It is grayish, bluish, or brownish, and is related to hauyne. Syn: *noselite.*

noseanite A feldspar- and olivine-free basalt that contains abundant nosean.

noseanolith An extrusive rock composed almost entirely of nosean.

noselite *nosean.*

noselitite An extrusive rock composed chiefly of nosean and a pyroxene or amphibole, or both.

nose-out A nose-shaped stratum as seen in outcrop.

notch [coast] A deep, narrow cut or hollow along the base of a sea cliff near the high-water mark, formed by undercutting due to wave erosion and/or chemical solution, and above which the cliff overhangs. See also: *nip.*

notch [geomorph] (a) A term used in the NE U.S. for a narrow passageway or short defile between mountains or through a ridge, hill, or mountain; a deep, close *pass.* Also, the narrowest part of such a passage. Cf: *gap; col.* (b) A breached opening in the rim of a volcanic crater.

notch filter *band-reject filter.*

notite A variety of palagonite (Hey, 1962, p.541).

notothyrial chamber The cavity in the umbo of the brachial valve of a brachiopod, bounded laterally by brachiophore bases (or homologues) or by posterior and lateral shell walls if brachiophore bases are absent. It corresponds to the *delthyrial chamber* of the pedicle valve.

notothyrial platform Umbonal thickening of the floor of the brachial valve of a brachiopod between brachiophore bases (or homologues).

notothyrium The median subtriangular opening in the brachial valve of a brachiopod, bisecting the dorsal cardinal area or pseudointerarea. Pl: *notothyria.* Cf: *delthyrium.*

noumeite A syn. of *garnierite,* esp. a dark-green unctuous variety of garnierite.

noup A Scottish term for a steep promontory.

nourishment [beach] The replenishment of a beach, either naturally (as by littoral transport) or artificially (as by the deposition of dredged materials).

nourishment [glaciol] A syn. of *accumulation* (the process).

novacekite A yellow secondary mineral of the autunite group: $Mg(UO_2)_2(AsO_4)_2 \cdot 9H_2O$.

novaculite (a) A very dense and hard, even-textured, light-colored, cryptocrystalline, siliceous sedimentary rock, similar to chert but characterized by dominance of microcrystalline quartz over chalcedony and by accessory minerals such as feldspar and garnet. It was formerly supposed to be consolidated siliceous slime, but is now considered to be a result of primary deposition of silica under geosynclinal conditions. Novaculite is used as a *whetstone* for sharpening cutting instruments. The term is little used outside of Arkansas and Oklahoma where it is found in lower Paleozoic strata. See also: *Arkansas stone; Washita stone.* Syn: *razor stone; Turkey stone; galactite.* (b) A term used in southern Illinois for an extensive *bedded chert* (J.E. Lamar, in Tarr, 1938, p.19). (c) A general name formerly used in England for certain argillaceous stones that served as whetstones.

novaculitic chert A generally gray chert that breaks into slightly rough, splintery fragments; it is less vitreous and somewhat coarser-grained than *chalcedonic chert*.

novakite A tetragonal mineral: $(Cu,Ag)_4As_3$.

nowackiite A mineral: $Cu_6Zn_3As_4S_{12-13}$.

nsutite A mineral: $Mn_{1-x}^{+4}Mn_x^{+2}O_{2-2x}(OH)_{2x}$. It was formerly called "gamma-MnO_2".

nubbin (a) One of the isolated bedrock knobs or small hills forming the last remnants of the crest of a mountain or mountain range that has succumbed to desert erosion (backwearing). The term was introduced by Lawson (1915) and extended by Cotton (1942) to include small remnants of spurs and ridges. (b) A residual boulder, commonly granitic, occurring on a desert dome or broad pediment (Stone, 1967, p.235).

nuclear age determination *radiometric dating*.

nuclear basin A postorogenic basin in a mobile belt; a contemporary *epieugeosyncline*.

nuclear clock *radioactive clock*.

nuclear fission *fission*.

nuclear fusion *fusion*.

nuclear geochemistry *isotope geology*.

nuclear geology *isotope geology*.

nuclear log *radioactivity log*.

nuclear magnetic resonance The selective absorption of electromagnetic radiation at the appropriate resonant frequency by nuclei undergoing precession in a strong magnetic field. Symbol: NMR.

nuclear magnetic resonance spectrometer An instrument for scanning and measuring the *nuclear magnetic resonance* spectrum of nuclei.

nuclear quadrupole resonance Resonance of an atomic nucleus whose electric charge distribution deviates from a spherical distribution. Symbol: NQR.

nuclear radiation Radiation of alpha or beta particles or of gamma rays from an atomic nucleus.

nuclear reaction Change in atomic nuclei, e.g. the union of heavy-hydrogen nuclei to form nuclei of helium. Syn: *reaction*.

nuclear resonance *Resonance* [phys] occurring when a nucleus is irradiated with gamma rays of exactly the same frequency as those which the nucleus naturally tends to radiate.

nuclear resonance magnetometer A type of magnetometer that measures total magnetic field intensity by means of the precession of magnetic nuclei, precession frequency being proportional to field intensity. In practice, only the *proton precession magnetometer* has been used.

nuclear snow gage Any type of gage using a radioactive source and a detector to measure, by the absorption of radiation, the water-equivalent mass of a snowpack.

nuclear twin-probe gage *profiling snow gage*.

nucleation The beginning of crystal growth at one or more points.

nucleoconch *embryonic apparatus*.

nucleogenesis The origin of the chemical elements of the universe.

nucleosynthesis The generation of elements from hydrogen nuclei or protons by nuclear processes under the high- temperature, high-pressure conditions common in the life of a star.

nucleus [phys sci] A small particle upon or around which other particles form and grow; e.g. a *crystallization nucleus* and a *condensation nucleus*.

nucleus [paleont] (a) The earliest-formed part of the shell or operculum of a gastropod. The term should not be used synonymously with *protoconch*. (b) The *madreporite* of an echinoderm.

nuclide A species of atom characterized by the number of neutrons and protons in its nucleus. See also: *radionuclide*.

nucule The female reproductive structure of a charophyte. It includes the oogonium and the outer protective cells.

nuculoid Any bivalve mollusk belonging to the order Nucu-loida, characterized by a taxodont, equivalve, isomyarian shell with closed margins.

nudibranch Any opisthobranch belonging to the order (or suborder) Nudibranchia, characterized chiefly by the absence of a shell in the adult stage and by the absence of gills, or their replacement by secondary gills.

nuée ardente A swiftly flowing and turbulent gaseous cloud, sometimes incandescent, erupted from a volcano and containing ash and other pyroclastics in its lower part. This lower part of the núee ardente is comparable to an *ash flow*, and the terms are sometimes used synonymously in this sense. Etymol: French, "glowing cloud". Syn: *Pelean cloud; glowing cloud*.

Nuevoleonian North American (Gulf Coast) stage: Lower Cretaceous (above Durangoan, below Trinitian).

nugget A large lump of placer gold or other metal. Cf: *heavy gold*.

nullah (a) A term used in the desert regions of India and Pakistan for a sandy river bed or channel, or a small ravine or gully, that is normally dry except after a heavy rain. (b) The small, intermittent, generally torrential stream that flows through a nullah.---Etymol: Hindi *nala*. Pron: *nala*. See also: *wadi; arroyo*. Syn: *nulla; nallah; nalla; nala*.

null hypothesis The assumption that no significant difference exists between two items or samples that are being compared statistically or that any observed difference is purely accidental and not due to a systematic cause.

nullipore A coralline alga, formerly thought to be an animal.

number of streams Total number of stream segments of specified order or orders in a given drainage basin. The symbol N_u refers to the total number of stream segments of given order *u* within a specified drainage basin. See also: *law of stream numbers*.

numerical aperture A measurement or indicator of a microscope's resolving power.

numerical taxonomy The use of statistics in classifying and analyzing fossils and their paleoecologic implications.

nummulite Any foraminifer belonging to the family Nummulitidae, characterized by a test that is usually planispiral. Its stratigraphic range is Upper Cretaceous to present. Adj: *nummulitic*. Var: *nummulitid*.

Nummulitic A syn. of *Paleogene*, used in Europe.

nummulitic limestone A *foraminiferal limestone* composed chiefly of nummulite shells; specif. the "Nummulite Limestone", a thick, distinctive, and widely distributed Eocene formation stretching from the Alps and northern Africa to China and eastern and southern Asia, composed esp. of the remains of the genus *Nummulites*.

nunakol A *nunatak* rounded by glacial erosion; a rounded "island" of rock in a glacier. Etymol: Eskimo. Syn: *rognon*.

nunatak An isolated hill, knob, ridge, or peak of bedrock that projects prominently above the surface of a glacier and is completely surrounded by glacier ice. Nunataks are common along the coast of Greenland. Etymol: Eskimo, "lonely peak". Swedish plural: *nunatakker*. Cf: *rognon; nunakol*. Also spelled *nunataq*.

nut An indehiscent, one-celled, and one-seeded hard and bony fruit, even if resulting from a compound ovary (Lawrence, 1951, p.762).

nutation The motion of the true axis of rotation of the Earth about its mean position with a principal term of about 18.6 years. See also: *differential nutation*.

nutational scanner An optical-mechanical line-scan system in which an oscillating plane mirror and a rotating prism permit full-frame image presentation from a fixed or hovering position. Thus, nutational scanners, e.g., with infrared detectors, do not depend on the forward motion of a fixed-wing aircraft to generate film-strip imagery. The nutational scanner can thus function as a ground instrument operating from a fixed tripod mounting and be used as a monitoring instrument at a

486

lcano observatory, for instance. The nutational scanner can
lso be operated from a hovering helicopter (Friedman, 1970,
39).

ut coal A size of bituminous coal that will pass through a 2-3
ch round mesh but not through a 0.75-1.50 inch round
esh (size standards vary regionally). Cf: *lump coal*. Syn:
hestnut coal.

utrient In oceanography, any inorganic or organic compound
used to sustain plant life; e.g. silicates for diatoms.

nymph (a) One of the narrow, thickened lunate processes or
platforms of many bivalve mollusks extending posteriorly from
the beak along the dorsal margin and serving for attachment
of ligament. Syn: *nympha*. (b) An immature stage in the life
cycle of an acarid arachnid; e.g. protonymph, deutonymph,
and tritonymph.

oasis [astrogeol] Any of numerous small, dark, roundish spots occurring at the intersection of canals on the planet Mars.

oasis [geog] A fertile, vegetated area in the midst of a desert where the water table has come close enough to the surface for wells and springs or seepages to exist, thus making it suitable for human habitation.

obcordate Said of a leaf that is deeply lobed at the base, rather than heart-shaped (*cordate*).

object glass *objective*.

objective The lens (or lenses) that gives an image of an object in the focal plane of a microscope's or telescope's *eyepiece*. Syn: *objective lens; object glass*.

objective lens *objective*.

objective synonym In taxonomy, any one of two or more synonyms based on the same type. Cf: *subjective synonym*.

oblate Flattened or depressed at the poles; e.g. "oblate pollen" whose equatorial diameters are much longer than the dimensions from pole to pole. Ant: *prolate*.

oblate spheroid A spheroid that is flattened at its poles.

obligate adj. Said of an organism that can grow only under certain restricted conditions. Cf: *facultative*. Syn: *obligative*.

oblique bedding An archaic syn. of *inclined bedding*, or bedding oblique to the principal surface of deposition; specif. *cross-bedding*.

oblique extinction *inclined extinction*.

oblique fault A fault, the strike of which is oblique to, rather than parallel or perpendicular to, the strike of the constituent rocks or dominant structure. Cf: *oblique-slip fault; strike fault; dip fault*. Syn: *diagonal fault*.

oblique joint *diagonal joint*.

oblique lamination (a) *cross-lamination*. (b) *transverse lamination*.

oblique photograph An aerial photograph taken with the camera axis intentionally inclined between the horizontal and the vertical (camera pointing down at an angle). It combines the ground view with the pattern obtained from a height. See also: *high-oblique photograph; low-oblique photograph*. Cf: *vertical photograph*. Syn: *oblique*.

oblique projection A projection that is not centered on a pole or on the equator and that does not use the equator or a meridian as a center line of orientation, or that has an axis inclined at an oblique angle to the equatorial plane; e.g. "oblique stereographic projection" or "oblique Mercator projection".

oblique section A slice through a foraminiferal test cut in a direction neither parallel to the axis of coiling nor normal to it.

oblique slip In a fault, movement or slip that is intermediate in orientation between the *dip slip* and the *strike slip*.

oblique-slip fault A fault, the slip of which is oblique to, rather than parallel or perpendicular to, the dip of the constituent rocks or dominant structure. Cf: *oblique fault*. Syn: *diagonal-slip fault*.

obovate Said of a leaf whose terminal end is broader than its basal end. Cf: *ovate*.

obovoid Said of a fruit whose terminal portion is broader than its basal portion. Cf: *ovoid*.

obruchevite A brown mineral of the pyrochlore group: $(Y,Na,Ca,U)(Nb,Ta,Ti,Fe)_2(O,OH)_7$.

obsequent [geomorph] Said of a geologic or topographic feature that does not resemble or agree with a *consequent* feature from which it developed at a later date; esp. said of a tilt-block mountain (or of a rift-block mountain) that was formerly the floor of the original valley (or graben) but that was left standing as a result of differential erosion, or said of a tilt-block valley (or of a rift-block valley) that occupies the site of the former mountain (or horst) after the original topography

was modified by differential erosion. Ant: *resequent*.

obsequent [streams] adj. Said of a stream, valley, or drainage system whose course or direction is opposite to that of the original consequent drainage. The term was proposed by Davis (1895, p.134). Etymol: "opposite to consequent". See also: *anaclinal*.---n. *obsequent stream*.

obsequent fault-line scarp A *fault-line scarp* that faces in the opposite direction as the original fault scarp (i.e. facing the upthrown block) or in which the structurally downthrown block is topographically higher than the upthrown block. Cf: *resequent fault-line scarp*.

obsequent stream A stream that flows in a direction opposite to that of an original consequent stream and that is a tributary to a subsequent stream developed along the strike of weak beds; e.g. a short stream flowing down the scarp slope of a cuesta, or a stream flowing in a direction opposite to that of the dip of the local strata or the tilt of the land surface. See also: *scarp stream; antidip stream; reversed consequent stream*. Syn: *obsequent; anticonsequent stream; inverted stream*.

obsequent valley A valley eroded by or containing an *obsequent stream*; a valley sloping in a direction opposite to that of the general dip of the strata.

observation well A special well drilled in a selected location for the purpose of observing parameters such as fluid levels and pressure changes, such as within an oil reservoir as production proceeds.

observed gravity Gravity value obtained by either relative or absolute measurements.

obsidian A black or dark-colored volcanic glass, usually of rhyolite composition characterized by conchoidal fracture. It is sometimes banded or has microlites. Usage of the term goes back as far as Pliny, who described the rock from Ethiopia. Obsidian has been used for making arrowheads, other sharp implements, jewelry, and art objects. Syn: *Iceland agate; hyalopsite; mountain mahogany*.

obsidian dating *obsidian hydration dating*.

obsidian hydration dating A method of calculating an age in years for an obsidian artifact or Holocene volcanic glass by determining the thickness of the hydration rim which has been produced by water vapor slowly diffusing into a freshly chipped surface and producing a hydrated layer or rind. It is applicable to glasses 200 to 200,000 years old. Syn: *hydration rind dating; obsidian dating*.

obsidianite A term, now obsolete, proposed by Walcott (1898) for a small, rounded, glassy, obsidian-like object now known as a *tektite*. Most stones originally described as "obsidianite" were later shown to be true obsidian and not tektites.

obstructed stream A stream whose valley has been blocked, as by a landslide, glacial moraine, sand dune, or lava flow; it frequently consists of a series of ponds or small lakes.

obstruction moraine A moraine formed where the movement of ice is obstructed, as by a ridge of bedrock.

obtuse Said of a leaf that is blunt or rounded.

obtuse bisectrix The *bisectrix* of the obtuse angle between the axes of a biaxial crystal. Cf: *acute bisectrix*.

obverse (a) Said of the aspect of a graptolite rhabdosome in which the sicula is most fully visible. (b) Pertaining to the *frontal* side of a bryozoan zooid or colony.---Cf: *reverse*.

occidental (a) Said of a gemstone that has an inferior quality (grade, luster, or value) or is an inferior variety; e.g. "occidental agate" (poorly marked and not very translucent), or "occidental chalcedony" (all but the quite translucent, gray to white chalcedony). (b) Misrepresenting a substitute as being the genuine gem it represents; e.g. "occidental turquoise" (or odontolite), or "occidental topaz" (or citrine). (c) Said of a gemstone found in any part of the world other than the Orient. --Cf: *oriental*.

occidental cat's-eye A syn. of quartz *cat's-eye*.

occipital condyle An articular surface on the bone along the back part of the head of a vertebrate and by which the skull articulates with the first cervical vertebra (atlas). See also: *condyle*.

occipital furrow (a) The transverse groove on the cephalon of a trilobite running from axial furrow to axial furrow and forming the posterior boundary of the glabella. (b) A groove in front of the rim along the posterior border of the prosoma of a merostome (TIP, 1955, pt.P, p.8).

occipital ring *neck ring*.

occludent margin The margin of scutum and tergum forming the aperture in a cirripede crustacean and occluding it with comparable margins of opposed scutum and tergum.

occlusion [chem] A syn. of *absorption* [chem]. Adj: *occluded*.

occlusion [meteorol] In a depression or low-pressure area, the closing in of a cold front over a warm front, which reduces the warm air to a line and moves it upward to the upper atmosphere, where it remains for some time.

occult mineral A mineral that might be expected to be present in a rock (as from the evidence of chemical analysis) but which is not identifiable, even with the aid of a microscope. Common examples are quartz and orthoclase in the glassy or cryptocrystalline groundmasses of certain lavas.

occupy To set a surveying instrument over a point for the purpose of making observations or measurements.

ocean The continuous salt-water body that surrounds the continents and fills the Earth's great depressions; also; one of its major geographic divisions. See also: *sea*.

ocean current (a) A permanent, nontidal, predominantly horizontal movement of the surface water of the ocean, and constituting part of the general ocean circulation. (b) Broadly, any current in the ocean (whether tidal or nontidal, permanent or seasonal, horizontal or vertical), characterized by a regularity usually as a continuous stream flowing along a definable path, or less commonly by a cyclic nature; it may be produced by wind stresses (drift current), long-wave motions (tidal current), or density gradients due to variations in temperature and salinity (density or geostrophic currents).

ocean-floor spreading *sea-floor spreading*.

ocean hole *blue hole*.

oceanic (a) Pertaining to those areas of the ocean that are deeper than the littoral and neritic. (b) Pertaining to the ocean.

oceanic climate The climate of islands and land areas bordering the ocean, characterized by only moderate temperature ranges and by the occurrence of maximum and minimum temperatures further away from summer and winter solstice, respectively, than in a *continental climate*. Syn: *marine climate; maritime climate*.

oceanic crust That type of the Earth's *crust* which underlies the ocean basins; it is equivalent to the *sima*, i.e. is characterized by the absence of the sialic layer. The oceanic crust is about 5 km thick; it has a density of 3.0 g/cm^3, and compressional seismic-wave velocities travelling through it exceed 6.2 km/sec. Cf: *continental crust*.

oceanic delta A delta built out into a tide-influenced sea and characterized by a concave delta plain (Moore, 1966, p.87).

oceanic ridge *mid-oceanic ridge*.

oceanic trench *trench* [marine geol].

oceanite A picritic basalt.

oceanization The conversion of continental crust into oceanic crust; *basification*.

oceanographic cast *hydrographic cast*.

oceanographic equator The zone of maximum temperature of the surface of the ocean; also defined as the zone in which the temperature of the surface of the ocean is higher than 28°C. Its position may vary seasonally, but it is always near to the geographic equator. Syn: *thermal equator*.

oceanography (a) The study of the ocean, including its physical, chemical, biologic, and geologic aspects. (b) In a nar-

rower sense, the study of the marine environment.----Syn: *oceanology*.

oceanology A syn. of *oceanography*, less commonly used.

ocellar Said of the texture, of an igneous rock (esp. one with nepheline), in which the phenocrysts are aggregates of smaller crystals (e.g. of biotite or acmite) arranged radially or tangentially around larger, euhedral crystals (e.g. of leucite or nepheline) or which form rounded eye-like branching forms. Also, said of a rock having such texture. See also: *ocellus* [petrology].

ocellus [paleont] (a) A minute simple eye in an arthropod; e.g. the only type of eye found in an arachnid, or the median visual organ located on the prosoma of a merostome, or an unpaired median eye common in some branchiopod and copepod crustaceans. Also, one of the elements of a *compound eye*. (b) A short hyaline process on the frustule in some diatoms (as in Auliscus).---Pl: *ocelli*.

ocellus [petrology] A phenocryst in an *ocellar* rock. Plural: *ocelli*.

ocher (a) An earthy, usually impure, pulverulent, and red, yellow, or brown iron oxide that is extensively used as a pigment; e.g. "yellow or brown ocher" (limonite) and "red ocher" (hematite). Also, any of various clays strongly colored by iron oxides. (b) A similar earthy and pulverulent metallic oxide used as a pigment; e.g. "antimony ocher" (stibiconite and cervantite), "lead ocher" (massicot and litharge), and "tungstic ocher" (tungstite and ferritungstite).---Cf: *umber; sienna*. Syn: *ochre*.

ocherous Pertaining to, containing, or resembling ocher; e.g. "ocherous iron ore", a red, powdery or earthy hematite. Var: *ochreous; ochrous*.

Ochoan North American provincial series: uppermost Permian (above Guadalupian, below Lower Triassic).

ochre Var. of *ocher*.

ochreous Var. of *ocherous*.

Ochrept In U.S. Dept. of Agriculture soil taxonomy, a suborder of the soil order *Inceptisol*, characterized by formation in a cold or temperate climate and by the presence of an ochric epipedon and a cambic horizon (SSSA, 1970). Cf: *Andept; Aquept; Plaggept; Tropept; Umbrept*.

ochric Pertaining to an *epipedon* that is thinner, lighter in color, and lower in content of organic matter than a *mollic* or an *umbric* epipedon (SSSA, 1970).

ochrous Var. of *ocherous*.

Ocoee A provincial series of the Precambrian in Virginia, Tennessee, North Carolina, and Georgia.

ocrite A group name for powdery ochers.

octactin A sponge spicule having six equidistant rays in one plane and two rays at right angles to them.

octahedral Pertaining to an *octahedron*.

octahedral borax A variety of *tincalconite* occurring in crystals that simulate octahedrons, as from the lagoons of Tuscany, Italy.

octahedral cleavage A type of crystal cleavage that occurs parallel to the faces of the octahedron; e.g. diamond.

octahedral coordination An atomic structure or arrangement in which an ion is surrounded by six ions of opposite sign, whose centers form the corners of an octahedron. An example is the structure of NaCl. Syn: *sixfold coordination*.

octahedral copper ore *cuprite*.

octahedral iron ore *magnetite*.

octahedral planes Those planes in a cubic crystal lattice having three equivalent Miller indices.

octahedrite [meteorite] The commonest *iron meteorite* containing 6-18% nickel in the metal phase and showing upon etching Widmanstatten structure due to the presence of intimate intergrowths (of plates of kamacite with narrow selvages of taenite) oriented parallel to the octahedral planes. Symbol: O. Cf: *hexahedrite; ataxite*.

octahedrite [mineral] A syn. of *anatase*. The term is a misnomer because anatase crystallizes in tetragonal dipyramids and not in octahedrons.

octahedron An isometric crystal form of eight faces which are equilateral triangles. Its indices are $\{111\}$ and its symmetry is $4/m\bar{3}\,2/m$. Adj: *octahedral*.

octane Any of the several isomeric liquid paraffin hydrocarbons having the formula C_8H_{18} including n-octane $CH_3(CH_2)_6$-CH_3 which is found in petroleum.

octaphyllite (a) A group of mica minerals that contain eight cations per ten oxygen and two hydroxyl ions. (b) Any mineral of the octaphyllite group, such as biotite; a trioctahedral clay mineral.---Cf: *heptaphyllite*.

octocoral Any anthozoan belonging to the subclass Octocorallia, characterized by exclusively polypoid forms with pinnate tentacles and by colonial growth. Their stratigraphic range is Silurian (questionably) to present. Syn: *alcyonarian*.

ocular [optics] *eyepiece*.

ocular [paleont] n. *ocular plate*.

ocular plate One of the outer circlet of primordial plates of the *apical system* of an echinoid, located at the aboral terminus of an ambulacrum and perforated by an ocular pore. Cf: *genital plate*. Syn: *ocular*.

ocular pore A perforation in an ocular plate of an echinoid for the passage of a *terminal tentacle*.

ocular sinus One of a pair of small and shallow sinuses at the sides of the aperture in the position of the eyes in *Nautilus*.

oculogenital ring A ring formed in echinoids by a circlet of ocular plates surrounding a circlet of genital plates in the center of the aboral surface at the apical end of ambulacral and interambulacral areas. It surrounds the periproct in regular echinoids, and it represents the initial plates of an echinoid skeleton. See also: *apical system*.

oculus A much-enlarged part of the pore structure of (usually triporate) pollen, consisting of a bulging, very thick protrusion of ektexine. Pl: *oculi*.

Oddo-Harkins rule A statement in geochemistry that, with four exceptions, the cosmic abundances of elements of even atomic number exceed those of adjacent elements of odd atomic number. This relationship was perceived by both Oddo and Harkins.

odd-pinnate Said of a pinnately compound leaf having an odd number of leaflets. Cf: *even-pinnate*.

odenite A variety of biotite supposed to contain a new element (odenium).

odinite A greenish-gray lamprophyre composed of labradorite and augite or diallage, sometimes hornblende, phenocrysts in a groundmass of fine lath-shaped or equigranular feldspar and a felty mesh of acicular hornblende crystals.

odograph An instrument that automatically plots the course and distance traveled by a vehicle and that draws directly on paper by electronic or photoelectric methods a continuous map of the route taken.

odometer An instrument attached to a wheel of a vehicle to count the number of turns made by the wheel and used to measure the approximate distance traveled as a function of the number of revolutions and the circumference of the wheel.

odometry Mechanical measurement of distances.

odontolite A fossil bone or tooth colored deep blue by iron phosphate (vivianite), and rarely green by copper, and resembling turquoise, such as that from the tusks of mammoths found in Siberia. It is cut and polished for jewelry. Syn: *bone turquoise; fossil turquoise*.

odontology The study of teeth, including their structure, development, and diseases.

oecostratigraphy Var. of *ecostratigraphy*.

oersted The cgs (centimeter gram second) unit of magnetic field intensity. Except in magnetized media, a magnetic field with an intensity H of one oersted has an induction B of one gauss.

offlap (a) The progressive offshore degression of the updip terminations of the sedimentary units within a conformable sequence of rocks (Swain, 1949, p.635), in which each successively younger unit leaves exposed a portion of the older unit on which it lies. Also, the successive contraction in the lateral extent of strata (as seen in an upward sequence) due to their being deposited in a shrinking sea or on the margin of a rising landmass. Example: the relation in Tennessee of Middle Ordovician Tellico Sandstone with respect to the underlying Farragut Limestone. Ant: *onlap*. Syn: *regressive overlap*. (b) The progressive withdrawal of a sea from the land. Cf: *regression*.

off-lying *offshore*.

off-reef Pertaining to the seaward margin or zone of a reef; e.g. the "off-reef facies" of reef talus, or the "off-reef floor" immediately surrounding a reef. Cf: *fore reef*. Also spelled: *offreef*.

offretite A zeolite mineral: $(K,Ca,Mg)_3Al_5Si_{13}O_{36}\cdot14H_2O$.

offset [cart] The small distance added (during construction of a map projection) to the length of meridians on each side of the central meridian in order to determine the top latitude of the constructed chart.

offset [coast] The migration of an upcurrent part of a shore to a position a little farther seaward than a downcurrent part; esp. the offset of a spit across a coastal inlet. Cf: *overlap*.

offset [geomorph] (a) A spur or minor branch from a range of hills or mountains. (b) A level terrace on a hillside.

offset [seis] (a) In seismic prospecting, the horizontal distance from a shothole to the line of profile, measured perpendicularly to the line. (b) In seismic refraction prospecting, the horizontal displacement, measured from the detector, of a point for which a calculated depth applies; in seismic reflection prospecting, the adjustment of a reflecting element from its position on a preliminary working profile to its true position in space.----v. To make such an adjustment of position or depth.

offset [paleont] A new corallite formed in corallum by budding; a corallite formed later than and from a protocorallite.

offset [drill] n. *offset well*.

offset [fault] In a fault, the horizontal component of displacement, measured parallel to the stroke of the fault. Cf: *normal shift*. Syn: *normal horizontal separation*.

offset [surv] (a) A short distance measured perpendicular to a traverse course or a surveyed line or principal line of measurement for the purpose of locating a point with respect to a point on the course or line; e.g. a perpendicular distance measured from a great-circle line to a parallel of latitude in order to locate a section corner on that parallel in the U.S. Public Land Survey system. (b) A jog in a survey line which has approximately the same direction both before and after passing the jog.

offset deposit A mineral deposit, esp of sulfides, formed partly by magmatic segregation and partly by hydrothermal solution, near the source rock.

offset line A supplementary line established close to and roughly parallel with the main survey line to which it is referenced by measured offsets. Offset lines are used where it is convenient to avoid obstructions, over which it would be difficult to make measurements, located along the main line.

offset ridge A ridge of resistant sedimentary rock that has been made discontinuous by faulting.

offset shotpoint In seismic shooting, the distance from the shotpoint to the nearest geophone or to the center of the nearest geophone group, or to any geophone. See also: *perpendicular offset; in-line offset*.

offset stream A stream displaced laterally or vertically by faulting.

offsetting dip *migrating dip*.

offset well An oil well drilled near the boundary of a property and opposite to a producing or completed well on an adjoining property, for the purpose of preventing the drainage of oil or gas to the earlier (productive) well. Syn: *offset*.

offshore (a) Situated off or at a distance from the shore; spe-f. said of the comparatively flat, always submerged zone of variable width extending from the breaker zone to the seaw-rd edge of the continental shelf where substantial movement f material is limited. The offshore zone is seaward of the *in-hore* zone or the *shoreface* (CERC, 1966, p.A43; and John-on, 1919, p.161), although it is often regarded (e.g. Shepard, 967, p.43) as the zone extending seaward from the low-ater shoreline. (b) Pertaining to a direction seaward or lak-ward from the shore; e.g. an *offshore* wind or one that blows way from the land, or an *offshore* current or one moving way from the shore.---Ant: *onshore*. See also: *nearshore*. yn: *off-lying*.

ffshore bar (a) *longshore bar*. (b) A catchall term used by ohnson (1919) for features now known as a *barrier beach* nd a *barrier island*.---The term is undesirable as it has been pplied both to a submerged feature (a *bar*) and an emergent eature (a *barrier*).

ffshore barrier *barrier beach*.

ffshore beach A syn. of *barrier beach*. The term was used by ilbert & Brigham (1902, p.306) for a long, narrow, low, andy beach with a belt of quiet water separating it from the ainland.

ffshore slope The frontal slope below the outer edge of the ave-built terrace.

ffshore terrace A wave-built terrace in the offshore zone, omposed of gravel and coarse sand. See also: *shoreface ter-ace*.

ffshore water Water that is adjacent to land and whose phys-al properties are influenced only slightly by continental con-itions. Ant: *inshore water*.

fftake *ground-water withdrawal*.

gdosymmetric Said of that *merohedral* crystal class, the gen-ral form of which has one eighth the number of equivalent aces of the corresponding holohedral form.

ghurd A term used in the Saharan region for a massive, ountainous dune, formed by some underlying rocky topogra-hic feature, and rising considerably above the general dune vel.

give A dark, curved, arcuate structure, one of a series re-eated periodically down a glacier, generally formed at the ase of an icefall, and resembling the pointed arch or rib cross a Gothic vault (the "arch" is convex downslope due to aster flow in the middle of the glacier); esp. *dirt-band ogive*. f: *Forbes band; wave ogive*.

id A suffix (derived from the Greek) meaning "like, having e form of". A rock name or geologic feature thus qualified e.g. granitoid, gneissoid) resembles but is not the same as e name or feature to which it is attached.

ikocryst In poikilitic fabric, the enclosing crystal.

il *petroleum*.

il accumulation *oil pool*.

il-base mud A *drilling mud* with clay particles suspended in il rather than in water.

il-cut mud A mixture of oil and drilling mud recovered in test-ng.

il field The surface limits of a petroleum-producing area which may correspond to an *oil pool* or may be circumscribed y political or legal limits.

il-field brine Connate waters found by the drill in rocks pene-rated at depth. They usually have a high concentration of cal-ium and sodium salts.

il pool A unique accumulation of petroleum whose limits are stablished by subsurface geologic factors. Cf: *oil field*. Syn: ool; *oil accumulation*.

il sand A term applied loosely to any porous stratum contain-ng petroleum or impregnated with hydrocarbons; specif. a andstone or unconsolidated sand formation from which oil is btained by drilled wells. See also: *gas sand; tar sand; sand drill*]. Cf: *water sand*.

oil seep The emergence of liquid petroleum at the surface as a result of the slow migration from its buried source through minute pores or fissure networks. Syn: *seepage; petroleum seep*.

oil shale A kerogen-bearing, finely laminated brown or black shale that will yield liquid or gaseous hydrocarbons on distilla-tion. Syn: *kerogen shale*.

oil show A partial syn. of *show*.

oil trap The accumulation of petroleum in a reservoir rock under such conditions that its migration and escape is pre-vented. Sealing can be effected by the abutment of imperme-able formations against the reservoir, by lateral variation with-in the beds to reduce permeability, or by the presence of water preventing downward migration. Syn: *trap* [*oil*].

oil-water contact *oil-water interface*.

oil-water interface The datum of a two-dimensional interface between oil and water. Syn: *oil-water contact*.

oil well A *well* from which petroleum is obtained or obtainable by pumping or by natural flow. Some statutes define the term on the basis of the gas-oil ratio.

ojo A term used in SW U.S. for a very small lake or a pond. Etymol: Spanish, "eye". Pron: *o-ho*.

okaite An ultramafic igneous rock composed chiefly of melilite and haüyne, with accessory biotite, perovskite, apatite, cal-cite, and opaque oxides. It resembles *turjaite* except that the feldspathoid is haüyne rather than nepheline.

okenite A whitish mineral: $CaSi_2O_4(OH)_2 \cdot H_2O$. Cf: *nekoite*.

old age [topog] The final stage of the cycle of erosion in the topographic development of a landscape or region in which the surface has been reduced by erosion almost to base level and the landforms are marked by simplicity of form and sub-dued relief. It is characterized by a few, large, meandering, widely spaced old streams flowing sluggishly across broad flood plains, separated by faintly swelling hills, and having dendritic distributaries, and by peneplanation. Cf: *senescence; senility*. Syn: *topographic old age*.

old age [coast] A stage in the development of a shore, shore-line, or coast characterized by a wide wave-cut platform, a faintly sloping sea cliff pushed far inland, and a coastal region approaching peneplanation. The stage is probably a theoreti-cal abstraction since it is doubtful whether stability of sea level is maintained long enough for the land to be reduced to the base level of wave erosion (about 600 ft or 185 m below sea level); the shoreline cycle is usually interrupted before this stage is reached.

old age [streams] The stage in the development of a stream at which erosion is decreasing in vigor and efficiency, and aggradation becomes dominant as the gradient is greatly re-duced. It is characterized by: a load that exceeds the stream's ability to carry it, and is therefore readily deposited; a very broad, shallow, open valley with gently sloping sides and a nearly level floor (flood plain) that may be 15 times the width of the meander belt; numerous oxbows, meander scars, levees, yazoos, bayons, and swamps and lakes on valley floors; a sluggish current; graded or mature tributaries, few in number; and slow erosion, effected chiefly by mass-wasting at valley sides.

Older Dryas n. A term used primarily in Europe for an interval of late-glacial time (centered about 11,500 years ago) fol-lowing the Bølling and preceding the Allerød, during which the climate as inferred from stratigraphic and pollen data in Denmark (Iversen, 1954) deteriorated favoring either expan-sion or retarded retreat of the waning continental and alpine glaciers; a subunit of the late-glacial Arctic interval, charac-terized by tundra vegetation.---adj. Pertaining to the late-gla-cial Older Dryas interval and to its climate, deposits, biota, and events.

Oldest Dryas n. A term used primarily in Europe for an inter-val of late-glacial time (centered about 13,000 years ago) preceding the Bølling, during which the climate as inferred from stratigraphic and pollen data in Denmark (Iversen, 1954) was colder than the succeeding Bølling; the oldest subunit of the late-glacial Arctic interval, characterized by tundra vegetation.---adj. Pertaining to the late-glacial Oldest Dryas interval and to its climate, deposits, biota, and events.

old-from-birth peneplain A term used by Davis (1922) for a peneplain presumably formed during an uplift of such extreme slowness over a long period of time that vertical corrasion was outpaced by valley-side grading and by general downwearing of the interstream uplands, thereby producing a landscape that will at once be "old" or that lacks any features characterizing youth or maturity; it is essentially a *primärrumpf*.

Oldham-Gutenberg discontinuity *Gutenberg discontinuity*.

oldhamite A pale-brown meteorite mineral: CaS.

old ice (a) Floating sea ice more than two years old (Armstrong et al, 1966, p.30). It may be as much as 3 m or more thick, and it shows features that are smoother than those in second-year ice. (b) A term formerly applied to sea ice that has survived at least one summer's melt and that shows features that are smoother than those in first-year ice; e.g. second-year ice and multi-year ice. (c) A term loosely applied to a deposit of ice in permafrost.

old lake (a) A lake in an advanced stage of filling by sediments or vegetation. (b) A eutrophic or dystrophic lake. See also: *aging.* (c) A lake whose shoreline exhibits an advanced stage of development.

oldland (a) Any ancient land; specif. an extensive area (as the Canadian shield) of ancient crystalline rocks reduced to low relief by long-continued erosion and from which the materials of later sedimentary deposits were derived. (b) A region of older land, projected above sea level behind a coastal plain, that supplied the material of which the coastal-plain strata were formed; the land adjoining a new land surface that has been brought above sea level for the last time. (c) A term proposed by Maxson & Anderson (1935, p.90) for the land surface of the old-age stage of the cycle of erosion, characterized by subdued relief. Maxson later adopted (1950, p.101) the earlier term *senesland* for this feature.

old mountain A mountain that was formed prior to the Tertiary period, esp. a *fold mountain* produced before the Alpine orogeny. Ant: *young mountain.*

Old Red Sandstone A thick sequence of nonmarine, predominantly red sedimentary rocks, chiefly sandstones, conglomerates, and shales, representing the Devonian System in parts of Great Britian and elsewhere in NW Europe. See Miller (1841). Abbrev. O.R.S.

old snow Fallen snow that has lost most traces of its original snow-crystal shapes, as *firn* or *settled snow*, through metamorphism, esp. equitemperature metamorphism. Ant: *new snow*. Syn: *firn snow.*

old Stone Age *Paleolithic.*

old stream A stream developed during the stage of *old age.*

ole A rock-name suffix suggested by Shand (1917) for rocks having dyad or triad minerals that are unsaturated.

olefin An unsaturated *aliphatic hydrocarbon*, empirical formula CnH_{2n}, which contains at least one double bond. The olefins form a series, analogous to the methane series, of which ethylene, formula C_2H_4, a sweet smelling gas present in common gas, is the lowest member. Also spelled: *olefine.*

olefine Var. of *olefin.*

olenellid Any trilobite belonging to the family Olenellidae, characterized generally by a subovate to elongate exoskeleton, the absence of dorsal sutures on the cephalon, numerous segments in the thorax, and well-developed pleural spines or acutely terminating, falcate distal portions (TIP, 1959, pt.O, p.191). They are known only from the Lower Cambrian.

oligist iron A syn. of *hematite.* Also spelled: *oligiste iron.*

Oligocene An epoch of the lower Tertiary period, after the Eocene and before the Miocene; also, the corresponding world-wide series of rocks. It is sometimes considered to be a period, when the Tertiary is designated as an era.

oligoclase A mineral of the plagioclase feldspar group with composition ranging from $Ab_{90}An_1$ to $Ab_{70}An_{30}$. It is common in igneous rocks of intermediate to high silica content.

oligoclasite A granular plutonic rock composed almost entirely of oligoclase. When the term was first defined, by Bombicci (1868), it was applied to a rock containing more orthoclase than oligoclase, i.e. what is now called *cavalorite* (Johannsen 1939, p.246). Syn: *oligosite.*

oligomictic [sed] Said of a clastic sedimentary rock composed of a single rock type, such as an orthoquartzitic conglomerate; also, said of the clasts of such a rock. Oligomictic rock are characteristic of stable conditions such as are found in epicontinental seas. Cf: *monomictic; polymictic.* Syn: *oligomictic.*

oligomictic [lake] Said of a lake that circulates only at very rare irregular intervals when abnormal cold spells occur; e.g. a lake of small or moderate area or of very great depth, or in a region of high humidity in which a small temperature difference between surface and bottom suffices to maintain stable stratification (Hutchinson, 1957, p.462). Cf: *polymictic.*

oligonite A variety of siderite (ferrous-carbonate mineral) containing up to 40% manganese carbonate. Syn: *oligon spar.*

oligopelic Said of a lake-bottom deposit that contains very little clay (Veatch & Humphrys, 1966, p.218).

oligophyre A light-colored diorite containing oligoclase phenocrysts in a groundmass of the same mineral.

oligosiderite A meteorite containing only a small amount of metallic iron. Cf: *holosiderite.*

oligosite *oligoclasite.*

oligostromatic Said of a plant part that is composed of only a few layers of cells. Cf: *monostromatic; distromatic; polystromatic.*

oligotrophic lake A lake that is characterized by a deficiency in plant nutrients and usually by abundant dissolved oxygen in the hypolimnion; its bottom deposits have relatively small amounts of slowly decaying organic matter and its water is often deep. Cf: *dystrophic lake; mesotrophic lake; eutrophic lake.*

oligotrophic peat Peat containing a small amount of plant nutrients. Cf: *mesotrophic peat; eutrophic peat.*

oligotrophy The quality or state of an *oligotrophic lake.*

olistoglyph A hieroglyph produced by sliding or interlaminar gliding (Vassoevich, 1953, p.61); specif. a *slide mark.*

olistolith An *exotic block* or other rock mass transported by submarine gravity sliding or slumping and included within the binder of an olistostrome. Term introduced by G. Flores in Beneo (1955, p.122).

olistostrome A sedimentary deposit consisting of a chaotic mass of intimately mixed heterogeneous materials (such as blocks and muds) that accumulated as a semifluid body by submarine gravity sliding or slumping of unconsolidated sediments. It is a mappable, lens-like stratigraphic unit lacking true bedding but intercalated among normally bedded sequences, as in the Tertiary basin of central Sicily. Term introduced by G. Flores in Beneo (1955, p.122). Etymol: Greek *olistomai*, "to slide", + *stroma*, "bed".

olivenite An olive-green, dull-brown, gray, or yellowish orthorhombic mineral: $Cu_2(AsO_4)(OH)$. Syn: *leucochalcite; wood copper.*

olivine (a) An olive-green, grayish-green, or brown orthorhombic mineral: $(Mg,Fe)_2SiO_4$. It comprises the isomorphous solid-solution series forsterite-fayalite. Olivine is a common rock-forming mineral of basic, ultrabasic, and low-silica igneous rocks (gabbro, basalt, peridotite, dunite); it crystallizes early from a magma, weathers readily at the Earth's surface, and metamorphoses to serpentine. (b) A name applied to a group of minerals forming the isomorphous system $(Mg,Fe, Mn,Ca)_2SiO_4$, including forsterite, fayalite, tephroite, and a hypothetical calcium orthosilicate. Also, any member of this system.---See also: *peridot; chrysolite.* Syn: *olivinoid.*

olivine basalt A group of basalts that contain olivine in addition to their other components; considered by some petrographers as a less-preferred syn. of *alkali olivine basalt.*

olivine leucitite *ugandite.*

olivine nephelinite An extrusive igneous rock differing in composition from nephelinite only by the presence of olivine. Syn: *nepheline basalt; ankaratrite.*

olivine rock *dunite.*

olivinite An olivine-rich ore-bearing igneous rock that also contains other pyroxenes and/or amphiboles.

olivinoid (a) An olivine-like substance found in meteorites. (b) *olivine.*

ollenite A type of hornblende schist characterized by abundant epidote, sphene, and rutile. Garnet is one of the accessories (Holmes, 1928, p.170).

olocyst Original misspelling of *holocyst.*

olsacherite A mineral: $Pb_2(SeO_4)(SO_4)$.

olshanskyite A mineral: $Ca_3B_4(OH)_{18}$.

olynthus (a) The first stage in the development of a sponge in which the initial functional aquiferous system has a single flagellated chamber. (b) Newly attached sponge larva resembling a vase in form and having a simple and asconoid body wall.

ombrogenous Said of peat deposits whose moisture content is dependent upon rainfall. Cf: *soligenous; topogenous.*

ombrophilous Said of a plant adapted to extremely rainy conditions. Cf: *ombrophobous.* Noun: *ombrophile.*

ombrophobous Said of an organism that cannot tolerate extremely rainy conditions. Cf: *ombrophilous.* Noun: *ombrophobe.*

ombrotiphic Pertaining to a short-lived pond whose water is derived from rainfall. Cf: *tiphic.*

omission The elimination or nonexposure of certain stratigraphic beds at the surface or in any specified section owing to disruption and displacement of the beds by faulting. Ant: *repetition.*

omission solid solution A crystal in which there is incomplete filling of particular atomic sites. Cf: *substitutional solid solution.* Syn: *defect-lattice solid solution.*

ommatidium One of the basic visual units of the compound eye of an arthropod. Pl: *ommatidia.*

omphacite A grass-green to pale-green, granular or foliated, vitreous, high-temperature, aluminous clinopyroxene mineral found as a common constituent in the rock eclogite; a variety of augite consisting of a solid solution containing jadeite and diopside components. In thin section, it is colorless, superficially resembling olivine.

omuramba A term used in central and NE South-West Africa for the clearly defined dry bed of an intermittent stream, carrying water sometimes only as a series of shallow lakes and vleis. Etymol: Bantu (Herero). Pl: *omirimbi.* Cf: *oshana.*

oncoid An algal biscuit that resembles an ancient oncolite.

oncolite A small, variously shaped (often spheroidal), concentrically laminated, calcareous sedimentary structure, resembling an oolith, and formed by the accretion of successive layered masses of gelatinous sheaths of blue-green algae. It is smaller than a *stromatolite* and generally does not exceed 10 cm in diameter. Syn: *onkolite.* Cf: *catagraphite.*

one face centered lattice A type of *centered lattice* in which the unit cell has one pair of faces centered, i.e. there are two lattice points per unit cell. If the (100) plane is centered, the symbol A is used; if the (010) plane is centered, the symbol B is used; and if the (001) plane is centered, the symbol C is used. In orthorhombic and monoclinic crystal lattices, all types are possible; in tetragonal crystal lattices, only the C type of centering is possible.

onegite A pale amethyst-colored sagenitic quartz penetrated by needles of goethite.

Onesquethawan North American stage: Lower and Middle Devonian (above Deerparkian, below Cazenovian).

one-year ice Sea ice of not more than one winter's growth, and a thickness of 70 cm to 2 m; it includes the "medium" and "thick" subdivisions of *first-year ice.*

onion-skin weathering A type of *spheroidal weathering* in which the successive shells of decayed rock so produced resemble the layers of an onion. Syn: *onion weathering.*

onkilonite A nepheline-leucite basalt that also contains olivine, augite, and perovskite, but no feldspar.

onkolite Var. of *oncolite.*

onlap (a) An *overlap* characterized by the regular and progressive pinching out, toward the margins or shores of a depositional basin, of the sedimentary units within a conformable sequence of rocks (Swain, 1949, p.635), in which the boundary of each unit is transgressed by the next overlying unit and each unit in turn terminates farther from the point of reference. Also, the successive extension in the lateral extent of strata (as seen in an upward sequence) due to their being deposited in an advancing sea or on the margin of a subsiding landmass. Ant: *offlap.* Cf: *overstep.* See also: *marine onlap.* Syn: *transgressive overlap.* (b) The progressive submergence of land by an advancing sea. Cf: *transgression.*

onofrite A mineral: $Hg(S,Se)$. It is a variety of metacinnabar containing selenium and is a source of selenium.

onokoid A small or microcrystalline, dense, nodular, pea-like body in ophthalmitic rocks (Niggli, 1954, p.191).

onset *arrival.*

onset-and-lee topography *stoss-and-lee topography.*

onshore (a) Pertaining to a direction toward or onto the shore; e.g. an *onshore* wind or one that blows landward from a sea or lake, or an *onshore* current or one moving toward the shore. (b) Situated on or near the shore, as *onshore* oil reserves.---Ant: *offshore.* Syn: *shoreside.*

Ontarian (a) Stage in New York State: Middle Silurian (middle and lower parts of Clinton Group). (b) An obsolete name for the Middle and Upper Ordovician in New York State.

ontogenetic stage Developmental stage in the growth of an individual organism.

ontogeny Development of an individual organism in its various stages from initiation to maturity. Adj: *ontogenetic.* Cf: *phylogeny.*

ontozone A term used by Henningsmoen (1961) for a biozone (rock strata) that signifies the range of a taxon.

onychium The distal subsegment of the tarsus carrying the claws, found in some arachnids but wanting in others. Pl: *onychia.*

onyx (a) A variety of chalcedony that is like *banded agate* in consisting of alternating bands of different colors (such as white and black, black and red, white and red, white and brown) but unlike it in that the bands are always straight and parallel. Onyx is used esp. in making cameos. Cf: *agate; sardonyx; jasponyx.* (b) A name applied incorrectly to dyed, unbanded, solid-colored (black, green, white) chalcedony; esp. *black onyx.* (c) *onyx marble.*---adj. (a) Parallel-banded; e.g. "onyx marble" and "onyx obsidian". (b) Of the color jet black.

onyx agate A banded agate with straight parallel alternating bands of white and different tones of gray.

onyx marble A hard, compact, dense, usually banded, generally translucent variety of calcite (or rarely of aragonite) resembling true onyx in appearance; esp. parallel-banded *travertine* capable of taking a good polish, and used as a decorative or architectural material or for small ornamental objects. It is usually deposited from cold-water solutions, often in the form of stalagmites and stalactites in caves. See also: *cave onyx.* Syn: *onyx; Mexican onyx; alabaster; oriental alabaster; Gibraltar stone; Algerian onyx.*

onyx opal Common opal with straight parallel markings.

oocast *oolicast.*

ooecium The *ovicell* or brood chamber in cheilostomatous bryozoans. Pl: *ooecia.*

oogonium In fungi and algae, the female gametangium consisting of a single cell.

ooid (a) An individual spherite of an oolitic rock; an *oolith.* The term has been used in preference to "oolith" to avoid confusion with "oolite". Syn: *ooide.* (b) A general, nongeneric term for a particle that resembles an oolith in outer appearance and size (Henbest, 1968, p.2). Cf: *pseudo-oolith.*---Adj: *ooidal.*

oolicast One of the small, subspherical openings found in an oolitic rock, produced by the selective solution of ooliths without destruction of the matrix. The term is inappropriate unless the opening is subsequently filled. See also: *oomold.* Syn: *oocast.*

oolicastic porosity The porosity produced in an oolitic rock by removal of the ooids and formation of oolicasts (Imbt & Ellison, 1947, p.369-370).

oolite (a) A sedimentary rock, usually a limestone, made up chiefly of ooliths cemented together. The rock was originally termed *oolith.* Syn: *roestone; eggstone.* (b) A term often used for *oolith,* or one of the ovoid particles of an oolite.---Etymol: Greek *oon,* "egg". Pron: o-o-lite, the first "o" as in "old", the second "o" an in "obey"; incorrectly pron. *oo-lite.* Cf: *pisolite.* Also spelled: *oölite.*

oolith One of the small, round (ovate, spherical, or oblate ellipsoidal) accretionary bodies in a sedimentary rock, resembling the roe of fish, and having diameters of 0.25 to 2 mm (commonly 0.5 to 1 mm). It is usually formed of calcium carbonate (but may be of dolomite, silica, iron oxide, pyrite, or other minerals) in successive concentric layers commonly around a nucleus (such as a shell fragment, an algal pellet, or a quartz-sand grain) in shallow, wave-agitated water, and often showing an internal radiating fibrous structure indicating outward growth or enlargement at the site of deposition. Ooliths are frequently formed by inorganic precipitation, although many noncalcareous ooliths are produced by replacement in which case they are less regular and spherical, and the concentric or radial internal structure is less well-developed, than in accretionary oolites. The term was originally used for a rock composed of ooliths (an *oolite*), and is sometimes so used today. Cf: *pisolith.* Also spelled: *oölith.* Syn: *ooid; oolite; ovulite.*

oolitic Pertaining to an oolite, or to a rock or mineral made up of ooliths; e.g. "oolitic ore", such as "oolitic phosphorite" and "oolitic ironstone" (in which iron oxide or iron carbonate has replaced the calcium carbonate of an oolitic limestone). Also spelled: *oölitic.*

oolitic limestone An even-textured limestone composed almost wholly of relatively uniform calcareous ooliths, with virtually no interstitial material. It is often an important oil reservoir (such as the Smackover Formation in Arkansas) and is also quarried for building purposes.

oolitic texture The texture of a sedimentary rock consisting largely of ooliths showing tangential contacts with one another.

oolitization The act or process of forming ooids or an oolitic rock. Also, the result of such action or process.

oolitoid A sedimentary particle similar in size and shape to an oolith, but lacking its internal structure (Bissell & Chilingar, 1967, p.162). Cf: *pseudo-oolith.*

ooloid A term used by Martin (1931, p.15) for a tiny, elliptically shaped, concretionary-like siliceous form constructed of thin concentric layers around a central siliceous mass. It is found singly or cemented together in irregularly shaped clusters embedded in silicified shells of bryozoans, brachiopods, etc., and even composing the entire pseudomorphic shell. See also: *beekite.*

oomicrite A limestone containing at least 25% ooliths and no more than 25% intraclasts and in which the carbonate-mud matrix (micrite) is more abundant than the sparry-calcite cement (Folk, 1959, p.14). It generally represents a mixing c two environments, such as where ooliths are swept from a ba into a muddy lagoon.

oomicrudite An oomicrite containing ooliths that are more than one millimeter in diameter.

oomold A spheroidal opening in a sedimentary rock or insoluble residue, produced by solution of an oolith. Adj: *oomoldic* See also: *oolicast.*

oopellet A spherical or subspherical grain displaying charac teristics of both an oolith and a pellet. "The internal part i pelletoidal, and thus may be ovoid in shape, but it has a accretionary coating, the thickness of all layers being equal t or slightly greater than the diameter of the pellet which the enclose" (Bissell & Chilingar, 1967, p.162). Cf: *superficial oo ith.*

oophasmic Said of a dolomite or recrystallized limestone tha contains vague but unmistakable traces of oolitic textur (Phemister, 1956, p.73).

oospararenite An oosparite containing sand-sized (mediu sand or coarse sand) ooliths; an oolitic sandstone.

oosparite A limestone containing at least 25% ooliths and n more than 25% intraclasts and in which the sparry-calcite ce ment is more abundant than the carbonate-mud matrix (mi crite) (Folk, 1959, p.14). It is common in environments c high wave or current energy, where the spar usually repre sents pore-filling cement. Cf: *pisosparite.*

oosparrudite An oosparite containing ooliths that are mor than one millimeter in diameter.

oosterboschite A mineral: $(Pd,Cu)_7Se_5$.

oovoid A void in the center of an incompletely replaced oolith.

ooze [geog] A piece of soft, muddy ground, such as a mud bank; a marsh, fen, or bog resulting from the flow of a sprin or brook.

ooze [marine geol] A pelagic sediment consisting of at leas 30% skeletal remains of pelagic organisms (either calcareou or siliceous), the rest being clay minerals. Grain size is ofte bimodal (partly in the clay range, partly in the sand or si range). Oozes are further defined by their characteristic orga nisms: *diatom ooze; foraminiferal ooze; globigerina ooze pteropod ooze; radiolarian ooze.* See also: *calcareous ooze siliceous ooze.*

ooze [sed] (a) A soft, soupy mud or slime, typically foun covering the bottom of a river, estuary, or lake. (b) Wet ear thy material that flows gently, or that yields easily to pressure.

oozy Pertaining to or composed of ooze; e.g. "oozy fraction of soils in which mineral grains are less than one micron i diameter.

opacite A general term applied to swarms of opaque, micro scopic grains in rocks, esp. in the groundmass of an igneou rock. Opacite is generally supposed to consist chiefly of mag netite dust. Cf: *viridite; ferrite [ign].*

opal A mineral (or mineral gel): $SiO_2 \cdot nH_2O$. It is an amor phous (colloidal) form of silica containing a varying proportio of water (as much as 20% but usually 3-9%) and occurring i nearly all colors. Opal is transparent to nearly opaque, an typically exhibits a definite and often marked iridescent play o color. It differs from quartz in being isotropic, having a lowe refractive index, and being softer and less dense. Opal usuall occurs massive and frequently pseudomorphous after othe minerals, and is deposited at low temperatures from silica bearing waters. It is found in cracks and cavities of igneou rocks, in flint-like nodules in limestones, in mineral veins, i deposits of hot and warm springs, in siliceous skeletons o various marine organisms (such as diatoms and sponges), i serpentinized rocks, in weathering products, and in mos chalcedony and flint. The transparent colored varieties exhib iting opalescence are valued as gemstones. Much so-calle opal gives weak X-ray patterns of cristobalite or tridymite Syn: *opaline.*

opal-agate A variety of banded opal having different shades o olor and being agate-like in structure, consisting of alternate ayers of opal and chalcedony. Cf: *agate opal.*

palescence A milky or somewhat pearly appearance or luster f a mineral, such as that shown by opal and moonstone.

paline n. (a) Any of several minerals related to or resembling pal; e.g. a pale-blue to bluish-white opalescent or girasol co-undum, or a brecciated impure opal pseudomorphous after erpentine. (b) *opal.* (c) An earthy form of gypsum. (d) A ock with a groundmass or matrix consisting of opal.---adj. Resembling opal, esp. in appearance; e.g. "opaline feldspar" (labradorite) or "opaline silica" (tabasheer).

palite An impure, colored variety of common opal; e.g. *myr-ckite.*

palized wood *silicified wood.*

pal jasper *jasper opal.*

pal phytolith A discrete, distinctively shaped, minute (less han 80 microns in diameter) *phytolith* or solid body of isotro-ic silica originally precipitated by terrestrial plants (sedges, eeds, some woods, and esp. grasses) as unwanted material r as reinforcement of cell structures. Such bodies may be ecent or fossil forms, and are often transported by wind and eposited in the ocean. Syn: *plant opal; grass opal.*

paque Said of a material that is impervious to visible light, or f a material that is impervious to radiant energy other than visible light, e.g. radiation. Cf: *transopaque; translucent; ransparent.*

opaque attritus Attritus that does not contain large quantities of transparent humic degradation matter. Cf: *translucent att-ritus.*

ppdalite A hypersthene-biotite granodiorite. It is named after Opdal, Norway. Cf: *farsundite.*

open bay An indentation between two capes or headlands, so broad and open that waves coming directly into it are nearly as high near its center as on adjacent parts of the open sea; a *bight.*

opencast mining *opencut mining.*

open channel A conduit in which water flows with a free sur-ace (ASCE, 1962).

open coast A coast exposed to the full action of waves and currents.

open-coast marsh A *salt marsh* found along an open coast. Cf: *coastal marsh.*

opencut mining Surficial mining, in which the ore is exposed to the sky by removing the overburden. Both coal and metalli-ferous ores (of iron, copper) are worked in this way. The term *quarrying* is reserved for building stone and sand and gravel deposits. Syn: *strip mining; opencast mining; openpit mining.*

open fault A fault, or a section of a fault, the two walls of which have become separated along the fault surface. Cf: *closed fault.*

open fold A fold, the limbs of which have been only moderate-y compressed. A fold can be described on a relative scale from open to closed. Ant: *closed fold.*

open form A crystal form whose faces do not enclose space, e.g. a trigonal prism. Cf: *closed form.*

open gash fracture *gash fracture.*

open hole (a) A well or borehole, or a portion thereof, that has not been cased at the depth referred to. (b) A borehole free of any obstructing object or material. (c) A borehole that s being drilled without cores (Nelson & Nelson, 1967, p.260).

open ice (a) Ice that is sufficiently broken up to permit pas-sage on navigable waters. (b) *broken ice.*

open joint *fissure.*

open lake (a) A lake that has an effluent; e.g. a *drainage lake.* Ant: *closed lake.* (b) A lake having *open* water (free of ice or emergent vegetation).

open-packed structure In crystal structure, a pattern of stack-ng of equal spheres in an orthogonal arrangement such that each sphere is in contact with six others. Cf: *open-packed structure.*

open pack ice Pack ice in which the concentration is 4/10 through 6/10 with many leads and polynyas; the floes are generally not in contact with one another. See also: *broken ice; scattered ice.*

open packing The manner of arrangement of uniform solid spheres packed as loosely as possible so that the porosity is at a maximum; e.g. *cubic packing.* Ant: *close packing.*

openpit mining *opencut mining.*

open rock Any stratum sufficiently open or porous to contain a significant amount of water or to convey it along its bed.

open sand A sand that is relatively porous and permeable, as contrasted with a *tight sand.* The term is used in petroleum geology.

open sound A sound (similar to a lagoon) with large openings between the protecting islands.

open-space structure A structure in a carbonate sedimentary rock, formed by the partial or complete occupation by internal sediments and/or cement (Wolf, 1965).

open structure A structure which, when represented on a map by contour lines, is not surrounded by closed contours. Ant: *closed structure.*

open system [**chem**] A chemical system in which, during the process under consideration, material is either added or re-moved. Cf: *closed system.*

open system [**permafrost**] A condition of freezing of the ground in which additional ground water is available either through free percolation or through capillary movement (Mul-ler, 1947, p.219), exemplified by the pingos of East Green-land. Ant: *closed system.*

open traverse A surveying traverse that starts from a station of known or adopted position but does not terminate upon such a station and therefore does not completely enclose a polygon. Cf: *closed traverse.*

open valley (a) A broad band of lowland between relatively straight and parallel valley sides, through which a stream swings from side to side in broad, open curves (Rich, 1914, p.469). (b) A strath produced by progressive widening of a valley by lateral stream cutting (Bucher, 1932, p.131).

open water [**ice**] A relatively large area of freely navigable water in an ice-filled region; specif. water in which floating ice is present in concentration less than 1/8 (or 1/10). Cf: *ice free; polynya.*

open water [**lake**] (a) Lake water that remains unfrozen or uncovered by ice during the winter. (b) Lake water that is free of emergent vegetation or artificial obstructions and of dense masses of submerged vegetation at very shallow depths.

open well (a) A well large enough (one meter or more in di-ameter) for a man to descend to the water level. See also: *combination well.* (b) An artificial pond formed where a large excavation into the zone of saturation has been filled with water to the level of the water table (Veatch & Humphrys, 1966, p.351).

openwork Said of a gravel with unfilled voids.

operational facies A term used by Krumbein & Sloss (1963, p.328) for stratigraphic *facies* designating lateral variations of any characteristic of a defined stratigraphic unit, occupying mutually exclusive areas bounded by arbitrarily (or preferably, quantitatively) determined limits, and usually comprising one or several lithosomes and biosomes that occur in vertical suc-cession or are intertongued.

operational unit A term used by Sloss et al (1949) for an arbi-trary stratigraphic unit that is distinguished by objective crite-ria for some practical purpose (such as regional facies map-ping or analysis); e.g. a unit delimited by easily recognizable and traceable markers, or a unit defined by the velocity of transmission of seismic or sonic energy. Its boundaries do not necessarily correspond with those of any conventional stratig-raphic unit. Syn: *parastratigraphic unit.*

operator variation A common effect of all experimental procedures (such as making visual estimates of particle sphericity and roundness) arising from two factors, "the one due to differences arising from a constant bias characteristic of each operator and the other due to inconsistent differences within one operator and among a group of operators" (Griffiths & Rosenfeld, 1954, p.74).

operculate adj. Having an operculum; e.g. said of a pollen grain having pore membranes with an operculum, or said of an archeopyle covered by a lid.---n. An operculate gastropod.

operculum [paleont] (a) A corneous or calcareous plate that develops on the posterior dorsal surface of the foot of a gastropod and that serves to close the aperture. (b) A generally uncalcified lamina or flap, hinged or pivoting on condyles, that closes the zooidal orifice in cheilostomatous bryozoans. (c) The valves (terga of scuta) and associated membranes forming an apparatus that guards the aperture of cirripede crustaceans. (d) A lid, usually disk-like and flat, and solid or composed of two parts, that closes an opening (such as the anus or genital opening) of an arachnid; a plate adjoining the appendages of genital segment of a merostome. (e) A lid-like covering of the calice in some solitary corals, formed of one or more independent plates. (f) A structure that may serve to close the pseudostome of chitinozoans. It may be external in position or sunken within the neck. (g) The flat pore-bearing base of the podoconus in nassellarian radiolarians; the central part of the astropyle of phaeodarian radiolarians.---Pl: *opercula*.

operculum [bot] A lid or cover in a protistan, e.g. in the fungi, it is part of the cell wall.

operculum [palyn] (a) A lid consisting of the plate or plates that originally filled the archeopyle of a dinoflagellate or the pylome of an acritarch. (b) A thicker central part of a pore membrane of a pollen grain, or a large section or cap of exine completely surrounded by a single circular colpus. For certain hilate spores and pollen, the operculum is a less well-defined lid of exine associated with the formation of the hilum.

opesiula One of the small notches or pores in a bryozoan cryptocyst for the passage of depressor muscles attached to the frontal membrane of some anascan cheilostomes. Syn: *opesiule*.

opesium The large opening below the frontal membrane in zooids of anascan cheilostomes (bryozoans), remaining after development of a cryptocyst. Pl: *opesia*.

ophicalcite A recrystallized limestone composed of calcite and serpentine, formed by dedolomitization of a siliceous dolomite. Cf: *forsterite marble*.

ophiocistioid Any quinqueradiate, free-living echinozoan belonging to the class Ophiocistioidea, having a depressed, dome-shaped body covered entirely or on one side only by plates. Their stratigraphic range is Lower Ordovician to Upper Silurian (possibly Middle Devonian, also).

ophiolite A group of mafic and ultramafic igneous rocks ranging from spilite and basalt to gabbro and peridotite, including rocks rich in serpentine, chlorite, epidote, and albite derived from them by later metamorphism, whose origin is associated with an early phase of the development of a geosyncline. The term was originated by Steinman in 1905 (Myashiro, 1968, p.826).

ophiolitic suite The association of ultramafic rocks with coarse-grained gabbro, coarse-grained diabase, volcanic rock, and red radiolarian chert in the Tethyan mountain system.

ophirhabd A sinuous oxea (sponge spicule).

ophite A general term for diabases which have retained their ophitic structure although the pyroxene is altered to uralite. The term was originated by Palasson in 1819.

ophitic Said of the holocrystalline, hypidiomorphic-granular texture of an igneous rock (esp. diabase) in which lath-shaped plagioclase crystals are partially or completely included in pyroxene crystals (typically augite). Also, said of a rock exhibiting ophitic texture (e.g. ophite) or, rarely, of a similar texture involving other pairs of minerals. The term *diabasic*, although generally considered synonymous, was distinguished from "ophitic" by Kemp (1900, p.158-159) who considered the latter as having an excess of augite over plagioclase while the former had a predominance of plagioclase, with augite filling the interstices. Cf: *poikilitic; poikilophitic*. Nonpreferred syn: *basiophitic; granitotrachytic*. Syn: *doleritic; gabbroid*.

ophiuroid Any asterozoan belonging to the subclass Ophiuroidea, characterized by slender, elongated arms that are distinct from the disc in almost all cases; e.g. starfish-like animals such as brittle stars and basket stars. Var: *ophiurid; ophiuran*.

ophthalmic ridge A longitudinal ridge above the compound eye of a merostome and extending forward and backward from it.

ophthalmite An *eyestalk* of a crustacean.

opisometer A *chartometer* consisting of a small toothed wheel geared to a pointer moving over a graduated recording dial, used for measuring distances on a map running the wheel along a given line which may be curved or irregular (such as one representing a stream, road, or railway).

opisthobranch Any marine gastropod belonging to the subclass Opisthobranchia, characterized by the reduction or absence of the shell.

opisthoclade A *clade* or bar in the ebridian skeleton that arises from an upper actine and is directed toward the posterior. In the triaene ebridian skeleton, it may rejoin the distal extremity of the rhabde. Cf: *proclade*.

opisthocline (a) Said of the hinge teeth (and in some genera, of the body of the shell) of a bivalve mollusk, sloping (from the lower end) in the posterior or backward direction. (b) Said of the growth lines that incline backward relative to the growth direction of a gastropod shell.---Cf: *prosocline*.

opisthodetic Said of a ligament of a bivalve mollusk situated wholly posterior to (or behind) the beaks. Cf: *amphidetic*.

opisthogyrate Said of the umbones (of a bivalve mollusk) curved so that the beaks point in the posterior or backward direction. Ant: *prosogyrate*. Syn: *opisthogyral*.

opisthoparian adj. Of or concerning a trilobite whose facial sutures extend backward from the eyes to the posterior margin of the cephalon; e.g. an "opisthoparian facial suture" that crosses a cheek, passes along the medial edge of the eye, and intersects the posterior border of the cephalon medial to the genal angle. Cf: *proparian*. Syn: *opisthoparous*.---n. An opisthoparian trilobite; specif. a trilobite of the order Opisthoparia including those in which the genal angles or genal spines are borne by the free cheeks.

opisthosoma The posterior part of the body of an arthropod; esp. the *abdomen* behind the *prosoma* of a merostome or following the fourth pair of legs of an arachnid.

opisthosome A dark fusiform body at the base of the body chamber of a chitinozoan, usually convex upward or even spherical, and commonly having a ragged appearance (as if burst open) and a longitudinally striate surface. It is not always present, and may at times be mistaken for a fold of the body-chamber wall. Cf: *prosome*.

opoka A porous, flinty, and calcareous sedimentary rock, with conchoidal or irregular fracture, consisting of fine-grained opaline silica (up to 90%), and hardened by the presence of silica of organic origin (silicified residues of radiolaria, sponge spicules, and diatoms). It differs from *gaize* in its absence of quartz grains and the rarity of glauconite, although a distinction is not drawn in Russia between the two rocks (P.G.H. Boswell in Allen, 1936, p.11). Etymol: Polish.

Oppelzone A term suggested as a syn. of *concurrent-range zone* (ISST, 1961, p.23). Named after Albert Oppel (1831-1865), German stratigrapher.

opposing wind A wind whose direction is opposite to that of ocean waves. Cf: *following wind*.

opposite In plant morphology, pertaining to the attachment of two parts, e.g. leaves, at opposite points at a node.

opposite tide The high tide at a corresponding place on the opposite side of the Earth accompanying a *direct tide.*

optalic metamorphism *caustic metamorphism.*

opthalmite Chorismite characterized by augen and/or other lenticular aggregates of minerals (Dietrich & Mehnert, 1961). Cf: *miarolithite.*

optical activity The property or ability of a mineral, e.g. quartz, to rotate the plane of polarization of light. Such a mineral is said to be optically active. Syn: *rotary polarization.*

optical axis In an optical system, the line passing through the nodal points of a lens.

optical calcite The type of calcite from which nicol prisms are made. It is usually *Iceland spar.*

optical center That point on the axis of an optical system at which light rays cross.

optical character A statement of the sign of uniaxial or biaxial crystals.

optical constant Any characteristic optical property of a crystal, e.g. index of refraction, optic angle.

optical crystallography That branch of crystallography which deals with the optical properties of a crystal, or *crystal optics.* Cf: *optical mineralogy.*

optical density *density* [optics].

optical emission spectrometry Chemical analysis performed by heating the sample to high temperatures whereupon the atoms emit light of definite wavelengths characteristic of the specific elements or molecules present (May & Cuttitta, 1967, p.130).

optical emission spectroscopy The observation of an optical emission spectrum and all processes of recording and measuring which go with it.

optical glass Glass that is suitable for use as prisms, lenses, and other optical purposes.

optically pumped magnetometer A type of magnetometer that measures total magnetic field intensity by means of the precession of magnetic atoms, with precession frequency proportional to field intensity. The magnetic atoms are usually gaseous rubidium, cesium, or helium, which are magnetized by optical pumping, i.e. by irradiation by circularly polarized light of suitable wavelength. Se also: cesium vapor magnetometer; rubidium vapor magnetometer.

optical microscope A microscope that utilizes visible light for illumination.

optical mineralogy The description of the original properties of minerals. Cf: *optical crystallography.*

optical oceanography That aspect of physical oceanography which deals with the optical properties of sea water and natural light in sea water.

optical path The path along which light rays travel through the *optical system* of a microscope or other optical apparatus. Syn: *path* [optics].

optical pyrometer A type of *pyrometer* that measures high temperature by comparing the intensity of light of a particular wavelength from the hot material with that of a filament of known temperature. It is used to determine the temperature of incandescent lavas.

optical rotation The angle of rotation, measured in degrees, of plane-polarized light as it passes through an optically active crystal.

optical square A small hand instrument used in surveying for accurately setting off a right angle by means of two plane mirrors placed at an angle of 45 degrees to each other or by means of a single plane mirror so placed that it makes an angle of 45 degrees with a sighting line.

optical system The lenses, prisms, and mirrors of an optical apparatus such as a microscope, through which goes the *optical path.* Syn: *optical train.*

optical train *optical system.*

optical twinning A type of twinning in quartz, the individuals of which are alternately right-handed and left-handed, e.g. Brazil twinning. Syn: *chiral twinning.*

optical wedge (a) A refracting prism of very small angle, inserted in an optical train to introduce a small bend in the ray path. It is used in the eyepiece of certain stereoscopes. (b) A strip of film or a glass plate used to reduce the intensity of light or radiation (gradually or in steps, as in determining the density of a photographic negative), and having a layer of neutral or colored substance varying progressively in transmittance with distance along the wedge; e.g. a *step wedge.*

optic angle The acute angle between the two optic axes of a biaxial crystal; its symbol is 2V. See also: *apparent optic angle.* Syn: *axial angle; optic-axial angle.*

optic-axial angle *optic angle.*

optic axis A direction in an anisotropic crystal along which there is no double refraction. In tetragonal and hexagonal crystals it is parallel to the threefold, fourfold, or sixfold symmetry axis; in orthorhombic, monoclinic, and triclinic crystals there are two optic axes, which are determined by the indices of refraction. See also: *primary optic axis; secondary optic axis.*

optic ellipse Any section through an ellipsoidal indicatrix.

optic indicatrix *indicatrix.*

optic normal The axis of a crystal that is perpendicular to the optic axis.

optimum A period of warmer and drier climate than that of the present; specif. the *climatic optimum* of post-Wisconsin age.

optimum moisture content The water content at which a specified compactive force can compact a soil mass to its maximum dry unit weight.

oral adj. (a) Being the surface on which the mouth of an invertebrate is situated, such as the upward-directed *actinal* surface of theca of an edrioasteroid. Also, relating to or located on an oral surface, or situated at, near, or toward the mouth or peristome (such as of an echinoderm); e.g. an "oral pole" representing the end theca containing the mouth in a cystoid. Ant: *aboral.* (b) Pertaining to the orifice (not the mouth) of a bryozoan zooid. (c) Toward the upper side of a conodont element.---n. An *oral plate* of an echinoderm.

oral disk The fleshy, more or less flattened wall closing off the upper or free end of the cylindrical column that forms the sides of a scleractinian coral polyp, its center containing the mouth. Cf: *basal disk.* Also spelled: *oral disc.*

oral margin The trace of the oral side of a conodont element in lateral (side) view. The term has also been used for the oral side itself.

oral membrane A sheet of cilia in the gullet of a tintinnid.

oral pinnule Any proximal pinnule of a crinoid, differentiated from distal pinnules in function or structure, or both.

oral plate Any of five interradially disposed plates forming a circlet surrounding or covering the mouth of an echinoderm. Syn: *oral.*

oral pole The end of a flask-shaped chitinozoan that includes the neck and the mouth. Cf: *aboral pole.*

oral side The upper side of a conodont element opposite that toward which the basal cavity opens. It commonly supports denticles, nodes, and ridges in compound and plate-like conodont elements. Cf: *aboral side.*

oral tooth One of the sharp triangular projections around the basal shell opening in phaeodarian radiolarians.

orangite A bright orange-yellow variety of thorite.

oranite A lamellar intergrowth of a potassium feldspar and a plagioclase near anorthite.

orate Said of a porate pollen grain having an internal opening (endopore or os) in the endexine.

O ray In uniaxial crystals, the ray which vibrates perpendicular to the optic axis; the *ordinary ray.* Cf: *E ray.* Syn: *ordinary wave.*

orbicular (a) Said of the structure of a rock containing numerous orbicules; also, said of a rock having orbicular structure. (b) Having the shape of an orbicule.----Syn: *nodular.*

orbiculate Said of a circular or disk-shaped leaf.

orbicule A more or less spherical body, from microscopic size to two or more centimeters in diameter, whose components are arranged in concentric layers. Cf: *spherulite* [*petrology*].

orbit [**waves**] The path of a water particle affected by wave motion, being almost circular in deep-water waves and almost elliptical in shallow-water waves.

orbit [**paleont**] A circular opening in the anterior part of the carapace of a decapod crustacean, enclosing the eyestalk.

orbite An igneous rock containing large phenocrysts of hornblende, or plagioclase and hornblende, in a groundmass having the composition of malchite.

orbitoid Any foraminifer belonging to the superfamily Orbitoidacea, characterized by test walls composed of radially arranged calcite crystals and by bilamellid septa. Their stratigraphic range is Cretaceous to present.

orbitolinid Any foraminifer belonging to the family Orbitolinidae and characterized by a relatively large conical test ranging from a high pointed cone to a broad shield or disc. Their stratigraphic range is Lower Cretaceous to Eocene.

orbitolite Any foraminifer belonging to the genus *Orbitolites* of the suborder Miliolina, and characterized by a discoidal test containing numerous small chambers in annular series. Its stratigraphic range is Upper Paleocene to Eocene.

orcelite A mineral: Ni$_2$As.

ordanchite An extrusive rock containing phenocrysts of sodic plagioclase, haüyne, hornblende, augite, and some olivine; an olivine-bearing haüyne tephrite.

order [**geomorph**] (a) *stream order.* (b) *basin order.*

order [**taxon**] A unit in taxonomy that contains one or more related families and ranking above "family" and below "class". In botany, the name of an order characteristically ends in -ales; e.g. Filicales.

order [**petrology**] In the C.I.P.W. classification of igneous rocks, the basic unit of the *class* [*petrology*].

order-disorder inversion *substitutional transformation.*

order-disorder polymorphs Two crystal substances of the same composition but of different atomic arrangement. In the higher-temperature or disordered form, two or more elements are randomly distributed over a particular set of atom sites; in the lower-temperature or ordered form, the atoms become ordered with respect to the same sites. The ordered form usually has lower symmetry.

order-disorder transformation *substitutional transformation.*

ordering *substitutional transformation.*

order in minerals The ordered substitution of one ion for another in the crystal structure, e.g. in microcline, in which one fourth of the Si positions are occupied by Al. See also: *short-range order; long-range order.*

order of crystallization The apparent chronologic sequence in which crystallization of the various minerals of an assemblage takes place, as evidenced mainly by textural features. Syn: *sequence of crystallization.*

ordinary chert A generally homogeneous *smooth chert* with an even fracture surface, approaching opacity, having slight granularity or crystallinity, and being of any color (chiefly white, gray, or brown, or sometimes mottled) (Ireland et al, 1947, p.1485). It has no distinctive characteristics that can be described.

ordinary coccolith One of the unmodified coccoliths in a coccolithophore exhibiting dimorphism.

ordinary lead *common lead.*

ordinary ray *O ray.*

ordinary tide level *mean tide level.*

ordinary wave *O ray.*

ordnance datum A name given to several horizontal datums to which heights have been referred on official maps of the British Ordnance Survey; specif. in Great Britain (but not Ireland)

the mean sea level at Newlyn in Cornwall. Abbrev: O.D.

ordonezite A brown tetragonal mineral: ZnSb$_2$O$_6$. Also spelled *ordoñ ezite.*

ordosite A dark-colored acmite syenite.

Ordovician The second earliest period of the Paleozoic er (after the Cambrian and before the Silurian), thought to hav covered the span of time between 500 and 430-440 millio years ago; also, the corresponding system of rocks. It i named after a Celtic tribe called the Ordovices. In the olde literature the Ordovician is sometimes known as the *Lower S lurian.* Se also: age of marine invertebrates. Obs syn: *Cham plainian.*

ore [**eco geol**] The naturally occurring material from which mineral, or minerals, of economic value can be extracte Also, the mineral(s) thus extracted. The term is generally b not always used to refer to metalliferous material, and is ofte modified by the name of the valuable constituent, e.g., "iro ore". See also:*mineral deposit; orebody; ore mineral.*

ore [**ign**] The term "ores" is sometimes applied collectively opaque accessory minerals, such as ilmenite and magnetite in igneous rocks.

ore beds Metal-rich layers in a sequence of sedimentar rocks.

ore block A section of an orebody, usually rectangular, that used for estimates of tonnage and quality of the orebody. Se also: *blocking out.*

ore blocked out Var. of *blocked-out ore,* which is a syn. *developed reserves.*

orebody A continuous, well-defined mass of material of suff cient *ore* content to make extraction economically feasible See also: *mineral deposit.*

ore channel A little-used term for the orebody or lode, inclu ing both gangue and economically valuable minerals. See als *lodestuff.* Syn: *lode country.*

ore chimney pipe [ore dep].

ore cluster A genetically related group of orebodies that ma have a common root or source rock but that may diffe structurally from one another.

ore control Any geologic feature (tectonic, lithologic, ge chemical, etc) considered to have influenced the formation ore. Cf: *metallotect.*

ore-forming fluid *mineralizer.*

oregonite A hexagonal mineral: Ni$_2$FeAs$_2$.

ore guide Any natural feature, e.g. alteration products, or certain structure or plant growth, known to be indicative of orebody. See also: *lithologic guide; stratigraphic guide.*

ore in sight *developed reserves.*

ore magma A term proposed by Spurr (1923) for a magm that may crystallize into an ore; the metallic facies of a solid fied magma. In a genetic sense, the distinction is made be tween a melt, which magma is, and a solution as fissure fil ings.

ore microscopy The study of ore minerals in polished sectio by reflecting microscope. Syn: *mineragraphy; mineralography.*

ore mineral The part of an *ore,* usually metallic, which is ec nomically desirable, as contrasted with the *gangue.*

orendite A porphyritic extrusive rock containing phlogopit phenocrysts in a nepheline-free reddish-gray groundmass leucite, sanidine, phlogopite, amphibole, and diopside; a phlo gopite-leucite trachyte. Its name is derived from Orend Butte, Wyoming.

ore of sedimentation *placer.*

oreography Var. of *orography.*

ore pipe pipe [ore dep].

ore roll Uranium or vanadium orebodies within sedimentar rock, esp. sandstone, that are discordant, forming s-shape or c-shaped cross sections (Shaw & Granger, 1965, p.241 Syn: *roll.*

ore shoot A large, elongate (pipelike or chimneylike) mass ore within a deposit (usually a vein), and representing th more valuable part of that deposit. Syn: *shoot.*

organ genus A genus name used for parts of fossil plants which can be classified in a family.

organic Pertaining or relating to a compound containing carbon, especially as an essential component. Organic compounds usually have hydrogen bonded to the carbon atom. Cf: *inorganic.*

organic bank *bank* [sed].

organic evolution *evolution.*

organic geochemistry That branch of chemistry which studies naturally occurring carbonaceous and biologically-derived substances which are of geological interest.

organic hieroglyph *bioglyph.*

organic lattice *growth lattice.*

organic mound *bioherm.*

organic reef A large or small sedimentary rock aggregate (see *reef*) of significant dimensions, usually well-developed vertically, erected by and composed almost exclusively of the remains of sedentary or colonial-type and sediment-binding organisms (mainly but not necessarily marine), chiefly corals and algae, less commonly crinoids, bryozoans, sponges, mollusks, and others, that lived their mature lives near but below the surface of the water (they may have some exposure at low tide), their exoskeletal hard parts remaining in place after death, and the deposit being firm enough to resist wave erosion; a *bioherm* of sufficient size to develop associated facies. An organic reef may also contain living organisms. See also: *coral reef; algal reef.* Cf: *organic bank.*

organic rock A sedimentary rock consisting primarily of the remains of living organisms (plant or animal), such as of material that originally formed part of the skeleton or tissues of an animal. Cf: *biogenic rock.*

organic soil A general term applied for a soil or a soil horizon that consists primarily of organic matter (or contains at least 30% organic matter), such as peat soils, muck soils, and peaty soil layers.

organic soil material Water-saturated soil material that contains a certain minimum percentage of organic carbon, depending on the clay content of the mineral fraction: for 50% or more clay, 14.7% or more carbon; and for less than 50% to no clay, 11.6% or more carbon. Soil material that has never been saturated with water must have 20.3% or more organic carbon (SSSA, 1970).

organic texture A sedimentary texture resulting from the activity of organisms (such as the secretion of skeletal material).

organic weathering Biologic processes and changes that assist in the breakdown of rocks; e.g. the penetrating and expanding force of roots, the presence of moss and lichen causing humic acids to be retained in contact with rock, and the work of animals (worms, moles, rabbits) in modifying surface soil. Syn: *biologic weathering.*

organized elements A term used by Claus & Nagy (1961) for circular, microscopic-sized organic particles observed in carbonaceous chondrites, having diameters about 4-30 microns, and resembling fossil algae, dinoflagellates, and other terrestrial microorganisms. They are believed by some to be extraterrestrial microfossils and by others to be terrestrial microbiologic contaminations or inorganic microstructures (such as mineral grains or sulfur droplets).

organogenic Said of a rock or sediment derived from organic substances; e.g. a crinoidal limestone (termed an "organogenic conglomerate" by Hadding, 1933) consisting almost exclusively of disks and plates detached from one another. Syn: *organogenous.*

organolite Any rock consisting mainly of organic material, esp. one derived from plants; e.g. coal, resin, and bitumen.

organometallic Describes a compound in which an atom of a metal is bound to an organic compound directly through a carbon atom. Cf: *metallo-organic.*

organosedimentary Pertaining to sedimentation as affected by organisms; e.g. said of a stromatolite, a sedimentary structure produced by the life processes of blue-green algae.

organotrophic Relating to the development and nourishment of living organs.

orido An Italian term for a gorge cut through a rock barrier holding a lake in a glaciated region, like those around Lago di Como in Italy.

orient [gem] (a) The minute play of color on, or just below, the surface of a gem-quality pearl, caused by diffraction and interference of light from the irregular edges of the overlapping crystals or platelets of aragonite that comprise the nacre of the pearl. (b) A pearl of great luster.

orient [surv] (a) To place or set a map so that the map symbols are parallel with their corresponding ground features. (b) To turn a plane table in a horizontal plane until all lines connecting positions on the plane-table sheet have the same azimuths as the corresponding lines connecting ground objects. (c) To turn a transit so that the direction of the zero-degree line of its horizontal circle is parallel to the direction it had in the preceding (or in the initial) setup or parallel to a standard line of reference.---Etymol: Latin *oriens,* from present participle of *oriri,* "to rise, come forth", used originally in connection with the rising of the Sun in the East. Medieval maps were drawn with the east at the top and the eastern side was the side "chiefly concerned in putting the map in agreement with the surroundings" (Cottingham, 1951, p.161).

oriental (a) Said of a genuine gemstone; e.g. "oriental ruby" (the true ruby), or "oriental sapphire" (the true sapphire), or "oriental turquoise". (b) Indicating a finer variety of a gem, or one having superior grade, luster, or value; e.g. "oriental chalcedony" (fine, translucent, gray to white chalcedony), or "oriental carnelian" (deep, bright red, translucent carnelian), or "oriental agate" (the most beautiful and translucent agate). When applied to minerals, the term is frequently used in the same sense as *precious,* implying the finest variety; e.g. "oriental garnet" (such as almandine), or "oriental opal" (such as fire opal). (c) Being corundum or sapphire but simulating another gem in color; e.g. "oriental amethyst", "oriental aquamarine", "oriental beryl", "oriental emerald", and oriental topaz" , all of which are varieties of corundum. (d) Said of a gem originating in the Orient.---Cf: *occidental.*

oriental alabaster *onyx marble.*

oriental amethyst (a) Violet to purple variety of sapphire. (b) Any amethyst of exceptional beauty.

oriental cat's-eye A syn. of chrysoberyl *cat's-eye.*

oriental chrysolite Greenish-yellow chrysoberyl.

oriental jasper *bloodstone.*

orientation [cryst] In describing crystal form and symmetry, the placing of the crystal so that its crystallographic axes are in the conventional position.

orientation [photo] (a) The establishment of the relationship in direction of a photograph when it correctly presents the perspective view of the ground directly in front of the observer or when images on the photograph appear in the same direction from the point of observation as do the corresponding map symbols. See also: *exterior orientation; interior orientation.* (b) The re-creation of natural terrain features at a miniature scale by the optical projection of overlapping photographs.

orientation [surv] The assignment or imposition of a definite direction in space; the act of establishing the correct relationship in direction, usually with reference to the points of the compass. Also, the state of being in such relationship.

orientation diagram In structural petrology, a general term for a point diagram or a contour diagram; a *fabric diagram.*

oriented Said of a specimen or thin section that is so marked as to show its exact, original position in space.

oriented core A core that can be positioned on the surface as it was in the borehole prior to extraction.

orientite A brown to black orthorhombic mineral: $Ca_2Mn^3_3$-$(SiO_4)_3(OH)$.

orifice (a) An opening in the zooid wall through which the lophophore and tentacles of a bryozoan are exserted. Cf: *aperture*. (b) An opening, in the upper part of a crustacean shell, containing the operculum. (c) Any major opening through the outer covering of an echinoderm. (d) An aperture or other opening in a foraminiferal test.

origin [surv] (a) A point in a coordinate system that serves as an initial in computing its elements or in prescribing its use; esp. the point defined by the intersection of coordinates axes and from which the coordinates are reckoned. The term has also been applied to the point to which the coordinate values of zero and zero are assigned (regardless of its position with reference to the axes) and to the point from which the computation of the elements of the coordinate system (projection) proceeds. Syn: *origin of coordinates*. (b) Any arbitrary zero or starting point from which a magnitude is reckoned on a scale or other measuring device.

original dip *primary dip*.

original horizontality The state of strata being horizontal or nearly horizontal at the time they were originally deposited. Se also: law of original horizontality.

original interstice An *interstice* that formed contemporaneously with the enclosing rock. Cf: *secondary interstice*. Syn: *primary interstice*.

original stream *consequent stream*.

original valley A valley formed by hypogene action or by epigene action other than that of running water (Geikie, 1898, p.347).

origofacies Facies of the primary sedimentary environment (Vassoevich, 1948); the sedimentary "facies" of most western authors (Teichert, 1958, p.2736).

ornament A pattern or collection of pictorial symbols (diagonal lines, plus signs, curlicues, wedges, etc.) printed over a color hue on a geologic map, distinguishing one cartographic unit from another of basically the same hue.

ornamental stone An attractive natural stone that is not practical for jewelry purposes but is useful for fashioning into ornamental and decorative objects, such as figurines, ash trays, and lamp bases; e.g. onyx marble, agate, or malachite.

ornamentation [paleont] The characteristic markings or patterns on the body of an animal; e.g. the external surface features of preserved hard parts (ridges, grooves, granules, spines, etc.) that may interrupt the smooth surface of a shell. See also: *sculpture* [paleont].

ornamentation [palyn] *sculpture*.

ornithocopros *guano*.

ornoite A variety of hornblende diorite.

orocline An orogenic belt with a change in trend (a curvature or a sharp bend), interpreted by Carey (1958) as a result of horizontal bending of the crust, or "deformation in plan". Syn: *geoflex*.

oroclinotath An orogenic belt, interpreted by Carey (1958) as having been subjected both to substantial bending and to substantial stretching.

orocratic Pertaining to a period of time in which there is much diastrophism. Cf: *pediocratic*.

orogen *orogenic belt*.

orogene *orogenic belt*.

orogenesis *orogeny*.

orogenetic A less preferred adj. of *orogeny*.

orogenic Adj. of *orogeny*. Cf: *orographic*.

orogenic belt A linear region that has been subjected to folding and other deformation during the *orogenic cycle*. Orogenic belts were *mobile belts* during their formative stages, and most of them later became mountain belts by postorogenic processes. Syn: *foldbelt; orogen; orogene*.

orogenic cycle The interval of time during which an originally *mobile belt* evolved into a stabilized *orogenic belt*, passing through *preorogenic, orogenic* and *postorogenic phases*. Syn: *tectonic cycle; geotectonic cycle*.

orogenic facies A term applied to a *geosynclinal facies* whe emphasizing its tectonic environment.

orogenic phase The median part of an *orogenic cycle* chara terized by the climax of crustal mobility and orogenic activit and by the formation of alpinotype structures. The orogen phase is commonly shorter than the *preorogenic* and *postoro genic phases* which precede and follow it, and may be les than a geologic period in length, although it is commonly pr longed by a succession of *pulsations*.

orogenic sediment Any sediment that is produced as the re sult of an orogeny or that is directly attributable to the oroger ic region in which it later becomes involved; e.g. a clast sediment such as flysch or molasse.

orogenic unconformity An *angular unconformity* produced l cally in a region affected by mountain-building movements.

orogeny Literally, the process of formation of mountains. Th term came into use in the middle of the 19th Century, whe the process was thought to include both the deformation rocks within the mountains, and the creation of the mountai ous topography. Only much later was it realized that the tw processes were mostly not closely related, either in origin time. Today, many geomorphologists and a few geologists us orogeny for the formation of mountainous topography; mo geologists regard this process as postorogenic and epeiroge ic. By present geological usage orogeny is the process b which structures within mountain areas were formed, includ ing thrusting, folding, and faulting in the outer and highe layers, and plastic folding, metamorphism, and plutonism the inner and deeper layers. This usage has practical advan tages; only in the very youngest, late Cenozoic mountains there any evident causal relation between rock structure an surface landscapes. Little such evidence is available for th early Cenozoic, still less for the Mesozoic and Paleozoic, an virtually none for the Precambrian--yet all the deformation structures are much alike, whatever their age, and are appr priately considered as products of orogeny. See also: *diastro phism*. Cf: *epeirogeny; tectogenesis; cymatogeny; morphotec tonics*. Syn: *orogenesis; mountain building; tectogenesis*. Ad *orogenic; orogenetic*.

orogeosyncline Kober's term for a geosyncline that later be came an area of orogeny (Glaessner & Teichert, 1947 p.588). See also: *mother geosyncline*.

orograph A machine used in making topographic maps an operated by being pushed across country to record both dis tances and elevations.

orographic [geog] (a) Pertaining to mountains, esp. in regar to their location and distribution. (b) Said of the precipitatio (or rainfall) that results when moisture-laden air encounters high barrier or is forced to rise over it, such as the precipita tion on the windward slopes of a mountain range facing steady wind from a warm ocean. Also, said of the lifting of a air current caused by its passage up and over a mountain. (c Pertaining to a *rain-shadow desert*.

orographic [tect] A term, now little used, for features relatin to mountain structure and topography. More explicit adjec tives such as *diastrophic, epeirogenic*, and *orogenic* are no preferred.

orographic desert *rain-shadow desert*.

orography (a) The branch of physical geography that deal with the disposition, character, formation, and structure o mountains and of chains, ranges, and systems of mountains (b) Broadly, the description or depiction of the relief of th Earth's surface or of a part of it, or the representation of suc relief on a map or model; the land features of a specified re gion.---Etymol: Greek *oros*, "mountain". Syn: *orology; oreo graphy*.

orohydrography A branch of hydrography dealing with th relations of mountains to drainage.

orology A syn. of *orography*; esp. the study of mountain build ing and mountain formation.

ometer An aneroid barometer that gives the approximate elevation above sea level of the place where the observation is made.

ometry The measurement of mountains.

ophilous Said of an organism that lives in subalpine conditions.

ophyte A plant growing in subalpine regions.

otath An orogenic belt, interpreted by Carey (1958) as having been substantially stretched in the direction of its length.

otvite A syenite composed of hornblende, biotite, plagioclase, nepheline, and cancrinite, with accessory sphene, ilmenite, and apatite.

piment A lemon-yellow to orange monoclinic mineral: As_2S_3. It is generally foliated or massive, and is frequently associated with realgar. Orpiment occurs as a deposit from some hot springs and as a sublimate from some volcanoes. Syn: *yellow arsenic*.

thembadism A term used in cartography as a synonym of *equivalence*. An "orthembadic projection" is an equal-area projection.

rthent In U.S. Dept. of Agriculture soil taxonomy, a suborder of the soil order *Entisol*, characterized by a fine earth fraction having textures of very fine sand or finer, by a coarse fragment content of 35% or more, and by an organic content that decreases with depth (SSSA, 1970). Cf: *Aquent; Arent; Fluvent; Psamment.*

rthid Any articulate brachiopod belonging to the order Orthida, characterized chiefly by an impunctate shell or one with endopuncta and by an open delthyrium and brachiophores. Their stratigraphic range is Lower Cambrian to Upper Permian.

rthid In U.S. Dept. of Agriculture soil taxonomy, a suborder of the soil order *Aridisol*, characterized by the presence of a cambic, calcic, petrocalcic, gypsic, or salic horizon or a duripan and by the absence of an argillic or natric horizon (SSSA, 1970). Cf: *Argid.*

rthite A syn. of *allanite*, applied esp. when occurring in slender prismatic or acicular crystals.

rtho- In petrology, a prefix that, when used with the name of a metamorphic rock, indicates that it was derived from an igneous rock, e.g. orthogneiss, orthoschist; it may also indicate the primary origin of a crystalline, sedimentary rock, e.g. orthoquartzite as distinguished from "metaquartzite".

rthoamphibole (a) A group name for amphiboles crystallizing in the orthorhombic system. (b) Any orthorhombic mineral of the amphibole group, such as anthophyllite, gedrite, and holmquistite.---Cf: *clinoamphibole.*

rthoandesite An andesite containing orthopyroxene. Syn: *anukitoid*. Cf: *sanukite.*

rthoantigorite A mineral of the serpentine group: $Mg_3Si_2O_5(OH)_4$. It is a six-layer orthorhombic form of antigorite.

rthoapsidal projection A map projection produced by means of the orthographic projection of a graticule from some solid body other than the sphere or spheroid.

rthoaxis In a monoclinic crystal, the lateral axis that has two-fold symmetry and/or is perpendicular to the mirror plane of symmetry; it is the *b* axis. Cf: *clinoaxis*. Syn: *orthodiagonal.*

rthobituminous Said of bituminous coal containing 87-89% carbon, analyzed on a dry, ash-free basis. Cf: *parabituminous; perbituminous.*

rthoceratite Any nautiloid belonging to the genus *Orthoceras*, characterized by the presence of three longitudinal furrows on the body chamber.

rthochamosite A mineral of the chlorite group: $(Fe^{+2},Mg, Fe^{+3})_6AlSi_3O_{10}(OH)_8$. It is the orthorhombic dimorph of chamosite.

rthochem An essentially normal precipitate formed by direct chemical action within a depositional basin or within the sediment itself, as distinguished from material transported in a solid state (Folk, 1959, p.7); e.g. aphanocrystalline calcareous ooze (micrite), intergranular cement, recrystallized sedimentary material, and replacement minerals such as dolomite. Adj: *orthochemical*. Cf: *allochem.*

orthochlorite (a) A group name for distinctly crystalline forms of chlorite (such as clinochlore and penninite). (b) A group name for chlorites conforming to the general formula: $(R_i^{+2}, R^{+3})_6(Si,Al)_4O_{10}(OH)$ (Hey, 1962, p.546).---Cf: *leptochlorite.*

orthochoanitic Said of a straight, cylindrical, retrochoanitic septal neck of a nautiloid that extends only a short distance to the preceding septum.

orthochronology *Geochronology* based on a standard succession of biostratigraphically significant faunas or floras, or based on irreversible evolutionary processes. Ideally, orthochronology is chronology based on a stratigraphic succession of species "where each successive species is the descendent of the one which immediately precedes it stratigraphically" (Teichert, 1958a, p.106). Cf: *parachronology*. See also: *biochronology.*

orthochrysotile A mineral of the serpentine group: $Mg_3Si_2O_5(OH)_4$. It is an orthorhombic form of chrysotile. Cf: *clinochrysotile.*

orthoclase (a) A colorless, white, cream-yellow, flesh-reddish, or grayish mineral of the alkali feldspar group: $KAlSi_3O_8$. It is the partly ordered, monoclinic modification of potassium feldspar and is dimorphous with *microcline*, being stable at higher temperatures; it usually contains some sodium in minor amounts. Ordinary or common orthoclase is one of the commonest rock-forming minerals; it occurs esp. in granites, acid igneous rocks, and crystalline schists, and is usually perthitic. Syn: *common feldspar; orthose; pegmatolite*. (b) A general term applied to any potassium feldspar that is or appears to be monoclinic; e.g. sanidine, submicroscopically twinned microcline, adularia, and submicroscopically twinned analbite.-- Cf: *plagioclase; anorthoclase.*

orthoclasite An orthoclase-bearing porphyritic extrusive rock, such as granite or syenite. The term is sometimes restricted to rocks containing more than 90 percent orthoclase.

orthocline (a) Said of the *inclination* of the cardinal area in either valve of a brachiopod lying on the continuation of the commissural plane. (b) Said of the hinge teeth (and in some genera, of the body of the shell) of a bivalve mollusk, oriented perpendicular or nearly perpendicular to the hinge axis. Syn: *acline* [*paleont*]. (c) Said of the growth lines that traverse the whorl at right angles to the growth direction of a gastropod shell.

orthocone A straight, slender nautiloid conch, resembling that of *Orthoceras*. Syn: *orthoceracone.*

orthoconglomerate A term used by Pettijohn (1957, p.256) for a conglomerate with an intact gravel framework, characterized by a mineral cement, and deposited by "ordinary" but highly turbulent water currents (either high-velocity streams or the surf); e.g. (orthoquartzitic conglomerate and *arkosic conglomerate*. It is strongly current-bedded and is associated with coarse cross-bedded sandstones. Cf: *paraconglomerate.*

orthocumulate A cumulate composed chiefly of one or more cumulus minerals plus the crystallization products of the intercumulus liquid. Cf: *mesocumulate.*

Orthod In U.S. Dept. of Agriculture soil taxonomy, a suborder of the soil order *Spodosol*, characterized by having less than six parts elemental iron to one part organic carbon, and an iron/carbon ratio of 0.2 or more, in its spodic horizon (SSSA, 1970). Cf: *Aquod; Ferrod; Humod.*

orthodiagonal n. A syn. of *orthoaxis.*

orthodolomite (a) A *primary dolomite*, or one formed by sedimentation. (b) A term used by Tieje (1921, p.655) for a dolomite rock so well-cemented that the particles are interlocking.

orthodome An old term for a monoclinic crystal form whose faces parallel the orthoaxis. Its indices are {h01}.

orthodont Said of the hinge of a bivalve mollusk in which the direction of the hinge teeth is parallel or nearly parallel to the cardinal margin.

orthodrome *great circle.*

orthodromic projection A map projection, derived from the gnomonic projection, in which angles are correct at two points and all great circles are straight lines.

orthoericssonite A mineral: $BaMn_2(FeO)Si_2O_7(OH)$.

orthofelsite *orthophyre.*

orthoferrosilite A mineral of the orthopyroxene group: (Fe, Mg)SiO_3; specif. a mineral consisting of the orthorhombic iron silicate $FeSiO_3$. See also: *ferrosilite.* Cf: *clinoferrosilite.*

orthogenesis Variation that follows a specific trend in successive generations of an organism, resulting in the evolution of a new type, independent of the effect of natural selection or other external factors. Cf: *rectilinear evolution.*

orthogeosyncline A geosyncline between continental and oceanic cratons, containing both volcanic (eugeosynclinal) and nonvolcanic (miogeosynclinal) belts (Stille, 1935, p.77-97). Syn: *primary geosyncline; geosynclinal couple.* See also: *eugeosyncline; miogeosyncline.*

orthogonal n. A curve that is everywhere perpendicular to the wave crests on a refraction diagram. Syn: *wave ray.*

orthogonal projection A projection in which the projecting lines (straight and parallel) are perpendicular to the plane of the projection; e.g. an *orthographic projection.*

orthograde Pertaining to the uniform distribution of dissolved oxygen in the hypolimnion of a lake, dependent only "on conditions at circulation and on subsequent physical events" (Hutchinson, 1957, p.603). Cf: *clinograde.*

orthogranite The original name for *kaligranite.*

orthographic projection (a) A perspective azimuthal map projection produced by straight parallel lines from a point at an infinite distance from the sphere to points on the sphere and perpendicular to the plane of projection. The largest area depicted is that of a hemisphere, and the projection is true to scale only at the center. The plane of projection may be perpendicular to the Earth's axis of rotation (polar orthographic projection, with the center at a pole) or parallel to the plane of some selected meridian (meridional orthographic projection, with the center on the equator). It is used for star charts and for pictorial world maps. (b) A similar projection used in optical mineralogical study of the origin of interference phenomena under the polarizing microscope, obtained by dropping perpendiculars from the poles (in the projection of the sphere) to the plane of projection which is normal to the north-south axis of the sphere. (c) *orthogonal projection.*

orthogyrate Said of the umbones (of a bivalve mollusk) curved so that each beak points neither anteriorly nor posteriorly but directly towards the other valve. Cf: *mesogyrate.*

orthohexagonal Pertaining to a set of crystallographic axes, e.g. in hexagonal or trigonal crystals, two of which have a fixed ratio.

orthohydrous (a) Said of coal containing 5-6% hydrogen, analyzed on a dry, ash-free basis. (b) Said of a maceral of normal hydrogen content, e.g. vitrinite.----Cf: *subhydrous; perhydrous.*

ortholignitous Said of coal containing 75-80% carbon, analyzed on a dry, ash-free basis.

ortholimestone A primary limestone, or one formed by sedimentation.

ortholith (a) A coccolith composed of one or very few crystals, as in the coccolithophorid *Braarudosphaera.* (b) An individual of the Ortholithae, a subdivision of the family Coccolithophoridae.-- -Cf: *heliolith.*

orthomagmatic stage The main stage in the crystallization of silicates from a typical magma; the stage during which as much as 90 percent of a magma may crystallize. Syn: *ortho-*

tectic stage.

orthometric correction A systematic correction that must be applied to a measured difference of elevation because of the fact that level surfaces at different elevations are not exactly parallel.

orthometric height The distance of a point above the geoid expressed in linear units measured along the plumb line at the point. Orthometric corrections are applied to measurements of precise leveling because level surfaces at different elevations are not parallel. Cf: *dynamic height; height.*

orthometric height Distance above sea level, measured along the plumb line in linear units.

orthomicrite A genetic term applied to unaltered or primary calcareous micrite (Chilingar et al, 1967, p.318). It includes *allomicrite* and *automicrite.* Cf: *pseudomicrite.*

orthomicrosparite A genetic term applied to microsparite that has developed by precipitation in open voids (Chilingar et al, 1967, p.228). Cf: *pseudomicrosparite.*

orthomimic feldspar A group of feldspars that by repeated twinning simulate a higher degree of symmetry with rectangular cleavages. Also spelled: *orthomic feldspar.*

orthomorphic projection *conformal projection.*

orthomorphism *conformality.*

orthophotograph A photographic copy, prepared from a photograph formed by a perspective projection, in which the displacements due to tilt and relief have been removed; a photograph that has been transformed to an orthographic projection.

orthophotomap A *photomap* made from an assembly of orthophotographs.

orthophotomosaic A uniform-scale, photographic mosaic consisting of an assembly of orthophotographs.

orthophyre An obsolete term for a porphyritic rock containing phenocrysts of orthoclase. Syn: *orthofelsite.*

orthophyric Said of the texture of the groundmass in certain holocrystalline porphyritic igneous rocks in which the feldspar crystals have quadratic or short, stumpy rectangular cross sections, rather than the lath-shaped outline observed in *trachytic* texture. Also, said of a groundmass with this texture, or of a rock having an orthophyric groundmass.

orthopinacoid *front pinacoid.*

orthopinakiolite A black orthorhombic mineral (Mg, $Mn^{+2})_2Mn^{33}BO_5$. It is a polymorph of pinakiolite.

orthopyroxene (a) A group name for pyroxenes crystallizing in the orthorhombic system and usually containing no calcium and little or no aluminum. (b) Any orthorhombic mineral of the pyroxene group, such as enstatite, bronzite, hypersthene, and orthoferrosilite.---Cf: *clinopyroxene.*

orthoquartzite A clastic sedimentary rock that is made up almost exclusively of quartz sand (with or without chert), that is relatively free of or lacks a fine-grained matrix, and that is derived by secondary silicification; a *quartzite* of sedimentary origin, or a "pure quartz sandstone". The term generally signifies a sandstone with more than 90-95% quartz and detrital chert grains that are well-sorted, well-rounded, and cemented primarily with secondary silica (sometimes with carbonate) in optical and crystallographic continuity with the grains. The rock is characterized by stable but scarce heavy minerals (zircon, tourmaline, magnetite), by lack of fossils, and by prominence of cross-beds and ripple marks. It commonly occurs as thin but extensive blanket deposits associated with widespread unconformities (e.g. an epicontinental deposit developed by an encroaching sea) and it represents intense chemical weathering of original minerals other than quartz, considerable transport and washing action before final accumulation (the sand may experience more than one cycle of sedimentation), and stable conditions of deposition (such as the peneplanation stage of diastrophism). Example: St. Peter Sandstone (Middle Ordovician) of midwestern U.S. The term as introduced by Tieje (1921, p.655) for a quartz sandstone whose interlocking particles were cemented by infiltration and pressure (in contrast to *paraquartzite*), and was used by rynine (1948, p.149) in contrast to *metaquartzite*, but the term is objectionable because it is an exception to the use of ortho-" for a metamorphic rock indicating an igneous origin and because "quartzite" is traditionally applied to quartzose rocks that break across instead of between grains. See also: *quartzose sandstone*. The term is essentially equivalent to *quartzarenite* and *quartzitic sandstone*. Syn: *sedimentary quartzite; orthoquartzitic sandstone*.

orthoquartzitic conglomerate A well-sorted, lithologically homogeneous, light-colored *orthoconglomerate* consisting of mature or supermature quartzose residues (chiefly vein quartz, chert, and quartzite, in fine to medium pebble size) that represent relatively stable material derived from eroded granitic or metamorphic terrain, with removal of finer material and less-stable lithologic types by weathering or by long transport. It is commonly interbedded with pure quartz sandstone. Syn: *quartz-pebble conglomerate*.

orthoquartzitic sandstone *orthoquartzite*.

orthorhombic system One of the six *crystal systems*, characterized by three axes of symmetry that are mutually perpendicular and of unequal length. Cf: *isometric system; tetragonal system; hexagonal system; monoclinic system; triclinic system*. Syn: *rhombic system*. Obsolete syn: *prismatic system*.

orthoscope A polarizing microscope in which light is transmitted by the crystal parallel to the microscope axis, in contrast to the *conoscope*, in which a converging lens and Bertrand lens are used.

orthose (a) A syn. of *orthoclase*, esp. yellow orthoclase. (b) An obsolete term introduced by Haüy (1801) for the feldspar group of minerals.---Etymol: French.

orthoselection The continuous action of natural selection in the same direction over a long period of time.

orthosilicate According to the now obsolete classification of silicates as oxyacids of silicon, a salt of the hypothetical orthosilicic acid, H_4SiO_4. Cf: *metasilicate*.

orthosite A light-colored coarse-grained igneous rock composed almost entirely of orthoclase. Although the rock is classified as plutonic, because of its texture, it almost always is hypabyssal.

orthosparite A sparite cement developed by physicochemical precipitation in open voids (Chilingar et al, 1967, p.320). Cf: *pseudosparite*.

orthostratigraphy Standard or "main" stratigraphy based on fossils identifying recognized biostratigraphic zones (such as trilobites in the Cambrian and graptolites in the Silurian) (Schindewolf, 1955, p.397). Cf: *parastratigraphy*.

orthostrophic Having harmonious coiling throughout; specif. said of the common gastropod shell in which the whorls are coiled on an erect cone so that the apex points backward rather than forward, and the spire is slightly to strongly elevated. Cf: *hyperstrophic*.

orthosymmetric Said of a crystal having orthorhombic symmetry.

orthotectic *magmatic*.

orthotectic stage *orthomagmatic stage*.

orthotectonics *alpinotype tectonics*.

orthotill A till formed by immediate release from the transporting ice, as by ablation and melting (Harland et al, 1966, p.231). Ant: *paratill*.

orthotriaene A sponge triaene in which the cladi are oriented close to 90 degrees to the rhabdome. Cf: *protriaene*.

Orthox In U.S. Dept. of Agriculture soil taxonomy, a suborder of the soil order *Oxisol*, characterized by being moist almost all or all of the time, by having a low to moderate organic-carbon content, and by having a mean annual soil temperature of

22°C or more (SSSA, 1970). Cf: *Aquox; Humox; Torrox; Ustox*.

orthozone A term suggested by Kobayashi (1944, p.742) to replace *zone* as defined by Oppel (1856-1858).

ortlerite An obsolete term for a porphyritic hypabyssal diorite resembling greenstone and containing minor accessory orthoclase. Its name is derived from the Ortles Alps, Italy. Cf: *suldenite*.

ortstein An old syn. of *hardpan*; the hardened B horizon of a podzol.

orvietite An extrusive rock composed of approximately equal amounts of plagioclase and sanidine, along with leucite, augite, minor biotite and olivine, and accessory apatite and opaque oxides.

oryctocoenosis That part of a thanatocoenosis that has been preserved as fossils. Var: *oryctocoenose*.

oryctognosy An obsolete syn. of *mineralogy*.

oryctology (a) An obsolete syn. of *mineralogy*. (b) A term used in the middle of the eighteenth century for the study of fossils, at that time including almost anything dug out of the ground. The term was restricted at the same time "fossil" was restricted and was later replaced by "paleontology" (Challinor, 1962, p.170).

os [glac] Anglicized spelling of the Swedish term *äs*, meaning *esker*. Pl: *osar*. Syn: *ose*.

os [palyn] *endopore*.

Osagian North American provincial series: Lower Mississippian (above Kinderhookian, below Meramecian). Syn: *Osagean*.

Osann's classification A pure chemical classification of igneous rocks which is also adaptable for metamorphic and sedimentary rocks. "The system is based on certain definite characteristics of the mineral combinations formed from the magmas, namely, on the combination of the alkalies with Al_2O_3 in definite proportions in the feldspars and feldspathoids, and on the union of lime with alumina in the anorthite molecule of the plagioclase, and with iron and magnesia in the ferromagnesian minerals. The rock is classified from the amounts of these combinations, the percentage of silica, the silica coefficient, and the ratio of soda to the sum of the alkalies" (Johannsen, 1931, v.1, p.68). See also: *Niggli's classification*.

osar Plural of *os*. The term is often mistakenly used as a singular noun.

osarizawaite A yellow mineral: $PbCuAl_2(SO_4)_2(OH)_6$.

osbornite A meteorite mineral: TiN.

oscillation [glac geol] *interstade*.

oscillation [stratig] A term used by Ulrich (1911) for the repeated transgressions and regressions of the seas in constantly shifting patterns, bringing about changes in the character of the sediments being deposited. Cf: *pulsation*.

oscillation cross ripple mark A *cross ripple mark* resulting from the concurrent or successive action of two sets of waves or from the intersection of a set of waves with a preexisting current ripple mark; e.g. *rectangular cross ripple mark* and *hexagonal cross ripple mark*. See also: *composite ripple mark*. Syn: *wave cross ripple mark; wave interference ripple mark*.

oscillation ripple *oscillation ripple mark*.

oscillation ripple mark A *symmetric ripple mark* with a sharp, narrow, and relatively straight crest between broadly rounded troughs, formed by the orbital or to-and-fro motion of water agitated by oscillatory waves on a sandy bottom at a depth theoretically shallower than wave base, the ripple remaining stationary. It may be found along a seacoast beyond the surf zone, or on the shallow bottom of a pond or lake whose water is not too strongly moved by currents. Cf: *current ripple mark*. Syn: *oscillation ripple; oscillatory ripple mark; wave ripple mark*.

oscillation theory A theory, proposed by Haarman, that cosmic energy produces the Earth's major tectonic features, and that secondary features are the result of gravitational sliding or compressional settling or subsidence.

oscillator quartz A natural or synthetic quartz crystal of sufficient quality and size to be used in the manufacture of oscillator plates.

oscillatory extinction *undulatory extinction.*

oscillatory ripple mark *oscillation ripple mark.*

oscillatory twinning Repeated, parallel twinning.

oscillatory wave A water wave in which the individual particles move in closed vertical orbits about a point with little or no change in position, although the wave form itself advances; e.g. an ocean wave in deep water. Cf: *wave of translation.* Syn: *wave of oscillation.*

oscillatory-wave theory A modern theory of the tides (replacing the *progressive-wave theory*) involving the assumption that the basic tidal movement in the open ocean consists of a system of waves oscillating within subdivided units of the ocean surface, each unit having a fixed node from which the height of tidal rise increases outward; the oscillations vary with the relative positions of the Earth, Moon, and Sun, together with the shape, size, and depth of the body of water within the unit. Progressive-wave movement is of secondary importance. Syn: *stationary-wave theory.*

osculum A large opening from the internal cavity of a sponge to the exterior and through which water leaves the sponge. Pl: *oscula.* Cf: *ostium; pore* [paleont]. Syn: *oscule.*

ose Var. of *os.* Pl: *oses.*

oshana A poorly defined stream channel in the flat-lying Ovamboland region of South-West Africa, containing water only during the highest floods and usually in the form of a chain of standing pools that quickly dry away. Cf: *omuramba.* Etymol: Afrikaans. Syn: *mulola.*

osmiridium A white to gray cubic mineral: (Ir,Os). It is a native alloy containing 25-40% osmium and 50-60% iridium, and is often found with platinum. The name has also been used as a syn. of *iridosmine.*

osmium A mineral: (Os,Ir).

osmosis The movement at unequal rates of solvent through a semipermeable membrane, which is usually separating the solvent, most often water, and a solution or a dilute solution and a more concentrated one, until the solutions on both sides of the membrane are equally strong. Cf: *dialysis.* See also: *electroosmosis.*

osseous amber Opaque or cloudy amber containing numerous minute bubbles. Syn: *bone amber.*

osseous breccia *bone breccia.*

ossicle Any of the numerous individual calcified elements or pieces of the skeleton of many echinoderms; e.g. a *plate.* The term is normally used for the larger of such elements. Syn: *ossiculum.*

ossiculum (a) *ossicle.* (b) *lithodesma.*---Pl: *ossicula.*

ossipite A coarse-grained variety of troctolite containing labradorite, olivine, magnetite, and a small amount of diallage. Also spelled: *ossypite.*

osteolite A massive, earthy mineral (apatite) consisting of an impure, altered calcium phosphate.

osteolith (a) A femur-shaped heterococcolith built up of lamellae as in the coccolithophorid *Ophiaster hydroideus.* (b) A fossil bone.

ostiole [geomorph] *tundra ostiole.*

ostiole [paleont] One of the small inhalant openings of a sponge; an *ostium.*

ostium Any opening through which water enters a sponge. The term is sometimes applied only to an opening larger than a *pore,* and it was used in the older literature as a synonym of *posticum.* Pl: *ostia.* Cf: *osculum.* Syn: *ostiole* [paleont].

ostracode Any aquatic crustacean belonging to the subclass Ostracoda, characterized by a bivalve, generally calcified carapace with a hinge along the dorsal margin. Most ostracodes

are of microscopic size (0.4-1.5 mm long) although freshwater forms up to 5 mm long and marine forms up to 30 mm long are known. Their stratigraphic range is Lower Cambrian present. Also spelled: *ostracod.*

ostracum A term used originally for the outer part of the calcareous wall of the shell of a bivalve mollusk, secreted at the edge of the mantle, but also applied by some later authors the entire calcareous wall. Cf: *hypostracum; periostracum.* P. *ostraca.*

ostraite A jacupirangite containing abundant green spinel.

Ostwald's rule The statement in phase studies that an unstable phase does not necessarily transform directly to the true stable phase, but rather it may first pass through successive intermediate phases, presumably due to lower activation energy barriers via that route.

osumilite A hexagonal mineral: $(K,Na)(Mg,Fe^{+2})_2(Al, Fe^{+3})_3(Si,Al)_{12}O_{30}.H_2O$. It is commonly mistaken for cordierite.

otavite A hexagonal mineral: $CdCO_3$. It is isostructural with calcite.

ottajanite A leucite tephrite having the chemical, but not mineralogic, composition of a *sommaite,* being composed of augite and leucite phenocrysts in a groundmass of calcic plagioclase, leucite, and augite, with some sanidine, nepheline, olivine, ore minerals, hornblende, biotite, and apatite. I name is derived from Ottajano, Vesuvius, Italy.

ottemannite A mineral: Sn_2S_3.

ottrelite A gray to black variety of chloritoid containing manganese.

ouachitite An olivine-free biotite monchiquite having a glass or analcime groundmass.

oued A var. of *wadi.* Etymol: French. Pron: *wed.* Pl: *oued, ouadi.*

ouenite A fine-grained igneous rock resembling eucrite and containing green augite, anorthite, and smaller amounts of hypersthene and olivine.

ouklip A term used in southern Africa for a conglomerate. Etymol: Afrikaans, "old rock".

oule A term used in the Pyrenees for *cirque.* Etymol: Spanish *olla,* "pot, kettle".

oulopholite *cave flower.*

outburst [glaciol] *glacier outburst flood.*

outcrop n. That part of a geologic formation or structure that appears at the surface of the Earth; also, bedrock that is covered only by surficial deposits such as alluvium. Cf: *exposure.* Syn: *crop (deprecated); cropping (deprecated); outcropping* ---v. To appear exposed and visible at the Earth's surface. The synonym *crop out* is preferred by the USGS (1958, p.164) to avoid confusion between the noun and verb senses of the word.

outcrop area The area (on a geologic map) shown as occupied by a particular rock unit.

outcrop curvature *settling* [mass move].

outcrop map A type of geologic map that shows the distribution and shape of actual outcrops, leaving blank those areas without outcrops. It often includes measured data for specific places, such as specimen or fossil collections, or strike and dip of beds.

outcropping n. *outcrop.*

outcrop spring *contact spring.*

outcrop water Rain and surface water which seeps downward through outcropping porous and fissured rock, fault planes, old shafts, or surface drifts.

outer bar A bar formed at the mouth of an ebb channel of a estuary. Ant: *inner bar.*

outer bark For stems and roots of dicotyledons and gymnosperms, a nontechnical term incorporating the periderm (a protective tissue of secondary origin, replacing the epidermis a the axis increases in girth) and the tissues of the axis isolated y (outside of) the vascular cambium. The technical term for ₁e outer bark is *rhytidome* (Esau, 1965, p.338).

uter beach The part of a sandy beach that is ordinarily dry, ₁eached only by the waves generated by violent storms. Cf: ₁ackshore.

uter core The outer or upper zone of the Earth's *core*, ex-₁nding to a depth of 5100 km, and including the *transition ₁one*; it is equivalent to the *E* and *F layers*. It is presumed to ₁e liquid because it sharply reduces compressional-wave ve-₁cities and does not transmit shear waves. Its density ranges ₁om 9 to 11 g/cm³. Cf: *inner core*.

uter epithelium The ectodermal epithelium adjacent to the ₁hell of a brachiopod and responsible for its secretion.

uter hinge plate Either of a pair of concave or subhorizontal ₁inge plates in the cardinalia of some brachiopods, separating ₁nner socket ridges and crural bases. Cf: *inner hinge plate*.

uter lamella The relatively thick mineralized shell layer of an ₁stracode, enclosed between thin chitinous layers, and serv-₁g to conceal and protect the soft parts of the body and ap-₁endages (TIP, 1961, pt.Q, p.53). Cf: *inner lamella*.

uter lip The abaxial (lateral) margin of the aperture of a gas-₁opod shell, extending from the suture to the foot of the colu-₁ella. Cf: *inner lip*.

uter mantle *upper mantle*.

uter mantle lobe The outer peripheral part of the mantle of a ₁rachiopod, separated by a mantle groove from an inner lobe, ₁nd responsible (in articulate brachiopods) for the secretion ₁f the primary shell layer (TIP, 1965, pt.H, p.149).

uter plate One of a pair of subvertical plates in the cardinalia ₁f pentameracean brachiopods, with the ventral surface fused ₁ the base of the brachial process and the dorsal edge at-₁ached to the floor of the valve (TIP, 1965, pt.H, p.149). Cf: ₁nner plate*.

uter side The portion of a conodont element on the convex ₁ide of the anterior-posterior midline. Ant: *inner side*.

utface *dip slope*.

utfall (a) The mouth of a stream or the outlet of a lake; esp. ₁e narrow end of a watercourse or the lower part of any body ₁f water where it drops away into a larger body. (b) The vent ₁r end of a drain, pipe, sewer, ditch, or other conduit that ₁arries waste water, sewage, storm runoff, or other effluent ₁to a stream, lake, or ocean.

utflow (a) The act or process of flowing out; e.g. the dis-₁harge of water from a river into the sea. Syn: *efflux*. (b) ₁/ater that flows out; e.g. ground-water seepage and stream ₁ater flowing out of a drainage basin. Also, the amount of ₁ater that has flowed out. (c) An *outlet* where water flows out ₁f a lake.

utflow cave A cave from which a stream issues, or is known ₁ have issued. Cf: *inflow cave; through cave*. Syn: *effluent ₁ave; cave of debouchure*.

utgassing The removal of occluded gases, usually by heat-₁g; e.g. the process involving the release of gases and water ₁apor from molten rocks, leading to the present formation of ₁e Earth's atmosphere and oceans.

utlet (a) The relatively narrow opening at the lower end of a ₁ke through which water is discharged into an outflowing ₁tream or other body of water. Syn: *outflow*. (b) A stream ₁owing out of a lake, pond, or other body of standing water; ₁lso, the channel through which such a stream flows. (c) The ₁wer end of a watercourse where its water flows into a lake ₁r sea; e.g. a channel, in or near a delta, diverging from the ₁ain river and delivering water into the sea. (d) A crevasse in ₁ levee.

utlet glacier A glacier issuing from an ice sheet or ice cap ₁hrough a mountain pass or valley, constrained to a channel ₁r path by exposed rock. Cf: *ice stream* [glaciol]; *glacial lobe; ₁istributary glacier*.

outlet head The place where water leaves a lake and enters an effluent.

outlier An area or group of rocks surrounded by outcrops of older age, e.g. an eroded geosynclinal trough, or a remnant on the downthrown side of a fault. Cf: *inlier*.

outline map A map that presents minimal geographic informa-tion, usually only coastlines, principal streams, major civil boundaries, and large cities, leaving as much space as possi-ble for the reception of particular additional data. See also: *base map*.

outside lateral muscle One of a pair of *protractor muscles* in some lingulid brachiopods, originating on the pedicle valve lat-erally to the central muscles, and extending posteriorly to in-sertions behind the *middle lateral muscles* on the brachial valve (TIP, 1965, pt.H, p.149).

outside pond A body of water enclosed or partly enclosed by sediment deposited by bifurcating distributaries in the outer-most extension of the Mississippi River delta; it is usually con-nected with the Gulf of Mexico. Cf: *inside pond*.

outwash [glac geol] (a) Stratified detritus (chiefly sand and gravel) removed or "washed out" from a glacier by meltwater streams and deposited in front of or beyond the terminal mo-raine or the margin of an active glacier. The coarser material is deposited nearer to the ice. Syn: *glacial outwash; over-wash; outwash drift*. (b) The meltwater from a glacier.

outwash [sed] Soil material washed down a hillside by rain-water and deposited upon more gently sloping land.

outwash apron *outwash plain*.

outwash cone A steeply sloping, cone-shaped accumulation of outwash deposited by meltwater streams at the margin of a shrinking glacier. Syn: *wash cone*.

outwash drift A deposit of *outwash*.

outwash fan A fan-shaped accumulation of outwash deposited by meltwater streams in front of the terminal moraine of a glacier. Coalescing outwash fans form an *outwash plain*.

outwash plain A broad, outspread, flat or gently sloping, alluvi-al sheet of outwash deposited by meltwater streams flowing in front of or beyond the terminal moraine of a glacier, and formed by coalescing *outwash fans*; the surface of a broad body of outwash. Cf: *valley train*. See also: *sand plain*. Syn: *apron; outwash apron; morainal apron; overwash plain; frontal apron; frontal plain; wash plain; marginal plain; sandur; mo-rainal plain*.

outwash-plain shoreline A prograding shoreline formed where the outwash plain in front of a glacier is built out into a lake or sea.

outwash terrace A dissected and incised valley train or bench-like deposit extending along a valley downstream from an out-wash plain or terminal moraine; a flat-topped bank of outwash with an abrupt outer face. Syn: *frontal terrace; overwash ter-race*.

outwash train *valley train*.

ouvarovite *uvarovite*.

ovary In a flower, the basal, enlarged part of the pistil, in which seeds develop.

ovate Said of a leaf whose basal end is broader than its termi-nal end. Cf: *obovate*.

oven (a) A rounded, sack-like, chemically weathered pit or hollow in a rock (esp. a granitic rock), having an arched roof and resembling an oven (Bell, 1894, p.358). Cf: *weather pit*. (b) *spouting horn*.

oven-dry soil A soil sample that has been dried at 105°C.

overbank deposit Fine-grained sediment (silt and clay) depos-ited from suspension on a flood plain by floodwaters that can-not be contained within the stream channel. See also: *flood-plain deposit*.

overburden [eco geol] Barren rock material, usually unconsol-idated, overlying a mineral deposit and which must be removed prior to mining. Syn: *top; baring. Caprock* and *capping* are synonyms that are usually used for consolidated material.

overburden [sed] (a) The upper part of a sedimentary deposit, compressing and consolidating the material below. (b) The loose soil, silt, sand, gravel, or other unconsolidated material overlying bedrock, either transported or formed in place; *regolith.*

overburdened stream *overloaded stream.*

overburden pressure (a) Vertical stress exerted by the weight of rock or soil; *geostatic pressure.* (b) *reservoir pressure.*

overconsolidation Consolidation (of sedimentary material) greater than that normal for the existing overburden; e.g. consolidation resulting from desiccation or from pressure of overburden that has since been removed by erosion. Ant: *underconsolidation.* Cf: *normal consolidation.*

overdeepened valley The degraded channel or valley of an alpine glacier, now occupied by an aggrading stream.

overdeepening The process by which an eroding glacier excessively deepens and broadens an inherited preglacial valley to a level below that of the subglacial surface. Cf: *oversteepening.*

overdraft Withdrawal of ground water in excess of replenishment.

overfall [eng] A place provided on a dam or weir for the overflow of surplus water such as from a canal or lock.

overfall [currents] A turbulent, disturbed surface of water (such as a breaking wave) caused by the meeting of strong currents, by winds moving against a current, or by a current setting over a submerged ridge or shoal; a *rip.* Term is usually used in the plural.

overfall [streams] An obsolete term for a waterfall.

overfit stream A *misfit stream* that is too large to have eroded the valley in which it flows, or whose flood plain is too small for the size of the stream. There is some doubt as to whether such a stream exists.

overflow v. To flow over the margin of; to cover with water.--- n. A flowing over the banks of a stream or river; an inundation.

overflow channel A channel or notch cut by the overflow waters of a lake, esp. the channel draining meltwater from a glacially dammed lake; an outlet of a proglacial lake. See also: *glacial drainage channel.* Syn: *spillway; glacial spillway; proglacial valley; sluiceway; crease.*

overflow ice Ice formed during high spring tides by water rising through cracks in the surface ice and then freezing (Swayne, 1956, p.104).

overflow spring A type of contact spring that develops where a permeable deposit dips beneath an impermeable mantle. Ground water overflows onto the land surface at the edge of the impermeable stratum.

overflow stream (a) A stream containing water that has overflowed the banks of a river or stream. Syn: *spill stream.* (b) An effluent from a lake, carrying water to a stream, sea, or another lake.

overfold An *overturned* fold.

overgrowth (a) Secondary material deposited in optical and crystallographic continuity around a crystal grain of the same composition, as in the diagenetic process of secondary enlargement. (b) A deposit of one mineral growing in oriented crystallographic directions on the surface of another mineral; e.g. hematite on quartz, or chalcopyrite on galena. See also: *mantle.*

overhand stoping A mining term for upward *magmatic stoping.*

overhang (a) *cliff overhang.* (b) A part of the mass of a salt dome that projects out from the top of the dome much like the cap of a mushroom.

overhanging ripple *rhomboid ripple mark.*

overhead stoping A mining term for upward *magmatic stoping.*

overite A pale-green to colorless mineral: $Ca_3Al_8(PO_4)_8$ $(OH)_6 \cdot 15H_2O$.

overland flow That part of surface runoff flowing over land surfaces toward stream channels; specif. *sheet flow* [geomorph]. After it enters a stream, it becomes a part of the total runoff (Langbein & Iseri, 1960). Syn: *unconcentrated flow.* Cf: *channel flow; streamflow.*

overlap [coast] The migration of an upcurrent part of a shore to a position that extends seaward beyond a downcurrent part; esp. the lapping over of an inlet by a spit. Cf: *offset.*

overlap [photo] The amount by which one aerial photograph covers the same area as covered by another (adjacent) photograph. It is usually expressed as a percentage. Also, the area so covered. See also: *end lap; side lap.*

overlap [stratig] (a) A general term referring to the extension of marine, lacustrine, or terrestrial strata beyond or over older underlying rocks whose edges are thereby concealed or "overlapped," and to the unconformity that commonly accompanies such a relation; esp. the relationship among conformable strata such that each successively younger stratum extends beyond the boundaries of the stratum lying immediately beneath. The term is often used in the sense of *onlap* (and "offlap"), and sometimes in the sense of *overstep* (as by De la Beche, 1832); because of such conflicting usage, Melton (1947, p.1869) and Swain (1949, p.634) urged that the term be abandoned. (b) *replacing overlap.*

overlap [fault] In a fault, the horizontal component of separation measured parallel to the strike. Cf: *gap (fault).* Obsolete syn: *stratigraphic heave; stratigraphic overlap.*

overlap fault (a) *thrust fault.* (b) A fault structure in which the displaced strata are doubled back upon themselves.

overlapping pair Two photographs taken at different camera stations in such a manner that part of one photograph shows the same terrain as shown on a part of the other photograph; e.g. *stereoscopic pair.*

overlapping spur *interlocking spur.*

overlay Graphic data on a transparent or translucent sheet to be superimposed on another sheet (such as a map or photograph) to show details not appearing, or requiring special emphasis, on the original; a *template.* Also, the medium or sheet containing an overlay.

overlie To lie over or be situated over or upon, or to occupy a higher position than. The term is usually applied to certain rocks (usually sedimentary or volcanic) resting or lying upon certain older rocks. Ant: *underlie.*

overload The amount of sediment that exceeds the ability of a stream to transport it and that is thereby deposited.

overloaded stream A stream that is so heavily loaded with sediment that its velocity is lessened and that is forced to deposit a part of its load; e.g. the Platte River in Nebraska. Syn: *overburdened stream.*

overpressure An occasionally used term signifying excessive pressure.

overprint [struc petrol] The development or superposition of metamorphic structures on original structures; the evidence of deformation in a rock fabric. Syn: *superprint; metamorphic overprint; imprint.*

overprint [geochron] A complete or partial disturbance of an isolated, radioactive system by thermal, igneous, or tectonic activities which results in loss or gain of radioactive or radiogenic isotopes and, hence, a change in the radiometric age that will be given by the disturbed system. See also: *updating; mixed ages; hybrid age.*

oversaturated Said of an igneous rock or magma that contains silica in excess of the amount required to form *saturated* minerals from the bases present. Syn: *silicic.* Cf: *undersaturated; unsaturated.*

oversteepened valley "An ice-free valley in which one side is higher and steeper than the other, a condition caused by the swing of a former glacier directed against that side" (Swayne, 1956, p.105).

oversteepened wall A *trough end* having an almost vertical slope due to glacial action.

oversteepening The erosive process by which an alpine glacier excessively steepens the sides of an inherited preglacial valley. Cf: *overdeepening*.

overstep n. (a) An *overlap* characterized by the regular truncation of older units of a complete sedimentary sequence by one or more later units of the sequence (Swain, 1949, p.635). The term, which is more commonly used in Great Britain than in U.S., refers to the progressive burial of truncated edges of underlying strata below an unconformity (esp. when an unconformity is not very obvious but is made evident by detailed mapping). Example: the overstep of Ordovician limestones over Precambrian rocks on the south margin of the Canadian Shield. Cf: *onlap*. See also: *strike-overlap; complete overstep; regional overstep*. (b) A stratum laid down on the upturned edges of underlying strata.---v. To transgress; e.g. an unconformable stratum that truncates the upturned edges of the underlying older rocks is said to "overstep" each of them in turn (except where the stratum and the underlying beds have the same strike).

overthrust A low-angle *thrust fault* of large scale, generally measured in miles. Cf: *underthrust fault*. Syn: *low-angle thrust; overthrust fault*.

overthrust block *overthrust nappe*.

overthrust fault *overthrust*.

overthrust nappe The body of rock that forms the hanging wall of a large-scale overthrust; a *thrust nappe*. Syn: *overthrust block; overthrust sheet; overthrust slice*.

overthrust sheet *overthrust nappe*.

overthrust slice *overthrust nappe*.

overturn The circulation, esp. in the fall and spring, of the layers of waters of a lake or sea, whereby surface water sinks and mixes with bottom water; it is caused by changes in density differences due to changes in temperature, and is esp. common wherever lakes are icebound in winter. See also: *turnover; circulation*.

overturned Said of a fold, or the limb of a fold, that has tilted beyond the perpendicular. Sequence of strata thus appears reversed. Such a fold may be called an *overfold*. Syn: *inverted [fold]; reversed*.

overturning The rising movement of bottom waters to the surface, either upwelling in the ocean or the slow, seasonal movement in lakes, known as *fall overturn*.

overwash [coast] (a) A mass of water representing the part of the uprush that runs over the berm crest (or other structure) and that does not flow directly back to the sea or lake. (b) The flow of water in restricted areas over low parts of barriers or spits, esp. during high tides or storms.

overwash [glac geol] *outwash*.

overwash mark A narrow, tongue-like ridge of sand formed by overwash on the landward side of a berm.

overwash pool A *tide pool* between a berm and a beach scarp which water enters only at high tide.

ovicell (a) The globular brood chamber in cheilostomatous bryozoans. Syn: *ooecium*. (b) A term used loosely and incorrectly for any skeletal structure that houses bryozoan larvae during their development.

ovoid Said of a fruit whose basal portion is broader than its terminal portion. Cf: *obovoid*.

ovulate Said of a gymnospermous megasporophyll in which the ovules are naked rather than enclosed in a pistil.

ovule In a seed plant, that body which, after fertilization, becomes the seed (Fernald, 1950, p.1579).

ovulite An individual spherite of an oolitic rock; an *oolith*.

owyheeite A steel-gray to silver-white mineral: $Ag_2Pb_5Sb_6S_{15}$. It occurs in metallic fibrous masses and acicular crystals.

Syn: *silver jamesonite*.

oxalite *humboldtine*.

oxammite A yellowish-white, transparent, orthorhombic mineral (ammonium oxalate): $(NH_4)_2C_2O_4 \cdot H_2O$.

oxbow (a) A closely looping stream meander resembling the U-shaped frame embracing an ox's neck, having an extreme curvature such that only a neck of land is left between two parts of the stream. Syn: *horseshoe bend*. (b) A term used in New England also for the land enclosed, or partly enclosed, within an oxbow (bend of a stream). (c) The abandoned, bow- or horseshoe-shaped channel of a former meander, left when the stream formed a cutoff across a narrow meander neck. See also: *cutoff meander*. Syn: *abandoned channel*. (d) *oxbow lake*.

oxbow lake The crescent-shaped, often ephemeral, body of standing water situated by the side of a stream in the abandoned channel (oxbow) of a meander after the stream formed a neck cutoff and the ends of the original bend were silted up. Examples are common along the banks of the Mississippi River, where they are often known as *bayous*. See also: *billabong*. Syn: *oxbow; loop lake; mortlake; moat; horseshoe lake; cutoff lake; crescentic lake*.

oxea A needle-shaped, monaxonic sponge spicule tapering to a sharp point at each end. Pl: *oxeas* or *oxeae*. Cf: *tornote*.

Oxfordian European stage: Upper Jurassic (above Callovian, below Kimmeridgian). Syn: *Divesian*.

oxic Said of a horizon of a mineral soil that is characterized by the almost complete lack of its original weatherable materials. It is at least 30cm thick.

oxidates Sediments composed of the oxides and hydroxides of iron and manganese, crystallized from acqueous solution. It is one of Goldschmidt's groupings of sediments or analogues of differentiation stages in rock analysis. Cf: *resistates; evaporates; reduzates; hydrolyzates*.

oxide A mineral compound characterized by the linkage of oxygen with one metallic element, such as cuprite, Cu_2O, rutile, TiO_2, or spinel, $MgAl_2O_4$. See also: *hydroxide*.

oxide-facies iron formation An *iron formation* in which the principal iron-rich minerals are oxides, typically hematite or magnetite (James, 1954, p.256-263). See also: *specular schist; itabirite*.

oxidite *shale-ball*.

oxidized zone An area of mineral deposits modified by surface waters, e.g. sulfides altered to oxides and carbonates. See also: *enrichment*. Cf: *sulfide zone; gossan; protore*.

oxidizing flame In blowpiping, the outer, almost invisible, and less intense part of the flame, in which oxygen may be added to the compound being tested. Cf: *reducing flame*.

Oxisol In U.S. Dept. of Agriculture soil taxonomy, a soil order characterized by the presence of either an oxic horizon within 2m, or of a continuous phase of plinthite within 30cm, of the surface. There is no underlying spodic or argillic horizon (SSSA, 1970). See also: *Aquox; Humox; Orthox; Torrox; Ustox*.

oxoferrite A variety of native iron with some FeO in solid solution.

oxyaster A stellate sponge spicule (aster) having acute, sharp rays.

oxybasiophitic Said of an ophitic rock that may be either *basiophitic* or *oxyophitic*.

oxycone A laterally compressed, coiled cephalopod shell with an acute periphery and a usually narrow or occluded umbilicus, as in *Oxynoticeras*.

oxygen deficit The difference between actual amount of dissolved oxygen in lake or sea water and the saturation concentration at the temperature of the water mass sampled.

oxygen demand *chemical oxygen demand*.

oxygen isotope fractionation Temperature-dependent *isotopic fractionation* of the oxygen-18/oxygen-16 isotope ratio in the carbonate shells of marine organisms which is used as an indication of water temperature at the time of deposition of the shell. See also: *carbonate thermometer*.

oxygen ratio The ratio of the number of atoms of oxygen in the basic oxides of a mineral or rock to the number of atoms of oxygen in SiO_2 (Johannsen, 1939, v.1, p.164). Syn: *acidity coefficient; acidity quotient; coefficient of acidity*.

oxyhexaster A hexaster whose simple terminal rays end in sharp points.

oxyhornblende *basaltic hornblende*.

oxylophyte A plant preferring or restricted to acid soil.

oxymagnite *maghemite*.

oxymesostasis The mesostasis (quartz, orthoclase, or micropegmatite) of an *oxyophitic* rock.

oxyophitic Said of the texture of an ophitic rock whose mesostasis is composed of quartz or orthoclase or both; also, said of an ophitic rock with such texture. Cf: *basiophitic; oxybasiophitic*. See also: *oxymesostasis*.

oxyphile *lithophile*.

oxysphere A term that was proposed as a replacement for *lithosphere*; that zone or layer of the Earth whose constituent rocks are 60% oxygen.

oxytylote A sponge spicule shaped like a common pin.

ozalid A print on light-sensitized material, made directly from a positive transparency and developed by a dry process that uses ammonia vapor.

Ozarkian A now obsolete term for the time represented by rocks formed between the Cambrian and the Canadian. It was so named after the Ozark uplift of Missouri.

ozarkite White massive *thomsonite* from Arkansas.

ozocerite A natural, brown to jet black paraffin wax consisting primarily of hydrocarbons. It occurs in irregular veins, is soluble in chloroform, has a variable melting point, and heating with a 20-30% solution of concentrated H_2SO_4 at 120° to 200° C bleaches it to yield ceresine. Varieties: *baikerite, celestialite; helenite; moldavite; pietricikite*. Syn: *ader wax; earth wax; fossil wax; mineral wax; native paraffin; ozokerite*.

ozokerite Original spelling of *ozocerite*.

P

paar A depression produced by the moving apart of crustal rocks rather than by subsidence within a crustal block. It is floored with upper-mantle igneous rocks and is essentially devoid of crustal material. Examples are the Gulf of California and the Dead Sea. Etymol: Hebrew.

pabstite A mineral: $Ba(Sn,Ti)Si_3O_9$.

pachnolite A colorless to white monoclinic mineral: $NaCaAl-$ .H_2O.

pachyodont Said of the dentition of a bivalve mollusk characterized by very large, heavy, blunt, thick, amorphous hinge teeth.

Pacific suite One of two large groups of igneous rocks, characterized by calcic and calc-alkalic rocks. Harker (1909) divided all Tertiary and Holocene igneous rocks of the world into two main groups (the *Atlantic suite* and the Pacific suite), the Pacific suite being so named because of the predominance of calcic and calc-alkalic rocks in the area of the circum-Pacific orogenic belt. Because there is such a wide variation in tectonic environments (and in their associated rock types) in the areas of Harker's Atlantic and Pacific suites, the terms are now seldom used to indicate kindred rock types. Cf: *mediterranean suite*. Syn: *anapeirean; circum-Pacific province*. See also: *andesite line*.

Pacific-type coastline A *concordant coastline*, esp. one as developed around the Pacific Ocean (e.g. the coastline of British Columbia) and reflecting the continuous linear trends of the circum-Pacific fold-mountain system. A "modified" Pacific-type coastline develops behind festoons of island arcs and adjacent foredeep trenches, such as the coastline of Asia. Ant: *Atlantic-type coastline*.

pack (a) *pack ice.* (b) *ice pack.*

packed biomicrite A *biomicrite* in which the skeletal grains make up over 50% of the rock. Cf: *sparse biomicrite.*

packed ice *close ice.*

packet texture A rarely used term for the close grouping of quartz crystals in pegmatite (Knopf, 1938, p.170).

pack ice (a) A term used in a broad sense to include any area of sea ice (other than fast ice) regardless of its form or disposition, composed of a heterogeneous mixture of ice of varying sizes and ages, and formed by the jamming or crushing together of pieces of floating ice; the mass may be either loosely or tightly packed but it covers the sea surface with little or no open water. See also: *drift ice.* (b) The ice material in an area of pack ice, or forming an ice pack.---Syn: *ice canopy; ice pack; pack.* "The terms 'pack ice' and 'ice pack' have been used indiscriminately for both the sea area containing floating ice, and the material itself" (Huschke, 1959, p.410).

packing [oil] *gravel packing.*

packing [sed] The manner of arrangement or spacing of the solid particles in a sediment or sedimentary rock, or of the atoms or ions in a crystal lattice; specif. the arrangement of clastic grains, entirely apart from any authigenic cement that may have crystallized between them. Cf: *fabric; compaction.*

packing density A measure of the extent to which the grains of a sedimentary rock occupy the gross volume of the rock in contrast to spaces between the grains, equal to the cumulative grain-intercept length along a traverse in a thin section (Kahn, 1956).

packing index The ratio of the ion volume to the volume of the unit cell in a crystal (Fairbairn, 1943).

packing proximity An estimate of the number of grains (in a sedimentary rock) that are in contact with their neighbors, equal to the total percentage of grain-to-grain contacts along a traverse measured on a thin section (Kahn, 1956).

packing radius Half the distance of closest approach of like atoms or ions in a crystal.

packsand A very fine-grained sandstone that is so loosely consolidated by a slight calcareous cement as to be readily cut by a spade.

packstone A term used by Dunham (1962) for a sedimentary carbonate rock whose granular material is arranged in a self-supporting framework, yet also contains some matrix of calcareous mud. Cf: *mudstone; grainstone; wackestone.*

paddle The flat distal part of the last prosomal appendage (toward the rear) in a merostome.

padmaragaya A light orange, reddish-yellow, or pinkish-orange variety of sapphire; a synthetic corundum of various shades of yellow or orange. Etymol: Sinhalese, "lotus color". Syn: *padmaradschah; padparadscha.*

padparadscha A syn. of *padmaragaya.* Also spelled: *padparadschah.*

paedogenesis *neoteny.*

paedomorphism *neoteny.*

paedomorphosis (a) *neoteny.* (b) Evolution as a result of modification in the immature growth stages.

pagoda stone (a) A Chinese limestone showing in section fossil orthoceratites arranged in pagoda-like designs. (b) An agate whose markings resemble pagodas. (c) *pagodite.*

pagodite Massive pinite or *agalmatolite* carved by the Chinese into miniature pagodas. Syn: *pagoda stone.*

paha A low, elongated, rounded glacial ridge or hill consisting mainly of drift, rock, or windblown sand, silt, or clay but capped with a thick cover of loess; found esp. in NE Iowa. Height varies between 10 and 30 m. Etymol: Dakota *pahà,* "hill". Pl: *paha; pahas.*

pahoehoe A type of lava flow having a glassy, smooth, and billowy or undulating surface; it is characteristic of Hawaii. It tends to be a basaltic, glassy, and porous type of lava. Cf: *aa.* Pron: pa-hó-e-hó-e. Etymol: Hawaiian. Obs. syn. *dermolith.* Syn: *ropy lava.*

Pahrump A provincial series of the Precambrian in California.

painite A mineral: $Ca_4BAl_{20}SiO_{38}$ (?).

paint A term used in SW U.S. for an earthy, pulverulent variety of cinnabar (Bureau of Mines, 1968, p. 788).

paint pot A type of *mud pot* containing multicolored mud. Also spelled: *paintpot; paint-pot.*

paired terrace One of two *stream terraces* that face each other at the same elevation from opposite sides of the stream valley and that represent the remnants of the same flood plain or valley floor. Cf: *unpaired terrace.* Syn: *matched terrace.*

paisanite A light-colored microgranitic hypabyssal igneous rock characterized by small sanidine and quartz phenocrysts, few in number, and aggregates of riebeckite in a groundmass of quartz and microperthite, frequently intergrown. Its name is derived from Paisano Pass, Texas. Syn: *ailsyte.*

pakihi A term used in New Zealand for a waterlogged gravel flat (Stamp, 1966, p.352). Etymol: Maori.

palae- Most terms beginning with the combining forms "palae-" and "palaeo-" are entered in this glossary under *pale-* or *paleo-*.

palaeocope Any ostracode belonging to the order Palaeocopida, characterized by a shell with a long straight dorsal margin and commonly with lobes, sulci, and ventral structures. Their stratigraphic range is Lower Ordovician to Middle Permian, with some questionably identified in present-day waters.

Palaeophyticum Plant life of the Paleozoic (Kobayashi, 1958). Cf: *Mesophyticum; Cainophyticum.*

palaetiology Explanation of past changes in the Earth's condition as being governed by the laws of cause and effect. Var: *paletiology.*

palagonite An altered tachylite, brown to yellow or orange and found in pillow lavas as interstitial material or amygdules.

palagonite tuff A pyroclastic rock consisting of angular fragments of palagonite.

palagonitization Formation of palagonite by hydration of tachylite.

palasome A syn. of *host*, used in economic geology. Syn: *palosome*. Cf: *paleosome; metasome*.

palate Part of a bryozoan avicularium occupied by the mandible. Syn: *rostrum*.

Palatinian orogeny *Pfälzian orogeny*.

palatinite An old term for basalts and diorites that contain orthopyroxenes.

paleic surface A smooth, preglacial erosion surface (Termier & Termier, 1963, p.411).

paleo- (a) A combining form denoting the attribute of great age or remoteness in regard to time (*Paleo*cene), or involving ancient conditions (*paleo*climate), or of ancestral origin, or dealing with fossil forms (*paleo*anthropic). Sometimes given as *pale-* before vowels (*pal*event). Also spelled: palaeo-; palaio-. (b) A prefix indicating pre-Tertiary origin, and generally altered character, of a rock to the name of which it is added; e.g. *paleo*picrite. By some writers the prefix has been further restricted to pre-Carboniferous rock (Holmes, 1928, p. 175).

paleoagrostology The study of fossil grasses.

paleoaktology Study of ancient nearshore and shallow-water environments.

paleoalgology The study of fossil algae. Syn: *paleophycology*.

paleoautochthon The original autochthon or basement of a tectonic region. It may be folded and faulted by later movements, but it has not been greatly displaced in a horizontal direction. Df: mesoautochthon; neoautochthon.

paleobiochemical Relating to ancient biochemical products, such as amino acids, fatty acids, and sugars, isolated from geological specimens and that have undergone little change since they were produced.

paleobiocoenosis An assemblage of organisms that lived together in the geologic past as an interrelated community. Syn: *paleocoenosis*.

paleobiology A branch of paleontology dealing with the study of fossils as organisms rather than features of historical geology.

paleobiotope A term sometimes used in paleoecology to designate a region of unspecified size which is characterized by essentially uniform environmental conditions and by a correspondingly uniform population of animals or plants or both. See also: *biotope* [*ecol*].

paleobotanic province A large region characterized and defined by similar fossil floras.

paleobotany The study of the plant life of the geologic past (Arnold, 1947, p.1). Syn: *phytopaleontology; paleophytology*.

Paleocene An epoch of the upper Tertiary period, after the Gulfian of the Cretaceous period and before the Eocene; also, the corresponding worldwide series of rocks.

paleochannel A remnant of a stream channel cut in older rock and filled by the sediments of younger overlying rock; a buried stream channel.

paleoclimate The climate of a given period of time in the geologic past. Syn: *geologic climate*.

paleoclimatologic map A paleogeographic map that depicts paleoclimatic data.

paleoclimatology The study of past climates (paleoclimates) throughout geologic time, and of the causes of their variations, either on a local or worldwide basis, and including temperature changes varying from a fraction of a degree (over a period of decades) to 8-12°C (over geologic time). It involves the interpretation of glacial deposits, fossils, and paleogeographic, isotopic, and sedimentologic data.

paleocoenosis *paleobiocoenosis*.

paleocrystic ice Old sea ice, esp. well-weathered polar ice, generally considered to be at least 10 years old; it is often found in floebergs and in the pack ice of the central Arctic Ocean.

paleocurrent An ancient current (generally of water) that existed in the geologic past and whose direction is inferred from the sedimentary structures and textures of the rocks formed at that time.

paleocurrent structure *directional structure*.

paleodepth The depth at which an ancient organism or group of organisms lived.

paleodrainage pattern A drainage pattern representing the distribution of a valley system as it existed at a given moment geologic time (Andresen, 1962).

paleoecology The study of the relationships between organisms and their environments, the death of organisms, and their burial and postburial history in the geologic past based on fossil faunas and floras and their stratigraphic position. See also: *ecology*.

paleoenvironment An environment in the geologic past.

paleoequator The position of the Earth's equator in the geologic past as defined for a specific geologic period and based on geologic evidence such as paleomagnetic measurements, oxygen-isotope ratios, fauna and flora, distribution of reefs, coal deposits, and tillites; e.g. the Ordovician paleoequator for North America, running from the southern tip of Baja California to the north end of Greenland. Paleoequators are great circles that were formerly normal to the axis of rotation but are now displaced and vary from continent to continent.

paleofluminology The study of ancient stream systems. Cf: *paleohydrology*.

Paleogene An interval of geologic time incorporating the Oligocene, Eocene, and Paleocene of the Tertiary; the lower Tertiary. When the Tertiary is designated as an era, then the Paleogene, together with the *Neogene*, may be considered to be its two periods. Syn: *Eogene; Nummulitic*.

paleogeographic event *palevent*.

paleogeographic map A map that shows the reconstructed physical geography at a particular time in the geologic past, including such information as the distribution of land and seas, the geomorphology of the land, the depth of the sea, the directions of currents in water and air, the distribution of bottom sediments, and the climatic belts. Cf: *paleotectonic map*.

paleogeographic stage *palstage*.

paleogeography The geography of ancient times; specif. the study and description of the physical geography of the geologic past, such as the historical reconstruction of the pattern of the Earth's surface or of a given area at a particular time in the geologic past, or the study of the successive changes of surface relief during geologic time. Syn: *paleophysiography*.

paleogeologic map A map that shows the areal geology of an ancient surface at some former time in the geologic past; esp. such a map of the surface immediately below a buried unconformity, showing the geology as it existed at the time the surface of unconformity was completed and before the overlapping strata were deposited. Paleogeologic maps were introduced by Levorsen (1933). Cf: *subcrop map*. Syn: *peel map*.

paleogeology A branch of geology that deals with geologic conditions and events in some former period of geologic time or with geologic features exposed at the surface during that time but now buried beneath rocks formed in a subsequent time.

paleogeomorphology A branch of geomorphology concerned with the recognition of ancient erosion surfaces and with the study of ancient topographies and topographic features that are now concealed beneath the surface or have been removed by erosion. Syn: *paleophysiography*.

paleogeophysics Geophysics in past geologic time; by extension, paleomagnetism, paleogeodynamics, etc.

paleohydrology (a) The study of the earliest uses and management of water. (b) The study of ancient hydrologic features preserved in rock. Cf: *paleofluminology*.

paleoichnology The study of trace fossils in the fossil state, as opposed to *neoichnology*. Syn: *palichnology*.

paleoisotherm The locus of points of equal temperature for some former period of geologic time.

paleokarst A rock or area that has been karstified and subsequently buried under later sediments (Monroe, 1970, p.13).

paleolatitude The latitude of a specific area on the Earth's surface in the geologic past; specif. distance measured in degrees from the paleoequator.

paleolimnology (a) The study of the past conditions and processes of ancient lakes; the interpretation of the accumulated sediments and the geomorphology and geologic history of ancient lake basins, most of which no longer contain lakes but are dry, flat areas that may occasionally be covered by water. (b) The study of the sediments and history of existing lakes.

paleolithic n. In archaeology, the first division of the *Stone age*, characterized by the appearance of man and man-made implements. Correlation of relative cultural levels with actual age (and, therefore, with the time-stratigraphic units of geology) varies from region to region; however, the age generally given for the Paleolithic more or less coincides with the Pleistocene. Cf: *Mesolithic; Neolithic*. See also: *lower Paleolithic; middle Paleolithic; upper Paleolithic; Eolithic*. Syn: *Old Stone age*.----adj. Pertaining to the Paleolithic.

paleolithologic map A paleogeologic map that shows lithologic variations at some buried horizon or within some restricted zone at a particular time in the geologic past.

paleomagnetic pole *virtual geomagnetic pole*.

paleomagnetic stratigraphy The use of natural remanent magnetization to identify stratigraphic units. It depends on the temporal variation of the ambient magnetic field, which is due to geomagnetic secular variation and reversals. Syn: *magnetic stratigraphy*.

paleomagnetism The study of natural remanent magnetization in order to determine the intensity and direction of the Earth's magnetic field in the geologic past.

paleometeoritics The study of variation of extraterrestrial debris as a function of time over extended parts of the geologic record, esp. in deep-sea sediments and possibly in sedimentary rocks, and for more recent periods in ice.

paleomorphology The *morphology* or study of form and structure of fossil remains (hard parts) in order to determine the original anatomy (soft parts) of an organism; e.g. the study of a brachiopod muscle scar whose depth may indicate the strength of the muscle.

paleomycology The study of fossil fungi.

paleontography The formal, systematic description of fossils. adj: *paleontographic*.

paleontologic facies A term recommended by Teichert (1958, 2734) to replace *biofacies* as used in stratigraphy, signifying the paleontologic characteristics of a sedimentary rock.

paleontologic species A morphologic species based on fossil specimens. It may include specimens that would be considered specifically distinct if living individuals could be observed.

paleontologist One who studies the fossilized remains of animals and/or plants (i.e. paleontology).

paleontology The study of life in past geologic periods, based on fossil plants and animals and including phylogeny, their relationships to existing plants and animals, and the chronology of the Earth's history. Cf: *neontology*. See also: *historical geology*.

paleopalynology A division of *palynology* concerned with the study of fossil spores and pollen. It is now interpreted broadly to include study of a wide range of fossil microscopic, usually organic bodies in addition to spores and pollen: animal remains such as chitinozoans, as well as fungal spores, dino-

flagellates, acritarchs, and other organisms resistant to acids and found in sedimentary rocks of all ages (nannofossils and diatoms are sometimes included). The usual criteria for inclusion are that the bodies be microscopic in size and composed of a resistant organic substance (usually sporopollenin, chitin, or pseudochitin) that results in the bodies being preserved in sedimentary rocks and available for separation by maceration from such rocks.

paleopedology The study of soils of past geologic ages, including determination of their ages.

paleophycology *paleoalgology*.

paleophyre A reddish porphyritic andesite intruded in Silurian strata in the Fichtelgebirge, German-Czechoslovakian border. Also spelled: *palaeophyre*.

paleophysiography (a) *paleogeomorphology*. (b) *paleogeography*.

Paleophytic A paleobotanic division of geologic time, signifying that time during which pteridophytes were abundant, between the development of algae and the appearance of the first gymnosperms. Cf: *Aphytic; Archeophytic; Eophytic; Mesophytic; Cenophytic*. Syn: *Pteridophytic*.

paleophytology An obsolete syn. of *paleobotany*.

paleoplain A term introduced by Hill (1900, p.5) for an ancient degradational plain that is now buried beneath later deposits.

paleopole A pole of the Earth, either magnetic or geographic, in past geologic time.

paleosalinity The salinity of a body of water in the geological past, as evaluated on the basis of chemical analyses of sediment or formation water.

paleosere A sequence of ecologic communities in the geologic past that led to a climax community; a *sere* in the geologic past.

paleoslope The direction of initial dip of a former land surface; esp. the regional slope of a large, ancient physiographic unit, such as a flood plain or a continental slope.

paleosol A buried soil horizon of the geologic past. When uncovered, it is said to be exhumed. See also: *dirt bed*. Syn: *buried soil; fossil soil*.

paleosome A geometric element of a composite rock or mineral deposit, appearing to be older than an associated younger rock element (or *neosome*); e.g. wall rock in a vein or replacement deposit, or the unaltered and relatively immobile pre-existing part of a migmatite. Sometimes used in place of the term *stereosome, metaster,* or *restite*. Cf: *host; palasome*.

paleostructure The geologic structure of a region or sequence of rocks at some former time in the geologic past; the structure of a paleogeologic area.

paleostructure map A map that shows, using thickness contour lines, the geologic structure of a lower boundary surface at the time that the upper boundary surface was formed as a horizontal plane (Levorsen, 1960, p.4).

paleotectonic map A map intended to show geologic and tectonic features as they existed at some time during the geologic past, rather than the sum of all the tectonics of the region, as portrayed on a general *tectonic map*. It is similar to a *paleogeographic map* but more emphasis is placed on the tectonic features than on the distribution of lands and seas, and a greater effort is commonly made to use factual rather than speculative data. Most of the paleotectonic maps now being made portray in much detail the well documented features in the cratonic areas, but show few features in the more intensely deformed areas. See also: *neotectonic map*.

paleotemperature [geol] The temperature at which a geologic process occurred in the past, such as the temperature at which a certain mineral developed during regional metamorphism.

paleotemperature [paleoclim] The mean climatic temperature at a given time or place in geologic history; esp. the paleoclimatic temperature of the sea.

paleothanatocoenosis A group of organisms buried together in the geologic past.

paleothermal Pertaining to or characteristic of warm climates of the geologic past; e.g. a "paleothermal fauna". Syn: *paleothermic*.

paleothermometry Measurement or estimation of paleotemperatures; esp. the determination of the temperature of a geologic-time unit based on the mass-spectrometric measurement of the abundance of oxygen isotopes in the carbon dioxide obtained from the carbonates of marine fossil shells. See also: *geologic thermometry*.

paleotopographic map A map that shows the relief of a surface of unconformity.

paleotopography The topographic relief of an area at a particular time in the geologic past; the topography of a paleogeologic area, such as the configuration of the surface of an unconformity at the time it was overlapped.

paleotypal Said of a fine-grained porphyritic igneous rock having the characteristics of altered extrusive or hypabyssal rocks such as those of pre-Tertiary age. This term and the term *cenotypal* were introduced to distinguish Tertiary and pre-Tertiary fine-grained igneous rocks; both are now obsolete.

paleovolcanic Said of extrusive rocks that are of pre-Tertiary age. Cf: *neovolcanic*.

paleowind An ancient wind of the geologic past. Its direction is recorded by distributions of volcanic ash falls, growth rates of colonial coral atolls, and orientation of sand dunes.

Paleozoic An era of geologic time, from the end of the Precambrian to the beginning of the Mesozoic. Obs syn: *Primary*.

paleozoology That branch of paleontology dealing with the study of subfossil and fossil animals.

palermoite A mineral: $(Li,Na)_2(Sr,Ca)Al_4(PO_4)_4(OH)_4$.

palette In a cave, a broad sheet or disc of calcite that is a solutional remnant. Syn: *shield*.

palevent A relatively sudden and short-lived paleogeographic happening, such as a short, static existence of a particular depositional environment, or a rapid geographic change separating two *palstages* (Wills, 1956, p.14). Syn: *paleogeographic event*.

pali [paleont] Plural of *palus*.

pali [geog] An Hawaiian term for a steep slope; e.g. the Nuuanu Pali, a steep-faced scarp on the NE side of Oahu.

palichnology Var. of *paleoichnology*.

paliform Resembling a *palus*; specif. "paliform lobes" of the septa in corals, formed by detached trabecular offsets from the inner edges of the septa, appearing in vertical succession, and differing from pali in not being formed as a result of substitution.

palimpsest [met] adj. Said of a structure (or texture) in a metamorphic rock in which remnants of the original structure (or texture) are preserved and, sometimes, megascopically visible. The term was first used by Sederholm (1891). Cf: *relict*.

palimpsest [streams] Said of a kind of drainage in which a modern, anomalous drainage pattern is superimposed upon an older one, clearly indicating different topographic and possibly structural conditions at the time of development.

palingenesis [paleont] Recapitulation, without change, by the young stages of an organism of the characteristics of their ancestors. See also: *recapitulation theory*.

palingenesis [petrology] Formation of a new magma by the melting of preexisting rock in situ. Considered by some workers as a syn. of *anatexis*; others apply the term to the formation of new rock by anatexis and/or metasomatism (Dietrich, 1960, p.50). Adj: *palingenetic*.

palingenetic [streams] *resurrected*.

palingenetic [petrology] Formed by or involving *palingenesis*. Syn: *palingenic*.

palinspastic map A name coined by Kay (1937) for a paleo-

geographic or paleotectonic map in which the features represented have been restored as nearly as possible to their original geographic positions, before the rocks of the crust were shortened by folding, or telescoped by thrusting.

palintrope A term used initially for the morphologically posterior sector of either valve (of some brachiopod shells) which was reflexed to grow anteriorly (mixoperipheral growth), but more recently for the curved surface of the shell, bounded by beak ridges and cardinal margin of nonstrophic shells (TIP, 1965, pt.H, p.149). It differs from a *planarea* in being curved in all directions.

palisade A picturesque, extended rock cliff or line of bold cliffs, rising precipitously from the margin of a stream or lake, esp. one consisting of basalt with columnar structure, as the Palisades along the Hudson River of New York and New Jersey. Term is usually used in the plural.

Palisade disturbance A time of deformation, or orogeny, supposed by Schuchert (1924) to have closed the Triassic Period in eastern North America and elsewhere. It is based on the block-faulted structure of the Upper Triassic Newark series of the Appalachian area, which was truncated before younger Mesozoic (mainly Cretaceous) strata were laid over it. The concept of a distinct orogeny at this time is dubious and has only local application at most.

palisade mesophyll *palisade tissue*.

palisade tissue A leaf tissue composed of long cylindrical chlorophyllous cells oriented normal to the lamina beneath the upper epidermis. Syn: *palisade mesophyll*.

palladium A soft, silver-white or steel-white, isometric mineral, the native metallic element Pd. It is one of the platinum metals, and it resembles and occurs with platinum, usually occurring in grains and frequently alloyed with platinum and iridium.

palladium amalgam *potarite*.

palladium gold *porpezite*.

pallasite [meteorite] A *stony-iron meteorite* composed essentially of large single glassy crystals of olivine embedded in a network of nickel-iron. Pallasites are believed to have been formed at the interface of the stony mantle and metal core of a layered planetoid. Syn: *pallas iron*.

pallasite [ign] Any ultramafic rock, whether of meteoric or terrestrial origin, which contains approximately 60 percent iron in the former, or more iron oxides than silica, in the latter; e.g. cumberlandite (Thrush, 1968, p.789).

pallasite shell A syn. of *lower mantle*, so named because its composition may be equivalent to that of a pallasite meteorite. Cf: *periodotite shell*.

pallial chamber *mantle cavity*.

pallial line A line or narrow band on the inner surface of a valve of a bivalve-mollusk shell, close to and more or less parallel with the margin, and marking the line of attachment of the marginal muscles of the mantle. It is typically distinguished by a groove or ridge and by a change in texture of shell material. Syn: *pallial impression*.

pallial sinus (a) An often conspicuous embayment or inward bend in the posterior and ventral part of the pallial line of a bivalve mollusk, marking the point of attachment of the siphon retractor muscles. See also: *sinus* [paleont]. (b) A mantle canal of a brachiopod.

pallium The *mantle* of a mollusk or brachiopod. Pl: *pallia*.

pallomancy A form of *dowsing* using a pendulum. Cf: *rhabdomancy*.

palmate Said of a leaf that is lobed or divided in a handlike fashion; digitate.

palmate venation In a leaf, a type of *net venation* in which the main veins branch out from the stalk apex like the fingers of a hand. Cf: *pinnate venation*.

palmierite A white hexagonal mineral: $(K,Na)_2Pb(SO_4)_2$. It is isostructural with kalistronite.

palmitic acid A long-chain, wax-like, fatty acid, formula

$_{16}H_{32}O_2$, present in numerous plant and animal fats as glycerides.

alosome Var. of *palasome*.

alp A reduced distal portion of the limb of a crustacean, usually only one of its rami, but may comprise both rami plus asis (TIP, 1969, pt.R, p.99). See also: *palpus*.

alpebral lobe One of two elevated portions of the fixed cheek a trilobite, extending laterally from the glabella to the upper ad inner margin of the visual surface of an eye.

alpi Plural of *palpus*.

alpus A term applied either to a *pedipalpus* (including pedialpal coxae) or more properly to one of the five segments following the coxa in an arachnid (TIP, 1955, pt.P, p.62). Pl: alpi. Adj: *palpal*. Se also: palp.

als Var. of *palsa*.

alsa A small, elliptical, dome-like *frost mound* containing at, commonly 3-6 m high and 2-25 m long, occurring in suarctic bogs of the tundra, esp. in Scandinavia, and often surrounded by shallow open water. Etymol: Swedish, "elliptical". : palsen (not "palses"). Syn: *pals; peat mound; peat humock*.

alsen Plural of *palsa*.

alstage A period of time when paleogeographic conditions are relatively static, or were changing gradually and progressively, with relation to such factors as sea level, surface reef, or distance from shore (Wills, 1956, p.14). Cf: *palevent*. n: *paleogeographic stage*.

aludal Pertaining to a marsh.

aludification *ulmification*.

aludous Pertaining or relating to marshes or marshy areas. n: *palustral*.

alus Any of several narrow, slender, vertical, calcareous mellae, plates, pillars, or other processes developed along e inner edge of certain entosepta of a coral and comprising e remnant part of a pair of exosepta joined at their inner argins. Pl: *pali*. See also: *paliform*.

alustral *paludous*.

alustrine Pertaining to material growing or deposited in a arsh or marsh-like environment.

alygorskite A chain-lattice clay mineral: $(Mg,Al)_2Si_4O_{10}$-$0H).4H_2O$. The term has also been used as a group name for htweight, tough, matted, fibrous clay minerals showing a onsiderable amount of substitution of aluminum for magnesin and characterized by distinctive rod-like shapes under the ectron microscope. Syn: *attapulgite*.

alyniferous Bearing pollen. The term in palynology usually refers to rocks or sediment samples that yield pollen, spores, or her palynomorphs on maceration.

alynofacies A term used in paleopalynology for an assemage of palynomorphs in a portion of a sediment, representing local environmental conditions and not typical of the gional palynoflora.

alynoflora The whole suite of palynomorphs from a given ck unit. The term *microflora* is sometimes used as a synnym but should be avoided as it better applies to assembla- s of extant microscopic algae and fungi.

alynology A branch of science concerned with the study of llen of seed plants and spores of other embryophytic plants, ether living or fossil, including their dispersal and applications in stratigraphy and paleoecology. Term suggested by de & Williams (1944, p.6). Etymol: Greek $\pi\alpha\lambda\nu\tau\omega$, I sprine, suggestive of $o\alpha\lambda\eta$, "fine meal" cognate with Latin *pollen*, ne flour, dust". See also: *paleopalynology; pollen analysis*.

alynomorph A microscopic, resistant-walled organic body und in palynologic maceration residues; a palynologic study ject. Palynomorphs include pollen, spores of many sorts, ritarchs, chitinozoans, dinoflagellate thecae and cysts, cer-

tain colonial algae, and other acid-insoluble microfossils. Cf: *sporomorph*.

palynostratigraphy The stratigraphic application of palynologic methods.

pamet A dry valley formed in glacial deposits on the outer part of Cape Cod, Mass.

pampa A vast, treeless, grassy plain of temperate regions, esp. as used in Argentina and adjacent parts of Uruguay. It is comparable to the prairies of North America, the steppes of the U.S.S.R., and the veld of South Africa.

pan [geomorph] (a) A shallow, natural depression or basin, esp. one containing a lake, pond, or other body of standing water; e.g. a shallow depression holding a temporary or permanent pool in a tidal marsh along the Atlantic coast of the U.S (b) A term used in South Africa for a hollow in the ground where the neck of a volcano formerly existed.

pan [salt] (a) A *salt pan*; specif. a term used in South Africa for a shallow, undrained, usually rounded depression or hollow occurring in an arid or semiarid region and holding water (received in the rainy season) that often evaporates in the dry season and leaves a salt deposit. Cf: *vloer*. (b) An artificial basin for producing salt by evaporation of salt water or brine. Also, a vessel for evaporating salt water or brine.

pan [soil] A hard, cement-like layer, crust, or horizon within or just beneath the surface soil, being strongly compacted, indurated, or very high in clay content, and usually impeding the movement of water and air and the growth of plant roots; specif. *hardpan*. See also: *genetic pan; pressure pan*.

pan [ice] (a) Shortened form of *pancake ice*. (b) An individual piece of pancake ice. (c) *ice pan*. (d) A large fragment of flat, relatively thin ice, having a diameter about 60 m, formed in a bay or fiord or along the shore and subsequently loosened to drift about the sea. Syn: *pan ice*.

panabase *tetrahedrite*.

panautomorphic *panidiomorphic*.

panautomorphic-granular *panidiomorphic-granular*.

pancake ice One or more small, predominantly circular pieces of newly formed sea ice (diameter varying from about 30 cm to about 3 m) with slightly raised rims caused by the pieces rotating and striking against one another; it often forms during the early fall in polar regions. See also: *lily-pad ice*. Syn: *pan [ice]; pancake*.

pancake-shaped bomb A type of volcanic bomb whose flattened shape is due to impact.

pandaite A mineral of the pyrochlore group: $(Ba,Sr)(Nb,Ti)(O, OH)_7$.

pandemic Said of conditions that occur over a broad geographic area and affect a major part of the population; also said of a widely dispersed population.

pandermite *priceite*.

panethite A meteorite mineral: $(Na,Ca,K)_2(Mg,Fe, Mn)_2(PO_4)_2$.

panfan A graded bedrock surface consisting of a series of coalescing pediments and representing the penultimate stage of an arid cycle of erosion. The synonymous term *pediplain* is preferred because the feature does not involve alluvial fans, although the term "panfan" was proposed by Lawson (1915, p.33) for a vast alluvial fan representing the end stage in the process of geomorphic development in a desert region.

Pangea A hypothetical supercontinent; supposed by many geologists to have existed at an early time in the geologic past, and to have combined all the continental crust of the Earth, from which the present continents were derived by fragmentation and movement away from each other by means of some form of continental displacement. During an intermediate stage of the fragmentation, between the existence of Pangea and that of the present widely separated continents, Pangea was supposed to have split into two large fragments, *Laurasia* on the north and *Gondwana* on the south. The proto-ocean around Pangea has been termed *Panthalassa*. Other geologists, while believing in the former existence of Laurasia and Gondwana, are reluctant to concede the existence of an original Pangea; in fact, the early (Paleozoic or older) history of continental displacement remains largely undeciphered.

Pang-Yang depression A large erosional basin with a flat bottom and steep sides, developed on a rocky plain or plateau (Stone, 1967, p.236). Type locality: Pang Yang, Burma.

panhole *solution pan.*

panicle A compound *inflorescence* with several main branches, each of which bears pedicelled flowers arranged along its axis; in grass panicles, the flowers are borne in spikelets on the pedicels (Fuller & Tippo, 1949, p.966).

panidiomorphic Said of an igneous rock, and of its texture, that is completely or predominantly *idiomorphic*; e.g. a *lamprophyric* texture. Syn: *panautomorphic.*

panidiomorphic-granular Said of the granular texture, of an igneous rock, characterized by a panidiomorphic fabric; also, said of a rock with such texture. Johannsen (1939, p.226) states that Rosenbusch, who first used the term in 1887, incorrectly applied it to the *granulitic* [*ign*] xenomorphic-granular texture of aplite, although he had defined it as involving idiomorphic components; this usage was corrected in the last edition of Rosenbusch's book, by its editor, Osann. Syn: *panautomorphic- granular.*

pan lake A lake occupying a shallow, natural depression, or pan.

panmixis The free interchange of genes within an interbreeding population.

panning A technique of prospecting for heavy metals, e.g. gold, by washing placer or crushed vein material in a pan. The lighter fractions are washed away, leaving the heavy metals behind, in the pan.

Pannonian European stage: lower Pliocene.

panplain (a) A term introduced by Crickmay (1933, p.344-345) for a very broad plain formed by the coalescence of several adjacent flood plains, each resulting from long-continued lateral erosion by meandering streams; it represents the end stage of an erosion cycle. Cf: *peneplain; plain of lateral planation.* Syn: *panplane.* (b) A very level plain with a general seaward inclination (Engeln, 1942).

panplanation The action or process of formation and development of a panplain; also the product resulting from such an action or process.

panplane *panplain.*

pantellerite A green to black extrusive rock (an alkalic rhyolite) characterized by acmite-augite or diopside, anorthoclase, and cossyrite phenocrysts in either a pumiceous, partly glassy, fine-grained holocrystalline trachytic, or microlitic groundmass composed of acmite and feldspar; plagioclase is typically absent and quartz, rare. Its name is derived from Pantelleria, an island in the Mediterranean Sea south of Sicily. Cf: *comendite.*

Panthalassa The hypothetical proto-ocean surrounding *Pangea,* supposed by some geologists to have combined all the oceans or areas of oceanic crust of the Earth at an early time in the geologic past, across which the present continents were gradually displaced to their present positions from the original protocontinent.

pantograph An instrument for copying a map or drawing on any predetermined scale (of reduction or enlargement), consisting of four bars hinged to form an adjustable parallelogram so that as one tracing stylus is moved over the material to be copied the other makes the desired copy. Syn: *pantagraph.*

pantometer An instrument for measuring all angles in determining elevations or distances.

pantonematic Said of a feather-like flagellum (as in Euglenophyta) provided throughout its length by a single row of tiny cilia.

papa A soft, bluish clay, mudstone, siltstone, or sandstone found in North Island, N.Z., and used for whitening fireplaces. Etymol: Polynesian.

papagoite A blue monoclinic mineral: $CaCuAlSi_2O_6(OH)_3$.

paper chromatography A chromatographic technique for separating components of a sample by moving it in a mixture of solution by gravity or capillarity through a paper substrate in such a way that the different components have different mobilities and thus become separated. The technique usually involves partition procedures (May & Cuttitta, 1967, p.116). See also: *chromatography.*

paper clay A fine-grained, white, kaolin-type clay with high retention and suspending properties and a very low content of free silica, used for coating or filling paper.

paper peat Thinly laminated peat. Syn: *leaf peat.*

paper shale A shale that easily separates on weathering into very thin, tough, uniform, somewhat flexible layers or laminae suggesting sheets of paper; it is often highly carbonaceous.

paper spar A crystallized variety of calcite occurring in thin lamellae or paper-like plates.

papery Descriptive of a fine-grained sedimentary rock that splits into laminae less than 2 mm in thickness (McKee & Weir, 1953, p.383).

papilla (a) A surficial mound associated with a pore in cystoids. (b) A minute scale-like ossicle or projection in ophiuroids. (c) *apical papilla.*---Pl: *papillae.*

papula A short protuberance of integument between ossicles of aboral or oral surface of an asteroid and functioning as an external gill. Pl: *papulae.*

papule A prolate to equant, somewhat rounded *glaebule* composed dominantly of clay minerals with a continuous and/or lamellar fabric and having sharp external boundaries (Brewer, 1964, p.274-275); e.g. a clay gall in soil material.

para- A prefix that, when used with a metamorphic rock name, indicates that the rock was derived from a sediment, e.g. paragneiss.

para-autunite An artificial mineral: $Ca(UO_2)_2(PO_4)_2$. It represents the complete dehydration product of autunite. Cf: *meta-autunite.*

parabiont Any one of the organisms involved in *parabiosis.*

parabiosis The condition in which members of two or more species maintain colonies close to one another without conflict. See also: *parabiont.*

parabituminous Said of bituminous coal containing 84-87% carbon, analyzed on a dry, ash-free basis. Cf: *perbituminous; orthobituminous.*

parabolic dune (a) A sand dune with a long, scoop-shaped form, convexly bowed in the downwind direction so that its horns point upwind (windward), and whose ground plan (when perfectly developed) approximates the form of a parabola. It is characteristically covered with sparse vegetation, and is often found along the coast where strong onshore winds are supplied with abundant sand. (b) A term used loosely as a syn. of *upsiloidal dune.*---Cf: *barchan.*

parabutlerite An orange orthorhombic mineral: $FeSO_4(OH).2H_2O$. Cf: *butlerite.*

paracelsian A pale-yellow orthorhombic mineral: $BaAl_2Si_2O_8$. It is dimorphous with celsian.

parachoma A ridge of dense calcite developed between adjacent foramina in some fusulinacean foraminiferal tests having multiple foramina (as in Verbeekinidae and Neoschwagerininae). Pl: *parachomata.* Cf: *choma.*

parachronology (a) Practical dating and correlation of stratigraphic units. (b) Geochronology based on fossils that supplement, or are used instead of, biostratigraphically significant fossils. Cf: *orthochronology.*

parachrysotile A mineral of the serpentine group: $Mg_3Si_2O_5(OH)_4$. It is a polymorph of chrysotile, distinct from clinochrysotile, orthochrysotile, and lizardite.

paraclase An obsolete term for a *fault*.

paraclavule An apparently monaxonic sponge spicule (microsclere) consisting of a short straight shaft pointed at one end and bearing an umbel at the other end. It resembles an amphidisc with one umbel missing.

paraclinal Said of a stream or valley that is oriented in a direction parallel to the fold axes of a region. Also said of a region having paraclinal streams. Term introduced by Powell (1874, p.50). Ant: *diaclinal*.

paracme The period in the phylogeny of a group of organisms that follows the *acme* and is marked by decadence or decline.

paraconformable Not really or not quite conformable; esp. said of strata exhibiting paraconformity.

paraconformity A term introduced by Dunbar & Rodgers (1957, p.119) for an obscure or uncertain *unconformity* in which no erosion surface is discernible or in which the contact is a simple bedding plane, and in which the beds above and below the break are parallel. This type of unconformity was formerly classed by Pirsson (1915, p.291-293) as a kind of *disconformity*, and is recognized in Great Britain as a *nonsequence* "of major time-significance" rather than as an unconformity (Challinor, 1967, p.261). Not to be confused with *paraunconformity*. Cf: *diastem*. Syn: *nondepositional unconformity*.

paraconglomerate A term proposed by Pettijohn (1957, p.261) for a conglomerate that is not a product of normal aqueous flow but deposited by such modes of mass transport as subaqueous turbidity flows (and slides) and glacier ice, that is characterized by a disrupted gravel framework (stones not generally in contact), that is often unstratified, and that is notable for a content of matrix greater than that of gravel-sized fragments (in many examples, pebbles form 10% or less of the rock). Examples include: tillites, pseudotillites, pebbly mudstones, and relatively structureless clay or shale bodies in which pebbles or cobbles are randomly distributed. Cf: *orthoconglomerate*. Syn: *conglomeratic mudstone*.

paracoquimbite A pale-violet rhombohedral mineral: $Fe_2(SO_4)_3.9H_2O$. It is dimorphous with coquimbite.

paracostibite A mineral: $CoSbS$.

paradamite A triclinic mineral: $Zn_2(AsO_4)(OH)$. It is isomorphous with tarbuttite and dimorphous with adamite.

paradelta A term proposed by Strickland (1940, p.10) for the landward or upper part of a delta, or that part undergoing degradation.

paradiagenetic Signifying a close relation with sedimentary diagenesis; e.g. "paradiagenetic movement", or deformation that is precrystalline in relation to spathization (Sander, 1951, p.52).

paradocrasite A mineral: $Sb_2(As,Sb)_2$.

para-ecology *taphonomy*.

paraffin base A crude oil which will yield large quantities of paraffin in the process of distillation. Cf: *naphthene base; mixed base*.

paraffin coal A type of light-colored, bituminous coal from which oil and paraffin is produced.

paraffin hydrocarbon Any of the hydrocarbons of the *methane series*.

paraffinic Pertaining or relating to a *paraffin* (i.e. *methane series*) *hydrocarbon* or paraffin wax.

paraffin series *methane series*.

paraffin wax A colorless, odorless, tasteless, amorphous solute of complex hydrocarbons with a high methane-series composition.

paraflagellar bosy A swelling near the base of a flagellum in some Euglenophyta. It possibly serves as a photoreceptor.

parafoliate An obsolete syn. of *foliate*.

paragaster The *cloaca* of a sponge.

paragenesis (a) The sequential order of mineral formation. (b) A characteristic association or occurrence of minerals.----

Syn: *mineral sequence; paragenetic sequence*.

paragenetic (a) Pertaining to paragenesis. (b) Pertaining to the genetic relations of sediments in laterally continuous and equivalent facies.

paragenetic sequence *paragenesis*.

parageosyncline (a) A geosyncline within a craton or stable area; an epeirogenic basin rather than an orogenic belt (Stille, 1935, p.77-97). Syn: *intrageosyncline*. (b) A contemporary oceanic depression marginal to the craton (Schuchert, 1923, p.151-260). Cf: *idiogeosyncline*.

paraglacial *periglacial*.

paragnath One of a pair of leafy lobes of the metastoma lying behind the mandibles in most crustaceans. Syn: *paragnathus*.

paragon A perfect diamond of 100 carats or more.

paragonite A yellowish or greenish mineral of the mica group: $NaAl_2(AlSi_3)O_{10}(OH)_2$. It corresponds to muscovite but with sodium instead of potassium, and it usually occurs in metamorphic rocks. Syn: *soda mica*.

paraguanajuatite A rhombohedral mineral: $Bi_2(Se,S)_3$.

parahilgardite A triclinic mineral: $Ca_2B_5O_8Cl(OH)_2$. It is dimorphous with hilgardite.

parahopeite A colorless triclinic mineral: $Zn_3(PO_4)_2.4H_2O$. It is dimorphous with hopeite.

parajamesonite A mineral: $Pb_4FeSb_6S_{14}$. It is dimorphous with jamesonite.

paralaurionite A white mineral: $Pb(OH)Cl$. It is dimorphous with laurionite.

paralectotype Any of the syntypes other than the one designated as lectotype. Syn: *lectoparatype*.

paraliageosyncline A geosyncline developing along a present-day continental margin, e.g. the Gulf Coast geosyncline (Kay, 1945).

paralic [coal] Said of coal deposits formed along the margin of the sea, as opposed to *limnic* coal deposits.

paralic [sed] By the sea, but nonmarine; esp. pertaining to intertongued marine and continental deposits laid down on the landward side of a coast or in shallow water subject to marine invasion, and to the environments (such as lagoonal or littoral) of the marine borders. Also said of basins, platforms, marshes, swamps, and other features marked by thick terrigenous deposits intimately associated with estuarine and continental deposits, such as deltas formed on the heavily alluviated continental shelves. Etymol: Greek *paralia*, "seacoast".

paralic swamp *marine swamp*.

paralimnion The littoral part of a lake, extending from the margin to the deepest limit of rooted vegetation. Adj: *paralimnetic*.

parallax [tides] The ratio of the mean radius of the Earth to the distance of a tide-producing body (usually the Moon), represented by the angle at the center of the Moon between a line to the center of the Earth and a line tangent to the Earth's surface. The term is used in regard to the variation in tide range or in tidal-current speed resulting from the continually changing distance of the Moon from the Earth.

parallax [surv] (a) The apparent displacement of the position of an object, with respect to a reference point or system, caused by an actual shift in the point of observation; e.g. "instrument parallax" in which an imperfect adjustment of a surveying instrument or a change in the position of the observer causes a change in the apparent position of an object with respect to the reference mark(s) of the instrument. (b) The difference in the apparent direction of an object as seen from two different points not on a straight line with the object (such as the apparent difference in position of a point on two consecutive photographs, or the apparent difference in direction between objects on the Earth's surface due to their difference in elevation); the angular distance between two straight lines drawn to an object from two different points of view.

parallax bar *stereometer.*

parallel (a) One of the imaginary circles on the surface of the Earth, parallel to the equator and to one another and connecting all points of equal *latitude*; a circle parallel to the primary great circle of a sphere or spheroid, or a closed curve approximating such a circle; an east-west line of constant latitude. Each parallel is a small circle except for the equator. (b) A line, corresponding to a parallel, drawn on a globe, map, or chart.---Cf: *meridian.* Syn: *parallel of latitude.*

parallel bedding *concordant bedding.*

parallel cleavage An obsolete syn. of *bedding-plane cleavage.*

parallel displacement fault A little-used term for a fault, the linear features of which that were parallel before displacement are still parallel afterwards.

parallel drainage pattern A drainage pattern in which the streams and their tributaries are regularly spaced and flow virtually parallel or subparallel to one another over a considerable area or in a number of successive cases. It is indicative of a region having a pronounced, uniform slope and a homogeneous lithology and rock structure.

parallelepiped A closed crystal form bounded by three pairs of parallelograms.

parallel evolution The development of similar forms by related but distinct phylogenetic lineages. See also:*parallelism.* Cf: *convergent evolution.*

parallel extinction A type of *extinction* in anisotropic crystals parallel to crystal outlines or traces of cleavage planes. Cf: *inclined extinction; undulatory extinction.*

parallel fold *concentric fold.*

parallel growth *parallel intergrowth.*

parallel intergrowth Intergrowth of two or more crystals in which one or more axes in each crystal are almost parallel. Syn: *parallel growth.*

parallelism The development or possession of similar characteristics by two or more related organisms as a result of similar environmental conditions. See also: *parallel evolution.* Cf: *convergence.*

parallelkanter An elongated *windkanter* having parallel faces or edges. Etymol: German *Parallelkanter,* "one having parallel edges".

parallel of latitude A *parallel* or line of latitude. Cf: *circle of latitude.*

parallel retreat of slope (a) The recession of a scarp or of the side of a hill or mountain (once the angle of slope is established) without change in declivity, the slope at any given time retreating parallel to its former positions. (b) The concept or principle of backwearing of a slope as proposed by W. Penck (1924).

parallel ripple mark A ripple mark with a relatively straight crest and an asymmetric profile; specif. a *current ripple mark.*

parallel roads A series of horizontal beaches or wave-cut terraces occurring parallel to each other at different levels on each side of a glacial valley, as those at Glen Roy in Scotland. Each beach ("road") represents a former shoreline that corresponds with a temporary level of overflow from a proglacial lake formed by ice-damming.

parallel section A slice through a foraminiferal test in a plane normal to the axis of coiling but not through the proloculus.

parallel shot In seismic prospecting, a test shot made with all the amplifiers connected in parallel and activated by a single geophone in order to check for lead, lag, polarity, and phasing in the amplifier to oscillograph circuits.

parallel texture A rock texture characterized by tabular to prismatic crystals oriented parallel to a plane (*plane parallel texture*) or to a line (*linear parallel texture*).

parallel twin A twinned crystal, the twin axis of which is parallel to the composition surface. Cf: *normal twin.*

parallel unconformity *disconformity.*

parallel venation In a leaf, a type of *venation* in which the main veins are parallel with each other and with the longitudinal axis of the leaf. Cf: *net venation.*

parallochthon Rocks that were brought from intermediate distances and deposited on or near an allochthonous mass during transit. See also: *mesoautochthon.* Also spelled: *paraallochthon.*

paramagnetic Having a small positive magnetic susceptibility. A paramagnetic mineral such as olivine, pyroxene, or biotite contains magnetic ions which tend to align along an applied magnetic field but which do not have a spontaneous magnetic order. Cf: *diamagnetic.* See also:*superparamagnetism.*

paramagnetic resonance *electron spin resonance.*

paramelaconite A black tetragonal mineral: $(Cu_{1-2x}^{+2}, Cu_{2x}^{+1})O_{1-x}.$

parameter [cryst] (a) Any of the axial lengths or interaxial angles that define a unit cell. Syn: *lattice constant.* (b) On a crystal face, the rational multiple of the axial length intercepted by a plane, which determines the position of the plane relative to the crystal lattice. (c) The proportions (x, y, z) of the unit cell axial lengths which define the position of an atom relative to any lattice point.

parameter [stat] (a) Any arbitrary numerical constant derived from a population or a probability distribution and characterizing by each of its particular values some particular member of a system; a quantity related to one or more variables in such a way that it remains constant for any specified set of values of the variable or variables. (b) An independent variable through functions of which other functions may be expressed. (c) Any measurable characteristic of a sample or population; any of a set of physical properties whose values determine the characteristics or behavior of a system.

parametric hydrology That branch of hydrology dealing with "the development and analysis of relationships among the physical parameters involved in hydrologic events and the use of these relationships to generate, or synthesize, hydrologic events" (Hofmann, 1965, p.120). Cf: *stochastic hydrology; synthetic hydrology.*

parametric latitude *reduced latitude.*

parametric sounding That form of electromagnetic *sounding* in which frequency is the variable. Cf: *geometric sounding.*

paramontroseite An orthorhombic mineral: VO_2.

paramorph A mineral whose internal form has changed, but without change in its composition or external form; this characteristic is known as *paramorphism.*

paramorphism The property of a mineral to change its internal structure without changing its external form or chemical composition. Such a mineral is called a *paramorph.* Syn: *allomorphism.*

paramoudra A flint nodule of exceptionally large size (up to a meter in length and one-third meter in diameter), shaped like a barrel, pear, or cylinder, standing erect in the chalk beds of NE Ireland and the eastern coast of England. It appears to be a gigantic fossil zoophyte allied to the sponges. Term introduced by Buckland (1817). Etymol: vernacular Irish. Pl: *paramoudras; paramoudrae.* Syn: *potstone.*

parapyla An accessory tubular aperture of the central capsule (in addition to astropyle) of a phaeodarian radiolarian. Pl: *parapylae.*

paraquartzite A term used by Tieje (1921, p.655) for a quartzite derived chiefly by contact metamorphism; thus, a variety of metaquartzite.

pararammelsbergite A mineral: $NiAs_2$. It is dimorphous with rammelsbergite.

para-ripple A term introduced by Bucher (1919, p.262-263) for a large and symmetric or nearly symmetric ripple having gentle surface slopes and "showing no assortment of grains". Cf: *metaripple.*

paraschoepite A mineral: $UO_3.2H_2O$ (?). It is closely related to *schoepite.*

parasitic [volc] Said of a volcanic cone, crater, or lava flow that occurs on the side of a larger cone; it is a subsidiary formation. Syn: *lateral; adventive.*

parasitic [ecol] Said of an organism that lives by *parasitism.*

parasitic ferromagnetism *weak ferromagnetism.*

parasitic fold *subsidiary fold.*

parasitism The relationship that exists when one organism derives its food and usually other benefits from another living organism without actually killing it, but usually causing it some harm. Cf: *commensalism; mutualism; symbiosis.* Adj: *parasitic.*

parastratigraphic unit *operational unit.*

parastratigraphy (a) Supplemental stratigraphy based on fossils other than those governing the prevalent *orthostratigraphy* (Schindewolf, 1955, p.397). (b) Stratigraphy based on operational units.

parastratotype Another section in the original locality where a stratotype was defined (Sigal, 1964).

parasymplesite A monoclinic mineral: $Fe_3(AsO_4)_2.8H_2O$. Cf: *symplesite.*

paratacamite A rhombohedral mineral: $Cu_2(OH)_3Cl$. It is dimorphous with atacamite.

parataxon A unit in taxonomy for those fossilized remains, esp. of animals, that are only part of the whole individual and that usually occur widely separated from other parts; e.g. a particular conodont. Cf: *form genus.* See also: *taxon.*

paratectonics *germanotype tectonics.*

paratellurite A tetragonal mineral: TeO_2. It is dimorphous with tellurite.

paratheca A wall of a scleractinian corallite, formed by closely spaced rows of dissepiments. Cf: *septotheca; synapticulotheca.*

paratill A till formed by ice-rafting in a marine or lacustrine environment; it includes deposits from ice floes and icebergs (Harland et al, 1966, p.232). Ant: *orthotill.*

para-time-rock unit A term introduced by Wheeler et al (1950, p.2364) for a working *time-stratigraphic unit* that is biostratigraphic and rock-stratigraphic in character and that therefore is intrinsically transgressive with respect to time; e.g. zone (lithizone, radiozone), stage, isobiolith, and isogeolith. It "approaches synchrony", whereas a true time-stratigraphic unit (such as system and series) expresses "absolute synchrony". Syn: *para-time-stratigraphic unit.*

paratype Any of the specimens, other than the holotype, upon which the original description of a species or subspecies is based. Nonpreferred syn: *cotype.*

paraunconformity A term, now obsolete, proposed by Crosby (1912, p.297) as a syn. of *disconformity.* Not to be confused with *paraconformity.* Syn: *parunconformity.*

parautochthonous [petrology] Said of a mobilized part of an *autochthonous* granite, moved higher in the crust or into a tectonic area of lower pressure, and characterized by variable and diffuse contacts with the country rocks.

parautochthonous [tect] Said of a rock unit that is intermediate in tectonic character between *autochthonous* and *allochthonous.*

paravane In seismic water shooting, a planning board used to keep a detector in a vertical position. In seismic water shooting, a device attached to the end of a towed line and so arranged that the device either, travels a path parallel to but offset from the path of the towing vessel or maintains a fixed depth below the surface, or both.

paravauxite A colorless mineral: $Fe^{+2}Al_2(PO_4)_2(OH)_2.10H_2O$. It has more water than vauxite and metavauxite.

parawollastonite A monoclinic mineral: $CaSiO_3$. It is dimorphous with wollastonite.

parchettite An extrusive rock similar in composition to leucite tephrite but containing more leucite and some orthoclase.

parenchyma [paleont] (a) The *mesohyle* of a sponge. (b) The endoplasm of a protozoan.

parenchyma [bot] A plant tissue composed of relatively unspecialized vacuolate, polyhedral cells with thin primary walls, often containing chloroplasts, and in some instances capable of meristematic activity, differentiation, or storage and support functions (Keeton, 1967, p.92-93).

parenchymalium One of the spicules of the interior of a hexactinellid sponge, excluding specialized dermal and gastral spicules. Pl: parenchymalia.

parenchymella Sponge larva composed of an envelope of uniflagellate cells surrounding more or less completely an internal mass of cells. See also: *parenchymula.*

parenchymula (a) The *planula* of a coelenterate. (b) An alternate, but not recommended, spelling of *parenchymella.*

parent The initial member of nuclide in a *radioactive series.* Cf: *daughter; end product.*

parental magma The magma from which a particular igneous rock solidified or from which another magma was derived. It is sometimes used as a syn. of *primary magma.*

parent material The unconsolidated material, mineral or organic, from which the solum or true soil develops. See also: *parent rock [soil]; residual material; transported soil material.*

parent rock [sed] *source rock.*

parent rock [soil] The rock mass from which a soil's *parent material* is derived.

parfacies A subfacies of a *diagenetic facies,* based on pH-Eh limits (Packham & Crook, 1960, p.400).

pargasite (a) A monoclinic mineral of the amphibole group: $NaCa_2Mg_4Al_3Si_6O_{22}(OH)_2$. Cf: *edenite.* (b) A green or blue-green variety of hornblende containing sodium and found in contact-metamorphosed rocks.

parichno A bundlelike strand of parenchyma that first parallels the vascular bundle in a lepidodendron leaf and then becomes indistinguishable from the mesophyll cells of the leaf.

paries The triangular middle part of a compartmental plate of a cirriped crustacean. Pl: *parietes.*

parietal (a) Pertaining to the walls of a part or cavity of an organism. (b) Said of a plant part that is peripheral in position or orientation.

parietal fold A *fold* or spirally wound ridge on the parietal region of a gastropod, projecting into the shell interior.

parietal gap An opening from the cloaca to the exterior of a lyssacine hexactinellid sponge, extending completely through the body wall and interrupting the regular skeletal framework.

parietal lip The part of the *inner lip* of a gastropod shell situated on the parietal region.

parietal pore (a) A *parietal tube* in a cirripede crustacean. (b) A gonopore in a cystoid.

parietal region The basal surface of a gastropod shell just within and immediately without the aperture.

parietal septum (a) A *longitudinal septum* in a cirripede crustacean. (b) A seldom preserved longitudinal wall extending inward in the posterior region of certain echinoderms.

parietal tube One of the longitudinal tubes in certain cirripede crustaceans situated between the inner and outer laminae of a compartmental plate and separated by longitudinal septa. Syn: *parietal pore.*

parietes Plural of *paries.*

pariety A name formerly applied to the *septum* of an archaeocyathid.

parisite A brownish-yellow secondary mineral: $(Ce,La)_2Ca(CO_3)_3F_2$. It is related to *synchysite.*

parivincular Said of a longitudinally elongated type of ligament of a bivalve mollusk, located posterior to the beaks and comparable to a cylinder split on one side with severed edges attached respectively along the dorsal margin of two valves.

park [karst] A local term for a solution-type sinkhole, used for such features on the Kaibab plateau of Arizona.

park [geomorph] (a) A term used in the Rocky Mountain region of Colorado and Wyoming for a wide, comparatively level, grassy, almost treeless open valley lying at a high altitude and walled in by wooded mountains; e.g. South Park in central Colorado. It is more extensive than a *hole*. Also, a level valley between mountain ranges. (b) A relatively large, open, grassy area surrounded by woodland, or interrupted by scattered clumps of trees and shrubby vegetation; e.g. a tropical grassland in Africa. Syn: *parkland.*

parkerite A bright-bronze mineral: $Ni_3(Bi,Pb)_2S_2$.

parkland A syn. of *park*, or a tract of grassy land with scattered trees.

parmal pore One of the pores piercing the shield of an acantharian radiolarian and bordered only by united branches of apophyses. Cf: *sutural pore.*

parna A term used in SE Australia for an eolian clay occurring in sheets. Etymol: an aboriginal word for "sandy and dusty ground".

parogenetic Formed previously to the enclosing rock; esp. said of a concretion formed in a different (older) rock from its present (younger) host. Term introduced by Bates (1938, p.91). Etymol: Greek *paros*, "before, formerly".

paroptesis A now obsolete term originated by Kinahan (1878) for the changes produced in rock by dry heat. Cf: *metapepsis.*

paroxysm Any sudden and violent action of physical forces occurring in nature, such as the explosive eruption of a volcano or the convulsive "throes" of an earthquake; specif. the most violent and explosive action during a volcanic eruption, sometimes leading to the destruction of the volcano and generally preceded and following by smaller explosions. Cf: *catastrophe.*

paroxysmal eruption *Vulcanian-type eruption.*

parricidal budding Formation of a new scleractinian coral polyp from the inner surface of a wedge-shaped fragment split off lengthwise from the parent.

parrot coal A syn. of *cannel coal*, so named because of the crackling noises it makes while burning.

parsettensite A copper-red mineral: $Mn_5Si_6O_{13}(OH)_8$ (?). It often contains appreciable aluminum and potassium.

parsonite A pale-yellow to pale-brown mineral: $Pb_2(UO_2)$-$(PO_4)_2.2H_2O.$

partial-duration flood A flood peak that exceeds a given base stage or discharge. Syn: *basic-stage flood; flood above a base.*

partial node That part (a point, line, or surface) of a standing wave where some characteristic of the wave field has a minimum amplitude differing from zero. Cf: *node.*

partial pediment (a) A term proposed by Mackin (1937, p.877) for a broadly planate, gravel-capped, interstream bench or terrace. (b) A broad, planate erosion surface formed by the coalescence of contemporaneous, valley-restricted benches developed at the same elevation in proximate valleys, and which would produce a pediment if uninterrupted planation were to continue at this level (Tator, 1953, p.52-53).

partial peneplain (a) *incipient peneplain.* (b) A planation surface intermediate in development between a berm (or a strath terrace) and a peneplain; a base-leveled area that need not be limited to the confines of a valley. It can cross divides on rocks of medium resistance or on decayed resistant rocks (Bascom, 1931, p.173). Syn: *local peneplain.*

partial pluton That part of a *composite intrusion* which represents a single intrusive episode.

partial range zone *local range zone.*

partial thermoremanent magnetization The thermoremanent magnetization acquired by cooling in an ambient field only over a restricted temperature interval, as opposed to the entire temperature range from Curie point to room temperature. Cf: *inverse thermoremanent magnetization.* Abbrev: *PTRM.*

particle A general term, used without restriction as to shape, composition, or internal structure, for a separable or distinct unit in a rock; e.g. a "sediment particle", such as a *fragment* or a *grain*, usually consisting of a mineral.

particle diameter The length of a straight line through the center of a sedimentary particle considered as a sphere; a common expression of *particle size.*

particle shape The spatial or geometric form of the particles in a sediment or rock; a fundamental property of a particle that determines the relation between its mass and surface area. It depends upon the *sphericity* and *roundness* of the particle, although the term is frequently applied to sphericity as distinguished from roundness. Syn: *grain shape.*

particle size The general dimensions (such as average diameter or volume) of the particles in a sediment or rock, or of the grains of a particular mineral that make up a sediment or rock, based on the premise that the particles are spheres or that the measurements made can be expressed as diameters of equivalent spheres. It is commonly measured by sieving, by calculating settling velocities, or by determining areas of microscopic images. See also: *particle diameter.* Syn: *grain size.*

particle-size analysis Determination of the statistical proportions or distribution of particles of defined size fractions of a soil, sediment, or rock; specif. *mechanical analysis.* Syn: *size analysis; size-frequency analysis.*

particle-size distribution The percentage, usually by weight and sometimes by number or count, of particles in each size fraction into which a powdered sample of a soil, sediment, or rock has been classified, such as the percentage of sand retained on each sieve in a given size range. It is the result of a particle-size analysis. Syn: *size distribution; size-frequency distribution.*

particle velocity The velocity with which an individual particle of a medium moves under the influence of wave motion. Cf: *group velocity; phase velocity.*

parting [ore dep] (a) A band or bed of waste material dividing mineral veins or beds. (b) *clay parting.*

parting [crystal] The breaking of a mineral along planes of weakness caused by deformation or twinning; e.g. garnet. Cf: *cleavage [mineral].*

parting [stratig] A lamina or very thin sedimentary layer, generally soft, following a surface of separation between thicker strata of different lithology; e.g. a *shale break* in sandstone, or a thin bed of shale or slate in a coal bed.

parting [struc geol] (a) A joint or fissure; specif. a plane or surface along which a hard rock (of uniform or nonuniform lithology) is readily separated or is naturally divided into layers, such as a *bedding-plane parting.* See also: *splitting.* (b) A plane of separation between individual layers of strata such as a thin shale bed.

parting cast A sand-filled tension crack produced by creep along the sea floor (Birkenmajer, 1959, p.111). Syn: *pseudo mud crack.*

parting lineation A faint or weakly defined, small-scale but distinct, primary sedimentary structure, consisting of a series of parallel ridges and grooves (a few millimeters wide and many centimeters long) formed parallel to the current, characteristically found on the bedding planes of horizontally laminated or thin-bedded sandstones, and indicative of formation in a fluvial environment or under shallow sheets of flowing water. See also: *parting-plane lineation; parting-step lineation.* Syn: *current lineation.*

parting-plane lineation A *parting lineation* on a laminated surface, consisting of subparallel, linear, shallow grooves and ridges of low relief, generally less than 1 mm (McBride & Yeakel, 1963).

parting-step lineation A *parting lineation* characterized by subparallel, step-like ridges where the parting surface cuts across several adjacent laminae (McBride & Yeakel, 1963).

partition In a cave, a solutional remnant of rock that spans a passage from floor to ceiling. Cf: *bridge; wall* [*speleo*].

partition coefficient The ratio of the molar concentration of a substance dissolved in two immiscible liquids, as described by the *Nernst distribution law.*

partitioning method *Lee configuration.*

partiversal Said of a series of local dips in different directions ranging through about 180° in compass direction, occurring at or near the end of a plunging anticlinal axis.

partridgeite *bixbyite.*

parunconformity Var. of *paraunconformity.*

parvafacies A term proposed by Caster (1934, p.19) for a body of rock that comprises the part of any *magnafacies* lying between designated time-stratigraphic planes or key beds traced across the magnafacies; a laterally limited or grading time-stratigraphic unit of different facies but formed at the same time; a heteropic facies of European usage. Etymol: Latin *parva*, "small", + *facies.*

parwelite A mineral: $(Mn,Mg)_5Sb(Si,As)_2O_{10-11}$.

Pasadenan orogeny The youngest of 30 or more short-lived orogenies during Phanerozoic time recognized by Stille (1936), in this case the middle of the Pleistocene, based on relations in southern California between Pliocene and lower Pleistocene strata, and the unconformably overlying upper Pleistocene; named for Pasadena, California. Syn: *Coast Range orogeny.*

pascoite A dark red-orange to yellow-orange mineral: $Ca_2V_6O_{17}.11H_2O$.

pass [coast] (a) A practically permanent channel through which a distributary on a delta flows to the sea; specif. a navigable channel in the Mississippi River delta. (b) A navigable channel connecting a body of water with the sea; e.g. a narrow opening between two closely adjacent islands or through a coastal obstruction such as a barrier reef, a barrier island, a bar, or a shoal. Cf: *inlet.* Syn: *passage.* (c) An expanse of open water in a marsh.

pass [geomorph] A natural passageway through high, difficult terrain; e.g. a break, depression, or other relatively low place in a mountain range, affording a passage across, or an opening in a ridge between two peaks, usually approached by a steep valley. Cf: *col; gap; notch.*

pass [streams] A river crossing; a ford. Syn: *passage.*

passage [coast] *pass.*

passage [speleo] In a cave, an opening between rooms of the cave. Partial syn: *aisle; corridor; crawl; squeeze.*

passage [streams] *pass.*

passage bed A stratum that is transitional in its lithologic or paleontologic character between rocks below and above or between rocks of two geologic systems; e.g. a deposit formed during the period of transition from one set of geographic conditions to another, such as a stratum of the Rhaetian Stage intermediate in character and position between the continental deposits of the Upper Triassic and the marine clays of the Lower Jurassic. See also: *transitional series.*

pass band In seismic profiling, that range of frequency in which transmission of signals is at its most efficient.

passing Transportation of sediment; e.g. *bypassing* and *total passing.*

passive earth pressure The maximum value of lateral *earth pressure* exerted by soil on a structure, occurring when the soil is compressed sufficiently to cause its internal shearing resistance along a potential failure surface to be completely mobilized; the maximum resistance of a vertical earth face to deformation by a horizontal force. Cf: *active earth pressure.*

passive fault block That wall of the fault which did not move; the autochthonous block.

passive fold A fold in which the mechanism of folding, either flow or slip, crosses the boundaries of the strata at random. It is characteristic of relatively ductile rocks.

passive glacier A glacier with sluggish movement, generally occurring in a continental environment at a high latitude, where accumulation and ablation are both small. Ant: *active glacier.*

passive method A construction method in permafrost areas by which the frozen ground near the structure is not disturbed or altered, and the foundations are provided with additional insulation to prevent thawing of the underlying ground. Ant: *active method.*

passive permafrost Permafrost that, having formed during an earlier colder period, will not, under present climatic conditions, refreeze after it is once disturbed or destroyed. Ant: *active permafrost.* Syn: *fossil permafrost.*

passive seismometer A seismometer that continuously monitors ground motions. Cf: *active seismometer.*

passive system An electromagnetic sensor which measures radiation emitted by the target.

passometer A pocket-size, watch-shaped instrument that registers the number of steps taken by a pedestrian. It is carried in an upright position attached to the body or to a leg. Cf: *pedometer.*

pass point A point whose horizontal and/or vertical position is determined from photographs by photogrammetric methods and which is intended for use as a supplemental *control point* in the orientation of other photographs.

paste The clay-like matrix of a "dirty" sandstone; e.g. the microcrystalline matrix of a graywacke, consisting of quartz, feldspar, clay minerals, chlorite, sericite, and biotite.

pastplain A plain that has been uplifted and dissected; thus "it is no longer a true plain" (Davis, 1890, p.88).

pat (a) A term used in Pakistan for an arid plain formed by deposits of fine, light-colored, muddy clay that accumulates upon evaporation of shallow pools of water. (b) A term used in Chota Nagpur, India, for a small, steep-sided plateau.--- Etymol: Sindhi.

patagium A spongy *veil* between arms in the skeleton of a spumellarian radiolarian. Pl: *patagia.*

patch [geog] A small, isolated piece of ground distinguished from that about it by its appearance or by the vegetation it bears.

patch [ice] (a) *ice patch.* (b) An irregular small mass of floating sea-ice fragments of any concentration.

patch reef (a) A small, subequidimensional or irregularly shaped, flat-topped organic reef, less extensive than a *platform reef,* forming a part of a reef complex. (b) A small, thick, generally unbedded lens of limestone or dolomite, more or less isolated and surrounded by rocks of unlike facies.---Cf: *reef patch.*

patella (a) The fourth segment in the pedipalpus or in the leg of an arachnid, following upon and forming the "knee" articulation with the femur (TIP, 1955, pt.P, p.62). (b) A joint forming the "knee" in the prosomal appendage of a merostome.-- Pl: *patellae* or *patellas.*

patellate Said of a low solitary corallite with sides expanding from apex at an angle of about 120 degrees. Cf: *trochoid; turbinate.*

paternoite A syn. of *kaliborite.* It was formerly regarded as a distinct mineral: $MgB_8O_{13}.4H_2O$.

paternoster lake One of a linear chain or series of small circular lakes occupying rock basins, usually at different levels, in a glacial valley, separated by morainal dams or riegels, but connected by streams, rapids, or waterfalls to resemble a rosary or a string of beads. Syn: *rock-basin lake; step lake; beaded lake.*

path [seis] The imaginary line along which a wave or ray travels; the course of travel between two points of a disturbance in an elastic medium. Syn: *ray path; wave path.*

path [optics] *optical path.*

patina [**palyn**] A thickening of the exine of spores that extends over approximately half of the surface, i.e. over the entire surface of one hemisphere. Adj: *patinate*.

patina [**geol**] (a) A colored film or thin outer layer produced on the surface of a rock or other material by weathering after long exposure; e.g. a *desert varnish*, or a case-hardened layer on a chert nodule. (b) Strictly, the greenish film formed naturally on copper and bronze after long exposure to a moist atmosphere, and consisting of a basic carbonate.

patination The quality or state of being coated with a patina, or the act or process of coating with a patina.

patronite A black mineral consisting of an impure vanadium sulfide whose exact composition is not known. It is mined as an ore of vanadium at Minasragra, Peru.

patterned ground A group term suggested by Washburn (1950, p.7-8) for certain well-defined, more or less symmetrical forms, such as circles, polygons, nets, steps, and stripes, that are characteristic of, but not necessarily confined to, surficial material (mantle) subject to intensive frost action. It is classified according to type of pattern and presence or absence of sorting. Patterned ground occurs principally in polar, subpolar, and arctic regions, but also includes features in tropical and subtropical areas. Previous terms more or less synonymous: *structure ground; soil structures; Strukturboden; frost-pattern soil; soil patterns*.

patterned sedimentation Sedimentation characterized by a systematic sequence of beds; e.g. recurrent sedimentation (interbedding, interdigitation, etc.), repetition of beds, or rhythmic or cyclic sedimentation.

pattern shooting In seismic prospecting, the firing of explosive charges arranged in a definite geometric pattern.

Patterson function *Patterson synthesis*.

Patterson map *Patterson projection*.

Patterson projection A projection of the *Patterson synthesis* on a section through a crystal. See also: *Patterson vectors*.

Patterson synthesis A type of Fourier synthesis whose coefficients are corrected for absorption and other factors; it is used in direct determination of crystal structure. See also: *Patterson vectors*. Syn: *Patterson function*.

Patterson vectors In analysis of crystal structure, the vectors of peaks relative to the origin in a *Patterson synthesis* or *Patterson projection*.

paulingite An isometric zeolite mineral consisting of an aluminosilicate of potassium, calcium, and sodium.

paulopost *deuteric*.

paurocrystalline Descriptive of an interlocking texture of a carbonate sedimentary rock having crystals whose diameters are in the range of 0.01-0.1 mm (Bissell & Chilingar, 1967, p.163).

paurograined Said of the texture of a carbonate sedimentray rock having clastic particles whose diameters are in the range of 0.01-0.1 mm (Bissell & Chilingar, 1967, p.163-164) or 0.004-0.1 mm (DeFord, 1946).

pavement A bare rock surface that suggests a paved road surface or other pavement, as in smoothness, hardness, horizontality, surface extent, or close packing of its units. Examples: boulder pavement; glacial pavement; desert pavement; limestone pavement; erosion pavement.

pavilion The portion of a faceted gemstone below the *girdle*. Cf: *crown*. Syn: *base*.

pavilion facet A main facet on the pavilion of any fashioned gemstone; e.g. a large facet extending from the girdle to the culet of a brilliant-cut gem, or a facet in the center row of facets on the pavilion of a step-cut gem.

pavonite A mineral: $AgBi_3S_5$.

pawdite A dark-colored, fine-grained, granular hypabyssal rock composed of magnetite, titanite, biotite, hornblende, calcic plagioclase, and traces of quartz. The plagioclase is abundant and approximates labradorite in composition, having

bytownite centers surrounded by oligoclase rims (Johannsen, 1939, p.273).

paxilla A pillar-like ossicle of an asterozoan, having a flattened summit bearing a tuft of spinelets or granules. Pl: *paxillae*.

paxillose (a) Resembling a little stake. (b) Bearing paxillae.

paxite A mineral: Cu_2As_3. It is probably orthorhombic.

pay adj. Said of a structure or stratum that contains a mineral deposit, e.g. pay gravel, pay streak; also, said of a mineral deposit or part of it that is especially profitable, e.g. pay ore. The term is colloquial.

pay zone The vertical interval(s) of the stratigraphic section of an oil field containing reservoir beds that will yield gas or petroleum in economic quantities.

peachblossom ore *erythrite*.

pea coal A size of anthracite that will pass through a 13/16 inch round mesh but not through a 9/16 inch round mesh. Cf: *broken coal; egg coal; stove coal; chestnut coal; buckwheat coal*.

peacock copper *peacock ore*.

peacock ore An iridescent copper mineral having a lustrous, tarnished surface exhibiting variegated colors, such as chalcopyrite and esp. *bornite*. Syn: *peacock copper*.

pea gravel Clean gravel, the particles of which are similar in size to that of peas.

pea grit A limestone containing calcareous pisoliths; a *pisolite*.

pea iron ore *pea ore*.

peak [**coast**] A headland or promontory; a jut of land.

peak [**geomorph**] (a) The more or less conical or pointed top of a hill or mountain; one of the crests of a mountain; a prominent summit or the highest point. (b) An individual mountain or hill taken as a whole, esp. when isolated or having a pointed, conspicuous summit.

peak diameter The dominant or maximum particle diameter as determined on a particle-size distribution curve.

peak discharge *flood peak*.

peak plain A high-level plain formed by a series of accordant summits, often explained as an uplifted and fully dissected peneplain. See also: *gipfelflur*. Syn: *summit plain*.

peak runoff The maximum rate of runoff at a given point or from a given area, during a specified period.

peak zone An informal *biostratigraphic zone* consisting of a body of strata characterized by the "exceptional abundance" of some taxon or taxa for which it is named (ACSN, 1961, art.20g) or representing the "maximum development" (but not necessarily the total range) of some taxon (ISST, 1961, p.23). The peak zone may represent one or more episodes of exceptional proliferation of a taxon, not only in number of individuals, but commonly in such respects as great lateral spread or dominance in the entire organic assemblage. The corresponding geologic-time unit is *hemera*. Cf: *assemblage zone; range zone*. Syn: *epibole; acme zone; flood zone; acrozone*.

pea ore A variety of pisolitic limonite or *bean ore* occurring in small, rounded grains or masses about the size of a pea. Syn: *pea iron ore*.

pearceite A black mineral: $Ag_{16}As_2S_{11}$. It may contain copper. Cf: *polybasite*.

pearl [**mollusk**] A dense, typically round, usually white or light-colored, calcareous concretion consisting of occasional layers of conchiolin and predominant layers of aragonite (or rarely, calcite) deposited concentrically about a foreign particle within or beneath the mantle of various marine and freshwater mollusks, either free from or attached to the shell.

pearlite *perlite*.

pearl mica *margarite* [mineral].

pearl opal *cacholong*.

pearl spar A crystalline carbonate mineral, such as ankerite, having a pearly luster; specif. *dolomite* [mineral]. Also spelled: *pearlspar*.

pearlstone *perlite*.

early *nacreous.*

ear shape A term used to describe the shape of the Earth.

ear-shape cut A variation of the *brilliant cut*, usually with 58 facets, having a pear-shaped girdle outline. See also: *pendeloque; briolette.*

eastone A rock whose texture resembles an aggregate of peas; a *pisolite.*

eat An unconsolidated deposit of semicarbonized plant remains of a watersaturated environment, such as a bog or fen, and of persistently high moisture content (at least 75%). It is considered an early stage or rank in the development of coal; carbon content is about 60% and oxygen content is about 30%. Structures of the vegetal matter can be seen. When dried, peat burns freely.

eat ball A *lake ball* containing an abundance of peaty fragments.

eat bed *peat bog.*

eat bog A *bog* in which peat has developed, under conditions of acidity, from the characteristic vegetation, esp. sphagnum. Syn: *peat moor; peat bed.*

eat breccia Peat that has been broken up and then redeposited by water. Syn: *peat slime.*

eat coal (a) A coal transitional between peat and brown coal or lignite. (b) Artificially carbonized peat that is used as a fuel.

eat flow A mudflow of peat produced in a peat bog by a *bog burst.* Cf: *bog flow.*

eat formation The decomposition of vegetable matter whose conditions intermediate between those of *moldering* and those of *putrefaction,* in stagnant water with small amounts of oxygen. Cf: *disintegration* [*coal*].

eat hummock *palsa.*

eat moor *peat bog.*

eat moss Moss from which peat has formed, usually *sphagnum moss.*

eat mound *palsa.*

eat ring A *nonsorted circle* in peat.

eat-sapropel Organic degradation matter that is transitional between peat and sapropel. Syn: *sapropel-peat.*

eat slime *peat breccia.*

eat soil An acid, humic soil that consists mainly of peat.

eat-to-anthracite theory A theory of coal formation as a process in which the progressive ranks of coal are indicative of the degree of *coalification* and, by inference, of the relative geologic age of the deposit. Peat, as the initial stage of coalification, is of recent geologic age; lignite, as an intermediate stage, is usually Tertiary or Mesozoic, and bituminous coal and anthracite, as the more advanced stages of coalification, are usually Carboniferous (Nelson & Nelson, 1967, p.271).

ebble [*gem*] (a) A rough gem occurring in the form of a pebble, as in a stream. (b) Transparent, colorless quartz, or *rock crystal*; e.g. "Brazilian pebble".

ebble [*part size*] (a) A general term for a small, roundish, esp. waterworn stone; specif. a rock fragment larger than a granule and smaller than a cobble, having a diameter in the range of 4-64 mm (1/6 to 2.5 in., or -2 to -6 phi units, or a size between that of a small pea and that of a tennis ball), being somewhat rounded or otherwise modified by abrasion in the course of transport. In Great Britain, the range of 10-50 mm has been used. The term had been formerly used to include fragments of cobble size; it is frequently used in the plural as a syn. of *gravel.* See also: *very coarse pebble; coarse pebble; medium pebble; fine pebble.* Syn: *pebblestone.* (b) A rock or mineral fragment in the soil, having a diameter in the range of 2-20 mm (Atterberg, 1905). The U.S. Bureau of Soils has used a range of 2-64 mm. Cf: *gravel.*

pebble armor A *desert armor* consisting of rounded pebbles, as on a *serir.*

pebble bed Any pebble conglomerate, esp. one in which the pebbles weather conspicuously and fall loose; e.g. the Bunter pebble beds of Devon and Somerset in England. Syn: *popple rock.*

pebble coal *ball coal.*

pebble conglomerate A consolidated rock consisting mainly of pebbles.

pebble dent A depression formed by a pebble on an unconsolidated sedimentary surface, represented by a downward curvature of laminae beneath the pebble.

pebble dike (a) A *clastic dike* composed largely of pebbles. (b) A tabular body containing sedimentary fragments in an igneous matrix, as from the Tintic district in Utah (Farmin, 1934); e.g. one whose fragments were broken from underlying rocks by gaseous or aqueous fluids of magmatic origin and injected upward into country rock and becoming rounded due to the milling and/or corrosive action of the hydrothermal fluids.

pebble gravel An unconsolidated deposit consisting mainly of pebbles.

pebble mosaic A *desert mosaic* consisting of pebbles.

pebble peat Peat that is formed in a semiarid climate by the accumulation of moss and algae, no more than 1/4 inch in thickness, under the surface pebbles of well-drained soils.

pebble phosphate A secondary phosphorite of either residual or transported origin, consisting of pebbles or concretions of phosphatic material, as in Florida; e.g. *land-pebble phosphate* and *river-pebble phosphate.*

pebble pup (a) A geologist's assistant. (b) A student of geology. (c) An inexperienced *rock hound.*

pebble size A term used in sedimentology for a volume greater than that of a sphere with a diameter of 4 mm (1/6 in.) and less than that of a sphere with a diameter of 64 mm (2.5 in.).

pebblestone *pebble* [*part size*].

pebbly Consisting, or containing appreciable or significant amounts, of pebbles; e.g. a "pebbly soil". Syn: *pebbled.*

pebbly mudstone A delicately laminated *conglomeratic mudstone* in which thinly scattered pebbles are embedded among somewhat distorted bedding planes. The term is advocated by Crowell (1957, p.1003) as a descriptive name without regard to manner of origin for a poorly sorted, till-like rock composed of dispersed pebbles in an abundant mudstone matrix. See also: *tilloid; pseudotillite.*

pebbly sand An unconsolidated sediment containing at least 75% sand and a conspicuous number of pebbles that does not exceed 25% of the total aggregate (William et al, 1942, p.343-344). Cf: *sandy gravel.*

pebbly sandstone (a) A consolidated *pebbly sand.* (b) A sandstone containing 10-20% pebbles (Krynine, 1948, p.141). Cf: *conglomeratic sandstone.* (c) A term used in Scotland for a conglomerate.

pecoraite A mineral of the serpentine group: $Ni_3Si_2O_5(OH)_4$.

pectinacean Any bivalve mollusk belonging to the superfamily Pectinacea, characterized by an orbicular, monomyarian, subequilateral shell with wing-like extensions from the hinge margin; e.g. a scallop.

pectinate Said of the spine connections of a cactus when there is a pattern of radiating, small, lateral spines.

pectinirhomb A specialized type of *pore rhomb* found in cystoids, consisting of a compact, rhomboidal structure of closely spaced, comb-like grooves. It is typically set in a distinct depressed area on thecal plates.

pectinite A fossil scallop shell.

pectolite A whitish or grayish monoclinic mineral: $NaCa_2$-$Si_3O_8(OH)$. It occurs in closely compacted masses of divergent or parallel fibers, commonly in cavities in basalts and scoriaceous lavas. Cf: *serandite.*

ped A naturally formed unit of soil structure, e.g. granule, block, crumb, aggregate. Cf: *clod.*

pedal elevator muscle A thin bundle of muscle fibers attached to the bivalve-mollusk shell in the umbonal cavity and serving to raise the foot.

pedalfer An old, general term for a soil in which there is a concentration of sesquioxides. It is the characteristic type of soil in a humid region. Cf: *pedocal.*

pedal gape An opening between margins of the shell of a bivalve mollusk for the protrusion of the foot.

pedality The physical nature of a soil as expressed by the features of its constituent peds.

pedal levator muscle A pedal muscle serving to retract the foot of a bivalve mollusk.

pedal muscle One of a pair of muscles connecting the foot of a mollusk to the interior surface of the shell; e.g. "pedal protractor muscle" serving to extend the foot of a bivalve mollusk, and "pedal retractor muscle" serving to retract the foot of a bivalve mollusk.

pedcal Var. of *pedocal.*

pedestal [geomorph] A relatively slender neck or column of rock capped by a wider mass of rock and produced by undercutting as a result of wind abrasion (as in the SW U.S.) or by differential weathering. See also: *pedestal rock.* Syn: *rock pedestal.*

pedestal [glaciol] *ice pedestal.*

pedestal rock (a) An isolated and residual or erosional mass of rock supported by or balanced on a *pedestal.* The term is also applied to the entire feature. See also: *balanced rock; mushroom rock.* Syn: *pedestal boulder.* (b) *perched block.*

pedia Plural of *pedion.*

pedial class That class in the triclinic system having symmetry 1 (no symmetry).

pedicel [paleont] (a) The greatly modified first segment of the abdomen in arachnids of the subclass Caulogastra, reaching its extreme development in spiders. (b) A small foot or footlike organ of an invertebrate, such as a tube foot of an echinoderm. (c) The area of attachment of the body of a tintinnid to the lorica. (d) A small or short stalk or stem in an animal; esp. a narrow basal part by which a larger part or body is attached, such as the *pedicle* of a brachiopod.

pedicel [bot] In an inflorescence, the stalk of an individual flower.

pedicellaria Any of various minute organs resembling forceps that are borne in large numbers on certain echinoderms; e.g. a minute stalked organ in the external integument of an echinoid and used for grasping or defending, or a minute pincerlike or valvate calcareous appendage on or in the skin, ossicles, or spines of an asteroid. Pl: *pedicellariae.*

pedicle A variably developed, cuticle-covered, fleshy or muscular appendage of a brachiopod, commonly attached to and protruding from the inner surface of the pedicle valve and serving to attach the animal to the substratum. See also: *pedicel* [paleont].

pedicle callist A localized thickening of secondary-shell layer in the apex of the pedicle valve of a brachiopod, representing the track of anterior migration of junction between pedicle epithelium and outer epithelium.

pedicle collar The complete or partial ring-like thickening of the inner surface of the ventral beak of a brachiopod, "continuous laterally with internal surface of deltidial plates, sessile, with septal support, or free anteriorly and secreted by anteriorly migrating outer epithelium at its junction with pedicle epithelium" (TIP, 1965, pt.H, p.149).

pedicle epithelium The ectodermal epithelium investing the pedicle of a brachiopod.

pedicle foramen The subcircular to circular perforation of shell, adjacent to the beak of the pedicle valve, through which the pedicle of a brachiopod passes. Cf: *pedicle opening.* Syn: *foramen.*

pedicle groove A commonly subtriangular groove dividing the ventral pseudointerarea medially and affording passage for th pedicle in many lingulids.

pedicle opening The variably shaped aperture in a brachiopod shell through which the pedicle emerges. Cf: *pedicle foramen.*

pedicle sheath An externally directed tube projecting posteriorly and ventrally from the ventral umbo of a brachiopod probably enclosing the pedicle in young stages of development of some shells with supra-apical pedicle opening (TIP, 1965, pt.H, p.150).

pedicle tube An internally directed tube of secondary shell of a brachiopod, continuous with margin of pedicle foramen and enclosing the proximal part of the pedicle (TIP, 1965, pt.H, p.150).

pedicle valve The valve of a brachiopod through which the pedicle commonly emerges. It is usually larger than the *brachial valve,* and it contains the teeth by which the hinging of the valves is accomplished. Syn: *ventral valve.*

pediment A broad, flat or gently sloping, rock-floored erosion surface or plain of low relief, typically developed by subaerial agents (including running water) in an arid or semiarid region at the base of an abrupt and receding mountain front or plateau escarpment, and underlain by bedrock (occasionally by older alluvial deposits) that may be bare but more often partly mantled with a thin and discontinuous veneer of alluvium derived from the upland masses and in transit across the surface. The longitudinal profile of a pediment is normally slightly concave upward, and its outward form may resemble a *bajada* (which continues the forward inclination of a pediment). The term was first applied to a landform by McGee (1897, p.92), although Gilbert (1877, p.130-131) first recognized and described the feature as a *hill of planation.* Etymol: from an architectural pediment, a triangular feature crowning a portico of columns in front of a Grecian-style building; in this sense, the term is not appropriate for a gently sloping surface commonly forming a broad approach to a mountain range. Cf: *rock fan; plain of lateral planation; peripediment.* Syn: *piedmont interstream flat; conoplain; rock pediment.*

pedimentation The action or process of formation and development of a *pediment* or pediments; also the product resulting from such an action or process. The two processes recognized as being most active in pediment formation are lateral planation by steep-gradient streams, and backwearing and removal of debris by rill wash and unconcentrated flow; the latter process appears to be the most widely accepted. Cf: *pediplanation.*

pediment dome *desert dome.*

pediment gap A term applied by Sauer (1930) to a broad opening formed by the enlargement of a *pediment pass.*

pediment pass A term applied by Sauer (1930) to a narrow, flat, rock-floored tongue extending back from a pediment and penetrating sufficiently along a mountain to join another pediment extending into the mountain front from the other side; the pediments are frequently at different levels. Cf: *pediment gap.*

pediocratic Pertaining to a period of time in which there is little diastrophism. Cf: *orocratic.*

pedion An open crystal form having only a single face, with no symmetrically equivalent face. Pl: *pedia.*

pedipalpus One of the second pair of cephalothoracic appendages that lie on each side of the mouth of an arachnid and that are subject to many variations in structure, such as being the largest and most conspicuous appendages in scorpions (ending in a powerful chela), stout and conspicuous appendages in whip scorpions and the order Phrynichida (but ending in a pointed joint), and the least conspicuous appendages in the order Architarbida. Pl: *pedipalpi.* See also: *palpus.* Syn: *pedipalp.*

pediplain A term proposed by Maxson & Anderson (1935, p.94) for an extensive, multiconcave, thinly alluviated, rock-it erosion surface formed in a desert region by the coalescence of two or more adjacent pediments and occasional desert domes, and representing the end result (the "peneplain") of the mature stage of the arid erosion cycle. Howard (1942) objected to the term because the surface to which it is applied is not wholly at the base of a slope and is not a "plain" in the true geomorphic sense. Cf: *pediplane; coalescing pediment.* Syn: *panfan; desert peneplain; desert plain.*

pediplanation (a) A general term for all the processes by which *pediplanes* are formed (Howard, 1942, p.11). (b) The action or process of formation and development of a *pediplain* pediments; *pedimentation* of regional magnitude, assisted by slope retreat. Also, the product resulting from such an action or process.

pediplane (a) A general term proposed by Howard (1942, p.11) for any planate erosion surface (such as a pediment or peripediment) produced in the piedmont area of an arid or semiarid region, either exposed or covered with a veneer of alluvium no greater than the depth of effective scour (the thickness that can be moved during floods). (b) A term sometimes used as a syn. of *pediplain.*

pedocal An old, general term for a soil in which there is an accumulation or concentration of carbonates, usually calcium carbonate. It is the characteristic type of soil in an arid or semiarid region. Cf: *pedalfer.* Also spelled: pedcal.

pedode A term proposed by Brewer (1964, p.271) for a spheroidal, discrete glaebule with a hollow interior, often with a drusy lining of crystals like that of a geode. It may have an outside layer of chalcedony.

pedogenesis *soil genesis.*

pedogenic Pertaining to soil formation.

pedogenics The study of the origin and development of soil; an aspect of *soil science.*

pedogeography The study of the geographic distribution of soils.

pedography The systematic description of soils; an aspect of *soil science.*

pedolith A surface formation that has undergone one or more pedogenic (soil-forming) processes (Dewolf, 1970).

pedologic age The relative maturity of a soil profile.

pedologic unit A soil considered without regard to its stratigraphic relations (ACSN, 1961, art.18b). Cf: *soil-stratigraphic unit.*

pedology A more formal syn. of *soil science.*

pedometer (a) A pocket-size, watch-shaped instrument that registers the linear distance a pedestrian covers by responding to his body motion at each step. It is carried in an upright position attached to the body or to a leg, and it can be adjusted to the length of the pace of the person carrying it. (b) A term formerly applied to an instrument now known as a *passometer.*

pedon The smallest unit or volume of soil that represents or exemplifies all the horizons of a soil profile. It is usually a horizontal, more or less hexagonal area of about one square meter, but may be larger. The term is part of the soil classification system of the National Cooperative Soil Survey.

pedorelic Said of a soil feature that is derived from a pre-existing soil horizon. Cf: *lithorelic.*

pedosphere That shell or layer of the Earth in which soil-forming processes occur.

pedotubule A soil feature consisting of skeleton grains (or skeleton grains plus plasma) and having a tubular external form (either single tubes or branching systems of tubes) characterized by relatively sharp boundaries and by a relatively uniform cross-sectional size and shape (circular or elliptical). Examples: *isotubule; striotubule; aggrotubule; granotubule.* The term was proposed by Brewer (1964, p.236). Cf: *rhizoconcretion.*

peduncle [paleont] (a) A narrow part by which some larger part or the whole body of an animal is attached, such as the pedicle of a brachiopod, the column of an echinoderm, or the basal portion of certain crustacean appendages. (b) The fleshy stalk-like portion of the body of certain cirripedes. (c) The mass of cyptoplasm projecting from a thecamoebian-test aperture, giving rise to pseudopodia; pseudopodial trunk.

peduncle [bot] In a plant, a stalk that bears an inflorescence or a strolibus.

peel map A *paleogeologic map*, so named because the formations overlying an unconformity are, in effect, "peeled off" and the preunconformity distribution of the underlying formations is uncovered and mapped (Levorsen, 1960, p.4).

peel-off time In seismic prospecting, the time correction to be applied to observed data to adjust them to a depressed reference datum.

peel thrust A sedimentary sheet peeled off a sedimentary sequence, essentially along a bedding plane. A series of peel thrusts may be imbricated above a décollement (Bucher, 1955).

peep-sight alidade An *alidade* used with a plane table, consisting of a rear (open) sight mounted on a straightedge.

peg adjustment The adjustment of a spirit-leveling instrument of the dumpy-level type in which the line of collimation is made parallel with the axis of the spirit level by means of two stable marks (pegs) the length of one instrument sight apart.

pegmatite An exceptionally coarse-grained (most grains one cm or more in diameter) igneous rock, with interlocking crystals, usually found as irregular dikes, lenses, or veins, esp. at the margins of batholiths. Although pegmatites having gross compositions similar to other rock types are known, their composition is generally that of granite; the composition may be simple or complex and may include rare minerals rich in such elements as lithium, boron, fluorine, niobium, tantalum, uranium, and rare earths. Pegmatites represent the last and most hydrous portion of a magma to crystallize and hence contain high concentrations of minerals present only in trace amounts in granitic rocks. The first use of the term "pegmatite" is attributed to Haüy (1822) who used it as a syn. of *graphic granite.* Cf: *pegmatoid; symplectite.* See also: *pegmatitic.* Syn: *giant granite.*

pegmatitic (a) Said of the texture of an exceptionally coarsely crystalline igneous rock; originally restricted to those rocks with graphic intergrowths, in which two components crystallized simultaneously and are mutually penetrating. Syn: *granophyric; graphic.* (b) Occurring in, pertaining to, or composed of *pegmatite.*

pegmatitic stage A stage in the normal sequence of crystallization of a magma containing volatiles, at which time the residual fluid is sufficiently enriched in volatile materials to permit the formation of coarse-grained rocks (i.e. pegmatites). The relative amounts of silicate and volatile materials in the fluid, the temperature range, and the relationship of these fluids to hydrothermal fluids are in dispute. Cf: *hydrothermal stage.*

pegmatitization Formation of, or replacement by, a pegmatite.

pegmatoid n. An igneous rock that has the coarse-grained texture of a *pegmatite* but that lacks graphic intergrowths and/or typically granitic composition.--adj. Said of the texture of a pegmatitic rock lacking graphic intergrowths and/or typical granitic composition.

pegmatolite *orthoclase.*

pegmatophyre *granophyre.*

pegmatophyric *granophyric.*

peg model A three-dimensional graphic method of depicting the structure and stratigraphic relations of an oil field. Vertical rods at scaled intervals represent wells, and correlative beds at scaled depths are joined by thread to represent formation surfaces.

pegostylite A speleothem that is formed by ascending waters.

523

peiroglyph A cross-cutting sedimentary structure, such as a sandstone dike (Vassoevich, 1953, p.37).

pelagic [oceanog] (a) Pertaining to the water of the ocean as an environment. See also: *epipelagic; mesopelagic*. (b) Said of marine organisms whose environment is the open ocean, rather than the bottom or shore areas. Pelagic organisms may be either *nektonic* or *planktonic*.

pelagic [lake] Pertaining to the deeper (10-20 m or more) part of a lake, characterized by deposits of mud or ooze and by the absence of aquatic vegetation.

pelagic deposits Deep-sea sediments without terrigenous material; they are either inorganic red clay or organic ooze. Cf: *terrigenous deposits; hemipelagic deposits*. Syn: *deep-sea sediments; abyssal deposits*.

pelagic limestone A fine-textured limestone formed chiefly by the accumulation, in relatively deep water, of the calcareous tests of floating organisms (esp. of post-Jurassic foraminifers). It is characteristic of geosynclinal belts.

pelagite A nodule or concretionary lump found in deep-sea deposits; specif. a *manganese nodule*. Syn: *halobolite*.

pelagochthonous Said of coal derived from a submerged forest or from driftwood.

pelagosite A term used by Revelle & Fairbridge (1957, p.258) for a superficial calcareous crust a few millimeters thick, generally white, gray, or brownish, with a pearly luster, formed in the intertidal zone by ocean spray and evaporation (alternate solution and evaporation), and composed of calcium carbonate accompanied by contents of magnesium carbonate, strontium carbonate, calcium sulfate, and silica that are higher than those found in normal limy sediments. It was originally described as an impure calcite mineral.

peldon An English term for a very hard, smooth, compact sandstone with conchoidal fracture, occurring in coal measures.

Peléan Adj. of *Pelée-type volcano* or eruption.

Peléan cloud A syn. of *nuée ardente*, so named because it is a characteristic type of eruption of Mt. Pelée.

pelecypod Any benthic aquatic mollusk belonging to the class Pelecypoda, characterized by a bilaterally symmetrical bivalve shell, a hatchet-shaped foot, and sheet-like gills. Syn: *lamellibranch*. Partial syn: *bivalve*.

peleeite A hypersthene-labradorite dacite similar to bandaite.

Pelée-type eruption A type of volcanic eruption characterized by conspicuous, gaseous clouds. Etymol: Mont Pelée, island of Martinique. Adj: Peléan. Cf: *Hawaiian-type eruption; Strombolian-type eruption; Vulcanian-type eruption*.

Pele's hair A naturally formed spun glass formed by blowing out during quiet fountaining of fluid lava, sometimes in association with *Pele's tears*. Its diameter is less than half a millimeter, but it can be as long as two meters in length. Etymol: Pele, Hawaiian goddess of fire. Syn: *lauoho o pele; filiform lapilli; capillary ejecta*.

Pele's tears [gem] Hawaiian name for a clear chalcedony or opal in cabochon cut.

Pele's tears [pyroclast] Small, solidified drops of volcanic glass behind which trail pendants of *Pele's hair*. They may be tear-shaped, spherical, or nearly cylindrical. Etymol: Pele, Hawaiian goddess of fire. Cf: *tear-shaped bomb*.

pelhamine A light gray-green serpentine mineral from Pelham, Mass. It may be altered chrysotile.

pelinite A term proposed by Searle (1912, p.148) for a hydrous aluminum silicate thought to be the true clay substance in clays other than the kaolins and considered to be an amorphous (colloidal) and plastic material of varying composition but of generally higher silica content than that in *clayite* and also with appreciable alkalies and/or alkaline earths.

pelionite A name proposed by W.F. Petterd for a bituminous coal resembling English cannel coal, occurring near Monte Pelion in Tasmania (U.S. Bureau of Mines, 1968, p.802).

pelite (a) A sediment or sedimentary rock composed of the finest detritus (clay- or mud-size particles); e.g. a mudstone or a calcareous sediment composed of clay, minute particles of quartz, or rock flour. The term is equivalent to the Latin-derived term, *lutite*. (b) A fine-grained sedimentary rock composed of more or less hydrated aluminum silicates with which are mingled other small particles of various minerals (Twenhofel, 1937, p.90); an aluminous sediment. (c) A term regarded by Tyrrell (1921, p.501-502) as the metamorphic derivative of lutite, such as the metamorphosed product of a siltstone or mudstone. "As commonly used, a pelite means an aluminous sediment metamorphosed, but if used systematically, it means a fine-grained sediment metamorphosed" (Bayly, 1968, p.230).---Etymol: Greek *pelos*, "clay mud". See also: *psammite; psephite*. Syn: *pelyte*.

pelitic (a) Pertaining to or characteristic of pelite; esp. said of a sedimentary rock composed of clay, such as a "pelitic tuff" representing a consolidated volcanic ash consisting of clay-size particles. (b) Said of a metamorphic rock derived from pelite; e.g. a "pelitic gneiss", a "pelitic hornfels", or a "pelitic schist", derived by metamorphism of an argillaceous or of a fine-grained aluminous sediment.---Cf: *argillaceous; lutaceous*.

pelitomorphic Pertaining to clay-size carbonate particles in limestone or dolomite rock. Also, said of a limestone or dolomite consisting of an aggregate of pelitomorphic particles or having a matrix of such particles.

pellet (a) A small, usually rounded aggregate of accretionary material, such as a lapillus or a fecal pellet; specif. a spherical to elliptical (commonly ovoid, sometimes irregularly shaped) homogeneous clast made up almost exclusively of clay-sized calcareous (micritic) material, devoid of internal structure, and contained in the body of a well-sorted carbonate rock. Folk (1959; 1962) suggested that the term apply to allochems less than 0.15-0.20 mm in diameter, the larger grains being referred to as *intraclasts*, although in some rocks it is impossible to draw a sharp division. Pellets appear to be mainly the feces of mollusks and worms; others include pseudo-ooliths and aggregates produced by gas bubbling, by algal "budding" phenomena, or by other intraformational reworking of lithified or semilithified carbonate mud. (b) A small rounded aggregate (0.1-0.3 mm in diameter) of clay minerals and fine quartz found in some shales and clays, separated from a matrix of the same materials by a shell of organic material, and ascribed to the action of water currents (Allen & Nichols, 1945).

pelleted limestone A limestone characterized by abundant pellets, such as some lower Paleozoic limestones whose major constituents are fecal pellets, or carbonate muds that display rounded or ellipsoidal aggregates of "grains of matrix" material. The adjectival term is sometimes given as "pellet", "pelletoid", "pelletoidal", "pelletal", and "pelletted".

pelletoid Pelleted, or containing abundant pellets; e.g. a "pelletoid limestone". Also spelled: *pelletoidal*.

pellet snow *graupel*.

pellicular envelope The delicate outer covering of soft parts in a tintinnid.

pellicular front The even front, developed only in pervious granular material, on which pellicular water depleted by evaporation, transpiration, or chemical action is regenerated by influent seepage (Tolman, 1937, p.593). Syn: *infiltration front; wetting front*.

pellicular water Water in layers more than one or two molecules thick that adheres to the surfaces of soil and rock particles in the zone of aeration. Layers more than a few microns in thickness are short-lived due to the requirement that free energy and capillary surface be at a minimum when moisture is at equilibrium (Smith, W.O., 1961, p.11). Syn: *adhesive water; film water; sorption water*. Cf: *pendular water; funicular* ater; attached ground water.

pell-mell structure A sedimentary structure characterized by absence of bedding in a coarse deposit of waterworn material; may occur where deposition is too rapid for sorting or where slumping has destroyed the layered arrangement.

pellodite A term used by Schuchert (1924, p.441) for a water-laid, sandy, varved clay, and by Pettijohn (1957, p.273) for the lithified equivalent of a varved clay. It is apparently a syn. of pelodite.

pelma (a) Entire crinoid column with attached cirri and holdfast structure, if present. (b) A pseudopore in a costa of a cribrimorph cheilostome (bryozoan).---Pl: pelmata.

pelmatozoan n. Any echinoderm, with or without a stem, that lives attached to a substrate.----adj. Said of an echinoderm having an attached mode of life. Var: pelmatozoic.----Cf: eleutherozoan.

pelmicrite A limestone consisting of a variable proportion of pellets and carbonate mud (micrite); specif. a limestone containing less than 25% intraclasts and less than 25% ooliths, with a volume ratio of pellets to fossils and fossil fragments greater than 3 to 1, and the carbonate-mud matrix more abundant than the sparry-calcite cement (Folk, 1959, p.14).

pelodite A term proposed by Woodworth (1912, p.78) for a lithified glacial rock flour, composed of glacial pebbles in a silty or clayey matrix, formed by redeposition of the fine fraction of a till. Syn: pellodite.

pelogloea Organic matter, mainly colloidal, occurring adsorbed on sedimentary particles in natural waters (Fox, 1957, p.384). Cf: leptopel.

pelolithic argillaceous.

pelophyte A lake-bottom deposit consisting mainly of fine, non-fibrous plant remains (Veatch & Humphrys, 1966, p.227). Cf: psephyte.

pelsparite A limestone consisting of a variable proportion of pellets and clear calcite (spar); specif. a limestone containing less than 25% intraclasts and less than 25% ooliths, with a volume ratio of pellets to fossils and fossil fragments greater than 3 to 1, and the sparry-calcite cement more abundant than the carbonate-mud matrix (micrite) (Folk, 1959, p.14).

peltate Said of a leaf, usually a shield-shaped leaf, that is attached to its stalk inside the margin (Lawrence, 1951, p.764).

pelyte Var. of pelite.

pen A British term variously used for a hill, mountain, highland, or headland. Etymol: Celtic.

peña A rock; a rocky point; a needle-like eminence; a cliff. Term is used in the SW U.S. Etymol: Spanish, "rock".

peñasco A term used in the SW U.S. for a projecting rock, esp. one isolated by the recession of a cliff or of a mountain slope. Etymol: Spanish, "large rock".

pencatite A recrystallized limestone containing periclase or brucite and calcite in approximately equal molecular proportions. It is formed by contact metamorphism of magnesian limestone. Cf: predazzite.

pencil cave A driller's term for hard, closely jointed shale that caves into a well in pencil-shaped fragments.

pencil cleavage Cleavage in which fracture produces long, slender pieces of rock. It is produced by the intersection of a direction of cleavage with the stratification of the rock, usually in weakly metamorphosed rocks.

pencil ganister A variety of ganister characterized by fine carbonaceous streaks or markings, and so called from the likeness of these to pencil lines. The carbonaceous traces are often recognizable as roots and rootlets of plants.

pencil gneiss A gneiss that splits into roughly cylindrical, pencil-like, quartz-feldspar crystal aggregates often mantled by mica flakes. Syn: stengel gneiss.

pencil ore Hard, fibrous masses of hematite that can be split up into thin rods.

pencil stone A compact pyrophyllite used for making slate pencils.

pendant [speleo] One of a series of connected projections from the ceiling of a cave. It is a solutional remnant. See also: horst [speleo]. Syn: rock pendant; solution pendant.

pendant [intrus rocks] A downward projection into an igneous intrusion of the country rock of its roof. Cf: cupola. Syn: roof pendant.

pendeloque A modification of the round brilliant cut, having an outline similar to that of the pear-shape cut, but with the narrower end longer and more pointed. Cf: briolette.

pendent [geomorph] Said of a landform that slopes steeply down (as a hillside) or overhangs (as a cliff).

pendent [paleont] Said of a graptoloid rhabdosome with approximately parallel stipes that hang below the sicula.

pendent terrace A connecting ribbon of sand that joins "an isolated point of rock with a neighboring coast" (Haupt, 1906, p.78).

pendletonite karpatite.

pendular water Capillary water ringing the contact points of adjacent rock or soil particles in the zone of aeration (Smith, W.O., 1961, p.2). Cf: funicular water; pellicular water; capillary condensation.

pendulum (a) A body so suspended from a fixed point as to swing freely to and fro under the combined action of gravity and momentum. Syn: physical pendulum. (b) A vertical bar so supported from below by a stiff spring as to vibrate to and fro under the combined action of gravity and the restoring force of the spring (Dobrin, 1952, p. 50-51). Syn: inverted pendulum.

pendulum level A leveling instrument in which the line of sight is automatically maintained horizontal by means of a built-in pendulum device (such as a horizontal arm and a plumb line at right angles to the arm).

penecontemporaneous Said of a geologic process, or resultant structure or mineral, occurring immediately after deposition but before consolidation of the enclosing rock. Cf: contemporaneous.

penecontemporaneous deformation contemporaneous deformation.

penecontemporaneous faulting A deformation occurring in soft rock, soon after the deposition of the strata involved, and caused by gravitational sliding or slump.

penecontemporaneous fold A fold that develops in sediment shortly after deposition.

peneloken pokelogan.

peneplain n. A term introduced by Davis (1889a, p.430) for a low, nearly featureless, and gently undulating or almost-plane land surface of considerable area which presumably has been reduced by the processes of long-continued subaerial erosion (primarily mass-wasting of and sheetwash on interstream areas of a mature landscape, assisted by stream erosion) almost to base level in the penultimate stage of a humid, fluvial geomorphic cycle; also, such a surface uplifted to form a plateau and subjected to dissection. The term has been extended in the literature of geomorphology to include surfaces produced by marine, eolian, and even glacial erosion; and many topographic surfaces interpreted as peneplains are actually pediplains, panplains, or plains of marine erosion. A peneplain may be characterized by gently graded and broadly convex interfluves sloping down to broad valley floors, by truncation of strata of varying resistance and structure, by accordant levels, and by isolated erosion remnants (monadnocks) rising above it. Etymol: Latin pene-, "almost", + plain. Cf: endrumpf; base-level plain; marine peneplain; rumpffläche. Syn: peneplane; base-level peneplain.---v. peneplane.

peneplanation The act or process of formation and development of a peneplain; esp. the decline and flattening out of hillsides during their retreat and the accompanying downwasting of divides and residual hills. The term "peneplaination" is never used.

peneplane n. A term suggested by Johnson (1916) to replace *peneplain* since the latter, when first introduced by Davis (1889a), was not intended to signify "almost a plain" (i.e. a region of nearly horizontal structure) but a region with almost a flat surface; despite a few exceptions, the suggestion has received little support.---v. To erode to a peneplain. Syn: *peneplain*.

penesaline (a) Said of an environment intermediate between normal marine and saline, characterized by evaporitic carbonates often interbedded with gypsum or anhydrite. According to Sloss (1953, p.145), the upper limit is characterized by a salinity of 352 parts per thousand (the maximum salinity at which sodium chloride is not precipitated) and the lower limit by a salinity high enough to be toxic to normal marine organisms. Cys (1963, p.162) proposed a lower-limit salinity of 72 parts per thousand which is "the lowest salinity at which calcium carbonate will be precipitated from solution by purely chemical processes with no influence of or catalyzing action by organisms or organic water". (b) Said of the water in the back-reef zone characterized by a salinity too great to sustain organisms (Lang, 1937, p.887).

penetration funnel An *impact crater*, generally funnel-shaped, formed by a small meteorite striking the Earth at a relatively low velocity and containing nearly all the impacting mass within it (Cassidy, 1968, p.117). Cf: *meteorite crater*.

penetration test A test to determine the relative densities of noncohesive soils, sands, or silts; e.g. the "standard penetration test" that determines the number of blows required by a standard weight, when dropped from a standard height (30 in. per blow), to drive a standard sampling spoon a standard penetration (12 in.), or the "dynamic penetration test" that determines the relative densities of successive layers by recording the penetration per blow or a specified number of blows. See also: *cone penetration test*.

penetration twin *interpenetration twin*.

penetrative Said of a texture of deformation that is uniformly distributed through the rock, without notable discontinuities, e.g. slaty cleavage. Ant: *nonpenetrative*. Cf: *continuous cleavage*.

penetrometer (a) An instrument for measuring the consistency of materials (such as soil, snow, asphalt, or coal) by indicating the pressure necessary to inject a rigid weight-driven rod or needle of specified shape to a specific depth. (b) An instrument that automatically records the depth and penetration rate of drilling.

penfieldite A white hexagonal mineral: $Pb_2(OH)Cl_3$.

penikkavaarite An intrusive rock composed chiefly of augite, barkevikite, and green hornblende in a feldspathic groundmass.

peninsula (a) An elongated body or stretch of land nearly surrounded by water and connected with a larger land area, usually by a neck or an isthmus. (b) A relatively large tract of land jutting out into the water, with or without a well-defined isthmus; e.g. the Italian peninsula.---Etymol: Latin *paeninsula*, "almost island". See also: *submarine peninsula*.

penitent [geomorph] A term used in the French Alps for an *earth pillar*.

penitent [glaciol] n. A syn. of *nieve penitente*.---adj. A term used to refer to a nieve penitente; e.g. "penitent ice" is a nieve penitente consisting mainly of ice, and "penitent snow" is one consisting mainly of snow.

penitent rock A variety of tor formed on rock with a dipping foliation, joint pattern, or bedding (Ackermann, 1962). Syn: *monk rock*.

penknife ice *candle ice*.

pennantite An orange, manganese-bearing mineral of the chlorite group: $(Mn,Al)_6(Si,Al)_4O_{10}(OH)_8$. Some ferric iron

may be present. It is isomorphous with thuringite.

pennate diatom A diatom having basically bilateral symmetr[y], a member of the diatom order Pennales. Cf: *centric diatom*.

pennine *penninite*.

penninite A mineral of the chlorite group: (Mg,Fe^{+2},Al) $(Si,Al)_4O_{10}(OH)_8$. It has an emerald, olive-green, pale-gree[n], or bluish color. Syn: *pennine*.

Pennsylvanian A period of the Paleozoic era (after the Missi[s]sippian and the Permian), thought to have covered the spa[ce] of time between 320 and 280 million years ago; also, the co[r]responding system of rocks. It is named after the state [of] Pennsylvania in which rocks of this age yield much coal. It [is] the approximate equivalent of the *Upper Carboniferous* of E[u]ropean usage. See also: *age of ferns*. Syn: *Carbonic*.

Penokean orogeny A time of deformation and granite e[mplacement during the Precambrian in Minnesota and Michig[an], dated radiometrically at about 1700 m.y. ago which occurre[d] between the formation of the Animikie ("Huronian") Seri[es] and the Keweenawan Series. It is the same as the *Hudsoni[an] orogeny* of the Canadian Shield. See also: *Killarney Revol[u]tion*.

peñon A high, rocky point. Etymol: Spanish, "large roc[k] rocky mountain". Obsolete syn: *peñol*.

penroseite A lead-gray mineral: $(Ni,Co,Cu)Se_2$. It may co[n]tain some lead and silver. Penroseite is structurally like pyr[ite] and occurs in radiating columnar masses. Syn: *blockite*.

pentactin A sponge spicule having five rays; specif. a hexact[in] with one ray suppressed.

pentagonal dodecahedron *pyritohedron*.

pentahydrite A triclinic mineral: $MgSO_4.5H_2O$. It is isostructu[r]al with chalcanthite.

pentahydroborite A mineral: $CaB_2O_4.5H_2O$ (?).

pentalith A coccolith formed of five crystal units diverging [at] 72 degrees.

pentameracean Any articulate brachiopod belonging to the s[u]perfamily Pentameracea, in general characterized by a larg[e,] strongly biconvex shell with a smooth, costellate, or costa[te] exterior and by a spondylium in the pedicle valve. Their str[a]tigraphic range is Upper (possibly Middle) Ordovician [to] Upper Devonian. Var: *pentamerid*.

pentamerous *quinqueradiate*.

pentane Any of three isomeric, low-boiling paraffin hydroca[r]bons, formula C_5H_{12}, found in petroleum and natural gas.

pentlandite A pale bronze-yellow to light-brown isometric mi[n]eral: $(Fe,Ni)_9S_8$. It is commonly intergrown with pyrrhoti[te,] from which it is distinguished by its octahedral cleavage a[nd] lack of magnetism. Pentlandite is the principal ore of nicke[l.] Syn: *folgerite; nicopyrite*.

penumbra (a) The partly shadowed region of an eclipse. ([b)] The outer, lighter region of a sunspot.----Cf: *umbra*.

Penutian North American stage: lower Eocene (above Bu[li]tian, below Ulatisian).

Peorian A term previously used as an interglacial stage be[tween the Iowan (earlier) and the Wisconsin (later) glaci[al] stages; now that the Iowan is classified as an early substa[ge] of the Wisconsin, the name "Peorian" is not used either for [a] distinct stage or substage. Named for exposures near Peori[a,] Ill.

peperino (a) An unconsolidated, gray tuff of the Italian Alban[o] hills, containing crystal fragments of leucite and other mine[r]als. (b) An indurated pyroclastic deposit containing fragmen[ts] of various sizes and types.

peperite A brecciaike material in marine sedimentary roc[k,] interpreted by some as a mixture of lava with sediment, or b[y] others as shallow intrusions of magma into the wet sediment.

pepino hill A less-preferred syn. of *mogote*, used in Puer[to] Rico.

pepper-and-salt texture Said of disseminated ores.

peracidite *silexite*.

peraeopod Var. of *pereiopod*.

peralkaline Said of an igneous rock in which the molecular proportion of aluminum oxide is less than that of sodium and potassium oxides combined; one of Shand's (1947) groups of igneous rocks, classified on the basis of the degree of aluminum-oxide saturation. Cf: *peraluminous; metaluminous; subaluminous*.

peraluminous Said of an igneous rock in which the molecular proportion of aluminum oxide is greater than that of sodium and potassium oxides combined; one of Shand's (1947) groups of igneous rocks, classified on the basis of the degree of aluminum-oxide saturation. Cf: *peralkaline; metaluminous; subaluminous*.

perbituminous Said of bituminous coal containing more than 5.8% hydrogen, analyzed on a dry, ash-free basis. Cf: *orthobituminous; parabituminous*.

percentage hypsometric curve Hypsographic curve using dimensionless parameters independent of an absolute scale of topographic features by relating the area enclosed between a given contour and the highest contour to the height of the given contour above a basal plane.

percentage log A *sample log* in which the percentage of each type of rock (except obvious cavings) present in each sample of cuttings is estimated and plotted. It is used in the Permian basin of the U.S. Cf: *interpretative log*.

percentage map A *facies map* that depicts the relative amount (thickness) of a single rock type in a given stratigraphic unit.

percent slope The direct ratio between the vertical distance and the horizontal distance for a given slope; e.g. a 3-meter rise in 10 meters horizontal distance would be a 30 *percent slope*.

perch A unit of length, varying locally in different countries, but by statute in Great Britain and U.S. equal to 16.5 ft. It was used extensively in the early public-land surveys and is equivalent in length to a *rod* or *pole*.

perched aquifer An aquifer containing *perched ground water*.

perched block (a) A large, detached rock fragment, generally of boulder size, believed to have been transported and deposited by a glacier, and lying in a conspicuous and relatively unstable or precariously poised position on a hillside. Cf: *erratic*. Syn: *perched boulder; perched rock; balanced rock; pedestal rock*. (b) A rock forming a glacier table in a glacier. (c) A rock capping an earth pillar.

perched boulder *perched block*.

perched glacial valley *hanging valley*.

perched ground water Unconfined ground water separated from an underlying main body of ground water by an unsaturated zone. See also: *perched aquifer; perching bed*. Syn: *perched water*.

perched lake A perennial lake whose surface level lies at a considerably higher elevation than those of other bodies of water, including aquifers, directly or closely associated with the lake; e.g. a lake on a bench that borders the shore of a larger lake.

perched rock *perched block*.

perched spring A spring whose source of water is a body of perched ground water.

perched stream (a) A stream or reach of a stream whose upper surface is higher than the water table and that is separated from the underlying ground water by an impermeable bed in the zone of aeration (Meinzer, 1923, p.57). (b) A stream flowing on an antecedent hillside along a graded valley of a higher-order stream into which it flows at nearly right angles (Horton, 1945, p.352).

perched water *perched ground water*.

perched water table The water table of a body of perched ground water. Syn: *apparent water table*.

perching bed A body of rock, generally stratiform, that sup-

ports a body of *perched ground water*. Its permeability is sufficiently low that water percolating downward through it is not able to bring water in the underlying unsaturated zone above atmospheric pressure. At a given place there may be two or more perching beds and bodies of perched ground water, separated from each other and from the main zone of saturation by unsaturated zones.

percolating water (a) A legal term for water that oozes, seeps, or filters through the soil without a definite channel in a course that is unknown or not discoverable. Cf: *underground stream*. (b) Water involved in percolation.

percolation Laminar flow of water, usually downward, by the force of gravity or under hydrostatic pressure, through small openings within a porous material. Also used as a syn. of "infiltration". Flow in large openings such as caves is not included. Cf: *infiltration*.

percolation rate The rate, expressed as either velocity or volume, at which water percolates through a porous medium.

percolation test A term used in sanitary engineering for a test to determine the suitability of a soil for the installation of a domestic sewage-disposal system. A hole is dug and filled with water and the rate of water-level decline measured; the dimensions of the hole and acceptable rate of decline vary from one jurisdiction to another.

percolation zone The area on a glacier or ice sheet where a limited amount of surface melting occurs, but the meltwater refreezes in the same snow layer and the whole thickness of the snow layer is not completely soaked or brought up to the melting temperature. The percolation zone may be bordered at higher altitudes by the dry snow line and at lower altitudes by the *saturation line*. Cf: *soaked zone*.

percrystalline A term, now obsolete, suggested by Cross et al (1906, p.694) for porphyritic rocks that are extremely crystalline with only a little glass, the ratio of crystals to glass being greater than 7 to 1.

percussion drilling *cable-tool drilling*.

percussion figure A pattern of radiating lines produced on a section of a crystal by a blow. Cf: *pressure figure*.

percussion mark A small, crescentic scar produced on a hard, dense pebble (esp. one of chert or quartzite) by a sharp blow, as by the violent impact of one pebble upon another; it may be indicative of high-velocity flow. Syn: *percussion scar*.

percussion scar *percussion mark*.

percylite A pale-blue mineral: $PbCuCl_2(OH)_2$ (?).

pereion The thorax of a malacostracan crustacean, exclusive of the somites bearing maxillipeds. It is usually provided with locomotory appendages (pereiopods). Pl: *pereia*. Syn: *pereon*.

pereionite A somite of the pereion of a malacostracan crustacean. Syn: *pereonite*.

pereiopod A locomotory thoracopod of a malacostracan crustacean; an appendage of the pereion. Syn: *peraeopod; pereopod; walking leg*.

pereletok A frozen layer of ground, between the *active layer* above and the *permafrost* below, that remains unthawed for one or several years. Etymol: Russian, "survives over the summer". Syn: *intergelisol*.

perennial lake A lake that retains water in its basin throughout the year, and that usually is not subject to extreme fluctuations in level.

perennially frozen ground *permafrost*.

perennial spring A spring that flows continuously, as opposed to an *intermittent spring* or a *periodic spring*.

perennial stream A stream or reach of a stream that flows continuously throughout the year and whose upper surface generally stands lower than the water table in the region adjoining the stream. Syn: *permanent stream; live stream*.

pereon Var. of *pereion*.

pereonite Var. of *pereionite*.

pereopod Var. of *pereiopod*.

perezone A depositional zone containing nonfossiliferous sediments and occurring mostly between low tide and low-lying land undergoing active erosion; it includes lagoons and brackish-water bays.

perfect crystal A crystal without lattice defects. It is an unattained ideal or standard.

perfect flower A flower having both stamens and carpels. Cf: *imperfect flower.*

perfect fractionation path On a phase diagram, a line or a path representing a crystallization sequence in which any crystal that has been formed remains inert, i.e. without altering its composition.

perfectly diffuse reflector A body that reflects radiant energy in such a manner that the reflected energy may be treated as if it were being radiated in accordance with *Lambert's law.* The energy reflected in any direction from a unit area of such a reflector varies as the cosine of the angle between the normal to the surface and the direction of the reflected radiant energy.

perfectly mobile component A component whose amount in a system is determined by its externally imposed chemical potential rather than by its initial amount in the system. Cf: *inert component; mobile component.* Syn: *boundary value component.*

perfect solution A solution that is ideal throughout its entire compositional range.

perfect stone (a) A *flawless* gemstone. (b) A colored stone in which small inclusions or structural faults are less undesirable (than in a flawless gemstone) and sometimes desirable (Shipley, 1951, p.171).

perfemic One of five classes in the C.I.P.W. (Cross, *et al,* 1902, p.585) chemical-mineralogic classification of igneous rocks and in which the ratio of salic minerals to femic is less than one to seven. Cf: *dofemic.*

perfoliate Said of a sessile leaf or bract that surrounds the stem so that the stem seems to pass through the leaf or bract.

perforate (a) Said of the wall between corallites of some colonies, characterized by the presence through it of many irregularly arranged small openings. Also, pertaining to Perforata, a division of corals whose skeleton has a porous texture. (b) Descriptive of foraminiferal-test walls punctured or pierced by numerous pores or small openings that are distinct from apertures, foramina, and canals. Perforate walls are esp. characteristic of calcareous hyaline tests. Also, pertaining to Perforata, a division of foraminifers whose shells have small perforations for the protrusion of pseudopodia. (c) Said of an echinoid tubercle with a small depression in the top for the ligament connecting a spine with the tubercle. (d) Said of a spiral mollusk shell having a permanently open umbilicus at the origin of the whorls.---Cf: *imperforate.*

perforated crust A type of *snow crust* containing pits and hollows produced by ablation.

perforation deposit A term suggested by Cook (1946) for an isolated kame consisting of material that accumulated in a vertical shaft piercing a glacier and affording no outlet for water at the bottom.

pergelation The formation of permafrost in the present and in the past (Bryan, 1946, p.640).

pergelic Said of a soil temperature regime in which the mean annual temperature is less than 0°C, and there is permafrost (SSSA, 1970).

pergelisol A term introduced by Bryan (1946) but later replaced by the syn. *permafrost.*

perhyaline In the C.I.P.W. (Cross, *et al,* 1906, p.694) textural classification of igneous rocks, those rocks in which the ratio of crystals to glassy material is greater than one to seven. This term is rarely used. Cf: *dohyaline.*

perhydrous (a) Said of coal containing more than 6% hydrogen, analyzed on a dry, ash-free basis. (b) Said of a maceral of high hydrogen content, e.g. exinite, resinite.----Cf: *orthohydrous; subhydrous.*

perianth A collective term for the corolla and the calyx of a flower when they are considered together or when they are not structurally differentiated.

periblain A kind of *provitrain* in which the cellular structure is derived from cortical material. Cf: *suberain; xylain.*

periblinite A variety of provitrinite characteristic of periblain and consisting of cortical tissue. Cf: *xylinite; suberinite; telinite.*

pericarp (a) The wall of a fruitified ovary. See also: *endocarp; mesocarp; epicarp.* (b) The wall of the capsule in mosses. (c) Improperly used of the protective husks surrounding certain fruits (Jackson, 1953, p.273).

periclase An isometric mineral: MgO. It is native magnesia but alters easily to brucite. Syn: *periclasite.*

periclinal [bot] Parallel to the surface or circumference of a plant organ. Cf: *anticlinal.*

periclinal [geol] Said of strata and structures that dip radially outward from, or inward towards, a center, to form a dome or a basin. Cf: *quaquaversal; centroclinal.*

pericline [mineral] A variety of albite elongated in the direction of the *b*-axis and often twinned with this axis as the twinning axis. It occurs in alpine veins as large milky-white opaque crystals. Pericline is probably an albitized oligoclase.

pericline [fold] A general term for a fold in which the dip of the beds has a central orientation; beds dipping away from a center form a *dome,* and beds dipping towards a center form a *basin.* The term is generally British in usage. See also: *centrocline; quaquaversal.*

pericline ripple mark A term used by Haaf (1959, p.22) for ripple mark arranged in an orthogonal pattern either parallel or transverse to the current direction and having a wavelength up to 80 cm and amplitude up to 30 cm.

pericline twin law A parallel twin law in triclinic feldspars, in which the twin axis is the crystallographic *b* axis and the composition surface is a rhombic section. It occurs alone or with the albite twin law

pericoel The space between the periphragm and endophragm in a cavate dinoflagellate cyst. Cf: *endocoel.*

pericolpate Said of pollen grains having more than three colpi, not meridionally arranged.

pericolporate Said of pollen grains having more than three colpi, not meridionally arranged, with at least part of the colpi provided with pores or transverse furrows.

pericycle In roots and stems, a layer (or layers) of cells immediately outside the phloem and inside the endodermic, from which branch roots develop (Fuller & Tippo, 1949, p.967).

pericyst *frontal shield.*

perideltaic Adjacent to or surrounding a delta.

perideltidium One of a pair of slightly raised triangular parts of the interarea of a brachiopod, flanking the pseudodeltidium or lateral to it, and characterized by vertical striae in addition to horizontal growth lines parallel to the posterior margin (TIP, 1965, pt.H, p.150).

periderm [paleont] The protein substances that compose the rhabdosome of a graptolite, consisting of an inner layer (fusellar tissue) with growth increments and an outer layer (cortical tissue) of concentric laminated material.

periderm [bot] A collective name for *cork, cork cambium,* and *phelloderm.* It is a major portion of what is commonly called bark of woody plants.

periderre A thinner layer located outside the *ectoderre* in the wall of a chitinozoan. Cf: *endoderre.*

peridot (a) A transparent to translucent and pale-, clear-, or yellowish-green gem variety of *olivine.* Syn: *peridote.* (b) A yellowish-green or greenish-yellow variety of tourmaline, approaching olivine in color. It is used as a semiprecious stone yn: *peridot of Ceylon.*

peridotite A general term for a coarse-grained plutonic rock composed chiefly of olivine with or without other mafic minerals such as pyroxenes, amphiboles, or micas, and containing little or no feldspars. Peridotites encompass the more specific forms saxonite, harzburgite, lherzolite, wehrlite, dunite. Accessory minerals of the spinel group are commonly present. Peridotite is commonly altered to serpentinite.

peridotite shell A syn. of the outer or *upper mantle,* so named because its composition may be peridotitic. Cf: *pallasite shell.*

periembryonic chamber An immature (nepionic) part of a foraminiferal test formed on the ventral side and partly surrounding the proloculus (as in Orbitolinidae).

perigean tide A tide of increased range (e.g. a spring tide) occurring monthly when the Moon is at or near the perigee of its orbit. Ant: *apogean tide.*

perigee That point on the orbit of an Earth satellite, e.g. the Moon, which is nearest to the Earth. Cf: *apogee.*

perigenic Said of a rock constituent or mineral formed at the same time as the rock of which it constitutes a part "but not at the specific location in which it is now found in that rock" (Lewis, 1964, p.875); e.g. said of a glauconite grain formed from an agglutinated clay pellet and subjected to "short" transportation prior to its final incorporation into the sediment.

periglacial (a) Said of the processes, conditions, areas, climates, and topographic features at the immediate margins of former and existing glaciers and ice sheets, and influenced by the cold temperature of the ice. (b) By extension, said of an environment in which frost action is an important factor, or of phenomena induced by a periglacial climate beyond the periphery of the ice. Syn: *cryergic; cryonival; paraglacial; subnival.*---Term introduced by Lozinski (1909).

periglacial geomorphology "The study of all processes and phenomena found in cold regions" (Hamelin & Cook, 1967, p.11). See also: *cryomorphology.*

perignathic girdle A continuous or discontinuous ring of internal processes around the peristomial opening of an echinoid, serving for the attachment of muscles supporting and controlling Aristotle's lantern. See also: *girdle* [paleont].

perigynous Said of a plant in which the stamens and petals are borne on a ring of the receptacle surrounding a pistil and usually adnate to the calyx. Cf: *epigynous; hypogynous.*

perilith *cored bomb.*

perilumen A raised inner border of a columnal articulum of a crinoid, developed as a smooth-topped, granulose, tuberculate, or vermiculate ridge or field surrounding a lumen.

perimagmatic Said of a hydrothermal mineral deposit located near its magmatic source. Cf: *apomagmatic; telemagmatic; cryptomagmatic.*

perine *perisporium.*

perinium A sometimes present, more or less sculptured outer coat of a pollen grain; *perisporium.* Pl: *perinia.* Adj: *perinate.*

period [geochron] (a) A geologic-time unit longer than an *epoch* [geochron] and a subdivision of an *era,* during which the rocks of the corresponding *system* were formed. It is the fundamental unit of the standard geologic time scale. (b) A term used informally to designate a length of geologic time; e.g. *glacial period.*

period [phys] (a) The interval of time required for the completion of a cyclic motion or recurring event, such as the time between two consecutive like phases of the tide or a current; the reciprocal of the *frequency.* See also: *wave period.* (b) Any portion or specified duration of time characterized in some particular manner.

periodic current *tidal current.*

periodic spring A spring that ebbs and flows periodically, apparently due to natural siphon action. Such springs issue mainly from carbonate rocks, in which solution channels form the natural siphons. It is distinguished from a geyser by its temperature--that of ordinary ground water--and general lack of gas emission. Syn: *ebbing-and-flowing spring.* Cf: *perennial spring; intermittent spring.*

periostracum The thin organic layer covering the exterior or the shell of brachiopods and many mollusks, such as the thin coat of chitinous material covering the calcareous part of the shell of a bivalve mollusk, or the outer horny shell layer of a gastropod, composed dominantly of conchiolin. Cf: *ostracum.*

peripediment A term proposed by Howard (1942, p.11) for the segment of a pediplane (a planate erosion surface produced in an arid, piedmont region) extending across the younger rocks or alluvium of a basin which is always beyond but adjacent to the segment (termed *pediment* by Howard) developed on the older upland rocks.

peripheral counter In structural petrology, an instrument used to prepare density contours for the marginal areas of a fabric diagram. It consists of a strip of plastic or similar material with a circular hole in either end, the area of each of which is equivalent to 1.0% of the area of the total diagram. Cf: *center counter.*

peripheral depression *ring depression.*

peripheral faults Arcuate faults bounding an elevated or depressed area such as a diapir. Partial syn: border fault. Cf: *arcuate fault.*

peripheral increase A type of *increase* (budding of corallites) characterized by offsets that arise in marginarial or coenenchymal tissue.

peripheral moraine A term proposed by Chamberlin (1879, p.14) and now considered an obsolete syn. of *recessional moraine.*

peripheral sink *rim syncline.*

peripheral stream A stream that flows parallel with the edge of a glacier, usually just beyond the moraine (Todd, 1902, p.39).

periphery The part of a gastropod shell or any particular whorl that is farthest from or most lateral to the axis of coiling; the outer margin of a coiled foraminiferal test.

periphract A continuous band, composed of muscles and of fibrous tissues (aponeuroses) providing means of linear attachment to the muscles, that encircles the body of a nautiloid (TIP, 1964, pt K, p. 57). Adj: *periphractic.* Syn: *annulus* [ceph].

periphragm The outer layer of a dinoflagellate cyst, usually carrying extensions in the form of spines, and projecting to the position of former thecal wall. It may have served as a support during the period of cyst formation. Cf: *ectophragm; endophragm.*

periphyton Micro-organisms that coat rocks, plants, and other surfaces on the bottom of lakes. Cf: *aufwuchs.*

peripolar space A three-sided pyramidal space resulting from formation of basal leaf cross in an acantharian radiolarian. Cf: *perizonal space.*

periporate Said of pollen grains having many pores scattered over the surface.

periproct The membranous area surrounding the anal opening of an echinoderm, such as the space in the CD interray of an echinoid containing the anus and covered in life by skin in which small plates are embedded. Adj: *periproctal.*

perisome The *body wall* of an invertebrate, esp. of an echinoderm. Syn: *perisoma.*

perispore The covering of a spore external to the exine; *perisporium.*

perisporium An additional wall layer external to the *exine* in certain spores and pollen. It is composed of thin and loosely attached sporopollenin and is therefore not usually encountered in palynomorphs. Syn: *perine; perinium; perispore.*

peristerite A gem variety of albite with blue or bluish-white luster characterized by sharp internal reflections of blue, green, and yellow; an inhomogeneous, unmixed sodic plagioclase with a composition ranging between An_2 (albite) and An_{24} (calcic oligoclase). It resembles moonstone, and is falsely called *moonstone* by jewelers.

peristome The region around the mouth in various invertebrates; e.g. the membranous area surrounding the mouth of an echinoderm (such as the space containing the oral plates at the summit of theca of a blastoid), or the edge of the aperture of the body chamber of a cephalopod, or the frontal depression above the mouth of a tintinnid or the raised rim around the aperture of a foraminiferal test, or the elevated rim surrounding the primary orifice in a cheilostomatous bryozoan or the free terminal portion of the zooid in a cyclostomatous bryozoan, or the part of the oral disk surrounding the mouth of a coral polyp.

peristomial Pertaining to the peristome; e.g. "peristomial ovicell" formed as a dilatation of the peristome of a bryozoan.

peristomice The opening at the outer extremity of the peristomie in some cheilostomatous bryozoans (TIP, 1953, pt́.G, p.13).

peristomie Tube-like extension of the peristome, projecting outward from the operculum-bearing orifice in some cheilostomatous bryozoans (TIP, 1953, pt.G, p.13).

peritabular Said of the surface features of a dinoflagellate cyst that originate immediately interior to the margins of reflected plate areas (as in *Areoligera* and *Eisenackia*). Cf: *intratabular; nontabular.*

perite An orthorhombic mineral: $PbBiO_2Cl$.

peritectic point *reaction point.*

peritreme The edge or margin of the aperture of a shell; e.g. the peristome of a gastropod.

peritrichous Said of a bacterial cell having flagella uniformly distributed over all the surface. Cf: *lophotrichous; monotrichous.*

perizonal space A four-sided region resulting from formation of basal leaf cross in an acantharian radiolarian. Cf: *peripolar space.*

perizonium An outer, silicified membrane on a diatom frustule formed during auxospore development and from which the new hypovalve and epivalve are produced.

perknide An informal term for any holocrystalline igneous rock composed almost entirely of dark-colored minerals; e.g. amphibolite, pyroxenite, peridotite. Cf: *perknite.*

perknite Any of a group of igneous rocks containing as their main components clinopyroxene and amphibole, with accessory orthopyroxene, biotite, iron ore, and little or no feldspar. Included in this group are pyroxenite and hornblendite. The term is not commonly used, having been replaced by "ultramafic rocks" and "ultramafites". Cf: *perknide.*

perlite A volcanic glass having the composition of rhyolite, perlitic texture, and a generally higher water content than obsidian. Syn: *pearlite; pearlstone.*

perlitic (a) Said of the texture of a glassy igneous rock that has cracked due to contraction during cooling, the cracks forming small spheruloids. (b) Pertaining to or characteristic of *perlite.*

permafrost Any soil, subsoil, or other surficial deposit, or even bedrock, occurring in arctic or subarctic regions at a variable depth beneath the Earth's surface in which a temperature below freezing has existed continuously for a long time (from two years to tens of thousands of years). This definition is based exclusively on temperature, and disregards the texture, degree of compaction, water content, and lithologic character of the material. Its thickness ranges from over 1000 m in the north to 30 cm in the south; it underlies about one-fifth of the world's land area. Etymol: *perman*ent + *frost.* See also: *tjaele.* Cf: *pereletok.* Term introduced by Muller (1947) who included as synonyms the terms "frozen ground" or "frozen soil" preceded by any of the following modifiers: "constantly", "eternally", "ever", "perennially", "permanently", "perpetu-

ally", and "stable". Syn: *pergelisol.*

permafrost island A small, shallow, isolated patch of perm‍ frost surrounded by unfrozen ground; occurs on protecte‍ north-facing slopes in regions of *sporadic permafrost.*

permafrost line A line on a map representing the border of t‍ arctic permafrost.

permafrost table The upper limit of permafrost, represented‍ an irregular surface dependent on local factors. Cf: *fr‍ table; frost line.* Syn: *pergelisol table.*

permanence of continents An hypothesis, propounded as‍ virtual dogma by many geologists during the 19th and ea‍ part of the 20th centuries, that the continents (and by implic‍ tion the intervening ocean basins) have been fixed in th‍ present positions throughout geologic time. Even these geo‍ gists found it necessary, in order to explain at least some‍ the intercontinental resemblances, to resort to borderlanc‍ isthmian links, and other supposed land features, now four‍ ered beneath the oceans. The hypothesis was severely cha‍ lenged later in the 20th Century by opposing hypotheses, su‍ as *continental displacement* and *oceanization,* and is now c‍ of favor in anything like its original form.

permanent axis The axis of the greatest moment of inertia‍ a rigid body, about which it can rotate in equilibrium.

permanent bench mark A readily identifiable, relatively perm‍ nent, recoverable *bench mark* that is intended to maintain‍ elevation with reference to an adopted datum without chan‍ over a long period of time and located where disturbing inf‍ ences are believed to be negligible. Abbrev: P.B.M. Cf: *te‍ porary bench mark.*

permanent extinction The *extinction* of a lake by destruction‍ the lake basin, such as due to deposition of sediments, er‍ sion of the basin rim, filling with vegetation, or catastroph‍ events.

permanent hardness *noncarbonate hardness.*

permanent icefoot An icefoot that does not melt complet‍ during the summer.

permanently frozen ground *permafrost.*

permanent magnet A *magnet* having a large, hard remane‍ magnetization.

permanent set The amount of permanent deformation of‍ material that has been stressed beyond its elastic limit. Sy‍ *set.*

permanent stream A syn. of *perennial stream.* The ter‍ should be avoided because a stream is not permanent (u‍ changing) in course, volume, or velocity.

permanent water A source of water that remains consta‍ throughout the year.

permanent wilting A degree of wilting from which a plant ca‍ recover only by adding water to the soil. Cf: *wilting point; te‍ porary wilting.*

permeability [magnet] The ratio of magnetic induction B to ‍ ducing field strength H. Syn: *magnetic permeability.*

permeability [geol] The property or capacity of a porous roc‍ sediment, or soil for transmitting a fluid without impairment‍ the structure of the medium; it is a measure of the relati‍ ease of fluid flow under unequal pressure. The customary u‍ of measurement is the *millidarcy.* Syn: *perviousness.* Adj: *pe‍ meable.* Ant: *impermeability.*

permeability coefficient The rate of flow of water in gallo‍ per day through a cross section of one square foot under‍ unit hydraulic gradient, at the prevailing temperature (*fie‍ permeability coefficient*) or adjusted for a temperature‍ 60°F. (Stearns, 1927, p.148). Cf: *capillary conductivity.* Sy‍ *hydraulic conductivity; coefficient of permeability.*

permeability trap An oil trap formed by lateral variation with‍ a reservoir bed sealing the contained hydrocarbon through‍ change of permeability.

permeameter An instrument for measuring the permeability‍ materials to fluids (gases or liquids).

ermeation [met] The intimate penetration of country rock by netamorphic agents, such as granitizing solutions, particularly f an already metamorphosed rock so that it becomes completely recrystallized (Read, 1931).

ermeation [grd wat] Penetration by passing through the interstices, as of a rock or soil, without causing physical change.

ermeation gneiss A gneiss formed as a result of or modified y the passage of geochemically mobile materials through or nto solid rock (Dietrich & Mehnert, 1961).

ermesothyridid Said of a brachiopod pedicle opening when ne foramen is located mostly within the ventral umbo (TIP, 965, pt.H, p.150). Cf: *submesothyridid*.

Permian The last period of the Paleozoic era (after the Carboniferous), thought to have covered the span of time between 280 nd 225 million years ago; also, the corresponding system of ocks. The Permian is sometimes considered to be part of the arboniferous, and sometimes is divided between the Carboniferous and Triassic. It is named after the province of Perm, JSSR, where rocks of this age were first studied. See also: *ge of amphibians*. Syn: *Dyassic*.

ermineralization A process of fossilization whereby the original hard parts of an animal have additional mineral material eposited in their pore spaces.

ermingeatite A mineral: Cu_3SbSe_4.

ermissive Said of a magmatic intrusion, and of the magma self, the emplacement of which is in spaces created by other orces than its own, e.g. orogenic forces. Cf: *aggressive*. Syn: *uctive*.

ermo-Carboniferous (a) The entire Permian and Carboniferus, considered as a single unit. (b) The Permian and Pennylvanian periods combined. (c) An age, or corresponding ock unit, transitional between the uppermost Pennsylvanian nd lowermost Permian.

ermutation Any different ordered subset, or arrangement, of given set of objects. See also: *combination [stat]*.

erofskite *perovskite*.

eroikic In the C.I.P.W. (Cross, *et al*, 1906, p.704) textural lassification of igneous rocks, those rocks in which the ratio f oikocrysts to xenocrysts is greater than seven to one. This erm is rarely used. Cf: *domoikic*.

erovskite A yellow, brown, or grayish-black mineral: $CaTiO_3$. t sometimes has cerium and other rare-earth metals. Cf: *larappite*. Syn: *perofskite*.

erpatic In the C.I.P.W. (Cross, *et al*, 1906, p.701) textural lassification of igneous rocks, those rocks in which the ratio f groundmass to phenocrysts is greater than seven to one. his term is rarely used. Cf: *dopatic*.

erpendicular A very steep slope or precipitous face, as on a nountain.

erpendicular offset That component of *offset shotpoint* which s the distance at right angles to the spread line. Cf: *in-line ffset*.

erpendicular separation The separation of a fault as measured at right angles to the fault plane.

erpendicular slip The component of the slip of a fault that is neasured perpendicular to the trace of the fault on any intersecting surface (Dennis, 1967, p.138).

erpendicular throw In a faulted bed, vein, or other surface, ne distance between two formerly adjacent points, measured erpendicular to the surface.

erpetual frost climate A type of *polar climate* having a mean emperature in the warmest month of less than $0°C$. Cf: *tunra climate*.

erpetually frozen ground *permafrost*.

erradial Having a meridional position; e.g. a "perradial suture" situated at the midline between two columns of ambularum in an echinoid, or a "perradial position" precisely along ome one of the radii of a crinoid, or a "perradial plane" ocupying a meridional position in acantharian radiolaria.

Perret phase That stage of a volcanic eruption that is characterized by the emission of much high-energy gas that may significantly enlarge the volcanic conduit.

perrierite A mineral: $(Ca,Ce,Th)_4(Mg,Fe)_2(Ti,Fe)_3Si_4O_{22}$.

perryite A meteorite mineral: $(Ni,Fe)_5,(Si,P)_2$.

persalic One of five classes in the C.I.P.W. (Cross, *et al*, 1902, p.585) chemical-mineralogic classification of igneous rocks and in which the ratio of salic minerals to femic is greater than seven to one. Cf: *dosalic*.

persemic In the C.I.P.W. (Cross, *et al*, 1906, p.701) textural classification of igneous rocks, those rocks in which the ratio of groundmass to phenocrysts is less than one to seven. This term is rarely used. Cf: *dosemic*.

persilicic A term proposed by Clarke (1908, p.357) to replace "acidic". Syn: *silicic*. Cf: *subsilicic; mediosilicic*.

personal error An *error*, either random or systematic, caused by an observer's personal habits in making observations, by his mental or physical reactions, or by his inability to perceive or measure dimensional values exactly.

perspective The appearance to the eye of objects in respect to their relative distance and positions. Also, a picture (or other representation) in perspective.

perspective center The point of origin or termination of bundles of rays directed to a point object such as a photographic image.

perspective plane Any plane containing the perspective center; its intersection with the ground always appears as a straight line on an aerial photograph.

perspective projection A projection of points by straight lines drawn through them from some given point to an intersection with the plane of projection; e.g. a photograph is formed by a perspective projection of light rays from a point within the lens. The point of projection (unless otherwise indicated) is understood to be within a finite distance from the plane of projection. Examples include: stereographic projection, orthographic projection, and gnomonic projection. Syn: *geometric projection*.

perthite A variety of alkali feldspar consisting of parallel or subparallel intergrowths in which the potassium-rich phase (usually microcline) appears to be the host from which the sodium-rich phase (usually albite inclusions) exsolved. The exsolved areas are visible to the naked eye, and typically form small strings, lamellae, blebs, films, or irregular veinlets. Cf: *antiperthite*.

perthitic (a) Said of a texture produced by parallel or subparallel intergrowths of sodium-rich feldspar (usually albite) occurring as small strings or irregular veinlets in a host of potassium-rich feldspar (usually microcline). Cf: *perthitoid*. (b) Pertaining to or characteristic of perthite.

perthitoid Said of perthitic texture produced by minerals other than the feldspars.

perthophyte A plant living on a dead plant or on the decaying portions of a live plant.

perthosite A light-colored syenite composed almost entirely of perthite, with less than three percent mafic minerals.

Peru saltpeter Naturally occurring sodium nitrate; *soda niter* occurring in Peru. Cf: *saltpeter*. Syn: *Peruvian saltpeter*.

pervalvar axis The axis connecting the midpoints of the two valves in a diatom frustule. Cf: *apical axis; transapical axis*.

perviousness *permeability [geol]*.

perxenic In the C.I.P.W. (Cross, *et al*, 1906, p.704) textural classification of igneous rocks, those rocks in which the ratio of oikocrysts to phenocrysts is less than one to seven. This term is rarely used. Cf: *doxenic*.

petal [paleont] An expanded, differentiated, petal-shaped segment of ambulacrum situated toward the apical system of an echinoid, characterized by tube feet more or less specialized for respiration and by typically unequal or enlarged pore pairs.

petal [bot] A member of the second set of floral leaves, i.e.

the set just internal to the sepals (Cronquist, 1961, p.879).

petalite A white, gray, or colorless monoclinic mineral: Li-AlSi$_4$O$_{10}$.

Petersen scale A system of measurement of the visible effects of wind on the seas, devised by Captain Petersen, a German sailing-ship master.

Peterson grab A type of *grab sampler* that encloses the ocean-bottom material in two semicylindrical buckets that rotate shut on a hinge when the sampler strikes the bottom.

petiole The stalk of a leaf.

Petoskey stone The state "rock" of Michigan (from Petoskey, Mich.): a Devonian colonial coral. Syn: *Petoskey agate.*

petra A term introduced by Swain (1958, p.2876) for "the rock materials produced in specific sedimentary organic environments"; a petrologic type. Pl: *petrai.* Etymol: Greek, a small mass of naturally occurring rock.

petralogy A term introduced by Pinkerton (1811) for what is known today as *petrology.*

petricole *petrocole.*

petricolous *endolithic.*

petrifaction A process of fossilization whereby organic matter is converted into a stony substance by the infiltration of water containing dissolved inorganic matter (e.g. calcium carbonate, silica) which replaces the original organic materials, sometimes retaining the original structure. Syn: *petrification.*

petrified moss A moss-like coating of *tufa* deposited on growing plants.

petrified rose *barite rosette.*

petrified wood *silicified wood.*

petroblastesis Formation of rocks chiefly as the result of crystallization of diffusing ions (Dietrich & Mehnert, 1961).

petrocalcic Said of a soil horizon that is characterized by an induration of calcium carbonate, sometimes with magnesium carbonate (SSSA, 1970).

petrochemistry The study of the chemical composition of rocks; it is an aspect of geochemistry, and is not equivalent to petroleum chemistry.

petroclastic rock *detrital rock.*

petrocole An organism that lives in rocky areas. Also spelled: *petricole.*

petrofabric *fabric* [struc petrol].

petrofabric analysis An equivalent term for *structural petrology,* used by Knopf and Ingerson (1938, p.13).

petrofabric diagram *fabric diagram.*

petrofabrics *structural petrology.*

petrofacies *petrographic facies.*

petrogenesis A branch of petrology that deals with the origin and formation of rocks, esp. igneous rocks. Cf: *lithogenesis.* Adj: *petrogenetic; petrogenic.* Syn: *petrogeny.*

petrogenetic grid A diagram whose coordinates are intensive parameters characterizing the rock-forming environment (e.g. pressure, temperature) on which are plotted equilibrium curves delimiting the stability fields of specific minerals and mineral assemblages.

petrogenic element An element that occurs mainly as an oxide, silicate, fluoride, or chloride, and is therefore a characteristic occurrence in ordinary rocks. It is one of H.S. Washington's two major groupings of elements of the lithosphere, now obsolete. Cf: *metallogenic element.*

petrogeny *petrogenesis.*

petrogeny's residua system The system NaAlSiO$_4$ - KAlSiO$_4$ - SiO$_2$, which represents a close approximation to the composition of many residual liquids from magmatic differentiation. The term was named by N.L. Bowen.

petrogeometry A syn. of *structural petrology,* according to Tomkeieff (1943, p.347).

petrographer One who does *petrography.*

petrographic Adj. of *petrography.*

petrographic facies Facies distinguished primarily on the basis

of appearance or composition without respect to their for■ boundaries, or mutual relations (Weller, 1958, p.627). The consist of actual large bodies of rock occurring in certa■ areas and in more or less restricted parts of the stratigraph■ section (e.g. "red-bed facies", "paralic facies", "geosynclin■ facies", "evaporite facies"), or they may consist of all rock of a single kind (e.g. "black-shale facies", "graywacke f■ cies"). See also: *facies* [*stratig*]. Cf: *stratigraphic facies.* Sy■ *petrofacies.*

petrographic microscope *polarizing microscope.*

petrographic period The extension in time of a rock associa■ tion.

petrographic province A broad area in which similar igneo■ rocks are considered to have been formed during the sam period of igneous activity. Syn: *magma province; comagma■ region; igneous province.*

petrography That branch of geology dealing with the descri■ tion and systematic classification of rocks, esp. igneous ar■ metamorphic rocks and esp. by means of microscopic exam■ nation of thin sections. Petrography is more restricted ■ scope than *petrology,* which is concerned with the origin, o■ currence, structure, and history of rocks. Adj: *petrographi■* Cf: *lithology.* See also: *petrographer.*

petroleum A naturally-occurring complex liquid hydrocarbo■ that may contain varying degrees of impurities (sulfur, nitro■ gen) which after distillation yields a range of combustib■ fuels, petrochemicals, and lubricants. Syn: *crude petroleum oil.*

petroleum coke A cokelike substance found in cavities of i■ neous intrusions into carbonaceous sediments (Tomkeie■ 1954).

petroleum geologist A geologist engaged in the exploration o production processes of hydrocarbon fuels. See also: *petro■ leum geology.*

petroleum geology That branch of economic geology which re■ lates to the origin, occurrence, migration, accumulation, an■ exploration for hydrocarbon fuels. Its practice involves the ap■ plication of geochemistry, geophysics, paleontology, structura■ geology, and stratigraphy to the problems of finding hydroca■ bons. See also: *petroleum geologist.*

petroleum seep *oil seep.*

petroleum series A complex hydrocarbon series that ma■ occur in liquid, gaseous or solid form. The gases usually ex■ clude methane when it occurs as marsh gas.

petrologen *kerogen.*

petrologic Adj. of *petrology.*

petrologic province *sedimentary petrologic province.*

petrologist One who does *petrology.*

petrology That branch of geology dealing with the origin, oc■ currence, structure, and history of rocks, esp. igneous an■ metamorphic rocks. Petrology is broader in scope than *petro■ graphy,* which is concerned with the description and classifi■ cation of rocks. Adj: *petrologic.* Cf: *lithology.* See also: *petro■ logist; petralogy.*

petromictic Said of a sedimentary deposit characterized by a■ assortment of metastable rock fragments; e.g. a "petromicti■ conglomerate" containing a mixture of pebbles or cobbles o■ plutonic, eruptive, sedimentary, and/or metamorphic rocks McElroy (1954, p.151) proposed the term "petromictic sand■ stone" to replace "greywacke" for certain Permian and Tri■ assic sedimentary rocks of New South Wales, having the gen■ eral granular composition of classical graywackes but which are light-colored, well-sorted, and mildly indurated, with a ma■ trix that may consist of an introduced mineral cement. Syn■ *petromict.*

petromorph A speleothem or cave formation that is exposed to the surface by erosion of the limestone in which the cav■ was formed.

petromorphology A syn. of *structural petrology,* according t■

◦mkeieff (1943, p.347).

◦trophilous *epilithic.*

◦trophysics Study of the physical properties of reservoir cks.

◦trosilex [petrology] (a) An old name for an extremely fine ystalline porphyry or quartz porphyry, and for the ground-ass of such porphyries; also, a finely crystalline aggregate w known to be devitrified glass. "It was practically a con-ssion by the older petrographers that they did not know of at the rock consisted" (Kemp, 1896, p.156). (b) A term plied by Lyell (1839, p.99) to igneous rocks. (c) An obso-te term formerly applied to clinkstone, to fusible hornstone, felsite, and to compact feldspar.

◦trosilex [sed] (a) An obsolete term applied to *flint* occurring the form of a rock mass, or as part of the rock, as distin-ished from detached nodular flint in chalk beds; also, a ck partly converted to flint. (b) A French term for *horn-one,* or flinty slate without slaty cleavage. (c) *amausite.*

◦trostratigraphy *lithostratigraphy.*

◦trotectonics *Structural petrology,* including or extending to alysis of the movements that produced the rock's fabric. n: *tectonic analysis.*

◦trous Said of a material that resembles stone in its hard-ss; e.g. petrous phosphates.

◦tschau twin law A rare type of parallel twin law in feldspar, aving a twin axis of [$\bar{1}$10].

◦ttersson theory An astronomical theory of climatic change, oposed by Sven Otto Pettersson (1848-1941), Swedish eanographer, in which climatic changes are related to tidal-rce cycles produced by systematic orbital variations of Earth d Moon about the Sun. Karlstrom (1961) integrates the ilankovitch and Pettersson theories as a possible explana-n for past glacial oscillations of both stage and lesser-rank vents.

◦tuntse *china stone.*

◦tunzyte *china stone.*

◦tzite A steel-gray to iron-black mineral: Ag_3AuTe_2.

◦zograph *regmaglypt.*

◦alzian orogeny One of the 30 or more short-lived orogenies uring Phanerozoic time identified by Stille, in this case at the d of the Permian. Syn: *Palatinian orogeny.*

◦nacellite *kaliophilite.*

◦haceloid Said of a fasciculate corallum having subparallel orallites.

◦hacoidal structure An infrequently used term for a lenticular etamorphic structure, e.g. *flaser structure, augen structure.*

◦hacolite [mineral] A variety of chabazite, characterized by olorless lenticular crystals.

◦nacolite [intrus rocks] Var. of *phacolith.*

◦hacolith A minor, concordant, doubly convex, and usually ranitic intrusion into folded strata. Var: *phacolite* [intrus cks].

◦haeodarian Any radiolarian belonging to the suborder Phae-darina, characterized mainly by a central capsule enclosed y a double-walled membrane.

◦hagotrophic Said of an organism that is nourished by the in-estion of solid matter.

◦haneric Said of the texture of a carbonate sedimentary rock esp. limestone) characterized by individual crystals or clastic ains whose diameters are greater than 0.01 mm (Bissell & hilingar, 1967, p.163). The term was proposed by DeFord 946) who used 0.004 mm as the limiting diameter. Cf: ohanic.

◦hanerite An igneous rock having *phaneritic* texture. Obs. syn: haneromere.

◦haneritic Said of the texture of an igneous rock in which the dividual components are distinguishable megascopically. lso, said of a rock having such texture (i.e. *phanerite*). Ant:

aphanitic. Syn: *phanerocrystalline; phenocrystalline; coarse-grained.*

phanerocryst *phenocryst.*

phanerocrystalline *phaneritic.*

phanerocrystalline-adiagnostic A nonpreferred syn. of *crypto-crystalline.*

phanerogam (a) A seed-bearing plant, as opposed to a spore-bearing plant, or *cryptogam.* Cf: *spermatophyte.* (b) A flower-ing plant in which the stamens and pistils are distinctly devel-oped (Jackson, 1953, p.279).

phanerogenic Of known origin; e.g. said of a species that is proved to descend from a known species found in an older geologic formation. Ant: *cryptogenic.*

phaneromere *phanerite.*

phaneromphalous Said of a gastropod shell with a completely open umbilicus. Cf: *anomphalous.*

phanerophyte A perennial plant whose overwintering buds are above the ground surface.

Phanerozoic That part of geologic time for which, in the corre-sponding rocks, the evidence of life is abundant, esp. of high-er forms. Cf: *Cryptozoic.*

phanoclastic rock An "even-grained or uniformly sized" clastic rock (Pettijohn, 1949, p.30).

phantom A bed or member that is missing from a given stra-tigraphic section although it elsewhere occupies a character-istic position in a sequence of similar age.

phantom bottom *deep scattering layer.*

phantom crystal A crystal (such as of quartz, calcite, or fluo-rite) within which an earlier stage of crystallization or growth is outlined by dust, tiny inclusions, or bubbles, e.g. serpentine containing a ghost or phantom of original olivine. Syn: *ghost crystal.*

phantom horizon In seismic reflection prospecting, a line so constructed that it is parallel to the nearest actual dip seg-ment everywhere along a profile.

pharetrone Any calcisponge having spicules that have the shape of a tuning fork.

pharmacolite A monoclinic mineral: $CaH(AsO_4).2H_2O$. It oc-curs in white or grayish silky fibers. Syn: *arsenic bloom.*

pharmacosiderite A green or yellowish-green mineral: $Fe_3-(AsO_4)_2(OH)_3.5H_2O$. It commonly occurs in cubic crystals. Syn: *cube ore.*

pharynx A differentiated part of the alimentary canal in many invertebrates; e.g. the tubular passageway between the mouth and gastrovascular cavity of an octocoral, or the internal oral tube in a phaeodarian radiolarian.

phase [chem] A homogeneous, physically distinct portion of matter in a nonhomogeneous, aqueous vapor.

phase [geochron] (a) A term approved by the 8th Internation-al Geologic Congress in Paris in 1900 for the geologic-time unit next in order of magnitude below *age,* during which the rocks of a *substage* (then referred to as a "zone") were formed; a *subage.* The term was seldom used and is now ob-solete in this usage. Syn: *episode (b); time (d).* (b) *moment* [stratig].

phase [glac geol] An informal subdivision of a *stage.* The term has been used by Flint (1957) for the deposits of various gla-cial Great Lakes at different levels and dates.

phase [seism] The onset of a displacement or oscillation on a seismogram indicating the arrival of a different type of seismic wave or of waves travelling along a different path. See also: *Airy phase; T-phase.*

phase [phys] A point or stage in the period to which any type of periodic motion (rotation, oscillation, etc.) has advanced with respect to a given initial point; expressed as an angular measure.

phase [ign] An interval in the development of a given process; esp. a chapter in the history of the igneous activity of a re-gion, such as the "volcanic phase" and major and minor "intrusive phases".

phase [sed] (a) A product of "deposition during a single fluctuation in the competency of the transporting agent" (Apfel, 1938). Such a subunit is probably a *lamina*. (b) A transitory or minor chance fluctuation in the velocity of a depositing current, resulting in the formation of a lamina.

phase [stratig] (a) A lithologic facies, esp. on a small scale, such as a minor variety within a dominant or normal facies, or a facies of short duration or local occurrence; e.g. the "marine phase" and the "fluviatile phase" of the Pocono Formation (Barrell, 1913, p.465). The term was used by McKee (1938, p.13-14) for a lateral subdivision (or facies) of a formation. (b) A term defined by Fenton & Fenton (1930, p.150) as "a local or regional aspect or condition of a stratum or group of strata, as determined both by original nature and secondary change; the latter being the determining factor"; e.g. a change arising from faulting, folding, secondary dolomitization, or erosion. (c) Characteristic strata repeated at various positions in a stratigraphic section that record the recurrence of a particular kind of environment (Allan, 1948, p.8). (d) A term that has been applied variously in stratigraphy to (see Weller, 1958, p.620): a part of a depositional cycle; a part of a cyclothem; sediments consisting of identical constituents deposited under relatively uniform conditions; a time-stratigraphic division; a type of lithologic development; a sedimentary province; and a major biotope.---Because the term has been used widely and vaguely in stratigraphy and because its technical application is not needed, "this word may well be left without restriction as to special meaning" (Moore, 1957a, p.1787).

phase [tect] The time of structural events in the development of a *system* [tect], usually named according to its geologic age, e.g. "the early Pennsylvanian phase of the Ancestral Rockies system" (Eardley, 1962, p. 10). Cf: *disturbance; pulsation* [tect]; *event*.

phase [surv] The apparent displacement of a surveying object or signal caused by one side being more strongly illuminated than the other and resulting in an error in sighting.

phase boundary *boundary line*.

phase diagram A graph designed to show the boundaries of the fields of stability of the various phases of a system. The coordinates are usually two or more of the intensive variables temperature, pressure, and compositions, but are not restricted to these. Syn: *equilibrium diagram*.

phase equilibria In physical chemistry, the study of those phases which, under specified conditions, may exist in equilibrium.

phase lag *tidal epoch*.

phase plate In a polarizing microscope, a plate of doubly refracting material, e.g. mica of a quarter-wave plate, that changes the relative phase of the polarized light's components.

phase response In seismology, the shift of phase with frequency, illustrated graphically. The amplitude-frequency response of a filter to the shape of pulses put through it will be different for different phase characteristics, leading to different phase distortion (Sheriff, 1968). Syn: *phase spectrum*.

phase rule The statement that for any system in equilibrium, the number of degrees of freedom is two greater than the difference between the number of components and the number of phases. It may be symbolically stated as $F = C-P+2$. Syn: *Gibbs phase rule*.

phase spectrum *phase response*.

phase velocity The velocity of advance with which an observable, individual wave or wave crest is propagated through a medium; the velocity of a point of constant phase. It is the product of wavelength and frequency. Symbol: c. Cf: *group velocity; particle velocity*.

phase-velocity method A procedure to determine phase velocity as a function of period, which used the differences in travel time, dt, of a particular seismic phase of period T at two stations separated by a distance $d\Delta$. Thus, $c(T) = d\Delta/d$ where c is the phase velocity.

phassachate A lead-colored agate.

phelloderm Secondary tissue of woody plant stems produce by the cork cambium on its inner surface as opposed to th cork, which is produced by the cork cambium on its oute surface (Cronquist, 1961, p.879). See also: *periderm*.

phellogen *cork cambium*.

phenacite *phenakite*.

phenakite A colorless, white, or very pale wine-yellow, pin blue, or brown glassy rhombohedral mineral: Be_2SiO_4. It sometimes confused with quartz. Phenakite is used as minor gemstone. Not to be confused with *fenaksite*. Sy *phenacite*.

phenetic system A classification system for living organism that is based on morphologic, anatomic, physiologic, or bic chemical criteria and does not reflect phylogeny, because th evolutionary history is unknown.

phengite (a) A variety of muscovite with a high silica conten (b) A transparent or translucent stone (probably crystallize gypsum) used by the ancients for windows.

phenhydrous Said of vegetal matter deposited under wate Cf: *cryptohydrous*.

phenicochroite *phoenicochroite*.

phenoclast One of the larger and more conspicuous frag ments in a sediment or sedimentary rock composed of var ous sizes of material, such as a cobble or pebble (*sphere clast*) embedded in a fine-textured matrix of a conglomerate or a fragment (*anguclast*) of a breccia.

phenoclastic rock A "nonuniformly sized" clastic rock contai ing phenoclasts (Pettijohn, 1949, p.30).

phenocryst A term suggested by J.P. Iddings, and wide used, for a relatively large, conspicuous crystal in a porphyr tic rock. The term *inset* has been suggested as a replace ment.

phenocrystalline *phaneritic*.

phenogenesis The development of the phenotype.

phenomenal gem A gemstone exhibiting an optical phenome non, such as asterism, chatoyancy, or play of color.

phenomenology The science that treats of the description an classification of phenomena.

phenoplast A large rock fragment (in a rudaceous rock) tha was plastic at the time of its incorporation in the matrix.

phenotype The visible characters of an organism that refle the interaction of genotype and environment.

phi Particle-size diameter, expressed as the negative loga rithm to the base 2 of the diameter in millimeters. Each intege of the phi grade scale is equal to a class limit of the Went worth grade scale. The negative values of phi are the value coarser than one millimeter, and as the phi unit increases th particle size decreases. Symbol: ϕ. Syn: *phi unit*.

phialine Said of an everted apertural rim (as on the neck of vial or bottle) of some foraminiferal tests.

phi deviation measure A graphic measure of dispersion of par ticle size shown on a plot of phi units.

phi grade scale A logarithmic transformation of the *Wentwort grade scale* in which the negative logarithm to the base 2 c the particle diameter (in millimeters) is substituted for the d ameter value (Krumbein, 1934); it has integers for the clas limits, increasing from -5 for 32 mm to +10 for 1/1024 mm The scale was developed specifically as a statistical device t permit the direct application of conventional statistical practi ces to sedimentary data. Syn: *phi scale*.

philippinite A tektite from Philippines. See also: *rizalite*.

philipstadite A monoclinic mineral of the amphibole group, ap proximately: $Ca_2(Fe,Mg)_5(Si,Al)_8O_{22}(OH)_2$.

phillipsite A white or reddish zeolite mineral: $(K_2,Na_2$

a) $Al_2Si_4O_{12}.4.5H_2O$. It sometimes contains no sodium, but ways contains considerable potassium. It commonly occurs complex (often cruciform) fibrous crystals, and makes up n appreciable part of the red-clay sediments in the Pacific cean.

hi mean diameter A *logarithmic mean diameter* obtained by sing the negative logarithms of the class midpoints to the ase 2. See also: *geometric mean diameter*.

hi scale *phi grade scale*.

hi unit *phi*.

hlebite Metamorphite or migmatite with roughly banded or eined appearance (Dietrich & Mehnert, 1961). Originally pro osed, without genetic connotation, to replace the term eined gneiss (Mehnert, 1968, p.17). Cf: *venite; arterite; omposite gneiss; injection gneiss*.

hleger corer A type of *gravity corer* that has a check valve r core catcher to retain the sample.

hloem The food-conducting tissue of vascular plants, con sting of various cell types such as sieve tubes, phloem arenchyma, fibers, companion cells, etc. Syn: *sieve tissue*.

hlogopite A magnesium-rich mineral of the mica group: $(Mg,Fe)_3AlSi_3O_{10}(OH,F)_2$. It is usually yellowish brown to rownish red or copper color, and usually occurs in crystalline mestones as a result of dedolomitization. Phlogopite is near iotite in composition, but contains little iron. Syn: *magnesia 1ica; amber mica; brown mica*.

hobotaxis Taxis in which an organism avoids a concentration r intensity of something. Cf: *strophotaxis; thigmotaxis*.

hoenicochroite A red mineral: Pb_2CrO_5. Syn: *phenicochroite; hoenicite; berezovite*.

holad Any bivalve mollusk belonging to the family Pholad lae, characterized by an equivalve shell of variable size, ommonly gaping open at the posterior end.

holerite A discredited name for a clay mineral identical with aolinite or one of the other kaolin minerals.

honolite (a) In the strictest sense, a group of fine-grained xtrusive rocks primarily composed of alkali feldspar (esp. odic orthoclase or sanidine), and nepheline as the main feld pathoid (Streckeisen, 1967, p.185); also, any rock in that roup; the extrusive equivalent of nepheline syenite. (b) In the roadest sense, any extrusive rock composed of alkali feld par, mafic minerals and any feldspathoid, such as nepheline, eucite, sodalite, etc. Syn: *clinkstone; klingstein; echodolite*.-- Streckeisen (1967, p.186) suggests that "phonolite" be pre eded by the name of the main feldspathoid mineral (e.g. leucite phonolite", "analcime phonolite", etc.). Etymol: Greek hone, "sound", in reference to the characteristic ringing ound emitted by a phonolite when struck with a hammer.

horogenesis The shifting or slipping of the Earth's crust rela ive to the mantle. See also: *drag-and-slippage zone*.

hosgenite A white, yellow, or grayish tetragonal mineral with damantine luster: $Pb_2Cl_2(CO_3)$. Syn: *horn lead; cromfordite*.

hosphate A mineral compound characterized by a tetrahedral onic group of phosphate and oxygen, PO_4^{-3}. An example is yromorphite, $Pb_5Cl(PO_4)_3$. Phosphorous, arsenic, and vana ium may substitute for each other in the tetrahedron. Cf: *ar senate; vanadate*.

hosphate rock *phosphorite*.

hosphatic Pertaining to or containing phosphates or phospho ic acid; said esp. of a sedimentary rock containing phosphate minerals, such as a "phosphatic limestone" produced by sec ndary enrichment of phosphatic material, or a "phosphatic hale" representing mixtures of primary or secondary phos hate and clay minerals.

hosphatic nodule A black (sometimes gray or brown), round d to irregular, earthy mass or "pebble" of variable size (di meter ranging from a few millimeters to more than 30 cm), having a hard shiny surface, often consisting of copro ites, corals, shells, bones, sand grains, mica flakes, or sponge spicules more or less enveloped in crusts of calcium phosphate, and occurring in marine strata (as in Permian beds of western U.S. or in the Cretaceous chalk of England) or forming presently along the sea floor (as off the coast of California). See also: *coprolite*.

phosphatite A sedimentary phosphatic rock composed of the mineral apatite in its various forms; e.g. phosphate rock.

phosphatization Conversion to a phosphate or phosphates; e.g. the diagenetic replacement of limestone, mudstone, or shale by phosphate-bearing solutions, producing phosphates of calcium, aluminum, or iron. Cf: *phosphorization*.

phosphide A mineral compound that is a combination of phos phorus with a metal. An example is schreibersite, $(Fe,Ni)_3P$.

phosphochalcite *pseudomalachite*.

phosphoferrite A white, pale-green, or yellow orthorhombic mineral: $(Fe,Mn)_3(PO_4)_2.3H_2O$. It is isomorphous with red dingite.

phosphophyllite A colorless or pale blue-green monoclinic mineral with perfect micaceous cleavage: $Zn_2(Fe,Mn)$-$(PO_4)_2.4H_2O$.

phosphorescence A type of *luminescence* in which the stimu lated substance continues to emit light after the external stim ulus has ceased; also, the light so produced. The duration of the emission is strongly temperature-dependent, and has a characteristic rate of decay. Cf: *fluorescence*.

phosphorite A sedimentary rock composed principally of phos phate minerals. Most commonly it is a bedded, marine rock composed of microcrystalline carbonate fluorapatite in the form of laminae, pellets, oolites, nodules, and skeletal and shell fragments. Aluminum and iron phosphate minerals are usually of secondary formation. Guano phosphorite is of com plex phosphate mineral composition. The term has also been applied to a sedimentary rock composed only of apatite and to an igneous rock containing appreciable apatite. See also: *brown rock; bone phosphate; pebble phosphate*. Syn: *phos phate rock; rock phosphate*.

phosphorization Impregnation or combination with phosphorus or a compound of phosphorus; e.g. the diagenetic process of *phosphatization*.

phosphorochalcite *pseudomalachite*.

phosphorroesslerite A monoclinic mineral: $MgH(PO_4).7H_2O$. It is isomorphous with roesslerite. Also spelled: *phosphorros slerite*.

phosphosiderite A pinkish-red monoclinic mineral: Fe-$PO_4.2H_2O$. It is dimorphous with strengite and isomorphous with metavariscite. Syn: *metastrengite*.

phosphuranylite A deep- to golden-yellow secondary mineral: $Ca(UO_2)_4(PO_4)_2(OH)_4.7H_2O$. It exhibits phosphorescence upon exposure to radium emanations.

photic zone *euphotic zone*.

photoalidade A photogrammetric instrument having a tele scopic alidade, a plateholder, and a hinged ruling arm, mount ed on a tripod frame, and used for plotting lines of direction and measuring vertical angles to selected features appearing on oblique and terrestrial photographs.

photoautotrophic *holophytic*.

pnotobase The length of the *air base* as represented on a photograph.

photoclinometer A device containing a sensitized graduated paper disk and a light source, used to determine the deflec tion from the vertical of a well bore. It hangs freely on a light cable, and upon being lowered to a determined depth, the energy source is activated to register a "burn" on the sensi tized disk.

photoclinometer log A *directional log* obtained by the use of a photoclinometer.

photoclinometry A technique for ascertaining slope informa tion from an image brightness distribution; it is used esp. for studying the amount of slope to a lunar crater wall or ridge by measuring the density of its shadow.

photoconductive Pertaining to a class of *radiation detector* that is based on a change of conductivity.

photoelasticity The property of a transparent, isotropic solid to become doubly refracting under tensile or compressive stress. This makes it possible to study stress-distribution patterns under the polariscope.

photogeologic anomaly Any systematic deviation of drainage, topography, vegetation, cultivation, photographic tone, or other photogeologic factor from the expectable norm in a given area.

photogeologic guide Any photographic element that assists in the interpretation of the geology of a given area.

photogeologic map A compilation of interpretations of a series of aerial photographs, including annotations of geologic features.

photogeology The identification, recording, and study of geologic features and structures by means of photography; specif. the geologic interpretation of aerial photographs and the presentation of the information so obtained. It now includes the interpretation of second-generation photographs obtained by photographing images recorded on television-type tubes (the images recording wavelengths outside the visible spectrum). See also: *aerogeology.*

photogeomorphology The study of landforms by means of aerial photographs.

photogoniometer A goniometer for measuring angles from the true perspective center to any point on a photograph.

photogrammetry The science or art of obtaining reliable measurements by means of photographs; specif. mapmaking and surveying with the aid of aerial and terrestrial photographs.

photograph A general term for a positive or negative picture obtained by photography. See also: *aerial photograph; oblique photograph; vertical photograph.*

photograph center The center of a photograph as indicated by the images of the fiducial marks of the camera. For a perfectly adjusted camera, it is identical to the *principal point* of the photograph.

photographic interpretation *photointerpretation.*

photograph nadir The point at which a vertical line through the perspective center of the camera lens pierces the plane of the photograph. Syn: *nadir point.*

photography The art or process of producing a positive or negative, and permanent or ephemeral, image of a subject directly or indirectly on a sensitized medium that is exposed to light or other form of radiant energy emitted or reflected from, or transmitted through, the subject.

photohydrology The science involving extraction of hydrologic data from aerial photographs.

photoindex A mosaic made by assembling individual photographs (with accompanying designations) into their proper relative positions and copying the assembly photographically at a reduced scale. It is not an *index map.* Also spelled: *photo index.*

photointerpretation The science of identifying and describing objects imaged on a photograph, such as deducing the topographic significance or the geologic structure of landforms on an aerial photograph. Also spelled: *photo interpretation.* Syn: *photographic interpretation.*

photomap An aerial photograph or a controlled mosaic of rectified photographs to which have been added a reference grid, scale, place names, marginal information, and other pertinent data or map symbols; e.g. an *orthophotomap.*

photometer An instrument for measuring the intensity of light. See also: *spectrophotometer; flame photometer.*

photometry (a) Study of ways and means to measure the intensity of light. (b) The art or process of using a *photometer* to measure the intensity of light.

photomicrograph A photographic enlargement of a microscop-

ic image such as a petrologic thin section; a type of *micr graph.* See also: *photomicrography.* Less-preferred syn: *micr photograph.*

photomicrography The preparation of *photomicrographs,* pe formed by the projection of the image through the eyepiece the microscope onto the photographic recording medium.

photomosaic *mosaic* [photo].

photoperiod The relative number of alternating daylight a dark hours in a 24-hour period. The photoperiod has a signi cant effect on the development of certain organisms.

photoreceptor A sense organ that undergoes specific stimul tion when exposed to light, such as (perhaps) a paraflagell body in some Euglenophyta.

photorelief map (a) A map consisting of a photograph of a r lief model of the area under study and showing salient phys cal features. (b) A diagrammatic map that simulates or give the impression of a photograph of a relief model of the are under study.---Also spelled: *photo-relief map.*

photostratigraphy A procedure of systematic field photograp in which photograph stations are so arranged that increasi detail is developed when the photographs are viewed in co secutive order (such as a distant view, a near view, on t outcrop, a close-up, and a photomicrograph).

phototaxis Taxis resulting from stimulation by light.

phototheodolite A ground-surveying instrument used in terre trial photogrammetry, combining the functions of a theodoli and a camera mounted on the same tripod.

phototopography The science of mapping and surveying which the detail is plotted entirely from photographs taken suitable ground stations; terrestrial photogrammetry.

phototriangulation The *extension* of horizontal and/or vertic control points by photogrammetric methods, whereby "th measurements of angles and/or distances on overlappir photographs are related into a spatial solution using the pe spective principles of the photographs" (ASP, 1966, p.1148 esp. *aerotriangulation.*

phototrophic *holophytic.*

photovoltaic Pertaining to a class of *radiation detector* th functions on a voltage change.

phragmites peat Peat that is derived mainly from the ree genus *Phragmites.*

phragmocone The part of a cephalopod conch consisting the camerae; e.g. the thin, conical, chambered, straight curved, internal shell of a belemnite, produced in front into very thin process resembling a blade or leaf, and fitted behir into a deep cavity in the anterior end of the guard.

phragmospore A plant spore having two or more septa; e.g. septate *fungal spore* that may have a chitinous wall and there fore be preserved as a palynomorph.

phreatic Said of a volcanic eruption or explosion of stean mud, or other material that is not incandescent; it is cause by the heating and consequent expansion of ground water du to an underlying igneous heat source. Cf: *phreatomagmatic.*

phreatic cycle The period of time during which the water tabl rises and then falls. It may be a daily, annual, or other cycl Syn: *cycle of fluctuation.*

phreatic gas A gas formed by the contact of atmospheric surface water with ascending magma. Cf: *juvenile* [water]; r surgent gas.

phreatic ground water *phreatic water.*

phreatic line *seepage line.*

phreaticolous Said of an organism or of the fauna inhabitir the interstices of mixtures of sand and gravel.

phreatic solution The solution action by ground water belo the water table. Cf: *vadose solution.*

phreatic surface *water table.*

phreatic water A term that originally was applied only to wate that occurs in the upper part of the zone of saturation unde water-table conditions (syn. of *unconfined ground water,* c ell water), but has come to be applied to all water in the ne of saturation, thus making it an exact syn. of *ground ater* (Meinzer, 1923, p.5). In 1939 Meinzer used "phreatic ater" (in the sense of "unconfined ground water") as a class plerotic water. Syn: *phreatic ground water.*

reatic-water discharge *ground-water discharge.*

reatic wave *ground-water wave.*

reatic zone *zone of saturation.*

reatomagmatic Said of a volcanic explosion that extrudes th magmatic gases and steam; it is caused by the contact the magma with ground water or with ocean water. Cf: *reatic.*

reatophyte A plant that obtains its water supply from the ne of saturation or through the capillary fringe and is char-cterized by a deep root system.

renotheca One of the thin, dense, diaphragm-like partitions at extend across chambers of a foraminiferal test at various ngles and in various parts of chamber (as in *Pseudofusuli-*a).

nthanite A term applied to dull cryptocrystalline siliceous cks, such as silicified shale, siliceous schist, Lydian stone asanite), and esp. *chert.* The term should be abandoned arr, 1938, p.25). Variant spellings: *phtanite; ptanite.*

hycobiont The algal partner or component of a lichen. Cf: ycobiont.

hyla The plural of *phylum.*

hyletic *phylogenetic.*

hyletic species A species based on the close genetic rela-onship of individuals.

hyllarenite A term used by Folk (1968, p.124) for a litharen-e composed chiefly of foliated, phyllosilicate-rich, metamor-hic rock fragments (such as of slate, phyllite, and schist). It ay have any particle size from silt through gravel, and any ay content, sorting, or rounding. See also: *subphyllarenite.*

hyllite [mineral] (a) A general term used by some French uthors for the scaly minerals, such as micas, chlorites, ays, and vermiculites. (b) A general term for minerals with a ayered crystal structure.

hyllite [petrology] An argillaceous rock commonly formed by egional metamorphism and intermediate in metamorphic rade between slate and mica schist. Minute crystals of seri-te and chlorite impart a silky sheen to the surfaces of cleav-ge (or schistosity). Cf: *phyllonite.*

hyllite-mylonite *phyllonite.*

hyllocarid Any malacostracan belonging to the subclass hyllocarida, characterized by a relatively large bivalve cara-ace which may or may not be hinged along the dorsal mar-in. Their stratigraphic range is Lower Cambrian to present.

hylloclade A somewhat flattened branch that functions as a eaf, e.g. in Christmas cactus. Cf: *phyllode.*

hyllode [paleont] The more or less depressed area of en-arged pores in the adoral part of an ambulacrum in an echin-id. It bears specialized podia. Cf: *bourrelet.*

hyllode [bot] A flattened and expanded petiole that functions s a leaf. Cf: *phylloclade.*

hyllofacies A facies differentiated on the basis of stratifica-on characteristics, esp. the stratification index (Kelley, 1956, .299).

hylloid Leaf-shaped, or resembling a leaf; esp. said of the ninor elements or endings of the saddles of an ammonoid uture.

hyllomorphic stage A term introduced by Dapples (1962) for he latest (most advanced) geochemical stage of diagenesis haracterized by authigenic development of micas, feldspars, nd chlorites at the expense of clays (unidirectional reac-ons). It follows the *locomorphic stage.* See also: *redoxomor-hic stage.*

hyllonite A rock that macroscopically resembles *phyllite* petrology] but that is formed by mechanical degradation

(mylonization) of initially coarser rocks (e.g. graywacke, granite, or gneiss). Silky films of recrystallized mica or chlo-rite smeared out along schistosity surfaces and formation by dislocation metamorphism are characteristic (Turner and Verhoogen, 1960). The term was originated by Sander (1911). Syn: *phyllite-mylonite.*

phyllonitization The processes of mylonitization and recrystalli-zation which together produce a phyllonite.

phyllopodium A broad, flat, leaf-like thoracic appendage of a crustacean. Pl: *phyllopodia.* Cf: *stenopodium.*

phyllosilicate A class or structural type of *silicate* character-ized by the sharing of three of the four oxygens in each tetra-hedron with neighboring tetrahedra, to form flat sheets; the Si:O ratio is 2:5. A example is the micas. Cf: *nesosilicate; sorosilicate; cyclosilicate; inosilicate; tectosilicate.* Syn: *layer silicate; sheet mineral; sheet silicate.*

phyllotaxy The arrangement of leaves or floral parts on their axis. It is generally expressed numerically by a fraction, the numerator representing the number of revolutions of a spiral made in passing from one leaf past each successive leaf to reach the leaf directly above the initial leaf, and the denomi-nator representing the number of leaves passed in the spiral thus made (Lawrence, 1951, p.765).

phyllotriaene A sponge *triaene* in which the cladi are expand-ed into flattened, sometimes digitate, leaf-like structures.

phyllovitrinite Vitrain in which vegetal structures are micro-scopically visible; *provitrain.*

phylogenetic The adj. of *phylogeny.* Syn: *phyletic.*

phylogenetic evolution Evolution within a single lineage.

phylogeny (a) The line, or lines, of direct descent in a given group of organisms, as opposed to the development of an in-dividual organism. Cf: *ontogeny.* (b) The study or history of such relationships.----Adj: *phylogenetic.*

phylum A major unit in the taxonomy of animals, ranking above "class" and below "kingdom". Equivalent to *division* in the classification of plants. Cf: *subkingdom.* Plural: *phyla.*

phyre A suffix which, in a rock name, signifies "porphyry".

physical exfoliation A type of *exfoliation* caused by physical forces, such as by the freezing of water that penetrated fine cracks in the rock or by the removal of overburden concealing deeply buried rocks.

physical geography That branch of *geography* which is the de-scriptive study of the Earth's surface as man's physical envi-ronment, dealing with the classification, form, and extent of the natural phenomena directly related to the exterior physical features and changes of the Earth, including land, water, and air. It differs chiefly from *geology* in that it considers the pres-ent rather than the past conditions of the Earth, and it is more inclusive than *geomorphology*, dealing not only with landforms but also climate, oceans, atmosphere, soils, geologic pro-cesses, natural resources, and sometimes the biogeographi-cal distribution of animal and plant life. In the 18th century, the term was applied in a broader sense, commonly including the races of men and their physical works on the Earth. Cf: *physiography.*

physical geology A broad division of *geology* that concerns it-self only with the processes and forces that have wrought changes on the morphology of the Earth, or on the constituent minerals, rocks, magmas, and core materials that are part of the Earth. It is not concerned with the dating of events or the evolutionary stages of the Earth's present forms. Cf: *histori-cal geology.*

physical landscape *natural landscape.*

physical oceanography The study of such physical aspects of the ocean as optical and acoustic properties; temperature; density; and currents, waves, and tides.

physical pendulum *pendulum* (a).

physical residue A *residue* formed by the mechanical weath-ering in place; e.g. a deposit of gravel resulting from the removal by water or wind of finer particles, as on the floor of a desert valley.

physical stratigraphy Stratigraphy based on the physical aspects of rocks (esp. the sedimentologic aspects); e.g. lithostratigraphy.

physical time A term used by Jeletzky (1956, p.682) to designate time as measured by any physical phenomenon or process (such as by radioactive decay of elements), and proposed by him to replace *absolute time* as used in the geologic sense.

physical weathering *mechanical weathering.*

physicogeographical Pertaining to *physical geography.*

physicochemical geology An archaic term for the applications of physical chemistry in petrology; in current usage, an aspect of geochemistry. Also spelled: *physico-chemical geology.*

physiofacies A term suggested by Moore (1949, p.17) for "the total inorganic characteristics of a sedimentary rock", or that part of lithofacies not represented by biofacies. The term is essentially identical with *lithofacies* as that term has been interpreted by some. Moore (1957a, p.1784-1785) later wrote that "the concept of physiofacies ... may well be forgotten". Cf: *physiotope.*

physiographic cycle *cycle of erosion.*

physiographic diagram A small-scale map showing landforms by the systematic application of a standardized set of simplified pictorial symbols that represent the appearances such forms would have if viewed obliquely from the air at an angle of about 45°. The first major map of this kind was published by Lobeck (1921). Syn: *morphographic map; landform map.*

physiographic feature A prominent or conspicuous *physiographic form* or noticeable part thereof (Mitchell, 1948, p.64). Cf: *topographic feature.*

physiographic form A *landform* considered with regard to its origin, cause, or history (Mitchell, 1948, p.64). Cf: *topographic form.*

physiographic geology A branch of geology that deals with topography; *geomorphology.* The term was previously used as a syn. of *physiography.*

physiographic pictorial map *trachographic map.*

physiographic province A region all parts of which are similar in geologic structure and climate and which has consequently had a unified geomorphic history; a region whose pattern of relief features or landforms differs significantly from that of adjacent regions (see Fenneman, 1914). Examples: the Valley and Ridge, the Blue Ridge, and Piedmont provinces in eastern U.S., and Basin and Range, the Rocky Mountains, and the Great Plains provinces in western U.S. Cf: *geologic province; geographic province; morphologic region; structural province.*

physiography Originally, a description of the physical nature (form, substance, arrangement, changes of real objects, or of natural features in their phenomenal or causal, as distinguished from theoretical, relationships; e.g. the "microscopical physiography of the rock-making minerals" (Rosenbusch, 1888). The term was introduced into geography in 1869 by Huxley (1877) for the study or description of "natural phenomena in general". The term later came to have a more restricted meaning, esp. in the U.S., for "a description of the surface features of the Earth, as bodies of air, water and land" (Powell, 1895, p.1), with an emphasis on mode of origin; i.e. it became synonymous with *physical geography*, and embraced geology, meteorology, and oceanography. Still later, esp. in the U.S., the term referred to a part of physical geography, namely the description and origin of landforms; in this sense, the term is obsolescent and is replaced by *geomorphology*, although there is a general assumption to regard "physiography" as the descriptive, and "geomorphology" as the interpretative, study of landforms. The term is also sometimes presently regarded as an integration of geomorphology, phytogeography, and soil science. See also *physiographic geology.* Etymol: Greek *physis*, "nature", *graphein*, " to write".

physiotope A term defined by Moore (1949, p.17) as "desig nation of all purely physiochemical elements of an enviro ment". but intended by him to represent the sedimentary env ronment of a *physiofacies.* Weller (1958, p.616) notes that the term is to be accepted as having a meaning similar "biotope" or "lithotope", it should be defined as an "area".

phytal zone The part of a lake bottom covered by water sha low enough to permit the growth of rooted plants. Cf: *aphyt zone.*

phyteral Vegetal matter in coal that is recognizable as mo phologic forms, e.g. cuticle, spore coats, or wax, as disti guished from the organic material forming the coal mass, *macerals.* See also: *muralite.*

phytoclast An organic particle similar to dispersed coal but roughly the same size as a mineral clast of the containir rock. Phytoclasts make up 0.1-0.5% of most shale and ar less abundant in sandstone; they occur widely in zeolitize and greenschist facies metasediments (Bostick, 1970, p.74).

phytocoenosis The entire plant population of a particular hab tat.

phytocollite A black, gelatinous, nitrogenous humic body o curring beneath or within peat deposits. *Dopplerite* may repr sent an accumulation of phytocollite concentrated by grour water (Swain, 1963, p. 105).

phytoecology The branch of ecology concerned with the rela tionships between plants and their environments. Cf: *zooeco ogy.*

phytogenic dam A natural dam consisting of plants and plar remains. Such dams may account for ponds and lakes in tur dra regions.

phytogenic dune Any dune in which the growth of vegetatic influences the form of the dune, as by arresting the drifting sand; e.g. a *foredune.*

phytogenic rock A *biogenic rock* produced by plants or direc ly attributable to the presence or activities of plants; e.g. alg deposits, peat, coal, some limestones, and lithified ooze co taining diatoms. Cf: *phytolith.* Syn: *phytogenous rock.*

phytogenous rock *phytogenic rock.*

phytogeography The branch of *biogeography* dealing with th geographic distribution of plants. Cf: *zoogeography.*

phytoleims Coalified remains of plants (Kristofovich, 194 p.138). See also: *meroleims; hololeims.*

phytolite A plant fossil. Syn: *phytolith* [paleont].

phytolith [paleont] (a) A stony (mineral) part of a living pla that secretes mineral matter; specif. *opal phytolith.* (b) Va of *phytolite.*

phytolith [sed] A *biolith* formed by plant activity or compose of plant remains; specif. *phytogenic rock.*

phytopaleontology *paleobotany.*

phytophagous Said of an organism that feeds on plants.

phytophoric Said of a rock that consists of plant remains.

phytoplankton The plant forms of *plankton*, e.g. diatoms. C *zooplankton.*

phytozoan *zoophyte.*

Piacentian A syn. of *Plaisancian.* Also spelled: *Piacenzian.*

pi axis In structural petrology, the normal to the great circl approximated by segments of a folded surface. It is usual written as "π axis", and is equivalent to *beta axis.* Syn: *pole.*

picacho A term used in the SW U.S. for a large, sharp pointed, isolated hill or mountain; a peak. Etymol: Spanish.

pi circle On a *pi diagram*, a girdle of points representing pole to folded surfaces.

pick In the interpretation of seismograph records, the selec tion of an event; also, any selected event on a seismic recorc Cf: *alternate pick.*

pickeringite A mineral: $MgAl_2(SO_4)_4.22H_2O$. It occurs in white to faintly colored fibrous masses. Syn: *magnesia alum*.

pickup *detector*.

picotite A dark-brown, chromium-bearing variety of hercynite (spinel). Much so-called picotite is ceylonite or magnesiochromite. Syn: *chrome spinel*.

picotpaulite A mineral: $TlFe_2S_3$.

picrite A dark-colored, generally hypabyssal rock containing abundant olivine along with pyroxene, biotite, possibly amphibole, and less than 10 percent plagioclase. The term was first used by Tschermak who applied it to a rock composed chiefly of olivine, titanaugite, and barkevikite, with or without biotite; later the term was used by Rosenbusch for a rock composed chiefly of olivine and augite, with or without hornblende and biotite (Streckeisen, 1967, p.176).

picritic Said of an olivine-rich igneous rock.

picrochromite An end member of the spinel group: $MgCr_2O_4$. It is produced synthetically. Syn: *magnesiochromite*.

picrocollite A hypothetical end member of the palygorskite group: $MgSi_3O_5(OH)_4.2H_2O$.

picrolite A term that has been applied to a fibrous or columnar variety of dark-green, gray, or brown serpentine mineral. It is now regarded as a syn. of *antigorite*.

picromerite A white mineral: $K_2Mg(SO_4)_2.6H_2O$. Syn: *schoenite*.

picropharmacolite A mineral: $H_2Ca_4Mg(AsO_4)_4.12H_2O$.

picurite *pitch coal*.

pi diagram In structural petrology, a plot of poles of planes of bedding or schistosity, which gives rise to a *pi circle*; a type of fabric diagram. Also spelled: π *diagram*. Cf: *beta diagram*.

piecemeal stoping *Magmatic stoping* in which only isolated blocks of roof rock are assimilated. Cf: *ring-fracture stoping*.

piedmont adj. Lying or formed at the base of a mountain or mountain range; e.g. a *piedmont* terrace or a *piedmont* pediment.---n. An area, plain, slope, glacier, or other feature at the base of a mountain; e.g. a foothill or a bajada. In the U.S., the Piedmont is a plateau extending from New Jersey to Alabama and lying east of the Appalachian Mountains. Etymol: from Piemonte, a region of NW Italy at the foot of the Alps.

piedmont alluvial plain *bajada*.

piedmont angle The sharp break of slope between a hill and a plain, such as the angle at the junction of a mountain front and the pediment at its base.

piedmont bench (a) An upfaulted alluvial fan or pediment surface at the base of a mountain, bounded on its outer side by a scarplet (piedmont scarp) (Sharp, 1954, p.23). Such features are displayed along the south sides of the San Gabriel and San Bernardino ranges in southern California. (b) *piedmont step*.

piedmont benchland One of several successions or systems of *piedmont steps*. Syn: *piedmonttreppe; piedmont stairway*.

piedmont bulb *expanded foot*.

piedmont flat *piedmont step*.

piedmont glacier A thick continuous sheet of ice at the base of a mountain range, resting on land, formed by the spreading out and coalescing of valley glaciers from the higher elevations of the mountains. Cf: *expanded-foot glacier*.

piedmont gravel Coarse gravel derived from high ground by mountain torrents and spread out on relatively flat ground where the velocity of the water is decreased.

piedmont interstream flat A term used by Tator (1949) for a planate rock surface along the east flank of the Colorado Front Range, and regarded by Tator (1953, p.47) as a syn. of *pediment*.

piedmontite *piemontite*.

piedmont lake An oblong lake occupying a partly overdeepened basin excavated from rock by, or dammed by a moraine of, a piedmont glacier.

piedmont pediment A term used by Davis (1930, p.154) for a pediment peripheral to, and along the base of, a mountainous area. Cf: *mountain pediment*.

piedmont plain *bajada*.

piedmont plateau A plateau lying between the mountains and the plains or the ocean; e.g. the plateau of Patagonia in southern Argentina and southern Chile, between the Andes and the Atlantic Ocean.

piedmont scarp A small, low cliff occurring in alluvium on a piedmont slope at the foot of, and essentially parallel to, a steep mountain range (as in the western U.S.), resulting from dislocation of the surface, esp. by faulting; term proposed by Gilbert (1928, p.34). See also: *fan scarp*. Syn: *scarplet*.

piedmont slope (a) *bajada*. (b) A gentle slope at the base of a mountain in a semiarid or desert region, composed of a pediment (upper surface of eroded bedrock) and a bajada (lower surface of aggradational origin).

piedmont stairway *piedmont benchland*.

piedmont step An extensive or regional, terrace- or bench-like feature sloping outward or down-valley (as in the Black Forest or Schwarzwald region of SW Germany), assumed by W. Penck (1924) to develop in response to a continually accelerated uplift of a rising or expanding dome. See also: *treppen concept; piedmont benchland*. Syn: *piedmont bench; piedmont flat*.

piedmonttreppe A syn. of *piedmont benchland*. Etymol: German *Piedmonttreppe*, "piedmont staircase".

piemontite A dark-reddish or reddish-brown, manganese-bearing mineral of the epidote group: $Ca_2(Al,Mn^{+3},Fe)_3$-$Si_3O_{12}(OH)$. Cf: *withamite*. Syn: *piedmontite; manganese epidote*.

pienaarite A sphene-rich malignite in which the feldspar is anorthoclase.

pier [eng] (a) An underground structural member that transmits a concentrated load to a stratum capable of supporting it without danger of failure or excessive settlement. Its ratio of base width to depth of foundation is usually about 1:4. Cf: *footing*. (b) A rectangular or circular column, usually of concrete or masonry, designed to support heavy concentrated loads from arches or the superstructure of a bridge.

pier [coast] (a) A long, narrow *wharf* extending out from the shore into the water, serving as a berthing or landing place for vessels or as a recreational facility. Cf: *jetty; quay*. Erroneous syn: *dock*. (b) A breakwater, groin, mole, or other structure used to protect a harbor or shore, and serving also as a promenade or landing place for vessels.

piercement dome *diapir*.

piercing fold *diapir*.

piercing point In a quaternary chemical system, the point at which a univariant curve (representing the compositions of liquids that can exist in equilibrium with three particular solid phases) and a ternary join intersect, at a point other than a ternary univariant point.

pier dam An engineering structure (such as a groin) built from shore to deepen a channel, to divert logs, or to influence the current. Syn: *wing dam*.

pierre-perdue Blocks of stone or concrete heaped loosely in the water to make a foundation. Etymol: French, "lost stone".

pierrepontite Iron-rich variety of tourmaline.

pierrotite A mineral: $Tl_2(Sb,As)_{10}S_{17}$.

piestic interval *potential drop*.

piestic water A term proposed by Meinzer (1939) as a syn. of *confined ground water* and one of two classes of *plerotic water*. It includes *hyperpiestic water, hypopiestic water*, and *mesopiestic water*. This classification is not commonly used in the U.S.

pietricikite A variety of *ozocerite*. Originally incorrectly spelled *zietrisikite*. See also: *neft-gil*.

piezocrystallization Crystallization of a magma under pressure, such as pressure associated with orogeny.

piezoelectric crystal A crystal, e.g. of quartz or tourmaline, that displays the *piezoelectric effect*. An nonconducting crystal lacking a center of symmetry may be piezoelectric.

piezoelectric detector In seismic prospecting, a type of detector which depends upon the piezoelectric effect by which an electric charge is produced on the faces of a properly cut crystal of certain materials, particularly quartz and Rochelle salt, when the crystal is strained. The detector is constructed from a pile of such crystals with intervening metal foil to collect the charge. An inertia mass is mounted on the top of the crystal stack which is included in a vacuum-tube circuit.

piezoelectric effect In certain crystals, the development of an electric potential in certain crystallographic directions when mechanical strain is applied, or, the development of a mechanical strain, hence vibration, when an electric potential is applied. Quartz and tourmaline are examples of naturally *peizoelectric crystals*.

piezogene Pertaining to the formation of minerals primarily under the influence of pressure (Kostov, 1961). Cf: *thermogene*.

piezoglypt *regmaglypt*.

piezomagnetism Stress dependence of magnetic properties. It is the inverse of *magnetostriction*.

piezometric contour *equipotential line*.

piezometric surface *potentiometric surface*.

pigeonite A mineral of the clinopyroxene group: $(Mg,Fe^{+2}, Ca)(Mg,Fe^{+2})Si_2O_6$. It is intermediate in composition between clinoenstatite and diopside, and has little calcium, little or no aluminum or ferric iron, and less ferrous iron than magnesium. Pigeonite is characterized optically by a small and variable axial angle ($2V = 0\text{-}30°$). It is found in basic igneous rocks at Pigeon Point in Minnesota. Cf: *augite*.

pigeon's-blood ruby A gem variety of ruby of the finest color quality, likened to that of the arterial blood of a freshly killed pigeon: intense, clear, medium- to medium-dark red to slightly purplish red. It is found almost exclusively in upper Burma. Also spelled: *pigeon-blood ruby*.

pigment minerals Those minerals whose economic value is as coloring agents. Red and yellow ochres are hydrated iron oxides; umber and sienna are aluminum silicated with iron and manganese. Minerals that may be used as white pigments are white clays, gypsum, talc, dolomite, and barytes (Nelson and Nelson, 1967).

pike A term used in England for any summit or top of a mountain or hill, esp. one that is peaked or pointed. Also, a mountain or hill having a peaked summit.

pikeite An obsolete name for an augite-bearing phlogopite peridotite. Its name is derived from Pike County, Arkansas, U.S.

pila Plural of *pilum*.

pilandite A hypabyssal rock containing abundant anorthoclase phenocrysts in a groundmass of the same mineral.

pilar A term used in the SW U.S. for a large pillar-like or projecting rock. Etymol: Spanish, "pillar".

pilate Said of spores and pollen having sculpture that is similar to that of *clavate* forms but that consists of smaller hair-like processes (pila) with more or less spherical knobs. Syn: *piliferous*.

pile A long, relatively slender structural foundation element (plate, post, plank, beam, board, etc.), usually made of timber, steel, or reinforced or prestressed concrete, that is driven or jetted into the ground or cast in place and that is used to support vertical or lateral loads, to form a wall to exclude water or soft material or to resist their pressure, to compact the surrounding ground, or rarely to restrain the structure from uplift forces. See also: *sheet pile*.

piliferous Bearing or producing hairs; e.g. said of pollen grains

bearing pila. The synonymous term *pilate* is preferred in palynology.

piling A structure or group of piles.

pilite (a) Actinolite pseudomorphous after olivine. (b) *tinder ore*.

pill An English term for a pool, and for a small stream or creek.

pillar [speleo] *stalacto-stalagmite*.

pillar [geomorph] A natural formation shaped like a pillar; specif. an *earth pillar* and a *rock pillar*.

pillar [paleont] (a) A tiny rod-like structure, larger and straighter than a *trabecula*, connecting discrete layers of sclerite in holothurians. (b) An elongate peg-like structure produced near the center or axis of coiling in certain foraminifera by thickening of the wall. The ends of the pillars appear as small bosses or nodes on the ventral side of the test.

pillar [struc geol] A joint block produced by columnar jointing.

pillar mining A method of mining coal and other bedded mineral deposits in which a network of passageways forms pillars of coal that are then extracted. Cf: *longwall mining*.

pillow breccia A deposit of pillows and fragments of lava in a matrix of tuff.

pillow lava A general term for those lavas displaying *pillow structure* and considered to have formed in a subaqueous environment; such a lava is usually basaltic or andesitic, esp. a spilite. Syn: *ellipsoidal lava*.

pillow structure [ign] A structure, observed in certain extrusive igneous rocks, that is characterized by discontinuous pillow-shaped masses ranging in size from a few centimeters to a meter or more in diameter (commonly between 30 and 60 cm). The pillows are close-fitting, the concavities of one matching the convexities of another. The spaces between the pillows are few and are filled either with material of the same composition as the pillows, with clastic sediments, or with scoriaceous material. Grain sizes within the pillows usually decrease toward the exterior. Pillow structures are considered to be the result of subaqueous deposition as evidenced by their association with sedimentary deposits, usually of deep-sea origin. See also: *pillow lava*.

pillow structure [sed] A primary sedimentary structure resembling the size and shape of a pillow; it is most characteristic of the basal parts of a sandstone overlying shale. See also: *ball-and-pillow structure*. Syn: *mammillary structure*.

pilotaxitic Said of the texture of the groundmass of a holocrystalline igneous rock in which lath-shaped microlites (usually of plagioclase) are arranged in a glass-free *felty* mesh, often aligned along the flow lines.

pilot channel One of a series of *cutoffs* for converting a meandering stream into a straight channel of greater slope. It is built only large enough to start flow along the new course, since erosion during floods is expected to create channels of adequate capacity.

pilum One of the small, spine-like rods on the exine of pollen and spores, characterized by rounded or swollen knob-like ends. Pl: *pila*.

pimelite An apple-green clay mineral of the montmorillonite group: $(Ni,Mg)_3Si_4O_{10}(OH)_2.4H_2O$.

pimple *pimple mound*.

pimple mound A term used along the Gulf Coast of eastern Texas and SW Louisiana for one of hundreds of thousands of low, flattened, rudely circular or elliptical domes composed of sandy loam that is coarser than, and entirely distinct from, the surrounding soil; the basal diameter varies from 3 m to more than 30 m, and the height from 30 cm to more than 2 m. Cf: *Mima mound*. Syn: *pimple*.

pimple plain A plain characterized by numerous and conspicuous *pimple mounds*.

pin [geomorph] An Irish term for a mountain peak. Etymol: Gaelic *beann* or *beinn*, "peak". See also: *ben*.

pin [sed] (a) A thin and irregular bed, band, or seam of iron-stone or other hard rock in the coal measures of south Wales. (b) A cylindrical nodule, usually of clay ironstone, in the coal measures of south Wales.

pin [surv] arrow.

pinacocyte One of the cells, generally flat, of the pinacoderm of a sponge.

pinacoderm An unstratified layer of cells (pinacocytes), other than the choanaderm, constituting the soft parts of a sponge and delimiting it from the external milieu; e.g. *endopinaco-derm* and *exopinacoderm*.

pinacoid An open crystal form consisting of two parallel faces. Adj: *pinacoidal*. Also spelled: *pinakoid*.

pinacoidal class That crystal class in the triclinic system having only a center of symmetry.

pinacoidal cleavage A type of crystal cleavage that occurs parallel to one of the crystal's pinacoidal surfaces; e.g. the [010] cleavage of gypsum.

pinakiolite A black mineral: $(Mg,Mn^{+2})_2Mn^{+3}BO_5$. It is polymorphous with orthopinakiolite.

pinakoid Var. of *pinacoid*.

pinch n. (a) A marked thinning or squeezing of a rock layer; e.g. a coming together of the walls of a vein, or of the roof and floor of a coal seam, so that the ore or coal is more or less completely displaced. See also: *nip [coal]*. (b) A thin place in, or a narrow part of, an orebody; the part of a mineral zone that almost disappears before it widens out in another place to form an extensive orebody, or swell. Cf: *make*. ----v. *pinch out*.

pinch-and-swell structure A structural condition commonly found in quartz veins and pegmatites in metamorphosed rocks, in which the vein is pinched at frequent intervals, leaving expanded parts between but without their being separated from one another (Ramberg, 1955).

pinch-out The termination or end of a stratum, vein, or other body of rock that narrows or thins progressively in a given horizontal direction until it disappears and the rocks it once separated are in contact; esp. a stratigraphic trap formed by the thinning out of a porous and permeable rock (sandstone) between two layers of impermeable rock (shale). The lithologic character of the stratum or rock is typically maintained to the feather edge of the layer. Cf: *shale-out*. See also: *wedge-out; nip-out*.

pinch out To taper or narrow progressively to extinction; to *thin out*.

pinfire opal Opal in which the patches (small pinpoints) of play of color are very small and close together and usually less regularly spaced than the color patches in *harlequin opal*.

pinger A battery-powered, low-energy source for an *echo sounder*.

pingo (a) A *frost mound*, esp. a relatively large conical mound of soil-covered ice (commonly 30-50 m high and up to 400 m in diameter), raised by hydrostatic pressure of water within or below the permafrost of Arctic regions (esp. Canada), and of more than one-year duration; an intrapermafrost ice-cored hill or mound. Its crest is sometimes ruptured or collapsed due to melting of the ice, thus forming a star-shaped crater; the term has also been applied to such a depression (Monkhouse, 1965, p.237). The mound itself often resembles a small volcano. The term was introduced for this feature by Porsild (1938, p.46). Pl: pingos (not "pinges"). See also: *ground-ice mound; hydrolaccolith*. (b) The term has been used in several related senses, as a conical hill or mound, or as a hill completely covered by an ice sheet but revealing its presence by surface indications (ADTIC, 1955, p.61). (c) *gull hummock*.---Etymol: Eskimo, "conical hill". Syn: *pingok*.

pingok Var. of *pingo*.

pingo-remnant A rimmed depression, as in northern Netherlands where it was previously regarded as a kettle (Gravenor

& Kupsch, 1959, p.62). It is formed by rupturing of a pingo summit resulting in exposure of the ice core to melting followed by partial or total collapse. See also: *fossil pingo*. Syn: *pseudokettle*.

pinguite *nontronite*.

pinhole chert Chert containing weathered pebbles pierced by minute holes or pores.

pinite A compact, fine-grained, usually amorphous mica (chiefly muscovite) of a dull-grayish, green, or brownish color, derived from the alteration of other minerals (such as cordierite, nepheline, scapolite, spodumene, and feldspar).

pink snow *red snow*.

pinna A primary subdivision of a pinnate leaf or frond; a leaflet. See also: *pinnule*.

pinnacle [geomorph] (a) A tall, very slender, tapering or pointed tower or spire-shaped pillar of rock, either isolated or at the summit of a mountain or hill; esp. a lofty peak. (b) A hill or mountain with a pointed summit.

pinnacle [reef] A small, pointed, isolated spire or column of rock or coral, either slightly submerged or awash; specif. a small reef patch, consisting of coral growing sharply upward (slopes vary from 45° to nearly vertical) within an atoll lagoon, often rising close to the water surface. Syn: *pinnacle reef; reef pinnacle; coral pinnacle; coral knoll*.

pinnacled iceberg An irregular iceberg shaped and weathered in such a way as to be topped with spires and pinnacles.

pinnacle reef *pinnacle [reef]*.

pinnate Said of a compound leaf whose leaflets are placed on either side of the rachis.

pinnate drainage pattern A *dendritic drainage pattern* in which the main stream receives many closely spaced, subparallel tributaries that join it at acute angles, resembling in plan a feather; it is believed to indicate unusually steep slopes on which the tributaries developed.

pinnate jointing *feather jointing*.

pinnate venation A type of *net venation* in which the secondary veins branch from the midrib in parallel pattern. Cf: *palmate venation*.

pinnoite A yellowish tetragonal mineral: $MgB_2O_4.3H_2O$.

pinnular A plate forming part of a pinnule of a crinoid.

pinnule [paleont] (a) One of several generally slender, unbifurcated, uniserial branchlets of the food-gathering system of a crinoid arm, typically borne on alternate sides of successive brachial plates. (b) A secondary branch of a plume-like organ, such as a digitate lateral branch of a tentacle of an octocoral polyp or one of the biserial branches of the cystoid *Caryocrinites*.---Syn: *pinnule*.

pinnule [bot] A subdivision of a *pinna*, or a secondary subdivision of a pinnate leaf or frond; a secondary leaflet.

pinolite A metamorphic rock containing magnesite (breunnerite) as crystals and as granular aggregates in a schistose matrix (phyllite or talc schist). It is so named because the magnesite inclusions resemble pine cones in shape (Holmes, 1928, p.184). It is found in Styria, Austria.

pintadoite A green mineral: $Ca_2V_2O_7.9H_2O$.

pinule (a) Var. of *pinnule*. (b) *pinulus*.

pinulus A sponge spicule (usually a pentactin or hexactin) in which one ray (such as the unimpaired one in a pentactin) is enlarged and projects either internally or externally from the sponge and bears numerous small oblique spines giving the spicule the appearance of a pine tree. Pl: *pinuli*. Syn: *pinule*.

pinwheel garnet *rotated garnet*.

pioneer In ecology, a community, species, flora, fauna, or individual that establishes itself in a barren area, initiating a new ecologic cycle. Cf: *climax*.

piotine *saponite*.

pipe [ore dep] A cylindrically shaped, more or less vertical orebody. The ore may be a vein deposit or diamond-bearing volcanic breccia. Syn: *ore pipe; ore chimney; chimney; neck; stock.*

pipe [intrus rocks] A discordant pluton of tubular shape.

pipe [sed] (a) A tubular cavity from several centimeters to a few meters in depth, formed esp. in calcareous rocks, and often filled with sand and gravel; e.g. a vertical joint or sinkhole in chalk, enlarged by solution of the carbonate material and filled with clastic material. See also: *sand pipe.* (b) *clastic pipe.*

pipe [volc] A vertical conduit through the Earth's crust below a volcano, through which magmatic materials have passed. It is usually filled with volcanic breccia and fragments of older rock. Cf: *plug [volc]; vent; diatreme.* Partial syn: *chimney [volc].* Syn: *breccia pipe.*

pipe [grd wat] *geyser pipe.*

pipe amygdule An elongate amygdule that occurs in a lava, towards the base of the flow, probably formed by the generation of gases or vapor from the underlying material.

pipe clay (a) A white to grayish-white, highly plastic clay practically free from iron, suitable for use in making tobacco pipes. The term has been extended to include any white-burning clay of considerable plasticity. Syn: *ball clay; cutty clay.* (b) A mass of fine clay, generally of lenticular form, forming the surface of bedrock and upon which the gravel of deep leads (old river beds) frequently rests.---Also spelled: *pipeclay.*

piperno A welded tuff characterized by *fiamme,* or flame structure. Such a rock is said to be pipernoid. Etymol: Italian.

pipernoid Said of the eutaxitic texture of certain extrusive igneous rocks in which dark patches and stringers occur in a light-colored groundmass. Also, said of a rock (i.e. piperno) exhibiting such texture.

pipe-rock A marine sandstone containing abundant scolites.

pipe-rock burrow *scolite.*

pipestone A pink or mottled argillaceous stone, carved by the Indians into tobacco pipes; esp. *catlinite.*

pipette analysis A kind of particle-size analysis of fine-grained sediment, made by removing samples from suspension with a pipette.

pipe vesicle Slender vertical cavities a few centimeters or tens of centimeters in length extending upward from the base of a lava flow. Most are formed by water vapor derived from the underlying wet ground that streamed upward into the lava.

piping Erosion by percolating water in a layer of subsoil, resulting in caving and in the formation of narrow conduits, tunnels, or "pipes" through which soluble or granular soil material is removed; esp. the movement of material, from the permeable foundation of a dam, by the flow or seepage of water along underground passages. See also: *water creep.* Syn: *tunnel erosion.*

pipkrake (a) A small, thin spike or needle-like crystal of ground ice, from 2.5-6 cm in length, formed just below, and growing perpendicular to, the surface of the soil in a region where the daily temperatures fluctuate across the freezing point. It is common in periglacial areas where it contributes to the sorting of material in patterned ground and to downslope movement of surface material. (b) A bundle, cluster, or tuft of pipkrakes.---Etymol: Swedish, "needle ice". Syn: *needle ice; feather ice; mush frost; spew frost.*

pi pole The trace of the *pi axis* in a stereographic projection.

piracy *capture* [streams].

pirate (a) *capturing stream.* (b) *pirate valley.*

pirated stream *captured stream.*

pirate stream *capturing stream.*

pirate valley A valley that appropriates the waters of another valley; a valley containing a capturing stream. Syn: *pirate.*

pirssonite A white to colorless orthorhombic mineral: $Na_2Ca(CO_3)_2.2H_2O$.

pisanite A blue mineral: $(Fe,Cu)SO_4.7H_2O$. It is a variety of melanterite containing copper.

pisiform Resembling the size and shape of a pea; e.g. a "pisiform concretion" (or pisolith).

pisolite [mineral] A variety of calcite or aragonite.

pisolite [sed] (a) A sedimentary rock, usually a limestone, made up chiefly of pisoliths cemented together; a coarse-grained oolite. Syn: *peastone; pea grit.* (b) A term often used for a *pisolith,* or one of the spherical particles of a pisolite.--Etymol: Greek *pisos,* "pea". Cf: *oolite.*

pisolite [volc] *mud ball.*

pisolith One of the small, round (spherical or ellipsoidal) accretionary bodies in a sedimentary rock, resembling a pea in size and shape (diameter: 2-10 mm), and constituting one of the grains that make up a pisolite. It is often formed of calcium carbonate, and is thought to have been produced (in some cases) by a biochemical algal-encrustation process. A pisolith is larger and less regular in form than an *oolith,* although it has the concentric and radial internal structure of an oolith. The term is sometimes used to refer to the rock made up of pisoliths. Syn: *pisolite [sed].*

pisolitic [sed] Pertaining to pisolite [sed], or to the texture of a rock made up of pisoliths or pea-like grains; e.g. "pisolitic bauxite" or "pisolitic limestone".

pisolitic [ign] Said of a tuff composed of accretionary lapilli or pisolites.

pisosparite A limestone containing at least 25% pisoliths and no more than 25% intraclasts and in which the sparry-calcite cement is more abundant than the carbonate-mud matrix (micrite) (Folk, 1959, p.22). Cf: *oosparite.*

pissasphalt Originally used by the ancients to describe what they considered a mixture of pitch and asphalt; now used as a syn. of *maltha.*

pistacite A syn. of *epidote,* esp. the pistachio-green variety rich in ferric iron. Also spelled: *pistazite.*

pistil In a flower, the female reproductive organ consisting of an ovary, style (when present), and stigma. It may consist of a single *carpel* (simple pistil) or of two or more carpels (compound pistil).

pistillate Said of a flower which has a pistil but no stamens. Cf: *staminate.*

piston corer An oceanographic *corer* containing a piston inside the cylinder which reduces friction by creating suction. There are several varieties, including the *Ewing corer,* the *Mackereth sampler,* and the *Kullenberg corer.* Cf: *gravity corer.*

piston organelle A mound-like structure rising from the floor of peristome in a tintinnid.

pit [drill] An excavation to hold water and drilling mud; specif. *slush pit.*

pit [geol] A small indentation or depression left on the surface of a rock particle (esp. of a clastic particle) as a result of some eroding or corrosive process, such as etching or differential solution.

pit [bot] A thin place on a cell wall. See also: *simple pit; bordered pit.*

pit and mound A sedimentary structure consisting of a small, raised, blister-like mound (1 mm high and 3-12 mm in diameter) that surrounds or contains at its summit a tiny, circular, crater-like central pit (up to one millimeter in diameter) simulating a rain print. It is formed during rapid settling of low-viscosity mud by gas bubbles or water currents escaping vertically upward through the mud and emerging at the surface. Term was introduced by Kindle (1916). Syn: *pit-and-mound structure.*

pitch [organic] A dark-colored, viscous to solid, nonvolatile fusible substance, consisting principally of hydrocarbons, that is the residue from the distillation of certain organic substances.

pitch [coast] An obsolete term for the tip of a piece of land uch as a cape, extending into a body of water.

pitch [speleo] A vertical shaft in a pothole.

pitch [struc geol] A syn. of *plunge*; for the geometry of folds, is the less preferred form. Syn: *rake*.

pitch [slopes] A steep place; a declivity.

pitchblende A massive, brown to black, and fine-grained (colloform), amorphous, or microcrystalline variety of *uraninite* found in hydrothermal sulfide-bearing veins and having a distinctive pitchy to dull luster. It contains a slight amount of radium, but thorium and the rare earths are generally absent. Syn: *pitch ore; nasturan.*

pitch coal (a) A brittle, lustrous bituminous coal or lignite, with conchoidal fracture. Syn: *bituminous lignite; glance coal; picurite; specular coal.* (b) A kind of *jet.*

pitches and flats flats and pitches.

pitching fold *plunging fold.*

pitch opal A yellowish to brownish inferior quality of common opal displaying a pitchy luster.

pitch ore (a) *pitchblende.* (b) *pitchy copper ore.*

pitch peat Peat that resembles asphalt.

pitchstone A volcanic glass, usually intrusive, with a waxy, dull, resinous pitchy luster rather than a bright, glassy luster. Its color and composition vary widely; it contains a higher percentage of water than does obsidian. Crystallites are detectable in thin section. Syn: *fluolite.*

pitchy copper ore A dark, pitch-like oxide of copper; a mixture of chrysocolla and limonite. Syn: *pitch ore.*

pitchy iron ore (a) *pitticite.* (b) *triplite.*

pit crater A volcanic *sink.*

pith Parenchymatous tissue occupying the central portion of a plant stem.

pith rays Primary bands of parenchyma cells extending from the pith to the pericycle in herbaceous and young woody stems (Fuller & Tippo, 1949, p.968).

pit lake A *kettle lake* in a pitted outwash plain.

piton A term commonly used for volcanic peaks, especially steep-sided domes, in the West Indies and other French-speaking regions.

pit run *bank gravel.*

pitted outwash plain An *outwash plain* marked by many irregular depressions such as kettles, shallow pits, and potholes; many are found in Wisconsin and Minnesota. See also: *kettle plain.* Syn: *pitted plain.*

pitted pebble A pebble having marked concavities not related to the texture of the rock in which it appears or to differential weathering (Kuenen, 1943). The depressions vary in size from minute pits caused by sand particles to cups a few centimeters across and a centimeter deep; they are common at the contacts between adjacent pebbles, and have been explained as the result of pressure-induced solution at points of contact. The term has also been applied to cobbles. Cf: *cupped pebble.*

pitted plain *pitted outwash plain.*

pitticite A brown to yellowish or reddish mineral found in reniform masses, consisting of a hydrous arsenate and sulfate of iron, and having a very variable composition. Syn: *pittizite; pitchy iron ore.*

pivotability A measure of roundness of sedimentary particles, expressed by the ease with which a particle can be dislodged from a surface or by the tendency of a particle to start rolling on a slope. The term was introduced by Shepard & Young (1961, p.198) who assigned the highest values of roundness to the particles that "could be most easily pivoted".

pivotal fault A partial syn. of *hinge fault.* Cf: *scissor fault.* See also: *trochoidal fault.*

placanticline A gentle, anticlinal-like uplift of the continental platform, usually asymmetric and without a typical outline. There is no corresponding synclinal-like structure. The term is used mainly in the Russian literature of the Volga-Urals region

(Shatsky, 1945). A corresponding term in western literature is *plains-type fold.*

placenta In a phanerogam or seed-bearing plant, the region which bears the ovules in an ovary, often the margin of the carpels; in a cryptogam or spore-bearing plant, the tissue from which the sporangia arise.

placer A surficial mineral deposit formed by mechanical concentration of mineral particles from weathered debris. The mechanical agent is usually alluvial but can also be marine, eolian, lacustrine, or glacial, and the mineral is usually a heavy metal such as gold. Cf: *lode.* Syn: *ore of sedimentation.*

placer claim A mining claim on a placer deposit in which a discovery has been made. Cf: *lode claim.*

placer mining The extraction and concentration of heavy metals from placers by various methods using running water. Cf: *hydraulic mining; drift mining.*

placic Said of a black to dark red soil horizon that is usually cemented with iron and is not very permeable (SSSA, 1970).

placochela A sponge chela in which both shaft and recurved ends are broadly expanded and flattened.

placolith A perforate coccolith having two shields connected by a central tube. See also: *tremalith.* Syn: *cyatholith.*

pladdy A term used in Northern Ireland for a "residual island drumlin awash at high tide" (Stamp, 1961, p.365).

pladorit A hornblende-bearing granitite or a magnesium- and mica-bearing hornblende granite (Johannsen, 1939, p.275).

plage A French term for a sandy beach, esp. at a seaside resort. The term is being supplanted by *lido.*

plaggen Said of an *epipedon* that is man-made by manuring and mixing, and that is more than 50cm in thickness (SSSA, 1970).

Plaggept In U.S. Dept. of Agriculture soil taxonomy, a suborder of the soil order *Inceptisol*, characterized by the presence of a plaggen epipedon (SSSA, 1970). Cf: *Andept; Aquept; Ochrept; Propept; Umbrept.*

plagiaplite An aplite composed chiefly of plagioclase (oligoclase to andesine), possibly green hornblende, and accessory quartz, biotite, and muscovite.

plagioclase (a) A group of triclinic feldspars of general formula: $(Na,Ca)Al(Si,Al)Si_2O_8$. At high temperatures it forms a complete solid-solution series from Ab ($NaAlSi_3O_8$) to An ($CaAl_2Si_2O_8$). The plagioclase series is arbitrarily subdivided and named according to increasing mole fraction of the An component: albite (An 0-10), oligoclase (An 10-30), andesine (An 30-50), labradorite (An 50-70), bytownite (An 70-90), and anorthite (An 90-100). The Al/*Si* ratio varies with increasing An content from 1:3 to 1:1. Plagioclases are one of the commonest rock-forming minerals, have characteristic twinning, and commonly display zoning. (b) A mineral of the plagioclase group; e.g. albite, anorthite, peristerite, and aventurine feldspar.---The term was introduced by Breithaupt (1847, p.490) who applied it to all feldspars having an oblique angle between the two main cleavages. Cf: *alkali feldspar; orthoclase.* Syn: *sodium-calcium feldspar.*

plagioclase-arenite A term used by McBride (1963, p.668) for an arkose containing more than 25% plagioclase, and by Folk (1968, p.124) for an arkose in which plagioclase is the main feldspar.

plagioclase rock *anorthosite.*

plagioclasite *anorthosite.*

plagiogranite A term commonly used by Russian petrologists for igneous rock having a low potassium content. It includes rocks ranging in composition from quartz diorite to trondhjemite. The original plagiogranite of Kruschov (1931) had the following average modal composition: 56% plagioclase, 27% quartz, 12% biotite, 5% amphibole.

plagiohedral class An early name for the *gyroidal class.*

plagionite A blackish lead-gray mineral: $Pb_5Sb_8S_{17}$.

plagiophyre A porphyritic igneous rock similar to an orthophyre but containing plagioclase rather than orthoclase phenocrysts.

plagiostome An asymmetrically placed aperture or pseudostome in a thecamoebian test (as in *Centropyxis* or *Plagiopyxis*).

plain (a) Broadly, any flat area, large or small, at a low elevation; specif. an extensive region of comparatively flat, smooth, and level or gently undulating land, having few or no prominent surface irregularities (hills, valleys) but sometimes having a considerable slope, and usually at a low elevation with reference to surrounding areas (local relief up to 60-150 m, although some, as the Great Plains of the U.S., are as much as 1000-1800 m above sea level). A plain may be either forested or bare of trees, and may be formed by deposition or by erosion. (b) A very extensive, broad tract of level or rolling, almost treeless country with a shrubby vegetation; a *prairie*. In Australia, "plain" implies treelessness. The term is usually used in the plural. (c) A region underlain by low-lying horizontal strata or characterized by horizontal structure, and which may be dissected into hills and valleys by stream erosion; Davis (1885) introduced this concept of a "plain" but the term should be used without regard to the underlying geologic structure.---Cf: *plateau* [geomorph].

plain of denudation A surface that has been reduced to or just above sea level by the agents of erosion (usually considered to be of subaerial origin); it is relatively flat but may be marked by residual hills of resistant rock rising somewhat above the general level. See also: *plain of marine denudation*.

plain of lateral planation An extensive, smooth, apron-like surface developed at the base of a mountain or escarpment (as at Book Cliffs, Utah) by the widening of valleys and the coalescence of flood plains as a result of *lateral planation*. It resembles the landscape form of a *pediment* and is often so called. Cf: *panplain*.

plain of marine denudation A plane or nearly plane surface worn down by the gradual encroachment of ocean waves upon the land, or a plane or nearly plane, imaginary surface representing such a plain after uplift and partial subaerial erosion. The concept of a "plain of marine denudation" was first recognized by Ramsay (1846); however, the term was often applied to a plain, produced by subaerial or nonmarine agents, that was subsequently submerged beneath the sea, a minor role being assigned to wave erosion. Syn: *plain of submarine denudation*.

plain of marine erosion A largely theoretical platform representing a plane surface of unlimited width, produced below sea level by the cutting away altogether of the land by marine processes acting over a very long period of stillstand; the ultimate abrasion platform, characteristic of a shoreline cycle during advanced old age. Cf: *plain of marine denudation; marine peneplain*. Syn: *marine plain; marine plane; sea plain; submarine plain*.

plain of submarine denudation *plain of marine denudation*.

plains-type fold An anticlinal or domelike structure of the continental platform that has no typical outline and for which there is no corresponding synclinal structure. It is associated with the vertical uplift of normal faulting. The corresponding term used in the Russian literature is *placanticline*. See also: *drape fold*.

plain tract The lower part of a stream, characterized by a low gradient and a wide flood plain. Cf: *mountain tract; valley tract*. See also: *deltaic tract*.

Plaisancian European stage: lower Pliocene (above Pontian of Miocene, below Astian). Also spelled: *Plaisanzian*. Syn: *Piacentian*.

plaiting A texture seen in some schists that results from the intersection of (relict) bedding planes with well developed cleavage planes. Syn: *gaufrage*.

Plait point A point representing conditions at which two conjugate phases become identical; a critical point.

plan A drawing, sketch, or diagram of any object or structure, made by horizontal projection upon a plane or flat surface; esp. a very large-scale and considerably detailed *map* of a small area, such as one showing underground mine workings.

planaas A Danish term for an outwash plain formed as a flat-topped delta in standing water between two walls of stagnant ice. Pron: pla-*nouse*.

planar Lying or arranged as a plane or in planes, usually implying more or less parallel planes, such as those of bedding or cleavage. It is a two-dimensional arrangement, in contrast to the one-dimensional *linear* arrangement.

planar cross-bedding (a) Cross-bedding in which the lower bounding surfaces are planar surfaces of erosion (McKee & Weir, 1953, p.385); it results from beveling and subsequent deposition. (b) Cross-bedding characterized by planar foreset beds.

planarea One of two flattened areas developed on either side of the posterior part of a brachiopod shell in place of the more common single median *interarea*. Cf: *palintrope*.

planar element A fabric element having two dimensions that are much greater than the third. Cf: *linear element; equant element*.

planar features Distinctive, multiple, closely spaced, parallel, plate-like, microscopic planes, distinct from cleavage planes, occurring in shock-metamorphosed minerals (particularly quartz, also feldspar) and regarded as unique and important indicators of shock metamorphism. The structures are characteristically multiple (often more than five distinct sets per grain) and are oriented parallel to specific planes in the host crystal lattice. They have been produced experimentally by shock pressures of approximately 80-250 kb. Syn: *shock lamellae*.

planar flow structure *platy flow structure*.

planate adj. Said of a surface that has been flattened or leveled by *planation*; e.g. a pediment is a nearly *planate* bedrock erosion surface.---v. To reduce to a plain or other flat surface.

planation (a) The process or processes of erosion whereby the surface of the Earth or any part of it is reduced to a fundamentally even, flat, or level surface; specif. *lateral planation* by a meandering stream. The term also includes erosion by waves and currents, and abrasion by glaciers or wind, in producing a flat surface. The term was originated by Gilbert (1877, p.126-127) who considered alluviation of the flattenend surface as part of the planation process; however, this condition is not necessary. (b) A broad term for the general lowering of the land; e.g. *peneplanation; panplanation; pediplanation; cryoplanation; altiplanation*.

planation stream piracy Capture effected by the lateral planation of a stream invading and diverting the upper part of a smaller stream.

planation surface *erosion surface*.

planchêite A blue mineral: $Cu_8Si_8O_{22}(OH)_4 \cdot H_2O$. Cf: *shattuckite*.

planchette (a) *circumferentor*. (b) *plane-table board*.

plane A general term for a two-dimensional geologic form that is without curvature; ideally, a perfectly flat or level surface, such as a bedding plane or a fault plane or an erosion surface produced in the ultimate stage of a geomorphic cycle. If suitably modified, it may refer to a curved plane, as curved bedding plane. Cf: *surface*.

plane bed A sedimentary bed "without elevations or depressions larger than the maximum size of the bed material" (Simons et al, 1961, p.vii). It is characteristic of the lower part of the upper flow regime.

plane coordinates (a) Two coordinates that represent the perpendicular distances of a point from a pair of axes that intersect at right angles, reckoned in the plane of those axes. (b) A coordinate system in a horizontal plane, used to describe the positions of points with respect to an arbitrary origin by means of plane coordinates. It is used in areas of such limited extent that the errors introduced by substituting a plane for the curved surface of the Earth are within the required limits of accuracy.---Syn: *rectangular coordinates; plane-rectangular coordinates.*

plane correction A correction applied to observed surveying data to reduce them to a common reference plane.

plane defect A type of crystal defect that occurs along the boundary plane (*lineage boundary* or *grain boundary*) of two regions of a crystal, or between two grains. See also: *stacking fault.*

plane fault A fault, the fault surface of which is planar rather than curved. Cf: *arcuate fault.*

plane group One of 17 two-dimensional patterns which can be produced by one asymmetric motif that is repeated by symmetry operations to produce a unit of pattern, which then is repeated by translation to build up an ordered pattern that fills any two-dimensional area. Cf: *space group.* Syn: *plane symmetry group.*

plane jet A flow pattern characteristic of *hyperpycnal inflow,* in which the inflowing water spreads as a parabola whose width is about three times the square root of the distance downstream from the mouth (Moore, p.87). Cf: *axial jet.*

plane of composition A term used by Cullison (1938, p.983) for the plane of contact between the part of an internal mold composed of material similar to the matrix in which the fossil shell was embedded and the material (such as secondary calcite) that partially filled the hollow interior of the shell. It defines the horizontal plane at the time the filling became hardened. Not to be confused with *composition plane.*

plane of contemporaneity A term used by Caster (1934, p.19 & 24) for the horizontal or nearly horizontal line between stratigraphic units (primarily formations) as seen in section; e.g. the line separating parvafacies belonging to the same magnafacies. Cf: *facies plane.*

plane of incidence A plane that contains an incident ray and the normal to the surface at the point of incidence.

plane of maximum shear stress Either of two planes that lie on opposite sides of the maximum principal stress axis at angles of 45° to it and that are parallel to the intermediate principal stress axis.

plane of mirror symmetry A symmetry element in a crystal that is a plane dividing the crystal into halves, one of which is the mirror image of the other. Syn: *mirror plane of symmetry; plane of symmetry; symmetry plane; reflection plane.*

plane of polarization *vibration plane.*

plane of saturation *water table.*

plane of stratification *bedding plane.*

plane of stretching A low-angle, normal fault caused by stretching of the solidified top of an igneous intrusion.

plane of symmetry [cryst] *plane of mirror symmetry.*

plane of symmetry [paleont] The plane that bisects a shell symmetrically.

plane of vibration *vibration plane.*

plane parallel texture The *parallel texture* of a rock in which the constituents are parallel to a plane but not to a line, as in *linear parallel texture.*

plane-polarized Said of a moving wave, e.g. of light, that has been polarized so that it vibrates in a single plane.

planerite A variety of coeruleolactite containing copper, or a variety of turquoise containing calcium.

plane strain A state of strain in which all displacements that arise from deformation are parallel to one particular plane.

plane stress A state of stress in which two of the principal stresses are always parallel to a given plane and are constant in the normal direction.

plane surveying Ordinary field and topographic surveying in which earth curvature is disregarded and all measurements are made or reduced parallel to a plane representing the surface of the Earth. The accuracy and precision of results obtained by plane surveying will decrease as the area surveyed increases in size. Cf: *geodetic surveying.*

plane symmetry group *plane group.*

planet (a) One of the nine celestial bodies of the solar system that revolve around the Sun in elliptical orbits and in the same direction. A planet shines only by reflected light. (b) A similar body in another solar system.

plane table A simple surveying instrument for graphically plotting the lines of a survey directly from field observations, consisting essentially of a small drawing board mounted on a tripod and fitted with a compass and a straightedge ruler (alidade) that is pointed at the object observed usually by means of a telescope or other sighting device. Also spelled: *planetable.*

plane-table board The drafting board of a plane-table instrument. Means are provided on the upper side for attachment of drawing paper, and a plate is threaded on the lower side for attachment to the tripod head. Syn: *planchette.*

plane-table map A map made by plane-table surveying methods. It includes maps made by complete field mapping on a base projection and by field contouring on a planimetric base map.

plane-tabling Plotting with a plane table; making use of a plane table.

planetary (a) Pertaining to the planets of the solar system. Cf: *terrestrial.* (b) Pertaining to the Earth as a whole.

planetary geology A science that applies geologic principles and techniques to the study of planets and their natural satellites. The term is frequently used as a syn. of *astrogeology.* Syn: *planetary geoscience.*

planetary wave A major, prominent atmospheric wave characterized by a long wavelength, a significant amplitude, and a velocity directed always to the west. Also, a similar free progressive wave in the model ocean, not yet convincingly demonstrated to exist there, but believed to be about 1600 km (1000 mi) long and to be caused by the gravitational attraction of the Sun and Moon on the Earth but largely governed by the depth of the water and by the Earth's rotational effects.

planetary wind Any wind system of the Earth's atmosphere which owes its existence and direction to solar radiation and to the rotation of the Earth, e.g. trade winds, mid-latitude westerlies.

planetesimal hypothesis *dust-cloud hypothesis.*

planetography The descriptive science of the physical features of planets.

planetoid *asteroid.*

planetology A term originally applied to the study and interpretation of surface markings of planets and their natural satellites, and later applied to the study of the condensed matter of the solar system, including planets, satellites, asteroids, meteorites, and interplanetary material. The term is frequently used as a syn. of *astrogeology,* and was redefined by Rankama (1962, p.519) as "the universal science that·studies the configuration and movements of matter and the accompanying energy transformations in planets, their natural satellites, and other cosmic bodies of a similar nature" in our solar system and other possible planetary systems.

planèze An erosional relief form consisting of a lava flow protecting the underlying volcanic cone. It may be a wedge-shaped unit on the slope of an erosionally dissected volcano, or a lava-capped plateau. Etymol: French, "lava plateau". Also spelled: *planeze.*

planimeter A mechanical instrument for measuring the area of any plane figure by means of a pointer or moving arm that traces its boundary or perimeter. It is used esp. for measuring irregular areas on a chart or map.

planimetric map A map that presents the relative horizontal positions only for the natural or cultural features represented. It is distinguished from a *topographic map* by the omission of relief in measurable form. Syn: *line map.*

planimetry (a) The measurement of plane surfaces; e.g. the determination of horizontal distances, angles, and areas on a map. (b) The plan details of a map; the natural and cultural features of a region (excluding relief) as shown on a map.

planisaic A photomap in which the planimetric detail is shown by overprints in colors. Cf: *toposaic.*

planispiral adj. Having the shell coiled in a single plane; esp. said of a gastropod shell formed by a spiral coiled in a single plane and ideally symmetric in that plane, and said of a coiled foraminiferal test with whorls of the coil in a single plane. Syn: *planospiral.*---n. A planispiral shell or test.

plankter An individual planktonic organism.

planktivorous Said of an organism that feeds on plankton.

plankton Aquatic organisms that drift or swim weakly. See also: *phytoplankton; zooplankton.* Adj: *planktonic.*

plankton bloom *water bloom.*

plankton equivalent A quantitative chemical relationship between one aspect of plankton and another, e.g. in phytoplankton, 1 mg carbon = 2.3 mg dry organic matter.

planktonic Said of that type of *pelagic* organism which floats; adj. of *plankton.* Cf: *nektonic.*

plankton snow *sea snow.*

planoconformity A term used by Crosby (1912, p.297) for the relation between conformable strata that are approximately uniform in thickness and sensibly parallel throughout.

plano-convex Flat on one side and convex on the other; e.g. said of a brachiopod shell having a flat brachial valve and a convex pedicle valve. Cf: *convexo-plane.*

planophyre A porphyritic rock characterized by *planophyric* texture.

planophyric A term, now obsolete, applied by Cross, et al (1906, p.703) to the texture of a porphyritic igneous rock in which the phenocrysts are arranged in layers or laminae in the groundmass; also, characteristic of or pertaining to a *planophyre.*

planorasion The process by which wind, working in conjunction with other erosional agents in a desert, "acts as an abrading and eroding agent" that works uphill (Hobbs, 1917, p.48); the process may grade a slope as steep as 4 degrees. See also: *antigravitational gradation.*

planosol An intrazonal, hydromorphic soil having a leached surface layer above a definite clay pan or hardpan. It is developed in a humid to subhumid climate.

planospiral Var. of *planispiral.*

plan-position indicator A radar-display device that consists of an oscilloscope (radarscope) equipped with a cathode-ray tube and a rotating antenna for scanning horizontally all or part of a complete circle and that presents visually or graphically in plan (map-like) position the range and direction (azimuth or bearing) of an object (such as a ship, building, cliff, or mountain) from which echoes are reflected in the form of spots of light whose brightness corresponds to the strength of the target signal detected by radar. The position of the radar itself is displayed in the center of the indicator. Abbrev: PPI.

plant Any member of the vegetable group of living organisms. Green plants are capable of manufacturing food from inorganic substances by photosynthesis. Plants lack the sensitivity of animals and are incapable of voluntary motion. Some tiny organisms share the characteristics of both plants and animals, e.g. Protista.

plant ash *inherent ash.*

plant indicator *indicator plant.*

plant opal *opal phytolith.*

planula The very young free-swimming larva of a coelenterate (such as of a coral polyp), consisting of an outer layer of cil-

iated ectoderm cells and an internal mass of endoderm cells. Pl: *planulae.* Syn: *parenchymula.*

planulate Said of a moderately evolute and compressed cephalopod shell with an open umbilicus and a bluntly rounded venter.

plash A shallow, standing, usually short-lived pool or small pond resulting from a flood, heavy rain, or melting snow; a puddle.

plasma [mineral] A faintly translucent or semitranslucent and bright-green, leek-green, or nearly emerald-green variety of chalcedony, sometimes having white or yellowish spots. The green color is attributed to chlorite. Cf: *bloodstone.*

plasma [soil] The part of a soil material that is capable of being, or that has been, moved, reorganized, and/or concentrated by soil-forming processes (Brewer & Sleeman, 1960) e.g. all mineral or organic material of colloidal size, and the relatively soluble material that is not bound up in *skeleton grains.* See also: *lac; floc.*

plaster conglomerate A conglomerate composed entirely of boulders derived from, and forming a wedge-like mass on the flank of, a partially exhumed monadnock.

plastering-on The addition of material to a ground moraine by the melting of ice at the base of a glacier (Gravenor & Kupsch, 1959, p.60). Cf: *lodgment.*

plaster stone *gypsum.*

plastic [biol] Having the capability of variation and phylogenetic change.

plastic [struc] Said of a body in which strain produces continuous, permanent deformation without rupture. Cf: *elastic.*

plastic deformation Permanent deformation of the shape or volume of a substance, without rupture, and which, once begun, is continuous without increase in stress. It is one of the processes of dynamothermal metamorphism. Cf: *elastic deformation; blastic deformation; clastic deformation.* Syn: *plastic flow; plastic strain.*

plastic equilibrium State of stress within a soil mass or a portion thereof that has been deformed to such an extent that its ultimate shearing resistance is mobilized (ASCE, 1958, term 263). Plastic equilibrium is "active" if it is obtained by an expansion of a soil mass, and "passive" if obtained by a compression of a soil mass.

plastic flow In structural geology, a syn. of *plastic deformation.*

plasticity index The water-content range of a soil at which it is plastic, defined numerically as the liquid limit minus the plastic limit.

plasticlast An intraclast consisting of calcareous mud that has been torn up while still soft (Folk, 1962). Cf: *protointraclast.*

plastic limit The water-content boundary of a sediment, e.g. a soil, between the plastic and semisolid states. It is one of the *Atterberg limits.* Cf: *liquid limit.*

plastic relief map A topographic map printed on plastic and then molded by heat and pressure into a three-dimensional form to emphasize the relief.

plastic shading Archaic syn. of *hill shading* (BNCG, 1966, p.31).

plastic strain *plastic deformation.*

plastic zone A region adjacent to the *rupture zone* of an explosion crater and at an increased distance from the shot site, differing from the rupture zone by having less fracturing and only small permanent deformations.

plastotype An artificial specimen molded or cast directly from a type specimen.

plastron The more or less inflated and enlarged adoral segment of the posterior interambulacral area of certain echinoids. Adj: *plastral.*

plat [cart] (a) A diagram drawn to scale, showing boundaries and subdivisions of a tract of land as determined by survey, together with all essential data required for accurate identifition and description of the various units shown and including one or more certificates indicating due approval. It differs from a map in that it does not necessarily show additional cultural, drainage, and relief features. (b) A precise and detailed plan or map representing a township, private land claim, mineral claim, or other surveyed area, and showing the actual or proposed divisions, special features, or uses of the land.---See also: plot.

plat [geog] An obsolete term for a plateau, tableland, or other expanse of open level land.

plate [geol] (a) A thin, flat, smooth rock fragment, such as a slab or a flagstone. (b) One of the large, nearly rigid, but still mobile segments or thin blocks involved in plate tectonics, with a thickness (50-250 km) that includes both crust and some part of the upper mantle.

plate [paleont] (a) Any discrete, normally flat or tabular ossicle in the skeleton of an echinoderm, composed of a single crystal of calcium carbonate. The term is sometimes used only for external plates, "but all calcareous bodies formed serve as framework of support for soft parts and constitute plates" (TIP, 1967, pt.S, p.113). (b) A structure consisting of inner and outer platforms and adjoining a portion of the axis of a plate-like conodont. The term is used incorrectly when referring to a platform. (c) A lamina that forms part of an animal body, such as a valve of a mollusk or crustacean, or an octocorallian sclerite too thick to be called a scale.

plate [snow] A snow crystal in the form of a flat, hexagonal plate..

plateau [geomorph] (a) Broadly, any comparatively flat area of great extent and elevation; specif. an extensive land region considerably elevated (more than 150-300 m in altitude) above the adjacent country or above sea level and commonly limited on at least one side by an abrupt descent, having a flat or nearly smooth surface but often dissected by deep valleys or canyons and surmounted by ranges of high hills or mountains, and having a large part of its total surface at or near the summit level. A plateau is usually higher and has more noticeable relief than a plain (it often represents an elevated plain), and is usually higher and more extensive than a mesa; it may be tectonic, residual, or volcanic in origin. See also: tableland. (b) A flat, upland region underlain by horizontal strata or characterized by horizontal structure, and which may be highly dissected; Davis (1885) introduced this concept of a "plateau" but the term should be used without regard to the underlying geologic structure.---Etymol: French. Pl: plateaus; plateaux.

plateau [marine geol] A broad, more or less flat-topped and well-defined elevation of the sea floor, generally over 200 m in elevation. Syn: submarine plateau.

plateau basalt An extensive, thick and smooth basaltic lava flow or successive flows of high-temperature, fluid basalt from fissure eruptions, accumulating to form a plateau, e.g. the Columbia-Snake Plateau of the northwestern U.S. Cf: shield basalt. Syn: flood basalt.

plateau eruptions Successive lava flows from fissures that spread in sheets over a large area. Cf: fissure eruption.

plateau glacier A highland glacier overlying a relatively flat mountain tract, and usually overflowing its edges in hanging glaciers. See also: ice plateau.

plateau gravel A sheet, spread, or patch of surficial gravel, often compacted, occupying a flat area on a hilltop, plateau, or other high region at a height above that normally occupied by a stream-terrace gravel. It may represent a continuous and extensive deposit that had been raised by earth movements and deeply dissected.

plateau mountain A pseudomountain produced by the dissection of a plateau; e.g. the Catskill Mountains, N.Y.

plateau plain An extensive plain surmounted by a sublevel summit area and bordered by escarpments (Hill, 1900, p.8).

Cf: mesa plain.

plate carrée projection A simple cylindrical projection with an evenly spaced network of horizontal parallels (spaced at their correct meridional distance from the equator) and vertical meridians (spaced at their correct equatorial distances). Only cardinal directions are true, and scale is true on all meridians and the standard parallel but is greatly distorted away from the center. If the equator is the standard parallel, the network consists of squares; any other standard parallel will produce rectangles with the north-south dimension the longer. The projection is neither equal-area nor conformal, and is used in geologic mapping for small areas and in geographic referencing for large-scale city maps. A modified version has the scale preserved along two parallels of latitude other than the equator. Etymol: French, "regular-square projection". Syn: equirectangular projection.

plate-equivalent Said of the part of the dinoflagellate-cyst wall judged to occupy a position equivalent to that occupied by a plate of the theca.

platelet A small ice crystal which, when united with other platelets, forms a layer of floating ice, esp. sea ice, and serves as seed crystals for further thickening of the ice cover. Platelets in sea ice retain their identity for some time because they are bounded by rows or layers of brine cells.

plate-like conodont element A conodont element having platforms or a greatly expanded basal cavity (cup).

plate-scale One of the oval to elliptic organic scales embedded in the surface layer of the periplast (cell membrane) of a coccolithophorid.

plate tectonics Global tectonics based on an Earth model characterized by a small number (10-25) of large, broad, thick plates (blocks composed of areas of both continental and oceanic crust and mantle) each of which "floats" on some viscous underlayer in the mantle and moves more or less independently of the others and grinds against them like ice floes in a river, with much of the dynamic activity concentrated along the periphery of the plates which are propelled from the rear by sea-floor spreading. The continents form a part of the plates and move with them, like logs frozen in the ice floes. Cf: raft tectonics.

platform [coast] A flat or gently sloping underwater erosional surface extending seaward or lakeward from the shore; specif. a wave-cut platform or an abrasion platform. See also: wave-built platform.

platform [geomorph] (a) A general term for any level or nearly level surface; e.g. a terrace or bench, a ledge or small space on a cliff face, a flat and elevated piece of ground such as a tableland or plateau, a peneplain, or any beveled surface. (b) A small plateau.

platform [paleont] (a) A laterally broadened structure or shelf along the inner side or the outer side of the anterior-posterior axis of a conodont. The term is also commonly used interchangeably with plate. (b) The muscle platform of a brachiopod. (c) The flat bottom or floor of the calyx of a coral.

platform [tect] That part of a continent which is covered by flat-lying or gently tilted strata, mainly sedimentary, which are underlain at varying depth by a basement of rocks that were consolidated during earlier deformations. Platforms are parts of the cratons.

platform beach A looped bar or ridge of sand and gravel formed on a wave-cut platform, as on Madeline Island along the Wisconsin shore of Lake Superior (Collie, 1901, p.212).

platform facies shelf facies.

platform reef An organic reef, generally small but more extensive than a patch reef, with a flat upper surface. Platform reefs are common off the coast of Australia. Cf: table reef.

platidiiform Said of a brachiopod loop consisting of descending branches from the cardinalia to the median septum, with only rudimentary prongs on the septum representing the as-

547

cending part of the loop (TIP, 1965, pt.H, p.150).

platina Crude native platinum.

platiniridium A silver-white cubic mineral: (Ir,Pt). It is a native alloy of iridium with platinum and other related metals.

platinite *platynite.*

platinum A very heavy, steel-gray to silvery- or grayish-white, isometric mineral, the native metallic element Pt, commonly containing palladium, iridium, iron, and nickel. It occurs as grains and nuggets in alluvial deposits (often associated with nickel sulfide and gold ores), and disseminated in basic and ultrabasic igneous rocks. Platinum is a highly corrosion-resistant, ductile, and malleable metal, and is the most abundant metal of the platinum group. Syn: *polyxene.*

platte A resistant knob of rock in a glacial valley or rising in the midst of an existing glacier, often causing a glacier to split near its snout. Etymol: German *Platte*, "slab". Pl: *platten.*

platting The action or process of mapping a surveyed area; the making of a plat.

plattnerite An iron-black mineral: PbO_2.

platy (a) Said of a sedimentary particle whose length is more than three times its thickness (Krynine, 1948, p.142). Cf: *acicular.* (b) Said of a sandstone or limestone that splits into laminae having thicknesses in the range of 2 mm to 10 mm (McKee & Weir, 1953, p.383).

platycone A coiled cephalopod shell with a flattened form without implication as to the width of the umbilicus or the shape of the venter.

platy flow structure An igneous rock structure of tabular sheets suggesting stratification. It is formed by contraction during cooling; the structure is parallel to the surface of cooling, and is commonly accentuated by weathering. Syn: *platy structure; linear flow structure; planar flow structure.*

platykurtic (a) Said of a frequency distribution that has a concentration of values about its mean less than for the corresponding normal distribution. (b) Said of a broad, flat-topped frequency distribution curve that is less peaked than the corresponding normal distribution curve.---Cf: *leptokurtic; mesokurtic.*

platynite An iron-black mineral: $PbBi_2(Se,S)_3$. It occurs in thin metallic plates resembling graphite. Syn: *platinite.*

platyproct Said of a sponge in which the exhalant surface is nearly or quite flat.

platy structure *platy flow structure.*

plauenite A potassic, plagioclase-rich syenite.

playa [coast] (a) A small, generally sandy land area at the mouth of a stream or along the shore of a bay. (b) A flat, alluvial coastland, as distinguished from a beach.---Etymol: Spanish, "beach, shore, strand, coast". Pron: *ply-ah.*

playa [geomorph] (a) A term used in SW U.S. for a dried-up, vegetation-free, flat-floored area composed of thin, evenly stratified sheets of fine clay, silt, or sand, and representing the bottom (lowermost or central) part of a shallow, completely closed or undrained, desert lake basin in which water accumulates (as after a rain) and is quickly evaporated, usually leaving deposits of soluble salts. It may be hard or soft, and smooth or rough. The term is also applied to the basin containing an expanse of playa. See also: *salina; alkali flat; salt flat; salt pan; salar; salada.* Syn: *dry lake; vloer; sebkha; kavir.* (b) A term that is often used for a *playa lake.*

playa basin *bolson.*

playa furrow A shallow but distinct indentation or trail left on a playa by a *playa scraper* moving across the still-moist surface.

playa lake A shallow, intermittent lake in an arid or semiarid region, covering or occupying a playa in the wet season but drying up in summer; an ephemeral lake that upon evaporation leaves or forms a playa. Syn: *playa.*

playa scraper An object (usually a cobble or boulder) that produces a *playa furrow.*

playfairite A mineral: $Pb_{16}Sb_{18}S_{43}$.

Playfair's law A generalized statement about the relation stream systems to their valleys in areas of uniform bedroc and structure which have been subject to stream erosion for long period of time; viz. that streams cut their own valle which are proportional in size to the streams they contai and that the stream junctions in these valleys are accorda in level (the latter is also known as the *law of accordant jun tions*). For a quantitative statement of a major part of Pla fair's law, see Horton (1945, p.293, eq.17). The law wa enunciated by John Playfair (1747-1819), professor of natur philosophy at Univ of Edinburgh: "Every river appears to co sist of a main trunk, fed from a variety of branches, each ru ning in a valley proportioned to its size, and all of them t gether forming a system of valleys, communicating with on another, and having such a nice adjustment of their decliv ties, that none of them join the principal valley, either on to high or too low a level; a circumstance which would be infi itely improbable, if each of these valleys were not the work the stream which flows in it" (Playfair, 1802, p.102).

play of color An optical phenomenon consisting of flashes of variety of prismatic colors, seen in rapid succession as ce tain minerals (esp. opal) or cabochon-cut gems are move about; e.g. opalescence. It is caused by diffraction of lig from innumerable, minute, regularly arranged, optically trans parent, uniform spherical particles of amorphous silica (an from the spaces between these particles), stacked in an o derly three-dimensional pattern that behaves like a diffractio grating. Cf: *fire; change of color.* Syn: *schiller.*

plaza A term used in the SW U.S. for the exceptionally wid floor of a flat, open valley; the flat bottom of a shallow can yon. Etymol: Spanish, "square, marketplace".

plazolite *hydrogrossular.*

pleat A longitudinal fold of retractile muscle fibers with assoc ated mesogloea on the side of a coral mesentery.

plectolophe A brachiopod lophophore in which each brachiur consists of a U-shaped side arm bearing a double row o paired filamentar appendages "but terminating distally in me dially placed plano-spire normal to commissural plane an bearing single row of paired appendages" (TIP, 1965, pt.H p.150). Cf: *deuterolophe; spirolophe.*

Pleistocene An epoch of the Quaternary period, after the Plio cene of the Tertiary and before the Holocene; also, the corre sponding worldwide series of rocks. When the Quaternary i designated as an era, the Pleistocene is considered to be period. Syn: *Ice Age; Great Ice Age; glacial epoch; Oiluvium.*

plenargyrite *matildite.*

pleochroic Said of a mineral that displays *pleochroism.*

pleochroic formula An expression of a crystal's pleochroism or the color of transmitted light. Cf: *absorption formula.*

pleochroic halo A minute sphere of color or darkening sur rounding and produced by a radioactive inclusion. Less-pre ferred syn: *radiohalo.*

pleochroic halo dating The increase in color darkening of th halo (*pleochroic halo*) of alpha-particle radiation damag around a zircon, monazite, xenotine, or apatite crystal is mea sured as a function of time and alpha-particle activity. How ever, so many variables (e.g. mica sensitivity to alpha radia tion, thermal annealing, color reversal) have been discovere that this dating method has only limited application.

pleochroism The ability or property of an anisotropic crystal t absorb differentially various wavelengths of transmitted light i various crystallographic directions, and thus to show differen colors in different directions. This property is more easily see under polarized light than by the naked eye. A mineral show ing pleochroism is said to be *pleochroic.* Cf: *bireflectance* Syn: *polychroism.* See also: *dichroism; trichroism.*

pleocrystalline An obsolete syn. of *holocrystalline.*

pleomere A somite of the abdomen of a malacostracan crus tacean. Syn: *pleonite*.

pleomorph *polymorph*.

pleomorphism *polymorphism*.

pleomorphous *polymorphic*.

pleon The abdomen of a malacostracan crustacean.

pleonaste *ceylonite*.

pleonastite An igneous rock similar in structure to diabase and composed of ceylonite, hercynite, and clinochlore surrounding corundum crystals (Thrush, 1968, p.836).

pleonite *pleomere*.

pleopod An abdominal limb of a crustacean; specif. any appendage of the pleon of a malacostracan crustacean, excluding caudal ramus and uropod. Syn: *pleopodite*.

pleosponge *archaeocyathid*.

pleotelson A structure of a malacostracan crustacean resulting from fusion of one or more abdominal somites (pleomeres) with telson.

plerotic water A syn. of *ground water* proposed by Meinzer (1939), including *piestic water* and *phreatic water*.

plesiostratotype A complementary stratotype (Sigal, 1964).

plessite A meteorite mineral consisting of an intimate, fine-grained intergrowth of kamacite and taenite. It occurs as triangular or polygonal areas in iron meteorites exhibiting Widmanstätten structure.

pleura (a) A laterally located part of the body of an invertebrate; e.g. a lateral part of the opisthosoma of a merostome. (b) A term used both as a synonym and a plural of *pleuron* of a trilobite and of a crustacean.---Pl: *pleurae*.

pleural angle The angle between two straight lines lying tangential to the last two whorls on opposite sides of a gastropod shell.

pleural furrow (a) A groove along the surface of a pleuron of a trilobite. (b) A groove crossing the pleura of a merostome.

pleuralia Sponge spicules (prostalia) on the sides of the body.

pleural spine A pointed or sharply rounded extension of the distal end of a pleuron of a trilobite. It is narrower than the medial part of the pleuron.

pleural suture The line of splitting apart in molting of the carapace of a decapod crustacean. It is present in all brachyurans.

pleurite *epimere*.

pleurocyst A secondary calcareous layer in some cheilostomatous bryozoans, spread inward over the *holocyst* from the region of the areolar pores.

pleuromyarian Said of a nautiloid in which the retractor muscles of the head-foot mass are attached to the shell along the lateral areas of the interior of the body chamber (TIP, 1964, pt.K, p.57). Cf: *dorsomyarian; ventromyarian*.

pleuron (a) One of the two lateral parts of each exoskeletal segment of a trilobite that extend outward from its axis; the portion of thoracic segment or pygidium that is lateral to the axial lobe (axis) of a trilobite. (b) An *epimere* of a crustacean.---Pl: *pleura*. Syn: *pleura*.

pleuston A community of organisms whose environment is the surface of the ocean; a marine type of *neuston*.

plexus [glac geol] An area, on a subglacial deposit, that encloses a giant's kettle (Stone, 1899).

plexus [evol] *evolutionary plexus*.

plica (a) One of the strong, sharp, parallel ridges and depressions involving the entire thickness of a bivalve shell (mollusk or brachiopod), extending radially from beak to shell margin, and appearing as corrugations on the inner as well as the outer surface of the shell; e.g. a major undulation of the commissure of a brachiopod, with crest directed dorsally, commonly but not invariably associated with the dorsal fold and the ventral sulcus (TIP, 1965, pt.H, p.150). A brachiopod plica is distinguished from a fold or sulcus by smaller amplitude and by occurrence to the sides of the midline. Syn: *plication [paleont]*. (b) A term used, irrespective of commissure,

for a small carina or fold in the surface of a brachiopod valve. --Pl: *plicae*.

plicated Adj. of *plication*. Syn: *crumpled*.

plication [paleont] A coarse radial corrugation in the surface of a bivalve-mollusk or brachiopod shell; specif. a *plica*.

plication [struc geol] Intense, small-scale folding. Adj: *plicated*. Cf: *crenulation*.

Pliensbachian European stage: Lower Jurassic (above Sinemurian, below Toarcian).

Plinian-type eruption *Vulcanian-type eruption*.

plinth A term suggested by Bagnold (1941, p.229) for the lower and outer part of a seif dune, beyond the slip-face boundaries, that has never been subjected to sand avalanches.

plinthite In a soil, a material consisting of a mixture of clay with quartz with other diluents, that is rich in sesquioxides and poor in humus and is highly weathered. It occurs as red mottles in a platy, polygonal, or reticulate pattern. Repeated wetting and drying changes plinthite to ironstone hardpan or irregular aggregates.

Pliocene An epoch of the Tertiary period, after the Miocene and before the Pleistocene; also, the corresponding worldwide series of rocks. It is sometimes considered to be a period, when the Tertiary is designated as an era.

pliomagmatic zone *eugeosyncline*.

pliothermic Pertaining to a period in geologic history characterized by more than average warmth of climate. Cf: *miothermic*.

plocoid Said of a massive scleractinian corallum in which corallites have separated walls and are united by costae, dissepiments, or coenosteum.

plot To place survey data upon a map or plat; to draw to scale. The term was formerly used in noun form as a syn. of *plat*.

plowshare A wedge-shaped feature developed on a snow surface by further ablation of *foam crust*. Syn: *ploughshare*.

plow sole A *pressure pan* representing a layer of soil compacted by repeated plowing to the same depth. Also spelled: *plowsole*. Syn: *hardpan*.

plucking [streams] *hydraulic plucking*.

plucking [glac geol] The process of glacial erosion by which sizable rock fragments, such as blocks, are loosened, detached, and borne away from bedrock by the freezing of water along joints and stratification surfaces with resulting removal of rock as the ice advanced. See also: *sapping*. Syn: *quarrying*.

pluck side The downstream, or lee, side of a roche moutonnée, roughened and steepened by glacial plucking. Ant: *scour side*.

plug [drill] n. A watertight or gastight seal installed in a borehole or well to prevent movement of fluids, such as a block cemented inside the casing or a cylindrical piece of wood inserted in the hole.---v. To insert a plug in a borehole; to fill in or seal off cracks, cavities, or other openings in the walls of a borehole.

plug [pat grd] (a) A cohesive, commonly vertical column of gravelly material with considerable fines, representing the continuance at depth of a sorted circle in a gravel beach, as on Victoria Island, Canada (Washburn, 1956, p.844). (b) A similar column-like feature occurring with a mud (nonsorted) circle (Bird, 1967, p.194).

plug [paleont] *umbilical plug*.

plug [volc] (a) A vertical, pipelike body of magma that represents the conduit to a former volcanic *vent*. Cf: *neck [volc]*; *plug [volc]*. (b) A crater filling of lava, the surrounding material of which has been removed by erosion.

plug [sed] A mass of sediment filling the part of a stream channel abandoned by the formation of a cutoff; e.g. a *clay plug* or a *sand plug*. See also: *valley plug*.

plug dome A volcanic dome characterized by an upheaved, consolidated conduit filling (Williams, 1932).

plugging (a) The act or process of stopping the flow of water, oil, or gas in strata penetrated by a borehole or well so that fluid from one stratum will not escape into another or to the surface; esp. the sealing up of a well that is dry and is to be abandoned. It is usually accomplished by inserting a plug into the hole, by sealing off cracks and openings in the sidewalls of the hole, by cementing a block inside the casing, or by capping the hole with a metal plate. (b) The act or process of drilling a borehole with a noncoring bit.

plugging back The act or process of cementing off a lower section of casing, or of blocking fluids below from rising in the casing to a higher section being tested.

plug reef A small triangular reef that grows with its apex pointing seaward through openings between linear shelf-edge reefs (Maxwell, 1968, p.101). Its outline is analogous with that of a sand ridge formed in the lower reach of a large river. Plug reefs are found off the coast of Australia where high tide range results in strong currents.

plum A clast embedded in a matrix of a different kind; esp. a pebble in a conglomerate.

plumalsite A mineral: $Pb_4Al_2(SiO_3)_7$.

plumasite A coarsely xenomorphic-granular hypabyssal rock of variable composition, being composed chiefly of corundum crystals enclosed in oligoclase grains.

plumb (a) Vertical. (b) A *plumb bob*.

plumbago A syn. of *graphite*. The term has also been applied to graphitic rock, to an impure graphite, and to graphitoid minerals such as molybdenite.

plumb bob A conical, metal device suspended by a cord, used to project a point vertically in space for short distances. Syn: *plumb*.

plumbic ocher *lead ocher*.

plumb line The line of force in the geopotential field; a continuous curve to which the direction of gravity is everywhere tangential.

plumboferrite A dark hexagonal mineral: $PbFe_4O_7$. Cf: *magnetoplumbite*.

plumbogummite (a) A mineral: $PbAl_3(PO_4)_2(OH)_5 \cdot H_2O$. (b) A group of isostructural minerals consisting of plumbogummite, gorceixite, goyazite, crandallite, florencite, and dussertite, and related to alunite and other sulfates isostructural with it.

plumbojarosite A mineral of the alunite group: $PbFe_6(SO_4)_4(OH)_{12}$.

plumbonacrite A mineral: $Pb_{10}(CO_3)_6(OH)_6O$ (?).

plumboniobite A dark-brown to black mineral of complex composition, consisting of a niobate of yttrium, uranium, lead, iron, and rare earths. It resembles samarskite, and may be a lead-bearing variety of samarskite.

plumbopyrochlore A mineral of the pyrochlore group: $(Pb,Y,U,Ca)_{2-x}Nb_2O_6(OH)$.

plumb point A nadir; specif. *ground nadir*.

plum-cake rock A quarryman's term applied in northern England to a breccia or breccia-conglomerate having sporadic fragments embedded in a preponderant matrix like plums in a pudding.

plume A flaw in a gem, as in an agate.

plume structure On the surface of a master joint, a ridgelike tracing in a plumelike pattern, usually oriented parallel to the upper and lower surfaces of the constituent rock unit. Syn: *plumose structure*. Less-preferred syn: feather fracture.

plumicome A sponge spicule (hexaster) in which the terminal branches are S-shaped and arranged in several tiers, forming a plume-like structure.

plumose mica A feathery variety of muscovite mica.

plumose spiculofiber A spiculofiber of an axinellid or ectyonine sponge skeleton, in which some or all of the component spicules face obliquely outward.

plumose structure *plume structure*.

plumosite An antimony-sulfide mineral having a feathery for e.g. jamesonite and boulangerite. Syn: *plumose ore*.

plum-pudding stone *puddingstone*.

plunge [struc geol] The inclination of a fold axis or oth geologic structure, measured by its departure from horizont It is mainly used for the geometry of folds. Cf: *dip*. Syn: *pi* [*struc geol*]; *rake*.

plunge [surv] v. (a) To set the horizontal cross wire of a the dolite in the direction of a grade when establishing a gra between two points of known level. (b) *transit*.

plunge basin A deep, relatively large hollow or cavity scour in the bed of a stream at the foot of a waterfall or cataract the force and eddying effect of the falling water. It is oft called a *plunge pool*. Cf: *pothole* [*streams*].

plunge line *breaker line*.

plunge point The line along which a plunging wave curls ov and collapses as it approaches the shore.

plunge pool (a) The water in a plunge basin. (b) A deep, c cular lake occupying a plunge basin after the waterfall h ceased to exist or the stream has been diverted. Syn: *wate fall lake*. (c) A small, deep *plunge basin*.

plunging breaker A type of *breaker* whose crest curls ov and collapses suddenly, with complete disintegration of t wave. Cf: *surging breaker; spilling breaker*.

plunging cliff A sea cliff bordering directly on deep water, ha ing a base that lies well below water level.

plunging fold A fold, the plunge of which is relatively stee Cf: *nonplunging fold; doubly plunging fold*. Syn: *pitching fold*.

pluricolumnal Two or more crinoid columnals attached to o another.

plush copper ore *chalcotrichite*.

plus sight *backsight*.

plutology The study of the interior of the Earth.

pluton (a) An igneous *intrusion*. (b) A body of rock formed metasomatic replacement. --The term originally signified on deep-seated or plutonic bodies of granitoid texture. See als *plutonism*.

plutonian Var. of *plutonic*.

plutonic (a) Pertaining to igneous rocks formed at gre depth. See also: *plutonic rock*. Cf: *hypabyssal*. (b) Pertaini to rocks formed by any process at great depth. ----Syn: *aby sal; plutonian; deep-seated; hypogene*.

plutonic breccia A breccia composed of angular fragments an older rock enclosed in younger plutonic rock.

plutonic cognate ejecta Pyroclastic fragments that were soli ified at depth but were brought to the surface by the erupti magma.

plutonic event A term proposed by Gilluly (1966, p.97) f "concentrations of radiometric dates" regarding named oroge nies, in his criticism of some of the criteria used for namin such orogenies.

plutonic metamorphism Deep-seated regional metamorphis at high temperatures and pressures, often accompanied b strong deformation; batholithic intrusion with accompanyin metasomatism, infiltration and injection phenomena (or, alte natively, differential fusion or anatexis) are characterist (Turner, 1948). Cf: *injection metamorphism*.

plutonic nodule *nodule* [ign].

plutonic rock A rock formed at considerable depth by crysta lization of magma or by chemical alteration. It is characterist cally medium- to coarse-grained, of granitoid texture.

plutonic water *Juvenile water* in, or derived from, magma at considerable depth, probably several kilometers. Cf: *magma tic water; volcanic water*.

plutonism (a) A general term for the phenomena associate with the formation of *plutons*. (b) The conception of the fo mation of the Earth by solidification of a molten mass. Th theory was promulgated by Hutton in the 18th Century.

plutonist A believer in the theory of plutonism as promulgated by Hutton. Ant: *neptunist*. Syn: *volcanist*.

plutonite A general term for a plutonic rock.

pluvial [geomorph] Said of a geologic episode, change, process, deposit, or feature resulting from the action or effects of rain; e.g. *pluvial* denudation, a landslide, or gully erosion and the consequent spreading out of the eroded material below. The term sometimes includes the fluvial action of rainwater flowing in a stream channel, esp. in the channel of an ephemeral stream.

pluvial [meteorol] (a) Pertaining to rain, or more broadly, to precipitation. Syn: *pluvious*. (b) Characterized by or regularly receiving abundant rain. Syn: *pluviose*.----Cf: *hyetal*.

pluvial [clim] Said of an episode of time, esp. one corresponding to a glacial age, characterized by abundant rainfall; also, said of such a climate.

pluvial lake A lake formed in a period of exceptionally heavy rainfall; specif. a lake formed in the Pleistocene epoch during a time of glacial advance, and now either extinct or existing as a remnant. Example: Lake Bonneville, a prehistoric lake in present Utah, eastern Nevada, and southern Idaho.

pluviofluvial Pertaining to the combined action of rainwater and streams; e.g. *pluviofluvial* denudation.

pluviograph (a) A self-registering *rain gage*. (b) A graph showing stream flow if all precipitation took the form of surface runoff (Nelson & Nelson, 1967).

pluviometer *rain gage*.

pluviometric coefficient The mean amount of rainfall at a given place and during a specific time interval, expressed as a percentage of the average rain fall (Swayne, 1956).

pneumatocyst A hollow area of a stipe which serves to keep some phaeophytes afloat.

pneumatogenic Said of a rock or mineral deposit formed by a gaseous agent. Cf: *hydatogenic; hydatopneumatogenic*.

pneumatolysis Alteration of a rock or crystallization of minerals by gaseous emanations from solidifying magma. Adj: *pneumatolytic* [petrology].

pneumatolytic [ore dep] *pneumatogenic*.

pneumatolytic [petrology] (a) Formed by *pneumatolysis*. (b) Applied to the surface products of gaseous emanations near volcanoes. (c) Applied to the stage of magmatic differentiation between the pegmatitic and hydrothermal stages. (d) Said of the effects of contact metamorphism adjacent to deep-seated intrusions.

pneumatolytic differentiation Magmatic differentiation by the process of *gaseous transfer*. Syn: *gaseous transfer differentiation*.

pneumatolytic metamorphism Contact metamorphism accompanied by strong metasomatism resulting from the chemical action of magmatic gases upon both the intrusive and country rock (Tyrrell, 1926).

pneumatolytic stage That stage in the cooling of a magma during which the solid and gaseous phases are in equilibrium.

pneumotectic Said of processes and products of magmatic consolidation affected to some degree by gaseous constituents of the magma.

pocket [eco geol] (a) A small, discontinuous occurrence or patch of ore, e.g. a mineralized cavity or crevice. See also: *pocket*. (b) A localized enrichment of an ore deposit. Syn: *belly*.

pocket [geog] (a) A *water pocket* in the bed of an intermittent stream. (b) A hollow or glen in a mountain.

pocket [coast] An enclosed or sheltered place along a coast, such as a reentrant between rocky cliffed headlands or a bight on a lee shore.

pocket [cryoped] The downfolded or sagging, convex part of a layer in a congeliturbate. Ant: *festoon*.

pocket [speleo] A concavity in the ceiling, walls, or floor of a cave; it is a solutional feature, the form of which is not structurally controlled, as it is, for instance, in a *ceiling cavity*.

pocket [grd wat] A colloquial term for a small body of ground water.

pocket [ice] *blind lead*.

pocket beach A small, narrow beach formed in a *pocket*, commonly crescentic in plan and concave toward the sea, and generally displaying well-sorted sands; a *bayhead beach*.

pocket lens *hand lens*.

pocket penitente A nieve penitente on the north side of whose base there is a water-filled depression in the ice, 30-60 cm deep, rounded or oval in outline and with perpendicular walls, at the bottom of which is a thin layer of debris that had absorbed solar heat and melted the ice (Workman, 1914, p.306). Such features are formed in late summer in the Himalayas.

pocket rock A term used in the SW U.S. for a desert boulder with a hard case of desert varnish.

pocket transit A small, compact surveyor's transit that fits in a pocket; specif. a *Brunton compass*.

pocket valley A valley whose head is enclosed by steep walls at the base of which underground water emerges as a spring. It is similar in shape to a *blind valley*. See also: *steephead*.

pocosen *pocosin*.

pocosin A local term along the Atlantic coastal plain south of Virginia for a swamp or marsh on a flat upland, bordering on or near the sea, in many places enclosing knobs or hummocks. Etymol: American Indian (Delaware). Syn: *pocoson; pocosen; dismal*.

pocoson *pocosin*.

pod [econ geol] An orebody of *podiform* shape.

pod [geomorph] A term used in the steppes of southern Russia for a very shallow depression as much as 10 km in diameter, containing an intermittent lake or lakes; it may indicate uneven loess deposition, preloess topography, deflation, or solution.

pod [meta] A term formerly used to describe certain bodies that are long in one dimension and short in two dimensions and are enclosed in schist with the long axis parallel to the schistosity.

podial pore A pore admitting the passage of a tube foot between the ambulacral plates of an echinoderm (such as of an edrioasteroid). See also: *tentacle pore*.

podiform Said of an orebody of an elongate, lenticular shape, e.g. chromite in alpine-type peridotite, or uranium-bearing cavities. Noun: roll.

podite A limb segment of an arthropod; e.g. a joint of a biramous appendage of a trilobite.

podium The cylindrical outer part of a tube foot of an echinoderm. The term has also applied to the tube foot itself. Pl: *podia*.

podoconus Internal cone within the central capsule of a nasselline radiolarian.

podocope Any ostracode belonging to the order Podocopida, characterized by a calcified shell with a curved dorsal margin or a straight dorsal margin shorter than the total length of the shell and by a muscle scar pattern consisting usually of a few secondary scars. Their stratigraphic range is Lower Ordovician to present.

podolite *carbonate-apatite*.

podomere An individual leg *segment* of an arthropod, connected by articulation with adjoining segments.

podophthalmite The distal segment of the eyestalk of a decapod crustacean, bearing the corneal surface of the eye. Cf: *basiophthalmite*.

podostyle A mass of cytoplasm that projects from the test aperture of monothalamous foraminifers and that gives rise to pseudopodia (TIP, 1964, pt.C, p.62).

Podsol Var. of *Podzol*.

Podzol A group of zonal soils having a surface layer of mats of organic material and thin horizons of organic minerals overlying gray, leached horizons and dark brown, illuvial horizons. It develops under coniferous or mixed forests or under heath, under a cool to temperate, moist climate. Also spelled: *Podsol*. Etymol: Russian *podsol*, "ash soil".

podzolization The process by which a soil becomes more acid due to the depletion of bases, and develops surface layers that have been leached of clay; the development of a *podzol*.

poêchore A climatic term for the part of the Earth's surface represented by steppes.

poecilitic The original, but now obsolete, spelling of *poikilitic*.

poeciloblast Var. of *poikiloblast*.

poeciloblastic Var. of *poikiloblastic*.

poenite A potassium spilite, such as those of Timor, formed by *adularization*.

Poikilitic An old term for the Permian and the Triassic.

poikilitic Said of the texture of an igneous rock in which small crystals of one mineral (e.g. plagioclase) are irregularly scattered without common orientation in a larger crystal of another mineral (e.g. pyroxene); also, said of the enclosed crystal. The larger crystal is typically anhedral and exhibits optical and crystallographic continuity; in hand specimen, this texture produces lustrous patches (*luster mottling*) due to reflection from cleavage planes. Originally spelled *poecilitic*. Cf: *ophitic; endoblastic*. Nonrecommended syn: *semipegmatitic*.

poikiloblast A large crystal (xenoblast) formed by recrystallization during metamorphism and containing numerous inclusions of small idioblasts. See also: *poikiloblastic*. Also spelled: *poeciloblast*.

poikiloblastic (a) Pertaining to a *poikiloblast*. (b) Said of a metamorphic texture due to the development, during recrystallization, of a new mineral around numerous relicts of the original minerals, thus simulating the poikilitic texture of igneous rocks. Cf: *helicitic*. (c) A metamorphic texture in which small idioblasts of one constituent lie within larger xenoblasts (Johannsen, 1931). Modern usage favors this meaning. Syn: *sieve texture*.----Also spelled: *poeciloblastic*.

poikilocrystallic A syn. of *poikilotopic*. The term was introduced by Phemister (1956, p.74).

poikilophitic Said of *ophitic* texture characterized by lath-shaped feldspar crystals completely included in large, anhedral pyroxene crystals (Schieferdecker, 1959, term 5255). Cf: *poikilitic*.

poikilotherm A poikilothermic organism; e.g. a frog. Syn: *heterotherm*.

poikilothermic Said of an organism having no internal mechanism for temperature regulation; having a body temperature that varies with the temperature of the environment; cold-blooded. Syn: *heterothermic*. Cf: *homoiothermic*.

poikilotope A large crystal enclosing smaller crystals of another mineral in a sedimentary rock showing poikilotopic fabric; e.g. a large calcite crystal enclosing smaller relics of incompletely replaced dolomite crystals in a dedolomitized rock, or a large gypsum crystal enclosing numerous grains of quartz and/or feldspar.

poikilotopic Said of the fabric of a crystalline sedimentary rock (recrystallized carbonate rock or chemically precipitated sediment) in which the constituent crystals are of more than one size and in which larger crystals enclose smaller crystals of another mineral. The term was proposed by Friedman (1965, p.651). Cf: *porphyrotopic*. Syn: *poikilocrystallic*.

point [coast] A tapering tract of relatively low land, such as a small *cape*, projecting from the shore into a body of water; specif. the tip section or extremity of such a projection, or the sharp outer end of any land jutting out into a body of water. Syn: *tongue*.

point [geomorph] A sharp, tapering or projecting, rocky prominence; esp. a peak of a mountain range.

point [gem] A unit of weight for diamonds and other gem-

stones, equal to 1/100 (0.01) part of a *carat*. A stone weighing 32/100 carat is called a 32-point stone or a 32 pointer.

point [surv] (a) One of the 32 precisely marked equidistant spots about the circumference of a circular card attached to compass that indicate the direction in which the various parts of the horizon lie; e.g. a *cardinal point*. The term is also applied to the angular distance of 11.25 degrees between two such successive points, and to the part of the horizon indicated precisely or approximately by a point of a compass card. (b) A *position* on a reference system determined by a survey and represented by a fix; e.g. a "point of observation".

point bar One of a series of low, arcuate ridges of sand and gravel developed on the inside of a growing meander by the slow addition of individual accretions accompanying migration of the channel toward the outer bank. Cf: *channel bar; meander scroll*. Syn: *meander bar*.

point-bar deposit A deposit consisting of a series of alternating *point bars* and intervening troughs.

point counter analysis A statistical method involving the estimation of the frequency of occurrence of an object, such as a fossil or mineral species, in a sample, determined by counting the number of times that object occurs at specified intervals throughout the sample. The analysis is commonly made with an automatic point counter attached to a microscope.

point defect A type of *crystal defect* occurring at a particular point and involving a particular atom in a lattice. See also: *interstitial defect; Frenkel defect; Schottky defect*.

point diagram A *fabric diagram* in which each item measured, e.g. preferred orientation of fabric elements, is represented by a point. Cf: *contour diagram*. Syn: *scatter diagram*.

point group *class* [cryst].

point maximum *maximum*.

point sample A sample of the sediment contained at a single point in a body of water. It is obtained either by an instantaneous sampler or a time-integrating sampler.

poised stream (a) A stream that is neither eroding nor depositing sediment. Cf: *graded stream*. (b) A stream that possesses stability from an engineering viewpoint.

Poiseuille's law A statement in physics that the velocity of flow of a liquid through a capillary tube varies directly as the pressure and the fourth power of the diameter of the tube and inversely as the length of the tube and the coefficient of viscosity (Webster, 1967, p.1751).

Poisson distribution A discrete, finite, and strongly skewed frequency distribution used for tests concerned with the number of occurrences per unit interval when the number of observations is large and the probability of occurrence is small. Named after Siméon D. Poisson (1781-1840), French mathematician.

Poisson's number The reciprocal of *Poisson's ratio*.

Poisson's ratio The ratio of the lateral unit strain to the longitudinal unit strain in a body that has been stressed longitudinally within its elastic limit. It is one of the *elastic constants*. Symbol: σ. See also: *Poisson's number; poisson's relation*.

Poisson's relation In experimental structural geology, a model of elastic behavior that takes *Poisson's ratio* as equal to 0.25, i.e. approximating many solids.

poitevinite A mineral: $(Cu,Fe,Zn)SO_4 \cdot H_2O$.

pokelogan A term of Algonquian origin for a marshy cove or inlet of a stream or lake. Sometimes the term, used primarily in N.E. and Wisconsin, is shortened to "logan". Syn: *bogan, peneloken*.

polar Relating or pertaining to the region of either or both of the two poles of the Earth.

polar area The part of a pollen grain poleward from the ends of the colpi and their associated structures.

polar-area index The ratio between the polar area of a pollen grain and its diameter.

polar axis (a) The primary axis of direction or the fixed refer-ice line from which the angle coordinate is measured in a system of polar coordinates; e.g. the axis of rotation of the earth. (b) An axis of symmetry that has different crystal faces opposite ends. (c) An imaginary line connecting the two poles of spores and pollen grains.

polar cap [astrogeol] An area at each pole of the planet Mars, covered with a white material believed to be solid carbon dioxide and varying in extent with the Martian seasons.

polar cap [struc petrol] *pole* [struc petrol].

polar cap [glaciol] (a) An ice sheet centered at one of the poles of the Earth, e.g. Antarctica. (b) A term incorrectly applied to the sea ice of the Arctic Ocean.

polar circle The *Arctic Circle* or the *Antarctic Circle*, two parallels of latitude to the equator and lying approx. 23°27′ from the poles of the Earth.

polar climate A type of climate of polar latitudes (above 66°33′), characterized by temperatures of 10° C and below. The two types of polar climate in Köppen's classification are *tundra climate* and *perpetual frost climate*. Cf: *frigid climate*.

polar convergence The line of *convergence* of polar and subpolar water masses in the ocean. It is indicated by a sharp change in water-surface temperatures.

polar coordinates (a) Two coordinates that represent the distance from a central point of reference (the pole or origin) along a line to a point whose position is being defined and the direction (angle) this line makes with a fixed line. (b) A coordinate system used to define the position of a point in space with respect to an arbitrarily chosen origin by means of three polar coordinates (two directions or angles and one distance).

polar desert A high latitude desert where the moisture present frozen in ice sheets, and thus is unavailable for plant growth (Stone, 1967, p.239). Syn: *arctic desert*.

polar firn Firn formed at low temperatures with no melting or liquid water present. Syn: *dry firn*. Cf: *melt firn*.

polar glacier A glacier whose temperature is below freezing to considerable depth, or throughout its mass, and on which there is no melting even in summer. See also: *high-polar glacier; temperate glacier; subpolar glacier*.

polar ice Any sea ice more than one year old and more than 3 thick. It is heavily hummocked and usually the thickest form of sea ice. Syn: *arctic pack*.

polarimeter An instrument for measuring the amount of polarization of light or the proportion of polarized light in a partially polarized ray caused by some property of a substance. Sometimes called a *polariscope*.

polarimetry The art or process of using a *polarimeter*.

polariscope Any of several optical instruments for observing the properties of polarized light or the effects produced on polarized light by various materials, e.g. the refraction of a crystal. It consists of a polarizer and an analyzer. Also used as a syn. of *polarimeter*.

polarite A mineral: Pd(Pb,Bi).

polarity epoch A period of time during which the Earth's magnetic field was predominantly or entirely of one polarity. The chronology of the epochs and their events forms the *geomagnetic polarity time scale*. See also: *polarity event*. Syn: *geomagnetic polarity epoch*.

polarity event A period of no more than about 100,000 years when the Earth's magnetic polarity was opposite to the predominant polarity of that *polarity epoch*. Syn: *geomagnetic polarity event*.

polarity reversal *geomagnetic reversal*.

polarization [elect] (a) The production of dipoles or higher order multipoles in a medium. (b) The state of an electric or magnetic field, i.e. linear polarization, elliptical polarization, and representing in time the terminus of the vector describing the field.

polarization [magnet] A nonrecommended syn. of *magnetization*.

polarization [optics] The modification of light so that its vibrations are restricted to a single plane. Polarized light is used in the study of thin sections of minerals and rocks, by the polarizing microscope.

polarization brush *isogyre*.

polarized light Light that has been changed by passage through a prism or other polarizer so that its transverse vibrations occur in a single plane, or in a circular or elliptical pattern. (In general, the term polarized light is taken to mean plane-polarized light). It is used in the polarizing microscope for optical analysis of minerals or rocks in thin section.

polarizer An apparatus for polarizing light; in a polarizing microscope, it may be the lower *Nicol prism* or the *polaroid*. Cf: *analyzer*. Partial syn: *nicol*.

polarizing angle The angle at which unpolarized light is incident upon a surface so that it acquires the maximum plane polarization. Syn: *Brewster angle*. See also: *Brewster's law*.

polarizing microscope A microscope that uses polarized light and a revolving stage for analysis of petrographic thin sections. Two prisms, one above and the other below the stage, polarize and analyze the light; the stage rotates about the line-of-sight axis. Syn: *petrographic microscope*.

polarizing prism *Nicol prism*.

polar lake A lake whose surface temperature never exceeds 4° C. Ant: *tropical lake*.

polar migration *polar wandering*.

polarograph An instrument for analyzing solutions by electrolysis (*polarography*) using a cathode consisting of falling mercury drops.

polarography An electrolytic technique for chemical analysis based on diffusion rates of ions to an electrode as a measure of the concentration of ions in the solution. A readily polarized electrode consisting of falling mercury drops is used to keep the ion concentration in the solution uniform. See also: *polarograph*.

polaroid A trademark name for a *polarizer* that consists of a sheet of cellulose that is impregnated with crystals of quinine iodosulphate. The crystals are aligned so that their optical axes are parallel, and they polarize light in two directions at right angles: in one direction most visible light is absorbed, and in the other, essentially white light is transmitted.

polar projection One of a group of projections that are centered on a pole of a sphere. Examples include: a *polar diagram*, and any of several azimuthal map projections (polar stereographic projection, polar gnomonic projection, and polar orthographic projection).

polar space A four-sided region resulting from formation of basal leaf cross in an acantharian radiolarian.

polar spine (a) One of the modified coccoliths located at the ends of nonmotile fusiform coccolithophores (such as *Calciosolenia*). (b) A spine normal to the equator, and defining one axis, of the shell in a radiolarian; e.g. a radial spine disposed according to the Müllerian law and marking a zone in an acantharian comparable to the polar zone of the terrestrial globe.

polar stereographic projection A *stereographic projection* generally on an equatorial plane, having its center located at one of the poles of the sphere. It is suitable for maps of the polar regions on planes cutting the Earth north of 60°, and it serves as the base of the Universal Polar Stereographic (UPS) Military Grid System for latitudes between 80° and 90°. It is also widely used in optical mineralogy and in structural geology. See also: *Wulff net*.

polar symmetry A type of crystal symmetry in which the two ends of the central crystallographic axis are not symmetrical. Such a crystal is said to display *hemimorphism*; hemimorphite is the characteristic example.

polar tubule One of the external cylinders occurring at opposed poles in the main axis of an elliptic shell of a spumellarian radiolarian.

polar view The view of a spore or pollen grain from more or less directly above one of the poles. See also: *amb*.

polar wandering Movement during geologic time of the poles of the Earth's rotation, and of the magnetic poles, the first suggested by shifts in the climatic zones, the second by paleomagnetic determinations. The extent of polar wandering during geologic time is difficult to evaluate, especially if the continents have been variously displaced, as this would leave no fixed points of reference. Possibly all indications of polar wandering can be accounted for by continental displacement, or perhaps both processes operated. Syn: *polar migration; Chandler motion*.

polder A generally fertile tract of flat, low-lying land (as in Netherlands and Belgium) reclaimed and protected from the sea, a lake, a river, or other body of water by the use of embankments, dikes, dams, or levees; e.g. a marsh that has been drained and brought under cultivation, or a lake that has been dried out by pumping. The term is usually reserved for coastal areas that are at or below sea level and that are constantly protected by an organized system of maintenance and defense. Etymol: old Flemish *poelen*, "to dig out". Syn: *polderland*.

polderization The creation of a polder or polders; esp. the draining and bringing under cultivation of a low-lying area reclaimed from the sea. See also: *empolder*. Syn: *empoldering*.

polderland *polder*.

pole [geog] Either extremity of an axis of a sphere or spheroid, or one of the two points of intersection of its surface and its axis; specif. a *geographic pole* of the Earth. The term is also applied more generally to the point of intersection of the surface of a sphere and the line drawn perpendicular through the center of any circle on the surface of the sphere.

pole [cryst] In crystallography, a line that is perpendicular to a crystal face and that passes through the center of the crystal. See also: *face pole*.

pole [paleont] (a) An end of the axis of coiling in planispirally coiled shells or tests, as in the fusulinids. (b) An end of theca in cystoids; e.g. "oral pole" containing the mouth, or "aboral pole" opposite the mouth and usually marking the end to which the column is attached.

pole [palyn] Either termination of the axis of a pollen grain or spore running from the center of the original tetrad to the center of the distal side of the grain, hence the center of both distal and proximal surfaces. The term is esp. useful for angiosperm pollen in which it is not apparent which is the proximal and which the distal surface.

pole [struc petrol] (a) The point, on the surface of a reference sphere, where a line (such as a crystal direction in a mineral), perpendicular to any plane passing through the center of the sphere, intersects the sphere. Also, the stereographic projection of such a point. (b) A center of concentration of points on a petrofabric diagram, representing axes of the fabric elements or the normals to specified planes. Syn: *polar cap*.

pole [surv] (a) A bar, staff, or rod used as a target in surveying; e.g. a range pole. (b) A unit of length measuring 16.5 ft or equivalent to a *perch* or *rod*. Also, a unit of area measuring 30.25 square yards or equal to a square perch or square rod. (c) The origin of a system of polar coordinates.

pole chain *Gunter's chain*.

pole coccolith A modified coccolith found at the flagellar and tail ends in flagellate coccolithophores exhibiting dimorphism (such as *Acanthoica*). Cf: *tail coccolith*.

pole-dipole array An electrode array used in lateral search in which one current electrode is placed at infinity while one current electrode and two potential electrodes in close proximity are moved across the structure to be investigated. The separation between the near current electrode and the closest po-

tential electrode is an integral number times the spacing b tween the potential electrodes. It is used in resistivity and duced polarization surveys and in drill hole logging.

pole-fleeing force A component of forces resulting from Earth's rotation that carry the crust away from the pole toward the equator. It is supposed by some geologists to ha been sufficiently great to have displaced free-moving co nental plates, an example cited being the closing up of Teth between the northern and southern continents of the Easte Hemisphere. Evidence for the effectiveness of such a force meager. Etymol: German, Polflucht. Cf: *Coriolis force*.

pole of inaccessibility *ice pole*.

pole-pole array An electrode array, used in lateral search or logging, in which one current electrode and one potent electrode are removed to infinity while the other current ele trode and the other potential electrode are kept in close pro imity and traversed across the structure.

polianite A syn. of *pyrolusite*, esp. in well-formed tetragor crystals.

polish An attribute of surface texture of a rock or partic characterized by the regularity of reflections and indicated high luster, glistening brightness, or the presence of highlig of reflected light, and produced by various agents; e.g. a de ert polish, a *glacial polish*, or the coating formed on a gastr ith. Syn: *gloss*.

polished section A section of rock or mineral which has be highly polished on one surface. It is used for study of opaq minerals by plane or polarized reflected light. Cf: *polished th section*. Partial syn: *microsection*.

polished surface *slickenside*.

polished thin section A thin section similar to that used in p trography but finished with a polished surface, not cover with a cover glass. It is useful for study by both transmitt and reflected light and by electron microprobe methods. C *polished section*. Partial syn: *microsection*.

polje A syn. of *interior valley*. Etymol: Serbo-Croatian, "fiel Var: *polya; polye*.

pollen The several-celled microgametophyte of seed plant enclosed in the microspore wall. Fossil pollen consists entire of the microspore wall or exine, from which the microgamet phyte itself was removed during or before lithification. T term "pollen" is a collective plural noun, and it is incorrect say "a pollen". See also: *pollen grain*.

pollen analysis (a) A branch of palynology dealing with t study of Quaternary (esp. late-Pleistocene and postglacia sediments by employing pollen diagrams and isopollen ma to show the relative abundance of various pollen types space and time; e.g. the identification and percentage dete mination of frequency of pollen grains of forest trees in pe bogs and lake beds as a means of dating fossil remains. It used as a geochronologic and paleoecologic tool, often in cc laboration with archaeology. (b) A term used prior to 1944 a synonym of what is now known as *palynology*.---Syn: *poll statistics*.

pollen diagram Any diagram of pollen abundance showir stratigraphic fluctuation; strictly, the graphical presentation relative abundances of various genera of pollen and spores successive levels of cores of Quaternary sediment studied pollen analysis. Syn: *pollen profile*.

pollen grain One of the dustlike particles of which *pollen* made up; a single unit of pollen. Syn: *grain* [palyn].

pollenite An igneous rock similar in composition to *tautirite* b containing olivine and having a glassy groundmass.

pollen mother cell A *mother cell* that is derived from the hyp dermis of the pollen sac of a seed plant and that gives rise meiosis to four cells, each of which develops into a polle grain. See also: *spore mother cell*.

pollen profile A vertical section of an organic deposit (such a a peat bog) showing the sequence of buried or fossil pollen;

ollen diagram.

ollen rain The total deposit of pollen (and spores) in a given ea and period of time, as estimated by study of sediment mples and by pollen-trapping devices.

ollen sac One of the pouches in a seed plant that contain the ollen; e.g. the *anther* of an angiosperm or flowering plant.

ollen spectrum One of the characteristic horizontal lines in a ollen diagram, showing the relative abundances (percent- es) of the various sorts of pollen and spores diagrammed in single sample analyzed from a single given level.

ollen statistics *pollen analysis*.

ollen sum A portion of the total pollen count (in pollen analy- s) from which certain sorts of pollen are excluded by defini- n. The most usual pollen sum excludes all nonarborescent ollen and some arborescent pollen as well. Where pollen ms are used, pollen abundances are calculated as ratios of ven sorts of pollen to the pollen sum, rather than to the raw tal count.

ollen symbol An arbitrary sign used in Quaternary pollen di- rams, representing a genus or other group of plants, and rving as an internationally understood identification of a line the pollen diagram.

ollen tube A more or less cylindric extension that develops m the wall of a pollen grain and that protrudes through one its apertures when the grain germinates on contact with the gmatic surface of flowering plants or the megasporangium gymnosperms. The tube acts as a haustorial (absorptive) gan in lower seed plants (such as cycads), but in flowering ants its primary function is to conduct the male nuclei to the cinity of the female gametophyte (embryo sac) to effect fer- ization.

ollination The fertilization of a seed plant; specif. the transfer pollen from a stamen or anther to an ovule or megasporan- um.

ollinium A large, coherent mass of pollen, usually the con- nts of a whole locule of an anther, shed in the mature stage a unit (as in the milkweed *Asclepias*). Pl: *pollinia*. Cf: *po- ad; massula*.

ollucite A colorless, transparent, zeolite mineral: (Cs, a)$_2$Al$_2$Si$_4$O$_{12}$.H$_2$O. It occurs massive or in cubes, and is sed as a gemstone. Syn: *pollux*.

ollution *contamination* [water].

ollux Obsolete syn. of *pollucite*.

olster A small, stunted, perennial cushion plant, usually iso- ted, that grows in a dense cushiony hummock capped with ncentric layers of moss or lichen which incorporate sandy t and small pebbles as they grow. It is esp. abundant on the out of Matanuska Glacier, Alaska, and it resembles a *gla- er mouse*. Etymol: German *Polster*, "cushion".

olya Var. of *polje*.

olyactin A sponge spicule having many rays diverging from a mmon center along more than four axes. Syn: *polyact*.

olyad A group of more than four mature pollen grains shed om the anther as a unit (as in *Acacia*). The grains within the olyad are usually in multiples of four. Cf: *pollinium; dyad; tet- d*.

olyargyrite A gray to black mineral: Ag$_{24}$Sb$_2$S$_{15}$ (?).

olyaxon A sponge spicule in which the rays grow along many es of development emanating from a central point.

olybasite An iron-black to steel-gray metallic-looking mineral: g,Cu)$_{16}$Sb$_2$S$_{11}$. Cf: *pearceite*.

olycentric Said of a corallite formed by a polyp retaining pol- tomodaeal condition permanently.

olychroism *pleochroism*.

olyclinal fold One of a group of adjacent folds, the axial sur- ces of which are oriented randomly, but which have similar rface axes.

olyconic projection (a) A map projection (neither conformal r equal-area) in which a series of right circular cones are

each tangent to the Earth's surface at successive latitudes, each parallel thus constructed serving as if it were the chosen standard parallel for a simple *conic projection*. All parallels (developed from the bases of the cones) are arcs of nonconcentric circles with their centers on the straight line generally representing the central meridian, all other meridians being curved lines drawn through the true divisions of the parallels. The scale along each parallel and along the central meridian is true, but it increases on the meridians with increasing distance from the central meridian. The projection is suitable for maps of small areas and for areas of great longitudinal extent (such as Chile). It is also used for the National Topographic Map Series published by the U.S. Geological Survey because the scale error within the standard 15-minute quadrangle is very small and because of the ease with which the projection can be constructed. (b) *modified polyconic projection*.

polycrase A black mineral: (Y,Ca,Ce,U,Th)(Ti,Nb,Ta)$_2$O$_6$. It is isomorphous with euxenite and occurs in granite pegmatites.

polycrystal An aggregate of crystals so assembled as to appear as one crystal. Cf: *glomerocryst*.

polycyclic A term favored by many geomorphologists in place of the syns. *multicyclic* or *multicycle*, esp. for a stream whose course reflects base-leveling to more than one former sea level.

polydemic Said of an organism that is native to several regions.

polydymite A mineral of the linnaeite group: Ni$_3$S$_4$. Much so-called polydymite is really violarite.

polye Var. of *polje*.

polygene adj. (a) *polygenetic*. (b) A term applied specif. by Naumann (1850, p.433) to an igneous rock composed of two or more minerals. Cf: *monogene; polymineralic; polymictic* [*sed*]. Syn: *polymere; polygenic*.

polygenetic (a) Resulting from more than one process of formation or derived from more than one source, or originating or developing at various places and times; e.g. said of a mountain range resulting from various orogenic episodes. (b) Consisting of more than one type of material, or having a heterogeneous composition; e.g. said of a conglomerate composed of materials from several different sources. --Cf: *monogenetic* [*geol*]. Syn: *polygene; polygenic*.

polygenic (a) *polygenetic*. (b) *polygene*.

polygenous *polygenetic* [geol].

polygeosyncline A geosynclinal-geoanticlinal belt along the continental margin and receiving sediments from a borderland on its oceanic side (Schuchert, 1923). Cf: *monogeosynlcine*. See also: *sequent geosyncline*.

polygon A form of horizontal patterned ground whose mesh is dominantly polygonal, as tetragonal, pentagonal, and hexagonal. Its formation is favored by intensive frost action. See also: *sorted polygon; nonsorted polygon*.

polygonal ground A form of patterned ground marked by polygonal or polygon-like arrangements of rock, soil, and vegetation, produced on a level or gently sloping surface by frost action; esp. a ground surface consisting of a large-scale network of ice-wedge polygons. Syn: *polygon ground; polygonal soil; Polygonboden; polygonal markings; cellular soil*.

polygonal karst A pattern of karst that is characteristic of tropical karst such as cone karst, with the surface completely divided into a polygonal network (Monroe, 1970).

polygonal soil *polygonal ground*.

polyhalite A mineral: K$_2$MgCa$_2$(SO$_4$)$_4$.2H$_2$O. It often occurs in fibrous masses of a brick-red color due to iron.

polyhedric projection A projection for large-scale topographic maps in which a small quadrangle on the sphere or spheroid is projected onto a plane trapezoid, the rectilinear parallels and meridians corresponding closely to arc distances on the sphere or spheroid.

polykinematic mélange A mélange that includes elements de-

555

rived from an earlier mélange.

polylithionite A mineral of the mica group: $KLi_2AlSi_4O_{10}(F, OH)_2$. It is related to lepidolite.

polymere An adjective applied by Rosenbusch (1898, p.17) to an igneous rock composed of two or more minerals; *polygene*.

polymetamorphic diaphthoresis Retrograde changes during a second phase of metamorphism that is clearly separated from a previous, higher-grade metamorphic period (Hsu, 1955).

polymetamorphism Polyphase or multiple metamorphism whereby two or more successive metamorphic events have left their imprint upon the same rocks. The superimposed metamorphism may be of a higher or lower grade than the earlier type. See also: *retrograde metamorphism; prograde metamorphism*. Syn: *superimposed metamorphism*.

polymict breccia A brecciated meteorite containing fragments of differing composition. Cf: *monomict breccia*.

polymictic [sed] (a) Said of a clastic sedimentary rock composed of many rock types, such as an arkose or graywacke, or a conglomerate with more than one variety of pebble; also, said of the clasts of such a rock. Polymictic rocks are characteristic of mobile (unstable) conditions such as are found in geosynclinal regions. Cf: *oligomictic*. (b) Said of a clastic sedimentary rock composed of more than one mineral species. Cf: *monomictic; polygene*.---Syn: *polymict*.

polymictic [lake] Said of a lake having no persistent thermal stratification; e.g. a lake of great area, moderate or little depth, in a region of low humidity or at great altitude. Cf: *oligomictic*.

polymignyte A black mineral: $(Ca,Fe^{+2},Y,Zr,Th)(Nb,Ti,Ta, Fe^{+3})O_4$. Also spelled: *polymignite*.

polymineralic Said of a rock composed of two or more minerals; esp. said of an igneous rock consisting of more than one essential mineral. Cf: *monomineralic; polygene*. Syn: *polymineral*.

polymodal distribution A frequency distribution characterized by two or more localized modes, each having a higher frequency of occurrence than other immediately adjacent individuals or classes. Cf: *bimodal distribution*.

polymodal sediment A sediment whose particle-size distribution shows one or more secondary maxima.

polymorph [cryst] A crystal form of a substance that displays *polymorphism*. Syn: *polymorphic modification; allomorph*.

polymorph [evol] An organism exhibiting *polymorphism*; also, one of the forms of such an organism. Syn: *pleomorph*.

polymorphic Said of a chemical substance that displays *polymorphism*, and said of the different crystal forms so displayed. Syn: *polymorphous; pleomorphous; allomorphic; allomorphous*.

polymorphic modification *polymorph*.

polymorphism [cryst] The characteristic of a chemical substance to crystallize in more than one form, e.g. rhombic and monoclinic sulfur. Such forms are called *polymorphs*. Adj: *polymorphic*. See also: *dimorphism; trimorphism; tetramorphism; polytypism; allotropy*.

polymorphism [evol] The existence of a species in several forms independent of sexual variations. Adj: *polymorphic*. See also: *polymorph [evol]*. Syn: *pleomorphism*.

polymorphous *polymorphic*.

polynigritite A type of *nigritite* that occurs finely dispersed in argillaceous rocks. Cf: *exinonigritite; humonigritite; keronigritite*.

polynya Any nonlinear-shaped opening enclosed in ice, esp. a large expanse of water, other than a *lead* [ice], surrounded by sea ice, but not large enough to be called *open water* [ice]; commonly found off the mouth of a large river. Pl: *polynyas; polynyi*. Etymol: Russian *polyn'ya*. See also: *shore polynya; recurring polynya*. Syn: *pool [ice]; glade [ice]; ice clearing; clearing*.

polyp A typical coelenterate individual with a hollow and tubular or columnar body terminating anteriorly in a central mouth surrounded by tentacles directed upward and being posterior closed and attached to the bottom (as in *Hydra*) or more less directly continuous with other individuals of a compound animal (as in most corals).

polypary The common investing structure or tissue in which coral polyps are embedded; a coral colony as a whole. Syn: *polyparium*.

polyphyletic Evolving from more than one ancestral stock. Cf: *monophyletic*.

polypide (a) The living portion (soft parts) of a bryozoan zooid; specif. the organs and tissues in a bryozoan autozooid that undergo periodic replacement, viz. tentacles, tentacle sheath, alimentary canal, associated musculature, and nerve ganglion. (b) An individual zooid of a bryozoan colony.

polyplacophoran A marine mollusk, originally considered a subclass of the *amphineurans*, but not elevated to a class (Polyplacophora), characterized by a protective girdle comprised of overlapping calcareous valves or plates, usually eight. Their stratigraphic range is Upper Cambrian to present. Syn: *chiton*. Cf: *aplacophoran*.

polyplicate Said of pollen grains (such as those of *Ephedra*) with multiple, longitudinal, linear thinnings in the exine that resemble, but are not, true colpi.

polyquartz A group term for $Al(PO_4)$, $Al(AsO_4)$, $B(PO_4)$, etc.

polyschematic Said of mineral deposits having more than one textural element. Cf: *monoschematic*.

polysomatic chondrule A chondrule consisting of several crystals. Cf: *monosomatic chondrule*.

polystomodaeal budding A type of budding in scleractinian corals in which more than three stomodaea are developed within a common tentacular ring. Examples: *intramural budding; circummural budding; circumoral budding*.

polystromatic Said of a plant part, e.g. an algal thallus, that is composed of many layers of cells. Cf: *monostromatic; distromatic; oligostromatic*.

polysynthetic twinning *Repeated twinning* of three or more individuals according to the same twin law and on parallel composition planes; e.g. albite twinning of plagioclase. It is often revealed megascopically by striated surfaces. Cf: *cyclic twinning*.

polythalamous Many-chambered; esp. said of a foraminifer or foraminiferal test composed of numerous chambers. See also: *multilocular*..

polytomy The division of a plant into more than two branches.

polytopism The independent origin of a species or other systematic group in more than one geographic area presumably as a result of identical change in scattered individuals of its ancestor. Cf: *monotopism*.

polytype A type of polymorph whose different possible forms are due to more than one possible mode of atomic packing; this property is *polytypism*. For example, in metals, hexagonal close packing has a sequence ABAB etc. along the sixfold axis, and face-centered cubic ABCABC etc. along the three-fold axis. Adj: *polytypic*.

polytypic [cryst] Adj. of *polytype*.

polytypic [paleont] (a) Said of a species consisting of subspecies that replace each other geographically. (b) Said of a taxon that contains two or more taxa of the next lower rank. Cf: *monotypic*.

polytypism The property of a mineral to crystallize in more than one form, due to more than one possible mode of atomic packing; a form of one-dimensional *polymorphism*. Such a mineral is a *polytype*.

polytypy *polytypism*.

polyvalent Said of foraminiferal specimens or individuals forming a vegetative and accidental association (probably due to crowding) with two or more embryonic apparatuses always c the same generation (microspheric or megalospheric) and of approximately the same age (TIP, 1964, pt.C, p.62).

polyxene *platinum.*

polyzoan *bryozoan.*

polyzoic Said of a habitat that supports a wide variety of animals.

polzenite A group of lamprophyres characterized by the presence of olivine and melilite; also, any rock in that group, e.g. *modlibovite, vesecite.*

pond (a) A natural body of standing fresh water occupying a small surface depression, usually smaller than a lake and larger than a pool. (b) A term frequently used interchangeably with *lake* and *pool* and applied indiscriminately to water bodies in various sections of the U.S. (c) A body of water formed in a stream by *ponding.* (d) A small, artificial body of water, used as a source of water. In Great Britain, the term usually refers only to a small body of standing water of artificial formation.

pondage land Land on which water is stored as dead water during flooding. It does not contribute to the downstream passage of flow. Syn: *flood fringe.*

ponding (a) The natural formation of a pond in a stream by an interruption of the normal streamflow, either by a transverse uplift whose rate of elevation exceeds that of the stream's erosion, or by a dam caused by landsliding, glacial deposition, volcanism, or strong flow of water from a side valley. (b) The artificial impoundment of stream water to form a pond.

pongo A term used in South America (esp. in Peru) for a canyon or gorge, esp. one cutting through a ridge or mountain range; also, a narrow and dangerous ford. Etymol: Quechua *puncu*, "door".

ponor A Yugoslavian term for a steep-sided hole in karstic topography that acts as a *swallow hole* at times, and at other times as a spring.

Pontian European stage: uppermost Miocene (above Sarmatian, below Plaisancian of Pliocene). It has also been regarded as lowermost Pliocene.

pontic Pertaining to sediments or facies deposited in comparatively deep and motionless water, such as an association of black shales and dark limestones deposited in a stagnant basin. Etymol: Greek *pontos*, "sea". Cf: *euxinic.*

ponzaite *ponzite.*

ponzite A feldspathoid-free trachyte containing augite and pyroxenes which may be rimmed with acmite or acmite-augite. Little or no biotite or amphibole forms phenocrysts. Its name is derived from Ponza Island, Italy. Var: *ponzaite.*

pool [coast] (a) *tide pool.* (b) *beach pool.*

pool [oil] *oil pool.*

pool [ice] (a) *polynya.* (b) A large *puddle.*

pool [water] (a) A small, natural body of standing water (usually fresh water); e.g. a stagnant body of water in a marsh, or a transient puddle in a depression following a rain, or a still body of water within a cave. (b) A small, quiet, and rather deep reach of a stream, as between two rapids or where there is very little current. See also: *plunge pool.* (c) A small or large body of impounded water, artificially confined above a dam or the closed gates of a lock.

pool spring A spring fed from deep pools, probably related to faults. A pool spring may develop the shape of a jug because of a peripheral platform that is developed over the water at the surface by vegetation and sediments blown in by the wind (Meinzer, 1923, p.55). Cf: *mound spring.*

poop shot *weathering shot.*

poorly graded (a) A geologic term for *poorly sorted.* (b) An engineering term pertaining to a *nongraded* soil or unconsolidated sediment in which all the particles are of about the same size or in which a continuous distribution of particle sizes from the coarsest to the finest is lacking.---Ant: *well-graded.*

poorly sorted Said of a clastic sediment or of a cemented detrital rock that is not *sorted* or that consists of particles of many sizes mixed together in an unsystematic manner so that no one size class predominates and that has a sorting coefficient in the range of 3.5 to 4.5 and higher. Based on the phi values associated with the 84 and 16 percent lines, Folk (1954, p.349) suggests sigma phi limits of 1.00-2.00 for poorly sorted material. Ant: *well-sorted.* Syn: *unsorted; assorted; nongraded; poorly graded.*

poort A term used in southern Africa for a mountain pass, esp. a water gap or a gorge cut by a river through a ridge or a range of hills or mountains. Etymol: Afrikaans, "gate".

popple rock An English term for *pebble bed.*

popular name *vernacular name.*

population [ecol] (a) All organisms occupying a certain area or environment. (b) All the individuals that developed from zygotes formed from the gametes of the same gene pool.

population [stat] Any theoretical group of items or samples all of which are capable of being measured statistically in one or more respects; all possible values of a variable, either finite or infinite, or continuous or discrete. Syn: *universe.*

porate Said of pollen grains having a pore or pores in the exine.

porcelain clay A clay suitable for use in the manufacture of porcelain; specif. *kaolin.* Syn: *porcelain earth.*

porcelainite (a) Var. of *porcellanite.* (b) *mullite.*

porcelain jasper A hard, naturally baked, impure clay (or *porcellanite*) which, because of its red color, had long been considered a variety of jasper.

porcelain stone *china stone* [ign].

porcelaneous Having the appearance of or resembling unglazed porcelain; e.g. said of a foraminiferal test having a calcareous, shiny, and commonly imperforate wall with a dull or matte white luster resembling that of porcelain in surface appearance and composed of microcrystalline calcite with c-axes tangential or more rarely radially arranged, or said of a rock (such as porcellanite) containing chert and carbonate impurities or containing a mixture of clay and a large but variable proportion of opaline silica. Syn: *porcellaneous; porcelanous; porcelainous; porcelanic.*

porcelaneous chert A hard, opaque to subtranslucent *smooth chert*, having a smooth fracture surface and a typically china-white appearance resembling chinaware or glazed porcelain (Ireland et al, 1947, p.1485).

porcelanite Var. of *porcellanite.*

porcellanite A hard, dense, siliceous rock having the texture, dull luster, hardness, fracture, or general appearance of unglazed porcelain; it is less hard, dense, and vitreous than chert. The term has been used for various kinds of rocks, such as: an impure chert, in part argillaceous and in part calcareous, or more rarely, sideritic (see also *siliceous shale*); an indurated or baked clay or shale with a dull, light-colored, cherty appearance, often found in the roof or floor of a burned-out coal seam (see also *porcelain jasper*); and a fine-grained, acidic tuff compacted by secondary silica (see also *hälleflinta*). Etymol: Italian *porcellana*, "porcelain". Syn: *porcelanite; porcelainite.*

pore [geol] A small to minute opening or passageway in a rock or soil; an *interstice.*

pore [paleont] (a) A small opening in an echinoderm; e.g. a *hydrospire pore* of a blastoid, or an opening from the exterior through the thecal plates of a cystoid, or a pit for the attachment of a ligament that fastens a spine to a tubercle in an echinoid. The term has also been used for a horizontal perforation (tube, canal, or slit) occupying parts of two adjoining thecal plates of a cystoid (TIP, 1967, pt.S, p.113). (b) One of numerous small openings from the exterior of a sponge; e.g. the terminus of a canal at any surface, or an opening surrounded by a single cell, or a smaller-sized opening serving for inward (incurrent) flow of water. Cf: *osculum; ostium; skeletal pore.*

pore [**palyn**] One of the external, more or less circular or slightly oval thinnings or openings in the exine of pollen grains, having dimensions in ratio of less than 2:1. Pores may occur by themselves or in association with colpi. Cf: *colpus*. See also: *germ pore*.

pore canal (a) A minute tubular passageway extending through the shell of an ostracode; e.g. *normal pore canal* and *radial pore canal*. (b) A perforation in a thecal plate of an echinoderm.

pore chamber A *dietella* of a cheilostomatous bryozoan.

pore diameter The diameter of a pore in a rock is measured as the diameter of the largest sphere that may be contained within it.

pore frame The raised edge around the area enclosing a minute opening in a radiolarian.

pore ice Ground ice that fills or partially fills pore spaces in the ground; it is formed by freezing of pore water in situ with no addition of water.

pore interconnection A constricted opening connecting pores in a pore system (Choquette & Pray, 1970, p.214). Syn: *pore throat*.

pore pair An ambulacral pore of an echinoid, divided by wall of stereome and through which a single tube foot passes.

pore plate The flat, pore-bearing base of the podoconus in a nassellarian radiolarian.

pore plug Minute, single, organic, microporous plates lying at the base of external openings in certain foraminifers (TIP, 1964, pt.C, p.62).

pore pressure *neutral stress*.

pore rhomb One of the rhombic or diamond-shaped structures on the surface of thecal plates of cystoids, consisting of a group of parallel, laterally directed perforations (tubes, grooves, slits) each end of which occupies parts of two adjacent plates (so that each plate of a pair bears one half of the rhomb). The ends may be exposed to the outside or covered by thin calcareous layers. Cf: *diplopore*. See also: *pectinirhomb*. Syn: *rhomb*.

pore space The open spaces in a rock or soil, considered collectively (Stokes & Varnes, 1955, p.112). Syn: *pore volume*.

pore system All the openings in a rock or sediment, considered as a unit (Choquette & Pray, 1970, p.214).

pore throat *pore interconnection*.

pore volume *pore space*.

pore water *interstitial water*.

pore-water pressure *neutral stress*.

poriferan *sponge*.

porocyte One of the large tubular cells that constitute the wall of the inhalant canals of some sponges, completely enclosing or surrounding a pore and capable of regulating its size by expansion or contraction.

porodic Said of a noncrystalline or amorphous substance. The term, now obsolete, was originally proposed as a syn. of "colloid". Syn: *porodine*.

porodine *porodic*.

porolith A coccolith in the form of a polygonal prism with an axial perforation; an axially perforated *prismatolith*. The term was introduced for the elements of the coccolithophorid *Thoracosphaera*, which electron-microscopic studies have shown to be both perforate and imperforate.

poros A coarse-grained limestone occurring in the Peloponnesus and extensively used as a building material by the ancient Greeks.

porosimeter An instrument that measures porosity.

porosity [**elect**] In electric log studies, the ratio of void volume to total volume of a porous medium. Symbol: ϕ.

porosity [**grd wat**] The property of a rock, soil, or other material of containing interstices. It is commonly expressed as a percentage of the bulk volume of material occupied by interstices, whether isolated or connected. Cf: *effective porosity*. Syn: *total porosity*. See also: *primary porosity; secondary porosity; porous*.

porosity curve *spontaneous-potential curve*.

porosity trap *stratigraphic trap*.

porous Having numerous interstices, whether connected or isolated. "Porous" usually refers to openings of smaller size than those of a *cellular* rock. Cf: *cavernous*. See also: *porosity* [*grd wat*].

porpezite A mineral consisting of a native alloy of palladium (5-10%) and gold. Syn: *palladium gold*.

porphyrin A large, complex, organic ring compound made up of, in addition to other rings, four substituted pyrrole rings. Chlorophyll is a porphyrin with magnesium coordinated in the center of the ring; heme (of hemoglobin) is a porphyrin with iron coordinated in the center. Porphyrins are found not only in plants, but also in carbonaceous shales, crude oils, coals, etc.

porphyrite An obsolete term synonymous with *porphyry*. The term was originally used to distinguish porphyry that contains plagioclase from porphyry that contains alkali feldspar phenocrysts.

porphyritic (a) Said of the texture of an igneous rock in which larger crystals (phenocrysts) are set in a finer groundmass which may be crystalline or glassy or both. Also, said of a rock with such texture or of the mineral forming the phenocrysts. The term is sometimes restricted to cases in which the phenocrysts and groundmass formed during two different crystallization generations.---See also: *hiatal*. (b) Pertaining to or resembling porphyry.

porphyro-aphanitic Said of the texture, of a porphyritic igneous rock (esp. an extrusive rock), having large megascopic phenocrysts in an aphanitic groundmass. Also, said of a rock with such texture.

porphyroblast A pseudoporphyritic crystal in a rock produced by thermodynamic metamorphism. Adj: *porphyroblastic*. Syn: *metacryst; pseudophenocryst*.

porphyroblastic (a) In current usage, pertaining to the texture of recrystallized metamorphic rock having large idioblasts of minerals possessing high form energy (e.g. garnet, andalusite) in a finer-grained crystalloblastic matrix. See also: *pseudoporphyroblastic*. (b) Originally, perteaining to a *pseudo-porphyritic* crystal of a rock produced by thermodynamic metamorphism. Such a crystal is called a *porphyroblast*.

porphyroclastic structure *mortar structure*.

porphyrocrystallic A syn. of *porphyrotopic*. The term was introduced by Phemister (1956, p.74).

porphyrogranulitic Said of ophitic texture characterized by large phenocrysts of feldspar and augite or olivine in a groundmass of smaller lath-shaped feldspar crystals and irregular augite grains; a combination of porphyritic and intergranular textures. Also, said of a rock having such texture.

porphyroid n. A blastoporphyritic or sometimes porphyroblastic metamorphic rock of igneous origin, or a feldspathic metasedimentary rock having the appearance of a porphyry It occurs in the lower grades of regional metamorphism.---adj. Said of or pertaining to such a rock.

porphyroid neomorphism A term introduced by Folk (1965 p.22) for *aggrading neomorphism* in which small crystals are converted to large ones by growth of a few large crystals in and replacing a static matrix; e.g. the replacement of an aragonite shell by calcite mosaic. Cf: *coalescive neomorphism*.

porphyroskelic Said of an arrangement in a soil fabric whereby the plasma occurs as a dense matrix in which skeleton grains are set in the manner of phenocrysts in a porphyritic rock (Brewer, 1964, p.170). Cf: *agglomeroplasmic; intertextic*.

porphyrotope A large crystal enclosed in a finer-grained matrix in a sedimentary rock showing porphyrotopic fabric; e.g. large dolomite crystal in finer-grained calcitic matrix.

porphyrotopic Said of the fabric of a crystalline sedimentary rock (recrystallized carbonate rock or chemically precipitate diment) in which the constituent crystals are of more than ne size and in which larger crystals are enclosed in a finer-ained matrix. The term was proposed by Friedman (1965, 649). Cf: *poikilotopic*. Syn: *porphyrocrystallic*.

orphyry An igneous rock of any composition that contains nspicuous phenocrysts in a fine-grained groundmass; a por-yritic igneous rock. The term was first applied to a purple-d rock quarried in Egypt and characterized by alkali feldspar enocrysts. The rock name descriptive of the groundmass omposition usually precedes the term, e.g. diorite porphyry. vn: *porphyrite*.

orphyry copper A copper deposit in which the copper-bearing inerals occur in disseminated grains and/or in veinlets rough a large volume of rock. The term was introduced be-use some of the first large copper deposits that were mined western USA occurred in porphyritic granodiorite and uartz monzonite. Today, the term implies a large low-grade sseminated copper deposit which may be also in schist, sili-ated limestone, and volcanic rocks, but quartz-bearing igne-us rocks are always in close association.

orta Part of the operculum in ascophoran cheilostomes ryozoans) that closes the anter. Cf: *vanna*.

orterfield North American stage: Middle Ordovician (lower ibdivision of older Mohawkian, above Ashby, below Wilder-ss) (Cooper, 1956).

orticus A distinctly asymmetric apertural flap in the tests of me planktonic foraminifers (such as *Ticinella* and *Praeglo-otruncana*). It was originally defined as being imperforate. : *portici*.

ortlandian European stage: uppermost Jurassic (above Kim-eridgian, below Berriasian of Cretaceous). Its upper part is metimes known as Purbeckian. Syn: *Tithonian*.

ortlandite A mineral: $Ca(OH)_2$. It occurs as hexagonal plates contact-metamorphic rocks and also in portland cement.

ortland stone (a) A yellowish-white, oolitic limestone from e Isle of Portland (a peninsula in southern England), used r building purposes. (b) A purplish-brown sandstone from ortland, Conn.

sepnyte A light-green to reddish-brown resin with a high 8%) oxygen content, found in plates and nodules in the eat Western mercury mine, Lake County, Calif. Also elled: *pošepnyite; posepnyte*.

osition (a) Data that define the location of a *point* with re-ect to a reference system in surveying. (b) The place occu-ed by a point on the surface of the Earth or in space. (c) e coordinates that define the location of a point on the eoid or spheroid. (d) A prescribed reading of the graduated rizontal circle of a direction theodolite to be used for the bservation on the initial station of a series of stations which e to be observed.

ositive [optics] Said of anisotropic crystals: of a uniaxial ystal in which the ordinary index of refraction is greater than e extraordinary index; and of biaxial crystal in which the in-rmediate index of refraction β is closer in value to α, and in hich Z is the acute bisectrix. Cf: *negative [optics]*.

ositive [photo] A photograph or *print* having approximately e same or similar rendition of tones as that of the original ibject (light for light and dark for dark); it is produced from negative. Cf: *diapositive*.

ositive [tect] n. An area of a craton that persistently tends to and higher than the surrounding area; a *positive element*.

ositive area *positive element*.

ositive birefringence *Birefringence* in which the velocity of e ordinary ray is greater than that of the extraordinary ray.

ositive confining bed The upper confining bed of an aquifer hose head is above the upper surface of the zone of satura-on, i.e. above the water table. Currently little used by hydro-eologists.

ositive element A large structural feature or portion of the arth's crust, characterized through a long period of geologic

time by repeated, progressive, or conspicuous upward move-ment (uplift, emergence), or by relative stability or a subsi-dence that is considerably less rapid or less frequent than those of adjacent *negative elements*. Syn: *positive area; ar-chibole; positive*.

positive elongation In a section of an anisotropic crystal, a *sign of elongation* that is parallel to the slower of the two plane-polarized rays. Cf: *positive elongation*.

positive estuary An estuary in which there is a measurable dilution of seawater by land drainage. Ant: *inverse estuary*. See also: *freshwater estuary*.

positive landform An upstanding topographic form, such as a mountain, hill, or plateau, or a volcanic feature formed by an excess of material (as a cinder cone). Ant: *negative landform*.

positive movement (a) An upward movement of the Earth's crust relative to an adjacent part of the crust, such as pro-duced by an uplift or by isostatic recovery; a positive move-ment of the land may result in a *negative movement* of sea level. (b) A relative rise of the sea level with respect to the land, such as produced by a *negative movement* of the Earth's crust or by an advance of the sea.

positive ore An orebody that has been exposed and developed on four sides, and for which tonnage and quality estimates have been made. Cf: *developed reserves; proved reserves*.

positive pole The north-seeking member of the *magnetic poles*. Cf: *negative pole; dipole field*. See also: *dipole field*.

positive shoreline A shoreline resulting from a negative move-ment of the land or by a positive movement of the sea level; a *shoreline of submergence*. Ant: *negative shoreline*.

posnjakite A mineral: $Cu_4(SO_4)(OH)_6 \cdot H_2O$.

possible ore A mineral deposit whose existence and extent is postulated on the basis of past geologic and mining experi-ence. Syn: *future ore; geological ore*. Cf: *inferred ore; indicat-ed ore; potential ore*. See also: *extension ore*.

possolan Var. of *pozzolan*.

post (a) An old English term, now largely obsolete, for a thick bed of sandstone or limestone. (b) A mass of slate traversed by so many joints as to be useless for building purposes.

postabdomen (a) The slender, attenuated, posterior part of the abdomen of a scorpion, composed of five segments and a telson modified as a poison gland; the narrow posterior part of the abdomen of a merostome. (b) The *telson* of a crustacean. (c) A joint succeeding the third segment (abdomen) of the shell of a nassellarian radiolarian.

postadaptation More perfect adjustment to an adaptive zone after an organism has entered it.

postcingular series The series of plates immediately behind the girdle of dinoflagellate theca, usually fewer in number and often larger in size than those of the *precingular series*.

Post-Classic n. In New World archaeology, the final cultural stage before the arrival of the European colonists; it follows the Classic.----adj. Pertaining to the Post-Classic.

postcollarette A fine membrane prolonging the collar and ap-pearing to close the pseudostome in some chitinozoans. It may have served for temporary closure or be a remnant of an attachment; it is commonly ragged or folded back and orna-mented by a filamentous network of clear lines.

poster Part of the orifice in ascophoran cheilostomes (bryo-zoans) that is proximal to the condyles and leads to the ascus. Cf: *anter*.

posterior adj. Situated toward the back of an animal, or at or toward the hinder part of the body (toward the tail, away from the head), as opposed to *anterior*; e.g. in a direction (in the plane of symmetry or parallel to it) toward the pedicle and away from the mantle cavity of a brachiopod, or in a direction (in the plane of bilateral symmetry) opposite the position of the head of a bivalve mollusk, or in a direction (typically api-cal) along the midline axis of a gastropod and opposite the head, or in a direction away from the head of an echinoderm (such as in a direction of a cystoid interambulacrum containing hydropore and in some forms also containing the periproct.---n. The hinder part or end of an animal; e.g. the part of a brachiopod shell occupied by the viscera and including the area nearest to the pedicle opening (the side defined by the position of the hinge line), or the end defined by the position of the pallial sinus of a bivalve mollusk, or the end opposite the aperture of a gastropod.

posterior margin The posterior part of the junction between edges of brachiopod valves. It may be a hinge line or a cardinal margin.

posterior oblique muscle One of a pair of muscles in discinacean brachiopods, originating posteriorly and laterally on the pedicle valve, and converging dorsally to insertions on the brachial valve between posterior adductor muscles (TIP, 1965, pt.H, p.150).

posterior side The back or rear end of a conodont; e.g. the concave side of cusp (the side facing in the direction toward which the tip of cusp points) in simple conodont elements, or the concave side of cusps and denticles in compound conodont elements, or the distal end of plate in many plate-like conodont elements (but orientation varies). Ant: *anterior side*.

Postglacial *Holocene*.

postglacial Pertaining to the time interval since the total disappearance of continental glaciers in middle latitudes or esp. from a particular area; e.g. "postglacial rebound". Partial syn: *subglacial*.

posthumous fold A kind of recurrent folding that occurs in younger sedimentary rocks overlying a buried fold belt. The term is little used.

posthumous structure Folds, faults, and other structural features in covering strata which revive or mimic the structure of older underlying rocks that are generally more deformed; for example, structures in Paleozoic platform deposits that revive structures or trends in the Precambrian basement.

posticum An obsolete term for the opening of an exhalant canal on the external or cloacal surface of a sponge. Pl: *postica*. See also: *ostium*.

postkinematic *posttectonic*.

postmagmatic An indefinite term applied generally to reactions or events occurring after crystallization of the bulk of the magma, and usually including the hydrothermal stage.

postmineral adj. In economic geology, said of a structural or other feature formed after mineralization. Cf: *premineral*.

Postnormapolles A group of Cretaceous and Cenozoic porate (unusually triporate) pollen without the usual pore apparatus or other features of the *Normapolles* group, from which it presumably derived.

postobsequent stream A strike stream developed after the obsequent stream into which it flows (Varney, 1921, p.198).

postorogenic Said of a geologic process or event, e.g. granitization, occurring after a period of orogeny; or said of a rock or feature so formed. Cf: *posttectonic*.

postorogenic phase The final phase of an orogenic cycle, following the climactic orogeny. In some belts the postorogenic rocks and structures are minor and do not obscure the orogenic structures, in others they nearly overwhelm them; in all belts, the postorogenic events have produced the present mountainous landscapes. Postorogenic structures are germanotype and epeirogenic. Postorogenic sediments are varied, although in Europe they are collectively called *molasse*. Portorogenic plutonic bodies include discordant granitic plutons and a wide array of hypabyssal intrusives. In many orogenic belts terrestrial volcanic rocks were spread widely over the deformed terrane during the postorogenic phase. Cf: *preorogenic phase; orogenic phase*.

postseptal passage An opening that interconnects all chamberlets of the same chamber of a foraminiferal test (as in Alveolinidae), located between the wall and septum at the rear of the chamber. Cf: *preseptal passage*.

post stone An English term for any fine-grained sandstone or limestone. Also spelled: *poststone*.

posttectonic Said of a geologic process or event occurring after any kind of tectonic activity; or said of a rock or feature so formed. Cf: *postorogenic*. Syn: *postkinematic*.

pot [geomorph] A general term for any hole, pit, or depression produced naturally in the ground (and often containing water) that suggests the shape or form of a large kettle or boiler; specif. any of various kinds of *potholes*.

pot [seis] A colloquial syn. of *seismic detector*.

pot [coal] (a) *pot bottom*. (b) *caldron bottom*.

potable water Water that is safe and palatable for human use; *fresh water* in which any concentrations of pathogenic organisms and dissolved toxic constituents have been reduced to safe levels, and which is, or has been treated so as to be, tolerably low in objectionable taste, odor, color, or turbidity and of a temperature suitable for the intended use.

potamic Pertaining to rivers or river navigation; e.g. "potamic transport", or transportation of sediments by river currents.

potamoclastic rock *fluvioclastic rock*.

potamogenic rock A sedimentary rock formed by precipitation from river water (Grabau, 1924, p.329).

potamography The description of rivers.

potamology The scientific study of rivers. Etymol: Greek *potamos*, "river".

potarite A silver-white isometric mineral: PdHg. It is a natural alloy of palladium and mercury. Syn: *palladium amalgam*.

potash (a) Potassium carbonate, K_2CO_3. (b) A term that is loosely used for potassium oxide, potassium hydroxide, or even for potassium in such deplorable expressions as potash feldspar.

potash alum (a) *alum*. (b) *kalinite*.

potash bentonite *potassium bentonite*.

potash feldspar A misnomer for *potassium feldspar*.

potash kettle *giant's kettle*.

potash lake An *alkali lake* whose waters contain a high content of dissolved potassium salts. Examples occur in the western part of the sandhills region of north-central Nebraska.

potash mica A potassium-rich mica; specif. *muscovite*.

potassic Said of a rock containing a significant amount of potassium.

potassium alum *alum*.

potassium-argon age method Determination of the age of a mineral or rock in years based on the known radioactive decay rate of potassium-40 to argon-40. Ordinarily, material to be dated must be older than 1 million years. Abbrev: K-Ar age method. Syn: *potassium-argon dating*.

potassium-argon dating *potassium-argon age method*.

potassium bentonite A potassium-bearing clay of the illite group, formed by alteration of volcanic ash; a *metabentonite* consisting of randomly interstratified layers of illite and montmorillonite with a ratio of 4 to 1 (potassium occupying about 80% of the exchangeable-cation positions of the mica portion). Syn: *K-bentonite; potash bentonite*.

potassium-calcium age method The determination of the age of a mineral or rock in years based on the known radioactive decay rate of potassium-40 to calcium-40. The method is not in common use as initially there is apt to be significant quantities of calcium-40 present.

potassium feldspar An alkali feldspar containing the Or molecule ($KAlSi_3O_8$); e.g. orthoclase, microcline, sanidine, and adularia. See also: *potash feldspar*. Syn: *K-feldspar*.

potassium-40 A radioactive isotope of potassium having a mass number of 40, a half-life of approximately 1.31 x 10⁹ years, and an atomic abundance of 0.000122 grams per gram of potassium. Potassium-40 decays by beta emission to calcium-40 and by electron capture to argon-40. Potassium-40 and its decay product argon-40 are commonly used to date geologic materials (*potassium-argon age method*).

potato stone A potato-shaped geode, esp. one consisting of a shell of hard silicified limestone with an internal lining of uartz crystals.

pot bottom A large boulder or concretion in the roof of a coal seam, having the rounded appearance of the bottom of an iron pot and easily detached. Cf: *caldron bottom; bell; kettle bottom.* Syn: *pot; potstone.*

potch A term used in Australia for an opal of inferior quality; it may be colorful but lacks the fine play of color.

pot clay (a) A refractory clay (fireclay) suitable for the manufacture of the melting pots in which glass is produced. (b) A clay bed associated with coal measures.

pot earth *potter's clay.*

potential With reference to ground water, the *fluid potential.*

potential barrier The resistance to change from one energy state to another in a chemical system, which must be overcome by *activation energy.*

potential density The *density* of a unit of water when it is raised by an *adiabatic* process to the surface, i.e., determined from in situ salinity and potential temperature. Cf: *in situ density.*

potential difference The difference in electric potential between two points that represents the work involved or energy released in the transfer of a given amount of electricity between them.

potential disturbance *disturbing potential.*

potential drop The difference in pressure between two equipotential lines. Syn: *piestic interval.*

potential electrode An electrode with which potential in the earth is measured.

potential gradient *hydraulic gradient.*

potential head *elevation head.*

potential of disturbing masses *disturbing potential.*

potential of random masses *disturbing potential.*

potential ore (a) As yet undiscovered mineral deposits. (b) A known mineral deposit for which recovery is not yet economically feasible.----Cf: *possible ore; inferred ore; indicated ore.* See also: *mineral resources; mineral reserves.*

potential temperature [meteorol] The temperature that a given unit of air would attain if it were reduced to a pressure of 1000 millibars without any heat transfer to or from it.

potential temperature [oceanog] The temperature of a unit of water, measured when it is raised by an *adiabatic* process to the surface. Cf: *in situ temperature.*

potential well In a field of force, a sharply defined area of minimum potential.

potentiometer An instrument for the precise measurement of electromotive forces by which a part of the voltage to be measured is balanced against that of a known electromotive force and computed therefrom.

potentiometric map A map showing the elevation of a potentiometric surface of an aquifer by means of contour lines or other symbols. Syn: *pressure-surface map.*

potentiometric surface An imaginary surface representing the static head of ground water and defined by the level to which water will rise in a well. The water table is a particular potentiometric surface. Syn: *piezometric surface; isopotential level; pressure surface.*

pothole [coast] A small, rounded, steep-sided depression or pit in a coastal marsh, containing water at or below low-tide level (Veatch & Humphrys, 1966, p.245). Syn: *rotten spot.*

pothole [glac geol] (a) *giant's kettle.* (b) A term applied in Michigan to a small pit depression (1-15 m deep), generally circular or elliptical, occurring in an outwash plain, a recessional moraine, or a till plain (Veatch & Humphrys, 1966, p.244).---Cf: *kettle.*

pothole [karst] (a) A funnel-shaped *doline.* (b) A vertical or steeply inclined shaft in limestone. Syn: *aven; cenote.*

pothole [speleo] A cave that is open upwards to the surface, shaftlike, usually in limestone.

pothole [geomorph] Any pot-shaped pit or hole.

pothole [salt] A term used in Death Valley, Calif., for a circular opening, about a meter in diameter, filled with brine and lined with salty crystals.

pothole [glaciol] *moulin.*

pothole [lake] A shallow depression, generally less than 10 acres, occurring between dunes on a prairie (as in Minnesota and the Dakotas), often containing an intermittent pond or marsh and serving as a nesting place for waterfowl.

pothole [streams] A smooth, roughly circular, bowl-shaped or cylindrical hollow, generally deeper than wide, formed in the rocky bed of a stream by the grinding action of a stone or stones, or of coarse sediment (sand, gravel, pebbles, boulders), whirled around and kept in motion by eddies or the force of the stream current in a given spot, as at a strong rapid or the foot of a waterfall. Cf: *plunge basin.* Syn: *pot; kettle; evorsion hollow; rock mill; churn hole; eddy mill; colk.*

pothole erosion *evorsion.*

potholer *caver.*

potholing *caving.*

pot lead Graphite used on the bottoms of racing boats.

potrero An elongate, island-like beach ridge, surrounded by mud flats and separated from the coast by a lagoon and barrier island, and made up of a series of accretionary dune ridges (Fisk, 1959, p.113); e.g. Potrero Lopeno, rising 10 m above the Laguna Madre Flats along the southern Texas coast.

Potsdam system A system of gravity values based on the determination of absolute gravity at Potsdam, Germany in 1906. Recent determinations show that the Potsdam value is about 14 milligals too high.

potstone [mineral] (a) A dark-green or dark-brown, impure steatite or massive talc, used in prehistoric times to make cooking pots and vessels. (b) A term used in Norfolk, Eng., for *paramoudra.*---Also spelled: *pot stone.*

potstone [coal] *pot bottom.*

potter's clay A plastic clay free from iron and devoid of fissility, suitable for modeling or making of pottery or adapted for use on a potter's wheel. It is white after burning. Syn: *potter's earth; pot earth; argil.*

poughite A mineral: $Fe_2(TeO_3)_2(SO_4).3H_2O$.

Poulter seismic method A type of *air shooting* in which the explosive is set on poles above the ground.

Pourtalès plan The arrangement of septa in some scleractinian corals (notably in the family Dendrophylliidae) characterized by much greater development of exosepta than that of entosepta. Named after Louis F. de Pourtalès (1824-1880), Swiss naturalist.

pow A Scottish term for a pool.

powder *powder snow.*

powder diffraction X-ray diffraction by a powdered, crystalline sample, commonly observed by the Debye-Scherrer camera method or by a recording diffractometer.

powder method A method of recording X-ray diffraction of a crystal from a powdered sample. See also: *powder pattern.*

powder pattern in the *powder method* of X-ray diffraction analysis, the display of lines made on film by the Debye-Scherrer method or on paper by a recording diffractometer. See also: *powder photograph.*

powder photograph The *powder pattern* made on film in the Debye-Scherrer method of X-ray diffraction analysis.

powder snow Dry fallen snow composed of loose snow crystals that accumulated under conditions of low temperature and that have not been compacted in any way. Cf: *sand snow; wild snow.* Syn: *powder.*

powdery avalanche *dry-snow avalanche.*

powellite A tetragonal mineral: $CaMoO_4$. It is isomorphous with scheelite and is a minor ore of molybdenum.

power efficiency The probablity of rejecting a statistical hypothesis when it is false. Syn: *power.*

power of crystallization *form energy.*

power reflection coefficient The square of the electric field intensity reflection coefficient [elect]. Syn: *reflection coefficient* [*elect*].

Poynting's law A special case of the *Clapeyron equation*, in which the fluid is removed as fast as it forms, e.g. under metamorphic stress, so that its volume may be ignored.

pozzolan Siliceous tuff, ash, or other material used in cement because when mixed with lime it hardens underwater. It is named after a leucitic tuff found near Puzzuoli, Itlay. Also spelled: *possolan; pozzolana; puzzolan; puzzuolana; puzzuolane.*

pradolina A syn. of *urstromtal*. Etymol: Polish, "ancient valley". Pl: *pradoliny.*

prairie (a) An extensive tract of level to rolling, generally treeless grassland in the temperate latitudes of the interior of North America (esp. in the Mississippi Valley region), characterized generally by a deep, fertile soil (suitable for wheat growing) and by a covering of tall, coarse grass and herbaceous plants. See also: *steppe; black prairie.* (b) One of a series of grassy *plains* into which the prairies proper of the Mississippi Valley region merge on the west, and whose treeless state is due to aridity. (c) A broad, low, wet, sandy, flat-bottomed, often water-covered, grass-grown tract or sink in the pinewoods of Florida.----Etymol: French, "meadow, grassland".

Prairie soil In early U.S. classification systems, a group of zonal soils having a surface horizon that is dark or grayish brown, which grades through brown soil into lighter-colored parent material. It is two to five feet thick, and develops under tall grass in a temperate and humid climate. Cf: *Reddish Prairie soil.* Syn: *Brunizem.*

prairillon A small prairie.

Prandtl number In fluid mechanics, a nondimensional parameter that is the ratio of kinematic viscosity to thermometric conductivity. It is named after Ludwig Prandtl, German physicist (d. 1953).

prase (a) A translucent and dull leek-green or light-grayish yellow-green variety of chalcedony. (b) Crystalline quartz containing a multitude of green hair-like crystals of actinolite.---Syn: *mother-of-emerald.*

prasinite A greenschist in which the proportions of the hornblende-chlorite-epidote assemblage are more or less equal (Holmes, 1928, p.189).

prasopal A green variety of common opal containing chromium. Syn: *prase opal.*

pratincolous Said of an organism that lives in meadows or low grassy areas.

Pratt isostasy The hypothesis of the mechanism of *isostasy*, proposed by G.H. Pratt, that postulates an equilibrium of crustal blocks of varying density; thus the topographically higher mountains would be of a lesser density than topographically lower units, and the depth of crustal material would be everywhere the same. Cf: *Airy isostasy.*

praya A beach or waterfront. Etymol: Portuguese *praia.*

preabdomen The enlarged anterior part of the abdomen of a scorpion, composed of seven segments; the broad anterior part of the opisthosoma of a merostome.

preadaptation Appearance of nonadaptive or inadaptive characters that later proves to be adaptive in a different or changed environment.

prealpine facies A geosynclinal facies characteristic of neritic areas, displaying thick limestone deposits and coarse terrigenous material, and resembling epicontinental platform sediments. It is generally overlain by flysch, as in the Alpine region.

Preboreal n. A term used primarily in Europe for an interval of postglacial time (from about 10,000 to 9000 years ago) following the Younger Dryas of the late-glacial Arctic interval and preceding the Boreal, during which the inferred climate was somewhat colder and wetter than the Boreal; a subunit of the (Blytt-Sernander climatic classification, characterized by

birch and pine vegetation. Also spelled: *Pre-Boreal.* Syn: *Subarctic.*---adj. Pertaining to the postglacial Preboreal interval and to its climate, deposits, biota, and events.

Precambrian All geologic time, and its corresponding rocks, before the beginning of the Paleozoic; it is equivalent to about 90% of geologic time. It has been divided according to several different systems, all of which use the presence of absence of evidence of life as a criterion. See also: *Azoic; Proterozoic.*

Precambrian W That part of the U.S. Geological Survey's purely temporal, four-fold division of the Precambrian corresponding to the time span 2600 million years ago and older. Cf: *Precambrian Z; Precambrian Y; Precambrian X.*

Precambrian X That part of the U.S. Geological Survey's purely temporal, four-fold division of the Precambrian corresponding to the time span 1700-2600 million years ago. Cf: *Precambrian Z; Precambrian Y; Precambrian W.*

Precambrian Y That part of the U.S. Geological Survey's purely temporal, four-fold division of the Precambrian corresponding to the time span 800-1700 million years ago. Cf: *Precambrian Z; Precambrian X; Precambrian W.*

Precambrian Z That part of the U.S. Geological Survey's purely temporal, four-fold division of the Precambrian corresponding to the time span 570-800 million years ago. Cf: *Precambrian Y; Precambrian X; Precambrian W.*

precession camera An X-ray diffraction camera used in the *Buerger precession method* for recording the diffractions of an individual crystal.

precession method *Buerger precession method.*

precession of the equinoxes A consequence of the precession of the Earth's spin axis wherein the intersection of the ecliptic with the celestial equator advances along the equator. It produces approximately a twenty-second difference in length between the tropical year and the sidereal year.

precingular archeopyle An *archeopyle* formed in a dinoflagellate cyst by loss of the middorsal plate of the precingular series.

precingular series The series of plates between the apical series and the girdle in dinoflagellate theca. Cf: *postcingular series.*

precious Said of the finest variety of a gem or mineral; e.g. "precious jade" (true jadeite that is wholly or partly deep green) or "precious scapolite" (gem-quality scapolite). See also: *oriental; precious stone.*

precious garnet (a) An unusually purplish and brilliant almandine. (b) An unusually reddish and brilliant pyrope.

precious metal A general term for gold, silver, or any of the minerals of the platinum group.

precious opal A gem variety of opal that exhibits a brilliant play of delicate colors; e.g. *white opal* and *black opal.* Cf: *common opal.*

precious serpentine A pale or dark oil-green, massive, translucent variety of the mineral serpentine.

precious stone (a) A gemstone that, owing to its beauty, rarity, durability, and hardness, has the highest commercial value and traditionally has enjoyed the highest esteem since antiquity; specif. diamond, ruby, sapphire, and emerald (and sometimes pearl, opal, topaz, and chrysoberyl). (b) Strictly, any genuine gem material.---Cf: *semiprecious stone.* See also: *precious.*

precipice A very steeply inclined, vertical, or overhanging wall or surface of rock; e.g. the high, steep face of a cliff.

precipitation [meteorol] The discharge of water (as rain, snow, hail, or sleet) from the atmosphere upon the Earth's surface. It is measured as a liquid regardless of the form in which it originally occurred; in this sense, it may be called *rainfall.*

precipitation excess The volume of water from precipitation that is available for direct runoff. Cf: *rainfall excess; abstraction [water]; effective precipitation.*

precipitation facies Facies characteristics that provide evidence of depositional conditions, as revealed mainly by sedimentary structures (such as cross-bedding and ripple marks) and by primary constituents (esp. fossils) (Sonder, 1956). Cf: *alimentation facies*.

precision (a) The degree of agreement or uniformity of repeated measurements of a quantity; the degree of refinement in the performance of an operation or in the statement of a result. It is exemplified by the number of decimal places to which a computation is carried and a result stated. Precision relates to the quality of the operation by which a result is obtained, as distinguished from *accuracy*, but it is of no significance unless accuracy is also obtained. (b) The deviation of a set of estimates or observations from their mean. (c) A term applied in surveying to the degree of perfection in the methods and instruments used when making measurements and obtaining results of a high order of accuracy.

precision depth recorder An *echo sounder* having an accuracy better than 1 in 3000. Abbrev: PDR.

Pre-Classic *Formative*.

preconsolidation pressure The greatest effective stress to which a soil has been subjected; the pressure exerted on unconsolidated sediment by overlying material (which may have been removed later by erosion), resulting in compaction. Syn: *prestress*.

precoxa An occasionally occurring limb segment proximal to the coxa of a crustacean.

precurrent mark A structure produced on the surface of unconsolidated sediment before the arrival of a turbidity current; e.g. an animal track.

predazzite A recrystallized limestone similar to *pencatite* but containing less brucite than calcite.

prediagenesis A term used by Chilingar et al (1967, p.322) for that part of *syngenesis* responsible for "those parts that were introduced subsequently" to *syndeposition* but "before the principal processes of diagenesis began"; e.g. internal sedimentation of clastic material.

prediluvian *antediluvian*.

preferred orientation In structural petrology, nonrandom orientation of planar or linear fabric elements.

pregeologic (a) Antedating reliable geologic data or theory. (b) Before the time when the surface of the Earth became generally similar to what it is today; e.g. "pregeologic time", or the part of geologic history that antedates the oldest rocks (about 3-5.5 b.y. ago).

preglacial (a) Pertaining to the time preceding a period of glaciation; specif. that immediately before the Pleistocene Epoch. (b) Said of material underlying glacial deposits; e.g. the loose sand and gravel lying beneath till in Iceland, where the term "preglacial drift" is (incorrectly) used.

prehistoric (a) Said of or pertaining to something in the past that is prior to the written records of man. (b) Pertaining to prehistory, i.e., the study of man during the time prior to his written records.

prehnite A pale-green, yellowish-brown, or white orthorhombic mineral: $Ca_2Al_2Si_3O_{10}(OH)_2$. It usually occurs in crystalline aggregates having a botryoidal or mammillary and radiating structure, and is commonly associated with zeolites in geodes, druses, fissures, or joints in altered igneous rocks.

pre-Imbrian (a) Pertaining to the oldest lunar topographic features and lithologic map units constituting a system of rocks that appear in the mountainous terrae and that are well displayed in the southern part of the visible lunar surface and over much of the reverse side. (b) Said of the stratigraphic period during which the pre-Imbrian System was developed.

preliminary waves The body waves of an earthquake, including both *P* waves and *S* waves.

premagadiniform Said of the loop, or of the early stages in the development of the loop, of a terebratellid brachiopod, marked by growth and completion of descending branches from both the cardinalia and the median septum and by the appearance of a tiny hood developing into a ring on the septum (TIP, 1965, pt.H, p. 151). Cf: *magadiniform*.

premineral adj. In economic geology, said of a structural or other feature extant before mineralization. Cf: *postmineral*.

preobrazhenskite A mineral: $Mg_3B_{11}O_{15}(OH)_9$.

preoccupied name In taxonomy, a name that is a junior homonym, i.e. unavailable for use because the same name was earlier applied to a different taxon.

preoral cavity The depression above the gullet in a tintinnid.

preorogenic phase The initial phase of an *orogenic cycle*, prior to the climactic orogeny. This phase is the time of formation of geosynclines, most of which are clearly divisible into eugeosynclinal (internal) and miogeosynclinal (external) parts, the first characterized by abundant submarine volcanism, the second by little magmatism and by carbonate-quartzite sedimentation. Preorogenic plutonic rocks include ultramafic bodies and rare early granitic plutons. Cf: *orogenic phase; postorogenic phase*.

prepollen Functional pollen grains that have haptotypic characters like those of spores, usually a trilete mark. They usually have also a colpus and such other pollenlike features as vesicles. Prepollen are typical of extinct primitive gymnosperms (mostly Mississippian to Permian).

preseptal passage An opening that interconnects all chamberlets of the same chamber of a foraminiferal test (as in Alveolinidae), located in the anterior part of the chamber. Cf: *postseptal passage*.

presque isle A promontory or peninsula extending into a lake, nearly or almost forming an island, its head or end section connected with the shore by a sag or low gap only slightly above water level, or by a strip of lake bottom exposed as a land surface by a drop in lake level (Veatch & Humphrys, 1966, p.246). Type example: Presque Isle (Mich.), extending into Lake Huron. Etymol: "presque" is French for "almost".

pressed amber *amberoid*.

pressolution *pressure solution*.

pressolved Said of a sedimentary bed or rock in which the grains have undergone pressure solution; e.g. "pressolved quartzite" whose toughness and homogeneity is due to a tightly interlocked texture of quartz grains subjected to pressure solution. Term was introduced by Heald (1956, p.22).

pressure The force exerted across a real or imaginary surface divided by the area of that surface; the force per unit area exerted on a surface by the medium in contact with it.

pressure altimeter An altimeter (such as an aneroid barometer) that indicates height by measuring differences in atmospheric pressure. Syn: *barometric altimeter*.

pressure altitude The altitude, in a standard atmosphere, at which a given pressure will be observed.

pressure arch A wave-like prominence, formed by pressure, on the surface of a glacier.

pressure box Jargon in experimental structural geology for an instrument that provides stress from one side.

pressure breccia *tectonic breccia*.

pressure bulb The zone in a loaded soil mass bounded by an arbitrarily selected isobar of stress (ASCE, 1958, term 277).

pressure burst A *rock burst* produced under stresses exceeding the elastic strength of the rock.

pressure cone *shatter cone*.

pressure decay The decline, usually gradual, from a temporary, abnormal pressure condition toward a normal pressure which is more nearly in balance with permanent or steady-state environmental conditions.

pressure dome *tumulus*.

pressure drag *pressure resistance*.

pressure figure A pattern resembling a six-rayed star, produced on a plate of mica by compression at a point. It is not as distinct as a *percussion figure*.

pressure fringe *pressure shadow*.

pressure gradient (a) The rate of variation (decrease) of pressure in a given direction in space at a fixed time, usually the horizontal component of that vector having a direction of the maximum increment of the pressure in a horizontal plane; in the ocean, it is caused by the vertical distribution of density (which depends on water temperature and salinity), by the slope of the sea surface with respect to the level surface, and by the difference at atmospheric pressure at the sea surface. (b) Loosely, the magnitude of the pressure gradient.----Cf: *hydraulic gradient.*

pressure head Hydrostatic pressure expressed as the height of a column of water that the pressure can support, expressed with reference to a specific level such as land surface. See also: *hydraulic gradient; static head; total head.*

pressure ice A general term for ice, esp. sea ice, whose surface in places has been deformed by stresses generated by wind, currents, or waves; it includes pieces of ice squeezed against the shore or each other, or forced upwards or downwards. Pressure ice may be "rafted", "hummocked", or "tented". See also: *deformed ice; rough ice; screw ice; pressure icefoot.* Ant: *level ice.*

pressure icefoot An icefoot formed along a shore by the freezing together of stranded *pressure ice.*

pressure melting Melting of ice in a place where its melting point is lowered by application of increased pressure; esp. the first part of the *regelation* process within a glacier, occurring at places where the overlying ice is especially thick.

pressure melting temperature The temperature at which ice can melt at a given pressure.

pressure pan An induced soil *pan* having a higher bulk density and a lower total porosity than the soil directly above or below it, produced as a result of pressure applied by normal tillage operations or by other artificial means (SSSA, 1965, p.341). Cf: *genetic pan.* Syn: *traffic pan.*

pressure penitente A nieve penitente composed of brilliantly white ice shaped into a slender ridge by lateral pressure of converging morainal streams and by melting of the adjacent debris-covered ice (Workman, 1914, p.316-317). Such features have broken upper surfaces and sharply inclined sides (resembling pointed cones, wedges, or pyramids), and usually occur on the lower parts of glaciers where morainal streams are strongly developed.

pressure pickup Any seismic detector which reacts to pressure variations.

pressure plateau An uplifted area of a thick, ponded lava flow, measuring up to three or four meters, the uplift of which is due to the intrusion of new lava from below that does not reach the surface. The lower part of such a flow may remain fluid for weeks.

pressure release The outward-expanding force of pressure which is released within rock masses by unloading, as by erosion of superincumbent rocks or by removal of glacial ice. It results in pulling away of the outer layers of the mass, especially in massive plutonic rocks, causing them to split into great shells or spalls; for example, in Yosemite Valley, California.

pressure-release jointing *Exfoliation* that occurs in once deeply buried rock that erosion has brought nearer the surface, thus releasing its confining pressure. See also: *sheeting structure.*

pressure resistance In fluid dynamics, a normal stress caused by acceleration of the fluid, which results in a decrease in pressure from the upstream to the downstream side of an object, and acting perpendicular to the boundary (Chow, 1957). Cf: *shear resistance.* Syn: *pressure drag.*

pressure ridge [seism] A seismic feature due to transverse pressure and shortening of the land surface; a *slice ridge.*

pressure ridge [volc] An elongate uplift of the congealing crust of a lava flow, probably due to the pressure of the underlying and still flowing lava.

pressure ridge [glac] (a) A ridge of glacier ice, produced [by] horizontal pressure associated with glacier flow. (b) *ic[e] pushed ridge.*

pressure ridge [ice] A rugged, irregular wall of broken float[ing] ice buckled upward by the lateral pressure of wind or curre[nt] forcing or squeezing one floe against another; it may be fre[sh] or weathered, and extend many kilometers in length and up [to] 30 m in height. Cf: *ice keel.* Syn: *ridge* [*ice*].

pressure shadow In structural petrology, an area adjoining [a] porphyroblast, characterized by a growth fabric rather than [a] deformation fabric, as seen in a section perpendicular to the [c-] axis of the fabric. Its sigmoid form indicates the direction [or] sense of the movement. Syn: *pressure fringe; strain shado[w;] stress shadow.*

pressure solution Solution (in a sedimentary rock) occurri[ng] preferentially at the contact surfaces of grains (crystal[s]) where the external pressure exceeds the hydraulic pressure [of] the interstitial fluid. It results in enlargement of the conta[ct] surfaces and thereby reduces pore space and tightly wel[ds] the rock. See also: *solution transfer.* Syn: *pressolution.*

pressure surface *potentiometric surface.*

pressure-surface map *potentiometric map.*

pressure texture *cataclastic texture.*

pressure tube A deep, slender, cylindrical hole formed in [a] glacier by the sinking of an isolated stone that has absorbe[d] more solar radiation than the surrounding ice (Mallet, 183[?,] p.326-327).

pressure wall A snow escarpment at the side of an avalanch[e] (ADTIC, 1955, p.63).

pressure wave *P wave.*

prestratigraphy A term proposed by Storey & Patters[on] (1959) for preliminary or introductory stratigraphy; *prostrati[g-] raphy.*

prestress *preconsolidation pressure.*

presuppression In seismic prospecting, the suppression of th[e] early events on a seismic record, for control of noise and r[e-] flections on that portion of the record.

prevailing current The ocean current most frequently observe[d] during a given period, as a month, season, or year.

previtrain The woody lenses in lignite that are equivalent to *[vi-]train* in coal of higher rank (Schopf, 1960, p.30).

Priabonian European stage: upper Eocene. It is believed [to] consist of Auversian and Bartonian.

priceite A snow-white earthy mineral: $Ca_4B_{10}O_{19}.7H_2O$ (?[).] Syn: *pandermite.*

priderite A red mineral: $(K,Ba)(Ti,Fe)_8O_{16}$.

prill An English term for a running stream.

primärrumpf A term proposed by W. Penck (1924) for a low, convex, rather featureless, erosional landscape or plain pro[-] duced by waxing uplift that proceeded so slowly with respe[ct] to the rate of denudation that there would be no actual n[et] rise of the surface or increase in its relief; an "expandi[ng] dome" that represents the universal and initial geomorph[ic] unit. Etymol: German *Primärrumpf,* "primary torso". Cf: *e[n-] drumpf; old-from-birth peneplain.* Syn: *primary peneplain.*

primary [eco geol] Said of a mineral deposit unaffected by s[u-] pergene enrichment.

primary [coast] Said of a youthful coast or shoreline whe[re] waves have not had time to produce notable effects and ha[v-] ing features that are produced chiefly by nonmarine agencie[s] (Shepard, 1937, p.605); e.g. coasts shaped by diastrophis[m,] volcanism, subaerial deposition, or land erosion. Cf: *seco[n-] dary.*

primary [metal] Said of a metal obtained directly from or[e.] Ant: *secondary* [*metal*]. Syn: *virgin.*

Primary A term applied in the early 19th century as equivale[nt] to *Primitive* or the period of time and associated rocks no[w] referred to as Precambrian. It was later extended to includ[e] the lower Paleozoic, and still later restricted to the whole [of] the Paleozoic Era. The term was abandoned in the late 19th entury in favor of *Paleozoic*. See also: *Secondary*.

imary allochthony In coal formation, accumulation of plant mains in a region that does not correspond to that in which e plants grew. Cf: *secondary allochthony*.

imary arc (a) A curved segment of elongated mountain ines that are the areas of the Earth's major and most recent ctonic activity, i.e. the *great-circle belts* (Strahler, 1963, 403). See also:*island arc*. (b) *internides*.

imary axial septulum A *primary septulum* in a foraminiferal st, representing an *axial septulum* observable in sagittal quatorial) section (as in *Lepidolina* and *Yabeina*).

imary basalt A presumed original magma from which all her rock types are supposedly obtained by various pro-sses.

imary clay A clay found in the place where it was formed; a *sidual clay*. Cf: *secondary clay*.

imary crater (a) An *impact crater* produced directly by the gh-velocity impact of a meteorite or other projectile; e.g. y of the lunar craters formed by collision of the Moon with arious-sized objects from space. Cf: *secondary crater*. (b) *ue crater*.

imary creep Elastic deformation that is time-dependent and sults from a constant differential stress acting over a long eriod of time. Cf: *secondary creep*. Syn: *transient creep*.

imary dip The slight dip of a bedded deposit assumed at its oment of deposition. Syn: *original dip; depositional dip; initial p*. Cf: *initial dip*.

imary dolomite A dense, finely textured (particle diameters ss than 0.01 mm) dolomite rock made up of crystals formed place by direct chemical or biochemical precipitation from eawater or lake water, recognized as a well-stratified (thinly minated) and wholly unfossiliferous dolomite associated with her primary sediments and commonly interbedded with hydrite, clay, and micritic limestone. Also, a similarly xtured dolomite rock made up of clastic particles formed by rect accumulation. Some authors consider the rock to be yndiagenetic (Fairbridge, 1967, p.66-67). See also: *dolomi-rite*. Syn: *orthodolomite*.

imary drilling Drilling of holes in solid rock in preparation for asting. Cf: *secondary drilling*.

imary fabric *apposition fabric*.

imary flat joint An approximately horizontal joint plane in ig-eous rocks. Syn: *L-joint*.

imary flowage Movement within an igneous rock that is still artly fluid (Cloos, 1946).

imary geosyncline Peyve & Sinitzyn's term for an *orthogeo-yncline* (1950).

imary gneiss A rock that exhibits planar or linear structures haracteristic of metamorphic rocks but that lacks observable ranulation or recrystallization and is therefore considered to e of igneous origin.

imary gneissic banding A kind of *banding* developed in cer-in igneous (plutonic) rocks of heterogeneous composition, roduced by the admixture of two magmas only partly misci-e or by magma intimately admixed with country rock into hich it has been injected along planes of bedding or folia-n. It is recognized by alternating layers of light and dark inerals. Cf: *gneissic structure*.

imary interstice *original interstice*.

imary lamella The first half whorl of each brachiopod spiral-m distal from its attachment to a crus (TIP, 1965, pt.H, 151).

imary layer The outer shell layer immediately beneath the eriostracum of a brachiopod, deposited extracellularly by co-mnar outer epithelium of the outer mantle lobe. It forms a ell-defined calcareous layer, devoid of cytoplasmic strands, most articulate brachiopods. Cf: *secondary layer*. Syn: *la-ellar layer*.

imary ligament The part of a ligament of a bivalve mollusk epresenting the original condition of structure, consisting of

periostracum and ostracum, but excluding secondary addi-tions.

primary magma A magma originating below the Earth's crust. It is sometimes used as a syn. of *parental magma*.

primary mineral A mineral formed at the same time as the rock enclosing it, by igneous, hydrothermal, or pneumatolytic processes, and that retains its original composition and form. Cf: *secondary mineral*.

primary optic axis One of two *optic axes* in a crystal that are perpendicular to the circular sections of the indicatrix and along which all light rays travel with equal velocity. Cf: *secon-dary optic axis*.

primary orogeny Orogeny that is characteristic of the *inter-nides* and that involves deformation, regional metamorphism, and granitization. Cf: *secondary orogeny*.

primary peneplain A syn. of *primärrumpf*, originally a German word with no adequate English equivalent. The term is unsa-tisfactory because the Davisian peneplain is developed by a different process and has different characteristics.

primary phase The only crystalline phase capable of existing in equilibrium with a given liquid; it is the first to appear on cooling from a liquid state, and the last to disappear on heat-ing to the melting point.

primary phase region On a phase diagram, the locus of all compositions having a common primary phase.

primary porosity [grd wat] The *porosity* that developed during the final stages of sedimentation or that was present within sedimentary particles at the time of deposition. "Primary po-rosity includes all predepositional and depositional porosity of a particle, sediment, or rock" (Choquette & Pray, 1970, p.249). Cf: *secondary porosity*.

primary precipitate A *precipitate* formed directly; e.g. an evaporite formed by evaporation of a saline solution, or a sediment formed as a result of a reaction between dissolved material and suspended clay or as a result of a change in acidity or a shift in the oxidation-reduction potential.

primary precipitate crystal *cumulus crystal*.

primary productivity In a body of water, the rate of photosyn-thetic carbon fixation by plants and bacteria forming the base of the food chain. See also: *productivity* [*lake*]; *production*.

primary rocks (a) Rocks all of whose constituents are newly formed particles that have never been constituents of pre-viously formed rocks and that are not the products of alter-ation or replacement, such as limestones formed by precipita-tion from solution; esp. igneous rocks formed directly by soli-dification from a magma. Cf: *secondary rocks*. (b) A "more appropriate" syn. of *primitive rocks* (Humble, 1843, p.210). The term in this usage is now obsolete.

primary sedimentary structure A syngenetic *sedimentary structure* determined by the conditions of deposition (mainly current velocity and rate of sedimentation) and developed be-fore the final consolidation of the rock in which it is found. It includes bedding in the broad sense (esp. the external form of the bed and its continuity and uniformity of thickness), bed-ding-plane markings (such as ripple marks and sole marks), and those deformational structures produced by preconsolida-tion movement due to unequal loading or to downslope sliding or slumping. Syn: *primary structure* [*geol*].

primary septulum A major partition of a chamberlet in a foraminiferal test; e.g. *primary axial septulum* and *primary transverse septulum*. Cf: *secondary septulum*.

primary spine The first-formed and usually largest spine of a plate of the corona of an echinoid. It is situated over the growth center of the plate except on a *compound plate*. Cf: *secondary spine*.

primary stratification Stratification developed when the sedi-ments were first deposited. Syn: *direct stratification*.

primary structure [**paleont**] Fine vacuoles or spaces in the wall of a tintinnid lorica. Cf: *secondary structure; tertiary structure*.

primary structure [geol] (a) A structure, in an igneous rock, that originated contemporaneously with the formation or emplacement of the rock, but before its final consolidation; e.g. pillow structure developed during the eruption of a lava, or banding developed during solidification of a magma. (b) *primary sedimentary structure.*----Cf: *secondary structure* [*geol*].

primary tectonite A tectonite whose fabric is *depositional fabric.* Most tectonites, however, are *secondary tectonites.* See also: *fusion tectonite.*

primary tissue Plant tissue derived directly by differentiation from an apical or intercalary meristem (Cronquist, 1961, p.880).

primary transverse septulum A *primary septulum* in a foraminiferal test, representing a *transverse septulum* that has a plane approximately normal to the axis of coiling and that is observable in axial section (as in *Lepidolina* and *Yabeina*).

primary type One of the fundamental types which determines the usage of a biologic name and upon which the understanding of a species is based; e.g. a holotype, lectotype, or syntype. Syn: *prototype.*

primary-type coal *banded ingredients.*

primary wall The first wall proper formed in a developing plant cell. It is the only wall in many types of cells (Esau, 1965, p.36-37).

primary wave *P wave.*

prime meridian An arbitrary meridian selected as a reference line having a longitude of zero degrees and used as the origin from which other longitudes are reckoned east and west to 180 degrees; specif. the *Greenwich meridian.* Local or national prime meridians are occasionally used. Syn: *zero meridian; initial meridian; first meridian.*

primeval Pertaining to the earliest ages of the Earth; e.g. said of a stream that has persisted along original drainage lines from earliest times and through the vicissitudes of geologic change, or said of lead that is associated with so little uranium (as in some meteorites) that the Pb-isotope composition has not changed appreciably in five billion years. See also: *primordial.*

primeval-fireball hypothesis "*big bang*" *hypothesis.*

primibrachial A plate of the proximal brachitaxis of a crinoid. It may or may not be an axillary, and it may or may not comprise part of theca. Syn: *primibrach.*

Primitive A name applied from the teachings of A.G. Werner in the 1790's to the group or series of rocks that were considered the first chemical precipitates derived from the ocean before emergence of land areas and that were believed to extend uninterruptedly around the world. The rocks included the larger intrusive igneous masses, all highly metamorphosed rocks, and roughly the rocks that later came to be known as Precambrian in age. See also: *Primary; Transition.*

primitive area U.S. National Forest land which is to be preserved in its natural state. The only changes permitted are those for fire prevention.

primitive circle That circle on a stereographic projection which is the intersection of the stereographic plane with the sphere of reflection; it is the sphere's equatorial circle.

primitive lattice A crystal lattice in which there are lattice points only at its corners. Cf: *centered lattice.* Syn: *simple lattice.*

primitive rocks A term applied by Lehmann (1756) to crystalline rocks devoid of fossils and rock fragments, and believed to be of chemical origin, having formed prior to the advent of life; also, a term for the rocks believed to have been first formed, being irregularly crystallized and aggregated without cement. They include gneiss, schist, primary limestone, and plutonic rocks such as granite. The term is obsolete as many of these rocks are found in all ages and formations. Cf: *secondary rocks.* Syn: *primary rocks.*

primitive unit cell *unit cell.*

primitive water Water imprisoned in the interior, in either mo-

lecular or dissociated form, since the formation of the Eart (Meinzer, 1923, p.31). Cf: *juvenile* [*water*].

primocryst A crystal in equilibrium with the magma where pr mary crystallization occurred. A primocryst becomes a *cum lus crystal* after settling out.

primordial Original, first in order or development, earliest, existing at or from the beginning; e.g. "primordial ocea basin" or "primordial potassium", or a "primordial magma postulated to be of basaltic composition. See also: *primeval.*

Primordial An obsolete term formerly applied to what is no called *Cambrian.* It was used by Joachim Barrande (179 1883), French paleontologist, for the oldest or lowest fossili erous strata as developed in Bohemia.

primordial plate One of the first plates formed following met morphosis in each plate system of an echinoid.

primordial valve A chitinous plate in certain cirripede crusta ceans, having a distinctive honeycomb appearance, and de veloping at incipient umbones of terga, scuta, and carina du ing metamorphosis (TIP, 1969, pt.R, p.100).

primordium The early cells in the differentiation of an organ.

principal axis [cryst] In a crystal, that crystallographic ax which is the most prominent. In the tetragonal and hexagon systems, it is the vertical or *c* axis; in the orthorhombi monoclinic, or triclinic systems, it is also usually the *c* axi although in monoclinic crystals such as epidote it may be th *b* axis.

principal axis [exp struc geol] In experimental structural ge ogy, a *principal axis of stress* or a *principal axis of strain.*

principal axis of strain One of the three axes corresponding the three axes of the body that were mutually perpendicula before deformation; also described as the axes of the stra ellipsoid. The longest or greatest is the axis of elongation, an the shortest or least is the axis of shortening. Syn: *strain axi principal axis.*

principal axis of stress One of the three mutually perpendic lar axes that are perpendicular to the *principal planes* stress. Syn: *stress axis; principal axis.*

principal direction One of two orthogonal directions at a point on a sphere or spheroid that remain orthogonal direc tions on the plane map of that point.

principal distance The perpendicular distance from the intern perspective center to the plane of a particular finished neg tive or print.

principal earthquake *main shock.*

principal focus The *focus* for a beam of incident rays parall to the axis of a lens or optical system. Syn: *focal point.*

principalia The main parenchymal megascleres (spicules) lyssacine hexactinellid sponges.

principal line [photo] The trace of the principal plane upon photograph; e.g. the line through the principal point (isocen er) and nadir of a tilted photograph.

principal line [geochem inst] *Spectral line* which is most easi excited or observed.

principal meridian A *central meridian* on which a rectangul grid is based; specif. one of a pair of coordinate axes (alon with the base level) used in the U.S. Public Land Survey sys tem to subdivide public lands in a given region, consisting of line extending north and south along the astronomic meridia passing through the initial point and along which standar township, section, and quarter-section corners are establ lished. The principal meridian is the line from which the sur vey of the township boundaries is initiated along the parallels.

principal plane (a) The vertical plane through the internal pe spective center (rear nodal point) and containing the perper dicular from that center to the plane of a tilted photograp (b) Any plane perpendicular to the axis of an optical syste and passing through its principal points.

principal plane of stress One of three mutually perpendicul planes, upon each of which the resultant stress is normal, i.

n which shear stress is zero. See also: *principal axis of stress.*

principal point (a) The foot of the perpendicular from the interior perspective center (rear nodal point) of a lens to the plane of the photograph; the geometric center of an aerial photograph, or the point where the optical axis of the lens meets the film plane in an aerial camera. Symbol: p. See also: *photograph center.* Syn: *center point.* (b) Either of two points on the axis of a lens such that a ray from any point of the object directed toward one principal point will emerge from the lens in a parallel direction but directed through the other principal point. When the initial and final media have the same index of refraction (such as for lenses used in air but not for oil-immersion systems as used in microscopes), the principal points coincide with the *nodal points.* (c) The point at which a principal visual ray intersects a perspective plane.

principal ray (a) The one ray within a bundle of incident rays that, upon entering an optical instrument from any given point of the object, passes through the optical center of the lens. (b) *principal visual ray.*

principal section In a uniaxial indicatrix, any plane passing through the optic axis.

principal shock *main shock.*

principal spine One of the large regularly placed spikes or needles in acantharian and spumellarian radiolarians.

principal stress A stress that is perpendicular to one of three mutually perpendicular planes that intersect at a point in a body on which the shearing stress is zero; a stress that is normal to a principal plane of stress. The three principal stresses are identified as least or minimum, intermediate, and greatest or maximum, in terms of sign. See also: *mean stress.*

principal visual ray A perpendicular extending from a station point to a perspective plane and theoretically passing exactly along the visual axis of a viewing eye. Syn: *principal ray.*

principle of uniformity *uniformitarianism.*

print A photographic copy made from a negative or from a transparency by photographic means, as by placing the negative or transparency in contact with a sensitized surface or by projecting the image on a screen or sensitized photographic medium and reproducing it. Cf: *positive [photo].*

prionodont Having a saw-like row of many simple and similar teeth; e.g. said of a hinge in a bivalve mollusk in which the teeth are developed in a direction transverse to the margin of the cardinal area, or said of an ostracode hinge resembling an adont hinge but distinguished by the presence of crenulations along elongate elevations and depressions. Cf: *taxodont.*

priorite A black mineral: $(Y,Ca,Th)(Ti,Nb)_2O_6$. It is isomorphous with aeschynite. Syn: *blomstrandine.*

priority In taxonomy, preference of a particular taxon because of having an earlier publication date or because of being generally accepted by most authors. See also: *law of priority.*

prior river A term applied in Australia to a river system that is older than the present system but postdates the *ancestral river.*

Prior's rules Two relationships regarding the basic chemical and mineralogic regularities in chondritic meteorites, established by Prior (1920): (1) the smaller the amount of nickel-iron in a chondrite, the higher the Ni/Fe ratio in the nickel-iron in a chondrite, the higher the FeO/MgO ratio in the ferromagnesian silicate minerals. Named after George T. Prior 1862-1936), English mineralogist.

prisere A sere that takes place in a barren area undisturbed by man's activities.

prism [cryst] A crystal form having three, four, six, eight, or twelve faces, with parallel intersection edges, and which is open only at the two ends of the axis parallel to the face intersection edges.

prism [sed] (a) A long, narrow, wedge-shaped sedimentary body whose width/thickness ratio is greater than 5 to 1, but less than 50 to 1 (Krynine, 1948, p.146); e.g. an alluvial fan adjacent to an escarpment, or one of the great conglomerates of the geologic record. It is typical of orogenic sediments formed during periods of intense crustal deformation, such as the arkoses found in fault blocks. Cf: *tabular; shoestring.* Syn: *wedge.* (b) *geosynclinal prism.*

prismatic (a) Said of a sedimentary particle whose length is 1.5 to 3 times its width (Krynine, 1948, p.142). Cf: *tabular.* (b) Pertaining to a sedimentary *prism.* (c) Pertaining to a crystallographic *prism.*

prismatic class That crystal class in the monoclinic system having symmetry $2/m$. Prisms of this system have four faces, are rhombic in cross section, and have as their axis either the *c* axis, the *a* axis, or any lattice row parallel to the *b* axis.

prismatic cleavage A type of crystal cleavage that occurs parallel to the faces of a prism, e.g. the [110] cleavage of amphibole.

prismatic compass A small *magnetic compass* held in the hand when in use and equipped with peep sights and a triangular glass prism so arranged that the magnetic bearing or azimuth of a line can be read (through the prism) from a circular graduated scale at the same time that the line is sighted over.

prismatic jointing *columnar jointing.*

prismatic layer (a) The middle layer of the shell of a bivalve mollusk, consisting essentially of prisms of calcium carbonate (calcite or aragonite). (b) The *secondary layer* in articulate brachiopods, secreted extracellularly as prismatic calcite.

prismatic structure *columnar jointing.*

prismatic system An old syn. of *orthorhombic system.*

prismatolith A coccolith constructed of polygonal prisms. It may be solid or axially perforated. See also: *porolith.*

prism crack A mud crack that develops in regular or irregular polygonal patterns on the surface of drying mud puddles and that breaks the sediment into prisms standing normal to bedding (Fischer, 1964, p.114).

prism level A type of dumpy level in which the level bubble can be viewed from the eyepiece end by means of an attached prism at the same time the rod is being read.

prismoid A solid body resembling a prism, having similar but unequal parallel, polygonal ends. It is a textural term used for sedimentary particles. Adj: *prismoidal.*

prismoidal Adj. of *prismoid.* It is a term used in sedimentary petrology, and is not to be confused with *prismatic*, which is a crystallographic term.

prism twin law A rare, normal twin law in monoclinic or triclinic feldspars, having a twin plane of (110) or (1$\bar{1}$0).

proancestrula The first-formed or basal part of the ancestrula of cyclostomatous and Paleozoic bryozoans.

probability A statistical measure (where zero is impossibility and one is certainty) of the likelihood of occurrence of an event.

probable error A quantity, or deviation from the mean, of such magnitude that the likelihood of its being exceeded in a set of observations or measurements is equal to the likelihood of its not being exceeded; a value about which one can assert with a probablity of 0.5 that it will not be exceeded in magnitude by the error of an estimate. Its value is that of the standard deviation (standard error) multiplied by ±0.6745, and it indicates the precision attained in a series of measurements. Probable error represents regular deviation within a determined distance on each side of the mean of a frequency curve (area under a normal frequency distribution curve equivalent to 50% of the distribution). It does not represent an error that is more apt to occur than an error of any other magnitude.

probable ore (a) A syn. of *indicated ore.* (b) A mineral deposit adjacent to a developed ore but not yet proven by development. Cf: *extension ore.*

probe n. Any measuring device that is placed in the environment to be measured, e.g. a potential electrode, a density probe in a drill hole, or oceanographic instruments that are lowered into the sea.

probertite A colorless monoclinic mineral: $NaCaB_5O_9.5H_2O$.

problematic fossil A natural object, structure, or marking in a rock, resembling a fossil but having a very doubtful organic nature or origin. Cf: *pseudofossil.* Syn: *quasi-fossil.*

problematicum A marking, object, structure, or other feature (in a rock) whose nature presents a problem, such as a doubtful sort of "fossil" that is probably of inorganic origin or whose organic nature is uncertain; esp. an undoubted organic remain (such as a trace fossil) with a more or less obscure nature. Pl: *problematica.*

proboscis A distal cylindric tube extending from an astropyle of a phaeodarian radiolarian.

Procellarian (a) Pertaining to lunar lithologic map units and topographic forms constituting, or closely associated with, the maria. Such features were formerly mapped as the Procellarum System, but are now considered a unit of the Imbrian System. (b) Said of the time interval during which the Procellarum Group was developed.

procephalic Pertaining to, forming, or situated on or near the front of the head; e.g. "procephalic lobe" or the anterior (preoral) part of a merostome embryo.

prochlorite *ripidolite.*

prochoanitic Said of a septal neck of a cephalopod directed forward (adorally, or toward the aperture). Ant: *retrochoanitic.*

prochronic Before time or creation. Ant: *diachronic.*

proclade A *clade* or bar in the ebridian skeleton that arises from the end of an upper actine and is directed toward the anterior or nuclear pole. Cf: *opisthoclade.*

procline Said of the ventral and anterior *inclination* of the cardinal area in the pedicle valve of a brachiopod, lying in the bottom right or second quadrant moving counterclockwise from the orthocline position (TIP, 1965, pt.H, p.60, fig.61).

Proctor Pertaining to or determined by a procedure designed by Ralph R. Proctor (1894-1962), U.S. civil engineer, to establish water content-density relationships of a remolded soil by application of compactive effort under standardized conditions; e.g. the "Proctor curve" (or compaction curve), the "Proctor compaction test", and the "Proctor density" of soil.

procumbent Said of a stem that trails or lies flat and that does not take root.

prod cast The cast of a prod mark, consisting of a short ridge that rises downcurrent and ends abruptly. Syn: *impact cast.*

prodelta The part of a delta that is below the effective depth of wave erosion, lying beyond the *delta front,* and sloping gently down to the floor of the basin into which the delta is advancing and where clastic river sediment ceases to be a significant part of the basin-floor deposits; it is entirely below the water level. Cf: *intradelta.*

prodelta clay The fine-grained river-borne material (very fine sand, silt, and clay) deposited as a broad fan on the floor of a sea or lake beyond the main body of a delta; the material in a bottomset bed.

prodissoconch The rudimentary or earliest-formed shell of a bivalve mollusk, secreted by the larva or embryo, and preserved at the tip of the beak of some adult shells.

prod mark A short tool mark oriented parallel to the current and produced by an object that plowed into and was then raised above the bottom; its longitudinal profile is asymmetric. The mark deepens gradually downcurrent where it ends abruptly (unlike a flute). Cf: *bounce mark.* Syn: *impact mark.*

producer An organism (e.g. most plants) that can form new organic matter from inorganic matter such as carbon dioxide, water, and soluble salts. Cf: *consumer.*

producing horizon A reservoir bed within the stratigraphic series of an oil province from which gas or liquid hydrocarbons can be obtained when penetrated by a well.

production (a) The growth of organisms in a lake. (b) A time-

rate unit of total amount of organisms grown. See also: *p mary productivity; productivity* [*lake*]; *yield* [*lake*].

productivity [biol] A general term for the organic fertility of body of water. The more precise terms *production* and *p mary production* are used.

productivity [lake] (a) The basic capacity of a lake to produc a particular organism. (b) The word often is used in place primary *productivity* and *production* [lake]. See also: *yie* [lake].

productoid Any articulate brachiopod belonging to the subc der Productidina, characterized by a pseudopunctate she having a flat or concave, rarely convex, brachial valve and convex pedicle valve. This group includes the largest ar most aberrant brachiopods yet known. Their stratigraph range is Lower Devonian to Upper Permian. Var: *productid.*

proepistome *interantennular septum.*

profile [geomorph] (a) The outline produced where the plar of a vertical section intersects the surface of the ground; e. the *longitudinal profile* of a stream, or the profile of a coast hill. Syn: *topographic profile.* (b) *profile section.*

profile [geophys] A graph or drawing that shows the variatic of one property such as elevation or gravity, usually as orc nate, with respect to another property, usually linear, such distance.

profile [seis] In seismic prospecting, the data recorded fro one shot point by a single line of receivers.

profile [palyn] *pollen profile.*

profile [struc petrol] A cross section of a homoaxial structure e.g. an s surface, drawn perpendicular to its axis. Syn: *te tonic profile; right section; normal cross section.*

profile [water] A vertical section of a water table or other pc tentiometric surface, or of a body of surface water.

profile line The top line of a *profile section,* representing th intersection of a vertical plane with the surface of the groune Cf: *line of section.*

profile method *two-dimensional method.*

profile of equilibrium [coast] The slightly concave slope of th floor of a sea or lake, taken in a vertical plane and extendin away from and transverse to the shoreline, being steepe near the shore, and having a gradient such that the amount sediment deposited by waves and currents is balanced by th amount removed by them; the transverse slope of a *grade shoreline.* The profile is easily disturbed by strong winds, larg waves, and exceptional high tides. Syn: *equilibrium profile graded profile.*

profile of equilibrium [streams] The longitudinal profile of graded stream or of one whose smooth gradient at every poir is just sufficient to enable the stream to transport the load c sediment made available to it. It is generally regarded as smooth, parabolic curve, gently concave to the sky, practica ly flat at the mouth and steepening toward the source. Sy equilibrium profile; graded profile.

profile section A diagram or drawing that shows along a give line the configuration or slope of the surface of the ground a it would appear if intersected by a vertical plane. The vertica scale is often exaggerated. See also: *line of section; profil line.* Syn: *profile.*

profiling Electrical exploration wherein the transmitter and re ceiver are moved in unison across structure and a profile c mutual impedance between transmitter and receiver is so ob tained. Cf: *sounding* [*elect*]. Syn: *lateral search.*

profiling snow gage A type of radioactive gage for measurin the water equivalent and density/depth distribution of a snow pack, consisting of a radioactive source and a radioactivit detector which move up and down in two adjacent vertica pipes surrounded by snow. Syn: *nuclear twin-probe gage.*

profluent stream A stream that is flowing copiously or smooth ly.

profundal adj. Pertaining to or existing in the deeper part of lake, below the limit of well developed zones of vegetation.

progenitor In biology, an ancestor.

proglacial Immediately in front of or just beyond the outer limits of a glacier or ice sheet, generally at or near its lower end; said of lakes, streams, deposits, and other features produced by or derived from the glacier ice.

proglacial lake A lake formed just beyond the frontal moraine of a retreating glacier and generally in direct contact with the ice. Cf: *glacier lake*. Syn: *ice-marginal lake*.

proglyph A hieroglyph consisting of a cast (Vassoevich, 1953, p.36); specif. a *groove cast*.

progradation The building forward or outward toward the sea of a shoreline or coastline (as of a beach, delta, or fan) by nearshore deposition of river-borne sediments or by continuous accumulation of beach material thrown up by waves or moved by longshore drifting. Ant: *retrogradation*. Cf: *advance*.

prograde metamorphism Metamorphic changes in response to higher pressure or temperature than that to which the rock last adjusted itself (Bayly, 1968); a type of *polymetamorphism*. Cf: *retrograde metamorphism*.

prograde motion The predominant apparent motion eastward of a body on the celestial sphere. Cf: *retrograde motion*.

prograding shoreline A shoreline that is being built forward or outward (into a sea or lake) by deposition and accumulation. Ant: *retrograding shoreline*.

program A plan for solution of a problem contained in a sequence of coded instructions for insertion into a mechanism (such as a computer) for automatic control of its operation.

progression *advance* [coast].

progressive fault *growth fault*.

progressive metamorphism Progressive change in the degree of metamorphism from lower to higher grade across a metamorphic terrane. The term may be applied to rocks in contact aureoles or to rocks traced through the different isograds or facies of regional metamorphism (Huang, 1962).

progressive overlap A general term used by Grabau (1906, p.569) for a "regular progressive" onward movement or spreading of the "zones of deposition", and including what are now known as onlap, offlap, and continental transgression.

progressive sand wave A term used by Bucher (1919, p.168) for a sand wave that migrates downcurrent. Ant: *regressive sand wave*.

progressive sorting Sorting in the downcurrent direction, resulting in a systematic downcurrent decrease in the mean grain size of the sediment (Pettijohn, 1957, p.541).

progressive wave A water wave, the wave form of which appears to move progressively. Cf: *standing wave*.

progressive-wave theory A former theory of the tides involving the formation of two tidal waves in the Southern Ocean, one following the Moon and the other on the opposite diameter of the Earth (Monkhouse, 1965, p.249); it is replaced by *oscillatory-wave theory*.

projected profile A diagram that includes only those features of a series of profiles, usually drawn along several regularly spaced and parallel lines on a map) that are not obscured by higher intervening ground (Monkhouse & Wilkinson, 1952); it gives a panoramic effect with a distant skyline, a middleground, and a foreground, and it represents an outline landscape-drawing showing only summit detail. Cf: *superimposed profile; composite profile*.

projection (a) A systematic, diagrammatic representation on a plane (flat) surface of three-dimensional space relations, produced by passing lines from various points to their intersection with a plane; esp. a *map projection*. (b) Any orderly method by which a projection is made; the process or operation of transferring a point from one surface to a corresponding position on another surface by graphical or analytical means, so that each point of one corresponds to one and only one point of the other.

projection net *net* [struc petrol].

prolapsed bedding A term used by Wood & Smith (1958,

p.172) for bedding characterized by a series of flat folds with near-horizontal axial planes contained entirely within a bed having undisturbed boundaries.

prolate Extended or elongated in the direction of a line joining the poles; e.g. "prolate pollen" whose equatorial diameters are much shorter than the dimensions from pole to pole. Ant: *oblate*.

proloculus The initial or first-formed chamber of a foraminiferal test, typically at the small end of a series or at the center of a coil. Pl: *proloculi*. Syn: *proloculum*.

proloculus pore A single circular opening in a proloculus, leading to the next-formed chamber of a foraminiferal test (as in fusulinids).

proluvium A complex, friable, deltaic sediment accumulated at the foot of a slope as a result of an occasional torrential washing of fragmental material. Adj: *proluvial*.

promontory (a) A high, prominent projection or point of land, or cliff of rock, jutting out boldly into a body of water beyond the coastline; a *headland*. Syn: *cobb; reach; ness; nook*. (b) A *cape*, either low-lying or of considerable height, with a bold termination. (c) A bluff or prominent hill overlooking or projecting into a lowland.

prong [geomorph] *spur*.

prong [streams] A term applied in the southern Appalachian Mountains to a fork or branch of a stream or inlet.

prong reef A wall reef that has developed irregular buttresses normal to its axis in both leeward and (to a smaller degree) seaward directions (Maxwell, 1968, p.99 & 101).

proostracum The anterior (adoral) horny or calcareous blade-like prolongation of the dorsal border of the phragmocone of belemnites and related cephalopods, forming a protecting shield over the visceral mass of the animal. Pl: *proostraca*.

propaedeutic stratigraphy *prostratigraphy*.

propagule The minimum number of individuals of a species required for the successful colonization of a habitable island (MacArthur & Wilson, 1967, p.190).

propane An inflammable, gaseous hydrocarbon, formula C_3H_8, of the methane series and that occurs naturally in crude petroleum and natural gas. It is also produced by cracking and is used primarily as a fuel and in the making of chemicals.

proparea One of a pair of flattened subtriangular halves of the pseudointerarea of a brachiopod, divided medially by various structures (such as homeodeltidium, intertrough, or pedicle groove).

proparian adj. Of or concerning a trilobite whose facial sutures extend outward from the eyes to the lateral margin of the cephalon; e.g. a "proparian facial suture" that crosses the dorsal surface of the cephalon, passes along the medial edge of the eye, and intersects the lateral border of the cephalon in front of or at the genal angle. Cf: *opisthoparian*. Syn: *proparous*.---n. A proparian trilobite; specif. a trilobite of the order Proparia in which the posterior branch of the facial suture cuts the lateral margin of the cephalon.

propeltidium An anterior sclerite of a segmented carapace of an arachnid, situated in front of the *mesopeltidium*.

property An attribute or characteristic of a material that may be measured quantitatively.

propodosoma A section of the body of an acarid arachnid, bearing the first and second pairs of legs. Cf: *metapodosoma*.

propodus The sixth or penultimate segment of the pereiopod of a malacostracan crustacean, bounded proximally by the carpus and distally by the dactylus. Pl: *propodi*. Syn: *propodite*.

proportional counter A radiation detector consisting of a gas-filled tube in which the intensity of the discharge pulses is proportional to the energy of the ionizing particles.

proportional limit The highest value of stress that a material can undergo before it loses its linear relationship between stress and strain, i.e. before it ceases to behave according to

Hooke's law.

propylite A hydrothermally altered andesite resembling a greenstone and containing calcite, chlorite, epidote, serpentine, quartz, pyrite, and iron ore. The term was first used by Richtofen in 1868. See also: *propylitization*.

propylitization A hydrothermal process involving the formation of a *propylite* by the introduction of, or replacement by, an assemblage of minerals including carbonates, epidote, quartz, and chlorite.

prorsiradiate Said of an ammonoid rib inclined forward (adorally) from the umbilical side toward the venter. Cf: *rursiradiate; rectiradiate*.

proseptum One of the initial partitions in the apical part of an ammonoid shell.

prosicula The proximal, initially formed part of the *sicula* of a graptolite, secreted as a single conical unit. Cf: *metasicula*.

prosiphon A small structure extending from the adapical part of the caecum to the wall of the protoconch of an ammonoid and having the form of a partial cone (TIP, 1959, pt.L, p.5).

prosobranch Any gastropod belonging to the subclass Prosobranchia, characterized in most cases by the presence of a shell, commonly with an operculum, and by the anterior position of the auricle with respect to the ventricle.

prosochete An *inhalant canal* of a sponge.

prosocline (a) Said of the hinge teeth (and in some genera, of the body of the shell) of a bivalve mollusk, sloping (from the lower end) in the anterior or forward direction. (b) Said of the growth lines that incline forward relative to the growth direction of a gastropod shell.---Cf: *opisthocline*.

prosodus A small canal of uniform diameter in a sponge, leading from an inhalant canal to a prosopyle of approximately the same cross-sectional area. Pl: *prosodi*. Cf: *aphodus*.

prosogyrate Said of the umbones (of a bivalve mollusk) curved so that the beaks point in the anterior or forward direction. Ant: *opisthogyrate*. Syn: *prosogyral*.

prosoma (a) The anterior part of the body of various invertebrates; esp. the *cephalothorax* of an arachnid or merostome. See also: *opisthosoma*. (b) The *prosome* of a copepod.

prosome (a) The anterior region of the body of a copepod crustacean, commonly behind by major articulation. See also: *urosome*. Syn: *prosoma*. (b) A structure within the neck of the body of a chitinozoan, extended to various intermediate positions, or even projecting beyond the collar. Its upper surface may be flat, convex, conical, or truncate, and an upper flange may lie against the pseudostome; its top may be marked by dark radial fibers, and the tubular area commonly has many dark annular rings. Cf: *opisthosome*.

prosopite A colorless mineral: $CaAl_2(F,OH)_8$.

prosopore The entrance opening of an inhalant canal of a sponge. Cf: *apopore*.

prosopyle Any aperture through which water enters a flagellated chamber of a sponge. Cf: *apopyle*.

prospect n. (a) An area that is a potential site of mineral deposits, based on preliminary exploration. (b) Sometimes, an area that has been explored in a preliminary way but has not given evidence of economic value. (c) An area to be searched by some investigative technique, e.g. geophysical prospecting, electrical prospecting. (d) A geologic or geophysical anomaly, especially one which is recommended for additional exploration.----A prospect is distinct from a *mine* in that it is nonproducing. See also: *prospecting*.

prospect [geophys] (a) An area that is being investigated by geophysical methods. (b) A geologic or geophysical anomaly, especially one which is recommended for additional exploration.

prospect hole A general term for any shaft, pit, adit, drift, tunnel, or drill hole made for the purpose of prospecting mineral-bearing ground; a *test hole* used for this purpose. More specific terms, such as *prospect shaft* and *prospect pit* are generally used.

prospecting [geophys] (a) Searching for a *prospect*. (b) *geo physical prospecting*.----Syn: *prospection*.

prospecting seismology The application of seismology to th exploration for natural resources, esp. gas and oil; *seism. exploration; applied seismology*.

prospection *prospecting*.

prospective ecospace As proposed by Valentine (1969 p.687), "the total *ecospace* that an organism or other ecolog cal unit may utilize if it is physically available". Cf: *realize ecospace*.

prospector An individual engaged in prospecting for valuab mineral deposits, generally working alone or in a small group and on foot with simple tools or portable detectors. The terr implies an individual searching for his own gain, rather tha an employee of a mining company.

prospect pit *prospect hole*.

prospect shaft *prospect hole*.

prostal A sponge spicule (megasclere) that protrudes fror the surface or beyond the body of a hexactinellid sponge. P *prostals* or *prostalia*.

prostratigraphy A term proposed by Schindewolf (1954) fc "preliminary stratigraphy" including lithologic and paleontolog ic studies without consideration of the time factor; the "ra material" for stratigraphy, consisting of local observation, de scription, and arrangement of strata, but not yet methodicall linked together by the concept of time. See also: *protostratic raphy; prestratigraphy; topostratigraphy*. Syn: *propaedeut stratigraphy*.

prosuture The line of junction of a proseptum with the walls c an ammonoid shell.

protactinium-ionium age method Calculation of an age i years for deep-sea sediments formed during the last 150,00 years, based on the assumption that the initial protactinium 231 to ionium (thorium-230) ratio for newly formed sediment has remained constant for the total section of sediment under study. The age depends on the gradual change wit time of the protactinium-231 to ionium ratio because of th difference in half-lives. See also: *uranium-series age methods* Syn: *protactinium-231 to thorium-230 age method; thorium 230 to protactinium-231 excess method*.

protactinium-231 to thorium-230 age method *protactinium-io nium age method*.

protalus rampart An arcuate ridge consisting of boulders an other coarse debris marking the downslope edge of an exist ing or melted snowbank. It is formed by the sliding or rollin downslope of frost-riven boulders over the snow (*winter-talu ridge*), or by the subsurface movement of finer material (probably over the permafrost table) due to solifluction. Syn *protalus*.

protaspis An early juvenile trilobite whose small, oval-lik exoskeleton is not yet divisible into articulated cephalon, tho rax, and pygidium; the earliest recognized stage in the devel opment of a trilobite. Pl: *protaspides*.

protaxis An antique term for the central axis of a mountai chain, supposedly consisting of the oldest rocks and struc tures; for example, an "Archean protaxis".

protected thermometer A *reversing thermometer* that is pro tected against hydrostatic pressure by a glass shell. Cf: *un protected thermometer*.

protectite A rock formed by the crystallization of a primar magma. See also: *anatexite, syntectite*.

protegulal node The apical, commonly raised portion of a adult brachiopod shell, representing the site of the protegulum and later growth up to the brephic stage (TIP, 1965, pt.H p.151).

protegulum The smooth, biconvex, first-formed shell of organ ic material (chitin or protein) of a brachiopod, secreted simul taneously by both mantles.

proteolite An old term for hornfels-type rock, introduced b Boase in 1832 and used by Bonney in 1886 for andalusit

‖rnfels. The term is now obsolete. Cf: *cornubianite; leptynol-* ‖.

‖oterobase A diabase in which the mafic mineral is primary ‖rnblende.

‖oterogenesis *neoteny.*

‖oteroglacial Pertaining to the earlier great ice age (Hansen, ‖94, p.128). Cf: *deuteroglacial.*

‖oterophytic *Archeophytic.*

‖oterosoma The anterior section of the body of an acarid ‖achnid, ending behind the second pair of legs.

‖oterozoic (a) The more recent division of the Precambrian. ‖: *Archeozoic.* Syn: *Algonkian; Agnotozoic.* (b) The entire ‖ecambrian.

‖othallus The *gametophyte* of a fern or other pteridophyte, ‖ually a flattened thallus-like structure attached to the soil. ‖ *prothalli.* Syn: *prothallium.*

‖otheca (a) The proximal part of graptoloid theca before it is ‖ferentiated from the succeeding theca. It is considered ‖uivalent to the *stolotheca* in those graptolites with more ‖an one type of theca. (b) A primary element of the wall of a ‖sulinid, comprising diaphanotheca and tectum.

‖otist A single-celled organism belonging to the kingdom Pro-‖ta, comprising organisms with both plant and animal affini-‖s, e.g. *protozoans,* bacteria, and some algae, fungi, and ‖uses. Syn: *protistan.*

‖otobitumen Any of the fats, oils, waxes, resins, etc. present ‖ unaltered or nearly unaltered plant and animal products ‖m which fossil bitumens are formed. (Tomkeieff, 1954). ‖e also: *labile, stabile.*

‖otoclase Leith's term for a rock possessing what he consid-‖ed to be primary cleavage, e.g. bedding planes in sedimen-‖y rock, formed concurrently with the rock (1905, p.12). Cf: ‖*etaclase.*

‖otoclastic Said of igneous rocks in which the earlier formed ‖nerals show granulation and deformation, the result of dif-‖ential flow of the magma before complete solidification.

‖otoconch (a) The first portion of the embryonic shell of a ‖phalopod, its preservation in fossil and in living forms being ‖certain. The term is sometimes applied to the first camera ‖hamber) of the shell, located at the apex of the phragmo-‖ne or at the center of the coil, and closed in an ammonoid ‖a proseptum. (b) The apical, usually smooth whorls of a ‖stropod shell, usually well-demarcated from the *teleoconch.* ‖e term applies to the fully formed embryonic shell of a gas-‖‖pod and should not be used synonymously with *nucleus,* ‖hough it has been restricted by some authors to the simple ‖p-shaped plate that constitutes the first shell rudiment (see ‖ight, 1941).

‖otocorallite The first-formed corallite of a colony.

‖otodolomite (a) A crystalline calcium-magnesium carbonate ‖h a disordered lattice in which the metallic ions occur in ‖ same crystallographic layers instead of in alternate layers ‖ in the dolomite mineral. (b) An imperfectly crystallized arti-‖ial material of composition near $CaMg(CO_3)_2$.

‖otoenstatite An artificial, unstable modification of $MgSiO_3$, ‖oduced by decomposition of talc by heating and convertible ‖ enstatite by grinding or by heating to a high temperature.

‖otoforamen The primary aperture of a foraminiferal test as-‖ciated with a fully developed or rudimentary tooth plate. Cf: ‖uteroforamen.

‖otogene An old term for a primary rock. Adj: *protogenous.* ‖ *deuterogene.* Syn: *protogine.*

‖otogenesis Reproduction by budding.

‖otogenic Said of an older crystalline rock believed to have ‖en formed by igneous activity.

‖otogenous Adj. of *protogene.*

‖otogine (a) Granitic rock occurring in the Alps that has gne-‖ic structure, contains sericite, chlorite, epidote, and garnet, ‖d shows evidence of composite origin or crystallization (or ‖rtial recrystallization) under stress after consolidation. Also

spelled: *protogene.* (b) *protogene.*

protointraclast A genetic term suggested by Bosellini (1966) for a limestone component that resulted from a premature attempt at resedimentation while still being in an unconsolidated and viscous or plastic state and that never existed as a free, clastic entity. Cf: *intraclast; plasticlast.*

protolith The unmetamorphosed rock from which a given metamorphic rock was formed by metamorphism. Syn: *parent rock.*

protomylonite (a) A mylonitic rock produced from contact-metamorphosed rock, with granulation and flowage being due to overthrusts following the contact surfaces between the intrusion and the country rock (Holmes, 1920). (b) A coherent crush breccia whose characteristically lenticular, megascopic particles faintly retain primary structures. It is a lower grade in the development of *mylonite* and *ultramylonite* (Waters & Campbell, 1935, p.479).

protonema The filamentous gametophyte stage of charophyte algae (stoneworts) and many bryophytes.

proton precession magnetometer A type of *nuclear resonance magnetometer* that accurately measures total magnetic intensity by the use of the precession of protons in a hydrogen-rich liquid about the magnetic field direction. The precession frequency is proportional to field strength. See also: *proton vector magnetometer.*

proton vector magnetometer A type of *proton precession magnetometer* with a system of auxiliary coils that permits measurement of horizontal intensity H or vertical intensity Z as well as total intensity F.

protonymph The first postembryonic stage in the arachnid order Acarida.

protophyte A single-celled plant. Cf: *metaphyte.*

protopod The proximal portion of a limb of a crustacean, consisting of coxa, basis, and sometimes precoxa, often fused to each other. Its distal edge generally bears the endopod and exopod. Syn: *protopodite; sympod.*

protopore A single fine opening or perforation in a foraminiferal test, rounded at least on the inner wall. Cf: *deuteropore.*

protoquartzite A well-sorted, quartz-enriched sandstone that lacks the well-rounded grains of an orthoquartzite; specif. a lithic sandstone intermediate in composition between subgraywacke and orthoquartzite, containing 75-95% quartz and chert, less than 15% detrital clay matrix, and 5-25% unstable materials in which the rock fragments exceed the feldspar grains in abundance (Pettijohn, 1954, p.364). It commonly forms shoestring sands. Examples: Venango Formation (Upper Devonian) of New York and Pennsylvania, and Hartshorne Sandstone (Pennsylvanian) of Oklahoma and Arkanasas. Pettijohn (1957, p.316) later used 10-25% unstable materials. The term was used by Krynine (1951) for a "cleaned-up" graywacke (matrix washed out), intermediate in composition between quartzose graywacke and orthoquartzite. Syn: *quartzose subgraywacke.*

protore The rock below the sulfide zone of *enrichment;* the primary, noneconomic material. See also: *oxidized zone; sulfide zone.*

protoscience Philosophic speculations concerning nature that were offered before the application of the scientific method.

protoseptum One of the six first-formed septa of a corallite. Cf: *metaseptum.*

protostratigraphy A term proposed by Henningsmoen (1961) for preliminary or introductory stratigraphy, including lithostratigraphy and biostratigraphy; *prostratigraphy.*

prototheca The roughly conical or cup-shaped structure constituting the embryonic exoskeleton of a coral.

prototype (a) An ancestral form; the most primitive form in a group of related organisms. Syn: *archetype.* (b) *primary type.*

protoxylem The first-formed, primary xylem of a plant. Cf: *metaxylem.*

protozoan A single celled organism belonging to the *protist* phylum Protozoa and characterized by the absence of tissues and organs. Some members of the phylum have both plant and animal affinities (flagellates); other members are characterized by their development of calcium carbonate and siliceous skeletons (radiolarians, foraminifers). Nonpreferred syn: *eozoan; protozoon.*

Protozoic (a) That part of Precambrian time for which, in the corresponding rocks, traces of life appear. Cf: *Azoic.* (b) The lower Paleozoic.----The term is obsolete.

protractor An instrument used in drawing and plotting, designed for laying out or measuring angles on a flat or curved surface, and consisting of a plate marked with units of circular measure.

protractor muscle A muscle that extends an organ or part; e.g. an *outside lateral muscle* or *middle lateral muscle* in some lingulid brachiopods, or a longitudinal fibril in connective tissue of the pedicle of some articulate brachiopods (TIP, 1965, pt.H, p.151). Cf: *retractor muscle.*

protriaene A sponge triaene in which the cladi curve away from the rhabdome, making an angle to the rhabdome that is noticeably greater than that of a normal tetraxon. Cf: *orthotriaene.*

protrusion A proposed term for a rock mass that has been tectonically intruded in the solid state; it is in contrast to an igneous intrusion (Lockwood, 1971). Adj: *protrusive.*

protrusive Tectonically intrusive; adj. of *protrusion.*

proustite A cochineal-red rhombohedral mineral: Ag_3AsS_3. It is isomorphous with pyrargyrite, and is a minor ore of silver. Cf: *xanthoconite.* Syn: *light ruby silver; light red silver ore.*

prove v. In economic geology, to establish, by drilling, trenching, underground openings, etc., that a given deposit of a valuable substance exists (and where), and that its grade or tenor and dimensions equal or exceed some specified amounts. See also: *proved reserves.*

proved ore *proved reserves.*

proved reserves Mineral reserves, esp. of crude oil, natural gas liquids, and natural gas, for which reliable quantity and quality estimates have been made. Cf: *prove; developed reserves; positive ore.* Syn: *proved ore.*

provenance A place of origin; specif. the area from which the constituent materials of a sedimentary rock or facies are derived. Also, the rocks of which this area is composed. Cf: *distributive province.* Syn: *provenience; source area; sourceland.*

provenience *provenance.*

province [geog] Any large area or region considered as a whole, all parts of which are characterized by similar features or history differing significantly from those of adjacent areas; specif. a geologic province or a physiographic province.

province [ecol] (a) A group of temporally and spatially associated plant or animal communities. (b) Part of a *region,* isolated and defined by its climate and topography and characterized by a particular group of organisms.

provincial alternation The overlapping of sedimentary petrologic provinces, caused by oscillation of the boundary between two provinces during time (Pettijohn, 1957, p.573-574).

provincial series A time-stratigraphic *series* recognized only in a particular region and involving a major division of time within a period; e.g. the Wolfcampian Series within the Permian System in west Texas and New Mexico.

provincial succession A succession of sedimentary petrologic provinces, produced by changes in provenance leading to mineral associations that change with time (Pettijohn, 1957, p.574).

provinculum A primitive taxodont hinge composed of minute teeth developed in some bivalve mollusks before the permanent dentition.

provitrain *Vitrain* in which some plant structure is microscopically visible. Cf: *euvitrain.* See also: *periblain; suberain; xylain.* Syn: *telain.*

provitrinite A variety of the maceral *vitrinite* characteristic of

provitrain and including the varieties periblinite, suberini and xylinite. Plant cell structure is visible under the micr scope. The term *telinite* has been proposed as a preferab synonym. Cf: *euvitrain.* Syn: *phyllovitrinite.*

prowersite An orthoclase- and biotite-rich minette. Its name derived from Prowers County, Colorado, U.S.A.

proximal [paleont] Next to or nearest the point of attachme or place of reference, a point conceived of as central, or t point of view. Examples in invertebrate morphology: "proxim direction" toward the dorsal pole or the mouth of a crinoid toward the mouth or center of disc of an asterozoan, or nea er to the proloculus in the growth of a foraminiferal she "proximal (first-formed) portion" of the rhabdosome of a gra tolite colony, nearest to the point of origin; "proximal ray" o sponge spicule directed inward from a bounding surface the sponge; "proximal shield" on the concave side of a plac ith; and "proximal side" toward the ancestrula or origin growth of a bryozoan colony. Ant: *distal.*

proximal [palyn] Said of the parts of pollen grains or spor nearest or toward the center of the original tetrad; e.g. said the side of a monocolpate pollen grain opposite the colpus, said of the side of a trilete spore provided with contact area Ant: *distal.*

proximal [sed] Said of a sedimentary deposit consisting coarse clastics and formed nearest the source area; e.g. "proximal turbidite" consisting of thick sandy varves. Cf: o tal.

proximale The noncirriferous topmost columnal of a crino typically distinguished by enlargement and permanent attac ment to the dorsal cup.

proximate admixture A term applied by Udden (1914) to *admixture* (in a sediment of several size grades) whose par cles are most similar in size to those of the dominant or ma mum grade; material in one of the two classes adjacent to t maximum histogram class.

proximate analysis The determination of compounds contain in a mixture; for coal, the determination of moisture, volat matter, fixed carbon, and ash. Cf: *ultimate analysis.*

proximate cyst A dinoflagellate cyst of nearly the same si as, and closely resembling, the motile theca. The ratio of t diameter of the main body to the total diameter of the cyst e ceeds 0.8. The term refers to the supposed proximity of t main cyst wall to the theca at the time of encystment. S also: *chorate cyst; proximochorate cyst.*

proximochorate cyst A dinoflagellate cyst having sutured o growths that more readily indicate the tabulate character. T ratio of the diameter of the main body to the total diameter the cyst is between 0.6 and 0.8. See also: *chorate cyst; pro mate cyst.*

proxistele The proximal region of a crinoid column near thec generally not clearly delimited from the mesistele. Cf: *d istele.*

proxy v. To substitute one ion or atom for another in a crys structure.----adj. Said of such a substituted ion or atom, or the mineral so formed.

proxying *diadochy.*

przhevalskite A bright greenish-yellow mineral: $Pb(UO_2 (PO_4)_2 \cdot 2H_2O.$

Psamment in U.S. Dept. of Agriculture soil taxonomy, a s order of the soil order *Entisol,* characterized by a texture loamy fine sand or coarser, and a coarse fragment content less than 35% (SSSA, 1970). Cf: *Aquent; Arent; Fluve Orthent.*

psammite (a) A clastic sediment or sedimentary rock co posed of sand-size particles; a sandstone. The term is equi lent to the Latin-derived term, *arenite.* (b) A term forme used in Europe for a fine-grained, fissile, clayey sandsto (as distinguished from a more siliceous and gritty sandstor in which "the component grains are scarcely distinguishal by the unassisted eye" (Oldham, 1879, p.44). (c) A term

rded by Tyrrell (1921, p.501-502) as the metamorphic de-
ative of arenite, such as the metamorphosed product of a
ndstone.---Etymol: Greek *psammos*, "sand". See also: *pse-
ite; pelite*. Syn: *psammyte*.

ammitic (a) Pertaining to or characteristic of psammite; *ar-
aceous*. Cf: *sandy*. (b) Said of a metamorphic rock derived
m a psammite; e.g. a "psammitic gneiss" or a "psammitic
hist" derived by metamorphism of an arenaceous sediment.

ammobiotic Said of an organism that lives in sand or sandy
eas.

ammofauna The animals associated with sandy substrates.

ammogenic dune A dune "caused by the effect of sand sur-
ces in trapping more sand" (Schieferdecker, 1959, term
48).

ammon In a body of fresh water, that part of the environent composed of sandy beach and bottom lakeward from
e water line.

ammophilic Said of an organism or of the fauna found in
nd. Noun: *psammophile*.

ammophyte A plant preferring sand or very sandy soil for
owth.

ammosere A sere that develops in a sandy environment.

ammyte Var. of *psammite*.

ephicity A term used by Mackie (1897, p.301) for the
oefficient of roundability" of a pebble- or sand-size mineral
agment, expressed as the ratio of specific gravity to hard-
ss (as measured in air) or the quotient of specific gravity
nus one divided by hardness (as measured in water).

ephite (a) A sediment or sedimentary rock composed of
rge fragments (coarser than sand) set in a matrix varying in
nd and amount; e.g. rubble, talus, breccia, glacial till, tillite,
ingle, gravel, and esp. conglomerate. The term is equivalent
the Latin-derived term, *rudite*. (b) A term regarded by Tyr-
ll (1921, p.501-502) as the metamorphic derivative of ru-
e, such as the metamorphosed product of a conglomerate
breccia.---Etymol: Greek *psephos*, "pebble". See also:
ammite; pelite*. Syn: *psephyte*.

ephitic (a) Pertaining to or characteristic of psephite. (b)
id of a metamorphic rock derived from a psephite.---Cf:
daceous; gravelly*.

ephonecrocoenosis A necrocoenosis of dwarf individuals.

ephyte [sed] Var. of *psephite*.

ephyte [lake] A lake-bottom deposit consisting mainly of
arse, fibrous plant remains (Veatch & Humphrys, 1966,
248). Cf: *pelphyte*.

eudatoll *pseudoatoll*.

eudoactin A ray-like arm or branch of a sponge spicule that
ntains no axial filament or axial canal.

eudoallochem An object resembling an *allochem* but proced in place within a calcareous sediment by a secondary
ocess such as recrystallization (Folk, 1959, p.7).

eudoaquatic Said of an organism living in moist or wet but
t truly aquatic conditions.

eudoatoll (a) An atoll that rises from the outer margin of
nless shoals; a reef platform encircling a shallow pool of
ater, rising up esp. at low tide from the continental shelf.
n: *bank atoll; shelf atoll; lagoon atoll*. (b) A ring-shaped
and or reef composed of material other than true coral-reef
nestone.---Syn: *pseudatoll*.

eudoautunite A pale-yellow to white mineral: $(H_3O)_4Ca_2(U-$
$_2)_2(PO_4)_4.5H_2O$ (?). It is not a member of the autunite
oup.

eudobed A group of nearly parallel plane surfaces that dip
current in climbing-ripple laminae and that are formed ei-
er by nondeposition or by erosion on the upcurrent sides of
grating superimposed ripple laminae: "between successive
eudobeds are sets of laminae that dip steeply in the oppo-
e direction, formed by deposition on the lee side of each
pple crest and resembling, in general, the foresets of tabular

planar cross-beds" (Edwin D. McKee in Middleton, 1965,
p.250).

pseudobedding [petrology] *pseudostratification*.

pseudobedding [sed] Bedding developed by concentration or
combining of ripple laminae representing the approach slopes
of ripple deposits (McKee, 1939, p.72); e.g. bedding due to
lee-side concentration in ripple cross-lamination. See also:
false bedding; pseudo cross-bedding.

pseudoboleite A mineral: $Pb_5Cu_4Cl_{10}(OH)_8.2H_2O$. Also
spelled: *pseudoboléite*.

pseudobomb *lava ball*.

pseudobreccia A partially and irregularly dolomitized limestone, characterized by a mottled appearance that gives the
rock a texture mimicking that of a breccia or by a weathered
surface that appears deceptively fragmental. It is produced diagenetically by selective grain growth in which localized, patchy, and irregularly shaped recrystallized masses of coarse
calcite (usually visible to the naked eye: 1-20 mm in diameter) are embedded in a lighter-colored and less-altered matrix
of calcareous mud. The boundaries between the "clasts" and
the matrix are indistinct or gradational. The term was introduced by Tiddeman in Strahan (1907, p.10-15), and used by
Dixon & Vaughan (1911, p.507) and Wallace (1913). Cf:
pseudopsephite. Syn: *recrystallization recrystallization breccia*.

pseudobrookite A brown or black orthorhombic mineral:
Fe_2TiO_5. It resembles brookite.

pseudocannel coal Cannel coal that contains much humic
matter. Syn: *humic-cannel coal*.

pseudocarina A perforate, ridge-like thickening of the peripheral part of a chamber wall of a foraminiferal test, situated
approximately in the plane of coiling.

pseudoceratite A Jurassic and Cretaceous ammonoid cephalopod explained as a reversionary or atavistic modification of a
normal ammonite and having a suture similar to that of a ceratite.

pseudoceratitic suture A type of suture in ammonoids that
approximates *ceratitic suture* in form but is not related to ceratites; specif. a suture in pseudoceratites.

pseudochamber A partly subdivided cavity of a foraminiferal
test (as in the family Tournayellidae), indicated by a slight
protuberance or an incipient septum.

pseudochitin A resistant organic substance, the exact chemical structure of which is uncertain, though it apparently consists of compounds of C-H-O-N. The behavior of pseudochitin
is similar to *chitin*, but by definition it does not yield a positive
chitin staining reaction. Various fossils, including graptolites
and chitinozoans, contain or consist mostly of this substance.

pseudochitinous Consisting of *pseudochitin*.

pseudochlorite (a) *swelling chlorite*. (b) *septechlorite*. (c) An
artificial product obtained by adsorbing magnesium salts on
montmorillonite and precipitating magnesium hydroxide between the layers of the mineral (Youell, 1960).

pseudochrysolite *moldavite* [tektite].

pseudocirque A term used by Freeman (1925) and recommended by Charlesworth (1957, p.244) for a feature that is
similar but not homologous to a glacial cirque. See also: *cirque*
[geomorph].

pseudocol A term proposed by Chamberlin (1894a) for a landform represented by a constriction of the valley of a stream
diverted by glacial ponding, formed by the cutting through of a
cover of drift and subsequent exposure of a former col; the
feature occurs in regions of reversed drainage along the border of ancient glacial formations, as along several segments
of the Ohio River valley.

pseudocolpus A colpus-like modification of the exine of pollen
grains, differing from a true *colpus* in that it is not a site of
pollen-tube emergence. Pl: *pseudocolpi*.

pseudoconcretion A subspherical, secondary sedimentary
structure resembling a true concretion but not formed by or-

derly precipitation of mineral matter in the pores of a sediment; e.g. an armored mud ball or certain algal structures.

pseudoconformity A term used by Fairbridge (1946, p.88) for a stratigraphic relationship that appears conformable but is characterized by nonaccumulation or deficiency of sediment, such as a slump gap in which an entire formation slipped away off the crest of a rising anticline or in which no trace of a hiatus is immediately apparent from the structure.

pseudoconglomerate A rock that resembles, or may easily be mistaken for, a true or normal (sedimentary) conglomerate; e.g. a crush conglomerate consisting of cemented fragments that had been rolled and rounded nearly in place by orogenic forces, or a sandstone packed with many rounded concretionary bodies, or an aggregate of rounded boulders produced in place by spheroidal weathering and surrounded by clayey material. Term introduced by Van Hise (1896, p.679). Cf: *pseudopsephite.*

pseudocotunnite A mineral: K_2PbCl_4 (?).

pseudo cross-bedding (a) An inclined bedding produced by deposition in response to ripple-mark migration and characterized by foreset beds that appear to dip into the current. See also: *pseudobedding.* (b) A structure resembling cross-bedding, caused by distortion-free slumping and sliding of a semiconsolidated mass of sediments (such as sandy shales).--Also spelled: *pseudocross-bedding.* Syn: *pseudo cross-stratification.*

pseudocruralium An excessive thickening of the secondary shell of a brachiopod, bearing dorsal adductor impressions, and elevated anteriorly above the floor of the valve.

pseudocrystal A substance that appears to be crystalline but that doesn't give a diffraction pattern that confirms it as truly crystalline.

pseudodeltidium A single and convex or flat plate affording variably complete cover of the delthyrium of a brachiopod but invariably closing the apical angle (subtended by the region of shell surface adjacent to umbo) when the pedicle foramen is supra-apical (located in ventral umbo away from the apex of the delthyrium) or absent and always dorsally enclosing the apical foramen (TIP, 1965, pt.H, p.151). Cf: *deltidium.*

pseudo-diffusion Mixing of thin superpositioned layers of slowly accumulated marine sediments by the action of water motion and/or subsurface organisms. This phenomenon can lead to serious errors in determining the rate of sedimentation if disturbed sediments are dated by carbon-14 or other radiometric methods (Bowen, 1966, p.208).

pseudofault A term coined by Palmer (1920, p.851) for a fault-like feature resulting from weathering along joint, shrinkage, or bedding planes.

pseudofibrous peat Peat that is fibrous in texture but that is plastic and incoherent. Cf: *fibrous peat; amorphous peat.*

pseudofjord A nonglaciated fjord-like valley.

pseudofossil A natural object, structure, or mineral of inorganic origin but which may resemble or be mistaken for a fossil. Cf: *problematic fossil.*

pseudogalena *sphalerite.*

pseudogeneric name A name applied as a generic name, but without being taxonomically valid, to a group of organisms that are not sufficiently known to permit valid classification.

pseudogley A densely packed, silty soil that is alternately waterlogged and rapidly dried out (Kubiéna, 1953).

pseudogradational bedding A structure in metamorphosed sedimentary rock in which the original textural graduation (coarse at the base, finer at the top) appears to be reversed, due to the formation of porphyroblasts in the finer-grained part of the rock (Schrock, 1948, p.426).

pseudogritty structure A type of *mortar structure* in which the larger relics are angular, due to fracture along cleavage planes. The term is not in common use.

pseudohexagonal Said of a crystal form, e.g. some orthorhombic forms, that simulate the hexagonal form.

pseudointerarea The somewhat flattened, posterior sector the shell of some inarticulate brachiopods "secreted by poste rior sector of mantle not fused with that of opposite valve (TIP, 1965, pt.H, p.151). Cf: *interarea.*

pseudokame *residual kame.*

pseudokarst A topography that resembles karst but that is n formed by the dissolution of limestone, usually a rough-su faced lava field in which ceilings of lava tubes have collapse Such an area is characterized by tunnels, tubes, stalactite and stalagmites of lava.

pseudokettle *pingo-remnant.*

pseudolagoon The shallow pool of water encirlced by a pse doatoll. Syn: *miniature lagoon.*

pseudolaueite A monoclinic mineral: $MnFe_2(PO_4)_2(OH)$ $78H_2O$. Cf: *laueite.*

pseudoleucite A pseudomorph after leucite, consisting of mixture of nepheline, orthoclase, and analcime, such as occ in certain syenites in Arkansas, Montana, and Brazil.

pseudomalachite A bright-green to blackish-green minera $Cu_5(PO_4)_2(OH)_4.H_2O$ (?). It resembles malachite in appea ance and it occurs in the oxidized zone of hydrothermal co per deposits. Syn: *dihydrite; phosphochalcite; phosphorocha cite; tagilite.*

pseudomicrite A genetic term applied to calcareous micri that has formed by secondary changes such as "degener tive" recrystallization (crystal diminution) of faunal and flo material (Chilingar et al, 1967, p.319). Cf: *orthomicrite.*

pseudomicroseism A microseism due to instrumental effects.

pseudomicrosparite A genetic term applied to microspari that has developed by recrystallization or by grain grow (Chilingar et al, 1967, p.228). Cf: *orthomicrosparite.*

pseudomonoclinic Said of a triclinic crystal form, e.g. that microcline, that simulates the monoclinic form.

pseudomorph A mineral whose outward crystal form is that another mineral species; it has developed by alteration, su stitution, incrustation, or paramorphism. A pseudomorph described as being "after" the mineral whose outward form has, e.g. quartz after fluorite (Dana, p.206). Se also: *pseud morphism.* Adj: *pseudomorphous.* Syn: *false form.* Partial sy *allomorph.*

pseudomorphism The process of becoming, and the conditi of being, a *pseudomorph.*

pseudomorphous Adj. of *pseudomorph.*

pseudomountain A term used by Tarr (1902) for a mounta formed by differential erosion, as contrasted with one pr duced by uplift; e.g. a *plateau mountain.*

pseudo mud crack A term used by Ksiazkiewicz (1958, pl.1 fig.2) for a sedimentary structure now known as a *parti cast.* See also: *false mud crack.*

pseudonodule A primary sedimentary structure consisting of ball-like mass of sandstone enclosed in shale or mudston characterized by a rounded base with upturned or inrolle edges, and resulting from the settling of sand into underlyir clay or mud which welled up between isolated sand masse The term was introduced by Macar (1948) who attributed th structure to horizontal displacement or vertical founderin See also: *ball-and-pillow structure; flow roll.* Syn: *sand roll.*

pseudo-oolith A spherical or roundish pellet or particle (gene ally less than 1 mm in diameter) in a sedimentary rock, exte nally resembling an oolith in size or shape but of seconda origin and amorphous or crypto- or micro-crystalline, an lacking the radial or concentric internal structure of an oolit e.g. a fecal pellet, a worn calcite grain, a shell fragment, or glauconite granule, or an oolith whose peripheral layers hav been resorbed or replaced. Cf: *oolitoid; ooid.* Also spelle *pseudoolith.* Syn: *false oolith.*

pseudo-ophitic Said of a texture of sedimentary rock gypsu that is formed by a diagenetic rather than a metamorphic pr cess, and that is characterized by large, platy selenite cry tals enclosing small, well formed euhedra and are probably ter origin than the matrix (Pettijohn, 1957).

seudo-orthorhombic Said of a monoclinic or triclinic crystal at approximates an orthorhombic crystal in lattice geometry ˌ crystal form.

seudophenocryst *porphyroblast.*

seudophite A general name for compact, massive chlorites ᵻsembling serpentines, in part clinochlore and in part pennin- ₑ (Hey, 1962, p.569).

seudoplankton Marine organisms other than plankton that ₑ usually benthonic but that are attached to floating vegeta- ɔn or otherwise drift about.

seudopluton An igneous rock mass that resembles a pluton ɹt lacks dike clusters and peripheral selvages, e.g. a *rheoig- ₗmbrite.*

seudopodium A temporary or semipermanent projection or ₊tractile process of the protoplasm of a cell (such as a uni- ₑllular organism) that serves for locomotion, attachment, ₁d food gathering and that changes in shape, character, and ɔsition with the activity of the cell. It may be lobose, filamen- us, bifurcating, or anastomosing. Examples: *axopodium; ₊ticulopodium; rhizopodium.* Pl: *pseudopodia.* Syn: *pseudo- ɔd.*

seudopore [paleont] (a) A tissue-filled lacuna in the calcifi- ₐtion of the outer zooidal wall in many bryozoans. (b) A pore the outer covering in various calcisponges, the covering ₑing formed by outgrowth from the peripheral part of the in- ₐlant canals.

seudopore [palyn] An esp. thin area in the *leptoma* of certain ɔniferous pollen (as in the families Cupressaceae and Taxa- ₑae).

seudoporphyritic [meta] With reference to metamorphic ₊ck, a syn. of *porphyroblastic* texture.

seudoporphyritic [ign] Said of the texture of an igneous rock which larger crystals have developed in a macrocrystalline ₊oundmass, but were formed, at least in part, after the rock ₀lidified (e.g. large orthoclase crystals in a granite). The ₊rm was first used by Lasaulx in 1875 (Johannsen, 1939, 230).

seudoporphyroblastic Pertaining to a structure resembling ₀rphyroblastic texture but that is due to processes other than ₊owth, e.g. to differential granulation.

seudopsephite The equivalent of *pseudobreccia* or *pseudo- ₒnglomerate* (Read, 1958).

seudopuncta A conical deflection of secondary shell of a ₊achiopod, with or without a *taleola*, pointing inwardly and ₒmmonly anteriorly so as to appear on the internal surface of ₑe valve as a tubercle. It may weather out in fossil shells, ₐving a tiny opening that may be mistaken for a *puncta* in ₊nctate shells. Pl: *pseudopunctae.* Syn: *pseudopunctum.*

seudopylome A prominent thickening of the wall at the oppo- ₊e (antapical) end of the vesicle in some acritarchs, resem- ₊ing the rim of a pylome. In some species (such as *Axis- ₊haeridium* and *Polyancistrodorus*), a central depression or ɔtch seemingly continues as a canal.

seudoraphe On the frustule of some pennate diatoms, a ₑar area on the valve between striae or costae.

seudo Rayleigh wave A *Rayleigh wave* that is a leaking ₊ode.

seudo ripple mark A term used by Kuenen (1948, p.372) for ₊bedding-plane feature resembling a ripple mark but attrib- ₊ed to lateral pressure caused by slumping (such as a mud- ₊ow structure imitating a ripple mark) or by local, small-scale ₊ctonic deformation (such as a corrugation on the cleavage ₊ce of slate). See also: *crinkle mark; creep wrinkle.*

seudorostrum The anterior part of the gnathothorax of a cu- ₊acean (malacostracan crustacean), formed by a pair of an- ₊rior and lateral parts of the cephalic shield projecting for- ₊ard and meeting medially in front of the true rostrum.

seudorutile A mineral: $Fe_2Ti_3O_9$. It is an oxidation product of ilmenite, and is common in beach sands. Cf: *arizonite [miner- al].*

pseudosaccus An ektexinous sac attached to a fossil spore (such as *Endosporites*), resembling the *vesicle* (saccus) of some pollen grains but typically not showing internal structure. The distinction between a pseudosaccus and a vesicle is rath- er slight.

pseudo section A display of resistivity and induced-polariza- tion data, obtained with the pole-dipole or dipole-dipole array, in which the observed data values are plotted in section at the intersections of lines drawn at 45 degrees from the mid-point of the current and potential electrode pairs (midpoint of po- tential pair and through the near current electrode for pole- dipole array); an artifice used to present all of the data from a sounding-profiling in one section. The vertical dimension of the pseudo section bears no simple relationship to the geolog- ic section.

pseudoseptum (a) A spine- or tooth-like skeletal projection in octocorals of the order Coenthecalia. Pseudosepta "bear no constant relationship with soft septa of polyps" (TIP, 1956, pt.F, p.174). (b) The plane of junction in a nautiloid between hyposeptal deposits of one septum and episeptal deposits on the preceding septum (TIP, 1964, pt.K, p.58).

pseudoskeleton A sponge skeleton consisting of foreign bod- ies not secreted by the sponge. Cf: *autoskeleton.*

pseudoslickenside In a tectonite, a surface resembling slick- enside, developed by rotation without any indication of the di- rection of movement.

pseudosparite A limestone consisting of relatively large, clear calcite crystals that have developed by recrystallization or by grain growth (Folk, 1959, p.33). Cf: *orthosparite.*

pseudospharolith A spherulite consisting of two minerals, one with parallel and one with inclined extinction, growing from the same center (Johannsen, 1939, p.193).

pseudospicule *spiculoid.*

pseudospondylium A cup-shaped chamber accommodating the ventral muscle field of a brachiopod and comprising an undercut callus (excessive thickening of secondary shell lo- cated on valve floor) contained between discrete dental plates. Cf: *spondylium.*

pseudostome (a) An aperture in a thecamoebian test from which pseudopodia protrude. It may be a simple opening or have definite structure. (b) An opening at the end of a chitino- zoan neck. It may be simple, bordered by a small lip, or have a tubular collar.

pseudostratification [glac geol] A concentric structure, resem- bling stratification, that occurs in till deposits overridden by ice, formed partly by plastering-on of layers of debris and partly by shearing of the till due to pressure of superincum- bent ice.

pseudostratification [ign] Apparent layering in some igneous rocks caused by structural features, esp. horizontal jointing, that have the appearance of stratification. Cf: *false stratifica- tion.* Syn: *pseudobedding.*

pseudostratification [struc geol] *sheeting structure.*

pseudosymmetry Apparent symmetry of a crystal, resembling that of another system; it is generally due to twinning.

pseudotachylite (a) A black rock that externally resembles *tachylite* and that occurs in irregularly branching veins. The material carries fragmental enclosures, and shows evidence of having been at high temperature. Miarolitic and spherulitic crystallization took place in the extremely dense base. It be- haves like an intrusive and has no structures related to local crushing (Shand, 1916, p.198). (b) A dense rock produced in the compression and shear conditions associated with intense and extensive fault movements, involving extreme mylonitiza- tion and/or partial melting. Similar rocks such as Sudbury breccias contain shock-metamorphic effects and may be in- jection breccias emplaced in fractures formed during meteor- ic impact.

pseudo telescope structure A term proposed by Blissenbach (1954, p.181) for an alluvial-fan structure resulting from slumping of unconsolidated deposits, such as that formed in a fan that has been cut by a series of small normal faults.

pseudotheca The false wall of a coral, formed by the thickening and fusion of the outer ends of septa.

pseudotill A nonglacial deposit resembling a glacial till.

pseudotillite A term proposed by Schwarzbach (1961) for a definitely nonglacial tillite-like rock, such as a *pebbly mudstone* formed on land by flow of nonglacial mud or deposited by a subaqueous turbidity flow; an indurated pseudotill. Harland et al (1966, p.233) urge the use of "pseudotillite" as an "unambiguous" and "negative" term for tillite-like rocks found to be nonglacial: "a deposit so named is ... likely to show positive characters which will lead to different nomenclature". The term is equivalent to *tilloid* as used by Pettijohn (1957, p.265).

pseudoumbilicus A deep depression, either wide or narrow, between the inner umbilical chamber walls of a trochospirally enrolled foraminiferal test where the sharply angled umbilical shoulder occurs (as in *Globorotalites*).

pseudounconformity A term used by Fairbridge (1946, p.88) for a stratigraphic relationship that appears unconformable but is characterized by superabundance or excess accumulation of sediment, such as due to submarine slumping penecontemporaneous with sedimentation off the sides of a rising anticline or dome.

pseudoviscous flow *secondary creep.*

pseudovitrinite A maceral of coal that is superficially similar to vitrinite but that is higher in reflectance from polished surfaces in oil immersion, has slitted structure, remnant cellular structures, uncommon fracture patterns, higher relief, and paucity or absence of pyrite inclusions (Benedict, 1968, p.125).

pseudovitrinoid Pseudovitrinites that occur in bituminous coals (Benedict, 1963, p.126).

pseudovolcanic Pertaining to a *pseudovolcano*; said of phenomena (such as a phreatic explosion, a bomb crater, or a Carolina bay) that give the impression of being volcanic in origin, but are actually of a different nature.

pseudovolcano A large crater or circular hollow believed not to be associated with recent volcanic activity; e.g. a crater that is possibly meteoritic in origin but is more probably the result of phreatic explosion or cauldron subsidence. Adj: *pseudovolcano.*

pseudowavellite *crandallite.*

psilate Said of the smooth walls of pollen and spores that lack sculpture. The term is usually applied to exines with pits less than one micron in diameter. Syn: *laevigate.*

psilomelane (a) A general field term for mixtures of manganese minerals or for a massive, hard, botryoidal, colloform, and heavy manganese oxide whose mineral composition is not specifically determined. Cf: *wad.* Syn: *manganomelane.* (b) A manganese-oxide mineral; specif. *romanechite.*

psilophyte An extinct *psilopsid*; a member of the Psilophytales, ranging from the Silurian to the Devonian.

psilopsid An early and primitive vascular plant corresponding to the division Psilopsida. It is generally without leaves but with spore-bearing organs at the stem tips. Such plants range from the Silurian. See also: *psilophyte.*

psittacinite *mottramite.*

Psychozoic n. A now obsolete term for the era in geologic time characterized and initiated by the appearance of man on Earth.

psychrometer A type of *hygrometer* that uses both a wet-bulb and a dry-bulb thermometer.

psychrophilic A syn. of *cryophilic.* Noun: *psychrophile.*

psychrophyte A plant adapted to arctic or alpine conditions.

psychrotolerant Said of an organism that lives at 0°C and tolerates temperatures above 20°C.

ptanite Var. of *phthanite.*

pterate chorate cyst A dinoflagellate *chorate cyst* characterized by a pronounced equatorial outgrowth in the form of so processes linked distally or in mesh-like fashion (as *Wanea*).

pteridophyte A fernlike, vascular plant that reproduces spores. Members of this division, which appeared in the D vonian, include lycopods, horsetails or scouring rushes, a ferns. Cf: *spermatophyte; cryptogam.*

Pteridophytic *Paleophytic.*

pteridosperm A gymnosperm with fernlike foliage and tr seeds borne on leaves, not in cones; a *seed fern.* It rang from the Devonian to the Mesozoic.

pterocavate Said of a dinoflagellate cyst having a pronounc equatorial pericoel (as in *Stephodinium*).

pteropod Any opisthobranch belonging to the order Pteropod which includes pelagic forms sometimes with shells. Tl shells are generally conical and composed of aragonite.

pteropod ooze An *ooze* whose skeletal remains are the tes of pteropods; it is a calcareous ooze.

pteropsid A vascular plant having an advanced leaf form, e. a fern, seed fern, gymnosperm, or angiosperm.

ptilolite *mordenite.*

ptycholophe A brachiopod lophophore in which the brachia a folded into one or more lobes in addition to median indent tion (TIP, 1965, pt.H, p.151).

ptychopariid Any trilobite belonging to the order Ptychopariid characterized generally by opistharian sutures, more th three segments in the thorax, and a simple glabella. The stratigraphic range is Lower Cambrian to Middle Permian.

ptygma Pegmatitic material within migmatite or gneiss, havi the appearance of disharmonic folds (Dietrich, 1959, p.35 The genesis of this type of folding remains controversial, wi hypotheses favoring both primary and secondary origins (D trich, 1960a, p. 140). Syn: *ptygmatic fold.*

ptygmatic fold *ptygma.*

ptygmatic injection A thin magmatic intrusion having conv lute form. According to Washburne (1943, p.495) it has be incorrectly called ptygmatic folding.

pubescent Said of a plant that is covered with soft, dow hairs.

public domain Land owned, controlled, or heretofore dispos of by the U.S. Federal government. It includes the land th was ceded to the government by the original thirteen State together with certain subsequent additions acquired by ce sion, treaty, and purchase. At its greatest extent, the pub domain occupied more than 1,820,000,000 acres. See als *public land.*

public land Land owned by a government, esp. a national go ernment; specif. the part of the U.S. *public domain* to whi title is still vested in the Federal government and that subject to appropriation, sale, or disposal under the gener laws.

public-land survey A survey of public lands; specif. the U. Public Land Survey system (USPLS) by which much of tl United States was surveyed and divided into a rectangu grid system using townships, sections, and fractions of se tions.

pucherite A reddish-brown orthorhombic mineral: $BiVO_4$.

pudding ball *armored mud ball.*

puddingstone (a) A popular name applied chiefly in Great Br ain to a *conglomerate* consisting of well-rounded pebbl whose colors are in such marked contrast with the abunda fine-grained matrix or cement that in section the ro suggests an old-fashioned pudding containing plums or ra sins. Example: the Hertfordshire Puddingstone (lower Eocen in England, composed of black or brown flint pebbles ceme ed by white silica, with or without brown iron hydroxide. Sy *plum-pudding stone.* (b) A siliceous rock cut into blocks furnace linings.---Also spelled: *pudding stone.*

puddle A small accumulation of meltwater in a depression or hollow on the surface of any form of ice, produced mainly by the melting of snow and ice, and in most cases fresh and potable.

puddle-core dam An earth dam in which the impervious core is constructed of puddled clay.

puddle wall A core wall of a dam, or an impervious cutoff in natural materials, made of puddled clay.

puerto A term used in the SW U.S. for a pass over or through an escarpment or mountain range. Etymol: Spanish.

puff n. A high spot or elevation in *gilgai* (a microrelief of heavy clay soil). Cf: *crab hole*.

puffing hole *blowhole* [coast].

pug *gouge*.

puglianite A coarse-grained igneous rock composed of euhedral augite, leucite, anorthite, sanidine, hornblende, and biotite.

pulaskite A light-colored, feldspathoid-bearing, granular or trachytoid alkali syenite composed chiefly of orthoclase, soda pyroxene, arfvedsonite, and nepheline.

pull-apart n. A precompaction sedimentary structure resembling *boudinage*, consisting of beds that have been stretched and torn apart into relatively short slabs, the intervening cracks being filled in from the top (or in some cases possibly from below) (Natland & Kuenen, 1951, p.89-90); e.g. stiff clay embedded in more mobile, water-soaked sand, or compact sandstone embedded in hydroplastic clayey rock.---adj. Said of a structure or bed characterized by pull-aparts.

pulmonate n. Any terrestrial or freshwater gastropod belonging to the subclass Pulmonata, characterized by the modification of the mantle cavity for air-breathing and by the presence of a shell, but, only rarely, of an operculum. (TIP, 1960, pt.I, p.I53).

pulp cavity An improper term for the *basal cavity* of a conodont.

pulpit rock *chimney rock* [geomorph].

pulps A term used by Allen & Day (1935, p.65) for "a fine, mealy, opaline silica", much like sand.

pulpy peat *sedimentary peat*.

pulsating spring *geyser*.

pulsation [stratig] (a) A term used by Grabau (1936a) for a long rhythm, conceived to be nearly the length of a geologic time period, representing a eustatic movement of sea level that resulted in simultaneous transgression and regression of widespread and semipermanent seas over whole continents. Cf: *oscillation*. (b) A distinct step or change in a series of rythmical or regularly recurring movements.

pulsation [tect] A minor time of deformation, or a subdivision of a more prolonged epoch of orogeny. Cf: *event; phase* [tect]; *disturbance*.

pulsation [clim] A series of rhythmic and widespread (if not worldwide) changes of climate that have occurred in the geologic past.

pulse In ecology, a sudden increase in the number of organisms or kinds or organisms, usually at regularly occurring intervals.

pulverite A sedimentary rock composed of silt- or clay-sized aggregates of constructional (nonclastic) origin, simulating in texture a lutite of clastic origin; e.g. a rock formed of diatom pustules. The term was introduced by Grabau (1911, p.1007). Syn: *pulveryte*.

pulverization *comminution*.

pulverulent Said of a mineral that may be easily powdered.

pulveryte *pulverite*.

pumice A light-colored, vesicular, glassy rock commonly having the composition of a rhyolite. It is often sufficiently buoyant to float on water and is economically useful as a lightweight aggregate and as an abrasive. The adjectival form, *pumiceous*, is usually applied to pyroclastic ejecta. Cf: *scoria*. Syn: *foam; volcanic foam; pumicite; pumice stone*.

pumice fall The descent of pumice from an eruption cloud by *air-fall deposition*. Cf: *ash fall*.

pumice flow For Japanese geologists, a syn. of *ash flow*.

pumiceous Said of the texture of a pyroclastic rock, e.g. pumice, characterized by numerous small cavities presenting a spongy, frothy appearance; finer than *scoriaceous*. Also, said of a rock exhibiting such texture.

pumilith A lithified deposit of volcanic ash.

pump A mechanical device for transferring either liquids or gases from one place to another, or for either compressing or attenuating gases.

pumpage (a) The quantity of water or other liquid pumped, as of ground water. (b) The act of pumping.

pumpellyite A greenish, epidote-like mineral: $Ca_4Al_4(Al,Fe^{+2}, Fe^{+3},Mg,Mn)_2Si_6O_{23}(OH)_3.2H_2O$. It is probably related to clinozoisite. See also: *chlorastrolite*. Syn: *zonochlorite; lotrite*.

pumpellyite-prehnite-quartz facies Rocks formed by burial metamorphism at moderate pressures (not exceeding 5000 bars) and low temperatures of 300°-400°C, i.e. in an environment of higher temperatures but essentially the same pressures as those of the laumontite-prehnite-quartz or zeolite facies (Winkler, 1967).

Pumpelly's rule The generalization that the axial surfaces of minor folds of an area are in accord with those of the major fold structures (Pumpelly, 1894, p.158).

puncta (a) One of the minute, closely spaced pores, perforations, or tubules extending perpendicularly from the inner to the outer surface of a brachiopod shell. The term is also used as a plural of *punctum*. See also: *endopuncta; exopuncta; pseudopuncta*. Syn: *punctum*. (b) Any of various thin places arranged in characteristic pattern in the frustule of pennate diatoms, being smaller and simpler than an *areola*; specif. the smallest structure on a diatom valve, such as one of the pores having diameters as small as 0.037 micron (but commonly 0.5 to 1.0 micron), occurring either scattered or in rows, and sometimes having fine porous plates at their inner extremity. (c) A hole in the external wall of a foraminiferal chamber.---Pl: *punctae*.

punctate Minutely pitted, or having minute dots, spots, or depressions, such as a "punctate leaf"; specif. said of a brachiopod or brachiopod shell possessing endopunctae. Cf: *impunctate*.

punctation The condition of being punctate.

punctum A small area marked off in any way from a surrounding surface; specif. a minute pit on the shell surface of a gastropod (not a tubule penetrating shell substance), or a *puncta* in the shell of a brachiopod. Pl: *puncta*.

punky Said of a semi-indurated rock, such as a leached limestone; esp. said of a tuff that is weakly welded.

pup A term used in Alaska for a small tributary stream.

puppet *loess doll*.

Purbeckian Stage in Great Britain: uppermost Jurassic (above Bononian, below Cretaceous). It is equivalent to upper Portlandian, and in France to Aquilonian.

pure coal An informal syn. of *vitrain*.

pure rotation *rotational strain*.

pure shear A particular example of irrotational strain or flattening in which the body is elongated in one direction and shortened at right angles to this as a consequence of differential displacements on two sets of intersecting planes. Cf: *simple shear*.

purga A violent arctic snowstorm. Etymol: Karelian, *purgu*, "snowstorm".

purgatory (a) A term used in New England for a long, deep, narrow, steep-sided cleft or ravine along a rugged coast, into which waves rush during a storm with great noise and violence; a rock chasm without a stream, often covered at the bottom with large, angular rocks, and difficult to traverse. (b) A swamp that is dangerous or difficult to traverse.

purl A swirling or eddying stream or rill, moving swiftly around obstructions; a stream making a soft, murmuring sound.

purple blende *kermesite.*

purple copper ore *bornite.*

purpurite A deep-red or reddish-purple mineral: $(Mn^{+3}, Fe^{+3})PO_4$. It is isomorphous with heterosite.

push [volc] *squeeze-up.*

push [hydraul] A force exerted directly by the wind upon the exposed sides of wave crests (Strahler, 1963, p.308). Cf: *drag.*

push moraine A broad, smooth, arc-shaped morainal ridge consisting of material mechanically pushed or shoved along by an advancing glacier. Examples are common in Netherlands and NW Germany. See also: *ice-pushed ridge.* Syn: *shoved moraine; push-ridge moraine; upsetted moraine; thrust moraine.*

push-pull wave *P wave.*

push wave *P wave.*

pustule A minute boss on an asterozoan ossicle, having a central depression in which a spine articulates.

pusule apparatus The sack-like vacuole in a dinoflagellate connected with the exterior by a slender canal opening into a flagellar pore.

Putnam anomaly *average level anomaly.*

putrefaction The decomposition of organic matter by slow distillation, in the presence of water, without air. Methane and other gaseous products (H_2, NH_3, H_2S) are formed. Cf: *disintegration* [coal]; *moldering; peat formation.*

puy A small, remnant volcanic cone; it is the French word for such structures in the Auvergne district of central France.

puzzolan Var of *pozzolan.*

p-veatchite A mineral: $Sr_2B_{11}O_{16}(OH)_5 \cdot H_2O$. It is dimorphous with veatchite, and has a space group $P2_1/m$.

P wave That type of seismic *body wave* which is propagated by alternating compression and expansion of material in the direction of propagation. It is the fastest of the seismic waves (traveling 6.0-6.7 km/sec in the crust and 8.0-8.5 km/sec in the upper mantle), and it is the type which carries sound. The P stands for primary; it is so named because it arrives before the S *wave* (secondary body wave). Syn: *longitudinal wave; irrotational wave; push wave; pressure wave; dilatational wave; primary wave; compressional wave; push-pull wave.*

pycnite A variety of topaz occurring in massive columnar aggregations.

pycnocline (a) A density gradient; esp. a vertical gradient marking a sharp change. Cf: *thermocline.* (b) A layer of water in the ocean, characterized by a rapid change of density with depth.

pycnogonid Any marine arthropod belonging to the subphylum Pycnogonida, resembling the *chelicerates* in the presence of one pair of chelae but lacking the well-developed abdomen. Their stratigraphic range is Lower Devonian to present.

pycnostromid An *algal biscuit* produced by *Pycnostroma.*

pycnotheca The dense, nonalveolate inner layer of the test wall of schwagerinid fusulinids, penetrated by septal pores, and wedged between tectum and keriotheca of antetheca.

pygidium A caudal structure or terminal body region of various invertebrates; esp. the posterior part or tail piece of an exoskeleton of a trilobite, consisting of several fused segments. Pl: *pygidia.* Adj: *pygidial.* Cf: *abdomen.*

pylome (a) A more or less circular opening in an acritarch, commonly closed by an operculum, and probably serving for emergence from a cyst. (b) A large opening in spumellarian radiolarians, commonly only in the outermost of concentric shells.

pyrabol *pyribole.*

pyrabole *pyribole.*

pyralmandite A garnet intermediate in chemical composition between pyrope and almandine.

pyralspite A group of garnets of formula: $M_3Al_2(SiO_4)_3$, where $M = Mg$, Fe^{+2}, or Mn^{+2}. It includes pyrope, almandine, and spessartine, and their intermediate forms.

pyramid [cryst] An open crystal form consisting of three, fo six, eight, or twelve nonparallel faces that meet at a point. (*dipyramid.* Adj: *pyramidal.*

pyramid [paleont] A large beak- or wing-like element of Ar totle's lantern in interambulacral position of an echinoid. S also: *demipyramid.*

pyramidal Having the symmetry of a *pyramid.*

pyramidal cleavage A type of crystal cleavage that occu parallel to the faces of a pyramid, e.g. the [011] cleavage scheelite.

pyramidal dune *star dune.*

pyramidal peak *horn.*

pyramidal system *tetragonal system.*

pyramid pebble *dreikanter.*

pyranometer An *actinometer* which measures the combine intensity of incoming direct solar radiation and diffuse sky r diation. It consists of a recorder and a radiation-sensing e ment which is mounted so that it views the entire sky (Mark 1969, p.364). Syn: *solarimeter.*

pyrargyrite A dark-red, gray, or black rhombohedral miner Ag_3SbS_3. It is isomorphous with proustite and polymorpho with pyrostilpnite, and is an important ore of silver. Syn: *da ruby silver; dark red silver ore.*

pyrene (a) A nucule or nutlet. (b) The small stone of a dru or similar fruit.

Pyrenean orogeny One of the 30 or more short-lived orogeni during Phanerozoic time identified by Stille, in this case duri the late Eocene, between the Bartonian and Ludian stages.

Pyrenean-type facies series A type of dynamothermal region metamorphism characteristic of the Pyrenees in which t pressure range is 3500-5000 bars. The mineral sequence, order of rising temperature, is staurolite - andalusite - sil manite - cordierite (Hietanen, 1967, p.193). Cf: *Idahoan-ty facies series.*

pyreneite *melanite.*

pyrgeometer An *actinometer* which measures the effecti terrestrial radiation. See also: *Ångström pyrgeometer.*

pyrheliometer An *actinometer* which measures the intensity direct solar radiation, consisting of a radiation sensing el ment enclosed in a casing which is closed except for a sm aperture, through which the direct solar rays enter, and a re corder unit (Marks, 1969, p.364). See also: *Ångström cor pensation pyrheliometer.*

pyribole A mnemonic term coined by Johannsen in 1911 in h classification of igneous rocks to indicate the presence of e ther or both a pyroxene and/or an amphibole. Var: *pyrabol pyrabol; pyrobol.* Etymol: *pyroxene + amphibole.*

pyric pond A pool of water that collects in a shallow hole sink formed as a result of fires and subsequent subsidence peat deposits, lignite, and coal beds.

pyrite A common, pale-bronze or brass-yellow, isometric mi eral: FeS_2. It is dimorphous with marcasite, and often co tains small amounts of other metals. Pyrite has a brilliant me tallic luster and an absence of cleavage, and has been mi taken for gold (which is softer and heavier). It common crystallizes in cubes (whose faces are usually striated), octa hedrons, or pyritohedrons, and it also occurs in shapeles grains and masses. Pyrite is the most widespread and abun ant of the sulfide minerals and occurs in all kinds of rock such as in nodules in sedimentary rocks and coal seams or a a common vein material associated with many different mi erals. Pyrite is an important ore of sulfur, less so of iron, an is burned in making sulfur dioxide and sulfuric acid; it sometimes mined for the associated gold and copper. Cf: p rites. Syn: *iron pyrites; fool's gold; mundic; common pyrites.*

pyrites (a) Any of various metallic-looking sulfides of whic pyrite ("iron pyrites") is the commonest. The term is use with a qualifying term that indicates the component meta e.g. "copper pyrites" (chalcopyrite), "tin pyrites" (stannite "white iron pyrites" (marcasite), "arsenical pyrites" (arseno-pyrite), "cobalt pyrites" (linnaeite), and "nickel pyrites" (millerite). When used popularly and without qualification, the term usually signifies *pyrite*. (b) An obsolete term for a stone that may be used for striking fire.

pyritization Introduction of, or replacement by pyrite; e.g. the replacement of the original material of the hard parts of certain fossil animals and plants by pyrite. Pyritization is a common process of hydrothermal alteration and often involves the introduction of fine-grained pyrite disseminated as specks in rock adjacent to veins.

pyritohedral Adj. of *pyritohedron*.

pyritohedron A crystal form that is a *dodecahedron* consisting of 12 pentagonal faces that are not regular. Its symmetry is $2/m\overline{3}$ and its indices are $\{210\}$. It is named after pyrite, which characteristically has this crystal form. Adj: *pyritohedral*. Cf: *rhombic dodecahedron*. Syn: *pentagonal dodecahedron; regular dodecahedron; pyritoid*.

pyritoid As a noun, a syn. of *pyritohedron*.

pyroaurite A gold-like or brownish rhombohedral mineral: $Mg_6Fe_2(CO_3)(OH)_{16}.4H_2O$. It is dimorphous with sjögrenite, and may contain up to 5% MnO.

pyrobelonite A fire-red to deep brilliant-red orthorhombic mineral: $PbMn(VO_4)(OH)$.

pyrobiolite An organic rock containing organic remains that have been altered by volcanic action.

pyrobitumen Any of the dark-colored, fairly hard, nonvolatile native substances composed of hydrocarbon complexes which may or may not contain oxygenated substances and are often associated with mineral matter. The nonmineral constituents are infusible, insoluble in water, and relatively insoluble in carbon disulfide. Pyrobitumens, upon heating, generally yield bitumens, i.e. decompose, rather than melt.

pyrobituminous Pertaining to substances which yield bitumens upon heating.

pyrobole *pyribole*.

pyrochlore (a) A pale-yellow, reddish, brown, or black isometric mineral: $(Na,Ca)_2(Nb,Ta)_2O_6(OH,F)$. It is isomorphous with microlite, with Nb greater than Ta, and it usually contains cerium and titanium. Pyrochlore occurs in pegmatites derived from alkalic igneous rocks and constitutes an ore of niobium. Syn: *pyrrhite*. (b) A group of minerals of the general formula: $A_2B_2O_6(O,OH,F)$, where A = Na, Ca, K, Fe^{+2}, U^{+4}, Sb^{+3}, Pb, Th, Ce, or Y, and B = Nb, Ta, Ti, Sn, Fe^{+3}, or W. It includes minerals such as pyrochlore, microlite, betafite, obruchevite, and pandaite.

pyrochroite A hexagonal mineral: $Mn(OH)_2$. It is white when fresh, but darkens upon exposure; it is very similar to brucite in appearance.

pyroclast An individual pyroclastic fragment or clast. It is usually classified according to size. Pl: *pyroclasts*. Cf: *pyroclastics*.

pyroclastic Pertaining to clastic rock material formed by volcanic explosion or aerial expulsion from a volcanic vent; also, pertaining to rock texture of explosive origin. It is not synonymous with the adjective "volcanic".

pyroclastic breccia *explosion breccia*.

pyroclastic flow A syn. of *ash flow*, but often used without the restriction of high temperature, in a more general, genetic sense.

pyroclastic rock A rock that is composed of materials fragmented by volcanic explosion. It is characterized by a lack of sorting; the size of the individual pyroclasts varies.

pyroclastics A general term for a deposit of *pyroclasts*. Syn: *tephra*.

pyroelectricity The simultaneous development, in any crystal lacking a center of symmetry, of opposite electric charges at opposite ends of a crystal axis, due to certain changes in temperature.

pyrogenesis A broad term encompassing the intrusion and ex-trusion of magma and its derivatives. Adj: *pyrogenic*.

pyrogenetic mineral An anhydrous mineral of an igneous rock, usually crystallized at high temperature in a magma containing relatively few volatile components.

pyrogenic Said of a process or of a deposit involving the intrusion and/or extrusion of magma. See also: *pyrogenesis*. Syn: *pyrogenetic; pyrogenous*.

pyrogenic rock An igneous rock.

pyrogenous A syn. of *pyrogenic*, originally used as a syn. of *igneous*.

pyrogeology A synonym of *volcanology* that was proposed by Grabau (1924).

pyrognomic Said of a metamict mineral which easily becomes incandescent when heated. The term is little used.

pyrolite A model proposed by Ringwood (Green and Ringwood, 1963) for the material of the upper mantle, composed of one part basalt to three parts dunite and consisting mainly of olivine and pyroxenes.

pyrolith A term proposed by Grabau (1904) for igneous rock.

pyrolusite A soft, iron-black or dark steel-gray, tetragonal mineral: MnO_2. It is the most important ore of manganese and is dimorphous with ramsdellite. Pyrolusite is generally massive or reniform, sometimes with a fibrous or radiate structure. Syn: *polianite; gray manganese ore*.

pyromagma A highly mobile lava, oversaturated with gases, that exists at shallower depths than *hypomagma*. Cf: *epimagma*.

pyromelane *brookite*.

pyromeride A devitrified rhyolite characterized by spherulitic texture; a nodular rhyolite.

pyrometamorphism Metamorphic changes taking place without the action of pressure or water vapor, at temperatures near the melting points of the component minerals; it is a local, intense type of *thermal metamorphism* resulting from the unusually high temperatures at the contact of a rock with magma, e.g. in xenoliths (Turner, 1948). Cf: *igneous metamorphism*.

pyrometasomatism The formation of contact-metamorphic mineral deposits at high temperatures by emanations issuing from the intrusive and involving replacement of enclosing rock with addition of materials (Lindgren, 1933).

pyrometer An instrument that measures high temperature, e.g. of molten lavas, by electrical or optical means. See also: *optical pyrometer; pyrometry*.

pyrometric cone *Seger cone*.

pyrometry The measurement of high temperatures by electrical or optical means, using a *pyrometer*. Its geologic application is to incandescent lavas.

pyromorphism An obsolete syn. of *thermal metamorphism*.

pyromorphite A green, yellow, brown, gray, or white mineral of the apatite group: $Pb_5(PO_4)_3Cl$. It is isomorphous with mimetite and vanadinite, and may contain arsenic or calcium. Pyromorphite is found in the oxidized zone of lead deposits, and is a minor ore of lead. Syn: *green lead ore*.

pyrope (a) The magnesium-aluminum end member of the garnet group, characterized by a deep fiery-red color: $(Mg,Fe)_3Al_2(SiO_4)_3$. It is rarely crystallized, and occurs in detrital deposits as rounded and angular fragments, or associated with olivine and serpentine in basic igneous rocks such as kimberlites. See also: *Cape ruby; Bohemian garnet*. Syn: *rock ruby*. (b) An obsolete name for a bright red gem, such as a ruby.

pyrophane (a) *fire opal*. (b) An opal (such as hydrophane) artificially impregnated with melted wax.

pyrophanite A blood-red mineral: $MnTiO_3$. It is isomorphous with ilmenite.

pyrophyllite A white, greenish, gray, or brown mineral: $AlSi_2O_5(OH)$. It resembles talc and occurs in a foliated form or in compact masses in quartz veins, granites, and esp. metamorphic rocks. Syn: *pencil stone*.

pyropissite An earthy, nonasphaltic pyrobitumen made up primarily of water, humic acid, wax (it is a source of montan wax), and silica. It is frequently found associated with brown coal which is then called pyropissitic brown coal. Syn: *wax coal.*

pyroretinite A type of retinite found in the brown coals of Aussig (Usti nad Labem), Bohemian Czechoslovakia.

pyroschist A schist or shale that has a sufficiently high carbon content to burn with a bright flame, or to yield volatile hydrocarbons, when heated.

pyrosmalite A colorless, pale-brown, gray, or grayish-green mineral: $(Mn,Fe)_4Si_3O_7(OH,Cl)_6$. Cf: *manganpyrosmalite.*

pyrosphere The zone of the Earth below the lithosphere, consisting of magma; it is equivalent to the *barysphere.* Syn: *magmosphere.*

pyrostibite *kermesite.*

pyrostilpnite A hyacinth-red monoclinic mineral: Ag_3SbS_3. It is polymorphous with pyrargyrite. Syn: *fireblende.*

pyroxene (a) A group of dark, rock-forming silicate minerals closely related in crystal form and composition and having the general formula: $ABSi_2O_6$, where A = Ca, Na, Mg, or Fe^{+2}, and B = Mg, Fe^{+3}, or Al, with silicon sometimes replaced in part by aluminum. It is characterized by a single chain of tetrahedra with a silicon:oxygen ratio of 1:3, by short, stout, stumpy prismatic crystals, and by good prismatic cleavage in two directions parallel to the crystal faces and intersecting at angles of about 87° and 93°; colors vary from white to dark green or black. Pyroxenes may crystallize in the orthorhombic, monoclinic, or triclinic systems; they constitute a common constituent of igneous rocks, and are analogous in chemical composition to the *amphiboles* (except that pyroxenes lack hydroxyls). (b) A mineral of the pyroxene group, such as enstatite, hypersthene, diopside, hedenbergite, acmite, jadeite, pigeonite, and esp. augite.---Etymol: Greek *pyros,* "fire", + *xenos,* "stranger", apparently so named from the mistaken belief that the pyroxenes "were only accidentally caught up in the lavas that contain them" (Challinor, 1967, p.205). Pron: *pie*-rok-seen or *peer*-ik-seen.

pyroxene alkali syenite In Tobi's (1971, p.202) classification of the charnockite suite, a quartz-poor (less than 20%) member characterized by the presence of microperthite.

pyroxene-hornfels facies Rocks formed in the high grades of thermal (contact) metamorphism at temperatures exceeding

550°C (to as high as 800°C) and at low pressures not exceeding about 3000 bars (Turner and Verhoogen, 1960 p.509). It is part of the *hornfels facies.* Cf: *hornblende-hornfels facies; albite-epidote-hornfels facies.* Syn: *K-feldspar-cordierite- hornfels facies.*

pyroxene monzonite In Tobi's (1971, p.202) classification of the charnockite suite, a quartz-poor member containing approximately equal amounts of microperthite and plagioclase; *mangerite.*

pyroxene-perthite Lamellar intergrowths of any of several pyroxenes, as with the feldspars.

pyroxene syenite In Tobi's (1971, p.202) classification of the charnockite suite, a quartz-poor member having more microperthite than plagioclase; a mangerite-syenite.

pyroxenide An informal term, used in the field, for any holocrystalline, medium- to coarse-grained igneous rock composed chiefly of pyroxene; e.g. a pyroxenite.

pyroxenite An ultramafic plutonic rock chiefly composed of pyroxene with accessory hornblende, biotite, or olivine. Syn: *pyroxenolite.*

pyroxenoid Any mineral chemically analogous to pyroxene but with the SiO_4-tetrahedra connected in rings of three rather than in long chains as in the pyroxenes; e.g. wollastonite and rhodonite.

pyroxenolite *pyroxenite.*

pyroxferroite A yellow mineral found in Apollo 11 lunar samples: $(Fe,Mn,Ca)SiO_3$. It is the iron analogue of pyroxmangite.

pyroxmangite A red or brown triclinic mineral: $(Mn,Fe,Ca,Mg)SiO_3$. It is a variety of rhodonite containing appreciable iron.

pyrrhite *pyrochlore.*

pyrrhotine *pyrrhotite.*

pyrrhotite A common reddish-brown to brownish-bronze hexagonal mineral: $Fe_{1-x}S$. It has a defect lattice in which some of the ferrous ions are lacking. Pyrrhotite is attracted by the magnet (but with varying intensity) and is darker and softer than pyrite; it is usually found massive and commonly associated with pentlandite, often containing as much as 5% nickel, in which case it is mined as a valuable ore of nickel. Syn: *pyrrhotine; magnetic pyrites; dipyrite.*

pythmic Pertaining to the bottom of a lake (Klugh, 1923, p.372).

Q

qanat A term used in Iran for an ancient, gently inclined, underground channel or conduit dug so as to conduct ground water by gravity from alluvial gravels and the foot of hills to an arid lowland; a horizontal well. Etymol: Arabic. Cf: *foggara; karez*. Syn: *kanat*.

Q-joint A partial syn. of *cross joint*, used for a cross joint that s perpendicular to flow structure.

quad Shortened form of *quadrangle*.

quadrangle (a) A four-sided tract of country bounded by parallels of latitude and meridians of longitude, used as an area unit in systematic mapping. The dimensions of a quadrangle are not necessarily the same in both directions, and its size and the scale at which it is mapped are determined by the prime purpose of the map. (b) A sheet representing a quadrangle.---Syn: *quad*.

quadrangle map A map of a quadrangle, the size being given in minutes or degrees; e.g. a 7.5-minute quadrangle map (scale: 1/24,000) having dimensions of 7.5 minutes in both latitude and longitude, the bounding parallels of latitude and meridians of longitude being integral multiples of 7.5 minutes. The U.S Geological Survey has also published 15-minute (scale:1/62,500) and 30-minute (scale: 1/125,000) quadrangle maps.

quadrant [paleont] The space in the interior of a rugose corallite, bounded by the cardinal septum and an alar septum or by the counter septum and an alar septum.

quadrant [surv] (a) An instrument formerly used in surveying and astronomy for measuring angles and altitudes, consisting of a graduated arc of 90 degrees (180 degrees in range) equipped with a sighting device and a movable index or vernier and usually a plumb line or spirit level for fixing the vertical or horizontal direction. It is now largely superseded by the sextant. (b) A quarter of a circle, an arc of 90 degrees, or an arc subtending a right angle at the center. Also, the area bounded by a quadrant and two radii.

quadratic An old syn. of *tetragonal*, as in "quadratic system".

quadrature [geophys] A component of an electrical or magnetic quantity that has a phase difference of one-quarter cycle as compared to the primary quantity.

quadrature [astron] Either of two points in the Moon's orbit about the Earth when the Moon is in its first or third quarters, or when a line from Earth to Moon makes a right angle with respect to the line from Earth to Sun. Cf: *syzygy*.

quagmire (a) A soft, wet, miry marsh or bog that gives under pressure. (b) *quaking bog*.

quake n. A syn. of *earthquake*; also, a *seismic event* on another planetary body.

quake sheet A well-defined bed resembling a slump sheet but produced by an earthquake and resulting in load casting without horizontal slip (Kuenen, 1958, p.20).

quaking bog A peat bog that is either actually floating or growing over water-saturated land such that it shakes or trembles when walked on. *Quagmire* is sometimes used as a synonym.

quality of snow An obsolescent term for the amount of ice in a snow sample, expressed as a percent of the weight of the sample, equal to one minus the free water content.

quantitative geomorphology The assignment of dimensions of mass, length, and time to all descriptive parameters of landform geometry and geomorphic processes, followed by the derivation of empirical mathematical relationships and formulation of rational mathematical models relating those parameters.

quantitative system *C.I.P.W. classification*.

quantum detector A semiconductor used in radiometers and line-scan systems to count the quantity of photons striking a sensitive element. A photoconductive, photovoltaic, or photoelectromagnetic method may be used. The photons striking the quantum detector interact with the crystal lattice, freeing electrons or holes.

quantum evolution Relatively rapid transition from one established type of biologic adaptation to another completely different type under the influence of some strong selection pressure.

quaquaversal adj. Said of strata and structures that dip outward in all directions away from a central point. It is an old term. Ant: *centroclinal*. The term has also been used as a syn. of *periclinal*.----n. A geologic structure, such as a dome or ridge, having a quaquaversal dip. Cf: *pericline*. Ant: *centrocline*.

quaquaversal fold *dome* [struc geol].

quar A Welsh term for sandstone.

quarfeloid A term coined by Johannsen in 1911 in his classification of igneous rocks to indicate the mineral combinations of quartz and feldspars and feldspars and feldspathoids (Johannsen, 1939, p.194).

quar ice A term used in Labrador for ice formed during the spring by meltwater running off the land onto an icefoot or fast ice, where it refreezes (ADTIC, 1955, p.64).

quarry A surficial mine; open workings; usually for building stone or gravel and sand.

quarrying [geomorph] (a) *plucking* [glac geol]. (b) *hydraulic plucking*.

quarrying [mining] The extraction of building stone or other valuable constituent, usually nonmetallic, from a surficial mine, or quarry. See also: *opencut mining*.

quarry sap *quarry water*.

quarry water Subsurface water retained in freshly quarried rock, called *greenstone*. Syn: *quarry sap*.

quarter post A post marking a corner at an extremity of a boundary of a quarter section of the U.S. Public Land Survey system. It is located midway between the controlling section corners, or 40 chains (0.5 mi) from the controlling section corner depending on location within the township.

quarter-quarter section A sixteenth of a normal *section* of the U.S. Public Land Survey system, representing a piece of land normally a quarter mile square and containing 40 acres as nearly as possible; a quarter section divided into four parts. It is usually identified as the northeast quarter, northwest quarter, southeast quarter, or southwest quarterarter of a particular quarter section and section.

quarter section A fourth of a normal *section* of the U.S. Public Land Survey system, representing a piece of land normally a half mile square and containing 160 acres as nearly as possible. It is usually identified as the northeast quarter, northwest quarter, southeast quarter, or southwest quarter of a particular section.

quarter-wave plate *mica plate*.

quartile Any one of three particle-size values (diameters) dividing a frequency distribution into four classes, obtained graphically from a cumulative curve by following the 25, 50, or 75 percent line to its intersection with the curve and reading the value on the diameter scale directly below the intersection; e.g. the first quartile (the 25 percentile) is the size such that 25% of the particles are larger than itself and 75% smaller, this size being larger than the third quartile (the 75 percentile) which is the size such that 75% of the particles are larger than itself and 25% smaller. Abbrev: Q.

quartz [ore dep] Any hard gold ore, or sometimes silver ore, that is either broken or in place, as distinguished from gravel or earth containing gold or silver.

quartz [mineral] (a) Crystalline silica, an important rock-forming mineral: SiO_2. It is, next to feldspar, the commonest mineral, occurring either in colorless and transparent hexagonal crystals (sometimes colored yellow, brown, purple, red, green, blue, or black by impurities) or in crystalline or cryptocrystalline masses. Quartz is the commonest gangue mineral of ore deposits, forms the major proportion of most sands, and has a widespread distribution in igneous (esp. granitic), metamorphic, and sedimentary rocks. It has a vitreous to greasy luster, a conchoidal fracture, an absence of cleavage, and a hardness of 7 on Mohs' scale (scratches glass easily, but cannot be scratched by a knife); it is composed exclusively of silicon-oxygen tetrahedra with all oxygens joined together in a three-dimensional network. Symbol: Q. Abbrev: qtz; qz. Etymol: German provincial *Quarz*. Cf: *tridymite; cristobalite; coesite; stishovite*. (b) A general term for a variety of noncrystalline or cryptocrystalline minerals having the same chemical composition as that of quartz, such as chalcedony, agate, and opal.

quartz andesite *dacite.*

quartzarenite A sandstone that is composed primarily of quartz; specif. a sandstone containing more than 95% quartz framework grains (excluding detrital chert grains) and having any amount of clay matrix and any sorting, rounding, texture, or hardness (Folk, 1968). McBride (1963, p.667), who included chert and quartzite in the 95% quartz content, coined the term as a contracted form of "quartz arenite", a term used by Gilbert (1954, p.294 & 316) for a mature sandstone containing more than 80% quartz, chert, and quartzite and less than 10% each of argillaceous matrix, feldspars, and unstable fine-grained rock fragments. The term is essentially equivalent to orthoquartzite.

quartz-banded ore A term used in Scandinavia for a metamorphosed *iron formation.*

quartz-bearing diorite A syn. of *quartz diorite*, although Streckeisen (1967, p.157) suggests that the term be restricted to diorite in which quartz constitutes 5 to 20 percent of the light-colored components.

quartz-bearing monzonite As recommended by Streckeisen (1967, p.157), any monzonite in which quartz constitutes from 5 to 20 percent of the light-colored components. In the most recent Soviet classification, it is a syn. of *quartz monzonite.*

quartz crystal *rock crystal.*

quartz diorite A group of plutonic rocks having the composition of diorite but with an appreciable amount of quartz, i.e. more than 20 percent of the light-colored constituents, according to Streckeisen (1967, p.157); also, any rock in that group; the approximate intrusive equivalent of *dacite*. Quartz diorite grades into *granodiorite* as the alkali feldspar content increases. Syn: *tonalite; quartz-bearing diorite.*

quartzfels *silexite.*

quartz felsite *quartz porphyry.*

quartz-flooded limestone A limestone characterized by an abundance of quartz particles that had been imported suddenly from a nearby source by wind or water currents, but that gradually die out upward and completely disappear within a few centimeters (Shrock, 1948, p.87).

quartz-free wacke A wacke with more than 90% unstable materials (Fischer, 1934).

quartz graywacke A term used by Gilbert (1954, p.294) for a graywacke containing abundant grains of quartz and chert and less than 10% each of feldspars and rock fragments. See also: *quartzose graywacke.*

quartzic *quartziferous.*

quartziferous Quartz-bearing. The term is applied to a rock (such as a limestone or syenite) that contains a minor proportion of quartz, to distinguish it from a variety (usually commoner) of the same rock that contains no quartz. Cf: *quartzose*. Syn: *quartzic.*

quartz index [petrology] A derived quantity (qz) in the Niggli system of rock classification, which may be either positive or negative, and is a valuable indicator of the minerals to be expected.

quartz index [sed] A term used by Dapples et al (1953, p.294 & 304) to indicate the mineralogic maturity of a sandstone by measuring the percentage of detrital quartz. It is expressed as the ratio of quartz and chert to the combined percentage of sodic and potassic feldspar, rock fragments, and clay matrix. The index is used as a basis for evaluating the degree of weathering of the source rock and the degree to which the sediment has been transported. Values for sandstones range between 3 and 19.

quartzine Fibrous chalcedony characterized by fibers having a positive crystallographic elongation (fibers parallel to the *c* axis). Syn: *quartzin.*

quartzite [met] A granoblastic metamorphic rock consisting mainly of quartz and formed by recrystallization of sandstone or chert by either regional or thermal metamorphism; *meta-quartzite.*

quartzite [sed] A very hard but unmetamorphosed sandstone consisting chiefly of quartz grains that have been so completely and solidly cemented (diagenetically) with secondary silica that the rock breaks across or through the individual grains rather than around them; an *orthoquartzite*. The cement grows in optical and crystallographic continuity around each quartz grain, thereby tightly interlocking the grains as the original pore spaces are completely filled with secondary enlargements developed on the grains. Skolnick (1965) believes that most sedimentary quartzites are compacted sandstones developed by pressure solution of quartz grains.

quartzitic sandstone A term used by Krynine (1940, p.51) for a sandstone that contains 100% quartz grains cemented with silica. The term is essentially equivalent to *orthoquartzite*. Cf: *quartzose sandstone.*

quartz latite *rhyodacite.*

quartz mengwacke A wacke with 10-33% unstable materials (Fischer, 1934).

quartz mine A mine in which the valuable constituent, e.g. gold, is found in veins rather than in placers. It is so named because quartz is the chief accessory mineral in such deposits.

quartz monzonite In U.S. usage, granitic rock in which quartz comprises 10-50% of the felsic constituents, and in which the alkali feldspar/total feldspar ratio is between 35% and 65%; the approximate intrusive equivalent of *rhyodacite*. With an increase in plagioclase and femic minerals, it grades into *granodiorite*, and with more alkali feldspar, into a granite. As introduced by Brögger in 1895, the term designated monzonite containing only small amounts of quartz, and it is still used in this sense by Soviet geologists. According to Tröger (1935, p.47), Lindgren, in 1900, changed the definition to apply to andesine-bearing granites. Now the term is applied by most British petrologists to granites with quartz constituting 20-60% of the light-colored components and with a plagioclase/total feldspar ratio of 35/65. Streckeisen (1967, p.167) recommends replacing the term with *quartz-bearing monzonite*. Syn: *adamellite.*

quartz norite In Tobi's (1971, p.202) classification of the charnockite suite, a member which contains plagioclase but no potassium feldspar.

quartzose Containing quartz as a principal constituent; esp. applied to sediments and sedimentary rocks (such as sands and sandstones) consisting chiefly of quartz. Cf: *quartziferous*. Syn: *quartzous; quartzy.*

quartzose arkose A term used by Hubert (1960, p.176-177) for a sandstone containing 50-85% quartz, chert, and metamorphic quartzite, 15-25% feldspars and feldspathic crystalline rock fragments, and 0-25% micas and micaceous metamorphic rock fragments. Cf: *quartzose graywacke.*

quartzose chert A vitreous, sparkly, shiny chert, which under high magnification shows a heterogeneous mixture of pyramids, prisms, and faces of quartz (Grohskopf & McCracken, 1949, pl.3), but also including chert in which the secondary quartz is largely anhedral. Also known as "drusy chert".

quartzose graywacke (a) A term used by Hubert (1960,

).176-177) for a sandstone containing 50-85% quartz, chert, and metamorphic quartzite, 15-25% micas and micaceous metamorphic rock fragments, and 0-25% feldspars and feldspathic crystalline rock fragments. Cf: *quartzose arkose*. (b) A term used by Krynine (1951) for a graywacke that has lost its micaceous constituents through abrasion and thus tends to approach an orthoquartzite. It is equivalent to *subgraywacke* of Folk (1954). See also: *quartz graywacke*.

quartzose sandstone (a) A well-sorted sandstone that contains (if pure) more than 95% clear quartz grains and a minor amount of matrix (less than 5%, or commonly lacking) and cement (Krumbein & Sloss, 1963, p.169-170). (b) A sandstone that contains at least 95% quartz, but is not cemented with silica (Krynine, 1940, p.51). Cf: *quartzitic sandstone*. (c) A sandstone that contains 99% quartz and quartz cement (Shrock, 1948a). (d) A sandstone that contains 90% quartz grains (Dunbar & Rodgers, 1957).---See also: *orthoquartzite*. Syn: *quartz sandstone*.

quartzose shale A green or gray shale composed dominantly of rounded quartz grains of silt size, commonly associated with highly mature sandstones (orthoquartzites), and representing the reworking of residual clays as transgressive seas encroached on old land areas (marked by relatively stable conditions with gentle rates of subsidence).

quartzose subgraywacke *protoquartzite*.

quartz-pebble conglomerate A term used by Krumbein & Sloss (1963, p.163) for *orthoquartzitic conglomerate*.

quartz porphyry A porphyritic extrusive or hypabyssal rock containing phenocrysts of quartz and alkali feldspar (usually orthoclase) in a microcrystalline or cryptocrystalline groundmass; a rhyolite. European petrologists called pre-Tertiary and Tertiary extrusive equivalents of granite "quartz porphyry" and post-Tertiary equivalents, "rhyolite" (Streckeisen, 1967, p.189). Syn: *quartz felsite*.

quartz sandstone *quartzose sandstone*.

quartz schist A schist whose foliation is due mainly to streaks and lenticles of nongranular quartz. Mica is present but in lesser quantities than in mica schist (Holmes, 1928, p.195).

quartz syenite A group of plutonic rocks having the characteristics of syenite but with a greater amount of quartz, i.e. 5 to 20 percent of the light-colored constituents, according to Streckeisen (1967, p.157); also, any rock in that group. Streckeisen also suggests that rocks of this group be referred to as "quartz-bearing syenites".

quartz topaz A frequently used but incorrect syn. of *citrine*. Cf: *topaz quartz*.

quartz trachyte *rhyolite*.

quartz wacke A gray to buff, moderately well-sorted, commonly fine-grained sandstone containing up to 90% quartz and chert, and with more than 10% argillaceous matrix (largely sericite and chlorite), less than 10% feldspar, and less than 10% rock fragments (bits of coal, shale, etc.) (Gilbert, 1954, p.292-293). Krumbein & Sloss (1963, p.172-173) give a lower limit of 15-20% matrix, and regard the rock as a slightly "cleaner" or "washed" graywacke such as one occurring under conditions of moderate subsidence in an unstable depositional area. The rock is equivalent to *subgraywacke* as originally defined by Pettijohn (1949, p.227 & 256). Term introduced by Fischer (1934) for a wacke with less than 10% unstable materials. Examples include many coal-measures sandstones of Pennsylvanian age (such as the Atokan Series sandstones). Also spelled: *quartzwacke*.

quartz wedge In an optical system such as a polarizing microscope, an elongate wedge of clear quartz that is used in analysis of a mineral's fast and slow vibration-plane traces, optical sign, and interference colors.

quartzy *quartzose*.

quasicratonic Said of or pertaining to a part of oceanic crust marginal to the continent that is considered to be formerly continental and that stretched and foundered during expansion. Syn: *semicratonic*.

quasi-equilibrium The state of balance or grade in a stream cross section, whereby "conditions of approximate equilibrium tend to be established in a reach of the stream as soon as a more or less smooth longitudinal profile has been established in that reach even though downcutting may continue" (Leopold & Maddock, 1953, p.51).

quasi-fossil *problematic fossil*.

quasi-geoid A nonequipotential surface in the vicinity of the geoid that is defined as the locus of points whose distances below the terrain are the normal heights.

quasi-instantaneous Geologically instantaneous; occurring within an interval of geologic time too small to be subdivided (Termier & Termier, 1956).

Quaternary The second period of the Cenozoic era (following the Tertiary), thought to cover the last two or three million years. It consists of two epochs (the Pleistocene and the Holocene). Its name was originally assigned as an era rather than as a period designation, with the epochs considered to be periods, and it is still sometimes used as such in the geologic literature. The Quaternary may also be incorporated into the Neogene, when the Neogene is designated as a period of the Tertiary era. See also: *age of man*.

quaternary sediment A sediment consisting of a mixture of four components or end members; e.g. a sediment with a clastic component (such as quartz), a secondary mineral (such as a clay mineral), a chemical component (such as calcite), and an organic residue.

quaternary system A chemical system having four principal components.

quay A wharf of solid construction, built roughly parallel to the shoreline and accommodating vessels on one side only. Pron: *key*. Cf: *pier*.

quebrada A term used in the SW U.S. for a ravine or gorge, esp. one that is usually dry but filled by a torrent during a rain; a *barranco*. Also, a stream or brook. Etymol: Spanish.

queenstownite *Darwin glass*.

queluzite An igneous rock composed chiefly of the mineral spessartite, occasionally with amphiboles, pyroxenes, or micas. Economically significant manganese ores are derived from this rock.

quenching In experimental petrology, the very rapid cooling of a heated charge in order to preserve certain physical-chemical characteristics of the high-temperature state which would be changed by slow cooling.

quenite A fine-grained, dark-colored hypabyssal rock composed of anorthite, chrome diopside, with less olivine and a small amount of bronzite.

quenselite A black monoclinic mineral: $PbMnO_2(OH)$ or $Pb_2Mn_2O_5.H_2O$.

quenstedtite A mineral: $Fe_2(SO_4)_3.10H_2O$.

quernstone An English term for a millstone, and (in Norfolk) used as a syn. of *carstone*.

Querwellen wave An obsolete syn. of *Love wave*.

quick [ore dep] Said of an economically valuable or productive mineral deposit, in contrast to a *dead* ground or area. An ore is said to be quickening as its mineral content increases. Syn: *alive*.

quick [mineral] A local term used in western U.S. for *quicksilver*.

quick [sed] Said of a sediment that, when mixed with or absorbing water, becomes extremely soft, incoherent, or loose, and is capable of flowing easily under load or by force of gravity; e.g. "quick clay" of glacial or marine origin, which, if disturbed, loses practically all of its shear strength and flows plastically.

quick [soil] (a) Said of a condition of soil by which a decrease in intergranular pressure allows water to flow upward with sufficient velocity to reduce significantly the soil's bearing capacity. (b) Said of a highly porous soil that readily absorbs heat.

quick clay A clay that loses all or nearly all its shear strength after being disturbed; a clay that shows no appreciable regain in strength after remolding.

quicksand (a) A thick mass or bed of fine sand (as at the mouth of a river or along a seacoast) that consists of smooth rounded grains with little tendency to mutual adherence and that is usually thoroughly saturated with water flowing upward through the voids, forming a soft, shifting, semiliquid, highly mobile mass that yields easily to pressure and tends to suck down and readily swallow heavy objects resting on or touching its surface. Syn: *running sand.* (b) An area marked by the presence of one or more such beds. (c) The loose, incoherent, or unstable sand found in a bed of quicksand.

quicksilver A term applied to *mercury* where it occurs as a native mineral or has been mined but not yet used (as in "flasks of quicksilver").

quickstone A consolidated rock that had flowed under the influence of gravity before lithification; a quick sediment that has become lithified.

quickwater The part of a stream characterized by a strong current.

quiet reach *stillwater.*

quilted surface A land surface characterized by broad, rounded, uniformly convex hills separating valleys that are comparatively narrow "like the seams by which a quilt is furrowed" (Davis, 1918, p.124).

quinary system A chemical system having five main components, e.g. Na_2O-CaO-K_2O-Al_2O_3-SiO_2.

quinqueloculine Having five externally visible chambers as a result of growth of foraminiferal test in varying planes about an elongate axis; specif. pertaining to the foraminifer *Quinqueloculina.*

quinqueradiate Said of radial symmetry of certain echinoderms characterized by five rays extending from the mouth. Syn: *pentamerous.*

quisqueite A highly sulfurous asphaltum; a black, brittle, lustrous, asphalt-like substance mostly composed of sulfur (37%) and carbon (43%) and accompanying the vanadium ores of Peru.

Q wave *Love wave.*

R

a A Norwegian term, used esp. in southern Norway, for a morainal ridge covered with a surface layer of large stones (Stamp, 1961, p.383); most of them are in or near the sea.

abbittite A pale greenish-yellow secondary mineral: $Ca_3Mg_3(UO_2)_2(CO_3)_6(OH)_4.18H_2O$.

abdolith Var. of *rhabdolith*.

ace [water] (a) A strong or rapid current of water flowing through a narrow channel or river, e.g. a *tide race*. (b) The constricted channel or river in which such a current flows. It may occur by the meeting of two tides, as near a headland separating two bays, or it may be artificial and used for an industrial purpose, such as conveying water to (headrace) or away from (tailrace) the waterwheel of a mill. Syn: *water race*.

ace [paleont] A group of organisms with similar characteristics but not sufficiently distinctive to be classified as a species or subspecies.

ace [sed] Small calcium-carbonate concretions commonly found in brick clay; bits of chalk set in a clayey matrix. Syn: *rance*.

rachis In a plant, the axis that bears flowers or leaflets.

radar (a) An electronic detection device or *active system* for locating or tracking a distant object by measuring elapsed circuit time of travel of ultrahigh-frequency radio waves (microwaves) of known propagation velocity emitted from a transmitter and reflected or scattered back by the object to or near the point of transmission in such a way that range, bearing, height, and other characteristics of the object may be determined. Radar operation is unaffected by darkness, but moisture in the form of fog, snow, rain, or heavy clouds may cause varying degrees of attenuation or reflection of the radio energy. The term is usually prefixed by a code letter indicating the frequency band for certain wavelength ranges; e.g. K-band (around 2 cm), X-band (around 4 cm), and P-band (around one meter). (b) A name applied to the method or technique of locating or tracking objects by means of radar, such as the observation and analysis of minute radio signals reflected from the objects and displayed in a radar system.--- Etymol: *radio detecting* (or *detection*) and *ranging*. See also: *monostatic radar; bistatic radar; scatterometry*.

radar imagery Imagery provided by scanning devices using microwave radiation.

radar shadow A no-return area extending in range from an object which is elevated above its surroundings. The object obstructs the radar beam, preventing illumination of the area behind it.

raddle A term used in Lancashire, Eng., for an ironstone nodule, a clay gall, or a deeply iron-stained rock.

radial [paleont] adj. (a) Belonging to or in the direction of a ray of an echinoderm. (b) A syn. of *ambulacral*; e.g. referring to the position of a line extending from the centrally placed mouth to the aboral end of any ambulacrum of a blastoid. (c) Directed outward from the center of the umbilicus of an ammonoid and at right angles to the axis of coiling and growth; transverse. (d) Said of the microstructure of hyaline calcareous foraminiferal tests consisting of calcite or aragonite crystals with *c*-axes perpendicular to the surface and exhibiting between crossed nicols a black cross with with concentric rings of color mimicking a negative uniaxial interference figure (TIP, 1964, pt.C, p.63).---n. A *radial plate* together with all structures borne by it.

radial [palyn] Pertaining to trilete-spore features associated closely with the arms of the laesura. Cf: *interradial*.

radial [photo] A line or direction from the center (principal point, isocenter, nadir point, or substitute center) to any point on a photograph.

radial array *azimuthal survey*.

radial assumption The assumption that on a nearly vertical photograph all displacements resulting from tilt and relief are radial from the principal point of the photograph. It is the basis of planimetric mapping by photogrammetric methods.

radial beam An internal rod usually connecting concentric lattice shells of spumellarian radiolarians.

radial canal (a) A canal extending radially from the ring canal beneath an ambulacrum of an echinoderm. It is closed at its outer end and bears rows of podia. The ambulacral system of echinoderms contains five radial canals. (b) One of the numerous minute canals, lined with choanocytes, radiating from the cloaca in some sponges and ending just below the surface of the sponge.

radial drainage pattern A drainage pattern in which consequent streams radiate or diverge outward, like the spokes of a wheel, from a high central area; it is best developed on the slopes of a young, unbreached domal structure or of a volcanic cone. Cf: *centripetal drainage pattern*. Syn: *centrifugal drainage pattern*.

radiale A *radial plate* of a crinoid. Pl: *radialia*.

radial facet A smooth or sculptured distal face of a radial plate of a crinoid, bearing marks of ligamentary or muscular articulation with the first primibrachial. It is lacking in radial plates that bear no arms.

radial faults Faults in a pattern of radiating from a common central point.

radial plate Any of various plates of an echinoderm; e.g. an undivided proximal plate of any crinoid, lying between the basal and brachial plates, or one of a circlet of plates on the sides of theca of a blastoid, or a prominent ossicle on the aboral surface of an asteroid. Syn: *radial; radiale; radius*.

radial pore canal One of a series of tubules or *pore canals* in an ostracode leading from the line of concrescence through the area of the duplicature to the free margin of the valve, usually housing sensory setae protecting the gape of the open carapace. Cf: *normal pore canal*.

radial shield One of a pair of relatively large ossicles adjacent to the base of an arm on the aboral surface of the disc in many ophiuroids.

radial spine A tangential rod or needle in the skeleton of an acantharian or phaeodarian radiolarian.

radial suture A suture in a heterococcolith corresponding to a radius in a circular coccolith or to a straight line drawn through the nearest focus or the line connecting the foci of an elliptic coccolith.

radial symmetry The condition, property, or state of having similar parts of an organism regularly arranged about a common central axis (as in a starfish); e.g. a type of symmetry exemplified in a flower that can be separated into two approximate halves by a longitudinal cut in any plane passing through the center of the flower. Cf: *bilateral symmetry*.

radial triangulation A triangulation procedure in which direction lines (radials) from the centers (principal points, isocenters, or nadir points) of overlapping vertical (or nearly vertical) or oblique photographs to control points imaged on the photographs are measured and used for horizontal-control extension by successive intersection and resection of such lines.

radial tube A centrifugal cylinder in an acantharian radiolarian.

radial zone The chamber portion of a foraminiferal test with essentially radial elements, situated between the marginal zone and the central complex (as in Orbitolinidae).

radiance Radiant flux per unit solid angle per unit area. Symbol: N.

radiant n. An organism or group of organisms, such as a species, that has arrived at its present geographic location as the result of dispersal from its main place of origin. Cf: *radiation* [evol].

radiant emittance Radiant flux emitted per unit area of a source. Symbol: W.

radiant energy Energy transferred by electromagnetic waves, measured in joules or ergs. Symbol: U.

radiant flux The rate of transfer of radiant energy, measured in watts or ergs per second. Symbol: P. Syn: *radiant power.*

radiant intensity Radiant flux per unit solid angle. Symbol: J.

radiant power *radiant flux.*

radiate aperture A foraminiferal-test opening consisting of numerous diverging slits (as in the superfamily Nodosariacea).

radiated Said of an aggregate of acicular crystals that radiate from a central point. Cf: *spherulitic.*

radiate mud crack A term used by Kindle (1926, p.73) for a mud crack that displays an incomplete radiate pattern and that lacks normal polygonal development.

radiating fault *radial fault.*

radiation [evol] In evolution, the dispersal of a group of organisms into different environments accompanied by divergent change in the evolutionary structure. Cf: *convergence; radiant.*

radiation [surv] A method of surveying in which points are located by a knowledge of their distances and directions from a central point.

radiation damage The damage done to a crystal lattice (or glass) by passage of fission particles or alpha-particles from the nuclear decay of a radioactive element residing in the lattice. The damage paths (*alpha-particle recoil tracks* or *fission tracks*) can be enlarged to microscopic size by suitable etching technques and used to determine an age for the material. High concentration of a radioactive element can cause destruction of the surrounding material.

radiation detector A device providing an electrical output that is a useful measure of incident radiation. Broadly divisible into two groups; thermal or sensitive to temperature changes; and photodetectors or sensitive to changes in photon flux incident on the detector. Typical thermal detectors are *thermocouples, thermopiles* and *thermistors;* also, *trimetal detectors.* Typical photodetectors are either *photovoltaic* (voltage changes) or *photoconductive* (conductivity change).

radiation log *radioactivity log.*

radiaxial Radially axial; e.g. "radiaxial calcite" in sedimentary rocks, occurring as cavity linings composed of subparallel individual crystals elongated normal to the cavity wall (Fischer, 1964, p.148).

radiciform Said of a root-like epithecal process (outgrowth) of a corallite wall, serving for fixation.

radicle [paleont] (a) A bryozoan root-like structure composed of one or more kenozooids. Syn: *rhizoid.* (b) An individual root-like branch of a crinoid radix.

radicle [bot] The lower portion of the hypocotyl which grows into the primary root of a seedling; the root-primordium of an embryo (Fuller & Tippo, 1949, p.970).

radii Plural of *radius.*

radioactivation analysis *activation analysis.*

radioactive Pertaining to or exhibiting *radioactivity.*

radioactive age determination *radiometric dating.*

radioactive chain *radioactive series.*

radioactive clock A *geochronometer* consisting of a radioactive isotope, e.g. carbon-14, rubidium-87, or potassium-40, whose decay constant is known and is low enough to be calibrated to time units, usually years. Radioactive clocks are the basis of absolute age determinations and the specific element being used is sometimes designated as a clock, e.g. carbon clock. Syn: *atomic clock; nuclear clock.*

radioactive constant A less-preferred syn. of *decay constant.*

radioactive dating *radiometric dating.*

radioactive decay Spontaneous, radioactive transformation of one nuclide to another, or of the energy state of the same nuclide. Syn: *decay; radioactive disintegration; disintegration.* Partial syn: *radioactivity.*

radioactive disintegration *radioactive decay.*

radioactive equilibrium In radioactivity, the condition of equilibrium in which the rate of decay of the parent isotope is exactly matched by the rate of decay of every intermediate daughter isotope. When equilibrium has been established, the concentrations of intermediate daughters remain virtually constant. Specimens which contain the natural radioactive elements and which have been in radioactive equilibrium for very long time are said to be in *secular equilibrium.*

radioactive half-life The time necessary for a radioactive substance to lose half its radioactivity during decay. Each radionuclide has a unique or characteristic half-life. Syn: *half-life; half-life period.*

radioactive heat Heat produced within a medium as a result of the absorption of radiation.

radioactive series A series or succession of *nuclides,* each of which becomes the next by radioactive decay, until a stable nuclide is achieved. See also: *parent; daughter; end product.* Syn: *radioactive chain.*

radioactive spring A spring whose water has a high and readily detectable radioactivity.

radioactivity (a) The spontaneous decay of the atoms of certain isotopes into new isotopes, which may be stable or undergo further decay until a stable isotope is finally created; *radioactive decay.* Radioactivity is accompanied by the emission of alpha particles, beta particles, and gamma rays and by the generation of heat. (b) A particular radiation component from a radioactive source, such as gamma radioactivity. (c) A *radionuclide,* such as a radioactivity produced in a bombardment. ----Adj: *radioactive.* Radioactivity is a form of *radiation.*

radioactivity anomaly An anomaly, or deviation from expected results, found when a survey of radioactivity reveals aberrant values. It is often used for mineral exploration.

radioactivity log A *geophysical log* that measures and records natural or artificially produced radiations from the rocks penetrated by a borehole or well. It is obtained by lowering and raising a sonde on a wire line and making in-situ measurements (continuously recorded at the surface) of the radioactive properties of the rocks in relation to depth in the hole. The measurements can be made in both open and cased holes filled with any fluid (including air). The two principal types are the *gamma-ray log* and the *neutron log.* Syn: *nuclear log; radiation log.*

radioassay An assay procedure involving the measurement of radiation intensity of a radioactive sample.

radioautograph *autoradiograph.*

radiocarbon Radioactive carbon, esp. *carbon-14,* but also carbon-10 and carbon-11.

radiocarbon age *carbon-14 age.*

radiocarbon dating *carbon-14 dating.*

radiochemistry The chemical study of irradiated and naturally occurring radioactive materials and their behavior. It includes their use in tracer studies and other chemical problems.

radioecology The branch of ecology concerned with the relationship between natural communities and radioactive material.

radiogenic Said of a product of a radioactive process, e.g. heat, lead.

radiogenic age determination *radiometric dating.*

radiogenic argon Argon occurring in rocks and minerals that is the result of in situ decay of potassium-40 since the formation of the Earth. Cf: *atmospheric argon; excess argon; inherited argon.*

radiogenic dating *radiometric dating.*

radiogenic isotope An isotope that was produced by the decay of a *radioisotope,* but which itself may or may not be radioactive.

radiogenic lead Stable, end-product lead (Pb-206, Pb-207, and Pb-208) occurring in rocks and minerals that is the result of in situ decay of uranium and thorium since the formation of the Earth. Cf: *primeval lead.* See also: *common lead.*

radiogenic strontium Strontium-87 occurring in rocks and min-als that is the direct result of in situ decay of rubidium-87 ...nce the formation of the Earth. See also: *common stronti-*...n.

...diogeology The study of the distribution patterns of radioac-...e elements in the Earth's crust, and the role of radioactive ...ocesses in geologic phenomena. The term was introduced ... the Russian geologist Vernadskiy.

...diograph A less-preferred syn. of *autoradiograph*.

...diohalo A little-used syn. of *pleochroic halo*.

...diohydrology The study of the hydrologic relationships of ex-...action, processing, and use (including use in hydrologic in-...stigations) of radioactive materials and disposal of the as-...ociated waste products.

...dioisotope A radioactive *isotope* of an element. More gener-...ly, a syn. of *radionuclide*. See also: *radiogenic isotope*. Syn: *...nstable isotope*.

...diolarian Any actinopod belonging to the subclass Radi-...aria, characterized mainly by a siliceous skeleton and a ma-...ne pelagic environment. Their stratigraphic range is Cam-...ian to present. In some classifications the radiolarians are ...ouped with the rhizopods.

...diolarian chert A well-bedded, cryptocrystalline radiolarite ...at has a well-developed siliceous cement or groundmass; it ...ppears to have accumulated in distinctly shallow water.

...diolarian earth A *siliceous earth* composed predominantly of ...e remains (lattice-like skeletal framework) of Radiolaria; the ...nconsolidated equivalent of *radiolarite*.

...diolarian ooze An *ooze* whose skeletal remains consist of ...e opaline silica tests of radiolarians; it is a siliceous ooze.

...diolarite [paleont] A fossil shell of the Radiolaria.

...diolarite [sed] (a) The comparatively hard, very fine-...ained, chert-like, homogeneous, consolidated equivalent of ...diolarian earth. (b) Indurated radiolarian ooze. (c) A term ...at is often applied as a syn. of *radiolarian earth*.

...diolite A spherulite composed of radially arranged acicular ...ystals.

...diolitic [ign] Said of the texture, of an igneous rock, charac-...rized by radial, fan-like groupings of acicular crystals, re-...embling sectors of spherulites.

...diolitic [sed struc] Said of limestones in which the compo-...ents radiate from central points, with the cement comprising ...ss than 50 percent of the total rock (Krumbein & Sloss, ...963, p.179).

...dioluminescence Luminescence that was stimulated by the ...npact of radioactive particles.

...diolysis Radiation-caused chemical decomposition.

...diometer A radiation-measuring instrument having substan-...ally equal response to a band of wavelengths, usually either ... the infrared or visible, but occasionally from ultraviolet to ...r infrared, microwave, radio, etc. Most radiometers measure ...e difference between the source radiation incident on the ...etector and a radiant energy (blackbody) reference.

...diometric Pertaining to measuring geologic time by the dis-...tegration rate of radioactive elements (i.e. *radiometric dat-...g*).

...diometric age An age expressed in years and calculated ...om the quantitative determination of radioactive elements ...nd their decay products. A common syn. is *absolute age*. ...yn: *isotopic age*.

...diometric age determination *radiometric dating*.

...diometric dating Calculating an age in years for geologic ...aterials by measuring the presence of a short-life radioac-...ve element, e.g. carbon-14, or by measuring the presence of ...long-life radioactive element plus its decay product, e.g. po-...ssium-40/argon-40. The term applies to all methods of age ...etermination based on nuclear decay of natural elements. ...yn: *isotopic age determination; radiometric age determina-...on; radioactive age determination; radioactive dating; radi-...genic age determination; radiogenic dating; nuclear age de-...ermination*.

radiometric line-scan system An image-forming system which produces radiometric imagery, i.e., the gray scale of the ima-ges produced is related to the incident flux striking the detec-tor by a known energy-transfer function, hence a scanning ra-diometer. Radiometric systems require an internal blackbody reference of known temperature and other special features. The gray scale of the images produced may be converted by means of amplitude level slicing or optical densitometry to ra-diation temperature isolines. If the images are also rectified to eliminate geometric distortion, thermal isoline maps of the Earth's surface can be derived (Friedman, 1970, p.35). Cf: *differential infrared line-scan system*.

radiometry The measurement of optical electromagnetic ra-diation.

radionuclide A radioactive *nuclide*. The term *radioisotope* is loosely used synonymously. See also: *radioactivity*.

radiosonde An airborne device (usually a balloon) used for the gathering and radio transmission of meteorologic data as it travels upward.

radio wave method Any electromagnetic exploration method wherein electromagnetic waves transmitted from radio broad-cast stations may be used as an energy source in determining the electrical properties of the Earth.

radiozone A para-time-rock unit representing a zone or suc-cession of strata established on common radioactivity criteria (Wheeler et al, 1950, p.2364).

radius (a) A ray of an echinoderm, such as any of five radiat-ing ossicles in Aristotle's lantern of an echinoid; esp. a *radial plate*. (b) The lateral part of a compartmental plate of a cirri-pede crustacean, overlapping the ala of an adjoining plate, and differentiated from the paries by change in direction of growth lines. (c) An imaginary radial line dividing the body of a radially symmetrical animal into similar parts.---Pl: *radii*.

radius of influence The radial distance from the center of a well to the edge of its area of influence.

radius ratio The radius of a cation divided by that of an ion. Relative ionic radii are pertinent to crystal lattice structure.

radix Root-like distal anchorage of a crinoid column.

radon-220 A radioactive, gaseous isotope of radon-219 and radon-222; it is a member of the thorium series and a daugh-ter of radium-224. Less-preferred syn: *thoron*.

Radstockian European stage: lower upper Upper Carbonifer-ous (above Staffordian, below Stephanian). It is equivalent to uppermost Westphalian.

radula A chitinous band or strip of horny material in nearly all univalve mollusks that bears numerous transverse rows of file- or rasp-like and usually very minute teeth on its dorsal surface, that can be protruded out through the mouth from its position on the floor of the digestive canal, and that serves to gather and tear up food and draw it into the mouth. Pl: *radu-lae*.

radulifer Hook-shaped or rod-like brachiopod crura that arise on the ventral side of the hinge plates and project toward the pedicle valve.

rafaelite A nepheline-free orthoclase-bearing hypabyssal rock that also contains analcime and calcic plagioclase.

raft [ign] A rock fragment caught up in a magma and drifting freely, more or less vertically.

raft [streams] An accumulation or jam of floating logs, drift-wood, dislodged trees, or other debris, formed naturally in a stream by caving of the banks, and acting as an impedance to navigation. See also: *raft lake*.

raft [coal] *float coal*.

raft breccia A breccia having fragments that remained unworn during transportation, as by an iceberg, a floe, or vegetation such as trees or seaweed (Norton, 1917, p.172).

rafted ice A form of pressure ice in which one floe overrides another as a result of *rafting* [ice].

rafting [geol] The transporting of land-derived rocks and other materials by floating ice (*ice-rafting*) or by floating organic material (as seaweed or logs) to places not reached by water currents.

rafting [ice] A form of deformation of floating ice whereby one piece of ice overrides another, thus creating *rafted ice*; most common in *new ice* and *young ice*. See also: *finger-rafting*.

raft lake A relatively short-lived body of water impounded along a stream by a *raft*; examples are commonly found in the Red River of Louisiana during times of high water.

raft tectonics A term used by Dickinson (1966, p.707) for tectonics based on the effects imposed on surficial plates ("rafts") of little strength by the motions of active undermasses to which the rafts are coupled and upon which they ride passively. It explains geometric patterns of folding and faulting in shallow Earth layers, particularly in Cenozoic orogenic belts. Cf: *plate tectonics*.

rag Any of various hard, coarse, rubbly or shelly rocks that weather with a rough irregular surface; e.g. a flaggy sandstone or limestone used as a building stone. The term appears in certain British stratigraphic names, as the Kentish Rag (a Cretaceous sandy limestone in East Kent). See also: *coral rag*. Syn: *ragstone*.

raglanite A nepheline syenite composed of oligoclase, nepheline, and corundum with minor amounts of mica, calcite, magnetite, and apatite. Cf: *craigmontite*.

ragstone *rag*.

raguinite A mineral: $TlFeS_2$.

rain [geol] A falling of numerous particles, such as the long unending deposition of pelagic matter to the bottom of the ocean. Also, the falling particles themselves, such as *pollen rain*.

rain [meteorol] Water in the Earth's atmosphere that is precipitated out as particles that have become too large (about 0.5 mm in diameter) to be held in suspension.

rainbeat *raindrop impact*.

rainbow Chromatic iridescence observed in drilling fluid that has been circulated in a well, indicating contamination or contact with fresh hydrocarbons.

rainbow quartz An *iris* quartz that exhibits the colors of the rainbow.

rain crust A type of *snow crust* formed by refreezing of surface snow that had been melted or wetted by liquid precipitation; it usually has a dimpled surface.

rain desert A desert in which rainfall is sufficient to maintain a sparse general vegetation. Cf: *runoff desert*.

raindrop impact The action of raindrops striking or falling upon the surface of the ground. Syn: *rainbeat*.

raindrop impression *rain print*.

raindrop imprint *rain print*.

rainfall (a) The quantity of water that is precipitated out in the atmosphere as rain, in a given period of time. (b) The liquid product of precipitation in whatever form. In this sense the term is synonymous with *precipitation*.

rainfall excess The volume of water from rainfall that is available for direct runoff (Langbein & Iseri, 1960). Cf: *abstraction* [*water*]; *precipitation excess*. Syn: *excess water*.

rainfall penetration The depth below the soil surface to which water from a given rainfall has been able to infiltrate.

rain forest A tropical forest where the annual rainfall is at least 100 inches. The region is characterized by tall, lush evergreen trees.

rain gage A device used to measure precipitation (melted snow, sleet, or hail as well as rain). It consists of a receiving funnel, a collecting vessel, and a measuring cylinder. See also: *pluviograph*. Syn: *pluviometer; hyetometer; snow gage*.

rain pillar A minor landform consisting of a column of soil or soft rock capped and protected by pebbles or concretions, produced by the differential erosion effected by the impact of falling rain (Stokes & Varnes, 1955, p.118).

rain print A small, shallow, circular to elliptical, and vertical or slanting depression or crater-like pit surrounded by a slightly

raised rim, formed in soft sediment (fine sand, silt, clay) or in the mud of a tidal flat by the impact of a falling raindrop, and sometimes preserved on the bedding planes of sedimentary rocks or as casts on the undersides of overlying sandstone beds. See also: *hail imprint; spray print*. Syn: *raindrop imprint; raindrop impression*.

rain shadow A region, usually very dry, on the lee side of a topographic obstacle (usually a mountain or mountain range) facing away from the prevailing winds, where the rainfall is noticeably less than on the windward side; e.g. the White Mountains in east-central California are in a rain shadow of the Sierra Nevadas.

rain-shadow desert A desert occurring on the lee side of a mountain or mountain range that deflects moisture-laden air upward on the windward side. See also: *middle-latitude desert*. Syn: *orographic desert*.

rainwash (a) The washing away of loose surface material by rainwater after it has reached the ground but before it has been concentrated into definite streams; specif. *sheet erosion*. Also, the movement downslope (under the action of gravity) of material loosened by rainwater. It occurs esp. in semiarid or scantily vegetated regions. Syn: *hillwash*. (b) The material that originates by the process of rainwash; material transported and accumulated, or washed away, by rainwater. (c) The rainwater involved in the process of rainwash.---Also spelled: *rain wash*.

rainwater Water that has fallen as rain and has not yet collected soluble matter from the soil, thus being quite soft.

rain wave train Overland flow in the form of wave trains or series of uniformly spaced waves and involving nearly all the runoff. Rain-wave trains are usually associated with heavy rains, esp. cloudbursts (Horton, 1945, p.313).

raised bank *natural levee*.

raised beach An ancient beach occurring above the present shoreline and separated from the present beach, having been elevated above the high-water mark either by local crustal movements (uplift) or by lowering of sea level, and often bounded by inland cliffs. Cf: *marine terrace*. See also: *strand line*.

raised bog An area of acid, peaty soil, especially that developed from moss, in which the center is relatively higher than the margins. Syn: *Hochmoor*.

raised peat bog A *highmoor bog* with a thick accumulation of peat in the center giving it a convex surface.

rake The inclination from the horizontal of ore shoots, lineation, slickensides, or other structures, measured on the plane of the associated veins, faults or foliation. It is also an obsolete syn. of *pitch* [struc geol] and *plunge*.

ralstonite A colorless, white, or yellowish mineral: $Na_xMg_xAl_{2-x}$ $(F,OH)_6 \cdot H_2O$. It occurs in octahedral crystals.

ram An underwater ledge or projection from an ice wall, ice front, iceberg, or floe, and usually caused by the more intensive melting and erosion of the unsubmerged part. Syn: *apron; spur*.

Raman effect Monochromatic light sent through a transparent substance undergoes energy transformations with some molecules of the substance which changes the frequency of some of the light resulting in the addition of certain lines (*Raman lines*) to the spectrum (*Raman spectrum*). Named in honor of Chandrasekhara V. Raman (1888-), Indian physicist.

Raman lines Added lines in the *Raman spectrum*.

Raman spectroscopy The observation of a *Raman spectrum* and all processes of recording and measuring that go with it.

Raman spectrum The characteristic spectrum observed when monochromatic light is scattered by a transparent substance. See also: *Raman effect*.

ramassis A local term in southern Louisiana for a mass of decomposed plant debris, dried plant remains, driftwood, and other flotsam occurring in coastal marshes, or flotants (Russell, 1942, p.96-97).

rambla A dry ravine, or the dry bed of an ephemeral stream. Etymol: Spanish, from Arabic *ramlah*, "sand".

ramdohrite A dark-gray mineral: $Ag_2Pb_3Sb_6S_{13}$. It is closely related to andorite and fizelyite.

ramentum A thin, chaffy scale on a leaf, e.g. on some ferns.

rami Plural of *ramus*.

rammell An English term for a rock containing a mixture of shale (or clay) and sand.

rammelsbergite A gray mineral: $NiAs_2$. It is dimorphous with pararammelsbergite and related to loellingite. Syn: *white nickel*.

ramose Consisting of or having branches; e.g. said of a bryozoan colony consisting of erect, round, or moderately flattened branches.

ramosite A basic scoria.

ramp [paleont] The abapically sloping surface of a gastropod whorl next below a suture.

ramp [struc geol] A thrust fault, the fault surface of which is steeply inclined, at least near the surface. See also: *ramp valley*.

ramp [snow] An accumulation of snow that forms an inclined plane between land or land ice and sea or shelf ice. Cf: *bridge snow*].

rampart [geomorph] (a) A narrow, wall-like ridge, 1-2 m high, built up by waves along the seaward edge of a reef flat, and consisting of boulders, shingle, gravel, or reef rubble, commonly capped by dune sand. (b) A wall-like ridge of unconsolidated material formed along a beach by the action of strong waves and currents.

rampart [volc] A crescentic or ringlike deposit of pyroclastics around the top of a volcano.

rampart [lake] *lake rampart*.

rampart wall A *rimming wall* formed along the outer or seaward margin of a terrace, as on various "high limestone" Pacific islands (Flint et al, 1953, p.1258).

ram penetrometer *ramsonde*.

ramp trough *ramp valley*.

ramp valley A valley that is bounded by high-angle thrust faults, or *ramps*. Syn: *ramp trough*.

ram resistance A hardness parameter of a snow layer, as measured by the resistance to penetration of a cone-shaped penetrometer or ramsonde, expressed in kilograms.

ramsayite *lorenzenite*.

ramsdellite An orthorhombic mineral: MnO_2. It is dimorphous with pyrolusite.

ramsonde A cone-tipped metal rod or tube that is driven downward into snow to measure its hardness. Syn: *ram penetrometer*. Also spelled: *ramsond*; *Rammsonde*.

ramule A bifurcating or nonbifurcating minor branch of a crinoid arm, differing from a pinnule in less regular occurrence and in some crinoids by the presence of pinnules on it.

ramus A projecting part or elongated process of an invertebrate; e.g. a branch of a crustacean limb, or the main branch of a crinoid arm. Pl: *rami*.

raña A Spanish term for a consolidated mudflow deposit containing angular blocks of rock of all sizes; e.g. a fanglomerate.

rance (a) A dull red marble, with blue-and-white markings, from Hainaut province in Belgium. Etymol: French. (b) *race* [sed].

rancholabrean Stage in southern California: upper Pleistocene (above Irvingtonian).

rancieite A mineral: $(Ca,Mn^{+2})Mn_4^{+4}O_9 \cdot 3H_2O$.

rand (a) An English term for the low, marshy border of a lake or of a river overgrown with reeds. (b) A term used in South Africa for a long, low, rocky ridge or range of hills often covered with scrub; e.g. Witwatersrand (popularly contracted to "The Rand"), a ridge consisting of a rich gold-bearing reef, 100 km long, situated near Johannesburg.

randannite (a) A dark variety of diatomaceous earth containing humic material, occurring in the Puy-de-Dôme (Randan) region of France. (b) An earthy form of opal.---Also spelled: *randanite*.

randkluft A crevasse at the head of a mountain glacier, separating the moving ice and snow from the surrounding rock wall of the valley where no ice apron is present. It may be enlarged where heat radiated from the rock wall causes the ice to melt. Etymol: German *Randkluft*, "rim crevice".

random error Any *error* that is wholly due to chance and does not recur; an *accidental error*. Ant: *systematic error*.

random line (a) A trial surveying line that is directed as closely as circumstances permit toward a fixed terminal point that cannot be seen from the initial point. (b) *random traverse*.

random process *stochastic process*.

random sample A subset of a statistical population in which each item has an equal and independent chance of being chosen; e.g. a sample chosen to determine (within defined limits) the average characteristics of an orebody.

random traverse A survey traverse run from one survey station to another station which cannot be seen from the first station in order to determine their relative positions. Syn: *random line*.

random variable A real-valued mathematical function or *variate*, arising from a mathematical process, that is defined over a sample space.

rang One of the units of subdivision in the C.I.P.W. classification of igneous rocks.

range [eco geol] An area in which a mineral-bearing formation outcrops, e.g. the "iron range" and "copper range" of the Lake Superior region; a mineral belt.

range [geomorph] (a) A *mountain range*. Also, a line of hills if the heights are comparatively low. (b) A term sometimes used in Australia for a single mountain. (c) Mountainous country; term usually used in the plural.

range [hydrog] An established or well-defined line or course whose position is known and along which soundings are taken in a hydrographic survey.

range [ecol] The geographic area over which an organism or group of organisms is distributed. Syn: *distribution*.

range [sed] A measure of the variability between the largest and smallest particle sizes of a sediment or sedimentary rock.

range [stat] The numerical difference between the highest and lowest values in any series.

range [stratig] *stratigraphic range*.

range [surv] (a) Any series of contiguous *townships* (of the U.S. Public Land Survey system) situated north and south of each other and numbered consecutively east and west from a principal meridian to which it is parallel (e.g. "range 3 east" indicates the third range or row of townships to the east from a principal meridian). Also, any series of contiguous sections similarly situated within a township. Abbrev (when citing specific location): R. Cf: *tier*. (b) The distance between any two points, usually an observation point and an object under observation; also, two points in line with the point of observation. Two or more objects in line are said to be "in range".

range [radioactivity] The distance that radiation penetrates a medium before its velocity becomes no longer detectable.

range chart A chart that records for a given area the local range zone (and often the peak zone) of each significant fossil taxon encountered in terms of genera and species.

range finder A *tachymeter* designed for finding the distance from a single point of observation to other points at which no instruments are placed. It uses the parallax principle, and is usually constructed to give a rapid mechanical solution of a triangle having the target at its apex and the range finder at one corner of its base. See also: *telemeter* [surv].

range line One of the imaginary boundary lines running north and south at six-mile intervals and marking the relative east and west locations of townships in a U.S. public-land survey; a meridional township boundary line. Cf: *township line*.

range midpoint *midrange.*

range pole A *range rod.* Syn: *ranging pole.*

range rod A straight, slender, wood or metal level rod, rounded or octagonal in section, 6-8 ft long and one inch or less in diameter, fitted with a sharp-pointed metal shoe, usually painted in one-foot bands of alternate contrasting colors of red and white, and used for sighting points and lines in surveying or for showing the position of a ground point. Syn: *ranging rod; range pole; lining pole; line rod; sight rod.*

range zone A formal *biostratigraphic zone* consisting of a body of strata comprising the total horizontal (geographic) and vertical (stratigraphic) range of occurrence of a specified taxon or group of taxa (ACSN, 1961, art.22). It comprises the rocks that contain the taxon whose name it bears. Range zones do not usually coincide with *assemblage zones* named for the same fossil. The term is more or less synonymous with certain usages of *biozone* (Hedberg, 1958, p.1888; and Teichert, 1958a, p.114-115), and is most commonly used (when unmodified) in referring to the interregional range of an individual taxon. Cf: *peak zone.* See also: *local range zone; concurrent-range zone.* Syn: *acrozone; zonite.*

rank [coal] A *coal classification* based on degree of metamorphism. Cf: *type* [coal]; *grade* [coal].

rank [bot] A vertical row of leaves.

rank [meta] *metamorphic rank.*

rankamaite A mineral: $(Na,K,Pb)_3(Ta,Nb,Al)_{11}(O,OH)_{30}$.

rankar A term used mostly in Europe and Asia for a soil whose humic layer lies directly on the parent rock, which is usually lime-deficient and siliceous.

rankinite A monoclinic mineral: $Ca_3Si_2O_7$. It is dimorphous with kilchoanite.

Ranney collector *collector well.*

ranquilite A mineral: $Ca(UO_2)_2Si_6O_{15}.12H_2O$ (?).

ransomite A sky-blue mineral: $CuFe_2(SO_4)_4.6H_2O$.

Raoult's law In its original sense, the statement that the partial vapor pressure of a solvent liquid is proportional to its mole fraction. It is now usually used, however, in a more general form to specify a model for the *ideal solution*: The statement that the activity of each component in a solution is equal to its mole fraction. It is obeyed by all solutions for the major component in sufficiently concentrated regions, and approximately by many over large compositional ranges.

rapakivi n. In the U.S., granite or quartz monzonite that is characterized by orthoclase phenocrysts (commonly ellipsoidal) that are mantled with plagioclase (oligoclase). The term was introduced in 1694 by Urban Hjarne to denote crumbly stone in certain weathered outcrops in Finland. In Scandinavia, the term is used to denote the youngest Precambrian granitic rocks in the Christiania district. Etymol: Finnish, "rotten stone". Syn: *wiborgite.*----adj. Said of volcanic as well as plutonic rocks having orthoclase phenocrysts that are mantled with plagioclase.

rapakivi texture Said of the texture observed in igneous and metamorphic rocks in which rounded crystals of potassium feldspar, a few centimeters in diameter, are surrounded by a mantle or rim of sodium feldspar in a finer-grained matrix, usually composed of quartz and colored minerals.

raphe (a) That portion of the funiculus of an ovule that is adnate to the integument, usually represented by a ridge. It is present in most anatropous ovules (Lawrence, 1951, p.767). (b) A vertical, unsilicified groove or cleft in the value of some pennate diatoms (Scagel, et al, 1965, p.630).

raphide A very thin, hair-like sponge spicule (monaxon or oxea).

rapid flow Water flow whose velocity exceeds the velocity or propagation of a long surface wave in still water (Middleton, 1966). Cf: *tranquil flow.* Syn: *shooting flow; supercritical flow.*

rapids (a) A part of a stream where the current is moving with a greater swiftness than usual and where the water surface is broken by obstructions but without a sufficient break in

slope to form a waterfall, as where the water descends over series of small steps. It commonly results from a sudde steepening of the stream gradient, from the presence of a r stricted channel, or from the unequal resistance of the su cessive rocks traversed by the stream. The singular for "rapid" is rarely used. See also: *cascade; cataract.* (b) swift, turbulent flow or current of water through a rapids.

raqqaite An extrusive rock having the composition of a pyro enite (Streckeisen, 1967, p.188).

rare In the description of coal constituents, less than 5% of particular constituent occurring in the coal (ICCP, 1963). C common; very common; abundant; dominant.

rarefaction *kataseism.*

ras A cape or headland. Etymol: Arabic *ra's,* "head".

rash Very impure coal, so mixed with waste material (cla slate, other argillaceous substances taken from the top bottom of the coal seam) as to be unsalable; a dark su stance intermediate in character between coal and shale; dirty coal. Not to be confused with *rashing.*

rashing A soft, friable, and flaky or scaly rock (shale or cla immediately beneath a coal seam, often containing mu carbonaceous material (numerous slickensided surfaces a streaks of coal), and readily mixed with the coal in mining. may also overlie or be interstratified with the coal. Term oft used in the plural. Not to be confused with *rash.*

rasorite *kernite.*

raspberry spar (a) *rhodochrosite.* (b) Pink tourmaline.

raspite A yellow or brownish-yellow monoclinic minera $PbWO_4$. It is dimorphous with stolzite.

Rassenkreis A polytypic species. Etymol: German *Rass* "race", plus *Kreis,* "cycle". Plural: *Rassenkreise.*

rasskar A Norwegian term for a cirque which has served "an old scree channel" and has been "carved upward weathering" (Termier & Termier, 1963, p.412).

raster lines Lines projected on an oscilloscope by a line-sca system to produce a TV-like image.

rate-of-change map A derived stratigraphic map that show the rate of change of structure, thickness, or composition of given stratigraphic unit (Krumbein & Sloss, 1963, p.484). It based on analysis of the contour lines on an initial ma (structure-contour map, isopach map, facies map, etc.).

rate of sedimentation The amount of sediment accumulated an aquatic environment over a given period of time, usua expressed as thickness of accumulation per unit tim Throughout geologic time, there appears to have been a pr gressive increase in the rate with decreasing age of the sec ments, with an overall average of about 22 cm of thicknes per 1000 years (Pettijohn, 1957, p.688). Syn: *sedimentatio rate.*

rathite A dark-gray mineral: $(Pb,Tl)_3As_5S_{10}.$.

rating curve In hydraulics, a *stage-discharge curve.*

ratio map A *facies map* that depicts ratio of thicknesses be tween rock types in a given stratigraphic unit; e.g. a "san shale ratio map" showing the ratio of sandstone thickness shale thickness in a given unit.

ratiometer An instrument used to measure the ratio of two p tential differences (Schieferdecker, 1959, term 3686).

rational formula In hydraulics, the expression of peak di charge (in cfs units) as equal to rainfall (in inches/hr) time drainage area (in acres) times a runoff coefficient dependir on drainage-basin characteristics (Chow, 1957).

rational horizon (a) A *celestial horizon;* e.g. a great circle 9 degrees from the zenith and constituting the equator of th horizon system of coordinates. (b) *actual horizon.*

rattlesnake ore A gray, black, and yellow mottled ore of ca notite and vanoxite, its spotted appearance resembling that a rattlesnake.

rattle stone A concretion composed of concentric laminae different composition, in which the more soluble layers hav been removed by solution, leaving the central part detache from the outer part, such as a concretion of iron oxide filled with loose sand that rattles on shaking. Syn: *klapperstein*. Also spelled: *rattlestone*.

rauenthalite A mineral: $Ca_3(AsO_4)_2.10H_2O$.

rauhaugite A *carbonatite* that contains ankerite.

rauk A Swedish term for a sea *stack*. Pl: *raukar*.

Rauracian Substage in Great Britain: Upper Jurassic (middle Lusitanian: above Argovian, below Sequanian).

rauvite A purplish- to bluish-black mineral: $Ca(UO_2)_2V_{10}O_{28}.16H_2O$.

ravelly ground Rock that breaks into small pieces when drilled and that tends either to slough or partly cave into the drill hole when the drill string is pulled or to bind the drill string by becoming wedged between the drill pipe and the borehole wall. Cf: *cavings*.

Ravenian Floral stage in Washington State: upper Eocene.

ravinated Said of a landform or landscape having or cut by ravines.

ravine (a) A small, narrow, deep, steep-sided depression, less precipitous than and not as grand as a *gorge*, smaller than a canyon but larger than a *gully*, and usually carved by running water; esp. the narrow excavated channel of a mountain stream. (b) A stream with a slight fall between rapids.--- Etymol: French, "mountain torrent".

ravinement [geomorph] (a) The formation of a ravine or ravines. (b) *gully erosion*.

ravinement [stratig] A term introduced by Stamp (1921, p.109) for "an irregular junction which marks a break in sedimentation", such as an erosion line occurring where shallow-water marine deposits have "scooped down into" (or "ravined") slightly eroded underlying beds; a small-scale disconformity caused by periodic invasions of the sea over a deltaic area. Etymol: French, "hollowing out (by waters), gullying".

raw Said of a mineral, fuel, or other material in its natural, unprocessed state, as mined.

raw humus *mor*.

raw map A *seismic map* whose z coordinate is time. Cf: *migrated map*.

ray [lunar] One of the long, relatively bright, almost white streaks, loops, or lines observed on the Moon's surface and appearing to radiate from a large, well-formed lunar crater, in some examples extending for hundreds of kilometers. Rays are brightest at high Sun angles and nearly visible at low Sun angles except for rough ground. They are believed to be formed in some way by fine-grained debris explosively ejected from craters either by impact or by volcanic activity.

ray [phys] A vector normal to the wave surface, indicating both the velocity and direction of propagation.

ray [paleont] (a) Any of the radiating divisions of the body of an echinoderm together with all structures borne by it; e.g. a segment of an echinoderm body that includes one ambulacral axis, or a radial plate or an arm of a crinoid. Also, a radial direction established by the position of an *ambulacrum*. (b) One of the primary subdivisions of a sponge spicule containing an axial filament or an axial canal.

ray [bot] (a) In a composite inflorescence, the corolla of a marginal flower. (b) A *vascular ray*. (c) In dasycladacean algae, a branch.

ray crater A large, relatively young lunar crater with visible rays; e.g. Copernicus.

Rayleigh scattering Scattering by particles small in size compared with the wavelengths being scattered, e.g., scattering of blue light by the atmosphere. Cf: *forward scatter; Mie theory*.

Rayleigh wave A type of *surface wave* having a retrograde, elliptical motion at the free surface. It is named after Lord Rayleigh, the English physicist who predicted its existence. See also: *pseudo Rayleigh wave; Rg wave*. Syn: *R wave*.

ray parameter A function p that is constant along a given seismic ray, defined as $p=rv^{-1} \sin i$, where r is the distance from the center O of the Earth, v is the velocity, and i is the

angle that the ray at a point P makes with the radius OP (Runcorn, 1967, p.1173).

ray path *path*.

razorback A sharp, narrow ridge, resembling the back of a razorback hog. "There is little or no implication as to geologic structure, hence the term is not quite so specific as *hogback*" (Stokes & Varnes, 1955, p.119).

razor stone *novaculite*.

Rb-Sr Rubidium-strontium, e.g. *rubidium-strontium age method*.

reach [geog] (a) A continuous and unbroken expanse or surface of water or land. Syn: *stretch*. (b) An unstated but specific distance; an interval.

reach [coast] (a) An arm of the sea extending up into the land; e.g. an estuary or a bay. (b) *promontory*.

reach [hydraul] (a) The length of a channel, uniform with respect to discharge, depth, area, and slope. (b) The length of a channel for which a single gage affords a satisfactory measure of the stage and discharge. (c) The length of a stream between two specified gaging stations.----See also: *test reach*.

reach [lake] (a) A relatively long and straight section of water along a lake shore; also, a narrow arm of the lake, reaching into the land. (b) A straight and narrow expanse of shore or land extending into a lake.

reach [streams] (a) A straight, continuous, or extended part of a stream, viewed without interruption (as between two bends) or chosen between two specified points; a straight section of a restricted waterway, much longer than a narrows. See also: *sea reach*. (b) The level expanse of water between locks in a canal.

reactance [elect] That part of the impedance of an alternating-current circuit that is due to either capacitance, inductance, or both. It is expressed in ohms (Webster, 1967).

reactance [seis] *acoustic reactance*.

reaction [radioactivity] *nuclear reaction*.

reaction border *reaction rim*.

reaction boundary *reaction line*.

reaction curve *reaction line*.

reaction line In a ternary system, a special case of the *boundary line*, along which one of the two crystalline phases present reacts with the liquid, as the temperature is decreased, to form the other crystalline phase. Syn: *reaction boundary; reaction curve*.

reaction pair Any two minerals, one of which is formed at the expense of the other by reaction with liquid; esp., any two adjacent minerals in a *reaction series*.

reaction point A (usually isobarically) invariant point on a liquidus diagram in which the composition of the liquid cannot be stated in terms of position quantities of all the solid phases in equilibrium at the point. In a binary system it is equivalent to an incongruent melting point, or *peritectic point*.

reaction principle The concept of a *reaction series*.

reaction rim A peripheral zone around one mineral that is composed of another mineral species and that represents the reaction of the earlier solidified mineral with the surrounding magma. Cf: *corrosion border; kelyphytic rim*. Syn: *reaction border*.

reaction series A series of minerals in which any early-formed mineral phase tends to react with the melt, later in the differentiation, to yield a new mineral further down in the series; e.g. early-formed crystals of olivine react with later liquids to form pyroxene crystals, and these in turn may further react with still later liquids to form amphiboles. There are two different series, a *continuous reaction series* and a *discontinuous reaction series*. This concept is frequently referred to as *Bowen's reaction series*, after N.L. Bowen, who first proposed it, or as the *reaction principle*. See also: *reaction pair*.

readvance (a) A new *advance* made by a glacier after it had receded from the position it reached in an earlier advance.

(b) A time interval during which a readvance occurred.

realgar A bright-red to orange-red monoclinic mineral: AsS. It occurs as nodules in ore veins and as a massive or granular deposit from some hot springs, and it is frequently associated with orpiment. Syn: *red arsenic; sandarac; red orpiment.*

realized ecological hyperspace *biospace.*

realized ecospace That portion of the *ecospace* actually utilized by an organism (Valentine, 1969, p.687). Cf: *prospective ecospace.*

realm (a) A portion of the Earth consisting of several *regions.* (b) A large *region.*

real time Transmission of data at the time of occurrence, with no delay.

reamer A rotary-drilling tool with a special bit used for enlarging, smoothing, or straightening a drill hole, or making the hole circular when the drill has failed to do so, or splitting a casing. Also, the bit of a reamer.

recapitulation theory A theory in biology stating that an organism passes through successive stages resembling its ancestors so that the ontogeny of the individual is a recapitulation of the phylogeny of its group. See also: *palingenesis* [*paleont*]. Syn: *Haeckel's law.*

recemented glacier *glacier remanié.*

Recent Holocene.

recentered projection A term preferred by the British National Committee for Geography (1966, p.33) to the synonym *interrupted projection.*

receptacle The apex of a pedicel (or peduncle) from which the organs of a flower grow out; also, the inflated tips of certain brown algae within which gametangia are borne (Fuller & Tippo, 1949, p.970).

receptaculitid Any of a group of early and middle Paleozoic fossils (Ordovician to Devonian, possibly Carboniferous, also) of uncertain systematic position, belonging to the family Receptaculitidae, characterized by ovoid, globose, or discoidal calcareous hard parts. The receptaculitids have been variously classified as calcareous algae, foraminifers, sponges, echinoderms, or an independent, extinct phylum.

receptor *geophone.*

recess [*geomorph*] An indentation into a surface bounded by a straight line; e.g. a cleft in a steep rock bank. See also: *reentrant.*

recess [*fold*] An area in which the axial traces of folds are concave toward the outer edge of the folded belt. Ant: *salient* [*fold*].

recession [*geomorph*] (a) The backward movement or retreat of an eroded escarpment; e.g. the slow wasting away of a cliff under the influence of weathering and erosion. (b) The moving back of a slope from a former position without a change in its angle. (c) The gradual upstream retreat of a waterfall.

recession [*coast*] (a) A continuing landward movement of a shoreline or beach undergoing erosion. Also, a net landward movement of the shoreline or beach during a specified period of time. Ant: *advance.* Cf: *retrogradation.* Syn: *retrogression.* (b) The withdrawal of a body of water (as a sea or lake), thereby exposing formerly submerged areas. The shoreline moves successively toward the water and away from higher land.

recession [*glaciol*] (a) A decrease in length of a glacier, resulting in a displacement backwards of the terminus caused when processes of ablation (usually melting and/or calving) exceed the speed of ice flow; normally measured in meters per year. Syn: *retreat; glacial retreat.* (b) An overall decrease in the volume of a glacier.----Syn: *glacial recession; glacier recession.*

recessional moraine An end moraine built during a temporary but significant halt or pause in the final retreat of a glacier. Also, a moraine built during a slight or minor readvance of the ice front during a period of general recession. Syn: *peripheral moraine; retreatal moraine; stadial moraine.*

recession curve A hydrograph showing the decrease of the runoff rate after rainfall or the melting of snow. Direct runoff and base runoff are usually given separate curves as they recede at different rates. The use of the term *depletion curve* in reference to the base-runoff recession is considered incorrect (Langbein & Iseri, 1960).

recessive Said of a characteristic of an organism that must be inherited from both parents if it is to be exhibited by offspring. A recessive character can be passed on to offspring without being exhibited by a parent.

recharge The processes involved in the absorption and addition of water to the zone of saturation. This does not include water reaching the belt of soil water or the intermediate belt. Also, the amount of water added. Syn: *intake; replenishment* [*grd wat*]; *ground-water replenishment; ground-water recharge; ground-water increment; increment.*

recharge area An area in which water is absorbed that eventually reaches the zone of saturation in one or more aquifers. Cf: *catchment area* [*grd wat*]; *discharge area.* Syn: *intake area.*

recharge basin A basin constructed in sandy material to collect water, as from storm drains, for the purpose of replenishing ground-water supply.

recharge well A well used to inject water into one or more aquifers in the process of artificial recharge. Syn: *injection well.*

reciprocal bearing *back bearing.*

reciprocal lattice A lattice array of points formed by drawing perpendiculars to each plane (hkl) in a crystal lattice through a common point as origin. Points are located on each perpendicular at a distance from the origin (000) inversely proportional to spacing of the specific lattice planes (hkl). The axes of the reciprocal lattice are the *a** axis, the *b** axis, and the *c** axis, which are perpendicular, respectively, to (100), (010), and (001) of the crystal lattice. The coordinates of each reciprocal lattice point are (*hkl*) or whole multiples (n*h*, n*k*, n*l*) in terms of the unit lengths a*, b*, and c*. Cf: *direct lattice.*

reciprocal leveling Trigonometric leveling in which vertical angles have been observed at both ends of the line in order to eliminate instrumental errors; e.g. leveling across a wide river by establishing a turning point on each bank of the river from one side and taking a backsight on each to determine the height of instrument on the other side. The mean of the differences in level represents the true difference.

reciprocal salt-pair diagram *Jänecke diagram.*

reciprocal strain ellipsoid In elastic theory, an ellipsoid of certain shape and orientation which under homogeneous strain is transformed into a set of orthogonal diameters of the sphere. Cf: *strain ellipsoid.*

reciprocity (a) The statement in electrical studies that the potential at a point A due to a point source at a point B is identical to the potential at point B due to a point source at point A (b) The statement that a transmitting coil at point A will produce an electric or magnetic field at point B equal to the field observed at point A when the transmitting coil is at point B.

reclined (a) Said of a graptoloid rhabdosome with stipes extending above the sicula and enclosing an angle less than 18 degrees between their dorsal sides. Cf: *reflexed; declined; deflexed.* (b) Said of a tabulate corallite growing and opening obliquely with respect to the surface of corallum (TIP, 1956, pt.F, p.250).

reclined fold *recumbent fold.*

recomposed granite (a) An arkose consisting of consolidate feldspathic residue (produced by surface weathering of a underlying granitic rock) that has been so little reworked an so little decomposed that upon cementation the rock look very much like the granite itself. It has a faint bedding, an ur usual range of particle sizes (unlike the even-grained or po phyritic texture of true granite), and a greater percentage

quartz than is normal for granite. Syn: *reconstructed granite*.
(b) A conglomerate that has been recrystallized by strong metamorphism into a rock that simulates granite, as in the Lake Superior region.---Cf: *meta-arkose*.

recomposed rock A rock produced in place by the cementation of the fragmental products of surface weathering; e.g. a recomposed granite. The term has been applied to a rock of intermediate character straddling an unconformable surface between the breccia of the lower formation and the conglomeratic base of the upper formation (Leith, 1923).

reconnaissance (a) A general, exploratory examination or survey of the main features (or certain specific features) of a region, usually conducted as a preliminary to a more detailed survey; e.g. an engineering survey in preparing for triangulation of a region. It may be performed in the field or office, depending on the extent of information available. (b) A rapid geologic survey made to gain a broad, general knowledge of the geologic features of a region.

reconnaissance map A map based on the information obtained in a reconnaissance survey. It may incorporate data obtained from other sources.

reconnaissance survey A preliminary survey, usually executed rapidly and at relatively low cost, prior to mapping in detail and with greater precision.

reconnoiter To make a reconnaissance of; esp. to make a preliminary survey of an area for geologic purposes.

reconsequent *resequent* [streams].

reconstitution The formation of new chemicals, minerals, or structures under the influence of metamorphism, e.g. development of mica from clay minerals, development of schistosity, etc.

reconstructed glacier *glacier remanié*.

reconstructed granite *recomposed granite*.

reconstructed stone A gem material made by the fusing or sintering together small particles of the genuine stone; e.g. amberoid and reconstructed turquoise. Cf: *synthetic stone*.

reconstructive transformation A type of crystal *transformation* that involves the breaking of either first- or second-order coordination bonds. It is usually a slow transformation. An example is quartz-tridymite. Cf: *dilatational transformation; dislacive transformation; rotational transformation; substitutional transformation*.

record (a) *geologic record*. (b) *stratigraphic record*.

recovery [mining] In mining, the percentage of valuable constituent derived from an ore, or of coal from a coal seam; the yield.

recovery [surv] A visit to a survey station to identify its mark or monument as authentic and in its original location and to verify or revise its description.

recovery [grd wat] (a) The rise in static water level in a well occurring upon the cessation of discharge from that well or a nearby well. (b) *ground-water withdrawal*.

recrystallization The formation of new, crystalline mineral grains in a rock, essentially in the solid state, under the influence of metamorphic (rather than diagenetic) processes. The new grains may have the same chemical and mineralogical composition as the original rock, or they may have a different composition, in which case the process is also called *neomineralization*.

recrystallization breccia *pseudobreccia*.

recrystallization calcite Patchy mosaics of calcite crystals, interrupting or replacing a finer-grained calcite fabric in sedimentary rocks (Leighton & Pendexter, 1962, p.60).

recrystallization flow *Flow* [exp struc geol] in which there is molecular rearrangement by solution and redeposition, solid diffusion, or local melting.

rectangular coordinates Two- or three-dimensional coordinates on any system in which the axes of reference intersect at right angles; *plane coordinates*. Also, a coordinate system using rectangular coordinates. Syn: *rectilinear coordinates*.

rectangular cross ripple mark An *oscillation cross ripple mark* consisting of two sets of ripples intersecting at right angles and enclosing a rectangular pit; it is formed by waves that oscillate at right angles to the direction of the original ripple mark.

rectangular drainage pattern A drainage pattern in which both the main streams and their tributaries display many right-angle bends and exhibit sections of approximately the same length; it is indicative of streams following prominent fault or joint systems that break the rocks into rectangular blocks. It is "more irregular" than the *trellis drainage pattern* as the side streams are not perfectly parallel and not necessarily as conspicuously elongated, and secondary tributaries need not be present (Zernitz, 1932, p.503). Examples are well developed along the Norwegian coast and in parts of the Adirondack Mountains. See also: *angulate drainage pattern; joint valley*. Syn: *lattice drainage pattern*.

rectangular joint structure *mural joint structure*.

rectification [eng] A new alignment to correct a deviation of a stream channel or bank.

rectification [coast] The simplification and straightening of the outline of an initially irregular and crenulate shoreline by marine erosion cutting back headlands and offshore islands, and by deposition of waste resulting from erosion or of sediment brought down by neighboring rivers.

rectification [elect] The conversion of alternating current to direct current.

rectification [photo] The *transformation* of a photograph onto a horizontal plane so as to remove or correct displacements (distortions in perspective) by tilt; e.g. projecting an oblique photograph onto a horizontal plane, the angular relation between the photograph and the plane being determined from known (ground) measurements. It is accomplished by printing the negative with a compensating tilt so that vertical and horizontal lines in the reproduction have the same appearance as in the original.

rectilinear coordinates *rectangular coordinates*.

rectilinear current *reversing current*.

rectilinear evolution Continued change of the same type and in the same direction within a line of descent over a considerable length of time. It is similar to *orthogenesis*, but without implying how the direction is determined and maintained.

rectilinear shoreline A long, relatively straight shoreline.

rectimarginate Said of a brachiopod having a plane (straight) anterior commissure (as in the terebratulaceans); also, said of such a commissure.

rectiradiate Said of an ammonoid rib in straight radial position, bending neither forward nor backward. Cf: *prorsiradiate; rursiradiate*.

rectorite A white clay-mineral mixture with a regular interstratification of two mica layers (pyrophyllite and vermiculite) and one or more water layers. Syn: *allevardite*.

recumbent fold An overturned fold, the axial surface of which is nearly horizontal. Syn: *reclined fold*.

recumbent ovicell *independent ovicell*.

recurrence horizon In peat bogs, the demarcation between older, more decomposed peat and younger material; a parting or horizon marking an abrupt change in lithology of a peat bog, reflecting climatic change. Syn: *grenz*.

recurrence interval (a) The average time interval between actual occurrences of a hydrological event of a given or greater magnitude. (b) In an annual flood series, the average interval in which a flood of a given size recurs as an annual maximum. (c) In a partial duration series, the average interval between floods of a given size, regardless of their relationship to the year or any other period of time. This distinction holds even though for large floods recurrence intervals are nearly the same on both scales.----(ASCE, 1962).

recurrent Said of an organism or group of organisms that reappears in an area from which it had been previously expelled; e.g., a fossil present in two different rock units separated by a unit or units in which it is absent.

recurrent folding A type of folding due to periodic deformation or subsidence and characterized by thinning or possibly disappearance of formations at the crest. Cf: *supratenuous fold.* Syn: *revived folding.*

recurring polynya A *polynya* that is found in the same region every year.

recurve A region produced by the successive landward extension of a spit.

recurved spit A spit whose outer end is turned landward by current deflection, by the opposing action of two or more currents, or by wave refraction. Syn: *hook; hooked spit.*

red algae A group of algae corresponding to the phylum Rhodophyta, that owes its reddish color to the presence of the pigment phycoerythrin. Its members may be filamentous, membranous, branched, or encrusting in form, and it has a worldwide distribution. Cf: *blue-green algae; brown algae; green algae.*

red antimony *kermesite.*

red arsenic *realgar.*

red beds Sedimentary strata deposited in a continental environment, composed largely of sandstone, siltstone, and shale, with locally thin units of conglomerate, limestone, or marl, that are predominantly red in color due to the presence of ferric oxide (hematite) usually coating individual grains; e.g. the Permian and Triassic sedimentary rocks of western U.S., and the Old Red Sandstone facies of the European Devonian. At least 60% of any given succession must be red before the term is appropriate, the interbedded strata being of any color (Hatch & Rastall, 1965, p.371). Also spelled: *redbeds.* Syn: *red rock.*

red clay A pelagic deposit that is fine-grained and bright to reddish brown or chocolate-colored, formed by the slow accumulation of material at depths generally greater than 3500 meters. It contains relatively large proportions of windblown particles, meteoric and volcanic dust, pumice, shark teeth, whale earbones, manganese nodules, and debris rafted by ice. The content of $CaCO_3$ ranges from 0-30%. Syn: *brown clay.*

red cobalt *erythrite.*

red copper ore *cuprite.*

Red Desert soil In early U.S. classification systems, a group of zonal soils having a light, friable, reddish brown surface over a heavy, reddish brown or red horizon, underneath which is an accumulation of lime. It is developed in deserts of tropical to warm-temperate climate. Cf: *Reddish Brown soil.*

reddingite A pinkish- or yellowish-white orthorhombic mineral: $(Mn,Fe)_3(PO_4)_2.3H_2O$. It is isomorphous with phosphoferrite.

Reddish-Brown Lateritic soil One of a zonal, lateritic group of soils developed from a mottled red parent material and characterized by a reddish-brown surface horizon and underlying red clay.

Reddish Brown soil A group of zonal soils having a reddish, light brown surface horizon overlying a heavier, more reddish horizon and a light-colored lime horizon. It is developed in warm, temperate to tropical, semiarid climate under shrub and short-grass vegetation. Cf: *Red Desert soil.*

Reddish Chestnut soil In early U.S. classification systems, a group of zonal soils having a thick surface horizon that varies from dark brown to reddish or pinkish, below which is a heavier, reddish-brown horizon and a carbonate accumulation. It is developed under mixed grasses with some shrubs, in a warm and temperate, semiarid climate. Cf: *Chestnut soil.*

Reddish Prairie soil In early U.S. classification systems, a group of zonal soils having a surface horizon that is acidic and of a dark, reddish brown color, and that grades through a heavier reddish soil to the parent material. It is developed under tall grass in a warm to temperate, humid to subhumid climate. Cf: *Prairie soil.*

reddle *red ocher.*

red earth A general term for the soil that is characteristic of tropical climate; it is leached, red in color, deep, and clayey. Syn: *red loam.*

redeposition Formation into a new accumulation, such as the deposition of sedimentary material that has been picked up and moved (reworked) from the place of its original deposition, or the solution and reprecipitation of mineral matter. See also: *resedimentation.*

red hematite A syn. of *hematite.* Cf: *brown hematite; black hematite.*

redingtonite A pale-purple mineral: $(Fe,Mg,Ni)(Cr,Al)(SO_4)_4.22H_2O.$

red iron ore *hematite.*

red lake A lake containing reddish water. The color may be due to iron-secreting bacteria, reddish plankton, ferrous-iron compounds in solution, or red clay held in suspension.

red lead *minium.*

red lead ore *crocoite.*

redledgeite A mineral: $Mg_4Cr_6Ti_{23}Si_2O_{61}(OH)_4$ (?).

red loam *red earth.*

red manganese A reddish manganese mineral; specif. rhodonite and rhodochrosite. Syn: *red manganese ore.*

Red Mediterranean soil A type of soil developed over either calcareous or noncalcareous parent material, with A, B, and C horizons. Its B horizon is red or yellow and clayey. The term is little used. Cf: *Brown Mediterranean soil.*

red mud A type of *mud* [marine geol] that is terrigenous and contains as much as 25% calcium carbonate. Its color is due to the presence of ferric oxide.

red ocher A red, often impure or clayey, earthy *hematite* used as a pigment. Syn: *reddle; ruddle.*

red ore A red-colored ore mineral; specif. hematite or metahewettite.

red orpiment *realgar.*

red oxide of copper *cuprite.*

red oxide of zinc *zincite.*

redoxomorphic stage A term introduced by Dapples (1962) for the earliest geochemical stage of diagenesis characterized by mineral changes primarily due to oxidation and reduction reactions (reversible reactions). It is typical of unlithified sediment and preceeds the *locomorphic stage.* See also: *phyllomorphic stage.*

Red Podzolic soil Formerly, one of a group of zonal soils that is now considered part of the classification *Red-Yellow Podzolic soil.*

red rock (a) *red beds.* (b) A driller's term for any reddish sedimentary rock.

redruthite *chalcocite.*

red schorl (a) *rubellite.* (b) *rutile.*

red silver ore A red silver-sulfide mineral; specif. "dark red silver ore" (pyrargyrite) and "light red silver ore" (proustite). Syn: *red silver.*

red snow A general name for reddish or pinkish snow colored by the presence of various red or pink microscopic algae (such as the species of *Sphaerella* and *Scotiella*) in the upper layers of snow in arctic and alpine regions. Cf: *green snow.* Syn: *pink snow.*

redstone A reddish sedimentary rock, such as a red-colored sandstone; specif. a deep-red, clayey sandstone or siltstone representing a flood-plain micaceous arkose, as in the Triassic deposits of Connecticut.

red tide A type of *water bloom* that is caused by dinoflagellates.

reduced latitude The angle at the center of a sphere tangent to a reference ellipsoid along the equator, between the plane of the equator and a radius to the point intersected on the sphere by a straight line perpendicular to the plane of the equator. Syn: *parametric latitude; geometric latitude.*

reduced mud *black mud.*

ducing flame In blowpiping, the blue, more intense part of e flame, in which oxygen in the compound being tested is rtly burned away. Cf: *oxidizing flame.*

duction [geomorph] The lowering of a land surface by erosion.

duction [grav] *gravity reduction.*

duction body A multicellular mass resulting from the disoranization of a sponge and capable of reorganizing into a onge with a functional aquiferous system.

duction index The rate of wear of a sedimentary particle bject to abrasion in the course of transportation, expressed s the difference between the mean weight of the particle bere and after transport divided by the product of mean weight fore transport and the distance traveled (Wentworth, 1931, 25). Abbrev: R.I. Cf: *durability index.*

duction sphere A white, leached, spheroidal mass produced a reddish or brownish sandstone by a localized reducing nvironment, commonly surrounding an organic nucleus or a ebble and ranging in size from a poorly defined speck to a rge perfect sphere more than 25 cm in diameter (Hamblin, 958, p.24-25); e.g. in the Jacobsville Sandstone of northern ichigan.

duction to sea level The application of a correction to a easured horizontal length on the Earth's surface, at any altide, to reduce it to its projected or corresponding length at a level.

duzates Sediments accumulated under reducing conditions nd thus characteristically rich in organic carbon and in iron ulphide minerals; coal and black shales are principal examles. It is one of Goldschmidt's groupings of sediments or anogues of differentiation stages in rock analysis. Cf: *resistes; evaporates; hydrolyzates; oxidates.*

d vitriol *bieberite.*

ed-Yellow Podzolic soil Any of a group of acidic, zonal soils aving a leached, light-colored surface layer and a subsoil ontaining clay and oxides of aluminum and iron, varying in olor from red to yellowish red to a bright yellowish brown. Its arent material is clayey but siliceous, and of variegated olor. It is developed under forest vegetation in a warm, temerate, or tropical and humid climate. See also: *Yellow Podolic soil; Red Podzolic soil.*

d zinc ore *zincite.*

ed cast A vertical and cylindrical cast of sand presumably presenting the filling of a mold left by a reed.

edmergnerite A colorless triclinic mineral: $NaBSi_3O_8$. It is e boron analogue of albite.

ed peat *telmatic peat.*

ef [ore dep] A provincial term for a metalliferous mineral eposit, esp. gold-bearing quartz, usually bedded (e.g. *saddle ef*).

ef [geomorph] A term formerly used widely, esp. in the estern U.S., for a projecting outcrop, or a jagged ridge of enerally upturned rock; e.g. a hogback.

ef [navigation] (a) A narrow chain, range, or ridge of rocks, sp. coral and sometimes sand, gravel or shells, elevated oove the surrounding bottom of the sea, lying at or near the urface of the water and sometimes visible at low tide, occuring either in a loose or firmly wave-resistant state, and danerous to surface navigation; specif. an elevation at a depth f 10 fathoms (formerly 6) or less, composed of rock or oral. Cf: *shoal.* (b) A bar of earth or sand submerged in a ver channel. See also: *sandbar.* (c) A ridge of sand, gravel, r rock in a large lake, as in the Great Lakes. (d) Loosely, ny area of partly submerged rock.

ef [sed] A ridge- or mound-like, layered, sedimentary rock tructure, or part thereof, built by and composed almost exlusively of the remains of sedentary organisms (esp. corals), nd usually enclosed in rock of differing lithology; specif. an rganic reef characterized by wave resistance and by topoaphic relief above surrounding contemporaneously deposited

sediment. Cf: *bank.*

reefal Pertaining to a reef and its integral parts, esp. to the carbonate deposits in and adjacent to a reef.

reef breccia A rock formed by the consolidation of limestone fragments broken off from a reef by the action of waves and tides. Cf: *reef-rock breccia.*

reef cap A deposit of fossil-reef material overlying or covering an island or mountain. Cf: *coral cap.*

reef cluster A group of reefs of wholly or partly contemporaneous growth, found within a circumscribed area or geologic province. See also: *hermatopelago.*

reef complex The solid reef proper (reef core) and the heterogeneous and contiguous fragmentary material derived from it by abrasion; the aggregate of reef, fore-reef, back-reef, and interreef deposits, bounded on the seaward side by the basin sediments and on the landward side by the lagoonal sediments (Nelson et al, 1962, p.249). Term introduced by Henson (1950, p.215-216) to include the reef and "all genetically(?) associated sediments".

reef conglomerate *reef talus.*

reef core The rock mass constructed in place, and within the rigid growth lattice formed, by reef-building organisms; the solid reef proper. See also: *reef wall.*

reef detritus Fragmental material derived from the erosion of an organic reef, as by boring animals detaching finer material carried away by waves, the coarser fragments forming a *talus apron.* See also:*reef talus.* Syn: *reef debris.*

reef edge The seaward margin of the reef flat, commonly marked by surge channels. Cf: *reef front.*

reef flank The part of the reef that surrounds, interfingers with, and locally overlies the reef core, often indicated by massive or medium beds of reef talus dipping steeply away from the reef core; the relatively narrow transitional zone where biologic forces of reef expansion contend with physical and biologic forces of reef destruction.

reef flat A stony platform of coral fragments and coral sand, formed on the inner or lagoonal side of a coral reef, generally dry at low tide, and possessing a horizontal surface diversified by dunes, shallow pools, irregular gullies, vegetation (esp. palms), and widely scattered colonies of the more hardy species of coral; the summit of the reef above low water.

reef front The upper part of the outer or seaward slope of a reef, extending to the *reef edge* from above the dwindle point of abundant living coral and coralline algae.

reef-front terrace A shelf- or bench-like eroded surface, sometimes veneered with organic growth, sloping seaward to a depth of 8-15 fathoms; e.g. the terrace in Bermuda.

reef knoll (a) A bioherm or fossil coral reef represented by a small, prominent, rounded hill, up to 100 m high, consisting of resistant reef material, being either a local exhumation of an original reef feature or a feature produced by later erosion; specif. a small, pinnacle-like or conical mass of coralline limestone, more or less circular in ground plan, often with a mushroom-shaped top, and commonly surrounded by rock of different lithology, as in the type area of the Craven district in Yorkshire, England. (b) A present-day reef in the form of a knoll; a small *reef patch* developed locally and upward rather than outward. The term was first used by Tiddeman (1890, p.600) for a reef feature that originated as a knoll. See also: *knoll reef; coral head.*

reef limestone A limestone consisting of the remains of active reef-building organisms, such as corals, sponges, and bryozoans, and of sediment-binding organic constituents, such as calcareous algae. See also: *coral-reef limestone.*

reef milk A very fine-grained matrix material of the back-reef facies, consisting of white, opaque microcrystalline calcite derived from abrasion of the reef core and reef flank.

reefoid Resembling a reef; e.g. "reefoid rocks" consisting of a ridge-like accumulation of hard, unstratified biogenic material at or near sea level.

reef patch A growth of coral formed independently on a shelf of less than 70 m depth in the lagoon of a barrier reef or of an atoll, varying in extent from an expanse measuring several kilometers across to a mushroom-shaped growth consisting of a single large colony (Kuenen, 1950, p.426). See also: *reef knoll; shoal reef.* Cf: *patch reef.*

reef pinnacle *pinnacle* [reef].

reef ring *atoll.*

reef rock A hard, resistant, unstratified rock consisting of sand, shingle, and the calcareous remains of reef-building organisms, cemented by calcium carbonate. Also spelled: *reef-rock.* Syn: *hermatolith.*

reef-rock breccia A *coral rag* "in which masses of coral retain the attitude and position of growth, and to which the varied animal and vegetal life of the reef contributes" (Norton, 1917, p.179). Cf: *reef breccia.*

reef segment A part of an organic reef lying between passes, gaps, or channels.

reef slope The face of a reef rising from the sea floor (Maxwell, 1968, p.106-107).

reef talus Massive inclined strata consisting of *reef detritus* deposited along the seaward margin of an organic reef; a reef-flank deposit. Syn: *reef conglomerate.*

reef tract A poorly defined offshore area in which reefs are found. The term is not recommended for scientific use.

reef tufa Drusy, prismatic, fibrous calcite deposited directly from supersaturated water upon the void-filling internal sediment of the calcite mudstone of a reef knoll (Bissell & Chilingar, 1967, p. 165). See also: *stromatactis.*

reef wall A wall-like upgrowth of living coral and the skeletal remains of dead coral and other reef-building organisms, reaching intertidal level where it acts as a partial barrier between adjacent environments (Henson, 1950, p.227); the *reef core.* See also: *wall reef.*

reefy (a) Containing reefs, such as a "reefy harbor". (b) Containing sedimentary material that resembles the material of a sedimentary reef.

reentrant adj. Reentering or directed inward; e.g. a *reentrant angle* in a coastline or on a twinned crystal.---n. A prominent, generally angular indentation into a landform; e.g. an inlet between two promontories along a coastline, or a transverse valley extending into an escarpment. Also spelled: *re-entrant.* Ant: *salient.* See also: *recess* [geomorph].

reentrant angle The angle between two plane surfaces on a solid, in which the external angle is less than 180°.

reevesite A mineral: $Ni_6Fe_2(OH)_{16}(CO_3).4H_2O$.

reference axis *fabric axis.*

reference ellipsoid *reference spheroid.*

reference level A *datum plane*; e.g. a standard level (in the study of underwater sound) to which sound levels can be related.

reference line Any line that serves as a reference or base for the measurement of other quantities; e.g. a *datum line.*

reference locality A locality containing a reference section, established to supplement the *type locality.*

reference plane *datum plane.*

reference section A rock section, or group of sections, designated to supplement the *type section,* or sometimes to supplant it (as where the type section is no longer exposed), and to afford a standard for correlation for a certain part of the geologic column; e.g. an auxiliary section of particular regional or facies significance, established through correlation with the type section, and from which lateral extension of the boundary horizons may be made more readily than from the type section. See also: *standard section.*

reference seismometer In seismic prospecting, a detector placed to record successive shots under similar conditions, to permit overall time comparisons. It is used in connection with the shooting of wells for velocity measurements.

reference spheroid A theoretical figure whose dimensions closely approach the dimensions of the geoid, and who[se] exact dimensions are determined by various considerations [of] the section of the Earth's surface concerned. Cf: *ellipso[id]* Syn: *reference ellipsoid.*

reference station A place where tidal constants previous[ly] have been determined and which is used as a standard for th[e] comparison of simultaneous observations at a second statio[n.] Also, a place where independent daily predictions are given [in] the tide and tidal-current tables, from which correspondi[ng] predictions are obtained for other stations by means of diffe[r]ences or factors (CERC, 1966, p.A26). Cf: *tide station.* Sy[n:] *standard point.*

referencing The process of measuring the horizontal ([and] slope) distances and directions from a survey station to nea[r]by landmarks, reference marks, and other permanent objec[ts] which can be used in the recovery or relocation of the station.

refikite A white, very soft mineral occurring in modern resin[.] $C_{20}H_{32}O_2$ (?). Syn: *reficite.*

reflectance The ratio of reflected radiant flux to incident rad[i]ant flux. Symbol: ρ. See also: *reflectivity.*

reflected buried structure The distortion of surface beds th[at] reflect a similar structural distortion of underlying formations.

reflected wave An elastic wave which has been reflected on[e] or more times at an interface or boundary between two elast[ic] media. Such a wave may become a *transformed wave.* It [is] indicated by such symbols as SS, SP, PSS, etc.

reflecting goniometer *reflection goniometer.*

reflecting horizon In seismic profiling of the ocean floor, [a] major layer of *reflection.* It may be either sedimentary (cher[t]) or igneous (basalt). Three layers are distinguished: *horizon* [A,] *horizon beta,* and *horizon B.*

reflection The return of a wave incident upon a surface to [its] original medium. See also: *law of reflection; total reflectio[n.]* Also, in seismic prospecting, the indication on a record [of] such reflected energy. Cf: *refraction; diffraction.*

reflection angle *Bragg angle.*

reflection coefficient [elect] *power reflection coefficient.*

reflection coefficient [seis] The ratio of the amplitude of th[e] reflected wave to that of the incident wave. The ratio of th[e] reflected energy to the incident energy is the reflection coeff[i]cient squared. Syn: *reflectivity; reflectance.*

reflection goniometer A *goniometer* that measures the angle[s] between crystal faces by reflection of a parallel beam of lig[ht] from the successive crystal faces. Cf: *contact goniomete[r;]* *two-circle goniometer.*

reflection plane *plane of mirror symmetry.*

reflection pleochroism *bireflectance.*

reflection profile A seismic profile obtained by designing th[e] spread geometry in such a manner as to enhance reflecte[d] energy. Cf: *refraction profile.* See also: *subbottom profile.*

reflection shooting A type of *seismic shooting* based on th[e] measurement of the travel times of waves which, originatin[g] from an artificially produced disturbance, have been reflecte[d] back to detectors from subsurface boundaries separatin[g] media of different elastic-wave velocities. Cf: *refraction shoo[t]ing.*

reflection spectrum The spectrum seen when incident wave[s] are selectively altered by a reflecting substance.

reflection twin A crystal twin whose symmetry is formed b[y] apparent mirror image across a plane. Cf: *rotation twin.*

reflectivity (a) *Reflectance* of a layer of material of such [a] thickness that there is no change of reflectance with increas[e] in thickness. It is a material property, as distinguished fro[m] reflectance which refers to the property of an actual sample[.] Symbol: ρ. (b) The numeric of albedo; a fundamental proper[ty] of a material that has a specular surface and is sufficient[ly] thick to be opaque.

reflectometer An apparatus for measuring the reflectivity of [a] substance, using some form of radiant energy such as light.

reflector A layer or horizon that reflects seismic waves.

flexed Said of a graptoloid rhabdosome with stipes extending above the sicula and their initial parts enclosing an angle as than 180 degrees between their dorsal sides and distal parts tending to the horizontal. Cf: *reclined; deflexed; defined.*

foliation A foliation that is subsequent to and oriented differently from an earlier foliation.

fraction The deflection of a ray of light or of an energy wave such as a seismic wave) due to its passage from one medium to another of differing optical densities, which changes its velocity. Cf: *reflection; diffraction.* See also: *single refraction; refringence.*

fraction angle *angle of refraction.*

fraction profile A seismic profile obtained by designing the read geometry in such a manner as to enhance refracted energy. Cf: *reflection profile.*

fraction shooting A type of seismic shooting based on the measurement of seismic energy as a function of time after the shot and of distance from the shot, by determining the arrival times of seismic waves which have traveled nearly parallel to the bedding in high-velocity layers, in order to map the depth to such layers (Sheriff, 1968). Cf: *reflection shooting.*

fractive index *index of refraction.*

fractive power *refractivity.*

fractivity The power a substance has to refract light. Such ability can be quantitatively expressed by the index of refraction. See also: *specific refractivity.* Syn: *refractive power; refringence.*

fractometer An apparatus for measuring the indices of refraction of a substance, either solid or liquid. Various types are designed for various substances; the chief type used for analysis of gems and minerals is the *Abbe refractometer.*

fractometry The measurement of indices of refraction, by means of a *refractometer.*

fractory adj. Said of an ore from which it is difficult or expensive to recover its valuable constituents.

fractory clay *fireclay.*

fringence *refractivity.*

efugian North American stage: Eocene and Oligocene (above Fresnian, below Zemorrian).

fugium (a) A sufficiently isolated area that underwent little environmental change, permitting a fauna or flora to persist locally long after it had been exterminated elsewhere. Syn: *asylum.* (b) A restricted area in which plants and animals persisted during a period of continental climatic change that made surrounding areas uninhabitable; esp. an ice-free or unglaciated area within or close to a continental ice sheet or upland ice cap, where hardy biotas eked out an existence during a glacial phase. It later served as a center of dispersal for the repopulation of surrounding areas after climatic readjustment. : *refugia.* Syn: *glacial refuge.*

g An extensive, nearly level, low desert plain from which the sand has been removed by the wind, leaving a sheet of coarse, smoothly angular, wind-polished gravel and small stones lying on an alluvial soil and strongly cemented by mineralized solutions to form a broad *desert pavement*; a *stony desert* of the plains, as in the Algerian Sahara and parts of American deserts. Etymol: Hamitic. Pl: *regs.* See also: *serir; hammada.* Syn: *gravel desert.*

egelation A two-fold process involving the melting of ice under excess pressure (*pressure melting*) and the refreezing of the derived meltwater upon release of that pressure. The term is sometimes restricted to the refreezing part of the process, but in some European literature it has been applied to often-repeated freezing and thawing (or *multigelation*).

egenerated anhydrite Anhydrite produced by the dehydration of gypsum that was itself formed by the hydration of anhydrite (Goldman, 1961).

egenerated crystal A large crystal that has grown in a mass of crushed material, such as mylonite. Se also: *integration.*

regenerated flow control Control of glacial drainage by modified morainal features, resulting from the readvance of a previously stagnant glacier (Gravenor & Kupsch, 1959, p.56).

regenerated glacier (a) *glacier remanié.* (b) A glacier that becomes active after a period of stagnation.

regenerated rock A clastic rock. The term "regenerirte Gesteine" was used by Zirkel (1866, p.3).

regeneration The renewal, regrowth, or restoration of a body or of a part, tissue, or substance of a body following injury or as a normal bodily process.

regime [geol] A regular or systematic pattern of occurrence or action, or a condition or style having widespread influence; e.g. a sedimentation regime, a tectonic regime, a periglacial regime, a hydrologic regime. See also: *flow regime.*

regime [glaciol] *balance.*

regime [streams] (a) The existence in a stream channel of a balance or grade between erosion and deposition over a period of years. (b) The condition of a stream with respect to the rate of its average flow as measured by the volume of water passing different cross sections in a specified period of time. In this unspecialized sense, the term is incorrectly used as a syn. of regimen.---Etymol: French régime.

regime channel The channel of a regime stream; a channel that is neither scoured nor filled.

regimen [water] The characteristic behavior and the total quantity of water involved in a drainage basin, determined by measuring such quantities as rainfall, surface and subsurface storage and flow, and evapotranspiration. Syn: *hydrologic regimen; water regimen.*

regimen [glaciol] *balance.*

regimen [lake] An analysis of the total quantity of water involved with a lake over a specified period of time (usually a year), including water losses (seepage, evaporation, transpiration, outflow, diversion) and water gains (precipitation, inflow, ground-water migration, water pumped or drained into the lake basin) (Veatch & Humphrys, 1966, p.172). Syn: *hydrologic regimen.*

regimen [streams] The flow characteristics of a stream; specif. the habits of an individual stream (including low flows and floods) with respect to such quantities as velocity, volume, form of and changes in the channel, capacity to transport sediment, and amount of material supplied for transportation. Cf: *regime [streams].*

regime stream A stream with a mobile (erodible) boundary, making at least part of its boundary from its transported load and part of its transported load from its boundary, carrying out the process at different places and times in a balanced or alternating manner that prevents unlimited growth or removal of the boundary (Blench, 1957, p.2). Cf: *graded stream.*

regime theory A theory of the formation of a channel in material carried and deposited by its stream.

region [geog] A very large expanse of land usually characterized or set apart by some aspect such as its being a political division or area of similar geography.

region [ecol] A major division of the Earth having distinctive climatic and topographic features and floral and faunal provinces. Cf: *realm.*

regional [geol] A general term applied in geology to a geologic feature or process that occurs over a relatively large area. It is contrasted with *local.*

regional [geophys] n. In geophysics, a contribution to the observed gravity or magnetic anomaly due to density irregularities at much greater depths than those of the possible structures, the location of which was the purpose of the survey (Nettleton, 1940, p.125). Cf: *residual [geophys].* Adj. said of such an anomaly or gradient, e.g. regional gravity.

regional correlation Correlation of rock units, major structures, or other geologic features over or across wide areas of the Earth's surface.

regional dip The nearly uniform inclination of strata over a

597

wide area, generally at a low angle, as in the Atlantic and Gulf coastal plains and parts of the Midcontinent region. Cf: *homocline*. Syn: *normal dip*.

regional geology The geology of any relatively large region, treated broadly and primarily from the viewpoint of the spatial distribution and position of stratigraphic units, structural features, and surface forms. Cf: *areal geology*.

regional gravity map A gravity map showing only gradual changes of gravity.

regional metamorphism A general term for metamorphism affecting an extensive region, as opposed to *local metamorphism* that is effective only in a restricted area. As introduced in the nineteenth century, the term covered only those changes due to deep burial metamorphism; it has, however, been used almost synonymously with *dynamothermal metamorphism* (Holmes, 1920). Cf: *dynamic metamorphism*.

regional metasomatism Metasomatic processes affecting extensive areas whereby the introduced material may be derived from partial fusion of the rocks involved from deep-seated magmatic sources. Cf: *contact metasomatism*.

regional overstep A term proposed by Swain (1949, p.634) for an *overstep* in which an unconformity occurs "widespread, but not universally, over very large parts of a craton (platform, shelf)".

regional snowline The level above which, averaged over a large area, snow accumulation exceeds ablation year after year. See also: *climatic snowline*. Cf: *snowline*.

regional unconformity An unconformity that extends continuously throughout an extensive region that may be nearly continent-wide in extent and that usually represents a relatively long period. Cf: *local unconformity*.

register mark A small figure, cross, circle, or other pattern at each corner of a map that is to be printed in more than one color. The accuracy of printing of each color is checked by synchronization of the register marks on each printing plate.

regmagenesis The diastrophic production of regional strike-slip displacements. Also spelled: *rhegmagenesis*. Adj: *regmatic*.

regmaglypt Any of various small, well-defined, characteristic indentations or pits on the surface of meteorites, frequently resembling the imprints of fingertips in soft clay. They are often polygonal, sometimes round, almond-shaped, or elliptic; their diameters range from a few millimeters to many centimeters. Syn: *piezoglypt; pezograph*.

regmatic Adj. of *regmagenesis*; said of strike-slip displacements, e.g. regmatic pattern.

regolith A general term for the entire layer or mantle of fragmental and loose, incoherent, or unconsolidated rock material, of whatever origin (residual or transported) and of very varied character, that nearly everywhere forms the surface of the land and overlies or covers the more coherent bedrock. It includes rock debris (weathered in place) of all kinds, volcanic ash, glacial drift, alluvium, loess and eolian deposits, vegetal accumulations. and soils. Term originated by Merrill (1897, p.299). Etymol: Greek *rhegos*, "blanket", + *lithos*, "stone". See also: *soil; lunar regolith*. Syn: *mantle; mantle rock; rock mantle; overburden; rhegolith*.

Regosol In early U.S. classification systems, one of an azonal group of soils that develops from deep, unconsolidated deposits and that has no definite genetic horizons.

regradation The formation by a stream of a new profile of equilibrium, as when the former profile, after *gradation*, became deformed by crustal movements.

regrading stream A stream that is simultaneously upbuilding (aggrading) and downcutting (degrading) along different parts of its profile.

regression [evol] (a) A hypothetical reversal in the direction of evolution that is sometimes used to explain certain paleontologic phenomena such as the extinction of the graptolites. (b) The trend exhibited by offspring, in respect to their inherited characteristics, away from specializations exhibited by the parents and toward the mean development of their biotype.

regression [stat] The fact that in associated variables on selecting one variable with a given value the second has (on the average) a less value and regresses toward the value for the mean of all variables of the class.

regression [stratig] The retreat or contraction of the sea from land areas, and the consequent evidence of such withdraw (such as enlargement of the area of deltaic deposition). Also any change (such as fall of sea level or uplift of land) that brings nearshore, typically shallow-water environments areas formerly occupied by offshore, typically deep-water conditions, or that shifts the boundary between marine and nonmarine deposition (or between deposition and erosion) toward the center of a marine basin. Ant: *transgression*. Cf: *offlap*.

regression [streams] The name given to the theory that some rivers have their sources on the rainier sides of mountain ranges and gradually erode their heads backward until the ranges are cut through.

regression analysis A statistical technique applied to paired data to determine the degree or intensity of mutual association of a dependent variable with one or more independent variables.

regression coefficient A coefficient in a regression equation; the slope of the regression line.

regression conglomerate A coarse sedimentary deposit formed during a retreat (recession) of the sea.

regression curve A curve that best fits particular data according to some principle. See also: *regression line*.

regression equation An experimentally determinable equation of a regression curve; e.g. an approximate, generally linear relation connecting two or more quantities and derived from the correlation coefficient.

regression line A *regression curve* that is a straight line; the line or curve from a family of curves that best fits the empirical relation between a dependent variable and an independent variable.

regressive diagenesis *hypergenesis*.

regressive overlap *offlap*.

regressive reef One of a series of nearshore reefs or bioherms superimposed on basinal deposits during the rising of a landmass or the lowering of the sea level, and developed more or less parallel to the shore (Link, 1950); e.g. the Capitan reef during the Permian in Texas and New Mexico. Cf: *transgressive reef*.

regressive ripple A term used by Jopling (1961) for an asymmetric ripple mark formed by a current but oriented in direction opposite to the general movement of current flow (steep side facing upcurrent).

regressive sand wave A term proposed by Bucher (1919, p.165) to replace *antidune* as used by Gilbert (1914, p.31). Ant: *progressive sand wave*.

regressive sediments Sediments deposited during the retreat or withdrawal of water from a land area or during the emergence of the land, and characterized by an offlap arrangement.

regular In paleontology, pertaining to an echinoid of the Regularia division having a more or less globular symmetrical shell with 20 meridional rows of plates and displaying an *endocyclic* test in which the periproct is located within the oculogenital ring. Cf: *irregular*.

regular dissepimentarium A *dissepimentarium* in rugose corals in which the dissepiments are developed only in spaces between major septa and minor septa.

regular dodecahedron *pyritohedron*.

regulation Artificial manipulation of stream flow.

regur One of a group of calcareous, intrazonal soils characterized by dark color, and a high clay content. It is formed from rocks low in silica, as on the volcanic Deccan plateau of dia. Etymol: Hindi. Syn: *black cotton soil.*

reh (a) A mixture of soluble sodium salts that rise by capillary action and effloresce at the surface of the ground in arid and semiarid regions in India (as in the Indo-Gangetic region). The term has also been used for a limy deposit (caliche) in the soil. (b) A saline soil rendered worthless for cultivation by efflorescence of reh.----Etymol: Hindi. See also: *usar.* Syn: *kallar.*

reibungsbreccia A syn. of *fold breccia.* Etymol: German *reibung,* "rubbing, friction".

Reichenbach's lamellae Thin platy inclusions of foreign minerals (usually troilite, schreibersite, or chromite) occurring in iron meteorites. Named after Karl von Reichenbach (1788-1869), German chemist.

reid mechanism *elastic rebound theory.*

reinerite A pale yellow-green mineral: $Zn_3(AsO_3)_2$. Not to be confused with renierite.

rejected recharge Water that infiltrates to the water table but then discharges because the aquifer is full to overflowing and cannot accept it.

rejuvenated Said of a structural feature, e.g. a fault scarp, along which the original stress has been renewed. Syn: *revived.*

rejuvenated fault scarp A fault scarp freshened by renewed movement along an old fault line after the initial scarp had been partly dissected or eroded. Syn: *revived fault scarp.*

rejuvenated stream A stream that, after having developed to maturity or old age, has reverted to the activities and forms of a more youthful stage as a result of rejuvenation. It may be characterized by entrenched meanders, stream terraces, and meander cusps. Syn: *revived stream.*

rejuvenated water Water returned to the terrestrial water supply as a result of compaction and metamorphism. It is divided into *water of compaction* and *metamorphic water.*

rejuvenation (a) The action of stimulating a stream to renewed erosive activity, as by uplift or by a drop of sea level; the renewal or restoration of youthful vigor in a stream that has attained maturity or old age. (b) The development or restoration of youthful features of a landscape or landform in an area previously worn down nearly to base level, usually caused by regional uplift or eustatic movements, followed by renewed downcutting by streams; a change in conditions of erosion, leading to the initiation of a new cycle of erosion. (c) The renewal of any geologic process, such as the reactivation of a fissure.---Syn: *revival.*

rejuvenation head A *knickpoint* resulting from rejuvenation or from an uplift.

relative abundance The number of individuals of a taxon in comparison with the number of individuals of other taxa in a certain area or volume. See also: *abundance; absolute abundance.*

relative age The *geologic age* of a fossil organism, rock, or geologic feature or event defined relative to other organisms, rocks, or features or events rather than in terms of years. Cf: *absolute age.*

relative aperture The diameter of the stop, diaphragm, or other physical element that limits the size of the bundle of rays traversing an optical instrument from a given point. It is expressed as a fraction of the focal length of the camera lens, with the symbol *f* being used instead of 1 as the numerator; e.g. a lens whose relative aperture is 1/4.5 of its focal length has a relative aperture of *f*/4.5 or *f*:4.5. See also: *f-number; speed.*

relative chronology *Geochronology* in which the time-order is based on superposition and/or fossil content rather than on an age expressed in years (*absolute chronology*).

relative consistency *consistency index.*

relative dating The proper chronological placement of a feature, object, or happening in the *geologic time scale* without reference to its absolute age.

relative density The ratio of the difference between the void ratio of a cohesionless soil in the loosest state and any given void ratio to the difference between its void ratios in the loosest and in the densest states (ASCE, 1958, term 296).

relative dispersion *dispersive power.*

relative fugacity The ratio of the fugacity in a given state to the fugacity in a defined standard state; *activity.*

relative gravity instruments Devices for measuring the differences in the gravity force or acceleration at two or more points. They are of two principal types: a static type in which a linear or angular displacement is observed or nulled by an opposing force (a gravimeter), and a dynamic type in which the period of oscillation is a function of gravity and is the quantity directly observed. In another type the gravity-field distortion is measured, e.g., the Eötvös torsion balance. Cf: *absolute gravity instruments.*

relative humidity The ratio, expressed as a percentage, of the actual amount of water vapor in a given volume of air to the amount that would be present if the air were saturated at the same temperature. See also: *saturation* [*meteorol*]. Cf: *absolute humidity; specific humidity.*

relative index of refraction An *index of refraction* that is the ratio of the velocity of light in one crystal to that of another crystal.

relative refractive index *relative index of refraction.*

relative relief (a) *local relief.* (b) Within a drainage basin, the ratio of *basin relief* to *basin perimeter.* Symbol: R_{hp}. Cf: *relief ratio.*

relative tilt The *tilt* of a photograph with reference to an arbitrary plane (not necessarily a horizontal plane), such as to that of the preceding or subsequent photograph in a strip.

relative time *Geologic time* determined by the placing of events in a chronologic order of occurrence; esp. time as determined by organic evolution or superposition. Cf: *absolute time; fossil time.*

relative time scale An uncalibrated *geologic time scale,* based on both layered rock sequences and the paleontologic evidence contained therein, giving the relative order for a succession of events. Cf: *biologic time scale; atomic time scale.*

relative water content *liquidity index.*

relaxation [**geophys**] In an elastic medium, the decrease of elastic resistance under applied stress, resulting in permanent deformation. See also: *relaxation oscillation.*

relaxation [**exp struc geol**] In experimental structural geology, the release of applied stress with time, due to any of various creep processes.

relaxation oscillation A cyclical motion resulting from applied force that repeatedly drives a system to a form of instability such that the system reverts essentially to its original state, by *relaxation.*

relaxation time The time required for a substance to return to its normal state after release of stress. Se also: *rheidity.*

release adiabat A curve or locus of points which defines the succession of states through which a mass, shocked to a high-pressure state, passes while monotonically returning to zero pressure. The process operates over a short time interval compared with the characteristic time for heat flow in the material.

released mineral A mineral formed during the crystallization of a magma as a consequence of an earlier phase failing to react with the liquid. Thus the failure of earlier formed olivine to react with the liquid portion of a magma to form pyroxene may result in the enrichment of the liquid in silica, which finally crystallizes as quartz, the "released mineral".

release fracture A fracture developed as a consequence of the relief of stress in one particular direction. The term is generally applied to a fracture formed when the maximum principal stress decreases sufficiently that it becomes the minimum principal stress; the fracture is an extension fracture oriented perpendicular to the then minimum principal stress direction.

release joint *sheeting structure.*

relic [geomorph] A landform that has survived decay or disintegration (such as an *erosion remnant*) or that has been left behind after the disappearance of the greater part of its substance (such as a *remnant island*). The term is sometimes used adjectively as a synonym of *relict*, but this usage is not recommended.

relic [meta] Var. of *relict.*

relic [sed] A vestige of a particle in a sedimentary rock, such as a trace of skeletal material in a carbonate rock or an incompletely recrystallized mineral in a diagenetic rock.

relict [geomorph] adj. Said of a topographic feature that remains after other parts of the feature have been removed or have disappeared; e.g. a "relict beach ridge" or a "relict hill". Cf: *relic; residual.* Syn: *relicted.*---n. A relict landform.

relict [paleont] n. A remnant of an otherwise extinct flora or fauna or kind of organism that has persisted since the extinction of the rest of the group.----Adj. Said of a remnant of an extinct group.

relict [meta] adj. Pertaining to a mineral, structure, or feature of a rock that represents those of an earlier rock and which persist in spite of processes tending to destroy it, e.g. metamorphism.----n. Such a mineral, structure, or other feature. See also: *stable relict; unstable relict.*----Also spelled: *relic.* Cf: *palimpsest.*

relict aperture One of the short radial slits around the umbilicus of a planktonic foraminiferal test that remain open when the umbilical parts of the equatorial aperture are not covered by succeeding chambers (as in Planomalinidae) or that, even when secondarily closed, the elevated apertural lips or flanges remain visible around the umbilicus (as in *Planomalina* and *Hastigerinoides*) (TIP, 1964, pt.C, p.63).

relict dike In a granitized mass, a tabular body of crystalloblastic texture that represents a dike that was emplaced prior granitization and that was relatively resistant to the granitization process (Goodspeed, 1955, p.146).

relict glacier A remnant of an older and larger glacier.

reliction The slow and gradual withdrawal or recession of the water in the sea, a lake, or a stream, leaving the former bottom as permanently exposed and uncovered dry land; it does not include seasonal fluctuations in water levels. Legally, the added land belongs to the owner of the adjacent land against which it abuts. Also, the land left uncovered by reliction. Cf: *dereliction; accretion.*

relict lake A lake that survives in an area formerly covered by the sea or a larger lake, or a lake that represents a remnant resulting from a partial extinction of an original body of water; a lake that has become separated from the sea by gentle uplift of the sea bottom.

relict mountain *mountain of circumdenudation.*

relict permafrost Permafrost that was formed in the past and persists in places where it could not form today (Hopkins et al, 1955).

relict sediment A sediment that had been deposited in equilibrium with its environment, but that is now unrelated to its present environment even though it remains unburied by later sediments; e.g. a land-laid or shallow-marine sediment occurring in deep water (as near the seaward edge of the continental shelf).

relict texture In mineral deposits, an original texture that remains after partial replacement.

relief [geomorph] (a) A term used loosely for the actual physical shape, configuration, or general unevenness of a part of the Earth's surface, considered with reference to variations of height and slope or to irregularities of the land surface; the elevations or differences in elevation, considered collectively, of a land surface. The term is frequently confused with *topography*, although the use of the two terms in the sense of surface configuration is "thoroughly established both in general speech and in technical geomorphological literature" in the

U.S. (C.D. Harris, in Stamp, 1961, p.454). Syn: *topograph... relief.* (b) The vertical difference in elevation between the hi... tops or mountain summits and the lowlands or valleys of ... given region. A region showing a great variation in elevatio... has "high relief", and one showing little variation has "low r... lief". See also: *local relief; available relief.*

relief [crystal optics] An apparently rough surface of a crys... section under the microscope. Such an appearance is assoc... ated with the relative indices of refraction of the crystal an... its mounting medium: the relief is positive if the refractiv... index of the mineral is greater than that of the medium, ... negative in the reverse case. Syn: *shagreen.*

relief feature *landform.*

relief limonite *Indigenous limonite* that is porous and cavern... ous in texture.

relief map A map that depicts the surface configuration or re... lief of an area by any method, such as by use of contour line... (contour map) and hachures, by hill shading (shaded-reli... map), by photography (photorelief map), by layer tinting, b... pictorial symbols (physiographic diagram), by molding plast... in three dimensions (plastic relief map), or by a combinatio... of these methods. Cf: *relief model.*

relief model A three-dimensional representation of the phys... cal features or relief of an area, in any size or medium, b... not necessarily constructed to true scale (the vertical scale... generally exaggerated to accentuate the relief). Cf: *reli... map.*

relief ratio Within a drainage basin, the ratio of *basin relief... basin length*; it is a measure of the overall steepness of th... basin and the intensity of erosion on its slopes. Symbol: R... Cf: *relative relief.*

relief shading *hill shading.*

relief well A well used to relieve excess hydrostatic pressur... as to reduce waterlogging of soil or to prevent blowouts o... the land side of levees during floods in the adjacent stream... Cf: *drainage well.*

Relizian North American stage: Miocene (above Saucesia... below Luisian).

remainder mountain *mountain of circumdenudation.*

remanent magnetization That component of a rock's *magnet... zation* that has a direction fixed relative to the rock and whic... is independent of moderate, applied magnetic fields such a... the Earth's magnetic field. Cf: *induced magnetization.* Se... also: *hysteresis; natural remanent magnetization.*

remanié [geol] adj. A French word meaning "reworked" o... "available for rehandling", and applied in geology to fragment... "derived from older materials", esp. to fossils removed fro... or washed out of an older bed and redeposited in a new one... Also applied to boulders in a glacial till, pebbles in a conglom... erate, country rock engulfed in a batholith, etc. Cf: *reworke... derived.* Anglicized version sometimes spelled: *remanie.*----n... A fragment of an older formation incorporated in a younge... deposit.

remanié [glaciol] A French term meaning "reworked" but ap... plied to glaciers that have been "recemented" or "recor... structed", as *glacier remanié*, or the Anglicized version *re... manié glacier.*

remnant *erosion remnant.*

remolded soil Soil that has had its natural internal structur... modified or disturbed by manipulation so that it lacks shea... strength and gains compressibility.

remolding Disturbance of the internal structure of clay or co... hesive soil.

remolding index The ratio of the modulus of deformation of ... soil in the undisturbed state to that of a soil in the remolde... state.

remolding sensitivity *sensitivity ratio.*

remolinite *atacamite.*

remote sensing (a) The measurement or acquisition of infor... mation of some property of an object or phenomenon, by ...

ecording device that is not in physical or intimate contact with the object or phenomenon under study. The technique employs such devices as the camera, lasers, infrared and ultraviolet detectors, microwave and radio frequency receivers, adar systems, etc. (b) The practice of data collection in the wavelengths from ultraviolet to radio regions. This restricted sense is the practical outgrowth from airborne photography. It sometimes called *rapid reconnaissance*.

enardite A yellow mineral: $Pb(UO_2)_4(PO_4)_2(OH)_4.7H_2O$.

endoll In U.S. Dept. of Agriculture soil taxonomy, a suborder f the soil order *Mollisol*, characterized by the absence of an rgillic or calcic horizon and the presence of 40% or more aCO_3 equivalent within or just below a mollic epipedon SSSA, 1970). Cf: *Alboll; Aquoll; Boroll; Udoll; Ustoll; Xeroll*.

endzina soil One of an intrazonal, calcimorphic group of soils aving a brown or black, friable surface horizon and a light ray or yellow, soft, calcareous underlying horizon. It is developed under grasses or grasses with forest, in a humid to emiarid climate.

enewed consequent stream *resequent stream*.

enierite A mineral: $Cu_3(Fe,Ge,Zn)(S,As)_4$. Not to be confused with reinerite.

eniform Kidney-shaped. Said of a crystal structure in which adiating crystals terminate in rounded masses; also said of nineral deposits having a surface of rounded, kidneylike hapes. Cf: *colloform; botryoidal*.

ensselaerite A soft, compact, fibrous talc pseudomorphous fter pyroxene and found in Canada and northern New York. t is harder than talc, takes a good polish, and is often made nto ornamental articles.

epeated reflection *multiple reflection*.

epeated twinning Crystal twinning that involves more than wo simple crystals; it may be *cyclic twinning* or *polysynthetic winning*. See also: *multiple twin*.

epetition The duplication of certain stratigraphic beds at the urface or in any specified section owing to disruption and lisplacement of the beds by faulting or intense folding. Ant: *mission*.

Repettian North American stage: lower Pliocene (above Delnontian, below Venturian).

epi A group term for "all lakes, ponds, or other \standing vater bodies related to sinks, or to subsidence, of land surace" (Veatch & Humphrys, 1966, p.264). Etymol: Greek.

eplacement [paleont] Substitution of inorganic matter for the riginal organic constituents of an organism during fossilizaion.

eplacement [meta] The process of practically simultaneous capillary solution and deposition by which a new mineral of artly or wholly differing chemical composition may grow in he body of an old mineral or mineral aggregate (Lindgren, 932).

eplacement [stratig] The gradual movement of the sea either oward or away from land areas, such as "marine replacement" (or transgression) and "continental replacement" (or egression).

eplacement dike A dike formed by gradual transformation of vall rock by solutions along fractures or permeable zones Goodspeed, 1955, p.146).

eplacing overlap A term, now obsolete, used by Grabau 1920, p.398) for a nonmarine overlap involving a receding horeline, occurring where continental sediments are deposited and progressively "replace" the corresponding and all but contemporaneous marine sediments into which they grade (as lescribed by Grabau, 1906, p.628-629). The misuse of *overap* in this sense for a facies change from marine to continenal sediments of the same age is confusing and "ungeological" Lovely, 1948, p.2295).

eplat (a) A French term for a horizontal surface (such as a bench, shelf, or shoulder), wider than a ledge, occurring along the steep side of a U-shaped valley (Stamp, 1961,

p.391). (b) A French term for a stretch of relatively horizontal land interrupting a slope.

replenishment [speleo] The stage in development of a cavern in which the presence of air in the passages allows the deposition of speleothems.

replenishment [grd wat] *recharge*.

repose imprint A term used by Kuenen (1957, p.232) for a sole mark formed by an animal lying on or taking cover in bottom sediment.

representative fraction The *scale* of a map, expressed in the form of a numerical fraction that relates linear distances on the map to the corresponding actual distances on the ground, measured in the same unit (centimeters, inches, feet); e.g. "1/24,000" indicated that one unit on the map represents 24,000 equivalent units on the ground. Abbrev: R.F. Syn: *natural scale*.

represo *charco*.

reproduction *synthetic stone*.

reptant (a) Creeping or prostrate; esp. said of a corallite with a creeping habit, growing attached along one side to some foreign body. Syn: *reptoid*. (b) Said of a bryozoan colony consisting of largely separate zooecial tubes that lie attached to the substrate.

reptation A syn. of *surface creep* (Scheidegger, 1961, p.290).

reptilian age An informal designation of the Mesozoic; *age of reptiles*.

resaca A term applied in SW U.S. to a long, narrow, meandering lake occupying the bed of a former stream channel; a series of connected bancos. Also, the dry channel or the former marshy course of a stream, now containing a resaca. Etymol: American Spanish, from Spanish *resacar*, "to redraw".

resection (a) A method in surveying by which the horizontal position of an occupied point is determined by drawing lines from the point to two or more points of known position. The most usual problem in resection is the *three-point problem* when three known positions are observed to locate the occupied station. Cf: *intersection*. (b) A method of determining a plane-table position by orienting along a previously drawn foresight line and drawing one or more rays through the foresight from previously located stations.

resedimentation (a) Sedimentation of material dervied from a preexisting sedimentary rock; *redeposition* of sedimentary material. (b) Mechanical deposition of material in cavities of postdepositional age, such as the deposition of carbonate muds and silts by internal mechanical erosion or solution of a limestone. (c) The general process of subaqueous, downslope movement of sediment under the influence of gravity, such as the formation of a turbidity-current deposit.

resedimented rock (a) A rock consisting of reworked sediments. (b) A turbidity-current deposit; e.g. a flysch or other similar graywacke, showing graded bedding, and alternating with shales in a thick sequence.

resequent [geomorph] Said of a geologic or topographic feature that resembles or agrees with a *consequent* feature but that developed from such a feature at a later date; esp. said of a tilt-block mountain (or of a rift-block mountain) that is similar in form to the original tilt block (or horst) but that is shaped by differential erosion after the original topography was destroyed and uplifted, or said of a tilt-block valley (or of a rift-block valley) that coincides with the site of the former valley (or graben) after the original topography was destroyed by erosion. Ant: *obsequent*.

resequent [streams] adj. Said of a stream, valley, or drainage system whose course or direction follows an earlier pattern but on a newer and lower surface, as in an area of ancient folding subjected to long-continued erosion. Etymol: re + *consequent*. Syn: *reconsequent*.---n. *resequent stream*.

resequent fault-line scarp A *fault-line scarp* that faces in the same direction as the original fault scarp (i.e. facing the downthrown block) or in which the structurally downthrown block is also topographically lower than the upthrown block. Cf: *obsequent fault-line scarp.*

resequent stream A stream that flows down the dip of underlying strata in the same direction as an original consequent stream but developed later at a lower level than the initial slope (as on formerly buried resistant strata) and generally tributary to a subsequent stream; e.g. a stream flowing down the back slope of a cuesta. Syn: *resequent; renewed consequent stream.*

resequent valley A valley eroded by or containing a *resequent stream*; a consequent valley produced by a new adjustment to structure at a deeper level.

reservoir [paleont] The enlarged posterior part of the gullet in some motile cells in protists such as Cryptophyceae and Euglenophyta.

reservoir [oil] A subsurface accumulation of crude oil or natural gas under adequate trap conditions.

reservoir [water] An artificial or natural storage place for water, such as a lake or pond, from which the water may be withdrawn as for irrigation, municipal water supply, or flood control.

reservoir [grd wat] *ground-water reservoir.*

reservoir gas-oil ratio The number of cubic feet of gas per barrel of oil originally in the reservoir. See also: *gas-oil ratio.*

reservoir pressure The pressure on fluids (water, oil, gas) in a subsurface formation, or the pressure under which fluids are confined in rocks; esp. the pressure within the solid structure of an oil pool, such as the *bottom-hole pressure* at the face of an oil-producing formation when the oil well is closed down.

reservoir rock In petroleum geology, any rock with adequate porosity or joint and fracture systems to contain liquid or gaseous hydrocarbons. Sandstones and limestones are the usually encountered reservoir beds, but accumulation in fractured igneous rocks is not unknown.

residual [ore dep] adj. Said of a mineral deposit formed by mechanical concentration, e.g. a placer, or by chemical concentration, e.g. bauxite, limonite.

residual [geomorph] adj. Said of a topographic or geologic feature (such as a rock, hill, mountain, or plateau) that represents a small part or trace of a formerly greater mass or area, and that is remaining above the surrounding surface which has been reduced by erosion. Cf: *relict.*---n. *erosion remnant.*

residual [geophys] n. In geophysics, that which is left after the regional has been subtracted; a field from which gross effects have been subtracted, in order to emphasize local anomalies. Cf: *regional* [geophys].----adj. Said of such an anomaly or gradient, e.g. residual gravity.

residual [stat] *residual error.*

residual [weath] adj. Pertaining to or constituting a *residue*; esp. said of material ultimately left after the weathering of rock in place, such as a *residual deposit* or a *residual soil.* Syn: *residuary; eluvial.*

residual anticline In salt tectonics, a relative structural high that is created as the result of the depression of two adjacent *rim synclines.* Syn: *residual dome.*

residual boulder *boulder of weathering.*

residual clay Extremely finely divided clay material formed in place by the weathering of rock, derived either by the chemical decay of feldspar and other rock minerals or by the removal of nonclay-mineral constituents by solution from a clay-bearing rock (such as an argillaceous limestone); a soil or a product of the soil-forming processes. Cf: *secondary clay.* Syn: *primary clay.*

residual compaction The difference between the amount of compaction that will occur ultimately for a given increase in applied stress, and that which has occurred at a specified time (Poland, et al, in press).

residual deposit (a) The *residue* formed by weathering in

place. (b) An ore deposit formed in clay by the conversion metallic compounds (as of manganese, iron, lead, or zin into oxidized forms by weathering at or near the Earth's su face.

residual dome *residual anticline.*

residual error The difference between any measured value a quantity in a series of observations (corrected for know systematic errors) and the computed value of the quantity o tained after the adjustment of that series. In practice, it is t residual errors that enter into a computation of probable errc Syn: *residual* [stat].

residual geosyncline *autogeosyncline.*

residual kame A ridge or mound of sand or gravel formed the denudation of glaciofluvial material that had been deposi ed in glacial lakes or on the flanks of hills of till (Gregor 1912, p.175). Syn: *pseudokame.*

residual liquid A term used for the volatile components of magma that remain in the magma chamber after much cry tallization has taken place.

residual liquor *rest magma.*

residual magma *ichor.*

residual map A stratigraphic map that displays the small-sca variations (such as local features in the sedimentary enviro ment) of a given stratigraphic unit (Krumbein & Sloss, 196 p.486). It is superimposed on the underlying pattern of a tren map.

residual material Unconsolidated or partly weathered pare material of a soil, presumed to have developed in place (b weathering) from the consolidated rock on which it lies; it the material from which soils are formed. See also: *residu soil.* Cf: *transported soil material; cumulose.*

residual mineral A mineral that has been concentrated place by weathering and leaching of rock; e.g. quartz.

residual rays *reststrahlen.*

residual sediment *resistate.*

residual soil A soil that developed from *residual material.* Th term is obsolete; the adjective "residual" is more correct applied to the parent material (USDA, 1957, p.766). Syn: *se entary soil.*

residual swelling "The difference between the original pre freezing level of the ground and the level reached by the se tling after the ground is completely thawed" (Muller, 1947 p.221).

residual valley An intervening trough between uplifted mour tains, as in the Basin and Range Province of western U.S (Gilbert, 1875, p.63).

residue (a) An accumulation of rock debris formed by weath ering and remaining essentially in place after all but the leas soluble constituents have been removed, usually forming comparatively thin surface layer concealing the unweathere or partly altered rock below; e.g. a soil. See also: *chemica residue; physical residue.* Syn: *residuum; residual deposit; e uvium.* (b) *insoluble residue.*

residuite The translucent *residuum* that occurs in clarain.

residuo-aqueous sand A term used by Sherzer (1910, p.627 for a sand containing water-rounded particles that were sub sequently subjected to weathering. Cf: *aqueo-residual sand.*

residuum [coal] In microscopic analysis of coals, the struc tureless groundmass of microscopically identifiable constitu ents, consisting of particles of one to two microns or less usually opaque, and of a dark color. It is the same as the lower range of fine *micrinite.* See also: *desmite; residuite.*

residuum [weath] *residue.*

resilience The ability of a material to store the energy of elas tic strain. This ability is measured in terms of energy per uni volume.

resilifer A spoon-shaped recess or process on the hinge plate of some bivalve mollusks (as in *Mactra*) to which the resilium is (was) attached or by which it is (was) supported. See also *chondrophore.* Syn: *resiliifer.*

resilium The internal ligament within the hinge line of a bivalve mollusk, compressed by the hinge plate when the valves are closed. It resembles cartilage in consistency but is in fact chitinous, although the term is applied irrespective of composition. Pl: *resilia*.

resin Any of various hard, brittle, solid or semisolid, usually transparent or translucent, and mainly amorphous substances formed esp. in plant secretions and obtained as exudates of recent or fossil origin (as from pine or fir trees, or from certain tropical trees) by the condensation of fluids on a loss of volatile oils. Resins are yellowish to brown with a characteristic luster; they are fusible and flammable, are soluble in ether and other organic solvents but not in water, and represent a complex mixture of terpenes, resin alcohols, and resin acids and their esters. See also: *fossil resin; mineral resin.* Syn: *natural resin.*

resin canal *resin duct.*

resin duct A long, narrow intercellular canal in wood, surrounded by one or more layers of parenchyma cells of the epthelium (Record, 1934, p.72). See also: *resin rodlet.* Syn: *resin canal.*

resinite A variety of *exinite* consisting of resinous compounds, often in elliptical or spindle-shaped bodies representing cell-filling matter or resin rodlets. Cf: *alginite; cutinite; sporinite.*

resin jack *rosin jack.*

resinoid A maceral group that includes the macerals in the resinite series.

resin opal A wax-, honey-, or ocherous-yellow variety of common opal with a resinous luster or appearance.

resinous coal Coal, usually younger coal, in which the attritus contains a large proportion of resinous material.

resinous luster The luster on the fractured surfaces of certain minerals (such as opal, sulfur, amber, and sphalerite) and rocks (such as pitchstone) that resemble the appearance of resin.

resin rodlet A resinous secretion, usually deposited in a *resin duct* by the surrounding epithelium.

resin tin *rosin tin.*

resistance [elect] The property that limits the steady electric current in a conductor; it is measured by the ratio of the applied constant electromotive force to the current.

resistance [seis] *acoustic resistance.*

resistates Sediments composed of chemically resistant minerals, enriched in weathering residues; thus highly quartzose sediments characteristically rich also in zircon, ilmenite, rutile, and, more rarely, cassiterite, monazite, and gold, and thus typified by high contents of Si, Ti, and Zr. It is one of Goldschmidt's groupings of sediments or analogues of differentiation stages in rock analysis. Cf: *hydrolyzates; oxidates; reduzates; evaporates.*

resistivity (a) *electrical resistivity.* (b) *thermal resistivity.*

resistivity curve One of several curves on an *electric log* (usually on the right side of the log), showing the varying resistivity to an electric current of the sequence of rock units opposite the traveling electrode. It is used to determine the nature of the strata penetrated and to indicate the content of gas, oil, or water enclosed in rock pores: formations containing salt water have low resistivities, those containing freshwater or oil have higher resistivities.

resistivity factor *formation factor.*

resistivity log An electric log consisting of resistivity curves; e.g. a *microlog,* a *normal log,* and a *lateral log.* See also: *wall-resistivity log.*

resistivity method Any electrical exploration method in which current is introduced in the ground by two contact electrodes and potential differences are measured between two or more other electrodes.

resistivity profile A survey by the resistivity method in which a contact array of electrodes is moved along profiles, lateral variation in and resistivity being portrayed.

resolution The ability of a given electrical method to resolve or clearly indicate each of the anomalies due to two adjacent structures.

resonance n. Induced oscillations of maximum amplitude produced in a physical system when an applied oscillatory stress and the natural oscillatory frequency of the system are the same. Because resonance is characteristic of the absorbing material, each type of resonance (e.g. *nuclear resonance, electron spin resonance,* and *nuclear magnetic resonance*) has given rise to analytical instrumentation.----adj. *resonant.*

resonant frequency The frequency at which maximum (or minimum) response of a system occurs.

resorbed reef A reef characterized by embayed margins and by the numerous isolated patches of reef that are closely distributed about the main mass (Maxwell, 1968, p.106-107). Resorbed reefs frequently rise from larger, submerged platforms and they are suggestive of restrictive growth or of degeneration of the reef mass.

resorption (a) The act or process of reabsorption or readsorption; specif. the partial or complete re-fusion or solution, by and in a magma, of previously formed crystals or minerals with which it is not in equilibrium or, owing to changes of temperature, pressure (depth), or chemical composition, with which it has ceased to be in equilibrium. "The term is often wrongly applied to immature crystals, and to crystals which have decomposition borders through change of pressure or otherwise" (Holmes, 1928, p.198). (b) The geologic capture of a mineral by another that is relatively fixed (Pryor, 1963).

resorption border A *corrosion border* representing partial resorption and recrystallization by a molten magma of previously crystallized minerals. Syn: *resorption rim.*

resorption rim A *resorption border* as seen in section.

responsivity In infrared detector terminology, the ratio of signal output to incident radiant flux, usually expressed as volts/watt (Bernard, 1970, p.56). Symbol R_v.

rest-hardening The increase of strength, with time, of a clay subsequent to its deposition, remolding, or modification by the application of shear stress.

resting spore A spore that remains dormant for a period before germination; e.g. a *chlamydospore,* or a desmid *zygospore* having thick cell walls and able to withstand adverse conditions (such as heat, cold, or drying out). See also: *statospore; cyst* [palyn].

restite A term used by Mehnert (1968, p.356) as an essentially nongenetic designation for all immobile or less mobile parts of migmatites during migmatization. Cf: *metaster; stereosome.* See also: *paleosome.*

rest magma The part of a magma that remains after many minerals have crystallized from it during a long series of differentiations. Syn: *residual liquor.*

restricted Said of tectonic transport or movement in which elongation of particles is transverse to the direction of movement. Cf: *unrestricted.*

restricted basin A depression in the ocean floor characterized by topographically restricted water circulation, often resulting in oxygen depletion. Syn: *silled basin; barred basin.*

reststrahlen Narrow bands of entranced reflectance that occur in transparent materials where either the refractive index is high, or when the absorption coefficient is large. Syn: *residual rays.*

resupinate Inverted or reversed in position; esp. referring to reversal in relative convexity of postbrephic brachiopod shells in which the convex pedicle valve becomes concave and the concave brachial valve becomes convex during successive adult stages of growth.

resurgence (a) The point where an underground stream reappears at the surface to become a surface stream. It is usually near the point where an impermeable stratum, underlying a rock such as limestone, intersects the surface. See also: *debouchure.* Syn: *rise; emergence; exsurgence.* (b) The rising again of a stream from an underground cave. Syn: *rising*. (c) *karst spring*.

resurgent [petrology] Said of magmatic water or gases that were derived from sources on the Earth's surface, from its atmosphere, or from country rock of the magma. Cf: *juvenile* [*water*].

resurgent [pyroclast] In the classification of pyroclastics, the equivalent of *accessory* ejecta. Cf: *juvenile*.

resurgent cauldron A *cauldron* (caldera) in which the cauldron block, following subsidence, has been uplifted, usually in the form of a structural dome (Smith & Bailey, 1968, p.613).

resurgent gas *resurgent vapor*.

resurgent vapor (a) Ground water volatilized by contact with hot rock. (b) Gas in magma, possibly derived from dissolved or assimilated country rock.----Syn: *resurgent gas*. Cf: *phreatic gas; juvenile* [*water*].

resurrected (a) Said of a surface, landscape, or feature (such as a mountain, peneplain, or fault scarp) that has been restored by *exhumation* to its previous status in the existing relief. Syn: *exhumed*. (b) Said of a stream that follows an earlier drainage system after a period of brief submergence had slightly masked the old course by a thin film of sediments. Syn: *palingenetic*.

resurrected-peneplain shoreline A shoreline of submergence formed where the sea rests against an inclined resurrected peneplain (Johnson, 1925, p.27); it may be remarkably straight for long distances.

retained water Water retained in a rock or soil after the gravity ground water has drained out. It is no longer ground water but has become vadose water. Most of it is held by molecular attraction, but part may be in isolated interstices or held by other, more or less obscure, forces, and part remains as water vapor in interstices from which water has drained (Meinzer, 1923, p.27-28).

retaining wall A thick wall designed to resist the lateral pressure (other than wind pressure) of the material behind it, as a bulkhead preventing an earth slide.

retard A permeable bank-protection structure situated at and parallel to the toe of a slope and projecting into a stream channel, designed to check stream velocity and induce silting or accretion.

retardation [tide] *lunitidal interval*.

retention The amount of water from precipitation that has not escaped as runoff or through evapotranspiration; "the difference between total precipitation and total runoff on a drainage area" (Nelson & Nelson, 1967).

reteporiform Said of bryozoans having reticulate colonies, as in the cheilostome *Retepora*.

retgersite A tetragonal mineral: $NiSO_4.6H_2O$.

reticle A system of wires, cross hairs, threads, dots, or very fine etched lines, placed in the eyepiece of an optical instrument (such as a surveyor's telescope) perpendicular to its principal focus, to define the line of sight of the telescope or to permit a specific pointing to be made on a target or signal or a reading to be made on a rod or scale. Syn: *reticule*.

reticulate [ore dep] Said of a vein or lode with netlike texture, e.g. *stockwork*.

reticulate [paleont] Said of evolutionary change that involves repeated intercrossing between a number of lines; specif. a change involving the complex recombination of genes from varied strains of a diversified interbreeding population.

reticulate [paleont] Said of a netted pattern of an invertebrate, or of one resembling a network or having the appearance or form of a net; e.g. a "reticulate layer" consisting of ornamental ridges at the surface of a foraminiferal test, or a "reticulate ornamentation" on the exterior of a brachiopod shell, commonly involving a node-like enlargement formed by the intersection of concentric rugae with radial costae or costellae.

reticulate [palyn] Said of pollen and spores having sculpture consisting of a more or less regular network of ridges.

reticulate [petrology] Said of a rock texture in which crystals are partially altered to a secondary mineral, forming a network that encloses the remnants of the original mineral. Syn: *mesh texture; reticulated; reticular*.

reticulated bar One of a group of slightly submerged sandbars in two sets both of which are diagonal to the shoreline, forming a crisscross pattern (Shepard, 1952, p.1909). Reticulated bars are observed in bays and lagoons on the inside of barrier islands.

reticule *reticle*.

reticulite *thread-lace scoria*.

reticulopodium A foraminiferal *pseudopodium* that bifurcates and anastomoses to form a network. Pl: *reticulopodia*.

retinalite A massive, honey-yellow or greenish serpentine mineral with a waxy or resinous luster; a variety of chrysotile.

retinasphalt A light-brown variety of retinite usually found with lignite.

retinite (a) A general term for a large group of fossil resins of variable composition (oxygen content generally 6-15%), characterized by the absence of succinic acid, and found in the younger (brown) coals or peat. (b) Any fossil resin of the retinite group, such as glessite, krantzite, muckite, and ambrite. (c) A general name applied to fossil resins.

retinosite A microscopic constituent of torbanite, consisting of translucent orange-red discs (Dulhunty, 1939).

retractor muscle A muscle that draws in an organ or part; e.g. "siphonal retractor muscle" serving to withdraw the siphon of a bivalve mollusk partly or wholly within the shell, or "pallial retractor muscle" withdrawing marginal parts of mantle within a bivalve-mollusk shell where there is no distinct line of muscle attachment, or an *anterior lateral muscle* in a lingulid brachiopod. Cf: *protractor muscle*.

retral Posterior, or situated at or toward the back; e.g. "retral processes" in foraminiferal tests, consisting of backward-pointing extensions of chamber cavity and enclosed protoplasm, located beneath external ridges on the chamber wall, and ending blindly at the chamber margins (as in *Elphidium*).

retreat A decrease in length of a glacier, resulting in a displacement up valley or up slope of the position of the terminus, caused when processes of ablation (usually melting and/or calving) exceed the speed of ice flow: normally measured in meters per year. Cf: *recession* [*glaciol*].

retreatal moraine *recessional moraine*.

retrochoanitic Said of a septal neck of a cephalopod directed backward (adapically). Ant: *prochoanitic*.

retrodiagenesis *hypergenesis*.

retrogradation The backward (landward) movement or retreat of a shoreline or of a coastline by wave erosion; it produces a steepening of the beach profile at the breaker line. Ant: *progradation*. Cf: *recession*.

retrograde boiling The separation of a gas phase in a cooling magma as a result of its residual enrichment in the dissolved gaseous components by progressive crystallization of the magma.

retrograde diagenesis *hypergenesis*.

retrograde metamorphism A type of *polymetamorphism* by which metamorphic minerals of a lower grade are formed at the expense of minerals which are characteristic of a higher grade of metamorphism, a readjustment necessitated by a change in physical conditions, e.g. lowering of temperature. Cf: *prograde metamorphism*. Syn: *diaphthoresis; retrogressive metamorphism*.

retrograde motion A period when a planet moves westward on the celestial sphere. For example, when the Earth is overtaking Jupiter in terms of their mutual angular motion about the Sun, Jupiter will be in retrograde motion before and after opposition. Cf: *prograde motion*.

retrograding shoreline A shoreline that is being moved backward by wave attack. Ant: *prograding shoreline*. Syn: *abrasion shoreline*.

retrogression [**coast**] A syn. of *recession* in regard to a beach.

retrogression [**evol**] The passage from a higher to a lower or from a more to a less specialized state or type of organization or structure during the development of an organism.

retrogressive erosion *headward erosion.*

retrogressive metamorphism *retrograde metamorphism.*

return flow Irrigation water not consumed by evapotranspiration but returned to its source or to another body of ground or surface water. Water discharged from industrial plants is also considered return flow (Langbein & Iseri, 1960). Syn: *waste water; return water.*

returns Those surface waves on the record of a large earthquake which have traveled around the Earth's surface by the long (greater than 180°) arc between epicenter to station, or which have passed the station and returned after traveling the entire circumference of the Earth.

return water *return flow.*

retusoid Said of spores, mostly Devonian, with prominent contact areas and curvaturae perfectae.

retzian A brown orthorhombic mineral: $Mn_2Y(AsO_4)(OH)_4$.

reverberation *singing.*

reversal *geomagnetic reversal.*

reverse (a) Said of the aspect of a graptolite rhabdosome in which the sicula is more or less concealed by crossing canal(s). (b) Pertaining to the basal side of an incrusting or freely growing bryozoan colony.---Cf: *obverse.*

reverse bearing *back bearing.*

reversed *overturned* [fold].

reversed consequent stream A consequent stream whose direction of flow is contrary to that normally consistent with the geologic structure; e.g. the part of a captured consequent stream between the escarpment slope and the elbow of capture. See also: *obsequent stream.*

reversed fault *thrust fault.*

reversed gradient A local gradient opposite to the general gradient; esp. a valley gradient at the downstream side of a glacially overdeepened valley.

reversed magnetization A natural remanent magnetization that is opposite to the present ambient geomagnetic field.

reversed polarity (a) A natural remanent magnetization opposite to the present ambient geomagnetic field direction. See also: *geomagnetic reversal.* (b) A configuration of the Earth's magnetic field with the magnetic positive pole, where field lines leave the Earth, located near the geographic north pole.---*normal polarity.*

reverse drag *dip reversal.*

reversed stream A stream whose direction of flow has been reversed, as by glacial action, landsliding, gradual tilting of a region, or capture.

reversed tide An oceanic tide that is out of phase with the apparent motions of the tide-producing body, so that low tide is directly under the tide-producing body and is accompanied by a low tide on the opposite side of the Earth. Cf: *direct tide.* Syn: *inverted tide.*

reversed zoning *Zoning* in a plagioclase crystal in which the core is more sodic than the rim. Cf: *normal zoning.* Syn: *inverse zoning.*

reverse fault *thrust fault.*

reverse-flowage fold A fold in which flow from deformation has thickened the anticlinal crests and thinned the synclinal troughs, contrary to the normal flow pattern of a *flow fold.*

reverse saddle A mineral deposit associated with the trough of a synclinal fold and following the bedding plane. Syn: *trough reef.* Cf: *saddle reef.*

reverse scarplet An *earthquake scarplet* facing in toward the mountain slope and enclosing a trench, produced by reversal of earlier movement along a fault (Cotton, 1958, p.165-166); examples are numerous in New Zealand. Syn: *earthquake rent.*

reverse similar fold A fold, the strata of which are thickened

on the limbs and thinned on the axes, contrary to the pattern of a *similar fold.*

reverse slip fault *thrust fault.*

reversible pendulum A pendulum that can swing around either of two knife edges placed in such a way that for both cases the period is the same; it is used in absolute-gravity determinations.

reversible process A thermodynamic process in which an infinitesimal change in the variables characterizing the state of the system can change the direction of the process.

reversing current A tidal current that flows in an alternating pattern of opposite directions for approximately equal lengths of time, with a slack period of no movement at each reversal. A reversing current occurs in estuaries, restricted channels, and inland bodies of water. Cf: *rotary current.* Syn: *rectilinear current.*

reversing dune A dune that tends to develop unusual height but migrates only a limited distance "because seasonal shifts in direction of dominant wind cause it to move alternately in nearly opposite directions" (McKee, 1966, p.10). Its general shape may resemble that of a barchan or a transverse dune, but it differs in the complexity of its internal structural orientation due to reversals in direction of the slip face.

reversing thermometer A mercury-in-glass thermometer used to measure temperatures of the sea at depth. The temperature is recorded when the thermometer is inverted; and the recording is maintained until it is once again upright. A *protected thermometer* and an *unprotected thermometer* are usually used as a pair. Reversing thermometers are attached to a *Nansen bottle.* See also: *thermometric depth.*

reversion A return toward an ancestral type or condition, such as the reappearance in an organism of an ancestral characteristic.

revet-crag A term proposed by Gilbert (1877, p.26) for one of a series of narrow, pointed outliers or ridges of eroded strata inclined like a revetment against a mountain spur.

revetment A facing of stone, concrete, or other material, built to protect an embankment (as of a stream or lake) or a shore structure from wave erosion.

revier A term applied in SW Africa to a deeply cut river bed that usually remains dry.

revival *rejuvenation.*

revived *rejuvenated.*

revived fault scarp *rejuvenated fault scarp.*

revived folding *recurrent folding.*

revived stream *rejuvenated stream.*

revolution A term formerly popular among geologists for a time of profound orogeny and other crustal movements, on a continentwide or even worldwide scale, the assumption being that such revolutions produced abrupt changes in geography, climate, and environment, hence were related to changes in the forms of life. Schuchert (1924) classed all orogenies at the close of geologic eras as revolutions, in contrast to *disturbances,* or orogenies within the eras. The basic premises of all these concepts are dubious, and the term revolution is little used now.

reworked Said of a sediment, fossil, rock fragment, or other geologic material that has been removed or displaced by natural agents from its place of origin and incorporated in recognizable form in a younger formation, such as a "reworked tuff" carried by flowing water and redeposited in another locality. Cf: *derived; remanié.*

reyerite A mineral possibly identical with truscottite.

Reynolds critical velocity *critical velocity* (b).

Reynolds number A numerical quantity used as an index to characterize the type of flow in a hydraulic structure in which resistance to motion is dependent upon the viscosity of the liquid in conjunction with the resisting force of inertia. It is the ratio of inertia forces to viscous forces, and is equal to the product of a characteristic velocity of the system (it may be the mean, surface, or maximum velocity) and a characteristic linear dimension, such as diameter or depth, divided by the kinematic viscosity of the liquid; all expressed in consistent units in order that the combinations will be dimensionless. The number is chiefly applicable to closed systems of flow, such as pipes or conduits where there is free water surface, or to bodies fully immersed in the fluid so the free surface need not be considered (ASCE, 1962).

rezbanyite A gray mineral: $Pb_3Cu_2Bi_{10}S_{19}$.

Rg wave A slow, short-period *Rayleigh wave* that travels only along a nonoceanic path. The subscript "g" refers to the possible importance of the granitic layer to its propagation. Cf: *Lg wave*.

rhabd (a) An obsolete term for *monaxon*. Syn: *rhabdus*. (b) *rhabdome*.

rhabdacanth A trabecula of a rugose coral in which the fibers are related to any number of separate, transient (shifting) centers of growth grouped around a main one (TIP, 1956, pt.F, p.235). Cf: *monacanth*.

rhabde The lower or axial branch in the triaene spicule of an ebridian skeleton.

rhabdite A syn. of *schreibersite*, esp. occurring in rods or needle-shaped crystals.

rhabdodiactin A seemingly monaxonic sponge spicule, formed by suppression of two of the axes of a hexactin which are preserved internally as an axial cross. See also: *diactin*.

rhabdoglyph A collective term used by Fuchs (1895) for a trace fossil consisting of a presumable worm trail appearing on the undersurface of flysch beds (sandstones) as a nearly straight bulge with little or no branching. Cf: *graphoglypt; vermiglyph*.

rhabdolith A minute (diameters average 3 microns), calcareous, spinose, rod- or club-like, supposedly perforate *coccolith* having a shield surmounted by a long stem. Rhabdoliths are found both at the surface and on the bottom of the ocean, and they have been classed as protozoans and as algae. Syn: *rabdolith*.

rhabdomancy A form of *dowsing* using a rod or twig. Cf: *pallomancy*.

rhabdome The long ray of a triaene sponge spicule. Syn: *rhabd*.

rhabdophane A brown, pinkish, or yellowish-white mineral: $(Ce,La)PO_4.H_2O$. It contains yttrium and rare-earth elements. Syn: *rhabdophanite*.

rhabdosome The skeleton of a graptolithine colony, composed of protein substances. Also, the entire colony of graptolites, or a colonial graptolite derived from a single individual.

Rhaetian European stage: uppermost Triassic (above Norian, below Hettangian of Jurassic). It is interpreted as lowermost Jurassic in some areas (as in France and Great Britain) or as a transitional stage between the Triassic and Jurassic. Syn: *Rhaetic*.

rhagon (a) The earliest developmental stage of a sponge with a functional aquiferous system having several flagellated chambers. Also, a sponge or sponge larva in such a stage. (b) A term used incorrectly as a syn. of *leucon*.---Adj: *rhagonoid*.

rhax A kidney-shaped sterraster (sponge spicule).

rhegmagenesis *regmagenesis*.

rhegolith Var. of *regolith*.

rheid A substance (below its melting point) which deforms by viscous flow during the time of applied stress at an order of magnitude at least three times that of elastic deformation under similar conditions.

rheid fold A fold, the strata of which have deformed by flow as if they were fluid. Cf: *flow fold*. Syn: *rheomorphic fold*.

rheidity *Relaxation time* of a substance, multiplied by 1000.

rhenium-osmium age method The determination of an age in years based on the known radioactive decay rate of rhenium-187 to osmium-187. The low crustal abundance of rhenium

limits the application of this method.

rheoglyph A hieroglyph produced by syngenetic deformation, such as by slumping (Vassoevich, 1953, p.55).

rheoignimbrite An ignimbrite, on the slope of a volcanic crater, that developed secondary flowage due to high temperatures.

rheologic settling Failure of a sediment under a stress load by plastic deformation or flow.

rheology The study of the deformation and flow of matter.

rheomorphic Said of a rock whose form and internal structure indicate a finite amount of flow in a plastic state; also, said of the phenomena causing such a rock. The term is generally applied to material that has been rendered plastic by magma. See also: *rheomorphism*.

rheomorphic fold *rheid fold*.

rheomorphic intrusion The injection of country rock that has become mobilized (rheomorphic) into the igneous intrusion that caused the rheomorphism. Such an intrusion usually resembles the metamorphosed country rock.

rheomorphism The process by which a rock becomes mobile by at least partial fusion that is commonly accompanied by, if not promoted by, addition of new material by diffusion (Larsen, 1961). Cf: *mobilization*.

rheopexy The accelerated gelation of a thixotropic sol by agitating it in some manner, e.g. stirring.

rheophile adj. Said of an organism that lives in or prefers flowing water.

rheotaxis Taxis resulting from mechanical stimulation by a stream of fluid, such as water. Cf: *rheotropism*.

rheotropism Tropism resulting from mechanical stimulation by a stream of fluid, such as water. Cf: *rheotaxis*.

rhexistasy The mechanical breaking up and transport of old soils or other surface residual materials (Erhart, 1955). Etymol: Greek *rhexis*, "act of breaking", + *stasis*, "condition of standing". Adj: *rhexistatic*. See also: *biorhexistasy; biostasy*.

rhinestone (a) An inexpensive and lustrous imitation of diamond, consisting of glass that has been backed with a thin leaf of metallic foil to simulate the brilliancy of a diamond. (b) Originally, a syn. of *rock crystal*.

rhinocanna *nasal tube*.

rhizic water A syn. of *soil water* proposed by Meinzer (1939) as one of three classes of *krematic water*.

rhizoclone A monocrepid desma (of a sponge) consisting of a straight or curved body bearing branching outgrowths along its entire length. See also: *megarhizoclone*.

rhizoconcretion A small, cylindrical or conical, usually branching or forked, concretion-like structure in a sedimentary rock, resembling a root of a tree. It may consist of material such as caliche or chert. Cf: *rhizocretion; pedotubule*. See also: *root sheath*. Syn: *rhizomorph; root cast*.

rhizocretion A term used by Kindle (1923, p.631) for a hollow concretion-like mass that had formed around the root of a living plant. Cf: *rhizoconcretion*.

rhizoid [paleont] adj. Resembling a root; e.g. "rhizoid spine" of a brachiopod, resembling a plant rootlet and serving for attachment either by entanglement or by extending along and cementing itself to a foreign surface.---n. A *radicle* of a bryozoan.

rhizoid [bot] A unicellular or multicellular, rootlike filament that attaches some nonvascular plants and gametophytes of some vascular plants to the substrate (Scagel, et al, 1965, p.630). Cf: *rhizome; rhizophore*.

rhizome An underground stem that lies horizontally and that is often enlarged in order to store food. Cf: *rhizoid; rhizophore*.

rhizomorine Any lithistid demosponge belonging to the suborder Rhizomorina and characterized by the presence of rhizoclones.

rhizomorph A term used by Northrop (1890) for a structure now known as a *rhizoconcretion*.

rhizophore In a club moss, a leafless and dicotomous root-bearing organ. Cf: *rhizoid; rhizome.*

rhizophytous Said of a sponge that is fastened to the substrate by branching extensions of the body.

rhizopod A protozoan belonging to the class Rhizopodea, generally characterized by lobose pseudopodia and by zoned protoplasm in shelled forms and protoplasm differentiated into endo- and ectoplasm in nonshelled forms. Cf: *actinopod.*

rhizopodial Said of a morphologic type or growth form in which the cell is somewhat amoeboid.

rhizopodium A bifurcating and anastomosing ectoplasmic pseudopodium that is typical of many foraminifers. Pl: *rhizopodia.* Syn: *rhizopod.*

rhizosphere The soil in the immediate vicinity of the plant roots in which the abundance or composition of the microbial population is affected by the presence of roots. Syn: *root zone [soil].*

Rhodanian orogeny One of the 30 or more short-lived orogenies during Phanerozoic time identified by Stille, in this case at the end of the Miocene.

rhodesite A mineral: $(Ca,Na_2,K_2)_8Si_{16}O_{40}.11H_2O$. Cf: *mountainite.*

rhodite A mineral consisting of a native alloy of rhodium (about 40%) and gold.

rhodizite A mineral: $CsAl_4Be_4B_{11}O_{25}(OH)_4$.

rhodochrosite A rose-red or pink to gray or brownish rhombohedral mineral: $MnCO_3$. It is isomorphous with calcite and siderite, and commonly contains some calcium and iron; it is a minor ore of manganese. Syn: *dialogite; manganese spar; raspberry spar.*

rhodolite A pink, rose, or purplish- to violetish-red garnet that is intermediate in chemical composition between pyrope and almandine, characterized by a lighter tone and a higher degree of transparency than either of the other two, and used as a gem.

rhodonite A pale-red, rose-red, or flesh-pink to brownish-red or red-brown triclinic mineral: $MnSiO_3$. It sometimes contains calcium, iron, magnesium, and zinc, and is often marked by black streaks and veins of manganese oxide. Rhodonite is used as an ornamental stone (esp. in Russia). Syn: *manganese spar; manganolite.*

rhodusite A monoclinic mineral of the amphibole group: $Na_2(Fe^{+2},Mg)_3Fe_2^{+3}Si_8O_{22}(OH)_2$. It is near riebeckite in chemical composition.

rhohelos A stream-crossed, nonalluvial marsh typical of filled lake areas.

rhomb [cryst] An oblique, equilateral parallelogram; in crystallography, a *rhombohedron.*

rhomb [paleont] (a) *pore rhomb.* (b) A six-sided, roughly equidimensional crystal composing some heterococcoliths.

rhombic (a) Adj. of *rhomb.* (b) Adj. of *orthorhombic.*

rhombic-dipyramidal class That crystal class in the orthorhombic system having symmetry $2/m\,2/m\,2/m$.

rhombic-disphenoidal class That crystal class in the orthorhombic system having symmetry 222.

rhombic dodecahedron A crystal form in the cubic system that is a *dodecahedron,* the faces of which are equal rhombs. Cf: *pyritohedron.*

rhombic-pyramidal class That crystal class in the orthorhombic system having symmetry mm2.

rhombic system *orthorhomic system.*

rhombochasm A term used by Carey (1958) for a parallel-sided gap in the sialic crust occupied by simatic crust, interpreted as due to spreading and separation. Cf: *sphenochasm.* See also: *rift [struc geol].*

rhomboclase A mineral: $HFe^{+3}(SO_4)_2.4H_2O$. It occurs in colorless rhombic plates with basal cleavage.

rhombohedral class That crystal class in the rhombohedral division of the hexagonal system having symmetry $\bar{3}$.

rhombohedral cleavage A type of crystal cleavage in the hexagonal system that occurs parallel to the faces of the rhombohedron, e.g. calcite.

rhombohedral iron ore (a) *hematite.* (b) *siderite* [mineral].

rhombohedral lattice A centered lattice of the hexagonal system in which the primitive unit cell is a rhombohedron. It may occur in crystal classes having one threefold axis. The unit cell contains three lattice points: one at the corners and two equally spaced along one long diagonal of the primitive hexagonal unit.

rhombohedral packing The "tightest" manner of systematic arrangement of uniform solid spheres in a clastic sediment or crystal lattice, characterized by a unit cell of six planes passed through eight sphere centers situated at the corners of a regular rhombohedron (Graton & Fraser, 1935). An aggregate with rhombohedral packing has the minimum porosity (25.95%) that can be produced without distortion of the grains. Cf: *cubic packing.* See also: *close packing.*

rhombohedral system A division of the *trigonal system* in which the basic unit cell is a rhombohedron.

rhombohedron A trigonal crystal form that is a parallelepiped whose six identical faces are rhombs. It is characteristic of the hexagonal system. Syn: *rhomb.*

rhomboid current ripple *rhomboid ripple mark.*

rhomboid ripple mark An aqueous *current ripple mark* characterized by diamond-shaped tongues of sand arranged in a reticular pattern resembling the scales of certain fish, each tongue (ranging from 12 to 25 mm in width and 25 to 50 mm in length) having one acute angle (formed by two steep sides) pointing downcurrent and another acute angle (formed by the gentle side extending into the reentrant angle of the steep sides of two tongues of the following) pointing upcurrent; it is extremely common on sand beaches where it forms during the final stages of backrush of each retreating wave. The sides are not more than 1 mm high. The term "rhomboid" applied to a ripple or ripple mark was introduced by Bucher (1919, p.153). Syn: *rhomboid current ripple; overhanging ripple.*

rhombolith *scapholith.*

rhomb-porphyry A porphyritic alkalic syenite containing phenocrysts of small augites, occasional olivine and anorthoclase or potassium oligoclase with rhombohedron-shaped cross sections in a groundmass composed chiefly of alkali feldspars. Var: *rhombenporphyry; rhombenporphyr.*

rhomb spar A *dolomite* mineral in rhombohedral crystals.

rhopaloid septum A rugose corallite septum characterized by distinctly thickened axial edge, appearing club-shaped in cross section (TIP, 1956, pt.F, p.250).

rhourd A pyramid-shaped sand dune, formed by the intersection of other dunes (Aufrere, 1931, p.376). Cf: *khurd.*

rhumb line A curved line on the surface of the Earth that crosses successive meridians at a constant oblique angle and that spirals around and toward the poles in a constant true direction but theoretically never reaches them; a straight line on a Mercator projection, representing a line of constant bearing or compass direction. Syn: *loxodrome; loxodromic curve; Mercator track.*

rhyacolite *sanidine.*

rhyncholite A fossil beak or part of a cephalopod jaw; specif. the calcified tip of a jaw of a Triassic nautiloid.

rhynchonelloid Any articulate brachiopod belonging to the order Rhynchonellida, characterized by a rostrate shell, a functional pedicle, and a delthyrium partially closed by deltidial plates. Their stratigraphic range is Middle Ordovician to present. Var: *rhynchonellid.*

rhyocrystal One of a group of idiomorphs arranged in "streamlines" (Lane, 1902, p.386).

rhyodacite A group of extrusive porphyritic igneous rocks intermediate in composition between dacite and rhyolite, with quartz, plagioclase, and biotite (or hornblende) as the main phenocryst minerals and a fine-grained to glassy groundmass composed of alkali feldspar and silica minerals; the extrusive equivalent of *granodiorite* or *quartz monzonite*. Also, any member of that group. Syn: *quartz latite; dellenite.*

rhyodiabasic Said of the ophitic texture of an igneous rock in which the plagioclase phenocrysts are more or less parallel; a nonrecommended term.

rhyolite A group of extrusive igneous rocks, generally porphyritic and exhibiting flow texture, with phenocrysts of quartz and alkali feldspar (esp. orthoclase) in a glassy to cryptocrystalline groundmass; also, any rock in that group; the extrusive equivalent of granite. Rhyolite grades into rhyodacite with decreasing alkali feldspar content and into trachyte with a decrease in quartz. Etymol: Greek *rhyo-*, from *rhyax*, "stream of lava". Syn: *liparite; quartz trachyte.* Cf: *quartz porphyry.*

rhyotaxitic texture *flow texture.*

rhythm A term often used in geology interchangeably with *cycle*, although Bissell (1964a, p.44) applied it to regular or measured movements in which time is considered. According to Challinor (1967, p.213), the term "suggests the recurrence, at more or less frequent and regular intervals, of one thing in particular, or an alternation, or a repetition of a sequence, on a rather small scale", although he noted that it is sometimes used for large-scale repetitions, such as Grabau's (1940) "rhythm of the ages".

rhythmic crystallization A phenomenon, observed in igneous rocks, in which different minerals crystallize in concentric layers, giving rise to orbicular texture.

rhythmic layering That type of *layering* in an igneous intrusion which is easily observable and in which there is repetition of zones of varying composition having the pattern abc, abc, etc. Cf: *cryptic layering.* See also: *zebra layering.*

rhythmic sedimentation The consistent repetition, through a sedimentary succession, of a regular sequence of two or more rock units organized in a particular order and indicating a frequent and predictable recurrence or pattern of the same sequence of conditions. It may involve only two components (such as interbedded laminae or varves of silt and clay), or broad changes in sediment character spanning whole systems (or longer intervals) and units up to hundreds of meters thick, or any sequence intermediate between these two extremes. See also: *cyclic sedimentation.*

rhythmic succession A succession of rock units showing continual and repeated changes of lithology. The term was used by Hudson (1924) for a continual repetition of a more or less complete suite comprising successive beds of certain kinds of sediments accompanied by an equally marked variation in the kind of fossils they contain.

rhythmic unit (a) *rhythmite.* (b) A layer or band of a rhythmically layered intrusive igneous rocks.

rhythmite An individual unit of a rhythmic succession or of beds developed by rhythmic sedimentation; e.g. a *cyclothem.* The term was used by Bramlette (1946, p.30) for the couplet of distinct types of sedimentary rock, or the graded sequence of sediments, that forms a unit in rhythmically bedded deposits. The term implies no limit as to thickness or complexity of bedding and it carries no time or seasonal connotation. See also: *laminite.* Syn: *rhythmic unit.*

rhytidome The technical term for *outer bark.*

ria (a) Any long, narrow, sometimes wedge-shaped inlet or arm of the sea (but excluding a fjord) whose depth and width gradually and uniformly diminish inland and which is produced by drowning due to submergence of the lower part of a narrow river valley or of an estuary; it is shorter and shallower than a *fjord*. Originally, the term was restricted to such an inlet produced where the trend of the coastal rock structure is at right angles to the coastline; it was later applied to any submerged land margin that is dissected transversely to the coastline. (b) Less restrictedly, any broad or estuarine river mouth, including a fjord, and not necessarily an embayment produced by partial submergence of an open valley.---See also: *estuary.* Etymol: Spanish *ría*, from *río*, "river".

ria coast A coast having several long, parallel *rias* extending far inland and alternating with ridge-like promontories; e.g. the coasts of SW Ireland and NW Spain. It is especially developed where the trend of the coastal structures is transverse to that of the coastline.

ria shoreline A shoreline characterized by numerous *rias* and produced by drowning due to partial submergence of a land margin subaerially dissected by numerous river valleys (Johnson, 1919, p.173).

rib [geomorph] A layer or dike of rock forming a small ridge on a steep mountainside.

rib [paleont] A radial or transverse fold upon a shell; e.g. a radial ornament on a brachiopod shell, or a *costa* on the shell of a bivalve mollusk, or a raised radial ridge on the coiled conch of a nautiloid.

rib [bot] A primary leaf vein.

rib and furrow A term used by Stokes (1953, p.17-21) for the bedding-plane expression of *micro cross-bedding*, consisting of small, transverse, arcuate markings (convex upcurrent) occurring in sets and confined to relatively long, parallel, narrow (3-5 cm wide) grooves oriented parallel to the current flow and separated from one another by very narrow and not altogether continuous ridges. It represents the eroded edges of upturned arcuate laminae. Syn: *rib-and-furrow structure.*

riband jasper *ribbon jasper.*

ribbed moraine One of a group of irregularly subparallel, locally branching, generally smoothly rounded and arcuate ridges that are convex in the downstream direction of a glacier but that curve upstream adjacent to eskers (J.A. Elson in Fairbridge, 1968, p.1217). They are most common in the continental ice sheets, and are abundant in the Arctic.

ribble *ripple till.*

ribbon [ore dep] adj. Said of a vein having alternating streaks differing in color, of ore with gangue or with country rock, e.g. ribbon quartz. Cf: *banded; book structure.*

ribbon [petrology] One of a set of parallel bands or streaks in a mineral or rock, e.g. ribbon jasper; when the lines of contrast are on a larger scale, the term *banding* is used. When occurring in slate, the structure is known as *slate ribbon.* Syn: *stripe.*

ribbon banding A *banding* produced in the bedding of a sedimentary rock by thin strata of contrasting colors, giving the rock an appearance suggesting bands of ribbons.

ribbon bomb A type of volcanic bomb that is elongate and flattened, and derived from ropes of lava.

ribbon diagram A single, continuous geologic cross section drawn in perspective along a curved or sinuous line.

ribbon injection A tonguelike igneous intrusion along the cleavage planes of a foliated rock.

ribbon jasper Beautifully banded jasper with parallel, ribbon-like stripes of alternating colors or shades of color (as of red, green, and esp. brown). Syn: *riband jasper.*

ribbon reef A linear reef within the Great Barrier Reef off the NE coast of Australia, having inwardly curved extremities, and forming a festoon along the precipitous edge of the continental shelf. They are variable in length (3-24 km), less so in width (300-470 m).

ribbon rock A rock characterized by a succession of thin layers of differing composition or appearance; e.g. a sedimentary rock consisting of gray shales interspersed with thin, varve-like seams of brown dolomite and lighter-colored limestone (Goldring, 1943), or a vein rock with narrow quartz bands separated by stripes of altered wall rock.

ribbon slate Slate produced by incomplete metamorphism of still clearly visible residual bedding planes that cut across the cleavage surface; slate characterized by varicolored ribbons.

rice coal One of the sizes of *buckwheat coal*, equivalent to number two in the series. It will pass through a 5/16 inch round mesh but not through a 3/16 inch round mesh. Cf: *barley coal.*

Richardson effect *thermionic emission.*

richellite A dubious mineral: $Ca_3Fe_{10}(PO_4)_8(OH,F)_{12}.nH_2O$. It occurs in amorphous yellow masses.

Richmondian North American stage: Upper Ordovician (above Maysvillian, below Lower Silurian).

richterite (a) A brown, yellow, or rose-red monoclinic mineral of the amphibole group: $(Na,K)_2(Mg,Mn,Ca)_6Si_8O_{22}(OH)_2$. (b) An end member of the amphibole group: $Na_2CaMg_5Si_8O_{22}(OH)_2$. Cf: *soda tremolite.*

Richter scale The range of numerical values of *earthquake magnitude*, devised in 1935 by the seismologist C.F. Richter. Very small earthquakes, or microearthquakes, can have negative magnitude values. In theory there is no upper limit to the magnitude of an earthquake. However, the strength of Earth materials produces an actual upper limit of slightly less than 9.

rickardite A deep-purple mineral: Cu_4Te_3.

ricolettaite A dark-colored syenite-gabbro containing calciclase as the plagioclase, along with olivine and augite.

rideau A small ridge or mound of earth, or a slightly elevated piece of ground. Etymol: French.

ridge [beach] (a) *beach ridge.* (b) A low mound that is sometimes found above the water level on the foreshore of a sand beach during low tide. See also: *runnel.*

ridge [geomorph] (1) A general term for a long, narrow elevation of the Earth's surface, usually sharp-crested with steep sides, occurring either as an independent hill or as part of a larger mountain or hill; e.g. an extended upland between valleys. A ridge is "generally less than" 8 km (5 mi) long (Eardley, 1962, p.6). (b) A term occasionally applied to a range of hills or mountains. (c) A top or upper part of a hill; a narrow, elongated crest of a hill or mountain.

ridge [marine geol] An elongate, steep-sided elevation of the ocean floor, having rough topography; Syn: *submarine ridge.*

ridge [paleont] An elevated body part of an animal, projecting from a surface; e.g. a relatively long narrow elevation of secondary shell of a brachiopod, or a *transverse ridge* on a crinoid. Also, an area separating adjacent pairs of ambulacral pores of a regular echinoid.

ridge [ice] *pressure ridge.*

ridge-and-ravine topography Hack's term for the *ridge-and-valley topography* of the Appalachians (1960).

ridge-and-valley topography A rolling topography characterized by a close succession of parallel or nearly parallel ridges and valleys, and resulting from the differential erosion of highly folded strata of varying resistances. Type region: Ridge and Valley region in the Appalachian Mountains, lying to the west of the Blue Ridge. Syn: *ridge-and-ravine topography.*

ridged ice Sea ice having readily observed surface features in the form of one or more *pressure ridges*; it is usually found in first-year ice. See also: *ropak; ridging.*

ridge fault A fault structure that is a set of two faults bounding a horst. Cf: *trough fault.*

ridge-top trench A trench, occasionally found at or near the crest of high, steep-sided mountain ridges, formed by the creep displacement of a large slab of rock along shear surfaces more or less parallel with the side slope of the ridge. Trenches are usually parallel with the crest of the ridge. See also: *sackungen.*

ridging A form of deformation of floating ice, caused by lateral pressure, whereby ice is forced or piled haphazardly one piece over another to form *ridged ice*. Cf: *tenting.*

riebeckite A blue or black monoclinic mineral of the amphibole group: $Na_2(Fe,Mg)_5Si_8O_{22}(OH)_2$. It occurs as a primary constituent in some acid or sodium-rich igneous rocks. See also: *crocidolite.*

Riecke's principle The statement in thermodynamics that solution of a mineral tends to occur most readily at points where external pressure is greatest, and that crystallization occurs most readily at points where external pressure is least. It is applied to recrystallization in metamorphic rocks with attendant change in mineral shapes. It is named after the German physicist E. Riecke (1845-1915) although it was actually discovered and described earlier by Sorby in 1863.

riedel shear A slip surface which develops during the early stage of shearing. Such shears are typically arranged en échelon, usually at inclinations of between $10°$ and $30°$ to the direction of relative movement (Riedel, 1929).

riedenite An igneous rock composed of large tabular biotite crystals in a granular groundmass of nosean, biotite, pyroxene, and small amounts of sphene and apatite.

riegel A low, transverse ridge or barrier of bedrock on the floor of a glacial valley, esp. common in the Alps; it separates a rock basin from the gently sloping valley bottom farther downstream. See also: *rock step.* Etymol: German *Riegel*, "crossbar". Syn: *rock bar; threshold; verrou.*

Riel discontinuity A seismic-velocity discontinuity noted in Alberta that may be equivalent to the *Conrad discontinuity.*

riffle (a) A natural shallows or other expanse of shallow bottom extending across a stream bed over which the water flows swiftly and the water surface is broken in waves by obstructions wholly or partly submerged; a shallow rapids of comparatively little fall. See also: *rift.* (b) An expanse of shallow water flowing over a riffle or at the head of a rapids. (c) A low bar or bedrock irregularity in a stream, resembling a riffle. (d) A wave of a riffle.---Syn: *ripple.*

riffle hollow A shallow depression in a stream bed, commonly 8-30 cm deep, produced by differential erosion of alternate layers of hard and soft rock (Bryan, 1920, p.192).

riffler *sample splitter.*

rift [speleo] A narrow and high passage in a cave, the shape of which is controlled by a joint or by a bedding or fault plane.

rift [geomorph] A narrow cleft, fissure, or other opening in rock (as in limestone), made by cracking or splitting.

rift [streams] A shallow or rocky place in a stream, forming either a ford or a rapids. The term is used in NE U.S. as a syn. of *riffle.*

rift fault A fault that bounds a rift valley.

rift lake *sag pond.*

rift trough *rift valley.*

rift valley [marine geol] The deep, central cleft in the crest of the *mid-oceanic ridge*, about 25-50 km in width, with a mountainous rather than flat floor. Syn: *central valley; median rift valley; mid-ocean rift.*

rift valley [struc geol] A syn. of *rift* [struc geol]. Cf: *graben.* Syn: *rift trough; fault trough.*

rift-valley lake *sag pond.*

rift zone (a) A system of crustal fractures; a *rift* [struc geol]. (b) In Hawaii, a zone of volcanic features associated with underlying dike complexes. Syn: *volcanic rift zone.*

rig [drill] (a) *drilling rig.* (b) An oil *derrick.*

rig [geomorph] An Iranian term for a sand dune occurring in the dasht (Fisher, 1950, p.267).

right bank The bank of a stream situated to the right of an observer who is facing downstream.

right-handed [cryst] Said of an optically active crystal which rotates the plane of polarization of light to the right. Cf: *left-handed.* Syn: *dextrorotatory.*

right-handed [paleont] *dextral.*

right-handed separation *right-lateral separation.*

right-lateral fault A fault, the displacement of which is *right-lateral separation.* Syn: *dextral fault; right-lateral slip fault; right-slip fault.*

right-lateral separation Movement of a *lateral fault* along which, in plan view, the side opposite the observer appears to have moved to the right. Cf: *left-lateral separation.* Syn: *right-handed separation.*

right-lateral slip fault *right-lateral fault.*

right section *profile* [struc petrol].

right-slip fault *right-lateral fault.*

right valve The valve lying on the right-hand side of a bivalve mollusk when the shell is placed with the anterior end pointing away from the observer, the commissure being vertical and the hinge being uppermost. Abbrev: RV. Ant: *left valve*.

right way up The state of strata where the present upward succession of layers is the original (normal) order of deposition. See also: *way up*. Syn: *right side up*.

rigidity The property of a material to resist applied stress that would tend to distort it, e.g. a fluid has zero rigidity.

rigidity modulus *modulus of rigidity*.

rigolet A term applied in the Mississippi River valley to a small stream, creek, or rivulet. Etymol: French *rigole*, "trench, small ditch, channel".

rig time The time (hours, days) during which a drilling rig is in use in actual drilling and other related operations (such as logging) not chargeable to the drilling contractor; the time devoted to the operator's, rather than the drilling contractor's, interest.

rijkeboerite A mineral: $Ba(Ta,Nb)_2(O,OH)_7$. It is the barium analogue of microlite.

rill [lunar] *rille*.

rill [beach] (a) A small, transient *runnel* carrying to the sea or a lake the water of a wave after breaking on a beach, esp. one formed following an outgoing tide. It may be 2-10 mm wide, 0.5 m or more long, and about 1 mm deep. (b) The minute stream or thin sheet of water flowing in a rill.

rill [speleo] A small channel formed by circulating water in the wall, floor, or ceiling of a cave.

rill [stream] (a) A very small brook or trickling stream of water; a streamlet or rivulet. (b) A small channel eroded by a rill, esp. in soil; one of the first and smallest channels formed by runoff, such as a *shoestring rill*. Syn: *rill channel*.

rill cast A term used by Dzulynski & Slaczka (1958, p.230) for a sole mark that is probably the same as a *furrow flute cast*.

rill channel A *rill* formed by running water; specif. a *shoestring rill*.

rille One of several relatively long (up to several hundred kilometers), narrow (1-2 km), trench- or crack-like valleys commonly occurring on the Moon's surface. Rilles may be extremely irregular with meandering courses ("sinuous rilles"), or they may be relatively straight depressions ("normal rilles"); they have relatively steep walls and usually flat bottoms. Rilles are essentially youthful features and apparently represent fracture systems originating in brittle material. Syn: *rill; rima*.

rillenstein A pattern of tiny solution grooves of about one millimeter or less in width, formed on the limestone surface of a karstic region. Etymol: German.

rill erosion The development of numerous, minute, closely spaced channels resulting from the uneven removal of surface soil by running water that is concentrated in streamlets of sufficient volume and velocity to generate cutting power. It is an intermediate process between *sheet erosion* and *gully erosion*. Cf: *channel erosion*. Syn: *rill wash; rilling; rillwork*.

rillet A little rill.

rill flow Surface runoff flowing in small, irregular channels too small to be considered rivulets.

rilling *rill erosion*.

rill mark (a) A small, dendritic channel, groove, or furrow formed on the surface of beach mud or sand by a wave-generated *rill* or by a retreating tide; esp. one formed on the lee side of a half-buried pebble, shell, or other obstruction, and usually showing an upstream bifurcation (a branching up the beach). (b) A small, dendritic channel formed by a small stream debouching on a sand flat or a mud flat; it shows a downslope bifurcation.---See also: *lobate rill mark*.

rillstone *ventifact*.

rill wash A syn. of *rill erosion*. Also spelled: *rillwash*.

rillwork *rill erosion*.

rim [glac geol] A ridge of morainal material, generally unbroken and of uniform height, surrounding a central depression (Gravenor & Kupsch, 1959, p.52).

rim [geomorph] The border, margin, edge, or face of a landform, such as the curved brim surrounding the top part of a crater or caldera; specif. the *rimrock* of a plateau or canyon.

rim [paleont] (a) One of the two flanges of a caneolith coccolith peripheral to the wall; e.g. the distal "upper rim" and the proximal "lower rim". (b) The outer, usually flange-like component of a *wheel* of a holothurian. It may be recurved, and the inner margin of its upper side is commonly denticulate or dentate. The rim is inclined to or within the plane of the wheel.

rim [ign] (a) *reaction rim*. (b) *kelyphytic rim*. (c) *corona* [ign].

rima A long, narrow aperture, cleft, or fissure; specif. a lunar *rille*. Pl: *rimae*.

rim cementation A term used by Bathurst (1958, p.21) for *secondary enlargement* in detrital sediments; e.g. the chemical deposition of calcium carbonate forming a single, completely enveloping rim on a grain of the same composition, as in a crinoidal sandstone where each grain (or crinoidal fragment) is a single crystal and is permeated by the calcite cement in lattice or optical continuity. Cf: *granular cementation*.

rime A deposit of rough ice crystals formed as a result of contact between the supercooled droplets of fog and a solid object at a temperature below $0°C$ (McIntosh, 1963, p.216).

rim gypsum Gypsum in thin films between anhydrite crystals, believed to have been introduced in solution rather than produced by replacement (Goldman, 1952, p.2).

rim height The maximum height of the rim of a crater above the original ground surface. See also: *lip height*.

rimmed kettle A morainal depression with raised edges (Gravenor & Kupsch, 1959, p.53).

rimming wall A steep, ridge-like erosional remnant of continuous layers of porous, permeable, poorly cemented, detrital limestones, believed to form under tropical or subtropical conditions (as on Okinawa and other Pacific islands) by surface-controlled secondary cementation of an original steep slope followed by differential erosion that brings the cemented zone into relief (Flint et al, 1953). See also: *rampart wall*.

rimpylite A group name for several green and brown hornblendes having high contents of $(Al,Fe)_2O_3$.

rim ridge A minor ridge of till defining the edge of a *moraine plateau* (Hoppe, 1952, p.5).

rimrock [eco geol] The bedrock forming or rising above the margin of a placer or gravel deposit. Also spelled: *rim rock*.

rimrock [geomorph] (a) An outcrop of a horizontal layer of resistant rock (as a lava flow) exposed on the edge of or overlying a plateau, butte, or mesa; it generally forms a cliff or ledge overlooking lower ground. (b) The edge or face of an outcrop of rimrock, esp. a cliff or a relatively vertical face of rock in the wall of a canyon. Syn: *rim*.

rimrocking Prospecting for carnotite, specifically on the Colorado Plateau, where the favorable beds, more or less flatlying, crop out in cliffs, or rims.

rimstone In a cave, a thin crustlike deposit of calcite that forms a ring around an overflowing basin or pool of water.

rimstone bar *rimstone dam*.

rimstone barrier *rimstone dam*.

rimstone dam A formation of rimstone that forms a pool or basin called a *rimstone pool*. Syn: *gour; rimstone barrier; rimstone bar; travertine dam; travertine terrace; terraced flowstone*.

rimstone pool The pool or basin of water that is formed of and bounded by a *rimstone dam*. Syn: *gour*.

rim syncline In salt tectonics, a local depression that develops as a border around a salt dome, as the salt in the underlying strata is displaced toward the dome. See also: *residual anticline*. Syn: *peripheral sink*.

rin A promontory. Also, a point of flat land running into the sea. Etymol: Celtic. Syn: *rinn; rhyn; rhinn.*

rincon (a) A term used in the SW U.S. for a square-cut recess or hollow in a cliff or a reentrant in the borders of a mesa or plateau. Cf: *cove.* (b) A term used in the SW U.S. for a small, secluded valley, and for a bend in a stream.--- Etymol: Spanish *rincón,* "inside corner, nook".

rindle An English syn. of *runnel.*

ring [geol] *ring structure.*

ring [paleont] The precursor to ascending branches of premagadiniform loop of terebratellid brachiopods, "consisting of thin circular ribbon, narrow ventrally and broadening dorsally to its attachment on median septum" (TIP, 1965, pt.H, p.152).

ring canal A hollow tube or *canal* forming a closed ring about the mouth of an echinoderm and from which radial canals branch.

ring complex An association of the ring-shaped igneous intrusive forms *ring dikes* and *cone sheets.*

ring current One of a system of electric currents in the equatorial region of the magnetosphere which causes depression of the magnetic field inside the region of the ring currents. During magnetic storms, the effect at the Earth's surface may be a few hundred gamma; at quiet times, it is approximately 20-40 gamma.

ring depression The annular, structurally depressed area surrounding the central uplift of a cryptoexplosion structure. Faulting and folding may be involved in its formation. Syn: *ring syncline; peripheral depression.*

ring dike A dike that is arcuate or roughly circular in plan and is vertical or inclined away from the center of the arc. It is associated with *cone sheets* to form *a ring complex.* Syn: *ring-fracture intrusion.*

ringer A thin bed of tough, tightly cemented, fine-grained sandstone that gives out a clear, resonant sound when struck with a hammer.

ring fault A steep-sided fault pattern that is cylindrical in outline and that is associated with *cauldron subsidence.* Syn: *ring fracture.*

ring fissure A roughly circular desiccation crack formed on a playa around a point source (generally a phreatophyte).

ring fracture *ring fault.*

ring-fracture intrusion *ring dike.*

ring-fracture stoping Large-scale *magmatic stoping* that is associated with *cauldron subsidence.* Cf: *piecemeal stoping.*

ring hill An isolated, till-covered hill in Lapland, which remained above the marine limit and is surrounded by a very pronounced ring of bedrock washed clear of material (Stephens & Synge, 1966, p.28).

ringing *singing.*

ringite An igneous rock formed by the mixing of silicate and carbonatite magmas.

ring mark A *saltation mark* produced by a fish vertebra, consisting of a ring-like ridge whose higher side is upcurrent; often the ring is incomplete, forming a semicircle that is concave downcurrent.

ring moor A *string bog* with concentric ridges.

ring ore *cockade ore.*

ring plain A lunar crater of exceptionally large diameter and with a relatively smooth interior. See also: *walled plain.*

ring-porous wood Wood in which the pores (vessels) of one part of an annual ring are distinctly different in size or number or both from those in the other part of the ring (Fuller & Tippo, 1949, p.970).

ring reef *atoll.*

ring silicate *cyclosilicate.*

ring structure A general term for an epigenetic structure with ring-shaped trace in plan; e.g. a ring dike, and a lunar crater resulting from a meteorite impact. Cf: *annulation.* Syn: *ring.*

ring syncline *ring depression.*

ringwall A bordering wall that encircles a mare or crater on the surface of the Moon, formed in part by the mountains and lesser eminences of lunarite.

ringwoodite A purple mineral: $(Mg,Fe)_2SiO_4$. It is a cubic dimorph of olivine.

rinkite *mosandrite.*

rinkolite *mosandrite.*

rinneite A colorless, pink, violet, or yellow rhombohedral mineral: NaK_3FeCl_6. It is isomorphous with chlormanganokalite. Also spelled: *rinneïte.*

Rinnental A syn. of *tunnel valley.* Etymol: German, "channel valley".

rio A term used in SW U.S. for a river or stream, usually a permanent stream. Etymol: Spanish *río.*

rip (a) A turbulent agitation of water, generally caused in the sea by the meeting of water currents or the interaction of currents and wind, or in a river or a nearshore region by currents flowing rapidly over an irregular bottom; an *overfall.* See also: *tide rip; current rip.* (b) An abbreviated form of *ripple* [current], often used in the plural.

ripa A legal term for the bank of a stream or lake (Veatch & Humphrys, 1966, p.268).

riparian (a) Pertaining to or situated on the bank of a body of water, esp. of a watercourse such as a river; e.g. "riparian land" situated along or abutting upon a stream bank, or a "riparian owner" who lives or has property on a riverbank. Cf: *riverain.* Syn: *ripicolous; riparial; riparious.* (b) *littoral.*

riparian water loss Discharge of water through evapotranspiration along a water-course, esp. water transpired by vegetation growing along the watercourse. Discharged water may be derived from the watercourse, adjacent ground water, and/or soil moisture.

rip channel A channel, often more than 2 m deep, carved on the shore by a rip current.

rip current A strong, narrow, surface or near-surface current of short duration (few minutes to an hour or two) and high velocity (up to 2 knots) flowing seaward from the shore through the breaker zone at nearly right angles to the shoreline, appearing as a visible band of agitated water returning to the sea after being piled up on the shore by incoming waves and wind; it consists of a *feeder current,* a *neck,* and a *head.* Cf: *undertow.* Often miscalled a *rip tide.*

ripe Said of peat that is in an advanced state of decay. Cf: *unripe.*

ripe snow Snow that is wet throughout so that additional melting can produce meltwater runoff. See also: *ripe-snow area.*

ripe-snow area The area of a drainage basin where coarsely crystalline snow is in a condition to discharge meltwater upon the addition of heat (as by rain); expressed in percent of drainage basin or in square kilometers. Abbrev: RSA. See also: *ripe snow.*

Riphean The most recent era of Precambrian time, as defined by Russian geologists. Approximately equivalent terms are *Sinian, Beltian,* and *Eocambrian.*

ripicolous *riparian.*

ripidolite A mineral of the chlorite group: $(Mg,Fe^{+2})_9$-$Al_6Si_5O_{20}(OH)_{16}$. The name is sometimes applied to *clinochlore.* Syn: *prochlorite; aphrosiderite.*

ripple [current] (a) A syn. of *capillary wave.* (b) The light ruffling of the surface of the water by a breeze.----Syn: *rip.*

ripple [sed struc] A very small ridge of sand resembling or suggesting a ripple of water and formed on the bedding surface of a sediment; specif. a *ripple mark,* or a small sand wave similar to a dune in shape but smaller in magnitude. Syn: *sedimentary ripple.*

ripple [streams] (a) A shallow reach of running water in a stream, roughened or broken into small waves by a rocky or uneven bottom. (b) *riffle.*

ripple bedding (a) A bedding surface characterized by ripple marks. (b) A term preferred by Hills (1963, p.10-11) to *cur-*

rent bedding when used for "the small-scale ripple-like bedding of rapidly deposited sand".

ripple biscuit A bedding structure produced by lenticular lamination of sand in a bay or lagoon (Moore, 1966, p.99).

ripple cross-lamination Small-scale cross-lamination formed by migrating current ripples developed during deposition, characterized by individual laminae whose thicknesses range between 0.08 cm (1/32 in.) and 0.3 cm (1/8 in.) (McKee, 1939, p.72). See also: *ripple lamina*. Syn: *rolling strata*.

ripple drift A term used by Sorby (1857, p.278) for a small-scale pattern of cross-lamination formed by deposition of sediment on both sides of a migrating ripple. See also: *drift bedding; climbing ripple*.

ripple height The vertical distance from crest to trough of a ripple on a ripple-marked surface. If the ripple is asymmetric, the height is measured from the trough adjacent to the steeper (downcurrent) slope. The term was used by Allen (1963, p.192). See also: *ripple-mark amplitude*.

ripple index The ratio of ripple-mark wavelength to ripple-mark amplitude. The ratio usually varies from 3 to 10 for ripples produced by water currents or waves and from 20 to 50 for ripples produced by wind. Syn: *ripple-mark index; vertical form index*.

ripple lamina An internal sedimentary structure formed in sand or silt by currents or waves, as opposed to a ripple mark formed externally on a surface. The term, as commonly used in the plural, "includes sets of laminae in incomplete ripple profiles and isolated ripple lenses, as well as series of superposed rippled layers" (McKee, 1965, p.66). McKee proposed: "ripple laminae-in-rhythm", a general term for all ripple structures superimposed in an orderly sequence; and "ripple laminae-in-phase", a general term for ripple laminae in which the crests of vertically succeeding laminae (as seen in sections parallel to the direction of current or wave motion) are directly above one another. See also: *ripple cross-lamination*. Syn: *ripple lamination*.

ripple load cast A load cast of a ripple mark that shows signs of penecontemporaneous deformation (caused by unequal loading, settling, and compaction) in the accentuation of its trough and crest and in the oversteepening of component laminae (Kelling, 1958, p.120-121).

ripple mark (a) An undulatory surface or surface sculpture consisting of alternating, subparallel, usually small-scale ridges and hollows of primary origin formed at the interface between a fluid and incoherent sedimentary material (esp. granular material such as loose sand) on land by wind action and subaqueously by currents or by the agitation of water in wave action, and trending at right angles or obliquely to the direction of flow of the moving fluid. It is no longer regarded as evidence solely of shallow water. (b) One of the small and fairly regular ridges, of various shapes and cross sections, produced on a ripple-marked surface; esp. a *ripple* preserved in consolidated rock as a structural feature of original deposition on a sedimentary surface and useful in determining the environment and order of deposition. The term was formerly restricted to *symmetric ripple marks*, but now includes *asymmetric ripple marks*. See also: *sand wave; wavemark*. Syn: *fossil ripple; ripple ridge*. (c) A corrugation on a snow surface, produced by wind.---The singular form may be used to denote general ripple structure (as well as a specific ripple), and the plural form to describe a particular example. Also spelled: *ripple-mark*.

ripple-mark amplitude The height of a ripple on a ripple-marked surface, measured as the vertical distance between the crest of the ripple and the adjacent trough; it is generally a centimeter or less. This use of the term "amplitude" is at variance with the convention used in physics and mathematics in which "amplitude" refers to displacement relative to a mean or equilibrium value. See also: *ripple height*.

ripple-mark index *ripple index*.

ripple-mark wavelength The horizontal distance from crest to crest (or trough to trough) of a ripple on a ripple-marked surface, measured parallel to the direction of current or wave propagation; it is generally several centimeters. Syn: *ripple spacing*.

ripple ridge A *ripple mark*.

ripple scour A shallow, linear trough with transverse ripple mark (Potter & Glass, 1958, pl.5).

ripple spacing *ripple-mark wavelength*.

ripple symmetry index A term used by Tanner (1960, p.481) to express the degree of symmetry of a ripple mark, defined as the ratio of the horizontal length of the gentle (upcurrent) side to that of the steep (downcurrent) side; an asymmetric ripple mark has an index greater than 1. Abbrev: RSI. See also: *horizontal form index*.

ripple till A till sheet containing low, winding, smooth-topped ridges, 6-15 m high and 200-3000 m long, lying at right angles to the direction of ice movement, and grouped into narrow belts up to 80 km long that are generally parallel to the direction of ice movement (F.K. Hare in Stamp, 1961, p.395); found in parts of northern Ontario. Syn: *ribble*.

rippling A surface characterized by ripple mark; a collective term for a series or occurrences of ripples or ripple marks.

riprap (a) A layer of large, durable, dense, specially selected and graded, broken rock fragments thrown together irregularly (as in deep water or on a soft bottom) to prevent erosion through wave action, tidal forces, or strong currents and thereby preserve the shape of a surface, slope, or underlying structure. It is used for irrigation channels, river-improvement works, spillways at dams, and shore protection (seawalls). Riprap may be cemented. (b) The stone used for riprap.

rip tide A popular, but improper, term used as a syn. of *rip current*. The usage is erroneous because a rip current has no relation to the tide. Also spelled: *riptide*.

rip-up (a) Said of a sedimentary structure formed by shale clasts (usually of flat shape) that have been "ripped off" by currents from a semiconsolidated mud deposit and transported to a new depositional site. (b) Said of a clast in a rip-up structure.

rise [streams] (a) A syn. of *resurgence*, or the point where an underground stream reappears at the surface. (b) *karst spring*.

rise [geomorph] (a) An upward slope in the land. (b) The top part of a hill or other landform that is higher than the surrounding ground.

rise [marine geol] A broad, elongate, and smooth elevation of the ocean floor. Syn: *swell [marine geol]*.

rise pit A pit through which an underground stream rises to the surface with a calm and steady flow.

riser The vertical or steeply sloping surface of one of a series of natural step-like landforms, as those of a glacial stairway or of successive stream terraces. Ant: *tread*.

rising (a) A syn. of *resurgence*, or a reappearance of an underground stream. (b) *karst spring*.

rising dune *climbing dune*.

rising tide That part of a tide cycle between low water and the following high water, characterized by landward or advancing movement of water. Also, an inflowing tidal river. Ant: *falling tide*. Syn: *flood tide*.

Riss (a) European stage: Pleistocene (above Mindel, below Würm). (b) The third glacial stage of the Pleistocene Epoch in the Alps, after the Mindel-Riss interglacial stage. See also: *Illinoian; Saale*.---Etymol: Riss River, Germany. Adj: *Rissian*.

Riss-Würm A term applied in the Alps to the third interglacial stage of the Pleistocene Epoch, after the Riss glacial stage and before the Würm. See also: *Sangamon*.

rithe An English term for a small stream. Syn: *rive*.

rivadavite A mineral: $Na_6MgB_{24}O_{40} \cdot 22H_2O$.

rive *rithe*.

river [coast] A term used in place names for an estuary, la

on, tidal river, inlet, or strait; e.g. York River, Va., and Indi-
n River, Fla.

ver [gem] A pure-white diamond of very high grade. See
so: *extra river*. Cf: *water*.

ver [streams] (a) A general term for a natural, freshwater,
urface *stream* of considerable volume and a permanent or
asonal flow, moving in a definite channel toward a sea,
ke, or another river; any large stream, or one larger than a
ook or a creek, such as the trunk stream and the larger
anches of a drainage system. (b) A term applied in New
ngland to a small watercourse which elswhere in the U.S. is
nown as a *creek*.

verain Pertaining to a riverbank; situated on or near a river.
he term has a wider meaning than *riparian*.

verbank The bank of a river.

ver bar A ridge-like accumulation of alluvium in the channel,
ong the banks, or at the mouth, of a river. It is commonly
mergent at low water and constitutes a navigational obstruc-
on.

ver-bar placer *bench placer*.

ver basin The entire area drained by a river and its tribu-
ries. Cf: *drainage basin*.

ver bed The channel containing or formerly containing the
ater of a river. Also spelled: *riverbed*.

ver bluff A bluff or steep hillslope or line of slopes proximal
a river bank. Cf: *river cliff*.

ver bottom The low-lying alluvial land along a river.

ver breathing Fluctuation of the water level of a river (ASCE,
962). Syn: *breathing*.

ver capture *capture* [streams].

ver cliff The steep *undercut slope* formed by the lateral ero-
on of a river. Cf: *river bluff*.

ver-delta marsh A brackish or freshwater marsh bordering
e mouth of a distributary stream.

ver-deposition coast A deltaic coast characterized by lobate
eaward bulges crossed by river distributaries and bordered
y lowlands (Shepard, 1948, p.72).

ver drift Rock material deposited by a river in one place after
aving been moved from another.

ver end The lowest point of a river with no outlet to the sea,
tuated where its water disappears by percolation or evapora-
on (Swayne, 1956, p.121).

iver engineering A branch of civil engineering that deals with
e control of rivers, their improvement, training, regulation,
nd flood mitigation.

iveret An obsolete syn. of *rivulet*.

iver flat An *alluvial flat* adjacent to a river; a bottom.

iver forecasting Forecasting the river stage and discharge, by
ydrology and meteorology, including research into forecast-
g methods. In some countries, the term hydrometeorology is
sed with this limited meaning (ASCE, 1962).

verhead The source or beginning of a river.

iver ice Ice formed on a river, regardless of observed loca-
on; ice carried by a river.

iverine (a) Pertaining to or formed by a river; e.g. a "riverine
ake" created by a dam across a river. (b) Situated or living
long the banks of a river; e.g. a "riverine ore deposit".

iverlet A little river.

iver morphology The study of the *channel pattern* and the
hannel geometry at several points along a river channel, in-
luding the network of tributaries within the drainage basin.
yn: *channel morphology; fluviomorphology; stream morpho-
gy*.

iver pattern A river *channel pattern*.

iver-pebble phosphate A term used in Florida for a trans-
orted, dark variety of *pebble phosphate* obtained from bars
nd flood plains of rivers. Cf: *land-pebble phosphate*. Syn:
iver pebble; river rock.

iver piracy *capture* [streams].

iver plain *alluvial plain*.

river profile The *longitudinal profile* of a river.

river rock A syn. used in South Carolina for *river-pebble phosphate*.

river run gravel Natural gravel as found in deposits that have
been subjected to the action of running water (Nelson, 1965,
p.373).

rivershed The drainage basin of a river.

riversideite A white mineral: $Ca_5Si_6O_{16}(OH)_2.2H_2O$.

river system A river and all of its tributaries. Syn: *water system*.

river terrace *stream terrace*.

river valley An elongate depression of the Earth's surface,
carved by a river during the course of its development.

riverwash (a) Soil material that has been transported and deposited by rivers. (b) An alluvial deposit in a river bed or flood
channel, subject to erosion and deposition during recurring
flood periods.

riviera A resort coastline much frequented by tourists, usually
having extensive sandy beaches and a mild climate. Type locality: the Riviera along the coast of the Mediterranean Sea
between Marseille, France, and La Spezia, Italy.

riving The splitting off, cracking, or fracturing of rock, esp. by
frost action. See also: *congelifraction*.

rivotite A mixture of malachite and stibiconite.

rivulet (a) A small stream; a brook or a runnel. (b) A small
river. (c) A streamlet developed by rills running down a steep
slope.---Obsolete syn: *riveret*.

rizalite A *philippinite* tektite from Rizal.

rizzonite A local variety of limburgite occurring on Monte Riz-
zoni, Italy.

r-meter A device that measures X-ray or gamma-ray intensity.
Syn: *roentgen meter; roentgenometer*.

road [coast] A *roadstead*. Term is usually used in the plural;
e.g. Hampton Roads, Va.

road [glac geol] One of a series of erosional terraces in a glacial valley, formed as the water level dropped in an ice-
dammed lake. See also: *parallel roads*.

road log A descriptive record of the route taken on a field trip
and of the geology observed along it.

roadstead An area of water near a shore, sheltered by a reef,
sandbank, or island, or an open anchorage, usually a shallow
indentation in the coast, where vessels may lie in relative
safety from winds and heavy seas; it is often outside, and less
sheltered than, a harbor. An "open roadstead" is unprotected
from the weather. Syn: *road*.

roaring sand A *sounding sand*, found on a desert dune, that
sets up a low roaring sound that sometimes can be heard for
a distance of 400 m. See also: *booming sand*.

robbery *capture* [streams].

robinsonite A mineral: $Pb_7Sb_{12}S_{25}$.

rocdrumlin *rock drumlin*.

roche moutonnée A small, elongate, protruding knob or hillock
of bedrock, most commonly granitic, so sculptured by a large
glacier as to have its long axis oriented in the direction of ice
movement, and characterized by an upstream (stoss or
scour) side that is gently inclined and smoothly rounded but
striated and by a downstream (lee or pluck) side that is
steep, rough, and hackly. It is usually a few meters in height,
length, and breadth. The term is commonly supposed to have
been originated by Saussure, but there is no record that he
ever used it (Longwell, 1933); Saussure (1786, par.1061,
p.512-513) did use "moutonnée", a French adjective meaning
"fleecy", ruffled, or curled", in describing an assemblage of
rounded knobs in the Alps because, when seen at a distance,
the rocks reminded him of "perruques moutonnées" (wigs
made of hair so curled as to resemble sheep's wool). The
term thus came to connote a resemblance between a single
knob of the character described and a grazing sheep. Much
later, the term was applied to a single glaciated knob so
roughened by plucking as to resemble a sheep's back in sur-

face texture as well as in general form, but this kind of surface is not generally regarded as essential. Pl: *roches moutonnées*. Pron: rosh moo-to-naa. See also: *moutonnée*. Syn: *sheepback; sheepback rock; sheep rock; whaleback; embossed rock*.

rochlederite The soluble resin extracted from *melanchym* by alcohol. See also: *melanellite*.

rock [geol] (a) Any naturally formed, consolidated or unconsolidated material (but not soil) composed of two or more minerals, or occasionally of one mineral, and having some degree of chemical and mineralogic constancy; also, a representative sample of such material. (b) Popularly, any hard, consolidated material derived from the Earth and usually of relatively small size. Partial syn: *stone* [geol].

rock [coast] A jagged, rocky coastline, esp. where dangerous to shipping.

rock [geomorph] (a) Any notable, usually bare peak, cliff, promontory, or hill considered as one mass; e.g. the *rock* of Gibraltar. (b) A rocky mass lying at or near or projecting above the surface of a body of water.

rock [gem] A slang term for a gem or a diamond.

rockallite A coarse-grained, mafic, alkalic granite composed of quartz, acmite, albite, and microcline.

rock asphalt *asphalt rock*.

rock association A group of igneous rocks within a petrographic province that are related chemically and petrographically, generally in a systematic manner such that chemical data for the rocks plot as smooth curves on variation diagrams. See also: *tribe*. Syn: *rock kindred; kindred; association* [petrology].

rock avalanche The very rapid downslope flowage of rock fragments, during which the fragments may become further broken or pulverized. Rock avalanches typically result from very large rockfalls and rockslides, and their patterns of displacement have led to the term *rock fragment flow* (Varnes, 1958). Characteristic features include chaotic distribution of large blocks, flow morphology and internal structure, relative thinness in comparison to large areal extent, high porosity, angularity of fragments, and lobate form.

rock baby An odd-shaped, protruding hill of sandy bedrock, produced by differential erosion in the desert region of the Henry Mountains, Utah (Hunt et al, 1953, p.175).

rock bar *riegel*.

rock basin A depression in solid rock, sometimes of great extent; esp. one formed by local erosion of the uneven floor of a cirque or glacial valley in a mountainous region, and usually containing a lake. See also: *glacial basin*.

rock-basin lake A glacial lake occupying a rock basin; e.g. a *paternoster lake*.

rock bench (a) A narrow valley-side niche developed during backweathering of weaker beds in a stratified rock section; a *structural bench* cut in solid rock. (b) A *wave-cut bench* produced on a rock surface.

rock bind An English term for a sandy shale or a banded or nonbanded siltstone.

rock bolt A bar, usually of steel, used in *rock bolting*. It is usually at least one meter in length and 2 cm in diameter, and it is provided with a device for expanding the leading end so that it is anchored firmly in rock. Rock bolts are classified according to the means by which they are secured or anchored in rock: expansion, wedge, grouted, and explosive. Also spelled: *rockbolt*. Syn: *roof bolt*.

rock bolting A method of securing or strengthening closely jointed or highly fissured rocks in mine workings, tunnels, or rock abutments by inserting and firmly anchoring *rock bolts* in predrilled holes that range in length from one meter to about 12 m and that are oriented perpendicular to the rock face or mine opening.

rock borer Any of certain bivalve mollusks that live in cavities they have bored into soft rock, concrete, or other material, usually by rotating the shell. Cf: *saxicavous*.

rockbridgeite A mineral: $Fe^{+2}Fe_6^{+3}(PO_4)_4(OH)_8$.

rock burst A sudden and often violent breaking, expulsion, burst of a mass of rock from the walls of a quarry, tunr mine, or other near-surface or underground opening, caus by failure of highly stressed rock and the very rapid or insta taneous release of accumulated strain energy, often with c sure of, or projection of broken rock into, the opening, a accompanied by ground tremors, rockfalls, and air concu sions. Rock bursts are not likely to occur until a depth about 1000 meters (about 3000 ft) below the surface reached. See also: *crush burst; strain burst; pressure bur* Also spelled: *rockburst*.

rock cave *shelter cave*.

rock cork *mountain cork*.

rock crystal Quartz that is transparent or nearly so and that usually colorless, having a low refractive index resulting in l brilliancy, and used for lenses, wedges, and prisms in optic instruments and for frequency control in electronics, or fas ioned into beads or other ornamental objects. It may or m not be in distinct crystals. Syn: *crystal; berg crystal; mounta crystal; quartz crystal; pebble*.

rock cycle A sequence of events leading to the formation, teration, destruction, and reformation of rocks as a result such processes as magmatism, erosion, transportation, dep sition, lithification, and metamorphism. One possible sequen in the cycle involves the crystallization of magma to form i neous rocks which may then be broken down to sediment a result of weathering, the sediments later being lithified a possible melted to form another magma as a result of met morphism.

rock-defended terrace (a) A river terrace protected from lat undermining by a projecting ledge or outcrop of resistant ro at its base (or at successively lower levels of the river). (b) marine terrace protected from wave erosion by a mass of r sistant rock at the base of the wave-cut cliff formed in th overlying coastal-plain sediments undergoing marine erosior -Syn: *rock-perched terrace*.

rock desert An upland desert area in which bedrock has bee exposed after the removal of sand and dust particles by win erosion, or in which bedrock is covered by a thin veneer coarse rock fragments; e.g. a *hammada*. Cf: *stony dese*. Syn: *rocky desert*.

rock doughnut A raised, rounded, annular ridge encircling weather pit, as one of those occurring on certain grani domes of central Texas (Blank, 1951). Syn: *doughnut*.

rock drift [mass move] *rock creep*.

rock drill (a) A machine for boring or making holes in roc such as by percussion (e.g. a jackhammer) effected by rec procating motion or by abrasion (e.g. a rotary drill) effecte by rotary motion. (b) A conical bit for drilling hard rock.

rock drumlin A smooth, streamlined hill, having a core of be rock usually veneered with a layer of till; it is modelled by gla cial erosion, and its long axis is parallel to the direction of ic movement. It is similar in outline and form to a true *drumli* but is generally less symmetrical and less regularly shape Syn: *false drumlin; rocdrumlin; rock drum; drumlinoid*.

rock fabric *fabric* [struc petrol].

rock face [geomorph] An exposed surface of rock in a wall c cliff.

rock failure *failure*.

rockfall (a) The relatively free falling or precipitous movemen of a newly detached segment of bedrock (usually massive homogeneous, or jointed) of any size from a cliff or other ve steep slope; it is the fastest-moving landslide and is most fr quent in mountain areas and during spring when there is re peated freezing and thawing. Movement may be straigh down, or a series of leaps and bounds down the slope; it i not guided by an underlying slip surface. (b) The mass c rock moving in or moved by a rockfall; a mass of fallen rocks --Also spelled: *rock fall*.

ckfall avalanche A rockfall that has turned into a flow, oc-rring only when large rockfalls and rockslides, involving mil-ns of metric tons, attain extremely rapid speeds; most com-on in a rugged mountainous region, as that which occurred 1903 at Frank, Alberta (McConnell & Brock, 1904).

ck fan An eroded, convex, fan-shaped bedrock surface hav-g its apex at the point where a mountain stream debouches on a piedmont slope, and occupying the zone where a pedi-ent meets the mountain slope (assuming that the mountain nt retreats as a result of lateral planation). According to hnson (1932), a *pediment* evolves from a coalescence of ck fans, although the term "rock fan" is often considered an uivalent of "pediment".

ck-fill dam A dam composed primarily of large, broken, and osely placed or pervious rocks with either an impervious re or an upstream facing.

ck-floor robbing A form of *sheetflood erosion* in which eetfloods remove crumbling debris from rock surfaces in sert mountains (Cotton, 1958, p.258).

ck flour Finely comminuted, chemically unweathered mate-al, consisting of silt- and clay-sized angular particles of rock-rming minerals, chiefly quartz, formed when rock fragments e pulverized while being transported or are crushed by the eight of superincumbent material. The term is most com-only applied to the very fine powder that is formed when ones embedded in a glacier or ice sheet abrade the underly-g rocks, and that is deposited as the matrix in till in outwash posits. Syn: *glacier meal; glacial meal; glacial flour; rock eal.*

ck flowage *flow* [exp struc geol].

ck-forming Said of those minerals that enter into the com-osition of a rock, and determine its classsification. The more portant rock-forming minerals include quartz, feldspars, icas, amphiboles, pyroxenes, olivine, calcite, and dolomite hallinor, 1967).

ck fragment flow *rock avalanche.*

ck generator *spinner magnetometer.*

ck glacier A mass of poorly sorted angular boulders and fine aterial cemented by interstitial ice a meter or so below the rface, occurring in high mountains in a permafrost area, d derived from a cirque wall or other steep cliff by frost ac-on. It has the general appearance and slow movement of a all valley glacier, varying from a few hundred meters to ore than a kilometer in length, and having a distal area arked by a series of transverse, arcuate, and rounded rid-es. When active, it may be 50 m thick with a surface move-ent (resulting from the flow of interstitial ice or from frost ction) of 0.5-2 m/yr. Rock glaciers are numerous in Alaska here they are classified in plan as lobate, tongue-shaped, or patulate. Cf: *rock stream; chrystocrene.* Syn: *talus glacier.*

ck-glacier creep A rapid *talus creep* of tongues of rock aste in a cold region, caused by the expansive force of the ternate freeze and thaw of ice in the interstices of the rock aste (Sharpe, 1938, p.31).

ck gypsum Massive, coarsely crystalline to earthy, finely ranular, often impure gypsum occurring in *gyprock.*

ck hole An Australian term for a *rock tank.*

ck hound (a) An amateur mineralogist. (b) A petroleum ex-loration geologist. ----See also: *pebble pup.*

cking stone A stone or boulder, often of great size, so finely oised upon its foundation (as on the side of a hill or cliff) at it can be rocked or slightly moved backward and forward ith little force (as with the hand) and still retain its original osition. It may be a glacial erratic or a rounded residual lock formed in place by weathering. See also: *logan stone; alanced rock; elephant rock.* Syn: *roggan.*

ck island (a) *meander core.* (b) A bedrock hill surrounded y alluvium in an aggraded stream valley.

ck kindred *rock association.*

ck leather *mountain leather.*

rock magnetism The study of the origins and characteristics of magnetization in rocks and minerals.

rock mantle *regolith.*

rock meal (a) *rock milk.* (b) *rock flour.* (c) *diatomaceous earth.*

rock mechanics The theoretical and applied science of the mechanical behavior of rocks, representing a "branch of me-chanics concerned with the response of rock to the force fields of its physical environment" (NAS-NRC, 1966, p.3).

rock milk (a) A soft, white, earthy or powdery variety of cal-cite occurring in caves or fissures of limestone or as an efflo-rescence. Syn: *rock meal; agaric mineral; bergmehl; fossil fa-rina.* (b) *lublinite.*

rock mill A *pothole* in a stream bed.

rock pedestal *pedestal* [geomorph].

rock pediment A *pediment* developed on a bedrock surface.

rock pendant *pendant.*

rock-perched terrace *rock-defended terrace.*

rock phospahte *phosphorite.*

rock pillar (a) A column of rock produced by differential weathering or erosion, as along a joint plane; a *hoodoo.* (b) In a cave, a pillar-type structure that is residual bedrock rather than a stalacto-stalagmite.

rock plane A term used by Johnson (1932) as a syn. of *pedi-ment;* the term is not recommended in this usage as there are many approximately planate rock surfaces that lack the areal extent and climatic restriction of a pediment.

rock platform (a) A *wave-cut platform* eroded on a rock sur-face. (b) A *high-water platform* eroded on a rock surface.

rock pool A *tide pool* formed along a rocky shoreline.

rock pressure (a) The pressure exerted by surrounding solids upon the supports of underground openings, including that due to the weight of the overlying material, residual unrelieved stresses, and pressures associated with swelling clays (Stokes & Varnes, 1955, p.125). (b) The compressive stress within the solid body of underground geologic material. (c) *geostatic pressure.*

rock river A very long and narrow *rock stream.*

rock ruby A fine red variety of garnet; specif. *pyrope.*

rock salt (a) Coarsely crystalline *halite* occurring as a mas-sive, fibrous, or granular aggregate, and constituting a nearly pure sedimentary rock that may occur in domes or plugs or as extensive beds resulting from evaporation of saline water. It is frequently stained by iron or mixed with fine-grained sedi-ments. (b) Artificially prepared salt in the form of large crys-tals or masses.

rock sea *felsenmeer.*

rock series *igneous rock series.*

rock shelter A cave that is formed by a ledge of overhanging rock, e.g. by erosion of an underlying area, as in limestone terrain of tropical areas, or as in a sea cave.

rock silk A silky variety of asbestos.

rockslide (a) A landslide involving a downward and usually sudden and rapid movement of newly detached segments of bedrock sliding or slipping over an inclined surface of weak-ness, as a surface of bedding, jointing, or faulting, or other preexisting structural feature. The moving mass is greatly de-formed and usually breaks up into many small independent units. Rockslides frequently occur in high mountain ranges, as the Alps or Canadian Rocky Mountains. (b) The mass of rock moving in or moved by a rockslide.---Also spelled: *rock slide.* Syn: *rock slip.*

rockslide avalanche A large rockslide that has turned into a flow, as that which occurred in 1925 along the valley of the Gros Ventre River, Wyo. (Alden, 1928).

rock slip *rockslide.*

rock soap *mountain soap.*

rock stack A rocky crag that has been uplifted from an old sea floor (Fairbanks, 1906, p.141).

rock step (a) A *knickpoint* produced by the outcrop of a resistant rock. (b) One of a series of ledges or other irregularities of gradient in the upper reaches of a hanging valley; an abrupt descent in the floor of a glacial valley.

rock-stratigraphic unit A stratigraphic unit having a substantial degree of lithologic homogeneity, consisting of a body of strata that is unified with respect to adjacent strata by possessing certain objective physical features observable in the field or subsurface, or consisting dominantly of a certain rock type or combination of rock types, and considered completely independent of time; "a subdivision of the rocks in the Earth's crust distinguished and delimited on the basis of lithologic characteristics" (ACSN, 1961, art.4). It is essentially the practical unit of general geologic work. Its name is binomial, consisting preferably of a geographic name (derived from an appropriate feature of the type area) combined with a descriptive lithologic term (usually the dominant rock type) or with the appropriate rank term alone. Rock-stratigraphic units in order of decreasing rank: group, formation, member, and bed. See also: *lithogenetic unit.* Syn: *lithostratigraphic unit; lithostratic unit; lithologic unit; rock unit; geolith.*

rock-stratigraphy *lithostratigraphy.*

rock stream An accumulation of rock fragments, typically boulders and large blocks of rock, occurring in a valley bottom or on a slope along which it moves or has moved slowly under its own weight or aided by frost action. It may be produced by solifluction or frost heaving, by washing away of finer materials in a bouldery deposit such as till, or by rock falling or sliding from valley walls. The feature was first described by Cross & Howe (1905) as a special landslide. Cf: *rock glacier.* Syn: *stone river; stone stream; rock river; block stream; boulder stream.*

rock stripe *stone stripe.*

rock tank A natural *tank* formed in rock by differential weathering or differential erosion. See also: *sand tank.* Syn: *rock hole.*

rock terrace A *stream terrace* produced on the side of a valley by erosion in horizontal beds of unequal resistance, composed of strong bedrock that is worn back less rapidly than the weaker beds above and below. Cf: *alluvial terrace.* Syn: *stream-cut terrace; cut terrace; erosion terrace.*

rock train A term suggested by Kendall & Wroot (1924, p.448) for the rock material in "process of transport at the sides and in the middle of a glacier" and "subject to the dynamic forces of the glacier".

rock type [coal] *banded ingredients.*

rock type [petrology] (a) In classifying rocks, one of the three major groups of rocks: igneous, sedimentary, metamorphic. (b) A particular kind of rock having a specific set of characteristics. It may be a general classification, e.g. a basalt, or a specific classification, e.g. a basalt from a particular area and having a unique description.

rock unit *rock-stratigraphic unit.*

rock waste Loose material resulting from weathering of rock by mechanical and chemical means; a syn. of *debris.*

rock weathering The chemical decomposition and mechanical disintegration of rocks in place, at or near the Earth's surface.

rockwood A name applied to a type of sedimentary iron ore mined at Rockwood, Tennessee.

rock wood *mountain wood.*

rock wreath *sorted circle.*

rocky desert *rock desert.*

Rocky Mountain orogeny A name proposed by W.H. White (1959) for a time of major folding and thrusting during Late Cretaceous and Paleocene time in the Rocky Mountains of eastern British Columbia and adjacent Alberta; it is broadly equivalent to the *Laramide orogeny* of western United States.

rod [paleont] (a) An elongate holothurian sclerite having a circular cross section, one or more axes, and an *eye* at its end. (b) A part of a heterococcolith having one dimension large and two much smaller.

rod [sed] A rod-like or prolate shape of a sedimentary particle, defined in *Zingg's classification* as having a width/length ratio less than 2/3 and a thickness/width ratio greater than 2/3. Syn: *roller.*

rod [surv] (a) A bar or staff for measuring, such as a graduated pole used as a target in surveying; specif. a *level rod.* (b) A unit of length equal to 16.5 ft. Also called a *perch; pole.*

rodding In metamorphic rocks, a linear structure in which the stronger parts, such as vein quartz or quartz pebbles have been shaped into parallel rods. Whether the structure has formed parallel to the direction of transport or parallel to the fold axes has been debated. Cf: *mullion.*

roddon A term used in East Anglia, Eng., for a natural levee built of sediment carried upstream by the tide rather than downstream by a river.

rodingite A medium- to coarse-grained, commonly calcium-enriched gabbroic rock containing, as essential minerals, grossular and diallage. Altered varieties also contain prehnite or serpentine or both.

rodite An obsolete syn. of *diogenite.*

rod level A spirit level attached to a level rod or stadia rod to assure a vertical position of the rod prior to instrument reading.

rodman One who uses or carries a surveyor's level rod; *chainman.* Syn: *rodsman.*

roeblingite A white mineral: $Pb_2Ca_7Si_6O_{14}(OH)_{10}(SO_4)_2$.

roedderite A meteorite mineral: $(Na,K)_2(Mg,Fe)_5Si_{12}O_{30}$.

roemerite A rust-brown to yellow mineral: $Fe^{+2}Fe_2^+(SO_4)_4.14H_2O$. Syn: *römerite.*

roentgenite *röntgenite.*

roentgen meter *r-meter.*

roentgenometer *r-meter.*

Roentgen ray A syn. of *X-ray.* Also spelled: *roentgen ray; röntgen ray.*

roesslerite A monoclinic mineral: $MgH(AsO_4).7H_2O$. It is isomorphous with phosphorroesslerite. Also spelled: *rösslerite.*

roestone A rock whose texture resembles the roe of fish; an oolite. Also spelled: *roe stone.*

rofla A term used by E. Desor for an extremely narrow, tortuous gorge, frequently formed by meltwater streams flowing from a glacier (Marr, 1900, p.172 & 314); e.g. the gorge of the Trient, near Vernayaz, Switzerland. Pl: *roflas.*

rogenstein An oolite in which the ooliths are united by argillaceous cement. Etymol: German *Rogenstein,* "roestone". Also spelled: *roggenstein.*

roggan *rocking stone.*

roggianite A mineral: $NaCa_6Al_9Si_{13}O_{46}.20H_2O$.

rognon (a) A small rocky peak or ridge surrounded by glacier ice in a mountainous region. Also, a similar peak projecting above the bed of a former glacier (ADTIC, 1955, p.67). Cf: *nunatak.* (b) A rounded nunatak (Lliboutry, 1958, p.264); *nunakol.*

roil A small section of a stream, characterized by swiftly flowing, turbulent water.

roily (a) Said of muddy or sediment-filled water. Cf: *turbid.* (b) Said of turbulent, agitated, or swirling water.

roll [ore dep] *ore roll.*

roll [sed] A primary sedimentary structure produced by deformation involving subaqueous slump or vertical foundering; e.g. a flow roll and a pseudonodule.

roll [coal mining] *horseback.*

rolled garnet *rotated garnet.*

roller [waves] A general term, usually meaning one of a series of massive, long-crested waves that roll in upon a coast (as after a storm), usually retaining its form until it reaches the beach or shoal. Cf: *comber.*

roller [sed] *rod.*

rolling beach The upper part of an accumulation at the base of a sea cliff of boulders and pebbles being ground to sand d finer particles (Shaler, 1895).

lling prairie A term used in Texas for "a plain of undulating rounded hilly relief" (Hill, 1900, p.7).

ling strata (a) *ripple cross-lamination.* (b) *wavy bedding.*

ling topography Any land surface having a gradual succession of low, rounded hills or undulations that impart a wave ect to the surface; esp. a land surface much varied by any small hills and valleys.

ll mark One of a series of similar tool marks following each her in a line parallel to the direction of the current, produced an object that was rolled along the bottom. Cf: *saltation* ark.

ll-off A measure of the rate at which the seismograph magication (or whatever response characteristic is being considered) decreases with frequency, for a specified interval of equency.

llover A syn. of *dip reversal,* used by Gulf Coast geologists.

ll-up structure *convolutional ball.*

manechite An iron-black to steel-gray mineral: $Mn^{+2}Mn_8^{+4}O_{16}(OH)_4$. Calcium, potassium, sodium, cobalt, d copper are sometimes present. Romanechite has a ownish-black streak, and commonly occurs massive, botryal, reniform, or stalactitic. It is an important ore of manganse. Syn: *psilomelane; black hematite.*

manzovite A dark-brown variety of grossular garnet.

meite A honey-yellow to yellowish-brown mineral occurring minute octahedrons: $(Ca,Fe,Mn,Na)_2(Sb,Ti)_2O_6(O,OH,F)$. so spelled: *romèite.*

merite *roemerite.*

nd A British term for a narrow *washland* that separates a oad from the river; it is connected with the river by an artifial passageway.

ngstockite A medium- to fine-grained plutonic rock composed of zoned plagioclase, orthoclase, some cancrinite, aute, mica, hornblende, magnetite, sphene, and apatite. The ck resembles essexite but contains less nepheline and has odic rather than calcic plagioclase.

ntgenite A wax-yellow to brown mineral: $Ca_2(Ce, a)_3(CO_3)_5F_3$. Syn: *roentgenite.*

ntgen ray Var. of *Roentgen ray.*

of [ore dep] The rock above an orebody.

of [intrus rocks] The country rock bordering the upper surce of an igneous intrusion. Cf: *floor [intrus rocks].*

of bolt A *rock bolt* used to support the roof of a mine or ine shaft.

of collapse *roof foundering.*

of control The scientific study of rock behavior when underined by mining operations, the systematic measurement of e movement of strata and the forces and stresses involved, d the most effective measures to prevent or reduce roof ovements (such as by using support systems and waste acking) (Nelson, 1965, p.378). Syn: *strata control.*

ofed dike A dike that has an upward termination, e.g. a eder.

ofed mud crack *vaulted mud crack.*

of foundering The collapse of the overlying rocks into an nderlying reservoir or magma, usually following the evacuaon of a large quantity of the magma. Less preferred syn: *roof ollapse.*

ofing slate Any slate whose qualities of finely developed leavage, compactness, and homogeneity make it suitable for plitting into the thin slabs required for roofing.

of pendant *pendant.*

oom A general term for an open area in a cave. Partial syn: hamber [speleo].

oom-and-pillar Said of a coral-reef structure characterized by terconnected and roofed-over surge channels.

ooseveltite A white monoclinic mineral: $BiAsO_4$.

ooster tail A plumelike form of water and sometimes spray hat occurs at the intersection of two crossing waves.

root [ore dep] (a) Syn. of *bottom.* (b) The conduit leading up through the basement to an ore deposit in the superjacent rocks.

root [paleont] An expanded, branching, tree-like extension at the distal end of a blastoid stem.

root [fold] The basal part of a fold nappe that was originally linked to its source, or *root zone.*

root [tect] According to the *Airy isostasy* hypothesis, the downward extension of lower-density crustal material as isostatic compensation for its greater mass and high topographic elevation. Syn: *downward bulge.* Cf: *antiroot.*

root cap A thimblelike mass of cells which fits over the growing tip of a rootlet and protects it.

root cast (a) A slender, tubular, near-vertical, and commonly downward-branching sedimentary structure formed by the filling of a tubular opening left by a root. (b) *rhizoconcretion.*

root clay An *underclay* characterized by the occurrence of fossil roots of coal plants. See also: *rootlet bed.*

rootless vent A source of lava that is not directly connected to a volcanic vent or magma source; it may be an accumulation of overflow or an outflow from an otherwise solidified lava flow.

rootlet bed A stratum characterized by the occurrence of fossil rootlets of plants; e.g. a *root clay* beneath a coal bed.

root level The place within a sediment at which plant roots are found in the living position.

root-mean-square The square root of the arithmetic mean of the squares of a set of numbers. Abbrev: *rms.*

root-mean-square deviation *standard deviation.*

root scar *root zone.*

root sheath A hollow *rhizoconcretion.*

root-tuft A tuft of subparallel, elongate spicules protruding from the base of a sponge and serving to fix it in the substrate.

root zone [soil] *rhizosphere.*

root zone [fault] That area in which a low-angle thrust fault becomes steeper and descends into the crust.

root zone [fold] The source or original attachment of the *root* of a nappe. Syn: *root scar.*

ropak A pinnacle or slab of sea ice standing vertically on edge, rising as high as 8 m above the surrounding ice, and representing an extreme formation of *ridged ice.* Etymol: Russian.

rope drilling *cable-tool drilling.*

ropy lava *pahoehoe.*

roquesite A tetragonal mineral: $CuInS_2$.

rosasite A pale- to bright-green or sky-blue mineral: $(Cu, Zn)_2CO_3(OH)_2$. It is a zinc-bearing malachite, and is a secondary mineral of hydrothermal origin.

roscherite A dark-brown monoclinic mineral: $(Ca,Mn,Fe)_3 Be_3(PO_4)_3(OH)_3.2H_2O$.

roscoelite A vanadium-bearing mineral of the mica group, having the approximate formula: $K(V,Al,Mg)_3Si_3O_{10}(OH)_2$. It is tan, grayish brown, or greenish brown, and occurs in minute scales or flakes in the cement of certain sandstones and in certain gold-quartz deposits.

rose [gem] n. (a) *rose cut.* (b) *rose diamond.* (c) A diamond so small that it can be cut little if at all.---adj. Said of a gem having a rose, pink, or lilac color; e.g. "rose topaz".

rose [sed] *rosette.*

rose [surv] (a) *compass rose.* (b) A chart showing true and magnetic courses.

rose cut An early style of cutting for a gemstone, now used primarily on small diamonds; it usually has a flat, unfaceted base and a somewhat dome-shaped top that is covered by a varied number of triangular facets and terminates in a point. Syn: *rose; rosette.*

rose diagram A circular or semicircular, star-shaped graph indicating values or quantities in several directions of bearing, consisting of radiating rays drawn proportional in length to the value or quantity; e.g. a wind rose, a current rose, a structural diagram for plotting strikes of planar features, a polar coordinate representation of a circular frequency distribution, or a "histogram" of orientation data.

rose diamond A rose-cut diamond. Syn: *rose*.

roselite A rose-red monoclinic mineral: $Ca_2(Co,Mg)(AsO_4)_2.2H_2O$. It is isomorphous with brandite and dimorphous with beta-roselite.

rosenbuschite A mineral: $(Ca,Na)_3(Zr,Ti)Si_2O_8F$.

Rosenbusch's law A statement of the sequence and crystallization of minerals from magmas, proposed in 1882 by the German geologist Harry Rosenbusch (d. 1914) and to which many exceptions have been taken.

rosenhahnite A mineral: $3CaSiO_3.H_2O$.

rose opal An opaque variety of common opal having a fine red color.

rose quartz A pink to rose-red and commonly massive variety of crystalline quartz often used as a gemstone or ornamental stone. The color is perhaps due to titanium, and is destroyed or becomes paler on exposure to strong sunlight. Syn: *Bohemian ruby*.

rosette [gem] *rose cut*.

rosette [paleont] (a) A delicate calcareous plate formed of metamorphosed basal plates, centrally located within radial pentagon in some free-swimming crinoids. (b) The group of five petal-shaped ambulacra on certain echinoids. (c) A cluster of parts in circular form; e.g. a discoaster. (d) A flower-shaped button within a hexagonal pore frame in a radiolarian skeleton.

rosette [sed] A radially symmetric, sand-filled, crystalline aggregate or cluster with a fanciful resemblance to a rose, formed in sedimentary rocks by barite, marcasite, or pyrite. See also: *barite rosette*. Syn: *rose*.

rosette plate A multiporous subcircular area in the vertical walls of cheilostomatous bryozoans for the passage of mesodermal fibers between adjacent zooids.

rosette texture A flowerlike or scalloped pattern of a mineral aggregate.

rosickyite A mineral: γ-S. It consists of native *sulfur* in the gamma crystal form. Syn: *gamma-sulfur*.

rosieresite A yellow to brown mineral consisting of a hydrous phosphate of lead, copper, and aluminum.

rosin jack A yellow variety of sphalerite. Syn: *resin jack*.

rosin tin A reddish or yellowish variety of cassiterite. Syn: *resin tin*.

Rosiwal analysis In petrography, a quantitative method of estimating the volume percentages of the minerals in a rock. Thin sections of a rock are examined with a microscope fitted with a micrometer which is used to measure the linear intercepts of each mineral along a particular set of lines. This method "is based on the assumption that the area of a mineral on an exposed surface is proportional to its volume in the rock mass" (Nelson & Nelson, 1967, p.320).

rosolite *landerite*.

ross A promontory. Etymol: Celtic.

Rossi-Forel scale One of the earthquake *intensity scales*, devised by the Italian geologist Michele Stefano De Rossi and the Swiss naturalist Francois Alphonse Forel, in 1883. It has a range of one to ten. It has been replaced by the *modified Mercalli scale*. Cf: *Richter scale*.

rossite A yellow mineral: $CaV_2O_6.4H_2O$.

rosslerite *roesslerite*.

rostellum A low projection between the anterior adductor muscle scars of the pedicle valve of some craniacean brachiopods to which the internal oblique muscles are attached.

rosterite *vorobyevite*.

rosthornite A brown to garnet-red variety of retinite with a low (4.5%) oxygen content, found in lenticular masses in coal.

rostral plate An anteriorly projecting, unpaired, movably articulated, median extension of the carapace of a malacostracan

crustacean; the *rostrum* of a trilobite or cirripede.

rostrate Having a rostrum; specif. said of a brachiopod with prominent beak of the pedicle valve projecting over a narro cardinal margin.

rostrolateral One of a pair of plates in certain cirriped crust ceans lying between the rostrum and lateral. Syn: *rost latus*.

rostrum (a) A part of an arachnid suggesting a bird's bill; e upper lip of a spider, or tube-like "beak" in the order Solpu ida, or anterior spike of the carapace of the order Eophry idae, similar to the rostrum of a lobster (TIP, 1955, pt. p.62). (b) The anteriorly projecting, unpaired, usually rig spine-like median extension of the carapace of a crustacea e.g. an unpaired plate adjacent to the scuta of a cirriped See also: *rostral plate*. (c) A small median ventral plate of th head region in a trilobite, immediately anterior to the hypo tome. Syn: *epistome; rostral plate*. (d) An elevation of th secondary shell on the inner surface of the brachial valve some craniacean brachiopods, in front of the anterior addu tor muscles, and consisting of a pair of low club-shape protuberances forming the seat of attachment for the brach protractor muscles. Also, the beak of articulate brachiopod (e) The grooved extension of any of many gastropod shel protecting the siphon; the attenuated extremity of the la whorl (of a gastropod shell) other than the siphonal canal, in *Tibia*. Also, the snout of gastropod mollusk when nonretra tile. (f) A pointed projection of peristome on the venter of a ammonoid. Also, the *guard* of a belemnite. (g) A spike-li prolongation of a bryozoan avicularium. Also, a distal part palate of an avicularium occupied by the mandible.---Pl: *rost* or *rostrums*.

rotaliid Any trochospirally, rather than planispirally, coile foraminifer, but not including trochospirally coiled plankton genera.

rotary current A tidal current that flows continually (general clockwise in the Northern Hemisphere) and changes directio progressively through 36° during a tide cycle, and that retur to its original direction after a period of 12.42 hours; occurs open ocean and along the coast where flow is not restricte Cf: *reversing current*.

rotary drill A rock drill that makes holes by a rotary actio the motive power being compressed air, electricity, or stear It is used extensively in oil-well drilling in which the holes m be up to 45 cm (18 in.) in diameter.

rotary drilling The commonest method of drilling, being a h draulic process consisting of a rotating drill pipe at the botto of which is attached a hard-toothed drill bit. The rotary moti is transmitted through the pipe from a rotary table at the su face: as the pipe turns, the bit loosens, or grinds a hole i the bottom material. During drilling, a stream of drilling mud in constant circulation down the pipe and out through the b from where it and the cuttings from the bit are forced back u the hole outside the pipe and into pits where the cuttings ar removed and the mud is picked up by slush pumps and force back down the pipe. The method is now used extensively exploration, esp. when cores are required. Cf: *cable-tool dri ing*.

rotary fault A partial syn. of *hinge fault*. Cf: *scissor fault*.

rotary polarization *optical activity*.

rotary table A chain- or gear-driven circular table or drivir mechanism (at the surface) that rotates the drill pipe and dr bit in rotary drilling, thereby permitting the pipe to slide dow ward as the hole deepens.

rotate Said of a flower whose parts are flat and radiating wheel-shaped.

rotated garnet A garnet crystal that shows evidence that has been rotated during metamorphic movement (Knopf an Ingerson, 1938). Syn: *rolled garnet; pinwheel garnet; spir garnet; snowball garnet*.

rotation [struc petrol] (a) *internal rotation*. (b) *external rotation*.

rotational bomb A pyroclastic bomb whose shape is formed by spiral motion or rotation during flight; rotation produces such shapes as spheroidal, tear-shaped or spindle-shaped bombs. See also: *fusiform bomb*.

rotational cylindroidal fold A cylindrical fold, the axial surface of which has been distorted by a subsequent or cross fold (Whitten, 1959).

rotational deflection The deflection of currents of air or water by rotation of the Earth, as stated in *Ferrel's law*.

rotational fault A partial syn. of *hinge fault*. Cf: *scissor fault*.

rotational flow Flow in which each fluid element rotates about its own mass center.

rotational landslide A landslide in which shearing takes place on a well defined, curved shear surface, concave upward in cross-section, producing a backward rotation in the displaced mass (Hutchinson, 1968). It may be single, successive (repeated up- and-down-slope), or multiple (as the number of slide components decreases).

rotational movement Apparent fault-block displacement in which the blocks have rotated relative to one another, so that alignment of formerly parallel features is disturbed. Cf: *translational movement*. See also: *rotary fault; rotational fault*. Less-preferred syn: *rotatory movement*.

rotational strain *Strain* in which the orientation of the axes of strain is changed. Cf: *irrotational strain*. Syn: *pure rotation*.

rotational transformation A type of crystal *transformation* that is a change from an ordered phase to a partially disordered phase by rotation of groups of atoms. It is usually a rapid process. Cf: *dilatational transformation; displacive transformation; reconstructive transformation; substitutional transformation*.

rotational wave *S wave*.

rotation axis *symmetry axis*.

rotation method A method of X-ray diffraction analysis using a rotating single crystal, monochromatic radiation, and a cylindrical film coaxial with the rotation axis of the crystal.

rotation twin A crystal twin whose symmetry is formed by apparent axial rotation of 180°. Cf: *reflection twin*.

rotatory dispersion In crystal optics, the breaking up of white light into colors by passing it through an optically active substance, such as quartz.

rotatory movement A less-preferred syn. of *rotational movement*.

rotatory reflection axis *rotoreflection axis*.

Rotliegende European series (esp. in Germany): Lower and Middle Permian (below Zechstein). It contains the Autunian and Saxonian stages. Obsolescent syn: *Rothliegende*.

rotoinversion axis A type of crystal symmetry element that combines a rotation of 60°, 90°, 120°, or 180° with inversion across the center. Syn: *symmetry axis of rotoinversion; symmetry axis of rotary inversion*.

rotoreflection axis A type of symmetry element that combines a rotation of 60°, 90°, 120°, or 180° with reflection across the plane perpendicular to the axis. Syn: *rotatory reflection axis*.

rotten ice Ice in which the grains or crystals are loosened, one from the other, forming a *honeycomb structure* [ice], due to melting along grain boundaries. See also: *candle ice*.

rotten spot *pothole* [coast].

rottenstone Any highly decomposed but still coherent rock; specif. a soft, friable, lightweight, earthy residue consisting of fine-grained silica and resulting from the decomposition of siliceous limestone (or of a highly shelly sandstone) whose calcareous material has been removed by the dissolving action of water. Cf: *tripoli*. Syn: *terra cariosa*.

rotula One of the five massive radial elements in ambulacral position at the top of Aristotle's lantern of an echinoid. Pl: *rotulae*.

rubaultite A mineral: $Cu_2(UO_2)_3(OH)_{10}.5H_2O$.

rougemontite A coarse-grained igneous rock composed of anorthite, titanaugite, and small amounts of olivine and iron ore. Its name is derived from Rougemont, Montreal, Canada.

rough n. An uncut gemstone.---adj. Pertaining to an uncut or unpolished gemstone; e.g. a *rough* diamond in its natural state.

rough ice An expanse of ice having an uneven surface caused by formation of *pressure ice* or by growlers frozen in place (ADTIC, 1955, p.68).

roughneck A driller's helper and general workman in a drilling crew; also, one who builds and repairs oil wells. Cf: *roustabout*.

roughness coefficient A factor in formulas for computing the average velocity of flow of water in a conduit or channel which represents the effect of roughness of the confining material upon the energy losses in the flowing water.

rounded Round or curving in shape; specif. said of a sedimentary particle whose original edges and corners have been smoothed off to rather broad curves and whose original faces are almost completely removed by abrasion (although some comparatively flat surfaces may be present), such as a pebble with a roundness value between 0.40 and 0.60 (midpoint at 0.500) and few (0-5) and greatly subdued secondary corners that disappear at roundness 0.60 (Pettijohn, 1957, p.59). The original shape is still readily apparent. Also, said of the *roundness class* containing rounded particles.

roundness The degree of abrasion of a clastic particle as shown by the sharpness of its edges and corners, expressed by Wadell (1932) as the ratio of the average radius of curvature of the several edges or corners of the particle to the radius of curvature of the maximum inscribed sphere (or to one-half the nominal diameter of the particle). The value is more conveniently computed from a plane figure (a projection or cross section); thus, roundness may be defined as the ratio of the average radius of curvature of the corners of the particle *image* to the radius of the maximum inscribed circle. A perfectly rounded particle (such as a sphere) has a roundness value of 1.0; less-rounded particles have values less than 1.0. The term has been used carelessly and should not be confused with *sphericity*: a nearly spherical particle may have sharp corners and be angular, while a flat pebble, far from spherical in shape, may be well-rounded. Cf: *flatness*. See also: *angularity; roundness class*.

roundness class An arbitrarily defined range of *roundness* values for the classification of sedimentary particles. Pettijohn (1957, p.58-59) recognizes five classes: *angular; subangular; subrounded; rounded; well-rounded*. Powers (1953) adds a sixth class: *very angular*. Syn: *roundness grade*.

roundness grade *roundness class*.

roundstone (a) A term proposed by Fernald (1929) for any naturally rounded rock fragment of any size larger than a sand grain (diameter greater than 2 mm), such as a boulder, cobble, pebble, or granule. Cf: *sharpstone*. (b) *cobblestone*.

roustabout A common laborer called upon to do any of the unskilled manual assignments in an oil field or refinery, or around a mine. Cf: *roughneck*.

routine A sequence of computer instructions for accomplishing a specific, well-defined, or limited task; e.g. a subdivision of a computer program.

routivarite A fine-grained igneous rock containing orthoclase, plagioclase, quartz, and garnet.

rouvillite A light-colored theralite composed predominantly of labradorite and nepheline, with small amounts of titanaugite, hornblende, pyrite, and apatite.

roweite A light-brown mineral: $H_2Ca(Mn,Zn)(BO_3)_2$.

rowlandite A massive dark-green mineral, approximately: $Y_3(SiO_4)_2(OH,F)$.

royal agate A mottled obsidian.

rozenite A mineral: $FeSO_4.4H_2O$.

R-tectonite A tectonite whose fabric is thought to have been caused by rotation. Cf: *S-tectonite; L-tectonite; B-tectonite.*

rubasse A crystalline variety of quartz stained a ruby red by numerous small scales or flecks of hematite distributed within it. Syn: *rubace; Mont Blanc ruby.*

rubble [geol] (a) A loose mass, layer, or accumulation of rough, irregular, or angular rock fragments broken from larger masses usually by physical (natural or artificial) forces, coarser than sand (diameter greater than 2 mm), and commonly but not necessarily poorly sorted; the unconsolidated equivalent of breccia. Cf: *gravel.* Syn: *rubblestone.* (b) *talus.* (c) Loose, angular, waterworn rock fragments, smaller than boulders, occurring along a beach. (d) *volcanic rubble.*

rubble [mining] (a) A loose covering of angular material overlying outcropping rock; the upper, more or less fragmentary or decomposed part of a rock stratum in a quarry. (b) Rough, irregular pieces of broken stone as it comes from the quarry; if large or massive, it is termed "block rubble".

rubble [ice] Fragments of floating or grounded sea ice in hard, roughly spherical blocks measuring 0.5-1.5 m in diameter, and resulting from the breaking up of larger ice formations. When afloat, commonly called *brash ice.* Syn: *rubble ice.*

rubble beach A beach composed of angular rock fragments or rubble.

rubble breccia (a) A breccia in which no matching fragments are parted by initial planes of rupture, and having fragments that are close-set and in touch (Norton, 1917, p.161). (b) A tectonic breccia characterized by prominent relative displacement of fragments and by some rounding (Bateman, 1950, p.133). Cf: *shatter breccia.*

rubble drift (a) A rubbly deposit (or *congeliturbate*) formed by solifluction under periglacial conditions; e.g. *head* and *coombe rock.* (b) A coarse mass of angular debris and large blocks set in an earthy matrix of glacial origin.

rubble island *debris island.*

rubble ore A term used in Brazil for iron ore found on the surface of itabirite and derived from it by "the breaking up of the thinner intercalated layers, more less completely freed from the associated siliceous elements by rain and wind action" (Derby, 1910, p.818). Cf: *sandy ore; canga.*

rubblerock *breccia* [geol].

rubblestone (a) A graywacke (Humble, 1843, p.224). (b) *rubble.*

rubble tract The part of the reef flat immediately behind and lagoonward of the reef front, paved with cobbles, pebbles, blocks, and other coarse reef fragments; when consolidated it forms a *reef breccia.*

rubellite A pink or red (pale rose red to deep ruby red), transparent, lithian variety of tourmaline, used as a gemstone. Syn: *red schorl.*

rubicelle A yellow or orange-red gem variety of spinel. See also: *ruby spinel.*

rubidium-strontium age method Determination of an age for a mineral or rock in years based on the known radioactive rate of decay of rubidium-87 to strontium-87. The method is most applicable to Mesozoic or older geologic material of high potassium-low calcium content. Abbrev: Rb-Sr age method. Syn: *rubidium-strontium dating.*

rubidium-strontium dating *rubidium-strontium age method.*

rubidium vapor magnetometer A type of *optically pumped magnetometer* that uses magnetic atoms of rubidium. Cf: *cesium vapor magnetometer.*

rubinblende A name applied to the red silver-sulfide minerals pyrargyrite, proustite, and miargyrite. Syn: *ruby blende.*

Rubrozem An ABC soil, the A layer of which is highly organic and low in bases, and the B layer of which is reddish to brown and prismatic.

ruby The red (purplish-red, blood-red, and light-red) variety of corundum, containing small amounts of chromium, used as a gemstone, and found esp. in the Orient (Burma, Ceylon, Thailand). Cf: *sapphire.*

ruby blende (a) A red, brownish-red, or reddish-brown transparent variety of sphalerite. Syn: *ruby zinc.* (b) *rubinblende.*

ruby copper Cuprous oxide; specif. *cuprite.* Syn: *ruby copper ore.*

ruby sand A red-colored beach sand containing garnets, as at Nome, Alaska.

ruby silver A red silver-sulfide mineral; specif. "dark ruby silver" (pyrargyrite) and "light ruby silver" (proustite). Syn: *ruby silver ore.*

ruby spinel A clear-red gem variety of magnesian spinel $MgAl_2O_4$, containing small amounts of chromium and having the color but none of the other attributes of true ruby. See also: *spinel ruby; balas ruby; rubicelle.*

ruby zinc A deep-red, transparent, zinc mineral; specif. *ruby blende* and *zincite.*

rudaceous Said of a sedimentary rock composed of a significant amount of fragments coarser than sand grains; pertaining to a rudite. The term implies no special size, shape, or roundness of fragments throughout the gravel range, and is broader than "pebbly", "cobbly", and "bouldery". Also said of the texture of such a rock. Term introduced by Grabau (1904, p.242). Cf: *psephitic.*

ruddle *red ocher.*

rudemark *rute-mark.*

rudist Any bivalve mollusk belonging to the superfamily Hippuritacea, characterized by an inequivalve shell, usually attached to a substrate, rarely free, and either solitary or gregarious, in reef-like masses. "Although the first rudists were only slightly inequivalve, their descendants very early became strongly so, with the two valves of individuals usually differing greatly from each other in size, shape, and shell wall structure" (TIP, 1969, pt.N, P.751). They are frequently found in association with corals. Their stratigraphic range is Upper Jurassic to Upper Cretaceous, and possibly into the Paleocene.

rudite A general name used for consolidated sedimentary rocks composed of rounded or angular fragments coarser than sand (granules, pebbles, cobbles, boulders, or gravel or rubble); e.g. conglomerate, breccia, and calcirudite. The term is equivalent to the Greek-derived term, *psephite*, and was introduced as *rudyte* by Grabau (1904, p.242) who used it with appropriate prefixes in classifying coarse-grained rocks (e.g. "autorudyte", "autosilicirudyte", "hydrorudyte", and "hydrocalcirudyte"). Etymol: Latin *rudus*, "crushed stone, rubbish, debris, rubble". See also: *lutite; arenite.*

rudyte Var. of *rudite.*

Rudzki anomaly A gravity anomaly calculated by replacing the surface topography by its mirror image within the geoid.

ruffle (a) A ripple mark produced by an eddy (Hobbs, 1917). (b) A roughness or disturbance of a surface, such as a ripple on a surface of water.

ruffled groove cast A *groove cast* with a feather pattern, consisting of a groove with lateral wrinkles that join the main cast in the downcurrent direction at an acute angle (Haaf, 1959, p.32). See also: *vibration mark.*

ruga A visceral fold or wrinkle; e.g. a concentric or oblique wrinkling of the external surface of a brachiopod shell, or a *growth ruga* on the shell of a bivalve mollusk. Pl: *rugae.*

ruggedness number A dimensionless number formed by the product of maximum basin relief and drainage density within a given drainage basin; it expresses the essential geometric characteristics of the drainage system, and implicitly suggests steepness of slope (Strahler, 1958, p.289). Symbol: N_r.

rugose Said of a seemingly wrinkled leaf whose venation appears to be impressed into the surface of the leaf.

rugose coral Any zoantharian belonging to the order Rugosa, characterized by calcareous corallites that may be solitary and cone-shaped or cylindrical, either curved or erect, or compound and branching or massive. Their stratigraphic range is Ordovician to Permian. Syn: *tetracoral.*

rugulate Said of sculpture of pollen and spores consisting of wrinkle-like ridges that irregularly anastomose.

ruin agate A brown variety of agate displaying on a polished surface markings that resemble or suggest the outlines of ruins or ruined buildings.

ruin marble A brecciated limestone that, when cut and polished, gives a mosaic effect suggesting the appearance of ruins or ruined buildings.

rule of constant proportion *constancy of relative proportions.*

rule of V's The outcrop of a formation that crosses a valley forms an acute angle (a V) that points in the direction in which the formation lies underneath the stream. The V points upstream where the outcrops of horizontal beds exactly parallel the topographic contours, where the beds dip upstream, or where the beds dip downstream at a smaller angle than the stream gradient; the V points downstream where the beds dip downstream at a larger angle than the stream gradient.

rumanite A brittle, yellow-brown to red (also black) variety of amber containing 1-3% sulfur and found in Rumania. Also spelled: *rumanite.*

rumpffläche A plain extending across a region underlain by massive or undifferentiated rocks; a term used "purely to express relief, with no implication as to position in the cycle of erosion" (W. Penck, 1953, p.420). The term has often been used to indicate a *peneplain.* Etymol: German *Rumpffläche,* "torso plain". Syn: *torso plain.*

run [ore dep] n. A ribbonlike, flatlying and irregular orebody following the stratification of the host rock.

run [intrus rocks] n. A branching or fingerlike extension of the feeder of an igneous intrusion. Runs usually spread laterally along several stratigraphic levels.

run [streams] n. A small, swiftly flowing watercourse; a brook or a small creek.

rundle Var. of *runnel.*

runic *graphic.*

runite A syn. of *graphic granite.* The term, first used by Pinkerton, was suggested by Johannsen (1939, p.273) as a replacement for "graphic granite".

runlet *runnel.*

runnel [beach] (a) A trough-like hollow, larger than that between ripple marks, formed landward of a *ridge* on the foreshore of a tidal sand beach by the action of tides or waves. It carries the water drainage off the beach as the tide retreats and is flooded as the tide advances. Cf: *rill.* (b) *swale.* (c) *trough.*

runnel [streams] (a) A little brook; a rivulet or a streamlet. Syn: *runlet; rundle; rindle.* (b) The channel eroded by a runnel.

running ground In mining, incoherent surface material; earth, soil, or rock that will not stand, esp. when wetted, and that tends to flow into mine workings. See also: *mudrush.*

running sand (a) A term used in petroleum geology for a suspension of sand particles in water or oil. Syn: *floating sand.* (b) *quicksand.*

running water Water that is flowing in a stream or that is not stagnant or brackish. Ant: *standing water.*

runoff [water] That part of precipitation appearing in surface streams. It is more restricted than streamflow as it does not include stream channels affected by artificial diversions, storage, or other works of man. With respect to promptness of appearance after precipitation, it is divided into *direct runoff* and *base runoff;* with respect to source, it is divided into *surface runoff, storm seepage,* and *ground-water runoff.* It is the same as *total runoff* used by other workers (Langbein & Iseri, 1960). Syn: *virgin flow.* Cf: *water yield.* See also: *runoff cycle.*

runoff [coal mining] The collapse of a pillar of coal in a mine; the pillar is said to have run off.

runoff coefficient The percentage of precipitation that appears as runoff. The value of the coefficient is determined on the basis of climatic conditions and physiographic characteristics of the drainage area and is expressed as a constant between

zero and one (Chow, 1964, p.20-8, 21-37). Symbol: *C.*

runoff cycle That part of the hydrologic cycle involving water between the moment of its precipitation onto land and its subsequent evapotranspiration or discharge through stream channels. See also *runoff* [water].

runoff desert An arid region in which local rain is insufficient to support any perennial vegetation except in drainage or runoff channels. Cf: *rain desert.*

runoff intensity The excess of rainfall intensity over infiltration capacity, usually expressed in inches in depth of rainfall per hour. Strictly, the volume of water derived from an area of land surface per hour. Symbol: Q. Syn: *runoff rate.*

runoff rate *runoff intensity.*

run-of-mine Said of the ore in its natural, unprocessed state; pertaining to ore just as it is mined.

runout *water yield.*

runup *uprush.*

runway The channel of a stream.

Rupelian European stage: middle Oligocene (above Tongrian, below Chattian). Syn: *Stampian.*

rupestral Said of an organism living among rocks or in rocky areas. Syn: *rupestrine; rupicolous; lithophilous; saxicolous; saxigenous.*

rupestrine *rupestral.*

rupicolous *rupestral.*

rupture *fracture* [exp struc geol].

rupture envelope *Mohr envelope.*

rupture strength The differential stress that a material sustains at the instant of rupture. The term is normally applied to the differential stress at rupture when deformation occurs at atmospheric confining pressure and room temperature.

rupture zone The region immediately adjacent to the boundary of an explosion crater, characterized by excessive in-place crushing and fracturing where the stresses produced by the explosion have exceeded the ultimate strength of the medium. Cf: *plastic zone.*

rursiradiate Said of an ammonoid rib inclined backward (adapically) from the umbilical area toward the venter. Cf: *prorsiradiate; rectiradiate.*

rusakovite A mineral: $(Fe,Al)_5(VO_4,PO_4)_2(OH)_9 \cdot 3H_2O$.

russellite A mineral: Bi_2WO_6.

rust An English term for a black shale discolored by ocher.

rust ball A lump of yellow iron ore found in chalk in Cambridgeshire, Eng.

rustumite A mineral: $Ca_4Si_2O_7(OH)_2$.

rusty gold Native gold that has a thin coat of iron oxide or silica that prevents it from amalgamating readily.

rute-mark (a) A type of polygonal ground found in arctic regions, consisting of a polygon enclosed by a row of stones. Etymol: Norwegian *rutemark,* "route mark". Syn: *rutmark; rutemark; rudemark.* (b) A crack in soil or mud, similar in form to a rute-mark.

rutherfordine A yellow secondary mineral: $(UO_2)(CO_3)$.

rutilated quartz Sagenitic quartz characterized by the presence of enclosed needle-like crystals of rutile. See also: *sagenite.* Syn: *Venus hairstone.*

rutile A usually reddish-brown (sometimes yellowish, deep-red, or black) tetragonal mineral: TiO_2. It is trimorphous with anatase and brookite, and often contains a little iron. Rutile forms prismatic crystals radiately grouped in other minerals (esp. quartz), and it occurs as a primary mineral in some acid igneous rocks (esp. those rich in hornblende), in metamorphic rocks, and as residual grains in sediments and beach sands. It is an ore of titanium. Syn: *red schorl.*

rutmark *rute-mark.*

rutterite A medium-grained, equigranular, dark-pink plutonic rock composed chiefly of microperthite, microcline, and albite, with small amounts of nepheline, biotite, amphibole, graphite, and magnetite.

ruware A term used in southern Africa for a low, flattish or gently domed, granitic pediment or outcrop of bare rock. It occurs where flat or gently dipping joint systems are prominent.

***R* wave** *Rayleigh wave.*

S

Saale The term applied in northern Europe to the third glacial stage of the Pleistocene Epoch, after the Elster glacial stage and before the Weichsel; equivalent to the *Riss* and *Illinoian* glacial stages.

Saalic orogeny One of the 30 or more short-lived orogenies during Phanerozoic time recognized by Stille, in this case early in the Permian between the Autunian and Saxonian stages.

sabach A term used in Egypt for a calcareous accumulation; specif. *caliche*. Syn: *sabath*.

sabakha Var. of *sebkha*.

Sabinas North American (Gulf Coast) provincial series: Upper Jurassic (above older Jurassic, below Coahuilan) (Murray, 1961).

Sabinian North American (Gulf Coast) stage: Eocene (above Midwayan, below Claibornian). It includes strata most commonly grouped as *Wilcoxian*.

sabkha Var. of *sebkha*. Kinsman (1969, p.832) proposed that the form "sabkha" be used, "with the connotation of a salt flat which is inundated only occasionally".

sabugalite A yellow secondary mineral: $HAl(UO_2)_4$-$(PO_4)_4 \cdot 16H_2O$.

sabulous Sandy or gritty; *arenaceous*. Syn: *sabulose; sabuline*.

sac A pouch within an animal or plant; e.g. *pollen sac* and *air sac*.

saccate Like or having the form of a sac or pouch; e.g. "saccate mantle canal" of a brachiopod, without terminal branches and not extending to the anterior and lateral periphery of the mantle, or "saccate pollen" containing vesicles. See also: *vesiculate*.

saccharoidal (a) Said of a granular or crystalline texture resembling that of loaf sugar; specif. said of the xenomorphic-granular texture typically developed in aplites, or said of the crystalline granular texture seen in some sandstones, evaporites, marbles, and dolomites in which the constituent crystals are well-developed and of approximately uniform size. (b) Said of a white (or nearly white), equigranular rock having a saccharoidal texture.----See also: *aplitic*. Syn: *sucrosic; sugary*.

saccus A wing-like extension or *vesicle* of the exine in gymnospermous pollen and prepollen. Pl: *sacci*.

Sackungen Deep-seated rock creep which has produced a *ridge-top trench* by gradual settlement of a slablike mass into an adjacent valley. The top of the settled slab is usually parallel to the crest line of the ridge (Zischinsky, 1969).

saddle [ore dep] *saddle reef*.

saddle [geomorph] (a) A flattish ridge connecting the summits of two higher elevations; a low point in the crest line of a ridge, commonly on a divide between the heads of streams flowing in opposite directions. (b) A broad, flat gap or pass, sloping gently on both sides, and resembling a saddle in shape; a *col*.

saddle [paleont] An element or undulation of a suture line in a cephalopod shell that forms an angle or curve whose convexity is directed forward or toward the aperture (or away from the apex). Ant: *lobe*.

saddle [struc geol] A low point, sag, or depression along the surface axis or axial trend of an anticline.

saddle [coal mining] A less preferred syn. of *kettle bottom*.

saddleback [geomorph] A hill or ridge having a concave outline along its crest.

saddle fold A type of flexure fold that has an additional flexure near its crest, at right angles to that of the parent fold and much larger in radius.

saddle reef A mineral deposit associated with crest of an anti-clinal fold and following the bedding plane, usually found in vertical succession, esp. the gold-bearing quartz beins of Australia. Syn: *saddle; saddle vein*. Cf: *reverse saddle*. See also: *reef* [*ore dep*].

saddle vein *saddle reef*.

safe yield A syn. of *economic yield* that is also applied to surface-water supplies. Use of the term is discouraged because the feasible rate of withdrawal depends on the location of wells in relation to aquifer boundaries and rarely can be estimated in advance of development.

safflorite A tin-white orthorhombic mineral: $CoAs_2$. It is isomorphous with loellingite and usually contains considerable iron.

sag [geomorph] (a) A saddle-like pass or gap in a ridge or mountain range. (b) A shallow depression in an otherwise flat or gently sloping land surface; a small valley between ranges of low hills or between swells and ridges in an undulating terrain.

sag [sed] (a) A depression in a coal seam. (b) A *sag structure*.

sag [struc geol] (a) A basin or downwarp of regional extent; a broad, shallow structural basin with gently sloping sides, such as the Michigan and Illinois basins. (b) A depression produced by downwarping of beds near a fault in a direction opposite to that of frictional drag. (c) *fault sag*.

sag-and-swell topography An undulating surface characteristic of till sheets, as the landscape of midwestern U.S.; it may include moraines, kames, kettles, and drumlins. Cf: *swell-and-swale topography*. Syn: *sag and swell*.

sag correction A *tape correction* applied to the apparent length of a level base line to counteract the sag in the measuring tape. It is the difference between the effective length of the tape (or part of the tape) when supported continuously throughout its length and when supported at a limited number of independent points. Base tapes usually are used with three or five points of support and hang in *catenaries* between adjacent supports.

sagenite (a) An acicular variety of rutile that occurs in reticulated twin groups of slender, needle-like crystals crossing at 60 degrees and that is often enclosed in quartz or other minerals. See also: *Venus hair*. (b) A crystal of sagenite. Also, a similar crystal of tourmaline, goethite, actinolite, or other minerals penetrating quartz. (c) Sagenitic quartz; esp. *rutilated quartz*.---Etymol: Latin *sagena*, "large fishing net".

sagenitic Containing or enclosing *acicular* minerals; esp. sagenitic quartz, a transparent, colorless to nearly colorless variety of quartz containing needlelike crystals of rutile, actinolite, goethite, tourmaline, or other minerals, regardless of their arrangement.

sagger A coarse *fireclay*, often forming the floor of a coal seam, so called because it is used for making saggers or protective boxes in which delicate ceramic pieces are placed while being baked. Etymol: corruption(?) of "safeguard". Also spelled: *saggar; seggar; sagre*.

sagittal Pertaining to or situated in the median anterior-posterior plane of a body having bilateral symmetry or any plane parallel thereto; e.g. a "sagittal plane" dividing an edrioasteroid into two similar halves, or a "sagittal axis" extending along the length of the frustule of a pennate diatom, or a "sagittal section" representing a slice of a foraminiferal test in an *equatorial* plane, or a "sagittal ring in certain radiolarians representing a hoop or band of variable size and shape reinforcing the latticed wall in a medial vertical plane and from which processes may extend, or a "sagittal triradiate" of a sponge having two mirror-imaged rays and a coplanar third ray pointing away from them along the axis of four-fold symmetry.

sagittate Said of a triangularly shaped leaf whose basal lobes point downward toward the petiole.

sag pond A small body of water occupying an enclosed depression or *sag* formed where active or recent fault movement has impounded drainage; specif. one of many ponds and small lakes along the San Andreas Fault in California. Also spelled: *sagpond*. Cf: *swag*. Syn: *fault-trough lake; rift lake; rift-valley lake*.

sag structure A general term for load casts and other related sedimentary structures.

sagvandite A carbonatite with a high content of eustatite and magnesite (Johannsen, 1939, p.278).

sahamalite A mineral: $(Mg,Fe)Ce_2(CO_3)_4$.

sahlinite A sulfur-yellow mineral: $Pb_{14}(AsO_4)_2O_9Cl_4$.

sahlite *salite*.

sai A term used in central Asia for a gravelly talus, a river bed filled with stones, and a dry wash (Stone, 1967, p.258), and also for a piedmont plain covered with patinated pebbles (Termier & Termier, 1963, p.413).

saif *seif dune*.

sailing ice *scattered ice*.

sainfeldite A mineral: $H_2Ca_5(AsO_4)_4.4H_2O$.

Saint Croixian *Croixian*.

Saint Venant substance A material that demonstrates *elasticoplastic* behavior: it behaves elastically below a yield stress, but deforms continuously under a constant stress equal to yield stress.

sakalavite A glassy andesite containing phenocrysts of plagioclase (intermediate in composition between calcic and sodic) and augite and quartz xenocrysts.

sakhaite A mineral: $Ca_{12}Mg_4(CO_3)_4(BO_3)_7Cl(OH)_2.H_2O$.

sakharovaite A mineral: $(Pb,Fe)(Bi,Sb)_2S_4$.

Sakmarian European stage: lowermost Permian (above Stephanian of Carboniferous, below Artinskian).

sakuraiite A mineral: $(Cu,Zn,Fe)_3(In,Sn)S_4$.

sal *sial*.

salada A term used in SW U.S. for a salt-covered plain where a lake has evaporated. Etymol: Spanish, feminine of *salado*, "salted, salty". See also: *playa*.

sal ammoniac An isometric mineral: NH_4Cl. It is a white, crystalline, volatile salt and occurs esp. as an encrustation around volcanoes. Syn: *salmiac*.

salaquifer *saline aquifer*.

salar A term used in SW U.S. and in the Chilean nitrate fields for a salt flat or for a salt-incrusted depression that may represent the basin of a salt lake. Etymol: Spanish, "to salt". Pl: *salares; salars*. See also: *playa*.

salband The *selvage* of an igneous mass or of a mineral vein. Etymol: German *Salband* or *Sahlband*.

salcrete A term suggested by Yasso (1966) for a thin, hard crust of salt-cemented sand grains, occurring on a marine beach that is occasionally or periodically saturated by saline water.

saléeite A lemon-yellow mineral of the autunite group: $Mg(UO_2)_2(PO_4)_2.8H_2O$.

salesite A bluish-green mineral: $Cu(IO_3)(OH)$.

salfemic One of five classes in the C.I.P.W. (Cross, *et al*, 1902, p.585) chemical-mineralogic classification of igneous rocks and in which the ratio of salic minerals to femic is less than five to three but greater than three to five. Cf: *dosalic; dofemic*.

salic [petrology] Said of certain light-colored silica-, or magnesium-rich minerals present in the norm of igneous rocks; e.g. quartz, feldspars, feldspathoids. Also, applied to rocks having one or more of these minerals as major components of the norm. Etymol: a mnemonic term derived from silicon + *al*uminum + *ic*. Cf: *femic; mafic; felsic*.

salic [soil] Pertaining to a soil horizon, at least 15cm thick, characterized by enrichment with secondary salts. It contains at least 2.0% salts.

salient [geomorph] adj. Projecting or jutting upward or outward; e.g. a *salient* point, or one formed by a conspicuous outward projection from the coast.---n. A landform that projects or extends outward or upward from its surroundings; e.g. a cape along a shoreline, or a spur from the side of a moun-

tain. Ant: *reentrant*.

salient [fold] An area in which the axial traces of folds ar convex toward the outer edge of the folded belt. Ant: *reces [fold]*.

saliferous Salt-bearing; esp. said of strata producing, contain ing, or impregnated with salt. See also: *saline*.

salina (a) A place where crystalline salt deposits are forme or found, such as a salt flat, a salt pan, a salada, or a sa lick; esp. a salt-incrusted *playa* or a *wet playa*. (b) A body c saline water, such as a salt pond, a salt lake, a salt well, salt spring, or a playa lake having a high concentration c salts. (c) *saltworks*. (d) *salt marsh*.---Etymol: Spanish, "sa pit, salt mine, saltworks". Anglicized equivalent: *saline*.

salinastone A general term proposed by Shrock (1948a p.127) for a sedimentary rock composed dominantly of salin minerals (which are usually precipitated but may be fragmer tal); e.g. anhydrock and gyprock.

saline n. (a) A natural deposit of common salt or of any othe soluble salt; e.g. an evaporite. See also: *salines*. (b) An angl cized form of *salina*. In this usage, a "saline" may refer t various features such as a playa, a salt flat, a salt pan, a sa marsh, a salt lake, a salt pond, a salt well, and a saltworks (c) *salt spring*. (d) A term used along the coast of Louisian for a body of water behind a barrier island.---adj. (a) Pertair ing to, consisting of, or containing salt, such as a "saline solu tion". The term is used to designate salt water (such as sea water) and also to signify a salinity appreciably greater tha that of seawater. See also: *saliferous; salty*. (b) Said of taste resembling that of common salt, esp. in describing th properties of a mineral.

saline-alkali soil A *salt-affected soil* whose content of ex changeable sodium is greater than 15% and which contain much soluble salts. Its pH value is usually less than 9.5. C *sodic soil; saline soil; nonsaline alkali soil*.

saline aquifer An aquifer containing saline water. Syn: *salaqu fer*.

saline deposit *evaporite*.

salinelle A *mud volcano* erupting saline mud.

saline residue *evaporite*.

salines (a) A general term for the naturally occurring solubl salts, such as common salt, sodium carbonate, sodium n trate, potassium salts, and borax. (b) A general term for sa mines, salt springs, salt beds, salt rock, and salt lands.

saline soil A nonalkali, *salt-affected soil* having a high conter of soluble salts. Its exchangeable-sodium percentage is les than 15, and its pH value is less than 8.5. Cf: *saline-alka soil; sodic soil*. See also: *Solonchak soil*.

saliniferous Said of a stratum that yields salt.

salinity The total quantity of dissolved salts in sea water, mea sured by weight in parts per thousand, with the following quar ifications: all the carbonate has been converted to oxide, a the bromide and iodide to chloride, and all the organic matte has been completely oxidized. Salinity is usually compute from some other factor, such as chlorinity.

salinity current A *density current* in the ocean, the flow c which is caused, controlled, or maintained by its relativel greater density due to excessive salinity.

salinity log *chlorine log*.

salinity meter *salinometer*.

salinization In a soil of an arid, poorly drained region, the ac cumulation of soluble salts by the evaporation of the water that bore them to the soil zone.

salinometer An instrument that is used to measure the salinit of sea water, e.g. by electrical conductivity methods. Syn: *sa linity meter*.

salite A mineral of the clinopyroxene group: $Ca(Mg,Fe)Si_2O_6$ It is a grayish-green to black variety of diopside with mor magnesium than iron. Syn: *sahlite*.

salitral A term used in Patagonia for a swampy place wher salts (esp. potassium nitrate) become encrusted in the dr eason. Etymol: Spanish, "saltpeter bed".

salitrite A lamprophyre composed chiefly of titanite and diopside with acmite, accessory apatite, microcline, and occasionally anorthoclase and baddeleyite.

salmiac *sal ammoniac.*

salmian *Tremadocian.*

salmonsite A buff-colored mineral: $Mn_9Fe_2^{+3}(PO_4)_8.14H_2O$ (?). Cf: *landesite.*

salnatron Crude sodium carbonate.

salopian European stage: Middle Silurian and part of Upper Silurian. The name was proposed for the Wenlockian and Lower Ludlovian, but is not now often used.

salpausselka A Finnish term for a steep recessional moraine, usually interpreted as a series of end moraines, as the one extending east-west across Finland.

salphingiform Shaped like a trumpet; e.g. said of a cyrtolith coccolith with a trumpet-shaped central structure (as in *Discosphaera tubifer*).

salsima According to Bemmelen (1949), the theoretical layer of the Earth's crust beneath the sial and above the Mohorovičić discontinuity that is considered to be of basaltic composition. Also spelled: *sialsima.* Cf: *sifema.*

salsuginous Said of a plant growing in soil or water with a high content of salts; i.e. a halophyte.

salt In geographic terminology, a *salt marsh*, esp. one flooded by the tide.

salt-affected soil A general term for a soil that is not suitable for the growth of crops because of an excess of salts, of exchangeable sodium, or both; e.g. *saline-alkali soil, saline soil, sodic soil.*

salt-and-pepper Said of a sand or sandstone consisting of a mixture of light- and dark-colored particles, such as a strongly cherty graywacke (Krynine, 1948, p.152) or a lighter-colored and speckled subgraywacke (Pettijohn, 1957, p.319); e.g. the Bow Island Sandstone of Cretaceous age in Alberta.

salt anticline A diapiric or piercement structure like a *salt dome*, but in which the salt core is linear rather than equidimensional, e.g. the salt anticlines in the Paradox basin of the central Colorado Plateau. Syn: *salt wall.*

saltation [evol] Sudden evolution of a new organism derived from older ones without the appearance of intermediate forms. This process is almost impossible genetically. See also: *saltatory evolution.*

saltation [sed] A mode of sediment transport in which the particles are moved progressively forward in a series of short intermittent leaps, jumps, hops, or bounces from a bottom surface; e.g. sand particles skipping downwind by impact and rebound along a desert surface, or bounding downstream under the influence of eddy currents that are not turbulent enough to sustain the particles in suspension and thereby return them to the stream bed at some distance downstream. It is intermediate in character between suspension and the rolling or sliding traction. Etymol: Latin *saltare,* "to jump, leap".

saltation load The part of the *bed load* that is bouncing along the stream bed or is moved, directly or indirectly, by the impact of bouncing particles.

saltation mark A tool mark made by an object proceeding along a saltatory path; e.g. a *ring mark*. Cf: *roll mark.*

saltatory evolution The theory of evolution by *saltation.*

salt bottom A flat piece of relatively low-lying ground encrusted with salt.

salt burst Rock destruction caused by soluble salts that enter pores and crystallize from nearly saturated solutions. In deserts, salt bursts may be due to crystallization pressure, to the volumetric expansion of salts in capillaries, and to hydration pressures of the entrapped salts (Winkler & Wilhelm, 1970).

salt corrie A cirque-like hollow, resembling a crater or caldera, produced by the solution of salt.

salt crust A salt deposit formed on an ice surface by crystal growth forcing salt out of young sea ice and pushing it upward.

salt-crystal cast A *crystal cast* formed by solution of a soluble salt crystal, followed by filling with mud or sand or by crystallization of a pseudomorph (such as calcite pseudomorphous after halite). See also: *hopper.*

salt-crystal growth The growth of salt crystals in openings in rock or soil, capable of exerting powerful stresses and producing granular disintegration in a dry climate. See also: *salt weathering.*

salt desert A desert with a saliferous soil; e.g. a *kavir.*

salt dome A diapiric or piercement structure with a central, nearly equidimensional *salt plug*, generally one to two kilometers or more in diameter, which has risen through the enclosing sediments from a mother salt bed 5 km to more than 10 km beneath the top of the plug. Many salt plugs have a *cap rock*, which is a concentration of the less soluble evaporite minerals originally associated with the salt. Most plugs have nearly vertical walls, but some overhang. The enclosing sediments are commonly turned up and complexly faulted next to a salt plug, and serve as reservoirs for oil and gas. Salt domes are characteristic features of the Gulf Coastal Plain in North America and the North German Plain in Europe, but occur in many other regions. Cf: *salt anticline.* See also: *salt tectonics.*

salt-dome breccia A breccia found in deep shale sequences and occurring as a dome-shaped mass in a broad zone surrounding a salt plug. It is believed to be a result of differential pressure caused by diapiric intrusion of salt into shale (Kerr & Kopp, 1958).

saltern (a) A *saltworks* where salt is produced by boiling or evaporation of salt or brine. (b) *salt garden.*

saltfield An area overlying a usually workable salt deposit of economic value.

salt flat The level, salt-encrusted bottom of a lake or pond that is temporarily or permanently dried up; e.g. the Bonneville Salt Flats west of Salt Lake City, Utah. See also: *playa; alkali flat.*

salt flower An *ice flower* forming on surface sea ice around a salt-crystal nucleus.

salt garden A large, shallow basin or pond where seawater is evaporated by solar heat. Syn: *saltern.*

salt glacier A gravitational flow of salt down the slopes of a salt plug, following the preexisting structure. It can be compared with the coulees of lava flows.

salt hill An abrupt hill of salt, with sinkholes and pinnacles at its summit (Thornbury, 1954, p.521).

saltierra A deposit of salt left by evaporation of a shallow, inland lake. Etymol: Spanish, "salt earth".

salting (a) A British term for the slightly higher part of a *salt marsh* flooded only by spring tides, containing little bare mud, and supporting grassy vegetation. Syn: *high marsh.* (b) A term used in parts of Great Britain for land regularly covered by the tide, as distinguished from a salt marsh.----The term is usually used in the plural.

salt lake An inland body of water situated in an arid or semi-arid region, having no outlet to the sea, and containing a high concentration of dissolved salts (principally sodium chloride). Examples include the Great Salt Lake in Utah, and the Dead Sea in the Near East. See also: *alkali lake; bitter lake.* Syn: *brine lake.*

salt lick (a) A place to which wild animals (such as buffalo) go to lick up salt lying on the surface of the ground; e.g. a swampy or boggy area surrounding a salt spring. (b) A term that has been used for a *salt spring* or a salt brook; this usage is improper because a lick is dry and a spring or brook is not. --Syn: *lick.*

salt marsh Flat, poorly drained land that is subject to periodic or occasional overflow by salt water, containing water that is brackish to strongly saline, and usually covered with a thick mat of grassy halophytic plants; e.g. a coastal marsh periodically flooded by the sea, or an inland marsh (or *salina*) in an arid region and subject to intermittent overflow by water containing a high content of salt. Cf: *tidal marsh; marine marsh.* See also: *low marsh; salting; sea marsh; open-coast marsh; tidal-delta marsh; salt-marsh plain.* Syn: *salt.*

salt-marsh plain A *salt marsh* that has been raised above the level of the highest tide and has become dry land.

salt meadow A meadow subject to overflow by salt water.

salt pan (a) An undrained, usually small and shallow, natural depression or hollow in which water accumulates and evaporates, leaving a salt deposit. Also, a shallow lake of brackish water occupying a salt pan. See also: *playa; pan[salt]; marsh pan.* (b) A large pan for recovering salt by evaporation.---Also spelled: *saltpan.*

saltpeter (a) Naturally occurring potassium nitrate; *niter.* Cf: *Chile saltpeter; Peru saltpeter; wall saltpeter.* (b) A speleologic term for earthy cave deposits of nitrate minerals.---Also spelled: *saltpetre.*

saltpeter earth A cave deposit containing calcium nitrate; it can be mined for saltpeter production.

salt pillow An embryonic salt dome rising from its source bed, still at depth.

salt pit (a) A pit in which seawater is received and evaporated and from which salt is obtained. Syn: *vat; wich.* (b) A body of salt water occupying a salt pit.---Also spelled: *saltpit.*

salt plug The salt core of a *salt dome.* It is nearly equidimensional, about one to two kilometers in diameter, and has risen through the enclosing sediments from a mother salt bed five to ten kilometers below.

salt polygon A surface of salt on a playa, having three to eight sides marked by ridges of material formed as a result of the expansive forces of crystallizing salt, and ranging in width from several centimeters to 30 m (Stone, 1967, p.244).

salt pond (a) A large or small body of salt water in a marsh or swamp along the seacoast. (b) An artificial pond used for evaporation in the production of salt from seawater.

salt prairie *soda prairie.*

salt ribbon A ribbon-like growth of salt from a network of small cracks (Brown, 1946).

saltspar Coarsely crystallized and cleavable halite.

salt spring A *mineral spring* whose water contains a large quantity of common salt; a spring of salt water. See also: *salt lick.* Syn: *saline; brine spring.*

salt stock A general term for a diapiric salt body of whatever shape.

salt table The flat upper surface of a salt stock along which ground-water solution leads to the formation of cap rock by freeing anhydrite (Goldman, 1952).

salt tectonics A general term for the study of the structure and mechanism of emplacement of *salt domes.* Syn: *halokinesis.*

salt wall *salt anticline.*

salt water A syn. of *sea water*, as the antonym of fresh water in general.

salt-water encroachment Displacement of fresh surface or ground water by the advance of salt water due to its greater density, usually in coastal and estuarine areas but also, as by movement of brine from beneath a playa lake toward wells discharging fresh water. Encroachment occurs when the total haed of the salt water exceeds that of adjacent fresh water. Syn: *encroachment [grd wat]; sea-water intrusion; intrusion [grd wat]; salt-water intrusion; sea-water encroachment.* See also: *Ghyben-Herzberg ratio.*

salt-water front The interface between fresh and salty water in a coastal aquifer or in an estuary. Under certain conditions, a similar front may by found inland.

salt-water intrusion *salt-water encroachment.*

salt-water underrun A type of density current occurring in a tidal estuary, due to the greater salinity of the bottom water

(ASCE, 1962).

salt-water wedge An intrusion, into an estuary or tidal rive dominated by freshwater circulation, of salty ocean water i the form of a wedge that underlies the freshwater and tha slopes slightly downward in the upstream direction, and that i characterized by a pronounced increase in salinity with dept

salt weathering The granular disintegration or fragmentation rock material effected by saline solutions or by *salt-cryst growth* (Wellman & Wilson, 1965). See also: *exsudation.*

salt wedge A wedge-shaped mass of salt water from an ocea or sea which intrudes the mouth and lower course of a rive The denser, salt water underlies the fresher, river water. E tent is regulated by river discharge and tides.

salt well A bored or driven well from which brine is obtaine See also: *brine pit.*

saltworks A building or group of buildings where salt is pro duced commercially, as by extraction from seawater or from the brine of salt springs. Syn: *salina; saltern.*

salty Pertaining to, containing, or resembling salt; *saline.*

samara A dry, indehiscent, usually one-seeded winged fru such as that of elm or maple.

samarskite A velvet-black to brown, commonly metamict min eral: $(Y,Ce,U,Ca,Fe,Pb,Th)(Nb,Ta,Ti,Sn)_2O_6$. It has a sple dent vitreous or resinous luster, and is found in granite peg matites. Syn: *ampangabèite; uranotantalite.*

samiresite A mineral that is possibly a variety of betafite con taining lead, but perhaps an independent species.

Sammelkristallization The action, depending on surface te sion or on total free surface energy, by which smaller grain become unstable in relation to larger grains and will eventuall be devoured by the larger grains (Barth, 1962, p.399). Ther is no English equivalent for this term. Etymol: German.

sample In statistics, the part or subset, of a statistical popula tion, that if properly chosen may be used to estimate parame ters.

sampleite A blue mineral: $NaCaCu_5(PO_4)_4Cl.5H_2O$.

sample log A log showing the rocks penetrated in drilling borehole or well, compiled by a geologist from information ob tained through microscopic study of drilling samples (core and cuttings) recovered at the surface, and plotted on a lo strip subdivided into units of depth. It shows the sequenc and characteristics of the strata penetrated in drilling an consists of colors and/or symbols with a written, abbreviate description of the lithology. See also: *interpretative log; pe centage log.*

sample splitter A device for separating dry incoherent materi (such as sediment) into truly representative samples of worl able size for laboratory study. Syn: *riffler.*

sampling In economic geology, the gathering of specimens ore for development of the orebody, esp. assay. The averag of many samples is used; representative sampling, therefor is crucial to accurate analysis or an orebody. The term is us ally modified to indicate the mode or locality, e.g. hand sam pling, mine sampling.

samsonite A steel-black monoclinic mineral: $Ag_4MnSb_2S_6$.

sanakite A glassy andesite composed of bronzite, augit magnetite, and a few large plagioclase and garnet crysta (Thrush, 1968, p.959).

sanbornite A white triclinic mineral: $BaSi_2O_5$.

sand [drill] (a) A driller's term applied loosely to any porou friable sedimentary rock or unconsolidated sediment contai ing oil, gas, or water; specif. an oil-producing sandstone unconsolidated sand formation. See also: *oil sand.* (b) Ro chips and other waste products produced by drilling action.

sand [eng] A rounded fragment having a diameter in t range of 0.074 mm (retained on U.S. standard sieve no.20 to 4.76 mm (passing U.S. standard sieve no.4). See als *coarse sand; medium sand; fine sand.*

sand [geomorph] (a) A tract or region of sand, such as sandy beach along the seashore, or a desert land. (b) andbank or a sandbar.----The term is usually used in the plural; e.g. "sea sands".

and [sed] (a) A rock fragment or detrital particle smaller an a granule and larger than a coarse silt grain, having a ameter in the range of 1/16 to 2 mm (62-2000 microns, or 0025-0.08 in., or 4 to -1 phi units, or a size between that at e lower limit of visibility of an individual particle with the unded eye and that of the head of a small wooden match), eing somewhat rounded by abrasion in the course of transrt. In Great Britain, the range of 0.1-1 mm has been used. ee also: *very coarse sand; coarse sand; medium sand; fine and; very fine sand.* (b) A loose aggregate of unlithified miner-al or rock particles of sand size; an unconsolidated or modrately consolidated sedimentary deposit consisting essentially medium-grained clastics. The material is most commonly omposed of quartz resulting from rock disintegration, and hen the term "sand" is used without qualification, a siliceous omposition is implied; but the particles may be of any miner-composition or mixture of rock or mineral fragments, such s "coral sand" consisting of limestone fragments. Also, a ass of such material, esp. on a beach or a desert or in a ream bed. (c) *sandstone.*

and [soil] (a) A term used in the U.S. for a rock or mineral article in the soil, having a diameter in the range of 0.05-2 m; prior to 1947, the range 1-2 mm was called "fine grav-". The diameter range recognized by the International Socie-of Soil Science is 0.02-2 mm. (b) A textural class of soil aterial containing 85% or more of sand, with the percentage silt plus 1.5 times the percentage of clay not exceeding 15; ecif. such material containing 25% or more of very coarse and, coarse sand, and medium sand, and less than 50% of he sand or very fine sand (SSSA, 1965, p.347). The term as also been used for a soil containing 90% or more of sand.) *sandy soil.*----See also: *very coarse sand; coarse sand; edium sand; fine sand; very fine sand.*

and apron A deposit of sand along the shore of a lagoon of a ef.

andar Plural of *sandur.*

andarac A syn. of *realgar.* Also spelled: *sandarach.*

and avalanche Movement of large masses of sand down a une face when the angle of repose is exceeded, or when the une is disturbed (Stone, 1967, p.245).

andbag In the roof of a coal seam, a deposit of glacial debris rmed by scour and fill subsequent to coal formation.

andbank (a) A submerged ridge of sand in the sea, a lake, a river, usually exposed during low tide; a sandbar. (b) A rge deposit of sand, esp. in a shallow area near the shore.

andbar A *bar* or low ridge of sand that borders the shore and built up, or near, to the water surface by currents in a river by wave action along the shore of a lake or sea. Syn: *sand ef.*

andblast (a) A stream of windblown sand driven against an xposed rock surface. (b) A gust of wind, carrying sand.

andblasting A type of *blasting* in which the particles are hard ineral grains (usually of quartz) of sand sizes. Syn: *sand-ast action.*

andblow A patch of coarse sandy soil denuded of vegetation wind action.

and boil A spring that bubbles through a river levee, with an ection of sand and water, as a result of water in the river eing forced through permeable sands and silts below the vee during flood stage. Syn: *blowout [grd wat].*

and-calcite A calcite crystal containing a large percentage of and-grain inclusions; a *sand crystal* of calcite. See also: *Fon-inebleau sandstone.*

and cay A British syn. of *sandkey.*

and cone [geomorph] A cone-shaped deposit of sand, pro-uced esp. in an alluvial cone.

and cone [glaciol] A low *debris cone* whose protective ve-eer consists of sand.

sand crystal A large euhedral or subhedral crystal (as of bar-ite, gypsum, and esp. calcite) loaded with detrital-sand inclu-sions (up to 60%), developed by growth in an incompletely cemented sandstone during or as a result of cementation. See also: *sand-calcite.*

sand dike A sedimentary dike consisting of sand that has been squeezed or injected upward into a fissure.

sand dome *dome* [beach].

sand drift (a) A general term for surface movement of wind-blown sands, occurring in deserts or along the shore. (b) An accumulation of sand formed in the lee of some fixed obstruc-tion, such as a rock or a bush, and usually smaller than a dune. See also: *sand shadow.*

sand drip A rounded or crescentic surface form on beach sand, resulting from the sudden absorption of overwash (Ros-alsky, 1949, p.12).

sand dune A *dune* consisting of loose sand piled or heaped up by the wind, commonly found along low-lying seashores above high-tide level, more rarely on the border of a large lake or river valley, as well as in various desert regions and generally where there is abundant, dry surface sand during some part of the year. See also: *sand hill.*

sanderite A mineral: $MgSO_4 \cdot 2H_2O$.

sand fall An accumulation of sand swept over a cliff or es-carpment (Stone, 1967, p.245). It may occur in a submarine canyon as well as on land.

sandfall The *slip face* on the lee side of a dune. Syn: *sandfall face.*

sand flag Fine-grained sandstone that is cleavable or splits up into flagstones.

sand flat A sandy *tidal flat* barren of vegetation. Cf: *mud flat.*

sand flood A vast body of sand moving or borne along a des-ert, as in the Arabian deserts.

sand flow [mass move] A flow of wet sand, as that along banks of noncohesive clean sand that is subject to scour and to repeated fluctuations in pore-water pressure due to rise and fall of the tide (Varnes, 1958, p.41).

sand flow [marine geol] In a submarine canyon, a discontin-uous movement of sand down the axis, in a series of slumps. See also: *sand fall.*

sand flow [pyroclast] (a) An obsolete syn. of *ash flow.* (b) A term applied by Fenner (1923) to an unsorted rhyolitic tuff in the vicinity of Mt. Katmai, Alaska..

sand gall *sand pipe.*

sand glacier (a) An accumulation of sand that is blown up the side of a hill or mountain and through a pass or saddle, and then spread out on the opposite side to form a wide fan-shaped plain. (b) A horizontal plateau of sand terminated by a steep talus slope.

sand hill A ridge of sand; esp. a *sand dune* in a desert region. See also: *chop hill.*

sandhills A region of *sand hills*, as in north-central Nebraska.

sand hole A small pit (7-8 mm in depth and a little less wide than deep) with a raised margin, formed on a beach by waves expelling air from a formerly saturated mass of sand; it resem-bles an impression made by a raindrop.

sand horn A pointed sand deposit extending from the shore into shallow water. Cf: *sand lobe.*

sandia An oblong, oval, or rounded mountain mass resembling a watermelon; e.g. the Sandia Mountains in New Mexico. Etymol: Spanish *sandía*, "watermelon".

sanding The accumulation or building up of sand, such as by the action of currents in filling a harbor.

sanding up The filling in or choking with sand, as in a well that produces sand mixed with oil and gas.

sandkey A small sandy island that is not notably elongate and that is parallel with the shore. Syn: *sand cay.*

sand lens A body of sand with the general form of a *lens*, thick in the middle and thinning toward the edges.

sand levee A *whaleback* in the desert.

sand line [drill] A *wire line* used to raise and lower a bailer or sand pump to remove cuttings from a borehole. Also spelled: *sandline*.

sand line [glac] An "easily overlooked" mark made by glacier ice, about 5-10 cm long, fine as a hair and similar to one of the marks made by the "finest sandpaper" (Campbell, 1865, p.4).

sand lobe A rounded sand deposit extending from the shore into shallow water. Cf: *sand horn*.

sand pavement A sandy surface derived from coarse-grained sand ripples, developed on the lower, windward slope of a dune or rolling sand area during a period of intermittent light, variable winds (E. Holm, 1957).

sand pipe A *pipe* formed in sedimentary rocks, filled with considerable sand and some gravel. Cf: *gravel pipe*. Syn: *sand gall*.

sand plain [glac geol] (a) A small *outwash plain* consisting chiefly of sand deposited by meltwater streams flowing from a glacier. (b) A term used in New England for an *esker delta*.

sand plain [geomorph] A sand-covered plain. The large sand plains in Western Australia have an uncertain origin: they may originate by deflation of sand dunes, the lower limit of erosion being governed by the ground-water level. Also spelled: *sandplain*.

sand plateau *esker delta*.

sand plug A mass of sand that fills the upper end of a stream channel abandoned by the formation of a chute cutoff.

sand pump A *bailer* fitted with a plunger for removing wet sand, mud, or silt from the bottom of a borehole.

sandr Var. of *sandur*.

sand reef *sandbar*.

sand ridge (a) A generic name for any low ridge of sand formed at some distance from the shore, and either submerged or emergent. Examples include a *longshore bar* and a *barrier beach*. (b) One of a series of long, wide, extremely low, parallel ridges believed to represent the eroded stumps of former longitudinal sand dunes, as in western Rhodesia. (c) A crescent-shaped landform found on a sandy beach; e.g. a *beach cusp*. (d) *sand wave*.

sand ripple A ripple composed of sand; esp. a *wind ripple* consisting of medium sand.

sand river A river that deposits much of its sand load along its middle course, to be subsequently removed by the wind; e.g. the Red River in Texas. Cf: *sand stream*.

sandrock (a) A field term for a sandstone that is not firmly cemented (Tieje, 1921, p.655). (b) A term used in southern England for a sandstone that crumbles between the fingers. (c) *sandstone*.

sand roll *pseudonodule*.

sand run (a) A fluid-like motion of dry sand. (b) A mass of dry sand in motion.

sand sea [desert] An extensive assemblage of sand dunes of several types in an area where a great supply of sand is present, and which is characterized by an absence of travel lines, or directional indicators, and by a wave-like appearance of dunes separated by troughs much as if storm sea waves were frozen into place. See also: *erg*.

sand sea [pyroclast] The flat, rain-smoothed plain of volcanic ash and other pyroclastics on the floor of a caldera.

sand shadow A lee-side accumulation of sand, as a small turret-shaped dune, formed in the shelter of, and immediately behind, a fixed obstruction, as clumps of vegetation. See also: *sand drift*.

sandshale A sedimentary deposit consisting of thin alternating beds of sandstone and shale.

sand-shale ratio A term introduced by Sloss et al (1949, p.100) for the ratio of the thickness or percentage of sandstone (and of conglomerate) to that of shale in a stratigraphic section, disregarding the amount of nonclastic material; e.g. a ratio of 3.2 indicates that the section contains an average of 3.2 m of sandstone per meter of shale. Cf: *clastic ratio*.

sand sheet A thin accumulation of coarse sand or fine gravel formed of grains too large to be transported by saltation characterized by an extremely flat or plain-like surface broke only by small sand ripples.

sand size A term used in sedimentology for a volume greater than that of a sphere with a diameter of 1/16 mm (0.0025 in. and less than that of a sphere with a diameter of 2 mm (0.0 in.).

sand snow Dry snow that has fallen at very cold temperature (usually below -25°C), having small and loosely compacte crystals; its surface has the consistency of dry sand. Cf: *wi snow; powder snow*.

sandspit A spit consisting chiefly of sand.

sand splay A *flood-plain splay* consisting of coarse sand part cles.

sand stalagmite A stalagmite that is developed on sand an that is partly composed of sand, cemented by calcite.

sandstone (a) A medium-grained, clastic sedimentary roc composed of abundant and rounded or angular fragments of sand size set in a fine-grained matrix (silt or clay) and mor or less firmly united by a cementing material (common silica, iron oxide, or calcium carbonate); the consolidate equivalent of sand, intermediate in texture between conglom erate and shale. The sand particles usually consist of quartz and the term "sandstone", when used without qualification, in dicates a consolidated clastic rock containing about 85-90° quartz sand (Krynine, 1940). The rock varies in color, bein commonly red, yellow, brown, gray, or white; it may be de posited by water or wind action, and primary features (sedi mentary structures and fossils) are common. Sandstones ma be classified according to rock and mineral composition of particles, mineralogic maturity, textural maturity, fluidity index diastrophism, primary structures, and type of cement (Kleir 1963). (b) A field term for any clastic rock containing individ ual particles that are visible to the unaided eye or slightl larger.---Abbrev: ss. Syn: *sand; sandrock*.

sandstone-arenite A term used by Folk (1968, p.124) for *sedarenite* composed chiefly of sandstone fragments.

sandstone cylinder *sandstone pipe*.

sandstone dike (a) A *clastic dike* composed of sandstone of lithified sand. (b) *stone intrusion*.

sandstone pipe A *clastic pipe* consisting of sandstone. It ma originate in various ways: gravitational foundering of sand int underlying water-saturated mud; filling of a spring vent; fillin of cavities caused by solution of underlying limestone or b volcanic explosions; or penecontemporaneous sag due to re moval of support by flowage. Syn: *sandstone cylinder*.

sandstone sill A tabular mass of sandstone that has been em placed by sedimentary injection parallel to the structure of bedding of preexisting rock in the manner of an igneous sil such as one injected at the mud-water interface by the under flow of a dense slurry.

sand streak A low, linear ridge formed at the interface of san and air or water, oriented parallel to the direction of flow, an having a symmetric cross section.

sand stream A small sand delta spread out at the mouth of gully, or a deposit of sand along the bed of a small creek formed by a torrential rain (Stephenson & Veatch, 1915 p.112). Cf: *sand river*.

sand stretch A striation worn in a rock surface by windblow sand.

sand strip A long, narrow ridge of sand extending for a lon distance downwind from each horn of a dune.

sand tank A *rock tank* filled with sand.

sand trap A *trap* for separating sand and other particles from flowing water and generally including means for ejectin them; e.g. such a device for separating heavy, coarse parti cles from the cuttings-laden fluid overflowing a drill collar.

sand tube A tubular *fulgurite* formed in sand.

sand tuff (a) A tuffaceous sandstone. (b) A tuff whose component fragments are in the size range of sand. This usage is obsolete.

sand twig A small, twig-like aggregate of sand that stands more or less upright on the surface of a sand dune undergoing wind scour, apparently forming around a root or stem of a plant exposed on the dune surface (Carroll, 1939, p.20-21).

sandur Icelandic term signifying "sand", but generally adopted for *outwash plain*. Pl: *sandar*. Syn: *sandr*.

sand volcano An accumulation of sand resembling a miniature volcano or low volcanic mound (maximum diameter is 15 m), generally situated on top of a slump sheet or the upper surface of a highly contorted layer of laminated sediments, and produced by the upward expulsion (under submarine conditions) of sand-laden water from slumped masses prior to normal deposition of overlying beds. Examples occur in County Clare, Ireland.

sandwash A sandy or gravel stream bed, devoid of vegetation, containing water only during a sudden and heavy rainstorm.

sand wave A large, ridge-like primary structure displayed on the upper surface of a sedimentary bed, having a shape somewhat resembling a water wave, and formed by high-velocity currents (of air or water) that move sand usually approximately normal to the direction of flow; e.g. a *dune* moving downcurrent and an *antidune* moving upcurrent. It is usually periodic and may be symmetric, asymmetric, or irregular in shape. The term has been used to describe sand ripples and sand dunes of all types and sizes, but it is usually restricted to very large and linear subaqueous sand dunes or sandbars formed on a stream bed; the smallest sand waves are known as *ripple marks* and a large sand wave as a *megaripple*. Syn: *sand ridge*.

sand wedge A body of sand shaped like a vertical wedge with the apex downward. In areas of patterned ground, esp. Antarctica, it forms by infilling of debris into winter contraction cracks.

sand-wedge polygon A *nonsorted polygon* formed by infilling of sand and gravel in intersecting fissures resulting from thermal contraction. Surface diameter and depth range from 15 to 30 cm (Bird, 1967, p.193). Syn: *tesselation*.

sandy [geomorph] n. A low stream terrace whose upper surface rises upstream where aggradation keeps pace with the growth of a downstream dam, as along Big Sandy River in eastern Kentucky (Shaw, 1911, p.489). Cf: *island hill; muddy*.

sandy [sed] adj. Pertaining to or containing sand, or consisting of, abounding in, or covered with sand; *arenaceous*. Cf: *sammitic*. Syn: *sanded*.

sandy breccia A term used by Woodford (1925, p.183) for a breccia containing at least 80% rubble and 10% sand, and no more than 10% of other material.

sandy chert Chert with oolith-like structures, formed when silica replaces cement or fills pore spaces in sandy beds and incorporates large, rounded sand grains into a cherty body or matrix. The perimeters of the sand grains are commonly resorbed, giving gradational contacts between them and the secondary silica.

sandy clay (a) An unconsolidated sediment containing 10-50% sand and having a ratio of silt to clay less than 1:2 (Folk, 1954, p.349). (b) An unconsolidated sediment containing 40-75% clay, 12.5-50% sand, and 0-20% silt (Shepard, 1954). (c) A soil containing 35-55% clay, 45-65% sand, and 0-20% silt (SSSA, 1965, p.347).

sandy clay loam A *clay loam* containing 20-35% clay, 45-80% sand, and less than 28% silt.

sandy conglomerate (a) A conglomerate containing 30-80% sand and having a ratio of sand to mud (silt + clay) greater than 9:1 (Folk, 1954, p.347); a consolidated sandy gravel. (b) A conglomerate containing more than 20% sand (Krynine, 1948, p.141).

sandy desert An area of sand accumulation in an arid region, usually having an undulating surface of dunes; an *erg* or *koum*.

sandy gravel (a) An unconsolidated sediment containing 30-80% gravel and having a ratio of sand to mud (silt + clay) greater than 9:1 (Folk, 1954, p.346); if the ratio is between 1:1 and 9:1, the sandy gravel is "muddy". (b) An unconsolidated sediment containing more particles of gravel size than of sand size, more than 10% sand, and less than 10% of all other finer sizes (Wentworth, 1922, p.390). (c) An unconsolidated sediment containing 50-75% sand and 25-50% pebbles (William et al, 1942, p.343-344). Cf: *pebbly sand*.

sandy loam A soil containing 43-85% sand, 0-50% silt, and 0-20% clay, or a soil containing at least 52% sand and no more than 20% clay and having the percentage of silt plus twice the percentage of clay exceeding 30, or a soil containing 43-52% sand, less than 50% silt, and less than 7% clay (SSSA, 1965, p.347); specif. such a soil containing at least 30% very coarse sand, coarse sand, and medium sand, and less than 25% very coarse sand and less than 30% fine sand or very fine sand. It is subdivided into coarse sandy loam, fine sandy loam, and very fine sandy loam. Sandy loam contains sufficient silt or clay to make the soil somewhat coherent. Cf: *loamy sand*.

sandy mud An unconsolidated sediment containing 10-50% sand and having a ratio of silt to clay between 1:2 and 2:1 (Folk, 1954, p.349).

sandy ore A term used in Brazil for iron ore found along the bottoms of valleys and derived from *rubble ore* "by the natural sluicing of the streams" (Derby, 1910, p.818).

sandy silt (a) An unconsolidated sediment containing 10-50% sand and having a ratio of silt to clay greater than 2:1 (Folk, 1954, p.349). (b) An unconsolidated sediment containing 40-75% silt, 12.5-50% sand, and 0-20% clay (Shepard, 1954). (c) An unconsolidated sediment containing more particles of silt size than of sand size, more than 10% silt, and less than 10% of all other sizes (Wentworth, 1922).

sandy siltstone (a) A consolidated *sandy silt*. (b) A siltstone containing more than 20% sand (Krynine, 1948, p.141).

sandy soil A soil containing a large amount of sand, such as one with more than 70% sand and less than 15% clay. Syn: *sand*.

Sangamon Pertaining to the third interglacial stage of the Pleistocene Epoch in North America, after the Illinoian glacial stage and before the Wisconsin. Etymol: Sangamon County, Ill. See also: *Riss-Würm*. Syn: *Sangamonian*.

sanidal Of or pertaining to the continental shelf.

sanidaster A spinose, rod-like, monaxonic sponge spicule (streptaster).

sanidine A high-temperature mineral of the alkali feldspar group: $KAlSi_3O_8$. It is a highly disordered, monoclinic form of orthoclase occurring in clear, glassy, often tabular crystals embedded in unaltered acid volcanic rocks (such as trachytes); it appears to be stable under equilibrium conditions above approximately $500°C$. Sanidine forms a complete solid-solution series with monalbite, and some sodium is always present. Syn: *glassy feldspar; ice spar; rhyacolite*.

sanidine nephelinite A fine- to medium-grained magnetite-rich murite composed chiefly of pyroxene, nepheline, and sanidine, with nosean sometimes present instead of nepheline, and occasionally with mica and acmite.

sanidinite An igneous rock composed almost entirely of sanidine. The term has also been applied to rocks composed of other alkali feldspars.

sanidinite facies According to Eskola (1939), rocks that have recrystallized at maximum temperatures ($600°-1000°C$ and at minimum pressures (less than 1000 bars) under conditions of pyrometamorphism (e.g. xenoliths in lavas, fragments of volcanic breccia, contact rocks of near-surface intrusions) approaching partial to complete fusion (Turner, 1948).

sanjuanite A mineral: $Al_2(PO_4)(SO_4)(OH).9H_2O$.

sanmartinite A mineral: $ZnWO_4$. The zinc is sometimes replaced by iron, manganese, or calcium.

sannaite An extrusive rock containing phenocrysts of barkevikite, pyroxene, and biotite (in order of decreasing abundance) in a fine-grained to dense groundmass of alkali feldspar, acmite, chlorite, calcite, and pseudomorphs of mica after nepheline.

Sannoisian *Tongrian.*

sansicl An unconsolidated sediment consisting of a mixture of *sand*, *silt*, and *clay*, in which no component forms 50% or more of the whole aggregate.

Sanson-Flamsteed projection A syn. of *sinusoidal projection.* Named after Nicolas Sanson (1600-1667), French geographer, and John Flamsteed (1646-1719), English astronomer.

santafeite A black orthorhombic mineral: $Na_2(Mn,Ca, Sr)_6Mn_3^{+4}(V,As)_6O_{28}.8H_2O$.

santite A mineral: $KB_5O_8.4H_2O$.

Santonian European stage: Upper Cretaceous (above Coniacian, below Campanian).

santorinite (a) A light-colored extrusive rock containing approximately 60-65 percent silica and calcic plagioclase (labradorite to anorthite) as the only feldspar. (b) A hypersthene andesite containing plagioclase crystals that have labradorite cores and sodic rims and a groundmass with microlites of sodic oligoclase. Its name is derived from Santorini (or Thera), Greece.

sanukite An andesite characterized by orthopyroxene as the mafic mineral, andesine as the plagioclase, and a glassy groundmass. Cf: *orthoandesite.*

sanukitoid *orthoandesite.*

sapanthracite Sapropelic coal of anthracitic rank; it is the highest stage in the *sapropelic series.* Cf: *humanthracite.*

sapanthracon Sapropelic coal of Carboniferous age; it is the fifth stage in the *sapropelic series.* Cf: *humanthracon.*

saphir d'eau A syn. of *water sapphire* (variety of cordierite). Etymol: French, "water sapphire".

saponite A soft, soapy, white or light-buff to bluish or reddish, trioctahedral, magnesium-rich clay mineral of the montmorillonite group: $(Ca/2,Na)_{0.33}(Mg,Fe)_3(Si_{3.67}Al_{0.33})O_{10}$-$(OH)_2.4H_2O$. It represents an end member in which the replacement of aluminum by magnesium in the octahedral sheets is essentially complete. Saponite occurs in masses and fills veins and cavities in serpentine and basaltic rocks; it has an unctuous feel and is somewhat plastic, but does not adhere to the tongue. Syn: *bowlingite; mountain soap; piotine; soapstone.*

sappare (a) *sapphire.* (b) *kyanite.*

sapperite A natural, pure white cellulose, formula $(O_6H_{10}O_5)_n$, which occurs in brown coal and fossil wood.

sapphire (a) Any pure, gem-quality corundum other than *ruby*; esp. the fine blue transparent variety of crystalline corundum of great value, containing small amounts of oxides of cobalt, chromium, and titanium, used as a gemstone, and found esp. in the Orient (Kashmir, Burma, Thailand, and Ceylon). Other colors, such as pink, purple, yellow, green, and orange, are applied to *fancy sapphire.* Syn: *sappare.* (b) Any gem from a corundum crystal.

sapphire quartz (a) A rare, opaque, indigo-blue variety of quartz colored by included nonparallel fibers of silicified crocidolite. Cf: *hawk's-eye.* Syn: *azure quartz; blue quartz; siderite.* (b) A term used in western U.S. for chalcedony having a light to pale sapphire-blue color.

sapphirine (a) A green or pale-blue mineral: $(Mg,Fe)_{15}(Al, Fe)_{34}Si_7O_{80}$. It is a principal constituent of certain high-grade silica-deficient metamorphic rocks and occurs usually in granular form. (b) A name applied to certain blue minerals such as hauyne and blue chalcedony.

sapping [geomorph] (a) The natural process of erosion along the base of a cliff by the wearing away of softer layers, thus removing the support for the upper mass which breaks off into large blocks falling from the cliff face. See also: *landslide*; *sapping.* Syn: *cliff erosion; undermining.* (b) *spring sapping.*

sapping [glac geol] (a) *basal sapping.* (b) Sometimes used as a syn. of *plucking.*

Saprist In U.S. Dept. of Agriculture soil taxonomy, a suborder of the soil order *Histosol*, characterized by water saturation for sufficient periods of time to make cultivation without artificial drainage difficult, and by a high content of well decomposed plant material. Bulk density of Saprists is 0.2 or more (SSSA, 1970). Cf: *Fibrist; Folist; Hemist.*

saprobe *saprophyte.*

saprobic *saprophytic.*

saprocol Indurated sapropel; it is the second stage in the *sapropelic series.* Also spelled: *saprokol.* Cf: *humocoll.*

saprodil A sapropelic coal of Tertiary age; it is the third stage in the *sapropelic series.* Cf: *humodil.*

saprodite Sapropelic coal of brown-coal rank; it is the fourth stage in the *sapropelic series.* Cf: *humodite.*

saprogen An organism that lives on dead organic matter and can cause its decay. Adj: *saprogenic.*

saprogenic Said of an organism that produces decay or putrefaction. Cf: *saprophytic.* Syn: *saprogenous.* Noun: *saprogen.*

saprohumolith series Organic materials and coals intermediate between the *sapropelite series* and the *humolith series*, with sapropelic materials predominating. Cf: *humosapropelic series.*

saprokol Var. of *saprocol.*

saprolite A soft, earthy, clay-rich, thoroughly decomposed rock formed in place by chemical weathering of igneous and metamorphic rocks. It often forms a thick (as much as 100 m) layer or cover, esp. in a humid and tropical or subtropical climate; the color is commonly some shade of red or brown. The term was proposed by Becker (1895). Cf: *geest; laterite.* Syn: *saprolith; sathrolith.*

sapromyxite *boghead coal.*

sapront *saprophyte.*

sapropel An unconsolidated, jelly-like ooze or sludge composed of plant remains, most often algae, macerating and putrefying in an anaerobic environment on the shallow bottoms of lakes and seas. It may be a source material for petroleum and natural gas. Cf: *dy; gyttja.*

sapropel-calc A sedimentary deposit in which the amount of calcareous-algae remains exceeds that of sapropel.

sapropel-clay A sedimentary deposit in which the amount of clay exceeds that of sapropel.

sapropelic Pertaining to or derived from *sapropel*; indicating high sulfate or reducing environment.

sapropelic coal Coal that is derived from organic residues (finely divided plant material, spores, algae) in stagnant or standing bodies of water. Putrefaction under anaerobic conditions rather than peatification is the formative process. The main types of sapropelic coal are cannel coal, boghead coal, and torbanite. Sapropelic coals are high in volatiles, generally dull, massive, and relatively uncommon. Cf: *humic coal.* See also: *sapropelite series.*

sapropelite *sapropelic coal.*

sapropelite series The organic materials of *sapropelic coal* of metamorphic rank: sapropel, saprocol, saprodil, saprodite, sapranthracon, and sapanthracite (Heim & Potonié, 1930, p.146). Cf: *humolith series; humosapropelic series; saprohumolith series.*

sapropel-peat *peat-sapropel.*

saprophilous *saprophytic.*

saprophyte A plant that lives on decayed or decaying organic matter. Syn: *saprobe; sapront.* Adj: *saprophytic.*

saprophytic Said of a plant that receives its nourishment from the products of decaying organic matter; i.e. a saprophyte. Syn: *saprophilous; saprobic; saprozoic.* Cf: *saprogenic.*

sapropsammite Sandy sapropel.

saprovitrinite Vitrinite in vitrain of sapropelic coal. Cf: *humovitrinite*.

saprozoic *saprophytic*.

saracen stone A syn. of *sarsen*. The term originally signified a pagan stone or monument. Syn: *Saracen's stone*.

sarcodictyum The outermost layer of cytoplasm in a radiolarian; a network of protoplasm on the surface of the calymma of a radiolarian.

sarcodina A subphylum of protozoans characterized mainly by their capability of forming pseudopodia. Among the members of the subphylum are rhizopods and actinopods.

sarcopside A mineral: $(Fe,Mn,Mg)_3(PO_4)_2$.

sarcotesta The fleshy, outer seed coat, e.g. of a cycad. Cf: *sclerotesta*.

sard A translucent brown, reddish-brown, or deep orange-red variety of chalcedony, similar to *carnelian* but having less-intense colors (darker and more brownish), and classed by some as a variety of carnelian. Syn: *sardius; sardine*.

sardic orogeny One of the 30 or more short-lived orogenies during Phanerozoic time identified by Stille, in this case near the end of the Cambrian.

sardius Original name for *sard*.

sardonyx A gem variety of chalcedony that is like *onyx* in structure but includes straight parallel brownish-red or reddish-brown bands of sard alternating with white, brown, black, or other colored bands of another mineral. The name is applied incorrectly to carnelian and sard.

Sargasso Sea A warm region of the open North Atlantic Ocean to the east and south of the Gulf Stream, characterized by a large mass of floating vegetation that is mainly sargasso (gulfweed), a seaweed (brown alga) of the genus *Sargassum*.

sarkinite A flesh-red monoclinic mineral: $Mn_2(AsO_4)(OH)$.

Sarmatian European stage: upper Miocene (above Tortonian, below Pontian).

sarmientite A yellow mineral: $Fe_2(AsO_4)(SO_4)(OH).5H_2O$. It is isomorphous with diadochite.

sarnaite A feldspathoid-bearing syenite composed of cancrinite and acmite. Its name is derived from Sarna, Sweden.

sarospatakite A micaceous clay mineral composed of mixed layers of illite and montmorillonite, found in Sárospatak, Hungary. Cf: *bravaisite*. Syn: *sarospatite*.

sarsden stone *sarsen*.

sarsen A large residual mass of stone left after the erosion of once continuous bed of which it formed a part; specif. one of the large, rounded, gray blocks or fragments of silicified sandstone or conglomerate strewn over the surface of the English chalk downs (esp. in Wiltshire) and far from any similar beds, being the only remnants of the former Tertiary (Eocene?) cover. Etymol: alteration of *Saracen*, a Moslem or "outlandish stranger". Syn: *sarsen stone; Saracen stone; sarsden stone; greywether; druid stone*.

sartorite A dark-gray monoclinic mineral: $PbAs_2S_4$.

sarule A *sceptrule* with a single, spinose, terminal outgrowth that resembles a brush.

saryarkite A white tetragonal mineral: $(Ca,Y,Th)_2Al_4(SiO_4,SO_4)_4(OH).9H_2O$.

sassolite A white or gray mineral consisting of native boric acid: $B(OH)_3$ or H_3BO_3. It usually occurs in small pearly scales as an incrustation or as tabular triclinic crystals around fumaroles or vents of sulfurous emenations. Syn: *sassoline*.

sastruga One of a series of irregular ridges up to two inches high, formed in a level or nearly level snow surface by wind erosion, often aligned parallel to the wind direction, with steep, concave or overhanging ends facing the wind; or cut into snow dunes previously deposited by the wind. Pl: *sastrugi*. Syn: *zastruga; skavl*. Cf: *erosion ridge; wind ridge*.

satelite Fibrous serpentine with a slight chatoyant effect, being pseudomorphous after asbestiform tremolite that has been silicified. It occurs in Tulare County, Calif.

satellite A secondary celestial body that revolves about another, primary body, e.g. the Moon about the Earth. The solar system also contains manmade satellites.

satellite geodesy Near earth satellites are used in navigational systems for position fixing, and are used in the determination of the geoid as well as in the determination of global geometric networks providing connections between the major land areas of the Earth.

satellitic crater *secondary crater*.

sathrolith *saprolite*.

satimolite A mineral: $KNa_2Al_4B_6O_{15}Cl_3.13H_2O$.

satin ice *acicular ice*.

satin spar (a) A white, translucent, fine fibrous variety of *gypsum*, characterized by chatoyancy or a silky luster. (b) A term used less correctly for a fine fibrous or silky variety of calcite or aragonite.---Syn: *satin stone*.

satpaevite A yellow mineral: $Al_{12}V_2^{+4}V_6^{+5}O_{37}.30H_2O$.

saturated [geol] (a) Said of a mineral that can form in the presence of free silica, i.e. that contains the maximum amount of combined silica. (b) Said of an igneous rock composed chiefly of these minerals, or of its magma. (c) Said of a rock having quartz in its norm.----Cf: *oversaturated; undersaturated; unsaturated*.

saturated [water] A condition in which the interstices of a material are filled with a liquid, usually water. It applies whether the liquid is under greater than or less than atmospheric pressure, so long as all connected interstices are full. See also: *zone of saturation*.

saturated permafrost Permafrost that contains no more ice than the ground could hold if the water were in the liquid state; permafrost in which all available pore spaces are filled with ice.

saturated surface *water table*.

saturated zone *zone of saturation*.

saturation [phys] In a nonlinear dependency of an observable upon a variable, for certain limiting values of the variable, the condition in which the observable does not change to a measurable degree.

saturation [meteorol] The maximum possible content of water vapor in the Earth's atmosphere, per unit volume and at a given temperature. See also: *dew point*.

saturation curve [soil] A curve showing the weight of solids per unit volume of a saturated soil mass as a function of water content.

saturation line [petrology] The line, on a variation diagram of an igneous rock series, that represents saturation with respect to silica. Rocks to the right of the line are oversaturated and those to the left, undersaturated.

saturation line [glaciol] The boundary on a glacier between the *soaked zone* and the *percolation zone*.

saturation magnetization The maximum possible magnetization of a material, i.e. alignment of all magnetic ions.

Saucesian North American stage: Oligocene and Miocene (above Zemorrian, below Relizian).

sauconite A trioctahedral, zinc-bearing clay mineral of the montmorillonite group: $Na_{0.33}Zn_3(Si_{3.47}Al_{0.53})O_{10}(OH)_2.4H_2O$. It represents an end member in which the replacement of aluminum by zinc in the octahedral sheets is essentially complete.

saucyite A glassy rhyolitic rock composed of large sanidine phenocrysts in a groundmass of orthoclase microlites and minute crystals of biotite, augite, sphene, zircon, and magnetite (Thrush, 1968, p.964).

sault A waterfall or rapids in a stream. Etymol: Latin *saltus*, past participle of *salire*, "to leap". Pron: *sue*.

sausage structure *boudinage*.

saussurite A tough, compact, and white, greenish, or grayish mineral aggregate consisting of a mixture of albite (or oligoclase) and zoisite or epidote, together with variable amounts of calcite, sericite, prehnite, and calcium-aluminum silicates other than those of the epidote group. It is produced by alteration of calcic plagioclase. Saussurite was originally thought to be a specific mineral.

saussuritization The replacement, esp. of plagioclase in basalts and gabbros, by a fine-grained aggregate of zoisite, epidote, albite, calcite, sericite, and zeolites. It is a metamorphic process and is frequently accompanied by chloritization of the ferromagnesian minerals.

savanna (a) An open, grassy, essentially treeless plain, esp. as developed in tropical or subtropical regions. Usually there is a distinct wet and dry season; what trees and shrubs are found there are drought-resistant. (b) The tall grass characteristic of a *savanna*. (c) Along the southeastern Atlantic Coast of the U.S. the term (often spelled *savannah*) is used for marshy alluvial flats with occasional clumps of trees.

Savic orogeny One of the 30 or more short-lived orogenies during Phanerozoic time identified by Stille, in this case in the late Oligocene, between the Chattian and Aquitanian Stages.

Savonius rotor current meter A sensor device used in oceanography for measuring the speed of a current. It utilizes an S-shaped rotor, or a pair of them with their axes perpendicular to each other.

saw-cut A large canyon that cuts across a terrace "with startling abruptness" so that it "cannot be seen until one has almost reached the edge" (Smith, 1898, p.469).

saw-toothed (a) *serrate*. (b) Descriptive of *hacksaw structure*.

saxicavous Said of an organism that bores into rock; *rock borer*.

saxicolous *rupestral*.

saxifragous Said of a plant that grows in rock crevices and promotes its splitting; i.e., a *chasmophyte*.

saxigenous *rupestral*.

Saxonian European stage: Middle Permian (above Autunian, below Thuringian).

Saxonian-type facies series A type of dynamothermal regional metamorphism for which the classical locality is the granulite area in Saxony. The pressure range of formation is extensive (2000-8000 bars) and the temperature range from 100° to 700°C, probably involving polymetamorphism. They differ little mineralogically from those formed in the *Barrovian-type facies series* except at the highest pressure and temperature values at which kyanite-orthoclase-bearing granulites are formed (Hietanen, 1967, p.201).

saxonite A peridotite composed chiefly of olivine and orthopyroxene. It is considered by some petrologists as a syn. of *harzburgite* and by others as distinct from harzburgite due to the presence of opaque oxide minerals in saxonite.

sborgite A triclinic mineral: $NaB_5O_8 \cdot 5H_2O$.

scabland Elevated, essentially flatlying, basalt-covered land with only a thin soil cover, sparse vegetation, and usually deep, dry channels (*channeled scabland*) scoured into the surface. An example is the Columbia lava plateau of eastern Washington which was widely and steeply eroded by glacial meltwaters. See also: *scabrock*.

scabrate Rough or scaly; esp. said of sculpture of pollen and spores consisting of more or less isodiametric projections less than one micron in diameter.

scabrock (a) An outcropping of *scabland*. (b) Weathered material of a scabland surface.

scacchite (a) A mineral: $MnCl_2$. (b) A name applied to various minerals, including monticellite, a doubtful selenide of lead, and a brick-red powdery fluoride containing rare earths.

scaglia A dark, very fine-grained, more or less calcareous shale typically developed in the Upper Cretaceous and lower Tertiary of the northern Apennines. Etymol: Italian, 'scale, chip". Pron: *skahl*-ya.

scalariform Pertaining to the ladderlike thickenings in the walls of certain xylem cells.

scale [cart] An indication of the proportion between the linear

distances on a map, chart, globe, model, photograph, or othe[r] drawing and the corresponding actual distances on the su[r]face being mapped. It may be expressed in the form of a d[i]rect or verbal statement using different units (e.g. "1 inch to [1] mile" or "1 inch = 1 mile" indicates that two points on th[e] ground exactly one mile apart will appear on the map exact[ly] one inch apart), a *representative fraction* or numerical rat[io] (e.g. "1/24,000" or "1:24,000" indicates that one unit on th[e] map represents 24,000 equivalent units on the ground), or [a] graphic measure (such as a bar or line marked off in fee[t,] miles, kilometers, etc., or subdivided at selected intervals[). The scale of a photograph is usually taken as the ratio of th[e] principal distance of the camera to the altitude of the camer[a] station above mean ground elevation. See also: *small-sca[le] map; large-scale map*. Syn: *map scale*.

scale [paleont] (a) A small plate-like structure attached [to] septal grooves and interseptal ridges in some rugose cora[l]lites, as in *Tryplasma*. Also, a thin and flat or nearly flat scle[r]ite in an octocoral. (b) *scaphocerite*. (c) A small, more [or] less flattened, rigid, and definitely circumscribed plate formin[g] part of the external body covering of certain vertebrates.

scale factor A multiplier for reducing a distance (obtaine[d] from a map) by computation or scaling to the actual distanc[e] on the datum of the map.

scalenohedral Having the form or symmetry of a *scalenoh[e]dron*.

scalenohedron A closed crystal form whose faces are scalen[e] triangles; a *hexagonal scalenohedron* has twelve faces, an[d] the *tetragonal scalenohedron* has eight. Adj: *scalenohedral*.

scaler An electronic instrument that counts the pulses from [a] radiation detector.

scaling A type of *exfoliation* that produces thin flakes, lam[i]nae, or scales.

scallop [speleo] British syn. of *flute*.

scallop [paleont] (a) Any of many marine bivalve mollusks [of] the family Pectinidae having a shell that is characteristical[ly] rather flat, radially ribbed, and marginally undulated, that ha[s] a single large adductor muscle, and that is able to swim b[y] opening and closing the valves. (b) One of the valves of th[e] shell of a scallop. Syn: *scallop shell*.

scallop [sed] *scalloping*.

scalloped upland "The region near or at the divide of an up[land] into which glacial cirques have cut from opposite sides[" (Stokes & Varnes, 1955, p.129). Cf: *fretted upland*.

scalloping A term used by Gruner et al (1941, p.1621-1622[) for a sedimentary structure of uncertain origin, superficial[ly] resembling oscillation ripple mark, and having a concave sid[e] that is always oriented toward the top of the bed. It may hav[e] formed by differential expansion or shrinkage of adjoinin[g] layers of mud before complete consolidation. Syn: *scallo[p] [sed]*.

scalped anticline *breached anticline*.

scaly Said of the texture of a mineral, esp. a mica, in whic[h] small plates break or flake off from the surface like scales.

scan n. A graphic or photographic depiction of the distributio[n] of radioactivity of a substance. See also: *autoradiograph; sk[i]agram*.

scandent Said of a graptoloid with stipes that grew ere[ct] along or enclosing the virgula.

scanner *line scanner*.

scanning electron microscope An *electron microscope* i[n] which a finely focused beam of electrons is electrically o[r] magnetically moved across the specimen to be examined[,] from point to point, again and again, and the reflected an[d] emitted electron intensity measured and displayed, sequentia[l]ly building up an image. The ultimate magnification and reso[lution is less than for the conventional *electron microscope* but opaque objects can be examined, and great depth of fiel[d] is obtained.

scanning radiometer An image-forming system consisting of adiometer, which by the use of a plane mirror rotating at 45° to the optical axis, can see a circular path normal to the instrument. When the system is moved forward at velocity V and at altitude H, a suitable V/H ratio may be established, so that consecutive circular scans are just touching. If the downward-pointing sectors are used and consecutively displayed as raster lines on an oscilloscope, a TV-like image may be produced. A scanning radiometer is distinguished from other image-forming line-scan systems which produce differential images by an internal blackbody reference and other features which enable it to produce radiometric images in which the gray scale is related to the infrared signal by a known energy transfer function.

scanoite An igneous rock similar to ghizite but containing normative nepheline.

scaphocerite A flattened plate on the second joint of the antennae of many crustaceans; e.g. a scale-like exopod of an antenna of a eumalacostracan. Syn: *scale* [paleont].

scapholith An elongate, diamond- or boat-shaped heterococcolith with a central area of parallel laths. Syn: *rhombolith*.

scaphopod Any benthic marine univalve mollusk belonging to the class Scaphopoda, characterized by an elongate body completely surrounded by mantle and a tubular calcareous shell open at both ends. Their stratigraphic range is Devonian to present.

scapolite (a) A group of minerals of general formula: $(Na,Ca,)_4[Al_3(Al,Si)_3Si_6O_{24}](Cl,F,OH,CO_3,SO_4)$. It consists of generally white or grayish-white minerals crystallizing in the dipyramidal class of the tetragonal system, and commonly forms an isomorphous series between marialite and meionite. Scapolite minerals characteristically occur in calcium-rich metamorphic rocks (associated with impure limestones) or in igneous rocks as the products of alteration of basic plagioclase feldspars. (b) A specific mineral of the scapolite group, intermediate in composition between marialite and meionite (Ma:Me from 2:1 to 1:3), containing 46-54% silica, and resembling feldspar when massive but having a fibrous appearance and a higher specific gravity. Syn: *wernerite*. (c) A member of the scapolite group, including scapolite, marialite, meionite, and mizzonite.

scapolitization Introduction of, or replacement by, scapolite. Plagioclase is commonly so replaced. The replacement may involve introduction of chlorine. Syn: *dipyrization*.

scar [geomorph] (a) A cliff, precipice, or other steep, rocky eminence or slope (as on the side of a mountain) where bare rock is well exposed to view; e.g. a limestone face in northern England. Originally, the term referred to a crack or breach; later, an isolated or protruding rock. Etymol: Old Norse *sker*, "skerry". Syn: *scaur; scaw*. (b) A rocky *shore platform*. (c) *landslide scar*. (d) *meander scar*.

scar [paleont] (a) *muscle scar*. (b) *cicatrix*.

scarbroite A white mineral: $Al_{14}(CO_3)_3(OH)_{36}$.

scarp (a) A line of cliffs produced by faulting or by erosion. The term is an abbreviated form of *escarpment*, and the two terms commonly have the same meaning, although "scarp" is more often applied to cliffs formed by faulting. See also: *fault scarp; erosion scarp*. (b) A relatively steep and straight, cliff-like face or slope of considerable linear extent, breaking the general continuity of the land by separating level or gently sloping surfaces lying at different levels, as along the margin of a plateau, mesa, terrace, or bench. A scarp may be of any height. The term should not be used for a slope of highly irregular outline. See also: *scarp slope*. (c) *beach scarp*. (d) A steep surface on the undisturbed ground around the periphery of a landslide, caused by movement of slide material away from the undisturbed ground; also, a similar but smaller feature on the disturbed material, produced by differential movements within the sliding mass.

scarped plain An area marked by a succession of faintly inclined or gently folded strata, as the eastern part of the Great Plains of the U.S. The inclination of strata has perceptible influence upon even the smaller elements of the topography.

scarped ridge *cuesta*.

scarp face *scarp slope*.

scarp-foot spring A spring that flows onto the land surface at or near the foot of an escarpment.

scarpland A region marked by a succession of nearly parallel cuestas separated by lowlands.

scarplet (a) A low, miniature scarp, varying in height from several centimeters to 6 m or more; specif. a *piedmont scarp*. Also, a small scarp formed on a wave-cut platform by the outcropping of resistant rocks. (b) *earthquake scarplet*.

scarp retreat The *slope retreat* of a scarp.

scarp slope (a) The relatively steeper face of a *cuesta*, facing in a direction opposite to the dip of the strata. Cf: *dip slope; back slope*. Syn: *scarp face; inface; front slope*. (b) A *scarp* or an *escarpment*.

scarp stream An *obsequent stream* flowing down a scarp, such as down the scarp slope of a cuesta.

scatter diagram [struc petrol] *point diagram*.

scatter diagram [stat] A graphic representation of paired measurements, usually along Cartesian axes, that aids in visualizing the relationships between two or more variables.

scattered gamma-ray log *gamma-gamma log*.

scattered ice An obsolete term for sea-ice concentration of 1/10 to 5/10; now replaced generally by *open pack ice* and *very open pack ice*. Syn: *sailing ice*.

scatterometry A method of using non-imaging *radar*, in which the radar scatterometer measures the variation of radar scattering coefficient with angle, wavelength and polarization. These variations may be used by geoscientists to discriminate between surfaces with different roughness and materials.

scavenger An organism that feeds on dead matter, refuse, or matter detrimental to the health of humans.

scavenger well A well located between a well, or group of wells, yielding usable water and a source of potential contamination; it is pumped (or allowed to flow) as waste to prevent the contaminated water from reaching the good wells. The most common application is in coastal areas, where scavenger wells are used to prevent salt water from reaching water-supply wells.

scaw (a) An obsolete term for a headland or promontory. (b) A steep, rocky slope; a *scar*.---Etymol: Scandinavian.

scawtite A colorless monoclinic mineral: $Ca_7Si_6O_{18}(CO_3).2H_2O$.

sceptrule A hexactinellid-sponge spicule (microsclere) that consists of one long ray with one end containing an axial cross and usually bearing various anaxial outgrowths. See also: *sarule; scopule; clavule; lonchiole*.

schafarzikite A red to brown tetragonal mineral: $FeSb_{2-x}(O,OH,H_2O)_4$.

schairerite A colorless rhombohedral mineral: $Na_{21}(SO_4)_7F_6Cl$. Cf: *galeite*.

schallerite A brown mineral: $(Mn,Fe)_8Si_6As(O,OH,Cl)_{26}$.

schalstein An altered tuff with shear structures; it is usually basic or calcareous (Holmes, 1928). Etymol: German. Cf: *adinole; spotted slate*.

schapbachite A syn. of *matildite*. The term has also been applied to an intimate intergrowth of matildite and galena and to a high-temperature polymorph of matildite.

schaurteite A mineral: $Ca_3Ge(SO_4)_2(OH)_6.3H_2O$.

scheelite A yellowish-white or brownish tetragonal mineral: $CaWO_4$. It is found in pneumatolytic veins associated with quartz, and fluoresces to show a blue color. Scheelite is isomorphous with powellite, and is an ore of tungsten.

schefferite A brown to black monoclinic mineral of the pyroxene group: $(Ca,Mn)(Mg,Fe,Mn)Si_2O_6$. It is a variety of diopside containing manganese and frequently much iron.

S-chert Stratigraphically controlled chert, occurring in beds (bedded chert) or in groups of nodules (nodular chert) distributed parallel to bedding (Dunbar & Rodgers, 1957, p.248).

schertelite A mineral: $(NH_4)_2MgH_2(PO_4)_2.4H_2O$.

scheteligite A mineral: $(Ca,Y,Sb,Mn)_2(Ti,Ta,Nb,W)_2O_6(O,OH)$.

Schiefer A general rock term referring to a rock's laminated or foliated structure, commonly used to describe rocks ranging from shale to schist, depending on its use as a modifying noun: e.g. Schieferton or argillaceous shale, Tonschiefer or slate. Adj: *schiefrig*. Etymol: German.

schiefrig Adj. of *Schiefer*.

schiller A syn. of *play of color*. Etymol: German. See also: *schillerization*.

schillerization The development of *schiller* or play of color due to the arrangement of minute inclusions in the crystal.

schiller spar A syn. of *bastite*. Also spelled: *schillerspar*.

schirmerite A mineral: $PbAg_4Bi_4S_9$.

schist A strongly foliated crystalline rock formed by dynamic metamorphism which can be readily split into thin flakes or slabs due to the well developed parallelism of more than 50% of the minerals present, particularly those of lamellar or elongate prismatic habit, e.g. mica, hornblende. The mineral composition is not an essential factor in its definition (American usage) unless specifically included in the rock name, e.g. quartz-muscovite schist. Varieties may also be based on general composition, e.g. calc-silicate schist, amphibolite schist, or on texture, e.g. spotted schist.

schist-arenite A light-colored sandstone containing more than 20% rock fragments derived from an area of regionally metamorphosed rocks (Krynine, 1940); specif. a lithic arenite having abundant fragments of schist. The term was applied by Krynine (1937, p.427) to the medium-grained clastic rocks of the Siwalik Series in northern India, averaging about 40% quartz, 15% feldspar, 35-40% schist and phyllite fragments, and 5-10% accessory materials.

schistic schistose.

schistoid adj. Resembling *schist*.

schistose Said of a rock displaying *schistosity*. Cf: *greissic*. Syn: *schistic*.

schistosity The foliation in schist or other coarse-grained, crystalline rock due to the parallel, planar arrangement of mineral grains of the platy, prismatic, or ellipsoidal types, usually mica. It is considered by some to be a type of *cleavage*. Adj: *schistose*.

schizocarp A dry fruit that, at maturity, splits apart into several one-seeded, indehiscent carpels.

schizochroal eye A trilobite eye whose external visual surface consists of numerous small convexities corresponding to underlying lenses. Cf: *holochroal eye*.

schizodont (a) Said of the dentition of a bivalve mollusk with one tooth (median of left valve) broad and divided into two equal parts (bifid), and characterized by coarse, variable, and amorphous teeth diverging sharply from beneath the beak. (b) Said of a subclass of amphidont hinges in ostracodes, having anterior tooth and socket of one valve both bifid, and a reverse arrangement of elevations and depressions in the opposed valve (TIP, 1961, pt.Q, p.54).

schizolite A light-red variety of pectolite containing manganese.

schizolophe A brachiopod lophophore indented anteriorly and medially to define a pair of brachia that bear a row of paired filamentar appendages, at least distally (TIP, 1965, pt.H, p.152).

schizomorphic A *deuteromorphic* crystal modified by cataclastic processes. The term is obsolete.

schizomycete An organism of the class Schizomycetes, a group of unicellular or noncellular organisms comprising the bacteria and variously classified with the fungi, with the blue-green algae, or separately. Cf: *myxomycete; eumycete*.

schizoporellid Said of bryozoans characterized by a median sinus at the proximal margin of the orifice, as in the cheilos-

tome family Schizoporellidae.

schizorhysis A *skeletal canal* in dictyonine hexactinellid sponges passing completely through the dictyonal framework and well as interconnecting laterally. It is covered by exopinacoderm. Pl: *schizorhyses*.

schlanite The soluble resin extracted from *anthracoxene* by ether. See also: *anthracoxenite*.

schlenter A term used in Africa for an imitation diamond.

schlieren In some igneous rocks, irregular streaks or masses that contrast with the rock mass but have shaded borders. They may represent segregations of dark or light minerals, or altered inclusions, elongated by flow. Etymol: German for flaw in glass due to a zone of abnormal composition. Singular *schliere*. Also spelled: *schliere*. Adj: *schlieric*. Cf: *flow layer*.

schlieren arch A term introduced by Balk (1937, p.56) for an intrusive igneous body with flow layers along its borders but poorly developed or absent in its interior. Cf: *schlieren dome*.

schlieren dome A term introduced by Balk (1937, p.56) for an intrusive igneous body more or less completely outlined by flow layers which culminate in one central area. Cf: *schlieren arch*.

Schlumberger An informal term designating an *electric log* obtained from a well. Named after Schlumberger Well Surveying Corporation, Houston (Tex.), a pioneer electric well surveying company. Commonly pron: slum-bur-jay.

Schlumberger array An electrode array in which two closely spaced potential electrodes are placed midway between two current electrodes.

schmeiderite A mineral: $(Pb,Cu)_2SeO_4(OH)_2(?)$.

Schmidt field balance An instrument that is both a *horizontal field balance* and a *vertical field balance* and consists of a permanent magnet pivoted on a knife edge. Cf: *torsion magnetometer*.

Schmidt net A coordinate system used to plot a *Schmidt projection*, used in crystallography for statistical analysis of data obtained esp. from universal-stage measurements and in structural geology for plotting azimuths as angles measured clockwise from north and about a point directly beneath the observer.

Schmidt projection A term used in crystallography and structural geology for a *Lambert azimuthal equal-area projection* of the lower hemisphere of a sphere onto the plane of a meridian. Named after Walter Schmidt (1885-1945), Austrian petrologist and mineralogist, who first used the projection in structural geology (Schmidt, 1925, p.395-399). See also: *Schmidt net*.

schmitterite A mineral: $(UO_2)TeO_3$.

schoderite An orange monoclinic mineral: $Al_2(PO_4)(VO_4) \cdot 8H_2O$.

Schoenflies notation An older system of describing crystal classes. Cf: *Hermann-Mauguin symbols*. Syn: *Schoenflies symbols*.

Schoenflies symbols Schoenflies notation.

schoenite A syn. of *picromerite*. Also spelled: *schönite*.

schoepite A yellow secondary mineral: $UO_3 \cdot 2H_2O$. See also: *metaschoepite; paraschoepite*. Syn: *epiianthinite*.

scholzite A colorless to white mineral: $CaZn_2(PO_4)_2 \cdot 2H_2O$.

schönfelsite A very dark-colored basalt containing approximately 28 percent bytownite, in the fine-grained groundmass along with apatite, magnetite, and bronzite, and phenocrysts of olivine and augite.

schorenbergite A tinguaitic hypabyssal rock containing noséan, or sometimes leucite, phenocrysts in a groundmass of leucite, nepheline, and acmite.

schorl (a) A term commonly applied to *tourmaline*, esp. to the black, iron-rich, opaque variety. (b) An obsolete term for any of several dark minerals other than tourmaline; e.g. hornblende.---Syn: *shorl; schorlite*.

schorlomite A black mineral of the garnet group: $Ca_3(Fe,Ti)_2(Si,Ti)_3O_{12}$. Cf: *melanite*.

schorl rock A term used in Cornwall, Eng., for a granular rock composed essentially of aggregates of needle-like crystals of black tourmaline (schorl) associated with quartz, and resulting from the complete tourmalinization of granite.

schorre A Dutch term for that part of a sandy beach covered by the sea only during spring tides.

schott Var. of shott.

schottky defect In a crystal lattice, the absence of an atom; it is a type of point defect. Cf: Frenkel defect; interstitial defect. Syn: defect lattice.

schreibersite A silver-white to tin-white, highly magnetic, tetragonal meteorite mineral: $(Fe,Ni)_3P$. It contains small amounts of cobalt and traces of copper, and tarnishes to brass yellow or brown. Schreibersite occurs in tables or plates as oriented inclusions in iron meteorites. Syn: rhabdite.

schriesheimite An amphibole peridotite that contains diopside. Its name is derived from Schriesheim, Germany.

schroeckingerite A greenish-yellow secondary mineral: $NaCa_3(UO_2)(CO_3)_3(SO_4)F.10H_2O$. It is an ore of uranium. Also spelled: schröckingerite. Syn: dakeite.

schrötterite An opaline variety of allophane rich in aluminum. Material from the type locality has been shown to be a mixture of glassy halloysite and earthy variscite.

schrund line A term introduced by Gilbert (1904, p.582) for "the base of the bergschrund at a late stage in the excavation of the cirque basin". The line separates the steeper slope of the cirque wall from the gentler, usually scalable, slope below.

schubnelite A mineral: $Fe_2(VO_4)_2.2H_2O$.

schuermann series A list of metals so arranged that the sulfide of one is precipitated at the expense of the sulfide of any lower metal in the series.

schuetteite A yellow mineral: $Hg_3(SO_4)_2O$.

schuilingite A blue mineral: $Pb_3Ca_6Cu_2(CO_3)_8(OH)_6.6H_2O$.

schultenite A colorless mineral: $PbHAsO_4$.

schulze's reagent An oxidizing mixture very commonly used in palynologic macerations, consisting of a saturated aqueous solution of $KClO_3$ and varying amounts of concentrated HNO_3 (Schulze, 1855). Named after Franz F. Schulze (1815-1873), German chemist. Syn: Schulze's mixture; Schulze's solution.

schungite shungite.

schuppen structure A syn. of imbricate structure [tect]. Etymol: German Schuppenstruktur. The German word is sometimes used in English geologic literature as schuppenstruktur.

schwagerinid Any fusulinid belonging to the subfamily Schwagerininae.

schwartzembergite A mineral: $Pb_6(IO_2)_2Cl_4O_2(OH)_2$.

schwatzite A variety of tetrahedrite containing mercury.

schweydar mechanical detector A seismic detector that detects and records refracted waves. It consists of a lead sphere suspended by a flat spring; an aluminum cone magnifies the sphere's motion by moving a bow around a spindle which carries a mirror. The motion is photographically recorded.

scientific hydrology Hydrologic study devoted principally to fundamental processes and relationships of the hydrologic cycle.

scientific name The formal Latin name of a taxon. Cf: vernacular name.

scientific stone A synthetic, reconstructed, or imitation gemstone. Syn: scientific gem.

scintillation [gem] The flashing, twinkling, or sparkling of light, or the alternating display of reflections, from the polished facets of a gemstone.

scintillation [radioactivity] A small flash of light produced by an ionizing agent (such as radioactive particles) in a phosphor or scintillator. See also: scintilloscope.

scintillation counter An instrument that measures ionizing radiation by counting individual scintillations of a substance. It consists of a phosphor and a photomultiplier tube that registers the phosphor's flashes. Compared with a Geiger-Müller counter, it is smaller and more sensitive. It is used in spectrometry as well as prospecting. Syn: scintillometer.

scintillation spectrometer An instrument for measuring a mass or energy spectrum, therein similar to a gamma-ray spectrometer, and determining their frequency distribution by the use of a scintillation counter.

scintillator Any transparent material (crystalline, liquid, or organic) which emits small flashes of light when bombarded by an ionizing agent such as radioactive particles. Syn: phosphor.

scintillometer scintillation counter.

scintilloscope An instrument that displays the scintillation of a substance on a screen. Also spelled: scintilliscope.

sciophyte A plant preferring growth in light of low intensity.

scissor fault A fault on which there is increasing offset or separation along the strike from an initial point of no offset, with reverse offset in the opposite direction. The separation is commonly attributed to a scissorlike or pivotal movement on the fault, whereas it is actually the result of uniform strike-slip movement along a fault across a synclinal or anticlinal fold. The term is much misunderstood; pivotal fault, hinge fault, rotary fault, and rotational fault are similarly used and their use should be questioned. See also: node; hinge line. Syn: differential fault.

scleracoma A collective term for the hard skeletal parts of radiolarians.

scleractinian Any zoantharian belonging to the order Scleractinia, characterized by solitary and colonial forms with calcareous exoskeletons secreted by the ectoderm. This order includes most post-Paleozoic and living corals; their stratigraphic range is Middle Triassic to present. Syn: hexacoral.

sclere A minute skeletal element; esp. a sponge spicule. See also: megasclere; microsclere.

sclerenchyma (a) Thick-walled strengthening tissue in a plant. It may consist of either elongate cells called fibers or shorter cells called stone cells. (b) sclerenchyme.

sclerenchyme (a) The calcareous tissue of rugose corallites, esp. the notably thickened parts of the skeleton (TIP, 1956, pt.F, p.250). (b) The vesicular skeletal structure between corallites in colonial coralla, such as the stony substance secreted by the coenenchyme of a scleractinian coral (Shrock & Twenhofel, 1953, p.133).---Cf: mesenchyme; stereome. Syn: sclerenchyma; scleroderm.

sclerite A hard chitinous or calcareous plate, piece, or spicule of an invertebrate; e.g. a hardened, chitinized cover forming part of the external skeleton of a merostome or arachnid, or a calcareous ossicle (anchor, hook, rod, wheel, or disc) of a holothurian, or a calcareous skeletal element of the mesogloea of an octocoral, irrespective of form, or a thickened line in the operculum, mandible, or frontal membrane of a bryozoan.

scleroblast (a) One of the cells of a sponge by which a spicule is formed; a mother cell of one or more sclerocytes. Also, a sclerocyte. (b) One of the ectodermal cells of octocorallian mesogloea that produce calcareous spicules; e.g. axoblast.

sclerocyte A cell that secretes all or part of a sponge spicule. Syn: scleroblast; spiculoblast.

scleroderm The hard sclerenchyme of the skeleton of a scleractinian coral.

sclerodermite (a) The center of calcification and surrounding cluster of calcareous (aragonitic) fibers making up a septum of a scleractinian coral. Sclerodermites are the apparent primary elements in septa and they are variously arranged in vertical series to make trabeculae. (b) A spine or plate of a holothurian. (c) The hard integument of an arthropod segment.

sclerometer An instrument used in mineral analysis to determine hardness by measuring the pressure required to scratch a polished surface of the material with a diamond point.

scleroseptum A calcareous radial septum of a coral.

sclerosome A continuous deposit in a calcareous sponge of nonspicular calcium carbonate that may form part or all of the skeleton.

sclerotesta The hard, bony coat of a seed, e.g. of a cycad. Cf: *sarcotesta*.

sclerotinite A variety of inertinite consisting of the sclerotia of fungi and that is characteristic of Tertiary lignites. Cf: *micrinite; fusinite; semifusinite*.

sclerotium In the myxomycetes, a hard plasmodial resting stage; in the eumycetes, a resting body composed of a hardened mass of hyphae, and frequently rounded in shape.

sclerotized Said of the covering of an invertebrate (esp. an arthropod) hardened by substances other than chitin.

scolecite A zeolite mineral: $CaAl_2Si_3O_{10}.3H_2O$. It usually occurs in delicate radiating groups of white fibrous or acicular crystals, and in some forms it shows a worm-like motion when heated.

scolecodont The fossil jaw, with denticles, of an *annelid*. They are composed of silica and chitin, the chitin being carbonized to a jet black during fossilization.

scolite Any of various tubular or vermiform trace-fossil structures found in Cambrian and Ordovician quartz-rich sandstones (and also in upper Precambrian rocks), consisting of narrow and vertical or usually straight tubes or tube fillings, about 0.2-1 cm in diameter, commonly but not always closely crowded, and generally flaring out into cup-like depressions at their tops. They are believed to be the fossil burrows of marine worms or phoronids, and are assigned to the "genus" *Scolithus* (properly *Skolithos*), a burrowing organism. See also: *worm tube*. Syn: *scolithus; pipe-rock burrow*.

scolithus *scolite*.

scopule A *sceptrule* in which the terminal outgrowths are a pair or ring of spines whose ends may be clubbed and bear rings of recurved teeth.

scopulite A rodlike or stemlike crystallite that terminates in brushes or plumes.

score *scoring*.

scoria [coal] *clinker* [coal].

scoria [volc] Vesicular, cindery, crust on the surface of andesitic or basaltic lava, the vesicular nature of which is due to the escape of volcanic gases before solidification; it is usually heavier, darker, and more crystalline than *pumice*. The adjectival form, *scoriaceous*, is usually applied to pyroclastic ejecta. Pl: *scoriae*. *Cinder* is sometimes used synonymously. See also: *thread-lace scoria*.

scoriaceous [pyroclast] Said of the texture of a coarsely *vesicular* pyroclastic rock (e.g. scoria), usually of andesitic or basaltic composition, and coarser than a *pumiceous* rock. The walls of the vesicles may be either smooth or jagged. Also, said of a rock exhibiting such texture. Syn: *scoriform; scorious*.

scoriaceous [sed] Said of a sedimentary rock whose surface is pitted and irregular like that of volcanic scoria; e.g. "scoriaceous limestone" produced by dissolution of nodules of a former nodular limestone.

scoriae The plural of *scoria*.

scoria tuff A deposit of fragmented scoria in a fine-grained tuff matrix.

scoriform *scoriaceous* [ign].

scorilite A volcanic glass (Hey, 1962, p.593).

scoring (a) The formation of parallel scratches, lines, or grooves in a bedrock surface by the abrasive action of rock fragments transported by a moving glacier. (b) A scratch, line, or groove produced by scoring. Syn: *score*.

scorious *scoriaceous*.

scorodite A pale leek-green or liver-brownish orthorhombic mineral: $FeAsO_4.2H_2O$. It is isomorphous with mansfieldite and represents a lesser ore of arsenic.

scorzalite A blue mineral: $(Fe^{+2},Mg)Al_2(PO_4)_2(OH)_2$. It is isomorphous with lazulite.

Scotch pebble A rounded fragment of agate, carnelian, cairngorm, or other varieties of quartz, found in the gravels of parts of Scotland, and used as a semiprecious stone.

Scotch topaz A yellow transparent variety of quartz resembling the color of *topaz*; specif. *cairngorm*.

Scotch-type volcano A volcanic form characterized by concentric cuestas and produced by cauldron subsidence (Guilcher, 1950).

scour [eng] An artificial current or flow of water for clearing away mud or other deposit from a stream bed; also, the structure built to produce such a current.

scour [geomorph] (a) The powerful and concentrated clearing and digging action of flowing air or water, esp. the downward erosion by stream water in sweeping away mud and silt on the outside curve of a bend, or during time of flood. (b) A place in a stream bed swept (scoured) by running water, generally leaving a gravel bottom. (c) *glacial scour*. (d) *tidal scour*.

scour [tides] *tidal scour*.

scour and fill [geomorph] A process of alternate excavation and refilling of a channel, as by a stream or the tides; esp. such a process occurring in time of flood when the volume and velocity of an aggrading stream are suddenly increased, causing the digging of new channels that become filled with sediment when the flood subsides. Cf: *cut and fill*.

scour and fill [sed struc] A sedimentary structure consisting of a small erosional channel that is subsequently filled; small-scale *washout*.

scour cast A sole mark consisting of a cast of a scour mark; specif. a *flute cast*.

scour channel A large groove-like erosional features produced in sediments by scour.

scour depression A crescentic hollow produced in a stream bed near the outside of a bend by water that scours below the grade of the stream (Bryan, 1920, p.191).

scour finger *flute cast*.

scouring The process of erosion by the action of flowing air, ice, or water, esp. *glacial scour*. See also: *scour*.

scouring velocity That velocity of water which is necessary to dislodge stranded solids from the stream bed.

scour lineation A smooth, low, narrow (2-5 cm wide) ridge formed on a sedimentary surface and believed to result from the scouring action of a current of water. It is characterized by symmetric ends so that the line of current movement, but not its direction, can be ascertained.

scour mark A *current mark* produced by the cutting or scouring action of a current of water flowing over the bottom; e.g. flute. See also: *transverse scour mark*. Syn: *scour marking*.

scour side The upstream, or stoss, side of a roche moutonnée smoothed, striated, and rounded by glacial abrasion. Ant: *pluck side*.

scourway A channel produced by a strong glacial stream near the margin of an ice sheet.

scrap mica Mica whose size, color, or quality is below specifications for sheet mica; e.g. flake mica, or the mica obtained as a product from the preparation of sheet mica, or the mica obtained from waste in fabricating sheet mica.

scratch *striation*.

scree (a) A term more commonly used in Great Britain as the loose equivalent of *talus* in each of its senses: broken rock fragments; a heap of such fragments; and the steep slope consisting of such fragments. Some authorities regard "scree" as the material (broken rock debris) that makes up the sloping land feature known as "talus"; others consider "scree" as a sheet of any loose, fragmental material lying on or mantling a mountain slope or hillside, and "talus" as that material accumulating specif. at the base of, and obviously derived from, a cliff or other projecting mass. Some local British synonyms of scree in the sense of a mass of loose stones include: *clitter; glitter; glyders; screef*. (b) A deposit of loose ngular material greater than 10 cm in diameter (Hatch & Rastall, 1913, p.21). (c) A loose stone or pebble, as on a hill-side. Syn: *glidder.*---Etymol: Old Norse *skritha,* "landslide, or the rock that slides away under the foot".

scree creep The gradual and steady downhill movement of individual large blocks of rock on a slope that is often gentle; it is most noticeable where the rocks are massive or well-jointed. See also: *talus creep.* Syn: *rock drift.*

screef *scree.*

screen An apparatus used to separate material according to size of its particles or to allow the passage of smaller parts while preventing that of larger (as in grading coal, ore, rock, or aggregate); it usually consists of a perforated plate or sheet, or of meshed wire or woven cloth, with regularly spaced circular holes of uniform size, mounted in a suitable frame. Cf: *sieve.*

screen analysis Determination of the particle-size distribution of a soil, sediment, or ore by measuring the percentage of the particles that will pass through standard screens of various sizes.

screening The operation of passing loose materials (such as gravel or coal) through a screen so that constituent particles are separated into defined sizes.

screw axis A type of crystal symmetry element that is a combination of a rotation of $360°/n$ with a translation of a/mn where a is a lattice period (usually the a, b, or c crystal axis), n may be 1, 2, 3, 4, or 6, and m is an integer between 0 and n.

screw dislocation A type of *line defect* in a crystal: a row of atoms along which a crystallographic plane seems to spiral. See also: *spiral growth.*

screw ice (a) Small ice fragments in heaps or ridges, produced by the crushing together of ice cakes. (b) A small formation of *pressure ice.*

scribing The process of preparing a map or other drawing for reproduction by cutting the detail to be shown into an opaque medium that coats a sheet of transparent plastic, using a scriber (an instrument holding one of a set of needles or blades of various diameters or cross-sectional shapes, sharpened to prescribed dimensions). The result of the process is a negative of the material to be reproduced.

scrobicule One of the smooth, shallow, depressed rings or trenches surrounding the bases of echinoid tubercles and serving for attachment of muscles of spines. See also: *areole.* Syn: *scrobicula; scrobiculus.*

scrobis septalis The inframarginal, asymmetric, sometimes deep indentation or concave surface of the apertural face of a foraminiferal test (as in *Alabamina*) (TIP, 1964, pt.C, p.63). Etymol: Latin. See also: *infundibulum.* Syn: *inframarginal sulcus.*

scroll (a) One of a series of crescentic deposits built by a stream on the inner bank of a shifting channel; e.g. a *flood-plain scroll.* (b) *meander scroll.*

scroll meander A *forced-cut meander* in which the building of meander scrolls on the inner bank is the cause of erosion on the outer bank of the meander (Melton, 1936, p.597). See also: *lacine meander.*

scrub Low-growing or stunted vegetation growing on poor soil or in semiarid regions and which sometimes forms inpenetrable masses.

sculpture [geomorph] (a) The carving out of surficial features of the Earth's surface by erosive agents, such as rain, running water, waves, glaciers, and wind. The term has been loosely applied to include also the processes of deposition and earth movement. Syn: *earth sculpture; land sculpture; glyptogenesis.* (b) A landform resulting from a modification or sculpturing of an existing form.

sculpture [paleont] Strongly developed *ornamentation* of preserved hard parts of an animal; e.g. the relief pattern on the surface of a gastropod shell.

sculpture [palyn] The external textural modifications (such as spines, warts, granules, pila, pits, streaks, and reticulations) of the exine of pollen grains and spores. It is always a feature of ektexine. Cf: *structure.* Syn: *ornamentation.*

scum A film, often putrid, that floats on a liquid, such as a stagnant pool. The film may be composed of soap, of precipitated calcium carbonate, etc., and, hence, not necessarily putrid.

scutum (a) One of a pair of opercular valves, adjacent to the rostrum in cirripede crustaceans, with adductor-muscle attachments. Cf: *tergum.* (b) A lateral *marginal spine,* generally broad and flat, overhanging the frontal area in some anascan cheilostomes (bryozoans).---Pl: *scuta.*

scyelite A coarse-grained ultramafic igneous rock characterized by poikilitic texture resulting from the inclusion of olivine crystals in crystals of other minerals, esp. amphiboles. Mica and some magnetite are also present. Its name is derived from Loch Scye, Scotland.

scyphozoan Any marine coelenterate belonging to the class Scyphozoa, characterized by the predominance of medusoid forms. Their stratigraphic range is Precambrian or Cambrian to present.

Scythian European stage: Lower Triassic (above Tataran of Permian, below Anisian). Syn: *Skythian; Werfenian.*

S-dolostone Stratigraphically controlled dolostone, occurring in extensive beds generally intertongued with limestone (Dunbar & Rodgers, 1957, p.238).

se In structural petrology, a fabric external to an inclusion. It may or may not be parallel to an s surface inside the grain. Cf: *si.*

sea [astrogeol] *mare.*

sea [oceanog] (a) An inland body of salt water. (b) A geographic division of an *ocean.* (c) An ocean area of wave generation.

sea [waves] A series of short-period, asymmetric waves generated by wind and that lies within its area of generation. Such a wave is called a *sea wave* or a *wind wave;* when it leaves its area of generation, it becomes *swell* [waves].

sea arch An opening through a headland, formed by wave erosion or solution (as by the enlargement of a sea cave, or by the meeting of two sea caves from opposite sides) and leaving a bridge of rock over the water. Syn: *marine arch; marine bridge; sea bridge; natural arch; natural bridge.*

sea ball A spherical mass of somewhat fibrous material of living or fossil vegetation (esp. algae), produced mechanically in shallow waters along a seashore by the compacting effect of wave movement. Cf: *lake ball.*

sea bank (a) *seashore.* (b) A *sandbank* adjacent to the sea. (c) *seawall.*

seabeach A beach lying along a sea or ocean.

seabeach placer *beach placer.*

seaboard (a) The strip of land bordering a seacoast. (b) *seacoast.*

sea bridge *sea arch.*

sea-captured stream A stream, flowing parallel to the seashore, that is cut in two as a result of marine erosion and that may enter the sea by way of a waterfall (Cleland, 1925).

sea cave A cleft or cavity in the base of a sea cliff, excavated where wave action has enlarged natural lines of weakness in easily weathered rock; it is usually at sea level and affected by the tides. Syn: *marine cave; sea chasm; cave.*

sea chasm A deep, narrow *sea cave.*

sea cliff A cliff or slope produced by wave erosion, situated at the seaward edge of the coast or the landward side of the wave-cut platform, and marking the inner limit of beach erosion. It may vary from an inconspicuous slope to a high, steep escarpment. Example: Gay Head at Cape Cod, Mass. See also: *wave-cut cliff; shore cliff.* Also spelled: *seacliff.* Syn: *cliff; marine cliff.*

sea coal An old British syn. of *bituminous coal,* named after coal washed ashore and used for fuel; the name was extended to mined coal, as well.

seacoast The coast adjacent to a sea or ocean. Syn: *seaboard*.

sea cucumber A *holothuroid* having a body shape resembling a cucumber, a flexible body wall, and able to creep along the sea floor.

sea-cut *marine-cut*.

sea fan *submarine fan*.

sea-floor spreading A hypothesis that the oceanic crust is increasing by convective upwelling of magma along the *mid-oceanic ridges* or *world rift system*, and a moving away of the new material at a rate of from one to ten centimeters per year. This movement provides the source of power in the hypothesis of *plate tectonics*. This hypothesis supports the *continental displacement* hypothesis. See also: *expanding Earth*. Syn: *ocean-floor spreading; spreading concept; spreading-floor hypothesis*.

sea-floor trench *trench* [marine geol].

sea-foam *sepiolite*.

seafront The land, buildings, or section of a town along a seashore or bordering a sea.

sea gate (a) A restricted passage leading or giving access to the sea. (b) A gate that protects against the sea.

sea gully *slope gully*.

sea ice (a) Any form of ice originating from the freezing of seawater (thus excludes icebergs). Ant: *land ice*. See also: *field ice*. (b) A mariner's term for any ice that is floating in the sea or that has drifted to the sea.

sea-ice shelf Sea ice floating in the vicinity of its formation and separated from fast ice (of which it may have been a part) by a tide crack or a family of such cracks.

seaknoll *knoll* [marine geol].

sea level A popular syn. of *mean sea level*.

sea-level datum A determination of *mean sea level* that has been adopted as a standard datum for heights or elevations, based on tidal observations over many years at various tide stations along the coasts; e.g. the Sea-Level Datum of 1929 used by the U.S. Coast and Geodetic Survey, the year referring to the last general adjustment of the level.

sealing The natural or artificial filling of the interstices or fractures in a rock formation to reduce their permeability to zero.

sealing-wax structure A term used by Fairbridge (1946, p.85 & 87) for a primary sedimentary flow structure produced by slumping, characterized by a lack of a sharply defined slip plane at the base or a contemporaneous erosion plane at the top, and occupying a zone of highly fluid contortion in otherwise normal sedimentary succession.

sea loch A *fjord* along the western Highland coast of Scotland.

seam [ore dep] A bed or vein or series of beds of ore, usually said of coal but also be said of metallic minerals.

seam [stratig] (a) A thin layer or stratum of rock separating two distinctive layers of different composition or greater magnitude. (b) Strictly, the line of separation between two different strata, resembling the seam between two parts of a garment.

seamanite A pale-yellow to wine-yellow orthorhombic mineral: $Mn_3(PO_4)(BO_3).3H_2O$.

sea marsh A *salt marsh* periodically overflowed or flooded by the sea. Syn: *sea meadow*.

sea mat A *bryozoan*, esp. an incrusting bryozoan.

sea meadow *sea marsh*.

sea meadows Those upper layers of the open ocean that have such an abundance of phytoplankton that they provide food for marine organisms. The term is usually used in the plural.

sea moat *moat* [marine geol].

seamount An elevation of the sea floor, 1000 m or higher, either flat-topped (called a *quyot*) or peaked (called a *seapeak*). Seamounts may be either discrete, or arranged in a

linear or random grouping, or may be connected at the bases and aligned along a ridge or rise.

sea mud Mud from the sea; specif. a rich, slimy deposit in salt marsh or along a seashore, sometimes used as a ma nure. Syn: *sea ooze*.

sea ooze *sea mud*.

seapeak A type of *seamount* that has a pointed summit.

sea peat A rare type of peat, formed from seaweeds.

sea plain *plain of marine erosion*.

seapoose A term used along the shore of Long Island, N.Y for a shallow inlet or tidal river. Etymol: Algonquian, akin Delaware *sepus*, "small brook".

seaquake An earthquake that occurs beneath the ocean an that can be felt on board a ship in the vicinity of the epicente Syn: *submarine earthquake*.

search coil A coil that is used in electromagnetic methods fo measuring an alternating magnetic field.

sea reach The *reach* of the lower course of a stream where approaches the sea.

sea rim The apparent horizon as actually observed at sea; th sea-level horizon.

searlesite A white mineral: $NaB(SiO_3)_2.H_2O$.

seascarp A relatively long, high, and rectilinear submarine cl or wall.

seashore (a) The narrow strip of land adjacent to or borderin a sea or ocean. Syn: *seaside*. (b) A legal term for all th ground between the ordinary tide levels; the *foreshore*.---Sy *seastrand*.

seashore lake A lake along the seashore, containing eithe freshwater or salt water, and isolated from the sea by a ba rier of sediment built by waves or by a river on a delta.

seaside *seashore*.

sea slick A smooth area on the surface of an ocean or boc of fresh water, caused by organic material, e.g. water bloom.

sea slope A slope of the land toward the sea.

sea snow The drifting descent of organic detritus in th ocean. Syn: *plankton snow; marine snow*.

seasonally frozen ground Ground that is frozen by low seasor al temperatures and remaining frozen only during the winter; corresponds to the *active layer* in permafrost regions. Syr *frost zone*.

seasonal recovery Recharge to ground water during and afte a wet season, with a rise in the level of the water table.

seasonal stream An *intermittent stream* that flows only durin a certain climatic season; e.g. a winterbourne.

sea stack *stack* [coast].

sea state A description of the roughness of the ocean su face, either numerical or in words. Syn: *state of the sea*.

seastrand *seashore*.

seat clay *underclay*.

seat earth A British term for a bed of rock underlying a coa seam, representing an old soil that supported the vegetatic from which the coal was formed; specif. *underclay*. A high siliceous seat earth is known as *ganister*. Also spelled: *sea tearth*. Syn: *seat rock; seat stone; spavin; coal seat*.

sea terrace *marine terrace*.

seat rock *seat earth*.

seat stone *seat earth*.

sea urchin An *echinoid* having a globular shape and a theca o calcareous plates, commonly with sharp movable spines.

seawall A wall or embankment of concrete, stone, or othe material built by man along a shore to resist encroachment the sea and to prevent wave erosion. Also spelled: *sea wa* Syn: *sea bank*.

sea wall (a) A long, steep-faced embankment of shingle o boulders (without gravel), built by powerful storm waves alon a seacoast at the high-water mark. (b) *seawall*.

sea water The water of the oceans, characterized by its salini ty and distinguished from the fresh water of lakes, streams and from rain water. *Salt water* is sometimes used synonym usly, as the antonym of fresh water in general. Also spelled: seawater; sea-water.

sea-water encroachment *salt-water encroachment.*

sea-water intrusion *salt-water encroachment.*

sea wave One of a series of waves known as *sea*. See also: *wind wave.*

seaworn Diminished or wasted away by the sea, as a *seaworn* shore.

sebastianite A plutonic rock composed of euhedral anorthite, biotite, and some augite and apatite, but without feldspathoids and quartz. This rock has been found as fragments in extrusive rocks of Monte Somma, Italy.

sebkha (a) A term used in northern Africa for a smooth, flat, usually saline plain, sometimes occupied after a rain by a marsh or a temporary shallow lake; a *playa* or the dry, salt-incrusted bed of a lake. (b) A term used in northern Africa for a salt marsh. (c) A term used on the Arabian peninsula for a salt flat or low salt-incrusted plain restricted to a coastal area, as along the Persian Gulf. Cf: *mamlahah.*---Etymol: Arabic *sabkhah*, "saline infiltration". See also: *shott*. The term has many variants, including: *sebka; sabkha; sabakha; sabka; sabkhah; sebcha; sebkra; sebja; sebjet; sebchet; sebkhat*. It is known in Iraq and Egypt as *sabkhat*, in Syria as *sebkaha*, and in the central Asian deserts as *sabkhet*.

secchi disc An instrument used to measure seawater *transparency* or clarity; a disc of variable color and diameter is lowered into the water, and an average is taken of the depth at which it disappears and at which it reappears when raised.

second (a) A unit of time equal to 1/60 of a minute or 1/3600 of an hour; specif. the cgs unit of time, originally equal to 1/86,400 part of the mean solar day but now defined as the duration of 9,192,631,770 cycles of frequency associated with the transition between two hyperfine levels of the fundamental state of the atom of cesium-133. Abbrev: sec; s (in physical tables). (b) A unit of angular measure equal to 1/60 of a minute of arc or 1/3600 of a degree. Abbrev: '. (c) An informal oceanographic term used to describe distance or depth, equal to about 1463 m (4800 ft) or the distance that sound will travel through seawater during one second of time.

secondary [eco geol] Said of a *supergene* mineral deposit or enrichment.

secondary [coast] Said of a mature coast or shoreline whose features are produced chiefly by present-day marine processes (Shepard, 1937, p.605); e.g. coasts shaped by wave erosion, marine deposition, or marine organisms. Cf: *primary.*

secondary A term applied in the early 19th century as a syn. of *Floetz*. It was later applied to the extensive series of stratified rocks separating the older *Primary* and the younger *Tertiary* rocks, and ranging from the Silurian to the Cretaceous; still later, it was restricted to the whole of the Mesozoic Era. The term was abandoned in the late 19th century in favor of *Mesozoic.*

secondary [metal] Said of metal obtained from scrap rather than directly from ore. Ant: *primary* [metal].

secondary allochthony In coal formation, accumulation of plant remains in a region characterized by erosion, transport, and resedimentation of coal masses previously deposited elsewhere in place. Cf: *primary allochthony.*

secondary arc *externides.*

secondary ash *extraneous ash.*

secondary axial septulum A *secondary septulum* in a fusulinid foraminiferal test, representing an *axial septulum* located between primary axial septula.

secondary clay A clay that has been transported from its place of formation and redeposited elsewhere. Cf: *residual clay; primary clay.*

secondary cleavage An old term for *cleavage* used by Leith 1905, p.11) to emphasize its development after consolidation of the rock, by deformation or metamorphism. See also: *metaclase.*

secondary consequent stream (a) A tributary of a subsequent stream, flowing parallel to or down the same slope as the original consequent stream; it is usually developed after the formation of a subsequent stream, but in a direction consistent with that of the original consequent stream. (b) A stream flowing down the flank of an anticline or syncline. Syn: *lateral consequent stream.*---Syn: *subconsequent stream.*

secondary consolidation Consolidation of sedimentary material, at essentially constant pressure, resulting from internal processes such as recrystallization.

secondary corner One of the minor convexities seen in the profile of a sedimentary particle, being numerous (15-30) for angular particles but quickly disappearing during abrasion and completely absent at a roundness value of 0.60 (Pettijohn, 1957, p.58-59).

secondary crater An *impact crater* produced by the relatively low-velocity impact of fragments ejected from a large *primary crater*; e.g. any of several lunar craters ("splash structures") formed by fragments thrown up from the Moon's surface as a result of violent primary impacts. Syn: *satellitic crater.*

secondary creep Deformation of a material under a constant differential stress, with the strain-time relationship as a constant. Cf: *primary creep.* Syn: *steady-state creep.*

secondary drilling Drilling of holes in rock that has been blasted loose during *primary drilling* but which is still too large for handling by shovels or for feeding to crushers.

secondary enlargement Deposition, around a clastic mineral grain, of material of the same composition as that grain and in optical and crystallographic continuity with it, often resulting in crystal faces characteristic of the original mineral (Pettijohn, 1957, p.119); e.g. the addition of a quartz overgrowth around a silica grain in sandstone, or the growth of new material around detrital nuclei such as calcite, feldspar, and tourmaline. Cf: *rim cementation.* Syn: *secondary growth.*

secondary flowage *gliding flow.*

secondary geosyncline (a) A geosyncline appearing at the culmination of or after geosynclinal orogeny, such as *exogeosyncline, epieugeosyncline, intradeep* (Peyve & Sinitzyn, 1950). (b) Haug's term for a *sequent geosyncline* (1900, p.617-711).

secondary glacier A small valley glacier that joins a larger trunk glacier as a *tributary glacier.*

secondary growth *secondary enlargement.*

secondary interstice An *interstice* that formed after the formation of the enclosing rock. Cf: *original interstice.*

secondary layer The inner shell layer of a brachiopod deposited by the outer epithelium median of the outer mantle lobes. It is secreted in most articulate brachiopods intracellularly as fibers bounded by cytoplasmic sheaths, or less commonly, extracellularly as prismatic calcite. Cf: *primary layer.* Syn: *fibrous layer; prismatic layer.*

secondary limestone Limestone deposited from solution in cracks and cavities of other rocks; esp. the limestone accompanying the salt and gypsum of the Gulf Coast salt domes.

secondary maximum A term used by Udden (1914) for a particle size of a sediment or rock, having greater frequency than other size ranges surrounding it but not greater than the modal diameter. It may be obtained graphically by locating the second highest peak of the frequency curve. A given sample may have more than one secondary maximum. Syn: *secondary mode.*

secondary mineral A mineral formed later than the rock enclosing it and usually at the expense of an earlier-formed, *primary mineral* and as a result of weathering, metamorphic, or solution activity.

secondary mode *secondary maximum.*

secondary optic axis One of two *optic axes* in a crystal along which all light rays travel with equal velocity. Secondary optic axes are close to but do not necessarily coincide with the *primary optic axes.*

secondary orogeny Orogeny that is characteristic of the *externides* and that involves intense deformation. Cf: *primary orogeny.*

secondary porosity The porosity developed in a rock formation subsequent to its deposition or emplacement, either through natural processes of dissolution or stress distortion, or artificially through acidization or the mechanical injection of coarse sand. Cf: *primary porosity.*

secondary production The organic matter that zooplankton herbivores produce within a given marine area of volume in a given time. Cf: *primary production.*

secondary reflection (a) *multiple reflection.* (b) *ghost.*

secondary rocks (a) Rocks composed of particles derived from the erosion or weathering of preexisting rocks, such as residual, chemical, or organic rocks formed of detrital, precipitated, or organically accumulated materials; specif. clastic sedimentary rocks. Cf: *primary rocks.* (b) A term applied by Lehmann (1756) to fossiliferous and stratified rocks, containing material eroded from the older *primitive rocks.*

secondary septulum A minor partition of a chamberlet in a foraminiferal test, reaching downward (adaxially) from spirotheca (as in Neoschwagenininae); e.g. *secondary axial septulum* and *secondary transverse septulum.* Cf: *primary septulum.*

secondary spine An intermediate-sized echinoid spine, appearing later than the *primary spine.*

secondary stratification Stratification developed when sediments already deposited are thrown into suspension and redeposited. Syn: *indirect stratification.*

secondary structure [paleont] Coarse structure, commonly between distinct laminae, in the wall of a tintinnid lorica. Cf: *primary structure; tertiary structure.*

secondary structure [geol] A structure that originated subsequently to the deposition or emplacement of the rock in which it is found, such as a fault, fold, or joint produced by tectonic movement; esp. an epigenetic *sedimentary structure,* such as a concretion or nodule produced by chemical action, or a sedimentary dike formed by infilling. Cf: *primary structure [geol].*

secondary succession An association of plants that develops after the destruction of all or part of the original plant community.

secondary sulfide zone *sulfide zone.*

secondary tectonite A tectonite whose fabric is *deformation fabric.* Most tectonites are of this type. Cf: *primary tectonite.*

secondary tissue Plant tissue developed from meristem, e.g. cambium.

secondary transverse septulum A *secondary septulum* in a foraminiferal test, representing a *transverse septulum* whose plane is approximately normal to the axis of coiling.

secondary twinning *deformation twinning.*

secondary vein In mining law, a vein discovered subsequent to the one on which the mining claim was based; an incidental vein. Cf: *discovery vein.*

secondary wall An inner cell-wall layer deposited upon the primary wall layer in a plant cell, and generally of different composition than the primary wall (Cronquist, 1961, p.881).

secondary wave *S wave.*

second boiling point The development of a gas phase from a liquid phase upon cooling. During the cooling, crystallization of large quantities of compounds low in or lacking volatile materials, e.g. feldspar, results in a sufficient increase in the concentration of volatile materials, e.g. water, in the residual liquid that finally the vapor pressure of this liquid becomes greater than the confining pressure and a gas phase develops, i.e. the liquid boils.

second bottom The first terrace above the normal flood plain (or *first bottom*) of a river.

second-class ore An ore that needs some preliminary treatment before it is of a sufficiently high grade to be acceptable for market. Syn: *milling ore.* Cf: *first-class ore.*

second-derivative map A *derivative map* of the second vertical derivative of a potential field such as the Earth's gravity o magnetic field.

second law of thermodynamics The statement concerning er tropy as a function of the state of the system, which says tha for all reversible processes, the change in entropy is equal the heat which the system exchanges with the outside wor divided by the absolute temperature. In irreversible processe the change in entropy is greater than the quotient of heat an temperature.

second-order leveling Spirit leveling that has less stringent re quirements than those of *first-order leveling,* in which line between bench marks established by first-order leveling ar run in only one direction using first-order instruments an methods (or other lines are divided into sections, over whic forward and backward runnings are to be made) and in whic the maximum allowable discrepancy is 8.4 mm times th square root of the length of the line (or section) in kilometer (or 0.035 ft times the square root of the distance in miles Cf: *third-order leveling.*

second-order pinacoid In a monoclinic or triclinic crystal, an $\{h01\}$ or $\{\bar{h}01\}$ pinacoid. Cf: *first-order pinacoid; third-orde pinacoid; fourth-order pinacoid.*

second-order prism A crystal form: in a tetragonal crystal, th $\{100\}$ prism; in a hexagonal crystal, the $\{11\bar{2}0\}$ prism; and an orthorhombic crystal, any $\{h01\}$ prism. Cf: *first-orde prism; third-order prism; fourth-order prism.* See also: *macro dome.*

second water The quality or luster of a gemstone next belo *first water,* such as that of a diamond that is almost perfec but contains slight flaws or turbid patches. Cf: *third water.*

second-year ice Sea ice that has survived only one summer melt; it is thicker and less dense than first-year ice and there fore stands higher out of the water, and any hummocks pres ent show greater weathering. Syn: *two-year ice; young pola ice.*

secretion [geol] A secondary structure formed of material de posited (from solution) within an empty cavity in any rock esp. a deposit formed on or parallel to the walls of the cavit the first layer being the outer one; e.g. a mineral vein, an am ygdule, or a geode. The space may be completely or on partly filled with mineral matter carried in by infiltrating solu tions. Cf: *concretion.*

secretion [paleont] The act or process by which animals an plants transform mineral material from solution into skeleta forms.

sectile Said of a mineral, the texture of which is tenaceou enough so that it can be cut with a knife; e.g. argentite.

section [geol] (a) An exposed surface or cut, either natura (such as a sea cliff or stream bank) or artificial (such as quarry face or road cut), through a part of the Earth's crust may be vertical or inclined. (b) A description, or graphic rep resentation drawn to scale, of the successive rock units or th geologic structure revealed by such an exposed surface or a they would appear if cut through by any intersecting plane such as a diagram of the geologic features or mine working penetrated in a shaft, borehole, trial pit, or drilled well; esp. *vertical section.* See also: *horizontal section; structure sec tion.----*Syn: *geologic section.*

section [petrology] *thin section.*

section [stratig] (a) *columnar section.* (b) *geologic section* (c) *type section.*

section [surv] (a) One of the 36 units of subdivision of *township* of the U.S. Public Land Survey system, representin a piece of land normally one square mile in area (containin 640 acres as nearly as possible), with boundaries conformin to meridians and parallels within established limits. Section within a normal township are numbered consecutively begin ning with number one in the northeast section and progressin west and east alternately with progressive numbers in th tiers to number 36 in the southeast section. See also: *frac*

onal section; half section; quarter section; quarter-quarter
ection. (b) The part of a continuous series of measured dif-
rences of elevation that is recorded and abstracted as a
nit. It always begins and ends on a bench mark.

ection-gage log caliper log.

ection line The boundary line of a section in surveying.

ection [stratig] (a) columnar section. (b) geologic section.
c) type section.

ection [surv] (a) One of the 36 units of subdivision of a
ownship of the U.S. Public Land Survey system, representing
 piece of land normally one square mile in area (containing
40 acres as nearly as possible), with boundaries conforming
 meridians and parallels within established limits. Sections
ithin a normal township are numbered consecutively begin-
ng with number one in the northeast section and progressing
est and east alternately with progressive numbers in the
ers to number 36 in the southeast section. See also: frac-
onal section; half section; quarter section; quarter-quarter
ection. (b) The part of a continuous series of measured dif-
rences of elevation that is recorded and abstracted as a
nit. It always begins and ends on a bench mark.

ection-gage log caliper log.

ection line The boundary line of a section in surveying.

ector graben A volcanic graben on the slope of a volcanic
one; a syn. of barranco [volc].

ecular (a) Said of a process or event lasting or persisting for
n indefinitely long period of time, e.g. secular variation; pro-
ressive rather than cyclic. (b) According to Langbein & Iseri
1960, p.13), any process, event, or change that is complete
ithin a century. Cf: climatic.

ecular equilibrium Long-term radioactive equilibrium of natu-
ally radioactive elements.

ecular movements Systematic, persistent movements of the
arth's crust, either upward or downward, which take place
lowly and imperceptibly, or day by day, over long periods of
eologic time.

ecular variation Changes in the Earth's magnetic field, mea-
ured in centuries. See also: westward drift. Syn: geomagnet-
c secular variation.

ecule A syn. of moment. Term suggested by Jukes-Browne
1903. p.37) for the duration of a biostratigraphic zone. Etym-
l: Latin seculum, "age".

ecundine dike A dike which has been intruded into hot coun-
ry rock. Pegmatites and aplites commonly occur in this
mode. See also: welded dike.

edarenite A term used by Folk (1968, p.124) for a litharenite
composed chiefly of sedimentary rock fragments; e.g. sandstone-arenite and shale-arenite. It may have any clay content,
orting, or rounding.

edentary Said of a sediment or soil that is formed in place,
without transportation, by the disintegration of the underlying
ock or by the accumulation of organic material.

edentary soil residual soil.

ederholmite A mineral: beta-NiSe.

edge peat carex peat.

edifluction The subaquatic or subaerial movement of material
n unconsolidated sediments, occurring in the primary stages
of diagenesis (Richter, 1952).

ediment (a) Solid fragmental material, or a mass of such
material, either inorganic or organic, that originates from
weathering of rocks and is transported by, suspended in, or
deposited by, air, water, or ice, or that is accumulated by
other natural agents, such as chemical precipitation from so-
ution or secretion by organisms, and that forms in layers on
he Earth's surface at ordinary temperatures in a loose,
unconsolidated form; e.g. sand, gravel, silt, mud, till, loess,
alluvium. (b) Strictly, solid material that has settled down
from a state of suspension in a liquid.---In the singular, the
term is usually applied to material held in suspension in water
or recently deposited from suspension. In the plural, the term

is applied to all kinds of deposits, and refers to essentially
unconsolidated materials; the plural usage as applied to
consolidated sedimentary rocks should be avoided (USGS,
1958, p.86). Cf: deposit.

sedimental Formed of or from sediment.

sedimentary adj. (a) Pertaining to or containing sediment;
e.g. a "sedimentary deposit" or a "sedimentary complex". (b)
Formed by the deposition of sediment (e.g. a "sedimentary
clay"), or pertaining to the process of sedimentation (e.g.
"sedimentary volcanism").---n. A sedimentary rock or deposit.

sedimentary ash extraneous ash.

sedimentary breccia A breccia formed by sedimentary processes; e.g. a talus breccia. It is usually (but not necessarily)
characterized by imperfect mechanical sorting of its materials
and either by the predominance of material from one local
source or by the presence of a marked variety of materials indiscriminately jumbled together (Wentworth, 1935, p.227).
Syn: sharpstone conglomerate.

sedimentary-contact shoreline A shoreline formed by the partial submergence of the slope left by the removal of weak
beds from one side of a straight sedimentary contact (Johnson, 1925, p.19).

sedimentary cover cover [tect].

sedimentary cycle cycle of sedimentation.

sedimentary differentiation The progressive separation (by
erosion and transportation) of a well-defined rock mass into
physically and chemically unlike products that are resorted
and deposited as sediments over more or less separate areas;
e.g. the segregation and dispersal of the components of an
igneous rock into sandstones, shales, limestones, etc.

sedimentary dike A tabular mass of sedimentary material that
cuts across the structure or bedding of preexisting rock in the
manner of an igneous dike and that is formed by the filling of
a crack or fissure either from below, above, or laterally by
forcible injection or intrusion of sediments under abnormal
pressure (such as by gas pressure or by the weight of overlying rocks, or by earthquakes), or from above by the simple
infilling of sediments; esp. a clastic dike. See also: sediment
vein.

sedimentary facies A term used by Moore (1949, p.32) for a
stratigraphic facies representing any areally restricted part of
a designated stratigraphic unit (or of any genetically related
body of sedimentary deposits) which exhibits lithologic and
paleontologic characters significantly different from those of
another part or parts of the same unit. It comprises "one of
any two or more different sorts of deposits which are partly or
wholly equivalent in age and which occur side by side or in
somewhat close neighborhood" (Moore, 1949, p.7).

sedimentary fault growth fault.

sedimentary injection injection [sed].

sedimentary insertion A term proposed by Challinor (1962,
p.177) for the emplacement of sedimentary material among
deposits or rocks already formed, such as by infilling, injection or intrusion, or localized solution subsidence.

sedimentary intrusion intrusion [sed].

sedimentary laccolith A term introduced by Raaf (1945) for an
intrusion of plastic sedimentary material (such as clayey salt
breccia) forced up under high pressure and penetrating parallel or nearly parallel to the bedding planes of the invaded formation, and characterized by a very irregular thickness.

sedimentary lag Delay between the formation of potential
sediment by weathering and its removal and deposition.

sedimentary mantle cover [tect].

sedimentary marble crystalline limestone.

sedimentary ore A sedimentary rock of ore grade; an ore deposit formed by sedimentary processes, e.g. saline residues,
phosphatic deposits, coal seams or iron ore formed by the
chemical reaction of iron in solution with calcareous material.

sedimentary peat Peat formed under water, usually lacustrine,
and consisting mainly of algae and related forms. Syn: lake

peat; pulpy peat; dredge peat.

sedimentary petrography The description and classification of sedimentary rocks. Syn: *sedimentography.*

sedimentary petrologic province An area underlain by sediments with a common provenance (Pettijohn, 1957, p.573). Cf: *dispersal shadow.*

sedimentary petrology The petrology of sediments and sedimentary rocks; esp. the study of the composition, characteristics, and origin of sedimentary deposits. Often miscalled "sedimentation".

sedimentary quartzite *orthoquartzite.*

sedimentary ripple *ripple* [sed struc].

sedimentary rock (a) A rock resulting from the consolidation of loose *sediment* that has accumulated in layers; e.g. a *clastic rock* (such as conglomerate, eolianite, or tillite) consisting of mechanically formed fragments of older rock transported from its source and deposited in water or from air or ice, or a *chemical rock* (such as rock salt or gypsum) formed by precipitation from solution, or an *organic rock* (such as certain limestones) consisting of the remains or secretions of plants and animals. The term is sometimes restricted by some authors to include only those rocks consisting of mechanically derived sediment; others extend the term to embrace all rocks other than purely igneous and completely metamorphic rocks, thereby including pyroclastic rocks composed of fragments blown from volcanoes and deposited on land or in water. Syn: *stratified rock; derivative rock.* (b) Less restrictedly, a general term for any unconsolidated and consolidated sedimentary material. This usage should be avoided, "notwithstanding the fact that geologists have agreed to call loose unconsolidated material 'rock'" (Challinor, 1967, p.224).

sedimentary structure A *structure* in a sedimentary rock, formed either contemporaneously with deposition (a *primary sedimentary structure*) or subsequently to deposition (a *secondary structure* [geol]).

sedimentary tectonics Folding and deformation in geosynclinal basins caused by geosynclinal subsidence and buckling of strata in the geosyncline. An example of a large anticline developed at depth in a geosyncline is the Cedar Creek anticline in the Williston Basin of North America (Gussow, 1970). Cf: *gravity orogenesis.*

sedimentary trap An area in which sedimentary material accumulates instead of being carried farther, as in an area between a high-energy and low-energy environment.

sedimentary tuff (a) A tuff containing a subordinate amount of nonvolcanic detrital material. (b) A deposit of reworked tuff and other detrital material.

sedimentary volcanism The expelling, extruding, issuing forth, or breaking through of overlying formations by a mixture of sediment (sand, clay, etc.), water, and gas, with gas under pressure furnishing the driving force. Also, the production of phenomena (such as sand volcanoes and mud volcanoes) much like those associated with volcanic activity; also, the phenomena so produced. Sedimentary volcanism may result from diapiric intrusion, volcanism in a fumarolic stage, escaping hydrocarbons from oil wells, oozing out of material (during thaw) in a permafrost region, or orogenic pressure release as during an earthquake.

sedimentation (a) The act or process of forming or accumulating sediment in layers, including such processes as the separation of rock particles from the material from which the sediment is derived, the transportation of these particles to the site of deposition, the actual deposition or settling of the particles, the chemical and other (diagenetic) changes occurring in the sediment, and the ultimate consolidation of the sediment into solid rock. (b) Less broadly, the process of *deposition* of sediment. (c) Strictly, the act or process of depositing sediment by mechanical means from a state of suspension in a liquid. (d) The accumulation of deposits of colluvium and alluvium derived from accelerated erosion of the soil. (e)

silting up. (f) A term often used erroneously for "sedimenta petrology".

sedimentation analysis Determination of the particle-size di tribution of a sediment by measuring settling velocities of d ferent size fractions.

sedimentation balance An apparatus used to measure the se tling rate of small particles dispersed in a liquid.

sedimentation curve An experimentally derived curve showir cumulatively the quantity of sediment deposited or remove from an originally uniform suspension in successive units time (Krumbein & Pettijohn, 1938, p.112-115).

sedimentation diameter A measure of particle size, equal the diameter of a hypothetical sphere of the same specif gravity and the same settling velocity as those of a given sed mentary particle in the same fluid; twice the *sedimentation r dius.* Cf: *equivalent radius; nominal diameter.*

sedimentation radius One half of the *sedimentation diameter.*

sedimentation rate *rate of sedimentation.*

sedimentation trend The direction in which sediments wer laid down.

sedimentation unit A layer or deposit resulting from one di tinct act of sedimentation, defined by Otto (1938, p.574) a "that thickness of sediment which was deposited under e sentially constant physical conditions"; the deposit made du ing a time period when the prevailing current has a mean ve locity and deposits some mean size, such as a cross-bedde layer of sand formed under conditions of essentially consta flow and sediment discharge. It is distinguished from like uni by changes in particle size and/or fabric indicating changes velocity and/or direction of flow.

sediment charge In a stream, the ratio of the weight or vo ume of sediment to the weight or volume of water passing given cross section per unit of time (ASCE, 1962).

sediment concentration The ratio of the dry weight of the sed ment in a water-sediment mixture (obtained from a stream c other body of water) to the total weight of the mixture. It usually expressed in percent (for high concentration values or in parts per million (for low concentration values).

sediment-delivery ratio The ratio of sediment yield of a drain age basin to the total amount of sediment moved by shee erosion and channel erosion; expressed in percent (Chow 1964, p.17-12).

sediment discharge The amount of sediment moved by stream in a given time, measured by dry weight or by volume the rate at which sediment passes a section of a stream. Syr *sediment-transport rate.*

sediment-discharge rating The ratio between the total dis charge of a stream and the discharge of its load.

sediment load The solid material (*load*) that is transported b a natural agent, esp. by a stream. The total sediment load o a stream is equal to *bed-material load* plus *wash load,* and i expressed as the dry weight of all sediment that passes given point in a given period of time.

sedimento-eustatism Worldwide changes in sea level pro duced by a change in the capacity of the ocean basins be cause of sediment accumulation. Cf: *diastrophic eustatism glacio-eustatism.* See also: *eustacy.*

sedimentogenesis The formation of sediments.

sedimentography *sedimentary petrography.*

sedimentology The scientific study of sedimentary rocks an of the processes by which they were formed; the description classification, origin, and interpretation of sediments.

sediment-production rate Sediment yield per unit of drainage area, derived by dividing the annual sediment yield by th area of the drainage basin; expressed as tons or acre-feet o sediment per square mile per year (Chow, 1964, p.17-11).

sediment station A vertical cross-sectional plane of a stream usually normal to the mean direction of flow, where sample of suspended load are collected on a systematic basis for de termining concentration, particle-size distribution, and other characteristics.

sediment transport The movement and carrying away of sediment by natural agents; esp. the conveyance of a stream load by suspension, saltation, solution, or traction.

sediment-transport rate *sediment discharge.*

sediment vein A *sedimentary dike* formed by the filling of a fissure from above with sedimentary material.

sediment yield The amount of material eroded from the land surface by runoff and delivered to a stream system.

sedovite A brown to reddish-brown mineral: $U(MoO_4)_2$.

seed [cryst] *seed crystal.*

seed [bot] The ripened ovule of a plant containing the *embryo.*

seed crystal A small, suitably oriented piece of crystal used in *crystal seeding.* Syn: *seed.*

seed fern *pteridosperm.*

seed leaf *cotyledon.*

seed plant *spermatophyte.*

Seelandian European stage: lowermost Paleocene.

seeligerite A mineral: $Pb_3(IO_3)Cl_3O$.

seep n. An area, generally small, where water, or another liquid such as oil, percolates slowly to the land surface. For water, it may be considered as a syn. of *seepage spring,* but it is used by some for flows too small to be considered a spring. Syn: *seepage* [water]. v. To move slowly through small openings of a porous material.

seepage [oil] *oil seep.*

seepage [water] (a) The act or process involving the slow movement of water or other fluid through a porous material such as soil. Cf: *influent seepage; effluent seepage.* (b) The amount of fluid that has been involved in seepage. (c) *seep.*

seepage face A belt along a slope, such as the bank of a stream, along which water emerges at atmospheric pressure and flows down the slope. The uppermost level at which flowing water emerges in the *seepage line* and represents an outcrop of a water table; above it is moist material representing an outcrop of the capillary fringe.

seepage lake (a) A *closed lake* that loses water mainly by seepage through the walls and floor of its basin. Cf: *drainage lake.* (b) A lake that receives its water mainly from seepage, as from irrigation waters in parts of western U.S.

seepage line The free-water surface of a zone of seepage. Syn: *line of seepage; phreatic line.* See also: *seepage face.*

seepage loss Loss of water by influent seepage from a stream, canal, or other body of surface water.

seepage spring This term may be used as a syn. of *filtration spring,* but is often limited to springs of small discharge (Meinzer, 1923, p.50). See also: *seep.* Syn: *weeping spring.*

seepage stress The force that is transferred from water flowing through a porous medium to the medium itself by means of viscous friction.

seepage velocity The rate at which seepage water is discharged through a porous medium per unit area of pore space perpendicular to the direction of flow.

Seger cone A small cone made in the laboratory of a mixture of clay and salt, that softens at a definite, known temperature. It is used in volcanology to determine the approximate temperature of a molten lava. Syn: *pyrometric cone.*

seggar Var. of *sagger.*

segment (a) One of the constituent parts into which an invertebrate body is naturally separated or divided; esp. any of the succeeding or repeated body parts of an arthropod, many of which are likely to be similar in form and function (such as one of the varying individual components of the thorax of a trilobite, connected by articulation with adjoining segments), or the somite of a crustacean. See also: *podomere; article; joint* [paleont]. (b) Any of the parts into which a heterococcolith naturally separates or divides.

segmentation The process by which a coastal lagoon is subdivided into smaller patches of water by the accumulation of transverse bars (Price, 1947).

segregated vein A fissure whose mineral filling is derived from the country rock by the action of percolating water. Syn: *exudation vein.* Cf: *infiltration vein.* See also: *lateral secretion.*

segregation [petrology] (a) *magmatic segregation.* (b) A concentration of crystals of a particular mineral (or minerals) that accumulated during an early stage of consolidation as a result of magmatic segregation.

segregation [sed] A secondary feature formed as a result of chemical rearrangement of minor constituents within a sediment after its deposition; e.g. a nodule of iron sulfide, a concretion of calcium carbonate, or a geode.

segregation banding A compositional banding in gneisses that is not primary in origin, but rather is the result of segregation of material from an originally homogeneous rock (Billings, 1954). Cf: *cleavage banding.*

segregation ice A syn. of *Taber ice.* Also known as *segregated ice.*

seiche (a) A free or standing-wave oscillation of the surface of water in an enclosed or semienclosed basin (as a lake, landlocked sea, bay, or harbor) that varies in period (depending on the physical dimensions of the basin) from a few minutes to several hours and in height from several centimeters to a few meters, that is initiated chiefly by local changes in atmospheric pressure aided by winds, tidal currents, and small earthquakes, and that continues pendulum fashion for a time after the cessation of the originating force; it usually occurs in the direction of longest diameter of the basin, but occasionally it is transverse. The term has also been applied to an oscillation superimposed upon the tidal waves of the open ocean. (b) A term used in the Great Lakes area for any sudden rise (whether oscillatory or not) in the water of a harbor or lake.---Etymol: French, supposedly from Latin *siccus,* "dry"; a term used locally to describe the occasional rise and fall of water at the narrow end of Lake Geneva, Switzerland, where the phenomenon was first observed. Pron: *saysh.* Cf: *internal seiche.*

seidozerite A brownish-red mineral: $Na_2(Zr,Ti,Mn)_2Si_2O_8F_2$.

seif dune A very large, sharp-crested, tapering *longitudinal dune* or chain of sand dunes, commonly found in the Sahara Desert; its crest in profile consists of a succession of peaks and cols, and it bears on one side a succession of curved slip faces produced by strong but infrequent cross winds that tend to increase its height and width. A seif dune may be as high as 100-200 m, it may range in length from 400 m to more than 100 km (300 km in Egypt). Etymol: Arabic *saif,* "sword"; the term originated in North Africa but is applied elsewhere to similar dunes of appreciably smaller size. Pron: *safe.* Syn: *seif; sif; saif; sword dune.*

seis A colloquial syn. of *seismic detector.*

seism *earthquake.*

seismic Pertaining to an earthquake or Earth vibration, including those that are artificially induced.

seismic activity *seismicity.*

seismic anisotropy The dependence of seismic velocity on the direction of propagation.

seismic area (a) An *earthquake zone.* (b) The region affected by a particular earthquake.

seismic belt An elongate *earthquake zone,* esp. the belts of the circum-Pacific, the Mediterranean and Trans-Atlantic, the Mid-Atlantic, and the mid-Indian belt.

seismic constant In building codes dealing with earthquake hazards, an arbitrarily set quantity of steady acceleration, in units of acceleration of gravity, that a building must withstand.

seismic detector Any instrument, e.g. *seismometer, seismograph, geophone,* that receives seismic impulses. Colloquial syn: *pot; seis.*

seismic discontinuity *discontinuity* [seism].

seismic efficiency The proportion of the total available strain energy which is radiated as seismic waves. The remainder goes into gravitational potential energy, friction, and heating.

seismic-electric effect The variation of resistivity with elastic deformation of rocks.

seismic event An *earthquake* or a somewhat similar transient Earth motion caused by an explosion. Syn: *event; quake.*

seismic exploration The use of seismic techniques, usually involving explosions, to map subsurface geologic structures with the aim of locating economic deposits; *prospecting seismology; applied seismology.*

seismic focus A little-used syn. of *focus* [seism].

seismic gradient The variation of seismic velocity with direction, e.g. with depth. Syn: *velocity gradient.*

seismic intensity The average rate of flow of seismic wave energy through a unit section perpendicular to the direction of propagation. See also: *sound intensity.*

seismicity The phenomenon of Earth movements; *seismic activity* (U.S. Naval Oceanographic Office, 1966, p.144).

seismic map A contour map constructed from seismic data. The z coordinate could be either time or depth: when it is time, the map is called a *raw map;* when depth, a *migrated map.* A depth map may be tied by well data.

seismic method A method of geophysical prospecting which uses the generation, reflection, refraction, detection, and analysis of elastic waves in the Earth.

seismic noise *microseism.*

seismic profile *profile* [seis].

seismic prospecting A type of geophysical prospecting that is based on analysis of elastic waves artificially generated in the Earth. See also: *seismic survey.*

seismic ratio The ratio of bulk modulus to density.

seismic ray The path along which seismic energy travels.

seismic record In geophysical prospecting, a photographic or magnetic record of reflected or refracted seismic waves; in earthquake seismology, a record of all seismic activity, including background noise and body and surface waves from both natural and artificial events.

seismic reflection *reflection* [seism].

seismic sea wave *tsunami.*

seismic shooting A method of geophysical prospecting in which elastic waves are produced in the Earth by the firing of explosives. See also: *shoot* [seis]; *reflection shooting; refraction shooting.*

seismic spread *spread* [seis].

seismic stringer A term for a thin, usually erratic *high-speed layer.*

seismic surge *tsunami.*

seismic survey The gathering of seismic data from an area; the initial phase of *seismic prospecting.*

seismic velocity The rate of propagation of an elastic wave, usually measured in km/sec. The wave velocity depends upon the type of wave, as well as the elastic properties and density of the Earth material through which it travels.

seismic wave (a) A general term for all elastic sea waves produced by earthquakes or generated artificially by explosions. It includes both *body waves* and *surface waves.* Obs. syn: *earth wave.* (b) A seismic sea wave, or tsunami.----Syn: *earthquake wave.*

seismogram The record made by a *seismograph.* Syn: *earthquake record.*

seismograph An instrument that records vibrations of the Earth, especially earthquakes. The movement of the frame of the instrument relative to its stationary pendulum is amplified and recorded; the record is a *seismogram.* Cf: *seismometer; vibrograph; seismic detector.*

seismography (a) The study of the theory of seismographs. (b) The description of earthquakes. This is an obsolete usage.

seismologist One who pursues the science of seismology; one who applies the methods or principles of seismology to his work, e.g. earthquake prediction, seismic exploration.

seismology The study of earthquakes; by extension, the study of the structure of the interior of the Earth via both natural and artificially generated seismic signals.

seismometer An instrument that detects Earth motions. It is the detector part of the seismograph system, and does not by itself contain a recording unit. Cf: *seismograph; seismic detector; geophone; seismoscope.*

seismometer plant The manner in which a seismic detector is placed on or in the Earth.

seismometer spread *spread* [seis].

seismometry The study of seismometers; the instrumental aspects of seismology.

seismoscope An instrument that merely indicates the occurrence of an earthquake. It is considered by some, however, to be the equivalent of a *seismometer.*

sejunction water Capillary water bounded by menisci and in static equilibrium in the soil above the capillary fringe. This water may occur dispersed or as a coherent body (Schieferdecker, 1959, term 0243). Cf: *funicular water.*

sekaninaite A violet variety of cordierite in which magnesium is largely replaced by ferrous iron: $(Fe,Mg)_2Al_4Si_5O_{18}$. Syn: *iron cordierite.*

selagite A mica trachyte characterized by abundant tabular biotite crystals in a holocrystalline groundmass of orthoclase and diopside, and possibly quartz and olivine.

selbergite A hypabyssal rock containing leucite, nosean, sanidine, acmite-augite, and biotite phenocrysts in a fine-grained groundmass of nepheline, alkali feldspar, and acmite. The phenocrysts comprise a higher percentage of the volume of the rock than does the groundmass. Its name is derived from Selberg, Germany.

Selbornian *Albian.*

Seldovian Floral stage in Alaska: Oligocene(?) and Miocene.

selection *natural selection.*

selective fusion The fusion of only a portion of a mixture such as a rock. The liquid portion will generally contain a larger proportion of the more fusible components than the parent material did.

selective weathering *differential weathering.*

selenate A mineral compound characterized by discrete $(SeO_4)^{\pm}$ groups. An example is montanite, $(BiO)_2$-$(TeO_4).2H_2O.$

selenide A mineral compound that is a combination of selenium with a more positive element or radical. A example is eucairite, CuAgSe.

seleniferous plant A plant that absorbs and retains large quantities of selenium from the soil. Syn: *selenophile.*

selenite The clear, colorless variety of *gypsum,* occurring (esp. in clays) in distinct, transparent monoclinic crystals or in large crystalline masses that easily cleave into broad folia. Syn: *spectacle stone.*

selenite butte A small tabular mound, rising 1-3 m above a playa, composed of lake sediments capped with a veneer of selenite formed by deflation of the playa or by the effects of rising ground water (Stone, 1967, p.246).

selenite plate *gypsum plate.*

selenizone A sharply defined spiral band of closely spaced crescentic growth lines or linear ridges generated by a narrow notch or slit in the outer lip of the aperture of a gastropod generally at the periphery of a shell whorl. It typically marks the positions of the notch or slit during earlier stages of growth.

selenochronology Chronology of the Moon.

selenodesy "Geodesy" of the Moon.

selenofault A term used by Fielder (1965, p.172) for a large-scale fault on the Moon's surface.

selenographic chart A map representing the surface of the Moon and on which positions are measured in latitude from the Moon's equator and in longitude from a reference meridian.

selenography (a) The science of the physical features of the

Moon; the observation and recording of lunar features. (b) The topography or physical geography of the Moon.

selenolite [mineral] A white mineral: SeO_2.

selenolite [sed] A sedimentary rock composed of gypsum or anhydrite.

selenology A branch of astronomy that deals with the Moon; the science of the Moon, including *lunar geology*.

selenomorphology "Geomorphology" of the Moon; the study of lunar landforms and their origin, evolution, and distribution.

selenophile *seleniferous plant*.

selenotectonics Tectonics of the Moon; the study of lunar structures and their movements as a result of the development of the Moon as a whole.

selen-tellurium A blackish-gray mineral: (Se,Te). The Te:Se ratio is nearly 3:2. It is probably an isomorphous mixture. Syn: *ondurasite*.

self-grown stream An *autogenetic* stream developed independently on an undisturbed land surface, diverging upstream in the manner of the branches of a tree (Willis, 1907, p.8).

self-inductance The property of an electric circuit which determines the e.m.f. induced in it by a change of the current flowing in it.

self-potential curve *spontaneous-potential curve*.

self-potential log *spontaneous-potential log*.

self-potential method An electrical exploration method in which one determines the spontaneous electrical potentials (*spontaneous polarization*) that are caused by electrochemical reactions associated with metallic mineral deposits. Syn: *spontaneous-potential method*.

self-reading rod *speaking rod*.

self-reversal Acquisition by a rock of a natural remanent magnetization opposite to the ambient magnetic field direction at the time of rock formation.

self-rising ground Puffy, irregular, surface or near-surface zone of certain playas, formed by the effects of capillary rise of ground water, and consisting of a thin clay crust underlain by loose, friable, and granular sediment (silt, clay, and salt) (Stone, 1967, p.246).

seligmannite A lead-gray orthorhombic mineral: $PbCuAsS_3$.

sellaite A colorless tetragonal mineral: MgF_2.

sellate Said of certain cephalopod sutures having a saddle. Ant: *asellate*.

selvage [ore dep] *gouge*.

selvage [ign] A marginal zone of a rock mass, having some distinctive feature of fabric or composition; specif. the chilled border of an igneous mass (as of a dike or lava flow), usually characterized by a finer grain or sometimes a glassy texture, such as the glassy inner margins on the pillows in pillow lava. Syn: *selvedge; salband*.

selvage [fault] The altered, clayey material found along a fault zone; *fault gouge*. Also spelled: *selvedge*.

selvedge *selvage*.

semianthracite Coal having a fixed-carbon content of between 86% and 92%. It is between bituminous coal and anthracite in metamorphic rank, although its physical properties more closely resemble those of anthracite.

semiarid Said of a type of climate in which there is slightly more precipitation (10-20 inches or 12-16 inches) than in an arid climate, and in which grasses are the charactertistic vegetation. In Thornthwaite's classification, the moisture index is between -20 and -40. Syn: *subarid*.

semibituminous coal Coal that ranks between bituminous coal and semianthracite; it is harder and more brittle than bituminous coal. It has a high fuel ratio and burns without smoke. Syn: *smokeless coal*. Cf: *metabituminous coal*.

semibolson A wide desert basin or valley that is drained by an intermittent stream flowing through canyons at each end and reaching a surface outlet (such as another stream, a lower basin, or the sea); its central playa is absent or poorly developed. It may represent a *bolson* where the alluvial fill reached

a level sufficient to permit occasional overflow across the lowest divide.

semibright coal A type of banded coal defined microscopically as consisting of between 80% and 61% of bright ingredients such as vitrain, clarain, and fusain, with clarodurain and durain composing the remainder. Cf: *semidull coal; bright coal; dull coal; intermediate coal*.

semicannel coal *lean cannel coal*.

semiconductor radiation detector A solid-state detector of ionizing radiation, fabricated from material such as germanium or silicon. Lithium is commonly diffused ("drifted") into the semiconductor to compensate for impurities. Energy of the detected radiation is determined by collecting the electrical change produced along the ionizing path. Semiconductor detectors usually yield one to two orders of magnitude greater energy resolution than scintillation-type detectors. See also: *silicon detector; lithium-drifted germanium detector*.

semicratonic *quasicratonic*.

semicrystalline *hypocrystalline*.

semidesert An area intermediate in character and often located between a desert and a grassland or woodland.

semidiurnal Said of a tide or tidal current having a period that is approximately equal to half a lunar day, or 12.42 solar hours.

semidull coal A type of banded coal defined microscopically as consisting mainly of clarodurain and durain, with from 40% to 21% bright ingredients such as vitrain, clarain, and fusain. Cf: *semibright coal; bright coal; dull coal; intermediate coal*.

semifusain A coal lithotype transitional between fusain and vitrain, but predominantly fusain. Cf: *fusovitrain*. Also spelled: *semifusite*. Syn: *vitrifusain; vitrofusain*.

semifusinite A maceral of coal whose optical properties are intermediate between those of vitrinite and those of fusinite. It has a well-defined woody structure.

semifusinoid Fusinite, the optical properties of which are transitional between those of fusinoid and those of associated xylinoids, vitrinoids, and anthrinoids (American Society for Testing and Materials, 1970, p.364).

semifusite Var. of *semifusain*.

semilogarithmic Said of graph or plotting paper or of a chart made on such paper having a logarithmic scale on one axis and an arithmetic scale or uniform spacing on the other axis usually at right angles to the first. Syn: *semilog*.

semiopal A loosely used term for common opal, hydrophane, and any partly dehydrated or impure opal, as distinguished from precious opal and fire opal. Syn: *hemiopal*.

semipegmatitic A term suggested by Lacroix in 1900 for a type of pegmatitic texture in which the grains of one mineral, enclosed in a larger crystal of another mineral, have variable extinction angles while the enclosing mineral has uniform extinction (Johannsen, 1939, p.232); a nonrecommended term synonymous with *poikilitic*.

semiperched ground water Unconfined ground water separated by a low-permeability, but saturated, bed from a body of confined water whose hydrostatic level is below the water table (Meinzer, 1923, p.41). Because of the gradation between unconfined and confined ground water, the term has little significance and is not in common use.

semiprecious stone Any gemstone other than a *precious stone*, or any gemstone of less commercial value than a precious stone; specif. a mineral that is less than 8 on Mohs' scale of hardness. A gemstone may also be regarded as semiprecious because of its comparative abundance, inferior brilliance, or unfamiliarity to the public, or due to the whims of fashion. This arbitrary classification is misleading as it does not recognize the fact, for example, that a poor-quality ruby may be far less costly than a fine specimen of jadeite.

semisplint coal A type of banded coal that is intermediate in composition and character between *bright-banded coal* and *splint coal*; it corresponds to *duroclarain; clarodurain*. It is defined quantitatively as having 20-30% opaque attritus and more than 5% *anthraxylon*.

semitropical *subtropical*.

sempatic In the C.I.P.W. (Cross, *et al*, 1906, p.701) textural classification of igneous rocks, those rocks in which the ratio of groundmass to phenocrysts is less than five to three but greater than three to five. This term is rarely used. Cf: *dopatic; dosemic*.

semseyite A gray to black mineral: $Pb_9Sb_8S_{21}$.

senaite A black mineral: $(Fe,Mn,Pb)TiO_3$. It occurs in rounded crystals and grains in diamond-bearing sands.

senarmontite A colorless, white, or gray isometric mineral: Sb_2O_3. It is polymorphous with valentinite.

Senecan North American provincial series: lower Upper Devonian (above Erian, below Chautauquan).

senescence [**geomorph**] The part of the cycle of erosion at which a landscape or region enters upon the stage of *old age*. Cf: *senility*.

senescence [**paleont**] Old age; esp. the later stages in the life cycle of a species or other group.

senescent (a) Pertaining to the stage of *senescence* of the cycle of erosion; esp. said of a landscape that is growing old or aging (such as one characterized by a pediplain). (b) Said of a lake that is nearing extinction, as from filling by the remains of aquatic vegetation.

senesland A term proposed by Davis (1932, p.429) for a land surface that has "lost the full measure of relief that characterizes maturity"; a land surface intermediate between a *matureland* and a peneplain (Maxson, 1950, p.101). Cf: *oldland*.

sengierite A yellowish-green mineral: $Cu(UO_2)_2(VO_4)_2.-8-10H_2O$.

senile Pertaining to the stage of *senility* of the cycle of erosion; esp. said of a landscape or topography that is approaching a base-level plain or the end of the erosion cycle, or of a stream approaching an ultimate stage that is seldom fully reached, characterized by a sluggish current and a tendency to meander through a peneplain of slight relief only a little above base level.

senility The stage of the cycle of erosion in which erosion of a land surface has reached a minimum, most of the hills have disappeared, and base level has been approached. Cf: *old age; senescence*.

senior In taxonomy, said of the earlier published of two synonyms or homonyms. Cf: *junior*.

Senonian European stage: Upper Cretaceous (above Turonian, below Danian). It includes: Coniacian, Santonian, Campanian, and Maestrichtian (although some authors omit Maestrichtian).

sensible horizon (a) The plane tangent to the Earth's surface at the observer's position; the *apparent horizon*. (b) *astronomic horizon*.

sensitivity [**soil**] The effect of remolding on the consistency of a clay or cohesive soil, regardless of the physical nature of the causes of the change. A "sensitive" clay is one whose shear strength is decreased to a very small fraction of its former value on remolding at constant moisture content. See also: *sensitivity ratio*.

sensitivity [**phys**] (a) The least change in an observed quantity that can be perceived on the indicator of a given instrument. (b) The displacement of the indicator of a recording unit of an instrument per unit of change of a measurable quantity.

sensitivity ratio A measure of the degree of *sensitivity* of a clay or cohesive soil. It is the ratio of the unconfined compressive strength of an "undisturbed" specimen to that of the same specimen at the same moisture content but in a remolded or "disturbed" state. The sensitivity ratio for most clays ranges between 2 and 4, although extrasensitive clays may have values greater than 8. Abbrev: St. Syn: *remolding sensitivity*.

sensu lato In a broad sense, esp. with reference to the name

of a taxon that is used in a more inclusive sense than the current description dictates. Etymol: Latin. Cf: *sensu stricto*.

sensu stricto In a narrow sense, esp. with reference to the name of a taxon that is used in restricted sense. Etymol: Latin. Cf: *sensu lato*.

sepal One of the separate parts of the calyx, the outermost organ of a flower, usually green and more or less of a leaflike texture.

separate *soil separate*.

separation On a fault, *apparent* rather than actual or net relative displacement measured in any given direction. Separation is two-dimensional and very often it is all that is ever seen in routine work with faults. Cf: *slip* [*struc geol*]. See also: *dip separation; strike separation*. Syn: *apparent relative movement*.

separation disc A breakage area in a filament of blue-green algae formed by the death of a cell.

separation layer *abscission layer*.

sepiolite A chain-lattice clay mineral: $Mg_4(Si_2O_5)_3(OH)_2.6H_2O$. It is an extremely lightweight, absorbent, soft, compact to fibrous, and white to light-gray or light-yellow material found chiefly in Asia Minor and used for making tobacco pipes, pipe bowls, cigar and cigarette holders, and ornamental carvings. Sepiolite occurs in veins with calcite, and in alluvial deposits formed from weathering of serpentine masses. Syn: *meerschaum; sea-foam*.

septa Plural of *septum*.

septal angle The angle between tangents drawn from the apex of a planispiral nautiloid shell to two successive septa and measured on a section made along the plane of symmetry (TIP, 1964, pt.K, p.58).

septal cycle All septa belonging to a single stage in ontogeny of a scleractinian corallite as determined by the order of appearance of septal groups (six protosepta comprising the first cycle and later-formed exosepta and entosepta in constantly arranged succession being introduced in sextants) (TIP, 1956, pt.F, p.250).

septal flap The extension of each lamella in the tests of foraminifers of the superfamily Rotaliacea, formed on the inner side of the chamber over the distal face of the previous chamber, and resulting in secondarily doubled septa (TIP, 1964, pt.C, p.63).

septal fluting One of the foldings, rufflings, wrinklings, or corrugations of septa (and antetheca) in a fusulinid test transverse to the axis of coiling, generally strongest in the lower (adaxial) part of septa and toward the poles and decreasing in intensity toward the top.

septal foramen (a) An intercameral *foramen* (opening) in the test of a foraminifer. It may be homologous with the aperture or be secondarily formed. (b) An opening in septum at the siphuncle of a cephalopod, allowing the passage of the siphuncular cord.

septal furrow (a) The narrow middorsal region of a nautiloid in which the mural part of the septum (attached to wall of conch) is lacking (TIP, 1964, pt.K, p.58). Syn: *dorsal furrow*. (b) An *external furrow* of a fusulinid.

septal groove A longitudinal groove on the outer surface of the wall of a corallite, corresponding in position to a septum on the inner surface of the wall. Cf: *interseptal ridge*.

septalial plate One of the *crural plates* forming the floor of the septalium of a brachiopod and united with the earlier-formed part of the median septum.

septalium A trough-like structure of the brachial valve of a brachiopod, situated between hinge plates (or homologues), consisting of septalial plates (or homologues) enveloping and buttressed by the median septum. It does not carry adductor muscles.

septal neck A short funnel- or tube-like flexure or extension of a septum along the siphuncle of a cephalopod. Syn: *neck* [*paleont*].

septal pore A small perforation in a septum (and antetheca) of the test of a fusulinid. Cf: *mural pore*.

septal tooth A small projection along the upper margin of a septum of a scleractinian coral, formed by extension of a trabecula beyond calcareous tissue connecting it with others.

septaria Plural of *septarium*.

septarian Said of the irregular polygonal pattern of internal cracks developed in septaria, closely resembling the desiccation structure of mud cracks; also said of the epigenetic mineral deposits which may occur as fillings of these cracks.

septarian boulder *septarium*.

septarian nodule *septarium*.

septarium (a) A large (8-90 cm in diameter), roughly spheroidal concretion, usually of an impure argillaceous carbonate (such as caly ironstone), characterized internally by irregular polyhedral blocks formed by a series of radiating cracks that extend from within outward and widen toward the center and that intersect a series of cracks concentric with the margins, the cracks invariably filled or partly filled by crystalline minerals (most commonly calcite) that cement the blocks together. Its formation involves the formation of an aluminous gel, case hardening of the exterior, shrinkage cracking due to dehydration of the colloidal mass in the interior, and vein filling. When worn on the surface, the veins sometimes weather in relief, thus producing a septate pattern. Cf: *melikaria*. Syn: *septarian nodule; septarian boulder; beetle stone; turtle stone*. (b) A crystal-lined crack or fissure in a septarium.---Pl: *septaria*.

septechlorite A name given to a group of chlorite-like minerals (amesite, cronstedtite, berthierine) with basal spacings of 7 angstroms. Syn: *pseudochlorite*.

septifer Said of brachiopod crura having the form of septa that descend directly from the dorsal side of the hinge plates to the floor of the brachial valve.

septotheca A wall of a scleractinian corallite, formed by thickened outer parts of septa along the axis of trabecular divergence. Cf: *synapticulotheca; paratheca*.

septula (a) A syn. of *septulum*. Pl: *septulae*. (b) Plural of *septulum*.

septulum (a) One of the small perforations (uniporous or multiporous) in the walls of cheilostomatous bryozoans for the passage of mesodermal fibers between adjacent zooids. Syn: *septule*. (b) A small septum; e.g. a ridge or small partition extending adaxially downward from the lower surface of spirotheca in the test of a fusulinid foraminifer so as to partially subdivide the chambers of the test.---Pl: *septula*. Syn: *septula*.

septum [paleont] (a) One of the transverse, internal, calcareous partitions dividing the shell of a cephalopod, such as a partition that divides the phragmocone of a nautiloid into camerae and that is attached to the inside of the wall of the conch. (b) One of several radially disposed, longitudinal, calcareous plates or partitions of a corallite, occurring between or within mesenterial pairs. It presumably supported the basal disk and lower wall of the polyp. Also, a thin radial noncalcareous partition (composed of soft tissue) dividing the gastrovascular cavity of an octocorallian polyp. (c) A partition or wall dividing a foraminiferal test interiorly into chambers, commonly formed by previous outer wall or apertural face. (d) A relatively long, narrow, commonly blade-like elevation of the secondary shell of a brachiopod; a median ridge in either valve of a brachiopod. It is indicated in articulates (within the underlying floor of a valve) by persistent high, narrow deflections of fibrous calcite originating near the primary layer, and in inarticulates by comparable deflections of shell lamellae. (e) The wall separating the two rows of thecae in a biserial graptoloid. (f) A radial, longitudinal, normally perforate plate connecting the inner and outer walls of an archaeocyathid. Formerly known as *pariety*. (g) A plate-like structure in an echinoid spine, radiating from the axial zone toward the anterior of the spine and seen in a cross section of the spine.---Pl: *septa*. Adj: *septal*.

septum [bot] (a) In algae, a crosswall, generally perpendicular to the length of the filament. (b) In a dinoflagellate cyst, a more or less membranous, linear projection that is perpendicular to the wall.

Sequanian Substage in Great Britain: Upper Jurassic (upper Lusitanian; above Rauracian Substage, below Kimmeridgian Stage). Syn: *Astartian*.

Sequanian river A river that rises in comparatively low ground and that decreases in volume in the summer owing to evaporation and vegetative absorption. Type river: the Seine in France.

sequence (a) A succession of geologic events, processes, or rocks, arranged in chronologic order to show their relative position and age with respect to geologic history as a whole. (b) A major informal rock-stratigraphic unit of greater than group or supergroup rank, traceable over large areas of a continent, and bounded by unconformities of interregional scope, such as in the cratonic interior of North America (Sloss, 1963); a geographically discrete succession of major rock units that were deposited under related environmental conditions (Silberling & Roberts, 1962). See also: *sub-sequence*. Syn: *stratigraphic sequence*. (c) A term, now obsolete, used by Moore (1933, p.54) for the rocks formed during an era; an *erathem*. (d) A faunal succession.

sequence of crystallization *order of crystallization*.

sequent geosyncline The constituent geosynclines of a *polygeosyncline*, separated from one another by the development of geanticlines (Schuchert, 1923). Syn: *secondary geosyncline (b)*.

sequential landform One of an orderly succession of smaller landforms that are developed by the erosion, weathering, and mass-wasting of larger *initial landforms*; it includes "erosional landforms" resulting from the progressive removal of earth materials, and "depositional landforms" resulting from the accumulation of the products of erosion. Cf: *ultimate landform; destructional*. Syn: *sequential form*.

serac A jagged pinnacle, sharp ridge, needle-like tower, or irregularly shaped block of ice on the surface of a glacier (commonly among intersecting crevasses, such as on an icefall), formed where the glacier is periodically broken as it passes over a steep slope. Etymol: French *sérac*, a solid curdy white cheese made in the Alps and sold in blocks made by pressing it into quadrangular molds, which glacial seracs resemble in appearance and shape. Both the French form and the anglicized version are in regular use. Pron: suh-*rahk*. Cf: *nieve penitente*.

seral The adj. of *sere*.

serandite A rose-red monoclinic mineral: $Na(Mn,Ca)_2-Si_3O_8(OH)$. Cf: *pectolite*.

sere A sequence of ecologic communities that succeed one another in development from the pioneer stage to climax. Adj: *seral*. See also: *paleosere*.

serendibite An indigo-blue to grayish blue-green mineral: $Ca_4(Mg,Fe,Al)_6(Al,Fe)_9(Si,Al)_6B_3O_{40}$.

serial homology The similarity that exists between members of a series of structures in an organism; e.g. the resemblance to each other of the vertebrae in the vertebral column.

serial sampling A method of gathering *samples* by a set pattern, such as a grid, to insure randomness.

seriate An *anisometric* texture in which the sizes of the crystals vary gradually or in a continuous series. Cf: *hiatal*.

sericite A white, fine-grained potassium mica occurring in small scales and flakes as an alteration product of various aluminosilicate minerals, having a silky luster, and found in various metamorphic rocks (esp. in schists and phyllites) or in the wall rocks, fault gouge, and vein fillings of many ore deposits. It is usually *muscovite* or very close to muscovite in composition, and may also include much illite.

sericitic sandstone A sandstone in which sericite (derived by decomposition of feldspar) intermingles with finely divided quartz and fills the voids between quartz grains.

sericitization A hydrothermal or metamorphic process involving the introduction of, or replacement by, sericite.

series [cart] *map series.*

series [geol] Any number of rocks, minerals, or fossils having characteristics, such as growth patterns, succession, composition, or occurrence, that make it possible to arrange them in a natural sequence.

series [ign] (a) *igneous rock series.* (b) A term that is often misused for a sequence of rocks resulting from a succession of eruptions or intrusions and that is usually preceded by an adjective such as "eruptive", "intrusive", or "volcanic" to indicate the origin of the rock. The term "group" should replace "series" in this usage (ACSN, 1961, art.9f).

series [stratig] (a) A time-stratigraphic unit generally classed next in rank below *system* and above *stage,* properly based on a clearly designated stratigraphic interval in a type area (although many series have come to be adopted quite generally without explicit indication of their limits) (ACSN, 1961, art.30); the rocks formed during an *epoch* of geologic time. Some series are recognized as worldwide units, others as *provincial series.* Most systems have been divided into three series but the number may vary from two to six. The term is not restricted to stratified rocks but may be applied to intrusive rocks in the same time-stratigraphic sense. Formal series names are binomial, usually consisting of a geographic name (with the adjectival ending "-an" or "-ian", although it is permissible to use the geographic name without any special ending, such as "Cincinnati Series") and the word "Series", the initial letter of both terms being capitalized. (b) A term that is often misused in a rock-stratigraphic sense for a succession of continuously deposited sedimentary rocks; e.g. an assemblage of formations (or group) or an assemblage of formations and groups (a supergroup), esp. in the Precambrian. The term "should no longer be so used" and should be replaced by "group" (ACSN, 1961, art.9f). See also: *subsequence.* (c) A term used in England for a rock-stratigraphic unit, generally for a large unit or esp. one throughout which the strata are conformable.

series [radioactivity] *radioactive series.*

series circuit An electrical circuit so connected that there is a single continuous path and all the current flows through each component.

serir A desert plain strewn with rounded pebbles, older than the gravel-covered *reg;* a *stony desert* from which the sand has been blown away, as in the Sahara of Libya and Egypt. Etymol: Arabic, "dry". Pl: *serir.* See also: *pebble armor; hammada.* Syn: *shore.*

Serozem soil Var. of *Sierozem.*

serpenticone A very evolute, many-whorled, coiled cephalopod shell with the whorls hardly overlapping, resembling a coiled snake or rope.

serpentine (a) A group of common rock-forming minerals having the formula: $(Mg,Fe)_3Si_2O_5(OH)_4$. Serpentines have a greasy or silky luster, a slightly soapy feel, and a tough, conchoidal fracture; they are usually compact but may be granular or fibrous, and are commonly green, greenish yellow, or greenish gray (sometimes brown, black, or white) and often veined or spotted with red, green, and white. Serpentines are always secondary minerals, derived by alteration of magnesium-rich silicate minerals (esp. olivines), and are found in both igneous and metamorphic rocks; they generally crystallize in the monoclinic system, but only as pseudomorphs. Translucent varieties are used for ornamental and decorative purposes, often as a substitute for jade. (b) A mineral of the serpentine group, such as chrysotile, antigorite, lizardite, parachrysotile, and orthochrysotile. (c) A term strictly applied in place of *chrysotile* as a species name, chrysotile becoming a variety and antigorite a separate species not included in the term "serpentine" (Hey, 1962, p.595).---Etymol: Latin *serpen-*

tinus, "resembling a serpent", from the mottled shades o green (on massive varieties) suggestive of the markings on serpent's skin.

serpentine asbestos *chrysotile.*

serpentine jade A variety of the mineral serpentine resemblin jade in appearance and used as an ornamental stone; speci *bowenite* [mineral].

serpentine marble *verd antique.*

serpentine rock *serpentinite.*

serpentine spit A spit that is extended in more than one direc tion due to variable or periodically shifting currents. Syn: *se pent spit.*

serpentine-talc A mineral: $Mg_6Si_6O_{15}(OH)_6$. It is intermediat in composition and physical characteristics between serpen tine and talc.

serpentinite A rock consisting almost wholly of serpentine group minerals, e.g. antigorite, chrysotile, or serpophite de rived from the alteration of previously existing ferromagnesia silicate minerals such as olivine and pyroxene. Accessor chlorite and talc may be present. Syn: *serpentine rock.*

serpentinization The process or state of hydrothermal alter ation (metasomatism) by which magnesium-rich silicate min erals (olivine, pyroxenes, amphiboles in dunites, peridotites and other ultrabasic rocks) are converted into or replaced b serpentine minerals forming serpentinite.

serpent kame A term introduced by Shaler (1889, p.549) for sinuous *esker* and known in New England also as an *India ridge.*

serpent stone A stone (usually a highly absorbent aluminou gem) formerly believed to be formed by snakes and to be eff cacious in drawing out poison; specif. *adder stone.*

serpierite A bluish-green mineral: $Ca(Cu,Zn)_4(SO_4)_2(OH)_6.3H_2O.$

serpophite A metacolloidal variety of the mineral serpentine.

serpulid Any annelid that belongs to the family Serpulidae an that characteristically builds a contorted calcareous or leather tube on a submerged surface. See also: *serpulid reef.*

serpulid reef A reef patch with a raised rim, characterized b cup-shaped central parts excavated by *serpulid* worms of the family Serpulidae.

serra (a) *sierra.* (b) A term used in Brazil for an elevate mountain zone supporting luxuriant vegetation.---Etymol: Por tuguese, "saw".

serrate [geomorph] adj. Said of topographic features that ar notched or toothed on the edge, or have a saw-edged profile e.g. *serrate* divide. Syn: *saw-toothed; serrated.---*n. A rock mountain summit having a serrate profile (Stone, 1967 p.246).

serrate [bot] Said of a leaf margin that is saw-toothed, and i which the toothlike projections point forward.

serrate [petrology] Said of saw-toothed contacts betwee minerals, usually resulting from replacement; e.g. the serrat texture of megacrysts in contact with plagioclase in igneou rocks.

serule *microsere.*

sesquan A *cutan* consisting of sesquioxides or hydroxides (o aluminum or iron) (Brewer, 1964, p.214).

sessile Said of a plant or animal that is permanently attache to a substrate and is not free to move about. Cf: *vagile.*

sessile cruralium Cruralium united with the floor of the bra chial valve of a brachiopod without intervention of the support ing median septum.

sessile spondylium Spondylium united with the floor of the pedicle valve of a brachiopod without intervention of the sup porting median septum.

seston Both living and nonliving organic material in water.

set [current] The compass direction toward which a current is flowing; a direction of flow. Syn: *current direction.*

set [stratig] A term introduced by McKee & Weir (1953 p.382-383) for "a group of essentially conformable strata o

ross-strata, separated from other sedimentary units by sur-
aces of erosion, nondeposition, or abrupt change in charac-
er"; it is composed of two or more consecutive beds of the
ame lithology, and is the smallest and most basic group unit.
ee also: coset.

et [exp struc geol] permanent set.

eta [paleont] A slender, typically rigid or bristly and springy
rgan or part of an invertebrate; e.g. a movable, spiny, whip-
ke component of a bryozoan vibraculum, homologous with
e operculum of an autozooid, or a chitinous bristle arising
om invagination of mantle groove of a brachiopod, or a hair-
ke process of and articulating with the external membrane of
crustacean. Pl: setae.

eta [bot] (a) The sporophyte stalk in the mosses and liver-
orts. (b) A stiff, short hair on a plant.----Syn: awn.

etaceous Said of a plant part that is covered with stiff, short
airs.

eter A Norwegian term for a wave-cut rock terrace.

ettled snow An old snow that has been strongly metamor-
hosed and compacted.

ettlement (a) The gradual downward movement of an engi-
eering structure, due to compression of the soil below the
oundation. See also: differential settlement. (b) The gradual
owering of the overlying strata in a mine, due to extraction of
he mined material. (c) The subsidence of surficial material
such as coastal sediments) due to compaction.

ettling [mass move] The sag in outcrops of layered strata,
aused by rock creep (Sharpe, 1938, p.33). See also: terminal
urvature. Syn: outcrop curvature.

ettling [sed] (a) The deposition of sediment. (b) A sediment
r precipitate.

ettling [snow] The vertical component of snowcreep.

ettling basin (a) An artificial basin or trap designed to collect
he suspended sediment of a stream before it flows into a res-
rvoir and thereby prevent the rapid siltation of the reservoir;
.g. a desilting basin. It is usually provided with means to
raw off the clear water. (b) An industrial sedimentation
tructure designed to remove pollutant materials from factory
ffluents.

ettlingite A hard, brittle, pale-yellow to deep-red hydrocarbon
H:C about 1.53) found in resinous drops on the walls of a
ead mine at Settling Stones in Northumberland, Eng. Syn:
Settling Stones resin.

ettling reservoir A reservoir consisting of a series of shallow
asins arranged in steps and connected by long conduits al-
owing the removal of only the clear upper layer of water in
ach basin.

ettling sand A driller's term for friable sandstone that caves
nto a well and settles around the drill bit.

ettling velocity The rate at which suspended solids subside
nd are deposited. Syn: fall velocity.

etup (a) The assembly and arrangement of the equipment
nd apparatus required for the performance of a surveying
peration; specif. a surveying instrument (transit or level)
laced in position and leveled, ready for taking measurements. (b) The actual physical placing of a leveling instrument
ver an instrument station. (c) instrument station. (d) The
orizontal distance from the fiducial mark on the front end of
surveyor's tape (or the part of a tape which is in use at the
ime), measured in a forward direction to the point on the
round mark or monument to which the particular measure is
eing made.---Also spelled: set-up.

Sevier orogeny A name proposed by R.L. Armstrong (1968)
or the well known deformations that occurred along the eastern edge of the Great Basin in Utah (eastern edge of Cordil-
eran miogeosyncline) during times intermediate between the
Nevadan orogeny farther west and the Laramide orogeny far-
her east, culminating early in the Late Cretaceous. During the
rogeny, the folding and eastward thrusting of the miogeosynclinal rocks over their foreland was largely completed.

sexine The outer division of the exine of pollen, more or less
equivalent to ektexine. Cf: nexine.

sexiradiate A sponge spicule in the form of six equidistant coplanar rays arising from a common center.

sextant A double-reflecting, hand instrument used for measuring the angular distance between two objects in the plane defined by the two objects and the point of observation, and
originally characterized by having an arc of 60 degrees (and a
range of 120 degrees) but now applied to similar instruments
regardless of range; esp. such an instrument used in navigation for measuring apparent altitudes of celestial bodies from
a moving ship or airplane, or used in hydrographic surveying
for measuring horizontal angles at a point in a moving boat
between shore objects. Cf: quadrant; astrolabe.

sexual dimorphism A condition in which the two sexes of an
organism are markedly dissimilar in appearance.

seybertite clintonite.

seyrigite A variety of scheelite containing molybdenum.

sferics The natural fluctuations of the Earth's electromagnetic
field. Their frequency is usually above one hertz, and they are
usually caused by lightning. Etymol: A short form of atmospherics. Also spelled: 'spherics.

s-fold A fold, the vertical cross section of which has the form
of the letter "s". It is a common type of fold in areas of overthrusting and intense folding.

shabka A desert landscape formed by wind erosion of alluvial
basins. Etymol: Arabic, "network, fiber".

shackanite An analcime trachyte containing rhombohedral
phenocrysts of feldspar.

shackhole shakehole.

shaded-relief map A map of an area whose relief is made to
appear three dimensional by the method of hill shading.

shading hill shading.

shadow weathering Mechanical weathering in which disintegration of rock occurs along the margin of sunlight and shade;
a kind of insolation, or the effect of sunlight on weathering of
rock.

shadow zone [dunes] wind shadow.

shadow zone [seis] (a) An area in which there is very little
penetration of acoustic waves. (b) A region 100°-140° from
the epicenter of an earthquake in which, due to refraction
from the low-velocity zone inside the core boundary, there is
no direct penetration of seismic waves.

shaft [speleo] A passage in a cave that is vertical or almost
so.

shaft [paleont] (a) The main part of the spine of an echinoid.
(b) The ridge- or stalk-like proximal part of the cardinal process of a brachiopod, supporting the myophore.

shagreen relief [crystal optics].

shake shakehole.

shakehole A caver's term for a doline or swallow hole, especially a doline that has been formed by subsidence rather than
by collapse. Syn: shake; shackhole.

shake wave S wave.

shaking prairie A term used in Louisiana to describe delta
land with a surface of matted vegetable material resting on
water, or waterlogged peat or sand, and which trembles when
walked on. Syn: trembling prairie.

shale A fine-grained, indurated, detrital sedimentary rock
formed by the consolidation (as by compression or cementation) of clay, silt, or mud, and characterized by finely stratified
(laminae 0.1-0.4 mm thick) structure and/or fissility that is
approximately parallel to the bedding (along which the rock
breaks readily into thin layers) and that is commonly most
conspicuous on weathered surfaces, and by a composition
with an appreciable content of clay minerals or derivatives
from clay minerals and with a high content of detrital quartz; a
thinly laminated or fissile claystone, siltstone, or mudstone. It
normally contains at least 50% silt, with 35% "clay or fine
mica fraction" and 15% chemical or authigenic materials

649

(Krynine, 1948, p.154-155). Shale is generally soft but sufficiently indurated so that it will not fall apart on wetting; it is less firm than argillite and slate, commonly has a splintery fracture and a smooth feel, and is easily scratched. Its color may be red, brown, black, gray, green, or blue. The term "shale" is regarded sometimes as a structural term with the significance of thin bedding or fissility and without implying any particular composition; it has been loosely applied to massive or blocky indurated silts and clays that are not laminated, to laminated silts and clays that are not indurated, to fine-grained and thinly laminated sandstones, and to slates (as in "marl slate"). A review of the origin and use of the term "shale" is given by Tourtelot (1960), who notes that the term originally meant a "laminated clayey rock" but historically has also been applied to the "general class of fine-grained rocks", and who states that the general trend prior to 1850 in the U.S. "seems to have been to use 'shale' for almost any clayey rock of Paleozoic age; afterwards the term came to be applied to many clayey rocks of all ages" (p.341). Etymol: Teutonic, probably Old English *scealu*, "shell, husk", akin to German *Schale*, "shell". Abbrev: sh.

shale-arenite A term used by Folk (1968, p.124) for a *sedarenite* composed chiefly of shale fragments.

shale-ball A meteorite partly or wholly converted to iron oxides by weathering. Syn: *oxidite*.

shale break A thin layer or *parting* of shale between harder strata or within a bed of sandstone or limestone.

shale crescent A term used by Shrock (1948, fig.86) for a crescent formed by the filling of a ripple-mark trough by shale.

shale ice A mass of thin and brittle plates of river or lake ice formed when sheets of skim ice break up into small pieces which are gathered into bunches.

shale line A reference line, usually straight, that can generally be traced along the extreme positive edges of the spontaneous-potential curve. It characterizes impermeable formations such as shales.

shale-out A *stratigraphic trap* formed by the *lateral variation* or facies change of a porous sandstone in which the clay constituents increase and the hard mineral grains are reduced until the porosity and permeability disappear and the bed grades to claystone or shale. Cf: *pinch-out*.

shale shaker A cylindrical sieve or vibrating table that receives the drilling mud on its return to the surface and screens out or removes the coarser drill cuttings for geologic examination. Syn: *shale screen*.

shaley Var. of *shaly*.

shalification The formation of shale; e.g. solution of calcium carbonate from a bed of impure limestone, leaving argillaceous material behind to form a thin layer of shale.

shaliness The quality of being shaly; specif. the property of clay-rich rocks (esp. calcareous shales) of splitting with concave or "shelly" surfaces roughly parallel to the bedding planes (Grabau, 1924, p.785).

shallow earthquake *shallow-focus earthquake.*

shallow-focus earthquake An earthquake whose focus occurs at a depth of less than 50, 60, or 70 km (there is no agreement). Most earthquake activity is of this type. Cf: *intermediate-focus earthquake; deep-focus earthquake.* Syn: *shallow earthquake; normal earthquake.*

shallow percolation Precipitation that moves downward and laterally toward streams. Syn: *storm seepage.* Cf: *deep percolation.*

shallows An indefinite term applied to a shallow place or area in a body of water, or to an expanse of shallow water; a *shoal.*

shallow scattering layer A stratified area of marine organisms over a continental shelf that scatter sound waves from an echo sounder. Cf: *deep scattering layer; surface scattering layer.*

shallow-water wave A wave on the surface of a body of water the wave length of which is 25 or more times the water depth and for which the water depth is an influence on the shape of the orbital and on the velocity. Cf: *deep-water wave; transitional-water wave.* Syn: *long wave [water].*

shallow well (a) A water well, generally dug by hand or by excavating machinery, or put down by driving or boring, that taps the shallowest aquifer in the vicinity. The water is generally unconfined ground water. (b) A well whose water level is shallow enough to permit use of a shallow-well (suction) pump, the practical lift of which is taken as 22 ft.----Cf: *water table well; deep well.*

shaly (a) Pertaining to, composed of, containing, or having the character or properties of shale; esp. readily split along closely spaced bedding planes, such as "shaly cleavage" (or "shaly structure" or "shaly parting") developed approximately parallel to the bedding. Also, said of a fine-grained, thinly laminated sandstone having the characteristic fissility of shale due to the presence of thin layers of shale, or said of a siltstone possessing bedding-plane fissility. (b) Said of bedding that consists of laminae varying in thickness between 2 mm and 10 mm (Payne, 1942); e.g. "shaly parting", a thin leaf of shale separating the layers of a thinly bedded rock.---Cf: *argillaceous.* Syn: *shaley; shelly.*

shandite A rhombohedral mineral: $Ni_3Pb_2S_2$.

Shand's classification A quantitative mineralogic classification of igneous rocks based on their silica content. This system was developed in 1927 by S.J. Shand (Shand, 1947).

shank [paleont] (a) The connection between flukes and (where present) the stock of an *anchor* of a holothurian. (b) The part connecting the eye and spear of a *hook* of a holothurian.

shank [fold] An obsolete syn. of *limb* of a fold.

shantung A monadnock in the process of burial by *huangho* deposits (Grabau, 1936, p.266). Type locality: Shantung rocky mass of northern China.

Shantung soil An early name for *Noncalcic Brown soil.* Var: Shantung Brown soil.

shape *particle shape.*

shape class The general group (oblate, prolate, or intermediate shapes) to which a pollen grain belongs in terms of the ratio between the equatorial diameter and the pole-to-pole dimension.

shapometer A device for measuring the shapes of sedimentary particles (Tester & Bay, 1931).

shard A vitric fragment in pyroclastics, having a characteristically curved surface of fracture.

sharkskin pahoehoe A type of pahoehoe, the surface of which displays innumerable tiny spicules or spines produced by escaping gas bubbles. Cf: *corded pahoehoe; elephant-hide pahoehoe; entrail pahoehoe; festooned pahoehoe; filamented pahoehoe; shelly pahoehoe; slab pahoehoe.*

shark-tooth projection A structure formed by the pulling or tearing apart of plastic lava into fine, sharp points several centimeters in length. Such projections may occur along the edge of a flow or along a slump scarp.

sharpite A greenish-yellow mineral: $(UO_2)(CO_3).H_2O$ (?).

sharp sand A sand composed of angular grains, nearly or wholly free from foreign particles (as of clay), and used in making mortar.

sharpstone (a) A collective term proposed by Shrock (1948a) for any rock fragment larger than a sand grain (diameter greater than 2 mm) and having angular edges and corners; a clastic constituent of rubble. Cf: *roundstone.* (b) A term used in Yorkshire (Eng.) for a fine-grained, nonargillaceous sandstone that breaks into angular fragments. Also spelled: *sharpstone.*

sharpstone conglomerate *sedimentary breccia.*

Shasta Provincial series in California: Lower Cretaceous.

shastaite An obsolete term originally applied to a dacite.

shastalite An obsolete term originally applied to unaltered glassy andesites as distinct from *weiselbergite*.

shatter belt A less-preferred syn. of *fault zone*.

shatter breccia A tectonic breccia composed of angular fragments that show little rotation (Bateman, 1950, p.133). Cf: *rubble breccia*. Syn: *crackle breccia*.

shatter cone A distinctively striated conical fragment of rock along which fracturing has occurred, ranging in length from less than a centimeter to several meters, generally found in nested or composite groups in the rocks of cryptoexplosion structures, and generally believed to have been formed by shock waves generated by meteorite impact (Dietz, 1959). Shatter cones superficially resemble cone-in-cone structure in sedimentary rocks; they are most common in fine-grained homogeneous rocks such as carbonate rocks (limestones, dolomites), but are also known from shales, sandstones, quartzites, and granites. The striated surfaces radiate outward from the apex in horsetail fashion; the apical angle varies but is close to 90 degrees. Syn: *shear cone; pressure cone*.

shatter-cone segment A part of an imcompletely developed shatter cone, consisting of a single curved, striated surface, generally 10-45 degrees of cross section, of a cone whose apical angle may range from 90 to 120 degrees (Manton, 1965, p.1021). Most shatter-coned rocks display only segments.

shatter coning A mode of rock failure characterized by the development of shatter cones (Manton, 1965, p.1021).

shattering The breaking up into highly irregular, angular blocks of a very hard rock that has been subjected to severe stresses; the fractures may cut across mineral grains and other structures in the rock.

shatter zone An area of randomly fissured or cracked rock that may be filled by mineral deposits, forming a network pattern of veins.

shattuckite A blue mineral: $Cu_5(SiO_3)_4(OH)_2$. Cf: *planchéite*.

shcherbakovite A monoclinic mineral: $(K,Na,Ba)_3(Ti,Nb)_2(Si_2O_7)_2$.

sheaf structure A bundled arrangement of crystals that is characteristic of certain fibrous minerals, e.g. stibnite.

shear A strain resulting from stresses that cause or tend to cause contiguous parts of a body to slide relatively to each other in a direction parallel to their plane of contact; specifically, the ratio of the relative displacement of these parts to the distance between them. It is the mode of failure of a body or mass whereby the portion of the mass on one side of a plane or surface slides past the portion on the opposite side. In geological literature the term refers almost invariably to strain rather than to stress. It is also used to refer to surfaces and zones of failure by shear, and to surfaces along which differential movement has taken place. Syn: *shear strain*.

shear cleavage A syn. of *slip cleavage*; it is a general term because there are many types of shear in deformation and metamorphism.

shear-cleavage fold *cleavage fold*.

shear cone *shatter cone*.

shear crack A *strain crack* in sea ice, caused by different but parallel forces acting simultaneously in different directions along a plane. Cf: *tension crack*.

shear drag *shear resistance*.

shear fold A similar fold, the mechanism of which is shearing or slipping along closely spaced planes parallel to the fold's axial surface. Syn: *slip fold; glide fold*. See also: *cleavage fold*.

shear fracture A fracture caused by compressive stress. See also: *shear joint*. Cf: *tension fracture*.

shear joint A joint that is a *shear fracture*; a potential plane of shear. Less-preferred syn: *slip joint*.

shear modulus *modulus of rigidity*.

shear moraine A debris-laden surface or zone found along the margin of any ice sheet or ice cap, dipping in toward the center of the ice sheet but becoming parallel to the bed at the base. Shear moraines are thought to separate stagnant ice from active ice and/or to represent surfaces along which debris is brought up to the surface from the bed. Cf: *shear plane [glaciol]*.

shear plane [exp struc geol] *shear surface*.

shear plane [glaciol] A planar surface in a glacier, usually laden with rock debris, attributed to discontinuous shearing or overthrusting. Apparent displacements on some shear planes may simply be due to differential ablation. Cf: *shear moraine*.

shear resistance In fluid dynamics, a tangential stress caused by the fluid viscosity and taking place along a boundary of the flow in the tangential direction of local motion (Chow, 1957). Cf: *pressure resistance*. Syn: *shear drag*.

shear slide A landslide, esp. a slump, produced by shear failure usually along a plane of weakness, as bedding or cleavage.

shear sorting Sorting of sediments in which the smaller grains tend to move toward the zone of greatest shear strain and the larger grains toward the zone of least shear. In sand dunes, it is characterized by a lamination of fine-grained dark minerals in the zone between moving and residual sand (Stone, 1967, p.246).

shear strain *shear*.

shear strength The internal resistance of a body to shear stress. See also: *cohesion*.

shear stress That component of *stress* which acts tangential to a plane through any given point on a body; any of the tangential components of the stress tensor. Symbol: τ. Cf: *normal stress*. Syn: *tangential stress*.

shear structure Any rock structure caused by shearing, e.g. crushing, crumpling, or cleavage.

shear surface A surface along which differential movement has taken place parallel to the surface. Syn: *shear plane*.

shear velocity The square root of the product of the acceleration due to gravity, the hydraulic mean depth of flow, and the slope of the energy grade line (ASCE, 1962).

shear wave *S wave*.

shear zone A tabular zone of rock that has been crushed and brecciated by many parallel fractures due to shear strain. Such an area is often mineralized by ore-forming solutions. See also: *sheeted-zone deposit*.

sheath [paleont] An investing cover or case of an animal body or body part; e.g. a tentacle sheath of a bryozoan, or a receptacle or container in radiolarians, or an expanded basal part of a bar, blade, or limb of a conodont element, or the thickened inner edge of the orifice of a cirripede crustacean, supporting the opercular membrane.

sheath [bot] (a) A tubular, enrolled part or organ of a plant, e.g. the lower part of a leaf in the grasses. (b) In some algae, a covering external to the cell wall.

shed A divide of land; e.g. a *watershed*.

sheen A subdued and often iridescent or metallic glitter that approaches but is just short of optical reflection and that modifies the surface luster of a mineral; e.g. the optical effect still visible in the body of a gem (such as tiger's-eye) after its silky surface appearance has been removed by polishing.

sheepback rock A term used as a syn. of *roche moutonnée* in reference to the fanciful resemblance of the landform to a sheep's back. Syn: *sheep rock; sheepback*.

sheep rock *sheepback rock*.

sheer A steep face of a cliff; a precipice.

sheet [ore dep] A term used in the Upper Mississippi lead-mining region of the U.S. for galena occurring in thin, continuous masses.

sheet [speleo] In a cave, a thin flowstone coating of calcite.

sheet [intrus rocks] A general term for a tabular igneous intrusion, esp. those that are concordant or only slightly discordant. In this general sense, the term *dike* is used for a vertical or steeply dipping tabular body, and the term *sill* for a horizontal or gently dipping one. Cf: *intrusive vein.*

sheet [**sed**] *blanket.*

sheet [**water**] *sheetflood.*

sheet crack A planar crack attributed to shrinkage of sediment due to dewatering (Fischer, 1964, p.148). It is commonly parallel to the bedding and filled with sparry calcite or mud.

sheet deposit A mineral deposit that is generally stratiform, more or less horizontal, and areally extensive relative to its thickness.

sheet drift An evenly spread deposit of glacial drift that did not significantly alter the form of the underlying rock surface.

sheeted Said of an igneous rock such as a granite that has undergone pressure-release jointing or exfoliation, sometimes giving it the appearance of being stratified.

sheeted fissures Fissures that are closely spaced.

sheeted-zone deposit A mineral deposit consisting of veins or lodes filling a zone of shear faulting, or *shear zone.*

sheet erosion Erosion in which thin layers of surface material are gradually removed more or less evenly from an extensive area of gently sloping land by broad, continuous sheets of running water rather than by streams flowing in well-defined channels; e.g. erosion that occurs when rain washes away a thin layer of topsoil. Cf: *channel erosion; rill erosion; gully erosion.* Syn: *sheetflood erosion; sheetwash; unconcentrated wash; rainwash; slope wash; surface wash.*

sheetflood A broad expanse of moving, storm-borne water that spreads as a thin, continuous, relatively uniform film over a large area in an arid region and that is not concentrated into well defined channels; its distance of flow is short and its duration is measured in minutes or hours. Sheetfloods usually occur before runoff is sufficient to promote channel flow, or after a period of sudden and heavy rainfall. Cf: *streamflood.* See also: *sheet flow.* Also spelled: sheet flood. Syn: *sheetwash; sheet* [*water*].

sheetflood erosion A form of *sheet erosion* caused by a sheetflood. See also: *rock-floor robbing.*

sheet flow [**geomorph**] An *overland flow* or downslope movement of water taking the form of a thin, continuous film over relatively smooth soil or rock surfaces and not concentrated into channels larger than rills. Cf: *streamflow.* See also: *sheetflood.* Also spelled: *sheetflow.*

sheet flow [**hydraul**] *laminar flow.*

sheet ground A term used in the Joplin district of Missouri for extensive, disseminated, low-grade zinc-lead deposits.

sheet ice Ice formed in a smooth, relatively thin layer by the rapid freezing of the surface layer of a body of water. Not to be confused with *ice sheet.*

sheeting *exfoliation.*

sheeting plane In igneous rocks, the primary cleavage plane or parting. It may or may not be related to the flow lineation.

sheeting structure The type of fracture or jointing formed by *pressure-release jointing* or exfoliation. Syn: *sheet structure; expansion joint; exfoliation joint; release joint; pseudostratification.*

sheet jointing *exfoliation.*

sheet line A line that limits a map sheet. Cf: *neat line.*

sheet mica Mica that is relatively flat and sufficiently free from structural defects to enable it to be punched or stamped into specified shapes for use by the electronic and electrical industries.

sheet mineral *phyllosilicate.*

sheet pile A *pile* with a generally flat cross section, meshed or interlocked with adjacent like members to form a diaphragm, wall, or bulkhead, and designed to resist mainly a lateral force or to reduce seepage. Syn: *sheeting pile.*

sheet sand *blanket sand.*

sheet silicate *phyllosilicate.*

sheet spar A sheet crack filled with spar (Fischer, 1964, p.148).

sheet structure *sheeting structure.*

sheetwash (a) A *sheetflood* occurring in a humid region. (The material transported and deposited by the water of sheetwash. (c) A term used as a syn. of *sheet flow* (a mov ment) and *sheet erosion* (a process).---Also spelled: *she wash.*

shelf [**geomorph**] (a) Bedrock or other solid rock beneath luvial soil or deposits; a flat-surfaced layer or stratum. (b) flat, projecting layer or ledge of rock, as on a slope. Sy *shelve.*

shelf [**marine geol**] *continental shelf.*

shelf [**paleont**] The subhorizontal part of the whorl surfac next to a suture in a gastropod shell, bordered on the sic toward the periphery of the whorl by a sharp angulation or ▮ a carina.

shelf [**tect**] As a sedimentary-tectonic feature, the part of stable, cratonic-type area of sedimentation, where it was bc dered by a more rapidly subsiding, more mobile basin of sec mentation, generally a geosyncline. The edges of som shelves were unstable, resulting in more disturbed sediment. tion than in the rest of the craton, and in slumps and slides shelf material into the basin.

shelf atoll *pseudoatoll.*

shelf break An obvious steepening of the gradient between th continental shelf and the continental slope. Cf: *shelf edge.*

shelf channel A shallow, somewhat discontinuous valley alor the continental shelf, e.g. the extension of the New York Ci harbor in the North Atlantic.

shelf edge The demarcation between the continental shelf an the continental slope, but not so obvious a change in gradie that it can be called a *shelf break.*

shelf facies A sedimentary facies that contains sediments pro duced in the neritic environment of the shelf seas marginal a very low-lying, stable land surface. It is also known as she ly facies in recognition of the importance of its characterist carbonate rocks and fossil shells. Cf: *geosynclinal facies.* Sy *platform facies; foreland facies.*

shelf ice (a) The ice of an ice shelf; a term introduced b Nordenskjold (1909, p.322) to describe the type of floatin freshwater ice formed in, or broken away from, the featur then known as an "ice barrier" but now referred to as an ic *shelf.* Syn: *barrier ice.* (b) An improper syn. of *ice shelf.*

shelf sea A shallow *marginal sea* situated on the continenta shelf, rarely exceeding 150 fathoms (275 m) in depth, e.g. th North Sea.

shelfstone A speleothem formed at the water's edge as a ho rizontally projecting ledge.

shell [**geol**] (a) The crust of the Earth. Also, any of the contir uous and distinctive concentric zones or layers composing th interior of the Earth (beneath the crust). The term was for merly used for what is now called the "mantle". Syn: *Eart shell.* (b) A thin and generally hard layer of rock; esp. such stratum encountered in drilling a well.

shell [**paleont**] (a) The hard, rigid outer covering of an anima commonly largely calcareous but in some cases chiefly c partly chitinous, siliceous, horny, or bony; e.g. the hard part of an ammonoid (including the protoconch and the conch, bu excluding the aptychus and the beaks or jaw structures) or c a cirripede crustacean (including compartmental plates, cal careous basis, and opercular valves). (b) A shell-bearing ani mal; esp. a shell-bearing mollusk.

shell [**sed**] A sedimentary deposit consisting primarily of ani mal shells.

shell hash A sediment layer composed of coquina (Russell 1968, p.76).

shell ice Thin ice originally formed on a sheet of water and re maining as an unbroken shell after the underlying water ha been withdrawn. Syn: *cat ice.*

shell marl A light-colored calcareous deposit formed on th bottoms of small freshwater lakes, composed largely of unce mented mollusk shells and precipitated calcium carbonate ong with the hard parts of minute organisms. See also: *dun*.

ell sand A loose aggregate that is largely composed of shell agments of sand size; specif. a marine sand containing 5% ell fragments (Hatch & Rastall, 1965, p.93), formed on a each by wave action.

elly [paleont] (a) Pertaining to the shell of an animal; chitious, siliceous, or testaceous. (b) Having a shell.

elly [sed] (a) Said of a sediment or sedimentary rock conining the shells of animals; e.g. a "shelly sand" composed of high proportion of loose shell fragments, or a "shelly limeone" composed chiefly of fossil shell fragments. (b) Said of nd abounding in or covered with shells, such as a "shelly eashore". (c) Var. of *shaly*.

elly facies A nongeosynclinal sedimentary facies that is ommonly characterized by abundant calcareous fossil shells, ominant carbonate rocks (limestones and dolomites), mature thoquartzitic sandstones, and paucity of shales. The term is equently used in reference to lower Paleozoic strata, as in e upper Mississippi Valley and the Great Lakes area. The cies is also known as *shelf facies* in recognition of the premed structural stability of the site of deposition.

elly formation A thin layer of hard rock encountered in drillg a well.

elly pahoehoe A type of pahoehoe, the surface of hich contains open tubes and blisters; its crust is 1-30 cm ick. Cf: *corded pahoehoe; elephant-hide pahoehoe; entrail ahoehoe; festooned pahoehoe; filamented pahoehoe; sharkin pahoehoe; slab pahoehoe*.

helter cave A cave that extends only a short way underround, and whose roof of overlying rock usually extends beond its sides. Syn: *rock cave*.

helter porosity A type of primary interparticle porosity defined y Choquette & Pray (1970, p.249) as the porosity "created y the sheltering effect of relatively large sedimentary partiles which prevent the infilling of pore space beneath them by ner clastic particles".

hergottite An achondritic stony meteorite composed mainly f pigeonite and maskelynite.

heridanite A talc-like mineral of the chlorite group: $(Mg, Al)_6(Si,Al)_4O_{10}(OH)_8$. It is pale greenish to nearly colorless.

herry topaz A brownish-yellow to yellow-brown variety of paz resembling the color of sherry wine. It is one of the ore valuable and important varieties of topaz.

herwoodite A blue-black tetragonal mineral: $Ca_3V_8O_{2}.15H_2O$.

heugh A Scottish term for a small ravine, esp. one containing stream. Syn: *sheuch*.

hield [tect] A large area of exposed basement rocks in a craon commonly with a very gently convex surface, surrounded y sediment-covered platforms; e.g. Canadian Shield, Baltic hield. The rocks of virtually all shield areas are Precambrian. yn: *continental shield, cratogene; continental nucleus*.

hield [eng] A metal (iron or steel) or wood framework or diahragm used in tunneling and mining, moved forward at the nd of the tunnel or adit in process of excavation, and used to upport the ground ahead of the lining. It is used esp. in vater-bearing or loose material.

hield [speleo] *palette*.

hield [paleont] (a) A protective cover or structure on an aninal, likened to or resembling a shield; e.g. an ossicle of an phiuroid, or the carapace of a crustacean, or a large scale n the head of a lizard. (b) A flat or curved, lateral outgrowth t one or more levels of a tangential rod or needle in the skelton of an acantharian radiolarian and forming by fusion the attice shell. (c) One of the discoidal elements of the placolith occolith.

hield basalt A basaltic lava flow that erupted from numerous, mall, and closely spaced shield-volcano vents and that coalesced to form a single unit. It is generally of smaller extent

than a *plateau basalt*.

shielding n. A grounded metallic enclosure intended to reduce noise capacitively or inductively coupled into a circuit.

shield volcano A volcano in the shape of a flattened dome, broad and low, built by flows of very fluid, basaltic lava; it is the characteristic type of volcano in Hawaii. Syn: *lava dome; basaltic dome*.

shift The relative displacement of the units affected by a fault but outside the fault zone itself; a partial syn. of *slip* [struc geol]. See also: *strike shift; dip shift*.

shifting [streams] The movement of the crest of a divide away from a more actively eroding stream (as on the steeper slope of an asymmetric ridge) toward a weaker stream on the gentler slope; the change in position of a divide (and of the stream channel) where one stream is abstracted by another. See also: *monoclinal shifting; leaping; creeping*. Syn: *migration*.

shifting [shores] The fluctuation or oscillation of sea level; the change in position of a shoreline.

shinarump Var. of *chinarump*.

shingle [beach] (a) Coarse, loose, well-rounded, and waterworn detritus or alluvial material of various sizes; esp. beach gravel composed of smooth and spheroidal or flattened (lenticular) fragments relatively free from finer material, consisting of pebbles, cobbles, and sometimes small boulders, generally measuring 20-200 mm in diameter (i.e. larger than a walnut), and occurring typically on the higher parts of a beach. Strictly, the term refers to beach pebbles and cobble of roughly the same size; more commonly, it includes any beach material coarser than ordinary gravel. The term is more widely used in Great Britain. (b) A place strewn with shingle; e.g. a *shingle beach*.---Etymol: probably Scandinavian, akin to *singel*, "coarse gravel that sings or crunches when walked on".

shingle [tect] *schuppe*.

shingle barchan A dune-like ridge of shingle formed at right angles to the beach in shallow water between troughs of sand (Cornish, 1898, p.639).

shingle beach A narrow beach, usually the first to form on a coastline having resistant bedrock and cliffs, composed of *shingle*, and commonly having a very steep slope on both its landward and seaward sides. Syn: *cobble beach; shingle*.

shingle-block structure An obsolete syn. of *imbricate structure* [tect].

shingle rampart A *rampart* of shingle built along the seaward edge of a reef.

shingle ridge A steeply sloping bank of shingle heaped up on and parallel with the shore.

shingle structure [eco geol] The arrangement of closely spaced veins en echelon in the manner of shingles on the roof.

shingle structure [sed] *imbricate structure*.

shingling *imbrication* [sed].

shingly Composed of or containing abundant shingle; e.g. a "shingly beach".

Shipek bottom sampler The commercial name of a popular type of *grab sampler*; it is of the clamshell variety.

shipping ore *first-class ore*.

shiver An old English term for soft and crumbly shale, or slate clay approaching shale. Etymol: Middle English *scifre*. Adj: *shivery*.

shiver spar A variety of calcite with slaty structure; specif. argentine. Syn: *slate spar*.

shoal adj. Having little depth; shallow.----n. (a) A relatively shallow place in a stream, lake, sea, or other body of water; a *shallows*. (b) A submerged ridge, bank, or bar producing a shoal, consisting of or covered by sand, mud, gravel, or other unconsolidated material, and rising from the bed of a body of water to near the surface so as to constitute a danger to surface navigation; specif. an elevation, or an area of such elevations, at a depth of 10 fathoms (formerly 6) or less, composed of material other than rock or coral. It may be exposed at low water. Cf: *reef.* (c) A rocky area on the sea floor within soundings. (d) A growth of vegetation on the bottom of a deep lake, occurring at any depth.----v. To become shallow gradually; to cause to become shallow; to fill up or block off with a shoal; to proceed from a greater to a lesser depth of water.

shoal breccia A breccia formed by the action of waves and tides on a shoal, and resulting from diastrophism or aggradation (Norton, 1917, p.177).

shoaling A bottom effect that describes, in terms of wave heights, the alteration of a wave as it proceeds from deep water into shallow water, and evidenced by an initial decrease in height of the incoming wave, followed by an increase in height as the wave arrives on the shore.

shoal reef (a) Any formation in which reef growth develops in irregular patches amidst submerged shoals of calcareous reef detritus derived from a large reef (Henson, 1950, p.227); it is smaller in area than a *bank reef.* See also: *reef patch.* (b) A term sometimes used as a syn. of *bank reef.*

shock n. *earthquake.*

shock breccia A fragmental rock formed by the action of shock waves; e.g. suevite formed by meteorite impact.

shock lamellae *planar features.*

shock-lithification The conversion of originally loose fragmental materials into coherent aggregates; e.g. instant rock, by the action of shock waves, such as by those generated by explosions or meteorite impacts (Short, 1966). It apparently involves such mechanisms as fracturing, compaction, and intergranular melting.

shock loading The process of subjecting material to the action of high-pressure shock waves generated by artificial explosions or by meteorite impact.

shock melting Fusion of material as a result of the high temperatures produced by the action of high-pressure shock waves.

shock-metamorphic facies A discredited term for an association of mineralogic features (such as multiple parallel planes and thetomorphic glasses) formed by a particular degree of shock metamorphism. As it implies a near approach to equilibrium, the term is not recommended; the term "stages of shock metamorphism" is preferred.

shock metamorphism The totality of observed permanent physical, chemical, mineralogic, and morphologic changes produced in rocks and minerals by the passage of transient high-pressure shock waves acting over short time intervals ranging from a few microseconds to a fraction of a minute (French, 1966). The only known natural mechanism for producing shock-metamorphic effects is the hypervelocity impact of large meteorites, but the term also includes identical effects produced by shock waves generated in small-scale laboratory experiments and in nuclear and chemical explosions. See also: *impact metamorphism.*

shock wave (a) A compressional wave formed whenever the speed of a body relative to a medium exceeds that at which the medium can transmit sound, having an amplitude that exceeds the elastic limit of the medium in which it travels, and characterized by a disturbed region of small but finite thickness within which very abrupt changes occur in the pressure, temperature, density, and velocity of the medium; e.g. the wave sent out through the air by the discharge of the shot initiating an explosion. It travels at supersonic velocities and is capable of vaporizing, melting, mineralogically transforming, or strongly deforming rock materials. See also: *blast wave.* (b) *blast* [geophys].

shock zone A volume of rock in or around an impact or explosion crater in which a distinctive shock-metamorphic deformation or transformation effect is present.

shoe stone A sharp-grained sandstone used as a whetstone by shoemakers and other leatherworkers. Also, the whetstone

so used.

shoestring A very long, relatively narrow, and generally straight sedimentary body whose width/thickness ratio is less than 5 to 1, and is usually on the order of 1 to 1 or even smaller (Krynine, 1948, p.147); e.g. a channel fill, a bar, dune, or a beach deposit. Cf: *prism.*

shoestring rill One of several long, narrow, uniform channels closely spaced and roughly parallel with one another, that merely score the homogeneous surface of a relatively steep slope of bare soil or weak, clay-rich bedrock, and that develop wherever overland flow is intense. Syn: *rill channel; shoestring gully.*

shoestring sand A shoestring composed of sand or sandstone usually buried in the midst of mud or shale; e.g. a buried sandbar or channel fill. Syn: *shoestring sandstone.*

shonkinite A dark-colored syenite composed chiefly of augite and orthoclase, and possibly containing olivine, hornblende, biotite, and nepheline. Its name is derived from Shonkin, the Indian name for the Highwood Mountains of Montana, U.S.A., where the rock was first found (Johannsen, 1939, p.280).

shoot [ore dep] *ore shoot.* Also spelled: *chute.*

shoot [seis] In seismic prospecting, to explore an area by the *seismic shooting* technique; to set off an explosion used in seismic shooting.

shoot [streams] (a) A place where a stream flows or descends swiftly. (b) A natural or artificial channel, passage, or trough through which water is moved to a lower level. (c) A rush of water down a steep place or a rapids.---Etymol: French *chute.* See also: *chute.*

shooting [seis] *seismic shooting.*

shooting-flow (a) *jet flow.* (b) *rapid flow.*

shooting-flow cast A term proposed by Wood & Smith (1958, p.169) for one of a series of "strong parallel ridges", representing the cast of a deep groove up to 10 cm deep, 30 cm wide, and 2 m long.

shooting star A visual *meteor* (meteoroid) appearing as a thin temporary streak or trace of light in the nighttime sky. It is not a true star. Syn: *falling star.*

shor A salt lake in Turkestan. Etymol: Russian.

shoran A precise electronic measuring system for indicating distance from a mobile (airborne or shipborne) station to each of two fixed ground stations simultaneously by recording (by means of cathode-ray screens) the time required for round-trip travel of electromagnetic pulses (radar signals or high-frequency radio waves) and thereby determining the position of the mobile station. Its range is effectively limited to line-of-sight distances (about 40 nautical miles). Shoran is used in control of aerial photography, geophysical (airborne) prospecting, offshore hydrographic surveys, and geodetic surveying for measuring long distances. Cf: *loran.* Etymol: *short range navigation.*

shore [desert] *serir.*

shore [coast] (a) The narrow strip of land immediately bordering any body of water, esp. a sea or a large lake; specif. the zone over which the ground is alternately exposed and covered by tides or waves, or the zone between high water and low water. The shore is seaward of the *coast*; its upper boundary is the landward limit of effective wave action at the base of the cliffs and its seaward limit is the low-water line. Subdivided into a foreshore and a backshore. See also: *beach; strand.* (b) The term is commonly used in the sense of the *shoreline* and of the *foreshore.* (c) A nautical term for land as distinguished from the sea.

shore cliff A cliff at the edge of a body of water or extending along the shore. See also: *sea cliff.*

shore drift *littoral drift.*

shore dune A sand dune produced by wind action on beach sands along a shore.

shoreface (a) The narrow, rather steeply sloping zone seaward or lakeward from the low-water shoreline, permanently covered by water, and over which beach sands and gravels actively oscillate with changing wave conditions (Johnson, 1919, p.161). The zone lies between the seaward limit of the shore and the more nearly horizontal surface of the *offshore* zone. The term "shore face" was originally used by Barrell (1912, p.385-386), in his study of deltas, for the relatively narrow slope developed by breaking waves and separating the subaerial plain from the subaqueous one below. Not to be confused with *beach face*. See also: *inshore*. (b) A relatively steep but short concave inner portion of the continental shelf (Price, 1954, p.81).

shoreface terrace A wave-built terrace in the shoreface region, composed of gravel and coarse sand swept from the wave-cut bench into deeper water. See also: *offshore terrace*.
Shore hardness scale An empirical scale of hardness of rocks, metals, ceramics, or other materials as determined by a Shore scleroscope which utilizes the height of rebound of a small standard object (such as a diamond-tipped hammer) dropped from a fixed height onto the surface of a specimen. Named after Albert F. Shore, 20th-century U.S. manufacturer, who proposed the technique in 1906.
shore ice (a) The basic form of *fast ice*, attached to the shore and, in shallow water, also grounded. (b) Sea ice that has been driven ashore and beached by wind, waves, currents, tides, or the pressure of adjacent ice. Cf: *stranded ice*. (c) Floating sea ice adjacent to the shore; it may or may not be attached to the shore.
shore-ice belt *icefoot*.
shoreland Land along a shore, or bordering a body of water.
shore lead A *lead* between pack ice and the shore, or between pack ice and an ice front. Formerly included what is now known as a *flaw lead*.
shoreline (a) The intersection of a specified plane of water with the shore or beach; it migrates with changes of the tide or of the water level. The term is frequently used in the sense of "high-water shoreline" or the intersection of the plane of mean high water with the shore or beach, or the landward limit of the intermittently exposed shore. Syn: *waterline; shore; strandline*. (b) The general configuration or outline of the shore.---The terms *shoreline* and *coastline* are often used synonymously, but there is a tendency to regard "coastline" as a limit fixed in position for a relatively long time and "shoreline" as a limit constantly moving across the beach.
shoreline cycle The succession of progressive changes or stages through which sequential forms of coastal features normally pass during the development of a shoreline from the time when the water first assumed its level and rested against the new shore to the time when the water can do no more work (either erosion or deposition).
shoreline-development ratio A ratio indicating the degree of irregularity of a lake shoreline, given as the length of the shoreline to the circumference of a circle whose area is equal to that of the lake (Veatch & Humphrys, 1966, p.289).
shoreline of depression A *shoreline of submergence* that implies an absolute subsidence of the land.
shoreline of elevation A *shoreline of emergence* that implies an absolute rise of the land. Not to be confused with *elevated shoreline*.
shoreline of emergence A shoreline resulting from the dominant relative *emergence* of the floor of an ocean or lake; the water surface comes to rest against the partially emerged land which is marked by marine-produced forms and structures. The shoreline is straight or gently curving, with no bays or promontories; it is simpler in outline than a *shoreline of submergence*, and is bordered by shallow water. The term carries no implication as to whether it is the land or the sea that has moved (Johnson, 1919, p.173). See also: *shoreline of elevation*. Syn: *emerged shoreline; negative shoreline*.
shoreline of submergence A shoreline resulting from the dominant relative *submergence* of a landmass; the water surface

comes to rest against the partially submerged land which is marked by subaerially produced forms and structures. The shoreline is characterized (in its youthful stage) by bays, promontories, offshore islands, spits, bars, cliffs, and other minor features; it is more irregular in outline than a *shoreline of emergence*, and is bordered by water of variable (and locally considerable) depth. The term carries no implication as to whether it is the land or the sea that has moved (Johnson, 1919, p.173). See also: *shoreline of depression*. Syn: *submerged shoreline; positive shoreline*.
shore platform A descriptive term for the horizontal or gently sloping surface produced along a shore by wave erosion; specif. a *wave-cut bench*. Also, sometimes used as a purely descriptive term for *wave-cut platform*. Syn: *scar*.
shore polynya A *polynya* between pack ice and the coast, or between pack ice and an ice front, and formed by a current or by wind.
shore reef *fringing reef*.
shoreside n. The margin of a shore.---adj. *onshore*.
shore terrace (a) A terrace produced by the action of waves and currents along the shore of a lake or sea; e.g. a *wave-built terrace*. (b) *marine terrace*.
shoreward Directed or moving toward the shore.
shorl *schorl*.
short *tender*.
short eruption rate *age-specific eruption rate*.
short-flame coal Coal that is low in volatiles. Syn: *lean coal; dry coal*. Cf: *long-flame coal*.
shortite A mineral: $Na_2Ca_2(CO_3)_3$.
short period A period of seismic activity that is less than six seconds in duration. Cf: *long period*.
short-range order A weak tendency for the random atoms in a random solid solution to order as the solution cools from the elevated temperature at which it was formed. Cf: *long-range order*.
short shot *weathering shot*.
short wave *deep-water wave*.
shoshonite A basaltic rock composed of olivine and augite phenocrysts in a groundmass of labradorite with orthoclase rims, olivine, augite, a small amount of leucite, and some dark-colored glass. Shoshonite grades into *absarokite* with an increase in olivine and into *banakite* with more sanidine. Its name is derived from the Shoshone River, Wyoming.
shot The explosive charge used in the shooting technique of seismic prospecting.
shot break In seismic prospecting, the electrical impulse which records the instant of explosion.
shot copper Small, rounded particles of native copper, molded by the shape of vesicles in basaltic host rock, and resembling shot in size and shape.
shotcrete *gunite*.
shot datum In seismic work, any convenient reference surface or plane to which calculations are referred, simulating a condition in which the charge is shot and its energy is recorded on the same reference surface. It is used in the preparation of time-depth charts.
shot depth In seismic work, the distance down the hole from the surface to the explosive charge. In the case of a small charge, it is measured to the middle or bottom of the charge, but in the case of a large charge, it is measured to both its top and its bottom. Syn: *effective shot depth; shot elevation*.
shot drill A core drill employed in rotary drilling in hard rock or very hard ground, using chilled iron or steel shot as the abrasive cutting medium. Not to be confused with *shothole drill*.
shot elevation The elevation of the explosive charge in the shot hole. The term is not to be confused with *shothole elevation*.
shothole In seismic prospecting, the borehole in which an explosive is placed for blasting in the seismic shooting technique.

shothole bridge An obstruction in a shot hole that prevents the explosive charge from going any deeper. It may be accidental or intentional.

shothole drill A drill (esp. a rotary drill) used for making a shothole. Not to be confused with *shot drill*.

shothole elevation The elevation of the top of the shot hole. The term is not to be confused with *shot elevation*.

shothole fatigue *hole fatigue*.

shot instant *time break*.

shot moment *time break*.

shotpoint (a) In seismic shooting, that point at which the explosive charge is fired. (b) The location of other sources of seismic energy. (c) The area surrounding a shothole or shotholes.

shot soil A warm-latitude soil containing *shot*; it is usually poorly drained and poorly aerated.

shott (a) A shallow and brackish or saline lake or marsh in southern Tunisia or on the plateaus of northern Algeria, usually dry during the summer; a playa lake, often many tens of kilometers in diameter. (b) A closed basin containing a shott; esp. the dried bed existing after the water has disappeared, characterized by salt deposits and frequently by absence of vegetation.---Etymol: Arabic *shatt*. See also: *sebkha*. Syn: *chott; schott*.

shoulder [glac geol] A bench on the side of a glaciated valley, occurring at the marked change of slope where the steep side of the inner, glaciated valley meets the much gentler slope above the level of glaciation. Cf: *alp*.

shoulder [geomorph] (a) A short, rounded spur projecting laterally from the side of a mountain or hill. (b) The sloping part of a mountain or hill below the summit. (c) *valley shoulder*.

shoulder [paleont] (a) The salient angulation of a gastropod-shell whorl parallel to the coiling and forming the abaxial edge of a subsutural ramp. (b) The ventral and lateral blunt angle of a whorl of an ammonoid shell (TIP, 1959, pt.L, p.5).---See also: *umbilical shoulder*.

shoulder [struc geol] A joint structure formed on the face of the joint by the intersection of plume-structure ridges with fringe joints.

shoved moraine *push moraine*.

show (a) A trace of oil or gas detected in a core, cuttings, or circulated drilling fluid, or interpreted from the electrical or geophysical logs run in a well. Partial syn: *oil show*. (b) A small particle of gold found in panning a gravel deposit.

shrinkage The decrease in volume of soil, sediment, fill, or excavated earth due to the reduction of voids by mechanical compaction, superimposed loads, natural consolidation, or drying.

shrinkage crack A small crack produced in fine-grained sediment or rock by the loss of contained water during drying or dehydration; e.g. a *desiccation crack* and a *syneresis crack*.

shrinkage index The numerical difference between the plastic limit of a material and its shrinkage limit.

shrinkage limit That moisture content of a soil below which a decrease in moisture content will not cause a decrease in volume, but above which an increase in moisture will cause an increase in volume.

shrinkage polygon *desiccation polygon*.

shrinkage pore A term used by Fischer (1964, p.116) for an irregular pore formed in muddy sediment by shrinkage (desiccation). It may become a bird's-eye (in a limestone) when filled with sparry calcite. Syn: *fenestra*.

shrinkage ratio The ratio of a volume change to the moisture-content change above the shrinkage limit.

shrub-coppice dune A small, streamlined dune that forms to the lee of bush-and-clump vegetation on a smooth surface of very shallow sand.

shuga An accumulation of spongy lumps of white sea ice, measuring a few centimeters across, and formed from grease ice or sludge, and sometimes from anchor ice rising to the surface (U.S. Naval Oceanographic Office, 1968, p.B36).

shungite A hard, black, amorphous, coal-like material containing over 98% carbon, found interbedded among Precambrian schists. It is probably the metamorphic equivalent of bitumen, but it may represent merely impure graphite. Syn: *schungite*.

shut-in A narrow, steep-sided gorge along the course of an otherwise wide and shallow stream valley.

shutterridge A ridge formed by vertical, lateral, or oblique displacement on a fault traversing a ridge-and-valley topography, with the displaced part of a ridge "shutting in" the adjacent ravine or canyon (Buwalda, 1937).

si In structural petrology, a fabric of crystal inclusions that have grown during metamorphism and that have an orderly internal structure. The "i" is actually a subscript referring to "internal". Cf: *se*.

sial A petrologic name for the upper layer of the Earth's crust, composed of rocks that are rich in silica and alumina; it is the source of granitic magma. It is characteristic of the *continental crust*. Etymol: an acronym for *si*lica + *al*umina. Adj: *sialic*. Cf: *sialma*. Syn: *sal; granitic layer*.

sialic Adj. of *sial*.

sialite *clay mineral*.

siallite (a) A group name for the kaolin clay minerals and allophane. (b) A rock composed of siallite minerals.

siallitic An old term used to describe weathered rock material consisting mainly of alumino-silicate clay minerals and highly leached of alkalies and alkaline earths (SSSA, 1970, p.14).

sialma A layer of the Earth's crust that is intermediate in both depth and composition between the *sial* and the *sima*. Etymol: an acronym for *si*lica + *al*umina + *ma*gnesia.

sialsima Var. of *salsima*.

siberite A violet-red or purplish, lithian variety of tourmaline; rubellite from Siberia.

sibirskite A mineral: $CaHBO_3$.

sibling species One of two or more species that are closely related, very similar morphologically, but reproductively isolated.

sicklerite A dark-brown mineral: $Li(Mn^{+2},Fe^{+3})PO_4$. It is isomorphous with ferrisicklerite.

sickle trough A flat-bottomed, crescent-shaped rock basin sculptured by a glacier. Syn: *skärtråg*.

sicula The chitinous skeleton secreted by the initial zooid of a graptolite colony, divisible into a conical *prosicula* and a distal *metasicula* that has growth increments commonly present throughout the remainder of the rhabdosome. Pl: *siculae*.

side (a) A slope of a mountain, hill, or bank; e.g. hill*side*. (b) A bank, shore, or other land bordering a body of water; e.g. sea*side*. (c) A geographic region; e.g. country*side*.

side canyon A ravine or other valley smaller than a canyon through which a tributary flows into the main stream.

side-centered lattice A type of *centered lattice* that is centered on the side faces only. It is found in orthorhombic crystals.

side lap The lateral *overlap* between aerial photographs of areas in adjacent parallel flight lines.

sideling A dialectal term for a slope.

side-looking airborne radar An airborne radar system in which a long, narrow, stabilized antenna, aligned parallel to the motion of an aircraft or satellite, projects radiation at right angles to the flight path. It enables extremely fine-resolution photography and mapping of the ground surface. Abbrev: SLAR. Syn: *side-looking radar*.

side moraine *lateral moraine*.

side pinacoid In an orthorhombic, monoclinic, or triclinic crystal, the {010} pinacoid. Cf: *front pinacoid; basal pinacoid*.

side plate One of numerous small plates of ambulacrum of an echinoderm, such as the broader of the floor plates in a cystoid; esp. a subquadrangular plate arranged along ambulacral borders of a blastoid and covering space between a lancet ate and adjacent deltoid and radial plates. See also: *ambu-cral.*

deraerolite *stony-iron meteorite.*

derazot A mineral: Fe_5N_2. Also spelled: *siderazote.*

dereal Pertaining to the stars.

derite [meteorite] A general name for *iron meteorites*, com-sed almost wholly of iron alloyed with nickel. Syn: *aerosid-ite.*

derite [mineral] (a) A rhombohedral mineral: $FeCO_3$. It is omorphous with magnesite and rhodochrosite, and common- contains magnesium and manganese. Siderite is usually llowish-brown, brownish-red, or brownish-black, but is metimes white or gray; it is often found in impure form in eds and nodules (of clay ironstone) in clays and shales and s a directly precipitated deposit partly altered into iron ox-es. Siderite is a valuable ore of iron. Syn: *chalybite; spathic on; sparry iron; rhombohedral iron ore; iron spar; siderose; hite iron ore.* (b) An obsolete syn. of *sapphire quartz.* (c) An osolete term formerly applied to various minerals, such as rnblende, pharmacosiderite, and lazulite.

derodot A variety of siderite (ferrous-carbonate mineral) ntaining calcium.

deroferrite A variety of native iron occurring as grains in pe-fied wood.

derogel A mineral consisting of truly amorphous $FeO(OH)$ nd occurring in some bog iron ores.

derolite *stony-iron meteorite.*

deronatrite An orange to straw-yellow mineral: Na_2 $e^{+3}(SO_4)_2(OH.3H_2O$.

deronitic texture In mineral deposits, a mesh of silicate min-als so shattered and pressed as to force out solutions and her volatiles.

derophile (a) Said of an element enriched in the metallic ther than in the silicate and sulfide phases of meteorites, nd that is probably enriched in the Earth's core relative to its antle and crust (in Goldschmidt's scheme of element parti-n in the solid Earth). Cf: *chalcophile; lithophile.* (b) Said of n element with a weak affinity for oxygen and sulfur and that readily soluble in molten iron.----(Rankama & Sahama,)50, p.88). Examples are: Fe, Ni, Co, P, Pt, Au.

derophile element An element that has a relatively weak af-nity for oxygen and sulfur and that is readily soluble in mol-n iron. It is concentrated in iron meteorites and presumably the Earth's inner core. In Goldschmidt's classification of imordial differentiation, siderophile elements include iron, ckel, cobalt, platinum metals, gold, tin, and tantalum.

derophyllite An iron-rich variety of biotite.

derophyre A stony-iron meteorite containing crystals of onzite and tridymite in a network of nickel-iron. Syn: *sidero-hyry.*

derose adj. Containing or resembling iron. The term was oposed to replace *"ferruginous"* when designating a form of on other than iron oxide; e.g. "siderose cement" consisting iron carbonate in a sandstone.----n. A syn. of *siderite* (fer-us-carbonate mineral). Also spelled: *sidérose.*

derosphere A term used for the iron *inner core* of the Earth.

derotil A mineral: $(Fe,Cu)SO_4.5H_2O$.

de shot A reading or measurement from a survey station to cate a point that is off the traverse or that is not intended to used as a base for the extension of the survey. It is usually ade to determine the position of some object that is to be own on a map.

de stream A *tributary* that receives its water from a drainage ea separate from that of the main stream into which it ows.

deswipe (a) A phenomenon of two cross reflections coming om a single seismograph, due to the almost simultaneous rival of reflection energy from both limbs of a syncline or om two nearby, steeply dipping fault scarps. (b) In refrac-n shooting, the lateral deflection of a minimum time path to

include a nearby, steeply dipping, high-velocity boundary such as a flank of a salt dome.

sidetracking The deliberate act or process of deflecting and drilling a borehole away from normal, straight course; e.g. drilling to the side of and beyond a piece of drilling equipment that is permanently lost in the hole, or deflecting a borehole so as to bypass obstructions in the hole, or drilling of another oil well beside a nonproducing well but making use of the upper part of the nonproducing well.

sidewall core A finger-sized cylinder of rock (2 cm in diame-ter and up to 5 cm in length) extracted by *sidewall sampling* from the walls of an uncased drill hole.

sidewall sampling The process of obtaining *sidewall cores*, usually by percussion (shooting hollow and retractable cylin-drical bullets into the walls). Syn: *sidewall coring.*

Siegenian European stage: Lower Devonian (above Gedinnian, below Emsian).

siegenite A mineral of the linnaeite group: $(Co,Ni)_3S_4$. It may contain copper or iron or both in appreciable amounts.

sienna Any of various brownish-yellow earthy substances that occur in limonite and that are used as pigments for oil stains and paints. It becomes orange red to reddish brown when burnt, and is generally darker and more transparent in oils than *ochers*. Named after Siena, a commune in Tuscany, Italy. Cf: *umber.*

Sierozem In early U.S. classification systems, a group of zonal soils having a brownish-gray surface horizon and a light-colored subsurface overlying a layer of carbonate accumula-tion and, sometimes, hardpan. It is developed under condi-tions of temperate to cool aridity, and under mixed shrub veg-etation. Etymol: Russian *serozem*, "gray earth". Also spelled: *Serozem; cerozem.* Syn: *Gray Desert soil; gray earth.*

sierra (a) A high range of hills or mountains, esp. one having jagged or irregular peaks that when projected against the sky resemble the teeth of a saw; e.g. the Sierra Nevada in Califor-nia. The term is often used in the plural, and is common in the SW U.S. and in Latin America. Syn: *serra.* (b) A mountainous region in a sierra.---Etymol: Spanish, "saw", from Latin *serra*, "saw".

sierranite A rock, consisting of onyx and chert, found in the Sierra Nevada of California.

sieve An apparatus used to separate soil or sedimentary ma-terial according to the size of its particles; it is usually made of brass with a wire-mesh cloth (having regularly spaced square holes of uniform diameters) spread across the base. Cf: *screen.*

sieve analysis Determination of the particle-size distribution in a soil, sediment, or rock by measuring the percentage of the particles that will pass through standard sieves of various sizes.

sieve deposition A term proposed by Hooke (1967, p.454) for the formation of coarse-grained lobate masses on an alluvial fan whose material is sufficiently coarse and permeable to permit complete infiltration of water before it reaches the toe of the fan.

sieve diameter The size (diameter) of a sieve opening (mesh) through which a given particle will just pass.

sieve lobe A coarse-grained lobate mass produced by *sieve deposition* on an alluvial fan.

sieve membrane A sievelike, partly closing membrane in the areolae of a locular-walled diatom. It may occur at an outer or inner position of the wall.

sieve plate [paleont] (a) A perforated diaphragm extending across the oscular end of the cloaca of a sponge. (b) A unilaminar and circular, subcircular, or polygonal perforate plate of a holothurian (TIP, 1966, pt.U, p.653). (c) A minute discoidal plate with numerous circular, triangular, and polygo-nal micropores arranged in concentric rows, contained in a pore canal of certain foraminifers. Also, a *trematophore.* (d) A flat, circular, porous plate in spumellarian radiolarians.

657

sieve plate [bot] In a plant, the perforated wall or wall portion between two sieve elements, e.g. in a sieve tube.

sieve texture A syn. of *poikiloblastic* texture.

sieve tissue *phloem.*

sieve tube A phloem tube formed from several sieve elements set end to end (Cronquist, 1961, p.881).

sieving The operation of shaking loose materials in a sieve so that the finer particles pass through the mesh bottom. It is the most common method of measuring particle sizes of sediments, esp. in the range 1/16 mm (very fine sand) to about 30 mm (coarse pebbles).

sif A syn. of *seit dune.* Pl: *siuf.*

sifema According to Bemmelen (1949), the theoretical ultrabasic layer underlying the sima; it is the equivalent of the sima of some authors, and of the ultrasima of others. Cf: *salsima.*

siferna A term that has been used for the sima, in a scheme in which the ultrasima is referred to as the sima (Schieferdecker, 1959, term 4547).

sight (a) An observation (such as of the altitude of a celestial body) taken for determining direction or position. Also, the data obtained by such an observation; e.g. a bearing or angle taken with a compass or transit when making a survey. (b) A device with a small aperture through which objects are seen and by which their directions are determined; e.g. an "open sight" of an alidade.

sight rod *range rod.*

sigillarid n. An arborescent club moss of the genus *Sigillaria* that occurs in Carboniferous deposits.----adj. Pertaining to *Sigillaria.*----Cf: *lepidodendrid.*

sigloite A triclinic mineral: $(Fe^{+3}, Fe^{+2})Al_2(PO_4)_2(O, OH) \cdot 8H_2O.$

sigma [paleont] A C-shaped, siliceous, monaxonic sponge spicule (microsclere). Cf: *sigmaspire.* Pl: *sigmata* or *sigmas.*

sigma [math] (a) *standard deviation.* (b) The sum of variables in a series. Symbol: $\Sigma.$

sigma phi The verbalized expression for $\sigma\varphi$ or the standard deviation (sorting) of a particle-size distribution computed in terms of phi units of the sample. It is a measure of *degree of sorting.*

sigmaspire An S-shaped, siliceous, smooth or spinose, monaxonic sponge spicule (microsclere); a *sigma* twisted in the form of a spiral of about one revolution. Cf: *spinispire.*

sigma-t *density* [oceanog].

sigmoidal dune An S-shaped, steep-sided, sharp-crested sand dune formed under the influence of alternating and opposing winds of roughly equal velocities (D. Holm, 1957); a transitional dune between a crescentic form and some of the dune complexes, being up to 50 m high, 1-2 km long, and 50-200 m wide.

sigmoidal fold A recumbent fold, the axial surface of which is so curved as to resemble the Greek letter sigma.

sigmoid distortion A distortion present in line-scan imagery such that straight lines cut obliquely appear as sigmoid curves; a type of *barrel distortion.*

signal [geophys] In geophysics, a desired physical impulse, e.g. a seismic signal. Cf: *noise.*

signal [surv] A natural or artificial object or structure located at or near a survey station and used as a sighting point or target for survey measurements; e.g. a flag on a pole, or a rigid structure erected over or close to a triangulation station.

signal correction In seismic analysis, a correction to eliminate the time differences between reflection times, resulting from changes in the outgoing signal from shot to shot.

signal effect In seismology, variation in arrival times of reflections recorded with identical filter settings, as a result of changes in the outgoing signal.

signal-to-noise ratio The comparison between the amplitude of the seismic signal (or pulse) and the amplitude of noise caused by seismic unrest and/or the seismic instruments. Abbrev: S/N.

signature (a) The trace recorded on a seismograph by th waves from an earthquake or other source of mechanic energy. (b) A graph of pressure versus time for points passe over by a wave.

significance level The probability that a stated statistical h pothesis will be rejected when in fact it is true.

significant wave A statistical term for a fictitious wave whos height and period are equal to the average height and averag period of the highest one-third of the actual waves that pass given point.

sign of elongation In hexagonal and tetragonal crystals of pri matic habit, the sign of the long crystallographic direction; negative sign of elongation indicates that the trace of the v bration plane of the fast component is parallel to the lor axis; a positive sign of elongation indicates that it is the slo component that is parallel to that axis. A crystal with a neg tive sign is said to be "length fast", and a crystal with a pos tive sign is termed "length slow". See also: *negative elong tion.*

sike (a) A British term for a small stream, esp. one that flow through flat or marshy ground and that is often dry in sun mer. (b) A British term for a gully, trench, drain, or hollow.- Syn: *syke.*

sikussak Very old, rough-surfaced sea ice trapped in a fjor as along the north coast of Greenland; it resembles glacier ic because snow accumulation contributes to its formation an perpetuation. According to Koch (1926, p.100), to be calle sikussak "the ice must be at least 25 years old". Etymo Eskimo, "very old ice". Cf: *fjord ice.*

sil *yellow ocher.*

silan A *cutan* consisting of silica in its various forms, esp. si or clay-size quartz and poorly crystalline chalcedony (Brewe 1964, p.216).

silcrete (a) A term suggested by Lamplugh (1902) for a co glomerate consisting of surficial sand and gravel cemente into a hard mass by silica. Examples occur in post-Creta ceous strata of the U.S (b) A siliceous *duricrust.* Syn: *billy.* Etymol: *sil*iceous + con*crete.* Cf: *calcrete; ferricrete.*

silex (a) The French term for *flint.* (b) Silica; esp. quart such as a pure or finely ground form for use as a filler. (c) A old term formerly applied to a hard, dense rock, such as ba salt or compact limestone.---Etymol: Latin, "hard stone, flir quartz". The term was used by Pliny for quartz.

silexite [ign] An igneous rock composed essentially of primar quartz (60-100 percent). The term was first used by Mille (1919, p.30) to include a quartz dike, segregation mass, c inclusion inside or outside its parent rock. Syn: *igneou quartz; peracidite; quartzfels.* Cf: *tarantulite.*

silexite [sed] The French term for *chert;* specif. chert occu ring in calcareous beds (Cayeux, 1929, p.554).

silica The chemically resistant dioxide of silicon: $SiO_2.$ It oc curs naturally in five crystalline polymorphs (the mineral quartz, tridymite, cristobalite, coesite, and stishovite), in cryp tocrystalline form (as chalcedony), in amorphous and hydrat ed forms (as opal), in less pure forms (such as sand, dia tomite, tripoli, chert, and flint), and combined in silicates a an essential constituent of many minerals.

silica coefficient In Ossan's chemical classification of igneou rocks, the number that expresses the ratio of the total silica a rock to the silica in the feldspars and metasilicates (Jo hannsen, 1939, p.61-82). Abbrev: K.

silica glass A glass or supercooled liquid consisting of pure c nearly pure silica, such as naturally occurring *lechatelierit* and artificially prepared *vitreous silica.* The term has been ap plied to impactites and to tektites.

silicalemma The three-layered organic membrane of a diatoi cell in which silica is deposited and which probably forms th basis of the organic skin of the mature diatom wall.

silicalite A term used by Wadsworth (1893, p.92) for any roc

...omposed of silica, such as quartz, jasper, or diatomaceous ...arth.

silica rock An industrial term for a rock that has a very high percentage (as much as 90%) of silicon dioxide, or SiO_2. It is used as a raw material. Cf: *silica sand.*

silica sand An industrial term for a sand that has a very high percentage of silicon dioxide, or SiO_2. It is a source of silicon and also has other industrial uses. Cf: *silica rock.*

silicastone A term suggested by Shrock (1948a, p.125) for any sedimentary rock composed of siliceous minerals.

silicate A compound whose crystal lattice contains SiO_4 tetrahedra, either isolated or joined through one or more of the oxygen atoms to form groups, chains, sheets, or three-dimensional structures with metallic elements. Silicates were once classified according to hypothetical oxyacids of silicon (see *metasilicate* and *orthosilicate*) but are now classified according to crystal structure (see *nesosilicate, sorosilicate, cyclosilicate, inosilicate, phyllosilicate, tectosilicate*).

silicated Said of a rock in which has occurred the process of *silication.*

silicate-facies iron formation An iron formation in which the principal iron minerals are silicates, such as greenalite, stilpnomelane, minnesotaite, and iron-rich chlorite (James, 1954, p.263-272).

silication The process of converting into or replacing by silicates, esp. in the formation of skarn minerals in carbonate rocks. Cf: *silicification.* Adj: *silicated.*

siliceous [petrology] Said of a rock containing abundant silica, esp. free silica rather than as silicates.

siliceous [ecol] *silicicolous.*

siliceous cyst A resting stage common in various yellow-green algae that is endogenous, flasklike or bottle-shaped, and from six to ten microns or rarely as large as 20 microns in size. It is composed of cellulose or pectin that is highly impregnated with silica, and it is closed by an organic plug.

siliceous earth A loose, friable, soft, porous, homogeneous, lightweight, very fine-grained, usually white sediment consisting chiefly of siliceous (opaline) material, having a dry earthy feel and appearance, and derived usually from the remains of organisms; e.g. *diatomaceous earth* and radiolarian earth. Cf: *tripoli.*

siliceous fireclay A fireclay composed mainly of fine white clay mixed with clean sharp sand.

siliceous limestone (a) A dense, dark, commonly thin-bedded limestone representing an intimate admixture of calcium carbonate and chemically precipitated silica that are believed to have accumulated simultaneously. It is found in geosynclinal associations. (b) A silicified limestone, bearing evidence of replacement of calcite by silica.

siliceous ooze Any *ooze* whose skeletal remains are siliceous, e.g. radiolarian ooze. Cf: *calcareous ooze.*

siliceous residue An *insoluble residue* chiefly composed of siliceous material, such as quartz or chert.

siliceous sandstone A sandstone cemented with quartz or cryptocrystalline silica; e.g. an orthoquartzite.

siliceous sediment A sediment composed of siliceous materials that may be fragmental, concretionary, or precipitated, and of either organic or inorganic origin; e.g. chert, novaculite, geyserite, and diatomaceous earth. Siliceous sediments may be formed by primary (chemical) deposition of silica or by secondary silicification and replacement.

siliceous shale A hard, fine-grained rock of shaly texture with an exceptional amount of silica (as much as 85%); it may have formed by silicification of normal shale (as by precipitation of silica derived from opal or devitrified volcanic ash) or by accumulation of organic material (such as diatom and radiolarian tests) at the same time as clay is deposited. Tarr (1938, p.20) prefers to describe such rock as *porcellanite* because it is not truly a shale.

siliceous sinter The white or nearly white, lightweight, porous, opaline variety of silica deposited as an incrustation by precipitation from the hot mineral waters of geysers and hot springs. The term has been applied loosely to any deposit made by a geyser or hot spring. Syn: *sinter; pearl sinter; geyserite; fiorite.*

siliceous sponge Any sponge having a skeleton composed of siliceous spicules.

silicic Said of a silica-rich igneous rock or magma. Although there is no firm agreement among petrologists, the amount of silica is usually said to constitute at least 65 percent or two-thirds of the rock. In addition to the combined silica in feldspars, silicic rocks generally contain free silica in the form of quartz. Granite and rhyolite are typical silicic rocks. The synonymous terms *acid* and *acidic* are used almost as frequently as "silicic". Syn: *oversaturated; persilicic.* Cf: *basic [geol]; intermediate; ultrabasic.*

siliciclastic Pertaining to clastic noncarbonate rocks "which are almost exclusively silicon-bearing, either as forms of quartz or as silicates" (Braunstein, 1961).

silicicolous Said of an organism living in siliceous soil. Syn: *siliceous [ecol].*

silicification [paleont] A process of fossilization whereby the original organic components of an organism are replaced by silica, either as quartz, chalcedony, or opal.

silicification [meta] The introduction of, or replacement by, silica, generally resulting in the formation of fine-grained quartz, chalcedony, or opal, which may both fill pores and replace existing minerals. The term covers all varieties of such processes as late diagenetic. Cf: *silication.* Adj: *silicified.* Syn: *silification.*

silicified Adj. of *silicification* [meta].

silicified wood A material formed by replacement of wood by silica in such a manner that the original form and structure of the wood is preserved. The silica is generally in the form of opal or chalcedony. Syn: *petrified wood; woodstone; agatized wood; opalized wood.*

silicilith (a) A term suggested by Grabau (1924, p.298) for a quartz (sedimentary) rock. Syn: *silicilyte.* (b) A sedimentary rock composed principally of the siliceous remains of organisms (Pettijohn, 1957, p.429); e.g. a diatomite.

silicinate An adjective restricted by Allen (1936, p.23) to designate the silica cement of a sedimentary rock.

siliciophite A mineral consisting of serpentine penetrated by opal.

silicle The short fruit of certain members of the crucifer family of herbs.

silicoflagellate Any chrysomonad protozoan belonging to the family Silicoflagellidae and characterized by a skeleton composed of siliceous rings and spines.

silicomagnesiofluorite A mineral: $Ca_4Mg_3Si_2O_5(OH)_2F_{10}$.

silicon detector A *semiconductor radiation detector* containing silicon rather than germanium.

silicon-oxygen tetrahedron A complex ion formed by four oxygen ions surrounding a silicon ion; with a negative charge of 4 units; the basic unit of the silicates. It is commonly written as SiO_4.

silicotelic *telechemic.*

silification *silicification.*

silique A simple, dry, dehiscent fruit, developed from two fused carpels which separate at maturity, leaving a persistent partition between (Fuller & Tippo, 1949, p.971).

silk Microscopically small, needle-like inclusions in a natural gem (such as ruby, sapphire, or garnet), from which subsurface reflections produce a whitish sheen resembling that of silk fabric. The inclusions are minute crystals of rutile.

silky A type of mineral luster characteristic of certain fibrous minerals, e.g. chrysotile.

sill [marine geol] (a) A submarine ridge or rise at a relatively shallow depth, separating a partly closed basin from another or from an adjacent sea; e.g. in the Straits of Gibraltar. (b) A ridge of bedrock or earth material at a shallow depth near the mouth of a fjord, separating the deep water of the fjord from the deep ocean water outside. Syn: *threshold.*

sill [intrus rocks] A tabular igneous intrusion that parallels the planar structure of the surrounding rock. Cf: *dike [intrus rocks]; sheet [intrus rocks].*

sillar (a) The deposit from an ash cloud or núee ardente that became indurated by recrystallization due to escaping gases rather than by welding, as is the case with *welded tuff*; it is a type of *ignimbrite.* (b) A nonwelded tuff.

silled basin *restricted basin.*

sillenite A cubic mineral: Bi_2O_3. It occurs in greenish and earthy or waxy masses, and is polymorphous with bismite. Also spelled: *sillénite.*

sillimanite (a) A brown, grayish, pale-green, or white orthorhombic mineral: Al_2SiO_5. It is trimorphous with kyanite and andalusite. Sillimanite occurs in long, slender, needle-like crystals often found in wisp-like or fibrous aggregates in schists and gneisses; it forms at the highest temperatures and pressures of a regionally metamorphosed sequence and is characteristic of the innermost zone of contact-metamorphosed sediments. Syn: *fibrolite.* (b) A group of aluminum-silicate minerals comprising sillimanite, kyanite, andalusite, dumortierite, topaz, and mullite.

silt [eng] Nonplastic or slightly plastic material exhibiting little or no strength when air-dried, consisting mainly of particles having diameters less than 0.074 mm (passing U.S. standard sieve no.200). Cf: *clay.*

silt [sed] (a) A rock fragment or detrital particle smaller than a very fine sand grain and larger than coarse clay, having a diameter in the range of 1/256 to 1/16 mm (4-62 microns, or 0.00016-0.0025 in., or 8 to 4 phi units; the upper size limit is approximately the smallest size that can be distinguished with the unaided eye), being somewhat rounded by abrasion in the course of transport. In Great Britain, the range of 0.01-0.1 mm has been used. See also: *coarse silt; medium silt; fine silt; very fine silt.* (b) A loose aggregate of unlithified mineral or rock particles of silt size; an unconsolidated or moderately consolidated sedimentary deposit consisting essentially of fine-grained clastics. It varies considerably in composition but commonly has a high content of clay minerals. The term is sometimes applied loosely to a sediment containing considerable sand- and clay-sized particles, and incorrectly to any clastic sediment (such as the muddy sediment carried or laid down by streams or by ocean currents in bays and harbors). (c) Sedimentary material (esp. of silt-sized particles) suspended in running or standing water; mud or fine earth in suspension.

silt [soil] (a) A soil-texture term used in the U.S. for a rock or mineral particle in the soil, having a diameter in the range of 0.002-0.05 mm; prior to 1937, the range was 0.005-0.05 mm. The diameter range recognized by the International Society of Soil Science is 0.002-0.02 mm. (b) *silt soil.*

siltage A mass of silt.

siltation *silting.*

silting The deposition or accumulation of silt that is suspended throughout a body of standing water or in some considerable portion of it; esp. the choking, filling, or covering with stream-deposited silt behind a dam or other place of retarded flow, or in a reservoir. The term often includes sedimentary particles ranging in size from colloidal clay to sand. Syn: *siltation.*

silting up The filling, or partial filling, with silt, as of a reservoir that receives fine-grained sediment brought in by streams and surface runoff. The term is often used synonymously with *sedimentation* without regard to any specific grain size.

siltite A term used by Kay (1951) for a *siltstone.*

silt load A *suspended load* consisting essentially of silt.

silt loam A soil containing 50-88% silt, 0-27% clay, and 0-50% sand; e.g. a soil containing at least 50% silt and 12-27% clay, or one containing 50-88% silt and less than 12%

clay (SSSA, 1965, p.347). Examples: the Newport silt loam in southern New England, and the Lebanon silt loam in Missouri.

silt shale A consolidated sediment consisting of no more than 10% sand and having a silt/clay ratio greater than 2:1 (Folk, 1954, p.350); a fissile siltstone.

silt size A term used in sedimentology for a volume greater than that of a sphere with a diameter of 1/256 mm (0.00016 in.) and less than that of a sphere with a diameter of 1/16 mm (0.0025 in.). See also: *dust size.*

silt soil A soil containing 80% or more of silt, and not more than 12% of clay and 20% of sand. Syn: *silt.*

siltstone An indurated or somewhat indurated silt having the texture and composition, but lacking the fine lamination or fissility, of shale; a massive *mudstone* in which the silt predominates over clay; a nonfissile silt shale. Pettijohn (1957, p.377) regards siltstone as a rock whose composition is intermediate between those of sandstone and shale and of which at least two-thirds is material of silt size; it tends to be flaggy, containing hard, durable, generally thin layers, and often showing various primary current structures. Syn: *siltite.*

silttil A friable, brownish to buff, open-textured silt containing a few small siliceous pebbles, representing a chemically decomposed and eluviated till that may originally have been clayey, and developed in an undulatory, well drained area, as the drift sheets in Illinois (Leighton & MacClintock, 1930, p.41). Pronounced as if spelled "silt-till". Cf: *mesotil; gumbotil.*

silty breccia A term used by Woodford (1925, p.183) for a breccia containing at least 80% rubble and 10% silt, and no more than 10% of other material.

silty clay (a) An unconsolidated sediment containing 40-75% clay, 12.5-50% silt, and 0-20% sand (Shepard, 1954). (b) An unconsolidated sediment containing more particles of clay size than of silt size, more than 10% silt, and less than 10% of all other coarser sizes (Wentworth, 1922). (c) A soil containing 40-60% clay, 40-60% silt, and 0-20% sand (SSSA, 1965, p.347).

silty clay loam A *clay loam* containing 27-40% clay, 60-73% silt, and less than 20% sand.

silty sand (a) An unconsolidated sediment containing 50-90% sand and having a ratio of silt to clay greater than 2:1 (Folk, 1954, p.349). (b) An unconsolidated sediment containing 40-75% sand, 12.5-50% silt, and 0-20% clay (Shepard, 1954). (c) An unconsolidated sediment containing more particles of sand size than of silt size, more than 10% silt, and less than 10% of all other sizes (Wentworth, 1922).

silty sandstone (a) A consolidated *silty sand.* (b) A sandstone containing more than 20% silt (Krynine, 1948, p.141).

Silurian A period of the Paleozoic, thought to have covered the span of time between 430-440 and 395 million years ago; also, the corresponding system of rocks. The Silurian is after the Ordovician and before the Devonian; however, the Ordovician is sometimes considered to be the lower part of the Silurian. It is named after the Silures, a Celtic tribe. See also: *age of fishes.*

silvanite *sylvanite.*

silver A soft, white, isometric mineral, the native metallic element Ag. It occurs in strings and veins in volcanic and sedimentary rocks and in the upper parts of silver-sulfide lodes, and is often associated with small amounts of gold, mercury, copper, lead, tin, platinum, and other metals. Silver is very ductile, malleable, and resistant to oxidation or corrosion (but tarnishes brown), and it has the highest thermal and electric conductivity of any substance. It is used for coinage, jewelry, and tableware, in photography, dentistry, and electroplating, and as a catalyst.

silver amalgam Naturally occurring *amalgam.*

silver Cape A *Cape diamond* having a very slight tint of yellow.

silver-copper glance *stromeyerite.*

silver glance *argentite.*

silver jamesonite owyheeite.

silver lead ore Galena containing more than one percent silver; argentiferous galena.

silvicolous Said of an organism that lives in wooded areas.

sima A petrologic name for the lower layer of the Earth's crust, composed of rocks that are rich in silica and magnesia; is the source of basaltic magma. It is equivalent to the ocenic crust and to the lower portion of the continental crust, underlying the sial. Etymol: an acronym for silica + magnesia. Adj: simatic. Cf: sialma. Syn: intermediate layer.

simatic Adj. of sima.

simetite A deep-red to light orange-yellow or brown variety of amber, having high contents of sulfur and oxygen and a low content of succinic acid, and occurring in the waters off Sicily.

similar fold A fold that is part of a pattern of incompetent beds in which the successive folds are congruent or resemble each other, with the limbs thinner than the axes. Cf: reverse similar fold; concentric fold.

simonellite A hydrocarbon mineral: $C_{15}H_{20}$.

simple coral solitary coral.

simple crater A meteorite impact crater of relatively small diameter, characterized by a uniformly concave-upward shape and a maximum depth in the center, and lacking a central uplift (Dence, 1968, p.171); e.g. Barringer Crater (Meteor Crater) in Coconino County, Ariz. Cf: complex crater.

simple cross-bedding Cross-bedding in which the lower bounding surfaces are nonerosional surfaces (McKee & Weir, 1953, p.385); it is formed by deposition alone.

simple cuspate foreland A foreland in which beach ridges, swales, and other symmetrical lines of growth are oriented parallel with both shores of the cusp (Johnson, 1919, p.325). Cf: complex cuspate foreland.

simple fold A single fold or flexure. Cf: compound fold.

simple lattice primitive lattice.

simple operculum A dinoflagellate operculum that is not at all, or only incompletely, divided by accessory archeopyle sutures. Cf: compound operculum.

simple ore An ore of a single metal. Cf: complex ore.

simple pit A pit [bot] in which there is no over-arching wall. Cf: bordered pit.

simple radiation electromagnetic radiation.

simple shear A particular type of strain that results in internal rotation of fabric elements and is caused by differential movements on one set of parallel planes. Cf: pure shear.

simple spit A spit, either straight or recurved, without the development of minor spits at its end or along its inner side. Cf: compound spit; complex spit.

simple stream A stream, generally small, whose drainage basin is "of practically one kind of structure and of one age" (Davis, 1889b, p.218).

simple trabecula A trabecula of a scleractinian coral, composed of a series of single sclerodermites. Cf: compound trabecula.

simple twin A twinned crystal composed of only two individuals in twin relation.

simple valley A valley that maintains a constant relation to the general structure of the underlying strata; e.g. a longitudinal valley, or a transverse valley. Term introduced by Powell 1874, p.50). Cf: complex valley; compound valley.

simplotite A dark-green monoclinic mineral: $CaV_4O_9 \cdot 5H_2O$.

simpsonite A hexagonal mineral: $Al_4(Ta,Nb)_3(O,OH,F)_{14}$.

simulated stone Any substance fashioned as a gemstone that imitates it in appearance: an imitation.

simulation The representation of a physical system by a device (such as a computer, model, or piece of equipment) that imitates the behavior of the system.

sincosite A leek-green tetragonal mineral: $CaV_2^{+4}(PO_4)_2 \cdot (OH)_4 \cdot 3H_2O$.

Sinemurian European stage: Lower Jurassic (above Hettangian, below Pliensbachian).

singing A resonance phenomenon of marine seismographs that is produced by short-path multiples in a water layer. Syn: reverberation; ringing.

singing sand sounding sand.

single cut A simplified brilliant cut consisting of 18 facets: a table, a culet, 8 bezel facets, and 8 pavilion facets.

single-cycle mountain A fold mountain that has been destroyed without reelevation of any important part (Hinds, 1943).

single-ended spread A type of seismic spread in which the shot point is located at one end of the arrangement of geophones.

single-grain structure A type of structure of a noncoherent soil in which there is no aggregation or orderly arrangement. It is characteristic of coarse-grained soils.

single-line stream A watercourse too narrow to depict (at the given scale on a map) by two lines representing the banks. Cf: double-line stream; split stream.

single refraction Refraction in an isotropic crystal, as opposed to the birefringence of an anisotropic crystal.

single shear A shear along one surface only.

single tombolo A single, simple bar connecting an island with the mainland or with another island. Cf: double tombolo; triple tombolo.

singular crystal form fixed form.

sinhalite A brown orthorhombic mineral: $MgAl(BO_4)$. It is structurally related to olivine.

Sinian An approximate equivalent of Riphean.

sinistral Pertaining, inclined, or spiraled to the left; specif. pertaining to the reversed or counterclockwise direction of coiling of some gastropod shells. A sinistral gastropod shell in apical view (apex toward the observer) has the whorls apparently turning from the right toward the left; when the shell is held so that the axis of coiling is vertical and the apex or spire is up, the aperture is open toward the observer to the left of the axis. Actually, the definition depends on features of soft anatomy: with genitalia on the left side of the head-foot mass, the soft parts and shell are arranged as in a mirror image of a dextral shell (TIP, 1960, pt.I, p.133). Ant: dextral. Syn: left-handed [paleont].

sinistral fault left-lateral fault.

sinistral fold An asymmetric fold, the long limb of which appears to have a leftward offset, viewed along its dip. Cf: dextral fold.

sinistral imbrication The condition in a heterococcolith in which each segment overlaps the one to the left when viewed from the center of the cycle. Ant: dextral imbrication.

sink [glac geol] An obsolete term for a depression in a terminal moraine.

sink [karst] Var. of sinkhole.

sink [desert] A slight, low-lying depression containing a central playa or saline lake with no outlet, as where a desert stream comes to an end or disappears by evaporation; e.g. Carson Sink in Nevada.

sink [volc] A circular or ellipsoidal depression on the flank of or near to a volcano, formed by collapse. It has no lava flows or rim surrounding it. Cf: collapse caldera. Syn: pit crater; volcanic sink.

sinkhole A partial syn. of doline; this usage is North American. Var. sink; lime sink; limestone sink; leach hole.

sinking [geol] subsidence.

sinking [currents] The downward movement of surface water, generally caused by converging currents or by a water mass that becomes denser than the surrounding water. Ant: upwelling. Syn: downwelling [currents].

sinking creek lost stream.

sink lake karst pond.

sinnerite A mineral: $Cu_{1.4}As_{0.9}S_2$. (?). Cf: luzonite.

sinoite A meteorite mineral: Si_2N_2O.

sinopite A brick-red, earthy, ferruginous clay mineral used by the ancients as a red paint.

sinople A blood-red or brownish-red (with a tinge of yellow) variety of quartz containing inclusions of hematite. Also spelled: *sinopal; sinopel.*

sinter A chemical sedimentary rock deposited as a hard incrustation on rocks or on the ground by precipitation from hot or cold mineral waters of springs, lakes, or streams; specif. *siliceous sinter* and *calcareous sinter* (travertine). The term is indefinite and shoIud be modified by the proper compositional adjective, although when used alone it usually signifies "siliceous" sinter. Etymol: German *Sinter*, "cinder". Cf: *tufa.*

sinuosity Ratio of the length of the channel or thalweg to the down-valley distance (Leopold & Wolman, 1957, p.53). Channels with sinuosities of 1.5 or more are called "meandering".

sinupalliate Said of a bivalve mollusk possessing a pallial line with a posterior embayment (pallial sinus). Cf: *integripalliate.*

sinus [coast] A bay of the sea.

sinus [paleont] (a) A curved, moderately deep groove, indentation, or reentrant in the outer lip of the aperture of a gastropod shell. It is progressively filled in as the shell grows and forms a distinct band, and is distinguished from the *slit* by nonparallel sides. (b) Any part of a transverse feature (apertural margin, rib, growth line) of a cephalopod, concave toward the aperture. (c) A major undulation or rounded depression along the commissure of a brachiopod (generally found on the pedicle valve), with the crest directed ventrally and commonly but not invariably associated with the ventral fold and the dorsal sulcus. The term is also used, irrespective of commissure, as a syn. of *sulcus*. (d) A slit at the proximal edge of orifice in some ascophoran cheilostomes (bryozoans). (e) A V-shaped indentation of blastoid ambulacrum along the margins of deltoid plates and radial plates. (f) A *pallial sinus* of a bivalve mollusk.

sinus [bot] The space or recess between two lobes or divisions of a leaf or other expanded organ (Lawrence, 1951, p.770).

sinusoidal projection An equal-area map projection representing the limiting form of the *Bonne projection*, using the equator as the standard parallel, and showing all parallels as equally spaced parallel straight lines drawn to scale. The meridians are sine curves, concaving toward the central meridian (a straight line, one half the length of, and at right angles to, the equator) along which the scale is true. The projection shows the entire globe but suffers from extreme distortion (shearing) in marginal zones at high latitudes; it is often used in atlases for the map of Africa. Syn: *Sanson-Flamsteed projection; Mercator equal-area projection.*

siphon [speleo] A passage in a cave system that connects with a *water trap*. Cf: *sump; conduit.*

siphon [paleont] (a) Either of a pair of posterior tube-like extensions of the mantle in many bivalve mollusks, serving for the passage of water currents; e.g. an inhalant ventral tube that conducts water to the mouth and gills and confines the current flowing into the mantle cavity, and an exhalant dorsal tube that carries away waste water and confines the current flowing from the mantle cavity. (b) An anterior channel-shaped prolongation of the mantle in many gastropods, serving for the passage of water to the gills, and often being protected by a grooved extension of the margin of the shell. (c) The swimming funnel of a cephalopod; the membranous siphuncle of a shelled cephalopod. (d) An internal tube extending inward from a foraminiferal aperture.

siphon [hydraul] A water conduit that is a U-shaped channel in which the water is in hydrostatic equilibrium. Also spelled *syphon*. Cf: *inverted siphon.*

siphonal canal The tabular or trough-like extension of the anterior (abapical) part of the apertural margin of a gastropod shell, serving for the shielding of the inhalant siphon.

siphonal deposit A calcareous accumulation within or along the siphuncle of a cephalopod, attaining considerable thickness in some nautiloids.

siphonal fasciole A band of abruptly curved growth lines near the foot of the columella of a gastropod, marking successive positions of the siphonal notch.

siphonal notch The narrow sinus of the apertural margin near the foot of the columella of a gastropod, serving for the protrusion of the inhalant siphon. It virtually separates the inner lip and the outer lip.

siphonoglyph A strongly ciliated groove extending down one side of the pharynx of a coral.

siphonostomatous Said of a gastropod shell with the apertural margin interrupted by a canal, spout, or notch for the protrusion of the siphon; e.g. said of various marine snails having the front edge of the aperture prolonged in the form of a channel for the protection of the siphon. Cf: *holostomatous.*

siphonous line An evolutionary trend in the green algae, characterized by multinucleate thalli and the absence of cell partitions. Cf: *tetrasporine line; volvocine line.*

siphonozooid A degenerate octocorallian polyp with reduced or no tentacles and commonly reduced septal filaments (thickened, convoluted edges of septa). It is usually much smaller than an *autozooid* and is supposed to serve to regulate the water supply of the colony.

siphuncle (a) A long and slender or thick membranous tube extending through all the camerae and septa from the protoconch to the base of the body chamber of a cephalopod shell and consisting of soft and shelly parts, including septal necks, connecting rings, calcareous deposits, and siphuncular cord. (b) The tubular or funnel-shaped shelly septal structures that ensheathe and support the siphuncle.

siphuncular cord The fleshy interior tissues of the siphuncle of a cephalopod.

sirloin-type ice A term used by Higashi (1958) and now regarded as a syn. of *Taber ice.*

siserskite *sysertskite.*

sismondine A magnesium-bearing chloritoid.

sismondinite A schist in which sismondine (magnesium chloritoid) is the chief constituent. The term was originated by Franchi in 1897.

sitaparite *bixbyite.*

site The area of land that has been inspected, selected, or investigated for a particular purpose; e.g. the spot where a drilling rig will be set up and where a borehole is to be drilled, or the place where a mine, dam, tunnel, or other engineering works is to be erected.

site investigation The collection of basic facts about, and the testing of, surface and subsurface features (including physical nature, thickness, geologic structure, and other engineering properties) at a site for the purpose of preparing suitable foundations and building designs for engineering structures.

sixfold coordination *octahedral coordination.*

sixth-power law A law asserting that the carrying power of a stream is proportional to the sixth power of its velocity; e.g. the stream flows twice as rapidly, the size of the particle carried may be increased 64 times. The law postulates complete transfer of kinetic energy from the water to the particle and makes no allowance for the effect of viscous drag.

size *particle size.*

size analysis *particle-size analysis.*

size distribution *particle-size distribution.*

size-frequency analysis *particle-size analysis.*

size-frequency distribution *particle-size distribution.*

sizing The arrangement, grading, or classification of particles according to size; e.g. the separation of mineral grains of a sediment into groups each of which has a certain range of size or maximum diameter, such as by sieving or screening.

sjögrenite A hexagonal mineral: $Mg_6Fe_2(CO_3)(OH)_{16}\cdot4H_2O$. It is dimorphous with pyroaurite. Also spelled: *sjogrenite.*

S-joint *longitudinal joint.*

skarn As used by Fennoscandian geologists, an old Swedish mining term for silicate gangue (amphibole, pyroxene, garnet, etc.) of certain iron-ore and sulfide deposits of Archean age, particularly those which have replaced limestone and dolomite. Its meaning has been generally expanded to include lime-bearing silicates of any geologic age derived from nearly pure limestone and dolomite with the introduction of large amounts of Si, Al, Fe and Mg (Holmes, 1920, p.211).

skartråg The Swedish term for *sickle trough*.

skauk A term introduced by Taylor (1951, p.620) for an extensive field of crevasses in a glacier.

skavl A Norwegian term for a large wind-eroded ridge of snow on a glacier. The term is generally equivalent to *sastruga*. Pl: *skavler*.

skedophyre A porphyritic rock characterized by *skedophyric* texture.

skedophyric A term, now obsolete, applied by Cross, et al 1906, p.703) to porphyritic igneous rocks in which the phenocrysts are more or less uniformly scattered throughout the groundmass; of, or pertaining to, a *skedophyre*.

skeletal (a) Pertaining to material derived from organisms and consisting of the hard parts secreted by the organisms or of the hard material around or within organic tissue. (b) A term used by Nelson et al (1962, p.234) to refer to a limestone that consists of, or owes its characteristics to, virtually in-place accumulation of skeletal matter (as distinguished from a fragmental limestone formed by mechanical transport), but regarded by Leighton & Pendexter (1962) as synonymous with "bioclastic", indicating faunal or floral fragments, or whole components of these organisms, that are not in their place of origin.

skeletal canal A canal-like cavity in a coherent skeletal framework of a sponge. It may or may not correspond to a canal of the aquiferous system. Examples: *amararhysis*; *apophysis*; *diarhysis*; *epirhysis*; *schizorhysis*; *surface groove*.

skeletal crystal growth Microscopic development of the outline or framework of a crystal, with incomplete filling in of the crystal faces. Crystals formed in this way are called skeleton crystals.

skeletal duplicature The outer exoskeletal layers or molted skin of a branchiopod crustacean shed during ecdysis. See also: *duplicature*.

skeletal fiber Any fiber-like structure of the sponge skeleton, such as a spiculofiber, a spicule tract, a sclerosomal trabecula, or a spongin fiber.

skeletal framework A coherent meshwork in a sponge, built of sclerosomal trabeculae, fused spicules, interlocking spicules, spongin-cemented spicules or sand grains, or spongin alone. Syn: *skeletal mesh; skeletal net*.

skeletal pore An opening between spicules or between skeletal fibers of the regular skeletal framework in a sponge; as distinct from larger openings (such as ostia or oscula) that interrupt the regular net and as distinguished from the true pores of the soft parts, with which the skeletal pores may not correspond.

skeletal residue An insoluble residue whose constituent material comprises less than 25% of the volume, and containing rhombohedral (dolomoldic) or spheroidal (oomoldic) openings (Ireland et al, 1947, p.1482-1483). Cf: *lacy residue*.

skeletal soil Lithosol.

skeletan A *cutan* consisting of skeleton grains adhering to the surface (Brewer, 1964, p.217); e.g. bleached sand and silt grains high in quartz and low in feldspar.

skeleton [paleont] The hard or bony structure that constitutes the framework supporting the softer parts of an animal and protecting or covering its internal organs; e.g. the mesh of spicules of a sponge, the shell of a brachiopod or mollusk, the chitinous covering of an arthropod, or the bones of a vertebrate. See also: *endoskeleton; exoskeleton*.

skeleton [bot] (a) The vascular system of a vascular plant. (b) A protective covering of a nonvascular plant, e.g. the frustule of a diatom.

skeleton grain A relatively stable and not readily translocated grain of a soil material, concentrated or reorganized by soil-forming processes (Brewer & Sleeman, 1960); e.g. a mineral grain, or a resistant siliceous or organic body larger than colloidal size. Cf: *plasma* [*soil*].

skeleton layer The structure formed at the bottom of sea ice while freezing, consisting of vertically oriented platelets of ice separated by layers of brine; this layer has almost no mechanical strength.

skerry A low, small, rugged and rocky island or reef; an isolated rock detached from the mainland, rising above sea level from a shallow strandflat, and covered by the sea during high tides or stormy weather. Examples occur along the coasts of Scotland and Scandinavia. Etymol: Scandinavian, akin to Old Norse *sker*. See also: *stack*.

skerry-guard A line, belt, or fringe of skerries, parallel to and extending along a coast for hundreds of kilometers, seemingly acting as a breakwater or "guard". The term is a common, but incorrect, translation of the Norwegian term *skjergaard*, "skerry enclosure", which properly refers to an area of calm water between a line of skerries and the mainland or enclosed by skerries.

sketch map An outline map drawn freehand from observation or from loose and uncontrolled surveys rather than from exact survey measurements, showing only the main features of an area. It preserves the essential space relationships, but does not truly preserve scale or orientation.

skewed projection Any standard projection, used in construction of maps or charts, that does not conform to a general north-south format with relation to the neat lines of the map or chart.

skewness (a) The quality, state, or condition of being distorted or lacking symmetry; esp. the quality or state of asymmetry shown by a frequency distribution that is bunched on one side of the average and tails out on the other side. It results from lack of coincidence of the mode, median, and arithmetic mean of the distribution. (b) A measure of asymmetry of a frequency distribution; specif. the quotient of the difference between the arithmetic mean and the mode divided by the standard deviation. The logarithm of this measure to the base ten has been used to express the asymmetry of the central half of the particle-size distribution of a sediment: $Q_1 Q_3 / (Md)^2$, where Q_1 and Q_3 are the particle diameters, respectively, at the 25% and 75% intersections on the cumulative frequency distribution, and Md is the median particle diameter. Positive skewness is defined for the longer slope of the plotted distribution in the direction of increasing variate values (mean greater than mode, or coarser particles exceed finer particles in a particle-size distribution); negative skewness is defined for the longer slope of the plotted distribution in the direction of decreasing variate values (mode greater than mean, or finer particles exceed coarser particles in a particle-size distribution). Several coefficients of skewness have been devised in an attempt to assign genetic significance to sediment distributions. Abbrev: Sk.---Cf: *kurtosis*.

skiagite A hypothetical end member of the garnet group: $Fe_3{}^{+2}Fe_2{}^{+3}(SiO_4)_3$.

skiagram An obsolete type of *scan* using X-ray shadows.

skialith A vague remnant of country rock in granite, obscured by the process of granitization.

skid boulder An isolated angular block of stone resting on the floor of a playa, derived from an outcrop near the playa margin, and associated with a trail or mark indicating that the boulder has recently slid across the mud surface, in some instances as much as 300 m.

Skiddavian Arenigian.

skim ice First formation of a thin layer of ice on the water surface. Syn: *skin* [*ice*].

skimming (a) Diversion of water from a stream or conduit by shallow overflow in order to avoid diverting sand, silt, or other debris carried as bottom load (Langbein & Iseri, 1960). (b) Withdrawal of fresh ground water from a thin body or lens floating on salt water by means of shallow wells or infiltration galleries.

skin *skim ice.*

skin depth *depth of penetration.*

skin effect [**drill**] The phenomenon in which alterations in permeability in the vicinity of a drill hole are caused by drilling and completion operations.

skin effect [**elect**] The concentration of alternating current in a conductor towards its exterior boundary.

skin friction [**eng geol**] (a) The frictional resistance developed between soil and an engineering structure. (b) The resistance of the ground to penetration by a pile, pipe, probing rod, etc.

skin friction [**hydraul**] In hydraulics, the friction between a fluid and the surface of solid moving through it (ASCE, 1962). Syn: *surface drag.*

skiodrome A term used in optical mineralogy for an orthographic projection of curves of equal velocity as they would appear on a sphere, assuming that the light source is at the center of the sphere.

skiou A facetious term used by Davis (1912, p.116) for a *morvan.*

skip cast The cast of a skip mark.

skip mark One of a linear series of regularly spaced and crescent-shaped tool marks produced by an object that intermittently impinged on or skipped along the bottom.

skjergaard See: *skerry-guard.*

skleropelite A term proposed by Salomon (1915) for argillaceous and allied rocks which have been indurated by low-grade metamorphism. It is more massive and dense than shale, and differs from slate by the absence of cleavage. Cf: *hornfels.*

sklodowskite A strongly radioactive, citron-yellow, orthorhombic secondary mineral: $Mg(UO_2)_2Si_2O_7.6H_2O$. It is isostructural with uranophane and cuprosklodowskite. Syn: *sklodovskite.*

skolite A scaly, dark-green variety of glauconite rich in aluminum and calcium and deficient in ferric iron.

skomerite A fine-grained, compact extrusive rock containing microscopic grains and crystals of augite, olivine, and phenocrysts of decomposed plagioclase (probably albite) in a groundmass of plagioclase, thought to be more calcic than the phenocrysts (Johannsen, 1939, p.280).

skutterudite A tin-white to silver-gray isometric mineral: $(Co, Ni)As_3$. It may contain considerable iron, and it represents a minor ore of cobalt and nickel. See also: *smaltite.*

skylight A submariner's term for a *polynya* or lead during the winter; it is covered by relatively thin ice (usually less than 1 m thick) and has a normally flat undersurface. Cf: *lake* [*ice*].

skystone *meteorite.*

Skythian Var. of *Scythian.*

slab A layer in, or the whole thickness of, a snowpack whose internal cohesion is large compared to its external adhesion to other snow layers or the ground. The absolute hardness of a snow slab may vary over several orders of magnitude. The characteristic identifying property of a snow slab is the ability to sustain elastic deformation under stress and hence the propagation of fractures. Cf: *wind slab.*

slab avalanche A snow avalanche that starts from a fracture line, in snow possessing a certain amount of cohesion. Cf: *loose-snow avalanche; wind-slab avalanche.*

slab jointing Jointing produced in rock by the formation of numerous cleaved or closely spaced parallel fissures dividing the rock into thin slabs.

slab pahoehoe A type of pahoehoe, the surface of which consists of a jumbled arrangement of plates or slabs of flow crust, presumably so arranged due to the draining away of the

underlying molten lava. Cf: *corded pahoehoe; elephant-hid pahoehoe; entrail pahoehoe; festooned pahoehoe; filament pahoehoe; sharkskin pahoehoe; shelly pahoehoe.*

slabstone A rock that readily splits into slabs; *flagstone.*

slack [**weath**] *grus.*

slack [**coast**] A hollow or depression between lines of sho dunes or in a sandbank or mudbank on a shore.

slack [**geomorph**] (a) A British term for a hollow or dip in th ground; e.g. a pass between hills, or a small and shallow va ley, or a depression in a hillside. (b) A British term for a so boggy piece of ground; a marsh or morass.

slack [**water**] *slack water.*

slack ice A syn. of *broken ice,* esp. if floating on slowly mo ing water.

slacking index *weathering index.*

slack tide *slack water.*

slack water (a) The condition of a tidal current or horizont motion of water when its velocity is very weak (less than 0. knot) or zero, esp. at the turn of the tide when there is a re versal between ebb current and flood current. Also, the inte val of time during which slack water occurs. Syn: *slack tid* (b) A quiet part of, or a still body of water in, a stream; e. on the inside of a bend, where the current is slight. Syn: *slac* [*water*].

slag [**sed**] A British term for a friable shale with many fossil

slag [**pyroclast**] A scoriaceous or cindery pyroclastic rock.

slaking (a) The crumbling and disintegration of the earth ma terials upon exposure to air or moisture; specif. the breakin up of dried clay or indurated soil when saturated with or im mersed in water, or the breaking up of clay-rich sedimentar rocks when exposed to air. (b) The disintegration of tunne walls in swelling clay due to inward movement and circumfe ential compression (Stokes & Varnes, 1955, p.137). (c) Th treating of lime with water to give hydrated (slaked) lime.

slang [**streams**] A term used in Vermont for a small stream.

slant drilling *directional drilling.*

slant range The direct distance from the radar antenna to th target.

slant well *directional well.*

slap A British term for a pass, notch, or gap between hills.

slash [**geog**] (a) A local term in eastern U.S. for a marsh or low swampy area overgrown with dense underbrush, an often covered by water. (b) An open or cutover tract in a fo est strewn with debris (logs, bark, branches, etc.), as fro logging or fire. Also, the debris in such a tract. Syn: *slashing.*

slash [**beach**] A term used in New Jersey for a wet or marsh *swale* between two parallel beach ridges.

slate (a) A compact, fine-grained, metamorphic rock forme from such rocks as shale and volcanic ash, which posses the property of fissility along planes independent of the orig nal bedding (slaty cleavage), whereby they can be parted int plates which are lithologically indistinguishable (Himus, 1954 (b) A coal miner's term for any shale accompanying coa also, sometimes the equivalent of *bone coal.*

slate clay A clay more or less transformed into slate; specif. fireclay occurring in the English coal measures.

slate ground A term used in southern Wales for a dark fissil shale, resembling slate.

slate ribbon A relict *ribbon* structure on the cleavage surfac of slate, in which varicolored and straight, wavy, or crumple stripes cross the cleavage surface.

slate spar *shiver spar.*

slat-flecked ice Ice swept clear of snow except for wind rip ples saturated with brine (ADTIC, 1955, p.73).

slatiness The quality of being slaty, such as a sedimentar rock splitting into thin layers or plates parallel to the beddin with essential regularity of surfaces similar to true slaty cleav age (Grabau, 1924, p.786).

slaty cleavage A pervasive, parallel foliation of fine-grained platy minerals (mainly chlorite and sericite) in a direction per endicular to the direction of compression, developed in slate r other homogeneous, sedimentary rock by deformation and w-grade metamorphism. Most slaty cleavage is also *axial- lane cleavage*. Syn: *flow cleavage*.

lavikite A greenish-yellow rhombohedral mineral: $MgFe_3^{+3}$-$SO_4)_4(OH)_3 \cdot 18H_2O$.

leugh British var. of *slough* in the sense of a small marsh.

lew A syn. of *slough*, esp. a wet spot not large enough to be alled a swamp or a marsh.

lice [stratig] An arbitrary informal division (either of uniform hickness or constituting some uniform vertical fraction) of an otherwise indivisible stratigraphic unit, distinguished for indiidual facies mapping or analysis.

lice [struc geol] Var. of *thrust slice*.

lice ridge A narrow linear ridge, a meter to 100 m high, repesenting a slice of rock squeezed up within a fault zone (esp. long a strike-slip fault). Syn: *fault-slice ridge; pressure ridge*.

lick n. A glassy smooth patch on an otherwise rippled water urface, occurring in coastal and inland waters. It is caused y a monomolecular layer of organic material that has the efect of reducing surface tension.

lickens Extremely fine-grained material, such as finely pulerized tailings discharged from hydraulic mines or a thin ayer of extremely fine silt deposited by a stream during flood.

lickenside A polished and smoothly striated surface that reults from friction along a fault plane. A surface bearing slickenside is said to be "slickensided". Etymol: English dialect, *slicken*, "slick, smooth". Cf: *striation* [fault]; Syn: *polished urface; mullion structure; slip-scratch*.

lickolite A field term proposed by Bretz (1940, p.338) for a vertically discontinuous striation produced by slippage and hearing and developed on strongly dipping bedding planes of imestone that forms the molding on the wall of a solution cavity. The structure resembles a slickenside but shows evidence of some solution and the development of incipient asymmetric stylolites oriented parallel to bedding.

lick spot On a soil, an alkali patch.

lide [mass move] (a) A mass movement or descent resulting rom failure of earth, snow, or rock under shear stress along one or several surfaces that are either visible or may reasonably be inferred; e.g. *landslide; snowslide; rockslide*. The moving mass may or may not be greatly deformed, and movement may be rotational or planar. A slide can result from lateral erosion, lateral pressure, weight of overlying material, accumulation of moisture, earthquakes, frost action, regional tilting, undermining, and human agencies. (b) The track of bare rock or furrowed earth left by a slide. See also: *landslide track*. (c) The mass of material moved in or deposited by a slide. See also: *slide-rock*. (d) A shortened form of *landslide*. (e) An earthflow in the Appalachian Plateau region of the U.S.

lide [fault] A term proposed by Fleuty (1964) for a fault formed in close connection with folding, and that is conformable with the fold limb or axial surface. It is accompanied by thinning and/or disappearance of the folded beds. Cf: *fold fault*.

lide cast The cast of a *slide mark*, commonly smoothly curved and less than a meter in length.

lide mark A scratch or *groove* left on a sedimentary surface by subaqueous sliding or slumping (Kuenen, 1957, p.251); it tends to be wider and shallower than a typical *drag mark*, and may be formed by sliding objects such as a mass of sediment, a plant mat, or a large soft-bodied animal. See also: *slide cast*. Syn: *olistoglyph*.

liding *gravitational sliding*.

lieve An Irish term for a mountain.

likke A Dutch term used also in France for a *tidal flat* or a *mud flat*, esp. one rich in decaying organic matter mixed with sand and crossed by tidal channels (Termier & Termier, 1963, p.414).

lim hole (a) An oil-driller's term for a diamond-drill borehole

having a diameter of 5 in. or less. (b) A drill hole of the smallest practicable size, drilled with less-than-normal-diameter tools (less than 7.875-inch rotary bit, thus a 6.25-inch bit or less), used primarily as seismic shotholes and structure tests and sometimes as stratigraphic tests (Williams & Meyers, 1964, p.375).

slip [coast] (a) An extension of navigable water into the space between adjacent structures (as piers) within which vessels can be berthed; a *dock*. (b) A sloping ramp extending into the water and serving as a landing place for vessels.

slip [geomorph] A narrow mountain pass; a defile.

slip [clay] A suspension of fine clay in water, having the consistency of cream, and used in the decoration of ceramic ware. See also: *slip clay*.

slip [cryst] *crystal gliding*.

slip [struc geol] (a) On a fault, the actual relative displacement along the fault plane of two formerly adjacent points on either side of the fault. Slip is three-dimensional, whereas *separation* is two-dimensional. See also: *actual*. Partial syn: *shift*. Syn: *actual relative movement; total displacement*. (b) A small fracture along which there has been some displacement.

slipband One of the parallel lines known as *Luders lines* or *Hartmann lines* in the crystalline grains of a material stressed behind its elastic limit that are visible only under a microscope and are produced by intracrystalline slip. See also: *deformation lamella*. Syn: *deformation lamellae*.

slip bedding A variety of *convolute bedding* produced by subaqueous sliding; esp. *slump bedding*. Cf: *glide bedding*.

slip block A separate rock mass that has "slid away from its original position and come to rest some way down the slope without being much deformed" (Kuenen, 1948, p.371).

slip clay An easily fusible clay containing a high percentage of fluxing impurities, used to produce a natural glaze on the surface of clayware. See also: *slip*.

slip cleavage A type of cleavage that is superposed on slaty cleavage or schistosity and this is characterized by finite spacing of cleavage planes (*spaced cleavage*) between which there occurs thin, tabular bodies of rock displaying a crenulated cross lamination. Syn: *crenulation cleavage; shear cleavage; strain- slip cleavage; close-joints cleavage*.

slip dike A dike that has been intruded along a fault plane.

slip face (a) The steeply sloping surface on the lee side of a dune, standing at or near the angle of repose of loose sand, and advancing downwind by a succession of slides wherever that angle is exceeded. Syn: *sandfall*. (b) The leeward surface of a sand wave, exhibiting foreset bedding. Syn: *slip slope*.--- Also spelled: *slipface*.

slip fold *shear fold*.

slip joint A less-preferred syn. of *shear joint*; it is descriptive and nongenetic.

slip-off slope The long, low, relatively gentle slope on the inside of a stream meander, produced on the downstream face of the meander spur by the gradual outward migration of the meander as a whole; located opposite to the *undercut slope*.

slip-off slope terrace A local terrace on the *slip-off slope* of a meander spur, formed by a brief halt during the irregular incision by a meandering stream.

slip plane [mass move] A planar *slip surface*.

slip plane [cryst] *glide plane*.

slip-scratch A term proposed by Challinor and Williams (1926) for the markings on a rock surface made by the movement over it of another rock mass, but of a type other than *slickenside*. Cf: *groove* [fault]; *striation* [fault]; *chattermark; mullion structure*.

slip sheet A stratum or rock unit on the limb of an anticline that, having become fractured at its base, has slid down and away from the anticline. It is a gravity-collapse structure.

slip slope The *slip face* of a sand wave.

slip surface [mass move] A landslide displacement surface, often slickensided, striated, and subplanar. It is best exhibited in argillaceous materials and in those materials which are highly susceptible to clay alteration when granulated. See also: *slip plane.* Syn: *landslide shear surface; gliding surface.*

slip surface [struc petrol] *flow surface.*

slip tectonite A tectonite whose deformation is along the most prominent s planes; a type of *S-tectonite* (Knopf and Ingerson, 1938, p.69).

slit A parallel-sided reentrant in the outer lip of the aperture of a gastropod shell, ranging from a shallow incision to a deep fissure as much as half a whorl in extent (TIP, 1960, pt.I, p.133). Cf: *sinus [paleont].*

slither Loose rubble or talus; angular debris.

slob (a) A dense accumulation of heavy sludge of sea ice. Syn: *slob ice.* (b) A term used in Newfoundland for soft snow or mushy ice.

slob land (a) *low marsh.* (b) A term used in Ireland for muddy ground or soil, or for a level tract of muddy ground; esp. alluvial land that has been reclaimed.

sloot *sluit.*

slope [stream] *stream gradient.*

slope [geomorph] (a) *gradient.* (b) The inclined surface of any part of the Earth's surface, as a *hillslope;* also, a broad part of a continent descending toward an ocean, as the Pacific *slope.*

slope correction A *tape correction* applied to a distance measured on a slope in order to reduce it to a horizontal distance, between the vertical lines through its end points; a correction for inclination of tape. Syn: *grade correction.*

slope curvature The rate of change of angle of slope with distance.

slope-discharge curve A graphic presentation of the discharge at a given gaging station, taking into account the slope of the water surface as well as the gage height.

slope element A curved part of a slope profile; a smooth concave or convex area of a slope or portion of the slope profile. Cf: *slope segment.*

slope facet *slope segment.*

slope gully A small, discontinuous submarine valley, usually formed by slumping along a fault scarp or the slope of a river delta. Syn: *sea gully.*

slope length (a) The linear distance between the top and bottom of a slope, measured on its surface. (b) *length of overland flow.*

slope map A map that shows the distribution of the degree of surface inclination; e.g. *isotangent map* and *isosinal map.*

slope-over-wall structure A geologic structure produced where coastal talus deposits are cut back by marine denudation to form a truncated deposit overlying a cliff face.

slope retreat The progressive recession of a scarp or of the side of a hill or mountain; suggested causes include backwearing (*parallel retreat of slope*) and downwearing. Syn: *slope recession; scarp retreat.*

slope sector The part of a slope element on which the slope curvature remains constant.

slope segment A rectilinear part of a slope or slope profile. Cf: *slope element.* Syn: *slope facet.*

slope sequence The part of a hillside surface that consists, in succession, of a waxing slope, a maximum slope segment, and a waning slope.

slope stability The quality of permanence or resistance of a natural or artificial slope or other inclined surface to failure by landsliding (including planar or rotational slides, flows, falls, and raveling). See also: *bank stability.*

slope unit (a) The smallest feature of a slope profile, consisting either of a slope element or of a slope segment. (b) A system consisting of a base level of denudation and the correlated slope lying above it (Penck, 1953, p.129). Slope units are separated by breaks of gradient. Etymol: English translation of German *Formsystem.*

slope wash (a) Soil and rock material that is or has bee transported down a slope by mass-wasting assisted by ru ning water not confined to channels. Cf: *colluvium.* (b) T process by which slope-wash material is moved; specif. *she erosion.*---Also spelled: *slopewash.*

slough [drill] Fragmentary rock material that has crumble and fallen away from the sides of a borehole (or mine wor ing). It either falls into and obstructs the hole or is wash out during circulation of the drilling mud.

slough [geog] (a) A small marsh; esp. a marshy tract lying a swale or other local, shallow, and undrained depression a piece of dry land, as on the prairies of the Midwest U. Also, a dry depression that becomes marshy or filled wi water. Syn: *slew; slue; sleugh.* (b) A large wetland, as swamp; e.g. in the Everglades of Florida. (c) A term use esp. in the Mississippi Valley for a creek or sluggish body water in a tidal flat, bottomland, or coastal marshland. (d) sluggish channel of water, such as a side channel of a rive in which water flows slowly through low, swampy ground, along the Columbia River, or a section of an abandoned riv channel, containing stagnant water and occurring in a floo plain or delta. Also, an indefinite term indicating a small lak a marshy or reedy pool or inlet, a bayou, a pond, or a sma and narrow backwater. Syn: *slew; slue.* (e) A small bay eastern England. (f) A piece of soft, miry, muddy, or wate logged ground; a place of deep mud, as a mudhole. The ter "slough" is an obsolete syn. of "mud" or mire.----Pron: *sloo.*

slough ice Slushy ice or snow.

slow ray In crystal optics, that component of light in any bire fringent crystal section which travels with the lesser veloci and has the higher index of refraction. Cf: *fast ray.*

slud [mass move] (a) The muddy material that has move downslope by solifluction. (b) Ground that behaves as a vi cous fluid, including material moved by solifluction as well by mechanisms not limited to gravitational flow (Muller, 194 p.221).---Etymol: a provincial English word for a soft, we slippery mass, as mud or mire.

slud [ice] *young ice.*

sludge [drill] Mud obtained from a drill hole in boring; mu from drill cuttings. The term has also been used for the *cu tings* produced by drilling.

sludge [sed] (a) A soft and soupy or muddy bottom deposi such as found on tideland or in a stream bed; specif. blac ooze on the bottom of a lake (Twenhofel, 1937, p.90). (b) semifluid, slushy, and murky mass or sediment of solid matte resulting from treatment of water, sewage, or industrial an mining wastes, and often appearing as local bottom deposit in polluted bodies of water.

sludge [ice] A dense, soupy accumulation of new sea ic formed during an early stage of freezing, and consisting c incoherent floating frazil crystals that may or may not b slightly frozen together; it forms a thin gluey layer and give the sea surface a steel-gray or leaden-tinted color. See also *ice gruel.* Syn: *slush; cream ice; sludge ice.*

sludge cake An accumulation of hardened sludge (sea ice strong enough to bear the weight of a man. See also: *sludg floe.*

sludge cast *furrow flute cast.*

sludge floe A large *sludge cake.*

sludge ice *sludge [ice].*

sludge lump An irregular mass of sludge (sea ice) shaped b strong wind action.

sludge pit *slush pit.*

sludging *solifluction.*

slue *slough.*

slug flow Movement of an isolated body of water, such as a body of gravity water moving downward in the zone of aera tion. The term is based on slang for a small amount of liquid such as a slug of whiskey.

sluggish Said of a stream in which the peaks of a flood form more slowly because of the decrease in the slope as the age of the stream system advances, or as the flow is reduced or retarded by the withdrawal or storage in upstream reaches.

sluice (a) A conduit or passage for carrying off surplus water at high velocity or fitted with a valve or gate for stopping or regulating the flow. (b) A gate, such as a *floodgate*. (c) A body of water flowing through or pent up behind a floodgate.

sluiceway *overflow channel.*

sluit An African term for a narrow, usually dry ditch, gully, or gulch, produced naturally by the washing of heavy rains in a large natural fissure; it is shallower than a ravine. Also, a similar watercourse produced artificially for irrigation or drainage. Etymol: Afrikaans, from Dutch *sloot*, "ditch". Syn: *sloot.*

slump [mass move] (a) A landslide characterized by a shearing and rotary movement of a generally independent mass of rock or earth along a curved slip surface (concave upward) and about an axis parallel to the slope from which it descends, and by backward tilting of the mass with respect to that slope so that the slump surface often exhibits a reversed slope facing uphill. Syn: *slumping.* (b) The sliding down of a mass of sediment shortly after its deposition on an underwater slope; esp. the downslope flowage of soft, unconsolidated marine sediments, as at the head or along the side of a submarine canyon. This is the "commonest usage in geology in Britain", although "*subaqueous slump* would be more precise" (Challinor, 1967, p.232). Syn: *slumping.* (c) The mass of material slipped down during, or produced by, a slump. See also: *slump block.*

slump [intrus rocks] A rarely used term for a preconsolidation movement in a cooling intrusion.

slump ball A relatively flattened mass of sandstone resembling a large concretion, measuring 2 cm to 3 m across, commonly thinly laminated with internal contortions and a smooth or lumpy external form, and formed by subaqueous slumping (Kuenen, 1948, p.369). Cf: *crumpled ball; spiral ball.* Syn: *snowball.*

slump basin A shallow basin near the base of a canyon wall and on a shale hill or ridge, formed by small, irregular slumps, and usually containing a short-lived lake (Worcester, 1939).

slump bedding A term applied loosely to any disturbed bedding; specif. a variety of *convolute bedding* produced by subaqueous slumping or lateral movement of newly deposited sediment. It is a type of *slip bedding.* Syn: *slurry bedding.*

slump block The mass of material torn away as a coherent unit during a *block slump.* It may be as long as 2 km and as thick as 300 m.

slump breccia A contorted sedimentary bed produced by subaqueous slumping and exhibiting brecciation.

slump fault (a) A gravity fault or *normal fault.* (b) A *growth fault.*

slump fold An intraformational fold produced by slumping of soft sediments, as at the edge of the continental shelf.

slumping The downward movement, such as sliding or settling, of a slump. Syn: *slump.*

slump mark A mark made by sand (wet or dry) avalanching down the lee side of a sand wave or dune.

slump overfold A fold consisting of hook-shaped masses of sandstone produced during slumping (Crowell, 1957, p.998).

slump scarp A low cliff or rim of thin solidified lava occurring along the margins of a lava flow and against the valley walls or around steptoes after the central part of the lava crust collapsed due to outflow of still molten underlying layers; the inward-facing cliff may be several meters high. Term introduced by Finch (1933), but Sharpe (1938, p.70) would prefer "lava subsidence scarp" or "lava slump scarp".

slump sheet A well-defined bed of limited thickness and wide horizontal extent, containing slump structures (Kuenen, 1948, p.373).

slump structure A genetic term for any sedimentary structure produced by subaqueous slumping; esp. a small-scale but complicated fold associated with convolute bedding, and formed by postdepositional sliding (such as when a relatively thick turbidite slumps relative to the more plastic and clayey bed immediately below).

slurry A very wet, highly mobile, semiviscous mixture or suspension of finely divided, insoluble matter; e.g. a muddy lake-bottom deposit having the consistency of a thick soup.

slurry bedding *slump bedding.*

slurry slump A slump in which the incoherent sliding mass is mixed with water and disintegrates into a quasi-liquid slurry (Dzulynski & Slaczka, 1958, p.217).

slush [drill] Liquid mud used in drilling a well.

slush [geog] A soft mud; mire.

slush [ice] *sludge.*

slush [snow] (a) Soft snow saturated and mixed with water, occurring on land or ice surfaces, or as a viscous floating mass in water after a heavy snowfall. (b) Frazil slush in a river.----Syn: *snow slush.* snow slush.

slush avalanche A rapid and far-reaching downslope transport of rock debris released by snow supersaturated with meltwater and marking the catastrophic opening of ice- and snow-dammed brooks to the spring flood.

slush ball The result of extremely compact accretion of snow, frazil, and ice particles. This is produced by either wind and wave action along the shore of lakes or in long stretches of very turbulent flows in rivers.

slush field An area of water-saturated snow having a soupy consistency and in which men and animals readily sink. If saturated to the surface, the snow is bluish; if not saturated to the surface, the snow is similar in color to the surrounding snow. Syn: *snow swamp.*

slushflow (a) A mudflow-like outburst of water-saturated snow along a stream course, commonly occurring in the Arctic after intense thawing produced more meltwater than can drain through the snow, and having a width generally several times greater than that of the stream channel (Washburn & Goldthwait, 1958). (b) A flow of clear slush on a glacier, as in Greenland.

slush pit A surface excavation or diked area to hold water, drilling mud, sludge, and discharged matter from drilling, such as a pit in which water (mixed with mud if necessary) can be stored for circulation through a drill hole or a pit in which the cuttings from drilling are received, trapped, and separated from the drilling mud. Also spelled: *slushpit.* Syn: *mud pit; sludge pit; slush pond; sump.*

slush pond [drill] *slush pit.*

slush pond [glaciol] A pool or lake containing slush, on the ablation surface of a glacier. It is especially common during summer.

small boulder A *boulder* having a diameter in the range of 256-512 mm (10-20 in., or -8 to -9 phi units).

small calorie *calorie.*

small circle A curve formed on the surface of a sphere by the intersection of any plane that does not pass through the center of the sphere; specif. a circle on the Earth's surface, the plane of which does not pass through the center of the Earth, such as any parallel of latitude other than the equator. Cf: *great circle.*

small-circle girdle *cleft girdle.*

small cobble A geologic term for a *cobble* having a diameter in the range of 64-128 mm (2.5-5 in., or -6 to -7 phi units).

smaller foraminifera An informal term generally used to designate those foraminifers that are studied with the aid of thin sectioning. Cf: *larger foraminifera.*

small-scale map A map drawn at a scale (in the U.S., smaller than 1/62,500) such that a large area can be covered showing only generalized detail; a map whose representative fraction has a large denominator (such as 1/250,000).

small spore A term that is sometimes used as if synonymous with *microspore*; more accurately, a term that includes pollen and prepollen, as well as spores other than megaspores. The term is therefore nearly synonymous with *miospore* but lacks its precise size definition.

small watershed A drainage basin that is "so small that its sensitivities to high-intensity rainfalls of short durations and to land use are not suppressed by the channel-storage characteristics" and in which "the effect of overland flow rather than the effect of channel flow is a dominating factor affecting the peak runoff" (Chow, 1957, p.379); its size may be a few acres to 1000 acres, or even up to 130 sq km (50 sq mi).

smaltite (a) A tin-white or pale-gray isometric mineral: (Co, Ni)As$_{3-x}$. It is a variety of *skutterudite*. Smaltite usually contains some iron, often occurs with cobaltite, and represents an ore of cobalt and nickel. Syn: *smaltine; tin-white cobalt; gray cobalt; white cobalt; speisscobalt.* (b) A term applied to undetermined, apparently isometric arsenides of cobalt or to a mixture of cobalt minerals.

smaragd *emerald* [mineral].

smaragdite A fibrous or thinly foliated green amphibole (near actinolite) pseudomorphous after pyroxene (such as omphacite) in rocks such as eclogite.

smectite (a) A disapproved name for the *montmorillonite* group of clay minerals. The name is in common use in England to designate dioctahedral (montmorillonite) and trioctahedral (saponite) clay minerals (and their chemical varieties) that possess swelling properties and high cation-exchange capacities. Syn: *montmorillonite-saponite.* (b) A term that was originally applied to fuller's earth and later to the mineral montmorillonite, and that has also been applied to certain clay deposits that are apparently bentonite and to a greenish variety of halloysite (Kerr & Hamilton, 1949, p.59).

smirnovite *thorutite.*

smithite A red monoclinic mineral: AgAsS$_2$.

smithsonite (a) A white or nearly white to yellow, gray, brown, or greenish mineral: ZnCO$_3$. It is a secondary mineral associated with sphalerite and often found as a replacement in limestone; it is commonly reniform, botryoidal, stalactitic, or granular, and is distinguished from hemimorphite by its effervescence with acids. Smithsonite is an ore of zinc. Syn: *dry-bone ore; calamine; zinc spar; szaskaite.* (b) A term sometimes used as a syn. of *hemimorphite.*

smokeless coal (a) *semibituminous coal.* (b) Any coal that burns without smoke; from semibituminous to superanthracitic in rank.

smokestone *smoky quartz.*

smoking crest The crest of a dune, along which sand grains are being winnowed.

smoky quartz A smoky-yellow, smoky-brown, or brownish-gray and often transparent variety of crystalline quartz sometimes used as a semiprecious gemstone. It often contains inclusions of both liquid and gaseous carbon dioxide. The color is probably due to some organic compound. Syn: *cairngorm; smokestone.*

smoky topaz A trade name for smoky quartz used for jewelry.

smolnitz Var. of *smonitza.*

smonitza A hydromorphic soil of Yugoslavia that is black to dark gray, has a surface that is leached of calcium carbonate, and is developed over a calcareous clay and underlying sand. Also spelled: *smolnitz.*

smooth chert A hard, dense, homogeneous chert (insoluble residue) characterized by a conchoidal to even fracture surface that is devoid of roughness and by a lack of crystallinity, granularity, or other distinctive structure (Ireland et al, 1947, p.1484). See also: *chalcedonic chert; ordinary chert; porcelaneous chert.* Cf: *granular chert; chalky chert.*

smooth phase The part of stream traction whereby a mass of sediment travels as a sheet with gradually increasing density from the surface downward (Gilbert, 1914, p.30-34). Cf: *dune phase; antidune phase.*

smothered bottom A term introduced by Shrock (1948,

p.307-308) for a sedimentary surface on which complete well-preserved, and commonly very fragile and delicate fossils were saved by an influx of mud that buried them instantly Such surfaces are common in sequences composed of alternating marine limestone and shale layers.

smythite A rhombohedral mineral: Fe$_3$S$_4$.

snake hole A horizontal (or nearly horizontal) borehole used for blasting, drilled approximately on a level with the floor of a quarry or under a boulder to be broken up.

snaking stream A winding, sinuous stream; a *meandering stream.*

snapper *grab sampler.*

Snell's law *law of refraction.*

snout [geog] A protruding mass of rock; a promontory.

snout [glaciol] The protruding lower extremity, leading edge or front of a glacier. Syn: *terminal face; terminus; glacier snout; glacier front; ice front; front* [glaciol].

snow (a) A form of ice composed of small, white or translucent, delicate, often branched or star-shaped hexagonal (tabular or prismatic) crystals of frozen water, formed directly by sublimation of atmospheric water vapor around solid nuclei at a temperature below the freezing point. The crystals grow while floating or falling to the ground, and are often agglomerated into snowflakes. (b) A consolidated mass of fallen snow crystals. (c) *snowfall.* (d) A region covered with permanent snow. The term is usually used in the plural, as the high *snows.*

snow avalanche An avalanche consisting of relatively pure snow, although considerable earth and rock material may also be carried downward. Syn: *snowslide.*

snowball A *slump ball* or a *spiral ball.* Term used by Hadding (1931, p.390) for a structure attributed to subaqueous sliding. The name is misleading as the ball did not roll down a slope picking up new layers of sediment (Kuenen, 1948, p.369).

snowball garnet *rotated garnet.*

snowbank A flat-topped mound of snow, often having a broad or long base.

snowbank glacier *nivation glacier.*

snow banner A stream of snow blowing off a mountain top, streaming out several miles from its source.

snow barchan A crescentic or horseshoe-shaped *snow dune* of windblown snow with the ends pointing downwind. Syn: *snow medano.*

snow blanket A surface accumulation of snow.

snowblink A bright, white glare in the sky near the horizon or on the underside of a cloud layer, produced by light reflected from a snow-covered surface (as a snowfield); brighter than *iceblink.* Also spelled: *snow blink.* Syn: *snow sky; snow sheen.*

snowbreak A protective barrier that shelters an area or object against drifting snow; e.g. planted trees along a road cut, or a fence on the windward side of a railroad track.

snowbridge An arch or layer of snow that has drifted across a crevasse in a glacier or a connecting splinter of ice or snow which allows a person or vehicle to cross a crevasse. Also spelled: *snow bridge.* Not to be confused with *bridge* [snow].

snow cap (a) A covering of snow on a mountain peak or ridge when no snow exists at lower elevations. (b) An accumulation of snow on the surface of a frozen lake.---Also spelled: snowcap.

snow concrete Snow that is compacted at low temperatures by heavy objects (as by a vehicle) and that sets into a tough substance of considerably greater strength than uncompressed snow. Syn: *snowcrete.*

snow cone A cone of snow, formed by fine snow sifting through a small opening.

snow cornice *cornice.*

snow course A line or series of connecting lines of regularly spaced observation stations (usually not fewer than 10) at which snow samples are taken for measuring depth, density, and water equivalent for forecasting subsequent runoff. See also: *snow survey*.

snow cover (a) All snow that has accumulated on the ground, including that derived from snowfall, drifting or blowing snow, avalanches, rain stored in the snow frozen or unfrozen, rime, and hoar. Syn: *snow mantle*. (b) The areal extent of ground partly or wholly covered with snow in a particular area, usually expressed as a percent of the total area. (c) The average depth of accumulated deposited snow on the ground in a particular area, usually expressed in centimeters.----Syn: *cover* [*snow*].

snowcreep The slow internal deformation of a snowpack resulting from the stress of its own weight and metamorphism of snow crystals; it usually involves shear parallel to the slope and compaction perpendicular to the slope. Cf: *snow glide; settling* [*snow*].

snowcrete *snow concrete.*

snow crust A firm or hard surface of snow overlying a layer of softer snow; it is formed by the melting and refreezing of surface snow, wind compaction, the freezing of rainwater on the surface, etc. Cf: *ice crust* [*glaciol*]; *film crust; wind crust; sun crust; rain crust.*

snow crystal A single ice crystal precipitated in the atmosphere or found in deposited snow. Cf: *snowflake.*

snow cushion An accumulation of snow, commonly deep, soft, and unstable, deposited in the lee of a cornice on a steep mountain slope.

snow density The weight of snow per unit volume, usually given in kg/m³ or mg/m³. It may also be defined as the ratio of the weight of a unit volume of snow to the weight of a unit volume of water, and the ratio of the volume of meltwater derived from a sample of snow to the original volume of the sample, but these two definitions are less preferred even though the same numerical results are obtained.

snowdrift (a) An accumulation, mound, or bank of snow piled together in a heap by the wind, usually in the lee of an obstruction or surface irregularity, and sometimes rising to heights of 30 m or more. (b) *drifting snow.*----Also spelled: *snow drift.*

snowdrift glacier *drift glacier.*

snow dune An accumulation of wind-transported snow resembling any of the many forms of sand dunes; e.g. *snow barchan.*

snow dust Fine snow crystals fragmented or driven by the wind.

snowfall (a) The deposition, on the ground or other surface, of snow precipitated out in the atmosphere. (b) The rate of deposition of snow.

snowfield (a) A broad, level, unbroken expanse of ground or ice covered with snow, relatively smooth and uniform in appearance, occurring usually at high latitudes or in mountainous regions above the snowline, and persisting throughout the year. (b) A region of permanent snow cover, as at the head of a glacier; the *accumulation area* of a glacier. (c) Any small glacier or accumulation of perennial snow and ice too small to be termed a glacier.

snowflake An aggregation or amalgamation of several single *snow crystals* that have become attached together while falling; it is the form in which snow reaches the ground. Syn: *flake* [*snow*].

snowflake obsidian An obsidian that contains white, gray, or reddish spherulites ranging in size from microscopic to a meter or more in diameter.

snowflush An accumulation of drifted snow, windblown soil, and wind-transported seeds on a lee slope, characteristically marked during the winter by a dark patch of soil (ADTIC 1955, p.75).

snow gage *rain gage.*

snow garland Snow festooned from trees or fences in the form of a rope, a few meters in length and 15-20 cm in diameter, formed and sustained near the melting point by the surface tension of thin water films bonding individual snow crystals.

snow glide The slow slip of a snowpack over the ground surface caused by the stress of its own weight. Cf: *snowcreep.*

snow grain (a) A mechanically separate particle of snow, often but not necessarily a single ice crystal. (b) A particle of *granular snow.*----Cf: *grain* [*glaciol*].

snow ice Ice that has been formed when snow slush, a mixture of snow and water, has frozen. It has a whitish appearance if air bubble inclusions are present.

snowline (a) A momentary line delimiting an area or altitude with complete snow cover, or in a zone of patchy snow the area or altitude of more than 50 percent snow cover. Syn: *transient snowline.* Cf: *climatic snowline.* (b) The line or altitude on land separating areas in which deposited snow disappears in summer from areas in which snow remains throughout the year; on glaciers it is identical to the *firn line.* Cf: *regional snowline; equilibrium line.* (c) The ever-changing extreme limit from the equator within which no snow reaches the ground or falls unmelted. Its position depends on such physical conditions as altitude and nearness to the sea. The term is applied esp. to the winter snowline in the Northern Hemisphere.----Also spelled: *snow line.*

snow mantle A term meaning an accumulation of snow and used as a syn. of *snow cover.*

snow medano *snow barchan.*

snowmelt The water resulting from the melting of snow. Also spelled: snow melt. Syn: *snow water.*

snow moisture *free water content.*

snow niche *nivation hollow.*

snowpack (a) Any snow cover. (b) The amount of annual accumulation of snow at higher elevations in mountains which provides water for hydroelectric power and irrigation.----Also spelled: *snow pack.*

snow patch An isolated mass of perennial snow and firn not large enough to be called a glacier.

snow-patch erosion *nivation.*

snow penitente A *nieve penitente* consisting of compacted snow.

snow pillow A device used to record the changing weight of the snow cover at a point; it consists of a fluid-filled bladder lying on the ground to which is connected a pressure transducer or a vertical pipe and float.

snowquake The sudden collapse of one or more layers of surface or subsurface snow, often accompanied by sound which may resemble that of a distant explosion (ADTIC, 1955, p.76). Syn: *snow tremor.*

snow resistograph An instrument for recording a hardness profile of a snow cover by recording the force required to move a blade up through the snow.

snow roller A cylinder or muff-shaped mass of moist, cohesive snow, formed and rolled along by the wind, esp. down a slope. It may be as large as 1.2 m in length and more than 2 m in circumference. Cf: *sun ball.*

snow sampler A snow-surveying instrument, in the form of a hollow tube, used to collect in situ a section or core of deposited snow, which is subsequently weighed to determine density and water equivalent.

snowshed A drainage basin primarily supplied by snowmelt.

snow sheen *snowblink.*

snow sky *snowblink.*

snowslide A *snow avalanche.* The term has also been used for a mass of downward-moving snow too small to be called a *snow avalanche.* Syn: *snowslip.*

snow sludge A soft elastic crust formed of fallen snow on a water surface.

snow slush *slush* [*snow*].

snow survey The process of determining the depth, density, and water equivalent of snow that has fallen on a particular area by sampling representative points along a *snow course*. Snow surveys made in the spring are used for forecasting subsequent snowmelt runoff.

snow swamp *slush field.*

snow tremor *snowquake.*

snow water *snowmelt.*

snub-scar A term used by Wentworth (1936) for a "push-off" end or edge (or lee-end pressure spall) characteristic of a glacial cobble.

soakaway A British term for a sink or depression in the Earth's surface, into which waters flow and naturally drain away.

soaked zone The area on a glacier where considerable surface melting occurs in summer, meltwater percolates through the whole mass of the snow layer bringing it all to the melting temperature. However, melting is not sufficient to remove all of the snow and snow persists at the surface at the end of summer. The soaked zone may be bordered at higher altitudes by the *saturation line* and at lower altitudes by the equilibrium line. Cf: *percolation zone.*

soap clay *bentonite.*

soap earth Massive talc; *steatite.*

soap hole A term used in Wyoming for a hole resulting from the wetting of the outcrop surface of bentonite.

soaprock (a) *soapstone* [mineral]. (b) *soapstone* [rock].

soapstone [mineral] (a) A mineral name applied to *steatite*, or massive talc. Syn: *soaprock.* (b) *saponite.* (c) A term loosely applied to much agalmatolite.

soapstone [rock] (a) A metamorphic rock of massive, schistose, or interlaced fibrous texture and soft, unctuous feel, composed essentially of talc with varying amounts of micas, chlorite, amphibole, pyroxenes, etc and derived from the alteration of ferromagnesian silicate minerals. (b) A miner's and driller's term for any soft, unctuous rock such as micaceous shale or sericitic schist. (c) A term used in Lancashire, NW England, for a smooth, fine-grained shale or mudstone (Rice, 1954).

soapy A type of mineral texture that is slippery, smooth, and soft; e.g. talc. Syn: *unctuous.*

socket A recess or depression along the hinge line of a bivalve for the reception of a projecting hinge tooth from the opposite valve; esp. a *dental socket* of a brachiopod.

socket ridge A linear elevation of secondary shell extending laterally from the cardinal process of a brachiopod and bounding the margin of dental sockets.

soda alum A cubic mineral of the alum group: $NaAl(SO_4)_2.12H_2O$. Cf: *mendozite.* Syn: *sodium alum.*

sodaclase *albite.*

soda feldspar A misnomer for *sodium feldspar.*

soda hornblende *arfvedsonite.*

soda lake An *alkali lake* whose waters contain a high content of dissolved sodium salts, chiefly sodium carbonate accompanied by sodium chloride and sodium sulfate. Examples occur in Mexico and Nevada. Syn: *natron lake.*

soda leucite A hypothetical sodium-rich variety of leucite, postulated as the original material of some pseudoleucites.

soda-lime feldspar A misnomer for *sodium-calcium feldspar.*

sodalite (a) A mineral of the feldspathoid group: $Na_4Al_3Si_3O_{12}Cl$. It is usually blue or blue-violet, but may be white, greenish, gray, pink, or yellow, and it occurs in various sodium-rich igneous rocks. (b) A group of bluish feldspathoid minerals containing sodium silicate, including sodalite, haüyne, nosean, and lazurite.

sodalithite An extrusive rock in which sodalite is the only light-colored mineral present, and olivine is absent.

sodalitite A coarse-grained igneous rock composed chiefly of sodalite, with smaller amounts of acmite, eudialyte, and alkali feldspar.

soda mica *paragonite.*

soda microcline A variety of microcline in which sodium re-

places potassium; specif. *anorthoclase.*

soda minette An alkalic minette containing alkali feldspar (specif. cryptoperthite), dark-brown mica, acmite, apatite and sphene.

soda niter A white or colorless, transparent, hexagonal mineral: $NaNO_3$. It is a deliquescent, soluble crystalline salt that occurs naturally, esp. in the crude form (as in caliche) i Chile, and that is associated with halite and sandy and claye material. Soda niter is a source of nitrates. Cf: *niter.* Syn: *n tratine; Chile saltpeter; Peru saltpeter.*

soda orthoclase *loxoclase.*

soda prairie A vast, level, barren tract of land covered with whitish efflorescence of sodium carbonate (natron), as i parts of SW U.S. and Mexico. Syn: *salt prairie.*

soda sanidine A mineral of the alkali feldspar group, containing 40-60% albite in solid solution.

soda straw A thin, tubular, hollow stalactite that maintains th diameter of a drop of water. Syn: *straw stalactite; soda straw tubular stalactite.*

soda tremolite A monoclinic mineral of the amphibole group $Na_2CaMg_5Si_8O_{22}(OH)_2$. It differs from tremolite in having so dium in place of half of the calcium. Cf: *richterite.*

soddyite A pale-yellow orthorhombic mineral: $(UO_2)_5Si_2 O_9.6H_2O$. Syn: *soddite.*

sodic soil A *salt-affected soil* having a high content of sodium Cf: *saline soil; saline-alkali soil.*

sodium alum *soda alum.*

sodium autunite A yellow mineral of the autunite group: $Na(U O_2)(PO_4).4H_2O$.

sodium-calcium feldspar A syn. of *plagioclase.* See also *soda-lime feldspar; lime-soda feldspar.*

sodium feldspar An alkali feldspar containing the Ab molecul $(NaAlSi_3O_8)$; specif. *albite.* See also: *soda feldspar.*

sodium illite *brammallite.*

sodium uranospinite A yellow-green to lemon and straw-yello secondary mineral: $(Na_2,Ca)(UO_2)_2(AsO_4)_2.5H_2O$.

soengei *sungei.*

sofar (a) A triangulation method of determining the location o an underwater explosion from points on shore. (b) A soun channel in the deep ocean which propagates acoustic wave for long distances with little attenuation.----Etymol: an acro nym; sound fixing and ranging. Also spelled: SOFAR.

soffione Steam-type fumaroles; the term was originally applie to boric-acid fumaroles of the Tuscany region of Italy. Etymo Italina *sòffio*, "puff" or "blast".

soft coal A syn. of *bituminous coal.* Cf: *hard coal;* a syn. *anthracite.*

softening Reduction of the hardness of water by removin hardness-forming ions (chiefly calcium and magnesium) b precipitation or ion exchange, or sequestering them as b combining them with substances, such as certain phosphates that form soluble but nonionized salts.

soft ground [eng] (a) Ground that is too moist or too yieldin to support weight and thereby allows an object to sink in. (b Rock that does not stand well and requires heavy timbering such as that about an underground opening.

soft ground [mining] In mining geology, that part of an or which can be mined without drilling and blasting. It is usuall the upper, weathered portion of the vein.

soft hail *graupel.*

soft magnetization Magnetization that is easily destroyed; spe cifically, remanent magnetization with a small coercivity. C hard magnetization.

soft mineral A mineral that is softer than quartz, i.e. less tha seven according to Moh's scale. Cf: *hard mineral.*

soft ore A term used in the Lake Superior region for an earth y, incoherent iron ore mainly composed of hematite or limo nite (goethite) and containing 45-60% iron. Cf: *hard ore.*

soft rock (a) A term used loosely for a sedimentary rock, a distinguished from an igneous or metamorphic rock. (b) A ock that is relatively nonresistant to erosion. (c) Rock that an be removed by air-generated hammers, but cannot be andled economically by pick. (d) A term used loosely by drilers for a post-Cretaceous sedimentary rock (typically unconolidated sandstone or shale) that is drilled relatively rapidly and that produces samples difficult to classify as to exact depth.----Ant: *hard rock.*

soft-rock geology A colloquial term for geology of sedimentary rocks, as opposed to *hard-rock geology.*

soft shore A shore composed of peat, muck, mud, or soft marl, or of marsh vegetation. Ant: *hard shore.*

soft water Water that lathers readily with ordinary soap; water containing not more than 60 mg/l of hardness-forming constituents expressed as $CaCO_3$ equivalent. Cf: *hard water; hardness* [*water*].

softwood The wood of a gymnosperm, lacking wood fibers. Actually, such wood may be either soft or hard. Cf: *hardwood.*

sogdianovite A mineral: $(K,Na)_2Li_2(Li,Fe,Al)_2ZrSi_{12}O_{30}$.

soggendalite A dark-colored dolerite containing abundant pyroxene. Cf: *mimosite.*

sogrenite A black organic material containing uranium.

sohngeite A mineral: $Ga(OH)_3$.

soil [lunar] *lunar regolith.*

soil [eng geol] All unconsolidated earthy material over bedrock. It is approximately equivalent to *regolith.*

soil [soil] (a) The natural medium for growth of land plants. (b) A term used in soil classification for the collection of natural bodies on the Earth's surface, in places modified or even made by man of earthy materials, containing living matter, and supporting or capable of supporting plants out-of-doors. The lower limit is normally the lower limit of biologic activity, which generally coincides with the common rooting of native perennial plants.----Etymol: Latin *solum,* "ground".

soil [lake] The bed or bottom of a lake.

soil association Two or more soils in a given geographic area that are distinguishable among themselves but that, on all but very detailed soil maps, are grouped together on the basis of their common characteristics, because of their intricate areal distribution. See also: *catena.* Syn: *soil complex.*

soil atmosphere The part of ground air that is in the soil and is similar to the air of the atmosphere but depleted or enriched in certain constituents, such as carbon dioxide. Cf: *subsurface air; ground air.*

soil binder A grass or other plant capable of preventing soil erosion by forming dense mats or roots; also, in an engineering sense, a fine material such as clay that gives cohesiveness to a coarse aggregate such as sand and gravel.

soil blister *frost mound.*

soil caliche Calcium carbonate leached from the topsoil in an arid to subhumid region (seasonal rainfall less than 30 in.) and concentrated in the lowermost part of the B horizon and the upper part of the C horizon in a developed zonal soil; it occurs as a continuous zone, several centimeters to a few meters thick, and is well cemented in the upper part. See also: *caliche.*

soil category One of any group of taxonomically related soils, according to any system of soil classification.

soil circle A term used loosely for any circular form of patterned ground developed on a soil surface, either sorted or nonsorted, and with or without vegetation. Syn: *earth circle.*

soil climate The moisture and temperature of a soil.

soil colloids The inorganic and organic matter in soils having very small particle size and a correspondingly large surface area per unit of mass. The term includes much, but not all, the clay and humus in soils.

soil complex *soil association.*

soil-cover complex (a) A group of similar areas in which the soils, slopes, litter, and vegetation cover have comparable physical characteristics (Chow, 1964, p.22-51). (b) The com-

bination of a specific soil and a specific vegetation cover, used as a parameter in estimating the runoff in a drainage basin (Chow, 1964, p.21-11).

soil creep The gradual and steady downhill movement of soil and loose rock material on a slope that may be very gentle but is usually steep. Syn: *surficial creep.*

soil discharge The release of water from the soil by evaporation and transpiration. The water may have been derived from the soil or from the zone of saturation by way of the capillary fringe. Syn: *soil evaporation.* See also: *vadose-water discharge.*

soil erosion Detachment and movement of topsoil, or soil material from the upper part of the profile, by the action of wind or running water, esp. as a result of changes brought about by human activity (such as unsuitable or mismanaged agricultural methods). It includes: rill erosion, gully erosion, sheet erosion, and wind erosion.

soil evaporation *soil discharge.*

soilfall A *debris fall* involving soil material.

soil family In early U.S. soil classification, a group of soils with a wider range in characteristics than those of a soil series but narrower than those of a soil subgroup. A group has restricted ranges in particle-size distribution in horizons of major biologic activity below plow depth, in mineralogy of the same horizons, in temperature regimes, and in thickness of the soil penetrable by roots.

soil fertility The status of a soil with respect to the amount and availability to plants of elements necessary for plant growth (SSSA, 1970, p.7). It is not synonymous with *soil productivity.*

soil flow *solifluction.*

soil fluction *solifluction.*

soil formation *soil genesis.*

soil-formation factors The natural conditions and substances that interact to produce a soil: parent material, climate, plants and other organisms, topography, and time.

soil genesis (a) The mode of origin of the soil, with special reference to the processes of soil-forming factors responsible for the development of the solum, or true soil, from unconsolidated parent material. (b) A division of soil science concerned with soil genesis.----Syn: *soil formation; pedogenesis.*

soil horizon A layer of a soil that is distinguishable from adjacent layers by characteristic physical properties such as structure, color, or texture, or by chemical composition, including content of organic matter, or degree of acidity or alkalinity. Soil horizons are generally designated by a capital letter, with or without a numerical annotation, e.g. A horizon, A_2 horizon. Syn: *horizon; soil zone.*

soil map A map showing the distribution of kinds of soil in relation to prominent physical and cultural features of the Earth's surface. Kinds of soil are expressed in terms of soil taxonomic units, such as series, or as phases of series. Maps showing single soil characteristics or qualities, such as slope, texture, depth fertility, or erodibility are not soil maps.

soil material The unit of study in pedography in which the characteristics being studied are relatively constant and whose size will vary "with the kind and extent of development of those characteristics" (Brewer, 1964, p.10).

soil mechanics The application of the laws and principles of mechanics and hydraulics to engineering problems dealing with the behavior and nature of soils, sediments, and other unconsolidated accumulations of solid particles produced by the mechanical disintegration and chemical decomposition of rocks, regardless of whether or not they contain an admixture of organic constituents (Terzaghi, 1943). It is the detailed and systematic study of the physical properties and utilization of soils, esp. in relation to highway and foundation engineering and to the study of other problems relating to soil stability.

soil moisture *soil water.*

soil-moisture tension *moisture tension.*

soil-moisture weathering Accelerated weathering of granite below an old soil line, often causing the steepening of margins of granitic inselbergs (Stone, 1967, p.249).

soil order A group of soils in the broadest category. For example, in early U.S. classification, the three soil orders are zonal soil, intrazonal soil, and azonal soil. Orders are divided into suborders and great soil groups. Names of soil orders, as well as of great soil groups, are usually but not invariably capitalized.

soil patterns Obsolete syn. of *patterned ground*.

soil phase A subdivision of any taxonomic unit in any category of the natural system of soil classification that is based upon any characteristic or combination of characteristics significant to use and management of the soils. Most phases are subdivisions of soil series.

soil physics The organized body of knowledge concerned with the physical characteristics of soils and with the methods and instruments used in determining these characteristics.

soil polygon A group term for forms of polygonal ground developed on a soil surface, frequently in permafrost areas but also in regions where contraction occurs (as in playa lakes and deserts), and occurring with or without a stone border. Diameter: a few millimeters to many tens of meters. The term is misleading because soil need not be present.

soil productivity The capacity of a soil, in situ, to produce a specified plant or sequence of plants under a specified system of management. (SSSA, 1970), p.13). The term is not synonymous with *soil fertility*.

soil profile A vertical section of a soil which displays all its horizons and its parent material.

soil reaction The degree of a soil's acidity or alkalinity, expressed by its pH value.

soil science The study of soil as a natural resource; soil formation, properties, classification, and mapping are included. See also: *pedogenics; pedography*. Syn: *pedology*. Obs syn: *agrology*.

soil separate A group of rock and mineral particles in the soil, obtained in separation (as in mechanical analysis), having equivalent diameters less than 2 mm, and ranging between specified size limits (from "very coarse sand" to "clay"). Cf: *coarse fragment*. Syn: *separate*.

soil series The basic unit in soil classification and more specific than a soil family; a group of soils having genetic horizons of similar characteristics and arrangement in the soil profile (or pedon), except for texture of the surface soil, and developed from a particular type of parent material.

soil slip *debris slide*.

soil solution Soil water considered as a solution of various salts, organic compounds, gases, etc., that are of significance to plant growth or to the consequences of flushing the soil solution to a body of ground or surface water.

soil stabilization Chemical or mechanical treatment designed to increase or maintain the stability of a soil mass or otherwise to improve its engineering properties (ASCE, 1958, term 337), such as by increasing its shear strength, reducing its compressibility, or decreasing its tendency to absorb water. Stabilization methods include physical compaction and treatment with cement, lime, and bitumen.

soil-stratigraphic unit A soil whose physical features and stratigraphic relations permit its consistent recognition and mapping as a stratigraphic unit (ACSN, 1961, art.18). It is formed essentially in place from underlying rock-stratigraphic units that may be of diverse composition and geologic age, and it may comprise one or more *pedologic units* or parts of units. The definition of a soil-stratigraphic unit should be based on as full knowledge as possible of its lateral variations and should be independent of concepts based on geologic history.

soil strip *soil stripe*.

soil stripe A *sorted stripe* whose texture is considerably finer than that of a *stone stripe*. Syn: *soil strip; earth stripe*.

soil structure [pat grd] A term formerly used (usually in the plural) by Sharp (1942, p.275) for *patterned ground*, but now discarded because the implied presence of humus and a soil profile may be absent in patterned ground. Syn: *structure soil*.

soil structure [soil] The combination or aggregation of primary soil particles into compound particles, or clusters of primary particles (peds), which are separated from adjoining peds by surfaces of weakness. Soil structure is classified on the basis of size, shape, and distinctness into classes, types, and grades respectively.

soil survey A general term for the systematic examination of soils in the field and in the laboratories, their description and classification, the mapping of kinds of soil, and the interpretation of soils for many uses, including their suitabilities or limitations for growing various crops, grasses, and trees, or for various engineering uses, and predicting their behavior under different management systems; for growing plants, and for engineering uses.

soil type A phase or subdivision of a soil series based primarily on texture of the surface soil to a depth at least equal to plow depth (about 6 inches). In Europe, the term is roughly equivalent to the term *great soil group*.

soil ulmin *humus*.

soil water Water in the belt of soil water. Syn: *soil moisture; rhizic water*.

soil-water belt *belt of soil water*.

soil-water zone *belt of soil water*.

soil zone *soil horizon*.

sol (a) A homogeneous suspension or dispersion of colloidal matter in a fluid (liquid or gas). (b) A completely mobile mud. ---A sol is in a more fluid form than a *gel*.

sola Plural of *solum*.

solar attachment An auxiliary instrument that may be attached to a surveyor's transit or compass for determining the true meridian directly from the Sun. Cf: *solar compass*.

solar compass A surveying instrumnet that attains a similar result as that of the *sun compass* and that permits the establishment and surveying of the astronomic meridian or astronomic parallel directly by observation on the Sun. It has been replaced by the *solar attachment* in combination with a transit.

solar constant The rate at which solar radiant energy is received outside the atmosphere on a surface normal to the incident radiation, at the Earth's mean distance from the Sun. The value of the mean solar constant is 1.94 gram calories per minute per square centimeter.

solarimeter *pyranometer*.

solar infrared *near infrared*.

solar lake A lake that has no connection to the sea and whose water temperature and salinity increase with depth.

solar salt Coarsely crystalline salt obtained by evaporating seawater or other brine by the heat of the sun. Syn: *bay salt*.

solar system The Sun and the celestial bodies which are in orbit around it, and the unknown region beyond the orbit of Pluto, the outermost planet.

solar tide The part of the tide caused solely by the tide-producing force of the Sun. Cf: *lunar tide*.

solar wind The motion of interplanetary plasma or ionized particles away from the Sun and towards the Earth, near which it interacts with the Earth's magnetic field (McIntosh, 1963, p.235).

sole [geol] (a) The relatively flat undersurface of a rock body or a vein; esp. the bottom of a sedimentary stratum. (b) The lowest part of a valley.

sole [mass move] The middle and lower, more gently inclined, portion of the shear surface of a landslide.

sole [fault] The underlying fault plane of a thrust nappe. See also: *sole fault; lubricating layer*. Syn: *sole plane*.

sole [glaciol] The lower part or basal ice of a glacier, often containing rock fragments and dirty in appearance, and often separated from clean ice by an abrupt boundary.

sole cast A *sole mark* preserved as a raised swelling or positive feature on the underside of a bed immediately overlying the upper surface of a finer-grained bed containing a primary sedimentary structure (depression).

soled boulder A stone with blunted corners and smoothed or flattened (and sometimes striated) sides, esp. one shaped by glacial grinding.

sole fault A low-angle thrust fault forming the *sole* of the thrust nappe; also, the basal main fault of an imbrication. Syn: *décollement fault; detachment fault; detachment thrust; basal thrust plane.*

sole injection An igneous intrusion that was emplaced along a thrust plane.

sole mark A general and descriptive term applied, without regard to origin or orientation, to a directional structure or to a small, wave-like, mainly convex irregularity or penetration found on the original underside or lower surface of a bed of sandstone or siltstone (rarely limestone) along the contact where it overlies a softer and finer-grained layer (such as shale). The term usually refers to a filling of a primary sedimentary structure (such as a crack, track, groove, or other depression) formed on the upper surface of the underlying mud by agents such as currents, organisms, and unequal loading, and preserved as a *sole cast* after the underlying material had consolidated and weathered away. Examples: load cast, flute cast, and groove cast. Syn: *sole marking.*

Solenhofen stone A *lithographic limestone* of Upper Jurassic age found at Solenhofen (Solnhofen), a village in Bavaria, West Germany. It is evenly and very thinly stratified and contains little clay.

sole plane Var. of *sole.*

solfatara A type of *fumarole*, the gases of which are characteristically sulfurous. Cf: *solfataric stage.* Etymol: the Solfatara volcano, Italy.

solfataric stage A late or decadent type of volcanic activity characterized by the emission of sulfurous gases from the vent. See also: *solfatara.* Cf: *fumarolic stage.*

solid Sedimentary material that is in solution or suspension but when freed of solvent or suspending medium has the form and properties of a solid. The term is usually used in the plural; e.g. *dissolved solids.*

solid diffusion Diffusion through a rock that remains solid throughout the process, particularly if it is metasomatic in nature (Challinor, 1967). Cf: *metamorphic diffusion.*

solid earth *lithosphere.*

solid flow Flow in a solid by rearrangement among or within the constituent particles. Cf: *liquid flow; viscous flow.*

solid geology A British term for *bedrock* geology.

solidification The process of becoming solid or hard; esp. the change from the liquid to the solid state on the cooling of a magma. The term *lithification* is more generally applied in the case of sedimentary rocks. See also: *consolidation.*

solid map A British term for a geological map showing the extent of *solid rock*, on the assumption that all surficial deposits, other than alluvium, are absent or removed (Nelson & Nelson, 1967, p.352). Cf: *drift map.*

solid rock A British term for consolidated bedrock.

solid solution A single crystalline phase which may be varied in composition within finite limits without the appearance of an additional phase. Syn: *mix-crystal; mixed crystal.*

solid-solution series *isomorphous series.*

solid stage That stage in the cooling of a magma during which the magma becomes completely solid, but while the magma is still present below.

solidus On a temperature-composition diagram, the locus of points in a system at temperatures above which solid and liquid are in equilibrium and below which the system is completely solid. In binary systems without solid solutions, it is a straight line; in binary systems with solid solutions, it is a curved line or a combination of straight and curved lines; like-wise, in ternary systems, it is a flat plane or a curved surface, respectively.

soliflual Said of debris resulting from solifluction (Baulig, 1957, p.927). Syn: *solifluidal.*

solifluction (a) The slow (normally 0.5-5.0 cm/yr), viscous, downslope flow of waterlogged soil and other unsorted and saturated surficial material; esp. the flow occurring at high elevations in regions underlain by frozen ground (not necessarily permafrost) acting as a downward barrier to water percolation, initiated by frost action and augmented by meltwater resulting from alternate freezing and thawing of snow and ground ice. The term was proposed by Andersson (1906, p.95-96) as "the slow flowing from higher to lower ground of masses of waste saturated with water", but as he did not state explicitly that it referred to flow over frozen ground, the term has been extended to include similar movement in temperate and tropical regions; also, it has been used as a syn. of *soil creep* although solifluction is generally more rapid. It is preferable to restrict the term to slow soil movement in periglacial areas. Syn: *soil flow; solifluxion; soil fluction; sludging.* (b) *subaqueous solifluction.*

solifluction lobe An isolated, tongue-shaped feature, up to 25 m wide and 150 m long, formed by more rapid solifluction on certain sections of the slope showing variations in gradient. It commonly has a steep front (15°-25°) and a relatively smooth upper surface. Syn: *solifluction tongue.*

solifluction mantle The unsorted, water-saturated, locally derived material moved downslope by solifluction. Syn: *flow earth.*

solifluction sediment A sediment that has resulted from solifluction.

solifluction sheet A broad deposit of *solifluction mantle*, occurring evenly across a wide slope.

solifluction slope A smooth curvilinear slope of 2° to 30° produced by solifluction or along which solifluction occurs.

solifluction step The flattish area at the front of a small solifluction lobe; the tread of a small, turf-banked terrace usually restricted to immediately above timberline.

solifluction stream A narrow, laterally confined, stream-like deposit of *solifluction mantle.*

solifluction stripe A form of *striped ground* associated with solifluction. The term was formerly used by Washburn (1947, p.94) as a syn. of *nonsorted stripe*, but solifluction may also be associated with sorted stripes.

solifluction terrace A low terrace or bench formed by solifluction at the foot of a slope; it may have a lobate margin reflecting uneven movement.

solifluction tongue *solifluction lobe.*

solifluidal *soliflual.*

solifluxion *solifluction.*

soligenous Said of peat deposits whose moisture content is dependent both on rainfall and on surface water. Cf: *ombrogenous; topogenous.*

solimixtion A term introduced by Rosauer (1957, p.65) for "a relatively homogeneous blending in the vertical plane of two different materials due to frost action with little or no macro-optical structures", as at the contact of two different layers in a loess profile.

soliqueous A term proposed by Leet & Leet (1965, p.620) to describe a state of matter, such as the materials in the Earth's mantle, that is neither solid, liquid, nor gaseous, but a mixture of all three; it is maintained and controlled by pressure, and it has neither a crystalline structure nor the chilled-liquid molecular arrangements of a glass.

solitaire A single diamond or sometimes other gem, set alone; a diamond *nonpareil.*

solitary coral A coral that does not form part of a colony; an individual corallite (of a polyp) that exists unattached to other corallites. Cf: *colonial.* Syn: *simple coral.*

solitary wave A wave consisting of a single elevation (above the water surface), its height not necessarily small compared to the depth, and neither followed nor preceded by another elevation or depression of the water surface.

Solod soil *Soloth soil.*

Solonchak soil One of an intrazonal, halomorphic group of soils containing much soluble salts and usually having a light color but no characteristic structure; a *saline soil*. It is developed under salt-tolerant vegetation, under conditions of a semi-arid or desert climate and poor drainage. Cf: *Solonetz soil; Soloth soil.*

Solonetz soil One of an intrazonal group of soils that is a black alkali soil formed by the leaching of salts from a saline or *Solonchak soil*. It has a characteristic columnar structure. See also: *Soloth soil.*

Soloth soil One of an intrazonal, halomorphic group of soils that has developed from saline material; a degraded, desalinized, decalcified *Solonetz soil*. It has a brown, friable surface layer, below which is a light-colored, leached horizon and an underlying dark horizon. Cf: *Solonchak soil*. Also spelled: *Solod soil.*

solstitial tide A tide occurring near the times of the solstices, when the Sun reaches a maximum north or south declination.

soluan A *cutan* consisting of crystalline salts, such as carbonates, sulfates, and chlorides of calcium, magnesium, and sodium (Brewer, 1964, p.216).

solubility The equilibrium concentration of a solute in a solution saturated with respect to that solute at a given temperature and pressure.

solubility product A syn. of *dissociation constant* that refers to a very slightly soluble compound.

solum The upper part of a soil profile, in which soil-forming processes occur. In a mature soil, the A and B horizons constitute the solum. Plural: *sola*. Syn: *true soil.*

solusphere That zone of the Earth in which water solutions affect geologic, chemical and life processes.

solution A process of chemical weathering by which rock material passes into solution; e.g. the dissolution and removal of the calcium carbonate in limestone or chalk by carbonic acid derived from rainwater containing carbon dioxide acquired during its passage through the atmosphere.

solution basin A shallow surface depression, either man-made or natural, produced by solution of surface material, or resulting from the settlement of a surface through the removal in solution of underlying material (such as salt or gypsum); specif. a *solution depression* in a karstic region.

solution breccia A *collapse breccia* formed where soluble material has been partly or wholly removed by solution, thereby allowing the overlying rock to settle and become fragmented; e.g. a breccia consisting of chert fragments gathered from a limestone whose carbonate material has been dissolved away. See also: *evaporite-solution breccia*. Syn: *ablation breccia.*

solution channel Tubular or planar channel formed by solution in carbonate-rock terranes, usually along joints and bedding planes. It is the main water carrier in carbonate rocks. Cf: *solution opening.*

solution collapse Collapse due to the solution of underlying rock, e.g. karstic activity. Cf: *solution subsidence.*

solution depression A general term for *solution basin* occurring in a karst region; a *doline*. Syn: *solution sink.*

solution facet A nearly plane face developed on the surface of a rock fragment (such as a pebble or boulder of limestone exposed above the land surface) by progressive solution (esp. by freely falling rain), and bounded by a narrow rim or raised edge (Scott, 1947, p.144). Examples occur on pebbles in the semiarid regions of New Mexico and southern Texas.

solution groove One of a series of continuous, subparallel furrows developed on an inclined or vertical surface of a soluble and homogeneous rock (such as the limestone walls of a cave) by the slow corroding action of trickling water.

solution lake (a) A syn. of *karst pond*. (b) A lake occupying a

basin formed by surface solution of bedrock.

solution load *dissolved load.*

solution mining The in-place dissolution of water-soluble mineral salts of an ore with a leaching solution; a type of *chemical mining.*

solution-morel Said of a pebble, feature, surface, or sculpturing characterized by a pattern (developed by solution) of branching ridges and furrows broken by a few pinnacles rising above the level of the ridges, and resembling the pattern of the morel mushroom (Scott, 1947, p.144). The pattern of solution-morel pebbles requires a convex surface, impure inhomogeneous limestone, and an arid to semiarid climate. See also: *morel basin.*

solution opening (a) An opening produced by direct solution by water penetrating pre-existing interstices. (b) An opening resulting from the decomposition of less soluble rocks by water penetrating pre-existing interstices, followed by solution and removal of the decomposition products. (c) *solution channel.*

solution pan In a region of karst topography such as a bare limestone plain, a very shallow, flat-bottomed basin with overhanging sides, formed by solution. It ranges in size from a few centimeters to several meters in diameter and from a centimeter to a meter in depth. Syn: *panhole; etched pothole; tinajita; kamenitza.*

solution pendant *pendant.*

solution pipe A vertical, cylindrical hole in carbonate rock, formed by solution and often without surface expression, that has become filled with detrital matter (Monroe, 1970).

solution plane Lines of chemical weakness in a crystal, along which solution tends to occur under certain physical circumstances, e.g. great pressure.

solution ripple A *flute* resembling an aqueous current ripple mark, formed on the walls, floor, or ceiling of a cave by solution.

solution sink *solution depression.*

solution subsidence Subsidence due to the solution of underlying rock, e.g. karstic activity. Cf: *solution collapse.*

solution transfer The process of *pressure solution* of detrital grains at points of contact, followed by chemical redeposition of the dissolved material on the less-strained parts of the grain surfaces. See also: *Riecke's principle.*

solution valley *karst valley.*

Solvan European stage: Middle Cambrian (above Caerfaian, below Menevian).

solvate A chemical compound consisting of a dissolved substance and its solvent, e.g. hydrated calcium sulfate.

solvation The chemical union of a dissolved substance and its dissolving liquid.

solvsbergite A fine-grained, holocrystalline, rarely porphyritic hypabyssal rock composed chiefly of sodic feldspar and a smaller amount of potassium feldspar, sodic pyroxene or amphibole, and little or no quartz. Cf: *lindoite.*

solvus The curved line in a binary system or the surface in a ternary system that separates a field of homogeneous solid solution from a field of two or more phases which may form from the homogeneous one by exsolution. Cf: *hypersolvus; subsolvus.*

somal unit A stratigraphic unit that intertongues laterally with its neighbor; e.g. lithosome or biosome.

somite One of the longitudinal series of body segments into which many animals (such as articulates and vertebrates) are more or less distinctly divided; esp. the basic embryologic unit of segmentation of the body of an arthropod, approximately equivalent (except where fusion of somites has occurred) to the part of the body covered by a single exoskeletal "body ring" (often divided into tergite and sternite) and bearing no more than one pair of limbs.

somma n. A circular or crescentic ridge that is steep on its inner side and represents the rim of an ancient volcanic crater caldera. Etymol: Mt. Somma, the ancient crater rim that surrounds Vesuvius. Syn: *somma ring.*----adj. Said of a volcanic crater with a central cone surrounded by a somma.

sommaite An essexite which contains leucite instead of nepheline and sanidine instead of orthoclase. Its name is derived from Monte Somma, Italy. Cf: *ottajanite.*

somma ring *somma.*

sonar An acronym that means sound navigation and ranging, a method used in oceanography to study the ocean floor.

sondalite A metamorphic rock consisting of cordierite, quartz, garnet, tourmaline, and kyanite (Holmes, 1928, p.213).

sonde A downhole device containing the measuring instrument in logging a well or borehole, lowered on a wire line; e.g. a circular container used in electric logging and in which the electrode devices are set, or a device containing a powerful neutron-emitting source used in neutron logging. Pron: *sahnd.* Syn: *tool.*

song of the desert The booming or roaring sound made by a sounding sand on a desert. Syn: *voices of the desert.*

sonic depth finder *echo sounder.*

sonic-layer depth The depth of the surface layer of the ocean into which acoustic waves are trapped by upward refraction.

sonic log A *geophysical log* made by an instrument, lowered and raised in the borehole or well, that continuously records as a function of depth the velocity (or interval time) of sound waves as they travel over short distances (often less than one meter) in the adjacent rocks. Sonic logs reflect lithologic changes and are used in correlation and formation evaluation (the velocity is related to porosity and to the nature of the liquid occupying the pores) and in showing the depth of fluid level in a well. Syn: *acoustic log; velocity log; continuous velocity log.*

sonic wave *acoustic wave.*

sonobuoy A buoy, either anchored or free, that contains a hydrophone which transmits a radio signal when it detects an acoustic wave in the water. It is used in seismic surveying to help determine locations.

sonograph A type of seismograph developed by Frank Rieber for the application of reflection methods to areas of complex geology and steeply dipping beds. The ordinary oscillograph traces are replaced by "sound tracks" of variable transparency on a moving picture film, and the analyzer adds up impulses which are in phase while the random effects tend to cancel one another.

sonolite A mineral: $Mn_9(SiO_4)_4(OH,F)_2$.

sonoprobe A type of echo sounder that generates sound waves and records their reflections from inequalities beneath a sedimentary surface.

Sonstadt solution A solution of mercuric iodide in potassium iodide that is used as a *heavy liquid;* its specific gravity is 3.2. Cf: *bromoform; Klein solution; Clerici solution; Sym: Thoulet solution.*

sooty chalcocite *sooty ore.*

sooty ore A black, pulverulent, impure variety of chalcocite, of supergene origin. Syn: *sooty chalcocite.*

sorbyite A mineral: $Pb_{17}(Sb,As)_{22}S_{50}$.

sordawalite *tachylite.*

soredium A mass of algal cells surrounded by fungus hyphae, extruded through the outer or upper cortex of a lichen.

sorensenite A mineral: $Na_4Be_2SnSi_6O_{16}(OH)_4$.

Soret effect A syn. of *thermodiffusion.* Var: *Soret action.*

sorkedalite An ultramafic, feldspathoid-free igneous rock resembling essexite or kjelsasite in composition but with high titanium, iron, and phosphorus contents.

sorosilicate A class or structural type of *silicate* characterized by the linkage of two SiO_4 tetrahedra by the sharing of one oxygen, with a Si:O ratio of 2:7. An example of a sorosilicate is hemimorphite, $Zn_4(Si_2O_7)(OH)\cdot H_2O$. Cf: *nesosilicate; cyclosilicate; inosilicate; phyllosilicate; tectosilicate.*

sorption water *pellicular water.*

sorted [pat grd] Said of a nongenetic group of patterned-ground features displaying a border of stones (including boulders) commonly surrounding or alternating with fines (including sand, silt, and clay). Ant: *nonsorted.*

sorted [part size] Said of an unconsolidated sediment or of a cemented detrital rock consisting of particles of essentially uniform size or of particles lying within the limits of a single grade; *graded.* See also: *well-sorted; moderately sorted; poorly sorted.*

sorted bedding A type of *graded bedding* in which only one particle size is present at each horizon within the bed, the size decreasing upward.

sorted circle A form of patterned ground "whose mesh is dominantly circular and has a sorted appearance commonly due to a border of stones surrounding finer material" (Washburn, 1956, p.827); developed singly or in groups. Diameter: a few centimeters to more than 3 m (some exceed 10 m); the stone border may be 35 cm high and 8-12 cm wide. See also: *debris island; stone pit.* Syn: *stone circle; stone ring; stone wreath; rock wreath; frost circle.*

sorted crack A form of nearly horizontal patterned ground consisting of a concentration of boulders along a straight line, as in the Swedish Caledonides.

sorted field *felsenmeer.*

sorted net A form of patterned ground "whose mesh is intermediate between that of a sorted circle and a sorted polygon and has a sorted appearance commonly due to a border of stones surrounding finer material" (Washburn, 1956, p.830). Diameter: a few centimeters to 3 m.

sorted polygon A form of patterned ground "whose mesh is dominantly polygonal and has a sorted appearance commonly due to a border of stones surrounding finer material" (Washburn, 1956, p.831); never developed singly. Diameter: a few centimeters to 10 m. Syn: *stone polygon; stone ring; stone net; stone mesh.*

sorted step A form of patterned ground "with a steplike form and a sorted appearance due to a downslope border of stones embanking an area of finer material upslope" (Washburn, 1956, p.833); formed in groups, rarely if ever singly. Dimensions: 1-3 m wide; up to 8 m long in downslope direction. See also: *stone garland; stone-banked terrace.*

sorted stripe One of the alternating bands of finer and coarser material comprising a form of patterned ground characterized by "a striped pattern and a sorted appearance due to parallel lines of stones and intervening strips of dominantly finer material oriented down the steepest available slope" (Washburn, 1956, p.836). It never forms singly, usually occurring as one of many evenly spaced, sometimes sinuous, stripes that often exceed 100 m in length on slopes as high as 30°. An individual stripe may be a few centimeters to 2 m wide, with the intervening area two to five times wider. See also: *soil stripe; stone stripe; striped ground.*

sorting (a) The dynamic process by which sedimentary particles having some particular characteristic (such as similarity of size, shape, or specific gravity) are naturally selected and separated from associated but dissimilar particles by the agents of transportation (esp. by the action of running water). (b) The result of sorting; the degree of similarity of sedimentary particles in a sediment. (c) A measure of sorting, or of the spread or range of the particle-size distribution on either side of an average.---Cf: *gradation* [part size].

sorting coefficient A *sorting index* developed by Trask (1932), being a numerical expression of the geometric spread of the central half of the particle-size distribution of a sediment, and defined as the square root of the ratio of the larger quartile, Q_1 (the diameter having 25% of the cumulative size-frequency distribution larger than itself), to the smaller quartile, Q_3 (the diameter having 75% of the cumulative size-frequency distribution larger than itself). It is indicative of the range of conditions present in the transporting fluid (velocities, turbulence, etc.) and to some extent indicative of the distances of transportation. A perfectly sorted sediment has a coefficient of 1.0; less perfectly sorted sediments have higher coefficients. The Trask sorting coefficient is not considered a useful particle-size measure and is no longer used by sedimentologists. Abbrev: So or So.

sorting index A measure of the degree of sorting or of uniformity of particle size in a sediment, usually based on the statistical spread of the frequency curve of particle sizes; e.g. *sorting coefficient* and *grading factor.*

sorus A cluster of sporangia on the leaf of a fern.

sótano In Mexico, a deep, vertical shaft in the limestone of a karst area, that may or may not lead to a cave (Monroe, 1970). Etymol: Spanish, "underground cellar".

soufriere A common name for a volcanic crater or area of solfataric activity, used especially in the West Indies and other French-speaking regions. Also spelled: *soufrière.*

sound (a) A relatively long, narrow waterway connecting two larger bodies of water (as a sea or lake with the ocean or another sea) or two parts of the same body, or an arm of the sea forming a channel between a mainland and an island; it is generally wider and more extensive than a *strait* [coast]. (b) A long, large, rather broad inlet of the ocean, generally extending parallel to the coast; e.g. Long Island Sound between New England and Long Island, N.Y. (c) A lagoon along the SE coast of the U.S.; e.g. Pamlico Sound, N.C. (d) A long bay or arm of a lake; a stretch of water between the mainland and a long island in a lake.

sound channel That region in a column of water in which the velocity of the sound waves changes from a pattern of decreasing due to depth to a pattern of increasing due to pressure, and in which the sound waves are trapped by increasing temperature above and increasing pressure below. Syn: *SOFAR.*

sounding [eng] Measuring the thickness of soil or depth to bedrock by driving a pointed steel rod into the ground or by using a penetrometer.

sounding [geophys] Any scientific investigation or penetration of the natural environment.

sounding [elect] Mapping of (nominally) horizontal interfaces by resistivity, induced polarization, or electromagnetics. It usually involves variation of electrode interval for resistivity and induced polarization sounding, but may involve variation of either frequency or coil separation in electromagnetics. See also: *parametric sounding; geometric sounding.* Cf: *profiling.*

sounding [oceanog] The measurement of water depth taken from ship by either an *echo sounder* or a *lead line.*

sounding line *lead line.*

sounding sand Sand, usually clean and dry, that emits a musical, humming, or crunching sound when disturbed, such as a desert sand when sliding down the slip face of a dune or a beach sand when it is stirred or walked over. Examples: *musical sand; booming sand; roaring sand; whistling sand.* Syn: *singing sand.*

sound intensity The average rate of flow of acoustic wave energy through a unit section perpendicular to the direction of propagation; average power transmission per unit area. Syn: *acoustic intensity; seismic intensity.*

sound ranging The location of a source of seismic energy by acoustic triangulation, i.e. by recording signals on receivers at known positions.

sound wave *acoustic wave.*

sour Said of crude oil or natural gas containing significant fractions of sulfur compounds. Cf: *sweet.*

source (a) The point of origin of a stream of water; the point at which a river rises or begins to flow. Syn: *fountain.* (b) A *headwater,* or one of the headwaters, of a stream; e.g. a fountainhead.

source area *provenance.*

source-bed concept The theory of sulfide ore genesis that postulates an original syngenetic deposition of sulfides i◼ particular stratigraphic horizon of a sedimentary environme◼ and their subsequent migration due to a rise in temperature the rock environment (Knight, 1957, p.808).

source bias An effect in which azimuthally dependent ◼ partures from standard traveltimes in the upper mantle ◼ neath the hypocenter result in consistent errors in the e◼ mated epicenter (Herrin & Taggart, 1968).

sourceland *provenance.*

source-receiver product In seismic prospecting, the prod◼ of the number of detectors per trace and the number of sh◼ holes fired simultaneously.

source region That extensive area of the Earth's surface o◼ which an *air mass* develops and acquires its distinctive ch◼ acteristics.

source rock [petrol] Sedimentary rock deposited together w◼ the organic material which under pressure, heat, and ti◼ was transformed to liquid or gaseous hydrocarbons. Sour◼ rock is usually shale and limestone.

source rock [sed] The rock from which fragments and oth◼ detached pieces have been derived to form a later (sedime◼ tary) rock. Syn: *parent rock; mother rock.*

source-rock index A term used by Dapples et al (1953, p.2◼ & 304) to indicate the extent of contribution to a sandstone◼ fragments from igneous and metamorphic rocks by measuri◼ the degree of mixing of "arkose" and "graywacke" types. It◼ expressed as the ratio of sodic and potassic feldspar (arko◼ tendency) to the total of assorted rock fragments plus mat◼ of clay and micas (graywacke tendency). The index indica◼ source-rock types regardless of subsequent depositional h◼ tory or diagenesis. Values greater than 3 indicate arkos◼ values less than 0.75 indicate graywacke.

south (a) The direction of the south terrestrial pole, or the ◼ rection to the right of one facing east or of one facing t◼ sunrise when the Sun is near one of the equinoxes. (b) T◼ cardinal point directly opposite to north. Abbrev: S (c) T◼ direction along any meridian toward that pole of the Ear◼ viewed from which the Earth's rotation is clockwise. (d) T◼ direction to the right when one faces the direction of revol◼ tion of the Earth around the Sun. (e) The point of the horiz◼ having an azimuth of 180 degrees measured clockwise fro◼ north.

South African ruby *Cape ruby.*

south geographic pole *south pole.*

southing A *latitude difference* measured toward the sou◼ from the last preceding point of reckoning; e.g. a linear di◼ tance southward from an east-west reference line.

south pole [geog] The *geographic pole* in the southern hem◼ sphere of the Earth at lat. 90°S, representing the souther◼ most point of the Earth or the southern extremity of its axis ◼ rotation. Also spelled *South Pole.* Syn: *south geographic pole.*

south pole [astron] The south *celestial pole* representing th◼ zenith of the heavens as viewed from the south geograph◼ pole.

souzalite A green mineral: $(Mg,Fe)_3(Al,Fe)_4(PO_4)_4(OH)_◼$ $2H_2O$.

sovite A *carbonatite* that contains calcite.

sowback A long, low hill or ridge shaped like the back of a f◼ male hog; e.g. *hogback; horseback; drumlin.*

sowneck A narrow divide between two expanses of lowlan◼ or a narrow boundary between two bodies of water, formed b◼ a gentle rise of ground.

spa (a) A *medicinal spring.* (b) A place where such spring◼ occur, often a resort area or hotel.----The name is derive◼ from that of a town in eastern Belgium where medicin◼ springs occur.

spaced cleavage In schist, the spacing or separation of cleav◼ age planes from a few millimeters to a microscopic scale◼ e.g. *slip cleavage.* It is a *nonpenetrative* texture. Cf: *continu◼ ous cleavage.*

ace geology *astrogeology.*

ace group In a crystal structure, one of 230 different ways arranging atoms in a homogeneous array.

ace lattice *crystal lattice.*

ace-time unit A stratigraphic unit whose lateral limits are termined by geographic coordinates and whose vertical ex-nt is measured in terms of geologic time (Wheeler, 1958).

ad An iron, brass, or tin nail, up to 5 cm long, with a hook eye at the head for suspending a plumb line, used to mark underground survey station (as in a mine or tunnel).

adaite A mineral: $MgSiO_2(OH)_2.H_2O$ (?).

all (a) A chip or fragment removed from a rock surface by eathering; esp. a small, relatively thin, commonly curved d sharp-edged slab or other piece of rock produced by ex-iation. (b) A similar rock fragment produced by chipping h a hammer; e.g. a piece of ore broken by spalling.

allation The ejection of atomic particles from a nucleus fol-wing the collision of an atom and a high-energy particle .g. a cosmic ray) which results in the formation of a differ-t isotope that is not a *fission* [isotope] product.

alling The chipping, fracturing, or fragmentation, and the ward and outward heaving, of rock caused by the interac-n of a shock (compressional) wave at a free surface, spe-. *exfoliation.*

almandite A garnet intermediate in chemical composition tween spessartine and almandine (almandite); a variety of essartine rich in iron.

an (a) The length of a time interval. (b) An informal desig-tion for a local geologic-time unit.

andite A garnet intermediate in chemical composition be-een spessartine and andradite; a variety of spessartine rich calcium and iron.

angolite A dark-green hexagonal mineral: $Cu_6Al-(SO_4)(OH)_{12}Cl.3H_2O.$

anish chalk A variety of steatite from the Aragon region of ain.

anish topaz (a) Any orange, orange-brown, or orange-red riety of quartz resembling the color of *topaz*; e.g. heat-eated amethyst. (b) A wine-colored or brownish-red citrine ccurring in Spain.

ar [mineral] A term loosely applied to any transparent or anslucent, light-colored, nonmetallic, usually readily cleava-e, and somewhat lustrous crystalline mineral, esp. one oc-urring as gangue in a metalliferous vein; e.g. calcspar and eland spar (calcite), fluorspar (fluorite), heavy spar (barite), feldspar. Obs. syn: *spath.*

ar [mining] A miner's term for a small clay vein in a coal am.

aragmite A collective term for the late Precambrian frag-ental rocks of Scandinavia, esp. the feldspathic sandstones the Swedish Jotnian, consisting mainly of coarse arkoses d subarkoses (characterized by high proportions of micro-ine), together with polygenetic conglomerates and gray-ackes. Etymol: Greek *sparagma*, "fragment, thing torn, ece".

arite (a) A descriptive term for the crystalline and clear, ansparent, or translucent interstitial component of limestone, onsisting of clean, relatively coarse-grained calcite (or ara-onite) that either accumulated during deposition or was in-oduced later as a cement. It is more coarsely crystalline an *micrite*, the grains having diameters that exceed 10 mi-rons (Folk, 1959) or 20 microns (Chilingar et al, 1967, 320). Syn: *sparry calcite; calcsparite.* (b) A limestone in hich the sparite cement is more abundant than the micrite atrix. Syn: *sparry limestone.*

ark spectrum The spectrum of light emitted by a substance, sually a gas or vapor, when an electric spark is passed rough it. The spectrum is representative of the ionized toms. Cf: *arc spectrum.*

arnacian European stage: upper upper Paleocene (above

Thanetian, below Ypresian of Eocene).

sparry (a) Pertaining to, resembling, or consisting of spar; e.g. "sparry vein" or "sparry luster". (b) Pertaining to sparite, esp. in allusion to the relative clarity both in thin section and hand specimen of the calcite cement; abounding with sparite, such as a "sparry rock".

sparry calcite Clean, coarse-grained calcite crystal; *sparite.*

sparry iron A syn. of *siderite* (ferrous-carbonate mineral).

sparry limestone (a) *sparite.* (b) A coarsely crystalline mar-ble.

sparse biomicrite A *biomicrite* in which the skeletal grains make up 10-50% of the rock. Cf: *packed biomicrite.*

spartalite *zincite.*

spasmodic turbidity current A single, rapidly developed turbidi-ty current, such as one initiated by a submarine earthquake. Cf: *steady turbidity current.*

spastolith A deformed oolith; e.g. a chamositic oolith that has been closely twisted or misshapen due to its soft condition at the time of burial (Pettijohn, 1957, p.97).

spate (a) A sudden flood on a river, caused by heavy rains or rapidly melting snow higher up the valley; a *freshet.* (b) A Scottish term for a flood.

spath An obsolete syn. of *spar* [mineral].

spathic Resembling spar, esp. in regard to having good cleav-age. Syn: *spathose.*

spathic iron Ferrous-carbonate mineral with good rhombohe-dral cleavage; specif. *siderite.* Syn: *spathic iron ore; spathose iron.*

spathization Widely distributed crystallization of sparry carbo-nates such as calcite and dolomite (Sander, 1951, p.3); de-velopment of relatively large sparry crystals that have good cleavage.

spathose *spathic.*

spatial dendrite A type of snow crystal somewhat like a *stellar crystal* except that branched arms are attached together in a random way instead of building a pattern of hexagonal sym-metry in a single plane.

spatial sediment concentration The sediment contained in a unit volume of flow used to measure *transport concentration.*

spatiography A science that deals with space beyond the Earth's atmosphere; esp. the description of the physical char-acteristics of the Moon and the planets (Webster, 1967, p.2184). The term is obsolete.

spatium A localized widening of an axial canal of a crinoid co-lumnal opposite interarticular sutures. Pl: *spatia.*

spatter [meteorite] Droplets on the surface of meteorites, often partly fused with the crust.

spatter [pyroclast] n. An accumulation of small pyroclastic fragments.----adj. Pertaining to the forms of such accumula-tions, e.g. *spatter cone,* spatter rampart. Syn: *driblet.*

spatter cone A low, steep-sided cone of *spatter* built up on a fissure or vent; it is usually of basaltic material. Syn: *volcanel-lo; agglutinate cone.*

spatulate Said of a leaf that is spoon-shaped.

spavin An English term for a hard, unstratified, sandy clay or mudstone underlying a coal seam; *seat earth.*

SP curve *spontaneous-potential curve.*

speaking rod A *level rod* with graduations designed to be read directly by the observer at the leveling instrument. Cf: *target rod.* Syn: *self-reading rod.*

spear The recurved part of a *hook* of a holothurian.

spear pyrites A form of *marcasite* in twin crystals showing reentrant angles that resemble the head of a spear. Cf: *cocks-comb pyrites.*

special creation The theory, strongly supported before the theory of evolution was generally accepted, that each species of organisms inhabiting the Earth was created fully formed and perfect by some divine process.

special erosion Erosion effected by agents (such as wind, waves, and glaciers) that are important only within strictly limited areas or that work with help from the agents of *normal erosion* (Cotton, 1958, p.38). "The modern tendency ... is to regard any distinction between 'normal' and 'special' agencies as unreal" (Stamp, 1961, p.340).

special-purpose map Any map designed primarily to meet specific requirements, and usually omitting or subordinating nonessential or less important information. Cf: *general-purpose map.*

speciation The production of new species of organisms from pre-existing ones during evolution.

species [**mineral**] A mineral, distinguished from others by its unique chemical and physical properties, but which may have *varieties.*

species [**paleont**] A group of organisms, either plant or animal, that may interbreed and produce fertile offspring that have similar structure, habits, and functions. As a basic unit in taxonomy, it ranks next below "genus". The name of a species is a binomen; e.g. *Nuculana diversa.* Adj: *specific.* Abbrev: *sp.* Plural: *species.* Cf: *ecospecies.*

species-group A group of species that replace one another geographically and may all be descended from a common stock, but that have characteristics that make them separate species. Syn: *gens; Artenkreis; collective species.*

specific The adj. of *species.*

specific absorption The capacity of water-bearing material to absorb liquid, after removal of gravity water; the ratio of the volume of water absorbed to the volume of the saturated material. It is equal to specific yield except when the water-bearing material has been compacted due to the weight of overlying rocks.

specific acoustic impedance *acoustic impedance.*

specific acoustic resistance *Acoustic resistance* referred to dimensionless units.

specific activity (a) The *activity* of a radioactive isotope, measured per unit weight of the element in the sample. (b) the *activity* per unit weight of a sample of radioactive material. (c) The *activity* per unit mass of a pure radionuclide.----(U.S. Naval Oceanographic Office, 1968).

specific capacity The rate of discharge of a water well per unit of drawdown, commonly expressed in gallons per minute per foot. It varies slowly with duration of discharge. If the specific capacity is constant except for the time variation, it is proportional to the hydraulic diffusivity of the aquifer.

specific character A particular characteristic that distinguishes one species from all other species of the same genus.

specific compaction The decrease in thickness of a deposit per unit of increase in applied stress during a specified period of time (Poland, et al, in press).

specific conductivity With reference to the movement of water in soil, a factor expressing the volume of transported water per unit of time in a given area.

specific discharge *Discharge* per unit area. It is often used to define the magnitude of a flood (ASCE, 1962).

specific energy The energy of water in a stream; it equals the mean depth plus velocity head of the mean velocity (ASCE, 1962).

specific expansion The increase in thickness of deposits per unit of decrease in applied stress (Poland, et al, in press).

specific-gravity liquid *heavy liquid.*

specific head The height of the *energy line* above the bed of a conduit (ASCE, 1962).

specific heat capacity The *heat capacity* of a system per unit mass, measured in calories/gm-°C.

specific humidity The mass of water vapor in a given mass of moist air, usually expressed in g/g or g/kg.

specific magnetization Magnetic moment per unit mass; magnetization divided by density.

specific name (a) The second term in a binomen. (b) A less preferred syn. of *binomen.*

specific permeability A factor expressing the permeability o[f] stream bed; it equals a constant times the square of rep[re]sentative pore diameter. Symbol: k. Syn: *intrinsic permeab[ili]ty.*

specific refractivity The *refractivity* of a substance divided [by] its density.

specific retention The ratio of the volume of water a giv[en] body of rock or soil will hold against the pull of gravity to [the] volume of the body itself. It is usually expressed as a perce[nt]age. Cf: *field capacity.*

specific rotation The angle of rotation of plane-polarized lig[ht] passing through a substance, measured in degrees per de[ci]meter for liquids and solutions and in degrees per millime[ter] for solids.

specific seismicity The square root of the energy, per un[it] area per unit time, released by the earthquakes of a given r[e]gion.

specific susceptibility Susceptibility divided by density; t[he] ratio of specific induced magnetization to the strength H [of] the magnetic field causing the magnetization. Syn: *mass su[s]ceptibility.*

specific tenacity The ratio of a material's tensile strength to [its] density.

specific unit compaction The compaction of deposits per un[it] of thickness per unit of increase in applied stress, during [a] specified time period. Ultimate specific unit compaction is a[t]tained when pore pressures in the aquitards have reached h[y]draulic equilibrium with pore pressures in contiguous aquifer[s;] at that time, specific unit compaction equals gross compress[i]bility of the system (Poland, et al, in press).

specific unit expansion The expansion per unit of thickne[ss] per unit decrease in applied stress. It is a net value if com[-]paction is occurring in parts of the interval being measure[d] during the period of decrease in applied stress (Poland, et a[l,] in press).

specific yield The ratio of the volume of water a given mass [of] saturated rock or soil will yield by gravity to the volume of th[e] mass. This ratio is stated as a percentage. Cf: *effective po[-]rosity; storage coefficient.*

specimen ore A particularly rich or well crystallized orebody.

speckstone An early name for talc or steatite. Etymol: Ge[r]man *Speckstein,* "bacon stone", alluding to its greasy fee[l.] See also: *bacon stone.*

spectacle stone *selenite.*

spectra pl. of *spectrum.*

spectral Pertaining to a *spectrum* [phys], e.g. *spectral line.*

spectral absorptance A term referring to the *absorptanc[e]* measured at a specified wavelength.

spectral line One component line in the array of intensit[y] values of a spectrum emitted by a source. See also: *principa[l] line.*

spectral log A *gamma-ray log* that records both the energy o[f] the gamma radiation and the relative intensities of gamm[a] rays emitted by strata penetrated in drilling. It permits a dis[-]tinction to be made in the gamma-ray emissions of various ra[-]dioactive elements with different wavelength energies. Sy[n:] *spectral gamma-ray log.*

spectral radiance Radiance per unit wavelength interval at [a] particular wavelength. Symbol: N_λ.

spectral radiant power *Radiant power* measured at a given, o[r] specified, wavelength.

spectra-zonal system A type of multi-band sensing system, fo[r] which the latter term is preferred. Cf: *multi-band system.* Sy[n:] *multi-zonal system.*

spectrochemical analysis Chemical analysis based on the spectrum of a substance, each substance having a character[-]istic spectrum. Syn: *spectrum analysis.*

spectrochemistry The branch of chemistry based on *spectro[-] chemical analysis.*

spectrocolorimeter Essentially an absorption *spectrophotomet[er]* used to measure the absorbance of solutions over an entire spectrum providing quantitative information about the composition of the solution. *Spectrophotometer* is often used as a syn. of spectrocolorimeter. See also: *colorimeter*.

spectrocolorimetry The art or process of using the *spectrocolorimeter* for the quantitative study of color.

spectrogram A map, photograph, or other picture of a spectrum, usually produced by a *spectrograph*.

spectrograph A *spectroscope* designed to map or photograph a spectrum.

spectrography The art or process of using a *spectrograph* to photograph or map a spectrum.

spectrometer A *spectroscope* designed to measure wavelengths and indices of refraction of rays in a spectrum.

spectrometry The art or process of using a *spectrometer* to measure spectra.

spectrophotometer A *photometer* for measuring and comparing the intensity of light in different parts of a spectrum as a function of wavelength. Common usage assigns this term mainly to analytical instruments which measure the characteristic absorption spectra of chemicals. See also: *flame photometer; spectrocolorimeter*.

spectrophotometry The art or process of using a *spectrophotometer* to measure the intensity of light in different parts of a spectrum as a function of wavelength.

spectroreflectometer An instrument for measuring and analyzing the *reflection spectrum* of a source.

spectroscope An instrument for producing and visually observing a spectrum.

spectroscopy The production and observation of a spectrum and all methods of recording and measuring, including the use of the *spectroscope*, that go with them.

spectrum [phys] n. (a) An array of visible light ordered according to its constituent wavelengths (colors) by being sent through a prism or diffraction grating. (b) An array of intensity values ordered according to any physical parameter, e.g. electromagnetic wave spectrum, energy spectrum, mass spectrum, and nuclear magnetic resonance spectrum. Pl: *spectra.*----adj. *spectral*.

spectrum [palyn] *pollen spectrum*.

spectrum analysis *spectrochemical analysis*.

specular coal *pitch coal*.

specular iron A syn. of *specularite*. Also called: *specular iron ore*.

specularite A brilliant, black or gray variety of *hematite* with a highly splendent metallic luster and often showing iridescence. It occurs in micaceous or foliated masses, or in tabular or disk-like crystals. Syn: *specular iron; gray hematite; iron glance*.

specular schist Metamorphosed *oxide facies iron formation* characterized by a high percentage of strongly aligned flakes of specular hematite.

specular surface One which is smooth with respect to the wavelength incident upon it.

speed In photogrammetry, the response or sensitivity of a photographic film, plate, or paper to light; also, the light-gathering power of a lens or optical system, expressed as the *relative aperture* of the lens. See also: *f-number*.

speisscobalt *smaltite*.

spelean Said of or pertaining to a feature in a cave.

speleochronology The dating or chronology of a cave's formation, or of its mineral deposits or filling. The dating may be either relative or absolute.

speleogen In a cave, any relief feature that is formed by solution, e.g. a scallop.

speleogenesis The process of cave formation.

speleologist A scientist engaged in *speleology*.

speleology The exploration and scientific study of caves, e.g. their genesis, morphology, and mineral deposits. See also: *caving; speleologist*.

speleothem Any secondary mineral deposit that is formed in a cave by the action of water. Syn: *cave formation; formation* [speleo].

spelunker *caver*.

spelunking *caving*.

spencerite (a) A pearly-white monoclinic mineral: $Zn_4(PO_4)_2(OH)_2.3H_2O$. (b) An artificial substance: (Fe, Mn)$_3$(C,Si).

spencite A dark-brown metamict mineral: (Y,Ca,La,Fe)$_5$(Si, B,Al)$_3$(O,OH,F)$_{13}$.

spergenite A name proposed by Pettijohn (1949, p.179 & 301) for a biocalcarenite that contains ooliths and fossil debris (such as bryozoan and foraminiferal fragments) and that has a quartz content not exceeding 10%. Type locality: Spergen Hill, situated a few miles SE of Salem (Washington County), Ind., where the Salem Limestone (formerly the Spergen Limestone) is found. Syn: *Bedford limestone; Indiana limestone*.

spermatophyte A vascular plant that produces seeds, e.g. a gymnosperm or angiosperm; a *seed plant*. Such plants range from the Carboniferous. Cf: *pteridophyte; phanerogam*.

spermatozoid In embryophytic plants, a free, usually ciliate haploid male reproductive cell produced by the antheridia of the gametophyte. Fusion of this cell with an egg produces a diploid zygote which develops into an embryo.

sperone A vesicular leucitite that contains small melanite crystals.

sperrylite A tin-white isometric mineral: $PtAs_2$. It is the only compound of platinum known to occur in nature.

spessartine The manganese-aluminum end member of the garnet group: $Mn_3Al_2(SiO_4)_3$. It has a brownish-red, orangy-brown, yellow-orange, orangy-red, or yellow-brown color, and it usually contains some iron, magnesium, and other elements in minor amounts. Spessartine is rather rare, and occurs in pegmatites and granites. Syn: *spessartite*.

spessartite [mineral] *spessartine*.

spessartite [petrology] A lamprophyre composed of green hornblende phenocrysts in a groundmass of sodic plagioclase, with accessory olivine, biotite, apatite, and opaque oxides.

spew frost *pipkrake*.

sphaeraster A many-rayed sponge spicule (euaster) in which the rays radiate from a prominent, solid, spherical centrum. Syn: *spheraster*.

sphaerite A light-gray or bluish mineral consisting of hydrous aluminum phosphate in globular concretions. It is perhaps the same as *variscite*. Syn: *spherite*.

sphaeroclone An *ennomoclone* in which six or more proximal arms radiate from one side of a frequently spherical centrum and terminate in cuplike zygomes.

sphaerocobaltite *spherocobaltite*.

sphaerocone A coiled, depressed, involute, globular cephalopod shell that has a small and quite or nearly occluded umbilicus and a round venter and that commonly opens out suddenly along the last whorl (as in *Sphaeroceras*).

sphaerolite A variant spelling of *spherulite* [petrology].

sphaerolitic *spherulitic*.

sphaeroplast A lorica-forming granule representing part of the shield-like mass formed during reproduction in tintinnids when a single cell divides into two theoretically equal parts (binary fission).

sphaerosiderite *spherosiderite*.

sphagnum atoll An *atoll moor* containing sphagnum.

sphagnum bog An acid, very wet, freshwater bog containing abundant sphagnum and which ultimately usually forms deposits of sphagnum peat (highmoor peat). See also: *balsam bog*.

sphagnum moss A moss of ths genus *Sphagnum*, often forming peat; *peat moss*.

sphagnum peat *highmoor peat*.

sphalerite A brown or black, sometimes yellow or white, isometric mineral: (Zn,Fe)S. It is dimorphous with wurtzite, and often contains manganese, arsenic, cadmium, and other elements. Sphalerite has a highly perfect dodecahedral cleavage and a resinous to adamantine luster. It is a widely distributed ore of zinc, commonly associated with galena in veins and other various deposits. Syn: *blende; zinc blende; jack; blackjack; steel jack; false galena; pseudogalena; mock ore; mock lead.*

Spharokrystal A spherulite composed of a single mineral species.

sphene A usually yellowish or brownish mineral: $CaTiSiO_5$. It often contains other elements such as niobium, chromium, fluorine, sodium, iron, manganese, and yttrium. Sphene occurs in wedge- or lozenge-shaped monoclinic crystals as an accessory mineral in granitic rocks and in calcium-rich metamorphic rocks. Syn: *titanite; grothite.*

sphenitite A sphene-rich jacupirangite. Syn: *sphenite.*

sphenochasm A triangular gap of oceanic crust separating two continental blocks and converging to a point; it is interpreted by Carey (1958) as having originated by the rotation of one of the blocks with respect to the other. A small-scale example is the Bay of Biscay. Cf: *rhombochasm.*

sphenoconformity A term used by Crosby (1912, p.297) for the relation between conformable strata that are thinner in one locality than in the other, though fully represented in both.

sphenoid An open crystal form having two nonparallel faces that are symmetrical to an axis of twofold symmetry. It occurs in monoclinic crystals of the sphenoidal class. Cf: *dome; disphenoid.*

sphenoidal class That crystal class in the monoclinic system having symmetry 2.

sphenolith [paleont] A coccolith having a prismatic base formed by radial elements surmounted by a cone.

sphenolith [intrus rocks] A wedgelike igneous intrusion, partly concordant and partly discordant.

sphenopsid n. A type of pteridophyte characterized by distinctly jointed sporophyte stems and a whorl pattern of leaves and sporangia. Sphenopsids are known mainly from Paleozoic deposits; the only living genus in this group is *Equisetum.* Syn: *horsetail.*

spheraster *sphaeraster.*

sphere A standard shape taken as a reference form in the analysis of sedimentary-particle shapes; the limiting shape assumed by many rock and mineral fragments upon prolonged abrasion, being a solid figure bounded by a uniformly curved surface, any point on which is equidistant from the center. It has the least surface area for a given volume and the greatest settling velocity in a fluid of any possible shape (under conditions of low velocity and of constant volume and density). See also: *spheroid* [sed].

sphere ore *cockade ore.*

spherical bomb *spheroidal bomb.*

spherical cap In gravimetry, part of a spherical shell limited by a circular cone with the apex in the center of the sphere.

spherical coordinates (a) Three coordinates that represent a distance and two angles in space, consisting of two polar coordinates in a plane and the angle between this plane and a fixed plane containing the primary axis of direction (polar axis). The term includes coordinates on any spherical surface or on any surface approximating a sphere (such as the surface of the Earth). (b) A system of polar coordinates in which the origin is the center of a sphere and the points all lie on the surface of the sphere; a coordinate system used to define the position of a point with reference to two great circles that form a pair of axes (such as longitude and latitude) at right angles to each other or with reference to an origin and a great circle through the point.

spherical triangle A *triangle* on the surface of a sphere, having sides that are arcs of three great circles.

spherical wave A seismic wave propagated from a point source whose front surfaces are concentric spheres (U.S

Naval Oceanographic Office, 1966, p.154).

spherical weathering *spheroidal weathering.*

sphericity The relation to each other of the various diamete (length, width, thickness) of a particle; specif. the degree which the shape of a sedimentary particle approaches that a sphere. True sphericity, as originally defined by Wad (1932), is the ratio of the surface area of a sphere of same volume as the particle to the actual surface area of particle. Due to the difficulty of determining the actual surfa area and volume of irregular solids, Wadell (1934) develop an operational definition expressed as the ratio of the tr nominal diameter of the particle to the diameter of a circu scribing sphere (generally the longest diameter of the par cle). A perfect sphere has a sphericity of 1.0; all oth objects have values less than 1.0. Not to be confused w *roundness* or *angularity.*

spherite [mineral] *sphaerite.*

spherite [sed] (a) A sedimentary rock composed of grav sized aggregates of constructional (nonclastic) origin, sim lating in texture a rudite of clastic origin; e.g. a rock form of volcanic bombs. The term was introduced by Grab (1911, p.1007). Syn: *spheryte.* (b) An individual spheric grain in a sedimentary rock, such as a concentric oolith in oolite or a radial *spherulite* in a limestone.

spheroclast A rounded *phenoclast,* such as a pebble or cobb of a conglomerate. Cf: *anguclast.*

spherocobaltite A peachblossom-red mineral: $CoCO_3$. It o curs in spherical masses. Syn: *sphaerocobaltite; cobaltoc cite.*

spheroid [geodesy] Any figure that generally differs little fro a sphere. In geodesy it is usually an *ellipsoid of revolution.* C *ellipsoid.*

spheroid [sed] A spherical, or equant or equiaxial, shape of sedimentary particle, defined in *Zingg's classification* as ha ing width/length and thickness/width ratios greater than 2: See also: *sphere.*

spheroidal (a) Having the shape of a spheroid. (b) Compose of spherules. (c) Said of the texture of a rock composed numerous spherules.

spheroidal bomb A rotational volcanic bomb in the shape an oblate spheroid. Syn: *spherical bomb.*

spheroidal jointing *spheroidal parting.*

spheroidal parting A series of concentric and spheroidal or e lipsoidal cracks produced about compact nuclei in fine grained, homogeneous igneous rocks. Each set of cracks more strongly developed during *spheroidal weathering.* Sy *spheroidal jointing.*

spheroidal symmetry *axial symmetry.*

spheroidal weathering A form of chemical weathering in whic concentric or spherical shells of decayed rock (varying in di ameter from 2 cm to 2 m) are successively loosened an separated from a usually well-jointed and fine-grained block rock (esp. of basalt) by water penetrating the intersectin joints or other fractures and attacking the block from all side at once. It commonly forms a rounded *boulder of decompos tion.* It is similar to the larger-scale *exfoliation* produced usu ally by mechanical weathering. See also: *spheroidal partin onion-skin weathering, concentric weathering; spherica weathering.*

spherop An equipotential surface in the normal gravity field o the Earth. A surface such that the spheropotential is constar and the normal gravity is perpendicular to it at every point. Se also: *spheropotential number.* Syn: *spheropotential surface.*

spherophyre An igneous rock in which the phenocrysts ar aggregations of crystals in the form of spherulites.

spheropotential The potential function of the normal gravit defined as either the external gravity potential of a level ellip soid approximating the Earth, or as a function consisting o the first few terms of some expansion of the *geopotential.*

spheropotential number The spheropotential difference be

en the mareograph *spherop* and the spherop through an
ervation point. The number, which is expressed in geopo-
ial units, is sometimes called the *normal geopotential
nber.*

eropotential surface *spherop.*

erosiderite A variety of siderite (ferrous-carbonate miner-
occurring in globular concretionary aggregates of blade-
crystals radiating from a center, generally in a clayey ma-
(such as those in or below beds of underclays associated
coal measures). It appears to be the result of weathering
water-logged sediments in which iron, leached out of sur-
e soil, is redeposited in a lower zone characterized by re-
ing conditions. Syn: *sphaerosiderite.*

erule A little sphere or spherical body; e.g. a "magnetic
erule" in a deep-sea sediment, or an object that appears
be an amygdule or a spherulite, or a small spherical
nge spicule.

erulite [petrology] A rounded or spherical mass of acicular
stals, commonly feldspar, radiating from a central point.
erulites may range from microscopic in size to several
timeters in diameter (Stokes & Varnes, 1955, p.140). Cf:
iole; spheruloid; orbicule. Also spelled: *sphaerolite.*

erulite [sed] (a) Any more or less spherical body or
rsely crystalline aggregate with a radial internal structure
anged around one or more centers, varying in size from
roscopic grains to objects many centimeters in diameter,
ned in a sedimentary rock in the place where it is now
nd; e.g. a minute particle of chalcedony in certain lime-
nes, or a large carbonate concretion or nodule in shale.
spherite. (b) A small (0.5-5 mm in diameter), spherical or
eroidal particle composed of a thin, dense calcareous
er layer with a sparry calcite core. It can originate by re-
stallization or by biologic processes.

erulitic Said of the texture of a rock composed of numer-
s spherulites; also, said of a rock containing spherulites. Cf:
iolitic; radiated. Syn: *globular; sphaerolitic.*

eruloid n. A spherule or nodule that lacks radial structure,
in perlitic lava (Nelson & Nelson, 1967, p.355). Cf: *spheru-
[petrology].*

eryte *spherite [sed].*

inctozoan Any calcisponge having a skeleton that consists
straight, curved, or branched series of hollow spheroidal
dies (TIP, 1955, pt.E, p.100).

cularite A sediment or rock composed principally of the
ceous spicules of invertebrates; esp. a *spongolite* com-
sed principally of sponge spicules. Syn: *spiculite; sponge-
cule rock.*

iculation The formation, or the form and arrangement, of
cules. Also, a spicular component (such as of a sponge).

icule (a) One of the numerous, often very minute calcare-
s or siliceous bodies, having very varied and often charac-
istic forms, occurring in and serving to stiffen and support
tissues of various invertebrates, and frequently found in
arine-sediment samples or in Paleozoic and Cretaceous
erts. Examples: a small, discrete, autochthonous skeletal
ment of a sponge, typically a needle-like rod or a fused
ster of such rods; a long, sharp, calcareous, variously
aped skeletal element (sclerite) of the mesogloea of an
tocoral; a discrete, commonly elongate or needle-like skel-
al element of many radiolarians; a scale-like calcareous
ject borne on the girdle of a primitive chiton; a small, irreg-
r, calcareous body secreted within the connective tissue of
body wall, mantle, and lophophore of a brachiopod; and a
ry minute, irregular, and cylindrical or radiate skeletal ele-
ent of an asterozoan. (b) The empty siliceous shell of a dia-
m.

icule tract A linear series or bundle of separate sponge spi-
les, usually held together by spongin.

iculin The chemically undetermined protein substance that
ms the axial filament of sponge spicules.

spiculite [petrology] A spindle-shaped belonite thought to have
formed by the coalescence of globulites.

spiculite [sed] *spicularite.*

spiculoblast *sclerocyte.*

spiculofiber A fiber-like structure built of a bundle of sponge
spicules held together by mutual interlocking, by fusion, by
spongin, or by sclerosome.

spiculoid A discrete autochthonous element of a sponge skeleton, resembling a spicule but made of organic material only.
Syn: *pseudospicule.*

spike The known amount of an isotope added to a sample to
determine the unknown amount present in analysis by *isotope
dilution.*

spilite An altered basalt, characteristically amygdaloidal or vesicular, in which the feldspar has been albitized and is usually
accompanied by chlorite, calcite, epidote, chalcedony, prehnite, or other low-temperature hydrous crystallization products
characteristic of a greenstone. Spilite often occurs as submarine lava flows and exhibits pillow structure. Adj: *spilitic.*

spilitic suite A group of extrusive and minor intrusive rocks
that characteristically have a high albite content. The group is
named for its type member, spilite.

spilitization Hydrothermal albitization of a basalt to form a
spilite.

spill bank A term used in Great Britain and India for a *natural
levee.*

spilling breaker A breaker whose crest collapses gradually
over a nearly flat bottom for a relatively long distance, forming
a foamy patch at the crest, the water spilling down continuously over the advancing wave front. See also: *comber.* Cf:
plunging breaker; surging breaker.

spilling point Var. of *spillpoint.*

spillpoint The point of maximum filling of a structural trap by
oil or gas. Syn: *spilling point.*

spill stream An *overflow stream* from a river.

spillway [eng] A passage or outlet through which surplus
water flows over, around, or in a rim or barrier (such as a
dam or natural obstruction).

spillway [glac geol] *overflow channel.*

spilosite A rock representing an early stage of the formation
of *adinole* or *spotted slate.*

spinach jade Dark-green nephrite.

spindle-shaped bomb A rotational volcanic bomb of fusiform
shape, with earlike projections at its ends. See also: *fusiform
bomb.* Syn: *almond-shaped bomb.*

spindle stage A single-axis *stage* of a microscope, consisting
of a liquid-filled cell in which the crystal is immersed, and may
be rotated 180°.

spine [paleont] A long or short, straight or curved, sharp or
blunt, solid or hollow projection of the shell surface, found on
various invertebrates; e.g. a radiolarian spicule, or a movable,
elongated, calcareous shaft mounted on, and articulating with,
a tubercle on the test of an echinoderm (such as an echinoid
or asteroid), or a cylindrical or elongated triangular projection
from the external shell surface of a brachiopod.

spine [bot] In a plant, a rigid, sharply pointed structure that
may be a modified leaf or leaf part, petiole, or stipule.

spine [volc] A pointed mass or monolith of solidified lava that
sometimes occurs over the throat of a volcano. It may be
formed by slow, forced extrusion of viscous lava, or it may
represent magma in the pipe that was exposed by weathering.

spinel (a) A mineral: $MgAl_2O_4$. The magnesium may be replaced in part by ferrous iron, and the aluminum by ferric iron.
Spinel has great hardness, usually forms octahedral crystals
(isometric system), varies widely in color (from colorless to
purplish red, greenish, and yellow to black), and is used as a
gemstone. It occurs typically as products of contact metamorphism of impure dolomitic limestones, and less commonly
as accessory minerals (original constituents) in basic igneous
rocks; it also occurs in alluvial deposits. (b) A group of miner-

als of general formula: AB_2O_4, where A represents magnesium, ferrous iron, zinc, or manganese, or any combination of them, and B represents aluminum, ferric iron, or chromium; specif. an isomorphous series of oxides, $(Mg,Fe,Zn,Mn)Al_2O_4$, consisting of spinel, hercynite, gahnite, and galaxite. (c) A member of the spinel group or spinel series. (d) A substance (such as a sulfide) that has a similar formula and the same crystal structure as a spinel. (e) An artificial substance, similar to the mineral spinel, that is used as a gemstone, as a refractory, or as instrument bearings; e.g. ferrospinel.---Syn: *spinelle; spinell; spinelite.*

spinellid A mineral of the spinel group.

spinellide A name applied to the spinel group.

spinellite A medium- to coarse-grained, hypidiomorphic-granular, titaniferous, magnetite-rich igneous rock containing a preponderance of spinel (up to 20%) (Johannsen, 1938, p.469).

spinel ruby A deep-red gem variety of spinel. The term is sometimes used inappropriately as a syn. of *ruby spinel.*

spinel twin law A twin law in crystals of the hexoctahedral, isometric system, e.g. spinel, having a twin axis of threefold symmetry with the twin plane parallel to one of the octahedron's faces.

spinispire A spinose, siliceous, monaxonic sponge spicule (microsclere) in the form of a spiral of more than one revolution. Cf: *sigmaspire.* Syn: *spiraster.*

spinneret A spinning organ, or an organ for producing threads of silk from the secretion of the silk glands of an arachnid; specif. an abdominal appendage of spiders, with spinning tubes at the end, or a special spinning organ on the movable finger (galea) of a chelicera of pseudoscorpions.

spinner magnetometer A type of magnetometer that continuously rotates the specimen whose remanent magnetization it is measuring, to produce an alternating voltage in a nearby coil by electromagnetic induction. Syn: *rock generator.*

spinose Spine-like, or full of or armed with spines; e.g. said of a foraminiferal test having fine elongate solid spines on its surface (as in *Hastigerinella*) with each spine comprising a single calcite crystal elongated along the c-axis.

S-P interval In earthquake seismology, the time interval between the first arrivals of longitudinal and transverse waves, which is a measure of the distance from the earthquake source.

spinulus A closed-end process extending from the surface of a valve of a diatom frustule.

spiracle [paleont] (a) A large, generally rounded opening near the adoral tip of a deltoid plate of a blastoid and excavated within it. It opens into the space enclosed by a hydrospire. (b) A *stigma* of an arachnid.

spiracle [volc] A fumarolic vent in a lava flow formed by a gaseous explosion in lava that is still fluid, probably due to generation of steam from underlying wet material. It is usually about one meter in diameter and up to five meters high, although in the northwestern U.S where spiracles are common, they may be 10 meters in diameter and 12 meters or more in height.

spiracular slit An elongate spiracle at the side of a blastoid ambulacrum, excavated in adjoining radial and deltoid plates.

spiraculate Having spiracles.

spiral angle The angle formed between two straight lines tangent to the periphery of two or more whorls on opposite sides of a gastropod shell. It is commonly determined by drawing tangents to the lowermost whorls of the spire. Syn: *spire angle.*

spiral ball A term used by Hadding (1931, p.389) for a sandstone body having a rolled-up, spiral structure due to lateral mass flowage of thin interbedded sands and shales. Cf: *slump ball.* Syn: *snowball.*

spiral canal The part of the canal system in the umbilical region of a foraminiferal test (as in *Elphidium*), parallel to and inside the lateral chamber margins.

spiral garnet *rotated garnet.*

spiral growth Growth of a crystal along a *screw dislocation.* may result in a *growth island* on the surface of the crystal.

spiralium A spiral brachidium; one of a pair of spirally coiled ribbon-like, calcareous supports for the deuterolophe or the spirolophe of certain brachiopods, composed of secondary shell. Pl: *spiralia.* Syn: *spire.*

spiral lamina The coiled or winding part of the lorica of a tintinnid.

spiral suture A line of contact between two whorls in the coiled test of a foraminifer.

spiral tracheid A *tracheid* in which the additional cell-wall material is deposited in a spiral. Cf: *annular tracheid.*

spiramen A median pore in the proximal wall of a bryozoan peristome, into the cavity of which it leads.

spiraster A spiral sponge spicule; e.g. a *streptosclere* or a spinispire.

spire (a) The adapical, visible, upper part of a spiral gastropod shell including the whole series of whorls except the last or body whorl. (b) A *spiralium* of a brachiopod. (c) *spirillum.*

spiriferoid Any articulate brachiopod belonging to the order Spiriferida, characterized generally by a spiral brachidium and a biconvex, rarely plano-convex, shell. Their stratigraphic range is Middle Ordovician to Jurassic. Var: *spiriferid; spirifer.*

spirilline Said of a foraminiferal test consisting of a planispiral nonseptate tube enrolled about a globular proloculus; specif. pertaining to the foraminiferal genus *Spirillina.*

spirillum (a) A helical or coiled morphologic form of a bacterial cell. Syn: *spire; spiril.* (b) Any bacterium of the genus *Spirillum* now restricted to elongated forms having tufts of flagella at one or both ends.---Pl: *spirilla.*

spirit level (a) A sensitive device for finding a horizontal line or plane, consisting of a closed glass tube or vial of circular cross section, its center line also forming a circular arc, nearly filled with a liquid of low viscosity (such as ether or alcohol), with enough free space being left for the formation of a bubble of air or gas that will always assume a position at the top of the tube. See also: *circular level.* Syn: *level [surv].* (b) An instrument using a spirit level to establish a horizontal line of sight.

spirit leveling A type of *leveling* using a spirit level to establish a horizontal line of sight. The term has been broadened to include leveling by means of other types of precise levels, such as a pendulum level.

spiroffite A red to purple monoclinic mineral: $(Mn,Zn)_2Te_3O_8$.

spirogyrate Said of the umbones of a bivalve mollusk, coiled outward from an anteriorly and posteriorly directed (sagittal) plane of symmetry.

spirolophe A brachiopod lophophore in which the brachia are spirally coiled and bear a single row of paired filamentary appendages. Cf: *deuterolophe; plectolophe.*

spirotheca The outer or upper spiral wall of the test in fusulinids.

spiroumbilical aperture An *interiomarginal aperture* in a foraminiferal test, extending from the umbilicus to the periphery and finally onto the spiral (dorsal) side where all whorls are visible.

spit (a) A small point or low tongue or narrow embankment of land commonly consisting of sand or gravel deposited by longshore drifting and having one end attached to the mainland and the other terminating in open water, usually the sea; a finger-like extension of the beach. (b) A relatively long, narrow shoal or reef extending from the shore into a body of water.

spitskop A term used in South Africa for a hill with a sharply pointed top. Etymol: Afrikaans. Cf: *tafelkop.*

s plane A planar fabric element, e.g. bedding; approximately the equivalent of *s surface.*

splash cup A concavity at the top of a stalagmite.

splash erosion The dislodgment and movement of soil particles es under the impact of falling raindrops.

splash zone The area, along a coast, affected by the splashing of seawater from breaking waves. Cf: *spray zone.*

splay *flood-plain splay.*

splaying crevasse A *crevasse* in a valley glacier that is parallel to the direction of flow in the center of the glacier but curves toward the margin downstream. Cf: *longitudinal crevasse; marginal crevasse; transverse crevasse.*

splaying out (a) The breakup of a fault into a number of minor faults. (b) The dying out of a fault by its dispersal into a number of minor faults. See also: *splays.*

splays A series of minor faults at the extremities of a major fault; the fault pattern formed by *splaying out.* It is associated with rifts.

splendent Said of mineral luster of the highest degree of intensity.

splent coal *splint coal.*

splint *splint coal.*

splint coal A type of banded coal that is hard, dull, blocky, and greyish black, with rough, uneven fracture and granular texture. It is defined quantitatively as having more than 5% anthraxylon and more than 30% opaque attritus. Syn: *splent coal; splint.* Cf: *semisplint coal.*

split A coal seam that is separated by a parting of other sedimentary rock and that cannot be mined as a single unit. Syn: *split coal; coal split.*

split coal *split.*

split spread A type of seismic spread in which the shot point is at the center of the arrangement of geophones. It is commonly used for continuous profiling and for dip shooting. Syn: *straddle spread; symmetric spread.*

split stream (a) A stream shown on a map by a single line and containing an island that divides the stream into two channels. (b) A *single-line stream* that divides, with the branches flowing into separate drainage areas.

splitting [paleont] In taxonomy, the practice of subdividing species and genera on the basis of more or less minute differences that may be of doubtful significance. A taxonomist known for his frequent splitting of taxa is called a "splitter". Cf: *lumping.*

splitting [sed] (a) Abrasion of a rock fragment resulting in the production of two or three subequal parts or grains. (b) The property or tendency of a stratified rock of separating along a plane or surface of *parting.* (c) The sampling of a large mass of loose material (such as of a sediment) by dividing it into two or more parts; e.g. *quartering.*

SP log *spontaneous-potential log.*

spodic Said of a soil horizon that is characterized by the illuvial accumulation of amorphous materials, specif. aluminum and organic carbon, sometimes with iron (SSSA, 1970).

spodosol In U.S. Dept. of Agriculture soil taxonomy, a soil order characterized by the presence of a spodic or placic horizon overlying a fragipan (SSSA, 1970). Suborders and great soil groups of this soil order have the suffix -od. See also: *Aquod; Ferrod; Humod; Orthod.*

spodumene A mineral of the clinopyroxene group: $LiAlSi_2O_6$. It occurs in white to yellowish-, purplish-, or emerald-green prismatic crystals often of great size esp. in granitic pegmatites. Spodumene is an ore of lithium. See also: *kunzite; hiddenite.* Syn: *triphane.*

spoil Waste material removed in mining, quarrying, dredging, or excavating; e.g. the overburden or nonore material removed in surface mining, or refuse earth, rock, and dirt material removed from a pond, drainage ditch, or other cut in engineering earthworks.

spoil bank A bank, mound, or other accumulation composed of spoil; e.g. a submerged embankment of waste earth material dredged from a channel and dumped along it. See also: *spoil heap.*

spoil ground An area where dredged or excavated material is deposited or dumped.

spoil heap A pile of refuse material from an excavation or mining operation; e.g. a pile of dirt removed from, and stacked at the surface of, a mine in a conical heap or in layered deposits, such as a tip heap from a coal mine. See also: *spoil bank.*

spoke A radial and typically flat component of a *wheel* of a holothurian, connecting the central part and the rim.

spondylium A trough-shaped or spoon-like plate serving for muscle attachment and accommodating a ventral muscle field in the posterior part of one or both valves of a brachiopod, composed of dental plates in various stages of coalescence, usually with a median septum. Pl: *spondylia.* Cf: *pseudospondylium.*

spondylium duplex A spondylium formed by convergence of dental plates and supported by variably developed median septum arising from the floor of the pedicle valve.

spondylium simplex A spondylium formed by convergence of dental plates and supported ventrally by a variably developed simple median septum.

spong A term used in the Pine Barrens, N.J., for a *cripple* without a growth of cedar or having flowing water only after a rain; sometimes defined as any lowland area where highbush blueberries grow. Pron: *spung.*

sponge A many-celled aquatic invertebrate belonging to the phylum *Porifera* and characterized by an internal skeleton composed most frequently of opaline silica and less commonly of calcium carbonate. Their stratigraphic range is Precambrian to present. Syn: *poriferan.*

sponge-spicule rock A lithified *spicularite* composed principally of sponge spicules.

spongework An entangled pattern or complex of irregular, interconnecting, tubular channels or cavities of various sizes (generally very small) produced randomly by solution in the walls of limestone caves and separated by amazingly intricate and perforated partitions and remnants of partitions. The relations are as complicated as those of the pores of a sponge. Syn: *anastomosis [speleo].*

spongin (a) A variety of collagen or scleroprotein (an insoluble fibrous protein) secreted by sponges and forming their skeletons. (b) A general term for any fibrous, organic skeletal material secreted by sponges. This usage is not recommended, although it is the sense in which the term has been used in all but the most recent literature.

spongioblast A cell that secretes spongin. Syn: *spongoblast; spongiocyte.*

spongocoel The *cloaca* of a sponge.

spongolite A sediment or rock composed principally of the remains of sponges; esp. a *spicularite.* Syn: *spongolith.*

spontaneous fission dating A method of determining the age in years of uranium minerals based on the known rate of spontaneous fission of uranium-238 to xenon.

spontaneous fission-track dating *fission-track dating.*

spontaneous generation An early concept in which living matter was thought to appear from dead material without the influence of outside or supernatural forces. Syn: *abiogenesis.*

spontaneous magnetization The magnetization within a domain in the absence of an applied magnetic field, due to spontaneous magnetic order caused by exchange forces.

spontaneous polarization Development of differences in static electrical potential between points in the Earth as a result of chemical reactions, differences in solution concentration, or the movement of fluids through porous media. See also: *self-potential method.*

spontaneous-potential curve A curve on an *electric log* (usually on the left side of the log), showing spontaneous potentials at different depths in the borehole, and representing small electromotive forces caused by infiltration (by the mud) of the reservoir rocks or possibly by an electrochemical reaction between mud and reservoir fluid. It is used to indicate the porosity and permeability of the rocks penetrated: a more or less straight line on the curve corresponds to shales and the maxima to the left corresponds to permeable strata. Syn: *SP curve; self-potential curve.*

spontaneous-potential log An electrical log consisting of a spontaneous-potential curve. Syn: *SP log; self-potential log.*

spontaneous-potential method *self-potential method.*

sporadic permafrost A region of dominantly unfrozen ground containing scattered areas of permafrost (*permafrost islands*); it occurs along the southern limits of regions where summer frost conditions are usual. Cf: *discontinuous permafrost; continuous permafrost.*

sporadosiderite A stony meteorite throughout which iron is more or less disseminated.

sporae dispersae Pollen and spores obtained by maceration of rocks, in contrast with those that have been found within the sporangia that bore them.

sporal Being, pertaining to, or having the special characteristics of a spore. The term is not in good or current usage in palynology.

sporangiospore A spore produced in a sporangium. The term is not in good or current usage in palynology.

sporangite A fossilized spore case of a plant. The term is not in good or current usage in palynology.

sporangium An organ within which spores are produced or borne; e.g. an organ in embryophytic plants in which spores are produced, such as a pollen sac of a gymnosperm and an anther of an angiosperm. Pl: *sporangia.* See also: *microsporangium; megasporangium.* Syn: *spore case.*

sporbo A term used in the San Joaquin Valley, Calif., for oolite. Pl: *sporbo.* Etymol: smooth-polished-round-black (*blue or brown*)-object (Galliher, 1931, p.257).

spore Any of a wide variety of minute unicellular reproductive bodies or cells that are often adapted to survive unfavorable environmental conditions and that are capable of developing independently into new organisms (directly if asexual or after union with another spore if sexual); e.g. one of the haploid, dispersed reproductive bodies of embryophytic plants, having a very resistant outer wall (exine), and frequently occurring as fossils from Silurian to Holocene.

spore case A case containing spores; a *sporangium.* The term is not in good or current usage in palynology.

spore coat *sporoderm.*

spore mother cell The *mother cell* in the microsporangium of a spore-bearing plant (as in the anther of an angiosperm, or in the pollen sac of other seed plants), which, by reduction division, produces the tetrad of haploid microspores. See also: *pollen mother cell.* Syn: *sporocyte.*

sporinite A variety of *exinite* consisting of spore exines that are generally compressed parallel to the stratification. Cf: *cutinite; alginite; resinite.*

sporite A coal microlithotype containing the macerals of the exinite group. Syn: *liptite.*

sporocyte *spore mother cell.*

sporoderm The wall of a spore or pollen grain, generally consisting of an outer layer (exine) and an inner layer (intine), and, when present, an extra third layer (perisporium) outside of the exine. Syn: *spore coat.*

sporogenous Producing or adapted to the production of spores, or reproducing by spores; e.g. "sporogenous tissue" in sporangium from which spore mother cells originate.

sporologic Of or pertaining to palynology. The term is sometimes used as a synonym of the more current term "palynologic".

sporomorph A fossil pollen grain or spore. Cf: *palynomorph.*

sporophitic Said of the ophitic texture, of an igneous rock, in which large pyroxene crystals enclose much smaller, widely separated plagioclase crystals (Walker, 1957, p.2); a form of poikilitic texture. Cf: *nesophitic.*

sporophyll A spore-bearing leaf.

sporophyte The individual or asexual generation of a plant th[a] produces spores; e.g. the diploid generation of an embryoph[y] tic plant, produced by fusion of egg and spermatozoid in low[e] vascular plants or by fusion of egg nucleus and the sperm n[u] cleus produced by the pollen of seed plants. Cf: *gametophyte.*

sporopollenin The very resistant and refractory organic su[b] stance of which the exine of spores and pollen is compose[d] and which gives the sporomorph its extreme durability durin[g] geologic time, being readily destroyed only by oxidation. It is [a] high-molecular-weight polymer of C-H-O (primarily [a] monocarboxylic or dicarboxylic fatty acids), but its exa[ct] structural composition has not yet been resolved. Als[o] spelled: *sporopollenine.*

sport An individual that exhibits a sudden spontaneous devi[a] tion beyond the normal limits of individual variation, usually a[s] a result of mutation.

spot correlation In seismology, the correlation of reflection[s] on isolated seismograms by noticing similarities in charact[er] and interval.

spot elevation (a) An elevation shown on a topographic ma[p] at a critical point (such as a break in slope or a summit alon[g] a stream) to supplement the map information given by con[-] tour lines and bench marks from which contours can be co[r] rectly drawn. (b) A point on a map or chart whose elevation [is] noted; a *spot height.*

spot height A predominately British term for a point, indicate[d] on a map, whose elevation above a given datum has bee[n] correctly measured on the ground but, in contrast to a benc[h] mark, is seldom indicated on the ground. See also: *spot el[e]vation.*

spot medallion *frost scar.*

spotted *maculose.*

spotted schist *spotted slate.*

spotted slate A shaly, slaty, or schistose argillaceous roc[k] whose spotted appearance is the result of incipient growth [of] porphyroblasts in response to contact metamorphism of low [to] medium intensity. Cf: *desmosite; Schalstein; spilosite; adino[le].* See also: *Fleckschiefer; Fruchtschiefer; Garbenschiefe[r;] Knotenschiefer.* Syn: *spotted schist; knotted schist; knotte[d] slate.*

spot test A method of testing a chemical solution by applyin[g] a drop of the solution to a paper impregnated with a sensiti[ve] reagent. The identity of the solution is indicated by the spec[i]fic color change.

spotty Said of a mineral deposit in which the valuable consti[t]uent occurs in concentrated lumps or nodules.

spout (a) A discharge or jet of water ejected with some vi[o]lence, either continuously (e.g. a spring) or periodically (e.[g.] a geyser). (b) A rush of water to a lower level; e.g. a wate[r]fall. (c) *waterspout.*

spouting horn A sea cave with a rearward or upward openin[g] through which water spurts or sprays after waves enter th[e] cave. Syn: *chimney; oven.*

spouty land Land so wet that it spouts water when walked o[n;] e.g. a marshland.

spray ice Ice formed from ocean spray blown along the sho[re] or upon floating ice.

spray print A small pit or depression similar to a *rain print* b[ut] formed by spray or windblown drops of water. Syn: *spray pit.*

spray ridge An ice formation on an icefoot, formed by th[e] freezing of wind-blown ocean spray.

spray zone The area along a coast affected by the spray [of] breaking waves. Cf: *splash zone.*

spread [seis] The layout of geophone groups from which da[ta] from a single shot are recorded simultaneously. It usua[lly] consists of 24 groups with 50-400 ft intervals between adj[a]cent group centers (Sheriff, 1968). Syn: *seismometer sprea[d;] seismic spread.*

spread [gem] The surface or width in proportion to the dep[th] of a cut stone, such as of a diamond.

spread [stream] A marsh or shallow body of water resulting from the expansion in width of a stream, as where a natural obstruction, aquatic vegetation, or sediment infilling chokes or impedes streamflow. Syn: *widespread*.

spread correction *normal moveout*.

spreading concept *sea-floor spreading*.

spreading-floor hypothesis *sea-floor spreading*.

spring A place where ground water flows naturally from a rock or the soil onto the land surface or into a body of surface water. Its occurrence depends on the nature and relationship of rocks, esp. permeable and impermeable strata, on the position of the water table, and on the topography.

spring alcove A term used along the Snake River Canyon in Oregon and Idaho for a Short, steep-sided, rather deep, amphitheater- or box-headed canyon formed by solutional spring sapping of basalt plateaus.

spring dome A descriptive term suggested by Williamson (1961) for a nondiastrophic limestone structure consisting of a circular or elliptical mound, usually with a central hollow or crater, believed to result from the expulsion of water from an underlying source (such as from a semirigid sediment).

spring-fed intermittent stream An *intermittent spring* that is generally produced by fluctuations of the water table whereby the stream channel stands a part of the time below and a part of the time above the water table (Meinzer, 1923, p.57). Cf: *surface-fed intermittent stream*.

spring-fed lake *spring lake*.

spring flood A flood occurring in the spring season.

springhead The *fountainhead* of a stream.

spring lake (a) A lake, usually of small size, that is created by the emergence of a spring or springs, esp. one having visibly flowing springs on its shore or springs rising from its bottom. (b) A lake that receives all or part of its waters directly from a spring.---Syn: *limnokrene; spring-fed lake*.

springlet (a) A little spring. (b) *streamlet*.

spring line A line of springs marking the intersection of the water table with the land surface, as at the foot of an escarpment.

spring mound A low (5-6 m high, 12 m across), roughly circular or elliptical mound of sand and silt, formed by a spring rising to the surface and depositing its load. See also: *mound spring*.

spring neck A long, narrow trench, commonly 60-90 cm wide and a few meters deep, formed by percolating water flowing toward the central level of a playa from a spring at the margin of the playa (Stone, 1967, p.249).

spring pit A small crater formed on a sand beach by ascending water, characterized by coarse sand in the center and finer sand around the edge, and measuring 30-60 cm across and about 15 cm deep (Quirke, 1930).

spring pot A shallow depression formed on the margin of a alluvial lake bed or a modern playa by spring flow, measuring 90-120 cm across and 60-90 cm deep (Stone, 1967, p.249).

spring range *mean spring range*.

spring sapping The erosion of a hillside around the fountainhead of a strongly flowing stream, causing small landslides and resulting in the retreat of the valley head. Syn: *springhead sapping*.

spring snow A coarse, granular, wet snow formed during spring by alternating freezing and thawing. Syn: *corn snow*. Cf: *granular snow*.

spring stream A stream that derives its normal runoff from a spring.

spring tide A tide occurring twice each month at or near the times of new moon (conjunction) and full moon (opposition) when the gravitational pull of the Sun reinforces (or acts in the same direction as) that of the Moon, and having an unusually large or increased tide range. Cf: *neap tide*. Syn: *syzygy tide*.

springwood In a woody stem, the portion of the annual ring formed in the early part of a growing season and consisting typically of cells which are larger than those formed later in the season (Fuller & Tippo, 1949, p.972). Also spelled: *spring wood*. Cf: *summerwood*. Syn: *early wood*.

spruit A term used in southern and eastern Africa for a small stream, esp. one that is usually dry but nourished by sudden floods. Etymol: Afrikaans, from Middle Dutch *spruten*, "to sprout".

spudder (a) A small drilling rig, used primarily to start a new well. (b) A drill used for drilling seismic shotholes in hard rock or gravel. (c) The special drill bit used to begin a borehole.

spudding in The beginning of actual drilling operations on a well or borehole (regardless of the type of equipment used); the first abrasion of the soil by the drill, or the first entrance of the drill into the ground; the preliminary boring of a well through earth material down to rock or other solid substrata. Syn: *spudding*.

spumellarian Any radiolarian belonging to the suborder Spumellina, characterized by a thick-walled central capsule perforated by fine, evenly distributed pores.

spur [ore dep] A small vein branching from a main one.

spur [eng] An artificial obstruction (such as a pier dam) built out from the bank of a stream channel for the purpose of deflecting the current or of protecting the shore from erosion.

spur [geomorph] (a) A subordinate ridge or lesser elevation that projects sharply at right angles to, or in a lateral direction from, the crest or side of a hill, mountain, or other high land surface; a small hill extending from a prominent range of hills or mountains. Syn: *prong*. (b) *meander spur*.

spur [marine geol] (a) A ridge or other prolongation of a terrestrial mountain range, extending from the shore onto or across the continental shelf or insular shelf; e.g. the Bahama Spur in the Atlantic Ocean. (b) A subordinate ridge projecting outward from a larger submarine feature or elevation.

spur [paleont] A dependent projection of the basal margin of a tergum of a cirripede crustacean.

spur [ice] *ram*.

spur-and-groove structure A comb-tooth structure common to all reef fronts, best developed on the windward side, consisting of grooves separated by seaward-extending spurs or ridges (Maxwell, 1968, p.110). Syn: *groove-and-spur structure*.

spur-end facet *triangular facet*.

spur furrow A groove on the outer surface of a tergum of a cirripede crustacean, extending along a spur to the apex. Syn: *spur fasciole*.

spurrite A light-gray mineral: $Ca_5(SiO_4)_2(CO_3)$.

squall A strong and sudden wind (either rotary or straight) caused by atmospheric intensity, and often accompanied by precipitation, thunder, and lightning. There are many local names for various types of squalls. See also: *line squall*.

squall line *line squall*.

squamiform load cast One of a group of crowded, lobate load casts overlapping downcurrent, resembling sagged flute casts but having an opposite orientation with regard to current direction.

squamose Said of a plant that is covered with small scales. Cf: *lepidote*.

squamula A small plate projecting subhorizontally in an eave-like manner from the wall of a tabulate corallite toward the axis, as in *Emmonsia*. Pl: *squamulae*.

square emerald cut An *emerald cut* with a square girdle outline but modified by corner facets.

square slide rule A slide rule used for the solution of right-angled triangle problems. Its upper scale is measured according to the square of the scaled distance and is identical with the scale on the slider. The labeling is in terms of distance so that the sum of two squares may be obtained.

squeaking sand A term used by Humphries (1966, p.135) for *whistling sand*.

squeeze [drill] v. To pump or inject cement, under high pressure, into a cavity or fracture on the outside of the casing in a borehole in order to seal off or recement a channeled area. See also: *squeeze job.*---n. The plastic movement of a soft rock in the walls of a borehole, in which the diameter of an opening in the wall is reduced.

squeeze [eng geol] (a) The gradual increase in load on tunnel or mine supports with some movement of ground around resisting support members. (b) The gradual closing of a mine working by the settling of the overlying strata. (c) The fracturing, crushing, or downward bending of the roof strata over a mine working; the gradual upheaval of the floor of a mine due to the weight of overlying strata. (d) A mine area (such as a section in a coal seam) undergoing a squeeze.

squeeze [speleo] In a cave, a *passage* that is traversable only with difficulty. Syn: *squeezeway.*

squeeze job The forcing of cement, under high pressure, upon the outside of the casing in a borehole in order to shut off leakage of undesirable fluids. It is usually a secondary cementation. See also: *squeeze [drill]; diesel squeeze.*

squeeze-up A small extrusion of viscous lava, or *toothpaste lava*, from a fracture or opening on the solidified surface of a flow, caused by pressure. It may take various forms, generally bulbous or linear, and may be from a few centimeters to almost a meter in height. It may be marked by vertical grooves. Syn: *push [volc].*

squeezeway *squeeze.*

squeezing ground Soil or rock that contains a large amount of clay and that advances slowly (flows plastically) with no perceptible increase in volume and without fracturing. Cf: *swelling ground.*

s surface In structural petrology, one of a set of parallel surfaces of discontinuity or mechanical inhomogeneity; a nongenetic term for a plane of movement, e.g. foliation, bedding. Cf: *s plane.*

stabile Resistant to chemical change, or decomposing with difficulty; e.g. "stabile protobitumen", a plant or animal product (such as wax, resin, spores, or leaf cuticle) that forms fossil carbonaceous deposits such as amber or cannel coal. Ant: *labile.*

stability [eng] The quality of permanence or resistance of a structure, slope, embankment, or other foundation to failure by sliding, overturning, collapsing, or other prevailing condition of stress; e.g. *bank stability* and *slope stability.*

stability [geochem] In thermodynamics, an equilibrium state of a system to which the system will tend to move from any other state under the same external conditions. Since it is never possible to examine all alternative states, assertions about the stability of real systems must always contain, at least implicitly, reference to the alternative states relative to which stability is claimed.

stability [tect] The quality or state of enduring or persisting without change of position or form, such as the condition of an undeformed part of the Earth's crust relative to mountain-making forces.

stability field The range of conditions within which a mineral or mineral assemblage is stable.

stability series A grouping of minerals arranged according to their persistence in nature; i.e. to their resistance to alteration or destruction by weathering of the parent rock, abrasion during transportation, and postdepositional solution (Goldich, 1938); e.g. olivine (least stable), augite, hornblende, biotite (most stable). The most stable minerals are those that tend to be at equilibrium at the Earth's surface.

stabilization [eng] *soil stabilization.*

stabilization [ecol] The characteristic of a climax, in which the greatest degree of adjustment between organisms and environment has been attained.

stabilized dune *anchored dune.*

stabilizing force The ordinary restoring force in an unstable gravimeter.

stable [sed] (a) Said of a constituent of a sedimentary rock that effectively resists further mineralogic change and the represents an end product of sedimentation (often resulting from more than one cycle or erosion and deposition); e. quartz, quartzite, chert, and accessory minerals such as zircon, rutile, muscovite, and tourmaline. (b) Said of a mature sedimentary rock (such as orthoquartzite) consisting of stable particles that are rounded or subrounded, well-sorted, an composed essentially of silica.

stable [tect] Said of an area or part of the Earth's crust that shows neither uplift nor subsidence or that is not readily deformed; e.g. a "stable shoreline" that is neither advancing nor receding.

stable [radioactivity] Said of a substance that is not spontaneously radioactive. Cf: *unstable [radioactivity].*

stable equilibrium A state of equilibrium of a body (such as pendulum) when it tends to return to its original position after being slightly displaced. Cf: *unstable equilibrium; dynamic equilibrium.*

stable gravimeter An instrument which uses a high order of optical and/or mechanical magnification so that an extremely small change in the position of a weight or associated property can be accurately measured.

stable isotope A nuclide that does not undergo *radioactive decay.* Cf: *unstable isotope.*

stable magnetization Remanent magnetization which does not change over geologic time, i.e., does not show magnetic viscosity. In practice it is nearly the same as *hard magnetization.*

stable relict A *relict* [meta] that was not only stable under the conditions of its formation but also under the newly imposed conditions of metamorphism. Cf: *unstable relict.*

stac A term used in the St. Kilda Islands (lying west of the Outer Hebrides, Scotland) for a *stack* consisting of hard igneous rock, sometimes eroded to or just below sea level.

stack [coast] A small, lofty, isolated, commonly steep-sided pillar-like rocky island or mass near a cliffy shore, detached from a headland by wave erosion assisted by weathering; especially one showing columnar structure and roughly horizontal stratification. Examples occur off the chalk cliffs of the Normandy coast. A stack may also form along the shore of a very large lake. See also: *skerry; stac; chimney; chimney rock.* Syn: *sea stack; marine stack; rauk.*

stack [geomorph] An upstanding, steep-sided mass of rock rising above its surroundings on all sides from a slope or hill.

stacking fault A type of *plane defect* in a crystal, caused by alternate stacking sequences between two parts of a closest packed crystal. Syn: *fault [cryst].*

stacking sequence A repetition of layers in the atomic structure of a crystal.

STADAN An acronym derived from the name *Satellite Tracking and Data Acquisition Network.*

stade A substage of a glacial stage marked by a glacial readvance; "a climatic episode within a glaciation during which a secondary advance of glaciers took place" (ACSN, 1961, art.40). Syn: *stadial.*

stadia (a) A surveying technique or method using a stadia rod in which distances from an instrument to the rod are measured by observing through a telescope the intercept on the rod subtending a small known angle at the point of observation, the distance to the rod being proportional to the rod intercept. The angle is usually defined by two fixed lines in the reticle of the telescope. (b) *stadia rod.* (c) An instrument used in a stadia survey; esp. an instrument with stadia hairs. -Pl: *stadias.* The term is also used as an adjective in such expressions as "stadia surveying", "stadia distance", and "stadia station".

stadia constant (a) The ratio by which the intercept on the stadia rod must be multiplied to obtain the distance to the rod. On most surveying instruments, it is 100. (b) The ratio between

hich the sum of the stadia intervals of all sights of a continu-
us series of measured differences of elevation is converted
the length of the series in kilometers.

adia hairs Horizontal cross hairs equidistant from the central
orizontal cross hair; esp. two horizontal parallel lines or
arks in the reticle of a transit telescope, arranged symmetri-
lly above and below the line of sight, and used in the stadia
ethod of surveying. Syn: *stadia wires*.

adia interval The length of stadia rod subtended between the
p and bottom cross hairs in the leveling instrument as seen
ojected against the face of the rod. Syn: *stadia intercept*.

adial adj. Pertaining to or formed during a *stade*.---n. *stade*.

adial moraine *recessional moraine*.

adia rod A graduated rod used with an instrument having
adia hairs to measure the distance from the observation
int to the place where the rod is positioned. Syn: *stadia*.

adia table A mathematical table from which may be found,
ith minimal computation, horizontal distance and the differce in elevation, knowing the stadia distance and the angle
sight.

adia traverse A surveying traverse (such as a transit tra-
rse or a traverse accomplished by plane-table methods) in
hich distances are measured by the stadia method.

adia wires *stadia hairs*.

adimeter A surveying instrument for determining the dis-
nce to an object of known height by measuring the angle
btended at the observer by the object.

affelite *francolite*.

aff gage A type of *gage* that is a staff, wall, pier, etc. conining a graduated scale used in gaging water-surface eleva-
on. Cf: *chain gage*.

affordian European stage: middle Upper Carboniferous
above Yorkian, below Radstockian). It is equivalent to part of
pper Westphalian.

age [geochron] An obsolete term for a geologic-time unit of
sser duration than *age* [geochron], during which the rocks
a *formation* [stratig] were formed. See also: *substage* [geo-
hron].

age [geomorph] (a) A period, or a phase in development, of
cycle of erosion in which the features of the landscape have
stinctive (although arbitrarily defined) characteristic forms
at distinguish them from similar features in other parts of
e cycle; e.g. the *stages* of youth, maturity, and old age in
e development of a stream or region. It is one of the factors
at determines the development of landforms in the Davisian
ycle of erosion, although Davis (1899) originally referred to
is factor as "time" (changed by later writers to "stage"). (b)
particular phase in the historical development of a geologic
ature; e.g. the Calumet *stage* of Lake Chicago.

age [glac geol] A time term for a major subdivision of a gla-
al epoch; specif. a major Pleistocene subdivision equated
ith a rock unit of formation rank. It includes glacial stage
nd interglacial stage. The American Commission on Stratigaphic Nomenclature (1961, art.31b) rejects this usage be-
ause of conflict with the definition of "stage" (as a time-stra-
graphic unit) and the requirement that stages be extended
eographically on the basis of time-equivalent criteria.

age [cryst] In a microscope or similar optical apparatus, the
mall platform on which the object to be studied is placed.
ee also: *universal stage; mechanical stage*.

age [stratig] (a) A time-stratigraphic unit next in rank below
eries and above *substage*, commonly based on a succession
biostratigraphic zones that are considered to approximate
osely time-equivalent deposits (ACSN, 1961, art.31); the
cks formed during an *age* of geologic time. Stages are the
asic working units of local time-stratigraphic correlation, and
e often used to relate the various kinds of minor stratigra-
ic units in one geologic section or area to those in another
th respect to time of origin. Most stage names are based on
ck-stratigraphic units, although preferably a stage should

have a geographic name not previously used in stratigraphic
nomenclature; the adjectival ending for the geographic name
is most commonly "-an" or "-ian", although it is permissible to
use the geographic name without any special ending, such as
"Claiborne Stage". (b) An informal term used to indicate "any
sort" of time-stratigraphic unit of approximate stage rank
"which is not a part of the standard hierarchy" of named time-
stratigraphic units (ISST, 1961, p.24-25). (c) A para-time-
rock unit consisting of two or more zones (Wheeler et al,
1950, p.2364). (d) A term used in England for a rock-stratigraphic unit.

stage [hydraul] The height of a water surface above an arbitrarily established datum plane. Syn: *gage height*.

stage-capacity curve A graphic illustration of the relationship
between the surface elevation of the water in a reservoir and
the volume of water (Langbein & Iseri, 1960, p.17). Cf: *stage-
discharge curve*.

stage-discharge curve A graphic illustration of the relationship
between gage height and volume of flowing water, expressed
as volume per unit of time (Langbein & Iseri, 1960, p.17-18).
Cf: *stage-capacity curve*. Syn: *rating curve; discharge-rating
curve*.

staghorn coral Any coral (esp. a scleractinian belonging to the
genus *Acropora*), characterized by large branching colonies
whose shape resembles antlers.

stagmalite A general term for both *stalactites* and *stalagmites*;
a partial syn. of *dripstone*. Obs syn: *dropstone*.

stagnant basin An isolated or barred basin containing essentially motionless water, rich in organic accumulations and noxious substances, but deficient in oxygen and capable of supporting only anaerobic organisms.

stagnant glacier *dead glacier*.

stagnant ice *dead ice*.

stagnation The condition or quality of water unstirred by a current or wave, or not running in a current or stream, or of a
glacier that has ceased to flow. Syn: *stagnancy*.

stagnation point On the surface of a solid immersed in a flowing fluid, that point at which the stream lines separate.

stagnicolous Said of an organism that lives in or frequents
stagnant water.

stagnum A small lake or pool of water lacking an outlet.

stained stone A gemstone whose color has been altered by
the use of a coloring agent, such as a dye, or by impregnation
with a substance, such as sugar, followed by either chemical
or heat treatment, which usually produces a permanent color;
e.g. green chalcedony. Cf: *heated stone*.

stainierite *heterogenite*.

staircase pond One of a sequent group of ponds (from a
dozen to a hundred) following the approximate axis of a
poorly developed watercourse on a sloping, thinly soil-mantled
flat in a high-altitude valley, and resulting from the "armoring"
and binding of naturally created bars by dense and rapidly
growing grass (Ives, 1941, p.287-290); e.g. in Albion Valley,
Colo.

stairway (a) *glacial stairway*. (b) *cirque stairway*.

stalactite [speleo] A conical or cylindrical speleothem that is
developed and hangs from the roof of a cave. It is deposited
by dripping water and is usually composed of calcium carbonate but may also be formed of metallic carbonates. Cf: *stalagmite; stagmalite; stalacto-stalagmite*.

stalactite [volc] A conical formation of lava hinging from the
roof or walls of a lava tunnel or other cavity and developed by
the dripping of fluid lava. It generally measures about 15-30
cm in length. Cf: *stalagmite* [*volc*].

stalacto-stalagmite A columnar deposit formed by the union of
a *stalactite* with its complementary *stalagmite*. Syn: *column*
[*speleo*]; *pillar* [*speleo*].

stalagmite [speleo] A conical speleothem that is developed
upwards from the floor of a cave by the action of dripping
water. It is usually composed of calcium carbonate but may

also be formed of metallic carbonates. Cf: *stalactite; stagmalite; stalacto-stalagmite.*

stalagmite [volc] A conical formation of lava that is built up from the floor of a cavity in a lava flow, and formed as a corresponding feature to a *stalactite* of lava. It generally measures up to 30 cm in height and up to 10 cm in diameter.

stalk That part of a plant by which a plant part is attached and supported, e.g. the petiole of a leaf, the stipe of an ovary, the peduncle of a fruit.

stamen That organ of a flower which produces pollen.

staminate flower Said of a flower which has a stamen but no carpel. Cf: *pistillate.*

Stampian *Rupelian.*

stamukha A fragment of sea ice stranded on a shoal or a shallows. Pl: stamukhas. Etymol: Russian.

stand [drill] Two or more lengths or connected joints of drill pipe or casing handled as a unit length as they are taken from a borehole and racked (set upright) in the derrick while changing the bit.

stand [tide] *stand of tide.*

standard atmosphere A hypothetical condition of the atmosphere described as a temperature of 15°C, a pressure of 1, 013.25 millibars, measured at mean sea level. It is used as a representative standard in various types of atmospheric analysis.

standard-cell method A method of studying the chemical relationships between rocks by calculating the number of various cations in the rock per 160 oxygen ions (Barth, 1948).

standard depth One of a series of depths (in meters) at which, by international agreement, physical measurements of sea water are to be taken.

standard deviation The square root of the average of the squares of the deviations about the mean of a set of data. It is a statistical measure of dispersion. Symbol: σ. Syn: *root-mean-square deviation; sigma.*

standard-deviation map A *vertical-variability* map, or *moment map,* that shows the degree of statistical dispersion of one lithologic type (in a given stratigraphic unit) about its mean position in the unit. Cf: *center-of-gravity map.*

standard Earth An Earth model in which each surface of *P* or *S* seismic velocity in the interior of the Earth is spherical, and encloses the same volume as the corresponding surface of equal velocity in the actual Earth (Runcorn, 1967, v.2, p. 1437).

standard error A measure of the accuracy of a sample mean as an estimator for the population mean; the standard deviation of the sampling distribution of a statistical parameter, or the standard deviation of a sample mean; the standard deviation divided by the square root of the number of observations of a sampled variate.

standard meridian (a) The meridian used for determining standard time. (b) A meridian of a map projection, along which the scale is as stated.

standard mineral A mineral whose presence in a rock is theoretically possible on the basis of certain chemical analyses. A standard mineral may or may not be actually present in the rock. See also: *norm.* Syn: *normative mineral.*

standard parallel (a) Any parallel of latitude that is selected as a standard axis on which to base a grid system; specif. one of a set of parallels of latitude (other than the base line) of the U.S. Public Land Survey system, passing through a selected township corner on a principal meridian, and on which standard township, section, and quarter-section corners are established. Standard parallels are usually at intervals of 24 miles north or south of the base line, and they are used to limit the convergence of range lines that intersect them from the south so that nominally square sections and townships can be layed out. Syn: *correction line.* (b) A parallel of latitude that is used as a control line in the computation of a map projection; e.g. the parallel of a normal-aspect conical

projection along which the principal scale is preserved. (c) parallel of latitude on a map or chart along which the scale as stated for that map or chart.

standard port British term for *reference station.*

standard rig A common, but now inappropriate, name for cable-tool drilling rig.

standard sea water *normal water.*

standard section A *reference section* showing as complete possible a sequence of all the strata in a certain area, in the correct order, and thus affording a standard for correlation. supplements (and sometimes supplants) the type section esp. for time-stratigraphic units.

standard state A condition in the rocks in which the pressu is the same in all directions at any point, and which is cause by the weight of the overlying rocks. Cf: *load.*

standing crop *biomass.*

standing floe A separate floe standing vertically or inclin and enclosed by rather smooth ice.

standing level The water level in a well (or other excavatio penetrating the zone of saturation, from which water is n being withdrawn, whether or not it is affected by withdrawa from nearby wells or other ground-water sources. Partial sy *static level.* Syn: *standing water level.*

standing oscillation *standing wave.*

standing water Surface water that has no perceptible flow a that remains essentially in place, such as the water of sor lakes and ponds; stagnant water, such as that enclosed marshes and swamps. Ant: *running water.*

standing water level *standing level.*

standing wave A water wave, the wave form of which osc lates vertically between two points or nodes, without progre sive movement. Syn: *standing oscillation; stationary wave.*

stand of tide The time during which there is no appreciab change in the height of the tide; it occurs at high water and low water, and is generally shorter when the tide range large and longer when the tide range is small. Syn: *tide stand; stand [tide].*

standstill *stillstand.*

stanfieldite A meteorite mineral: $Ca_4(Mg,Fe,Mn)_5(PO_4)_6$.

stannic Relating to or containing tin in its tetravalent state. C *stannous; stanniferous.*

stanniferous Relating to or containing tin, e.g. stanniferou ore. Cf: *stannic; stannous.*

stannite (a) A steel-gray or iron-black tetragonal mineral: Cu $FeSnS_4$. Zinc often replaces iron. Stannite has a metal luster and usually occurs in granular masses in tin-bearir veins, associated with cassiterite. Syn: *tin pyrites; bell-me ore.* (b) Impure cassiterite.---Syn: *stannine.*

stannoidite A mineral: $Cu_5(Fe,Zn)_2SnS_8$.

stannopalladinite A mineral: Pd_3Sn_2.

stannous Relating to or containing tin in its bivalent state. C *stannic; stanniferous.*

stantienite A black variety of retinite having a very high (23% oxygen content. Syn: *black amber.*

star n. (a) A rayed figure in a crystal, consisting of two more intersecting bands of light radiating from a bright cente and observed best under strong illumination; an optical ph nomenon caused by reflected light from inclusions or cha nels, and brought to sharp lines in gem materials by caboche cutting. Stars usually have four, six, or twelve rays, but three five-, seven-, or nine-rayed stars occur, or are possible due the absence of inclusions in a portion of the stone. See als *asterism.* (b) A gemstone showing such a figure. (c) *st facet.* (d) *star cut.*---adj. Said of a mineral, crystal, or ge stone that exhibits asterism; e.g. "star agate". Syn: *asteri ed.*

star cluster A system of gravitationally interacting stars, nu bering from hundreds to millions of stars, sharing a commo evolution. Two fairly distinct types are recognized. Possessi a spherical, or halo, distribution with respect to the center

Galaxy are the globar clusters, which are highly centrally ndensed, very massive, and relatively old (billions of years). stributed in the galactic plane, especially within the spiral ns, are the open or galactic clusters. They are very much s massive, show no central condensation, are sometimes young as a few million years, and are relatively enriched in e heavy elements. Syn: *galactic cluster.*

ir cut A general term that refers to any brilliant-cut stone ose pavilion facets present a star effect when viewed ough the table. Syn: *star.*

ir dune An isolated hill of sand, its base resembling in plan several-pointed star, and its sharp-crested ridges converg- from the basal points to a central peak that may be as h as 100 m above the surrounding plain; it tends to remain ed in place for centuries in an area where the wind blows m all directions. Syn: *pyramidal dune; heaped dune.*

ir facet One of the eight triangular facets between the main zel facets and bounding the table of a round brilliant-cut m. Syn: *star.*

ringite A mineral: $(Fe,Sn,Ta,Nb)_6O_{12}$.

rkeyite A mineral: $MgSO_4.4H_2O$. Syn: *leonhardtite.*

ir-pattern spread A type of seismic spread in which the ophones or shot points are arranged along the diagonals of ve- or six-pointed star or hexagon.

ir quartz An asteriated variety of quartz containing within crystal whitish or colored radiations along the diametral nes. The asterism is due to the inclusion of submicroscopic edles of some other mineral arranged in parallel orientation.

ir ruby A semiopaque to semitransparent asteriated variety ruby with normally six chatoyant rays.

ir sapphire A semiopaque to semitransparent asteriated va- ty of sapphire with normally six rays resulting from the esence of microscopic crystals (such as rutile needles) in rious orientations within the gemstone. Syn: *asteria.*

rstone (a) An *asteria;* esp. a star sapphire. (b) Less cor- ctly, any asteriated stone, including even petrified wood ntaining small star-like figures in its more transparent parts.

irved basin A sedimentary basin in which the rate of subsi- nce is more rapid than the rate of sedimentation. Sediment ckness is greater at the margins than at the center (Adams al, 1951).

rved ripple mark *incomplete ripple mark.*

issfurtite A massive variety of *boracite* from Stassfurt, Ger- ny, sometimes having a subcolumnar structure and resem- ng a fine-grained white marble or granular limestone.

issfurt salt Potassium salt from deposits in Stassfurt, Ger- ny.

ite-line fault A tongue-in-cheek term for the discontinuity of ologic structures appearing at the borders of geologic maps adjacent geographic areas, due to differences in interpreta- n.

ite of the sea *sea state.*

itic field An electric or magnetic field in which the field antity is unidirectional and time-invariant.

itic granitization Formation of a granitic rock by metasoma- m in the absence of compressive force.

itic head The height above a standard datum of the surface a column of water or other liquid that can be supported by static pressure at a given point. It is the sum of the *eleva- n head* and the *pressure head,* the velocity head being neg- ble under conditions to which Darcy's law can be applied. e also: *head[hydraul]; hydrostatic head.*

itic level (a) *hydrostatic level.* (b) *standing level.* (c) That er level of a well that is not being affected by withdrawal ground water.

itic metamorphism (a) A variety of regional metamorphism ught about by the action of heat and solvents at high hostatic) pressures, the latter being due to a superincum- it load and not induced by orogenic deformation, i.e. shear- stress (Daly, 1917). See also: *load metamorphism.* Cf:

thermal metamorphism; geothermal metamorphism. (b) Ac- cording to Grabau (1932), the equivalent of diagenesis, a group of processes which are no longer considered to be part of metamorphism.

static modulus A *modulus of elasticity* that is produced by a very slow application of load.

static pressure Pressure that is "standing" or stabilized due to the fact that it has attained the maximum possible from its source, and is not being diminished by loss.

static rejuvenation A kind of *rejuvenation* resulting from a de- crease in stream load, an increase in runoff (owing to in- creased rainfall), or an increase in stream volume through acquisition of new drainage; it involves neither uplift of the land nor eustatic lowering of sea level.

static zone A term suggested for the zone below the lowest point of discharge, i.e. below the *zone of discharge,* suppos- edly where there is little or no water movement. This concept is inaccurate as there is substantial movement below this level in both surface- and ground-water bodies.

station [geophys] A position at which a geophysical observa- tion is made.

station [surv] (a) A definite point on the Earth's surface whose position and location has been or will be determined by surveying methods; e.g. *triangulation station* and *instrument station.* It may or may not be marked on the ground. (b) A length of 100 ft, measured on a traverse along a given line, which may be straight, broken, or curved.

stationary block A mass of relatively undeformed rocks be- neath an overthrust. Cf: *autochthon.*

stationary field A nonvarying physical field, e.g. a magnetic field, either artificial or natural.

stationary flow *steady flow.*

stationary mass In some seismometers, a heavy weight, ei- ther suspended or supported, that tends to remain quiescent during an earthquake. Syn: *steady mass.*

stationary wave *standing wave.*

stationary-wave theory *oscillatory-wave theory.*

station error *deflection of the vertical.*

station pole A pole, rod, or staff for marking stations in surveying; e.g. a range rod and a level rod. Also known as a "station rod" or "station staff".

statistical lithofacies A *lithofacies* that grades laterally into its neighbors and whose boundaries are vertical arbitrary-cutoff planes (Weller, 1958, p.633). It is a mappable unit and is the most commonly recognized and used kind of facies. Cf: *inter- tongued lithofacies.*

statistics (a) The pure and applied science of devising, apply- ing, and developing techniques such that the uncertainty of numerical inferences may be calculated. (b) The art of reduc- ing numerical data and their interrelationships to comprehen- sible summaries or parameters. (c) Numbers describing data taken from any sampled population.

statobiolith A term introduced by Sander (1967, p.326) for a rock composed mainly of the remains of sessile reef- or shoal-building organisms in their positions of growth

statoblast A hard-shelled reproductive body formed in many freshwater bryozoans (such as those in the class Phylactolae- mata) and enclosed in a chitinous envelope in the parent body. It generally serves to preserve the species in winter, and develops into a new individual in the spring. See also: *hibernaculum.*

statoscope A sensitive barometer used in aerial photography for measuring altitude differences between adjacent air sta- tions by recording small differences in atmospheric pressure.

statospore A *resting spore;* e.g. a siliceous thick-walled resis- tant *cyst* formed within the frustules of various chiefly marine centric diatoms and characterized by two overlapping convex valves and by absence of a girdle, or an intracellular cyst in various algae of the division Chrysophyta.

statute mile A measure of distance used on land, equal to

5280 ft, 1760 yd, 1609.35 m, 1.61 km, 880 fathoms, 80 chains, 320 rods, or 8000 links, and roughly equivalent to 0.87 international nautical mile. It is usually referred to as *mile*.

stauractin A spicule of hexactinellid sponges in the form of four coplanar rays at approximately right angles to one another.

staurodisc A hexactinellid-sponge spicule (microsclere) in the form of two interpenetrating amphidiscs at right angles to one another. Cf: *hexadisc*.

staurolite A brownish to black orthorhombic mineral: (Fe, Mg)$_2$Al$_g$Si$_4$O$_{23}$(OH). It is often twinned so as to resemble a cross (six-sided prisms intersecting at 90° and 60°), and it is a common constituent in rocks (such as mica schists and gneisses) that have undergone medium-grade metamorphism. Syn: *staurotide; cross-stone; grenatite; fairy stone*.

stauroscope A type of polariscope that is used to determine the position of vibration-plane traces on a crystal for the accurate measurement of angles of extinction.

staurotide *staurolite*.

staurotite A type of mica schist characterized by porphyroblasts of staurolite. It is usually garnetiferous.
The term was originated by Cordier in 1868.

stead A term used in south Wales for very thin bands of ironstone in coal measures.

steady flow In hydraulics, flow that remains constant in magnitude or in direction of the velocity vector. Cf: *unsteady flow*. Syn: *stationary flow*.

steady mass *stationary mass*.

steady-state creep *secondary creep*.

steady-state stream *graded stream*.

steady-state theory A model of the Universe as a stationary, expanding world that does not change in appearance through space and time. It has no beginning or end, and has a constant density of matter which is continuously and spontaneously created throughout space (Rogers, 1966, p.15). Syn: *continuous-creation hypothesis*.

steady turbidity current A persistent turbidity current, such as one produced where a stream heavily laden with sediment flows into a body of deep standing water. Cf: *spasmodic turbidity current*.

steam vent A type of hot spring from which superheated steam is rapidly and violently expelled.

steatite (a) A compact, massive, fine-grained, fairly homogeneous rock consisting chiefly of talc but usually containing much other material; an impure talc-rich rock. See also: *soapstone* [*rock*]. (b) A term originally used as an alternative mineral name for *talc*, often restricted to grayish-green or brown massive talc that can be easily carved into ornamental objects. Syn: *lardite; lard stone; soapstone* [*mineral*]; *soap earth*. (c) *steatite talc*.

steatite talc A relatively pure or high-grade variety of talc suitable for use in electronic insulators. It is the purest commercial form of talc, although it is usually not all talc. Syn: *steatite*.

steatitization Introduction of, or replacement by, talc or steatite; esp. the act or process of hydrothermal alteration of ultrabasic rocks which finally results in the formation of a talcose rock (such as steatite, soapstone, or relatively pure concentrations of talc).

S-tectonite A tectonite whose fabric is dominated by planar surfaces of formation or deformation, e.g. slate. Cf: *L-tectonite; R-tectonite; B-tectonite*. See also: *slip tectonite*.

steel galena Galena having a fine-grained texture resulting from mechanical deformation or from incipient transformation to anglesite.

steel jack A colloquial syn. of *sphalerite*.

steel ore A name given to various iron ores (esp. to siderite) because they were readily used for making steel.

steenstrupine A dark-brown to black mineral: (La,Ca,Na)(Al, Fe,Mn)(Si,P)(O,OH,F)$_4$ (?).

steep dip In seismology, an observed reflecting horizon whi◄ must be plotted in true spatial position (migrated) instead ◄ directly under a point halfway between the shot point and r◄ ceiver.

steephead A nearly vertical, semicircular wall (at the head ◄ a *pocket valley*) at the base of which springs emerge (S◄ lards & Gunter, 1918, p.27).

steep-to Said of a shore, coast, bank, shoal, or other coast◄ feature that has a precipitous, almost vertical, slope.

Stefan's Law The statement that the total energy emitted ◄ the form of heat radiation per unit time from unit area of ◄ blackbody is proportional to the fourth power of its absolu◄ temperature (Stefan, 1879, p.391-428).

stegidium The convex plate closing the gap between the d◄ thyrial plate and the brachial valve of a spiriferoid brachiop◄ and consisting of a series of concentric layers deposited ◄ the outer epithelium associated with the atrophying pedic◄ migrating dorsally (TIP, 1965, pt.H, p.153).

steigerite A canary-yellow mineral: AlVO$_4$.3H$_2$O.

steilwand A syn. of *gravity slope*. Etymol: German *Steilwan◄* "steep wall" (Penck, 1924).

steinkern Rock material consisting of consolidated mud ◄ sediment that filled the hollow interior of a fossil shell (su◄ as a bivalve shell) or other organic structure. Also, the fos◄ thus formed after dissolution of the mold. Etymol: Germ◄ *Steinkern*, "stone kernel". Syn: *internal cast; endocast*.

stele In a plant, the primary vascular structure of a stem ◄ root, together with the tissues (such as the pith) which m◄ be enclosed (Cronquist, 1961, p.873). Syn: *central cylinder*.

stell An English term for a brook.

stellar coal *stellarite*.

stellar crystal A common and beautiful type of snow crys◄ having the shape of a flat hexagonal star, often with intrica◄ branches. Cf: *spatial dendrite*.

stellarite A variety of *albertite* from Stellarton, Nova Scotia.◄ is fusible, burning with a bright smoky flame and droppi◄ sparks, thus the origin of the name. Syn: *stellar coal*.

stellate Said of an aggregate of crystals in a starlike arrang◄ ment; e.g. wavellite.

stellerite An orthorhombic mineral: CaAl$_2$Si$_7$O$_{18}$.7H$_2$O. It is◄ variety of stilbite.

stem [**drill**] *drill stem*.

stem [**paleont**] A narrow structure by which a sessile anin◄ is made fast; e.g. the *column* of an echinoderm.

stemflow Water from precipitation that reaches the ground ◄ running down the trunks of trees or plant stems. Cf: *throug◄ fall*.

stemming (a) The material (sand, clay, limestone) that fills◄ shothole after the explosive charge has been inserted. It ◄ packed between the charge and the outer end of the shothe◄ and is used to prevent the explosive from "blowing" out alo◄ the hole. (b) The act of inserting the stemming into a sh◄ thole.---See also: *tamping*.

stem stream *trunk stream*.

stenecious Said of an organism that can adjust or survive ◄ only a limited range of environments.

stengel gneiss A syn. of *pencil gneiss*. Etymol: German.

stenhuggarite A bright yellow, tetragonal mineral: Ca$_2$Fe$^+$ Sb$^{+5}_2$(As^{+3}O$_3$)$_4$O$_3$.

stenobathic Said of a marine organism that tolerates only◄ narrow range of depth. Cf: *eurybathic*.

stenobiontic A term used as the English equivalent of a Ru◄ sian term applied to an organism that requires a stable a◄ uniform environment.

stenohaline Said of a marine organism that tolerates only◄ narrow range of salinities. Cf: *euryhaline*.

stenonite A monoclinic mineral: (Sr,Ba,Na)$_2$Al(CO$_3$)F$_5$.

stenoplastic Having limited capacity for modification and a◄ aptation to new environmental conditions; incapable of maj◄ evolutionary differentiation. Cf: *euryplastic*.

enopodium A slender, elongate limb of a crustacean, comsed of rod-like segments. Pl: *stenopodia*. Cf: *phyllopodium*.

enoproct Said of a sponge in which the cloaca is cylindrical.

nosiphonate Said of nautiloids with relatively narrow siuncles. Cf: *eurysiphonate*.

nothermal Said of a marine organism that tolerates only a rrow range of temperature. Cf: *eurythermal*.

notopic (a) Said of an organism that has little adaptability changes in the environment. (b) Said of an organism that is stricted to one habitat or to a relatively few habitats.----Cf: *enotropic*.

notropic Said of an organism that has a narrow range of erance for variable environmental conditions. Cf: *stenotop-*

entorg A well-defined felsenmeer, usually on the crest and nks of an esker, and often striped by former shorelines and ach ridges (Stamp, 1961, p.430). Etymol: Swedish, "stone arketplace".

ep [coast] (a) The nearly horizontal section, not necessarily rmanent, that roughly divides the beach from the shoreface, cated just seaward of the low-tide shoreline, and in many ses marked by the presence of coarse sediment (such as avel on a sand beach). Syn: *toe*. (b) An abrupt downward flection composed of coarse sand or gravel that marks the eakpoint of relatively small waves and swells prevailing durg the summer.

ep [pat grd] A form of patterned ground characteristic of oderate slopes, and having a step-like form; it is transitional tween a *polygon* (upslope) and a *stripe* (downslope). A ep typically develops as a lobate solifluction terrace with the wer border convex downslope. See also: *sorted step; nonrted step*. Obsolete syn: *terracette*. Syn: *lobate soil*.

ep [geomorph] (a) A *canyon bench* greatly broadened by osion, as those characteristic of the high plateaus of westn U.S.; a step-like landform on a hillside or valley slope that otherwise smoothly rising. (b) *rock step*.

ep-and-platform topography A landscape developed in a reon in which the rocks dip very gently in one direction over a rge area and are composed of alternating layers of hard and ft material, and characterized by a sequence of lowlands d cuestas (Marbut, 1896, p.31-32).

epanovite A mineral oxalate: $NaMgFe^{+3}(C_2O_4)_3.8-9H_2O$. *: zhemchuzhnikovite*.

ep cline An irregular or broken cline.

ep cut A style of cutting for a gemstone in which long, raight, usually narrow, four-sided facets form in a series or w parallel to the girdle and decrease in length as they reede from both above and below the girdle, giving the appearce of steps. The number of rows, or steps, may vary, alough it is usually three on the crown and three on the paviln. Different shapes of step cuts are described by their oute, such as rectangular step cut or square step cut. Cf: *emrald cut; brilliant cut; mixed cut*. See also: *baguette*. Syn: *ap cut*.

ep delta One of a series of deltas built in a body of water hose level was alternately standing and falling, the upper elta being the oldest (Dryer, 1910, p.259); examples are nuerous in the Finger Lakes region of New York.

ep fault (a) One of a set of parallel, closely spaced faults ver which the total displacement is distributed. Cf: *fault ne*. Syn: *multiple fault; distributive fault*. (b) One of a series f low-angle thrust faults in which the fault planes step both wn and laterally in the stratigraphic section to lower glide anes. Step faulting is due to variation in the competence of e beds in the stratigraphic section (Jones, P.B., 1971).

ep fold An abrupt downward flexure of horizontal strata; it is monoclinal structure.

ep function A mathematical expression whose value remains nstant within stated intervals but changes from one interval the next.

Stephanian European stage: Upper Carboniferous (Upper Pennsylvanian; above Westphalian, below Sakmarian of Permian).

stephanite An iron-black orthorhombic mineral: Ag_5SbS_4. It is an ore of silver. Syn: *brittle silver ore; black silver; goldschmidtine*.

stephanocolpate Said of pollen grains having more than three colpi, meridionally arranged.

stephanocolporate Said of pollen grains having more than three colpi, meridionally arranged and provided with pores.

stephanolith A crown- or star-shaped coccolith.

stephanoporate Said of pollen grains having more than three pores, disposed on the equator.

step lake A lake occupying one of a series of rock basins on a glacial stairway; e.g. a *paternoster lake*.

stepout *stepout time*.

step-out A well drilled at a distance from a producing oil or gas well in an effort to extend the productive limits of a field.

stepout correction *normal moveout*.

stepout time In seismic prospecting, the time differential in arrivals of a given peak or trough of a reflected or refracted event for successive detector positions on the Earth's surface. This difference gives information on the dip of the reflecting or refracting horizon in the Earth. Syn: *stepout; angularity; moveout*.

steppe An extensive, treeless grassland area in southeastern Europe and Asia developing in the semiarid mid-latitudes of that region. They are generally considered drier than the *prairie* which develops in the subhumid mid-latitudes of the U.S.

stepped crescent One of a succession of crescentic scarps (several centimeters to a meter high at the center) along the course of an arroyo, resembling together a stair with broad steps and shallow risers (Sharpe, 1938, p.25).

stepped plain A plain that has a succession of step-like levels, "like the steps of a staircase belonging to one flight of stairs" (Schwarz, 1912, p.95). Syn: *klimakotopedion*.

stepping method A rapid method of determining differences in elevations by the use of stadia hairs. The method is now replaced by one using the Beaman stadia arc.

stepping stone An island used by a species that spread (or is spreading) from one region to another (McArthur & Wilson, 1967, p.191).

step tablet *step wedge*.

step terrace (a) A man-made terrace having several step-like levels along which cultivation is done. (b) A terrace with a vertical drop on the downhill side.

steptoe An isolated protrusion of bedrock, e.g. the summit of a hill or mountain, in a lava flow. Syn: *dagala; kipuka*.

step vein In economic geology, a vein that alternately conforms with and cuts through the country-rock bedding.

step wedge An *optical wedge* whose transparency diminishes in discrete, graduated adjacent steps from one end to the other. Cf: *gray scale*. Syn: *step tablet*.

step zone A zone along a shoreline, located slightly below mean sea level and characterized by sediments that are coarser than those above (on the beach) or below (on the shoreface).

stercorite A white mineral: $HNa(NH_4)(PO_4).4H_2O$. It is native *microcosmic salt*.

stereochemistry *crystal chemistry*.

stereocomparator A stereoscope for accurately measuring the three space coordinates of the image of a point on a photograph; it is used in making topographic measurements by the comparison of stereoscopic photographs.

stereogenetic The adj. form of *stereosome*. Also spelled: *stereogenic*.

stereogram [geol] A graphic diagram on a plane surface, giving a three-dimensional representation, such as projecting a set of angular relations; e.g. a block diagram of geologic structure, or a *stereographic projection* of a crystal.

stereogram [photo] A *stereoscopic pair* of photographs (or of other perspective views) correctly oriented and mounted for viewing with a stereoscope. Syn: *stereograph.*

stereograph (a) A stereometer with a pencil attachment, used to plot topographic detail from a properly oriented stereoscopic pair (stereogram). (b) *stereogram* [photo].

stereographic net *net* [struc petrol].

stereographic projection (a) A perspective, conformal, azimuthal map projection of a complete hemisphere in which meridians and parallels are projected onto a tangent plane, with the point of projection on the surface of the sphere diametrically opposite to the point of tangency of the projecting plane. Any point of tangency may be selected (at a pole, on the equator, or a point in between). Any circle on the sphere is represented as a true circle on the projection except for great circles passing through the point of tangency and projecting as straight lines. Meridians and parallels are circular arcs that intersect at right angles. Scale exaggeration increases radially and systematically outward from the center so that the shapes of large areas are somewhat distorted (areas at a great distance from the center are exaggerated). It is the only azimuthal projection that is conformal. Stereographic projections are much used for maps of a hemisphere and are useful in showing geophysical relations (such as patterns of islands arcs, mountain arcs, and their associated earthquake epicenters). (b) A similar projection used in optical mineralogy and structural geology, made on an equatorial plane (passing through the center of the sphere) with the point of projection at the south pole. Syn: *stereogram* [geol].--See also: *polar stereographic projection.*

stereome (a) The calcareous exoskeletal deposit forming the plates of an echinoderm. Also, the calcareous tissue in the mesodermal endoskeleton of a living echinoderm. Cf: *stroma.* (b) The more or less dense calcareous skeletal deposit generally covering and thickening various parts of a scleractinian corallite. Cf: *sclerenchyme.*---Also spelled: *stereom.*

stereometer A device used to measure heights of Earth features on a stereoscopic pair of aerial photographs, containing a micrometer movement by means of which the separation of two index marks can be changed to measure parallax difference. Syn: *parallax bar.*

stereometric map A relief map made by the application of the stereoscopic principle to aerial or terrestrial photographs. Syn: *stereotopographic map.*

stereonet A term used in structural geology for a *Wulff net.*

stereopair *stereoscopic pair.*

stereophotogrammetry Photogrammetry utilizing stereoscopic equipment and methods.

stereophytic Said of a sedimentary rock of organic origin (such as coral rocks and some algal limestones) that has been "built up directly, from the beginning, as a quite solid material" (Tyrrell, 1926, p.234).

stereoplasm Gelated protoplasm; specif. the relatively solid axis of granular-reticulose pseudopodia in foraminifers, surrounded by a granular fluid outer portion (rheoplasm). It is noted in foraminifers *Peneroplis* and *Elphidium*, but is not visible in most agglutinated types (TIP, 1964, pt.C, p.64).

stereoscope An optical instrument for obtaining from two overlapping photographs, diagrams, or other perspective views the mental impression of a three-dimensional image by means of two lenses used one with each eye and set in such a manner (as the approximate distance apart of the human eyes, or about 5-6 cm) as to produce the effect of light rays originating at a single point.

stereoscopic fusion The mental process by which two perspective views or images (one for each eye) are combined and brought to a focus on the retina of each eye in such a manner as to give the impression of a three-dimensional effect or model.

stereoscopic image The mental impression of a three-dimensional model that results from stereoscopic fusion (su as from viewing two overlapping perspective photographs o stereoscopic pair). Syn: *stereoscopic model.*

stereoscopic model A syn. of *stereoscopic image.* Al spelled: *stereomodel.*

stereoscopic pair An *overlapping pair* of photographs th when properly oriented and used with a stereoscope, gives three-dimensional view of the area of overlap. See also: *ste ogram; anaglyph; vectograph.* Syn: *stereopair.*

stereoscopic principle The formation of a single, three-mensional image by simultaneous vision with both eyes of tw photographic images of the same terrain taken from differe camera stations.

stereoscopic radius The limiting distance at which an obje can be seen in stereoscopic relief; it is about 450 m (1500 with unaided human vision.

stereoscopic vision Simultaneous vision with both eyes which the mental impression of depth and distance is o tained, usually by means of two different perspectives of object (such as two photographs of the same area taken fro different camera stations); the viewing of an object in thr dimensions. Syn: *spectroscopy; stereovision.*

stereoscopy (a) The science and art that deals with the use simultaneous vision with both eyes for observation of a pair overlapping photographs or other perspective views, and w the methods by which such viewing is performed and su effects are produced. (b) *stereoscopic vision.*

stereosome The part of a chorismite that remained solid at times during the chorismitization process. Adj: *stereogenet* Cf: *metaster; restite.* See also: *paleosome.*

stereosphere A term that was originally proposed for the i nermost shell of the mantle, but is also used as equivalent the lithosphere. Cf: *chalcosphere.*

stereostatic *geostatic.*

stereotheca The inner layer of a thecal plate of a cystoid. It thicker than *epitheca.*

stereotopographic map *stereometric map.*

stereotriangulation A triangulation procedure that uses stereoscopic plotting instrument to establish horizontal and/ vertical control data by means of successive orientation stereoscopic pairs of photographs in a continuous strip. Se also: *bridging.*

stereozone An area of dense skeletal deposits in a corallit generally peripheral or subperipheral in position.

steric Pertaining to phenomena involving molecular dime sions or arrangements in space; e.g. a steric sea level resul ing from variations in the density (as by changes in wat temperature or salinity) of a column of seawater withou change in mass.

sterigma A minute, spore-bearing process of the basidiomy cete fungi.

sternal pore A pore in radiolarian skeletons of the subfami Trissocyclinae, directly below the vertical spine and part framed by the tertiary-lateral bars. It is not present on spec mens having a vertical pore.

sternbergite A dark-brown or black mineral: $AgFe_2S_3$. It oc curs in tabular crystals or soft flexible laminae. Syn: *flexib silver ore.*

Sternberg's law The decline in size of a clastic particle trans ported downstream is proportional to the weight of the particl in water and to the distance it has traveled, or to the wor done against friction along the bed: $W = W_0e^{-as}$, where W the weight at any distance s, W_0 is the initial weight of th particle, and a is the coefficient of size reduction (Pettijoh 1957, p.530). This relation was observed by H. Sternber (1875), 19th century German engineer(?).

sternite The ventral part of a somite of an arthropod; e.g. th chitinous plate forming the ventral cover of an abdominal o thoracic segment of an insect, or the sclerotized plate formin the ventral cover of a somite of an arachnid or merostome, o

e sclerotized ventral surface of a single somite of a crusta-
ean.

ernum The ventral surface of the body, of a single tagma, or
a somite of an arthropod; the whole ventral wall of the tho-
x of an arthropod (such as of an arachnid). Pl: *sterna.*

eroid Any one of several complex hydrocarbons occurring
aturally in living organisms and having a polycyclic structure
e same as found in the sterols. The fatty acids found in bile
nd vitamin D are examples of steroids.

erraster A sponge spicule (euaster) of globular or kidney
hape with a granular surface, formed ontogenetically by ex-
ansion of a centrum to engulf all but the tips of the rays.

errettite A syn. of *kolbeckite.* It was formerly described as a
drous phosphate of aluminum: $Al_6(PO_4)_4(OH)_6.5H_2O$.

erryite A mineral: $Pb_{12}(Sb,As)_{10}S_{27}$.

evensite A mineral: $Mg_3Si_4O_{10}(OH)_2$. Syn: *aphrodite.*

ewartite (a) A brownish-yellow mineral: $Mn_3(PO_4)_2.4H_2O$. It
sually occurs in minute crystals and tufts of fibers in pegma-
tes. (b) A steel-gray, ash-rich, fibrous variety of bort conining iron, having magnetic properties, and found in the dia-
ond mines of Kimberley, South Africa.

ibianite *stibiconite.*

ibiconite A mineral: $Sb_3O_6(OH)$. It is usually yellowish to
halky white, and occurs as an alteration product of stibnite.
yn: *stibianite.*

ibiocolumbite An orthorhombic mineral: $Sb(Nb,Ta)O_4$. It is
omorphous with stibiotantalite.

ibiopalladinite A silver-white to steel-gray isometric mineral:
d_3Sb.

ibiotantalite A brown or yellowish orthorhombic mineral:
b$(Ta,Nb)O_4$. It is isomorphous with stibiocolumbite.

ibium An ancient name for *stibnite* used (as in Egypt) as a
osmetic for painting the eyes.

ibnite A lead-gray mineral: Sb_2S_3. It has a brilliant metallic
ster, differs from galena by ease of fusion, and often con-
ains gold and silver. Stibnite occurs in massive forms and in
rismatic orthorhombic crystals that show highly perfect
leavage and are striated vertically. It is the principal ore of
ntimony. Syn: *antimonite; antimony glance; gray antimony;
tibium.*

ichtite A lilac-colored rhombohedral mineral: Mg_6Cr_2-
$CO_3)(OH)_{16}.4H_2O$. It is dimorphous with barbertonite and
nay contain some iron.

tick-slip A jerky, sliding-type motion associated with fault
novement in laboratory experiments. It may be a mechanism
n shallow-focus earthquakes.

ticky limit The lowest water content at which a soil will ad-
ere to a metal blade drawn across the surface of the soil
nass (ASCE, 1958, term 353). Cf: *sticky point.*

ticky point A condition of consistency at which a soil material
arely fails to adhere to a foreign object; specif. the moisture
ontent of a well-mixed, kneaded soil material that barely fails
o adhere to a polished nickel or stainless-steel surface when
he shearing speed is 5 cm/sec (SSSA, 1965, p.349). Cf:
ticky limit.

tictolith Chorismite with spotted appearance (Dietrich &
Mehnert, 1961).

tiff clay Clay of low plasticity.

tiffening limit The water content (expressed as percent by
weight of dry soil) at which a thoroughly stirred thixotropic soil
till flows under its own weight in a test tube of 11 mm diame-
er after exactly one minute of rest (Mielenz & King, 1955,
.223).

tigma [paleont] An opening of an arachnid; esp. an opening
of an air-conveying tube forming the respiratory system of an
arachnid or an opening into a saccular breathing organ (book
ung) occurring in many arachnids. Pl: *stigmata.* Syn: *spiracle
paleont].*

tigma [bot] That part of a pistil (usually at the tip of the
style) which receives pollen grains and on which they germi-

nate.

stigmaria The subaerial portions of such Carboniferous plants
as sigillarids and lepidodendrons; the form genus *Stigmaria.*

stilbite (a) A zeolite mineral: $NaCa_2Al_5Si_{13}O_{36}.14H_2O$. It occurs in sheaf-like aggregates of crystals and also in radiated
masses. Syn: *desmine.* (b) A term used by German mineralogists as a syn. of *heulandite.*

stilleite A mineral: ZnSe

stillstand (a) A condition of stability, or of remaining stationary, with reference to the Earth's interior or to sea level, applicable to an area of land, as a continent or island; e.g. an
unvarying base level of erosion between periods of crustal
movement, or an undisturbed sea level relative to that of the
land. (b) A period of time during which there is a stillstand.--
-Syn: *standstill; stand.*

stillwater (a) A reach of a stream that is so nearly level as to
have no visible current or other motion. Syn: *quiet reach.* (b)
A sluggish stream, the water of which appears to be marked
by little or no agitation.---Also spelled: *still water.*

stillwellite A rhombohedral mineral: $(Ce,La,Ca)BSiO_5$.

stilpnomelane A black or greenish-black mineral, approximately: $K(Fe,Mg,Al)_3Si_4O_{10}(OH)_2.xH_2O$. It occurs in mica-
like plates, fibrous forms, and velvety bronze-colored incrustations. Syn: *chalcodite.*

stinkquartz A variety of quartz that emits a fetid odor when
struck.

stinkstein A syn. of *stinkstone.* Etymol: German *Stinkstein.*

stinkstone A stone that emits an odor on being struck or
rubbed; specif. a *bituminous limestone* (or brown dolomite)
that gives off a fetid smell (owing to decomposition of organic
matter) when rubbed or broken. It may emit a "sweet-and-
sour" smell if the carbonate rock is rich in organic-phosphatic
material. See also: *anthraconite.* Syn: *stinkstein.*

stipe [paleont] One branch of a branched graptolite colony
(rhabdosome), made up of a series of overlapping tubes (thecae). In an unbranched rhabdosome, a stipe is the entire
graptolithine colony.

stipe [bot] The stalk of a pistil or other small organ when axile
in origin; also, the petiole of a fern leaf (Lawrence, 1951,
p.771). Adj: *stipitate.*

stipitate Said of a plant having a *stipe* or special stalk.

stipoverite *stishovite.*

stipule In a leaf, a basal appendage of a petiole. There are
usually two stipules on a complete leaf.

stishovite A tetragonal mineral: SiO_2. It is a high-pressure,
extremely dense (4.35 g/cm^3) polymorph of quartz, produced
under static conditions at pressures above about 100 kb and
found naturally associated with coesite and only in shock-metamorphosed quartz-bearing rocks such as those from Barringer Crater (Meteor Crater), Ariz., and the Ries basin, Germany. Its occurrence provides a criterion for meteorite impact. Stishovite has a closely packed rutile-type structure in
which the silicon has a coordination number of 6 (instead of 4
as in quartz and coesite); it forms at higher pressures than
coesite and is apparently less stable at lower pressures after
formation. Syn: *stipoverite.*

stistaite A mineral: SnSb.

stochastic hydrology That branch of hydrology involving the
"manipulation of statistical characteristics of hydrologic variables with the aim of solving hydrologic problems, using the
stochastic properites of the events" (Hofmann, 1965, p.120).
Cf: *parametric hydrology; synthetic hydrology.*

stochastic process A process in which the dependent variable
is random (so that prediction of its value depends on a set of
underlying probabilities) and the outcome at any instant is not
known with certainty. Ant: *deterministic process.* Syn: *random
process.*

stock [ore dep] A rarely used term for a chimneylike orebody;
a syn. of *pipe.*

stock [paleont] The terminal bar of an *anchor* of a holothurian, perpendicular to the shank and of varying shape.

stock [**intrus rocks**] An igneous intrusion that is less than 40 sq mi in surface exposure, is usually but not always discordant, and resembles a batholith except in size. Cf: *boss*.

stockade Piling that serves as a breakwater.

stockwork A mineral deposit in the form of a network of veinlets diffused in the country rock; a deposit of *reticulate* or *chambered* texture. Syn: *network deposit; stringer lode*.

stoichiometric With reference to a compound or a phase, pertaining to the exact proportions of its constituents specified by its chemical formula. It is generally implied that a stoichiometric phase does not deviate measurably from its ideal composition.

stoichiometric coefficient One of the numerical coefficients in an equation that specify the combining proportions of the reactants and products of the reaction described.

Stoke's formula A formula first published in 1849 for computing the undulations of the *compensated geoids* from gravity anomalies.

stokesite A colorless orthorhombic mineral: $CaSnSi_3O_9 \cdot 2H_2O$.

Stokes' law A formula that expresses the rates of settling of spherical particles in a fluid: $V = Cr^2$, where V is velocity (in cm/sec), r is the particle's radius (in cm), and C is a constant relating relative densities of fluid and particle, acceleration due to gravity, and the viscosity of the fluid. It is named after Sir George Stokes, British mathematician and physicist. Cf: *impact law*.

stolidium The thin marginal extension of one or both valves of certain brachiopods (as in the superfamily Stenoscismatacea) forming a narrow to broad frill protruding at a distinct angle to the main contour of the shell (TIP, 1965, pt.H, p.153).

stolon [**paleont**] (a) A slender internal thread-like tubule from which graptolithine thecae appear to originate, and lying within the common canal. (b) A creeping and ribbon-like or membranous basal expansion from which certain octocorallian polyps arise (as those of Stolonifera and Telestacea). (c) A slender tube of kenozooids bearing autozooids of ctenostomatous bryozoans along its length. (d) A small, short, calcareous, tube-like projection serving as a prolonged connection between chambers in the test of an orbitoid foraminifer.

stolotheca A type of graptolithine theca (tube) that encloses the main stolon and proximal parts of three new thecae (*autotheca* and *bitheca*, and daughter stolotheca). It is probably secreted by immature autothecal zooid, constituting in effect the proximal part of autotheca. Stolotheca is equivalent to *protheca* of graptoloids.

stolzite A tetragonal mineral: $PbWO_4$. It is isomorphous with wulfenite and dimorphous with raspite.

stoma A pore in the epidermis of a plant through which gases are exchanged. Plural: *stomata*.

stomach stone *gastrolith*.

stomatal coccolith One of the modified coccoliths surrounding the flagellar field in flagellate coccolithophores exhibiting dimorphism.

stomodaeum An esophagus-like tubular passageway or pharynx leading from the mouth of a coral polyp to the gastrovascular cavity (TIP, 1956, pt.F, p.251). Pl: *stomodaea*. Adj: *stomodaeal*.

stomostyle Thickened outer membrane invaginated in cytoplasm of the apertural region of some foraminifers and from which the mass of cytoplasm emerges giving rise to pseudopodia.

stone [**geol**] Any small piece of *rock* [geol] or mineral.

stone [**meteorite**] A *stony meteorite*.

stone [**gem**] (a) A cut-and-polished natural gemstone; a gem or a precious stone. (b) A term often used incorrectly for an artificial reproduction of, or a substitute for, a gem.

stone [**sed**] (a) One of the larger fragments in a variable matrix of a sedimentary rock; a phenoclast or a megaclast. (b) Crushed or naturally angular rock particles that will pass a 3-inch sieve (76 mm) and be retained on U.S. standard sie no.4 (4.76 mm).

stone [**eco geol**] (a) Ore before processing. (b) Quarried, pr pared rock.

Stone Age In archaeology, a cultural level that was origina the first division of the *three-age system*, and was subs quently divided into the *Paleolithic*, *Mesolithic*, and *Neolith* It is characterized by the use of materials other than met e.g. stone, wood, or bone, for technical purposes. Correlatic of relative cultural levels with actual age (and, therefore, wi the time-stratigraphic units of geology) varies from region region; e.g. this oldest cultural level has been discovered exist in recent times.

stone band *dirt band* [coal].

stone-banked terrace A *sorted step* whose steep slope is bo dered by stones. The term should be reserved for a terrac like feature that lacks a regular pattern and is not a well-de fined form of patterned ground (Washburn, 1956, p.833). C *stone garland*.

stone bind An English term for interbedded layers of san stone and shale, or for a rock (such as siltstone) intermedia between a sandstone and a mudstone (Arkell & Tomkeiet 1953, p.115).

stone-bordered strip *stone stripe*.

stone bubble *lithophysa*.

stone canal A calcified, typically short tube or *canal* leadin from madreporite to ring canal in the water-vascular syste of an echinoderm.

stone cells In a vascular plant, thick-walled, isodiametri strengthening cells of sclerenchyma.

stone circle [**pat grd**] *sorted circle*.

stone clunch A term used in England for a very hard under clay (*clunch*) with interbeds of sand.

stone coal *anthracite*.

stone eye *stone intrusion*.

stone field *felsenmeer*.

stone gall A clay concretion found in certain sandstones. Als spelled: *stonegall*.

stone garland A *sorted step* consisting of a tongue-shape mass of fine material enclosed on the downslope side by crescentic stony embankment; similar to but smaller than stone-banked terrace. Syn: *stone semicircle; garland*.

stone guano A secondary deposit formed by leaching of guan and consequent enrichment in insoluble phosphates.

stone intrusion An irregular, bulbous, sometimes much distort ed mass, or a vertical to steeply dipping sheet, of sandstone occurring within a coal seam or penetrating it (frequently from top to bottom), and always connected with a similar sand stone in the roof or in higher strata. The British usage of the term *intrusion* for such a mass of sedimentary rock in a coa seam is not recommended (BSI, 1964, p.10). Syn: *stone eye sandstone dike*.

stone lace *stone lattice*.

stone land Legally, an area that is economically valuable fo its rock, e.g. granite or sandstone, etc. Cf: *mineral lands*.

stone lattice A *honeycomb structure* produced on a rock face in a desert by sandblast that "pecks away at the softer places and leaves the harder ones in relief" (Hobbs, 1912, p.205) Syn: *stone lace*.

Stoneley wave A type of *guided wave* that is propagated along an internal surface of discontinuity; an interface wave. Cf *channel wave*.

stone line A broken line of angular and subangular rock frag ments, paralleling a sloping topographic surface and lying just above the parent material of a soil at a depth of a few meters below that surface (Sharpe, 1938, p.24); it outcrops in natura and artificial cuts. Syn: *carpedolith*.

stone mesh *sorted polygon*.

stone net *sorted polygon*.

stone packing "A frost structure exclusively occurring on pebble beaches in arctic areas, consisting of a large, flat-lying boulder surrounded by a cluster of edgewise-lying flat stones, in an arrangement resembling the petals of a rose" (Schieferdecker, 1959, term 2403).

stone peat The dark, compacted peat at the bottom of a bog.

stone pit A shallow *sorted circle*, less than a meter in diameter, consisting of a floor of isolated and dominantly circular stones (without finer material) surrounded by thick vegetation; term introduced by Lundqvist (1949, p.336).

stone pitch *Pitch* that is as hard as stone.

stone polygon *sorted polygon*.

stone reef A longshore bar whose upper 3-4 m has been solidly cemented by calcium carbonate derived from organic material. Examples occur off the coast of Brazil near Recife.

stone ring A syn. of *sorted circle* and *sorted polygon*; the term refers to the circular or polygonal border of stones surrounding a central area of finer material.

stone river A term used in the Falkland Islands for a *rock stream* formed by solifluction. Syn: *stone run*.

stone run *stone river*.

stone semicircle *stone garland*.

stone stream *rock stream*.

stone stripe A *sorted stripe* consisting of coarse rock debris, and occurring between wider stripes of finer material. Cf: *soil stripe*. Syn: *stone-bordered strip; rock stripe*.

stone wall *hogback*.

stoneware clay A clay suitable for manufacture of stoneware (ceramic ware fired to a hard, dense condition and with an absorption of less than 5%), used for items such as crocks, jugs, and jars. It possesses good plasticity, fusible minerals, and a long firing range.

stonewort *charophyte*.

stone wreath *sorted circle*.

stony desert A desert area whose surface has been deflated, leaving a concentration of coarse fragments after the removal of sand and dust particles, as a gravel-strewn *reg* or a pebble-strewn *serir*; a desert surface covered with *desert armor*. Cf: *rock desert*.

stony-iron meteorite A general name for relatively rare meteorites containing large (at least 25%) and approximately equal amounts (by weight) of both nickel-iron and heavy basic silicates (such as pyroxene and olivine); e.g. *pallasite* and *mesosiderite*. Syn: *stony-iron; siderolite; iron-stony meteorite; lithosiderite; syssiderite; aerosiderolite; sideraerolite*.

stony meteorite A general name for meteorites consisting largely or entirely of silicate minerals (chiefly olivine, pyroxene, and plagioclase); e.g. *chondrite* and *achondrite*. Stony meteorites resemble ultramafic rocks in composition, and they constitute more than 90% of all meteorites seen to fall. Syn: *stone; aerolite; meteoric stone; meteorolite; asiderite; brontolith*.

stool stalagmite A form of stalagmite that widens and flattens at its top. It is formed by the growth of shelfstone around a partially submerged stalagmite. Syn: *mushroom stalagmite; lily pad*.

stop A dam or weir.

stopbank An Australian term for a levee.

stope n. In a mine, the underground excavation formed by *stoping*, or working in a series of steps, usually in a vertical or steeply inclined orebody.

stoping [intrus rocks] *magmatic stoping*.

stoping The extraction of ore from the Earth in an underground mine by working horizontally in a series of levels or steps. The process is distinct from working in a shaft or tunnel or in a room in a horizontal drift, although the term is used in a general sense to mean the extraction of ore. See also: *stope*.

storage (a) Artificially impounded water, in surface or subsurface reservoirs, for future use. Also, the amount of water so impounded. (b) Water naturally detained in a drainage basin, e.g. ground water, depression storage, and channel storage.

storage coefficient (a) For surface waters such as a reservoir, a coefficient expressing the relationship of the surface area to the mean annual flow that supplies it. (b) For an aquifer, the volume of water released from storage in a vertical column of 1.0 sq ft when the water table or other piezometric surface declines 1.0 ft. In an unconfined aquifer, it is approximately equal to the specific yield (Theis, 1938, p.894). Syn: *coefficient of storage*.

storage curve *capacity curve*.

storage ratio The net available storage of a body of water divided by the mean annual flow (Langbein & Iseri, 1960, p.18).

storage-required frequency curve A graphic illustration of the frequency with which storage equal to or greater than selected amounts will be required to maintain selected rates of regulated flow (Langbein & Iseri, 1960, p.18).

storied Said of a landform or landscape characterized by two or more adjacent levels; e.g. a *storied* peak plain with summit levels at different altitudes. See also: *two-story cliff*.

storis A floating mass of closely crowded icebergs and floes, esp. the remnants of the thickest polar ice drifting from the Arctic Ocean into the North Atlantic between Spitsbergen and Greenland. Pl: storis. Etymol: Danish, "large ice".

storm [geophys] A disturbance or considerable variation of a geophysical field, e.g. a magnetic storm.

storm [meteorol] (a) A general term for any meteorologic disturbance of the atmosphere, e.g. precipitation (rainstorm), wind (windstorm). (b) On the Beaufort wind scale, a wind force of eleven.

storm beach (a) A low, rounded ridge of coarse materials (coarse gravel, cobbles, boulders) piled up by very powerful storm waves behind or at the inner margin of a beach, above the level reached by normal high spring tides or by ordinary waves. Syn: *storm terrace*. (b) A beach as it appears immediately after an exceptionally violent storm, characterized either by removal or deposition of beach materials.

storm berm A low ridge along a beach, marking the limit of wave action during storms. See also: *winter berm*.

storm cusp A transient *cusp* measuring 70-120 m between the crescentic tips and developed during a period of relatively heavy seas. Term introduced by Evans (1938).

storm delta *washover*.

stormflow *direct runoff*.

storm icefoot An icefoot produced by the breaking of a heavy sea and consequent freezing of the wind-driven spray.

storm microseism A long-period (25+ sec) microseism caused by ocean waves.

storm roller A term used by Chadwick (1931) for a wave-formed sedimentary structure "mismentioned" as a concretion. The feature is now regarded as a *flow roll* formed by load deformation.

storm runoff *direct runoff*.

storm seepage The *runoff* [water] infiltrating the surface soil and moving toward streams as ephemeral, shallow, perched ground water above the main ground-water level. It is usually considered part of direct runoff (Langbein & Iseri, 1960). Syn: *subsurface storm flow; subsurface runoff; subsurface flow; shallow percolation; interflow*. Cf: *surface runoff; ground-water runoff*.

storm surge An abnormal, sudden rise of sea level along an open coast during a storm, caused primarily by onshore-wind stresses, or less frequently, by atmospheric pressure reduction, resulting in water piled up against the coast. It is most severe when accompanied by a high tide. Erroneous syn: *tidal wave; storm tide*. Syn: *surge* [waves]; storm wave; hurricane surge; hurricane wave.

storm terrace *storm beach*.

storm tide An erroneous syn. of *storm surge*.

storm water *direct runoff*.

storm wave *storm surge*.

stoss Said of the side or slope of a hill, knob, or prominent rock facing the direction from which an advancing glacier or ice sheet moved; facing the upstream side of a glacier, and most exposed to its abrasive action. Etymol: German *stossen*, "to push, thrust". Ant: *lee* [*glac geol*].

stoss-and-lee topography An arrangement, in a strongly glaciated area, of small hills or prominent rocks having gentle slopes on the stoss side and somewhat steeper, plucked slopes on the lee side; this arrangement is the reverse of *crag and tail*. Syn: *onset-and-lee topography*.

stottite A brown tetragonal mineral: $FeGe(OH)_6$.

stove coal A size of anthracite that will pass through a 2 7/16 inch round mesh but not through a 1 5/8 inch round mesh. Cf: *broken coal; egg coal; chestnut coal; pea coal; buckwheat coal.*

straat A term used in the Kalahari region of southern Africa for the trough between dunes, often floored with clayey sand, and generally 100-150 m wide. Etymol: Afrikaans, "street". Pl: *straate.*

straddle spread *split spread.*

straight suture An externally visible line of articular contact perpendicular to the longitudinal axis of adjoined crinoid ossicles.

strain Change in the shape or volume of a body as a result of *stress*, defined as the ratio of the change to the original shape or volume; a change in relative configuration of the particles of a substance. See also: *rotational strain; homogeneous strain.*

strain axis *principal axis of strain.*

strain break A fracture of rock in a quarry that occurs when the compressive stress is locally relieved by the quarrying process, e.g. a *lift joint.*

strain burst A minor *rock burst* in which increased pressure at the face of a mine working resulted in splinters, flakes, and fragments of pillars and walls being projected into the workings with explosive violence. Cf: *crush burst.*

strain crack A *crack* in sea ice, caused by stresses that exceed the elastic limit of the ice; e.g. *shear crack* and *tension crack.*

strain ellipse In two-dimensional analysis of rock deformation, the imaginary ellipse whose half axes are the greatest and least principal strains.

strain ellipsoid A geometric representation of the strain of a homogeneous body by homogeneous strain, or of heterogeneous stress at a particular point. Its axes are the principal axes of strain. Cf: *reciprocal strain ellipsoid.* Syn: *deformation ellipsoid.*

strain-energy density function The work required to impart elastic (reversible) displacements within a body, either adiabatically or isothermally.

strain field The state of strain at any point within the volume of material of interest.

strain gage A general term for devices with which mechanical strain can be measured, commonly by an electrical signal, e.g. a wire strain gage.

strain hardening The behavior of a material whereby each additional increment of strain requires an additional increment of differential stress.

strain recrystallization Recrystallization in which a deformed mineral changes to a mosaic of undeformed crystals of the same mineral, e.g. strained to unstrained calcite (Folk, 1965, p.15).

strain seismometer A seismometer that is designed to detect deformation of the ground by measuring relative displacement of two points.

strain shadow (a) *undulatory extinction.* (b) *pressure shadow.*

strain-slip cleavage *slip cleavage.*

strait [**coast**] A relatively narrow waterway connecting two larger bodies of water, as the Strait of Gibraltar linking the Atlantic Ocean with the Mediterranean sea; a small *channel.*

The term is often used in the plural. Cf: *sound.*

strait [**geog**] (a) A neck of land. (b) An obsolete term for a gorge.

strand [**coast**] (a) A syn. of *shore* and *beach*; the land bordering any large body of water, esp. the beach of a sea or an arm of the ocean, or the bank of a large river. (b) An Anglo-Saxon term for the narrow strip of land lying between high water and low water, being alternately exposed and covered by the tide.

strand [**streams**] A British term for a stream or current, and for a channel.

strand crack A fissure at the junction of a sheet of inland ice, an ice piedmont, or an ice rise with an ice shelf, the latter being subject to the rise and fall of the tide (Armstrong & Roberts, 1958, p.96).

stranded ice Floating ice that is deposited on the shore by retreating high water. Cf: *grounded ice; shore ice.*

stranded icefoot An icefoot formed by the stranding of ice floes or small icebergs along a shore; it may be built upward by freezing of spray or breaking seas.

strandflat (a) Any *wave-cut platform*; esp. a low, flat, very wide (up to 65 km) platform extending for many hundreds of kilometers off the rocky coast of western Norway, either partly submerged or standing slightly above the present sea level as a result of isostasy, and supporting thousands of stacks, skerries, and other small islands. (b) A discontinuous shelf of land inside a fjord, reaching to about 30 m in height, having a rounded and dissected form.---Sometimes spelled: *strand flat.*

strandline (a) The ephemeral line or level at which a body of standing water, as the sea, meets the land; the *shoreline*, esp. a former shoreline now elevated above the present water level. (b) A beach, esp. one raised above the present sea level.---Also spelled: *strand line.* See also: *raised beach.*

strand mark Any inorganic sole mark on a sedimentary surface along the shore (Clarke, 1918). Cf: *undertow mark.* Syn: *strand marking.*

strand plain A prograded shore built seaward by waves and currents, and continuous for some distance along the coast (Cotton, 1958, p.431). Syn: *foreland.*

stranskiite A blue triclinic mineral: $Zn_2Cu(AsO_4)_2$.

strashimerite A mineral: $Cu_8(AsO_4)_4(OH)_4.5H_2O$.

strata Plural of *stratum.*

strata-bound Said of a mineral deposit confined to a single stratigraphic unit. The term can refer to a *stratiform* deposit or to a randomly oriented orebody contained within a single stratigraphic unit. Cf: *bedded.*

strata control *roof control.*

stratal Pertaining to a stratum or strata; e.g. "stratal dip" or "stratal unit".

strata time Geologic time estimated from the thickness of strata and the rate of deposition (Kobayashi, 1944a, p.476).

strategic materials Materials that are vital to the security of a nation, but that must be procured entirely or in large part from foreign sources because the available domestic production will not meet that nation's requirements in time of war; e.g. *strategic minerals.* Cf: *critical materials.*

strategic minerals Minerals that are considered to be *strategic materials*; e.g. aluminum-, antimony-, chromium-, mercury-, and tin-bearing minerals were some of the "strategic minerals" during World War II.

strath [**erosion**] (a) An extensive, undissected, terrace-like remnant of a broad, flat valley floor that has undergone dissection following uplift; e.g. a continuous river terrace along a valley wall, interrupted in its development during the mature stage of a former erosion cycle. Bucher (1932) prefers the term *strath terrace* for this feature. Bascom (1931) proposed that "strath" be replaced by *berm.* (b) A broad, flat valley bottom formed in bedrock and resulting from degradation, "first by lateral stream cutting and later by whatever additional processes of degradation may be involved" (Bucher, 1932 p.131); a level valley floor representing a local base level. It is usually covered by a veneer of alluvium, and is wider (or larger) and flatter than a *glen*. Etymol: Scottish, from the Gaelic *strath*, "valley". Syn: *incipient peneplain*.

strath [marine geol] An elongate, broad, and steep-sided depression on the continental shelf, usually glacial in origin. It is often deeper on its nearshore side.

strath lake A small lake occupying the upper part of a long narrow valley that has been cut off by natural levees of alluvium built by an inflowing stream, commonly at the head of the lake.

strath stage The stage in the peneplanation of a region when the main streams have carved broad, flat-floored valleys that are graded to the same regional base level.

strath terrace (a) A term used by Bucher (1932) for an extensive remnant of a *strath* (i.e. a flat valley bottom) that belonged to a former erosion cycle and that has undergone dissection by a rejuvenated stream following uplift. The term is synonymous with *strath* and *berm* as used by other writers. Cf: *fillstrath terrace*. (b) A *strath* (i.e. a remnant) of greater extension than that of a narrow ribbon along one valley (Engeln, 1942, p.222).

strath valley (a) A stream valley characterized by the development of a flat valley bottom (strath) resulting from degradation (Bucher, 1932, p.131). (b) A valley abandoned by a stream whose course was dislocated (Engeln, 1942, p.224).

stratic Pertaining to or designating the order or sequence of strata; stratigraphic, or pertaining to stratigraphy. Grabau (1924, p.821) referred to a disconformity as a "stratic unconformable relation".

straticulate Characterized by numerous, very thin, parallel layers (separable or not), either of sedimentary deposition (as a bed of clay) or of deposition from solution (as in a stalagmite or banded agate).

straticulation The formation of a straticule or straticules.

straticule A French term for a thin sedimentary layer, or *lamina*.

stratification [sed] (a) The formation, accumulation, or deposition of material in layers; specif. the arrangement or disposition of sedimentary rocks in strata. (b) A structure produced by deposition of sediments in strata; a stratified formation, or stratum. It may be due to differences of texture, hardness, cohesion or cementation, color, internal structure, and mineralogic or lithologic composition. (c) The state of being stratified; a term describing a layered or bedded sequence, or signifying the existence of strata.---The term should be restricted to sedimentary rocks or deposits, although some authors have broadened the term to include "any somewhat horizontal layering in an igneous intrusion" (Challinor, 1967, p.238). See also: *bedding; layering*.

stratification [lake] The state of a body of water consisting of two or more horizontal layers of differing characteristics; esp. the arrangement of the waters of a lake into layers of differing densities. See also: *density stratification; thermal stratification*.

stratification [snow] Layering in a snow, firn, or ice mass; it is caused by sedimentation of different kinds of snow, by sedimentation of rock dust during summer periods, or by the development of layers of depth hoar at times of rapid changes in temperature.

stratification index A measure of the "beddedness" of a stratigraphic unit, expressed as the number of beds in the unit per 100 feet of section (Kelley, 1956, p.295). It is determined by multiplying the number of beds times 100, and dividing by the unit's thickness in feet. Syn: *beddedness index*.

stratification plane *bedding plane*.

stratified Formed, arranged, or laid down in layers or strata; esp. said of any layered sedimentary rock or deposit. See also: *bedded*.

stratified cone A less-preferred syn. of *stratovolcano*.

stratified drift Fluvioglacial drift consisting of sorted and layered material deposited by a meltwater stream or settled from suspension in a body of quiet water adjoining the glacier. Cf: *till*. Obsolete syn: *washed drift; modified drift*.

stratified estuary An estuary in which salinity increases with depth as well as along its length. An estuary is "highly stratified" if there is a density discontinuity separating surface river flow and bottom seawater; it is "slightly stratified" if the amount of increase in salinity with depth is not significant. Ant: *vertically mixed estuary*.

stratified lake A lake exhibiting *stratification* of its waters.

stratified rock A rock displaying stratification. The term is virtually synonymous with *sedimentary rock*, although some sedimentary rocks (such as tillite) are without internal stratification. The term is sometimes, but not generally, applied to layered igneous rocks.

stratified volcanic cone A less-preferred syn. of *stratovolcano*.

stratified volcano A less-preferred syn. of *stratovolcano*.

stratiform [ore dep] Said of a layered mineral deposit, of either sedimentary or igneous origin. Cf: *strata-bound [ore dep]*; *bedded [ore dep]*.

stratiform [sed struc] Having the form of a layer, bed, or stratum; consisting of roughly parallel bands or sheets, such as a "stratiform intrusion".

stratify To lay down or arrange in strata.

stratigrapher A geologist who studies or specializes in stratigraphy.

stratigraphic break *break [stratig]*.

stratigraphic classification The arbitrary but systematic arrangement, zonation, or partitioning of the sequence of rock strata of the Earth's crust into units with reference to any or all of the many different characters, properties, or attributes which the strata may possess (Hedberg, 1958, p.1881-1882).

stratigraphic code A usefully comprehensive, yet concisely stated formulation of generally accepted views on stratigraphic principles, procedures, and practices, designed to obtain the greatest possible uniformity in applying such principles, etc.; specif. "a systematic collection of rules of formal stratigraphic classification and nomenclature" (ACSN, 1961, art.3). It is applicable to all kinds (sedimentary, igneous, metamorphic) of rocks.

stratigraphic column *geologic column*.

stratigraphic control [ore dep] In ore deposition, the influence of stratigraphic features, e.g. ore minerals selectively replacing calcareous beds. Cf: *structural control*.

stratigraphic control [stratig] The degree of understanding of the stratigraphy of an area; the body of knowledge that can be used to interpret the stratigraphy or geologic history of an area.

stratigraphic correlation The process by which stratigraphic units in two or more separated areas are demonstrated or determined to be laterally similar in character or mutually correspondent in stratigraphic position, as based on geologic age (time of formation), lithologic characteristics, fossil content, or any other property of a stratum; *correlation* in the usual or narrowest sense. Unless otherwise stated, the term usually means *time-correlation*, or the determination of equivalence (in time) of stratigraphic units. See also: *lithologic correlation*.

stratigraphic cutoff *cutoff [stratig]*.

stratigraphic facies Facies distinguished primarily on the basis of form, nature of boundaries, and mutual relations, to which appearance and composition are subordinated (Weller, 1958, p.627). These facies are all stratigraphic bodies of one kind or another; they may occur in vertical succession and have boundaries that are more or less horizontal stratigraphic planes (e.g. systems, formations, biostratigraphic zones, and lithostromes), or they may be laterally intergrading parts of some kind of a stratigraphic unit and separated at more or less arbitrary vertical cutoff boundaries (e.g. lithofacies), or they may bear both lateral and vertical relations to each other and have irregular boundaries (e.g. the "magnafacies" of Caster, 1934). See also: *facies* [stratig]. Cf: *petrographic facies*.

stratigraphic geology *stratigraphy*.

stratigraphic guide In mineral exploration, a rock unit known to be associated with an ore. Cf: *lithologic guide*. See also: *ore guide*.

stratigraphic heave An obsolete syn. of *gap* [fault] and of *overlap* [fault].

stratigraphic interval The distance between the corresponding parts of any two strata in a sedimentary sequence, measured in a direction at right angles to the bedding (or its projection in space). The term refers not to the rocks, but to their thickness.

stratigraphic leak The deposition of sediments and/or fossils of a younger age within or under rocks of an older age. Such a deposit is said to be laid down in intraposition (Foster, 1966). Var: *stratigraphic leakage*. See also: *intrapositional deposit*.

stratigraphic map A map that shows the areal distribution, configuration, or aspect of a stratigraphic unit or surface. It involves a span of geologic time. Examples include isopach map, structure-contour map, facies map, and vertical-variability map.

stratigraphic overlap An obsolete syn. of *overlap* [fault].

stratigraphic paleontology The study of fossils and of their distribution in various geologic formations based on the stratigraphic relations (conditions and order of deposition) of the sedimentary rocks in which they are contained. Cf: *biostratigraphy*.

stratigraphic range The distribution or spread of any given species, genus, or other taxonomic group of organisms through geologic time, as indicated by its distribution in strata whose geologic age is known. Also, the persistence of a fossil organism through the stratigraphic sequence. Syn: *range; geologic range; time-rock span*.

stratigraphic record The *geologic record* based on or derived from a study of the stratigraphic sequence; the rocks arranged chronologically as in a geologic column. Syn: *record*.

stratigraphic section *geologic section*.

stratigraphic separation *stratigraphic throw*.

stratigraphic sequence A chronologic succession of sedimentary rocks from older below to younger above, essentially without interruption; e.g. a *sequence* of bedded rocks of interregional scope and bounded by unconformities.

stratigraphic test A hole drilled to determine stratigraphic information (thickness, lithology, sequence, porosity, permeability) of the rock penetrated or to locate the position of a key bed. It is frequently drilled to penetrate a potentially productive zone. Cf: *structure test*. Syn: *strat test*.

stratigraphic throw The thickness of the strata that originally separated two beds brought into contact at a fault. Syn: *stratigraphic separation*.

stratigraphic trap The sealing of a reservoir bed as the result of lithologic changes rather than through structural trapping conditions. See also: *shale-out; pinch-out*. Syn: *porosity trap*.

stratigraphic unconformity *disconformity*.

stratigraphic unit A stratum or body of strata recognized as a unit in the classification of the rocks of the Earth's crust with respect to any specific rock character, property, or attribute (ISST, 1961, p.18) or for any purpose such as description, mapping, and correlation. Rocks may be classified stratigraphically on the basis of lithology (rock-stratigraphic units), fossil content (biostratigraphic units), age (time-stratigraphic units), or properties (such as mineral content, radioactivity, seismic velocity, electric-log character, chemical composition) in categories for which formal nomenclature is lacking for units. A geologic-time unit is not a stratigraphic unit.

stratigraphy (a) The branch of geology that deals with the definition and description of major and minor natural divisions of

rocks (mainly sedimentary, but not excluding igneous and m... tamorphic) available for study in outcrop or from subsurfac... and with the interpretation of their significance in geologic h... tory; specif. the geologic study of the form, arrangement, g... ographic distribution, chronologic succession, classificatio... and esp. correlation and mutual relationships of rock stra... (and other associated rock bodies) "in normal sequence w... respect to any or all of the characters, properties, and att... butes which rocks may possess" (ISST, 1961, p. 18). It ther... by involves interpretation of these features of rock strata... terms of their origin, occurrence, environment, thickness, ... thology, composition, fossil content, age, history, paleoge... graphic conditions, relation to organic evolution, and relati... to other geologic concepts. Syn: *stratigraphic geology*. (... The arrangement of strata, esp. as to geographic position a... chronologic order of sequence. (c) The sum of the characte... istics studied in stratigraphy; the part of the geology of a... area or district pertaining to the character of its stratifi... rocks. (d) A term that is sometimes used to signify the stu... of *historical geology*.

stratofabric The arrangement of strata in any body of stratifi... rock, "from the dimensions of a thin section to those of... sedimentary basin" (Fischer, 1964, p.148).

stratose Arranged in strata.

stratosphere In oceanography, the waters of the ocean belo... the thermocline. Cf: *troposphere*.

stratotype A specifically bounded *type section* of rock stra... to which a time-stratigraphic unit is ascribed, consisting des... rably of a complete and continuously exposed and deposite... sequence of correlatable strata, and extending from a readi... identifiable basal boundary to a readily identifiable top boun... ary. It may consist of "a single designated type section or... may be a composite stratotype, formed by coordination ... several separate sections which together, completely, b... without duplication of the intervals to be accepted as typ... represent the desired interval" (ISST, 1965, p.1701). C... *boundary stratotype*.

stratous Composed of strata.

stratovolcano A volcano that is constructed of alternatin... layers of lava and pyroclastics. Viscous, acidic lava may flo... from fissures radiating from a central vent, from which pyr... clastics are ejected. Syn: *composite volcano*. Less-preferre... syn: *stratified volcano; bedded volcano; composite cone; stra... tified cone; stratified volcanic cone.*

strat test Slang equivalent of *stratigraphic test*.

stratum A tabular or sheet-like mass, or a single and distin... *layer*, of homogeneous or gradational sedimentary materia... (consolidated rock or unconsolidated earth) of any thickness... visually separable from other layers above and below by a dis... crete change in the character of the material deposited or b... a sharp physical break in deposition, or by both; a sedimenta... ry *bed*. The term should be restricted to sedimentary materia... and is generally considered to be synonymous with "bed", al... though it has been defined as a stratigraphic unit that may b... composed of a number of beds (Dana, 1895, p.91), as a laye... greater than 1 cm in thickness and constituting a part of ... bed (Payne, 1942, p.1724), and as a general term that in... cludes both "bed" and "lamination" (McKee & Weir, 1953... p.382). The term is more frequently used in its plural form... *strata*. Cf: *lamina*.

stratum plain A plain having a *stripped structural surface*. Ex... amples in the Colorado Plateau region of the U.S. commonl... form isolated buttes and mesas, or bench- or terrace-lik... areas along valley sides, but some have considerable area... extent. See also: *dip plain; cut plain*. Syn: *structural plai... stripped plain; stripped structural plain*.

stratum spring *contact spring*.

Straumanis camera method In X-ray diffraction analysis, ... method of mounting film in a cylindrical X-ray camera to al... low for recording of both front and back reflections on bot...

des of the exit and entry ports, enabling the determination of m diameter from the measurements. Cf: *Wilson technique.*

straw stalactite *soda straw.*

stray A lenticular rock formation encountered unexpectedly (accidentally) in drilling an oil or gas well and separated by a short interval from a more persistent formation by a sharp change in lithology or hardness; e.g. *stray sand.*

stray current An electric current that is introduced in the earth by leakage of currents from cultural installations.

stray sand A *stray* consisting of sandstone.

streak [geog] A long, narrow, irregular strip or stretch of land water.

streak [min] The color of a mineral in its powdered form, usually obtained by rubbing the mineral on a *streak plate* and observing the mark it leaves. Streak is an important characteristic in mineral analysis; it is sometimes different from the color of the sample, and is generally constant for the same mineral.

streak [sed] (a) A comparatively small and flattish or elongate sedimentary body, visibly differing from the adjacent rock, but without the sharp boundaries typical of a lens or layer (Stokes & Varnes, 1955, p.143). (b) A long, narrow body of sand, perhaps representing an old shoreline; a shoestring. (c) The outcropping edge of a coal bed.

streaked-out ripples A term used to describe sedimentary flame structure.

streaking *mineral streaking.*

streak plate In mineral analysis, a piece of unglazed porcelain used for rubbing a sample to obtain its powder color, or *streak.* It has a hardness of about seven.

stream [glaciol] (a) *ice stream* [glaciol]. (b) A stream of melt-water.

stream [streams] (a) Any body of running water, great or small (from a large river to a small rill), moving under gravity flow to progressively lower levels in a relatively narrow but clearly defined channel on the surface of the ground, in a subterranean cavern, or beneath or in a glacier; esp. such a body flowing in a natural channel. It is a mixture of water and of dissolved, suspended, or entrained matter. Cf: *river.* (b) A term popularly applied to a *brook* (as in Maine) or to a small river. (c) The water flowing in a stream. (d) A term used in quantitative geomorphology interchangeably with *channel.*

stream action *fluviation.*

stream azimuth Orientation of the mean line of a stream from head to mouth, measured in degrees from some arbitrary direction, generally north (Strahler, 1954, p.346). Symbol: α.

stream bed The channel containing or formerly containing the water of a stream. Also spelled: *streambed.*

stream-built terrace *alluvial terrace.*

stream capture *capture* [streams].

stream channel The hollow bed where a natural stream of water runs or may run; the long, narrow, sloping, trough-like depression shaped by the concentrated flow of a stream and covered continuously or periodically by water. Syn: *streamway.*

stream current (a) A relatively narrow, deep, well-defined, fast-moving ocean current; e.g. the Gulf Stream. A *drift current* deflected by an obstruction, such as a shoal or land, may form a stream current. Syn: *stream.* (b) A steady current in a terrestrial stream or river.

stream-cut terrace *rock terrace.*

stream deposition The accumulation of any transported rock particles on the bed or adjoining flood plain of a stream, or on the floor of a body of standing water into which a stream empties.

stream-entrance angle *axil angle.*

stream erosion The progressive removal by a stream of bedrock, overburden, soil, or other exposed matter from the surface of its channel, as by hydraulic action, corrasion, and corrosion.

streamflood A flood of water in an arid region, characterized by the "spasmodic and impetuous flow" of a *sheetflood* but confined to a definite, shallow channel that is normally dry (Davis, 1938).

streamflow A type of *channel flow*, applied to that part of surface runoff traveling in a stream whether or not it is affected by diversion or regulation. Also spelled: *stream flow.* Cf: *sheet flow* [geomorph]; *overland flow.*

streamflow depletion The amount of water that flows into a given land area minus the amount that flows out of that area.

streamflow wave A traveling wave caused by a sudden increase of water flow (ASCE, 1962).

stream frequency Ratio of the number of streams of all orders within a drainage basin to the area of that basin; a measure of topographic texture. Symbol: F. Syn: *channel frequency.*

stream gaging Measurement of the velocity of a stream of water in a channel or open conduit and of the cross-sectional area of the water, in order to determine discharge. See also: *chemical gaging.* Syn: *gaging.*

stream gold Gold occurring in alluvial placers.

stream gradient The angle between the water surface (of a large stream) or the channel floor (of a small stream) and the horizontal, measured in the direction of flow; the "slope" of the stream. Symbol: S. See also: *law of stream gradients.* Syn: *stream slope.*

stream-gradient ratio Ratio of the gradient of a stream channel of a given order to that of a stream of the next higher order in the same drainage basin. Symbol: R_s. Syn: *channel-gradient ratio.*

streamhead The source or beginning of a stream.

streaming flow [glaciol] Glacier flow in which the ice moves without cracking or breaking into blocks, such as where the walls and bottom are relatively smooth for a long distance.

streaming flow [hydraul] *tranquil flow.*

streaming potential *electrofiltration potential.*

stream length The length of a stream segment of a given order u. Symbol: L_u. See also: *law of stream lengths.* Syn: *channel length.*

stream-length ratio Ratio of the mean stream length of a given order to the mean stream length of the next lower order within a specified drainage basin (Horton, 1945, p.296). Symbol: R_L.

streamlet A small stream. Syn: *springlet.*

stream line An imaginary line connecting a series of fluid particles in a moving fluid so that, at a given instant, the velocity vector of every particle on that line is tangent to it (Middleton, 1938, p.35-36).

streamline flow *laminar flow.*

stream load (a) The solid material that is actually transported by a stream, either as visible sediment (carried in suspension, or moved along the stream bed by saltation and traction) or in chemical or colloidal solution. (b) The actual quantity or amount of such material at any given time or passing a given point in a given period of time, and expressed as a weight or volume per unit time.---Material in solution is sometimes excluded in the usage of the term. See also: *suspended load; bed load; dissolved load.*

stream morphology *river morphology.*

stream numbers *number of streams.*

stream order The designation by a dimensionless integer series (1, 2, 3, etc.) of the relative position of stream segment(s) in a drainage-basin network: the smallest, unbranched tributaries, terminating at an outer point, are designated order 1; the junction of two first-order streams produces a stream segment of order 2; the junction of two second-order streams produces a stream segment of order 3; and so on. The order of the single trunk segment of the stream terminating in the single flow exit of the drainage basin is therefore the highest integer and determines the order of the drainage basin. Symbol u refers to an order number; symbol k refers to the highest stream order in a given basin. See also: *basin order.* Syn: *channel order.*

stream piracy *capture* [streams].

stream profile The *longitudinal profile* of a stream.

stream robbery *capture* [streams].

stream segment The portion of a stream extending between specified tributary junctions. Syn: *channel segment*.

stream sink That point at which a stream disappears underground. Cf: *swallet*.

stream slope *stream gradient*.

stream terrace One of a series of level surfaces in a stream valley, flanking and more or less parallel to the stream channel, originally occurring at or below, but now above, the level of the stream, and representing the dissected remnants of an abandoned flood plain, stream bed, or valley floor produced during a former stage of erosion or deposition. See also: *alluvial terrace; rock terrace; meander terrace; inset terrace*. Syn: *terrace; river terrace*.

stream tin Cassiterite occurring in the form of rolled or water-worn fragments or pebbles in alluvial or placer deposits or on bedrock along valleys or streams, such as that resulting from the wearing away of pneumatolytic veins of tin associated with acid rocks. Cf: *lode tin*. Syn: *alluvial tin*.

stream transportation The movement in and by a stream of weathered or eroded rock material in chemical solution, in turbulent suspension, or by rolling, dragging, or bouncing along the stream bed.

streamtube A passage in a cave that is or has been completely filled with flowing water, and whose surfaces contain scallops.

stream underflow Percolating water in the permeable bed of a stream and flowing parallel to the stream (ASCE, 1962).

stream valley An elongate depression of the Earth's surface, carved by a stream during the course of its development.

stream velocity The rate of flow, measured by distance per time unit, e.g. ft/sec.

streamway (a) The current of a stream. (b) *stream channel*.

street A part of a bare desert floor that forms a gap separating chains of sand dunes. See also: *straat*.

strengite A pale-red orthorhombic mineral: $FePO_4.2H_2O$. It is isomorphous with variscite and dimorphous with phosphosiderite, and may contain some manganese.

strength A term used in experimental structural geology that is meaningful only when all the environmental conditions of the experiment are specified; in general, the ability to withstand differential stress, measured in the same units as is stress.

streptaster A sponge spicule (microsclere) having the form of a modified aster in which the rays do not arise from a common center but radiate from an axis; e.g. a streptosclere, a sanidaster, or a discorhabd. Cf: *euaster*.

streptosclere A siliceous sponge spicule (streptaster) in which long ray-like spines are given off in spiral succession about a central axis, and which intergrade with simple euasters. Syn: *spiraster*.

streptospiral Said of a foraminiferal test coiled like a ball of twine.

stress In a solid, the force per unit area, acting on any surface within it, and variously expressed as pounds or tons per square inch, or dynes or kilograms per square centimenter; also, by extension, the external pressure which creates the internal force. The stress at any point is mathematically defined by nine values: three to specify the normal component and six to specify the shear component, relative to three mutually perpendicular reference axes. Cf: *strain*. See also:*normal stress; shear stress*.

stress axis *principal axis of stress*.

stress difference The difference between the greatest and least of the three principal stresses.

stress ellipsoid A geometric representation of the state of stress at a point that is defined by the three mutually perpendicular principal stresses and their intensities.

stress field The state of stress, either homogeneous or varyi from point to point and through time, in a given domain.

stress mineral A term suggested by Harker (1918) for a m eral such as chlorite, chloritoid, talc, albite, epidote, amp boles, kyanite, etc. whose formation in metamorphosed roc is favored by shearing stress. The term is now obsolete. (*antistress mineral*.

stress shadow *pressure shadow*.

stress-strain curve The plot of conventional strain in perce shortening or elongation as the abscissa, and true longitudir differential stress, i.e. the difference between the greatest a least principal stresses, as the ordinate. Syn: *stress-strain agram*.

stress-strain diagram *stress-strain curve*.

stress tensor A notation of the state of stress at a point whi in the most general case involves nine components, each which is associated with one of the three orthogonal referen axes. Three components are those normal to each of the re erence axis (normal stress); the remaining six are parallel the axes (shear stress).

stretch A *reach* of water or land.

stretched Said of a structure or texture produced by dynam metamorphism in which the long direction of most of the co stituents are parallel, and are stretched and broken in this rection, e.g. stretch-pebble conglomerate (Johannsen, 193 p.230). The term should not be confused with the concept lineation.

stretch fault *stretch thrust*.

stretching In metamorphic rocks, the elongation of miner grains, gas bubbles, or other features; a type of lineation. (*mineral streaking*.

stretch modulus *Young's modulus*.

stretch thrust A little-used term for a reverse fault formed shear in the middle limb of an overturned fold. It is actual caused by compression. Syn: *stretch fault*.

strewn field (a) A restricted geographic area within which specific group of tektites is found. Examples include weste Czechoslovakia, the southern half of Australia, the Ivo Coast, and southern U.S. (Texas and Georgia). (b) *dispersi ellipse*.

stria [geol] *striation*.

stria [paleont] One of a series of very fine radiating, conce tric, or parallel grooved lines or threads on the surface some shells (esp. on otherwise smooth shells); e.g. one the parallel, small to minute grooves or channels on the su face of a nautiloid conch separated by *lirae* and not easily di cernible with the naked eye, or a very narrowly incised sha low groove on a gastropod shell. Pl: *striae*.

striae [cryst] (a) Parallel lines on the surface of a crystal, i dicative of a vacillation between two crystal forms. (b) Para lel, straight lines on the cleavage planes of a crystal such plagioclase, calcite, or corundum, indicative of polysynthe cleavage.----Syn: *striations*.

striate [geol] Var. of *striated*.

striate [palyn] (a) Said of spores and pollen having a streake sculpture characterized by multiple, usually parallel groove and ribs in the exine; specif. referring to Striatiti. (b) *striated*.

striated Adj. of *striation* [geol]. Var: *striate* [geol].

striated ground *striped ground*.

striation [geol] (a) A superficial scratch, a tiny furrow, or thread-like line inscribed on a rock surface or rock fragme by a geologic agent (such as glaciers, torrents, mass mov ments, and faulting) and usually occurring as one of a seri of parallel or subparallel lines; e.g. *glacial striation*. Syn: *stri scratch*. (b) The condition of being striated; the disposition striations.----Adj: *striate; striated*.

striation [glac geol] *glacial striation*.

striation [mineral] One of the very shallow, parallel depre sions or narrow bands on the cleavage face of a minera caused by oscillation in growth between differently oriente

striation [sed] A short, narrow, and curved or straight mark of small dimensions (1-2 mm deep and many centimeters long), nearly cut (as by torrents) below a sedimentary surface on which it appears; it is smaller and narrower than a *groove*. See also: *drag mark*.

striation [fault] One of a series of hairlike, parallel, usually straight scratches or smooth furrows developed on a rock surface by tectonic forces, as the abrasion of one projecting rock against another during fault movement; the striations often indicate the direction of movement. Cf: *slickenside; groove; slip-scratch; mullion structure.* Syn: *fault striae.*

striation cast The cast of a striation produced on a sedimentary surface; it is usually found on the underside of a thin siltstone or fine sandstone bed interlayered with mudstone. Cf: *groove cast.* Syn: *microgroove cast.*

striatiti Abundant upper Paleozoic and lower Mesozoic pollen with very characteristic striate sculpture in the exine of the body of the pollen grain, the grooves and ribs usually oriented perpendicular to the axes of the vesicles (if these are present). They are presumably pollen of conifers and gnetaleans.

stricture A contraction between successive shell joints of the skeleton of a nasselline radiolarian.

striding compass A compass on a theodolite for orientation.

striding level (a) A spirit level so mounted that it can be placed above and parallel with the horizontal axis of a surveying instrument and so supported that it can be used for precise leveling of the horizontal axis of the instrument or for measuring any remaining inclination of the horizontal axis. (b) A demountable spirit level that can be attached to the telescope tube to level the line of sight.

strigovite (a) A dark-green mineral of the chlorite(?) group: $Fe_3^{+2}(Al,Fe^{+3})_3Si_3O_{11}(OH)_7$. (b) A hypothetical end member of the chlorite group: $(Mg,Fe)_2Al_2Si_2O_7(OH)_4$.

strike [eco geol] n. The discovery of a mineral deposit, esp. a sudden or unexpected one. v. To suddenly or unexpectedly discover or reach a mineral deposit, e.g. "to strike oil".

strike [struc geol] n. The direction or trend that a structural surface, e.g. a bedding or fault plane, takes as it intersects the horizontal. See also: *attitude.* Cf: *trend; trace.* Syn: *line of strike.*

strike fault A fault that strikes parallel with the strike of the strata involved. Cf: *dip fault; oblique fault.*

strike fold *longitudinal fold.*

strike joint A joint, the strike of which is approximately parallel to the bedding or cleavage of the constituent rock. Cf: *dip joint.*

strike-overlap A term proposed by Melton (1947, p.1870) for truncation of sedimentary rocks below unconformities, esp. for a slow, extremely low-angle regional truncation below a regional unconformity. The term is essentially synonymous with *overstep* if it is assumed that "in most bodies of marine, or interfingering marine and nonmarine rock, angular unconformities eventually pass downdip into disconformities, which in turn disappear farther out in the basin" (Swain, 1949, p.634).

strike separation In a fault, the distance of *separation* of two formerly adjacent beds on either side of the fault surface, measured parallel to the strike of the fault. Cf: *dip separation; strike slip.*

strike-separation fault *lateral fault.*

strike shift In a fault, the *shift* or relative displacement of the rock units parallel to the strike of the fault, but outside the fault zone itself; a partial syn. of *strike slip.*

strike-shift fault *strike-slip fault.*

strike slip In a fault, the component of the movement or slip that is parallel to the strike of the fault. Cf: *dip slip; strike separation; oblique slip.* Syn: *horizontal displacement; horizontal separation.* Partial syn: *strike shift.*

strike-slip fault A fault, the actual movement of which is parallel to the strike of the fault. Cf: *dip-slip fault; lateral fault.* See also: *transcurrent fault.* Syn: *strike-shift fault.*

strike stream A *subsequent stream* that follows the strike of the underlying strata.

strike valley A *subsequent valley* eroded in, and developed parallel to the strike of, underlying weak strata; a valley containing a *strike stream.*

string [drill] (a) A set of well-drilling tools and drilling equipment (including the rig), used esp. for cable-tool drilling; specif. the *drill string.* (b) The casing, tubing, or pipe, of one size, used in a well. See also: *flow string.*

string [ore dep] *stringer.*

string bog A linear Pleistocene periglacial muskeg or moor with an undulating surface, occurring in the boreal needle-tree forest zone of the northern hemisphere (esp. western Siberia and the Hudson Bay area), and characterized by shallow water-filled depressions and festoons of lenticular ridges (up to 1 m high) consisting of floating fen or moss vegetation. Its origin is controversial, but it may be related to collapse and melting of permafrost. See also: *ring moor.*

stringer [ore dep] A mineral veinlet or filament occurring in a discontinuous pattern in the host rock. See also: *stringer lode.* Syn: *string.*

stringer [stratig] A thin sedimentary bed.

stringer lode A zone of shattered host rock containing a network of *stringers*; a *stockwork.*

string galvanometer A form of the moving-coil galvanometer in which the coil is a single wire of fine silvered quartz, suspended between the poles of an electromagnet, whose deflection is observed with a microscope when direct or alternating current is passed through the wire.

striotubule A *pedotubule* composed of skeleton grains and plasma that are not organized into recognizable aggregates but exhibit a basic fabric with a semiellipsoidal directional arrangement related to the external form, with the walls approximately tangential to the semiellipsoid (Brewer, 1964, p.241).

strip [photo] *flight strip.*

strip [ice] A long narrow area of pack ice about 1 km or less in width, usually composed of small fragments detached from the main mass of ice, and run together under the influence of wind, swell, or current; it is more limited than a *belt.*

stripe [pat grd] One of the alternating bands of fine and coarse surficial material, or of rock or soil and vegetation-covered ground, comprising a form of patterned ground characteristic of slopes steeper than those of *steps.* It is usually straight, but may be sinuous or branching, and is probably the result of solifluction acting in conjunction with other processes, such as rillwork. See also: *sorted stripe; nonsorted stripe; contraction stripe.*

stripe [met] *ribbon.*

striped ground A form of patterned ground marked by alternating *stripes* produced on a sloping surface by frost action. Syn: *striped soil; striated ground.*

striped soil *striped ground.*

stripe hummock A *nonsorted stripe*, probably closely related to an earth hummock, but formed on sloping ground.

strip log (a) A *graphic log* plotted on a *log strip*, commonly on a scale of 100 ft to the inch or of 100 ft to two inches. (b) Mounted cores or cuttings from a borehole.

strip mining *opencut mining.*

stripped bedding plane The exposed top side of a resistant stratum that forms a *stripped structural surface* when extended over a considerable area.

stripped illite *degraded illite.*

stripped structural surface An *erosion surface* developed during the mature stage of an erosion cycle in an area underlain by horizontal or gently inclined strata of unequal resistance, the overlying softer beds having been removed by erosion so as to expose the more or less smooth surface of a resistant stratum that has served as a local base level and thereby controlled the depth of erosion; specif. the surfaces produced on a *structural plateau*, a *stratum plain*, and a *structural terrace*. Syn: *stripped surface*.

stripped surface *stripped structural surface*.

stripping the Earth *layer stripping*.

strip thrust An obsolete syn. of *décollement*.

strip width The average dimension, measured normal to the flight line, of a series of neat models in a flight strip. It is generally equal to the width between flight lines.

strobilus A conelike collection of sporophylls, e.g. in club mosses.

strobilus theory A theory of plant evolution which states that the sporophyte of a vascular plant derives from a primitive structure resembling the strobilus of a sporophyll.

stroma (a) The supporting framework of an animal organ, such as organic tissue in the mesodermal endoskeleton of a living echinoderm. Cf: *stereome*. (b) A compact mass of fungous cells, or of mixed host and fungous cells, in or on which spores or sporocarps are formed.---Pl: *stromata*.

stromatactis Open-space sedimentary structures characterized by horizontal or nearly flat bottoms and by irregular upper surfaces, formed by the filling of original cavities with internal sediments and/or sparry-calcite cement, as in the central part of a reef core (Chilingar et al, 1967, p.321); sometimes called *reef tufa*. They have been variously interpreted as cavities caused by the burial and decay of soft-bodied but rigid frame-building organisms (known as *Stromatactis* according to Lowenstam, 1950), although they may represent syngenetic voids in calcareous sediments.

stromatite A chorismite with two or more textural elements arranged in essentially parallel layers, flat or folded. Syn: *stromatolith*.

stromatocerque An arcuate projection from one side of the opening of a cyst of a yellow-green alga, so that the cyst is asymmetrical.

stromatolite A term that has been generally applied to a variously shaped (often domal), laminated, calcareous sedimentary structure formed in a shallow-water environment under the influence of a mat or assemblage of sediment-binding blue-green algae that trap fine (silty) detritus and precipitate calcium carbonate and that commonly develop colonies or irregular accumulations of a constant shape but with little or no microstructure. It has a variety of gross forms, from near-horizontal to markedly convex, columnar, and subspherical. Stromatolites were originally considered animal fossils, and although they are still regarded as "fossils" because they are the products of organic growth, they are not fossils of any specific organism but consist of associations of different genera and species of organisms that can no longer be recognized and named or that are without organic structures. "The various generic names given to these structures are apparently invalid, since the names apply only to various forms assumed by the accumulated entrapped sediment and may not be directly related to specific organisms" (Pettijohn. 1957, p.221). Term introduced by Kalkowsky (1908) as *stromatolith*. Cf: *oncolite*.

stromatolith [intrus rocks] A complex, sill-like igneous intrusion that is interfingered with sedimentary strata (Foye, 1916).

stromatolith [migma] *stromatite*.

stromatolith [sed] *stromatolite*.

stromatology A term, now obsolete, proposed to embrace "the history of the formation of the stratified rocks" (Page, 1859, p.340).

stromatoporoid Any one of a group of invertebrates of uncertain systematic position currently placed (TIP, 1956, pt.F, p.127) in the hydrozoan order Stromatoporoidea, characterized by a calcareous skeleton and colonial, massive, sheet-like, or dendroid growth. The stromatoporoids have been variously classified as algae, foraminifers, sponges, hydrozoan other coelenterates, and bryozoans and may include representatives of more than one of these groups. Their stratigraphic range is Cambrian to Cretaceous. Var: *stromatoporid*.

strombite A petrified gastropod shell of the genus *Strombus*.

Strombolian-type bomb A general type of volcanic bomb produced from lava that is less fluid than that of a *Hawaiian-type bomb* and that is usually larger and pear-shaped.

Strombolian-type eruption A type of volcanic eruption characterized by fire fountains of fluid, basaltic lava from a central crater. Etymol: Stromboli, Lipari Islands of Italy. Cf: *Hawaiian-type eruption; Pelée-type eruption; Vulcanian-type eruption*.

stromeyerite A dark steel-gray orthorhombic mineral with blue tarnish: CuAgS. Syn: *silver-copper glance*.

stromoconolith A layered igneous intrusion that is either conical or funnel-shaped (Tomkeieff, 1961).

strong Said of large or important mineral veins or faults.

strong motion *Ground motion* that is sufficiently strong to be of interest in engineering seismology.

strongyle A rod-shaped sponge spicule (monaxon) with both ends blunt. Syn: *strongyl*.

strongylote Said of a sponge spicule having one end rounded.

strontianite A pale-green, white, gray, or yellowish orthorhombic mineral of the aragonite group: $SrCO_3$. It frequently occurs in limestones, and is a source of strontium chemicals.

strontioborite A mineral: $SrB_8O_{13} \cdot 2H_2O$ (?).

strontioginorite A mineral: $(Sr,Ca)_2B_{14}O_{23} \cdot 8H_2O$. It is a strontian variety of ginorite.

strontium-apatite A pale-green to yellowish-green mineral of the apatite group: $(Sr,Ca)_5(PO_4)_3(OH,F)$.

strophic Said of a brachiopod shell in which the true hinge line is parallel to the hinge axis. Cf: *nonstrophic*.

strophomenid Any articulate brachiopod belonging to the order Strophomenida, characterized chiefly by a plano- to concavo-convex shell that may be resupinate or geniculate. Their stratigraphic range is Lower Ordovician to Lower Jurassic.

strophotaxis Taxis in which an organism tends to turn in response to some external stimulus. Cf: *phobotaxis; thigmotaxis*.

structural Said of or pertaining to features that are the result of crustal folding or faulting.

structural adjustment A term proposed by Salisbury (1904, p.710) for the rearrangement of the drainage of an area so as to conform to the geologic structure; esp. the flowing of streams along the strike of the strata. Cf: *topographic adjustment*.

structural analysis A term preferred by Turner and Weiss (1963, p.6) for *structural petrology*.

structural basin *basin* [struc geol].

structural bench A bench representing the resistant edge of a *structural terrace* that is being reduced by erosion (Cotton, 1958, p.94-95). Syn: *rock bench*.

structural closure *closure*.

structural control In ore deposition, the influence of structural features, e.g. ore minerals filling fractures. Cf: *stratigraphic control*.

structural crystallography That branch of crystallography which deals with crystal structure.

structural datum *datum horizon*.

structural depression A hollow in the land surface caused by structural deformation of the Earth's crust.

structural diagram A figure illustrating the spatial array of orientation of objects.

structural dome *dome* [struc geol].

structural fabric *fabric* [struc geol].

structural feature A feature produced by deformation or displacement of the rocks, such as a fold or fault. For such features the more colloquial term *structure* (used as a specific noun) has been long in use, and is now being accepted by ditors and purists.

structural geology The branch of geology that deals with the form, arrangement, and internal structure of the rocks, and especially with the description, representation, and analysis of structures, chiefly on a moderate to small scale. The subject is similar to *tectonics*, but the latter is generally used for the broader regional or historical phases.

structural high *high* [struc geol].

structural lake *tectonic lake.*

structural landform A landform developed by erosion and controlled by the structure of the rocks. Cf: *tectonic landform.*

structural low *low* [struc geol].

structural map *structure-contour map.*

structural nose *nose.*

structural petrology Study of the internal structure or fabric of a rock, commonly with the aim of clarifying the rocks' deformational history. Various terms are used synonymously: *petrofabrics; petrofabric analysis; structural analysis; petreometry; petromorphology; fabric analysis; microtectonics.* See also: *petrotectonics.*

structural plain *stratum plain.*

structural plateau A plateau-like landform with a *stripped structural surface.* Syn: *stripped structural plateau.*

structural province A region whose geologic-structure features differ significantly from those of adjacent regions. It is generally very similar in extent to a *physiographic province.*

structural relief (a) The vertical distance between stratigraphically equivalent points at the crest of an anticline and in the trough of an adjacent syncline. (b) More generally, the difference in elevation between the highest and lowest points of a bed or stratigraphic horizon in a given region.

structural terrace (a) A local, shelf- or step-like flattening in more steeply inclined and otherwise uniformly dipping strata, composed of a synclinal bend above and followed by an anticlinal bend at a lower level. (b) A terrace-like landform controlled by the structure of the underlying rocks; esp. a terrace produced by the more rapid erosion of weaker strata lying on more resistant rocks in a formation with horizontal bedding. Cf: *structural bench.*

structural trap The containment of oil or gas within a reservoir bed as the result of flexure or fracture.

structural unconformity *angular unconformity.*

structural valley A valley that owes its origin or form to the underlying geologic structure. Cf: *tectonic valley.*

structure [petrology] (a) A megascopic feature of a rock mass or rock unit, generally seen or studied best in the outcrop rather than in hand specimen or thin section, and representing a discontinuity or major inhomogeneity; one of the larger morphologic features of a rock mass, such as columnar structure due to fracture, blocky features in lava, platy parting, slaty cleavage, foliation, bedding, certain concretions, and even fossils. The term is also applied to the appearance, or to a smaller-scale feature, of a heterogeneous rock in which the structure or composition is different in neighboring parts; e.g. banded structure, orbicular structure. The term *texture* is generally used for the smaller features or particles composing a rock, and "structure" for those features that indicate the way the rock is organized or made up of its component parts; although the two terms are often used interchangeably, they should not be considered synonymous, even though some textures may parallel major structural features of a rock. See also: *sedimentary structure.* (b) The sum total of such features; the arrangement of a rock mass with respect to such features.

structure [geomorph] A comprehensive term for the assemblage of rocks upon which erosive agents are, and have been, at work; the terrane underlying a landscape. The term indicates the product of all constructional agencies, and includes the arrangement and disposition of the rocks, their nature and mode of aggregation, and even the initial forms prior to erosion.

structure [mineral] (a) The form assumed by a mineral; e.g. bladed, columnar, tabular, or fibrous structures. (b) *crystal structure.*

structure [palyn] (a) The internal complexity in the makeup of the ektexine of pollen grains and spores, usually consisting of rodlets (columellae) that may be branched and more or less fused laterally. Cf: *sculpture.* (b) A term that is sometimes, but less desirably, used to describe major morphologic characteristics of spores, esp. those of the Paleozoic.

structure [struc geol] (a) The general disposition, attitude, arrangement, or relative positions of the rock masses of a region or area; the sum total of the *structural features* of an area, consequent upon such deformational processes as faulting, folding, and igneous intrusion. (b) A term used in petroleum geology for any physical arrangement of rocks (such as an anticline, unconformity, or cavity) that may lead to the accumulation of oil or gas; in this sense it is synonymous with *structural feature,* although such usage is not recommended (USGS, 1958, p.168).

structure contour A *contour* that portrays a structural surface such as formation boundaries or a fault. Syn: *subsurface contour.* See also: *structure-contour map.*

structure-contour map A map that portrays subsurface configuration by means of *structure contour* lines. See also: *contour map; tectonic map.* Syn: *structural map; structure map.*

structure ground A term used by Antevs (1932, p.48) but now replaced by its syn. *patterned ground.*

structure map *structure-contour map.*

structure-process-stage The name given to the Davisian principle (Davis, 1899) that the development of all landforms in the cycle of erosion is a function of three basic factors: geologic structure, geomorphic process, and stage of development. Davis originally referred to "structure, process, and time", but later writers have changed this to "structure-process-stage".

structure section A *vertical section* that shows the observed geologic structure on a vertical or nearly vertical surface, or, more commonly, one that shows the inferred geologic structure as it would appear on a vertical plane cutting through a part of the Earth's crust. The vertical scale is often exaggerated.

structure soil *soil structure* [pat grd].

structure test A generally shallow hole drilled to determine geologic structure only, although other types of information may be acquired during drilling. It is frequently drilled to a structural datum that is normally short of the known or expected producing zone or zones. Cf: *stratigraphic test.*

structure type A group of crystals having the same atomic structure, i.e. having the constituent atoms arranged in a geometrically analogous way. An example is the NaCl structure type, in which equal numbers of cations and anions occur in sixfold coordination; it includes sylvite, periclase, and galena.

struggle for existence The natural process by which members of a population compete automatically for a limited supply of vital necessities resulting in *natural selection.*

struggle for the divide Contention between two adjacent river systems to extend their drainage basins.

Strukturboden A term formerly used for what is now known as *patterned ground.* Etymol: German, "structure ground" or "structure soil".

Strunian European stage: uppermost Devonian, transitional into Carboniferous. Syn: *Etroeungtian.*

strunzite A straw-yellow monoclinic mineral: $MnFe_2(PO_4)_2(OH)_2.8H_2O$. It is polymorphous with laueite.

strut thrust An obsolete term for a fault initiated by the shearing of a strut, or competent bed.

struvite A colorless to yellow or pale-brown orthorhombic mineral: $Mg(NH_4)(PO_4).6H_2O$.

stubachite An altered diallage peridotite characterized by the presence of tremolite, talc, serpentine, magnetite, pyrite, and breunnerite.

Student's t test A statistical test used to compare sample and population means when the variances are unknown. Named after Student, pseudonym of William S. Gosset, 20th century Irish statistician. Syn: *t test*.

stuffed mineral A mineral having extra ions of a foreign element within the large interstices of its structure; e.g. garnet with an extra cation.

stuffing box A chamber or space in a borehole, filled with a soft, compressible packer (such as rubber) to maintain a fluid-tight joint and prevent leakage along a drill pipe where it enters the casing at the collar of the hole.

stuffing-box casinghead *bradenhead.*

sturtite A black mineral: $Mn_3{}^{+2}Fe^{+3}Si_4O_{11}(OH)_3.10H_2O$ (?).

stützite A mineral: $Ag_{5-x}Te_3$. It was formerly regarded as identical with empressite.

stylaster Any one of a group of hydrozoans belonging to the order Stylasterina, characterized by a calcareous skeleton and by sexual individuals that remain attached to the colony. Cf: *hydroid; millepore.*

style [paleont] (a) A sponge spicule (monaxon) with one blunt end and one pointed end; e.g. *tylostyle.* (b) A tubule that arises from the galea in phaeodarian radiolarians. (c) A central calcareous process in certain pores of a stylaster coral. (d) The *telson* of a crustacean.

style [bot] A cylindrical, centrally located flower organ that connects the stigma to the ovary.

styliform columella A solidly fused and longitudinally projecting coral *columella.* It is fused to scleractinian entosepta by secondary stereome.

styliform cyrtolith A cyrtolith coccolith with a long spinose central structure; e.g. a pole coccolith.

stylocerite A rounded or spiniform process on the outer part of antennular peduncle in some decapod crustaceans.

stylolite (a) A thin seam or a surface or contact usually occurring in "pure" or homogeneous carbonate rocks (certain limestones, dolomites, marble, bedded siderites, calcareous shales) and more rarely in sandstones and quartzites, marked by an irregular and interlocking or mutual interpenetration of the two sides, the columns, pits, and teeth-like projections on one side fitting into their counterparts on the other. As usually seen in cross section, it resembles somewhat a suture or the tracing of a stylus. The seam is characterized by a concentration of insoluble constituents (clay, carbon, sand, iron oxides, etc.) of the rock, and it is commonly parallel to the bedding. Stylolites are supposedly formed diagenetically in consolidated rock by differential vertical movement under pressure, accompanied by solution. See also: *microstylolite; suture joint.* Syn: *stylolite seam.* (b) A small, short (less than a millimeter to more than 30 cm in length, but commonly less than a centimeter), straight, vertically grooved column (of the same material as the rock in which it occurs) fitting into a corresponding socket in a stylolitic seam and being highly inclined or at right angles to the bedding plane. It often results from the slipping under vertical pressure of a part capped by a fossil shell through adjacent parts not so capped.---Term introduced by Klöden (1828, p.28). Etymol: Greek *stylos,* "pillar", + *lithos,* "stone". Obsolete syn: *epsomite; crystallite; toenail; devil's toenail; crowfoot; lignilite.*

stylolitic Pertaining to a stylolite, such as "stylolitic seam" or "stylolitic column"; said of a sedimentary surface penetrated by sinous, peaked films of clayey (and often organic) matter.

stylotypite A syn. of *tetrahedrite.* The name has been applied esp. to tetrahedrite containing considerable silver.

Styrian orogeny One of the 30 or more short-lived orogenies during Phanerozoic time identified by Stille, in this case in the Miocene, between the Burdigalian and Aquitanian Stages.

suanite A mineral: $Mg_2B_2O_5$.

subactive volcano *dormant volcano.*

subaerial Occurring beneath the atmosphere or in the ope air; esp. said of conditions and processes (such as erosio that exist or operate on or immediately adjacent to the lar surface, or of features and materials (such as eolian depo its) that are formed or situated on the land surface. Befor eolian processes were recognized as significant, "subaeria was considered the same as "fluvial" (Johnson, 1919, p.166 Cf: *subaqueous; subterranean.* See also: *surficial.*

subaerial bench A term used by Lawson (1915) for a nonall viated, concave-upward pediment.

subaerialism The doctrine that the landscape and its lan forms are formed chiefly by subaerial agents (esp. rainwash and processes.

subage A seldom-used term for a geologic-time unit short than an *age,* during which the rocks of the corresponding *su stage* were formed. It is usually characterized by the occu rence of some specific phenomenon, such as the depositio of loess. Syn: *episode (b); time (d); phase [geochron]* (*a).*

subalkalic (a) Said of an igneous rock that contains no alka minerals other than feldspars. (b) Formerly used to describ an igneous rock of the Pacific suite.

suballuvial bench A term used by Lawson (1915, p.34) for th outward or basinward extension of a pediment, covered b alluvium as the basin slowly filled (the thickness increasir basinward to several hundred meters), and exhibiting a co vex-upward longitudinal profile. Cf: *concealed pediment.*

subalpine *montane.*

subalpine peat *hill peat.*

subaluminous Said of an igneous rock in which there is littl or no excess of aluminum oxide over that required to for feldspars or feldspathoids; one of Shand's (1947) groups o igneous rocks, classified on the basis of the degree of alum num-oxide saturation. Cf: *peralkaline; peraluminous; metalu minous.*

subangular Somewhat angular, free from sharp angles but n smoothly rounded; specif. said of a sedimentary particl showing definite effects of slight abrasion, retaining its origin general form, and having faces that are virtually untouche and edges and corners that are rounded off to some exter such as a glacial boulder with numerous (10-20) seconda corners and a roundness value between 0.15 and 0.25 (mi point at 0.200) (Pettijohn, 1957, p.59), or one with one-thi of its edges smooth (Krynine, 1948, p.142). Also, said of th *roundness class* containing subangular particles.

subaquatic plant A hydrophyte that is not a *submerged aqua ic plant.*

subaqueous Said of conditions and processes that exist or op erate, or of features and deposits that are formed or situate in or under water or beneath the surface of water, esp. freshwater (as in a lake or stream). Cf: *subaerial.* See als *submarine; underwater.*

subaqueous sand dune *dune* [stream].

subaqueous till *berg till.*

subarctic [geog] Pertaining or relating to the regions direct adjacent to the arctic circle or to areas that have characteri tics such as climate, vegetation, and animals, similar to thes regions.

Subarctic n. *Preboreal.*

subarkose A sandstone that does not have enough feldspar t be classed as an *arkose,* or a sandstone that is intermedia in composition between arkose and pure quartz sandston Examples of quantitative definitions: an arkosic sandstor containing 75-95% quartz and chert, less than 15% detrit clay matrix, and 5-25% unstable materials in which the fel spar grains exceed the rock fragments in abundance (Pett john, 1954, p.364); a sandstone containing 5-25% feldspa and igneous rock fragments and less than 10% micas ar metamorphic rock fragments, and having any clay conten orting, and rounding (Folk, 1954, p.354); a sandstone containing 5-25% feldspar, 65-95% quartz, quartzite, and chert, and less than 10% fine-grained rock fragments (McBride, 1963, p.667); and a sandstone with 75-95% quartz and metamorphic quartzite and a content (5-25%) of feldspar and fragments of gneiss and granite that exceeds that of all other fine-grained rock fragments (including chert), regardless of clay content or texture (Folk, 1968, p.124). Pettijohn (1957, p.322) later used 10-25% unstable fragments so that a subarkose might have as little as 5% feldspar. The rock is roughly equivalent to *feldspathic arenite* of Gilbert (1954). Syn: *feldspathic quartzite; feldspathic sandstone.*

subartesian Said of confined ground water that is under sufficient pressure to rise above the water table but not to the land surface; mesopiestic water.

subatlantic n. A term used primarily in Europe for a period of postglacial time (approximately the last 2500 years, or from 500 B.C. to the present) following the Subboreal, during which the inferred climate became generally milder and wetter; a subunit of the Blytt-Sernander climatic classification, characterized by beech and linden vegetation. It corresponds roughly with the last half of the Little Ice Age. Also spelled: *Sub-Atlantic.*---adj. Pertaining to the postglacial Subatlantic interval and to its climate, deposits, biota, and events. Also spelled: *Sub-Atlantic.*

subbase Another base or underlying support placed below that which ordinarily forms the base; specif. a layer of material (earth, rock, etc.) placed between the base course and the *subgrade*, designed to give additional support or to form a porous layer (such as the first layer of large stone laid down in constructing a road, airstrip, or other graded surface).

subbentonite *metabentonite.*

subbituminous A coal A type of *subbituminous coal* having 10,500 or more, but less than 13,000 BTU per pound.

subbituminous B coal A type of *subbituminous coal* having 9,500 or more, but less than 10,500 BTU per pound.

subbituminous C coal A type of *subbituminous coal* having 8,300 or more, but less than 9,500 BTU per pound.

subbituminous coal A black coal intermediate in rank between lignite and bituminous coals, or in some classifications, the equivalent of *black lignite.* It is distinguished from lignite by higher carbon and lower moisture contents. Further classification of subbituminous coal is made on the basis of calorific value. See also: *subbituminous A coal; subbituminous B coal; subbituminous C coal; lignite.* Cf: *gloss coal.*

subboreal (a) Said of a climate that is very cold or approaching frigidity. (b) Pertaining to a biogeographic zone that approaches a boreal climatic condition. (c) Pertaining to the Subboreal postglacial period, and to the climate of such a period. Also spelled: *sub-Boreal.*

Subboreal n. A term used primarily in Europe for an interval of postglacial time (from about 4500 to 2500 years ago) following the Atlantic and preceding the Subatlantic, during which the inferred climate became generally cooler and drier periodically warm with cooler intervals, and rather dry but with considerable variation of humidity); a subunit of the Blytt-Sernander climatic classification, characterized by oak, ash, and linden vegetation. It corresponds roughly with the first half of the Little Ice Age. Also spelled: *Sub-Boreal.*---adj. Pertaining to the postglacial Subboreal interval and to its climate, deposits, biota, and events.

subbottom profile A type of *reflection profile* in which the reflectors lie beneath the bottom of the ocean.

subcannel coal Cannel coal of brown-coal to subbituminous rank. Cf: *metacannel coal; lean cannel coal.*

subcapillary interstice An *interstice* sufficiently smaller than a *capillary interstice* that the molecular attraction of its walls reaches across the entire opening. Water held in it by adhesive forces is immovable except by forces in excess of pressures commonly found in subsurface water. "The conditions existing in these interstices are, however, only very imperfectly understood and are largely a matter of speculation" (Meinzer, 1923, p.19). Cf: *supercapillary interstice.*

subchela The distal prehensile or grasping part of a crustacean limb formed by folding dactylus against propodus or dactylus and propodus against carpus. Pl: *subchelae.*

subclass A group of orders, within a class, that have characteristics distinct from those of other orders in the class.

subconsequent stream (a) *secondary consequent stream.* (b) An obsolete syn. of *subsequent stream.*

subcontinent (a) A division or part of a continent having characteristics that distinguish it from the rest of the continent, e.g. the Indian subcontinent. This subdivision is typically based on geologic or geomorphic characteristics. (b) A large land mass, such as Greenland or Antarctica, that is smaller than any of the seven recognized continents.

subcortical crypt An inhalant aquiferous cavity lying beneath a cortex in a sponge and differentiated from a canal by virtue of its larger size and distinctive shape.

subcrevasse channel A shallow channel eroded in subglacial material by a stream flowing along the bottom of a crevasse that completely penetrated a glacier (Leighton, 1959, p.340).

subcritical flow *tranquil flow.*

subcrop (a) An occurrence of strata in contact with the undersurface of an inclusive stratigraphic unit that succeeds an important unconformity on which overstep is conspicuous; a "subsurface outcrop" that describes the areal limits of a truncated rock unit at a buried surface of unconformity. (b) An area within which a formation occurs directly beneath an unconformity.---The term, in common use in petroleum geology, appears to have been used first by Swesnick (1950, p.401) at the suggestion of Thom H. Green.

subcrop map A geologic map that shows the distribution of formations that have been preserved and remain covered beneath a given stratigraphic unit or immediately underlying an unconformity; properly, a map of an area where the overlapping formation is still present. The term "may be considered a generalization" of the term *paleogeologic map* (Krumbein & Sloss, 1963, p.448). Cf: *supercrop map.*

subdelta A small delta, forming a part of a larger delta or complex of deltas.

subdermal space A *vestibule* of a sponge.

subdiabasic Said of the texture, of an igneous rock, that is similar to ophitic texture except that the augite of the groundmass is not optically continuous but is divided into granular aggregates.

subdivide A drainage divide between the tributaries of a main stream; a subordinate divide.

subdrainage Drainage from beneath, either natural or artificial.

subdrift topography Topography of a bedrock surface underlying unconsolidated glacial drift.

subduction The process of one crustal block descending beneath another, by folding or faulting or both. The concept was originally used by Alpine geologists. See also: *zone of subduction.*

subduction zone An elongate region along which a crustal block descends relative to another crustal block, e.g. the descent of the Pacific plate beneath the Andean plate along the Andean trench. See also:*subduction.*

subdued Said of a landform or landscape that is marked by a broadly rounded form and by moderate height, as if produced by long-continued weathering and erosion; esp. said of a mountain in the stage of senescence in a cycle of erosion, sufficiently worn down to have lost its peaks and cliffs, and having its moderately steep slopes covered with its own detritus. Cf: *feral.*

subepoch A term proposed by Sutton (1940, p.1402) for a geologic-time unit representing the first division of an *epoch.* It is applied only to a few portions of geologic time. Cf: *subseries.*

subera A little used term referring to a "portion of an *era* comprised of two or more periods" (Sutton, 1940, p.1410).

suberain A kind of *provitrain* in which the cellular structure is derived from corky material. Cf: *periblain; xylain.*

suberinite A variety of provitrinite characteristic of suberain and consisting of corky tissue. Cf: *periblinite; xylinite; telinite.*

subfabric Any part of the *fabric* of a rock; the spatial array of any particular *fabric element.*

subface The basal or lower surface of a stratigraphic unit.

subfacies A subdivision of a facies, such as of a broadly defined sedimentary facies, or of a metamorphic facies based on compositional differences rather than pressure-temperature relations.

subfamily A group of genera, within a family, with characteristics that set them apart from the other genera. In zoology, the name of a subfamily characteristically has the ending -inae; e.g. Cytheredeinae.

subfeldspathic (a) Said of a mature lithic wacke (or lithic graywacke) in which quartz grains and fragments of siliceous and argillaceous rocks predominate, and feldspars make up less than 10% of the rock and may be altogether lacking (Gilbert, 1954, p.302-303). Such rocks have also been called *subgraywackes.* (b)Said of a mature lithic arenite containing abundant quartz grains and fragments of the more stable rocks (such as cherts), and less than 10% feldspar grains (Gilbert, 1954, p.304 & 307).

subfluvial Situated or formed at the bottom of a river; e.g. a "subfluvial deposit".

subfossil n. A fossil that is younger than what would be considered typical fossil age but not strictly recent either. It may have lost its organic components but not yet have had them replaced.----adj. Applied to an organism that would be considered a "subfossil".

subgelisol A layer of unfrozen ground (*talik*) occurring beneath the permafrost.

subgenus A group of species, within a genus, with characteristics that distinguish it from other groups in the genus. The name of a subgenus, when used with the name of the genus, is placed in parentheses after the genus and is followed by the name of the species, if it is used; it does not count as one of the words in a binomen or trinomen; e.g. *Palaeneilo (Koenenia) emarginata.* Cf: *cenospecies.*

subglacial (a) Formed or accumulated in or by the bottom parts of a glacier or ice sheet; said of meltwater streams, till, moraine, etc. Syn: *infraglacial.* (b) Pertaining to the area immediately beneath a glacier, as *subglacial* eruption or *subglacial* drainage. (c) *postglacial.*

subgrade A layer, stratum, or surface immediately beneath some principal surface; specif. a layer of material (earth, rock, etc.) that is leveled off to receive the foundation of an engineering structure (such as the soil or natural ground that is prepared and compacted to support a structure and that lies directly below a road, pavement, building, airfield, or railway). Cf: *subbase.*

subgraphite *meta-anthracite.*

subgraywacke (a) A term introduced by Pettijohn (1949, p.227 & 255-256) for a sedimentary rock that has less feldspar and more and better-rounded quartz grains than graywacke; specif. a sandstone containing 15-85% quartz and chert, 15-75% detrital clay matrix (chiefly sericite and chlorite), less than 10-15% feldspar, and an appreciable (5%) quantity of rock fragments. This rock, as originally defined, is equivalent to *quartz wacke* of Krumbein & Sloss (1963), to *low-rank graywacke* of Krynine (1948), to *lithic graywacke* of Pettijohn (1954), and to the *subfeldspathic* wackes (lithic wacke and lithic graywacke) of Gilbert (1954). (b) A term redefined by Pettijohn (1957, p.316-320) for the most common type of sandstone, intermediate in composition between graywacke and orthoquartzite, containing less than 75% quartz and chert (commonly 30-65%), less than 15% detrital

clay matrix, and an abundance (more than 25%) of unstab[le] materials (feldspar grains and rock fragments) in which th[e] rock fragments (at least 15%) exceed the feldspars, and hav[ing] voids and/or mineral cement (esp. carbonates) exceedin[g] the amount of clay matrix. The rock is lighter-colored and be[t]-ter-sorted, and has less matrix, than *graywacke,* and com[mon]ly forms great lenticular bodies as a result of paralic sed[i]mentation from normal subaqueous currents. Example: th[e] Oswego Sandstone (Upper Ordovician) of central Pennsylva[n]ia. (c) A term used by Folk (1954, p.354) for a sedimentar[y] rock that does not have enough rock fragments to be classe[d] as a graywacke; specif. a sandstone with 5-25% micas an[d] metamorphic rock fragments and less than 10% feldspars an[d] igneous rock fragments, and having any clay content or sor[t]ing. This rock is equivalent to *quartzose graywacke* of Krynin (1951).

subgroup A formally differentiated assemblage of formation[s] within a *group* (ACSN, 1961, art.9d).

subhedral (a) Said of an individual mineral crystal, in an igne[ous] ous rock, that is partly faced or incompletely bounded by it[s] own crystal faces and partly bounded by surfaces forme[d] against preexisting crystals. (b) Said of a crystal, in a sed[i]mentary rock (such as a calcite crystal in a recrystallize[d] dolomite), characterized by partially developed crystal face[s] (c) Said of the shape of a subhedral crystal.----Intermediat[e] between *euhedral* and *anhedral.* The term was proposed, i[n] reference to igneous-rock components, by Cross et al (190[6], p.698) in preference to the synonymous terms *hypidiomorph[ic]* and *hypautomorphic* (as these were originally defined).

subhedron A subhedral crystal. The term was introduced b[y] Cross et al (1906) in reference to an igneous-rock compo[nent] nent (crystal) only partly bounded by the crystal faces char[-] acteristic of the mineral species to which the crystal belong[s]. Pl: *subhedrons; subhedra.*

Subhercynian orogeny One of the 30 or more short-lived oro[-] genies during Phanerozoic time recognized by Stille, in thi[s] case in the Late Cretaceous, between the Turonian and Seno[-] nian stages.

subhumid Said of a climate type that is transitional betwee[n] humid and subarid types according to quantity and distributio[n] of precipitation. In Thornthwaite's classification, the moistur[e] index is between zero and -20.

subhydrous (a) Said of coal containing less than 5% hydro[-] gen, analyzed on a dry, ash-free basis. (b) Said of a macera[l] of low hydrogen content, e.g. fusinite.----Cf: *orthohydrous[;] perhydrous.*

subida A rock-floored belt produced by wind scour and "po[-] tentially reaching to the base of a mountain range" (Stone 1967, p.250). Etymol: Spanish, "ascent, acclivity".

subidioblast *hypidioblast.*

subidiomorphic *hypidiomorphic.*

subimposed Said of a subterranean stream that becomes [a] surface stream, as when the roof of a limestone cavern fall[s] in. An obsolete term, originally proposed by Russell (1898b p.246).

subirrigation Irrigation of plants with water delivered to th[e] roots from underneath, either naturally or artificially.

subjacent [geomorph] Being lower, but not lying directl[y] below; e.g. "hills and subjacent valleys".

subjacent [intrus rocks] Said of an igneous intrusion, general[-] ly a discordant one, with no known floor, and that presumabl[y] enlarges downward to an unknown depth.

subjacent [stratig] Said of a stratum situated immediatel[y] under a particular higher stratum or below an unconformity Ant: *superjacent.* See also: *underlying.*

subjacent karst Karst that is developed beneath a surface for[-] mation of more resistant rock. Syn: *subterranean karst; bur[-] ied karst.*

subjective synonym In taxonomy, any one of two or mor[e] synonyms based on different types, but considered as refer[-] ring to the same taxon by those who regard them as synonyms. Cf: *objective synonym*.

subjoint A minor joint associated with a major joint, either divergent or parallel in arrangement.

subkingdom A group of organisms, within a kingdom, having characteristics distinct from other groups in the kingdom. It is sometimes considered synonymous with *phylum* and sometimes ranked above it.

sublacustrine Existing or formed beneath the waters, or on the bottom, of a lake; e.g. a "sublacustrine channel" eroded in the lake bed by a surface stream before the lake was there or by a strong current within the lake.

sublimate A solid that has been deposited by a gas or vapor; in volcanology it refers to such a deposit made by a volcanic gas, e.g. metals around the mouth of a fumarole.

sublimation [**ore dep**] The process of ore deposition, usually sulfur, by vapors, the volatilization and transportation of minerals followed by their deposition at reduced temperatures and pressures. Sublimation deposits are generally associated with fumarolic activity.

sublimation [**chem**] The process by which a solid substance vaporizes without passing through a liquid stage. Cf: *evaporation*.

sublimation loss Loss of water through the direct evaporation of ice and snow on lakes or from any body of ice or snow.

sublitharenite (a) A term introduced by McBride (1963, p.667) for a sandstone that does not have enough rock fragments to be classed as a *litharenite*, or a sandstone that is intermediate in composition between litharenite and pure quartz sandstone; specif. a sandstone with 5-25% fine-grained rock fragments, 65-95% quartz, quartzite, and chert, and less than 10% feldspar. (b) A term used by Folk (1968, p.124) for a sandstone, regardless of clay content or texture, with 75-95% quartz and metamorphic quartzite and a content (5-25%) of fine-grained volcanic, metamorphic, and sedimentary rock fragments (including chert) that exceeds that of feldspar and fragments of gneiss and granite.

sublithistid Said of a sponge containing desmoids.

sublithographic Said of a limestone whose texture approaches the exceedingly fine grain of lithographic limestone. Also, said of the texture of such a rock.

sublittoral Said of that part of the *littoral* zone that is between low tide and a depth of about 100 m; a syn. of *neritic*.

submarginal channel A channel formed by a meltwater stream flowing near the ice margin but also cutting across spurs or "behind small outlying hills" (Rich, 1908, p.528).

submarginal moraine *lodge moraine*.

submarginal ring A prominent ring of thick plates exposed on both oral and aboral sides of theca of cyclocystoids and representing their "most conspicuous and best-preserved feature" (TIP, 1966, pt.U, p.201). Cf: *marginal ring*.

submarine bar A *longshore bar* that is always submerged, or never exposed above the water level even by low tides.

submarine barchan A large-scale, straight-crested, lunate asymmetric ripple mark on the sea floor, ranging in length from centimeters to 100 m or more. Examples occur in the shallow-water areas of the Bahamas.

submarine canyon (a) A steep-sided, V-profile trench or valley winding along the continental shelf or continental slope, having tributaries and resembling unglaciated, river-cut land canyons. (b) A general term for all valleys of the deep-sea floor. Syn: *submarine valley*.

submarine cone *submarine fan*.

submarine delta *submarine fan*.

submarine earthquake *seaquake*.

submarine fan A terrigenous, cone- or fan-shaped deposit located seaward of large rivers and submarine canyons. Syn: *submarine cone; abyssal cone; abyssal fan; subsea apron; deep-sea fan; submarine delta; sea fan; fan* [*marine geol*]; *cone* [*marine geol*].

submarine geology *geological oceanography*.

submarine geomorphology That aspect of geological oceanography which deals with the relief features of the ocean floor and with the forces that modify them.

submarine meadow A grassland consisting of marine plants such as turtle grass.

submarine plain A syn. of *plain of marine erosion*. Term is not recommended because some of these plains have been uplifted. Syn: *submarine platform*.

submarine plateau *plateau* [marine geol].

submarine platform *submarine plain*.

submarine ridge *ridge* [marine geol].

submarine spring A large offshore emergence of fresh water, usually associated with a coastal karst area but sometimes with lava beds.

submarine valley *submarine canyon*.

submarine volcano A volcano on the ocean floor, characteristically of tholeiitic basalt. See also: *volcanic island*.

submarine weathering *halmyrolysis*.

submask geology The geology of the surface underlying a cover of alluvium, glacial drift, windblown sand, low-angle overthrust sheets, or water (as under shallow lakes and bays) (Kupsch, 1956).

submature [**geomorph**] Said of a topographic feature that has passed through the stage of youth but is not completely matured; e.g. a *submature* shoreline characterized by the cutting back of headlands and by the near closing of baymouths by bars, thus simplifying an earlier intricately embayed shoreline (Cotton, 1958, p.456).

submature [**sed**] Pertaining to the second stage of textural maturity (Folk, 1951); said of a clastic sediment intermediate in character between an *immature* and a *mature* sediment, characterized by little or no clayey material and by poorly sorted and angular grains. Example: a clean "submature sandstone" containing less than 5% clay and commonly occurring in stream channels. Cf: *supermature*.

submeander A small meander contained within the banks of the main channel (Langbein & Iseri, 1960, p.19); it is associated with relatively low discharges.

submerged aquatic plant A hydrophyte the main part of which grows below the surface of the water. Cf: *subaquatic plant*.

submerged coastal plain The continental shelf representing the seaward continuation of a coastal plain on the land. Syn: *coast shelf*.

submerged contour A contour on the bed of a lake or reservoir, joining points of equal elevation where the elevation is related to a datum (usually mean sea level) used for mapping adjacent land (BNCG, 1966, p.14). Cf: *isobath* [*oceanog*].

submerged forest Forest remains, e.g. stumps still rooted in peaty soil, seen at low tide or found below sea level indicating a relative rise in sea level or a relative subsidence of the coast.

submerged land A legal term for the land at the bottom of a lake, or the land covered by water when the lake is at its mean high-water level or at a level set by court decree (Veatch & Humphrys, 1966, p.324).

submerged shoreline *shoreline of submergence*.

submerged valley A *drowned valley*, such as a ria.

submergence A change in the relative levels of water and land such that the water level is higher (in relation to the land) than it was before the change and areas formerly dry land become inundated; it results either from a sinking of the land or from a rise of the water level. Ant: *emergence*.

submesothyridid Said of a brachiopod pedicle opening when the foramen is located mainly in the delthyrium and partly in the ventral umbo (TIP, 1965, pt.H, p.153). Cf: *permesothyridid*.

submetallic Said of mineral lusters that are classified indeterminately between *metallic* and *nonmetallic*. Chromite, for example, has a metallic to submetallic luster.

subnival *periglacial.*

subnormal-pressure surface A potentiometric surface that is below the upper surface of the zone of saturation (Meinzer, 1923, p.39). The term is not in general use among hydrogeologists. Cf: *normal-pressure surface; artesian-pressure surface.*

subophitic Said of the ophitic texture of an igneous rock in which the feldspar crystals are approximately the same size as the pyroxene and are only partially included by them.

suborder A unit within an order that contains a single distinctive family or a group of families that are distinct from other families in the order. It is sometimes considered equivalent to *superfamily* and sometimes as the next higher rank.

suboutcrop *blind apex.*

subperiod A geologic-time unit that is a portion of a *period* [geochron], but longer than an *epoch* [geochron], during which the rocks of the corresponding *subsystem* were formed.

subpermafrost water Ground water in the unfrozen ground beneath the permafrost.

subphyllarenite A *phyllarenite* containing 3-25% rock fragments.

subphylum A group of classes, within a phylum, that is distinct from other groups in the phylum.

subpolar glacier A glacier on which there is some surface melting during the summer but which is below freezing in temperature throughout most of its mass. Cf: *polar glacier.*

subrounded Partially rounded; specif. said of a sedimentary particle showing considerable but incomplete abrasion and an original general form that is still discernible, and having many of its edges and corners noticeably rounded off to smooth curves, such as a cobble with a reduced number (5-10) of secondary corners, a considerably reduced area of the original faces, and a roundness value between 0.25 and 0.40 (midpoint at 0.315) (Pettijohn, 1957, p.59), or one with two-thirds of its edges smooth (Krynine, 1948, p.142). Also, said of the *roundness class* containing subrounded particles.

subsea apron *submarine fan.*

subseptate Having imperfect or partial septum or septa; e.g. having slight protuberances or incipient septa that form pseudochambers in a foraminiferal test (as in the family Tournayellidae).

sub-sequence A term applied by R.C. Moore (1958, p.80) to a Precambrian rock division (often called a *series* in Canada) that "cannot be correlated from one region to another". The term requires a hyphen in order to distinguish it from the noun "subsequence". See also: *sequence.*

subsequent [geomorph] Said of a geologic or topographic feature that followed in time the origin and development of a *consequent* feature or of the system of which the subsequent feature is a part; e.g. a *subsequent* ridge (such as one formed by differential erosion of a consequent ridge), or a *subsequent* mountain (such as one representing the remains of a preexisting plateau), or a *subsequent* waterfall (such as one developed where a downcutting stream encounters rocks of varying hardness).

subsequent [streams] adj. Said of a stream, valley, or drainage system that is developed independently of, and subsequent to, the original relief of a land area, as by shifting of divides, stream capture, or adjustment to rock structure. The concept was originally discussed by Jukes (1862, p.393-395).---n. *subsequent stream.*

subsequent divide A divide between two subsequent streams.

subsequent fold *cross fold.*

subsequent stream A tributary that has developed its valley (mainly by headward erosion) along a belt of underlying weak rock and is therefore adjusted to the regional structure; esp. a stream that flows approximately in the direction of the strike of the underlying strata and that is subsequent to the formation of a consequent stream of which it is a tributary. Obsolete syn: *subconsequent stream.* Syn: *subsequent; strike stream; longitudinal stream.*

subsequent valley A valley eroded by or containing a *subsequent stream*; a structurally controlled valley, developed after a consequent valley into which it passes. See also: *strike valley; longitudinal valley.*

subsere (a) A secondary ecologic succession that arises on a denuded area following an ecologic climax. (b) A seral community which is prevented from reaching ecological climax by a temporary interference, e.g. by human activity.

subseries A term proposed by Sutton (1940, p.1402) for the first division of a series, representing the rocks formed during a *subepoch.*

subsidence (a) A local mass movement that involves principally the gradual downward settling or sinking of the solid Earth's surface with little or no horizontal motion and that does not occur along a free surface (not the result of a landslide or failure of a slope). The movement is not restricted in rate, magnitude, or area involved. Subsidence may be due to: natural geologic processes such as solution, erosion, oxidation, thawing, lateral flow, or compaction of subsurface materials; earthquakes, slow crustal warping, and volcanism (withdrawal of fluid lava beneath a solid crust); or man's activity such as removal of subsurface solids, liquids, or gases and wetting of some types of moisture-deficient loose or porous deposits. See also: *cauldron subsidence.* Syn: *land subsidence; bottom subsidence.* (b) A sinking of a large part of the Earth's crust relative to its surrounding parts, such as the formation of a rift valley or the lowering of a coast due to tectonic movements.----Syn: *sinking.*

subsidence caldera *collapse caldera.*

subsidence/head-decline ratio The ratio between land subsidence and the hydraulic head decline in the coarse-grained beds of the compacting aquifer system (Poland, et al, in press).

subsidence theory A theory of coral-atoll and barrier-reef formation according to which upward reef growth kept pace uninterruptedly over a long period with slow subsidence of a volcanic island, forming first a fringing reef that became a barrier reef and later an atoll when the island was completely submerged; it accounts most satisfactorily for the majority of Pacific Ocean reefs. Theory was proposed by Charles Darwin in 1842. Cf: *glacial-control theory; antecedent-platform theory.*

subsidiary fold A minor fold that has the same relationship to a major fold as does a *drag fold*, but for which origin by drag is uncertain or unlikely. Syn: *parasitic fold.* Cf: *minor fold.*

subsidiary fracture A less-preferred syn. of *tension fracture.*

subsilicic A term proposed by Clarke (1908, p.357) to replace *basic* [geol]. Cf: *persilicic; mediosilicic.*

subsoil (a) A syn. of *B horizon*, in a soil profile having distinct horizons. (b) The soil below the *surface soil*; this is an older meaning.----Cf: *topsoil.*

subsoil ice *ground ice.*

subsoil weathering A term used by Davis (1938) for the chemical decomposition that produces spheroidal boulders (beneath the regolith in granitic areas of the desert) by percolation of water along joints followed by exposure and exfoliation.

subsolidus A chemical system that is below its melting point, and in which reactions may occur in the solid state.

subsolvus Said of those granites, syenites, and nepheline syenites that are characterized by both potassium feldspar and plagioclase (Bowen and Tuttle, 1958, p.129). Cf: *solvus; hypersolvus.*

subspeciation Division into or formation of subspecies.

subspecies A group of individuals, within a species, that is defined by a set of minor structural peculiarities. Such groups that are geographically isolated from one another are geographic subspecies; groups separated in geologic time are chronologic subspecies. The name of a subspecies is a trinomen; e.g. *Bollia americana zygocornis.* Cf: *variety; ecotype.*

substage [geochron] An obsolete term for a geologic-time unit of lesser duration than *stage* [geochron], during which the rocks of a *member* were formed.

substage [glac geol] A time term for a subdivision of a glacial stage during which there was a secondary fluctuation of glacial advance and retreat; specif. a Pleistocene subdivision equated with a rock unit of member rank, such as the "Tazewell substage" of the Wisconsin stage.

substage [optics] In a microscope, an attachment for holding polarizers or other attachments below the stage.

substage [stratig] A time-stratigraphic unit next in rank below *stage*; the rocks formed during a *subage* of geologic time. The unit is considered a formal subdivision of a stage (ISST, 1961, p.25; and ACSN, 1967, p.1865). The frequently used synonym *zone* is not recommended (ISST, 1961, p.13).

substitute Any substance represented to be, or used to imitate, a more valuable or better-known gemstone; e.g. plastic, glass, doublet, synthetic ruby, or natural spinel all could be substitutes for natural ruby.

substitution [chem] *ionic substitution.*

substitutional transformation A type of crystal *transformation* of a disordered phase (a substitutional solid solution) to an ordered phase. It is usually a slow transformation. Cf: *dilatational transformation; displacive transformation; reconstructive transformation; rotational transformation.* Syn: *order-disorder transformation; ordering.*

substitution solid solution A crystal in which a particular atomic site can be occupied by any of two or more elements. Cf: *omission solid solution.*

substrate [ecol] The substance, base, or nutrient on (or medium in) which an organism lives and grows, or the surface to which a fixed organism is attached; e.g. soil, rocks, water, and leaf tissues, or perhaps a gel for the accumulation and preservation of prebiologic organic matter. Syn: *substratum.*

substratum [ecol] *substrate.*

substratum [soil] Any layer lying beneath the solum or true soil. It is applied to both parent materials and to other layers unlike the parent material, below the B horizon or subsoil.

subsurface n. (a) The zone below the surface whose geologic features, principally stratigraphic and structural, are interpreted on the basis of drill records and various kinds of geophysical evidence. (b) Rock and soil materials lying near, but not exposed at, the Earth's surface; esp. the soil occurring immediately above the subsoil.----adj. Formed or occurring beneath a surface, esp. beneath the Earth's surface. Cf: *surficial.* See also: *subterranean.*

subsurface air Gas in interstices in the zone of aeration that open directly or indirectly to the surface and that is therefore at or near atmospheric pressure. Its composition is generally similar though not identical to that of the atmosphere (Meinzer, 1923, p.21). Cf: *soil atmosphere; natural gas; included gas; ground air.*

subsurface contour A syn. of *structure contour,* used to distinguish it from a surface or topographic contour.

subsurface drainage The removal of surplus water from within the soil by natural or artificial means, such as by drains placed below the surface to lower the water table below the root zone.

subsurface flow *storm seepage.*

subsurface geology Geology and correlation of rock formations, structures, and other features beneath the land or seafloor surface as revealed or inferred by exploratory drilling, underground workings, and geophysical methods. Ant: *surface geology.* Syn: *underground geology.*

subsurface ice *ground ice.*

subsurface map A map depicting geologic data or features below the Earth's surface; esp. a plan of mine workings, or a structure-contour map of an underground ore deposit, coal seam, or key bed.

subsurface perched stream Vadose water flowing toward the

water table in fractures in solution openings. There may be solid rock instead of a true zone of aeration beneath such a stream, and it could be considered simply as gravity ground water on its way to the water table by the easiest route. According to Meinzer (1923, p.22), "Water in transit from the surface to a zone of saturation may for convenience be regarded as passing through a zone of aeration or may be regarded as an irregular and perhaps temporary projection of the zone of saturation".

subsurface runoff *storm seepage.*

subsurface storm flow *storm seepage.*

subsurface water Water in the lithosphere in solid, liquid, or gaseous form. It includes all water beneath the land surface and beneath bodies of surface water. Syn: *subterranean water; underground water; ground water (b).*

subsystem [chem] Any part of a system that may be treated as an independent system. The substances which designate the subsystem must be components of it.

subsystem [stratig] A time-stratigraphic unit of higher than series rank, proposed for part of a system (ACSN, 1961, art.29c); the rocks formed during a *subperiod* of geologic time. The term was proposed for Mississippian or Pennsylvanian rocks to harmonize American stratigraphic classification with European, in which these units are regarded as parts of the Carboniferous System.

subtalus buttress The convex-upward rock surface developed under a rising talus slope as the cliff above it weathers back (Howard, 1942, p.27).

subterposition The state of being placed beneath something else; specif. the order in which strata are disposed in descending sequence. The term is obsolete.

subterrane n. The bedrock beneath a surficial deposit or below a given geologic formation. Syn: *subterrain.* ----adj. *subterranean.*

subterranean adj. Formed or occurring beneath the Earth's surface, or situated in the Earth. Cf: *subaerial.* See also: *subsurface.* Syn: *subterrestrial; subterrane.*

subterranean cutoff Diversion of a surface stream by the development of a shorter underground course across a meander neck.

subterranean ice *ground ice.*

subterranean karst *subjacent karst.*

subterranean stream A body of subsurface water flowing through a cave or a group of communicating caves, as in a karstic region. Cf: *underground stream.*

subterranean stream piracy Capture and diversion through an underground channel to an adjacent entrenched stream of a surface stream perched upon soluble rocks. It occurs in areas where limestone or other soluble rock lies above the base level of erosion.

subterranean water (a) A syn. of *subsurface water.* (b) A syn. of *ground water* in less preferred usage.

subterrestrial *subterranean.*

subtropical Said of the climate of the subtropics, which borders that of the tropics and is intermediate in character between tropical and temperate, though more like the former than the latter. Syn: *semitropical.*

subtuberant Said of a topographic high that is caused by igneous intrusion.

subulate Said of a leaf that tapers to an apical point from its base.

subvective system An organ system for transporting food particles to the mouth of an echinoderm. It cannot be separated morphologically from the *ambulacral system* (TIP, 1966, pt.U, p.155).

subvolcanic *hypabyssal* [ign].

subvolcano A term used for a small intrusion not far from the surface, tapering downward.

subweather velocity The velocity of a seismic wave in a bed immediately underlying the low-velocity weathered layer or zone. This velocity is distinctly greater than that in the weathered zone. Cf: *weathering velocity.*

subzone (a) A subdivision of a *biostratigraphic zone*, next lower in rank than the zone itself (ACSN, 1961, art.20e). It is a biostratigraphic unit that is usually recognizable across a continent, and it may include any number of *zonules.* (b) A subdivision of an informal stratigraphic zone of any kind (ISST, 1961, p.29).

succession [ecol] The progressive change in a biologic population as a result of the response of the members of the population to the environment.

succession [stratig] (a) A number of rock units or a mass of strata that succeed one another in chronologic order; e.g. an inclusive stratigraphic sequence involving any number of stages, series, or systems, or parts thereof, such as shown graphically in a geologic column or seen in actual superposition in an exposed section. (b) The chronologic order of rock units.

successional speciation The gradual evolution from one species to another, eventually leading to its replacement by that species.

succinic acid A crystalline dicarboxylic acid, formula HOOC-CH$_2$CH$_2$COOH, a constituent of wood bark and occurring in the amber group (but not in the retinite group) of fossil resins.

succinite (a) An old name for *amber*, esp. that mined in East Prussia or recovered from the Baltic Sea. (b) A light-yellow, amber-colored variety of grossular garnet.

sucrosic A syn. of *saccharoidal.* The term is commonly applied to idiotopic dolomite rock. Syn: *sucrose.*

suctive *permissive.*

sudburite An augite-bearing hypersthene basalt characterized by pillow structure and that also contains bytownite and magnetite. Its name is derived from Sudbury District, Ontario, Canada.

Sudetic orogeny One of the 30 or more short-lived orogenies during Phanerozoic time identified by Stille, in this case between the Early and Late Carboniferous.

sudoite A dioctahedral aluminum-rich mineral of the chlorite group: (Al,Fe,Mg)$_{4-5}$(Si,Al)$_4$O$_{10}$(OH)$_8$.

Suess effect The specific radioactivity of modern wood due to C-14 is slightly lower than expected due to dilution of atmospheric carbon dioxide by nonradioactive carbon from the burning of fossil fuels by industry.

Suess torsion balance A double variometer torsion balance with visual observation, which replaced the single balance.

suevite A grayish or yellowish fragmental rock (depositional breccia) that is associated with meteorite impact craters and that contains both shock-metamorphosed rock fragments and glassy inclusions that occur typically as aerodynamically shaped bombs. It closely resembles a tuff breccia or pumiceous tuff but is of nonvolcanic origin and can be distinguished by the presence of shock-metamorphic effects. The term was originally applied to material from the Ries basin, Germany, but is now used to designate similar brecciated material (impactites) found at other meteorite impact structures. See also: *trass.*

suffosion The bursting out on the surface in little eruptions of highly mobile or water-saturated material; esp. a destructive process operating under periglacial conditions in which underground water, resulting from partial melting of ground ice, exerts upward pressure and bursts through a hard dried upper skin to deposit a mound of mud, clay, sand, and/or boulders. Suffosional forms due to corrasion of underground water include dimpling, pits, blind valleys, shafts, and cavern entrances.

suffosion knob A *frost mound* (Muller, 1947, p.222).

sugar iceberg An iceberg consisting of porous glacier ice that is formed at very low temperatures.

sugarloaf A conical or conoidal hill or mountain comparatively bare of timber, resembling the shape of a loaf of sugar; e.g. Sugarloaf Mountain, Maine.

sugar sand A sandstone that breaks up into granules, resembling sugar.

sugar snow *depth hoar.*

sugar stone Compact, white to pink datolite from the Michigan copper district (Pough, 1967, p.140).

sugary *saccharoidal.*

suicidal stream A stream that rises in desert mountains and that loses its small amount of water by evaporation and infiltration soon after reaching the desert plain below (Stone, 1967, p.250).

suite [ign] A set of apparently comagmatic igneous rocks.

suite [stratig] A term used by Caster (1934, p.18) for a body of rocks intermediate between monothem (formation) and member, consisting of several intimately related members bracketed together; esp. a repeated sequence of such closely associated strata.

sukulaite A mineral: Sn$_2$Ta$_2$O$_7$.

sulcal notch A ventral indentation in the margin of some apical archeopyles in a dinoflagellate cyst, corresponding to the *sulcal tongue.*

sulcal plate One of the plates of the ventral furrow region in dinoflagellates possessing a theca. The plates are subdivided as to left or right, and anterior or posterior position.

sulcal tongue An extension of the operculum of a dinoflagellate cyst, in the position of the first apical plate of the theca. It is normally bordered by those parts of the archeopyle suture corresponding to thecal structures separating the first apical plate from the first precingular plate, the last precingular plate, and the intervening anterior sulcal plate. Cf: *sulcal notch.*

sulcate Pertaining to a sulcus, or scored with furrows, grooves, or channels, esp. lengthwise; e.g. said of a form of alternate folding in brachiopods with the brachial valve bearing a median sulcus and an anterior commissure median sinus (TIP, 1965, pt.H, p.153). Ant: *uniplicate.* Syn: *sulcated.*

sulcus [paleont] (a) A major longitudinal depression in the surface of either valve of a brachiopod, externally concave in transverse profile and radial from the umbo, and usually median in position. It is typically associated with the *fold.* See also *sinus.* (b) An elongate, shallow, vertical depression in the lateral surface of ostracode valve, extending from the dorsal region toward the venter. (c) A radial depression of the surface of the shell of a bivalve mollusk. (d) A longitudinal groove on the venter of an ammonoid shell.----Pl: *sulci.*

sulcus [palyn] (a) A relatively broad, longitudinal furrow in the exine of pollen grains; a *colpus.* The term is usually only applied to the distal furrow of monocolpate pollen grains. (b) A longitudinal posterior groove lying on the surface of dinoflagellate thecae and containing the trailing flagellum.

suldenite A hornblende andesite differing from *ortlerite* in being andesitic rather than dioritic.

sulfate A mineral compound characterized by the sulfate radical SO$_4$. Anhydrous sulfates, such as barite, BaSO$_4$, have divalent cations linked to the sulfate radical; hydrous and basic sulfates, such as gypsum, CaSO$_4$·2H$_2$O, contain water molecules. Cf: *chromate.*

sulfide A mineral compound characterized by the linkage of sulfur with a metal or semimetal, such as galena, PbS, or pyrite, FeS$_2$. See also: *sulfosalt.*

sulfide enrichment *enrichment.*

sulfide-facies iron formation An *iron-formation* consisting essentially of pyritic, carbonaceous slate.

sulfide zone An area of enrichment of sulfide deposits that have not yet been oxidized by near-surface waters. Cf: *oxidized zone; protore.* Syn: *secondary sulfide zone.*

sulfoborite A mineral: Mg$_3$B$_2$(SO$_4$)(OH)$_2$.4H$_2$O. Also spelled: *sulphoborite.*

sulfohalite *sulphohalite.*

sulfosalt A type of *sulfide* in which both a metal and a semimetal are present, forming a double sulfide, e.g. enargite, Cu_3AsS_4.

sulfur (a) An orthorhombic mineral, the native nonmetallic element S. It occurs in yellow crystals or in masses or layers often associated with limestone, gypsum, and other minerals (as in volcanic regions, in salt domes, or with hot springs) or combined esp. in sulfides and sulfates. Sulfur exists in several allotropic forms, including the ordinary yellow orthorhombic alpha form stable below 95.5°C and the pale-yellow monoclinic crystalline beta form. See also: *rosickyite*. Syn: *sulphur; brimstone*. (b) A mining term used for iron sulfide (pyrite) occurring in coal seams and with Wisconsin and Missouri zinc ores.

sulfur bacteria Anaerobic bacteria that obtain the oxygen needed in metabolism by reducing sulfate ions to hydrogen sulfide or elemental sulfur. Accumulations of sulfur in this way are *bacteriogenic* ore deposits. Cf: *iron bacteria*.

sulfur ball [pyroclast] A sulfurous mud skin that formed on a bubble of hot volcanic gas and became firm on contact with the air.

sulfur ball [coal] A pyritic impurity in coal, occurring as a spheroidal or irregular mass.

sulfur-mud pool *mud pot*.

sulfur spring A spring containing sulfur water.

sulfur water Generally, water containing enough hydrogen sulfide to smell and taste. Except for the hydrogen sulfide, it may not differ in mineral content from ordinary potable water, or may qualify as saline water. In either case, it is usually considered a mineral water.

sullage Mud and silt deposited by flowing water.

sulphatite Free sulfuric acid (H_2SO_4) found in some waters.

sulphoborite *sulfoborite*.

sulphohalite A mineral: $Na_6(SO_4)_2FCl$. Syn: *sulfohalite*.

sulphophile element An element which occurs preferentially in a mineral that is oxygen-free, e.g. fluorine or chlorine, i.e. as a sulfide, selenide, telluride, arsenide, antimonide, etc. The term incorporates the chalcophile elements and some of the siderophile elements of Goldschmidt's classification. The term is rarely used. Syn: *thiophile element*.

sulphur *sulfur*.

sulphur ore A mining term used for both pyrite and native sulfur.

sulvanite A bronze-yellow isometric mineral: Cu_3VS_4. Not to be confused with *sylvanite*.

sumacoite A magnetite-rich extrusive rock containing abundant phenocrysts of plagioclase and augite and rare olivine in a groundmass of andesine, orthoclase, nepheline, and hauyne. Its name is derived from Sumaco crater, Ecuador.

summation method In seismology, a correction of the arrival times of reflected waves for the time they spend in the low-velocity zone. On a continuous, reversed, and interlocked reflection profile between two holes shot at reasonable depths below the low-velocity zone, the low-velocity time under each geophone is equal to one half the sum of the first arrival times received at that geophone from both shot holes minus the average high-velocity time, which is obtained by subtracting the uphole time from the first arrival time at the geophone at the shothole in the opposite end of the profile.

summer balance The change in mass of a glacier from the maximum value in a certain year to the following minimum value of that year; sometimes called *apparent ablation* or (erroneously) *net ablation*. Cf: *winter balance*.

summer berm A berm built on the backshore by the uprush of waves during the summer. Cf: *winter berm*.

summer season In glaciology, that period of a year when the balance of a glacier decreases from a maximum value for the year to a minimum value for the year. This is a period when, on the average, ablation exceeds accumulation. Cf: *winter season*. Syn: *ablation season*.

summer surface An observable or measurable horizon (e.g. *dirt band* [glaciol]) in a glacier marking at each point the time of minimum mass of the glacier in a year's time. See also: *balance year; net balance*.

summerwood The denser, smaller-celled, later-formed portion of each annual ring in a woody stem. Also spelled: *summer wood*. Cf: *springwood*. Syn: *late wood*.

summit (a) The top, or the highest point or level, of an undulating land feature, as of a hill, mountain, volcano, or rolling plain; a peak. See also: *crest*. (b) Loosely, a divide or pass; e.g. Donner Summit, Calif.

summit concordance Equal or nearly equal elevation of ridgetops or mountain summits over a region. The concordance is commonly thought to indicate the existence of an ancient erosion plain of which only scattered patches are preserved. See also: *accordant summit level; even-crested ridge*. Syn: *accordance of summit levels; concordance of summit levels*.

summit graben A *volcanic graben* on the summit of a volcanic cone, more or less rectangular or triangular. Cf: *sector graben*.

summit level [eng] The highest point of a road, railroad, or canal; the highest of a series of elevations over which a road or canal is carried.

summit level [geomorph] The elevation of a summit plane. See also: *accordant summit level*.

summit plain *peak plain*.

summit plane The plane passing through a series of accordant summits. See also: *gipfelflur*.

sump [drill] A slush pit. Syn: *sump hole; sump pit*.

sump [geog] (a) A subterranean hole into which the drainage water of an area is collected and subsequently used either for irrigation or for wild-fowl conservation. (b) A dialectal term for a swamp or morass, and for a stagnant pool or puddle of dirty water. (c) An English term for a cove or a muddy inlet.

sump [speleo] (a) A pool of water in a cave, the exit or extension of which lies underneath it. (b) *water trap*.----Cf: *siphon*.

sun ball A ball of snow formed by downhill rolling of masses of snow loosened by the thawing action of the sun on a snow-covered slope. Cf: *snow roller*.

sun compass A navigational compass that determines the direction of the astronomic meridian mechanically and instantaneously from an observation on the Sun. It is used to establish direction esp. in high latitudes. Cf: *solar compass*.

sun crack A crack in sediment or rock, formed by the drying action of the Sun's heat; esp. a *mud crack*.

sun crust A type of *snow crust* formed by refreezing of surface snow that had been melted by the sun; it is usually thin and has a smooth surface.

sun cup A cuspate hollow or depression in a snow surface, formed during sunny weather by complex ablation processes. In some environments, sun cups may grow into *nieve penitente*.

sundtite *andorite*.

sungei A marine channel cutting across the Aroe Islands north of Australia, representing a Quaternary antecedent stream that was invaded by the Flandrian Transgression (Fairbridge, 1951). Etymol: Malay, "river, large stream". Syn: *soengei*.

sunken caldera *collapse caldera*.

sunken island A high relief feature of a lake basin, such as a basin divide or the crest of a knob, covered by a shallow depth of water; it was "never originally above water level" and is therefore "not due to subsidence" (Veatch & Humphrys, 1966, p.317). Cf: *blind island*.

sunken stream *lost stream*.

sun opal *fire opal*.

sun spike *nieve penitente*.

sunspot A relatively dark area on the Sun's surface representing lower temperature and consisting of a dark central umbra surrounded by a penumbra which is intermediate in brightness between the umbra and the surrounding surface of photosphere (NASA, 1966, p.47).

sunstone An *aventurine feldspar*, usually a brilliant, translucent variety of oligoclase that emits a reddish or golden billowy reflection from minute scales or flakes of hematite spangled throughout and arranged parallel to planes of repeated twinning. Cf: *moonstone*. Syn: *heliolite*.

suolunite A mineral: $Ca_2H_2Si_2O_7.H_2O$.

superanthracite *meta-anthracite*.

supercapillary interstice An *interstice* sufficiently larger than a *capillary interstice* that surface tension will not hold water far above a free water surface. Water moving in it, as by *supercapillary percolation*, may develop currents and eddies (Meinzer, 1923, p.18). Cf: *subcapillary interstice*.

supercapillary percolation Percolation through *supercapillary interstices*.

supercooling The process of lowering the temperature of a phase or assemblage below the point or range at which a phase change should occur at equilibrium, i.e. making the system metastable by lowering the temperature. It generally refers to a liquid taken below its liquidus temperature. Cf: *superheating*. Syn: *undercooling*.

supercritical Said of a system that is at a temperature higher than its critical temperature; also, said of the temperature itself.

supercritical flow *rapid flow*.

supercrop map A geologic map that shows the distribution of stratigraphic units lying immediately above a given rock body or a surface at a given time. Cf: *subcrop map*. Syn: *worm's-eye map*.

superface The top or upper surface of a stratigraphic unit.

superfacies A large-scale stratigraphic facies, generally consisting of two or more subordinate facies; e.g. a laterally equivalent and contrasting part of a formation, within which smaller-scale, laterally equivalent, and contrasting parts are recognized.

superfamily A unit in taxonomy that contains one or several families and that is considered equivalent to a *suborder* or between "suborder" and "family" in rank.

superficial Pertaining to, or lying on, occurring in, or affecting only, a surface or surface layer; e.g. "superficial weathering" of a rock surface, or a "superficial structure" formed in a sedimentary rock or sediment by surface creep. The term is used esp. in Great Britain, but the synonymous term *surficial* is more generally applied in the U.S.

superficial deposit *surficial deposit*.

superficial fold *décollement fold*.

superficial moraine *surficial moraine*.

superficial oolith An oolith with an incomplete or single layer; specif. an oolith in which the thickness of the accretionary coating is less than the radius of the nucleus (Beales, 1958, p.1863). Cf: *oopellet*.

superfluent lava flow A flow of lava issuing from a summit crater and streaming down the flanks of the volcano (Dana, 1890); an obsolete term. Cf: *effluent lava flow; interfluent lava flow*.

supergene Said of a mineral deposit or *enrichment* formed by descending solutions; also, said of those solutions and of that environment. Cf: *hypogene; mesogene*. Syn: *hypergene*.

superglacial Carried upon, deposited from, or pertaining to the top surface of a glacier or ice sheet; said of meltwater streams, till, drift, etc.

supergroup A formally named assemblage of related *groups*, or of formations and groups, having significant lithologic features in common (ACSN, 1961, art.9e). Cf: *megagroup*.

superheating (a) The addition of more heat than necessary to complete a given phase change. (b) In a magma, the addition of more heat than necessary to cause complete melting. The

temperature increase above liquidus is called the superhea(c) The process of increasing heat beyond that point at whic a phase or assemblage changes at equilibrium, i.e. to a meta stable state in the sense analogous to *supercooling*.

superimposed [stratig] Said of rocks that are layered or strati fied.

superimposed [streams] Said of a stream or drainage syster let down from above by erosion through the formations o which it was developed onto rocks of different structure lyin unconformably beneath. The term was first applied by Powe (1875, p.165-166) to the valley thus formed, although Ma (1866, p.443-444) earlier discussed the concept. Syn: *super posed; inherited; epigenetic*.

superimposed drainage Drainage by streams that were estab lished on a preexisting surface now eroded and whose cours es are unrelated to the geologic structures and underlying rocks on which they are now developed.

superimposed fan A term proposed by Blissenbach (1954 p.180-181) for a newly deposited alluvial fan that has a steep er gradient than the older fan upon which it is developed; results from tectonic movements that initiate a new stage o deposition.

superimposed fold A syn. of *cross fold*. Var: *superposed fold*.

superimposed ice Ice formed when meltwater percolate down through a snowpack on a glacier and refreezes at th base of the snowpack, or as it is trapped on a lower horizo of reduced permeability such as a firn-ice boundary. This ic appears at the surface of a glacier in the *superimposed ic zone*.

superimposed ice stream Ice from a tributary glacier tha rests upon the surface of a larger glacier but does not sin down into it, such as where a tributary glacier flows onto th surface of a trunk glacier. Cf: *inset ice stream*.

superimposed ice zone The area on a glacier or ice shee where surface melting occurs, much of the meltwater is refro zen at the base of the snowpack as *superimposed ice*, an the snowpack is removed exposing superimposed ice at th surface. The superimposed ice zone is bordered at higher al titudes by the *firn line*, and at lower altitudes by the equilib rium line.

superimposed metamorphism *polymetamorphism*.

superimposed profile A diagram on which a series of profiles drawn along several regularly spaced and parallel lines on a map are placed one on top of the other (Monkhouse & Wilkin son, 1952); it may emphasize such features as accordan summit levels and erosion platforms. Cf: *composite profile projected profile*.

superimposed stream A stream that was established on a new surface and that maintained its course despite different preex isting lithologies and structures encountered as it erodec downward into the underlying rocks. Syn: *superinducec stream*.

superimposed valley A valley eroded by or containing a *super imposed stream*.

superimposition [stratig] *superposition*.

superimposition [streams] The establishment, originally on a cover of rocks now removed by erosion, of a stream or drain age system on existing underlying rocks independently of their structure. Gilbert (1877, p.144) recognized superimposition from an unconformable cover of sediments, from a surface o alluvium, and from a surface produced by planation. Syn: *epigenesis*.

superincumbent Said of a *superjacent* layer, esp. one that is situated so as to exert pressure.

superindividual An aggregate of crystal grains whose lattices are not uniformly oriented but are less divergent from each other than from neighboring grains (Knopf and Ingerson 1938).

superinduced stream *superimposed stream*.

superjacent Said of a stratum situated immediately upon or over a particular lower stratum or above an unconformity. Ant: *subjacent*. See also: *superincumbent*.

superlattice The crystal lattice of an alloy, in which the repeat unit cell is a multiple of that representing the disordered solid solution. Syn: *superstructure*.

supermature Pertaining to the fourth and last stage of textural maturity (Folk, 1951); said of a *mature* clastic sediment whose well-sorted grains have become subrounded to well-rounded, such as a clay-free "supermature sandstone" whose sand-size quartz grains have an average roundness that exceeds 0.35 and that is presumed to form mainly as dune sands. Cf: *immature; submature*.

superparamagnetism The *paramagnetic* behavior of an assembly of extremely small particles of ferromagnetic or ferrimagnetic minerals.

superperiodicity A crystal-lattice period in an ordered superlattice which is a simple multiple of a corresponding direction in the disordered sublattice.

superposed A term introduced by McGee (1888) as a shortened form of *superimposed* (in regard to streams and drainage systems).

superposed fold Var. of *superimposed fold*.

superposition (a) The actual order in which rocks are placed or accumulated in beds one above the other, the highest bed being the youngest. (b) The process by which successively younger sedimentary layers are deposited on lower and older layers; also, the state of being superposed.---See also: *law of superposition*. Syn: *superimposition; supraposition*.

superprint *overprint*.

supersaturated permafrost Permafrost that contains more ice than the ground could possibly hold if the water were in the liquid state.

supersaturated solution A solution which contains more of the solute than is normally present when equilibrium is established between the saturated solution and undissolved solute.

superstage A time-stratigraphic unit of higher than stage rank (Schindewolf, 1959, p.68).

superstratum An overlying stratum.

superstructure [cryst] *superlattice*.

superstructure [tect] The upper structural layer in an orogenic belt, subjected to relatively shallow or near-surface deformational processes, in contrast to an underlying and more complexly deformed and metamorphosed *infrastructure*. Also spelled: *suprastructure*.

supersystem A time-stratigraphic unit of higher than system rank (ISST, 1961, p.27). See also: *erathem*.

superterranean Occurring on or above the Earth's surface. Syn: *superterrene; superterrestrial*.

superzone An assemblage of two or more informal stratigraphic zones of any kind (ISST, 1961, p.29). The term is used in cases where intercontinental correlation is rough and free-ranging or where detailed zonation cannot be carried out. Syn: *megazone*.

supplementary contour A dashed or dotted contour line drawn, at less than the regular interval between intermediate contours, to increase the topographic expression of an area, such as in an area of extremely low relief. Syn: *auxiliary contour*.

suppressed Marked or affected by, or showing, suppression, or being vestigial to the degree of not being evident superficially but whose presence in ancestral forms may be indicated by other features; e.g. said of aborted conodont denticles that could not develop into mature structures owing to crowded conditions along the growing edge of the structure.

suppressor In seismic apparatus, a circuit component which facilitates the recognition of early reflections by reducing the burst of energy associated with the refracted first arrival.

supra-anal plate The dorsal, posteriorly produced part of the telson of a branchiopod crustacean.

supracrustal Said of rocks that overlie the basement.

supraembryonic area Circular apical area over the megalospheric proloculus in some foraminifers of the family Orbitolinidae.

supragelisol *suprapermafrost layer*.

suprageneric name The name of any taxon above the level of genus, i.e. "family", "order", "class", "phylum", "kingdom".

supraglacial Situated or occurring at or immediately above the surface of a glacier or ice sheet; said of meltwater streams, till, drift, etc.

supralithion Animals that swim above rock but that are dependent on it as their source of food.

supralittoral Pertaining to the shore area marginal to the *littoral* zone, just above high-tide level. Syn: *supratidal*.

suprapelos Animals that swim above soft mud but that are dependent on it as their source of food.

suprapermafrost layer The layer of ground (or soil) above the permafrost, consisting of the active layer and, wherever present, taliks and the pereletok. Syn: *supragelisol*.

suprapermafrost water Ground water existing above the impervious permafrost table (Muller, 1947, p.222).

supraposition *superposition*.

suprapsammon Animals that swim above sand but that are dependent on it as their source of food.

suprastructure Var. of *superstructure*.

supratenuous fold A pattern of fold in which there is thickening at the synclinal troughs and thinning at the anticlinal crests. It is formed by differential compaction around the basement morphology. See also: *compaction fold*.

supratidal *supralittoral*.

suranal plate One of the first-formed and largest plates of the periproctal system of an echinoid, often filling the central and anterior parts of the periproct, but not recognizable in many echinoids. Syn: *suranal*.

surf (a) The wave activity in the *surf zone*. (b) A collective term for *breakers*.

surface (a) The exterior or outside part of the solid or liquid Earth; the top of the ground or the exposed part of a rock formation. (b) A two-dimensional boundary between geologic features or structures, such as a *bedding surface* or a *fault surface*, or an imaginary surface such as the *axial surface* of a fold; usually an internal boundary, rather than one occurring on the outside of a feature (Challinor, 1967, p.246). It need not be flat. Cf: *plane*.

surface anomaly A geophysical anomaly caused by an irregularity at the Earth's surface or in the near-surface zone which interferes with geophysical measurements; *surface interference*.

surface axis *axial trace*.

surface conductivity Conduction along the surfaces of certain mineral grains due to excess ions in the diffuse layer of adsorbed cations.

surface correction A correction of a geophysical measurement to remove the influence of varying surface elevation and of surface anomalies.

surface creep The slow downwind advance of large sand grains along a surface by impact of smaller grains in saltation, as in the shifting or movement of a sand dune. See also: *surficial creep*. Syn: *creeping; reptation*.

surface curve *surface profile*.

surface density (a) The density of the surface material within the range of the elevation differences of the gravitational surface. Both the Bouguer correction and the terrain correction depend on the density of the surface material. (b) A quantity (as mass and electricity) per unit area distributed over a surface.

surface detention *detention*.

surface drag *skin friction*.

surface drainage The removal, or the prevention of entry into the soil, of unwanted water from the surface of the ground by natural or artificial means, such as by grading and smoothing the land to remove barriers and to fill in depressions, by terracing or digging ditches, or by diverting runoff from adjacent areas to natural waterways.

surface factor *fineness factor.*

surface-fed intermittent stream An *intermittent stream* that receives water from some surface source, generally the melting of snow in a mountainous area. Cf: *spring-fed intermittent stream.*

surface forces Any of the forces acting over the surface of a body of material, measured in units of force per unit area; this force per unit area is stress. Cf: *body forces.*

surface geology (a) Geology and correlation of rock formations, structures, and other features as seen at the Earth's surface. Ant: *subsurface geology.* (b) *surficial geology.*

surface groove A *skeletal canal* in the form of a groove on the surface of the skeletal framework of a sponge and generally corresponding to an exhalant canal of the soft parts.

surface hoar A type of hoar consisting of leaf- or plate-shaped ice crystals formed directly on a snow surface by condensation from vapor. Cf: *depth hoar; ice flower.*

surface interference An anomaly in a geophysical measurement caused by a surficial or near-surface irregularity; a *surface anomaly.*

surface moraine *surficial moraine.*

surface of concentric shearing In flexural folding, the bedding plane along which slip occurs. Var: *concentric shearing surface.*

surface of no strain A surface along which the original configuration of an array of points remains unchanged after deformation of the body in which it occurs. In two-dimensional models sometimes referred to as the *neutral axis.* Syn: *neutral surface.*

surface of unconformity The surface of contact between two groups of rocks displaying an unconformable relationship, such as a buried surface of erosion or of nondeposition separating younger strata from underlying older rocks. Syn: *unconformity; hard ground.*

surface phase In metamorphism, a thin layer of rock having properties which may differ from those of the volume phases on either side (Turner and Verhoogen, 1960, p.461). Phase in this context is used in the geochemical sense. Syn: *volume phase.*

surface pipe The first length of *casing* set in a well, generally used to shut off and protect shallow freshwater sands from contamination by deeper saline waters. It also serves as a conductor of the drilling mud through loose, near-surface formations during deeper drilling. Syn: *surface string.*

surface profile The longitudinal profile assumed by the surface of a stream of water in an open channel. See also: *backwater curve.* Syn: *surface curve.*

surface runoff The *runoff* that travels over the soil surface to the nearest surface stream; runoff of a drainage basin that has not passed beneath the surface since precipitation. The term is misused when applied in the sense of direct runoff (Langbein & Iseri, 1960). Cf: *storm seepage; ground-water runoff.*

surface scattering layer An area of marine organisms near the ocean surface that scatter sound waves from an echo sounder. Cf: *shallow scattering layer; deep scattering layer.*

surface-ship gravimeter An instrument designed to produce useful gravity observations aboard a surface ship underway. Due to large rotations and accelerations encountered, a complex system is required to stabilize the gravity sensor and by filtering to distinguish between inertial forces and the gravity value sought. Accuracies attained range from 1 to 5 milligals depending largely on quality of the navigation and state of the sea.

surface slope The inclination of the water surface expressed as change of elevation per unit of slope length; the sine of the angle which the water surface makes with the horizontal. The

tangent of that angle is ordinarily used; no appreciable error results except for the steeper slopes (ASCE, 1962).

surface soil The upper five to eight inches of a soil; that region or depth of a soil which is tilled. Cf: *subsoil.* Partial syn: *topsoil.*

surface texture The aggregate of the minute or minor, genetic features (roughness) of the surfaces of sedimentary particles, independent of size, shape, or roundness; e.g. polish, frosting, and striations.

surface velocity The velocity of a seismic wave in the Earth's surface layer.

surface wash *sheet erosion.*

surface water [oceanog] A *water mass* of varying salinity and temperature, occurring at the ocean surface or up to 300 meters depth. Cf: *intermediate water; deep water; bottom water.*

surface water [water] All waters on the surface of the Earth, including fresh and salt water, ice and snow.

surface wave [seis] A *seismic wave* that travels along the surface of the Earth, or parallel to the Earth's surface. Surface waves include *Rayleigh waves, Love waves,* and *coupled waves.* See also: *hydrodynamic waves.* Cf: *body wave; guided wave.* Obs syn: *circumferential wave; long wave; large wave.* Syn: *L wave.*

surface wave [water] (a) A progressive gravity wave in which the particle movement is confined to the upper limits of a body of water; strictly, a gravity wave whose celerity is a function of wavelength only. (b) *deep-water wave.*

surf base The depth at which a wave begins to peak and, under storm conditions, to break; it is usually 10-20 meters. Cf: *wave base.* Syn: *surge base.*

surf beat A wind-generated ocean wave, traveling shoreward, caused by interference of two different wind-wave trains, and usually associated with the onset of heavy surf. It has a long wave period (1-10 min), and a beat frequency and amplitude of a few centimeters.

surficial Pertaining to, situated at, or formed or occurring on, a surface, esp. the surface of the Earth. Cf: *subsurface.* See also: *subaerial.* Syn: *superficial.*

surficial creep A syn. of *soil creep.* See also: *surface creep.*

surficial deposit Unconsolidated and residual, alluvial, or glacial deposits lying on bedrock or occurring on or near the Earth's surface; it is generally unstratified and represents the most recent of geologic deposits. Syn: *surface deposit; superficial deposit.*

surficial geology Geology of *surficial deposits,* including soils; the term is sometimes applied to the study of bedrock at or near the Earth's surface. See also: *surface geology.*

surficial moraine A moraine, such as a lateral or medial moraine, in transit on the surface of a glacier. Syn: *superficial moraine; surface moraine.*

surf ripple A general term proposed by Kuenen (1950, p.292) for a ripple mark formed on a sandy beach by wave-generated currents in the surf zone.

surfusion An obsolete term proposed by Fournet (1844) for a condition under which the fusing points of substances are lowered to temperatures much below the points at which they usually solidify (see Zittel, 1901, p.342).

surf zone The area bounded by the landward limit of wave uprush and the farthest seaward breaker. Syn: *surf; breaker zone.*

surge [glaciol] The period of very rapid flow of a *surging glacier;* also, the displacement or advance of ice resulting from the very rapid flow. Syn: *glacier surge; catastrophic advance.*

surge [waves] (a) *storm surge.* (b) Horizontal oscillation of water with a comparatively short period, accompanying a seiche.

surge [hydraul] In fluid flow, long-interval variations in velocity and pressure that are not necessarily periodic and may even be transient.

surge base *surf base.*

surge channel A transverse channel that cuts across the outer edge of an organic reef and in which the water level rises and falls as the result of wave and tidal action.

surging breaker A type of *breaker* that peaks up and then surges onto the beach face without spilling or plunging. It forms over a very steep bottom gradient. Cf: *plunging breaker; spilling breaker.*

surging glacier A glacier which alternates periodically between brief periods (usually one to four years) of very rapid flow called surges and longer periods (usually 10 to 100 years) of near stagnation. During a *surge* [glaciol], a large volume of ice from an ice reservoir area is rapidly (up to several meters per hour) displaced downstream into an ice receiving area, and the affected portion of the glacier is chaotically crevassed. Only in exceptional cases does the glacier advance beyond its former limit. In the interval between surges the ice reservoir is slowly replenished by accumulation and normal ice flow and the ice in the receiving area is greatly reduced by ablation.

sursassite A mineral: $Mn_5Al_4Si_5O_{21}.3H_2O$ (?).

survey v. To determine and delineate the form, extent, position, boundaries, value, or nature of a tract of land, coast, harbor, or the like, esp. by means of linear and angular measurements and by the application of the principles of geometry and trigonometry.---n. (a) The orderly and exacting process of examining, determining, finding, and delineating the physical or chemical characteristics of the Earth's surface, subsurface, or internal constitution by topographic, geologic, geophysical, or geochemical measurements; esp. the act or operation of making detailed measurements for determining the relative positions of points on, above, or beneath the Earth's surface. (b) The associated data or results obtained in a survey; a map or description of an area obtained by surveying. (c) An organization engaged in making a survey; e.g. a government agency such as the U.S. Geological Survey or the U.S. Coast and Geodetic Survey.

surveying (a) The art of making a survey; specif. the applied science that teaches the art of making such measurements as are necessary to determine the area of any part of the Earth's surface, the lengths and directions of the boundary lines, and the contour of the surface, and of accurately delineating the whole on paper. (b) The act of making a survey; the occupation of one that surveys.

surveyor's compass A surveying instrument used for measuring horizontal angles; specif. one designed for determining a magnetic bearing of a line of sight by the use of a sighting device, a graduated horizontal circle, and a pivoted magnetic needle. The surveyor's compass used on the early land surveys in U.S. had a pair of peep sights to define the line of sight and was usually mounted on a Jacob's staff. See also: *circumferentor.* Syn: *land compass.*

surveyor's cross A simple surveying instrument used in establishing right angles, consisting of two bars forming a right-angled cross with sights at each end.

surveyor's level A *leveling instrument* consisting of a telescope (with cross hairs) and a spirit level mounted on a tripod, revolving on a vertical axis, and having leveling screws that are used to adjust the instrument to the horizontal.

surveyor's measure A system of measurement used in land surveying, having the surveyor's chain (one chain = 4 rods = 66 feet = 100 links = 1/80 mile) as a unit.

surveyor's rod *level rod.*

susannite A rhombohedral mineral: $Pb_4(SO_4)(CO_3)_2(OH)_2$. It is dimorphous with leadhillite.

susceptibility [elect] The ratio of the electric polarization to the electric intensity in a polarized dielectric.

susceptibility [mag] The ratio of *induced magnetization* to the strength H of the magnetic field causing the magnetization. See also: *susceptibility anisotropy.* Syn: *magnetic susceptibility; volume susceptibility.*

susceptibility anisotropy In minerals of low crystal symmetry or in rocks with planar or linear fabric: magnetic *susceptibility* which is not perfectly parallel with the inducing magnetic field, because it depends on direction and induced magnetization. Syn: *magnetic anisotropy.*

suspended current A *turbidity current,* in a body of standing water, that is not in contact with the bottom (as on a slope or where the current overrides denser underlying water) and that continues in suspension (Dzulynski & Radomski, 1955).

suspended load (a) The part of the total *stream load* that is carried for a considerable period of time in suspension, free from contact with the stream bed; it consists mainly of mud, silt, and sand. (b) The material collected in, or computed from samples collected with, a suspended-load sampler.--- Syn: *suspension load; suspensate; silt load; wash load.*

suspended-load sampler A device that collects a sample of water with its sediment load without separating the sediment from the water.

suspended water *vadose water.*

suspensate *suspended load.*

suspension (a) A mode of sediment transport in which the upward currents in eddies of turbulent flow are capable of supporting the weight of undissolved sediment particles and keeping them indefinitely held in the body of the surrounding fluid (such as silt in water or dust in air). Cf: *traction.* (b) The state of a substance in such a mode of transport; also, the substance itself.

suspension current *turbidity current.*

suspension feeder An animal that feeds by selecting microorganisms and detritus suspended in the surrounding water.

suspension flow Flow of a water-sediment mixture in which the sediment is maintained in suspension by the combination of water turbulence and relatively low settling velocities of particles. It commonly occurs in a turbidity current.

suspension load *suspended load.*

suspensive lobe The visible external part of the umbilical lobe of an ammonoid suture on the exposed part of a whorl, comprising the portion from which auxiliary lobes spring (TIP, 1959, pt.L, p.6).

suspensor A structure in the embryo sporophyte of higher plants, which attaches or forces the embryo into gametophyte or endosperm tissue (Fuller & Tippo, 1949, p.973).

sussexite [mineral] A white mineral: $(Mn,Mg)BO_2(OH)$. It is isomorphous with szaibelyite.

sussexite [petrology] A porphyritic tinguaitic rock composed chiefly of nepheline and acmite and free of appreciable feldspar. Its name is derived from Sussex County, New Jersey, U.S.A.

sustained runoff *base runoff.*

sutural Pertaining to a suture, or corresponding to sutures in position; e.g. "sutural supplementary apertures" in a foraminiferal test.

sutural element One of the major undulations of a suture of a cephalopod shell, such as a lobe or a saddle.

sutural pore One of the pores bordered by sutures along the meeting branches of two or more adjacent radial spines in an acantharian radiolarian. Cf: *parmal pore.*

suture [paleont] (a) The line of junction of a septum of a cephalopod's shell with the inner surface of the shell wall. It is commonly more or less undulated or plicated, and is visible (as in an ammonoid) only when the shell wall is removed. (b) The line of contact between two whorls of a gastropod shell. It is typically a spiral on the outer surface, and also on the inner surface around the axis of coiling where the umbilicus is not closed. (c) A plane of junction between adjacent plates of an echinoderm. Also, the boundary line on the surface marking the area of contact between the adjacent plates. (d) A line of demarcation between two fused or partly fused limb segments or body somites of a crustacean; e.g. a line or seam at the juncture of two compartmental plates of a cirripede, or a weak or uncalcified narrow seam between parts of a trilobite exoskeleton which may separate at time of molting or after death. (e) A line of contact between two chambers or two whorls of a foraminiferal test. It may be reflected on the outer wall by a groove- or ridge-like feature. (f) The boundary between segments of heterococcoliths.

suture [bot] On a fruit, the line of dehiscence or splitting.

suture [palyn] The line along which the laesura of an embryophytic spore opens on germination; loosely, the *laesura*. See also: *commissure*.

sutured Said of the texture found in igneous, metamorphic, and sedimentary rocks in which mineral grains or irregularly shaped crystals interfere with their neighbors, producing interlocking, irregular contacts without interstitial spaces, resembling the sutural structures in bones. Also, said of the crystal contacts in rock with this texture. Syn: *consertal*.

suture joint A very small *stylolite*.

svabite A colorless, yellowish-white, or gray mineral of the apatite group: $Ca_5(AsO_4)_3F$. It may contain phophorus, lead, magnesium, or manganese.

svanbergite A colorless to yellow, rose, or reddish-brown rhombohedral mineral: $SrAl_3(PO_4)(SO_4)(OH)_6$. It is isomorphous with corkite, hinsdalite, and woodhouseite.

Svecofennian A division of the Proterozoic of the Baltic Shield.

sviatonossite An andradite syenite in which the pyroxene is acmite-augite and the plagioclase is oligoclase.

swab n. A piston-like device equipped with an upward-opening check valve and provided with flexible rubber suction cups, lowered into a borehole or casing by means of a wire line for the purpose of cleaning out drilling mud or of lifting oil.---v. To draw out oil from a well by means of a swab; to clean with a swab.

swag (a) A shallow pocket or closed depression in flat or gently rolling terrain, often filled with water (as in the bottomlands of the lower Mississippi Valley). (b) A shallow water-filled hollow produced by subsidence resulting from underground mining.----Cf: *sag pond*.

swale (a) A slight depression, sometimes swampy, in the midst of generally level land. (b) A shallow depression in an undulating ground moraine due to uneven glacial deposition. (c) A long, narrow, generally shallow, trough-like depression between two *beach ridges*, and aligned roughly parallel to the coastline. Syn: *low; furrow; slash; runnel*.

swallet (a) *swallow hole*. (b) An area in which water disappears underground, as into alluvium; it differs from a swallow hole in that it is not a depression.----Cf: *stream sink*.

swallow hole (a) A closed depression or *doline* into which all or part of a stream disappears underground. Partial syn: *swallet; ponor*. (b) An underwater outlet of a lake.

swamp A watersaturated area, intermittently or permanently covered with water, having shrub- and tree-type vegetation, essentially without peatlike accumulation. Cf: *marsh; bog*.

swamp ore *bog iron ore*.

swamp theory *in-situ theory*.

swarm [seism] *earthquake swarm*.

swarm [intrus rocks] *dike swarm*.

swarm earthquakes *earthquake swarm*.

swartzite A green monoclinic mineral: $CaMg(UO_2)(CO_3)_3.12H_2O$.

swash bar (a) A small, transitory bar built above the stillwater level by wave uprush, and forming a tiny lagoon on its landward side (King & Williams, 1949, p.81). (b) A bar over which the sea washes. Syn: *swash*.

swash channel (a) A narrow sound or secondary channel of water lying within a sandbank or between a sandbank and the shore, or passing through or shoreward of an inlet or river bar. Syn: *swash; swashway; swatch*. (b) A channel cut on an open shore by flowing water in its return to the parent body.

swash mark A thin, delicate, wavy or arcuate (convex landward) line or very small ridge on a beach, marking the farthest advance of wave uprush, and consisting of fine sand, mica flakes, bits of seaweed, and other debris. Syn: *wave line; wavemark; debris line*.

swash pool A shallow pool of water formed behind a swash bar.

swashway *swash channel*.

swash zone The sloping part of the beach that is alternately covered and uncovered by the uprush of waves, and where longshore movement of water occurs in a zigzag (upslope downslope) manner.

swatch A British syn. of *swash channel*. Syn: *swatchway*.

S wave That type of seismic *body wave* which is propagated by a shearing motion of material, so that there is oscillation perpendicular to the direction of propagation. It does not travel through liquids, or through the outer core of the Earth. Its speed is 3.0-4.0 km/sec in the crust and 4.4-4.6km/sec in the upper mantle. The *S* stands for secondary; it is so named because it arrives later than the *P wave* (primary body wave). Syn: *shear wave; secondary wave; rotational wave; tangential wave; equivoluminal wave; distortional wave; transverse wave; shake wave*.

swedenborgite A colorless to wine-yellow mineral: $NaBe_4SbO_7$.

Swedish mining compass A compass in which a magnetic needle is suspended on a jewel and a stirrup so that it can rotate about both a horizontal and a vertical axis.

sweeping The progressive down-valley movement or shift of a system of meanders. See also: *wandering*. Syn: *sweep*.

sweepstakes route A biogeographic dispersal path which constitutes a formidable obstacle to the migration of plants and animals but which is occasionally conquered. The term implies that the odds against crossing the barrier are as great as those against someone winning a lottery, i.e., great but not impossible.

sweet Said of crude oil or natural gas that contains little or no sulfur compounds. Cf: *sour*.

sweet water *fresh water*.

swell [ore dep] An enlarged place in, or part of, an orebody, as opposed to a *pinch*.

swell [eng geol] The increase in volume of soil or hard rock upon excavation or absorption of water; the tendency of soil to increase in volume on being removed from their natural compacted state.

swell [waves] (a) One of a series of regular, long-period somewhat flat-crested waves that has traveled out of its generating area. See also: *ground swell*. (b) The slow and regular undulation of the surface of the open ocean; a series of unbroken waves.----Ant: *sea*.

swell [marine geol] *rise* [marine geol].

swell [struc geol] A general, imprecise term for *dome* and *arch*.

swell *horseback*.

swell-and-swale topography A low-relief, undulating landscape characteristic of the ground moraine of a continental glacier exhibiting gentle slopes and well-rounded hills interspersed with shallow depressions. Cf: *sag-and-swell topography*.

swelling The increase in volume of surficial deposits due to frost action. See also: *residual swelling*.

swelling chlorite A chlorite-like mineral, found in clays, that behaves like a chlorite on heating but has its basal spacing expand on glycerol treatment. It contains incomplete hydroxide (brucite or gibbsite) layers, and might be regarded as a special interlayering of chlorite with smectite or vermiculite (Martin Vivaldi & MacEwan, 1960). See also: *corrensite*. Syn: *pseudochlorite*.

swelling clay Clay that is capable of absorbing large quantities of water; e.g. bentonite.

swelling ground Soil or rock that contains a large amount of clay and that advances (flows plastically) principally because of volume expansion, as when wetted. Cf: *squeezing ground.*

swelling pressure The pressure exerted by a clay or shale when it absorbs water in a confined space.

swimming leg The hindmost prosomal appendage of a merostome, serving as a swimming organ.

swimming stone *floatstone* [mineral].

swinestone *anthraconite.*

swinging The steady lateral movement of a meander belt from one side of the valley floor to the other. See also: *wandering.* Syn: *swing.*

swinging dip *migrating dip.*

swing mark A circular or semicircular sedimentary structure formed by wind action on an anchored plant stem or root that is swept to and fro, or round and round, across a sandy surface.

swither A colloquial term used in the Wisconsin lead-mining region for an offshoot or branch of a main lode.

switzerite A mineral: $(Mn,Fe)_3(PO_4)_2 \cdot 4H_2O$.

sword dune *seif dune.*

sychnodymite *carrollite.*

sycon A sponge or sponge larva in which separate flagellated chambers open directly (without the intervention of exhalant canals) into a central cloaca lined with pinacoderm. Cf: *ascon; leucon.* Adj: *syconoid.*

syenide An informal term, for field use, applied to any holocrystalline, medium- to coarse-grained igneous rock containing one or more feldspars and, generally, biotite or hornblende. The dark minerals comprise less than half of the rock. The term includes syenites and light-colored diorites.

syenite A group of plutonic rocks containing alkali feldspar (usually orthoclase, microcline, or perthite), a small amount of plagioclase (less than in *monzonite*), one or more mafic minerals (esp. hornblende), and quartz, if present, only as an accessory; also, any rock in that group; the intrusive equivalent of *trachyte.* With an increase in the quartz content, syenite grades into *granite.* Its name is derived from Syene, Egypt, where the rock was quarried in ancient times.

syenodiorite A group of plutonic rocks intermediate in composition between syenites and diorites, containing both alkali feldspar (usually orthoclase) and plagioclase feldspar, usually more of the former; also, any rock in that group. Generally considered a syn. of *monzonite,* but may also include both monzonite and rocks intermediate between monzonites and diorites (Streckeisen, 1967, p.170).

syenogabbro A plutonic rock differing in composition from a gabbro by the presence of orthoclase.

syke Var. of *sike.*

sylvanite A steel-gray, silver-white, or brass-yellow monoclinic mineral: $(Au,Ag)Te_2$. It often occurs in implanted crystals resembling written characters. Not to be confused with *sulvanite.* Syn: *silvanite; graphic tellurium; white tellurium; yellow tellurium; goldschmidtite.*

sylvine *sylvite.*

sylvinite A mixture of halite and sylvite, mined as a potassium ore; a rock that contains chiefly impure potassium chloride.

sylvite A white or colorless isometric mineral: KCl. It is the principal ore of potassium. Sylvite occurs in cubes or crystalline masses or as a saline residue, and it has a sharper taste than that of halite. Syn: *sylvine; leopoldite.*

symbiosis The relationship that exists between two different organisms that live in close association without either being harmed. Cf: *parasitism; mutualism; commensalism.* Adj: *symbiotic.*

symbiotic The adj. of *symbiosis.*

symbol A diagram, design, letter, color hue, abbreviation, or other graphic device placed on maps and charts, which by convention, usage, or reference to a legend is understood to signify or represent a specific characteristic, feature, or

object, such as structural data, rock outcrops, or mine openings.

symmetrical fold A fold, the limbs of which have about the same angle of dip relative to the axial surface, which is essentially vertical. Cf: *asymmetric fold.* Syn: *normal fold.*

symmetric bedding Bedding characterized by lithologic types or facies that follow each other in a "retracing" arrangement illustrated by the sequence 1-2-3-2-1-2-3-2-1. Cf: *asymmetric bedding.*

symmetric ripple mark A *ripple mark* having a symmetric profile in cross section, being similarly shaped on both sides of the crest which, in plan view, is predominantly straight; specif. *oscillation ripple mark.* Ant: *asymmetric ripple mark.*

symmetric spread *split spread.*

symmetry [cryst] The repeat pattern of similar crystal faces that indicates the ordered internal arrangement of a crystalline substance.

symmetry [paleont] Correspondence in size, shape, and relative position of organs or parts that are on opposite sides of a dividing line or median plane or that are distributed about a center or axis (Webster, 1967, p.2317); e.g. *bilateral symmetry* and *radial symmetry.*

symmetry [struc petrol] A symmetry of fabric around a plane or an axis, as displayed in a hand specimen or a fabric diagram.

symmetry axis In a crystal, an imaginary line about which the crystal may be rotated, during which there may be two, three, four, or six repetitions of its appearance (lines, angles, or faces); it is one of the symmetry elements. Syn: *axis of symmetry; rotation axis; symmetry axis of rotation.*

symmetry axis of rotary inversion *rotoinversion axis.*

symmetry axis of rotation *symmetry axis.*

symmetry center *center of symmetry.*

symmetry elements The axes, plane, and center of symmetry, by which crystal symmetry can be described. There are 32 possible arrangements of the elements of symmetry; each arrangement is a crystal class. Syn: *elements of symmetry.*

symmetry operations Various movements of a crystal that bring it into coincidence with its original position; these are rotation about an axis, reflection across a plane, inversion, and rotary inversion (Dana, 1959, p.17).

symmetry plane *plane of mirror symmetry.*

symmetry principle The statement that if a geometric configuration is the result of a particular cause, such as a vector field, then the symmetry of the cause cannot be less (lower) than the symmetry of the resulting configuration (Paterson & Weiss, 1961, p.845). The principle is used in structural petrology.

symmict Said of a structureless sedimentation unit, as in a varved clay or a graded series, composed of material in which coarse- and fine-grained particles are mixed to a greater extent in the lower part due to rapid flocculation. Also, said of the sedimentary structure so formed. Syn: *symminct.*

symmictite [ign] A term used by Sederholm (1924, p.148) for a homogenized eruptive breccia composed of a mixture of country rock and intrusive rock.

symmictite [sed] *diamictite.*

symmicton *diamicton.*

symminct Var. of *symmict.*

symmixis Flocculation induced in sediments by certain electrolytes (esp. sodium chloride), resulting in the mixing of silt and clay particles and the formation of a nearly homogeneous or nonlaminated clay.

symon fault A syn. of *horseback,* named after such a structure in the Coalbrookdale coalfield of England that was originally thought to be a large fault.

sympatric Said of populations occupying the same area without losing their identity as a result of interbreeding. Noun: *sympatry.* Cf: *allopatric.*

sympatric species Related species whose geographic distributions overlap or coincide.

symphrattism A term proposed by Grabau (1904) for regional or dynamothermal metamorphism. The term is now obsolete. Cf: *aethoballism*.

symphytium A single plate formed in certain brachiopods by fusion of deltidial plates dorsally or anteriorly from the pedicle foramen and lacking a median line of junction.

symplectic Said of a rock texture produced by the intimate intergrowth of two different minerals, sometimes restricted to those of secondary origin. One of the minerals may assume a vermicular habit. Also, said of a rock exhibiting such texture or of the intergrowth itself, i.e. *symplectite*. Also spelled: *symplektic; symplectitic; symplektitic*. Cf: *dactylitic*.

symplectite An intimate intergrowth of two different minerals, sometimes restricted to those of secondary origin; also, a rock (igneous or thermally metamorphosed) characterized by *symplectic* texture. Also spelled: *symplektite*. Cf: *pegmatite*. Nonpreferred syn: *implication*.

symplesite A blue, bluish-green, or pale-indigo triclinic mineral: $Fe_3(AsO_4)_2.8H_2O$. Cf: *parasymplesite; ferrisymplesite*.

sympod The *protopod* of a crustacean. Syn: *sympodite*.

sympodium A plant stem made up of a series of branches that grow on each other. It gives the effect of a simple stem.

symptomatic mineral *diagnostic mineral.*

synadelphite A black mineral: $(Mn,Mg,Ca,Pb)_4(AsO_4)(OH)_5$.

synaeresis Var. of *syneresis*.

synangium (a) An aggregate fern sporangium that forms a series of locules. (b) An anther in the plant genus *Ephedra* (Jackson, 1953, p.374).

synantectic Said of a primary mineral formed by the reaction of two other minerals, as in the formation of a reaction rim. See also: *deuteric*.

synantexis *Deuteric* alteration.

synapticula (a) One of numerous, small, conical or cylindrical, transverse, calcareous, spine-like projections (rods or bars) that extend between and connect the opposed faces of adjacent septa of some corals and that perforate the mesenteries between them. A "compound synapticula" consists of a broad bar formed by fusion of opposed ridges on adjacent septa. The term is also used as a plural of *synapticulum*. (b) An anaxial bar of secondarily secreted silica connecting adjacent spicules in hexactinellid sponges. (c) A rod-like structure extending between septa of an archaeocyathid.---Pl: *synapticulae*.

synapticulotheca The porous outer wall of a scleractinian corallite, formed by union of one or more rings of synapticulae along the axis of trabecular divergence. Cf: *septotheca; paratheca*.

synapticulum A coral synapticula. Pl: *synapticula*.

syncarpous Said of a plant ovary having two or more united carpels; also, said of partially united pistils within a flower (Lawrence, 1951, p.772). Cf: *apocarpous*.

synchisite *synchysite*.

synchronal *synchronous*.

synchrone (a) A zone representing equal time. (b) A stratigraphic surface on which every point has the same geologic age; a *time plane*.

synchroneity The state of being *synchronous* or simultaneous; coincident existence, formation, or occurrence of geologic events or features in time, such as "glacial synchroneity". Syn: *synchronism*.

synchronism *synchroneity*.

synchronogenic Formed in the same part of geologic time (R.C. Moore, 1958, p.21); e.g. said of rocks possessing identical or nearly identical geologic ages. Cf: *syntopogenic*.

synchronous Occurring, existing, or formed at the same time; contemporary or simultaneous. The term is applied to rock surfaces on which every point has the same geologic age, such as the boundary between two ideal time-stratigraphic units in continuous and unbroken succession. It is also ap-

plied to growth (or depositional) faults and to plutons emplaced contemporaneously with orogenies. Cf: *isochronous; diachronic*. Syn: *synchronal; synchronic*.

synchysite A mineral: $(Y,Ce)Ca(CO_3)_2F$. It is related to parisite. Syn: *synchisite*.

synclinal n. An obsolete form of *syncline*. adj. Pertaining to syncline.

synclinal axis *trough surface*.

syncline A fold, the core of which contains the stratigraphically younger rocks; it is concave upward. Ant: *anticline*. See also: *synform; synclinal*.

synclinorium A composite synclinal structure of regional extent composed of lesser folds. Cf: *anticlinorium*. See also *geosyncline*. Pl: *synclinoria*.

syncolpate Said of pollen grains in which the colpi join, normally near the pole.

syndeposition A term used by Chilingar et al (1967, p.322) for that part of *syngenesis* comprising "processes responsible for the formation of the sedimentary framework". Cf: *prediagenesis*.

syndepositional fold A fold structure that forms contemporaneously with sedimentation. It is a feature associated with sedimentary tectonics.

syndiagenesis A term used by Bissell (1959) for the sedimentational, prediastrophic phase of diagenesis, including alterations occurring during transportation (halmyrolysis) and during deposition of sediments, and continuing through the early stages of compaction and cementation but ending before that of deep burial (less than 100 m). It is characterized by the presence of large amounts of interstitial or connate water that is expelled only very slowly and by extreme variations in pH and Eh. It is equivalent to *early diagenesis*. See also: *epidiagenesis; anadiagenesis*. Cf: *syngenesis*. Adj: *syndiagenetic*.

syndromous load cast A term used by Haaf (1959, p.48) for an elongate, shallow load cast having sharp creases that combine to form a dendritic pattern, the junctures always occurring downcurrent.

syneclise A negative or depressed structure of the continental platform; it is of broad, regional extent (tens to hundreds of thousands of square kilometers) and is produced by slow crustal downwarp during the course of several geologic periods. The term is used mainly in the Russian literature, e.g. the Caspian syneclise. Ant: *anteclise*.

synecology The study of the relationships between communities and their environments. Cf: *autecology*.

syneresis The spontaneous separation or throwing off of a liquid from or by a gel or flocculated colloidal suspension during aging, resulting in shrinkage and in the formation of cracks, pits, mounds, cones, or craters. Syn: *synaeresis*.

syneresis crack A *shrinkage crack* formed by the spontaneous throwing off of water by a gel during aging.

syneresis vug A vug formed by syneresis, esp. in a sedimentary carbonate rock.

synform A synclinal-type structure, the stratigraphic sequence of which is unknown. Cf: *syncline*. Ant: *antiform*.

syngenesis (a) A term introduced by Fersman (1922) for the formation, or the stage of accumulation, of unconsolidated sediments in place, including the changes affecting detrital particles still in movement in the waters of the depositional basin. The term is in dispute by Russian geologists (Dunoyer de Segonzac, 1968, p.170): some would apply it to initial diagenesis (designating exchange phenomena between fresh sediment and the sedimentary environment), others would extend it to all the transformations undergone by a sediment before its compaction. The term is equivalent to *early diagenesis* as used in the U.S. Cf: *syndiagenesis*. (b) A term used by Chilingar et al (1967, p.322) for the "processes by which sedimentary rock components are formed simultaneously and penecontemporaneously"; it includes *syndeposition* and *prediagenesis*.

syngenetic [ore dep] Said of a mineral deposit formed contemporaneously with the enclosing rocks. Cf: *epigenetic; xplogenetic.* Syn: *idiogenous.*

syngenetic [sed] (a) Said of a primary sedimentary structure (such as a ripple mark) formed contemporaneously with the deposition of the sediment. (b) Pertaining to sedimentary syngenesis.---Cf: *epigenetic.*

syngenetic karst Karst that has developed simultaneously with the lithification of dune sands, or eolian calcarenite (Monroe, 1970).

syngenite A colorless or white monoclinic mineral: $K_2Ca(SO_4)_2.H_2O$.

synglyph A hieroglyph formed contemporaneously with sedimentation (Vassoevich, 1953, p.33).

synkinematic *syntectonic.*

synneusis Said of the texture of a rock in which some of the crystal components are aggregated in clusters. This texture differs from *glomeroporphyritic* and *cumulophyric* in that it need not involve a porphyritic rock or have the mineral clusters form the phenocrysts. Etymol: Greek, "to swim together". Cf: *gregaritic.*

synonym In taxonomy, a scientific name that is rejected in favor of another name because of being incorrect (e.g. in form, in spelling) or because of the evidence of the priority of the other name. See also: *synonymy.*

synonymy (a) In taxonomy, the relationship existing between two or more different names that have been applied to the same taxon. (b) Also, a list of *synonyms* that have been applied to a particular taxon.

synoptic Pertaining to simultaneously existing, meteorologic conditions that together give a description of the weather; also, said of a weather map or chart that shows such conditions.

synoptic oceanography Continuous gathering and reporting of simultaneous oceanographic data. It has become more important and feasible with the development of satellites. Syn: *hydropsis.*

synorogenic Said of a geologic process or event occurring during a period of orogenic activity; or said of a rock or feature so formed. In an orogenic belt, synorogenic sediments are preserved mainly in the external parts of the belt; they include flysch as strictly defined, accumulating in narrow, deep, rapidly subsiding troughs, as well as broad sheets of deposit spread far from their orogenic sources and onto the foreland. Synorogenic plutonic rocks include concordant granites which have replaced and partly disrupted the rocks of the internal parts of the foldbelts. Cf: *syntectonic.*

synrhabdosome A colony of graptolites made up of rhabdosomes; e.g. a growth association of biserial graptoloid rhabdosomes attached distally by their nemas around a common center (TIP, 1955, pt.V, p.7).

synsedimentary fault *growth fault.*

syntactic Recommended adj. of *syntaxy.*

syntactic growth *syntaxy.*

syntaphral Descriptive of a type of tectonics involving gravity sliding of unconsolidated sediments toward the axis of a geosyncline (Carey, 1963, p.A6). Cf: *diataphral; apotaphral.*

syntaxial Adj. of *syntaxy.*

syntaxial rim An optically oriented crystal overgrowth of a detrital grain, developed during diagenesis.

syntaxic Adj. of *syntaxy.*

syntaxis A sheaf-like pattern, as shown on a map, of mountain ranges converging at a common center. Ant: *virgation.*

syntaxy Crystallographically oriented intergrowth of two alternating, chemically identical substances that crystallized simultaneously, and whose corresponding edges or axes are in the ratio of small integers. Donnay proposed abandoning the strict requirement of chemical identity in such intergrowths (1953). Adj: *syntactic; syntaxic; syntaxial.* Cf: *topotaxy; epitaxy.* Syn: *syntactic growth.*

syntectic The adj. of *syntexis.*

syntectite A rock formed by syntexis. See also: *anatexite; protectite.*

syntectonic Said of a geologic process or event occurring during any kind of tectonic activity; or said of a rock or feature so formed. Cf: *synorogenic.* Syn: *synkinematic.*

syntexis (a) The formation of magma by melting of two or more rock types and assimilation of country rock; *anatexis* of two or more rock types. (b) Modification of the composition of a magma by *assimilation.*----(Dietrich & Mehnert, 1961). Adj: *syntectic.*

synthetic [gem] adj. Said of a substance produced artificially, such as a gemstone (ruby, sapphire) made by the Verneuil process, or a diamond produced by subjecting a carbonaceous material to extremely high temperature and pressure.--n. Something produced by synthesis; a *synthetic stone.*

synthetic [fault] Pertaining to minor normal faults that are of the same orientation as the major fault with which they are associated. Ant: *antithetic.* Syn: *homothetic [fault].*

synthetic antenna The effective antenna produced by storing and comparing the *doppler signals* received while the aircraft travels along its flight path. This synthetic antenna is many times longer than the physical antenna, thus sharpening the effective beam width and improving azimuth resolution.

synthetic group A rock-stratigraphic unit consisting of two or more formations that are associated because of similarities or close relationships between their fossils or lithologic characters (Weller, 1960, p.434). Cf: *analytic group.*

synthetic hydrology A catchall term for new techniques in hydrologic analysis involving the generation of hydrologic information or sequences of hydrologic events by means other than direct measurement or observation. It has been suggested that the term be abandoned and replaced by the terms *parametric hydrology* and *stochastic hydrology* (Hofmann, 1965, p.119).

synthetic ore A term used by the U.S. Bureau of Mines for material that is the equivalent of, or better than, natural ore, that can be put to the same uses, and is produced by means other than ordinary concentration, calcining, sintering, or nodulizing. (U.S. Bureau of Mines, 1968, p.1113-1114).

synthetic stone A man-made stone that has the same physical, optical, and chemical properties, and the same chemical composition, as the genuine or natural stone that it reproduces. Many gem materials have been made synthetically as a scientific experiment, but only corundum, spinel, emerald, rutile, garnet, sphene, and strontium titanate have been made commercially and cut as gemstones for the jewelry trade. Cf: *reconstructed stone.* Syn: *synthetic; reproduction.*

syntopogenic Formed in the same or similar place, or denoting similar conditions of origin (R.C. Moore, 1958, p.21); e.g. said of sedimentary rocks deposited in identical or nearly identical environments in marine waters or on land. Cf: *synchronogenic.*

syntype Any of the specimens upon which the description of a species or subspecies is based when no holotype has been designated. Nonpreferred syn: *cotype.*

synusia A subdivision of an ecologic community or habitat having a characteristic life pattern or uniform conditions. Plural: *synusiae.* Adj: *synusial.*

syphon Var. of *siphon.*

syrinx A tube of secondary shell located medially on the ventral side of the delthyrial plate of certain brachiopods (as in *Syringothyris*) and split along its ventral and anterior surface. Pl: *syringes* or *syrinxes.*

sysertskite A variety of iridosmine containing 50-80% osmium (or 20-50% iridium). Syn: *siserskite.*

syssiderite An obsolete syn. of *stony-iron meteorite.*

system [chem] (a) Any portion of the material universe which can be isolated completely and arbitrarily from the rest for consideration of the changes which may occur within it under varied conditions. (b) A conceptual range of compositions defined by a set of components in terms of which all compositions in the system can be expressed, e.g. the system $CaO-MgO-SiO_2$. Ricci has proposed to distinguish this sense from the ordinary thermodynamic one by capitalizing the word when it is used in the present sense.

system [geol] A group of related natural features, objects, or forces; e.g. a *drainage system* or a *mountain system*.

system [cryst] *crystal system.*

system [stratig] (a) A major time-stratigraphic unit of worldwide significance, representing the fundamental unit of time-stratigraphic classification of Phanerozoic rocks, extended from a type area or region and correlated mainly by its fossil content (ACSN, 1961, art.29); the rocks formed during a *period* of geologic time. It is next in rank above *series* and below *erathem*. The systems commonly recognized in the U.S. (in increasing order of age): Quaternary, Tertiary, Cretaceous, Jurassic, Triassic, Permian, Pennsylvanian, Mississippian, Devonian, Silurian, Ordovician, and Cambrian. Internationally, there is considerable divergence of opinion with respect to classification and nomenclature, and to boundaries for almost all systems. Although the basis for definition of a system should be a specifically designated and delimited type or reference sequence of strata, most of the systems in current use today lack adequate type or reference sections and were not established according to any systematic plan for geochronologic division of the Earth's strata as a whole (ISST, 1961, p.27). Systems in the Precambrian have only local significance or have not been placed in widely accepted orderly successions, and do not serve as the fundamental units of time-stratigraphic classification. (b) An informal term sometimes used locally for a large rock-stratigraphic unit that does not coincide with, and is somewhat larger in scope than, the formal or standard system of time-stratigraphic classification; e.g. "Karroo system" of Africa and "Hokonui system" of New Zealand. The informal term "sequence" is suggested for this usage (ISST, 1965, p.1698).

system [struc geol] (a) In structural geology, a group of related structures such as faults, joints, or dikes. (b) In tectonics, it is sometimes used for an individually named, or distinct orogenic belt, as in "Appalachian system". In this context, the term might be confused with the formally named stratigraphic systems. See also: *phase.*

systematic error Any *error* that persists and cannot be considered as due entirely to chance, or an error that follows some definite mathematical or physical law or pattern and that can be compensated, at least partly, by the determination and application of a correction; e.g. an error whose magnitude changes in proportion to known changes in observational conditions, such as an error caused by the effects of temperature or pressure on a measuring instrument or on the object to be measured. Ant: *random error.* See also: *constant error; instrument error.*

systematic joints Joints that occur in sets or patterns. They cross other joints, their surfaces are flat or only broadly curved, they are oriented perpendicular to the boundaries of the constituent rock unit, and the structures on their faces are oriented. Cf: *nonsystematic joints.*

systematics Water diffused in the atmosphere or the ground, including soil water.

syzygial (a) Pertaining to syzygy; e.g. a "syzygial pair" consisting of two crinoid ossicles joined by syzygy. (b) Pertaining to zygosis in sponges.

syzygy [paleont] (a) Ligamentary articulation of crinoid plates with opposed joint faces bearing numerous fine culmina that radiate from axial canal, the culmina meeting one another instead of fitted into crenellae. It forms a single segment and allows very slight mobility of joined ossicles in all directions. (b) The segment formed by syzygy.

syzygy [astron] Either of two points in the Moon's orbit about the Earth when the Moon is new (in conjunction) or full (in opposition), or when the Sun, Moon, and Earth are in a near straight-line configuration. Etymol: Greek *syzygos*, "yoked together". Cf: *quadrature.*

syzygy tide *spring tide.*

szaibelyite A white to yellowish acicular mineral occurring in nodular masses: $MgBO_2(OH)$. It is isomorphous with sussexite. Also spelled: *szájbelyite.* Syn: *ascharite.*

szaskaite *smithsonite.*

szmikite A monoclinic mineral: $MnSO_4.H_2O$.

szomolnokite A yellow or brown monoclinic mineral: $FeSO_4.H_2O$.

T

taaffeite A lilac mineral: $BeMgAl_4O_8$. It resembles mauve-colored spinel.

tabasheer Translucent to opaque and white to bluish-white opaline silica of organic origin (deposited within the joints of the bamboo), valued in the East Indies as a medicine and used in native jewelry. Var: *tabaschir; tabashir*.

tabbyite A variety of solid *asphalt* found in veins in Tabby Canyon, Utah.

tabella One of several small subhorizontal plates in the central part of a corallite, forming part of an *incomplete tabula*. Pl: *tabellae*.

taber ice A seam, lens, or layer of ground ice, usually pure, formed by the drawing in of water to a growing ice crystal as the ground became frozen. Named in honor of Stephen Taber (1882-1963), American geologist. Syn: *segregation ice; sirloin-type ice*.

tabetification The process of forming a talik (Bryan, 1946, p.640).

tabetisol *talik*.

table [geomorph] (a) The flat summit of a mountain, as one capped with horizontal masses of basalt. (b) A term used in the western U.S. for *tableland*.

table [gem] (a) The large facet that caps the crown of a faceted gemstone. In the standard round brilliant, it is octagonal in shape and is bounded by eight star facets. (b) *table diamond*.

table cut (a) An early style of fashioning diamonds, in which opposite points of an octahedron were ground down to squares to form a large culet and a larger table, and the remaining parts of the eight octahedral faces were polished. (b) A term sometimes used loosely to describe any one of the variations of the *bevel cut*, provided it has the usual large table of that cut.

table diamond A relatively flat, table-cut diamond. Syn: *table gem*].

table iceberg *tabular iceberg*.

tableknoll *guyot*.

tableland (a) A general term for a broad, elevated region of land with a nearly level or undulating surface of considerable extent; e.g. South Africa. Syn: *continental plateau*. (b) A plateau bordered by abrupt cliff-like edges rising sharply from the surrounding lowland; a mesa.

tablemount *guyot*.

table mountain A mountain having a comparatively flat summit and one or more precipitous sides. See also: *mesa*.

table reef A small, isolated, flat-topped organic reef, with or without islands, that does not enclose a lagoon. Cf: *platform reef*.

tablet (a) A tabular crystal. (b) A table-cut gem.

tabula (a) One of the transverse and nearly plane or upwardly convex or concave partitions (septa) of a corallite, extending to the outer walls or accompanying only the central part of the corallite. See also: *complete tabula; incomplete tabula*. (b) A subhorizontal perforate plate in the intervallum of an archaeocyathid, extending from one septum to another or in some genera supplanting the septa. (c) A six-sided heterococcolith, with two dimensions equal and the third smaller.---Pl: *tabulae*.

tabular (a) Said of a slab-like feature having a plane surface and two dimensions that are much larger or longer than the third (such as an igneous dike or a tablet-like orebody), or of a geomorphic feature having a flat surface (such as a plateau). (b) Said of the shape of a sedimentary body whose width/thickness ratio is greater than 50 to 1, but less than 1000 to 1 (Krynine, 1948, p.146); e.g. a graywacke formation in a geosynclinal deposit. Cf: *blanket; prism*. (c) Said of a sedimentary particle whose length is 1.5-3 times its thickness

(Krynine, 1948, p.142). Cf: *prismatic*. (d) Said of a crystal form having two prominent parallel faces that give it a broad, flat appearance, e.g. that of wollastonite.

tabula rasa theory A theory according to which during the Pleistocene Epoch the entire Scandinavian peninsula became covered with ice and its fauna and flora were completely destroyed, and subsequent immigration from central Europe, England, and Siberia produced an entirely new biota (Dahl, 1955, p.1500).

tabular berg *tabular iceberg*.

tabular cross-bedding Cross-bedding in which the cross-bedded units are bounded by planar, essentially parallel surfaces, forming a tabular body; e.g. *torrential cross-bedding*.

tabular dissepiment A nearly flat plate extending across an entire scleractinian corallite or confined to its axial part.

tabular iceberg A flat-topped iceberg which may be very large (up to 160 km long and more than 500 m thick), with cliff-like sides. Tabular icebergs are usually detached from an ice shelf and are numerous in the Antarctic. See also: *ice island*. Syn: *tabular berg; table iceberg*.

tabularium The axial part of the interior of a corallite in which tabulae are developed. Cf: *marginarium*.

tabular spar *wollastonite*.

tabular structure The structure of a mineral or rock whereby it tends to separate into plates or laminae.

tabulate adj. (a) Having tabulae; specif. said of a coral characterized by prominent tabulae. (b) Having plates; e.g. said of phaeodarian radiolarians having smooth plates, or said of dinoflagellate theca having armored plates.---n. Any zoantharian belonging to the order Tabulata. Their stratigraphic range is Middle Ordovician to Permian, possibly Triassic to Eocene, also.

tacharanite A mineral: $(Ca,Mg,Al)(Si,Al)O_3 \cdot H_2O$.

tacheometer *tachymeter*.

tachygenesis The extreme crowding and eventual loss of those primitive phylogenetic stages which are represented early in the life of an individual.

tachygraphometer A tachymeter with alidade for surveying. Syn: *tacheographometer*.

tachyhydrite A yellowish mineral: $CaMg_2Cl_6 \cdot 12H_2O$. Syn: *tachydrite; tachhydrite*.

tachylite A volcanic glass that may be black, green, or brown because of abundant crystallites. It is formed from basaltic magma and is commonly found as chilled margins of dikes, sills, or flows. Syn: *tachylyte; hyalobasalt; basalt glass; jaspoid; basalt obsidian; sordawalite; wichtisite*. Cf: *hyalomelane; hydrotachylite; sideromelane*. See also: *pseudotachylite*.

tachymeter A surveying instrument designed for use in the rapid determination from a single observation of the distance, direction, and elevation difference of a distant object; esp. a transit or theodolite with stadia hairs, or an instrument in which the base line for distance measurements is an integral part of the instrument. Tachymeters include *range finders* with self-contained bases, although range finders do not usually afford the means for the determination of elevation. Syn: *tacheometer*.

tachymetry A method of rapid surveying using the tachymeter; e.g. the stadia method of surveying used in U.S.

tachytely A phylogenetic' phenomenon characterized by a rapid temporary spurt in evolution that occurs as populations shift from one major zone of adaptation to another; episodic evolution. Cf: *bradytely; horotely; lipogenesis*.

Taconian orogeny Var. of *Taconic orogeny*, used to distinguish it from other Taconic features, e.g. Taconic allochthone, Taconic thrust.

Taconic orogeny An orogeny in the latter part of the Ordovician period, named for the Taconic Range of eastern New York State and well developed through most of the northern Appalachians in U.S. and Canada. As with many other orogenies, it can be strictly defined in places at a point in the Late

Ordovician by its relation to limiting fossiliferous strata, or it can be extended to include many pulsations which occurred from place to place from early in the Ordovician to early in the Silurian; perhaps it could most properly be regarded as an orogenic era in the sense of Stille. Taconic plutonic rocks are less voluminous than the Acadian plutonic rocks of the same region, but are widely distributed; they yield radiometric ages of about 400 to 450 m.y. Also spelled: *Taconian orogeny*.

taconite A local term used in the Lake Superior iron-bearing district of Minnesota for any bedded ferruginous chert or variously tinted jaspery rock, esp. one that enclosed the Mesabi iron ores (granular hematite); an unleached *iron formation* containing magnetite, hematite, siderite, and hydrous iron silicates (greenalite, minnesotaite, and stilpnomelane). The term is specifically applied to this rock when the iron content, either banded or disseminated, is high enough (25%) so that natural leaching can convert it into a low-grade iron ore (50-60% iron). (b) Since World War II, a low-grade, magnetite-quartz *iron formation* suitable for the artificial concentration of magnetite and from which pellets containing 62-65% iron can be manufactured. Also spelled: *taconyte*.

tactite A rock of complex mineralogical composition formed by contact metamorphism and metasomatism of carbonate rocks.

tactoid A spindle-shaped body, e.g. in vanadium pentoxide sol, visible under a polarizing microscope.

tactosol A sol that contains tactoids which are in a spontaneous, parallel arrangement.

tadjerite A black, semiglassy, chondritic stony meteorite composed of bronzite and olivine.

tadpole nest A small, irregular *cross ripple mark* characterized by an erratic and polygonal or cell-like pattern formed by the intersection of two sets of ripples (approximately at right angles to each other) and formerly believed to have been made by tadpoles. The height of the ripple mark is considerably greater than in the equivalent form associated with transverse ripple mark. Term introduced by Hitchcock (1858, p.121-123).

taele (a) Older form of the Norwegian term *tele*. (b) Anglicized version of the Swedish term *tjäle*, and used as a syn. of *tjaele*.

taenia An irregularly bent small plate in the intervallum of an archaeocyathid (TIP, 1955, pt.E, p.7).

taeniolite A white or colorless mineral of the mica group: $KLiMg_2Si_4O_{10}F_2$. Syn: *tainiolite*.

taenite A meteorite mineral consisting of the face-centered cubic gamma-phase of a *nickel-iron* alloy, with a nickel content varying from about 27% up to 65%. It occurs in iron meteorites as lamellae or strips flanking bands of *kamacite*.

tafelberg A term used in South Africa for a mesa or a table mountain; a large *tafelkop*. Etymol: Afrikaans.

tafelkop A term used in South Africa for an isolated hill with a flat top; a butte. Etymol: Afrikaans. Cf: *spitskop*.

tafone (a) A Corsican dialect term for one of the natural cavities in a *honeycomb structure*, formed by *cavernous weathering* on the face of a cliff in a dry region or along the seashore. The hole or recess may reach a depth of 10 cm, and is explained as due to solution of free salts in crystalline rock (granite, gneiss, quartz) following heating by insolation. (b) A granitic or gneissic block or boulder hollowed out by cavernous weathering.---Pl: *tafoni*.

tafrogenesis Var. of *taphrogenesis*.

Tagg's method A method of interpretation of resistivity sounding data obtained with a Wenner array over a two-layered Earth.

Taghanican North American stage: uppermost Middle Devonian (above Tioughniogan, below Fingerlakesian).

tagilite *pseudomalachite*.

tagma A major division of the body of an arthropod, each composed of several somites; e.g. cephalon, thorax, and abdomen. Pl: *tagmata*.

tahitite A feldspathoid-bearing trachyandesite that contain haüyne phenocrysts and generally more sodic plagioclas than orthoclase. Its name is derived from Tahiti.

tahoma (a) A generic term indicating a high snowy or glacier clad mountain in the Pacific NW U.S., as Mount Hood. (b The V-shaped residual ridge between two cirques on a glacial ly carved volcanic cone in the Pacific NW U.S., as one o several on Mount Rainier (Russell, 1898a, p.382-383).-- Etymol: An approximation of one of several Indian names fo Mount Rainier; presumably the higher the mountain, the long er the second syllable was drawn out, as "ta-hoooom-ah".

taiga A swampy area of coniferous forest sometimes foun lying between tundra and steppe regions. Etymol: Russian.

tail [coast] (a) A bar or barrier formed behind a small islan or a skerry. Syn: *trailing spit; banner bank*. (b) The outermos part of a projecting bar.

tail [glac geol] An accumulation or streak of till taperin down-valley from the lee side of a knob of resistant rock. Se also: *crag and tail*.

tail [paleont] Any of various backwardly directed and usuall posterior parts on the body of an invertebrate, esp. if it is a tenuated.

tail [sed] The rear part of a turbidity current, which is les dense than the *nose* and moves more slowly.

tail [streams] (a) The downstream section of a pool o stream. (b) The comparatively calm water after a current or reach of rough water.

tail coccolith A modified coccolith found at the end opposit the flagellar field in flagellate coccolithophores exhibiting di morphism (such as *Calciopappus*). Cf: *pole coccolith*.

tail dune A dune that accumulates on the leeward side of a obstacle and that tapers away gradually for a distance of u to 1 km. Cf: *head dune*.

tail fan *caudal fan*.

tailing adj: *leggy*.

tail-land A term proposed by Dryer (1899, p.273) for a mean der lobe that slopes regularly from an elevated mainland o meander neck to a low point.

tail water The water downstream from a structure, as below dam.

taimyrite A nosean-bearing trachyte. It is named after the Ta myr River, Siberia, U.S.S.R.

Taimyr polygon An *ice-wedge polygon*, so-called from its oc currence in northern Siberia.

tainiolite *taeniolite*.

tala A term introduced by Berkey & Morris (1924, p.105) for broad structural basin of internal drainage, formed in the Gob Desert by subsidence or warping and bounded by inconspicu ous divides or mountain ranges. Cf: *gobi*. Etymol: Mongoliar "open steppe-country".

talc (a) An extremely soft, whitish, greenish, or grayis monoclinic mineral: $Mg_3Si_4O_{10}(OH)_2$. It has a characteristi soapy or greasy feel and a hardness of 1 on Mohs' scale, an it is easily cut with a knife. Talc is a common secondary min eral derived by alteration (hydration) of nonaluminous magne sium silicates (such as olivine, enstatite, and tremolite) i basic igneous rocks or by metamorphism of dolomite rocks and it usually occurs in foliated, granular, or fibrous masses Talc is used as a filler, coating pigment, dusting agent, and i ceramics, rubber, plastics, lubricants, and talcum powder Originally spelled: *talck*. See also: *steatite*. (b) A talcose rock a rock consisting mainly of talc, such as steatite and soap stone. (c) A thin sheet of muscovite mica.

talcite (a) A massive variety of talc. (b) *damourite*.

talcoid n. A mineral: $Mg_3Si_5O_{12}(OH)_2$. It is possibly a mixtur of talc and quartz.---adj. Resembling talc; e.g. "talcoi schist".

talcose (a) Pertaining to or containing talc; e.g. "talcos granite". (b) Resembling talc; e.g. a "talcose rock" that ft and soapy to the touch.

lc schist A schist in which talc, in association with mica and quartz, is the dominant schistose material (Holmes, 1928, 223).

lc slate An impure, hard, slaty variety of talc, a little harder an French chalk; "indurated talc".

leola A cylinder or rod of granular, nonfibrous calcite in the ial region of some *pseudopunctate* of brachiopods. Pl: *tale-ae.*

let A term used in the High Atlas of Morocco for a dried-out rrential gully (Termier & Termier, 1963, p.415). Etymol: Ber-er.

lik A Russian term for a layer of unfrozen ground above, thin, or beneath the permafrost; occurs in regions of *dis-ontinuous permafrost.* It may be permanent or temporary. ee also: *subgelisol.* Syn: *tabetisol.*

lmessite *arsenate-belovite.*

nakhite A mineral: $Cu_9(Fe,Ni)_8S_{16}$.

lpatate (a) A surficial rock formed by the cementing action calcium carbonate on sand, soil, or volcanic ash, and partly quivalent to caliche. See also: *tepetate.* (b) A poor, thin soil onsisting of partly decomposed volcanic ash that is more or ss consolidated.---Etymol: American Spanish, from Nahuatl Aztec) *tepetatl,* "stone matting". Syn: *talpetate.*

lus [geol] (a) *talus slope.* (b) Rock fragments of any size or hape (usually coarse and angular) derived from and lying at e base of a cliff or very steep, rocky slope. Also, the out-ard sloping and accumulated heap or mass of such loose roken rock, considered as a unit, and formed chiefly by avitational falling, rolling, or sliding. See also: *scree.* Syn: bble.---Etymol: French *talu,* later *talus,* "a slope", originally the military sense of fortification for the outside of a ram-art or sloping wall whose thickness diminishes with height; om Latin *talutium,* a gossan zone or slope indicative of gold probably of Iberian origin). Pl: *taluses.*

lus [reef] *reef talus.*

lus apron A poorly sorted but well-bedded accumulation of ef detritus, usually much larger in volume than the parent ef, with a surface inclination up to 40°.

lus breccia A breccia formed by the accumulation and con-olidation of talus.

lus cave A cave formed accidentally by a fall of talus.

lus cone A small, cone-shaped or apron-like landform at the ase of a cliff and consisting of poorly sorted talus that has ccumulated episodically by mass-wasting. Also, a similar eature of fluvial origin and tapering up into a gully. Maximum eight is 300 m.

lus creep The slow downslope movement of talus, either the ock fragments on a talus slope or the mass of fragments oving as a unit; it becomes more rapid during frequent eeze and thaw. See also: *rock creep; rock-glacier creep.*

lus fan *alluvial fan.*

lus glacier A *rock glacier* consisting of loose debris on a teep slope.

lus slope A steep, concave slope formed by an accumula-on of loose rock fragments; esp. such a slope at the base of cliff or steep slope and formed by the coalescence of *talus ones;* the surface profile of an accumulation of talus. It has constant angle of repose close to 35°, although the upper art of the slope may be steeper. Etymologically, "talus slope" s a tautology as the term "talus" originally signified a slope, lthough often used in the U.S. for the material composing the lope. See also: *scree.* Syn: *talus; debris slope.*

lus spring A spring occurring at the base of a talus slope nd originating from water falling upon or seeping into the lope.

luvium A term introduced by Wentworth (1943) for a detrital over consisting of talus and colluvium; the fragments vary om large blocks to silt.

lweg Var. of *thalweg.*

tamaraite A dark-colored hypabyssal rock resembling a basalt in appearance and containing augite, hornblende, biotite, nepheline, plagioclase, orthoclase, minor accessories, and secondary cancrinite and analcime.

tamarugite A colorless mineral: $NaAl(SO_4)_2.6H_2O$. It was originally named *lapparentite.*

tamping (a) The act or an instance of filling up a drill hole above a blasting charge with moist, loose material (mud, clay, earth, sand) in order to confine the force of the explosion to the lower part of the hole. (b) The material used in tamping.--See also: *stemming.*

tang A Scottish term for a low, narrow cape.

tangeite A syn. of *calciovolborthite.* Also spelled: *tangueite; tangeïte.*

tangent n. (a) A straight line that touches, but does not tran-sect, a given curve or surface at one and only one point; a line that touches a circle and is perpendicular to its radius at the point of contact. (b) The part of a traverse included be-tween the point of tangency (the point in a line survey where a circular curve ends and a tangent begins) of one curve and the point of curvature (the point in a line survey where a tan-gent ends and a circular curve begins) of the next curve. (c) A great-circle line that is tangent to a parallel of latitude at a township corner in the U.S Public Land Surveys system. (d) A term sometimes applied to a long straight line of a traverse whether or not the termini of the line are points of curve. (e) The ratio of the length of the leg opposite an acute angle in a right-angled triangle to the length of the leg adjacent to the angle.---adj. Said of a line or surface that meets a curve or surface at only one point.

tangential cross-bedding Cross-bedding in which the foreset beds appear in section as smooth arcs meeting the underlying surface at low angles; it implies deposition by wind. Cf: *angu-lar cross-bedding.*

tangential ray A sponge-spicule ray that lies approximately parallel to the surface of the sponge.

tangential section (a) A slice through part of a foraminiferal test parallel to the axis of coiling or growth but not through the proloculus. (b) A section of zoarium of a bryozoan, cut at right angles to the zooecial tubes. (c) A section of a cylindri-cal organ (such as a stem) cut lengthwise and at right angles to the radius of the organ.

tangential stress *shear stress.*

tangential wave *S wave.*

tangent screw A very fine, slow-motion screw giving a tangen-tial movement for making the final setting to a precision surveying instrument (such as for completing the alignment of sight on a theodolite or transit by gentle rotation of the read-ing circle about its axis).

tangi A term used in Baluchistan (an arid region in southern Asia) for a narrow, transverse gorge or cleft through which a stream penetrates a longitudinal ridge. Etymol: Persian *tang,* "narrow".

tangiwai A term used by the Maoris of New Zealand for *bow-enite.* Syn: *tangiwaite; tangawaite.*

tangle sheet Mica with intergrowths of crystals or laminae re-sulting in books that split well in some places but tear to pro-duce a large proportion of partial films (Skow, 1962, p.170).

tangue Calcareous mud occurring in the shallow bays along the coast of Brittany (NW France), consisting in part of a flu-violacustrine silt reworked by recent marine transgressions and in part of finely powdered molluscan shell material trans-ported and deposited by the tides. It contains 25-60% calcium carbonate, is coherent (even when dry), and has a permeabil-ity similar to that of fine sand. Pron: *tong.*

tank (a) A term applied in SW U.S. to a natural depression or cavity in impervious rocks (usually crystalline) in which rain-water, floodwater, snowmelt, and seepage are collected and preserved during the greater part of the year. See also: *rock tank; charco.* (b) A natural or artificial pool, pond, or small lake occupying a tank; esp. an artificial reservoir for supplying water for livestock. (c) A term used in Ceylon and the drier parts of peninsular India for an artificial pond, pool, or lake formed by building a mud wall across the valley of a small stream to retain the monsoon rainwater.

tannbuschite A rarely used term applied to dark-colored nepheline basalt.

tantalite A black mineral: $(Fe,Mn)(Ta,Nb)_2O_6$. It is isomorphous with *columbite* and dimorphous (orthorhombic phase) with tapiolite; it occurs in pegmatites, and is the principal ore of tantalum.

tanteuxenite A black or brown mineral: $(Y,Ce,Ca)(Ta,Nb, 0)_2(O,OH)_6$. It is a variety of euxenite with Ta largely or almost entirely substituting for Nb. Syn: *delorenzite; eschwegeite*.

tanzanite A sapphire-blue gem variety of zoisite that exhibits strong pleochroism.

tape [ore dep] A thin band or streak or ribbon of ore.

tape [surv] A continuous ribbon or strip of steel, invar, dimensionally stable alloys, specially made cloth, or other suitable material, having a constant cross section and marked with linear graduations, used by surveyors in place of a *chain* for the measurement of lengths or distances.

tape correction A quantity applied to a taped distance to eliminate or reduce errors due to the physical condition of the tape and the the manner in which it is used; e.g. a correction based on the length, temperature, tension, or alignment of the tape. See also: *sag correction; slope correction*.

tapeman *chainman*.

tapetum Tissue of nutritive cells in the sporangium of embryophytic plants, digested during development of the spores. In angiosperms, it is the inner wall of the anther locules and provides nutritive substances for the developing pollen. Pl: *tapeta*.

taphocoenose *thanatocoenosis*.

taphocoenosis *thanatocoenosis*.

taphoglyph A hieroglyph representing the impression of the body of a dead animal (Vassoevich, 1953, p.72).

taphonomy The branch of paleoecology concerned with the manner of burial and the origin of plant and animal remains. It includes *Fossildiagenese* and *biostratonomy*. Syn: *para-ecology*.

taphrogenesis *taphrogeny*. Also spelled: *Tafrogenesis*.

taphrogenic Adj. of *taphrogeny*.

taphrogeny A general term for the formation of rift phenomena, characterized by high-angle or block faulting and associated subsidence. Etymol: Greek, *taphre*, "trench". Also spelled: *tafrogeny*. Adj: *taphrogenic*. Syn: *taphrogenesis*.

taphrogeosyncline A geosyncline developed as a rift or trough between faults (Kay, 1945). Cf: *aulacogen*.

taping The operation of measuring distances on the ground by means of a surveyor's tape. Cf: *chaining*.

tapiolite A mineral: $Fe(Ta,Nb)_2O_6$. It is isomorphous with mossite and dimorphous (tetragonal phase) with tantalite; it occurs in pegmatites or detrital deposits, and is an ore of tantalum.

tapoon A subsurface dam built in a dry wash, either to increase recharge to nearby well or to impound water for direct use. In the latter case, a pipe is run from the dam to the point of use.

tar A thick, brown to black, viscous organic liquid, free of water, which is obtained by condensing the volatile products of the destructive distillation of coal, wood, oil, etc. It has a variable composition, depending on the temperature and material used to obtain it.

taramellite A brownish-red or reddish-brown mineral: $Ba_4(Fe, Mg)Fe^+_2{}^3TiSi_8O_{24}(OH)_2$.

taramite A black monoclinic mineral of the amphibole group, approximately: $(Ca,Na,K)_3Fe_5(Si,Al)_8O_{22}(OH)_2$.

taranakite A yellowish-white clayey mineral: $KAl_3(PO_4)_3-$

$(OH).9H_2O$.

Tarannon European stage: Lower Silurian (above Llandoverian, below Wenlockian).

tarantulite A plutonic rock containing more than 50 percen quartz, less alkali feldspar, over half of which is orthoclas and the rest albite, and up to five percent dark minerals. Th term was proposed by Johannsen (1939, p.283) as a subs tute for *alaskite-quartz*. The rock is transitional between *ala kite* and *silexite*. Its name is derived from Tarantula Sprin Nevada.

tarapacaite A yellow mineral: K_2CrO_4.

tarasovite A mixed-layer clay mineral, related to hydromica.

taraspite A mottled, compact dolomite rock from Taras Switzerland, used for decorative purposes.

tarbuttite A colorless or pale-yellow, brown, red, or green t clinic mineral: $Zn_2(PO_4)(OH)$. It is isomorphous with parad mite.

tar coal Resinous, bitumen-rich brown coal.

tare A sudden jump in reading between observations, usual applied to gravimeters but occasionally to gravity pendulums.

target [photo] (a) The distinctive marking or instrumentatio of a ground point to aid in its identification on a photograph. is a material marking so arranged and placed on the groun as to form a distinctive pattern over a geodetic or other co trol-point marker, on a property corner or line, or at the pos tion of an identifying point above an underground facility feature (ASP, 1966, p.1156). (b) The image pattern on an ae rial photograph of the actual mark or target placed on th ground prior to photography.

target [surv] The vane or sliding sight on a surveyor's lev rod; a device, object, or point upon which sights are made.

target rod A *level rod* with an adjustable target that is move into position by the rodman in accordance with signals give by the man at the leveling instrument and that is read and re corded by the rodman when it is bisected by the line of coll mation. Cf: *speaking rod*.

tarn (a) A landlocked pool or small lake such as those tha occur within tracts of swamp, marsh, bog, or muskeg Michigan (Davis, 1907, p.116), and on moors in northern En land. In unrestricted usage, "any small lake may be called tarn" (Veatch & Humphrys, 1966, p.326). (b) A relative small and deep, steep-banked lake or pool amid high moun tains, esp. one occupying an ice-gouged rock basin amid gl ciated mountains. (c) *cirque lake*.---Etymol: Icelandic.

tarnish The altered condition, in both color and luster, of th surface of a mineral in comparison with its interior. Tarnish characteristic of copper-bearing minerals.

tar pit An area in which an accumulation of natural bitumen is exposed at the land surface, forming a trap into which an mals (esp. vertebrates) fall and sink, their hard parts bein preserved in the bitumens.

tar sand A type of *oil sand* or sandstone from which the ligh er fractions of crude oil have escaped, leaving a residual as phalt to fill the interstices.

tarsus The distal part of the limb of an arthropod; e.g. the las segment (sometimes subsegmented) of a leg of an arachnid or a joint of the distal part of a prosomal appendage of a mer ostome. Pl: *tarsi*.

Tartarian Var. of *Tatarian*.

tasco A fireclay from which melting pots are made. Als spelled: *tasko*.

tasmanite [coal] An impure coal, transitional between canne coal and oil shale. Syn: *combustible shale; yellow coal; Mer sey yellow coal; white coal*.

tasmanite [ign] An intrusive rock similar in composition to ijo ite but containing zeolites in place of nepheline. Among th zeolites are natrolite, thomsonite, phillipsite, and hydrargillite It is named after Tasmania.

tasmanites An informal term for members of the genu *Tasmanites*, a large, spherical palynomorph with a thick per

rate wall and probably representing the resting body of cer-
_in green algae. Its fossils (ranging from Ordovician to Ceno-
ɔic) are usually classed with the acritarchs; certain organic-
ch shales (also known as tasmanite, as in Australia) contain
1ormous numbers of these fossils.

atarian European stage: Upper Permian (above Kazanian,
ɛlow Scythian of Triassic). Syn: _Tartarian; Chideruan._

tarskite A mineral: $Ca_6Mg_2(SO_4)_2(CO_3)_2Cl_4(OH)_4.7H_2O$.

uactin A spicule of hexactinellid sponges, having three co-
anar rays of which two lie along the same axis.

urite A sodic rhyolite containing acmite and differing from
ɔmendite in having a spherulitic or granophyric groundmass.

utirite An igneous rock composed of potassium feldspar, an-
ɛsine, nepheline, and amphibole, with abundant accessory
ɔhene. Cf: _pollenite._

utochron A graph showing the variation of ground tempera-
ɪre with depth for a definite period of time. Air temperature
ɪutochrons may be plotted above the ground surface.

utonym A binomen or trinomen in which the term designat-
g the genus is the same as that for the species or sub-
ɔecies; e.g. _Troglodytes troglodytes._ See also: _tautonymy._

utonymy A rule of binomial nomenclature by which a
ɔecies having a name identical to the name of the genus to
hich it belongs automatically becomes the type species of
ɪat genus. See also: _tautonym._

utozonal faces Faces of a crystal that occur in the same
ɔne.

ɪvistockite _carbonate-apatite._

ɪvolatite An igneous rock containing large phenocrysts of
ɪucite, hauyne, acmite-augite, and garnet in a microphyric
ɪoundmass of the same minerals with interstitial leucite,
ɪrthoclase, hauyne, labradorite, acmite-augite, biotite, garnet,
ɪd nepheline.

ɪvorite A yellow mineral: $LiFe^{+3}(PO_4)(OH)$.

ɪwite A plutonic rock, similar in composition to _ijolite,_ con-
ɪining 30 to 60 percent mafic minerals, but with sodalite in
lace of nepheline as the dominant sodium feldspathoid.

ɪwmawite A yellow or green to dark-green variety of epidote
ɪontaining chromium and found in Tawmaw, upper Burma.

ɪxa The plural of _taxon._

ɪxichnic Said of a dolomite rock in which the original texture
r structure of a limestone is preserved or distinguishable.
ɛrm introduced by Phemister (1956, p.74).

ɪxis [paleont] A definite linear series of plates in any part of
ɪe crown of a crinoid; e.g. anitaxis and brachitaxis. Pl: _taxes._

ɪxis [ecol] A movement or orientation of an organism orient-
d with respect to a source of stimulation. Cf: _tropism._

ɪxite A general term for a volcanic rock that appears to be
lastic because of its mixture of materials of varying texture
nd structure from the same flow. See also: _ataxite; eutaxite._

ɪxodont adj. Said of the dentition of a bivalve mollusk char-
ɪcterized by numerous, short, subequal hinge teeth forming a
ɪontinuous row, with some or all of the teeth transverse to the
ɪargin of the hinge. The term is virtually synonymous with
ɪrionodont.---n. A taxodont mollusk; specif. a bivalve mollusk
ɪf the order Taxodonta, having numerous and unspecialized
ɪinge teeth and also having equally developed adductor mus-
les. Cf: _heterodont._

ɪxon In taxonomy, a unit of any rank, such as a particular
ɪpecies, family, or class; also, the name applied to that unit.
ɪ taxon may be designated by a formal Latin name or by a
ɛtter, number, or other symbol. Plural: _taxa; taxons._ See also:
ɪarataxon.

ɪxonomy The theory and practice of classifying plants and
ɪnimals. At one time "taxonomy" and _systematics_ were used
ɪore or less synonymously. The main taxonomic units are, in
ɪrder of increasing rank, species, genus, family, order, class,
ɪhylum, kingdom. Cf: _classification._

ɪayloran North American (Gulf Coast) stage: Upper Creta-
ɛeous (above Austinian, below Navarroan).

taylorite [clay] An obsolete name first applied by Knight
(1897) to the rock subsequently designated _bentonite._ Named
after William Taylor, who made the first commercial ship-
ments of the clay from the Rock Creek district, Wyo.

taylorite [mineral] A white mineral: $(K,NH_4)_2SO_4$. It is a variety of arcanite containing ammonia, and occurs in compact
lumps in the guano beds on certain offshore Peruvian islands.

tazheranite A cubic mineral: $(Zr,Ca,Ti)O_2$.

tcheremkhite An algal sapropelic deposit found in the vicinity
of Cheremkhovo, USSR, and which has been interpreted as
an aggregation of peaty matter washed from other deposits
(Twenhofel, 1939, p.434).

T-chert Tectonically controlled chert, occurring in irregular
masses related to fractures and orebodies (Dunbar & Rodgers, 1957, p.248).

Tchornozem Var. of _Chernozem._

T.D. curve _traveltime curve._

_t_ direction In structural petrology, the direction of movement
in a glide plane. See also: _t axis._

T-dolostone Tectonically controlled dolostone, occurring in irregular masses related to fracture systems (Dunbar & Rodgers, 1957, p.239).

TΔT process A method of obtaining or measuring the vertical
velocity of sound through subsurface sediments by use of the
reflection seismograph technique.

teallite A black or blackish-gray mineral: $PbSnS_2$.

tear fault A very steep to vertical fault associated with a low-
angle overthrust fault and occurring in the hanging wall. It
strikes perpendicular to the strike of the overthrust; displacement may be horizontal, and there may be a scissor effect. It
is considered by some to be a type of strike-slip fault.

tear-shaped bomb A rotational volcanic bomb shaped like a
teardrop and having an ear at its constricted end; it ranges in
size from 1 mm to more than 1 cm in length. Cf: _Pele's tears._

technical scale A standard of fifteen minerals by which the
hardness of a mineral may be rated. The scale includes, from
softest to hardest and numbered one to fifteen: talc; gypsum;
calcite; fluorspar; apatite; orthoclase; pure quartz glass;
quartz; topaz; garnet; fused zircon; corundum; silicon-carbide;
boron carbide; and diamond. Cf: _Mohs' scale._

tectate Said of a pollen grain whose ektexine has an outer
surface supported by more or less complicated inner structure
usually consisting of columellae that support the tectum.

tectine An albuminoid (proteinaceous) organic substance in
the wall of some foraminifers and having the appearance of,
but chemically distinct from, chitin (TIP, 1964, pt.C, p.64).

tectite _tektite._

tectocline _geotectocline._

tectofacies A lithofacies that is interpreted tectonically. The
term was introduced by Sloss et al (1949, p.96) for "a group
of strata of different tectonic aspect from laterally equivalent
strata", and was defined by Krumbein & Sloss (1951, p.383)
as "laterally varying tectonic aspects of a stratigraphic unit".
The term appears to be of "very limited practical value" because generally the nature of a tectofacies is noted "only
after the area of the tectofacies has been outlined on the
basis of other considerations" (Weller, 1958, p.635). Not to
be confused with _tectonic facies._ See also: _facies [stratig]._

tectogene (a) A long, relatively narrow unit of downfolding of
sialic crust considered to be related to mountain-building processes. The term was proposed by Haarmann (1926, p.107)
as a substitute for _orogene._ Syn: _geotectogene._ (b) the downfolded portion of an orogene (Hess, 1938). Syn: _downbuckle._

tectogenesis A syn. of _orogenesis,_ in its present meaning of
the process by which mountainous areas are formed, e.g.
folding or thrusting, without implication of the formation of
mountainous topography.

tectomorphic A _deuteromorphic_ crystal modified by magmatic
corrosion. The term is obsolete.

tectonic Said of or pertaining to the forces involved in, or the resulting structures or features of, *tectonics.* Syn: *geotectonic.*

tectonic analysis *petrotectonics.*

tectonic axis A *fabric axis* used as reference for the movement symmetry of a deformed rock.

tectonic breccia A *breccia* formed as a result of crustal movements and produced by lateral or vertical pressure or by tension. It is usually developed from brittle rocks subjected to pressures that were insufficient to render them plastic. The two major varieties are *fault breccia* and *fold breccia.* See also:*crush breccia.* Syn: *dynamic breccia; pressure breccia.*

tectonic conglomerate *crush conglomerate.*

tectonic creep Slow, apparently continuous movement of a fault.

tectonic cycle (a) The cycle that relates the larger structural features of the Earth's crust to gross crustal movements and to the kinds of rocks that form in the various stages of development of these features; *orogenic cycle.* (b) A *geosynclinal cycle* of three stages: peneplanation (widespread deposition on a relatively stable flat surface), geosynclinal (deposition during subsidence), and orogenic (postgeosynclinal uplift commonly marked by faulting, after folding and magmatic intrusion in the geosyncline) (Krynine, 1941).

tectonic denudation The stripping during deformation of an underbody, such as basement or other competent rock by the movement of an upper stratified layer over it. During movement of rootless masses of the upper rocks by gravity tectonics, the surface of the underbody is laid bare in places. Cf: *décollement.*

tectonic earthquake An earthquake due to faulting rather than to volcanic activity. The term is little used. Cf: *volcanic earthquake.*

tectonic enclave A body of rock that has become detached or isolated from its source by tectonic forces. Cf: *tectonic inclusion.*

tectonic fabric *deformation fabric.*

tectonic facies A collective term for rocks that owe their present characteristics mainly to tectonic movements; e.g. mylonites and some phyllites. This concept was introduced by Sander (1912). Not to be confused with *tectofacies.*

tectonic flow *tectonic transport.*

tectonic framework The combination or relationship in space and time of subsiding, stable, and rising tectonic elements in sedimentary provenance and depositional areas. Var: *framework* [tect].

tectonic gap *lag fault.*

tectonic inclusion A body of rock that has become detached or isolated from its source by tectonic disruption, and that is enclosed or included in the surrounding rock, e.g. a boudin. The term was used by Rast (1956, p.401) to replace *boudin,* a term that is often misused for any such inclusion. Cf: *tectonic enclave.*

tectonic lake A lake occupying a basin produced mainly by crustal movements; e.g. a lake resulting from the uplift above sea level of a submarine basin formed by differential marine sedimentation, or a lake caused by the impoundment of a drainage system as a result of upwarping, or a lake occupying a graben (such as Lake Baikal in Russia). Syn: *structural lake.*

tectonic land Linear fold ridges and volcanic islands which had an ephemeral existence in the internal parts of an orogenic belt during the early or geosynclinal phase. Kay (1951) compares them with modern island arcs, and proposes that their existence probably accounts for many of the features formerly ascribed to *borderlands.*

tectonic landform A landform produced by earth movements. Cf: *structural landform.*

tectonic lens A body of rock similar to a *boudin* but interpreted as having been formed by distortion of a continuous incompetent layer enclosed between competent layers, rather

than vice versa as for a boudin.

tectonic line A major, extensive fault, the displacement which is both lateral and vertical, and which cuts across delineates orogenic belts.

tectonic map A map that portrays the architecture of th upper part of the Earth's crust. It is similar to a *structure-co tour* map, which primarily shows folds, faults, structure co tours and the like, that also appear on a tectonic map, but th latter also presents some indication of the ages and kinds rocks from which the structures were made, as well as the historical development. Cf: *paleotectonic map.*

tectonic moraine An aggregation of boulders incorporated the base or sole of an overthrust mass. It is often mistake for a conglomerate as it has a local concordant relation to th associated strata (Pettijohn, 1957, p.281).

tectonic overpressure The pressure exceeding the load pres sure at times during metamorphism, by amounts which de pend on the strength of the rocks (1000-2000 bars) (Clark 1961).

tectonic profile *profile* [struc petrol].

tectonic rotation *Internal rotation* of a tectonite in the directio of transport.

tectonics A branch of geology dealing with the broad arch tecture of the upper part of the Earth's crust, that is, the re gional assembling of structural or deformational features, study of their mutual relations, their origin, and their historica evolution. It is closely related to *structural geology,* with whic the distinctions are blurred, but tectonics generally deals wit the larger, and structural geology with the smaller, features Adj: *tectonic.* Syn: *geotectonics.*

tectonic style The total character of a group of related struc tures that distinguishes them from other groups of structures in the same way that the style of a building or an art objec distinguishes it from others of different periods or influences.

tectonic transport In structural petrology, *componental move ment* of a rock mass during its deformation. Syn: *tectoni flow.*

tectonic unmixing Mechanical segregation of minerals due t recrystallization in a shear zone (Knopf and Ingerson, 1938).

tectonic valley A valley that is produced mainly by crusta movements, such as by faulting or folding. Cf: *structural va ley.*

tectonism A less preferred syn. of *diastrophism.*

tectonite Any rock whose fabric reflects the history of its de formation; a rock whose fabric clearly displays coordinate geometric features that indicate continuous solid flow durin formation (Turner and Weiss, 1963, p.39). Also spelled: *tek tonite.*

tectonization A term sometimes used as a generalized syn onym of orogenesis, diastrophism, etc. Thus, deformed rock in orogenic belts are said to have been *tectonized.* This usag is not recommended.

tectonized Adj. of *tectonization.*

tectono-eustatism *diastrophic eustatism.*

tectonophysics A branch of geophysics that deals with th forces responsible for movements in, and deformation of, th Earth's crust.

tectonosphere The zone or layer of the Earth above the leve of isostatic equilibrium, in which crustal or tectonic move ments originate. It is equivalent to the *crust,* and is define petrologically as the sial, salsima, and sima layers.

tectono-stratigraphic unit A mixture of rock-stratigraphic unit resulting from tectonic deformation; e.g. a mélange.

tectorium The internal lining of a foraminiferal chamber (as ir fusulinids), composed of dense calcite formed at or near the same time as that in which the tunnel in the test was excavat ed. It may include the *lower tectorium* and the *upper tecto rium.* Pl: *tectoria.*

tectosequent Said of a surface feature that reflects the under lying geologic structure. Ant: *morphosequent.*

tectosilicate A class or structural type of *silicate* characterized by the sharing of all four oxygens of the SiO_4 tetrahedra with neighboring tetrahedra, and a Si:O ratio of 1:2. Quartz, SiO_2, is an example. Cf: *nesosilicate; sorosilicate; cyclosilicate; inosilicate; phyllosilicate.* Syn: *framework silicate.*

tectosome A term used by Sloss (in Weller, 1958, p.625) for "body of strata indicative of uniform tectonic conditions"; the sedimentary rock record of a uniform tectonic environment or of a tectotope. The term replaces *tectotope* as that term was originally defined.

tectosphere A term for a layer or shell of the Earth that has been variously equated with the lithosphere, the asthenosphere, and the tectonosphere.

tectostratigraphic Pertaining to facies aspects determined by tectonic conditions and influences; said of an interpretative (rather than an objective) stratigraphic facies characterized lithologically "in whatever way is considered to be most significant tectonically" (Weller, 1958, p.630).

tectotope An area of uniform tectonic environment. The term was originally defined by Sloss et al (1949, p.96) as a "stratum or succession of strata with characteristics indicating accumulation in a common tectonic environment", but was used by Krumbein & Sloss (1951, p.381) to designate a tectonic environment. Sloss (in Weller, 1958, p.616) later regarded the term as referring to an area, rather than to a stratigraphic body or an environment, and notes that it "is an almost pure abstraction dependent upon interpretation" of a *tectosome.* Weller (1958, p.636) considers the term "superfluous" because tectonic areas are "generalized and extensive" and not subject to the differentiation possible in the consideration or description of sedimentary environments".

tectum [paleont] (a) The thin, dense or dark, outermost layer of the spirotheca in fusulinids. Cf: *diaphanotheca.* (b) Marginal prolongation of a chamber in trochospirally coiled foraminiferal tests, making sutures of the dorsal (spiral) side more inclined than those of the ventral (umbilical) side. This usage is not recommended because of prior adoption of the term for fusulinids.---Pl: *tecta.*

tectum [palyn] (a) The surface of tectate pollen grains. (b) A term sometimes unfortunately used to designate a projecting flap of exine associated with the laesura on a fossil spore.

teepleite A mineral: $Na_2BO_2Cl.2H_2O$.

teggoglyph *load cast.*

tegillum An umbilical covering in a planktonic foraminiferal test (as in *Globotruncana* and *Rugoglobigerina*), comprising an extension from a chamber comparable to a highly developed apertural lip but extending across the umbilicus, thus completely covering the primary aperture (main opening of the test) and attached at its farther margin or at the tegillum of an earlier chamber. It may have small openings along its margin or be pierced centrally. Pl: *tegilla.*

tegmen [paleont] The oral surface of an echinoderm body; strictly the calcareous adoral (ventral) part of crinoid theca roofing the dorsal cup and situated above the origin of free arms of occupying space between them. It may include calcareous ambulacral and interambulacral plates or be composed entirely of soft tissue (Beerbower, 1968, p.400).

tegmen [bot] (a) The inner coat of a seed, previously the secundine of an ovule. Syn: *endopleura.* (b) The *glume* of a grass.----(Jackson, 1953, p.378).

teilchron A term proposed by Arkell (1933) for the locally recognizable time span of a taxonomic entity. It is synonymous with *teilzone* as defined by Pompeckj (1914), but also represents a geologic-time unit corresponding to the spatial (or biostratigraphic) usage of "teilzone".

teilzone (a) A time term introduced in the German literature as "Teilzone" by Pompeckj (1914) to designate the local duration of existence of a species. Syn: *teilchron.* (b) A spatial term used as a biostratigraphic equivalent of teilzone (or teilchron) and applied to the strata at a specific locality through which some particular fossil actually ranges; e.g. the total observed vertical dimension (range) in space of a given taxonomic entity at a specific locality (Wheeler, 1958a, p.641 & 654). The term in this usage "seems unnecessary" and is replaced by *local range zone* (ACSN, 1961, art.22g).---Etymol: German *Teilzone*, "part of zone".

teineite A blue mineral: $CuTeO_3.2H_2O$.

tejon A term used in the SW U.S. for a solitary, disk-shaped eminence separated by erosion from the mass of which it was originally a part. Etymol: Spanish *tejón*, "round gold ingot". Pron: tay-*hone.* Cf: *huerfano.*

tektite A small (usually walnut-sized), rounded, pitted, jet-black to olive-greenish or yellowish body of silicate glass of nonvolcanic origin, found usually in groups in several widely separated areas of the Earth's surface and bearing no relation to the associated geologic formations. Most tektites have uniformly high silica (68-82%) and very low water contents (average 0.005%): their composition is unlike that of obsidian and more like that of shale. They have various shapes (teardrop, dumbbell, canoe) strongly suggesting modelling by aerodynamic forces, and they average a few grams in weight (largest weighs 3.2 kg). Tektites are believed to be of extraterrestrial origin (e.g. moon splash, formed as gravity-escaping ejecta following large lunar impacts), or alternatively the product of large hypervelocity meteorite impacts on terrestrial rocks. Term proposed by Suess (1900) who believed they were meteorites which at one time underwent thorough melting. Etymol: Greek tektos, "molten". Syn: *tectite; obsidianite.*

tektite field *strewn field.*

tektonite Var. of *tectonite.*

telain A syn. of *provitrain.* It is used in the names of transitional coal lithotypes, e.g. clarotelain. Also spelled: *telite.*

tele A Norwegian term for *frozen ground*, but often used erroneously as a syn. of *permafrost.* Syn: *taele.*

telechemic Said of the earliest minerals to crystallize during solidification of a magma; e.g. zircon, apatite, corundum. Syn: *silicotelic.*

teleconnection Identification and correlation of a series of varves, esp. over great distances or even worldwide, in order to construct a uniform time scale for a part of the Pleistocene Epoch.

telemagmatic Said of a hydrothermal mineral deposit located far from its magmatic source. Cf: *apomagmatic; perimagmatic; cryptomagmatic.* See also: *telemagmatic.*

Telemark snow Any hard snow or snow crust thick enough to support a skier, but soft enough on top to permit controlled turns on skis; it is wet old snow having grains appreciably smaller than those of spring snow. Named after Telemark, a region in southern Norway. Also spelled: *telemark snow.*

telemeter A surveying instrument for measuring the distance of an object from an observer; e.g. a telescope with stadia hairs in which the angle subtended by a short base of known length is measured. See also: *range finder.*

telemetry The automatic transmittal of data to a point remote from the sensing instrument. In widespread usage, the term is restricted to transmittal by electromagnetic propagation.

teleoconch The entire gastropod shell exclusive of the *protoconch.*

teleodont Said of the dentition of a bivalve mollusk (e.g. *Venus*) having differentiated cardinal teeth and lateral teeth, similar to *diagenodont* dentition but with additional elements giving rise to a more complicated hinge.

telescoped Said of ore deposits that represent varying depth of plutonic formation but that appear in a much abbreviated or superimposed sequence.

telescope structure A term proposed by Blissenbach (1954, p.180) for an alluvial-fan structure "characterized by younger fans with flatter gradients spreading out from between fan mesas of older fans with steeper gradients"; e.g. in the Santa Catalina Mountains, Ariz.

telescopic alidade An *alidade* used with a plane table, consisting of a telescope mounted on a straightedge ruler, fitted with level bubble, scale, and vernier to measure angles, and calibrated to measure distances.

teleseism An earthquake that is distant from the recording station.

teleseismology That aspect of seismology which deals with records made at long distances from the point of origin of the impulses. Cf: *engysseismology.*

telethermal Said of a hydrothermal mineral deposit formed at shallow depth and at mild temperatures, with little or no wallrock alteration. Also, said of that environment. See also: *telemagmatic.* Cf: *hypothermal; mesothermal; epithermal; xenothermal; leptothermal.*

teleutospore A *fungal spore* developed in the final stage of the life cycle of rust fungi and whose thick walls may be composed of chitin. Such spores may occur as microfossils in palynologic preparations. Cf: *urediospore.* Syn: *teliospore.*

telinite (a) A variety of provitrinite characteristic of vitrain and consisting of cell-wall material. (b) A suggested, preferable syn. of *provitrinite.*

teliospore *teleutospore.*

telite Var. of *telain.*

tellurbismuth *tellurobismuthite.*

telluric Pertaining to the Earth, esp. the depths of the Earth.

telluric bismuth *tetradymite.*

telluric current *Earth current.*

telluric method An electrical exploration method in which the Earth's natural electric field is measured at two or more stations simultaneously and a quantitative estimate of the geoelectric section obtained thereby.

telluric ocher *tellurite.*

telluric water Water formed by the combination of hydrogen with the oxygen of the atmosphere at high temperature and pressure (Swayne, 1956, p.137). Cf: *juvenile water.*

telluride A mineral compound that is a combination of tellurium with a metal. An example is hessite, Ag_2Te.

tellurite A white or yellowish orthorhombic mineral: TeO_2. It is dimorphous with paratellurite. Syn: *telluric ocher.*

tellurium A silvery-white to brownish-black mineral, the native semimetallic element Te. It is occasionally found in native form (in pyrites and sulfur, or from the fine dust of gold-telluride ores) but more often combined (esp. with metals in tellurides).

tellurium glance *nagyagite.*

tellurobismuthite A pinkish mineral: Bi_2Te_3. It is often intergrown with tetradymite. Syn: *tellurbismuth.*

telluroid A surface near the terrain which is the locus of points in which the spheropotential coincides with the geopotential of corresponding points on the terrain.

tellurometer A rugged, lightweight, portable electronic device that measures ground distances precisely by determining the velocity of a phase-modulated, continuous, microwave radio signal transmitted between two instruments operating alternately as master station and remote station. It has a range up to 65 km (35-40 miles). Etymol: *Tellurometer,* a trade name for a distance-measuring system produced by Tellurometer Ltd., Cape Town, Union of South Africa. Cf: *geodimeter.*

telmaro A term proposed by Veatch & Humphrys (1966, p.326) for a river traversing a peat marsh or peat swamp. Etymol: Greek *telma,* "marsh".

telmatic peat A lowmoor peat developed in very shallow water and consisting mainly of reeds. Syn: *reed peat.*

telmatology The study of wet lands, e.g. marshy areas and swamps.

teloclarain A transitional lithotype of coal characterized by the presence of telinite, but more of other macerals than of telinite. Cf: *clarotelain.* Also spelled: *teloclarite.*

teloclarite Var. of *teloclarain.*

telodurain A coal lithotype transitional between durain and te-

lain, but predominantly durain. Cf: *durotelain.*

telofusain A coal lithotype transitional between fusain and te[lain, but predominantly fusain. Cf: *fusotelain.*

telogenetic A term proposed by Choquette & Pray (197[p.220-221) for the period during which long-buried carbona[rocks are affected significantly by processes related to weath[ering and subaerial and subaqueous erosion. Also applied [the porosity that develops during the telogenetic stage. C[*eogenetic; mesogenetic.*

telome In a plant that is thought of as a branched axis, eac[ultimate branch or division of the axis; the fundamental stru[tural unit of a plant (Darrah, 1939, p.10). See also: *telon[theory.*

telome theory A theory in botany that describes the enti[plant body as a branched axis, each division or branch havin[a certain function. See also: *telome.*

telson (a) The last somite of the body of a crustacean, bea[ing the anus and commonly the caudal furca. Syn: *postabd[men; style* [paleont]. (b) A dorsal, postanal extension of th[body of an arachnid, articulated to the last abdominal se[ment; a postanal spine or plate of a merostome. (c) A term[nal or anal segment of a trilobite. The term is sometimes i[correctly used for a spine mounted on the terminal or one [the near-terminal segments of a trilobite and directed pos[eriorly along the midline, such as the first macrospine on th[posterior part of the thorax in certain Olenellidae (TIP, 195[pt.O, p.126).

temblor A syn. of *earthquake.* Etymol: Spanish, a "trembling[Also spelled: *tremblor.*

temperate Said of a temperature that is moderate or mild. Th[term is also used to describe temperatures of the middle la[itudes, whether moderate or not.

temperate glacier A glacier of the type characteristic of th[temperate zone, in which at the end of the ablation seaso[the firn and ice of which it consists are near the melting poi[(Ahlmann, 1933). Its temperature is in fact close to $0°$[throughout, except during the winter, when its upper part [frozen to a depth of several meters. Examples: almost all th[glaciers in Scandinavia and the Alps, and in the U.S. outsid[of northern Alaska. Cf: *polar glacier.*

temperature A basic property of a system in thermal equilit[rium, measured by various arbitrary, empirical scales base[on changes in volume, electrical resistance, thermal electr[motive force, or length. Systems in thermal equilibrium wit[each other have the same temperature. See also: *thermome[ter.*

temperature coefficient A number equal to the ratio of a mea[sured temperature change and a simultaneous measure[change in some other physical property such as solubility, vo[ume, electrical resistance, length. Coefficients vary dependin[on use.

temperature compensator An instrument that corrects for th[effect of temperature on a physical measurement.

temperature gradient *thermal gradient.*

temperature-gradient metamorphism A process of modifica[tion of ice crystals in deposited snow, characterized by vapo[transfer under strong vapor pressure and temperature gradi[ents, and resulting in the growth of complex-shaped crystals[usually with stepped or layered surfaces, termed *depth hoa[* Syn: *constructive metamorphism.* Cf: *equitemperature meta[morphism.*

temperature log A *geophysical log* that graphically shows th[relation of temperature to depth in a borehole or well. It i[useful in detecting the top of heat-producing cement pumpe[outside casing and (together with gamma-ray logs) in showin[temperatures in cased holes. Syn: *thermal log.*

temperature-salinity diagram A plotting of temperature an[salinity in a column of water, thus identifying the water mass[es within it, its stability, and σ_T value. Syn: *T-S diagram.*

temperature survey Measurement of temperature in drill hole[

to an absolute accuracy of about 0.05°C and to a precision (relative accuracy) of about 0.005°C. Maps of isotherm surfaces can be constructed which help to detect or interpret anomalies in geologic structure or subsurface ground-water conditions. The shapes of isotherm surfaces may be related to differences in the thermal conductivities of rocks, differences in heat flow, ground-water disturbances, etc.

temperature zone A general term for a region characterized by a relatively uniform temperature or temperature range. It may refer to a region of a particular *climatic classification* (a latitudinal division) or to a particular temperature belt on a mountainside (a vertical or altitudinal division). Cf: *climatic zone.*

templet Var. of *template.* The term is used widely in photogrammetry for a transparent celluloid overlay made over an aerial photograph and showing the center (generally the principal point of the photograph) and all radial lines from the center through images of control points as well as showing the azimuth lines connecting the center with images of points that show on the photograph and are themselves the centers of other photographs.

temporal transgression A name given by Wheeler & Beesley (1948, p.75) to the principle that rock units and unconformities vary in age from place to place.

temporary base level Any *base level,* other than the actual level of the ocean (sea level), below which a large or small land area cannot be reduced, for the time being, by ordinary erosion; e.g. a level locally controlled by a particularly resistant stratum in a stream bed, or the surface of a lake in an interior basin, or (for a tributary) the level of the main stream into which the tributary flows. Cf: *ultimate base level.* Syn: *local base level.*

temporary bench mark A supplementary *bench mark* of less enduring character than a *permanent bench mark,* intended to serve for only a comparatively short period of time (such as a few years); e.g. an intermediate bench mark established at a junction of sections of a continuous series of measured differences of elevation for the purpose of holding the end of a completed section and serving as an initial point from which the next section is run. It elevation determination may not be precise. Examples include a chiseled square or cross on masonry, a nail and washer in the root of a tree, a spike or screw in a pole, and a bolt on a bridge. Abbrev: T.B.M.

temporary extinction The *extinction* of a lake by the loss of water only (such as due to climatic changes), the lake basin remaining intact to receive water at some future time.

temporary hardness *carbonate hardness.*

temporary lake *intermittent lake.*

temporary plankton *meroplankton.*

temporary stream *intermittent stream.*

temporary wilting A degree of wilting from which a plant can recover by decreasing its rate of transpiration and without adding water to the soil. Cf: *permanent wilting; wilting point.*

tenacity The property of the particles or molecules of a substance to resist separation; *tensile strength.*

tender A general descriptor for a material that fractures under stress rather than absorbing it by plastic deformation; i.e. that lacks tensile strength. Cf: *tough.* Syn: *short.*

tendril A vinelike, coiling plant part that helps support the stem.

tenebrescence In optics, the absorption of light by a crystal under the influence of radiation.

tengerite A mineral: $CaY_3(CO_3)_4(OH)_3.3H_2O$ (?). The name was applied originally to a supposedly beryllium-yttrium carbonate.

tennantite A blackish lead-gray isometric mineral: $(Cu, Fe)_{12}As_4S_{13}$. It is isomorphous with tetrahedrite, and sometimes contains zinc, silver, or cobalt replacing part of the copper. It is an important ore of copper. Syn: *fahlore; gray copper ore.*

tenor *grade* [ore dep].

tenorite A triclinic mineral: CuO. It occurs in minute, shining, steel- or iron-gray scales, in black powder, or in black earthy masses, generally in the oxidized (weathered) zones or gossans of copper deposits. Tenorite is an ore of copper. Syn: *melaconite; black copper.*

tensile Said of or pertaining to a substance or force that is undergoing or exerting tension.

tensile strength The maximum applied tensile stress that a body can withstand before failure occurs. Syn: *tenacity.*

tensile stress A *normal stress* that tends to cause separation across the plane on which it acts. Cf: *compressive stress.*

tension A state of stress in which tensile stresses predominate; stress that tends to pull a body apart.

tension crack An *extension fracture* caused by tensile stress. Cf: *shear crack.*

tension fault A general and nonrecommended term for any fault in which crustal tension is assumed to be a factor. This would include most normal faults. Cf: *compression fault.* Syn: *extensional fault.* Partial syn: *extension fault.*

tension fracture A minor fracture in a rock that develops perpendicular to the direction of greatest tension, e.g. at right angles to a fold axis. Cf: *shear fracture.* See also: *tension joint; extension fracture.* Syn: *subsidiary fracture.*

tension gash A short tension fracture along which the walls have been pulled apart. They may be open or filled, and commonly have an en echelon pattern. They may occur diagonally in fault zones, or perpendicular to the cleavage in boudinage.

tension joint A joint that is a *tension fracture.*

tension zone *ecotone.*

tentacle Any of various elongate and flexible processes that have various functions and that are borne by invertebrates; e.g. a movable, commonly simple but rarely forked, tubular extension of soft integument rising from the oral disk of a coral polyp and closed terminally at the tip and serving primarily for food getting, or a tube foot of an asterozoan or arm of a crinoid, or a short, slender, arm-like sensory appendage surrounding the mouth and extending anteriorly in front of the head of a cephalopod, or one of the numerous small ciliated processes borne on the arms of a brachiopod or the lophophore of a bryozoan and used primarily for food getting.

tentacle pore A term more commonly used for *podial pore* of ophiuroids (TIP, 1966, pt.U, p.30).

tentacle sheath A thin, delicate, membranous, introverted part of the body wall of a bryozoan, enclosing the tentacles when the polypide is retracted.

tentaculitid An invertebrate animal characterized by radial symmetry, a small, conical shell with transverse rings of variable size and spacing, longitudinal striae, an embryonic chamber with a bluntly pointed apex, and by the presence of small pores in the shell wall. The tentaculitids belong to the order Tentaculitida and are questionably assigned to the mollusks. Their stratigraphic range is Lower Ordovician to Upper Devonian, the oldest representatives belonging to the genus *Tentaculites.* Var: *tentaculite.*

tented ice Sea ice deformed by *tenting.*

tent hill A term used in Australia for a butte or flat-topped hill resembling a canvas tent; it often is capped by resistant rock from a former plateau surface. Cf: *tepee butte.*

tenting The vertical displacement upward of sea ice under lateral pressure to form a flat-sided arch over a cavity between the raised ice and the water beneath; a type of *ridging.*

tenuitas A thin area in the exine of pollen and spores, usually germinal in function, as the annular tenuitas of *Classopollis.* A tenuitas is a less distinct feature than a colpus or pore.

tepee butte A conical hill or knoll resembling an American Indian tepee; esp. an isolated, residual hill formed by a capping of resistant rock that protects the underlying softer material from erosion, as one of the limestone hills in the Pierre Shale of Colorado, or one of the sandstone-capped hills in the Painted Desert, Ariz. Cf: *tent hill*.

tepee structure A disharmonic sedimentary structure consisting of a fold which in cross section resembles a chevron or an "inverted depressed V" or the profile of a peaked dwelling of the North American Indians (Newell et al, 1953, p.126). It is believed to be a diagenetic structure formed during the hydration of anhydrite. See also: *enterolithic*.

tepetate (a) An evaporite consisting of a calcareous crust coating solid rocks on or just beneath the surface of an arid or semiarid region; a deposit of *caliche*. (b) A term used in Mexico for a volcanic tuff, or a secondary volcanic or chemical nonmarine deposit, very commonly calcareous (Brown & Runner, 1939, p.338).---Etymol: Mexican Spanish, from Nahuatl (Aztec) *tepetatl*, "stone matting". See also: *talpatate*.

tephra A general term for all *pyroclastics* of a volcano.

tephrite A group of extrusive rocks, of basaltic character, primarily composed of calcic plagioclase, augite, and nepheline or leucite as the main feldspathoids, with accessory sodic sanidine; also, any member of that group; the extrusive equivalent of *theralite*. With the addition of olivine, the rock would then be called a *basanite*.

tephritoid n. A term proposed by Bucking in 1881, but never adopted, for a group of rocks intermediate in composition between basalts and tephrites (Johannsen, 1939, p.283), i.e. having the chemical composition of a tephrite but with a soda-rich glassy groundmass in place of nepheline.----adj. Said of a tephrite-like rock.

tephrochronology The dating of layers of volcanic ash in order to establish a sequence of geologic and archaeologic events.

tephroite A mineral of the olivine group: Mn_2SiO_4. It occurs with zinc and manganese minerals.

terebrataliiform Said of the loop, or of the growth stage in the development of the loop, of a dallinid brachiopod (as in *Terebratalia*), consisting of long descending branches with connecting bands to the median septum, then recurving into ascending branches that meet in transverse band (TIP, 1965, pt.H, p.154). The terebrataliifrom loop is morphologically similar to the *terebratelliform* loop.

terebratellacean n. Any terebratuloid belonging to the superfamily Terebratellacea, characterized by a long brachial loop. Their stratigraphic range is Upper Triassic to present.----adj. Said of a terebratuloid having a long brachial loop, or of its shell.

terebratellid Any terebratuloid belonging to the suborder Terebratellidina, characterized by a loop that develops in connection with both cardinalia and median septum. Their stratigraphic range is Lower Devonian to present.

terebratelliform Said of the loop, or of the growth stage in the development of the loop, of a terebratellid brachiopod (as in the subfamily Terebratellinae), consisting of long descending branches with connecting bands to the median septum, then recurving into ascending branches that meet in transverse band (TIP, 1965, pt.H, p.154). The terebratelliform loop is morphologically similar to the *terebrataliiform* loop.

terebratulacean Any terebratuloid belonging to the superfamily Terebratulacea, characterized by the development of the cardinal process and outer hinge plates and by the absence of inner hinge plates or hinge plates. Their stratigraphic range is Upper Triassic to present.

terebratuliform Said of a short, typically U- or W-shaped loop found in most terebratulacean brachiopods.

terebratuliniform Said of a short brachiopod loop in which crural processes are fused medially to complete a ring- or box-like apparatus (TIP, 1965, pt.H, p.154).

terebratuloid Any articulate brachiopod belonging to the order Terebratulida, characterized chiefly by a punctate shell with a teardrop-shaped outline, pointed at the posterior end. Their stratigraphic range is Lower Devonian to present. Var: *terebratulid*.

tergal fold *epimere*.

tergite The dorsal plate or dorsal part of the covering of a segment of an articulate animal; e.g. the sclerotized dorsal surface of a single crustacean somite, or a hardened chitinous plate on the dorsal surface of an arachnid body segment, or a plate forming the dorsal cover of a merostome somite.

tergum (a) One of a pair of opercular valves, adjacent to the carina in cirripede crustaceans, lacking adductor-muscle attachments. Cf: *scutum*. (b) The back or dorsal surface of the body of an animal.---Pl: *terga*.

terlinguaite A yellow monoclinic mineral: Hg_2ClO.

terminal curvature (a) The sharp, local change in the dip of strata or cleavage near a fault, caused by the drag of the downthrown side against the fault plane; *drag*. (b) *terminal creep*.----The expression is not commonly used in the U.S.

terminal face The lower extremity, or *snout*, of a glacier.

terminal moraine (a) An end moraine, extending across a glacial valley as an arcuate or crescentic ridge, that marks the farthest advance or maximum extent of a glacier; the outermost end moraine of a glacier. It is formed at or near a more-or-less stationary edge, or at a place marking the cessation of an important glacial advance. (b) A term sometimes used as a syn. of *end moraine*.

terminal plate A single plate at the end of an arm of an asteroid and appearing very early in ontogeny. Syn: *terminal*.

terminal tentacle A terminal podium of a radial vessel of the water-vascular system of an echinoid, extending through an *ocular pore*.

terminal velocity The limiting velocity reached asymptotically by a particle falling under the action of gravity in a still fluid (ASCE, 1962).

terminator The line separating the illuminated and the unilluminated parts of a celestial body (a planet, the Moon, etc.); the dividing line between day and night as observed from a distance.

terminus The lower margin or extremity of a glacier; the *snout*.

termitarium A mound of mud built by termites, ranging up to 4 m in height, and commonly found in the lateritic soil belts of the tropics and subtropics. Pl: *termitaria*. Syn: *termite mound; anthill*.

ternary diagram A triangular diagram that graphically depicts the composition of a three-component mixture or ternary system.

ternary feldspar Any feldspar containing more than 5% of a third component; e.g. anorthoclase, soda sanidine, and potassian oligoclase or potassian andesine.

ternary sediment A sediment consisting of a mixture of three components or end members; e.g. a sediment with one clastic component (such as feldspar) and two chemical components (such as calcite and quartz), or an aggregate containing sand, silt, and clay.

ternary system A system having three components, e.g. CaO-Al_2O_3 - SiO_2.

ternovskite A monoclinic mineral of the amphibole group: $Na_2(Mg,Fe^{+2})_3Fe_2^{+3}Si_8O_{22}(OH)_2$. It is near riebeckite in chemical composition.

terra A bright upland or mountainous region on the surface of the Moon, characterized by a lighter color than that of a *mare*, by relatively high albedo, and by a rough texture formed by large intersecting or overlapping craters. It may represent a remnant of an ancient lunar surface, sculptured largely by impact of meteorites; it may also be attributed to igneous and volcanic activity from within the Moon. Etymol: Latin, "earth". Pron: *ter-uh*. Pl: *terrae*. Syn: *continent*.

terra cariosa A syn. of *rottenstone*. Etymol: Latin, "rotten earth".

terrace [soil] A horizontal or gently sloping, artificial ridge or embankment of earth built along the contours of a sloping hillside for the purpose of conserving moisture, reducing ero-

...ion, or controlling runoff.

errace [coast] (a) A narrow, gently sloping, constructional coastal strip extending seaward or lakeward, and veneered by a sedimentary deposit; esp. a *wave-built terrace*. See also: *marine terrace*. (b) Loosely, a stripped wave-cut platform that has been exposed by uplift or by lowering of the water level; an elevated wave-cut bench.

errace [geomorph] (a) Any long, narrow, relatively level or gently inclined surface, generally less broad than a plain, bounded along one edge by a steeper descending slope and along the other by a steeper ascending slope; a large bench or step-like ledge breaking the continuity of a slope. The term is usually applied to both the lower or front slope (the riser) and the flattish surface (the tread), and it commonly denotes a valley-contained, aggradational form composed of unconsolidated material as contrasted with a *bench* eroded in solid rock. A terrace commonly occurs along the margin and above the level of a body of water, marking a former water level; e.g. a *stream terrace*. (b) A term commonly but incorrectly applied to the deposit underlying the tread and riser of a terrace, esp. the alluvium of a stream terrace; "this deposit ... should more properly be referred to as a fill, alluvial fill, or alluvial deposit, in order to differentiate it from the topographic form" (Leopold et al, 1964, p.460). (c) *structural terrace*.

errace [marine geol] A bench-like structure on the ocean floor.

errace cusp *meander cusp*.

erraced flowstone *rimstone dam*.

erraced flute cast A flute cast with external sculpturing resembling differentially weathered laminae, but "in reality, a cast of differentially eroded laminations in the underlying shale and unrelated to internal structure of the cast" (Pettijohn & Potter, 1964, p.347).

erraced pool One of the shallow, rimmed pools on a reef surface, circular around the point where the water reaches the surface, produced by the growth of lime- and silica-secreting algae, and "arranged in successively lower terraces as sections of a circle around the overflow from the higher level" (Kuenen, 1950, p.431-432).

Terrace epoch An obsolete term formerly applied informally to the earlier part of the Holocene Epoch characterized by the general formation of stream terraces in the drift-filled valleys of the regions glaciated during the preceding Pleistocene Epoch. Syn: *Terracian*.

errace flight A series of terraces resembling a series of stairs, formed by the swinging meanders of a degrading stream that continuously excavates its valley.

errace meander A meander formed by the incision of a free meander associated with a former valley floor whose remnants now form a terrace (Schieferdecker, 1959, term 1492).

errace placer *bench placer*.

errace plain A well-developed stream terrace that represents a narrow but "true" plain (Tarr, 1902, p.88).

errace slope The scarp or bluff below the outer edge of a terrace; the front or face of a terrace.

erracette [mass move] A small ledge, bench, or step-like form produced on the surface of a slumped soil mass along a steep grassy slope or hillside, varying from several centimeters to 1.5 m in height and averaging a meter in width, and developed as a result of small landslides and subsequent backward tilting of the soil surface. See also: *sheep track; catstep*.

erracette [pat grd] An obsolete syn. of *step*.

Terracian *Terrace epoch*.

Terracing (a) The formation of terraces, as by the shrinkage of a glacier. (b) A terraced structure, feature, or contour.

erra-cotta A fired or kiln-burnt clay of a peculiar brownish-red or yellowish-red color, used for statuettes, figurines, and vases, and for ornamental work on the exterior of buildings.

Also, an object made of terra-cotta. Etymol: Italian, "baked earth".

terra-cotta clay A general term applied loosely to any fine-textured, fairly plastic clay that acquires a natural vitreous skin in burning and that is used in the manufacture of terra-cotta. It is characterized by low shrinkage, freedom from warping, strong bonding, and absence of soluble salts.

terradynamics The study of projectile penetration of natural earth materials (Colp, 1967, p.38).

terrae Plural of *terra*.

terrain Var. of *terrane*.

terrain correction A correction applied to observed values obtained in geophysical surveys in order to remove the effect of variations to the observations due to the topography in the vicinity of the sites of observation. (Heiland, 1940) Syn: *topographic correction*.

terrain profile recorder *airborne profile recorder*.

terra Lemnia (a) *Lemnian bole*. (b) A clay, perhaps cimolite (Dana, 1892, p.689).

terra miraculosa *bole*.

terrane An obsolescent term applied to a rock or group of rocks and to the area in which it outcrops. The term is used in a general sense and does not necessarily imply a specific rock unit or group or rock units. Var: terrain.

terraqueous zone That part of the lithosphere that is penetrated by water.

terra rosa Var. of *terra rossa*.

terra rossa A reddish-brown, residual soil found as a mantle over limestone bedrock, typically in the karst areas around the Adriatic Sea, under conditions of Mediterranean-type climate. Also spelled: *terra rosa*. Etymol: Italian, "red earth".

terra roxa A deep, porous, reddish purple soil of the Paraná plateau of eastern Brazil. It is developed from diabase and has a high content of humus. The term is not to be confused with *terra rossa*. Etymol: Portuguese, "purple soil".

terra verde A syn. of *green earth*. Etymol: Italian.

terreplein An embankment of earth with a broad level top.

terrestrial (a) Pertaining to the Earth. Cf: *planetary*. (b) Pertaining to the Earth's dry land.

terrestrial deposit (a) A sedimentary deposit laid down on land above tidal reach, as opposed to a marine deposit, and including sediments resulting from the activity of glaciers, wind, rainwash, or streams; e.g. a lake deposit, or a *continental deposit*. (b) Strictly, a sedimentary deposit laid down on land, as opposed to one resulting from the action of water; e.g. a glacial or eolian deposit. (c) A sedimentary deposit formed by springs or by underground water in rock cavities.---Cf: *terrigenous deposit*.

terrestrial equator The *equator* on the Earth's surface.

terrestrial latitude Latitude on the Earth's surface.

terrestrial longitude Longitude on the Earth's surface.

terrestrial magnetism *geomagnetism*.

terrestrial meridian A *meridian* on the Earth's surface; specif. an *astronomic meridian*.

terrestrial peat Peat that is developed above the water table.

terrestrial planet A planet similar to the Earth in terms of size and mean density and possession of derived oxidizing atmospheres, specif. Mercury, Venus, Earth, and Mars. Sometimes Pluto is included.

terrestrial pole *geographic pole*.

terrestrial radiation The infrared radiation emitted from the Earth's surface (including the oceanic surface). Cf: *counter-radiation*. See also: *effective terrestrial radiation*. Syn: *eradiation; Earth radiation*.

terre verte A syn. of *green earth*. Etymol: French.

terrigenous deposits Shallow marine sediments consisting of material eroded from the land surface. Cf: *hemipelagic deposits; pelagic deposits*.

territorial sea The coastal waters (and the accompanying seabed) under the jurisdiction of a maritime nation or state, usually measured from the mean low-water mark or from the seaward limit of a bay or river mouth and, as originally defined under international law, extending 3 nautical miles (about 5.6 km) outward to the *high seas*, although the U.S. officially accepts a 6-mile limit and attempts have been made to extend it as much as 15 miles (28 km). See also: *marginal sea.* Syn: *territorial waters.*

territorial waters (a) The surface waters under the jurisdiction of a nation or state, including *inland waters* and *marginal sea.* (b) *territorial sea.*

Tertiary The first period of the Cenozoic era (after the Cretaceous of the Mesozoic era and before the Quaternary), thought to have covered the span of time between 65 and three to two million years ago. It is divided into five epochs: the Paleocene, Eocene, Oligocene, Miocene, and Pliocene. The name was originally assigned as an era rather than a period designation; in this sense, it may be considered to have either four periods (Eocene, Oligocene, Miocene, Pliocene) or two (the Paleogene and Neogene), with the Pleistocene and Holocene included in the Neogene.

tertiary structure Very coarsely irregular shell material in the lorica of a tintinnid. Cf: *primary structure; secondary structure.*

Tertiary-type ore deposit An *epithermal* ore deposit, generally associated with volcanism.

tertschite A mineral: $Ca_4B_{10}O_{19} \cdot 20H_2O$.

teschemacherite A yellowish to white mineral: $(NH_4)HCO_3$.

teschenite A granular hypabyssal rock containing calcic plagioclase, augite, sometimes hornblende, and a small amount of biotite, with interstitial analcime. It is of darker color than *bogusite* and is distinguished from theralite by the presence of analcime in place of nepheline.

tesselation A geomorphic feature resembling a mosaic pattern, as a *sand-wedge polygon* or the fractured surface of the salty crust or certain dry lakes in Australian deserts. Also spelled: *tessellation.*

tessera The singular form of *tesserae.*

tesserae A syn. of *felder.* Its singular form is *tessera.*

tesseral system An obsolete syn. of *isometric system.*

test [paleont] (a) The external shell, secreted exoskeleton, or other hard or firm covering or supporting structure of many invertebrates, such as the plates of the coronal, apical, periproctal, and peristomial systems of an echinoid; esp. a gelatinous, calcareous, or siliceous foraminiferal shell composed of secreted platelets or solid wall, or formed of agglutinated foreign particles, or a combination of these. A test may be enclosed within an outer layer of living tissue, such as a protozoan shell enclosed in cytoplasm. (b) The *theca* of a dinoflagellate.

test [oil] Any procedure for sampling the content of an oil reservoir; e.g. a drill-stem test or a wire-line test. Cf: *A test well.*

testa The seed coat of a flowering plant developed from the *integument* of the ovule.

testaceous Having or consisting of a shell; specif. pertaining to the test of an invertebrate.

tester A service-company representative who supervises drill-stem testing operations.

test hole (a) A general term for any type of hole, pit, shaft, etc., dug or drilled for subsurface reconnaissance. More specific terms, e.g. *test pit, trial pit,* are generally used. (b) Such a hole used for investigating mineral-bearing ground; a *prospect hole.*

test pit *test hole.*

test reach A *reach,* esp. one between two gaging stations, that is long enough to be used in the determination of slope.

test well [oil] An *exploratory well* drilled to determine the presence and commercial value of oil in an unproven area. Cf: *test* [oil].

test well [water] A well drilled in search for water; e.g. a well drilled adjacent to a lake to determine the relationship be-

tween the ground-water level and the lake level.

tetartohedral Said of that crystal class in a system, the general form of which has only one fourth the number of equivalent faces of the corresponding form in the *holohedral* class of the same system. Cf: *merohedral.*

tetartohedron Any crystal form in the tetartohedral class of a crystal system.

tetartoid An isometric, closed crystal form having 12 faces that correspond to one fourth of the faces of a hexoctahedron. A tetartoid may be right-handed or left-handed.

tetartoidal class That crystal class of the isometric system having symmetry 23.

Tethys A sea, similar to the present Mediterranean Sea, which existed for long periods of geologic time between the northern and southern continents of the Eastern Hemisphere, along the general course of the Alpine-Himalayan orogenic belt; a composite geosyncline out of which many of the structures of the present orogenic belt were formed. Its principal history was between the Hercynian and Alpine orogenies, or from the Permian into the early Tertiary.

tetraclone A four-armed desma of a sponge built about a tetraxial crepis; e.g. a *trider.*

tetracoral *rugose coral.*

tetracrepid Said of a desma (of a sponge) with a tetraxial crepis.

tetractin A sponge spicule having four rays. Syn: *tetract.*

tetrad A symmetric grouping of four embryophytic spores (or pollen grains) that result from meiotic division of one mother cell. A number of pollen types regularly remain in united tetrads as mature pollen when shed by the pollen sacs (as in the fossil *Classopollis* or in the living *Rhododendron*). Cf: *dyad; polyad.*

tetrad scar *laesura.*

tetradymite A pale steel-gray mineral: Bi_2Te_2S. It occurs usually in foliated masses in auriferous veins, often with tellurobismuthite. Syn: *telluric bismuth.*

tetraene A sponge spicule with one long ray and four short rays at one end.

tetragonal dipyramid A crystal form of eight faces consisting of two tetragonal pyramids repeated across a mirror plane of symmetry. A cross section perpendicular to the unique four-fold axis is ideally square. Its indices are $\{h0l\}$ or $\{hhl\}$ in most tetragonal crystals, also $\{hkl\}$ or $\{khl\}$ in class $4/m$.

tetragonal dipyramidal class That crystal class in the tetragonal system having symmetry $4/m$.

tetragonal disphenoid A crystal form consisting of four faces, ideally isosceles triangles, in which the unique $\bar{4}$-axis joins those two edges which are at right angles. Its indices are $\{hhl\}$, $\{h\bar{h}l\}$, in $\bar{4}2m$, or $\{h0l\}$, $\{hhl\}$, or $\{hkl\}$ in $\bar{4}4$. Cf: tetrahedron; orthorhombic disphenoid.

tetragonal-disphenoidal class That crystal class in the tetragonal system having symmetry 4.

tetragonal prism A crystal form of four equivalent faces parallel to the symmetry axis that are, ideally, square in cross section. Its indices are $\{100\}$ or $\{110\}$ with symmetry $4/m \, 2/m \, 2/m$, or $\{hk0\}$ in $4/m$.

tetragonal pyramid A crystal form consisting of four equivalent faces, ideally isosceles triangles, in a pyramid that is square in cross section. Its indices are $\{h0l\}$ and $\{hhl\}$ in symmetry 4mm, also $\{hkl\}$ in symmetry 4.

tetragonal-pyramidal class That crystal class in the tetragonal system having symmetry 4.

tetragonal-scalenohedral class That crystal class in the tetragonal system having symmetry $\bar{4} \, 2m$.

tetragonal scalenohedron A *scalenohedron* of eight faces, with symmetry $\bar{4} \cdot 2m$ and indices $\{hkl\}$. It resembles a disphenoid Cf: *hexagonal scalenohedron.*

tetragonal system One of the six *crystal systems,* characterized by a fourfold rotation or rotatory inversion axis, and in which the crystals are referred to three mutually perpendicular axes, the vertical one of which is of unequal length relative to the two horizontal axes. Cf: *isometric system; hexagonal system; orthorhombic system; monoclinic system; triclinic system.* Syn: *quadratic system; pyramidal system.*

tetragonal-trapezohedral class That crystal class in the tetragonal system having symmetry 422.

tetragonal trapezohedron A crystal form of eight faces, each of which is a trapezium. Its indices are $\{hkl\}$ in symmetry 422, and it may be right- or left-handed.

tetragonal tristetrahedron *deltohedron.*

tetrahedral Having the symmetry or form of a *tetrahedron.*

tetrahedral coordination An atomic structure or arrangement in which an ion is surrounded by four ions of opposite sign, whose centers form the points of a tetrahedron around it. It is typified by SiO_4. Syn: *fourfold coordination.*

tetrahedral hypothesis A corollary of the hypothesis of a *contracting Earth*; as the Earth cooled and contracted it assumed a broadly tetrahedral form, with the continents on the nodes and the ocean basins on the sides. The original concept was fanciful, and neither it nor the contraction hypothesis can be reconciled with modern concepts of Earth evolution.

tetrahedral radius The radius of a cation when in tetrahedral coordination.

tetrahedrite A steel-gray to iron-black isometric mineral: $(Cu, Fe)_{12}Sb_4S_{13}$. It is isomorphous with tennantite, and often contains zinc, lead, mercury, cobalt, nickel, or silver replacing part of the copper. Tetrahedrite commonly occurs in characteristic tetrahedral crystals associated with copper ores. It is an important ore of copper and sometimes a valuable ore of silver. Syn: *fahlore; gray copper ore; panabase; stylotypite.*

tetrahedron A crystal form in cubic crystals having symmetry $4\,3m$ or 23. It is a four-faced polyhedron, each face of which is a triangle. It is descriptive of the silicon-oxygen structure of silicates. Adj: *tetrahedral.*

tetrahexahedron An isometric crystal form having 24 faces that are isosceles triangles and that are arranged four to each side of a cube. Its indices are $\{hk0\}$ and its symmetry is $4/m\,\overline{3}\,2/m$. Syn: *tetrakishexadron.*

tetrakalsilite A mineral: $(K,Na)AlSiO_4$. It is a structural variety of kalsilite with an *a*-axis of about 20 angstroms. Cf: *trikalsilite.*

tetrakishexadron *tetrahexahedron.*

tetramorph One of four crystal forms displaying *tetramorphism.*

tetramorphism That type of *polymorphism* in which there occur four crystal forms, known as *tetramorphs.* Adj: *tetramorphous.* Cf: *dimorphism; trimorphism.*

tetramorphous Adj. of *tetramorphism.*

tetrasporine line An evolutionary series in the green algae, ranging from the palmelloid type to the filamentous-growth type. Cf: *siphonous line; volvocine line.*

tetratabular archeopyle An *apical archeopyle* formed in a dinoflagellate cyst by the loss of four plates.

tetraxon A sponge spicule in which the rays grow along four axes arranged as the diagonals of a tetrahedron.

Texas tower A radar-equipped offshore platform, mounted on the continental shelf or a shoal, and designed in part to provide oceanographic and meteorological data. Etymol: *texas*, a structure on the deck of a steamboat.

textulariid Any agglutinated foraminifer belonging to the family Textulariidae. Their stratigraphic range is Carboniferous to present.

textural maturity A type of sedimentary *maturity* in which a sand approaches the textural end product to which it is driven by the formative processes that operate upon it. It is defined in terms of uniformity of particle size and perfection of rounding and depends upon the stability of the depositional site and the input of modifying wave and current energy; it is independent of mineral composition (Folk, 1951). A sandstone may pass sequentially through four stages of textural maturity: im-

mature, submature, mature, and supermature. Cf: *compositional maturity; mineralogic maturity.*

texture [petrology] The general physical appearance or character of a rock, including the geometric aspects of, and the mutual relations among, the component particles or crystals; e.g. the size, shape, and arrangement of the constituent elements of a sedimentary rock, or the crystallinity, granularity, and fabric of the constituent elements of an igneous rock. The term is applied to the smaller (megascopic or microscopic) features as seen on a smooth surface of a homogeneous rock or mineral aggregate. The term *structure* is generally used for the larger features of a rock. "Texture" should not be used synonymously with "structure", although some textural features such as foliation or flow texture may parallel major structural features of a rock.

texture [geomorph] *topographic texture.*

texture [soil] The physical nature of the soil according to the relative proportions of elements of various composition and their particle size, e.g. sand, clay, and silt and their various mixtures.

texture ratio Ratio of the greatest number of streams crossed by a contour line within a drainage basin to the length of the upper basin perimeter intercept (Smith, 1950, p.656); a measure of *topographic texture.* Symbol: T.

tey A muddy islet near the mouth of the Rhône River (Reclus, 1872, p.358).

thalassic (a) Pertaining to the deep ocean. (b) Pertaining to the seas and gulf.----The term is not commonly used.

thalassocratic (a) Adj. of *thalassocraton.* (b) Said of a period of high sea level in the gologic past. Cf: *epeirocratic.*

thalassocraton A *craton* that is part of the oceanic crust. The concept of cratonic areas in oceanic crust is now outmoded. Cf: *hedreocraton; epeirocraton.* Adj: *thalassocratic.*

thalassogenesis A Russian term synonymous with *basification.*

thalassoid A lunar *mare basin* not filled, or only partly filled, with mare material; e.g. the Nectaris Basin.

thalassophile element An element that is relatively more abundant in sea water than it is in normal continental waters, e.g. chlorine, bromine, iodine, sodium, and boron. The term is rarely used.

thalenite A flesh-red or pink mineral: $Y_2Si_2O_7$. Cf: *thortveitite; yttrialite.*

thallite A yellowish-green *epidote.*

thallogen A plant whose growth is not restricted to an apical growing point.

thallophyte A nonvascular plant without differentiated roots, stems, or leaves. Algae and fungi are thallophytes. Cf: *bryophyte.*

thallus The body of certain simple plants such as algae, seaweeds, and liverworts, that is characterized by having relatively little cellular differentiation and no true roots, stems, or leaves. It is flattened in shape, perhaps lobed or ribbonlike.

thalweg [geomorph] The line of continuous maximum descent from any point on a land surface; e.g. the line of greatest slope along a valley floor, or the line crossing all contour lines at right angles, or the line connecting the lowest points along the bed of a stream. Etymol: German *Talweg,* "valley way".

thalweg [coast] *midway.*

thalweg [grd wat] A subsurface ground-water stream percolating beneath and generally in the same direction as the course of a surface stream or valley.

thalweg [streams] (a) The line connecting the lowest or deepest points along a stream bed or valley, whether under water or not; the *longitudinal profile* of a stream or valley; the line maximum depth. Syn: *valley line.* (b) The median line of a stream; the *valley axis.* (c) *channel line.*---Etymol: German *Thalweg* (later *Talweg*), "valley way". Pron: tal-veg. Syn: *talweg.*

thamnasterioid Said of a massive corallum characterized by absence of corallite walls and by confluent septa that join neighboring corallites together, with a pattern of septa resembling lines of force in a magnetic field.

thanatocenosis Var. of *thanatocoenosis*.

thanatocoenosis A group of dead organisms (or fossils) that may represent the *biocoenosis* of an area or the biocoenosis plus the thanatocoenosis of another environment; all of the fossils present at a particular place in a sediment. The term was introduced by the German hydrobiologist Wasmund in 1926. Var: *thanatocenosis; thanatocoenose; thanatocenose*. Plural: *thanatocoenoses*. Etymol: Greek *thanatos*, "death" +*koinos*, "general, common". Syn: *death assemblage; taphocoenose; taphocoenosis*.

thanatotope The total area in which the dead specimens of a taxon or taxa are deposited.

Thanetian European stage: uppermost Paleocene (above Montian, below Ypresian of Eocene).

thaumasite A white mineral: $Ca_3Si(OH)_6(CO_3)(So_4).12H_2O$.

thaw v. To go from a frozen state, as ice, to a liquid state; to melt.---n. The end of a frost, when the temperature rises above the freezing point, and ice or snow undergo melting. Also, the transformation of ice or snow to water.

thaw depression A hollow in the ground resulting from subsidence following the unequal melting of ground ice in a permafrost region. See also: *cave-in lake*. Syn: *thermokarst depression*.

thaw hole A vertical *hole* [ice] in sea ice, formed where a surface puddle melts through to the underlying water.

thaw lake [permafrost] *cave-in lake*.

thaw lake [glaciol] A pool of water formed on the surface of a large glacier by accumulation of meltwater.

thaw sink A closed *thaw depression* with subterranean drainage, believed to have originated as a cave-in lake (Hopkins, 1949).

theca (a) Sac-like echinoderm skeleton consisting of calcareous plates and enclosing the body and internal organs; e.g. the dorsal cup of the calyx of a crinoid. The term is generally applied to all fossilized parts, and includes ambulacra but excludes the column or stem and appendages such as free arms and brachioles. (b) An individual tube or cup that housed a zooid of a graptolite colony. (c) The external skeletal deposit of a coelenterate, such as the calcareous *wall* enclosing a corallite and presumably the sides of a coral polyp. (d) The sometimes resistant-walled coat or external covering, formed of numerous plates, of the nonencysted or actively swimming stage of the life cycle of some dinoflagellates. Syn: *test* [paleont].---Pl: *thecae*.

thecal plate Any of numerous calcareous plates that form an element in theca of an echinoderm. It is usually distinguished from a plate of an ambulacrum or arm.

thecamoebian Any one of a group of, usually freshwater, testaceous protozoans, the fossil members of which belong to the orders Arcellinida and Gromida and to part of the suborder Allogromiina of the foraminifers.

Theis curve Log-log plot of drawdown or recovery of head against time used in an aquifer test based on the *Theis equation*.

Theis equation An equation relating drawdown or recovery of ground-water head to rate of withdrawal or addition of water and to the hydraulic characteristics of the aquifer (Theis, 1935, p.520). See also: *Theis curve*.

thenardite A white or brownish orthorhombic mineral: Na_2SO_4. It occurs in masses or crusts often in connection with salt lakes.

theodolite A precision surveyor's instrument that can be rotated on a horizontal base so as to be sighted first upon one point and then upon another and that is used for measuring angular distances in both vertical and horizontal planes. It consists of a telescope so mounted on a tripod as to swivel vertically in supports secured to a revolvable table carrying a vernier for reading horizontal angles (azimuths), and usually

includes a compass, a spirit level, and an accurately graduated circle for determining vertical angles (altitudes). See also: *transit*.

theory A *hypothesis* that is supported to some extent by experimental or factual evidence but that has not been so conclusively proved as to be generally accepted as a law; e.g. the "theory of continental drift."

theralite A group of mafic plutonic rocks composed of calcic plagioclase, feldspathoids, and augite, with lesser amounts of sodic sanidine and sodic amphiboles and accessory olivine; also, any rock in that group; the intrusive equivalent of *tephrite*. Theralite grades into nepheline monzonite with an increase in the alkali feldspar content, into gabbro as the feldspathoid content diminishes, and into diorite with both fewer feldspathoids and increasingly sodic plagioclase. The term was defined by Rosenbusch in 1887.

therm Any of several units of quantity of heat such as a calorie, a kilogram calorie, 1000 kilogram calories, or 100,000 British thermal units.

thermal [glac] n. A syn. of *interglacial stage*.

thermal [geophys] Of, or pertaining to, or caused by heat. Syn: *thermic*.

thermal [meteorol] n. A vertically moving current of air that is caused by differential heating of the ground below it.

thermal analysis The study of chemical and/or physical changes in materials as a function of temperature, i.e. the heat evolved or absorbed during such changes. See also: *differential thermal analysis*. Syn: *thermoanalysis*.

thermal anomaly A pattern of thermal-energy distribution at the Earth's surface, which is anomalous to that from adjoining areas (e.g., as recorded by an airborne infrared line-scan system or as measured on the ground by thermistors, thermocouples, radiometers, etc.). Recommended usage is restricted to heat-flow anomalies that are persistent, and greater than the normal heat flow of the Earth rather than for diurnal or seasonal thermal variations, ephemeral thermal differences resulting from variations in thermal characteristics, albedo or emissivity of surficial materials, thermal variations resulting from differential solar warming of topographic slopes, or thermal variations resulting from differential irradiance of the surface because of meteorological conditions (Friedman, 1970).

thermal aureole *aureole*.

thermal band A general term for middle-infrared spectral wavelengths which are transmitted through atmospheric transmission windows between 3 and 15 or 20 micrometers where the far infrared commences.

thermal bar A boundary separating regions of a lake having considerable differences in surface temperature; commonly occurring in the spring and autumn in large lakes in temperate climates. In the spring, water shoreward of the thermal bar is commonly warmer than the main body of the lake, and in the autumn the shoreward water is colder.

thermal capacity *heat capacity*.

thermal conduction *heat conduction*.

thermal conductivity (a) The time rate of transfer of heat by conduction, through unit thickness, across unit area for unit difference of temperature. (b) A measure of the ability of a material to conduct heat. Typical values of thermal conductivity for rocks range from 3 to 15 millicalories/cm-sec-°C. Rocks with abundant quartz have high thermal conductivities. Poorly consolidated sediments have lower thermal conductivities (Jaeger, 1965, p.7).

thermal demagnetization A technique of partial *demagnetization* by heating the specimen to a temperature T then cooling to room temperature in a nonmagnetic space; this destroys a partial thermoremanent magnetization for that temperature interval but leaves unaffected a partial thermoremanent magnetization for temperature intervals above T. Cf: *alternating field demagnetization; chemical demagnetization*.

thermal detector A major type of infrared detector, whose operation is based on the change of a physical property such as the resistance or thermoelectric force of the detector as its own temperature is changed by incident radiation. Examples of thermal detectors are the thermocouple, and the thermistor bolometer (Bernard, 1970, p.61).

thermal diffusivity *Thermal conductivity* divided by the product of the density and specific heat capacity of the material. In rock, the common range of values is from 0.005 to 0.025 cm^2/sec (Jaeger, 1965, p.7).

thermal energy yield In volcanology, the thermal energy equivalent of a volcanic eruption, based on the thermal characteristics and volume of the volcanic products (Yokoyama, 1956-57, p.75-97).

thermal equator *oceanographic equator.*

thermal exfoliation A type of *exfoliation* caused by the heating of rock during the day and its rapid cooling at night.

thermal expansion An increase in the linear dimensions of a solid or in the volume of a fluid as a result of an increase in temperature.

thermal fracture The disintegration of, or the formation of a fracture or crack in, a rock as a result of sudden temperature changes. It occurs where rock-forming minerals have varying coefficients of expansion, where a cliff face receives and loses radiation rapidly, or where there is a quick drop of temperature after sundown. Also, the result of such a process.

thermal gradient The rate of change of temperature with distance. When applied to the Earth, the term *geothermal gradient* is used. Syn: *temperature gradient.*

thermal head A temperature difference that causes a flow of heat energy by conduction, or heat transfer by convection.

thermal imaging A procedure in which an image-forming system, equipped with a thermal or quantum detector sensitive to infrared radiation (generally in the 3-14 μm wavelength region) converts an infrared signal to a voltage output which is amplified and recorded on film or magnetic tape, or viewed directly. The result, when printed as a film positive, commonly is a black-and-white image in which the gray scale steps represent increments in the radiant flux from the object imaged. In geology and geophysics, the images obtained are generally of a portion of the Earth's surface.

thermal inertia A measure of the rate of heat transfer from the surface of a substance; the square root of the product of thermal conductivity, bulk density, and specific heat of the substance. Symbol: β. The reciprocal is often used and is called the *thermal parameter.*

thermal infrared The preferred term for the middle wavelength ranges of the infrared region, extending roughly from 3 micrometers at the end of the near infrared, to about 15 or 20 micrometers where the far infrared commences.

thermal lineament A linear feature, usually observed on an infrared or thermographic image, representing actual linear distribution of thermal energy at the Earth's surface at a point in time. Less preferred syn: *thermal linear.*

thermal linear n. A nonrecommended colloquialism and synonym for the preferred term *thermal lineament.*

thermal log *temperature log.*

thermal maximum A term proposed by Flint & Deevey (1951) as a substitute for *climatic optimum* and *Altithermal*, and then redefined by Deevey & Flint (1957) to designate secondary warming trends separated by cooler climatic phases within the *Hypsithermal*. The utility of the term is limited by the difficulties of distinguishing between the primacy of either temperature or precipitation changes that separately or in differing combinations could have produced the changes recorded in most paleoclimatic records and by similar secondary climatic oscillations that are possibly controlled primarily by atmospheric humidity changes (Karlstrom, 1956). Also spelled: *Thermal Maximum.*

thermal metamorphism A type of metamorphism resulting in chemical reconstitution controlled by temperature and in-

fluenced to a lesser extent by the confining pressure (as a function of depth); there is no simultaneous deformation (Turner, 1948). See also: *pyrometamorphism.* Cf: *geothermal metamorphism; static metamorphism; load metamorphism.* Syn: *thermometamorphism.* Obs. syn: *pyromorphism.*

thermal parameter The reciprocal of *thermal inertia.*

thermal prospecting A system of geophysical prospecting based on the measurement of underground temperatures or gradients and the relationship of their irregularities to geologic structure. Cf: *geothermal prospecting.*

thermal resistivity The reciprocal of thermal conductivity.

thermal shock Failure of a material, esp. a brittle material, due to the thermal stress of rapidly rising temperature.

thermal spring A spring whose water temperature is appreciably higher than the local mean annual atmospheric temperature. A thermal spring may be a *hot spring* or a *warm spring* (Meinzer, 1923, p.54).

thermal stratification The *stratification* of a lake produced by changes in temperature at different depths and resulting in horizontal layers of differing densities. See also: *density stratification.*

thermal stress Stress in a body caused by a local temperature gradient within the body.

thermal structure An arrangement of metamorphic zones of increasing metamorphic grade in some distinct, structural pattern which produces, for example, a thermal anticline or a thermal dome. Such features are associated with orogenesis and are produced by a localized heat source, possibly anatexis (Winkler, 1967).

thermal unit *therm.*

thermal water Water, generally of a spring or geyser, whose temperature is appreciably above the local mean annual air temperature.

thermic [geophys] *thermal.*

thermic [soil] Said of a soil temperature regime in which the mean annual temperature is at least 15°C but less than 22°C, with a variation of more than 5°C between summer and winter measurements, at 50cm depth (SSSA, 1970). Cf: *isothermic.*

thermionic emission The emission of electrons from a hot cathode, as in a vacuum tube. Syn: *Richardson effect.*

thermistor A thermally sensitive resistor employing a semiconductor with a large negative resistance-temperature coefficient and used as an electrical thermometer.

thermistor chain A chain that is towed astern, carrying instruments to measure seawater temperatures.

thermite An old name for any fossil combustible substance.

thermoanalysis *thermal analysis.*

thermobarometer *hypsometer.*

thermocline [phys sci] A temperature gradient; esp. a vertical gradient that marks a sharp change of a property and that has values appreciably greater than those of the gradients in adjacent areas. Cf: *pycnocline.*

thermocline [oceanog] A vertical, negative gradient of temperature that is characteristic of a layer of ocean water; also, the layer in which it occurs. A thermocline may be either seasonal or permanent. Cf: *halocline.* See also: *discontinuity layer.*

thermocline [lake] (a) The horizontal plane in a thermally stratified lake located at the depth where temperature decreases most rapidly with depth. (b) The horizontal layer of water characterized by a rapid decrease of temperature and increase of density with depth; sometimes arbitrarily defined as the layer in which the rate of temperature decrease with depth is equal to at least 1°C per meter. This is an older and less preferred definition, and one that is often used in engineering literature. Syn: *metalimnion.*

thermocouple A thermoelectric couple used to measure temperature differences and hence thermal radiation.

thermodiffusion Diffusion of material in solution due to a temperature gradient. Syn: *Soret effect.*

thermodynamic equilibrium constant *equilibrium constant.*

thermodynamic potential Any thermodynamic function of state, an extremum of which is a necessary and sufficient criterion of equilibrium for a system under specified conditions. Thus, for example, the *Gibbs free energy* is the thermodynamic potential for a system at constant pressure and temperature, the *Helmholtz free energy* is the thermodynamic potential for a system at constant temperature and volume. See also: *free energy*.

thermodynamic process A change in any macroscopic property of a thermodynamic system.

thermodynamics The mathematical treatment of the relation of heat to mechanical and other forms of energy.

thermoelastic effect A fall in temperature under tension or a rise in temperature under compression during elastic deformation.

thermoerosional niche A niche resulting from undercutting produced by bank erosion in north-flowing Arctic rivers during the limited summer flow, esp. in Alaska and Siberia where niches as wide as 8 m have been formed in one year (Hamelin & Cook, 1967, p.123).

thermogene Pertaining to the formation of minerals primarily under the influence of temperature (Kostov, 1961). Cf: *piezogene*.

thermogenesis The rise in temperature in a body from reactions in that body, as by oxidation, or the decay of radioactive elements.

thermogram (a) The continuous trace of air temperature made by a self-recording thermometer. (b) *thermographic image*.

thermograph (a) A self-recording thermometer. (b) *thermographic image*.

thermographic image A two-dimensional thermal image produced by an infrared line-scan image-forming system. It is not synonymous with *infrared photograph*, which records, among other wavelengths, the near infrared. Syn: *infrared image; thermogram; thermograph*.

thermography (a) *differential thermal analysis*. (b) A term that has been suggested (Williams, R., 1972) as a replacement for the phrase "thermal infrared imagery", in a way analogous to the term "photography". No confusion would then exist between "photograph" (a record of reflected solar energy) and "thermograph" (a record of emitted thermal energy) as there now exists between infrared photography and infrared imagery.

thermogravimetric analysis A syn. of *thermogravimetry*. Symbol: TGA.

thermogravimetry A method of analysis which measures the loss or gain of weight by a substance as the temperature of the substance is raised or lowered at a constant rate. Symbol: TG. Syn: *thermogravimetric analysis* (Symbol: TGA).

thermohaline circulation Vertical movement of sea water that is generated by density differences caused by the combined effect of variations in temperature and salinity. Syn: *thermohaline convection*.

thermohaline convection *thermohaline circulation*.

thermokarst (a) Karst-like topographic features produced in a permafrost region by the local melting of ground ice and the subsequent settling of the ground. Cf: *glaciokarst*. (b) A region marked by *thermokarst topography*. (c) The process of formation of a thermokarst topography.---Syn: *cryokarst*.

thermokarst depression *thaw depression*.

thermokarst lake *cave-in lake*.

thermokarst mound A residual, polygonal hummock, bordered by depressions that were formed by the melting of ground ice in a permafrost region.

thermokarst topography An irregular land surface containing cave-in lakes, bogs, and accumulation of snow, and caverns, pits, and other small depressions, formed in a permafrost region by the melting of ground ice; in exterior appearance, it resembles the uneven *karst topography* formed by the solution of limestone.

thermolabile Said of a material that is decomposable by heat.

thermomer A relatively warm period within the Pleistocene Epoch, such as an *interstade* (Lüttig, 1965, p.582). Ant: *kryomer*.

thermometamorphism *thermal metamorphism*.

thermometer An instrument for measuring *temperature*. There are several types: a liquid-in-glass thermometer that is based on the increase in volume of the liquid in a capillary tube in response to an increase in temperature; a resistance thermometer that utilizes the change in electrical resistance of such substances as platinum in response to change in temperature; and a constant-volume gas thermometer that is based on the expansion of a gas whose volume is kept constant and whose pressure change in response to change in temperature is measured (Sears, 1958, p.498-499).

thermometric depth The depth, in meters, at which a pair of *reversing thermometers* are inverted.

thermometric leveling A type of indirect leveling in which elevations above sea level are determined from observed values of the boiling point of water. Cf: *barometric leveling*.

thermometry The measurement of temperature.

thermonatrite A white mineral: $Na_2CO_3.H_2O$. It is found in some lakes and alkali soils, and as a saline residue.

thermo-osmosis Osmosis occurring under the influence of a temperature difference between fluids on either side of a semipermeable membrane, with movement from the warmer to the cooler side, as when water flows in small openings from the warmer to the cooler parts of a soil mass.

thermophilic Said of an organism that prefers high temperatures, esp. bacteria that thrive in temperatures between 115 and 175°F. Noun: *thermophile*.

thermophyte A plant preferring high temperatures for growth.

thermopile An instrument for measuring thermal radiation, consisting of a number of thermocouples, either connected in series (if e.m.f. is measured) or in parallel (if electric current is measured), as a means of increasing the sensitivity beyond that possible with a single thermocouple (McIntosh, 1963 p.256).

thermoremanence *thermoremanent magnetization*.

thermoremanent magnetization Remanent magnetization acquired as a rock cools in a magnetic field from above the Curie point down to room temperature. It is very stable and exactly parallel to the ambient field at time of cooling. Abbrev: TRM. Syn: *thermoremanence*.

thermuticle *porcelanite*.

thesocyte An amoebocyte in a sponge, filled with inclusions of stored by-products (metabolites) of protoplasmic activity.

Thetis hairstone A variety of *hairstone* containing or penetrated by tangled or ball-like inclusions of green fibrous crystals of hornblende, asbestos, and esp. actinolite. Also spelled: *Thetis's-hairstone*.

thetomorph A glass or glassy phase, commonly of quartz or feldspar (maskelynite), produced by solid-state alteration of an originally crystalline mineral by the action of shock waves and retaining the form and original textures (such as fractures, twin lamellae, and grain boundary shapes) of the preexisting mineral or grain.

thetomorphic Pertaining to a thetomorph; e.g. "thetomorphic silica glass" having a refractive index of 1.46 and retaining the morphology of the host quartz from which it was formed by shock. Term introduced by Chao (1967a, p.211-212). Etymol: Greek *thetos*, "adopted", + *morphe*, "form". Cf: *diaplectic*.

thick bands In banded coal, vitrain bands from 5.0 to 50.0 mm thick (Schopf, 1960, p.39). Cf: *thin bands; medium bands; very thick bands*.

thick-bedded (a) A relative term applied to uniformly occurring sedimentary beds having great extent from one surface to another and variously defined as more than 6.4 cm (2.5 in.) to more than 100 cm (40 in.) in thickness; specif. said of a bed whose thickness is in the range of 60-120 cm (2-4 ft), a bed greater than 120 cm being "very thick-bedded" (McKee, & Weir, 1953, p.383). (b) A term used in sandstone quarrying to describe strata whose seams are more than 90 cm (3 ft) apart (AIME, 1960, p.333).---Cf: *thin-bedded; medium-bedded.*

thickness [paleont] (a) The greatest distance between the two valves of a brachiopod shell at right angles to the length and width. It is equal to the *height* in biconvex, plano-convex, and convexo-plane shells. (b) The distance between the inner and outer surfaces of the wall of a bivalve-mollusk shell. Also, a term used to denote the shell measurement (of a bivalve mollusk) commonly called *inflation.*

thickness [geol] The extent of a tabular unit from its bottom boundary surface to its top surface, usually measured along a line normal to these bounding surfaces; e.g. the distance at right angles between the hanging wall and the footwall of a lode, or the distance of a stratigraphic unit measured at right angles to the bedding surface in a vertical cross section, parallel to the dip direction and normal to the strike. See also: *true thickness; apparent thickness.*

thickness contour *isopach.*

thickness line *isopach.*

thickness map *isopach map.*

thief formation A rock formation that causes excessive fluid loss in drilling operations.

thigmotaxis Taxis in response to mechanical or tactile stimuli. Cf: *strophotaxis; phobotaxis.*

thill A British term for the floor of a coal mine or coal seam; hence, *underclay.* The term is also used for a thin stratum of fireclay.

thin bands In banded coal, vitrain bands from 0.5 to 2.0 mm thick (Schopf, 1960, p.39). Cf: *medium bands; thick bands; very thick bands.*

thin-bedded (a) A relative term applied to uniformly occurring sedimentary beds having little extent from one surface to another and variously defined as less than 30 cm (1 ft) to less than 1 cm (0.4 in.) in thickness; specif. said of above thickness is in the range of 5-60 cm (2 in. to 2 ft.), a bed less than 5 cm but more than 1 cm thick being "very thin-bedded" (McKee & Weir, 1953, p.383). (b) Said of a shale that is easy to split (Alling, 1945, p.753). (c) A term used in sandstone quarrying to describe strata that are less than 90 cm (3 ft) but more than several centimeters apart (AIME, 1960, p.333).---Cf: *thick-bedded; medium-bedded.*

thinic Pertaining to a sand dune (Klugh, 1923, p.374).

thin-layer chromatography An essentially adsorptive chromatographic technique for separating components of a sample by moving it in a mixture or solution through a uniformly thin deposit of adsorbent on rigid supporting plates in such a way that the different components have different mobilities and thus become separated (May & Cuttitta, 1967, p.116). Symbol: TLC. See also: *chromatography.*

thinolite (a) A pale-yellow to light-brown variety of calcite, often terminated at both ends by pyramids. It may be pseudomorphous after gaylussite. (b) *thinolitic tufa.*

thinolitic tufa A tufa deposit consisting in part of layers of delicate, interlaced, prismatic, skeletal crystals of thinolite (up to 20 cm long and 1 cm thick), occurring exposed in the desert basins of NW Nevada, as in the dome-like masses along the shore of the extinct Lake Lahontan where it overlies *lithoid tufa* and underlies *dendroid tufa.* Syn: *thinolite.*

thin out To grow progressively thinner in one direction until extinction. The term is applied to a stratum, vein, or other body of rock that becomes gradually less thick so that its upper and lower surfaces approach each other and eventually meet and the layer or rock disappears. The thinning may be original or due to truncation at an acute angle beneath an unconformity. Syn: *pinch out; wedge out.*

thin section A fragment of rock or mineral mechanically

ground to a thickness of approximately 0.03 mm, polished and mounted between glasses as a microscope slide. This reduction renders most rocks and minerals transparent or translucent, thus making it possible to study their optical properties. Syn: *section [petrology].*

thin-skinned structure A term used by Rodgers (1963) for an interpretation that folds and faults of miogeosynclinal and foreland rocks in an orogenic belt involve only the surficial strata, and lie on a *décollement* beneath which the structure differs; it is also called the *no-basement interpretation.* Proposed examples are in the Valley and Ridge and Plateau provinces of the Appalachian belt, and in the Jura Mountains. *Thick-skinned structure* is a contrasting interpretation of the same structures.

thiophile element *sulphophile element.*

thiospinel A general term for minerals with the spinel structure having the general formula: AR_2S_4.

third-law entropy The difference in entropy between a substance at some finite temperature and at absolute zero, as defined by the *third law of thermodynamics,* and determined from calorimetric measurements by integration of the relationship $dS = Cp \, d \ln T$, where S = entropy, Cp = heat capacity at constant pressure, and T = absolute temperature.

third law of thermodynamics The statement that the entropy of any perfect crystalline substance becomes zero at the absolute zero of temperature. See also: *third-law entropy.*

third-order leveling Spirit leveling that does not attain the quality of *second-order leveling,* in which lines are not extended more than 30 miles (48.3 km) from lines of first- or second-order leveling and must close upon lines of equal or higher order of accuracy and in which the maximum allowable discrepancy is 12 mm times the square root of the length of the line in kilometers (or 0.05 ft times the square root of the distance in miles). It is used for subdividing loops of first- and second-order leveling and in providing local vertical control for detailed surveys. Cf: *first-order leveling.*

third-order pinacoid In a triclinic crystal, any $\{hk0\}$ or $\{\bar{h}k0\}$ pinacoid, with symmetry $\bar{1}$. Cf: *first-order pinacoid; second-order pinacoid; fourth-order pinacoid.*

third-order prism A crystal form: in a tetragonal crystal, a $\{hk0\}$ prism, with symmetry $4m$, $\bar{4}$, or 4; in a hexagonal crystal, a $\{hki0\}$ prism, with symmetry $6/m$, 6, or $\bar{3}$; and in orthorhombic and monoclinic crystals, any $\{hk0\}$ prism. Cf: *first-order prism; second-order prism; fourth-order prism.*

third water The quality or luster of a gemstone next below *second water,* such as that of a diamond that is clearly imperfect and contains flaws.

thixotropic clay A clay that displays *thixotropy,* i.e. weakens when disturbed and strengthens when left undisturbed. Syn: *false body.*

thixotropy The property of certain colloidal substances, e.g. a *thixotropic clay,* to weaken or change from a gel to a sol when shaken but to increase in strength upon standing.

tholeiite A group of basalts primarily composed of plagioclase (approximately An50), pyroxene (esp. augite or subcalcic augite), and iron oxide minerals as phenocrysts in a glassy groundmass or intergrowth of quartz and alkali feldspar; also, any rock in that group. Little or no olivine is present. The term was first used in 1840 by Steininger (Johannsen, 1939, p.283). Its name is derived from Tholei, in Germany (Lower Saxony). Cf: *alkali basalt.*

tholoid *volcanic dome.*

thomsenolite A white monoclinic mineral: $NaCaAlF_6 \cdot H_2O$.

thomsonite A zeolite mineral: $NaCa_2Al_5Si_5O_{20} \cdot 6H_2O$. It has considerable replacement of CaA by NaSi, and sometimes contains no sodium. It usually occurs in masses of radiating crystals. Syn: *ozarkite.*

thonstein An obsolete name for a porphyry tuff or felsite tuff (Thrush, 1968, p.1136).

thoracic Pertaining to, located within, or involving the thorax;

e.g. a "thoracic limb" attached to any somite of the thorax of a crustacean. Syn: *thorasic*.

thoracomere A somite of the thorax of a crustacean.

thoracopod A limb of any thoracic somite of a crustacean; a thoracic limb, such as a maxilliped or a pereiopod. Syn: *thoracopodite*.

thorax (a) The central tagma of the body of an arthropod, consisting of several generally moveable segments; e.g. the nearly always limb-bearing tagma between the cephalon and the abdomen of a crustacean, or the middle part of an exoskeleton of a trilobite, extending between the cephalon and the pygidium and consisting of several freely articulated segments, or the middle of the three chief divisions of the body of an insect. See also: *cephalothorax*. Syn: *trunk* [*paleont*]. (b) The second joint of the shell of a nasselline radiolarian.---Pl: *thoraxes* or *thoraces*.

thorbastnaesite A brown mineral: $Th(Ca,Ce)(CO_3)_2F_2.3H_2O$. It occurs as an accessory mineral in iron-rich albitites and in selvages of veinlets and stockworks.

thoreaulite A brown mineral: $SnTa_2O_7$.

thorianite A mineral: ThO_2. It is isomorphous with uraninite, often contains rare-earth metals and uranium, and is strongly radioactive.

thorite A brown, black, or sometimes orange-yellow tetragonal mineral: $ThSiO_4$. It is isostructural with thorogummite, strongly radioactive, and usually metamict, and may contain as much as 10% uranium. Thorite resembles zircon and occurs as a minor accessory mineral of granites, syenites, and pegmatites. It is dimorphous with huttonite.

thorium-lead age method Calculation of an age in years for geologic material based on the known radioactive rate of thorium-232 to lead-208. It is part of the more inclusive *uranium-thorium-lead age method* in which the parent-daughter pairs are considered simultaneously.

thorium series The radioactive series of thorium, beginning with Th-232 as parent.

thorium-230 to protactinium-231 deficiency method Calculation of an age in years for fossil coral, shell, or bone 10,000 to 250,000 years old, based on the growth of uranium daughter products from uranium isotopes that enter the carbonate or phosphate material shortly after its formation or burial. The age depends on measurement of the thorium-230 to uranium-234 and protactinium-231 to uranium-235 activity ratios which vary with time. See also: *uranium-series age methods; ionium-deficiency method*.

thorium-230 to protactinium-231 excess method *protactinium-ionium age method*.

thorium-230 to thorium-232 age method *ionium-thorium age method*.

thorn A short, sharply pointed triangular or conical surface extension in the skeleton of a spumellarian radiolarian.

Thornthwaite's classification of climate A *climate classification* (formulated by the U.S. agricultural climatologist Warren Thornthwaite) that is based on ratios of precipitation to evaporation. Five humidity provinces are distinguished: perhumid, humid, subhumid, semiarid, and arid. Cf: *Köppen's classification of climate*.

thorogummite A secondary mineral: $Th(SiO_4)_{1-x}(OH)_{4x}$. It is isostructural with thorite and may contain as much as 31.4% uranium. Syn: *mackintoshite*.

thoron A less-preferred syn. of *radon-220*.

thorosteenstrupine A dark-brown to nearly black mineral: $(Ca,Th,Mn)_3Si_4O_{11}F.6H_2O$.

thoroughfare (a) A tidal channel or creek providing an entrance to a bay or lagoon behind a barrier or spit. Syn: *thorofare*. (b) A navigable waterway, as a river or strait; esp. one connecting two bodies of water.

thortveitite A grayish-green mineral: $(Sc,Y)_2Si_2O_7$. It is a source of scandium. Cf: *thalenite*.

thorutite A black mineral: $(Th,U,Ca)Ti_2(O,OH)_6$. Syn: *smirnovite*.

Thoulet solution *Sonstadt solution*.

thread (a) A thin stream of water. (b) The middle of a stream. (c) The line along the surface of a stream connecting points of maximum current velocity. Cf: *channel line*.

thread-lace scoria A *scoria* in which the vesicle walls have collapsed and are represented only by a network of threads. Syn: *reticulite*.

three-age system In archaeology, the original classification scheme of relative prehistoric time, including the *Stone Age*, *Bronze Age*, and *Iron Age*. Since its formulation in the early nineteenth century, it has been expanded by the division of the Stone Age into the *Paleolithic*, *Mesolithic*, and *Neolithic* and by the addition of the *Copper Age* between the Neolithic and Bronze Age (Bray and Trump, 1970, p.231).

three array An electrode array used in lateral search (profiling) in which one current electrode is placed at infinity while one current electrode and two potential electrodes are in close proximity and moved across the structure to be investigated. When the separation between the potential electrodes equals the separation between the near current electrode and the closest potential electrode, the three array is reduced to the pole-dipole array. It is used in resistivity and induced polarization surveys and in drill hole logging.

threefold coordination *triangular coordination*.

three-layer structure A type of *layer structure* having three unit layers to the full repeat unit; e.g. some phlogopites, which have three tetrahedral layers per unit along the c axis. Such micas are usually hexagonal. Cf: *two-layer structure*.

threeling *trilling*.

three-mile limit The seaward limit of the *territorial sea* or *marginal sea* of 3 nautical miles (about 5.6 km), the one-time range of a cannon shot.

three-phase inclusion An inclusion in a gemstone consisting of tiny crystals, gas, and liquid. Cf: *two-phase inclusion*.

three-point method A geometric method of calculating the dip and strike of a structural surface from three points of varying elevation along the surface.

three-point problem (a) The problem of locating the horizontal geographic position of a point of observation from data comprising two observed horizontal angles subtended by three known sides of a triangle (or situated between three points of known position). It is solved analytically by trigonometric calculation in triangulation, mechanically by means of a three-arm protractor, or graphically by trial-and-error change in the orientation of the board in plane-table surveying. Cf: *two-point problem*. (b) A name applied to the method of solving the three-point problem in plane-table surveying, commonly by taking backsights on three previously located stations.---See also: *resection*.

three-swing cusp A *meander cusp* formed by three successive swings of a meander belt, the scar produced by the third swing meeting the point of the cusp formed by the first two swings (Lobeck, 1939, p.240-241); it may be seen on the edge of a rock-defended terrace.

threshold [*geochem*] The entrance, boundary, or beginning of a new domain; the lowest detectable value; the point at which a process or effect commences.

threshold [*glac geol*] *riegel*.

threshold [*speleo*] That part of a cave which received daylight.

threshold [*marine geol*] *sill* [*marine geol*].

threshold of detectability The minimum detectable ascending radiant flux attributable to the geothermal flux in areas of surficial thermal anomalies, in distinction to the ascending radiant flux arising from diurnal and seasonal temperature variations, differences in emissivity, albedo, and thermal characteristics of surface materials, and differential solar warming of topographic slopes. The practical threshold of detectability for airborne infrared line-scan systems was estimated in 1970 at

738

pproximately the equivalent of 200 microns cal·cm⁻² convective heat transfer at the Earth's surface. Note that the threshold of detectability is dependent on terrestrial or hydrographic surface conditions and is not synonymous with instrument detectivity.

threshold pressure *yield stress.*

threshold velocity The minimum velocity at which wind or water, in a given place and under specified conditions, will begin to move particles of soil, sand, or other material.

throat plane A plane passed through the centers of the spheres in a layer of rhombohedral packing.

through cave A cave, into and out of which a stream flows or is known to have flown. Cf: *inflow cave; outflow cave.*

throughfall Water from precipitation that falls through the plant cover directly onto the ground or that drips onto the ground from branches and leaves. Cf: *interception; stemflow.*

through glacier A double-ended glacier, consisting of two valley glaciers situated in a single depression, from which they flow in opposite directions. A "through-glacier system" is a body of glacier ice consisting of interconnected through glaciers that may lie in two or more drainage systems. Cf: *transection glacier.*

through valley A flat-floored depression or channel eroded across a divide by glacier ice or by meltwater streams; a valley excavated by a through glacier.

throw (a) On a fault, the amount of vertical displacement. Cf: *heave.* See also: *upthrow; downthrow.* (b) The vertical component of the net slip.

throwing clay Clay plastic enough to be shaped on a potter's wheel.

throwout Fragmental material ejected from an impact or explosion crater during formation and redeposited on or outside the crater lip. Cf: *fallout [crater]; fallback.*

thrust (a) An overriding movement of one crustal unit over another, as in thrust faulting. (b) *thrust fault.*

thrust block *thrust nappe.*

thrust fault A fault with a dip of 45° or less in which the hanging wall appears to have moved upward relative to the footwall. Horizontal compression rather than vertical displacement is its characteristic feature. Cf: *normal fault.* Syn: *reverse fault; reverse slip fault; thrust slip fault; thrust.* Partial syn: *overthrust; contraction fault; overlap fault.*

thrust moraine (a) A moraine produced by the overriding and pushing forward, by a regenerated glacier, of dead ice and its deposits (Gravenor & Kupsch, 1959). (b) *push moraine.*

thrust nappe The body of rock that forms the hanging wall of a large-scale thrust fault, the fault surface of which is either horizontal or of very low angle; an *overthrust nappe.* Syn: *thrust sheet; thrust plate; thrust slice; thrust block.*

thrust outlier *klippe.*

thrust plane The *thrust surface* of a thrust fault, when the surface is planar.

thrust plate *thrust nappe.*

thrust pond A small, shallow, roughly circular pond on the floor of a slightly inclined mountain valley, and bordered by a raised rim of thick soil resulting from ice push and supporting a very dense growth of coarse alpine grass (Ives, 1941, p.290); e.g. in the high parts of the Rocky Mountains in Colorado.

thrust scarp The scarp along the forward edge of a thrust nappe.

thrust sheet *thrust nappe.*

thrust slice A syn. of *thrust nappe.* Var: *slice.*

thrust slip fault *thrust fault.*

thrust surface The surface, usually a plane, along which thrust faulting occurs. See also: *thrust plane.*

thucolite A brittle, jet-black mixture or complex of organic matter (hydrocarbons) and uraninite, with some sulfides, occurring esp. in gold conglomerates (as in those of the Witwatersrand of South Africa) or in pegmatites (as in Canada). It

may contain up to 48% thorium in the ash.

thufa An Icelandic term for *earth hummock.* Pl: *thufur* (not "thufurs").

Thulean province A region of Tertiary volcanic activity (basalt flood) including Iceland and most of Britain and Greenland. Etymol: Greek *Thule,* "the northernmost part of the habitable world."

thulite A pink, rose-red, or purplish-red variety of zoisite containing manganese and used as an ornamental stone.

thumper A device for generating seismic waves, in the *weight-dropping* method. It is a device which drops a three-ton weight from a ten-foot elevation.

thumping *weight dropping.*

thunder egg A popular term for a small, geodelike body of chalcedony, opal, or agate that has weathered out of the welded tuffs of central Oregon.

Thuringian European stage: Upper Permian (above Saxonian, below Triassic).

thuringite An olive- or pistachio-green mineral of the chlorite group: $(Fe^{+2},Fe^{+3},Al,Mg)_6(Al,Si)_4O_{10}(OH)_8$. Some manganese may be present. It is isomorphous with pennantite.

thurm A term used in Nova Scotia for a ragged and rocky headland swept by the sea. Syn: *thurm cap.*

Thyssen gravimeter An early gravity meter of the unstable equilibrium type.

tibia (a) The fifth segment of a typical leg or pedipalpus of an arachnid, following upon the patella which may be completely fused with it (TIP, 1955, pt.P, p.63). (b) A joint of the distal part of a prosomal appendage of a merostome.---Pl: *tibiae.*

tickle (a) Any narrow passage connecting larger bodies of water. (b) A term used in the Gulf of St. Lawrence region for an inlet of the sea into a lagoon.

tidal basin A dock or basin (in a tidal region) in which water is maintained at a desired level by means of a gate; it is filled at high tide by water that is retained and then released at low tide.

tidal bedding Sedimentary bedding caused by tides in a tidal channel, a tidal flat, or a marsh; esp. bedding produced where currents of high tides are stronger than those of low tides flowing in the opposite direction, as where a layer of coarse sediments deposited by a high tide is not destroyed by the low tide.

tidal bench mark A durable bench mark fixed rigidly in stable ground and set to reference a tide staff at a tide station.

tidal bore *bore* [tide].

tidal bulge The tidal effect on the side of the Earth nearest to the Moon, where lunar attraction is greatest. Cf: *antipodal bulge.*

tidal channel (a) A major channel followed by the tidal currents, extending from offshore well into a tidal marsh or a tidal flat. (b) *tidal inlet.*

tidal compartment The part of a stream that "intervenes between the area of unimpeded tidal action and that in which there is a complete cessation or absence of tidal action" (Carey & Oliver, 1918, p.8).

tidal constant A parameter of a tide that, for a given locality, usually remains constant. *Tide amplitude* and *tidal epoch* are harmonic tidal constants; *tide range* is a disharmonic constant.

tidal correction In gravity observations, a correction that is applied to remove the effect of Earth tides. The value of gravity at any point varies in a cyclical manner during the course of a day due to the changing positions of the Sun and the Moon relative to the area being investigated. The tidal correction commonly is included in the drift correction and may be determined by a series of observations at a fixed base station.

tidal creek A relatively small tidal inlet or estuary. Syn: *creek.*

tidal current The periodic horizontal movement of ocean water associated with the vertical rise and fall of the tides and resulting from the gravitational attraction of the Moon and Sun upon the Earth. In the open ocean, its direction rotates 360° on a diurnal or semidiurnal basis; in coastal areas, however, topography influences its direction. Incorrect syn: *tide*. British syn: *tidal stream*. Syn: *periodic current*.

tidal cycle *tide cycle*.

tidal datum A *chart datum* based on a phase of the tide.

tidal day The interval between two consecutive high waters of the tide at a place, averaging 24 hours and 51 minutes. Cf: *lunar day*.

tidal delta A delta formed at the mouth of a tidal inlet on both the seaward and lagoon sides of a barrier island or baymouth bar by changing tidal currents that sweep sand in and out of the inlet.

tidal-delta marsh A *salt marsh* found around distributary patterns of tidal rivers inside a tidal inlet.

tidal divide A divide between two adjacent tidal channels.

tidal efficiency The ratio of the fluctuation of water level in a well to the tidal fluctuation causing it, expressed in the same units such as feet of water. Symbol: C. Cf: *barometric efficiency*.

tidal epoch A *tidal constant* representing the phase difference or time elapsing between the Moon's transit over a fixed meridian and the ensuing high tide. Syn: *epoch* [*tide*]; *phase lag*.

tidal flat An extensive, nearly horizontal, marshy or barren tract of land that is alternately covered and uncovered by the rise and fall of the tide, and consisting of unconsolidated sediment (mostly mud and sand). It may form the top surface of a deltaic deposit. See also: *tidal marsh; mud flat*. Syn: *tide flat*.

tidal flushing The removal of sediment, as from an estuary, by the ebb and flow of tidal currents that are stronger, more constant, or more prevalent than the incoming river flow.

tidal friction The frictional effect of the tides, resulting in dissipation of energy, esp. along the bottoms of shallow seas where the tidal epoch is lengthened and the Earth's rotational velocity is retarded, thereby slowly increasing the length of the day and theoretically causing the Moon to recede and accelerate over the course of geologic time.

tidal glacier *tidewater glacier*.

tidal inlet Any *inlet* through which water flows alternately landward with the rising tide and seaward with the falling tide; specif. a natural inlet maintained by tidal currents. Syn: *tidal outlet; tidal channel*.

tidalite A sediment deposited by tidal tractive currents, by an alternation of tidal tractive currents and tidal suspension deposition, or by tidal slack-water suspension sedimentation. Tidalites occur both in the intertidal zone and in shallow, subtidal, tide-dominated environments. See also: *intertidalite*.

tidal marsh A low, flat marsh bordering a coast (as in a shallow lagoon or a sheltered bay), formed of mud and of the resistant mat of roots of salt-tolerant plants, and regularly inundated during high tides; a marshy *tidal flat*. Cf: *salt marsh*.

tidal outlet *tidal inlet*.

tidal pool *tide pool*.

tidal prism The volume of water that flows in or out of a harbor or estuary with the movement of the tide, and excluding any freshwater flow; it is computed as the product of the tide range and the area of the basin at midtime level, or as the difference in volume at mean high water and at mean low water.

tidal range *tide range*.

tidal resonance theory George Drawin's postulation of the cause of continental drift; the Moon separated from the Earth, leaving the Pacific Ocean as a scar, and the consequent shortage of continental crust caused a global tension, resulting in drifting (Fairbridge, 1966, p.83).

tidal river A river whose lower part for a considerable distance is influenced by the tide of the body of water into which it flows; the movement of water in and out of an estuary or other inlet as a result of the alternating rise and fall of the

tide. Syn: *tidal stream*.

tidal scour The downward and sideward erosion of the sea floor by powerful tidal currents resulting in the removal of in shore sediments and the formation of deep channels and holes. Syn: *scour*.

tidal stand *stand of tide*.

tidal stream (a) *tidal river*. (b) A British syn. of *tidal current*.

tidal swamp A swamp partly covered during high tide by the backing up of a freshwater river (Stephenson & Veatch, 1915, p.37). Cf: *upland swamp*.

tidal water *tidewater*.

tidal wave An erroneous syn. of both *storm surge* and *tsunami*.

tidal wedge A tidal channel that is narrower and shallower a the downstream end.

tide (a) The rhythmic, alternate rise and fall of the surface (o water level) of the ocean and bodies of water connected with the ocean (as estuaries and gulfs), occurring twice a day over most of the Earth, and resulting from the gravitational attraction of the Moon (and in lesser degree of the Sun) acting unequally on different parts of the rotating Earth. (b) An incorrect syn. of *tidal current*. (c) *earth tide*. (d) *atmospheric tide*.

tide amplitude A *tidal constant* representing one-half of the *tide range*; the elevation of tidal high water above mean sea level.

tide crack A *crack* [ice], usually parallel to the shore, at the junction line between an immovable icefoot or ice wall and fast ice, and caused by the rise and fall of the tide which moves the fast ice upward and downward.

tide curve A graphic record of the height (rise and fall) of the tide, with time as abscissa and tide height as ordinate. Syn: *marigram*.

tide cycle A period that includes a complete set of tide conditions or characteristics, such as a tidal day or a lunar month. Syn: *tidal cycle*.

tide flat A syn. of *tidal flat*. Also spelled: *tideflat*.

tide gage A device for measuring the height (rise and fall) of the tide; esp. an instrument automatically making a continuous graphic record of tide height versus time. See also: *tide staff; marigraph*.

tideland (a) The coastal area that is alternately covered and uncovered by the ebb and flow of the ordinary daily tides; land that is sometimes covered by *tidewater* during a flood tide. (b) Land that underlies the ocean beyond the low-water mark, but within the territorial waters of a nation. Term is often used in the plural.

tide pole *tide staff*.

tide pool A pool of water, as in rock basin, left on a beach or reef by an ebbing tide. Syn: *tidal pool*.

tide race A type of *race* [current] caused by a greater tide range at one end of the channel than at the other end.

tide range A *tidal constant* representing the difference in height between consecutive high water and low water at a given place; it is twice the *tide amplitude*. Cf: *mean range*. Syn: *tidal range*.

tide rip A *rip* produced by the meeting of opposing tides (as where tidal currents are converging and sinking), or by a tidal current suddenly entering shallow water.

tide staff A *tide gage*, either fixed or portable, consisting of a long, vertical, graduated rod or board, from which the height of the tide can be read directly at any time. Syn: *tide pole*.

tide station A place where tide observations are obtained. Cf: *reference station*.

tidewater (a) Water that overflows the land during a flood tide; water that covers the *tideland*. Also, stream water that is affected by the rise and fall of the tide. Syn: *tidal water*. (b) A broad term for the seacoast, or low-lying coastal land traversed by tidewater streams.

tidewater glacier A glacier that terminates in the sea, where it usually ends in an ice cliff from which icebergs are disharged. Syn: *tidal glacier*.

tie (a) A survey connection from a point of known position to point whose position is desired. (b) A survey connection to lose a survey on a previously determined point.

tie bar *tombolo*.

tied island An island connected with the mainland or with another island by a tombolo. Syn: *tombolo island*.

tie-in In geophysics, the relating of a new station or value to one already established.

tie line [chem] A line at constant temperature that connects any two phases that are in equilibrium at the temperature of the tie line. See also: *conjugation line*. Syn: *conode*.

tie line [surv] A line measured on the ground to connect some object to a survey; e.g. a line joining opposite corners of a four-sided figure, thereby enabling its area to be checked by triangulation.

tiemannite A dark-gray or nearly black mineral: HgSe.

tienshanite A mineral: $Na_2BaMnTiB_2Si_6O_{20}$.

tie point A point to which a tie is made; esp. a point of closure of a survey either on itself or on another survey.

tier Any series of contiguous *townships* (of the U.S. Public Land Survey system) situated east and west of each other and numbered consecutively north and south from a base line. Also, any series of contiguous sections similarly situated within a township. Cf: *range*.

tierra blanca A Spanish term for "white ground" or "white earth", and applied to white calcareous deposits such as tufa, caliche, and chalky limestone.

tiff A sparry mineral. The term is applied to calcite in SW Missouri and to barite in S Missouri.

tiger's-eye A chatoyant, translucent to semitranslucent, yellowish-brown or brownish-yellow gem and ornamental variety of quartz pseudomorphous after crocidolite whose fibers (penetrating the quartz) are changed to iron oxide (limonite); silicified crocidolite stained yellow or brown by iron oxide. Upon heating, the limonite turns to hematite and produces a red to brownish-red sheen. Cf: *hawk's-eye; cat's-eye [mineral]*. Syn: *tiger-eye; tigereye; tigerite*.

tight fold *closed fold*.

tight sand A sand whose interstices are filled with finer grains or the matrix material, thus effectively destroying porosity and permeability. The term is used in petroleum geology. Cf: *open sand*. Syn: *close sand*.

tikhonenkovite A monoclinic mineral: $SrAlF_4(OH).H_2O$.

tilaite A dark-colored plutonic rock having crystalline-granular texture and containing abundant green diopside and olivine in a small amount of calcic plagioclase; a rock intermediate in composition between a peridotite and a gabbro.

tilasite A violet-gray monoclinic mineral: $CaMg(AsO_4)F$. It is isomorphous with isokite.

tile ore A red or brownish earthy variety of cuprite often mixed with red iron oxide.

tilestone An English term for a flagstone (flaggy sandstone) used for roofing.

till [glac geol] Unsorted and unstratified drift, generally unconsolidated, deposited directly by and underneath a glacier without subsequent reworking by water from the glacier, and consisting of a heterogeneous mixture of clay, sand, gravel, and boulders varying widely in size and shape. Cf: *stratified drift*. See also: *moraine*. Syn: *boulder clay; glacial till; ice-laid drift*.

till [soil] An agricultural term originally applied in Scotland to a stiff, hard clay subsoil, generally impervious and unstratified, often containing gravel and boulders; an extremely poor or unproductive soil.

till ball An *armored mud ball* whose core is made of till, occurring in certain Pleistocene glacial deposits.

till billow An undulating or swelling accumulation of glacial drift irregularly disposed with regard to the direction of movement of the ice (Chamberlin, 1894b, p.523).

till crevasse filling A ridge of unstratified morainal material deposited in a crack of a wasting glacier and left standing after the ice melted; term introduced by Gravenor (1956, p.10). Cf: *crevasse filling*.

tilleyite A white mineral: $Ca_5(Si_2O_7)(CO_3)_2$.

tillite A consolidated or indurated sedimentary rock formed by lithification of glacial till, esp. pre-Pleistocene till (such as the Late Carboniferous tillites in South Africa and India).

tilloid A term introduced by Blackwelder (1931a, p.903) for a till-like deposit of "doubtful origin", but redefined by Pettijohn (1957, p.265) as a nonglacial *conglomeratic mudstone* (also known as *geröllton*) varying from "a chaotic unassorted assemblage of coarse materials set in a mudstone matrix to a mudstone with sparsely distributed cobbles" (such as a sedimentary deposit resulting from extensive slides or flows of mud on the margin of a geosyncline). Harland et al (1966, p.251) urge the use of "tilloid" as a nongenetic term for a rock resembling tillite in appearance but whose origin is in doubt or unknown. Cf: *pebbly mudstone; pseudotillite*.

till plain An extensive area, with a flat to undulating surface, underlain by till, which is commonly covered by ground moraines and subordinate end moraines; such plains occupy parts of Indiana, Illinois, and Iowa.

till-shadow hill A glacial hill, without a core of resistant rock, that has a gentle south slope on which till thickens but does not form a well-developed tail (Coates, 1966); examples occur in central New York State.

till sheet A sheet, layer, or bed of till, without reference to its topographic expression. It may form ground moraine.

tillstone A boulder or other stone in a deposit of glacial till. Also spelled: *till stone*.

till tumulus A low, stony mound representing the immature nucleus of a drumlin (Chamberlin, 1894b, p.522-523).

till wall A ridge consisting of morainal material squeezed upward into a crevasse by the pressure of overlying ice (Gravenor & Kupsch, 1959, p.58).

tilly Composed or having the nature of glacial till; e.g. *tilly land*.

tilt (a) The angle at the perspective center between the plumb line and the perpendicular from the interior perspective center to the plane of the photograph. See also: *relative tilt*. (b) The lack of parallelism (or the angle) between the plane of the photograph from a downward-pointing aerial camera and the horizontal plane (normal to the plumb line) of the ground.

tilt angle The inclination of the major axis of an ellipse of polarization, measured from the horizontal.

tilt block A *fault block* that has become tilted, perhaps by rotation on a hinge line. Syn: *tilted fault block*.

tilt-block basin *fault-angle valley*.

tilted fault block *tilt block*.

tilted iceberg A tabular iceberg that has become unbalanced due to melting or calving, so that its flat top is inclined.

tilted photograph An aerial photograph taken with a camera whose plane of film is not parallel (horizontal) with the plane of the ground at the time of exposure.

tilth The physical condition of a soil relative to its fitness for the growth of a specified plant or sequence of plants.

tiltmeter An instrument that measures slight changes in the tilt of the Earth's surface, usually in relation to a liquid-level surface or to the rest position of a pendulum. It is used in volcanology and in earthquake seismology.

timazite A greenstonelike, metamorphically altered hornblende-biotite andesite occurring in the Timok Valley of Siberia (Johannsen, 1939, p.284). The term was originated by Breithaupt in 1861, and is now obsolete.

timberline The elevation (as on a mountain) or latitudinal limits (on a regional basis) at which tree growth stops. Syn: *tree line*.

time (a) Measured or measurable duration; a nonmaterial dimension of the universe, representing a period or interval during which an action, process or condition exists or continues.

See also: *geologic time*. (b) A reference point from which du- ration is measured; e.g. the instant at which a seismic event occurs relative to a chosen reference time such as a shot in- stant. (c) A reckoning of time, or a system of reckoning dura- tion. (d) An informal term proposed by the International Sub- commission on Stratigraphic Terminology (1961, p.13 & 25) for the geologic-time unit next in order of magnitude below *age* [geochron] (a), during which the rocks of a *substage* (or of any time-stratigraphic unit of lesser rank than a *stage*) were formed. It is a syn. of *subage*; *episode* (b); *phase* [geo- chron] (a). (e) Any division of geologic chronology, such as "Paleozoic time" or "Miocene time".

time at shot point *uphole time*.

time break An indication on a seismic record of the instant of detonation of a shot or charge. Cf: *time signal; timing line*. Syn: *shot instant; shot moment*.

time-correlation The demonstration or determination of age equivalence or mutual time relations (such as contemporane- ity of origin) of stratigraphic units in two or more separated areas. It is the oldest kind of *stratigraphic correlation*, and is established mainly by fossil content.

time delay *time lag*.

time-depth chart A graphical expression of the functional rela- tion between the velocity function and the times observed in the seismic method of geophyscial exploration. It permits the time increments to be converted to the corresponding depths. Syn: *time-depth curve*.

time-depth curve *time-depth chart*.

time-distance curve *traveltime curve*.

time-distance graph *traveltime curve*.

time domain Transmission of a single or repetitive pulse of electromagnetic energy, frequently of square waveform, and reception of electromagnetic energy, as a function of time, in the time interval after the transmitted waveform has been turned off. It is used with induced electrical polarization and electromagnetic methods.

time gradient (a) In the reflection seismic methods applied to dipping reflectors, the reciprocal of apparent velocity which varies with the spread from shot point to detector. (b) In seis- mic prospecting, the rate of change of traveltime with depth.

time lag In refraction seismic interpretation, the departure from normal traveltime for a path from source to detector in a low-velocity layer with abnormally long arrival time. Also, rela- tive to seismic prospecting, a time lag may be due to other factors such as phase shifts in filtering, or shothole fatigue. Syn: *time delay*.

time lead In a method of interpretation of refraction seismic records (used in salt-dome exploration) in which arrival times are plotted against shot-to-detector distances, the departure from the traveltime curve indicating a high-speed segment of the paths. It is proportional to the horizontal extent of the high-speed segment.

time line (a) A line indicating equal age in a geologic cross section or correlation diagram; e.g. a line separating two time-stratigraphic units. (b) A rock unit represented by a time line; e.g. an intraformational conglomerate formed by suba- queous slump and turbidity flows of but a few hour's duration.

time mark A minute mark from which *arrival times* of various seismic phases are measured.

time of concentration *concentration time*.

time-parallel (a) Said of a surface that is parallel to or that closely approximates a synchronous surface and that involves a geologically insignificant amount of time, such as the sur- face of a rapidly transgressed unconformity. (b) Said of a stratum bounded by time-parallel surfaces.

time plane A stratigraphic horizon identifying an *instant* in geologic time. Syn: *synchrone*.

time-rock span *stratigraphic range*.

time-rock unit *time-stratigraphic unit*.

time scale *geologic time scale*.

time series A series of statistical data collected at regular in- tervals of time; a frequency distribution in which the indepen- dent variable is time.

time signal A signal transmitted by telegraph or radio that in- dicates an exact point in time; specif. in seismology, such signal indicating the time of explosion in a shothole. Cf: *time break*.

time standard Any category of physical or biologic phenomena or processes by which segments of time can be measured or subdivided; e.g. radioactive decay of elements, orderly evolu- tion of forms of life, rotation of the Earth on its axis, revolution of the Earth around the Sun, and human artifacts. All such time standards are "partial" with respect to time in the ab- stract (Jeletzky, 1956, p.681).

time-stratigraphic facies A stratigraphic facies that is recog- nized on the basis of the amounts of geologic time during which sedimentary deposition and nondeposition occurred; a facies that is a laterally segregated, statistical variant of a stratigraphic interval and whose boundaries (vertical surfaces or arbitrary cutoffs) extend from the bottom to the top of the interval (Wheeler, 1958, p.1060).

time-stratigraphic unit A stratigraphic unit based on geologic age or time of origin, having boundaries that are everywhere of the same age or time value and thus ideally independent of lithology, fossil content, or conditions of origin; "a subdivision of rocks considered solely as the record of a specific interval of geologic time" (ACSN, 1961, art.26). It is a material unit or body of strata based on actual sections or sequences of rock and it represents the preserved record or rocks formed during an arbitrary interval of finite geologic time that extended from the beginning to the ending of its deposition or intrusion. Al- though the unit is fundamentally independent of all physical objective properties of rocks, the interpretation of such prop- erties in terms of time "provides us with the only available technique of determining approximate isochronous surfaces and thereby geographically extending approximate working boundaries for time-stratigraphic units" (Hedberg, 1958, p.1890). Dunbar & Rodgers (1957, p.293) regard time-stratig- raphic units as groupings of biostratigraphic units because "away from their individual type regions, systems, series, and stages can be delimited (defined) only by their fossils". The magnitude of the unit is measured by the length of the time interval to which its rocks correspond, not by their thickness. Time-stratigraphic units in order of decreasing rank: erathem, system, series, stage, and substage. Cf: *para-time-rock unit; geologic-time unit*. Syn: *chronostratigraphic unit; chronostratic unit; chronolithologic unit; time-rock unit; chronolith*.

time-stratigraphy A syn. of *chronostratigraphy*. Term used by Wheeler (1958) for abstract or subjective stratigraphy involv- ing three-dimensional entities (which delineate all interpreted deposition, nondeposition, post-deposition erosional removal or combinations thereof) that are related to stratal relation- ships and that are defined in a framework consisting of two lateral space dimensions and a vertical time dimension.

time tie The identification of seismic events on different rec- ords by their arrival times, when they possess common ray paths.

time-transgressive *diachronous*.

time-transitional Said of a rock unit including within itself an important geologic time plane and thus consisting of strata that belong to two adjacent time-stratigraphic units (such as systems).

time unit *geologic-time unit*.

time value The amount or interval of geologic time represent- ed by or involved in producing a stratigraphic unit, an uncon- formity, the range of a fossil, or any geologic feature or event. See also: *hiatus (b)*.

timing line One of a series of marks or lines placed on seis- mic records at precisely determined intervals of time (usually at intervals of 0.01 or 0.005 sec) for the purpose of measurg the time of events recorded. The timing mechanism commonly includes an accurate tuning fork for the determination f small time intervals. Cf: *time break*.

imiskamian A division of the Archeozoic of the Canadian hield. Also spelled: *Timiskaming*.

imiskaming Var. of *Timiskamian*.

n (a) A bluish-white mineral, the native metallic element Sn.) A term used loosely to designate cassiterite and concenates containing cassiterite with minor amounts of other minrals.

naja (a) A term used in SW U.S. for a *water pocket* develped below a waterfall, esp. when partly filled with water. (b) term used loosely in New Mexico for a temporary pool, or a spring too feeble to form a stream.---Etymol: Spanish, large earthen jar". Pron: te-*na*-ha.

najita *solution pan*. Etymol: Spanish, diminutive of *tinaja*, water jar".

inakste A mineral: $K_2NaCa_2TiSi_7O_{19}(OH)$.

incal An old name for crude *borax* formerly obtained from ibetan-lake shores and deposits and once the chief source of oric compounds.

incalconite A colorless to dull-white rhombohedral mineral: $a_2B_4O_7.5H_2O$. It is one of the principal ores of boron. Syn: *nohavite; octahedral borax*.

ind A Norwegian term for a glacial *horn* that is detached from he main mountain range by the lateral recession of cirques utting through an upland spur between two glacial troughs Thornbury, 1954, p.373). Syn: *monument*.

inder ore An impure variety of jamesonite. Syn: *pilite*.

inguaite A hypabyssal igneous rock, being a textural variety of phonolite, typically found in dikes, and characterized by onspicuous acicular crystals of acmite arranged in radial or riss-cross patterns in the groundmass. The phenocrysts are quigranular feldspar and nepheline crystals. The name is deived from the Tingua Mountains (Serra de Tingua) near Rio le Janeiro, Brazil. Adj: *tinguaitic*. Cf: *muniongite*.

in ore *cassiterite*.

in pyrites *stannite*.

insel A flagellum having a central axis from which extend many fine short hairs (mastigonemes) in one or two rows along its length.

instone *cassiterite*.

int *hypsometric tint*.

inticite A creamy-white mineral: $Fe_6(PO_4)_4(OH)_6.7H_2O$.

intinaite A mineral: $Pb_5(Sb,Bi)_8S_{17}$.

intinnid A ciliate protozoan belonging to the family Tintinnidae and characterized by a lorica that is almost always inflated in he oral region.

in-white cobalt *smaltite*.

inzenite A yellow monoclinic mineral: $(Ca,Mn,Fe)_3Al_2$-$(BO_3)(Si_4O_{12})(OH)$. It is a variety of axinite rich in manganese.

Tioughniogan North American stage: Middle Devonian (above Cazenovian, below Taghanican).

tiphic Pertaining to a pond or ponds. Cf: *ombrotiphic*.

tiphon *diapirism*.

tirodite A monoclinic mineral of the amphibole group: $(Mg,Mn)_8Si_8O_{22}(OH)_2$.

tirs In Morocco, a black soil that is similar to a Vertisol. Etymol: Arabic.

tissue An aggregate of cells into a structural unit that performs a particular function.

titanaugite A variety of augite rich in titanium and occurring in basaltic rocks: $Ca(Mg,Fe,Ti)(Si,Al)_2O_6$.

titanic iron ore A syn. of *ilmenite*. Also called: *titaniferous iron ore*.

titanite *sphene*.

titanomaghemite A general term applied to an abnormal titanium-bearing magnetite with varying cation vacancies in an oxygen framework of spinel structure.

titanomagnetite (a) A titaniferous variety of magnetite: Fe-$(Fe^{+2},Fe^{+3},Ti)_2O_4$. It is strictly a homogeneous cubic solid solution of ilmenite in magnetite. (b) A term loosely used for mixtures of magnetite, ilmenite, and ulvöspinel.

Tithonian Southern European equivalent of *Portlandian*.

title box *cartouche*.

tjaele A syn. of *frozen ground*. The term has been used erroneously as a syn. of *permafrost* (Bryan, 1951). Etymol: Swedish *tjäle*, "frozen ground". Syn: *tjäle; taele*.

tjosite A dark-colored, porphyritic, nepheline-bearing igneous rock intermediate in composition between a syenite and a lamprophyre. Its name is derived from Tjose, Norway.

toadback marl A term used in Lancashire, Eng., for unlaminated marl with lumpy fracture. Cf: *beechleaf marl*.

toad's-eye tin A reddish or brownish variety of cassiterite occurring in botryoidal or reniform shapes that display an internal concentric and fibrous structure. Syn: *toad's-eye*.

toadstone A fossilized object, such as a fish tooth or palatal bone, that was thought to have formed within a toad and was frequently worn as a charm or an antidote to poison.

toadstool rock *mushroom rock*.

Toarcian European stage: Lower Jurassic (above Pliensbachian, below Bajocian).

tobacco jack A miner's term for *wolframite*.

tobacco rock A term used in SW U.S. for a favorable host rock (for uranium) characterized by light yellow or gray color and by brown limonite stains.

tobermorite A mineral: $Ca_5Si_6O_{16}(OH)_2.4H_2O$.

todorokite A mineral: $(Mn,Ca,Mg)Mn_3^{+4}O_7.H_2O$. It may contain some barium and zinc.

toe [drill] The bottom of a drill hole (esp. one used for blasting), as distinguished from its open end.

toe [coast] *step*.

toe [mass move] The lower, usually curved, margin of the disturbed material of a landslide pushed over onto the undisturbed slope; it is most distant from the place of origin. Cf: *foot*.

toe [fault] The leading edge of a thrust nappe.

toe [volc] *lava toe*.

toe [slopes] The lowest part of a slope or cliff; the downslope end of an alluvial fan.

toellite *tollite*.

toenail (a) A curved joint intersecting a sheet structure, usually along the strike or sometimes differing from it in strike by 45° or more. (b) Obsolete syn. of *stylolite*.

toernebohmite *törnebohmite*.

toeset The forward part of a tangential foreset bed.

toe slope *wash slope*.

toe-tap flood plain The outer end of a meander lobe, built by a stream as it meanders down-valley.

toft A British term for an isolated hill, knoll, or other eminence in a flat region, esp. one suitable for a homesite.

toise An old French unit of length used in early geodetic surveys and equal to 6 French feet, 6.396 U.S. feet, or 1.949 meters.

tokeite A dark-colored olivine basalt similar in composition to schönfelsite and in which the plagioclase is labradorite.

tollite A seldom-used term applied to a hypabyssal rock containing phenocrysts of hornblende, andesine, and some biotite, white mica, quartz, orthoclase, and feldspar. No dark-colored minerals occur in the groundmass. Syn: *toellite*.

tolt A term used in Newfoundland for an isolated peak rising abruptly from a plain.

tombarthite A mineral: $Y_4(Si,H_4)_4O_{12-n}(OH)_{4+2n}$.

tombolo A sand or gravel bar (or spit) that connects an island with the mainland or with another island. Etymol: Italian, "sand dune"; from Latin *tumulus*, "mound". The Italian term refers to the dune or mound atop the bar but the "failure" of the popular mind to appreciate the independent origin of the bars and the dunes which surmount them ... resulted in the application of the term ... to the bars themselves" (Johnson, 1919, p.312). Pl: *tombolos*. Syn: *connecting bar; tie bar; tying bar*.

tombolo cluster *complex tombolo.*

tombolo island *tied island.*

tombolo series *complex tombolo.*

tomite *boghead coal.*

tonalite A syn. of *quartz diorite*; sometimes restricted to quartz diorite in which biotite and hornblende are the main mafic minerals. Its name is derived from Tonale Pass, northern Italy. See also: *adamellite*.

Tonawandan Stage in New York State: middle Middle Silurian.

Tongrian European stage: lower Oligocene (above Ludian of Eocene, below Rupelian). Syn: *Lattorfian; Sannoisian*.

tongue [coast] (a) A *point* or long, low, narrow strip of land projecting from the mainland into the sea or other body of water. (b) *inlet*.

tongue [oceanog] An extension of one type of water into water of differing salinity or temperature, e.g. salt water into the mouth of a river.

tongue [intrus rocks] A branch or offshoot of a larger intrusive body. See also: *epiphysis*. Syn: *apophysis*.

tongue [stratig] n. (a) A minor rock-stratigraphic unit of limited geographic extent, being a subdivision of a formation and similar in rank to a *member*, and disappearing laterally (usually by facies change) in one direction; "a member that extends outward beyond the main body of a formation" (ACSN, 1961, art.7). Cf: *lentil*. (b) A lateral extension of a formation, wedging out in one direction between strata of a different kind and passing in another direction into a thicker body of similar rock type.---v. To thin laterally to disappearance.

tongue [volc] A lava flow that is an offshoot from a larger flow; it may be as much as several kilometers in length.

tongue [glaciol] A long narrow extension of the lower part of a glacier, either on land or afloat. Cf: *glacier tongue afloat; glacial lobe*. Syn: *glacier tongue*. Nonpreferred syn: *ice tongue*.

tongue [ice] A projection of the *ice edge* up to several kilometers in length, caused by wind and current.

tongue [streams] *meander lobe.*

tonnage-volume factor In economic geology, the number of cubic feet in a ton of ore.

tonsbergite A red igneous rock that is sometimes porphyritic and resembles larvikite, the feldspar being represented by orthoclase and andesine.

tonstein A compact argillaceous rock containing the clay mineral kaolinite in a variety of forms together with occasional detrital and carbonaceous material, commonly occurring as a thin band in a Carboniferous coal seam (or locally in the roof of a seam), and often used as an aid in correlating European strata of Westphalian age. Etymol: German *Tonstein*, "claystone".

tool A *sonde* used in logging a well or borehole.

tool mark A *current mark* produced by the impact against a muddy bottom of a solid object swept along by the current, and sometimes preserved as a cast on the underside of the overlying bed. The mark may be produced by an object in continuous contact with the bottom (e.g. a groove or a striation) or in intermittent contact with the bottom (e.g. a skip mark or a prod mark). The engraving "tools" include: shell fragments, sand grains, pebbles, fish bones, seaweed, and wood chips. Syn: *tool marking*.

toolpusher One who operates, or assumes the responsibility of operating, a drill machine; the general supervisor of drilling operations; the head driller.

tooth (a) Any of various usually hard and sharp horny, chitinous, or calcareous projections of an invertebrate that functions like or resembles the vertebrate jaws, such as one of the numerous and minute horny processes on the radula of a gastropod. (b) A tooth-like process on the margin of a bivalve shell; specif. *hinge tooth*. (c) A calcareous rod located in the pyramid of Aristotle's lantern of an echinoid. Its upper end uncalcified. (d) A projection in the aperture of a foraminiferal test. It may be simple or complex, and single or multiple.

toothpaste lava Viscous lava that is extruded as a *squeeze up*.

tooth plate An internal, apertural modification of a foraminiferal test, commonly consisting of a contorted plate that extends from the aperture through chamber to previous septal foramen. One side may be attached to the chamber wall or base attached to the proximal border of foramen, the opposite side being free and folded (TIP, 1964, pt.C, p.64).

top [ore dep] A quarrymen's syn. of *overburden*.

top [mass move] The highest point of contact between the disturbed material of a landslide and the scarp face along which it moved. Cf: *tip*.

top [gem] *crown.*

top [stratig] The uppermost surface of a geologic formation where it is first encountered during drilling, usually characterized by the first appearance of a distinctive feature (such as a marked change in lithology or the occurrence of a guide fossil). It is often determined by a distinctive configuration of an electric log, and it is widely used in correlation.

topaz (a) A white, orthorhombic mineral: $Al_2SiO_4(F,OH)_2$. It occurs as a minor constituent in highly siliceous igneous rocks and in tin-bearing veins as translucent or transparent prismatic crystals and as masses, and also as rounded water-worn pebbles. Topaz has a hardness of 8 on Mohs' scale. (b) A colorless, yellow, brown, reddish, pink, or light-blue transparent topaz used as a gemstone. (c) A yellow quartz that resembles topaz in appearance, such as smoky quartz turned yellow by heating; specif. *false topaz* and *Scotch topaz*. See also: *Spanish topaz*. (d) A term used for a greenish-yellow to orange-yellow mineral resembling topaz in appearance, such as "oriental topaz" (a yellow corundum).

topazfels *topazite.*

topazite A hypabyssal rock composed almost entirely of quartz and topaz. Syn: *topaz rock; topazfels; topazoseme; topazogene.*

topazogene *topazite.*

topazolite A greenish-yellow to yellow-brown variety of andradite garnet, having the color and transparency of topaz.

topazoseme *topazite.*

topaz quartz Topaz-colored quartz; specif. *citrine*. Cf: *quartz topaz.*

topaz rock *topazite.*

top conglomerate A conglomerate formed of gravels lying at the top of a stratum, and not separated from it by an erosional surface (Twenhofel, 1939, p.203-204).

tophus A syn. of *tufa*. Etymol: Latin. Pl: *tophi*.

topocentric horizon *apparent horizon.*

topocline A cline related to a geographic zone and usually unrelated to any ecologic condition.

topogenous Said of peat deposits whose moisture content is dependent on surface water. Cf: *ombrogenous; soligenous*.

topographic (a) Pertaining to *topography*. (b) Surveying or representing the topography of a region; e.g. a "topographic survey" or a "topographic map".---Syn: *topographical*.

topographic adjustment The condition existing where the gradient of a tributary is harmonious with that of the main stream. Cf: *structural adjustment*.

topographic adolescence *adolescence.*

topographic contour *contour* [cart].

topographic correction [cart] Correction of errors in a topographic map.

topographic correction [grav] *terrain correction.*

topographic deflection of the vertical An expression used to indicate that the deflection of the vertical has been computed from the topography. This method may be used when it is not possible to compare astronomic and geodetic positions directly.

topographic depression *closed depression.*

topographic desert Deserts of low rainfall because of their location in the middle of a continent, far from the ocean, or on the lee side of high mountains, cut off from prevailing winds.

topographic divide A drainage *divide.*

topographic expression The effect achieved by shaping and spacing contour lines on a map so that topographic features can be interpreted with the greatest ease and fidelity.

topographic feature A prominent or conspicuous *topographic form* or noticeable part thereof (Mitchell, 1948, p.80). Cf: *physiographic feature.*

topographic form A *landform* considered without regard to its origin, cause, or history (Mitchell, 1948, p.80). Cf: *physiographic form.*

topographic grain The *grain* (alignment and direction) of the topographic-relief features of a region.

topographic high (a) A knoll, hill, mountain, or other elevated land mass, esp. if relatively isolated. (b) An area of relatively high elevation in an oil field. Cf: *geologic high.*----The term is applied regardless of the age of the material composing the high feature. Syn: *high.*

topographic infancy *infancy.*

topographic license The freedom to adjust, add, or omit contour lines, within allowable limits, in order to attain the best topographic expression; it does not permit the adjustment of contours by amounts that significantly impair their accuracy.

topographic low An area of relatively low elevation in an oil field, regardless of the age of the material composing the feature. Cf: *geologic low.* Syn: *low.*

topographic map A map on a sufficiently large scale showing, in detail, selected man-made and natural features of a part of a land surface, including its relief (generally by means of contour lines) and certain physical and cultural features (vegetation, roads, drainage, etc.). Its distinguishing characteristic is the portrayal of the position (horizontal and vertical), relation, size, shape, and elevation of the features of the area. Topographic maps are frequently used as base maps. Cf: *planimetric map.*

topographic maturity *maturity* [topog].

topographic old age *old age* [topog].

topographic profile *profile* [geomorph].

topographic relief *relief* [geomorph].

topographic survey A survey that determines the configuration (relief) of the Earth's surface (ground) and the location of natural and artificial features thereon. Also, an organization making such a survey.

topographic texture Disposition, grouping, or average size of the topographic units composing a given topography; usually restricted to a description of the relative spacing of drainage lines in stream-dissected regions. See also: *coarse topography; fine topography; texture ratio.*

topographic unconformity (a) The relationship between two parts of a landscape or two kinds of topography that are out of adjustment with one another, due to an interruption in the ordinary course of the erosion cycle of a region; e.g. a lack of harmony between the topographic forms of the upper and lower parts of a valley, due to rejuvenation. (b) A land surface exhibiting topographic unconformity.

topographic youth *youth* [topog].

topography (a) The general configuration of a land surface or any part of the Earth's surface, including its relief and the position of its natural and man-made features. See also: *geomorphy.* Cf: *relief.* Syn: *lay of the land.* (b) The natural or physical surface features of a region, considered collectively as to form; the features revealed by the contour lines of a map. In nongeologic usage, the term includes man-made features (such as are shown on a topographic map). (c) The art or practice of accurately and graphically delineating in detail, as on a map or chart or by a model, selected natural and man-

made surface features of a region. Also, the description, study, or representation of such features. Cf: *chorography.* (d) Originally, the term referred to the detailed description of a particular place or locality (such as a city, parish, or tract of land) as distinguished from the general geography of a country or other large part of the world, and also to the science or practice of such a description; this usage is practically obsolete.---Etymol: Greek *topos*, "place", + *graphein*, "to write".

topology (a) The analytical, detail study of minor landforms, requiring fairly large scales of mapping (Matthes, 1912, p.336). Cf: *topometry.* (b) The topographic study of a particular place; specif. the history of a region as indicated by its topography (Webster, 1967, p.2411).

topometry The art, process, or science of making large-scale, high-precision maps (1:20,000 or larger) upon which geomorphic features are "measured in with mathematical accuracy, practically nothing being 'sketched in' by eye" (Matthes, 1912, p.338). Cf: *topology.*

toposaic A photomap on which topographic or terrain-form lines are shown. Cf: *planisaic.*

toposequence A sequence of kinds of soil in relation to location on a topographic slope, from top to bottom of the slope or vice versa.

topostratigraphic unit A term proposed by Jaanusson (1960, p.218) for a "convenient regional stratigraphic unit" consisting of a combined rock- and bio-stratigraphic unit.

topostratigraphy Preliminary or introductory stratigraphy, including lithostratigraphy and biostratigraphy; *prostratigraphy.*

topotactic Adj. of *topotaxy.*

topotaxial Adj. of *topotaxy.*

topotaxy Strong preferred orientation of a crystal aggregate, produced by transformation of a polymorph. It occurs in displacive transformations with no breaking of primary interatomic bonds, as in low-to-high quartz transformation. The degree of topotaxy is low in reconstructive transformations. Adj: *topotactic; topotaxial.* Cf: *epitaxy; syntaxy.*

topotype A specimen of a particular species that comes from the type locality of that species.

topozone A syn. of *local range zone.* The term was proposed by Moore (1957) for a paleontologically defined horizon or zone recognizable in a single locality.

top-reef deposit A sedimentary deposit on the reef flat.

topset A *topset bed.*

topset bed One of the nearly horizontal layers of fine-grained sediments deposited on the top surface of an advancing delta and continuous with the landward alluvial plain; it truncates or covers the edges of the seaward-lying *foreset bed.* See also: *bottomset bed.* Also spelled: *top-set bed.* Syn: *topset.*

topsoil (a) A presumably fertile soil used to cover areas of special planting. (b) A syn. of *surface soil.* (c) A syn. of *A horizon.* (d) The dark-colored upper portion of a soil, varying in depth according to soil type.----Cf: *subsoil.*

tor A high, isolated, craggy hill, pinnacle, or rocky peak; or a pile of rocks, much-jointed and usually granitic, exposed to considerable weathering, and often assuming peculiar or fantastic shapes, as the granite rocks standing as prominent masses on the sides and tops of hills in Devon and Cornwall, England. Linton (1955) suggests that a tor is a residual mass of bedrock resulting from subsurface rotting through the action of acidic ground water penetrating along joint systems, followed by mechanical stripping of loose material. Periglacial processes may also be important in the formation of tors. Etymol: Celtic(?). See also: *core-stone.*

torbanite Essentially synonymous with *boghead coal*, but often considered as a highly carbonaceous oil shale. It is named from its type locality, Torbane Hill, in Scotland. Cf: *cannel coal; wollongongite.* Syn: *kerosene shale; bitumenite.*

torbernite A green, radioactive, tetragonal mineral: $Cu(UO_2)_2(PO_4)_2 \cdot 8\text{-}12H_2O$. It is isomorphous with autunite. Torbernite is commonly a secondary mineral and occurs in tabular crystals or in foliated form. Syn: *chalcolite; copper uranite; cuprouranite; uran-mica.*

torch peat A waxy, resinous peat derived mainly from pollen.

tordrillite A light-colored rhyolite that lacks mafic minerals and has the same chemical composition as alaskite. It is named after the Tordrillo Mountains, Alaska, U.S.A.

törnebohmite A greenish mineral: $Ce_3Si_2O_8(OH)$. Syn: *toernebohmite.*

tornote A monaxonic sponge spicule having abruptly pointed ends. Cf: *oxea.*

toroid A mold, commonly consisting of sand, of a circular scour pit made in firm, shallow-water sediments (such as hard mud) by an eddy or whirlpool in flowing water. It has a characteristic swirled shape like a folded bun, but with a homogeneous internal structure and texture.

torose load cast One of a group of elongate load casts that pinch and swell along their trends and that may terminate downcurrent in bulbous, tear-drop, or spiral forms (Crowell, 1955, p.1360).

torque The effectiveness of a force that tends to rotate a body; the product of the force and the perpendicular distance from its line of action to its axis.

torrent (a) A violent and rushing stream of water; e.g. a flooded river, or a rapidly flowing stream in a mountain ravine, or a stream suddenly raised by heavy rainfall or rapid snowmelt and descending a steep slope. Also, any similar stream as of lava. (b) A mountain channel that is intermittently filled with rushing water at certain times or seasons.---Adj: *torrential.*

torrential cross-bedding A variety of *angular cross-bedding* resulting from rapid deposition (as under desert conditions of concentrated rainfall, strong winds, and playa-lake deposition), exhibiting uniform cross-beds of coarser material that meet fine, horizontally laminated beds at an acute angle both above and below. It is essentially *tabular cross-bedding.*

torrential plain An early term used by McGee (1897) for a feature now known as a *pediment.*

torrent tract *mountain tract.*

Torrert In U.S. Dept. of Agriculture soil taxonomy, a suborder of the soil order *Vertisol,* characterized by formation in an arid region and the continuous presence of wide, deep surface cracks (SSSA, 1970). Cf: *Udert; Ustert; Xerert.*

torreyite A mineral: $(Mg,Mn,Zn)_7(SO_4)(OH)_{12}.4H_2O$. Cf: *mooreite.* Syn: *delta-mooreite.*

torricellian chamber In a cave, a room or passage that is filled with air but below water level and sealed by water. Its pressure is below atmospheric pressure and its air-water contact surface is higher than that of the adjacent, free air-water surface (Monroe, 1970).

Torridonian A division of the upper Precambrian in Scotland.

Torrox In U.S. Dept. of Agriculture soil taxonomy, a tentative suborder of the soil order *Oxisol,* characterized by formation in a torric soil moisture regime (SSSA, 1970). Cf: *Aquox; Humox; Orthox; Ustox.*

torsion The state of stress produced by two force couples of opposite moment acting in different but parallel planes about a common axis.

torsion balance *Eötvös torsion balance.*

torsion coefficient The resistance of a material to torsional stress, measured as the work necessary to overcome it, in c.g.s. units.

torsion crack A *crack* [ice] in sea ice, produced by the twisting of the ice beyond its elastic limit.

torsion fault *wrench fault.*

torsion head That part of a torsion balance from which the filament or wire is suspended; a rotary cap, often graduated in degrees, atop the vertical tube supporting a torsion suspension (Weld, 1937).

torsion magnetometer An instrument that is both a *horizontal field balance* and a *vertical field balance* and consists of a suspended permanent magnet. Cf: *Schmidt field balance.*

torsion modulus *modulus of rigidity.*

torsion period The natural period of oscillation of the suspended system in a torsion balance.

torsion seismometer A seismometer designed to detect the horizontal component of an earthquake by the torsion of vertical, suspended thread.

torsion wire The filament or wire supporting the beam in a torsion balance or gravity meter.

torso mountain A mountain rising above a peneplain; a *monadnock.*

torso plain *rumpffläche.*

torticone A cephalopod shell coiled in an irregular three-dimensional spiral with progressive twisting of the conch (like most gastropods) as distinguished from one coiled in a plane spiral. Syn: *trochoceroid.*

tortoise *camel back.*

Tortonian European stage: Miocene (above Helvetian, below Sarmatian).

tortuga A colloquial syn. of *geophone.* Etymol: Spanish, "turtle".

tortuosity [elect] The inverse ratio of the length of a rock specimen to the length of the equivalent path of electrolyte within it.

tortuosity [hydraul] The ratio of the actual length of a river channel, measured along the middle of the main channel, to the axial length of the river (ASCE, 1962).

tortuous flow *turbulent flow.*

torus (a) An invagination or protuberance of exine more or less paralleling the laesura of a spore. Cf: *kyrtome.* (b) The thickening of the closing membrane in a bordered pit.

toryhillite A plutonic rock containing sodaclase, nepheline, pyroxene, garnet, iron ore, apatite, and calcite. Sodalite may be present, and potassium feldspar is absent.

tosca (a) A term used in Patagonia for a white deposit of calcium carbonate occurring in the loess of the pampas. (b) A term used in Mexico for various rocks, such as clayey vein matter, a talc seam, and soft, decomposed porphyry. (c) A soft coral limestone, used in Puerto Rico for masonry, road surfacing, and as fertilizer.---Etymol: Spanish feminine of *tosco,* "rough, coarse, unpolished".

toscanite An extrusive rock intermediate in composition between rhyolite and dacite and containing sanidine, plagioclase (intermediate between calcic and sodic), and accessory hypersthene, biotite, apatite, and opaque oxides in a silica-rich glassy groundmass.

tosudite A dioctahedral mineral consisting of interlayered chlorite and montmorillonite.

total absorptance A term referring to the *absorptance* measured over all wavelenths. It was formerly called "total absorptivity".

total displacement A syn. of *slip* [struc geol]. Cf: *normal displacement.*

total field The vector sum or combination of all components of a field under consideration, e.g. a magnetic or gravitational field. Syn: *total intensity* [geophys].

total hardness *hardness* [water].

total head The sum of the *elevation head, pressure head,* and *velocity head* of a liquid. For ground water, however, the velocity-head component is generally negligible.

total intensity [geophys] *total field.*

total intensity [mag] The magnitude of the intensity of the magnetic field, symbolized by F; it is one of the *magnetic elements.* Syn: *total magnetic intensity.*

total magnetic intensity *total intensity* [mag].

total passing Transportation of all sediment across an area without any being deposited. Cf: *bypassing.*

total porosity *porosity* [grd wat].

total reflection *Reflection* in which all of the incident wave is returned.

total rock *whole rock.*

total runoff *runoff* [water].

total slip *net slip*.

total time correction The sum of all corrections applied to a reflection traveltime in seismic prospecting to express the time as that to a selected datum plane. The main corrections are those for the low-velocity layer and the so-called elevation correction to datum.

touchstone A black, flinty, siliceous stone (such as a silicified shale, a siliceous slate, or a variety of quartz closely allied to or grading into chert or jasper) whose smoothed surface was formerly used to test the purity or fineness of alloys of gold and silver by comparing the steak left on the stone when rubbed by the metal with that made by an alloy of predetermined composition. Syn: *Lydian stone; basanite; flinty slate*.

tough A general descriptor for a material that is able to absorb stress by plastic deformation, i.e. that has tensile strength. Cf: *tender*.

tourmaline (a) A group of minerals of general formula: (Na, Ca)(Mg,Fe^{+2},Fe^{+3},Al,Li)$_3$Al$_6$(BO$_3$)$_3$Si$_6$O$_{18}$(OH)$_4$. It sometimes contains fluorine in small amounts. (b) Any of the minerals of the tourmaline group, such as buergerite, elbaite, and dravite.---Tourmaline occurs in 3-, 6-, or 9-sided prisms (usually vertically striated) or in compact or columnar masses; it is commonly found as an accessory mineral in granitic pegmatites, and is widely distributed in acid igneous rocks, metamorphic rocks (such as gneisses), and clay slates. Its color varies greatly and gives a basis for naming the varieties; when transparent and flawless, it may be cut into gems. See also: *schorl*. Syn: *turmaline*.

tourmalinization Introduction of, or replacement by, tourmaline.

tourmalite A rock composed almost entirely of tourmaline and quartz with a mottled appearance and a texture ranging from dense to granular to schistose. It is of secondary origin, resulting from metasomatic and pneumatolytic effects along the margins of igneous intrusions (Johannsen, 1939, p.22).

Tournaisian European stage: Lower Carboniferous (Lower Mississippian; above Famennian of Devonian, below Viséan).

towan A coastal sand dune in Cornwall, Eng. Cf: *tywyn*.

tower A very high rock formation or peak marked by precipitous sides; e.g. Devils Tower, Wyo.

tower karst A type of karst that is characterized by isolated, steep-sided limestone hills that may be flat-topped and that are surrounded by detrital material. Etymol: German, *Turmkarst*.

towhead A low alluvial island or shoal in a river; esp. a sandbar covered with a growth of cottonwoods or young willows.

township The unit of survey of the U.S. Public Land Survey system, representing a piece of land that is bounded on the east and west by meridians approximately six miles apart (exactly six miles at its south border) and on the north and south by parallels six miles apart, and that is normally subdivided into 36 *sections*. Townships are located with reference to the initial point of a principal meridian and base line, and are normally numbered consecutively north and south from a base line (e.g. "township 14 north" indicates a township in the 14th *tier* north of a base line). The term "township" is used in conjunction with the appropriate *range* to indicate the coordinates of a particular township in reference to the initial point (e.g. "township 3 south, range 4 west" indicates the particular township which is the 3rd township south of the base line and the 4th township west of the principal meridian controlling the surveys in that area). Abbrev (when citing specific location): T. See also: *fractional township*.

township line One of the imaginary boundary lines running east and west at six-mile intervals and marking the relative north and south locations of townships in a U.S. public-land survey. Cf: *range line*.

toxa A siliceous, monaxonic sponge spicule (microsclere) with a central arch and broadly recurved extremities that taper to a point, the whole resembling an archer's bow. Pl: *toxae* or *toxas*.

T phase A seismic *phase* applied to a wave period of 1 sec or less which travels through the ocean with the speed of sound in water. It is occasionally identified on the records of earthquakes in which a larger part of the path from epicenter to station is across the deep ocean.

T plane *glide plane*.

trabecula [paleont] (a) A rod or pillar of radiating calcareous fibers forming a skeletal element in the structure of the septum and related components of a coral. See also: *simple trabecula; compound trabecula*. (b) A branch separating the fenestrae in reticulate cheilostomatous bryozoans. (c) One of the individual anastomosing filaments of a hexactinellid sponge that form a web in which the flagellated chambers (lined by choanocytes) are suspended and that form the pinacoderm. Also, any rod- or beam-like skeletal element of a sponge other than a ray or branch of a single spicule; esp. such a structure of sclerosome. (d) A tiny rod-like structure, smaller and less regular than a *pillar*, connecting layers of sclerite in holothurians.---Pl: *trabeculae*.

trabecula [bot] (a) In the wood of a gymnosperm, a small bar extending across the lumina of the ordinary tracheid from one tangential wall to another. It has no known function. (b) In a moss, a transverse bar of the teeth of the peristome. (c) In some dinoflagellate cysts, a narrow, solid rod distally connecting processes. (d) In the lycopod order Isoetales, a plate forming a partial septum in the microsporangium. (e) In the lycopod *Selaginella*, the lacunar tissue between the cortex and the central bundle.----Plural: *trabeculae*.

trabecular columella A spongy *columella* in scleractinian corals, formed of trabeculae loosely joined with synapticulae or paliform lobes.

trabecular linkage The connection between corallite centers in scleractinian corals, reflecting in the hard parts the *indirect linkage* of stomodaea.

trabecular network (a) A network of sponge trabeculae made of sclerosome. (b) The cellular web of a hexactinellid sponge.

trabecular chorate cyst A dinoflagellate *chorate cyst* possessing trabeculae (e.g. *Cannosphaeropsis*).

trace [geochem] n. A concentration of a substance that is too minute for determination but that is nevertheless detectable.

trace [seis] The record made by a recording device such as a seismometer on paper or film.

trace [meteorol] A quantity of precipitation that is insufficient to be measured by a gage.

trace [paleont] A sign, evidence, or indication of a former presence; specif. a mark left behind by an extinct animal, such as a *trace fossil*.

trace [struc geol] The intersection of a geological surface with another surface, e.g. the trace of bedding on a fault surface, or the trace of a fault or outcrop on the ground. Cf: *trend; strike*.

trace analysis Computation of correct seismic reflection times for all seismic recording geophones, and the plotting of these reflection times in a true reflecting position with respect to each other and the energy source.

trace element (a) An element that is not essential in a mineral but that is found in small quantities in its structure or adsorbed on its surfaces. Although not quantitatively defined, it is conventionally assumed to be less than 1.0% of the mineral. Syn: *accessory element; guest element*. (b) An element that occurs in minute quantities in plant or animal tissue and that is essential physiologically. Syn: *minor element; microelement*.

trace fossil A sedimentary structure consisting of a fossilized track, trail, burrow, tube, boring, or tunnel resulting from the life activities (other than growth) of an animal, such as a mark made by an invertebrate moving, creeping, crawling, feeding, hiding, browsing, running, or resting on or in soft sediment at the time of its accumulation. It is often preserved as a raised or depressed form in a sedimentary rock. Many trace fossils were formerly assumed to be bodily preserved plants or animals. Syn: *ichnofossil; trace; vestigiofossil; lebensspur.*

tracer Any substance that is used in a process to trace its course, specif. radioactive material introduced into a chemical, biological, or physical reaction.

trace slip In a fault, that component of the net slip which is parallel to the trace of an index plane, such as bedding, on the fault plane. See also: *trace-slip fault.*

trace-slip fault A fault, the net slip of which is *trace slip*, or slip parallel to the trace of the bedding or other index plane.

tracheid An elongate cell in xylem that has pitted walls and that is used for both strength and water conduction.

tracheophyte A *vascular plant*; variously classified as incorporating either pterophytes and spermatophytes or psilopsids, lycopsids, spenopsids, and pteropsids.

trachographic map A map using perspective symbols to show local relief and average slope of the Earth's surface, after the style of Erwin Raisz (1959). Syn: *physiographic pictorial map.*

trachyandesite An extrusive rock, intermediate in composition between trachyte and andesite, with acid plagioclase, alkali feldspar, and one or more mafic minerals (biotite, amphibole, or pyroxene). Although "trachyte" has been considered a syn. of *latite*, Streckeisen (1967, p.185) suggests that they can be distinguished by considering trachyandesites as including latites plus latite-andesites.

trachybasalt An extrusive rock intermediate in composition between a trachyte and a basalt, characterized by the presence of both calcic plagioclase and sanidine, along with augite, olivine, and possibly minor analcime or leucite. According to Streckeisen (1967, p.185), the term has been variously defined as synonymous with *latite*, intermediate between latite and basalt, and intermediate between trachyte and basalt, but he recommends applying it to latite and latite-basalt. The term was also used, by Rosenbusch, as a syn. of *trachydolerite* (Johannsen, 1939, p.284).

trachydiscontinuity A term proposed by Sanders (1957, p.293) for an unconformity characterized by an irregular surface. Cf: *leurodiscontinuity.*

trachydolerite (a) An alkalic basalt composed of orthoclase or anorthoclase, along with labradorite and a small amount of feldspathoids. (b) A rock intermediate in composition between trachyte and basalt, and, in this sense, a syn. of *trachybasalt*, as used by Rosenbusch.

trachyophitic Said of the ophitic texture of an igneous rock in which the feldspar crystals enclosed by pyroxene have a parallel or subparallel alignment; a form of nesophitic texture (Walker, 1957, p.2).

trachyostracous Thick-shelled; esp. said of a thick-shelled gastropod.

trachyte A group of fine-grained, generally porphyritic, extrusive rocks having alkali feldspar, minor mafic minerals (biotite, hornblende, or pyroxene) as the main components, and possibly a small amount of acid plagioclase; also, any member of that group; the extrusive equivalent of *syenite*. Trachyte grades into latite as the alkali feldspar content decreases, and into rhyolite with an increase in quartz. Etymol: Greek *trachys*, "rough", in reference to the fact that rocks of this group are rough to the touch.

trachytic (a) Said of igneous rocks in which the lath-shaped feldspar microlites of the groundmass have a subparallel arrangement corresponding to the flow lines of the nearly consolidated magma or lava. Cf: *trachytoid texture; orthophyric.* (b) Pertaining to or composed of *trachyte.*

trachytoid texture The texture of a phaneritic extrusive igneous rock in which the microlites of a mineral, not necessarily feldspar, in the groundmass have a subparallel or randomly divergent alignment. Cf: *trachytic.*

track [paleont] (a) A fossil structure consisting of a mark left in soft material by the foot of a bird, reptile, mammal, or other animal. Cf: *trail.* (b) *muscle track.*

track [photo] The actual path of an aircraft over the Earth's surface.

tract [geog] A region or area of land that may be precisely or indefinitely defined.

tract [streams] A part of a stream, such as a *mountain tract* or a *valley tract.*

traction [sed] A mode of sediment transport in which the particles are swept along (on, near, or immediately above) and parallel to a bottom surface by rolling, sliding, dragging, pushing, or *saltation*; e.g. boulders tumbling along a stream bed, or sand carried by the wind over a desert surface or moved by waves and currents on a beach. The term was introduced into geology by Gilbert (1914, p.15) for the entire complex process of carrying material along a stream bed. Cf: *suspension.*

traction [exp struc geol] The stress component in any given direction; a vector quantity. It may be normal, tensile, compressive, or shear.

traction load The *bed load* moved by stream traction. Also spelled: *tractional load.*

tractive current A current, in standing water, that transports sediment along and in contact with the bottom, as in a stream (Passega, 1957, p.1961). Cf: *turbidity current.* Syn: *traction current.*

tractive force In hydraulics, drag or shear developed on the wetted area of the stream bed, acting in the direction of flow. As measured per unit wetted area, unit tractive force equals the specific weight of water times hydraulic radius times slope of the channel bed (Chow, 1957). See also: *critical tractive force.*

trade-wind desert *tropical desert.*

trade winds A major system of tropical winds moving from the subtropical highs to the equatorial trough. It is northeasterly in the Northern Hemisphere and southeasterly in the Southern Hemisphere. Cf: *antitrades.*

trafficability The quality or suitability of the soil or a terrain to permit passage, such as the cross-country movement of military troops; esp. the capacity of the soil to support moving vehicles.

traffic pan A *pressure pan* produced by the passage of machines, such as tractors.

trail [glac geol] "A line or belt of rock fragments picked up by glacial ice at some localized outcrop and left scattered along a more or less well defined tract during the movement and melting of a glacier" (Stokes & Varnes, 1955, p.154). See also: *train.*

trail [mass move] *congeliturbate.*

trail [paleont] (a) A fossil structure consisting of a trace or sign of the passing of one or many animals; esp. a more or less continuous marking left by an organism moving over the bottom, such as a *worm trail.* Cf: *track.* (b) The extension of either valve of a brachiopod shell anterior to the geniculation (or anterior to the "visceral disc" or the part of the shell posterior to the geniculation) (TIP, 1965, pt.H, p.154 & 155).

trail [sed] An obsolete term used by Fisher (1866) for material that fills furrows in SE England.

trailed fault A fault that has become deflected by a later fault.

trailing spit *tail* [coast].

trail of a fault Crushed material along the fault surface that is used as an indication of the direction of displacement. Such material can be a source of mineral deposits (*drag ore*).

train [glac geol] A term applied to a long, narrow glacial deposit extending for a long distance, such as a *valley train* or a *boulder train.* See also: *trail; rock train.*

train [phys] (a) A series of reflections on a seismograph record. (b) A succession of repetitive events, such as physical oscillations.

trajectory [seism] The path of a seismic wave; a line representing the locus of points determined by experiment or computation. Syn: *curve [seism]*.

trajectory [exp struc geol] The curve that a moving body in a field, or a characteristic of a field, describes in space.

tranquil flow Water flow whose velocity is less than that of a long surface wave in still water (Middleton, 1966). Cf: *rapid flow*. Syn: *subcritical flow; streaming flow*.

transapical axis An axis of a pennate diatom frustule that is perpendicular to both the *apical axis* and the *pervalvar axis*.

transceiver An instrument that both transmits and receives radio-frequency waves.

transcontinental geophysical survey A comprehensive geological and geophysical study from coast to coast across a continent; specif. the study of a band 4° wide (about 440 km) centered on lat. 37°N, extending across the U.S and offshore into the Atlantic and Pacific oceans. Abbrev: TGS.

transcurrent fault A large-scale *strike-slip fault* in which the fault surface is steeply inclined. Syn: *transverse thrust*.

transect n. In ecology, a sample area chosen as the basis for studying the characteristics of a particular assemblage of organisms.

transection glacier A glacier that fills an entire valley system, concealing the divides between the valleys. Cf: *through glacier*.

transfer A single process occurring continuously in space-time in which erosion is followed by transportation and deposition of sediment (Wilson, 1959).

transfer impedance The complex ratio of voltage at one pair of terminals to the current at another pair in a four-terminal network.

transfer percentage For any element, the ratio of the amount present in sea water to the amount supplied to sea water during geologic time by weathering and erosion, multiplied by 100.

transfluence The flowing of glacier ice through a breach made by the headward growth of cirques on both sides of a mountain ridge.

transformation [chem] In phase studies, a syn. of *inversion*.

transformation [isotope] *transmutation*.

transformation [cryst] The change from one crystal polymorph to another, by one of several processes: dilatation, displacement, reconstruction, rotation, or substitution. See also: *dilatational transformation; displacive transformation; reconstructive transformation; rotational transformation; substitutional transformation*. Syn: *inversion [cryst]*.

transformation *granitization*.

transformation [photo] The process of projecting a photograph (mathematically, graphically, or photographically) from its plane onto another plane by translation, rotation, and/or scale change. See also: *rectification*.

trans-formational breccia A term used by Landes (1945) for a breccia occurring in a vertical body and cutting through a stratigraphic section, and believed to have been produced by collapse, such as above a dissolved salt bed.

transformation twin A crystal twin developed by a transformation from a higher to a lower symmetry; e.g. Dauphiné twinning in the transformation from high quartz to low.

transformed wave A *reflected wave* that has been transformed from a *P* wave to an *S* wave or vice versa, one or more times by one or more reflections. It is indicated by such symbols as PS, SP, PSS, etc.

transform fault A strike-slip fault characteristic of midoceanic ridges and along which the ridges are offset. Analysis of transform faults is based on the concept of sea-floor spreading.

transformism A theory that explains the origin of granite as a result of *granitization;* opposed to *magmatism*. A proponent of this theory is called a *transformist*.

transformist A proponent of the theory of *transformism*. Syn:

granitizer; antimagmatist.

transfusion A complex series of processes for the generation of potassium-rich, ultrabasic magmas involving highly energized emanations rich in alkalis and other materials to react with rocks of the Earth's crust encountered in their ascent from deep-seated sources. The products of reaction are presumed to range from essentially solid, metasomatically altered rocks to completely liquid magmas. The term is now being used in a much broader sense to include the entry and exit of any gaseous or hydrothermal fluid into solid rocks to produce such igneous rocks as granite. Cf: *granitization*.

transgression [stratig] (a) The spread or extension of the sea over land areas, and the consequent evidence of such advance (such as strata deposited unconformably on older rocks, esp. where the new marine deposits are spread far and wide over the former land surface). Also, any change (such as rise of sea level or subsidence of land) that brings offshore, typically deep-water environments to areas formerly occupied by nearshore, typically shallow-water conditions, or that shifts the boundary between marine and nonmarine deposition or between deposition and erosion) outward from the center of a marine basin. Ant: *regression*. Cf: *continental transgression; onlap*. Syn: *invasion; marine transgression*. (b) A term used mostly in Europe for discrepancy in the boundary lines of continuous strata; i.e. *unconformity*.---Syn: *transgress*.

transgressive [intrus rocks] Said of a minor igneous intrusion, e.g. a sill, that cuts across strata rather than confining itself to a single stratum.

transgressive overlap *onlap*.

transgressive reef One of a series of nearshore reefs or bioherms superimposed on back-reef deposits of older reefs during the sinking of a landmass or a rise of the sea level, and developed more or less parallel to the shore (Link, 1950). Cf: *regressive reef*.

transgressive sediments Sediments deposited during the advance or encroachment of water over a land area or during the subsidence of the land, and characterized by an onlap arrangement.

transient [elect] A voltage or current pulse that is usually of short time duration and that may be repetitive or nonrepetitive.

transient [evol] A subdivision of a species whose members varied with time; comparable to a subspecies for a subdivision of a species in space.

transient beach A beach whose sand is removed by storm waves but is quickly restored by longshore-current deposition.

transient creep *primary creep*.

transient methods Resistivity or electromagnetic methods in which the transmitted waveform is a transient.

transient response The secondary electric or magnetic field excited in a geologic structure, or model of a geologic structure, upon excitation by a transient.

transient snowline *snowline*.

transient strain A less precise, loosely used syn. of *creep recovery*.

transit n. (a) A *theodolite* in which the telescope can be reversed (turned end for end) in its supports without being lifted from them by rotating it 180 degrees or more about its horizontal transverse axis. Syn: *transit theodolite*. (b) The act of reversing the direction of a telescope (of a transit) by rotation about its horizontal axis.---v. To reverse the direction of a telescope (of a transit) by rotating it 180 degrees about its horizontal axis. Syn: *plunge*.

Transition A name, now obsolete, applied by Jameson (1808) from the teachings of A.G. Werner in the 1790's to the group or series of rocks occurring between the older and more crystalline *Primitive* rocks and the younger and better stratified *Floetz*, and roughly corresponding to the upper Precambrian and to the lower Paleozoic strata now assigned to the Cambrian, Ordovician, and Silurian. The rocks, consisting of dikes and sills, thick graywackes, and thoroughly indurated limestones, were considered to have been the first orderly deposits formed from the ocean during the passage (transition) of the Earth from its chaotic state to its habitable state; they were laid down with original steep dip and contained the first traces of organic remains, and were believed to extend uninterruptedly around the world.

Transitional *Mesolithic.*

transitional series A series of *passage beds*; esp. a large and widespread series of beds that are transitional in the character of their fossils.

transitional-water wave A wave that is moving from deep water to shallow water, i.e. is transitional between a *deep-water wave* and a *shallow-water wave*; its wavelength is more than twice but less than 25 times the depth of the water, and the wave orbitals are beginning to be influenced by the bottom. Syn: *intermediate wave.*

transition point *inversion point.*

transition temperature *inversion point.*

transition zone (a) A region within the *upper mantle* bordering the lower mantle, at a depth of 410-1000 km, characterized by a rapid increase in density of about 20% and an increase in seismic-wave velocities; it is equivalent to the *C layer.* (b) A region within the *outer core*, transitional to the inner core; the *F layer.*

transit line Any line of a traverse that is projected, either with or without measurement, by the use of a transit or other device; an imaginary straight line between two transit stations.

transitory frozen ground Ground that is frozen by a sudden drop of temperature and that remains frozen for a short period, usually hours or days (Muller, 1947, p.223).

transit theodolite *transit.*

transit traverse A surveying traverse in which the angles are measured with a transit or theodolite and the lengths with a metal tape. It is usually executed for the control of local surveys.

translation A shift in position without rotation. When applied to plastic deformation, it refers to the movement of one block of atoms past another.

translational Pertaining to or said of a uniform movement in one direction, without rotation.

translational fault A fault in which there has been translational movement and no rotational component of movement; dip in the two walls remains the same. It can be strictly applied only to segments of faults (Dennis, 1967). Syn: *translatory fault.*

translational movement Apparent fault-block displacement in which the blocks have not rotated relative to one another, so that features that were parallel before movmement remain so afterwards. Cf: *rotational movement.* See also: *translational fault.* Less-preferred syn: *translatory movement.*

translational slide A major landslide classification group involving the downslope displacement of soil-rock material on a surface which is roughly parallel to the general ground surface, in contrast to rockfalls and rotational landslides. The term includes such diverse landslide types as rock slides, block glides, slab or flake slides, debris slides, mudflows, earthflows, and rapid-flow failures such as liquefaction slides, including loess flows and quick clay slides.

translation gliding *crystal gliding.*

translation lattice *crystal lattice.*

translation plane *gliding plane.*

translation vector In tectonics, a term used by Bhattacharji (1958, p.626) for the vector representing the direction and the net displacement of material from a reference point; the sum of the vectors for compression and for flow.

translatory fault *translational fault.*

translatory movement A less-preferred syn. of *translational movement.*

translucent Said of a mineral that is capable of transmitting

light, but is not *transparent.* Cf: *opaque.*

translucent attritus Attritus consisting mainly of transparent humic degradation matter, with minor quantities of opaque materials. Cf: *opaque attritus.* Syn: *humodurite.*

translucent humic degradation matter *Humic degradation matter* that is transparent and of the same deep red color as anthraxylon; humic degradation matter that is less than 14 microns in width, measured perpendicular to the bedding. Abbrev: THDM.

translunar Pertaining to phenomena, or being the space, beyond the Moon's orbit about the Earth. Cf: *cislunar.*

transmedian muscle One of a pair of muscles in some lingulid brachiopods, anterior to the umbonal muscle. One muscle originates on the left side of the pedicle valve and rises dorsally to be inserted on the right side of the brachial valve whereas the other muscle originates on the right side of the pedicle valve and is inserted on the left side of the brachial valve (TIP, 1965, pt.H, p.154).

transmissibility coefficient In an aquifer, the rate of flow of water, in gallons per day, at the prevailing water temperature through each vertical strip of the aquifer 1 ft wide having a height equal to the thickness of the aquifer and under a unit hydraulic gradient (Theis, 1938, p.894). Syn: *coefficient of transmissibility.*

transmission capacity In a column of soil of unit cross section, the volume of water that flows per unit of time, with a hydraulic gradient unity or with a hydraulic head equal to the length of the soil column (Horton, 1945, p.308).

transmission coefficient The complex ratio of electric field intensity transmitted beyond to that incident upon an interface.

transmission constant An expression of the ability of a permeable medium to transmit a fluid under pressure. As applied to ground water, the discharge in cubic feet per minute through each square foot of cross-sectional area under a 100-percent hydraulic gradient (Tolman, 1937, p.564).

transmission window *infrared atmospheric transmission window.*

transmissivity [remote sensing] Internal *transmittance* of a layer of a material such that the path of the radiation is of unit length, and under conditions in which the boundary of the material has no influence (Nicodemus, 1971, p.299).

transmissivity [hydraul] In an aquifer, the rate at which water of the prevailing kinematic viscosity is transmitted through a unit width under a unit hydraulic gradient. Though spoken of as a property of the aquifer, it embodies also the saturated thickness and the properties of the contained liquid.

transmissometer An instrument that measures the capability of a fluid to transmit light; esp. one that measures the turbidity of water by determining the percent transmission of a light beam. See also: *turbidimeter.*

transmittance The ratio of the transmitted flux to the incident flux. Symbol: τ. See also: *transmissivity.*

transmutation [isotope] The transformation of one element into another. Radioactive decay is the spontaneous transmutation of one element to another, and artificial transmutation can be accomplished by bombardment of atoms with high-speed particles. Syn: *transformation.*

transmutation [evol] The change from one species to another.

transopaque Said of a mineral that is *transparent* in one part of the visible spectrum and *opaque* in another; e.g. goethite, hematite (Salisbury & Hunt, 1968).

transparency [oceanog] The ability of sea water to transmit light; the depth to which water is transparent may be measured by the use of a *Secchi disc.*

transparency [photo] A photographic print or positive image on a clear base (glass or film), intended to be viewed by transmitted light, either black and white or in color; a *diapositive.*

transparent [seis] Pertaining to a region of the Earth that transmits seismic waves with little distortion and alteration.

The antonym "opaque" is not often used.

transparent [min] Said of a mineral that is capable of transmitting light, and through which an object may be seen. Cf: *translucent; opaque; transopaque.*

transpiration The process by which water absorbed by plants, usually through the roots, is evaporated into the atmosphere from the plant surface. Cf: *guttation.*

transport [struc petrol] *tectonic transport.*

transport [sed] A syn. of *transportation.* The term is favored in British usage (Stamp, 1961, p.458), and often occurs in combined terms such as *sediment transport* and *mass transport.*

transportation A phase of sedimentation concerned with the actual movement, shifting, or carrying away by natural agents (such as flowing water, ice, wind, or gravity) of sediment or of any loose, broken, or weathered material, either as solid particles or in solution, from one place to another (over a short or long distance) on or near the Earth's surface; e.g. the drifting of sand along a seashore under the influence of currents, the creeping movement of rocks on a glacier, and the conveyance of mud and dissolved salts by a stream. Syn: *transport [sed].*

transportation velocity *nonsilting velocity.*

transport concentration In a stream, the rate of flow of sediment passing through a given cross-sectional area perpendicular to the flow, compared with the rate of flow of the suspension of water and sediment passing through the same area (ASCE, 1962). See also: *spatial sediment concentration.*

transported Said of material, such as clay or overburden, that has been carried by natural agents from its former site to another place on or near the Earth's surface.

transported soil material *Parent material* that has been moved and redeposited from the site of its parent rock. The adjective "transported" is also applied to the soil formed from such a parent material. Cf: *residual material.*

transporting erosive velocity That velocity of water in a channel which both maintains silt in movement and scours the bed. Cf: *noneroding velocity.*

transposed hinge A hinge in bivalve mollusks in which certain hinge teeth present in one valve occupy positions of teeth usually found in the other valve.

transposition structure A primary sedimentary structure resulting from hydroplastic or fluid flow after deposition and sometimes after consolidation of the sediment (Hills, 1963, p.30).

transverse [ore dep] Said of a vein or lode that is oriented across the bedding of the host rock.

transverse [geomorph] Said of an entity that is extended in a crosswise direction; esp. said of a topographic feature that is oriented at right angles to the general strike of a region. Ant: *longitudinal.*

transverse band The connecting lamella joining the posterior ends of the ascending branches of a brachiopod loop.

transverse bar A slightly submerged sand ridge that extends more or less at right angles to the shoreline. It has been described as a "giant sand wave" and a "plateau-like sandbar".

transverse basin *exogeosyncline.*

transverse coastline *discordant coastline.*

transverse crevasse A *crevasse* developed across a glacier, roughly perpendicular to the direction of ice movement, and convex on the downstream side. Cf: *marginal crevasse; splaying crevasse.*

transverse dune A strongly asymmetrical sand dune elongated perpendicularly to the direction of the prevailing winds, having a gentle windward slope and a steep leeward slope standing at or near the angle of repose of sand; it generally forms in areas of sparse vegetation.

transverse electric electromagnetic wave An electromagnetic wave in which there is no axial component of electric field. Cf: *transverse magnetic electromagnetic wave.*

transverse fault A fault that strikes obliquely or perpendicular to the general structural trend of the region.

transverse flagellum A flagellum, often ribbon-like in form, encircling the body of a dinoflagellate in a nearly transverse plane, usually lodged in a deep encircling groove (girdle), and arising from the anterior pole near the proximal end of the girdle.

transverse fold *cross fold.*

transverse furrow An equatorial thinning in the exine of a dicotyledonous pollen grain, usually occurring at the equator, and always running perpendicular to a meridional colpus. Syn: *colpus transversalis.*

transverse joint *cross joint.*

transverse lamination Lamination of cleavage transverse to bedding. Syn: *oblique lamination.*

transverse magnetic electromagnetic wave An electromagnetic wave in which there is no axial component of magnetic field. Cf: *transverse electric electromagnetic wave.*

transverse Mercator projection A cylindrical conformal map projection, equivalent to the regular *Mercator projection* turned (transversed) 90° in azimuth, so that the cylinder is tangent along a given meridian (or any pair of opposing meridians) rather than along the equator. The central meridian is represented by a straight line and is divided truly; all other meridians and all parallels (except the equator, if shown) are curved lines intersecting at right angles. Lines of constant direction (rhumb lines) are also curved lines. The projection is designed to minimize scale error or variation along a narrow zone by using a great circle, centrally located to the area to be mapped, as the "theoretical equator". It is used for maps of areas extending short distances from the central meridian, for charts of polar regions, and as a worldwide standard for plotting military maps. A special case of the projection is used as the basis of the Universal Transverse Mercator (UTM) grid. See also: *Gauss projection.*

transverse profile *cross profile.*

transverse projection A projection in which the projection axis is rotated 90° in azimuth; e.g. "transverse Mercator projection" or "transverse polyconic projection". Syn: *inverse projection.*

transverse resistivity Resistivity of rock measured across the direction of bedding. Cf: *longitudinal resistivity.*

transverse ridge A generally denticulate fulcral elevation on an articular face of an ossicle of a crinoid ray, disposed perpendicularly or slightly oblique to the axis extending from the dorsal toward the ventral side.

transverse ripple mark A ripple mark formed approximately at right angles to the direction of the current, such as one controlled by longshore currents; its profile may be asymmetric or symmetric.

transverse scour mark A *scour mark* whose long axis is transverse to the main direction of the current. The regular spacing of such marks may lead to confusion with ordinary transverse ripple marks. See also: *current-ripple cast.*

transverse section *cross section.*

transverse septulum A minor partition of a chamber in a foraminiferal test, oriented transverse to the axis of coiling and observable in sagittal (equatorial) and parallel sections. See also: *primary transverse septulum; secondary transverse septulum.*

transverse septum One of a series of septa dividing the parietal tubes of a cirripede crustacean into a series of cells, oriented normal to a longitudinal septum and parallel to the basis (TIP, 1969, pt.R, p.103).

transverse thrust *transcurrent fault.*

transverse valley (a) A valley having a direction at right angles to the general strike of the underlying strata; a *dip valley.* (b) A valley that cuts across a ridge, range, or chain of mountains or hills at right angles (Conybeare & Phillips, 1822, p.xxiv).---Cf: *longitudinal valley.* Syn: *cross valley.*

transverse wave *S wave.*

trap [eng] A device for separating sediment from flowing water; e.g. a *sand trap*.

trap [speleo] *water trap*.

trap [cryst] An imperfection in a crystal structure that may trap a mobile electron, usually temporarily.

trap [oil] *oil trap*.

trap [ign] Any dark-colored, fine-grained, nongranitic, hypabyssal or extrusive rock such as a basalt, peridotite, diabase, or fine-grained gabbro; also, applied to any such rock used in building roads.----Etymol: Swedish *trappa*, "stair, step" in reference to the stairstep appearance created by the abrupt termination of successive flows. Syn: *trapp; trap rock; trappide*.

trap cut *step cut*.

trap-door fault A circular fault, hinged at one edge; it is an *intrusion displacement* structure that is common in the Little Rockies of Montana.

trap efficiency The ability of a reservoir to trap and retain sediment, expressed as the percent of incoming sediment (sediment yield) retained in the basin.

trapezohedral Said of those crystal classes in the tetragonal and hexagonal systems in which the general form is a trapezohedron.

trapezohedron (a) An isometric crystal form of 24 faces, each face of which is ideally a four-sided figure having no two sides parallel, or a trapezium. Syn: *tetragonal trisoctahedron; leucitohedron; icositetrahedron*. (b) A crystal form consisting of six, eight, or twelve faces, half of which above are offset from the other half below. Each face is, ideally, a trapezium. The tetragonal and hexagonal forms may be right- or left-handed.

trapezoidal projection A map projection in which equally spaced straight parallels and straight converging meridians divide the area into trapezoids.

trapp *trap* [ign].

trappide *trap* [ign].

trapshotten gneiss A gneiss injected with flinty crush rock (pseudotachylitic). The term was originated by King and Foote in 1864.

trash ice Broken or crumbled ice mixed with water. Syn: *trash*.

trash line A line on a beach, marking the farthest advance of high tide, and consisting of debris (Pettijohn & Potter, 1964, p.350). Cf: *debris line*.

traskite A mineral: $Ba_9Fe_2Ti_2Si_{12}O_{36}(OH,Cl,F)_6 \cdot 6H_2O$.

trass A pumiceous tuff resembling pozzolan that is used in the production of hydraulic cement.

traveled Removed from the place of origin, as by streams or wind, and esp. by glacier ice, as a *traveled* stone; a syn. of *erratic*. Also spelled: *travelled*.

traveling beach A beach that is continually moving in one general direction under the influence of floods.

traveling dune *wandering dune*.

traveltime The time required for a wave train to travel from its source to a point of observation. In seismic exploration, the time elapsed between the detonation of the charge in the shothole and the arrival of the first waves of interest at the detector. Traveltime may be plotted against distance in a *traveltime curve*.

traveltime curve In seismology, a plot of wave-train *traveltime* against corresponding distance along the Earth's surface from the source to the point of observation. Syn: *time-distance graph; time-distance curve; T.D. curve; T-X graph; hodochrone; hodograph*.

traverse [geol] (a) A line across a thin section or other sample along which grains of various minerals are counted or measured. (b) A vein or fissure in a rock, running obliquely and in a transverse direction.

traverse [surv] n. (a) A sequence or system of measured lengths and directions of straight lines connecting a series of surveyed points (or stations) on the Earth's surface, obtained by or from field measurements, and used in determining the relative positions of the points (or stations). (b) *traverse sur-*

vey. (c) A line surveyed across a plot of ground.----v. T make a traverse; to carry out a traverse survey.

traverse map A map made from a traverse survey.

traverse survey A survey in which a series of lines joined en to end are completely determined as to length and direction these lines being often used as a basis for triangulation. It i used esp. for long narrow strips of land (such as for railroads and for underground surveys. Syn: *traverse*.

traverse table A mathematical table listing the lengths of th two sides opposite the oblique angles for each of a series o right-angled plane triangles as functions of every degree o angle (azimuth or bearing) and of all lengths of the hypote nuse from 1 to 100. Traverse tables are used in computing la titudes and departures in surveying and courses in navigation.

travertine (a) A hard, dense, finely crystalline, compact o massive but often concretionary limestone, of white, tan, o cream color, often having a fibrous or concentric structur and splintery fracture, formed by rapid chemical precipitatio of calcium carbonate from solution in surface and ground wa ters, as by agitation of stream water or by evaporation aroun the mouth or in the conduit of a spring (esp. of a hot spring) It also occurs in limestone caves where it forms banded de posits of dripstone, flowstone, and rimstone in the form of sta lactites, stalagmites, and other cave deposits; and also as vein filling, along faults, and in soil crusts. The spongy or les compact variety is *tufa*. The term is also regarded as a miner al name for a variety of calcite or aragonite. See also: *ony marble*. Syn: *calcareous sinter; calc-sinter*. (b) A term some times applied to any cave deposit of calcium carbonate. (c) A term used inappropriately as a syn. of *kankar*.---Etymol: Ital ian tivertino, from the old Roman name of Tivoli, a town nea Rome where travertine forms an extensive deposit. Syn: *trav ertin*.

travertine dam *rimstone dam*.

travertine terrace *rimstone dam*.

tread The flat or gently sloping surface of one of a series o natural step-like landforms, as those of a glacial stairway o of successive stream terraces; a bench level. Ant: *riser*.

treanorite *allanite*.

treated stone A gemstone that has been heated, stained, o coated, or one that has been treated with X-rays, radium, o in a cyclotron, to improve or otherwise alter its color. Also, stone that has been treated to disguise flaws, such as an opa whose cracks have been filled with oil or other liquid.

trechmannite A red rhombohedral mineral: $AgAsS_2$.

tree A woody, perennial plant having a single main stem.

tree agate A *moss agate* whose dendritic markings resembl trees.

tree line *timberline*.

Tree of Life A figurative reference to the branching pattern o evolution, often represented by a fanciful picture of a tree.

tree ore A high-grade uranium ore consisting of buried *carbo trash* that has been replaced or enriched with uranium-bearin solutions.

tree pollen A syn. of *arborescent pollen*. Abbrev: TP.

tree ring *growth ring*.

tree-ring chronology *dendrochronology*.

trellis drainage pattern A drainage pattern characterized b parallel main streams intersected at or near right angles b their tributaries which in turn are fed by elongated secondar tributaries parallel to the main streams, resembling in plan th stems of a vine on a trellis. It is commonly developed wher the beveled edges of alternating hard and soft rocks outcro in parallel belts, as in a rejuvenated folded-mountain region o in a maturely dissected belted coastal plain of tilted strata; is indicative of marked structural control emphasized by sub sequent and secondary consequent streams. Examples ar well displayed in the Appalachian Mountains region. Cf: *faul trellis drainage pattern; rectangular drainage pattern*. Syn *trellised drainage pattern; grapevine drainage pattern; espalie*

752

drainage pattern.

trema An orifice, occurring singly or in series, in the outer wall of some gastropod shells for excretory functions. Pl: *tremata.*

tremadocian European stage: Lower Ordovician (above Dolgellian of Cambrian, below Arenigian). In Great Britain, the stage has been traditionally classed as uppermost Cambrian, although its fauna has greater affinities with the Ordovician. Syn: *Salmian.*

tremalith A centrally and minutely perforate coccolith, such as a placolith or a rhabdolith. The term is sometimes restricted to *placolith* only. Cf: *discolith.* Syn: *trematolith.*

trematolith *tremalith.*

trematophore A perforated plate over the aperture of the test in some miliolid foraminifers. Syn: *sieve plate.*

trembling prairie *shaking prairie.*

tremblor Var. of *temblor.*

tremocyst A porous calcareous layer of the frontal wall in some cheilostomatous bryozoans, developed evenly over the *olocyst.*

tremolite A white to dark-gray monoclinic mineral of the amphibole group: $Ca_2Mg_5Si_8O_{22}(OH)_2$. It has varying amounts of iron, and may contain manganese and chromium. Tremolite occurs in long blade-shaped or short stout prismatic crystals and also in columnar, fibrous, or granular masses or compact aggregates, generally in metamorphic rocks such as crystalline dolomitic limestones and talc schists. It is used as a form of asbestos insulation. Cf: *actinolite.*

tremopore A pseudopore in the tremocyst of a bryozoan, becoming tubular as thickening proceeded.

tremor A minor earthquake, esp. a foreshock or an aftershock. Syn: *earth tremor; earthquake tremor.*

tremor tract In coal mining, an area of complex folding, faulting, and gliding of coal seams and associated rocks. It may be formed by seismic shocks during the deposit's semicompacted state (Nelson & Nelson, 1967, p.387).

trempealeauan North American stage: uppermost Cambrian (above Franconian, below Lower Ordovician).

trench [geomorph] (a) A long, straight, relatively narrow, U-shaped valley or depression between two mountain ranges, often occupied by parts of two or more streams alternately draining the depression in opposite directions. Syn: *trough.* (b) A narrow, steep-sided canyon, gully, or other depression eroded by a stream. (c) Any long, narrow cut or excavation produced naturally in the Earth's surface by erosion or tectonic movements. Also, a similar feature produced artificially, such as a ditch.

trench [marine geol] A narrow, elongate depression of the deep-sea floor, with steep sides and oriented parallel to the trend of the continent and between the continental margin and the abyssal hills. Such a trench is about 2 km deeper than the surrounding ocean floor, and may be thousands of kilometers long. Cf: *foredeep; trough.* Syn: *oceanic trench; marginal trench; sea-floor trench.*

trend [paleont] In evolutionary paleontology, the evolution of a specific structure or morphologic characteristic within a group, esp. in taking an overall view of a large group, such as an order or class; e.g., the evolution of the form of the septal suture, from simple to complex, in tracing the ammonoids as a group from the Devonian to the Triassic.

trend [stat] (a) The direction or rate of increase or decrease in magnitude of the individual members of a time series of data when random fluctuations of individual members are disregarded; the general movement over a sufficiently long period of time of some statistical progressive change. (b) *trend line.*

trend [struc geol] A general term for the direction or bearing of the outcrop of a geological feature of any dimension, such as a layer, vein, ore body, fold, or orogenic belt. Cf: *strike; trace.* Syn: *direction.*

trend line [stat] A straight line or other statistical curve expressing the best empirical relationship between two variables (commonly a regression line) or showing the tendency of some function to increase or decrease over a period of time. Syn: *trend.*

trend map A stratigraphic map that displays the relatively systematic, large-scale features of a given stratigraphic unit, such as of those indicating broad postdepositional structural and erosional changes or those controlled by regional deposition (Krumbein & Sloss, 1963, p.485-488). Cf: *residual map.*

trend surface analysis A statistical method for fitting and evaluating the degree of fit of a set of data (usually contoured data) to a calculated mathematical surface of linear, quadratic, or higher degree.

Trentonian North American stage: Middle Ordovician (above Wilderness, below Edenian); it is equivalent to the upper part of *Mohawkian.* It has also been regarded as a substage (upper Mohawkian Stage), above the Blackriverian Substage.

trepostome Any ectoproct bryozoan belonging to the order Trepostomata and characterized by tubular zooecia with distinct endozone and exozone and a terminal aperture. Adj: *trepostomatous.*

treppen concept The concept that, on a surface reduced to old age by streams and then uplifted, the rejuvenated streams develop second-cycle valleys first near their mouths and that these young valleys are extended headward to form *piedmont steps.* Etymol: German *Treppen,* "stair steps".

treptomorphism *isochemical metamorphism.*

trevalganite A tourmaline granite containing large phenocrysts of a pink feldspar and/or quartz in a groundmass of schorl rock.

trevor A dark-brown granitic rock from Wales, used to make curling stones.

trevorite A black or brownish-black mineral of the magnetite series in the spinel group: $NiFe_2O_4$.

triactin A sponge spicule having three rays. Syn: *triact.*

triad Said of a symmetry axis that requires a rotation of $120°$ to repeat the crystal's appearance. It refers to *threefold symmetry.* Cf: *diad.*

triaene (a) A tetraxon in which three similar rays differ from the fourth; specif. an elongated sponge spicule with one long ray (the rhabdome) and three similar short rays (the cladi), which are sometimes branched or otherwise modified, divergent at one end. See also: *dichotriaene; phyllotriaene; discotriaene.* (b) The initial four-rayed or trident-like spicule of the ebridian skeleton.

triakisoctahedron *trisoctahedron.*

triakistetrahedron *trigonal tristetrahedron.*

trial pit *test hole.*

triangle An ordinarily plane figure bounded by three straight-line sides and having three internal angles. Cf: *spherical triangle.*

triangle closure The amount by which the sum of the three measured angles of a triangle fails to equal exactly 180 degrees plus the spherical excess (the amount by which the sum of the three angles of a triangle on a sphere exceeds 180 degrees); the *error of closure* of triangle. Also known as "closure of triangle".

triangular coordination An atomic structure or arrangement in which an ion is surrounded by three ions of opposite sign. Syn: *threefold coordination.*

triangular diagram A method of plotting compositions in terms of the relative amounts of three materials or components, involving a triangle in which each apex represents a pure component. The perpendicular distances of a point from each of the three sides (in an equilateral triangle) will then represent the relative amounts of each of the three materials represented by the apexes opposite those sides.

triangular facet A physiographic feature, having a broad base and an apex pointing upward; specif. the face on the end of a *faceted spur*, usually the small remnant of a dissected fault plane at the base of a block mountain. A triangular facet may also form by wave erosion of a mountain front or by glacial truncation of a spur. Syn: *spur-end facet*.

triangular organelle A small sensory structure near the peristome of a tintinnid.

triangular texture In mineral deposits, texture produced by exsolution or replacement in which the exsolved or replacement mineral crystals are arranged in a triangular way, following the crystallographic directions of the host mineral.

triangulate To divide into triangles; esp. to use, or to survey, map, or determine by, triangulation. Etymol: back-formation from *triangulation*.

triangulation (a) A trigonometric operation for finding the directions and distances to and the coordinates of a point by means of bearings from two fixed points a known distance apart; specif. a method of surveying in which the stations are points on the ground at the vertices of a chain or network of triangles, whose angles are measured instrumentally, and whose sides are derived by computation from selected sides or base lines the lengths of which are obtained by direct measurement on the ground or by computation from other triangles. Triangulation is generally used where the area surveyed is large and requires the use of geodetic methods. Cf: *trilateration*. (b) The network or system of triangles into which any part of the Earth's surface is divided in a trigonometric survey.

triangulation net A *net* or series of adjoining triangles covering an area in such a manner that the lengths and relative directions of all lines forming the triangles can be computed successively from a single base line; arcs of triangulation connected together to form a system of loops or circuits extending over an area. See also: *base net*.

triangulation station A surveying *station* whose position is determined by triangulation. It is usually a permanently marked point that has been occupied (such as one identified by a bench mark), as distinguished from secondary points such as church spires, chimneys, water tanks, and prominent summits located by intersection. See also: *trigonometric point*.

triangulation tower An engineering structure used to elevate the line of sight above intervening obstacles (such as trees and topographic features), usually consisting of two separate towers built one within the other, the central one supporting the theodolite and the outer one supporting the observing platform; e.g. the Bilby steel tower consisting of two steel tripods that are demountable and easily erected.

Trias *Triassic*.

Triassic The first period of the Mesozoic era (after the Permian of the Paleozoic era, and before the Jurassic), thought to have covered the span of time between 225 and 195-190 million years ago; also, the corresponding system of rocks. The Triassic is so named because of its threefold division in the rocks of Germany. Syn: *Trias*.

triaxial compression test A test in which a cylindrical specimen or rock encased in an impervious membrane is subjected to a confining pressure and then loaded axially to failure. See also: *unconfined compression text*. Cf: *extension test*. Syn: *compression test*.

triaxial extension test A test in which a cylindrical specimen or rock encased in an impervious membrane is subjected to a confining pressure and the axial load is decreased to failure. Cf: *triaxial compression test*. Syn: *extension test*.

triaxial state of stress A stress system in which none of the principal stresses is zero.

triaxon A siliceous sponge spicule in which six rays or their rudiments grow along three mutually perpendicular axes.

tribe A subdivision of the *rock association*.

tributary n. (a) A stream feeding, joining, or flowing into a larger stream (at any point along its course) or into a lake. Ant: *distributary*. Syn: *tributary stream; affluent; feeder; side stream; contributory*. (b) A valley containing a tributary

stream.---adj. Serving as a tributary.

tributary glacier A glacier that flows into a larger glacier. Se also: *secondary glacier*.

tributary stream *tributary*.

tricentric Said of a corallite formed by a polyp retaining tristo modaeal condition permanently.

trichalcite *tyrolite*.

trichite [paleont] A hair-like siliceous sponge spicule occurrin in fascicles.

trichite [petrology] A straight or curved hairlike crystallite usually black. Trichites occur singly or radially arranged i clusters and are found in glassy igneous rocks.

trichobothrium A sensory hair arising from the center of disk-like membrane on the legs or pedipalpi of an arachni and serving for perception of air currents (TIP, 1955, pt.F p.63). Also, a sensory organ consisting of one or more suc hairs together with its supporting structures. Pl: *trichobothria*.

trichotomocolpate Said of monocolpate pollen grains in whic the colpus is triangular and may simulate a trilete laesura Syn: *trichotomosulcate*.

trichotomosulcate *trichotomocolpate*.

trichroic Said of a mineral that displays *trichroism*.

trichroism *Pleochroism* of a crystal that is indicated by thre different colors. A mineral showing trichroism is said to be *tr chroic*. Cf: *dichroism*.

trickle A small, thin, slowly moving stream; a rill. Syn: *tricklet*.

triclinic system One of the six *crystal systems*, characterize by a onefold axis of symmetry, and having three unequal axe that intersect obliquely.

tricolpate Said of pollen grains having three meridionally ar ranged colpi which are not provided with pores. Tricolpate pol len are typical of dicotyledonous plants, and they first ap peared in the fossil record in the Aptian-Albian interval (uppe part of Lower Cretaceous).

tricolporate Said of pollen grains having three colpi which ar provided with pores or other, usually equatorial modifications.

tricranoclone An *ennomoclone* with three proximal rays.

trider A *tetraclone* consisting of three similar arms differin from the fourth.

tridymite A mineral: SiO_2. It is a high-temperature polymorp of *quartz*, and usually occurs as minute, thin, tabular, white o colorless crystals or scales in cavities in acidic volcanic rock (such as trachyte and rhyolite). Tridymite is stable betwee 870° and 1470°C, it has an orthorhombic structure (alpha-tr dymite) at low temperatures and a hexagonal structure (beta tridymite) at higher temperatures. Cf: *cristobalite*.

trifilar suspension A construction used in some gravity meter in order to increase their sensitivity. It consists of a disk sup ported by a helical spring at its center and three equall spaced wires at its circumference (Heiland, 1940, p.131).

trigonal dipyramid A crystal form of six faces, ideally isoscele triangles, consisting of two trigonal pyramids repeated acros a mirror plane of symmetry. It is trigonal in cross section, an its indices are $|h0l|$ in symmetry $\bar{6}\,m2$, also $|hhl|$ in $\bar{6}$ an 32, and $|hkl|$ in $\bar{6}$.

trigonal dipyramidal class That crystal class in the hexagona system having symmetry $\bar{6}$.

trigonal prism A crystal form of three faces that are parallel t a threefold symmetry axis. Its indices are $|100|$ in symmetr $\bar{6}\,m2$, $\bar{6}$, $3m$, and 3; or $|110|$ in symmetry $\bar{6}$, 32, and 3; o $|hk0|$ in $\bar{6}$ and 3.

trigonal pyramid A crystal form consisting of three faces, ide ally isosceles triangles, in a pyramid with a triangular cros section. Its indices are $|h0l|$ in symmetry $3m$, and $|h0l|$, $|hhl|$ and $|hkl|$ in symmetry 3.

trigonal-pyramidal class That class in the rhombohedral divi sion of the hexagonal system having symmetry 3.

trigonal-scalenohedral class *hexagonal-scalenohedral class*.

trigonal system A crystal system of threefold symmetry that i often considered as part of the *hexagonal system* since th attice may be either hexagonal or rhombohedral. See also: *rhombohedral system.*

trigonal-trapezohedral class That crystal class in the rhombohedral division of the hexagonal system having symmetry 32.

trigonal trapezohedron A crystal form of six faces, having a threefold axis and three twofold axes, but neither mirror planes nor a center of symmetry. The top and bottom trigonal pyramids are rotated less than 30° about c with respect to each other. It may be either right-handed or left-handed. Its indices are {hkl}.

trigonal trisoctahedron *trisoctahedron.*

trigonal tristetrahedron A *tristetrahedron* whose faces are triangular rather than quadrilateral, as in the *deltohedron.* Syn: *triakistetrahedron.*

trigoniid Any bivalve mollusk belonging to the family Trigoniidae, characterized by a variably shaped and ornamented shell, generally with opisthogyrate umbones and with the ornamentation of the posterior portion of the valves differing from that of the flank.

trigonite A yellow to brown monoclinic mineral: $MnPb_3H(AsO_3)_3$. It occurs in triangular wedge-shaped crystals.

trigonododecahedron An obsolete syn. of *deltohedron.*

trigonometric leveling A type of indirect leveling in which differences of elevation are determined by means of observed vertical angles combined with measured or computed horizontal or inclined distances. Syn: *vertical angulation.*

trigonometric point A fixed point determined with great accuracy in the triangulation method of surveying; a *triangulation station* being the vertex of the triangle. Shortened form: *trig point.*

trigonometric survey A survey accomplished by triangulation and by trigonometric calculation of the elevations of points of observation. It is generally preliminary to a topographic survey, and is performed after careful measurement of a base line and of the angles made with this line by the lines toward points of observation.

trihedron A geometric form composed of three planes that meet at a central point, e.g. the trigonal pyramid crystal form.

trikalsilite A hexagonal mineral: (K,Na)AlSiO. It is a structural variety of kalsilite with an *a*-axis of 15 angstroms. Cf: *tetrakalsilite.*

trilateration A method of surveying in which the lengths of the three sides of a series of touching or overlapping triangles are measured (usually by electronic methods) and the angles are computed from the measured lengths. Cf: *triangulation.*

trilete adj. Said of an embryophytic spore and some pollen having a laesura consisting of a three-pronged mark somewhat resembling an upper-case "Y". Cf: *monolete.*---n. A trilete spore. The usage of this term as a noun is improper.

trill *trilling.*

trilling A cyclic crystal twin consisting of three individuals. Cf: *twoling; fourling; fiveling; eightling.* Syn: *threeling; trill.*

trilobite Any marine arthropod belonging to the class Trilobita, characterized by a three-lobed, ovoid to subelliptical exoskeleton divisible longitudinally into axial and side regions and transversely into anterior, middle, and posterior regions. Their stratigraphic range is Lower Cambrian to Permian.

triloculine Having three chambers; specif. said of a foraminiferal test having three externally visible chambers, resembling *Triloculina* in form and chamber plan.

trimaceral Said of a coal microlithotype consisting of three macerals. Cf: *monomaceral; bimaceral.*

trimerite A salmon-colored mineral: Be(Ca,Mn)(SiO₄).

trimerous Said of radial symmetry of certain echinoderms (such as edrioasteroids) characterized by three primary rays extending from the mouth, each of two lateral rays giving off two branches.

trimetal detector A *radiation detector* device.

trimetric projection A projection based on representation of a spherical triangle by a plane triangle whose sides are lines of zero distortion, and in which the three spatial axes are represented as unequally inclined to the plane of projection (equal distances along the axes are drawn unequal).

trimetrogon A system of compiling map data from aerial photographs involving an assembly of three cameras arranged systematically at fixed angles and simultaneously taking photographs (one vertical photograph, and two oblique right and left photographs along the flight line at 60 degrees from the vertical) at regular intervals over the area being mapped. Etymol: originally equipped with wide-angle Metrogon lenses.

trimline A sharp boundary line delimiting the maximum extent of the margins of a glacier that has receded from an area. It usually coincides with the upper limit of unweathered rock on a valley wall or a nunatak; but the trimline of a long-extinct glacier may be marked by a sharp change in the age, constitution, or density of vegetation.

trimming The elimination of spurs that jut across a widening stream valley, effected by lateral erosion where a stream flows against and undercuts the sides of the spurs in going around meanders.

trimorph One of three crystal forms displaying *trimorphism.*

trimorphism That type of *polymorphism* in which there occur three crystal forms, known as *trimorphs.* Adj: *trimorphous.* Cf: *dimorphism; tetramorphism.*

trimorphous Adj. of *trimorphism.*

trinacrite A tuff having the composition of palagonite (Hey, 1962, p.629).

Trinitian North American (Gulf Coast) stage: Lower Cretaceous (above Nuevoleonian, below Fredericksburgian).

trinomen A name of a plant or animal that consists of three words, the first designating the genus, the second the species, and the third the subspecies; e.g. *Odontochile micrurus clarkei.* Syn: *trinomial.*

trinomial n. A syn. of *trinomen.*

trioctahedral Pertaining to a layered-mineral structure in which all possible octahedral positions are occupied. Cf: *dioctahedral.*

triode (a) The initial triradial spicule (in which one ray is atrophied) of the ebridian skeleton. (b) A *triradiate* (sponge spicule). Also spelled: *triod.*

tripartite method A method of determining the apparent surface velocity and direction of propagation of microseisms or earthquake waves by determining the times at which a given wave passes three separated points.

tripestone (a) A concretionary variety of anhydrite composed of contorted plates suggesting pieces of tripe. (b) A stalactite resembling intestines. (c) A variety of barite.---Also spelled: *tripe stone.*

triphane *spodumene.*

triphylite A grayish-green or bluish-gray orthorhombic mineral: $Li(Fe^{+2},Mn^{+2})PO_4$. It is isomorphous with lithiophilite.

triple point An invariant point at which three phases coexist in a unary system. When not otherwise specified, it usually refers to the coexistence of solid, liquid, and vapor of a pure substance.

triple stomodaeal budding A type of budding in scleractinian corals similar to *tristomodaeal budding* in which the three stomodaea invariably form a triangle and only one interstomodaeal couple of mesenteries occurs between each pair of neighboring stomodaea.

triplet An assembled gemstone made by cementing together three parts of which the top and bottom are usually genuine and the middle is a colored substitute. Cf: *doublet.*

triple tombolo Three separate bars connecting an island (usually of large extent and close to the shore) with the mainland. Cf: *single tombolo; double tombolo.*

triplite A dark-brown mineral: $(Mn,Fe,Mg,Ca)_2(PO_4)(F,OH)$. Syn: *pitchy iron ore.*

triploblastic Said of the structure of animals having three layers (ectoderm, mesoderm, and endoderm).

triploidite A yellowish to reddish-brown mineral: $(Mn,Fe)_2$-$(PO_4)(OH)$. It is isomorphous with wolfeite.

tripod (a) A sponge spicule having three equal rays that radiate as if from the apex of a pyramid. (b) A stool-shaped shell formed from divergent rods united at a common center in nasselline radiolarians.

tripoli (a) A finely divided, very porous, lightweight, friable, and white, gray, pink, red, buff, or yellow siliceous (largely chalcedonic) sedimentary rock occurring confined to the Earth's surface usually in powdery or earthy masses and resulting from the weathering (leaching, hydration) of chert or of siliceous limestone (Tarr, 1938, p.27). It has a harsh, rough feel, and is used for the polishing of metals and stones. (b) A term used for an incompletely silicified limestone from which the carbonate has been leached; *rottenstone*. (c) A term that was originally, but is now incorrectly, applied to a *siliceous earth* that closely resembles tripoli; specif. *diatomaceous earth*, such as the type deposit of northern Africa. See also: *tripolite*.

tripoli-powder *diatomaceous earth*.

tripolite A term that is properly applied as a syn. of *diatomaceous earth*, in reference to the material from the north African location of Tripoli. It has also been used, less correctly, as a syn. of *tripoli* (a residual product consisting of nondiatomaceous silica).

triporate Said of pollen grains having three pores, usually disposed at 120 degrees from each other in the equator.

trippkeite A blue-green mineral: $CuAs_2O_4$. It has excellent prismatic cleavage that permits crystals to be broken into flexible fibers.

tripuhyite A greenish-yellow to dark-brown mineral: $FeSb_2O_6$. Syn: *flajolotite*.

triradiate n. A sponge spicule in the form of three coplanar rays radiating from a common center. Syn: *triode*.

triradiate crest In a trilete spore, the three-rayed raised figure on the proximal surface caused by the intersection of the contact areas. Syn: *triradiate ridge*.

triserial Arranged in, characterized by, or consisting of three rows or series; specif. said of the chambers of a foraminiferal test arranged in three columns or in a series of three parallel or alternating rows, such as a trochospiral test with three chambers in each whorl.

trisoctahedron An isometric crystal form of 24 faces, each of which is an isosceles triangle. Its indices are $\{hhk\}$. Syn: *triakisoctahedron; trigonal trisoctahedron*.

tristetrahedron An isometric crystal form having 12 faces that are either triangular (*trigonal tristetrahedron*) or quadrilateral (*deltohedron*). Its indices are $\{hkk\}$ and its symmetry is $\bar{4}3m$ or 23.

tristomodaeal budding A type of budding in scleractinian corals in which three stomodaea are developed within a common tentacular ring and either occur in series or form a triangle, and two interstomodaeal couples of mesenteries are located between the original and each new stomodaeum. See also: *triple stomodaeal budding*.

tritium A radioactive isotope of hydrogen having two neutrons and one proton in the nucleus.

tritium dating Calculation of an age in years by measuring the concentration of radioactive hydrogen-3 (tritium) in a substance, usually water. Maximum possible age limit is about 30 years; however, the method also provides a means of tracing subsurface movements of water and determining its velocities.

tritomite A dark-brown mineral: $(Ce,La,Y,Th,Zr)_5(Si,B)_3(O, OH,F)_{13}$ (?).

tritonymph The third developmental stage in the arachnid order Acarida.

trituration *comminution*.

trivariant Pertaining to a chemical system having three degrees of freedom, i.e. having a variance of three.

trivium (a) The three anterior ambulacra of an echinoid. (b) The part of an asterozoan containing three rays, excluding the bivium. The term is not recommended as applied to asterozoans (TIP, 1966, pt.U, p.30).---Pl: *trivia*. Cf: *bivium*.

trochanter (a) The second segment of a pedipalpus or leg of an arachnid, so articulated to coxa and femur as to permit motion of the entire leg in any direction (TIP, 1955, pt.P, p.63). It corresponds physiologically to vertebrate hip articulation. (b) A joint of the proximal part of a prosomal appendage of a merostome.

trochiform Shaped like a top; e.g. said of a gastropod shell (e.g. of *Trochus*) with a flat base, evenly conical sides, and not highly acute spire.

trochite A wheel-shaped joint of the stem of a fossil crinoid.

trochoceroid A synonym (esp. in the older literature) of *toricone*.

trochoid adj. (a) Said of a horn-shaped solitary corallite with sides regularly expanding from apex at an angle of about 40 degrees. Cf: *turbinate; patellate*. (b) Said of a foraminiferal test with spirally or helically coiled chambers, evolute on one side of the test and involute on the opposite side. Syn: *trochospiral*.---n. A trochoid corallite or foraminiferal test.

trochoidal fault A type of *pivotal fault*, the pivotal point of which has also moved or slipped along the fault surface (Nelson & Nelson, 1967).

trocholophe A brachiopod lophophore disposed as a ring surrounding the mouth, bearing a single row of unpaired (or more rarely, a double row of paired) filamentar appendages (TIP, 1965, pt.H, p.154).

trochospiral Said of a foraminiferal test with spirally coiled chambers; *trochoid*.

troctolite A gabbro that is composed chiefly of calcic plagioclase (e.g. labradorite) and olivine with little or no pyroxene. Syn: *forellenstein*.

troegerite A lemon-yellow mineral: $(UO_2)_3(AsO_4)_2.12H_2O$. Syn: *trögerite*.

Tröger's classification A quantitative mineralogic classification of igneous rocks proposed by E. Tröger in 1935.

troglobiont A troglodyte, esp. one living in the lightless water of caves.

troglobite Any organism living in caves, underground streams, or associated solution cavities. Cf: *troglodyte; troglophile; trogloxene*.

troglodyte Any organism that lives in a cave or rock shelter. Adj: *troglodytic*. Syn: *troglobiont*. Cf: *troglobite; troglophile; trogloxene*.

troglophile Any organism that completes its life cycle in a cave but that also occurs in certain environments outside the cave. Cf: *troglodyte; trogloxene; troglobite*.

trogloxene Any organism that regularly or accidentally enters a cave but that must return to the surface to maintain its existence. Cf: *troglophile, troglodite; troglobite*.

trogschluss A syn. of *trough end*. Etymol: German *Trogschluss*.

trogtalite An isometric mineral: $CoSe_2$. It is dimorphous with hastite.

troilite A hexagonal mineral present in small amounts in almost all meteorites: FeS. It is a variety of pyrrhotite with almost no ferrous-iron deficiency.

trolleite A mineral: $Al_5(PO_4)_4(OH)_4$.

trona A gray-white or yellowish-white monoclinic mineral: $Na_2(CO_3).Na(HCO_3).2H_2O$. It occurs in fibrous or columnar layers and masses in saline residues. Trona is a source of sodium compounds. Syn: *urao*.

trondhjemite A light-colored plutonic rock primarily composed of sodic plagioclase (esp. oligoclase), quartz, sparse biotite, and little or no alkali feldspar; a leuco-quartz diorite with oligoclase as the sole feldspar. Its name is derived from Trondhjem, Norway. Also spelled: *trondjemite; trondheimite*.

troostite A mineral: $(Zn,Mn)_2SiO_4$. It is a reddish variety of willemite containing manganese and occurring in large crystals.

Tropept In U.S. Dept. of Agriculture soil taxonomy, a suborder of the soil order *Inceptisol*, characterized by formation in an isomesic temperature regime and by the possible presence of an ochric epipedon with a cambic horizon, or an umbric or mollic epipedon, but no plaggen epipedon (SSSA, 1970). Cf: *Andept; Aquept; Ochrept; Plaggept; Umbrept.*

trophic Of or pertaining to nutrition.

trophic level A stage of nourishment representing one of the segments of the food chain.

trophism Nutrition involving metabolic exchanges in the tissues.

trophocyte An amoebocyte of a sponge that serves as a nourishing cell of an oocyte, embryo, or gemmule.

trophogenic Said of the upper or illuminated zone of a lake in which photosynthesis converts inorganic matter to organic matter. Cf: *tropholytic.*

tropholytic Said of the deeper part of a lake in which organic matter tends to be dissimilated. Cf: *trophogenic.*

tropic n. The area of the Earth falling between the latitudes of the Tropics of Cancer and Capricorn.----adj. Pertaining to features, climate, vegetation and animals characteristic of the tropic region.

tropical (a) Said of a climate characterized by high temperature and humidity and by abundant rainfall. An area of tropical climate borders that of *equatorial* climate. (b) Pertaining to the tropic regions.

tropical cyclone An intense *cyclone* that forms over the ocean in the tropics, and varies from 25 to 600 miles in diameter. It moves first westward, then northeastward in the Northern Hemisphere and southeastward in the Southern Hemisphere; its wind velocity is moderate at its fringe area but increases to as much as 150 mph at its center. Cf: *extratropical cyclone.* Syn: *hurricane; typhoon.*

tropical desert A hot, dry desert lying within lat. 15° to 30° north and south of the equator, more specifically, near the Tropic of Cancer or Capricorn, where subtropical high pressure air masses prevail, creating an area of very low, sporadic rainfall. See also: *west coast desert.* Syn: *low-latitude desert; trade-wind desert.*

tropical lake A lake whose surface temperature is constantly above 4°C. Ant: *polar lake.*

tropical spine A radial spine disposed according to the Müllerian law and marking a zone in an acantharian radiolarian comparable to the tropical zone of the terrestrial globe.

tropic current A twice-monthly type of tidal current that occurs in conjunction with the maximum declinations of the Moon.

Tropic of Cancer The parallel of lat. approx. 23° north of the equator and which indicates the northernmost latitude of the Sun's vertical rays.

Tropic of Capricorn The parallel of lat. approx. 23° south of the equator and which indicates the southernmost latitude of the Sun's vertical rays.

tropic tide A tide occurring twice monthly when the Moon is at its maximum declination north or south of the Earth's equator (when the Moon in nearly above the tropics of Cancer and Capricorn), and displaying the greatest diurnal inequality. Cf: *equatorial tide.*

tropism An involuntary orientational movement or growth in which an organism or one of its parts turns or curves as a positive or negative response to a stimulus. It may be indistinguishable from *taxis* in motile organisms.

tropophilous Said of an organism that has adapted physiologically to periodic changes in the environment, e.g. seasonal changes.

tropophyte A plant adapted to seasonal changes in moisture and temperature; e.g. the deciduous trees of temperate and tropical regions.

troposphere In oceanography, the waters of the ocean above the thermocline. Cf: *stratosphere.*

tropotaxis Taxis resulting from the simultaneous comparison of stimuli of different intensity that are acting on separate ends of the organism.

trottoir A narrow, organic, intertidal reef construction, composed of either a solid mass or a simple crust covering a rocky substratum, separating the shoreline from the sea in the same manner that a sidewalk separates the street from the adjoining houses. Etymol: French, "sidewalk".

trough [beach] A small linear depression formed just offshore on the bottom of a sea or lake and on the landward side of a *longshore bar.* It is generally parallel to the shoreline, and is always under water. A trough may be excavated by the extreme turbulence of wave and current action in the zone where breakers collapse. Inappropriate syn: *runnel; low.*

trough [geomorph] (a) Any long and narrow depression in the Earth's surface, such as one between hills or with no surface outlet for drainage; esp. a broad, elongated, U-shaped valley, such as a *glacial trough* or a *trench.* (b) The channel in which a stream flows.

trough [marine geol] An elongate depression of the sea floor that is wider and shallower than a *trench,* with less steeply dipping sides. Troughs and trenches are gradational forms; a trough may develop from a trench by becoming filled with sediment.

trough [paleont] (a) The furrow on the posterior part of the pedicle valve of an inarticulate brachiopod beneath the apex which provides space for the pedicle (Moore et al, 1952, p.207). (b) *hinge trough.*

trough [fault] *graben.*

trough [folds] A line that connects the lowest points of a fold; the *axis of a syncline.* Cf: *trough plane.* Syn: *trough line.*

trough [sed] *geosynclinal trough.*

trough banding A mineralogic layering or alignment across the bottom of a magma chamber, probably produced by movements during cooling.

trough cross-bedding Cross-bedding in which the lower bounding surfaces are curved surfaces of erosion (McKee & Weir, 1953, p.385); it results from channeling and subsequent deposition. See also: *festoon cross-bedding.* Syn: *crescent-type cross-bedding.*

trough end The steep, semicircular rock wall forming the abrupt head or end of a glacial trough. See also: *oversteepened wall.* Syn: *trough wall; trogschluss.*

trough fault A fault structure that is a set of two faults bounding a graben. Cf: *ridge fault.*

trough-in-trough Said of a cross profile depicting two or more glaciations, each of which shaped its own trough-like valley, esp. where a steep-sided inner trough lies within a wider trough with a flatter bottom.

trough line *trough.*

trough plane *trough surface.*

trough reef *reverse saddle.*

trough surface A surface that connects the *troughs* of the beds of a syncline. Syn: *trough plane; synclinal axis.*

trough valley *U-shaped valley.*

trough wall *trough end.*

troutstone A troctolite characterized by a speckled appearance due to dark spots of olivine in a groundmass of light-colored feldspar.

trowlesworthite A coarse-grained plutonic rock characterized by the presence of red orthoclase, tourmaline, fluorite, and quartz. According to Johannsen (1939, p.285), it may represent a vein in granite that has undergone pneumatolysis.

trudellite An amber-yellow mineral: $Al_{10}(SO_4)_3Cl_{12}-(OH)_{12}.30H_2O.$

true azimuth The *azimuth* measured clockwise from true north through 360 degrees.

true bearing The *bearing* expressed as a horizontal angle between a geographic meridian and a line on the Earth; esp. a horizontal angle measured clockwise from true north. Cf: *magnetic bearing.*

true cleavage A quarryman's term for the dominant cleavage in a rock, e.g. slaty cleavage, to distinguish it from minor or *false cleavage.* Geologically, the term is misleading and should be avoided.

true crater A primary depression, formed by impact or explosion, before modification by slumping or by deposition of ejected material; the crater prior to fallback of debris. "The true crater is defined as the boundary between the loose, broken fallback material and the underlying material that has been crushed as fractured but has not experienced significant vertical displacement" (Nordyke, 1962, p.3447). Cf: *apparent crater.* Syn: *primary crater.*

true dip A syn. of *dip,* used in comparison with *apparent dip.* Syn: *full dip.*

true folding Folding due to lateral compression. Ant: *false folding.* Syn: *buckle folding.*

true granite A syn. of *two-mica granite,* used by Rosenbusch.

true homology *homology.*

true horizon (a) A *celestial horizon.* Also, the horizon at sea. (b) A horizontal plane passing through a point of vision or a perspective center. The *apparent horizon* approximates the true horizon only when the point of vision is very close to sea level.

true north The direction from any point on the Earth's surface toward the geographic north pole; the northerly direction of any geographic meridian or of the meridian through the point of observation. It is the universal zero-degree (or 360-degree) mapping reference. True north differs from *magnetic north* by the amount of magnetic declination at the given point. Syn: *geographic north.*

true resistivity The resistivity of a locally homogeneous medium. Cf: *apparent resistivity.*

true soil *solum.*

true thickness The *thickness* of a stratigraphic unit or other tabular body, measured at right angles to the direction of extension of the unit or body. See also: *vertical.* Cf: *apparent thickness.*

truffite Nodular masses of woody lignite occurring within Cretaceous lignite of France. It is so named because of its truffle-like odor.

trug A term used in Devon, Eng., for red limestone.

Truman gravimeter One of the earliest practical field gravity meters of the unstable equilibrium type. (Nettleton, 1940, p.32).

trumpet log *microlaterolog.*

trumpet valley A narrow valley or gorge that cuts through the central part of a lobe (in the morainal landscape of a former piedmont glacier) and opens out into a broad funnel as it reaches the glaciofluvial sand and gravel cone or fan of the lower piedmont. Examples are numerous along the northern Alpine foothills in Bavaria.

truncate v. In crystal structure, to replace the corner of a crystal form with a plane. Such a crystal form is said to be truncated.

truncated [geomorph] Said of a landform (such as a headland or mountain) or of a geologic structure that has been abbreviated by *truncation*; esp. said of a conical eminence (such as a volcano) whose apical part has been replaced by a plane section parallel to the land surface.

truncated [soil] Said of a soil profile in which upper horizons are missing.

truncated spur A spur (as an interlocking spur) that formerly projected into a preglacial valley and that was partially worn away or beveled by a moving glacier that widened and straightened the valley. See also: *faceted spur.*

truncation [geomorph] An act or instance of cutting or breaking off the top or end of a geologic structure or landform, as by erosion. Cf: *beveling.*

truncation [paleont] The natural loss, in life, of the apical portion of a nautiloid shell.

trunk [paleont] The *thorax* of an arthropod; esp. the postcephalic part of the body of a crustacean.

trunk [streams] The principal channel of a system of tributaries; the channel of a trunk stream.

trunk glacier A central or main valley glacier formed by the union of several tributary glaciers. See also: *dendritic glacier.*

trunk stream A *main stream* having an axial or central position in a drainage system. Syn: *stem stream.*

truscottite A mineral: $(Ca,Mn)_2Si_4O_9(OH)_2$. It is perhaps identical with reyerite.

trustedtite A mineral: Ni_3Se_4.

tscheffkinite *chevkinite.*

tschermakite (a) A mineral of the amphibole group: $Ca_2Mg_3(Al,Fe^{+3})_2(Al_2Si_6)O_{22}(OH,F)_2$. Not to be confused with *Ca-Tschermak molecule.* (b) A gray-white feldspar (albite?) containing some magnesium but no calcium, from Bamble, Norway. (c) A plagioclase feldspar (oligoclase or albite) with composition ranging from $Ab_{95}An_5$ to $Ab_{80}An_{20}$.

Tschermak molecule *Ca-Tschermak molecule.*

tschermigite A mineral of the alum group: $(NH_4)Al(SO_4)_2 \cdot 12H_2O$. Syn: *ammonia alum; ammonium alum.*

Tschernosem Var. of *Chernozem.*

Tschernosiom Var. of *Chernozem.*

T-S diagram *temperature-salinity diagram.*

tsilaisite Manganese-rich variety of tourmaline.

tsingtauite A granite that contains only feldspar as phenocrysts. The feldspar is microperthite and sodic plagioclase.

tsumebite An emerald-green mineral: $Pb_2Cu(PO_4)(SO_4)(OH)$. It was previously regarded to be: $Pb_2Cu(PO_4)(OH)_3 \cdot 3H_2O$.

tsunami A gravitational sea wave produced by any large-scale, short-duration disturbance of the ocean floor, principally by a shallow submarine earthquake, but also by submarine earth movement, subsidence, or volcanic eruption, characterized by great speed of propagation (up to 950 km/hr), long wavelength (up to 200 km), long period (varying from 5 min to a few hours, generally 10-60 min), and low observable amplitude on the open sea although it may pile up to great heights (30 m or more) and cause considerable damage on entering shallow water along an exposed coast, often thousands of kilometers from the source. Etymol: Japanese, "harbor wave". Pron: (t)soo-*nah*-me. Pl: *tsunamis; tsunami.* Adj: *tsunamic.* Erroneous syn: *tidal wave.* Syn: *seismic sea wave; seismic surge; earthquake sea wave; tunami.*

t test *Student's t test.*

tube [speleo] A passage in a cave that is elliptical to almost circular in cross section, and is smooth-sided. See also: *half-tube.*

tube [paleont] (a) The central cylinder connecting the two shields of a placolith coccolith. (b) One of the siphons of a bivalve mollusk.

tube foot One of numerous small tentacular, flexible or muscular, extensible, and cylindrical organs of echinoderms, being the end of a branch of the water-vascular system and serving for grasping, adhesion, locomotion, respiration, feeding, or combination of these.

tubercle (a) One of the small, rounded, knob-like structures on the outer surface of the test plates of an echinoid, bearing a spine which is movably articulated with it. (b) A low rounded prominence of intermediate size on the surface of an ostracode valve, commonly along the free margin (TIP, 1961, pt.Q, p.55). See also: *eye tubercle.* (c) Any fine, low, rounded protuberance on either surface of a brachiopod valve, irrespective of origin (TIP, 1965, pt.H, p.154). (d) A small, rounded, moderately prominent elevation on the surface of a gastropod shell (TIP, 1960, pt.I, p.133).

tubercle texture In mineral deposits, a texture in which gangue is replaced by automorphic minerals. Cf: *atoll texture.*

tuberose Said of a mineral whose form is irregular, rootlike shapes or branches.

tube well (a) *driven well.* (b) *tubular well.*----Also spelled: *tube well.*

tubing (a) A small-diameter (2-4 in.), removable pipe, set and sealed in a well inside a large-diameter casing placed in a producing horizon, through which oil and gas are produced (brought to the surface from the reservoir). (b) The act or process of setting and sealing a tubing into a well. (c) The tube lining of a borehole.

tubular spring A gravity or an artesian spring whose water issues from rounded openings, such as lava tubes or solution channels.

tubular stalactite *soda straw.*

tubular well General term for a drilled, driven, or bored well of circular cross section, and of a depth that is relatively great compared to diameter. Syn: *tube well.*

tubule An irregular, hollow, twig-like, calcareous concretion characteristic of loess deposits.

tubulospine A foraminiferal chamber produced radially into a long hollow extension (as in *Schackoina*).

tufa A chemical sedimentary rock composed of calcium carbonate, formed by evaporation as a thin, surficial, soft, spongy, cellular or porous, semifriable incrustation around the mouth of a hot or cold calcareous spring or seep, or along a stream carrying calcium carbonate in solution, and exceptionally as a thick, bulbous, concretionary or compact deposit in a lake or along its shore. It may also be precipitated by algae or bacteria. The hard, dense variety is *travertine.* The term is rarely applied to a similar deposit consisting of silica. The term is not to be confused with *tuff.* Etymol: Italian *tufo.* Cf: *sinter.* Syn: *calcareous tufa; calc-tufa; tuft; petrified moss; tophus.*

tufaceous Pertaining to or like tufa. Not to be confused with *tuffaceous.*

tuff A compacted pyroclastic deposit of volcanic *ash* and dust that may or may not contain up to 50% sediments such as sand or clay. The term is not to be confused with *tufa.* Adj: *tuffaceous.*

tuffaceous Said of sediments containing up to 50% *tuff.*

tuff ball *mud ball.*

tuffeau A term used in France for tufa, micaceous chalk, and soft, very porous, extremely coarse limestone made up of bryozoan fragments.

tuffisite A term proposed by Cloos (1941) for pipes formed of fragmented country rock in Swabia, southwest Germany. Syn: *intrusive tuff.*

tuffite A term used in Germany for a tuff containing both pyroclastic and detrital material, but predominantly pyroclastics.

tufflava An extrusive rock containing both pyroclastic and lava-flow characteristics, so that it is considered to be an intermediate form between a lava flow and a *welded-tuff* type of *ignimbrite.* Whether or not it is actually a genetically distinct type of rock is a matter of debate. Also spelled: *tuffolava; tuff lava; tuflava.* Cf: *ignispumite.*

tuffolava Var. of *tufflava.*

tuffstone A sandstone that contains pyroclastics of sand-grain size.

tuft A term used in England for any porous or soft stone, such as the sandstone in the Alston district of Cumberland, and the rock now known as *tufa* (Arkell & Tomkeieff, 1953, p.121).

tugtupite A mineral: $Na_4BeAlSi_4O_{12}Cl$. It is related to sodalite.

tuhualite A mineral: $(Na,K)_2(Fe^{+2},Fe^{+3},Al)_3Si_7O_{18}(OH)_2$.

tulare A syn. of *tule land.* Etymol: Spanish, "tule field".

tule *tule land.*

tule land A local term given in the Sacramento River valley (Calif.) to a large tract of overflowed land (or *flood basin*) on which the tule, a variety of bulrush, is the dominant or characterized native plant. The land is often referred to as *tule* or "the tules". Etymol: Spanish. Syn: *tulare.*

tumescence The swelling of a volcanic edifice due to accumulation of magma in the reservoir. It may or may not be followed by eruption. Syn: *bulge; inflation.*

tump (a) A mound, hummock, hillock, or other small rise of ground. (b) A clump of vegetation, such as trees, shrubs, or grass; esp. one forming a small, dry island in a marsh or swamp.

tumuli Plural of *tumulus.*

tumulus [paleont] A secondary deposit on the chamber floor of a foraminiferal test, appearing in cross section as more or less a symmetric node with a rounded summit. Pl: *tumuli.*

tumulus [volc] A doming or small mound on the crust of a lava flow, caused by pressure due to the difference in rate of flow between the cooler crust and the more fluid lava below. Unlike a *blister*, it is a solid structure. Pl: *tumuli.* Syn: *pressure dome.*

tunami *tsunami.*

tundra A treeless, level or gently undulating plain characteristic of arctic and subarctic regions. It usually has a marshy surface which supports a growth of mosses, lichens and numerous low shrubs and is underlain by a dark, mucky soil and permafrost.

tundra climate A type of *polar climate* having a mean temperature in the warmest month of between 0° and 10°C. Cf: *perpetual frost climate.*

tundra crater A circular or shapeless "island" of silt (without vegetation) found on a tundra and formed during the period of thawing by the forced rise of silt to the surface and its pouring out like lava onto the surface.

tundra ostiole A *mud circle* in the tundra soil of northern Quebec. Syn: *ostiole* [geomorph].

tundra peat Peat occurring in subarctic areas and derived from mosses, heaths, and birch and willow trees.

tundra polygon *ice-wedge polygon.*

Tundra soil In early U.S. classification systems, a group of zonal soils having dark brown, highly organic upper horizons and grayish lower horizons. It is developed over a permafrost substratum in the tundra, under conditions of cold, humidity, and poor drainage.

tundrite A mineral: $(Ce,La,Nd)_2Ti(Si,P)(O,OH)_7.4H_2O$.

tunellite A colorless monoclinic mineral: $SrB_6O_{10}.4H_2O$.

tungomelane A variety of psilomelane containing tungsten.

tungstate A mineral compound characterized by the radical WO_4, in which the six-valent tungsten ion and the four oxygens form a flattened square rather than a tetrahedron. An example of a tungstate is wolframite, $(Fe,Mn)WO_4$. Tungsten and molybdenum may substitute for each other. Cf: *molybdate.*

tungsten (a) An obsolete term formerly applied to tungsten minerals such as scheelite and wolframite. (b) A metallic element with atomic number 74. Syn: *wolfram.*

tungstenite A mineral: WS_2. it occurs in small, dark, lead-gray folia or scales. The term is "sometimes erroneously given as a translation from German or Russian for *wolframite*" (Fleischer, 1966, p.1317).

tungstic ocher (a) *tungstite.* (b) *ferritungstite.*---Syn: *wolfram ocher.*

tungstite An earthy mineral: $WO_3.H_2O$. It occurs in yellow or yellowish-green pulverulent masses. Syn: *tungstic ocher; wolframine.*

tungusite A mineral: $Ca_4Fe_2Si_6O_{15}(OH)_6$.

tuning-fork spicule A sponge spicule (triradiate) in which two of the rays are subparallel and at approximately 180 degrees to the third ray. These spicules are found in the class Calcarea.

tunnel [speleo] *natural tunnel.*

tunnel [paleont] A low slit-like opening representing a resorbed area at the base of septa in the central part of the test in many fusulinids and serving to facilitate communication between adjacent chambers.

tunnel cave *natural tunnel.*

tunneldale A syn. of *tunnel valley.* Etymol: Danish, "tunnel valley".

tunnel erosion *piping.*

tunneling (a) A form of failure occurring in earth dams and embankments in which a tunnel is created when cracks, developed in the structure under dry conditions, collapse internally when brought into sudden contact with water. It starts at the wet face or upstream side and proceeds downstream. (b) The operation of excavating, driving, and lining tunnels.

tunnel lake A glacial lake occupying a *tunnel valley.*

tunnel valley A shallow trench cut in drift and other loose material by a subglacial stream not loaded with coarse sediment. See also: *ice-walled channel.* Syn: *tunneldale; Rinnental.*

turanite An olive-green mineral: $Cu_5(VO_4)_2(OH)_4$ (?).

turbation Churning, stirring, mixing, or other modifications of a sediment or soil by agents not determined. The term is generally preceded by a prefix denoting agent, if known; e.g. congeliturbation and bioturbation.

turbid Stirred or disturbed up, such as by sediment; not clear or translucent, being opaque with suspended matter, such as of a sediment-laden stream flowing into a lake; cloudy or muddy in physical appearance, such as of a feldspar containing minute inclusions. Cf: *roily.*

turbidimeter An instrument for measuring or comparing the turbidity of liquids in terms of the reduction in intensity of a light beam passing through the medium. See also: *transmissometer.*

turbidimetry The determination and measurement of the amount of suspended or slow-settling matter in a liquid; the measurement of the decrease in intensity of a light beam passed through a medium. Cf: *nephelometry.*

turbidite A sediment or rock deposited from, or inferred to have been deposited from, a turbidity current. It is characterized by graded bedding, moderate sorting, and well-developed primary structures in the sequence noted in the Bouma cycle.

turbidity (a) The state, condition, or quality of opaqueness or reduced clarity of a fluid, due to the presence of suspended matter. (b) A measure of the ability of suspended material to disturb or diminish the penetration of light through a fluid.

turbidity current A *density current* in water, air, or other fluid, caused by different amounts of matter in suspension, such as a dry-snow avalanche or a descending cloud of volcanic dust; specif. a bottom-flowing current laden with suspended sediment, moving swiftly (under the influence of gravity) down a subaqueous slope and spreading horizontally on the floor of the body of water, having been set and/or maintained in motion by locally churned- or stirred-up sediment that gives the water a density greater than that of the surrounding or overlying clear water. Such currents are known to occur in lakes, and are believed to have produced the submarine canyons notching the continental slope. They appear to originate in various ways, such as by storm waves, tsunamis, earthquake-induced sliding, tectonic movement, over-supply of sediment, and heavily charged rivers in spate with densities exceeding that of sea water. The term was introduced by Johnson (1939, p.27), and is applied to a current due to turbidity, not to one showing that property. See also: *turbidity flow; suspended current.* Cf: *tractive current.* Syn: *suspension current.*

turbidity fan A local fan-shaped area of turbid water at the mouth of a stream flowing into a lake or adjacent to an eroding bank of a lake (Veatch & Humphrys, 1966, p.334).

turbidity flow A tongue-like flow of dense, muddy water moving down a slope; the flow of a *turbidity current.*

turbidity limestone A limestone indicating resedimentation by turbidity currents (Bissell & Chilingar, 1967, p.168).

turbidity size analysis A kind of particle-size analysis based upon the amount of material in turbid suspension, the turbidity decreasing as the particles settle.

turbinate (a) Said of a horn-shaped solitary corallite with sides expanding from apex at an angle of about 70 degrees. Cf: *trochoid; patellate.* (b) Shaped like a top; e.g. said of a spiral gastropod shell with a generally rounded base, a broadly conical spire, and whorls decreasing rapidly from base to apex, or said of a protist shaped like a cone with the point down.

turbodrilling A system of drilling in which the drill bit is directly rotated by a turbine that is attached to the end of the drill pipe at the bottom of the hole and that is driven by drilling mud pumped under high pressure. It was developed in the U.S.S.R. for drilling deep oil wells.

turboglyph A current-produced hieroglyph (Vassoevich, 1953, p.36 & 65); specif. a *flute cast.*

turbulence *turbulent flow.*

turbulence spectrum *eddy spectrum.*

turbulent diffusion *eddy diffusion.*

turbulent flow Water flow in which the flow lines are confused and heterogeneously mixed. It is typical of flow in surface-water bodies. Cf: *laminar flow; mixed flow.* Syn: *turbulence; tortuous flow.*

turbulent flux *eddy flux.*

turbulent velocity That velocity of water in a stream above which the flow is turbulent, and below which it may be either laminar or turbulent. Cf: *laminar velocity.*

turf Peat that has been dried for use as a fuel.

turf-banked terrace A *nonsorted step* whose riser is covered by vegetation and whose tread is composed of fine soil. The term should be reserved for an irregular, terrace-like feature that is not a clearly defined form of patterned ground (Washburn, 1956, p.835). Syn: *turf garland.*

turf garland *turf-banked terrace.*

turgite A red fibrous mineral: $Fe_2O_3.nH_2O$. It is equivalent to hematite with adsorbed water or to an iron ore intermediate between hematite and goethite (hematite being in excess). It occurs as a ferruginous cement in sandstones. Syn: *hydrohematite.*

turjaite A dark-colored plutonic foidite containing a sodium feldspathoid (nepheline) and 60 to 90 percent mafic minerals, esp. melilite, its presence distinguishing turjaite from *melteigite.* Its name is derived from Turja, on the Kola Peninsula, USSR. Cf: *okaite.*

turkey-fat ore A popular name used in Arkansas and Missouri for smithsonite colored yellow by greenockite. Syn: *turkey ore.*

Turkey stone (a) A very fine-grained siliceous rock (containing up to 25% calcite) quarried in central Turkey and used as a whetstone; *novaculite.* Syn: *Turkey slate.* (b) *turquoise.*

turlough An Irish term for a winter lake that is dry or marshy in summer. Also, the ground or hollow periodically flooded to form a turlough. Etymol: Gaelic *turloch,* "dry lake".

turma An artificial suprageneric grouping of form genera of fossil spores and pollen (mostly pre-Cenozoic) based on morphology. It is subdivided into other groups such as "subturmae" and "infraturmae". The system is not governed by the International Code of Botanical Nomenclature. Pl: *turmae.* See also: *ante-turma.*

turmaline Var. of *tourmaline.* The mineral name was originally spelled: "turmalin".

turning point (a) A surveying point on which a level rod is held, after a foresight has been made on it, and before the differential-leveling instrument is moved to another station so that a backsight may be made on it to determine the height of instrument after the resetting; a point of intersection between survey lines, such as the intervening point between two bench marks upon which rod readings are taken. It is established for the purpose of allowing the leveling instrument to be moved forward (alternately leapfrogging with the rod) along the line of survey without a break in the series of measured differences of elevation. Abbrev: T.P. (b) A physical object representing a turning point, such as a steel pin or stake driven in the ground.

turnover [ecol] (a) The process by which some species become extinct and are replaced by other species (MacArthur and Wilson, 1967, p.191). (b) The number of animal generations which replace each other during a given length of time Thorson, 1957, p.491).

turnover [struc geol] *dip reversal.*

turnover [lake] A period (usually in the fall or spring) of uniform vertical temperature when vertical convective circulation occurs in a lake; the time of an *overturn.* See also: *circulation.*

Turonian European stage: Upper Cretaceous, or Middle Cretaceous of some authors (above Cenomanian, below Coniacian).

turquoise A semitranslucent to opaque triclinic mineral: $CuAl_6(PO_4)_4(OH)_8.5H_2O$. It is isomorphous with chalcosiderite. Turquoise is blue, bluish green, greenish blue, or yellowish green; when sky blue it is valued as the most important of the nontransparent gem materials. It usually occurs as reniform masses with a botryoidal surface in the zone of alteration of aluminum-rich igneous rocks (such as trachytes). Syn: *turquois; Turkey stone; calaite.*

urrelite An asphaltic shale found in Texas.

urriculate Turreted, or furnished with or as if with turrets; esp. said of a gastropod shell (e.g. of *Turritella)* with an acutely or highly conical spire composed of numerous flat-sided whorls.

turtleback An extensive, smooth, and curved topographic surface, apparently unique to the Death Valley (Calif.) region, that resembles the carapace of a turtle or the nose of a large, elongate dome with an amplitude up to a few thousand meters. Turtlebacks were first mapped, described, and named by Curry (1938).

turtle stone A large, flattened, oval *septarium* released from its matrix and so weathered and eroded that the interior vein-filled system of cracks may be seen. Its form has a rough resemblance to that of a turtle (it was formerly believed to be a petrified turtle) and its polished surface bears a fanciful resemblance to the back of a turtle. Such concretions are abundant in the Devonian shales of eastern North America. Also spelled: *turtlestone.* Formerly called: *beetle stone.*

tusculite A melilite leucitite that contains small quantities of pyroxene, ilmenite, and feldspar. Its name is derived from Tusculum, Italy.

tussock A dense tuft of grass or grasslike plants usually forming one of many firm hummocks in the midst of a marshy or boggy area.

tussock-birch-heath polygon A *vegetation polygon* characterized by the assemblage indicated (Hopkins & Sigafoos, 1951, p.52-53, 87-92); permafrost appears essential for its formation. Diameter: 2-4.6 m.

tussock ring A *nonsorted circle* consisting of tussocks surrounding a patch of bare soil.

Tuttle lamellae Planes of inclusions in quartz, oriented randomly rather than with reference to the enclosing crystal. Cf: *Boehm lamellae.*

tutvetite A light-reddish trachytoid rock composed chiefly of albite, microcline, a decomposed mafic mineral (possibly acmite), and accessory pyrite and possibly anatase and nordenskioldite. Its name is derived from Tutvet, Norway.

tuxtlite A clinopyroxene intermediate in composition between jadeite and diopside: $NaCaMgAlSi_4O_{12}$. It is a pea-green variety of jadeite containing magnesium and calcium and found in Tuxtla, SE Mexico. Syn: *diopside-jadeite.*

tuya A flat-topped and steep-sided volcano that erupted into a lake thawed in a glacier by the volcano's heat. Examples of tuyas occur in northern British Columbia.

tveitasite A dark-colored, medium- to fine-grained contact igneous rock composed chiefly of a clinopyroxene (acmite-diopside) and an alkali feldspar (orthoclase, cryptoperthite, microperthite, albite), with accessory sphene, apatite, pyroxene, and possibly nepheline and calcite. The rock is probably a hybrid (Johannsen, 1939, p.285).

***T* wave** A short-period (0.5 sec), acoustic wave in the sea.

twenty-degree discontinuity The break in the traveltime curve of seismic *P* waves, originally defined as occurring at an angular distance of about $20°$ and at a depth of 413 km. Later studies determined the angular distance to be $15°$. The term is also written as "20° discontinuity".

twig *divining rod.*

twilight zone *disphotic zone.*

twin An intergrowth of two or more single crystals of the same mineral in a mathematically describable manner, so that some lattices are parallel whereas others are in reversed position. The symmetry of the two parts may be reflected about a common plane, axis, or center. See also: *twinning.* Syn: *twin crystal; twinned crystal.*

twin axis The crystal axis about which one individual of a twin crystal may be rotated (usually $180°$) to bring it into coincidence with the other individual. It cannot be coincident with the axes of twofold, fourfold, or sixfold axes of symmetry. Cf: *twin plane; twin center.* Syn: *twinning axis.*

twin center The crystal point about which the individuals of a twin may be symmetrically arranged. Cf: *twin plane; twin axis.*

twin crystal *twin.*

twin gliding *Crystal gliding* that results in the formation of crystal twins.

twin law A definition of a twin relationship in a given mineral or mineral group, specifying the twin axis, center, or plane, defining the composition surface or plane if possible, and giving the type of twin.

twinned crater A lunar surface feature consisting of two craters with overlapping rims.

twinned crystal *twin.*

twinning The development of a twin crystal by growth, by transformation, or by gliding.

twinning axis *twin axis.*

twinning displacement Displacement in a crystal due to twin gliding.

twinning plane *twin plane.*

twinnite A mineral: $Pb(Sb,As)_2S_4$.

twin plane The common plane across which the individual components of a crystal twin are symmetrically arranged or reflected. It is parallel to a possible crystal face but not to a plane of symmetry of a single crystal. Cf: *twin axis; twin center.* Syn: *twinning plane.*

twin shell Shell of a spumellarian radiolarian with a median transverse constriction.

two-circle goniometer A *goniometer* that measures the azimuth and polar angles for the pole of each face on a crystal by reflection of a parallel beam of light from a cross slit. The two circle angles are readily plotted in stereographic or gnomonic projection for indexing in any crystal system. Cf: *contact goniometer; reflection goniometer.*

two-cycle coast A coast characterized by *two-story cliffs* (Cotton, 1926, p.426).

two-cycle valley A valley produced by rejuvenation, as by headward erosion or by differential earth movements, and characterized by a *valley-in-valley* cross profile. Syn: *two-story valley.*

two-dimensional method A simplified method for calculating the effect on gravity of geological structures in section, in which these structures are assumed to be infinitely long at right angles to the section. Syn: *profile method.*

two-layer structure A type of *layer structure* having two unit layers to the full repeat unit; e.g. most muscovite, which has two tetrahedral layers per unit along the c axis. Cf: *three-layer structure.*

twoling A crystal twin consisting of two individuals. Cf: *trilling; fourling; fiveling; eightling.*

two-mica granite A granite containing both dark (biotite) and light (muscovite) mica. This granite was called *true granite* by Rosenbusch and *binary granite* by Keyes. Cf: *aplogranite; granitelle.*

two-phase inclusion An angular cavity in a gemstone consisting of a gas bubble in a liquid; the cavity may or may not coincide with a possible crystal form of the host mineral. Examples occur in corundum. Cf: *three-phase inclusion.*

two-point problem A problem in plane-table surveying of determining the position of a point with the known factor being the length of one line that does not include the point to be located. Cf: *three-point problem* [*surv*].

two-story cliff A sea cliff consisting of an ancient, uplifted cliff of a former shoreline cycle separated from a lower-lying cliff of a later cycle by a narrow wave-cut bench. Syn: *two-storied cliff.*

two-story valley *two-cycle valley.*

two-sweep cusp A *meander cusp* formed by the sweep of two successive meanders migrating downstream while the stream remains on the same side of the flood plain (Lobeck, 1939, p.241).

two-swing cusp A *meander cusp* formed by two successive swings of a meander belt, the scar produced by the first swing intersecting the scar of the second swing in such a way that a Y-shaped feature results, with the handle of the Y pointing either upstream or downstream (Lobeck, 1939, p.240-241).

two-year ice *second-year ice.*

T-X graph *traveltime curve.*

tychite A white isometric mineral: $Na_6Mg_2(CO_3)_4(SO_4)$.

tychopotamic Said of an aquatic organism adapted to living chiefly in still, fresh water. Cf: *autopotamic; eupotamic.*

tying bar *tombolo.*

tylaster A small tylote aster (sponge spicule).

Tyler standard grade scale A *grade scale* for the particle-size classification of sediments and soils, devised by the W.S. Tyler Company of Cleveland, Ohio; it is based on the square root of 2, with the midpoint values of each size class being simple whole numbers or common fractions. It is used as specifications for sieve mesh.

tylosis A proliferation of the protoplast of a ray or wood parenchyma cell through a pit-pair into the lumen of an adjacent inactive tracheary element, where it may or may not divide (Record, 1934, p.68) Cf: *tylosoid.*

tylosoid A tylosislike intrusion of a parenchyma cell into an intercellular space. It differs from a *tylosis* in that it does not pass through the cavity of a pit (Record, 1934, p.70).

tylostyle A *style* (sponge spicule) in which the blunt end is swollen or knobbed. Syn: *tylostylus.*

tylote n. A slender, elongate sponge spicule (monaxon) with a knob at both ends.---adj. Said of a sponge spicule with the ends of the rays knobbed or swollen.

tylotoxea A rod-like sponge spicule tapering to a sharp point at one end and to a knob at the other.

tympanoid Said of a squat, drum-shaped scleractinian corallite.

Tyndall figure A small hollow in the shape of a round or hexagonal disk or a hexagonal star, partly filled with water, oriented parallel to the basal plane within an ice crystal, and formed through melting by radiation absorbed at points of defect in the ice lattice. Named after John Tyndall (1820-1893), British physicist. Syn: *Tyndall flower; Tyndall star.*

type [**coal**] A *coal classification* based on the constituent plant materials. Cf: *rank* [*coal*]; *grade* [*coal*].

type [**taxon**] A lower taxonomic unit upon which a higher taxonomic unit is based, being the standard of reference for determining the exact application of the scientific name of that higher taxon and usually being considered as most closely ex-

emplifying the higher taxon. The type of a species is a single specimen; the type of a genus is a single species.

type [**petrology**] *rock type.*

type area An area containing the type locality, within which the diagnostic relations of the type section are widely represented; e.g. an area in which good exposures (of weakly consolidated rocks) are evanescent (ACSN, 1961, art.13i). Many early definitions of stratigraphic units indicate a type area without specifying a type section. Syn: *type region.*

type-boundary section *boundary stratotype.*

type concept A basic principle of binomial nomenclature stating that each binomen and its description be associated with a preserved specimen or specimens, each generic name and its description with a named species, and each taxon of a higher rank with a definite member of a lower taxon included within it.

type curves Curves of electrical parameters, computed for simple subsurface models, with which observed curves are compared in interpretation.

type fossil A term occasionally used as a syn. of *index fossil.*

type genus That genus upon which the description of a family subfamily, or superfamily is based; the type of a family, subfamily, or superfamily.

type locality (a) The place at which a stratigraphic unit (such as a formation or series) is typically displayed and from which it derives its name. It contains the type section, and is contained within the type area. Cf: *reference locality.* (b) The place where a geologic feature (such as an ore occurrence, a particular kind of igneous rock, or the type specimen of a fossil species or subspecies) was first or originally recognized and described.

type material All of the specimens upon which the description of a new species is based. Syn: *hypodigm.*

type region *type area.*

type section The original sequence of strata as described for a given locality or area. It serves as an objective standard with which spatially separated parts of the stratigraphic unit may be compared, and it is preferably in an area where the unit shows maximum thickness and is completely exposed (or at least shows top and bottom). Type sections for rock-stratigraphic units can never be changed (ACSN, 1961, art.13h) there is only one type section, although there may be more than one typical section. Cf: *reference section.* See also: *stratotype.*

type species That species upon which the original description of a genus or subgenus is largely or entirely based; the type of a genus or subgenus. Syn: *genotype.*

type specimen The single specimen upon which the original description of a particular species or subspecies is based. The type specimen may be a *holotype, neotype,* or *lectotype.*

typhoon A *tropical cyclone,* esp. in the western Pacific. Etymol: Chinese, "great wind".

typochemical element An element that is characteristically present in a mineral, though it is not essential to its composition.

typomorphic mineral A mineral that is typically developed in only a narrow range of temperature and pressure. The term was originated by Becke. Cf: *critical mineral; index mineral.*

tyretskite A mineral: $Ca_3B_8O_{13}(OH)_4$ (?).

tyrolite A mineral: $Cu_5Ca(AsO_4)_2(CO_3)(OH)_4.6H_2O$. Syn: *trichalcite.*

tyrrellite A mineral: $(Cu,Co,Ni)_3Se_4$.

tysonite *fluocerite.*

tyuyamunite An orthorhombic mineral: $Ca(UO_2)_2(VO_4)_2.5-8H_2O$. It is an ore of uranium, and occurs in yellow incrustations as a secondary mineral. Syn: *calciocarnotite.*

U

ubac A mountain slope so oriented as to receive the minimum available amount of light and warmth from the Sun during the day; esp. a northward-facing slope of the Alps. Etymol: French dialect, "shady side". Cf: *adret*.

uehebe A low volcanic cone composed of pyroclasts which are mostly accidental.

Udalf In U.S. Dept. of Agriculture soil taxonomy, a suborder of the soil order *Alfisol*, characterized by formation in a udic moisture regime and in a mesic or warmer temperature regime. Udalfs are generally brown (SSSA, 1970). Cf: *Aqualf; Boralf; Ustalf; Xeralf.*

Udden grade scale A logarithmic *grade scale* devised by Johan A. Udden (1859-1932), U.S. geologist; it uses 1 mm as the reference point and progresses by the fixed ratio of 1/2 in the direction of decreasing size and of 2 in the direction of increasing size, such as 0.25, 0.5, 1, 2, 4 (Udden, 1898). See also: *Wentworth grade scale.*

Udert In U.S. Dept. of Agriculture soil taxonomy, a suborder of the soil order *Vertisol*, characterized by formation in a humid region so that surface cracks do not remain open for more than two or three months (SSSA, 1970). Cf: *Torrert; Ustert; Xerert.*

udic Said of a soil moisture regime that is characterized by not being completely dry for either 90 consecutive days or for 0 consecutive days in the 90-day period following summer solstice, when soil temperature at a depth of 50 cm is above °C (SSSA, 1970).

Udoll In U.S. Dept. of Agriculture soil taxonomy, a suborder of the soil order *Mollisol*, characterized by formation in a udic moisture regime and by the lack of a calcic or gypsic horizon (SSSA, 1970). Cf: *Alboll; Aquoll; Boroll; Rendoll; Ustoll; Xeroll.*

Udult In U.S. Dept. of Agriculture soil taxonomy, a suborder of the soil order *Ultisol*, characterized by a low to moderate organic-carbon content, argillic horizons that are reddish or yellowish, and by formation in a udic soil moisture regime (SSSA, 1970). Cf: *Aquult; Humult; Ustult; Xerult.*

ugandite An extrusive rock containing leucite, augite, and abundant olivine in a soda-rich glassy groundmass; *olivine leucitite.*

ugrandite A group name for the calcium garnet minerals uvarovite, grossular, and andradite.

uhligite (a) A black isometric mineral consisting of a titanate and zirconate of calcium and aluminum. (b) An amorphous variscite or fischerite.

uintahite A black, shiny asphaltite, with a brown streak and conchoidal fracture, which is soluble in turpentine; it occurs primarily in veins in the Uinta Basin, Utah. See also: *wurtzilite.* Syn: *gilsonite; uintaite.*

uintaite *uintahite.*

Uinta structure A basement diapir or doming or upwarp in the form of a regional, flattened anticlinal flexure on which denudation has exposed the basement-rock core. It is named after the Uinta Mountains of northeastern Utah.

uklonskovite A mineral: $NaMg(SO_4)(OH).2H_2O$.

ukrainite A monzonite with less than 20 percent quartz.

Ulatisian North American stage: middle Eocene (above Penutian, below Narizian).

ulexite A white triclinic mineral: $NaCaB_5O_9.8H_2O$. It forms rounded reniform masses of extremely fine acicular crystals and is usually associated with borax in saline crusts on alkali flats in arid regions. Syn: *boronatrocalcite; natroborocalcite; cotton ball.*

uliginous Said of an organism living in wet or swampy ground.

ullmannite A steel-gray to black mineral: NiSbS. It usually contains a little arsenic. Syn: *nickel-antimony glance.*

ulmain A kind of *euvitrain* that consists completely of ulmin but that is not precipitated from solution, as is *collain.*

ulmic acid *ulmin.*

ulmification The process of peat formation. See also: *ulmin.* Syn: *paludification.*

ulmin Vegetable-degradation material occurring in coal as amorphous, brown to black substance of gel and that is insoluble in alkaline solution. It is abundant in peat and lignite, and forms vitrinite. Syn: *ulmic acid; humin; carbohumin; humogelite; fundamental jelly; jelly; gélose; fundamental substance; vegetable jelly.*

ulminite A variety of euvitrinite characteristic of ulmain and consisting of jellified but not precipitated plant material. Cf: *collinite.*

ulrichite [**mineral**] A syn. of *uraninite*; specif. the original unoxidized UO_2.

ulrichite [**petrology**] A dark-colored hypabyssal rock composed of large phenocrysts of alkali feldspar, soda pyroxene, and amphibole, and smaller accessory olivine phenocrysts in a groundmass of feldspar, pyroxene, and amphibole; an olivine-bearing tinguaite.

Ulsterian North American provincial series: Lower Devonian (above Cayugan of Silurian, below Erian).

ultimate analysis The determination of the elements in a compound; for coal, the determination of carbon, hydrogen, sulfur, nitrogen, ash, and oxygen. Cf: *proximate analysis.*

ultimate base level The lowest possible *base level*; for a stream, it is sea level, projected inland as an imaginary surface beneath the stream. Cf: *temporary base level.* Syn: *general base level.*

ultimate bearing capacity The average load per unit of area required to produce failure by rupture of a supporting soil mass. See also: *bearing capacity.*

ultimate landform The theoretical landform produced near the end of a cycle of erosion. Cf: *initial landform; sequential landform.* Syn: *ultimate form.*

ultimate shear strength The maximum shearing stress (i.e., half the differential stress) corresponding with the *ultimate strength.*

ultimate strength The maximum differential stress that a material can sustain under the conditions of deformation. Beyond this point, rock *failure* occurs. See also: *ultimate shear strength.*

Ultisol In U.S. Dept. of Agriculture soil taxonomy, a soil order characterized by the presence of an argillic horizon having a base saturation of less than 35% at a pH of 8.2. It has a mean annual soil temperature of at least $8°C$ (SSSA, 1970). Suborders and great soil groups of the Ultisols have the suffix ult. See also: *Aquult; Humult; Udult; Ustult; Xerult.*

ultrabasic Said of an igneous rock having a low silica content (less than that of a basic rock). Percentage delimitations are arbitrary and vary with different petrologists. The term is frequently used interchangeably with *ultramafic*; although most ultrabasic rocks are also ultramafic, there are some exceptions; e.g. monomineralic rocks composed of pyroxenes are ultramafic but are not ultrabasic because of their high SiO_2 content; a monomineralic rock composed of anorthite would be considered ultrabasic (SiO_2 = 43.2 percent) but not ultramafic. "Ultrabasic" is one subdivision of a widely used system for classifying igneous rocks on the basis of silica content; the other subdivisions are *acidic, basic,* and *intermediate.* Cf: *silicic.*

ultrabasite *diaphorite.*

ultramafic Said of an igneous rock composed chiefly of mafic minerals, e.g. monomineralic rocks composed of hypersthene, augite, or aegerine. Cf: *ultrabasic.*

ultramafite An ultramafic rock.

ultramarine A syn. of *lazurite.* The term is also applied to artificial lazurite and to artificial compounds allied to lazurite; e.g.

the brilliant blue pigment formerly made by powdering lapis lazuli and characterized by the durability of its color.

ultrametagranite A granite formed as a result of extremely high-grade metamorphism, possibly involving partial remelting. Cf: *metagranite*.

ultrametamorphism Metamorphic processes at the extreme upper range of temperatures and pressures, at which partial to complete fusion of the affected rocks takes place and magma is produced. The term was originated by Holmquist in 1909.

ultramicroearthquake An earthquake having a magnitude of zero or less on the Richter scale. Such a limit is arbitrary, and may vary according to the user. Cf: *microearthquake; major earthquake*.

ultramylonite An ultra-crushed variety of *mylonite* in which primary structures and porphyroclasts have been entirely obliterated so that the rock becomes homogeneous and dense with little, if any, parallel structure (Quensel, 1916). Cf: *protomylonite*. Syn: *flinty crush rock*.

ultraplankton The smallest plankton; they are five microns and smaller. Cf: *nannoplankton; microplankton; macroplankton; megaloplankton*.

ultrasima The supposedly ultrabasic layer of the Earth below the sima, immediately below the Mohorovicic discontinuity.

ultraviolet absorption spectroscopy The observation of an *absorption spectrum* in the ultraviolet frequency region and all processes of recording and measuring which go with it. The absorption of radiant energy by the outer electrons of atoms or molecules and their subsequent transition to higher energy levels produce the ultraviolet absorption spectrum.

ultraviolet filter An optical filter for use when taking pictures from high altitudes. It cuts down the ultraviolet rays, but allows all visible light to pass; hence, no increase in exposure time is required.

ultravulcanian A type of volcanic eruption characterized by violent, gaseous explosions of lithic dust and blocks, with little if any incandescent scoria. It is commonly observed during the opening or reopening of a volcanic vent. Its type occurrence is the explosion of Krakatoa in 1883.

ulvospinel A mineral of the spinel group: Fe_2TiO_4. It usually occurs as fine exsolution lamellae, intergrown with magnetite. Syn: *ulvite*.

umangite a dark-red mineral: Cu_3Se_2.

umbel (a) An umbrella-like structure consisting of multiple recurved teeth attached to the tip of a ray or pseudoactin of a sponge spicule (such as of an amphidisc). (b) A sponge spicule consisting of a single shaft with an umbel at one end; e.g. a paraclavule or one type of a clavule.

umbelliferous (a) Said of a tabulate corallum having corallites arranged like ribs of an umbrella, growing outward in whorls. (b) Producing umbels.

umber A naturally occurring chestnut-brown to liver-brown earth that is darker than *ocher* and *sienna* and that consists of manganese oxides as well as hydrated ferric oxide, silica, alumina, and lime. It is highly valued as a permanent paint pigment, and is used either in the greenish-brown natural state ("raw umber") or in the dark-brown or reddish-brown calcined state ("burnt umber").

umbilical area The inner part or surface of a whorl of a cephalopod shell, separating the umbilical shoulder from the umbilical seam. It is called an "umbilical wall" if it rises somewhat vertically from the spiral plane and "umbilical slope" if it rises gently (TIP, 1959, pt.L, p.6).

umbilical lobe The large primary lobe of a suture of an ammonoid, centered on or near the umbilical seam, and forming part of both external suture and internal suture.

umbilical perforation The vacant space or opening around the axis of coiling and connecting the umbilici on opposite sides of a cephalopod shell.

umbilical plug (a) The deposit of secondary skeletal or shell material in the axis or umbilical region of certain coiled foraminiferal tests (e.g. in *Rotalia*). (b) The calcareous deposit filling the umbilicus of a cephalopod.---Syn: *plug* [paleont].

umbilical seam The helical line of junction or overlap of adjacent whorls of a coiled cephalopod (nautiloid or ammonoid) conch. Syn: *umbilical suture*.

umbilical shoulder (a) The part of a cephalopod shell bordering the umbilicus and forming its outer margin; e.g. the strongly bent part of a whorl of a nautiloid shell between the flank and the inner part of the umbilical area. (b) The angulation of whorls at the margin of and within the umbilicus of gastropod shell (Moore et al, 1952, p.289). (c) The part of foraminiferal test bordering the umbilicus.---See also: *shoulder* [paleont].

umbilical suture (a) A continuous line separating successive whorls as seen in the umbilicus of phaneromphalous gastropod shells. (b) An *umbilical seam* of a cephalopod.

umbilical tooth One of the projections forming a triangular modification of the apertural lip of a foraminiferal test, with those of successive chambers in forms (e.g. *Globoquadrina*) with umbilical aperture giving a characteristic serrate border to the umbilicus.

umbilicus (a) A cavity or depression in the center of the base of a spiral shell of a univalve mollusk; e.g. the cavity (typically a conical opening) formed around the central axis of a spiral gastropod shell between faces of adaxial walls of whorl where these do not coalesce to form a solid columella, or an external depression centered around the axis of coiling of cephalopod shell and formed by the diminishing width of whorls toward the axis (such as the depression near the center of whorls of a coiled nautiloid conch). (b) A circular depression or pit in the axis of a coiled foraminiferal test; e.g. the closed, shallow, axial depressed area formed by curvature of overlapping chamber walls in involute forms, or the space formed between inner margins of the walls of chambers belonging to the same whorl of the test.---Pl: *umbilici*.

umbo (a) The "humped" part of the shell of a bivalve mollusk, or the elevated and relatively convex part of a valve surrounding the point of maximum curvature of the longitudinal dorsal profile and extending to the beak when not coinciding with it. The term is often used synonymously with *beak*, but with most shells two distinct terms are needed. (b) The relatively convex, apical part of either valve of a brachiopod, just anterior to or containing the beak. It is usually swollen in the pedicle valve of a productoid, but a pit or rarely a blister-like elevation in the brachial valve (Muir-Wood & Cooper, 1960, p.8). (c) The apical part of either valve of the bivalved carapace of a crustacean; e.g. the point on the plate from which successive growth increments extend in a cirripede. (d) A blunt prominence on the frontal wall or ovicell in certain cheilostomatous bryozoans. (e) A central, round, elevated structure in discoidal foraminiferal tests. It is commonly due to lamellar thickening and may occur on one or both sides of the test. (f) A central projection on the thecal plate of an echinoderm, representing part of its ornamentation.---Pl: *umbones* or *umbos*. Syn: *umbone*.

umbonal angle (a) The approximate angle of divergence of the posterior/dorsal and anterior/dorsal parts of the longitudinal profile of bivalve-mollusk shells; specif. the angle of divergence of umbonal folds in pectinoid shells. (b) The angle subtended at the umbo of a brachiopod by the region of the shell surface adjacent to the umbo.

umbonal chamber One of a pair of posteriorly and laterally located cavities in either valve of a brachiopod, bounded in the pedicle valve by dental plates and shell walls and limited medially in the brachial valve by crural plates (or homologues) and shell walls.

umbonal fold The ridge originating at the umbo of a pectinoid shell and setting off the body of the shell from the auricle.

umbonal muscle A single muscle occurring in some lingulid brachiopods, thought to be homologous with the posterior adductor muscles, and consisting of two bundles of fibers, posteriorly and slightly asymmetrically placed (TIP, 1965, pt.H, p.154).

umbonate Having or forming an umbo; e.g. having an umbo on one or both sides of an enrolled foraminiferal test. Also, said of a foraminifer bearing a convex elevation in the center.

umbone A syn. of *umbo* [paleont]. The term "umbones" is the usual plural for "umbo".

umbra (a) The completely shadowed region of an eclipse. (b) The inner, darker region of a sunspot.----Cf: *penumbra*.

umbracer dune A *lee dune* tapering to a point downwind, formed under constant wind direction commonly behind a clump of bushes or a prominent bedrock obstacle (Melton, 1940, p.120). See also: *wind-shadow dune*.

umbrafon dune A *lee dune* developed to the leeward of a source or area of loose sand where the sand supply is constantly replenished (Melton, 1940, p.122); e.g. a dune on the lee side of a stream flood plain or landward from a sandy beach. Syn: *lee-source dune; source-bordering lee dune*.

Umbrept In U.S. Dept. of Agriculture soil taxonomy, a suborder of the soil order *Inceptisol*, characterized by formation in a cold or temperate climate and by the presence of an umbric epipedon. Umbrepts may, however, have a mollic or anthropic epipedon (SSSA, 1970). Cf: *Andept; Aquept; Ochrept; Plaggept; Tropept*.

umbric Pertaining to an *epipedon* that is similar to a *mollic* epipedon except for having a base saturation of less than 50%, measured at a pH of 7 (SSSA, 1970). Cf: *ochric*.

umohoite A black to bluish-black mineral: $(UO_2)MoO_4.4H_2O$.

umptekite A syenite composed chiefly of microperthite and soda amphibole, with accessory sphene, apatite, and opaque oxides, and occasionally small amounts of interstitial nepheline; a sodic syenite resembling pulaskite. Its name is derived from Umptek, Kola Peninsula, U.S.S.R.

unaka (a) A term proposed by Hayes (1899, p.22) for a large residual mass rising above a peneplain that is less advanced than one having a *monadnock*, and sometimes displaying on its surface the remnants of a peneplain older than the one above which it rises; an erosion remnant of greater size and height than a monadnock. (b) A group or sprawling mass of monadnocks, often occurring near the headwaters of stream systems where erosion has not yet reduced the area to the level of a peneplain (Lobeck, 1939, p.633).---Type locality: Unaka Mountains of eastern Tennessee and western North Carolina.

unakite A metamorphosed igneous rock composed of abundant epidote, with subordinate pink orthoclase and quartz and minor opaque oxides, apatite, and zircon. At its type locality, the Unaka Range on the border of Tennessee and North Carolina, it appears to have been derived from a hypersthene syenite (akerite) (Johannsen, 1931a, p.59).

unarmored Said of naked dinoflagellates (such as those of the order Gymnodiniales) lacking a plate-constructed theca or cell wall and enclosed by a thin, structureless pellicle. Ant: *armored*.

unary system A chemical system that has only one component. Syn: *unicomponent system*.

unavailable moisture *unavailable water*.

unavailable water Water that cannot be utilized by plants because it is held in the soil by adsorption or other forces; water in the soil in an amount below the wilting point. Syn: *unavailable moisture*.

unbalanced force A force that is not opposed by another force acting along the same line in the opposite sense of direction; an unbalanced force causes translation of a body.

uncinate A hexactinellid-sponge spicule (diactinal monaxon) covered on all sides with short thorn-like spines directed toward one end. Cf: *cleme*.

uncompahgrite A plutonic rock composed chiefly of melilite, along with pyroxene, opaque oxides, perovskite, apatite, calcite, anatase, melanite, and occasionally phlogopite; a member of the *melilitolite* group.

unconcentrated flow *overland flow*.

unconcentrated wash *sheet erosion*.

unconfined aquifer An aquifer having a water table; an aquifer containing unconfined ground water. Syn: *water-table aquifer*.

unconfined compression test A special condition of a *triaxial compression test* in which no confining pressure is applied. Syn: *crushing test*.

unconfined ground water Ground water that has a free water table, i.e. water not confined under pressure beneath relatively impermeable rocks. Ant: *confined ground water*. Syn: *phreatic water; nonartesian ground water; free ground water; unconfined water*.

unconfined water *unconfined ground water*.

unconformability The quality, state, or condition of being unconformable, such as the relationship of unconformable strata; *unconformity*.

unconformable Said of strata or stratification exhibiting the relation of unconformity to the older underlying rocks; not succeeding the underlying rocks in immediate order of age or not fitting together with them as parts of a continuous whole. In the strict sense, the term is applied to younger strata that do not "conform" in position or that do not have the same dip and strike as, those of the immediately underlying rocks. Also, said of the contact between unconformable rocks. Cf: *conformable; discordant*.

unconformity (a) A substantial break or gap in the geologic record where a rock unit is overlain by another that is not next in stratigraphic succession, such as an interruption in the continuity of a depositional sequence of sedimentary rocks or a break between eroded igneous rocks and younger sedimentary strata. It results from a change that caused deposition to cease for a considerable span of time, and it normally implies uplift and erosion with loss of the previously formed record. An unconformity is of longer duration than a *diastem*. (b) The structural relationship between rock strata in contact, characterized by a lack of continuity in deposition, and corresponding to a period of nondeposition, weathering, or esp. erosion (either subaerial or subaqueous) prior to the deposition of the younger beds, and often (but not always) marked by absence of parallelism between the strata; strictly, the relationship where the younger overlying stratum does not "conform" to the dip and strike of the older underlying rocks, as shown specif. by an angular unconformity. Cf: *conformity*. Syn: *unconformability; transgression* [stratig]. (c) *surface of unconformity*.---Common types of unconformities recognized in U.S.: *nonconformity; angular unconformity; disconformity; paraconformity*. Since the essential feature of an unconformity, as understood in Great Britain, is structural discordance rather than a time gap, the British do not recognize disconformity and paraconformity as unconformities. For an historical study of unconformities, see Tomkeieff (1962).

unconformity iceberg An iceberg consisting of two or more distinct layers or lenses that differ in composition and are separated by surface discontinuities; e.g. an iceberg in which crevassed glacier ice is overlain successively by a layer of silt and a layer of firn.

unconformity trap A stratigraphic trap associated with an unconformity.

unconsolidated material (a) A sediment that is loosely arranged or unstratified, or whose particles are not cemented together, occurring either at the surface or at depth. (b) Soil material that is in a loosely aggregated form.

uncontrolled mosaic An aerial *mosaic* formed solely by matching detail of overlapping photographs without spatial or directional adjustments to control points.

uncovers *dries*.

unctuous *soapy.*

undation theory A theory proposed by van Bemmelen (1933) that explains the structural and tectonic features of the Earth's crust by vertical upward and downward movements caused by waves that are generated by deep-seated magma. Cf: *blister hypothesis.*

undaturbidite A term proposed by Rizzini & Passega (1964, p.71) for a sediment formed from a suspension produced by violent storms; a deposit intermediate between an ordinary wave deposit and a turbidite. Cf: *fluxoturbidite.*

underclay A layer of fine-grained detrital material, usually clay, lying immediately beneath a coal bed or forming the floor of a coal seam. It represents the old soil in which the plants (from which the coal was formed) were rooted, and it commonly contains fossil roots (esp. of the genus *Stigmaria*). It is often a *fireclay*, and some underclays are commercial sources of fireclay. Syn: *underearth; seat earth; seat clay; root clay; thill; warrant; coal clay.*

underclay limestone A thin, dense, nodular, relatively unfossiliferous *freshwater limestone* underlying coal deposits, so named because it is closely related to underclay.

undercliff [geomorph] A terrace or subordinate cliff along a coast, formed of material fallen from the cliff above; the lower part of a cliff whose upper part underwent landsliding.

undercliff [sed] A term used in southern Wales for a shale forming the floor of a coal seam.

underconsolidation Consolidation (of sedimentary material) less than that normal for the existing overburden; e.g. consolidation resulting from deposition that is too rapid to give time for complete settling. Ant: *overconsolidation.*

undercooling *supercooling.*

undercurrent A current of water flowing beneath a surface current at a different speed or in a different direction; e.g. the Mediterranean Undercurrent off Gibraltar. See also: *equatorial undercurrent.*

undercut A reentrant in the face of a cliff, produced by undercutting.

undercutting The removal of material at the base of a steep slope or cliff or other exposed rock by the erosive action of falling or running water (such as a meandering stream), of sand-laden wind in the desert, or of waves along the coast.

underearth (a) A hard fireclay forming the floor of a coal seam; *underclay.* (b) The soil beneath the Earth's surface. (c) The depths of the Earth.

underfit stream A *misfit stream* that appears to be too small to have eroded the valley in which it flows; a stream whose volume is greatly reduced or whose meanders show a pronounced shrinkage in radius. It is a common result of drainage changes effected by capture, by glaciers, or by climatic variations.

underflow (a) The movement of ground water in an *underflow conduit*; the flow of water through the soil or a subsurface stratum, or under a structure. (b) The rate of discharge of ground water through an underflow conduit. (c) The water flowing beneath the bed or alluvial plain of a surface stream, generally in the same direction as, but at a much slower rate than, the surface drainage; esp. the water flowing under a dry stream channel in an arid region.

underflow conduit A permeable deposit that underlies a surface stream channel, that is more or less definitely limited at its bottom and sides by rocks of relatively low permeability, and that contains ground water moving in the same general direction as the stream above it (Meinzer, 1923, p.43). See also: *underflow.*

underground ice *ground ice.*

underground stream A body of water flowing as a definite current in a distinct channel below the surface of the ground, usually in an area characterized by joints or fissures. Legally, such a stream discoverable by men without scientific instruments. Application of the term to ordinary aquifers is incor-

rect. Cf: *subterranean stream; percolating water.*

underground water (a) A syn. of *ground water.* (b) A syn. of *subsurface water* in less preferred usage.

underhand stoping A mining term for downward *magmatic* stoping.

underlay In metal mining, the extension of a vein or ore deposit beneath the surface; also, the inclination of a vein or ore deposit from the vertical, that is, *hade.* Syn: *underlie.*

underlie [stratig] v. To lie or be situated under, to occupy lower position than, or to pass beneath. The term is usually applied to certain rocks over which certain younger rocks (usually sedimentary or volcanic) are spread out. Ant: *overlie.*

underlie [mining] *underlay.*

underloaded stream A stream that carries less than a full load of sediment and that erodes its bed.

undermass Harder rock material (or the basement) lying beneath the *cover mass*, characterized by a more complex or intensely deformed structure; the material below the surface of an angular unconformity. See also: *compound structure.*

undermelting The melting from below of any floating ice (Huschke, 1959, p.601).

undermining The action of wearing away supporting material, as the *undermining* of a cliff by stream erosion; *sapping.*

underplight A substratum, once consisting of soft mud, that preserves the form of an overlying thin layer of sand or gravel that has been contorted by alternate freezing and thawing (Spurrell, 1887).

undersaturated (a) Said of an igneous rock consisting of *unsaturated* minerals, e.g. feldspathoids and olivine. (b) Said of a rock whose norm contains feldspathoids and olivine, or olivine and hypersthene. Cf: *critically undersaturated; over-saturated; saturated.*

undersaturated permafrost Permafrost that contains less ice than the ground could hold if the water were in the liquid state.

underthrust fault A type of thrust fault in which it is the lower rock mass that has been actively moved under the upper passive rock mass. An underthrust may be difficult to distinguish from an *overthrust.*

undertow The seaward, return flow near the bottom of a sloping beach of water that was carried onto the shore by waves. Cf: *rip current.*

undertow mark A channeled structure on a sedimentary surface, believed to have been made by currents dragging heavy objects in very shallow water adjacent to a beach (Clarke, 1918). Cf: *strand mark.* Syn: *undertow marking.*

underwater gravimeter An instrument capable of measuring gravity when lowered to the sea bottom from a stationary surface vessel; it is leveled and read in a few minutes by remote control and has an accuracy of about 0.1 milligal.

underwater ice Ice formed below the surface of a body of water; e.g. *anchor ice.*

undisturbed Said of a soil sample in which the material has been subjected to so little human disturbance (such as by boring tools or excessive handling) that it is suitable for all laboratory tests and thereby for approximate in-situ determinations of such physical properties as strength, consolidation and permeability characteristics.

undivided Said of a surface, landscape, or area that has no noticeable feature separating the drainage of neighboring streams.

undulate Said of a margin of a leaf or petal that is wavy, i.e. up and down, not in and out (Lawrence, 1951, p.774).

undulating fold A minor fold with rounded apexes; a fold whose beds are bent into alternate elevations and depressions.

undulation [geomorph] (a) A landform having a wavy outline or form; e.g. a desert sand deposit similar to a *whaleback* but shorter and lacking the definite form of the whaleback (Stone, 1967, p.252). (b) A rippling or scalloped land surface, having wavy outline or appearance, or resembling waves in form; e.g. a rolling prairie.

undulation [geodesy] The separation, or height of the geoid above or below the *reference spheroid*.

undulatory extinction A type of *extinction* that occurs successively in adjacent areas, as the microscope's stage is turned. Cf: *parallel extinction; inclined extinction*. Syn: *strain shadow; oscillatory extinction; wavy extinction*.

uneven fracture A general type of mineral fracture that is rough and irregular.

ungaite A general term suggested for oligoclase-bearing dacite. Its name is derived from Unga Island, Kamchatka, U.S.S.R.

ungemachite A colorless to yellowish rhombohedral mineral: $Na_8K_3Fe(SO_4)_6(OH)_2 \cdot 10H_2O$.

unglaciated Said of a land surface that has not been modified by the action of a glacier or an ice sheet; "never-glaciated". Cf: *deglaciation*.

uniaxial Said of a crystal having only one optic axis, e.g. a tetragonal or hexagonal crystal. Cf: *biaxial*.

uniclinal shifting *monoclinal shifting*.

unicline An obsolete syn. of *monocline*.

unicomponent system *unary system*.

uniform channel In hydraulics, a channel having a uniform cross section and a constant roughness and slope (ASCE, 1962).

uniform development The production of a landscape where the rate of uplift is equal to the rate of downward erosion, characterized by constant relief and straight slopes. Cf: *accelerated development; declining development*.

uniform flow Flow of a current of water in which there is neither convergence nor divergence.

uniformitarian n. A believer in the doctrine of uniformitarianism.--- adj. Pertaining to the doctrine of uniformitarianism.

uniformitarianism (a) The fundamental principle or doctrine that geologic processes and natural laws now operating to modify the Earth's crust have acted in the same regular manner and with essentially the same intensity throughout geologic time, and that past geologic events can be explained by phenomena and forces observable today; the classical concept that "the present is the key to the past". The doctrine does not imply that any change has a uniform rate, and does not exclude minor local catastrophes. The term was originated by Lyell (1830), who applied it to a concept expounded by Hutton (1788). Cf: *catastrophism*. Syn: *actualism; principle of uniformity*. (b) The logic and method by which geologists attempt to reconstruct the past using the principle of uniformitarianism.

uniformity coefficient A numerical expression of the variety in particle sizes in mixed natural soils, defined as the ratio of the sieve size through which 60% (by weight) of the material passes to the sieve size that allows 10% of the material to pass. It is unity for a material whose particles are all of the same size, and it increases with variety in size (as high as 30 for heterogeneous sand).

uniform plane wave A plane wave in which the electric and magnetic vectors are of constant intensity across a plane of constant phase.

uniform strain *homogeneous strain*.

unilateral Said of a stream or drainage system in which "all tributaries come in from the north, while the south walls of the main valleys are practically unbroken" (Rich, 1915, p.145).

unilobite A descriptive term for a trace fossil consisting of a one-lobed (unilobate) trail. About 80 percent of all invertebrate tracks are unilobites. The term is seldom used.

unilocular Containing a single chamber or cavity; e.g. said of a single-chambered foraminifer. Syn: *monothalamous*.

unimodal sediment A sediment whose particle-size distribution shows no secondary maxima; e.g. a modern beach gravel.

uninverted relief A topographic configuration that reflects the underlying geologic structure, as where mountains mark the sites of anticlines and valleys mark the sites of synclines. Ant: *inverted relief*.

uniplicate Said of a form of alternate folding in brachiopods with the pedicle valve bearing a median sulcus and an anterior commissure median plica (TIP, 1965, pt.H, p.155). Ant: *sulcate*.

uniserial Arranged in, characterized by, or consisting of a single row or series; e.g. a "uniserial arm" of a primitive crinoid composed of brachial plates arranged in a single row with or without subparallel sutures, or a "uniserial ambulacrum" of an echinoid with pore pairs in a single longitudinal row, or a "uniserial test" of a foraminifer whose chambers are arranged in a single linear or curved series, or a "uniserial rhabdosome" of a graptoloid consisting of a single row of thecae. Cf: *biserial*.

unit cell The fundamental parallelepiped that forms a *crystal lattice* by regular repetition in space. It is sometimes called the *primitive unit cell*.

unit character A natural characteristic that is dependent on the presence or absence of a single gene.

unit circle In a gnomonic projection, the circle that is the projection of the equatorial plane of the sphere of projection. Its radius gives the scale used in plotting the projection.

unit coal Pure coal, free of moisture and noncoal mineral matter, prepared for analysis. Unit coal is expressed by the equation: unit coal $= 1.00 - (W + 1.08A + 0.55S)$, in which W = water, A = ash, and S = sulfur.

unit compaction The compaction per unit thickness of the compacting deposits.

unit compaction/head-decline ratio The ratio between the compaction per unit thickness of the compacting deposits and the head decline in the coarse-grained beds of the compacting aquifer system. If the observed head decline is a direct measure of increase in applied stress, the ratio equals specific unit compaction (Poland, et al, in press).

unit dry weight *dry unit weight*.

unit form A crystal form in a system other than the cubic, having intercepts on the chosen crystal axes that define the axial ratio. Unit forms have Miller indices $\{111\}$, $\{110\}$, $\{011\}$, $\{101\}$.

unit weight A term applied esp. in soil mechanics to the weight per unit of volume, such as grams per cubic centimeter; the density of a material. Symbol: γ. See also: *dry unit weight; effective unit weight; wet unit weight*.

univalve adj. Having or consisting of one valve only. Cf: *bivalve*. Syn: *univalved*.---n. (a) A univalve animal; specif. a mollusk with a univalve shell, such as a gastropod, a cephalopod, or a scaphopod. (b) A shell of a univalve animal; specif. a mollusk shell consisting of one piece.

univariant Said of a chemical system having one degree of freedom; said of an equilibrium system in which the arbitrary variation of more than one physical condition will result in the disappearance of one of the phases.

universal stage A *stage* of three, four, or five axes attached to the rotating stage of a polarizing microscope that enables the thin section under study to be tilted about two horizontal axes at right angles. It is used for optical study of low-symmetry minerals or for determining the orientation of any mineral relative to the section surface and edge directions. Syn: *U-stage; Fedorov stage*.

universe In statistics, a syn. of *population*.

unloading The removal by denudation of overlying material.

unmatched terrace *unpaired terrace*.

unmixing [chem] A syn. of *exsolution* that is also applied to the separation of immiscible liquids.

unmixing [sed] Segregation and concentration of sedimentary material during diagenesis.

unoriented [geol] Said of a rock or other geologic specimen whose original position in space, when collected, is unknown or not definitely ascertained.

unoriented [surv] Said of a map or surveying instrument whose internal coordinates are not coincident with corresponding directions in space.

unpaired terrace A *stream terrace* with no corresponding terrace on the opposite side of the stream valley, usually produced by a meandering stream swinging back and forth across a valley. See also: *meander terrace*. Cf: *paired terrace*. Syn: *unmatched terrace*.

unprotected thermometer A *reversing thermometer* that is not protected against hydrostatic pressure. Cf: *protected thermometer*.

unrestricted Said of tectonic transport or movement in which elongation of particles is parallel to the direction of movement. Cf: *restricted*.

unripe Said of peat that is in an early stage of decay, and in which original plant structures are visible. Cf: *ripe*.

unroofed anticline *breached anticline*.

unsaturated Said of a mineral that does not form in the presence of free silica; e.g. nepheline, leucite, olivine, feldspathoids. Cf: *undersaturated; saturated; oversaturated*.

unsaturated flow The flow of water in an undersaturated soil by capillary action and gravity.

unsaturated zone *zone of aeration*.

unsorted *poorly sorted*.

unstable [sed] (a) Said of a constituent of a sedimentary rock that does not effectively resist further mineralogic change and that represents a product of rapid erosion and deposition (as in a region of tectonic activity and high relief); e.g. feldspar, pyroxene, hornblende, and various fine-grained rock fragments. (b) Said of an immature sedimentary rock (such as graywacke) consisting of unstable particles that are angular to subrounded, poorly to moderately sorted, and composed of feldspar grains or rock fragments.---Cf: *labile* [geol].

unstable [radioactivity] Said of a spontaneously radioactive substance. Cf: *stable* [radioactivity].

unstable equilibrium A state of equilibrium from which a chemical system or a body (such as a pendulum) will depart in response to the slightest perturbation. Cf: *stable equilibrium*.

unstable gravimeter *astatic gravimeter*.

unstable isotope A syn. of *radioisotope*. Cf: *stable isotope*.

unstable relict A *relict* [meta] that is unstable under the newly imposed conditions of metamorphism but persists in a perhaps altered but still recognizable form owing to the low velocity of transformation. A more preferable term would be *metastable relict*. Cf: *stable relict*. See also: *armored relict*.

unstable remanent magnetization *viscous magnetization*.

unsteady flow In hydraulics, flow that changes in magnitude or direction with time. Cf: *steady flow*. Syn: *nonsteady flow*.

unstratified Not formed or deposited in strata; specif. said of *massive* rocks or sediments with an absence of layering, such as granite or glacial till.

unstuck Pertaining to the surfaces of contact between adjacent strata in a metamorphic terrane due to very marked disharmonious folding (Whitten, 1959).

unweathered *fresh* [weath].

Unwin's critical velocity *critical velocity* (e).

upbank thaw A thaw or marked rise of temperature occurring at hill or mountain level while the frost is unbroken in the valley below.

upbuilding The building up of a sedimentary deposit, as by a stream or in the ocean. Cf: *aggradation*.

upcoast Said of the coastal direction generally trending toward the north (CERC, 1966, p.A39). Ant: *downcoast*.

upconcavity The persistent downstream decrease in gradient as seen on the channel profiles of most streams.

updating A change, most frequently a decrease, in the radiometric age of a rock caused by a complete or partial disturbance of the isolated, radioactive system by thermal, igneous,

or tectonic activities which results in loss, usually of daughter products, from the system of radiogenic isotopes (only rarely the gain of radioactive isotopes). See also: *hybrid age; mixed ages; overprint* [geochron].

updip A direction that is upwards and parallel to the dip of a structure or surface. Cf: *downdip*.

updip block The rocks on the *upfaulted* side of a fault. Cf: *downdip block*.

updrift The direction opposite that of the predominant movement of littoral materials.

upfaulted Said of the rocks on the updip side of a fault, or the *updip block*. Cf: *downfaulted*.

upgrading *aggradation*.

uphole A borehole drilled at an upward angle in a direction pointed above the horizontal plane of the drill's swivel head (the mechanism that rotates and advances the drill string).

uphole shooting In seismic exploration, the setting off of successive shots in a shothole at varying depths in order to determine velocities and velocity variation of the materials forming the hole walls.

uphole time In seismic exploration, the time required for the seismic impulse to travel from a charge in a shothole to the surface. Syn: *time at shot point*.

upland (a) A general term for high land or an extensive region of high land, esp. far from the coast or in the interior of a country. Sometimes used synonymously with *fastland*. (b) The higher ground of a region, in contrast with a valley, plain, or other low-lying land; a plateau. (c) The elevated land above the low areas along a stream or between hills; any elevated region from which rivers gather drainage. Also, an area of land above flood level, or not reached by storm tides.----Ant: *lowland*.

upland plain A relatively level area of land lying at a considerable altitude; esp. a high-lying erosion surface.

upland swamp A swamp that "probably" occupies the site of a former shallow sound or coastal lagoon which has become land "through uplift and retreat of the sea" (Stephenson & Veatch, 1915, p.37). Cf: *tidal swamp*.

uplift [eng] The force that tends to raise an engineering structure or its foundation relative to its surroundings. This may be due to pressure of adjacent ground, surface water, or plastic soil on the base of a structure, or to movements produced by lateral forces such as wind.

uplift [tect] A structurally high area in the crust, produced by positive movements that raise or upthrust the rocks, as in a dome or arch. Cf: *depression*.

upper Pertaining to rocks or strata that are normally above those of earlier formations of the same subdivision of rocks. The adjective is applied to the name of a time-stratigraphic unit (system, series, stage) to indicate position in the geologic column and corresponds to *late* as applied to the name of the equivalent geologic-time unit; e.g. rocks of the Upper Jurassic System were formed during the Late Jurassic Period. The initial letter of the term is capitalized to indicate a formal subdivision (e.g. "Upper Devonian") and is lowercased to indicate an informal subdivision (e.g. "upper Miocene"). The informal term may be used where there is no formal subdivision of a system or of a series. Cf: *lower; middle*.

upper break *head*.

Upper Carboniferous In European usage, the approximate equivalent of the *Pennsylvanian*. Cf: *Lower Carboniferous*.

upper keriotheca The abaxial (upper) part of *keriotheca* in the wall of a fusulinid, characterized by fine alveolar structure (as in *Schwagerina*). Cf: *lower keriotheca*.

upper mantle That part of the *mantle* which lies above a depth of about 1000 km and has a density of 3.40 g/cm³, in which P-wave velocity is about 8.10 km/sec and S-wave velocity is about 4.7 km/sec. It is presumed to be peridotitic in composition. It is sometimes referred to as the *asthenosphere*, and includes the *transition zone*; it is equivalent to the B and yers. Syn: *outer mantle; peridotite shell.*

upper Paleolithic n. The third and most recent division of the *Paleolithic,* characterized by *Homo sapiens* and the appearance of man in Australia and the Americas. Cf: *lower Paleolithic; middle Paleolithic.*----adj. Pertaining to the upper Paleolithic.

upper plate The hanging wall of a fault. Cf: *lower plate.*

upper tectorium The abaxial secondary layer of spirotheca in the wall of a fusulinid, next above the tectum (as in *Profusulinella*). Cf: *tectorium; lower tectorium.*

upright fold A fold having an essentially vertical axial surface; *vertical fold.*

uprush The advance of water up the foreshore of a beach, following the breaking of a wave. Cf: *backwash.* Syn: *runup.*

upsetted moraine *push moraine.*

upside-down channel A channel scar on the roof or ceiling of a cave, presumably formed under phreatic conditions. Syn: *ceiling channel; ceiling meander.*

upsiloidal dune A general term for a U-shaped or V-shaped dune whose form is concave toward the wind; e.g. a *parabolic dune.*

upslope n. A slope that lies upward; uphill.---adj. In an upward direction, or ascending; e.g. an *upslope* ripple that climbed a sloping surface.

upstream Toward, at, or from a point nearer the source of a stream; in a direction from which a stream or glacier is flowing. Similarly, *upriver.*

upthrow adj. A syn. of *upthrown,* e.g. an *upthrow fault.* n. (a) the upthrown side of a fault; *upthrow side.* (b) The amount of upward vertical displacement of a fault.---Cf: *downthrow; heave.*

upthrow fault *upthrown side.*

upthrown Said of that side of a fault that appears to have moved upward, compared with the other side. Cf: *downthrown.* Syn: *upthrow (adj).*

upthrown block *upthrow side.*

upthrown side Var. of *upthrow side.*

upthrow side The upthrown side of a fault; an *upthrow.* Syn: *upthrow fault; upthrown side; upthrown block.*

up-to-basin fault A term used in petroleum geology for a fault whose upthrown side is toward the basin. An "up-to-coast fault" is one whose upthrown side is toward the coast. Syn: *up-to-the-basin fault.*

upwarping The upwards *warping* [tect] or uplift of a regional area of the Earth's crust, usually as the result of the release of isostatic pressure, e.g. melting of an ice sheet. Cf: *downwarping.*

upwelling [currents] The rising of cold, heavy subsurface water toward the surface, esp. along the western coasts of continents (as along the coast of southern California); the displaced surface water is transported away from the coast by the action of winds parallel to it or by diverging currents. Upwelling may also occur in the open ocean where cyclonic circulation is relatively permanent, or where southern trade winds cross the equator. Ant: *sinking [currents].*

upwelling [volc] The relatively quiet eruption of lava and volcanic gases, without much force.

uraconite A name that has been used for various uranium sulfates, but that "lacks specific meaning and should be abandoned" (Frondel et al, 1967, p.44).

uralborite A mineral: $CaB_2O_4.2H_2O$.

Uralian Stage in Russia: uppermost Carboniferous (above Gzhelian, below Sakmarian of Permian).

Uralian emerald (a) Emerald from near Sverdlovsk in the Ural Mountains, U.S.S.R. (b) *demantoid.*

uralite A green, generally fibrous or acicular variety of secondary amphibole (hornblende or actinolite) occurring in altered rocks and pseudomorphous after pyroxene (such as augite).

uralite diabase *uralitite.*

uralitite A term suggested for a diabase that contains augite

altered to uralite. Syn: *uralite diabase.*

uralitization The development of amphibole from pyroxene; specif. a late-magmatic or metamorphic process of replacement whereby uralite results from alteration of primary pyroxene. Also, the alteration of an igneous rock in which pyroxene is changed to amphibole; e.g. the alteration of gabbro to greenstone by pressure metamorphism.

uralolite A mineral: $CaBe_3(PO_4)_2(OH)_2.4H_2O$.

Ural-type glacier *drift glacier.*

uramphite A bottle-green to pale-green mineral: $(NH_4)(UO_2)(PO_4).3H_2O$.

uraninite A black, velvety-brownish, steel-gray, or greenish-black, strongly radioactive, octahedral or cubic mineral, essentially UO_2, but usually partly oxidized. It is the chief ore of uranium, and is isomorphous with thorianite. Uraninite often contains impurities such as thorium, radium, the cerium and yttrium metals, and lead; when heated, it often yields a gas consisting chiefly of helium. It occurs in veins of lead, tin, and copper minerals and in sandstone-type deposits, and is a primary constituent of igneous rocks, granites, and pegmatites. See also: *pitchblende.* Syn: *ulrichite; coracite.*

uranite A general term for a mineral group consisting of uranyl phosphates and arsenates of the autunite (lime uranite) and torbernite (copper uranite) type.

uranium-isotope age *uranium-uranium age.*

uranium-lead age method Calculation of an age in years for geologic material based on the known radioactive decay rate of uranium-238 to lead-206 and uranium-235 to lead-207. It is part of the more inclusive *uranium-thorium-lead age method* in which the parent-daughter pairs are considered simultaneously.

uranium ocher *gummite.*

uranium series The radioactive series of uranium, beginning with U-238 as parent.

uranium-series age methods Calculation of an age in years for Quaternary materials based on the general finding that the decay products uranium-234, thorium-230, and protactinium-231 in natural materials are commonly in disequilibrium with their parent isotopes, uranium-238 and uranium-235, either deficient or in excess. The age is determined from the measured activity ratios of these isotopes. See also: *ionium-thorium age method; ionium-excess method; ionium-deficiency method; thorium-230 to protactinium-231 deficiency method; protactinium-ionium age method; uranium-234 age method.*

uranium-thorium-lead age method Calculation of an age in years for geologic material, usually zircon, based on the known radioactive decay rate of uranium-238 to lead-206, uranium-235 to lead-207, and thorium-232 to lead-208 whose ratios give three independent ages for the same sample. The determined lead-207 to lead-206 ratio can be converted into a fourth age (*lead-lead age*). The method is most applicable to minerals that are Precambrian in age. Whether all four possible dates are concordant or discordant is useful for evaluating the results of this method, used alone or in comparison with other methods, and in determining if the initially-closed system has been disturbed. Partial syn: *uranium-lead age method; thorium-lead age method.* Syn: *uranium-thorium-lead dating.*

uranium-thorium-lead dating *uranium-thorium-lead age method.*

uranium-uranium age An age in years calculated from the ratio of uranium-235 to uranium-238, a by-product of the *uranium-thorium-lead age method.* Syn: *uranium-isotope age.*

uranium-234 age method The calculation of an age in years for fossil coral or shell (limited to those formed during the last million years), based on the assumption that the initial uranium-234 to uranium-238 ratio is known for the fossil. The change in this ratio is directly related to passage of time as the two isotopes have very different half-lives. See also: *uranium-series age methods.* Syn: *uranium-234 excess method;*

uranium-234 to uranium-238 age method; uranium-238 to uranium-234 disequilibrium method.

uranium-234 excess method uranium-234 age method.

uranium-234 to uranium-238 age method uranium-234 age method.

uranium-238 to uranium-234 disequilibrium method uranium-234 age method.

uran-mica A uranite, esp. torbernite.

uranocher A general name used chiefly for uranium sulfates (such as uranopilite) and for some uranium oxides. Also spelled: uranochre.

uranocircite A yellow-green mineral of the autunite group: $Ba(UO_2)_2(PO_4)_2.8H_2O$.

uranophane A strongly radioactive, lemon- to straw-yellow or orange-yellow, orthorhombic secondary mineral: $Ca(UO_2)_2Si_2O_7.6H_2O$. It is isostructural with sklodowskite and cuprosklodowskite, and dimorphous with beta-uranophane. Syn: uranotile.

uranopilite A yellow secondary mineral: $(UO_2)_6(SO_4)(OH)_{10}.12H_2O$.

uranosphaerite An orange-red or brick-red secondary mineral: $Bi_2U_2O_9.3H_2O$. Also spelled: uranospherite.

uranospinite A green to yellow secondary mineral of the autunite group: $Ca(UO_2)_2(AsO_4)_2.10H_2O$. It is isomorphous with zeunerite.

uranotantalite samarskite.

uranothallite liebigite.

uranothorianite A variety of thorianite containing uranium; an intermediate member in the uraninite-thorianite isomorphous series.

uranothorite A variety of thorite containing uranium.

uranotile A syn. of uranophane. Also spelled: uranotil.

urao trona.

urbainite An ilmenitite that contains 10-20 percent rutile and 3-5 percent sapphirine. Its name is derived from St. Urbain, Quebec, Canada.

urban geology The application of geologic knowledge and principles for future planning and management of high-density urban and urbanizing areas and their surroundings. It includes geologic studies for physical planning, waste disposal, land use, water-resources management, and the full range of usable raw materials. See also: environmental geology.

urediospore A yellow, orange, or reddish fungal spore of brief vitality, whose thin walls may be composed of chitin. Such spores may occur as microfossils in palynologic preparations. Cf: teleutospore. Syn: uredospore.

ureilite An achondritic stony meteorite composed essentially of olivine and clinobronzite, with accessory amounts of nickel-iron, troilite, diamond, and graphite. It is the only achondrite with an appreciable amount of nickel-iron.

ureyite A meteorite mineral of the pyroxene group: $NaCrSi_2O_6$. Syn: kosmochlor; cosmochlore.

Uriconian A division of the Precambrian in Great Britain.

uropod Either of the flattened leaf-like appendages of the last abdominal segment of various crustaceans that with the telson forms the caudal fan; e.g. limb of the sixth abdominal somite of a eumalacostracan, or one of the last three abdominal appendages of an amphipod. The term is sometimes applied to any abdominal appendage of a crustacean. Syn: uropodite.

urosome The part of the body of a copepod crustacean behind the major articulation that marks the posterior boundary of a prosome. Syn: urosoma.

ursilite A lemon-yellow mineral: $(Ca,Mg)_2(UO_2)_2Si_5O_{14}.9-10H_2O$.

urstromtal A wide, shallow, trench-like valley or depression excavated by a temporary meltwater stream flowing parallel to the front margin of a continental ice sheet, esp. one of the east-west depressions across northern Germany; a large-scale overflow channel. Etymol: German Urstromtal, "ancient

river valley". Pl: urstromtäler. Syn: pradolina.

urtite A light-colored member of the ijolite series that is composed chiefly of nepheline and 0-30% mafic minerals, esp. acmite and apatite. Cf: melteigite.

usamerite A term proposed by Boswell (1960, p.157) for rock comparable to the type graywacke and characterized b size grades ranging from gravel to sand, by poor sorting wi a "substantial" quantity of matrix, and by variable rock an mineral fragments that are predominantly angular to subangu lar. Etymol: United States of America + ite.

usar Barren, saline land in India, characterized by reh. Ety mol: Hindi.

U-shaped dune A dune having the form of the letter "U", i open end facing upwind.

U-shaped valley A valley having a pronounced parabolic cros profile suggesting the form of a broad letter "U", with stee parallel walls and a broad, nearly flat floor; specif. a valle carved by glacial erosion, such as a glacial trough. Cf: V shaped valley. Syn: U-valley; trough valley.

usovite A mineral: $Ba_2MgAl_2F_{12}$.

Ussherian Pertaining to the biblical chronology compiled b James Ussher (d.1656), Irish archbishop, who calculate from studies of the Scriptures that the Earth was formed o 26 October 4004 B.C. at 9:00 A.M.

ussingite A reddish-violet mineral: $Na_2AlSi_3O_8(OH)$.

U-stage universal stage.

Ustalf In U.S. Dept. of Agriculture soil taxonomy, a suborde of the soil order Alfisol, characterized by formation in a usti moisture regime and in a mesic or warmer temperature re gime. Ustalfs are brown or red (SSSA, 1970). Cf: Aqualf; Bc ralf; Udalf; Xeralf.

ustarasite A gray mineral: $Pb(Bi,Sb)_6S_{10}$.

Ustert In U.S. Dept. of Agriculture soil taxonomy, a suborde of the soil order Vertisol, characterized by wide, deep surfac cracks that are open for more than three months of the yea but that are not open continuously. Usterts form in a isohyperthermic temperature regime (SSSA, 1970). Cf: To rert; Udert; Xerert.

Ustoll In U.S. Dept. of Agriculture soil taxonomy, a suborde of the soil order Mollisol, characterized by formation in a usti soil moisture regime and in a mesic or warmer soil tempera ture regime. A Ustoll may have a calcic, petrocalcic, or gyps ic horizon (SSSA, 1970). Cf: Alboll; Aquoll; Boroll; Rendol. Udoll; Xeroll.

Ustox In U.S. Dept. of Agriculture soil taxonomy, a suborde of the soil order Oxisol, characterized by formation in an usti soil moisture regime and by a mean annual soil temperatur of 15°C or more (SSSA. 1970). Cf: Aquox; Humox; Orthox Torrox.

Ustult In U.S. Dept. of Agriculture soil taxonomy, a suborde of the soil order Ultisol, characterized by a low to moderat organic-carbon content, formation in an ustic soil moisture re gime, and by brownish or reddish colors (SSSA, 1970). Cf Aquult; Humult; Udult; Xerult.

utahite (a) jarosite. (b) natrojarosite.

utahlite A syn. of variscite, esp. that found in compact, nodu lar masses in Utah.

utricle (a) A bladdery, indehiscent fruit having one or a few seeds and a thin, membranous pericarp. (b) In the stonewor family Clavatoraceae, a hull or envelope commonly calcifie around the spiraled oogonium.

uvala A syn. of karst valley. Etymol: Serbo-Croatian.

U-valley U-shaped valley.

uvanite A brownish-yellow mineral: $U_2V_6O_{21}.15H_2O$ (?).

uvarovite The calcium-chromium end member of the garne group, characterized by an emerald-green color: $Ca_3Cr_2(SiO_4)_3$. It may have considerable amounts of alumina Syn: uwarowite; ouvarovite.

uvite A variety of tourmaline rich in calcium and magnesium.

uwarowite uvarovite.

uzbekite volborthite.

V

vacancy A vacant site in a crystal structure, due to the absence of an atom or ion from its normal structural position. Syn: *hole* [*cryst*].

vacuity *degradation vacuity.*

vacuole [*paleont*] A cavity in the cytoplasm of a cell of a plant or protozoan, often containing a watery solution enclosed by a membrane, and performing various functions such as digestion (food vacuole) and hydrostatic relation (contractile vacuole); e.g. one of the irregularly shaped *alveoles* in a foraminiferal-test wall. Also, the globular fluid inclusion or droplet enclosed in a vacuole.

vacuole [*petrology*] A syn. of *vesicle*; such usage is usually French.

vacuum-tube voltmeter A voltage-measuring instrument using electronic circuitry to obtain high impedance across the measuring probes so that very little current is drawn. Abbrev: *VTVM.*

vadose solution The solution action by vadose water above the level of the water table. Cf: *phreatic solution.*

vadose water Water of the zone of aeration. Syn: *kremastic water; suspended water; wandering water.*

vadose-water discharge The release, by evaporation, of water not originating in the zone of saturation. It may be in the form of *vegetal discharge* or *soil discharge.*

vadose zone *zone of aeration.*

vaesite An isometric mineral with pyrite structure: NiS_2.

vagile Said of a plant or animal that is free to move about. Cf: *sessile.*

vake The French term for *wacke* or soft, compact, mixed clay-like material with a flat, even fracture, commonly associated with basaltic rocks.

vakite A rock composed predominantly of vake. The term is not recommended.

val A longitudinal, synclinal valley in the folded Jura Mountains of the European Alps. Etymol: French, "narrow valley". Pl: *vaux.* Cf: *cluse.* See also: *combe.* Syn: *vallon.*

Valanginian European stage: Lower Cretaceous (above Berriasian, below Hauterivian).

valbellite A fine-grained, black hypabyssal rock containing bronzite, olivine, hornblende, and magnetite; a weigelith in which bronzite is present instead of enstatite and which contains more iron ore.

vale (a) A lowland, usually containing a stream; e.g. the depression between two parallel cuestas. It often forms the wider and flatter part of a valley. (b) A rift valley or tectonic valley; e.g. the Vale of Arabia. (c) A poetic var. of *valley*, esp. one that is relatively broad and flat.

valencianite A variety of adularia from Guanajuato, Mexico.

Valentian *Llandoverian.*

valentinite A white orthorhombic mineral: Sb_2O_3. It is polymorphous with senarmontite. Syn: *antimony bloom; white antimony.*

valid (a) Said of a taxon that meets all the requirements of the rules of nomenclature; e.g. that is neither a synonym nor a homonym of an older name. (b) Said of the publication of a taxon that meets the requirements of the rules of nomenclature, esp. in regard to availability, reproducibility, and purpose of publication.

valleriite A mineral: $2(Fe,Cu)_2S_2.3(Mg,Al)(OH)_2$.

valleuse A French term for a *hanging valley*, as on the chalk cliffs of France.

vallevarite A light-colored monzonitic igneous rock composed chiefly of andesine, microcline, and antiperthite, with small quantities of diopside, biotite, and apatite.

valley [*geomorph*] (a) Any low-lying land bordered by higher ground; esp. an elongate, relatively large, gently sloping depression of the Earth's surface, commonly situated between two mountains or between ranges of hills or mountains, and often containing a stream with an outlet. It is usually developed by stream erosion, but may be formed by faulting. (b) A broad area of generally flat land extending inland for a considerable distance, drained or watered by a large river and its tributaries; a river basin. Example: the Mississippi Valley.--- Etymol: Latin *vallis.* Syn: *vale; dale.*

valley [*marine geol*] A wide, low-relief depression of the ocean floor with gently sloping sides, as opposed to a submarine canyon.

valley axis A term used by Woodford (1951, p.803a) to replace *thalweg*, signifying "the surface profile along the center line of the valley".

valley bottom *valley floor.*

valley braid *anabranch.*

valley bulge *bulge.*

valley drift Outwash material constituting a *valley train.*

valley fill The unconsolidated sediment deposited by any agent so as to fill or partly fill a valley.

valley flat (a) The low or nearly level ground lying between valley walls and bordering a stream channel; esp. the small plain at the bottom of a narrow, steep-sided valley. Howard (1959, p.239) recommends that the term be applied noncommitally to a flat surface that cannot be identified with certainty as a flood plain or terrace. Syn: *flat.* (b) A bedrock surface produced by lateral erosion, commonly veneered with the alluvium of a *flood plain* (Thornbury, 1954, p.130).

valley floor The comparatively broad and flat bottom of a valley; it may be excavated and represent the level of a former erosion cycle, or it may be buried under a thin cover of alluvium. Syn: *valley bottom; valley plain.*

valley-floor basement The gently sloping, degraded bedrock underlying the *valley-floor side strip* and the valley floor (flood plain) proper, developed in a humid climate by lateral extension of the valley floor at the expense of the enclosing slopes, and covered with slowly creeping soil and flood-plain deposits (Davis, 1930).

valley-floor divide A divide in a valley; a dividing height located between two parts of the same valley, each part draining to a different river basin.

valley-floor increment The loose material coming to and lying upon the valley floor (Malott, 1928b, p.12).

valley-floor side strip The narrow and level to very slightly concave surface between the wash slope and the valley floor proper (the flood plain), produced by degradation and recession of the valley-side slope. See also: *valley-floor basement.*

valley glacier An *alpine glacier*; a glacier flowing down between the walls of a mountain valley in all or part of its length. Deprecated syn: *ice stream* [*glaciol*].

valley head The upper part of a valley.

valley-head cirque A cirque formed at the head of a valley. Cf: *hanging cirque.*

valley iceberg An iceberg eroded in such a manner that a large U-shaped slot, which may be awash, extends through the ice, separating pinnacles or slabs of ice. Syn: *drydock glacier.*

valley-in-valley (a) Said of the condition, structure, or cross profile of a valley form whose side is marked by a *valley shoulder* separating a steep-sided, youthful valley below from a more widely opened, older valley above. (b) Pertaining to a *two-cycle valley.*

valley line *thalweg* [*streams*].

valley-loop moraine *loop moraine.*

valley meander One of a series of curves of a *meandering valley.*

valley-moraine lake A glacial lake formed in a valley by the damming action of a recessional moraine produced by a mountain glacier. Cf: *drift-barrier lake.*

valley of elevation A syn. of *anticlinal valley.* The term was introduced in 1825 by Buckland (1829, p.123).

valley of subsidence A syn. of *synclinal valley*. The term was used by Hitchcock (1841, p.178).

valley plain (a) A continuous flood plain (Cotton, 1940). (b) *valley floor.*

valley-plain terrace A term used by Cotton (1940, p.28-29) for the remnant of a formerly continuous flood plain or valley floor; it would include the features now known as a *strath terrace* and a *fillstrath terrace.*

valley plug A local constriction in a stream channel, which may be formed by any of several types of channel obstructions and may cause rapid deposition. See also: *plug* [*sed*].

valley profile The *longitudinal profile* of a valley.

valley shoulder A bedrock surface made in a *valley-in-valley* form, representing the sharp angle or break in slope between the side or floor of the upper, older valley and the side of the lower, newer valley. It is a remnant of the valley floor formed during a previous erosion cycle, marking the former base level of erosion, and extending across rocks of varying lithology. Syn: *shoulder.*

valley-side moraine *lateral moraine.*

valley-side slope (a) A measure, generally expressed in degrees, of the steepest inclination of the side of a valley in stream-eroded topography. Maximum slope is measured at intervals along the valley walls on the steepest parts of the contour orthogonals running from divides to adjacent stream channels. Symbol: θ. Syn: *ground slope.* (b) The surface between a drainage divide and the valley floor. Syn: *valley side; valley wall.*

valley sink In a karst area, an elongate, narrow chasm or depression that is a solutional rather than an erosional feature. Cf: *karst valley.*

valley spring A type of depression spring issuing from the side of a valley at the outcrop of the water table.

valley storage (a) The volume of water in a body of water below the water-surface profile. (b) The natural storage capacity or volume of water of a stream in flood that has overflowed its banks; it includes both the water within the channel and the water that has overflowed.----(ASCE, 1962).

valley system A valley and all of its tributary valleys.

valley tract The middle part of a stream, characterized by a moderate gradient and a fairly wide valley. Cf: *mountain tract; plain tract.*

valley train A long, narrow body of outwash, deposited by meltwater streams far beyond the terminal moraine or the margin of an active glacier and confined within the walls of a valley below the glacier; it may or may not emerge from the mouth of the valley to join an outwash plain. See also: *gravel train; valley drift.* Syn: *outwash train.*

valley wall *valley-side slope.*

valley wind A daytime *anabatic wind* moving up a valley or mountain slope. Cf: *mountain wind.*

vallon A syn. of *val.* Etymol: French, "small valley".

Valmeyeran Provincial series in Illinois: Lower and Upper Mississippian (equivalent to Osagian and Meramecian elsewhere).

value In economic geology, (a) the valuable constituents of an ore; (b) their percentage in an orebody, or *assay grade;* (c) their quantity in an orebody, or *assay value.*

valval plane The plane of division between the valves of a diatom frustule; it parallels the valves.

valvate Said of leaves or petals in the bud that meet at the edges without overlapping and that open like valves.

valve (a) One of the distinct and usually movably articulated pieces that make up the shell of certain invertebrates; e.g. one of the two convexly curved (rarely flat or concave) calcareous plates that constitute the shell of a bivalve mollusk and are articulated along a dorsal hinge line, or one of the two halves of the carapace of a crustacean, divided by articulation along the middorsal line (such as an opercular plate of a cirri-

pede), or one of the two curved chitino-phosphatic or calcareous plates that form the shell of a brachiopod and that surround and lie, respectively, above and below the soft parts. (b) One of the two silicified pieces or encasing membranes forming the top or bottom surface of a diatom frustule; e.g. epivalve and hypovalve.

valve mantle The large marginal flange of the valve of a diatom frustule.

valverdite A rounded or lenticular glass object containing crystalline inclusions, found near Del Rio in Val Verde County, Texas. It is probably weathered obsidian.

valvular Resembling or having the function of a valve in the body of an invertebrate; e.g. "valvular pyramid" of a cystoid or edrioasteroid, composed of several more or less triangular plates covering the anus or a gonopore.

van A term used in the French Alps for *cirque* (Schieferdecker, 1959, term 1667).

vanadate A mineral compound characterized by pentavalent vanadium and oxygen in the anion. An example is vanadinite, $Pb_5Cl(VO_4)_3$. Cf: *arsenate; phosphate.*

vanadinite A red, yellow, or brown mineral of the apatite group: $Pb_5(VO_4)_3Cl$. It is isomorphous with pyromorphite, and commonly contains arsenic or phosphorus. Vanadinite often forms globular masses encrusting other minerals in lead mines, and is an ore of vanadium and lead.

vanado-magnetite *coulsonite.*

vanalite A bright-yellow mineral: $NaAl_8V_{10}O_{38}.30H_2O$.

vandenbrandeite A blackish-green mineral: $CuUO_4.2H_2O$.

vandendriesscheite A yellow or amber-orange mineral: $PbU_7O_{22}.12H_2O$.

van der Kolk method A test used in refractometry to determine the index of refraction of a mineral relative to that of the liquid medium in which it is immersed. When an obstacle blocks the light rays used for illumination, its shadow appears on the same side as itself when the mineral grain has a relatively higher refractive index, and on the opposite side when the mineral's refractive index is relatively lower than that of the medium.

Vandyke brown *black earth* [*coal*]. Etymol: its use by the 17th Century Flemish painter Van Dyck.

vane [*geophys*] In many geophysical instruments, a part, the resistance to movement of which in a gas, fluid, or magnetic field, tends to retard or damp the vibration of a suspended or balanced system.

vane [*surv*] (a) The target of a level rod. (b) One of the sights of a compass or quadrant.

vane test An in-place test to determine the shear strength of cohesive soils and other soft deposits (such as clays and silts), in which a rod affixed at the end with four thin, flat, radial blades (vanes) projecting at 90-degree intervals is forced into the soil and rotated, and the torque required to shear the soil (or the resistance to rotation of the rod) is determined as a measure of the shear strength.

vanna Part of the operculum in ascophoran cheilostomes (bryozoans) that closes the poster. Cf: *porta.*

vanoxite A black mineral: $V_4^{+4}V_2^{+5}O_{13}.8H_2O$ (?).

van't Hoff equation An equation giving the temperature dependence of the *equilibrium constant* of a reaction: $d \ln K/dT = \Delta H°/RT^2$, where K=equilibrium constant, T=absolute temperature, $\Delta H°$=enthalpy change for the hypothetical reaction with all substances in their standard states, and R=gas constant.

vanthoffite A colorless mineral: $Na_6Mg(SO_4)_4$.

van't Hoff law The statement in phase studies that, when a system is in equilibrium, of the two opposed interactions, the endothermic one is promoted by raising the temperature, and the exothermic one, by lowering it.

vanuralite A citron-yellow mineral: $Al(UO_2)_2(VO_4)_2$-$(OH).11H_2O$.

vanuranylite A bright-yellow mineral: $[(H_{30}),Ba,Ca,K)]_{1,6}$-

$(UO_2)_2(VO_4)_2.4H_2O(?)$.

vapor The gaseous phase of a known liquid or solid. The term sometimes refers to a gaseous phase of a substance below its critical temperature, i.e. formed by pressure.

vaporization *evaporation.*

vara Any of various old Spanish units of length used in Latin America and SW U.S., equal in different localities to between 31 and 34 inches; e.g. a unit equal to 33.3333 inches in Texas, to 33.372 inches in California, to 33.00 inches in Arizona and New Mexico, and to 32.9931 inches and 32.9682 inches (among others) in Mexico. For other values, see ASCE (1954, p.169-170).

variability [paleont] The quality or attribute of an organism that causes it to exhibit variation.

variability [grd wat] The ratio of the difference between maximum and minimum discharge of a spring to its average discharge, expressed as a percentage.

variable (a) Any measurable or changeable statistical quality or quantity; e.g. *independent variable* and *dependent variable.* See also: *attribute; variate.* (b) A quantity that can assume any of a given set of values at different stages in a computer program.

variable-area method A method of recording seismic impulses in which the area of exposure of a photosensitive film or paper is proportional to the intensity of the seismic impulse.

variable-density method A method of displaying the deflections of a seismometer in which the photographic density is proportional to the signal amplitude.

variance [chem] *degrees of freedom.*

variance [stat] The square of the standard deviation. Symbol: σ^2.

variant An individual exhibiting *variation.*

variate A quantitative *variable;* e.g. a *random variable.*

variation Divergence in the structural or functional characteristics of an organism from those that are considered typical of the group to which it belongs. See also: *variant.*

variation diagram A diagram constructed by plotting the chemical compositions of rocks in an igneous rock series in order to show the genetic relationships and the nature of the processes that have affected the series. The weight percent of SiO_2 is usually plotted as the abscissa and other major oxides, individually, as the ordinates. Syn: *Harker diagram.*

variegated Said of a sediment or sedimentary rock (such as red beds or sandstone) showing variations of colors or tints in irregular spots, streaks, blotches, stripes, or reticulate patterns. Cf: *mottled.*

variegated copper ore *bornite.*

varietal mineral A mineral that is either present in considerable amounts in a rock or characteristic of the rock; a mineral which distinguishes one variety of rock from another. Syn: *distinctive mineral; characterizing accessory mineral.*

variety [mineral] In gemology, a mineral that is a type of the mineral *species,* distinguished by color or other optical phenomenon or characteristic: e.g. emerald and aquamarine are varieties of beryl.

variety [taxon] A group of individuals, within a species, that differs from other groups in the species in some conspicuous way. The term "variety" was originally used synonymously with *subspecies* but is now seldom used and is not currently accepted as a formal unit in taxonomy.

varigradation A term used by McGee (1891, p.261-267) for the process by which all streams of progressively increasing volume tend constantly, in a degree varying inversely with the volume, to depart slightly from the normal gradients.

variole A pea-size spherule, usually composed of radiating crystals of plagioclase or pyroxene. This term is generally applied only to such spherical bodies in basic igneous rock, e.g. variolite. Cf: *spherulite.*

variolitic Said of the texture of a rock, esp. a basic igneous rock, composed of pea-size spherical bodies (varioles) in a finer-grained groundmass. Cf: *spherulitic.*

variometer An instrument for measurement of temporal variation of a magnetic element, using the torque on a permanent magnet in a uniform magnetic field.

Variscan orogeny The late Paleozoic orogenic era of Europe, extending through the Carboniferous and Permian. By current usage, it is synonymous with the *Hercynian orogeny.* Cf: *Armorican orogeny; Altaides.*

variscite A yellow-green or soft-green orthorhombic mineral: $AlPO_4.2H_2O$. It is isomorphous with strengite and dimorphous with metavariscite. Variscite is a popular material for cabochons and various kinds of carved objects, and is often used as a substitute for turquoise. See also: *sphaerite.* Syn: *utahlite.*

varix (a) One of the transverse elevations of the surface of a gastropod shell that is more prominent than a costa and that represents a halt in growth during which a thickened outer lip was developed (TIP, 1960, pt.I, p.134). (b) A thickening of an ammonoid shell marked on an internal mold by a transverse groove (Moore et al, 1952, p.366).---Pl: *varices.*

varlamoffite A mineral: $(Sn,Fe)(O,OH)_2$. It is perhaps fine-grained cassiterite.

varnish *desert varnish.*

varnsingite A coarse-grained, light-colored hypabyssal rock containing albite (over 50 percent), pyroxene, sphene, magnetite, apatite, and secondary epidote, prehnite, chlorite, amphibole, and muscovite.

varulite A dull olive-green mineral: $(Na_2,Ca)(Mn^{+2}, Fe^{+2})_2(PO_4)_2$. It is isomorphous with hühnerkobelite.

varve A sedimentary bed or lamina or sequence of laminae deposited in a body of still water within one year's time; specif. a thin pair of graded glaciolacustrine layers seasonally deposited (usually by meltwater streams) in a glacial lake or other body of still water in front of a glacier. A glacial varve normally includes a lower "summer" layer consisting of relatively coarse-grained, light-colored sediment (usually sand or silt) produced by rapid melting of ice in the warmer months, which grades upward into a thinner "winter" layer, consisting of very fine-grained (clayey), often organic, dark sediment slowly deposited from suspension in quiet water while the streams were ice-bound. Counting and correlation of varves have been used to measure the ages of Pleistocene glacial deposits. Etymol: Swedish *varv,* "layer" or "periodical iteration of layers" (Geer, 1912, p.242).

varved clay A distinctly laminated lacustrine sediment consisting of varves; specif. the upper, fine-grained, "winter" layer of a glacial varve. Syn: *varve clay.*

varvite An indurated rock consisting of ancient varves.

varvity The property of being varved; the seasonal and alternating lamination in varves.

varzea A term used in Brazil and Portugal for an alluvial flood plain or the bank of a river; also, a field or a level tract of land, esp. one that is sowed and cultivated. Etymol: Portuguese *várzea.*

vascular bundle In a vascular plant, a strand of xylem and phloem.

vascular plant A plant having vessels in the conducting tissues of the stele, and having structural differentiation into roots, stem, and leaves. Most terrestrial vegetation is vascular. Syn: *tracheophyte.*

vascular ray A ribbonlike aggregate of cells extending radially in stems through xylem and, often, phloem (Fuller & Tippo, 1949, p.975).

vascular tissue In vascular plants, conducting tissue composed of xylem and phloem.

vase Freshwater silt deposited in estuaries along the Atlantic coast of Europe and Africa, consisting of a mixture of sandy and pulverulent grains of quartz, calcite, clay minerals, and diatom shells, with a binder of *algon* (Bourcart, 1941). Etymol: French, "slime, mud". Pron: *vaaz.*

vashegyite A white, yellow, or rust-brown mineral: $2Al_4(PO_4)_3(OH)_3.27H_2O$ (?).

vat (a) *salt pit.* (b) A term used in SW U.S. for a dried and incrusted margin around a water hole.

vaterite A rare hexagonal mineral: $CaCO_3$. It is trimorphous with calcite and aragonite, and consists of a relatively unstable form of calcium carbonate.

vaterite-A Artificial calcite.

vaterite-B Artificial vaterite.

Vauclusian spring A *karst spring* formed where an underground stream has been actively eroding in limestone and emerges at the foot of a steep valley wall where the limestone overlies a layer of impervious rock. It is named after the Fontaine de Vaucluse in southern France, a spring whose water issues from large, ramifying solution channels in limestone. Syn: *gushing spring.*

vaughanite A term suggested by Kindle (1923a, p.370) for a pure, dense, homogeneous, dove-colored, fine-textured limestone that breaks with a smooth and more or less pronounced conchoidal fracture, that contains relatively few fossils, and that typically has a white, chalky appearance on weathered surfaces. Named after T. Wayland Vaughan (1870-1952), U.S. paleontologist.

vaugnerite A dark-colored, coarse-grained hypabyssal rock containing abundant biotite, along with green hornblende, white feldspar, and quartz. Microscopic grains of hornblende, biotite, plagioclase, and quartz, with accessory orthoclase, apatite, magnetite, pyrite, and sphene.

vault [geomorph] A structure in the Earth's crust, resembling or suggesting a vault or an arched room; e.g. a cavern or a volcanic crater.

vault [paleont] (a) The part of blastoid theca above the dorsal region (from aboral tips of ambulacra to dorsal pole). (b) An arched covering of calcareous plates between crinoid arms.

vaulted mud crack A raised mud crack on a playa, shaped like an inverted "V", and formed by salts and clayey material rising by capillary action through the narrow mud crack (Stone, 1967, p.252). Syn: *roofed mud crack.*

vauquelinite A green to brownish-black mineral: $Pb_2Cu(CrO_4)(PO_4)(OH)$. It is isomorphous with fornacite.

vaux Pl. of *val.*

vauxite A blue triclinic mineral: $Fe^{+2}Al_2(PO_4)_2(OH)_2.7H_2O$. It has less water than metavauxite and paravauxite.

vayrynenite A mineral: $MnBe(PO_4)(OH,F)$.

V-bar A *cuspate bar* whose seaward angle is fairly sharp, as where a secondary spit trails abruptly back toward the shore from the point of a primary spit.

V-coal Microscopic coal particles that are predominantly vitrain and clarain, as found in miners' lungs. Cf: *F-coal; D-coal.*

veatchite A white mineral: $Sr_2B_{11}O_{16}(OH)_5.H_2O$. It is dimorphous with p-veatchite, and has a space group A2/a. Veatchite was originally thought to be a hydrous calcium borate.

Vectian *Aptian*

vectograph A picture, photograph, or lantern slide composed of two superimposed stereoscopic images that polarize light in planes at right angles to each other, giving a three-dimensional effect when viewed through polarizing spectacles whose lens axes are at right angles. See also: *stereoscopic pair.*

vector structure *directional structure.*

veenite A mineral: $Pb_2(Sb,As)_2S_5$.

vegetable jelly *ulmin.*

vegetal discharge The release, through the transpiration of plants, of water derived either from the zone of areation or from the zone of saturation by way of the capillary fringe. See also: *vadose-water discharge.*

vegetation arabesque *vegetation polygon.*

vegetation coast A coast that is being extended seaward by the growth of vegetation, such as the mangrove trees along the coasts of Florida.

vegetation polygon A small *nonsorted polygon* whose fissured borders are emphasized by thick vegetation (usually moss, lichen, or willows) and whose center consists of fine-textured material or a mixture of fines and stones. Diameter: about 1 m. See also: *lichen polygon; tussock-birch-heath polygon.* Syn: *vegetation arabesque.*

vegetation stripe (a) A syn. of *nonsorted stripe.* (b) A *sorted stripe* emphasized by vegetation (Sigafoos, 1951, p.289).

vegetative reproduction Plant reproduction in which progeny have the same genetic composition as the parent plant. Various methods of vegetative reproduction include cuttings made from stems, roots, and even leaves; underground rhizomes such as "eyes" from the potato; leaf specializations such as the formation of "plantlets" on leaf tips; in some bryophytes, by gemmae; by posterior decay, in which two plants arise from one by separation of two branches upon the death of the main plant body. Syn: *budding.*

veil [cryst] An aggregate of minute bubbles creating a whitish or cloudlike appearance in quartz.

veil [paleont] A variously formed web- or net-like film in a radiolarian; e.g. *patagium.*

vein [ore dep] An epigenetic mineral filling of a fracture in a host rock, in tabular or sheetlike form, often with associated replacement of the host rock; a mineral deposit of this form and origin. Cf: *lode.*

vein [bot] One of the vascular bundles of a plant. See also: *venation.*

vein [intrus rocks] A thin, sheetlike igneous intrusion into a crevice.

vein [ice] (a) A narrow water channel within land ice; also, the stream of water flowing through such a channel. (b) A narrow lead or lane in pack ice.

vein [streams] (a) A narrow waterway or channel in rock or earth. Also, a stream of water flowing in such a channel. (b) An archaic term for the flow or current of a stream.

vein bitumen Any one of the black or dark-brown bitumens which give off a pitchy odor, burn readily with a smoky flame, and occupy fissures in rocks or less frequently form basin-shaped deposits on the surface (Nelson & Nelson, 1967).

veindike A pegmatitic intrusion that has the characteristics of both a vein and a dike. Also spelled: *vein-dike; vein dike.*

veined gneiss A composite gneiss with irregular layering. The term is generally used in the field and has no genetic implications (Dietrich, 1960, p.50). Cf: *venite; arterite; phlebite; composite gneiss.*

veinlet A small, irregular, tonguelike igneous intrusion. Syn: *stringer.*

vein quartz A rock composed chiefly of sutured quartz crystals of pegmatitic or hydrothermal origin and of variable size.

vein system An assemblage of veins of a particular area or age or fracture system, usually inclusive or more than one *lode.*

Vela Uniform A research program, sponsored by the Advanced Research Projects Agency of the U.S. Dept. of Defense, which had the objective of improving the capability of detecting underground nuclear explosions and of discriminating them from earthquakes.

veld An open grassland area of South Africa. The number of terms incorporating "veld", e.g. *bushveld*, denotes the diversity of the area regarding elevation, vegetation, soil, etc. Also spelled: *veldt.*

velocity *seismic velocity.*

velocity coefficient A numerical factor always less than unity, which expresses the ratio between the actual velocity issuing from an orifice or other hydraulic structure or device and the theoretical velocity which would exist if there were no friction losses due to the orifice, structure, or device. The square of the velocity coefficient is a measure of the efficiency of a structure as a waterway. It is a dimensionless number (ASCE, 1962).

velocity discontinuity *discontinuity* [seism].

velocity distribution The relationship between seismic velocity and depth.

velocity gradient [**seis**] *seismic gradient.*

velocity gradient [**hydraul**] The rate of change of velocity with respect to distance normal to the direction of flow (ASCE, 1962).

velocity head The energy of flow expressed as the vertical distance through which a fluid would fall in order to attain the given velocity; the height the kinetic energy of a liquid is capable of lifting the liquid above a given point. See also: *total head.*

velocity-head coefficient A correction factor applied to the velocity head of the mean velocity to correct for nonuniformity of velocity in a cross section. The factor is 1.0 where velocities are identical across a section and greater than 1.0 where velocities vary across a section (ASCE, 1962).

velocity/height ratio In airborne surveying, the quotient of apparent ground speed of an aircraft and the aircraft altitude above the terrain. In airborne image-forming systems of the line-scan type, e.g. scanning radiometers, the forward film transport rate is set to match the velocity/height ratio to eliminate linear distortion of the image or undue separation of the consecutive scan lines. Abbrev: *V/H ratio.*

velocity log *sonic log.*

velocity meter A seismometer that is used to record vibrations of a period very close to its own free period.

velocity profile A seismic shooting setup or spread used to record reflections over a large range of shot-to-geophone distances, which are used to determine seismic velocity from the time-distance relationship.

velu A term used in the Maldive Islands of the Indian Ocean for the lagoon of a faro.

velum A sail- or frill-like structure along the distal part of the valve of an ostracode, typically developed as the double-walled outfold of the carapace. Pl: *vela.*

velvet copper ore *cyanotrichite.*

venanzite A holocrystalline porphyritic extrusive rock composed of olivine and phlogopite phenocrysts in a fine-grained groundmass of these minerals plus melilite, leucite, and magnetite. Syn: *euktolite.*

venation The arrangement of the vascular bundles (*veins*) in a plant. See also: *parallel venation; net venation.*

veneer (a) A thin but extensive layer of sediments covering an older geologic formation or surface; e.g. a *veneer* of alluvium covering a pediment. (b) A weathered or otherwise altered coating on a rock surface; e.g. desert varnish.

Vening Meinesz zone A belt of negative gravity anomalies, generally related to island arcs and/or oceanic deeps. Syn: *negative strip.*

venite Migmatite the mobile portion(s) of which were formed by exudation (secretion) from the rock itself (Dietrich & Mehnert, 1961). Originally proposed, along with the term *arterite*, to replace *veined gneiss* with terms of genetic connotation (Mehnert, 1968, p.17). Cf: *phlebite; composite gneiss.*

vent The opening at the Earth's surface through which volcanic materials are extruded; also, the channel or conduit through which they pass. Cf: *neck* [*volc*]; *pipe* [*volc*]. Partial syn: *feeder* [volc]; *chimney* [volc].

venter [**paleont**] (a) The outer and convex part of the shell of a curved or coiled cephalopod or gastropod, or the peripheral wall of a cephalopod whorl comprising the part of the shell radially farthest from the protoconch; the underside of a nautiloid and of its conch, distinguished generally by the hyponomic sinus and often by a conchal furrow (TIP, 1964, pt.K, p.59). (b) The median region of the shell of a productoid brachiopod, situated between the valve surfaces on either side of the median sector of the shell (TIP, 1965, pt.H, p.155). (c) The belly region or the lower part of the carapace of an ostracode.---Cf: *dorsum.*

venter [**bot**] In the female gametophyte of certain plants, the enlarged base of an archegonium within which the egg develops.

ventifact A general term introduced by Evans (1911) for any stone or pebble shaped, worn, faceted, cut, or polished by the abrasive or sandblast action of windblown sand, generally under desert conditions; e.g. a *dreikanter*. See also: *windkanter*. Syn: *glyptolith; rillstone; wind-worn stone; wind-cut stone; wind-polished stone; wind-grooved stone; wind-scoured stone; wind-shaped stone.*

venting The escape through the Earth to the atmosphere of gases or radioactive products from an underground high-explosive or nuclear detonation.

ventral (a) Pertaining or belonging to or situated near or on the abdominal (belly) or lower surface of an animal or of one of its parts that is opposite the back; e.g. in the direction toward the pedicle valve from the brachial valve of a brachiopod, or pertaining to the region of the shell of a bivalve mollusk opposite the hinge where the valves open most widely, or pertaining to the side of the stipe on which the thecal apertures of a graptoloid are situated or to the inferior (commonly the umbilical or apertural) side of a foraminiferal test. (b) Referring to the direction or side of an echinoderm toward or containing the mouth, normally upward; adoral or oral.---Ant: *dorsal.*

ventrallite An alkalic igneous rock containing more potassium feldspar than calcic plagioclase and containing nepheline as the essential feldspathoid.

ventral lobe The main adapical lobe or inflection of a suture on the venter of a cephalopod shell. See also: *external lobe.* Cf: *dorsal lobe.*

ventral process A medially located excessive thickening of secondary shell of a brachiopod underlying the pseudodeltidium and projecting dorsally to fit between lobes of the cardinal process.

ventral shield An ossicle of secondary origin on the oral side of an arm in an ophiuroid. Cf: *dorsal shield.*

ventral valve The *pedicle valve* of a brachiopod.

ventromyarian Said of a nautiloid in which the retractor muscles of the head-foot mass are attached to the shell along the interior areas of the body chamber adjacent to, or coincident with, its ventral midline (TIP, 1964, pt.K, p.59). Cf: *dorsomyarian; pleuromyarian.*

Venturian North American stage: middle Pliocene (above Repettian, below Wheelerian).

Venus hair Extremely slender, needle-like, or acicular crystals of reddish-brown or yellow rutile, forming tangled swarms of inclusions in quartz. See also: *sagenite.*

Venus hairstone A variety of *hairstone* penetrated by Venus hair; *rutilated quartz.* Also spelled: *Venus's-hairstone.*

verd antique A type of marble that is mainly massive serpentine, commonly with veinlets of calcium and magnesium carbonates. It is capable of being polished (ASTM, 1970, p.462). Also spelled: *verde antique.* Syn: *serpentine marble.*

verdelite A green variety of tourmaline.

verdite A green rock, consisting chiefly of impure fuchsite and clayey matter, used as an ornamental stone.

vergence The direction of overturning or of inclination of a fold. The term is a translation of the German *Vergenz*, "overturn", coined by Stille (1930, p.379) for the direction in which a geologic structure or family of structures is facing. Cf: *regard; facing.*

verglas A thin film or layer of clear, hard, smooth ice on a rock surface, formed as a result of a frost following rain or snowmelt or by rime. Etymol: French. Cf: *black ice.*

verite A black extrusive rock containing phlogopite or biotite, augite, and olivine crystals in a glassy groundmass; a variety of lamproite. Its name is derived from the town of Vera, Spain.

vermeil (a) An orangy-, yellowish-, or brownish-red garnet.

Syn: *vermilion.* (b) A reddish-brown to orangy-red gem corundum; ruby. (c) An orangy-red spinel.---Syn: *vermeille.*

vermicular quartz Quartz occurring in worm-like forms intergrown with or penetrating feldspar, as in myrmekite.

vermiculated Said of a stone, carbonate sediment, or any corroded geologic feature that has the appearance of having been eaten into by worms.

vermiculite (a) A group of platy or micaceous clay minerals closely related to chlorite and montmorillonite, and having the general formula: $(Mg,Fe,Al)_3(Al,Si)_4O_{10}(OH)_2.4H_2O$. The minerals are derived generally from the alteration of micas (chiefly biotite and phlogopite), and they vary widely in chemical composition. They are characterized by marked exfoliation when heated above 150°C; at high temperature, their granules greatly expand into long worm-like threads that entrap air and produce a lightweight and highly water-absorbent material that is used as an insulator and as an aggregate in concrete and plaster. Vermiculites differ from montmorillonites in that the characteristic exchangeable cation is Mg^{+2}, the lattice expands only to a limited degree (hydration and dehydration is limited to two layers of water), and they have higher layer charges per formula unit (0.6-0.9) and higher cation-exchange capacities. Vermiculites are found mostly with basic rocks such as dunite and pyroxenite. (b) Any mineral of the vermiculite group, such as maconite, jefferisite, and lennilite.

vermiform Worm-like or having the form of a worm; e.g. "vermiform problematica" consisting of long, thin, and more or less cylindrical tubes.

vermiglyph A collective term used by Fuchs (1895) for a trace fossil consisting of a presumable worm trail appearing on the undersurface of flysch beds (mostly sandstones) as a thread-like, unbranched, and irregular relief form a few millimeters wide with a straight or variously winding course. Cf: *graphoglypt; rhabdoglyph.*

vermilion (a) *cinnabar.* (b) An orange-red garnet; *vermeil.*---Syn: *vermillion.*

vernacular name In biologic nomenclature, the common name of a plant or animal as opposed to its formal Latin name; e.g. sugar maple is the vernacular name of *Acer saccharum.* Syn: *popular name.* Cf: *scientific name.*

vernadite A name used in the U.S.S.R. for a mineral of supposed composition $MnO_2.nH_2O$.

vernadskite *antlerite.*

Verneuil process A method developed by Auguste V.L. Verneuil (1856-1913), French mineralogist and chemist, for the manufacture of large, synthetic crystals of corundum and spinel in which an alumina powder of the desired composition is melted in an oxyhydrogen flame, producing a series of drops that build up the boules of the synthetic gems.

vernier A short, uniformly divided, auxiliary scale that slides along the primary scale of a measuring device and that is used to measure accurately fractional parts of the smallest divisions of the primary scale or to obtain one more significant figure of a particular measurement. It is graduated such that the total length of a given number of divisions on a vernier is equal to the total length of one more or one less than the same number of divisions on the primary scale; parts of a division are determined by observing what line on the vernier coincides with a line on the measuring device. Named after Pierre Vernier (1580-1637), French mathematician.

vernier compass A surveyor's compass with a vernier, used for measuring angles without the use of the magnetic needle by means of a compensating adjustment made for magnetic variation.

Verona earth A naturally occurring iron silicate from Verona, Italy; specif. *celadonite.* Syn: *veronite.*

verplanckite A mineral: $Ba_2(Mn,Fe,Ti)Si_2O_6(O,OH,Cl,F)_2.3H_2O$.

verrou A syn. of *riegel.* Etymol: French, "bolt".

verrucate Warty, or covered with wart-like knobs or elevations; e.g. said of spores and pollen having sculpture consisting of wart-like projections. Syn: *verrucose.*

versant (a) The slope or side of a mountain or mountain chain. (b) The general slope of a region; e.g. the Pacific versant of the U.S.

verst A Russian unit of distance equal to 0.6629 mile or 1.067 km.

vertebra (a) One of the bony or cartilaginous elements that together make up the spinal column of a vertebrate. (b) One of a fused pair of opposite ambulacral plates of an asterozoan, articulating with neighboring vertebrae by ball-and-socket joints.---Pl: *vertebrae.*

vertebrate paleontology The branch of paleontology dealing with fossil vertebrates.

vertex The culmination or high point of a feature. In older tectonic writing, the term is used for the apex or nucleus of a continental or other large structure; example, the Angara Shield, supposed to have been the vertex of Asia.

vertical [geophys] adj. Said of a direction that is perpendicular to a *horizontal* plane.

vertical [photo] n. *vertical photograph.*

vertical [stratig] Said of the direction at right angles to the strike or to the direction of extension of the strata, as in the measurement of *true thickness.* The term has also been used to indicate a direction at right angles to the surface of the land, as in the measurement of *apparent thickness.* Cf: *lateral.*

vertical accretion Upward growth of a sedimentary deposit; e.g. settling of sediment from suspension in a stream subject to overflow. Cf: *lateral accretion.*

vertical-accretion deposit *flood-plain deposit.*

vertical angle An angle in a vertical plane; the angle between the horizontal and an inclined line of sight, measured on a vertical circle either upward or downward from the horizon. One of the directions that form a vertical angle in surveying is usually either the direction of the vertical (the angle being termed the *zenith distance*) or the line of intersection of the vertical plane in which the angle lies with the plane of the horizon (the angle being termed the altitude). Cf: *horizontal angle.*

vertical angulation *trigonometric leveling.*

vertical axis The line through the center of a theodolite or transit about which the alidade rotates. Cf: *horizontal axis.*

vertical circle Any great circle of the celestial sphere passing through the zenith.

vertical collimator A telescope so mounted that its collimation axis may be made to coincide with the vertical (or direction of the plumb line). It may be used for centering a theodolite on a high tower exactly over a station mark on the ground. See also: *collimator.*

vertical control A series of measurements taken by surveying methods for the determination of elevation with respect to an imaginary level surface (usually mean sea level) and used as fixed references in positioning and correlating map features.

vertical corrasion Corrasion of the stream bed, causing a deepening of the channel.

vertical dip slip *vertical slip.*

vertical erosion *downcutting.*

vertical exaggeration [cart] (a) A deliberate increase in the vertical scale of a relief model, plastic relief map, block diagram, or cross section, while retaining the horizontal scale, in order to make the model, map, diagram, or section more clearly perceptible. (b) The ratio expressing vertical exaggeration; e.g. if the horizontal scale is one inch to one mile and the vertical scale is one inch to 2000 ft, the vertical exaggeration is 2.64. Abbrev: V.E.

vertical exaggeration [photo] The apparent increase in the relief as seen in a stereoscopic image.

vertical fault A fault, the dip of which is 90 degrees. Cf: *horizontal fault.*

vertical field balance An instrument that measures the vertical component of the magnetic field by means of the torque that the field component exerts on a horizontal permanent magnet. The two most common types are the *Schmidt field balance* and the *torsion magnetometer*. Cf: *horizontal field balance*.

vertical fold A fold having an esentially vertical axial surface; a *symmetrical fold*.

vertical form index A term used by Bucher (1919, p.154) for a ratio now known as *ripple index*. Cf: *horizontal form index*.

vertical gradiometer An instrument for measuring the vertical gradient of gravity.

vertical intensity The vertical component of the vector magnetic field intensity; it is one of the *magnetic elements*, and is symbolized by Z. It is usually considered positive if downward, negative if upward. Cf: *horizontal intensity*.

vertical interval The difference in vertical height between two points on a land surface; specif. *contour interval*. Abbrev: V.I. Cf: *horizontal equivalent*. Syn: *vertical distance*.

vertical limb A graduated arc attached to a surveying instrument and used to measure vertical angles.

vertical-loop method An inductive electromagnetic method in which the transacting coil has its plane vertical (i.e., axis horizontal).

vertically mixed estuary An estuary in which there is no measurable variation in salinity with depth, although salinity may increase laterally from the head to the mouth of the estuary; occurs where tidal currents are very strong relative to river flow. Ant: *stratified estuary*. Syn: *vertically homogeneous estuary*.

vertical photograph An aerial photograph made with the camera axis vertical (camera pointing straight down) or as nearly vertical as possible in an aircraft. Cf: *oblique photograph*. Syn: *vertical*.

vertical pore A pore surrounding the vertical spine in the back of the lattice shell in the radiolarian skeletons of the subfamily Trissocyclinae.

vertical section (a) A natural or artificial, vertical exposure of rocks or soil. (b) A *section* or diagram representing a vertical segment of the Earth's crust either actually exposed or as it would appear if cut through by any intersecting vertical plane; e.g. a *columnar section* or a *structure section*.

vertical seismograph An instrument that detects, magnifies, and records the vertical component of the ground motion of an earthquake. See also: *Galitzin-type seismograph*.

vertical separation In a fault, the vertical component of the dip slip. Cf: *horizontal separation*.

vertical shift In a fault, the vertical component of the shift.

vertical slip In a fault, the vertical component of the net slip; it equals the vertical component of the dip slip. Cf: *horizontal slip*. Syn: *vertical dip slip*.

vertical tectonics The tectonics of Pratt-type isostasy, in which vertical movements of crustal units of varying specific gravity cause topographic relief.

vertical-variability map A stratigraphic map that depicts the relative vertical positions, thicknesses, and number of occurrences of specific rock types in a sequence of strata or within a designated stratigraphic unit; e.g. a "number-of-sandstones map" indicating the number of discrete sandstone units in a given stratigraphic body, or a "limestone mean-thickness map" indicating the average thickness of limestone units in a given stratigraphic body. The map gives information about the internal geometry of the stratigraphic unit in terms of a designated rock component or property, or it may show the degree of differentiation of the unit into subunits of different lithologic types. Cf: *facies map*. See also: *center-of-gravity map; standard-deviation map; multipartite map; interval-entropy map*.

vertical-velocity curve A graphic presentation of the relationship between depth and velocity, at a given point along a vertical line, of water flowing in an open channel or conduit (ASCE, 1962). Syn: *mean velocity curve; depth-velocity curve*.

verticil One *whorl* of similar body parts arranged like the spokes of a wheel along an axis; e.g. a circle of flowers about a point on an axis. Adj: *verticillate*.

verticillate Arranged in or having *verticils*; e.g. having successive whorls of branches arranged like the spokes of a wheel. Syn: *whorled*.

Vertisol In U.S. Dept. of Agriculture soil taxonomy, a soil order containing at least 30% clay. It is found in regions having pronounced dry seasons, and is characterized by the presence of deep, wide surface cracks and gilgai, when dry (SSSA, 1970). Suborders and great soil groups of this soil order have the suffix -ert. See also: *Torrert; Udert; Ustert; Xerert*.

very angular A term used by Powers (1953) to describe a sedimentary particle with a roundness value between 0.12 and 0.17 (midpoint at 0.14). Also, said of the *roundness class* containing very angular particles. Cf: *angular*.

very close pack ice Pack ice in which the concentration approaches 10/10; the floes are tightly packed with very little, if any, seawater visible. See also: *close ice*.

very coarsely crystalline Descriptive of an interlocking texture of a carbonate sedimentary rock having crystals whose diameters are in the range of 1-4 mm (Folk, 1959).

very coarse pebble A geologic term for a *pebble* having a diameter in the range of 32-64 mm (1.3-2.5 in., or -5 to -6 phi units) (AGI, 1958).

very coarse sand (a) A geologic term for a *sand* particle having a diameter in the range of 1-2 mm (1000-2000 microns, or zero to -1 phi units). Also, a loose aggregate of sand consisting of very coarse sand particles. (b) A soil term used in the U.S. for a *sand* particle having a diameter in the range of 1-2 mm. Obsolete syn: *fine gravel*.

very common In the description of coal constituents, 10-30% of a particular constituent occurring in the coal (ICCP, 1963). Cf: *rare; common; abundant; dominant*.

very fine clay A geologic term for a *clay* particle having a diameter in the range of 1/4096 to 1/2048 mm (0.24-0.5 microns, or 12 to 11 phi units). Also, a loose aggregate of clay consisting of very fine clay particles.

very finely crystalline Descriptive of an interlocking texture of a carbonate sedimentary rock having crystals whose diameters are in the range of 0.004-0.016 mm (Folk, 1959).

very fine pebble A term used by AGI (1958) as a syn. of *granule*.

very fine sand (a) A geologic term for a *sand* particle having a diameter in the range of 0.062-0.125 mm (62-125 microns, or 4 to 3 phi units). Also, a loose aggregate of sand consisting of very fine sand particles. Syn: *flour sand*. (b) A soil term used in the U.S. for a *sand* particle having a diameter in the range of 0.05-0.10 mm. (c) Soil material containing 85% or more of sand-size particles (percentage of silt plus 1.5 times the percentage of clay not exceeding 15) and 50% or more of very fine sand (SSSA, 1965, p.347).

very fine silt A geologic term for a *silt* particle having a diameter in the range of 1/256 to 1/128 mm (4-8 microns, or 8 to 7 phi units). Also, a loose aggregate of silt consisting of very fine silt particles.

very large boulder A *boulder* having a diameter in the range of 2048-4096 mm (80-160 in., or -11 to -12 phi units).

very open pack ice Pack ice in which the concentration is 1/10 through 3/10 with water predominating over ice; the floes are loose and widely spaced. See also: *scattered ice*.

very thick bands In banded coal, vitrain bands exceeding 50.0 mm in thickness (Schopf, 1960, p.39). Cf: *thin bands; medium bands; thick bands*.

vesbite An extrusive rock intermediate in composition between melilitite and leucitite, being composed of leucite, melilite, acmite-augite, and accessory apatite and opaque oxides.

vesecite A monticellite-bearing *polzenite* that also contains olivine, melilite, phlogopite, and nepheline. Cf: *modlibovite*.

vesicle [paleont] (a) A plant or animal structure having the general form of a membranous cavity; e.g. the space enclosed in the interior of a corallite. (b) A term incorrectly applied to a *dissepiment* in a corallite.

vesicle [palyn] An expanded, bladdery projection of ektexine extending beyond the main body of a pollen grain and typically showing more or less complex internal structure. One or more vesicles are characteristic of many gymnospermous (especially coniferous) genera, in which they tend to reduce the specific gravity of, and give buoyancy to, the pollen grains. Cf: *pseudosaccus*. Syn: *wing; bladder; air sac; saccus*.

vesicle [ign pet] A cavity of variable shape in a lava formed by the entrapment of a gas bubble during solidification of the lava. Syn: *vacuole*.

vesicle cylinder A cylindrical zone in a lava, in which there are abundant vesicles, probably formed by the generation of steam from underlying wet material. This feature occurs in the lavas of the northwestern U.S

vesicular [paleont] (a) Pertaining to or containing vesicles in a plant or animal. (b) Having dissepiments in corals. This usage is not recommended.

vesicular [petrology] Said of the texture of a rock, esp. a lava, characterized by abundant vesicles formed as a result of the expansion of gases during the fluid stage of the lava. Less preferred syn: *cellular*. Cf: *scoriaceous*.

vesicularity The condition of being vesicular. Cf: *vesiculation*.

vesiculate Possessing vesicles; e.g. said of *saccate* pollen.

vesiculation (a) The process of forming vesicles. (b) The arrangement of vesicles in a rock. Cf: *vesicularity*.

vesignieite A greenish mineral: $BaCu_3(VO_4)_2(OH)_2$. Also spelled: *vesigniéite*.

vessel A xylem tube formed from several vessel segments (modified tracheids with imperfect or no end walls) set end to end (Cronquist, 1961, p.883).

vestibulate Having or resembling a vestibule or vestibulum.

vestibule (a) An inhalant cavity, other than a canal, of the aquiferous system of a sponge, located close to the surface and receiving water from one or more ostia. Syn: *subdermal space*. (b) The distal (near-surface) part of the zooecium in a cryptostome bryozoan, often delimited basally by hemisepta. Syn: *vestibulum*. (c) The space between the duplicature and the calcareous wall (outer lamella) composing the externally visible shell of an ostracode. (d) A subcylindrical prolongation of the pedicle valve dorsal of the brachial valve of a brachiopod (TIP, 1965, pt.H, p.155).

vestibulum (a) The space between the external opening (*exopore*) in the ektexine and the internal opening (*endopore*) in the endexine of a pollen grain with a complex porate structure. The openings in ektexine and endexine are about the same size. Cf: *atrium* [palyn]. (b) The *vestibule* in a cryptostome bryozoan.---Pl: *vestibula*.

vestige A small and imperfectly developed or degenerate bodily part or organ that is a remnant of one more fully developed in an earlier stage in the life cycle of the individual, in a past generation, or in closely related forms. Adj: *vestigial*.

vestigial Of, pertaining to, or being a *vestige*.

vestigiofossil *trace fossil*.

vestured pits In certain dicotyledonous woods, bordered pits having punctations or a sievelike appearance resulting from the presence of minute but highly refractive outgrowths from the free surfaces of the secondary wall (Record, 1934, p.22-23).

Vesulian Stage in Great Britain: Middle Jurassic (above Bajocian, below Bathonian).

vesuvian (a) *vesuvianite*. (b) A mixture of calcite and hydromagnesite. (c) *leucite*.

Vesuvian garnet An early name for *leucite* whose crystal form resembles that of garnet.

vesuvianite A mineral: $Ca_{10}Mg_2Al_4(SiO_4)_5(Si_2O_7)_2(OH)_4$. It is usually brown, yellow, or green, sometimes contains iron and fluorine, and is commonly found in contact-metamorphosed limestones. Syn: *idocrase; vesuvian*.

Vesuvian-type eruption *Vulcanian-type eruption*.

vesuvite A tephrite containing abundant leucite.

veszelyite A greenish-blue mineral: $(Cu,Zn)_3PO_4(OH)_3.2H_2O$. Syn: *arakawaite*.

vey A French term used esp. in Normandy for a *tidal flat*.

vibetoite According to Johannsen (1931, p.149), an igneous rock described as containing 8.5 percent apatite, 14.2 percent primary calcite, and 76.9 percent dark-colored minerals.

vibraculum A specialized zooid (heterozooid) in a colony of cheilostomatous bryozoans, having an operculum in the form of a long seta slung between condyles. Pl: *vibracula*.

vibration gravimeter A device which affords a measurement of gravity by observation of the period of transverse vibration of a thin wire tensioned by the weight of a known mass, useful for observations at sea.

vibration magnetometer A type of magnetometer for individual rock specimens that uses the alternating voltage generated by relative vibration of the specimen and a coil. Syn: *Foner magnetometer*.

vibration mark A term used by Dzulynski & Slaczka (1958, p.234) for a sedimentary structure representing a modified groove consisting of crescentic depressions (concave upcurrent) presumed to result from the unsteady inscribing action of a solid object moved by the current. Cf: *chevron mark; chattermark*. See also: *ruffled groove cast*.

vibration meter *vibrograph*.

vibration plane In optics, a plane of polarized light, including the directions of propagation and vibration. Syn: *plane of polarization; plane of vibration*.

vibrograph An instrument that records other than seismic vibrations of the Earth; e.g. quarry blasts. Cf: *seismograph*. Syn: *vibration meter; vibrometer*.

vibrometer *vibrograph*.

vicarious (a) Having the function of a substitute; e.g. "vicarious avicularium" replacing a bryozoan autozooid in a series. (b) Pertaining to or being closely related kinds of organisms that occur in similar environments or as fossils in corresponding strata but in distinct and often widely separated areas. Also, characterized by the presence of or consisting of such organisms.

vicinal face A crystal face that modifies a normal crystal face, which it closely approximates in angle.

Vickers hardness test The testing of metals by indentation with a pyramidal diamond point, and measuring the area of indentation.

Vicksburgian North American (Gulf Coast) stage: Oligocene (above Jacksonian, below Chickasawhay).

vicoite An extrusive rock composed of leucite, sodic sanidine, calcic plagioclase, and augite; a leucite shoshonite.

vidicon (a) A storage-type electronically scanned photoconductive television camera tube, which often has a response to radiations beyond the limits of the visible region. (b) An image-plane scanning device.

Villafranchian European stage: lower Pleistocene (above Astian of Pliocene, below middle Pleistocene). It is the terrestrial equivalent (in France and Italy) of the marine *Calabrian*. Before 1948, it was used for the latest division of the Pliocene.

villamaninite A black isometric mineral: $(Cu,Ni,Co,Fe)(S,Se)_2$.

villiaumite A carmine to colorless isometric mineral: NaF.

vincularian Said of a rigid, unjointed bryozoan colony having cylindrical branches in which the orifices occur on all aspects of the surface.

Vindobonian European stage: middle Miocene.

vinogradovite A white to colorless mineral: $(Na,Ca,K)_4Ti_4AlSi_6O_{23}.2H_2O$.

vintlite A porphyritic hypabyssal rock composed of labradorite r bytownite and hornblende phenocrysts in a fine-grained groundmass of feldspar, hornblende, and quartz; a hornblende diorite.

violaite A highly pleochroic mineral of the clinopyroxene group: $Ca(Mg,Fe)(SiO_3)_2$.

violan A translucent, massive, fine-blue or bluish-violet variety of diopside containing MnO and Mn_2O_3. Syn: *violane*.

violarite A violet-gray mineral of the linnaeite group: Ni_2FeS_4.

virgal One of an articulated series of more or less rod-shaped or cylindrical ossicles forming a structure that extends outward from an ambulacral plate of an asterozoan. Pl: *virgals* or *virgalia*.

virgation [geomorph] A sheaf-like pattern, as shown on a map, of mountain ranges diverging from a common center. Ant: *syntaxis*.

virgation [fault] A divergent, branchlike pattern of fault distribution. The term is used in the Russian literature.

virgation [fold] A fold pattern in which the axial surfaces diverge or fan out from a central bundle.

virgella A spine developed during formation of the metasicula of a graptolithine. It is embedded in the wall of the metasicula and projects freely from the apertural margin.

Virgilian North American provincial series: uppermost Pennsylvanian (above Missourian, below Wolfcampian of Permian).

virgin *primary* [metal].

virgin clay Fresh clay, as distinguished from that which has been fired.

virgin flow Streamflow unaffected by artificial obstructions, storage, or other works of man in the stream channels or drainage basin; a syn. of *runoff* [water].

Virglorian Anisian.

virgula A hollow thread-like prolongation of the apex of the prosicula of a graptolite; the axial support of various graptolites. It is homologous with *nema*, and the term is used where the prolongation is enclosed within a scandent rhabdosome (biserial rhabdosomes) or incorporated in the dorsal wall (as in monograptids).

viridine A grass-green variety of andalusite containing manganese. Syn: *manganandalusite*.

viridite (a) A general term formerly applied to the indeterminable or obscure green alteration products (such as chlorite and serpentine) occurring in scales and threads in the groundmasses of porphyritic rocks. Cf: *opacite; ferrite*. (b) An iron-rich chlorite containing considerable ferric iron.

virtual geomagnetic pole A conventional form of expressing a measured remanent magnetization; the pole location of the dipole magnetic field for which the field direction at the rock's location is parallel to its measured remanent magnetization. Syn: *paleomagnetic pole*.

visceral Pertaining to or located on or among the internal organs of a body; e.g. "visceral area" of a brachiopod shell enclosing the body cavity, or "visceral skeleton" developed within a crinoid body and consisting of spicules or a calcareous network, or "visceral hump" consisting of the part of the body of a mollusk behind the head and above the foot in which the digestive and reproductive organs are concentrated.

viscoelastic *elasticoviscous*.

viscometer An instrument used to measure viscosity. Syn: *viscosimeter*.

viscometry The measurement of viscosity. Syn: *viscosimetry*.

viscosimeter *viscometer*.

viscosimetry *viscometry*.

viscosity The property of a substance to offer internal resistance to flow; its *internal friction*. Specifically, the ratio of the rate of shear stress to the rate of shear strain. This ratio is known as the *coefficient of viscosity*. See also: *Newtonian liquid*.

viscosity coefficient A numerical factor that measures the interval resistance of a fluid to flow; it equals the shearing force in dynes/sq cm transmitted from one fluid plane to another

that is 1 cm away, and generated by the difference in fluid velocities of 1 cm/sec in the two planes. The greater the resistance to flow, the larger the coefficient (ASCE, 1962). Syn: *absolute viscosity; dynamic viscosity*.

viscous creep Inelastic, time-dependent strain in which the rate of strain is constant at constant differential stress.

viscous damping A type of *damping*, such that the damping force is directly proportional to the velocity.

viscous flow A syn. of *Newtonian flow*. Cf: *liquid flow; solid flow*.

viscous magnetization A component of magnetization which behaves as remanent magnetization during the time needed for laboratory measurement and like induced magnetization during geologic time, thus showing *magnetic viscosity*. Syn: *viscous remanent magnetization; unstable remanent magnetization*.

viscous remanent magnetization *viscous magnetization*. Abbrev: *VRM*.

viscous stress The resistive force of water. It is proportional to the speed of the current, but acts opposite to its direction of flow (U.S. Naval Oceanographic Office, 1966, p.175).

Viséan European stage: Lower Carboniferous (lowermost Upper Mississippian; above Tournaisian, below lower Namurian). Also spelled: *Visean*.

viséite A mineral: $NaCa_5Al_{10}(SiO_4)_3(PO_4)_5(OH)_{14}.16H_2O$ (?). It is regarded as a zeolite with structure analogous to analcime but with some vacant lattice positions in the $(Al,Si,P)O_2$ net (Hey, 1962, 17.6.8). Also spelled: *viseite*.

vishnevite A sulfate-bearing feldspathoid mineral of the cancrinite group: $(Na,K,Ca)_{6-8}(Al_6Si_6O_{24})(SO_4,CO_3).nH_2O$, where n varies between 1 and 5.

Vishnu A division of the Archeozoic.

visible horizon *apparent horizon*.

visor The more or less inclined, overhanging surface directly above the wave-cut notch in a sea cliff, most commonly found in limestones in a tropical region.

vitalism The theory that some internal force or driving energy of organisms exerts a directional effect that more or less determines how variation and evolution will proceed. Cf: *holism*.

viterbite [mineral] A mixture of allophane and wavellite(?).

viterbite [petrology] An extrusive rock composed chiefly of sodic sanidine and large leucite phenocrysts, with smaller quantities of calcic plagioclase, augite, biotite, apatite, and opaque oxides; a leucite phonolite. Its name is derived from Viterbo, Italy.

vitr- A prefix which, in a rock name, suggests its glassy character. Cf: *hyalo-*.

vitrain A coal *lithotype* characterized by brilliant, vitreous luster, black color, and cubical cleavage with conchoidal fracture. Vitrain bands or lenticles are amorphous, usually 3-5 mm thick, and their characteristic microlithotype is *vitrite*. Cf: *clarain; durain; fusain*. See also: *euvitrain; provitrain*. Syn: *pure coal*.

vitreous [geol] A type of luster resembling that of glass. Quartz, for example, has a vitreous or *glassy* luster.

vitreous [paleont] Said of a hyaline foraminifer having the appearance and luster of glass.

vitreous copper *chalcocite*.

vitreous silver *argentite*.

vitric Said of pyroclastic material that is characteristically glassy, i.e. contains more than 75% glass.

vitric tuff A tuff that consists predominantly of volcanic glass fragments. Cf: *crystal-vitric tuff*.

vitrification Formation of a glassy or noncrystalline substance. Syn: *vitrifaction*.

vitrifusain *semifusain*.

vitrinertite A coal microlithotype that contains a combination of vitrinite and inertinite totalling at least 95%, and containing more of each than of exinite. It occurs in high-ranking bituminous coals.

vitrinite (a) An oxygen-rich maceral group that is characteristic of vitrain and is composed of humic material associated with peat formation. It includes *provitrinite* and *euvitrinite* and their varieties. Cf: *inertinite; exinite.* (b) A microlithotype that contains a combination of the macerals vitrinite and inertinite totalling at least 95% and containing more of each than of exinite.

vitrinization A process of *coalification* in which vitrain is formed. Cf: *incorporation; fusinization.*

vitrinoid Vitrinite that occurs in bituminous caking coals and that has a reflectance of 0.5-2.0% (American Society for Testing and Materials, 1970, p.466). Cf: *xylinoid; anthrinoid.*

vitriol peat Peat that contains abundant iron sulfate.

vitriphyric A term applied by Cross et al (1906, p.703) to the texture of a porphyritic igneous rock in which the groundmass is microscopically glassy. Cf: *vitrophyric.*

vitrite A coal microlithotype that contains a combination of collinite and telinite (i.e., vitrinite), totalling at least 95%. Cf: *vitrain.*

vitroclarain A transitional lithotype of coal characterized by the presence of vitrinite, but more of other macerals than of vitrinite. Cf: *clarovitrain.* Also spelled: *vitroclarite.*

vitroclarite Var. of *vitroclarain.*

vitroclastic Pertaining to a pyroclastic rock structure characterized by crescentically or triangularly fragmented bits of glass; also, said of a rock having such a structure.

vitrodurain A coal lithotype transitional between durain and vitrain, but predominantly durain. Cf: *durovitrain.*

vitrofusain A syn. of *semifusain.* Also spelled: *vitrofusite.*

vitrofusite Var. of *vitrofusain.*

vitrophyre Any porphyritic igneous rock having a glassy groundmass. Its composition is similar to that of a rhyolite. Adj: *vitrophyric.* Cf: *felsophyre; granophyre.* Syn: *glass porphyry.*

vitrophyric A term used by Vogelsang in 1872 and applied by Cross et al (1906, p.703) to the texture of porphyritic igneous rocks in which the groundmass is megascopically glassy. Cf: *vitriphyric.*

vitrophyride A porphyritic volcanic glass; suggested for field use only (Johannsen, 1939, p.287).

vitroporphyric Said of the texture of a porphyritic igneous rock having large intratelluric phenocrysts in a glassy groundmass.

vivianite A mineral: $Fe_3(PO_4)_2.8H_2O$. It is colorless, blue, or green when unaltered, but grows darker on exposure; it occurs as monoclinic crystals, fibrous masses, or earthy forms in copper, tin, and iron ores and in clays, peat, and bog iron ore. Syn: *blue iron earth; blue ocher.*

vladimirite A mineral: $Ca_5H_2(AsO_4)_4.5H_2O$.

vlasovite A colorless monoclinic mineral: $Na_2ZrSi_4O_{11}$.

vlei A Dutch word used in the Middle Atlantic States and in southern Africa for a shallow lake or marshy area, esp. one developed in the poorly drained valley of an intermittent stream. Syn: *vley; vly.*

vley *vlei.*

vloer A term used in South Africa for a flat surface of caked mud with a high salt content and generally destitute of vegetation; a *playa.* It has a more irregular shape, greater horizontal development, and less depth than a *pan,* and usually has an outlet. Etymol: Afrikaans, "floor".

vltavite *moldavite* [tektite].

vly *vlei.*

voe A term used in the Orkney and Shetland islands of Scotland for a narrow bay, creek, or inlet, or a narrow gully cut in a cliff. Etymol: Scandinavian.

vogesite A lamprophyre composed of hornblende phenocrysts in a groundmass of orthoclase and hornblende. Augite, diopside, olivine, and plagioclase feldspar also may be present.

voglite An emerald-green to bright grass-green mineral: Ca_2-$Cu(UO_2)CO_3)_4.6H_2O$ (?).

voices of the desert *song of the desert.*

void *interstice.*

voidal concretion A large, tube-like, iron-oxide concretion with a central hollow or cavity and a hard, dense limonitic rim, commonly found in sands and sandstones and in some clays. It appears to be a product of weathering (oxidation) of a sideritic concretion.

void ratio The ratio of the volume of void space to the volume of solid substance in any material consisting of voids and solid material, such as in a soil sample, a sediment, or sedimentary rock. Symbol: *e.* Syn: *voids ratio.*

volatile (a) adj. Readily vaporizable. (b) n. A syn. of *volatile component.*

volatile combustible *volatile matter.*

volatile component A material, such as water or carbon dioxide, in a magma whose vapor pressures are sufficiently high for them to be concentrated in any gaseous phase. Syn: *volatile; volatile flux.*

volatile flux *volatile component.*

volatile matter In coal, those substances, other than moisture, that are given off as gas and vapor during combustion. Standardized laboratory methods are used in analysis. Syn: *volatiles; volatile combustile.*

volatiles *volatile matter.*

volatile transfer *gaseous transfer.*

volborthite A green or yellow mineral: $Cu_3(VO_4)_2.3H_2O$. It may contain some calcium and barium, and it represents a principal ore of vanadium. Syn: *uzbekite.*

volcan (a) The component of the Earth's crust made up of volcanoes and various hypabyssal rocks (Makiyama, 1954, p.146). (b) *volcano.*

volcanello (a) A small, active cone within the central crater of a volcano, e.g. Mount Nuevo in Vesuvius. (b) A *spatter cone.*---Etymol: Italian, "small volcano".

volcanic (a) Pertaining to the activities, structures, or rock types of a volcano. (b) A syn. of *extrusive.*

volcanic accident A departure from the normal cycle of erosion, caused by the outbreak of volcanic activity.

volcanic arc *island arc.*

volcanic arenite A term used by Gilbert (1954, p.308) for a lithic arenite composed chiefly of volcanic detritus and having a low quartz content. It is a common rock among Tertiary and Mesozoic sediments around the Pacific Basin. Folk (1968, p.124) used "volcanic-arenite" for a litharenite composed chiefly of volcanic rock fragments, and having any clay content, sorting, or rounding; for more detailed specification, terms such as "basalt-arenite" and "andesite-arenite" may be used.

volcanic ash *ash* [volc].

volcanic ball *lava ball.*

volcanic blowpiping *gas fluxing.*

volcanic breccia (a) A pyroclastic rock that consists of angular volcanic fragments that are larger than 2 mm in diameter and that may or may not have a matrix. Cf: *agglomerate.* (b) A rock that is composed of accidental or nonvolcanic fragments in a volcanic matrix. Syn: *alloclastic breccia; lava breccia.*

volcanic butte An isolated hill or mountain resulting from the differential weathering or erosion and consequent exposure of a volcanic neck or of a narrow, vertical igneous intrusion into overlying weaker rock; e.g. Ship Rock, N. Mex. Cf: *mesabutte.*

volcanic chain A linear arrangement of a number of volcanoes, apparently associated with a controlling geologic feature.

volcanic clay *bentonite.*

volcanic cloud *eruption cloud.*

volcanic cluster A group of volcanic vents in apparent random arrangement, in contrast to an arrangement over a fissure or other structural control.

volcanic cone A conical hill of lava and/or pyroclastics that is built up around a volcanic vent. It may be intersected by dikes. Syn: *cone* [*volc*].

volcanic conglomerate A water-deposited conglomerate containing over 50% volcanic material, esp. coarse pyroclastics.

volcanic debris *volcanic rubble.*

volcanic dome A steep-sided, rounded extrusion of highly viscous lava squeezed out from a volcano, and forming a dome-shaped or bulbous mass of congealed lava above and around the volcanic vent. Portions of older lavas may be elevated by the pressure of the new lava rising from below. The structure generally develops inside a volcanic crater or on the flank of a large volcano, and is usually much fissured and brecciated (Williams, 1932). Cf: *lava dome.* Syn: *tholoid; dome volcano; cumulo-dome; cumulo-volcano.*

volcanic dumpling *lava ball.*

volcanic earthquake An earthquake associated with volcanic rather than tectonic forces. Cf: *tectonic earthquake.*

volcanic flow drain *lava tube.*

volcanic foam *pumice.*

volcanic focus The subterranean seat or center of volcanism of a region, or of a volcano.

volcanic gases Volatile matter released during a volcanic eruption that was previously dissolved in the magma. Water vapor forms about 90% of the gases; other constituents include carbon gases, esp. carbon dioxide; sulfur gases, esp. sulfur dioxide at high temperatures and hydrogen sulfide at low temperatures; hydrogen chloride; nitrogen as a free element, and others (Krauskopf, 1967). See also: *gas phase.*

volcanic glass A natural glass produced by the cooling of molten lava or a liquid fraction of it, too rapidly to permit crystallization. Examples are obsidian, pitchstone, sideromelane, and the glassy mesostasis of many extrusive rocks.

volcanic graben A straight-walled collapse structure on the summit or flanks of a volcanic cone. See also: *summit graben; sector graben.*

volcanic gravel A pyroclastic deposit, the individual clasts of which are in the size range 0.5 cm or smaller, but still distinguishable by the naked eye. The finer fractions are also called *volcanic sand.* Cf: *block* [*volc*]; *cinder; lapilli.*

volcanic graywacke *volcanic wacke.*

volcanic harbor A natural harbor formed by the sea breaking through a gap in the rim of a volcanic crater; e.g. that of St. Paul Rocks in the southern Atlantic Ocean.

volcanic island A *submarine volcano* that has been sufficiently built up to be exposed above the ocean's water level. Syn: *island volcano.*

volcanicity *volcanism.* Also spelled: *vulcanism.*

volcaniclastic Pertaining to a clastic rock containing volcanic material in whatever proportion, and without regard to its origin or environment.

volcanic mud A mixture of water and volcanic ash, either just erupted and hot or already cooled. The mixture may form a mud flow down the slope of the volcano.

volcanic plain An extensive lava flow or fall of volcanic ash that forms a flat surface and covers topographic irregularities.

volcanic rain *eruption rain.*

volcanic rent A great volcanic depression, bordered by fissures that are usually concentric in plan, caused by the pressure of magmatic injection or by the overloading of cone material on a weak substratum.

volcanic rift zone *rift zone* (b).

volcanic rock (a) A generally finely crystalline or glassy igneous rock resulting from volcanic action at or near the Earth's surface, either ejected explosively or extruded as lava; e.g. basalt. The term includes near-surface intrusions that form a part of the volcanic structure. See also: *volcanics.* Cf: *plutonic rock.* Syn: *volcanite* [*petrology*]. (b) A general term proposed by Read (1944) to include the effusive rocks and associated intrusive rocks; they are dominantly basic. Cf: *neptunic rock; plutonic rock.*

volcanic rubble The unconsolidated equivalent of volcanic breccia. Syn: *volcanic debris.*

volcanics Those igneous rocks that have reached or nearly reached the Earth's surface before solidifying. The common use of the term for *volcanic rocks* should be avoided (USGS, 1958, p.86).

volcanic sand A pyroclastic deposit, the individual clasts of which are in the size range 2-5 mm; the finer fractions of *volcanic gravel.*

volcanic sandstone An indurated deposit of rounded, water-worn pyroclastic fragments and a subordinate amount of non-volcanic detritus.

volcanic seismology The study of earthquakes related to volcanoes, in order to determine their location and depth, to locate their magmatic reservoirs, and to predict future eruptions.

volcanic sink *sink* [*volc*].

volcanic wacke A term used by Gilbert (1954, p.303) for a lithic wacke composed chiefly of detritus derived from intermediate (andesitic) and basic (basaltic) volcanic rocks and having a low quartz content. It is a common rock among Tertiary and Mesozoic sediments deposited in orogenic belts bordering the Pacific Ocean. Syn: *volcanic wacke.*

volcanic water Water in, or derived from, magma at the Earth's surface or at a relatively shallow level; *juvenile* [*water*] of volcanic origin. Cf: *plutonic water.*

volcanism The processes by which magma and its associated gases rise into the crust and are extruded onto the Earth's surface and into the atmosphere. Also spelled: *vulcanism.* Syn: *volcanicity.*

volcanist An obsolete syn. of *volcanologist.* Also spelled: *vulcanist.* Syn: *plutonist.*

volcanite [*mineral*] An old mineral name suggested as a synonym of pyroxene and of a variety of sulfur containing less than one percent selenium.

volcanite [*petrology*] (a) An extrusive rock composed chiefly of anorthoclase, andesine, and augite phenocrysts in a glassy groundmass containing feldspar and augite microlites. (b) A syn. of *volcanic rock.* Also spelled: *vulcanite* [*petrology*].

volcano (a) A vent in the surface of the Earth through which magma and associated gases and ash erupt; also, the form or structure, usually conical, that is produced by the ejected material. (b) Any eruption of material, e.g. mud, that resembles a magmatic volcano.----Obsolete var: *vulcano.* Pl: *volcanoes.* Etymol: the Roman diety of fire, Vulcan.

volcanogenic Of volcanic origin, e.g. volcanogenic sediments.

volcano-karst A terrain of freshly erupted volcanic materials (esp. pyroclastics such as certain tuffs and agglomerates that contain unstable minerals) that have been corroded by rainwater to develop microrelief forms resembling limestone karst. Term introduced by Naum et al (1962) as "vulcanokarstul".

volcanologist One who studies or does *volcanology.* Obsolete syn: *volcanist.*

volcanology The branch of geology that deals with volcanism, its causes and its phenomena. Also spelled: *vulcanology.* See also: *volcanologist.* Less-preferred syn: *pyrogeology.*

volcano shoreline A roughly circular, steeply-sloping shoreline formed where fragmental volcanic materials or flows of lava occur in a coastal location, or where an active volcano projects above the water surface, building a cone upward and outward by continued addition of ejected materials.

volcano-tectonic depression A large-scale depression that is usually linear and that is controlled by both tectonic and volcanic processes (van Bemmelen, 1932). An example is the Toba trough in northern Sumatra.

volhynite A quartz-bearing kersantite composed of plagioclase, hornblende, and sometimes biotite phenocrysts in a groundmass of quartz, feldspar, and abundant chlorite. Its name is derived from Volhynia, U.S.S.R.

volkonskoite A bluish-green, chromium-bearing clay mineral of the montmorillonite group; specif. a variety of nontronite containing chromium. Syn: *volchonskoite; wolchonskoite.*

volkovskite A mineral: $(Ca,Sr)B_6O_{10}.3H_2O.$

voltage gradient *electric field intensity.*

voltaite A dull oil-green to brown or black mineral: $K_2Fe_5^{+2}Fe_4^{+3}(SO_4)_{12}.18H_2O$ (?).

volt-second *weber.*

voltzite A yellowish or reddish material consisting of wurtzite plus an organometallic compound of zinc.

volume control *gain control.*

volume elasticity *bulk modulus.*

volume law *Lindgren's volume law.*

volume magnetization *magnetization.*

volume phase *surface phase.*

volume susceptibility *susceptibility* [mag].

volumetric analysis Quantitative chemical analysis where the amount of a substance in a solution is determined by adding a fixed volume of a standard solution to the prepared sample until a reaction occurs. The amount of standard solution needed to produce the desired reaction indicates the amount of substance in the original sample.

volumetric shrinkage The decrease in volume, expressed as a percentage of the soil mass when dried, of a soil mass when the water content is reduced from a given percentage to the shrinkage limit (ASCE, 1958, term 411).

volution A *whorl* of a spiral shell. Syn: *volute.*

volvocine line An evolutionary series of green algae, exemplified by a series of colonial forms with cells not arranged in a filament. Cf: *siphonous line; tetrasporine line.*

volynskite A mineral: $AgBiTe_2.$

von Baer's law The principle stated by Karl E. von Baer (1792-1876), Estonian embryologist and geologist, according to which the rotation of the Earth causes an asymmetrical, lateral erosion of stream beds.

von Kármán constant A dimensionless number which relates the mixing length to the flow condition in turbulent flow (Middleton, 1965, p.252). It is symbolized by k in formulas for velocity and sediment distribution in turbulent flow.

von Schmidt wave *head wave.*

vonsenite A coal-black orthorhombic mineral: $(Fe^{+2},Mg)_2-Fe^{+3}BO_5.$ It is isomorphous with ludwigite.

Von Sterneck-Askania pendulum A device for measuring the vertical component of gravity, characterized by the use of four pendulums in a single case (Jakosky, 1950, p. 277).

von Wolff's classification A quantitative chemical-mineralogic classification of igneous rock proposed in 1922 by F. von Wolff.

vorobyevite A rose-red, purplish-red, or pinkish gem variety of beryl containing cesium. Appreciable amounts of sodium and other alkalies may be present. Syn: *morganite; rosterite; worobieffite; vorobievite.*

vortex [struc geol] A vertical, cylindrical fold formed in incompetent rock by late-stage deformation during deep-zone orogeny (Wynne-Edwards, 1957, p.643).

vortex [hydraul] A fluid flow that has a revolving motion in which the stream lines are concentric circles, and in which the total head for each stream line is the same.

vortex cast *flute cast.*

Vraconian European stage: uppermost Lower Cretaceous or lowermost Upper Cretaceous.

vrbaite A gray-black to dark-red orthorhombic mineral: $Tl_4Hg_3Sb_2As_8S_{20}.$

vredenburgite (a) *beta-vredenburgite.* (b) *alpha-vredenburgite.*

V-shaped valley A valley having a pronounced cross profile suggesting the form of the letter "V", characterized by steep sides and short tributaries; specif. a young, narrow valley resulting from downcutting by a stream. The "V" becomes broader as the amount of wasting increases. Cf: *U-shaped valley.* Syn: *V-valley.*

V's, rule of *rule of V's.*

V-terrace A triangular-shaped terrace, commonly formed in a long, narrow arm of an old lake, having one side built against an even coast and the opposite angle pointing toward open water (Gilbert, 1890, p.58).

vug [petrology] A small cavity in a vein or in rock, usually lined with crystals of a different mineral composition from the enclosing rock. Etymol: Cornish *vooga*, "underground chamber, cavern, cavity". Adj: vuggy. Cf: *druse; geode.* Syn: *bughole.* Var: *vugh.*

vug [oil] A term used in petroleum geology for any opening in a rock, from the size of a small pea upwards. See also: *vuggy porosity.*

vuggy Pertaining to a *vug* or having numerous vugs.

vuggy porosity In petroleum geology, porosity induced by the presence of *vugs* [oil]; it is usually used with reference to limestones.

vugh (a) A relatively large and usually irregular void in a soil material, not normally interconnected with other voids of comparable size (Brewer, 1964, p.189). Cf: *vesicle* [soil]. (b) Var. of *vug* [petrology].

vulc- For most terms beginning with the combining form "volc-", this is a variant spelling.

Vulcanian-type eruption A type of volcanic eruption characterized by periodic explosive events. Also spelled: *Vulcano-type eruption.* Syn: *Vesuvian-type eruption; paroxysmal eruption; Plinian-type eruption.*

vulcanite [mineral] An orthorhombic mineral: $CuTe.$

vulcanite [petrology] *volcanite* [petrology].

vulcanorium A structure transitional between a mid-oceanic ridge and an island arc, e.g. the Arctic vulcanorium between the Nansen and Amundsen deeps in the Arctic Ocean (Runcorn, 1967, p.161).

Vulcano-type eruption Var. of *Vulcanian-type eruption.*

vulpinite A scaly, granular, grayish-white variety of anhydrite. It sometimes has an admixture of silica.

vulsinite An extrusive rock similar to trachyte, being composed chiefly of alkali feldspar (sanidine) and also calcic plagioclase and augite.

V-valley *V-shaped valley.*

vysotskite A tetragonal mineral: $(Pd,Ni)S.$

W

wacke (a) A "dirty" sandstone that consists of a mixed variety of angular and unsorted or poorly sorted mineral and rock fragments and of an abundant matrix of clay and fine silt; specif. an impure sandstone containing more than 10% argillaceous matrix (Gilbert, 1954, p.290). The term is used for a major category of sandstone, as distinguished from *arenite*. (b) A term used by Fischer (1934) for a clastic sedimentary rock in which the grains are almost evenly distributed among the several size grades; e.g. a sandstone consisting of sediment "poured in" to a basin of deposition at a comparatively rapid rate without appreciable selection or reworking by currents after deposition, or a mixed sediment of sand, silt, and clay in which no component forms more than 50% of the whole aggregate. (c) A term commonly used as a shortened form of *graywacke*. This usage is not recommended. (d) Originally, a term applied to a soft and earthy variety of basalt, or to the grayish-green to brownish-black clay-like residue resulting from the partial chemical decomposition in place of basalts, basaltic tuffs, and related igneous rocks. Syn: *vake*.-Etymol: German *Wacke*, an old provincial mining term signifying a large stone or "stoniness" in general.

wackestone A term used by Dunham (1962) for a mud-supported carbonate sedimentary rock containing more than 10% grains (particles with diameters greater than 20 microns); e.g. a calcarenite. The use of this term is discouraged because it is apt to lead to confusion with terrigenous "wacke-type" rocks. Cf: *mudstone; packstone*.

wad [coast] A Dutch term for *tidal flat*. Pl: *wadden*. The spelling "wadd" is incorrect.

wad [mineral] (a) A massive, amorphous, earthy, dark-brown or black mineral substance consisting chiefly of an impure mixture of manganese oxides and other oxides, with varying amounts of other minerals (such as copper, cobalt, and silica) and 10-20% water. It is commonly very soft, soiling the hand, but is sometimes hard and compact, and it has a low apparent specific gravity. Wad generally occurs in low, damp, marshy areas as a result of decomposition of manganese minerals. Cf: *psilomelane*. Syn: *bog manganese; black ocher; earthy manganese*. (b) A general term applied to hydrated oxides of manganese (or of manganese and other metals) whose true natures are unknown or which have variable and uncertain compositions and some at least of which may be amorphous. (c) An English dialectal term for graphite.

wadden Plural of the Dutch term *wad*, meaning a "tidal flat". The form "waddens" is incorrect.

waddy Var. of *wadi*.

wadeite A mineral: $K_2CaZr(SiO_3)_4$.

wadi (a) A term used in the desert regions of SW Asia and northern Africa for a stream bed or channel, or a steep-sided and bouldery ravine, gully, or valley, or a dry wash, that is usually dry except during the rainy season, and that often forms an oasis. (b) The intermittent and torrential stream that flows through a wadi and ends in a closed basin. (c) A shallow, usually sharply defined, closed basin in which a wadi terminates.---Etymol: Arabic. Variant plurals: *wadis; wadies; wadian; widan*. See also: *arroyo; nullah*. Syn: *wady; waddy; oued; widiyan*.

wady Var. of *wadi*.

wagnerite A yellow, red, or greenish monoclinic mineral: $Mg_2(PO_4)F$. Ferrous iron and calcium may be present.

wairakite A zeolite mineral: $CaAl_2Si_4O_{12}\cdot2H_2O$. It is isostructural with analcime.

wairauite A mineral: $CoFe$.

wake dune A sand dune occurring on the leeward side of a larger dune, and trailing away in the direction of the wind.

wakefieldite A mineral: YVO_4.

walchowite A honey-yellow variety of retinite containing a little nitrogen, found in brown coal at Walchow in Moravia, Czechoslovakia.

walker's earth A syn. of *fuller's earth*. Etymol: German *Walkererde*.

walking beam A rigid, oscillating bar or beam used to activate the cable in cable-tool drilling by alternately raising and dropping the drill bit in the hole. It is mounted on a fulcrum and connected to a power source at one end and carrying the string of tools at the other end.

walking leg (a) A prosomal appendage of a merostome, serving for walking. (b) A *pereiopod* of a malacostracan crustacean.

walking out A simple method of correlation by which stratigraphic units are traced from place to place along continuous outcrops.

wall [eng] An engineering structure that serves to hold back pressure (such as of water or sliding earth); e.g. a seawall and a retaining wall.

wall [speleo] (a) The side of a cave passage. (b) A series of *columns* along a joint crack that have fused together into a solid mass. Cf: *partition*.

wall [mining] The side of a lode, or of mine workings. Cf: *footwall; hanging wall*.

wall [paleont] (a) An external layer surrounding internal parts of an invertebrate; e.g. a skeletal deposit, formed in various corallites, that encloses the column of a scleractinian polyp and unites the outer edges of septa, such as septotheca, paratheca, and synapticulotheca, or the part of a cephalopod conch comprising the external shell. (b) The raised margin in a caneolith.

wall [fault] The rock mass on a particular side of a fault, e.g. hanging wall, footwall. See also: *wall rock [fault]*. Syn: *fault wall*.

Wallace's line The hypothetical boundary that separates the distinctly different floras and faunas of Asia and Australia. It is usually drawn between the islands of Bali and Lombok, through the Strait of Macassar, between Borneo and Celebes, and south of the Philippines.. Named after Alfred Russel Wallace (1822-1913), English naturalist. Cf: *Weber's line*.

Wallachian orogeny One of the 30 or more short-lived orogenies during Phanerozoic time identified by Stille, in this case at the end of the Pliocene.

walled lake A lake bordered along its shore by *lake ramparts* or low, wall-like ridges composed of boulders and cobbles.

walled plain A large lunar crater characterized by a broad, nearly level floor and filled by dark material similar to that which forms the floors of the maria. It is not as deep as a cup-shaped crater. See also: *ring plain*. Syn: *cirque [lunar]*.

wallisite A mineral: $PbTl(Cu,Ag)As_2S_5$.

wall niche *meander niche*.

wallongite *wollongongite*.

wallow In geomorphology, a depression or area, often filled with water or mud, that suggests a place where animals have wallowed or that resembles a wallow; e.g. a *kommetje*.

wallpaper effect The swelling and blocking action (upon the introduction of water) shown by interstitial clay minerals of the expanding-lattice type.

wall reef A linear, steep-sided coral reef constructed on a *reef wall*.

wall-resistivity log A *resistivity log* that measures the formation factors of porous formations. It measures thickness, mud resistivity in-situ, and borehole diameter. Examples include microlog and contact log.

wall rock [eco geol] The rock enclosing a vein; the *country rock* of a vein deposit.

wall rock [intrus rocks] *country rock*.

wall rock [fault] The rock mass comprising the *wall* of a fault.

wall-rock alteration Alteration of country rock adjacent to hydrothermal ore deposits by the fluids responsible for the formation of the deposits; also, the alteration products themselves.

wall saltpeter Naturally occurring calcium nitrate; *nitrocalcite* occurring on walls of limestone caves. Cf: *saltpeter*.

wall-sided glacier A glacier that is on a steep slope and not laterally confined by valley walls.

walpurgite A yellow to yellow-orange mineral: Bi_4-$(UO_2)(AsO_4)_2O_4 \cdot 3H_2O$ (?). Syn: *waltherite*.

walstromite A mineral: $BaCa_2Si_3O_9$.

waltherite A mineral formerly believed to be a poorly defined carbonate of bismuth, but now known to be identical to *walpurgite*.

wandering The slow and winding, compound movement consisting of the *sweeping* of meanders and the *swinging* of a meander belt.

wandering dune A sand dune, such as a barchan, that is slowly shifted more or less as a unit in the leeward direction of the prevailing winds, and that is characterized by insufficient vegetation to anchor it. Cf: *anchored dune*. Syn: *migrating dune; traveling dune*.

wandering water *vadose water*.

waning development *declining development*.

waning slope The lower part of a hillside surface, tending to become concave below the *constant slope*, having an angle that decreases continuously downslope as the hillside stretches to the valley floor or other local base level (Wood, 1942). Cf: *wash slope*. Ant: *waxing slope*. Syn: *concave slope*.

want *nip* [coal].

warden A term used in south Wales for a strong massive sandstone associated with coal.

wardite A light bluish-green mineral: $NaAl_3(PO_4)_2(OH)_4 \cdot 2H_2O$.

wardsmithite A mineral: $Ca_5MgB_{24}O_{42} \cdot 30H_2O$.

warm front The sloping *front* between an advancing warm air mass and a colder air mass over which it moves. Its passage is usually preceded by considerable precipitation, whereas there is little or no precipitation afterwards. Cf: *cold front*.

warm loess Continental loess composed of desert dust, such as that formed in the inland basins and steppes encircling the modern deserts of central Asia between lat. 52°N and 56°N. Cf: *cold loess*.

warm spring A *thermal spring* whose temperature is appreciably above the local mean annual atmospheric temperature, but below that of the human body (Meinzer, 1923, p.54). Cf: *hot spring*.

warp [sed] (a) An English provincial term for the fine mud and silt held in suspension in waters artificially introduced over low-lying land. (b) A general term for a bed or layer of sediment deposited by water; e.g. an estuarine clay, or the alluvium laid down by a tidal river.

warp [mass move] *congeliturbate*.

warp [tect] n. A slight flexure or bend of the Earth's crust, either upwards or downwards, and usually on a broad or regional scale. See also: *warping* [tect].

warped fault A fault, usually a thrust fault, that has been folded; a partial syn. of *folded fault*.

warping [sed] (a) The flooding at high tides of low-lying land near an estuary or tidal river by water loaded with fine mud and silt (warp), the water remaining until the warp is deposited, and then allowing it to run off clear during low tides; a means of fertilizing or of raising the general level of large, low tracts, as the conversion of a lagoon or tidal flat into a marsh. Cf: *colmatage*. (b) The filling up of hollows or the choking of a channel with warp.

warping [tect] The slight flexing or bending of the Earth's crust on a broad or regional scale, either upwards (*upwarping*) or downwards (*downwarping*); the formation of a *warp* [tect].

warpland Low-lying land that has been built up or fertilized by the process of *warping*. Also spelled: *warp land*.

warrant A term used in England for a particularly "hard and

tough" *underclay* (Nelson, 1965, p.503). The term "possibly has the sense of a token or guarantee of the presence of coal" (Arkell & Tomkeieff, 1953, p.124).

warrenite (a) A general term for gaseous and liquid bitumens consisting of a mixture of paraffins, isoparaffins, etc.; a variety of petroleum rich in paraffins. (b) A pink variety of smithsonite containing cobalt. (c) A name applied to a mineral that may be owyheeite or jamesonite.

warwickite A dark-brown to dull-black orthorhombic mineral: $(Mg,Fe)_3Ti(BO_4)_2$.

wash [eco geol] An alluvial placer.

wash [geomorph] (a) Erosion effected by wave action. (b) The wearing away of soil by runoff water, as in gullying or sheet erosion; *rainwash*.

wash [coast] (a) A piece of land that is washed by the action of a sea or river, or that is alternately covered and uncovered by a sea or river; e.g. a sandbank or mudbank, or an area of such banks, alternately submerged and exposed by the tide. (b) The shallowest part of a river, estuary, or arm of the sea. (c) A bog, fen, or marsh.

wash [speleo] *drift* [speleo].

wash [sed] (a) Loose or eroded surface material (such as gravel, sand, silt) collected, transported, and deposited by running water, as on the lower slopes of a mountain range; esp. coarse alluvium. (b) A fan-shaped deposit, as an *alluvial fan* or an *alluvial cone*, or a mound of detritus below a cliff opening. (c) *downwash*.

wash [streams] (a) A term applied in the western U.S. (esp. in the arid and semiarid regions of the SW) to the broad, shallow, gravelly or stony, normally dry bed of an intermittent stream, often situated at the bottom of a canyon; it is occasionally filled by a torrent of water. Syn: *dry wash; washout*. (b) A shallow body of water; esp. a shallow creek.

wash-and-strain icefoot An icefoot formed from ice casts and slush, and attached to a shelving beach between the high- and low-water marks.

washboard moraine (a) One of several small, parallel, regularly spaced ridges that are oriented transverse to the ice movement in a general sense and that collectively resemble a washboard. They are abundant in north-central U.S. and in the western plains of Canada. (b) A subglacial disintegration feature formed at the base of a thrust plane in glacier ice by the periodic recession and readvance of a glacier which pushes previously deposited ground moraine into a ridge (Gravenor & Kupsch, 1959, p.54). Examples are common on the swell-and-swale topography of Alberta and Saskatchewan.

wash cone *outwash cone*.

washing (a) Erosion or wearing away by the action of waves or running water. (b) The selective sorting, and removal, of fine-grained sediment by water currents. Cf: *winnowing*.

washings Material abraded or transported by the action of water.

Washitan North American (Gulf Coast) stage: Lower and Upper Cretaceous (above Fredericksburgian, below Woodbinian).

Washita stone A porous, uniformly textured *novaculite* found in the Ouachita (Washita) River region and used esp. for sharpening woodworking tools.

washland An embanked, low-lying land bordering a river or estuary, usually part of the natural flood plain, over which floodwaters are allowed to flow periodically in order to control high water levels in the river. See also: *rond*.

wash load The part of the total *sediment load* (of a stream) composed of material that is usually supplied from bank erosion or some external upstream source (such as overland flow); it is the finer part of the load, or the part that the streamflow can easily carry in large quantities. The term is essentially synonymous with *suspended load*. Cf: *bed-material load*.

washout [geomorph] (a) The washing out or away of earth

materials as a result of a flood or a sudden and concentrated downpour, often causing extensive scouring and bank caving. (b) A place where part of a road or railway has been washed away by the waters of a freshet or local flood.

washout [**sed struc**] A channel or channel-like feature produced in a sedimentary deposit by the scouring action of flowing water and later filled with the sediment of a younger deposit. Cf: *channel cast*. Syn: *scour and fill; cut and fill*.

washout [**streams**] A narrow channel or gully cut in the surface of the ground by a swiftly flowing stream during and after a heavy rainfall; a *wash*.

washout [**mining**] *horseback*.

washover (a) Material deposited by the action of overwash; specif. a small delta built on the landward side of a bar or barrier, separating a lagoon from the open sea, produced by storm waves breaking over low parts of the bar or barrier and depositing sediment in the lagoon. Cf: *blowover*. Syn: *wave delta; storm delta*. (b) The process by which a washover is formed.

washover crescent A term used by Tanner (1960, p.484) for a shallow-water, barchan-like ripple mark between crescentic depressions a centimeter deep, having horns pointing downcurrent and a plane of symmetry parallel to the current direction.

washover fan A fan-like deposit consisting of sand washed onto the shore during a storm. Syn: *washover apron*.

wash plain (a) *outwash plain*. (b) An *alluvial plain* composed of coarse alluvium.

wash slope The lower, gentle slope of a hillside, lying at the foot of an escarpment or steep rock face and usually covered by an accumulation of talus; it is less steep than the *gravity slope* above and often consists of alluvial fans or pediments. Term introduced by Meyerhoff (1940). Cf: *waning slope*. See also: *foot slope*. Syn: *haldenhang; basal slope; toe slope*.

washy Easily washing out or eroding, such as a *washy* hillside.

wastage [**geomorph**] A general term for the denudation of the Earth's surface. See also: *mass-wasting*.

wastage [**glaciol**] *ablation* [glaciol].

waste [**geol**] Loose material resulting from weathering by mechanical and chemical means, and moved down sloping surfaces or carried short distances by streams to the sea; esp. *rock waste*.

waste bank A bank or other accumulation composed of waste material; e.g. a bank where excess earth excavated during the digging of a ditch is dumped parallel to it.

waste-disposal well *waste-injection well*.

waste-injection well A well used for the injection of waste water or other fluids into the subsurface. Because the wastes rarely degrade to an innocuous condition but are simply stored, they are not truly disposed of; hence, this term is preferred to *waste-disposal well*.

waste mantle Weathered (disintegrated and decomposed) rock material that overlies or covers bedrock.

waste plain (a) *alluvial plain* (b) *bajada*.

waste rock In mining, rock that must be broken and disposed of in order to gain access to or upgrade the ore; valueless rock that must be removed in mining. Syn: *muck; mullock*.

waste stream The loose rock debris in transit to the sea or to rock-rimmed desert basins, consisting wholly of debris or of debris and water in varying proportions (Grabau, 1924, p.541).

waste water (a) *return flow*. (b) Seepage of water from a ditch or reservoir.

wasting The gradual destruction or wearing away of a landform or surface by natural processes, including removal by wind, gravity, and rill wash, but excluding stream erosion; e.g. of a glacier by melting, or of rocks by weathering. See also: *mass-wasting; backwasting*. Cf: *wearing*.

water [**gem**] The quality, transparency, limpidity, or luster of a precious stone or pearl, and esp. of a diamond. Cf: *river* [gem].

water [**geog**] (a) A British term for lake, pond, pool, or other body of standing fresh water. (b) A Scottish term for a stream and for a stream bank or the land abutting a stream.

water agate *enhydros*.

water balance *hydrologic budget*.

water-balance equation *hydrologic equation*.

water-bearing An adj. used to describe bodies of rock that contain or yield water.

water bed A term used in the upper Mississippi valley for a bed of coarse gravel or pebbles occurring in the lower part of the upper till.

water biscuit *algal biscuit*.

water bloom An aquatic growth of algae in such concentration as to cause discoloration of the water. See also: *red tide*. Syn: *plankton bloom; bloom* [oceanog].

water-bound Said of a soil or a macadam road surface that is consolidated or held together by filling of the voids with water. Also spelled: *waterbound*.

water-break (a) A place in a stream where the surface of the water is broken by bottom irregularities. (b) *breakwater*.

water budget *hydrologic budget*.

water capacity The maximum amount of water a rock or soil can hold.

water color The apparent color of the surface waters of the ocean. Detrital, organic, or dissolved material in the water may affect its color.

water content [**sed**] Amount of water contained in a porous sediment or sedimentary rock, generally expressed as the ratio of the weight of the water in the sediment to that of the sediment when dried, multiplied by 100. See also: *moisture content*.

water content [**snow**] This term is not recommended because it has been used for both *water equivalent* and *free water content*, two different concepts.

watercourse (a) A natural, well-defined channel produced wholly or in part by a definite flow of water, and through which water flows continuously or intermittently. Also, a ditch, canal, aqueduct, or other artificial channel for the conveyance of water to or away from a given place, as for the draining of a swamp. (b) A stream or current of water. Legally, a natural stream arising in a given drainage basin but not wholly dependent for its flow on surface drainage in its immediate area, flowing in a channel with a well-defined bed between visible banks or through a definite depression (as a ravine or swamp) in the surrounding land, having a definite and permanent or periodic supply of water (the stream may be intermittent), and usually, but not necessarily, having a perceptible current in a particular direction and discharging at a fixed point into another body of water. (c) A legal right permitting the use of the flow of a stream (esp. of one flowing through one's land) or the receipt of water discharged upon land belonging to another.

water creep The movement of water under or around a structure (such as a dam) built on a permeable foundation. See also: *piping*.

water crop *water yield*.

water cupola A vaulted uprising of the surface of the ocean above a submarine volcanic explosion; the initial effect of a submarine explosion on the water surface. It immediately precedes the eruption of volcanic gases and ejectamenta. Syn: *water fountain*.

water cushion Water pumped into the drill pipe during a drill-stem test to retard fill-up and prevent collapse of the pipe under sudden pressure changes.

water cycle *hydrologic cycle*.

water equivalent The amount (or depth) of water that would result from the complete melting of a sample of deposited snow. Not to be confused with *free water content*. See also:

water content [*snow*].

water eye A small, shallow depression formed as a result of chemical weathering in crystalline rock (Russell, 1968, p.94).

water-faceted stone *aquafact*.

waterfall (a) A perpendicular or very steep, unsupported descent of the water of a stream; it occurs where the stream course is markedly or suddenly interrupted, as by the presence of a transverse outcrop of resistant rock overhanging softer rock which has been eroded, or along the edge of a plateau or cliffed coast. See also: *cascade; cataract.* Syn: *fall.* (b) An obsolete term for a riffle or rapids in a swift stream. (c) A falling away of the ground such that water may be drained off.

waterfall lake *plunge pool.*

waterfinder (a) One who seeks sources of water supply, esp. a *dowser.* (b) Any instrument purported to indicate the presence of water (e.g. *divining rod*; compass).

water-fit A Scottish term for the mouth of a river. Syn: *water-foot.*

waterflood A sweeping flood of water.

water fountain *water cupola.*

water gap A deep, narrow, low-level pass penetrating to the base of and across a mountain ridge, and through which a stream flows; esp. a narrow gorge or ravine cut through resistant rocks by an antecedent stream. Example: Delaware Water Gap, Penna. Cf: *wind gap.*

water gate A Scottish term for a natural watercourse.

waterhead The *headwater* of a stream.

water hemisphere That half of the Earth containing the bulk (about six-sevenths) of the ocean surface; it is mostly south of the equator with New Zealand as its approx. center. Cf: *land hemisphere.*

water-holding capacity The smallest value to which the water content of a soil can be reduced by gravity drainage (ASCE, 1958, term 414).

water hole [ice] A hole in the surface of a mass of ice.

waterhole [geog] (a) A natural hole, hollow, or small depression that contains water; esp. in an arid or semiarid region. (b) A spring in the desert. (c) A natural or artificial pool, pond, or small lake.

water horizon *aquifer.*

water humus Humus formed from organic deposits, allochthonous and autochthonous and including both plant and animal matter, in rivers, lakes, and seas.

water-laid Deposited in or by water; sedimentary.

water level [phys sci] (a) The level assumed by the surface of a body of standing water. (b) *sea level.*

water level [oil] The surface below which rock pores are virtually saturated with water and above which there is an exploitable concentration of hydrocarbons; e.g. oil-water interface, gas-water interface. Syn: *water surface; water table* [oil]; *edge-water line.*

water level [surv] An instrument that shows the level by means of the surface of water in a trough or in the legs of a U-tube.

water level [grd wat] *water table.*

water leveling A type of leveling in which relative elevations are obtained by observing heights with respect to the surface of a body of still water (as of a lake).

water-level mark (a) A small, horizontal, wave-cut "terrace" on an inclined surface of unconsolidated sediment, marking a former water level. (b) *watermark.*

water-level weathering In coastal areas, a lateral widening of a pool of water, due to the alternate wetting and drying in and around it which causes the banks to retreat. By this process, beaches are created that are unrelated to the stands of the sea, esp. in porous or readily eroded rock (Russell, 1968, p.94).

water lime (a) *hydraulic lime.* (b) A limestone from which hydraulic lime is made; *hydraulic limestone.*---Also spelled: *waterlime.*

waterline [phys sci] The common boundary between the water surface and any immersed structure.

waterline [coast] (a) The migrating line of contact between land and sea; the *shoreline.* (b) The actual line of contact, at a given time, between the standing water of a lake or sea and the bordering land. (c) The *limit of backrush* where waves are present on a beach.---Also spelled: *water line.*

waterline [grd wat] *water table.*

waterlogged Said of an area in which water stands near, at, or above the land surface, so that the roots of all plants except hydrophytes are drowned and the plants die.

water mass A synonym of *water type*; also, a mixture of two or more water types. See also: *central water; equatorial water; intermediate water; deep water; bottom water; surface water* [oceanog].

water mouth A Scottish term for the mouth of a river.

water of capillarity *capillary water.*

water of compaction *Rejuvenated water* originating from the destruction of interstices by compaction of sediments.

water of crystallization Water in a crystal structure that is chemically combined but may be driven off by heat; molecular water, e.g. in gypsum: $CaSO_4.2H_2O$.

water of dehydration Water that has been set free from its chemically combined state. Cf: *water of crystallization.*

water of dilation *water of supersaturation.*

water of hydration Water that is chemically combined in a crystalline substance to form a hydrate, but that may be driven off by heat.

water of imbibition (a) The amount of water a rock can contain above the water table. (b) *water of saturation.*

water of retention That part of the interstitial water in a sedimentary rock which remains in the interstices under capillary pressure and under conditions of unhindered flow; usually called connate water.

water of saturation The amount of water that can be absorbed by water-bearing material without dilation of that material. Syn: *water of imbibition.*

water of supersaturation Water in excess of that required for saturation; water in sedimentary materials that are inflated or dilated, such as plastic clay or flowing mud whose particles are not in contact and are separated by water. Syn: *water of dilation.*

water opal (a) *hyalite.* (b) Any transparent precious opal.

water opening Any break in sea ice which reveals the water; e.g. a *lead.*

water parting A term suggested by Huxley (1877, p.18) to replace *watershed* in the original meaning of that term (i.e. a drainage *divide*).

water-plasticity ratio *liquidity index.*

water pocket A small, bowl-shaped depression on a bedrock surface, where water may gather; esp. a *water hole* in the bed of an intermittent stream, formed at the foot of a cliff by the action of falling water when the stream is in the flood stage. Syn: *tinaja.*

waterpower The power of moving or falling water, once used to drive machinery directly, as by a water wheel, but now more commonly used to generate electricity by means of a power generator coupled to a turbine through which the water passes. Cf: *hydroelectric power; hydropower; white coal* [*water*].

water quality The fitness of water for use, being affected by physical, chemical, and biological factors.

water race A *race* or watercourse.

water regimen *regimen* [water].

water reserve (a) An area of land set aside for feeding streams that are used for water supply. (b) A general term for a quantity or source of water regarded as a supplemental or reserve supply.

water resources A general term referring to the occurrence, replenishment, movement, discharge, quantity, quality, and availability of water.

water-rolled Said of round and smooth sedimentary particles that have been rolled about by water.

waters The marine *territorial waters* of a nation or state.

water sand A porous sand with high or total water content. Cf: *oil sand.*

water sapphire (a) A light-colored blue sapphire. (b) An intense-blue variety of cordierite occurring in waterworn masses in certain river gravels (as in Ceylon) and sometimes used as a gemstone. Syn: *saphir d'eau.* (c) A term applied to waterworn pebbles of topaz, quartz, and other minerals from Ceylon.

watershed (a) A term used in Great Britain for a drainage *divide.* (b) A *drainage basin.*---Etymol: probably German *Wasserscheide,* "water parting". The original and "correct" meaning of the term "watershed" signifies a "water parting" or the line, ridge, or summit of high ground separating two drainage basins. However, the usage of the term, esp. in the U.S. and by several international agencies, has been changed to signify the region drained by, or contributing water to, a stream, lake, or other body of water. The term, when used alone, is ambiguous, and unless the context happens to suffice without aid from the word itself, "the uncertainty of meaning entailed by this double usage makes the term undesirable" (Meinzer, 1923, p.16).

watershed area The total area of the watershed above the discharge-measuring points. Symbol: A. Cf: *basin area.*

watershed leakage Seepage or flowage of water underground from one drainage basin to an outlet in a neighboring drainage basin or directly to the sea.

watershed line A drainage *divide.*

watershed management Administration and regulation of the aggregate resources of a drainage basin for the production of water and the control of erosion, streamflow, and floods. Also includes the operational functions.

water sky Dark or gray streaks or patches in the sky near the horizon or on the underside of low clouds, indicating the small amount of light reflected from water features in the vicinity of sea ice; darker than *land sky.*

watersmeet A meeting place of two streams.

waterspace The ecologic and social interplay (unity) among land, water, and social institutions having identical boundaries in time and space (Padfield & Smith, 1968).

watersplash A shallow ford in a stream.

water spreading Artificial recharge of ground water by spreading water over an absorptive area. Generally broadened to include all methods of artificial recharge involving surficial structures or shallow furrows, pits, or basins, as opposed to injection of water through wells or deep pits or shafts.

waterstead An English term for a stream bed.

water stone A mineral name that has been applied to moonstone, hyalite, enhydros, and jade.

waterstone [geol] An English term applied to a stratum whose surface has a watery appearance (like watered silk) and generally understood to express the water-bearing quality of the rock (Woodward, 1887, p.227); specif. the Waterstones, certain flaggy micaceous sandstones and marls in the Keuper of the English Midlands, from which some water is available. The term should not be used in place of "aquifer" (Stamp, 1966, p.485).

water supply A source or volume of water available for use; also, the system of reservoirs, wells, conduits, treatment facilities, etc., required to make the water available and usable. Syn: *water system.*

water surface *water level* [oil].

water system (a) *river system.* (b) *water supply.*

water table [oil] *water level* [oil].

water table [grd wat] The surface between the *zone of satura-* tion and the *zone of aeration;* that surface of a body of unconfined ground water at which the pressure is equal to that of the atmosphere. Syn: *waterline* [grd wat]; *water level* [grd wat]; ground-water table; ground-water surface; plane of saturation; saturated surface; level of saturation; phreatic surface; ground-water level; free-water elevation; free-water surface.

water-table aquifer *unconfined aquifer.*

water-table cement *ground-water cement.*

water-table divide *divide* [grd wat].

water-table map A map that shows the upper surface of the zone of saturation, by means of contour lines.

water-table mound *ground-water mound.*

water-table rock Rock cemented at or near the level of the water table, e.g. beachrock exposed by a retreating shoreline, rock outcropping along eroding stream banks, *ground-water cement*; a specific type of hardpan (Russell, 1968).

water-table stream Concentrated flow of ground water at the level of the water table in a structure or mass of rock having high permeability.

water-table well A well tapping unconfined ground water. Its level may, but does not necessarily, lie at the level of the water table. Cf: *artesian well; nonflowing well; shallow well.*

water tagging The introduction of foreign substances (tracers) into water to detect its movement by measurement of the subsequent location and distribution of the tracers.

water trap A chamber or part of a cave system that is filled with water, due to the dipping of the roof or ceiling below the water level. Cf: *siphon.* Syn: *trap* [speleo]. Partial syn: *sump.*

water type A body of sea water having characteristic temperature and salinity; *water mass* is sometimes used synonymously.

water-vascular system A hydrostatic circulatory system of canals or vessels peculiar to echinoderms, containing a watery fluid analogous to blood, and used to control the movement of tube feet and perhaps also functioning in excretion and respiration. It consists of a stone canal, a ring canal, radial canals, and tube feet. See also: *ambulacral system.*

water vein (a) Ground water in a crevice or fissure in dense rock. (b) A term popularly applied to any body of ground water, in part because dowsers commonly describe water as occurring in veins. Hence, the term is little used among hydrologists.

waterway (a) A way or channel, either natural (as a river) or artificial (as a canal or a grassed waterway), for conducting the flow of water or by which water may escape. (b) A navigable body or stretch of water available for passage; a watercourse.

water well (a) A *well* that extracts water from the zone of saturation or that yields useful supplies of water. (b) A well that obtains ground-water information or that replenishes ground water. (c) A well drilled for oil but yielding only water.

water witch (a) A device for determining the presence of water, usually electrically. Cf: *divining rod.* (b) *dowser.*----Nonpreferred syn: *witch.*

water witching *dowsing.*

waterwork A tank, dock, canal lock, levee, seawall, or other engineering structure built in, for, or as a protection against water.

waterworn Smoothed or polished by the action of water.

water yield The *runoff* from a drainage basin; precipitation minus evapotranspiration (Langbein & Iseri, 1960). Syn: *water crop; runout.*

wath A dialectal term for a *ford* in a stream.

watt A syn. of *tidal flat.* Pl: *watten.* Etymol: German *Watt.*

wattenschlick Tidal or intertidal mud. Etymol: German *Wattenschlick,* "tidal-flats mud".

wattevillite A colorless mineral: $Na_2Ca(SO_4)_2 \cdot 4H_2O$ (?). It occurs in hair-like monoclinic crystals. Also spelled: *wattevilleite.*

Waucoban North American provincial series: Lower Cambrian (above Precambrian, below Albertan). Syn: *Waucobian; Georgian.*

wave [seism] A *seismic wave.*

wave [water] An oscillatory movement of water manifested by an alternate rise and fall of a surface in or on the water.

wave age The state of development of a wind-generated, water-surface wave, expressed as the ratio of wave velocity to wind velocity (measured at about 8 m above stillwater level).

wave base The depth at which wave action no longer stirs the sediments; it is usually about 10 meters. Cf: *surf base.* Syn: *wave depth.*

wave-built Constructed or built up by the action of lake or sea waves, assisted by their currents. The term is widely used in regard to *marine-built* features. Cf: *wave-cut.*

wave-built platform A syn. of *wave-built terrace.* The term is inconsistent because a platform is usually regarded as an erosional feature.

wave-built terrace A gently sloping coastal surface entirely constructed at the seaward or lakeward edge of a wave-cut platform by sediment brought by rivers or derived from the wave-cutting and drifted along the shore or across the platform and deposited in the deeper water beyond. See also: *marine terrace; beach plain.* Syn: *wave-built platform; built terrace.*

wave cross ripple mark *oscillation cross ripple mark.*

wave-current ripple mark A longitudinal *compound ripple mark* in which the material forming the crest is believed to have accumulated by the oscillation produced by wave action on a preexisting transverse (current) ripple mark (Straaten, 1953a; and Kelling, 1958, p.124).

wave-cut Carved or cut away by the action of lake or sea waves, assisted by their currents. The term is widely used in regard to *marine-cut* features. Cf: *wave-built.*

wave-cut bench A level to gently sloping, narrow surface or platform produced by wave erosion, extending outward from above the base of the wave-cut cliff and occupying all of the shore zone and part or all of the shoreface (Johnson, 1919, p.162); it is developed mainly above water level by the spray and splash of storm waves aided by subaerial weathering and rainwash. The bench may be bare, freshly worn rock or it may be temporarily covered by a beach; it may end abruptly or grade into the abrasion platform. See also: *wave-cut platform.* Syn: *shore platform; beach platform; high-water platform.*

wave-cut cliff A cliff, esp. a *sea cliff*, produced by the breaking away of rock fragments after horizontal and landward undercutting by waves.

wave-cut notch A *notch* produced along the base of a sea cliff by wave erosion.

wave-cut pediment A *wave-cut platform* formed by erosion of a fault-scarp shoreline (Hinds, 1943, p.792). The term is not recommended.

wave-cut plain *wave-cut platform.*

wave-cut platform (a) A theoretically horizontal, but actually uniformly and gently sloping surface produced by wave erosion, extending far out to the sea or lake from the base of the wave-cut cliff. It represents both the wave-cut bench and the abrasion platform. Syn: *wave-cut terrace; cut platform; erosion platform; wave platform; shore platform; wave-cut plain; strandflat.* (b) A term sometimes used more restrictedly as a syn. of *abrasion platform.*

wave-cut terrace A syn. of *wave-cut platform.* The term is inconsistent because a terrace is usually regarded as a constructional feature.

wave delta *washover.*

wave depth *wave base.*

wave drift The net translation of water in the direction of wave movement, caused by the open orbital motion of water particles with the passage of each surface wave.

wave energy The capacity of waves to do work. The energy of a wave system is theoretically proportional to the square of the wave height, and the actual height of the waves (being a relatively easily measured parameter) is a useful index to wave energy: a high-energy coast is characterized by breaker heights greater than 50 cm and a low-energy coast is characterized by breaker heights less than 10 cm. Most of the wave energy along equilibrium beaches is used in shoaling and in sand movement. See also: *coastal energy.*

wave erosion *marine abrasion.*

wave-etched shoreline A relatively straight shoreline made irregular by differential wave erosion acting on coastal materials of varying resistance.

wave forecasting The theoretical determination of future wave characteristics, usually from observed or predicted meteorological phenomena such as wind velocities, duration, and fetch. Cf: *wave hindcasting.*

wave front [seism] A curve denoting the position of a traveling seismic disturbance at successive times; the surface of equal time elapse from the point of detonation to the position of the resulting outgoing signal at any given time after the charge has been detonated. Also spelled: *wavefront.*

wave front [optics] In optics, the locus of all the points reached by light that is sent outward in all directions from a point. In an isotropic medium, the wave front is a sphere; if the light is constrained to a beam, the wave front will be a plane surface. Cf: *wave normal.* Syn: *wave surface.*

wave-front chart A diagram used in seismology, consisting of a series of lines showing equal times from the point of detonation.

wave generation The creation and growth of waves by natural or mechanical means, as by a wind blowing over a water surface for a certain period of time.

wave guide A region, usually a layer, in the atmosphere, ocean, or solid Earth that tends to channel seismic energy.

wave hindcasting The calculation of ocean-wave characteristics that probably occurred for some past situation at a given time and place, based on historic synoptic wind charts giving direction, velocity, and duration of winds. Cf: *wave forecasting.*

wave interference ripple mark *oscillation cross ripple mark.*

wave line *swash mark.*

wavellite A white to yellow, green, or black orthorhombic mineral: $Al_3(PO_4)_2(OH)_3 \cdot 5H_2O$. It occurs usually in small hemispherical aggregates exhibiting a strongly developed internal radiating structure. See also: *fischerite.*

wavemark [coast] *swash mark.*

wavemark [sed] A *ripple mark* produced by wave action during the period of deposition.

wave meter An instrument for measuring and recording wave heights.

wave normal In optics, the line at a given point perpendicular to a plane that is tangent to the surface of a light wave at that point. Cf: *wave front [optics].*

wave of oscillation *oscillatory wave.*

wave of translation A water wave in which the individual particles of water are significantly displaced in the direction of wave travel. Cf: *oscillatory wave.*

wave ogive A curved undulation in the surface of a glacier forming an arc convex downslope, usually repeated periodically downstream and often merging into a *dirt-band ogive* formed at the base of certain icefalls. Cf: *Forbes band; dirt band [glaciol].* Syn: *glacier wave.*

wave path *path.*

wave platform *wave-cut platform.*

wave pole A device for measuring the heights and periods of water-surface waves, consisting of a graduated, weighted vertical pole below which a disk is suspended at a depth sufficiently deep for the wave motion associated with deep-water waves to be negligible. Syn: *wave staff.*

wave ray *orthogonal.*

wave refraction (a) The process by which a water wave, moving in shallow water as it approaches the shore at an angle, tends to be turned from its original direction. The part of the wave advancing in shallower water moves more slowly than the part still advancing in deeper water, causing the wave crests to become more nearly, but rarely exactly, parallel to the shoreline. (b) The bending of wave crests by currents.

wave ripple mark *oscillation ripple mark.*

wave spectrum (a) The description of wave energy with respect to frequency by mathematical function. The square of the wave height is related to the potential energy of the surface of the sea. (b) A graph that shows, for a region of the ocean, the distribution of wave height with respect to frequency.

wave staff *wave pole.*

wave steepness The ratio of wave height of a water wave to its wavelength. A wave with a ratio of 1/25 to 1/7 has "great" steepness; one with a ratio of less than 1/100 has "low" steepness. Syn: *steepness.*

wave surface *wave front* [optics].

wave velocity That velocity at which a wave train moves forward; also, a loosely used term for *phase velocity* or for *group velocity.*

wave wash The erosion of shores or embankments by the lapping or breaking of waves; esp. the erosion of levees during flood periods.

wave-worn Smoothed or polished by, or showing attrition from, wave action.

wavy bedding Bedding characterized by undulatory bounding surfaces. Syn: *rolling strata.*

wavy extinction *undulating extinction.*

wax A solid, noncrystalline hydrocarbon of mineral origin such as ozocerite and paraffin wax and which is composed of the fatty acid esters of the higher hydrocarbons.

waxing development *accelerated development.*

waxing slope The upper part of a hillside surface, tending to become convex by being rounded off immediately above an escarpment, having an angle that increases continuously downslope as the hillside is worn back (Wood, 1942). Ant: *waning slope.* Syn: *convex slope.*

wax opal Yellow opal with a waxy luster.

waxy A type of mineral luster that is soft like that of wax, as seen in chalcedony, for example.

waylandite A white mineral: $(Bi,Ca)Al_3(PO_4,SiO_4)_2(OH)_6$.

way up The upward direction, orientation, or position of a succession of strata. See also: *right way up.*

W-chert Chert nodules formed by weathering (Dunbar & Rodgers, 1957, p.249).

W-dolostone Dolostone produced by weathering (Dunbar & Rodgers, 1957, p.239).

weak ferromagnetism *Antiferromagnetism* in which the opposing atomic magnetic moments do not cancel perfectly, so that there is a weak, spontaneous macroscopic magnetization. An example of a mineral displaying weak ferromagnetism is α Fe_2O_3, hematite. Syn: *parasitic ferromagnetism.*

weal A descriptive field term for one of the crisscrossing raised bands, 5-7.5 cm wide, occurring on a more or less evenly patterned sedimentary surface (Donaldson & Simpson, 1962, p.74). The bands in cross section are almost semicircular.

wear The reduction of size or change of shape of clastic fragments by one or more of the mechanical processes of abrasion, impact, or grinding (Wentworth, 1931, p.24-25). See also: *wearing.*

wearing The gradual destruction of a landform or surface by friction or attrition. Cf: *wasting.* See also: *backwearing; downwearing; wear.*

weather [meteorol] n. The condition of the Earth's atmosphere, specif. its temperature, barometric pressure, wind velocity, humidity, clouds, and precipitation.----adj. A syn. of *windward.*

weather [geol] v. To undergo changes, such as the discoloration, softening, crumbling, or pitting of rock surfaces, brought about by exposure to the atmosphere and its agents. See also: *weathering.*

weather chart *weather map.*

weather coal Brown coal that has been weathered and displays bright colors.

weathered ice Sea ice that has undergone a gradual elimination of surface irregularities by thermal and mechanical processes of removal and addition of material; ice whose hummocks and pressure ridges are smoothed and rounded.

weathered iceberg An iceberg that has undergone prolonged ablation, which generally gives it a very irregular but rounded shape.

weathered layer In seismology, that zone of the Earth that is immediately below the surface, characterized by low wave velocities.

weathering The destructive process or group of processes constituting that part of erosion whereby earthy and rocky materials on exposure to atmospheric agents at or near the Earth's surface are changed in character (color, texture, composition, firmness, or form), with little or no transport of the loosened or altered material; specif. the physical disintegration and chemical decomposition of rock that produce an in-situ mantle of waste and prepare sediments for transportation. Most weathering occurs at the surface, but it may take place at considerable depths, as in well-jointed rocks that permit easy penetration of atmospheric oxygen and circulating surface waters. Some authors restrict weathering to the destructive processes of surface waters occurring below 100°C and 1 kb; others broaden the term to include biologic changes and the corrasive action of wind, water, and ice. Syn: *demorphism; clastation.*

weathering correction In explosion seismology, a time correction applied to reflection and refraction data to correct for the travel time of the observed signals in the low-velocity or weathered layer. Syn: *low-velocity correction.*

weathering escarpment An escarpment developed where gently dipping sedimentary rocks of varying resistance are subjected to degradation; the term is not too appropriate because mass-wasting, sheetwash, and stream erosion "are equally or more important" than weathering (Thornbury, 1954, p.71-72).

weathering front The interface of fresh and weathered rock; a term proposed by Mabbutt (1961) to replace *basal surface.*

weathering index A measure of the weathering characteristics of coal, according to a standard laboratory procedure. Syn: *slacking index.*

weathering map In seismic prospecting, a map on which the low-velocity layer (weathering layer) is plotted and contoured to show areal variations.

weathering out The exposing of relatively resistant rock as the surrounding softer rock is reduced by weathering.

weathering-potential index A measure of the degree of susceptibility to weathering of a rock or mineral, computed from a chemical analysis, and expressed as the mol-percentage (the percentage of any constituent divided by its molecular weight) ratio of the sum of the alkalies and alkaline earths (less combined water) to the total mols present exclusive of water (Reiche, 1943, p.66).

weathering rind An outer crust or layer on a pebble, boulder, or other rock fragment, formed by weathering.

weathering shot In seismic shooting, detonation of a small explosive charge in the weathering or low-velocity layer to determine the characteristics of that layer by refraction data. It is seldom used now, however, since the information is derived from uphole shooting times and from first-arrival reflection times of a reflection shot. Syn: *short shot; poop shot.*

weathering velocity That velocity with which a seismic *P* wave travels through the low-velocity or weathered layer. Cf: *subweather velocity.*

weather map A chart that is used to show temperature, pressure, precipitation, wind direction and velocity, air masses, and fronts of a given area. Syn: *weather chart.*

weather pit A shallow depression on the flat or gently sloping summit of large exposures of granite or granitic rocks (as in the Sierra Nevada, Calif.), attributed to strongly localized solvent action of impounded water (Matthes, 1930, p.64); diameter is 30-45 cm, and depth ranges up to 15 cm. Cf: *rock tank; oven.*

weather shore A shore lying to the windward or in the direction from which the wind is blowing, and thereby exposed to strong wave action. Ant: *lee shore.*

weber The mks (meter kilogram second) unit of magnetic flux. Weber $=10^8$ *maxwell.* Syn: *volt-second.*

weberite A pale-gray mineral: Na_2MgAlF_7.

Weber number The relationship of the forces of inertia to those of surface energy, expressed as the product of density times velocity of flow squared times length divided by surface energy. It is important in the movement of water in porous media and capillaries (Chow, 1964, p.7-5).

Weber's line A hypothetical boundary between the Asian and Australasian biogeographic regions. It generally coincides with the Australian-Papuan shelf and is sometimes used in preference to *Wallace's line.* Named after Max Weber (d. 1937), German zoologist.

websterite [**mineral**] *aluminite.*

websterite [**ign**] A pyroxenite composed chiefly of ortho- and clinopyroxene.

weddellite A mineral (calcium oxalate): $CaC_2O_4.2H_2O$. It is found as small isolated crystals in urinary calculi and in mud at the bottom of Weddell Sea, Antarctica. Cf: *whewellite.*

wedge [**optics**] (a) *optical wedge.* (b) *quartz wedge.*

wedge [**paleont**] A five-sided part of a heterococcolith, having two dimensions subequal and the third dimension small at one edge and approaching zero at the other edge.

wedge [**stratig**] (a) The shape of a stratum, vein, or intrusive body that thins out; specif. a wedge-shaped sedimentary body, or *prism.* (b) *sand wedge.*

wedge ice *foliated ground ice.*

wedge-out n. The edge or line of *pinch-out* of a lensing or truncated rock formation.

wedge out v. To become progressively thinner or narrower to the point of disappearance; to *thin out.*

wedge theory A corollary of the contracting Earth theory which supposes that shrinking of the crust breaks it into wedge-shaped blocks, which are uplifted and laterally compressed along their margins, resulting in two-sided orogenic structures.

wedgework The action of rock disintegration by the wedgelike insertion of agents such as roots and esp. ice. See also: *frost wedging.* Also, the results of wedgework action.

wedging The splitting, breaking, or forcing apart of a rock as if by a wedge, such as by the growth of salt or mineral crystals in interstices; specif. *frost wedging.*

weedia A type of stromatolite consisting of algal crusts that are nearly flat or essentially parallel to the bedding and appearing in cross section as a branching network of bedding planes that join in an irregular manner (Pettijohn, 1957, p.222 & 399).

weeksite A yellow orthorhombic mineral: $K_2(UO_2)_2(Si_2O_5)_3.4H_2O$.

weeping rock A porous rock from which water oozes.

weeping spring A spring of small yield; a syn. of *seepage spring.*

Wegener hypothesis *continental displacement.*

wegscheiderite A triclinic mineral: $Na_5(CO_3)(HCO_3)_3$.

wehrlite [**mineral**] A mineral: BiTe. It is a native alloy of bismuth and tellurium, and was earlier formulated Bi_2Te_3. Syn: *mirror glance.*

wehrlite [**ign**] A peridotite composed chiefly of olivine and clinopyroxene with common accessory opaque oxides.

weibullite A mineral: $Pb_4Bi_6S_9Se_4$.

Weichsel The term applied in northern Europe to the fourth and last glacial stage of the Pleistocene Epoch, after the Saale glacial stage; equivalent to the *Würm* and *Wisconsin.* Adj: *Weichselian.*

weigelith An amphibole- and enstatite-bearing peridotite.

weight dropping A method used in seismic prospecting, in which a heavy weight is dropped to create a source of reflected waves. See also: *thumper.* Syn: *thumping.*

weighting A statistical method for expressing the relative importance of different measurements for a set of data; the purposeful addition of statistical bias.

weilerite A mineral: $BaAl_3(AsO_4)(SO_4)(OH)_6$ (?).

weilite A mineral: $CaHAsO_4$.

weinschenkite (a) A white mineral: $YPO_4.2H_2O$. Syn: *churchite.* (b) A dark-brown variety of hornblende low in ferrous iron and high in ferric iron, aluminum, and water.

weir (a) A small dam in a stream, designed to raise the water level or to divert its flow through a desired channel; e.g. a "leaping weir". (b) A notch in a levee, dam, embankment, or other barrier across or bordering a stream, through which the flow of water is regulated; e.g. a "wasteweir".

weisbachite A variety of anglesite containing barium.

weiselbergite An altered glassy basalt characterized by labradorite, augite, and iron-ore phenocrysts in a groundmass of plagioclase and augite microlites and interstitial glass. Cf: *shastalite.*

Weisenboden *meadow soil.*

Weissenberg pattern The pattern of x-ray diffraction spots obtained from a single crystal using a Weissenberg camera and monochromatic radiation, by a moving-film method which enables unambiguous indexing of all diffractions for any properly oriented single crystal.

weissite A bluish-black mineral: Cu_5Te_3.

welded contact Any intimate, closely fitting contact between two bodies of rock that have not been disrupted tectonically; e.g. a contact between two parallel limestone beds separated by a paraconformity. The term does not imply a preliminary softening by heat.

welded dike A *secundine dike* whose boundaries have become obscured by continued mineral growth of the granitic country rock into the intrusion.

welded texture A texture of pyroclastic rocks, especially those derived from ash flows and núees ardentes, that is formed by the heat and pressure of still-plastic particles as they are deposited.

welded tuff A pyroclastic rock that has been indurated by the combined action of the heat retained by particles, the weight of overlying material, and hot gases. It contains rhyolite and obsidian and appears banded or streaky. The term includes both air-fall and *ignimbrite* deposits. Cf: *sillar.* Syn: *tuff lava.*

welding (a) Consolidation of sediments (esp. of clays) by pressure resulting from the weight of superincumbent material or from earth movement, characterized by cohering particles brought within the limits of mutual molecular attraction as water is squeezed out of the sediments (Tyrrell, 1926, p.196). (b) The diagenetic process whereby discrete crystals and/or grains become attached to each other during compaction, often involving pressure solution and solution transfer (Chilingar et al, 1967, p.322).

welinite A mineral: $(Mn^{+4},W)_{1-x}(Mn^{+2},W,Mg)_{3-y}Si(O,OH)_7$.

well [**eng**] A hollow cylinder of reinforced concrete, steel, timber, or masonry constructed in a pit or hole in the ground that reaches to hardpan or bedrock and used as a support for a bridge or building. Also, the pit or hole in which the well is built.

well [gem] The small, dark, nonreflecting spot in the center of a fashioned stone, esp. in that of a colorless diamond cut improperly (too thick).

well [oil] A borehole or shaft sunk into the ground for the purpose of obtaining oil and/or gas from an underground source or of introducing water or gas under pressure into an underground formation. See also: *oil well; gas well.*

well [water] (a) An artificial excavation (pit, hole, tunnel), generally cylindrical in form and often walled in, sunk (drilled, dug, driven, bored, or jetted) into the ground to such a depth as to penetrate water-yielding rock or soil and to allow the water to flow or to be pumped to the surface; a *water well.* (b) A term originally applied to a natural spring or to a pool formed by or fed from a spring; esp. a mineral spring. (c) A term used chiefly in the plural form for the name of a place where mineral springs are located or of a health resort featuring marine or freshwater activities; a spa.

well-bedded Said of a *bedded* rock whose beds are numerous and very clearly defined.

well bore The hole made by a well. Also spelled: *wellbore.*

well casing (a) The *casing* or tubular lining of a bored or drilled well. (b) The apparatus used in sinking a well (esp. an oil well).

well cuttings The *cuttings* produced in the process of drilling a well.

well-data system A system designed for computer storage and retrieval of well data, including programs necessary to update the file.

well-graded (a) A geologic term for *well-sorted.* (b) An engineering term pertaining to a *graded* soil or unconsolidated sediment with a continuous distribution of particle sizes from the coarsest to the finest, in such proportions that the successively smaller particles almost completely fill the spaces between the larger particles.---Ant: *poorly graded.*

wellhead The source from which a stream flows; the place in the ground where a spring emerges.

wellhole (a) The pit or shaft of a well. (b) A large-diameter (about 15 cm) vertical hole used in a quarry or opencast pit for taking heavy explosive charges in blasting.

well log A *log* obtained from a well, showing such information as resistivity, radioactivity, spontaneous potential, and acoustic velocity as a function of depth; esp. a lithologic record of the rocks penetrated.

well logging (a) The act or process of making or recording a well log. (b) A general term for the various methods or techniques in which a subsurface formation is analyzed by the making of a well log.

well point A hollow vertical tube, rod, or pipe terminating in a perforated pointed shoe and fitted with a fine-mesh screen, connected with others in parallel by way of a header pipe to a drainage pump, and driven into an excavation to remove underground water or to lower the water table and thereby minimize flooding during construction and strengthen the ground.

well record A concise statement of the available data regarding a well, such as a *scout ticket*; a full history or day-by-day account of a well, from the day the well was surveyed to the day production ceased.

well-rounded Said of a sedimentary particle whose original faces, edges, and corners have been destroyed by abrasion and whose entire surface consists of broad curves without any flat areas; specif. said of a particle with no secondary corners and a roundness value between 0.60 and 1.00 (Pettijohn, 1957, p.59). The original shape is suggested by the present form of the particle. Also, said of the *roundness class* containing well-rounded particles.

well sample A sample of well cuttings, usually one showing the lithologic character of a single stratum penetrated by the well.

well shooting In seismic prospecting, a method of determining the average velocity as a function of depth by lowering a geo-

phone into a hole and recording energy from shots fired from surface shotholes (Sheriff, 1968).

well site The *location* selected for the sinking of a well.

wellsite A zeolite mineral: $(Ba,Ca,K_2)Al_2Si_3O_{10}.3H_2O.$

well-sorted Said of a *sorted* sediment that consists of particles all having approximately the same size and that has a sorting coefficient less than 2.5. Based on the phi values associated with the 84 and 16 percent lines, Folk (1954, p.349) suggests sigma phi limits of 0.35-0.50 for well-sorted material. Ant: *poorly sorted.* Syn: *well-graded.*

wellspring The *fountainhead* of a stream.

wellstrand A Scottish term for a stream flowing from a spring.

well ties The comparison of seismic datum points with geologic datum points at well locations; it is the measure of the reliability of the seismic map.

well water Water obtained from a well; water from the zone of saturation or from a perched aquifer; *phreatic water.*

weloganite A mineral: $Sr_5Zr_2(CO_3)_9.4H_2O.$

welt A nongenetic germ used by Bucher (1933) for a raised part of the Earth's crust of any size with a distinct linear development. Cf: *furrow.*

wenkite A mineral: $(Ba,Ca)_9Al_9Si_{12}O_{42}(OH)_5(SO_4)_2.$

Wenlockian European stage: Middle Silurian (above Tarannon, below Ludlovian).

wennebergite A quartz-bearing porphyry containing orthoclase, biotite, and quartz phenocrysts in a microlitic, chloritic groundmass of apatite and sphene.

Wenner array An electrode array in which the four electrodes are in-line and equally spaced, and in which the outer pair is used to inject current into the ground while the inner pair is used to measure potential differences.

Wentworth grade scale An extended version of the *Udden grade scale,* adopted by Chester K. Wentworth (1891-1969), U.S. geologist, who modified the size limits for the common grade terms but retained the geometric interval or constant ratio of 1/2 (Wentworth, 1922). The scale ranges from clay particles (diameter less than 1/256 mm) to boulders (diameter greater than 256 mm). It is the grade scale generally used by North American sedimentologists. See also: *phi grade scale.*

Werfenian *Scythian.*

wermlandite A mineral: $Ca_2Mg_{14}(Al,Fe)_4(CO_3)(OH)_{42}.29H_2O.$

Wernerian adj. Of or relating to Abraham G. Werner (1749-1817), German geologist and mineralogist, who classified minerals according to their external characters, advocated the theory of *neptunism,* and postulated a worldwide age sequence of rocks based on their lithology. Also, said of one who is a great, but dogmatic, teacher of geology.---n. An adherent of Wernerian beliefs; a *neptunist.*

wernerite A syn. of common *scapolite,* a specific mineral of the scapolite group intermediate between meionite and marialite.

werneritite A light-colored igneous rock composed almost entirely of scapolite (Thrush, 1968, p.1231).

wesselite A hypabyssal rock containing anomite, barkevikite, titanaugite, haüyne, and nepheline.

west (a) The general direction of sunset; the direction toward the left of one facing north. (b) The place on the horizon where the Sun sets when it is near one of the equinoxes. (c) The cardinal point directly opposite to east. Abbrev: W. (d) The direction opposite to that of the Earth's daily rotation and its revolution around the Sun. (e) The point of the horizon having an azimuth of 270 degrees measured clockwise from north.

west coast desert A *coastal desert* found on the western edge of continents and occurring in the *tropical-desert* latitude, i.e. near the Tropic of Cancer or Capricorn. However, the temperature fluctuation, both annually and daily, is much less than for inland tropical deserts (Strahler, 1963, p.335).

westerwaldite An extrusive rock composed of phenocrysts of

serpentinized olivine, in some cases with augite rims, in a groundmass of labradorite, sanidine, augite, and biotite, with interstitial nepheline.

westgrenite A mineral: $(Bi,Ca)(Ta,Nb)_2O_6(OH)$.

westing A *departure* (difference in longitude) measured to the west from the last preceding point of reckoning; e.g. a linear distance westward from the north-south (vertical) grid line that passes through the origin of the grid.

Westphal balance In mineral analysis, a balance used to determine specific gravity of the *heavy liquid*. Syn: *beam balance*.

Westphalian European stage: Upper Carboniferous (Middle Pennsylvanian; above upper Namurian, below Stephanian).

westward drift A component of the *secular variation* of the Earth's magnetic field; the movement is about $0.2°$ per year of the broad-scale departures of the actual geomagnetic field from an ideal dipole field.

wet analysis A method of estimating the effective diameters of particles smaller than 0.06 mm by mixing the sample in a measured volume of water and checking its density at intervals with a sensitive hydrometer (Nelson, 1965, p.512).

wet assay Any type of assay procedure that involves liquid as a means of separation. Cf: *dry assay*.

wet avalanche *wet-snow avalanche*.

wet beach The lower part of a beach that is covered by ordinary wave water. Ant: *dry beach*.

wet blasting Abrasion or attrition effected by the impact of water against an exposed surface; e.g. the formation of an *aquafact* by wave action.

wet-bulb temperature The lowest temperature to which air can be cooled by the evaporation of water into it. During the process, the heat required for the evaporation is supplied by the cooling of the air.

wet chemical analysis Any of the methods for chemical determinations using water or other liquids as part of the process.

wet gas A natural gas containing liquid hydrocarbons. Cf: *dry gas*.

wet playa A playa that is soft under foot, having a thin and puffy surface that is coated with white efflorescent salts indicating the active discharge of near-surface ground water by evaporation (Thompson, 1929); a *salina*. It is underlain by loose granular silt, salt crystals, and moist clay. Cf: *dry playa*. Syn: *moist playa*.

wet snow Deposited snow that contains appreciable liquid water. Cf: *dry snow*.

wet-snow avalanche An avalanche composed of damp snow or wet snow, and caused by a sudden spring thaw which releases downslope a single blanket-like mass of heavy snow. Due to friction, it is the slowest-moving of the snow avalanches, and it can stop suddenly when its momentum is lost. Cf: *damp-snow avalanche*. Syn: *wet avalanche; ground avalanche*.

wettability The ability of a liquid to form a coherent film on a surface, due to the dominance of molecular attraction between the liquid and the surface over the cohesive force of the liquid itself.

wetted perimeter (a) The length of the wetted contact between a stream of flowing water and its containing conduit or channel, measured in a plane at right angles to the direction of flow. (b) The length of the perimeter of a conduit below the water surface. (c) The entire perimeter of a conduit flowing full. The wetted perimeter is used when computing the *hydraulic radius* (ASCE, 1962).

wetting front *pellicular front*.

wet unit weight The *unit weight* of soil solids plus water per unit of total volume of soil mass, irrespective of the degree of saturation. Syn: *mass unit weight*.

whaleback (a) A large mound or hill having the general shape of a whale's back, esp. a smooth elongated ridge of desert sand having a rounded crest and ranging widely in size (about

300 km long, 1-3 km wide, and perhaps 50 m high). It forms a coarse-grained platform or pedestal built up and left behind by a succession of longitudinal (seif) dunes along the same path. Syn: *sand levee*. (b) A rounded, elongated rock mass commonly granite, found in tropical areas associated with tors. (c) A *roche moutonnée*, often of granitic composition, as those in Canada and Finland.

wharf Any structure (as a *pier* or *quay*) built out from the shore and serving as a berthing place for vessels.

wheel A holothurian sclerite in the form of a vehicular wheel consisting of a *hub*, a *rim*, and *spokes*.

Wheelerian North American stage: upper Pliocene (above Venturian, below Hallian).

wheelerite A yellowish variety of retinite that is soluble in ether and that fills fissures in, or is thinly interbedded with, lignite beds in northern New Mexico.

wheel ore The mineral *bournonite* esp. when occurring in wheel-shaped twin crystals.

wherryite A pale-green mineral: $Pb_4Cu(CO_3)(SO_4)_2(Cl,OH)O$ (?).

whetstone Any hard, fine-grained, naturally occurring, usually siliceous rock suitable for sharpening cutting instruments (such as razors, knives, and mechanic's tools); e.g. *novaculite*.

whewellite A white or colorless monoclinic mineral (calcium oxalate): $CaC_2O_4.H_2O$. It occurs as a warty and somewhat opaline incrustation on marble. Cf: *weddellite*.

whiplash A smooth-surfaced flagellum (without mastigonemes) of some algae and protozoans, having a long rigid basal part and a short thinner distal region.

whipstock (a) A long, slender, wedge-shaped, steel device with a concave groove along its inclined face, dropped into or placed in an oil well and used during drilling to deflect and guide the drill bit away from the vertical (such as to one side of an obstruction) and toward the direction in which the inclined grooved surface is facing. (b) The procedure or technique of using a whipstock to drill a directional well. Syn: *whipstocking*.

whirl ball A spindle-shaped, tubular, ellipsoidal, or spherical mass of fine sandstone embedded in silt, its long axis being vertical or steeply inclined. It is attributed to vortices in mudflows.

whirlpool A body of water moving rapidly in a circular path of relatively limited radius. It may be produced by a current's passage through an irregular channel or by the meeting of two opposing currents. Cf: *eddy; maelstrom*.

whirl zone A zone of transition between a slump sheet and the overlying strata.

whistling sand A *sounding sand*, often found on a beach, that gives rise to a high-pitched note when stepped on or struck with the hand, the sound apparently resulting from the translation of grain over grain. Syn: *squeaking sand; musical sand*.

Whitbian Stage in Great Britain: upper Lower Jurassic (above Domerian, below Yeovilian).

white agate A term sometimes applied to white or whitish chalcedony.

white alkali Sodium sulfate or other salt that may develop as a crust over an *alkali soil*. Cf: *black alkali*.

white antimony *valentinite*.

white band A layer in a glacier consisting of ice that is white and opaque because it contains many small air bubbles. Cf: *blue band* [*glaciol*].

white-bedded phosphate A term used in Tennessee for a phosphatic limestone characterized by partial replacement of calcite by calcium phosphate, and by a matrix consisting of cryptocrystalline quartz. It occurs in regular bands alternating with thinner beds of chert. Cf: *brown rock; hard-rock phosphate*.

whitebody A body which reflects all incident wavelengths totally. In natural materials this is possible only over a finite wavelength range. Cf: *blackbody; graybody.*

whitecap The white froth on the crest of a wave; it is caused by wind.

white chert A light-colored *chert,* or chert proper, as distinguished from the dark variety or *black chert.*

white clay *kaolin.*

white coal [coal] *tasmanite.*

white coal [water] A fanciful term meaning *waterpower, hydroelectric power,* or *hydropower,* from the French term 'houille blanche''.

white cobalt (a) *cobaltite.* (b) *smaltite.*

white copperas (a) *goslarite.* (b) *coquimbite.*

whitedamp A term for carbon monoxide in coal mines. Cf: *blackdamp; afterdamp; firedamp.*

white earth A siliceous, earthy material that is used as a pigment in paint.

white feldspar *albite.*

white garnet (a) A translucent variety of grossular garnet, sometimes resembling white jade in appearance. (b) *leucite.*

white gold A pale alloy of gold that resembles silver or platinum; esp. gold alloyed with a high proportion of nickel or palladium to give it a white color, with or without other alloying metals (such as tin, zinc, or copper).

white ice (a) Sea ice of not more than one winter's growth, and a thickness of 30-70 cm. The term "thin" *first-year ice* is synonymous. (b) Coarsely granular, porous glacier ice formed by compaction of snow and appearing white. Cf: *black ice; blue ice.*

white iron ore A syn. of *siderite* (ferrous-carbonate mineral).

white iron pyrites *marcasite* [mineral].

white lead ore *cerussite.*

white mica A light-colored mica; specif. *muscovite.*

white mundic *arsenopyrite.*

white nickel (a) *nickel-skutterudite.* (b) *rammelsbergite.*

white olivine *forsterite.*

white opal A form of *precious opal* with a predominant body color of any light color, as distinguished from *black opal*; e.g. a pale bluish-white gem variety of opal.

whiteout The diffusion of daylight by multiple reflection between fallen snow and overcast clouds, so that the horizon and surface features are impossible to discern.

white pyrite *marcasite* [mineral].

white pyrites (a) *arsenopyrite.* (b) *marcasite* [mineral].

Whiterock North American (California, Nevada, Oklahoma) stage: lowermost Middle Ordovician (above Lower Ordovician, below Marmor) (Cooper, 1956).

white sand Quartzitic sand, pure enough to resist heat, used in steel furnaces.

white sapphire The colorless or clear, pure variety of crystallized corundum.

white schorl *albite.*

white stone A clear, colorless imitation gem, resembling the diamond.

white tellurium (a) *sylvanite.* (b) *krennerite.*

white top In coal mining, light gray shale that occurs above a coal seam and that is gradational between the coal and the overlying, darker shale roof. It may be arenaceous, and is usually unlaminated.

white trap A term used in the Midland Valley of Scotland for an intrusive igneous rock, usually of basic composition, that has been bleached at the contact with coal or other carbonaceous rock. It is created where gaseous hydrocarbons and carbon dioxide, resulting from the local breakdown of the sedimentary organic matter, invade the igneous body at a late stage during its cooling and convert the ferromagnesian and feldspar minerals into a mixture of carbonates and clay minerals.

white vitriol *goslarite.*

whitleyite An achondritic stony meteorite of the *aubrite* class, containing fragments of black chondrite.

whitlockite A mineral: $Ca_9(Mg,Fe)H(PO_4)_7$. Its formula was formerly given as: $Ca_3(PO_4)_2$.

whole rock Adj. used in analytical geology to indicate that a portion of rock rather than individual minerals was examined. In the rubidium-strontium age method the rock may have remained a closed system for rubidium and strontium isotopes whereas the constituent minerals did not. Thus, a calculated age for the whole rock would give the apparent age of formation whereas the individual minerals might give discordant ages. This whole-rock, closed-system feature does not hold true for all isotopic systems. Syn: *total rock.*

whorl (a) One of the turns of a spiral or coiled shell; specif. a single complete turn through 360 degrees of a gastropod shell, of a cephalopod conch, or of a foraminiferal test. See also: *body whorl.* Syn: *volution.* (b) An arrangement of two or more anatomical parts or organs of one kind in a circle around the same point on an axis; e.g. a circle of equally spaced branches around the stem of a plant, arranged like the spokes of a wheel. Syn: *verticil.*

whorl coccolith One of the modified coccoliths forming a whorl about the naked pole in nonmotile coccolithophores exhibiting dimorphism (such as *Ophiaster*).

whorled In plant morphology, pertaining to attachment of three or more parts, e.g. leaves, at a single node. Cf: *alternate; opposite; acyclic.*

whorl height The height of an ammonoid whorl measured at right angles to the maximum width (the horizontal distance between points located between ribs or spines on opposite whorl sides), comprising the distance from the middle of the venter to the middle of the dorsum plus the depth of the impressed area (TIP, 1959, pt.L, p.6). In practice, the "oblique whorl height" is commonly used, consisting of the distance from the umbilical seam to the middle of the venter.

whorl section A transverse section of a cephalopod whorl.

whorl side The *flank* of a cephalopod conch; esp. the lateral wall of an ammonoid whorl between the umbilical seam and venter.

wiborgite *rapakivi.*

wich A term used in England for a damp meadow or a marshy place, esp. where salt is found or has been worked; also, a salt pit. Syn: *wych.*

Wichita orogeny A name used by van der Gracht (1931) for the first major phase of deformation in the Wichita and Ouachita orogenic belts of southern Oklahoma. In the Wichita belt it is dated by adjacent strata as early Pennsylvanian (Morrow); in the Ouachita belt, it includes a Mississippian phase that produced the great flysch body of the Stanley and Jackfork formations.

wichtisite *tachylite.*

wick A Scottish term for a small inlet or creek.

wickenburgite A mineral: $Pb_3Al_2CaSi_{10}O_{24}(OH)_6$.

wickmanite A mineral: $MnSn(OH)_6$.

widespread *spread.*

wide water A local term applied in northern Michigan to a wide, shallow expanse of water backed up behind a natural dam or produced by a widening in the course of a stream.

widiyan Var. of *wadi* used in the north African deserts (Stone, 1967, p.264).

Widmanstatten structure A triangular pattern observed on polished and etched surfaces of iron meteorites (octahedrites), composed of parallel bands or plates of kamacite bordered by taenite and intersecting one another in two, three, or four directions. The kamacite bands, arranged parallel to the octahedral planes in the host taenite, are produced by exsolution from an originally homogeneous taenite crystal. As the bands become finer (thinner), the nickel content increases. Named after Aloys B. Widmanstatten (1753?-1849), Austrian mineralogist who discovered the structure in 1808. Also spelled: *Widmanstätten structure.* Syn: *Widmanstatten figure; Widmanstatten pattern.*

width On a brachiopod, the maximum dimension measured perpendicular to the plane of symmetry (at right angles to the length and thickness or height).

Wiechert-Gutenberg discontinuity *Gutenberg discontinuity.*

Wien's displacement law (a) The statement that when the temperature of a radiating blackbody increases, the wavelength corresponding to maximum radiance decreases in such a way that the product of the absolute temperature and wavelength is constant (Wien, 1894, p.132-165). (b) The statement that the wavelength of the most intense radiation emitted by a blackbody is inversely proportional to the absolute temperature of the body (Swenson et al, 1971, p.299).

wiggle stick *divining rod.*

wightmanite A colorless triclinic mineral: $Mg_9B_2O_{12}.8H_2O$.

wiikite A poorly defined mineral high in niobium, tantalum, titanium, and yttrium, found to be variable mixtures of euxenite and obruchevite.

Wilcoxian *Sabinian.*

wildcat n. *wildcat well.* adj. Said of a risky or unproven venture---a company, mine, or well---in the mineral resources industry.

wildcat well An *exploratory well* drilled for oil or gas on a geologic structure without sufficient geologic information, or in an unproven territory or in a horizon that has never produced or is not known to be productive in the general area, regardless of how good or how poor the prospects may be. Syn: *wildcat.*

Wilderness North American stage: Middle Ordovician (above Porterfield, below Trentonian; it includes uppermost Black River and Rockland rocks) (Cooper, 1956). See also: *Blackriverian.*

wilderness An area or tract of land that is uncultivated and uninhabited by man.

wilderness area An area set aside by government for preservation of natural conditions for scientific or recreational purposes; e.g. an area of a forest land or of a barren plain. See also: *natural area.*

wildflysch A type of *flysch* facies representing a mappable stratigraphic unit displaying large and irregularly sorted blocks and boulders resulting from tectonic fragmentation, and twisted, contorted, and confused beds resulting from slumping or sliding under the influence of gravity. It consists of shales, clays, and sandstones, often extremely coarse-grained. The term was first applied by Kaufmann (1886) in the Alps.

wild land Uncultivated land, or land that is unfit for cultivation; e.g. a wasteland or a desert.

wild river (a) A river whose shores and waters remain essentially in a virgin condition, unmodified by man. (b) A torrential river.

wild snow Deposited snow of low density, less than 30 kg/m; newly fallen snow that is fluffy or powdery and unstable, accumulating in a lightweight mass, and falling only during a dead calm at low temperatures. Cf: *sand snow; powder snow.* Syn: *dust snow.*

wilkeite A rose-red or yellow mineral of the apatite group, containing hydroxyl, and in which the phosphate is partly replaced by carbonate, sulfate, or silicate: $Ca_5(SiO_4,PO_4,SO_4)_3(O,OH,F)$.

wilkmanite A mineral: Ni_3Se_4.

willemite A rhombohedral mineral: Zn_2SiO_4. It is a minor ore of zinc and commonly contains manganese. Willemite varies in color from white or greenish yellow to green, reddish, and brown; it exhibits an intense bright-yellow fluorescence in ultraviolet light.

willemseite A talc-like mineral: $(Ni,Mg)_3Si_4O_{10}(OH)_2$.

williamsite A massive, yellowish- or apple-green, impure variety of antigorite resembling jade in appearance and used for decorative purposes. It usually contains specks of chromite.

willyamite A pseudocubic mineral: $(Co,Ni)SbS$, with Co greater than Ni.

wilsonite [mineral] A purplish-red mineral consisting of an aluminosilicate of magnesium and potassium, and representing an altered scapolite.

wilsonite [pyroclast] A tuff composed of fragments of pumice and andesite in a matrix of vitric and granular material (Holmes, 1928, p.240).

Wilson technique In x-ray diffraction analysis, a method of mounting film in a cylindrical x-ray powder camera which enlarges the area for recording back reflection diffractions on both sides of the entry port. Cf: *Straumanis camera method.*

wilting coefficient *wilting point.*

wilting percentage *wilting point.*

wilting point The point at which the water content of the soil becomes too low to prevent the *permanent wilting* of plants. As originally introduced and, to a certain extent, today, the point at which a soil-water deficiency produces any degree of wilting is the wilting point. Syn: *wilting coefficient; wilting percentage.* Cf: *temporary wilting.*

wiluite (a) A green variety of grossular garnet. (b) A greenish variety of vesuvianite.

wind (a) Naturally moving air, of any direction or velocity. (b) More specifically, a meteorologic term for that component of air which moves parallel to the Earth's surface. Its direction and velocity can be measured.

wind abrasion A process of erosion whereby windblown particles (of rock material or snow) scour and wear away exposed surfaces of any kind. Syn: *wind corrasion.*

wind avalanche *dry-snow avalanche.*

wind corrasion *wind abrasion.*

wind crust A type of *snow crust* formed by the packing of previously fallen snow into a hard layer by wind action. Cf: *wind slab.*

wind current *drift* [oceanog].

wind-cut stone *ventifact.*

wind-deposition coast A coast built out into the sea by sand dunes advancing in the direction of the prevailing winds; generally found on the lee side of a sandy neck of land.

wind drift (a) *drift* [oceanog]. (b) "That portion of the total vector drift of sea ice from which the effects of the current have been subtracted" (Baker et al, 1966, p.183). (c) The average direction of the wind over a period of time.

wind-driven current *drift* [oceanog].

wind erosion Detachment, transportation, and deposition of loose topsoil by wind action, esp. in duststorms in arid or semiarid regions or where a protective mat of vegetation is inadequate or has been removed. See also: *deflation.*

wind-faceted stone *windkanter.*

wind gap (a) A shallow notch in the crest or upper part of a mountain ridge, usually at a higher level than a water gap. (b) A former *water gap*, now abandoned (as by piracy) by the stream that formed it; a pass that is not occupied by a stream. Syn: *dry gap.*---Syn: *air gap; wind valley.*

wind-grooved stone *ventifact.*

windkanter A *ventifact*, usually highly polished, bounded by one or more, curved or nearly flat, smooth faces or facets ending or intersecting in one or more sharp edges or angles. The faces may be cut at different times, as when the wind changes seasonally or the pebble is undermined and turned over on its flattened face thereby permitting another face to be cut. Etymol: German *Windkanter*, "one having wind edges". See also: *einkanter; zweikanter; dreikanter; parallelkanter.* Syn: *faceted pebble; wind-faceted stone.*

window [geomorph] The opening under a natural bridge (Gregory, 1917, p.134).

window [paleont] An opening in the skeleton of an ebridian, such as a "lower window" between the opisthoclades, a "middle window" between the mesoclades and the actines, and an "upper window" between the proclades.

window [tect] An eroded area of a thrust sheet that displays the rocks beneath the thrust sheet. Syn: *fenêtre, fenster.*

window [**river**] A part of a river surrounded by river ice, remaining unfrozen during all or part of the winter, and caused by local inflow of warm water or by turbulence or a strong current.

wind packing The compaction of snow by wind action.

wind polish *desert polish.*

wind-polished stone A *ventifact* having a *desert polish.*

wind ridge A ridge of snow formed by the deposition of blowing snow at right angles to the direction of the prevailing wind. Its lee side is the steeper. Cf: *sastruga.*

wind-rift dune A sand dune produced in a shrub-covered area by a strong wind of constant direction, typically marked by a gap or "rift" at the very tip or downwind end of a hairpin-shaped sand rim (the hairpin or elongated chevron is opened toward the wind), and extending up to 2 km in length and about 100 m in width (Melton, 1940, p.129-130); it is usually found along a seashore, rarely on a desert. The term is also applied to the "doublet" of parallel sand ridges resulting from the formation of the "rift". The spelling "windrift" is not recommended owing to possible confusion with "wind-drift".

wind ripple [**sed**] One of many wave-like, asymmetric undulations produced on a sand surface by the saltatory movement of particles by air currents (wind) and occasionally found in eolian rocks; it is generally longer and of smaller height than an aqueous ripple mark, but is similar in having a steep lee side (facing downcurrent) and a gentle windward side (facing upcurrent). See also: *sand ripple; granule ripple.* Cf: *antiripple.*

wind ripple [**snow**] One of a series of wave-like formations on a snow surface, lying at right angles to the wind direction, and formed as snow grains are moved along by the wind.

windrow (a) A low bank, heap, or other accumulation of material, formed naturally by the wind (as a drift of snow) or the tide (as a pile of beach shells), or artificially (as a ridge of construction material along a road or building site). (b) Part of a *slick* that has broken up into a narrower and shorter band or bands at wind speeds greater than about 7 knots, its long axis always oriented along the wind direction.

windrow ridge A term used by Tanner (1960, p.482) for a shallow-water ripple mark that is parallel with and directly beneath a windrow on a water surface, that consists of a straight, tapered ridge becoming narrower and shorter in the downwind direction, and that cuts regularly across a preexisting ripple mark.

wind scale A numerical scale for expressing the different degrees of wind velocity, in a manner suitable for easy communication and plotting on a weather map. The *Beaufort wind scale* is the most used.

wind scoop A saucer-like depression in the snow near an obstruction (such as a tree or rock), caused by the eddying action of the deflected wind.

wind-scoured basin *deflation basin.*

wind-scoured stone *ventifact.*

wind set-up (a) The vertical rise of the still-water level on the leeward side of a body of water, caused by the force of wind on the surface of the water; the difference between the leeward and windward sides of the form. It is a type of *meteorologic tide.* Syn: *wind tide.*

wind shadow The area in the lee of an obstacle where air motion is not capable of moving material (such as sand in saltation), thereby trapping it when it falls there; the zone that is gradually filled with sand drift during the formation of a sand dune, and which determines the shape of the dune. Syn: *shadow zone.*

wind-shadow dune A longitudinal *umbracer dune.*

wind-shaped stone *ventifact.*

wind slab A layer of snow that is packed tightly by the wind as the snow is being deposited; if it adheres poorly to underlying snow it may fail rapidly when broken at any point, and thus is one possible cause of a *wind-slab avalanche.* Cf: *wind crust;*

slab. Also spelled: *windslab.*

wind-slab avalanche An avalanche started by the dislodging or slipping of a rigid *wind slab* from the underlying snow. Cf: *slab avalanche; loose-snow avalanche.*

windsorite A light-colored quartz monzonite with a minor amount of biotite. It is named after Windsor, Vermont, U.S.A.

wind stress The force per unit area of the wind acting on a water surface to produce waves and currents; its magnitude depends on wind speed, air density, and roughness of water surface.

wind sweep The trough-shaped part of the windward slope of an advancing dune, up which the main wind currents pass.

wind tide A syn. of *wind set-up,* used for a lake, reservoir, or smaller body of water.

wind valley *wind gap.*

windward adj. (a) Said of the side (as of a shore or reef) located toward the direction from which the wind is blowing; facing the wind, such as the "windward slope" of a dune, up which sand moves by saltation. (b) Said of a tide moving toward the direction from which the wind is blowing.----n. The part or side (as of a hill or shore) from which the wind is blowing; the side facing the wind. Also, the direction from which the wind is blowing, or the direction opposite to that toward which the wind is blowing. Ant: *leeward.*

wind wave A wind-generated wave; a *sea wave,* or part of *sea.*

wind-worn stone *ventifact.*

wineglass valley A valley resembling in cross section a goblet or a tulip champagne glass: it flares broadly open at its upper end (where it has a cup-shaped or a wide, funnel-shaped head), then narrows sharply to form a constricted, gorge-like lower section (the stem of the wineglass), and flares open again on a spreading alluvial fan (the base of the wineglass). The valley commonly forms at right angles to a fault scarp in an arid region. Syn: *goblet valley; hourglass valley.*

wing [**geomorph**] The forward extending, outer end of a dune; a *horn.*

wing [**paleont**] A solid or fenestrated extension from the side wall of the shell of a nasselline radiolarian.

wing [**palyn**] *vesicle.*

wing bar A sandbar that partly crosses the entrance to a bay or the mouth of a river.

wing dam *pier dam.*

winged headland A headland having spits extending from both sides in opposite directions. It may be produced by waves that are unable to move material to the bayhead. Syn: *winged beheadland.*

Winkler method In oceanography, a chemical method of determining the amount of dissolved oxygen in sea water.

winnowing The selective sorting, or removal, of fine particles by wind action, leaving the coarser grains behind. The term is often applied to removal by or sorting in water, but the term *washing* is more appropriate for such a process.

winter balance The change in mass of a glacier from the minimum value at the beginning of a balance year to the following maximum value; sometimes called *apparent accumulation* or (erroneously) *net accumulation.*

winter berm A berm built on the backshore by the uprush of large storm waves during the winter; it is landward of, and somewhat higher than, the *summer berm.* See also: *storm berm.*

winterbourne A regular *bourne* that breaks out every year at the same spot in the floor of a dry valley; specif. one that flows only or chiefly in winter, when the water table rises above the valley floor, as in the chalk regions of southern England.

winter ice A term formerly used for sea ice of less than one winter's growth, and a thickness varying from 15 cm to 3.7 m (12 ft); it is now being replaced by the term *first-year ice.*

winter moraine (a) A moraine "pushed up" along the *winter* *line* during a single winter. (b) A minor end moraine formed under water; examples are found in Sweden (Gravenor & Kupsch, 1959, p.54).

winter season In glaciology, that period of a year when the balance of a glacier increases to the maximum for the year. This is the part of the year when, on the average, accumulation exceeds ablation. Cf: *summer season.* Syn: *accumulation season.*

winter-talus ridge A wall-like *protalus rampart* formed on the floor of a cirque by rapid frost action that dislodged boulders from a snowbank-occupied cirque wall before the summer heat melted the snow across which the talus had rolled. Syn: *nivation ridge.*

wire adj. A syn. of *capillary*; said of native metals, e.g. wire silver.

wire line A general term for any rope made of steel wires twisted together to form the strands and used with a drill machine to hoist drill pipes, casing, and other borehole-drilling equipment; esp. a steel *cable.* See also: *sand line.* Also spelled: *wireline.*

wire-line coring Cutting and removing of a core sample (of soft sandstone or shale) with the drill bit still in place and without withdrawing and dismantling the drill pipes, as by raising the core in a retractable core barrel and lowering the same or an alternate barrel into place inside the drill pipe.

wire-line test A procedure for measuring the potential productivity of an individual oil reservoir by means of a tool lowered into a borehole by a wire line, in which a sample of fluid (rather than a rate of flow) is obtained. It is faster than *drill-stem test* and is used in unconsolidated, noncarbonaceous, sandy deposits.

wire strain gage An instrument consisting of a fine wire used to indicate minute changes in strain by detecting corresponding changes in electrical resistance via elongation of the wire.

wiry A syn. of *capillary*; said of native metals.

Wisconsin Pertaining to the fourth glacial stage (and the last definitely ascertained, although there appear to be others) of the Pleistocene Epoch in North America, following the Sangamon interglacial stage; it began about 85,000 ± 15,000 years ago and ended about 7,000 years ago. Substages, in order of age (oldest first): Iowan, Cary, Mankato, Valders, and Cochrane. See also: *Würm.* Syn: *Wisconsinan.*

Wisconsinan (a) The uppermost Pleistocene stage in Illinois and Wisconsin. (b) *Wisconsin.*

wiserite A mineral: $Mn_4B_2O_5(OH,Cl)_4$.

witch A nonpreferred syn. of *water witch.*

witching stick *divining rod.*

withamite A red to yellow variety of epidote containing a little manganese and occurring in andesites in Glencoe, Scotland. Cf: *piemontite.*

withdrawal The act of removing water from a source for use; also, the amount removed.

witherite A yellowish- or grayish-white orthorhombic mineral of the aragonite group: $BaCO_3$.

witness butte *butte témoin.*

witness corner A monumented survey point near a *corner* and usually on a line of the survey, established as a reference mark where the true corner is inaccessible or cannot be monumented or occupied; e.g. a post set near the corner of a mining claim, with the distance and direction of the true corner indicated thereon.

witness mark A physical structure (such as a post, rock, stake, or tree) placed at a known distance and direction from a property corner, instrument, or other survey station, to aid in its recovery and identification; e.g. a blazed tree on the bank of a river, indicating the corner which is at the intersection of some survey line with the center line of the river and therefore cannot be marked directly.

witness point A monumented station on a line of the survey, used to perpetuate an important location more or less remote from, and without special relation to, any regular corner.

witness rock *zeuge.*

witness tree *bearing tree.*

wittichenite A steel-gray to tin-white mineral: Cu_3BiS_3.

wittite A lead-gray mineral: $Pb_5Bi_6(S,Se)_{14}$.

wobbling of the pole An expression sometimes used to describe the period polar motion identified by a Chandler term (approximately a 14-month period) and an annual term.

wodanite A variety of biotite containing titanium.

wodginite A black mineral: $(Ta,Nb,Sn,Mn,Fe)_{16}O_{32}$.

woebourne A *bourne* that is regarded in some English localities as appearing only when some disaster was about to happen.

woehlerite *wöhlerite.*

wöhlerite (a) A yellow or brown mineral: $NaCa_2(Zr,Nb)Si_2O_8(O,OH,F)$. Cf: *lavenite.* (b) A name for organic matter in carbonaceous chondrites.---Syn: *woehlerite.*

wolchonskoite *volkonskoite.*

wold A range of hills produced by differential erosion from inclined sedimentary rocks; a *cuesta.*

Wolfcampian North American provincial series: lowermost Permian (above Virgilian of Pennsylvanian, below Leonardian).

wolfeite A mineral: $(Fe,Mn)_2(PO_4)(OH)$. It is isomorphous with triploidite.

wolfram (a) *wolframite.* (b) The metallic element *tungsten.*

wolframine (a) *tungstite.* (b) *wolframite.*

wolframite (a) A brownish or grayish-black mineral: $(Fe,Mn)WO_4$. It is isomorphous with and intermediate between huebnerite and ferberite, and occurs in monoclinic crystals (commonly twinned so as to imitate orthorhombic tabular forms) and in granular masses or columnar aggregates (as in pneumatolytic veins near granite masses and associated with tin ores). Wolframite is the principal ore of tungsten. See also: *tungstenite.* Syn: *wolfram; wolframine; tobacco jack.* (b) A name applied to an isomorphous mineral series consisting of the end members huebnerite and ferberite and of wolframite.

wolfram ocher *tungstic ocher.*

wolframoixiolite A mineral: $(Nb,W,Ta,Fe,Mn)_3O_6$.

wolfsbergite *chalcostibite.*

wolgidite A leucitite that contains leucite, magnophorite, diopside, and minor amounts of olivine and phlogopite.

wollastonite A triclinic mineral: $CaSiO_3$. It is dimorphous with parawollastonite. Wollastonite is found in contact-metamorphosed limestones, and occurs usually in cleavable masses or sometimes in tabular twinned crystals; it may be white, gray, brown, red, or yellow. It is not a pyroxene. Symbol: Wo. Syn: *tabular spar.*

Wollaston prism In an optical system, a double-image prism consisting of two right-angled calcite prisms that produce two perpendicular beams of plane-polarized light.

wollongite *wollongongite.*

wollongongite A coal-like shale similar to *torbanite.* It is named from its type locality, Wollongong, New South Wales, Australia. Also spelled: *wollongite; wallongite.*

wolsendorfite A red or orange-red orthorhombic mineral: $(Pb,Ca)U_2O_7.2H_2O$.

wood Technically, *xylem*; more popularly, the hard, fibrous xylem of trees and shrubs (Fuller & Tippo, 1949, p.976).

wood agate A term used for *agatized wood*, esp. agate formed by petrifaction of wood.

Woodbinian North American (Gulf Coast) stage: Upper Cretaceous (above Washitan, below Eaglefordian).

wood coal *woody lignite.*

wood copper A fibrous variety of *olivenite.*

woodenite An extrusive rock containing olivine and augite in an alkalic, brown, glassy groundmass. Its chemical composition is similar to that of absarokite.

wood hematite A finely radiated variety of hematite exhibiting alternate bands of brown or yellow or varied tints.

woodhouseite A colorless rhombohedral mineral: $CaAl_3$(PO_4)(SO_4)(OH)$_6$. It is isomorphous with svanbergite, corkite, and hinsdalite.

wood iron ore A fibrous variety of limonite from Cornwall, Eng.

wood opal A variety of common opal that has filled the cavities in, and replaced the organic matter of, wood and that often preserves the original features of the wood. See also: *opalized wood*. Syn: *xylopal; lithoxyl*.

woodruffite A mineral: $(Zn,Mn)_2Mn_5O_{12}.4H_2O$.

woodstone *silicified wood*.

wood tin A nodular or reniform, massive, brownish variety of cassiterite having a concentric structure of radiating fibers resembling dry wood in appearance. Syn: *dneprovskite*.

woodwardite A bluish mineral: $Cu_4Al_2(SO_4)(OH)_{12}.2-4H_2O$ (?).

woodyard *forest bed*.

woody lignite Lignite that shows the fibrous structures of wood. Cf: *earthy lignite*. Syn: *xyloid lignite; xyloid coal; wood coal; board coal; bituminous wood*.

woody peat *fibrous peat*.

woody plant A vascular and usually perennial plant with a large development of xylem.

wool An English term for a sandy shale or shaly flagstone with irregular curly bands or bedding.

woolpack A term used in Shropshire, Eng., for a concretionary, ball-like mass of crystalline limestone occurring in the Wenlockian limestones; a *ballstone*.

Worden gravimeter A compact, small, temperature-compensated gravity meter in which a system is held in unstable equilibrium about an axis, so that an increase in the gravitational pull on a mass at the end of a weight arm causes a rotation opposed by a sensitive spring. The meter weighs 5 pounds and has a sensitivity of better than 0.1 mgal.

work v. To undergo gradual movement, such as heaving, sliding, or sinking; said of rock materials.

workings The system of openings or excavations made in mining or quarrying for the purpose of exploitation; esp. the area where the ore is actually mined.

World Data Centers Centers for the collection, exchange, and general availability of data from various geophysical disciplines, e.g. solid-earth geophysics, solar-terrestrial geophysics, oceanography, glaciology, meteorology, tsunamis. They were originally established for the International Geophysical Year, but are being continued under the auspices of the International Council of Scientific Unions (ICSU).

world geodetic system Any system which connects the major continental geodetic datums and land masses into a unified Earth-centered network.

world point A term used by Kobayashi (1944, p.745) for a single restricted outcrop regarded as representative of a geologic province or part of the world.

world rift system A major tectonic element of the Earth, consisting of midoceanic ridges and their associated trenches, such as those along the Mid-Atlantic Ridge. It is believed to be the locus of tensional splitting and upwelling of magma that has resulted in *sea-floor spreading*. Cf: *rift* [*struc geol*].

world time *Geologic time* as indicated by the life range of a single cosmopolitan fossil species (Kobayashi, 1944, p. 745).

worm boring *worm tube*.

worm cast (a) A sinuous fossil trail of a worm, preserved as a sand cast on the bedding plane of an arenaceous rock. (b) *worm casting*.----Also spelled: *wormcast*.

worm casting A cylindrical mass of earth or mud excreted by an earthworm. Syn: *erpoglyph; worm cast*.

worm's-eye map (a) A term applied to what is more formally known as a *supercrop map* or a *lap-out map*, in reference to the pattern of formations that would be visible to an observer looking upward at the bottom of the rocks overlying a given surface. (b) A map showing overlap of sediments, or of progressive transgressions of a sea over a given surface.

worm trail A marking in a fossiliferous rock, formed by the passage of an extinct worm or worms.

worm tube (a) A fossilized tubular structure built by a marine worm and preserved in the top of a bed that had been exposed for some time as the sea floor; e.g. a *scolite*. Syn: *worm boring*. (b) A membranous tube, usually of calcium carbonate or particles of mud or sand, built on a submerged surface by a marine worm.

worobieffite *vorobyevite*.

wrench fault A *lateral fault* in which the fault surface is more or less vertical. Syn: *basculating fault; torsion fault*.

Wright biquartz wedge *biquartz plate*.

wrinkle ridge A sinuous, irregular, segmented, apparently smooth elevation occurring within the borders of a mare region of the Moon's surface and characterized by dike-like outcrops, crest-top craters, and longitudinal rifts. Wrinkle ridges are up to 35 km wide and 100 m high, and may extend for hundreds of kilometers. They probably originated in fissure eruptions or from volcanic activity along fractures. Syn: *mare ridge*.

wrist *carpus*.

W-shaped valley A valley having an inverted and faintly pan-shaped cross profile suggesting the form of the letter "W", such as the valley of a river having the highest parts of its flood plain immediately near both banks (Lane, 1923).

wulfenite A yellow, orange, or bright orange-yellow or orange-red (sometimes grayish or green) tetragonal mineral: $PbMoO_4$. It is isomorphous with stolzite. Wulfenite occurs in tabular crystals and in granular masses, and represents an ore of molybdenum. Syn: *yellow lead ore*.

Wulff net (a) A coordinate system used in crystallography to plot a *polar stereographic projection* with conservation of equal angles, such as for plotting angular relations obtained from universal-stage measurements. (b) *stereonet*.---Named after Georgij V. von Wulff (1862-1925), Polish(?) crystallographer, who introduced the net (Wulff, 1902).

Würm (a) European stage: uppermost Pleistocene (above Riss, below Holocene). (b) The fourth glacial stage of the Pleistocene Epoch in the Alps, after the Riss-Würm interglacial stage. See also: *Wisconsin; Weichsel*.---Etymol: Würm, a lake in Germany. Adj: *Würmian*.

wurtzilite A black, massive, sectile, infusible, asphaltic pyrobitumen, closely related to *uintahite*, but insoluble in turpentine, and derived from the metamorphosis of petroleum. It is found in veins in Uinta County in Utah.

wurtzite A brownish-black hexagonal mineral: $(Zn,Fe)S$. It is dimorphous with sphalerite. Wurtzite occurs in hemimorphic pyramidal crystals, or in radiating needles and bundles within lamellar sphalerite. Many polymorphs with slight variants on the wurtzite structure are known, and separate names proposed for some of these are "superfluous and not generally accepted" (Hey, 1962, 3.4.3).

wüstite A mineral: FeO. Artificially prepared specimens are characteristically deficient in iron. Also spelled: *wustite*. Syn: *iozite*.

W wave An archaic term for a surface wave that returns through the antipode of the epicenter to the detector.

wyartite A violet-black secondary mineral: $Ca_3U^{+4}(UO_2)_6$(CO_3)$_2$(OH)$_{18}.3-5H_2O$. It was erroneously called *ianthinite*.

wych Var. of *wich*.

wye level A leveling instrument having a removable telescope, with attached spirit level, supported in Y-shaped rests, in which it may be rotated about its longitudinal (or collimation) axis, and from which it may be lifted and reversed, end for end, for testing and adjustment. Cf: *dumpy level*. Syn: *Y level*.

wyomingite A pink hypabyssal lamproite containing phlogopite phenocrysts in a fine-grained groundmass of leucite and diopside; a phlogopite-leucite phonolite. Its name is derived from the state of Wyoming, U.S.A.

X

x-acline B twin law A complex twin law in feldspars with twin axis parallel to (010) and composition plane (100).

xalostocite *landerite.*

xanthiosite A yellow mineral: $Ni_3(AsO_4)_2$.

xanthite A yellowish to yellowish-brown variety of vesuvianite.

xanthochroite *greenockite.*

xanthoconite A cochineal-red, orange-yellow, or brown monoclinic mineral: Ag_3AsS_3. Cf: *proustite.*

xanthophyllite *clintonite.*

xanthosiderite *goethite.*

xanthoxenite A wax-yellow monoclinic mineral: Ca_4Fe_2-$(PO_4)_4(OH)_2.3H_2O$.

x-Carlsbad twin law A complex twin law in feldspar, having a twin axis at right angles to [001] and a composition plane (100). It is supposedly a true interpretation of the *acline-B twin law.*

xenoblast A mineral of low form energy which has grown during metamorphism without development of its characteristic crystal faces and the texture produced thereby. It is a type of *crystalloblast.* The term was originated by Becke (1903). Cf: *idioblast; hypidioblast.* Syn: *allotrioblast.*

xenocryst A crystal resembling a phenocryst in igneous rock that is foreign to the body of rock in which it occurs. See also: *disomatic.* Syn: *chadacryst.*

xenogenous A little-used syn. of *epigenetic.* Cf: *idiogenous; hysterogenous.*

xenoikic A term suggested by Cross, et al (1906, p.704) for a variety of poikilitic texture in which the relative proportions between enclosing crystals (oikocrysts) and enclosed crystals (chadacrysts) is less than 5 to 3 but greater than 3 to 5. The term is now obsolete.

xenolith An inclusion in an igneous rock to which it is not genetically related. Cf: *autolith.* Syn: *exogenous inclusion; accidental inclusion.*

xenology The dating of early events in the chronology of the planetary system on the basis of excess xenon-129 in meteorites. Xenon-129 is a decay product of radioactive iodine-129 which has a half-life of approximately 17 million years.

xenomorphic (a) Said of the texture or fabric of an igneous rock having or characterized by crystals not bounded by their own crystal faces and which have their form impressed upon them by preexisting adjacent mineral crystals. Also, said of an igneous rock with xenomorphic texture. Cf: *idiomorphic; hypidiomorphic.* (b) An obsolescent syn. of *anhedral.*----The term *xenomorphisch* was proposed by Rohrbach (1885, p.17-18) originally to describe in an igneous rock the individual mineral crystals (now known as anhedral crystals) whose mutual growths have prevented the assumption of outward crystal form. Current usage tends to apply "xenomorphic" to an igneous-rock texture or fabric characterized by anhedral crystals. Syn: *allotriomorphic; anidiomorphic; leptomorphic.*

xenomorphic-granular Said of the granular texture, of an igneous rock, characterized by a xenomorphic fabric; also said of the rock with such a texture. Syn: *allotriomorphic-granular.*

xenomorphism The state or condition of special sculpture at the umbonal region of the unattached valve of a bivalve mollusk, resembling the configuration of the substratum onto which the attached valve is or was originally fixed. It is known in the Anomiidae, Gryphaeidae, and Ostreidae, and on the right valves in oysters and on the left valves in Anomia. Erroneous syn: *allomorphism.*

xenothermal Said of a hydrothermal mineral deposit formed at a high temperature but at shallow depth; also, said of that environment. Cf: *telethermal; epithermal; mesothermal; hypothermal; leptothermal.*

xenotime A brown, yellow, or reddish tetragonal mineral:

YPO_4. It is isostructural with zircon, and often contains erbium, cerium, and other rare earths, as well as thorium, uranium, aluminum, calcium, beryllium, zirconium, or other elements. Xenotime occurs as an accessory mineral in granites and pegmatites.

xenotopic Said of the fabric of a crystalline sedimentary rock in which the majority of the constituent crystals are anhedral. Also, said of the rock (such as an evaporite, a chemically deposited cement, or a recrystallized limestone or dolomite) with such a fabric. The term was proposed by Friedman (1965, p.648). Cf: *idiotopic; hypidiotopic.*

Xeralf In U.S. Dept. of Agriculture soil taxonomy, a suborder of the soil order *Alfisol,* characterized by formation in a xeric moisture regime and by brown or red colors (SSSA, 1970). Cf: *Aqualf; Boralf; Udalf; Ustalf.*

xerarch adj. Said of an ecologic succession (i.e. a sere) that develops under *xeric* conditions. Cf: *mesarch; hydrarch.* See also: *xerosere.*

Xerert In U.S. Dept. of Agriculture soil taxonomy, a suborder of the soil order *Vertisol,* characterized by formation in a Mediterranean climate and by a once yearly opening and closing of its wide, open surface cracks (SSSA, 1970). Cf: *Torrert; Udert; Ustert.*

xeric [ecol] Said of a habitat characterized by a low or inadequate supply of moisture; also, said of an organism or group of organisms existing in such a habitat. Cf: *mesic; hydric.* See also: *xerarch.*

xeric [soil] Said of a soil moisture regime that is characteristic of the cool, moist winter and warm, dry summer of a Mediterranean climate (SSSA, 1970).

xerochore A climatic term for the part of the Earth's surface represented by waterless deserts.

xerocole *xerophilous.*

Xeroll In U.S. Dept. of Agriculture soil taxonomy, a suborder of the soil order *Mollisol,* characterized by formation in a xeric soil moisture regime. A Xeroll may have a calcic, petrocalcic, or gypsic horizon, or a duripan (SSSA, 1970). Cf: *Alboll; Aquoll; Boroll; Rendoll; Udoll; Ustoll.*

xeromorphic (a) Said of a plant characterized by the morphology of a xerophyte. (b) Said of conditions favorable for the growth of xerophilous organisms.

xerophile (a) *xerophyte.* (b) *xerophilous.*

xerophilous Said of an organism adapted to dry conditions. Syn: *xerophile; xerocole.* Cf: *xerophobous.*

xerophobous Said of a plant that cannot tolerate dry conditions. Cf: *xerophilous.*

xerophyte A plant adapted to dry conditions; a desert plant. Cf: *xerotherm; hydrophyte; mesophyte.* Syn: *eremophyte; xerophile.*

xerophytization Adaptation, esp. in the development of a species, to conditions of low moisture supply, i.e. to xeric conditions.

xerosere A sere that develops under extremely dry (i.e. xeric) conditions; a *xerarch* sere. Cf: *hydrosere; mesosere.*

xerotherm A plant adapted to hot dry conditions. Cf: *xerophyte.*

xerothermic Said of a hot, dry climate; also, pertaining to the climate of the Xerothermic postglacial interval.

Xerothermic n. A term used to designate a postglacial interval of both warmer and drier climate. It has been used as a syn. of *Altithermal* (or of Long Drought), as an interval of time distinct from and later than the Altithermal, and as a constituent subunit of the *Hypsithermal* (equivalent to *Subboreal*). These differences in part reflect dating uncertainties as well as inherent difficulties in distinguishing between moisture and temperature changes as inferable from many types of paleoclimatic records. Syn: *Xerothermal.*

Xerult In U.S. Dept. of Agriculture soil taxonomy, a suborder of the soil order *Ultisol,* characterized by a low to moderate organic-carbon content, formation in a xeric soil moisture regime, and by brownish or reddish colors (SSSA, 1970). Cf: *Aquult; Humult; Udult; Ustult.*

xiphosuran Any merostome belonging to the subclass Xiphosura, characterized by a trilobate dorsal shield. Horseshoe crabs are included in this group. Cf: *eurypterid.*

xonotlite A pale-pink, white, or gray mineral: $5CaSiO_3.H_2O$.

x-pericline twin law A complex twin law in feldspar. The *Carlsbad B twin law* is now considered to be equivalent to the x-pericline twin law.

X-ray Non-nuclear electromagnetic radiation of very short wavelength, in the interval of 0.1-100 angstroms, i.e. between that of gamma rays and ultraviolet radiation (NASA, 1966, p.51). Also spelled: *x-ray*. Syn: *Roentgen ray.*

X-ray diffraction The phenomenon of the apparent bending of X-rays when passing near opaque objects; also, the study of this phenomenon. The diffraction depends on the crystal structure (grating effect) of the substance.

X-ray diffraction pattern The characteristic interference pattern of lines obtained when X-rays are diffracted by a substance. A single crystal substance gives rise to a pattern of spots called a *Laue pattern;* and a powdered substance, a pattern of rings called an *X-ray powder pattern*. From the X-ray diffraction pattern of a substance, crystal structure and unit cell dimensions can be determined.

X-ray emission spectroscopy The qualitative study of a substance by exciting its characteristic X-ray spectrum and measuring the wavelengths present. See also: *X-ray fluorescence spectroscopy.*

X-ray fluorescence spectroscopy A type of *X-ray emission spectroscopy* in which the characteristic X-ray spectrum of a substance is produced by using X-rays of short wavelength to induce the substance to emit X-rays of a longer wavelength.

X-ray powder diffraction The phenomenon of the apparent bending of X-rays when passing through a powdered substance giving rise to an interference pattern of rings.

X-ray powder pattern The interference pattern of rings seen on photographic film when X-rays are diffracted by a powdered substance. See also: *X-ray diffraction pattern.*

X-ray scattering The phenomenon of changes in direction of X-ray transmission by interaction of the waves with objects or with the transmitting medium due to reflection, refraction, or diffraction.

X-ray spectrograph An instrument for producing, recording, and analyzing an *X-ray spectrum* by reflecting X-rays from a given sample, measuring the angle of diffraction, and thence determining the wavelengths of the X-rays. Sometimes called an *X-ray spectrometer.*

X-ray spectrometer *X-ray spectrograph.*

X-ray spectroscopy The observation of an *X-ray spectrum* and all processes of recording and measuring which go with it.

X-ray spectrum The characteristic spectrum of X-rays emitted when a substance is bombarded with cathode rays due to the diffraction, or grating, effect of that substance. The spectrum is usually recorded as a picture.

x twin law A normal twin law in feldspar.

xylain A kind of *provitrain* in which the cellular structure is derived from woody material. Cf: *periblain; suberain.*

xylem In vascular plants, a complex tissue that is involved in water conduction, storage, and strengthening. Types of cells commonly found in xylem include tracheids (for conduction and strengthening), vessels (for conduction), fibers (for strengthening), and parenchyma (for storage). Syn: *wood.*

xylinite A variety of provitrinite characteristic of xylain and consisting of xylem or lignified tissue. Cf: *suberinite; periblinite; telinite.*

xylinoid Vitrinite that occurs in noncaking subbituminous coals and lignite and that has a reflectance of less than 0.5% (American Society for Testing and Materials, 1970, p.466). Cf: *vitrinoid; anthrinoid.*

xylith A type of lignite that is composed almost entirely of anthraxylon (Parks, 1951, p.30).

xyloid coal *woody lignite.*

xyloid lignite *woody lignite.*

xylopal *wood opal.*

xylotile A delicately fibrous mineral, approximately: $(Mg, Fe^{+2})_3Fe_2^{+3}Si_7O_{20}.10H_2O$. It is a serpentine mineral derived from alteration of asbestos or chrysolite.

xylovitrain *euvitrain.*

Y

yaila A term used in central Kurdistan (of eastern Turkey) for a small, grassy, upland plain.

yakatagite A name proposed by Miller (1953, p.26) for a "conglomeratic sandy mudstone" from Yakataga, SE Alaska. It is poorly indurated, tillite-like, glaciomarine sedimentary rock containing angular gravel-sized fragments.

yamaskite A medium- to fine-grained basalt containing hornblende, titanaugite, a small amount of anorthite, and accessory biotite and iron ores; an amphibole-bearing jacupirangite.

yamatoite Hypothetical end member of the garnet group: $Mn_3V_2(SiO_4)_3$.

yardang (a) A long, irregular, sharp-crested, undercut ridge between two round-bottomed troughs, carved on a plateau or unsheltered plain in a desert region by wind erosion, and consisting of soft but coherent deposits (such as clayey sand); it lies in the direction of the dominant wind, and may be up to 6 m high and 40 m wide. Syn: *yarding; jardang.* (b) A landscape form produced in a region of limestone or sandstone by infrequent rains combined with wind action, and characterized by "a surface bristling with a fine and compact lacework of sharp ridges pitted by corrosion" (Stone, 1967, p.254).--- Etymol: ablative of Turki *yar*, 'steep bank''.

yardang trough A long, shallow, round-bottomed groove, furrow, trough, or corridor excavated in the desert floor by wind abrasion, and separating two yardangs.

yardarm carina One of oppositely placed *carinae* of a rugose coral that give the appearance of yardarms along a mast to cross sections of a septum. Cf: *zigzag carina.*

yarding Var. of *yardang.*

Yarmouth Pertaining to the second interglacial stage of the Pleistocene Epoch in North America, after the Kansan glacial stage and before the Illinoian. Etymol: Yarmouth, a town in Iowa. See also: *Mindel-Riss.* Syn: *Yarmouthian.*

yaroslavite A mineral: $Ca_3Al_2F_{10}(OH)_2 \cdot H_2O$.

yatalite A pegmatitic rock composed chiefly of uralite, albite, magnetite, sphene, and some quartz.

Yavapai A provincial series of the Precambrian in Arizona.

yavapaiite A monoclinic mineral: $KFe(SO_4)_2$.

yazoo (a) *yazoo stream.* (b) *deferred junction.*

yazoo stream A tributary that flows parallel to the main stream for a considerable distance before joining it at a *deferred junction*; esp. such a stream forced to flow along the base of a natural levee formed by the main stream. Type example: Yazoo River in western Mississippi, joining the Mississippi River at Vicksburg. Also spelled: *Yazoo stream.* Syn: *yazoo; Yazoo-type tributary; deferred tributary.*

yeatmanite A brown mineral: $(Mn,Zn)_{16}Sb_2Si_4O_{29}$.

yellow arsenic *orpiment.*

yellow coal *tasmanite* [coal].

yellow copperas *copiapite.*

yellow copper ore *chalcopyrite.*

yellow earth [mineral] Impure *yellow ocher.*

yellow earth [sed] Loess of northern (mainland) China.

yellow-green algae A group of algae corresponding to the division Chrysophyta, that owes its yellowish green to golden brown color to chromatophores of that range of pigmentation. Such algae usually have a cell wall composed of overlapping halves. Cf: *blue-green algae; green algae; brown algae; red algae.*

yellow ground Oxidized kimberlite of yellowish color found at the surface of diamond pipes (e.g. South Africa), above the zone of *blue ground.*

yellow lead ore *wulfenite.*

yellow ocher (a) A mixture of limonite usually with clay and silica, used as a pigment. See also: *yellow earth* [mineral].

Syn: *sil.* (b) A soft, earthy, yellow variety of limonite or of goethite.

yellow ore A yellow-colored ore mineral; specif. carnotite and chalcopyrite.

Yellow Podzolic soil Formerly, one of a group of zonal soils that is now considered part of the classification *Red-Yellow Podzolic soil.*

yellow pyrites *chalcopyrite.*

yellow quartz *citrine.*

yellow substance Dissolved organic matter in sea water; commonly carbohydrate-humic acids.

yellow tellurium *sylvanite.*

yenite *ilvaite.*

yentnite A coarse-grained granitic rock originally thought to contain scapolite, plagioclase, and biotite. The scapolite was later discovered to be quartz, and the name was withdrawn (Johannsen, 1939, p.288).

Yeovilian Stage in Great Britain: uppermost Lower Jurassic (above Whitbian, below Aalenian).

yield [exp struc geol] v. To undergo permanent deformation as a result of applied stress.

yield [lake] n. (a) The amount of water that can be taken continuously from a lake for any economic purpose. (b) The amount of organic matter (plant and animal) produced by a lake, either naturally or under management.

yield point *yield stress.*

yield strength A syn. of *yield stress*; the stress at which a material begins to undergo permanent deformation.

yield stress The differential stress at which permanent deformation first occurs in a material. Syn: *yield point; yield strength; threshold pressure.*

Y level *wye level.*

Y-mark A trilete *laesura* on embryophytic spores and some pollen, consisting of a three-pronged mark somewhat resembling an upper-case "Y". It is commonly also a commissure or suture along which the spore germinates. The term is also applied to analogous marks, which are not laesurae, on pollen grains.

Ynezian North American stage: lower Paleocene (above Upper Cretaceous, below Bulitian).

yoderite A purple mineral: $(Mg,Al)_8Si_4(O,OH)_{20}$.

yogoite An obsolete term originally applied to a syenite that contains approximately equal amounts of orthoclase, plagioclase, and augite.

yoked basin *zeugogeosyncline.*

yoke-pass *joch.*

Yorkian European stage: lower Upper Carboniferous (above Lanarkian, below Staffordian). It is equivalent to part of lower Westphalian.

yosemite A portion of a glacial valley, esp. in the Sierra Nevada of California, that is deeply U-shaped, with sheer walls, hanging troughs, and a wide almost level floor, and hence resembles the Yosemite Valley, Calif.

yoshimuraite An orange-brown mineral: $(Ba,Sr)_2TiMn_2(SiO_4)_2(PO_4,SO_4)(OH,Cl)$.

young [geomorph] Pertaining to the stage of *youth* of the cycle of erosion; esp. said of a stream that has not developed a profile of equilibrium, and of its valley. Syn: *youthful.*

young [struc geol] v. To *face*, in the sense "to present the younger aspect" of one formation toward another; e.g. if formation A "youngs" toward formation B, then B is younger than A unless some fold, fault, unconformity, or intrusion intervenes. Term coined by Bailey (1934, p.469) and used "as a record of observation and not of stratigraphic deduction".

young coastal ice Sea ice in the initial stage of fast-ice formation, consisting of nilas or young ice of local origin, and having a width varying from a few meters to 100-200 m from the shoreline.

Younger Dryas n. A term used primarily in Europe for an interval of late-glacial time (centered about 10,500 years ago) following the Allerød and preceding the Preboreal, during which the climate as inferred from stratigraphic and pollen data in Denmark (Iversen, 1954) deteriorated favoring either expansion or retarded retreat of the waning continental and alpine glaciers; the youngest subunit of the late-glacial Arctic interval, characterized by birch and park-tundra vegetation.-- adj. Pertaining to the late-glacial Younger Dryas interval and to its climate, deposits, biota, and events.

young ice Newly formed, flat sea ice in the transition stage between nilas and first-year ice; it is 10-30 cm thick (formerly given as 5-20 cm). Includes: *gray ice; gray-white ice.* Syn: *slud* [*ice*]; *fresh ice.*

younging A colloquial syn. of *facing* [struc geol].

young lake A lake developed during the stage of *youth.* See also: *aging.*

youngland The land surface, with its plateaus and valleys, of the youthful stage of the cycle of erosion (Maxson & Anderson, 1935, p.90).

young mountain A mountain that was formed during the Tertiary and Quaternary periods, esp. a *fold mountain* produced during the last great period of folding (i.e. the Alpine orogeny). Ant: *old mountain.*

young polar ice *second-year ice.*

Young's modulus A *modulus of elasticity* in tension or compression, involving a change of length. See also: *elastic compliance.* Syn: *stretch modulus.*

young stream A stream developed during the stage of *youth.*

youth [**topog**] The first stage of the cycle of erosion in the topographic development of a landscape or region in which the original surface or structure is still the dominant feature of the relief and the landforms are being accentuated or are tending toward complexity. It is characterized by: a few, small, widely spaced young streams; broad, flat-topped interstream divides and upland surfaces, only little modified by erosion; partly developed or poorly integrated drainage systems, with numerous swamps and shallow lakes; and rapid and progressive increase of local relief, with sharp landforms, steep and irregular slopes, and a surface well above sea level. Cf: *infancy.* Syn: *topographic youth.*

youth [**coast**] A stage in the development of a shore, shoreline, or coast characterized by an ungraded profile of equilibrium. For a shoreline of submergence: an irregular or crenulate outline, vigorous wave action, formation of sea cliffs and associated erosional forms, a steep offshore profile, and the presence of bays, promontories, offshore islands, spits, bars, and other minor irregularities. For a shoreline of emergence: a usually straight and simple outline, larger waves breaking well offshore, smaller waves coming to land to produce a nip or low cliff, and the formation of barrier beaches, lagoons, and marshes. See also: *primary* [*coast*].

youth [**streams**] The first stage in the development of a stream, at which it has just entered upon its work of erosion and is increasing in vigor and efficiency, being able everywhere to erode its channel and having not reached a graded condition. It is characterized by: an ability to carry a load greater than the load it is actually carrying; active and rapid downcutting, forming a deep, narrow, steep-walled, V-shaped valley (gorge or canyon) with a steep and irregular gradient and rocky outcrops; numerous waterfalls, rapids, and lakes; a swift current and clear water; a few, short, straight tributaries; an absence of flood plains as the stream occupies all or nearly all of the valley floor; and an ungraded bed.

youthful Pertaining to the stage of *youth* of the cycle of erosion; esp. said of a topography or region, and of its landforms (such as a plain or plateau), having undergone little erosion or being in an early stage of development. Cf: *infantile.* Syn: *young; juvenile.*

Ypresian European stage: lowermost Eocene (above Thanetian of Paleocene, below Cuisian). Syn: *Londinian.*

Y-shaped valley A valley having a cross profile suggesting the form of the letter "Y", such as a rejuvenated valley in which the grade of the river has recently been increased by uplift of the headwaters (Lane, 1923).

-yte A suffix used by Grabeau for rock names. Cf: *-lite.*

Y-tombolo A tombolo consisting of two embankments that extend shoreward from an island or seaward from the mainland and that unite "to form a single ridge before the connection is completed" (Johnson, 1919, p.315); there is a body of water between the prongs of the "Y".

yttrialite An olive-green mineral: $(Y,Th)_2Si_2O_7$. Cf: *thalenite.*

yttrocerite A violet-blue variety of yttrofluorite containing cerium.

yttrocolumbite A mineral: $(Y,U,Fe)(Nb,Ta)O_4$. Cf: *yttrotantalite.*

yttrocrasite A black mineral: $(Y,Th,U,Ca)_2Ti_4O_{11}$ (?).

yttrofluorite A mineral: $(Ca,Y)F_{2-3}$. It is a variety of fluorite containing yttrium.

yttrotantalite A black or brown mineral: $(Y,U,Fe)(Ta,Nb)O_4$. Cf: *yttrocolumbite.*

yttrotungstite A mineral: $Y_2W_5O_{18}.4H_2O$. It may contain a little thorium.

yugawaralite A zeolite mineral: $CaAl_2Si_6O_{16}.4H_2O$.

yukonite An obsolete term originally assigned to an igneous rock intermediate in composition between a tonalite and an aplite. It is named after the Yukon River, Alaska, U.S.A

Z

zanjón A Puerto Rican term for a *corridor* [karst]. Etymol: Spanish, "deep ditch".

zanoga A glacial *cirque* (Engeln, 1942, p.447).

zap crater An informal syn. of *micrometeorite crater*.

zaphrentid A simple coral with marked pinnate septal arrangement and lacking dissepiments or axial structures.

zaratite An emerald-green mineral: $Ni_3(CO_3)(OH)_4.4H_2O$. It occurs in secondary incrustations or compact masses. Syn: *emerald nickel*.

zastruga Var. of *sastruga*. Pl: *zastrugi*.

zavaritskite A mineral: BiOF.

zawn An English term for a sandy cove in a cliff (Robson & Nance, 1959, p.39) or a little inlet of the sea (Stamp, 1966, p.495).

zeasite An opal, formerly an old name for "fire opal" but now applied to "wood opal".

zebra dolomite A term used in the Leadville district of Colorado for an altered dolomite rock that shows conspicuous banding (generally parallel to bedding) consisting of light-gray coarsely textured layers alternating with darker finely textured layers. See also: *zebra rock*.

zebra layering *Rhythmic layering* in which dark and light bands alternate, reflecting changes in the amounts of pyroxene and plagioclase.

zebra limestone A term used by Fischer (1964, p.135) for a limestone banded by parallel sheet cracks filled with calcite.

zebra rock (a) A term used in the Colville district of NE Washington State for a dolomite that shows narrow banding consisting of black layers (indicative of organic matter) alternating with white, slightly coarse-grained, and somewhat vuggy layers. See also: *zebra dolomite*. (b) A term used in Western Australia for a banded quartzose rock of Cambrian age.

Zechstein European series (esp. in Germany): Upper Permian (above Rotliegende). It contains the Thuringian Stage.

zeilleriid Said of a long brachiopod loop (as in the superfamily Zeilleriacea) not attached to the dorsal septum in adult.

zellerite A lemon-yellow secondary mineral: $Ca(UO_2)-(CO_3)_2.5H_2O$.

zemannite A mineral: $(Zn,Fe)_2(TeO_3)_3Na_xH_{2-x}.yH_2O$.

Zemorrian North American stage: Oligocene and Miocene (above Refugian, below Saucesian).

zenith The point on the celestial sphere that is directly above the observer and directly opposite to the *nadir* [geodesy]. In a more general sense, the term denotes the stretch of sky overhead.

zenithal projection *azimuthal projection*.

zeolite (a) A generic term for a large group of white or colorless (sometimes red or yellow) hydrous aluminosilicates that are analogous in composition to the feldspars, with sodium, calcium, and potassium as their chief metals (rarely barium or strontium), that have a ratio of (Al + Si) to nonhydrous oxygen of 1:2, that are characterized by their easy and reversible loss of water of hydration and by their ready fusion and intumescence when strongly heated under the blowpipe, and that occur as secondary minerals (derived by hydrothermal alteration of various aluminosilicates such as feldspars and feldspathoids) filling cavities and coating cracks and joint planes in basaltic lavas and less frequently in granite and gneiss, and as authigenic minerals in sedimentary rocks. (b) Any of the minerals of the zeolite group, including natrolite, heulandite, analcime, chabazite, stilbite, mesolite, scolecite, phillipsite, laumontite, mordenite, clinoptilolite, erionite, harmotome, and other less important minerals, as well as minerals not yet classified. (c) Any of various silicates that are processed natural materials (such as glauconite) or artificial granular sodium aluminosilicates used in the base-exchange method of water softening and as gas adsorbents or drying agents. The term now includes such diverse groups of compounds as sulfonated organics or basic resins, which act in a similar manner to effect either cation or anion exchange.---Etymol: Greek *zein*, "to boil".

zeolite facies Metamorphic rocks formed in the zone of transition from diagenesis to metamorphism, estimated to take place at load pressures of 2000-3000 bars and at temperatures of 200°-300°C (with 4000 bars at 400°C as the possible extreme limit). Dynamothermal regional metamorphism must, however, be responsible for the formation of the zeolitic and associated index minerals. The term was established by Fyfe et al (1958, p.215). Syn: *laumontite-prehnite-quartz facies*.

zeolitic ore deposits Ore deposits, particularly of native copper in basalts, which have zeolites as distinctive, though not necessarily abundant, gangue minerals.

zeolitization Introduction of, or replacement by, a mineral (or minerals) of the zeolite group. This process occurs chiefly in rocks containing calcic feldspars or feldspathoids and is sometimes associated with copper deposits.

zeophyllite A white rhombohedral mineral: $Ca_4Si_3O_7(OH)_4F_2$. It sometimes contains iron.

zerdeb A term used in Algeria for an interdune hollow floored with calcareous tufa (Capot-Rey, 1945, p.397).

zero curtain A layer of ground, between the active layer and the permafrost, where a nearly constant temperature of 0°C persists for a considerable period (up to 115 days/yr) during the freezing and thawing of overlying ground (Muller, 1947, p.224).

zero distortion Conformality of shape on a map projection, in which the scale is preserved either along a line (the arc of a circle) or at a point.

zero-energy coast A coast characterized by average breaker heights of 3 cm or less. Cf: *low-energy coast*.

zero-length spring A special type of gravimeter spring so constructed that the total length is proportional to the applied force. It is used in astatic, or unstable, instruments.

zero-length spring gravimeter *LaCoste-Romberg gravimeter*.

zero meridian *prime meridian*.

zeuge A tabular mass of resistant rock left standing on a pedestal of softer rocks, resulting from differential erosion by the scouring effect of windblown sand in a desert region; it may vary in height from 2 m to 50 m. Etymol: German *Zeuge*, "witness". Pl: *zeugen*. Pron: *tzoy-geh*. See also: *mushroom rock*. Syn: *witness rock*.

zeugenberg A syn. of *butte témoin*. Etymol: German *Zeugenberg*, "witness hill".

zeugogeosyncline A parageosyncline with an adjoining uplifted area also in the craton, receiving clastic sediments; an intracratonic trough (Kay, 1945). Syn: *yoked basin*. Cf: *autogeosyncline*.

zeunerite A green secondary mineral of the autunite group: $Cu(UO_2)_2(AsO_4)_2.10-16H_2O$. It is isomorphous with uranospinite.

zeylanite *ceylonite*.

zhemchuzhnikovite A green mineral oxalate: $NaMg(Al, Fe^{+3})(C_2O_4)_3.8H_2O$. Cf: *stepanovite*.

zietrisikite Incorrect spelling of *pietricikite*, a variety of ozocerite.

zigzag carina One of not quite oppositely placed *carinae* of a rugose coral on the two sides of a septum. Cf: *yardarm carina*.

zigzag cross-bedding *chevron cross-bedding*.

zigzag fold An *accordian fold*, the limbs of which are of unequal length. Cf: *chevron fold*. See also: *knee fold*.

zigzag ridge A continuous ridge that trends first in one direction, then in another, each alternating part being roughly parallel; it is formed by the converging of two ridges in one direction, each ridge converging with a different ridge in the opposite direction. Zigzag ridges are produced in the Appalachian Mountains by truncation of plunging folds.

zigzag watershed A drainage divide through which rivers have broken by headward erosion, the divide retaining its original position between the drainage basins.

zinalsite A clay mineral: $Zn_7Al_4(SiO_4)_6(OH)_2.9H_2O$ (?).

zinc A bluish-white mineral, the native metallic element Zn. The occurrence of native zinc is unconfirmed, although it has been reported to have been found in basalt and also in auriferous sands in Victoria, Australia.

zincaluminite A light-blue mineral: $Zn_6Al_6(SO_4)_2(OH)_{26}.5H_2O$.

zinc blende *sphalerite.*

zinc bloom *hydrozincite.*

zincite A deep-red to orange-yellow, brittle mineral: (Zn, Mn)O. It is an ore of zinc, such as that occurring in New Jersey where it is associated with franklinite and willemite. Syn: *red zinc ore; red oxide of zinc; ruby zinc; spartalite.*

zinckenite *zinkenite.*

zinc-melanterite A monoclinic mineral: $(Zn,Cu,Fe)SO_4.7H_2O$.

zincobotryogen A mineral: $(Zn,Mg,Mn)Fe(SO_4)_2(OH).7H_2O$.

zincocopiapite A mineral: $ZnFe_4(SO_4)_6(OH)_2.18H_2O$.

zincrosasite A mineral: $(Zn,Cu)_2(CO_3)(OH)_2$. It is a variety of rosasite with zinc greater than copper.

zincsilite A mineral: $Zn_3Si_4O_{10}(OH)_2.nH_2O$. It is the aluminum-free end member of the montmorillonite-sauconite series.

zinc spar *smithsonite.*

zinc spinel *gahnite.*

zinc vitriol *goslarite.*

Zingg's classification A classification of pebble shapes, devised by Theodor Zingg (1905-), Swiss meteorologist and engineer, based on the graphical representation of the diameter ratio of intermediate (width) to maximum (length) plotted against the diameter ratio of minimum (thickness) to intermediate (Zingg, 1935). The classification distinguishes four shape classes: *spheroid; disk; blade; rod.*

zinkenite A steel-gray orthorhombic mineral: $Pb_6Sb_{14}S_{27}$. Syn: *zinckenite.*

zinnwaldite A mineral of the mica group: $K_2(Li,Fe,Al)_6(Si, Al_8O_{20}(OH,F)_4$. It is a pale-violet, yellowish, brown, or dark-gray variety of lepidolite containing iron, and is the characteristic mica of greisens.

zippeite A powdery or earthy and orange-yellow to bright-yellow mineral, approximately: $(UO_2)_2(SO_4)(OH)_2.4H_2O$.

zircon A mineral: $ZrSiO_4$. It occurs in tetragonal prisms, has various colors (brown, green, pale blue, red, orange, golden yellow, grayish, and colorless), and is a common accessory mineral in siliceous igneous rocks, crystalline limestones, schists, and gneisses, in sedimentary rocks derived therefrom, and in beach and river placer deposits. It is the chief ore of zirconium, and is used as a refractory; when cut and polished, the colorless varieties provide exceptionally brilliant gemstones. Syn: *zirconite; hyacinth; jacinth.*

zirconite Gray or brownish *zircon.*

zirconolite A mineral: $CaZrTi_2O_7$. Cf: *zirkelite.* Syn: *blakeite.*

zircosulfate A mineral: $Zr(SO_4)_2.4H_2O$.

zirkelite [mineral] A mineral: $(Ca,Fe,Th,U)_2(Ti,Nb,Zr)_2O_7$ (?) (Frondel et al, 1967, p.49). Cf: *zirconolite.*

zirkelite [ign] An altered basaltic glass.

zirklerite A mineral: $(Fe,Mg,Ca)_9Al_4Cl_{18}(OH)_{12}.14H_2O$ (?).

zittavite A type of lustrous, black lignite. It is harder and more brittle than dopplerite.

zoantharian Any anthozoan belonging to the subclass Zoantharia, characterized by paired mesenteries. They are or may not have a calcareous exoskeleton. Their known stratigraphic range is Ordovician to present.

zoarium (a) The collective skeletal parts (zooecia) of an entire bryozoan colony. It is composed of calcite and/or chitin. (b) A colony of colonial bryozoans.---Pl: *zoaria.* Syn: *zooarium.*

ZoBell bottle A sterilized bottle used for collection of sea water for bacteriological analysis.

zobtenite A gabbro-gneiss characterized by augen of diallage surrounded by streams of uralite and embedded in granular epidote and plagioclase (saussurite)(Holmes, 1928, p.242). Cf: *flaser gabbro.*

zodiacal dust *cosmic dust.*

zoecium Var. of *zooecium.*

zoichnic Said of a dolomite or recrystallized limestone in which animal fossils, though partly destroyed by recrystallization, are still recognizable in outline or by traces of internal structure (Phemister, 1956, p.74). Cf: *zoophasmic.*

zoisite An orthorhombic mineral of the epidote group: $Ca_2-Al_3Si_3O_{12}(OH)$. It often contains appreciable ferric iron, and is white, gray, brown, green, or rose red in color. Zoisite occurs in metamorphic rocks (esp. schists formed from calcium-rich igneous rocks) and in altered igneous rocks, and is an essential constituent of saussurite. Cf: *clinozoisite.*

zona *zone* [palyn].

zonal axis *zone axis.*

zonal equation The statement that if a given crystal face (*hkl*) belongs to a zone with the axis [*uvw*], then $hu+kv+lw=0$.

zonal guide fossil A *guide fossil* that makes possible the identification of a specific biostratigraphic zone and that gives its name to the zone. It need not necessarily be either restricted to the zone or found throughout every part of it. Syn: *zonal fossil; zone fossil; zonal index fossil.*

zonal profile *composite profile.*

zonal soil In early U.S. classification systems, one of the *soil orders* that embraces soils that have well-developed characteristics that presumably reflect the influence of the agents of soil genesis, esp. climate and living organisms (chiefly plants); also, any soil belonging to the zonal order. Cf: *intrazonal soil; azonal soil.* Syn: *mature soil.*

zonal structure [ore dep] In mineral deposits, a *banded* pattern due to slight compositional differences established during formation of the material.

zonal structure [cryst] *zoning.*

zonal theory A theory of hypogene mineral-deposit formation and spatial distribution patterns of mineral sequences, based on the changes in a mineral-bearing fluid as it migrates away from a magmatic source (Park & MacDiarmid, 1964, p.157); See also: *zoning of ore deposits.*

zonate Said of spores possessing a zone.

zonation The condition of being arranged or formed in zones; e.g. the distribution of distinctive fossils, more or less parallel to the bedding, in biostratigraphic zones.

zone [geog] A term used generally, even vaguely, for an area or region of latitudinal character, more or less set off from surrounding areas by some special or distinctive characteristics; e.g. any of the five great belts or encircling regions into which the Earth's surface is divided with respect to latitude and temperature, viz. the torrid zone, the two temperate zones, and the two frigid zones.

zone [geol] (a) A regular or irregular belt, layer, band, or strip of earth materials (such as rock or soil), disposed horizontally, vertically, concentrically, or otherwise, and characterized as distinct from surrounding parts by some particular property or content; e.g. a part of the Earth's crust (such as the "zone of saturation" or the "zone of fracture"), or a "structural zone" characterized by folding of different styles and periods. Also, a series of zones. (b) A body of rock thin in one dimension as compared to others; e.g. "fault zone", "mineralized zone", or "conglomeratic zone".

zone [cryst] *crystal zone.*

zone [ecol] (a) Part of a biogeographic region characterized by uniform climatic conditions, fauna, and flora and usually by a sloping, band-formed area. (b) An area characterized by the dominance of a particular organism.

zone [palyn] An annular, more or less equatorial extension of a spore, having varying equatorial width and being as thick as or thinner than the spore wall. It is much thinner than a *cingulum*. The term is also used in a general sense for any equatorial extension of the spore wall. Cf: *flange; corona; auricula; crassitude.* Syn: *zona.*

zone [meta] *aureole.*

zone [stratig] (a) A *biostratigraphic zone.* The term "zone" was given independent and definitely established status by Oppel (1856-1858) who defined it as a belt of strata identified paleontologically (irrespective of thickness or lithology), marked in any one place by the joint and persistent occurrences of two or more species with different but restricted vertical ranges (within the strata) and by the absence of species that may characterize the preceding or succeeding zones (see Teichert, 1958a, p.109; and Wheeler, 1958a, p.654). Oppel's "zone" is equivalent to Buckman's (1902) *faunizone* to Kobayashi's (1944) *orthozone*, and to the present-day usage of *concurrent-range zone.* The unmodified term "zone" does not define a formal biostratigraphic unit "because it has been used indiscriminately for several different concepts [such as assemblage and range] and does not distinguish between them" and because "it is used in other kinds of stratigraphic classification" (ACSN, 1961, art.20a). Appropriately modified, the term may be used as an informal biostratigraphic unit indicating "a body of strata unified in a general way by paleontologic features but for which there is insufficient need or insufficient information to justify designation as a formal named unit" (ISST, 1961, p.22); e.g. "the zone of crustacean remains" or "the second bryozoan zone". (b) A general, informal working term for a stratigraphic unit of any kind, esp. any rock stratum or body of strata "which it is useful to recognize as a unit because of its being characterized by some unifying property, attribute, or content" (ISST, 1961, p.18-19 &29), or that is set off as distinct from surrounding parts by certain general characteristics and that may include all or parts of a bed, a member, a formation, or even a group (ACSN, 1961, art.4g); e.g. "lithologic zone", "(heavy-)mineral zone", "fossil zone", "marine zone", "calcareous zone", "tar zone", "(oil-) producing zone", "coal-bearing zone", or any body of strata more or less unified with respect to the feature named. The term should always be preceded by a modifying word to indicate the kind of zone to which reference is made. (c) A term used as a formal stratigraphic unit in certain categories of stratigraphic classification "where other special terms are not available" (ISST, 1961, p.19). (d) A term approved by the 8th International Geological Congress in Paris in 1900 for a time-stratigraphic unit next in rank below stage. It is suggested that the term *substage* be used for such a unit in order to avoid confusion between a chronozone and a biostratigraphic zone or other kinds of zones (ISST, 1961, p.13 & 25). (e) *chronozone.* (f) A term used by Wheeler et al (1950, p.2364) for the fundamental para-time-rock unit, consisting of "a succession of strata possessing in common distinguishable characteristics which are useful in establishing a relative space-time relationship", such as lithozone, radiozone, and esp. faunizone and florizone. (g) A term sometimes used to designate the time span of strata, or (as in Germany) the survival time of a species or other taxonomic group. This usage is "careless, confusing, and unnecessary; the term is not a time term" (Dunbar & Rodgers, 1957, p.300).

zone axis That line or crystallographic direction through the center of a crystal which is parallel to the intersection edges of the crystal faces defining the *crystal zone.* Syn: *zonal axis.*

zone-breaking species A fossil species that is confined to a biostratigraphic zone in certain areas but transgresses the boundaries of that zone at other places (Arkell, 1933, p.32).

zone fossil A fossil characteristic of a zone; a *zonal guide fossil.*

zone of ablation *ablation area.*

zone of accumulation [soil] *B horizon.*

zone of accumulation [snow] (a) *accumulation area.* (b) In respect to an avalanche, a syn. of *accumulation zone.*

zone of aeration A subsurface zone containing water under pressure less than that of the atmosphere, including water held by capillarity; and containing air or gases generally under atmospheric pressure. This zone is limited above by the land surface and below by the surface of the *zone of saturation*, i.e. the *water table.* The zone is subdivided into the *belt of soil water*, the *intermediate belt*, and the *capillary fringe.* Syn: *vadose zone; unsaturated zone; zone of suspended water.*

zone of capillarity *capillary fringe.*

zone of cementation The layer of the Earth's crust below the *zone of weathering*, in which percolating waters cement unconsolidated deposits by the deposition of dissolved minerals from above. Syn: *belt of cementation.*

zone of deposition "The area in which continental glaciers deposit materials derived from the *zone of erosion.* It is usually covered with drift and has the general aspect of a plain" (Stokes & Varnes, 1955, p.164).

zone of discharge A term suggested for that part of the zone of saturation having a means of horizontal escape. Cf: *static zone.*

zone of erosion "The area from which continental glaciers have removed material by erosion. It is mostly a bare rock surface" (Stokes & Varnes, 1955, p.164). Ant: *zone of deposition.*

zone of flow [interior Earth] *zone of plastic flow.*

zone of flow [glaciol] The inner, mobile main body or mass of a glacier, in which most of the ice flows without fracture. Cf: *zone of fracture.*

zone of flowage *zone of plastic flow.*

zone of fracture [interior Earth] The upper part of the Earth's crust which is brittle, i.e. in which deformation is by fracture rather than by plastic flow; that region of the crust in which fissures can exist. Cf: *zone of plastic flow; zone of fracture and plastic flow.* Syn: *zone of rock fracture.*

zone of fracture [glaciol] The outer, rigid part of a glacier, in which the ice is much fractured. Cf: *zone of flow.*

zone of fracture and plastic flow That region of the Earth's crust which is intermediate in depth and pressure between the *zone of fracture* and the *zone of plastic flow*, and in which deformation of the weaker rocks is by plastic flow, and of the stronger rocks, by fracture.

zone of illuviation *B horizon.*

zone of intermittent saturation A term applied by Monkhouse (1965) to the temporary zone of saturation formed in the soil by infiltration from rainfall or snowmelt at a rate in excess of that at which the water can move downward to the main water table.

zone of leaching *A horizon.*

zone of mobility *asthenosphere.*

zone of plastic flow That part of the Earth's crust which is under sufficient pressure to prevent fracturing, i.e. is ductile, so that deformation is by flow. Cf: *zone of fracture; zone of fracture and plastic flow.* Syn: *zone of flow; zone of rock flowage; zone of flowage.*

zone of rock flowage *zone of plastic flow.*

zone of rock fracture *zone of fracture.*

zone of saturation A subsurface zone in which all the interstices are filled with water under pressure greater than that of the atmosphere. Although the zone may contain gas-filled interstices or interstices filled with fluids other than water, it is still considered *saturated.* This zone is separated from the *zone of aeration* (above) by the *water table.* Syn: *saturated zone; phreatic zone.*

zone of soil water *belt of soil water.*

zone of suspended water *zone of aeration.*

zone of weathering The superficial layer of the Earth's crust above the water table that is subjected to the destructive agents of the atmosphere, and in which soils develop. Cf: *zone* of cementation. Syn: *belt of weathering.*

zone symbol The symbol of the zone axis of a crystal in terms of the crystal lattice, e.g. the symbols for the zone axis of a series of (*hk*0) faces would be [001]. Cf: *indices of lattice row.*

zone time A syn. of *moment.* The term was suggested by Kobayashi (1944, p.742) for the average duration in years of a biostratigraphic zone in any given geologic system, ranging between about 300,000 and 5 million years. Cf: *instant.*

zoning [crystal] A variation in the composition of a crystal from core to margin, due to a separation of the crystal phases during its growth by loss of equilibrium in a continuous reaction series. The higher-temperature phases of the isomorphic series form the core, with the lower-temperature phases toward the margin. Cf: *armoring.* See also: *normal zoning; reversed zoning.* Syn: *zonal structure.*

zoning [meta] The development of areas of metamorphosed rocks which may exhibit zones in which a particular mineral or suite of minerals is predominant or characteristic, reflecting the pressure and temperature conditions of formation, the original rock composition, duration of metamorphism or whether or not material was added during metamorphism (Stokes and Varnes, 1955).

zoning of ore deposits Spatial distribution patterns of mineral types; paragenetic sequences, either syngenetic or epigenetic (Park & MacDiarmid, 1964, p. 160); See also: *zonal theory.* Syn: *mineral zoning.*

zonite A term proposed by Henningsmoen (1961) to replace *range zone.*

zonochlorite A syn. of *pumpellyite* occurring in green pebbles of banded structure (as in the Lake Superior region). It was previously thought to be an impure prehnite.

zonolimnetic Pertaining to a definite depth zone, in a body of fresh water, inhabited by planktonic animals.

zonotrilete Said of a trilete spore characterized by an equatorial zone or other thickening.

zonule The smallest recognized subdivision of a *biostratigraphic zone,* generally consisting of a single stratum or small thickness of strata (ACSN, 1961, art.20f); a locally recognizable biostratigraphic unit, such as in a sedimentary basin or similar restricted area of sedimentation. It may be a subdivision of a *subzone,* or it may be distinguished as a minor component of a zone without subdividing the zone into subzones. The term was defined by Fenton & Fenton (1928, p.20-22) as the strata or stratum that contains a *faunule* or *florule,* its thickness and area being limited by the vertical and horizontal range of that faunule or florule. Wheeler (1958a, p.649 & 654) regards zonules as tongues of biosomes or as "vertically contiguous units, each of which is characterized (unless barren) by one or more taxonomic entities, without consideration of their total ranges ... or total concurrent limits, and each of which differs in its taxonomic constitution from the immediately subjacent and superjacent zonules".

zooarium Var. of *zoarium.*

zoobenthos Animal forms of the benthos.

zoochore A plant whose seeds or spores are distributed by living animals.

zooeciule A small zooecium with orifice but not all parts of a bryozoan autozooid.

zooecium The calcareous or chitinous skeleton of a bryozoan zooid, consisting of tubular walls and various internal structures. Pl: *zooecia.* Adj: *zooecial.* Syn: *zoecium.*

zooecology The branch of ecology concerned with the relationships between animals and their environment. Cf: *phytoecology.*

zoogenic rock A *biogenic rock* produced by animals or directly attributable to the presence or activities of animals; e.g. shell limestone, coral reefs, guano, and lithified calcareous ooze. Cf: *zoolith.* Syn: *zoogenous rock.*

zoogenous rock *zoogenic rock.*

zoogeography The branch of *biogeography* dealing with the geographic distribution of animals. Cf: *phytogeography.*

zooid A more or less independent animal produced by other than direct sexual methods and therefore having an equivocal individuality; any individual of a colony, irrespective of its morphologic specifications, such as an octocorallian polyp, or a single bryozoan individual (including polypide and zooecium), or a soft-bodied graptolite individual inhabiting a theca.

zoolite An animal fossil. Syn: *zoolith.*

zoolith [paleont] Var. of *zoolite.*

zoolith [sed] A *biolith* formed by animal activity or composed of animal remains; specif. *zoogenic rock.*

zoophasmic Said of a dolomite or recrystallized limestone that contains vague but unmistakable traces of the former presence of animal fossils (Phemister, 1956, p.74). Cf: *zoichnic.*

zoophyte An invertebrate animal that resembles a plant in forming branching colonies attached to a substrate. Syn: *phytozoan.*

zooplankton The animal forms of *plankton,* e.g. jellyfish. They consume the *phytoplankton.*

zootrophic *heterotrophic.*

zooxanthella A symbiotic, unicellular, yellow-brown protistan in the endoderm of hermatypic coral polyps.

Z phenomenon The possible time lag (a few seconds or less) between the issuance of *P* and *S* waves from an earthquake locus (Runcorn, 1967).

Zuloagan North American (Gulf Coast) stage: Upper Jurassic (above older Jurassic, below LaCasitan; it is equivalent to European Oxfordian) (Murray, 1961).

zunyite A mineral: $Al_{13}Si_5O_{20}(OH,F)_{18}Cl$. It occurs in a minute transparent tetrahedral crystals.

zussmanite A mineral: $K(Fe,Mg,Mn)_{13}(Si,Al)_{18}O_{42}(OH)_{14}$.

zvyagintsevite A mineral: $(Pd,Pt)_3(Pb,Sn)$.

zweikanter A *windkanter* or stone having two faces intersecting in two sharp edges. Etymol: German *Zweikanter,* "one having two edges". Pl: *zweikanters; zweikanter.*

Zwischengebirge A syn. of *median mass.* Etymol: German.

zygolith A coccolith in the form of an elliptic ring with a crossbar arching slightly or strongly upward (e.g. distally), and bearing a knob or short spine.

zygolophe A brachiopod lophophore in which each brachium consists of a straight or crescentic side arm bearing two rows of paired filamentar appendages (TIP, 1965, pt.H, p.155).

zygome An articulatory structure of a desma in a sponge.

zygomorphic Said of an organism or organ that is bilaterally symmetric or capable of division into essentially symmetric halves by only one longitudinal plane passing through the axis. Cf: *actinomorphic.*

zygosis The interlocking of sponge desmas, without fusion, by means of zygomes.

zygospore A *resting spore* of various nonvascular plants (such as desmids), produced by sexual fusion of two protoplasts. It often has a thick, resistant wall and can therefore occur as a palynomorph.

zygous basal plate One of the two large plates of the basalia of a blastoid, located in the right posterior (BD) or left anterior (DA) position and formed by fusion of a pair of antecedent small basal plates comparable to *azygous basal plate* in the AB interray (TIP, 1967, pt.S, p.350).

Bibliography

This bibliography presents the citations for references used in the text. Some corporate authors appear in abbreviated form; these abbreviations, with the full form by which they may be located in the bibliography, are here given:

ACSN: American Commission on Stratigraphic Nomenclature.

ADTIC: U.S. Air Force. Air University. Aerospace Studies Institute. Arctic, Desert, Tropic Information Center.

AGI: American Geological Institute.

AIME: American Institute of Mining, Metallurgical, and Petroleum Engineers.

ASCE: American Society of Civil Engineers.

ASP: American Society of Photogrammetry.

ASTM: American Society for Testing and Materials.

BNCG: British National Committee for Geography.

BSI: British Standards Institution.

CERC: U.S. Army. Coastal Engineering Research Center.

DOD: U.S. Department of Defense.

ESSA: U.S. Environmental Science Services Administration.

ICCP: International Committee for Coal Petrology.

ICZN: International Commission of Zoological Nomenclature.

ISST: International Subcommission on Stratigraphic Terminology.

NASA: National Aeronautics and Space Administration.

NAS-NRC: National Academy of Sciences-National Research Council.

SSSA: Soil Science Society of America.

TIP: Moore, Raymond C., ed. (1953-) *Treatise on invertebrate paleontology.*

USDA: U.S. Department of Agriculture.

USGS: U.S. Geological Survey.

Ackermann, Ernst (1951) *Geröllton!* Geologische Rundschau, v.39, p.237-239.

Ackermann, Ernst (1962) *Büssersteine—Zeugen vorzeitlicher Grundwasserschwankungen.* Zeitschrift für Geomorphologie, n.F., Bd.6, p.148-182.

Adams, John A.S., and Weaver, Charles E. (1958) *Thorium-to-uranium ratios as indicators of sedimentary processes—example of concept of geochemical facies.* American Association of Petroleum Geologists. Bulletin, v.42, p.387-430.

Adams, John E., and others (1951) *Starved Pennsylvanian Midland Basin.* American Association of Petroleum Geologists. Bulletin, v.35, p.2600-2607.

Agassiz, Louis (1866) *Geological sketches.* Boston: Ticknor & Fields. 311p.

Ager, Derek V. (1963) *Principles of paleoecology: an introduction to the study of how and where animals and plants lived in the past.* New York: McGraw-Hill, Inc.

Ahlmann, H.W. (1933) *Scientific results of the Swedish-Norwegian Arctic Expedition in the summer of 1931, Pt.8.* Geografiska Annaler, v.15, p.161-216, 261-295.

Alden, W.C. (1928) *Landslide and flood at Gros Ventre, Wyoming.* American Institute of Mining and Metallurgical Engineers. Transactions, v.76, p.347-361. (Technical publication no. 140).

Allan, Robin S. (1948) *Geological correlation and paleoecology.* Geological Society of America. Bulletin, v.59, p.1-10.

Allen, Eugene T., and Day, Arthur L. (1935) *Hot springs of the Yellowstone National Park.* Microscopic examinations by Herbert E. Merwin. Carnegie Institution of Washington. Publication 466. 525p.

Allen, John R.L. (1960) *Cornstone.* Geological Magazine, v.97, p.43-48.

Allen, John R.L. (1963) *Asymmetrical ripple marks and the origin of water-laid cosets of cross-strata.* Liverpool and Manchester Geological Journal, v.3, p.187-236.

Allen, Victor T. (1936) *Terminology of medium-grained sediments. With notes by P.G.H. Boswell.* National Research Council. Division of Geology and Geography. Annual report for 1935-1936, appendix I, exhibit B. 23p. (Its Committee on Sedimentation. Report, exhibit B).

Allen, Victor T., and Nichols, Robert L. (1945) *Clay-pellet conglomerates at Hobart Butte, Lane County, Oregon.* Journal of Sedimentary Petrology, v.15, p.25-33.

Alling, Harold L. (1943) *A metric grade scale for sedimentary rocks.* Journal of Geology, v.51, p.259-269.

Alling, Harold L. (1945) *Use of microlithologies as illustrated by some New York sedimentary rocks.* Geological Society of America. Bulletin, v.56, p.737-755.

Allum, J.A.E. (1966) *Photogeology and regional mapping.* New York: Pergamon Press. 107p.

American Commission on Stratigraphic Nomenclature (1957) *Nature, usage, and nomenclature of biostratigraphic units.* American Association of Petroleum Geologists. Bulletin, v.41, p.1877-1889. (Its Report 5).

American Commission on Stratigraphic Nomenclature (1959) *Application of stratigraphic classification and nomenclature to the Quaternary.* American Association of Petroleum Geologists. Bulletin, v.43, p.663-675. (Its Report 6).

American Commission on Stratigraphic Nomenclature (1961) *Code of stratigraphic nomenclature.* American Association of Petroleum Geologists. Bulletin, v.45, p.645-660.

American Commission on Stratigraphic Nomenclature (1965) *Records of the Stratigraphic Commission for 1963-1964.* American Association of Petroleum Geologists. Bulletin, v.49, p.296-300. (Its Note 31).

American Commission on Stratigraphic Nomenclature (1967) *Records of the Stratigraphic Commission for 1964-1966.* American Association of Petroleum Geologists. Bulletin, v.51, p.1862-1868. (Its Note 34).

American Geological Institute. Data Sheet Committee (1958) *Wentworth grade scale.* Geo-Times, v.3, no.1, p.16. (Its Data sheet 7).

American Institute of Mining, Metallurgical, and Petroleum Engineers (1960) *Industrial minerals and rocks (nonmetallics other than fuels).* 3rd ed. New York: American Institute of Mining, Metallurgical, and Petroleum Engineers. 934p.

American Society for Testing and Materials (1970) *Annual book of ASTM standards, part 33: glossary of ASTM definitions and index to ASTM standards.* Philadelphia: American Society for Testing and Materials. 706p.

American Society of Civil Engineers. Soil Mechanics and Foundations Division. Committee on Glossary of Terms and Definitions in Soil Mechanics (1958) *Glossary of terms and definitions in soil mechanics; report of the committee.* R.E. Fadum, chairman. Its Journal, 1958, no.SM4, pt.1. 43p. (American Society of Civil Engineers. Proceedings, v.84, paper 1826).

American Society of Civil Engineers. Surveying and Mapping Division. Committee on Definitions of Surveying Terms (1954) *Definitions of surveying, mapping, and related terms.* Charles B. Breed, chairman. New York: American Society of Civil Engineers. 202p. *(Manuals of engineering practice, no.34).*

American Society of Photogrammetry. Committee on Nomenclature (1966) *Definitions of terms and symbols used in photogrammetry.* Editor: Robert D. Turpin. In: American Society of Photogrammetry. *Manual of photogrammetry,* v.2, p.1125-1161. 3rd ed. Falls Church (Va.): American Society of Photogrammetry. 1199p.

Amiran, D.H.K. (1950-1951) *Geomorphology of the central Negev highlands.* Israel Exploration Journal, v.1, p.107-120.

Anderson, Duwayne M., and others (1969) *Bentonite debris flows in northern Alaska.* Science, v.164, p.173-174.

Andersson, J. Gunnar (1906) *Solifluction, a component of subaerial denudation.* Journal of Geology, v.14, p.91-112.

Andreev, P.F., and others (1968) *Transformation of petroleum in nature.* Translated from the Russian edition by Robert B. Gaul and Bruno C. Metzner. Translation editors: E. Barghoorn and S. Silverman. New York: Pergamon Press. 468p. *(International series of monographs in earth sciences, v.29).*

Andresen, Marvin J. (1962) *Paleodrainage patterns: their mapping from subsurface data, and their paleogeographic value.* American Association of Petroleum Geologists. Bulletin, v.46, p.398-405.

Antevs, Ernst V. (1932) *Alpine zone of Mt. Washington Range.* Auburn (Me.): Merrill & Weber. 118p.

Antevs, Ernst V. (1948) *Climatic changes and pre-white man.* Utah. University. Bulletin, v.38, no.20, p.168-191. *(The Great Basin, with emphasis on glacial and postglacial times, 3).*

Antevs, Ernst V. (1953) *Geochronology of the Deglacial and Neothermal ages.* Journal of Geology, v.61, p.195-230.

Apfel, Earl T. (1938) *Phase sampling of sediments.* Journal of Sedimentary Petrology, v.8, p.67-68.

Arkell, William J. (1933) *The Jurassic System in Great Britain.* Oxford: Clarendon Press. 681p.

Arkell, William J. (1956) *Jurassic geology of the world.* New York: Hafner. 806p.

Arkell, William J., and Tomkeieff, Sergei I. (1953) *English rock terms, chiefly as used by miners and quarrymen.* London: Oxford University Press. 139p.

Arkley, Rodney J., and Brown, Herrick C. (1954) *The origin of Mima mound (hogwallow) microrelief in the far western states.* Soil Science Society of America. Proceedings, v.18, p.195-199.

Armstrong, R.L. (1958) *Sevier orogenic belt in Nevada and Utah*. Geological Society of America. Bulletin, v.79, p.429-458.

Armstrong, Terence, and Roberts, Brian (1956) *Illustrated ice glossary*. [Pt. 1]. Polar Record, v.8, no.52, p.4-12.

Armstrong, Terence, and Roberts, Brian (1958) *Illustrated ice glossary*. Pt. 2. Polar Record, v.9, no.59, p.90-96.

Armstrong, Terence, and others (1966) *Illustrated glossary of snow and ice*. Cambridge (Eng.): Scott Polar Research Institute. 60p.

Arnold, Chester A. (1947) *An introduction to paleobotany*. New York: McGraw-Hill, Inc. 433p.

Association of Engineering Geologists (1969) *Definition of engineering geology*. Its News letter, v.12, no.4, p.3.

Atterberg, Albert (1905) *Die rationelle Klassifikation der Sande und Kiese*. Chemiker-Zeitung, Jahrg.29, p.195-198.

Aubouin, Jean (1965) *Geosynclines*. Amsterdam: Elsevier. 335p.

Aufrère, L. (1931) *Le cycle morphologique des dunes*. Annales de Géographie, v.40, no. 226, p.362-385.

Back, William (1966) *Hydrochemical facies and ground-water flow patterns in northern part of Atlantic Coastal Plain*. U.S. Geological Survey. Professional Paper 498-A. 42p.

Bagnold, Ralph A. (1941) *The physics of blown sand and desert dunes*. London: Methuen. 265p.

Bagnold, Ralph A. (1956) *The flow of cohesionless grains in fluids*. Royal Society of London. Philosophical Transactions, ser.A, v.249, p.235-297.

Bailey, Edward B. (1934) *West Highland tectonics: Loch Leven to Glen Roy*. Geological Society of London. Quarterly Journal, v.90, p.462-525.

Baker, Alfred A. (1959) *Imprisoned rocks: a process of rock abrasion*. Victorian Naturalist, v.76, p.206-207.

Baker, B.B., and others (1966) *Glossary of oceanographic terms*. 2nd ed. Washington, D.C.: U.S. Naval Oceanographic Office. 204p. (U.S. Naval Oceanographic Office. Special Publication 35).

Baker, Herbert A. (1920) *On the investigation of the mechanical constitution of loose arenaceous sediments by the method of elutriation, with special reference to the Thanet beds of the southern side of the London Basin*. Geo-logical Magazine, v.57, p.321-332, 363-370, 411-420, 463-467.

Balk, Robert (1937) *Structural behavior of igneous rocks (with special reference to interpretations by H. Cloos and collaborators)*. Geological Society of America. Memoir 5. 177p.

Ballard, Thomas J., and Conklin, Quentin E. (1955) *The uranium prospector's guide*. New York: Harper. 251p.

Barrell, Joseph (1907) *Geology of the Marysville mining district, Montana: a study of igneous intrusion and contact metamorphism*. U.S. Geological Survey. Professional Paper 57. 178p.

Barrell, Joseph (1912) *Criteria for the recognition of ancient delta deposits*. Geological Society of America. Bulletin, v.23, p.377-446.

Barrell, Joseph (1913) *The Upper Devonian delta of the Appalachian geosyncline*. American Journal of Science, ser.4, v.36, p.429-472.

Barrell, Joseph (1917) *Rhythms and the measurements of geologic time*. Geological Society of America. Bulletin, v.28, p.745-904.

Barth, Thomas F.W. (1948) *Oxygen in rocks: a basis for petrographic calculations*. Journal of Geology, v.56, p.50-60.

Barth, Thomas F.W. (1959) *Principles of classification and norm calculations of metamorphic rocks*. Journal of Geology, v.67, no.2, p.135-152.

Barth, Thomas F.W. (1962) *Theoretical petrology*. 2d ed. New York: John Wiley & Sons, 416p.

Barton, Donald C. (1929) *The Eötvös torsion balance method of mapping geologic structure*. American Institute of Mining, Metallurgical, and Petroleum Engineers. [Transactions, v.18], Geophysical Prospecting, p.416-479.

Bascom, Florence (1931) *Geomorphic nomenclature*. Science, v.74, p.172-173.

Bastin, Edson S. (1909) *Chemical composition as a criterion in identifying metamorphosed sediments*. Journal of Geology, v.17, p.445-472.

Bateman, Alan M. (1950) *Economic mineral deposits*. 2nd ed. New York: Wiley. 916p.

Bates, Robert L. (1938) *Occurrence and origin of certain limonite concretions*. Journal of Sedimentary Petrology, v.8, p.91-99.

Båth, Markus (1966) *Earthquake seismology*. Earth-Science Reviews, v.1, no.1, p.69-86.

Bathurst, R.G.C. (1958) *Diagenetic fabrics in some British Dinantian limestones*. Liverpool and Manchester Geological Journal, v.2, pt.1, p.11-36.

Baulig, Henri (1956) *Vocabulaire franco-anglo-allemand de géomorphologie*. Paris: Soc. Ed. Belles Lettres. 229p.

Baulig, Henri (1957) *Peneplains and pediplains.* Translated from the French by C.A. Cotton. Geological Society of America. Bulletin, v.68, p.913-929.

Baumhoff, Martin A., and Heizer, Robert F. (1965) *Postglacial climate and archaeology in the Desert West.* In: Wright, H.E., jr., and Frey, David G., eds. *The Quaternary of the United States,* p.697-707. Princeton (N.J.): Princeton Univ. Press. 922p.

Bayly, Brian (1968) *Introduction to petrology.* Englewood Cliffs (N.J.): Prentice-Hall. 371p.

Beales, Francis W. (1958) *Ancient sediments of Bahaman type.* American Association of Petroleum Geologists. Bulletin, v.42, p.1845-1880.

Beasley, Henry C. (1914) *Some fossils from the Keuper Sandstone of Alton, Staffordshire.* Liverpool Geological Society. Proceedings, v.12, p.35-39.

Becke, F. (1903) *Über Mineralbestand und Struktur der Kristallinischen Schiefer.* Comptes Rendus, Congrès Géologique International, 9th, Vienna.

Becke, F. (1913) *Ueber Minerallesband und Struktur der krystallinischen Schiefer.* Akademie der Wissenschaften in Wien. Denkschriften, v.75, p.1-53.

Becker, George F. (1895) *Reconnaissance of the gold fields of the southern Appalachians.* U.S. Geological Survey. Annual Report, 16th, pt.3, p.251-331.

Becker, Hans (1932) *Report on some work on sediments done in Germany in 1931.* National Research Council. Committee on Sedimentation. Report, 1930-1932, p.82-89. (National Research Council. Bulletin, no.89).

Beerbower, James R. (1964) *Cyclothems and cyclic depositional mechanisms in alluvial plain sedimentation.* Kansas. State Geological Survey. Bulletin 169, v.1, p.31-42.

Beerbower, James R. (1968) *Search for the past; an introduction to paleontology.* 2nd ed. Englewood Cliffs (N.J.): Prentice-Hall. 512p.

Bell, Robert (1894) *Pre-Paleozoic decay of crystalline rocks north of Lake Huron.* Geological Society of America. Bulletin, v.5, p.357-366.

Bemmelen, R.W. van (1933) *The undation theory of the development of the Earth's crust.* International Geological Congress. 16th, Washington, D.C. Proceedings, v.2, p.965-982, [1935].

Bemmelen, R.W. van (1949) *The geology of Indonesia,* Vol. 1A. The Hague: Government Printing Office. 732p.

Benedict, L.G., and others (1968) *Pseudovitrinite in Appalachian coking coals.* Fuel, v.47, p.125-143.

Beneo, E. (1955) *Les résultats des études pour la recherche pétrolifère en Sicile.* World Petroleum Congress. 4th, Rome, 1955. Proceedings, sec.1, p.109-124.

Berkey, Charles P., and Morris, Frederick K. (1924) *Basin structures in Mongolia.* American Museum of Natural History. Bulletin, v.51, p. 103-127.

Bernard, B. (1970) *ABC's of infrared.* Indianapolis: Howard W. Sames & Co.

Berry, L.G., and Mason, Brian (1959) *Mineralogy: concepts, descriptions, determinations.* San Francisco: W.H. Freeman & Co. 630p.

Berthelsen, Asger (1970) *Globulith: a new type of intrusive structure, exemplified by metabasic bodies in the Moss area, SE Norway.* Norges Geologiske Undersökelse, [Publikasjoner]. No. 266 (Årbok 1969).

Bhattacharji, Somdev (1958) *Theoretical and experimental investigations on crossfolding.* Journal of Geology, v.66, p.625-667.

Billings, B.H. (1963) *Optics.* In: *American Institute of Physics Handbook.* 2d ed.

Billings, Marland P. (1954) *Structural geology.* 2nd ed. Englewood Cliffs (N.J.): Prentice-Hall. 514p.

Birch, A.F. (1952) *Elasticity and constitution of the Earth's interior.* Journal of Geophysical Research, v.57, p.227-286.

Bird, J. Brian (1967) *The physiography of Arctic Canada, with special reference to the area south of Parry Channel.* Baltimore: Johns Hopkins Press. 336p.

Birkenmajer, Krzysztof (1958) *Oriented flowage casts and marks in the Carpathian flysch and their relation to flute and groove casts.* Acta geologica Polonica, v.8, p.139-148.

Birkenmajer, Krzysztof (1959) *Classification of bedding in flysch and similar graded deposits.* Studia geologica Polonica, v.3, p.81-133.

Bissell, Harold J. (1959) *Silica in sediments of the Upper Paleozoic of the Cordilleran area.* In: Ireland, Hubert A., ed. *Silica in sediments —a symposium.* Tulsa: Society of Economic Paleontologists and Mineralogists. 185p. (Its Special Publication, no.7, p.150-185).

Bissell, Harold J. (1964) *Ely, Arcturus, and Park City groups (Pennsylvanian-Permian) in eastern Nevada and western Utah.* American Association of Petroleum Geologists. Bulletin, v.48, p.565-636.

Bissell, Harold J. (1964a) *Patterns of sedimentation in Pennsylvanian and Permian strata of* part of the eastern Great Basin. Kansas. State Geological Survey. Bulletin 169, v.1, p.43-56.

Bissell, Harold J., and Chilingar, George V. (1967) Classification of sedimentary carbonate rocks. In: Chilingar, George V., and others, eds. Carbonate rocks; origin, occurrence and classification, p.87-168. Amsterdam: Elsevier. (Developments in Sedimentology 9A, ch.4).

Black, Robert F. (1966) Comments on periglacial terminology. Biuletyn peryglacjalny, no.15, p. 329-333.

Blackwelder, Eliot (1931) Desert plains. Journal of Geology, v.39, p.133-140.

Blackwelder, Eliot (1931a) Pleistocene glaciation in the Sierra Nevada and Basin ranges. Geological Society of America. Bulletin, v.42, p. 865-922.

Blank, Horace R. (1951) "Rock doughnuts", a product of granite weathering. American Journal of Science, v.249, p.822-829.

Blench, Thomas (1957) Regime behaviour of canals and rivers. London: Butterworths. 138p.

Blissenbach, Erich (1954) Geology of alluvial fans in semiarid regions. Geological Society of America. Bulletin, v.65, p.175-189.

Bold, Harold C. (1967) Morphology of plants. 2d ed. New York: Harper & Row.

Bonney, Thomas George (1886) The anniversary address of the President. "Metamorphic" rocks. Geological Society of London. Quarterly Journal, v.42, (Proc.), p.38-115.

Bosellini, A. (1966) Protointraclasts: texture of some Werfenian (Lower Triassic) limestones of the Dolomites (northeastern Italy). Sedimentology, v.6, p.333-337.

Bostick, Neely H. (1970) Measured alteration of organic particles (phytoclasts) as an indicator of contact and burial metamorphism in sedimentary rocks. Geological Society of America. Abstracts with Programs, v.2, no.2, p.74.

Boswell, P.G.H. (1960) The term graywacke. Journal of Sedimentary Petrology, v.30, p.154-157.

Bouma, Arnold H. (1962) Sedimentology of some flysch deposits; a graphic approach to facies interpretation. Amsterdam: Elsevier. 168p.

Bourcart, Jacques (1939) Essai d'une définition de la vase des estuaires. Académie des Sciences, Paris. Comptes rendus hebdomadaires des séances, v.209, no.14, p.542-544.

Bourcart, Jacques (1941) Essai de définition des vases des eaux douces. Académie des Sciences, Paris. Comptes rendus hebdomadaires des séances, v.212, no.11, p.448-450.

Bowen, R. (1966) Oxygen isotopes as climatic indicators. Earth-science Reviews, v.2, p.199-224.

Boydell, H.C. (1926) A discussion of metasomatism and the linear "force of growing crystals". Economic Geology, v.21, p.1-55.

Bradley, William C. (1963) Large-scale exfoliation in massive sandstones of the Colorado Plateau. Geological Society of America. Bulletin, v.74, p.519-527.

Bradley, Wilmot H. (1930) The behavior of certain mud-crack casts during compaction. American Journal of Science, ser.5, v.20, p. 136-144.

Bradley, Wilmot H. (1931) Origin and microfossils of the oil shale of the Green River Formation of Colorado and Utah. U.S. Geological Survey. Professional Paper 168. 58p.

Bramlette, Milton N. (1946) The Monterey Formation of California and the origin of its siliceous rocks. U.S. Geological Survey. Professional Paper 212. 57p.

Branco, W., and Fraas, E. (1905) Das kryptovulcanische Becken von Steinheim. Akademie der Wissenschaften, Berlin. Physikalische Abhandlungen, Jahrg. 1905, Abh.1. 64p.

Braunstein, Jules (1961) Calciclastic and siliciclastic. American Association of Petroleum Geologists. Bulletin, v.45, p.2017.

Bray, Warwick, and Trump, David (1970) The American Heritage guide to archaeology. New York: American Heritage Press. 269p.

Breithaupt, August (1847) Handbuch der mineralogie. Vol. 3. Dresden & Leipzig: Arnoldische Buchhandlung. 496p.

Bretz, J Harlen (1929) Valley deposits immediately east of the channeled scabland of Washington. Journal of Geology, v.37, p.393-427, 505-541.

Bretz, J Harlen (1940) Solution cavities in the Joliet Limestone of northeastern Illinois. Journal of Geology, v.48, p.337-384.

Bretz, J Harlen (1942) Vadose and phreatic features of limestone caverns. Journal of Geology, v.50, p.675-811.

Brewer, Roy (1964) Fabric and mineral analysis of soils. New York: Wiley. 470p.

Brewer, Roy, and Sleeman, J.R. (1960) Soil structure and fabric: their definition and description. Journal of Soil Science, v.11, p.172-185.

Brigham, Albert P. (1901) A text-book of geology. New York: Appleton. 477p.

British National Committee for Geography (1966) Glossary of technical terms in cartography. London: The Royal Society. 84p.

British Standards Institution (1964) *Glossary of mining terms; section 5: Geology.* London: British Standards Institution. 18p. (B.S. 3618).

Brongniart, Alexandre (1823) *Macigno.* Dictionnaire des sciences naturelles, v.27, p.297-504.

Brown, John S. (1943) *Suggested use of the word microfacies.* Economic Geology, v.38, p.325.

Brown, Roland W. (1946) *Salt ribbons and ice ribbons.* Washington Academy of Sciences. Journal, v.36, p.14-16.

Brown, Victor J., and Runner, Delmar G. (1939) *Engineering terminology; definitions of technical words and phrases.* 2nd ed. Chicago: Gillette. 439p.

Bryan, Andrew Bonnell (1937) *Gravimeter design and operation.* Geophysics, v.2, no.4, p.301-308.

Bryan, Kirk (1920) *Origin of rock tanks and charcos.* American Journal of Science, 4th ser., v.50, p.186-206.

Bryan, Kirk (1923a) *Erosion and sedimentation in the Papago country, Arizona, with a sketch of the geology.* U.S. Geological Survey. Bulletin 730-B, p.19-90.

Bryan, Kirk (1923b) *Geology and ground-water resources of Sacramento Valley, California.* U.S. Geological Survey. Water-supply Paper 495. 285p.

Bryan, Kirk (1940) *Gully gravure, a method of slope retreat.* Journal of Geomorphology, v.3, p.89-107.

Bryan, Kirk (1946) *Cryopedology, the study of frozen ground and intensive frost-action, with suggestions on nomenclature.* American Journal of Science, v.244, p.622-642.

Bryan, Kirk (1951) *The erroneous use of tjaele as the equivalent of perennially frozen ground.* Journal of Geology, v.59, p.69-71.

Bucher, Walter H. (1919) *On ripples and related sedimentary surface forms and their paleogeographic interpretation.* American Journal of Science, ser.4, v.47, p.149-210, 241-269.

Bucher, Walter H. (1932) *"Strath" as a geomorphic term.* Science, v.75, p.130-131.

Bucher, Walter H. (1933) *The deformation of the Earth's crust.* Princeton: Princeton University Press. 518p.

Bucher, Walter H. (1952) *Geologic structure and orogenic history of Venezuela; text to accompany the author's geologic tectonic map of Venezuela.* Geological Society of America. Memoir 49. 113p.

Bucher, Walter H. (1955) *Deformation in orogenic belts.* Geological Society of America. Special Paper 62, p.343-368.

Bucher, Walter H. (1963) *Cryptoexplosion structures caused from without or from within the Earth? ("astroblemes" or "geoblemes"?).* American Journal of Science, v.261, p.597-649.

Bucher, Walter H. (1965) *Role of gravity in orogenesis.* Geological Society of America. Bulletin, v.67, no.10, p.1295-1318.

Buckland, William (1817) *Description of the Paramoudra, a singular fossil body that is found in the Chalk of the North of Ireland; with some general observations upon flints in chalk, tending to illustrate the history of their formation.* Geological Society of London. Transactions, v.4, p.413-423.

Buckland, William (1829) *On the formation of the Valley of Kingsclere and other valleys by the elevation of the strata that enclose them; and on the evidences of the original continuity of the basins of London and Hampshire.* Geological Society of London. Transactions, ser.2, v.2, p.119-130.

Buckland, William (1829a) *On the discovery of coprolites, or fossil faeces, in the Lias at Lyme Regis, and in other formations.* Geological Society of London. Transactions, ser.2, v.3, p.223-236.

Buckman, Sydney S. (1902) *The term "hemera".* Geological Magazine, dec.4, v.9, p.554-557.

Buckman, Sydney S. (1893) *The Bajocian of the Sherborne district: its relation to subjacent and superjacent strata.* Geological Society of London. Quarterly Journal, v.49, p.479-522.

Burt, Frederick A. (1928) *Melikaria: vein complexes resembling septaria veins in form.* Journal of Geology, v.36, p.539-544.

Buwalda, John P. (1937) *Shutterridges, characteristic physiographic features of active faults.* Geological Society of America. Proceedings, 1936, p.307.

Cady, Gilbert H. (1921) *Coal resources of District IV.* Illinois. State Geological Survey. Cooperative Mining Series, Bulletin 26. 247p.

Cairnes, Delorme D. (1912) *Some suggested new physiographic terms.* American Journal of Science, v.34, p.75-87.

Campbell, John F. (1865) *Frost and fire; natural engines, tool-marks & chips with sketches taken at home and abroad by a traveller.* Vol. 2. Philadelphia: Lippincott. 519p.

Campbell, Marius R. (1896) *Drainage modifica-*

tions and their interpretation. Journal of Geology, v.4, p.567-581, 657-678.

Capot-Rey, R. (1945) *Dry and humid morphology in the Western Erg.* Geographical Review, v.35, p.391-407.

Carey, Alfred E., and Oliver, F.W. (1918) *Tidal lands; a study of shore problems.* London: Blackie. 284p.

Carey, S. Warren (1958) *A tectonic approach to continental drift.* In: Carey, S. Warren, convener. *Continental drift; a symposium.* Hobart: University of Tasmania, Geology Dept. p.177-355.

Carey, S. Warren, convener (1963) *Syntaphral tectonics and diagenesis; a symposium.* Hobart: University of Tasmania, Geology Dept. 190p.

Carozzi, Albert V. (1957) *Micro-mechanisms of sedimentation in epicontinental environment.* Geological Society of America. Bulletin, v.68, p.1706-1707.

Carozzi, Albert V., and Textoris, Daniel A. (1967) *Paleozoic carbonate microfacies of the eastern stable interior (U.S.A.).* Leiden: Brill. 41p. *(International sedimentary petrographical series, v.11).*

Carroll, Dorothy (1939) *Movement of sand by wind.* Geological Magazine, v.76, p.6-23.

Cassidy, William A. (1968) *Meteorite impact structures at Campo del Cielo, Argentina.* In: French, Bevan M., and Short, Nicholas M., eds. *Shock metamorphism of natural materials,* p.117-128. Baltimore: Mono Book Corp. 644p.

Caster, Kenneth E. (1934) *The stratigraphy and paleontology of northwestern Pennsylvania.* Pt. 1. Bulletins of American Paleontology, v.21, no.71. 185p.

Cayeux, Lucien (1929) *Les roches sedimentaires de France; roches siliceuses.* Paris: Imprimerie Nationale. 774p.

Chadwick, George H. (1931) *Storm rollers.* Geological Society of America. Bulletin, v.42, p.242.

Chadwick, George H. (1939) *Geology of Mount Desert Island, Maine.* American Journal of Science, v.237, p.355-363.

Chadwick, George H. (1948) *Ordovician "dinosaur-leather" markings (exhibit).* Geological Society of America. Bulletin, v.59, p.1315.

Challinor, John (1962) *A dictionary of geology.* 1st ed. New York: Oxford University Press. 235p.

Challinor, John (1967) *A dictionary of geology.* 3rd ed. New York: Oxford University Press. 298p.

Challinor, J., and Williams, K.E. (1926) *On some curious marks on a rock surface.* Geological Magazine, v.63, no.746, p.341-343.

Chamberlin, Thomas C. (1879) *Annual report of the Wisconsin Geological Survey for the year 1878.* Madison: Wisconsin Geological Survey. 52p.

Chamberlin, Thomas C. (1883) *Terminal moraine of the second glacial epoch.* U.S. Geological Survey. Annual Report, 3rd. p.291-402.

Chamberlain, Thomas C. (1888) *The rock-scorings of the great ice invasions.* U.S. Geological Survey. Annual Report, 7th, v.3, p.147-248.

Chamberlain, Thomas C. (1893) *The horizon of drumlin, osar and kame formation.* Journal of Geology, v.1, p.255-267.

Chamberlin, Thomas C. (1894a) *Pseudo-cols.* Journal of Geology, v.2, p.205-206.

Chamberlin, Thomas C. (1894b) *Proposed genetic classification of Pleistocene glacial formations.* Journal of Geology, v.2, p.517-538.

Chamberlin, Thomas C. (1897) *The method of multiple working hypotheses.* Journal of Geology, v.5, p.837-848.

Chao, Edward C.-T. (1967) *Shock effects in certain rock-forming minerals.* Science, v.156, p.192-202.

Chao, Edward C.-T. (1967a) *Impact metamorphism.* In: Abelson, P.H., ed. *Researches in geochemistry,* v.2, p.204-233. New York: Wiley. 663p.

Charlesworth, John K. (1957) *The Quaternary era, with special reference to its glaciation.* 2v. London: Arnold. 1700p.

Chayes, F. (1956) *Petrographic modal analysis.* New York: Wiley.

Chebotarev, I.I. (1955) *Metamorphism of natural waters in the crust of weathering.* [Pt.] 1-3. Geochimica et cosmochimica acta, v.8, p.22-48, 137-170, 198-212.

Chilingar, George V. (1957) *Classification of limestones and dolomites on basis of Ca/Mg ratio.* Journal of Sedimentary Petrology, v.27, p.187-189.

Chilingar, George V., and others (1967) *Diagenesis in carbonate rocks.* In: Larsen, Gunnar, and Chilingar, George V., eds. *Diagenesis in sediments.* Amsterdam: Elsevier. 551p. *(Developments in sedimentology 8, p.179-322).*

Chinner, G.A. (1966) *The significance of the aluminum silicates in metamorphism.* Earth-Science Reviews, v.2.

Choquette, Philip W. (1955) *A petrographic study of the "State College" siliceous oölite.* Journal of Geology, v.63, p.337-347.

Choquette, Philip W., and Pray, Lloyd C. (1970) *Geologic nomenclature and classification of porosity in sedimentary carbonates.* American Association of Petroleum Geologists. Bulletin, v.54, no.2, p.207-250.

Chow, Ven Te, chairman (1957) *Report of the Committee on Runoff, 1955-1956.* American Geophysical Union. Transactions, v.38, p.379-384.

Chow, Ven Te, ed. (1964) *Handbook of applied hydrology; a compendium of water-resources technology.* New York: McGraw-Hill. 1418p.

Church, W.R. (1968) *Eclogites.* In: Hess, H.H., and Poldervaart, Arie, eds. *Basalts — the Poldervaart treatise on rocks of basaltic composition,* v.2, p.755-798. New York: Interscience Publishers.

Clapp, Charles H. (1913) *Contraposed shorelines.* Journal of Geology, v.21, p.537-540.

Clarke, James W. (1958) *The bedrock geology of the Danbury quadrangle.* Connecticut. State Geological and Natural History Survey. Quadrangle report no.7. 47p.

Clarke, John I. (1966) *Morphometry from maps.* In: Dury, G.H., ed. *Essays in geomorphology,* p.235-274. New York: American Elsevier. 404p.

Clarke, John M. (1911) *Observations on the Magdalen Islands.* New York State. State Museum. Bulletin 149, p.134-155.

Clarke, John M. (1918) *Strand and undertow markings of Upper Devonian time as indications of the prevailing climate.* New York State. State Museum. Bulletin 196, p.199-238.

Claus, George, and Nagy, Bartholomew (1961) *A microbiological examination of some carbonaceous chondrites.* Nature, v.192, p.594-596.

Cleland, Herdman F. (1910) *North American natural bridges, with a discussion of their origin.* Geological Society of America. Bulletin, v.21, p.313-338.

Cleland, Herdman F. (1916) *Geology; physical and historical.* New York: American Book Co. 718p.

Cleland, Herdman F. (1925) *Geology; physical and historical.* New York: American Book Co. 718p.

Cloos, Ernst (1946) *Lineation, a critical review and annotated bibliography.* Geological Society of America. Memoir 18. 122p.

Cloos, Hans (1939) *Hebung, Spaltung, Vulkanismus; Elemente einer geometrischen Analyse indischer Grossformen.* Geologische Rundschau, v.30, no.4a, p.405-527.

Cloos, Hans (1941) *Bau und Tätigkeit von Tuffschloten: Untersuchungen an dem schwäbischen Vulkan.* Geologische Rundschau, v.32 p.709-800.

Close, U., and McCormick, E. (1922) *Where the mountains walked.* National Geographic Magazine, v.41, p.445-464.

Cloud, Preston E., jr. (1957) *Nature and origin o. atolls; a progress report.* Pacific Science Congress. 8th, Philippines, 1953. Proceedings v.3A, p.1009-1036.

Cloud, Preston E., jr., and Barnes, Virgil E. (1957) *Early Ordovician sea in central Texas.* Geological Society of America. Memoir 67, v.2 p.163-214.

Cloud, Preston E., jr., and others (1943) *Stratigraphy of the Ellenburger Group in centra. Texas—a progress report.* Texas. University Publication 4301, p.133-161.

Clough, Charles Thomas, et al. (1909) *The cauldron-subsidence of Glen Coe, and the associated igneous phenomena.* Geological Society of London. Quarterly Journal, v.65, p.611-678

Coates, Donald R. (1966) *Glaciated Appalachian Plateau: till shadows on hills.* Science, v.152, p.1617-1619.

Coats, Robert R. (1968) *The Circle Creek Rhyolite, a volcanic complex in northern Elko County, Nevada.* Geological Society of America. Memoir 116, p.69-106.

Coffey, George N. (1909) *Clay dunes.* Journal of Geology, v.17, p.754-755.

Cohee, George V. (1962) *Stratigraphic nomenclature in reports of the U.S. Geological Survey.* Washington: U.S. Geological Survey. 35p.

Coleman, Alice (1952) *Selenomorphology.* Journal of Geology, v.60, p.451-460.

Collie, George L. (1901) *Wisconsin shore of Lake Superior.* Geological Society of America. Bulletin, v.12, p.197-216.

Colp, John L. (1967) *Terradynamics: a study of projectile penetration of natural earth materials.* Geological Society of America. Special Paper, no.115, p.38).

Comstock, T.B. (1878) *An outline of general geology, with copious references.* Ithaca: University Press. 82p.

Conybeare, W.D., and Phillips, William (1822) *Outlines of the geology of England and Wales, with an introductory compendium of the general principles of that science, and comparative views of the structure of foreign countries.* Part 1. London: William Phillips. 470p.

Cook, John H. (1946) *Kame complexes and perforation deposits.* American Journal of Science, v.244, p.573-583.

Coombs, D.S. (1961) *Some recent work on the lower grades of metamorphism.* The Australian Journal of Science, v.24, no.5, p.203-215.

Cooper, Byron N. (1945) *Industrial limestones and dolomites in Virginia; Clinch Valley district.* Virginia. Geological Survey. Bulletin 66. 259p.

Cooper, Byron N., and Cooper, Gustav A. (1946) *Lower Middle Ordovician stratigraphy of the Shenandoah Valley, Virginia.* Geological Society of America. Bulletin, v.57, p.35-113.

Cooper, Gustav A. (1956) *Chazyan and related brachiopods.* Smithsonian Miscellaneous Collections, v.127, pt.1-2. 1245p.

Cooper, John R. (1943) *Flow structure in the Berea Sandstone and Bedford Shale of central Ohio.* Journal of Geology, v.51, p.190-203.

Cooper, William S. (1958) *Terminology of post-Valders time.* Geological Society of America. Bulletin, v.69, p.941-945.

Cornish, Vaughan (1898) *On sea-beaches and sandbanks.* Geographical Journal, v.11, p.628-651.

Cornish, Vaughan (1899) *On kumatology.* Geographical Journal, v.13, p.624-628.

Correns, Carl W. (1950) *Zur Geochemie der Diagenese; I: Das Verhalten von $CaCO_3$ und SiO_2.* Geochimica et Cosmochimica Acta, v.1, p.49-54.

Cottingham, Kenneth (1951) *The geologist's vocabulary.* Scientific Monthly, v.72, p.154-163.

Cotton, Charles A. (1916) *Fault coasts in New Zealand.* Geographical Review, v.1, p.20-47.

Cotton, Charles A. (1922) *Geomorphology of New Zealand.* Pt.1. Wellington: Dominion Museum. 462p. (New Zealand Board of Science and Art. Manual no.3).

Cotton, Charles A. (1940) *Classification and correlation of river terraces.* Journal of Geomorphology, v.3, p.27-37.

Cotton, Charles A. (1942) *Climatic accidents in landscape-making.* Christchurch (N.Z.): Whitcombe & Tombs. 354p.

Cotton, Charles A. (1948) *Landscape as developed by the processes of normal erosion.* 2nd ed. New York: Wiley. 509p.

Cotton, Charles A. (1950) *Tectonic scarps and fault valleys.* Geological Society of America. Bulletin, v.61, p.717-757.

Cotton, Charles A. (1958) *Geomorphology; an introduction to the study of landforms.* 7th ed., rev. Christchurch (New Zealand): Whitcombe and Tombs. 505p.

Crickmay, Colin H. (1933) *The later stages of the cycle of erosion; some weaknesses in the theory of the cycle of erosion.* Geological Magazine, v.70, p.337-347.

Cronquist, Arthur (1961) *Introductory botany.* New York: Harper & Row.

Crosby, William O. (1912) *Dynamic relations and terminology of stratigraphic conformity and unconformity.* Journal of Geology, v.20, p.289-299.

Cross, Whitman, and Howe, Ernest (1905) *Description of the Silverton quadrangle.* U.S. Geological Survey. Geologic Atlas, folio no. 120. 34p.

Cross, Whitman, and others (1902) *A quantitative chemico-mineralogical classification and nomenclature of igneous rocks.* Journal of Geology, v.10, p.555-690.

Cross, Whitman, and others (1906) *The texture of igneous rocks.* Journal of Geology, v.14, p.692-707.

Crowell, John C. (1955) *Directional-current structures from the Prealpine Flysch, Switzerland.* Geological Society of America. Bulletin, v.66, p.1351-1384.

Crowell, John C. (1957) *Origin of pebbly mudstones.* Geological Society of America. Bulletin, v.68, p.993-1009.

Crowell, John C. (1964) *Climatic significance of sedimentary deposits containing dispersed megaclasts.* In: Nairn, A.E.M., ed. *Problems in palaeoclimatology,* p.86-99. New York: Interscience. 705p.

Cullison, James S. (1938) *Origin of composite and incomplete internal moulds and their possible use as criteria of structure.* Geological Society of America. Bulletin, v.49, p.981-988.

Cumings, Edgar R. (1930) *List of species from the New Corydon, Kokomo, and Kenneth formations of Indiana, and from reefs in the Mississinewa and Liston Creek formations.* Indiana Academy of Science. Proceedings, v.39, p.204-211.

Cumings, Edgar R. (1932) *Reefs or bioherms?* Geological Society of America. Bulletin, v.43, p.331-352.

Cumings, Edgar R., and Shrock, Robert R. (1928) *Niagaran coral reefs of Indiana and adjacent states and their stratigraphic relations.* Geological Society of America. Bulletin, v.39, p. 579-619.

Curry, H. Donald (1938) *"Turtleback" fault surfaces in Death Valley, Calif.* Geological Society of America. Bulletin, v.49, p.1875.

Curry, H. Donald (1954) *Turtlebacks in the central Black Mountains, Death Valley, California.*

California. Division of Mines. Bulletin 170, ch.4, [pt.]7, p.53-59.

Cuvillier, Jean (1951) *Corrélations stratigraphiques par microfaciès en Aquitaine occidentale*. Avec la collaboration de V. Sacal. Leiden: E.J. Brill. 23p. 90 plates.

Cys, John M. (1963) *A new definition of the penesaline environment*. Compass, v.40, p.161-163.

Dahl, Eilif (1955) *Biogeographic and geologic indications of unglaciated areas in Scandinavia during the glacial ages*. Geological Society of America. Bulletin, v.66, p.1499-1519.

Dale, T. Nelson (1923) *The commercial granites of New England*. U.S. Geological Survey. Bulletin 738. 488p.

Daly, Reginald A. (1902) *The geology of the northeast coast of Labrador*. Harvard College. Museum of Comparative Zoology. Bulletin 38, Geological Series 5, p.205-270.

Daly, Reginald A. (1917) *Metamorphism and its phases*. Geological Society of America. Bulletin, v.28, p.375-418.

Damon, Paul E. (1968) *Potassium-argon dating of igneous and metamorphic rocks with applications to the Basin Ranges of Arizona and Sonora*. In: Hamilton, E.I., and Farquhar, R.M., eds. *Radiometric dating for geologists*, p.1-71. New York: Interscience Publishers. 506p.

Dana, James D. (1873) *On some results of the Earth's contraction from cooling including a discussion of the origin of mountains and the nature of the Earth's interior*. American Journal of Science, v.5, p.423-443.

Dana, James D. (1874) *A text-book of geology*. 2nd ed. New York: Ivison, Blakeman, Taylor. 358p.

Dana, James D. (1890) *Characteristics of volcanoes . . .* New York: Dodd, Mead. 399p.

Dana, Edward S. (1892) *The system of mineralogy of James Dwight Dana, 1837-1868; descriptive mineralogy*. 6th ed. New York: Wiley. 1134p.

Dana, James D. (1895) *Manual of geology; treating of the principles of the science with special reference to American geological history*. 4th ed. New York: American Book Co. 1088p.

Dapples, Edward C. (1962) *Stages of diagenesis in the development of sandstones*. Geological Society of America. Bulletin, v.73, p.913-933.

Dapples, Edward C., and others (1953) *Petrographic and lithologic attributes of sandstones*. Journal of Geology, v.61, p.291-317.

Darling, F. Fraser, and Boyd, J. Morton (1964) *The Highlands and Islands*. London: Collins. 336p.

Darrah, William Culp (1939) *Principles of paleobotany*. Leiden, Holland: Chronica Botanica Co. 239p.

Daubrée, Gabriel Auguste (1879) *Études synthétiques de géologie expérimentale*. Paris: Dunod. 828p.

Davis, Charles A. (1907) *Peat, essays on its origin, uses, and distribution in Michigan*. Michigan. Geological Survey. Annual Report 1906, p.93-395.

Davis, William M. (1885) *Geographic classification, illustrated by a study of plains, plateaus, and their derivatives*. American Association for the Advancement of Science. Proceedings, v.33, p.428-432.

Davis, William M. (1889a) *Topographic development of the Triassic formation of the Connecticut Valley*. American Journal of Science, 3rd ser., v.37, p.423-434.

Davis, William M. (1889b) *The rivers and valleys of Pennsylvania*. National Geographic Magazine, v.1, p.183-253.

Davis, William M. (1890) *The rivers of northern New Jersey, with notes on the classification of rivers in general*. National Geographic Magazine, v.2, p.81-110.

Davis, William M. (1894) *Physical geography in the university*. Journal of Geology, v.2, p.66-100.

Davis, William M. (1895) *The development of certain English rivers*. Geographical Journal, v.5, p.127-146.

Davis, William M. (1897) *Current notes on physiography*. Science, n.s., v.6, p.22-24.

Davis, William M. (1899) *The geographical cycle*. Geographical Journal, v.14, p.481-504.

Davis, William M. (1902) *Base level, grade, and peneplain*. Journal of Geology, v.10, p.77-111.

Davis, William M. (1909) *Geographical essays*. Boston: Ginn. 777p.

Davis, William M. (1911) *The Colorado Front Range; a study in physiographic presentation*. Association of American Geographers. Annals, v.1, p.21-83.

Davis, William M. (1912) *Relation of geography to geology*. Geological Society of America. Bulletin, v.23, p.93-124.

Davis, William M. (1918) *A handbook of northern France*. Cambridge: Harvard Univ. Press. 174p.

Davis, William M. (1922) *Peneplains and the geographical cycle*. Geological Society of America. Bulletin, v.33, p.587-598.

Davis, William M. (1925a) *The undertow myth.* Science, v.61, p.206-208.

Davis, William M. (1925b) *A Roxen lake in Canada.* Scottish Geographical Magazine, v.41, p.65-74.

Davis, William M. (1927) *The rifts of southern California.* American Journal of Science, 5th ser., v.13, p.57-72.

Davis, William M. (1930) *Rock floors in arid and in humid climates.* Journal of Geology, v.38, p.1-27, 136-158.

Davis, William M. (1932) *Piedmont benchlands and Primärrümpfe.* Geological Society of America. Bulletin, v.43, p.399-440.

Davis, William M. (1933) *Granite domes of the Mojave Desert, California.* San Diego Society of Natural History. Transactions, v.7, no.20, p.211-258.

Davis, William M. (1938) *Sheetfloods and streamfloods.* Geological Society of America. Bulletin, v.49, p.1337-1416.

Deane, Roy E. (1950) *Pleistocene geology of the Lake Simcoe district, Ontario.* Canada. Geological Survey. Memoir 256. 108p.

Deevey, Edward S., jr., and Flint, Richard Foster (1957) *Postglacial hypsithermal interval.* Science, v.125, p.182-184.

DeFord, Ronald K. (1946) *Grain size in carbonate rock.* American Association of Petroleum Geologists. Bulletin, v.30, p.1921-1928.

De la Beche, Henry T. (1832) *Geological manual.* Philadelphia: Carey & Lea. 535p.

Dence, M.R. (1968) *Shock zoning at Canadian craters: petrography and structural implications.* In: French, Bevan M., and Short, Nicholas M., eds. *Shock metamorphism of natural materials,* p.169-184. Baltimore: Mono Book Corp. 644p.

Dennis, John G., ed. (1967) *International tectonic dictionary; English terminology.* American Association of Petroleum Geologists. Memoir 7. 196p.

Derby, Orville A. (1910) *The iron ores of Brazil.* In: International Geological Congress. 11th, Stockholm, 1910. *The iron ore resources of the world,* v.2, p.813-822. Stockholm Generalstabens Litografiska Anstalt. 2v. 1068p.

DeSitter, L.U. (1954) *Gravitational sliding tectonics—an essay on comparative structural geology.* American Journal of Science, v.252, p.321-344.

DeVoe, J.R., and Spijkerman, J.J. (1966) *Mössbauer spectrometry.* Analytical Chemistry, v.38, no.5, p.382R-393R.

DeWiest, Roger J.M. (1965?) *Geohydrology.* New York: Wiley. 366p.

Dewolf, Y. (1970) *Les argiles à siles: paléosols ou pédolithes.* l'Association française pour l'étude du Quaternaire. Bulletin, no.2-3, p.117-119.

Dickinson, William R. (1966) *Structural relationships of San Andreas fault system, Cholame Valley and Castle Mountain Range, California.* Geological Society of America. Bulletin, v.66, p.707-725.

Dickinson, William R., and Vigrass, Laurence W. (1965) *Geology of the Suplee-Izee area, Crook, Grant, and Harney counties, Oregon.* Oregon. Department of Geology and Mineral Resources. Bulletin, no.58, 110p.

Dietrich, R.V. (1959) *Development of ptygmatic features within a passive host during partial anatexis.* Beiträge zur Mineralogie und Petrologie, v.6, no.6, p.357-365.

Dietrich, R.V. (1960) *Nomenclature of migmatitic and associated rocks.* GeoTimes, v.4, no.5, p.36-37, 50-51.

Dietrich, R.V. (1960a) *Genesis of ptygmatic features.* International Geological Congress, 21st, Copenhagen, 1960. Report, Norden. Proceedings, sec.14, p.138-148.

Dietrich, R.V. (1963) *Banded gneisses of eight localities.* Norsk Geologisk Tidsskrift, v.43, no.1, p.89-119.

Dietrich, R.V. (1969) *Hybrid rocks.* Science, v.163, p.557. [book review].

Dietrich, R.V., and Mehnert, K.R. (1961) *Proposal for the nomenclature of migmatites and associated rocks.* In: Sorensen, H., ed. *Symposium on migmatite nomenclature.* International Geological Congress, 21st, Copenhagen, 1960. Report, pt.26, sec.14, p.56-67.

Dietz, Robert S. (1959) *Shatter cones in cryptoexplosion structures (meteorite impact?).* Journal of Geology, v.67, p.496-505.

Dietz, Robert S. (1960) *Meteorite impact suggested by shatter cones in rock.* Science, v.131, p.1781-1784.

Dietz, Robert S. (1961) *Astroblemes.* Scientific American, v.205, p.50-58.

Dietz, Robert S. (1963) *Wave-base, marine profile of equilibrium, and wave-built terraces: a critical appraisal.* Geological Society of America. Bulletin, v.74, p.971-990.

Dietz, Robert S., and Holden, John C. (1966) *Miogeoclines (miogeosynclines) in space and time.* Journal of Geology, v.74, no.5, pt.1, p.566-583.

Dixon, E.E.L., and Vaughan, Arthur (1911) *The

Carboniferous succession in Gower (Glamorganshire), with notes on its fauna and conditions of deposition. Geological Society of London. Quarterly Journal, v. 67, p.477-571.

Dobrin, Milton B. (1952) *Introduction to geophysical prospecting.* New York: McGraw-Hill. 435p.

Donaldson, Douglas, and Simpson, Scott (1962) *Chomatichnus, a new ichnogenus, and other trace-fossils of Wegber Quarry.* Liverpool and Manchester Geological Journal, v.3, p.73-81.

Donnay, Gabrielle, & Donnay, J.D.H. (1953) *The crystallography of bastnaesite, parisite, roentgenite, and synchisite.* American Mineralogist, v.38, nos.11&12, pp.932-963.

Dorr, John Van N., II, and Barbosa, A.L. de Miranda (1963) *Geology and ore deposits of the Itabira district, Minas Gerais, Brazil.* U.S. Geological Survey. Professional Paper 341-C. 110p.

Dott, Robert H., jr. (1964) *Wacke, graywacke and matrix—what approach to immature sandstone classification?* Journal of Sedimentary Petrology, v.34, p.625-632.

Draper, John W. (1847) *On the production of light by heat.* Philosophical Magazine, ser.3, v.30, p.345-360.

Drew, G. Harold (1911) *The action of some denitrifying bacteria in tropical and temperate seas, and the bacterial precipitation of calcium carbonate in the sea.* Marine Biological Association of the United Kingdom. Journal, v.9, p.142-155.

Drewes, Harold (1959) *Turtleback faults of Death Valley, California: a reinterpretation.* Geological Society of America. Bulletin, v.70, p.1497-1508.

Dryer, Charles R. (1899) *The meanders of the Muscatatuck at Vernon, Ind.* Indiana Academy of Science. Proceedings, 1898, p.270-273.

Dryer, Charles R. (1901) *Certain peculiar eskers and esker lakes of northeastern Indiana.* Journal of Geology, v.9, p.123-129.

Dryer, Charles R. (1910) *Some features of delta formation.* Indiana Academy of Science. Proceedings, 1909, p.255-261.

Duff, P. McL. D., and Walton, Ewart K. (1962) *Statistical basis for cyclothems: a quantitative study of the sedimentary succession in the East Pennine Coalfield.* Sedimentology, v.1, p.235-255.

Dulhunty, J.A. (1939) *The torbanites of New South Wales, Pt. 1.* Royal Society of New South Wales. Journal and Proceedings, 1938, v.72, p.179-198.

Dunbar, Carl O., and Rodgers, John (1957) *Principles of stratigraphy.* New York: Wiley. 356p.

Dunham, Robert J. (1962) *Classification of carbonate rocks according to depositional texture.* American Association of Petroleum Geologists. Memoir 1, p.108-121.

Dunn, Joseph Avery (1942) *Granite and magmation and metamorphism.* Economic Geology, v.37, no.3, p.231-238.

Dunoyer de Segonzac, G. (1968) *The birth and development of the concept of diagenesis (1866-1966).* Earth-Science Reviews, v.4, p. 153-201.

DuToit, A. (1920) *The Karroo dolerites of South Africa.* Geological Society of South Africa. Transactions, v.23, p.1-42.

DuToit, A. (1937) *Our wandering continents.* New York: Hafner. 379p.

Dzulynski, Stanislaw, and Radomski, Andrzej (1955) *Pochodzenie śladów wleczenia na tle teorii pradów zawiesinowych.* Acta geologica Polonica, v.5, p.47-66.

Dzulynski, Stanislaw, and Sanders, John E. (1962) *Current marks on firm mud bottoms.* Connecticut Academy of Arts and Sciences. Transactions, v.42, p.57-96.

Dzulynski, Stanislaw, and Slaczka, Andrzej (1958) *Directional structures and sedimentation of the Krosno beds (Carpathian flysch).* Polskie Towarzystwo Geologiczne. Rocznik, t.28, p. 205-260.

Dzulynski, Stanislaw, and others (1959) *Turbidites in flysch of the Polish Carpathian Mountains.* Geological Society of America. Bulletin, v.70, p.1089-1118.

Eakin, Henry M. (1916) *The Yukon-Koyukuk region, Alaska.* U.S. Geological Survey. Bulletin 631. 88p.

Eardley, Armand J. (1962) *Structural geology of North America.* 2nd ed. New York: Harper & Row. 743p.

Eaton, Joseph E. (1929) *The by-passing and discontinuous deposition of sedimentary materials.* American Association of Petroleum Geologists. Bulletin, v.13, p.713-761.

Eaton, Joseph E. (1951) *Inadvisability of restricting terms that designate the relative.* Geological Society of America. Bulletin, v.62, p.77-80.

Eckis, Rollin (1928) *Alluvial fans of the Cucamonga district, southern California.* Journal of Geology, v.36, p.225-247.

Ehrenberg, Christian G. (1854) *Mikrogeologie; das Erde und Felsen schaffende wirken des unsichtbar kleinen selbstständigen Lebens auf der Erde.* Leipzig: L. Voss. 347p.

Eisenack, Alfred (1931) *Neue Mikrofossilien des baltischen Silurs.* Pt.1. Palaeontologische Zeitschrift, Bd.13, p.74-118.

Elliott, R.E. (1965) *A classification of subaqueous sedimentary structures based on rheological and kinematical parameters.* Sedimentology, v.5, p.193-209.

Elton, C.S. (1927) *The nature and origin of soil-polygons in Spitsbergen.* Geological Society of London. Quarterly Journal, v.83, p.163-194.

Emery, K.O. (1948) *Submarine geology of Bikini atoll.* Geological Society of America. Bulletin, v.59, p.855-859.

Emiliani, Cesare (1955) *Pleistocene temperatures.* Journal of Geology, v.63, p.538-578.

Emmons, W.H. (1933) *On the mechanism of the deposition of certain metalliferous lode systems associated with granitic batholiths.* In: The Committee on the Lindgren Volume, ed. *Ore deposits of the western states.* New York: American Institute of Mining and Metallurgical Engineers. 797p.

Engelhardt, Wolf von, and Stoffler, D. (1968) *Stages of shock metamorphism in crystalline rocks of the Ries basin, Germany.* In: French, Bevan M., and Short, Nicholas M., eds. *Shock metamorphism of natural materials,* p.159-168. Baltimore: Mono Book Corp. 644p.

Engeln, Oskar D. von (1942) *Geomorphology; systematic and regional.* New York: Macmillan. 655p.

Erhart, Henri (1955) *"Biostasie" et "rhexistasie"; esquisse d'une théorie sur le rôle de la pédogenèse en tant que phénomène géologique.* Académie des Sciences, Paris. Comptes rendus hebdomadaires des séances, t.241, no.18, p.1218-1220.

Erhart, Henri (1956) *La genèse des sols; esquisse d'une théorie géologique et géochimique: biostasie et rhexistasis.* Paris: Masson. 90p.

Esau, Katherine (1965) *Plant anatomy.* 2d ed. New York: Wiley.

Eskola, Pentti (1915) *Om sambandet mellan kemisk och mineralogisk sammansättning hos Orijävitraktens metamorfa bergarter.* With an English summary of the contents. Finland, Commission Géologique. Bulletin, v.8, no.44, 145p.

Eskola, Pentti (1938) *On the esboitic crystallization of orbicular rocks.* Journal of Geology, v. 46, p.448-485.

Eskola, Pentti (1939) *Die metamorphen Gesteine.* In: Barth, T.F.W., Correns, Carl W., and Eskola, Pentti. *Die Entstehung der Gesteine, ein Lehrbuch der Petrogenese.* Berlin: Julius Springer. 422p.

Eskola, Pentti (1948) *The problem of mantled gneiss domes.* Geological Society of London. Quarterly Journal, v.104, p.461.

Eskola, Pentti (1961) *Granitentschung bei Orogenese und Epeirogenese.* Geologische Rundschau, v.50, p.105-113.

Evans, J.W. (1911) *Dreikanter.* Geological Magazine, no.565, dec.5, v.8, p.334-335.

Evans, Oren F. (1938) *The classification and origin of beach cusps.* Journal of Geology, v.46, p.615-627.

Evitt, William R. (1963) *A discussion and proposals concerning fossil dinoflagellates, hystrichospheres, and acritarchs.* Pt.2. National Academy of Sciences, U.S.A. Proceedings, v.49, p.298-302.

Eythórsson, Jón (1951) *Jökla-mýs.* Journal of Glaciology, v.1, p.503.

Fairbairn, Harold W. (1943) *Packing in ionic minerals.* Geological Society of America. Bulletin, v.54, p.1305-1374.

Fairbanks, Harold W. (1906) *Practical physiography.* Boston: Allyn & Bacon. 542p.

Fairbridge, Rhodes W. (1946) *Submarine slumping and location of oil bodies.* American Association of Petroleum Geologists. Bulletin, v.30, p.84-92.

Fairbridge, Rhodes W. (1951) *The Aroe Islands and the continental shelf north of Australia.* Scope, v.1, no.6, p.24-29.

Fairbridge, Rhodes W. (1954) *Stratigraphic correlation by microfacies.* American Journal of Science, v.252, p.683-694.

Fairbridge, Rhodes W. (1958) *What is a consanguineous association?* Journal of Geology, v.66, p.319-324.

Fairbridge, Rhodes W., ed. (1966) *The encyclopedia of oceanography.* New York: Reinhold. 1021p. *(Encyclopedia of earth sciences series, v.1).*

Fairbridge, Rhodes W. (1967) *Phases of diagenesis and authigenesis.* In: Larsen, Gunnar, and Chilingar, George V., eds. *Diagenesis in sediments.* Amsterdam: Elsevier. 551p. *(Developments in sedimentology 8, p.19-89).*

Fairbridge, Rhodes W., ed. (1968) *The encyclopedia of geomorphology.* New York: Reinhold. 1295p. *(Encyclopedia of earth sciences series, v.3).*

Fairchild, Herman L. (1904) *Geology under the planetesimal hypothesis of Earth origin.* Geological Society of America. Bulletin, v.15, p.243-266.

Fairchild, Herman L. (1913) *Pleistocene geology of New York State.* Geological Society of America. Bulletin, v.24, p.133-162.

Fairchild, Herman L. (1932) *New York moraines.* Geological Society of America. Bulletin, v.43, p.627-662.

Farmin, Rollin (1934) *"Pebble dikes" and associated mineralization at Tintic, Utah.* Economic Geology, v.29, p.356-370.

Fay, Albert H. (1918) *A glossary of the mining and mineral industry.* Washington: Government Printing Office. 754p. (U.S. Bureau of Mines. Bulletin 95).

Fenneman, Nevin M. (1909) *Physiography of the St. Louis area.* Illinois. State Geological Survey. Bulletin, v.12. 83p.

Fenneman, Nevin M. (1914) *Physiographic boundaries within the United States.* Association of American Geographers. Annals, v.4, p.84-134.

Fenner, C.N. (1948) *Incandescent tuff flows in southern Peru.* Geological Society of America. Bulletin, v.59, p.879-893.

Fenton, Carroll L., and Fenton, Mildred A. (1928) *Faunule and zonule.* American Midland Naturalist, v.11, p.1-23. *(Ecologic interpretations of some biostratigraphic terms,* 1).

Fenton, Carroll L., and Fenton, Mildred A. (1930) *Zone, subzone, facies, phase.* American Midland Naturalist, v.12, p.145-153. *(Ecologic interpretations of some biostratigraphic terms,* 2).

Fernald, F.A. (1929) *Roundstone, a new geologic term.* Science, n.s., v.70, p.240.

Fernald, Merritt L. (1950) *Gray's manual of botany.* 8th ed. New York: American Book Company.

Fersman, Aleksandr E. (1922) *Geokhimiia Rossii.* St. Petersburg: Nauchnoe Khimichesko-Tekhnicheskoe izdatel'stvo. ca. 210p.

Fielder, Gilbert (1965) *Lunar geology.* Chester Springs (Penna.): Dufour. 184p.

Finch, R.H. (1933) *Slump scarps.* Journal of Geology, v.41, p.647-649.

Finsterwalder, Richard (1950) *Some comments on glacier flow.* Journal of Glaciology, v.1, no.7, p.383-388.

Fischer, Alfred G. (1964) *The Lofer cyclothem of the Alpine Triassic.* Kansas. State Geological Survey. Bulletin 169, v.1, p.107-149.

Fischer, Alfred G. (1969) *Geological time-distance rates: the Bubnoff unit.* Geological Society of America. Bulletin, v.80, p.549-551.

Fischer, Alfred G., and others (1954) *Arbitrary cut-off in stratigraphy.* American Association of Petroleum Geologists. Bulletin, v.38, p.926-931.

Fischer, Georg (1934) *Die Petrographie der Grauwacken.* Preussische Geologische Landesanstalt. Jahrbuch, 1933, Bd.54, p.320-343.

Fisher, O. (1866) *On the warp (of Mr. Trimmer) —its age and probable connexion with the last geological events.* Geological Society of London. Quarterly Journal, v.22. p.553-565.

Fisher, W.B. (1950) *The Middle East; a physical, social, and regional geography.* London: Methuen. 514p.

Fisk, Harold N. (1959) *Padre Island and the Laguna Madre Flats, coastal south Texas.* Coastal Geography Conference, 2nd, Baton Rouge (La.), 1959. [Proceedings], p.103-151.

Flanders, Phyllis L., and Sauer, Fred M. (1960) *A glossary of geoplosics: the systematic study of explosion effects in the Earth.* Kirtland Air Force Base (N.Mex.): U.S. Air Force Special Weapons Center. 34p. (U.S. Air Force. Special Weapons Center. Technical note 60-20).

Flawn, Peter T. (1953) *Petrographic classification of argillaceous sedimentary and low-grade metamorphic rocks in subsurface.* American Association of Petroleum Geologists. Bulletin, v.37, p.560-565.

Fleischer, Michael (1966) *Index of new mineral names, discredited minerals, and changes of mineralogical nomenclature in volumes 1-50 of* The American Mineralogist. American Mineralogist, v.51, p.1247-1357.

Fleuty, M.J. (1964) *Tectonic slides.* Geological Magazine, v.101, no.5, p.452-456.

Flint, D.E., and others (1953) *Limestone walls of Okinawa.* Geological Society of America. Bulletin, v.64, p.1247-1260.

Flint, Richard Foster (1928) *Eskers and crevasse fillings.* American Journal of Science, v.15, p.410-416.

Flint, Richard Foster (1957) *Glacial and Pleistocene geology.* New York: Wiley. 553p.

Flint, Richard Foster, and Deevey, Edward S., jr. (1951) *Radiocarbon dating of late-Pleistocene events.* American Journal of Science, v.249, p.257-300.

Flint, Richard Foster, and others (1960a) *Symmictite: a name for nonsorted terrigenous sedimentary rocks that contain a wide range of particle sizes.* Geological Society of America. Bulletin, v.71, p.507-509.

Flint, Richard Foster, and others (1960b) *Diamictite, a substitute term for symmictite.* Geological Society of America. Bulletin, v.71, p.1809.

Folk, Robert L. (1951) *Stages of textural maturity in sedimentary rocks.* Journal of Sedimentary Petrology, v.21, p.127-130.

Folk, Robert L. (1954) *The distinction between grain size and mineral composition in sedimentary-rock nomenclature.* Journal of Geology, v.62, p.344-359.

Folk, Robert L. (1959) *Practical petrographic classification of limestones.* American Association of Petroleum Geologists. Bulletin, v.43, p.1-38.

Folk, Robert L. (1962) *Spectral subdivision of limestone types.* American Association of Petroleum Geologists. Memoir 1, p.62-84.

Folk, Robert L. (1965) *Some aspects of recrystallization in ancient limestones.* Society of Economic Paleontologists and Mineralogists. Special Publication no.13, p.14-48.

Folk, Robert L. (1968) *Petrology of sedimentary rocks.* Austin (Tex.): Hemphill's Book Store. 170p.

Forgotson, James M., jr. (1954) *Regional stratigraphic analysis of the Cotton Valley Group of Upper Gulf Coastal Plain.* American Association of Petroleum Geologists. Bulletin, v.38, p.2476-2499.

Forgotson, James M., jr. (1957) *Nature, usage, and definition of marker-defined vertically segregated rock units.* American Association of Petroleum Geologists. Bulletin, v.41, p.2108-2113.

Forgotson, James M., jr. (1960) *Review and classification of quantitative mapping techniques.* American Association of Petroleum Geologists. Bulletin, v.44, p.83-100.

Foster, Norman H. (1966) *Stratigraphic leak.* American Association of Petroleum Geologists. Bulletin, v.50, no.12, p.2604-2611.

Fournet, J. (1844) *Sur l'état de surfusion du quartz dans les roches éruptives et dans les filons métallifères.* Académie des Sciences, Paris. Comptes rendus, v.18, p.1050-1057.

Fournet, J. (1847) *Résults d'une exploration des Vosges.* Société géologique de France. Bulletin, series 2, v.4, p.220-254.

Fowler, George M., and Lyden, Joseph P. (1932) *The ore deposits of the Tri-State district (Missouri-Kansas-Oklahoma).* American Institute of Mining and Metallurgical Engineers. Transactions, v.102, p.206-251.

Fowler, Henry W. (1937) *A dictionary of modern English usage.* 1st ed, reprinted with corrections. London: Oxford University Press. 742p.

Foye, James Clark (1916) *Are the "batholiths" of the Haliburton-Bancroft area, Ont., correctly named?* Journal of Geology, v.24, p.783-791.

Fox, Denis L. (1957) *Particulate organic detritus.* Geological Society of America. Memoir 67, v.1, p.383-389.

Freeman, Otis W. (1925) *The origin of Swimming Woman Canyon, Big Snowy Mountains, Montana, an example of a pseudo-cirque formed by landslide sapping.* Journal of Geology, v.33, p.75-79.

French, Bevan M. (1966) *Shock metamorphism of natural materials.* Science, v.153, p.903-906.

Friedman, Gerald M. (1965) *Terminology of crystallization textures and fabrics in sedimentary rocks.* Journal of Sedimentary Petrology, v.35, p.643-655.

Friedman, Jules D. (1970) *The airborne infrared scanner as a geophysical research tool.* Optical Spectra, June, p.35-44.

Frondel, Judith W., and others (1967) *Glossary of uranium- and thorium-bearing minerals.* 4th ed. U.S. Geological Survey. Bulletin 1250. 69p.

Frye, John C., and Willman, Harold B. (1960) *Classification of the Wisconsinan Stage in the Lake Michigan glacial lobe.* Illinois. State Geological Survey. Circular 285. 16p.

Fuchs, Theodor (1895) *Studien über Fucoiden und Hieroglyphen.* Akademie der Wissenschaften, Vienna. Mathematisch-Naturwissenschaftliche Classe. Denkschriften, v.62, p.369-448.

Fuller, Harry J., and Tippo, Oswald (1954) *College botany.* revised edition. New York: Holt, Rinehart & Winston.

Fuller, Myron L. (1914) *The geology of Long Island, N.Y.* U.S. Geological Survey. Professional Paper 82. 231p.

Fyfe, W.S., Turner, F.J., & Verhoogen, J. (1958) *Metamorphic reactions and metamorphic facies.* Geological Society of America. Memoir 73. 259p.

Galkiewicz, T. (1968) *Geocosmology.* International Geological Congress. 23rd, Prague, 1968. Report, sec.13, p.145-149.

Galliher, E. Wayne (1931) *Collophane from Miocene brown shales of California.* American Association of Petroleum Geologists. Bulletin, v.15, p.257-269.

Galloway, Jesse J. (1922) *Value of the physical characters of sand grains in the interpretation of the origin of sandstones.* Geological Society of America. Bulletin, v.33, p.104-105.

Galton, Francis (1889) *On the principle and*

methods of assigning marks for bodily effi-ciency. Nature, v.40, p.649-653.

Garland, George D. (1971) *Introduction to geophysics: mantle, core, and crust.* Philadelphia: W.B. Saunders. 420p.

Gault, D.E., and others (1968) *Impact cratering mechanics and structures.* In: French, Bevan M., and Short, Nicholas M., eds. *Shock metamorphism of natural materials,* p.87-100. Baltimore: Mono Book Corp. 644p.

Geer, Gerard de (1912) *A geochronology of the last 12,000 years.* International Geological Congress. 11th, Stockholm, 1910. Report, v.1, p.241-253.

Geikie, James (1898) *Earth sculpture, or the origin of land-forms.* New York: Putnam. 397p.

Gilbert, Charles M. (1954) *Sedimentary rocks.* In: Williams, Howel, and others. *Petrography; an introduction to the study of rocks in thin sections,* p.249-384. San Francisco: Freeman. 406p.

Gilbert, Grove Karl (1875) *Report on the geology of portions of Nevada, Utah, California, and Arizona, examined in the years 1871 and 1872.* U.S. Geographical Surveys West of the One Hundredth Meridian. Report, v.3, pt.1, p.17-187.

Gilbert, Grove Karl (1876) *The Colorado Plateau province as a field for geological study.* American Journal of Science, ser.3, v.12, p.16-24, 85-103.

Gilbert, Grove Karl (1877) *Report on the geology of the Henry Mountains.* Washington: U.S. Geographical and Geological Survey of the Rocky Mountain Region. 160p.

Gilbert, Grove Karl (1885) *The topographic features of lake shores.* U.S. Geological Survey. Annual Report, 5th, p.69-123.

Gilbert, Grove Karl (1890) *Lake Bonneville.* U.S. Geological Survey. Monograph 1. 438p.

Gilbert, Grove Karl (1898) *A proposed addition to physiographic nomenclature.* Science, n.s., v.7, p.94-95.

Gilbert, Grove Karl (1899) *Ripple-marks and cross-bedding.* Geological Society of America. Bulletin, v.10, p.135-140.

Gilbert, Grove Karl (1904) *Systematic asymmetry of crest lines in the high Sierra of California.* Journal of Geology, v.12, p.579-588.

Gilbert, Grove Karl (1914) *The transportation of débris by running water.* U.S. Geological Survey. Professional Paper 86. 263p.

Gilbert, Grove Karl (1928) *Studies of Basin Range structure.* U.S. Geological Survey. Professional Paper 153. 92p.

Gilbert, Grove Karl, and Brigham, Albert P.

(1902) *An introduction to physical geography.* New York: Appleton. 380p.

Giles, Albert W. (1918) *Eskers in the vicinity of Rochester, N.Y.* Rochester Academy of Science. Proceedings, v.5, p.161-240.

Gilluly, James (1966) *Orogeny and geochronology.* American Journal of Science, v.264, p.97-111.

Glaessner, M.F., and Teichert, C. (1947) *Geosynclines: a fundamental concept in geology.* 2 parts. American Journal of Science, v.245, p.465-482, p.571-591.

Glen, J.W. (1955) *The creep of polycrystalline ice.* Royal Society of London. Proceedings, ser.A, v.228, no.1175, p.519-538.

Glock, Waldo S. (1928) *An analysis of erosional forms.* American Journal of Science, ser.5, v.15, p.471-483.

Glock, Waldo S. (1932) *Available relief as a factor of control in the profile of a landform.* Journal of Geology, v.40, p.74-83.

Goldich, Samuel S. (1938) *A study in rock weathering.* Journal of Geology, v.46, p.17-58.

Goldman, Marcus I. (1921) *Lithologic subsurface correlation in the "Bend" series of north-central Texas.* U.S. Geological Survey. Professional Paper 129, p.1-22.

Goldman, Marcus I. (1933) *Origin of the anhydrite cap rock of American salt domes.* U.S. Geological Survey. Professional Paper 175-D, p.83-114.

Goldman, Marcus I. (1952) *Deformation, metamorphism, and mineralization in gypsum-anhydrite cap rock, Sulphur salt dome, Louisiana.* Geological Society of America. Memoir 50. 169p.

Goldman, Marcus I. (1961) *Regenerated anhydrite redefined.* Journal of Sedimentary Petrology, v.31, p.611.

Goldring, Winifred (1943) *Geology of the Coxsackie quadrangle, New York.* New York State. State Museum. Bulletin, no. 332. 374p.

Goldschmidt, V.M. (1937) *The principles of distribution of chemical elements in minerals and rocks.* Chemical Society (of London). Journal, v.1937, p.655-673.

Goldschmidt, V.M. (1954) *Geochemistry.* Oxford: Clarendon Press.

Goldsmith, Richard (1959) *Granofels: a new metamorphic rock name.* Journal of Geology, v.67, no.1, p.109-110.

Goldthwait, James W. (1908) *Intercision, a peculiar kind of modification of drainage.* School Science and Mathematics, v.8, p.129-139.

Goode, Harry D. (1969) *Geoevolutionism: a step*

beyond catastrophism and uniformitarianism. Geological Society of America. Abstracts with programs for 1969, pt.5, p.29.

Goodspeed, George Edward (1948) *Origin of granites.* In: Gilluly, J. *Origin of granite.* Geological Society of America. Memoir 28. p.55-78.

Goodspeed, George Edward (1955) *Relict dikes and relict pseudodikes.* American Journal of Science, v.253, no.3, p.146-161.

Grabau, Amadeus W. (1903) *Paleozoic coral reefs.* Geological Society of America. Bulletin, v.14, p.337-352.

Grabau, Amadeus W. (1904) *On the classification of sedimentary rocks.* American Geologist, v.33, p.228-247.

Grabau, Amadeus W. (1905) *Physical characters and history of some New York formations.* Science, v.22, p.528-535.

Grabau, Amadeus W. (1906) *Types of sedimentary overlap.* Geological Society of America. Bulletin, v.17, p.567-636.

Grabau, Amadeus W. (1911) *On the classification of sand grains.* Science, n.s., v.33, p.1005-1007.

Grabau, Amadeus W. (1920) *Principles of salt deposition.* 1st ed. New York: McGraw-Hill. 435p. (*Geology of the non-metallic mineral deposits other than silicates,* v.1).

Grabau, Amadeus W. (1920a) *General geology.* Boston: Heath. 864p. (*A textbook of geology,* pt.1).

Grabau, Amadeus W. (1924) *Principles of stratigraphy.* 2nd ed. New York: A.G. Seiler. 1185p.

Grabau, Amadeus W. (1932) *Principles of stratigraphy.* 3d ed. New York: A.G. Seiler. 1185p.

Grabau, Amadeus W. (1936) *The Great Huangho Plain of China.* Association of Chinese and American Engineers. Journal, v.17, p.247-266.

Grabau, Amadeus W. (1936a) *Oscillation or pulsation.* International Geological Congress. 16th, Washington, 1933. Report, v.1, p.539-553.

Grabau, Amadeus W. (1940) *The rhythm of the ages; Earth history in the light of the pulsation and polar control theory.* Peking: Henri Vetch. 561p.

Graton, Louis C., and Fraser, Horace J. (1935) *Systematic packing of spheres, with particular relation to porosity and permeability.* Journal of Geology, v.43, p.785-909.

Gravenor, Conrad P. (1956) *Air photographs of the plains region of Alberta.* Research Council of Alberta. Preliminary Report 56-5. 35p.

Gravenor, Conrad P., and Kupsch, W.O. (1959)

Ice-disintegration features in western Canada. Journal of Geology, v.67, p.48-64.

Gray, Henry H. (1955) *Stratigraphic nomenclature in coal-bearing rocks.* Geological Society of America. Bulletin, v.66, p.1567-1568.

Green, D.H., and Ringwood, A.E. (1963) *Mineral assemblages in a model mantle composition.* Journal of Geophysical Research, v.68, p.937-945.

Greensmith, J. Trevor (1957) *The status and nomenclature of stratified evaporites.* American Journal of Science, v.255, p.593-595.

Gregory, Herbert E. (1917) *Geology of the Navajo country; a reconnaissance of parts of Arizona, New Mexico, and Utah.* U.S. Geological Survey. Professional Paper 93. 161p.

Gregory, J.W. (1912) *The relations of kames and eskers.* Geographical Journal, v.40, p.169-175.

Gressly, Amanz (1838) *Observations géologiques sur le Jura Soleurois.* [Pt.1]. Allgemeine Schweizerische Gesellschaft für die gesammten Naturwissenschaften. Neue Denkschriften, v.2, [pt.6]. 112p.

Griffiths, John C., and Rosenfeld, Melvin A. (1954) *Operator variation in experimental research.* Journal of Geology, v.62, p.74-91.

Grim, Ralph E. (1953) *Clay mineralogy.* New York: McGraw-Hill. 384p.

Grim, Ralph E. (1968) *Clay mineralogy.* 2nd ed. New York: McGraw-Hill. 596p.

Grim, Ralph E., and others (1937) *The mica in argillaceous sediments.* American Mineralogist, v.22, p.813-829.

Grohskopf, John G., and McCracken, Earl (1949) *Insoluble residues of some Paleozoic formations of Missouri, their preparation, characteristics and application.* Missouri. Division of Geological Survey and Water Resources. Report of Investigations, no.10. 39p.

Grossman, William L. (1944) *Stratigraphy of the Genesee group of New York.* Geological Society of America. Bulletin, v.55, p.41-75.

Grout, Frank F. (1932) *Petrography and petrology, a textbook.* New York: McGraw-Hill. 522p.

Grubenmann, Ulrich (1904) *Die kristallinen Schiefer.* Berlin: Gebrüder Borntraeger. 2 vol.

Grubenmann, Ulrich, and Niggli, P. (1924) *Die Gesteinsmetamorphose: I. Allgemeiner Teil.* Berlin: Gebrüder Borntraeger. 539p.

Gruner, John W. (1950) *An attempt to arrange silicates in the order of reaction energies at relatively low temperatures.* American Mineralogist, v.35, p.137-148.

Gruner, John W., and others (1941) *Structural geology of the Knife Lake area of northeastern*

Minnesota. Geological Society of America. Bulletin, v.52, p.1577-1642.

Guilcher, André (1950) *Définition d'un type de volcan "écossais".* Association de Géographes Français. Bulletin, no.206-207, p.2-11.

Gulliver, Frederick P. (1896) *Cuspate forelands.* Geological Society of America. Bulletin, v.7, p.399-422.

Gulliver, Frederick P. (1899) *Shoreline topography.* American Academy of Arts and Sciences. Proceedings, v.34, p.149-258.

Gümbel, Carl Wilhelm von (1868) *Geognostische Beschreibung des ostbayerischen Grenzgebirges oder des bayerischen und oberpfälzer Waldgebirges.* Gotha: Justus Perthes. 968p. (Bavaria. Bayerisches Oberbergamt Geognostische Abteilung. *Geognostische Beschreibung des koenigreichs Bayern,* v.2).

Gümbel, Carl Wilhelm von (1874) *Die paläolithischen Eruptivgesteine des Fichtelgebirges (als vorläufige Mittheilung).* Munich. 50p.

Gussow, William Carruthers (1958) *Metastasy or crustal shift.* Alberta Society of Petroleum Geologists. Journal, v.6, p.253-257.

Haaf, Ernst ten (1959) *Graded beds of the northern Apennines.* PhD thesis, Rijks University of Groningen. 102p.

Haarmann, E. (1926) *Tektogenese oder Gefugebildung statt Orogenese oder Gebirgsbildung.* Deutsche Geologische Gesellschaft. Zeitschrift, v.78, p.105-107.

Haarmann, E. (1930) *Die Oszillationstheorie; eine erklarung der krustenbewegungen von erder und mond.* Stuttgart: F. Enke, 260p.

Hack, J.T. (1960) *Interpretation of erosional topography in humid temperate regions.* American Journal of Science, v.258a, p.80-97.

Hadding, Assar (1931) *On subaqueous slides.* Geologiska Föreningen, Stockholm. Förhandlingar, Bd.53, no.387, p.377-393.

Hadding, Assar (1933) *On the organic remains of the limestones; a short review of the limestone forming organisms.* Acta universitatis Lundensis, n.s., Bd.29, pt.2, no.4. 93p. *(The pre-Quaternary sedimentary rocks of Sweden,* 5).

Haeckel, Ernst (1862) *Die Radiolarien (Rhizopoda radiaria); eine Monographie.* Berlin: Georg Reimer. 572p.

Hall, James (1843) *Geology of New York; part IV, comprising the survey of the fourth geological district.* Albany: Carroll and Cook. 683p.

Hall, James (1859) *Description and figures of the organic remains of the lower Helderberg Group and the Oriskany Sandstone.* New York Geological Survey. Paleontology, v.3. 532p.

Hamblin, W. Kenneth (1958) *The Cambrian sandstones of northern Michigan.* Michigan. Geological Survey Division. Publication 51. 149p.

Hamelin, Louis-Edmond (1961) *Périglaciaire du Canada: idées nouvelles et perspectives globales.* Cahiers de géographie de Québec, v.5, no.10, p.141-203.

Hamelin, Louis-Edmond, and Clibbon, Peter (1962) *Vocabulaire périglaciaire bilingue (français et anglais).* Cahiers de géographie de Québec, v.6, no.12, p.201-226.

Hamelin, Louis-Edmond, and Cook, Frank A. (1967) *Le périglaciaire par l'image; illustrated glossary of periglacial phenomena.* Quebec: Presses de l'Université Laval. 237p. (Centre d'Etudes Nordiques. Travaux et documents 4).

Hammer, W. (1914) *Das Gebiet der Bündnerschiefer im tirolischen Oberinntal.* Kaiserliche Köningliche Geologische Reichsanstalt (Austria). Jahrbuch, v.64, p.443-567.

Hansen, A.M. (1894) *The glacial succession in Norway.* Journal of Geology, v.2, p.123-143.

Harbaugh, John W. (1962) *Geologic guide to Pennsylvanian marine banks, southeast Kansas.* Kansas Geological Society. Guidebook; field conference, 27th, p.13-67.

Hardy, Clyde T., and Williams, James S. (1959) *Columnar contemporaneous deformation.* Journal of Sedimentary Petrology, v.29, p.281-283.

Harker, Alfred (1909) *The natural history of igneous rocks.* New York: Macmillan. 377p.

Harker, Alfred (1918) *The anniversary address of the President.* Geological Society of London. Quarterly Journal, v.74, p.i-lxxx.

Harker, Alfred (1939) *Metamorphism.* 2d ed. London: Methuen. 362p.

Harland, W.B., and others (1966) *The definition and identification of tills and tillites.* Earthscience Reviews, v.2, p.225-256.

Harrassowitz, H. (1927) *Anchimetamorphose, das Gebiet zwischen Oberflächen- und Tiefenumwandlung der Erdrinde.* Oberhessischen Gesellschaft für Natur- und Heilkunde zu Giessen. Naturwissenschaftliche Abteilung. Bericht, 12, p.9-15.

Harrington, Horacio J. (1946) *Las corrientes de barro ("mud-flows") de "El Volcán", quebrada de Humahuaca, Jujuy.* Sociedad Geológica Argentina. Revista, t.1, p.149-165.

Harris, Stanley E., jr. (1943) *Friction cracks and the direction of glacial movement.* Journal of Geology, v.51, p.244-258.

Hartshorn, J.H. (1958) *Flowtill in southeastern*

Massachusetts. Geological Society of America. Bulletin, v.69, p.477-482.

Harvey, Roger D. (1931) *Glacial chutes from the Peruvian Cordillera.* American Journal of Science, 5th ser., v.31, p.220-231.

Harvie-Brown, J.A. (1910) *"Caledonia rediviva".* Scottish Geographical Magazine, v.26, p.93-94.

Hatch, F.H., and Rastall, R.H. (1913) *The petrology of the sedimentary rocks; a description of the sediments and their metamorphic derivatives.* London: George Allen. 425p.

Hatch, F.H., and Rastall, R.H. (1965) *Petrology of the sedimentary rocks.* 4th ed. revised by J. Trevor Greensmith. London: Thomas Murby. 408p. *(Textbook of petrology, v.2).*

Haug, Emile (1900) *Les géosynclinaux et les aires continentales; contribution à l'étude des transgressions et des régressions marines.* Société Géologique de France. Bulletin, ser.3, v.28, p.617-711.

Haug, Émile (1907) *Les phénomènes géologiques.* Paris: Librairie Armand Colin. 538p. (His Traité de géologie, 1).

Haupt, Arthur W. (1953) *Plant morphology.* New York: McGraw-Hill.

Haupt, Lewis M. (1883) *The topographer, his instruments and his methods.* Philadelphia: Henry Carey Baird. 247p.

Haupt, Lewis M. (1906) *Changes along the New Jersey coast.* New Jersey. Geological Survey. Annual report of the State Geologist, 1905, p.27-95.

Haüy, Réné J. (1801) *Traité de minéralogie.* T. 2. Paris: Chez Louis. 444p.

Hawkes, Herbert E., and Webb, John S. (1962) *Geochemistry in mineral exploration.* New York: Harper & Row. 415p.

Hayes, Charles W. (1899) *Physiography of the Chattanooga district, in Tennessee, Georgia, and Alabama.* U.S. Geological Survey. Annual Report, 19th, pt.2, p.1-58.

Heald, Milton T. (1956) *Cementation of Simpson and St. Peter sandstones in parts of Oklahoma, Arkansas, and Missouri.* Journal of Geology, v.64, p.16-30.

Hedberg, Hollis D. (1958) *Stratigraphic classification and terminology.* American Association of Petroleum Geologists. Bulletin, v.42, p.1881-1896.

Hedberg, Hollis D. (1961) *The stratigraphic panorama (an inquiry into the bases for age determination and age classification of the Earth's rock strata).* Geological Society of America. Bulletin, v.72, p.499-517.

Heiland, Carl August (1940) *Geophysical exploration.* New York: Prentice-Hall. 1013p.

Heim, Arnold (1908) *Über rezente und fossile subaquatische Rutschungen und deren lithologische Bedeutung.* Neues Jahrbuch für Mineralogie, Geologie und Paläontologie, 1908, Bd. 2, p.136-157.

Heim, Arnold, and Potonié, Robert (1932) *Beobachtungen über die Entstehung der tertiären Kohlen (Humolithe und Saprohumolith) in Zentralsumatra.* Geologische Rundschau, v.23, no.3,4, p.145-172.

Henbest, Lloyd G. (1952) *Significance of evolutionary explosions for diastrophic division of Earth history.* Journal of Paleontology, v.26, p.299-318.

Henbest, Lloyd G. (1968) *Diagenesis in oolitic limestones of Morrow (Early Pennsylvanian) age in northwestern Arkansas and adjacent Oklahoma.* U.S. Geological Survey. Professional Paper 594-H. 22p.

Hendricks, C. Leo (1952) *Correlations between surface and subsurface sections of the Ellenburger group of Texas.* Texas. University. Bureau of Economic Geology. Report of Investigations, no.11. 44p.

Henningsmoen, Gunnar (1961) *Remarks on stratigraphical classification.* Norges Geologiske Undersökelse, no.213, p.62-92.

Henson, F.R.S. (1950) *Cretaceous and Tertiary reef formations and associated sediments in Middle East.* American Association of Petroleum Geologists. Bulletin, v.34, p.215-238.

Herrick, Clarence L. (1904) *The clinoplains of the Rio Grande.* American Geologist, v.33, p.376-381.

Herrin, Eugene, and Taggart, James (1968) *Source bias in epicenter determinations.* Seismological Society of America. Bulletin, v.58, p.1791-1796.

Hess, H.H. (1938) *Gravity anomalies and island arc structures, with particular reference to the West Indies.* American Philosophical Society. Proceedings, v.79, p.71-96.

Hesse, Richard (1924) *Tiergeographie auf ökologischer Grundlage.* Jena: G. Fischer. 613p.

Hesse, Richard, and others (1937) *Ecological animal geography.* New York: Wiley. 597p.

Hey, Max H. (1962) *An index of mineral species & varieties arranged chemically, with an alphabetical index of accepted mineral names and synonyms.* 2nd rev. ed., reprinted with corrections. London: British Museum (Natural History). 728p.

Hey, Max H. (1963) *Appendix to the second edi-*

tion of *An index of mineral species and varieties arranged chemically.* London: British Museum (Natural History). 135p.

Hietanan, Anna (1967) *On the facies series in various types of metamorphism.* Journal of Geology, v.75, no.2, p.187-214.

Higashi, Akira (1958) *Experimental study of frost heaving.* U.S. Army. Snow, Ice & Permafrost Research Establishment. Research Report 45. 20p.

Hill, Robert T. (1891) *The Comanche series of the Texas-Arkansas region.* Geological Society of America. Bulletin, v.2, p.503-528.

Hill, Robert T. (1900) *Physical geography of the Texas region.* U.S. Geological Survey. Topographic Atlas, folio 3. 12p.

Hills, Edwin S. (1940) *The lunette, a new land form of aeolian origin.* Australian Geographer, v.3, no.7, p.15-21.

Hills, E. Sherbon (1963) *Elements of structural geology.* New York: Wiley. 483p.

Himus, Godfrey W. (1954) *A dictionary of geology.* Baltimore: Penguin Books. 153p.

Hind, Wheelton, and Howe, John A. (1901) *The geological succession and palaeontology of the beds between the Millstone Grit and the limestone-massif at Pendle Hill and their equivalents in certain other parts of Britain.* Geological Society of London. Quarterly Journal, v.57, p.347-404.

Hinds, Norman E.A. (1943) *Geomorphology, the evolution of landscape.* New York: Prentice-Hall. 894p.

Hitchcock, Edward (1841) *Elementary geology.* 2nd ed. New York: Dayton & Saxton. 346p.

Hitchcock, Edward (1843) *The phenomena of drift, or glacioaqueous action in North America between the Tertiary and alluvial periods.* Association of American Geologists and Naturalists. Reports, 1st-3rd mtngs, p.164-221.

Hitchcock, Edward (1844) *Report on ichnolithology or fossil footmarks, with description of several new species and the coprolites of birds, and of a supposed footmark from the valley of Hudson River.* American Journal of Science, v.47, p.292-322.

Hitchcock, Edward (1858) *Ichnology of New England; a report on the sandstone of the Connecticut Valley, especially its footmarks.* Boston: William White. 220p.

Hobbs, William H. (1901) *The Newark system of the Pomperaug Valley, Connecticut.* U.S. Geological Survey. Annual Report, 21st, pt.3, p.7-160.

Hobbs, William H. (1907) *On some principles of seismic geology.* Beiträge zur Geophysik, v.7 p.219-292.

Hobbs, William H. (1911a) *Characteristics o existing glaciers.* New York: Macmillan. 301p

Hobbs, William H. (1911b) *Repeating patterns ir the relief and in the structure of the land.* Geological Society of America. Bulletin, v.22 p.123-176.

Hobbs, William H. (1912) *Earth features and their meaning; an introduction to geology for the student and the general reader.* New York: Macmillan. 506p.

Hobbs, William H. (1917) *The erosional ana degradational processes of deserts, with especial reference to the origin of desert depressions.* Association of American Geographers. Annals, v.7, p.25-60.

Hobbs, William H. (1921) *Studies of the cycle of glaciation.* Journal of Geology, v.29, p.370-386.

Hoffmeister, J.E., and Ladd, H.S. (1944) *The antecedent-platform theory.* Journal of Geology, v.52, p.388-402.

Hofmann, Walter, chairman (1965) *Parametric hydrology and stochastic hydrology; report of the Committee on Surface Water Hydrology.* American Society of Civil Engineers. Proceedings, Hydraulics Division. Journal, v.91, no. HY6, p.119-122.

Hollingworth, S.E., Taylor, J.H., & Kellaway, G.A. (1950) *Large-scale superficial structures in the Northampton Ironstone Field.* Geological Society of London. Quarterly Journal, v.100, p.1-44.

Holm, Donald A. (1957) *Sigmoidal dunes; a transitional form.* Geological Society of America. Bulletin, v.68, p.1746.

Holm, Esther A. (1957) *Sand pavements in the Rub' al Khali.* Geological Society of America. Bulletin, v.68, p.1746.

Holmes, Arthur (1920) *The nomenclature of petrology.* 1st ed. London: Thomas Murby. 284p.

Holmes, Arthur (1928) *The nomenclature of petrology.* 2nd ed. London: Thomas Murby. 284p.

Holmes, Arthur (1959) *A revised geological time-scale.* Edinburgh Geological Society. Transactions, v.17, p.183-216.

Holmes, Arthur (1965) *Principles of physical geology.* 2d ed. New York: Ronald Press. 1288p.

Hooke, Roger LeB. (1967) *Processes on arid-region alluvial fans.* Journal of Geology, v.75, p.438-460.

Hopkins, David M. (1949) *Thaw lakes and thaw sinks in the Imuruk Lake Area, Seward Penin-*

sula, Alaska. Journal of Geology, v.57, p.119-131.

Hopkins, David M., and others (1955) *Permafrost and ground water in Alaska.* U.S. Geological Survey. Professional Paper 264-F, p.113-146.

Hopkins, David M., and Sigafoos, R.S. (1951) *Frost action and vegetation patterns on Seward Peninsula, Alaska.* U.S. Geological Survey. Bulletin 974-C, p.51-100.

Hoppe, Gunnar (1952) *Hummocky moraine regions, with special reference to the interior of Norbotton.* Geografiska Annaler, v.34, p.1-71.

Horberg, Leland (1954) *Rocky Mountain and continental Pleistocene deposits in the Waterton region, Alberta, Canada.* Geological Society of America. Bulletin, v.65, p.1093-1150.

Horton, Robert E. (1932) *Drainage basin characteristics.* American Geophysical Union. Transactions, v.13, p.350-361.

Horton, Robert E. (1945) *Erosional development of streams and their drainage basins; hydrophysical approach to quantitative morphology.* Geological Society of America. Bulletin, v.56, p.275-370.

Hörz, F., Hartung, J.B., and Gault, D.E. (1971) *Micrometeorite craters on lunar rock surfaces.* Journal of Geophysical Research, v.76, p.5770-5798.

Howard, Arthur D. (1942) *Pediment passes and pediment problem.* Journal of Geomorphology, v.5, p.1-31, 95-136.

Howard, Arthur D. (1959) *Numerical systems of terrace nomenclature; a critique.* Journal of Geology, v.67, p.239-243.

Howard, W.V., and David, M.W. (1936) *Development of porosity in limestones.* American Association of Petroleum Geologists. Bulletin, v.20, no.11, p.1389-1412.

Howell, Jesse V. (1922) *Notes on the pre-Permian Paleozoics of the Wichita Mountain area.* American Association of Petroleum Geologists. Bulletin, v.6, p.413-425.

Hsu, Kenneth Jinghwa (1955) *Monometamorphism, polymetamorphism and retrograde metamorphism.* American Journal of Science, v.253, no.4, p.237-239.

Huang, W.H., and Walker, R.M. (1967) *Fossil alpha-particle recoil tracks: a new method of age determination.* Science, v.155, p.1103-1106.

Huang, W.T. (1962) *Petrology.* New York: McGraw-Hill. 480p.

Hubert, John F. (1960) *Petrology of the Fountain and Lyons formations, Front Range, Colorado.* Colorado School of Mines. Quarterly, v.55, no.1, p.1-242.

Hudson, George H. (1909) *Some items concerning a new and an old coast line of Lake Champlain.* New York State. State Museum. Bulletin 133, p.159-163.

Hudson, George H. (1910) *Joint caves of Valcour Island—their age and their origin.* New York State. State Museum. Bulletin 140, p.161-196.

Hudson, R.G. (1924) *On the rhythmic succession of the Yoredale Series in Wensleydale.* Yorkshire Geological Society. Proceedings, n.s., v.20, p.125-135.

Humble, William (1843) *Dictionary of geology and mineralogy, comprising such terms in botany, chemistry, comparative anatomy, conchology, entomology, palaeontology, zoology, and other branches of natural history, as are connected with the study of geology.* 2nd ed., with additions. London: Henry Washbourne. 294p.

Humphries, D.W. (1966) *The booming sand of Korizo, Sahara, and the squeaking sand of Gower, S. Wales: a comparison of the fundamental characteristics of two musical sands.* Sedimentology, v.6, p.135-152.

Hunt, Charles B., and others (1953) *Geology and geography of the Henry Mountains region, Utah.* U.S. Geological Survey. Professional Paper 228. 234p.

Huschke, Ralph E., ed. (1959) *Glossary of meteorology.* Boston: American Meteorological Society. 638p.

Hutchinson, G. Evelyn (1957) *A treatise on limnology.* Vol.1. New York: Wiley. 1015p.

Hutchinson, J.N. (1967) *The free degradation of London Clay cliffs.* Geotechnical conference on shear strength properties of natural soils and rocks, Oslo. Proceedings, 1, p.113-118.

Hutchinson, J.N. (1968) *Mass movement.* In: Fairbridge, R.W., ed. *Encyclopedia of geomorphology,* p.688-696. New York: Reinhold. 1295p.

Hutchinson, J.N., & Bhandari, R.K. (1971) *Undrained loading, a fundamental mechanism of mudflows and other mass movements.* Géotechnique, v.21, no.4, pp.353-358.

Hutton, James (1788) *Theory of the Earth; or an investigation of the laws observable in the composition, dissolution, and restoration of land upon the globe.* Royal Society of Edinburgh. Transactions, v.1, p.209-304.

Huxley, Thomas H. (1862) *The anniversary address.* Geological Society of London. Quarterly Journal, v.18, p.xl-liv.

Huxley, Thomas H. (1877) *Physiography: an introduction to the study of nature.* London: Macmillan. 384p.

Hyde, H.A., and Williams, D.A. (1944) *The right word.* Pollen Analysis Circular, no.8, p.6.

Illing, Leslie V. (1954) *Bahaman calcareous sands.* American Association of Petroleum Geologists. Bulletin, v.38, p.1-95.

Imbrie, John, and Buchanan, Hugh (1965) *Sedimentary structures in modern carbonate sands of the Bahamas.* Society of Economic Paleontologists and Mineralogists. Special Publication, no.12, p.149-172.

Imbt, William C., and Ellison, Samuel P., jr. (1947) *Porosity in limestone and dolomite petroleum reservoirs.* Drilling and Production Practice, 1946, p.364-372.

Ingle, James C. (1966) *The movement of beach sand; an analysis using fluorescent grains.* New York: Elsevier. 221p. (*Developments in sedimentology, 5*).

International Commission on Zoological Nomenclature (1964) *International code of zoological nomenclature, adopted by the XV International Congress of Zoology.* N.R. Stoll, Editorial Committee, chairman. 2nd ed. London: International Trust for Zoological Nomenclature. 176p.

International Committee for Coal Petrology, Nomenclature subcommittee (1963). *International handbook of coal petrography.* 2d ed. Paris: Centre National de la Recherche Scientifique.

International Subcommission on Stratigraphic Terminology (1961) *Stratigraphic classification and terminology.* Edited by Hollis D. Hedberg. International Geological Congress. 21st, Copenhagen, 1960. Report, pt.25. 38p.

International Subcommission on Stratigraphic Terminology (1965) *Definition of geologic systems.* American Association of Petroleum Geologists. Bulletin, v.49, p.1694-1703. (American Commission on Stratigraphic Nomenclature. Note 32).

Ireland, H.A., and others (1947) *Terminology for insoluble residues.* American Association of Petroleum Geologists. Bulletin, v.31, p.1479-1490.

Ireland, Herbert O. (1969) *Foundations for heavy structures.* Reviews in Engineering Geology, v.2, p.1-15.

Irwin, J.S. (1926) *Faulting in the Rocky Mountain region.* American Association of Petroleum Geologists. Bulletin, v.10, no.2, p.105-129.

Isacks, Bryan, and others (1968) *Seismology and the new global tectonics.* Journal of Geophysical Research, v.73, p.5855-5899.

Iversen, Johs. (1954) *The late-glacial flora of Denmark and its relation to climate and soil.* Denmark. Geologiske Undersögelse. Danmarks geologiske undersögelse, ser.2, no.80, p.87-119.

Ives, Ronald L. (1941) *Tundra ponds.* Journal of Geomorphology, v.4, p.285-296.

Jaanusson, Valdar (1960) *The Viruan (Middle Ordovician) of Öland.* Uppsala. University. Geological Institutions. Bulletin, v.38, p.207-288.

Jacks, G.V., and others, eds. (1960) *Multilingual vocabulary of soil science.* 2nd ed., revised. New York(?): United Nations, Food and Agriculture Organization, Land & Water Development Division. 430p.

Jackson, Benjamin D. (1950) *A glossary of botanic terms with their derivation and accent.* 4th ed. New York: Hafner. 481p.

Jackson, J.R. (1834) *Hints on the subject of geographical arrangement and nomenclature.* Royal Geographical Society. Journal, v.4, p.72-88.

Jaeger, John C. (1965) *Application of the theory of heat conduction to geothermal measurements,* ch.2. In: *Terrestrial heat flow.* American Geophysical Union. Geophysical Monograph No.8, p.7-23.

Jakosky, J.J. (1950) *Exploration geophysics.* 2d ed. Los Angeles: Trija.

James, Harold L. (1954) *Sedimentary facies of iron-formation.* Economic Geology, v.49, p.235-293.

Jameson, Robert (1808) *Elements of geognosy.* Edinburgh: William Blackwood. 368p. (His *System of mineralogy, v.3*).

Jamieson, J.A., and others (1963) *Infrared physics and engineering.* New York: McGraw-Hill.

Jeletzky, Jurjiz A. (1956) *Paleontology, basis of practical geochronology.* American Association of Petroleum Geologists. Bulletin, v.40, p.679-706.

Jeletzky, Jurjiz A. (1966) *Comparative morphology, phylogeny, and classification of fossil Coleoidea.* University of Kansas. Paleontological Contributions, art.7. 162p.

Jennings, Jesse D. (1968) *Prehistory of North America.* New York: McGraw-Hill. 391p.

Johannsen, Albert (1917) *Suggestions for a quantitative mineralogical classification of igneous rocks.* Journal of Geology, v.25, p.63-97.

Johannsen, Albert (1931) *Introduction, textures, classifications and glossary.* 1st ed. Chicago:

University of Chicago Press. 267p. *(A descriptive petrography of the igneous rocks, v.1.)*

Johannsen, Albert (1931a) *The quartz-bearing rocks.* 1st ed. Chicago: University of Chicago Press. *(A descriptive petrography of the igneous rocks, v.2.)*

Johannsen, Albert (1938) *The feldspathoid rocks; the peridotites and perknites.* Chicago: University of Chicago Press. 523p. *(A descriptive petrography of the igneous rocks, v.4.)*

Johannsen, Albert (1939) *Introduction, textures, classifications and glossary.* 2nd ed. Chicago: University of Chicago Press. 318p. *(A descriptive petrography of the igneous rocks, v.1.)*

Johnson, Douglas W. (1916) *Plains, planes, and peneplanes.* Geographical Review, v.1, p.443-447.

Johnson, Douglas W. (1919) *Shore processes and shoreline development.* New York: Wiley. 584p.

Johnson, Douglas W. (1925) *The New England-Acadian shore line.* New York: Wiley. 608p.

Johnson, Douglas W. (1932) *Rock fans of arid regions.* American Journal of Science, 5th ser., v.23, p.389-416.

Johnson, Douglas W. (1939) *The origin of submarine canyons; a critical review of hypotheses.* New York: Columbia University Press. 126p.

Johnson, Paul H., and Bhappu, Roshan B. (1969) *Chemical mining—a study of leaching agents.* New Mexico. State Bureau of Mines and Mineral Resources. Circular 99. 10p.

Jones, Owen T. (1937) *On the sliding or slumping of submarine sediments in Denbigshire, North Wales, during the Ludlow period.* Geological Society of London. Quarterly Journal, v.93, p.241-283.

Jones, Paul H. (1969) *Hydrology of Neogene deposits in the northern Gulf of Mexico basin.* Louisiana Water Resources Research Institute. Bulletin GT-2, 105p.

Jones, P.B. (1971) *Folded faults and sequence of thrusting in Alberta foothills.* American Association of Petroleum Geologists. Bulletin, v.55, no.2, p.292-306.

Joplin, Germaine A. (1968) *A petrography of Australian metamorphic rocks.* New York: American Elsevier. 262p.

Jopling, Alan V. (1961) *Origin of regressive ripples explained in terms of fluid-mechanic processes.* U.S. Geological Survey. Professional Paper 424-D, art.299, p.15-17.

Judson, S. Sheldon, jr. (1953) *Geology of the San Jon site, eastern New Mexico.* Smithsonian Miscellaneous Collections, v.121, no.1. 70p.

Jukes, J. Beete (1862) *On the mode of formation of some of the river-valleys in the south of Ireland.* Geological Society of London. Quarterly Journal, v.18, p.378-403.

Jukes-Browne, Alfred J. (1903) *The term "hemera".* Geological Magazine, dec.4, v.10, p.36-38.

Jurine, Louis (1806) *Réflexions sur la nécessité d'une nouvelle nomenclature en géologie, et l'exposé de celle qu'il propose.* Journal des Mines, v.19, p.367-378.

Kahn, James S. (1956) *The analysis and distribution of the properties of packing in sand-size sediments.* 2 parts. Journal of Geology, v.64, p.385-395, 578-606.

Kalkowsky, Ernst (1880) *Ueber die Erforschung der archäischen Formationen.* Neues Jahrbuch für Mineralogie, Geologie und Palaeontologie, Jahrg.1880, Bd.1, p.1-28.

Kalkowsky, Ernst (1908) *Oolith und Stromatolith im norddeutschen Buntsandstein.* Deutsche Geologische Gesellschaft. Zeitschrift, Bd.60, p.68-125.

Karcz, Iaakov (1967) *Harrow marks, current-aligned sedimentary structures.* Journal of Geology, v.75, p.113-121.

Karlstrom, Thor N.V. (1956) *The problem of the Cochrane in late Pleistocene chronology.* U.S. Geological Survey. Bulletin 1021-J, p.303-331.

Karlstrom, Thor N.V. (1961) *The glacial history of Alaska; its bearing on paleoclimatic theory.* New York Academy of Sciences. Annals, v.95, art.1, p.290-340.

Karlstrom, Thor N.V. (1966) *Quaternary glacial record of the North Pacific region and worldwide climatic changes.* In: Blumenstock, David I., ed. *Pleistocene and post-Pleistocene climatic variations in the Pacific area,* p.153-182. Honolulu: Bishop Museum Press. 182p.

Kaufmann, Franz J. (1886) *Emmen- und Schlierengegenden nebst Umgebungen bis zur Brünigstrasse und Linie Lungern-Grafenort geologisch aufgenommen und dargestellt.* Beiträge zur geologischen Karte der Schweiz, v.24, pt.1. 608p. with atlas of 30 plates.

Kay, G.F. (1916) *Gumbotil, a new term in Pleistocene geology.* Science, v.44, p.637-638.

Kay, G. Marshall (1937) *Stratigraphy of the Trenton Group.* Geological Society of America. Bulletin, v.48, p.233-302.

Kay, G. Marshall (1945) *North American geosynclines: their classification* [abstract]. Geological Society of America. Bulletin, v.56, p. 1172.

Kay, G. Marshall (1945a) *Paleogeographic and palinspastic maps.* American Association of Petroleum Geologists. Bulletin, v.29, p.426-450.

Kay, G. Marshall (1947) *Geosynclinal nomenclature and the craton.* American Association of Petroleum Geologists. Bulletin, v.31, no.7, p.1287-1293.

Kay, G. Marshall (1951) *North American geosynclines.* Geological Society of America. Memoir 48. 143p.

Keeton, William T. (1967) *Biological science.* New York: Norton.

Keith, Arthur (1895) *Description of the Knoxville sheet.* U.S. Geological Survey. Geologic Atlas, folio 16. 6p.

Keith, Mackenzie L., and Degens, Egon T. (1959) *Geochemical indicators of marine and freshwater sediments.* In: Abelson, Philip H., ed. *Researches in geochemistry,* p.38-61. New York: Wiley. 511p.

Keller, Walter D. (1958) *Argillation and direct bauxitization in terms of concentrations of hydrogen and metal cations at surface of hydrolyzing aluminum silicates.* American Association of Petroleum Geologists. Bulletin, v.42, p.233-245.

Keller, Walter D. (1963) *Diagenesis in clay minerals—a review.* National Conference of Clays and Clay Minerals. 11th, Ottawa, 1962. Clays and clay minerals; proceedings, p.136-157. *(International series of monographs on earth sciences, v.13).*

Kelley, Vincent C. (1956) *Thickness of strata.* Journal of Sedimentary Petrology, v.26, p.289-300.

Kelling, Gilbert (1958) *Ripple-mark in the Rhinns of Galloway.* Edinburgh Geological Society. Transactions, v.17, p.117-132.

Kemp, James F. (1896) *A handbook of rocks, for use without the microscope.* New York: J.F. Kemp. 176p.

Kemp, James F. (1900) *Handbook of rocks for use without the microscope.* New York: Van Nostrand. 185p.

Kemp, James F. (1934) *A handbook of rocks for use without the microscope, with a glossary of the names of rocks and other lithological terms.* 5th ed. New York: Van Nostrand. 300p.

Kemp, James F. (1940) *A handbook of rocks (for use without the petrographic microscope).* 6th ed., completely revised & edited by Frank F. Grout. New York: Van Nostrand.

Kendall, Percy F. (1902) *A system of glacier-lakes in the Cleveland Hills.* Geological Society of London. Quarterly Journal, v.58, p.471-571.

Kendall, Percy F., and Bailey, E.B. (1908) *The glaciation of East Lothian south of the Garleton Hills.* Royal Society of Edinburgh. Transactions, v.46, p.1-31.

Kendall, Percy F., and Wroot, Herbert E. (1924) *Geology of Yorkshire; an illustration of the evolution of northern England.* Vol. 1. Vienna: Hollinek Brothers. 660p.

Kennedy, J.F. (1963) *The mechanics of dunes and antidunes in erodible-bed channels.* Journal of Fluid Mechanics, v.16, p.521-544.

Keroher, Grace C., and others (1966) *Lexicon of geologic names of the United States for 1936-1960.* Parts 1-3. U.S. Geological Survey. Bulletin 1200. 4341p.

Kerr, Paul F., and Hamilton, P.K. (1949) *Glossary of clay mineral names.* New York: Columbia University. 66p. (American Petroleum Institute. Project 49: Clay Mineral Standards. Preliminary Report no.1).

Kerr, Paul F., and Kopp, Otto C. (1958) *Salt-dome breccia.* American Association of Petroleum Geologists. Bulletin, v.42, p.548-560.

Kerr, Richard C., and Nigra, John O. (1952) *Eolian sand control.* American Association of Petroleum Geologists. Bulletin, v.36, p.1541-1573.

Kerr, Washington C. (1881) *On the action of frost in the arrangement of superficial earthy material.* American Journal of Science, v.21, p.345-358.

Kesseli, John E. (1941) *The concept of the graded river.* Journal of Geology, v.49, p.561-588.

Kessler, Paul (1922) *Über Lochverwitterung und ihre Beziehungen zur Metharmose (Umbildung) der Gesteine.* Geologische Rundschau, v.12, p.237-270.

Keyes, Charles R. (1910) *Deflation and the relative efficiencies of erosional processes under conditions of aridity.* Geological Society of America. Bulletin, v.21, p.565-598.

Keyes, Charles R. (1913) *Antigravitational gradation.* Science, n.s., v.38, p.206.

Kieffer, S.W. (1971) *Shock metamorphism of the Coconino sandstone at Meteor Crater, Arizona.* Journal of Geophysical Research, v.76, p.5449-5473.

Kinahan, George Henry (1878) *Manual of the geology of Ireland.* London: C.K. Paul. 444p.

Kindle, Edward M. (1916) *Small pit and mound structures developed during sedimentation.* Geological Magazine, dec.6, v.3, p.542-547.

Kindle, Edward M. (1917) *Recent and fossil ripple-mark.* Canada. Geological Survey. Museum Bulletin no.25. 121p. (Canada. Department of Mines. Geological Series, no.34).

Kindle, Edward M. (1923) *Range and distribution of certain types of Canadian Pleistocene concretions.* Geological Society of America. Bulletin, v.34, p.609-648.

Kindle, Edward M. (1923a) *Nomenclature and genetic relations of certain calcareous rocks.* Pan-American Geologist, v.39, p.365-372.

Kindle, Edward M. (1926) *Contrasted types of mud cracks.* Royal Society of Canada. Proceedings and Transactions, ser.3, v.20, sec.4, p.71-75.

King, Clarence (1878) *Systematic geology.* Washington: U.S. Government Printing Office. 803p. (U.S. Geological Exploration of the Fortieth Parallel. Report, v.1).

King, Cuchlaine A.M., and Williams, W.W. (1949) *The formation and movement of sand bars by wave action.* Geographical Journal, v.113, p.70-85.

King, Lester C. (1948) *A theory of bornhardts.* Geographical Journal, v.112, p.83-87.

King, Lester C. (1959) *Denudational and tectonic relief in south-eastern Australia.* Geological Society of South Africa. Transactions, v.62, p.113-138.

King, Lester C. (1962) *The morphology of the Earth; a study and synthesis of world scenery.* New York: Hafner. 699p.

King, Philip (1959) *The evolution of North America.* Princeton: Princeton University Press. 190p.

Kinsman, David J.J. (1969) *Modes of formation, sedimentary associations, and diagnostic features of shallow-water and supratidal evaporites.* American Association of Petroleum Geologists. Bulletin, v.53, p.830-840.

Kirwan, Richard (1794) *Elements of mineralogy.* Vol. 1. 2nd ed. London: J. Nichols. 510p.

Klein, George de Vries (1963) *Analysis and review of sandstone classifications in the North American geological literature, 1940-1960.* Geological Society of America. Bulletin, v.74, p.555-575.

Klöden, Karl Friedrich von (1828) *Beiträge zur mineralogischen und geognostischen Kenntniss der Mark Brandenburg.* [Pt.1]. Berlin: W. Dieterici. 108p.

Klugh, A. Brooker (1923) *A common system of classification in plant and animal ecology.* Ecology, v.4, p.366-377.

Knight, C.L. (1957) *Ore genesis—the source bed concept.* Economic Geology, v.52, no.7, p.808-817.

Knight, James B. (1941) *Paleozoic gastropod genotypes.* Geological Society of America. Special Paper 32. 510p.

Knight, Wilbur C. (1897) *"Mineral soap".* Engineering and Mining Journal, v.63, p.600-601.

Knight, Wilbur C. (1898) *Bentonite.* Engineering and Mining Journal, v.66, p.491.

Knopf, Eleanora Bliss, and Ingerson, Earl (1938) *Structural petrology.* Geological Society of America. Memoir 6. 270p.

Knox, Alexander (1904) *Glossary of geographical and topographical terms.* London: Edward Stanford. 432p. (*Stanford's compendium of geography and travel; suppl. vol.*).

Knutson, Ray M. (1958) *Structural sections and the third dimension.* Economic Geology, v.53, p.270-286.

Kobayashi, Teiichi (1944) *An instant in the Phanerozoic Eon and its bearings on geology and biology.* Imperial Academy, Tokyo. Proceedings, v.20, p.742-750. (*Concept of time in geology, 3*).

Kobayashi, Teiichi (1944a) *On the major classification of the geological age.* Imperial Academy, Tokyo. Proceedings, v.20, p.475-478. (*Concept of time in geology, 1*).

Koch, Lauge (1926) *Ice cap and sea ice in north Greenland.* Geographical Review, v.16, p.98-107.

Koppejan, A.W., van Wanelen, B.M., and Weinberg, L.J.H. (1948) *Coastal flow slides in the Dutch province of Zeeland.* International Conference on Soil Mechanics and Foundation Engineering. 2nd, Rotterdam, 1948?. Proceedings, 5, p.89-96.

Kostov, Ivan (1961) *Genesis of kyanite in quartz veins.* International Geology Review, v.3, no.8, p.645-651.

Krasheninnikov, G.F. (1964) *Facies, genetic types, and formations.* International Geology Review, v.6, p.1242-1248.

Krauskopf, Konrad B. (1967) *Introduction to Geochemistry.* New York: McGraw-Hill. 721p.

Krishtofovich, A.N. (1945) *The mode of preservation of plant fossils and their bearing upon the problem of coal formation.* Akademiya Nauk SSSP. Izvestiya, Seriya Geologicheskaya, no.2, p.136-150.

Krivenko, V., and Lapchik, T. (1934) *On the petrography of the crystalline rocks of the* rapids of the Dnipro. Ukrainskii Naukovo-Doslidchii Geologichnii Institute. Trudy, v.5, no.2, p.67-77.

Kruger, Fredrick C. (1946) Structure and metamorphism of the Bellows Falls quadrangle of New Hampshire and Vermont. Geological Society of America. Bulletin, v.57, p.161-205.

Krumbein, William C. (1934) Size frequency distributions of sediments. Journal of Sedimentary Petrology, v.4, p.65-77.

Krumbein, William C. (1955) Composite end members in facies mapping. Journal of Sedimentary Petrology, v.25, p.115-122.

Krumbein, William C., and Libby, Willard G. (1957) Application of moments to vertical variability maps of stratigraphic units. American Association of Petroleum Geologists. Bulletin, v.41, p.197-211.

Krumbein, William C., and Pettijohn, Francis J. (1938) Manual of sedimentary petrography. New York: Appleton. 549p.

Krumbein, William C., and Sloss, Laurence L. (1951) Stratigraphy and sedimentation. San Francisco: Freeman. 497p.

Krumbein, William C., and Sloss, Laurence L. (1963) Stratigraphy and sedimentation. 2nd ed. San Francisco: Freeman. 660p.

Krynine, Paul D. (1937) Petrography and genesis of the Siwalik Series. American Journal of Science, ser.5, v.34, no.204, p.422-446.

Krynine, Paul D. (1940) Petrology and genesis of the Third Bradford Sand. Pennsylvania. State College. Mineral Industries Experiment Station. Bulletin 29. 134p.

Krynine, Paul D. (1941) Differentiation of sediments during the life history of a landmass. Geological Society of America. Bulletin, v.52, p.1915.

Krynine, Paul D. (1945) Sediments and the search for oil. Producers Monthly, v.9, p.12-22.

Krynine, Paul D. (1948) The megascopic study and field classification of sedimentary rocks. Journal of Geology, v.56, p.130-165.

Krynine, Paul D. (1951) Reservoir petrography of sandstones. In: Payne, Thomas G., and others. Geology of the Arctic Slope of Alaska, sheet 2. U.S. Geological Survey. Oil and Gas Investigations, map OM 126. 3 sheets.

Ksiazkiewicz, Marian (1958) Submarine slumping in the Carpathian flysch. Polskie Towarzystwo Geologicne. Rocznik, t.28, p.123-151.

Kubiëna, Walter Ludwig (1953) The soils of Europe. London: Thomas Murby. 317p.

Kuenen, Philip H. (1943) Pitted pebbles. Leidsche geologische mededeelingen, v.13, p.189-201.

Kuenen, Philip H. (1947) Water-faceted boulders. American Journal of Science, v.245, p.779-783.

Kuenen, Philip H. (1948) Slumping in the Carboniferous rocks of Pembrokeshire. Geological Society of London. Quarterly Journal, v.104, p.365-385.

Kuenen, Philip H. (1950) Marine geology. New York: Wiley. 568p.

Kuenen, Philip H. (1953) Significant features of graded bedding. American Association of Petroleum Geologists. Bulletin, v.37, p.1044-1066.

Kuenen, Philip H. (1957) Sole markings of graded graywacke beds. Journal of Geology, v.65, p.231-258.

Kuenen, Philip H. (1958) Experiments in geology. Geological Society of Glasgow. Transactions, v.23, p.1-28.

Kulp, J. Laurence (1961) Geologic time scale. Science, v.133, p.1105-1114.

Kümmel, Henry B. (1893) Some rivers of Connecticut. Journal of Geology, v.1, p.371-393.

Kupsch, Walter O. (1956) Submask geology in Saskatchewan. International Williston Basin Symposium. 1st, Bismarck (N.D.), 1956. Williston Basin Symposium, p.66-75.

Lacroix, A. (1922) Minéralogie de Madagascar. v.2: minéralogie appliquée; lithologie. Paris: Augustin Challamel. 694p.

Lahee, Frederic H. (1923) Field geology. 2nd ed. New York: McGraw-Hill. 649p.

Lahee, Frederic H. (1961) Field geology. 6th ed. New York: McGraw-Hill. 926p.

Lamar, John E. (1928) Geology and economic resources of the St. Peter Sandstone of Illinois. Illinois. State Geological Survey. Bulletin no.53. 175p.

Lamb, H.H. (1970) Volcanic dust in the atmosphere. Royal Society of London, Philosophical Transactions, Series A, Mathematics and Physical Sciences, v.266, p.471.

Lambe, Thomas W. (1953) The structure of inorganic soil. American Society of Civil Engineers. Proceedings, v.79, separate no.315. 49p.

Lamont, Archie (1957) Slow anti-dunes and flow marks. Geological Magazine, v.94, p.472-480.

Lamplugh, G.W. (1895) The crush-conglomerates of the Isle of Man. With a petrographical appendix by W.W. Watts. Geological Society of London. Quarterly Journal, v.51, p.563-599.

Lamplugh, G.W. (1902) "Calcrete". Geological Magazine, n.s., dec.4, v.9, no.462, p.575.

Landes, Kenneth K. (1945) The Mackinac Brec-

cia. Michigan. Geological Survey Division. Publication 44, Geological Series 37, ch.3, p.123-153.

Landes, Kenneth K. (1957) *Chemical unconformities.* Geological Society of America. Bulletin, v.68, p.1759.

Lane, Alfred C. (1903) *Porphyritic appearance of rocks.* Geological Society of America. Bulletin 14, p.369-379.

Lane, Alfred C. (1923) *Communication.* Journal of Geology, v.31, p.348.

Lane, Alfred C. (1928) *Isontic?* Science, v.68, p.37.

Lang, Walter T.B. (1937) *The Permian formations of the Pecos Valley of New Mexico and Texas.* American Association of Petroleum Geologists. Bulletin, v.21, p.833-898.

Langbein, W.B., and Iseri, Kathleen T. (1960) *General introduction and hydrologic definitions.* U.S. Geological Survey. Water-supply Paper 1541-A. 29p. *(Manual of hydrology: pt.1, p.1-29).*

Larsen, Esper Signius (1938) *Some new variation diagrams for groups of igneous rocks.* Journal of Geology, v.46, p.505-520.

Larsen, Gunnar, and Chilingar, George V., eds. (1967) *Diagenesis in sediments.* Amsterdam: Elsevier. 551p. *(Developments in sedimentology, 8).*

Larsen, Ole, & Sörensen, H. (1960) *Principles of classification and norm calculations of metamorphic rocks: a discussion.* Journal of Geology, v.68, no.6, p.681-683, 1960.

Lasius, Georg S.O. (1789) *Beobachtungen über die Harzgebirge, nebst einem Profilnisse, als ein Beytrag zur mineralogischen Naturkunde.* Hannover: Helwingischen Hofbuchhandlung. 559p. in 2 v.

Lasky, Samuel Grossman (1950) *How tonnage and grade relations help predict ore reserves.* Engineering and Mining Journal, v.151, no.4, p.81-85.

Lawrence, George H.M. (1951) *Taxonomy of vascular plants.* New York: Macmillan.

Lawson, Andrew C. (1885) *Report on the geology of the Lake of the Woods region, with special reference to the Keewatin (Huronian?) belt of the Archean rocks.* Canada. Geological Survey. Annual Report, 1:CC. 151p.

Lawson, Andrew C. (1894) *The geomorphogeny of the coast of northern California.* California. University. Department of Geology. Bulletin, v.1, no.8, p.241-271.

Lawson, Andrew C. (1904) *The geomorphogeny of the upper Kern Basin.* California. University,

Berkeley. Department of Geology. Bulletin, v.3, p.291-376.

Lawson, Andrew C. (1913) *The petrographic designation of alluvial-fan formations.* California. University, Berkeley. Department of Geology. Bulletin, v.7, p.325-334.

Lawson, Andrew C. (1915) *The epigene profiles of the desert.* California. University, Berkeley. Department of Geology. Bulletin, v.9, p.23-48.

Lawson, Andrew C., and others (1908) *The California earthquake of April 18, 1906; report of the State Earthquake Commission.* Vol. 1. Carnegie Institution of Washington. Publication no.87. 451p.

Lee, Charles A. (1840) *The elements of geology, for popular use; containing a description of the geological formations and mineral resources of the United States.* New York: Harper. 375p.

Lee, Willis T. (1900) *The origin of the débris-covered mesas of Boulder, Colo.* Journal of Geology, v.8, p.504-511.

Lee, Willis T. (1903) *The canyons of northeastern New Mexico.* Journal of Geography, v.2, p.63-82.

Lees, George Martin (1952) *Foreland folding.* Geological Society of London. Quarterly Journal, v.108, p.1-34.

Leet, L. Don, and Leet, Florence J. (1965) *The Earth's mantle.* Seismological Society of America. Bulletin, v.55, p.619-625.

Legget, Robert F. (1962) *Geology and engineering.* 2nd ed. New York: McGraw-Hill. 884p.

Legrand, H.E. (1965) *Patterns of contaminated zones of water in the ground.* Water Resources Research, v.1, no.1, p.83-95.

Lehmann, Johann Gottlob (1756) *Versuch einer Geschichte von Flötz-Gebürgen, betreffend deren Entstehung, Lage, darinne befindliche Metallen, Mineralien und Fossilien, gröstentheils aus eigenen Wahrnehmungen, chymischen und physicalischen Versuchen, und aus denen Grundsätzen der Natur-Lehre hergeleitet, und mit nöthigen Kupfern versehen.* Berlin: F.A. Lange. [329]p.

Leighton, Morris M. (1959) *Stagnancy of the Illinoian glacial lake east of the Illinois and Mississippi rivers.* Journal of Geology, v.67, p.337-344.

Leighton, Morris M., and MacClintock, Paul (1930) *Weathered zones of the drift-sheets of Illinois.* Journal of Geology, v.38, p.28-53.

Leighton, Morris W., and Pendexter, C. (1962) *Carbonate rock types.* American Association of Petroleum Geologists. Memoir 1, p.33-61.

Leith, Charles K. (1905) *Rock cleavage.* U.S. Geological Survey. Bulletin 239. 216p.

Leith, Charles K. (1923) *Structural geology.* Revised ed. New York: Holt. 390p.

Leith, Charles K., and Mead, Warren J. (1915) *Metamorphic geology; a text-book.* New York: Holt. 337p.

Leopold, Luna B., and Langbein, Walter B. (1962) *The concept of entropy in landscape evolution.* U.S. Geological Survey. Professional Paper 500-A, p.1-20.

Leopold, Luna B., and Maddock, Thomas, jr. (1953) *The hydraulic geometry of stream channels and some physiographic implications.* U.S. Geological Survey. Professional Paper 252. 57p.

Leopold, Luna B., and Wolman, M.G. (1957) *River channel patterns—braided, meandering, and straight.* U.S. Geological Survey. Professional Paper 282-B, p.39-85.

Leopold, Luna B., and others (1964) *Fluvial processes in geomorphology.* San Francisco: Freeman. 522p.

Leopold, Luna B., and others (1966) *Channel and hillslope processes in a semiarid area, New Mexico.* U.S. Geological Survey. Professional Paper 352-G, p.193-253.

Lesevich, Vladimir V. (1877) *Opyt kriticheskogo izsledovaniia osnovonachal pozitivnoi filosofii.* St. Petersburg: M. Stasiulevich. 295p.—Also in: Lesevich, V.V. (1915) *Sobranie sochinenii,* v.1, p.189-454. Moscow: IU.V. Leontovich.

Leverett, Frank (1903) *Glacial features of lower Michigan.* Journal of Geology, v.11, p.117-118.

Levin, Ernest M., and others (1964) *General discussion of phase diagrams for ceramists.* Margie K. Reser, ed. Columbus (Ohio): American Ceramic Society. 601p.

Levorsen, A.I. (1933) *Studies in paleogeology.* American Association of Petroleum Geologists. Bulletin, v.17, p.1107-1132.

Levorsen, A.I. (1943) *Discovery thinking.* American Association of Petroleum Geologists. Bulletin, v.27, p.887-928.

Levorsen, A.I. (1960) *Paleogeologic maps.* San Francisco: Freeman. 174p.

Lewis, Douglas W. (1964) *"Perigenic"; a new term.* Journal of Sedimentary Petrology, v.34, p.875-876.

Lilly, H.D. (1966) *Late Precambrian and Appalachian tectonics in the light of submarine exploration on the Great Bank of Newfoundland and in the Gulf of St. Lawrence; preliminary views.* American Journal of Science, v.264, p.569-574.

Lindgren, Waldemar (1912) *The nature of replacement.* Economic Geology, v.7, p.521-535.

Lindgren, Waldemar (1928) *Mineral deposits.* 3d ed. New York: McGraw-Hill Book Company. 1049p.

Lindgren, Waldemar (1933) *Mineral deposits.* 4th ed. New York: McGraw-Hill. 930p.

Lindsey, Alton A., and others (1969) *Natural areas in Indiana and their preservation.* Lafayette: Indiana Natural Areas Survey, Department of Biological Sciences, Purdue University. 594p.

Link, Theodore A. (1950) *Theory of transgressive and regressive reef (bioherm) development and origin of oil.* American Association of Petroleum Geologists. Bulletin, v.34, p.263-294.

Linton, D.L. (1955) *The problem of tors.* Geographical Journal, v.121, p.470-487.

Lliboutry, Louis (1958) *Studies of the shrinkage after a sudden advance, blue bands and wave ogives on Glaciar Universidad (central Chilean Andes).* Journal of Glaciology, v.3, no.24, p.261-272.

Lobeck, Armin K. (1921) *A physiographic diagram of the United States.* Chicago: Nystrom. map. 1:3,000,000.

Lobeck, Armin K. (1924) *Block diagrams and other graphic methods used in geology and geography.* New York: Wiley. 206p.

Lobeck, Armin K. (1939) *Geomorphology; an introduction to the study of landscapes.* New York: McGraw-Hill. 731p.

Lockwood, J.P. (1971) *Sedimentary and gravity-slide emplacement of serpentinite.* Geological Society of America. Bulletin, v.82, p.919-936.

Loewinson-Lessing, Franz J. (1899) *Note sur la classification et la nomenclature des roches éruptives.* International Geological Congress. 7th, St. Petersburg, 1897. Compte rendu, p.53-71.

Logan, William Edmund (1863) *Report on the geology of Canada.* Canada. Geological Survey. Report of Progress to 1863.

Lohest, M.J. Maximilien, and others (1909) *Compte rendu de la session extraordinaire de la Société Géologique de Belgique tenue à Eupen et à Bastogne les 29, 30 et 31 août et 1, 2 et 3 septembre 1908.* Société Géologique de Belgique. Annales, t.35, p.351-414.

Lombard, Augustin (1963) *Laminites: a structure of flysch-type sediments.* Journal of Sedimentary Petrology, v.33, p.14-22.

Long, Albert E. (1960) *A glossary of the diamond-drilling industry.* U.S. Bureau of Mines. Bulletin 583. 98p.

Longwell, Chester R. (1933) *Meaning of the term "roches moutonnées".* American Journal of Science, v.225, p.503-504.

Longwell, Chester R. (1951) *Megabreccia developed downslope from large faults.* American Journal of Science, v.249, p.343-355.

Longwell, Chester R., Knopf, Adolph, and Flint, R.F. (1969) *Physical geology.* [3rd] ed. New York: Wiley. 685p.

Lovely, H.R. (1948) *Onlap and strike-overlap.* American Association of Petroleum Geologists. Bulletin, v.32, p.2295-2297.

Lovén, Sven L. (1874) *Etudes sur les Echinoïdées.* K. Svenska Vetenskaps-Akademien, Stockholm. Handlingar, ser.4, Bd.11, no.7, p.1-91. 53 plates.

Lovering, T.S. (1963) *Epigenetic, diplogenetic, syngenetic, and lithogene deposits.* Economic Geology, v.58, no.3, p.315-331.

Lowe, Charles H., jr. (1961) *Biotic communities in the Sub-Mogollon region of the inland Southwest.* Arizona Academy of Science. Journal, v.2, p.40-49.

Lowenstam, Heinz A. (1950) *Niagaran reefs of the Great Lakes area.* Journal of Geology, v.58, p.430-487.

Lozinski, W.V. (1909) *Das Sandomierz-Opalower Lössplateau.* Globus, v.96, p.330-334.

Lucia, Floyd J. (1962) *Diagenesis of a crinoidal sediment.* Journal of Sedimentary Petrology, v.32, p.848-865.

Lundqvist, G. (1949) *The orientation of the block material in certain species of flow earth.* Geografiska Annaler, v.31, p.335-347.

Lüttig, Gerd (1965) *Interglacial and interstadial periods.* Journal of Geology, v.73, p.579-591.

Lyell, Charles (1830) *Principles of geology.* Vol. 1. London: John Murray. 511p.

Lyell, Charles (1833) *Principles of geology.* Vol. 3. London: John Murray. 398p. plus 109p. of appendix.

Lyell, Charles (1838) *Elements of geology.* London: John Murray. 543p.

Lyell, Charles (1839) *Elements of geology.* 1st American, from the 1st London ed. Philadelphia: J. Kay. 316p.

Lyell, Charles (1840) *Principles of geology.* 6th ed. London: John Murray. 3v.

Lyell, Charles (1854) *Principles of geology; or, the modern changes of the Earth and its inhabitants considered as illustrative of geology.* 9th rev. ed. New York: Appleton. 834p.

Lynch, Edward J. (1962) *Formation evaluation.* New York: Harper & Row. 422p.

Mabbutt, J.A. (1961) *"Basal surface" or "weathering front".* Geologists' Association. Proceedings, v.72, p.357-358.

Macar, P. (1948) *Les pseudo-nodules du Famennien et leur origine.* Société Géologique de Belgique. Annales, v.72, p.47-74.

MacArthur, Robert H., and Wilson, Edward O. (1967) *The theory of island biogeography.* Princeton (N.J.): Princeton University Press. 203p. (Monographs in population biology).

MacBride, Thomas H. (1910) *Geology of Hamilton and Wright cos.* Iowa. Geological Survey. Volume 20, p.97-149.

MacDonald, Gordon Andrew (1953) *Pahoehoe, aa, and block lava.* American Journal of Science, v.251, no.3, p.169-191.

MacEwan, Douglas M.C. (1947) *The nomenclature of the halloysite minerals.* Mineralogical Magazine, v.28, p.36-44.

Mackie, William (1897) *On the laws that govern the rounding of particles of sand.* Edinburgh Geological Society. Transactions, v.7, p.298-311.

Mackin, J. Hoover (1937) *Erosional history of the Big Horn Basin, Wyoming.* Geological Society of America. Bulletin, v.48, p.813-893.

Mackin, J. Hoover (1948) *Concept of the graded river.* Geological Society of America. Bulletin, v.59, p.463-511.

Mackin, J. Hoover (1950) *The down-structure method of viewing geologic maps.* Journal of Geology, v.58, p.55-72.

Mackinder, Halford J. (1919) *Democratic ideals and reality; a study in the politics of reconstruction.* London: Constable. 272p.

MacNeil, Francis S. (1954) *Organic reefs and banks and associated detrital sediments.* American Journal of Science, v.252, p.385-401.

Makiyama, Jiro (1954) *Syntectonic construction of geosynclinal neptons.* Kyoto. University. College of Science. Memoirs, ser.B, v.21, p.115-149.

Mallet, Robert (1838) *On the mechanism of glaciers, being an attempt to ascertain the causes and effects of their peculiar, and in part, unobserved, motions.* Geological Society of Dublin. Journal, v.1, p.317-335.

Malott, Clyde A. (1928a) *An analysis of erosion.* Indiana Academy of Science. Proceedings, v.37, p.153-163.

Malott, Clyde A. (1928b) *The valley form and its development.* Indiana University Studies, v.15, no.81, p.3-34.

Mann, C. John (1970) *Isochronous, synchronous,*

and coetaneous. Journal of Geology, v.78, p.749-750.

Manton, W.I. (1965) *The orientation and origin of shatter cones in the Vredefort Ring.* New York Academy of Sciences. Annals, v.123, art.2, p.1017-1049.

Marbut, Curtis F. (1896) *Physical features of Missouri.* Missouri. Geological Survey. Volume 10, p.11-109.

Marks, Robert W. (1969) *The new dictionary and handbook of aerospace.* New York: Praeger. 531p.

Marr, John E. (1900) *The scientific study of scenery.* London: Methuen. 368p.

Marr, John E. (1901) *The origin of moels, and their subsequent dissection.* Geographical Journal, v.17, p.63-69.

Marr, John E. (1905) *The anniversary address of the President.* Geological Society of London. Quarterly Journal, v.61, p.xlvii-lxxxvi.

Marschner, Hannelore (1969) *Hydrocalcite (CaCO$_3$.H$_2$O) and nesquehonite (MgCO$_3$.3H$_2$O) in carbonate scales.* Science, v.165, p.1119-1121.

Martin, Henry G. (1931) *Insoluble residue studies of Mississippian limestones in Indiana.* Indiana. Department of Conservation. Publication no.101. 37p.

Martin Vivaldi, J.L., and MacEwan, D.M.C. (1960) *Corrensite and swelling chlorite.* Clay Minerals Bulletin, v.4, no.24, p.173-181.

Mason, Brian (1958) *Principles of geochemistry.* New York: Wiley.

Matthes, François É. (1912) *Topology, topography and topometry.* American Geographical Society. Bulletin, v.44, p.334-339.

Matthes, François É. (1930) *Geologic history of the Yosemite Valley.* U.S. Geological Survey. Professional Paper 160. 137p.

Matthes, François É. (1939) *Report of Committee on Glaciers, April 1939.* American Geophysical Union. Transactions, 20th annual meeting (1939), pt.4, p.518-523.

Maw, George (1866) *On subaerial and marine denudation.* Geological Magazine, v.3, p.439-451.

Mawe, John (1818) *A new descriptive catalogue of minerals, consisting of more varieties than heretofore published, and intended for the use of students, with which they may arrange the specimens they collect.* 3rd ed. London: Longman, Hurst, Rees, Orme, and Brown. 96p.

Maxey, George B. (1964) *Hydrostratigraphic units.* Journal of Hydrology, v.2, p.124-129.

Maxson, John H. (1940) *Fluting and faceting of rock fragments.* Journal of Geology, v.48, p.717-751.

Maxson, John H. (1940a) *Gas pits in non-marine sediments.* Journal of Sedimentary Petrology, v.10, p.142-145.

Maxson, John H. (1950) *Physiographic features of the Panamint Range, California.* Geological Society of America. Bulletin, v.61, p.99-114.

Maxson, John H., and Anderson, George H. (1935) *Terminology of surface forms of the erosion cycle.* Journal of Geology, v.43, p.88-96.

Maxwell, W.G.H. (1968) *Atlas of the Great Barrier Reef.* New York: Elsevier. 258p.

May, Irving, and Cuttitta, Frank (1967) *New instrumental techniques in geochemical analysis.* In: Abelson, Philip H., ed. *Researches in geochemistry,* p.112-142. New York: Wiley. 663p.

McBride, Earle F. (1962) *Flysch and associated beds of the Martinsburg Formation (Ordovician), central Appalachians.* Journal of Sedimentary Petrology, v.32, p.39-91.

McBride, Earle F. (1962a) *The term graywacke.* Journal of Sedimentary Petrology, v.32, p.614-615.

McBride, Earle F. (1963) *A classification of common sandstones.* Journal of Sedimentary Petrology, v.33, p.664-669.

McBride, Earle F., and Yeakel, Lloyd S. (1963) *Relationship between parting lineation and rock fabric.* Journal of Sedimentary Petrology, v.33, p.779-782.

McCammon, Richard B. (1968) *The dendrograph: a new tool for correlation.* Geological Society of America. Bulletin, v.79, p.1663-1670.

McConnell, Richard G., and Brock, R.W. (1904) *Report on the great landslide at Frank, Alberta.* Canada. Department of the Interior. Annual report 1902-1903, pt.8, appendix. 17p.

McElroy, C.T. (1954) *The use of the term "greywacke" in rock nomenclature in N.S.W.* Australian Journal of Science, v.16, p.150-151.

McGee, WJ (1888) *The geology of the head of Chesapeake Bay.* U.S. Geological Survey. Annual Report, 7th, p.537-646.

McGee, WJ (1891) *The Pleistocene history of northeastern Iowa.* U.S. Geological Survey. Annual Report, 11th, pt.1, p.189-577.

McGee, WJ (1897) *Sheetflood erosion.* Geological Society of America. Bulletin, v.8, p.87-112.

McGee, WJ (1908) *Outlines of hydrology.* Geological Society of America. Bulletin, v.19, p.193-200.

McIntosh, D.H., ed. (1963) *Meteorological glos-*

sary. London: Her Majesty's Stationery Office. 288p.

McIver, N.L. (1961) *Upper Devonian marine sedimentation in the central Appalachians.* PhD thesis, Johns Hopkins Univ. 347p.

McKee, Edwin D. (1938) *The environment and history of the Toroweap and Kaibab formations of northern Arizona and southern Utah.* Carnegie Institution of Washington. Publication 492. 268p.

McKee, Edwin D. (1939) *Some types of bedding in the Colorado River delta.* Journal of Geology, v.47, p.64-81.

McKee, Edwin D. (1954) *Stratigraphy and history of the Moenkopi Formation of Triassic age.* Geological Society of America. Memoir 61. 133p.

McKee, Edwin D. (1965) *Experiments on ripple lamination.* Society of Economic Paleontologists and Mineralogists. Special Publication no.12, p.66-83.

McKee, Edwin D. (1966) *Structures of dunes at White Sands National Monument, New Mexico (and a comparison with structures of dunes from other selected areas).* Sedimentology, v.7, p.1-69.

McKee, Edwin D., and Weir, Gordon W. (1953) *Terminology for stratification and cross-stratification in sedimentary rocks.* Geological Society of America. Bulletin, v.64, p.381-389.

McKinstry, Hugh Exton (1948) *Mining geology; including a glossary of mining and geological terms.* New York: Prentice-Hall. 680p.

McNitt, James R. (1965) *Review of geothermal resources,* ch. 9 in: *Terrestrial heat flow.* American Geophysical Union. Geophysical Monograph No.8, p.240-266.

McWolfinvan, C., ed. (1971) *Orismologic progeny.* Bethesda (Md.)/Washington, D.C.: Rebecca Elizabeth Wolf and Matthew Joseph Sullivan.

Mead, Daniel Webster (1919) *Hydrology, the fundamental basis of hydraulic engineering.* 1st ed. New York: McGraw-Hill. 647p.

Mechtly, E.A. (1964) *International system of units: physical constants and conversion factors.* National Aeronautics and Space Administration: George C. Marshall Space Flight Center. 19p.

Medlicott, H.B., and Blanford, W.T. (1879) *A manual of the geology of India.* Pt.1-2. 2v. Calcutta: Geological Survey of India. 817p.

Mehnert, K.R. (1968) *Migmatites and the origin of granitic rocks.* Amsterdam: Elsevier. 393p.

Meier, Mark F. (1961) *Mass budget of South Cascade Glacier, 1957-60.* U.S. Geological Survey. Professional Paper 424-B, p.206-211.

Meinzer, Oscar E. (1923) *Outline of groundwater hydrology, with definitions.* U.S. Geological Survey. Water-supply Paper 494. 71p.

Meinzer, Oscar E. (1939) *Discussion of question No. 2 of the International Commission on Subterranean Water: definitions of the different kinds of subterranean water.* American Geophysical Union. Transactions, pt.4, p.674-677.

Meinzer, Oscar E. (1942) *Hydrology.* New York: Dover. 712p.

Mellor, Joseph W. (1908) *A note on the nomenclature of clays.* English Ceramic Society. Transactions, v.8, p.23-30.

Melton, Frank A. (1936) *An empirical classification of flood-plain streams.* Geographical Review, v.26, p.593-609.

Melton, Frank A. (1940) *A tentative classification of sand dunes; its application to dune history in the southern High Plains.* Journal of Geology, v.48, p.113-173.

Melton, Frank A. (1947) *Onlap and strike-overlap.* American Association of Petroleum Geologists. Bulletin, v.31, p.1868-1878.

Merriam, C. Hart (1890) *Results of a biological survey of the San Francisco mountain region and desert of the Little Colorado in Arizona.* U.S. Department of Agriculture. Division of Ornithology and Mammalogy. North American fauna, no.3. 136p.

Merriam, Daniel F. (1962) *Late Paleozoic limestone "buildups" in Kansas.* Kansas Geological Society. Guidebook; field conference, 27th, p.73-81.

Merriam, Daniel F. (1963) *The geologic history of Kansas.* Kansas. State Geological Survey. Bulletin 162. 317p.

Merrill, George P. (1897) *A treatise on rocks, rock-weathering and soils.* New York: Macmillan. 411p.

Meyerhoff, Howard A. (1940) *Migration of erosional surfaces.* Association of American Geographers. Annals, v.30, p.247-254.

Middleton, Gerard V., ed. (1965) *Primary sedimentary structures and their hydrodynamic interpretation.* Society of Economic Paleontologists and Mineralogists. Special Publication no.12. 265p.

Mielenz, Richard C., and King, Myrle E. (1955) *Physical-chemical properties and engineering performance of clays.* California. Division of Mines. Bulletin 169, p.196-254. (National Conference on Clays and Clay Technology. 1st, Berkeley (Calif.), 1952. Proceedings.)

Milankovitch, Milutin (1920) *Théorie mathéma-*

tique des phénomènes thermiques produits par la radiation solaire. Paris: Gauthier-Villars. 338p.

Miller, Don J. (1953) *Late Cenozoic marine glacial sediments and marine terraces of Middleton Island, Alaska.* Journal of Geology, v.61, p.17-40.

Miller, Hugh (1841) *The Old Red Sandstone; or, new walks in an old field.* Edinburgh: J. Johnstone. 275p.

Miller, Hugh (1883) *River-terracing: its methods and their results.* Royal Physical Society of Edinburgh. Proceedings, v.7, p.263-306.

Miller, William J. (1919) *Pegmatite, silexite, and aplite of northern New York.* Journal of Geology, v.27, no.1, p.28-54.

Milner, Henry B. (1922) *The nature and origin of the Pliocene deposits of the county of Cornwall and their bearing on the Pliocene geography of the south-west of England.* Geological Society of London. Quarterly Journal, v.78, p.348-377.

Milner, Henry B. (1940) *Sedimentary petrography, with special reference to petrographic methods of correlation of strata, petroleum technology and other economic applications of geology.* 3rd ed. London: Thomas Murby. 666p.

Milton, Daniel J. (1969) *Astrogeology in the 19th century.* Geotimes, v.14, no.6, p.22.

Mitchell, Hugh C. (1948) *Definitions of terms used in geodetic and other surveys.* Washington: U.S. Government Printing Office. 87p. (U.S. Coast and Geodetic Survey. Special Publication no.242).

Miyashiro, Akiho (1961) *Evolution of metamorphic belts.* Journal of Petrology, v.2, no.3, p.277-311.

Monkhouse, F.J. (1965) *A dictionary of geography.* Chicago: Aldine. 344p.

Monkhouse, F.J., and Wilkinson, H.R. (1952) *Maps and diagrams; their compilation and construction.* London: Methuen. 330p.

Monroe, Watson H. (1970) *A glossary of karst terminology.* U.S. Geological Survey. Water-supply Paper 1899-K.

Moore, David G., and Scruton, Philip C. (1957) *Minor internal structures of some recent unconsolidated sediments.* American Association of Petroleum Geologists. Bulletin, v.41, p.2723-2751.

Moore, Derek (1966) *Deltaic sedimentation.* Earth-Science Reviews, v.1, p.87-104.

Moore, P. Fitzgerald (1958) *Nature, usage, and definition of marker-defined vertically segre-gated rock units.* American Association of Petroleum Geologists. Bulletin, v.42, p.447-450.

Moore, Raymond C. (1926) *Origin of inclosed meanders on streams of the Colorado Plateau.* Journal of Geology, v.34, p.29-57.

Moore, Raymond C. (1933) *Historical geology.* New York: McGraw-Hill. 673p.

Moore, Raymond C. (1936) *Stratigraphic classification of the Pennsylvanian rocks of Kansas.* Kansas. State Geological Survey. Bulletin 22. 256p.

Moore, Raymond C. (1949) *Meaning of facies.* Geological Society of America. Memoir 39, p.1-34.

Moore, Raymond C., ed. (1953-   ) *Treatise on invertebrate paleontology.* Prepared under the guidance of the Joint Committee on Invertebrate Paleontology. Lawrence (Kans.): Univ. of Kansas Press and the Geological Society of America. Issued in various separate parts identified by letter.

Moore, Raymond C. (1957) *Minority report.* In: American Commission on Stratigraphic Nomenclature. *Nature, usage, and nomenclature of biostratigraphic units,* p.1888. American Association of Petroleum Geologists. Bulletin, v.41, p.1877-1889. (Its Report 5).

Moore, Raymond C. (1957a) *Modern methods of paleoecology.* American Association of Petroleum Geologists. Bulletin, v.41, p.1775-1801.

Moore, Raymond C. (1958) *Introduction to historical geology.* 2nd ed. New York: McGraw-Hill. 656p.

Moore, Raymond C., and others (1952) *Invertebrate fossils.* 1st ed. New York: McGraw-Hill. 766p.

Moore, W.G. (1967) *A dictionary of geography; definitions and explanations of terms used in physical geography.* 3rd ed. New York: Praeger. 246p.

Morlot, Adolphe von (1847) *Ueber Dolomit und seine künstliche Darstellung aus Kalkstein.* Naturwissenschaftliche Abhandlungen, gesammelt und durch Subscription hrsg. von Wilhelm Haidinger, v.1, p.305-315.

Moss, J.H. (1951) *Early man in the Eden valley.* Philadelphia: University of Pennsylvania. University Museum Monograph. 92p.

Mueller, Ivan Istvan, and Rockie, John D. (1966) *Gravimetric and celestial geodesy; a glossary of terms.* New York: Ungar. 129p.

Muir-Wood, Helen M., and Cooper, G. Arthur (1960) *Morphology, classification and life habits of the Productoidea (Brachiopoda).*

Geological Society of America. Memoir 81. 447p.

Müller, German (1967) *Diagenesis in argillaceous sediments.* In: Larsen, Gunnar, and Chilingar, George V., eds. *Diagenesis in sediments.* Amsterdam: Elsevier. 551p. *(Developments in sedimentology 8,* p.127-177).

Müller, Johannes (1858) *Ueber die Thalassicollen, Polycystinen und Acanthometren des Mittelmeeres.* Akademie der Wissenschaften, Berlin. Abhandlungen (physikalische), 1858, p.1-62.

Muller, P.M., and Sjogren, W.L. (1968) *Mascons: lunar mass concentrations.* Science, v.161, p.680-684.

Muller, Siemon W., compiler (1947) *Permafrost or permanently frozen ground and related engineering problems.* Ann Arbor (Mich.): Edwards. 231p.

Muller, Siemon Wm. (1965) *Superposition of strata: part 1. Physical criteria for determining top and bottom of beds.* American Geological Institute: Data Sheet no.10.

Murchison, Roderick I. (1839) *The Silurian System.* Pt. 1. London: John Murray. 576p.

Murray, A.N. (1930) *Limestone oil reservoirs of the northeastern United States and Ontario, Canada.* Economic Geology, v.25, no.5, p.452-469.

Murray, Grover E. (1961) *Geology of the Atlantic and Gulf Coast Province of North America.* New York: Harper. 692p.

Murray, Grover E. (1965) *Indigenous Precambrian petroleum?* American Association of Petroleum Geologists. Bulletin, v.49, p.3-21.

Murray, R.C. (1960) *Origin of porosity in carbonate rocks.* Journal of Sedimentary Petrology, v.30, no.1, p.59-84.

Murzaevs, E., and Murzaevs, V. (1959) *Slovar' mestnykh geograficheskikh terminov.* Moscow: Geografizdat. 303p.

Myashiro, A. (1968) *Metamorphism of mafic rocks.* In: Hess, H.H., and Poldervaart, Arie, eds. *Basalts; the Poldervaart treatise on rocks of basaltic composition,* v.2, p.483-862. New York: Interscience.

Nabholz, Walther K. (1951) *Beziehungen zwischen Fazies und Zeit.* Eclogae geologicae Helvetiae, v.44, p.131-158.

National Academy of Sciences—National Research Council. Committee on Rock Mechanics (1966) *Rock-mechanics research; a survey of United States research to 1965, with a partial survey of Canadian universities.* Washington, D.C. 82p. (Its Publication 1466).

National Aeronautics and Space Administration (1966) *Short glossary of space terms.* Its publication: SP-1.

National Aeronautics and Space Administration, Goddard Space Flight Center (1971) *NASA Earth resources data users' handbook.* (Document #71SD4249)

Natland, Manley L., and Kuenen, Philip H. (1951) *Sedimentary history of the Ventura Basin, California, and the action of turbidity currents.* Society of Economic Paleontologists and Mineralogists. Special Publication no.2, p.76-107.

Naum, T., and others (1962) *Vulcanokarstul din Masivul Calimanului (Carpatii Orientali).* Bucharest. Universitatea. Analele; seria stiintele naturii: geologie-geografie, anul 11, no.32, p.143-179.

Naumann, Carl Friedrich (1850) *Lehrbuch der Geognosie.* Bd. 1. Leipzig: Wilhelm Engelmann. 1000p.

Naumann, Carl Friedrich (1858) *Lehrbuch der Geognosie.* Bd. 1. 2nd ed. Leipzig: Wilhelm Engelmann. 960p.

Nayak, V.K. (1970) *Geoplanetology: a new term for geology of the planets including the Moon.* Geological Society of America. Bulletin, v.81, p.1279.

Nelson, A. (1965) *Dictionary of mining.* New York: Philosophical Library. 523p.

Nelson, A., and Nelson, K.D. (1967) *Dictionary of applied geology, mining and civil engineering.* New York: Philosophical Library. 421p.

Nelson, Henry F., and others (1962) *Skeletal limestone classification.* American Association of Petroleum Geologists. Memoir 1, p.224-252.

Nettleton, Lewis Lomax (1940) *Geophysical prospecting for oil.* 1st ed. New York: McGraw-Hill. 444p.

Neumann, A. Conrad (1966) *Observations on coastal erosion in Bermuda and measurements of the boring rate of the sponge, Cliona lampa.* Limnology and Oceanography, v.11, no.1, p.92-108.

Nevin, Charles Merrick (1949) *Principles of structural geology.* 4th ed. New York: Wiley. 410p.

Newell, Norman D., and others (1953) *The Permian reef complex of the Guadalupe Mountains region, Texas and New Mexico—a study in paleoecology.* San Francisco: Freeman. 236p.

Nicodemus, F.E. (in press) *Radiometric nomenclature.* In: *Radiometry targets, backgrounds,* and atmospheres of proposed mil standard for infrared terms and definitions by Philco-Ford Corporation. Philco-Ford Corporation.

Niggli, Paul (1954) *Rocks and mineral deposits.* English translation by Robert L. Parker. San Francisco: Freeman. 559p.

Nikolaev, V.A. (1959) *Some structural characteristics of mobile tectonic belts.* International Geology Review, v.1, p.50-64.

Noble, Levi F. (1941) *Structural features of the Virgin Spring area, Death Valley, California.* Geological Society of America. Bulletin, v.52, p.941-999.

Nockolds, S.R. (1934) *The production of normal rock types by contamination and their bearing on petrogenesis.* Geological Magazine, v.71, p.31-39.

Nordenskjöld, Otto (1909) *Einige Beobachtungen über Eisformen und Vergletscherung der antarktischen Gebiete.* Zeitschrift für Gletscherkunde, Bd.3, p.321-334.

Nordyke, Milo D. (1962) *Nuclear craters and preliminary theory of the mechanics of explosive crater formation.* Journal of Geophysical Research, v.66, p.3439-3459.

Norris, D.K. (1958) *Structural conditions in Canadian coal mines.* Canada. Geological Survey. Bulletin 44. 54p.

Northrop, John I. (1890) *Notes on the geology of the Bahamas.* New York Academy of Sciences. Transactions, v.10, p.4-22.

Norton, W.H. (1917) *A classification of breccias.* Journal of Geology, v.25, p.160-194.

Ogilvie, Ida H. (1905) *The high-altitude conoplain; a topographic form illustrated in the Ortiz Mountains.* American Geologist, v.36, p.27-34.

Ohio. Legislative Service Commission (1969) *Preservation of natural areas, and Report of Committee to Study Natural Areas.* Project officer: John P. Bay. Columbus: Ohio Legislative Service Commission. 20p. (Its Staff Research Report no.89).

Oldham, Thomas (1879) *Geological glossary for the use of students.* Edited by R.D. Oldham. London: Stanford. 62p.

Oppel, Albert (1856-1858) *Die Juraformation Englands, Frankreichs und des südwestlichen Deutschlands; nach ihren einzelnen Gliedern einigetheilt und verglichen.* Stuttgart: Ebner & Seubert. 857p. (Württembergische Naturwissenschaftliche Jahreshefte. Jahrg.12, p.121-132, 313-556; 13, p.141-396; 14, p.128-291).

Oriel, Steven S. (1949) *Definitions of arkose.* American Journal of Science, v.247, p.824-829.

Osborne, Robert H. (1969) *Undergraduate instruction in geomathematics.* Journal of Geological Education, v.17, p.120-124.

Otterman, Joseph, and Bronner, Finn E. (1966) *Martian wave of darkening: a frost phenomenon?* Science, v.153, p.56-60.

Otto, George H. (1938) *The sedimentation unit and its use in field sampling.* Journal of Geology, v.46, p.569-582.

Packer, R.W. (1965) *Lag mound features on a dolostone pavement.* Canadian Geographer, v.9, p.138-143.

Packham, G.H. (1954) *Sedimentary structures as an important factor in the classification of sandstones.* American Journal of Science, v.252, p.466-476.

Packham, G.H., and Crook, Keith A.W. (1960) *The principle of diagenetic facies and some of its implications.* Journal of Geology, v.68, p.392-407.

Padfield, Harland I., and Smith, Courtland L. (1968) *Water and culture: new decision rules for old institutions.* Rocky Mountain Social Science Journal, v.5, no.2, p.23-32.

Page, David (1859) *Handbook of geological terms and geology.* Edinburgh: William Blackwood. 416p.

Palmer, Allison R. (1965) *Biomere—a new kind of biostratigraphic unit.* Journal of Paleontology, v.39, p.149-153.

Palmer, Leroy A. (1920) *Desert prospecting.* Engineering and Mining Journal, v.110, p.850-853.

Park, Charles F., jr. (1959) *The origin of hard hematite in itabirite.* Economic Geology, v.54, p.573-587.

Park, Charles F., jr., and MacDiarmid, Roy A. (1970) *Ore deposits.* 2d ed. San Francisco: Freeman. 522p.

Park, James (1914) *A text-book of geology; for use in mining schools, colleges, and secondary schools.* London: Charles Griffin. 598p.

Parks, B.C. (1951) *Petrography of American lignites.* Economic Geology, v.46, no.1, p.23-50.

Passega, R. (1957) *Texture as characteristic of clastic properties.* American Association of Petroleum Geologists. Bulletin, v.41, p.1952-1984.

Paterson, Mervyn Silas, and Weiss, Lionel E. (1961) *Symmetry concepts in the structural analysis of deformed rocks.* Geological Society of America. Bulletin, v.72, no.6, p.841-882.

Pavlov, Aleksei P. (1889) *Geneticheskie tipy materikovykh obrazovanii lednikovoi i poslelednikovoi epokhi*. Geologicheskogo Komiteta (St. Petersburg). Izvestiia, v.7, p.243-261.

Payne, Thomas G. (1942) *Stratigraphical analysis and environmental reconstruction*. American Association of Petroleum Geologists. Bulletin, v.26, p.1697-1770.

Peacock, M.A. (1931) *Classification of igneous rock series*. Journal of Geology, v.39, p.54-67.

Pearn, William C. (1964) *Finding the ideal cyclothem*. Kansas. State Geological Survey. Bulletin 169, v.2, p.399-413.

Peltier, Louis C. (1950) *The geographic cycle in periglacial regions as it is related to climatic geomorphology*. Association of American Geographers. Annals, v.40, p.214-236.

Pelto, Chester R. (1954) *Mapping of multicomponent systems*. Journal of Geology, v.62, p.501-511.

Penck, Albrecht (1900) *Geomorphologische Studien aus der Herzegowina*. Deutscher und. Oesterreichischer Alpenverein. Zeitschrift, Bd. 31, p.25-41.

Penck, Albrecht (1919) *Die Gipfelflur der Alpen*. Akademie der Wissenschaften, Berlin. Sitzungsberichte, Jahrg. 1919, p.256-268.

Penck, Walther (1942) *Die morphologische Analyse; ein Kapitel der physikalischen Geologie*. Stuttgart: J. Engelhorns. 283p. (Geographische Abhandlungen, 2 Reihe, Hft.2).

Penck, Walther (1953) *Morphological analysis of land forms; a contribution to physical geology*. Translated by Hella Czech and Katharine Cumming Boswell. New York: St. Martin's Press. 429p.

Pepper, James F., and others (1954) *Geology of the Bedford Shale and Berea Sandstone in the Appalachian basin*. U.S. Geological Survey. Professional Paper 259. 111p.

Pepper, T.B. (1941) *The Gulf underwater gravimeter*. Geophysics, v.6, no.1, p.34-44.

Pettijohn, Francis J. (1949) *Sedimentary rocks*. New York: Harper. 526p.

Pettijohn, Francis J. (1957) *Sedimentary rocks*. 2nd ed. New York: Harper. 718p.

Pettijohn, Francis J. (1954) *Classification of sandstones*. Journal of Geology, v.62, p.360-365.

Pettijohn, Francis J., and Potter, Paul E. (1964) *Atlas and glossary of primary sedimentary structures*. New York: Springer-Verlag. 370p.

Peyve, A.V., and Sinitzyn, V.M. (1950) *Certains problèmes fondamentaux de la doctrine des géosynclinaux*. Akademiya Nauk SSSR. Izvestiya, Seriya Geologicheskaya, v.4, p.28-52.

Phemister, James (1956) *Petrography*. In: Great Britain. Geological Survey. *The limestones of Scotland; chemical analyses and petrography*. Edinburgh: H.M.S.O. 150p. (Its Memoirs; special reports on the mineral resources of Great Britain, v.37, p.66-74).

Philipsborn, H. von (1930) *Zur chemisch-analytischen Erfassung der isomorphen Variation gesteinbildender Minerale. Die Mineralkomponenten des Pyroxengranulits von Hartmannsdorf (Sa.)*. Chemie der Erde, v.5, p.233-253.

Phillips, William (1818) *A selection of facts from the best authorities, arranged so as to form an outline of the geology of England and Wales*. London: William Phillips. 250p.

Picard, M. Dane (1953) *Marlstone—a misnomer as used in Uinta Basin, Utah*. American Association of Petroleum Geologists. Bulletin, v.37, p.1075-1077.

Pinkerton, John (1811) *Petralogy; a treatise on rocks*. 2 v. London: White, Cochrane. 599p. and 656p.

Pirsson, Louis V. (1896) *A needed term in petrography*. Geological Society of America. Bulletin, v.7, p.492-493.

Pirsson, Louis V. (1915) *Physical geology*. New York: Wiley. 404p. (*A text-book of geology*, pt.1).

Playfair, John (1802) *Illustrations of the Huttonian theory of the Earth*. Edinburgh: Wm. Creech. 528p.

Plumley, W.J., and others (1962) *Energy index for limestone interpretation and classification*. American Association of Petroleum Geologists. Memoir 1, p.85-107.

Poland, J.F., Lofgren, B.E., and Riley, F.S. (in press) *Glossary of selected terms useful in studies of the mechanics of aquifer systems and land subsidence due to fluid withdrawal*. U.S. Geological Survey. Water-supply Paper 2025.

Poldervaart, Arie, and Parker, Alfred B. (1964) *The crystallization index as a parameter of igneous differentiation in binary variation diagrams*. American Journal of Science, v.262, p.281-289.

Pompeckj, Josef Felix (1914) *Die Bedeutung des schwäbischen Jura für die Erdgeschichte*. Stuttgart: Schweizerbartsche. 64p.

Porsild, A.E. (1938) *Earth mounds in unglaciated*

Arctic northwestern America. Geographical Review, v.28, p.46-58.

Post, Lennart von (1924) *Ur de sydsvenska skogarnas regionala historia under postarktist tid.* Geologiska Föreningen, Stockholm. Förhandlingar, Bd.46, p.83-128.

Potter, Paul E., and Glass, Herbert D. (1958) *Petrology and sedimentation of the Pennsylvanian sediments in southern Illinois—a vertical profile.* Illinois. State Geological Survey. Report of Investigations 204. 60p.

Pough, Frederick H. (1967) *The story of gems and semiprecious stones.* Irvington-on-Hudson: Harvey House. 142p.

Powell, John Wesley (1873) *Some remarks on the geological structure of a district of country lying to the north of the Grand Cañon of the Colorado.* American Journal of Science, 3rd ser., v.5, p.456-465.

Powell, John Wesley (1874) *Remarks on the structural geology of the Valley of the Colorado of the West.* Philosophical Society of Washington. Bulletin, v.1, p.48-51.

Powell, John Wesley (1875) *Exploration of the Colorado River of the West and its tributaries.* Washington: Government Printing Office. 291p.

Powell, John Wesley (1895) *Physiographic features.* National Geographic Society. National Geographic Monographs, v.1, no.2, p.33-64.

Power, F. Danvers (1895) *A glossary of terms used in mining geology. Adelaide:* Australasian Institute of Mining Engineers. 69p.

Powers, M.C. (1953) *A new roundness scale for sedimentary particles.* Journal of Sedimentary Petrology, v.23, p.117-119.

Powers, R.W. (1962) *Arabian Upper Jurassic carbonate reservoir rocks.* American Association of Petroleum Geologists. Memoir 1, p.122-192.

Prentice, J.E. (1956) *The interpretation of flow-markings and load-casts.* Geological Magazine, v.93, p.393-400.

Price, W. Armstrong (1947) *Equilibrium of form and forces in tidal basins of coast of Texas and Louisiana.* American Association of Petroleum Geologists. Bulletin, v.31, p.1619-1663.

Price, W. Armstrong (1954) *Dynamic environments—reconnaissance mapping, geologic and geomorphic, of continental shelf of Gulf of Mexico.* Gulf Coast Association of Geological Societies. Transactions, v.4, p.75-107.

Prior, George T. (1920) *The classification of meteorites.* Mineralogical Magazine, v.19, no. 90, p. 51-63.

Pryor, Edmund J. (1963) *Dictionary of mineral technology.* London: Mining Publications. 437p.

Pumpelly, Raphael, Wolf, J.E., and T. Nelson Dale (1894) *Geology of the Green Mountains in Massachusetts; part III, Mount Greylock: its areal and structural geology,* by T. Nelson Dale. U.S. Geological Survey. Monograph 23.

Purdy, Ross C. (1908) *Qualities of clays suitable for making paving brick.* Illinois. State Geological Survey. Bulletin no.9, p.133-278.

Pustovalov, Leonid V. (1933) *Geokhimicheskie fatsii, ikh znachenie v obshchei i prikladnoi geologii.* Problemy Sovetskoi geologii, 1933, [ser.1], t.1, no.1, p.57-80.

Quensel, P. (1916) *Zur Kenntnis der Mylonitbildung, erläutert an Material aus dem Kebnekaisegebiet. Uppsala. Universitet.* Geologiska Institut. Bulletin, v.15, p.91-116.

Quirke, Terence T. (1930) *Spring pits, sedimentation phenomena.* Journal of Geology, v.38, p.88-91.

Raaf, J.F.M. de (1945) *Notes on the geology of the southern Rumanian oil district with special reference to the occurrence of a sedimentary laccolith.* Geological Society of London. Quarterly Journal, v.101, p.111-134.

Raistrick, Arthur, and Marshall, Charles E. (1939) *The nature and origin of coal and coal seams.* London: English Universities Press Ltd. 282p.

Raisz, Erwin Josephus (1959) *Landform maps—a method of preparation.* U.S. Office of Naval Research, Geography Branch. Part one of Final Report, Contract Nonr 2339(00). 23p.

Ramberg, Hans (1955) *Natural and experimental boudinage and pinch-and-swell structures.* Journal of Geology, v.63, no.6, p.512-526.

Ramsay, Andrew C. (1846) *On the denudation of South Wales and the adjacent countries of England.* Great Britain. Geological Survey. Memoirs, v.1, ch.2, p.297-335.

Rankama, Kalervo (1962) *Planetology and geology.* Geological Society of America. Bulletin, v.73, p.519-520.

Rankama, Kalervo (1967) *Megayear and gigayear: two units of geological time.* Nature, v.214, p.634.

Rankama, Kalervo, and Sahama, Th.G. (1950) *Geochemistry.* Chicago: University of Chicago Press. 912p.

Ransome, Frederick L., and Calkins, Frank C. (1908) *The geology and ore deposits of the Coeur d'Alene district, Idaho.* U.S. Geological Survey. Professional Paper 62. 203p.

Rast, Nicholas (1956) *The origin and significance of boudinage.* Geological Magazine, v.93, p.401-408.

Read, H.H. (1931) *The geology of central Sutherland (east-central Sutherland and south-western Caithness). (Explanation of sheets 108 and 109).* Geological Survey of Scotland. Memoir. 238p.

Read, H.H. (1944) *Meditation on granite, pt.2.* Geologists' Association. Proceedings, v.55, p.45-93.

Read, H.H. (1958) *A centenary lecture; Stratigraphy in metamorphism.* Geologists' Association. Proceedings, v.69, pt.2, p.83-102.

Rechard, Paul A., and McQuisten, Richard, compilers (1968) *Glossary of selected hydrologic terms.* University of Wyoming. Water Resources Research Institute. Water Resources Series, no.1, 54p.

Reclus, Elisée (1872) *The Earth: a descriptive history of the phenomena of the life of the globe.* Translated by B.B. Woodward. Edited by Henry Woodward. New York: Harper. 573p.

Record, Samuel James (1934) *Identification of the timbers of temperate North America, including anatomy and certain physical properties of wood.* New York: Wiley.

Reiche, Parry (1938) *An analysis of cross-lamination; the Coconino Sandstone.* Journal of Geology, v.46, p.905-932.

Reiche, Parry (1943) *Graphic representation of chemical weathering.* Journal of Sedimentary Petrology, v.13, p.58-68.

Reiche, Parry (1945) *A survey of weathering processes and products.* Albuquerque: University of New Mexico Press. 87p. (New Mexico. Univ. Publications in Geology, no.1).

Reid, H.F. (1924) *Antarctic glaciers.* Geographical Review, v.14, p.603-614.

Renevier, E., and others (1882) *Rapport du comité suisse sur l'unification de la nomenclature.* International Geological Congress. 2nd, Bologna, 1881. Compte rendu, p.535-548.

Reusch, Hans (1894) *The Norwegian Coast Plain.* Journal of Geology, v.2, p.347-349.

Revelle, Roger, and Fairbridge, Rhodes W. (1957) *Carbonates and carbon dioxide.* Geological Society of America. Memoir 67, v.1, p.239-295.

Rice, C.M. (1945) *Dictionary of geological terms.* 1st ed. Ann Arbor (Mich.): Edwards Brothers. 461p.

Rice, C.M. (1954) *Dictionary of geological terms.* 2d ed. Ann Arbor (Mich.): Edwards Brothers.

Rich, John L. (1908) *Marginal glacial drainage features in the Finger Lake region.* Journal of Geology, v.16, p.527-548.

Rich, John L. (1914) *Certain types of stream valleys and their meaning.* Journal of Geology, v.22, p.469-497.

Rich, John L. (1915) *Notes on the physiography and glacial geology of the northern Catskill Mountains.* American Journal of Science, 4th ser., v.39, p.137-166.

Rich, John L. (1938) *A mechanism for the initiation of geosynclines and geo-basins.* Geological Society of America. Proceedings, p.106-107.

Rich, John L. (1950) *Flow markings, groovings, and intra-stratal crumplings as criteria for recognition of slope deposits, with illustrations from Silurian rocks of Wales.* American Association of Petroleum Geologists. Bulletin, v.34, p.717-741.

Rich, John L. (1951) *Geomorphology as a tool for the interpretation of geology and Earth history.* New York Academy of Sciences. Transactions, ser.2, v.13, p.188-192.

Richter, C.F. (1958 *Elementary seismology.* San Francisco: Freeman. 768p.

Richter, Rudolf (1952) *Fluidal-Textur in Sediment-Gesteinen und über Sedifluktion überhaupt.* Hessisches Landesamt für Bodenforschung, Wiesbaden. Notizblatt, F.6, H.3, p.67-81.

Richtofen, Ferdinand von (1868) *The natural system of volcanic rocks.* California Academy of Sciences. Memoirs, v.1, pt.2, 94p.

Riecke, Eduard (1894) *Über das Gleichgewicht zwischen einem festen, homogen deformirten Körper und einer flüssigen Phase, insbesondere über die Depression des Schmelzpunctes durch einseitige Spannung.* Gesellschaft der Wissenschaften zu Göttingen. Mathematisch-Physikalische Klasse. Nachrichten, 1894, no.4, p.278-284.

Riedel, W. (1929) *Zur Mechanik geologischer Brucherscheinungen.* Zentralblatt fuer Mineralogie, Geologie, und Palaeontologie B, p.354-368.

Ries, Heinrich (1937) *Economic geology.* 7th ed. New York: Wiley. 720p.

Rittenhouse, Gordon (1943) *The transportation and deposition of heavy minerals.* Geological Society of America. Bulletin, v.54, p.1725-1780.

Rigby, J. Keith (1959) *Possible eddy markings in the Shinarump Conglomerate of northeastern Utah.* Journal of Sedimentary Petrology, v.29, p.283-284.

Rivière, André (1952) *Expression analytique*

générale de la granulométrie des sédiments meubles. Société Géologique de France. Bulletin, ser.6, v.2, p.155-167.

Rizzini, A., and Passega, R. (1964) Évolution de la sédimentation et orogenèse, vallée du Santerno, Apennin septentrional. In: Bouma, A.H., and Brouwer, A., eds. Turbidites. Amsterdam: Elsevier. 264p. (Developments in sedimentology 3, p.65-74).

Roberts, George (1839) An etymological and explanatory dictionary of the terms and language of geology; designed for the early student, and those who have not made great progress in that science. London: Longman, Orme, Brown, Green, & Longmans, 183p.

Roberts, Ralph Jackson (1951) Geology of the Antler Peak quadrangle, Nevada. U.S. Geological Survey. Quadrangle Map (GQ10), scale 1:62,500, with text.

Robinove, Charles J. (1963) Photography and imagery—a clarification of terms. Photogrammetric Engineering, v.29, p.880-881.

Robson, J., and Nance, R. Morton (1959) Geological terms used in S.W. England. Royal Geological Society of Cornwall. Transactions, 1955-1956, v.19, pt.1, p.33-41.

Rodgers, John (1963) Mechanics of Appalachian foreland folding in Pennsylvania and West Virginia. American Association of Petroleum Geologists. Bulletin, v.47, p.1527-1536.

Rodgers, John (1967) Chronology of tectonic movements in the Appalachian region of North America. American Journal of Science, v.265, p.408-427.

Roger, Henry Darwin (1860) On the distribution and probable origin of the petroleum, or rock oil, of western Pennsylvania, New York, and Ohio. Royal Philosophical Society of Glasgow. Proceedings, v.4, p.355-359.

Rogers, Harold H., compiler (1966) Glossary of terms frequently used in cosmology. New York: American Institute of Physics. 16p.

Rohrbach, Carl E.M. (1885) Ueber die Eruptivgesteine im Gebiete der schlesisch-mährischen Kreideformation. Tschermak's mineralogische und petrographische Mitteilungen, N.F., Bd.7, p.1-63.

Roques, M. (1961) Nomenclature de J. Jung et M. Roques pour certains types de migmatites. In: Sörensen, H., ed., Symposium on migmatite nomenclature, p.68. International Geological Congress, 21st, Copenhagen, 1960. Report, pt.26, sec.14.

Rosalsky, Maurice B. (1949) A study of minor beach features at Fire Island, Long Island, New York. New York Academy of Sciences. Transactions, ser.2, v.12, p.9-16.

Rosauer, Elmer A. (1957) Climatic conditions involved in glacial loess formation. Dr. Dissertation, University of Bonn. 105p.

Rosenbusch, Harry (1887) Mikroskopische Physiographie der massigen Gesteine. 2nd ed. Stuttgart: Schweizerbart'sche Verlagshandlung (E. Koch). 877p. (His Mikroskopische Physiographie der Mineralien und Gesteine, Bd.2).

Rosenbusch, Harry (1888) Microscopical physiography of the rock-making minerals: an aid to the microscopical study of rocks. Translated & abridged by Joseph P. Iddings. New York: Wiley. 333p.

Rosenbusch, Harry (1898) Elemente der Gesteinlehre. Stuttgart: Schweizerbart'sche. 546p.

Ross, Clarence S., and Shannon, Earl V. (1926) The minerals of bentonite and related clays and their physical properties. American Ceramic Society. Journal, v.9, p.77-96.

Rousseau, Jacques (1949) Modifications de la surface de la toundra sous l'action d'agents climatiques. Revue Canadienne Géographie, v.3, p.43-51.

Rubey, William W., and Hubbert, M. King (1959) Role of fluid pressure in mechanics of overthrust faulting. Geological Society of America. Bulletin, v.70, no.2, p.167-206.

Runcorn, S.K., and others, eds. (1967) International dictionary of geophysics. New York: Pergamon Press. 2 volumes. 1728p.

Russell, Israel C. (1885) Existing glaciers of the U.S. U.S. Geological Survey. Annual Report, 5th, p.303-355.

Russell, Israel C. (1898a) Glaciers of Mount Rainier. U.S. Geological Survey. Annual Report, 18th, pt.2, p.349-415.

Russell, Israel C. (1898b) Rivers of North America. New York: Putnam. 327p.

Russell, Richard J. (1942) Flotant. Geographical Review, v.32, p.74-98.

Russell, Richard J., ed. (1968) Glossary of terms used in fluvial, deltaic, and coastal morphology and processes. Coastal Studies Institute, Louisiana State University. Technical Report No. 63. (reproduced by the Clearinghouse for Federal Scientific & Technical Information, Springfield, Virginia 21151.)

Ruxton, Bryan P., and Berry, Leonard (1959) The basal rock surface on weathered granitic rocks. Geologists' Association. Proceedings, v.70, p.285-290.

Salisbury, John W., and Hunt, Graham R. (1968) *Martian surface materials: effect of particle size on spectral behavior.* Science, v.161, no.3839, p.365-366.

Salisbury, Rollin D., and others (1902) *The glacial geology of New Jersey.* New Jersey. Geological Survey. Final Report 5. 802p.

Salisbury, Rollin D. (1904) *Three new physiographic terms.* Journal of Geology, v.12, p.707-715.

Salomon, W. (1915) *Die Definitionen von Grauwacke, Arkose und Ton.* Geologische Rundschau, Leipzig, VI, p.398-404.

Sander, Bruno (1911) *Über Zusammenhänge zwischen Teilbewegung und Gefüge in Gesteinen.* Tschermaks Mineralogische und Petrographische Mitteilungen, v.30, p.281-314.

Sander, Bruno (1912) *Über tektonische Gesteinsfazies.* Austria. Geologische Reichsanstalt. Verhandlungen, 1912, no.10, p.249-257.

Sander, Bruno (1930) *Gefügekunde der Gesteine mit besonderer Berücksichtigung der Tektonite.* Vienna: Julius Springer. 352p.

Sander, Bruno (1936) *Beiträge zur Kenntnis der Anlagerungsgefüge (rhythmische Kalke und Dolomite aus der Trias).* [Pt.] 1-2. Mineralogische und petrographische Mitteilungen, Bd. 48, p.27-139, 141-209.

Sander, Bruno (1951) *Contributions to the study of depositional fabric.* Translated by Eleanora Bliss Knopf. Tulsa (Okla.): American Association of Petroleum Geologists. 207p.

Sander, N.J. (1967) *Classification of carbonate rocks of marine origin.* American Association of Petroleum Geologists. Bulletin, v.51, p.325-336.

Sanders, John E. (1957) *Discontinuities in the stratigraphic record.* New York Academy of Sciences. Transactions, ser.2, v.19, p.287-297.

Sanders, John E. (1963) *Concepts of fluid mechanics provided by primary sedimentary structures.* Journal of Sedimentary Petrology, v.33, p.173-179.

Sauer, Carl O. (1930) *Basin and Range forms in the Chiricahua area.* California. University, Berkeley. Publications in Geography, v.3, no.6, p.339-414.

Saussure, Horace-Benedict de (1786) *Voyages dans les Alpes, précédés d'un essai sur l'histoire naturelle des environs de Genève.* Vol. 2. Genève: Barde, Manget. 641p.

Scagel, R.F., and others (1965) *An evolutionary survey of the plant kingdom.* Belmont, California: Wadsworth.

Scheidegger, Adrian (1961) *Theoretical geomorphology.* Berlin: Springer-Verlag. 333p.

Scheidegger, Adrian E. (1963) *Principles of geodynamics.* 2nd ed. New York: Academic Press. 362p.

Schenck, Hubert G., and Muller, Siemon W. (1941) *Stratigraphic terminology.* Geological Society of America. Bulletin, v.52, p.1419-1426.

Schermerhorn, L.J.G. (1966) *Terminology of mixed coarse-fine sediments.* Journal of Sedimentary Petrology, v.36, p.831-835.

Schieferdecker, A.A.G., ed. (1959) *Geological nomenclature.* Gorinchem: Royal Geological and Mining Society of the Netherlands. 523p.

Schiffman, A. (1965) *Energy measurements in the swash-surf zone.* Limnology and Oceanography, v.10, p.255-260.

Schindewolf, Otto H. (1950) *Grundlagen und Methoden der paläontologischen Chronologie.* 3rd ed. Berlin-Nikolassee: Naturwissenschaftlichen Verlag vorm. Gebrüder Borntraeger. 152p.

Schindewolf, Otto H. (1954) *Über einige stratigraphische Grundbegriffe.* Roemeriana, H.1, p.23-38.

Schindewolf, Otto H. (1957) *Comments on some stratigraphic terms.* American Journal of Science, v.255, p.394-399.

Schindewolf, Otto H. (1959) *On certain stratigraphic fundamentals.* International Geology Review, v.1, no.7, p.62-70.

Schipman, Henry, jr. (1968) *Buffalo rings.* National Parks Magazine, v.42, no.254, p.14-15.

Schlanger, Seymour O. (1957) *Dolomite growth in coralline algae.* Journal of Sedimentary Petrology, v.27, p.181-186.

Schmidt, Volkmar (1965) *Facies, diagenesis, and related reservoir properties in the Gigas Beds (Upper Jurassic), northwestern Germany.* Society of Economic Paleontologists and Mineralogists. Special Publication no.13, p.124-168.

Schmidt, Walter (1925) *Gefügestatistik.* Tschermak's mineralogische und petrographische Mitteilungen, N.F., Bd.38, p.392-423.

Schmoll, Henry R., and Bennett, Richard H. (1961) *Axiometer—a mechanical device for locating and measuring pebble and cobble axes for macrofabric studies.* Journal of Sedimentary Petrology, v. 31, p.617-622.

Schofield, W. (1920) *Dumb-bell islands and peninsulas on the coast of South China.* Liverpool Geological Society. Proceedings, v.13, p.45-51.

Schopf, James M. (1960) *Field description and*

sampling of coal beds. U.S. Geological Survey. Bulletin 1111-B, p.25-70.

Schuchert, Charles (1910) Paleogeography of North America. Geological Society of America. Bulletin, v.20, p.427-606.

Schuchert, Charles (1923) Sites and natures of the North American geosynclines. Geological Society of America. Bulletin, v.34, p.151-260.

Schuchert, Charles (1924) Historical geology. 2nd, rev. ed. New York: Wiley. 724p. (A textbook of geology, pt.2).

Schulze, Franz (1855) Bemerkungen über das Vorkommen wohlerhaltener Cellulose in Braunkohle und Steinkohle. Akademie der Wissenschaften, Berlin. Bericht über die zur Bekanntmachung geeigneten Verhandlungen, 5 Nov 1855, p.676-678.

Schumm, Stanley A. (1956) Evolution of drainage systems and slopes in badlands at Perth Amboy, New Jersey. Geological Society of America. Bulletin, v.67, p.597-646.

Schwarz, E.H.L. (1912) South African geology. Glasgow: Blackie. 200p.

Schwarzbach, Martin (1961) Das Klima der Vorzeit; eine Einführung in die Paläoklimatologie. 2nd ed. Stuttgart: Ferdinand Enke. 275p.

Scott, Harold W. (1947) Solution sculpturing in limestone pebbles. Geological Society of America. Bulletin, v.58, p.141-152.

Scott, William B. (1922) Physiography, the science of the abode of man. New York: Collier. 384p.

Scrivenor, J.B. (1921) The physical geography of the southern part of the Malay peninsula. Geographical Review, v.11, p.351-371.

Searle, Alfred B. (1912) The natural history of clay. New York: Putnam. 176p.

Searle, Alfred B. (1923) Sands and crushed rocks. Vol. 1. London: Henry Frowde and Hodder & Stoughton. 475p. (Oxford technical publications).

Sears, F.W. (1958) Mechanics, wave motion and heat. Reading (Mass.): Addison-Wesley. 664p.

Sederholm, J.J. (1891) Studien über Archäis die Eruptingesteine aus dem sudwestlichen Finnland (auch Diss.). Tschermaks Mineralogische und Petrographische Mitteilungen, v.12, p.97-142.

Sederholm, J.J. (1907) Om granit och gneis deras uppkomst, uppträdande och utbredning inom urberget i Fennoskandia. Finland. Commission Géologique. Bulletin, no.23, 110p.

Sederholm, J.J. (1924) Granit-gneisproblemen belysta genom iakttagelser i Åbo-Ålands skär-

gard. Geologiska Föreningen, Stockholm. Förhandlingar, v.46, p.129-153.

Sellards, Elias H., and Gunter, Herman (1918) Geology between the Apalachicola and Ocklocknee rivers in Florida. Florida. Geological Survey. Annual Report, 10th-11th, p.9-56.

Shackleton, Robert M. (1958) Downward-facing structures of the Highland Border. Geological Society of London. Quarterly Journal, v.113, p.361-392.

Shaler, Nathaniel S. (1889) The geology of Cape Ann, Mass. U.S. Geological Survey. Annual Report, 9th, p.529-611.

Shaler, Nathaniel S. (1890) General account of the fresh-water morasses of the United States, with a description of the Dismal Swamp district of Virginia and North Carolina. U.S. Geological Survey. Annual Report, 10th, pt.1, p.255-339.

Shaler, Nathaniel S. (1895) Beaches and tidal marshes of the Atlantic coast. National Geographic Society. National Geographic Monographs, v.1, no.5, p.137-168.

Shand, S.J. (1916) The pseudotachylyte of Parijs (Orange Free State), and its relation to "trap-shotten-gneiss" and "flint-crush-rock". Geological Society of London. Quarterly Journal, v.72, p.198-220.

Shand, S.J. (1947) Eruptive rocks, their genesis, composition, classification, and their relation to ore deposits with a chapter on meteorites. 3rd ed. London: Thomas Murby. 488p.

Sharp, John Van Alstyne, and Nobles, Laurence Hewit (1953) Mudflow of 1941 at Wrightwood, Southern California. Geological Society of America. Bulletin, v.64, no.5, p.547-560.

Sharp, Robert P. (1942) Soil structures in the St. Elias Range, Yukon Territory. Journal of Geomorphology, v.5, p.274-301.

Sharp, Robert P. (1954) Physiographic features of faulting in southern California. California. Division of Mines. Bulletin 170, ch.5, [pt.] 3, p.21-28.

Sharpe, Charles F.S. (1938) Landslides and related phenomena; a study of mass-movements of soil and rock. New York: Columbia University Press. 136p.

Shatsky, Nikolai S. (1945) Outlines of the tectonics of Volga-Urals petroleum region and adjacent parts of the west slope of the southern Urals. In: Mather, Kirtley F., ed. Source book in geology 1900-1950, p.263-267. Cambridge, Massachusetts: Harvard University Press.

Shatsky, Nikolai S. (1955) O proiskhozhdenii Pachelmskogo progiba. Moskovskoe Obshchestvo Ispytatelei Prirody. Biulleten'; otdel geologicheskii, n.s., v.30, no.5, p.3-26.

Shatsky, Nikolai S., and Bogdanoff, A.A. (1957) *Explanatory note on the tectonic map of the U.S.S.R. and adjoining countries.* Moscow: State Scientific and Technical Publishing House. (English translation in International Geology Review, 1959, v.1, p.1-49).

Shatsky, Nikolai S., and Bogdanoff, A.A. (1960) *La carte tectonique internationale de l'Europe au 2 500 000ᵉ.* Akademiya Nauk S.S.S.R., Izvestiya, Seriya Geologicheskaya, no.4.

Shawe, D.R. & Granger, H.C. (1965) *Uranium ore rolls—an analysis.* Economic Geology, v.60, no.2, p.240-250.

Shaw, Eugene W. (1911) *Preliminary statement concerning a new system of Quaternary lakes in the Mississippi basin.* Journal of Geology, v.19, p.481-491.

Shepard, Francis P. (1937) *Revised classification of marine shorelines.* Journal of Geology, v.45, p.602-624.

Shepard, Francis P. (1948) *Submarine geology.* New York: Harper. 348p.

Shepard, Francis P. (1952) *Revised nomenclature for depositional coastal features.* American Association of Petroleum Geologists. Bulletin, v.36, p.1902-1912.

Shepard, Francis P. (1954) *Nomenclature based on sand-silt-clay ratios.* Journal of Sedimentary Petrology, v.24, p.151-158.

Shepard, Francis P. (1967) *The Earth beneath the sea.* Revised ed. Baltimore: Johns Hopkins Press. 242p.

Shepard, Francis P., and Cohee, George V. (1936) *Continental shelf sediments off the mid-Atlantic States.* Geological Society of America. Bulletin, v.47, p.441-457.

Shepard, Francis P., and Dill, Robert F. (1966) *Submarine canyons and other sea valleys.* Chicago: Rand McNally. 381p.

Shepard, Francis P., and Young, Ruth (1961) *Distinguishing between beach and dune sands.* Journal of Sedimentary Petrology, v.31, p.196-214.

Sheriff, R.E. (1968) *Glossary of terms used in geophysical exploration.* Geophysics, v.33, p.181-228.

Sherzer, William H. (1910) *Criteria for the recognition of the various types of sand grains.* Geological Society of America. Bulletin, v.21, p.625-662, 775-776.

Shipley, Robert M. (1951) *Dictionary of gems and gemology including ornamental, decorative and curio stones.* Assisted by Anna McC.

Beckley, Edward Wigglesworth, and Robert M. Shipley, jr. 5th ed. Los Angeles: Gemological Institute of America. 261p.

Short, Nicholas M. (1966) *Shock-lithification of unconsolidated rock materials.* Science, v.154, p.382-384.

Shrock, Robert R. (1947) *Loiponic deposits.* Geological Society of America. Bulletin, v.58, p.1228.

Shrock, Robert R. (1948) *Sequence in layered rocks, a study of features and structures useful for determining top and bottom or order of succession in bedded and tabular rock bodies.* New York: McGraw-Hill. 507p.

Shrock, Robert R. (1948a) *A classification of sedimentary rocks.* Journal of Geology, v.56, p.118-129.

Shrock, Robert R., and Twenhofel, William H. (1953) *Principles of invertebrate paleontology.* 2nd ed. New York: McGraw-Hill. 816p.

Shvetsov, Mikhail S. (1960) *K voprosu o diageneze.* In: Natsional'nyi komitet geologov Sovetskogo Soiuza. *Voprosy sedimentologii; doklady sovetskikh geologov k VI Mezhdunarodnomu kongressu po sedimentologii,* p.153-161. Moscow: Gos. nauchno-tekh. izd'vo lit'ry po geologii i okhrane nedr. 215p.

Siegel, R., and Howell, J.R. (1968) *Thermal radiation heat transfer. v.1.* National Aeronautics and Space Administration. SP-164.

Sigafoos, R.S. (1951) *Soil instability in tundra vegetation.* Ohio Journal of Science, v.51, p.281-298.

Sigal, Jacques (1964) *Une thérapeutique homéopathique en chronostratigraphie: les parastratotypes (ou prétendus tels).* France. Bureau de Recherches Géologiques et Minières. Dépt. d'Information Géologique. Bulletin trimestriel, année 16, no.64, p.1-8.

Silberling, Norman J., and Roberts, Ralph J. (1962) *Pre-Tertiary stratigraphy and structure of northwestern Nevada.* Geological Society of America. Special Paper 72. 58p.

Simons, D.B., and others (1961) *Flume studies using medium sand (0.45 mm).* U.S. Geological Survey. Professional Paper 1498-A. 76p.

Skolnick, Herbert (1965) *The quartzite problem.* Journal of Sedimentary Petrology, v.35, p.12-21.

Skow, Milford L. (1962) *Mica; a materials survey.* U.S. Bureau of Mines. Information Circular 8125. 241p.

Sloane, Richard L., and Kell, T.R. (1966) *The fabric of mechanically compacted kaolin.* In: National Conference on Clays and Clay Minerals. 14th, Berkeley (Calif.), 1965. Proceedings, p.289-296. New York: Pergamon Press. 443p. (International series of monographs on earth sciences, v.26).

Sloss, Laurence L. (1953) *The significance of evaporites.* Journal of Sedimentary Petrology, v.23, p.143-161.

Sloss, Laurence L. (1963) *Sequences in the cratonic interior of North America.* Geological Society of America. Bulletin, v.74, p.93-113.

Sloss, Laurence L., and Laird, Wilson M. (1947) *Devonian System in central and northwestern Montana.* American Association of Petroleum Geologists. Bulletin, v.31, p.1404-1430.

Sloss, Laurence L., and others (1949) *Integrated facies analysis.* Geological Society of America. Memoir 39, p.91-123.

Smirnov, V.I. (1968) *The sources of the ore-forming fluid.* Economic Geology, v.63, p.380-389.

Smit, David E., and Swett, Keene (1969) *Devaluation of "dedolomitization".* Journal of Sedimentary Petrology, v.39, p.379-380.

Smith, John T., jr. (1968) *Glossary of aerial photographic terms.* In: Smith, John T., jr., and Anson, Abraham, eds. *Manual of color aerial photography,* p.489-509. Falls Church (Va.): American Society of Photogrammetry. 550p.

Smith, Kenneth G. (1950) *Standards for grading texture of erosional topography.* American Journal of Science, v.248, p.655-668.

Smith, R.A., Jones, F.E., and Chasmar, R.P. (1968) *The detection and measurement of infra-red radiation.* 2d ed. Oxford: Clarendon Press.

Smith, Robert L., and Bailey, Roy A. (1968) *Resurgent cauldrons.* Geological Society of America. Memoir 116, p.613-662.

Smith, William S.T. (1898) *A geological sketch of San Clemente Island.* U.S. Geological Survey. Annual Report, 18th, pt.2, p.459-496.

Smith, W.O. (1961) *Mechanism of gravity drainage and its relation to specific yield of uniform sands.* U.S. Geological Survey. Professional Paper 402-A. 12p.

Smyth, Henry L. (1891) *Structural geology of Steep Rock Lake, Ont.* American Journal of Science, ser.3, v.42, p.317-331.

Soil Science Society of America (1965) *Glossary of soil science terms.* Its Proceedings, v.29, no.3, p.330-351.

Soil Science Society of America (1970) *Glossary of soil science terms.* Madison (Wisc.): Its publication. 27p.

Sonder, Richard A. (1956) *Mechanik der Erde;*

Elemente und Studien zur tektonischen Erdgeschichte. Stuttgart: Schweizerbartsche. 291p.

Sorby, Henry Clifton (1857) *On the physical geography of the Tertiary estuary of the Isle of Wight.* Edinburgh New Philosophical Journal, n.s., v.5, p.275-298.

Sorby, Henry Clifton (1863) *Über Kalkstein-Geschiebe mit Eindrücken.* Neues Jahrbuch für Mineralogie, Geologie, und Palaeontologie, Jahrg. 1863, p.801-807.

Sorby, Henry Clifton (1879) *The anniversary address of the President.* Geological Society of London. Proceedings, session 1878-1879, p.39-95. (Geological Society of London. Quarterly Journal, 1879, v.35).

Spate, O.H.K. (1954) *India and Pakistan; a general and regional geography.* With a chapter on Ceylon by B.H. Farmer. London: Methuen. 827p.

Speers, Elmer C. (1957) *The age relation and origin of common Sudbury breccia.* Journal of Geology, v.65, p.497-514.

Spencer, Arthur C. (1917) *The geology and ore deposits of Ely, Nevada.* U.S. Geological Survey. Professional Paper 96. 189p.

Spencer, Edgar W. (1969) *Introduction to the structure of the Earth.* New York: McGraw-Hill. 597p.

Spieker, Edward M. (1956) *Mountain-building chronology and nature of geologic time scale.* American Association of Petroleum Geologists. Bulletin, v.40, no.8, p.1769-1815.

Spotts, J.H., and Weser, O.E. (1964) *Directional properties of a Miocene turbidite, California.* In: Bouma, A.H., and Brouwer, A., eds. *Turbidites.* Amsterdam: Elsevier. 264p. (Developments in sedimentology 3, p.199-221).

Sproule, J.C. (1939) *The Pleistocene geology of the Cree Lake region, Saskatchewan.* Royal Society of Canada. Transactions, ser.3, v.33, sec.4, p.101-109.

Spurr, Josiah E. (1923) *The filling of fissure veins.* Engineering & Mining Journal, v.116, Aug.25, p.329-330.

Spurr, Josiah E. (1944) *The Imbrian Plain region of the Moon.* Lancaster (Penna.): Science Press. 112. (His *Geology applied to selenology,* v.1).

Spurrell, F.C.J. (1887) *A sketch of the history of the rivers and denudation of West Kent, etc.* Geological Magazine, dec.3, v.4, no.273, p.121-122.

Stach, E. (1965) *Basic principles of coal petrology; macerals, microlithotypes and some effects on coalification.* In: Murchison (, Duncan) & Westoll (, T. Stanley), eds. *Coal and coal-bearing strata*, p.3-17. New York: American Elsevier. 418p.

Stamp, L. Dudley (1921) *On cycles of sedimentation in the Eocene strata of the Anglo-Franco-Belgian basin*. Geological Magazine, v.58, p.108-114, 146-157, 194-200.

Stamp, L. Dudley, ed. (1961) *A glossary of geographical terms*. New York: Wiley. 539p.

Stamp, L. Dudley, ed. (1966) *A glossary of geographical terms*. 2nd ed. London: Longmans, Green. 539p.

Stearns, Harold T., and Macdonald, Gordon A. (1942) *Geology and ground-water resources of the island of Maui, Hawaii*. Hawaii. Division of Hydrography. Bulletin, no.7. 344p.

Stefan, Joseph (1879) *Über die Beziehung zwischen der Wärmestrahlung und der Temperatur*. Akademie der Wissenschaften in Wein. Sitzungsberichte, v.79, pt.2, p.391-428.

Steno, Nicolaus (1669) *Nicolai Stenonis de solido intra solidum naturaliter contento dissertationis prodromus*. Florence. 79p.—See also: Steno, Nicolaus (1916) *The prodromus of Nicolaus Steno's dissertation concerning a solid body enclosed by process of nature within a solid*. English version with introduction & explanatory notes by John Garrett Winter. Foreword by William H. Hobbs. New York: Macmillan. 119p. (Michigan. Univ. Studies; humanistic series, v.11, pt.2, p.165-283).

Stephens, N., and Synge, F.M. (1966) *Pleistocene shorelines*. In: Dury, G.H., ed. *Essays in geomorphology*. p.1-51. New York: American Elsevier. 404p.

Stephenson, Lloyd W., and Veatch, J.O. (1915) *Underground waters of the Coastal Plain of Georgia*. U.S. Geological Survey. Water-supply Paper 341. 539p.

Sternberg, H. (1875) *Untersuchungen über Längen- und Querprofil geschiebeführender Flüsse*. Zeitschrift für Bauwesen, v.25, p.483-506.

Stevens, Rollin E., and Carron, Maxwell K. (1948) *Simple field test for distinguishing minerals by abrasion pH*. American Mineralogist, v.33, p.31-50.

Stewart, Harris B., jr. (1956) *Contorted sediments in modern coastal lagoon explained by laboratory experiments*. American Association of Petroleum Geologists. Bulletin, v.40, p.153-161.

Stille, Hans (1930) *Über Einseitigkeiten in der germanotypen Tektonik Nordspaniens und Deutschlands*. Gesellschaft der Wissenschaften, Göttingen. Mathematisch-Physikalische Klasse. Nachrichten, 1930, H.3, p.379-397.

Stille, Hans (1935) *Der derzeitige tektonische Erdzustand*. Preussische Academie der Wissenschaften, Physikalisch-mathematische Klasse. Sitzungsberichte, v.13, p.179-219.

Stille, Hans (1936) *Present tectonic state of the Earth*. American Association of Petroleum Geologists. Bulletin, v.20, p.849-880.

Stille, Hans W. (1940) *Einführung in den Bau Amerikas*. Berlin: Gebrüder Borntraeger.

Stillwell, F.L. (1918) *The metamorphic rocks of Adelie Land*. Australasian Antarctic Expedition, 1911-1914, under the leadership of Sir Douglas Mawson. Scientific reports, Series A, Geography, physiography, glaciology, oceanography, and geology. v.3, pt.1, p.7-230.

Stoces, Bohuslav, and White, Charles H. (1935) *Structural geology, with special reference to economic deposits*. London: Macmillan. 460p.

Stockdale, Paris B. (1939) *Lower Mississippian rocks of the east-central interior*. Geological Society of America. Special Paper 22. 248p.

Stockwell, C.H. (1964) *Fourth report on structural provinces, orogenics, and time-classification of rocks of the Canadian Precambrian Shield*. In: *Age determinations and geological studies, part 2*. Canada. Geological Survey. Geological studies. Paper 64-17.

Stokes, William L. (1947) *Primary lineation in fluvial sandstones, a criterion of current direction*. Journal of Geology, v.55, p.52-54.

Stokes, William L. (1953) *Primary sedimentary trend indicators as applied to ore finding in the Carrizo Mountains, Arizona and New Mexico*. Pt. 1. Salt Lake City: University of Utah. 48p. (U.S. Atomic Energy Commission. Report RME-3043, pt.1).

Stokes, William L., and Judson, S. Sheldon (1968) *Introduction to geology, physical and historical*. Englewood Cliffs (N.J.): Prentice-Hall. 530p.

Stokes, William L., and Varnes, David J. (1955) *Glossary of selected geologic terms, with special reference to their use in engineering*. Denver: Colorado Scientific Society. 165p. (Colorado Scientific Society. Proceedings, v.16).

Stone, George H. (1899) *The glacial gravels of Maine and their associated deposits*. U.S. Geological Survey. Monograph 34. 499p.

Stone, Richard O. (1967) *A desert glossary*. Earth-science Reviews, v.3, p.211-268.

Storey, Taras P., and Patterson, John R. (1959) *Stratigraphy—traditional and modern con-*

cepts. American Journal of Science, v.257, p.707-721.

Störmer, Leif (1966) *Concepts of stratigraphical classification and terminology.* Earth-Science Reviews, v.1, p.5-28.

Straaten, L.M.J.U. van (1951) *Longitudinal ripple marks in mud and sand.* Journal of Sedimentary Petrology, v.21, p.47-54.

Straaten, L.M.J.U. van (1953) *Megaripples in the Dutch Wadden Sea and in the basin of Arcachon (France).* Geologie en mijnbouw, n.s., v.15, p.1-11.

Straaten, L.M.J.U. van (1953a) *Rhythmic patterns on Dutch North Sea beaches.* Geologie en mijnbouw, n.s., v.15, p.31-43.

Strahan, Aubrey (1907) *The country around Swansea; being an account of the region comprised in sheet 247 of the map.* London: H.M.S.O. 170p. (Great Britain. Geological Survey. Memoirs; England and Wales: the geology of the South Wales coal-field, pt.8).

Strahler, Arthur N. (1946) *Geomorphic terminology and classification of land masses.* Journal of Geology, v.54, no.1, p.32-42.

Strahler, Arthur N. (1952a) *Dynamic basis of geomorphology.* Geological Society of America. Bulletin, v.63, p.923-938.

Strahler, Arthur N. (1952b) *Hypsometric (area-altitude) analysis of erosional topography.* Geological Society of America. Bulletin, v.63, p.1117-1142.

Strahler, Arthur N. (1954) *Quantitative geomorphology of erosional landscapes.* International Geological Congress. 19th, Algiers, 1952. Comptes rendus, sec.13, pt.3, fasc.15, p.341-354.

Strahler, Arthur N. (1958) *Dimensional analysis applied to fluvially eroded landforms.* Geological Society of America. Bulletin, v.69, p.279-299.

Strahler, Arthur N. (1963) *The earth sciences.* New York: Harper & Row. 681p.

Strahler, Arthur N. (1964) *Quantitative geomorphology of drainage basins and channel networks.* In: Chow, Ven Te, ed. *Handbook of applied hydrology; a compendium of water-resources technology,* sec.4, p.39-76. New York: McGraw-Hill. 29 sections.

Streckeisen, Albert L. (1967) *Classification and nomenclature of igneous rocks (final report of an inquiry).* Neues Jahrbuch für Mineralogie. Abhandlungen, v.107, no.2, p.144-214.

Strickland, Cyril (1940) *Deltaic formation with special reference to the hydrographic proc-esses of the Ganges and the Brahmaputra.* New York: Longmans, Green. 157p.

Stringfield, Victor T. (1966) *Hydrogeology—definition and application.* Ground Water, v.4, no.4, p.2-4.

Ström, K. (1966) *Geophysiography.* Atlas, v.2, no.1, p.8-9.

Suess, Franz E. (1900) *Die Herkunft der Moldavite und verwandter Gläser.* Austria. Geologische Reichsanstalt. Jahrbuch, Bd.50, H.2, p.193-382.

Suess, Hans E., and Urey, Harold C. (1956) *Abundances of the elements.* Reviews of Modern Physics, v.28, p.53-74.

Suggate, R.P. (1965) *The definition of "interglacial".* Journal of Geology, v.73, p.619-626.

Sullwold, Harold H., jr. (1959) *Nomenclature of load deformation in turbidites.* Geological Society of America. Bulletin, v.70, p.1247-1248.

Sullwold, Harold H., jr. (1960) *Load-cast terminology and origin of convolute bedding: further comments.* Geological Society of America. Bulletin, v.71, p.635-636.

Sutton, Arle H. (1940) *Time and stratigraphic terminology.* Geological Society of America. Bulletin, v.51, p.1397-1412.

Svensson, Nils-Bertil (1968) *Lake Lappajärvi, central Finland: a possible meteorite impact structure.* Nature, v.217, p.438.

Swain, Frederick M. (1949) *Onlap, offlap, overstep, and overlap.* American Association of Petroleum Geologists. Bulletin, v.33, p.634-636.

Swain, Frederick M. (1958) *Organic materials of early Middle Devonian, Mt. Union area, Pennsylvania.* American Association of Petroleum Geologists. Bulletin, v.42, p.2858-2891.

Swain, Frederick M. (1963) *Geochemistry of humus.* In: Breger, Irving A., ed. *Organic geochemistry.* New York: Macmillan. 658p. (Earth science series 16, p.87-147).

Swain, Frederick M., and Prokopovich, N. (1954) *Stratigraphic distribution of lipoid substances in Cedar Creek Bog, Minnesota.* Geological Society of America. Bulletin, v.65, p.1183-1198.

Swann, David H., and Willman, Harold B. (1961) *Megagroups in Illinois.* American Association of Petroleum Geologists. Bulletin, v.45, p.471-483.

Swayne, J.C. (1956) *A concise glossary of geographical terms.* London: George Philip & Son. 164p.

Swenson, H.N., and Woods, J.E. (1971) *Physical science for liberal arts students.* 2d ed. New York: Wiley.

Swesnick, Robert M. (1950) *Golden Trend of*

south-central Oklahoma. American Association of Petroleum Geologists. Bulletin, v.34, p.386-422.

Taber, Stephen (1943) Perennially frozen ground in Alaska: its origin and history. Geological Society of America. Bulletin, v.54, p.1433-1548.

Tanner, William F. (1960) Shallow water ripple mark varieties. Journal of Sedimentary Petrology, v.30, p.481-485.

Tanton, Thomas L. (1944) Conchilites. Royal Society of Canada. Transactions, ser.3, v.38, sec.4, p.97-104.

Tarr, Ralph S. (1902) The physical geography of New York. New York: Macmillan. 397p.

Tarr, Ralph S. (1914) College physiography. Published under the editorial direction of Lawrence Martin. New York: Macmillan. 837p.

Tarr, Ralph S., and Engeln, Oskar D. von (1926) New physical geography. New York: Macmillan. 689p.

Tarr, Ralph S., and Martin, Lawrence (1914) Alaskan glacier studies of the National Geographic Society in the Yakutat Bay, Prince William Sound and lower Copper River regions. Washington: The National Geographic Society. 498p.

Tarr, William A. (1938) Terminology of the chemical siliceous sediments. National Research Council. Division of Geology and Geography. Annual report for 1937-1938, appendix A, exhibit A, p.8-27. (Its Committee on Sedimentation. Report, exhibit A).

Tator, Ben A. (1949) Valley widening processes in the Colorado Rockies. Geological Society of America. Bulletin, v.60, p.1771-1783.

Tator, Ben A. (1953) Pediment characteristics and terminology; part II. Association of American Geographers. Annals, v.43, p.47-53.

Taylor, Garvin L., and Reno, Duane H. (1948) Magnetic properties of "granite" wash and unweathered "granite". Geophysics, v.13, p.163-181.

Taylor, T. Griffith, ed. (1951) Geography in the twentieth century. New York: Philosophical Library. 630p.

Teall, J.J.H. (1887) On the origin of certain banded gneisses. Geological Magazine, v.4, no.11, p.484-493.

Teall, J.J.H. (1903) On dedolomitisation. Geological Magazine, n.s., dec.4, v.10, p.513-514.

Tebbutt, Gordon E., and others (1965) Lithogenesis of a distinctive carbonate rock fabric. Wyoming. University. Contributions to Geology, v.4, no.1, p.1-13.

Teichert, Curt (1958) Concept of facies. American Association of Petroleum Geologists. Bulletin, v.42, p.2718-2744.

Teichert, Curt (1958a) Some biostratigraphic concepts. Geological Society of America. Bulletin, v.69, p.99-119.

Termier, Henri, and Termier, Geneviève (1956) La notion de migration en paléontologie. Geologische Rundschau, v.45, p.26-42.

Termier, Henri, and Termier, Geneviève (1963) Erosion and sedimentation. Translated by D.W. Humphries and Evelyn E. Humphries. New York: Van Nostrand. 433p.

Terzaghi, Karl C. von (1943) Theoretical soil mechanics. New York: Wiley. 510p.

Tester, Allen C., and Bay, Harry X. (1931) The shapometer; a device for measuring the shapes of pebbles. Science, n.s., v.73, p.565-566.

Theis, Charles V. (1935) The relation between the lowering of the piezometric surface and the rate and duration of discharge of a well using ground-water storage. American Geophysical Union. Transactions, 1935, p.519-524.

Thomas, G.E. (1962) Grouping of carbonate rocks into textural and porosity units for mapping purposes. American Association of Petroleum Geologists. Memoir 1, p.193-223.

Thomas, Horace D. (1960) Misuse of "bioclastic limestone". American Association of Petroleum Geologists. Bulletin, v.44, p.1833-1834.

Thompson, David G. (1929) The Mohave Desert region, California; a geographic, geologic, and hydrologic reconnaissance. U.S. Geological Survey. Water-supply Paper 578. 759p.

Thornbury, William D. (1954) Principles of geomorphology. New York: Wiley. 618p.

Thornton, Charles Perkins, and Tuttle, Orville Frank (1960) Chemistry of igneous rocks—pt. 1, differentiation index. American Journal of Science, v.258, no.9, p.664-668.

Thorson, Gunnar (1957) Bottom communities (sublittoral or shallow shelf). Geological Society of America. Memoir 67, v.1, p.461-534.

Thrush, Paul W., comp. (1968) A dictionary of mining, mineral, and related terms. Compiled and edited by Paul W. Thrush and the Staff of the Bureau of Mines. Washington: U.S. Bureau of Mines. 1269p.

Tiddeman, R.H. (1890) On concurrent faulting and deposit in Carboniferous times in Craven, Yorkshire, with a note on Carboniferous reefs. British Association for the Advancement of Science. Report, 1889, 59th, p.600-603.

Tieje, Arthur J. (1921) Suggestions as to the

description and naming of sedimentary rocks. Journal of Geology, v.29, p.650-666.

Titkov, Nikolai, and others (1965) *Mineral formation and structure in the electrochemical induration of weak rocks.* New York: Consultants Bureau. 74p.

Todd, James E. (1902) *Hydrographic history of South Dakota.* Geological Society of America. Bulletin, v.13, p.27-40.

Todd, James E. (1903) *Concretions and their geological effects.* Geological Society of America. Bulletin, v.14, p.353-368.

Tolman, Cyrus Fisher (1937) *Ground water.* New York: McGraw-Hill. 593p.

Tomkeieff, Sergei I. (1943) *Megatectonics and microtectonics.* Nature, v.152, p.347-349.

Tomkeieff, Sergei I. (1946) *James Hutton's "Theory of the Earth", 1795.* Geologists' Association. Proceedings, v.57, p.322-328.

Tomkeieff, Sergei I. (1954) *Coals and bitumens and related fossil carbonaceous substances; nomenclature and classification.* London: Pergamon Press. 122p.

Tomkeieff, Sergei I. (1961) *Alkalic ultrabasic rocks and carbonatites of the U.S.S.R.* International Geology Review, v.3, no.9, p.739-758.

Tomkeieff, Sergei I. (1962) *Unconformity—an historical study.* Geologists' Association. Proceedings, v.73, p.383-417.

Törnebohm, A.E. (1880-1881) *Några ord om granit och gneis.* Geologiska Föreningen, Stockholm. Förhandlingar, v.5, p.233-248.

Tourtelot, Harry A. (1960) *Origin and use of the word "shale".* American Journal of Science, v.258-A (Bradley volume), p.335-343.

Tower, Walter S. (1904) *The development of cutoff meanders.* American Geographical Society. Bulletin, v.36, p.589-599.

Trask, Parker D. (1932) *Origin and environment of source sediments of petroleum.* Houston: American Petroleum Institute. 323p.

Tröger, Walter Ehrenreich (1935) *Spezielle Petrographie der Eruptivgesteine: Ein Nomenklatur-Kompendium.* Berlin: Verlag der Deutschen Mineralogischen Gesellschaft e.V. 360p.

Troll, Carl (1944) *Strukturböden, Solifluktion und Frostklimate der Erde.* Geologische Rundschau, Bd.34, p.545-694.

Trowbridge, Arthur C. (1921) *The erosional history of the Driftless Area.* Iowa. University. Studies in Natural History, v.9, no.3. 127p. (Iowa. University. Studies, 1st ser., no.40).

Trümpy, Rudolf (1955) *Wechselbeziehungen zwischen Palaogeographie und Deckenbau.* Naturforschende Gesellschaft in Zürich, Vierteljahrschrift, pt.C, p.217-231.

Trueman, A.E. (1923) *Some theoretical aspects of correlation.* Geologists' Association. Proceedings, v.34, p.193-206.

Turner, Francis J. (1948) *Mineralogical and structural evolution of the metamorphic rocks.* Geological Society of America. Memoir 30. 342p.

Turner, Francis J., and Verhoogen, Jean (1951) *Igneous and metamorphic petrology.* 1st ed. New York: McGraw-Hill.

Turner, Francis J., and Verhoogen, Jean (1960) *Igneous and metamorphic petrology.* 2d ed. New York: McGraw-Hill. 694p.

Turner, Francis J., and Weiss, Lionel E. (1963) *Structural analysis of metamorphic tectonites.* New York: McGraw-Hill. 545p.

Tuttle, Orville Frank, and Bowen, Norman Levi (1958) *Origin of granite in the light of experimental studies in the system $NaAlSi_3O_8$-$KAlSi_3O_8$-SiO_2-H_2O.* Geological Society of America. Memoir 74. 153p.

Twenhofel, William H. (1937) *Terminology of the fine-grained mechanical sediments.* National Research Council. Division of Geology and Geography. Annual report for 1936-1937, appendix I, exhibit F, p.81-104. (Its Committee on Sedimentation. Report, exhibit F).

Twenhofel, William H. (1939) *Principles of sedimentation.* New York: McGraw-Hill. 610p.

Twenhofel, William H. (1950) *Principles of sedimentation.* 2nd ed. New York: McGraw-Hill. 673p.

Tyrrell, George W. (1921) *Some points in petrographic nomenclature.* Geological Magazine, v.58, p.494-502.

Tyrrell, George W. (1926) *The principles of petrology; an introduction to the science of rocks.* London: Methuen. 349p.

Tyrrell, George W. (1950) *The principles of petrology; an introduction to the science of rocks.* London: Methuen; New York: E.P. Dutton.

Tyrrell, J. Burr (1904) *Crystosphenes or buried sheets of ice in the tundra of northern America.* Journal of Geology, v.12, p.232-236.

Tyrrell, J. Burr (1910) *"Rock glaciers" or chrystocrenes.* Journal of Geology, v.18, p.549-553.

Tyrrell, J. Burr, and Dowling, D.B. (1896) *Report on the country between Athabasca Lake and Churchill River with notes on two routes travelled between the Churchill and Saskatchewan rivers.* Ottawa: Geological Survey of Canada. 120p. Canada. Geological Survey. Annual report 1895, n.s., v.8, report D.

Udden, Johan A. (1898) *The mechanical composition of wind deposits.* Rock Island (Ill.): Augustana Library Publications, no.1. 69p.

Udden, Johan A. (1914) *Mechanical composition of clastic sediments.* Geological Society of America. Bulletin, v.25, p.655-744.

Ulrich, Edward O. (1911) *Revision of the Paleozoic systems.* Geological Society of America. Bulletin, v.22, p.281-680.

Umbgrove, J.H.F. (1933) *Verschillende Typen van tertiaire Geosyclinalen in den Indischen Archipel.* Leidsche Geologische Mededelingen, v.5, no.1, p.33-43.

Underwood, Lloyd Bradish (1957) *Rebound problem in the Pierre Shale at Oahe Dam, Pierre, South Dakota, Pt. 1.* (abstr.). Geological Society of America. Bulletin, v.68, no.12, pt.2, p.1807-1808.

U.S. Arctic, Desert, Tropic Information Center (1955) *Glossary of arctic and subarctic terms.* Maxwell Air Force Base (Ala.): Air University, Research Studies Institute. 90p. (ADTIC Publication A-105).

U.S. Army. Coastal Engineering Research Center (1966) *Shore protection, planning and design.* 3rd ed. Washington, D.C.: Government Printing Office. 580p. (includes 5 appendixes). (Its Technical report no.4).

U.S. Department of Agriculture (1957) *Soil.* Washington: Government Printing Office. 784p. (Its *Yearbook of agriculture,* 1957).

U.S. Department of Defense. Army Topographic Command (1969) *Glossary of mapping, charting, and geodetic terms.* 2nd ed. Washington, D.C.: U.S. Department of Defense. 281p.

U.S. Environmental Science Services Administration (1968) *ESSA.* Washington: U.S. Government Printing Office. [8]p.

U.S. Federal Committee on Research Natural Areas (1968) *A directory of research natural areas on Federal lands of the United States of America.* Washington: U.S. Government Printing Office. 129p.

U.S. Geological Survey (1958) *Suggestions to authors of the reports of the United States Geological Survey.* 5th ed. Washington, D.C.: U.S. Government Printing Office. 255p.

U.S. Geological Survey (1970) *Geologic time—the age of the Earth.* Washington, D.C.: U.S. Government Printing Office. 20p.

U.S. Naval Oceanographic Office (1968) *Ice observations.* Rev. 2nd ed. Washington: U.S. Naval Oceanographic Office. 42p. (H.O. publication no.606-d).

Valentine, James W. (1969) *Patterns of taxonomic and ecological structure of the shelf benthos during Phanerozoic time.* Palaeontology, v.12, pt.4, p.684-709.

Van Hise, Charles R. (1896) *Principles of North American pre-Cambrian geology.* U.S. Geological Survey. Annual Report, 16th, pt.1, p.571-843.

Van Hise, Charles R. (1904) *A treatise on metamorphism.* U.S. Geological Survey. Monograph 47. 1286p.

Van Hise, Charles R., and Leith, Charles K. (1911) *The geology of the Lake Superior region.* U.S. Geological Survey. Monograph 52. 641p.

Van Riper, Joseph E. (1962) *Man's physical world.* New York: McGraw-Hill. 637p.

Vanserg, Nicholas (1952) *How to write geologese.* Economic Geology, v.52, p.220-223.

Van Tuyl, Francis M. (1916) *The origin of dolomite.* Iowa. Geological Survey. Volume 25, p.251-421.

Varnes, David J. (1958) *Landslide types and processes.* National Research Council. Highway Research Board. Special Report 29, p.20-47.

Varney, W.D. (1921) *The geological history of the Pewsey Vale.* Geologists' Association. Proceedings, v.32, p.189-205.

Vassoevich, Nikolai B. (1948) *Evolyutsiya predstavlenii o geologicheskikh fatsiyakh.* Leningrad: Gostoptekhizdat. ?p. (Vsesouiznyi Neftianoi Nauchno-Issledovatel'skiy Geologo-Rezvedochnyi Institut. Litologicheskii sbornik, no.1).

Vassoevich, Nikolai B. (1953) *O nekotorykh flishevykh teksturakh (znakakh).* L'vovskoe Geologicheskoe Obshchestvo. Trudy; geologicheskaia seriia, no.3, p.17-85.

Vassoevich, Nikolai B. (1965) *Vernadskiy's views on the origin of oil.* International Geology Review, v.7, no.3, p.507-517.

Veatch, J.O., and Humphrys, C.R. (1966) *Water and water use terminology.* Kaukauna (Wisc.): Thomas Printing & Publishing Co. 381p.

Vening Meinesz, F.A. (1955) *Plastic buckling of the Earth's crust: the origin of geosynclines.* Geological Society of America. Special Paper 62, p.319-330.

Vistelius, Andrei B. (1967) *Studies in mathematical geology.* New York: Consultants Bureau. 294p.

Vitaliano, Dorothy B. (1968) *Geomythology.* Indiana. Univ. Folklore Institute. Journal, v.5, p.5-30.

Vogt, J.H.L. (1905) *Über anchi-eutektische und anchi-monomineralische Eruptivgesteine.* Norsk geologisk tidsskrift, v.1, no.1, paper 2. 33p.

von Bernewitz, M.W. (1931) *Handbook for prospectors.* 2d ed. New York: McGraw-Hill.

Wadell, Hakon A. (1932) *Volume, shape, and roundness of rock particles.* Journal of Geology, v.40, p.443-451.

Wadell, Hakon A. (1934) *Shape determinations of large sedimental rock fragments.* Pan-American Geologist, v.61, p.187-220.

Wadsworth, Marshman E. (1893) *A sketch of the geology of the iron, gold, and copper districts of Michigan.* Michigan. Geological Survey. Report of the State Geologist for 1891-1892, p.75-174.

Wager, Lawrence R. (1968) *Rhythmic and cryptic layering in mafic and ultramafic plutons.* In: Hess, H.H., and Poldervaart, Arie, eds. *Basalts; the Poldervaart treatise on rocks of basaltic composition,* v.2, p.483-862. New York: Interscience.

Wager, Lawrence R., and Brown, G.M. (1967) *Layered igneous rocks.* San Francisco: Freeman. 588p.

Wager, Lawrence R., and Deer, W.A. (1939) *The petrology of the Skaergaard intrusion, Kangerdlugssuaq, east Greenland.* Meddelelser om Grönland, v.105, no.4. 352p. (*Geological investigations in east Greenland,* pt.3).

Wahlstrom, Ernest E. (1948) *Optical crystallography.* New York: Wiley. 206p.

Walcott, R.H. (1898) *The occurrence of so-called obsidian bombs in Australia.* Royal Society of Victoria. Proceedings, n.s., v.11, pt.1, p.23-53.

Walker, Frederick (1957) *Ophitic texture and basaltic crystallization.* Journal of Geology, v.65, no.1, p.1-14.

Walker, T.L. (1902) *The geology of Kalahandi state, central provinces.* India, Geological Survey. Memoirs, v.33, part 3.

Wallace, Robert C. (1913) *Pseudobrecciation in Ordovician limestones in Manitoba.* Journal of Geology, v.21, p.402-421.

Walther, Johannes (1893-1894) *Einleitung in die Geologie als historische Wissenschaft; Beobachtungen über die Bildung der Gesteine und ihrer organischen Einschlusse.* Jena: G. Fischer. 1055p.

Walton, Ewart K. (1956) *Limitations of graded bedding: and alternative criteria of upward sequence in the rocks of the Southern Uplands.* Edinburgh Geological Society. Transactions, v.16, p.262-271.

Walton, John (1940) *An introduction to the study of fossil plants.* London: Adam & Charles Black. 188p.

Wanless, Harold R., and Weller, J. Marvin (1932) *Correlation and extent of Pennsylvanian cyclothems.* Geological Society of America. Bulletin, v.43, p.1003-1016.

Ward, F. Kingdon (1923) *From the Yangtze to the Irrawaddy.* Geographical Journal, v.62, p.6-20.

Ward, W.H. (1953) *Glacier bands; conference on terminology.* Journal of Glaciology, v.2, no.13, p.229-232.

Washburn, A.L. (1947) *Reconnaissance geology of portions of Victoria Island and adjacent regions, arctic Canada.* Geological Society of America. Memoir 22. 142p.

Washburn, A.L. (1950) *Patterned ground.* Revue Canadienne Géographie, v.4, no.3-4, p.5-59.

Washburn, A.L. (1956) *Classification of patterned ground and review of suggested origins.* Geological Society of America. Bulletin, v.67, p.823-865.

Washburn, A.L., and Goldthwait, R.P. (1958) *Slushflows.* Geological Society of America. Bulletin, v.69, p.1657-1658.

Washburne, Chester W. (1943) *Some wrong words.* Journal of Geology, v.51, p.495-497.

Waters, Aron Clement, and Campbell, Charles Duncan (1935) *Mylonites from the San Andreas fault zone.* American Journal of Science, 5th ser., v.29, no.174, p.473-503.

Waterschoot van der Gracht, W.A.J.M. van (1931) *The Permo-Carboniferous orogeny in the south-central United States.* K. Akad. Wetensch. Amsterdam Verh., Afd. Natuurk., Deel 27, no.3.

Wayland, E.J. (1920) *Some facts and theories relating to the geology of Uganda.* Uganda. Geological Dept. Pamphlet no.1. 52p.

Wayland, E.J. (1934) *Peneplains and some other erosional platforms.* Uganda. Geological Survey Dept. Annual Report and Bulletin, 1933, Bull. no.1, p.77-79.

Weast, Robert C., ed. (1970) *Handbook of chemistry and physics.* 51st ed. Cleveland: The Chemical Rubber Co.

Weatherley, A.H., ed. (1967) *Australian inland waters and their fauna; eleven studies.* Canberra: Australian National University Press. 287p.

Weaver, J.E., and Clements, F.E. (1938) *Plant ecology.* 2nd ed. New York: McGraw-Hill. 601p.

Webb, Robert W. (1936) *Kern Canyon fault,*

southern Sierra Nevada. Journal of Geology, v.44, p.631-638.

Webster's third new international dictionary of the English language; unabridged (1967) Philip B. Gove, ed. Springfield (Mass.): G. & C. Merriam. 2662p.

Weeks, L.G. (1952) Factors of sedimentary basin development that control oil occurrence. American Association of Petroleum Geologists. Bulletin, v.36, p.2071-2124.

Weeks, L.G. (1959) Geologic architecture of circum-Pacific. American Association of Petroleum Geologists. Bulletin, v.43, no.2, p.350-380.

Wegener, Alfred (1912) The origin of the continents and oceans—English translation, 1924. New York: Dutton.

Welch, Paul S. (1952) Limnology. 2nd ed. New York: McGraw-Hill. 538p.

Weld, LeRoy Dougherty, ed. (1937) Glossary of physics. 1st ed. New York: McGraw-Hill. 255p.

Weller, J. Marvin (1930) Cyclical sedimentation of the Pennsylvanian Period and its significance. Journal of Geology, v.38, p.97-135.

Weller, J. Marvin (1956) Argument for diastrophic control of late Paleozoic cyclothems. American Association of Petroleum Geologists. Bulletin, v.40, p.17-50.

Weller, J. Marvin (1958) Stratigraphic facies differentiation and nomenclature. American Association of Petroleum Geologists. Bulletin, v.42, p.609-639.

Weller, J. Marvin (1958a) Cyclothems and larger sedimentary cycles of the Pennsylvanian. Journal of Geology, v.66, p.195-207.

Weller, J. Marvin (1960) Stratigraphic principles and practice. New York: Harper. 725p.

Weller, J. Marvin, and others (1942) Stratigraphy of the fusuline-bearing beds of Illinois. Illinois. State Geological Survey. Bulletin 67, p.9-34.

Wellman, H.W., and Wilson, A.T. (1965) Salt weathering, a neglected geological erosive agent in coastal and arid environments. Nature, v.205, p.1097-1098.

Wells, Francis G. (1949) Ensimatic and ensialic geosynclines [abstract]. Geological Society of America. Bulletin, v.60, p.1927.

Wells, John W. (1944) Middle Devonian bone beds of Ohio. Geological Society of America. Bulletin, v.55, p.273-302.

Wells, John W. (1947) Provisional paleoecological analysis of the Devonian rocks of the Columbus region. Ohio Journal of Science, v.47, p.119-126.

Wentworth, Carl M. (1967) Dish structure, a primary sedimentary structure in coarse turbidites. American Association of Petroleum Geologists. Bulletin, v.51, p.485.

Wentworth, Chester K. (1922) A scale of grade and class terms for clastic sediments. Journal of Geology, v.30, p.377-392.

Wentworth, Chester K. (1922a) A method of measuring and plotting the shapes of pebbles. U.S. Geological Survey. Bulletin 730-C, p.91-114.

Wentworth, Chester K. (1922b) The shapes of beach pebbles. U.S. Geological Survey. Professional Paper 131-C, p.75-83.

Wentworth, Chester K. (1925) Chink-faceting; a new process of pebble-shaping. Journal of Geology, v.33, p.260-267.

Wentworth, Chester K. (1931) Pebble wear on the Jarvis Island beach. Washington University Studies, n.s., Science and Technology, no.5, p.11-37.

Wentworth, Chester K. (1935) The terminology of coarse sediments. With notes by P.G.H. Boswell. National Research Council. Division of Geology and Geography. Committee on Sedimentation. Report for 1932-1934, p.225-246. (National Research Council. Bulletin, no.98).

Wentworth, Chester K. (1936) An analysis of the shapes of glacial cobbles. Journal of Sedimentary Petrology, v.6, p.85-96.

Wentworth, Chester K. (1943) Soil avalanches on Oahu, Hawaii. Geological Society of America. Bulletin, v.54, p.53-64.

Wentworth, Chester K., and Williams, Howel (1932) The classification and terminology of the pyroclastic rocks. National Research Council. Division of Geology and Geography. Committee on Sedimentation. Report for 1930-1932, p.19-53. (National Research Council. Bulletin, no.89)

Wheeler, Harry E. (1958) Time-stratigraphy. American Association of Petroleum Geologists. Bulletin, v.42, p.1047-1063.

Wheeler, Harry E. (1958a) Primary factors in biostratigraphy. American Association of Petroleum Geologists. Bulletin, v.42, p.640-655.

Wheeler, Harry E. (1964) Baselevel, lithosphere surface, and time-stratigraphy. Geological Society of America. Bulletin, v.75, p.599-609.

Wheeler, Harry E., and Beesley, E. Maurice (1948) Critique of the time-stratigraphic concept. Geological Society of America. Bulletin, v.59, p.75-85.

Wheeler, Harry E., and Mallory, V. Standish (1956) Factors in lithostratigraphy. American

Association of Petroleum Geologists. Bulletin, v.40, p.2711-2723.

Wheeler, Harry E., and others (1950) *Stratigraphic classification.* American Association of Petroleum Geologists. Bulletin, v.34, p.2361-2365.

White, Charles A. (1870) *Report on the geological survey of the State of Iowa.* Vol. 1. Des Moines: Mills & Co. 391p.

White, David (1915) *Some relations in origin between coal and petroleum.* Washington Academy of Sciences. Journal, v.5, no.6, p.189-212.

White, Donald E. (1957) *Thermal waters of volcanic origin.* Geological Society of America. Bulletin, v.68, no.12, pt.1, p.1637-1657.

White, W.H. (1959) *Cordilleran tectonics in British Columbia.* American Association of Petroleum Geologists. Bulletin, v.43, p.60-100.

Whitney, J.D. (1888) *Names and places; studies in geographical and topographical nomenclature.* Cambridge (Eng.): Cambridge University Press. 239p.

Whitten, E.H. Timothy (1959) *A study of two directions of folding: the structural geology of the monadhliath and mid-Strathspey.* Journal of Geology, v.67, no.1, p.14-47.

Wickman, F.E. (1966) *Repose period patterns of volcanoes. Part 1: Volcanic eruptions regarded as random phenomena.* Arkiv for Mineralogi och Geologi, v.4, no.7, p.291-301.

Wiegel, Robert L. (1953) *Waves, tides, currents and beaches: glossary of terms and list of standard symbols.* Berkeley (Calif.): Council on Wave Research, The Engineering Foundation. 113p.

Wien, Willy (1894) *Temperatur und Entropie der Strahlung.* Annalen der Physik, ser.2, v.52, p.132-165.

Willard, Bradford (1930) *Conglomerite, a new rock term.* Science, v.71, p.438.

Williams, B.J., and Prentice, J.E. (1957) *Slump-structures in the Ludlovian rocks of North Herefordshire.* Geologists' Association. Proceedings, v.68, p.286-293.

Williams, Henry S. (1893) *The elements of the geological time-scale.* Journal of Geology, v.1, p.283-295.

Williams, Henry S. (1895) *Geological biology; an introduction to the geological history of organisms.* New York: Henry Holt. 395p.

Williams, Henry S. (1901) *The discrimination of time-values in geology.* Journal of Geology, v.9, p.570-585.

Williams, Howard R., and Meyers, Charles J. (1964) *Oil and gas terms; annotated manual of legal, engineering, tax words and phrases.* San Francisco: Matthew Bender. 449p.

Williams, Howel (1932) *The history and character of volcanic domes.* California. University. Dept of Geological Sciences. Bulletin, v.21, no.5, p.51-146.

Williams, Howel (1941) *Calderas and their origin.* University of California, Department of Geological Sciences. Bulletin, v.25, no.6, p.239-346.

Williams, M.Y. (1936) *Frost circles.* Royal Society of Canada. Transactions, ser.3, v.30, p.129-132.

Williams, Richard S., jr. (1972) *Thermography.* Photogrammetric Engineering, v.38, no.8.

Williamson, Iain A. (1961) *Spring domes developed in limestone.* Journal of Sedimentary Petrology, v.31, p.288-291.

Willis, Bailey (1893) *The mechanics of Appalachian structure.* U.S. Geological Survey. Annual Report, 13th, pt.2, p.211-281.

Willis, Bailey (1903) *Physiography and deformation of the Wenatchee-Chelan district, Cascade Range.* U.S. Geological Survey. Professional Paper 19, p.41-97.

Willis, Bailey (1907) *Geographic history of Potomac River.* U.S. Geological Survey. Water-supply Paper 192, p.7-22.

Willis, Bailey (1928) *Dead Sea problem: rift valley or ramp valley?* Geological Society of America. Bulletin, v.39, p.490-542.

Willis, Bailey (1938) *Asthenolith (melting spot) theory.* Geological Society of America. Bulletin, v.49, no.4, p.603-614.

Willman, Harold B., and others (1942) *Geology and mineral resources of the Marseilles, Ottawa, and Streator quadrangles.* Illinois. State Geological Survey. Bulletin 66. 388p.

Wills, Leonard J. (1956) *Concealed coalfields; a palaeogeographical study of the stratigraphy and tectonics of mid-England in relation to coal reserves.* London: Blackie. 208p.

Wilmarth, M. Grace (1938) *Lexicon of geologic names of the United States (including Alaska).* Parts 1-2. U.S. Geological Survey. Bulletin 896. 2396p.

Wilson, Gilbert (1953) *Mullion and rodding structures in the Moine series of Scotland.* Geologists' Association. Proceedings, v.64, p.118-151.

Wilson, John A. (1959) *Transfer, a synthesis of stratigraphic processes.* American Association of Petroleum Geologists. Bulletin, v.43, p.2861-2862.

Wilson, J. Tuzo (1950) *An analysis of the pattern*

and possible cause of young mountain ranges and island arcs. Geological Association of Canada. Proceedings, v.3, p.141-166.

Winchell, Newton H., and Winchell, Horace V. (1891) *The iron ores of Minnesota, their geology, discovery, development, qualities and origin, and comparison with those of other iron districts.* Minnesota. Geological and Natural History Survey. Bulletin no.6. 430p.

Winkler, E.M., and Wilhelm, E.J. (1970) *Salt burst by hydration pressures in architectural stone in urban atmosphere.* Geological Society of America. Bulletin, v.81, p.567-572.

Winkler, H.G.F. (1967) *Petrogenesis of metamorphic rocks.* 2d ed. New York: Springer-Verlag. 237p.

Wolf, Karl H. (1960) *Simplified limestone classification.* American Association of Petroleum Geologists. Bulletin, v.44, p.1414-1416.

Wolf, Karl H. (1965) *Littoral environment indicated by open-space structures in algal limestones.* Palaeogeography, Palaeoclimatology, Palaeoecology, v.1, p.183-223.

Wolfe, C. Wroe, and others (1966) *Earth and space science.* Boston: Heath. 630p.

Wood, Alan (1935) *The origin of the structure known as guilielmites.* Geological Magazine, v.72, p.241-245.

Wood, Alan (1941) *"Algal dust" and the finer-grained varieties of Carboniferous limestone.* Geological Magazine, v.78, p.192-200.

Wood, Alan (1942) *The development of hillside slopes.* Geologists' Association. Proceedings, v.53, p.128-138.

Wood, Alan, and Smith, Alec James (1958) *The sedimentation and sedimentary history of the Aberystwyth Grits (upper Llandoverian).* Geological Society of London. Quarterly Journal, v.114, p.163-195.

Woodford, Alfred O. (1925) *The San Onofre Breccia; its nature and origin.* California. University, Berkeley. Dept. of Geological Sciences. Bulletin, v.15, no.7, p.159-280.

Woodford, Alfred O. (1951) *Stream gradients and Monterey sea valley.* Geological Society of America. Bulletin, v.62, p.799-851.

Woodward, Horace B. (1887) *Geology of England and Wales: with notes on the physical features of the country.* 2d ed. London: G. Philip. 670p.

Woodward, Horace B. (1894) *The Jurassic rocks of Britain.* Vol. 4. London: Her Majesty's Stationery Office. 628p. (Great Britain. Geological Survey. Memoir).

Woodworth, Jay B. (1894a) *Postglacial eolian action in southern New England.* American Journal of Science, ser.3, v.47, p.63-71.

Woodworth, Jay B. (1894b) *Some typical eskers of southern New England.* Boston Society of Natural History. Proceedings, v.26, p.197-220.

Woodworth, Jay B. (1901) *Pleistocene geology of portions of Nassau County and Borough of Queens.* New York. State Museum. Bulletin, v.48, p.618-670.

Woodworth, Jay B. (1912) *Geological expedition to Brazil and Chile, 1908-1909.* Harvard College. Museum of Comparative Zoology. Bulletin, v.56, no.1. 137p.

Woolnough, W.G. (1927) *The duricrust of Australia.* Royal Society of New South Wales. Journal and Proceedings, v.61, p.24-53.

Worcester, Philip G. (1939) *A textbook of geomorphology.* New York: Van Nostrand. 565p.

Workman, W. Hunter (1914) *Nieve penitente and allied formations in Himalaya, or surface-forms of névé and ice created or modelled by melting.* Zeitschrift für Gletscherkunde, Bd. 8, p.289-330.

Wright, John K. (1944) *The terminology of certain map symbols.* Geographical Review, v. 34, p.653-654.

Wright, John K. (1947) *Terrae incognitae: the place of the imagination in geography.* Association of American Geographers. Annals, v.37, p.1-15.

Wright, William B. (1914) *The Quaternary ice age.* London: Macmillan. 464p.

Wright, William B. (1926) *Stratigraphical diachronism in the Millstone Grit of Lancashire.* British Association for the Advancement of Science. Report, 94th, p.354-355.

Wulff, Georgij V. von (1902) *Untersuchungen im Gabiete der optischen Eigenschaften isomorpher Krystalle.* Zeitschrift für Krystallographie und Mineralogie, Bd.36, p.1-28.

Wyckoff, Ralph Dewey (1941) *The Gulf gravimeter.* Geophysics, v.6, no.1, p.13-33.

Wyllie, Peter J. (1966) *Experimental petrology: an indoor approach to an outdoor subject.* Journal of Geological Education, v.14, no.3, p.93-97.

Wynne-Edwards, Hugh R. (1957) *Structure of the Westport concordant pluton in the Grenville, Ontario.* Journal of Geology, v.65, no.6, p.639-649.

Yaalon, Dan H. (1965) *Microminerals and micromineralogy.* Clay Minerals, v.6, no.1, p.71.

Yasso, Warren E. (1966) *Heavy mineral concentration and sastrugi-like deflation furrows in* a beach salcrete at Rockaway Point, New York. Journal of Sedimentary Petrology, v.36, p.836-838.

Yokoyama, Izumi (1956-57) Energetics and active volcanoes. Tokyo University. Earthquake Research Institute. Bulletin, v.35, pt.1, p.75-97.

Youell, R.F. (1960) An electrolytic method for producing chlorite-like substances from montmorillonite. Clay Minerals Bulletin, v.4, no.24, p.191-195.

Young, Alfred P. (1910) On the glaciation of the Navis Valley in North Tirol. Geological Magazine, v.7, p.244-258.

Zemansky, M.W. (1957) Heat and thermodynamics. New York: McGraw-Hill. 484p.

Zernitz, Emilie R. (1932) Drainage patterns and their significance. Journal of Geology, v.40, p.498-521.

Zingg, Theodor (1935) Beitrag zur Schotteranalyse; die Schotteranalyse und ihre Anwendung auf die Glattalschotter. Schweizerische mineralogische und petrographische Mitteilungen, Bd.15, p.39-140.

Zirkel, Ferdinand (1866) Lehrbuch der Petrographie. Bd. 1. Bonn: Adolph Marcus. 607p.

Zirkel, Ferdinand (1876) Microscopical petrography. U.S. Geological Exploration of the Fortieth Parallel. Report, v.6, 297p. (U.S. Army. Engineer Dept. Professional Paper, no. 18).

Zirkel, Ferdinand (1893) Lehrbuch der Petrographie. Bd. 1. 2nd ed. Leipzig: Wilhelm Engelmann. 845p.

Zischinsky, Ulf (1969) Über Sackungen. Rock Mechanics, v.1, p.30-52.

Zittel, Karl A. von (1901) History of geology and palaeontology to the end of the nineteenth century. Translated by Maria M. Ogilvie-Gordon. London: Walter Scott. 562p.